Biology

SIXTH EDITION

Eldra P. Solomon
University of South Florida

Linda R. Berg
St. Petersburg College

Diana W. Martin
Rutgers University

BROOKS/COLE
THOMSON LEARNING™

Australia • Canada • Mexico • Singapore • Spain
United Kingdom • United States

Dedication

To our families, friends, and colleagues
who gave freely of their love, support, knowledge,
and time as we labored over this revision of Biology . . .

Especially to . . .

Lauryn, Haley, Ariel, Eryn, Alexandra, Giliah, Justin, Bryce, Joshua, and Meriah
Alan, Jennifer, and Corey
Chuck and Margaret

BROOKS/COLE
THOMSON LEARNING™

Executive Editor: Nedah Rose
Developmental Editors: Melanie Cann, Anne Scanlan-Rohrer
Editorial Assistant: Alyssa White
Technology Project Managers: Keli Amann, Natalie Peretti
Marketing Manager: Ann Caven
Production Manager: Charlene Catlett Squibb
Project Editor: Bonnie Boehme
Text Designers: Paul Fry, Kim Menning
Art Director: Paul Fry
Photo Researcher: Nancy Lubars
Copy Editor: Sue Reilly
Illustrators: Elizabeth Morales, Rolin Graphics
Cover Designer: Larry Didona
Cover Image: ©Konrad Wothe, Minden Pictures
Cover Printer: Lehigh Press
Compositor: TechBooks
Printer: R. R. Donnelley Willard

Printed in the United States of America
2 3 4 5 6 7 05 04 03 02

For more information about our products, contact us at:
Thomson Learning Academic Resource Center
1-800-423-0563

For permission to use material from this text, contact us by:
Phone: 1-800-730-2214
Fax: 1-800-730-2215
Web: http://www.thomsonrights.com

Library of Congress Control Number: 2001095366
BIOLOGY, sixth edition
ISBN: 0-03-033503-5

Asia
Thomson Learning
60 Albert Street, #15-01
Albert Complex
Singapore 189969

Australia
Nelson Thomson Learning
102 Dodds Street
South Melbourne, Victoria 3205
Australia

Canada
Nelson Thomson Learning
1120 Birchmount Road
Toronto, Ontario M1K 5G4
Canada

Europe/Middle East/Africa
Thomson Learning
Berkshire House
168-173 High Holborn
London WC1 V7AA
United Kingdom

Latin America
Thomson Learning
Seneca, 53
Colonia Polanco
11560 Mexico D.F.
Mexico

Spain
Paraninfo Thomson Learning
Calle/Magallanes, 25
28015 Madrid, Spain

About the Cover The slow wingbeat of the gyrfalcon *(Falco rusticolus)*, a bird of prey found in arctic and subarctic regions from northern Alaska to Greenland to northern Eurasia, is caught on camera. The gyrfalcon is the largest species of falcon, with wingspans to 1.2 m, or 4 ft. It flies deceptively fast for a bird of its size and feeds on birds (such as ptarmigan), lemmings and other small mammals, and insects. Females lay three to four eggs on a cliff ledge, on the bare rock, or in a huge nest of sticks added to year after year by many generations of breeding gyrfalcons. Three color morphs occur: dark, gray, and white *(shown)*. *(Konrad Wothe/Minden Pictures)*.

About the Authors

Eldra P. Solomon has written several leading college-level textbooks in biology and in human anatomy and physiology. Her books have been translated into more than ten languages. Dr. Solomon earned an M.S. from the University of Florida and an M.A. and Ph.D. from the University of South Florida. She is an adjunct professor and member of the Graduate Faculty at the University of South Florida. She taught biology and nursing students for more than 20 years.

Dr. Solomon is a bio-psychologist as well as a biologist. As a psychologist, she specializes in health psychology, Post-traumatic Stress Disorder, dissociative disorders, and stress management. Her research has focused on the relationships among stress, emotions, and health. Dr. Solomon has served as Clinical Director of the Center for Mental Health Education, Assessment, and Therapy since 1992.

Dr. Solomon has been recognized nationally and internationally multiple times. She was an Invited Scientist to the XVth Congress of Scientific Investigation, sponsored by Interamerican University of Puerto Rico, where she presented the Plenary Session, "Mind–Body Connections—An Introduction to Psychoneuro-Immunology." She has been profiled more than twenty times in leading publications, including *Who's Who in America, Who's Who in Science and Engineering, Who's Who in Medicine and Healthcare, Who's Who in American Education, Who's Who of American Women,* and *Who's Who in the World.*

Linda R. Berg is an award-winning teacher and textbook author. She received a B.S. in science education, an M.S. in botany, and a Ph.D. in plant physiology from the University of Maryland. Her research focused on the evolutionary implications of steroid biosynthetic pathways in various organisms.

Dr. Berg taught at the University of Maryland at College Park for 17 years and is presently an Adjunct Professor at St. Petersburg College in Florida. During her career, she has taught introductory courses in biology, botany, and environmental science to thousands of students. At the University of Maryland, she received numerous teaching and service awards. Dr. Berg is also the recipient of many national and regional awards, including the National Science Teachers Association Award for Innovations in College Science Teaching, the Nation's Capital Area Disabled Student Services Award, and the Washington Academy of Sciences Award in University Science Teaching.

During her career as a professional science writer, Dr. Berg has authored or co-authored several leading college science textbooks. Her writing reflects her teaching style and love of science.

Diana W. Martin is the Director of General Biology, Division of Life Sciences, at Rutgers University, New Brunswick Campus. She received an M.S. at Florida State University, where she studied the chromosomes of related plant species to understand their evolutionary relationships. She earned a Ph.D. at the University of Texas at Austin, where she studied the genetics of the fruit fly, *Drosophila melanogaster*, and then conducted postdoctoral research at Princeton University. She has taught General Biology and other courses at Rutgers for more than 20 years and has been involved in writing textbooks since 1988. She is immensely grateful that her decision to study biology in college has led to a career that allows her many ways to share her excitement about all aspects of biology.

Preface

The science of biology is truly one of the most dynamic and rapidly expanding human endeavors. Knowledge in the biological sciences is relevant not only to students' personal lives but also to the future of humanity and of all life on Earth. As biologists have studied life, we have gained a greater understanding of life processes and have become more aware of our interdependence with the vast diversity of organisms with which we share our planet. With new advances in biological research, our lives have become healthier, safer, more comfortable, and also more challenging.

We want students to gain a better appreciation of Earth's diverse organisms, their remarkable adaptations to the environment, and their evolutionary and ecological relationships. We also want students to understand the dynamic way that science works and to appreciate the contributions of scientists whose discoveries not only contribute to our expanding knowledge of biology but also help shape and protect the future of our planet.

PHILOSOPHY AND APPROACH

***BIOLOGY* IS A BOOK FOR STUDENTS.** Since the earliest edition of *Biology,* we have always focused primarily on making the numerous facts and principles of biology interesting and accessible to the student. To that end, we take great care to present the student with a text that is enjoyable to read, with a well-developed art program and sound learning aids.

- **A student-oriented writing style.** One of our principal goals for *Biology* has always been to share with beginning biology students our sense of excitement about the biological sciences. We have worked hard to cultivate a writing style that respects the student's intelligence and presents concepts in a straightforward, accessible way. Special discussions, highlighted with a *Making the Connection* icon, are included to spark student interest and facilitate the integration of concepts. *Focus On* boxes explore issues that are of particular relevance to students' lives (such as the effects of smoking), as well as provide a forum for discussing certain key topics in more detail (such as mammalian cloning). *Tables* (many of them illustrated) and *Sequence Summaries* organize and summarize material presented in the text.

MAKING THE CONNECTION What is the evolutionary relationship between prokaryotic cells and the more complex cells of eukaryotes? Mitochondria and chloroplasts have provided valuable insights because these organelles have been shown to have many prokaryote features. For example, although most of the DNA in eukaryotic cells resides in the nucleus, both mitochondria and chloroplasts (as well as other plastids) have DNA molecules in their inner compartments. These DNA molecules code for a small number of the proteins found in these organelles. These proteins are synthesized by mitochondrial or chloroplast ribosomes, which are similar to the ribosomes of prokaryotes. Most of the mitochondrial and chloroplast proteins, however, are coded for by nuclear genes, manufactured on free ribosomes in the cytosol, and then transported to their appropriate locations within.

The existence of a separate set of ribosomes and DNA molecules in mitochondria and chloroplasts and their similarity in size to many bacteria, along with
teristics, provide support for the e
20–7 and 24–23). According to t
chloroplasts evolved from prokar
residence inside larger cells and
function as autonomous organism
Evolutionary biologists are se

EUKARYOT
A CYTOSKE

We have seen that t
cells. When we watch
they frequently chan
about. The shapes o
mined in large part b
tein fibers (Fig. 4–21
port, the cytoskelet
transport of material
The cytoskeleton
Its framework is mad
tubules, microfilame
termediate filaments
fibers formed from be
be rapidly assembled

rounding cells; most of this water diffuses back to the xylem to be transported upward. This water movement decreases the hydrostatic pressure inside the sieve tubes at the sink. Thus, phloem unloading at the sink is as follows:

Sugar is transported out of sieve tube member ⟶ water diffuses out of sieve tube member and into xylem ⟶ hydrostatic pressure decreases within sieve tube

Thus, the pressure-flow hypothesis explains the movement of dissolved sugar in phloem by means of a pressure gradient. It is the difference in sugar concentrations between the source and

Features, such as the *Career Visions* interviews that conclude each part, help students become aware of the many doors that a biology degree can open by sharing the experiences of people not much older than the students who have found fulfilling careers in which they use their knowledge of biology. We hope that the combined effect of an engaging writing style and interesting features will captivate students, encouraging them to continue their study of biology.

- **An art program that is fully integrated with the text.** The purpose of an art program is to reinforce and expand on concepts discussed in the text. Numerous photographs, both alone and in combination with line art, help students understand concepts by connecting the "real" to the "ideal." The line art, substantively revised and augmented in this edition, uses devices such as ***orientation icons*** on "serial" figures to help the student put the detailed figures into the proper broad context. Symbols and colors are used consistently throughout the book to avoid confusion and help students connect concepts. For example, the same four colors and shapes are used consistently throughout the book to represent guanine, cytosine, adenine, and thymine.

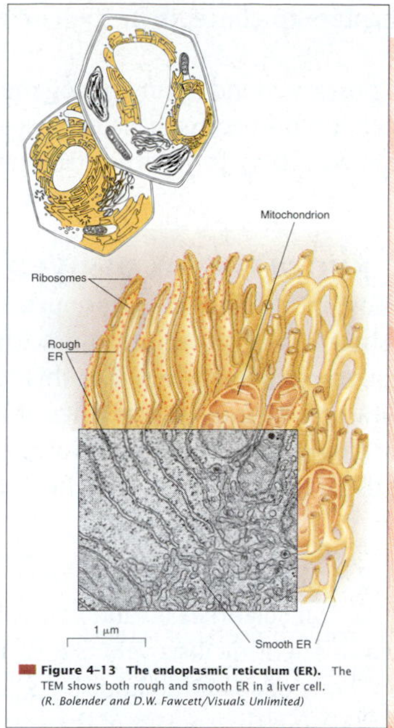

Figure 4-13 The endoplasmic reticulum (ER). The TEM shows both rough and smooth ER in a liver cell. *(R. Bolender and D.W. Fawcett/Visuals Unlimited)*

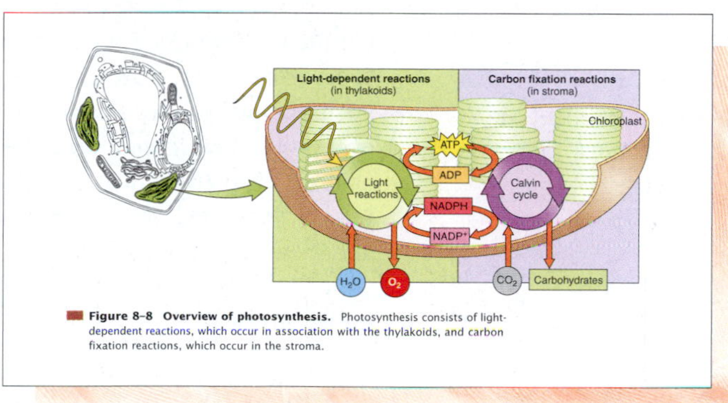

Figure 8-8 Overview of photosynthesis. Photosynthesis consists of light-dependent reactions, which occur in association with the thylakoids, and carbon fixation reactions, which occur in the stroma.

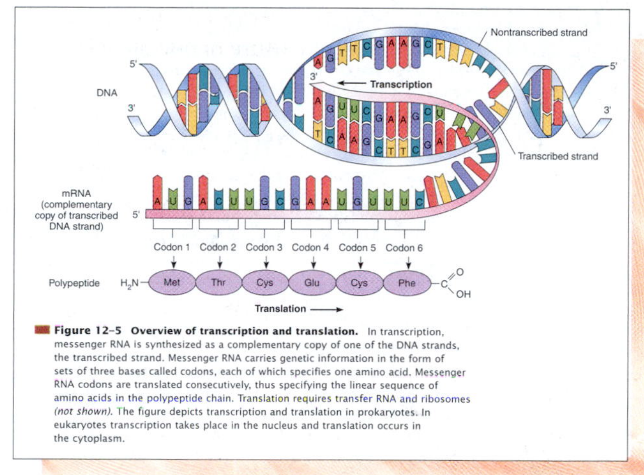

Figure 12-5 Overview of transcription and translation. In transcription, messenger RNA is synthesized as a complementary copy of one of the DNA strands, the transcribed strand. Messenger RNA carries genetic information in the form of sets of three bases called codons, each of which specifies one amino acid. Messenger RNA codons are translated consecutively, thus specifying the linear sequence of amino acids in the polypeptide chain. Translation requires transfer RNA and ribosomes *(not shown)*. The figure depicts transcription and translation in prokaryotes. In eukaryotes transcription takes place in the nucleus and translation occurs in the cytoplasm.

- **Tried-and-true learning aids.** Learning the principles of biology is a challenging endeavor. To help the student achieve mastery of the subject, the text contains a repertoire of learning aids that consistently rate approval from both professors and students. Here we highlight five learning tools that have significant pedagogical effectiveness:

  1. *Learning Objectives* in the chapter introduction indicate, in behavioral terms, what the student must be able to do to demonstrate mastery of the material in the chapter.
  2. *Concept-Statement Heads* introduce each section, previewing and summarizing the key idea or ideas that will be discussed in that section. The headings and subheadings provide a concise outline of the chapter's contents; chapter outlines may also be found on our Web site (http://www.info.brookscole.com/solomonbergmartin).
  3. *A Summary with Key Terms* in outline form at the end of each chapter provides a review of the material. Selected key terms are boldfaced in the summary, enabling students to study vocabulary words within the context of related concepts.
  4. Three types of *End-of-Chapter Questions* provide students with the opportunity to evaluate their understanding of the material in the chapter. The ***Post-Test*** consists of multiple-choice questions, some of which are based on the recall of important terms, whereas others challenge students to integrate their knowledge. Answers to the Post-Test questions are provided in Appendix A. The Post-Test is followed by a series of ***Review Questions*** that assess comprehension by asking the student to explain, compare and contrast, or illustrate important concepts. Finally, a series of ***You Make the Connection*** questions encourages critical thinking and active learning. To answer these questions, the student must apply the concepts just learned

to new situations or relate concepts learned in previous chapters to concepts in the current chapter.

5. A *Glossary* at the end of the book, the most comprehensive glossary found in any biology text, provides precise definitions of terms. The Glossary is especially useful because it is extensively cross-referenced and includes pronunciations. The vertical tan bar along the margin facilitates rapid access to the Glossary.

AN UNDERSTANDING OF HOW THE BODY OF SCIENTIFIC KNOWLEDGE IS DERIVED IS CRITICAL FOR SCIENTISTS AND NONSCIENTISTS ALIKE. Most introductory survey courses of biology seek to familiarize the student with the basic concepts and vocabulary of biology, of course. But equally important is helping the student understand *how* that body of knowledge evolved. For this reason, *Biology* seeks to give the student insight into what science is, how scientists work, the roles of the many scientists who have contributed to our current understanding of biology, and how scientific knowledge impacts daily life. Two features of the sixth edition clearly reflect this emphasis:

- A *"Process of Science"* icon, embedded at relevant points in the text, highlights discussions about the work that scientists do and helps students to understand that the process of scientific discovery is ongoing and indefinite. Examples of how the body of scientific knowledge has been derived are given frequently and within the context of the particular subject being discussed.
- *On the Cutting Edge* boxes present exciting research areas, such as how light regulates gene expression (see Chapter 36). The abstract for the cutting edge box is formatted like a much-abbreviated science report, which further reinforces the scientific process.

Process of Science The goal of systematics is to determine the evolutionary relationships, or **phylogeny** (literally, "production of phyla"), among and between species and higher taxa. Once these relationships are established, systematists can use these evolutionary relationships to build classifications based on common ancestry. Once phylogenies are determined, they can be used to help answer other questions in biology. For example, phylogenies can help us understand evolutionary patterns that might provide clues to the origin and spread of HIV and other pathogens. Phylogenies are also useful in predicting characteristics of newly discovered species. Systematists may classify a new species bas
specific characters that it shares with other organisms in
ticular grouping. They might then infer that the new s
shares other characters with organisms in that group.

As you read the following sections, it is important
member that phylogenies are testable hypotheses. They ar
ported or falsified by the available data. Thus, systematics is
namic science that changes as biologists discover new s
and use ever more sophisticated techniques to investigat
evolutionary relationships among organisms.

Process of Science ON THE CUTTING EDGE

How Light Regulates Gene Expression

HYPOTHESIS: The transcription factor PIF3 is involved in the mechanism of phytochrom
METHOD: To test the hypothesis, researchers used molecular techniques to determin
promoter sequence and, if so, whether phytochrome interacts with the DN
RESULTS: PIF3 binds to a DNA sequence that is present in several light-regulated ge
reversibly to the PIF3-promoter complex.
CONCLUSION: PIF3 plays a crucial role in phytochrome signal transduction.

For several years, Dr. Peter H. Quail and colleagues at the University of California, Berkeley, and the U.S.Department of Agriculture Western Regional Research Center in Albany, California, have actively worked on the mechanism that links phytochrome action to light-responsive genes. They

DNA. They found, however,
corresponding to a phytoch
present only in the samples
light prior to electrophores
phytochrome in the Pfr for

AN OVERVIEW OF *BIOLOGY*, SIXTH EDITION

Three themes provide the structural foundation for *Biology*: the transmission of biological information, the evolution of life, and the flow of energy through living systems. As we introduce the concepts of modern biology, we explain how these themes are connected and how life depends on them.

Educators present the major topics of an introductory biology course in a variety of orders. For this reason, we have carefully designed the eight parts so that they do not depend heavily on preceding chapters and parts. The eight parts and their 55 chapters can be presented in any number of sequences with pedagogical success. Chapter 1, which introduces the student to the major principles of biology, provides an adequate springboard for future discussions, whether the professor prefers to start with the "big picture" and work down, or vice versa.

In this edition, as in previous editions, we examined every line of every chapter for its accuracy and currency, and we made a serious attempt to update every topic and verify all new material. The following brief survey of the eight parts of *Biology* provides a general overview of changes made to the sixth edition.

Part 1: The organization of life

The five chapters that comprise Part 1 seek to provide the student with basic background knowledge. We begin Chapter 1 with a discussion of some current areas of biological research, including genetically modified crops, the Human Genome Project, and stem cells. We then introduce the main themes of the book—evolution, energy transfer, and information transfer. Chapter 1 examines several other fundamental concepts: the similarities of all living things; the organization of life on individual and ecological levels; the diversity of life and evolution; how biologists classify organisms; and how science works. Chapters 2 and 3, which focus on the molecular level of organization, establish the foundations in chemistry necessary for an understanding of biological processes. Chapters 4 and 5 focus on the cellular level of organization. Several recent advances in cell biology are introduced, such as our new understanding of mitochondrial function (including information on mitochondrial mutations, free radicals, and apoptosis), and a new hypothesis of Golgi complex function.

Part 2: Energy transfer through living systems

Because all living cells require energy for life processes, the flow of energy through living systems—that is, capturing energy and converting it to usable forms—is a basic theme of *Biology*. Chapter 6 examines how cells capture, transfer, store, and use energy. Chapters 7 and 8, which can be taught in either order, discuss the metabolic adaptations by which organisms obtain and use energy through cellular respiration and photosynthesis.

Part 3: The continuity of life: genetics

The seven chapters of Part 3 explore the science of genetics, giving the student the tools needed to comprehend the important new findings reported almost daily. This unit begins with a discussion of mitosis and meiosis (Chapter 9), providing a foundation for the consideration of Mendelian genetics and related patterns of inheritance in Chapter 10. We then turn our attention to the flow of information in cells, beginning with the structure and replication of DNA (in Chapter 11), followed by a discussion of RNA and protein synthesis (Chapter 12). Gene regulation is discussed in Chapter 13, and in Chapter 14, we focus on genetic engineering. These chapters provide the necessary foundation for an exploration of the human genome in Chapter 15. (The new title of this chapter, "The Human Genome," reflects the chapter's greater emphasis on the molecular aspects of human genetics.) In Chapter 16, we introduce the role of genes in development, emphasizing studies on specific model organisms that have led to spectacular advances in this field.

Part 4: The continuity of life: evolution

Although evolution as the cornerstone of biology is explored throughout the book, this unit delves into the subject in-depth; we provide the history behind the discovery of the theory of evolution, the mechanism by which it occurs, and the methods by which it is studied and tested. Chapter 17 introduces Darwinian evolution and presents several kinds of evidence that support evolution. In Chapter 18, we examine evolution at the population level. Chapter 19 describes the evolution of new species and discusses aspects of macroevolution. Chapter 20 summarizes the evolutionary history of life on Earth. In Chapter 21 we recount the evolution of the primates, including humans. Many topics and examples have been added to the sixth edition, such as the role of chance in evolution, MN blood groups as an example of genetic equilibrium, the evolution of mammalian ear bones from reptilian jaw bones, and recent fossil discoveries of human ancestors.

Part 5: The diversity of life

As we revised Part 5 of *Biology*, we struggled to find a workable balance between the classical evolutionary systematics approach and the newer phylogenetic systematics (cladistics) approach. On the basis of input from reviewers, we decided that for this sixth edition of *Biology*, we would take a moderate stance, including both approaches in our discussions and in the art portraying phylogenetic relationships.

We use an evolutionary framework to discuss each group of organisms, presenting traditional and current hypotheses of how groups of organisms are related. By focusing on evolutionary relationships and the structural and functional adaptations of each group of organisms, we avoid the traditional parade through the phyla characteristic of many biology textbooks.

In Chapter 22, we discuss *why* organisms are classified and provide insight into the scientific process of deciding *how* they are classified. Chapter 23, devoted to the viruses and bacteria, compares the three domains and discusses viroids, prions, and emerging viruses. Chapter 24 describes the protists, including a new section on recent changes in the classification of protists, and Chapter 25 describes the fungi. Chapters 26 and 27 present the members of the plant kingdom. In Chapters 28 through 30, which cover the diversity of animals, we introduce recent classification approaches, including the subdivision of the protostomes into Lophotrochozoa and Ecdysozoa branches.

Part 6: Structure and life processes in plants

Part 6 introduces students to the fascinating organisms we know as plants. It stresses the relationships between structure and function in plant cells, tissues, organs, and individual organisms. In Chapter 31, plant structure, growth, and differentiation are introduced. Chapters 32 through 34 discuss the structural and physiological adaptations of leaves, stems, and roots. Chapter 35 describes reproduction in flowering plants, including asexual reproduction, flowers, fruits, and seeds. Chapter 36 focuses on growth responses and regulation of growth. The sixth edition strives to present the explosion of knowledge in plant biology, particularly at the molecular level, that has occurred in the last few years. Some of the new topics include the role of cell walls in signal transduction; the discovery of the gene involved in abscission zone development; the functions of calcium, nickel, sodium, and silicon in plants; self-incompatibility as a mechanism to prevent self-pollination; and phytochrome mutants in *Arabidopsis*.

Part 7: Structure and life processes in animals

In Part 7, we emphasize the structural, functional, and behavioral adaptations that help animals meet environmental challenges. We use a comparative approach to examine how various animal groups have solved similar and diverse problems. In Chapter 37, we discuss the basic tissues and organ systems of the animal body, homeostasis, and how animals regulate their body temperature. Chapter 38 focuses on body coverings, skeletons, and muscles. In Chapters 39 through 41, we discuss neural signaling, neural regulation, and sensory reception. In Chapters 42 through 49, we compare how different animal groups carry on specific life processes, such as internal transport, internal defense, gas exchange, digestion, reproduction, and development. Each chapter in this part considers the human adaptations for the life processes being discussed. The unit ends with a discussion of behavioral adaptations in Chapter 50.

Research on receptors and signal transduction continues to provide knowledge of how many animal functions are carried out on a cellular level. Because we are convinced that this information is best delivered within a context, many research findings on receptors and signal transduction have been integrated throughout Part 7 (as well as in other relevant chapters in the book). Reflecting recent research findings, we have added new material on neural plasticity, the neurophysiology of learning, the role of dendritic cells in immune responses, DNA vaccines, mechanisms of human immunodeficiency virus infection, mechanisms of hormone action, and reproductive physiology.

Part 8: The interactions of life: ecology

Part 8 focuses on the dynamics of populations, communities, and ecosystems and on the application of ecological principles to disciplines such as conservation biology. Chapters 51 through 54 give the student an understanding of the ecology of populations, communities, ecosystems, and the biosphere, whereas the final chapter (55) focuses on environmental problems caused by humans. In the sixth edition, we have made a concerted effort to interweave ecological theory and the scientific process by giving clear, concrete examples of ecological studies to illustrate conceptual points. Some of the topics new to Part 8 include Connell's research on barnacles; experiments at the Hubbard Brook Experimental Forest; Vitousek's study of negative human effects on the global nitrogen cycle; Wallace and Meyers' research on aquatic–terrestrial linkages; and an overview of the hundreds of recent research studies on declining amphibian populations.

A COMPREHENSIVE PACKAGE FOR LEARNING AND TEACHING

To further facilitate learning and teaching, a carefully designed supplement package is available. In addition to the usual print resources, we are pleased to present student multimedia tools that have been developed in conjunction with the text.

Additional Resources for Students

- **New! Interactive Student CD.** Packaged with the textbook at no additional charge, this CD delivers interactive exercises for each of the 55 chapters in the book, as well as "mini-simulations" for each part. ("Mini-simulations" simulate the laboratory by allowing the student to establish parameters and predict outcomes.) Many of these interactive exercises and mini-simulations are supported by animations derived from art that is in the text; text art is also the basis for many of the chapter-specific exercises. Icons placed in the margins of the text notify the student that a tutorial or simulation on the student CD addresses the topic that the student is reading in the text.
- **Study Guide to Accompany *Biology,* Sixth Edition,** by Ronald S. Daniel of California State Polytechnic University, Pomona; Sharon C. Daniel of Orange Coast College; and Ronald L. Taylor. This study guide provides the student with extensive opportunities to review chapter concepts. Multiple-choice study questions, coloring book exercises, vocabulary-building exercises, and many other types of active-learning tools are provided to suit different cognitive learning styles.
- ***A Problem-Based Guide to Basic Genetics*** by Donald Cronkite of Hope College. This brief guide provides students with a systematic approach to solving genetics problems, along with numerous solved problems and practice problems.
- ***Process of Science: Discovering Biology.*** This interactive CD-ROM allows students to explore the discoveries of some of the most important concepts in biology. In the *Interactive Investigations* mode, students retrace the steps of scientists' experiments and discoveries using the scientific process as a road map. An *Investigator's Notepad* allows students to track their progress through each investigation, pose new questions to pursue, or initiate a discussion with the instructor or other students via an Internet connection. *Concept Tutorials* provide students with essential background information in general biology for the course and the investigations.

Additional Resources for the Entire Biology Community

- **Web site.** The Web site to accompany *Biology*, sixth edition, gives students access to additional quizzes through the QT (Quizzing and Testing) feature and additional activities to support each chapter. Professors can access WebTutor, a course-management tool; the instructor's manual; the image bank; the PowerPoint™ lecture slides; and the test bank through this Web site as well.

Additional Resources for Instructors

- **The *Instructor's Resource Manual,*** by Lisa Shimeld of Crafton Hills College, features lecture outlines, research and discussion topics, teaching suggestions, and additional suggested readings. A section in each chapter called "Connections and Resources" helps instructors integrate all of the supplements offered with *Biology* into their teaching programs.
- **PowerPoint™ Lectures,** by Anthony Moss of Auburn University, are available. A slide presentation to accompany each chapter in the book is available on the instructor's resource side of the *Biology*, sixth edition, Web site (http://www.info.brookscole.com/solomonbergmartin).
- **The *Test Bank,*** by Wendy Ryan of Kutztown University of Pennsylvania, includes approximately 2200 multiple-choice questions, as well as more than 150 short-answer questions.

Many of the "fact recall" multiple-choice questions from previous editions have been replaced with multiple-choice questions that require higher-order thinking on the part of the student. *The Computerized Test Bank*, available for both Windows™ and Macintosh® platforms, houses these questions and has a feature that allows automatic test generation.

- **The *Instructor's Resource CD-ROM*** provides instructors with all the line art from the text in formats suitable for hard-copy printouts (e.g., on paper or on acetate), incorporation into PowerPoint™ presentations, and incorporation into Web sites. The CD-ROM is available for both Windows™ and Macintosh® platforms.
- A set of 250 **Overhead Transparencies** is available. The images in this set, chosen by the authors based on their own classroom experience, have been manipulated to ensure maximum visibility.
- **WebCT** is an Internet course-management tool that allows professors to create course-specific Web sites. Using a simple point-and-click interface, anyone familiar with today's Internet browsers can create a fully functional Web site that can include timed on-line quizzes, image archives, bulletin boards, chat rooms, an on-line syllabus and lecture notes, and more.
- ***Plain Talk About Using the Web in Teaching*** by Arthur L. Buikema, Jr., and Brian Ward of Virginia Polytechnic Institute and State University, is a basic handbook on intelligently designing a course that fully integrates computer technology and the World Wide Web.
- **A Laboratory Manual** written by Russell V. Skavaril, Mary M. Finnen, and Steven M. Lawton, all of Ohio State University, and an accompanying **Laboratory Instructor's Manual** are available. Also available is a **Laboratory Manual** written by Carolyn Eberhard of Cornell University. Additionally, professors may build a laboratory manual customized to their students' needs using our custom publication service.
- ***Handbook on Teaching Undergraduate Science Courses: A Survival Training Manual,*** by Gordon E. Uno of the University of Oklahoma, is a complete guide for new instructors or those who are looking for new ideas, and it is the ideal, everything-you-need-to-know companion for teaching assistants.
- ***Student-Active Science: Models of Innovation in College Science Teaching,*** edited by Ann P. McNeal and Charlene D'Avanzo of Hampshire College, is a collection of articles by college faculty and administrators from across the country who came together to attend a conference on inquiry-based science teaching that was sponsored by the National Science Foundation. At the forefront of innovative biology teaching, these attendees offer insights on how to encourage student activities that are investigative and collaborative, and that foster higher-order thinking skills.
- ***Thinking Toward Solutions: Problem-Based Learning Activities in General Biology,*** by Deborah Allen and Barbara Duch of the University of Delaware, presents complex, open-ended problems that cover all aspects of biology from cells to ecology. Addressing real-world applications of biology with questions of ethics, economics, and daily living, the problems foster critical thinking, cooperative learning, and problem-solving skills. An accompanying **Instructor's Manual** provides detailed, practical suggestions on how to use the problems in any size course and with a variety of teaching styles. **Thinking Toward Solutions** also has an Internet component, which offers links to resources that assist students in solving each problem.

Introductory biology is an exciting yet demanding course. It is important to have a text that is comprehensive, yet readable and that emphasizes (and re-emphasizes) key concepts. It is our hope that our love for and excitement about biology is clearly conveyed and that *Biology*, sixth edition, will prove to be an invaluable resource for students beginning their study of this remarkable science.

ELDRA P. SOLOMON

LINDA R. BERG

DIANA W. MARTIN

Acknowledgments

The development and production of the sixth edition of *Biology* required extensive interaction and cooperation among the authors and many individuals in our home and professional environments. We appreciate the valuable input and support from editors, colleagues, students, family, and friends.

We are grateful to the editorial and production staff at the erstwhile Harcourt College Publishers for their help and support throughout this project. We thank our publisher, Emily Barrosse, and Executive Editor, Nedah Rose, for their support, enthusiasm, and ideas. Our Developmental Editors Melanie Cann and Anne Scanlan-Rohrer efficiently guided the project through the revision process, providing us with many thoughtful reviews and useful suggestions.

We are grateful to our very talented Art Editor, Elizabeth Morales, for her new and exciting ideas for reconceptualizing and improving the art. We thank Photo Editor Nancy Lubars for helping us find outstanding photographs to enhance the text. We thank freelance journalist Mary Catherine Hager, who interviewed recent biology graduates and wrote the excellent career interviews for this edition.

We appreciate the help of our Project Editor, Bonnie Boehme, who expertly guided the project through the complexities of production. We thank Senior Art Director Paul Fry for coordinating the art program and design. All of these dedicated professionals and many others at Harcourt provided the skill and attention needed to produce *Biology*, sixth edition. We thank them for their help and support throughout this project.

We thank our families and friends for their understanding, support, and encouragement as we struggled through many revisions and deadlines. We especially thank Dr. Amy Solomon, Dr. Kathleen M. Heide, Alan and Jennifer Berg, and Dr. Charles Martin and Margaret Martin for their input and support. We greatly appreciate the expert help of Mary Kay Hartung of Florida Gulf Coast University, who (whenever we had difficulty finding resources or ran out of time) rapidly located and sent us needed research studies from the Internet. We also thank immunologist Dr. Susan Pross of the University of South Florida, College of Medicine, who has reviewed the immunology chapter for every edition of *Biology*.

Our colleagues and students who have used our book have provided valuable input by sharing their responses to the fifth edition of *Biology*. We thank them and ask again for their comments and suggestions as they use this new edition. We can be reached through the Internet (http://www.info.brookscole.com/solomonbergmartin) or through our editors at Brooks/Cole, a division of Thomson Learning. We express our thanks to the many biologists who have read the manuscript during various stages of its development and provided us with valuable suggestions for improving it. Sixth edition reviewers include:

Julie H. Aires *(Florida Community College, Jacksonville),* David E. Alexander *(University of Kansas),* Susan Bandoni Muench *(State University of New York, Geneseo),* James Barron *(Ohio University, Lancaster),* Lisa G. Bates *(Florida Community College at Jacksonville),* David Begun *(University of Texas at Austin),* Nancy Boury *(Iowa State University),* Robert S. Boyd *(Auburn University),* Arthur L Buikema, Jr. *(Virginia Polytechnic Institute and State University),* Ruth Buskirk *(University of Texas at Austin),* Warren R. Buss *(University of Northern Colorado),* David M. Carlberg *(California State University, Long Beach),* Robert E. Cleland *(University of Washington),* Jim Colbert *(Iowa State University),* Linda T. Collins *(University of Tennessee at Chattanooga),* Mark S. Condon *(Dutchess Community College),* Amy Cook *(East Carolina University),* Rebecca A. Cook *(Lambuth University),* Elizabeth A. Cowles *(Eastern Connecticut State University),* James T. Cronin *(University of North Dakota),* Kenneth J. Curry *(University of Southern Mississippi),* Anne Cusic *(University of Alabama at Birmingham),* Stan Dalton *(Jones County Community College),* Peter J. Davies *(Cornell University),* Jonathan R. Day *(Pennsylvania State University),* John V. Dean *(DePaul University),* Daniel V. DerVartanian *(University of Georgia),* Laura DiCaprio *(Ohio University),* David Dilcher *(University of Florida),* Ernest F. DuBrul *(University of Toledo),* Janice Edgerly-Rooks *(Santa Clara University),* David H. Evans *(University of Florida),* Robert C. Evans *(Rutgers University, Camden),* Dale Fast *(St. Xavier University),* Craig S. Feibel *(Rutgers University),* Steven K. Fisher *(University of California, Santa Barbara),* Kathy Foltz *(University of California, Santa Barbara),* Bruce Fowles *(Colby College),* James French *(Rutgers University),* Gary Galbreath *(Northwestern University),* Daniel L. Gebo *(Northern Illinois University),* George W. Gilchrist *(Clarkson University),* Gene Godbold *(University of Alabama, Huntsville),* Wayne Goodey *(University of British Columbia),* H. Thomas Goodwin *(Andrews University),* John S. Graham *(Bowling Green State University),* Thaddeus A. Grudzien *(Oakland University),* Jeff Hardin *(University of Wisconsin, Madison),* Michael B. Harvey *(East Tennessee State University),* Wiley Henderson *(Alabama A&M University),* Donna Hoefner *(Delaware County Community College),* Luke Holbrook *(Rowan University),* Robert Holmberg *(Athabasca University),* Joan E. N. Hudson *(Sam Houston State University),* Alice C. Jacklet *(State University of New York, Albany),* Paul Jarrell *(Pasadena City College),* Thomas C. Kane *(University of Cincinnati),* Alan J. Katz *(Illinois State University),* Joanne M. Kilpatrick *(Auburn University at Montgomery),* Loren W. Knapp *(University of South Carolina),* Dan E. Krane *(Wright State University),* James B. Kring *(Roane State Community College),* Zhi-Chun Lai *(Pennsylvania State University),* Joe Leverich *(St. Louis University),* Graeme Lindbeck *(Valencia Community College),* Marion B. Lobstein *(Northern Virginia Community College, Manassas Campus),* Jennifer Lundmark *(California State University, Sacramento),* Ronald H. Matson *(Kennesaw State University),* Tim McDowell *(East Tennessee State University),* James E. Mickle *(North Carolina State University),* Anthony Moss *(Auburn University),* Alison M. Mostrom *(University of the Sciences in Philadelphia),* Alan Muchlinski *(California State University, Los Angeles),* Darrel L. Murray *(University of Illinois at Chicago),* Patrick M. Muzzall *(Michigan State University),* William H. Nelson *(Morgan State University),* Dan Nickrent *(Southern Illinois University),* Stephen F. Norton *(East Carolina University),* Daniel M. Pavuk *(Bowling Green State University),* David Pennock *(Miami University of Ohio),* Chris E. Petersen *(College of DuPage),* Gregory W. Phillips *(Blinn College, Brenham),* Richard E. Phillips *(University of Minnesota),* Shirley Porteus-Gafford *(Fresno City College),* Susan Pross *(University of South Florida College of Medicine),* Jerry Purcell *(San Antonio College),* Eric Ribbens *(Western Illinois University),* Laurel Roberts *(University of Pittsburgh),* Wayne Rowley *(Iowa State University),* Lori S. Rynd *(Pacific University),* Mimi M. Sayed *(Michigan State University),* Louis A. Scala *(Monmouth University),* Karen Schumaker *(University of Arizona),* David Seigler *(University of Illinois),* Mark Sheridan *(North Dakota State University),* James Siedow *(Duke University),* Deborah Smith *(University of Kansas),* Nancy G. Solomon *(Miami University of Ohio),* Bruce Stallsmith *(University of Alabama, Huntsville),* Martha L. Stauderman *(University of San Diego),* Marshall D. Sundberg *(Emporia State University),* Walter K. Taylor *(University of Central Florida),* Sylvia D. Torti *(University of Utah),* Nathan Tublitz *(University of Oregon),* Mary S. Tyler *(University of Maine),* Thomas C. Vogelmann *(University of Wyoming),* Fred Wasserman *(Boston University),* Judy Watts *(Cleveland State Community College),* Michael Weber *(Carleton University),* Mary E. White *(Southeastern Louisiana University),* Stephen Yazulla *(State University of New York, Stony Brook),* Robert Yost *(Indiana University/Purdue University at Indianapolis)*

Reviewers of Previous Editions

Faculty

Dawn Adams *(Baylor University),* James Adams *(Dalton College),* John H. Adler *(Michigan Technical University),* Surinder Aggarwal *(Michigan State University),* Julie Aires *(Florida Community College, Jacksonville),* Venita Allison *(Southern Methodist University),* Sylvester Allred *(Northern Arizona University),* Jane Aloi *(Saddleback College),* Marvin Alvarez *(University of South Florida),* David Asch *(Youngstown State University),* Edward Ashworth *(Purdue University),* Sonya Baird *(University of Georgia),* Susan Bandoni *(State University of New York, Geneseo),* Penny Bauer *(Colorado State University),* Lester Bazinet *(Community College of Philadelphia),* Chris Beard *(Carnegie Museum of Natural History),* J.T. Beatty *(University of British Columbia),* Ed Bedecarrax *(City College of San Francisco),* David Begun *(University of Texas at Austin),* Vincent Bellis *(East Carolina University),* David Benner *(Eastern Tennessee State University),* Todd Bennethum *(Purdue University),* Gerald Bergtrom *(University of Wisconsin, Milwaukee),* Dorothy Berner *(Temple University),* Charles Biggers *(University of Memphis),* William L. Bischoff *(University of Toledo),* Del Blackburn *(Clark College),* Gary Booth *(Brigham Young University),* Nicole Bournias *(California State University at San Bernardino),* George Bowes *(University of Florida),* Barry Bowman *(University of California at Santa Cruz),* George Boyajian *(University of Pennsylvania),* Robert Boyd *(Auburn University),* Dean Bratis *(Delaware Community College),* W.H. Breazeale, Jr. *(Francis Marion College),* Anne Britt *(University of California, Davis),* William Brooks *(Florida Atlantic University),* George Brown *(Iowa State University),* Gary Brusca *(Humboldt State University),* Arthur Buikema, Jr. *(Virginia Tech),* Vicki Cameron *(Ithaca College),* David Carlberg *(California State University, Long Beach),* David Carr *(University at Maryland at College Park),* Barry Chess *(Pasadena City College),* W. Dennis Clark *(Arizona State University),* Keith Clay *(Indiana University),* William Cohen *(University of Kentucky),* Gary Cole *(University of Texas at Austin),* Bruce Condon *(Seattle Pacific University),* Mark Condon *(Dutchess Community College),* Joyce Corban *(Wright State University),* Jeffrey Corden *(The Johns Hopkins University),* Robert Cordero *(St. Joseph's University),* William Cordes *(Loyola University),* Harry O. Corwin *(University of Pittsburgh),* David Cotter *(Georgia College),* Kenneth Curry *(University of Southern Mississippi),* Anne Cusic *(University of Alabama, Birmingham),* Henry Daniell *(Auburn University),* Thomas Davis *(University of New Hampshire),* Jonathan Day *(Pennsylvania State University),* Patricia DeLeon *(University of Delaware),* Jean DeSaix *(University of North Carolina at Chapel Hill),* Laura DiCaprio *(Ohio University),* Stephen J. Dina *(St. Louis University),* Linda Dion *(University of Delaware),* Peter Dixon *(University of California, Irvine),* Penny Dobbins *(Syracuse University),* Andrew Dobson *(University of Rochester),* Warren Dolphin *(Iowa State University),* Rob Dorit *(Harvard University),* Lee C. Drickamer *(Williams College* and *Southern Illinois University at Carbondale),* Peter Ducey *(State University of New York, Cortland),* Inge Eley *(Hudson Valley Community College),* John Evans *(Memorial University of Newfoundland),* Robert Evans *(Rutgers University, Camden),* Sharon Eversman *(Montana State University),* Guy Farish *(Adams State College),* Millicent Ficken *(University of Wisconsin, Milwaukee),* Milton Fingerman *(Tulane University),* David Firmage *(Colby College),* Steven K. Fisher *(University of California, Santa Barbara),* Malinda Fitzgerald *(Christian Brothers University),* Jim Florini *(Syracuse University),* Bruce Fowles *(Colby College),* James French *(Rutgers University, New Brunswick),* David Fromson *(California State University, Fullerton),* Bernard Frye *(University of Texas at Arlington),* Douglas Gaffin *(University of Oklahoma),* Michael Gaines *(University of Kansas),* Gary Galbreath *(Northwestern University),* Darrell Galloway *(Ohio State University),* Dan Gebo *(Northern Illinois University),* Patricia Gensel *(University of North Carolina, Chapel Hill),* Robert P. George *(University of Wyoming),* Michael Ghedotti *(University of Kansas),* Florence Gleason *(University of Minnesota),* Elizabeth A. Godrick *(Boston University),* David Goldstein *(Wright State University),* Paul Goldstein *(University of Texas at El Paso),* Nels Granholm *(South Dakota State University),* Judith Goodenough *(University of Massachusetts, Amherst),* Wayne Goodey *(University of British Columbia),* Thomas Goodwin *(Andrews University),* Edward J. Greding, Jr. *(Del Mar College),* Katharine B. Gregg *(West Virginia Wesleyan College),* Peter Gregory *(Cornell University),* Floyd Grimm *(Harford Community College),* Mark Gromko *(Bowling Green University),* Thaddeus Grudzien *(Oakland University),* David Hale *(Texas A&M University),* Thomas Hanson *(Temple University),* Alexander Harcourt *(University of California at Davis),* Paul K. Hayes *(University of Bristol),* James Hayward *(Andrews University),* Steven Heidemann *(Michigan State University, East Lansing),* Jean Heitz *(University of Wisconsin at Madison),* Louis Held *(Texas Tech University),* Jean Helgeson *(Collin County Community College),* Charles Henry *(University of Connecticut),* Fritz Hertel *(University of California, Los Angeles),* Martinez Hewlett *(University of Arizona),* Linden Higgins *(University of Texas at Austin),* Andrew Hill *(Yale University),* Betsy Hirsch *(University of Minnesota),* Helmut Hirsch *(State University of New York, Albany),* Ricky Hirschorn *(Hood College),* Carl Hoagstrom *(Ohio Northern University),* Dan Hoffman *(Bucknell University),* Rebecca Holburton *(University of Mississippi),* Robert Holmberg *(Athabasca University),* Jill VanWort Hood *(University of Texas at Arlington),* Linda Hsu *(Seton Hall University),* Joan Hudson *(Sam Houston State University),* Stephen Hudson *(Furman University),* Pat Humphrey *(Ohio University),* Robert Hurst *(Purdue University),* Gerard Iwantsch *(Fordham University),* Alice Jacklet *(State University of New York, Albany),* Mark Jacobs *(Swarthmore College),* Charles Janson *(State University of New York, Stony Brook),* Dan Johnson *(Eastern Tennessee State University),* Randal Johnston *(University of Calgary),* Claudia Jolls *(East Carolina University),* Kenneth Kardong *(Washington State University),* Richard Karp *(University of Cincinnati),* Glenn Kasparian *(Brookhaven Community College),* Donald Keefer *(Loyola College, Maryland),* Phil Keeting *(West Virginia University),* Tasneem Khaleel *(Montana State University),* M.A.Q. Khan *(University of Illinois, Chicago),* William Kimbel *(Institute of Human Origins),* Robert Kitchin *(University of Wyoming),* Ross Koning *(Eastern Connecticut State University),* Robert W. Korn *(Bellarmine College),* Dan E. Krane *(Wright State University),* William Kroen *(Wesley College),* Paul Kugrens *(Colorado State University),* Paul Lago *(University of Mississippi),* Vaughn A. Langman *(Louisiana State University),* Ralph Larson *(San Francisco State University),* Virginia Latta *(Jefferson State College),* Brenda Leicht *(University of Iowa),* Joe Leverich *(St. Louis University),* Harvey Lillywhite *(University of Florida),* Roger M. Lloyd *(Florida Community College, Jacksonville),* Miriam Lobstein *(Northern Virginia Community College),* Heather Lorimer *(Youngstown State University),* Victor Lotrich *(University of Delaware),* Karl Maddox *(Miami University of Ohio),* Sharook Madon *(Pace University, Westchester),* Charles Mallery *(University of Miami),* Arthur Mange *(University of Massachusetts, Amherst),* Dennis Matthews *(University of New Hampshire),* James Mauseth *(University of Texas at Austin),* Jeffrey May *(Marshall University),* Henry M. McHenry *(University of California, Davis),* Roger McMacken *(The Johns Hopkins University),* Michael Meighan *(University of California, Berkeley),* Lee Meserve *(Bowling Green State University),* Joseph Michalewicz *(Holy Family College),* Ann Mickle *(LaSalle University),* Roger Milkman *(University of Iowa),* Lillian Miller *(Florida Community College),* Charles Mims *(University of Georgia),* Manuel Molles

(University of New Mexico), Marion Monahan *(Immaculata College),* John Moner *(University of Massachusetts, Amherst),* Russell Monson *(University of Colorado, Boulder),* Darrell Moore *(East Tennessee State University),* James Moore *(University of California, San Diego),* Robert E. Moore *(Montana State University),* Edward Morgan *(Temple Junior College, Texas),* Michael Morgan *(University of Wisconsin, Green Bay),* Jim Morrone *(Louisiana State University),* Anthony G. Moss *(Auburn University),* Alison Mostrom *(Philadelphia College of Pharmacy and Science),* Debbie Mueler *(Cardinal Stritch College),* James Murray *(University of Virginia),* John Murray *(University of Pennsylvania),* Richard Myers *(Southwest Missouri State University),* Thomas L. Naples *(Delaware Community College),* William H. Nelson *(Morgan State University),* Anne Penney Newton *(Temple Junior College, Texas),* Frank G. Nordlie *(University of Florida),* David O. Norris *(University of Colorado, Boulder),* Gary Ogden *(Moorpark College),* Carolyn Ogren *(Parkland College),* John Olsen *(Rhodes College),* Beulah Parker *(North Carolina State University),* Glenn R. Parsons *(University of Mississippi),* Robert Patterson *(San Francisco State University),* Greg Paulson *(Shippensburg University),* Jerome Perry *(North Carolina State University),* Greg Phillips *(Blinn Bryan College),* Ronald Phillips *(Seattle Pacific University),* Ruth Pitkin *(Shippensburg University),* Thomas Pitzer *(Florida International University),* Jeanne S. Poindexter *(The Public Health Research Institute of the City of New York, Inc.),* Dave Polcyn *(California State University at San Bernardino),* Trevor Price *(University of California, San Diego),* Susan Pross *(University of South Florida),* Jerry Purcell *(San Antonio College),* James Pushnik *(Chico State University),* Richard Racusen *(University of Maryland),* Peggy Redshaw *(Austin College),* Arthur Repak *(Quinnipiac College),* Florence Ricciuti *(Albertus Magnus College),* Robert Roberson *(Arizona State University),* Laurel Roberts *(University of Pittsburgh),* Martin Roeder *(Florida State University),* Rodney A. Rogers *(Drake University),* John Romeo *(University of South Florida),* Marvin J. Rosenberg *(California State University, Fullerton),* Jean Saillant *(University of Michigan, Dearborn),* Ted Sargent *(University of Massachusetts, Amherst),* Mimi A. Sayed *(Michigan State University),* Carl Schlicting *(Pennsylvania State University, University Park),* John A. Schmidt *(Ohio State University),* Edward Schneider *(University of California, Santa Barbara),* Janet L. Schottel *(University of Minnesota),* Brian Schwartz *(Columbus State University),* Kathleen Scott *(Rutgers University, New Brunswick),* William A. Searcy *(University of Miami),* Duane W. Sears *(University of California, Santa Barbara),* David Seigler *(University of Illinois),* Mary Colavito Shepanski *(Santa Monica College),* Mark Sheridan *(North Dakota State University),* Lisa Shimeld *(Crafton Hills College),* David Shomay *(University of Illinois, Chicago),* Jane Shoup *(Purdue University at Calumet),* J. Kenneth Shull, Jr. *(Appalachian State University),* Paul Small *(Eureka College),* Bruce Smith *(Brigham Young University),* Dennis M. Smith *(Wellesley College),* Phillip Snider *(University of Houston),* Richard C. Snyder *(University of Washington),* Bruce Stallsmith *(University of Massachusetts, Boston),* Karen Steudel *(University of Wisconsin, Madison),* Charles L. Stevens *(University of Pittsburgh),* Robert Stockhouse *(Pacific University),* Gerald Summers *(University of Missouri),* Marshall Sundberg *(Louisiana State University),* Daryl Sweeney *(University of Illinois at Urbana, Champaign),* Chris Tarp *(Contra Costa College),* Walter Taylor *(University of Central Florida),* Robert M. Timm *(University of Kansas),* Ian Tizard *(Texas A&M University),* Kenneth Thomulka *(Philadelphia College of Pharmacy),* John Tudor *(St. Joseph's University),* Gordon Uno *(University of Oklahoma),* Frederick Utech *(Carnegie Museum of Natural History),* John Utley *(University of New Orleans),* Joseph W. Vanable *(Purdue University),* Steve Vessey *(Bowling Green State University),* Darrel Vodopich *(Baylor University),* Jack Waber *(West Chester State University),* Charles Walcott *(Cornell University),* Elizabeth Waldorf *(Mississippi Gulf Community College),* Eileen Walsh *(Westchester Community College),* Fred Wasserman *(Boston University),* Jacqueline F. Webb *(Villanova University),* David Whetstone *(Jacksonville State University, Alabama),* Matthew White *(Ohio University),* John Whitmarsh *(University of Illinois, Urbana),* Donald Whitmore *(University of Texas, Arlington),* Varley Wiedeman *(University of Louisville),* Leslie Williams *(Mountain View College),* David Wilson *(University of Miami),* Lawrence Winship *(Hampshire College),* Dwayne Wise *(Mississippi State University),* Steven Woeste *(Scholl College),* Clarence Wolfe *(North Virginia Community College),* Drew H. Wolfe *(Hillsborough Community College),* David Woodruff *(University of California, San Diego),* Roger Young *(Texas A&M University),* John L. Zimmerman *(Kansas State University),* William Zimmerman *(Amherst College)*

Students

Felisha Avery *(Mountain View College, Texas Women's University),* Leslie Baker *(Eastfield College, Texas Tech University),* Sreddevi Chittineni *(University of Delaware),* Chris Churchman *(Furman University, North Lake Community College),* Beverly Cimino *(Montgomery County Community College),* Alan Cohen *(University of Michigan, Ann Arbor),* Christopher David Colson *(Ohio University),* Jana A. Damphousse *(Oakland University),* Karen Davis *(Tarrant County Junior College),* Anjali Cherise D'Souza *(North Lake Community College, University of Texas at Austin),* Estelle S. D'Souza *(University of Toledo),* Kimberly Dunham *(University of Delaware),* Katharine Edmund *(University of Michigan, Ann Arbor),* Cory Fajardo *(East Carolina University),* Michael A. Fox *(North Lake Community College, Parker Chiropractic College),* Michele France *(University of Delaware),* Hannah A. M. Gilkenson *(University of Michigan, Ann Arbor),* Shuaib A. Gill *(Wayne State University),* Beth Glaze *(Orange Coast College),* Lindsay Goodman *(Eastfield College, University of Texas at Austin),* Kelly B. Hall *(North Lake Community College, Texas A&M University),* Cory Hinchman *(Orange Coast College),* Stacy Hirth *(Ohio University),* Shelli Hornberger *(Mountain View College, Stephen F. Austin University),* William DeVaughn Hunt *(East Carolina University),* Rebekah L. Hunter *(Eastern Michigan University),* James Patrick Jarvis *(East Carolina University),* Jennifer Ray Jones *(North Lake Community College, Texas Tech University),* Jeffrey L. Kacsandi *(Ohio University),* Michael Kane *(Delaware County Community College),* Jenny Kerekles *(University of Michigan, Ann Arbor),* Evelyn Knox *(East Carolina University),* Elizabeth Kucera *(Ohio University),* Mike Tien Minh Le *(Orange Coast College),* Nancy Lee *(Orange Coast College),* Charles W. Luce *(East Carolina University),* Nasser Mahaud *(Delaware County Community College),* Emedio Marchozzi *(Montgomery County Community College),* Joe Matthews *(Delaware County Community College),* Glenda McCourt *(Delaware County Community College),* Marianna J. McSweeney *(University of Delaware),* Beth L. Measamer *(Brookhaven College),* Jami Miller *(Ohio University),* Kimberly S. Miller *(Eastfield Community College, Abilene Christian University),* Michael Jason Miller *(East Carolina University),* Ngoc Quang Nguyen *(University of North Texas, Texas Women's University),* Mark Nolan *(University of Texas at Arlington),* Anthony Orlando *(University of Michigan, Ann Arbor),* Stacy Peebles *(Tarrant County Junior College),* Rick Poce *(Delaware County Community College),* Mary A. Radlick *(Oakland University),* Cynthia Rainey *(Texas A&M University),* Tanga M. Ray *(University of Delaware),* Gus Reese *(University of Texas at Arlington),* Renee Sandora *(Eastern Michigan University),* Jennifer Schklair *(Mountain View Community College),* Heather Slater *(Montgomery County Community College),* Katherine Strafford *(Ohio University),* Theresa Tidd *(Delaware County Community College),* Nicholas John Urbanczyk *(University of Michigan, Dearborn),* Travis Vaughn *(University of Delaware),* Jill Wauldron *(Oakland University),* Irene Wedderien *(Orange Coast College),* Alex M. Zadeh *(Orange Coast College)*

We would also like to thank the Introductory Biology Students at Ohio University and Montgomery County Community College.

To The Student

Biology is one of the most varied and challenging subjects one can study. The thousands of students we have taught have differed in their life goals and learning styles. Some have had excellent backgrounds in science, others poor ones. Regardless of their backgrounds, it is common for students taking their first college biology course to find that they must work harder than they expected. You can make the task easier by using approaches to learning that have been successful for a broad range of our students over the years.

MAKE A STUDY SCHEDULE

Many college professors suggest that students study three hours for every hour spent in class. This major investment in study time is one of the main differences between high school and college. To succeed academically, college students must learn how to manage their time effectively. The actual number of hours you spend studying biology will vary depending on your course load and personal responsibilities, such as work schedules and family commitments.

The most successful students are often those who are best organized. It is helpful to make a detailed daily calendar at the beginning of the semester. Mark off the hours that you are in each class, along with transportation time to and from class if you are a commuter. After you have received course syllabi, add to your calendar the dates of all exams, quizzes, papers, and reports. It also helps to add an entry one week before each major exam or assignment so it does not catch you unaware. Now add your work schedule and other personal commitments to your calendar. Making a calendar helps you find convenient study times.

Many of our successful biology students set aside two hours a day to study biology, as opposed to a weekly marathon session for eight or ten hours during the weekend (it rarely happens). Put your study hours into your daily calendar, and make every effort to adhere to your schedule.

DETERMINE WHETHER THE PROFESSOR EMPHASIZES TEXT MATERIAL OR LECTURE NOTES

Some professors test almost exclusively on material covered in lecture, whereas others rely on their students' learning most, or even all, of the content in assigned chapters. Try to ascertain from your professor what his or her requirements are, because the way you study will vary according to how and on what you are being tested.

How to study when professors test lecture material

If lectures are the main source of examination questions, your lecture notes must be as complete and organized as possible. Before going to class, skim over the chapter, identifying key terms and examining the main figures, so that you will be able to take effective lecture notes. You should spend no more than one hour on this.

You should rewrite (or type) your notes within 24 hours after class. Before rewriting, however, read the notes and make marginal notes about anything that is not clear. Then read the corresponding material in your text. Highlight or underline any sections that clarify questions you had in your notes. Read the *entire* chapter, including parts that are not covered in lecture. This extra information will give you breadth and help you form key concepts.

After reading the text, you are ready to rewrite your notes, incorporating relevant material from the text. It also helps to use the Glossary to help define unfamiliar terms.

Many students think that developing a set of flash cards of key terms and concepts is a good way to study, but in our experience, flash cards produce mixed results. They are not effective when the student tries to second-guess the professor. ("She won't ask this, so I won't make a flash card of it.") Flash cards are also a hindrance when they are relied on exclusively. Studying flash cards instead of reading the text is a bit like reading the first page of each chapter in a mystery novel. It's hard to fill in the missing parts because you are learning the facts in an unconnected way. Flash cards are a useful tool, however, to help you learn scientific terminology. They are portable and can be used at times when other studying is not possible, for example, when riding a bus.

How to study when professors test material in the book

If the assigned readings in the text are going to be tested, you must use your text intensively. After reading the chapter introduction, read through the list of Learning Objectives for the chapter. These objectives are written in behavioral terms; that is, you must "do" something in order to demonstrate mastery. The objectives provide you with a concrete set of goals for the chapter. You will revisit the Learning Objectives later in the study process.

Read the chapter *actively*. Many students read and study passively. An active learner always has questions in mind and is constantly making connections. For example, there are many processes that must be understood in biology. Do not try to blindly memorize these; instead, think about causes and effects, so that every process becomes a story. Eventually you will see that many processes are connected by common elements.

You will probably have to read each chapter two or three times before mastering the material. The second and third times through will be much easier than the first because you will be reinforcing concepts that you have already partially learned.

After reading the chapter, you should write a four- to six-page chapter outline by using the subheads as the body of the outline (first-level heads are boldface and all caps; second-level heads are boldface but not all caps). Flesh out your outline by adding important concepts and boldface terms with definitions. You will use this outline when preparing for the exam.

Now it is time to test yourself. Answer the Post-Test questions, and check your answers in the Appendix. Write answers to each of the Review questions and "You Make the Connection" questions. Finally, go back to the Learning Objectives and try to answer them. If your professor has told you that some or all of the exam will be short-answer or essay format, you should write the answer for each Learning Objective. Remember that this is a self-test. If you do not know an answer to a question, find it in the text. If you can't find the answer, use the Index.

LEARN THE VOCABULARY

One stumbling block for many students is the necessity of learning a great deal of terminology. In fact, it would be much more difficult to learn and communicate if we did not have this terminology, for words are really tools for thinking. Learning terminology generally becomes easier if you realize that most biological terms are modular. They are composed of mostly Latin and Greek roots; once you learn many of these, you will find you have a good idea of the meaning of a new word even before it is defined. For this reason, we have included an Appendix on Understanding Biological Terms. To be sure you understand the precise definition of a term, you will want to use the Index and Glossary. The more you use biological terms in speech and writing, the more comfortable you will be.

FORM A STUDY GROUP

Active learning is facilitated if you do some of your studying in a small group. In a study group, the roles of teacher and learner should be interchangeable: the best way to be sure you understand is to teach. A study group allows you to be challenged in a nonthreatening environment and can provide some emotional support. Study groups are effective learning tools when combined with individual study of text and lecture notes. If, however, you and other members of your study group have not prepared for your meetings by studying individually in advance, the study session can be a waste of time.

PREPARE FOR THE EXAM

Your calendar tells you it is now one week before your first biology exam. If you have been following these suggestions, you are well prepared and will need only some last-minute reviewing. No all-nighters will be required.

During the week prior to the exam, spend two hours each day actively studying your lecture notes or chapter outlines. It helps many students to read these notes out loud (most people listen to what they say!). Begin with the first lecture/chapter covered on the exam, and continue in the order on the lecture syllabus. Stop when you have reached the end of your two-hour study period. The following day, begin where you stopped the previous day. When you reach the end of your notes, start at the beginning and study them a second time. The material should be *very* familiar to you by the second or third time around. At this stage, use your textbook only to answer questions or clarify important points.

The night before the exam, do a little light studying, eat a nutritious dinner, and get a full night's sleep. You will arrive in class on exam day with a well-rested body (and brain) and the self-confidence that goes with being well prepared.

Contents Overview

Contents

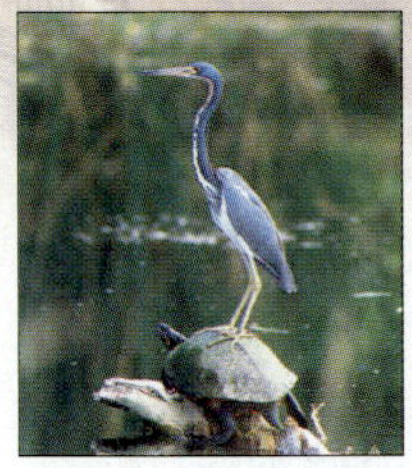

PART 1
The Organization of Life 1

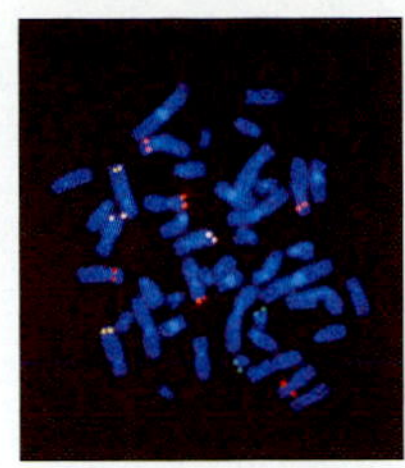

PART 3
The Continuity of Life: Genetics *197*

PART 6
Structure and Life Processes in Plants 673

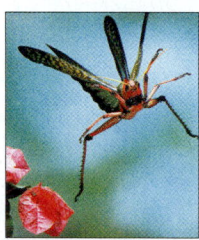

PART 8
The Interactions of Life: Ecology 1135

Part

1

Biologists study the details of living organisms and all of their interactions. The Louisiana heron *(Hydranassa tricolor)* and the Florida red-bellied turtle *(Chrysemys nelsoni)* on which it has perched are complex organisms that make their home in the Florida cypress swamp. *(Mark J. Thomas/Dembinsky Photo Associates)*

The Organization of Life

Chapters

1

A View of Life

Human genome research. Researchers carrying out DNA sequencing at the Joint Genome Institute (JGI). The JGI, a collaboration between three of the U.S. Department of Energy's National Laboratories, is part of the international Human Genome Organization (HUGO) that aims to complete the mapping of the entire human genome, the 3 billion base pairs that form the genetic code, by 2003. HUGO will provide the information needed to discover the genetic cause of many diseases. Computers in the laboratory sit between the large automated sequencers. *(David Parker/Science Photo Library/Photo Researchers, Inc.)*

LEARNING OBJECTIVES

After you have studied this chapter you should be able to

1. Define biology and discuss its applications to human life and society.
2. Distinguish between living and nonliving things by describing the features that characterize living organisms.
3. Relate metabolism to homeostasis and give specific examples of these life processes.
4. Summarize the importance of information transfer to living systems, giving specific examples.
5. Construct a hierarchy of biological organization, including levels of an individual organism and ecological levels.
6. Demonstrate the binomial system of nomenclature using several specific examples, and classify an organism (e.g., a human) in its domain, kingdom, phylum, class, order, family, genus, and species.
7. Contrast the six kingdoms of living organisms and cite examples of each group.
8. Give a brief overview of the theory of evolution and explain why it is the principal unifying concept in biology.
9. Apply the theory of natural selection to any given adaptation, suggesting a logical explanation of how the adaptation may have evolved.
10. Contrast the roles of producers, consumers, and decomposers and cite examples of their interdependence.
11. Design an experiment to test a given hypothesis, using the procedure and terminology of the scientific method.

This is an exciting time to begin your study of **biology,** the science of life. Almost daily, biologists are making remarkable new discoveries about ourselves—the human species—and about the millions of other organisms with which we share our planet. Applications of this basic research have provided us with the technology to transplant kidneys, livers, and hearts, manipulate genes, treat many diseases, and increase world food production. Biology has been a powerful force in providing us with the quality of life that most of us enjoy.

Whatever your college major or career goals, a knowledge of biological concepts is a vital tool for understanding our world and for meeting many of the personal, societal, and global challenges that confront us. Among these challenges are the expanding human population, diminishing natural resources, decreasing biodiversity, and diseases such as cancer, Alzheimer's disease, malaria, and acquired immunodeficiency syndrome (AIDS). Meeting these challenges will require the combined efforts of biologists and other scientists, politicians, and biologically informed citizens. The debate over genetically modified (GM) crops is an example of the importance of consumer understanding. Scientists have promoted the development of GM crops as a means of increasing yields to meet the needs of growing populations. However, citizens in many countries are fearful of these new methods. In India, farmers recently destroyed fields of GM cotton. In the United States, fear of these new products has kept some off grocery shelves. Consumer demands will affect the course of future research.

Two other remarkable areas of biological investigation, genome research and stem cell studies, have potential applications that could transform health care and many other aspects of our lives during the 21st century. The **Human Genome Project** is mapping the complete set of genes that make up human genetic material (the genome). In 2000 a group of publicly funded scientists led by Francis Collins and a team of privately funded researchers led by J. Craig Venter announced that each had mapped and completed a working draft of the human genome. Their work has been heralded as a brilliant achievement, a big step toward deciphering the "book of life." Locating the genes is just the first step, however. Scientists need to refine what has been done and determine which genes do what and how they function. Detailed analyses of other primate genomes must also be carried out. The Human Genome Project also encompasses the next level of research, studying the proteins for which the genes code. Proteins make up the structural framework of an organism and also serve as enzymes that regulate the biochemical reactions essential to life. Genome research will contribute to the new science of gene therapy and will provide knowledge that will be applied to treatment, and even prevention, of many human disorders.

Stem cells are another exciting focus of biological research. **Stem cells** are cells that are unspecialized and have the capacity to divide, giving rise to more stem cells *and* to one or more specialized types of cell. Stem cells permit the body to repair both normal wear and tear and injury. For example, stem cells in the skin continuously divide, and some differentiate to replace skin cells that are constantly being worn off. Similarly, stem cells in the lining of the intestine reproduce themselves to replace cells that wear out. Stem cells in the bone marrow differentiate to restore the various types of blood cells. Humans and other mammals have an estimated 20 major types of stem cells.

Some tissues in the body, for example the heart, the brain, and the islets of the pancreas that produce insulin, have few or no stem cells and cannot effectively repair themselves. This is why research on embryonic stem (ES) cells is important. ES cells can give rise to all of the types of cells in the body. If ES cells could be cultured in the laboratory and then implanted in damaged hearts, brains, spinal cords, or livers, they could potentially provide the cells needed for repairing injuries or for reversing the ravages of disease processes. Some researchers are confident that stem cells will eventually be used to treat disorders like Parkinson's disease, Alzheimer's disease, multiple sclerosis, diabetes, and spinal cord injury.

In the 1970s and 1980s biologists worked out techniques for culturing ES cells from mouse embryos. Human ES cells were first grown in laboratory cultures in 1998 by two different research teams, one headed by James A. Thomson at the University of Wisconsin and the other led by John D. Gearhart of Johns Hopkins Medical Institutions. Gearhart's team obtained human ES cells from aborted embryos, whereas Thomson's group obtained human ES cells from very early embryos produced in fertility clinics. (After embryos are produced in laboratory glassware and one is implanted in a patient, the unused embryos can be frozen for future implantations, discarded, or donated for this type of ES cell research.) The ethical questions posed by new technology in genetics and developmental biology represent one group of issues that demand a biologically informed society.

The technology being developed for stem cell research is related to the techniques used in cloning animals. **Cloning** is the process of producing genetically identical cells or organisms by asexual reproduction of a single cell or organism. Cloning techniques have been combined with genetic engineering procedures to produce transgenic animals that can produce needed human proteins such as blood clotting factors. Cloning can also be used to improve livestock, to produce animal tissues and organs for transplant into humans, and to reproduce endangered species, pulling them back from the brink of extinction. For example, in 2000 an Iowa cow gave birth to the first cloned endangered species, a baby gaur (an oxlike animal native to India and southeast Asia). In the future, cloning may be used to save cheetahs, giant pandas, ocelots, and many other animals from extinction. In this way cloning can be an important tool in maintaining world biodiversity. The future applications of genetic research, stem cell studies, and cloning research have the potential to enhance our health and well-being and the well-being of our planet.

LIFE DEPENDS ON EVOLUTION, TRANSMISSION OF INFORMATION, AND ENERGY

This book is a starting point for your exploration of biology. It will provide you with the basic knowledge and the tools to become a part of this fascinating science and a more informed member of society. In this first chapter we introduce three basic themes of biology: **evolution of life, transmission of information,** and **flow of energy through living systems.** Scientists have accumulated a wealth of evidence showing that the diverse life forms on our planet are related and that organisms have evolved through time from earlier life forms. The process of evolution is the framework for the science of biology and a major theme of this book.

The process of evolution, as well as the survival and function of every organism, depends on the orderly transmission of information. At the molecular level, instructions for producing and maintaining each living organism and each new generation are encoded in the **deoxyribonucleic acid (DNA)** molecules that make up the genes. At higher levels, the activities of organisms are coordinated by many forms of chemical signaling. Animals also use chemical, as well as behavioral, signals to communicate with one another. For example, many female animals release chemical substances that attract males.

Energy is required to maintain the precise order that characterizes living systems. Maintaining the chemical transactions and cellular organization essential to life requires a continuous input of energy. We begin our study of biology by developing a more precise understanding of the fundamental characteristics of living systems.

LIFE CAN BE DEFINED IN TERMS OF THE CHARACTERISTICS OF ORGANISMS

We can easily recognize that an oak tree, a butterfly, and a lamb are living, whereas a rock is not. Despite their diversity, the organisms that inhabit our planet share a common set of characteristics

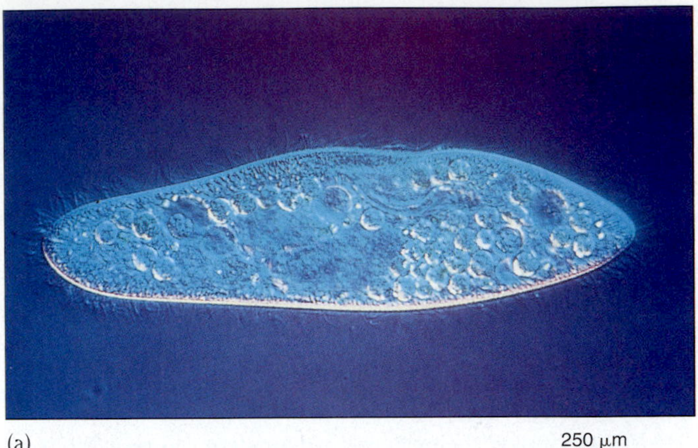

Figure 1–1 Unicellular and multicellular life forms. **(a)** Unicellular organisms are generally smaller than multicellular organisms and consist of one intricate cell that performs all the functions essential to life. Ciliates, such as this *Paramecium,* move about by beating their hairlike cilia. **(b)** Multicellular organisms, such as this African buffalo *(Syncerus caffer)* and the plants on which it grazes, may consist of billions of cells specialized to perform specific functions. *(a, Mike Abbey/Visuals Unlimited; b, McMurray Photography)*

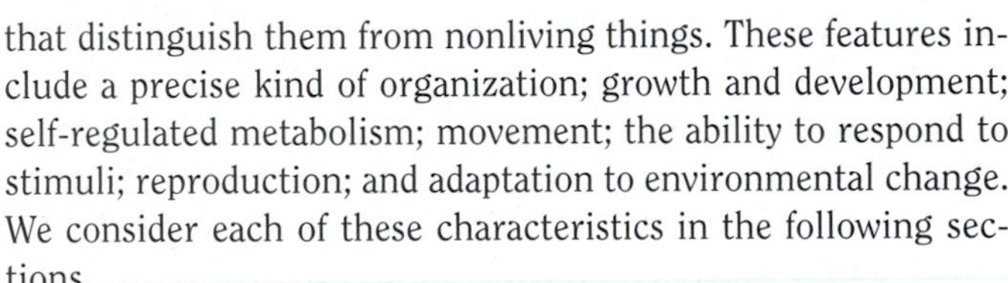
that distinguish them from nonliving things. These features include a precise kind of organization; growth and development; self-regulated metabolism; movement; the ability to respond to stimuli; reproduction; and adaptation to environmental change. We consider each of these characteristics in the following sections.

Organisms are composed of cells

Living organisms are highly organized, and, as discussed later in this chapter, we can identify a hierarchy of biological organization. As expressed in the **cell theory,** one of the fundamental unifying concepts of biology, all organisms are composed of basic units called **cells.** Three German scientists are credited with the cell theory. Matthias Schleiden (in 1838) and Theodor Schwann (in 1839) were the first to report that plants and animals consist of groups of cells. Later, physician Rudolph Virchow observed cells dividing and giving rise to daughter cells. In 1855, Virchow proposed that new cells are formed only by the division of previously existing cells. In other words, cells do not arise by spontaneous generation from nonliving matter, a belief that had prevailed for centuries.

Although they vary greatly in size and appearance, all organisms are composed of these small building blocks. Some of the simplest life forms, such as protozoa, are *unicellular,* meaning that each consists of a single cell (Fig. 1–1). In contrast, the body of a lamb or a maple tree is made of billions of cells. In such complex *multicellular* organisms, life processes depend on the coordinated functions of component cells that may be organized to form tissues, organs, and organ systems.

Organisms grow and develop

Some nonliving things appear to grow. For example, salt crystals form and enlarge in a supersaturated salt solution. However, this is not growth in the biological sense. **Biological growth** may involve an increase in the size of individual cells of an organism or in the number of cells, or both (Fig. 1–2). Growth may be uni-

Figure 1–2 Biological growth. The young African elephant *(Loxodonta africana)* eats and grows until it reaches the adult size of its parents, the largest living land animals. These elephants were photographed in Kenya. *(McMurray Photography)*

form in the various parts of an organism, or it may be greater in some parts than in others, causing the body proportions to change as growth occurs.

Some organisms, most trees, for example, continue to grow throughout their lives. Many animals have a defined growth period that terminates when a characteristic adult size is reached. One of the remarkable aspects of the growth process is that each part of the organism typically continues to function as it grows.

Living organisms develop as well as grow. **Development** includes all the changes that take place during the life of an organism. Each human, like many other organisms, begins life as a fertilized egg that then grows and develops. The structures and body form that develop are exquisitely adapted to the functions the organism must perform.

Organisms regulate their metabolic processes

Within all organisms, chemical reactions and energy transformations occur that are essential to nutrition, growth and repair of cells, and conversion of energy into usable forms. The sum of all the chemical activities of the organism is its **metabolism.** Metabolic processes occur continuously in every living organism, and they must be carefully regulated to maintain **homeostasis,** a balanced internal state. When enough of some cellular product has been made, its manufacture must be decreased or turned off. When a particular substance is needed, cellular processes that produce it must be turned on. These **homeostatic mechanisms** are self-regulating control systems that are remarkably sensitive and efficient.

The regulation of glucose (a simple sugar) concentration in the blood of complex animals is a good example of a homeostatic mechanism. Most cells require a constant supply of glucose, which they break down to obtain energy. The circulatory system delivers glucose and other nutrients to all the cells. When the concentration of glucose in the blood rises above normal limits, it is stored in the liver and in muscle cells. When the concentration begins to fall (between meals), stored nutrients are converted to glucose so that the concentration in the blood returns to normal levels. When glucose becomes depleted, we also feel hungry and restore nutrients by eating.

Movement is a basic property of cells

Organisms move as they interact with the environment, and in fact the living material within their cells is in continuous motion. In some organisms, locomotion results from the slow oozing of the cell (a process called *amoeboid motion*). Other organisms beat tiny hairlike extensions of the cell called **cilia** or longer structures called **flagella** (Fig. 1–3). Some bacteria move by means of rotating flagella.

Most animals move very obviously. They wiggle, crawl, swim, run, or fly by contracting muscles. A few animals, such as sponges, corals, and oysters, have free-swimming larval stages but do not move from place to place as adults. Even though these adults, described as **sessile,** remain firmly attached to some surface, they may have cilia or flagella. These structures beat rhythmically, moving the surrounding water that brings food and other necessities to the organism.

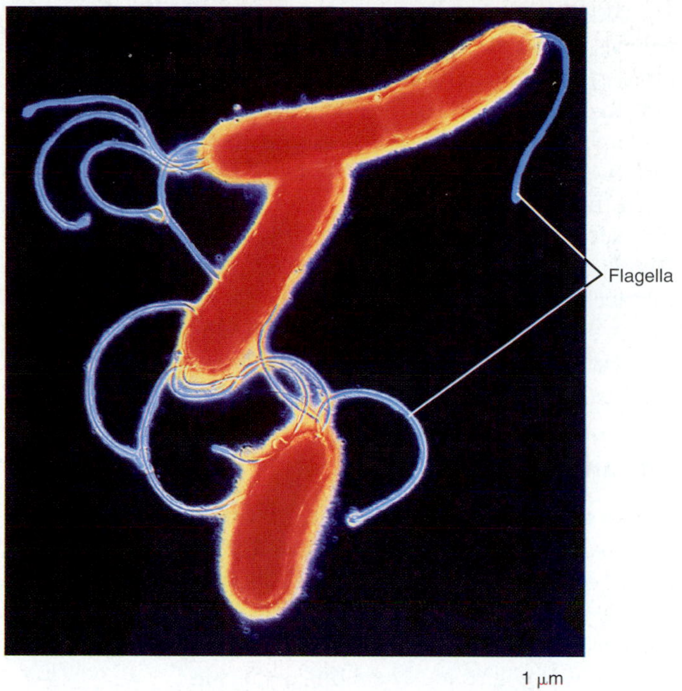

Figure 1–3 Biological movement. These bacteria *(Helicobacter pylori)*, equipped with flagella for locomotion, have been linked to stomach ulcers. The photograph is a color-enhanced scanning electron micrograph. *(A.B. Dowsett/Science Photo Library/Photo Researchers, Inc.)*

Although plants move more slowly than most animals, they do move. For example, plants orient their leaves to the sun and grow toward light. In some plants, such as the Venus flytrap, movement is obvious, even dramatic (described in the next section).

Organisms respond to stimuli

All forms of life respond to **stimuli,** physical or chemical changes in their internal or external environment. Stimuli that evoke a response in most organisms are changes in the color, intensity, or direction of light; changes in temperature, pressure, or sound; and changes in the chemical composition of the surrounding soil, air, or water.

In simple organisms, the entire individual may be sensitive to stimuli. Certain unicellular organisms, for example, respond to bright light by retreating. In complex animals such as polar bears and humans, certain cells of the body are highly specialized to respond to specific types of stimuli. For example, cells in the retina of the eye respond to light.

Although their responses may not be as obvious as those of animals, plants do respond to light, gravity, water, touch, and other stimuli. Many plant responses involve different rates of growth of various parts of the plant body. A few plants, such as the Venus flytrap of the Carolina swamps (Fig. 1–4), are remarkably sensitive to touch and can catch insects. Their leaves are hinged along the midrib, and they have a scent that attracts insects. Trigger hairs on the leaf surface detect the arrival of an insect and stimulate the leaf to fold. When the edges come together, the hairs interlock, preventing escape of the prey. The leaf then secretes enzymes that kill and digest the insect. The Venus flytrap is usually found in soil that is deficient in nitrogen. The plant obtains part of the nitrogen required for its growth from the insect prey it "eats."

Organisms reproduce

At one time worms were thought to arise spontaneously from horsehair in a water trough, maggots from decaying meat, and frogs from the mud of the Nile. Thanks to the work of several scientists, including Francesco Redi in the 17th century and Louis Pasteur in the 19th century, we now know that an organism can come only from previously existing organisms.

In simple organisms such as amoebas, reproduction may be **asexual,** that is, without the fusion of egg and sperm to form a fertilized egg (Fig. 1–5). When an amoeba has grown to a certain size, it reproduces by splitting in half to form two new amoebas. Before an amoeba divides, its hereditary material (set of genes) duplicates, and one complete set is distributed to each new cell. Except for size, each new amoeba is similar to the parent cell. The only way that variation occurs among asexually reproducing organisms is by genetic **mutation**, a permanent change in the genes.

In most plants and animals, **sexual reproduction** is carried out by the production of specialized egg and sperm cells that fuse to form a fertilized egg. The new organism develops from the fertilized egg. Offspring produced by sexual reproduction are the product of the interaction of various genes contributed by the mother and the father. Such genetic variation provides raw material for the vital processes of evolution and adaptation.

Populations evolve and become adapted to the environment

The ability of a population to evolve (change over time) and adapt to its environment enables it to survive in a changing world. **Adaptations** are traits that enhance an organism's ability to survive in a particular environment. They may be structural, physiological, behavioral, or a combination of all three (Fig. 1–6). The long, flexible tongue of the frog is an adaptation for catching insects, the feathers and lightweight bones of birds are adaptations for flying, and the thick fur coat of the polar bear is an adaptation for surviving frigid temperatures. Every biologically successful organism is a complex collection of coordinated adaptations produced through evolutionary processes.

(a)

(b)

Figure 1–4 Plants respond to stimuli. **(a)** Hairs on the leaf surface of the Venus flytrap (*Dionaea muscipula*) detect the touch of an insect, and the leaf responds by folding. **(b)** The edges of the leaf come together and interlock, preventing the fly's escape. The leaf then secretes enzymes that kill and digest the insect. *(David M. Dennis/Tom Stack & Associates)*

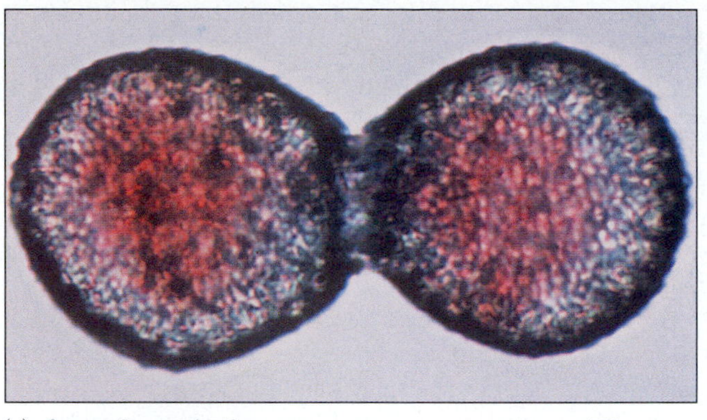

(a) Asexual reproduction

(b) Sexual reproduction

Figure 1–5 Asexual and sexual reproduction. **(a)** Asexual reproduction in *Difflugia,* a unicellular amoeba. In asexual reproduction, one individual gives rise to two or more offspring that are similar to the parent. **(b)** A pair of tropical flies mating. In sexual reproduction, typically two parents each contribute a gamete (sperm or egg). Gametes fuse to produce the offspring, which is a combination of the traits of both parents. *(a, Visuals Unlimited/Cabisco; b, L.E. Gilbert, Biological Photo Service)*

Figure 1–6 Adaptations. These Burchell's zebras *(Equus burchelli),* photographed at Ngorongoro Crater in Tanzania, are behaviorally adapted to position themselves to watch for lions and other predators. Stripes are thought to be an adaptation for visually protecting themselves against predators. They serve as camouflage or to break up form when spotted from a distance. The zebra stomach is adapted for feeding on coarse grass passed over by other grazers, an adaptation that helps it survive when food is scarce. *(McMurray Photography)*

INFORMATION MUST BE TRANSMITTED WITHIN AND BETWEEN INDIVIDUALS

For an organism to grow, develop, carry on self-regulated metabolism, move, respond, and reproduce, it must have precise instructions. The information an organism needs to carry on these life processes is coded and delivered in the form of chemical substances and electrical impulses.

DNA transmits information from one generation to the next

Humans give birth only to human babies, not to giraffes or rose bushes. In organisms that reproduce sexually, each offspring is a combination of the traits of its parents. In 1953, James Watson and Francis Crick worked out the structure of deoxyribonucleic acid, more simply known as DNA. This chemical substance makes up the **genes,** the units of hereditary material. The work of Watson and Crick led to the understanding of the genetic code that transmits information from generation to generation. This code works somewhat like our alphabet; it can spell an amazing variety of instructions for making organisms as diverse as bacteria, frogs, and redwood trees. The genetic code is a dramatic example of the unity of life because it is used to specify instructions for making every living organism (Fig. 1–7).

Information is transmitted by chemical and electrical signals

Genes control the development and functioning of every organism. DNA contains the "recipes" for making all the **proteins** needed by the organism. Proteins are large molecules that are important in determining the structure and function of cells and tissues. Brain cells are different from muscle cells in large part because they have different types of proteins. Some proteins are important in communication within and among cells. Certain proteins on the surface of a cell serve as markers so that other cells "recognize" them. Other cell surface proteins serve as receptors that combine with chemical messengers.

Cells use proteins and many other types of ions and molecules to communicate with one another. In a multicellular

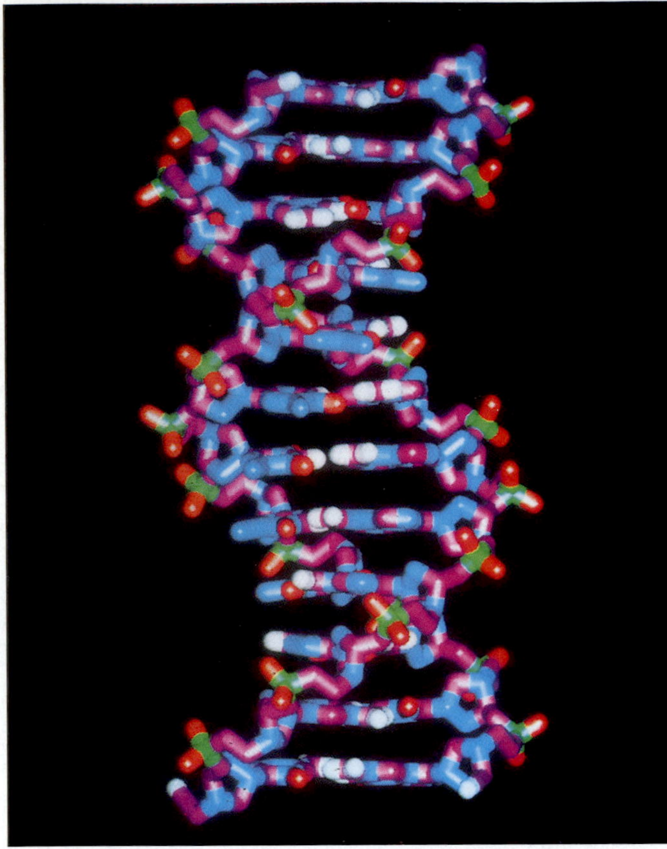

Figure 1–7 DNA. An organism's ability to transmit information from one generation to the next is essential to the continuity of life. In all organisms, the hereditary material is DNA. This computer graphic of a small (12 base-pair) segment of DNA shows its double-helix configuration. The DNA molecule consists of two chains of atoms twisted into a helix. Each chain consists of an outer sugar-phosphate backbone from which nucleotide bases project. The bases are paired in a complementary way. The sequence of bases makes up the genetic code. The atoms are color-coded as follows: carbon, blue; oxygen, red; nitrogen, purple; phosphorus, green; hydrogen, white. *(Jon Wilson/ Science Photo Library/Photo Researchers, Inc.)*

organism, chemical compounds secreted by cells help regulate growth, development, and metabolic processes in other cells. The mechanisms involved in **cell signaling** are complex, often involving multistep biochemical sequences, and cell signaling is currently an area of intense research. A major focus has been the transfer of information among cells of the immune system. A better understanding of how cells communicate promises new insights into how the body protects itself against disease organisms. Learning to manipulate cell signaling may lead to new methods of delivering drugs into cells and new treatments for cancer and other diseases. Examples of cell signaling are discussed throughout this book.

Hormones are molecules that function as chemical messengers that transmit information from one part of an organism to another. A hormone can signal cells to produce or secrete a certain protein or other substance.

Many organisms use electrical signals to transmit information. Most animals have nervous systems that transmit information by way of both electrical impulses and chemical compounds known as **neurotransmitters.** Information transmitted from one part of the body to another is important in regulating life processes. In complex animals, the nervous system transmits signals from sensory receptors such as the eyes and ears to the brain, giving the animal information about its outside environment.

Information must also be transmitted from one organism to another. Mechanisms for this type of communication include release of chemicals, visual displays, and sounds. Typically, organisms use a combination of several types of communication signals. For example, a dog may signal aggression by growling, using a particular facial expression, and positioning its ears back. Many animals perform complex courtship rituals in which they display parts of their bodies, often elaborately decorated, to attract a mate.

BIOLOGICAL ORGANIZATION IS HIERARCHICAL

Whether we study a single complex organism or the world of life as a whole, we can identify a hierarchy of biological organization (Fig. 1–8). At every level, structure and function are precisely coordinated. One way to study a particular level is by looking at its components. For example, biologists can learn about cells by studying atoms and molecules. Learning about a structure by studying its parts is known as **reductionism.** However, the whole is more than the sum of its parts. Each level has **emergent properties,** characteristics not found at lower levels. For example, populations have emergent properties such as population density, age structure, and birth and death rates. The individuals that make up a population lack these characteristics.

Organisms have several levels of organization

The **chemical level,** the most basic level of organization, includes atoms and molecules. An **atom** is the smallest unit of a chemical element (fundamental substance) that retains the characteristic properties of that element. For example, an atom of iron is the smallest possible amount of iron. Atoms combine chemically to form **molecules.** Two atoms of hydrogen combine with one atom of oxygen to form a single molecule of water. Although composed of two types of atoms that are gases, water is a liquid with very different properties, an example of emergent properties.

At the **cellular level** many different types of atoms and molecules associate with one another to form cells. However, a cell is much more than a heap of atoms and molecules. Its emergent properties make it the basic structural and functional unit of life, the simplest component of living matter that can carry on all the activities necessary for life. Every cell is surrounded by a **plasma membrane** that regulates the passage of materials between the cell and its surrounding environment. All cells have specialized

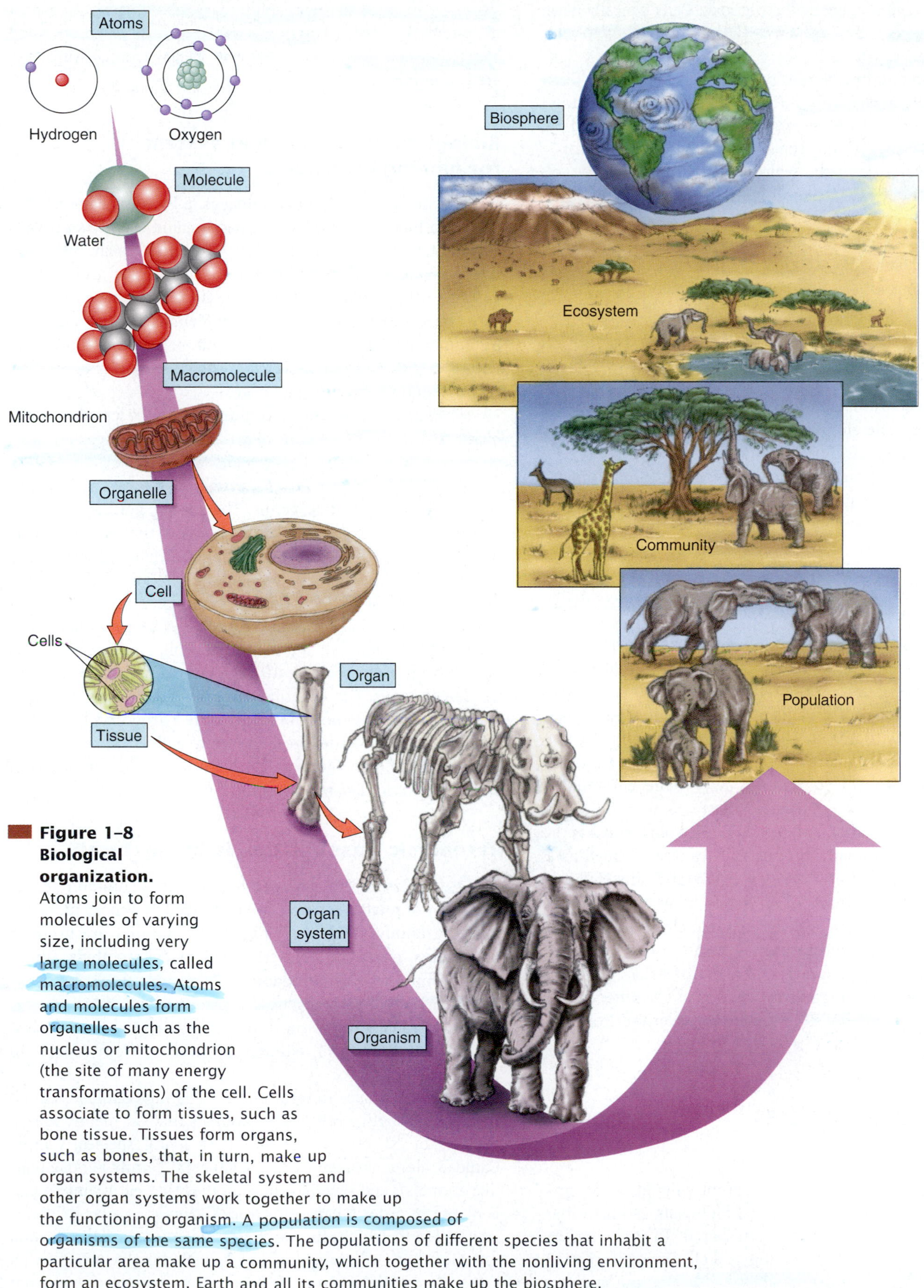

Figure 1–8 Biological organization. Atoms join to form molecules of varying size, including very large molecules, called macromolecules. Atoms and molecules form organelles such as the nucleus or mitochondrion (the site of many energy transformations) of the cell. Cells associate to form tissues, such as bone tissue. Tissues form organs, such as bones, that, in turn, make up organ systems. The skeletal system and other organ systems work together to make up the functioning organism. A population is composed of organisms of the same species. The populations of different species that inhabit a particular area make up a community, which together with the nonliving environment, form an ecosystem. Earth and all its communities make up the biosphere.

molecules that contain genetic instructions. Cells typically have internal structures called **organelles** that are specialized to perform specific functions.

Two fundamentally different types of cells are known. Bacteria are **prokaryotic cells.** Their cells lack membrane-bounded organelles. All other organisms are characterized by their **eukaryotic cells.** Unlike the structurally simpler prokaryotic cells, eukaryotic cells typically contain a variety of membrane-bounded organelles, including a **nucleus** that houses DNA.

During the evolution of multicellular organisms, cells associated to form **tissues.** For example, most animals have muscle tissue and nervous tissue, and plants have epidermis, a tissue that serves as a protective covering. In most complex organisms, tissues organize into functional structures called **organs,** such as the heart and stomach in animals and roots and leaves in plants. In animals, each major group of biological functions is performed by a coordinated group of tissues and organs called an **organ system.** The circulatory and digestive systems are examples of organ systems. Functioning together with great precision, organ systems make up a complex, multicellular **organism.** Again, emergent properties are evident. An organism is much more than its component organ systems.

Several levels of ecological organization can be identified

Organisms interact to form still more complex levels of biological organization. All the members of one species that live in the same geographic area at the same time make up a **population.** The populations of organisms that inhabit a particular area and interact with one another form a **community.** A community can consist of hundreds of different types of organisms. As populations within a community evolve, the community changes.

A community together with its nonliving environment is referred to as an **ecosystem.** An ecosystem can be as small as a pond (or even a puddle) or as vast as the Great Plains of North America or the Arctic tundra. All of Earth's ecosystems together are known as the **biosphere.** The biosphere includes all of Earth that is inhabited by living organisms—the atmosphere, the hydrosphere (water in any form), and the lithosphere (Earth's crust). The study of how organisms relate to one another and to their physical environment is called **ecology** (derived from the Greek *oikos,* meaning "house").

MILLIONS OF SPECIES HAVE EVOLVED

About 1.8 million species of extant (currently living) organisms have been scientifically identified, and biologists estimate that several million more remain to be discovered. To study life, we need a system for organizing, naming, and classifying its myriad forms. **Systematics** is the field of biology that studies the diversity of organisms and their evolutionary relationships. An aspect of systematics, called **taxonomy,** is the science of naming and classifying organisms. Biologists who specialize in classification are **taxonomists.**

Biologists use a binomial system for naming organisms

In the 18th century Carolus Linnaeus, a Swedish botanist, developed a hierarchical system of naming and classifying organisms that, with some modification, is still used today. The basic unit of classification is the **species.** A species is a group of organisms with similar structure, function, and behavior; in nature they breed only with each other. Members of a species have a common gene pool and share a common ancestry. Closely related species are grouped together in the next higher level of classification, the **genus** (pl. *genera*).

The Linnaean system of naming species is referred to as the **binomial system of nomenclature** because each species is assigned a two-part name. The first part of the name is the genus, and the second part, the **specific epithet,** designates a particular species belonging to that genus. The specific epithet is often a descriptive word expressing some quality of the organism. It is always used together with the full or abbreviated generic name preceding it. The generic name is always capitalized; the specific epithet is generally not capitalized. Both names are always italicized or underlined. For example, the dog, *Canis familiaris* (abbreviated *C. familiaris*), and the timber wolf, *Canis lupus (C. lupus),* belong to the same genus. The cat, *Felis catus,* belongs to a different genus. The scientific name of the American white oak is *Quercus alba,* whereas the name of the European white oak is *Quercus robur*. Another tree, the white willow, *Salix alba,* belongs to a different genus. The scientific name for our own species is *Homo sapiens* ("wise man").

Taxonomic classification is hierarchical

Just as species may be grouped together in a common genus, a number of related genera can be grouped in a more inclusive group, a **family** (Table 1–1; Fig. 1–9). Families are grouped into **orders,** orders into **classes,** and classes into **phyla.** Biologists group phyla into **kingdoms**, and kingdoms are assigned to **domains.** Each formal grouping at any given level is a **taxon.** Note that each taxon is more inclusive than the taxon below it. Together they form a hierarchy ranging from species to domain.

Consider a specific example. The family Canidae, which includes all doglike carnivores (animals that eat mainly meat), consists of 12 genera and about 34 living species. Family Canidae, along with family Ursidae (bears), family Felidae (catlike animals), and several other families that eat mainly meat, is placed in order Carnivora. Order Carnivora, order Primates (the order to which chimpanzees and humans belong), and several other orders belong to class Mammalia (mammals). Class Mammalia is grouped with several other classes that

TABLE 1–1 Classification of Domestic Cat, Human, and White Oak

Category	Cat	Human	White Oak
Domain	Eukarya	Eukarya	Eukarya
Kingdom	Animalia	Animalia	Plantae
Phylum	Chordata	Chordata	Anthophyta
Subphylum	Vertebrata	Vertebrata	None
Class	Mammalia	Mammalia	Dicotyledones
Order	Carnivora	Primates	Fagales
Family	Felidae	Hominidae	Fagaceae
Genus and specific epithet	*Felis catus*	*Homo sapiens*	*Quercus alba*

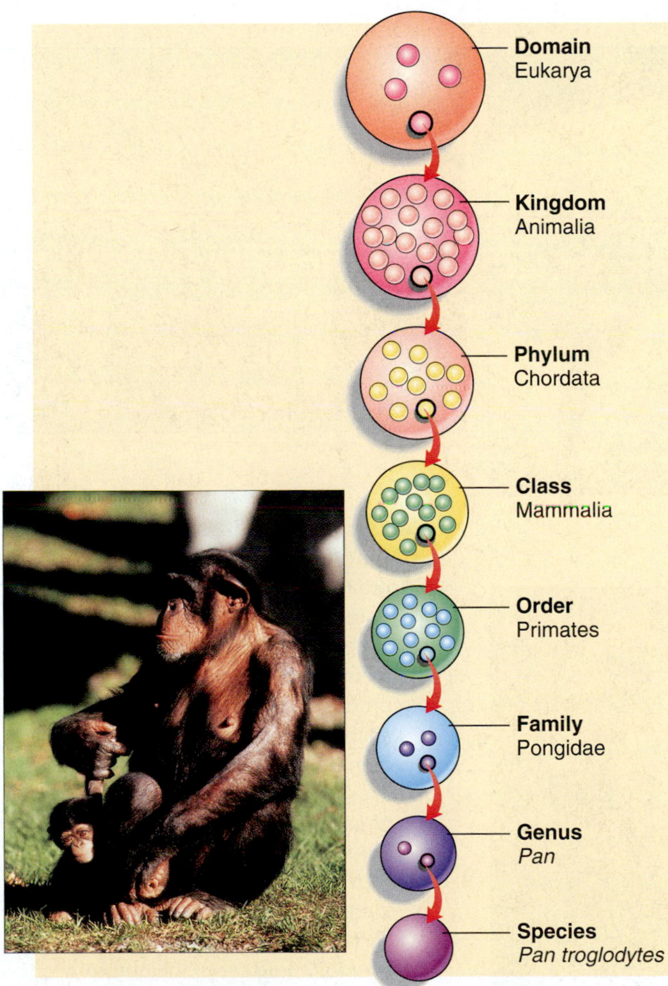

Figure 1–9 Classification of the chimpanzee *(Pan troglodytes)*. As illustrated by this example, the classification scheme used by biologists is hierarchical. In the series of taxonomic categories from species to domain, each category is more general and more inclusive than the one below it. *(T. Whittaker/ Dembinsky Photo Associates)*

include fishes, amphibians, reptiles, and birds in subphylum Vertebrata. The vertebrates belong to phylum Chordata, which is part of kingdom Animalia. Animals are assigned to domain Eukarya.

Organisms can be assigned to three domains and six kingdoms

Systematics has itself evolved as new molecular techniques have been developed. As researchers report new data, the classification of organisms changes. Although biologists do not agree on how organisms are related or on how to classify them, many biologists now assign organisms to three domains and six kingdoms.

Bacteria are unicellular prokaryotic cells; they differ from all other organisms in that they are **prokaryotes.** Two distinct groups have been recognized among the prokaryotes, and biologists assign them to two domains: **Archaea** and **Eubacteria** (also called *Bacteria*). The **eukaryotes**, organisms with eukaryotic cells, are classified in domain **Eukarya.**

In the system of classification used in this book, every organism is also assigned to one of six **kingdoms** (Fig. 1–10). Two kingdoms correspond to the prokaryotic domains: **Archaebacteria** corresponds to domain Archaea, and Eubacteria corresponds to domain Eubacteria. Kingdom **Protista** consists of protozoa, algae, water molds, and slime molds. These are unicellular or simple multicellular organisms. Some protists are adapted to carry out **photosynthesis,** the process in which light energy is converted to the chemical energy of food molecules. Kingdom **Fungi** is composed of the yeasts, mildews, molds, and mushrooms. These organisms do not photosynthesize. They obtain their nutrients by secreting digestive enzymes into food and then absorbing the predigested food.

Members of kingdom **Plantae** are complex multicellular organisms adapted to carry out photosynthesis. Among characteristic plant features are the *cuticle* (a waxy covering over aerial parts that reduces water loss), *stomata* (tiny openings in stems and leaves for gas exchange), and multicellular *gametangia* (organs that protect developing reproductive cells). Kingdom Plantae includes both nonvascular plants (mosses) and vascular plants (ferns, conifers, and flowering plants).

Kingdom **Animalia** is made up of multicellular organisms that must eat other organisms for nourishment. Complex animals exhibit considerable tissue specialization and body organization. These characters have evolved along with motility, complex sense organs, nervous systems, and muscular systems.

A more detailed presentation of the kingdoms can be found in Chapters 22 through 30, and classification is summarized in Appendix C. We refer to these groups repeatedly throughout this book as we consider the many kinds of challenges faced by living organisms and the various adaptations that have evolved in response to these challenges.

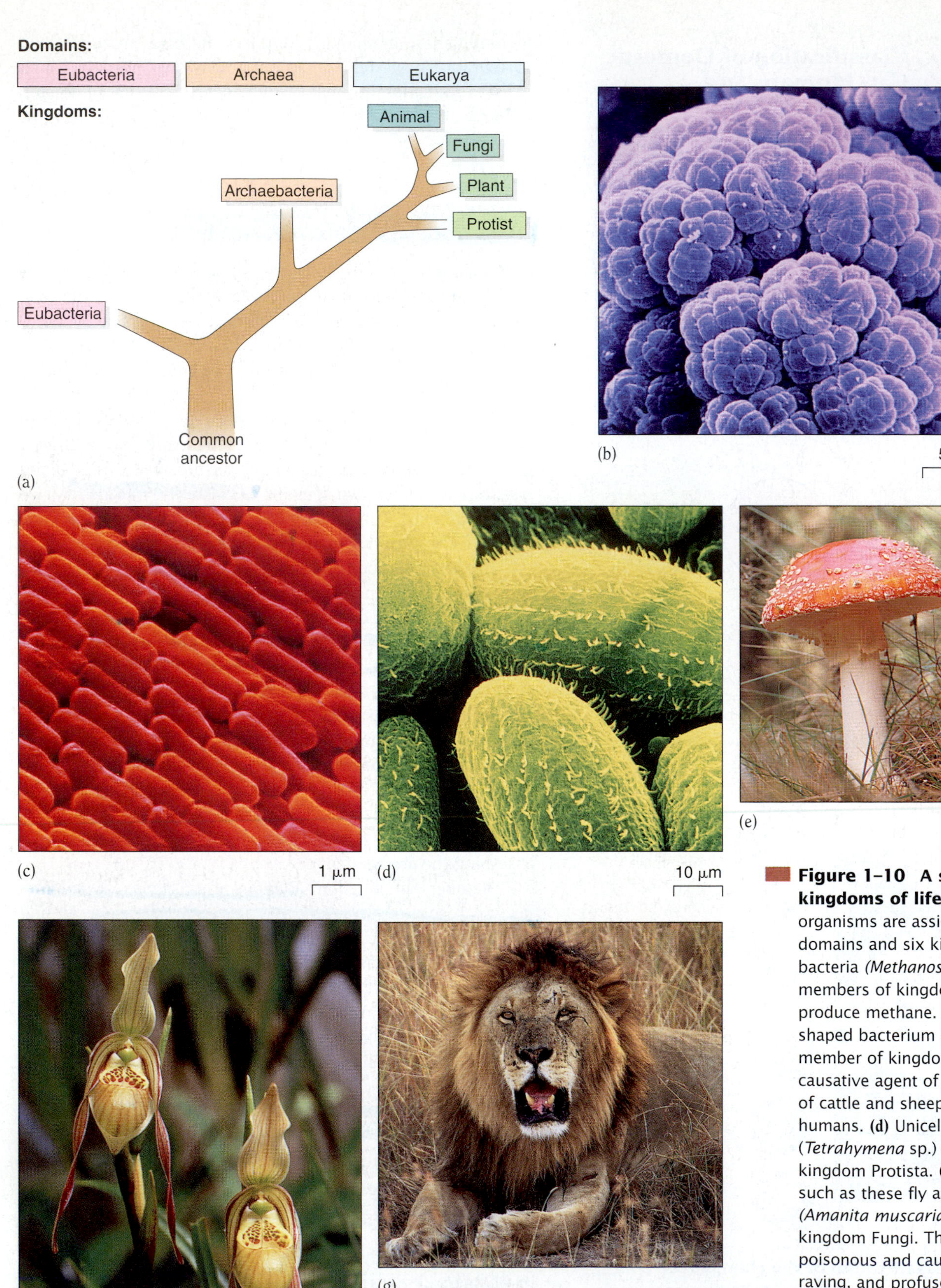

Figure 1–10 A survey of the kingdoms of life. **(a)** In this book organisms are assigned to three domains and six kingdoms. **(b)** These bacteria *(Methanosarcina mazei)*, members of kingdom Archaebacteria, produce methane. **(c)** The large rod-shaped bacterium *Bacillus anthracis*, a member of kingdom Eubacteria, is the causative agent of anthrax, a disease of cattle and sheep that can infect humans. **(d)** Unicellular protozoa (*Tetrahymena* sp.) are classified in kingdom Protista. **(e)** Mushrooms, such as these fly agaric mushrooms *(Amanita muscaria)*, belong to kingdom Fungi. The fly agaric is poisonous and causes delirium, raving, and profuse sweating when ingested. **(f)** The plant kingdom claims many beautiful and diverse forms such as the lady's slipper *(Phragmipedium caricinum)*. **(g)** Among the fiercest members of the animal kingdom, lions *(Panthero leo)* are also among the most sociable. The largest of the big cats, lions live in prides (groups). *(b, R. Robinson/Visuals Unlimited; c, CNRI/Science Photo Library/Photo Researchers, Inc.; d, David M. Phillips/Visuals Unlimited; e, Ulf Sjostedt/FPG International; f, John Arnaldi; g, McMurray Photography)*

EVOLUTION IS THE PRIMARY UNIFYING CONCEPT OF BIOLOGY

The theory of **evolution,** which explains how populations of organisms have changed over time, has become the greatest unifying concept of biology. Some element of an evolutionary perspective is present in every specialized field within biology. Biologists try to understand the structure, function, and behavior of organisms and their interactions with one another by considering them in light of the long, continuing process of evolution. Although evolution is discussed in depth in Chapters 17 through 21, we present a brief overview here to give you the background necessary to understand other aspects of biology.

Species adapt in response to changes in their environment

Every organism is the product of complex interactions between environmental conditions and the genes of its ancestors. If every individual of a species were exactly like every other, any change in the environment might be disastrous to all, and the species would become extinct. Adaptation to changes in the environment involves changes in populations rather than in individual organisms. Such adaptations are the result of evolutionary processes that occur over time and involve many generations.

Natural selection is an important mechanism by which evolution proceeds

Although the concept of evolution had been discussed by philosophers and naturalists through the ages, Charles Darwin and Alfred Wallace first brought the theory of evolution to general attention and suggested a plausible mechanism, **natural selection,** to explain it. In his book *The Origin of Species by Natural Selection,* published in 1859, Darwin synthesized many new findings in geology and biology. He presented a wealth of evidence that the present forms of life descended, with modifications, from previously existing forms. Darwin's book raised a storm of controversy in both religion and science, some of which still lingers.

Darwin's theory of evolution has helped shape the biological sciences to the present day. His work generated a great wave of scientific observation and research that has provided much additional evidence that evolution is responsible for the great diversity of organisms present on our planet. Even today, the details of the process of evolution are a major focus of investigation and debate.

Darwin based his theory of natural selection on the following four observations: (1) Individual members of a species show some variation from one another. (2) Organisms produce many more offspring than will survive to reproduce (Fig. 1–11). (3) Organisms compete for necessary resources such as food, sunlight, and space. Individuals who happen to have characteristics that permit them to effectively use resources are more likely to survive to reproductive maturity and are more likely to leave offspring. (4) The survivors that reproduce pass their adaptations for survival on to their offspring. Thus, the best adapted individuals of a population leave, on average, more offspring than do other individuals. Because of this differential reproduction, a greater proportion of the population becomes adapted to the prevailing environmental conditions. The environment *selects* the best adapted organisms for survival.

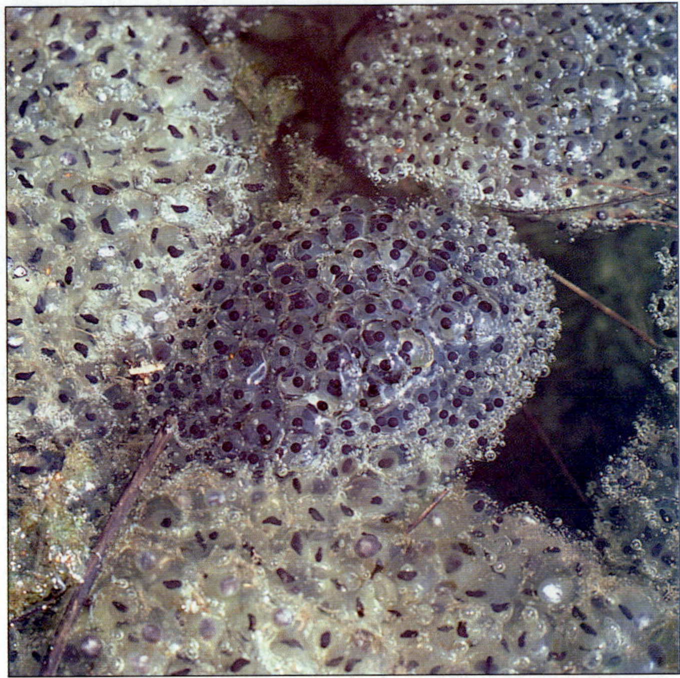

Figure 1-11 Egg masses of the wood frog *(Rana sylvatica)*. Many more eggs are produced than can possibly develop into adult frogs. Random events are largely responsible for determining which of these developing frogs will hatch, reach adulthood, and reproduce. However, certain traits possessed by each organism also contribute to its probability for success in its environment. Not all organisms are as prolific as the frog, but the generalization that more organisms are produced than survive is true throughout the living world. *(J. Serrao/Photo Researchers, Inc.)*

Darwin did not know about DNA or understand the mechanisms of inheritance. We now understand that most variations among individuals are a result of different varieties of genes that code for each characteristic. The ultimate source of these variations is random **mutations,** chemical or physical changes in DNA that persist and can be inherited. Mutations modify genes; by this process they provide the raw material for evolution.

Populations evolve as a result of selective pressures from changes in the environment

All the genes present in a population make up its **gene pool.** By virtue of its gene pool, a population is a reservoir of variation.

(a)

(b)

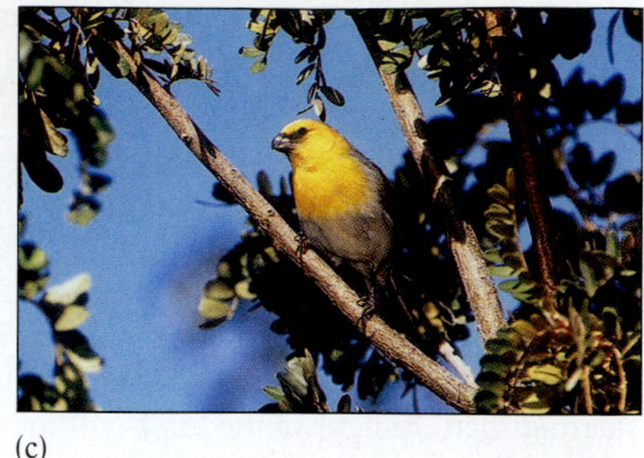

(c)

Figure 1–12 Adaptation and diversification in Hawaiian honeycreepers. **(a)** The bill of this Akiapolaau male *(Hemignathus munroi)* is adapted for extracting insect larvae from bark. The lower mandible (jaw) is used to peck at and pull off bark, while the upper mandible and tongue remove the prey. **(b)** 'I'iwi *(Vestiaria coccinea)* in 'Ohi'a blossoms. The bill is adapted for feeding on nectar in tubular flowers. **(c)** Palila *(Loxiodes bailleui)* in mamane tree. This finch-billed honeycreeper feeds on immature seeds in pods of the mamane tree. It also eats insects, berries, and young leaves. All three species shown here are endangered. *(a – c, Jack Jeffrey, Inc.)*

Natural selection acts on individuals within a population. Selection favors individuals with genes that specify traits that enable them to cope effectively with pressures exerted by the environment. These organisms are most likely to survive and produce offspring. As these successful organisms pass on their genetic recipe for survival, their traits become more widely distributed in the population. Over time, as organisms continue to change (and as the environment itself changes, bringing different selective pressures), the members of the population become better adapted to their environment and less like their ancestors.

As members of a population adapt to pressures exerted by the environment and exploit new opportunities for finding food or safety, or avoiding predators, the population diversifies and new species may evolve. The Hawaiian honeycreepers, a group of related birds, are a good example. When honeycreeper ancestors first reached Hawaii, few other birds were present and so there was little competition. Honeycreepers moved into a variety of food zones and various types of bills evolved (Figure 1–12; also see discussion in Chapter 19 and Figure 19–17). Some honeycreepers now have long, curved bills, adapted for feeding on nectar from tubular flowers. Others have short, thick bills for foraging for insects, and still others are adapted for eating seeds.

LIFE DEPENDS ON A CONTINUOUS INPUT OF ENERGY

A continuous input of energy from the sun enables life to exist. Every activity of a living cell or organism requires energy. Whenever energy is used to perform biological work, some is converted to heat and dispersed into the environment.

Energy flows through cells and organisms

Recall that all the energy transformations and chemical processes that occur within an organism are referred to as metabolism. Energy is necessary to carry on the metabolic activities essential for growth, repair, and maintenance. Each cell of an organism requires nutrients. Some nutrients are used as fuel for **cellular respiration,** a process during which some of the energy stored in the nutrient molecules is released for use by the cells (Fig. 1–13). This energy can be used for cellular work or for synthesis of needed materials such as new cellular components. All cells carry on cellular respiration.

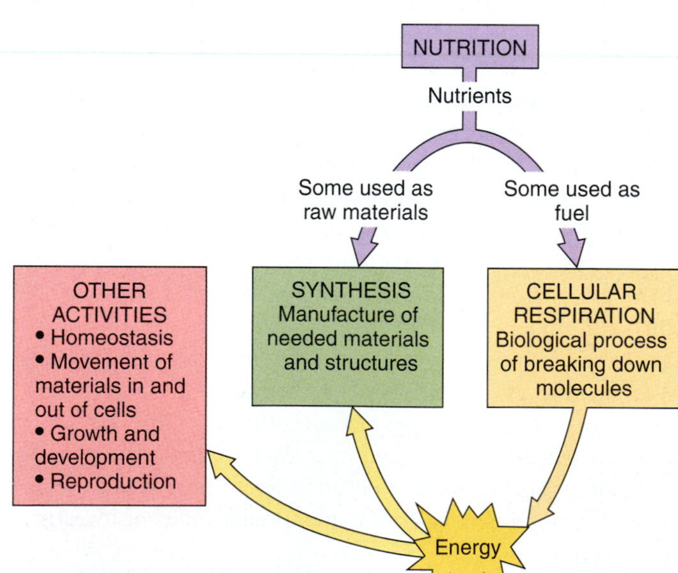

Figure 1–13 Relationships of metabolic processes. These processes occur continuously in the cells of living organisms. Some of the nutrients in food are used to synthesize needed materials and cell parts. Other nutrients are used as fuel for cellular respiration, a process that releases energy stored in food. This energy is needed for synthesis and for other forms of cellular work.

Energy flows through ecosystems

Like individual organisms, ecosystems depend on a continuous input of energy. A self-sufficient ecosystem contains three types of organisms—producers, consumers, and decomposers—and has a physical environment appropriate for their survival. These organisms depend on each other and on the environment for nutrients, energy, oxygen, and carbon dioxide. However, there is a one-way flow of energy through ecosystems. Organisms can neither create energy nor use it with complete efficiency. During every energy transaction, some energy is dispersed into the environment as heat and is no longer available to the organism (Fig. 1–14).

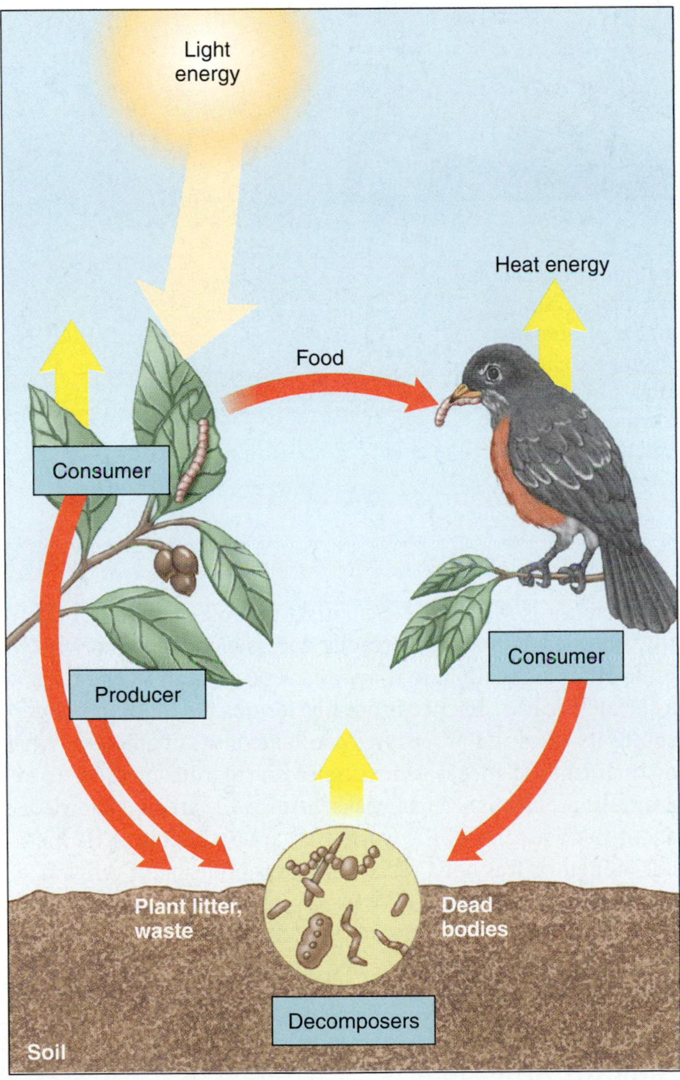

Figure 1–14 Energy flow. Continuous energy input from the sun operates the biosphere. During photosynthesis, producers use the energy from sunlight to make complex molecules from carbon dioxide and water. Consumers obtain energy, carbon, and other needed materials when they eat producers or consumers that have eaten producers. Wastes and dead organic material supply decomposers with energy and carbon. During every energy transaction some energy is lost to biological systems as it is dispersed as heat.

Producers, or **autotrophs,** are plants, algae, and certain bacteria that can produce their own food from simple raw materials. Most of these organisms use sunlight as an energy source and carry out photosynthesis, in which complex molecules are synthesized from carbon dioxide and water. The light energy is transformed into chemical energy, which is stored within the chemical bonds of the food molecules produced. Oxygen, which is required not only by plant cells but also by the cells of most other organisms, is produced as a byproduct of photosynthesis:

Carbon dioxide + Water + Light energy →
Sugars (food) + Oxygen

Animals are **consumers,** or **heterotrophs,** that is, organisms that depend on producers for food, energy, and oxygen. Consumers obtain energy by breaking down sugars and other food molecules originally produced during photosynthesis. Recall that the biological process of breaking down sugars and other fuel molecules is known as cellular respiration. When chemical bonds are broken during cellular respiration, their stored energy is made available for life processes:

Sugars (and other food molecules) + Oxygen →
Carbon dioxide + Water + Energy

Consumers contribute to the balance of the ecosystem. For example, consumers produce carbon dioxide needed by producers. The metabolism of consumers and producers helps maintain the life-sustaining mixture of gases in the atmosphere.

Bacteria and fungi are **decomposers,** heterotrophs that obtain nutrients by breaking down wastes, dead leaves, and bodies of dead organisms. In their process of obtaining energy, decomposers make the components of wastes and dead organisms available for reuse. If decomposers did not exist, nutrients would remain locked up in dead bodies, and the supply of elements required by living systems would soon be exhausted.

BIOLOGY IS STUDIED USING THE SCIENTIFIC METHOD

Biologists work in laboratories and out in the field (Fig. 1–15). Their investigations range from the study of molecular biology and viruses to the interactions of the communities of our biosphere. Perhaps you will decide to become a research biologist and help unravel the complexities of the human brain; discover new hormones that cause plants to flower; identify new species of animals or bacteria; or develop new stem cell strategies to treat cancer, AIDS, or heart disease. Or perhaps you will choose to enter an applied field of biology such as environmental science, dentistry, medicine, pharmacology, or veterinary medicine.

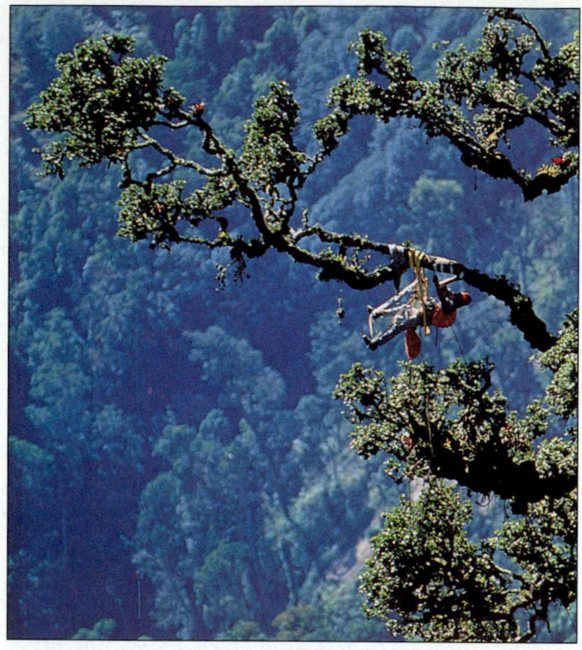

(a)

(b)

Figure 1–15 Biologists at work. **(a)** This biologist studying the rainforest canopy in Costa Rica is part of an international effort to study and preserve tropical rain forests. Researchers study the interactions of organisms and the effects of human activities on the rain forests. **(b)** This researcher is working on the Human Genome Project at the Sanger Centre in Cambridge, England. She is monitoring a beaker of a yeast-tryptone culture medium that is being prepared for use in growing yeasts or bacteria containing human DNA. As the organisms in the culture medium reproduce, they clone the human DNA. *(a, Mark Moffett/Minden Pictures; b, James King-Holmes/Science Photo Library/Photo Researchers, Inc.)*

A number of interesting careers in the biological sciences are discussed in the Career Visions sections of this book.

Biology is a **science.** The word *science* comes from a Latin word meaning "to know." Science is a way of thinking and a method of investigating the world around us in a systematic manner. Science enables us to uncover ever more about the world we live in and leads us to an expanded appreciation of our universe.

The **process of science** is investigative, dynamic, and often controversial. It changes over time as it is influenced by cultural, social, and historical contexts as well as by the personalities of scientists themselves. The observations made, the range of questions posed, and the design of experiments depend on the creativity of the individual scientist. In contrast, the **scientific method** involves a series of ordered steps and is a framework used by most scientists.

Using the scientific method, scientists make careful observations, ask critical questions, and develop **hypotheses,** which are testable statements. Based on their hypotheses, scientists make predictions that can be tested, and test their predictions by making further observations or by performing experiments (Fig. 1–16). They interpret the results of their experiments and draw conclusions from them. Even results that do not support the hypothesis may be valuable and may lead to new hypotheses. If the results support a hypothesis, a scientist may use them to generate related hypotheses.

Science is systematic. Scientists organize, and often quantify, knowledge, making it readily accessible to all who wish to build on its foundation. In this way science is both a personal and a social endeavor. Science is not mysterious. Anyone who understands its rules and procedures can take on its challenges. What distinguishes science is its insistence on rigorous methods to examine a problem. Science seeks to give us precise knowledge about those aspects of the world that are accessible to its methods of inquiry. It is not a replacement for philosophy, religion, or art. Being a scientist does not prevent one from participating in other fields of human endeavor, just as being an artist does not prevent one from practicing science.

Science requires systematic thought processes

Two types of systematic thought processes used by scientists are deduction and induction. With **deductive reasoning,** we begin with supplied information, called **premises**, and draw conclusions on the basis of that information. Deduction proceeds from general principles to specific conclusions. For example, if we accept the premise that all birds have wings, and the second premise that sparrows are birds, we can conclude deductively that sparrows have wings. Deduction helps us discover rela-

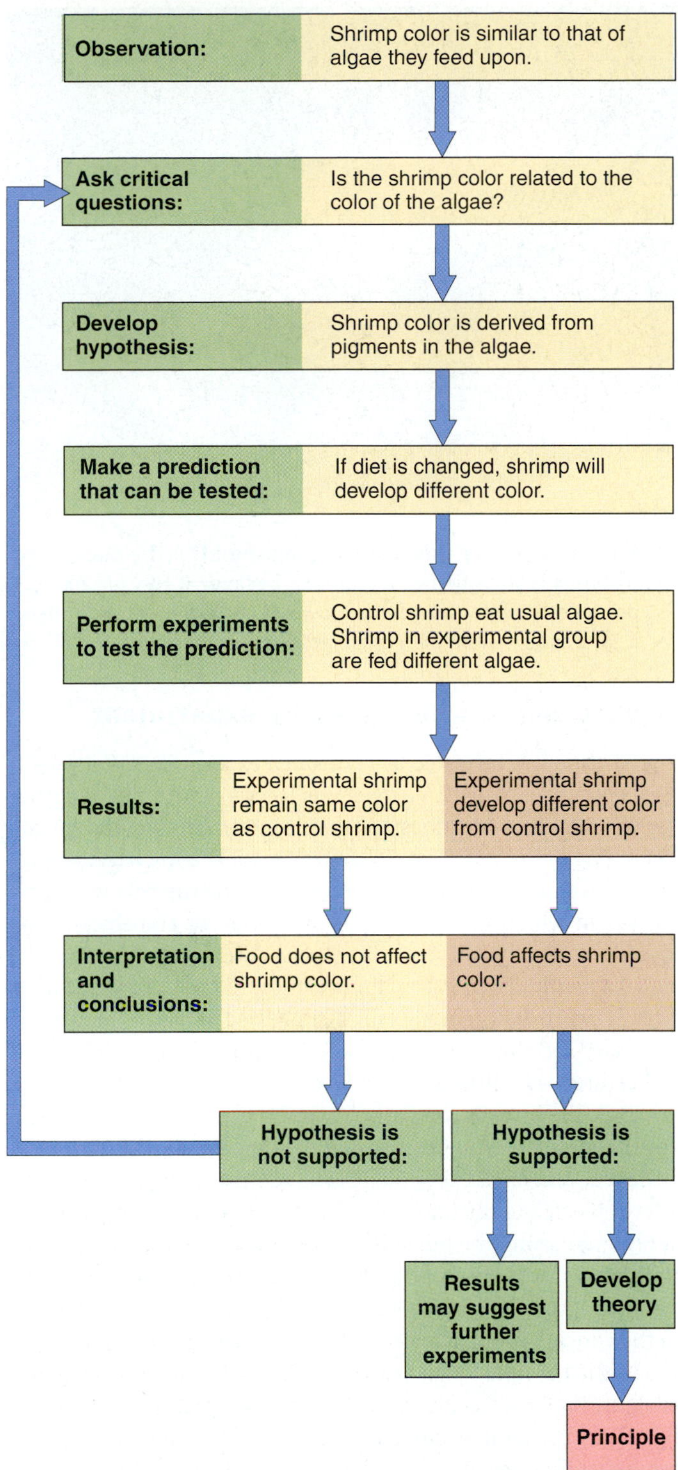
Observation:
Shrimp color is similar to that of algae they feed upon.
Ask critical questions:
Is the shrimp color related to the color of the algae?
Develop hypothesis:
Shrimp color is derived from pigments in the algae.
Make a prediction that can be tested:
If diet is changed, shrimp will develop different color.
Perform experiments to test the prediction:
Control shrimp eat usual algae. Shrimp in experimental group are fed different algae.
Results:
Experimental shrimp remain same color as control shrimp.
Experimental shrimp develop different color from control shrimp.
Interpretation and conclusions:
Food does not affect shrimp color.
Food affects shrimp color.
Hypothesis is not supported:
Hypothesis is supported:
Results may suggest further experiments
Develop theory
Principle

Figure 1–16 The scientific method. Scientists use the scientific method as a framework for their research.

tionships among known facts. The **hypothetico-deductive approach** emphasizes the use of deductive reasoning to test hypotheses.

Scientists also use a **hypothetico-inductive approach** that focuses on discovering new general principles. **Inductive reasoning** is the opposite of deduction. We begin with specific observations and draw a conclusion or discover a general principle. For example, if we know that sparrows have wings and are birds, and we know that robins, eagles, pigeons, and hawks have wings and are birds, we might induce that all birds have wings. In this way, the inductive method can be used to organize raw data into manageable categories by answering the question: What do all these facts have in common?

A weakness of inductive reasoning is that conclusions generalize the facts to all possible examples. We go from many observed examples to all possible examples when we formulate the general principle. This is known as the **inductive leap.** Without it, we could not arrive at generalizations. However, we must be sensitive to exceptions and to the possibility that the conclusion is not valid. For example, the kiwi bird of New Zealand does not have functional wings! The generalizations in inductive conclusions come from the creative insight of the human mind, and creativity, however admirable, is not infallible.

Scientists make careful observations and ask critical questions

Chance and luck are often involved in recognizing a phenomenon or problem, but significant discoveries are usually made by those who are in the habit of looking critically at nature. Necessary technology for investigating the problem must also be available. In 1928 the British bacteriologist Alexander Fleming observed that one of his bacterial cultures had become invaded by a blue mold. He almost discarded it, but before he did, he noticed that the area contaminated by the mold was surrounded by a zone where bacterial colonies did not grow well.

The bacteria were disease organisms of the genus *Staphylococcus,* which can cause boils and skin infections. Anything that could kill them was interesting! Fleming saved the mold, a variety of *Penicillium* (blue bread mold). It was subsequently discovered that the mold produced a substance that slowed reproduction of the bacterial population but was usually harmless to laboratory animals and humans. The substance was penicillin, the first antibiotic.

We may wonder how many times the same type of mold grew on the cultures of other bacteriologists who failed to make the connection and simply threw away their contaminated cultures. Fleming benefited from chance, but his mind was prepared to make observations and formulate critical questions, and his pen was prepared to publish them. Still, it was left to others to develop the practical applications. Although Fleming recognized the potential practical benefit of penicillin, he did not develop the chemical techniques needed to purify it, and more than 10 years passed before the drug was put to significant use.

In 1939 Sir Howard Florey and Ernst Boris Chain developed chemical procedures to extract and produce the active agent penicillin from the mold. Florey took the process to laboratories in the United States, and penicillin was first produced to treat wounded soldiers in World War II. In 1945, Fleming, Florey, and Chain shared the Nobel Prize in Medicine.

A hypothesis is a testable statement

In the early stages of an investigation, a scientist typically thinks of many possible hypotheses and hopes that the right one is among them. He or she then decides which, if any, could and should be subjected to experimental test. Why not test them all? Time and money are important considerations in conducting research. We must establish priority among the hypotheses to decide which to test first. Fortunately, some guidelines do exist. A good hypothesis exhibits the following:

1. It is reasonably consistent with well-established facts.
2. It is capable of being tested; that is, it should generate definite predictions, whether the results are positive or negative. Test results should also be repeatable by independent observers.
3. It is falsifiable, which means it can be proved false.

A hypothesis cannot really be proved true, but in theory (though not necessarily in practice) a well-stated hypothesis can be proved false. If one believes in an unfalsifiable hypothesis (e.g., the existence of invisible and undetectable angels), it must be on grounds other than scientific ones.

Consider the following hypothesis: All female mammals (animals that have hair and produce milk for their young) bear live young. The hypothesis was based on the observations that dogs, cats, cows, lions, and humans all are mammals and all bear live young. Consider further that a new species, species X, was identified as a mammal. Biologists predicted that females of species X would bear live young. When a female of the new species gave birth to offspring, this supported the hypothesis. Yet it did not really *prove* the hypothesis.

Before the Southern Hemisphere was explored, most individuals would probably have accepted the hypothesis without question, because all known furry, milk-giving animals did, in fact, bear live young. But it was discovered that two Australian animals (the duck-billed platypus and the spiny anteater) had fur and produced milk for their young but laid eggs (Fig. 1–17). The hypothesis, as stated, was false no matter how many times it had previously been supported. As a result, biologists either had to consider the platypus and the spiny anteater as nonmammals or had to broaden their definition of mammals to include them. (They chose the latter.)

A hypothesis is not true just because some of its predictions (the ones we happen to have thought of or have thus far been able to test) have been shown to be true. After all, they could be true by coincidence. Failure to observe a predicted outcome does not make a hypothesis false, but neither does it show that the hypothesis is true.

A prediction is a logical consequence of a hypothesis

A hypothesis is an abstract idea, so there is no way to test it directly. But hypotheses suggest certain logical consequences, that is, observable things that cannot be false if the hypothesis is true. On the other hand, if the hypothesis is, in fact, false, other definite predictions should disclose that. As used here, then, a **prediction** is a deductive, logical consequence of a hypothesis. It does not have to be a future event.

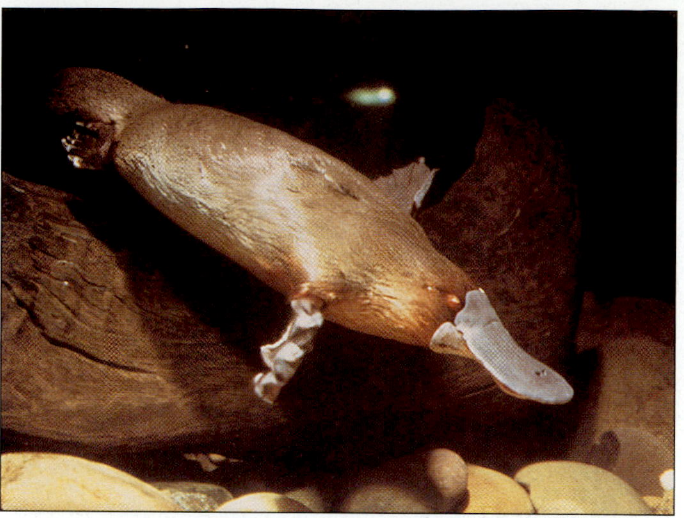

Figure 1–17 Is this animal a mammal? The duck-billed platypus is classified as a mammal because it has fur and produces milk for its young. However, unlike most mammals, it lays eggs. *(Tom McHugh/Photo Researchers, Inc.)*

Predictions can be tested by experiment

A prediction can be tested by controlled experiments. Early biologists observed that the nucleus was the most prominent part of the cell, and they hypothesized that it might be essential for the well-being of the cell. They predicted that if the nucleus were removed from the cell, the cell would die. Experiments were performed in which the nucleus of a unicellular amoeba was removed surgically with a microloop. After this surgery, the amoeba continued to live and move but it did not grow, and after a few days it died. These results suggested that the nucleus is necessary for the metabolic processes that provide for growth and cell reproduction (Fig. 1–18).

But, the investigators asked, what if the operation itself and not the loss of the nucleus caused the amoeba to die? They performed a *controlled* experiment in which two groups of amoebas were subjected to the same operative trauma. However, in the **experimental group** the nucleus was removed, whereas in the **control group** it was not. An experimental group ideally differs from a control group only with respect to the variable being studied. In the control group, a microloop was inserted into each amoeba and pushed around inside the cell to simulate the removal of the nucleus; then the needle was withdrawn, leaving the nucleus inside. Amoebas treated with such a sham operation recovered and subsequently grew and divided, but the amoebas without nuclei died. This experiment provided data that it was the removal of the nucleus and not simply the operation that caused the death of the amoebas. The data supported the hypothesis that the nucleus is essential for the well-being of the cell.

In scientific studies, care must be taken to avoid **bias.** For example, to prevent bias, most medical experiments today are carried out in a **double-blind** fashion. When a drug is being tested, one group of patients is given the new medication, while a second similar group of patients (the control group) is given a placebo, a harmless starch pill similar in size, shape, color, and

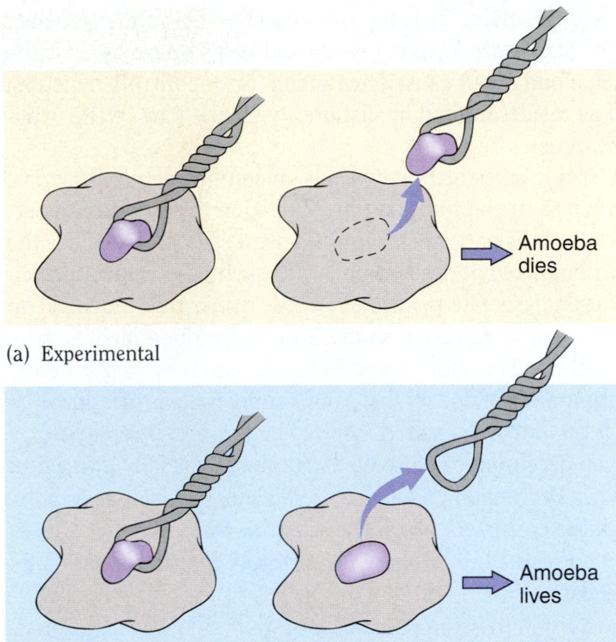

Figure 1–18 Testing a prediction. An early controlled experiment tested the prediction that if the nucleus is removed from a cell, the cell would die. The data gathered from this and similar experiments provided support for the hypothesis that the nucleus is essential for the well-being of the cell. **(a)** When its nucleus is surgically removed with a microloop, the amoeba dies. **(b)** Control amoebas subjected to similar surgical procedures (including insertion of a microloop), but without actual removal of the nucleus, do not die.

taste to the pill being tested. This is a double-blind study because neither the patient nor the physician knows who is getting the experimental drug and who is getting the placebo. The pills or treatments are coded in some way, and only after the experiment is over and the results are recorded is the code broken. Not all experiments can be so neatly designed; for one thing, it is often difficult to establish appropriate controls.

Scientists interpret the results of experiments and make conclusions

Scientists gather data in an experiment, interpret their results, and then formulate conclusions. For example, in the amoeba experiment described earlier, investigators concluded that the nucleus was essential for the well-being of the cell.

One reason for inaccurate conclusions is **sampling error.** Because *all* cases of what is being studied cannot be observed or tested (scientists cannot study every amoeba), we must be content with a sample, or subset, of them. Yet how can we know whether that sample is truly representative of whatever we are studying? In the first place, if the sample is too small, it may be different owing to random factors. A study with only two, or even nine, amoebas might not yield reliable data that could be generalized to other amoebas. This problem can usually be solved by using large numbers of subjects and applying the mathematics of statistical analysis (Fig. 1–19).

We must also ensure that the sample is typical of the group that we intend to study. Scientists use statistical techniques to ensure that there is no consistent bias in the way that experimental samples are chosen.

Even if a conclusion is based on results from a carefully designed experiment, it is still possible that new observations or results from other experiments can challenge the conclusion. If we test a large number of cases, we are more likely to draw accurate scientific conclusions. The scientist seeks to state with confidence that any specific conclusion has a certain statistical probability of being correct.

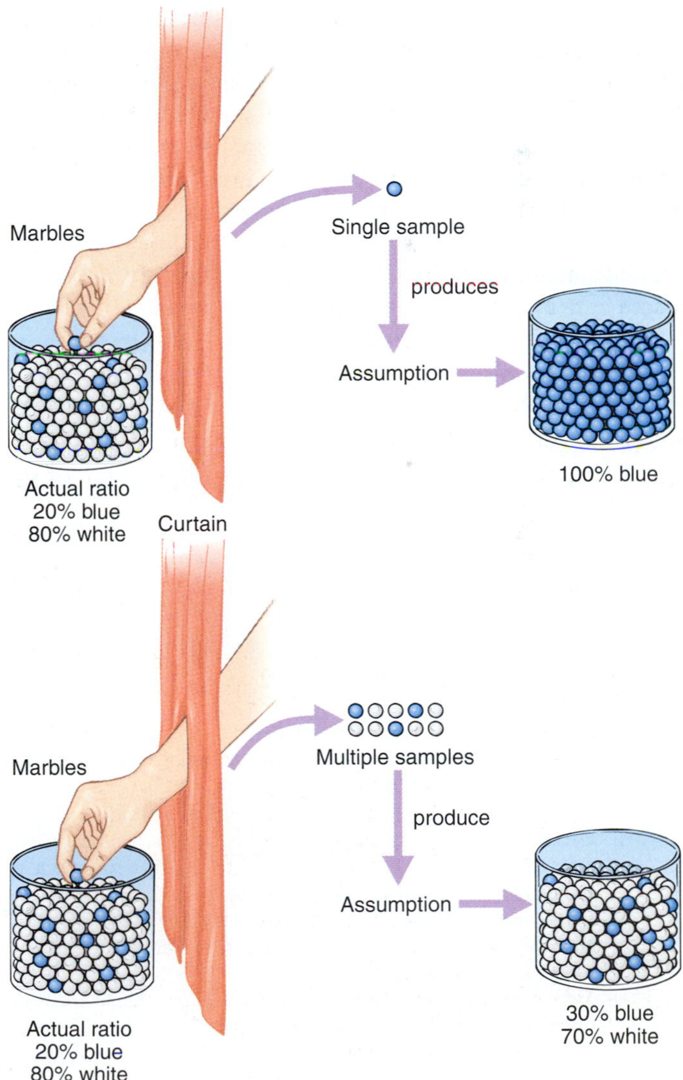

Figure 1–19 Statistical probability. Taking a single sample can result in sampling error. If the only marble sampled is blue, we might assume that all the marbles are blue. The greater the number of samples we take of an unknown, the more likely we can make valid assumptions about it.

Experiments must also be replicated. When researchers publish their findings in a scientific journal, they typically include a description of their methods and procedures so that other scientists can repeat the experiments. When the findings are replicated, the conclusions are, of course, strengthened.

A well-supported hypothesis may lead to a theory

Nonscientists often use the word *theory* incorrectly to refer to a hypothesis. A **theory** is actually an integrated explanation of a number of hypotheses, each supported by consistent results from many observations or experiments. A theory relates data that previously appeared to be unrelated. A good theory grows, building on additional facts as they become known. It predicts new facts and suggests new relationships among phenomena. It may even suggest practical applications.

A good theory, by showing the relationships among classes of facts, simplifies and clarifies our understanding of natural phenomena. As Einstein wrote, "In the whole history of science from Greek philosophy to modern physics, there have been constant attempts to reduce the apparent complexity of natural phenomena to simple, fundamental ideas and relations."

A theory that, over a long period, has withstood repeated testing and is almost universally accepted by scientists is referred to as a **scientific principle.** The term **law** is sometimes used for a principle judged to be of great basic importance, such as the law of gravity.

Science has ethical dimensions

Researchers who publish their work in scientific journals describe their experiments in sufficient detail to be independently performed by others. This permits objective observers to detect errors or bias in the original study and helps guard against the occasional odd result caused by random or uncontrolled factors, as well as results tainted by dishonesty on the part of the original researcher.

Scientific investigation depends on commitment to practical ideals such as truthfulness and the obligation to communicate results. Honesty is particularly important in science. Consider the great (though temporary) damage done whenever an unprincipled or even desperate researcher, whose career might depend on publication of a research study, knowingly disseminates false data. Until the deception is uncovered, researchers might devote many thousands of dollars and hours of precious professional labor to futile lines of research inspired by erroneous reports.

Such deception can also be dangerous, especially in medical research. Fortunately, science tends to be self-correcting through the consistent use of the scientific process itself. Sooner or later, someone's experimental results are bound to cast doubt on false data.

In addition to being ethical about their own work, scientists must face many broad ethical issues surrounding areas such as stem cell research, cloning, human and animal experimentation, and applications of genetic engineering. For example, some of the stem cells that show the greatest potential for use in treating human disease are found in early embryos. The cells can be taken from 5- or 6-day-old embryos and then cultured in laboratory glassware. Such cells could be engineered to treat failing hearts or brains harmed by stroke, injury, Parkinson's disease, or Alzheimer's disease. They could save the lives of burn victims and perhaps be engineered to treat specific cancers. Society will need to determine whether the potential benefits of stem cell research outweigh the ethical risks of using these cells.

SUMMARY WITH KEY TERMS

- I. **Biology** is the study of life. Basic themes of biology include the evolution of life, the transmission of information, and the flow of energy through organisms.
- II. A living organism is able to grow and develop, carry on self-regulated metabolism, move, respond to stimuli, and reproduce. Species evolve and adapt to their environment.
  - A. All living organisms are composed of one or more **cells.**
  - B. Organisms grow by increasing the size and number of their cells.
  - C. **Metabolism** refers to all the chemical activities that take place in the organism, including the chemical reactions essential to nutrition, growth and repair, and conversion of energy to usable forms. **Homeostasis** is the tendency of organisms to maintain a constant internal environment.
  - D. Movement, though not necessarily locomotion, is characteristic of living organisms. Some organisms use tiny extensions of the cell, called **cilia,** or longer **flagella** to move from place to place. Other organisms are **sessile** and remain rooted to some surface.
  - E. Organisms respond to **stimuli,** physical or chemical changes in their external or internal environment.
  - F. In **asexual reproduction,** offspring are typically identical to the single parent; in **sexual reproduction,** offspring are typically the product of the fusion of gametes, and genes are typically contributed by two parents.
  - G. Populations evolve and become adapted to their environment. **Adaptations** are traits that increase an organism's ability to survive in its environment.
- III. Organisms transmit information chemically, electrically, and behaviorally.
  - A. DNA, which makes up the **genes,** contains the instructions for the development of an organism and for carrying out life processes.
    1. Information encoded in **DNA** is transmitted from one generation to the next.
    2. DNA codes for **proteins,** which are important in determining the structure and function of cells and tissues.
  - B. **Hormones,** chemical messengers that transmit messages from one part of an organism to another, are one type of **cell signaling.**
  - C. Many organisms use electrical signals to transmit information; most animals have nervous systems that transmit electrical impulses and release **neurotransmitters.**
- IV. Biological organization is hierarchical.
  - A. A complex organism is organized at the **chemical, cellular, tissue, organ,** and **organ system** levels.
  - B. The basic unit of ecological organization is the **population.** Various populations form **communities;** a community and its physical environment are referred to as an **ecosystem**; all of Earth's communities and ecosystems together make up the **biosphere.**

V. Millions of species have evolved. A **species** is a group of organisms with similar structure, function, and behavior that, in nature, breed only with each other. Members of a species have a common **gene pool** and share common ancestry.
   A. Taxonomic classification is hierarchical; it includes **species, genus, family, order, class, phylum, kingdom,** and **domain.** Each grouping is referred to as a **taxon.**
   B. Biologists use a **binomial system of nomenclature** in which the name of each species includes a genus name and a **specific epithet.**
   C. Bacteria have **prokaryotic cells;** all other organisms have **eukaryotic cells.**
   D. Organisms can be classified into three domains: **Archaea, Eubacteria,** and **Eukarya,** and six kingdoms: **Archaebacteria, Eubacteria, Protista** (protozoa, algae, water molds, and slime molds), **Fungi** (molds and yeasts), **Plantae,** and **Animalia.**

VI. **Evolution** is the process by which populations change over time in response to changes in the environment.
   A. **Natural selection,** the mechanism by which evolution proceeds, favors individuals with traits that enable them to cope with environmental changes. These individuals are most likely to survive and to produce offspring.
   B. Charles Darwin based his theory of natural selection on his observations that individuals of a species vary; organisms produce more offspring than survive to reproduce; individuals that are best adapted to their environment are more likely to survive and reproduce; as successful organisms pass on their hereditary information, their traits become more widely distributed in the population.
   C. The source of variation in a population is random **mutation.**

VII. Activities of living cells require energy; life depends on continuous energy input from the sun.
   A. During **photosynthesis** plants, algae, and certain bacteria use the energy of sunlight to synthesize complex molecules from carbon dioxide and water.
   B. All cells carry on **cellular respiration,** a biochemical process in which they capture the energy stored in nutrients by producers. Some of that energy is then used to synthesize needed materials or to carry out other cell activities.
   C. A self-sufficient ecosystem includes **producers,** or **autotrophs,** which make their own food; **consumers,** which eat producers or organisms that have eaten producers; and **decomposers,** which obtain energy by breaking down wastes and dead organisms. Consumers and decomposers are **heterotrophs,** which are organisms that depend on producers as an energy source and for food and oxygen.

VIII. The **process of science** is a dynamic approach to investigation. The **scientific method** is a framework that scientists use in their work; it includes observing, recognizing a problem or stating a critical question, developing a hypothesis, making a prediction that can be tested, performing experiments, interpreting results, and drawing conclusions that support or falsify the hypothesis.
   A. Deductive reasoning and inductive reasoning are two categories of systematic thought processes that are used in the scientific method. **Deductive reasoning** proceeds from general principles to specific conclusions and helps us discover relationships among known facts. **Inductive reasoning** begins with specific observations and draws conclusions from them. Inductive reasoning helps us discover general principles.
   B. A **hypothesis** is a testable statement about the nature of an observation or relationship.
   C. A properly designed scientific experiment includes both a **control group** and an **experimental group,** and must be as free as possible from bias. The experimental group differs from a control group only with respect to the variable being studied.
   D. When a number of related hypotheses have been supported by conclusions from many experiments, scientists may develop a **theory** based on them. A well-established and tested theory may be referred to as a **scientific principle.**
   E. Science has important ethical dimensions.

POST-TEST

1. Metabolism (a) is the sum of all the chemical activities of an organism (b) results from an increase in the number of cells (c) is characteristic of plant and animal kingdoms only (d) refers to chemical changes in an organism's environment (e) does not take place in producers
2. Homeostasis (a) is the tendency of organisms to maintain a constant internal environment (b) generally depends on the action of cilia (c) is the long-term response of organisms to changes in their environment (d) occurs at the ecosystem level, not in cells or organisms (e) may be sexual or asexual
3. Structures used by some organisms for locomotion are (a) cilia and nuclei (b) flagella (c) nuclei (d) cilia and sessiles (e) cilia and flagella
4. The splitting of an amoeba into two is best described as an example of (a) locomotion (b) neurotransmission (c) asexual reproduction (d) sexual reproduction (e) metabolism
5. Cells (a) are the building blocks of living organisms (b) always have nuclei (c) are not found among the bacteria (d) answers a, b, and c are correct (e) answers a and b only are correct
6. An increase in the size or number of cells best describes (a) homeostasis (b) biological growth (c) chemical level of organization (d) asexual reproduction (e) adaptation
7. DNA (a) makes up the genes (b) transmits information from one species to another (c) cannot be changed (d) is a neurotransmitter (e) is produced during cellular respiration
8. Cellular respiration (a) is a process whereby sunlight is used to synthesize cellular components with the release of energy (b) occurs in heterotrophs only (c) is carried on by both autotrophs and heterotrophs (d) causes chemical changes in DNA (e) occurs in response to environmental changes
9. Which of the following is a correct sequence of levels of biological organization? (a) cellular, organ, tissue, organ system (b) chemical, cellular, organ, tissue (c) chemical, cellular, tissue, organ (d) tissue, organ, cellular, organ system (e) chemical, cellular, population, species
10. Which of the following is a correct sequence of levels of biological organization? (a) organism, population, ecosystem, community (b) organism, population, community, ecosystem (c) population, biosphere, ecosystem, community (d) species, population, ecosystem, community (e) ecosystem, population, community, biosphere
11. Protozoa are assigned to kingdom (a) Protista (b) Fungi (c) Archaebacteria (d) Animalia (e) Plantae
12. Yeasts and molds are assigned to kingdom (a) Protista (b) Fungi (c) Archaebacteria (d) Animalia (e) Plantae
13. In the binomial system of nomenclature, the first part of an organism's name designates the (a) specific epithet (b) genus (c) class (d) kingdom (e) phylum
14. Which of the following is a correct sequence of levels of classification (a) genus, species, family, order, class, phylum, kingdom (b) genus, species, order, phylum, class, kingdom (c) genus, species, order, family, class, phylum, kingdom (d) species, genus, family, order, class, phylum, kingdom (e) species, genus, order, family, class, kingdom, phylum

15. Darwin suggested that evolution takes place by (a) mutation (b) changes in the individuals of a species (c) natural selection (d) interaction of hormones (e) homeostatic responses to each change in the environment
16. A testable statement is a (an) (a) theory (b) hypothesis (c) principle (d) inductive leap (e) critical question
17. Ideally, an experimental group differs from a control group (a) only with respect to the hypothesis being tested (b) only with respect to the variable being studied (c) by being less subject to bias (d) because it is less vulnerable to sampling error (e) because its subjects are more reliable

REVIEW QUESTIONS

1. Contrast a living organism with a nonliving object.
2. In what ways might the metabolisms of an oak tree and a tiger be similar? Relate these similarities to the biological themes of transmission of information, energy, and evolution.
3. What would be the consequences if an organism's homeostatic mechanisms failed? Explain your answer.
4. What components do you think might be present in a balanced forest ecosystem? In what ways are consumers dependent on producers? On decomposers? Include energy considerations in your answer.
5. Why do you suppose that the binomial system of nomenclature has survived for more than 200 years and is still used by biologists?
6. How might you explain the sharp claws and teeth of tigers in terms of natural selection?
7. What is meant by a "controlled" experiment?
8. Make a prediction and devise a suitably controlled experiment to test each of the following hypotheses: (a) A type of mold found in your garden does not produce an effective antibiotic. (b) The rate of growth of a bean seedling is affected by temperature. (c) Estrogen alleviates the symptoms of Alzheimer's disease in elderly women.

YOU MAKE THE CONNECTION

1. How might a firm understanding of evolutionary processes be helpful to a biologist who is doing research in (a) animal behavior, (b) ecology, or (c) the development of a vaccine against human immunodeficiency virus, the virus that causes AIDS?
2. If you could influence U. S. policy on stem cell research, what position would you take? Explain. What would be your position on the use of genetically modified crops to increase world food supply or the use of genetically modified animals for producing drugs such as insulin and blood-clotting factors?

RECOMMENDED READINGS

Brown, K. "Seeds of Concern." *Scientific American,* Vol. 284, No. 4, Apr. 2001. What are the concerns and what is the evidence regarding genetically modified crops?

Cohen, J. "Can Cloning Help Save Beleaguered Species?" *Science,* Vol. 276, 30 May, 1997. A brief discussion of the benefits and concerns of cloning endangered species.

Lanza, R.P., B.L. Dresser, and P. Damiani. "Cloning Noah's Ark," *Scientific American,* Vol. 283, Nov., 2000. A description of the process and technology used to clone animals on the brink of extinction, and a discussion of some of the attendant controversy.

Mirsky, S., and J. Rennie. "What Cloning Means for Gene Therapy," *Scientific American,* Vol. 276, Jun., 1997. A special report about the benefits of combining cloning technology with other biotechnologies.

Moore, J.A. *Science as a Way of Knowing: The Foundations of Modern Biology*. Harvard University Press, Cambridge, 1993. An account of scientific thought as related to the history of modern biology.

Pedersen, R.A. "Embryonic Stem Cells for Medicine." *Scientific American,* Vol. 280, Apr., 1999. A special report on the potential and ethics of stem cell research.

Pennisi, E. "Human Genome: Finally, the Book of Life and Instructions for Navigating It." *Science,* Vol. 288, 30 Jun., 2000. A brief description of the announcement of the "completion" of the Human Genome Project and its implications.

Science, Vol. 277, 25 Jul., 1997. Special Issue: Human-Dominated Ecosystems. The authors examine the global consequences of human activity on several ecosystems.

Science, Vol. 287, 25 Feb., 2000. Special Section: Stem Cell Research and Ethics. Several articles discuss the state of the art, financial considerations, global issues, and ethical issues.

Scientific American, Vol. 281, Dec., 1999. End of the Millennium Special Issue. Several articles of interest to the beginning biology student, including "Deciphering the Code of Life," which discusses genome research.

Velander, W.H., H. Lubon, and W.N. Drohan. "Transgenic Livestock as Drug Factories," *Scientific American,* Vol. 276, Jan., 1997. A discussion of some exciting new medical applications of genetic techniques.

- Visit our Web site at **http://www.info.brookscole.com/solomonbergmartin** for links to chapter-related resources on the World Wide Web. Additional on-line materials relating to this chapter can also be found on our Web site.

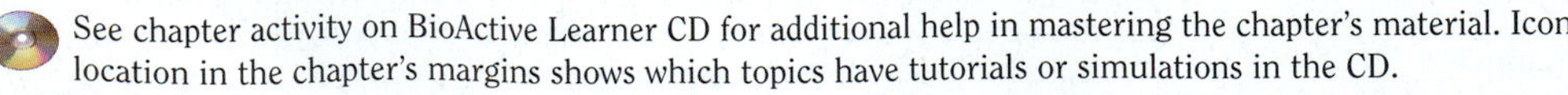

See chapter activity on BioActive Learner CD for additional help in mastering the chapter's material. Icon location in the chapter's margins shows which topics have tutorials or simulations in the CD.

2

Atoms and Molecules: The Chemical Basis of Life

LEARNING OBJECTIVES

After you have studied this chapter you should be able to

1. Name the principal chemical elements in living things and give an important function of each.
2. Compare the physical properties (mass and charge) and the locations of electrons, protons, and neutrons.
3. Distinguish between the atomic number and the mass number of an element.
4. Define the terms *electron orbital* and *electron shell.* Relate electron shells to principal energy levels.
5. Explain how the number of valence electrons of an atom is related to its chemical properties.
6. Distinguish among covalent bonds, hydrogen bonds, and ionic bonds. Compare them in terms of the mechanisms by which they form and their relative bond strengths.
7. Explain how cations and anions form and how they interact.
8. Distinguish between the terms *oxidation* and *reduction* and relate these processes to the transfer of energy.
9. Draw a simple ball-and-stick model of a water molecule, indicating the regions of partial positive and partial negative charge. Show how hydrogen bonds form between adjacent water molecules and explain how these are responsible for many of the properties of water.
10. Contrast acids and bases and discuss their properties.
11. Convert the hydrogen ion concentration (moles per liter) of a solution to a pH value. Describe how buffers help minimize changes in pH.
12. Describe the composition of a salt and explain why salts are important in organisms.

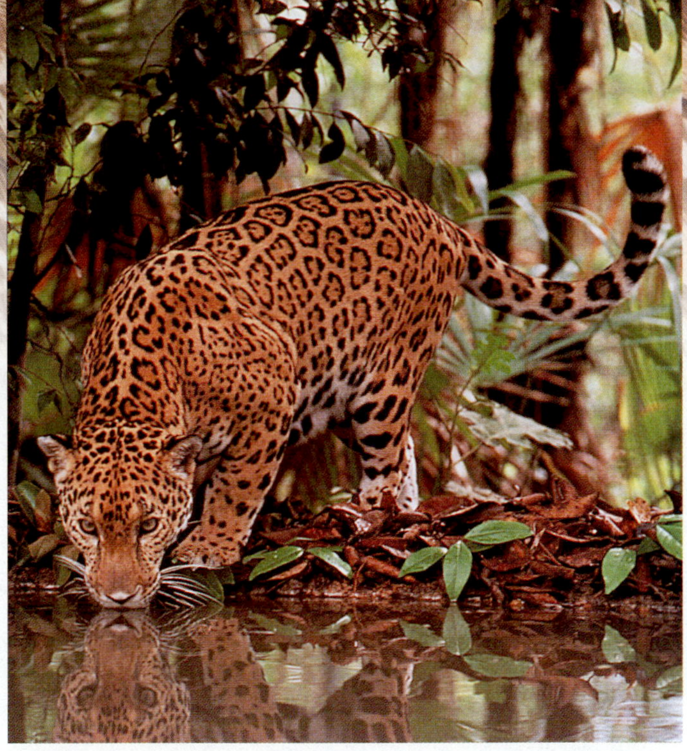

A jaguar, the largest cat in the Western Hemisphere, pauses to drink water from a rainforest stream. Water is a basic requirement for all life. *(Frans Lanting/Minden Pictures)*

Knowledge of chemistry is essential if we are to understand organisms and how they function. This jaguar and the plants of the tropical rain forest, as well as abundant unseen insects and microorganisms, share fundamental similarities in their chemical composition and basic metabolic processes. These chemical similarities provide strong evidence for the evolution of all organisms from a common ancestor and explain why much of what biologists learn from studying bacteria or rats in laboratories can be applied to other organisms, including humans. Furthermore, the basic chemical and physical principles governing organisms are not unique to living things, for they apply to nonliving systems as well.

Today much attention is given to the Human Genome Project, introduced in Chapter 1. The success of this exciting initiative relies heavily on biochemistry and **molecular biology,** the chemistry and physics of the molecules that constitute living things. A biochemist might investigate the precise interactions among a cell's atoms and molecules that maintain the energy flow essential to life, and a molecular biologist might study how proteins interact with deoxyribonucleic acid (DNA) in ways that control the expression of certain genes. However, an understanding of chemistry is essential to *all* biologists. An evolutionary biologist might study evolutionary relationships by comparing the DNA of different types of organisms. An ecologist might study how energy is transferred among the organisms living in an estuary or monitor the biological effects of changes in the salinity of the water. A botanist might study unique compounds produced by plants, and might even be a "chemical prospector," seeking new sources of medicinal agents.

In this chapter we lay a foundation for understanding how the structure of atoms determines the way they form chemical bonds to produce complex compounds. Most of our discussion centers around small, simple substances known as **inorganic compounds.** Among the biologically important groups of inorganic compounds are water, many simple acids and bases, and simple salts. We pay particular attention to water, the most abundant substance on Earth's surface and in organisms, and we examine how its unique properties affect living things as well as their nonliving environment. In Chapter 3 we extend our discussion to **organic compounds,** carbon-containing compounds that are generally large and complex. In all but the simplest organic compounds two or more carbon atoms are bonded to each other to form the backbone, or skeleton, of the molecule.

TABLE 2–1 Functions of Elements that Make Up Two Representative Organisms

Element (Chemical Symbol)	Human: % of Total Mass	Nonwoody Plant: % of Total Mass	Functions
Oxygen (O)	65	78	Required for cellular respiration; present in most organic compounds; component of water
Carbon (C)	18	11	Forms backbone of organic molecules; each carbon atom can form four bonds with other atoms
Hydrogen (H)	10	9	Present in most organic compounds; component of water; hydrogen ion (H^+) is involved in some energy transfers
Nitrogen (N)	3	*	Component of proteins and nucleic acids; component of chlorophyll in plants
Calcium (Ca)	1.5	*	Structural component of bones and teeth; calcium ion (Ca^{2+}) is important in muscle contraction, conduction of nerve impulses, and blood clotting; associated with plant cell wall
Phosphorus (P)	1	*	Component of nucleic acids and of phospholipids in membranes; important in energy transfer reactions; structural component of bone
Potassium (K)	*	*	Potassium ion (K^+) is a principal positive ion (cation) in interstitial (tissue) fluid of animals; important in nerve function; affects muscle contraction; controls opening of stomata in plants
Sulfur (S)	*	*	Component of most proteins
Sodium (Na)	*	*	Sodium ion (Na^+) is a principal positive ion (cation) in interstitial (tissue) fluid of animals; important in fluid balance; essential for conduction of nerve impulses; important in photosynthesis in plants
Magnesium (Mg)	*	*	Needed in blood and other tissues of animals; activates many enzymes; component of chlorophyll in plants
Chlorine (Cl)	*	*	Chloride ion (Cl^-) is principal negative ion (anion) in interstitial (tissue) fluid of animals; important in water balance; essential for photosynthesis
Iron (Fe)	*	*	Component of hemoglobin in animals; activates certain enzymes

*The asterisk indicates that these elements represent less than 1% of the total mass. Other elements found in very small (trace) amounts in animals, plants, or both include iodine (I), manganese (Mn), copper (Cu), zinc (Zn), cobalt (Co), fluorine (F), molybdenum (Mo), selenium (Se), boron (B), silicon (Si), and a few others.

ELEMENTS ARE NOT CHANGED IN NORMAL CHEMICAL REACTIONS

Elements are substances that cannot be broken down into simpler substances by ordinary chemical reactions. Scientists have assigned each element a **chemical symbol:** usually the first letter or first and second letters of the English or Latin name of the element. For example, O is the symbol for oxygen, C for carbon, H for hydrogen, N for nitrogen, and Na for sodium (Latin *natrium*).

Just four elements—oxygen, carbon, hydrogen, and nitrogen—are responsible for more than 96% of the mass of most organisms. Others, such as calcium, phosphorus, potassium, and magnesium, are also consistently present but in smaller quantities. Some elements, such as iodine and copper, are known as *trace elements* because they are required only in minute amounts. Table 2–1 lists the elements that make up two representative organisms, a human and a typical nonwoody plant (such as grass), and briefly explains why each is important.

ATOMS ARE THE FUNDAMENTAL PARTICLES OF ELEMENTS

An **atom** has been traditionally defined as the smallest portion of an element that retains its chemical properties. Atoms are much smaller than the tiniest particle visible under a light microscope. By scanning tunneling microscopy, with magnifications as high as $\times$ 5 million, researchers have been able to photograph the positions of some large atoms in molecules.

Physicists have discovered a number of subatomic particles, but for our purposes we need consider only three: electrons, protons, and neutrons. An **electron** is a particle that carries a unit of negative electrical charge; a **proton** carries a unit of positive charge; and a **neutron** is an uncharged particle. In an electrically neutral atom, the number of electrons is equal to the number of protons.

Clustered together, protons and neutrons compose the **atomic nucleus.** Electrons, however, have no fixed locations and move rapidly through the mostly empty space surrounding the atomic nucleus.

An atom is uniquely identified by its number of protons

Each kind of element has a fixed number of protons in the atomic nucleus. This number, called the **atomic number,** is written as a subscript to the left of the chemical symbol. Thus $_1H$ indicates that the hydrogen nucleus contains one proton, and $_8O$ means that the oxygen nucleus contains eight protons. It is the atomic number, the number of protons in its nucleus, that determines an atom's identity and defines the element.

The **periodic table** (Fig. 2–1 and Appendix B) is a chart in which elements are arranged in order by atomic number. As will become evident later in this chapter, the periodic table is an extremely useful device because it allows us to simultaneously correlate a great many of the relationships among the various elements.

Figure 2–1 includes representations of the **electron configurations** of several elements important in organisms. These *Bohr models,* which show the electrons arranged in a series of concentric circles around the nucleus, are convenient to use but inaccurate. The space outside the nucleus is actually extremely large compared to the nucleus, and, as we will see, electrons do not actually circle the nucleus in fixed concentric pathways.

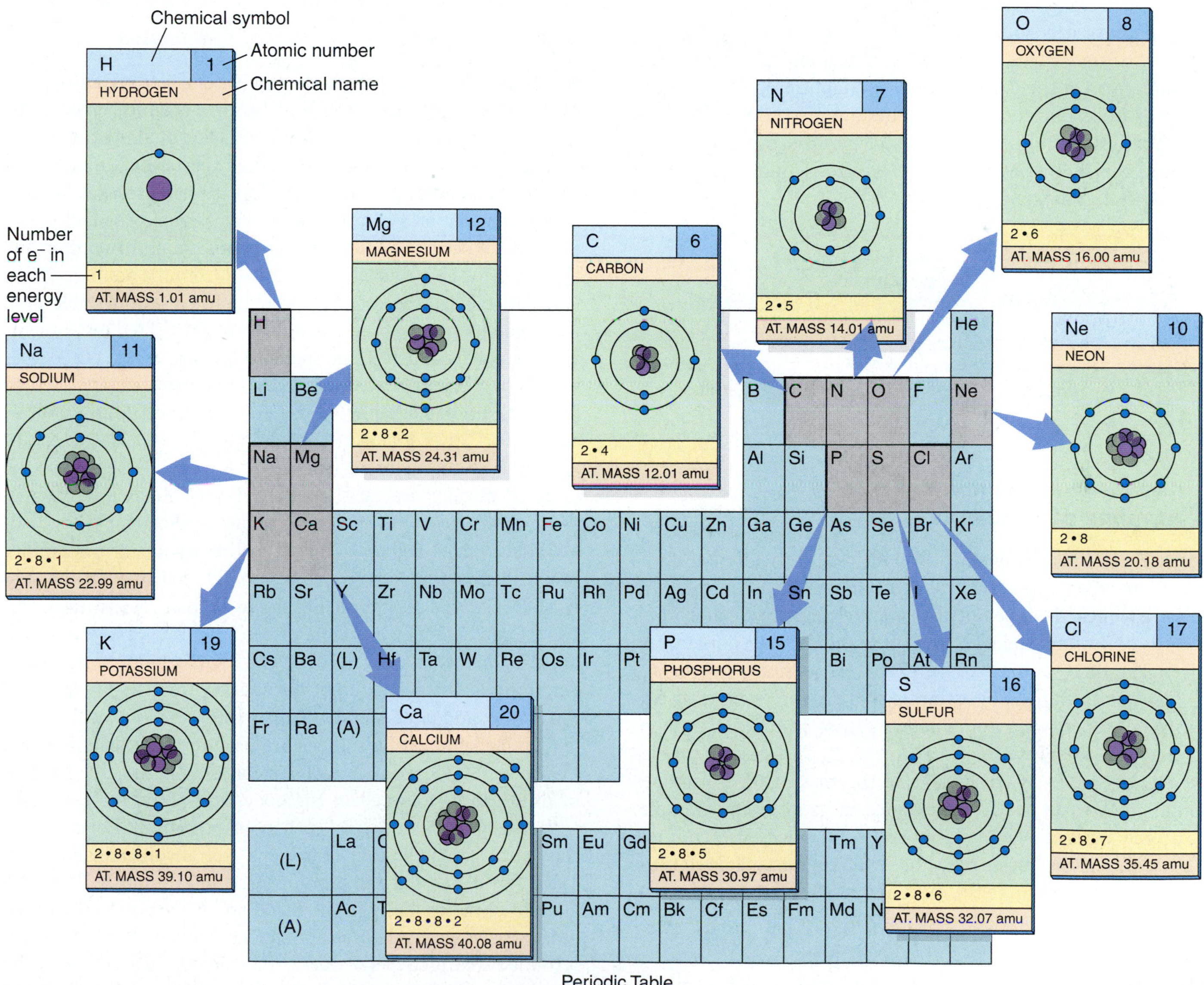

Figure 2–1 The periodic table. Note the Bohr models depicting the electron configuration of atoms of some biologically important elements. Although the Bohr model does not depict electron configurations accurately, it is commonly used because of its simplicity and convenience. A complete periodic table is given in Appendix B.

Protons plus neutrons determine atomic mass

The mass of a subatomic particle is exceedingly small, much too small to be conveniently expressed in grams or even micrograms.[1] Such masses are expressed in terms of the **atomic mass unit (amu),** also called the **dalton** in honor of John Dalton, who formulated an atomic theory in the early 1800s. One amu is equal to the approximate mass of a single proton or a single neutron. Protons and neutrons make up almost all the mass of an atom. The mass of a single electron is only about 1/1800 the mass of a proton or neutron.

The **atomic mass** of an atom is a number that indicates approximately how much matter it contains compared with another atom. This value is determined by adding the number of protons to the number of neutrons and expressing the result in atomic mass units or daltons.[2] The mass of the electrons is ignored because it is so small. The atomic mass number is indicated by a superscript to the left of the chemical symbol. The common form of the oxygen atom, with eight protons and eight neutrons in its nucleus, has an atomic number of 8 and a mass of 16 atomic mass units. It is indicated by the symbol $^{16}_{8}O$.

The characteristics of protons, electrons, and neutrons are summarized in the following table:

Particle	Charge	Approximate Mass	Location
Proton	Positive	1 amu	Nucleus
Neutron	Neutral	1 amu	Nucleus
Electron	Negative	Approx. 1/1800 amu	Outside nucleus

Isotopes of an element differ in number of neutrons

Most elements consist of a mixture of atoms with different numbers of neutrons and thus different masses. Such atoms are called **isotopes.** Isotopes of the same element have the same number of protons and electrons; only the number of neutrons varies. The three isotopes of hydrogen, $^{1}_{1}H$ (ordinary hydrogen), $^{2}_{1}H$ (deuterium), and $^{3}_{1}H$ (tritium), contain zero, one, and two neutrons, respectively. Bohr models of two isotopes of carbon, $^{12}_{6}C$ and $^{14}_{6}C$, are illustrated in Figure 2–2. The mass of an element is expressed as an average of the masses of its isotopes (weighted by their relative abundance in nature). For example, the atomic mass of hydrogen is not 1.0 amu, but 1.0079 amu, reflecting the natural occurrence of small amounts of deuterium and tritium in addition to the more abundant ordinary hydrogen.

[1] Tables of commonly used units of scientific measurement are printed inside the back cover of this text.

[2] Unlike weight, mass is independent of the force of gravity. For convenience, however, we consider mass and weight to be equivalent. Atomic weight has the same numerical value as atomic mass, but it has no units.

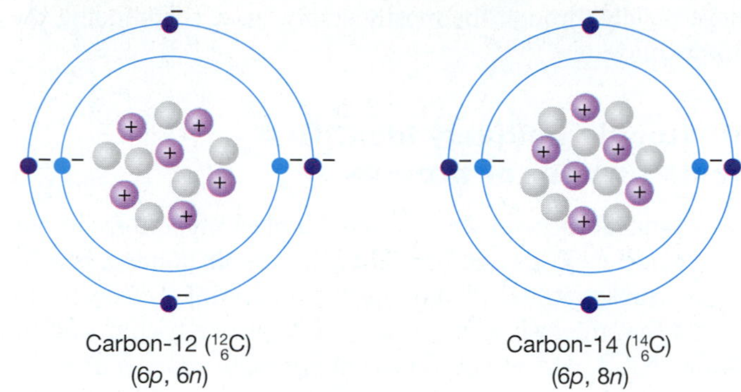

Figure 2–2 Isotopes differ in atomic mass. Carbon-12 ($^{12}_{6}C$) is the most common isotope of carbon. Its nucleus contains six protons and six neutrons, so its atomic mass is 12. Carbon-14 ($^{14}_{6}C$) is a rare radioactive carbon isotope. Because it contains eight neutrons, its atomic mass is 14.

Because they have the same number of electrons, all isotopes of a given element have essentially the same chemical characteristics. However, some isotopes are unstable and tend to break down, or decay, to a more stable isotope (usually becoming a different element). Such unstable isotopes are termed **radioisotopes** because they emit radiation when they decay. For example, the radioactive decay of $^{14}_{6}C$ occurs as a neutron decomposes to form a proton and a fast-moving electron, which is emitted from the atom as a form of radiation known as a beta (β) particle. The resulting stable atom is the common form of nitrogen, $^{14}_{7}N$. Sophisticated instruments allow scientists to detect and measure β particles and other types of radiation. Radioactive decay can also be detected by a method known as **autoradiography,** in which radiation causes the appearance of dark silver grains in photographic film (Fig. 2–3).

Because the different isotopes of a given element have the same chemical characteristics, they are essentially interchangeable in molecules. Molecules containing radioisotopes are usually metabolized and/or localized in the organism in a similar way to their nonradioactive counterparts, and they can be substituted. For this reason, radioisotopes such as ^{3}H (tritium), ^{14}C, and ^{32}P are extremely valuable research tools used in areas such as dating fossils (see Fig. 17–10), tracing biochemical pathways, determining the sequence of genetic information in DNA (see Fig. 14–9), and understanding sugar transport in plants.

In medicine, radioisotopes are used for both diagnosis and treatment. The location and/or metabolism of a sugar, hormone, or drug can be followed in the body by labeling the substance with a radioisotope such as carbon-14 or tritium. For example, the active component in marijuana (tetrahydrocannabinol, or THC) can be labeled and administered intravenously. Then the amount of radioactivity in the blood and urine can be measured at successive intervals. Results of such measurements have determined that for several weeks this compound remains in the blood, and products of its metabolism can be detected in the urine. Radioisotopes are also used to test thyroid gland function, to provide images of blood flow in the arteries supplying the heart muscle, and to study many other aspects of body function and chemistry. Because radiation can interfere with cell division,

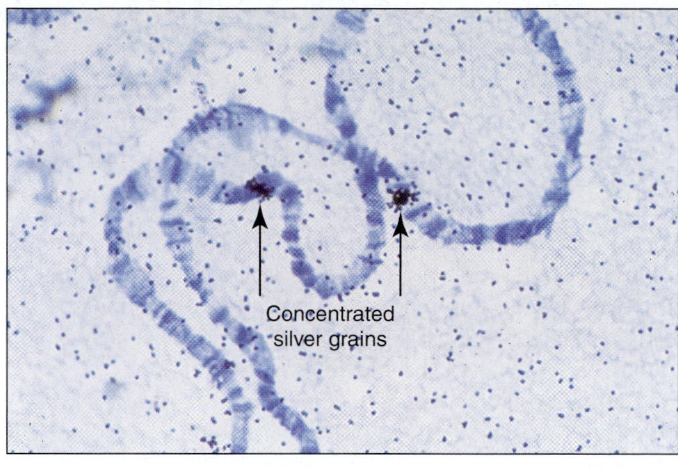

Figure 2–3 Autoradiography. The chromosomes of the fruit fly, *Drosophila melanogaster,* shown in this light micrograph (LM) have been covered with photographic film in which silver grains *(dark spots)* are produced when tritium (^{3}H) that has been incorporated into DNA undergoes radioactive decay. The concentrations of silver grains *(arrows)* mark the locations of specific DNA molecules. *(Peter J. Bryant/Biological Photo Service)*

radioisotopes have been used therapeutically in the treatment of cancer (a disease often characterized by rapidly dividing cells).

Electrons move in orbitals corresponding to energy levels

Electrons move through characteristic regions of three-dimensional space, termed **orbitals.** Each orbital contains a maximum of two electrons. Because it is impossible to know an electron's position at any given time, orbitals are most accurately depicted as "electron clouds," shaded areas whose density is proportional to the probability that an electron is present there at any given instant. The energy of an electron depends on the orbital it occupies. Electrons in orbitals with similar energies, said to be at the same **principal energy level,** make up an **electron shell.** These are illustrated in Figure 2–4.

In general, electrons in a shell distant from the nucleus have greater energy than those in a shell close to the nucleus. This is because energy is required to move a negatively charged electron farther away from the positively charged nucleus. The most energetic electrons, known as **valence electrons,** are said to occupy the **valence shell.** The valence shell is represented as the outermost concentric ring in a Bohr model.

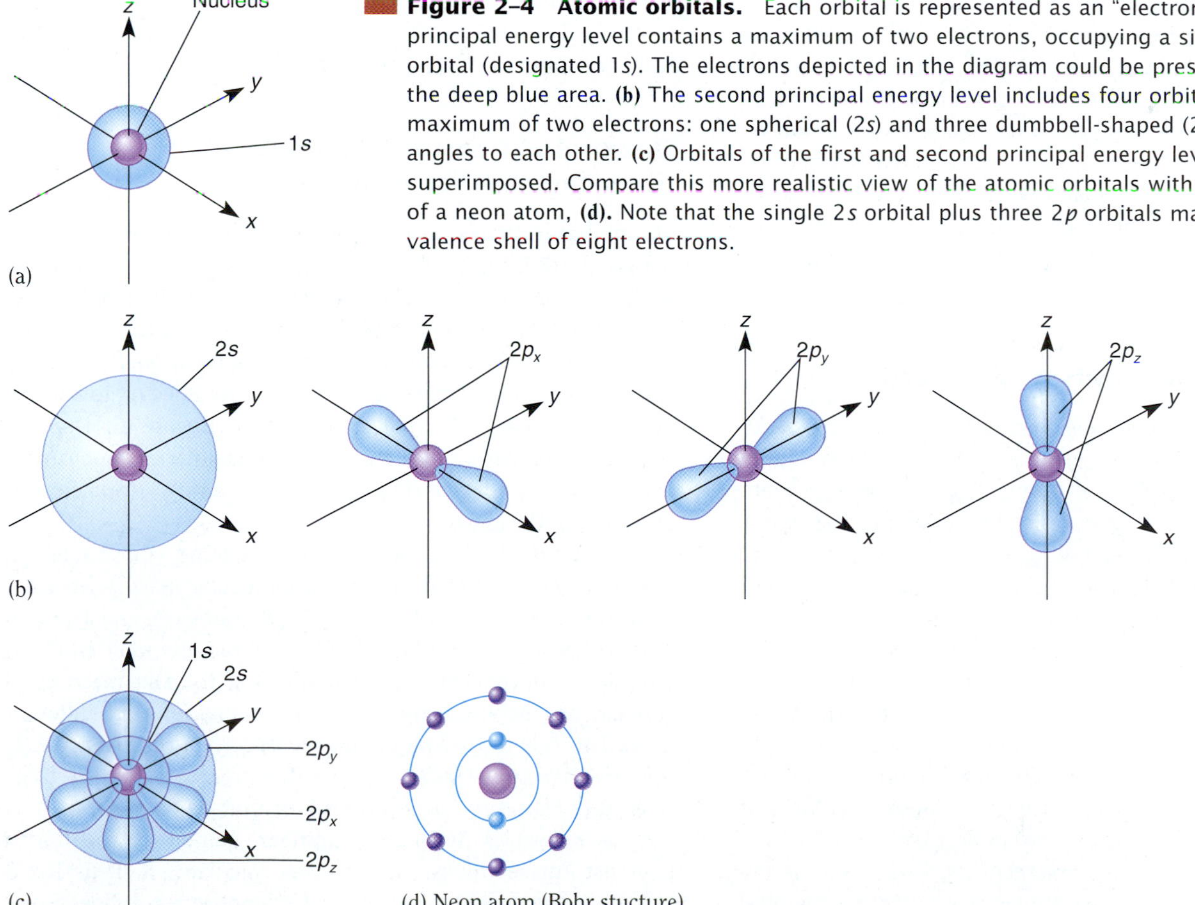

Figure 2–4 Atomic orbitals. Each orbital is represented as an "electron cloud." **(a)** The first principal energy level contains a maximum of two electrons, occupying a single spherical orbital (designated 1*s*). The electrons depicted in the diagram could be present anywhere within the deep blue area. **(b)** The second principal energy level includes four orbitals, each with a maximum of two electrons: one spherical (2*s*) and three dumbbell-shaped (2*p*) orbitals at right angles to each other. **(c)** Orbitals of the first and second principal energy levels are shown superimposed. Compare this more realistic view of the atomic orbitals with the Bohr structure of a neon atom, **(d).** Note that the single 2*s* orbital plus three 2*p* orbitals make up neon's full valence shell of eight electrons.

An electron can move to an orbital farther from the nucleus by receiving more energy, or it can give up energy and sink to a lower energy level in an orbital nearer the nucleus. Changes in electron energy levels are important in energy conversions in organisms. For example, during photosynthesis light energy absorbed by chlorophyll molecules causes electrons to move to a higher energy level (see Fig. 8–3).

VALENCE ELECTRONS PARTICIPATE IN CHEMICAL REACTIONS

The chemical behavior of an atom is determined primarily by the number and arrangement of its valence electrons. The valence shell of hydrogen or helium is full (i.e., stable) when it contains two electrons. The valence shell of any other atom is full when it contains eight electrons. When the valence shell is not full, the atom tends to lose, gain, or share electrons to achieve a full outer shell. The valence shells of all isotopes of an element are identical; this is why they have similar chemical properties and can substitute for each other in chemical reactions (e.g., tritium can substitute for ordinary hydrogen).

Elements that fall into the same vertical column (said to belong to the same *group*) of the periodic table have similar chemical properties because their valence shells have similar tendencies to lose, gain, or share electrons. For example, chlorine and bromine, included in a group commonly known as the *halogens,* are highly reactive. Because their valence shells have seven electrons, they tend to gain an electron in chemical reactions. By contrast, hydrogen, sodium, and potassium each have a single valence electron, which they tend to give up or share with another atom. Helium (He) and neon (Ne) belong to a group referred to as the "noble gases." They are quite unreactive because their valence shells are full. Note the incomplete valence shells of some of the elements important in organisms, including carbon, hydrogen, oxygen, and nitrogen, in Figure 2–1, and compare them with the full valence shell of neon in Figure 2–4*d*.

Atoms form compounds and molecules

Two or more atoms may combine chemically. When atoms of *different* elements combine, the result is a chemical compound. A **chemical compound** consists of atoms of two or more different elements combined in a fixed ratio. For example, water is a chemical compound composed of hydrogen and oxygen in a ratio of 2 : 1. Common table salt, sodium chloride, is a chemical compound made up of sodium and chlorine in a 1 : 1 ratio.

When two or more atoms combine chemically, units called **molecules** can be formed. For example, when two atoms of oxygen combine chemically, a molecule of oxygen is formed. Water is a molecular compound, with each molecule consisting of two atoms of hydrogen and one of oxygen. However, as we shall see, not all compounds are made up of molecules. Sodium chloride is an example of a compound that is not molecular.

Recall from Chapter 1 that emergent properties become evident as a system becomes more complex. Accordingly, a chemical compound has unique properties that are not merely the sum of the properties of its component atoms. For example, at room temperature water is a liquid, whereas hydrogen and oxygen are gases.

Simplest, molecular, and structural chemical formulas give different information

A **chemical formula** is a shorthand expression that describes the chemical composition of a substance. Chemical symbols indicate the types of atoms present, and subscript numbers indicate the ratios among the atoms. There are several types of chemical formulas, each providing specific kinds of information.

In a **simplest formula** (also known as an *empirical formula*), the subscripts give the smallest whole-number ratios for the atoms present in a compound. For example, the simplest formula for hydrazine is NH_2, indicating that there is a 1:2 ratio of nitrogen to hydrogen. (Note that when a single atom of a type is present, the subscript number 1 is never written.)

In a **molecular formula,** the subscripts indicate the actual numbers of each type of atom per molecule. The molecular formula for hydrazine is N_2H_4, which indicates that each molecule of hydrazine consists of two atoms of nitrogen and four atoms of hydrogen. The molecular formula for water, H_2O, indicates that each molecule consists of two atoms of hydrogen and one atom of oxygen.

A **structural formula** shows not only the types and numbers of atoms in a molecule but also their arrangement. From the molecular formula for water, H_2O, you would not know whether the atoms were arranged H—H—O or H—O—H. The structural formula, H—O—H, settles the matter, indicating that the two hydrogen atoms are attached to the oxygen atom.

One mole of any substance contains the same number of units

The **molecular mass** of a compound is the sum of the atomic masses of the component atoms of a single molecule; thus, the molecular mass of water, H_2O, is (hydrogen: 2 × 1 amu) + (oxygen: 1 × 16 amu), or 18 amu. (Owing to the presence of isotopes, atomic mass values are not whole numbers. However, for easy calculation each atomic mass value has been rounded off to a whole number.) Similarly, the molecular mass of glucose ($C_6H_{12}O_6$), a simple sugar that is a key compound in cellular metabolism, is (carbon: 6 × 12 amu) + (hydrogen: 12 × 1 amu) + (oxygen: 6 × 16 amu), or 180 amu.

The amount of an element or compound whose mass in grams is equivalent to its atomic or molecular mass is termed 1 **mole (mol).** Thus, 1 mol of water is 18 grams (g), and 1 mol of glucose has a mass of 180 g. As we shall see, the mole is an extremely useful concept because it allows us to make meaningful comparisons between atoms and molecules of very different mass. This is because *1 mol of any substance always has exactly the same number of units,* whether they are small atoms or large molecules. The very large number of units in a mole, 6.02×10^{23}, is known as **Avogadro's number,** named for the Italian physicist Amadeo Avogadro, who first calculated it. Thus 1 mol (180 g) of glucose contains 6.02×10^{23} molecules, as does 1 mol

(2 g) of molecular hydrogen (H_2). Although it is impossible to count atoms and molecules individually, this fact allows a scientist to "count" them simply by weighing a sample. Molecular biologists usually deal with smaller values, either millimoles (a mmol is one-thousandth of a mole) or micromoles (a μmol is one-millionth of a mole).

The mole concept also allows us to make useful comparisons among solutions. A 1-molar solution, represented by 1 *M*, contains 1 mol of that substance dissolved in 1 liter (L) of solution. For example, we can compare 1 L of a 1 *M* solution of glucose with 1 L of a 1 *M* solution of sucrose (table sugar, a larger molecule). They differ in the mass of the dissolved sugar (180 g and 340 g, respectively), but they each contain 6.02×10^{23} sugar molecules.

Chemical equations describe chemical reactions

During any moment in the life of an organism, be it a bacterial cell, a mushroom, or a butterfly, many complex chemical reactions are taking place. Chemical reactions, for example, the reaction between glucose and oxygen, can be described by means of chemical equations:

$$\underset{\text{Glucose}}{C_6H_{12}O_6} + \underset{\text{Oxygen}}{6\ O_2} \longrightarrow \underset{\text{Carbon dioxide}}{6\ CO_2} + \underset{\text{Water}}{6\ H_2O} + \text{Energy}$$

In a chemical equation, the **reactants** (the substances that participate in the reaction) are generally written on the left side, and the **products** (the substances formed by the reaction) are written on the right side. The arrow means "yields" and indicates the direction in which the reaction tends to proceed.

Chemical compounds react with each other in quantitatively precise ways. The numbers preceding the chemical symbols or formulas (known as *coefficients*) indicate the relative number of atoms or molecules reacting. For example, 1 mol of glucose burned in a fire or metabolized in a cell reacts with 6 mol of oxygen to form 6 mol of carbon dioxide and 6 mol of water.

Many reactions can proceed simultaneously in the reverse direction (to the left) as well as in the forward direction (to the right); at **dynamic equilibrium** the rates of the forward and reverse reactions are equal (see Chapter 6). Reversible reactions are indicated by double arrows:

$$\underset{\text{Carbon dioxide}}{CO_2} + \underset{\text{Water}}{H_2O} \rightleftharpoons \underset{\text{Carbonic acid}}{H_2CO_3}$$

In this example, the arrows are drawn in different lengths to indicate that when the reaction reaches equilibrium there will be more reactants (CO_2 and H_2O) than product (H_2CO_3).

ATOMS ARE JOINED BY CHEMICAL BONDS

The atoms of a compound are held together by forces of attraction called **chemical bonds.** Each bond represents a certain amount of chemical energy. **Bond energy** is the energy necessary to break a chemical bond. The valence electrons dictate how many bonds an atom can participate in. The two principal types of strong chemical bonds are covalent bonds and ionic bonds.

In covalent bonds electrons are shared

Covalent bonds involve the sharing of electrons between atoms in a way that results in each atom having a filled valence shell. A compound consisting mainly of covalent bonds is called a **covalent compound.** A simple example of a covalent bond is the joining of two hydrogen atoms in a molecule of hydrogen gas, H_2. Each atom of hydrogen has one electron, but two electrons are required to complete its valence shell. The hydrogen atoms have equal capacities to attract electrons, so neither donates an electron to the other. Instead, the two hydrogen atoms share their single electrons so that each of the two electrons is attracted simultaneously to the two protons in the two hydrogen nuclei. The two electrons thus whirl around both atomic nuclei, joining the two atoms.

A simple way of representing the electrons in the valence shell of an atom is to use dots placed around the chemical symbol of the element. Such a representation is called the *Lewis structure* of the atom, named for G.N. Lewis, who developed this type of notation. In a water molecule, two hydrogen atoms are covalently bonded to an oxygen atom:

$$H\cdot + H\cdot + \cdot\ddot{\underset{\cdot\cdot}{O}}\cdot \longrightarrow H:\ddot{\underset{\cdot\cdot}{O}}:H$$

Oxygen has six valence electrons; by sharing electrons with two hydrogen atoms, it completes its valence shell of eight. At the same time each hydrogen atom obtains a complete valence shell of two. (Note that in the structural formula H—O—H, each pair of shared electrons constitutes a covalent bond, represented by a solid line. Unshared electrons are usually omitted in a structural formula.)

The carbon atom has four electrons in its valence shell, all of which are available for covalent bonding:

$$\cdot\dot{\underset{\cdot}{C}}\cdot$$

When one carbon and four hydrogen atoms share electrons, a molecule of methane, CH_4, is formed:

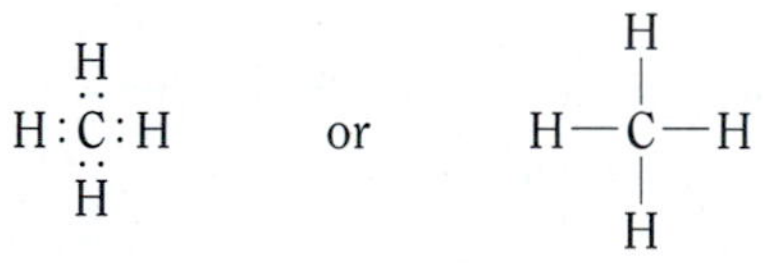

$$\begin{matrix} & H & \\ H & \ddot{\underset{\cdot\cdot}{C}} & H \\ & H & \end{matrix} \quad \text{or} \quad \begin{matrix} & H & \\ & | & \\ H- & C & -H \\ & | & \\ & H & \end{matrix}$$

Lewis structure — Structural formula

The nitrogen atom has five electrons in its valence shell. Recall that each orbital can hold a maximum of two electrons. Usually two electrons occupy one orbital, leaving three available for sharing with other atoms:

$$\cdot\ddot{\underset{\cdot}{N}}\cdot$$

When a nitrogen atom shares electrons with three hydrogen atoms, a molecule of ammonia, NH_3, is formed:

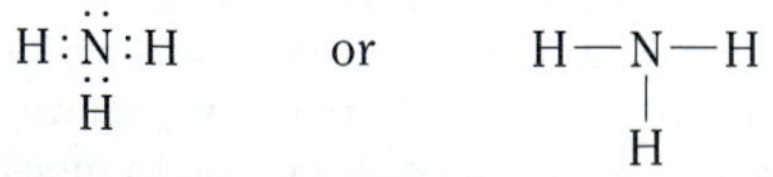

$$H:\ddot{\underset{\cdot\cdot}{N}}:H \quad \text{or} \quad \begin{matrix} H- & N & -H \\ & | & \\ & H & \end{matrix}$$
$$\quad\ H$$

Lewis structure — Structural formula

When one pair of electrons is shared between two atoms, the covalent bond is referred to as a **single covalent bond** (Fig. 2–5*a*). Two oxygen atoms may achieve stability by forming covalent bonds with one another. Each oxygen atom has six electrons in its outer shell. To become stable, the two atoms share two pairs of electrons, forming molecular oxygen (Fig. 2–5*b*). When two pairs of electrons are shared in this way, the covalent bond is called a **double covalent bond,** which is represented by two parallel solid lines. Similarly, a **triple covalent bond** (represented by three parallel solid lines) is formed when three pairs of electrons are shared between two atoms.

The number of covalent bonds usually formed by the atoms commonly present in biologically important molecules is summarized as follows:

Atom	Symbol	Covalent Bonds
Hydrogen	H	1
Oxygen	O	2
Carbon	C	4
Nitrogen	N	3
Phosphorus	P	5
Sulfur	S	2

The function of a molecule is related to its shape

In addition to being composed of atoms with certain properties, each kind of molecule has a characteristic size and a general overall shape. Although the shape of a molecule may change (within certain limits), the functions of molecules in living cells are dictated largely by their geometric shapes. A molecule that consists of two atoms, for example, is linear. Molecules composed of more than two atoms may have more complicated shapes. The geometric shape of a molecule provides the optimal distance between the atoms to counteract the repulsion of electron pairs.

When an atom forms covalent bonds with other atoms, the orbitals in the valence shell may become rearranged in a process known as **orbital hybridization,** thereby affecting the shape of the resulting molecule. For example, when four hydrogen atoms combine with a carbon atom to form a molecule of methane (CH_4), the hybridized valence shell orbitals of the carbon form a geometric structure known as a *tetrahedron*, with one hydrogen atom present at each of its four corners (Fig. 2–6; see Fig. 3–2*b*).

Covalent bonds can be nonpolar or polar

Atoms of different elements vary in their affinity for electrons. **Electronegativity** is a measure of an atom's attraction for shared electrons in chemical bonds. Very electronegative atoms such as oxygen, nitrogen, fluorine, and chlorine are sometimes called "electron-greedy." When covalently bound atoms have similar electronegativities, the electrons are shared equally, and the covalent bond is described as **nonpolar.** The covalent bond of the hydrogen molecule is nonpolar, as are the covalent bonds of molecular oxygen and methane.

In a covalent bond between two different elements, such as oxygen and hydrogen, the electronegativities of the atoms may be different. If so, electrons are pulled closer to the atomic nucleus of the element with the greater electron affinity (in this case, oxygen). A covalent bond between atoms that differ in elec-

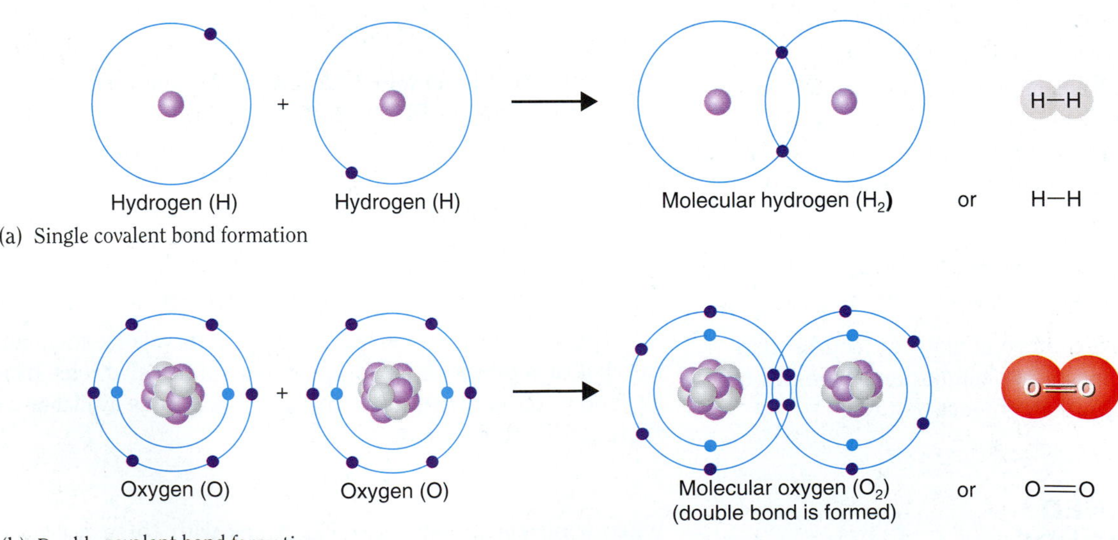

Figure 2–5 Electron sharing in covalent compounds. **(a)** Two hydrogen atoms achieve stability by sharing electrons, thereby forming a molecule of hydrogen. In the structural formula shown on the right, the straight line between the hydrogen atoms represents a single covalent bond. **(b)** In molecular oxygen, two oxygen atoms share two pairs of electrons, forming a double covalent bond.

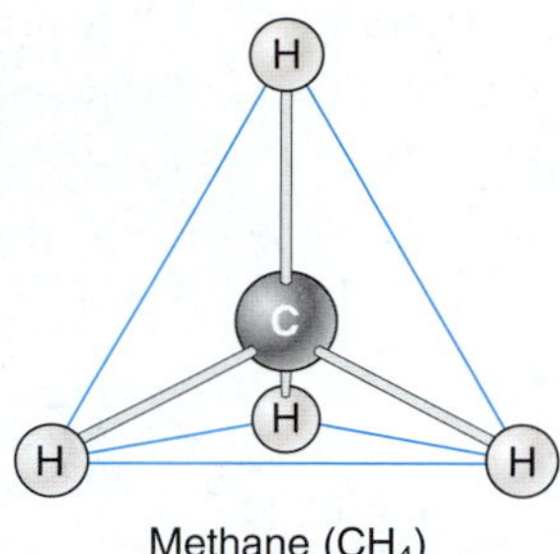

Figure 2-6 Orbital hybridization in methane. The four hydrogens are located at the corners of a tetrahedron owing to hybridization of the valence shell orbitals of carbon.

tronegativity is called a **polar covalent bond.** Such a bond has two dissimilar ends (or poles), one with a partial positive charge and the other with a partial negative charge. Each of the two covalent bonds in water is polar because there is a partial positive charge at the hydrogen end of the bond and a partial negative charge at the oxygen end, where the "shared" electrons are more likely to be found.

Covalent bonds differ in their degree of polarity, ranging from those in which the electrons are exactly shared (as in the nonpolar hydrogen molecule) to those in which the electrons are much closer to one atom than to the other (as in water). Oxygen is quite electronegative and forms polar covalent bonds with carbon, hydrogen, and many other atoms. Nitrogen is also strongly electronegative, although less so than oxygen.

A molecule with one or more polar covalent bonds can be polar even though it is electrically neutral as a whole. This is because a **polar molecule** has one end with a partial positive charge and another end with a partial negative charge. One example is water (Fig. 2–7). The polar bonds between the hydrogens and the oxygen are arranged in a V shape, rather than linearly. The oxygen end therefore constitutes the negative pole of the molecule, and the end with the two hydrogens is the positive pole.

Ionic bonds form between cations and anions

Some atoms or groups of atoms are not electrically neutral. A particle with one or more units of electrical charge is called an **ion.** An atom becomes an ion if it gains or loses one or more electrons. An atom with one, two, or three electrons in its valence shell tends to lose electrons to other atoms. Such an atom then becomes positively charged because its nucleus contains more protons than the number of electrons orbiting around the nucleus. These positively charged ions are termed **cations.** Atoms with five, six, or seven valence electrons tend to gain electrons from other atoms and become negatively charged **anions.**

The properties of ions are quite different from those of the electrically neutral atoms from which they were derived. For example, although chlorine gas is a poison, chloride ions (Cl^-) are essential to life (see Table 2–1). Because their electrical charges provide a basis for many interactions, cations and anions are involved in energy transformations within the cell, the transmission of nerve impulses, muscle contraction, and many other life processes (Fig. 2–8).

A group of covalently bonded atoms can also become an ion *(polyatomic ion)*. Unlike a single atom, a group of atoms can lose or gain protons (derived from hydrogen atoms) as well as electrons. Therefore, a group of atoms can become a cation if it loses one or more electrons or gains one or more protons. A group of atoms becomes an anion if it gains one or more electrons or loses one or more protons.

An **ionic bond** forms as a consequence of the attraction between the positive charge of a cation and the negative charge of an anion. An **ionic compound** is a substance consisting of anions and cations bonded together by their opposite charges.

A good example of how ionic bonds are formed is the attraction between sodium ions and chloride ions. A sodium atom has one electron in its valence shell. It cannot fill its valence shell by obtaining seven electrons from other atoms, for it would then have a large unbalanced negative charge. Instead, it gives up its single valence electron to a very electronegative atom, such as chlorine, which acts as an electron acceptor (Fig. 2–9). Chlorine cannot give up the seven electrons in its valence shell, because it

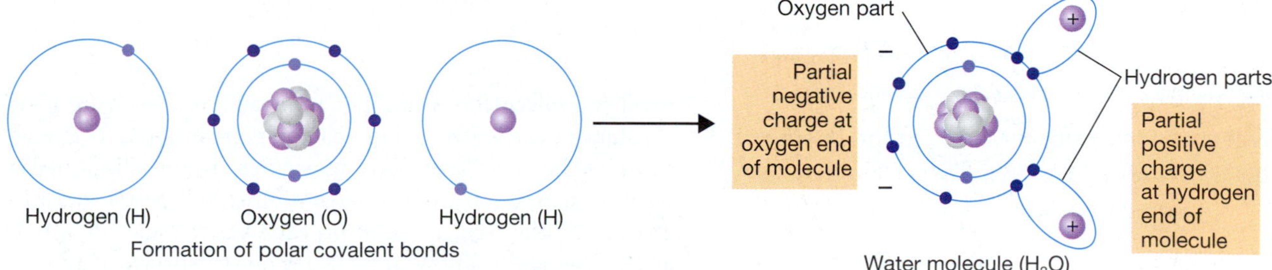

Figure 2-7 Water, a polar molecule. Note that the electrons tend to stay closer to the nucleus of the oxygen atom than to the hydrogen nuclei. This results in a partial negative charge on the oxygen portion of the molecule and a partial positive charge at the hydrogen end. Although the water molecule as a whole is electrically neutral, it is a polar covalent compound.

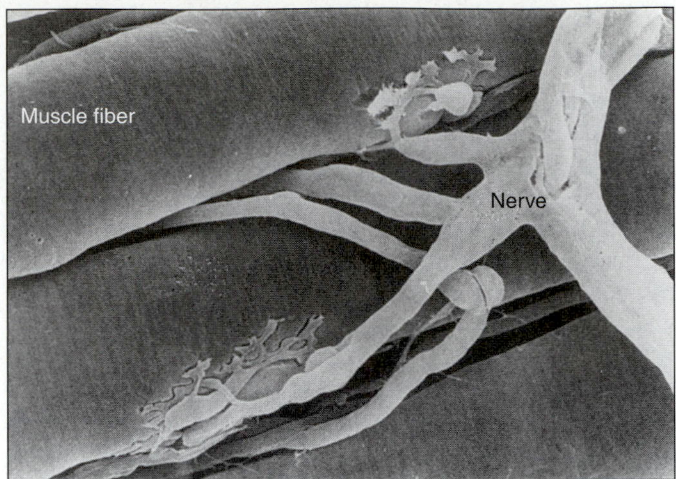

Figure 2–8 Ions and biological processes. Sodium, potassium, and chloride ions are essential for this nerve cell to stimulate these muscle fibers, initiating contraction. Calcium ions in the muscle cell are required for muscle contraction. *(D.W. Fawcett)*

would then have a large positive charge. Instead it strips an electron from an electron donor (sodium in this example) to complete its valence shell.

When sodium reacts with chlorine, sodium's valence electron is transferred completely to chlorine. Sodium becomes a cation, with one unit of positive charge (Na^+). Chlorine becomes an anion, a chloride ion with one unit of negative charge (Cl^-). These ions attract each other as a result of their opposite charges. They are held together by this electrical attraction in ionic bonds to form NaCl, sodium chloride,[3] or common table salt.

The term *molecule* does not adequately explain the properties of ionic compounds such as NaCl. When NaCl is in its solid crystal state, each ion is actually surrounded by six ions of opposite charge. The simplest formula, NaCl, indicates that sodium ions and chloride ions are present in a 1 : 1 ratio, but in the actual crystal, no discrete molecules composed of one Na^+ and one Cl^- ion are present.

Compounds joined by ionic bonds, such as sodium chloride, have a tendency to *dissociate* (separate) into their individual ions when placed in water:

$$\underset{\text{Sodium chloride}}{NaCl} \xrightarrow{\text{in } H_2O} \underset{\text{Sodium ion}}{Na^+} + \underset{\text{Chloride ion}}{Cl^-}$$

In the solid form of an ionic compound (i.e., in the absence of water), the ionic bonds are very strong. Water, however, is an

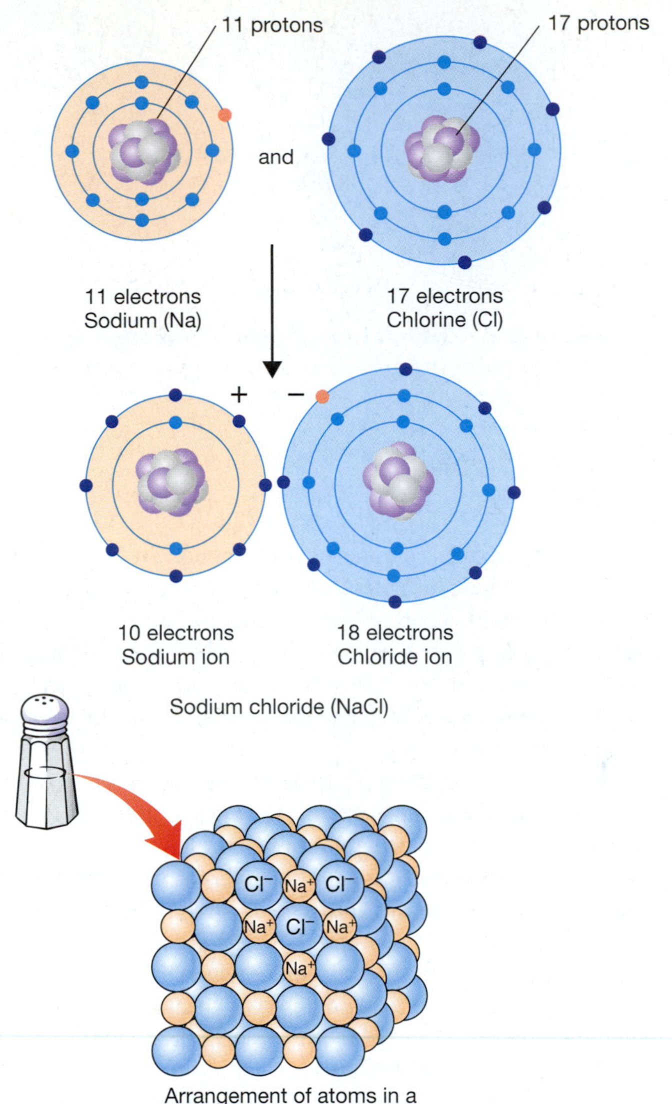

Figure 2–9 Ionic bonding. Sodium becomes a positively charged ion when it donates its single valence electron to chlorine, which has seven valence electrons. With this additional electron, chlorine completes its valence shell and becomes a negatively charged chloride ion. These sodium and chloride ions are attracted to one another by their unlike electrical charges, forming the ionic compound sodium chloride.

excellent **solvent;** as a liquid it is capable of dissolving many substances, particularly those that are polar or ionic. This is because of the polarity of water molecules. The localized partial positive charge (on the hydrogen atoms) and partial negative charge (on the oxygen atom) on each water molecule attract and surround the anions and cations on the surface of an ionic solid. As a result, the solid dissolves. A dissolved substance is referred to as a **solute.** In solution, each cation and anion of the ionic compound is surrounded by oppositely charged ends of the water molecules (Fig. 2–10). This process is known as **hy-**

[3] In both covalent and ionic binary compounds (*binary* denotes compounds consisting of two elements), the element having the greater attraction for electrons is named second, and an *-ide* ending is added to the stem name, such as sodium chloride and hydrogen fluoride. The *-ide* ending is also used to indicate an anion, as in chloride (Cl^-) and hydroxide (OH^-).

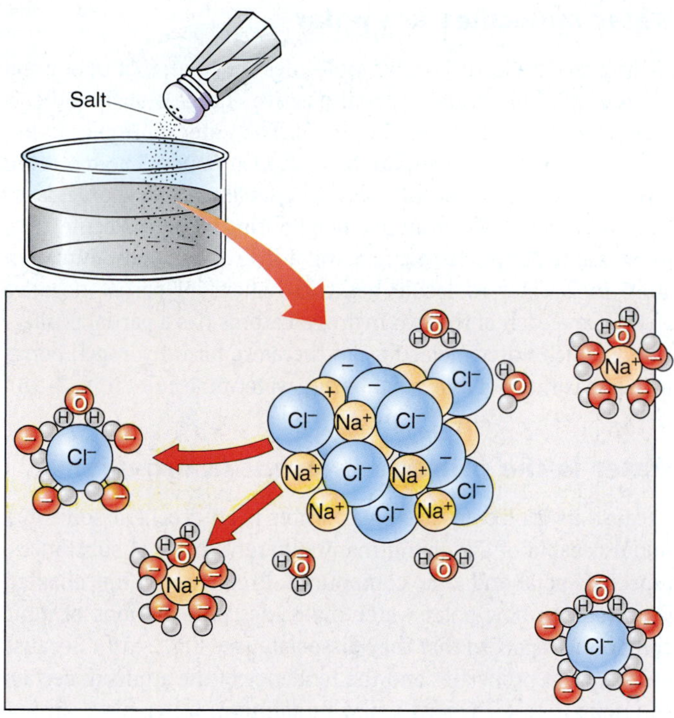

Figure 2–10 Hydration of an ionic compound. When the crystal of NaCl is added to water, the sodium and chloride ions are pulled apart as the partial negative ends of the water molecules are attracted to the positive sodium ions, and the partial positive ends of the water molecules are attracted to the negative chloride ions. When the NaCl is dissolved, each Na^+ and Cl^- is surrounded by water molecules electrically attracted to it.

dration. Hydrated ions still interact with each other to some extent, but the transient ionic bonds formed are much weaker than those in a solid crystal.

Hydrogen bonds are weak attractions

Another type of bond important in organisms is the **hydrogen bond.** When hydrogen combines with oxygen (or with another relatively electronegative atom such as nitrogen), it acquires a partial positive charge because its electron spends more time closer to the electronegative atom. Hydrogen bonds tend to form between an atom with a partial negative charge and a hydrogen atom that is covalently bonded to oxygen or nitrogen (Fig. 2–11). The atoms involved may be in two parts of the same large molecule or in two different molecules. Water molecules interact with each other extensively through hydrogen bond formation.

Hydrogen bonds are readily formed and broken. Although individually relatively weak, hydrogen bonds are collectively strong when present in large numbers. Furthermore, they have a specific length and orientation. As we will see in Chapter 3, these features are very important in determining the three-dimensional structure of large molecules such as DNA and proteins.

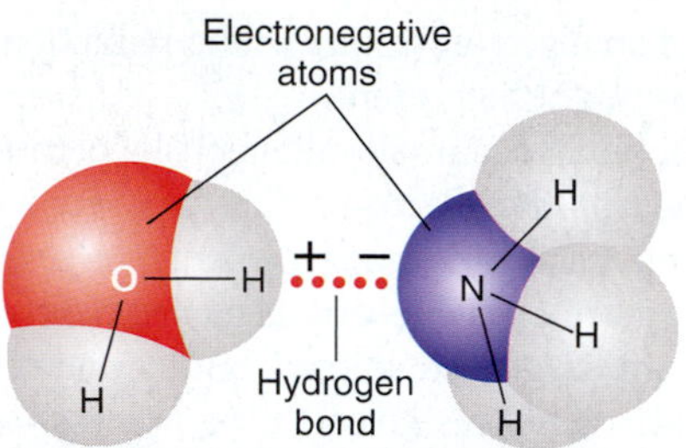

Figure 2–11 Hydrogen bonding. A hydrogen bond (generally indicated by a dotted line) can form between two molecules with regions of unlike partial charge. A hydrogen atom in a water molecule has a partial positive charge because of its polar covalent bond with oxygen. Nitrogen is strongly electronegative and, in molecules like ammonia (NH_3), has a partial negative charge because of its polar covalent bonds with hydrogen. Here, the nitrogen atom of the ammonia molecule is joined by a hydrogen bond to a hydrogen atom of a water molecule.

ELECTRONS AND THEIR ENERGY ARE TRANSFERRED IN REDOX REACTIONS

Many of the energy conversions that go on in a cell involve reactions in which an electron is transferred from one substance to another. This is because the transfer of an electron also involves the transfer of the energy of that electron. Such an electron transfer is called an *oxidation-reduction,* or **redox reaction.** Both cellular respiration (Chapter 7) and photosynthesis (Chapter 8) are essentially redox processes.

Rusting, which is the combination of iron (symbol Fe) with oxygen, is a simple illustration of oxidation and reduction:

$$4\,Fe + 3\,O_2 \longrightarrow \underset{\text{Iron (III) oxide}}{2\,Fe_2O_3}$$

Oxidation and reduction always occur together, but initially we will discuss them separately. **Oxidation** is a chemical process in which an atom, ion, or molecule loses electrons. In rusting, each iron atom becomes oxidized as it loses three electrons.

$$4\,Fe \rightarrow 4\,Fe^{3+} + 12e^-$$

The e^- is a symbol for an electron; the + superscript in Fe^{3+} represents an electron deficit. (When an atom loses an electron, it acquires one unit of positive charge from the excess of one proton. In our example, each iron atom loses three electrons and acquires three units of positive charge.)

You will recall that the oxygen atom is very electronegative, able to remove electrons from other atoms. In this reaction, oxygen gains electrons from iron.

$$3\,O_2 + 12e^- \rightarrow 6\,O^{2-}$$

Oxygen becomes reduced when it accepts electrons from the iron. **Reduction** is a chemical process in which an atom, ion, or molecule *gains* electrons. (The term *reduction* refers to the fact that the gain of an electron results in the reduction of any positive charge that might be present.)

Redox reactions occur simultaneously because one substance must accept the electrons that are removed from the other. In a redox reaction, one component, the *oxidizing agent,* accepts one or more electrons and becomes reduced. Oxidizing agents other than oxygen are known, but oxygen is such a common one that its name was given to the process. Another reaction component, the *reducing agent,* gives up one or more electrons and becomes oxidized.

In our example there was a complete transfer of electrons from iron (the reducing agent) to oxygen (the oxidizing agent). Similarly, in Figure 2–9, an electron was transferred from sodium (the reducing agent) to chlorine (the oxidizing agent).

Electrons are not easily removed from covalent compounds unless an entire atom is removed. In cells, oxidation often involves the removal of a hydrogen *atom* (an electron plus a proton that "goes along for the ride") from a covalent compound; reduction often involves the addition of the equivalent of a hydrogen atom (see Chapter 6).

WATER IS ESSENTIAL TO LIFE

A large part of the mass of most organisms is water. In human tissues the percentage of water ranges from 20% in bones to 85% in brain cells; about 70% of our total body weight is water. As much as 95% of a jellyfish and certain plants is water. Water is the source, through photosynthesis (see Chapter 8), of the oxygen in the air we breathe, and its hydrogen atoms become incorporated into many organic compounds. Water is also the solvent for most biological reactions and a reactant or product in many chemical reactions.

Water is not only important as an internal constituent of organisms but also is one of the principal environmental factors affecting them. Many organisms live in the ocean or in freshwater rivers, lakes, or puddles. Water's unique combination of physical and chemical properties is considered to have been essential to the origin of life, as well as to the continued survival and evolution of life on Earth (Fig. 2–12).

Water molecules are polar

As discussed previously, water molecules are polar; that is, one end of each molecule bears a partial positive charge and the other a partial negative charge (see Fig. 2–7). The water molecules in liquid water and in ice associate by hydrogen bonds. The hydrogen atom of one water molecule, with its partial positive charge, is attracted to the oxygen atom of a neighboring water molecule, with its partial negative charge, forming a hydrogen bond. An oxygen atom in a water molecule has two regions of partial negative charge, and each of the two hydrogen atoms has a partial positive charge. Each water molecule can therefore form hydrogen bonds with a maximum of four neighboring water molecules (Fig. 2–13).

Water is the principal solvent in organisms

Because its molecules are polar, water is an excellent solvent, a liquid capable of dissolving many different kinds of substances, especially polar and ionic compounds. Previously in this chapter, we discussed how polar water molecules pull the ions of ionic compounds apart so that they dissociate (see Fig. 2–10). Because of its solvent properties and the tendency of the atoms in certain compounds to form ions when in solution, water plays an important role in facilitating chemical reactions. Substances that interact readily with water are said to be **hydrophilic** ("water-loving"). Examples include table sugar (sucrose, a polar compound) and table salt (NaCl, an ionic compound), which dissolve readily in water. Not all substances in organisms are hydrophilic, however. Many **hydrophobic** ("water-fearing") substances found in living things are especially important because of their ability to form associations or structures that are not disrupted or dissolved by water. Examples, to be discussed more fully in Chapter 3, include fats and other nonpolar substances.

Hydrogen bonding makes water cohesive and adhesive

Water molecules have a strong tendency to stick to each other; that is, they are **cohesive.** This is due to the hydrogen bonds among the molecules. Because of the cohesive nature of water molecules, any force exerted on part of a column of water will be transmitted to the column as a whole. The major mechanism of

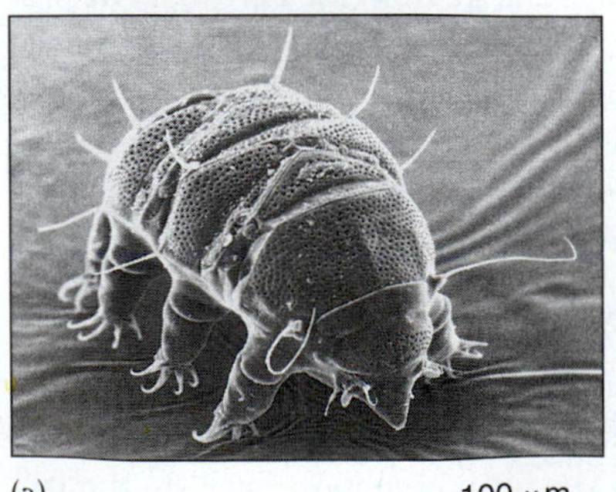

(a) 100 μm

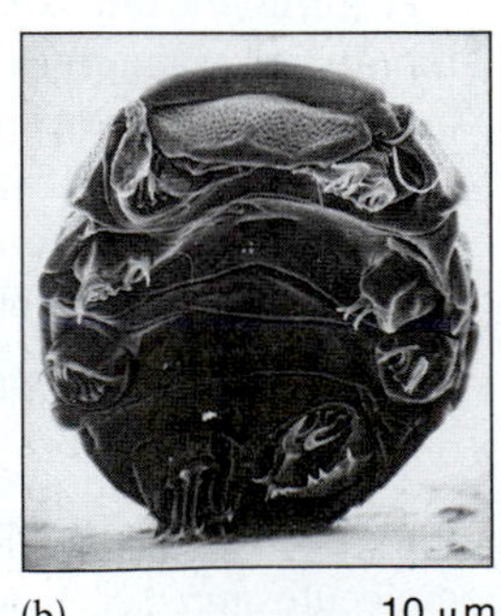

(b) 10 μm

Figure 2–12 Tardigrade. **(a)** Commonly known as "water bears," tardigrades such as these members of the genus *Echiniscus* are small (less than 1.2 mm long) animals that normally live in moist habitats, such as thin films of water on mosses. **(b)** When subjected to desiccation, tardigrades assume a barrel-shaped form known as a *tun,* remaining in this state, motionless but alive, for as long as 100 years. When rehydrated they assume their normal appearance and activities. *(a, Diane R. Nelson; b, Robert O. Schuster, courtesy of Diane R. Nelson)*

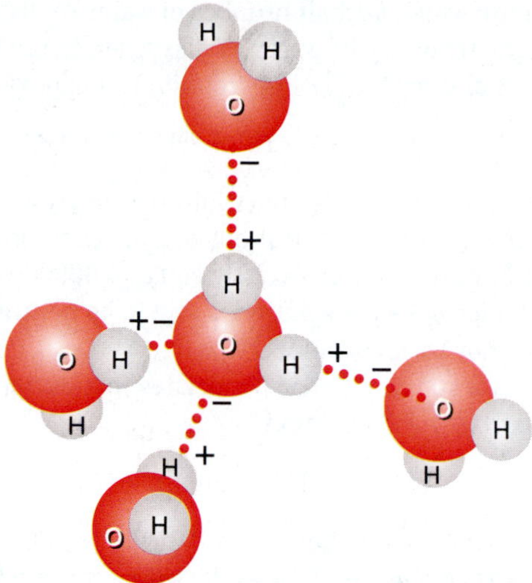

Figure 2–13 Hydrogen bonding of water molecules. Each water molecule can form hydrogen bonds *(dotted lines)* with as many as four neighboring water molecules.

water movement in plants (see Chapter 33) depends on the cohesive nature of water. Water molecules also stick to many other kinds of substances, most notably those with charged groups of atoms or molecules on their surfaces. These **adhesive** forces explain how water makes things wet.

A combination of adhesive and cohesive forces accounts for the phenomenon of **capillary action,** which is the tendency of water to move in narrow tubes, even against the force of gravity (Fig. 2–14). For example, water moves through the microscopic spaces between soil particles to the roots of plants by capillary action.

Water has a high degree of **surface tension** because of the cohesiveness of its molecules, which have a much greater attraction for each other than for molecules in the air. Thus, water molecules at the surface crowd together, producing a strong layer as they are pulled downward by the attraction of other water molecules beneath them (Fig. 2–15).

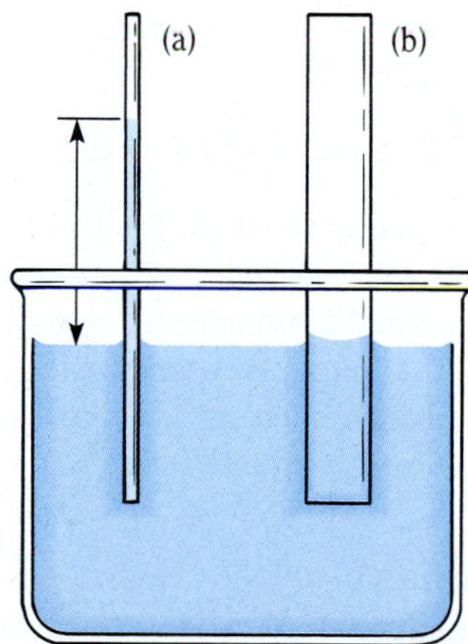

Figure 2–14 Capillary action. **(a)** In a narrow tube, adhesive forces attract water molecules to the glass wall of the tube. Other water molecules inside the tube are then "pulled along" by cohesive forces, which are due to hydrogen bonds between the water molecules. **(b)** In the wider tube, a smaller percentage of the water molecules line the glass wall. Because of this, the adhesive forces are not strong enough to overcome the cohesive forces of the water beneath the surface level of the container, and water in the tube rises only slightly.

Water helps maintain a stable temperature

Raising the temperature of a substance involves adding heat energy to make its molecules move faster, that is, to increase the **kinetic energy** (energy of motion) of the molecules (see Chapter 6). The term **heat** refers to the *total* amount of kinetic energy in a sample of a substance; **temperature** is a measure of the *average* kinetic energy of the particles. Water has a high **specific heat;** that is, the amount of energy required to raise the temperature of water is quite large. A **calorie** (cal) is a unit of heat energy (equivalent to 4.184 joules [J]) that equals the amount of heat required to raise the temperature of 1 g of water 1° Celsius (C). The specific heat of water is therefore 1 cal/g of water per degree Celsius. Most other common substances such as metals, glass, and ethyl alcohol have much lower specific heat values. The specific heat of ethyl alcohol, for example, is 0.59 cal/g/1° C (2.46 J/g/1° C).

The high specific heat of water results from the hydrogen bonding of its molecules. Some of the hydrogen bonds holding the water molecules together must first be broken to permit the molecules to move more freely. Much of the energy added to the system is used up in breaking the hydrogen bonds, and only a portion of the heat energy is available to speed the movement of the water molecules (thereby increasing the temperature of the water). Conversely, when liquid water changes to ice, additional

Figure 2–15 Surface tension of water. Hydrogen bonding between water molecules is responsible for the surface tension of water, which is strong enough to support these water striders, and causes the dimpled appearance of the surface. Although water striders are denser than water, these insects can walk on the surface of a pond because fine hairs at the ends of their legs spread their weight over a large area. *(Dennis Drenner)*

hydrogen bonds must be formed, liberating a great deal of heat into the environment.

Because so much heat input is required to raise the temperature of water (and so much heat is lost when the temperature is lowered), the ocean and other large bodies of water have relatively constant temperatures. Thus, many organisms living in the ocean are provided with a relatively constant environmental temperature. The properties of water are crucial in stabilizing temperatures on the Earth's surface. Although surface water is only a thin film relative to Earth's volume, the quantity is enormous compared to the exposed land mass. This relatively large mass of water resists both the warming effect of heat and the cooling effect of low temperatures.

Hydrogen bonding causes ice to have unique properties with important environmental consequences. Liquid water expands as it freezes because the hydrogen bonds joining the water molecules in the crystalline lattice keep the molecules far enough apart to give ice a density about 10% less than the density of liquid water (Fig. 2–16*c*). When ice has been heated enough to raise its temperature above 0° C (32° F), these hydrogen bonds among the water molecules are broken, freeing the molecules to slip closer together. The density of water is greatest at 4° C, above which water begins to expand again as the speed of its molecules increases. As a result, ice floats on the denser cold water.

This unusual property of water has been important in enabling life as we know it to appear, survive, and evolve on Earth. If ice had a greater density than water, it would sink; eventually all ponds, lakes, and even the ocean would freeze solid from the bottom to the surface, making life impossible. When a deep body of water cools, it becomes covered with floating ice. The ice insulates the liquid water below it, retarding freezing and permitting a variety of organisms to survive below the icy surface.

The high water content of organisms helps them maintain relatively constant internal temperatures. Such minimizing of temperature fluctuations is important because biological reactions can take place only within a relatively narrow temperature range.

Because its molecules are held together by hydrogen bonds, water has a high **heat of vaporization.** To change 1 g of liquid water into 1 g of water vapor, 540 cal of heat are required. The heat of vaporization of most other common liquid substances is much less. As a sample of water is heated, some molecules are moving much faster than others (i.e., they have more heat energy). These faster moving molecules are more likely to escape the liquid phase and enter the vapor phase (see Fig. 2–16*a*). When they do, they take their heat energy with them (thus lowering the temperature of the sample, a process called **evaporative cooling**). For this reason the human body can dissipate excess heat as sweat evaporates from the skin, and a leaf can keep cool in the bright sunlight as water evaporates from its surface.

ACIDS ARE PROTON DONORS; BASES ARE PROTON ACCEPTORS

Water molecules have a slight tendency to **ionize,** that is, to dissociate into hydrogen ions (H^+) and hydroxide ions (OH^-).[4] In pure water, a small number of water molecules ionize. This slight tendency of water to dissociate is reversible as hydrogen ions and hydroxide ions reunite to form water:

$$HOH \rightleftharpoons H^+ + OH^-$$

Because each water molecule splits into one hydrogen ion and one hydroxide ion, the concentrations of hydrogen ions and hydroxide ions in pure water are exactly equal (0.0000001 or 10^{-7} mol/L for each ion). Such a solution is said to be neutral, that is, neither acidic nor basic (alkaline).

An **acid** is a substance that dissociates in solution to yield hydrogen ions (H^+) and an anion.

$$\text{Acid} \rightarrow H^+ + \text{Anion}$$

An acid is a proton *donor*. (Recall that a hydrogen ion, or H^+, is nothing more than a proton.) An acidic solution has a hydrogen ion concentration that is higher than its hydroxide ion concentration. Acidic solutions turn blue litmus paper red and have a sour taste. Hydrochloric acid (HCl) and sulfuric acid (H_2SO_4) are examples of inorganic acids. Lactic acid ($CH_3CHOHCOOH$) from sour milk and acetic acid (CH_3COOH) from vinegar are two common organic acids.

A **base** is defined as a proton *acceptor*. Most bases are substances that dissociate to yield a hydroxide ion (OH^-) and a cation when dissolved in water. A hydroxide ion can act as a base by accepting a proton (H^+) to form water. Sodium hydroxide (NaOH) is a common inorganic base.

$$NaOH \rightarrow Na^+ + OH^-$$

$$OH^- + H^+ \rightarrow H_2O$$

Some bases do not dissociate to yield hydroxide ions directly. For example, ammonia (NH_3) acts as a base by accepting a proton from water, producing an ammonium ion (NH_4^+) and releasing a hydroxide ion.

$$NH_3 + H_2O \rightarrow NH_4^+ + OH^-$$

A basic solution is one in which the hydrogen ion concentration is lower than the hydroxide ion concentration. Basic solutions such as household ammonia turn red litmus paper blue and feel slippery to the touch. In later chapters we encounter a number of organic bases, such as the purine and pyrimidine bases that are components of nucleic acids.

pH is a convenient measure of acidity

The degree of a solution's acidity is generally expressed in terms of **pH,** defined as the negative logarithm (base 10) of the hydrogen ion concentration (expressed in moles per liter):

$$pH = -\log_{10}[H^+]$$

[4] The H^+ immediately combines with a negatively charged region of a water molecule, forming a hydronium ion (H_3O^+). However, by convention, H^+, rather than the more accurate H_3O^+, is used.

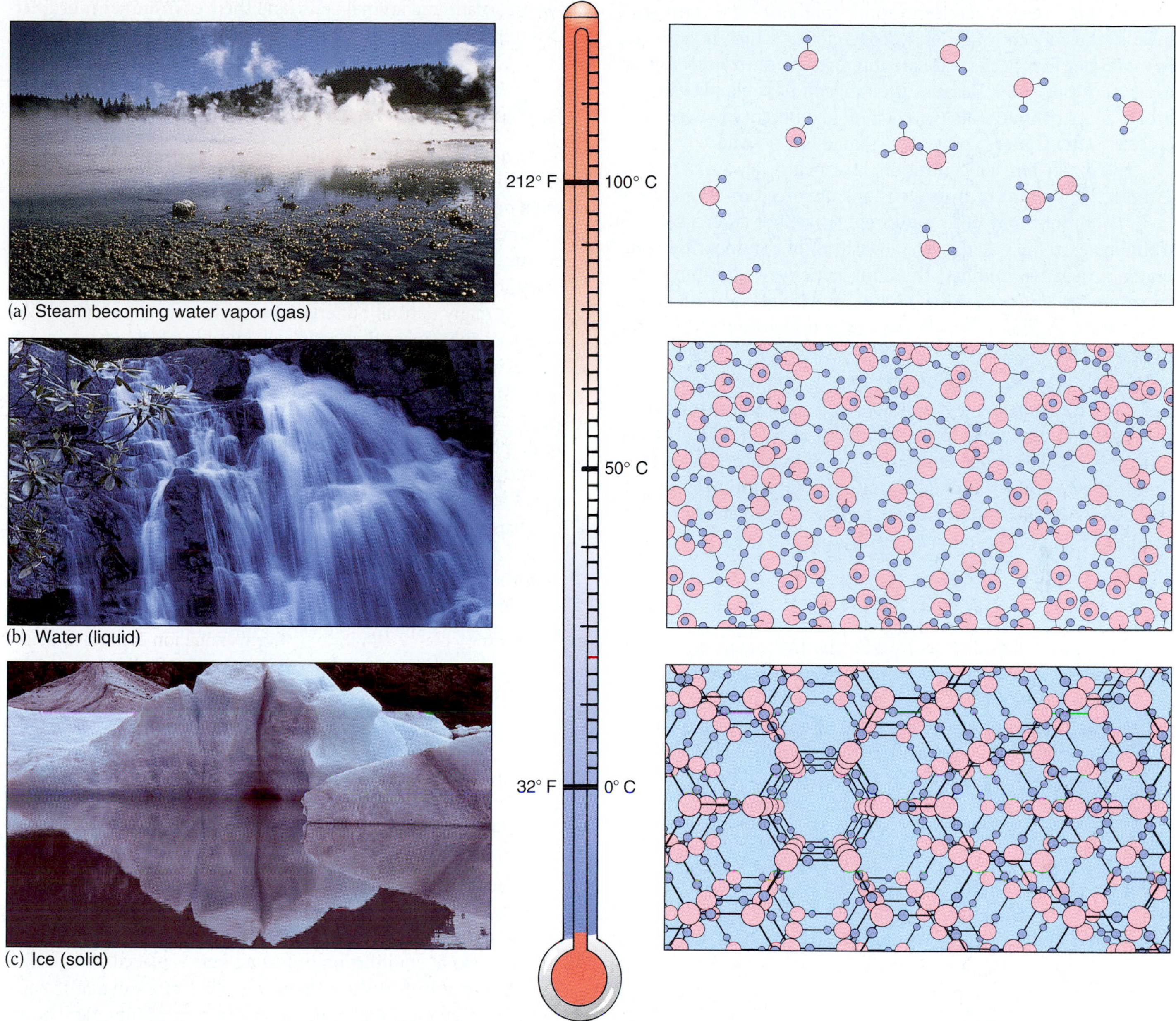

Figure 2–16 Three forms of water. **(a)** When water boils, as in this hot spring at Yellowstone National Park, many hydrogen bonds are broken, causing steam, consisting of minuscule water droplets, to form. If most of the remaining hydrogen bonds are subsequently broken, the molecules begin to move more freely as water vapor (a gas). **(b)** Water molecules in a liquid state continually form, break, and re-form hydrogen bonds with each other. **(c)** In ice, each water molecule participates in four hydrogen bonds with adjacent molecules, resulting in a regular, evenly distanced crystalline lattice structure. Because the water molecules move apart slightly as the hydrogen bonds form, water expands as it freezes; thus ice floats on water. *(a, Woodbridge Wilson/National Park Service; b, Gary R. Bonner; c, Barbara O'Donnell/Biological Photo Service)*

The brackets refer to concentration; therefore, the term $[H^+]$ means "the concentration of hydrogen ions," which is expressed in moles per liter because we are interested in the *number* of hydrogen ions per liter. Because the range of possible pH values is broad, a logarithmic scale (with a tenfold difference between successive units) is more convenient than a linear scale.

Hydrogen ion concentrations are nearly always less than 1 mol/L. One gram of hydrogen ions dissolved in 1 L of water (a 1 *M* solution) may not sound impressive, but such a solution would be extremely acidic. The logarithm of a number less than one is a negative number; thus, the *negative* logarithm corresponds to a *positive* pH value. (Solutions with pH values less than zero can be produced but do not occur under biological conditions.)

Whole number pH values are easy to calculate (Table 2–2). For instance, consider our example of pure water, which has a hydrogen ion concentration of 0.0000001 (10^{-7}) mol/L. The logarithm is -7. The negative logarithm is 7; therefore, the pH is 7.

If the hydrogen ion concentration of a solution is known, the hydroxide ion concentration can be easily calculated. The product of the hydrogen ion concentration and the hydroxide ion concentration is 1×10^{-14}:

$$[H^+][OH^-] = 1 \times 10^{-14}$$

In pure (freshly distilled) water, the hydrogen ion concentration is 10^{-7}; therefore, the hydroxide concentration is also 10^{-7}. Such a solution, in which the concentrations are equal, is called a **neutral solution. Acidic solutions** (those with an excess of hydrogen ions) have pH values smaller than 7. For example, the hydrogen ion concentration of a solution with pH 1 is ten times that of a solution with pH 2. **Basic solutions** (those with an excess of hydroxide ions) have pH values greater than 7.

The pH values of some common substances are shown in Figure 2–17. Although some very acidic compartments exist within cells (see Chapter 4), most of the interior of an animal or plant cell is neither strongly acidic nor strongly basic but instead an essentially neutral mixture of acidic and basic substances. Although certain bacteria are adapted to life in extremely acidic environments (see Chapter 23), a substantial change in pH is incompatible with life for most cells (Fig. 2–18). The pH of most types of plant and animal cells (and their environment) ordinarily ranges from around 7.2 to 7.4.

Buffers minimize pH change

Many homeostatic mechanisms operate to maintain appropriate pH values. For example, the pH of human blood is about 7.4 and must be maintained within very narrow limits. Should the blood become too acidic (e.g., as a result of respiratory disease), coma and death may result. Excessive alkalinity can result in overexcitability of the nervous system and even convulsions. Organisms contain many natural buffers. A **buffer** is a substance or combination of substances that resists changes in pH when an acid or base is added. A buffering system includes a weak acid or a weak base. A weak acid or weak base does not ionize completely. That is, at any given instant only a fraction of the molecules are ionized; most are not dissociated.

One of the most common buffering systems is found in the blood of vertebrates (see Chapter 44). Carbon dioxide, produced as a waste product of cellular metabolism, enters the blood, the main constituent of which is water. The carbon dioxide reacts with the water to form carbonic acid, a weak acid that dissociates to yield a hydrogen ion and a bicarbonate ion. The buffering system is described by the following expression:

$$\underset{\text{Carbon dioxide}}{CO_2} + \underset{\text{Water}}{H_2O} \rightleftharpoons \underset{\text{Carbonic acid}}{H_2CO_3} \rightleftharpoons H^+ + \underset{\text{Bicarbonate ion}}{HCO_3^-}$$

As indicated by the double arrows, all the reactions are reversible. Because carbonic acid is a weak acid, undissociated molecules are always present, as are all the other components of the system. The expression describes the system when it is at dynamic equilibrium, that is, when the rates of the forward and reverse reactions are equal and the relative concentrations of the components are not changing. A system at dynamic equilibrium tends to stay at equilibrium unless a stress is placed on it, which causes it to shift to reduce the stress until a new dynamic equilibrium is attained. A change in the concentration of any of the components is one such stress. Therefore, the system can be

TABLE 2–2 Relationship among pH, Hydrogen Ion, and Hydroxide Ion Concentrations

Substance	$[H^+]$*	log $[H^+]$	pH	$[OH^-]$†
Gastric juice	0.01, 10^{-2}	−2	2	10^{-12}
Pure water, neutral solution	0.0000001, 10^{-7}	−7	7	10^{-7}
Household ammonia	0.00000000001, 10^{-11}	−11	11	10^{-3}

* $[H^+]$ = hydrogen ion concentration (mol/L)

† $[OH^-]$ = hydroxide ion concentration (mol/L)

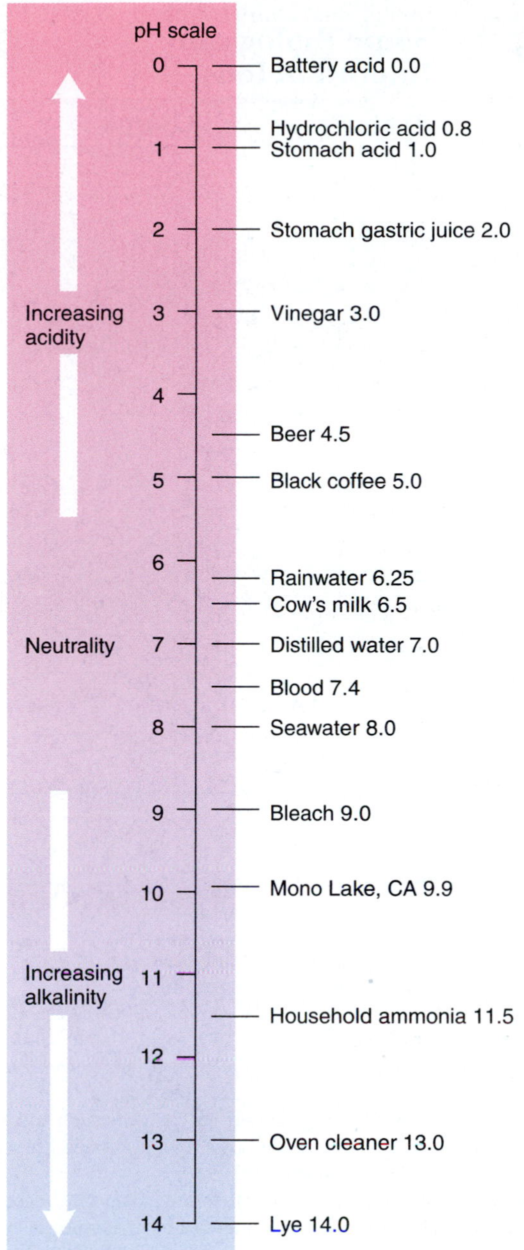

Figure 2–17 pH values of some common solutions. A neutral solution (pH 7) has equal concentrations of H^+ and OH^-. Acidic solutions, which have a higher concentration of H^+ than OH^-, have pH values lower than 7; pH values higher than 7 characterize basic solutions, which have an excess of OH^-.

Figure 2–18 Acid rain damage. Oxides of sulfur emitted from fossil fuel power plants and other industries, and oxides of nitrogen from automobile exhaust, are converted in the moist atmosphere into acids that become dispersed over wide areas. Unlike unpolluted rain (average pH 5.6), the pH of acid rain has been measured at 4.2 and even lower. Plants, such as these trees photographed in the Black Forest of Germany, may be damaged when the resulting increase in soil acidity causes certain minerals, particularly calcium ions, to leach out of the soil. *(Hans Reinhard/Bruce Coleman)*

"shifted to the right" by adding reactants or removing products. Conversely, it can be "shifted to the left" by adding products or removing reactants. Hydrogen ions are the important products to consider in this system.

The addition of excess hydrogen ions temporarily shifts the system to the left, as they combine with the bicarbonate ions to form carbonic acid. Eventually a new dynamic equilibrium is established; at this point the hydrogen ion concentration is similar to the original concentration, and the product of the hydrogen ion and hydroxide ion concentrations is restored to the equilibrium value of 1×10^{-14}.

If hydroxide ions are added, they combine with the hydrogen ions to form water, effectively removing a product and thus shifting the system to the right. More carbonic acid then ionizes, replacing the hydrogen ions that were removed.

Organisms contain many weak acids and weak bases, thus maintaining an essential reserve of buffering capacity and avoiding pH extremes.

An acid and a base react to form a salt

When an acid and a base are mixed together in water, the H^+ of the acid unites with the OH^- of the base to form a molecule of water. The remainder of the acid (an anion) combines with the remainder of the base (a cation) to form a salt. For example, hydrochloric acid reacts with sodium hydroxide to form water and sodium chloride:

$$HCl + NaOH \rightarrow H_2O + NaCl$$

A **salt** is a compound in which the hydrogen ion of an acid is replaced by some other cation. Sodium chloride, NaCl, is a compound in which the hydrogen ion of HCl has been replaced by the cation Na^+.

When a salt, an acid, or a base is dissolved in water, its dissociated ions can conduct an electrical current; these substances are called **electrolytes.** Sugars, alcohols, and many other substances do not form ions when dissolved in water; they do not conduct an electrical current and are referred to as **nonelectrolytes.**

Cells and extracellular fluids (such as blood) of animals and plants contain a variety of dissolved salts that are the source of the many important mineral ions essential for fluid balance and acid-base balance. The concentrations and relative amounts of the various cations and anions are kept remarkably constant. Any marked change results in impaired cellular functions and may lead to death. Nitrate and ammonium ions from the soil are the important nitrogen sources for plants. In animals, nerve and muscle function, blood clotting, bone formation, and many other aspects of body function depend on ions. Sodium, potassium, calcium, and magnesium are the chief cations present; chloride, bicarbonate, phosphate, and sulfate are important anions (Table 2–3).

TABLE 2–3 Some Biologically Important Ions

Name	Formula	Charge
Sodium	Na^+	1+
Potassium	K^+	1+
Hydrogen	H^+	1+
Magnesium	Mg^{2+}	2+
Calcium	Ca^{2+}	2+
Iron	Fe^{2+} or Fe^{3+}	2+ [iron(II)] or 3+ [iron(III)]
Ammonium	NH_4^+	1+
Chloride	Cl^-	1−
Iodide	I^-	1−
Carbonate	CO_3^{2-}	2−
Bicarbonate	HCO_3^-	1−
Phosphate	PO_4^{3-}	3−
Acetate	CH_3COO^-	1−
Sulfate	SO_4^{2-}	2−
Hydroxide	OH^-	1−
Nitrate	NO_3^-	1−
Nitrite	NO_2^-	1−

SUMMARY WITH KEY TERMS

I. The chemical composition and metabolic processes of all organisms are very similar. The physical and chemical principles that govern nonliving things also govern organisms.

II. Organisms are made up of small, simple, **inorganic compounds** as well as large, complex, carbon-containing **organic compounds.**

III. An **element** is a substance that cannot be decomposed into simpler substances by normal chemical reactions. Four elements—carbon, hydrogen, oxygen, and nitrogen—make up 96% or more of an organism's mass.

IV. Each **atom** is composed of a **nucleus** containing **protons** and **neutrons.** In the space outside the nucleus, **electrons** move rapidly in **orbitals** that correspond to energy levels.
   A. An atom is identified by its number of protons **(atomic number).**
   B. Atoms of the same element with different numbers of neutrons (different **atomic masses**) are **isotopes.** Some isotopes are radioactive **(radioisotopes).**
   C. In an electrically neutral atom, the number of protons equals the number of electrons.
   D. The chemical properties of an atom are determined chiefly by the number and arrangement of its most energetic electrons, known as **valence electrons.**

V. Different atoms are joined by chemical bonds to form **compounds.** A **chemical formula** gives the types and relative numbers of atoms in a substance. A **simplest formula** gives the smallest whole number ratio of the component atoms.
   A. One **mole** (the atomic or molecular mass in grams) of any substance contains 6.02×10^{23} atoms, molecules, or ions. This number is known as **Avogadro's number.**
   B. **Covalent bonds** are strong, stable bonds formed when atoms share **valence electrons,** forming **molecules.** When covalent bonds are formed, the orbitals of the valence electrons may become rearranged in a process known as **orbital hybridization.** A **molecular formula** gives the actual numbers of each type of atom in a molecule; a **structural formula** shows their arrangement.
      1. Covalent bonds are **nonpolar** if the electrons are shared equally between the two atoms.
      2. Covalent bonds are **polar** if one atom is more **electronegative** (has a greater affinity for electrons) than the other.
   C. An **ionic bond** is formed between a positively charged **cation** and a negatively charged **anion.** Ionic bonds are strong in the absence of water but relatively weak in aqueous solution.
   D. **Hydrogen bonds** are relatively weak bonds formed when a hydrogen atom with a partial positive charge is attracted to an atom (usually oxygen or nitrogen) with a partial negative charge already bonded to another molecule or in another part of the same molecule.

VI. **Oxidation** and **reduction (redox)** reactions are chemical processes in which electrons (and their energy) are transferred from a reducing agent to an oxidizing agent.

VII. Water accounts for a large part of the mass of most organisms. It is important in many chemical reactions within living things and has unique properties that also affect the environment.
   A. Water is a **polar molecule** because one end has a partial positive charge and the other has a partial negative charge.
   B. Because its molecules are polar, water is an excellent **solvent** for ionic or polar **solutes.**
   C. Water molecules are **cohesive** because they form hydrogen bonds with each other; they are also **adhesive** through hydrogen bonding to substances with ionic or polar regions.
   D. Water has a high **specific heat,** which helps organisms maintain a relatively constant internal temperature; this property also helps keep the ocean and other large bodies of water at a constant temperature.
   E. Water has a high **heat of vaporization.** Molecules entering the vapor phase carry a great deal of heat, which accounts for **evaporative cooling.**

F. The fact that ice is less dense than liquid water makes the aquatic environment less extreme than it would be if ice sank to the bottom.

G. Water has a slight tendency to **ionize,** that is, to dissociate to form hydrogen ions (protons, H^+) and hydroxide ions (OH^-).

VIII. **Acids** are proton (H^+) donors; **bases** are proton acceptors. Many bases dissociate in solution to yield hydroxide ions, which then accept protons.

A. The **pH scale** is a logarithmic expression of the hydrogen ion concentration of a solution. A **neutral solution** has a pH of 7; an **acidic solution** has a pH below 7; and a **basic solution** has a pH greater than 7.

B. A buffering system is based on a weak acid or a weak base. A **buffer** resists changes in the pH of a solution when acids or bases are added.

C. A **salt** is a compound in which the hydrogen atom of an acid is replaced by some other cation. Salts provide the many mineral ions essential for life functions.

POST-TEST

1. Which of the following elements is *mismatched* with its properties or function? (a) carbon—forms the backbone of organic compounds (b) nitrogen—component of proteins (c) hydrogen—very electronegative (d) oxygen—can participate in hydrogen bonding (e) all of the above are correctly matched
2. Which of the following applies to a neutron? (a) positive charge and located in an orbital (b) negligible mass and located in the nucleus (c) positive charge and located in the nucleus (d) uncharged and located in the nucleus (e) uncharged and located in an orbital
3. $^{32}_{15}P$, a radioactive form of phosphorus, has (a) an atomic number of 32 (b) an atomic mass of 15 (c) an atomic mass of 47 (d) 32 electrons (e) 17 neutrons
4. Which of the following facts allows you to determine that atom A and atom B are isotopes of the same element? (a) they each have six protons (b) they each have four neutrons (c) in each, the sum of their electrons and neutrons is 14 (d) they each have four valence electrons (e) they each have a mass number of 14
5. Sodium and potassium behave similarly in chemical reactions. This is because (a) they have the same number of neutrons (b) each has a single valence electron (c) they have the same atomic mass (d) they have the same number of electrons (e) they have the same number of protons
6. The orbitals comprising an atom's valence electron shell (a) are arranged as concentric spheres (b) contain the atom's least energetic electrons (c) may change shape when covalent bonds are formed (d) never contain more than one electron each (e) more than one of the above is correct
7. Which of the following bonds and properties are correctly matched? (a) ionic bonds—strong only if the participating ions are hydrated (b) hydrogen bonds—responsible for bonding oxygen and hydrogen to form a single water molecule (c) polar covalent bonds—can occur between two atoms of the same element (d) covalent bonds—may be single, double, or triple (e) hydrogen bonds—stronger than covalent bonds
8. In a redox reaction (a) energy is transferred from a reducing agent to an oxidizing agent (b) a reducing agent becomes oxidized as it accepts an electron (c) an oxidizing agent accepts a proton (d) a reducing agent donates a proton (e) the electrons in an atom move from its valence shell to a shell closer to its nucleus
9. A solution with a pH of 2 has a hydrogen ion concentration that is ________________ the hydrogen ion concentration of a solution with a pH of 4. (a) 1/2 (b)1/100 (c) two times (d) ten times (e) one hundred times
10. The high heat of vaporization of water accounts for (a) evaporative cooling (b) the fact that ice floats (c) the fact that heat is liberated when ice forms (d) the cohesive properties of water (e) capillary action
11. NaOH and HCl react to form Na^+, Cl^-, and water. Which of the following statements is true? (a) Na^+ is an anion, and Cl^- is a cation (b) Na^+ and Cl^- are both anions (c) a hydrogen bond can form between Na^+ and Cl^- (d) Na^+ and Cl^- are electrolytes (e) Na^+ is an acid, and Cl^- is a base
12. Which of the following statements is true? (a) the number of individual particles (atoms, ions, or molecules) contained in one mole varies depending on the substance (b) Avogadro's number is the number of particles contained in one mole of a substance (c) Avogadro's number is 10^{23} particles (d) one mole of ^{12}C has a mass of 12 g (e) both b and d are true

REVIEW QUESTIONS

1. What is the relationship between molecules and compounds? Are all compounds composed of molecules?
2. What are the ways an atom or molecule can become an anion or a cation?
3. What is a radioisotope? Why is it able to substitute for an ordinary (nonradioactive) atom of the same element in a molecule? What are some of the ways radioisotopes are used in biological research?
4. How do ionic and covalent bonds differ?
5. Why does water form hydrogen bonds? List some of the properties of water that result from hydrogen bonding. How do these properties contribute to the role of water as an essential component of organisms?
6. How can weak forces, such as hydrogen bonds, have significant effects in organisms?
7. A solution has a hydrogen ion concentration of 0.01 mol/L. What is its pH? What is its hydroxide ion concentration? How does this solution differ from one with a pH of 1?
8. Why are buffers important in organisms? Give a specific example of how a buffering system works.
9. Differentiate clearly among acids, bases, and salts.
10. Why must oxidation and reduction occur simultaneously? Why are redox reactions important in some energy transfers?
11. Describe a reversible reaction that is at chemical equilibrium. What would be the consequences of adding or removing a reactant or a product?

YOU MAKE THE CONNECTION

1. Element A has two electrons in its valence shell (which is complete when it contains eight electrons). Would you expect element A to share, donate, or accept electrons? What would you expect of element B, which has four valence electrons, and element C, which has seven?
2. A hydrogen bond formed between two water molecules is only about 1/20 as strong as a covalent bond between hydrogen and oxygen. In what ways would the physical properties of water be different if these hydrogen bonds were stronger (e.g., 1/10 the strength of covalent bonds)?
3. Consider the following reaction (in water).

$$HCl \rightarrow H^+ + Cl^-$$

Name the reactant(s) and product(s). Does the expression indicate that the reaction is reversible? Could HCl be used as a buffer?

RECOMMENDED READINGS

Atkins, P.W. *Periodic Kingdom.* Basic Books, Harper Collins, New York, 1995. In this imaginative work the periodic table is described as a landscape inhabited by elements whose properties are determined by the region in which they reside.

Ball, P. *Life's Matrix: A Biography of Water*. Farrar, Straus, & Giroux, New York, 2000. This comprehensive overview of the unique properties of water also provides insight into the process of science by examining some of the major scientific controversies of recent years.

Bettelheim, F.A., W.H. Brown, and J. March. *Introduction to General, Organic, and Biochemistry,* 6th ed. Harcourt College Publishers, Philadelphia, 2001. A very readable reference text for those who would like to know more about the chemistry basic to life.

Hedin, L.O., and G.E. Likens. "Atmosphere Dust and Acid Rain." *Scientific American,* Vol. 275, No. 6, Dec. 1996. This article examines the idea that recent reductions in basic compounds attached to dust particles in the atmosphere may be increasing the environmental damage done by acid rain.

Scerri, E.R. "The Periodic Table and the Electron." *American Scientist,* Vol. 85, No.5, 1997. This article argues that the chemical properties of atoms are only approximately explained by their electron configurations.

Suter, R.B. "Walking on Water." *American Scientist*, Vol. 87, Vol. 2, 1999. The author studies how fishing spiders use the properties of water to their advantage as they propel themselves over the surface.

Zimmer, C. "Wet, Wild and Weird." *Discover,* Dec. 1992. Computer simulations illustrate the many ways water molecules can interact through hydrogen bonding.

- Visit our Web site at **http://www.info.brookscole.com/solomonbergmartin** for links to chapter-related resources on the World Wide Web. Additional on-line materials relating to this chapter can also be found on our Web site.

See chapter activity on BioActive Learner CD for additional help in mastering the chapter's material. Icon location in the chapter's margins shows which topics have tutorials or simulations in the CD.

3

The Chemistry of Life: Organic Compounds

LEARNING OBJECTIVES

After you have studied this chapter you should be able to

1. Distinguish between organic and inorganic compounds.
2. Describe the properties of carbon that make it the central component of organic compounds.
3. Distinguish among the three principal types of isomers.
4. Identify the major functional groups present in organic compounds and describe their properties.
5. Compare the functions and chemical compositions of the major groups of organic compounds: carbohydrates, lipids, proteins, and nucleic acids.
6. Distinguish among monosaccharides, disaccharides, and polysaccharides. Compare storage polysaccharides with structural polysaccharides.
7. Distinguish among fats, phospholipids, and steroids; describe the composition, characteristics, and biological functions of each.
8. Sketch the structure of an amino acid. Explain how amino acids are grouped into classes based on the characteristics of their side chains.
9. Distinguish among the levels of organization of protein molecules.
10. Describe the components of a nucleotide. Name some nucleic acids and discuss the importance of these compounds in living organisms.

The Atlantic bay scallop, *Argopecten irradians*, like all other living things, contains many types of organic molecules. *(Paul A. Zahl/Photo Researchers, Inc.)*

Both inorganic and organic forms of carbon occur widely in nature. The scallop in the photograph contains **organic compounds,** in which carbon atoms are covalently bonded to each other to form the backbone of the molecule. Some very simple carbon compounds are considered inorganic if the carbon is not bonded to another carbon or to hydrogen. Inorganic carbon is represented by the scallop's shell, composed mainly of calcium carbonate. Organic compounds are so named because at one time it was thought that they could be produced only by living (organic) organisms. In 1928 the German chemist Friedrich Wühler synthesized urea, a metabolic waste product. Since that time, scientists have learned to synthesize many organic molecules and have discovered organic compounds not found in any organism.

Organic compounds are extraordinarily diverse; in fact, more than 5 million have been identified. There are many reasons for this diversity. **Hydrocarbons,** organic compounds consisting only of carbon and hydrogen, can be produced in a wide variety of three-dimensional shapes. Furthermore, the carbon atom can form bonds with a greater number of different elements than any other type of atom. The addition of chemical groups containing atoms of other elements, especially oxygen, nitrogen, phosphorus, and sulfur, can profoundly change the properties of an organic molecule.

Further diversity is provided by the fact that a great many organic compounds found in organisms are extremely large **macromolecules.** Cells construct these from simpler modular subunits. For example, protein molecules are built from smaller compounds called *amino acids.*

In this chapter we focus on some of the major groups of organic compounds important in organisms, including carbohydrates, lipids, proteins, and nucleic acids (DNA and RNA). Organic compounds are the main structural components of cells and tissues. They participate in and regulate metabolic reactions, transmit information, and provide energy for life processes. Evolution involves chemical changes in the organic compounds produced by organisms.

CARBON ATOMS FORM AN ENORMOUS VARIETY OF STRUCTURES

Carbon has unique properties that permit formation of the carbon backbones of the large, complex molecules essential to life (Fig. 3–1). Because a carbon atom has four valence electrons, it can complete its valence shell by forming a total of four covalent bonds (see Fig. 2–2). Each bond can link it to another carbon atom or to an atom of a different element. Carbon is particularly well suited to serve as the backbone of a large molecule because carbon-to-carbon bonds are strong and not easily broken. However, they are not so strong that it would be impossible for cells to break them. Carbon-to-carbon bonds are not limited to single bonds (based on sharing of one electron pair). Two carbon atoms can share two electron pairs with each other, forming double bonds:

$$>C{=}C<$$

In some compounds, triple carbon-to-carbon bonds are formed:

$$-C{\equiv}C-$$

As seen in Figure 3–1, carbon chains can be unbranched or branched, and carbon atoms can also be joined into rings. Rings and chains are joined in some compounds.

The molecules in the cell are analogous to the components of a machine. Each component has a shape that permits it to fill certain roles and to interact with other components (often with a

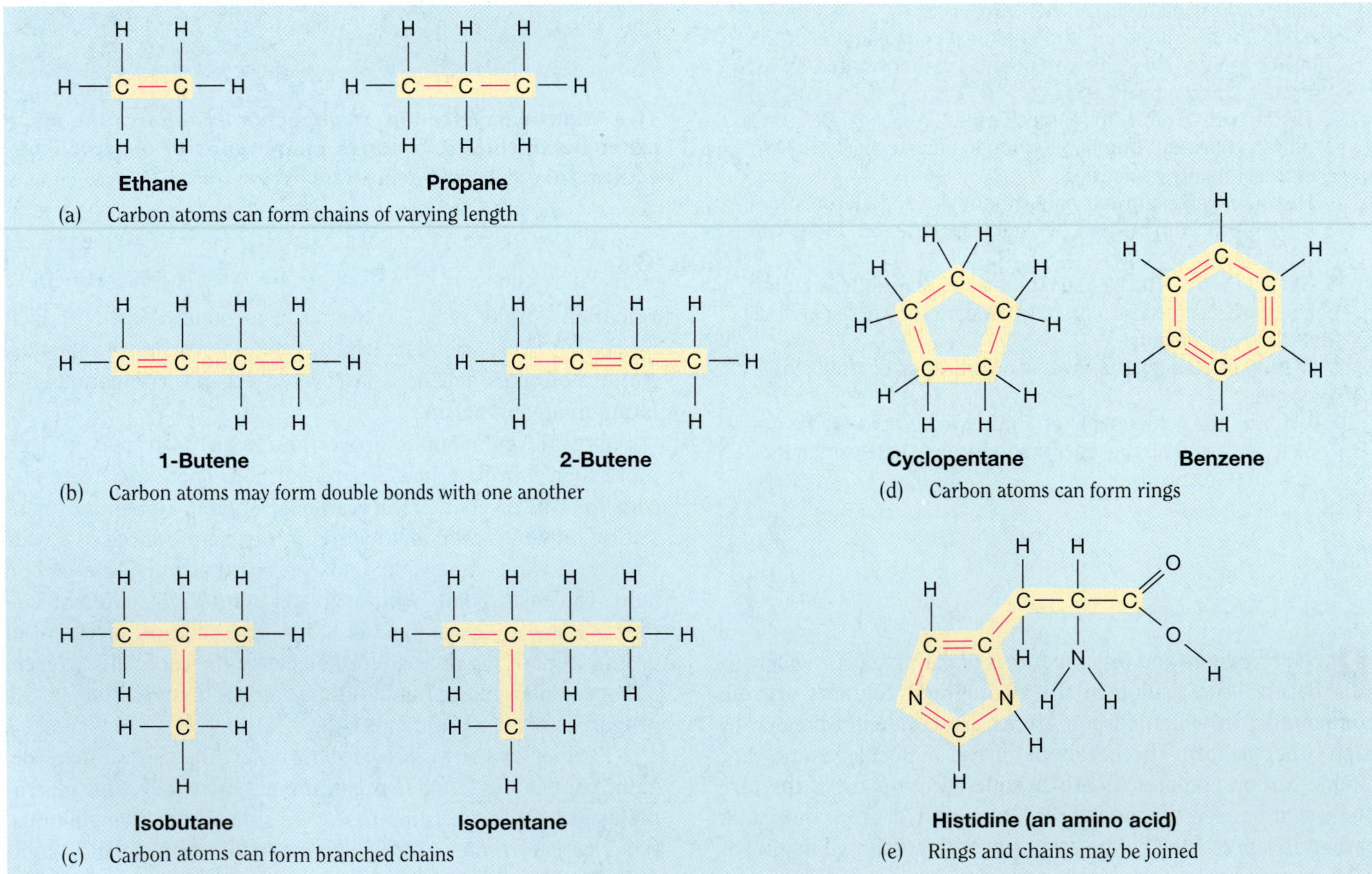

Figure 3–1 Organic molecules. Note that each carbon atom forms four covalent bonds, producing a wide variety of shapes.

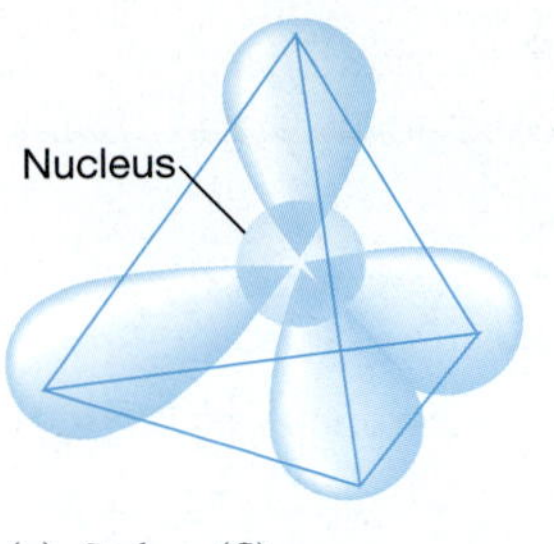

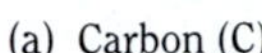
(a) Carbon (C)

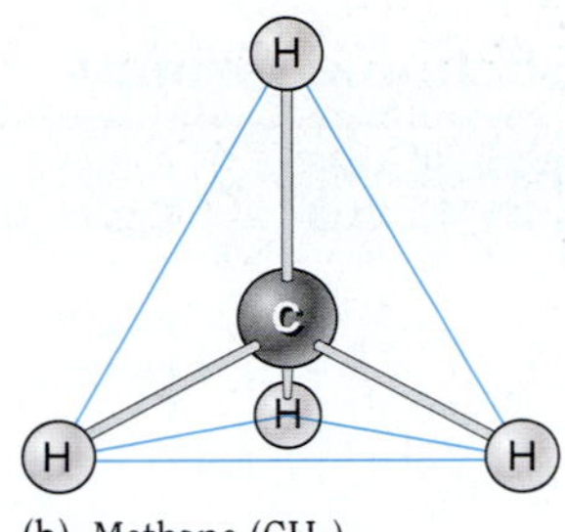

(b) Methane (CH_4)

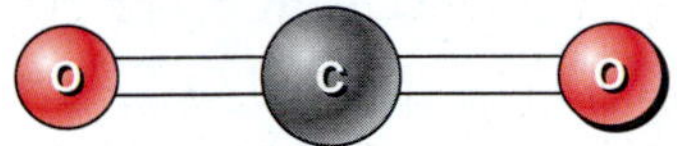

(c) Carbon dioxide (CO_2)

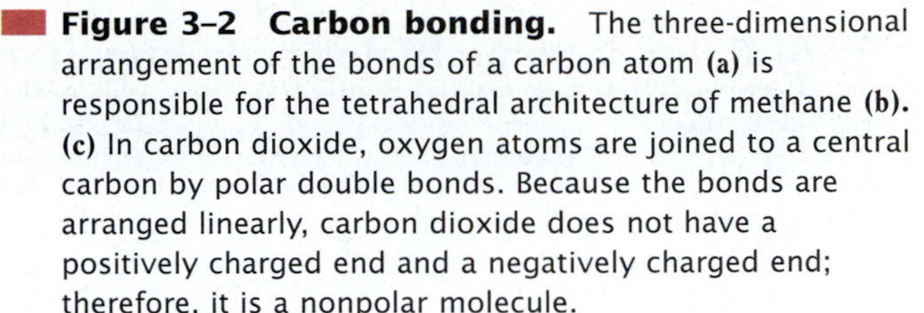
Figure 3–2 Carbon bonding. The three-dimensional arrangement of the bonds of a carbon atom **(a)** is responsible for the tetrahedral architecture of methane **(b)**. **(c)** In carbon dioxide, oxygen atoms are joined to a central carbon by polar double bonds. Because the bonds are arranged linearly, carbon dioxide does not have a positively charged end and a negatively charged end; therefore, it is a nonpolar molecule.

complementary shape). Similarly, the shape of a molecule is important in determining its biological properties and function. Carbon atoms are able to link to each other and to other atoms to produce a wide variety of three-dimensional molecular shapes. This is because the four covalent bonds of carbon do not form in a single plane. Instead, as discussed in Chapter 2, the valence electron orbitals become elongated and project from the carbon atom toward the corners of a tetrahedron (Fig. 3–2). The structure is highly symmetrical, with an angle of about 109.5 degrees between any two of these bonds. Keep in mind that, for simplicity, many of the figures in this book are drawn as two-dimensional graphic representations of three-dimensional molecules. Even the simplest hydrocarbon chains, such as those seen in Figure 3–1, are not actually straight but have a three-dimensional zigzag structure.

Generally, there is freedom of rotation around each carbon-to-carbon single bond. This property permits organic molecules to be flexible and to assume a variety of shapes, depending on the extent to which each single bond is rotated. Double and triple bonds do not permit rotation, so regions of a molecule with such bonds tend to be inflexible.

One reason for the great number of possible carbon-containing compounds is the fact that the same components usually can link together in more than one pattern, generating an even wider variety of molecular shapes.

ISOMERS HAVE THE SAME MOLECULAR FORMULA, BUT DIFFERENT STRUCTURES

Compounds with the same molecular formulas but different structures and thus different properties are called **isomers.** Isomers do not have identical physical or chemical properties and may have different common names. Cells can distinguish between isomers. Usually, one isomer is biologically active and the other is not. Three types of isomers are structural isomers, geometric isomers, and enantiomers.

Structural isomers are compounds that differ in the covalent arrangements of their atoms. For example, Figure 3–3*a* illustrates two structural isomers with the molecular formula C_2H_6O. Similarly, there are two structural isomers of the four-carbon hydrocarbon butane (C_4H_{10}), one with a straight chain and the other with a branched chain (isobutane). Large compounds have more possible structural isomers. There are only two structural isomers of butane, but there can be up to 366,319 isomers of $C_{20}H_{42}$.

Geometric isomers are compounds that are identical in the arrangement of their covalent bonds but different in the spatial arrangement of atoms or groups of atoms. Geometric isomers are present in some compounds with carbon-to-carbon double bonds. Because double bonds are not flexible like single bonds, atoms joined to the carbons of a double bond cannot rotate freely about the axis of the bonds. These *cis-trans* isomers may be drawn as shown in Figure 3–3*b*. The designation *cis* (Latin, "on this side") indicates that the two larger components are on the same side of the double bond. If they are on opposite sides of the double bond, the compound is designated a *trans* (Latin, "across") isomer.

Enantiomers are molecules that are mirror images of one another (Fig. 3–3*c*). Recall that the four groups bonded to a single carbon atom are arranged at the vertices of a tetrahedron. If the four bonded groups are all different, the central carbon is described as asymmetrical. Figure 3–3*d* illustrates that the four groups can be arranged about the asymmetrical carbon in two different ways that are mirror images of each other. The two molecules are enantiomers if they cannot be superimposed on one another no matter how they are rotated in space. Although enantiomers have similar chemical properties and most of their physical properties are identical, cells recognize the difference in shape, and usually only one form is found in organisms.

The existence of isomers is not the only source of variety among organic molecules. The addition of various combinations of atoms, known as *functional groups,* can generate a vast array of molecules with differing properties.

TABLE 3–1 Some Biologically Important Functional Groups

Functional Group	Structural Formula	Class of Compounds Characterized by Group	Description
Hydroxyl	R— OH	Alcohols 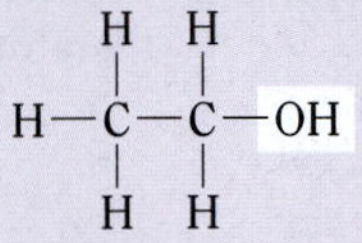 H—C(H)(H)—C(H)(H)—OH Ethanol	Polar because electronegative oxygen attracts covalent electrons
Carbonyl	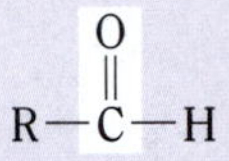 R—C(=O)—H	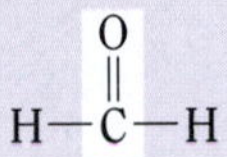 Aldehydes H—C(=O)—H Formaldehyde	Carbonyl group carbon is bonded to at least one H atom; polar because electronegative oxygen attracts covalent electrons
	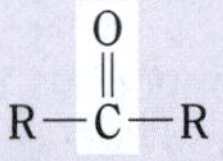 R—C(=O)—R	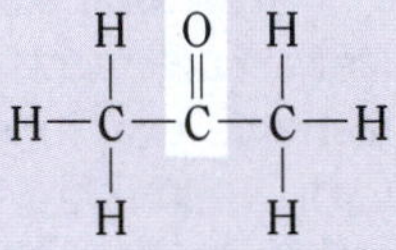 Ketones H—C(H)(H)—C(=O)—C(H)(H)—H Acetone	Carbonyl group carbon is bonded to two other carbons; polar because electronegative oxygen attracts covalent electrons
Carboxyl	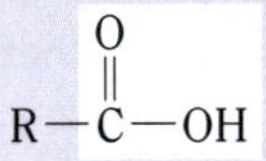 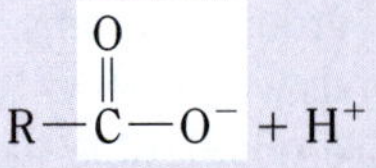 R—C(=O)—OH Nonionized R—C(=O)—O^- + H^+ Ionized	Carboxylic acids (organic acids) 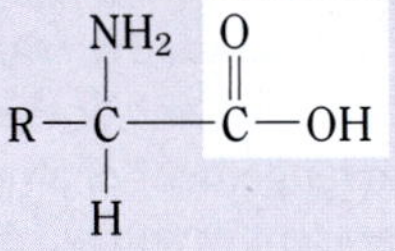 R—C(NH_2)(H)—C(=O)—OH Amino acid	Weakly acidic; can release an H^+ ion
Amino	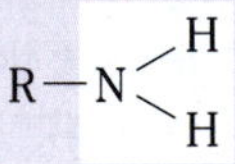 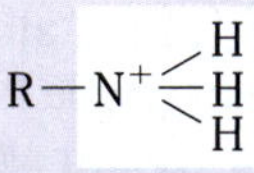 R—N(H)(H) Nonionized R—N^+(H)(H)(H) Ionized	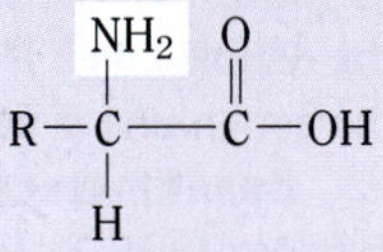 Amines R—C(NH_2)(H)—C(=O)—OH Amino acid	Weakly basic; can accept an H^+ ion
Methyl	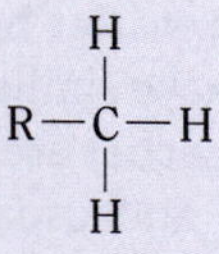 R—C(H)(H)—H	Component of many organic compounds 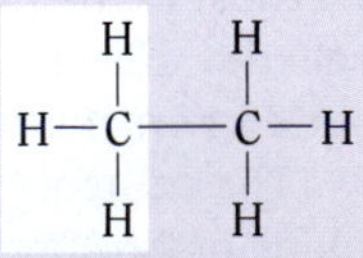 H—C(H)(H)—C(H)(H)—H Ethane	Hydrocarbon; nonpolar

TABLE 3–1 *Continued*

Functional Group	Structural Formula	Class of Compounds Characterized by Group	Description
Phosphate	$R-O-\overset{\overset{O}{\parallel}}{\underset{\underset{OH}{\vert}}{P}}-OH$ Nonionized $^*R-O-\overset{\overset{O}{\parallel}}{\underset{\underset{O^-}{\vert}}{P}}-O^- + 2\,H^+$ Ionized	Organic phosphates $HO-\overset{\overset{O}{\parallel}}{\underset{\underset{OH}{\vert}}{P}}-O-R$ Phosphate ester (as found in ATP)	Weakly acidic; one or two H^+ ions can be released
Sulfhydryl	$R-SH$	Thiols $H-\overset{\overset{H}{\vert}}{\underset{\underset{SH}{\vert}}{C}}-\overset{\overset{H}{\vert}}{\underset{\underset{NH_2}{\vert}}{C}}-\overset{\overset{O}{\parallel}}{C}-OH$ Cysteine	Helps stabilize internal structure of proteins

FUNCTIONAL GROUPS CHANGE THE PROPERTIES OF ORGANIC MOLECULES

Because covalent bonds between hydrogen and carbon are nonpolar, hydrocarbons lack distinct charged regions. For this reason, hydrocarbons are insoluble in water and tend to cluster together, through **hydrophobic interactions.** "Water fearing," the literal meaning of the term *hydrophobic,* is somewhat misleading. Hydrocarbons can interact with water, but much more weakly than the water molecules cohere to each other through hydrogen bonding. Hydrocarbons do interact weakly with each other, but the main reason for hydrophobic interactions is that they are driven together in a sense, having been excluded by the hydrogen-bonded water molecules.

However, the characteristics of an organic molecule can be changed dramatically by replacing one of the hydrogens with a group of atoms known as a **functional group.** Functional groups help determine the types of chemical reactions in which the compound participates. Most functional groups readily form associations, such as ionic and hydrogen bonds, with other molecules. Polar and ionic functional groups are **hydrophilic** because they associate strongly with polar water molecules.

The properties of the major classes of biologically important organic compounds—carbohydrates, lipids, proteins, and nucleic acids—are largely a consequence of the types and arrangement of functional groups they contain. When we know what kinds of functional groups are present in an organic compound, we can predict its chemical behavior. As you read the rest of this section please refer to Table 3–1 for the complete structural formulas of these groups, as well as additional information. Note that the symbol R is used to represent the *remainder* of the molecule of which each functional group is a part.

The **hydroxyl group** (abbreviated R—OH) must not be confused with the hydroxide ion (OH^-) discussed in Chapter 2. The hydroxyl group is polar owing to the presence of a strongly electronegative oxygen atom. If a hydroxyl group replaces one of the hydrogens of a hydrocarbon, the resulting molecule can have significantly altered properties. For example, ethane (see Fig. 3–1*a*) is a hydrocarbon that is a gas at room temperature. If a hydrogen atom is replaced by a hydroxyl group, the resulting molecule is ethyl alcohol, or ethanol, which is found in alcoholic beverages (Fig. 3–3*a*). Ethanol is somewhat cohesive because the polar hydroxyl groups of adjacent molecules interact; it is therefore liquid at room temperature. Unlike ethane, ethyl alcohol can dissolve in water because the polar hydroxyl groups interact with the polar water molecules.

The **carbonyl group** consists of a carbon atom that has a double covalent bond with an oxygen atom. This double bond is polar because of the electronegativity of the oxygen; thus, the carbonyl group is hydrophilic. The position of the carbonyl

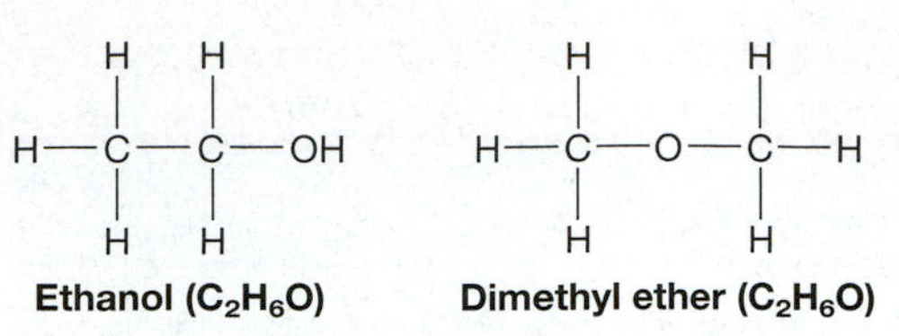

(a) Structural isomers

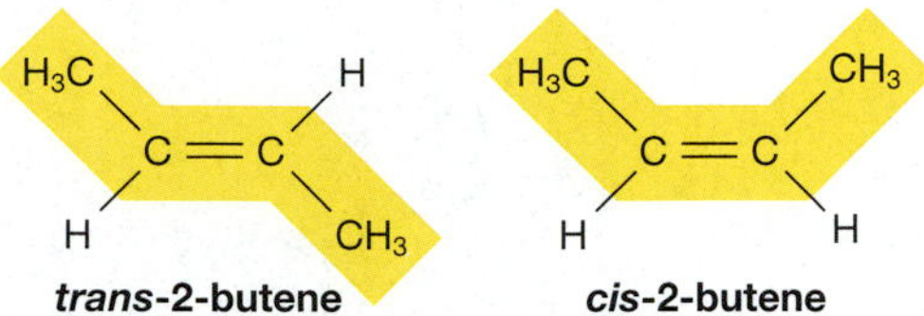

(b) Geometric isomers

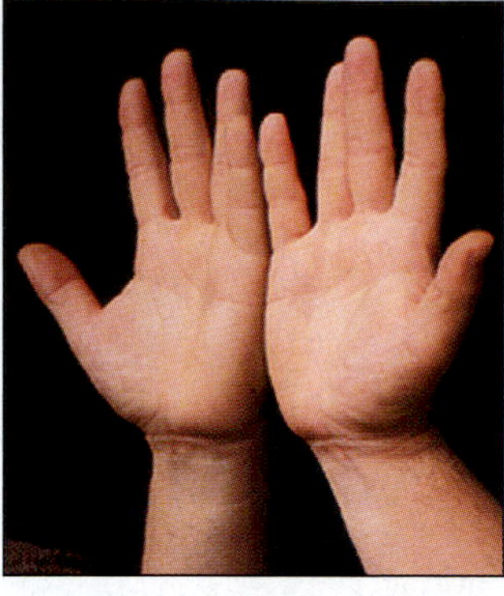
(c) Enantiomers are mirror images

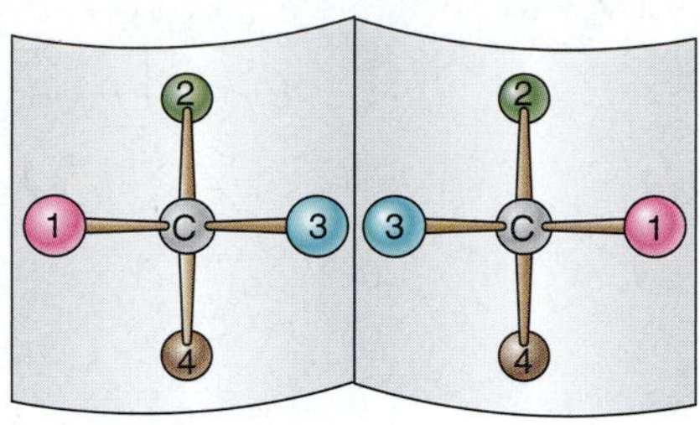
(d) Enantiomers

Figure 3–3 Isomers. Isomers have the same molecular formula, but their atoms are arranged differently. **(a)** Structural isomers differ in the covalent arrangement of their atoms. **(b)** Geometric, or *cis-trans,* isomers have identical covalent bonds but differ in the order in which groups of atoms are arranged in space. **(c and d)** Enantiomers are isomers that are mirror images of one another. The central carbon is asymmetrical because it is bonded to four different groups. Because of their three-dimensional structure the two figures cannot be superimposed no matter how they are rotated. *(c, Dennis Drenner)*

group in the molecule determines the class to which the molecule belongs. An **aldehyde** has a carbonyl group positioned at the end of the carbon skeleton (abbreviated R—CHO); a **ketone** has an internal carbonyl group (abbreviated R—CO—R).

The **carboxyl group** (abbreviated R—COOH) consists of a carbon atom joined by a double covalent bond to an oxygen atom, and by a single covalent bond to another oxygen, which is in turn bonded to a hydrogen atom. Two electronegative oxygen atoms in such close proximity establish an extremely polarized condition, which can cause the hydrogen atom to be stripped of its electron and released as a hydrogen ion (H^+). The carboxyl group then has one unit of negative charge (R—COO^-):

$$R{-}C{\overset{\displaystyle O}{\underset{\displaystyle O{-}H}{\lessgtr}}} \longrightarrow R{-}C{\overset{\displaystyle O}{\underset{\displaystyle O^-}{\lessgtr}}} + H^+$$

Carboxyl groups are weakly acidic; only a fraction of the molecules ionize in this way. This group can therefore exist in one of two hydrophilic states: ionic or polar. Carboxyl groups are essential constituents of amino acids.

An **amino group** (abbreviated R—NH_2) includes a nitrogen atom covalently bonded to two hydrogen atoms. Amino groups are weakly basic because they are able to accept a hydrogen ion (proton), thus acquiring a unit of positive charge. Amino groups are components of amino acids and of nucleic acids.

A **phosphate group** (abbreviated R—PO_4H_2) is weakly acidic. The attraction of electrons by the oxygen atoms can result in the release of one or two hydrogen ions, producing ionized forms with one or two units of negative charge. Phosphates are constituents of nucleic acids and certain lipids.

The sulfhydryl group (abbreviated R—SH), consisting of an atom of sulfur covalently bonded to a hydrogen atom, is found in molecules called *thiols.* As we will see, amino acids that contain a sulfhydryl group can make important contributions to the structure of proteins.

The **methyl group** (abbreviated R—CH_3), a common hydrocarbon group, is nonpolar.

MANY BIOLOGICAL MOLECULES ARE POLYMERS

Many biological molecules such as proteins and nucleic acids are very large, consisting of thousands of atoms. Such giant molecules are known as **macromolecules.** Most macromolecules are **polymers,** produced by linking small organic compounds called **monomers** (Fig. 3–4). Just as all the words in this book have been written by arranging the 26 letters of the alphabet in various combinations, monomers can be grouped together to form an almost infinite variety of larger molecules. Just as we use different words to convey information, cells use different molecules to convey information. The thousands of different complex organic compounds present in organisms are constructed from about 40 small, simple monomers. For example, the 20 monomers called amino acids can be linked end-

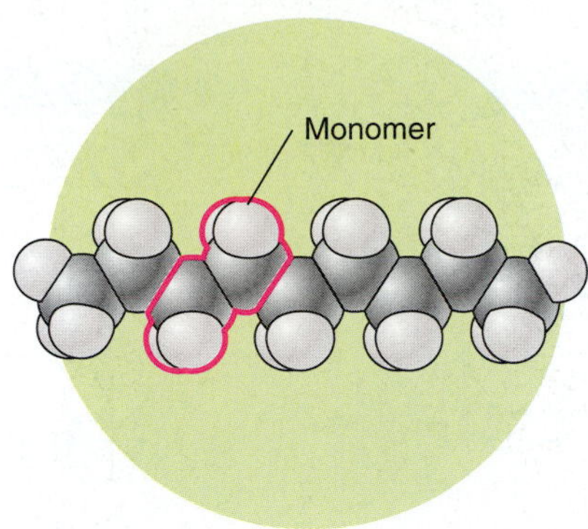

Figure 3–4 A simple polymer. This small polymer of polyethylene is formed by linking two-carbon ethylene (C_2H_4) monomers. One such monomer is outlined in red. The structure is represented by a space-filling model, which accurately depicts the actual three-dimensional shape of the molecule.

to-end in countless ways to form the polymers we know as proteins.

Polymers can be degraded to their component monomers by **hydrolysis** (which means "to break with water"). In a reaction regulated by a specific enzyme,[1] a hydrogen from a water molecule attaches to one monomer, and a hydroxyl from water attaches to the adjacent monomer (Fig. 3–5).

The synthetic process by which monomers are covalently linked is called **condensation.** Because the *equivalent* of a molecule of water is removed during the reactions that combine monomers, the term *dehydration synthesis* is sometimes used to describe the process (see Fig. 3–5). However, in biological systems the synthesis of a polymer is not simply the reverse of hydrolysis, even though the net effect is the opposite of hydrolysis. Synthetic processes such as condensation require energy and are regulated by different enzymes.

In the following sections we examine carbohydrates, lipids, proteins, and nucleic acids. Our discussion begins with the smaller, simpler forms of these compounds and extends to the linking of these monomers to form macromolecules.

[1] An enzyme is a biological catalyst that accelerates a specific chemical reaction (see Chapter 6). Most enzymes are proteins.

CARBOHYDRATES ARE USED FOR FUEL AND STRUCTURAL MATERIALS

Sugars, starches, and cellulose are **carbohydrates.** Sugars and starches serve as energy sources for cells; cellulose is the main structural component of the walls that surround plant cells. Carbohydrates contain carbon, hydrogen, and oxygen atoms in a ratio of approximately one carbon to two hydrogens to one oxygen $(CH_2O)_n$. The term *carbohydrate,* meaning "hydrate (water) of carbon," reflects the 2:1 ratio of hydrogen to oxygen, the same ratio found in water (H_2O). Carbohydrates contain one sugar unit *(monosaccharides),* two sugar units *(disaccharides),* or many sugar units *(polysaccharides).*

Monosaccharides are simple sugars

Monosaccharides typically contain from three to seven carbon atoms. In a monosaccharide, a hydroxyl group is bonded to each carbon except one; that carbon is double-bonded to an oxygen atom, forming a carbonyl group. If the carbonyl group is at the end of the chain, the monosaccharide is an aldehyde; if the carbonyl group is at any other position, the monosaccharide is a ketone. (By convention, the numbering of the carbon skeleton of a sugar begins with the carbon at or nearest the carbonyl end of the open chain.) The large number of polar hydroxyl groups, plus the carbonyl group, gives a monosaccharide hydrophilic properties.

Figure 3–6 shows simplified, two-dimensional representations of some common monosaccharides. The simplest carbohydrates are the three-carbon sugars (trioses): glyceraldehyde and dihydroxyacetone. Ribose and deoxyribose are common pentoses, sugars that contain five carbons; they are components of nucleic acids (DNA, RNA, and related compounds). Glucose, fructose, galactose, and other six-carbon sugars are called **hexoses.** (Note that the names of carbohydrates typically end in *-ose.*)

Glucose ($C_6H_{12}O_6$), the most abundant monosaccharide, is used as an energy source in most organisms. During cellular respiration (see Chapter 7), cells oxidize glucose molecules, converting the stored energy to a form that can be readily used

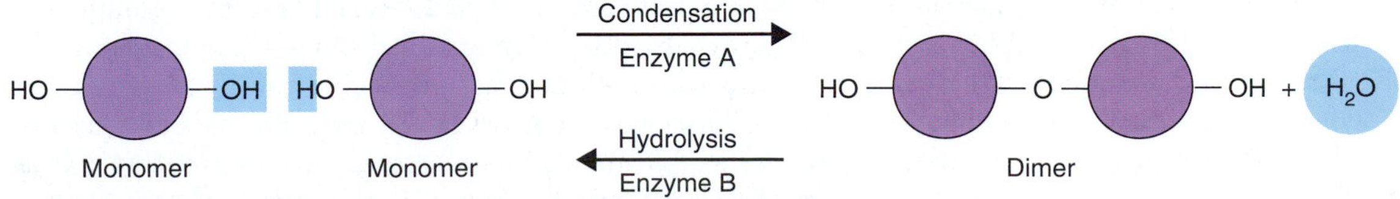

Figure 3–5 Condensation and hydrolysis reactions. Joining two monomers gives a dimer; incorporation of additional monomers produces a polymer. Note that these reactions are catalyzed by different enzymes.

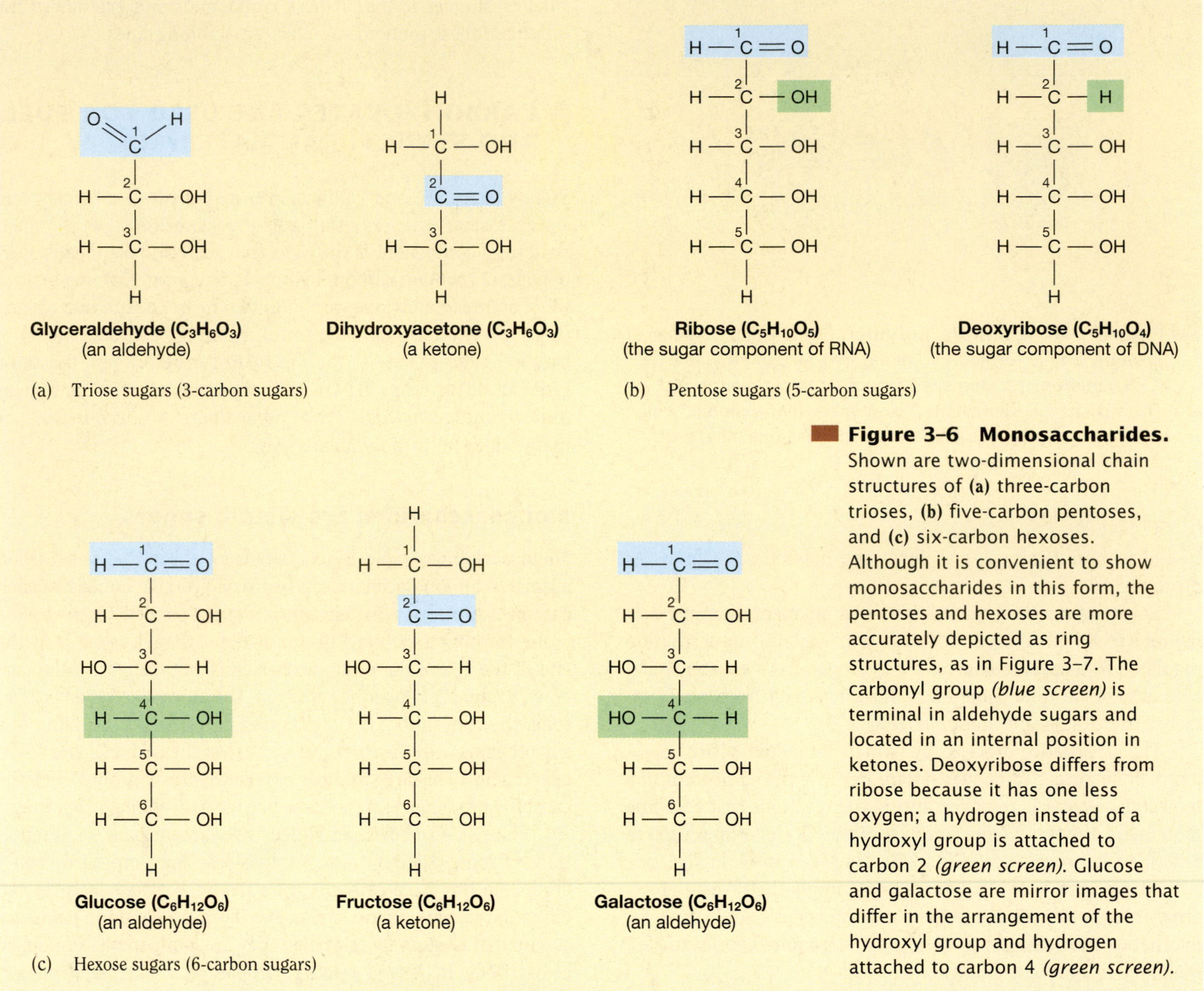

Figure 3–6 Monosaccharides. Shown are two-dimensional chain structures of **(a)** three-carbon trioses, **(b)** five-carbon pentoses, and **(c)** six-carbon hexoses. Although it is convenient to show monosaccharides in this form, the pentoses and hexoses are more accurately depicted as ring structures, as in Figure 3–7. The carbonyl group *(blue screen)* is terminal in aldehyde sugars and located in an internal position in ketones. Deoxyribose differs from ribose because it has one less oxygen; a hydrogen instead of a hydroxyl group is attached to carbon 2 *(green screen).* Glucose and galactose are mirror images that differ in the arrangement of the hydroxyl group and hydrogen attached to carbon 4 *(green screen).*

for cellular work. Glucose is also used as a component in the synthesis of other types of compounds such as amino acids and fatty acids. So important is glucose in metabolism that its concentration is carefully kept at a homeostatic (relatively constant) level in the blood of humans and other complex animals (see Chapter 47).

Glucose and fructose are structural isomers: They have identical molecular formulas, but their atoms are arranged differently. In fructose (a ketone) the double-bonded oxygen is linked to a carbon within the chain rather than to a terminal carbon as in glucose (which is an aldehyde). Because of these differences, the two sugars have different properties. For example, fructose tastes sweeter than glucose.

Glucose and galactose are both hexoses and aldehydes. However, they are mirror images (enantiomers) because they differ in the arrangement of the atoms attached to asymmetrical carbon atom 4.

The "stick" formulas in Figure 3–6 give a clear but somewhat unrealistic picture of the structures of some common monosaccharides. As has been discussed, molecules are not two-dimensional; in fact, the properties of each compound depend largely on its three-dimensional structure. Thus, three-dimensional formulas are helpful in understanding the relationship between molecular structure and biological function. Molecules of glucose and other pentoses and hexoses in solution are actually rings, rather than extended straight carbon chains.

Glucose in solution (as in the cell) typically exists as a ring of five carbons and one oxygen. It assumes this configuration when its atoms undergo a rearrangement, permitting a covalent bond to connect carbon 1 to the oxygen attached to carbon 5 (Fig. 3–7). When glucose forms a ring, two isomeric forms are possible, differing only in the orientation of the hydroxyl (—OH) group attached to carbon 1. When this hydroxyl group is on the same side of the plane of the ring as the

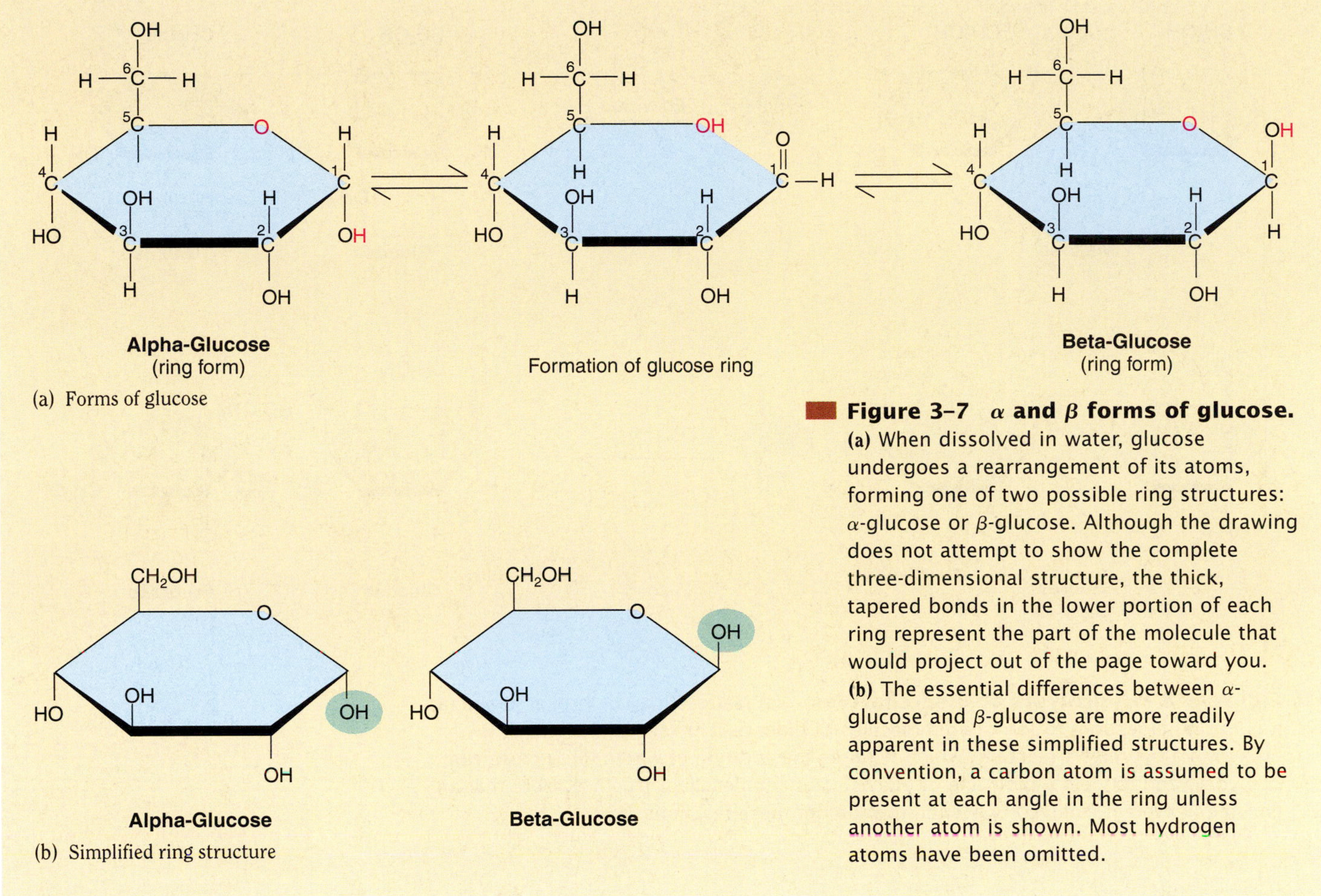

Figure 3–7 α and β forms of glucose. **(a)** When dissolved in water, glucose undergoes a rearrangement of its atoms, forming one of two possible ring structures: α-glucose or β-glucose. Although the drawing does not attempt to show the complete three-dimensional structure, the thick, tapered bonds in the lower portion of each ring represent the part of the molecule that would project out of the page toward you. **(b)** The essential differences between α-glucose and β-glucose are more readily apparent in these simplified structures. By convention, a carbon atom is assumed to be present at each angle in the ring unless another atom is shown. Most hydrogen atoms have been omitted.

—CH_2OH side group, the glucose is designated beta (β)-glucose. When it is on the side (with respect to the plane of the ring) opposite the —CH_2OH side group, the compound is designated alpha (α)-glucose.

Disaccharides consist of two monosaccharide units

A disaccharide (two sugars) contains two monosaccharide rings joined by a **glycosidic linkage,** consisting of a central oxygen covalently bonded to two carbons, one in each ring (Fig. 3–8). The glycosidic linkage of a disaccharide generally forms between carbon 1 of one molecule and carbon 4 of the other molecule. The disaccharide maltose (malt sugar) consists of two covalently linked α-glucose units. Sucrose, common table sugar, consists of a glucose unit combined with a fructose unit. Lactose (the sugar present in milk) is composed of one molecule of glucose and one of galactose.

As shown in Figure 3–8, a disaccharide can be hydrolyzed, that is, split by the addition of water, into two monosaccharide units. During digestion, maltose is hydrolyzed to form two molecules of glucose:

Maltose + Water ⟶ Glucose + Glucose

Similarly, sucrose is hydrolyzed to form glucose and fructose:

Sucrose + Water ⟶ Glucose + Fructose

Polysaccharides can store energy or provide structure

The most abundant carbohydrates are the **polysaccharides,** a group that includes starches, glycogen, and cellulose. A polysaccharide is a macromolecule consisting of repeating units of simple sugars, usually glucose. Although the precise number of sugar units varies, thousands of units are typically present in a single molecule. The polysaccharide may be a single long chain or a branched chain. Because they are composed of different isomers and because the units may be arranged differently, polysaccharides vary in their properties. Those that can be easily broken down to their subunits are well suited for energy storage, whereas the macromolecular three-dimensional architecture of others makes them particularly well suited to form stable structures.

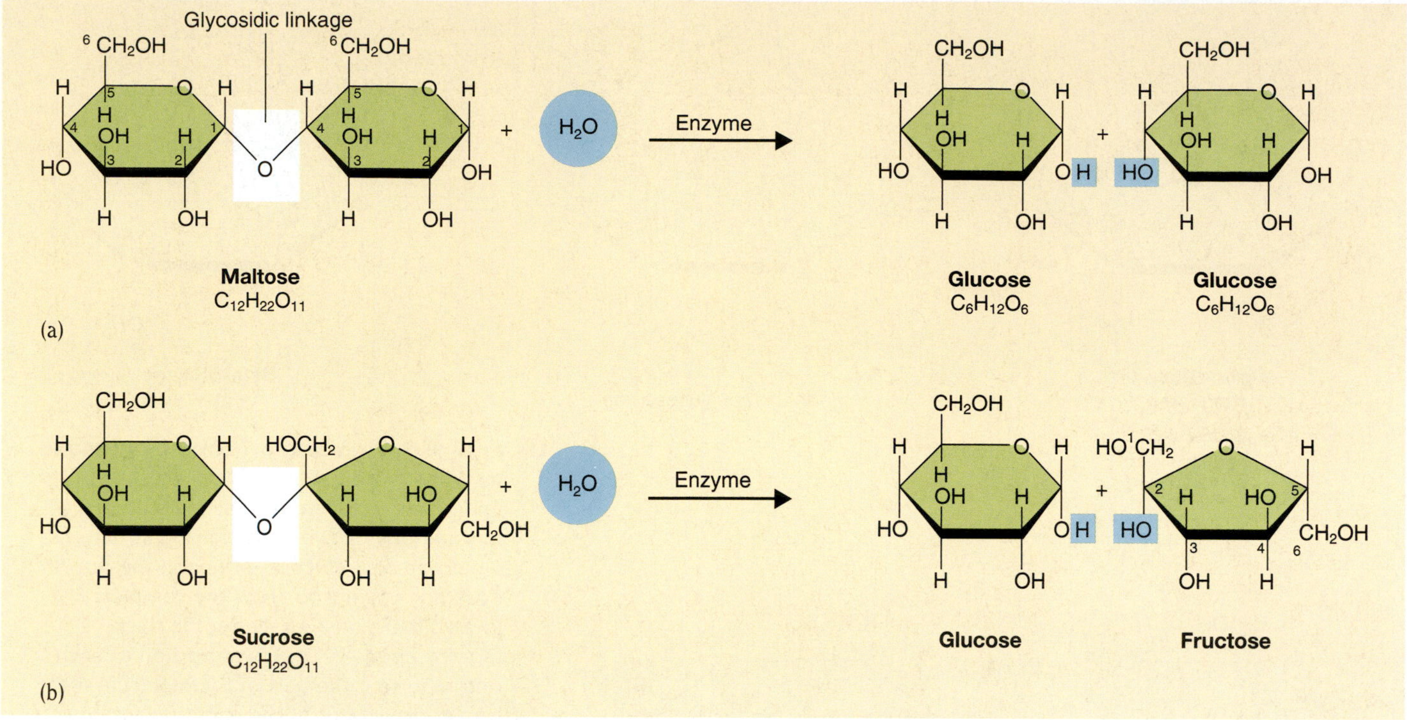

Figure 3–8 Hydrolysis of disaccharides. **(a)** Maltose may be broken down (as it is during digestion) to form two molecules of glucose. The glycosidic linkage is broken in a hydrolysis reaction, which requires the addition of water. **(b)** Sucrose can be hydrolyzed to yield a molecule of glucose and a molecule of fructose. Note that an enzyme (a protein catalyst) is needed to promote these reactions.

Starch is the main storage carbohydrate of plants

Starch, the typical form of carbohydrate used for energy storage in plants, is a polymer consisting of α-glucose subunits. These monomers are joined by α 1—4 linkages, which means that carbon 1 of one glucose is linked to carbon 4 of the next glucose in the chain (Fig. 3–9). Starch occurs in two forms: amylose and amylopectin. Amylose, the simpler form, is unbranched. Amylopectin, the more common form, usually consists of about 1000 glucose units in a branched chain.

Plant cells store starch mainly as granules within specialized organelles called **amyloplasts** (Fig. 3–9*a*). When energy is needed for cellular work, the plant can hydrolyze the starch, releasing the glucose subunits. Humans and other animals that eat plant foods have enzymes to hydrolyze starch.

Glycogen is the main storage carbohydrate of animals

Glycogen (sometimes referred to as *animal starch*) is the form in which glucose is stored as an energy source in animal tissues. It is similar in structure to plant starch but more extensively branched and more water-soluble. Glycogen is stored mainly in liver and muscle cells.

Cellulose is a structural carbohydrate

Carbohydrates are the most abundant group of organic compounds on Earth, and **cellulose** is the most abundant carbohydrate; it accounts for 50% or more of all the carbon in plants (Fig. 3–10). Cellulose is a structural carbohydrate. Wood is about half cellulose, and cotton is at least 90% cellulose. Plant cells are surrounded by strong supporting cell walls consisting mainly of cellulose.

Cellulose is an insoluble polysaccharide composed of many glucose molecules joined together. The bonds joining these sugar units are different from those in starch. Recall that starch is composed of α-glucose subunits, joined by α 1—4 glycosidic linkages. Cellulose contains β-glucose monomers joined by β 1—4 linkages. These bonds cannot be split by the enzymes that hydrolyze the α linkages in starch. Humans, like most organisms, do not have enzymes that can digest cellulose and therefore cannot use it as a nutrient. Because cellulose remains fibrous , it helps keep the digestive tract functioning properly.

Some microorganisms can digest cellulose to glucose. In fact, cellulose-digesting bacteria live in the digestive systems of cows and sheep, enabling these grass-eating animals to obtain nourishment from cellulose. Similarly, the digestive systems of termites contain microorganisms that digest cellulose (see Fig. 24–5*a*).

Amyloplasts

(a)

100 µm

Figure 3–9 Starch, a storage polysaccharide. **(a)** Starch *(stained purple)* is stored in specialized organelles, called amyloplasts, in these cells of a buttercup root. **(b)** Starch is composed of α-glucose molecules joined by glycosidic bonds. At the branch points are bonds between carbon 6 of the glucose in the straight chain and carbon 1 of the glucose in the branching chain. **(c)** Starch is made up of highly branched chains; the arrows indicate the branch points. Each chain is actually in the form of a coil or helix, stabilized by hydrogen bonds between the hydroxyl groups of the glucose subunits. *(a, Ed Reschke)*

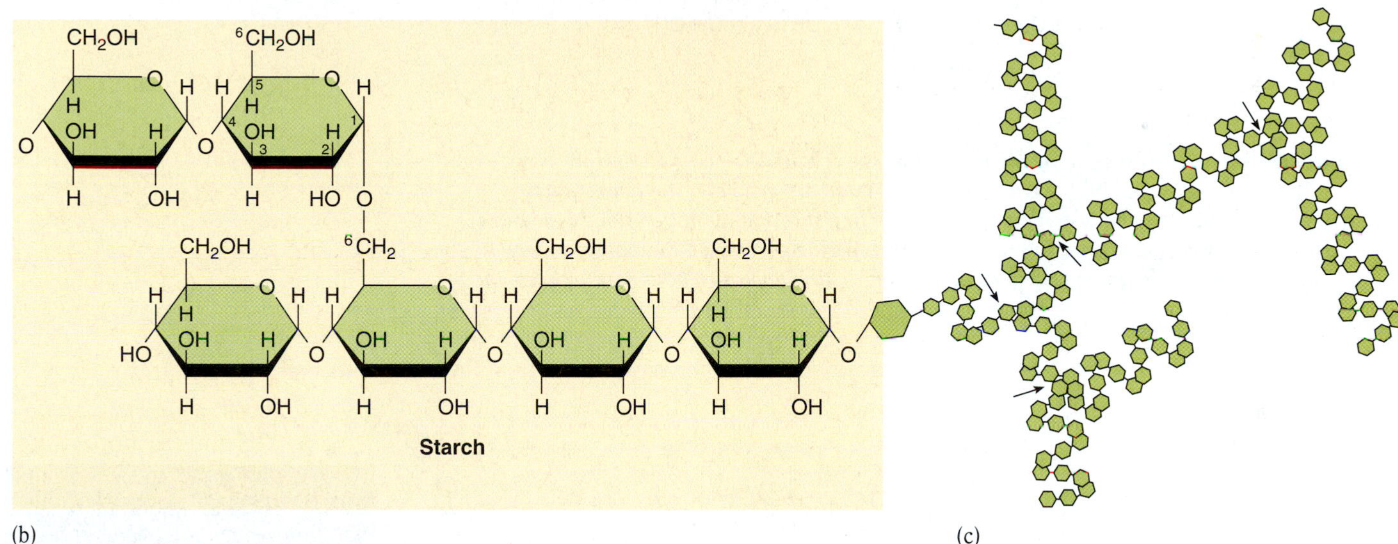

Cellulose molecules have characteristics that make them well suited for a structural role. The β-glucose subunits are joined in a way that allows extensive hydrogen bonding among different cellulose molecules. Thus, cellulose molecules aggregate in long bundles of fibers, as shown in Figure 3–10*a*.

Some modified and complex carbohydrates have special roles

Many derivatives of monosaccharides are important biological molecules. Some form important structural components. The amino sugars galactosamine and glucosamine are compounds in which a hydroxyl group (—OH) is replaced by an amino group (—NH_2). Galactosamine is present in cartilage, a constituent of the skeletal system of vertebrates. *N*-acetyl glucosamine (NAG) subunits, joined by glycosidic bonds, compose **chitin,** a main component of the external skeletons of insects, crayfish, and other arthropods (Fig. 3–11), and of the cell walls of fungi. Chitin forms very tough structures because, as in cellulose, its molecules interact through multiple hydrogen bonds. Some chitinous structures, such as the shell of a lobster, are further hardened by the addition of calcium carbonate.

Carbohydrates may also be combined with proteins to form **glycoproteins,** compounds present on the outer surface of cells other than bacteria. Some of these carbohydrate chains allow cells to adhere to one another, while others provide protection. Most proteins secreted by cells are glycoproteins. These include the major components of mucus, a complex protective material secreted by the mucous membranes of the respiratory and

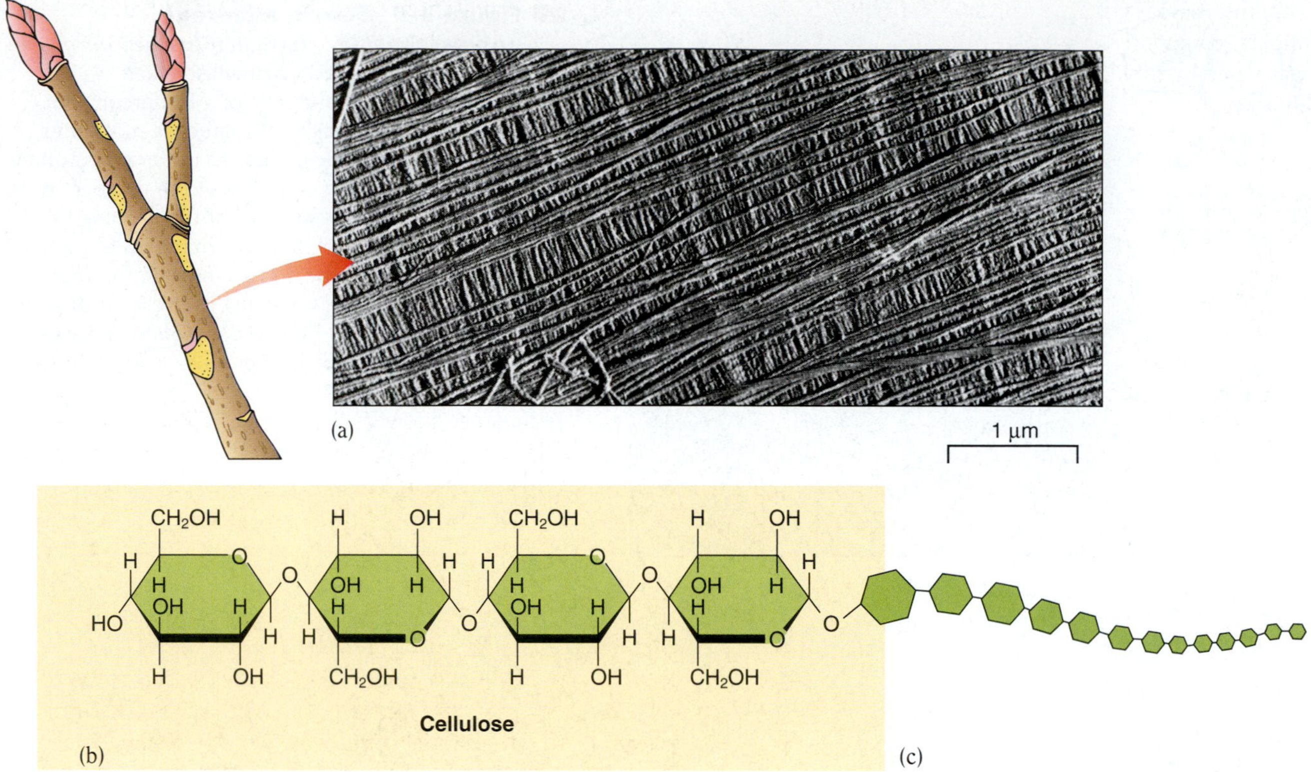

Figure 3-10 Cellulose, a structural polysaccharide. **(a)** An electron micrograph of cellulose fibers from a cell wall. The fibers visible in the photograph consist of bundles of cellulose molecules, interacting through hydrogen bonds. **(b** and **c)** The cellulose molecule is an unbranched polysaccharide composed of approximately 10,000 β-glucose units joined by glycosidic bonds. *(a, Omikron/Photo Researchers, Inc.)*

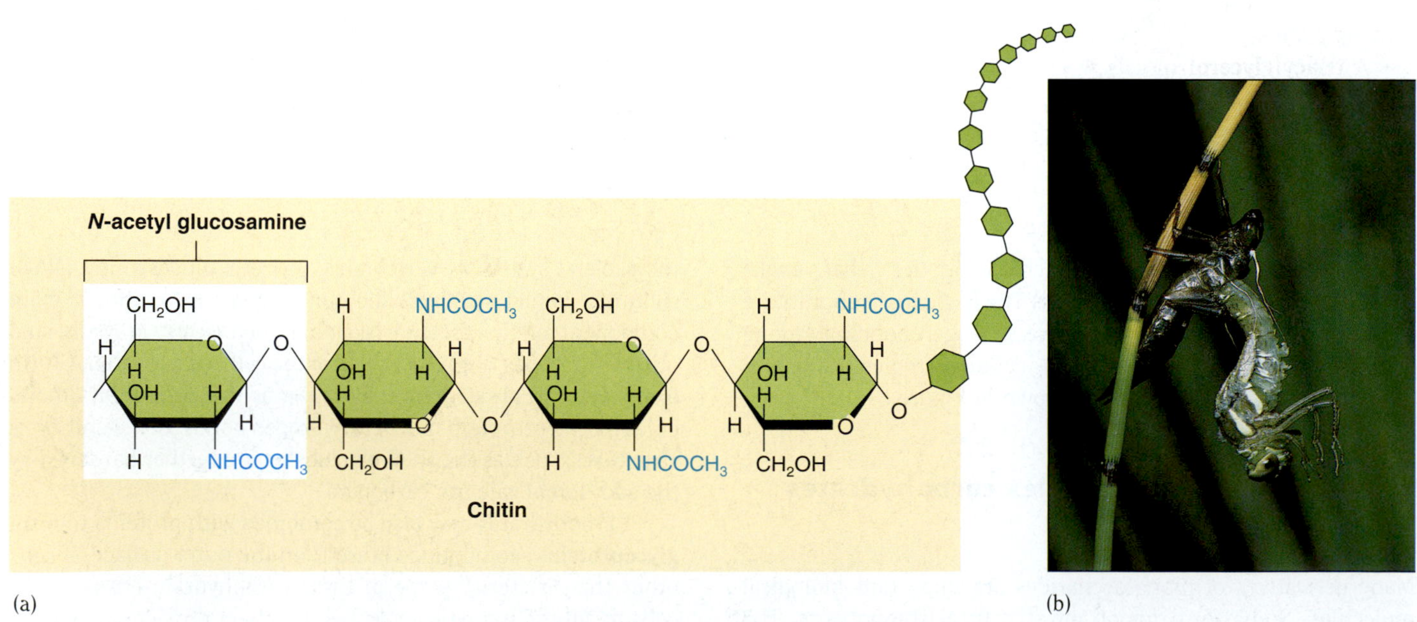

Figure 3-11 Chitin, a structural polysaccharide. **(a)** Chitin is a polymer composed of *N*-acetyl glucosamine (NAG) subunits. **(b)** Chitin is an important component of the exoskeleton (outer covering) that this dragonfly is shedding. *(b, Dwight R. Kuhn)*

digestive systems. Carbohydrates can combine with lipids to form **glycolipids,** compounds present on the surfaces of animal cells that are thought to allow cells to recognize and interact with one another.

LIPIDS ARE FATS OR FATLIKE SUBSTANCES

Lipids are a heterogeneous group of compounds that are defined, not by their structure, but rather by the fact that they are soluble in nonpolar solvents (such as ether and chloroform) and are relatively insoluble in water. Lipid molecules have these properties because they consist mainly of carbon and hydrogen, with few oxygen-containing functional groups. Hydrophilic functional groups typically contain oxygen atoms; therefore lipids, with little oxygen, tend to be hydrophobic. Among the biologically important groups of lipids are fats, phospholipids, carotenoids (orange and yellow plant pigments), steroids, and waxes. Some lipids are used for energy storage; others serve as structural components of cellular membranes; and some are important hormones.

Triacylglycerol is formed from glycerol and three fatty acids

The most abundant lipids in living organisms are **triacylglycerols,** also known as *triglycerides.* These compounds, commonly known as fats, are an economical form of reserve fuel storage because, when metabolized, they yield more than twice as much energy per gram as do carbohydrates. Carbohydrates and proteins can be transformed by enzymes into fats and stored within the cells of adipose (fat) tissue of animals and in some seeds and fruits of plants.

A **triacylglycerol** molecule consists of glycerol joined to three fatty acids (Fig. 3–12). **Glycerol** is a three-carbon alcohol that contains three hydroxyl (—OH) groups, and a **fatty acid** is a long, unbranched hydrocarbon chain with a carboxyl group (—COOH) at one end. A triacylglycerol molecule is formed by a series of three condensation reactions. In each reaction the equivalent of a water molecule is removed as one of the glycerol's hydroxyl groups reacts with the carboxyl group of a fatty acid, resulting in the formation of a covalent linkage known as an **ester linkage** (Fig. 3–12*b*). The first reaction yields a **monoacylglycerol** ***(monoglyceride),*** the second a **diacylglycerol** ***(diglyceride),*** and the third a triacylglycerol. During digestion triacylglycerols are hydrolyzed to produce fatty acids and glycerol (see Chapter 45). Diacylglycerol is an important molecule used for sending signals within the cell (see Chapter 47).

Saturated and unsaturated fatty acids differ in physical properties

About 30 different fatty acids are commonly found in lipids, and they typically have an even number of carbon atoms. For example, butyric acid, present in rancid butter, has four carbon atoms. Oleic acid, with 18 carbons, is the most widely distributed fatty acid in nature and is found in most animal and plant fats.

Saturated fatty acids contain the maximum possible number of hydrogen atoms. Palmitic acid, a 16-carbon fatty acid, is a common saturated fatty acid (Fig. 3–12*c*). Fats high in saturated fatty acids, such as animal fat and solid vegetable shortening, tend to be solid at room temperature. This is because even electrically neutral, nonpolar molecules can develop transient regions of weak positive charge and weak negative charge. This occurs as the constant motion of their electrons causes some regions to have a temporary excess of electrons, whereas others have a temporary electron deficit. These slight opposite charges result in attractions, known as **van der Waals forces,** between adjacent molecules. Although van der Waals interactions are individually weak, they can be strong when many occur among long hydrocarbon chains.

Unsaturated fatty acids include one or more adjacent pairs of carbon atoms joined by a double bond. Therefore they are not fully saturated with hydrogen. Fatty acids with one double bond are called **monounsaturated fatty acids,** whereas those with more than one double bond are **polyunsaturated fatty acids.** Oleic acid is a monounsaturated fatty acid, and linoleic acid is a common polyunsaturated fatty acid (Fig. 3–12*d* and *e*). Fats containing a high proportion of monounsaturated or polyunsaturated fatty acids tend to be liquid at room temperature. This is because each double bond produces a bend in the hydrocarbon chain that prevents it from aligning closely with an adjacent chain, thereby limiting van der Waals interactions.

At least two unsaturated fatty acids (linoleic acid and arachidonic acid) are essential nutrients that must be obtained from food because the human body cannot synthesize them. However, the amounts required are small, and deficiencies are rarely seen. There is no dietary requirement for saturated fatty acids.

Phospholipids are components of cellular membranes

Phospholipids belong to a group of lipids, called **amphipathic lipids,** in which one end of each molecule is hydrophilic and the other end is hydrophobic. The two ends of a phospholipid differ both physically and chemically. A phospholipid consists of a glycerol molecule attached at one end to two fatty acids, and at the other end to a phosphate group linked to an organic compound such as choline. The organic compound usually contains nitrogen (Fig. 3–13*a*). (Note that phosphorus and nitrogen are absent in triacylglycerols.) The fatty acid portion of the molecule (containing the two hydrocarbon "tails") is hydrophobic and not soluble in water. However, the portion composed of glycerol, phosphate, and the organic base (the "head" of the molecule) is ionized and readily water-soluble. The amphipathic properties of these lipid molecules cause them to form lipid bilayers in aqueous (watery) solution, making them uniquely suited to function as the fundamental components of cell membranes, structures that are discussed in Chapters 4 and 5 (Fig. 3–13*b*).

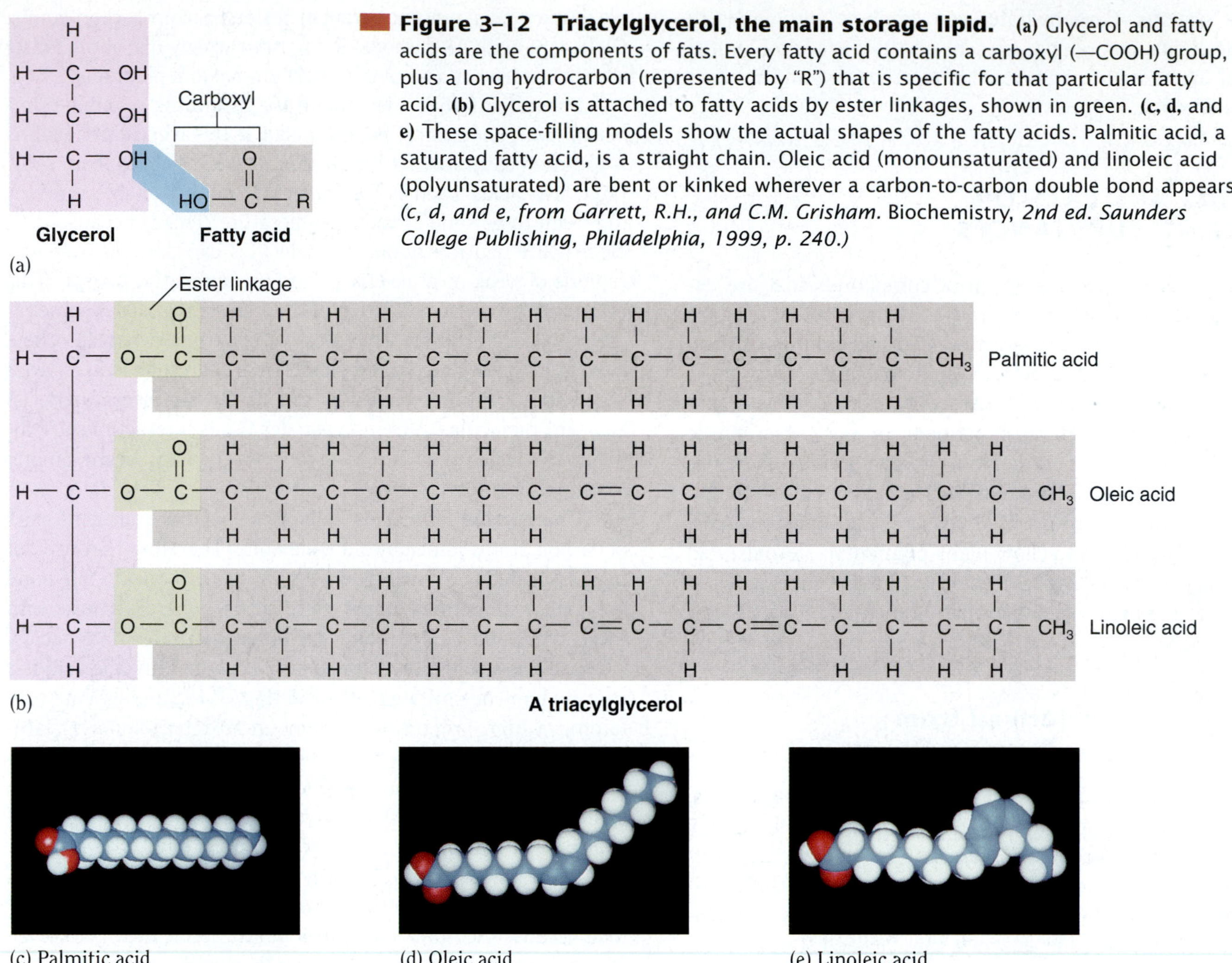

Figure 3–12 Triacylglycerol, the main storage lipid. **(a)** Glycerol and fatty acids are the components of fats. Every fatty acid contains a carboxyl (—COOH) group, plus a long hydrocarbon (represented by "R") that is specific for that particular fatty acid. **(b)** Glycerol is attached to fatty acids by ester linkages, shown in green. **(c, d,** and **e)** These space-filling models show the actual shapes of the fatty acids. Palmitic acid, a saturated fatty acid, is a straight chain. Oleic acid (monounsaturated) and linoleic acid (polyunsaturated) are bent or kinked wherever a carbon-to-carbon double bond appears. *(c, d, and e, from Garrett, R.H., and C.M. Grisham.* Biochemistry, *2nd ed. Saunders College Publishing, Philadelphia, 1999, p. 240.)*

Carotenoids and many other pigments are derived from isoprene units

The orange and yellow plant pigments called **carotenoids** are classified with the lipids because they are insoluble in water and have an oily consistency. These pigments, found in the cells of all plants, play a role in photosynthesis. Carotenoid molecules, such as β-carotene, and many other important pigments, consist of five-carbon hydrocarbon monomers known as **isoprene units** (Fig. 3–14).

Most animals can convert carotenoids to vitamin A, which can then be converted to the visual pigment **retinal.** Eyes are found in three different lines of animals—the mollusks, insects, and vertebrates—and all three groups have the same compound involved in the process of light reception, retinal, a molecule that is apparently uniquely fitted for this role.

Notice that carotenoids, vitamin A, and retinal all have a pattern of double bonds alternating with single bonds. The electrons that make up these bonds can move about relatively easily when light strikes the molecule. Such molecules are pigments that tend to be highly colored because these mobile electrons cause them to strongly absorb light of certain wavelengths and reflect other wavelengths.

Steroids contain four rings of carbon atoms

A **steroid** consists of carbon atoms arranged in four attached rings; three of the rings contain six carbon atoms, and the fourth contains five (Fig. 3–15). The length and structure of the side chains that extend from these rings distinguish one steroid from another. Like carotenoids, steroids are synthesized from isoprene units.

Among the steroids of biological importance are cholesterol, bile salts, reproductive hormones, and cortisol and other hormones secreted by the adrenal cortex. Cholesterol is an essential structural component of animal cell membranes, but when there is excess cholesterol in the blood it forms plaques on artery walls, leading to an increased risk of heart attack (see Chapter 42). Plant cell membranes contain molecules similar to

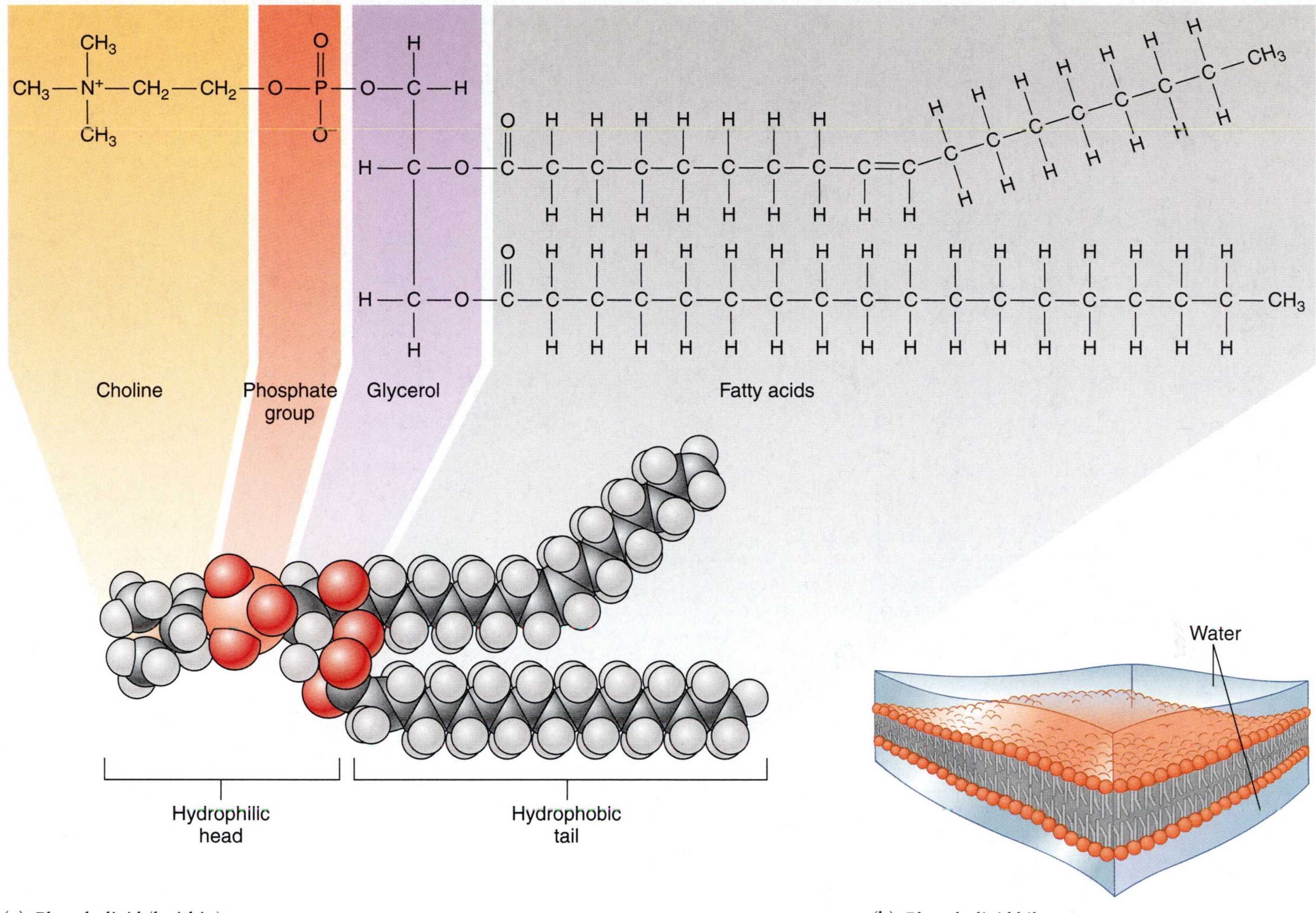

(a) Phospholipid (lecithin)

(b) Phospholipid bilayer

Figure 3–13 A phospholipid and a phospholipid bilayer. **(a)** A phospholipid consists of a hydrophobic tail, made up of two fatty acids, and a hydrophilic head, which includes a glycerol bonded to a phosphate group, which is in turn bonded to an organic group that can vary. Choline is the organic group in lecithin (or phosphatidylcholine), the molecule shown. The upper fatty acid in the figure is monounsaturated; it contains one double bond that produces a characteristic bend in the chain. **(b)** Phospholipids form lipid bilayers in which the hydrophilic heads interact with water and the hydrophobic tails are in the bilayer interior.

cholesterol. Interestingly, certain of these plant steroids are able to block the absorption of cholesterol by the intestine. Bile salts emulsify fats in the intestine so that they can be enzymatically hydrolyzed. Steroid hormones regulate certain aspects of metabolism in a variety of animals, including vertebrates, insects, and crabs.

Some chemical mediators are lipids

Animal cells secrete chemicals that permit them to communicate with each other or to regulate their own activities. Some chemical mediators are produced by the modification of fatty acids that have been removed from membrane phospholipids. These include prostaglandins, which have varied roles, including promoting inflammation and smooth muscle contraction. Certain hormones, such as the juvenile hormone of insects, are also fatty acid derivatives (see Chapter 47).

PROTEINS ARE THE MOST VERSATILE CELLULAR COMPONENTS

In June 2000 two groups of scientists, one a government-supported international consortium known as the Human Genome Project, and the other a private company, made a landmark announcement that created great excitement, not only in

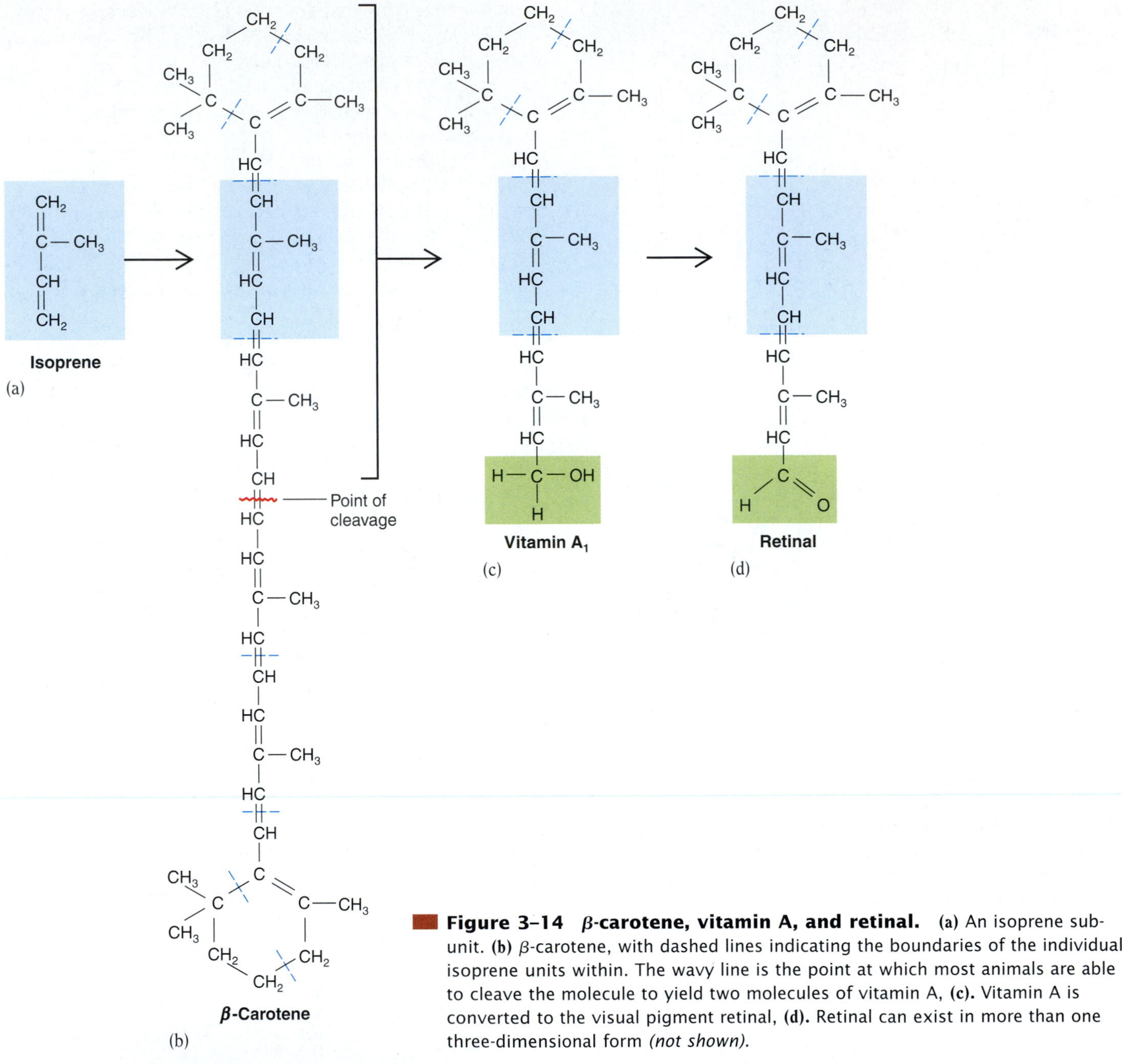

Figure 3–14 β-carotene, vitamin A, and retinal. **(a)** An isoprene subunit. **(b)** β-carotene, with dashed lines indicating the boundaries of the individual isoprene units within. The wavy line is the point at which most animals are able to cleave the molecule to yield two molecules of vitamin A, **(c).** Vitamin A is converted to the visual pigment retinal, **(d).** Retinal can exist in more than one three-dimensional form *(not shown).*

the scientific community but also among the public. In an immense undertaking they had succeeded in sequencing virtually all the genetic information in a human cell. (See Chapter 15 for additional discussion of the Human Genome Project.)

Some might think that sequencing human genes is the end of the story, but it is actually only the beginning. Most genetic information is used to specify the structure of **proteins,** and it has been predicted that most of the 21st century will be devoted to understanding the functioning of these extraordinarily versatile macromolecules that are of central importance in the chemistry of life. In a real sense, proteins are involved in virtually all aspects of metabolism because most **enzymes** (molecules that regulate the thousands of different chemical reactions that take place in an organism) are proteins. Proteins can be assembled into a variety of shapes, allowing them to serve as major structural components of cells and tissues. For this reason, growth and repair, as well as maintenance of the organism, depend on these compounds. As shown in Table 3–2, proteins serve in a great many other specialized capacities.

The protein constituents of a cell are the clues to its lifestyle. Each cell type contains characteristic forms, distributions, and amounts of protein that largely determine what the cell looks like and how it functions (see Fig. 16–2). A muscle cell contains large amounts of the proteins myosin and actin, which are responsible for its appearance as well as for its ability to contract. The protein hemoglobin, found in red blood

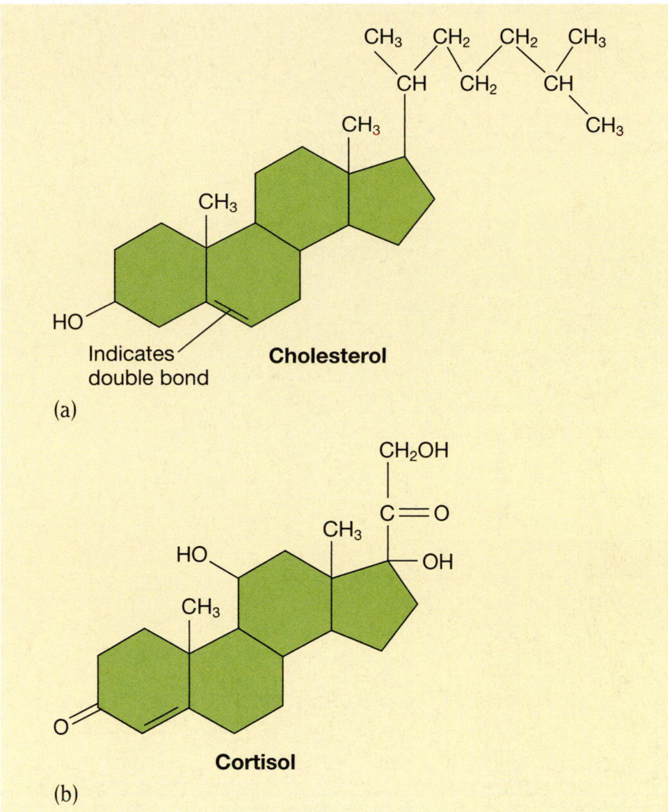

Figure 3–15 Steroids. Four attached rings—three six-carbon rings and one with five carbons—make up the fundamental structure of a steroid *(shown in green).* Note that some carbons are shared by two rings. In these simplified structures, a carbon atom is present at each angle of a ring; the hydrogen atoms attached directly to the ring have not been drawn. **(a)** Cholesterol is an essential component of animal cellular membranes. **(b)** Cortisol is a steroid hormone secreted by the adrenal glands. Notice that cortisol differs from cholesterol in its attached functional groups.

cells, is responsible for the specialized function of oxygen transport.

Amino acids are the subunits of proteins

Amino acids, the constituents of proteins, have an amino group ($—NH_2$) and a carboxyl group (—COOH) bonded to the same asymmetrical carbon atom, known as the **alpha carbon.** There are about 20 amino acids commonly found in proteins, each uniquely identified by the variable side chain (R group) bonded to the α carbon (Fig. 3–16). Glycine, the simplest amino acid, has a hydrogen atom as its R group; alanine has a methyl ($—CH_3$) group.

Amino acids in solution at neutral pH are mainly dipolar ions. This is generally how amino acids exist at cellular pH. Each carboxyl group (—COOH) donates a proton and becomes dissociated ($—COO^-$), whereas each amino group ($—NH_2$) accepts a proton and becomes $—NH_3^+$ (Fig. 3–17). Because of the ability of their amino and carboxyl groups to accept and release protons, amino acids in solution resist changes in acidity and alkalinity and so are important biological buffers.

The amino acids are grouped in Figure 3–16 by the properties of their side chains. These broad groupings actually include amino acids with a fairly wide range of properties. Amino acids classified as having *nonpolar* side chains tend to have hydrophobic properties, whereas those classified as *polar* are more hydrophilic. An acidic amino acid has a side chain that contains a carboxyl group. At cellular pH the carboxyl group is dissociated, giving the R group a negative charge. A basic amino acid becomes positively charged when the amino group in its side chain accepts a hydrogen ion. Acidic and basic side chains are ionic at cellular pH and therefore hydrophilic.

In addition to the 20 common amino acids, some proteins have unusual ones. These rare amino acids are produced by the modification of common ones after they have become part of a protein. For example, lysine and proline may be converted to hydroxylysine and hydroxyproline after they have been incorporated into collagen. These amino acids can form cross links between the peptide chains that make up collagen. Such cross links are responsible for the firmness and great strength of the collagen molecule, which is a major component of cartilage, bone, and other connective tissues.

TABLE 3–2 Some of the Major Classes of Proteins and Their Functions

Protein Types	Some Functions and Examples
Enzymes	Each is responsible for catalyzing a specific chemical reaction
Structural proteins	Strengthen and protect cells and tissues (e.g., collagen strengthens animal tissues)
Storage proteins	Store nutrients; particularly abundant in eggs (e.g., ovalbumin in egg white) and seeds (e.g., zein in corn kernels)
Transport proteins	Transport specific substances between cells (e.g., hemoglobin transports oxygen in red blood cells) Move specific substances (e.g., ions, glucose, amino acids) across cell membranes
Regulatory proteins	Some are protein hormones (e.g., insulin) Some control the expression of specific genes
Motile proteins	Participate in cellular movements (e.g., actin and myosin are essential for muscle contraction)
Protective proteins	Proteins that defend against foreign invaders (e.g., antibodies play a key role in the immune system)

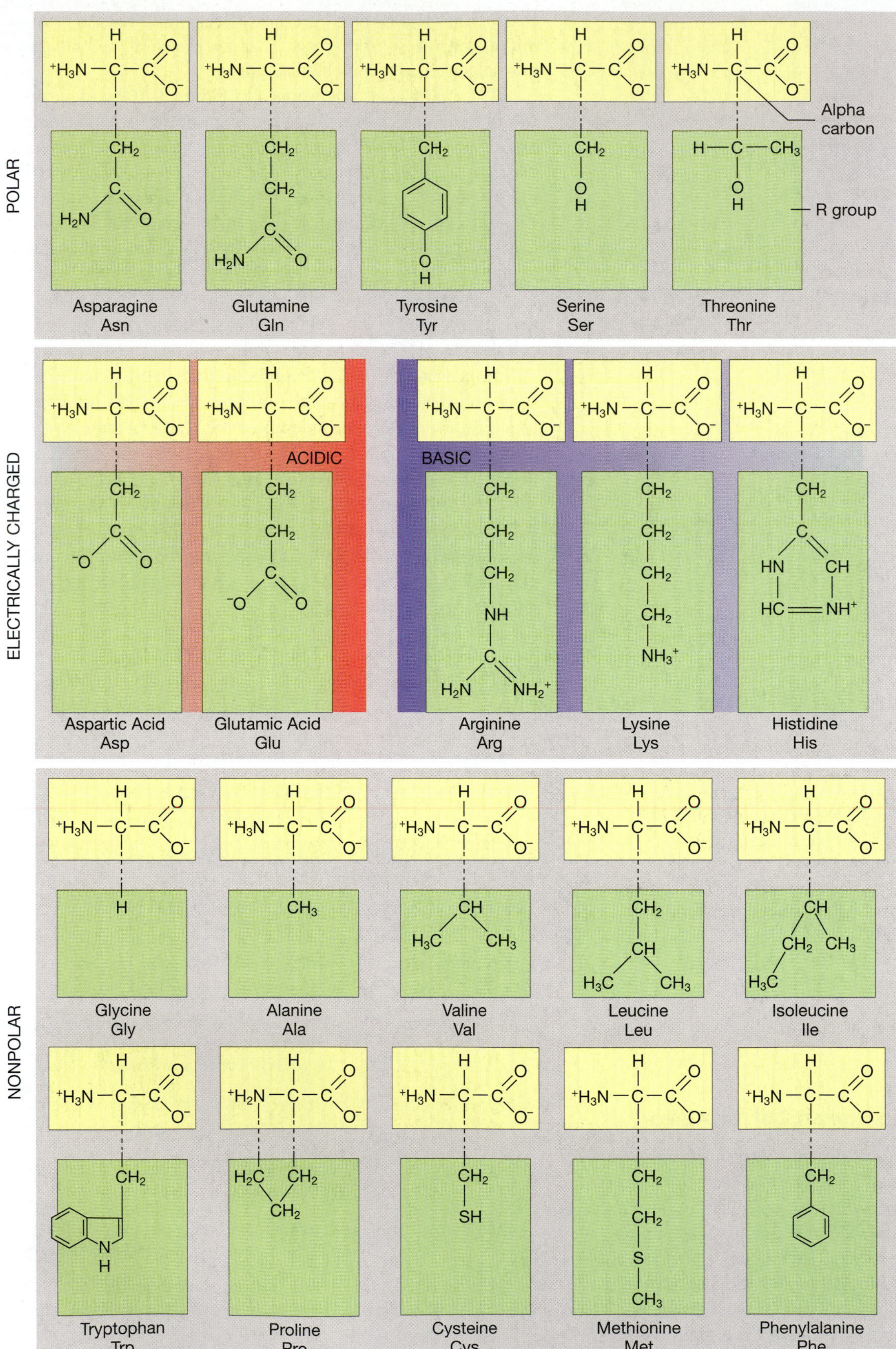

Figure 3-16 The 20 common amino acids. Amino acids designated *polar* have relatively hydrophilic side chains, whereas the side chains of those referred to as *nonpolar* are relatively hydrophobic. Carboxyl groups and amino groups are electrically charged at cellular pH; therefore, acidic and basic amino acids are hydrophilic. The three-letter symbols are the conventional abbreviations for the amino acids.

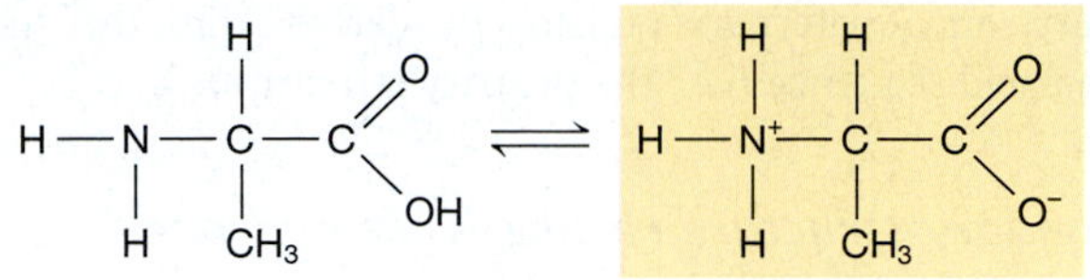

Ionized form

Figure 3–17 An amino acid at pH 7. In living cells, amino acids exist mainly in their ionized form, as dipolar ions.

With some exceptions, bacteria and plants can synthesize all their needed amino acids from simpler substances. If the proper raw materials are available, the cells of humans and animals can manufacture some, but not all, of the biologically significant amino acids. Those that animals cannot synthesize and so must obtain from the diet are known as **essential amino acids.** Animals differ in their biosynthetic capacities; what is an essential amino acid for one species may not be for another. The essential amino acids for humans include isoleucine, leucine, lysine, methionine, phenylalanine, threonine, tryptophan, valine, histidine, and (in children) arginine.

Peptide bonds join amino acids

Amino acids combine chemically with one another by a condensation reaction that bonds the carboxyl carbon of one molecule to the amino nitrogen of another (Fig. 3–18). The covalent carbon-to-nitrogen bond linking two amino acids together is called a **peptide bond.** When two amino acids combine, a **dipeptide** is formed; a longer chain of amino acids is a **polypeptide.** A protein consists of one or more polypeptide chains. Each polypeptide has a free amino group at one end and a free carboxyl group (belonging to the last amino acid added to the chain) at the opposite end. The other amino and carboxyl groups of the amino acid monomers (except those in side chains) are part of the peptide bonds. The complex process by which polypeptides are synthesized is discussed in Chapter 12.

A polypeptide may contain hundreds of amino acids joined in a specific linear order. The backbone of the polypeptide chain includes the repeating sequence

$$N—C—C—N—C—C—N—C—C$$

plus all other atoms *except those in the R groups.* The R groups of the amino acids extend from this backbone.

An almost infinite variety of protein molecules is possible, differing from one another in the number, types, and sequences of amino acids they contain. The 20 types of amino acids found in proteins may be thought of as letters of a protein alphabet; each protein is a very long sentence made up of amino acid letters.

Proteins have four levels of organization

The polypeptide chains making up a protein are twisted or folded to form a macromolecule with a specific *conformation,* or

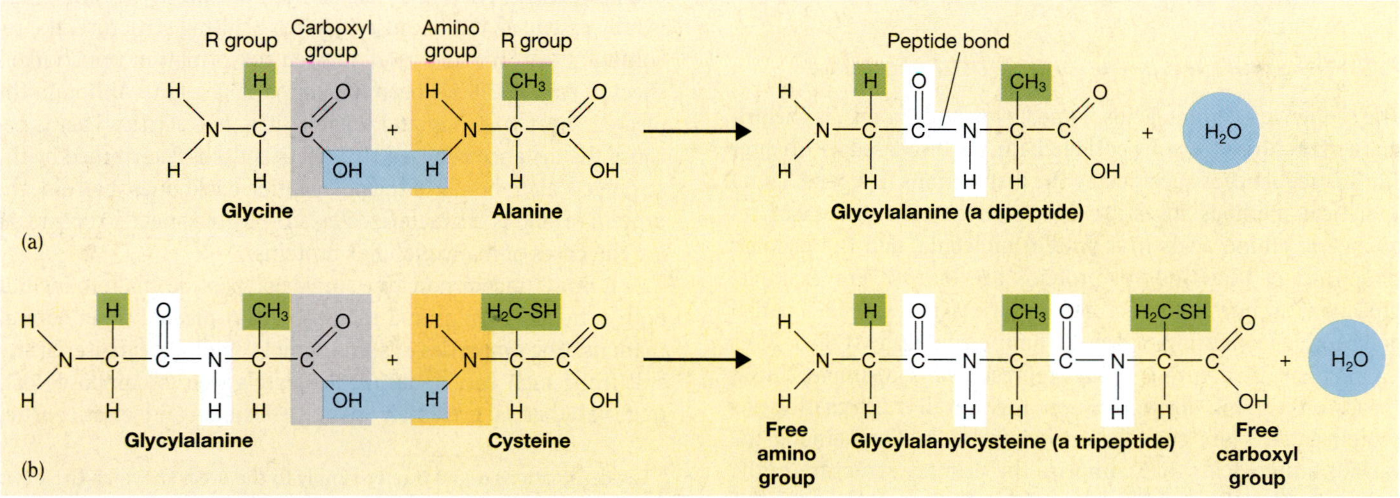

Figure 3–18 Peptide bonds. **(a)** A dipeptide is formed by a condensation reaction, that is, by the removal of the equivalent of a water molecule from the carboxyl group of one amino acid and the amino group of another amino acid. The resulting peptide bond is a covalent, carbon-to-nitrogen bond. Note that the carbon is also part of a carbonyl group, and that the nitrogen is also covalently bonded to a hydrogen. **(b)** The carboxyl group of the dipeptide reacts with the amino group of a third amino acid to form a chain of three amino acids (a tripeptide, or small polypeptide). Additional amino acids can be added to form a long polypeptide chain with a free amino group at one end and a free carboxyl group at the other.

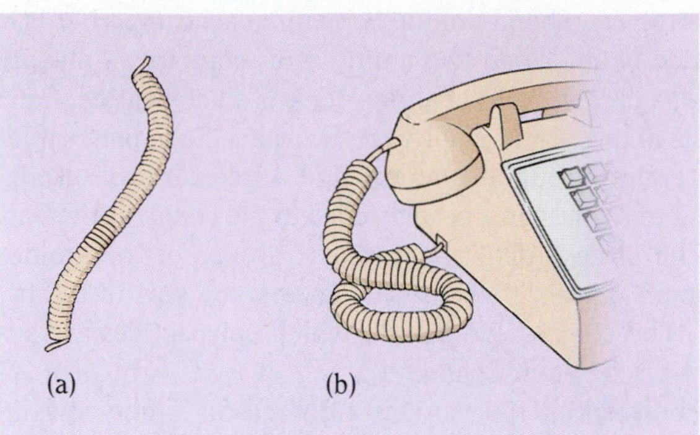

Figure 3–19 Levels of protein structure. A telephone cord provides a familiar analogy for the levels of protein structure. **(a)** Secondary structure, in this case analogous to an α-helix. **(b)** Tertiary structure, in which interactions among side chains cause the molecule to fold back on itself.

three-dimensional shape. Some polypeptide chains form long fibers. **Globular** proteins are tightly folded into compact, roughly spherical shapes. There is a close relationship between a protein's conformation and its function. For example, a typical enzyme is a globular protein with a unique shape that permits it to catalyze a specific chemical reaction. Similarly, the shape of a protein hormone enables it to combine with receptors on its target cell (the cell the hormone acts on).

Four main levels of protein organization can be recognized: primary, secondary, tertiary, and quaternary. An analogy for secondary and tertiary structure is depicted in Figure 3–19.

Primary structure is the amino acid sequence

The sequence of amino acids, joined by peptide bonds, is the **primary structure** of a polypeptide chain. As discussed in Chapter 12, this sequence is specified by the instructions in a gene. Using analytical methods investigators can determine the exact sequence of amino acids in a protein molecule, and the primary structures of thousands of proteins are known. For example, glucagon, a hormone secreted by the pancreas, is a small polypeptide, consisting of only 29 amino acid units (Fig. 3–20).

Primary structure is always represented in a simple, linear, "beads-on-a-string" form. However, the overall conformation of a protein is far more complex, involving interactions among the various amino acids that comprise the primary structure of the molecule. Therefore, the higher orders of structure—secondary, tertiary, and quaternary—ultimately derive from the specific amino acid sequence (i.e., the primary structure).

Secondary structure results from hydrogen bonding involving the backbone

Some regions of a polypeptide exhibit **secondary structure,** which is highly regular. A common secondary structure in protein molecules is the **α-helix,** a region where a polypeptide chain forms a uniform helical coil (Fig. 3–21*a*). The helical structure is determined and maintained by the formation of hydrogen bonds between the backbones of the amino acids in successive turns of the spiral coil. Each hydrogen bond forms between an oxygen with a partial negative charge and a hydrogen with a partial positive charge. The oxygen is part of the remnant of the carboxyl group of one amino acid; the hydrogen is part of the remnant of the amino group of the fourth amino acid down the chain. Thus 3.6 amino acids are included in each complete turn of the helix. Every amino acid in an α-helix is hydrogen-bonded in this way.

The α-helix is the basic structural unit of some fibrous proteins that make up wool, hair, skin, and nails. The elasticity of these fibers is due to a combination of physical factors (the helical shape) and chemical factors (hydrogen bonding). Although hydrogen bonds maintain the helical structure, these bonds can be broken, allowing the fibers to stretch under tension (like a telephone cord). When the tension is released, the fibers recoil and hydrogen bonds reform. This is why human hairs can stretch to some extent and then snap back to their original length.

Another type of secondary structure is the **β-pleated sheet**[2] (Fig. 3–21*b*). The hydrogen bonding in a β-pleated sheet takes place between different polypeptide chains, or different regions of a polypeptide chain that has turned back on itself. Each chain is fully extended, but because each has a zigzag structure, the resulting "sheet" has an overall pleated conformation (much like a sheet of paper that has been folded to make a fan). Although the pleated sheet is strong and flexible, it is not elastic. This is because the distance between the pleats is fixed, determined by the strong covalent bonds of the polypeptide backbones. Fibroin, the protein of silk, is characterized by a β-pleated sheet structure, as are the cores of many globular proteins.

It is not uncommon for a single polypeptide chain to include both α-helical regions and regions with β-pleated sheet conformations. The properties of some complex biological materials result from such combinations. A spider's web is composed of a material that is extremely strong, flexible, and elastic. Once

[2] The designations α and β refer simply to the order in which these two types of secondary structures were discovered.

Figure 3–20 Primary structure of a polypeptide. Glucagon is a very small polypeptide made up of 29 amino acids. The linear sequence of amino acids is indicated by ovals containing their abbreviated names (see Fig. 3–16).

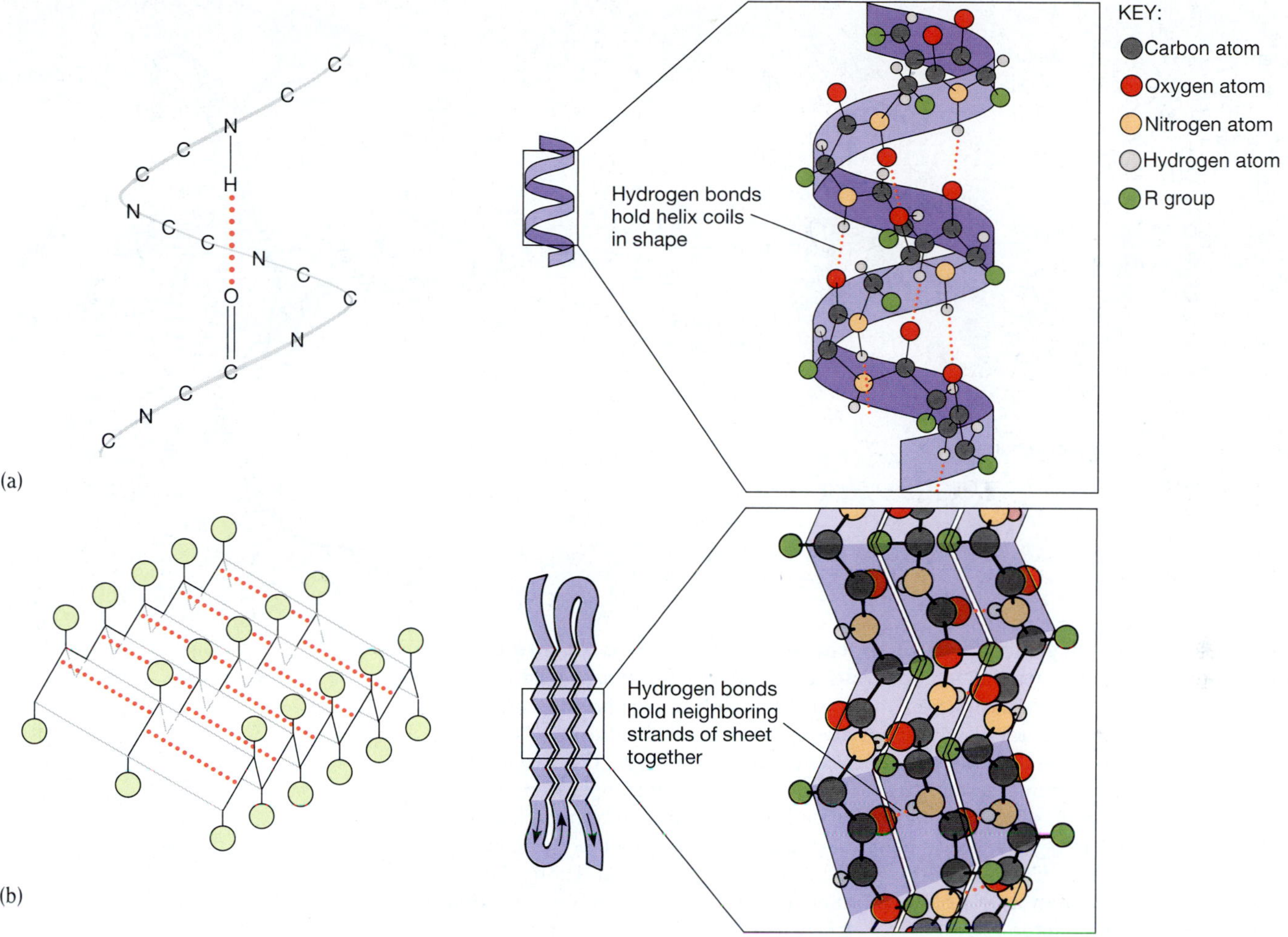

Figure 3–21 Secondary structure of a protein. **(a)** Note that the R groups project out from the sides of the α-helix. (The R groups have been omitted in the simplified diagram at left.) **(b)** A β-pleated sheet forms when a polypeptide chain folds back on itself *(arrows)*; half the R groups project above the sheet and the other half project below it.

again we see function and structure working together, as these properties derive from the fact that it is a composite of proteins with α-helical conformations (providing elasticity) and others with β-pleated sheet conformations (providing strength).

Tertiary structure depends on interactions among side chains

The **tertiary structure** of a protein molecule is the overall shape assumed by each individual polypeptide chain (Fig. 3–22). This three-dimensional structure is determined by four main factors that involve interactions among R groups (side chains) belonging to the same polypeptide chain. These include both weak interactions (hydrogen bonds, ionic bonds, and hydrophobic interactions) and strong covalent bonds.

1. Hydrogen bonds form between R groups of certain amino acid subunits.
2. Ionic attraction can occur between an R group with a unit of positive charge and one with a unit of negative charge.
3. Hydrophobic interactions result from the tendency of nonpolar R groups to be excluded by the surrounding water and therefore to associate in the interior of the globular structure.
4. Covalent bonds known as disulfide bonds or disulfide bridges (—S—S—) may link the sulfur atoms of two cysteine subunits belonging to the same chain. A disulfide bridge forms when the sulfhydryl groups of two cysteines react; the two hydrogens are removed, and the two sulfur atoms that remain become covalently linked.

Quaternary structure results from interactions among polypeptides

Many functional proteins are composed of two or more polypeptide chains, interacting in specific ways to form the biologically

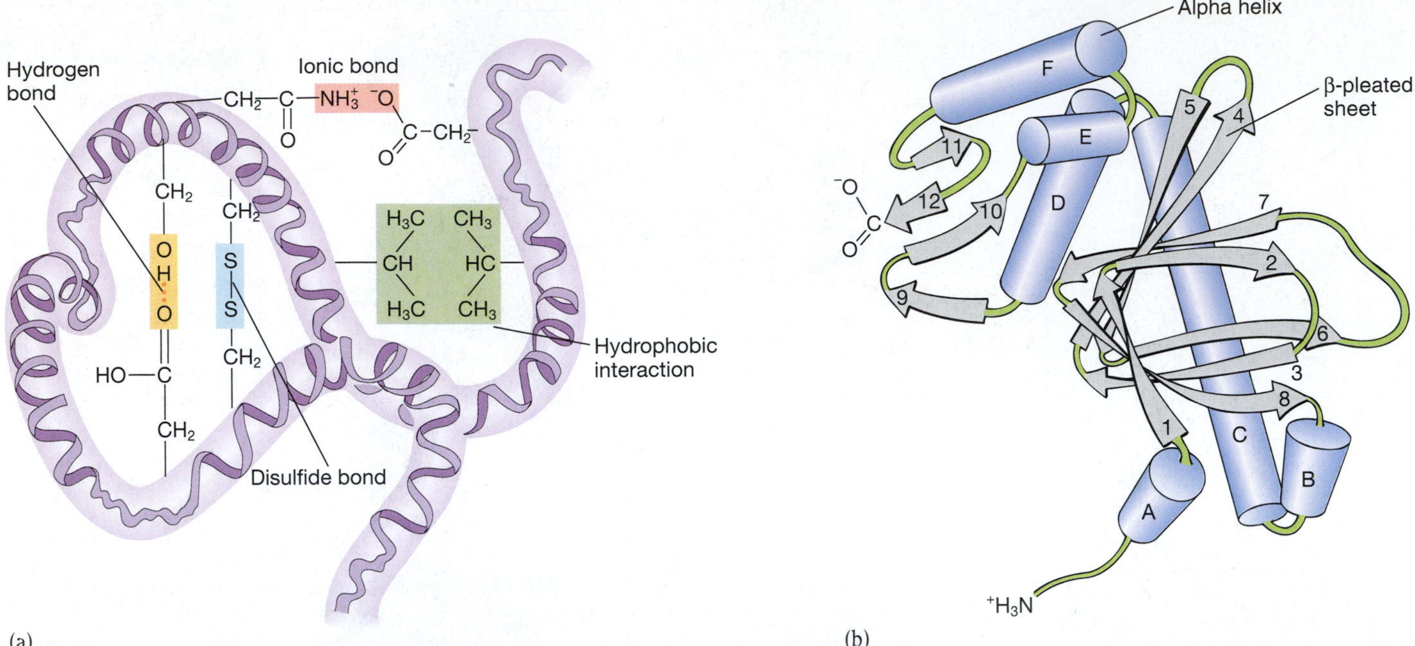

Figure 3–22 Tertiary structure of a protein. (a) Disulfide bonds, hydrogen bonds, hydrophobic interactions, and ionic attractions between R groups hold the parts of the molecule in the designated shape. (b) Schematic drawing of the tertiary structure of a polypeptide. α-Helical regions are represented as blue tubes lettered A through F; β-pleated sheets are the gray arrows numbered 1 through 12. Green lines represent connecting regions. Although the molecule seems very complicated, it is a single polypeptide chain, starting at the amino end *(bottom left)* and terminating at the carboxyl end *(upper left)*. Most of the bends and foldbacks that give the molecule its overall conformation (tertiary structure) are stabilized by R-group interactions. This polypeptide is a subunit of a DNA-binding protein (known as CAP) from the bacterium *Escherichia coli* (see Chapter 13).

active molecule. **Quaternary structure** is the resulting three-dimensional architecture of these polypeptide chains (each with its own primary, secondary, and tertiary structure). The same types of interactions that produce secondary and tertiary structure can also contribute to quaternary structure; these include hydrogen bonding, ionic bonding, hydrophobic interactions, and disulfide bridges.

For example, a functional antibody molecule consists of four polypeptide chains joined by disulfide bridges (see Chapter 43). Disulfide bridges are a common feature of proteins, such as antibodies, that are secreted from cells; these strong bonds stabilize the molecules in the extracellular environment.

Hemoglobin, the protein in red blood cells that is responsible for oxygen transport, is an example of a globular protein with quaternary structure (Fig. 3–23*a*). Hemoglobin consists of 574 amino acids arranged in four polypeptide chains: two identical chains called *alpha chains* and two identical chains called *beta chains*.

Collagen, mentioned previously, has a fibrous type of quaternary structure that well suits it to function as the major strengthener of animal tissues. It consists of three polypeptide chains, wound about each other and bound by cross links between their amino acids (Fig. 3–23*b*).

The amino acid sequence of a protein determines its conformation

Under defined experimental conditions in vitro (i.e, outside a living cell), a polypeptide can be demonstrated to spontaneously undergo folding processes that allow it to attain its normal, functional conformation. For example, in 1996 researchers at the University of Illinois at Urbana—Champaign conducted an experiment in which they completely unfolded myoglobin, a polypeptide that stores oxygen in muscle cells, and then used sophisticated technology to track the refolding process. They found that within a few fractions of a microsecond the molecule had coiled up to form α-helices, and formation of the tertiary structure was completed within 4 microseconds.

This and other types of evidence support the widely held conclusion that amino acid sequence is the ultimate determinant of protein conformation. However, because conditions in vivo (in the cell) are quite different from defined laboratory conditions, proteins do not necessarily always fold spontaneously. On the contrary, in recent years it has been learned that proteins known as **molecular chaperones** mediate the folding of certain proteins. Chaperones are thought to make the folding process more orderly and efficient and to prevent partially folded proteins from becom-

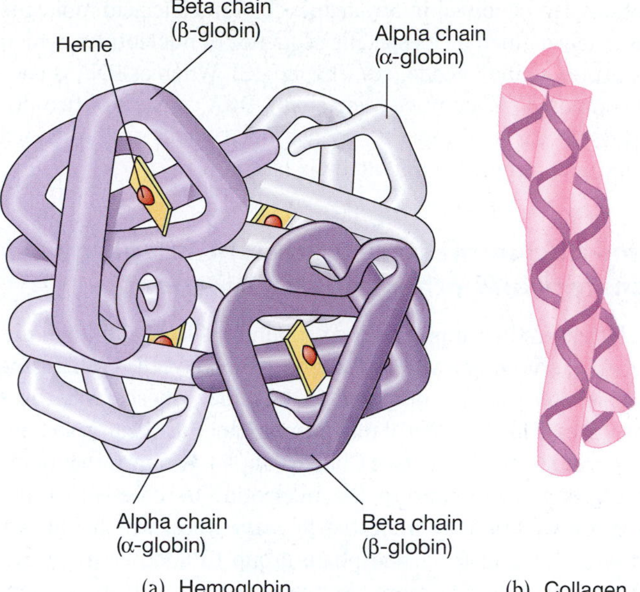

Figure 3-23 Quaternary structure. (a) Hemoglobin, a globular protein, consists of four polypeptide chains, each joined to an iron-containing molecule, a heme. (b) Collagen, a fibrous protein, is a triple helix consisting of three long polypeptide chains. *(b, from Tobin, A.J. and J. Dusheck.* Asking About Life, *2nd ed. Harcourt College Publishers, Philadelphia, 2001, p. 66)*

ing inappropriately aggregated. However, there is no evidence that chaperones actually dictate the folding pattern. For this reason, the existence of chaperones is not an argument against the idea that amino acid sequence determines conformation.

Protein conformation determines function

The overall structure of a protein helps determine its biological activity. A single protein may have more than one distinct structural region, each with its own function. Many proteins are modular, consisting of two or more globular regions, called *domains,* connected by less compact regions of the polypeptide chain. Each domain may have a different function. For example, a protein might have one domain that attaches it to a membrane and another that allows it to act as an enzyme.

The biological activity of a protein can be disrupted by a change in amino acid sequence that results in a change in conformation. For example, the genetic disease known as *sickle cell anemia* is due to a mutation that causes the substitution of the amino acid valine for glutamic acid at position 6 (the sixth amino acid from the amino end) in the beta chain of hemoglobin. The substitution of valine (which has a nonpolar side chain) for glutamic acid (which has a charged side chain) makes the hemoglobin less soluble and more likely to form crystal-like structures. This alteration of the hemoglobin affects the red blood cells, changing them to the crescent or sickle shapes that characterize this disease (see Fig. 15–9).

The biological activity of a protein may be affected by changes in its three-dimensional structure. When a protein is heated, subjected to significant pH changes, or treated with any of a number of chemicals, its structure can become disordered and the coiled peptide chains can unfold to give a more random conformation. This unfolding, which is mainly due to the disruption of hydrogen bonds and ionic bonds, is typically accompanied by a loss of normal function. Such changes in shape and the accompanying loss of biological activity are termed **denaturation** of the protein. For example, a denatured enzyme would lose its ability to catalyze a chemical reaction. An everyday example of denaturation occurs when we fry an egg. The consistency of the egg white protein, known as *albumin,* changes to a solid. Denaturation generally cannot be reversed (you can't "unfry" an egg). However, under certain conditions, some proteins have been denatured and have returned to their original shape and biological activity when normal environmental conditions have been restored.

 Protein conformation is studied through a variety of methods

The architecture of a protein can be ascertained directly through sophisticated types of analysis, such as the x-ray diffraction studies discussed in Chapter 11. Because these studies are tedious and costly, efforts have been made to develop alternative approaches, which rely heavily on the enormous databases generated by the Human Genome Project and related efforts, many of which are accessible through the Internet. Today a protein's primary structure can be determined rapidly through the application of genetic engineering techniques (see Chapter 14), or by the use of sophisticated technology such as mass spectrometry. A variety of efforts are being made to effectively use these amino acid sequence data to predict a protein's higher levels of structure: secondary, tertiary, and quaternary. As we have seen, side chains can interact in relatively predictable ways, such as through ionic and hydrogen bonds. In addition, regions with certain types of side chains appear more likely to form α-helices or β-pleated sheets. Very complex computer programs are used to make such predictions, but these are imprecise because of the many possible combinations of folding patterns.

Computers are an essential part of yet another strategy. Once the amino acid sequence of a polypeptide has been determined, researchers use computers to search databases to find polypeptides with similar sequences. If the conformations of any of those polypeptides or portions have already been determined by x-ray diffraction or other techniques, this information can be extrapolated to make similar correlations between amino acid sequence and three-dimensional structure for the protein under investigation. These predictions are becoming increasingly reliable as more information is added to the databases on a daily basis.

DNA AND RNA ARE NUCLEIC ACIDS

Nucleic acids transmit hereditary information and determine what proteins a cell manufactures. There are two classes of nucleic acids found in cells: **ribonucleic acids (RNAs)** and **deoxyribonucleic**

acids (DNAs). DNA comprises the genes, the hereditary material of the cell, and contains instructions for making all the proteins, as well as all the RNA, needed by the organism. RNA is required as a direct participant in the complex process in which amino acids are linked to form polypeptides. Some types of RNA, known as **ribozymes,** can even act as specific biological catalysts. Like proteins, nucleic acids are large, complex molecules. The name *nucleic acid* reflects the fact that they are acidic and were first identified, by Friederich Miescher in 1870, in the nuclei of pus cells.

Nucleic acids consist of nucleotide subunits

Nucleic acids are polymers of **nucleotides,** molecular units that consist of (1) a five-carbon sugar, either **ribose** (in RNA) or **deoxyribose** (in DNA); (2) one or more phosphate groups, which make the molecule acidic; and (3) a nitrogenous base, a ring compound that contains nitrogen. The nitrogenous base may be either a double-ringed **purine** or a single-ringed **pyrimidine** (Fig. 3–24).

DNA commonly contains the purines adenine (A) and guanine (G), the pyrimidines cytosine (C) and thymine (T), the sugar deoxyribose, and phosphate. RNA contains the purines adenine and guanine, and the pyrimidines cytosine and uracil (U), together with the sugar ribose, and phosphate.

The molecules of nucleic acids are made of linear chains of nucleotides, which are joined by **phosphodiester linkages,** each consisting of a phosphate group and the covalent bonds that attach it to the sugars of adjacent nucleotides (Fig. 3–25). Note that each nucleotide is defined by its particular base and that nucleotides can be joined in any sequence. A nucleic acid molecule is uniquely defined by its specific sequence of nucleotides, which constitutes a kind of code (see Chapter 12). Whereas RNA is usually composed of one nucleotide chain, DNA consists of two nucleotide chains held together by hydrogen bonds and entwined around each other in a double helix (see Fig. 1–7).

Some nucleotides are important in energy transfers and other cellular functions

In addition to their importance as subunits of DNA and RNA, nucleotides serve other vital functions in living cells. **Adenosine triphosphate (ATP),** composed of adenine, ribose, and three phosphates (Fig. 3–26), is of major importance as the primary energy currency of all cells (see Chapter 6). The two terminal phosphate groups are joined to the nucleotide by covalent bonds. These are traditionally indicated by wavy lines, which indicate that ATP can transfer a phosphate group to another molecule, making that molecule more reactive (see Fig. 6–6). In this way ATP is able to donate some of its chemical energy. Most of the readily available chemical energy of the cell is associated with the phosphate groups of ATP.

A nucleotide may be converted to an alternative form with specific cellular functions. ATP, for example, is converted to cyclic adenosine monophosphate **(cyclic AMP)** by the enzyme adenylyl cyclase (Fig. 3–27). Cyclic AMP regulates certain cellular functions and is important in the mechanism by which some hormones act (see Chapters 13, 39, and 47).

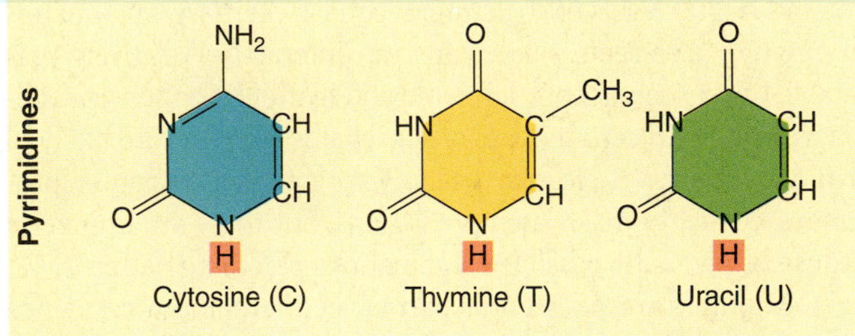

(a)

Purines

Adenine (A)

Guanine (G)

(b)

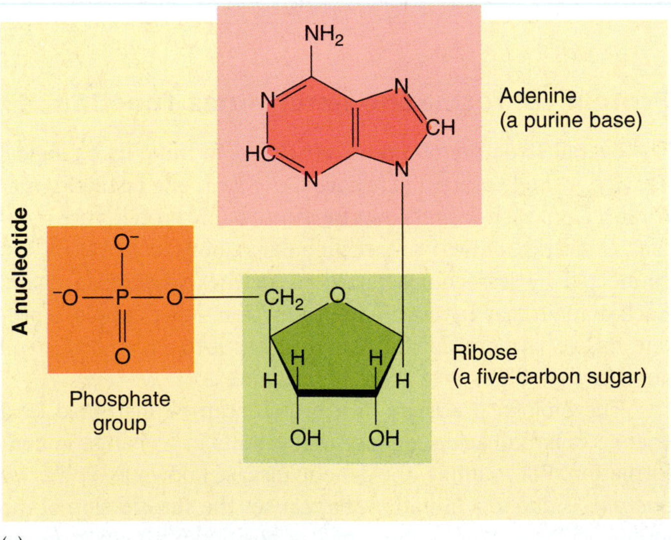

(c)

Figure 3–24 Nucleotides. **(a)** The three major pyrimidine bases found in nucleotides are cytosine, thymine (in DNA only), and uracil (in RNA only). **(b)** The two major purine bases found in nucleotides are adenine and guanine. The hydrogens indicated by the boxes are removed when the base is attached to a sugar. **(c)** A nucleotide, adenosine monophosphate (AMP).

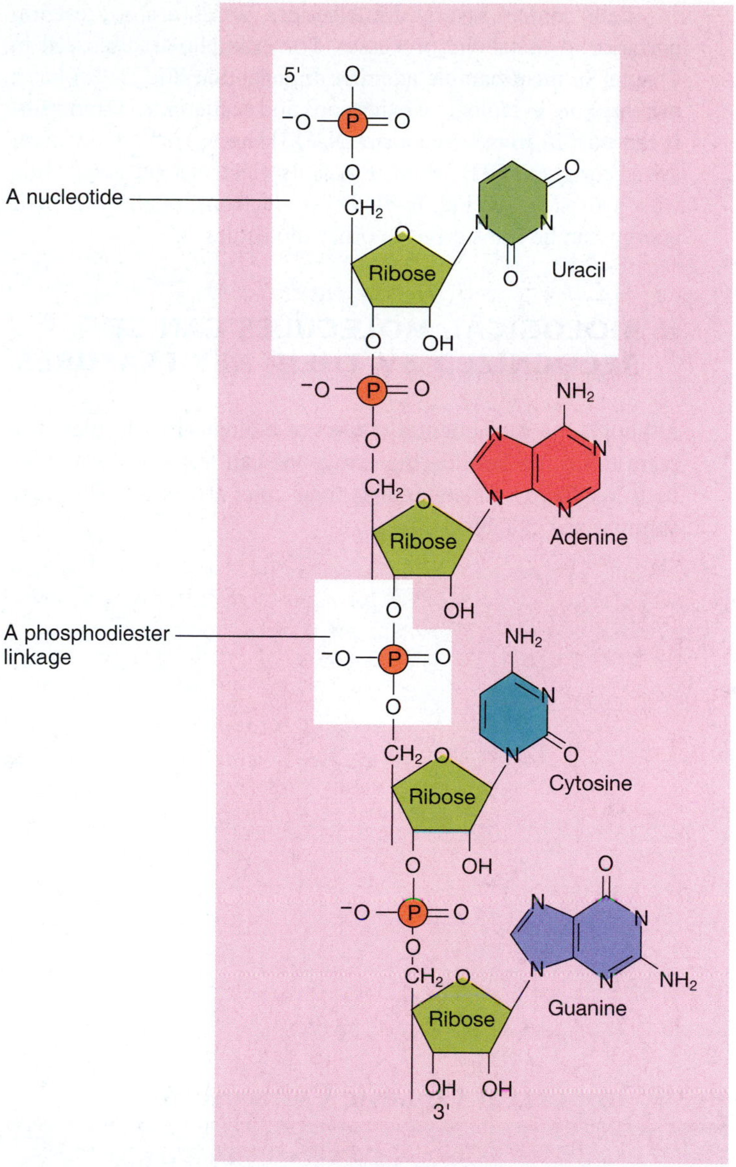

Figure 3-25 RNA. Nucleotides, each with a specific base, are joined by phosphodiester linkages.

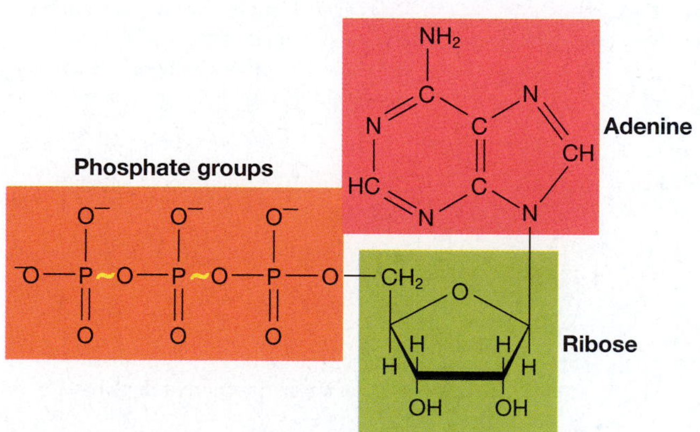

Figure 3-26 Adenosine triphosphate, a nucleotide. The two terminal phosphate groups are joined by covalent bonds (indicated by *wavy lines*) that permit the phosphates to be transferred to other molecules, making them more reactive.

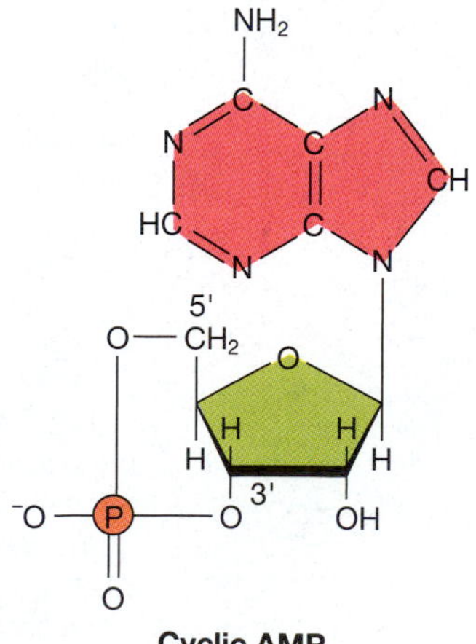

Figure 3-27 Cyclic adenosine monophosphate (cAMP). The single phosphate is part of a ring connecting two regions of the ribose.

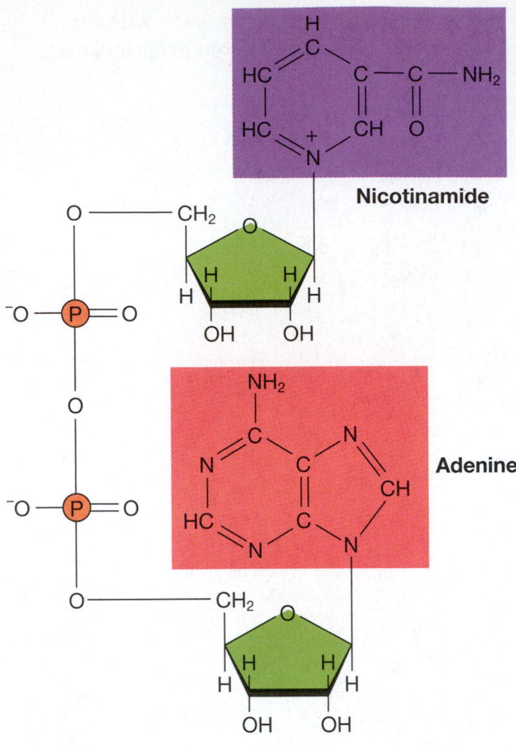

Figure 3–28 NAD$^+$, an important hydrogen (electron) acceptor. The nicotinamide portion of the molecule accepts hydrogen and becomes reduced in the process. The resulting reduced molecule, known as *NADH,* is an electron donor.

Cells contain several dinucleotides, which are of great importance in metabolic processes. For example, as discussed in Chapter 6, **nicotinamide adenine dinucleotide** (Fig. 3–28) has a primary role in biological oxidations and reductions within cells. It can exist in an oxidized form **(NAD$^+$)** that is converted to a reduced form **(NADH)** when it accepts electrons (in association with hydrogen; see Fig. 6–8). These electrons, along with their energy, can be transferred to other molecules.

BIOLOGICAL MOLECULES CAN BE RECOGNIZED BY THEIR KEY FEATURES

Although the fundamental classes of biological molecules may seem to form a bewildering array, one can learn to distinguish them readily by understanding their chief attributes. These are summarized in Table 3–3.

TABLE 3–3 Some of the Groups of Biologically Important Organic Compounds

Class of Compounds	Component Elements	Description	How to Recognize	Principal Function in Living Systems
Carbohydrates	C, H, O	Contain approximately 1 C:2 H:1 O (but make allowance for loss of oxygen and hydrogen when sugar units are linked)	Count the carbons, hydrogens, and oxygens.	Cellular fuel; energy storage; structural component of plant cell walls; component of other compounds such as nucleic acids and glycoproteins
		1. Monosaccharides (simple sugars). Mainly five-carbon (pentose) molecules such as ribose or six-carbon (hexose) molecules such as glucose and fructose	Look for the ring shapes:	Cellular fuel; components of other compounds
		2. Disaccharides. Two sugar units linked by a glycosidic bond, e.g., maltose, sucrose	Count sugar units	Components of other compounds; form of sugar transported in plants
		3. Polysaccharides. Many sugar units linked by glycosidic bonds, e.g., glycogen, cellulose	Count sugar units	Energy storage; structural components of plant cell walls

continued

TABLE 3–3 *Continued*

Class of Compounds	Component Elements	Description	How to Recognize	Principal Function in Living Systems
Lipids	C, H, O (sometimes N, P)	Contain much less oxygen relative to carbon and hydrogen than do carbohydrates		Energy storage; cellular fuel, components of cells; thermal insulation
		1. Neutral fats. Combination of glycerol with one to three fatty acids. Monoacylglycerol contains one fatty acid; diacylglycerol contains two fatty acids; triacylglycerol contains three fatty acids. If fatty acids contain double carbon-to-carbon linkages (C═C), they are unsaturated; otherwise they are saturated	Look for glycerol at one end of molecule: H H—C—O— H—C—O— H—C—O— H	Cellular fuel; energy storage
		2. Phospholipids. Composed of glycerol attached to one or two fatty acids and to an organic base containing phosphorus	Look for glycerol and side chain containing phosphorus and nitrogen.	Components of cell membranes
		3. Steroids. Complex molecules containing carbon atoms arranged in four attached rings. (Three rings contain six carbon atoms each, and the fourth ring contains five.)	Look for four attached rings:	Some are hormones, others include cholesterol, bile salts, vitamin D, components of cell membranes
		4. Carotenoids. Orange and yellow pigments; consist of isoprene units	Look for isoprene units. H CH$_3$ $H_2C{=}C{-}C{=}CH_2$	Retinal (important in photo-reception) and vitamin A are formed from carotenoids
Proteins	C, H, O, N. usually S	One or more polypeptides (chains of amino acids) coiled or folded in characteristic shapes.	Look for amino acid units joined by C—N bonds.	Serve as enzymes; structural components; muscle proteins; hemoglobin.
Nucleic acids	C, H, O, N, P	Backbone composed of alternating pentose and phosphate groups, from which nitrogenous bases project. DNA contains the sugar deoxyribose and the bases guanine, cytosine, adenine, and thymine. RNA contains the sugar ribose and the bases guanine, cytosine, adenine, and uracil. Each molecular subunit, called a *nucleotide,* consists of a pentose, a phosphate, and a nitrogenous base.	Look for a pentose-phosphate backbone. DNA forms a double helix.	Storage, transmission, and expression of genetic information

SUMMARY WITH KEY TERMS

I. An **organic compound** is made up of carbon covalently linked to carbon or to hydrogen. The major groups of biologically important organic compounds are carbohydrates, lipids, proteins, and nucleic acids.

II. The properties of carbon atoms make them extraordinarily versatile, able to form the backbones of the large variety of organic compounds essential to life.
   A. Each carbon atom can form four covalent bonds with four other atoms; these can be single, double, or triple bonds.
   B. Carbon forms covalent bonds with a greater number of different elements than does any other type of atom. Carbon atoms can form straight or branched chains or can join into rings.

III. **Isomers** are compounds with the same molecular formula but different structures.
   A. **Structural isomers** differ in the covalent arrangements of their atoms.
   B. **Geometric isomers,** or *cis-trans* isomers, differ in the spatial arrangements of their atoms.
   C. **Enantiomers** are isomers that are mirror images of each other. Cells can distinguish between these configurations.

IV. Organic compounds are made up of specific **functional groups** with characteristic properties.
   A. **Hydrocarbons,** organic compounds consisting of only carbon and hydrogen, are nonpolar and **hydrophobic.**
   B. Polar and ionic functional groups interact with each other and dissolve in water.
   C. Partial charges on atoms at opposite ends of a bond are responsible for the polar property of a functional group. **Hydroxyl** and **carbonyl** groups are polar.
   D. **Carboxyl** and **phosphate** groups are acidic, becoming negatively charged when they release hydrogen ions. The **amino** group is basic, becoming positively charged when it accepts a hydrogen ion.

V. Long chains of similar organic compounds linked together are called **polymers.** Large polymers such as polysaccharides, proteins, and DNA are referred to as **macromolecules.**

VI. **Carbohydrates** contain carbon, hydrogen, and oxygen in a ratio of approximately one carbon to two hydrogens to one oxygen.
   A. **Monosaccharides** are simple sugars such as glucose, fructose, and ribose.
   B. Two monosaccharides can be joined by a **glycosidic linkage,** forming a **disaccharide** such as maltose or sucrose.
   C. Most carbohydrates are **polysaccharides,** long chains of repeating units of a simple sugar.
      1. Carbohydrates are typically stored in plants as **starch** and in animals as **glycogen.**
      2. The cell walls of plants are composed mainly of the polysaccharide **cellulose.**

VII. **Lipid** molecules are composed mainly of hydrocarbon-containing regions, with few oxygen-containing (polar or ionic) functional groups. Lipids have a greasy or oily consistency and are relatively insoluble in water.
   A. **Triacylglycerol,** the main storage form of fat in organisms, consists of a molecule of **glycerol** combined with three **fatty acids.**
      1. **Monoacylglycerols** and **diacylglycerols** contain one and two fatty acids, respectively.
      2. A fatty acid can be either **saturated** with hydrogen, or **unsaturated.**
   B. **Phospholipids** are structural components of cellular membranes.
   C. **Steroid** molecules contain carbon atoms arranged in four attached rings. Cholesterol, bile salts, and certain hormones are important steroids.

VIII. **Proteins** are large, complex molecules made of simpler subunits, called **amino acids,** joined by **peptide bonds.**
   A. Proteins are the most versatile class of biological molecules, serving a variety of functions, such as **enzymes,** structural components, and cellular regulators.
   B. Proteins are composed of various linear sequences of 20 different amino acids. Two amino acids combine to form a **dipeptide.** A longer chain of amino acids is a **polypeptide.**
      1. All amino acids contain an amino group and a carboxyl group, but vary in their side chains. The side chains of amino acids dictate their chemical properties.
      2. Amino acids generally exist as dipolar ions at cellular pH and serve as important biological buffers.
   C. Four levels of organization can be distinguished in protein molecules.
      1. **Primary structure** is the linear sequence of amino acids in the polypeptide chain.
      2. **Secondary structure** is a regular conformation, such as an **α-helix** or a **β-pleated sheet;** it is due to hydrogen bonding between elements of the uniform backbone of the polypeptide.
      3. **Tertiary structure** is the overall shape of the polypeptide chains, as dictated by chemical properties and interactions of the side chains of specific amino acids. Hydrogen bonding, ionic bonds, hydrophobic interactions, and disulfide bridges contribute to tertiary structure.
      4. **Quaternary structure** is determined by the association of two or more polypeptide chains.

IX. The **nucleic acids DNA** and **RNA** store and transfer information that governs the sequence of amino acids in proteins and ultimately the structure and function of the organism.
   A. Nucleic acids are composed of long chains of **nucleotide** subunits, each composed of a two-ring **purine** or one-ring **pyrimidine** nitrogenous base, a five-carbon sugar (**ribose** or **deoxyribose**), and a phosphate group.
   B. **ATP** (**adenosine triphosphate**) is a nucleotide of special significance in energy metabolism. **NAD^+** is also involved in energy metabolism through its role as an electron (hydrogen) acceptor in biological oxidations.

POST-TEST

1. Which of the following is generally considered an inorganic form of carbon? (a) CO_2 (b) C_2H_4 (c) CH_3COOH (d) b and c (e) all of the above are inorganic
2. Carbon is particularly well suited to be the backbone of organic molecules because (a) it can form both covalent bonds and ionic bonds (b) its covalent bonds are very irregularly arranged in three-dimensional space (c) its covalent bonds are the strongest chemical bonds known (d) it can bond to atoms of a large number of other elements (e) all of the bonds it forms are polar
3. The structures depicted in the right-hand column are

```
   CH3 CH3            H   CH3
    |   |             |   |
H — C — C — H     H — C — C — H
    |   |             |   |
    H   H            CH3  H
```

(a) enantiomers (b) different views of the same molecule (c) geometric (*cis-trans*) isomers (d) both geometric isomers and enantiomers (e) structural isomers

4. Which of the following are generally hydrophobic? (a) polar molecules and hydrocarbons (b) ions and hydrocarbons (c) nonpolar molecules and ions (d) polar molecules and ions (e) none of the above
5. Which of the following is a nonpolar molecule? (a) water (H_2O) (b) ammonia (NH_3) (c) methane (CH_4) (d) ethane (C_2H_6) (e) more than one of the above
6. Which of the following functional groups normally acts as an acid? (a) hydroxyl (b) carbonyl (c) sulfhydryl (d) phosphate (e) amino
7. The synthetic process by which monomers are covalently linked is called (a) hydrolysis (b) isomerization (c) condensation (d) glycosidic linkage (e) ester linkage
8. A monosaccharide designated as an aldehyde sugar contains (a) a terminal carboxyl group (b) an internal carboxyl group (c) a terminal carbonyl group (d) an internal carbonyl group (e) a terminal carboxyl group and an internal carbonyl group
9. Structural polysaccharides typically (a) have extensive hydrogen bonding between adjacent molecules (b) are much more hydrophilic than storage polysaccharides (c) have much stronger covalent bonds than do storage polysaccharides (d) consist of alternating α-glucose and β-glucose subunits (e) form helical structures in the cell
10. Fatty acids are components of (a) phospholipids and carotenoids (b) carotenoids and triacylglycerol (c) steroids and triacylglycerol (d) phospholipids and triacylglycerol (e) carotenoids and steroids
11. Saturated fatty acids are so named because they are saturated with (a) hydrogen (b) water (c) hydroxyl groups (d) glycerol (e) double bonds
12. Which pair of amino acid side groups would be most likely to associate with each other by an ionic bond?
    1. $—CH_3$
    2. $—CH_2—COO^-$
    3. $—CH_2—CH_2—NH_3^+$
    4. $—CH_2—CH_2—COO^-$
    5. $—CH_2—OH$

    (a) 1 and 2 (b) 2 and 4 (c) 1 and 5 (d) 2 and 5 (e) 3 and 4
13. Which of the following levels of protein structure may be affected by hydrogen bonding? (a) primary and secondary (b) primary and tertiary (c) secondary, tertiary, and quaternary (d) primary, secondary, and tertiary (e) primary, secondary, tertiary, and quaternary
14. Each phosphodiester linkage in DNA or RNA includes a phosphate joined by covalent bonds to (a) two bases (b) two sugars (c) two additional phosphates (d) a sugar, a base, and a phosphate (e) a sugar and a base

REVIEW QUESTIONS

1. Marble is composed of the carbon-containing compound calcium carbonate ($CaCO_3$). Should calcium carbonate be considered an organic molecule? Why or why not?
2. What are some of the ways that the features of carbon-to-carbon bonds influence the stability and three-dimensional structure of organic molecules?
3. Draw pairs of simple sketches comparing two (a) structural isomers, (b) geometric isomers, and (c) enantiomers.
4. Sketch the following functional groups: methyl, amino, carbonyl, hydroxyl, carboxyl, and phosphate. Classify each as one of the following: nonpolar, polar, acidic, or basic.
5. What features related to hydrogen bonding give storage polysaccharides, such as starch and glycogen, very different properties from structural polysaccharides, such as cellulose and chitin?
6. Draw a structural formula of a simple amino acid and identify the carboxyl group, amino group, and R group.
7. How does the primary structure of a polypeptide influence its secondary and tertiary structures? How can the conformation of a protein be disrupted?
8. Compare the functions of proteins and nucleic acids.

YOU MAKE THE CONNECTION

1. Like oxygen, sulfur forms two covalent bonds. However, sulfur is far less electronegative. In fact, it is approximately as electronegative as carbon. How would the properties of the various classes of biological molecules be altered if you were to replace all the oxygen atoms with sulfur atoms?
2. In what ways are all species alike biochemically? How do species differ from one another biochemically?

RECOMMENDED READINGS

Bettelheim, F.A., W.H. Brown, and J. March. *Introduction to General, Organic and Biochemistry,* 6th ed. Harcourt College Publishers, Philadelphia, 2001. A very readable reference text for those who would like to know more about the chemistry basic to life.

Ezzell, C. "Beyond the Human Genome." *Scientific American,* Vol. 283, No. 1, Jul. 2000. A perspective on the emerging field of proteomics. An outgrowth of the Human Genome Project, proteomics is concerned with understanding the structures and interactions among the thousands of proteins in a cell.

Garrett, R.H., and C.M. Grisham. *Biochemistry,* 2nd ed. Saunders College Publishing, Philadelphia, 1999. A comprehensive, advanced biochemistry text.

Richards, F.M. "The Protein Folding Problem." *Scientific American,* Vol. 264, No. 1, Jan. 1991. A discussion of the mechanisms involved when a protein folds into its biologically active shape.

Strauss, E. "How Proteins Take Shape." *Science News,* 6 Sept. 1997. Chaperones help proteins fold efficiently.

- Visit our Web site at **http://www.info.brookscole.com/solomonbergmartin** for links to chapter-related resources on the World Wide Web. Additional on-line materials relating to this chapter can also be found on our Web site.

See chapter activity on BioActive Learner CD for additional help in mastering the chapter's material. Icon location in the chapter's margins shows which topics have tutorials or simulations in the CD.

4

Organization of the Cell

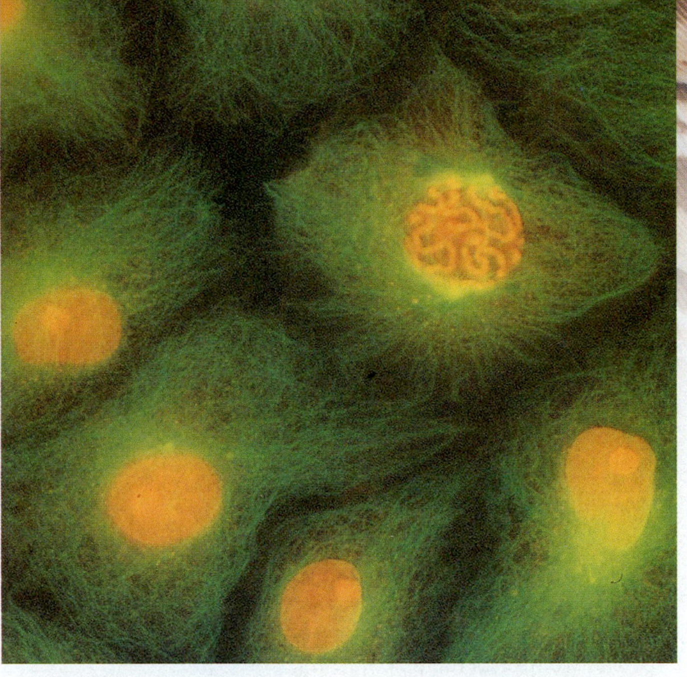

Microtubules, key components of the cytoskeleton. The cells shown here were stained with fluorescent antibodies (specific proteins) that bound to proteins associated with DNA *(orange)* and to a protein (tubulin) in microtubules *(green)*. This type of microscopy, known as confocal fluorescence microscopy, shows the extensive distribution of microtubules in these cells. *(Courtesy of Dr. John M. Murray, Department of Cell and Developmental Biology, University of Pennsylvania)*

LEARNING OBJECTIVES

After you have studied this chapter you should be able to

1. Explain why the cell is considered the basic unit of life and discuss some of the implications of the cell theory.
2. Compare and contrast the general characteristics of prokaryotic and eukaryotic cells.
3. Compare size relationships among different cells and cell structures, and explain why the relationship between surface area and volume of a cell is important in determining cell size limits.
4. Describe the structure of the nucleus and relate its structure to its function in eukaryotic cells.
5. Distinguish between smooth and rough endoplasmic reticulum in terms of both structure and function and discuss the relationship between the endoplasmic reticulum and other internal membranes in the cell.
6. Trace the path of proteins synthesized in the rough endoplasmic reticulum as they are subsequently processed, modified, and sorted by the Golgi complex and then transported to specific destinations.
7. Describe the functions of lysosomes.
8. Compare the functions of chloroplasts and mitochondria and discuss ATP synthesis by each of these organelles.
9. Explain the importance of the cytoskeleton to the cell and describe the structures of the major types of polymers that make up the cytoskeleton.
10. Relate the structural features of cilia and flagella to the way in which these organelles are able to move.

Cells are dramatic examples of the underlying unity of all living things. When we examine a variety of diverse organisms, ranging from simple bacteria to the most complex plants and animals, we find striking similarities at the cellular level. This is a reflection of the evolution of cells from a common ancestor, as well as the fact that living things have many common needs. Careful studies of shared cellular features help us trace the evolutionary history of various groups of organisms and furnish powerful evidence that all organisms alive today had a common origin.

Each cell is a microcosm of life. It is the smallest unit that can carry out all activities we associate with life. When provided with essential nutrients and an appropriate environment, some cells can be kept alive and growing in the laboratory for many years. By contrast, no isolated part of a cell is capable of sustained survival. Composed of a vast array of inorganic and organic ions and molecules including water, salts, carbohydrates, lipids, proteins, and nucleic acids, most cells have all the physical and chemical components needed for their own maintenance, growth, and division. Genetic information is stored in deoxyribonucleic acid (DNA) molecules and is faithfully replicated and passed on to each new generation of cells during cell division. Information in DNA codes for specific proteins that in turn determine cell structure and function. In this chapter and those that follow we discuss how cells use many of the chemical materials introduced in Chapters 2 and 3.

Cells exchange materials and energy with the environment. All living cells need one or more sources of energy, but a cell rarely obtains energy in a form that is immediately usable. Cells convert energy from one form to another, and that energy is used to carry out various activities, ranging from mechanical work to chemical synthesis. Cells convert energy to a convenient form, usually chemical energy stored in adenosine triphosphate (ATP) (see Chapter 3). Although the specifics vary, the basic strategies cells use for energy conversion are very similar. The chemical reactions that convert energy from one form to another are essentially the same in all cells, from bacteria to those of complex plants and animals.

Cells are the building blocks of complex multicellular organisms. Although they are basically similar, cells are also extraordinarily diverse and versatile. They can be modified in a variety of ways to carry out specialized functions. As technology advances, cell biologists have increasingly sophisticated tools to use in their search to better understand the structure and function of cells. For example, investigation of the cytoskeleton (cell skeleton), currently an active and exciting area of research, has been greatly enhanced by advances in microscopy. In the chapter opening illustration, *confocal fluorescence microscopy* was used to show the extensive distribution of microtubules in cells. Microtubules, key components of the cytoskeleton, help maintain cell shape, function in cell movement, and function in transport of materials within the cell.

THE CELL IS THE BASIC UNIT OF LIFE

Two German scientists, botanist Matthias Schleiden in 1838 and zoologist Theodor Schwann in 1839, were the first to point out that all plants and animals are composed of cells. Later, Rudolf Virchow, a German scientist and professor of pathology, observed cells dividing and giving rise to daughter cells. In 1855, Virchow proposed that new cells are formed only by the division of previously existing cells.

The work of Schleiden, Schwann, and Virchow gave rise to the **cell theory,** the unifying concept that cells are the basic living units of organization and function in all organisms and that all cells come from other cells. About 1880 another biologist, August Weismann, added an important corollary to Virchow's concept by pointing out that the ancestry of all the cells alive today can be traced back to ancient times. Evidence that all presently living cells have a common origin is provided by the basic similarities in their structures and in the molecules of which they are made.

CELL ORGANIZATION AND SIZE PERMIT HOMEOSTASIS

Recall from Chapter 1 that **homeostasis** is the process of maintaining an internal environment that is appropriate and supportive to life. Cells experience constant changes in their environments, including deviations in salt concentration, pH, and temperature. For its biochemical mechanisms to function, the cell must work continuously to restore appropriate conditions.

Organization is basically similar in all cells

Every cell must be able to separate its contents from the external environment. This function is performed by the **plasma membrane,** a structurally distinctive surface membrane that surrounds all cells, whether bacteria, algae, or human cells. The cells must also be able to accumulate materials and energy stores and exchange materials with the environment. These activities are highly regulated, and the plasma membrane must serve as an extremely selective barrier. By making the interior of the cell an enclosed compartment, the plasma membrane permits the chemical composition of the cell to be quite different from that outside the cell.

Typically, cells have internal structures, called **organelles,** specialized to carry on life activities such as converting energy to usable forms, synthesizing needed compounds, and manufacturing structures essential to function and reproduction. Each cell has genetic instructions coded in its DNA, which is concentrated in a limited region of the cell.

Cell size is limited

Although their sizes vary over a wide range (Fig. 4–1), most cells are microscopic, and very small units are required to measure them and their internal structures. The basic unit of linear measurement in the metric system (see inside back cover) is the meter (m), which is just a little longer than a yard. A millimeter (mm) is 1/1000 of a meter and is about as long as the bar enclosed in parentheses (-). The micrometer (μm) is the most convenient unit for measuring cells. A bar 1 μm long is far too short to be seen with the unaided eye, for it is 1/1,000,000 (one-millionth) of a meter, or 1/1000 of a millimeter, long. Most of us have difficulty thinking about units that are too small to see, but it is helpful to remember that a micrometer has the same relationship to a millimeter that a millimeter has to a meter (1/1000).

As small as it is, the micrometer is actually too large to measure most cellular components. For these purposes we use the nanometer (nm), which is 1/1,000,000,000 (one-billionth) of a meter, or 1/1000 of a micrometer. To mentally move down to the world of the nanometer, recall that a millimeter is 1/1000 of a meter, a micrometer is 1/1000 of a millimeter, and a nanometer is 1/1000 of a micrometer.

A good light microscope allows us to see most types of bacterial cells, and some specialized algae and animal cells are large enough to be seen with the naked eye. A human egg cell, for example, is about 130 μm in diameter, or approximately the size of the period at the end of this sentence. The largest cells are birds' eggs, but they are atypical because both the yolk and the egg white consist of food reserves. The functioning part of the cell is a small mass on the surface of the yolk.

Why are most cells so small? If you consider what a cell must do to grow and survive, it may be easier to understand the reasons for its small size. A cell must take in food and other materials and must rid itself of waste products generated by metabolic reactions. Everything that enters or leaves a cell must pass through its plasma membrane. The plasma membrane contains specialized "pumps" and "gates" that selectively regulate the passage of materials into and out of the cell. The plasma membrane must be large enough relative to the cell volume to keep up with the demands of regulating the passage of materials. This means that a critical factor in determining cell size is the ratio of its surface area to its volume (Fig. 4–2).

As a cell becomes larger, its volume increases at a greater rate than its surface area (i.e., its plasma membrane), which

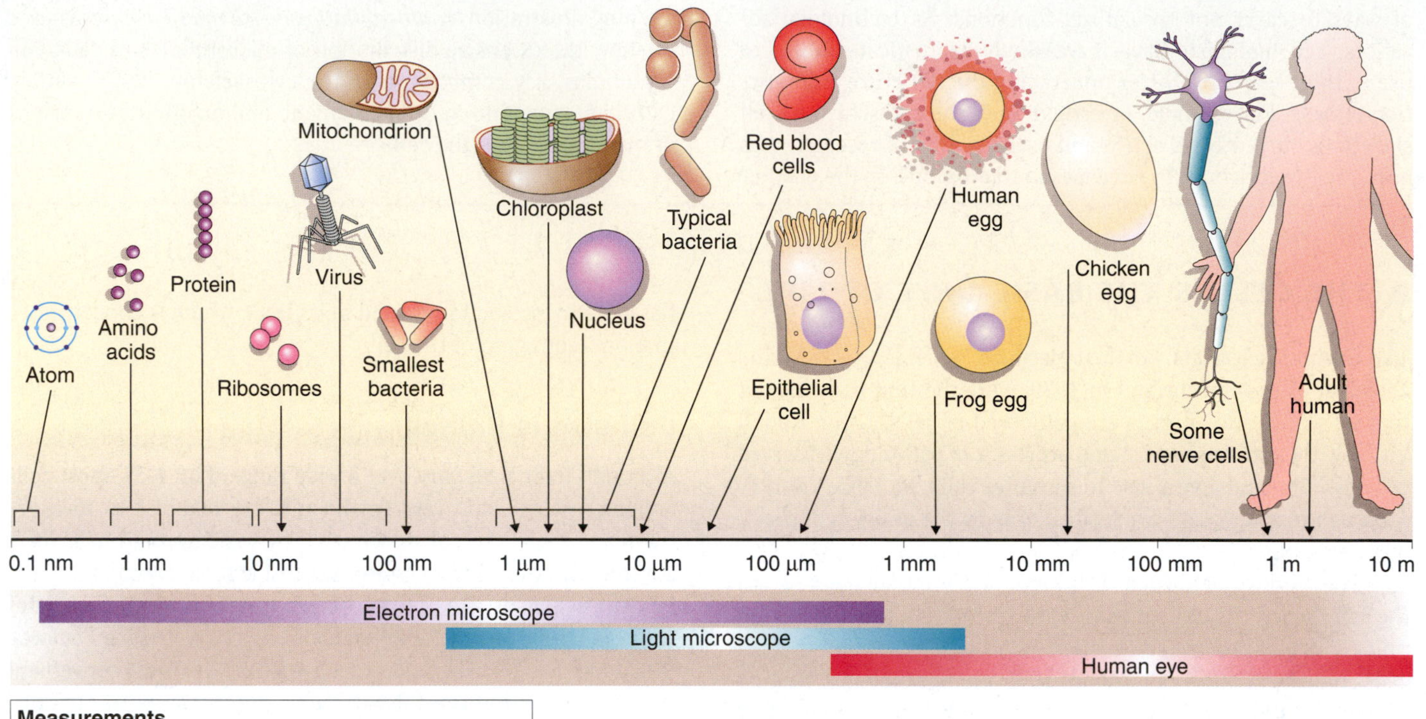

Measurements		
1 meter	=	1000 millimeters (mm)
1 millimeter	=	1000 micrometers (μm)
1 micrometer	=	1000 nanometers (nm)

Figure 4–1 Biological size and cell diversity. Relative size from chemical to organismic levels is most conveniently compared using a logarithmic scale (multiples of ten). The prokaryotic cells of bacteria typically range in size from less than 1 to 10 μm long; their small size enables them to grow and divide rapidly. Eukaryotic cells are typically 10 to 100 μm in diameter; most are between 10 and 30 μm. The nuclei of animal and plant cells range from about 3 to 10 μm in diameter. Mitochondria are about the size of small bacteria, whereas chloroplasts are usually larger, about 5 μm long. Ova (egg cells) are among the largest cells. Although microscopic, some nerve cells are very long, specialized to transmit messages from one part of the body to another. The cells shown here are not drawn to scale.

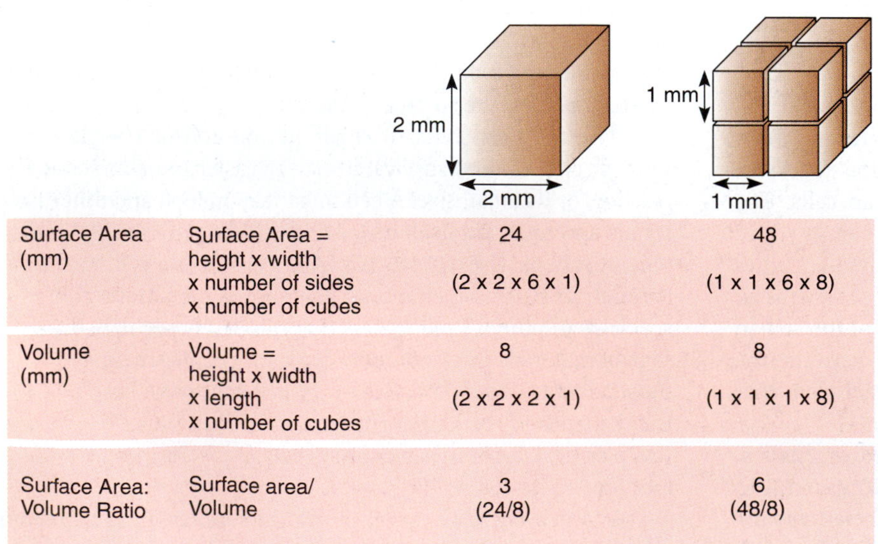

		2 mm cube	1 mm cubes
Surface Area (mm)	Surface Area = height x width x number of sides x number of cubes	24 (2 x 2 x 6 x 1)	48 (1 x 1 x 6 x 8)
Volume (mm)	Volume = height x width x length x number of cubes	8 (2 x 2 x 2 x 1)	8 (1 x 1 x 1 x 8)
Surface Area: Volume Ratio	Surface area/ Volume	3 (24/8)	6 (48/8)

Figure 4–2 Surface area-to-volume ratio. The surface area of a cell must be large enough relative to its volume to permit adequate exchange of materials with the environment. Although their volumes are the same, eight small cells have a much greater surface area (plasma membrane) in relation to their total volume than does one large cell. In the example shown, the ratio of the total surface area to total volume of eight 1-mm cubes is double the surface-to-volume ratio of the single large cube.

effectively places an upper limit on cell size. Above some critical size, the number of molecules required by the cell could not be transported into the cell fast enough to sustain its needs. In addition, the cell would not be able to regulate its concentration of various ions or efficiently export its wastes.

Of course, not all cells are spherical or cuboidal. Some very large cells have relatively favorable ratios of surface area to volume because of their shapes. In fact, some variations in cell shape represent a strategy for increasing the ratio of surface area to volume. For example, many large plant cells are long and thin, which increases their surface-to-volume ratio. Some cells, such as epithelial cells, have finger-like projections of the plasma membrane, called **microvilli,** that significantly increase the surface area used to absorb nutrients and other materials (see Fig. 45–10*c*).

Another reason that cells are small is that, once inside, molecules must be transported to the locations where they are converted into other forms. Because cells are small, the distances molecules travel within them are relatively short, which speeds up many cellular activities.

Cell size and shape are related to function

The sizes and shapes of cells are related to the functions they perform. Some cells, such as the amoeba and the white blood cell, can change their shape as they move about. Sperm cells have long, whiplike tails, called *flagella,* for locomotion. Nerve cells possess long, thin extensions that permit them to transmit messages over great distances. The extensions of some nerve cells in the human body may be as long as 1 m. Other cells, such as certain epithelial cells, are almost rectangular in shape and are stacked much like building blocks to form sheetlike structures.

CELLS ARE STUDIED BY A COMBINATION OF METHODS

Process of Science

One of the most important tools used to study cell structures has been the microscope. In fact, cells were not described until 1665, when Robert Hooke examined a piece of cork using a compound microscope (a microscope with more than one lens) he had made. In his book *Micrographia,* published in 1665, Hooke drew and described what he saw. Hooke chose the term *cell* because the tissue reminded him of the small rooms that monks lived in during that period. What Hooke saw were not actually living cells, but the walls of dead cork cells (Fig. 4–3*a*). Not until much later was it realized that the interior enclosed by the walls is the important part of living cells.

A few years later, inspired by Hooke's work, the Dutch naturalist Anton van Leeuwenhoek viewed living cells with small lenses that he made. Owing to technical problems in building them, early compound microscopes magnified objects only about 30 times. Leeuwenhoek did not build compound microscopes. He was highly skilled at grinding lenses and was able to magnify images more than 200 times. Among his important discoveries were bacteria, protists, blood cells, and sperm cells. Leeuwenhoek was among the first scientists to report cells in animals.

Leeuwenhoek was a merchant and was not formally trained as a scientist, but his skill, curiosity, and diligence in sharing his discoveries with scientists at the Royal Society of London brought an awareness of microscopic life to the scientific world. He did not share his techniques, however, and it was not until more than a century later, in the late 19th century, that microscopes were sufficiently developed for biologists to seriously focus their attention on the study of cells.

Light microscopes are used to study stained or living cells

The **light microscope (LM),** the type used by most students, consists of a tube with glass lenses at each end. (Because it contains several lenses, the light microscope is referred to as a compound microscope.) Visible light passes through the specimen being observed and through the lenses. Light is refracted (bent) by the lenses, magnifying the image.

Two features of a microscope determine how clearly a small object can be viewed: magnification and resolving power. **Magnification** is the ratio of the size of the image seen with the microscope to the actual size of the object. The best light microscopes usually magnify an object no more than 1000 times. **Resolution,** or **resolving power,** is the capacity to distinguish fine detail in an image. This is defined as the minimum distance between two points at which they can both be seen separately rather than as a single, blurred point. Resolving power depends on the quality of the lenses and the wavelength of the illuminating light. As the wavelength decreases, the resolution increases. The visible light used by light microscopes has wavelengths ranging from about 400 nm (violet) to 700 nm (red); this limits the resolution of the light microscope to details no smaller than the diameter of a small bacterial cell (about 1 μm).

By the early 20th century, refined versions of the light microscope, as well as certain organic compounds that specifically stain different cell structures, became available. These enabled biologists to discover that cells contain a number of different internal structures, the organelles. The contribution of organic chemists in the development of biological stains was essential to this understanding, because the interior of many cells is transparent. Most of the methods used to prepare and stain cells for observation, however, also kill them in the process.

More recently, sophisticated types of light microscopes have been developed that use interfering waves of light to enhance the internal structures of cells. With *phase contrast* and *Nomarski differential interference microscopes,* some internal structures can be seen in unstained living cells (Fig. 4–3*d* and *e*). One of the most striking things that can be observed with these microscopes is that living cells contain numerous internal structures that are constantly changing shape and location.

Fluorescence microscopes are used to detect the locations of specific molecules in cells. Fluorescent stains (like paints that glow under black light) are molecules that absorb light energy of one wavelength and then release some of that energy as light of a longer wavelength. One such stain binds specifically to DNA molecules and emits green light after absorbing ultraviolet light.

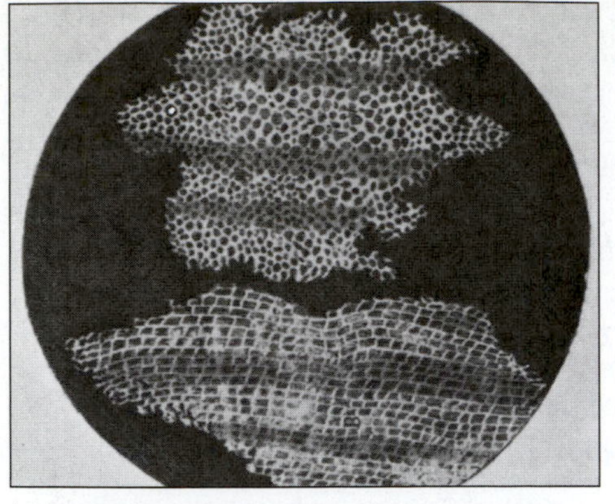

(a)

Figure 4–3 Viewing cells with various types of microscopes. **(a)** Using a crude microscope that he constructed, Robert Hooke looked at a thin slice of cork and drew what he saw. More sophisticated microscopes and techniques permit biologists to view cells in more detail. Unstained epithelial cells from the skin of human cheek are compared using **(b)** bright field (transmitted light), **(c)** dark field, **(d)** phase contrast, and **(e)** Nomarski differential interference microscopy. The phase contrast and differential interference microscopes enhance detail by increasing the differences in optical density in different regions of the cells. *(a, from Hooke's* Micrographica, *1665; b–e, Jim Solliday/Biological Photo Service)*

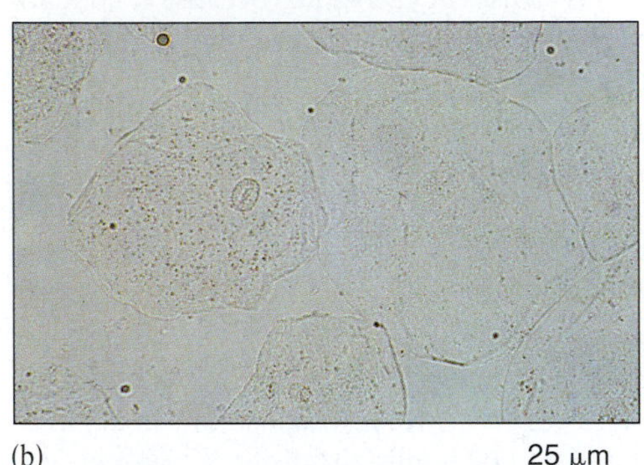

(b)

25 μm

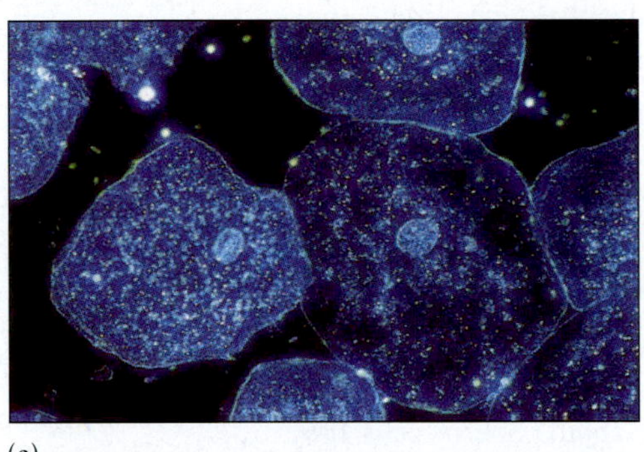

(c)

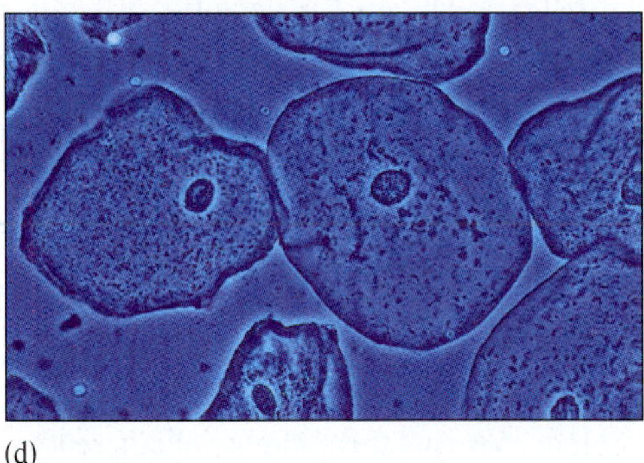

(d)

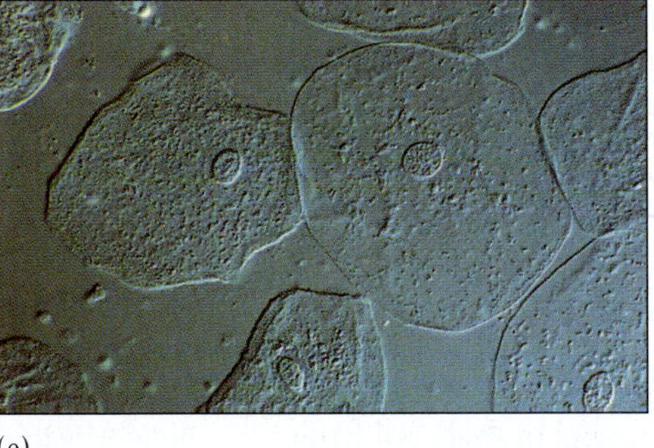

(e)

Cells can be stained, and the location of the DNA can be determined, by observing the source of the green fluorescent light within the cell.

Some fluorescent stains can be chemically bonded to *antibodies,* protein molecules important in internal defense. The antibody can then bind to a highly specific region of a molecule in the cell. A single type of antibody molecule binds to only one type of structure, such as a part of a specific protein or some of the sugars in a specific polysaccharide. Purified fluorescent antibodies known to bind to a specific protein isolated from a cell can be used to determine where that protein is located. Powerful computer imaging methods have allowed the development of the *confocal fluorescence microscope,* which greatly improves the resolution of structures labeled by fluorescent dyes (Fig. 4–4 and the micrograph in the chapter introduction).

Cell biologists are developing new techniques for viewing cells using computers, lasers, and photodetectors. Computer-based image processing synthesizes multiple images to produce three-dimensional views.

Electron microscopes provide a high-resolution image that can be greatly magnified

Even with improved microscopes and techniques for staining cells, ordinary light microscopes can distinguish only the gross details of many cell parts. In most cases all that can be seen clearly is the outline of a structure and its ability to be stained by some dyes and not by others. Not until the development of the **electron microscope (EM),** which came into wide use in the 1950s, were researchers able to study the fine details, or **ultrastructure,** of cells.

Whereas light microscopes magnify an object no more than about 1000 times, the electron microscope can magnify it

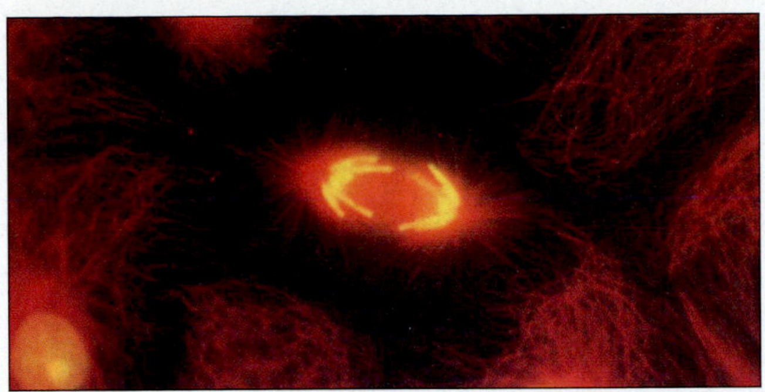

Figure 4–4 Confocal fluorescence micrograph of cultured animal cells. The cell in the center is dividing. The DNA of the chromosomes is yellow; the microtubules are red. The colors are from the fluorescent stain. *(Courtesy of Dr. John M. Murray, Department of Cell and Developmental Biology, University of Pennsylvania)*

250,000 times or more. And while the best light microscopes have about 500 times more resolution than the human eye, the electron microscope multiplies the resolving power by more than 10,000 (Fig. 4–5). This is because electrons have very short wavelengths, on the order of about 0.1 to 0.2 nm. Although such resolution is difficult to achieve with biological material, it can be approached when isolated molecules such as proteins and DNA are examined.

The image formed by the electron microscope cannot be seen directly. The electron beam itself consists of energized electrons, which, because of their negative charge, can be focused by electromagnets just as images are focused by glass lenses in a light microscope (see Fig. 4–5*b*). For **transmission electron microscopy (TEM),** the specimen is embedded in plastic and then cut into extraordinarily thin sections (50 to 100 nm thick) with a glass or diamond knife. A section is then placed on a small metal grid. The electron beam passes through the specimen and then falls onto a photographic plate or a fluorescent screen that works much like a television screen. When you look at TEMs in this chapter (and elsewhere), keep in mind that each represents only a thin cross section of a cell.

To reconstruct a three-dimensional view of the cell, it is necessary to study many consecutive sectional views (called *serial sections*) through the object. To understand the enormity of such a task, try imagining what it would be like to reconstruct an image of the contents of your home from a set of hundreds of consecutive 5-cm sections.

Special methods using antibody molecules that have very tiny gold particles bound to them allow the detection of specific molecules in electron microscope images. The dense gold particles block the electron beam and identify the location of the proteins recognized by the antibodies as precise black spots on the electron micrograph.

In another type of electron microscope, the **scanning electron microscope (SEM),** the electron beam does not pass through the specimen. Instead, the specimen is coated with a thin film of gold or some other metal. When the electron beam strikes various points on the surface of the specimen, secondary electrons are emitted whose intensity varies with the contour of the surface. The recorded emission patterns of the secondary electrons give a three-dimensional picture of the surface (see Fig. 4–5*c*). The SEM provides information about the shape and external features of the specimen that cannot be obtained with the TEM.

Note that the LM, TEM, and SEM are focused by similar principles. A beam of light or an electron beam is directed by the condenser lens onto the specimen and is magnified by the objective lens and the eyepiece in the light microscope or by the objective lens and the projector lens in the TEM. The TEM image is focused onto a fluorescent screen, and the SEM image is viewed on a type of television screen. Lenses in the electron microscopes are actually magnets that bend the beam of electrons.

Process of Science — Cell fractionation procedures permit study of cell components

The EM is a powerful tool for studying cell structure, but it has limitations. The methods used to prepare cells for electron microscopy kill them and may alter their structure. Furthermore, electron microscopy provides few clues about the functions of organelles and other cell components. To determine what organelles actually do, researchers have to be able to purify different parts of cells so that they can be studied by physical and chemical methods.

Cell fractionation procedures are methods for purifying organelles. Generally, cells are broken apart as gently as possible, and the mixture, referred to as the cell extract, is subjected to centrifugal force by spinning in a device called a **centrifuge.** An ultracentrifuge, a very powerful centrifuge, can spin at speeds exceeding 100,000 revolutions per minute (rpm), generating a centrifugal force of 500,000 $\times$ G (a G is equal to the force of gravity). Centrifugal force separates the extract into two fractions: a pellet and a supernatant. The *pellet* that forms at the bottom of the tube contains heavier materials, such as nuclei, packed together. The *supernatant,* the liquid above the pellet, contains lighter particles, dissolved molecules, and ions (Fig. 4–6).

The supernatant can be centrifuged again at a higher speed to obtain a pellet that contains the next heaviest cell components, for example, mitochondria and chloroplasts. In **differential centrifugation,** the supernatant is spun at successively higher speeds, permitting various cell components to be separated on the basis of their different sizes and densities.

Cell components in the resuspended pellets can be further purified by **density gradient centrifugation.** In this procedure, the resuspended pellet is placed in a layer on top of a density gradient, usually made up of a solution of sucrose (table sugar) and water. The concentration of sucrose is highest at the bottom of the tube and decreases gradually so that it is lowest at the top. Because the densities of organelles differ, each will migrate during centrifugation and form a band at the position in the gradient where its own density equals that of the sucrose solution.

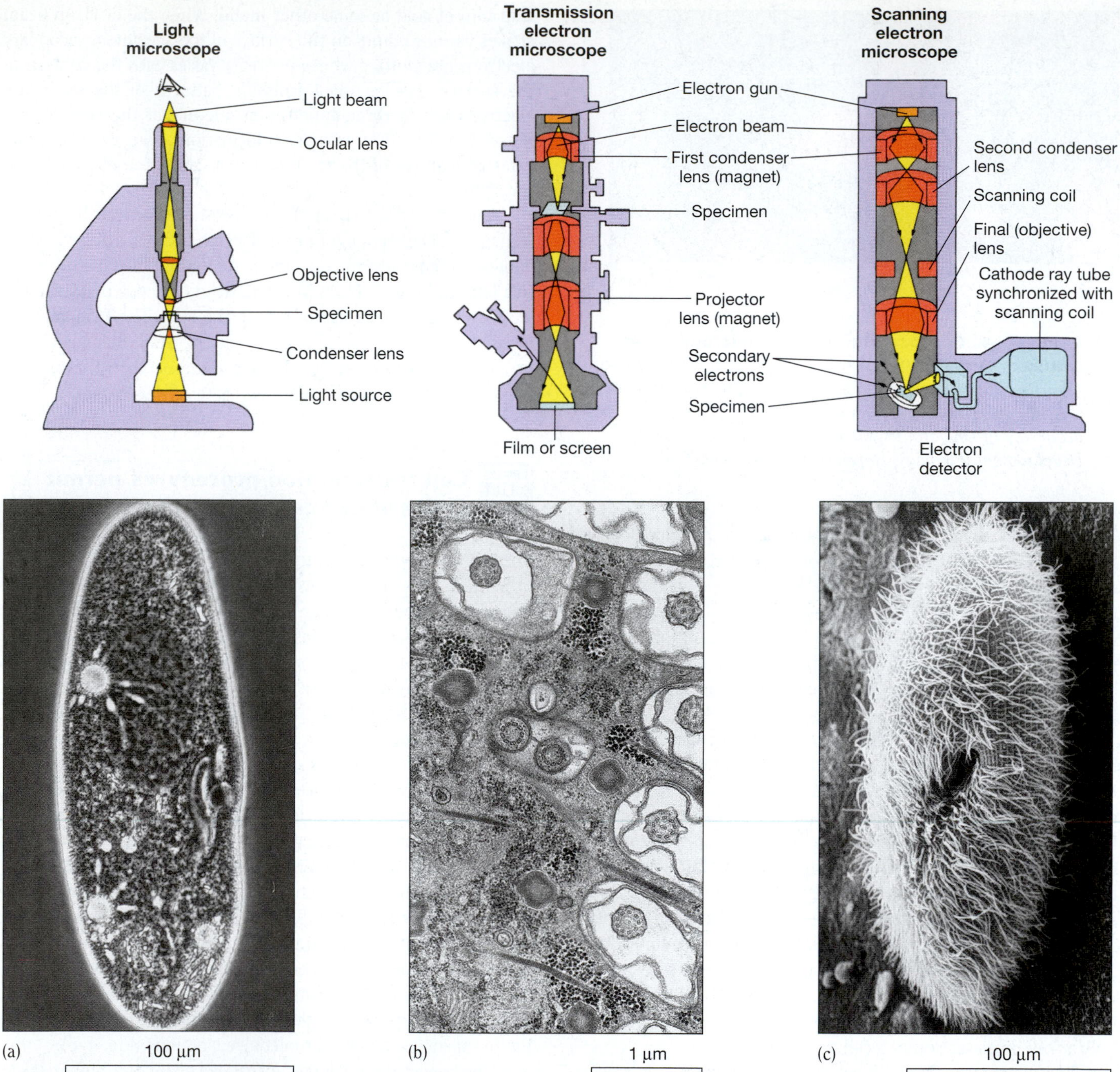

Figure 4-5 Comparison of light and electron microscopy. Distinctive images of cells, such as the protist *Paramecium* shown in the photomicrographs, are provided by three types of microscopes. **(a)** A phase-contrast light microscope can be used to view stained or living cells, but at relatively low resolution. **(b)** The transmission electron microscope (TEM) produces a high-resolution image that can be greatly magnified. Because of the high magnification, only a small part of the *Paramecium* is shown in the photograph. **(c)** The scanning electron microscope (SEM) is used to provide a clear view of surface features. *(Photos courtesy of T.K. Maugel/University of Maryland)*

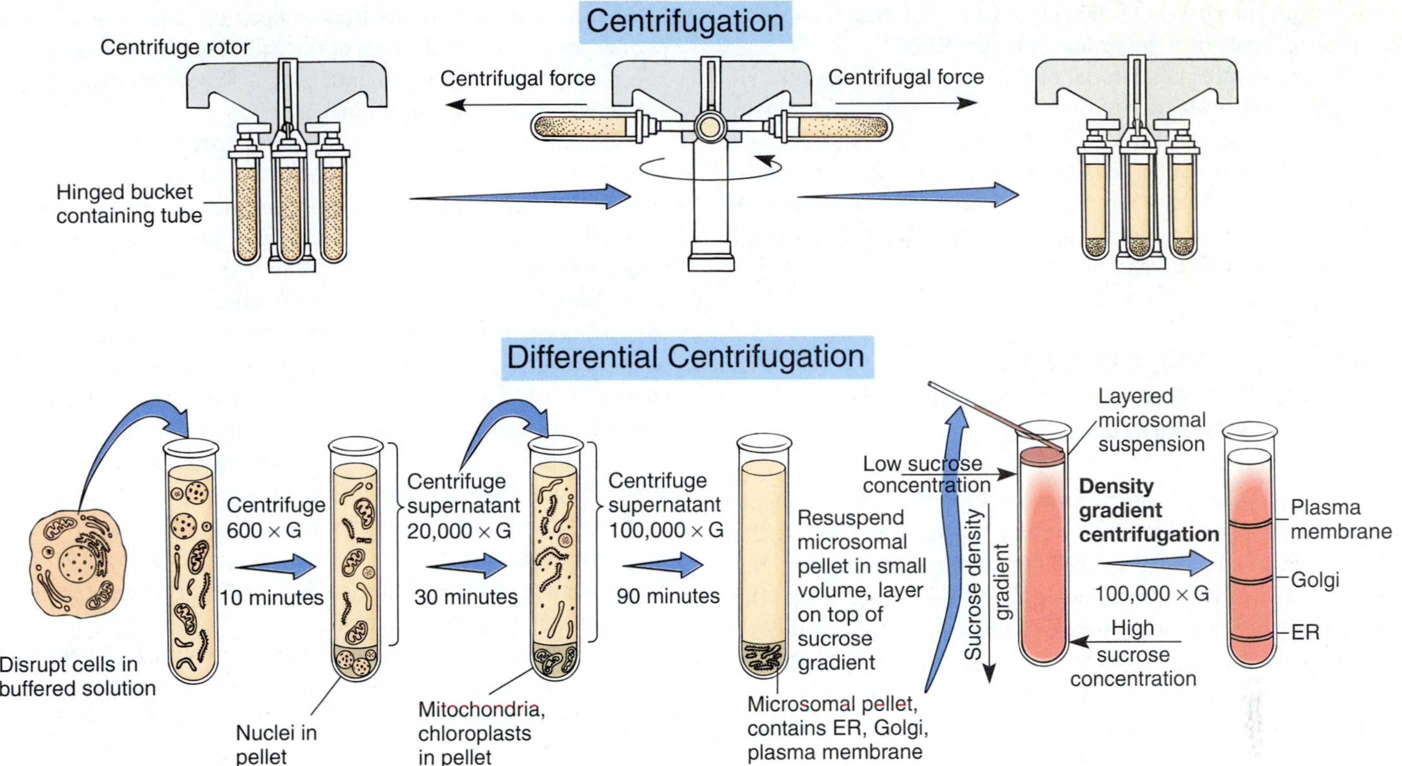

Figure 4–6 Cell fractionation. Differential centrifugation permits cell biologists to separate cell structures into various fractions by spinning the suspension at increasing revolutions per minute. Membranes and organelles from the resuspended pellets can then be further purified by density gradient centrifugation, shown as the last step in the figure. G is the force of gravity. ER is the endoplasmic reticulum.

Purified organelles can be examined to determine what kinds of proteins and other molecules they might contain, as well as the nature of the chemical reactions that take place within them. Cell biologists often use a combination of experimental approaches to study the functions of cellular structures.

PROKARYOTIC CELLS ARE STRUCTURALLY SIMPLER THAN EUKARYOTIC CELLS

Recall from Chapter 1 that two basic types of cells are known: **prokaryotic** and **eukaryotic.** Bacteria are prokaryotic cells. All other known organisms consist of eukaryotic cells. A major difference between prokaryotic and eukaryotic cells is that the DNA of prokaryotic cells is not enclosed in a nucleus. In fact the term *prokaryotic* means "before the nucleus." In the prokaryotic cell, the DNA is located in a limited region of the cell called a **nuclear area,** or **nucleoid,** which is not enclosed by a membrane (Fig. 4–7). Other types of internal membrane–bounded organelles are also absent in prokaryotic cells. These cells are typically smaller than eukaryotic cells. In fact, the average prokaryotic cell is only about one-tenth the diameter of the average eukaryotic cell.

Like eukaryotic cells, prokaryotic cells have a plasma membrane that confines the contents of the cell to an internal compartment. In some prokaryotic cells the plasma membrane may be folded inward to form a complex of membranes along which many of the cell's metabolic reactions take place. Most prokaryotic cells have **cell walls,** which are extracellular structures that enclose the entire bacterium, including the plasma membrane. Many prokaryotes have **flagella** (sing., *flagellum*), long fibers that

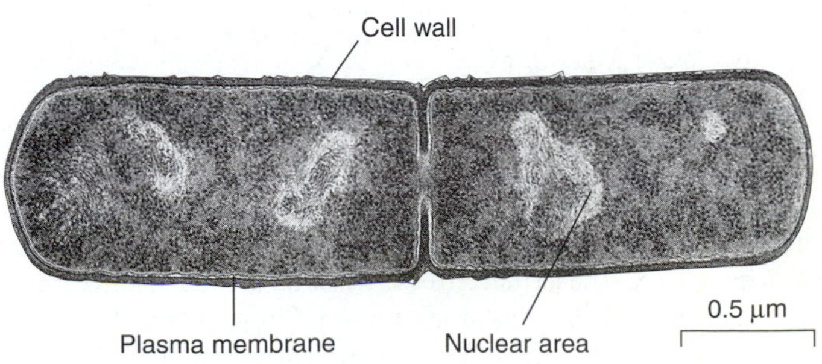

Figure 4–7 TEM of a prokaryotic cell. This bacterium, *Bacillus subtilis,* which is dividing, has a prominent cell wall surrounding the plasma membrane. The nuclear areas are clearly visible. *(Courtesy of A. Ryter)*

project from the surface of the cell. Prokaryotic flagella operate like propellers and are important in locomotion.

The dense internal material of the bacterial cell contains **ribosomes,** small complexes of ribonucleic acid (RNA) and protein that synthesize polypeptides (discussed later in this chapter). The ribosomes of prokaryotic cells are smaller than those found in eukaryotic cells. Prokaryotic cells also contain storage granules that hold glycogen, lipid, or phosphate compounds. (Prokaryotes are discussed in more detail in Chapter 23.)

EUKARYOTIC CELLS ARE CHARACTERIZED BY MEMBRANE-BOUNDED ORGANELLES

Eukaryotic cells are characterized by highly organized membrane-bounded organelles. The most prominent of these is the *nucleus,* which contains the hereditary material DNA. In fact, the name eukaryote means "true nucleus." Table 4–1 summarizes the types of organelles typically found in eukaryotic cells. Some organelles may be found only in specific cells. For example, chloroplasts, structures that trap sunlight for energy conversion, are found only in cells that carry on photosynthesis, such as certain plant cells. The many specialized organelles of eukaryotic cells solve some of the problems associated with large size, permitting eukaryotic cells to be considerably larger than prokaryotic cells. Eukaryotic cells also differ from prokaryotic cells in having a supporting framework, or cytoskeleton, important in maintaining shape and transporting materials within the cell.

Early biologists thought that the cell consisted of a homogeneous jelly, which they called *protoplasm.* With the electron microscope and other modern research tools, perception of the environment within the cell has been greatly expanded. We now know that the cell is highly organized and complex (Figs. 4–8 to 4–11). The eukaryotic cell has its own control center, internal transportation system, power plants, factories for making needed materials, packaging plants, and even a "self-destruct" system. Biologists refer to the part of the cell outside the nucleus as **cytoplasm** and the part of the cell within the nucleus as **nucleoplasm.** Various organelles are suspended within the fluid component of the cytoplasm, which is referred to as the **cytosol.** Therefore, the term *cytoplasm* includes both the cytosol and all the organelles other than the nucleus.

Membranes divide the cell into compartments

Membranes have unique properties that enable membranous organelles to carry out a wide variety of functions. For example, cell membranes never have free ends; therefore, a membranous organelle always contains at least one enclosed internal space or compartment. These membrane-bounded compartments allow certain cell activities to be localized within specific enclosed regions of the cell. Reactants that are located in only a small part of the total cell volume are far more likely to come in contact, and the rate of the reaction can be dramatically increased. Membrane-bounded compartments keep certain reactive compounds away from other parts of the cell that might be adversely affected by them. Compartmentalizing also permits many different activities to go on simultaneously.

Membranes also allow the storage of energy. The membrane provides a barrier that is analogous to a dam on a river. A difference in the concentration of some substance on the two sides of a membrane is a form of stored energy or *potential energy* (see Chapter 6). As particles of the substance move across the membrane from the side of higher concentration to the side of lower concentration, the cell can convert some of this potential energy to the chemical energy of ATP molecules. This process of energy conversion (discussed in Chapters 7 and 8) is a basic mechanism that cells use to capture and convert energy to sustain life.

Membranes also serve as important work surfaces. For example, many chemical reactions in cells are carried out by enzymes that are bound to membranes. Because the enzymes that carry out successive steps of a series of reactions are organized close together on a membrane surface, certain series of chemical reactions can occur more rapidly.

In a eukaryotic cell several types of membranes are generally considered to be part of the internal membrane system, or **endomembrane system.** Look at Figures 4–8 and 4–9 and notice how membranes divide the cell into many compartments, including the cell itself (bounded by the plasma membrane), the nucleus, endoplasmic reticulum (ER), Golgi complexes, lysosomes, and vesicles and vacuoles. Although it is not internal, the plasma membrane is also included because of its participation in the activities of the endomembrane system. (Mitochondria and chloroplasts are also separate compartments but are not generally considered part of the endomembrane system because they function independently of other membranous organelles.)

Some organelles have direct connections between their membranes and compartments. Others transport materials in **vesicles,** small membrane-bounded sacs formed by "budding" from the membrane of another organelle. Vesicles also carry materials from one organelle to another. Through a complex series of steps, a vesicle can form as a "bud" from one membrane and then be transported to another membrane to which it fuses, thus delivering its contents into another compartment.

The cell nucleus contains DNA

Typically, the **nucleus** is the most prominent organelle in the cell. It is usually spherical or oval in shape and averages 5 μm in diameter. Owing to its size and the fact that it often occupies a relatively fixed position near the center of the cell, some early investigators guessed long before experimental evidence was available that the nucleus served as the control center of the cell (see *Focus On:* Acetabularia *and the Control of Cell Activities* on page 86). Most cells have one nucleus, although there are exceptions.

The **nuclear envelope** consists of two concentric membranes that separate the nuclear contents from the surrounding cytoplasm (Fig. 4–12 on page 85). These membranes are separated by about 20 to 40 nm. At intervals the membranes come together to form **nuclear pores,** which consist of protein complexes. Nuclear

TABLE 4–1 Eukaryotic Cell Structures and Their Functions

Structure	Description	Function
Cell Nucleus		
Nucleus	Large structure surrounded by double membrane; contains nucleolus and chromosomes	Information in DNA is transcribed in RNA synthesis; specifies cellular proteins
Nucleolus	Granular body within nucleus; consists of RNA and protein	Site of ribosomal RNA synthesis; ribosome subunit assembly
Chromosomes	Composed of a complex of DNA and protein known as chromatin; condense during cell division, becoming visible as rodlike chromosomes	Contain genes (units of hereditary information) that govern structure and activity of cell
Cytoplasmic Organelles		
Plasma membrane	Membrane boundary of cell	Encloses cellular contents; regulates movement of materials in and out of cell; helps maintain cell shape; communicates with other cells (also present in prokaryotes)
Endoplasmic reticulum (ER)	Network of internal membranes extending through cytoplasm	Synthesizes lipids and modifies many proteins; origin of intracellular transport vesicles that carry proteins
Smooth	Lacks ribosomes on outer surface	Lipid biosynthesis; drug detoxification
Rough	Ribosomes stud outer surface	Manufacture of many proteins destined for secretion or for incorporation into membranes
Ribosomes	Granules composed of RNA and protein; some attached to ER, some free in cytosol	Synthesize polypeptides in both prokaryotes and eukaryotes
Golgi complex	Stacks of flattened membrane sacs	Modifies proteins; packages secreted proteins; sorts other proteins to vacuoles and other organelles
Lysosomes	Membranous sacs (in animals)	Contain enzymes to break down ingested materials, secretions, wastes
Vacuoles	Membranous sacs (mostly in plants, fungi, algae)	Store materials, wastes, water; maintain hydrostatic pressure
Peroxisomes	Membranous sacs containing a variety of enzymes	Site of many diverse metabolic reactions
Mitochondria	Sacs consisting of two membranes; inner membrane is folded to form cristae and encloses matrix	Site of most reactions of cellular respiration; transformation of energy originating from glucose or lipids into ATP energy.
Plastids (e.g., chloroplasts)	Double-membraned structure enclosing internal thylakoid membranes; chloroplasts contain chlorophyll in thylakoid membranes	Site of photosynthesis; chlorophyll captures light energy; ATP and other energy-rich compounds are formed and then used to convert CO_2 to glucose
Cytoskeleton		
Microtubules	Hollow tubes made of subunits of tubulin protein	Provide structural support; have role in cell and organelle movement and cell division; components of cilia, flagella, centrioles, basal bodies
Microfilaments	Solid, rodlike structures consisting of actin protein	Provide structural support; play role in cell and organelle movement and cell division
Intermediate filaments	Tough fibers made of protein	Help strengthen cytoskeleton; stabilize cell shape
Centrioles	Pair of hollow cylinders located near nucleus; each centriole consists of nine microtubule triplets (9×3 structure)	Mitotic spindle forms between centrioles during animal cell division; may anchor and organize microtubule formation in animal cells; absent in most plants
Cilia	Relatively short projections extending from surface of cell; covered by plasma membrane; made of two central and nine pairs of peripheral microtubules ($9 + 2$ structure)	Movement of some single-celled organisms; used to move materials on surface of some tissues
Flagella	Long projections made of two central and nine pairs of peripheral microtubules ($9 + 2$ structure); extend from surface of cell; covered by plasma membrane	Cellular locomotion by sperm cells and some unicellular eukaryotes

pores regulate the passage of materials between nucleoplasm and cytoplasm. Just how materials are transported through nuclear pores and how the process is regulated are active areas of research.

Most of the cell's DNA is located inside the nucleus. Recall from Chapter 3 that DNA molecules are composed of sequences of nucleotides called **genes,** which contain the chemically coded instructions for producing the proteins needed by the cell. The nucleus controls protein synthesis by transcribing and then sending RNA molecules to the cytoplasm (where proteins are manufactured).

DNA is associated with proteins, forming a complex known as **chromatin,** which appears as a network of granules and

(Text continues on page 85)

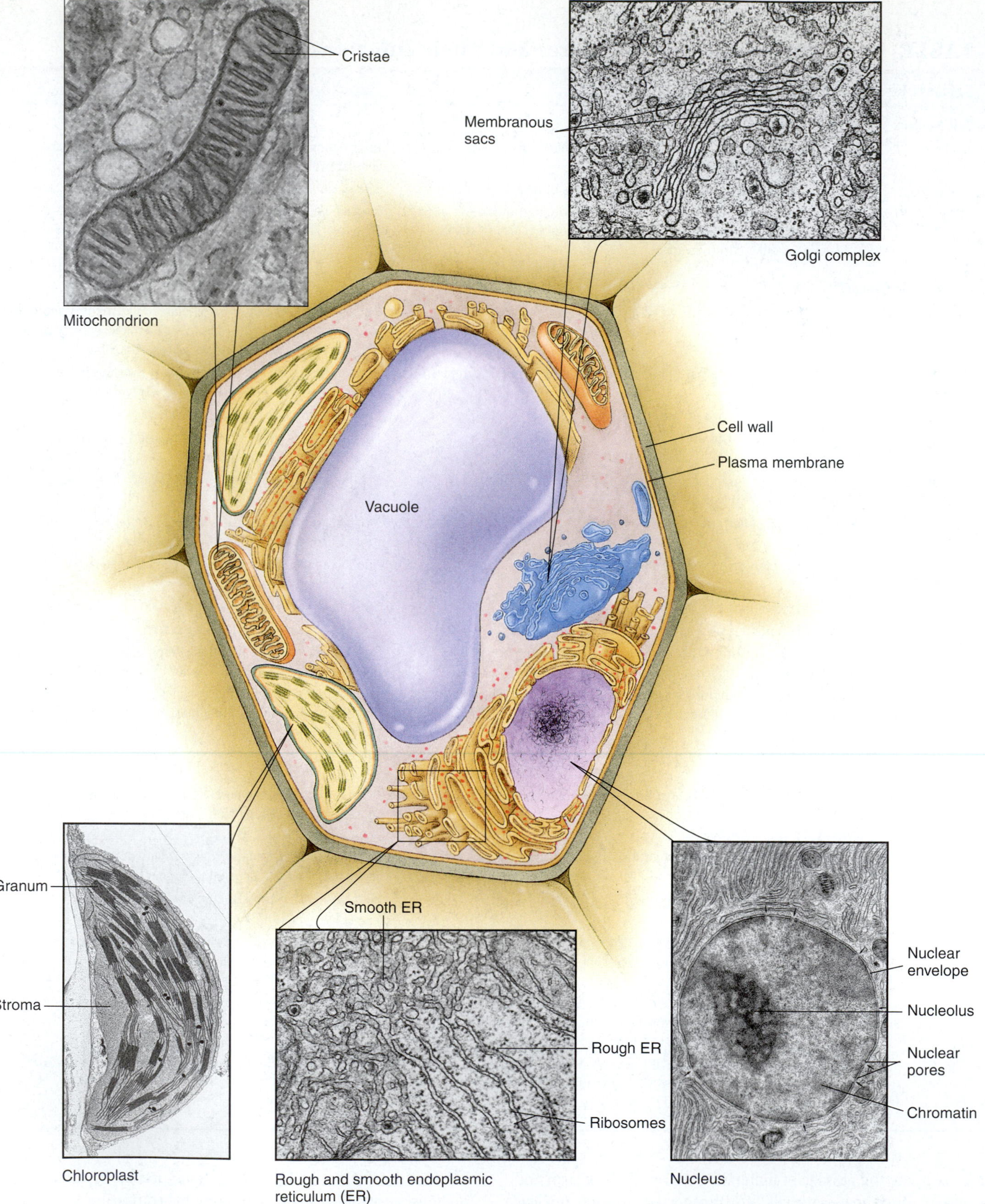

Figure 4-8 Composite diagram of a plant cell. The TEMs show certain structures or areas of the cell. Some plant cells do not have all the organelles shown. For example, leaf and stem cells that carry on photosynthesis contain chloroplasts, whereas root cells do not. Chloroplasts, a cell wall, and prominent vacuoles are characteristic of plant cells. Many of the other organelles, such as the nucleus, mitochondria, and endoplasmic reticulum (ER), are also found in protist, fungal, and animal cells. *(Clockwise from top left: D.W. Fawcett; D.W. Fawcett and R. Bolender; D.W. Fawcett/Visuals Unlimited; R.Bolender and D.W. Fawcett; E.H. Newcomb and W.P. Wergin, Biological Photo Service)*

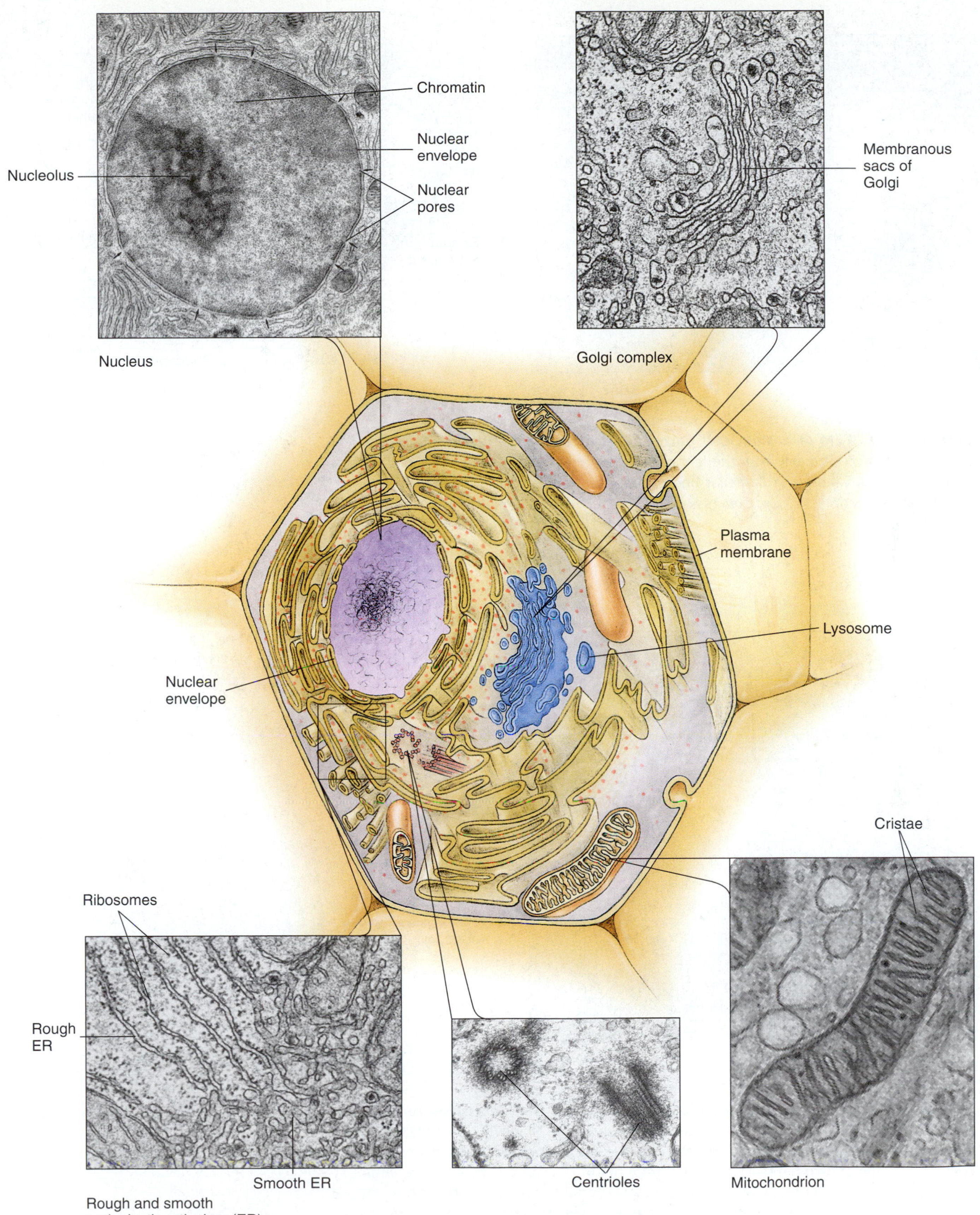

Figure 4–9 Composite diagram of an animal cell. This generalized animal cell is shown in realistic context surrounded by adjacent cells that cause it to be slightly compressed. The TEMs show the structure of various organelles. Depending on the cell type, certain organelles may be more or less prominent. *(Clockwise from top left: D.W. Fawcett; D.W. Fawcett and R. Bolender; D.W. Fawcett; B.F. King, Biological Photo Service; R. Bolender and D.W. Fawcett/Visuals Unlimited)*

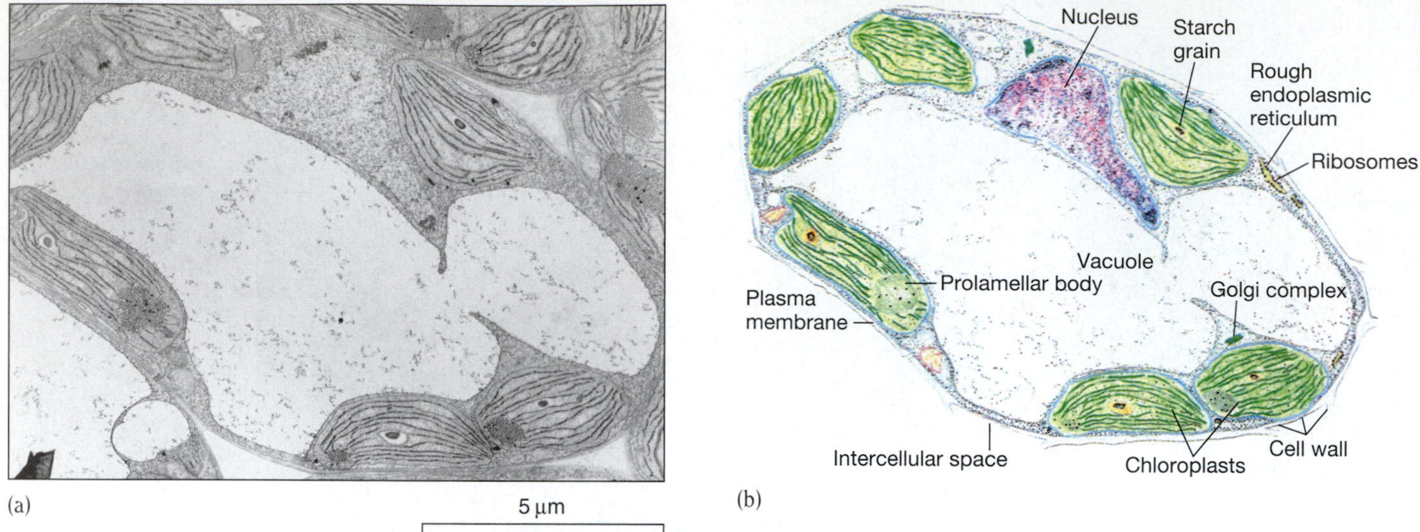

Figure 4–10 TEM of a plant cell paired with interpretive drawing. Most of this cross section of a cell from the leaf of a young bean plant *(Phaseolus vulgaris)* is dominated by a vacuole. Prolamellar bodies are membranous regions typically seen in developing chloroplasts. *(a, Courtesy of Dr. Kenneth Miller, Brown University)*

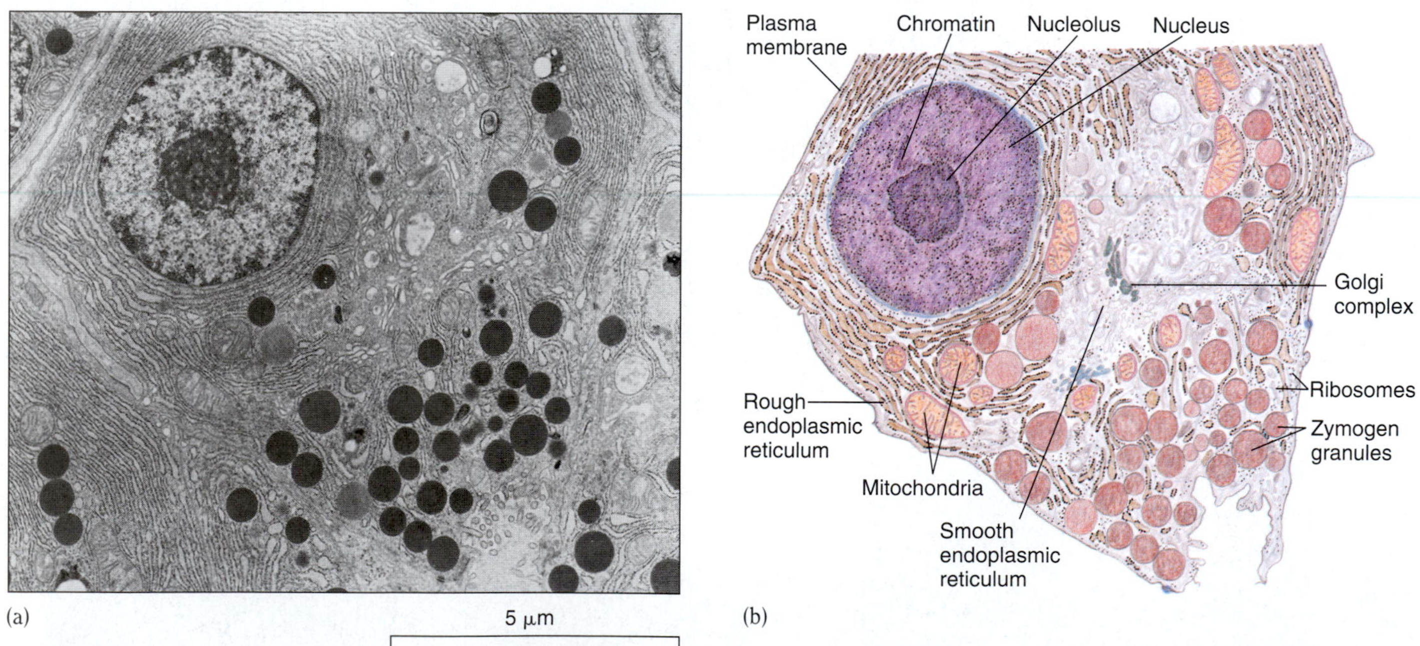

Figure 4–11 TEM of a human pancreas cell paired with interpretive drawing. Most of the structures of a typical animal cell are present. However, like most cells, this one has certain structures associated with its specialized functions. Pancreas cells such as the one shown here secrete large amounts of digestive enzymes. The large, dark, circular bodies in the TEM **(a)** and the corresponding structures in the drawing **(b)** are zymogen granules containing inactive enzymes. When released from the cell, they catalyze chemical reactions such as the breakdown of peptide bonds of ingested proteins in the intestine. Most of the membranes visible in this section are part of the rough endoplasmic reticulum, an organelle specialized to manufacture protein. *(a, Dr. Susumu Ito, Harvard Medical School)*

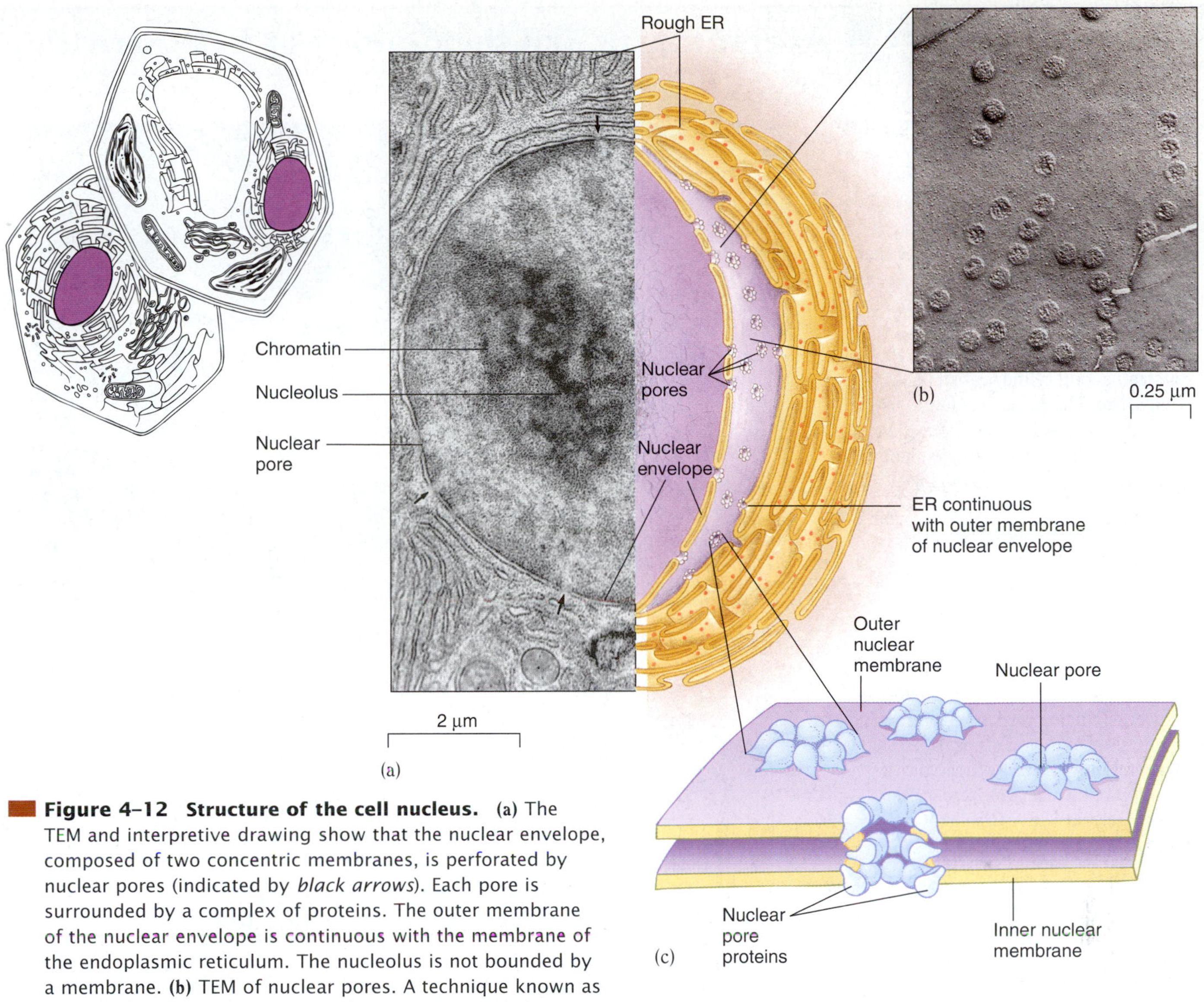

Figure 4-12 Structure of the cell nucleus. **(a)** The TEM and interpretive drawing show that the nuclear envelope, composed of two concentric membranes, is perforated by nuclear pores (indicated by *black arrows*). Each pore is surrounded by a complex of proteins. The outer membrane of the nuclear envelope is continuous with the membrane of the endoplasmic reticulum. The nucleolus is not bounded by a membrane. **(b)** TEM of nuclear pores. A technique known as freeze-fracture was used to split the membrane. **(c)** The nuclear pores, which are made up of proteins, form channels between the nucleoplasm and cytoplasm. *(a, D.W. Fawcett; b, R. Kessel-G. Shih/Visuals Unlimited)*

strands in cells that are not dividing. Although chromatin appears disorganized, it is not. Because DNA molecules are extremely long and thin, they must be packed inside the nucleus in a very regular fashion. In dividing cells, the chromatin condenses and becomes visible as distinct threadlike structures called **chromosomes.** If the DNA in the 46 chromosomes of one human cell could be stretched end to end, it would extend for 2 m.

Most nuclei have one or more compact structures called **nucleoli** (sing., *nucleolus*). A nucleolus is *not* membrane-bounded and usually stains differently from the surrounding chromatin. Each nucleolus contains a nucleolar organizer, made up of chromosomal regions containing instructions for making the type of RNA in ribosomes. This ribosomal RNA is synthesized in the nucleolus. The proteins needed to make ribosomes are synthesized in the cytoplasm and imported into the nucleolus. Ribosomal RNA and proteins are then assembled into ribosomal subunits that leave the nucleus through the nuclear pores.

Ribosomes manufacture proteins

Visible as small granules, ribosomes are molecular structures consisting of RNA and protein. Each eukaryotic ribosome is actually a knot of three RNA strands in association with about 75 different proteins. Free ribosomes are suspended in the cytosol. Other ribosomes are associated with certain internal membranes within the cell. Each ribosome has two main components: a large subunit and a small subunit. In addition to specific ribosomal RNAs, each ribosome contains specific proteins. Ribosomes, which contain the enzyme necessary to form peptide bonds (see Chapter 3), are tiny manufacturing plants that assemble proteins.

(Text continues on page 88)

FOCUS ON Process of Science *Acetabularia* and the Control of Cell Activities

To the romantically inclined, the little seaweed *Acetabularia* resembles a mermaid's wineglass, although the literal translation of its name, "vinegar cup," is somewhat less elegant (Fig. A). In the 19th century, biologists discovered that this marine eukaryotic alga consists of a single cell. At about 5 cm (2 in) in length, *Acetabularia* is small for a seaweed but gigantic for a cell. It consists of a rootlike **holdfast,** a long cylindrical **stalk,** and a cuplike **cap.** The nucleus is found in the holdfast, about as far away from the cap as it can be. Because it is a single giant cell, *Acetabularia* is easy to manipulate.

Figure A Light micrograph of *Acetabularia*. *(L. Sims/Visuals Unlimited)*

Regeneration Experiments Demonstrated that the Cap Shape Is under the Control of Something in the Stalk or the Holdfast

If the cap of *Acetabularia* is removed experimentally, another one grows after a few weeks. Such a response, common among simple organisms, is called **regeneration.** This fact attracted the attention of investigators, especially J. Hämmerling and J. Brachet, who became interested in whether a relationship exists between the nucleus and the physical characteristics of the alga. Because of its great size, *Acetabularia* could be subjected to surgery that would be impossible with smaller cells. During the 1930s and 1940s these researchers performed brilliant experiments that in many ways laid the foundation for much of our modern knowledge of the nucleus. Two species were used for most experiments: *Acetabularia mediterranea,* which has a smooth cap, and *Acetabularia crenulata,* which has a cap divided into a series of finger-like projections.

The kind of cap that is regenerated depends on the species of *Acetabularia* used in the experiment. As you might expect, *A. crenulata* regenerates a "cren" cap, and *A. mediterranea* regenerates a "med" cap. But it is possible to graft together two capless algae of different species. Through this union, they regenerate a common cap that has characteristics intermediate between those of the two species involved (Fig. B).

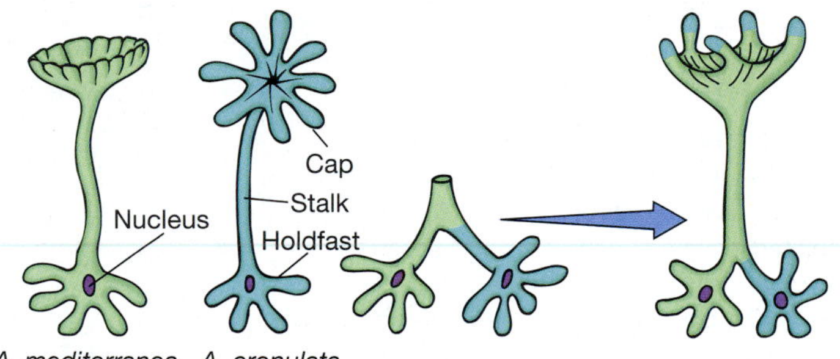

Figure B

Thus, it is clear that something about the lower part of the cell controls cap shape.

Stalk Exchange Experiments Indicated that Short-Term Control Can Be Exerted by the Stalk, but Long-Term Control Is in the Holdfast

It is possible to attach a section of *Acetabularia* to a holdfast that is not its own by telescoping the cell walls of the two into one another. In this way the stalks and holdfasts of different species may be intermixed.

First, we take *A. mediterranea* and *A. crenulata* and remove their caps. Then we sever the stalks from the holdfasts. Finally, we exchange the parts (Fig. C). What happens? Not, perhaps, what you would expect! The caps that regenerate are characteristic not of the species donating the holdfasts but of those donating the stalks!

However, if the caps are removed once again, this time the caps that regenerate are characteristic of the species that donated the holdfasts. This continues to be the case no matter how many more times the regenerated caps are removed.

From all these results Hämmerling and Brachet deduced that the ultimate control of the *Acetabularia* cell is

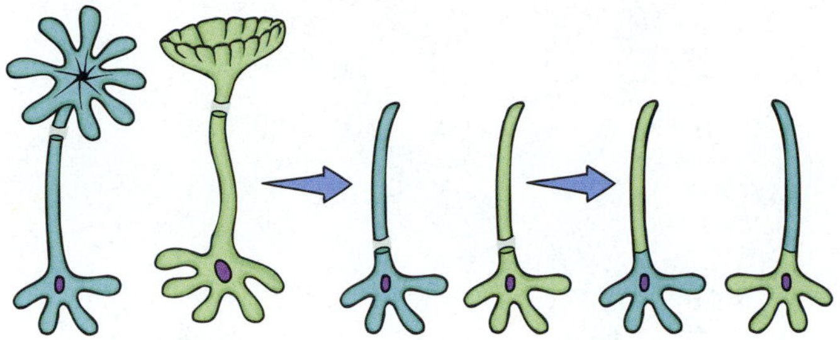

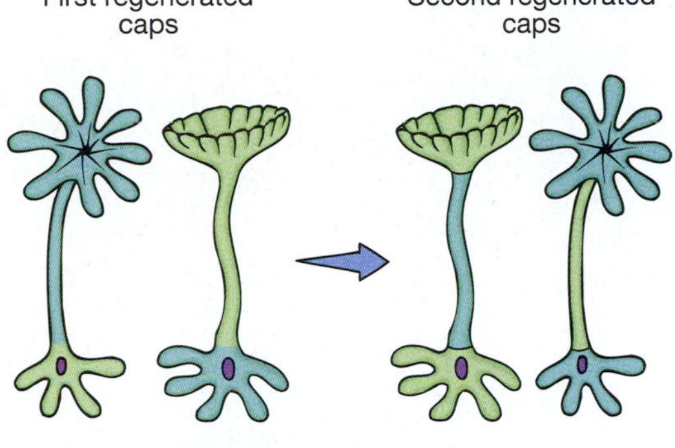

Figure C

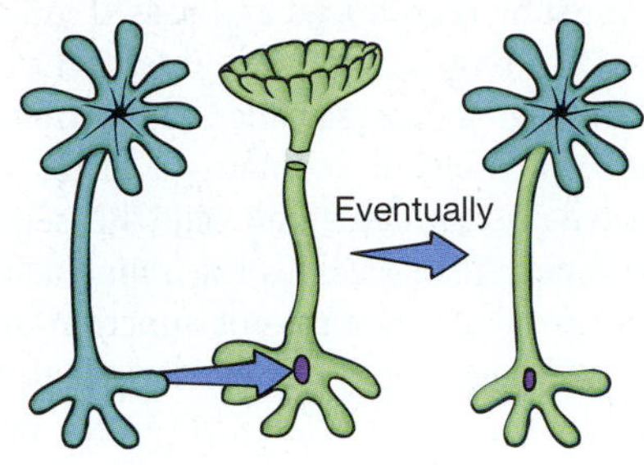

Figure E

associated with the holdfast. Because there is a time lag before the holdfast appears to take over, they hypothesized that it produces some temporary cytoplasmic messenger substance whereby it exerts its control. They further hypothesized that initially the grafted stalks still contain enough of the substance from their former holdfasts to regenerate a cap of the former shape. But this still leaves us with the question of what it is about the holdfast that accounts for its apparent control. An obvious suspect is the nucleus.

Nuclear Exchange Experiments Demonstrated that the Nucleus Is the Ultimate Source of Information for the Control of Cellular Activities

If the nucleus is removed and the cap cut off, a new cap regenerates (Fig. D). *Acetabularia,* however, can usually regenerate only once without a nucleus. If the nucleus of another species is now inserted and the cap is cut off once again, a new cap regenerates that is characteristic of the species of the nucleus (Fig. E)! If two kinds of nuclei are inserted, the regenerated cap is intermediate in shape between those of the species that donated the nuclei.

As a result of these and other experiments, biologists began to develop some basic ideas about the control of cell activities. The control of the cell exerted by the holdfast is attributable to the nucleus that is located there. Further, the nucleus is the apparent source of some "messenger substance" that can temporarily exert control but is limited in quantity and cannot be produced without the nucleus (Fig. F). This information helped provide a starting point for research on the role of nucleic acids in the control of all cells.

These ideas have been extended in our modern view of information flow and control in the cell. We now know that the nucleus of eukaryotes controls the cell's activities because it contains DNA, the ultimate source of biological information. DNA can pass on its information to successive generations because it is able to precisely replicate itself. The information in the DNA is used to specify the sequence of amino acids in all the proteins of the cell. To carry out its mission, DNA uses ribonucleic acid (RNA) as a cytoplasmic messenger substance.

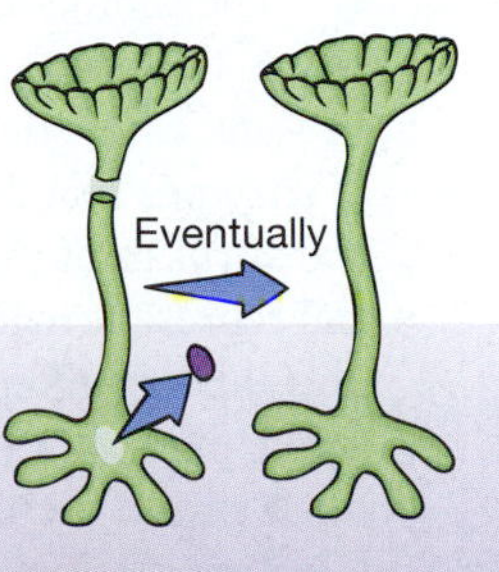

Figure D

The characteristics of the cell are governed by the messenger substance, and therefore ultimately by the nucleus.

Messenger substance

The nucleus produces the messenger substance

Figure F

MAKING THE CONNECTION How does a cell store and use genetic information? The cell stores information in the form of DNA. When a cell divides, the information stored in DNA must be reproduced and passed intact to the two daughter cells. DNA has the unique ability to make an exact duplicate of itself through a process called **replication.**

The DNA molecule contains a linear sequence of components called *nucleotides.* In all cells this sequence of nucleotides serves as a code that specifies the amino acid sequence (primary structure) in proteins. Proteins function as structural components of cells and as **enzymes,** molecules that catalyze and regulate virtually every chemical reaction that takes place in the cell. By specifying the structure of enzymes and other proteins, DNA directs the metabolism of the cell.

The cell uses a second nucleic acid, RNA, as a messenger. The sequence of bases in DNA that codes for a specific protein is copied as a sequence of bases called **messenger RNA.** This process is known as **transcription.** In eukaryotic cells, DNA is stored in the nucleus and transcription occurs there.

Messenger RNA leaves the nucleus through nuclear pores and attaches to ribosomes. Two other types of RNA are also involved in protein synthesis: ribosomal RNA (rRNA) and transfer RNA (tRNA). Transfer RNA delivers amino acids to the ribosomes for assembly into proteins. The complex process of assembling a chain of specific amino acids using the code in messenger RNA is called **translation.** After proteins are manufactured, they may be transported to the Golgi complex, where they are modified and sent on to other locations in the cell. (Protein synthesis is discussed in detail in Chapter 12.)

The endoplasmic reticulum is a major manufacturing center

One of the most prominent features in the electron micrograph in Figure 4–11 is a maze of parallel internal membranes that encircle the nucleus and extend into many regions of the cytoplasm. This complex of membranes, the **endoplasmic reticulum (ER),** forms a network that makes up a significant part of the total volume of the cytoplasm in many cells. A higher-magnification TEM of the ER is shown in Figure 4–13. Remember that a TEM represents only a thin cross section of the cell, so there is a tendency to interpret the ER as a series of tubes. In fact, many ER membranes consist of a series of tightly packed and flattened, saclike structures that form interconnected compartments within the cytoplasm.

The internal space enclosed by the membranes is called the **ER lumen.** In most cells the ER lumen forms a single internal compartment that is continuous with the space between outer and inner membranes of the nuclear envelope (see Fig. 4–12). The compartment formed between the two nuclear membranes is thus connected to the ER lumen. The membranes of other organelles are not directly connected to the ER and appear to form distinct and separate compartments within the cytoplasm.

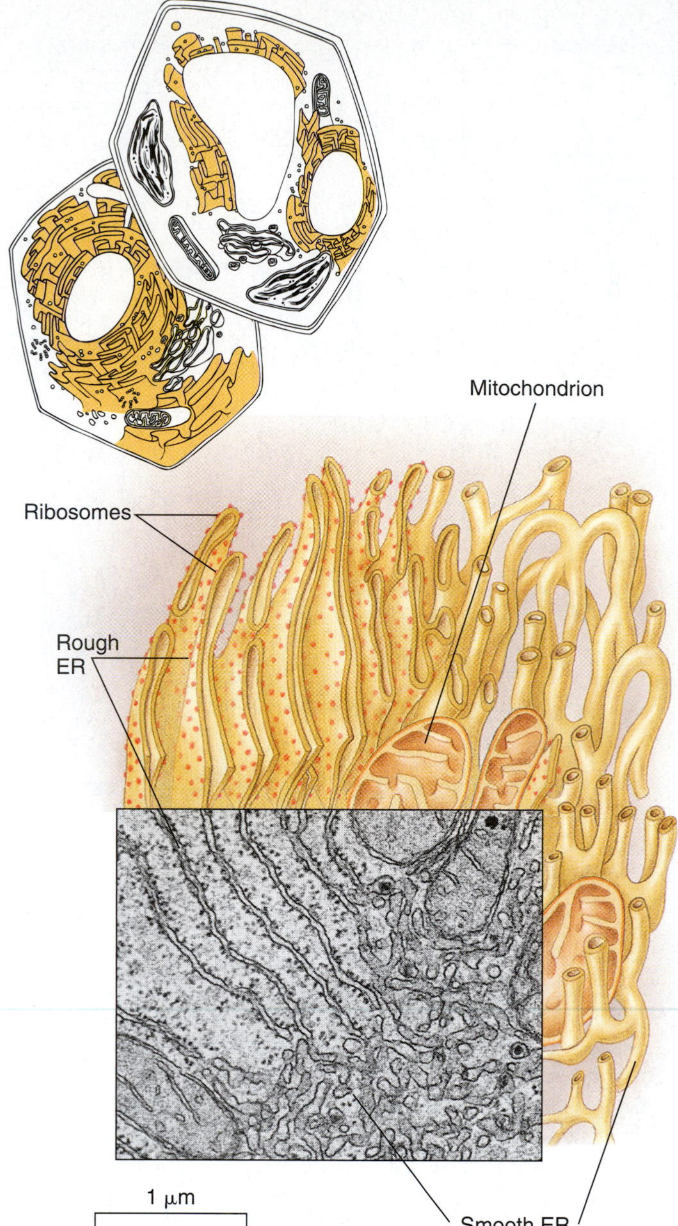

Figure 4–13 The endoplasmic reticulum (ER). The TEM shows both rough and smooth ER in a liver cell. *(R. Bolender and D.W. Fawcett/Visuals Unlimited)*

The ER membranes and lumen contain a large variety of enzymes that catalyze many different types of chemical reactions. In some cases the membranes serve as a framework for systems of enzymes that carry out sequential biochemical reactions. The two surfaces of the membrane contain different sets of enzymes and represent regions of the cell with different synthetic capabilities, just as different regions of a factory are used to make different parts of a particular product. Still other enzymes are located within the ER lumen.

Two distinct regions of the ER can be distinguished in TEMs: rough ER and smooth ER. Although these regions have different functions, their membranes are connected and their internal spaces are continuous. **Rough ER** has ribosomes attached to it and consequently appears rough in electron micrographs. Notice in Figure 4–13 that one membrane face (the lumen side) appears to be bare, while the other membrane face (the cytosolic side) is studded with ribosomes that appear as dark particles.

The rough ER plays a central role in the synthesis and assembly of proteins. Many proteins that are exported from the cell (such as digestive enzymes) and those destined for other organelles are synthesized on ribosomes attached to the ER membrane. The ribosome forms a tight seal with the ER membrane. A tunnel within the ribosome connects to an ER pore, or *translocon.* Proteins are transported through the tunnel and the pore in the ER membrane into the ER lumen. In the ER lumen, proteins may be modified by enzymes that add complex carbohydrates or lipids to them. Other enzymes, called **molecular chaperones,** in the ER lumen catalyze the efficient folding of proteins into proper conformations. The proteins are then transferred to other compartments by small **transport vesicles,** which bud off the ER membrane and then fuse with the membrane of some target compartment.

Smooth ER is more tubular and does not have ribosomes bound to it, so its outer membrane surfaces appear smooth. The smooth ER is the primary site of phospholipid, steroid, and fatty acid metabolism. While the smooth ER may be a minor membrane component in some cells, extensive amounts of smooth ER are present in others. For example, extensive smooth ER is present in human liver cells, where it synthesizes and processes cholesterol and other lipids and serves as a major detoxification site. Enzymes located along the smooth ER of liver cells break down toxic chemicals such as carcinogens (cancer-causing agents). These compounds are then converted to water-soluble products that can be excreted.

The Golgi complex processes and packages proteins

The **Golgi complex** (also known as the *Golgi body* or *Golgi apparatus*) was first described in 1898 by the Italian microscopist Camillo Golgi, who found a way to specifically stain that organelle. In many cells the Golgi complex consists of stacks of flattened membranous sacs called **cisternae** (sing., *cisterna*). In certain regions, cisternae may be distended because they are filled with cellular products (Fig. 4–14). Each of the flattened sacs has an internal space, or lumen. However, unlike the ER, most of these internal spaces of the Golgi complex and the membranes that form them are not continuous. Hence a Golgi complex contains a number of separate compartments, as well as some that are interconnected.

Each Golgi stack has three areas referred to as *cis* and *trans faces,* with a *medial* region between. Typically, the *cis* face is located nearest the nucleus and functions to receive materials from transport vesicles from the ER. The *trans* face, nearest to the plasma membrane, packages molecules in vesicles and transports them out of the Golgi.

In a cross-sectional view like that in the TEM in Figure 4–14, many of the ends of the sheetlike layers of Golgi membranes are distended, an arrangement that is characteristic of well-developed Golgi complexes in many types of cells. In some animal cells the Golgi complex is often located at one side of the nucleus; in other animal cells and in plant cells there are many

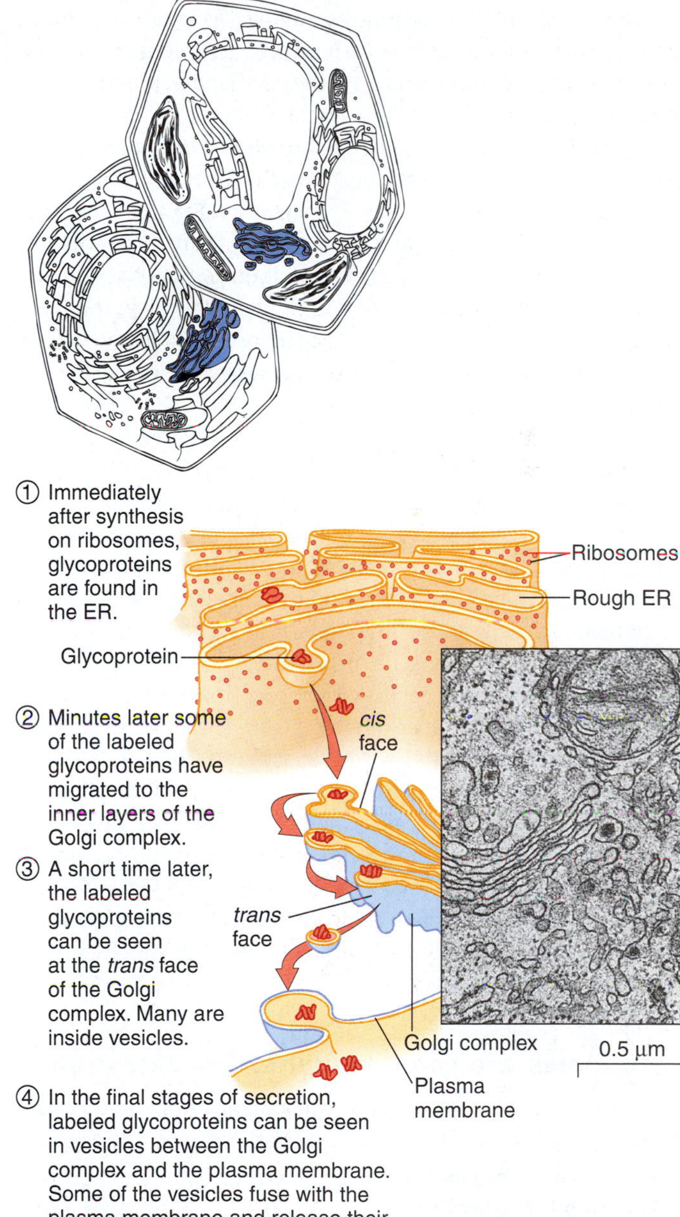

① Immediately after synthesis on ribosomes, glycoproteins are found in the ER.
Ribosomes
Rough ER
Glycoprotein
② Minutes later some of the labeled glycoproteins have migrated to the inner layers of the Golgi complex.
cis face
③ A short time later, the labeled glycoproteins can be seen at the trans face of the Golgi complex. Many are inside vesicles.
trans face
Golgi complex
0.5 μm
Plasma membrane
④ In the final stages of secretion, labeled glycoproteins can be seen in vesicles between the Golgi complex and the plasma membrane. Some of the vesicles fuse with the plasma membrane and release their contents outside the cell.

Figure 4–14 TEM and interpretive drawing of the Golgi complex. Glycoproteins are transported from the rough ER to the Golgi, where they are modified. This diagram shows the passage of glycoproteins through the Golgi complex during the secretory cycle of a mucus-secreting goblet cell that lines the intestine. Mucus is a complex mixture of covalently linked proteins and carbohydrates. *(D.W. Fawcett and R. Bolender)*

Golgi complexes, usually consisting of separate stacks of membranes dispersed throughout the cell. Cells that secrete large amounts of glycoproteins have large numbers of Golgi stacks. Golgi complexes of plant cells produce extracellular polysaccharides that are used as components of the cell wall.

Process of Science The Golgi complex functions principally to process, sort, and modify proteins. Cell biologists have studied the function of the Golgi complex by radioactively labeling newly manufactured amino acids or carbohydrates and observing their movement. Glycoproteins are synthesized and are first located in the rough ER and only later in the Golgi complex. The proteins are transported from the rough ER to the *cis* face of the Golgi complex in small transport vesicles formed from the ER membrane. Until recently, researchers thought that glycoprotein molecules released into the Golgi complex became enclosed in new vesicles that shuttle them from one compartment to another within the Golgi. A competing hypothesis, now the focus of research, holds that the cisternae themselves may move from *cis* to *trans* positions. The vesicles may move backward to recycle materials.

However proteins are moved through the Golgi complex, while there they are modified in different ways, resulting in the formation of complex biological molecules. For example, the carbohydrate part of a glycoprotein (first added to proteins in the rough ER) may be modified. In some cases the carbohydrate component may be a "sorting signal," a kind of zip code that routes the protein to a specific organelle.

Glycoproteins are packaged in secretory vesicles in the *trans* region. These vesicles pinch off from the Golgi membrane and transport their contents to a specific destination. Vesicles transporting products for export from the cell fuse with the plasma membrane. The vesicle membrane becomes part of the plasma membrane, and the glycoproteins are secreted from the cell. Other vesicles may store glycoproteins for secretion at a later time, while still others are routed to various organelles of the endomembrane system. In animal cells, the Golgi complex also manufactures lysosomes.

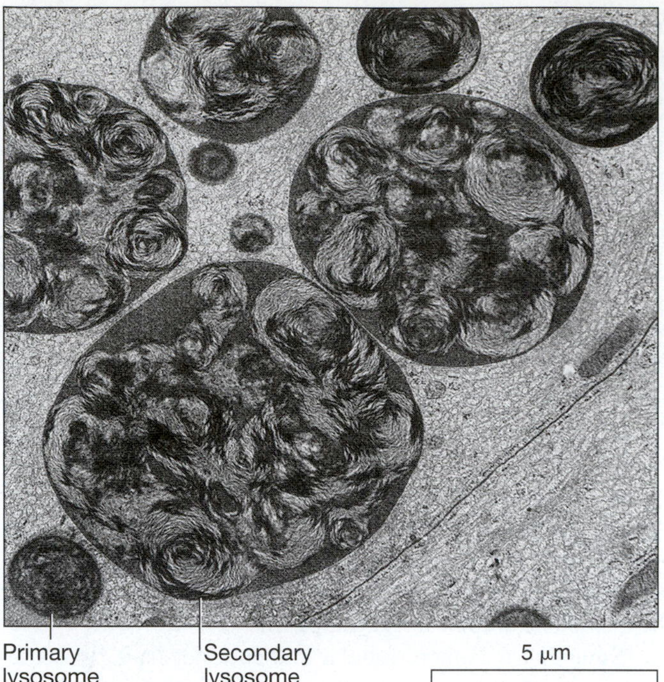

Figure 4–15 Lysosomes. The dark vesicles in this TEM are lysosomes. These compartments separate powerful digestive enzymes from the rest of the cell. Primary lysosomes bud off from the Golgi complex. After a lysosome encounters material to be digested, it is known as a secondary lysosome. The large vesicles shown here are secondary lysosomes containing various materials being digested. *(Don Fawcett/Photo Researchers, Inc.)*

Lysosomes are compartments for digestion

Small sacs of digestive enzymes called **lysosomes** are dispersed in the cytoplasm of most eukaryotic cells (Fig. 4–15). The enzymes in these organelles break down complex molecules—lipids, proteins, carbohydrates, and nucleic acids—that originate inside or outside the cell. About 40 different digestive enzymes have been identified in lysosomes; most are active under rather acidic conditions (pH 5). Under most normal conditions, the lysosome membrane confines its enzymes and their actions. Some forms of tissue damage have been related to "leaky" lysosomes. The powerful enzymes and low pH maintained by the lysosome provide an excellent example of the importance of separating functions within the cell into different compartments.

Primary lysosomes are formed by budding from the Golgi complex. Their hydrolytic enzymes are synthesized in the rough ER. As these enzymes pass through the lumen of the ER, sugars are attached to each molecule, identifying it as bound for a lysosome. This signal permits the Golgi complex to appropriately sort the enzyme to the lysosomes rather than export it from the cell.

Lysosomes degrade bacteria or debris ingested by scavenger cells. The ingested matter is enclosed in a vesicle formed from part of the plasma membrane. One or more primary lysosomes fuse with the vesicle containing the ingested material, forming a larger vesicle called a *secondary lysosome.* In the secondary lysosome the powerful enzymes come in contact with the ingested molecules and degrade them into their components. Under some conditions lysosomes break down organelles so that their components can be recycled or used as an energy source.

In certain genetic diseases of humans, known as *lysosomal storage diseases,* one of the normally present digestive enzymes is lacking. Its substrate (substance that the enzyme would normally break down) accumulates in the lysosomes, ultimately interfering with cellular activities. An example is Tay-Sachs disease (see Chapter 15), in which a normal lipid cannot be broken down in brain cells. The lipid accumulates in the cells, resulting in mental retardation and death.

Peroxisomes metabolize small organic compounds

Peroxisomes are membrane-bounded organelles containing enzymes that catalyze an assortment of metabolic reactions in which hydrogen is transferred from various compounds to oxygen (Fig. 4–16). During these reactions, hydrogen peroxide (H_2O_2), a substance toxic to the cell, is produced as a byproduct. Peroxisomes contain catalase, an enzyme that splits hydrogen peroxide, rendering it harmless.

Peroxisomes are found in large numbers in cells that synthesize, store, or degrade lipids. In plant seeds, specialized peroxisomes, called *glyoxysomes,* contain enzymes that convert stored fats to sugars. The sugars are used by the young plant as an energy source and as a component needed to synthesize other compounds. Animal cells lack glyoxysomes and cannot convert fatty acids into sugars.

When yeast cells grow in an alcohol-rich medium, they manufacture a large number of peroxisomes. These peroxisomes contain an enzyme that degrades the alcohol. Peroxisomes in human liver and kidney cells detoxify ethanol, the alcohol in alcoholic beverages.

Vacuoles are large, fluid-filled sacs with a variety of functions

Although lysosomes have been identified in almost all kinds of animal cells, their occurrence in plant and fungal cells is open to debate. Many of the functions carried out in animal cells by lysosomes are performed in plant cells by a large, single, membrane-bounded sac referred to as the **vacuole.** The vacuolar membrane, part of the endomembrane system, is referred to as the **tonoplast.** The term *vacuole,* which means "empty," refers to the fact that these organelles have no internal structure. Although the terms vacuole and vesicle are sometimes used interchangeably, vacuoles are usually larger structures, sometimes produced by the merging of many vesicles. As discussed earlier, some biologists now define *vesicle* as a membrane-enclosed structure that holds cargo.

Vacuoles play a significant role in plant growth and development. Immature plant cells are generally small and contain numerous small vacuoles. As water accumulates in these vacuoles, they tend to coalesce, forming a large central vacuole. A plant cell increases in size mainly by adding water to this central vacuole.

As much as 90% of the volume of a plant cell may be occupied by a large central vacuole containing water, as well as stored food, salts, pigments, and metabolic wastes (see Figs. 4–8 and 4–10). The vacuole may serve as a storage compartment for inorganic compounds and for molecules such as proteins in seeds. Plants lack organ systems for disposing of toxic metabolic waste products. Wastes may be recycled in the vacuole, or they may aggregate and form small crystals inside the vacuole. Compounds that are noxious to herbivores (animals that eat plants) may also be stored in some plant vacuoles as a means of defense. Plant vacuoles are lysosome-like in their ability to break down unneeded organelles and other cellular components. The vacuole is also important in maintaining hydrostatic (turgor) pressure in the plant cell.

Vacuoles have numerous other functions and are also present in many types of animal cells and in single-celled protists. Most protozoa have **food vacuoles,** which fuse with lysosomes so that the food they contain can be digested (Fig. 4–17). Many protozoa also have **contractile vacuoles,** which remove excess water from the cell (see Chapter 24).

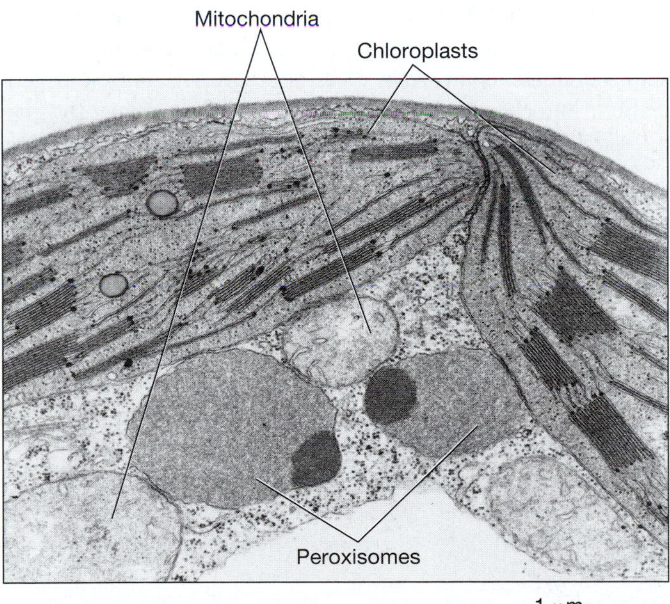

Figure 4–16 Peroxisomes. In this TEM of a tobacco leaf cell, peroxisomes can be seen in close association with chloroplasts and mitochondria. These organelles may cooperate in carrying out some metabolic processes. *(E.H. Newcomb and S.E. Frederick/Biological Photo Service)*

Mitochondria and chloroplasts are energy-converting organelles

When a cell obtains energy from its environment, it is usually in the form of chemical energy in food molecules (such as glucose) or in the form of light energy. These types of energy must be converted to forms that can be used more conveniently by cells. Some energy conversions go on in the cytosol, but other types take place in mitochondria and chloroplasts, organelles that are specialized to facilitate conversion of energy from one form to another. Chemical energy is most commonly stored in ATP. Recall from Chapter 3 that the chemical energy of ATP can be used to drive a variety of chemical reactions in the cell.

Figure 4–18 summarizes the main activities that take place in mitochondria, found in almost all eukaryotic cells (including algae and plants), and in chloroplasts, found only in algae and certain plant cells. In addition to their central roles in energy metabolism, mitochondria and chloroplasts contain small

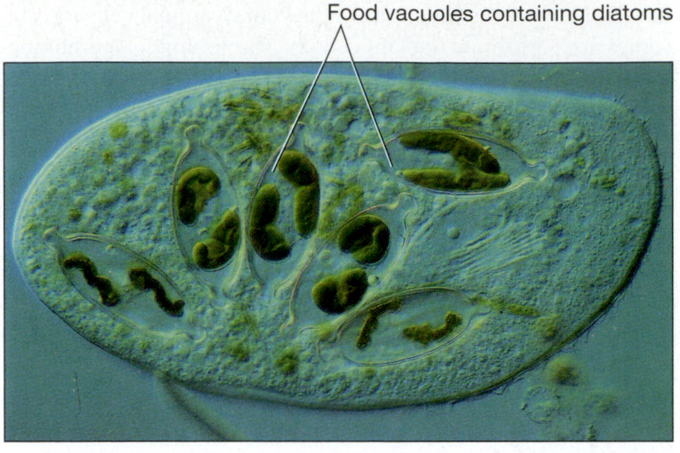

Figure 4–17 LM of food vacuoles. This protozoon, *Chilodonella,* has ingested many small, photosynthetic protists called diatoms *(dark areas)* that have been enclosed in food vacuoles. From the number of diatoms scattered about its cell, one might judge that *Chilodonella* has a rather voracious appetite. *(M.I. Walker/Photo Researchers, Inc.)*

amounts of DNA that code for some of their proteins. These interesting organelles grow and reproduce themselves.

Mitochondria make ATP through cellular respiration

Virtually all eukaryotic cells (plant, animal, fungal, and protist) contain complex organelles called **mitochondria** (sing., *mitochondrion*). These organelles are the site of **aerobic respiration,** a process that includes most of the reactions that convert the chemical energy present in certain foods to ATP (see Chapter 7). Aerobic respiration requires oxygen and results in the release of carbon atoms from food molecules as carbon dioxide (a waste product).

Mitochondria are most numerous in cells that are very active and therefore have high energy requirements. More than 1000 mitochondria have been counted in a single liver cell! Mitochondria vary in size, ranging from 2 to 8 μm in length, and are capable of changing size and shape rapidly. Mitochondria usually give rise to other mitochondria by growth and subsequent division.

Each mitochondrion is bounded by a double membrane, which forms two *different* compartments within the organelle: the intermembrane space and the matrix (Fig. 4–19; see Chapter 7 for more detailed descriptions of mitochondrial structure). The **intermembrane space** is the compartment formed between the outer and inner mitochondrial membranes. The **matrix,** the compartment enclosed by the inner mitochondrial membrane, contains enzymes that break down food molecules and convert their energy to other forms of chemical energy.

The **outer mitochondrial membrane** is smooth and allows many small molecules to pass through it. By contrast, the **inner mitochondrial membrane** has numerous folds and strictly regulates the types of molecules that can move across it. The folds, called **cristae** (sing., *crista*), extend into the matrix. Cristae greatly increase the surface area of the inner mitochondrial membrane, providing a surface for the chemical reactions that transform the chemical energy in food molecules into the energy of ATP. The membrane contains the complex series of enzymes and other proteins needed for these reactions.

In a mammalian cell, each mitochondrion has 5 to 10 identical, circular molecules of DNA, accounting for up to 1% of the total DNA in the cell. Mutations in mitochondrial DNA have been linked with certain genetic diseases, including a form of young adult blindness, and certain types of progressive muscle degeneration. Mitochondrial DNA mutates far more frequently than nuclear DNA, and an accumulation of mutations may interfere

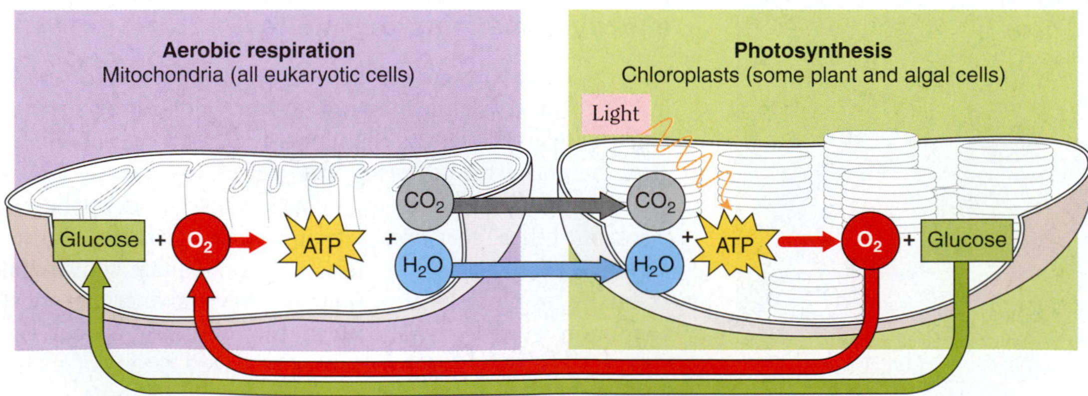

Figure 4–18 Cellular respiration and photosynthesis. Cellular respiration takes place in the mitochondria of eukaryotic cells. In this process, some of the chemical energy in glucose is transferred to ATP. Photosynthesis, which is carried out in chloroplasts in some plant and algal cells, converts light energy to ATP and to other forms of chemical energy. This energy is used to synthesize glucose from carbon dioxide and water.

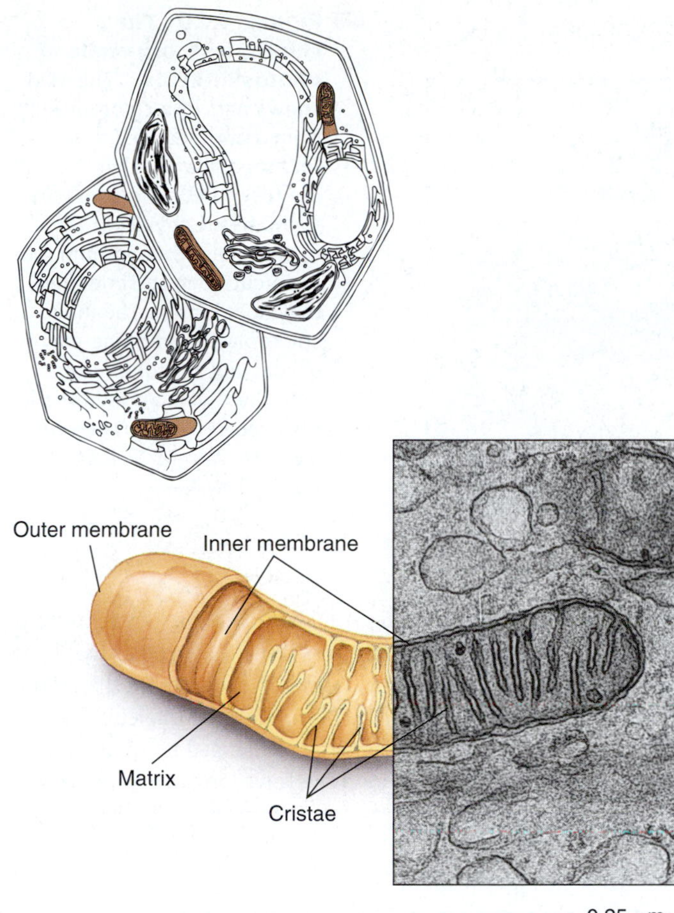

Figure 4–19 Mitochondria. Aerobic respiration takes place within mitochondria. Cristae are evident in the TEM as well as in the drawing. The drawing shows the relationship between the inner and outer mitochondrial membranes. *(D.W. Fawcett)*

with mitochondrial function. A diminished capacity to generate energy may contribute to the aging process.

Mitochondria also affect health and aging by leaking electrons. These electrons form **free radicals,** which are toxic, highly reactive compounds that have unpaired electrons. These electrons bond with other compounds in the cell, interfering with normal function.

Mitochondria play an important role in programmed cell death, or **apoptosis.** Unlike **necrosis,** which is uncontrolled cell death that causes inflammation and damages other cells, apoptosis is a normal part of development and maintenance. For example, during the metamorphosis of a tadpole to a frog, the cells of the tadpole tail must die. During human development, the hand is webbed until, through apoptosis, the tissue between the fingers is destroyed. Cell death also occurs in the adult. For example, cells in the upper layer of human skin and in the intestinal wall are continuously destroyed and replaced by new cells.

Mitochondria can initiate cell death in several different ways. For example, they can interfere with energy metabolism or activate enzymes that mediate cell destruction. When a mitochondrion is injured, large pores open in its membrane, and cytochrome *c*, a protein important in energy production, is released into the cytoplasm. Cytochrome *c* triggers apoptosis by activating a group of enzymes known as **caspases** that cut up vital compounds in the cell. Inappropriate initiation or inhibition of apoptosis may contribute to a variety of diseases including cancer, acquired immunodeficiency syndrome (AIDS), and Alzheimer's disease. Pharmaceutical companies are developing drugs that block apoptosis. However, cellular dynamics are extremely complex, and blocking apoptosis could lead to a worse fate, including necrosis.

Chloroplasts convert light energy to chemical energy through photosynthesis

Certain plant and algal cells carry out a complex set of energy conversion reactions known as **photosynthesis** (see Chapters 1 and 8). **Chloroplasts** are organelles that contain **chlorophyll,** a green pigment that traps light energy for photosynthesis. Chloroplasts also contain a variety of yellow and orange light-absorbing pigments known as **carotenoids** (see Chapter 3). A unicellular alga may have only a single large chloroplast, whereas a leaf cell may have 20 to 100. Chloroplasts tend to be somewhat larger than mitochondria, with lengths typically ranging from about 5 to 10 μm or longer.

Chloroplasts are typically disc-shaped structures and, like mitochondria, have a complex system of folded membranes (Fig. 4–20; see Chapter 8 for more detailed descriptions of structure). Two membranes, separated by a small space, separate the chloroplast from the cytosol. The inner membrane encloses a fluid-filled space called the **stroma,** which contains enzymes responsible for producing carbohydrates from carbon dioxide and water, using energy trapped from sunlight. A system of internal membranes, suspended in the stroma, consists of an interconnected set of flat, disclike sacs called **thylakoids.** The thylakoids are arranged in stacks called **grana** (sing., *granum*).

The thylakoid membranes enclose a third, innermost compartment within the chloroplast, called the **thylakoid lumen.** The thylakoid membranes, in which chlorophyll is found, are similar to the inner mitochondrial membranes in that they are involved in the formation of ATP. Energy absorbed from sunlight by the chlorophyll molecules is used to excite electrons; the energy in these excited electrons is then used to form ATP and other molecules that can transfer chemical energy. This chemical energy is used to produce carbohydrates from carbon dioxide and water in the stroma.

Chloroplasts belong to a group of organelles known as **plastids** that produce and store food materials in cells of plants and algae. All plastids develop from **proplastids,** precursor organelles found in less specialized plant cells, particularly in growing, undeveloped tissues. Depending on the special functions a cell will eventually have, its proplastids can mature into a variety of specialized mature plastids. These are extremely versatile organelles; in fact, under certain conditions even mature plastids can convert from one form to another.

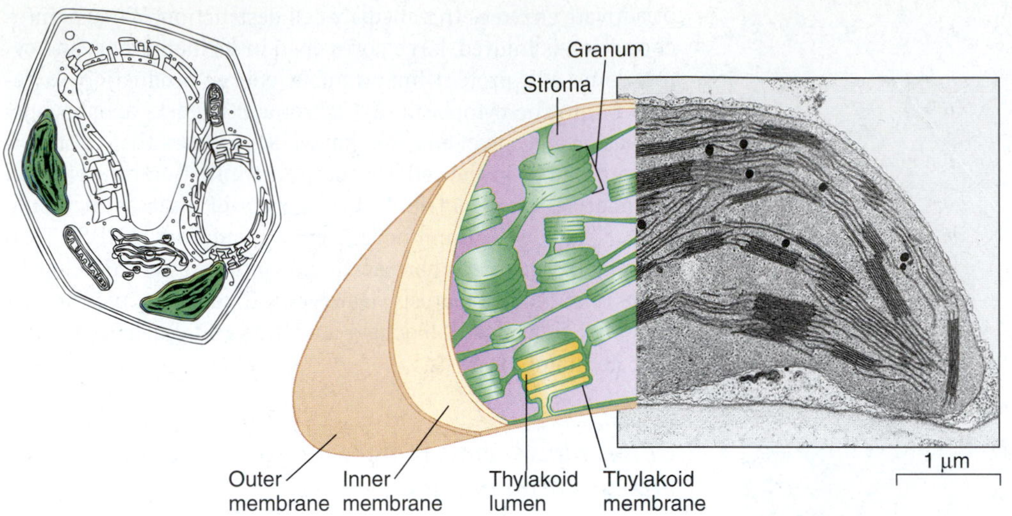

Figure 4–20 The chloroplast, organelle of photosynthesis. The TEM shows part of a chloroplast from a corn leaf cell. Chlorophyll and other photosynthetic pigments are found in the thylakoid membranes. One granum has been cut open to show the thylakoid lumen. The inner chloroplast membrane may or may not be continuous with the thylakoid membrane *(as shown). (E.H. Newcomb and W.P. Wergin/Biological Photo Service)*

Chloroplasts are produced when proplastids are stimulated by exposure to light. **Chromoplasts** contain pigments that give certain flowers and fruits their characteristic colors; these attract animals that serve as pollinators or as seed dispersers. **Leukoplasts** are unpigmented plastids; they include **amyloplasts** (see Fig. 3–9), which store starch in the cells of many seeds, roots, and tubers (e.g., white potatoes).

MAKING THE CONNECTION What is the evolutionary relationship between prokaryotic cells and the more complex cells of eukaryotes? Mitochondria and chloroplasts have provided valuable insights because these organelles have been shown to have many prokaryote features. For example, although most of the DNA in eukaryotic cells resides in the nucleus, both mitochondria and chloroplasts (as well as other plastids) have DNA molecules in their inner compartments. These DNA molecules code for a small number of the proteins found in these organelles. These proteins are synthesized by mitochondrial or chloroplast ribosomes, which are similar to the ribosomes of prokaryotes. Most of the mitochondrial and chloroplast proteins, however, are coded for by nuclear genes, manufactured on free ribosomes in the cytosol, and then transported to their appropriate locations within.

The existence of a separate set of ribosomes and DNA molecules in mitochondria and chloroplasts and their similarity in size to many bacteria, along with other prokaryote-like characteristics, provide support for the **endosymbiont theory** (see Figs. 20–7 and 24–23). According to this theory, mitochondria and chloroplasts evolved from prokaryotic organisms that took up residence inside larger cells and eventually lost the ability to function as autonomous organisms.

Evolutionary biologists are sequencing mitochondrial genes from protists to determine the mitochondrion's evolution from a bacterium to a highly specialized organelle. Evidence suggests that during this process, some genes moved from the early mitochondrion to the nucleus. One such gene codes for a molecular chaperone known as *chaperonin 60,* which helps other proteins fold appropriately. Some investigators think that protists that have retained this gene in their mitochondria evolved earlier than those in which the chaperonin 60 gene is found in the nucleus. The endosymbiont theory is currently being challenged by some investigators who argue that mitochondria evolved in eukaryotic cells.

EUKARYOTIC CELLS CONTAIN A CYTOSKELETON

We have seen that there are many different sizes and shapes of cells. When we watch cells growing in the laboratory, we see that they frequently change shape and that many types of cells move about. The shapes of cells and their ability to move are determined in large part by the **cytoskeleton,** a dense network of protein fibers (Fig. 4–21). In addition to providing mechanical support, the cytoskeleton functions in cell movement, in the transport of materials within the cell, and in cell division.

The cytoskeleton is highly dynamic and constantly changing. Its framework is made of three types of protein filaments: **microtubules, microfilaments** (also known as actin filaments), and **intermediate filaments.** Both microfilaments and microtubules are fibers formed from beadlike, globular protein subunits, which can be rapidly assembled and disassembled. Intermediate filaments are made from fibrous protein subunits and are more stable than microtubules and microfilaments.

Microtubules are hollow cylinders

Microtubules, the thickest filaments of the cytoskeleton, are about 25 nm in outside diameter and up to several micrometers in length. In addition to playing a structural role in the formation of the cytoskeleton, these extremely adaptable structures are involved in the movement of chromosomes during cell division. They serve as tracks for several other kinds of intracellular move-

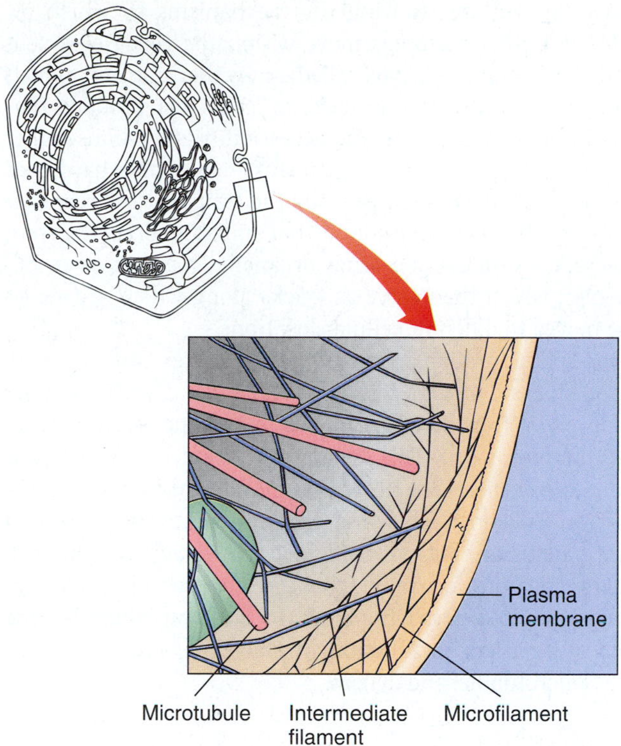

Figure 4–21 The cytoskeleton. Eukaryotic cells contain a cytoskeleton consisting of networks of several types of fibers, including microtubules, microfilaments, and intermediate filaments. The cytoskeleton contributes to the shape of the cell, anchors organelles, and sometimes rapidly changes shape during cellular locomotion.

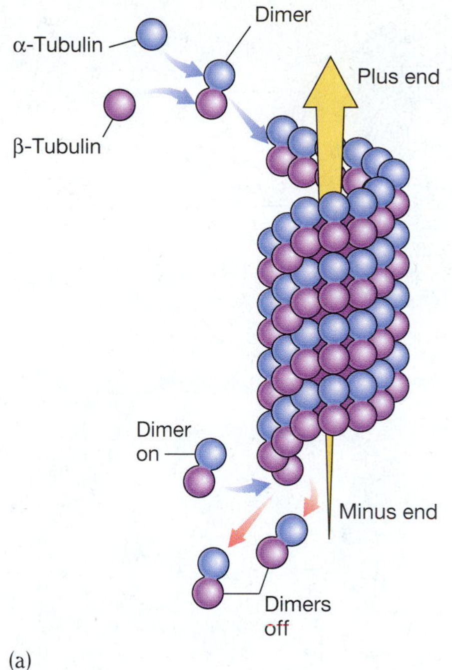

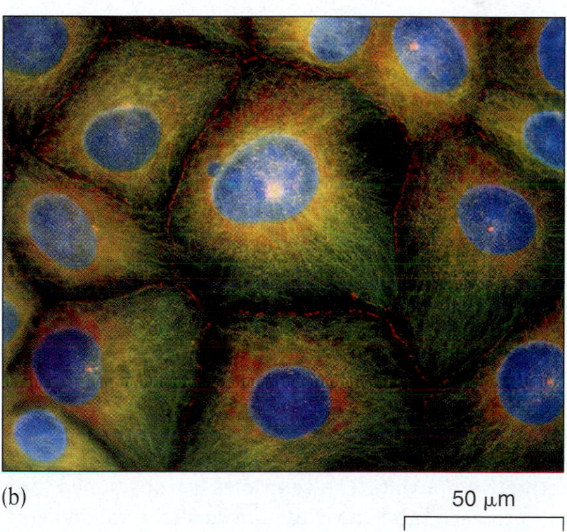

Figure 4–22 Organization of microtubules. **(a)** Microtubules are manufactured in the cell by adding dimers of α-tubulin and β-tubulin to an end of the hollow cylinder. Notice that the cylinder has polarity. The end shown at the top of the figure is the fast-growing or plus end; the opposite end is the minus end. Each turn of the spiral requires 13 dimers. **(b)** Confocal fluorescence LM showing microtubules in green. A microtubule-organizing center *(pink dot)* is visible beside or over most of the cell nuclei (*blue*). *(b, Nancy Kedersha)*

ment and are the major structural components of cilia and flagella—specialized structures used in some cell movements.

Microtubules consist of two very similar proteins: ***α*-tubulin** and ***β*-tubulin.** These proteins combine to form a dimer (recall from Chapter 3 that a dimer forms from the association of two similar, simpler units, referred to as *monomers*). A microtubule elongates by the addition of tubulin dimers (Fig. 4–22). Microtubules are disassembled by the removal of dimers, which can then be recycled to form microtubules in other parts of the cell. Each microtubule has polarity, and its two ends are referred to as *plus* and *minus.* The plus end elongates more rapidly.

For microtubules to act as a structural framework or participate in cell movement, they must be anchored to other parts of the cell. In nondividing cells, the minus ends of microtubules appear to be anchored in regions called **microtubule-organizing centers (MTOCs).** In animal cells, the main MTOC is the cell center or **centrosome,** a structure that is important in cell division. In almost all animal cells, the centrosome contains two structures called **centrioles** (Fig. 4–23). Some other cells also contain centrioles. These structures, which are oriented within the centrosome at right angles to each other, are known as *9 × 3 structures;* they consist of nine sets of three attached microtubules arranged to form a hollow cylinder. The centrioles are duplicated before cell division and may play a role in some types of microtubule assembly. Most plant cells and fungal cells have an MTOC but lack centrioles. This suggests either that centrioles are not essential to most microtubule assembly processes or that alternative assembly mechanisms are possible.

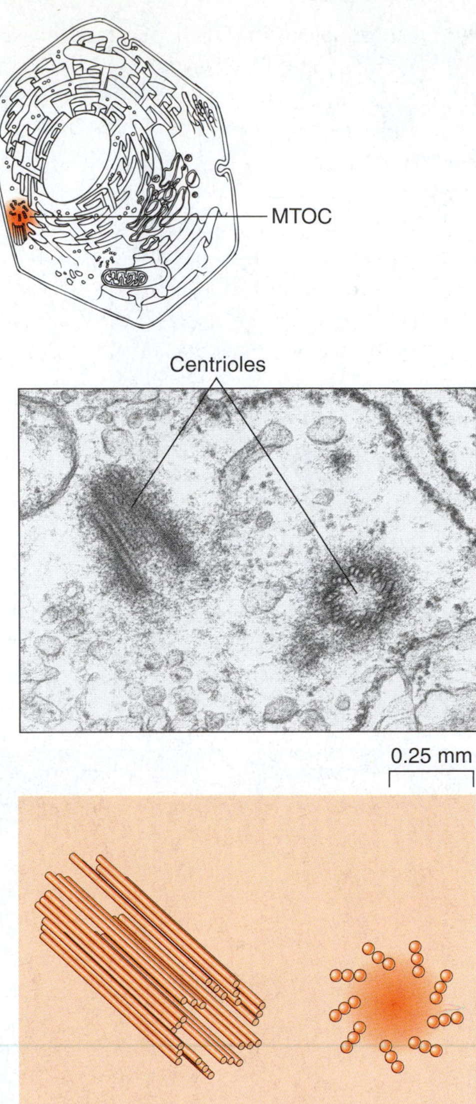

Figure 4–23 Centrioles. The TEM is paired with an interpretive drawing. The centrioles are positioned at right angles to each other, near the nucleus of a nondividing animal cell. Note the 9 × 3 arrangement of microtubules. The centriole on the left has been cut longitudinally and the one on the right transversely. *(B.F. King/Biological Photo Service)*

The ability of microtubules to assemble and disassemble rapidly can be seen during cell division, when much of the cytoskeleton appears to break down (see Chapter 9). Many of the tubulin subunits organize into a structure called the **spindle,** which serves as a framework for the orderly distribution of chromosomes when the cell divides.

Several **microtubule-associated proteins (MAPs)** have been identified and classified into two groups: structural MAPs and motor MAPs. Structural MAPs may help regulate microtubule assembly, and they cross-link microtubules to other cytoskeletal polymers. Motor MAPs use ATP energy to produce movement.

Investigators are studying the mechanisms by which organelles and other materials move within the cell. Nerve cells typically have long extensions called *axons* that transmit signals to other nerve cells, muscle cells, or cells that produce hormones. Because of its length and accessibility and because other cells use similar transport mechanisms, researchers have used the axon as a model system to study the transport of organelles within the cell. They have found that mitochondria, transport and secretory vesicles, and other organelles may attach to microtubules, which then serve as tracks along which organelles can be moved to different cellular locations.

One motor protein, named *kinesin,* moves organelles toward the plus end of a microtubule (Fig. 4–24). *Dynein,* another motor protein, transports organelles in the opposite direction, toward the minus end. This dynein movement is referred to as *retrograde transport.* In 1997, Claire Waterman-Storer of the University of Pennsylvania and her colleagues reported that dynein is necessary but not sufficient for retrograde transport. These investigators determined that a protein complex called *dynactin* is required. Dynactin binds to both microtubules and dynein and appears to function in transport, linking the organelle, microtubule, and dynein.

Cilia and flagella are composed of microtubules

Projecting from surfaces of many cells are thin, movable structures important in cell movement. If a cell has one, or only a few, of these appendages and if they are long (typically about 200 μm) relative to the size of the cell, they are called **flagella** (sing., *flagellum*). If the cell has many short (typically 2–10 μm long) appendages, they are called **cilia** (sing., *cilium*). Both cilia and flagella are used by cells to move through a watery environment, and cilia can be used to pass liquids and particles across the cell surface. These structures are commonly found on unicellular

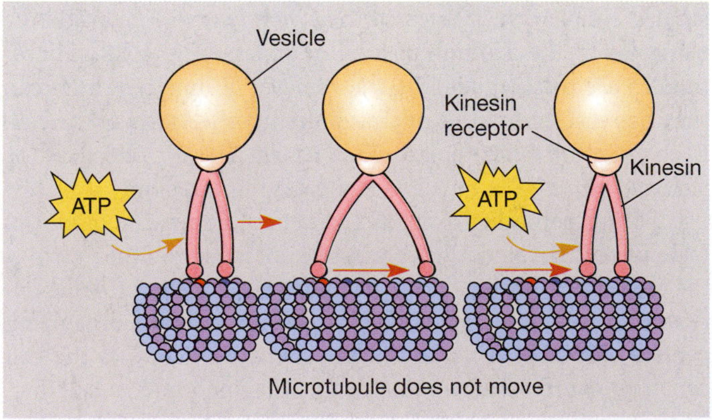

Figure 4–24 Hypothetical model of a kinesin motor. A kinesin molecule attaches to a specific receptor on the vesicle. Energy from ATP allows the kinesin molecule to change its conformation and "walk" along the microtubule, carrying the vesicle along.

and small multicellular organisms. In animals and certain plants, flagella serve as the tails of sperm cells. In animals cilia commonly occur on the surfaces of cells that line internal ducts of the body (e.g., respiratory passageways).

Eukaryotic cilia and flagella are structurally alike (but different from bacterial flagella). Each consists of a slender, cylindrical stalk covered by an extension of the plasma membrane. The core of the stalk contains a group of microtubules arranged so that there are nine attached pairs of microtubules around the circumference and two unpaired microtubules in the center (Fig. 4–25). This *9 + 2* arrangement of microtubules is characteristic of virtually all eukaryotic cilia and flagella, another example of the unity of organisms that reflects their common origin.

The microtubules move by sliding in pairs past each other. The sliding force is generated by dynein proteins, which are attached to the microtubules like small arms. These proteins use the energy from ATP to power the cilia or flagella. The dynein proteins (arms) on one pair of tubules change their shape and "walk" along the adjacent microtubule pair. Thus, the microtubules on one side of a cilium or a flagellum extend farther toward the tip than those on the other side. This sliding of microtubules translates into a bending motion (Fig. 4–25*b*). Cilia typically move like oars, alternating power and recovery strokes and exerting a force that is parallel to the cell surface. A flagellum moves like a whip, exerting a force perpendicular to the cell surface.

Each cilium or flagellum is anchored in the cell by a **basal body,** which has nine sets of three attached microtubules in a cylindrical array (a 9×3 structure). The basal body appears to be the organizing structure for the cilium or flagellum when it first begins to form. However, experiments have shown that as growth proceeds, the tubulin subunits are added much faster to the tips of the microtubules than to the base. Basal bodies and centrioles appear to be functionally related as well as structurally similar. In fact, centrioles are typically found in the cells of organisms that are capable of producing flagellated or ciliated cells; these include animals, certain protists, a few fungi, and a few plants. Both basal bodies and centrioles replicate themselves.

Microfilaments consist of intertwined strings of actin

Microfilaments, also called *actin filaments,* are flexible, solid fibers about 7 nm in diameter. Each microfilament consists of two intertwined polymer chains of beadlike **actin** molecules (Fig. 4–26). Actin filaments are cross-linked with one another and with other proteins by linker proteins. They form bundles of fibers that provide mechanical support for various cell structures. In many cells, a network of microfilaments can be seen in the cytosol just inside the plasma membrane.

In muscle cells, actin is associated with another protein, **myosin,** to form fibers that generate the forces involved in muscle contraction (see Chapter 38). In nonmuscle cells, actin can

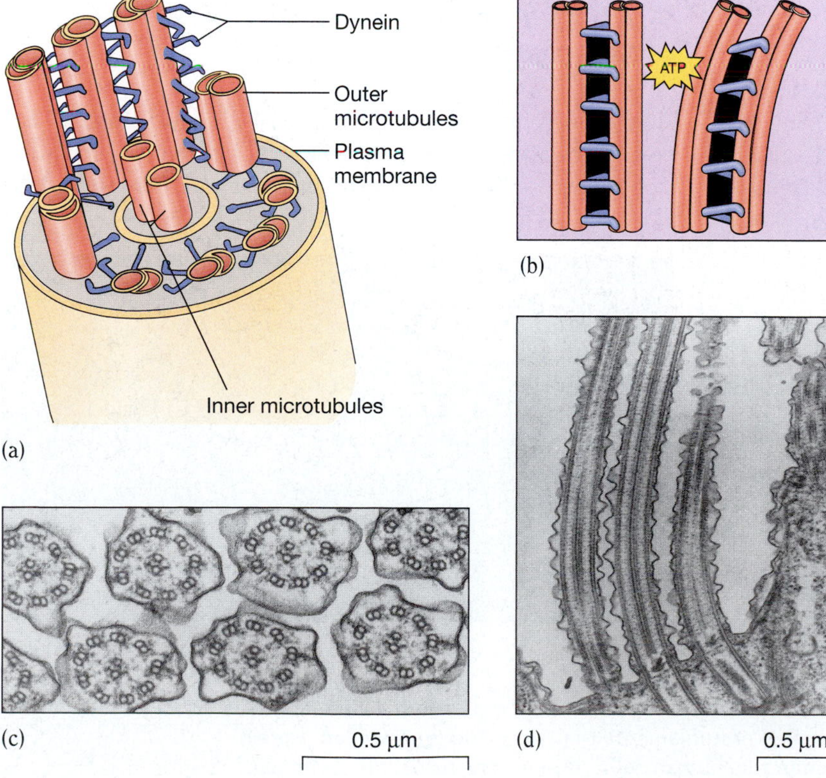

Figure 4–25 Structure of cilia. A cilium (or flagellum) contains microtubules in a 9 + 2 arrangement. **(a)** This three-dimensional representation shows nine attached microtubule pairs (doublets) arranged in a cylinder, with two unattached microtubules in the center. The "arms" are made of dynein, a motor protein that uses energy from ATP to bend the cilia by "walking" up and down the neighboring pair of microtubules. The dynein arms, shown widely spaced for clarity, are actually much closer together along the longitudinal axis. **(b)** The dynein arms move the microtubules by forming and breaking cross bridges on the adjacent microtubules, so that one tubule "walks" along its neighbor. **(c)** TEM of cross sections through cilia showing the 9 + 2 arrangement of microtubules. **(d)** TEM of a longitudinal section of three cilia of the protist *Tetrahymena,* an organism often used in genetic research. Some of the interior microtubules are visible. *(c, d, W.L. Dentler/ Biological Photo Service)*

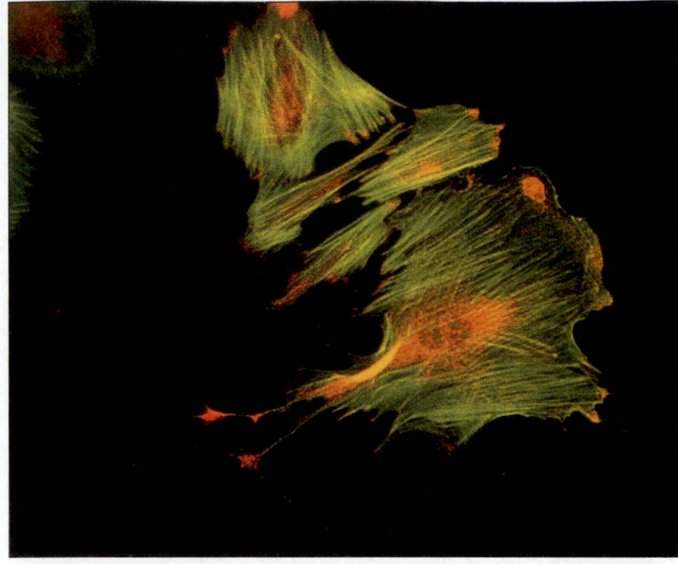

Figure 4-26 Microfilaments **(a)** Many bundles of aggregated microfilaments *(green)* are evident in this confocal fluorescence LM of fibroblasts (cells found in connective tissue). **(b)** An individual microfilament consists of two intertwined strings of beadlike actin molecules. *(a, Nancy Kedersha/ImmunoGen, Inc.)*

also associate with myosin, forming contractile structures that are involved in various cell movements. Actin filaments themselves cannot contract, but they can generate movement by rapidly assembling and disassembling. Actin filaments associated with myosin are involved in transient functions, such as cell division in animals, in which contraction of a ring of actin associated with myosin causes the constriction of the cell to form two daughter cells (see Chapter 9). Certain organelles in the giant axons of the squid move along microfilaments. A type of myosin appears to be the motor for this transport.

As mentioned earlier in the chapter, some types of cells have microvilli, projections of the plasma membrane that increase the surface area of the cell for transporting materials across the plasma membrane. Microvilli contain bundles of microfilaments, which extend and retract as a result of the building and breaking down of these microfilaments.

Intermediate filaments help stabilize cell shape

Intermediate filaments are tough fibers, typically 8 to 10 nm in diameter (Fig. 4–27). They vary widely in protein composition and size among different cell types and different organisms. Intermediate filaments function to strengthen the cytoskeleton and stabilize cell shape. They are abundant in parts of a cell that may be subject to mechanical stress applied from outside the cell. Certain proteins cross-link intermediate filaments with other types of filaments and mediate interactions between them.

Intermediate filaments are responsible for formation of a sheath called the **nuclear lamina,** just inside the nuclear envelope. They are important in regulating the timing of the disorganization and initiating the reorganization of the nucleus during division. In humans, more than 50 genes have been identified that code for proteins that assemble into intermediate filaments. Certain mutations in these genes weaken the cell and have been linked to several diseases.

AN EXTRACELLULAR MATRIX SURROUNDS MOST CELLS

Most eukaryotic cells are surrounded by a **glycocalyx,** or **cell coat,** formed by polysaccharide side chains of proteins and lipids that are part of the plasma membrane. Certain molecules of the

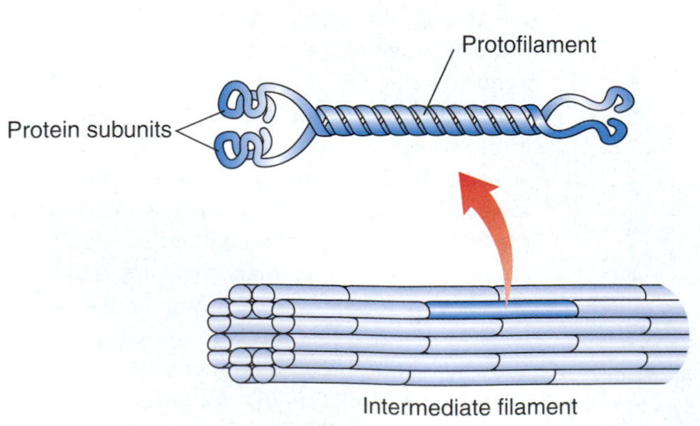

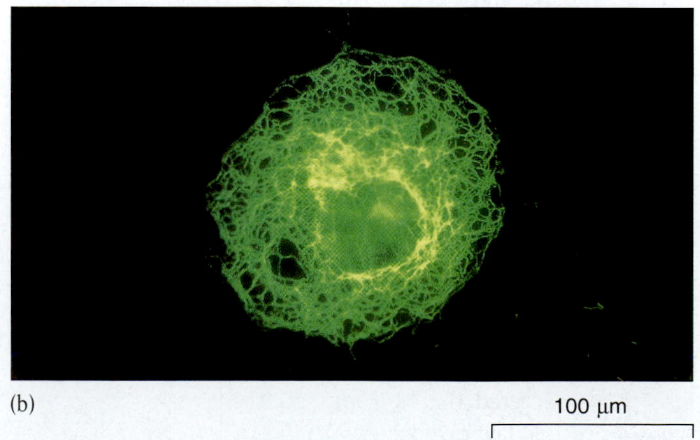

Figure 4-27 Intermediate filaments. **(a)** Intermediate filaments are flexible rods about 10 nm in diameter. Each intermediate filament consists of components called protofilaments that are made up of coiled protein subunits. **(b)** Intermediate filaments are stained green in this human cell. *(b, K.G. Murti/Visuals Unlimited)*

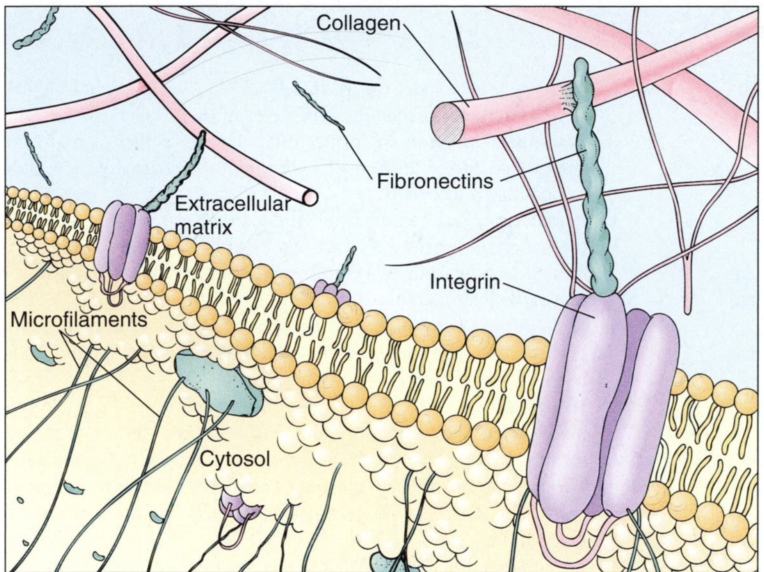

Figure 4–28 Extracellular matrix (ECM). Fibronectins, glycoproteins of the ECM, bind to integrins and other receptors in the plasma membrane.

cell coat permit cells to recognize one another, to make contact, and in some cases to form adhesive or communicating associations. Other molecules of the cell coat contribute to the mechanical strength of multicellular tissues.

Many animal cells are also surrounded by an **extracellular matrix (ECM)** which they secrete. It consists of a gel of carbohydrates and fibrous proteins (Fig. 4–28). The main structural protein in the ECM is **collagen,** which forms very tough fibers. Certain glycoproteins of the ECM, called **fibronectins,** bind to protein receptors that extend from the plasma membrane. The main membrane receptors for the ECM are **integrins.** These proteins activate many **cell signaling** pathways and help regulate a variety of cell functions. They appear to be important in cell movement and in organizing the cytoskeleton so that cells assume a definite shape. In many types of cells, integrins anchor the external ECM to the microfilaments of the internal cytoskeleton. When these cells are not appropriately anchored, apoptosis results. Cancer cells apparently lose this requirement to be anchored to the ECM.

Most bacteria, fungi, and plant cells are surrounded by a cell wall and proteins. Plant cells are surrounded by thick cell walls that contain multiple layers of the polysaccharide **cellulose** (see Fig. 3–10). Other polysaccharides in the plant cell wall form cross links between the bundles of cellulose fibers. Each cellulose fiber layer runs in a different direction from the adjacent layer, giving the cell wall great mechanical strength.

A growing plant cell secretes a thin, flexible *primary cell wall,* which can stretch and expand as the cell increases its size (Fig. 4–29). After the cell stops growing, either new wall material is secreted that thickens and solidifies the primary wall or multiple layers of a *secondary cell wall* with a different chemical composition are formed between the primary wall and the plasma membrane. Wood is made mainly of secondary cell walls. Between the primary cell walls of adjacent cells lies the **middle lamella,** a layer of gluelike polysaccharides called *pectins.* The middle lamella causes the cells to adhere tightly to one another. (See Chapter 31 discussion of the ground tissue system for additional information on plant cell walls.)

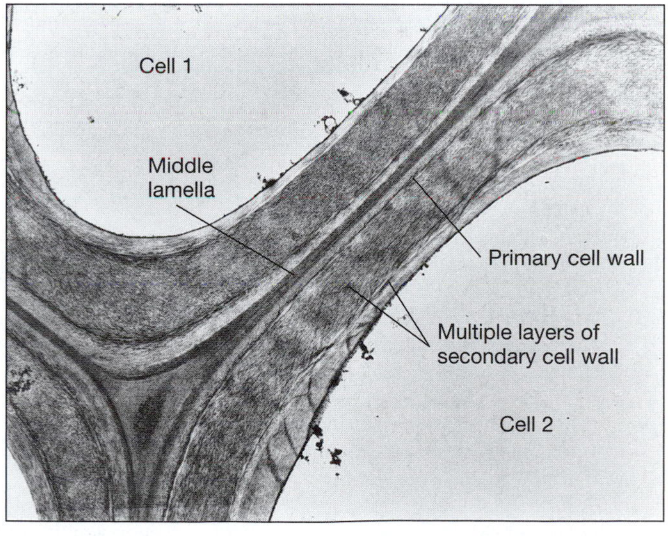

Figure 4–29 Plant cell walls. The cell walls of two adjacent plant cells are labeled in this TEM. The cells are cemented together by the middle lamella, a layer of gluelike polysaccharides called pectins. A growing plant cell first secretes a thin primary wall that is flexible and can stretch as the cell grows. The thicker layers of the secondary wall are secreted inside the primary wall after the cell stops elongating. *(Biophoto Associates)*

SUMMARY WITH KEY TERMS

I. The **cell** is considered the basic unit of life because it is the smallest self-sufficient, self-replicating unit of living material.

II. The **cell theory** states that organisms are composed of cells and that all cells arise by division of preexisting cells.

III. Every cell is surrounded by a **plasma membrane** that forms a cytoplasmic compartment. The plasma membrane serves as a selective barrier between the cell and its surrounding environment.
   A. Cells have genetic instructions coded in **DNA.**
   B. Cells have internal structures called **organelles** that are specialized to carry on specific functions.
   C. A critical factor in determining cell size is the ratio of the plasma membrane (surface area) to the cell's volume; the plasma membrane must be large enough to regulate the passage of materials into and out of the cell.
   D. Cell size and shape are related to function and are limited by the need to maintain homeostasis.

IV. Biologists have learned about cellular structure by studying cells with **light** and **electron microscopes** and by using a variety of chemical methods.
   A. The electron microscope has superior **resolving power,** enabling investigators to see details of cell structures not observable with conventional microscopes.
   B. Cell biologists use **cell fractionation** techniques and other biochemical methods to gain information about the function of cellular structures.

V. **Prokaryotic cells** are bounded by a plasma membrane but have little or no internal membrane organization. They have a **nuclear area** rather than a membrane-bounded nucleus. Prokaryotes typically have a **cell wall** and **ribosomes** and may have propeller-like **flagella.**

VI. **Eukaryotic cells** have a membrane-bounded **nucleus** and **cytoplasm,** which is organized into organelles; the fluid component of the cytoplasm is the **cytosol.**
   A. Plant cells differ from animal cells in that they have rigid cell walls, **plastids,** and large **vacuoles;** cells of most plants lack centrioles.
   B. Membranes divide the cell into membrane-bounded compartments; this allows cells to conduct specialized activities within small areas of the cytoplasm, concentrate molecules, and organize metabolic reactions. A system of interacting membranes forms the **endomembrane system.** Small membrane-bounded sacs, called **vesicles,** transport materials between compartments.
   C. The nucleus, the control center of the cell, contains genetic information coded in DNA.
      1. The nucleus is bounded by a **nuclear envelope** consisting of a double membrane perforated with **nuclear pores** that communicate with the cytoplasm.
      2. DNA in the nucleus associates with protein to form **chromatin.** During cell division, chromatin condenses and the **chromosomes** become visible.
      3. The **nucleolus** is a region in the nucleus that is the site of ribosomal RNA synthesis and ribosome assembly.
   D. The **endoplasmic reticulum (ER)** is a network of folded internal membranes in the cytosol.
      1. **Rough ER** is studded along its outer surface with ribosomes that manufacture proteins. Proteins synthesized on rough ER can be transferred to other cell membranes or secreted from the cells by **transport vesicles,** formed by membrane budding.
      2. **Smooth ER** is the site of lipid synthesis and detoxifying enzymes.
   E. The **Golgi complex** consists of stacks of flattened membranous sacs called **cisternae** that process, sort, and modify proteins synthesized on the ER. The Golgi adds carbohydrates and lipids to proteins and can route proteins, by way of **secretory vesicles,** to the plasma membrane for export from the cell, or to other destinations. The Golgi complex also manufactures lysosomes.
   F. **Lysosomes** function in intracellular digestion; they contain enzymes that break down both worn-out cell structures and substances taken into cells.
   G. **Peroxisomes** are membrane-bounded sacs containing enzymes that catalyze a variety of reactions in which hydrogen peroxide is formed as a byproduct. Peroxisomes contain catalase, an enzyme that splits hydrogen peroxide.
   H. Vacuoles are important in plant growth and development. Many protists have **food vacuoles** and **contractile vacuoles.** Vacuoles may be formed by the merging of many vesicles.
   I. **Mitochondria,** the sites of aerobic respiration, are double-membraned organelles in which the inner membrane is folded, forming **cristae** that increase the surface area of the membrane. The cristae and the compartment enclosed by the inner membrane, the **matrix,** contain enzymes for the reactions of aerobic respiration.
      1. Mitochondria contain DNA that codes for some of its proteins.
      2. Electrons that leak out of mitochondria form free radicals that disrupt cell function.
      3. Mitochondria play an important role in **apoptosis,** or programmed cell death.
   J. Cells of algae and plants contain plastids; **chloroplasts,** the sites of **photosynthesis,** are double-membraned plastids.
      1. Typically, the inner membrane encloses a fluid-filled space, the **stroma. Grana,** stacks of disclike sacs called **thylakoids,** are suspended in the stroma.
      2. **Chlorophyll,** the green pigment that traps light energy during photosynthesis, is found in the thylakoid membranes.

VII. The **cytoskeleton** is a dynamic internal framework made of microtubules, microfilaments, and intermediate filaments. The cytoskeleton provides structural support and functions in various types of cell movement, including transport of materials in the cell.
   A. **Microtubules** are hollow cylinders assembled from subunits of the protein **tubulin.**
      1. In cells that are not dividing, the minus ends of microtubules appear to be anchored in **microtubule-organizing centers (MTOCs).** The main MTOC of animal cells is the **centrosome,** which usually contains two **centrioles.** Each centriole has a 9×3 arrangement of microtubules.
      2. **Microtubule-associated proteins (MAPs)** include structural MAPs and motor MAPs. Two motor MAPs are kinesin and dynein.
      3. **Cilia** and **flagella** function in cell movement. Each consists of a $9 + 2$ arrangement of microtubules, and each is anchored in the cell by a **basal body** that has a 9×3 organization of microtubules.
   B. **Microfilaments,** or **actin filaments,** formed from subunits of the protein **actin,** are important in cell movement.
   C. **Intermediate filaments** are stable structures that strengthen the cytoskeleton and stabilize cell shape.

VIII. Most cells are surrounded by a **glycocalyx,** or **cell coat,** formed by polysaccharides extending from the plasma membrane.
   A. Many animal cells are also surrounded by an **extracellular matrix (ECM)** consisting of carbohydrates and protein. **Fibronectins** are glycoproteins of the ECM that bind to **integrins,** receptor proteins in the plasma membrane.
   B. Most bacteria, fungi, and plant cells are surrounded by a cell wall made of carbohydrates and protein. Plant cells secrete **cellulose** and other polysaccharides to form rigid cell walls.

POST-TEST

1. The ability of a microscope to reveal fine detail is known as (a) magnification (b) resolving power (c) cell fractionation (d) scanning electron microscopy (e) phase contrast
2. A plasma membrane is characteristic of (a) all cells (b) prokaryotic cells only (c) eukaryotic cells only (d) animal cells only (e) eukaryotic cells except for plant cells
3. Detailed information about the shape and external features of a specimen can best be obtained by using a (a) differential centrifuge (b) fluorescence microscope (c) transmission electron microscope (d) scanning electron microscope (e) light microscope
4. In eukaryotic cells, DNA is found in (a) chromosomes (b) chromatin (c) mitochondria (d) answers a, b, and c are correct (e) answers a and b only are correct
5. Which of the following structures would *not* be found in prokaryotic cells? (a) cell wall (b) ribosomes (c) nuclear area (d) nucleus (e) propeller-like flagellum
6. Which of the following is/are most closely associated with protein synthesis? (a) ribosomes (b) smooth ER (c) mitochondria (d) microfilaments (e) lysosomes
7. Which of the following is/are most closely associated with the breakdown of ingested material? (a) ribosomes (b) smooth ER (c) mitochondria (d) microfilaments (e) lysosomes
8. Which of the following are most closely associated with photosynthesis? (a) basal bodies (b) smooth ER (c) cristae (d) thylakoids (e) MTOCs
9. A 9 + 2 arrangement of microtubules best describes (a) cilia (b) centrosomes (c) basal bodies (d) microfilaments (e) microvilli
10. Which sequence most accurately describes information flow in the eukaryotic cell? (a) DNA in nucleus → messenger RNA → ribosomes → protein synthesis (b) DNA in nucleus → ribosomal RNA → mitochondria → protein synthesis (c) RNA in nucleus → messenger DNA → ribosomes → protein synthesis (d) DNA in nucleus → messenger RNA → Golgi complex → protein synthesis (e) DNA in nucleus → messenger RNA → smooth ER → protein synthesis
11. Which sequence most accurately describes glycoprotein processing in the eukaryotic cell? (a) smooth ER → transport vesicle → *cis* region of Golgi → *trans* region of Golgi → plasma membrane or other organelle (b) rough ER → transport vesicle → *cis* region of Golgi → *trans* region of Golgi → plasma membrane or other organelle (c) rough ER → transport vesicle → *trans* region of Golgi → *cis* region of Golgi → plasma membrane or other organelle (d) rough ER → nucleus → *cis* region of Golgi → *trans* region of Golgi → plasma membrane or other organelle (e) smooth ER → transport vesicle → *cis* region of Golgi → chloroplast
12. Which of the following is/are part of the cytoskeleton? (a) microfilaments (b) lysosomes (c) peroxisomes (d) ribosomes (e) endoplasmic reticulum
13. Which of the following function(s) in cell movement? (a) microtubules (b) cristae (c) grana (d) smooth ER (e) rough ER
14. Which of the following is/are *not* associated with mitochondria? (a) cristae (b) aerobic respiration (c) apoptosis (d) free radicals (e) thylakoids
15. The extracellular matrix (a) consists mainly of myosin and RNA (b) projects to form microvilli (c) houses the centrioles (d) contains fibronectins that bind to integrins (e) has an elaborate system of cristae

REVIEW QUESTIONS

1. Describe the basic needs of all living things and explain how a cell is able to meet these needs. How does the cell theory help us understand how living things function?
2. Compare and contrast prokaryotic and eukaryotic cells.
3. Draw a chloroplast and a mitochondrion. Label the membranes and their compartments.
4. Describe the functions of each of the following: (a) ribosomes (b) smooth endoplasmic reticulum (c) Golgi complex (d) microtubules
5. Trace the path of a protein from its site of synthesis to its final destination for the following: (a) a secreted protein (b) a protein found inside a lysosome (c) a protein associated with the plasma membrane
6. Contrast microfilaments and microtubules. Compare their structures and the different roles they play in cell structure and function.
7. Describe plant cell walls. How are they formed?
8. Label the diagrams of the animal and plant cells. How is the structure of each organelle related to its function? Use Figures 4–8 and 4–9 to check your answers.

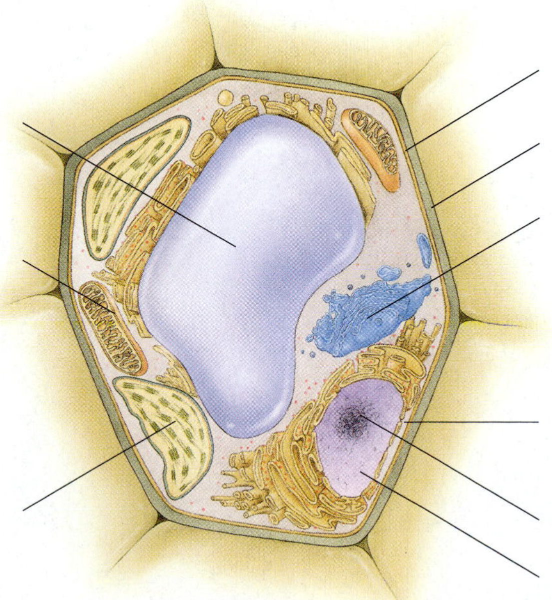

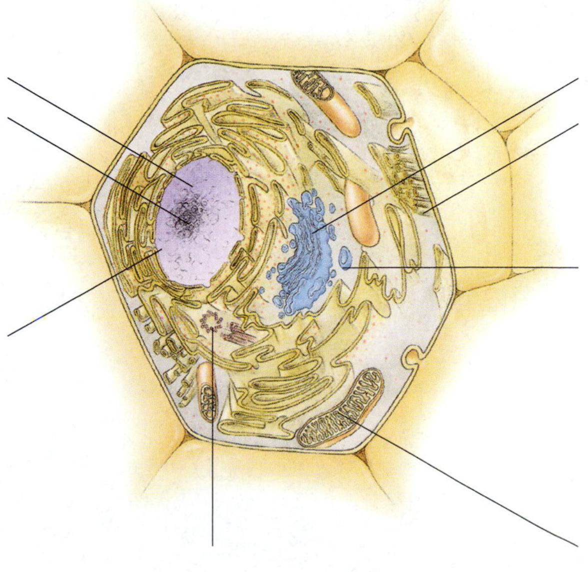

YOU MAKE THE CONNECTION

1. Why does a eukaryotic cell need both membranous organelles and fibrous cytoskeletal components?
2. Describe three examples illustrating the correlation between cell structure and function. (*Hint:* Think of mitochondrial structure.)
3. The *Acetabularia* experiments described in this chapter suggest that DNA is much more stable in the cell than is messenger RNA. Is this advantageous or disadvantageous to the cell? Why? How can *Acetabularia* live for a few days after its nucleus is removed?

RECOMMENDED READINGS

Brown, G.C. "Symbionts and Assassins." *Natural History,* Jul.-Aug. 2000. A highly readable account of the function of mitochondria, including a discussion of their role in apoptosis.

Duke, R.C., D.M. Oljcius, and J. Ding-E Young. "Cell Suicide in Health and Disease." *Scientific American,* Vol. 275, No. 6, Dec. 1996. A discussion of the role of apoptosis in regulation and disease.

Horwitz, A.F. "Integrins and Health." *Scientific American,* Vol. 276, No. 5, May 1997. Integrins are proteins in the plasma membrane that link microtubules of the cytoskeleton to the extracellular matrix.

Ingber, D. "The Architecture of Life." *Scientific American,* Vol. 278, No. 1, Jan. 1998. A discussion of the principles that guide patterns and architecture from cytoskeleton to skeleton.

Special Issue: Frontiers in Cell Biology: The Cytoskeleton. *Science,*Vol. 279, No. 5350, Jan. 1998. Several articles and an editorial discuss cytoskeletal structure and function.

Tobin, A.J., and R.E. Morel. *Asking About Cells.* Saunders College Publishing, Philadelphia, 1997. A clearly written, excellent introduction to cell biology.

Tran, P.B., and R.J. Miller. "Apoptosis: Death and Transfiguration." *Science & Medicine,* Vol. 6, No. 3, May/Jun. 1999. Mitochondria play an important role in programmed cell death.

Wallace, D.C. "Mitochondrial DNA in Aging and Disease." *Scientific American,* Vol. 277, No. 2, Aug. 1997. Mutations in mitochondrial DNA have been linked to several human diseases.

- Visit our Web site at **http://www.info.brookscole.com/solomonbergmartin** for links to chapter-related resources on the World Wide Web. Additional on-line materials relating to this chapter can also be found on our Web site.

See chapter activity on BioActive Learner CD for additional help in mastering the chapter's material. Icon location in the chapter's margins shows which topics have tutorials or simulations in the CD.

5

Biological Membranes

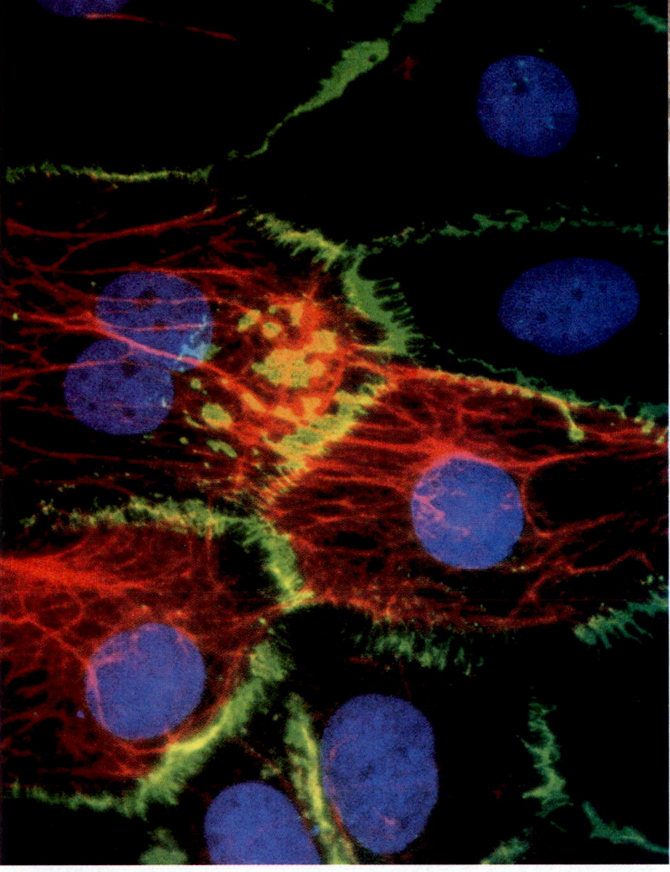

Cadherins. The human skin cells shown in this LM were grown in culture and stained with fluorescent antibodies. Cadherins, a group of membrane proteins, are seen as green belts around each cell in this sheet of cells. The nuclei appear as blue spheres; myosin in the cells appears red. *(Nancy Kedersha)*

LEARNING OBJECTIVES

After you have studied this chapter you should be able to

1. Evaluate the importance of membranes to the cell, emphasizing their various functions.
2. Make a detailed sketch of the fluid mosaic model of cell membrane structure.
3. Explain how the properties of the lipid bilayer are responsible for many of the physical properties of a cell membrane.
4. Explain how the various classes of membrane proteins associate with the lipid bilayer and discuss the different roles of membrane proteins.
5. Contrast the physical processes of simple diffusion and osmosis with the carrier-mediated physiological processes by which materials are transported across cell membranes.
6. Solve simple problems involving osmosis; for example, predict whether cells will swell or shrink under various osmotic conditions.
7. Summarize the main ways that small hydrophilic molecules can move across membranes.
8. Differentiate between the processes of facilitated diffusion and active transport and discuss the ways in which energy is supplied to active transport systems.
9. Compare endocytotic and exocytotic transport mechanisms.
10. Describe information transfer across the plasma membrane.
11. Describe the structures and compare the functions of desmosomes, tight junctions, gap junctions, and plasmodesmata.

To carry out the many chemical reactions necessary to sustain life, a cell must maintain an appropriate internal environment. The plasma membrane that surrounds every cell physically separates the cell from the outside world and defines the cell as a distinct entity. The plasma membrane helps maintain a life-supporting internal environment by regulating the passage of materials into and out of the cell. Many biologists view the evolution of biological membranes as an essential step in the origin of life. One can argue further that membranes made the evolution of complex cells possible because the extensive internal membranes of eukaryotes form multiple compartments with unique environments for highly specialized activities.

Cell biologists have found that biological membranes are not inanimate barriers; they are complex and dynamic structures made from lipid and protein molecules that are in constant motion. The unusual properties of membranes allow them to perform many functions in addition to defining the cell as a compartment and regulating passage of materials. These functions include participating in many chemical reactions, transmitting signals and information between the environment and the interior of the cell, and acting as an essential part of energy transfer and storage systems (see Chapters 7 and 8).

One exciting area of cell membrane research focuses on membrane proteins. Many proteins associated with the plasma membrane are enzymes. Others function in transport of materials or in transfer of information. Still others are important in connecting cells to one another to form tissues. Investigators are studying how membrane proteins function in health and disease. Membrane proteins known as **cadherins** are responsible for calcium-dependent adhesion between cells that form multicellular sheets, for example, the epithelium that makes up human skin (*see LM above*). An absence of these proteins has been observed when cells become malignant. Recent experimental evidence suggests that certain cadherins mediate the way that cells adhere in the early embryo and so are important in development.

In this chapter, we first consider what is known about the composition and structure of biological membranes. We survey how various materials, ranging from simple to complex molecules, and even particles, move across membranes. We then consider how information can cross the plasma membrane through a signal relay system. Finally, specialized structures that permit interactions between membranes of different cells are examined. Although much of our discussion centers on the structure and functions of plasma membranes, many of the concepts are also applicable to other cellular membranes.

BIOLOGICAL MEMBRANES ARE LIPID BILAYERS WITH ASSOCIATED PROTEINS

Long before the development of the electron microscope, it was known that membranes are composed of both lipids and proteins. Work by researchers in the 1920s and 1930s had provided clues that the core of cell membranes is composed of lipids, mostly phospholipids (see Chapter 3).

Phospholipids form bilayers in water

Phospholipids are primarily responsible for the physical properties of biological membranes. This is because certain phospholipids have unique attributes, including features that allow them to form bilayered structures. A phospholipid contains two fatty acid chains linked to two of the three carbons of a glycerol molecule (see Fig. 3–13). The fatty acid chains make up the nonpolar, hydrophobic ("water-fearing") portion of the phospholipid. Bonded to the third carbon of the glycerol is a negatively charged, hydrophilic ("water-loving") phosphate group, which in turn is linked to a polar, hydrophilic organic group. Molecules of this type, which have distinct hydrophobic and hydrophilic regions, are called **amphipathic** molecules. All lipids that make up the core of biological membranes have amphipathic characteristics.

Because one end of each phospholipid associates freely with water and the opposite end does not, the most stable orientation for them to assume in water results in the formation of a bilayer structure (Fig. 5–1*a*). This arrangement allows the hydrophilic heads of the phospholipids to be in contact with the aqueous medium, while their oily tails, the hydrophobic fatty acid chains, are buried in the interior of the structure away from the water molecules.

Amphipathic properties alone do not predict the ability of lipids to associate as a bilayer. Shape is also important. Phospholipids tend to have uniform widths; their roughly cylindrical shapes, together with their amphipathic properties, are responsible for bilayer formation. In summary, phospholipids form bilayers because the molecules have (1) two distinct regions, one strongly hydrophobic and the other strongly hydrophilic (making them strongly amphipathic), and (2) cylindrical shapes that allow them to associate with water most easily as a bilayer structure.

Many common detergents are amphipathic molecules, each containing a single hydrocarbon chain (like a fatty acid) at one end and a hydrophilic region at the other. These molecules are roughly cone-shaped, with the hydrophilic end forming the broad base and the hydrocarbon tail leading to the point. Because of their shapes, these molecules do not associate as bilayers but instead tend to form spherical structures in water (see Fig. 5–1*b*). Detergents are able to "solubilize" oil because the oil molecules associate with the hydrophobic interiors of the spheres.

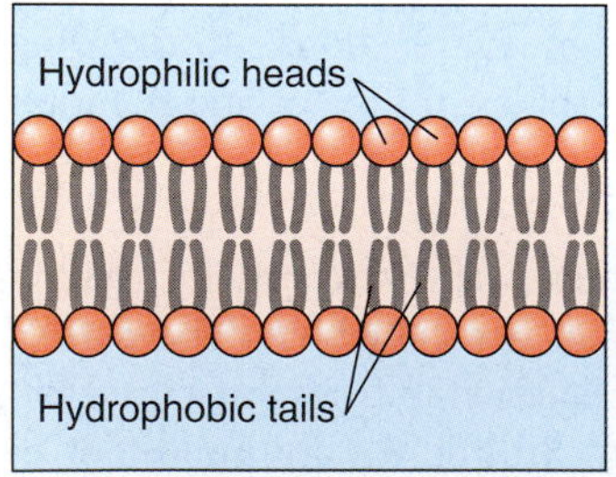

(a) Phospholipids in water

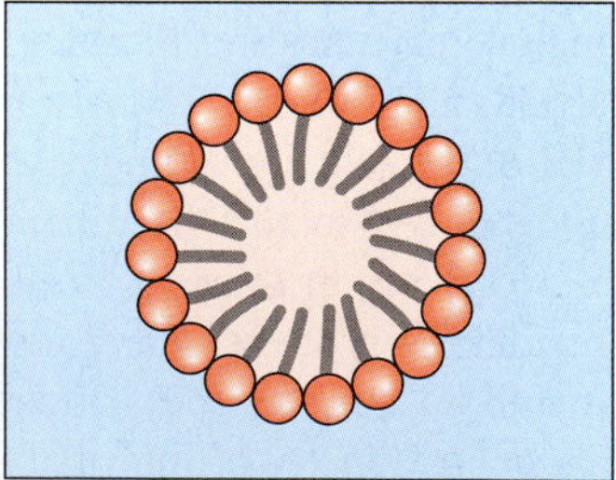

(b) Detergent in water

Figure 5–1 Lipid membranes. The ability of lipids to associate in water depends on their amphipathic properties and their shapes. (a) Phospholipids associate as bilayers in water because they are roughly cylindrical amphipathic molecules. The hydrophobic fatty acid chains associate with each other and are not exposed to the water. The hydrophilic phospholipid heads are in contact with the water. (b) Detergent molecules are roughly cone-shaped amphipathic molecules that associate in water as spherical structures.

Process of Science — Current data support a fluid mosaic model of membrane structure

By examining the plasma membrane of the mammalian red blood cell and comparing the surface area of the membrane with the total number of lipid molecules per cell, early investigators were able to calculate that the phospholipids are probably arranged so that the membrane is no more than two phospholipid molecules thick. These findings, together with other data, led H. Davson and J.F. Danielli in 1935 to propose a model in which they envisioned a membrane as a kind of "sandwich" consisting of a *lipid bilayer* (a double layer of lipid) between two protein layers (Fig. 5–2*a*). This very useful model had a great influence on the direction of membrane research for more than 20 years. Models are important in the scientific process; good ones not only explain the available data but are testable. That is, scientists can use the model to help them develop hypotheses that can be tested experimentally (see Chapter 1).

With the development of the electron microscope in the 1950s, cell biologists were able to see the plasma membrane for the first time. One of their most striking observations was how uniform and thin the membranes are. The plasma membrane is no more than 10 nm thick. The electron microscope revealed a three-layered structure, something like a railroad

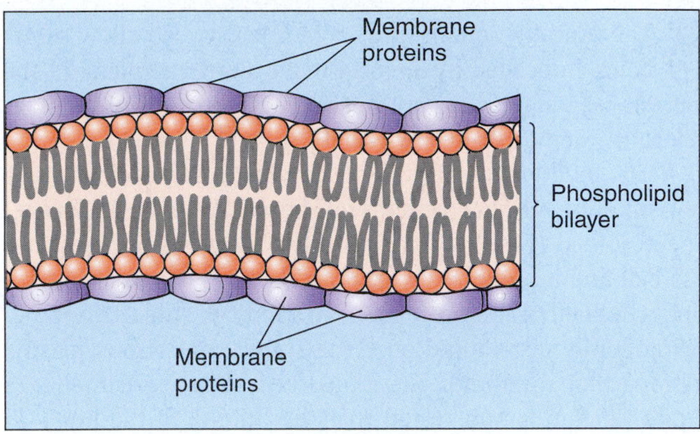

(a) The Davson-Danielli "sandwich" model

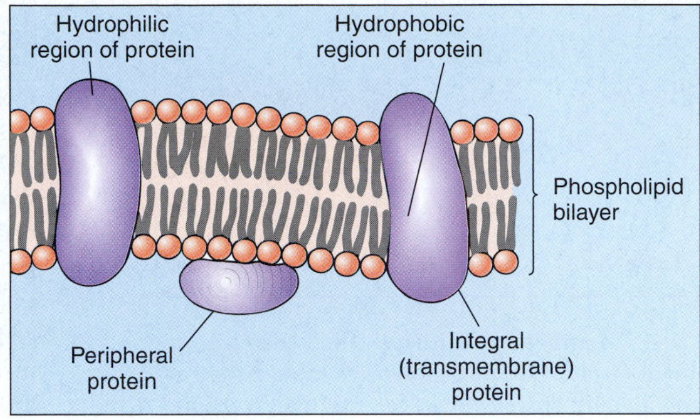

(b) Fluid mosaic model

Figure 5–2 Two models of membrane structure. **(a)** According to the Davson-Danielli model, the membrane is a sandwich of phospholipids spread between two layers of protein. Although accepted for more than 20 years, this model was shown to be incorrect. **(b)** According to the fluid mosaic model, a cell membrane is made up of a fluid lipid bilayer with a constantly changing "mosaic pattern" of associated proteins.

track, with two dark layers separated by a lighter layer (Fig. 5–3). Their findings seemed to support the protein-lipid-protein sandwich model.

During the 1960s, a paradox emerged regarding arrangement of the proteins. It was widely assumed that membrane proteins were uniform and had shapes that would allow them to lie like thin sheets on the membrane surface. When proteins from membranes were purified by cell fractionation, however, they were found to be far from uniform; in fact, they varied widely in composition and size. Some proteins were much larger than investigators had imagined. How could these proteins be arranged to fit within a surface layer of a membrane less than 10 nm thick? At first, some investigators attempted to address these objections by modifying the model with the hypothesis that the proteins on the membrane surfaces were a flattened, extended form, perhaps a β-pleated sheet (see Chapter 3).

Cell interior
Plasma membrane
Outside cell
0.1 μm

Figure 5–3 TEM of the plasma membrane of a mammalian red blood cell. The plasma membrane separates the cytoplasm *(darker region)* from the external environment *(lighter region)*. The hydrophilic heads of the phospholipids are seen as the parallel dark lines, while the hydrophobic tails are visible as the light zone between them. *(Omikron/Photo Researchers, Inc.)*

Other cell biologists found that instead of having sheetlike structures, many membrane proteins are rounded, or globular. Studies of a number of individual membrane proteins showed that one region (or domain) of the molecule could always be found on one side of the bilayer, while another part of the protein might be located on the opposite side. It appeared that, rather than forming a thin surface layer, many membrane proteins extend completely through the lipid bilayer. Thus, membranes appear to contain many different types of proteins of different shapes and sizes that are associated with the bilayer in a mosaic pattern.

In 1972, S.J. Singer and G.L. Nicolson proposed a model of membrane structure that represented a synthesis of the known properties of biological membranes. According to their **fluid mosaic model,** a cell membrane consists of a fluid bilayer of phospholipid molecules in which the proteins are embedded or otherwise associated, much like the tiles in a mosaic picture. This mosaic pattern is not static, however, because the positions of the proteins are constantly changing as they move about like icebergs in a fluid sea of phospholipids. This model has provided great impetus to research; it has been repeatedly tested and has been shown to accurately predict the properties of many kinds of cell membranes. Figure 5–2*b* depicts the plasma membrane of a eukaryotic cell; prokaryotic plasma membranes are discussed in Chapter 23.

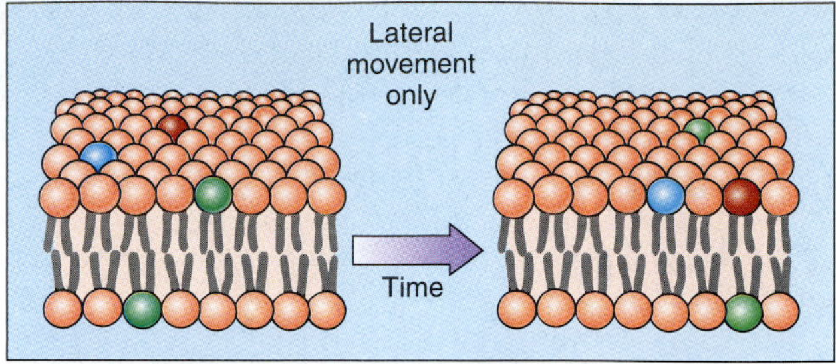

Figure 5-4 Membrane fluidity. The ordered arrangement of phospholipid molecules makes the cell membrane a liquid crystal. However, the hydrocarbon chains are in constant motion, allowing each molecule to move laterally on the same side of the bilayer.

Biological membranes are two-dimensional fluids

An important physical property of phospholipid bilayers is that they behave as *liquid crystals.* The bilayers are crystal-like in that the lipid molecules form an ordered array with the heads on the outside and fatty acid chains on the inside; they are liquid-like in that, despite the orderly arrangement of the molecules, their hydrocarbon chains are in constant motion. Thus, molecules are free to rotate and can move laterally within their single layer (Fig. 5–4). Such movement gives the bilayer the property of a *two-dimensional fluid.* Under normal conditions this means that a single phospholipid molecule can travel laterally across the surface of a eukaryotic cell in seconds.

Process of Science

The fluid-like qualities of lipid bilayers also allow molecules embedded in them to move along the plane of the membrane (as long as they are not anchored in some way). This was elegantly demonstrated by David Frye and Michael Ediden in 1970. They conducted experiments in which they followed the movement of membrane proteins on the surface of two cells that had been joined (Fig. 5–5). When the plasma membranes of a mouse cell and a human cell are fused, within minutes at least some of the membrane proteins from each cell migrate and become randomly distributed over the single continuous plasma membrane that surrounds the joined cells. The experiments of Frye and Ediden demonstrated that the fluidity of the lipids in the membrane allows many of the proteins to move, producing an ever-changing configuration.

If a membrane is to function properly, its lipids must be in a state of optimal fluidity. The structure of a membrane is weakened if its lipids are too fluid. On the other hand, it has been shown that many membrane functions, such as the transport of certain substances, are inhibited or cease if the lipid bilayer is too rigid. At normal temperatures, cell membranes are fluid. However, the motion of the fatty acid chains is slowed at low temperatures. If the temperature decreases to a critical point, the membrane is converted to a more solid gel state.

Certain properties of membrane lipids have significant effects on the fluidity of the bilayer. Recall from Chapter 3 that molecules are free to rotate around single carbon-to-carbon covalent bonds. Because most of the bonds in hydrocarbon

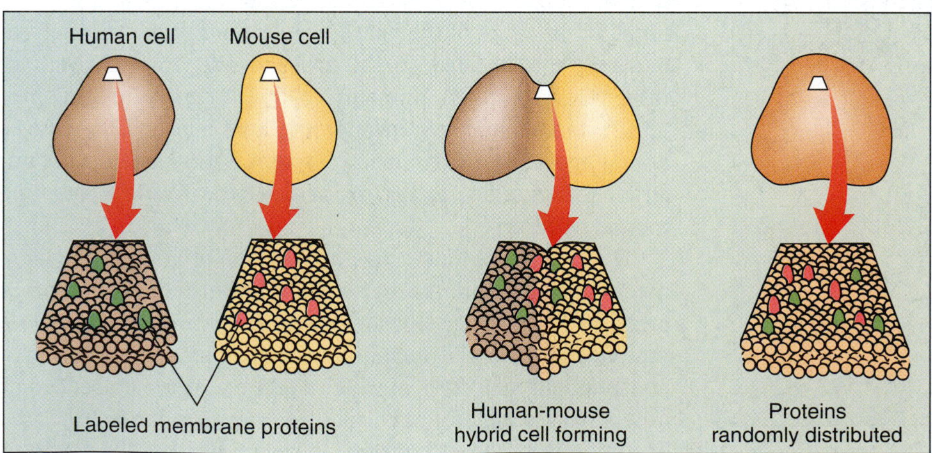

Figure 5-5 Mobility of membrane proteins. An elegant series of experiments by Frye and Ediden demonstrated that at least some membrane proteins are highly mobile entities in a two-dimensional fluid. Membrane proteins of mouse cells and human cells were labeled with fluorescent dye markers in two different colors. When the plasma membranes of a mouse cell and a human cell were fused, mouse proteins were observed migrating to the human side and human proteins to the mouse side. After a short time mouse and human proteins became randomly distributed on the cell surface.

chains are single bonds, the chains themselves can undergo rapid twisting motions that increase as the temperature increases.

The fluid state of the membrane depends on its component lipids. You have probably noticed that when melted butter is left at room temperature, it solidifies. Vegetable oils, however, remain liquid at room temperature. Recall from our discussion of fats in Chapter 3 that animal fats like butter are high in saturated fatty acids that lack double bonds. In contrast, a vegetable oil may be polyunsaturated with most of its fatty acid chains having two or more double bonds. Double bonds produce "bends" in the molecules that prevent the hydrocarbon chains from coming close together and interacting through van der Waals forces (see Chapter 3). In this way unsaturated fats lower the temperature at which oil or membrane lipids solidify.

Many organisms have regulatory mechanisms that allow them to maintain their membranes in an optimally fluid state. For example, some organisms that are unable to maintain a constant internal temperature can compensate for temperature changes by altering the fatty acid content of their membrane lipids. When grown at colder temperatures, such organisms are found to have relatively high proportions of unsaturated fatty acids in their membrane lipids.

Some membrane lipids have the ability to help stabilize membrane fluidity within certain limits. One such "fluidity buffer" is cholesterol, a steroid found in animal cell membranes. A cholesterol molecule is largely hydrophobic but is slightly amphipathic owing to the presence of a single hydroxyl group (see Fig. 3–15*a*). This hydroxyl group associates with the hydrophilic heads of the phospholipids; the hydrophobic remainder of the cholesterol molecule fits between the fatty acid hydrocarbon chains (Fig. 5–6).

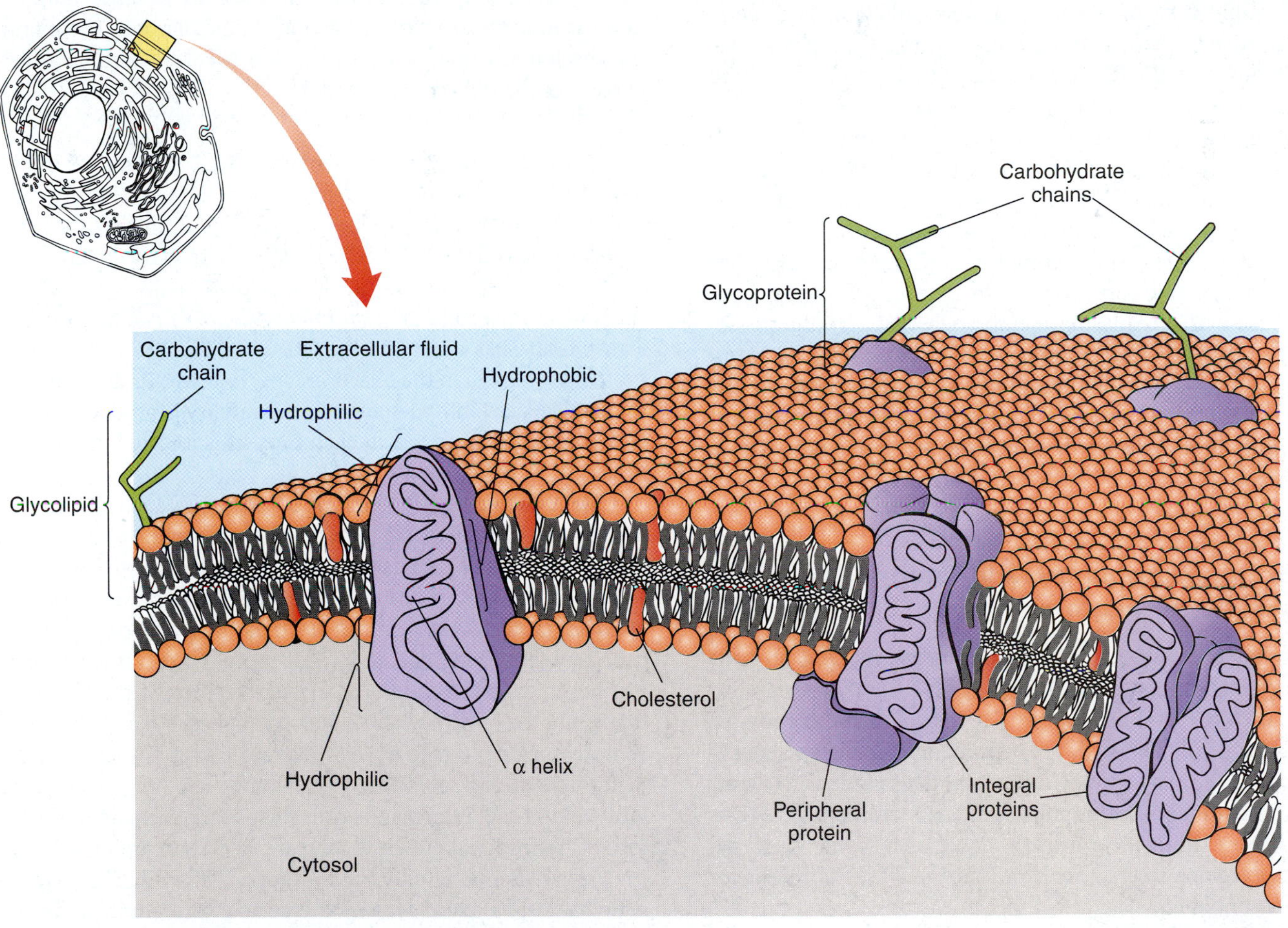

Figure 5–6 Structure of the plasma membrane according to the fluid mosaic model. Although the lipid bilayer consists mainly of phospholipids, other lipids such as cholesterol are present. Peripheral proteins are loosely associated with the bilayer, while integral proteins are tightly bound. The integral proteins shown here are transmembrane proteins that extend through the bilayer. They have hydrophilic regions on both sides of the bilayer, connected by a membrane-spanning α-helix with hydrophobic amino acid side chains *(not shown)*. Glycolipids (carbohydrates attached to lipids) and glycoproteins (carbohydrates attached to proteins) are exposed on the extracellular surface; they play roles in cell recognition and adhesion.

At low temperatures the cholesterol molecules act as "spacers" between the hydrocarbon chains, restricting van der Waals interactions that would promote solidifying. Cholesterol also helps prevent the membrane from becoming weakened or unstable at higher temperatures. This is because the cholesterol molecules interact strongly with the portions of the hydrocarbon chains closest to the phospholipid head. This interaction restricts motion in these regions. Plant cells have steroids other than cholesterol that carry out similar functions.

Biological membranes fuse and form closed vesicles

Lipid bilayers, particularly those in the liquid-crystalline state, have additional important physical properties. Bilayers tend to resist forming free ends; as a result, they are self-sealing and under most conditions spontaneously round up to form closed vesicles. Lipid bilayers are also flexible, allowing cell membranes to change shape without breaking. Finally, under appropriate conditions lipid bilayers have the ability to fuse with other bilayers.

Membrane fusion is an important cellular process. When a vesicle fuses with another membrane, both membrane bilayers and their compartments become continuous. Various transport and secretory vesicles form from and also merge with membranes of the ER and Golgi complex, allowing materials to be transferred from one compartment to another. A secretory vesicle can fuse with the plasma membrane when a product is secreted from the cell. This process is known as *exocytosis.* In *endocytosis,* large molecules are brought into the cell from the outside by the formation of vesicles from the plasma membrane. Both exocytosis and endocytosis are discussed later in this chapter.

Membrane proteins include integral and peripheral proteins

The two major classes of membrane proteins, integral proteins and peripheral proteins, are defined by how tightly they are associated with the lipid bilayer (see Fig. 5–6). **Integral membrane proteins** are firmly bound to the membrane. Cell biologists usually can release them only by disrupting the bilayer with detergents. These proteins are amphipathic. Their hydrophilic regions extend out of the cell or into the cytoplasm, while their hydrophobic regions interact with the fatty acid tails of the membrane phospholipids.

Some integral proteins do not extend all the way through the membrane. Many others, called **transmembrane proteins,** extend completely through the membrane. Some span the membrane only once, while others wind back and forth as many as 24 times. The most common kind of transmembrane protein is an α-helix (see Chapter 3) with hydrophobic amino acid side chains projecting out from the helix into the hydrophobic region of the lipid bilayer.

Peripheral membrane proteins are not embedded in the lipid bilayer. They are located on the inner or outer surfaces of the plasma membrane, usually bound to exposed regions of integral proteins by noncovalent interactions. Peripheral proteins can be easily removed from the membrane without disrupting the structure of the bilayer.

Proteins are oriented asymmetrically across the bilayer

Process of Science One of the most remarkable demonstrations that proteins are actually embedded in the lipid bilayer comes from freeze-fracture electron microscopy (Fig. 5–7), which enables investigators to literally see the membrane from "inside out." When cellular membranes are examined in this way, numerous particles are observed on the fracture faces. These particles are clearly integral membrane proteins because they are never seen in freeze-fractured artificial lipid bilayers. These findings profoundly influenced Singer and Nicolson in their development of the fluid mosaic model.

When the two sides of a membrane are compared (as in Fig. 5–7), large numbers of particles are found on one side and very few on the other. This does not necessarily mean that there are more proteins on one side of the membrane than on the other but rather that most are more firmly attached to a given side. Thus, the protein molecules are *asymmetrically oriented.* Each side of a membrane has different characteristics because each type of protein is oriented in the bilayer in only one way. Proteins are not randomly placed into membranes; asymmetry is produced by the highly specific way in which each protein is inserted into the bilayer.

Membrane proteins that will become part of the inner surface of the plasma membrane are manufactured by free ribosomes and move to the membrane through the cytoplasm. Membrane proteins that will be associated with the cell's outer surface are manufactured like proteins destined to be exported from the cell. As discussed in Chapter 4, these proteins are initially formed by ribosomes on the rough ER. They pass through the ER membrane into the lumen, where sugars are added, making them **glycoproteins.** Only a part of each protein passes through the ER membrane, so each completed protein has some regions that are located in the ER lumen and other regions that remain in the cytosol. Enzymes that attach the sugars to certain amino acids on the protein are found only in the lumen of the ER. Thus, carbohydrates can be added only to the parts of proteins that are located in that compartment.

Follow the vesicle budding and membrane fusion events that are part of the transport process from top to bottom in Figure 5–8. You can see that the same region of the protein that protruded into the ER lumen is also transferred to the lumen of the Golgi complex. There additional enzymes further modify the carbohydrate chains. Within the Golgi complex, the glycoprotein is sorted and directed to the plasma membrane. The modified re-

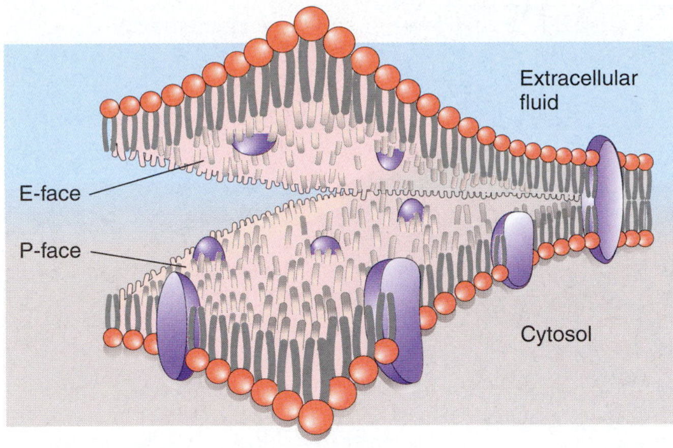

(a)

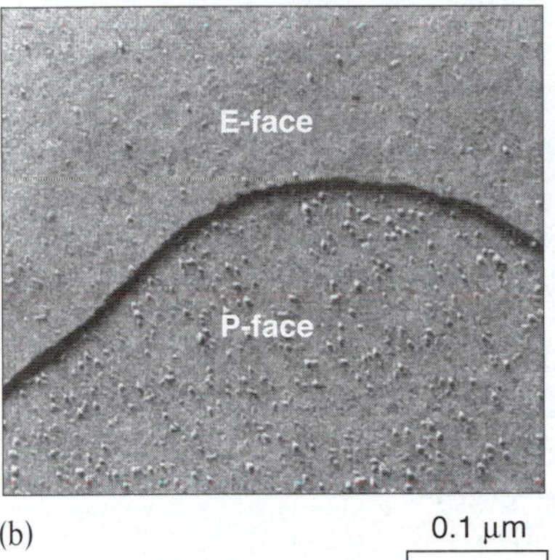

(b) 0.1 μm

Figure 5–7 Asymmetry of the plasma membrane. **(a)** In the freeze-fracture method, the path of membrane cleavage is along the hydrophobic interior of the lipid bilayer, resulting in two complementary fracture faces: (1) an inner half-membrane presenting the P-face (or protoplasmic face), from which project most of the membrane proteins, and (2) a relatively smooth, outer half-membrane presenting the E-face (or external face), which shows fewer protein particles. In a good fracture, particles are visible on both of the inside faces of the fractured membrane, as shown here. These particles are transmembrane proteins inserted into the lipid bilayer. Freeze-fractured bilayers of lipids alone do not have particles on the fracture planes. **(b)** A freeze-fracture TEM. Notice the greater number of proteins on the P-face of the membrane. *(b, D.W. Fawcett)*

gion of the protein remains inside a membrane compartment of a secretory vesicle as it buds from the Golgi complex. When the secretory vesicle fuses with the plasma membrane, the carbohydrate chain becomes the part of the membrane protein that extends to the exterior of the cell surface.

ER lumen ⟶ transport vesicle ⟶ vesicles in Golgi (transport to successive compartments) ⟶ secretory vesicle ⟶ plasma membrane

Membrane proteins function in transport, in information transfer, and as enzymes

Why should a membrane such as the plasma membrane illustrated in Figure 5–6 require so many different proteins? This diversity reflects the multitude of activities that take place in or on the membrane. Generally, plasma membrane proteins fall into several broad functional categories. Some membrane proteins anchor the cell to its substrate. For example, integrins, described in Chapter 4, attach to the extracellular matrix while simultaneously binding to microfilaments inside the cell (Fig. 5–9*a*). They also serve as receptors, or docking sites, for proteins of the extracellular matrix.

Many membrane proteins are involved in the transport of molecules across the membrane. Some form channels that selectively allow the passage of specific ions or molecules (Fig. 5–9*b*). Other proteins form pumps that use ATP to actively transport solutes across the membrane (Fig. 5–9*c*).

Certain membrane proteins are enzymes that catalyze reactions that take place near the cell surface (Fig. 5–9*d*). In some membranes, for example, mitochondrial or chloroplast membranes, enzymes may be organized in a sequence to regulate a series of reactions such as occurs in cellular respiration or photosynthesis.

Some membrane proteins are receptor proteins that receive information from other cells. For example, cells receive hormonal signals from endocrine cells. This information may be transmitted to the cell interior by *signal transduction,* discussed later in this chapter (Fig. 5–9*e*).

Membrane proteins can serve as identification tags that permit other cells to recognize them. Cells that recognize one another may connect to form a tissue. Human cells have distinctive major histocompatibility complex (MHC) receptors that identify them as part of a particular individual. Human cells recognize the surface proteins, or antigens, of bacterial cells as foreign. Antigens stimulate immune defenses that destroy the bacteria (Fig. 5–9*f*). Some membrane proteins form junctions between adjacent cells (Fig. 5–9*g*). These proteins may also serve as anchoring points for networks of cytoskeletal elements.

CELL MEMBRANES ARE SELECTIVELY PERMEABLE

Whether a membrane permits a substance to pass through it depends on the size and charge of the substance and on the composition of the membrane. A membrane is said to be *permeable* to a given substance if it permits that substance to pass through, and impermeable if it does not. A **selectively permeable membrane** allows some but not other substances to pass through it readily. In general, biological membranes are most permeable to small molecules and to lipid-soluble substances able to pass through the hydrophobic interior of the bilayer.

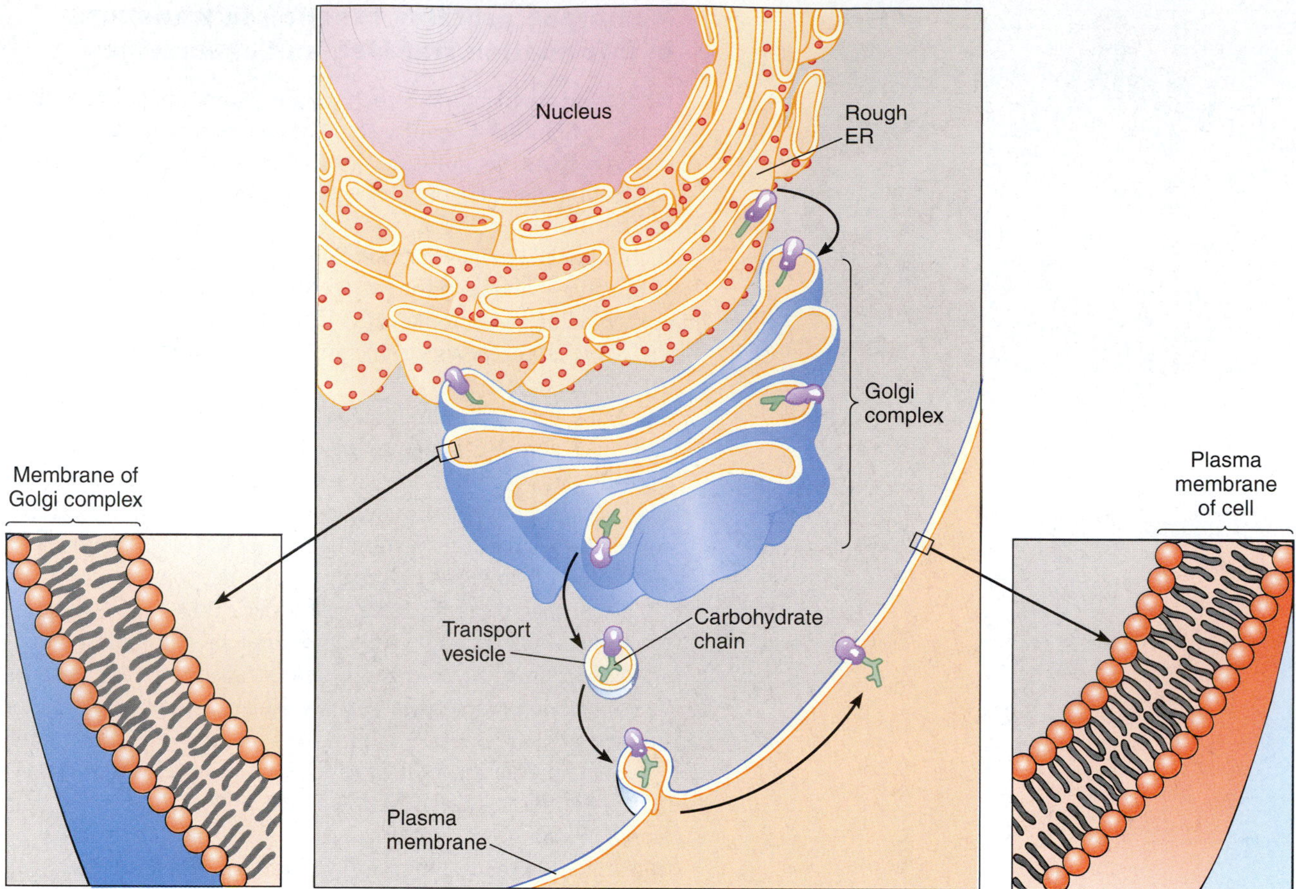

Figure 5-8 Formation of a plasma membrane protein. The surface of the rough ER membrane that faces the lumen of the rough ER also faces the lumen of the Golgi complex and vesicles. When a vesicle fuses with the plasma membrane, its inner surface becomes the extracellular surface of the plasma membrane. Note that the orientation of a protein in the plasma membrane is also a consequence of the pathway of its synthesis and transport in the cell. Carbohydrates added to proteins in the ER and then modified in the Golgi complex are associated with the extracellular surface of the plasma membrane.

Although they are polar (and therefore not lipid-soluble), water molecules can rapidly cross a fluid lipid bilayer. They can pass through a membrane in part because they are small enough to pass through gaps that occur as a fatty acid chain momentarily moves out of the way. Gases such as oxygen, carbon dioxide, and nitrogen; small polar molecules like glycerol; plus larger, nonpolar (hydrophobic) substances such as hydrocarbons also move through the lipid bilayer rapidly. Slightly larger polar molecules, such as glucose, and charged ions of any size pass through the bilayer much more slowly.

Although the bilayer is relatively impermeable to ions, cells must be able to move ions, as well as large and small polar molecules such as amino acids and sugars, across membranes. The permeability of membranes to those substances is due primarily to the activities of specialized membrane proteins. Biological membranes surrounding cells, nuclei, vacuoles, mitochondria, chloroplasts, and other organelles are selectively permeable to different types of molecules.

In response to varying environmental conditions or cellular needs, a plasma membrane may be a barrier to a particular substance at one time and actively promote its passage at another time. Changes in cell volume or ion distribution can critically affect cellular functions. Even routine activity, such as taking in nutrients, can cause chemical imbalance. By regulating chemical traffic across its plasma membrane, a cell can control its volume and its internal ionic and molecular composition, which can be quite different from that on the outside.

In the nonliving world, materials move passively by physical processes such as diffusion. In living cells, some particles cross the bilayer by diffusion. However, complex adaptations have evolved that rapidly move materials in and out of the cell. These physiological transport mechanisms, including active transport, exocytosis, and endocytosis (discussed later in this chapter), require a direct expenditure of metabolic energy by the cell.

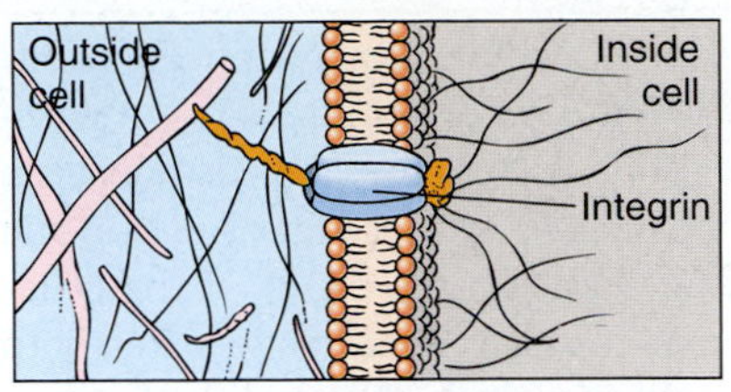

(a) **Anchoring cell.** Some membrane proteins, for example, integrins, anchor the cell to the extracellular matrix; they also connect to microfilaments within the cell.

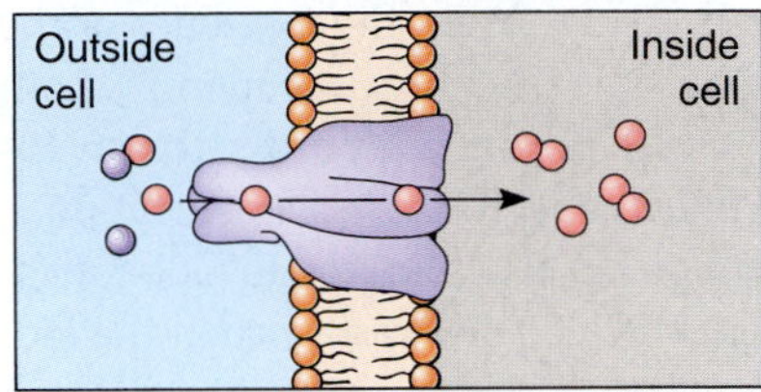

(b) **Passive transport.** Certain proteins form channels that allow selective passage of ions or molecules.

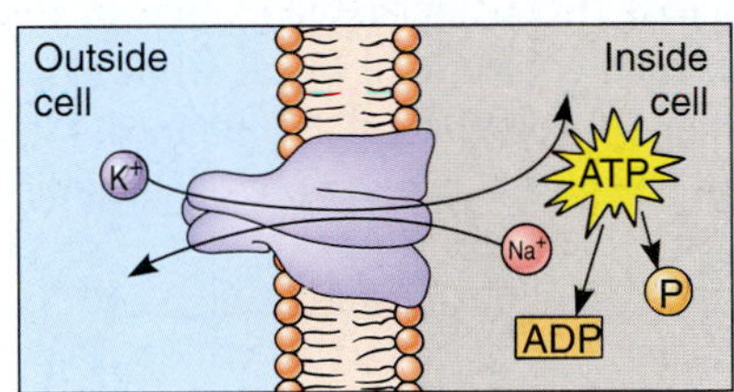

(c) **Active transport.** Some transport proteins pump solutes across the membrane, a process that requires a direct input of energy.

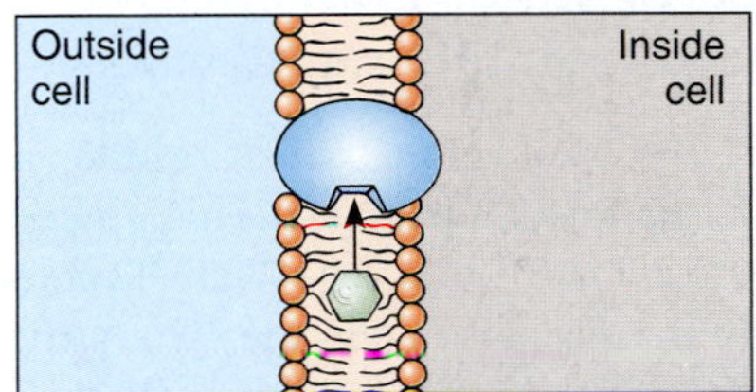

(d) **Enzymatic activity.** Many membrane-bound enzymes catalyze reactions that take place within or along the surface of the membrane.

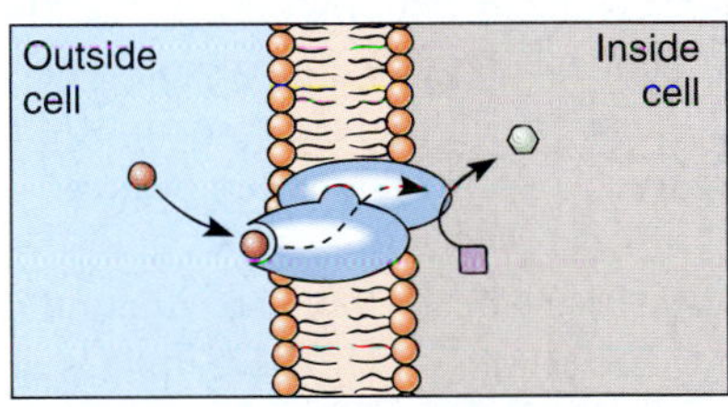

(e) **Signal transduction.** Some receptors bind with signal molecules such as hormones and transmit information into the cell by signal transduction.

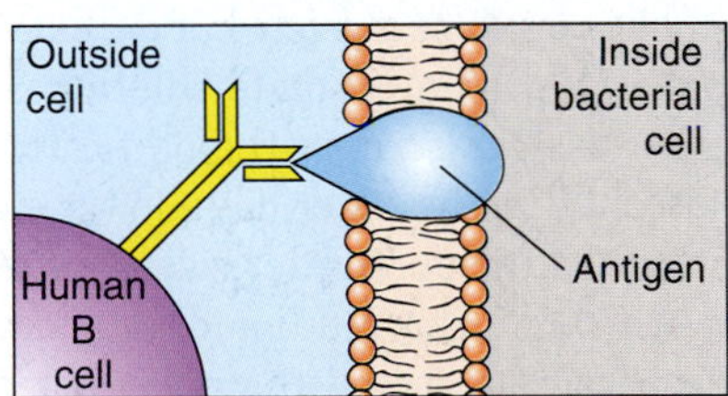

(f) **Cell recognition.** Some receptor proteins function as identification tags. For example, bacterial cells have surface proteins, or antigens, that human cells recognize as foreign. Antigens stimulate immune defenses that destroy the bacteria.

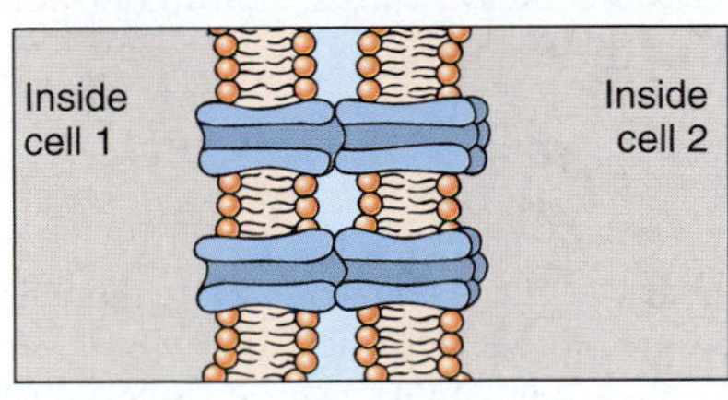

(g) **Junction between cells.** Cell adhesion proteins attach membranes of adjacent cells.

Figure 5-9 Functions of membrane proteins. Membrane proteins function **(a)** in anchoring the cell, **(b)** in passive transport, **(c)** in active transport, **(d)** as enzymes, **(e)** to transmit information into the cell by signal transduction, **(f)** in cell recognition, and **(g)** in forming junctions between cells.

Random motion of particles leads to diffusion

Some substances pass into or out of cells and move about within cells by simple **diffusion,** a physical process based on random motion. All atoms and molecules possess kinetic energy, or energy of motion, at temperatures above absolute zero (0 Kelvin, −273° Celsius, or −459.4° F). Matter may exist as a solid, liquid, or gas, depending on the freedom of movement of its constituent particles. The particles of a solid are closely packed, and the forces of attraction between them allow them to vibrate but not to move around. In a liquid the particles are farther apart; the intermolecular attractions are weaker, and the particles move about with considerable freedom. In a gas the particles are so far apart that intermolecular forces are negligible; molecular movement is restricted only by the walls of the container that encloses the gas. This means that atoms and molecules in liquids and gases move in a kind of "random walk," changing directions as they collide.

Although the movement of the individual particles is undirected and unpredictable, we can nevertheless make predictions about the behavior of groups of particles. If the particles (atoms, ions, or molecules) are not evenly distributed, then at least two regions exist: one with a higher concentration of particles and the other with a lower concentration. Such a difference in the concentration of a substance from one place to another establishes a **concentration gradient.**

In the phenomenon of diffusion, the random motion of particles results in their net movement "down" their own concentration gradient (from the region of higher concentration to the one of lower concentration). This does not mean that individual particles are prohibited from moving "against" the gradient. However, because there are initially more particles in the region of high concentration, it logically follows that more particles move randomly from there into the low-concentration region than vice versa.

Diffusion can occur rapidly over very short distances. The rate of diffusion is determined by the movement of the particles, which in turn is a function of their size and shape, their electrical charges, and the temperature. As the temperature rises, particles move faster and the rate of diffusion increases.

Particles of different substances in a mixture diffuse independently of each other. If particles are not added to or removed from the system, a state of **equilibrium,** a condition of no net change in the system, is ultimately reached. At equilibrium the particles are uniformly distributed.

More commonly in organisms, equilibrium is never attained. For example, carbon dioxide continually forms within a human cell as sugars and other molecules are metabolized during the process of aerobic respiration. Carbon dioxide readily diffuses across the plasma membrane but then is rapidly removed by the blood. This limits the opportunity for the molecules to reenter the cell, so a sharp concentration gradient of carbon dioxide molecules always exists across the membrane. Two special cases of diffusion are dialysis and osmosis.

How does kidney dialysis depend on diffusion? Dialysis is the diffusion of a solute across a selectively permeable membrane. To demonstrate dialysis, one can fill a cellophane bag[1] with a sugar solution and immerse it in a beaker of pure water (Fig. 5–10). If the cellophane membrane is permeable to sugar as well as to water, both sugar and water molecules will pass through it and the concentrations of sugar molecules in the water on the two sides of the membrane will eventually become equal. Subsequently, both solute and water molecules will continue to cross the membrane, but there will be no net change in their concentrations.

Dialysis has an important practical application in kidney dialysis, a process used to cleanse the blood of wastes when the kidneys do not function properly. Waste products in the form of small molecules diffuse readily across the artificial membrane in the dialysis apparatus. Wastes are removed from the blood, while blood cells, blood proteins, and other large molecules are retained. (See *Focus On: Kidney Disease* in Chapter 46 for a more detailed discussion of dialysis.)

Osmosis is diffusion of water (solvent) across a selectively permeable membrane

The selective permeability of cell membranes results in another special kind of diffusion called **osmosis.** This process involves the movement of *solvent* (in this case, water) molecules through a selectively permeable membrane. The water molecules pass freely in both directions, but, as in all types of diffusion, *net* movement is from the region where the water molecules are more concentrated to the region where they are less concentrated. Most solute molecules cannot diffuse freely through selectively permeable membranes of the cell.

The principles involved in osmosis can be illustrated using an apparatus called a U-tube (Fig. 5–11). The U-tube is divided into two sections by a selectively permeable membrane that allows solvent (water) molecules to pass freely but excludes solute molecules (e.g., sugar, salt). A water/solute solution is placed on one side, and pure water is placed on the other. The side containing the solute dissolved in the water has a lower effective concentration of water than the pure water side. This is because the solute particles, which are charged (ionic) or polar, interact with the partial electrical charges on the polar water molecules. Many of the water molecules are thus "bound up" and no longer free to diffuse across the membrane.

[1] Cellophane is often used as an "artificial membrane." It is made from cellulose and can be formed into a thin sheet that allows the passage of water molecules. Such membranes can be constructed with varying permeability to different solutes and can be quite different from biological membranes in their permeability. (When cellophane is used to package foods, it is coated to make it impermeable to air and water.)

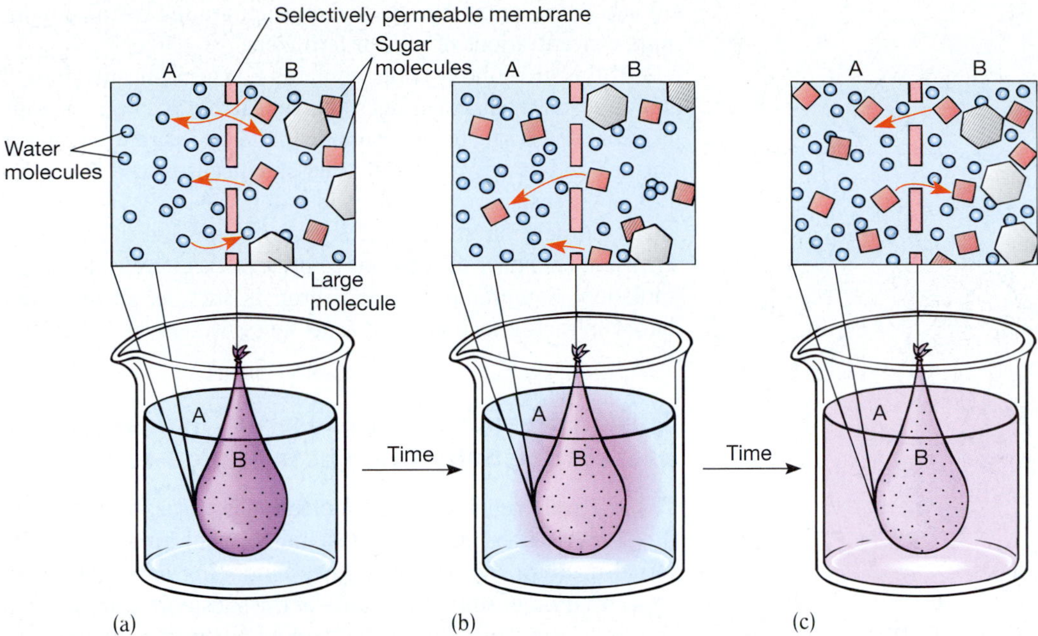

Figure 5–10 Dialysis. **(a)** A cellophane bag, filled with a mixture of sugar, water, and large molecules such as proteins, is immersed in a beaker of pure water. The cellophane acts as a selectively permeable membrane, permitting passage of the sugar and water molecules *(arrows)* but preventing passage of larger molecules. **(b)** The arrows indicate net movement of sugar molecules through the membrane into the water of the beaker. **(c)** Eventually the sugar becomes distributed equally between the two compartments. Although sugar and water molecules continue to diffuse back and forth *(arrows),* net movement is zero.

Because of the difference in effective water concentration, there is net movement of water molecules from the pure water side (with a high effective concentration of water) to the water/solute side (with a lower effective concentration of water). As a result, the fluid level drops on the pure water side and rises on the water/solute side. Because the solute molecules do not diffuse across the membrane, equilibrium is never attained. Net movement of water continues, and the fluid level continues to rise on the side containing the solute. The weight of the rising column of fluid eventually exerts enough pressure to stop further changes in fluid levels, although water molecules continue to pass through the selectively permeable membrane in both directions.

We define the **osmotic pressure** of a solution as the tendency of water to move into that solution by osmosis. In our U-tube example, we could measure the osmotic pressure by inserting a piston on the water/solute side of the tube and measuring how much pressure must be exerted by the piston to prevent the rise of fluid on that side of the tube. A solution with a high solute concentration has a low effective water concentration and a high osmotic pressure; conversely, a solution with a low solute concentration has a high effective concentration of water and a low osmotic pressure.

Two solutions may be isotonic, or one may be hypertonic and the other hypotonic

Dissolved in the fluid compartment of every living cell are salts, sugars, and other substances that give that fluid a specific osmotic pressure. When a cell is placed in a fluid with exactly the same osmotic pressure, no net movement of water molecules occurs, either into or out of the cell. The cell neither swells nor shrinks. Such a fluid is said to be **isotonic,** of equal solute concentration, to the fluid within the cell (Table 5–1). Normally, our blood plasma (the fluid component of blood) and all our other body fluids are isotonic to our cells; they contain a concentration of water equal to that in the cells. A solution of 0.9% sodium chloride (sometimes called *physiological saline*) is isotonic to the cells of humans and other mammals. Human red blood cells placed in 0.9% sodium chloride neither shrink nor swell (Fig. 5–12*a*).

If the surrounding fluid has a concentration of dissolved substances greater than the concentration within the cell, it has a higher osmotic pressure than the cell and is said to be **hypertonic** to the cell. Because the hypertonic solution has a lower effective water concentration, a cell placed in such a solution shrinks as it loses water by osmosis. Human red blood cells

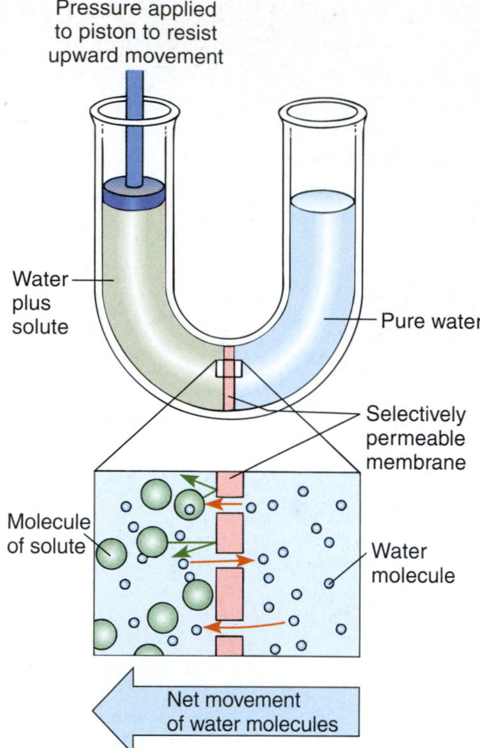

Figure 5–11 Osmosis. The U-tube contains pure water on the right and water plus a solute on the left, separated by a selectively permeable membrane. Water molecules are able to cross the membrane in both directions *(red arrows)*. Solute molecules are unable to cross *(green arrows)*. The fluid level rises on the left and falls on the right because net movement of water *(blue arrow)* is to the left. The force that must be exerted by the piston to prevent the rise in fluid level is equal to the osmotic pressure of the solution.

placed in a solution of 1.3% sodium chloride shrink and are said to be *crenated* (Fig. 5–12*b*). If a cell that has a cell wall is placed in a hypertonic medium, it loses water to its surroundings. Its contents shrink, and the plasma membrane separates from the cell wall, a process known as **plasmolysis** (Fig. 5–13). Plasmolysis occurs in plants when the soil or water around them contains high concentrations of salts or fertilizers.

If the surrounding fluid contains a lower concentration of dissolved materials than does the cell, it has a lower osmotic pressure and is said to be **hypotonic** to the cell; water then enters the cell and causes it to swell. Red blood cells placed in a solution of 0.6% sodium chloride gain water, swell (Fig. 5–12*c*), and may eventually burst. Many cells that normally live in hypotonic environments have adaptations to prevent excessive water accumulation. For example, certain protists such as *Paramecium* have contractile vacuoles that expel excess water (see Fig. 24–6).

Turgor pressure is the internal hydrostatic pressure usually present in walled cells

The relatively rigid cell walls of plant cells, algae, bacteria, and fungi enable these cells to withstand, without bursting, an external medium that is very dilute, containing only a very low concentration of solutes. Because of the substances dissolved in the cytoplasm, the cells are hypertonic to the outside medium (conversely, the outside medium is hypotonic to the cytoplasm). Water moves into the cells by osmosis, filling their central vacuoles and distending the cells. The cells swell, building up a pressure, termed **turgor pressure,** against the rigid cell walls (Fig. 5–13*a*). The cell walls can be stretched only slightly, and a steady state is reached when their resistance to stretching prevents any further increase in cell size and thereby halts the net movement of water molecules into the cells (although, of course, molecules continue to move back and forth across the plasma membrane). Turgor pressure in the cells is an important factor in providing support for the body of nonwoody plants. Thus, lettuce becomes limp in a salty salad dressing, and a picked flower wilts from lack of water.

Carrier-mediated transport requires special integral membrane proteins

The movement of water through the plasma membrane by osmosis is sometimes insufficient to account for the rapid bulk passage of water that occurs under some conditions. Recently,

TABLE 5–1 Osmotic Terminology

Solute Concentration in Solution A	Solute Concentration in Solution B	Tonicity	Direction of Net Movement of Water
Greater	Less	A hypertonic to B B hypotonic to A	B to A
Less	Greater	B hypertonic to A A hypotonic to B	A to B
Equal	Equal	A and B are isotonic to each other	No net movement

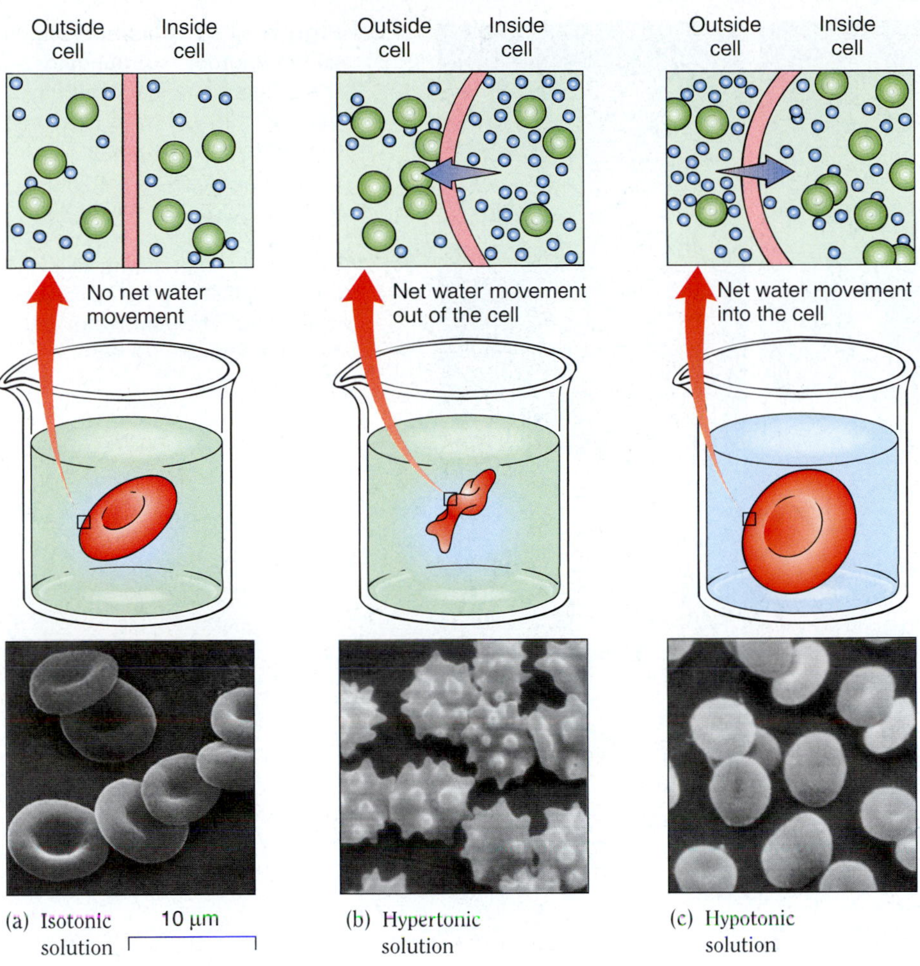

Figure 5-12 How animal cells respond to osmotic pressure differences. **(a)** When a cell is placed in an isotonic solution, water molecules pass in and out of the cell, but the net movement is zero. **(b)** When a cell is placed in a hypertonic solution, there is a net movement of water out of the cell *(arrow)*, and the cell becomes dehydrated, crenated (shrunken), and may die. **(c)** When a cell is placed in a hypotonic solution, there is a net movement of water molecules into the cell *(arrow)*, causing the cell to swell or even burst. *(SEMs of human red blood cells courtesy of Dr. R.F. Baker, University of Southern California Medical School)*

investigators have identified **aquaporin-1,** an integral membrane protein that functions as a gated water channel. In certain types of cells, for example red blood cells, or under certain environmental conditions in plant cells, aquaporin-1 permits the rapid passage of water. The channel formed by aquaporin-1 allows the passage of water molecules, but it excludes protons. In some cells, the number of these channels apparently changes in response to specific signals. Other aquaporins, the focus of current research, appear to regulate the passage of other molecules.

Cells also must continually acquire essential polar nutrient molecules such as glucose and amino acids. The lipid bilayer of the plasma membrane is relatively impermeable to most large polar molecules. This is advantageous to cells for a number of reasons. Most of the compounds required in metabolism are polar, and the impermeability of the plasma membrane prevents their loss by diffusion.

Ions play important roles in many physiological processes. Some ions, such as calcium ions, are used in cell signaling. Changes in their cytoplasmic concentration trigger changes in a number of cellular processes (such as muscle contraction; see Chapter 38). As a cell controls the influx and efflux of ions, it is able to directly or indirectly control many metabolic activities. Because of their electric charges, ions cannot cross a lipid bilayer by simple diffusion.

To transport ions and nutrients through membranes, systems of carrier proteins apparently evolved early in the origin of cells. This transfer of solutes by proteins located within the membrane is termed **carrier-mediated transport.** The two forms of carrier-mediated transport—facilitated diffusion and carrier-mediated active transport—differ in their capabilities and energy sources.

Facilitated diffusion occurs down a concentration gradient

In all processes in which substances move across membranes by passive diffusion, the net transfer of those molecules from one side to the other occurs as a result of a concentration gradient (see Fig. 6–4). If the membrane is permeable to a substance, there is net movement from the side of the membrane where it is more highly concentrated to the side where it is less concentrated. Such a gradient across the membrane is actually a form of stored energy. A concentration gradient can be established as a result of many different processes that take place in cells. The stored energy of the concentration gradient is released when molecules move from a region of high concentration to one of

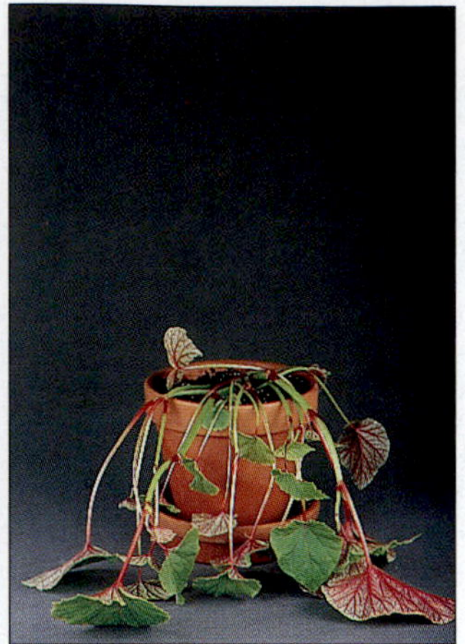
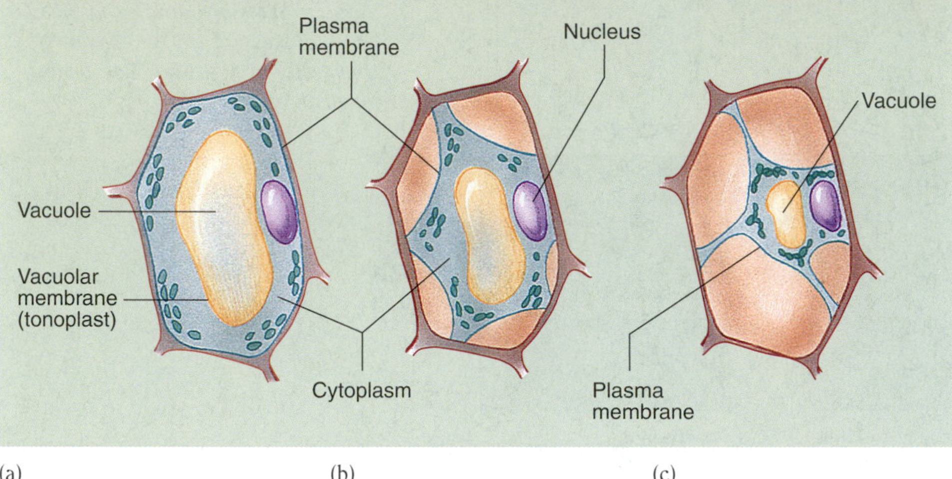

(a) (b) (c)

Figure 5–13 Plasmolysis. **(a)** In hypotonic surroundings, the vacuole of a plant cell fills, but the rigid cell walls prevent the cell from expanding. The cells of this healthy begonia plant are turgid. **(b and c)** When the begonia plant is exposed to a hypertonic solution, its cells become plasmolyzed as they lose water. The plant wilts and eventually dies. *(Dennis Drenner)*

low concentration; movement down a concentration gradient is therefore spontaneous. (These types of energy and spontaneous processes are discussed in greater detail in Chapter 6.)

In the type of transport known as **facilitated diffusion,** the membrane may be made permeable to a solute, such as an ion or a polar molecule, by a specific carrier or **transport protein** (Fig. 5–14). An important example of a transport protein that works by facilitated diffusion is glucose permease, a transmembrane protein that transports glucose into red blood cells. These cells keep the internal concentration of glucose low by immediately adding a phosphate group to entering glucose molecules, converting them to highly charged glucose phosphates that cannot pass back through the membrane. Because glucose phosphate is a different molecule, it does not contribute to the glucose concentration gradient. Thus, a steep concentration gradient for glucose is continually maintained, and glucose rapidly diffuses into the cell, only to be immediately changed to the phosphorylated form.

The mechanism of facilitated diffusion for glucose has been studied by using **liposomes,** artificial vesicles surrounded by

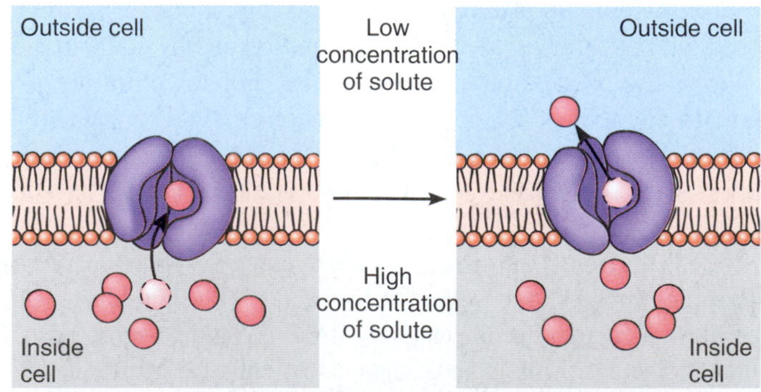

Figure 5–14 Facilitated diffusion. A transport protein in the membrane binds a solute particle. The transport protein changes its shape, opening a channel through the membrane. A specific solute can be transported from the inside of the cell to the outside or from the outside to the inside, but net movement is always from a region of higher solute concentration to a region of lower concentration. Facilitated diffusion requires the potential energy of a concentration gradient.

phospholipid bilayers. The phospholipid membrane of a liposome does not permit the passage of glucose unless researchers introduce the transport protein glucose permease into the membrane. Glucose permease and similar transport proteins temporarily bind to the molecules they transport. This mechanism appears to be similar to the way an enzyme binds with its substrate, the molecule on which it acts (see Chapter 6). In addition, as in enzyme action, binding apparently changes the shape of the transport protein. This change allows the glucose molecule to be released on the inside of the cell. According to this model, when the glucose is released into the cytoplasm, the transport protein reverts to its original shape and is available to bind another glucose molecule on the outside of the cell.

Another similarity to enzyme action is that transport proteins become saturated when the transported molecule is at high concentration. This may be because there are a finite number of transport molecules available and they operate at a defined maximum rate. When the concentration of solute molecules to be transported reaches a certain level, all of the transport molecules are working at their maximum rate.

Some carrier-mediated active transport systems "pump" substances against their concentration gradients

Although adequate amounts of some substances can be transported across cell membranes by diffusion, a cell often needs to move solutes against a concentration gradient. Many substances are required by the cell in concentrations higher than those outside the cell. These molecules are moved across cellular membranes by **carrier-mediated active transport** mechanisms. Because this active transport requires that particles be "pumped" from a region of low concentration to a region of high concentration (i.e., *against a concentration gradient*), transport must be coupled to an energy source such as ATP.

One of the most striking examples of an active transport mechanism is the **sodium-potassium pump** found in virtually all animal cells (Fig. 5–15). The pump is a group of specific proteins in the plasma membrane that uses energy in the form of ATP to exchange sodium ions on the inside of the cell for potassium ions on the outside of the cell. The exchange is unequal, so that usually only two potassium ions are imported for every three sodium ions exported. Because these particular concentration gradients involve ions, an electrical potential (separation of electrical charges) is generated across the membrane, and we say that the membrane is polarized.

Both sodium and potassium ions are positively charged, but because there are fewer potassium ions inside relative to the sodium ions outside, the inside of the cell is negatively charged relative to the outside. The unequal distribution of ions establishes an **electrical gradient** that drives ions across the plasma membrane. The action of sodium-potassium pumps helps maintain a charge separation across the plasma membrane. The separation of charges across a plasma membrane is called a **membrane potential.** Because there is both an electrical charge difference and a concentration difference on the two sides of the membrane, the gradient is referred to as an **electrochemical gradient.** Such gradients store energy (like water stored behind a dam) that can be used to drive other transport systems. So important is the electrochemical gradient produced by these pumps that some cells (e.g., nerve cells) expend 70% of their total energy just to power this one transport system.

Sodium-potassium pumps (as well as all other ATP-driven pumps) are transmembrane proteins that extend entirely through the membrane. By undergoing a series of conformational changes (i.e., changes in shape), the pumps are able to exchange sodium for potassium across the plasma membrane. Unlike facilitated diffusion, at least one of the conformational changes in the pump cycle requires energy, which is provided by ATP. The shape of the pump protein changes as a phosphate group (from ATP) first binds to it and is subsequently removed later in the pump cycle.

The use of electrochemical potentials for energy storage is not confined to the plasma membranes of animal cells. Plant and fungal cells use ATP-driven plasma membrane pumps to transfer protons from the cytoplasm of their cells to the outside. Removal of positively charged protons from the cytoplasm of these cells results in a large difference in the concentration of protons, such that the outside of the cells is relatively positively charged and the inside of the plasma membrane is relatively negatively charged. The energy stored in these electrochemical gradients can be made available to do many kinds of cell work.

Other proton pumps can be used in "reverse" to synthesize ATP. Bacteria, mitochondria, and chloroplasts use energy from food or light to establish proton concentration gradients (see Chapters 7 and 8). When the protons diffuse through the proton carriers from a region of high proton concentration to one of low concentration, ATP is synthesized. These electrochemical gradients form the basis for the major energy conversion systems in virtually all cells.

Ion pumps have other important roles. For example, they are instrumental in the ability of an animal cell to equalize the osmotic pressures of its cytoplasm and its external environment. If an animal cell does not control its internal osmotic pressure, its contents will become hypertonic relative to the exterior. Water will enter the cell by osmosis, causing it to swell and possibly burst (see Fig. 5–12*c*). By controlling the ion distribution across the membrane, the cell is able to indirectly control the movement of water, for when ions are pumped out of the cell, water leaves by osmosis.

Linked cotransport systems indirectly provide energy for active transport

The electrochemical concentration gradients generated by the sodium-potassium pump (and other pumps) provide sufficient energy to power the active transport of other essential substances. In these systems a transport protein can **cotransport** the required molecules *against* their concentration gradient, while sodium, potassium, or hydrogen ions move *down* their gradient.

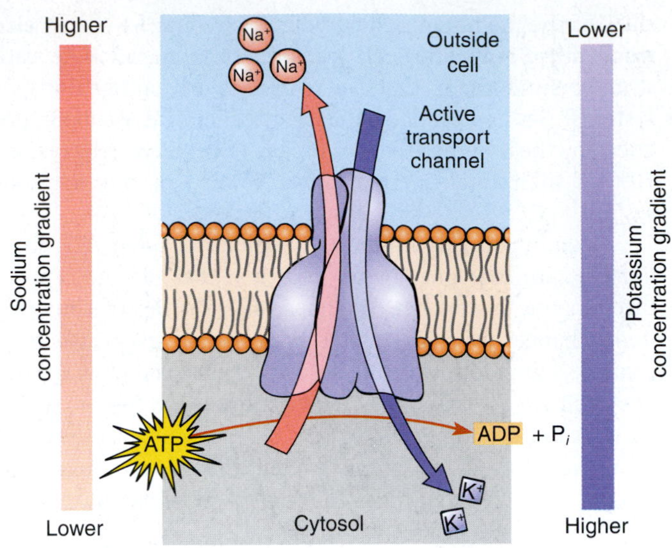

(a)

Figure 5-15 The sodium-potassium pump. **(a)** The sodium-potassium pump is an active transport system that requires energy from ATP. Each complete pumping cycle uses one molecule of ATP; three sodium ions are exported, and two potassium ions are imported. **(b)** A model illustrating the sodium-potassium pumping cycle.

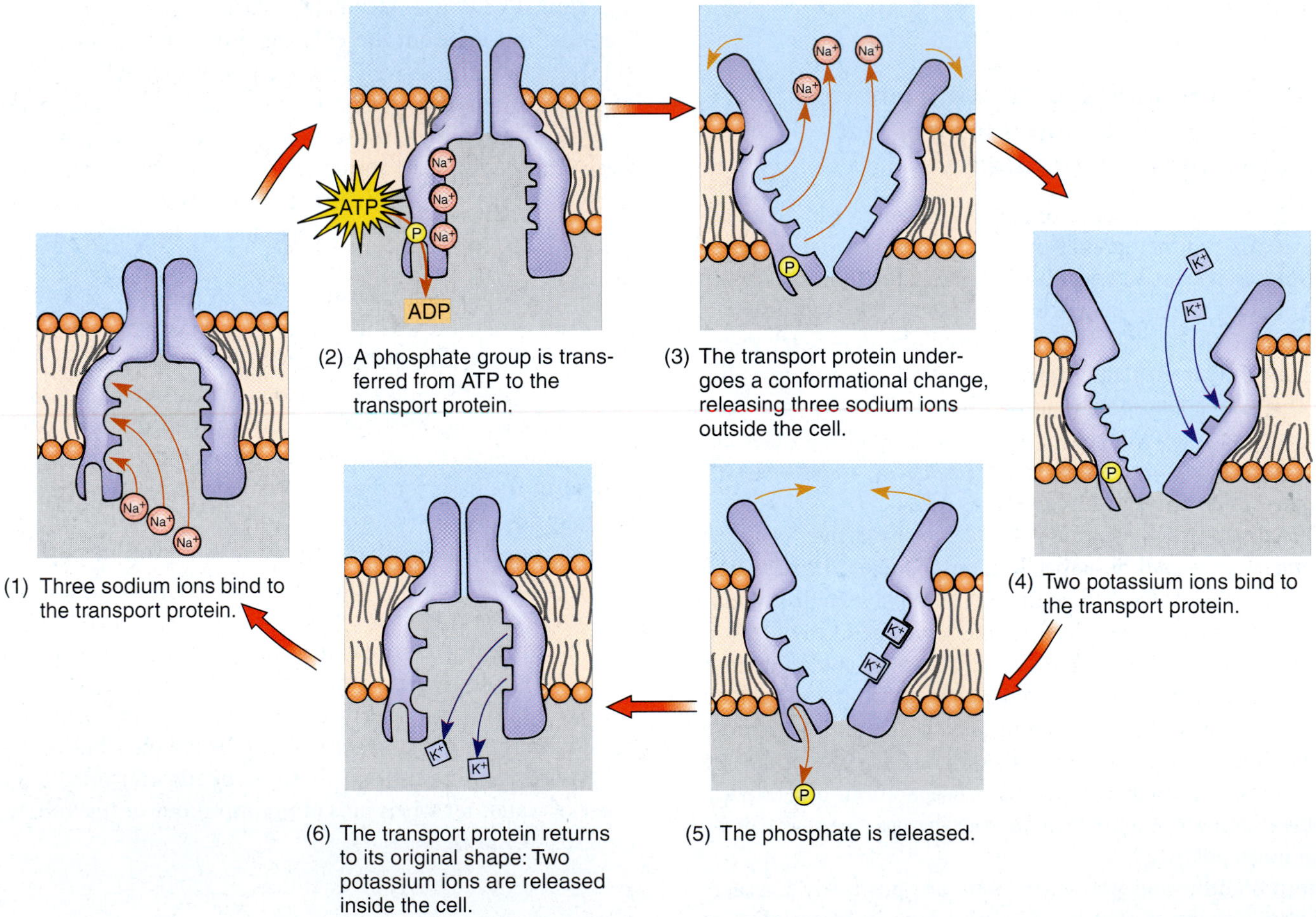

(1) Three sodium ions bind to the transport protein.

(2) A phosphate group is transferred from ATP to the transport protein.

(3) The transport protein undergoes a conformational change, releasing three sodium ions outside the cell.

(4) Two potassium ions bind to the transport protein.

(5) The phosphate is released.

(6) The transport protein returns to its original shape: Two potassium ions are released inside the cell.

(b)

Energy from ATP may be used indirectly in this process. ATP can be used to produce the ion gradient; the energy of this gradient is then used to drive the active transport of a required substance against its gradient.

In some cells more than one system may work to transport a given substance. For example, the transport of glucose from the intestine to the blood occurs through a thin sheet of epithelial cells that line the intestine. The surface that is exposed to the intestine has many **microvilli** (sing., *microvillus*), fingerlike extensions that effectively increase the surface area of the membrane available for absorption (see Fig. 45–10*c*). The glucose transport protein on that region of the cell surface is part of an active transport system for glucose that is "driven" by the cotransport of sodium. The sodium concentration inside the cell is kept low by an ATP-requiring sodium-potassium pump that transports sodium out of the cell and into the blood. Because of its high concentration inside the cell (relative to the blood), glucose can be transported to the blood by facilitated diffusion.

What are the signals that target each transport protein to its appropriate region in the plasma membrane? Some of the current research in cell biology focuses on understanding mechanisms such as those that allow the cell to place different transport proteins in separate regions of the same plasma membrane.

Facilitated diffusion is powered by a concentration gradient; active transport requires another energy source

It is a common misconception that diffusion, whether simple or facilitated, is somehow "free of cost" and that only active transport mechanisms require energy. Because diffusion always involves net movement of a substance down its concentration gradient, we say that the concentration gradient "powers" the process. However, energy is required to do the work of establishing and maintaining the gradient. Think back to the example of facilitated diffusion of glucose. The cell maintains a steep concentration gradient (high outside, low inside) by phosphorylating the glucose molecules once they enter the cell. One ATP molecule is spent for every glucose molecule phosphorylated (not to mention such additional costs as the energy required to make the enzymes that carry out the reaction).

An active transport system can work *against* a concentration gradient, pumping materials from a region of low concentration to a region of high concentration. The energy stored in the concentration gradient is not only unavailable to the system but actually works against it. For this reason some other source of energy must be provided. As we have seen, in many cases ATP energy is used directly. In a cotransport system, energy is provided by a concentration gradient for some other substance (e.g., an ion). ATP may be required indirectly to power the pump that produces the ion gradient.

To summarize, both diffusion and active transport require energy. The energy for diffusion is provided by a concentration gradient for the substance being transported. Active transport requires some other, usually more direct, expenditure of metabolic energy.

Process of Science: The patch clamp technique has revolutionized the study of ion channels

Because ions cannot cross a lipid bilayer by simple diffusion, every membrane of every cell contains numerous *ion channels.* Movement of ions across a membrane can result in a charge difference, or electrical gradient. If the cell is large enough, this charge difference (usually expressed in millivolts, mV) can be measured by using two microelectrodes connected to an extremely sensitive oscilloscope or voltmeter (Fig. 5–16*a*). One of the microelectrodes is inserted into the cell and the other is placed just outside the plasma membrane. Although valuable, these techniques have serious limitations, for they cannot be used on smaller cells and do not provide information on the function of individual ion channels.

In the mid 1970s Erwin Neher and Bert Sakmann developed a method, known as the **patch clamp technique,** that allows researchers to study single ion channels of very small cells. In this technique, the tip of a micropipette is tightly sealed to a patch of membrane so small that it generally contains only a single ion channel (Fig. 5–16*b*). The flow of ions through the channel can be measured using an extremely sensitive recording device.

The patch clamp technique enables neurobiologists to study the action of a single channel over time. They have found that the current flow is intermittent and corresponds to the opening and closing of the ion channel. The permeability of the channel affects the magnitude of the current. This technique has been modified in many ways and has been applied to studies of the roles of ion channels in a wide range of cellular processes in both plants and animals. For example, studies of single ion channels enabled researchers to demonstrate that the genetic disease cystic fibrosis (see Chapter 15) is caused by a defect in a specific type of chloride ion channel. Because of the far-reaching implications of their work, Neher and Sakmann were awarded a Nobel Prize in 1991.

In exocytosis and endocytosis, large particles are transported by vesicles or vacuoles

In both simple and facilitated diffusion, and in carrier-mediated active transport, individual molecules and ions pass through the plasma membrane. Larger quantities of material, such as particles of food and even whole cells, must also be moved into or out of cells. Such work requires that cells expend energy directly, making it a form of active transport.

In **exocytosis,** a cell ejects waste products or specific secretion products such as hormones by the fusion of a vesicle with the plasma membrane (Fig. 5–17). Exocytosis results in the in-

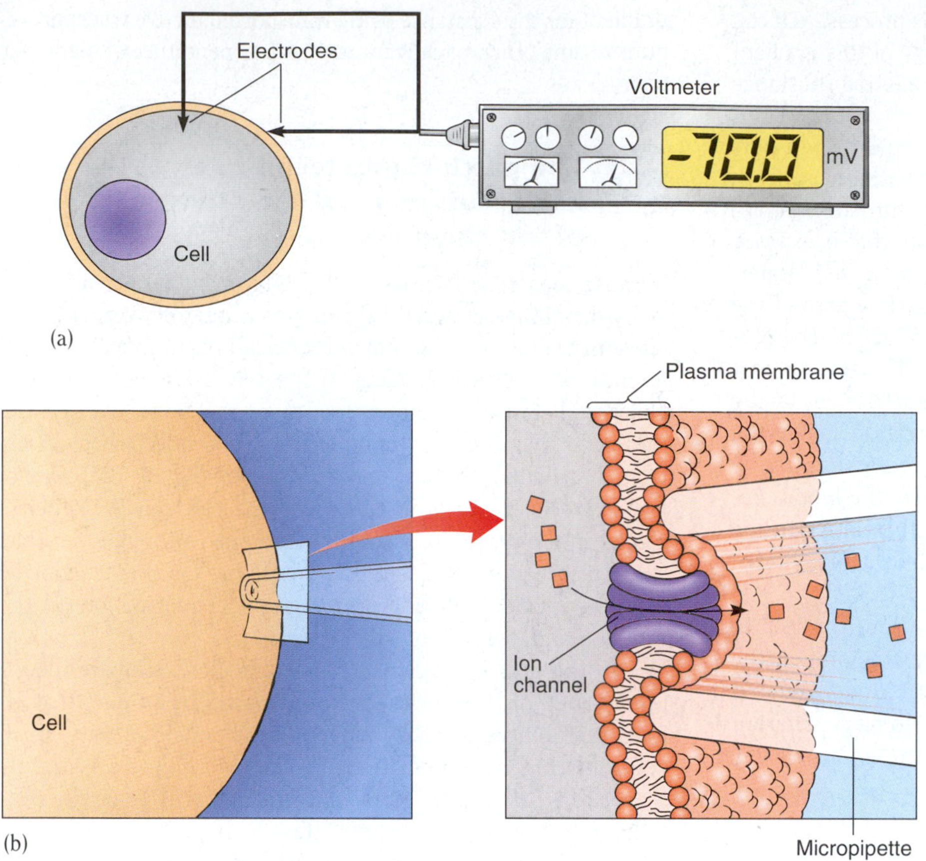

Figure 5–16 Patch clamp technique. **(a)** The difference in electrical charge across a membrane can be measured using microelectrodes and a voltmeter. **(b)** A micropipette can be used to form a tight seal with a patch of plasma membrane. The membrane can be pulled away from the rest of the cell, and the flow of ions through a single ion channel can be studied.

corporation of the membrane of the secretory vesicle into the plasma membrane, as well as the release of the contents of the vesicle from the cell. This is also the primary mechanism by which plasma membranes grow larger.

In **endocytosis,** materials are taken into the cell. Several types of endocytotic mechanisms operate in biological systems. These mechanisms include phagocytosis, pinocytosis, and receptor-mediated endocytosis. In **phagocytosis** (literally, "cell eating"), the cell ingests large solid particles such as bacteria and food (Fig. 5–18). Phagocytosis is a mechanism used by certain protists and by several types of vertebrate white blood cells to ingest particles, some of which are as large as an entire bacterium. During ingestion, folds of the plasma membrane enclose the particle, which has bound to the surface of the cell, forming a large membranous sac, or vacuole. When the membrane has encircled the particle, it fuses at the point of contact. The vacuole then fuses with lysosomes, and the ingested material is degraded.

In the form of endocytosis known as **pinocytosis** ("cell drinking"), the cell takes in dissolved materials. Tiny droplets of fluid are trapped by folds in the plasma membrane (Fig. 5–19), which pinch off into the cytosol as tiny vesicles. The liquid contents of these vesicles are then slowly transferred into the cytosol; the vesicles may become progressively smaller, to the point that they appear to vanish.

In a third type of endocytosis, called **receptor-mediated endocytosis,** specific molecules combine with *receptor proteins* embedded in the plasma membrane. A molecule that binds specifically to a receptor is called a **ligand.** The receptors are concentrated in *coated pits,* depressed regions on the cytoplasmic surface of the plasma membrane. Each pit is coated by a layer of a protein, called *clathrin,* found just below the plasma membrane. After a ligand binds with a receptor, the coated pit forms a *coated vesicle* by endocytosis.

Cholesterol in the blood is taken up by animal cells by receptor-mediated endocytosis. Much of the receptor-mediated endocytosis pathway was detailed through studies by M. Brown and J. Goldstein on the receptor for low-density lipoprotein (LDL), a primary cholesterol carrier in blood. In 1986 these investigators were awarded the Nobel Prize for their pioneering work. Their findings have important medical implications because cholesterol that remains in the blood instead of entering the cells can become deposited in the artery walls, increasing the risk of heart attack.

In Figure 5–20 the uptake of an LDL particle is shown. Seconds after the vesicle moves into the cytoplasm, the coating dissociates from it, leaving an uncoated vesicle, called an *endosome,* free in the cytoplasm. The endosome typically forms two vesicles. One contains the receptors, the other the LDL particle. The receptors are returned to the plasma membrane, where they are recycled. The remaining vesicle fuses with a lysosome, and its contents are then digested and released into the cytosol.

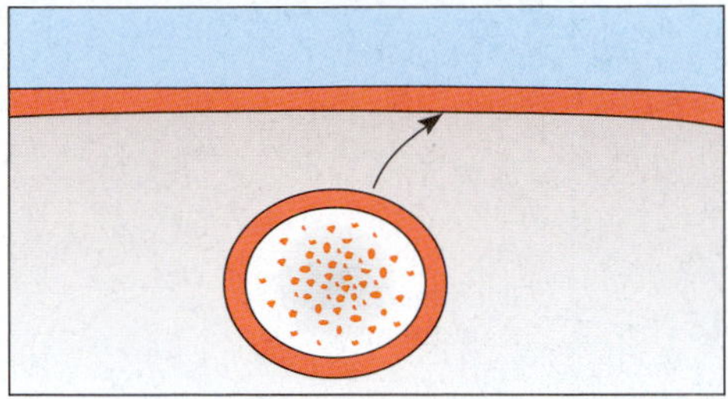

1 A vesicle approaches the plasma membrane,

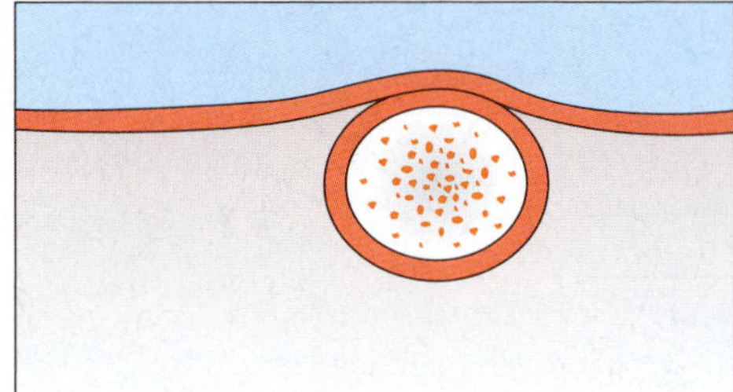

2 fuses with it, and

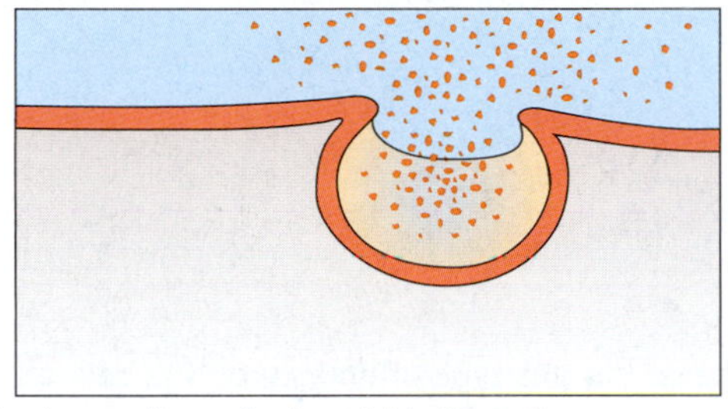

3 releases its contents outside the cell.

(a)

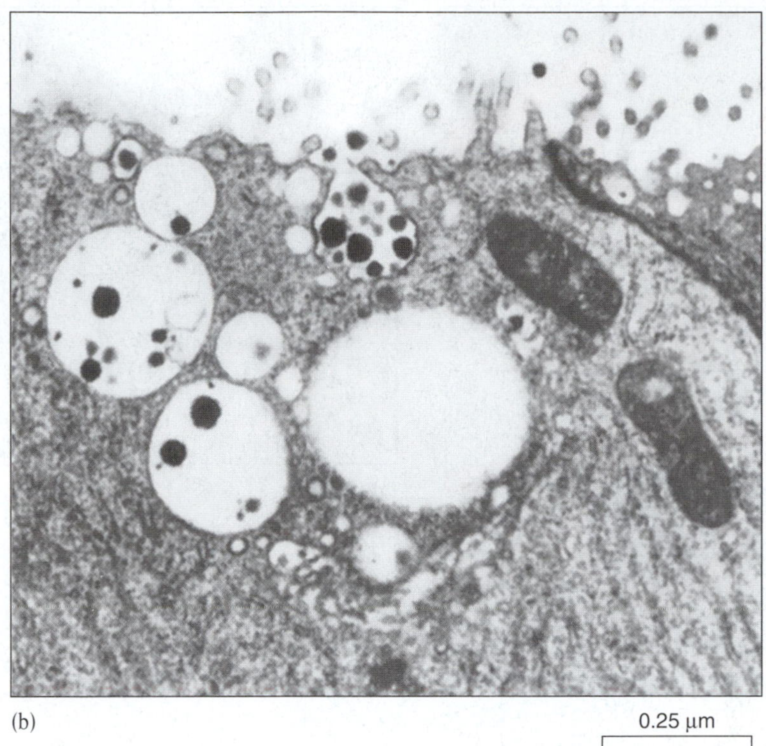

(b)

Figure 5–17 Exocytosis. **(a)** The process of exocytosis. **(b)** TEM showing exocytosis of the protein components of milk by a mammary gland cell. *(b, A. Ichikawa/from D.W. Fawcett)*

Ligand binds to receptors in coated pits of plasma membrane ⟶ coated vesicle forms by endocytosis ⟶ coating detaches from vesicle ⟶ endosome divides:
⟶ One vesicle returns receptors to plasma membrane, where they are recycled
⟶ Other portion of endosome fuses with lysosome ⟶ contents are digested and released into the cytosol

The recycling of LDL receptors to the plasma membrane through vesicles illustrates a problem common to all cells that employ endocytotic and exocytotic mechanisms. In cells that are constantly involved in secretion, an equivalent amount of membrane must be returned to the interior of the cell for each vesicle that fuses with the plasma membrane; if it is not, the cell surface will keep expanding even though the growth of the cell itself may be arrested. A similar situation exists for cells that use endocytosis. A type of phagocytic cell known as a macrophage, for example, ingests the equivalent of its entire plasma membrane in about 30 minutes, requiring an equivalent amount of recycling or new membrane synthesis for the cell to maintain its surface area.

CELL MEMBRANES TRANSFER INFORMATION

Cells communicate with one another and with their external environment mainly by means of proteins that are located in cellular membranes or by chemical compounds that are secreted to the outside of the cell. Unicellular bacteria, protists, and fungi communicate with other members of their species by secreting chemical compounds. For example, when food is scarce, the amoeba-like cellular slime mold *Dictyostelium* secretes **cyclic adenosine monophosphate (cAMP)** (see Fig. 24–20). This chemical compound diffuses through the cell's environment and induces nearby slime molds to come together and form a multicellular slug-shaped colony. Another example is the chemical communication between yeast cells that permits recognition between similar mating types.

About a billion years ago, when cells began to associate to form multicellular organisms, elaborate **cell signaling** systems evolved. The development and functioning of complex organisms require precise internal communication, as well as effective responses to the outside environment. In plants and animals, hormones serve as important chemical signals between

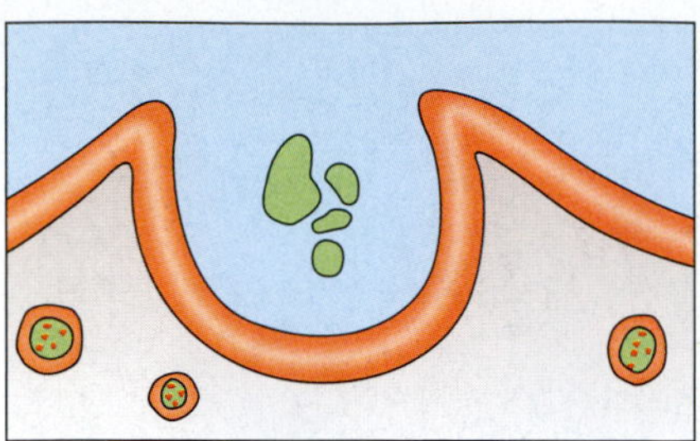

1 Folds of the plasma membrane surround the particle to be ingested, forming a small vacuole around it.

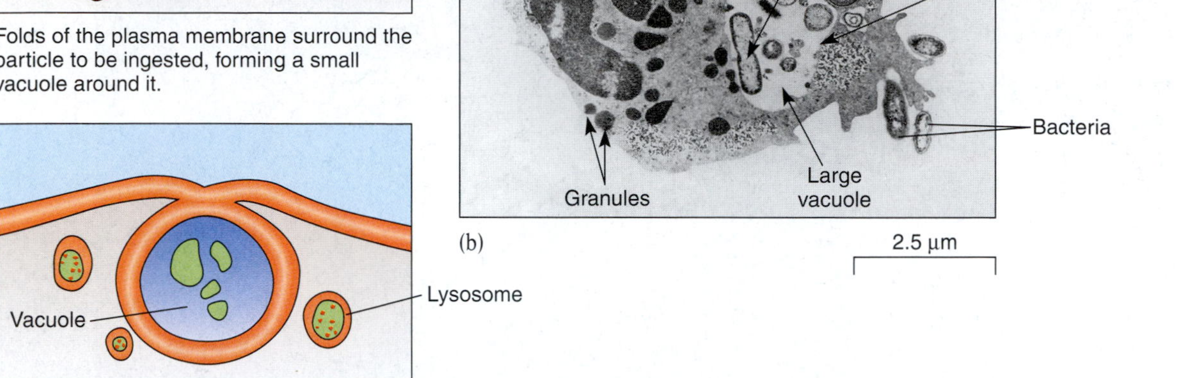

2 The vacuole then pinches off inside the cell.

(b)

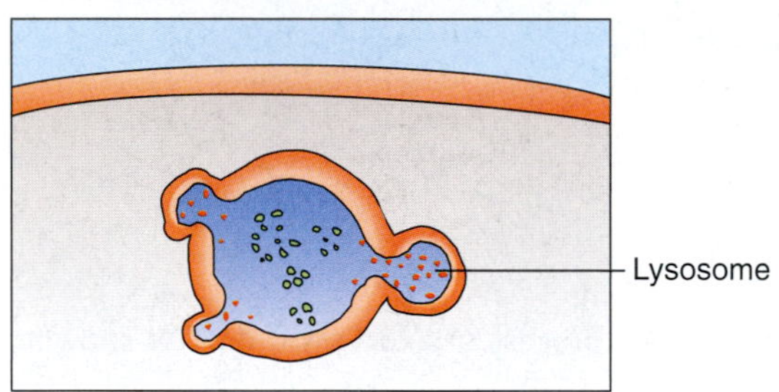

3 Lysosomes may fuse with the vacuole and pour their potent enzymes onto the ingested material.

(a)

Figure 5-18 Phagocytosis. In this type of endocytosis, a cell ingests relatively large solid particles. **(a)** The process of phagocytosis. **(b)** The white blood cell (known as a *neutrophil*) shown in this TEM is phagocytizing bacteria. The vacuoles contain bacteria that have already been ingested, while other bacteria are still outside the cell. The granules in the cytosol contain digestive enzymes. *(b, D.W. Fawcett)*

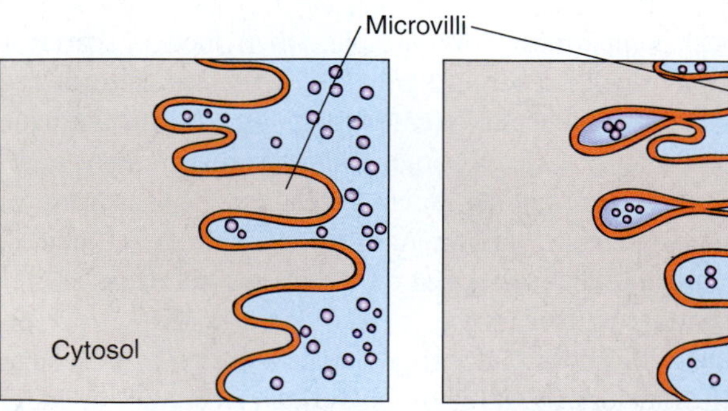

1 Tiny droplets of fluid are trapped by folds of the plasma membrane.

2 These pinch off into the cytosol as small fluid-filled vesicles.

3 The contents of these vesicles are then slowly transferred to the cytosol.

Figure 5-19 Pinocytosis or "cell-drinking."

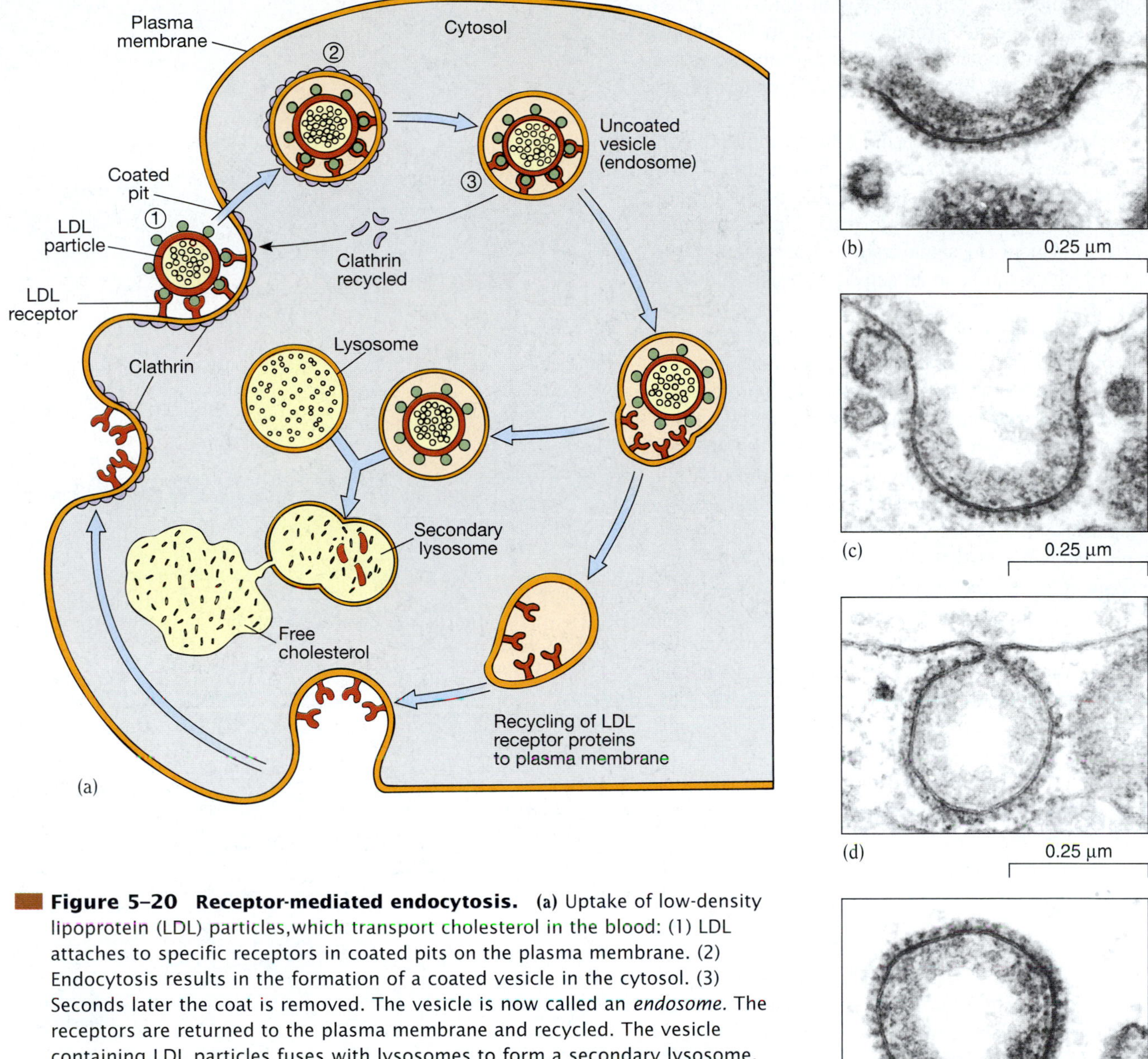

Figure 5–20 Receptor-mediated endocytosis. (a) Uptake of low-density lipoprotein (LDL) particles,which transport cholesterol in the blood: (1) LDL attaches to specific receptors in coated pits on the plasma membrane. (2) Endocytosis results in the formation of a coated vesicle in the cytosol. (3) Seconds later the coat is removed. The vesicle is now called an *endosome.* The receptors are returned to the plasma membrane and recycled. The vesicle containing LDL particles fuses with lysosomes to form a secondary lysosome. Hydrolytic enzymes then digest the cholesterol from the LDL particles for use by the cell. (b to e) Series of TEMs showing the formation of a coated vesicle from a coated pit. *(b to e, from Perry, M.M., and A.B. Gilbert,* J. Cell. Sci. *39:257–272, 1979)*

various cells and organs. Animals have evolved nervous systems in which neurons transmit information electrically and chemically.

The process of information transfer includes (1) **synthesis and release** of the signaling molecules; (2) **transport to target cells,** the cells that can respond to the signal; (3) **reception** of the information by target cells; (4) **signal transduction,** the process in which the cell converts an extracellular signal into an intracellular signal that affects some cell function; (5) **cellular response** to the information; and (6) **termination** of signaling by destruction of the signaling molecules.

Signaling molecules may be synthesized by neighboring cells or by specialized tissues some distance away from the target cells. Signaling molecules may find their way to target cells by diffusion, by transport in neurons, or via the circulatory system.

Reception is the process of receiving the signal. Target cells are typically equipped with receptor proteins that bind with signaling molecules. The receptor proteins for many chemical signals are specific proteins located in the plasma membrane. A signaling molecule such as a hormone or other regulatory molecule that binds with a receptor protein is called a ligand.

Many regulatory molecules transmit information to the cell interior without physically crossing the plasma membrane. These signal molecules rely on systems of interacting integral membrane proteins to transmit the information by signal transduction. Each component of a signal transduction system acts as a relay "switch," which can be in an activated ("on") state or an inactive ("off") state.

The first component in a signal transduction system is typically the receptor, which may be a transmembrane protein with a domain (a structural/functional component of a protein) exposed on the extracellular surface. A receptor generally has at least three domains. The external domain is a docking site for a signaling molecule. A second domain extends through the plasma membrane, and a third domain is a "tail" that extends into the cytoplasm.

In a typical signaling pathway, when the ligand binds with the receptor, it activates the receptor. Binding of the ligand changes the shape of the receptor tail that extends into the cytoplasm. The activated receptor changes the conformation of a second protein, which then becomes activated. The signal may be relayed through a sequence of proteins (Fig. 5–21). Ultimately these interactions result in the activation of a specific enzyme bound to the membrane. That enzyme may itself catalyze the production of large numbers of intracellular signaling molecules, or it may activate intracellular enzymes. In this way the original signal received by the receptor protein is amplified many times, and the metabolism of the cell may be profoundly altered.

The ligand that acts as a signaling molecule is sometimes referred to as the *first messenger.* Some ligand-receptor complexes bind to and activate specific integral membrane proteins, referred to as **G proteins.** In 1994 Alfred G. Gilman and Martin Rodbell were awarded a Nobel Prize for their research on G proteins. These proteins are so named because the active form is bound to **guanosine triphosphate (GTP),** a molecule similar to ATP but containing the base guanine instead of adenine. G proteins catalyze the hydrolysis of GTP to guanosine diphosphate (GDP), a process that releases energy.

In a complex sequence of events, a G protein relays the message from the receptor to an enzyme that catalyzes the production of a **second messenger.** Often, the enzyme adenylyl cyclase

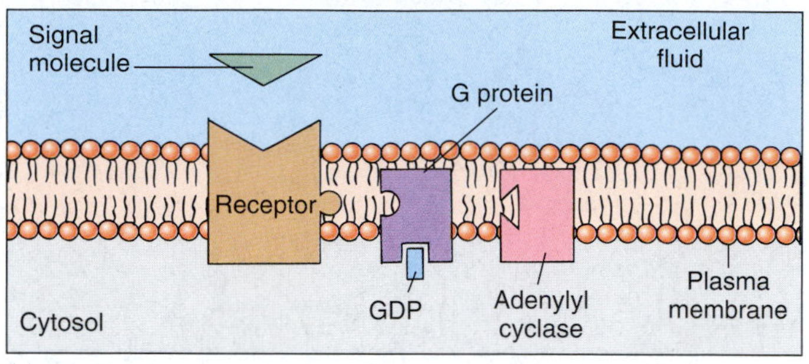

(a)

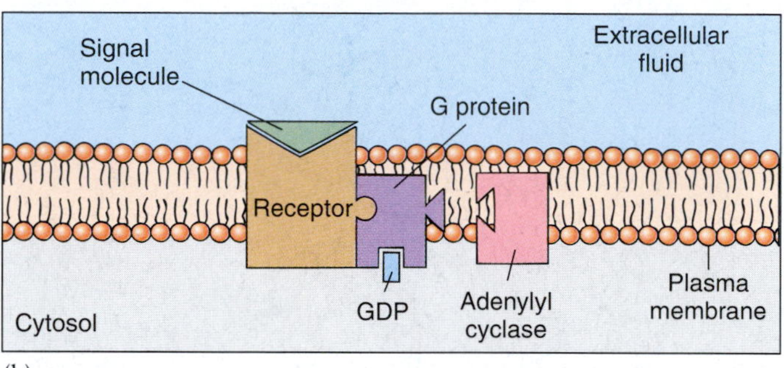

(b)

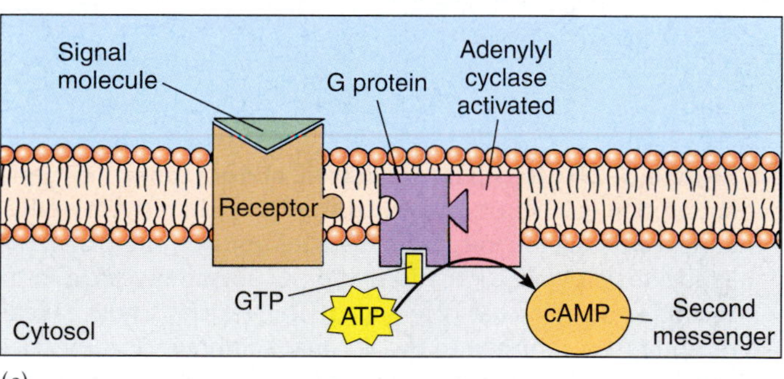

(c)

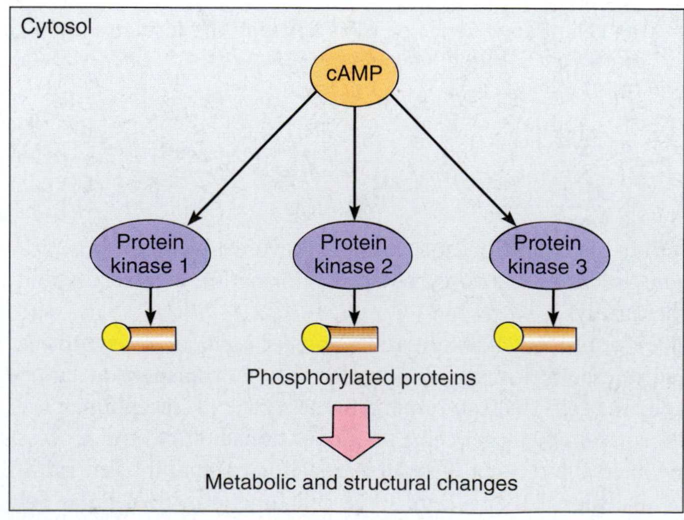

(d)

Figure 5–21 Transfer of information across the plasma membrane. **(a)** A signal molecule binds with a receptor in the plasma membrane. **(b)** The signal molecule-receptor complex activates a G protein. **(c)** G protein activates an enzyme that catalyzes the production of a second messenger such as cyclic AMP (cAMP). **(d)** cAMP then activates one or more enzymes such as protein kinases. The enzymes may phosphorylate proteins, which then alter the activity of the cell in some way.

catalyzes the formation of the second messenger cyclic AMP. Typically, the second messenger activates **protein kinases,** enzymes that activate specific proteins by transferring phosphate groups to them from ATP. This sequence of reactions, beginning with the binding of the signaling molecule to the receptor, leads to a change in some cell function.

G proteins are involved in a number of important signal transductions, including the action of many hormones (see Chapter 47). Some G proteins regulate channels that allow ions to cross the plasma membrane, and still others play important roles in the senses of sight, smell, and taste (see Chapter 41).

Ras proteins, a group of GTP-binding proteins that function somewhat like G proteins, are thought to be important in signal transduction necessary for many cell activities. Fibroblasts (a type of connective tissue cell) require the presence of two *growth factors,* epidermal growth factor and platelet-derived growth factor, for DNA synthesis. However, when investigators inactivated Ras proteins by injecting antibodies that bind to them into the fibroblasts, the growth factors were no longer effective. Data from this and similar experiments led to the conclusion that Ras proteins are important in signal transduction involving growth factors.

Cell biologists have demonstrated that when certain ligands bind to integrins (transmembrane proteins that connect the cell to the extracellular matrix) in the plasma membrane, certain signal transduction pathways are activated. Growth factors also turn on signaling pathways, and it is thought that growth factors and certain molecules of the extracellular matrix may modulate each other's messages. Integrins also respond to information received from inside the cell. This inside-out signaling affects how selective integrins are with respect to the molecules to which they bind and how strongly they bind to them.

JUNCTIONS ARE SPECIALIZED CONTACTS BETWEEN CELLS

Cells in close contact with each other typically develop specialized intercellular junctions. These structures may allow neighboring cells to form strong connections with each other, prevent passage of materials, or establish rapid communication between adjacent cells. Several types of junctions are found connecting animal cells, including anchoring junctions (desmosomes and adhering junctions), tight junctions, and gap junctions. Plant cells are connected by plasmodesmata.

Anchoring junctions connect cells of an epithelial sheet

Adjacent epithelial cells, such as those found in the outer layer of the skin, are so tightly bound to each other by anchoring junctions that strong mechanical forces are required to separate them. Cadherins, transmembrane proteins shown in the chapter opening photograph, are important components of these junctions. Anchoring junctions do not affect the passage of materials between adjacent cells. Two common types of anchoring junctions are desmosomes and adhering junctions.

Desmosomes are points of attachment between cells (Fig. 5–22). They hold cells together at one point like a rivet or a spot weld. As a result, cells can form strong sheets, and substances can still pass freely through the spaces between the

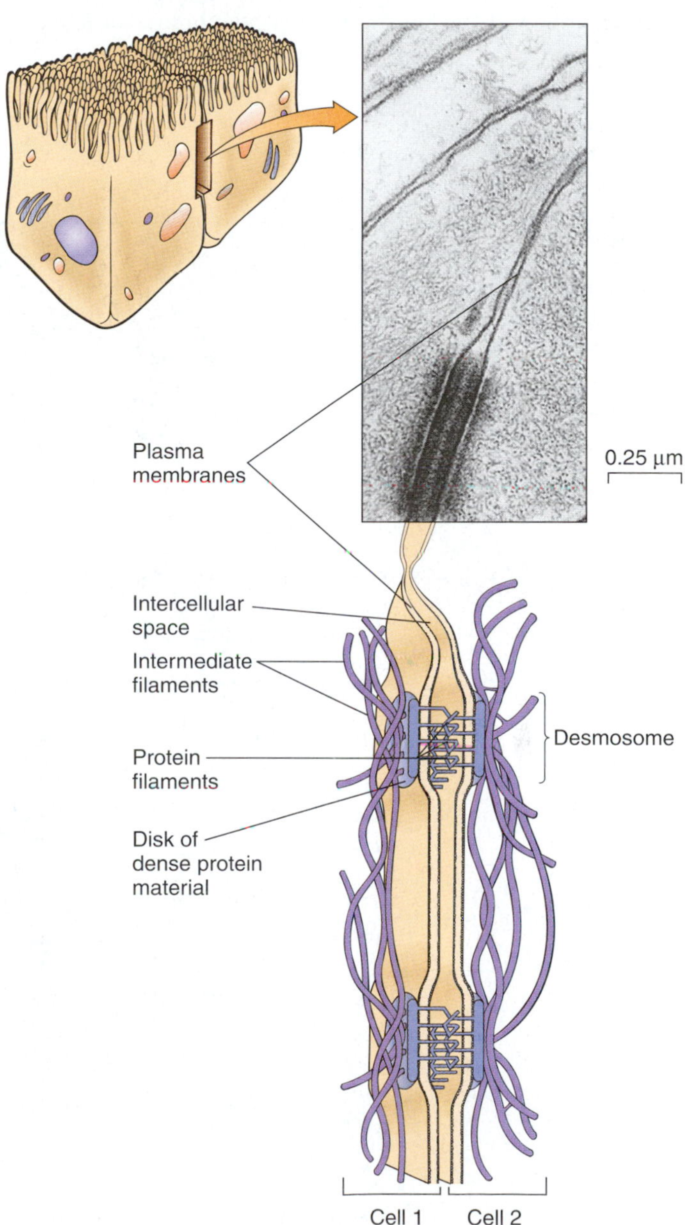

Figure 5–22 Desmosomes. The dense structure in the TEM is a desmosome. Each desmosome consists of a pair of button-like disks associated with the plasma membranes of adjacent cells, plus the intercellular protein filaments that connect them. Intermediate filaments in the cells are attached to the disks and are connected to other desmosomes. *(D.W. Fawcett)*

plasma membranes. Each desmosome is made up of regions of dense material associated with the cytosolic sides of the two plasma membranes, plus protein filaments that cross the narrow intercellular space between them. Desmosomes are anchored to systems of intermediate filaments inside the cells. Thus the intermediate filament networks of adjacent cells are connected so that mechanical stresses are distributed throughout the tissue.

Adhering junctions cement cells together. Cadherins, an important component of these junctions, form a continuous adhesion belt around each cell. These junctions connect to microfilaments that zip up the plasma membranes of adjacent cells.

Tight junctions seal off intercellular spaces between some animal cells

Tight junctions are literally areas of tight connections between the membranes of adjacent cells. These connections are so tight that no space remains between the cells; substances cannot leak between the cells. TEMs of tight junctions show that in the region of the junction the membranes of the two cells are in actual contact with each other, held together by proteins linking the two cells (Fig. 5–23).

Cells connected by tight junctions seal off body cavities. For example, tight junctions between cells lining the intestine prevent substances in the intestine from entering the body or the

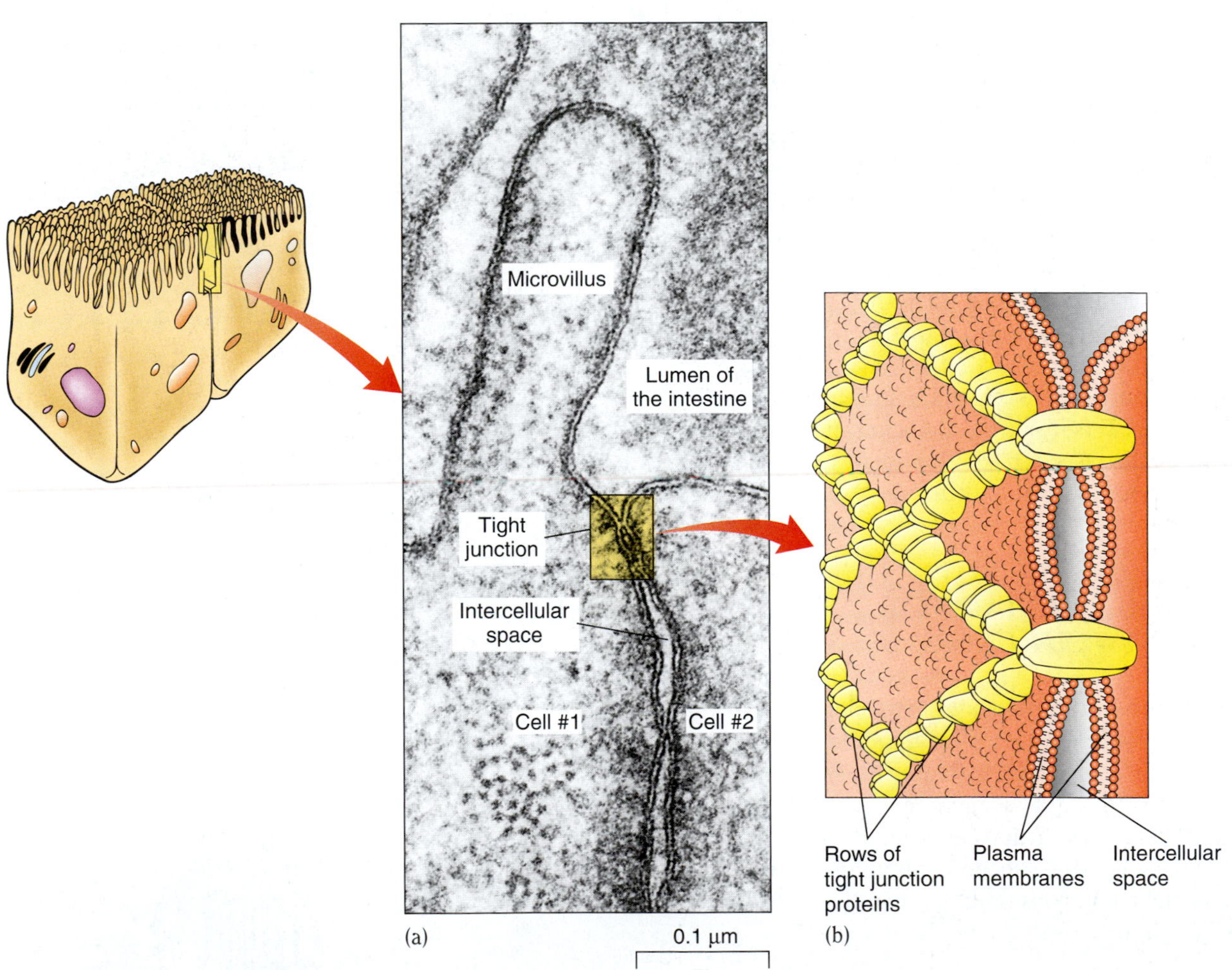

Figure 5–23 Tight junctions. These junctions prevent passage of materials through spaces between cells. Tight junctions occur at the points of contact between two cells and extend completely around the cells. **(a)** This TEM shows points of fusion between the plasma membranes of adjacent cells lining the intestine. One tight junction is marked by the box. **(b)** The diagram shows that a tight junction is formed by linkages between rows of proteins of adjacent cells. These proteins are tightly packed in rows that seal off the intercellular space. *(a, G.E. Palade)*

blood by passing around the cells. The sheet of cells thus acts as a selective barrier. Food substances must be transported across the plasma membranes and through the intestinal cells before they enter the blood. This arrangement helps prevent toxins and other unwanted materials from entering the blood, and also prevents nutrients from leaking out of the intestine.

Gap junctions permit transfer of small molecules and ions

The **gap junction** is like the desmosome in that it bridges the space between cells; however, the space it spans is somewhat narrower (Fig. 5–24). Gap junctions also differ in that they are communicating junctions. They not only connect the membranes but also contain channels connecting the cytoplasm of adjacent cells. A gap junction consists of a hexagonal array of proteins (connexins) on adjacent cells that form a channel, about 1 to 1.5 nm in diameter. Small inorganic molecules (e.g., ions) and some biological molecules (e.g., derivatives of ATP) can pass through the channels, but larger molecules are excluded. When appropriate marker substances are injected into one of a group of cells connected by gap junctions, the marker passes rapidly into the adjacent cells but does not enter the space between the cells.

Cells are able to control the passage of materials through gap junctions by opening and closing the channels (Fig. 5–24*d*). There is evidence that the open and closed states are regulated mainly by the intracellular concentrations of certain ions, such as Ca^{2+}.

Gap junctions provide for rapid chemical and electrical communication between cells. Cells in the pancreas, for example, are linked together by gap junctions in such a way that if one of a group of cells is stimulated to secrete insulin, the signal is passed through the junctions to the other cells in the cluster, ensuring a coordinated response to the initial signal. Gap junctions allow some nerve cells to be electrically coupled. Heart muscle cells are linked by gap junctions that permit the flow of ions necessary to synchronize contractions.

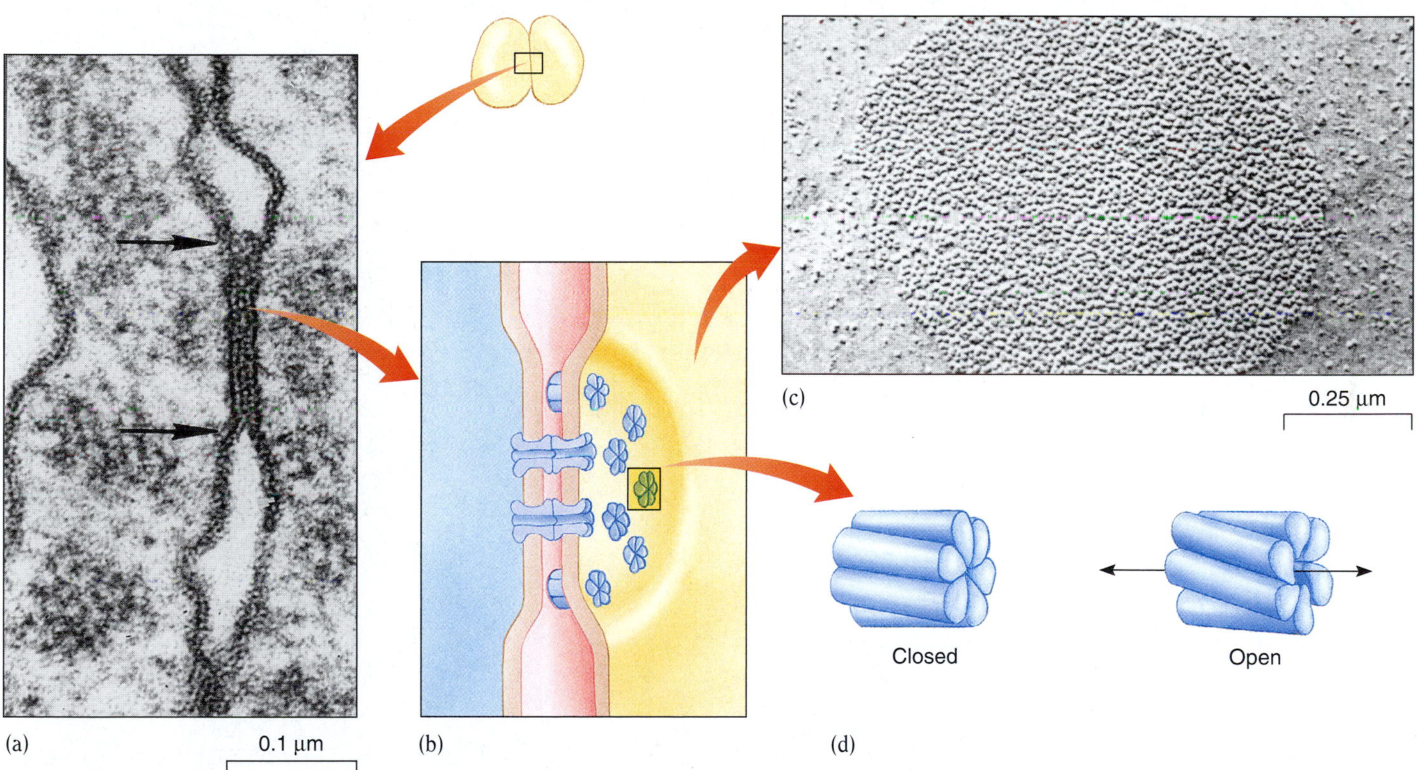

Figure 5–24 Gap junctions. These connections permit transfer of small molecules and ions between adjacent cells. **(a)** A TEM of a gap junction *(between the arrows).* **(b)** Model of a gap junction based on electron microscopic and x-ray diffraction data. The two membranes contain cylinders composed of six protein subunits. Two cylinders from opposite membranes are joined to form a pore about 1.5 to 2 nm in diameter connecting the cytoplasmic compartments of the two cells. **(c)** Freeze-fracture replica of the P-face of a gap junction between two ovarian cells of a mouse, showing the numerous protein particles present. **(d)** Model illustrating how a gap junction pore might open and close. *(a, D.W. Fawcett; c, E. Anderson,* J. Morphol. *156:339–366, 1978)*

Plasmodesmata allow movement of certain molecules and ions between plant cells

Plant cells do not need desmosomes for strength because they have cell walls. However, these same walls would isolate the cells, preventing them from communicating. For this reason, plant cells require connections that are functionally equivalent to the gap junctions of some animal cells. **Plasmodesmata** (sing., *plasmodesma*) are 20- to 40-nm-wide channels through adjacent cell walls connecting the cytoplasm of neighboring cells (Fig. 5–25). The plasma membranes of adjacent cells are therefore continuous with each other through the plasmodesmata. Most plasmodesmata contain a cylindrical membranous structure, called the *desmotubule*, which also runs through the opening and connects the ER of the two adjacent cells.

Plasmodesmata generally allow molecules and ions, but not organelles, to pass through the openings from cell to cell. The movement of ions through the plasmodesmata allows for a very slow type of electrical signaling in plants.

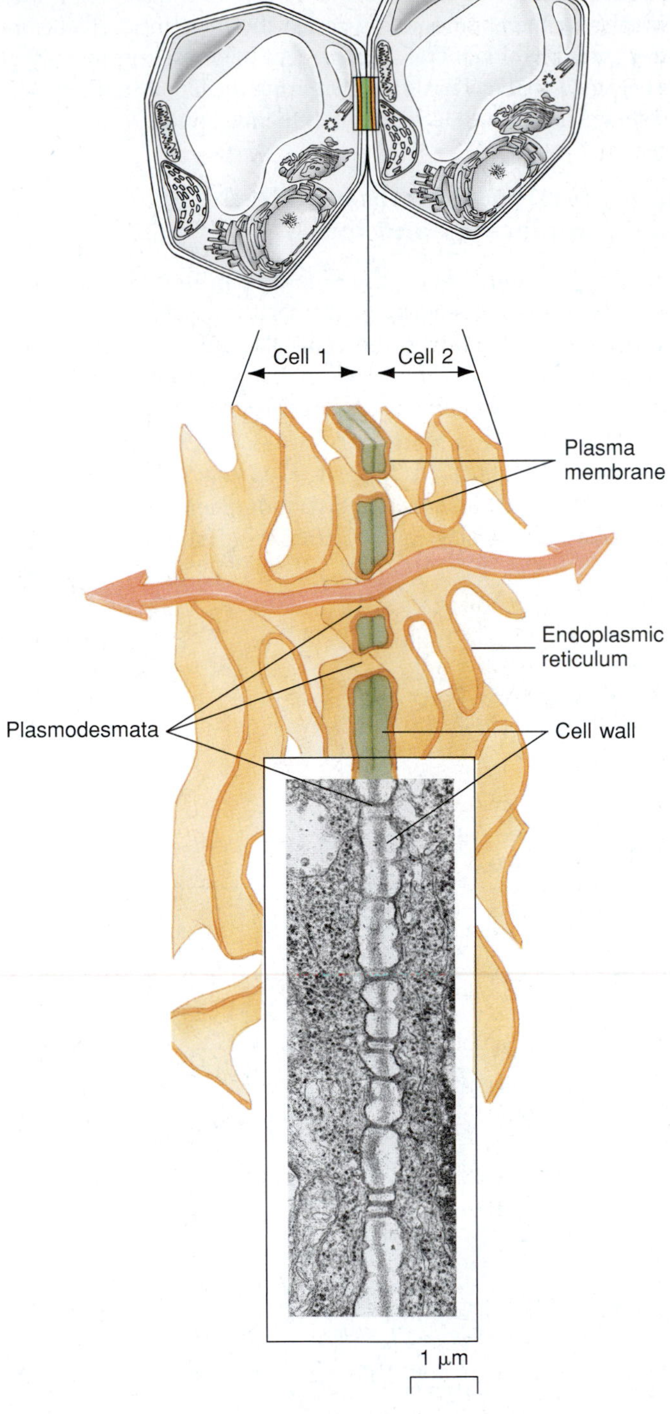

Figure 5–25 Plasmodesmata. TEM and line art of cytoplasmic channels through the cell walls of adjacent plant cells *(wide arrows)* that allow passage of water, ions, and small molecules. The channels are lined with the fused plasma membranes of the two adjacent cells. *(E.H. Newcomb, Biological Photo Service)*

SUMMARY WITH KEY TERMS

I. Cell membranes are complex structures that (1) physically separate the interior of the cell from the extracellular environment and (2) form compartments within the cells of eukaryotes that allow them to perform complex functions. Membranes have many different structural and functional roles.
   A. They regulate the passage of materials.
   B. They receive information that permits the cell to sense and respond to changes in its environment.
   C. They contain specialized structures that allow specific contacts and communications with other cells.
   D. They participate in and serve as surfaces for biochemical reactions.

II. According to the **fluid mosaic model,** membranes consist of a fluid phospholipid bilayer in which a variety of proteins are embedded.
   A. The phospholipid molecules are **amphipathic:** they have hydrophobic and hydrophilic regions.
   B. The lipid bilayer is arranged in such a way that the hydrophilic heads of the phospholipids are at the two surfaces of the bilayer and their hydrophobic fatty acid chains are in the interior.
   C. In almost all biological membranes, the lipids of the bilayer are in a fluid or liquid-crystalline state, which allows the molecules to move rapidly in the plane of the membrane.
   D. **Integral membrane proteins** are embedded in the bilayer with their hydrophilic surfaces exposed to the aqueous environment and their hydrophobic surfaces in contact with the hydrophobic interior of the bilayer. **Transmembrane proteins** are integral proteins that extend completely through the membrane.
   E. **Peripheral membrane proteins** are associated with the surface of the bilayer, usually bound to integral proteins, and are easily removed without disrupting the structure of the membrane.
   F. Membrane proteins, lipids, and carbohydrates are asymmetrically positioned with respect to the bilayer so that one side of the membrane has a different composition and structure than the other.
   G. Many materials are transported from one part of the cell to another in vesicles that bud from various cell membranes and then fuse with some other membrane.
   H. Membrane proteins have various functions including transport of materials, acting as enzymes or receptors, cell recognition, and structurally linking cells together.

III. Biological membranes are selectively permeable membranes; that is, they allow the passage of some substances but not others.
   A. **Diffusion** is the net movement of a substance down its **concentration gradient** (from a region of greater concentration to one of lower concentration).
      1. **Dialysis** is the diffusion of a solute across a selectively permeable membrane.
      2. **Osmosis** is a kind of diffusion in which molecules of water pass through a selectively permeable membrane from a region where water has a higher effective concentration to a region where its effective concentration is lower.
      3. The **osmotic pressure** of a solution is determined by its concentration of dissolved substances (solutes). Cells regulate their internal osmotic pressures to prevent shrinking or bursting.
      4. An **isotonic** solution has an equal solute concentration to another fluid, for example, the fluid within the cell. When placed in a **hypertonic** solution, one that has a greater solute concentration than the cell, cells lose water to the surroundings; plant cells undergo **plasmolysis,** a process in which the plasma membrane separates from the cell wall. When cells are placed in a **hypotonic** solution, one with a lower concentration of dissolved materials relative to the cell, water enters the cells and causes them to swell.
      5. Plant cells can withstand high internal hydrostatic pressure because their cell walls prevent them from expanding and bursting. When water moves into cells by osmosis, it fills the central vacuoles. The cells swell, building up **turgor pressure** against the rigid cell walls.
   B. In **carrier-mediated transport** special membrane proteins move ions or molecules across a membrane.
      1. Some substances pass through membranes by **facilitated diffusion,** a form of carrier-mediated transport that uses the energy of a concentration gradient for the substance being transported and that cannot work against the gradient.
      2. In **carrier-mediated active transport** the cell expends metabolic energy to move ions or molecules across a membrane *against* a concentration gradient. The **sodium-potassium pump** uses ATP to pump sodium ions out of the cell and potassium ions into the cell.
   C. In **cotransport** an ATP-powered pump such as the sodium-potassium pump transports ions or some other solute and indirectly powers the transport of other solutes by maintaining a concentration gradient.
   D. The **patch clamp technique** allows researchers to study single ion channels.
   E. In **exocytosis,** the cell ejects waste products or secretes substances such as hormones or mucus by fusion of vesicles with the plasma membrane.
   F. In **endocytosis** materials such as food may be moved into the cell; a portion of the plasma membrane envelops the material, enclosing it in a vesicle or vacuole that is then released inside the cell.
      1. In **phagocytosis,** the plasma membrane encloses a particle such as a bacterium or protist, forms a vacuole around it, and moves it into the cell.
      2. In **pinocytosis,** the cell takes in dissolved materials by forming tiny vesicles around droplets of fluid trapped by folds of the plasma membrane.
      3. In **receptor-mediated endocytosis, ligands** bind to specific receptors in coated pits along the plasma membrane. These pits, coated by the protein clathrin, form coated vesicles by endocytosis.

IV. In **signal transduction,** a receptor converts an extracellular signal into an intracellular signal that causes some change in the cell.
   A. Signal transduction typically involves a series of molecules that relay information from one to another.
   B. The signaling pathway often involves activation of **G proteins** by binding of a ligand to a receptor; a second messenger such as **cyclic AMP;** and **protein kinases,** enzymes that activate specific proteins by phosphorylating them. The function of the phosphorylated protein is then altered.

V. Cells in close contact with one another may develop intercellular junctions.
   A. Anchoring junctions include **desmosomes** and **adhering junctions.** Desmosomes spot weld adjacent animal cells together. Desmosomes and adhering junctions are found between cells that form a sheet of tissue.
   B. **Tight junctions** seal membranes of adjacent animal cells together, preventing substances from moving through the spaces between the cells.
   C. **Gap junctions** are protein complexes that form channels in membranes, allowing communication between the cytoplasm of adjacent animal cells.
   D. **Plasmodesmata** are channels connecting adjacent plant cells. Openings in the cell walls allow the plasma membranes and cytoplasm to be continuous, thus permitting certain molecules and ions to pass from cell to cell.

POST-TEST

1. Which of the following statements is *not* true? Biological membranes (a) are composed partly of amphipathic lipids (b) have hydrophobic and hydrophilic regions (c) are typically in a fluid state (d) are made mainly of lipids and of proteins that lie like thin sheets on the membrane surface (e) function in signal transduction
2. According to the fluid mosaic model, membranes consist of (a) a lipid-protein sandwich (b) mainly phospholipids with scattered nucleic acids (c) a fluid phospholipid bilayer in which proteins are embedded (d) a fluid phospholipid bilayer in which carbohydrates are embedded (e) a protein bilayer that behaves as a liquid crystal.
3. Transmembrane proteins (a) are peripheral proteins (b) are receptor proteins (c) extend completely through the membrane (d) extend along the surface of the membrane (e) are secreted from the cell
4. Which of the following is *not* a function of the plasma membrane? (a) transports materials (b) helps to structurally link cells together (c) manufactures proteins (d) anchors the cell to the extracellular matrix (e) has receptors that relay signals
5. Which of the following processes requires the cell to expend metabolic energy directly (e.g., from ATP)? (a) active transport (b) facilitated diffusion (c) dialysis (d) osmosis (e) simple diffusion
6. Which of the following is an example of carrier-mediated transport? (a) simple diffusion (b) facilitated diffusion (c) dialysis (d) osmosis (e) osmosis when a cell is in a hypertonic solution
7. Transport of sodium ions by sodium-potassium pumps is an example of (a) active transport (b) pinocytosis (c) dialysis (d) exocytosis (e) facilitated diffusion
8. Which of the following statements is *not* true of the patch clamp technique? (a) It allows researchers to study single ion channels (b) A micropipette is tightly sealed to a patch of membrane that contains a single ion channel (c) The technique can be used to study ion channels in animal cells, but not in plant cells (d) It helped researchers understand the correspondence between electrical current flow and opening of ion channels (e) It was developed by Neher and Sakmann
9. A cell takes in dissolved materials by forming tiny vesicles around fluid droplets trapped by folds of the plasma membrane. This process is (a) active transport (b) pinocytosis (c) receptor-mediated endocytosis (d) exocytosis (e) facilitated diffusion
10. When plant cells are in a hypotonic medium, they (a) undergo plasmolysis (b) build up turgor pressure (c) wilt (d) carry on dialysis (e) lose water to the environment
11. Which sequence most accurately describes receptor-mediated endocytosis? (a) ligand binds to receptors in coated vesicle → vesicle enters cytosol by cotransport mechanisms → clathrin accumulates around vesicle (b) ligands bind to receptors in coated pit → pit forms coated vesicle by endocytosis → clathrin coating detaches from vesicle (c) ATP binds to receptors in coated vesicle → vesicle enters cytosol by facilitated diffusion → protein coat dissolves (d) ligand binds to receptors in coated pit → pit forms coated vesicle by phagocytosis → coating detaches from vesicle (e) clathrin binds to receptors in coated pit → pit forms coated vesicle by endocytosis → protein coating forms around vesicle
12. In signal transduction (a) an extracellular signal is converted to an intracellular signal (b) a signal is relayed through a series of molecules in the membrane (c) signal molecules are destroyed before target cells can respond to the signal (d) answers a, b, and c are correct (e) answers a and b only
13. Anchoring junctions that hold cells together at one point like a spot weld are (a) tight junctions (b) microfilaments (c) desmosomes (d) gap junctions (e) plasmodesmata
14. Junctions that permit the transfer of water, ions, and molecules between adjacent plant cells are (a) tight junctions (b) microfilaments (c) desmosomes (d) gap junctions (e) plasmodesmata

REVIEW QUESTIONS

1. What molecules are responsible for the physical properties of a cell membrane?
2. Illustrate how a transmembrane protein might be positioned in a lipid bilayer. How do the hydrophilic and hydrophobic regions of the protein affect its orientation?
3. Describe the pathway used by cells to place carbohydrates on plasma membrane proteins. Explain why this pathway results in the carbohydrate groups being exposed on only one side of the lipid bilayer.
4. What is the source of energy for diffusion? State a rule for predicting the movement of particles along their concentration gradient. Is the rule different for facilitated diffusion compared with simple diffusion?
5. Distinguish between osmosis and dialysis.
6. Predict the consequences if a plant cell were to be placed in a relatively (a) isotonic, (b) hypertonic, or (c) hypotonic environment. How would you modify your predictions for an animal cell?
7. What are some of the functions of the plasma membrane? Discuss the nature of the proteins that carry out those functions and explain how their properties make them especially adapted for their functions.
8. Identify a common energy source for active transport. In what ways are facilitated diffusion and carrier-mediated active transport similar? In what ways do they differ?
9. Draw a diagram illustrating how membrane lipid bilayers fuse during the processes of exocytosis and endocytosis. Is one the exact reverse of the other? Why or why not?
10. Discriminate between the processes of phagocytosis and pinocytosis.
11. How are desmosomes and tight junctions functionally similar? How do they differ? Do they share any structural similarities?
12. What is the justification for considering gap junctions and plasmodesmata to be functionally similar? How do they differ structurally?
13. Label the diagram of a typical plasma membrane. Use Figure 5–6 to check your answers.

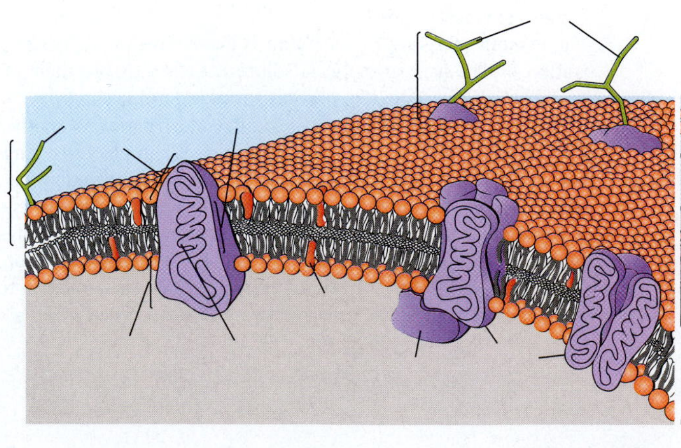

YOU MAKE THE CONNECTION

1. Why can't larger polar molecules and ions cross a lipid bilayer? Would it be advantageous to the cell if they could?
2. Most cells do not actively transport water, yet water is essential to life. How, then, are cells able to control their water content?
3. You prepare a salad with dressing in the morning but find that it is limp and unappetizing by lunch time. Why?
4. Most adjacent living cells in a plant are connected by plasmodesmata. On the other hand, only certain adjacent animal cells are associated through gap junctions. Why?

RECOMMENDED READINGS

Bayley, H. "Building Doors into Cells." *Scientific American,* Vol. 277, No. 3, Sept. 1997. Investigators are using recombinant DNA technology to create artificial pores in cell membranes. The technique has many clinical applications, including delivery of drugs.

Lasic, D.D. "Liposomes," *Science & Medicine,* Vol. 3, No. 3, May/Jun. 1996. Liposomes, artificial vesicles that can be produced commercially, are being investigated as vehicles for delivering drugs to specific cell types in the body.

Linder, M.E., and A. Gilman. "G Proteins." *Scientific American,* Vol. 267, No. 1, Jul. 1992. Pioneers in cell-signaling research discuss the many roles of G proteins.

Neher, E., and B. Sakmann. "The Patch Clamp Technique." *Scientific American,* Vol. 266, No. 3, Mar. 1992. The developers of the patch clamp technique discuss its varied applications.

Rothman, J.E., and L. Orci. "Budding Vesicles in Living Cells." *Scientific American,* Vol. 274, No. 3, Mar. 1996. The authors discuss the exciting process of discovering how cells form transport vesicles.

Scott, J.D. and T. Pawson. "Cell Communication: The Inside Story." *Scientific American,* Vol. 282, No. 6, Jun. 2000. A readable account of the process of signal transduction.

Vogel, S. "Dealing Honestly with Diffusion." *The American Biology Teacher,* Vol. 56, No. 7, Oct., 1994. An explanation of why most macroscopic phenomena attributed to diffusion actually have other explanations. This article emphasizes the fact that diffusion is rapid only over extremely short distances.

Zimmer, C. "Frozen Assets." *Natural History,* Dec. 1999/Jan. 2000. Considers the role of osmosis in protecting plant cells from freezing temperatures.

- Visit our Web site at **http://www.info.brookscole.com/solomonbergmartin** for links to chapter-related resources on the World Wide Web. Additional on-line materials relating to this chapter can also be found on our Web site.

See chapter activity on BioActive Learner CD for additional help in mastering the chapter's material. Icon location in the chapter's margins shows which topics have tutorials or simulations in the CD.

CAREER VISIONS

Pharmaceutical Sales Representative

JULIE HUANG

***Julie Huang** is a pharmaceutical sales representative for Janssen Pharmaceutica, a division of Johnson & Johnson, in Titusville, New Jersey. A native of the Chicago area, Julie graduated with a B.S. in biology and a minor in chemistry from the University of Illinois at Champaign-Urbana in 1997. After graduating, she moved to New York City and earned an M.S. in Biology from New York University in 1998. Julie began working for Janssen in early 1999. In her first position there, she earned a fourth-in-sales ranking among Janssen employees in the United States. Now working in Manhattan, Julie represents medications geared for the elderly. She enjoys the challenges and benefits of working independently and interacting with various medical professionals, and she welcomes the opportunities she has to apply her biology background.*

What led you to major in biology and minor in chemistry?

I come from a big family of physicians, and my original intent was to go to medical school. I was also very much interested in biology research. Because I also enjoy chemistry, I eventually found myself one course shy of a minor, so I took that course.

Did you consider a career as a research biologist?

I had actually wanted to get into an M.D.-Ph.D. program, which covers both medicine and research. Then I decided that I wanted to pursue research first. Medical school was not out of the question, but I just wanted to focus on one thing at a time. So I entered the graduate program at New York University.

What type of research did you do for your master's degree?

As an undergraduate, I had done research in the neurosciences, and I decided I wanted to go into that field. My project investigated pain in rats. It involved a drug that had not been approved by the U.S. Food and Drug Administration for pain. Through the research I identified different neural fibers that caused acute versus chronic pain.

How did you make the leap between basic research and pharmaceutical sales?

I just kind of stumbled on pharmaceutical sales as I was looking for a job, when I was finishing up my master's degree. I attended a job fair put on by NYU, with the intent of pursuing a research position. Janssen had a booth at the fair I attended, although they were looking for someone in marketing and sales, not research. I didn't think that was what I was looking for, but they called me back for an interview. And that's how I came to work at Janssen.

What sort of products do you sell at Janssen?

I'm in an eldercare position. My responsibilities are to make sure our products are represented at nursing homes, and to call on physicians who are focused on geriatrics. This is actually a relatively new division at Janssen. We have a pretty small sales force, but we know that elderly people will make up a greater percentage of the population in the United States as the baby boomers age. We sell an anti-psychotic, which we promote for elderly patients with dementia and psychosis; a patch used for chronic pain; and a drug that is a proton pump inhibitor for heartburn and for ulcers.

Do you have direct professional contact with physicians? With patients?

I don't have direct contact with patients, but I do with physicians. We know that physicians prescribe, so that's who we're out to find. But in eldercare we're more about educating and not so much about pushing our products; that comes secondary to what we do. I interact with physicians to determine their prescribing habits and learn their thoughts about certain types of medications versus ours, provide them with accredited continuing education programs, and consult with them on how to introduce our products to their patients.

Did you receive any business training?

Yes, though not in sales at first. I received four weeks of at-home training in general biology, things like anatomy, then four more weeks of intensive training, this time on the products that I was to sell. I've also gone through some very intensive selling seminars, after first spending time on the job.

Do you think your knowledge of the sciences makes learning about these products, or your ability to speak with physicians, any easier?

Oh, definitely. When we had the home study, I think a lot of people had difficulty with it; they didn't really understand a lot of the information that was given. To me, the at-home training was very straightforward, because that's my field. My background helps me assimilate product information, and it also helps me with study presentations. That's another big part of my job—to look at studies, find out what the key points are, and make those easier for a physician to understand. I feel that one of my strengths is my ability to understand scientific studies and to help others understand them.

What do you like best about your job?

Understanding studies and presenting them, interacting with physicians, and the flexibility of the work.

What advice can you offer biology students interested in pharmaceutical sales?

I would have to say that you should really be sure you like the job. It's a difficult transition; I had a hard time when I first started. It can be very tiring, following up on physicians. Be sure that you like that interaction and that you're willing to be an advocate for both the physicians (and their patients) and your company. You have to represent both interests.

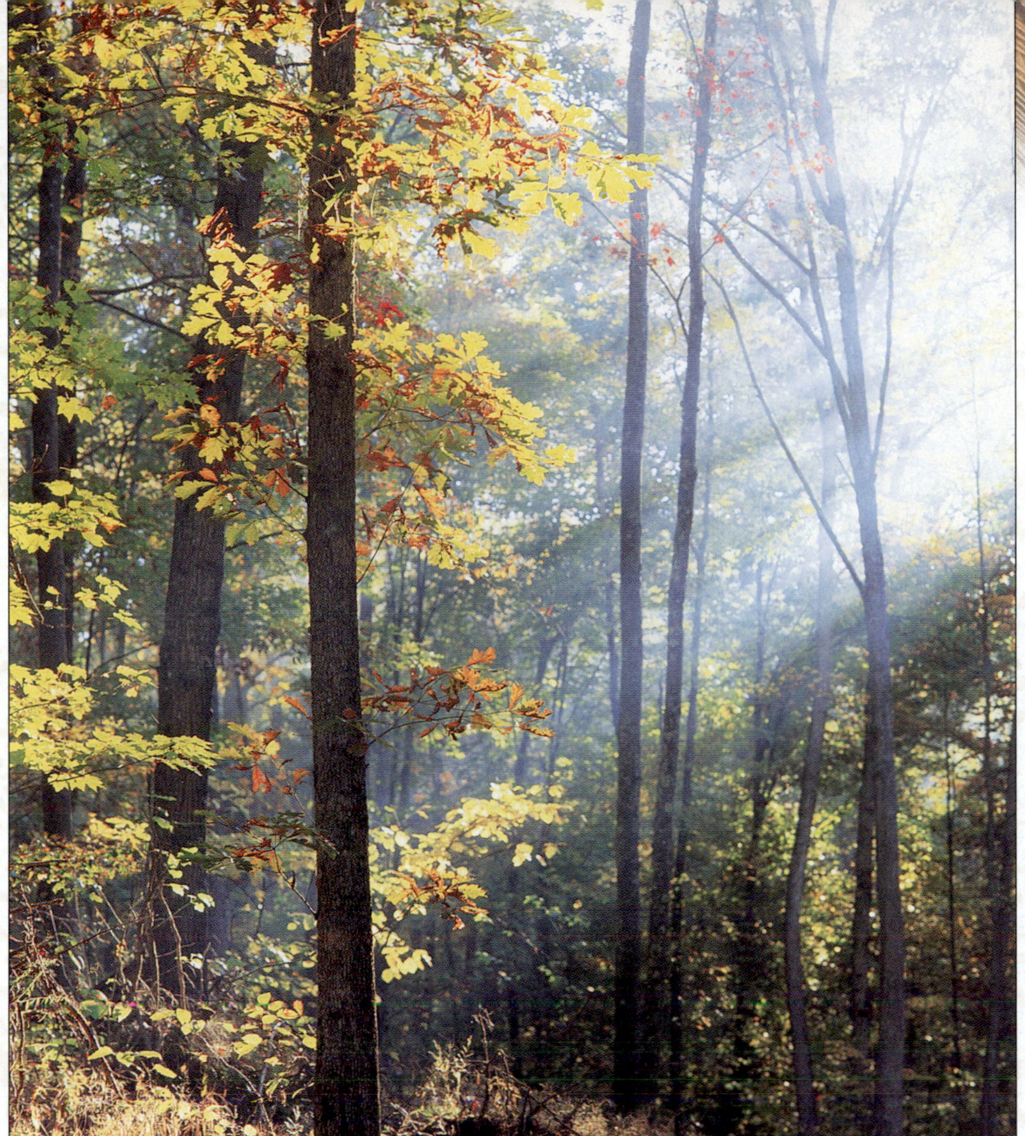

Sunlight filters through a forest clearing in West Virginia. Almost all life ultimately depends on solar energy. *(Michael Hubreich/ Photo Researchers, Inc.)*

Part

2

Energy Transfer Through Living Systems

Chapters

6

Energy and Metabolism

Black-tailed prairie dog *(Cynomys ludovicianus).* The chemical energy produced by photosynthesis and stored in seeds and leaves is transferred to the black-tailed prairie dog as it eats. *(Barbara Gerlach/Visuals Unlimited)*

LEARNING OBJECTIVES

After you have studied this chapter you should be able to

1. Define *energy,* emphasizing how it is related to work and to heat.
2. Use examples to contrast potential energy and kinetic energy.
3. State the first and second laws of thermodynamics and discuss the implications of these laws as they relate to organisms.
4. Discuss how changes in free energy in a reaction are related to changes in entropy and enthalpy.
5. Compare the energy dynamics of a reaction at equilibrium with the dynamics of a reaction not at equilibrium.
6. Distinguish between exergonic and endergonic reactions and give examples of how they may be coupled.
7. Explain how the chemical structure of ATP allows it to transfer a phosphate group. Discuss the central role of ATP in the overall energy metabolism of the cell.
8. Relate the transfer of electrons (or hydrogen atoms) to the transfer of energy.
9. Explain how an enzyme lowers the required energy of activation for a reaction.
10. Describe specific ways enzymes are regulated.

All living things require energy because life processes involve work. It may seem obvious that cells need energy to grow and reproduce, but even nongrowing cells need energy simply to maintain themselves. The sun is the ultimate source of almost all the energy that powers life; this **radiant energy** flows from the sun as electromagnetic waves. Plants and other photosynthetic organisms capture about 0.02% of the sun's energy that reaches Earth. In the process of photosynthesis, plants convert radiant energy to **chemical energy** in the bonds of organic molecules. The chemical energy captured by photosynthesis and stored in seeds and leaves is transferred to animals, such as the black-tailed prairie dog in the photograph, when they eat. Plants, animals, and other organisms need the energy stored in these organic molecules, and they commonly use the process of cellular respiration to break them apart and convert their energy to more immediately usable forms.

The first law of thermodynamics states that energy cannot be created or destroyed. A corollary to this in biological terms is that cells have no way to produce new energy. Thus, energy must be captured from the environment, temporarily stored, and then used to perform biological work. However, according to the second law of thermodynamics, not all of the captured energy can be used for work; at every step, some inevitably becomes converted to heat and is dispersed back into the environment.

Cells obtain energy in many forms, but seldom can that energy be used directly to power cellular processes. For this reason, cells have mechanisms that convert energy from one form to another. Because most of the components of these energy-conversion systems evolved very early in the history of life, many aspects of energy metabolism tend to be very similar in a wide range of organisms.

This chapter focuses on some of the basic principles that govern how cells capture, transfer, store, and use energy. We discuss the functions of adenosine triphosphate (ATP) and other molecules used in energy conversions, including those that transfer electrons in oxidation-reduction (redox) reactions. We also pay particular attention to the essential role of enzymes in cellular energy dynamics. In Chapter 7 we will explore some of the main metabolic pathways used in cellular respiration, and in Chapter 8 we will discuss the energy transformations of photosynthesis. The flow of energy in ecosystems is discussed in Chapter 53.

BIOLOGICAL WORK REQUIRES ENERGY

Energy, one of the most important concepts in biology, can be understood in the context of **matter,** which is anything that has mass and takes up space. **Energy** can be defined as the capacity to do **work,** which is any change in the state or motion of matter.

Biologists generally express energy in units of work **(kilojoules, kJ)** or units of heat energy **(kilocalories, kcal).** (**Heat energy** is thermal energy that flows from an object with a higher temperature—the heat source—to an object with a lower temperature—the heat sink.) One kilocalorie equals 4.184 kilojoules. Because heat energy cannot do cellular work, the kilojoule is the unit preferred by most biologists today. However, we will use both units because references to the kilocalorie are common in the scientific literature.

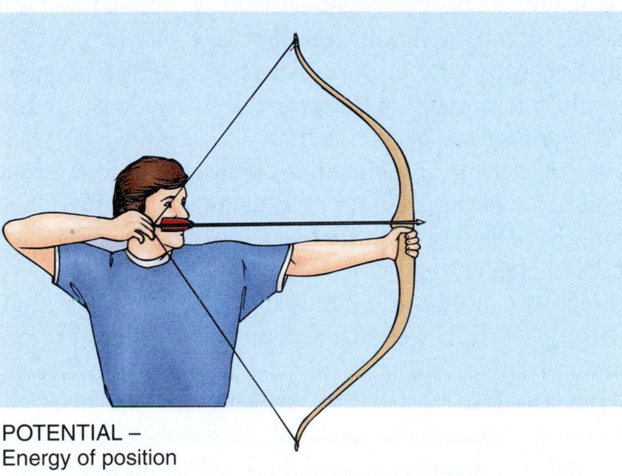

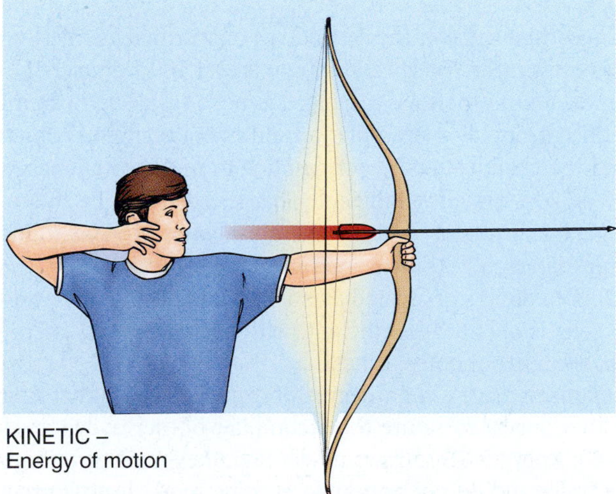

Figure 6–1 Potential versus kinetic energy. The potential chemical energy released by cellular respiration is converted to kinetic energy in the muscles, which do the work of drawing the bow. The potential energy stored in the drawn bow is transformed into kinetic energy as the bowstring pushes the arrow toward its target.

Many of the activities performed by an organism are examples of **mechanical energy,** which is energy in the movement of matter. At this very moment you are expending considerable energy to carry out such activities as breathing and circulating your blood. However, these forms of mechanical energy are the consequence of cellular activities. For example, the cells of the heart muscle use a great deal of energy to contract, thereby pumping the blood through your body. As we will see, however, not all of the work of cells is mechanical. A great deal of it is chemical. For example, heart muscle cells expend energy to synthesize the proteins required for contraction. Energy can be converted to many different forms, including not only mechanical and chemical energy but also heat energy, electrical energy, nuclear energy, and radiant energy.

Organisms carry out conversions between potential energy and kinetic energy

When an archer draws a bow, **kinetic energy,** which is energy of motion, is used and work is done (Fig. 6–1). The resulting tension in the bow and string represents stored energy, or **potential energy.** Potential energy is the capacity to do work owing to position or state. When the string is released, this potential energy is converted to kinetic energy in the motion of the bow, which propels the arrow.

Most of the actions of an organism involve a complex series of energy transformations that occur as kinetic energy is converted to potential energy or as potential energy is converted to kinetic energy. For example, potential energy derived from chemical energy of food molecules is converted to kinetic energy in the muscles of the archer.

TWO LAWS OF THERMODYNAMICS GOVERN ENERGY TRANSFORMATIONS

All the activities of our universe, from the life and death of cells to the life and death of stars, are governed by **thermodynamics,** which is the study of energy and its transformations. When considering thermodynamics, scientists use the term *system* to refer to an object that is being studied, whether a cell, an organism, or planet Earth. The rest of the universe other than the system being studied is known as the *surroundings.* A **closed system** is one that does not exchange energy or matter with its surroundings, whereas an **open system** is one that can exchange matter and energy with its surroundings (Fig. 6–2). Biological systems are open systems.

There are two laws about energy that apply to all things in the universe. These are known as the first and second laws of thermodynamics.

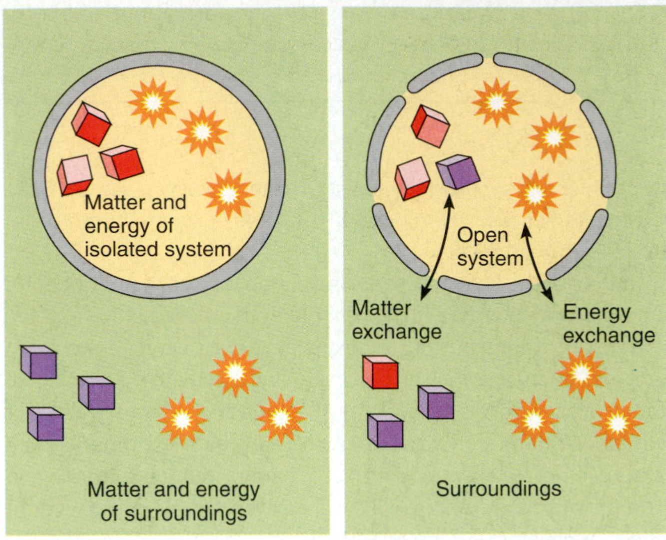

Figure 6–2 Closed and open systems. **(a)** Matter and energy are not exchanged between a closed system and its surroundings. **(b)** Matter and energy are exchanged between an open system and its surroundings. *(Adapted from Tobin, A.J., and R.E. Morel.* Asking About Cells. *Harcourt College Publishers, Philadelphia, 1997)*

The total energy in the universe does not change

According to the **first law of thermodynamics,** energy cannot be created or destroyed, although it can be transferred or changed from one form to another. As far as we know, the energy present in the universe at its formation, approximately 15 to 20 billion years ago, equals the amount of energy present in the universe today.[1] This is all the energy that can ever be present in the universe. Similarly, the energy of any system and its surroundings is constant. A system may absorb energy from its surroundings, or it may give up some energy to its surroundings, but the total energy content of that system and its surroundings is always the same.

As specified by the first law of thermodynamics, then, organisms cannot create the energy that they require to live. Instead, they must capture energy from the environment to use for biological work, a process involving the transformation of energy from one form to another. In photosynthesis, for example, plants absorb the radiant energy of the sun and convert it into the chemical energy contained in the bonds of carbohydrate molecules. Similarly, some of that chemical energy may later be transformed by the plant to do various types of cellular work, or some animal that eats the plant might convert it to the mechanical energy of muscle contraction or some other needed form.

[1] Technically, mass is a form of energy, and so we should say that the total mass-energy of the universe is a constant. Energy can be produced from mass (recall Einstein's famous equation $E = mc^2$). This is the basis behind the energy generated by the sun and stars. More than 4 billion kilograms of matter per second are converted to energy in our sun.

The entropy of the universe is increasing

As each energy transformation occurs, some of the energy is converted to heat energy that is then given off into the cooler surroundings. This energy can never again be used by any organism for biological work; it is lost, at least from a biological point of view. However, it is not really gone from a thermodynamic point of view because it still exists in the surrounding physical environment. For example, the use of food to enable us to walk or run does not destroy the chemical energy that was once present in the food molecules. After we have performed the task of walking or running, the energy still exists in the surroundings as heat.

The **second law of thermodynamics** can be stated as follows: When energy is converted from one form to another, some usable energy, that is, energy available to do work, is converted into a less usable form, heat, that disperses into the surroundings (see Fig. 53–1). As a result, the amount of usable energy available to do work in the universe decreases over time.

It is important to understand that the second law of thermodynamics is consistent with the first law; that is, the total amount of energy in the universe is *not* decreasing with time. However, the total amount of energy in the universe that is available to do work is decreasing over time.

Less-usable energy is more diffuse, or disorganized. **Entropy *(S)*** is a measure of this disorder or randomness; organized, usable energy has a low entropy, whereas disorganized energy, such as heat, has a high entropy.

Entropy is continuously increasing in the universe in all natural processes. It may be that at some time, billions of years from now, all energy will exist as heat uniformly distributed throughout the universe. If that happens, the universe will cease to operate because no work will be possible. Everything will be at the same temperature, so there will be no way to convert the thermal energy of the universe into usable mechanical energy.

As a consequence of the second law of thermodynamics, no process requiring an energy conversion is ever 100% efficient because much of the energy is dispersed as heat, resulting in an increase in entropy. For example, an automobile engine, which converts the chemical energy of gasoline to mechanical energy, is between 20% and 30% efficient. That is, only 20% to 30% of the original energy stored in the chemical bonds of the gasoline molecules is actually transformed into mechanical energy; the other 70% to 80% is dissipated as waste heat. Energy utilization in our cells is about 40% efficient, with the remaining energy given to the surroundings as heat.

Organisms have a high degree of organization, and at first glance they appear to refute the second law of thermodynamics (Fig. 6–3). As organisms grow and develop, they maintain a high level of order and do not appear to become more disorganized. However, organisms are able to maintain their degree of order over time only with the constant input of energy from their surroundings. That is why plants must photosynthesize and animals must eat. Although the order within organisms might tend to increase temporarily, the total entropy of the universe (organisms plus surroundings) will increase over time.

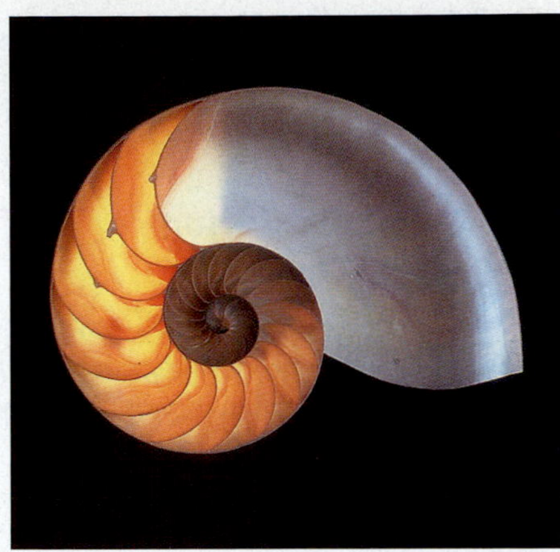

Figure 6–3 Cutaway view of the shell of the chambered nautilus (*Nautilus* sp.). The nautilus is a marine animal that lives in the outermost chamber of its shell. As it grows, the nautilus builds a new wall (septum) at the smaller end of the shell, thereby closing it off. The highly organized structure of this coiled shell is developed and maintained only by the constant input of energy from the small, shrimplike animals the nautilus eats. Thus, the high degree of organization of this animal, and of all living things, does not refute the second law of thermodynamics. *(Charles D. Winters)*

MAKING THE CONNECTION Is diffusion related to entropy? In Chapter 5 we saw that randomly moving particles can diffuse down their own concentration gradient (Fig. 6–4). That is, although the movements of the individual particles are random, net movement of the group of particles seems to be directional. What provides energy for this seemingly directed process? A **concentration gradient,** with a region of higher concentration and another region of lower concentration, is an orderly state. A cell must expend energy to produce a concentration gradient. Because work must be done to produce this order, the concentration gradient is a form of potential energy. As the particles move about randomly, disorder increases.

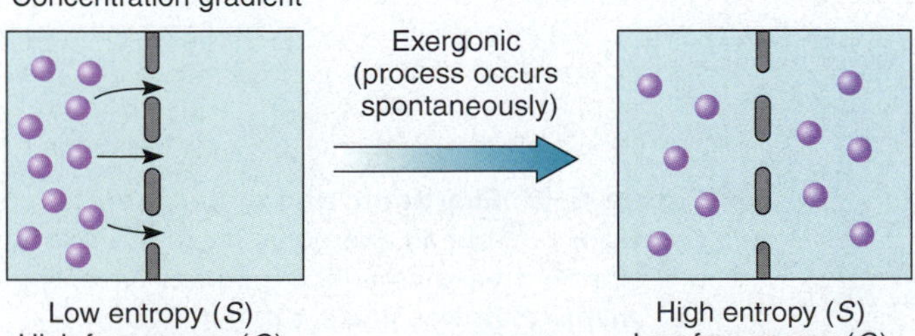

Figure 6–4 Entropy and diffusion. The tendency of entropy to increase can be used to produce work, in this case, diffusion. *(Left)* A concentration gradient is a form of potential energy. *(Right)* When molecules are evenly distributed, they have high entropy.

In cellular respiration and photosynthesis, the potential energy stored in a concentration gradient of hydrogen ions (H^+) can be transformed into chemical energy in adenosine triphosphate (ATP) as the hydrogen ions pass through a membrane down their concentration gradient. This important concept, known as **chemiosmosis,** will be discussed further in Chapters 7 and 8.

METABOLIC REACTIONS INVOLVE ENERGY TRANSFORMATIONS

The chemical reactions that enable an organism to carry on its activities—to grow, move, maintain and repair itself, reproduce, and respond to stimuli—together make up its metabolism. **Metabolism** was defined in Chapter 1 as the sum of all the chemical activities that take place in an organism. An organism's metabolism consists of many intersecting series of chemical reactions, or pathways, which are of two main types: anabolism and catabolism. **Anabolism** refers to the various pathways in which complex molecules are synthesized from simpler substances, such as the linking of amino acids to form proteins. **Catabolism** includes the pathways in which larger molecules are broken down into smaller ones, such as the degradation of starch to form monosaccharides.

As we will see, these changes involve not only alterations in the arrangement of atoms but also various energy transformations. Catabolism and anabolism are complementary processes; catabolic pathways involve an overall release of energy, some of which is used to power the anabolic pathways, which have an overall energy requirement. In the following sections we will discuss how to predict whether a particular chemical reaction requires energy or releases it.

Enthalpy is the total potential energy of a system

In the course of any chemical reaction, including the metabolic reactions of a cell, chemical bonds break, and new and different bonds may form. Every specific type of chemical bond has a certain amount of **bond energy,** defined as the energy required to break that bond. The total bond energy is essentially equivalent to the total potential energy of the system, a quantity known as **enthalpy *(H)*.**

Free energy is energy that is available to do cellular work

Entropy and enthalpy are related by a third type of energy, termed **free energy *(G)*,** which is the amount of energy available to do work under the conditions of a biochemical reaction.[2] Free energy, the only kind of energy that can do cellular work, is the aspect of thermodynamics of greatest interest to a biologist.

[2] The *G* that designates free energy comes from J. Willard Gibbs, one of the founders of thermodynamics.

Enthalpy, free energy, and entropy are related by the following equation:

$$H = G + TS$$

in which H is enthalpy, G is free energy, S is entropy, and T is the absolute temperature of the system, expressed in degrees Kelvin. Disregarding temperature (T) for the moment, enthalpy (the total energy of a system) is equal to free energy (the usable energy) and entropy (the unusable energy).

A rearrangement of the equation shows that entropy and free energy are related inversely; as entropy increases, the amount of free energy decreases:

$$G = H - TS$$

If we assume that entropy is zero, the free energy is simply equal to the total potential energy (enthalpy); an increase in entropy reduces the amount of free energy.

What is the significance of the temperature (T)? Remember that as the temperature increases, there is an increase in random molecular motion that contributes to disorder and multiplies the effect of the entropy term.

Chemical reactions involve changes in free energy

Biologists need ways to analyze the role of energy in the many reactions that comprise metabolism. Although the total free energy of a system (G) cannot be effectively measured, the equation $G = H - TS$ can be extended to predict whether any particular chemical reaction will release energy or require an input of energy. This is because *changes* in free energy can be measured. We use the Greek letter delta (Δ) to denote any change that occurs in the system between its initial state before the reaction and its final state after the reaction. To express what happens with respect to energy in a chemical reaction, the equation becomes:

$$\Delta G = \Delta H - T\Delta S$$

Notice that the temperature does not change; it is held constant during the reaction. Thus the change in free energy (ΔG) during the reaction is equal to the change in enthalpy (ΔH) minus the product of the absolute temperature (T) multiplied by the change in entropy (ΔS). ΔG and ΔH are expressed in kilojoules or kilocalories per mole; ΔS is expressed in kilojoules or kilocalories per degree.

Free energy decreases during an exergonic reaction

An **exergonic reaction** releases energy and is said to be a spontaneous or a "downhill" reaction, from higher to lower free energy (Fig. 6–5*a*). Because the total free energy in its final state is less than the total free energy in its initial state, ΔG is a negative number for exergonic reactions.

The term *spontaneous* may give the false impression that such reactions are always instantaneous. In fact, spontaneous reactions do not necessarily occur readily; some are extremely slow. This is because energy, known as activation energy, is required to initiate every reaction, even a spontaneous one. Activation energy will be discussed later in the chapter.

Free energy increases during an endergonic reaction

An **endergonic reaction** is a reaction in which there is a gain of free energy (Fig. 6–5*b*). Because the free energy of the products is greater than the free energy of the reactants, ΔG has a positive value. Such a reaction cannot take place in isolation. Instead, it must occur in such a way that energy can be supplied from the surroundings. Of course, many energy-requiring reactions take place in cells, and as we will see, metabolic mechanisms have evolved that supply the energy needed to "drive" these nonspontaneous cellular reactions in a particular direction.

Free energy changes depend on the concentrations of reactants and products

According to the second law of thermodynamics, any process that increases entropy can do work. Differences in concentration of a substance, for example, between two different parts of a cell, represent a more orderly state than when the substance

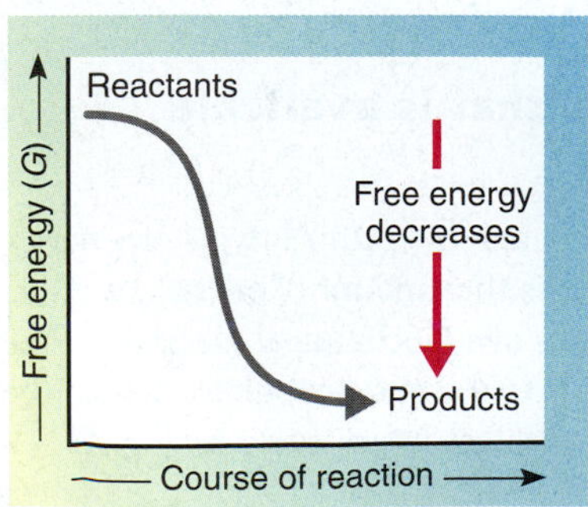

(a) Exergonic reaction (spontaneous; energy-releasing)

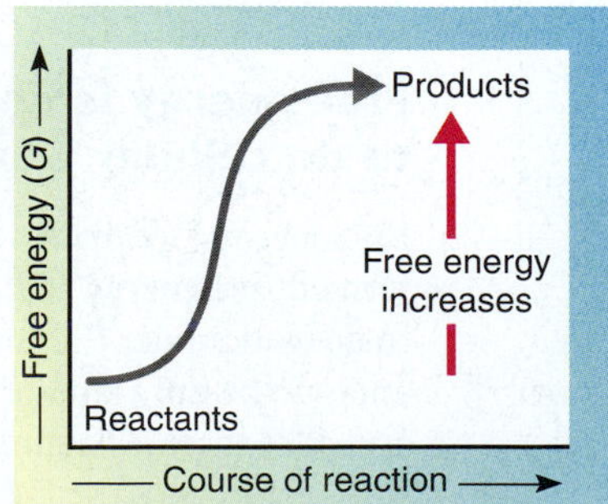

(b) Endergonic reaction (not spontaneous; energy-requiring)

Figure 6–5 Exergonic and endergonic reactions. **(a)** In an exergonic reaction there is a net loss of free energy. The products have less free energy than was present in the reactants, and the reaction proceeds spontaneously. **(b)** In an endergonic reaction there is a net gain in free energy. The products have more free energy than was present in the reactants. An endergonic reaction occurs only if energy is supplied by an exergonic reaction.

is diffused homogeneously throughout the cell (recall the discussion of concentration gradients earlier in the chapter). We have seen that free energy changes in any chemical reaction depend mainly on the difference in bond energies (enthalpy, *H*) between reactants and products. Free energy also depends on *concentrations* of both reactants and products. The change in molecules from a more concentrated to a less concentrated state increases entropy because it is movement from a more orderly to a less orderly state.

In most biochemical reactions there is little intrinsic free energy difference between reactants and products. Such reactions are reversible, a fact that is indicated by drawing double arrows ($\rightleftharpoons$) between the reactants and the products.

$$A \rightleftharpoons B$$

At the beginning of a reaction, only the reactant molecules (A) may be present. As the reaction proceeds, the concentration of the reactant molecules decreases, and the concentration of the product molecules (B) increases. As the concentration of the product molecules increases, they may have enough free energy to initiate the reverse reaction. The reaction thus proceeds in both directions simultaneously; if undisturbed it could eventually reach a state of **dynamic equilibrium,** in which the rate of the reverse reaction is equal to the rate of the forward reaction. At equilibrium there is no net change in the system; every forward reaction is balanced by a reverse reaction.

At a given temperature and pressure, each reaction has its own characteristic equilibrium. For any given reaction, chemists can perform experiments and calculations to determine the relative concentrations of reactants and products present at equilibrium. If the reactants have much greater intrinsic free energy than the products, the reaction goes almost to completion; that is, it reaches equilibrium at a point at which most of the reactants have been converted to products. Reactions in which the reactants have much less intrinsic free energy than the products reach equilibrium at a point where very few of the reactant molecules have been converted to products.

If we increase the initial concentration of A, then the reaction will "shift to the right," and more A will be converted to B. A similar effect can be obtained if B is removed from the reaction mixture. The reaction always shifts in the direction that reestablishes equilibrium, so that the proportions of reactants and products characteristic of that reaction at equilibrium are restored. The opposite effect occurs if the concentration of B is increased or if A is removed; here the system "shifts to the left." The actual free energy change that occurs during a reaction is defined mathematically to include these effects, which are a consequence of the relative initial concentrations of reactants and products.

Cells manipulate the relative concentrations of reactants and products of almost every reaction. Cellular reactions are virtually never at equilibrium. By displacing their reactions far from equilibrium, cells are able to supply energy to endergonic reactions and direct their metabolism in accordance with their needs (see an example of this in the Chapter 5 discussion of facilitated diffusion of glucose in red blood cells).

Cells drive endergonic reactions by coupling them to exergonic reactions

Many metabolic reactions, such as protein synthesis, are anabolic and endergonic. Because an endergonic reaction cannot take place without an input of energy, endergonic reactions are coupled to exergonic reactions. In **coupled reactions,** the thermodynamically favorable exergonic reaction provides the energy required to drive the thermodynamically unfavorable endergonic reaction. The endergonic reaction can proceed only if it absorbs free energy released by the exergonic reaction to which it is coupled.

Consider the free energy change, ΔG, in the following reaction:

$$(1)\ A \longrightarrow B \qquad \Delta G = +20.9 \text{ kJ/mol } (+5 \text{ kcal/mol})$$

Because ΔG has a positive value, we know that the product of this reaction has more free energy than the reactant. This is an endergonic reaction. It is not spontaneous and does not take place without an energy source.

By contrast, consider the following reaction:

$$(2)\ C \longrightarrow D \qquad \Delta G = -33.5 \text{ kJ/mol } (-8 \text{ kcal/mol})$$

The negative value of ΔG tells us that the free energy of the reactant is greater than the free energy of the product. This exergonic reaction proceeds spontaneously.

We can sum up Reactions 1 and 2 as follows:

(1) A ⟶ B	$\Delta G = +20.9$ kJ/mol (+5 kcal/mol)
(2) C ⟶ D	$\Delta G = -33.5$ kJ/mol (−8 kcal/mol)
Overall	$\Delta G = -12.6$ kJ/mol (−3 kcal/mol)

Because thermodynamics considers the overall changes in these two reactions, which show a net negative value of ΔG, the two reactions taken together are exergonic.

The fact that we can write reactions this way is a useful bookkeeping device, but it does not mean that an exergonic reaction can mysteriously transfer energy to an endergonic "bystander" reaction. However, these reactions can be coupled if their pathways are altered such that they are linked by a common intermediate. Reactions 1 and 2 might be coupled by an intermediate *(I)* in the following way:

(3) A + C ⟶ I	$\Delta G = -8.4$ kJ/mol (−2 kcal/mol)
(4) I ⟶ B + D	$\Delta G = -4.2$ kJ/mol (−1 kcal/mol)
Overall	$\Delta G = -12.6$ kJ/mol (−3 kcal/mol)

Note that Reactions 3 and 4 are sequential. Thus the reaction pathways have changed, but overall the reactants (A and C) and products (B and D) are the same, and the free energy change is the same.

Generally, for each endergonic reaction occurring in a living cell, there is a coupled exergonic reaction to drive it. Often, the exergonic reaction involves the breakdown of ATP. We now examine specific examples of the role of ATP in energy coupling.

ATP IS THE ENERGY CURRENCY OF THE CELL

In all living cells, energy is temporarily packaged within a remarkable chemical compound called **adenosine triphosphate (ATP),** which holds readily available energy for very short periods of time. We may think of ATP as the energy currency of the cell. When you work to earn money, you might say that your energy is symbolically stored in the money you earn. The energy the cell requires for immediate use is temporarily stored in ATP, which is like cash. When you earn extra money, you might deposit some in the bank; similarly, a cell might deposit energy in the chemical bonds of lipids, starch, or glycogen. Moreover, just as you dare not make less money than you spend, so too the cell must avoid energy bankruptcy, which would mean its death. Finally, just as you (alas) do not keep what you make very long, so too the cell continuously spends its ATP, which must be replaced immediately.

ATP is a nucleotide consisting of three main parts: adenine, a nitrogen-containing organic base; ribose, a five-carbon sugar; and three phosphate groups, identifiable as phosphorus atoms surrounded by oxygen atoms (Fig. 6–6). Notice that the phosphate groups are bonded to the end of the molecule in a series, rather like three cars behind a locomotive, and, like the cars of a train, they can be attached and detached.

ATP donates energy through the transfer of a phosphate group

When the terminal phosphate is removed from ATP, the remaining molecule is **adenosine diphosphate (ADP)** (see Fig. 6–6). If the phosphate group is not transferred to another molecule, it is released as inorganic phosphate (P_i). This is an exergonic reaction. ATP is sometimes called a "high-energy" compound because the hydrolysis reaction that releases a phosphate has a relatively large $-\Delta G$. (Calculations of the free energy of ATP hydrolysis vary somewhat, but range between about -28 and -37 kJ/mol, or -6.8 to -8.7 kcal/mol.)

$$\text{(5)}\quad \text{ATP} + H_2O \longrightarrow \text{ADP} + P_i \qquad \Delta G = -32 \text{ kJ/mol (or } -7.6 \text{ kcal/mol)}$$

Reaction 5 can be coupled to endergonic reactions in cells. Consider the following endergonic reaction in which the disaccharide sucrose is formed from two monosaccharides, glucose and fructose.

$$\text{(6)}\quad \text{Glucose} + \text{Fructose} \longrightarrow \text{Sucrose} + H_2O \qquad \Delta G = +27 \text{ kJ/mol (or } +6.5 \text{ kcal/mol)}$$

With a free energy change of -32 kJ/mol (-7.6 kcal/mol), the hydrolysis of ATP in Reaction 5 can drive Reaction 6, but only

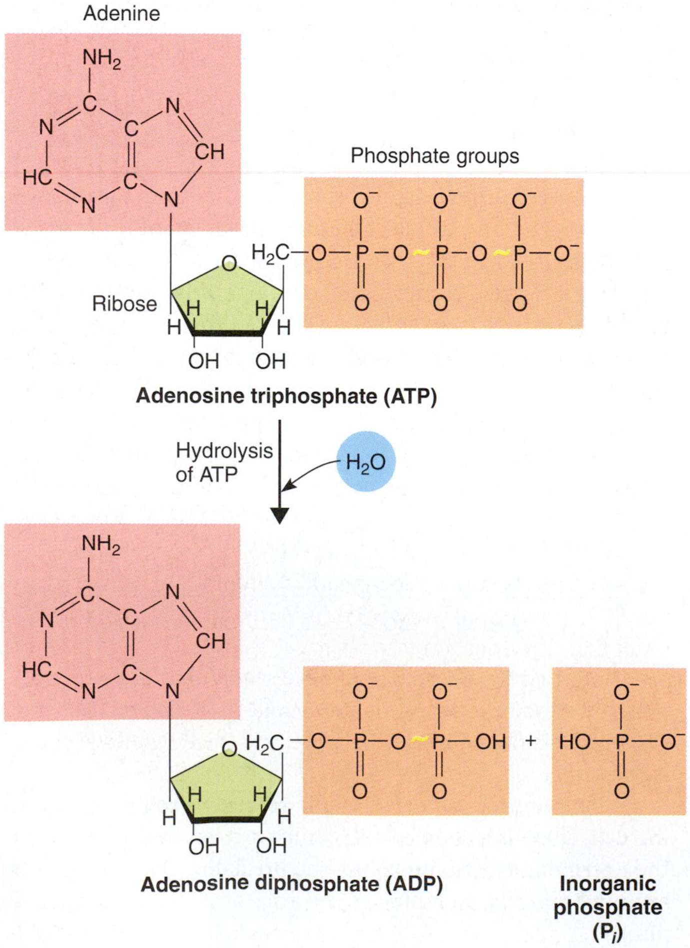

Figure 6–6 ATP and ADP The energy currency of all living things, ATP is composed of adenine, ribose, and three phosphate groups. The hydrolysis of ATP, an exergonic reaction, yields ADP and inorganic phosphate. (Yellow wavy lines indicate unstable bonds. These bonds permit the phosphates to be transferred to other molecules, making them more reactive.)

if the reactions can be coupled through a common intermediate. The following series of reactions is a simplified version of an alternative pathway used by some bacteria.

(7) Glucose + ATP ⟶ Glucose-P + ADP

(8) Glucose-P + Fructose ⟶ Sucrose + P_i

Reaction 7 is a **phosphorylation reaction,** one in which a phosphate group is transferred to some other compound. Glucose is phosphorylated to form glucose phosphate (glucose-P), the intermediate that links the two reactions. Glucose-P, which corresponds to "I" in Reactions 3 and 4, reacts exergonically with fructose to form sucrose. For energy coupling to work in this way, Reactions 7 and 8 must occur in sequence. It is convenient to summarize the reactions in the following way:

(9) Glucose + Fructose + ATP ⟶ Sucrose + ADP + P_i

$\Delta G = -5$ kJ/mol (-1.2 kcal/mol)

When encountering an equation written in this way, remember that it is actually a summary of a series of reactions and that transitory intermediate products (in this case, glucose-P) are sometimes not shown.

ATP links exergonic and endergonic reactions

We have just discussed how the transfer of a phosphate group from ATP to some other compound can be coupled to endergonic reactions in the cell. Conversely, adding a phosphate group to adenosine monophosphate (AMP; forming ADP) or to ADP (forming ATP) requires coupling to exergonic reactions in the cell.

AMP + P_i + Energy ⟶ ADP

ADP + P_i + Energy ⟶ ATP

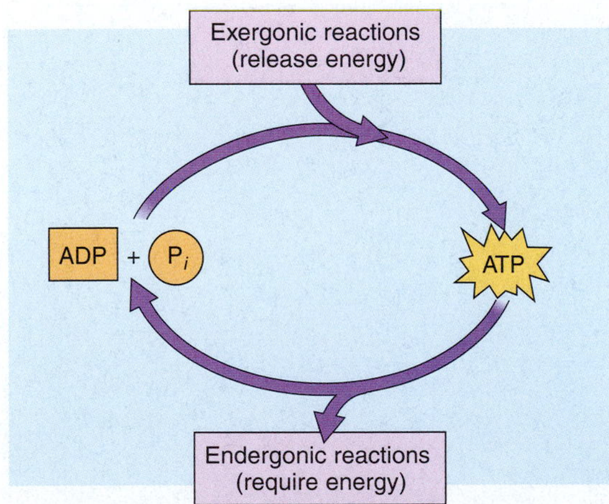

Figure 6–7 ATP links exergonic and endergonic reactions. Because ATP couples many exergonic and endergonic reactions, it is an important link between catabolic reactions, which are exergonic, and anabolic reactions, which are endergonic. Exergonic reactions *(top)* supply the energy to make ATP from ADP, and the hydrolysis of ATP supplies the energy to drive endergonic reactions *(bottom).*

Thus ATP occupies an intermediate position in the metabolism of the cell and is an important link between exergonic reactions, which are generally components of **catabolic pathways,** and endergonic reactions, which are generally part of **anabolic pathways** (Fig. 6–7).

THE CELL MAINTAINS A VERY HIGH RATIO OF ATP TO ADP

The cell maintains a ratio of ATP to ADP far from the equilibrium point. ATP is constantly formed from ADP and inorganic phosphate as nutrients are broken down in cellular respiration or as the radiant energy of sunlight is trapped in photosynthesis. At any point in time, a typical cell contains more than ten ATP molecules for every ADP molecule. The fact that the cell maintains the ATP concentration at such a high level (relative to the concentration of ADP) makes its hydrolysis reaction even more strongly exergonic and more able to drive the endergonic reactions to which it is coupled.

Although the cell maintains a high ratio of ATP to ADP, large quantities of ATP cannot be stored in the cell. The concentration of ATP is always very low, less than 1 mmol/L. In fact, studies suggest that a bacterial cell has no more than a 1-second supply of ATP. Thus, ATP molecules are used almost as quickly as they are produced. A human at rest uses about 45 kg (100 lb) of ATP each day, but the amount present in the body at any given moment is less than 1 g (0.035 oz). Every second in every cell, an estimated 10 million molecules of ATP are made from ADP and phosphate, and an equal number of ATPs transfer their phosphate groups along with their energy to whatever chemical reactions may require them.

CELLS TRANSFER ENERGY BY REDOX REACTIONS

We have seen that cells can transfer energy through the transfer of a phosphate group from ATP. Energy can also be transferred through the transfer of electrons. As discussed in Chapter 2, **oxidation** is the chemical process in which a substance loses electrons, whereas **reduction** is the complementary process in which a substance gains electrons. Because electrons released during an oxidation reaction cannot exist in the free state in living cells, every oxidation reaction must be accompanied by a reduction reaction, in which the electrons are accepted by another atom, ion, or molecule. Oxidation and reduction reactions are often called **redox reactions** because they occur simultaneously. The substance that becomes oxidized gives up energy as it releases electrons, and the substance that becomes reduced receives energy as it gains electrons.

Redox reactions often occur in a series as electrons are transferred from one molecule to another. These electron transfers, which are equivalent to energy transfers, are an essential part of cellular respiration, photosynthesis, and many other chemical reactions. Redox reactions, for example, release the energy stored in food molecules so that ATP can be synthesized using that energy.

Most electron carriers transfer hydrogen atoms

Generally it is not easy to remove one or more electrons from a covalent compound; it is much easier to remove a whole atom. For this reason, redox reactions in cells usually involve the transfer of a hydrogen atom rather than just an electron. A hydrogen atom contains an electron, plus a proton that does not participate in the oxidation/reduction.

When an electron, either singly or as part of a hydrogen atom, is removed from an organic compound, it takes with it some of the energy stored in the chemical bond of which it was a part. That electron, along with its energy, is transferred to an acceptor molecule. An electron progressively loses free energy as it is transferred from one acceptor to another.

One of the most frequently encountered acceptor molecules is **nicotinamide adenine dinucleotide (NAD^+).** When NAD^+ becomes reduced, it temporarily stores large amounts of free energy. Here is a generalized equation showing the transfer of hydrogen from a compound we call X to NAD^+:

$$\underset{\text{Oxidized}}{XH_2 + NAD^+} \longrightarrow \underset{\text{Reduced}}{X + NADH + H^+}$$

Note that the NAD^+ becomes reduced when it combines with hydrogen. NAD^+ is an ion with a net charge of +1. When two electrons and one proton are added, the charge is neutralized and the reduced form of the compound, **NADH,** is produced (Fig. 6–8). (Although the correct way to write the reduced form of NAD^+ is NADH + H^+, for simplicity we will present the reduced form as NADH in this and succeeding chapters.) Some of the energy stored in the bonds holding the hydrogen atoms to molecule X has been transferred by this redox reaction and is temporarily held by NADH. When NADH transfers the electrons to some other molecule, some of their energy is transferred. This energy is usually then transferred through a complex series of reactions that ultimately result in the formation of ATP (see Chapter 7).

Nicotinamide adenine dinucleotide phosphate ($NADP^+$) is a hydrogen acceptor that is chemically similar to NAD^+ but with an extra phosphate group. Unlike NADH, the reduced form of $NADP^+$ (abbreviated **NADPH**) is not involved in ATP synthesis. Instead, the electrons of NADPH are used more directly to provide energy for certain reactions, including certain essential reactions of photosynthesis (see Chapter 8).

Other important hydrogen acceptors or electron acceptors include **flavin adenine dinucleotide (FAD)** and the **cytochromes.**

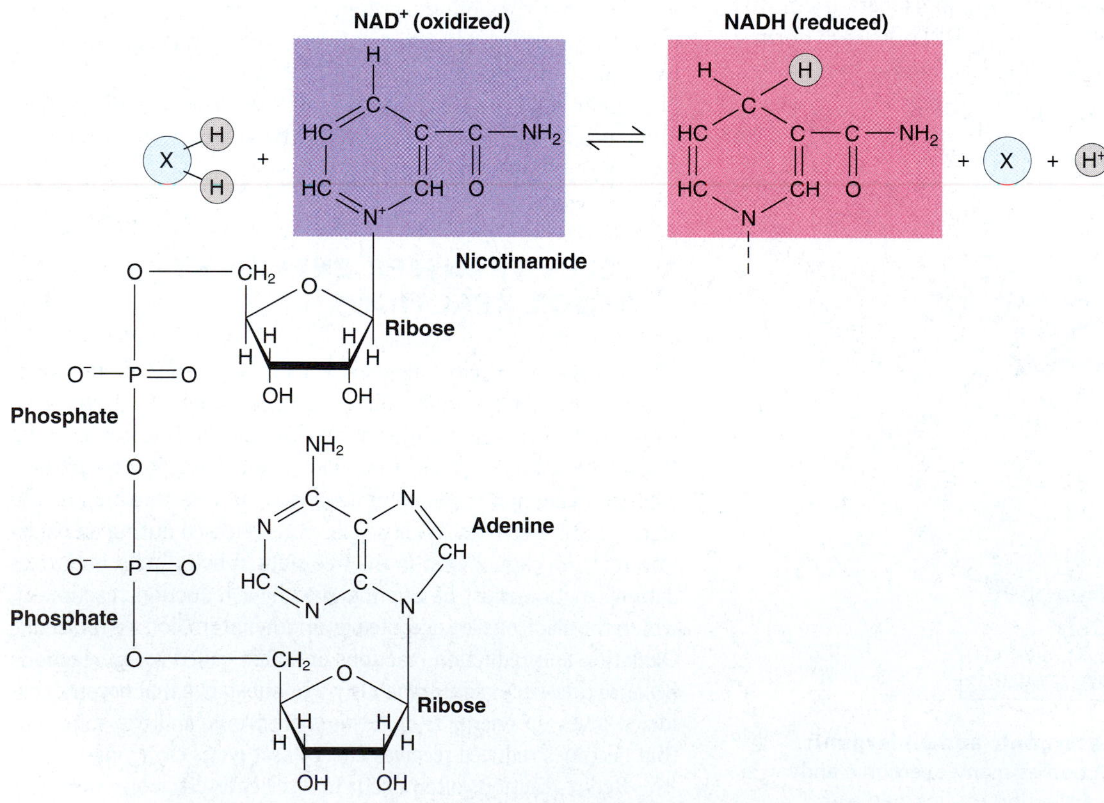

Figure 6–8 NAD^+. NAD^+ consists of two nucleotides, one with adenine and one with nicotinamide, that are joined at their phosphate groups. The oxidized form (NAD^+, *purple screen at top*) becomes reduced (NADH, *pink screen*) by the transfer of two electrons and one proton from another organic compound (XH_2), which becomes oxidized (to X) in the process.

FAD is a nucleotide that accepts hydrogen atoms and their electrons; its reduced form is **$FADH_2$.** The cytochromes are proteins that contain iron; the iron component accepts electrons from hydrogen atoms and then transfers these electrons to some other compound. Like NAD^+ and $NADP^+$, FAD and the cytochromes are electron-transfer agents. Each can exist in a **reduced state,** in which it has more free energy, or in an **oxidized state,** in which it has less. Each is an essential component of many redox reaction sequences in cells.

ENZYMES ARE CHEMICAL REGULATORS

The principles of thermodynamics help us predict whether a reaction can occur, but they tell us nothing about the speed of the reaction. The breakdown of glucose, for example, is an exergonic reaction, yet a glucose solution keeps virtually indefinitely in a bottle if kept free of bacteria and molds and not subjected to high temperatures or strong acids or bases. Cells cannot wait for centuries for glucose to break down, nor can they use extreme conditions to cleave glucose molecules. Cells regulate the rates of chemical reactions with **enzymes,** which are protein **catalysts** that affect the speed of a chemical reaction without being consumed by the reaction.[3]

Cells require a steady release of energy, and they must be able to regulate that release to meet metabolic energy requirements. Metabolism generally proceeds by a series of steps such that a molecule may go through as many as 20 or 30 chemical transformations before it reaches some final state. Even then, the seemingly completed molecule may enter yet another chemical pathway and become totally transformed or consumed to release energy. The changing needs of the cell require a system of flexible metabolic control. The key directors of this control system are enzymes.

The catalytic ability of some enzymes is truly remarkable. For example, hydrogen peroxide (H_2O_2) breaks down extremely slowly if the reaction is uncatalyzed, but a single molecule of the enzyme **catalase** brings about the decomposition of 40 million molecules of hydrogen peroxide per second! Catalase, which has the highest catalytic rate known for any enzyme, protects cells because hydrogen peroxide is a poisonous substance produced as a byproduct of some cellular reactions. The bombardier beetle uses the enzyme catalase as a defense mechanism (Fig. 6–9).

Figure 6–9 Catalase as a defense mechanism. When threatened, a bombardier beetle uses the enzyme catalase to decompose hydrogen peroxide. The oxygen gas formed in the decomposition ejects water and other chemicals with explosive force. Because the reaction releases a great deal of heat, the water comes out as steam. (A wire attached by a drop of adhesive to the beetle's back immobilizes it. His leg was prodded with the dissecting needle on the left to trigger the ejection.) *(Thomas Eisner and Daniel Aneshansley/Cornell University)*

All reactions have a required energy of activation

All reactions, whether exergonic or endergonic, have an energy barrier known as the **energy of activation (E_A),** or **activation energy.** The energy barrier is the energy required to break the existing bonds and begin the reaction. In a population of molecules of any kind, some have a relatively high kinetic energy, while others have a lower energy content. Only molecules with a relatively high kinetic energy are likely to react to form the product.

Even a strongly exergonic reaction, one that releases a substantial quantity of energy as it proceeds, may be prevented from proceeding by the activation energy required to begin the reaction. For example, molecular hydrogen and molecular oxygen can react explosively to form water:

$$2\,H_2 + O_2 \longrightarrow 2\,H_2O$$

This reaction is spontaneous, yet hydrogen and oxygen can be safely mixed as long as all sparks are kept away. This is because the required energy of activation for this particular reaction is relatively high. A tiny spark provides the activation energy that allows a few molecules to react. Their reaction liberates so much heat that the rest react, producing an explosion. Such an explosion occurred on the space shuttle *Challenger* on January 28, 1986 (Fig. 6–10). The failure of a rubber O-ring to seal properly caused the liquid hydrogen in the tank attached to the shuttle to leak and start burning. When the hydrogen tank ruptured a few seconds later, the resulting force caused the nearby oxygen tank to burst as well, mixing hydrogen and oxygen and igniting a huge explosion.

An enzyme lowers a reaction's activation energy

As do all catalysts, enzymes affect the rate of a reaction by lowering the energy needed to initiate the reaction. An enzyme greatly reduces the activation energy (E_A) necessary to initiate a

[3] In recent years scientists have learned that protein enzymes are not the only cellular catalysts; some types of RNA molecules have catalytic activity as well (see Chapter 12).

Figure 6–10 The space shuttle *Challenger* explosion. This disaster resulted from an explosive exergonic reaction between hydrogen and oxygen. All seven crew members died in the accident on January 28, 1986. *(AP Photo/Bruce Weaver)*

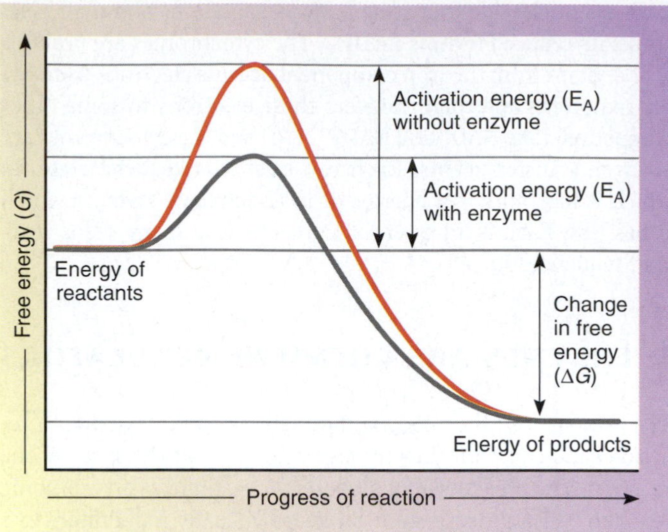

Figure 6–11 Activation energy and enzymes. An enzyme speeds up a reaction by lowering its activation energy (E_A). In the presence of an enzyme, reacting molecules require less kinetic energy to complete a reaction.

chemical reaction (Fig. 6–11). If molecules need less energy to react because the activation barrier is lowered, a larger fraction of the reactant molecules reacts at any one time. As a result, the reaction proceeds more quickly.

Although an enzyme lowers the activation energy for a reaction, it has no effect on the overall free energy change. That is, an enzyme can only promote a chemical reaction that could proceed without it. No catalyst can cause a reaction to proceed in a thermodynamically unfavorable direction or can influence the final concentrations of reactants and products if the reaction goes to equilibrium. Enzymes simply speed up reaction rates.

An enzyme works by forming an enzyme-substrate complex

An uncatalyzed reaction depends on random collisions among reactants. Because of its ordered structure, an enzyme is able to reduce this reliance on random events and thereby control the reaction. The enzyme is thought to accomplish this by forming an unstable intermediate complex with the **substrate,** the substance on which it acts. When the **enzyme-substrate complex,** or **ES complex,** breaks up, the product is released; the original enzyme molecule is regenerated and is free to form a new ES complex.

$$\text{Enzyme} + \text{Substrate(s)} \longrightarrow \text{ES complex}$$
$$\text{ES complex} \longrightarrow \text{Enzyme} + \text{Product(s)}$$

The enzyme itself is not permanently altered or consumed by the reaction and can be reused.

As shown in Figure 6–12*a,* every enzyme contains one or more **active sites,** regions to which the substrate binds, forming the ES complex. The active sites of some enzymes are grooves or cavities in the enzyme molecule, formed by amino acid side chains. The active sites of most enzymes are located close to the surface. During the course of a reaction, substrate molecules occupying these sites are brought close together and react with one another.

The shape of the enzyme does not seem to be exactly complementary to that of the substrate. When the substrate binds to the enzyme molecule, it causes a change, known as **induced fit,** in the shape of the enzyme molecule (Fig. 6–12*b*). Usually the shape of the substrate also changes slightly, in a way that may distort its chemical bonds. The proximity and orientation of the reactants, together with strains in their chemical bonds, facilitate the breakage of old bonds and the formation of new ones. Thus the substrate is changed into product, which moves away from the enzyme. The enzyme is then free to catalyze the reaction of more substrate molecules to form more product molecules.

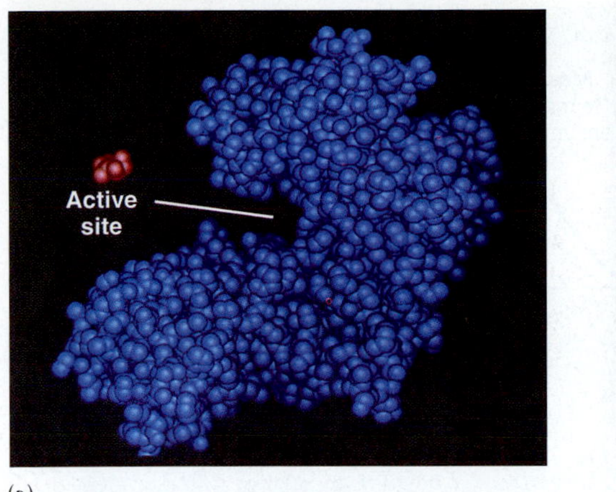

(a)

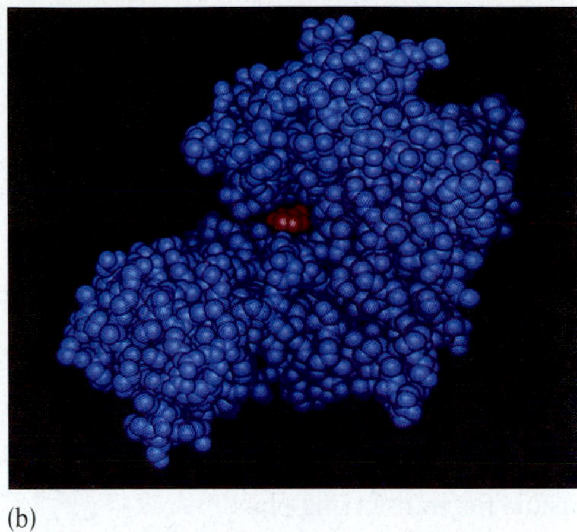

(b)

Figure 6–12 Active site and induced fit. **(a)** Computer graphic model of the enzyme hexokinase *(blue)* and its substrate, glucose *(red)*, before forming an ES complex. The active site of the enzyme is the furrow where glucose will bind. **(b)** The binding of glucose to the active site of hexokinase changes the shape of the enzyme, a phenomenon known as induced fit. Hexokinase, which is involved in cellular respiration, catalyzes the transfer of a phosphate group from ATP to glucose. *(Courtesy of Thomas A. Steitz)*

Most enzyme names end in *-ase*

Enzymes are usually named by the addition of the suffix *-ase* to the name of the substrate. The enzyme sucrase, for example, splits sucrose into glucose and fructose. A few enzymes retain traditional names that do not end in *-ase;* some of these end in *-zyme*. For example, lysozyme (from the Greek *lysis,* "to dissolve") is an enzyme found in tears and saliva; this enzyme breaks down bacterial cell walls. Other examples of enzymes with traditional names include pepsin and trypsin, which break internal peptide bonds in proteins.

Enzymes are specific

Enzymes catalyze virtually every chemical reaction that takes place in an organism. Because there is a close relationship between the shape of the active site and the shape of the substrate, the majority of enzymes are highly specific. Most are capable of catalyzing only a few closely related chemical reactions or, in many cases, only one particular reaction. For example, the enzyme urease, which decomposes urea to ammonia and carbon dioxide, attacks no other substrate. The enzyme sucrase splits only sucrose; it does not act on other disaccharides, such as maltose or lactose. A few enzymes are specific only to the extent that they require the substrate to have a certain kind of chemical bond. For example, lipase, secreted by the pancreas, splits the ester linkages connecting the glycerol and fatty acids of a wide variety of fats.

Enzymes that catalyze similar reactions are classified into groups, although each particular enzyme in the group may catalyze only one specific reaction. The six classes of enzymes that biologists recognize are described in Table 6–1. Each class is divided into many subclasses. For example, sucrase, mentioned above, is referred to as a *glycosidase* because it cleaves a glycosidic linkage (see Chapter 3). Glycosidases are a subclass of the hydrolases.

Many enzymes require cofactors

Some enzymes consist only of protein. For example, the enzyme pepsin, which is secreted by the animal stomach and digests dietary protein by breaking certain peptide bonds, is exclusively a protein molecule. Other enzymes have two components: a protein referred to as the **apoenzyme** and an additional chemical component called a **cofactor.** Neither the apoenzyme nor the cofactor alone has catalytic activity; only when the two are combined

TABLE 6–1 Some Important Classes of Enzymes

Enzyme Class	Function
Oxidoreductases	Catalyze oxidation-reduction reactions
Transferases	Catalyze the transfer of a functional group from a donor molecule to an acceptor molecule
Hydrolases	Catalyze hydrolysis reactions
Isomerases	Catalyze conversion of a molecule from one isomeric form to another
Ligases	Catalyze certain reactions in which two molecules are joined in a process coupled to the hydrolysis of ATP
Lyases	Catalyze certain reactions in which double bonds are formed or broken.

does the enzyme function. A cofactor may be inorganic, or it may be an organic molecule.

Some enzymes require a specific metal ion as a cofactor. Two very common inorganic cofactors are magnesium ions and calcium ions. Most of the trace elements, such as iron, copper, zinc, and manganese, all of which organisms require in very small amounts, function as cofactors.

An organic, nonpolypeptide compound that binds to the apoenzyme and serves as a cofactor is called a **coenzyme.** Most coenzymes are carrier molecules that transfer electrons or part of a substrate from one molecule to another. Some examples of coenzymes have already been introduced in this chapter. NADH, NADPH, and $FADH_2$ are coenzymes; they transfer electrons. ATP functions as a coenzyme; it is responsible for transferring phosphate groups. Yet another coenzyme, **coenzyme A,** is involved in the transfer of groups derived from organic acids. Most vitamins, which are organic compounds that an organism requires in small amounts but cannot synthesize itself, are coenzymes or components of coenzymes (see Table 45–3).

Enzymes are most effective at optimal conditions

Enzymes generally work best under certain narrowly defined conditions, such as appropriate temperature, pH, and ion concentration. Any departure from optimal conditions adversely affects enzyme activity.

Each enzyme has an optimal temperature

Most enzymes have an optimal temperature, at which the rate of reaction is fastest. For human enzymes, the temperature optima are near the human body temperature (35° to 40° C). Enzymatic reactions occur slowly or not at all at low temperatures. As the temperature increases, molecular motion increases, resulting in more molecular collisions. The rates of most enzyme-controlled reactions therefore increase as the temperature increases, within limits (Fig. 6–13*a*). High temperatures rapidly denature most enzymes. The molecular conformation (three-dimensional shape) of the protein becomes altered as the hydrogen bonds responsible for its secondary, tertiary, and quaternary structures are broken. Because this inactivation is usually not reversible, activity is not regained when the enzyme is cooled.

Most organisms are killed by even a short exposure to high temperature; their enzymes are denatured, and they are unable to continue metabolism. A few remarkable exceptions to this rule exist. Certain species of bacteria can survive in the waters of hot springs, such as those in Yellowstone Park, where the temperature is almost 100° C; these organisms are responsible for the brilliant colors in the terraces of the hot springs (Fig. 6–14). Still other bacteria live at temperatures much above that of boiling water, near deep-sea vents, where the extreme pressure keeps water in its liquid state (see Chapter 23 and *Focus On: Life Without the Sun* in Chapter 53).

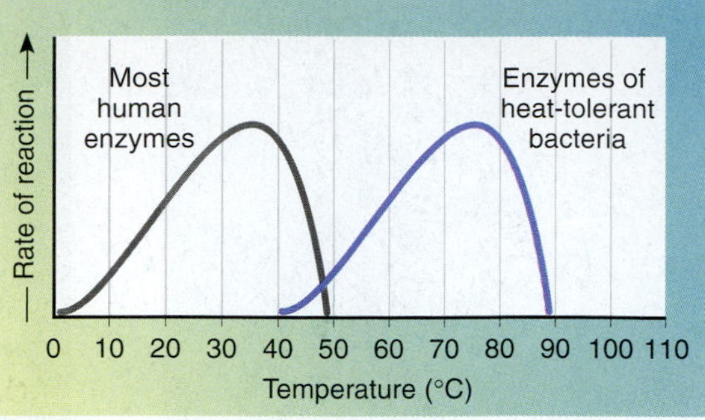

(a)

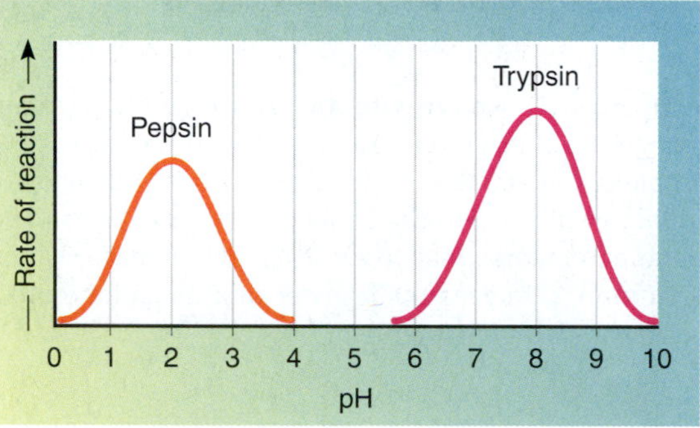

(b)

Figure 6–13 Effect of temperature and pH on enzyme activity. Substrate and enzyme concentrations are held constant in the reactions illustrated. **(a)** Generalized curves for the effect of temperature on enzyme activity. As temperature increases, enzyme activity increases until it reaches an optimal temperature. Enzyme activity abruptly falls after it exceeds the optimal temperature because the enzyme, being a protein, denatures. **(b)** Enzyme activity is very sensitive to pH. Pepsin is a protein-digesting enzyme in the very acidic stomach juice. Trypsin, secreted by the pancreas into the slightly basic small intestine, digests polypeptides.

Each enzyme has an optimal pH

Most enzymes are active only over a narrow pH range and have an optimal pH, at which the rate of reaction is fastest. The optimal pH for most human enzymes is between 6 and 8. (Recall from Chapter 2 that buffers minimize pH changes in cells so that the pH is maintained within a narrow limit.) Pepsin, a protein-digesting enzyme secreted by cells lining the stomach, is remarkable in that it works only in a very acid medium, optimally at pH 2 (Fig. 6–13*b*). In contrast, trypsin, a protein-splitting enzyme secreted by the pancreas, functions best under the slightly basic conditions found in the small intestine.

The activity of an enzyme may be markedly changed by any alteration in pH, which in turn alters charges on the enzyme. Changes in charge affect the ionic bonds that contribute to tertiary and quaternary structure, thus changing the protein's conformation and activity. Many enzymes become inactive, and usu-

Figure 6–14 Yellowstone National Park's Grand Prismatic Spring. The world's third largest spring, about 61 m (200 ft) in diameter, the Grand Prismatic Spring teems with heat-tolerant bacteria. The rings around the perimeter, where the water is slightly cooler, get their distinctive colors from the various kinds of bacteria living there. *(From Smith, R.B., and L.J. Siegel.* Windows into the Earth: The Geologic Story of Yellowstone and Grand Teton National Parks. *Oxford University Press, Oxford, 2000)*

ally irreversibly denatured, when the medium is made very acidic or very basic.

Enzymes are organized into teams in metabolic pathways

Enzymes play an essential role in reaction coupling because they usually work in sequence, with the product of one enzyme-controlled reaction serving as the substrate for the next. We can picture the inside of a cell as a factory with many different assembly (and disassembly) lines operating simultaneously. An assembly line is composed of a number of enzymes. Each enzyme carries out one step, such as changing molecule A into molecule B. Then molecule B is passed along to the next enzyme, which converts it into molecule C, and so on. Such a series of reactions is referred to as a **metabolic pathway.**

$$\mathrm{A} \xrightarrow{\text{Enzyme 1}} \mathrm{B} \xrightarrow{\text{Enzyme 2}} \mathrm{C}$$

Each of these reactions is theoretically reversible, and the fact that it is catalyzed by an enzyme does not change that fact. An enzyme does not itself determine the direction of the reaction it catalyzes. However, the overall reaction sequence is portrayed as proceeding from left to right. You will recall that if there is little intrinsic free energy difference between the reactants and products for a particular reaction, its direction will be determined mainly by the relative concentrations of reactants and products.

In biological pathways, both intermediate and final products are often removed and converted to other chemical compounds. Such removal drives the sequence of reactions in a particular direction. Let us assume that Reactant A is being constantly supplied and that its concentration remains constant. Enzyme 1 converts Reactant A to Product B. The concentration of B is always lower than the concentration of A because B is removed as it is converted to C in the reaction catalyzed by Enzyme 2. If C is removed as quickly as it is formed (perhaps by leaving the cell), the entire reaction pathway is "pulled" toward C.

The cell regulates enzymatic activity

Enzymes regulate the chemistry of the cell, but what controls the enzymes? One mechanism depends simply on controlling the amount of enzyme produced. A specific gene directs the synthesis of each type of enzyme. The gene, in turn, may be switched on by a signal from a hormone or by some other type of cellular product. When the gene is switched on, the enzyme is synthesized. The amount of enzyme present then influences the rate of the reaction.

If the pH and temperature are kept constant (as they are in most cells), the rate of the reaction can be affected by the substrate concentration or by the enzyme concentration. If an excess of substrate is present, the enzyme concentration is the rate-limiting factor. The initial rate of the reaction is then directly proportional to the concentration of enzyme (Fig. 6–15*a*).

If the enzyme concentration is kept constant, the rate of an enzymatic reaction is proportional to the concentration of substrate present. Substrate concentration is the rate-limiting factor at lower concentrations; the rate of the reaction is therefore directly proportional to the substrate concentration. However, at higher substrate concentrations the enzyme molecules become saturated with substrate, that is, substrate molecules are bound to all available active sites of enzyme molecules. In this situation, increasing the substrate concentration does not increase the reaction rate (Fig. 6–15*b*).

The product of one enzymatic reaction may control the activity of another enzyme, especially in a complex sequence of enzymatic reactions. For example, consider the following metabolic pathway:

$$\mathrm{A} \xrightarrow{\text{Enzyme 1}} \mathrm{B} \xrightarrow{\text{Enzyme 2}} \mathrm{C} \xrightarrow{\text{Enzyme 3}} \mathrm{D} \xrightarrow{\text{Enzyme 4}} \mathrm{E}$$

A different enzyme catalyzes each step, and the final product E may inhibit the activity of Enzyme 1. When the concentration of E is low, the sequence of reactions proceeds rapidly. However, an increasing concentration of E serves as a signal for Enzyme 1 to slow down and eventually to stop functioning. Inhibition of Enzyme 1 stops the entire reaction sequence. This type of enzyme regulation, in which the formation of a product inhibits an earlier reaction in the sequence, is called **feedback inhibition** (Fig. 6–16).

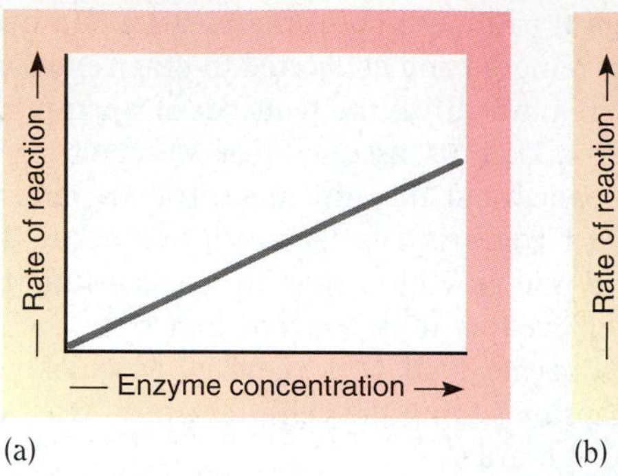

(a)

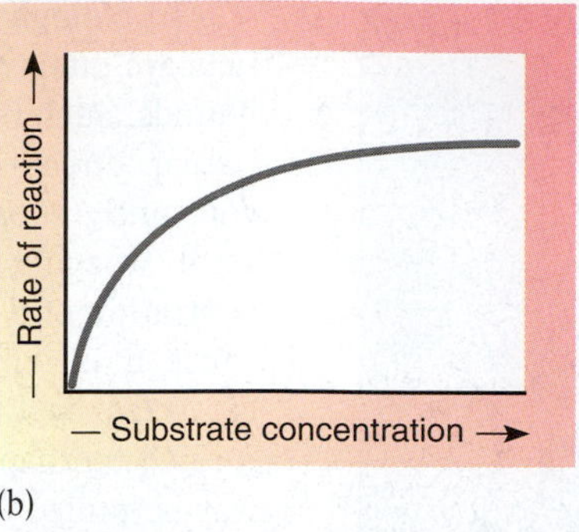

(b)

Figure 6–15 Effect of enzyme concentration and substrate concentration on the rate of a reaction. **(a)** In this example, the rate of reaction is measured at different enzyme concentrations, and an excess of substrate is present at all times. (Temperature and pH are kept at a constant level.) The rate of the reaction is therefore directly proportional to the enzyme concentration. **(b)** In this example, the rate of reaction is measured at different substrate concentrations, and enzyme concentration, temperature, and pH are constant. If the substrate concentration is relatively low, then the reaction rate is directly proportional to substrate concentration. However, higher substrate concentrations do not increase the reaction rate because the enzyme molecules become saturated with substrate.

Another important method of enzymatic control focuses on the activation of enzyme molecules. In their inactive form, the active sites of the enzyme are inappropriately shaped, so that the substrates do not fit. Among the factors that influence the shape of the enzyme are pH, the concentration of certain ions, and the addition of phosphate groups to certain amino acids in the enzyme.

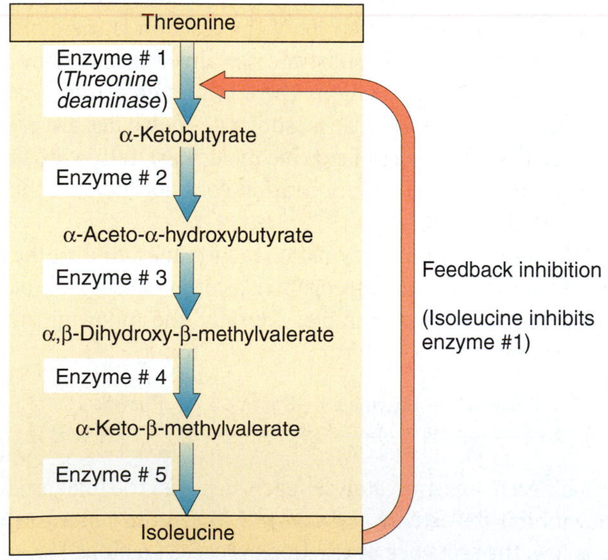

Figure 6–16 Feedback inhibition. Bacteria synthesize the amino acid isoleucine from the amino acid threonine. The isoleucine synthetic pathway involves five steps, each catalyzed by a different enzyme. When enough isoleucine accumulates in the cell, the isoleucine inhibits threonine deaminase, the enzyme that catalyzes the first step in this pathway.

Some enzymes possess a receptor site, called an **allosteric site,** on some region of the enzyme molecule other than the active site. (The word *allosteric* means "another space.") When a substance binds to an enzyme's allosteric site, the conformation of the enzyme's active site is changed, thereby modifying the enzyme's activity. Substances that affect enzyme activity by binding to allosteric sites are called **allosteric regulators.** Some allosteric regulators are inhibitors that keep the enzyme in its inactive shape. Other allosteric regulators are activators that result in an enzyme with a functional active site.

The enzyme **cyclic AMP-dependent protein kinase** is an allosteric enzyme with a regulator that is a protein that binds reversibly to the allosteric site and inactivates the enzyme. Protein kinase is in this inactive form most of the time (Fig. 6–17). When protein kinase activity is needed, the compound cyclic AMP (cAMP; see Fig. 3–27) contacts the enzyme-inhibitor complex and removes the inhibitory protein, thereby activating the protein kinase. Activation of protein kinases by cAMP is an important aspect of the mechanism of action of certain hormones (see Chapters 5 and 47).

Enzymes can be inhibited by certain chemical agents

Most enzymes may be inhibited or even destroyed by certain chemical agents. Enzyme inhibition may be reversible or irreversible. **Reversible inhibition** occurs when an inhibitor forms weak chemical bonds with the enzyme. Reversible inhibition can be competitive or noncompetitive.

In **competitive inhibition,** the inhibitor competes with the normal substrate for binding to the active site of the enzyme (Fig. 6–18*a*). Usually a competitive inhibitor is structurally similar to the normal substrate and so fits into the active site

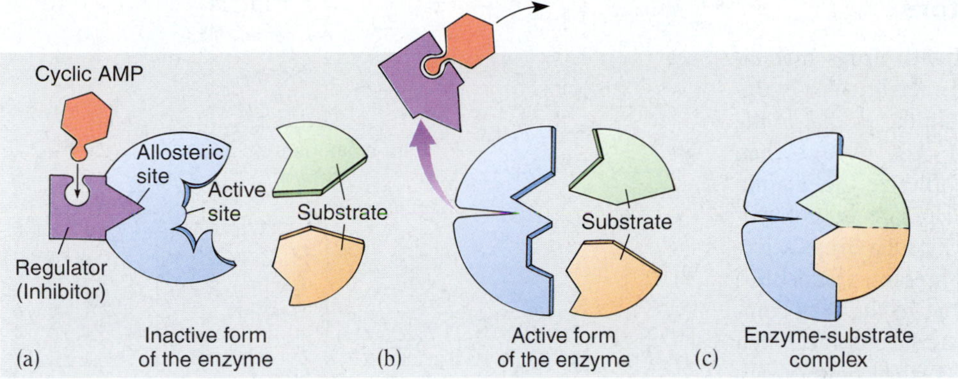

Figure 6–17 Allosteric enzyme. **(a)** The enzyme protein kinase is inhibited by a regulatory protein that binds reversibly to its allosteric site. When the enzyme is in this inactive form, the shape of the active site is modified so that the substrate cannot combine with it. **(b)** Cyclic AMP removes the allosteric inhibitor and activates the enzyme. The substrate can then combine with the active site **(c)**.

and combines with the enzyme. However, it is not similar enough to substitute fully for the normal substrate in the chemical reaction, and the enzyme cannot attack it to form product molecules. A competitive inhibitor occupies the active site only temporarily and does not permanently damage the enzyme. In competitive inhibition, an active site is occupied by the inhibitor part of the time and by the normal substrate part of the time. If the concentration of the substrate is increased relative to the concentration of the inhibitor, the active site will usually be occupied by the substrate. Competitive inhibition is demonstrated experimentally by the fact that increasing the substrate concentration reverses competitive inhibition.

In **noncompetitive inhibition,** the inhibitor binds with the enzyme at a site other than the active site (Fig. 6–18*b*). Such an inhibitor inactivates the enzyme by altering its shape so that the active site cannot bind with the substrate. Many important noncompetitive inhibitors are metabolic substances that regulate enzyme activity by combining reversibly with the enzyme. Noncompetitive inhibition has some features in common with allosteric inhibition, discussed previously.

In **irreversible inhibition,** an inhibitor permanently inactivates or destroys an enzyme when it combines with one of its functional groups, either at the active site or elsewhere. Many poisons are irreversible enzyme inhibitors. For example, heavy metals such as mercury and lead bind irreversibly to and denature many proteins, including enzymes. Certain nerve gases poison the enzyme acetylcholinesterase, which is important to the function of nerves and muscles. Cytochrome oxidase, one of the enzymes that transports electrons in cellular respiration, is especially sensitive to cyanide. Death results from cyanide poisoning because cytochrome oxidase is irreversibly inhibited and can no longer transfer electrons from its substrate to oxygen.

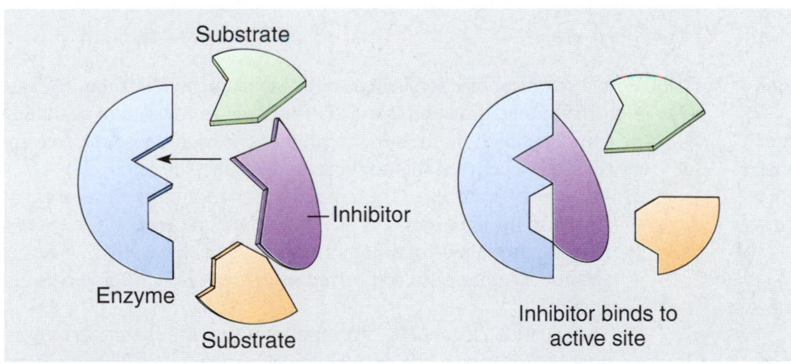

(a) Competitive inhibition

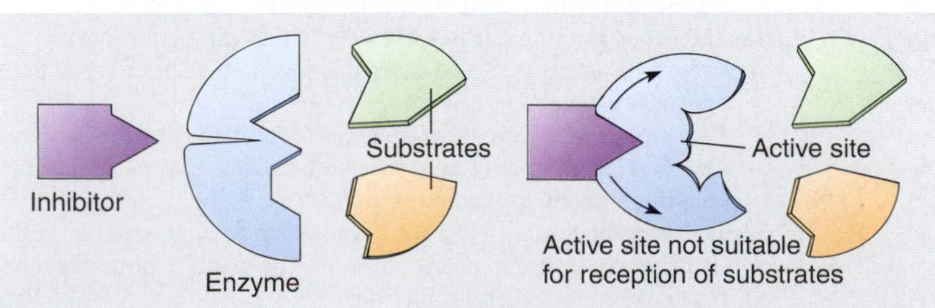

(b) Noncompetitive inhibition

Figure 6–18 Competitive and noncompetitive inhibition. **(a)** In competitive inhibition, the inhibitor competes with the normal substrate for the active site of the enzyme. A competitive inhibitor occupies the active site only temporarily. **(b)** In noncompetitive inhibition, the inhibitor binds with the enzyme at a site other than the active site, altering the shape of the enzyme and thereby inactivating it. Noncompetitive inhibition may be reversible.

Some drugs are enzyme inhibitors

Many bacterial infections are treated with drugs that directly or indirectly inhibit bacterial enzyme activity. For example, sulfa drugs have a chemical structure similar to that of the nutrient **para-aminobenzoic acid (PABA)** (Fig. 6–19). When PABA is available, microorganisms can synthesize the vitamin **folic acid,** which is necessary for growth. Humans do not synthesize folic acid from PABA, and that is why sulfa drugs selectively affect bacteria. When a sulfa drug is present, competitive inhibition occurs within the bacterium—that is, the drug competes with PABA for the active site of the bacterial enzyme. When bacteria use the sulfa drug instead of PABA, they synthesize a compound that cannot be used to make folic acid. Therefore, the bacterial cells are unable to grow.

Penicillin and related antibiotics irreversibly inhibit a bacterial enzyme called transpeptidase. This enzyme is responsible for establishing some of the chemical linkages in the bacterial cell wall. Susceptible bacteria cannot produce properly constructed cell walls and are prevented from multiplying effectively. Human cells do not have cell walls and do not employ this enzyme. Thus, except for individuals allergic to it, penicillin is harmless to humans. Unfortunately, during the years since it was introduced, resistance to penicillin has evolved in many bacterial strains. The resistant bacteria fight back with an enzyme of their own, penicillinase, which breaks down the penicillin and renders it ineffective. Because bacteria evolve at such a rapid rate, drug resistance is a growing problem in medical practice (see section in Chapter 17 on infectious disease organisms evolving resistance to drugs). Although new antibacterial drugs are constantly under development, certain serious infections, such as tuberculosis, are becoming increasingly difficult to treat.

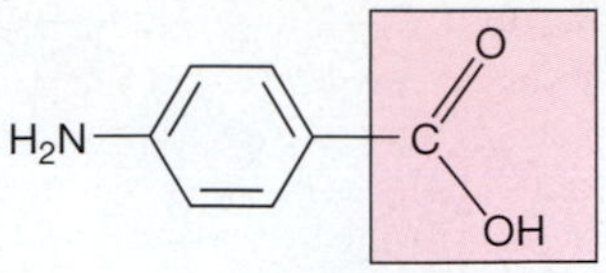

Figure 6–19 Para-aminobenzoic acid and sulfonamides. Sulfa drugs owe their antibiotic properties to their similarity in structure to para-aminobenzoic acid (PABA), a precursor in the synthesis of folic acid. Sulfa drugs block the synthesis of folic acid, an important vitamin necessary for growth. Animals, including humans, obtain folic acid in their diets, but many bacteria must synthesize it.

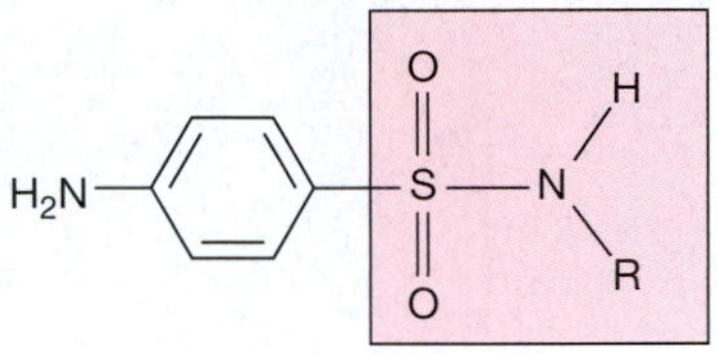

SUMMARY WITH KEY TERMS

I. **Energy** can be defined as the capacity to do work (expressed in **kilojoules, kJ**).
   A. All life depends on a continuous input of energy. Most producers capture **radiant energy** during photosynthesis and incorporate some of it into the chemical bonds of organic compounds. Some of this **chemical energy** then becomes available to consumers and decomposers.
   B. All forms of energy are interconvertible.
      1. **Potential energy** is stored energy; **kinetic energy** is energy of motion.
      2. Energy can be conveniently measured as **heat energy,** thermal energy that flows from an object with a higher temperature to an object with a lower temperature; the unit of heat energy is the **kilocalorie (kcal),** which is equal to 4.184 kilojoules. Heat energy cannot do cellular work.

II. The **first law of thermodynamics** states that energy cannot be created or destroyed but can be transferred and changed in form. The **second law of thermodynamics** states that disorder (entropy) in the universe is continuously increasing.
   A. The first law explains why organisms cannot produce energy but must continuously capture it from the surroundings.
   B. The second law explains why no process requiring energy is ever 100% efficient. In every energy transaction, some energy is dissipated as heat, which contributes to **entropy.**

III. When a chemical reaction is in a state of **dynamic equilibrium,** the rate of change in one direction is exactly the same as the rate of change in the opposite direction; the system can do no work because the **free energy** difference between the reactants and products is zero.
   A. As entropy increases, the amount of free energy decreases, as shown in the equation $G = H - TS$, in which G is the free energy, H is the **enthalpy** (total potential energy of the system), T is the absolute temperature (expressed in degrees Kelvin), and S is entropy.
   B. The equation $\Delta G = \Delta H - T\Delta S$ indicates that the change in free energy (ΔG) during a chemical reaction is equal to the change in enthalpy (ΔH) minus the product of the absolute temperature (T) multiplied by the change in entropy (ΔS).

IV. A spontaneous reaction releases free energy that can perform work.
   A. Free energy decreases in an **exergonic reaction.** Exergonic reactions are spontaneous.
   B. Free energy increases in an **endergonic reaction.** The input of free energy required to drive an endergonic reaction may be supplied by **coupling** it to an exergonic reaction.

V. **Adenosine triphosphate (ATP)** is the immediate energy currency of the cell; it generally transfers energy through the transfer of its terminal phosphate group to acceptor molecules.
   A. ATP is formed by the **phosphorylation** of **ADP,** an endergonic process that requires an input of energy.

B. ATP is the common cellular link between exergonic and endergonic reactions and between **catabolism** (degradation of large complex molecules into smaller, simpler molecules) and **anabolism** (synthesis of complex molecules from simpler molecules).

VI. Energy can be transferred in **oxidation-reduction (redox) reactions.**

A. A substance that becomes oxidized gives up one or more electrons (and energy) to a substance that becomes reduced. Electrons are typically transferred as part of hydrogen atoms.

B. **NAD^+** and **$NADP^+$** accept electrons as part of hydrogen atoms and become reduced to form NADH and NADPH, respectively. These electrons (along with some of their energy) can be transferred to other acceptors.

VII. An **enzyme** is a biological **catalyst;** it greatly increases the speed of a chemical reaction without being consumed.

A. An enzyme lowers the **activation energy,** the kinetic energy necessary to get a reaction going.

B. An **active site** of an enzyme is a three-dimensional region where **substrates** come into close contact and thereby react more readily. A substrate binds to an active site, causing an **induced fit,** in which the shapes of the enzyme and substrate change slightly.

C. Some enzymes consist of an **apoenzyme** (its protein component) and a **cofactor.**

1. Most inorganic cofactors are metal ions.
2. A **coenzyme** is an organic cofactor; many coenzymes transfer electrons or part of a substrate from one molecule to another.

D. Enzymes work best at specific temperature and pH conditions.

E. A cell can regulate enzymatic activity by controlling the amount of enzyme produced and by regulating metabolic conditions that influence the shape of the enzyme.

1. Some enzymes have **allosteric sites,** noncatalytic sites to which a substance can bind, changing the enzyme's activity.
2. Allosteric enzymes are subject to **feedback inhibition,** in which the formation of an end product inhibits an earlier reaction in the sequence.

F. Certain chemical substances inhibit most enzymes. Inhibition may be reversible or irreversible.

1. **Reversible inhibition** occurs when an inhibitor forms weak chemical bonds with the enzyme. Reversible inhibition may be **competitive,** in which the inhibitor competes with the substrate for the active site, or **noncompetitive,** in which the inhibitor binds with the enzyme at a site other than the active site.
2. **Irreversible inhibition** occurs when an inhibitor combines with an enzyme and permanently inactivates it.

POST-TEST

1. According to the first law of thermodynamics (a) energy may be changed from one form to another but is neither created nor destroyed (b) much of the work an organism does is mechanical work (c) the disorder of the universe is increasing (d) free energy is available to do cellular work (e) a cell is in a state of dynamic equilibrium
2. According to the second law of thermodynamics (a) energy may be changed from one form to another but is neither created nor destroyed (b) much of the work an organism does is mechanical work (c) the disorder of the universe is increasing (d) free energy is available to do cellular work (e) a cell is in a state of dynamic equilibrium
3. In thermodynamics, ____________ is a measure of the amount of disorder in the system. (a) bond energy (b) catabolism (c) entropy (d) enthalpy (e) work
4. The ____________ energy of a system is that part of the total energy available to do cellular work. (a) activation (b) bond (c) kinetic (d) free (e) heat
5. A reaction that requires a net input of free energy is described as (a) exergonic (b) endergonic (c) spontaneous (d) both a and c (e) both b and c
6. A reaction that releases energy is described as (a) exergonic (b) endergonic (c) spontaneous (d) both a and c (e) both b and c
7. A spontaneous reaction is one in which the change in free energy *(ΔG)* has a ____________ value. (a) positive (b) negative (c) positive or negative (d) none of these (ΔG has no measurable value)
8. To drive a reaction that requires an input of energy (a) an enzyme-substrate complex must form (b) the concentration of ATP must be decreased (c) the activation energy must be increased (d) some reaction that yields energy must be coupled to it (e) some reaction that requires energy must be coupled to it
9. Which of the following reactions could be coupled to an endergonic reaction with $\Delta G = +3.56$ kJ/mol? (a) $A \rightarrow B$, $\Delta G = +6.08$ kJ/mol (b) $C \rightarrow D$, $\Delta G = +3.56$ kJ/mol (c) $E \rightarrow F$, $\Delta G = 0$ kJ/mol (d) $G \rightarrow H$, $\Delta G = -1.22$ kJ/mol (e) $I \rightarrow J$, $\Delta G = -5.91$ kJ/mol
10. Consider the reaction: Glucose + 6 $O_2 \rightarrow$ 6 CO_2 + 6 H_2O ($\Delta G = -2880$ kJ/mol). Which of the following statements about this reaction is *not* true? (a) the reaction is spontaneous in a thermodynamic sense (b) a small amount of energy (activation energy) must be supplied to start the reaction, which then proceeds with a release of energy (c) the reaction is exergonic (d) the reaction can be coupled to an endergonic reaction (e) the reaction can be coupled to an exergonic reaction
11. The kinetic energy required to initiate a reaction is called (a) activation energy (b) bond energy (c) potential energy (d) free energy (e) heat energy
12. A biological catalyst that affects the rate of a chemical reaction without being consumed by the reaction is a(an) (a) product (b) cofactor (c) coenzyme (d) substrate (e) enzyme
13. The region of an enzyme molecule that combines with the substrate is the (a) allosteric site (b) reactant (c) active site (d) coenzyme (e) product
14. Which inhibitor binds to the active site of an enzyme? (a) noncompetitive inhibitor (b) competitive inhibitor (c) irreversible inhibitor (d) allosteric regulator (e) PABA
15. In the following reaction series, which enzyme(s) is/are most likely to have an allosteric site to which the end product E binds?

$$A \xrightarrow{\text{Enzyme 1}} B \xrightarrow{\text{Enzyme 2}} C \xrightarrow{\text{Enzyme 3}} D \xrightarrow{\text{Enzyme 4}} E$$

(a) enzyme 1 (b) enzyme 2 (c) enzyme 3 (d) enzyme 4 (e) enzymes 3 and 4

REVIEW QUESTIONS

1. You exert tension on a spring and then release it. Explain how these actions relate to work, potential energy, and kinetic energy.
2. Life is sometimes described as a constant struggle against the second law of thermodynamics. How do organisms succeed in this struggle without violating the second law?
3. Consider the free energy change in a reaction in which enthalpy decreases and entropy increases. Is ΔG zero, or does it have a positive value or a negative value? Is the reaction endergonic or exergonic?
4. Why do coupled reactions typically have common intermediates? Give a generalized example involving ATP. Why is ATP able to serve as an important link between exergonic and endergonic reactions?
5. What is activation energy? What effect does an enzyme have on activation energy?
6. Give the function of each of the following: (a) active site of an enzyme (b) coenzyme (c) allosteric site
7. Describe three factors that influence how an enzyme functions.
8. Label the diagram depicting the progress of a chemical reaction with these labels: energy of reactants, energy of products, activation energy with enzyme, activation energy without enzyme, change in free energy. Use Figure 6–11 to check your answers.

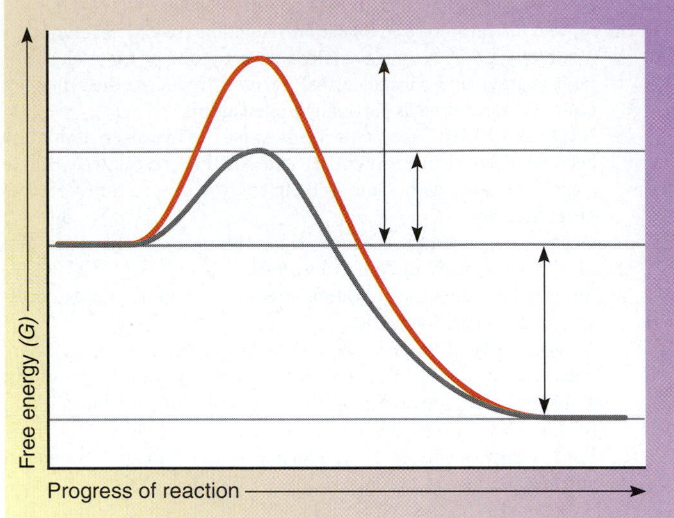

YOU MAKE THE CONNECTION

1. Reaction 1 and Reaction 2 happen to have the same free energy change: $\Delta G = -41.8$ kJ/mol (-10 kcal/mol). Reaction 1 is at equilibrium, but Reaction 2 is far from equilibrium. Is either reaction capable of performing work? If so, which one?
2. You are doing an experiment in which you are measuring the rate at which succinate is converted to fumarate by the enzyme succinic dehydrogenase. You decide to add a little malonate to make things interesting. You observe that the reaction rate slows markedly and conclude that malonate must be acting as an inhibitor. Design an experiment that will help you decide whether malonate is acting as a competitive inhibitor or a noncompetitive inhibitor.
3. Based on what you have learned in this chapter, explain why an extremely high fever (above 105° F or 40° C) is often fatal.

RECOMMENDED READINGS

Adams, S. "No Way Back." *New Scientist,* Oct. 1994. Examines the second law of thermodynamics.

Atkins, P.W. *The Second Law.* W.H. Freeman, San Francisco, 1984. A basic, understandable introduction to thermodynamics with an extensive section devoted to its biological implications.

Deeth, R. "Chemical Choreography." *New Scientist,* Jul. 1997. Reports on the use of computer modeling to study how enzymes function in cells.

Hinrichs, R.A., and M. Kleinbach. *Energy: Its Use and the Environment,* 3rd ed. Harcourt College Publishers, Philadelphia, 2002. This introductory text focuses on the physical principles behind energy use and its effects on the environment.

Tobin, A.J., and R.E. Morel. *Asking About Cells.* Saunders College Publishing, Philadelphia, 1997. A readable cell biology text with excellent coverage of cellular energetics

- Visit our Web site at **http://www.info.brookscole.com/solomonbergmartin** for links to chapter-related resources on the World Wide Web. Additional on-line materials relating to this chapter can also be found on our Web site.

See chapter activity on BioActive Learner CD for additional help in mastering the chapter's material. Icon location in the chapter's margins shows which topics have tutorials or simulations in the CD.

7

How Cells Make ATP: Energy-Releasing Pathways

LEARNING OBJECTIVES

After you have studied this chapter you should be able to

1. Write a summary reaction for aerobic respiration, showing which reactant becomes oxidized and which becomes reduced.
2. List and give a brief overview of the four stages of aerobic respiration, indicate where each stage takes place in a eukaryotic cell, and add up the energy captured (as ATP, NADH, and $FADH_2$) in each stage.
3. Draw a diagram illustrating chemiosmosis and explain (1) how a gradient of protons is established across the inner mitochondrial membrane and (2) the process by which the proton gradient drives ATP synthesis.
4. Summarize how the products of protein and lipid catabolism enter the same metabolic pathway that oxidizes glucose.
5. Compare and contrast aerobic and anaerobic pathways used by cells to extract free energy from nutrients; include the mechanism of ATP formation, the final electron acceptor, and the end products.
6. Summarize the basic similarities of alcohol and lactate fermentation.

Female gerenuks. Gerenuks *(Litocranius walleri)* live in the dry brush country of East Africa, where they browse on leaves, fruits, and flowers of thorny trees and shrubs. *(Renee Lynn/ Photo Researchers, Inc.)*

Cells are tiny factories that process materials on the molecular level, through thousands of metabolic reactions. Cells exist in a dynamic state and are continuously building up and breaking down the many different cellular constituents. As you learned in Chapter 6, metabolism has two complementary components: **catabolism,** which releases energy by splitting complex molecules into smaller components, and **anabolism,** the synthesis of complex molecules from simpler building blocks. Anabolic reactions produce proteins, nucleic acids, lipids, polysaccharides, and other complex molecules that help to maintain the cell or the organism of which it is a part. Most anabolic reactions are endergonic and require ATP or some other energy source to drive them.

Every organism must extract energy from the organic food molecules that it either manufactures by photosynthesis or captures from the environment. The gerenuks in the photograph, for example, eat the leaves of thorny shrubs and trees to obtain energy. In gerenuks and other complex animals, food is first broken down by the digestive system. During digestion, proteins are split into their component amino acids, carbohydrates are digested to simple sugars, and fats are split into glycerol and fatty acids. These nutrients are then absorbed into the blood and transported to all the cells. Each cell converts the energy in the chemical bonds of nutrients to chemical energy stored in ATP in a process known as **cellular respiration.** (The term *cellular respiration* is used to distinguish these cellular processes from *organismic respiration,* the exchange of oxygen and carbon dioxide with the environment by animals that have special organs, such as lungs or gills, for gas exchange.)

Cellular respiration may be either aerobic or anaerobic. **Aerobic** respiration requires molecular oxygen (O_2), whereas **anaerobic** pathways, which include anaerobic respiration and fermentation, do not require oxygen. Most cells use aerobic respiration, which is by far the most common pathway and the main subject of this chapter. All three pathways—aerobic respiration, anaerobic respiration, and fermentation—are exergonic and release free energy.

AEROBIC RESPIRATION IS A REDOX PROCESS

Most eukaryotes and prokaryotes carry out **aerobic** respiration, a form of cellular respiration requiring oxygen. During aerobic respiration, nutrients are catabolized to carbon dioxide and water. Most cells of plants, animals, protists, fungi, and bacteria use aerobic respiration to obtain energy from glucose, which enters the cell though a specific transport protein in the plasma membrane (see discussion of facilitated diffusion in Chapter 5). The overall reaction pathway for the aerobic respiration of glucose is summarized as follows:

$$C_6H_{12}O_6 + 6\ O_2 + 6\ H_2O \longrightarrow 6\ CO_2 + 12\ H_2O + \text{Energy (in the chemical bonds of ATP)}$$

Note that water is shown on both sides of the equation; this is because it is a reactant in some reactions and a product in others.

For purposes of discussion, the equation for aerobic respiration can be simplified to indicate that there is a net yield of water:

$$\overbrace{C_6H_{12}O_6 + 6\ O_2 \rightarrow 6\ CO_2}^{\text{Oxidation}} + 6\ H_2O + \text{Energy (in the chemical bonds of ATP)}$$

(Reduction: $O_2 \rightarrow H_2O$)

If we analyze this summary reaction, it appears that CO_2 is produced by the removal of hydrogen atoms from glucose. Conversely, water appears to be formed as the hydrogen atoms are accepted by oxygen. Because the transfer of hydrogen atoms is equivalent to the transfer of electrons, this is a redox process in which glucose becomes **oxidized** and oxygen becomes **reduced** (see Chapter 6).

The products of the reaction would be the same if the glucose were simply placed in a test tube and burned in the presence of oxygen. However, if cells were to burn glucose, its energy would be released all at once as heat, which not only would be unavailable to the cell but also would actually destroy it. For this reason, cells do not transfer hydrogen atoms directly from glucose to oxygen. Aerobic respiration is a redox process in which electrons associated with the hydrogen atoms in glucose are transferred to oxygen in a series of about 30 steps (Fig. 7–1). During this process, the free energy of the electrons is coupled to ATP synthesis.

AEROBIC RESPIRATION HAS FOUR STAGES

The chemical reactions of the aerobic respiration of glucose can be grouped into four stages (Fig. 7–2, Table 7–1, and summary equations at the end of this chapter). In eukaryotes, the first stage (glycolysis) takes place in the cytosol, and the remaining stages take place inside mitochondria. Most bacteria also carry out these processes, but because these cells lack mitochondria, the reactions of aerobic respiration occur in the cytosol and in association with the plasma membrane.

1. **Glycolysis.** A six-carbon glucose molecule is converted to two, three-carbon molecules of pyruvate,[1] and ATP and NADH are formed (see Chapter 6 for a review of ATP and NADH).[2]
2. **Formation of acetyl coenzyme A.** Each pyruvate enters a mitochondrion and is oxidized to a two-carbon group (acetate) that combines with coenzyme A, forming acetyl coenzyme A. NADH is produced, and carbon dioxide is released as a waste product.

[1] Pyruvate and many other compounds in cellular respiration exist as anions at the pH found in the cell. They sometimes associate with H^+ to form acids. For example, pyruvate forms pyruvic acid. In some textbooks these compounds are presented in the acid form.

[2] Although the correct way to write the reduced form of NAD^+ is NADH + H^+, for simplicity we will present the reduced form as NADH throughout the chapter.

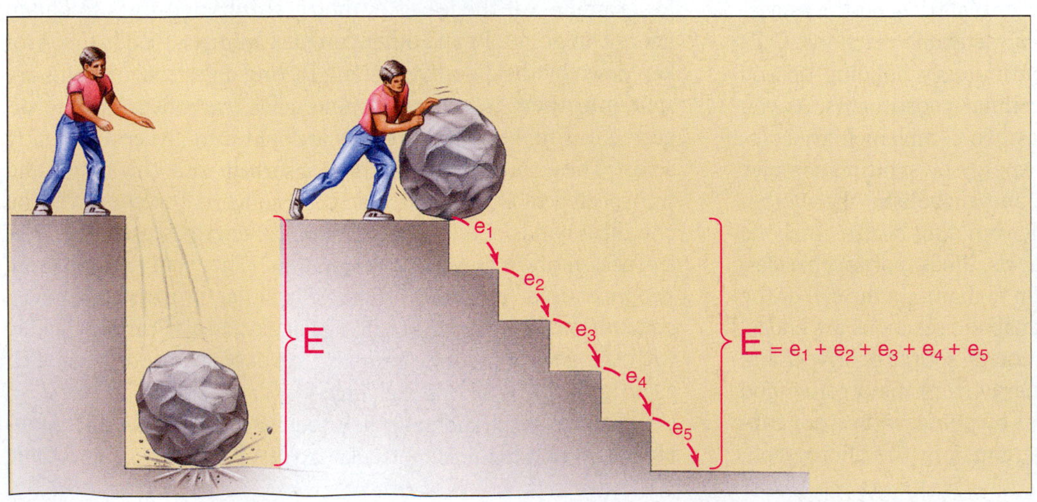

Figure 7–1 Changes in free energy. The release of energy from a glucose molecule is analogous to the liberation of energy by a falling object. The total energy released *(E)* is the same whether it occurs all at once or in a series of steps.

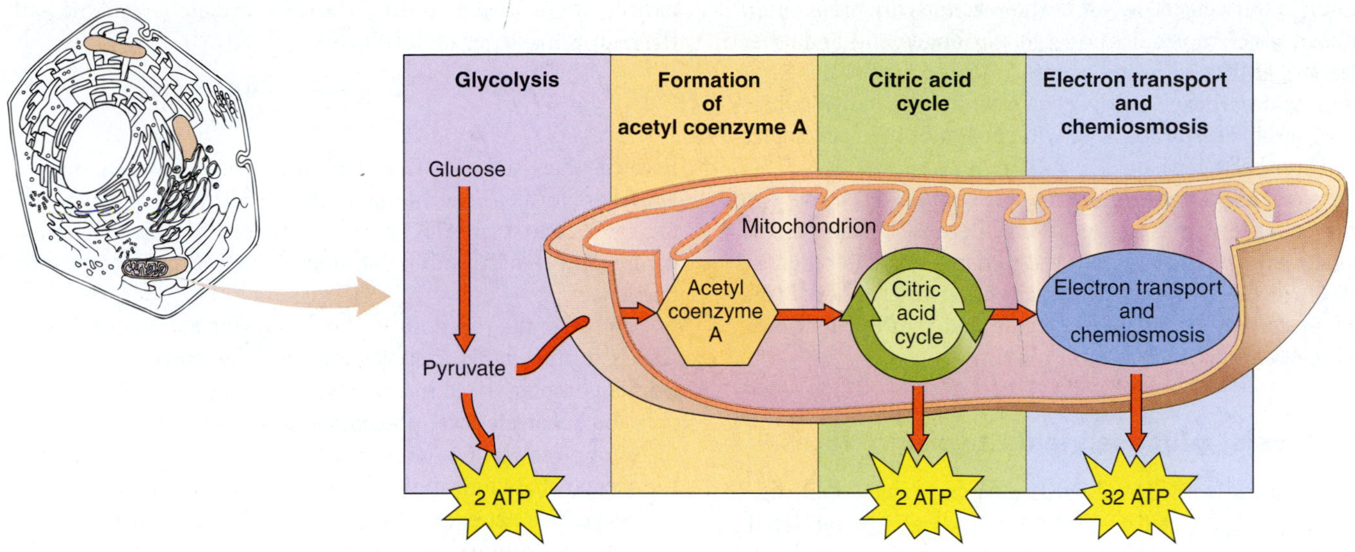

Figure 7–2 The four stages of aerobic respiration. In eukaryotic cells, much of aerobic respiration occurs in mitochondria. Glycolysis, the first stage of aerobic respiration, occurs in the cytosol. Pyruvate, the product of glycolysis, enters a mitochondrion, where cellular respiration continues with the formation of acetyl CoA, the citric acid cycle, and electron transport/chemiosmosis. Most ATP is synthesized by chemiosmosis.

3. **The citric acid cycle.** The acetate group of acetyl coenzyme A combines with a four-carbon molecule (oxaloacetate) to form a six-carbon molecule (citrate). In the course of the cycle, citrate is recycled to oxaloacetate, and carbon dioxide is released as a waste product. Energy is captured as ATP and the reduced, high-energy compounds NADH and $FADH_2$ (see Chapter 6 for a review of $FADH_2$).
4. **The electron transport chain and chemiosmosis.** The electrons removed from glucose during the preceding stages are transferred from NADH and $FADH_2$ to a chain of electron acceptor compounds. As the electrons are passed from one electron acceptor to another, some of their energy is used to move hydrogen ions (protons) across the inner mitochondrial membrane, forming a proton gradient. In a process known as chemiosmosis, to be described later, the energy of this proton gradient is used to produce ATP.

Most reactions involved in aerobic respiration are one of three types: dehydrogenations, decarboxylations, and those that we will informally categorize as preparation reactions. **Dehydrogenations** are reactions in which two hydrogen atoms (actually, two electrons plus one or two protons) are removed from the substrate and transferred to NAD^+ or FAD. **Decarboxylations** are reactions in which part of a carboxyl group (—COOH) is removed from the substrate as a molecule of CO_2. The carbon dioxide we exhale with each breath is derived from decarboxylations

TABLE 7–1 Summary of Aerobic Respiration

Phase	Summary	Some Starting Materials	Some End Products
1. Glycolysis (in cytosol)	Series of reactions in which glucose is degraded to pyruvate; net profit of 2 ATPs; hydrogen atoms are transferred to carriers; can proceed anaerobically	Glucose, ATP, NAD^+, ADP, P_i	Pyruvate, ATP, NADH
2. Formation of acetyl CoA (in mitochondria)	Pyruvate is degraded and combined with coenzyme A to form acetyl CoA; hydrogen atoms are transferred to carriers; CO_2 is released	Pyruvate, coenzyme A, NAD^+	Acetyl CoA, CO_2, NADH
3. Citric acid cycle (in mitochondria)	Series of reactions in which the acetyl portion of acetyl CoA is degraded to CO_2; hydrogen atoms are transferred to carriers; ATP is synthesized	Acetyl CoA, H_2O, NAD^+, FAD, ADP, P_i	CO_2, NADH, $FADH_2$, ATP
4. Electron transport and chemiosmosis (in mitochondria)	Chain of several electron transport molecules; electrons are passed along chain; released energy is used to form a proton gradient; ATP is synthesized as protons diffuse down the gradient; oxygen is final electron acceptor	NADH, $FADH_2$, O_2, ADP, P_i	ATP, H_2O, NAD^+, FAD

that occur in our cells. The rest of the reactions are preparation reactions in which molecules undergo rearrangements and other changes so that they can subsequently undergo further dehydrogenations or decarboxylations. As we examine the individual reactions of aerobic respiration, we will encounter many examples of these three basic types.

In following the reactions of aerobic respiration, it helps to do some bookkeeping as you go along. Because glucose is the starting material, it is useful to express changes on a per glucose basis. We will be paying particular attention to changes in the number of carbon atoms per molecule and to steps in which some type of energy transfer takes place.

In glycolysis, glucose yields two pyruvates

Glycolysis comes from Greek words meaning "sugar-splitting," which refers to the fact that the sugar glucose is metabolized. Glycolysis does not require oxygen and can proceed under aerobic or anaerobic conditions. Figure 7–3 shows a simple summary of glycolysis, in which a glucose molecule comprising six carbons is converted to two molecules of **pyruvate,** a three-carbon molecule. Some of the energy in the glucose is captured; there is a net yield of two ATP molecules and two NADH molecules. The reactions of glycolysis take place in the cytosol, where the necessary reactants, such as ADP, NAD^+, and inorganic phosphates, float freely and are used as needed.

The glycolysis pathway consists of a series of reactions, each of which is catalyzed by a specific enzyme (Fig. 7–4). Glycolysis is divided into two major phases: The first includes endergonic reactions that require ATP, while the second includes exergonic reactions that yield ATP and NADH.

The first phase of glycolysis requires an investment of ATP

The first phase of glycolysis is sometimes referred to as the "energy-investment" phase. Glucose is a relatively stable molecule and is not easily broken down. In two separate **phosphorylation reactions,** a phosphate group is transferred from ATP to the sugar. The resulting phosphorylated sugar (fructose-1,6-bisphosphate) is less stable and is broken enzymatically into two, three-carbon molecules, dihydroxyacetone phosphate and glyceraldehyde-3-phosphate (G3P); the dihydroxyacetone phosphate is enzymatically converted to G3P, so the products at this point in glycolysis are two molecules of G3P. We may summarize this portion of glycolysis as follows:

$$\underset{\text{Six-carbon compound}}{\text{Glucose}} + 2\ \text{ATP} \longrightarrow \underset{\text{Three-carbon compound}}{2\ \text{G3P}} + 2\ \text{ADP}$$

The second phase of glycolysis yields NADH and ATP

Each G3P is converted to pyruvate. In the first step of this process each G3P is oxidized by the removal of two electrons (as part of two hydrogen atoms). These immediately combine with the hydrogen carrier molecule, NAD^+:

$$\underset{\text{Oxidized}}{\text{NAD}^+} + \underset{\text{(From G3P)}}{2\ \text{H}} \longrightarrow \underset{\text{Reduced}}{\text{NADH}} + \text{H}^+$$

Because there are two G3P molecules for every glucose, two NADH are formed. The energy of the electrons carried by NADH can be used to form ATP later. The process by which this is accomplished is discussed in conjunction with the electron transport chain.

In two of the reactions leading to the formation of pyruvate, ATP is formed when a phosphate group is transferred to ADP from a phosphorylated intermediate (see Fig. 7–4). This process is called **substrate-level phosphorylation.** Note that in the first phase of glycolysis two molecules of ATP are consumed, but in the second phase four molecules of ATP are produced. Thus, glycolysis yields a net energy profit of *two* ATPs per glucose.

We may summarize the second phase of glycolysis as follows:

$$2\ \text{G3P} + 2\ \text{NAD}^+ + 4\ \text{ADP} \longrightarrow 2\ \text{Pyruvate} + 2\ \text{NADH} + 4\ \text{ATP}$$

Pyruvate is converted to acetyl CoA

In eukaryotes, the pyruvate molecules formed in glycolysis enter the mitochondria, where they are converted to **acetyl coenzyme A (acetyl CoA).** These reactions occur in the cytosol of aerobic prokaryotes. In this series of reactions, pyruvate undergoes a process known as **oxidative decarboxylation.** First, a carboxyl group is removed as carbon dioxide, which diffuses out of the cell (Fig. 7–5 on page 160). Then the remaining two-carbon fragment is oxidized, and NAD^+ accepts the electrons removed during the oxidation. Finally, the oxidized two-carbon fragment, an acetyl group, becomes attached to **coenzyme A,** yielding acetyl CoA. *Pyruvate dehydrogenase,* the enzyme that catalyzes these reactions, is an enormous multienzyme complex consisting of 72 polypeptide chains!

Recall from Chapter 6 that coenzyme A transfers groups derived from organic acids. In this case, coenzyme A transfers an acetyl group, which is related to acetic acid. Coenzyme A is manufactured in the cell from one of the B vitamins, pantothenic acid.

The overall reaction for the formation of acetyl coenzyme A is the following:

$$2\ \text{Pyruvate} + 2\ \text{NAD}^+ + 2\ \text{CoA} \longrightarrow 2\ \text{Acetyl CoA} + 2\ \text{NADH} + 2\ \text{CO}_2$$

Note that the original glucose molecule has now been partially oxidized, yielding two acetyl groups and two CO_2 molecules. The electrons removed have reduced NAD^+ to NADH. At this point in aerobic respiration, four NADH molecules have been formed as a result of the catabolism of a single glucose molecule: two during glycolysis and two during the formation of acetyl CoA from pyruvate. Keep in mind that these NADH molecules will be used later (during electron transport) to form additional ATP molecules.

(text continues on page 160)

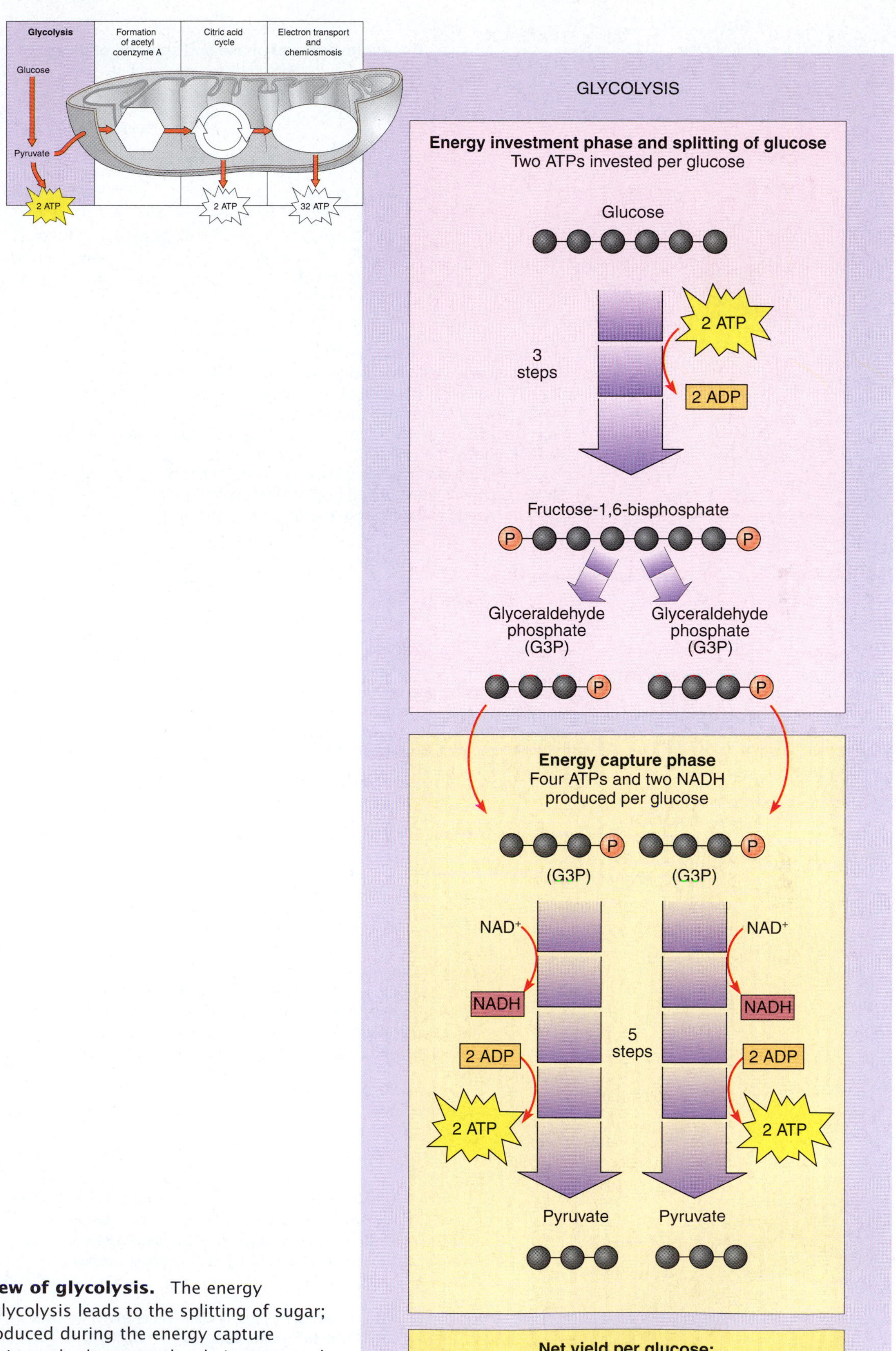

Figure 7–3 Overview of glycolysis. The energy investment phase of glycolysis leads to the splitting of sugar; ATP and NADH are produced during the energy capture phase. During glycolysis, each glucose molecule is converted to two pyruvates, with a net yield of two ATP molecules and two NADH molecules.

Energy investment phase and splitting of glucose
Two ATPs invested per glucose

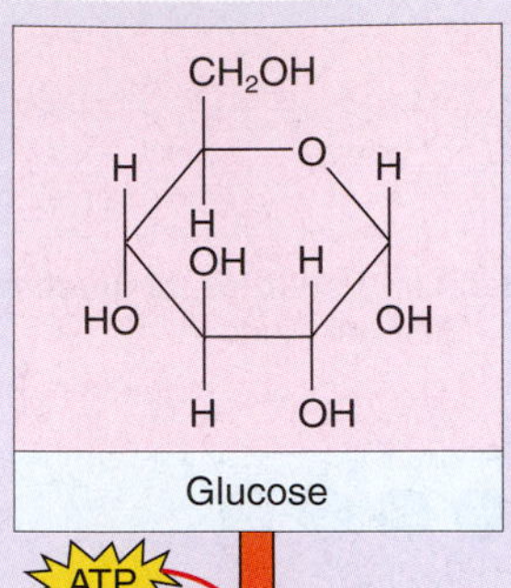

Glucose

ATP

Hexokinase

ADP

CH_2O—P

Glucose-6-phosphate

(1) Glycolysis begins with a preparation reaction in which glucose receives a phosphate group from an ATP molecule. The ATP serves as a source of both phosphate and the energy needed to attach the phosphate to the glucose molecule. (Once the ATP is spent, it becomes ADP and joins the ADP pool of the cell until turned into ATP again.) The phosphorylated glucose is known as glucose-6-phosphate. (Note the phosphate attached to its carbon atom 6.) Phosphorylation of the glucose makes it more chemically reactive.

Phosphoglucoisomerase

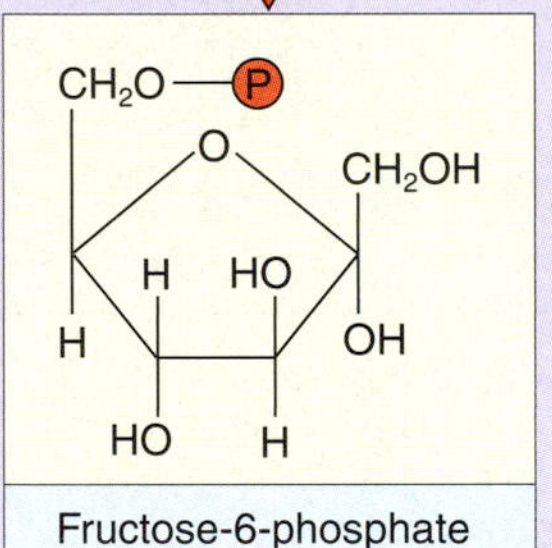

Fructose-6-phosphate

(2) Glucose-6-phosphate undergoes another preparation reaction, the rearrangement of its hydrogen and oxygen atoms. In this reaction glucose-6-phosphate is converted to its isomer, fructose-6-phosphate.

ATP

Phosphofructokinase

ADP

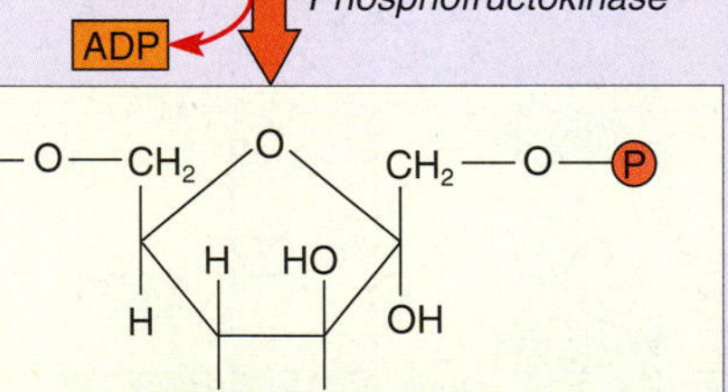

Fructose-1,6-bisphosphate

(3) Next, another ATP donates a phosphate to the molecule, forming fructose-1,6-bisphosphate. So far, two ATP molecules have been invested in the process without any being produced. Phosphate groups are now bound at carbons 1 and 6, and the molecule is ready to be split.

Aldolase

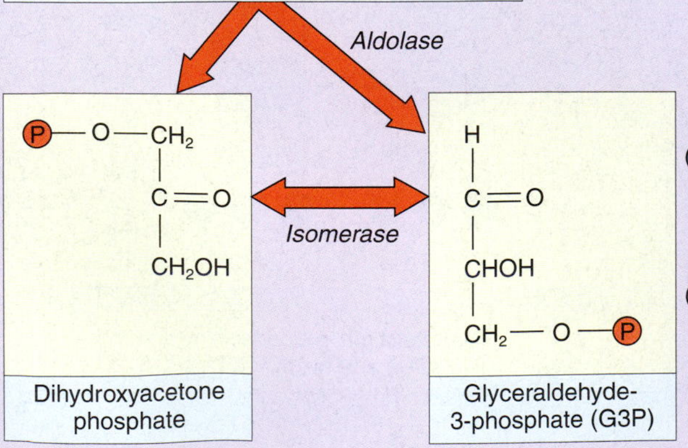

Dihydroxyacetone phosphate

Glyceraldehyde-3-phosphate (G3P)

(4) Fructose-1,6-bisphosphate is then split into two 3-carbon sugars, glyceraldehyde-3-phosphate (G3P) and dihydroxyacetone phosphate.

(5) Dihydroxyacetone phosphate is enzymatically converted to its isomer, glyceraldehyde-3-phosphate, for further metabolism in glycolysis.

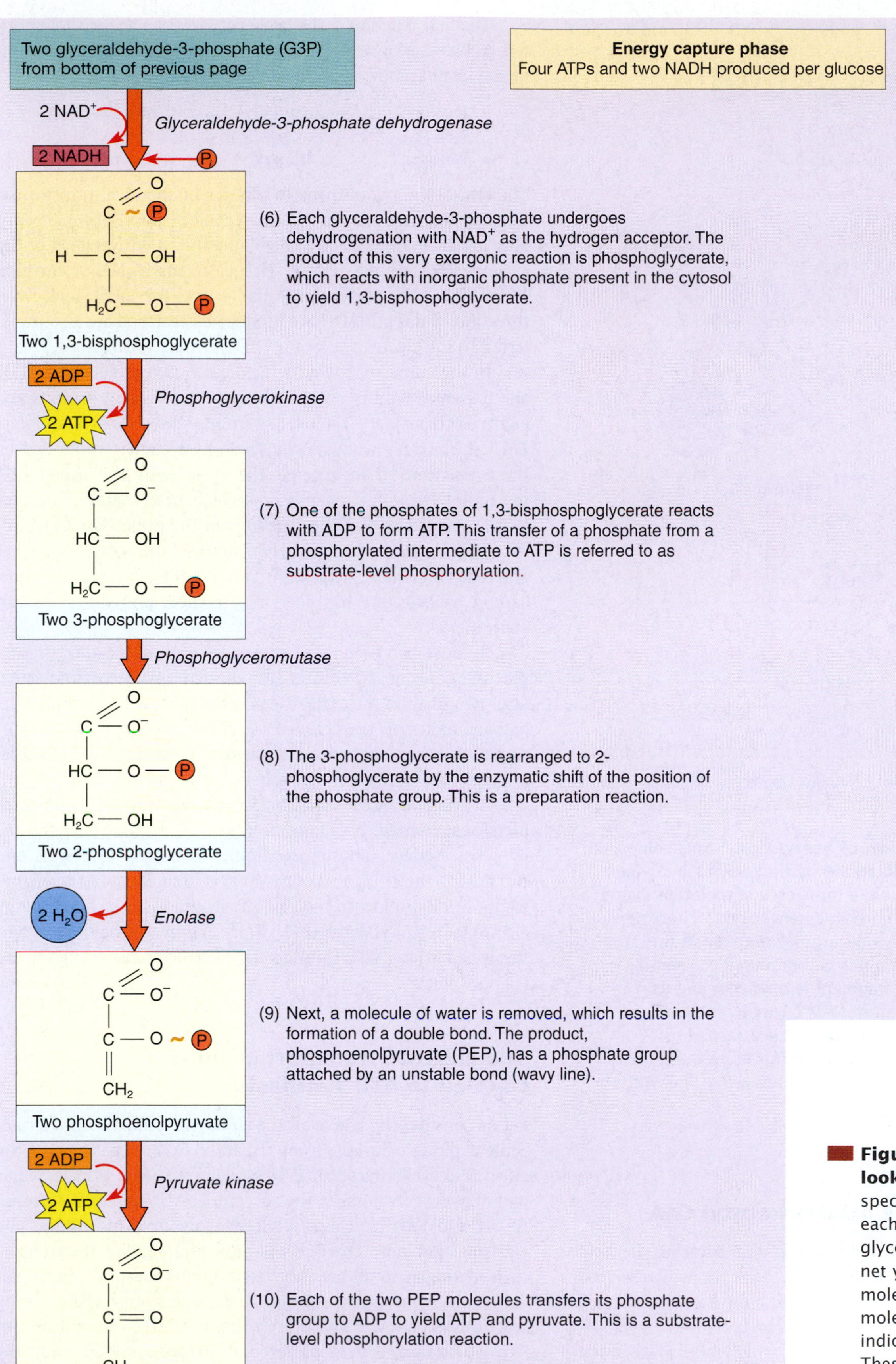

Two glyceraldehyde-3-phosphate (G3P) from bottom of previous page
Energy capture phase
Four ATPs and two NADH produced per glucose
2 NAD⁺
2 NADH
Glyceraldehyde-3-phosphate dehydrogenase
Two 1,3-bisphosphoglycerate
(6) Each glyceraldehyde-3-phosphate undergoes dehydrogenation with NAD⁺ as the hydrogen acceptor. The product of this very exergonic reaction is phosphoglycerate, which reacts with inorganic phosphate present in the cytosol to yield 1,3-bisphosphoglycerate.
2 ADP
2 ATP
Phosphoglycerokinase
Two 3-phosphoglycerate
(7) One of the phosphates of 1,3-bisphosphoglycerate reacts with ADP to form ATP. This transfer of a phosphate from a phosphorylated intermediate to ATP is referred to as substrate-level phosphorylation.
Phosphoglyceromutase
Two 2-phosphoglycerate
(8) The 3-phosphoglycerate is rearranged to 2-phosphoglycerate by the enzymatic shift of the position of the phosphate group. This is a preparation reaction.
2 H₂O
Enolase
Two phosphoenolpyruvate
(9) Next, a molecule of water is removed, which results in the formation of a double bond. The product, phosphoenolpyruvate (PEP), has a phosphate group attached by an unstable bond (wavy line).
2 ADP
2 ATP
Pyruvate kinase
Two pyruvate
(10) Each of the two PEP molecules transfers its phosphate group to ADP to yield ATP and pyruvate. This is a substrate-level phosphorylation reaction.

Figure 7–4 A detailed look at glycolysis. A specific enzyme catalyzes each of the reactions in glycolysis. Note that there is a net yield of two ATP molecules and two NADH molecules. (Yellow wavy lines indicate unstable bonds. These bonds permit the phosphates to be transferred to other molecules, in this case ADP.)

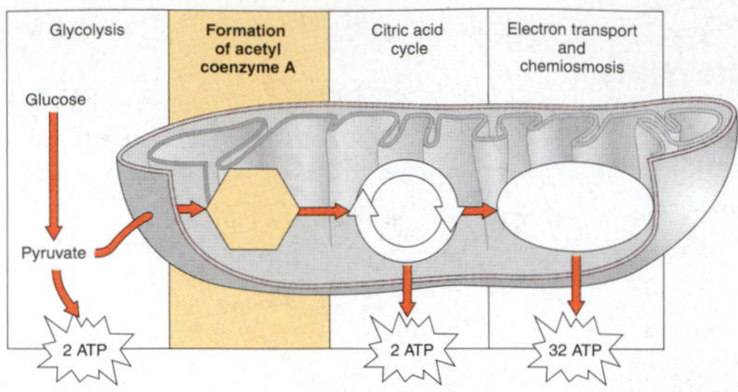

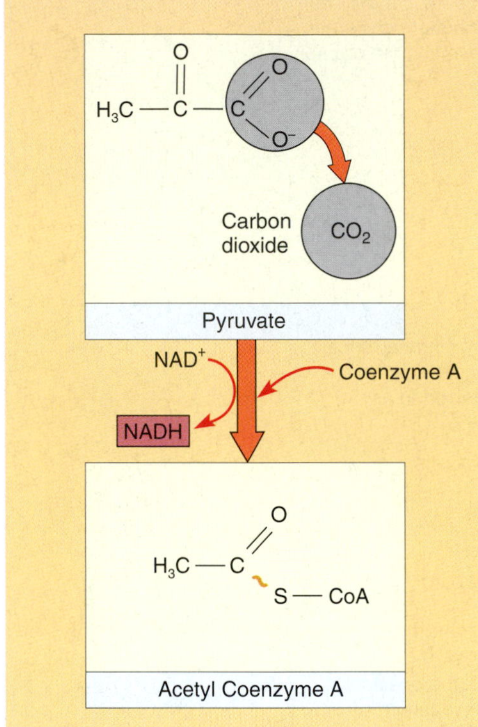

Figure 7–5 Formation of acetyl CoA. This complex series of reactions is catalyzed by the enzyme pyruvate dehydrogenase. Pyruvate, a three-carbon molecule that is the end product of glycolysis, enters the mitochondrion and undergoes oxidative decarboxylation. First, the carboxyl group is split off as carbon dioxide. Then the remaining two-carbon fragment is oxidized, and its electrons are transferred to NAD^+. Finally, the oxidized two-carbon group, an acetyl group, is attached to coenzyme A. Coenzyme A has a sulfur atom that forms a very unstable bond, shown as a yellow wavy line, with the acetyl group.

The citric acid cycle oxidizes acetyl CoA

The **citric acid cycle** is also known as the **tricarboxylic acid (TCA) cycle** and as the **Krebs cycle,** after Hans Krebs, the chemist who assembled the accumulated contributions of many scientists and worked out the details of the cycle in the 1930s. A simple summary of the citric acid cycle, which takes place in the matrix of the mitochondria, is given in Figure 7–6. The eight steps of the citric acid cycle are shown in Figure 7–7. A specific enzyme catalyzes each reaction.

The first reaction of the cycle occurs when acetyl CoA transfers its two-carbon acetyl group to the four-carbon acceptor compound **oxaloacetate,** forming **citrate,** a six-carbon compound:

Oxaloacetate	+	Acetyl CoA	$\longrightarrow$	Citrate	+	CoA
Four-carbon compound		Two-carbon compound		Six-carbon compound		

The citrate then goes through a series of chemical transformations, losing first one and then a second carboxyl group as CO_2. Most of the energy made available by the oxidative steps of the cycle is transferred as energy-rich electrons to NAD^+, forming NADH. For each acetyl group that enters the citric acid cycle, three molecules of NADH are produced. Electrons are also transferred to the electron acceptor FAD, forming $FADH_2$.

In the course of the citric acid cycle, two molecules of CO_2 and the equivalent of eight hydrogen atoms (eight protons and eight electrons) are removed, forming three NADH and one $FADH_2$. You may wonder why more hydrogen is generated by these reactions than entered the cycle with the acetyl CoA molecule. These hydrogen atoms come from water molecules that are added during the reactions of the cycle. The CO_2 produced accounts for the two carbon atoms of the acetyl group that entered the citric acid cycle. At the end of each cycle, the four-carbon oxaloacetate has been regenerated, and the cycle can continue.

Because two acetyl CoA molecules are produced from each glucose molecule, two cycles are required per glucose molecule. After two turns of the cycle, the original glucose has lost all of its carbons and may be regarded as having been completely consumed. To summarize, the citric acid cycle yields 4 CO_2, 6 NADH, 2 $FADH_2$, and 2 ATP per glucose molecule.

At the end of the citric acid cycle, glucose has been completely catabolized. Only four molecules of ATP have been formed by substrate-level phosphorylation: two during glycolysis and two during the citric acid cycle. At this time, most of the energy of the original glucose molecule is in the form of high-energy electrons in NADH and $FADH_2$. Their energy will be used to synthesize additional ATP through the electron transport chain and chemiosmosis.

The electron transport chain is coupled to ATP synthesis

Let us consider the fate of all the electrons removed from a molecule of glucose during glycolysis, acetyl CoA formation, and the citric acid cycle. Recall that these electrons were transferred as part of hydrogen atoms to the acceptors NAD^+ and FAD, forming NADH and $FADH_2$. These reduced compounds now enter the **electron transport chain,** where the high-energy electrons of their hydrogen atoms are shuttled from one acceptor to another. As the electrons are passed along in a series of exergonic redox reactions, some of their energy is used to drive the synthesis of ATP, which is an endergonic process. Because ATP synthesis (by phosphorylation of ADP) is coupled to the redox reactions in the electron transport chain, the entire process is known as **oxidative phosphorylation.**

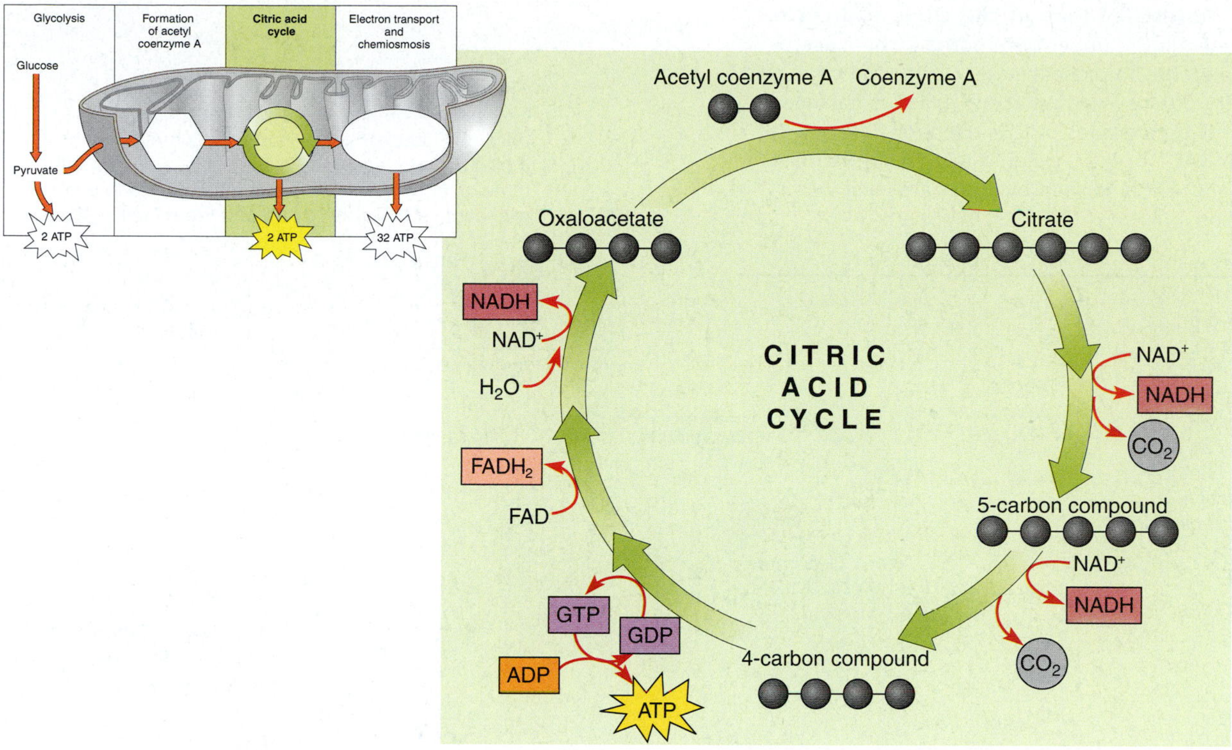

Figure 7–6 Overview of the citric acid cycle. For every glucose, two acetyl groups enter the citric acid cycle *(top)*. Each two-carbon acetyl group combines with a four-carbon compound, oxaloacetate, to form the six-carbon compound citrate. Two CO_2 molecules are removed, and energy is captured as one ATP, three NADH, and one $FADH_2$ per acetyl group (or two ATPs, six NADH, and two $FADH_2$ per glucose molecule).

The electron transport chain transfers electrons from NADH and $FADH_2$ to oxygen

The electron transport chain is a series of electron carriers embedded in the inner mitochondrial membrane of eukaryotes and in the plasma membrane of aerobic prokaryotes. Like NADH and $FADH_2$, each carrier can exist in an oxidized form or a reduced form. Electrons pass down the electron transport chain in a series of redox reactions that works much like a bucket brigade, the old-time chain of people that passed buckets of water from a stream to a building that was on fire. In the electron transport chain, each acceptor molecule is alternately reduced as it accepts electrons and oxidized as it gives up electrons. The electrons entering the electron transport chain have a relatively high energy content. They lose some of their energy at each step as they pass along the chain of electron carriers (just as some of the water spills out of the bucket as it is passed from one person to another).

Members of the electron transport chain include the flavoprotein *flavin mononucleotide (FMN),* the lipid *ubiquinone* (also called *coenzyme Q* or *CoQ*), several *iron-sulfur proteins,* and a group of closely related iron-containing proteins called *cytochromes.* Each of the electron carriers has a different mechanism for accepting and passing electrons. As cytochromes accept and donate electrons, for example, the charge on the iron atom, which is the electron carrier portion of the cytochromes, alternates between Fe^{2+} (reduced) and Fe^{3+} (oxidized).

The electron transport chain has been isolated and purified from the inner mitochondrial membrane as four large, distinct protein complexes, or groups, of acceptors (Fig. 7–8). *Complex I (NADH-ubiquinone oxidoreductase)* accepts electrons from NADH molecules that were produced during glycolysis, the formation of acetyl CoA, and the citric acid cycle. *Complex II (succinate-ubiquinone reductase)* accepts electrons from $FADH_2$ molecules that were produced during the citric acid cycle. Complexes I and II both produce the same product, reduced ubiquinone, which is the substrate of *complex III (ubiquinone-cytochrome c oxidoreductase).* That is, complex III accepts electrons from reduced ubiquinone and passes them on to cytochrome *c. Complex IV (cytochrome c oxidase)* accepts electrons from cytochrome *c* and uses these electrons to reduce molecular oxygen, forming water in the process. The electrons simultaneously unite with protons from the surrounding medium to form hydrogen, and the chemical reaction between hydrogen and oxygen produces water.

Because oxygen is the final electron acceptor in the electron transport chain, organisms that respire aerobically require oxygen. What happens when cells that are strict aerobes are deprived of oxygen? When no oxygen is available to accept them, the last

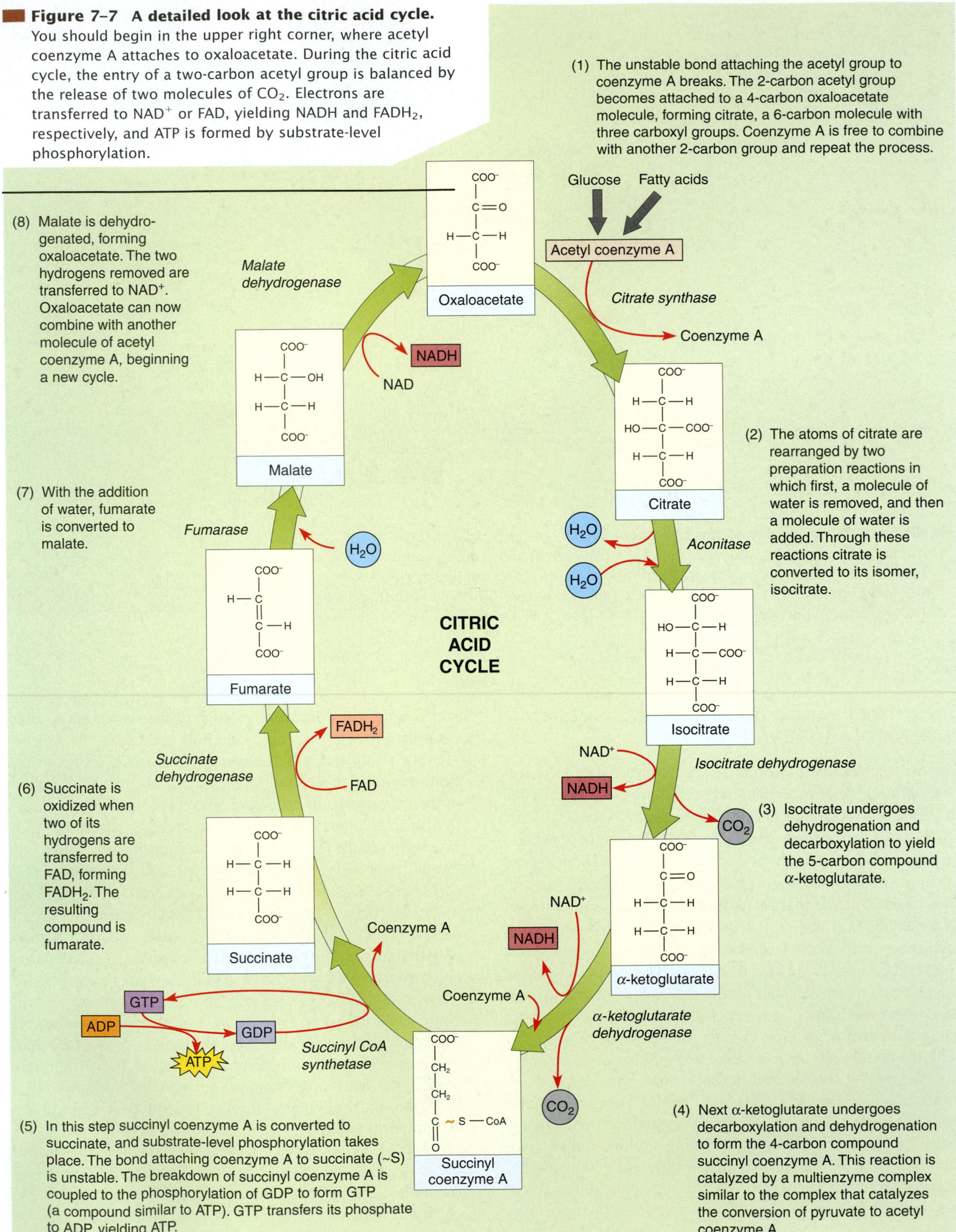

(1) The unstable bond attaching the acetyl group to coenzyme A breaks. The 2-carbon acetyl group becomes attached to a 4-carbon oxaloacetate molecule, forming citrate, a 6-carbon molecule with three carboxyl groups. Coenzyme A is free to combine with another 2-carbon group and repeat the process.
Glucose
Fatty acids
Acetyl coenzyme A
Citrate synthase
Coenzyme A
Oxaloacetate
Citrate
(2) The atoms of citrate are rearranged by two preparation reactions in which first, a molecule of water is removed, and then a molecule of water is added. Through these reactions citrate is converted to its isomer, isocitrate.
H_2O
Aconitase
H_2O
Isocitrate
NAD^+
Isocitrate dehydrogenase
NADH
CO_2
(3) Isocitrate undergoes dehydrogenation and decarboxylation to yield the 5-carbon compound α-ketoglutarate.
α-ketoglutarate
NAD^+
NADH
Coenzyme A
α-ketoglutarate dehydrogenase
CO_2
(4) Next α-ketoglutarate undergoes decarboxylation and dehydrogenation to form the 4-carbon compound succinyl coenzyme A. This reaction is catalyzed by a multienzyme complex similar to the complex that catalyzes the conversion of pyruvate to acetyl coenzyme A.
Succinyl coenzyme A
Coenzyme A
Succinyl CoA synthetase
GTP
ADP
GDP
ATP
(5) In this step succinyl coenzyme A is converted to succinate, and substrate-level phosphorylation takes place. The bond attaching coenzyme A to succinate (~S) is unstable. The breakdown of succinyl coenzyme A is coupled to the phosphorylation of GDP to form GTP (a compound similar to ATP). GTP transfers its phosphate to ADP, yielding ATP.
Succinate
Succinate dehydrogenase
FAD
$FADH_2$
(6) Succinate is oxidized when two of its hydrogens are transferred to FAD, forming $FADH_2$. The resulting compound is fumarate.
Fumarate
CITRIC ACID CYCLE
Fumarase
H_2O
(7) With the addition of water, fumarate is converted to malate.
Malate
Malate dehydrogenase
NAD
NADH
(8) Malate is dehydrogenated, forming oxaloacetate. The two hydrogens removed are transferred to NAD^+. Oxaloacetate can now combine with another molecule of acetyl coenzyme A, beginning a new cycle.

Figure 7–7 A detailed look at the citric acid cycle. You should begin in the upper right corner, where acetyl coenzyme A attaches to oxaloacetate. During the citric acid cycle, the entry of a two-carbon acetyl group is balanced by the release of two molecules of CO_2. Electrons are transferred to NAD^+ or FAD, yielding NADH and $FADH_2$, respectively, and ATP is formed by substrate-level phosphorylation.

cytochrome in the chain is stuck with its electrons. When that occurs, each acceptor molecule in the chain remains stuck with electrons (i.e., each is reduced), and the entire chain is blocked all the way back to NADH. Because oxidative phosphorylation is coupled to electron transport, no further ATPs are produced by way of the electron transport chain. Most cells of complex organisms cannot live long without oxygen because the small amount of ATP they produce by glycolysis alone is insufficient to sustain life processes.

Lack of oxygen is not the only factor that interferes with the electron transport chain. Some poisons, including cyanide, inhibit the normal activity of the cytochromes. Cyanide binds tightly to the iron in the last cytochrome in the electron transport chain (cytochrome a_3), making it unable to transport electrons on to oxygen. This blocks the further passage of electrons through the chain, halting ATP production.

Although the flow of electrons in electron transport is usually tightly coupled to the production of ATP, some organisms are able to uncouple the two processes to produce heat (see *Focus On: Electron Transport and Heat*).

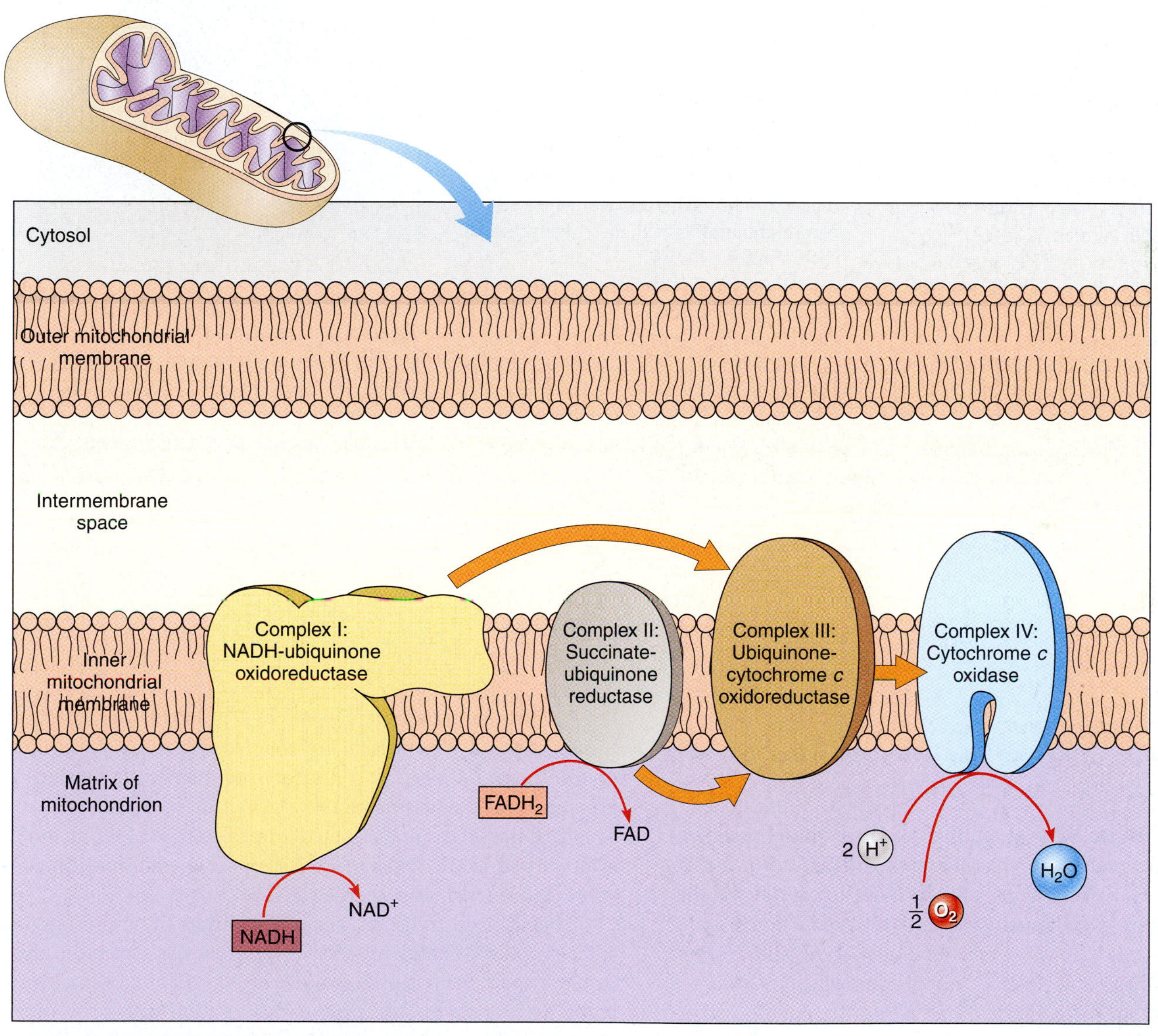

Figure 7–8 Overview of the electron transport chain. Electrons fall to successively lower energy levels as they are passed along the four complexes of the electron transport chain, which are located in the inner mitochondrial membrane. (The orange arrows indicate the pathway of electrons.) The carriers within each complex are alternately reduced and oxidized as they accept and donate electrons. The terminal acceptor is oxygen; one of the two atoms of an oxygen molecule (written as $\frac{1}{2}$ O_2) accepts two electrons, which are added to two protons from the surrounding medium to produce water.

FOCUS ON Electron Transport and Heat

What is the source of our body heat? Essentially, it is a byproduct of various exergonic reactions, especially those involving the electron transport chains in our mitochondria. Some cold-adapted animals, hibernating animals, and newborn animals are able to produce unusually large amounts of heat by uncoupling electron transport from ATP production. These animals have **adipose tissue** (tissue in which fat is stored) that is brown. The brown color comes from the large number of mitochondria found in the brown adipose tissue cells. The inner mitochondrial membranes of these mitochondria contain an *uncoupling protein* that produces a passive proton channel through which protons flow into the mitochondrial matrix. As a consequence, most of the energy of glucose is converted to heat rather than to chemical energy in ATP.

Certain plants, which are not generally considered "warm" organisms, also have the ability to produce large amounts of heat. Skunk cabbage *(Symplocarpus foetidus)*, for example, lives in North American swamps and wet woodlands and generally flowers during February and March when the ground is still covered with snow *(see figure)*. Its uncoupled mitochondria generate large amounts of heat, enabling the plant to melt the snow and attract insect pollinators by vaporizing certain odiferous molecules into the surrounding air. The flower temperature of skunk cabbage is 15° to 22° C (59° to 72° F) when the air surrounding it is −15° to 10° C (5° to 50° F). Skunk cabbage flowers maintain this temperature for two weeks or more. Other plants, such as splitleaf philodendron *(Philodendron selloum)* and sacred lotus *(Nelumbo nucifera)*, also generate heat when they bloom and maintain their temperatures within precise limits.

Some plants generate as much or more heat per gram of tissue than animals in flight, which have long been considered the greatest heat producers in the living world. The European plant lords-and-ladies *(Arum maculatum)*, for example, produces 0.4 joules (0.1 cal) of heat per second per gram of tissue, whereas a hummingbird in flight produces 0.24 J (0.06 cal) per second per gram of tissue.

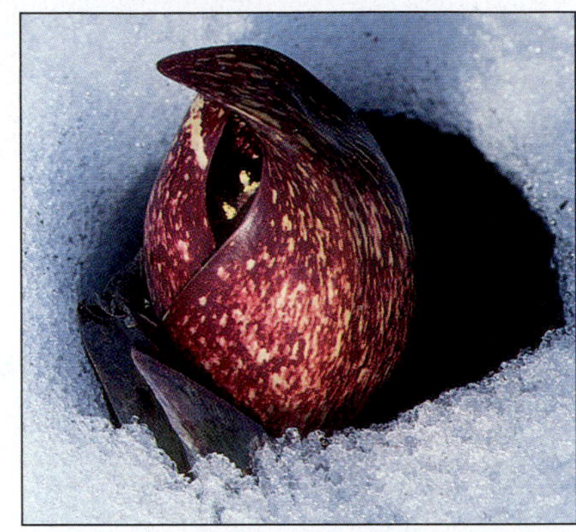

Skunk cabbage *(Symplocarpus foetidus)*. This plant not only produces a significant amount of heat when it flowers but also regulates its temperature within a specific range. *(Leonard Lee Rue III /Earth Scenes)*

The chemiosmotic model explains the coupling of ATP synthesis to electron transport in aerobic respiration

For decades scientists were aware that oxidative phosphorylation occurs in mitochondria, and many experiments had shown that the transfer of two electrons from each NADH to oxygen (via the electron transport chain) usually results in the production of up to three ATP molecules. However, for a long time, the connection between ATP synthesis and electron transport remained a mystery.

Then, in 1961 Peter Mitchell proposed the **chemiosmotic model,** based on his experiments with bacteria. Because the respiratory electron transport chain is located in the plasma membrane of an aerobic bacterial cell, the bacterial plasma membrane can be considered comparable to the inner mitochondrial membrane. Mitchell demonstrated that if bacterial cells are placed in an acidic environment (that is, an environment with a high hydrogen ion, or proton, concentration), the cells synthesized ATP even if electron transport was not taking place. On the basis of these and other experiments, Mitchell proposed that electron transport and ATP synthesis are coupled by means of a proton gradient across the inner mitochondrial membrane in eukaryotes (or across the plasma membrane in bacteria). His model was so radical that it was not accepted immediately. By 1978 so much evidence had accumulated in support of the chemiosmotic model that Peter Mitchell was awarded a Nobel Prize.

The electron transport chain establishes the proton gradient; some of the energy released as electrons pass down the electron transport chain is used to move protons (H^+) across a membrane. In eukaryotes the protons are moved across the inner mitochondrial membrane into the intermembrane space, that is, the space between the inner and outer mitochondrial membranes (Fig. 7–9). Hence the inner mitochondrial membrane separates a space with a higher concentration of protons (the intermembrane space) from a space with a lower concentration of protons (the mitochondrial matrix).

Protons are moved across the inner mitochondrial membrane by three of the four electron transport complexes (com-

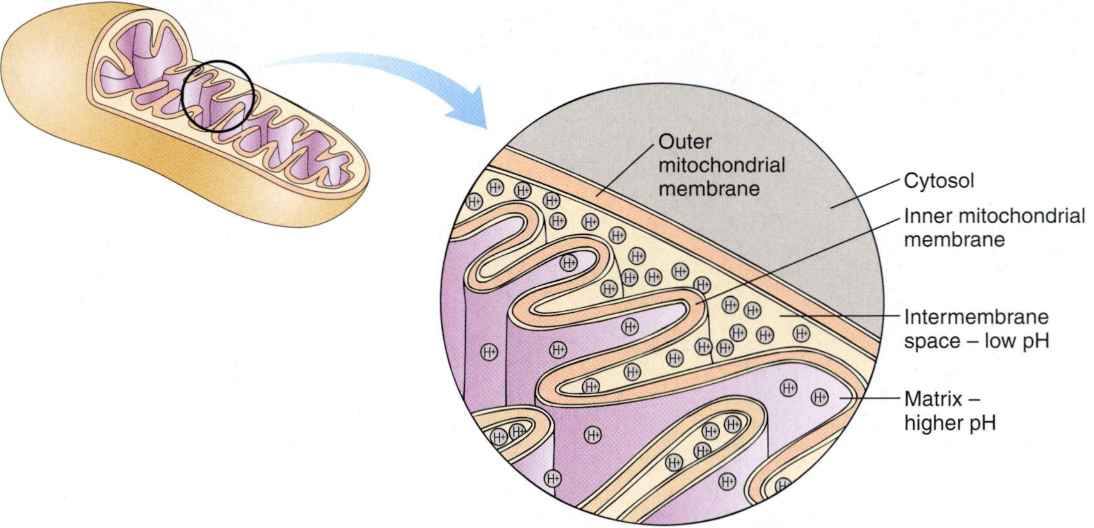

Figure 7–9 Accumulation of protons (H^+) within the intermembrane space. As electrons move down the electron transport chain, the electron transport complexes move protons (H^+) from the matrix to the intermembrane space, creating a proton gradient. The high concentration of H^+ in the intermembrane space lowers the pH.

plexes I, III, and IV) (Fig. 7–10). The result is a proton gradient across the inner mitochondrial membrane. Like water behind a dam, the proton gradient is a form of potential energy that can be harnessed to provide the energy for ATP synthesis.

Diffusion of protons from the intermembrane space, where they are highly concentrated, through the inner mitochondrial membrane to the matrix of the mitochondrion is limited to specific channels formed by a fifth enzyme complex, **ATP synthase,** a transmembrane protein. Portions of these complexes project from the inner surface of the membrane (the surface that faces the matrix) and are visible by electron microscopy (Fig. 7–11). Diffusion of the protons down their gradient, through the ATP synthase complex, is exergonic because the entropy of the system increases. This exergonic process provides the energy for ATP production, although the exact mechanism by which ATP synthase catalyzes the phosphorylation of ADP is still incompletely understood. In 1997 Paul Boyer and John Walker shared the Nobel Prize for Chemistry for the discovery that ATP synthase functions in an unusual way. Experimental evidence strongly suggests that ATP synthase acts like a highly efficient molecular motor: During the production of ATP from ADP and inorganic phosphate, a central structure of ATP synthase rotates, possibly in response to the force of protons moving through the enzyme complex. The rotation apparently alters the conformation of the catalytic subunits in a way that allows ATP synthesis.

Chemiosmosis is a fundamental mechanism of energy coupling in cells; it allows exergonic redox processes to drive the endergonic reaction in which ATP is produced by phosphorylating ADP. In photosynthesis (see Chapter 8), ATP is produced by a comparable process.

AEROBIC RESPIRATION OF ONE GLUCOSE YIELDS A MAXIMUM OF 36 TO 38 ATPs

Let us now review where biologically useful energy is captured in aerobic respiration and calculate the total energy yield from the complete oxidation of glucose. Figure 7–12 summarizes the arithmetic involved.

1. In glycolysis, glucose is activated by the addition of phosphates from 2 ATP molecules and converted ultimately to 2 pyruvates + 2 NADH + 4 ATPs, yielding a net profit of 2 ATPs.
2. The 2 pyruvates are metabolized to 2 acetyl CoA + 2 CO_2 + 2 NADH.
3. In the citric acid cycle the 2 acetyl CoA molecules are metabolized to 4 CO_2 + 6 NADH + 2 $FADH_2$ + 2 ATPs.

Because the oxidation of NADH in the electron transport chain yields up to 3 ATPs per molecule, the total of 10 NADH molecules can yield up to 30 ATPs. The 2 NADH molecules from glycolysis, however, yield either 2 or 3 ATPs each. This is because certain types of eukaryotic cells must expend energy to shuttle the NADH produced by glycolysis across the mitochondrial membrane (discussed shortly). Prokaryotic cells lack mitochondria; hence they have no need to shuttle NADH molecules. For this reason, bacteria are able to generate 3 ATPs for every NADH, even those produced during glycolysis. Thus, the maximum number of ATPs formed using the energy from NADH is 28 to 30.

The oxidation of $FADH_2$ yields 2 ATPs per molecule (recall that $FADH_2$ enters the electron transport chain at a different location than NADH), so the 2 $FADH_2$ molecules produced in the citric acid cycle yield 4 ATPs.

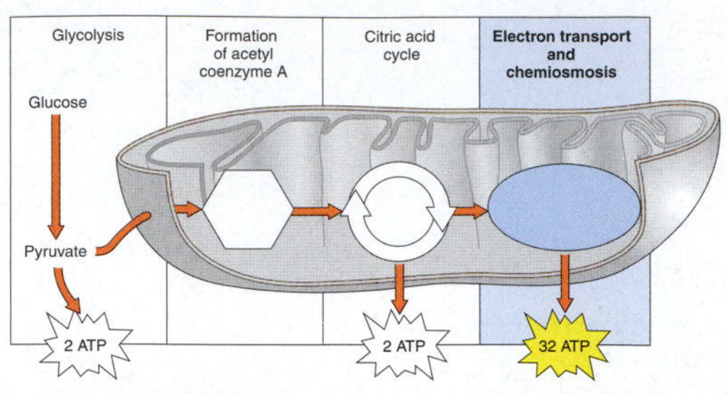

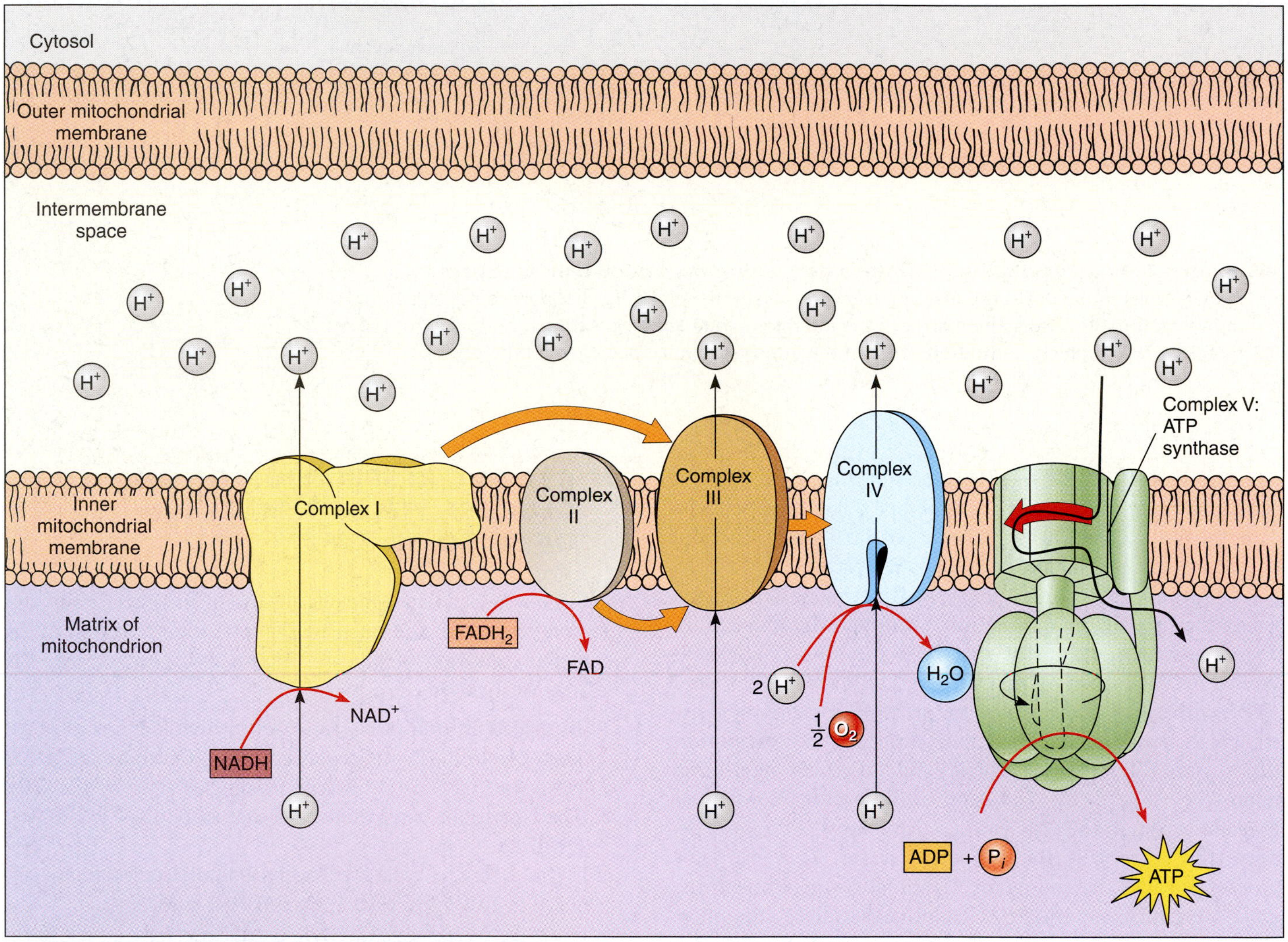

Figure 7–10 A detailed look at electron transport and chemiosmosis. The electron transport chain in the inner mitochondrial membrane includes three proton pumps that are located in three of the four electron transport complexes. (The orange arrows indicate the pathway of electrons.) The energy released during electron transport is used to transport protons (H^+) from the mitochondrial matrix to the intermembrane space, where a high concentration of protons accumulates. The protons are prevented from diffusing back into the matrix except through special channels in ATP synthase in the inner membrane. The flow of the protons through ATP synthase provides the energy for generating ATP from ADP and P_i. In the process, the inner part of ATP synthase rotates *(thick red arrow)* like a motor.

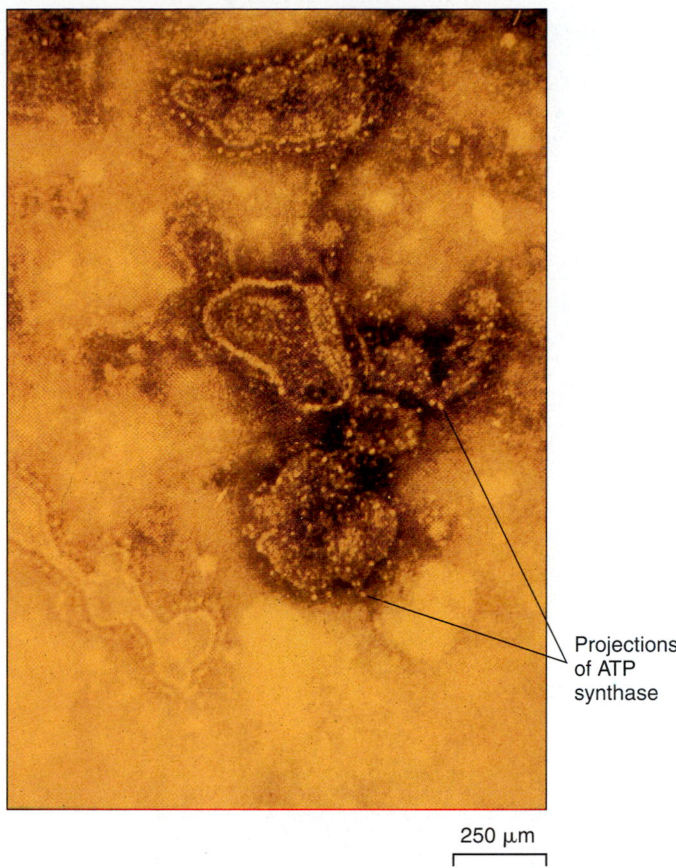

Figure 7–11 Inner mitochondrial membrane. This TEM shows hundreds of projections of ATP synthase complexes along the surface of the inner mitochondrial membrane. *(R. Bhatnagar/Visuals Unlimited)*

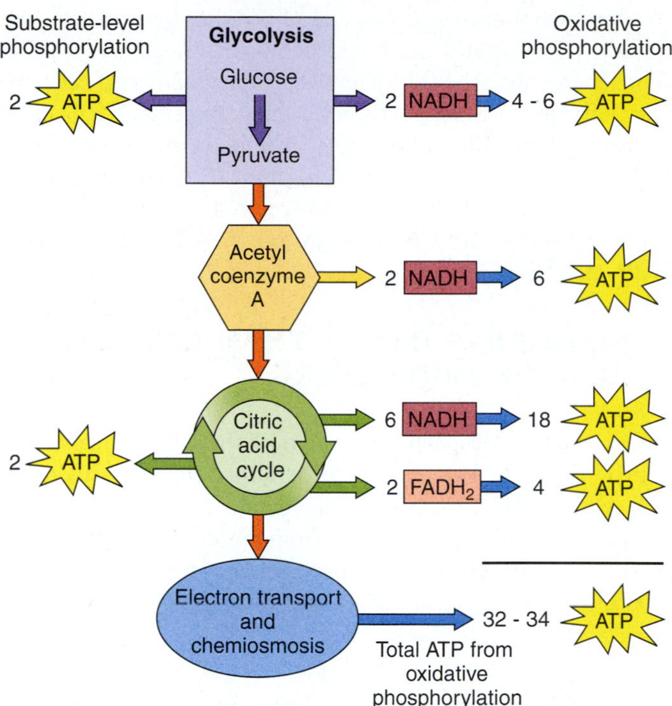

Figure 7–12 Energy yield from the complete oxidation of glucose by aerobic respiration. A maximum of 36 to 38 ATPs are produced per glucose molecule. Of these ATPs, four are produced by substrate-level phosphorylation, and the remainder by oxidative phosphorylation (that is, electron transport and chemiosmosis).

4. Summing all the ATPs (2 from glycolysis, 2 from the citric acid cycle, and 32 to 34 from electron transport and chemiosmosis), we see that the complete aerobic metabolism of one molecule of glucose yields a maximum of 36 to 38 ATPs. Note that most of the ATPs are generated by oxidative phosphorylation, which involves the electron transport chain and chemiosmosis. Only 4 ATPs are formed by substrate-level phosphorylation in glycolysis and the citric acid cycle.

We can analyze the efficiency of the overall process of aerobic respiration by comparing the free energy captured as ATP to the total free energy in a glucose molecule. You will recall from Chapter 6 that, although heat energy cannot power biological reactions, it is convenient to measure energy as heat. This can be done through the use of a calorimeter, an instrument that measures the heat of a reaction. A sample is placed in a compartment surrounded by a chamber of water. As the sample burns (becomes oxidized), the temperature of the water rises, providing a measure of the heat released during the reaction.

When 1 mol of glucose is burned in a calorimeter, some 686 kcal (2870 kJ) are released as heat. The free energy temporarily held in the phosphate bonds of ATP is about 7.6 kcal (31.8 kJ) per mole. When 36 to 38 ATPs are generated during the aerobic respiration of glucose, the free energy trapped in ATP amounts to 7.6 kcal/mol × 36, or about 274 kcal (1146 kJ) per mole. Thus, the efficiency of aerobic respiration is 274/686, or about 40%. (By comparison, a steam power plant has an efficiency of 35% to 36% in converting its fuel energy into electricity.) The remainder of the energy in the glucose is released as heat.

Mitochondrial shuttle systems harvest the electrons of NADH produced in the cytosol

The inner mitochondrial membrane is not permeable to NADH, which is a large molecule. Therefore the NADH molecules produced in the cytosol during glycolysis cannot diffuse into the mitochondria to transfer their electrons to the electron transport chain. Unlike ATP and ADP, NADH does not have a carrier protein to transport it across the membrane. Instead, several systems have evolved to transfer just the *electrons* of NADH, not the NADH molecules themselves, into the mitochondria.

In liver, kidney, and heart cells, a special shuttle system transfers the electrons from NADH through the inner mitochondrial membrane to an NAD^+ molecule in the matrix. These electrons are transferred to the electron transport chain in the inner mitochondrial membrane, and up to three molecules of ATP are produced per pair of electrons.

In skeletal muscle, brain, and some other types of cells, another type of shuttle operates. Because this shuttle requires more energy than the shuttle in liver, kidney, and heart cells, the electrons are at a lower energy level when they enter the electron transport chain. They are accepted by ubiquinone rather than by NAD^+ and so generate a maximum of 2 ATP molecules per pair of electrons. This is why the number of ATPs produced by aerobic respiration of 1 molecule of glucose in skeletal muscle cells is 36 rather than 38.

NUTRIENTS OTHER THAN GLUCOSE ALSO PROVIDE ENERGY

Many organisms depend on nutrients other than glucose as a source of energy. Humans and many other animals usually obtain more of their energy by oxidizing fatty acids than by oxidizing glucose. Amino acids derived from protein digestion are also used as fuel molecules. Such nutrients are transformed into one of the metabolic intermediates that are fed into glycolysis or the citric acid cycle (Fig. 7–13).

Amino acids are metabolized by reactions in which the amino group ($—NH_2$) is first removed, a process called **deamination.** In mammals and some other animals, the amino group is converted to urea (see Fig. 46–2) and excreted, but the carbon chain is metabolized and eventually is used as a reactant in one of the steps of aerobic respiration. The amino acid alanine, for example, undergoes deamination to become pyruvate, the amino acid glutamate is converted to α-ketoglutarate, and the amino acid aspartate yields oxaloacetate. Pyruvate enters aerobic respiration as the end product of glycolysis, and α-ketoglutarate and oxaloacetate both enter aerobic respiration as intermediates in the citric acid cycle. Ultimately, the carbon chains of all the amino acids are metabolized in this way.

Each gram of lipid in the diet contains 9 kcal (38 kJ), more than twice as much energy as 1 g of glucose or amino acids, which have about 4 kcal (17 kJ) per gram. Lipids are rich in energy because they are highly reduced; that is, they have many hydrogen atoms and few oxygen atoms. When completely oxidized in aerobic respiration, a molecule of a six-carbon fatty acid generates up to 44 ATPs (compared with 36 to 38 ATPs for a molecule of glucose, which also has 6 carbons).

Both the glycerol and fatty acid components of a triacylglycerol (see Chapter 3) are used as fuel; phosphate is added to glycerol, converting it to G3P or another compound that enters glycolysis. Fatty acids are oxidized and split enzymatically into two-carbon acetyl groups that are bound to coenzyme A; that is, fatty acids are converted to acetyl CoA. This process, which occurs in the mitochondrial matrix, is called **beta-oxidation,** or **β-oxidation.** Acetyl CoA molecules formed by β-oxidation enter the citric acid cycle.

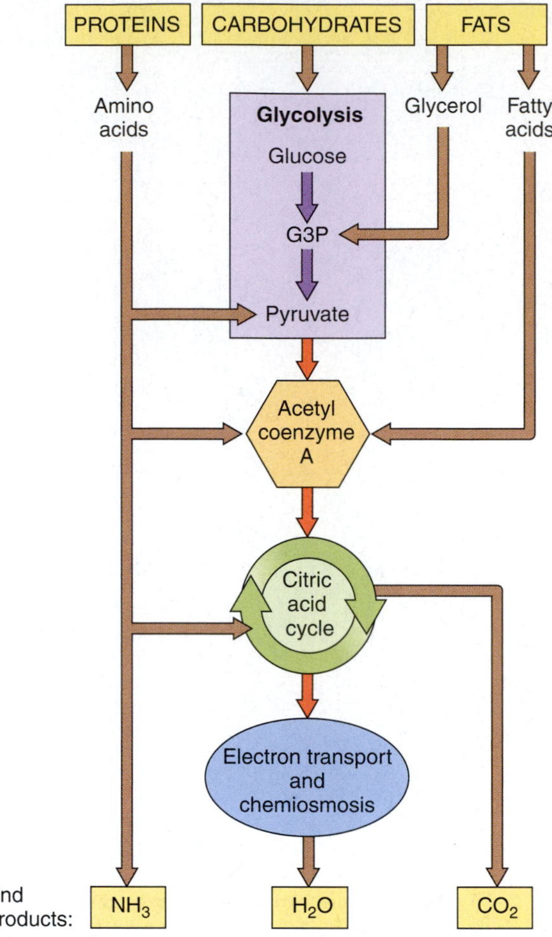

Figure 7–13 Energy from carbohydrates, proteins, and fats. Products of the catabolism of carbohydrates, proteins, and fats enter glycolysis or the citric acid cycle at various points. This diagram is greatly simplified and illustrates only a few of the principal catabolic pathways.

CELLS REGULATE AEROBIC RESPIRATION

Aerobic respiration requires a steady input of fuel molecules and oxygen. Under normal conditions these materials are adequately provided and do not affect the rate of respiration. Instead, the rate of aerobic respiration is regulated by how much ADP and phosphate are available. In a resting muscle cell, for example, ATP synthesis continues until most of the ADP has been converted to ATP. At this point oxidative phosphorylation slows considerably. Because electron flow is tightly coupled to oxidative phosphorylation, the flow of electrons also slows, which in turn slows down the citric acid cycle.

When ATP transfers energy to power an energy-requiring process like muscle contraction, many molecules of ATP are hydrolyzed. The ADP molecules produced can then accept phosphate to become ATP once again; aerobic respiration speeds up until most of the ADP has again been converted to ATP.

The control of most metabolic pathways is exerted on a particular enzyme that catalyzes a reaction early in the pathway (see Fig. 6–16 and discussion of feedback inhibition in Chapter 6). The regulated enzyme is usually inhibited by the presence of the end product of the pathway. One of the important control points in aerobic respiration in mammals is phosphofructokinase, an

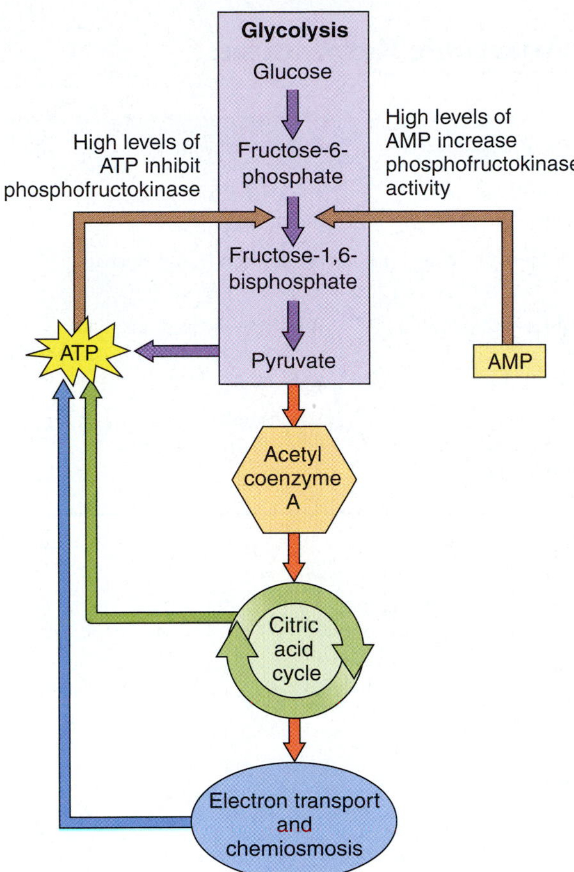

Figure 7–14 Regulation of aerobic respiration. When enough ATP accumulates in the cell *(left)*, the ATP inhibits phosphofructokinase, an enzyme that catalyzes an early step in glycolysis. Alternatively, when ATP levels are low and AMP levels are correspondingly high *(right)*, phosphofructokinase activity is high. These regulatory controls affect the rate of glycolysis (and therefore of aerobic respiration), matching it to the cell's demands for energy.

enzyme that catalyzes an early reaction of glycolysis (Fig. 7–14; also see Fig. 7–4). The active site of phosphofructokinase binds ATP and fructose-6-phosphate. However, the enzyme is inhibited by the presence of very high levels of ATP and activated by the presence of AMP (adenosine monophosphate, a molecule formed when two phosphates are removed from ATP). Therefore, this enzyme is inactivated when ATP levels are high and activated when they are low.[3] Phosphofructokinase possesses different allosteric sites for both enzyme inhibitors (in this case, ATP) and enzyme activators (in this case, AMP).

When respiration produces more ATP than the cell currently needs, some of the excess ATP binds to the allosteric inhibitor site of phosphofructokinase, changing its conformation so that it is no longer active. Thus, glycolysis and aerobic respiration slow down, and less ATP is produced. As excess ATP is used by the cell, ADP and AMP are produced. Now the allosteric inhibitor site of phosphofructokinase is no longer occupied by ATP. Instead, the allosteric activator sites are filled with AMP. Thus the enzyme becomes activated and respiration proceeds, generating more ATP.

[3] Other materials, including citrate, also affect the activity of phosphofructokinase.

ANAEROBIC RESPIRATION AND FERMENTATION DO NOT REQUIRE OXYGEN

Anaerobic respiration, which does not use oxygen as the final electron acceptor, is performed by some types of bacteria that live in such anaerobic environments as waterlogged soil, stagnant ponds, or animal intestines. As in aerobic respiration, electrons are transferred in anaerobic respiration from glucose to NADH; they then pass down an electron transport chain that is coupled to ATP synthesis by chemiosmosis. However, an inorganic substance such as nitrate (NO_3^-) or sulfate (SO_4^{2-}) replaces molecular oxygen as the terminal electron acceptor. The end products of this type of anaerobic respiration are carbon dioxide, one or more reduced inorganic substances, and ATP. One representative type of anaerobic respiration, summarized below, is part of the biogeochemical cycle known as the **nitrogen cycle** (see Chapter 53).

$$C_6H_{12}O_6 + \underset{\text{Potassium nitrate}}{12\ KNO_3} \longrightarrow 6\ CO_2 + 6\ H_2O + \underset{\text{Potassium nitrite}}{12\ KNO_2} + \underset{\text{(in the chemical bonds of ATP)}}{\text{Energy}}$$

Certain other bacteria, as well as some fungi, regularly use **fermentation,** an anaerobic pathway that does not involve an electron transport chain. During fermentation only two ATPs are formed per glucose (by substrate-level phosphorylation during glycolysis). One might expect that a cell that obtains energy from glycolysis would produce pyruvate, the end product of glycolysis. However, this cannot happen because every cell has a limited supply of NAD^+, and NAD^+ is required for glycolysis to continue. If virtually all NAD^+ becomes reduced to NADH during glycolysis, then glycolysis stops, and no more ATP can be produced.

In fermentation, NADH molecules transfer their hydrogen atoms to organic molecules, thus regenerating the NAD^+ needed to keep glycolysis going. The resulting, relatively reduced, organic molecules (commonly, alcohol or lactate) tend to be toxic to the cells and are essentially waste products. Table 7–2 summarizes features of aerobic respiration, anaerobic respiration, and fermentation.

Alcohol fermentation and lactate fermentation are inefficient

Yeasts are **facultative anaerobes.** These eukaryotic, unicellular fungi have mitochondria and carry out aerobic respiration when oxygen is available but switch to **alcohol fermentation** when deprived of oxygen (Fig. 7–15*a*). They have enzymes that decarboxylate pyruvate, releasing carbon dioxide and forming a two-carbon compound called **acetaldehyde.** NADH produced during glycolysis transfers hydrogen atoms to acetaldehyde, reducing it to form

TABLE 7–2 A Comparison of Aerobic Respiration, Anaerobic Respiration, and Fermentation

	Aerobic Respiration	Anaerobic Respiration	Fermentation
Immediate Fate of Electrons in NADH	Transferred to an electron transport chain	Transferred to an electron transport chain	Transferred to an organic molecule
Terminal Electron Acceptor of Electron Transport Chain	O_2	Inorganic substances such as NO_3^- or SO_4^{2-}	No electron transport chain
Reduced Product(s) Formed	Water	Relatively reduced inorganic substances	Relatively reduced organic compounds (commonly, alcohol or lactate)
Mechanism of ATP Synthesis	Oxidative phosphorylation/ chemiosmosis; also substrate-level phosphorylation	Oxidative phosphorylation/ chemiosmosis; also substrate-level phosphorylation	Substrate-level phosphorylation only (during glycolysis)

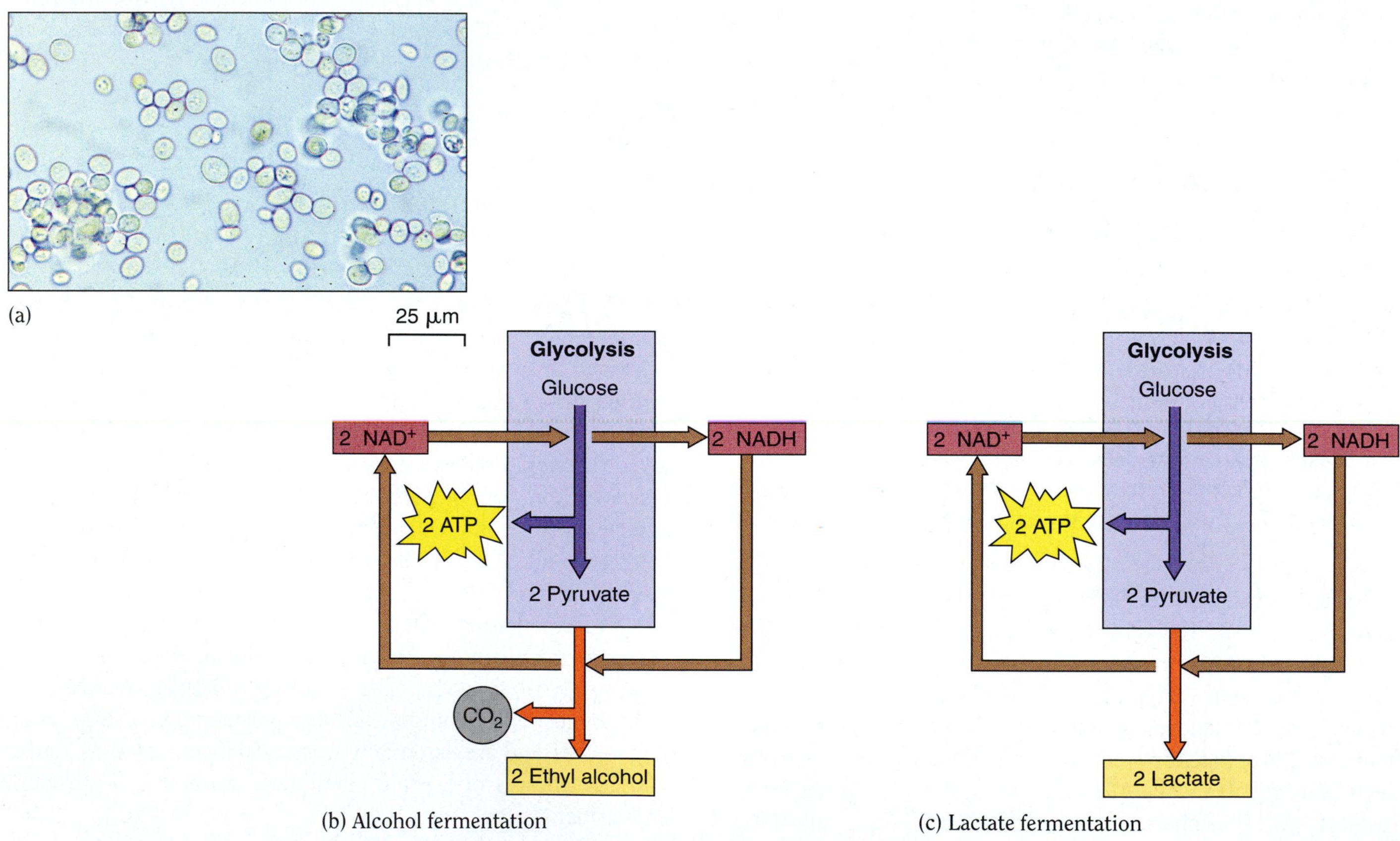

Figure 7–15 Fermentation. **(a)** Light micrograph of live brewer's yeast *(Saccharomyces cerevisiae)*. Yeast cells possess mitochondria and carry on aerobic respiration when O_2 is present. In the absence of O_2, yeasts carry on alcohol fermentation. **(b, c)** Glycolysis is the first part of fermentation pathways. **(b)** In alcohol fermentation, CO_2 is split off, and the two-carbon compound ethyl alcohol is the end product. **(c)** In lactate fermentation, the final product is the three-carbon compound lactate. In both alcohol and lactate fermentation, there is a net gain of only two ATPs per molecule of glucose. Note that the NAD^+ used during glycolysis is regenerated during both alcohol fermentation and lactate fermentation. *(a, Dwight R. Kuhn)*

ethyl alcohol (Fig. 7–15*b*). Alcohol fermentation is the basis for the production of beer, wine, and other alcoholic beverages. Yeast cells are also used in baking to produce the carbon dioxide that causes dough to rise; the alcohol evaporates during baking.

Certain fungi and bacteria perform **lactate (lactic acid) fermentation.** In this alternative pathway, NADH produced during glycolysis transfers hydrogen atoms to pyruvate, reducing it to form **lactate** (Fig. 7–15*c*). The ability of some bacteria to produce lactate is exploited by humans, who use these bacteria to make yogurt and to ferment cabbage for sauerkraut.

Lactate is also produced during strenuous activity in the muscle cells of humans and other complex animals. If the amount of oxygen delivered to muscle cells is insufficient to support aerobic respiration, the cells shift briefly to lactate fermentation. The shift is only temporary, however, and oxygen is required for sustained work. As lactate accumulates in muscle cells, it contributes to fatigue and muscle cramps.[4] About 80% of the lactate is eventually exported to the liver, where it is used to regenerate more glucose for the muscle cells. The remaining 20% of the lactate is metabolized in muscle cells in the presence of oxygen. This explains why you continue to breathe heavily after you have stopped exercising: The additional oxygen is needed to oxidize lactate, thereby restoring the muscle cells to their normal state.

Although humans can use lactate fermentation to produce ATP for only a few minutes, a few animals can live without oxygen for much longer periods. The red-eared slider can remain under water for as long as two weeks (Fig. 7–16). During this time, it is relatively inactive and therefore does not need to expend a great deal of energy. It relies on lactate fermentation for ATP production.

Both alcohol fermentation and lactate fermentation are highly inefficient because the fuel is only partially oxidized. Alcohol, the end product of fermentation by yeast cells, can be burned and can even be used as automobile fuel; obviously, it contains a great deal of energy that the yeast cells are unable to extract using anaerobic methods. Lactate, a three-carbon compound, contains even more energy than the two-carbon alcohol. In contrast, all available energy is removed during aerobic respiration because the fuel molecules become completely oxidized to CO_2. A net profit of only 2 ATPs is produced by the fermentation of one molecule of glucose, compared with up to 36 to 38 ATPs when oxygen is available.

The inefficiency of fermentation necessitates a large supply of fuel. For example, skeletal muscle cells, which often metabolize anaerobically for short periods, store large quantities of glucose in the form of glycogen. To perform the same amount of work, a cell engaged in fermentation must consume up to 20 times more glucose or other carbohydrate per second than a cell using aerobic respiration.

[4] Why exercise causes fatigue and muscle cramps is incompletely understood at this time, but these conditions may be related to lactate buildup, insufficient oxygen, and/or the depletion of fuel molecules.

Figure 7–16 Long-term use of lactate fermentation for ATP production. The red-eared slider *(Trachemys scripta)* can stay submerged for as long as two weeks. *(Cleveland P. Hickman, Jr./Visuals Unlimited)*

SUMMARY WITH KEY TERMS

I. Metabolism, the total of all the chemical reactions that occur in cells, has two complementary components.
   A. **Catabolism** releases energy by breaking down complex molecules into simpler molecules. Catabolic pathways include aerobic respiration, anaerobic respiration, and fermentation.
   B. **Anabolism** is the synthesis of complex molecules from simpler building blocks. Protein synthesis is an example of anabolism.

II. During **aerobic respiration,** a fuel molecule such as glucose is oxidized to form carbon dioxide and water.
   A. Aerobic respiration is a redox process in which electrons are transferred from glucose (which becomes **oxidized**) to oxygen (which becomes **reduced**).
   B. Up to 36 to 38 ATPs are produced per molecule of glucose.

III. The chemical reactions of aerobic respiration occur in four stages: glycolysis, formation of acetyl CoA, the citric acid cycle, and the electron transport chain/chemiosmosis.
   A. During **glycolysis,** which occurs in the cytosol, a molecule of glucose is degraded to two molecules of **pyruvate.**
      1. Two ATP molecules (net) are produced by **substrate-level phosphorylation** during glycolysis.
      2. Four hydrogen atoms are removed from the fuel molecule and used to produce two NADH.
   B. The two pyruvate molecules each lose a molecule of carbon dioxide, and the remaining acetyl groups each combine with **coenzyme A,** producing two molecules of **acetyl CoA.** One NADH is produced as each pyruvate is converted to acetyl CoA.

C. Each acetyl CoA enters the **citric acid cycle** by combining with a four-carbon compound, **oxaloacetate,** to form **citrate,** a six-carbon compound. Two acetyl CoA molecules enter the cycle for every glucose molecule.
   1. For every two carbons that enter the cycle as part of an acetyl CoA molecule, two leave as carbon dioxide.
   2. For every acetyl CoA, hydrogen atoms are transferred to three NAD^+ and one FAD; only one ATP is produced by substrate-level phosphorylation.

D. Hydrogen atoms (or their electrons) removed from fuel molecules are transferred from one electron acceptor to another down an **electron transport chain** located in the mitochondrial inner membrane.
   1. The electron transport chain is organized as four protein complexes. Complex I accepts electrons from NADH molecules, and complex II accepts electrons from $FADH_2$ molecules; both of these complexes produce reduced ubiquinone. Complex III accepts electrons from reduced ubiquinone and passes them on to cytochrome *c*. Complex IV accepts electrons from cytochrome *c* and uses these electrons to reduce molecular oxygen, forming water in the process.
   2. According to the **chemiosmotic model,** some of the energy of the electrons in the electron transport chain is used to establish a proton gradient across the inner mitochondrial membrane.
   3. The diffusion of protons through the membrane from the intermembrane space to the mitochondrial matrix (through channels formed by the enzyme **ATP synthase**) provides the energy needed to synthesize ATP.
   4. The coupling of ATP synthesis to the redox reactions in the electron transport chain is known as **oxidative phosphorylation.**

IV. Organic nutrients other than glucose are converted to appropriate compounds and fed into glycolysis or the citric acid cycle.
   A. Amino acids are **deaminated,** and their carbon skeletons are converted to metabolic intermediates of aerobic respiration.
   B. Both the glycerol and fatty acid components of lipids are oxidized as fuel. Fatty acids are converted to acetyl CoA molecules by the process of **β-oxidation.**

V. In **anaerobic respiration,** electrons are transferred from fuel molecules to an electron transport chain; the final electron acceptor is an inorganic substance such as nitrate or sulfate, not molecular oxygen.

VI. **Fermentation** is an anaerobic process that does not use an electron transport chain. There is a net gain of only two ATPs per glucose; these are produced during glycolysis. To maintain the supply of NAD^+ essential for glycolysis, hydrogen atoms are transferred from NADH to an organic compound derived from the initial nutrient.
   A. Yeast cells carry out **alcohol fermentation,** in which **ethyl alcohol** and carbon dioxide are the final waste products.
   B. Certain fungi, bacteria, and animal cells carry out **lactate fermentation,** in which hydrogen atoms are added to pyruvate to form **lactate,** a waste product.

Summary Reactions for Aerobic Respiration

Summary reaction for the complete oxidation of glucose:

$$C_6H_{12}O_6 + 6\ O_2 + 6\ H_2O \longrightarrow 6\ CO_2 + 12\ H_2O + \text{Energy (36 to 38 ATP)}$$

Summary reaction for glycolysis:

$$C_6H_{12}O_6 + 2\ \text{ATP} + 2\ \text{ADP} + 2\ P_i + 2\ NAD^+ \longrightarrow 2\ \text{Pyruvate} + 4\ \text{ATP} + 2\ \text{NADH} + H_2O$$

Summary reaction for the conversion of pyruvate to acetyl CoA:

$$2\ \text{Pyruvate} + 2\ \text{Coenzyme A} + 2\ NAD^+ \longrightarrow 2\ \text{Acetyl CoA} + 2\ CO_2 + 2\ \text{NADH}$$

Summary reaction for the citric acid cycle:

$$2\ \text{Acetyl CoA} + 6\ NAD^+ + 2\ \text{FAD} + 2\ \text{ADP} + 2\ P_i + 2\ H_2O \longrightarrow 4\ CO_2 + 6\ \text{NADH} + 2\ FADH_2 + 2\ \text{ATP} + 2\ \text{CoA}$$

Summary reactions for the processing of the hydrogen atoms of NADH and $FADH_2$ in the electron transport chain:

$$\text{NADH} + 3\ \text{ADP} + 3\ P_i + \tfrac{1}{2}\ O_2 \longrightarrow NAD^+ + 3\ \text{ATP} + H_2O$$

$$FADH_2 + 2\ \text{ADP} + 2\ P_i + \tfrac{1}{2}\ O_2 \longrightarrow \text{FAD} + 2\ \text{ATP} + H_2O$$

Summary Reactions for Fermentation

Summary reaction for lactate fermentation:

$$C_6H_{12}O_6 \longrightarrow 2\ \text{Lactate} + \text{Energy (2 ATP)}$$

Summary reactions for alcohol fermentation:

$$C_6H_{12}O_6 \longrightarrow 2\ CO_2 + 2\ \text{Ethyl alcohol} + \text{Energy (2 ATP)}$$

POST-TEST

1. The process of splitting larger molecules into smaller ones is an aspect of metabolism called (a) anabolism (b) fermentation (c) catabolism (d) oxidative phosphorylation (e) chemiosmosis
2. The synthetic aspect of metabolism is referred to as (a) anabolism (b) fermentation (c) catabolism (d) oxidative phosphorylation (e) chemiosmosis
3. A chemical process during which a substance gains electrons is called (a) oxidation (b) oxidative phosphorylation (c) deamination (d) reduction (e) dehydrogenation
4. The pathway through which glucose is degraded to pyruvate is referred to as (a) aerobic respiration (b) the citric acid cycle (c) the oxidation of pyruvate (d) alcohol fermentation (e) glycolysis
5. The reactions of ____________ take place within the cytosol of eukaryotic cells. (a) glycolysis (b) oxidation of pyruvate (c) the citric acid cycle (d) chemiosmosis (e) the electron transport chain
6. Before pyruvate enters the citric acid cycle, it is decarboxylated, oxidized, and combined with coenzyme A, forming acetyl CoA, carbon dioxide, and one molecule of (a) NADH (b) $FADH_2$ (c) ATP (d) ADP (e) $C_6H_{12}O_6$
7. In the first step of the citric acid cycle, acetyl CoA reacts with oxaloacetate to form (a) pyruvate (b) citrate (c) NADH (d) ATP (e) CO_2
8. Dehydrogenase enzymes remove hydrogen atoms from fuel molecules and transfer them to acceptors such as (a) O_2 and H_2O (b) ATP and FAD (c) NAD^+ and FAD (d) CO_2 and H_2O (e) CoA and pyruvate
9. Which of the following is a major source of electrons for the electron transport chain? (a) H_2O (b) ATP (c) NADH (d) ATP synthase (e) coenzyme A
10. In the process of ____________, electron transport and ATP synthesis are coupled by a proton gradient across the inner mitochondrial membrane. (a) chemiosmosis (b) deamination (c) anaerobic respiration (d) glycolysis (e) decarboxylation
11. Which of the following is a common energy flow sequence in aerobic respiration, starting with the energy stored in glucose? (a) glucose → NADH → pyruvate → ATP (b) glucose → ATP → NADH → electron transport chain (c) glucose → NADH → electron transport chain → ATP (d) glucose → oxygen → NADH → water (e) glucose → $FADH_2$ → NADH → coenzyme A
12. Which multiprotein complex in the electron transport chain is responsible for reducing molecular oxygen? (a) complex I (NADH-ubiquinone oxidoreductase) (b) complex II (succinate-ubiquinone reductase) (c) com-

plex III (ubiquinone-cytochrome *c* oxidoreductase) (d) complex IV (cytochrome *c* oxidase) (e) complex V (ATP synthase)

13. A net profit of only 2 ATPs can be produced anaerobically from the ____________ of one molecule of glucose, compared with a maximum of 38 ATPs produced in ____________. (a) fermentation; anaerobic respiration (b) aerobic respiration; fermentation (c) aerobic respiration; anaerobic respiration (d) dehydrogenation; decarboxylation (e) fermentation; aerobic respiration
14. When deprived of oxygen, yeast cells obtain energy by fermentation, producing carbon dioxide, ATP, and (a) acetyl CoA (b) ethyl alcohol (c) lactate (d) pyruvate (e) citrate
15. During strenuous muscle activity, the pyruvate in muscle cells may accept hydrogen from NADH to become ____________. (a) acetyl CoA (b) ethyl alcohol (c) lactate (d) pyruvate (e) citrate

REVIEW QUESTIONS

1. What is the specific role of oxygen in most cells? What happens when cells that can only respire aerobically are deprived of oxygen?
2. Mitochondria are often referred to as the "power plants" of the cell. Justify this with a specific explanation.
3. Refer to Figure 7–7, the diagram of the steps in the citric acid cycle. Look at each reaction and, without reading the description, determine if it is a dehydrogenation, decarboxylation, or preparation reaction.
4. Sketch a mitochondrion and indicate the locations of the electron transport chain and the proton gradient that drives ATP production.
5. Why is each of the following essential to chemiosmotic ATP synthesis? (a) electron transport chain (b) proton gradient (c) ATP synthase complex
6. Explain the roles of the following in aerobic respiration: (a) NAD^+ and FAD (b) oxygen
7. Sum up how much energy (as ATP) is made available to the cell from a single glucose molecule by the operation of (1) glycolysis, (2) the formation of acetyl CoA, (3) the citric acid cycle, and (4) the electron transport chain.
8. Trace the fate of hydrogen atoms removed from glucose during glycolysis when oxygen is present in muscle cells; compare this to the fate of hydrogen atoms removed from glucose when the amount of available oxygen is insufficient to support aerobic respiration.
9. Compare the ATP yields of aerobic respiration and fermentation.
10. Label the ten blank lines in the figure. Use Figure 7–2 to check your answers.

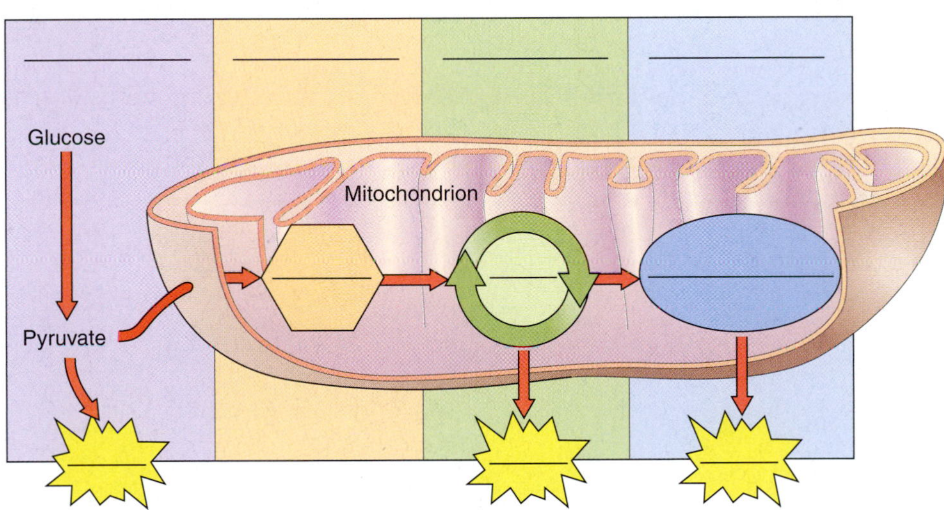

YOU MAKE THE CONNECTION

1. The reactions of glycolysis are identical in *all* organisms—bacteria, protists, fungi, plants, and animals—that obtain energy from glucose catabolism. What does this universality suggest about the evolution of glycolysis?
2. How are the endergonic reactions of the first phase of glycolysis coupled to the hydrolysis of ATP, which is exergonic? How are the exergonic reactions of the second phase of glycolysis coupled to the endergonic synthesis of ATP and NADH?
3. Why is the citric acid cycle also called the Krebs cycle? Why is it known as the tricarboxylic acid cycle? (*Hint:* Examine the chemical structures pictured in Fig. 7–7.)
4. What is the role of the inner mitochondrial membrane in the coupling of electron transport and ATP synthesis?
5. Based on what you have learned in this chapter, explain why a schoolchild can run 17 miles per hour in a 100-meter dash, but a trained athlete can run only about 11.5 miles per hour in a 26-mile marathon.

RECOMMENDED READINGS

Fackelmann, K. "Power Failure." *Science News,* Vol. 152, 27 Sept. 1997. Defects in mitochondria can lead to several different human diseases.

Garrett, R.H., and C.M. Grisham. *Biochemistry,* 2nd ed. Saunders College Publishing, Philadelphia, 1999. A comprehensive biochemistry text with good coverage of cellular respiration and related aspects of metabolism.

Offner, S. "A Plain English Map of the Human Glycolysis Enzymes." *The American Biology Teacher,* Vol. 61, No. 6, June 1999. Every enzyme is a protein coded for by a gene somewhere on the chromosomes. To illustrate this point, the author shows locations on the human chromosomes of the ten enzymes involved in glycolysis.

Seymour, R.S. "Plants That Warm Themselves." *Scientific American,* Vol. 276, No. 3, March 1997. Some plants produce a significant amount of heat when they flower, and a few precisely regulate their temperature.

Tobin, A.J., and R.E. Morel. *Asking About Cells.* Saunders College Publishing, Philadelphia, 1997. Contains a readable and detailed account of cellular energy metabolism.

- Visit our Web site at **http://www.info.brookscole.com/solomonbergmartin** for links to chapter-related resources on the World Wide Web. Additional on-line materials relating to this chapter can also be found on our Web site.

See chapter activity on BioActive Learner CD for additional help in mastering the chapter's material. Icon location in the chapter's margins shows which topics have tutorials or simulations in the CD.

8

Photosynthesis: Capturing Energy

These blue lupines *(Lupinus hirsutus)* and the trees behind them are photoautotrophs. They use CO_2 as a carbon source and light as an energy source. Photographed in Southern Michigan. *(Skip Moody/Dembinsky Photo Associates)*

LEARNING OBJECTIVES

After you have studied this chapter you should be able to

1. Explain the relationship between a wavelength of light and its energy and describe the physical properties of light.
2. Describe what can happen to an electron in a biololgical molecule such as chlorophyll when a photon of light energy is absorbed.
3. Diagram the internal structure of a chloroplast and explain how its components interact and facilitate the process of photosynthesis.
4. Write a summary reaction for photosynthesis, showing the origin and fate of each substance involved.
5. Distinguish between the light-dependent reactions and carbon fixation reactions of photosynthesis.
6. Describe photosynthesis as a redox process and contrast cyclic and noncyclic electron transport pathways.
7. Explain how a proton (H^+) gradient is established across the thylakoid membrane and how this gradient functions in ATP synthesis.
8. Summarize the three phases of the Calvin cycle and indicate the roles of ATP and NADPH in the process.
9. Discuss how the C_4 pathway increases the effectiveness of the Calvin cycle in certain types of plants.

This chapter deals with photosynthesis, which is the first step in the flow of energy through most living things. To help you appreciate the significance of photosynthesis, consider that all organisms—from microscopic bacteria to dolphins to palm trees—can be classified on the basis of two aspects of their nutrition: their carbon source and their energy source. Carbon atoms are required for the carbon skeletons of an organism's organic molecules, and as we have seen in Chapters 6 and 7, energy powers all life processes, from growth, to movement, to repair of worn or injured tissues.

Organisms obtain carbon in one of two ways. **Autotrophs** (from the Greek *auto,* "self," and *trophos,* "nourishing") are able to carry out carbon fixation; they use carbon dioxide (CO_2) as a carbon source. **Heterotrophs** (from the Greek *heter,* "other," and *trophos,* "nourishing") cannot fix carbon; they use organic molecules produced by other organisms as the building blocks from which they synthesize the carbon compounds they need.

Organisms obtain energy in one of two ways. **Phototrophs** are photosynthetic organisms that use light as their energy source. In contrast, **chemotrophs** use organic compounds, such as glucose, or inorganic substances, such as iron, nitrate, ammonia, or sulfur, as sources of energy. Chemotrophs typically obtain energy from these materials by oxidation–reduction reactions (see discussion of redox reactions in Chapters 6 and 7).

By combining both aspects of nutrition—that is, carbon and energy requirements—we find that all organisms fall into one of four groups:

1. Green plants *(see photograph),* algae, and certain bacteria, are **photoautotrophs** (that is, both phototrophs and autotrophs). Photoautotrophs are uniquely capable of absorbing and converting light energy into stored chemical energy of organic molecules by the process of **photosynthesis.** Photoautotrophs use light energy to make ATP and other molecules that temporarily hold chemical energy but are unstable and cannot be stockpiled in the cell. Their energy drives the anabolic pathway by which a photosynthetic cell synthesizes stable organic molecules from the simple inorganic compounds, CO_2 and water. These organic compounds are used not only as starting materials to synthesize all the other organic compounds the photosynthetic organism needs (e.g., complex carbohydrates, amino acids, and lipids) but also for energy storage. Glucose

and other carbohydrates produced during photosynthesis are relatively reduced compounds that can be subsequently oxidized by aerobic respiration or by some other catabolic pathway (see Chapter 7).

2. A few bacteria, known as nonsulfur purple bacteria, are **photoheterotrophs** (i.e., both phototrophs and heterotrophs). Photoheterotrophs are able to use light energy but unable to carry out carbon fixation. Photoheterotrophs must obtain carbon from organic compounds (as "food").
3. **Chemoautotrophs** are bacteria that are both chemotrophs and autotrophs. These bacteria obtain their energy from the oxidation of reduced inorganic molecules such as hydrogen sulfide (H_2S), nitrite (NO_2^-), or ammonia (NH_3). Some of this energy is then used to carry out carbon fixation.
4. All animals, fungi, and most bacteria are **chemoheterotrophs;** that is, they are both chemotrophs and heterotrophs. Chemoheterotrophs use preformed organic molecules as a source of both energy and carbon. Plants and other photosynthetic organisms produce almost all of the preformed organic molecules used by chemoheterotrophs.

From this discussion, it is clear that photosynthesis is the process that captures the vast majority of the energy used by living organisms. Photosynthesis not only sustains plants and other photoautotrophs but also indirectly supports almost all animals and other chemoheterotrophs in the biosphere. Each year plants and other photosynthetic organisms convert CO_2 into billions of tons of organic molecules. The chemical energy stored in these molecules fuels the metabolic reactions that sustain almost all life.

LIGHT IS COMPOSED OF PARTICLES THAT TRAVEL AS WAVES

Most life on our planet depends on light, either directly or indirectly, and so it is important to understand the nature of light and how it permits photosynthesis to occur. Visible light represents a very small portion of a vast, continuous range of radiation called the electromagnetic spectrum (Fig. 8–1). All radiation in this spectrum travels as waves. A **wavelength** is the distance from one wave peak to the next. At one end of the electromagnetic spectrum are gamma rays, which have very short wavelengths measured in fractions of nanometers, or nm. (One nanometer equals 10^{-9} m—that is, one billionth of a meter.) At the other end of the spectrum are radio waves, with wavelengths so long they can be measured in kilometers. The portion of the electromagnetic spectrum from 380 to 760 nm is called the visible spectrum because humans can see it. The visible spectrum includes all the colors of the rainbow (Fig. 8–2); violet has the shortest wavelength, and red has the longest.

Light behaves not only as waves do but also as particles. Light is composed of small particles, or packets, of energy called **photons.** The energy of a photon is inversely proportional to its wavelength; shorter wavelength light has more energy per photon than does longer wavelength light.

Why does photosynthesis depend on light detectable by the human eye (visible light) rather than on some other wavelength of radiation? We can only speculate on the answer. Perhaps it is because radiation within the visible light portion of the spectrum excites certain types of biological molecules, moving electrons into higher energy levels. Radiation with wavelengths longer than those of visible light does not possess enough energy to excite these biological molecules. Radiation with wavelengths shorter than those of visible light is so energetic that it disrupts the bonds of many biological molecules. Thus, visible light has just the right amount of energy to be useful in photosynthesis.

When a molecule absorbs a photon of light energy, one of its electrons becomes energized, which means that the electron shifts from a lower energy atomic orbital to a high-energy orbital that is more distant from the atomic nucleus. One of two things then hap-

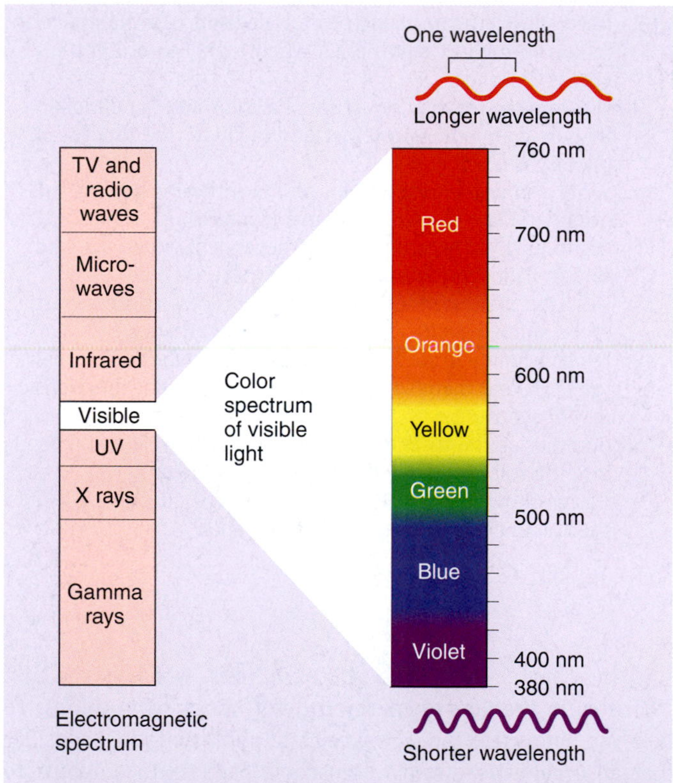

Figure 8–1 The electromagnetic spectrum. Waves in the electromagnetic spectrum have similar properties but different wavelengths and methods of production. Radio waves are the longest (and least energetic) waves, with wavelengths as long as 20 km. Gamma rays are the shortest (and most energetic) waves. Visible light represents a small fraction of the electromagnetic spectrum and consists of a mixture of wavelengths ranging from approximately 380 to 760 nm. The energy from visible light is used in photosynthesis.

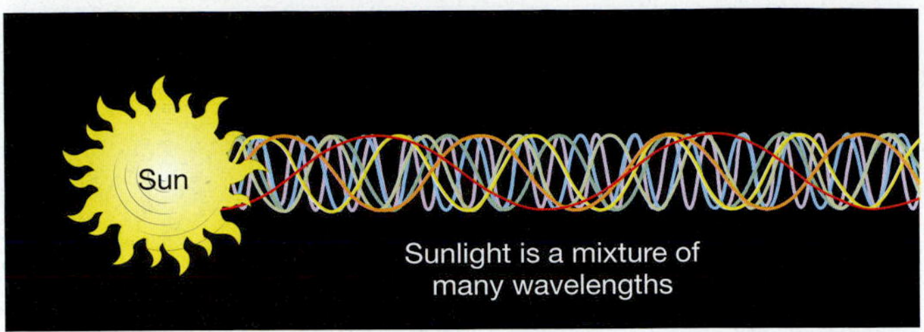
Sun
Sunlight is a mixture of many wavelengths

Figure 8–2 Visible radiation emitted from the sun. Electromagnetic radiation from the sun includes ultraviolet radiation and visible light of varying colors and wavelengths.

pens, depending on the atom and its surroundings (Fig. 8–3). The atom may return to its **ground state,** which is the condition in which all its electrons are in their normal, lowest energy levels. When an electron returns to its ground state, its energy is dissipated as heat or as an emission of light of a longer wavelength than the absorbed light; this emission of light is called **fluorescence.** Alternatively, the energized electron may leave the atom and be accepted by an electron acceptor molecule, which becomes reduced in the process; this is what occurs in photosynthesis.

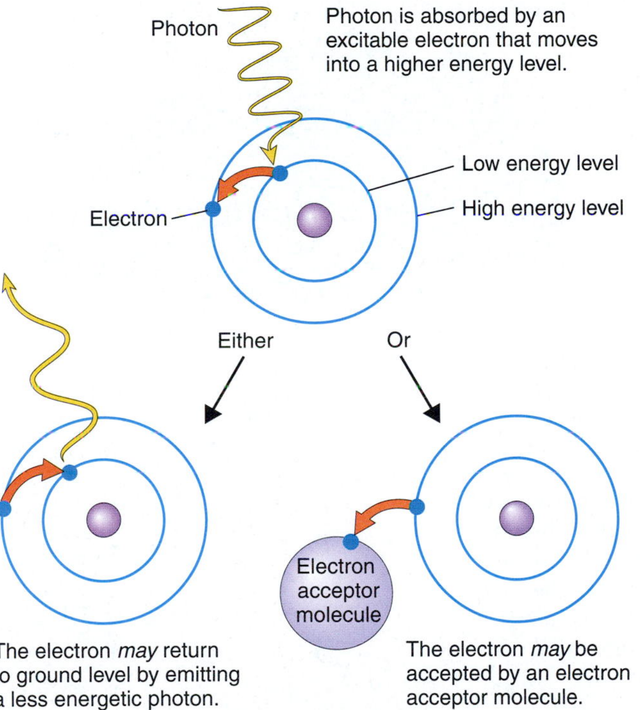
Photon
Photon is absorbed by an excitable electron that moves into a higher energy level.
Low energy level
Electron
High energy level
Either
Or
Electron acceptor molecule
The electron *may* return to ground level by emitting a less energetic photon.
The electron *may* be accepted by an electron acceptor molecule.

Figure 8–3 Interactions between light and atoms or molecules. *(Top)* When a photon of light energy strikes an atom, or a molecule of which the atom is a part, the energy of the photon may push an electron to an orbital farther from the nucleus (i.e., into a higher energy level). *(Lower left)* If the electron returns to the lower, more stable energy level, the energy may be released as a less energetic, longer-wavelength photon, known as *fluorescence (shown),* or as heat. *(Lower right)* If the appropriate electron acceptors are available, the electron may leave the atom. During photosynthesis, an electron acceptor captures the energetic electron and passes it to a chain of acceptors.

Now that we have an understanding of some of the properties of light, we will consider the cellular location where light is used for photosynthesis.

PHOTOSYNTHESIS IN EUKARYOTES TAKES PLACE IN CHLOROPLASTS

When a section of leaf tissue is examined under the microscope, we can see that the green pigment, **chlorophyll,** is not uniformly distributed in the cell but is confined to organelles called chloroplasts (Fig. 8–4). In plants, chloroplasts are located mainly in the cells of the **mesophyll,** a tissue inside the leaf. Each mesophyll cell has 20 to 100 chloroplasts.

The chloroplast, like the mitochondrion, is bounded by outer and inner membranes. The inner membrane encloses a fluid-filled region called the **stroma,** which contains most of the enzymes required to produce carbohydrate molecules. Suspended in the stroma is a third system of membranes that forms an interconnected set of flat, disclike sacs called **thylakoids.** The thylakoid membrane encloses a fluid-filled interior space, the **thylakoid lumen.** In some regions of the chloroplast, thylakoid sacs are arranged in stacks called **grana** (sing., *granum*). Each granum looks something like a stack of coins, with each "coin" being a thylakoid. Some thylakoid membranes extend from one granum to another. These membranes, like the inner mitochondrial membrane (see Chapter 7), are involved in ATP synthesis. (Photosynthetic prokaryotes have no chloroplasts, but thylakoid membranes are often arranged around the periphery of the cell as infoldings of the plasma membrane.)

Chlorophyll is found in the thylakoid membrane

Thylakoid membranes contain several kinds of pigments, which are substances that absorb visible light. Different pigments absorb light of different wavelengths. Chlorophyll, the main pigment of photosynthesis, absorbs light primarily in the blue and red regions of the visible spectrum. Green light is not appreciably absorbed by chlorophyll. Plants usually appear green because some of the green light that strikes them is scattered or reflected.

A chlorophyll molecule (Fig. 8–5) has two main parts, a complex ring and a long side chain. The ring structure, called a *porphyrin ring,* is made up of joined smaller rings composed of carbon and nitrogen atoms; it is the porphyrin ring that absorbs the light energy. The porphyrin ring of chlorophyll is strikingly similar to the heme portion of the red pigment hemoglobin in red blood cells. However, unlike heme, which contains an atom of iron in the center of the ring, chlorophyll contains an atom of magnesium in that position. The chlorophyll molecule also contains a long, hydrocarbon side chain that makes the molecule extremely nonpolar.

All chlorophyll molecules in the thylakoid membrane are associated with specific **chlorophyll-binding proteins.** Biologists have identified about 15 different chlorophyll-binding proteins. Each

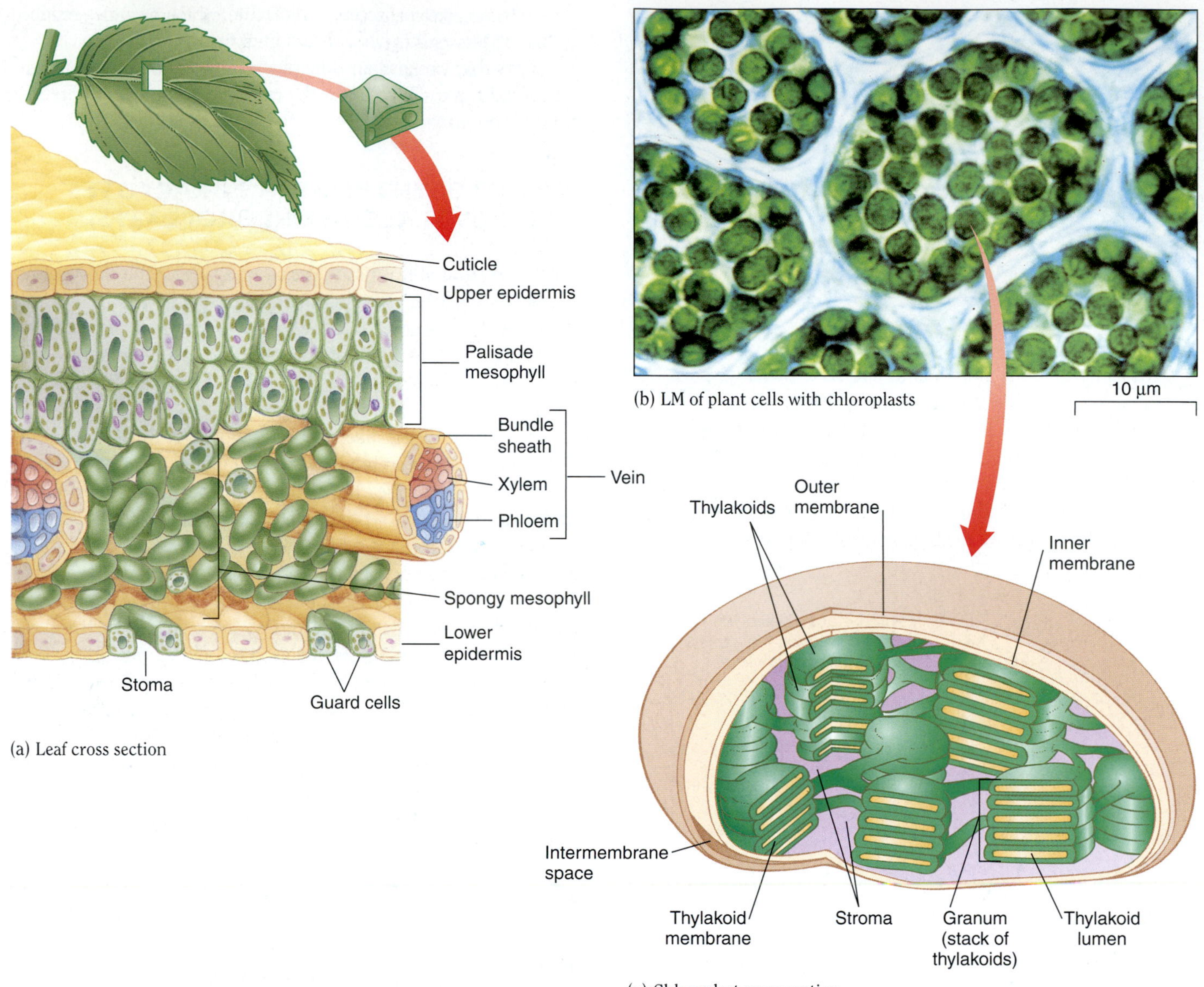

Figure 8–4 Where photosynthesis occurs in the plant. **(a)** A leaf cross section reveals that the mesophyll is the photosynthetic tissue. CO_2 enters the leaf through tiny pores called stomata (sing., stoma), and H_2O is carried to the mesophyll in veins. **(b)** LM of plant cells with numerous chloroplasts. **(c)** Chloroplast structure. The pigments necessary for the light-capturing reactions of photosynthesis are part of thylakoid membranes, whereas the enzymes for the synthesis of carbohydrate molecules are in the stroma. *(b, M. Eichelberger/Visuals Unlimited)*

thylakoid membrane is filled with precisely oriented chlorophyll molecules and chlorophyll-binding proteins, an arrangement that permits transfer of energy from one molecule to another.

There are several kinds of chlorophyll. The most important is **chlorophyll *a*,** the pigment that initiates the light-dependent reactions of photosynthesis. **Chlorophyll *b*** is an accessory pigment that also participates in photosynthesis. It differs from chlorophyll *a* only in a functional group on the porphyrin ring: The methyl group ($-CH_3$) in chlorophyll *a* is replaced in chlorophyll *b* by a terminal carbonyl group ($-CHO$). This difference shifts the wavelengths of light absorbed and reflected by chlorophyll *b,* making it yellow-green, whereas chlorophyll *a* is bright green.

Chloroplasts also have other accessory photosynthetic pigments, such as **carotenoids,** which are yellow and orange (see Fig. 3–14). Carotenoids absorb different wavelengths of light than chlorophyll does and so broaden the spectrum of light that provides energy for photosynthesis. Chlorophyll may be excited by light directly or indirectly, by energy passed to it from accessory pigments that have become excited by light. When a carotenoid

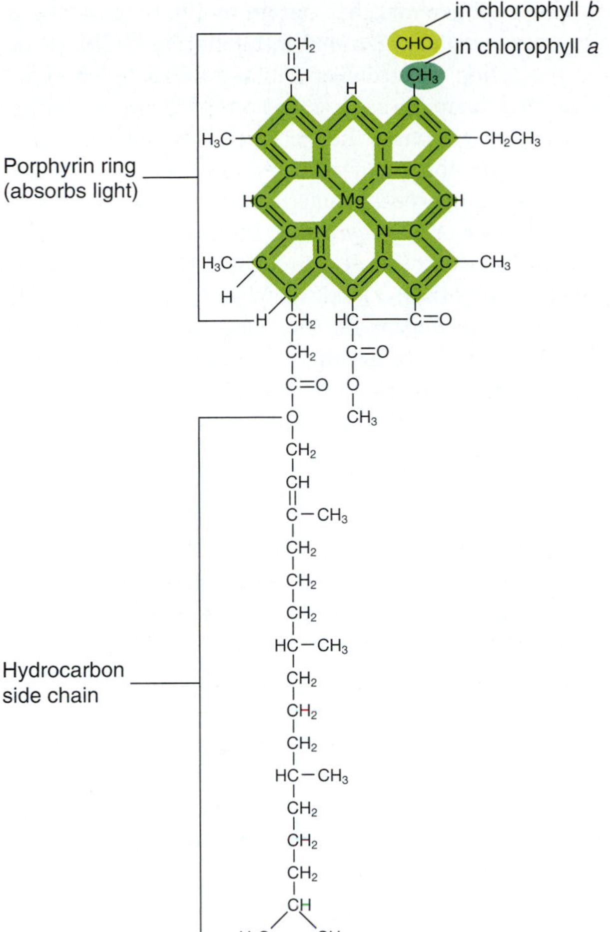

Figure 8-5 Chlorophyll structure. Chlorophyll consists of a porphyrin ring and a hydrocarbon side chain. The porphyrin ring, with a magnesium atom in its center, contains a system of alternating double and single bonds; these are commonly found in molecules that strongly absorb visible light. At the top right corner of the diagram, the methyl group (—CH_3) distinguishes chlorophyll *a* from chlorophyll *b*, which has a carbonyl group (—CHO) in this position.

molecule is excited, its energy can be transferred to chlorophyll *a*. Carotenoids also have an indispensable role in protecting chlorophyll and other parts of the thylakoid membrane from excess light energy that could easily damage the photosynthetic components. (High light intensities often occur in nature.)

Chlorophyll is the main photosynthetic pigment

As we have seen, the thylakoid membrane contains more than one kind of pigment. An instrument called a spectrophotometer is used to measure the relative abilities of different pigments to absorb different wavelengths of light. The **absorption spectrum** of a pigment is a plot of its absorption of light of different wavelengths. Figure 8–6*a* shows the absorption spectra for chlorophylls *a* and *b*.

The relative effectiveness of these different wavelengths of light in photosynthesis is given by an **action spectrum** of photosynthesis. To obtain an action spectrum, the rate of photosynthesis is measured at each wavelength for leaf cells or tissues exposed to monochromatic light (light of one wavelength) (Fig. 8–6*b*).

Process of Science

The first action spectrum was obtained in one of the classic experiments in biology. In 1883 the German biologist T.W. Engelmann carried out an experiment that took advantage of the shape of the chloroplast in a species of *Spirogyra,* a green

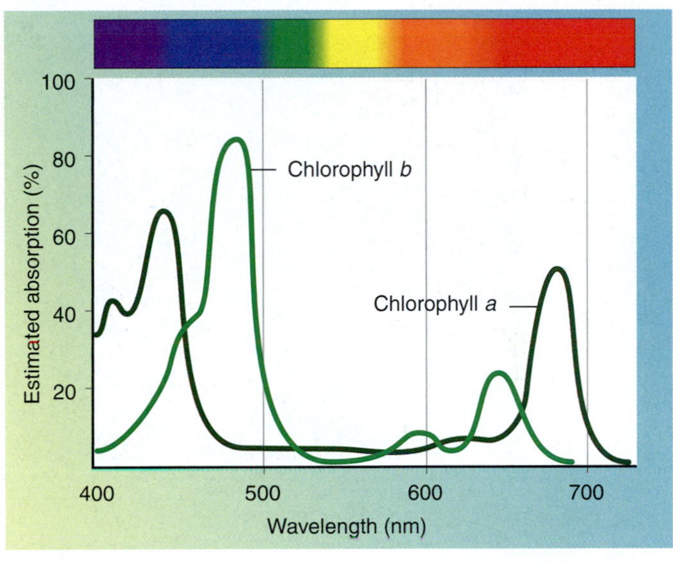

(a)

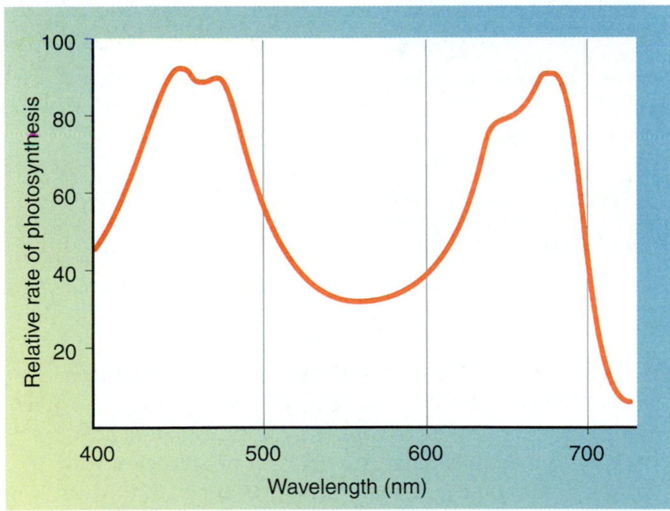

(b)

Figure 8–6 The absorption spectra for chlorophylls *a* and *b* and the action spectrum for photosynthesis. **(a)** Chlorophylls *a* and *b* absorb light mainly in the blue (422 nm to 492 nm) and red (647 nm to 760 nm) regions of visible light. **(b)** The action spectrum of photosynthesis shows how effective various wavelengths of light are in powering photosynthesis. Many plant species, including crop plants, have action spectra for photosynthesis that resemble the generalized action spectrum shown here.

alga that occurs as long, filamentous strands in freshwater habitats, especially slow-moving or still waters (Fig. 8–7*a*). The individual cells of *Spirogyra* each contain a long, spiral-shaped, emerald-green chloroplast embedded in the cytoplasm. Engelmann exposed these cells to a color spectrum produced by passing light through a prism. He reasoned that if chlorophyll were indeed responsible for photosynthesis, then it would take place most rapidly in the areas where the chloroplast was illuminated by the colors most strongly absorbed by chlorophyll.

Yet how could photosynthesis be measured in those technologically unsophisticated days? Engelmann knew that photosynthesis produces oxygen and that certain motile bacteria are attracted to areas of high oxygen concentration (Fig. 8–7*b*). He determined the action spectrum of photosynthesis by observing that the bacteria swam toward the portions of *Spirogyra* located in the red and blue regions of the spectrum. How did Engelmann know bacteria were not simply attracted to red or blue light? As a control, Engelmann exposed bacteria to the spectrum of visible light in the absence of *Spirogyra*. The bacteria showed no preference for any particular wavelength of light. Because the action spectrum of photosynthesis closely matched the absorption spectrum of chlorophyll, Engelmann concluded that chlorophyll in the chloroplasts (and not another compound in another organelle) is responsible for photosynthesis. Numerous studies using sophisticated instruments have since confirmed Engelmann's conclusions.

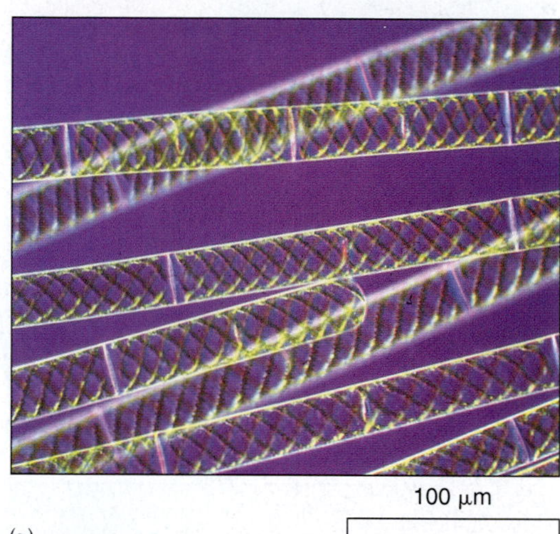

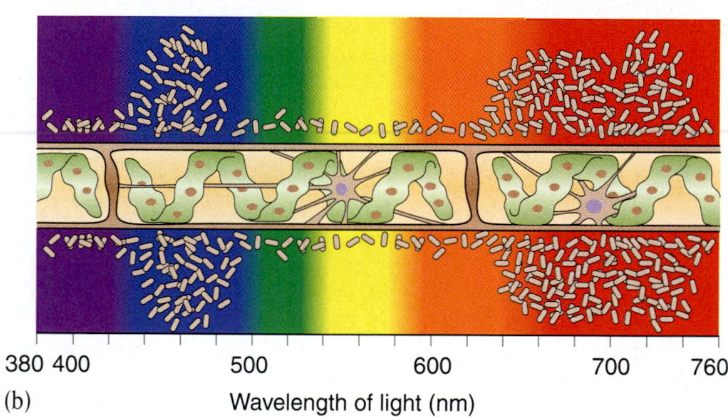

Figure 8–7 The first action spectrum of photosynthesis. **(a)** LM of filaments of *Spirogyra* sp., the green alga that Engelmann used in his classic experiment. **(b)** Engelmann illuminated a filament of *Spirogyra* with light that had been passed through a prism. In this way, different parts of the filament were exposed to different wavelengths of light. He estimated the rate of photosynthesis indirectly by observing the movement of motile aerobic bacteria toward the portions of the algal filament emitting the most oxygen. Watching through a microscope, Engelmann observed that the bacteria aggregated most densely along the cells in the red and, to a lesser extent, blue portions of the spectrum. This suggested that blue light and red light work most effectively for photosynthesis. *(a, T.E. Adams/Visuals Unlimited)*

The action spectrum of photosynthesis does not parallel the absorption spectrum of chlorophyll exactly (Fig. 8–6). This difference occurs because accessory pigments, such as carotenoids, transfer some of the energy of excitation produced by green light to chlorophyll molecules. The presence of these accessory photosynthetic pigments can be demonstrated by chemical analysis of almost any leaf, although it is obvious in temperate climates when leaves change color in the fall. Toward the end of the growing season, chlorophyll breaks down (and its magnesium is stored in the permanent tissues of the tree), leaving orange and yellow accessory pigments in the leaves.

PHOTOSYNTHESIS IS THE CONVERSION OF LIGHT ENERGY TO CHEMICAL BOND ENERGY

During photosynthesis, a cell uses light energy captured by chlorophyll to power the synthesis of carbohydrates. The overall reaction of photosynthesis can be summarized as follows:

$$\underset{\text{Carbon dioxide}}{6\,CO_2} + \underset{\text{Water}}{12\,H_2O} \xrightarrow[\text{Chlorophyll}]{\text{Light energy}} \underset{\text{Glucose}}{C_6H_{12}O_6} + \underset{\text{Oxygen}}{6\,O_2} + \underset{\text{Water}}{6\,H_2O}$$

The equation is typically written in the form given above, with H_2O on both sides, because water is a reactant in some reactions and a product in others. However, because there is no net yield of H_2O, we can simplify the summary equation of photosynthesis for purposes of discussion:

$$6\,CO_2 + 6\,H_2O \xrightarrow[\text{Chlorophyll}]{\text{Light}} C_6H_{12}O_6 + 6\,O_2$$

When we analyze this process, it appears that hydrogen atoms are transferred from H_2O to CO_2 to form carbohydrate, and so we recognize it as a **redox reaction.** As you learned in Chapter 6, in a redox reaction one or more electrons, usually as part of one or more hydrogen atoms, are transferred from an electron donor (a reducing agent) to an electron acceptor (an oxidizing agent).

$$6\ CO_2 + 6\ H_2O \xrightarrow[\text{Chlorophyll}]{\text{Light}} C_6H_{12}O_6 + 6\ O_2$$

(Reduction: $CO_2 \rightarrow C_6H_{12}O_6$; Oxidation: $H_2O \rightarrow O_2$)

When the electrons are transferred, some of their energy is transferred as well. However, the summary equation of photosynthesis is somewhat misleading because no direct transfer of hydrogen atoms actually occurs. The summary equation describes what happens but not how it happens. The "how" is much more complex and involves multiple steps, many of which are redox reactions.

The reactions of photosynthesis are divided into two parts: the light-dependent reactions (the *photo* part of photosynthesis) and the carbon fixation reactions (the *synthesis* part of photosynthesis). Each set of reactions occurs in a different part of the chloroplast: the light-dependent reactions in association with the thylakoids, and the carbon fixation reactions in the stroma (Fig. 8–8).

ATP and NADPH are the products of the light-dependent reactions: an overview

Light energy is converted to chemical energy in the **light-dependent reactions,** which are associated with the thylakoids. The light-dependent reactions begin as chlorophyll captures light energy, which causes one of its electrons to move to a higher energy state. The energized electron is transferred to an acceptor molecule and is replaced by an electron from H_2O. When this happens, H_2O is split and molecular oxygen is released (Fig. 8–9). Some of the energy of the energized electrons is used to phosphorylate **adenosine diphosphate (ADP),** forming **adenosine triphosphate (ATP).** In addition, the coenzyme **nicotinamide adenine dinucleotide phosphate ($NADP^+$)** becomes reduced, forming NADPH.[1] The products of the light-dependent reactions, ATP and NADPH, are both needed in the endergonic carbon fixation reactions.

Carbohydrates are produced during the carbon fixation reactions: an overview

The ATP and NADPH molecules produced during the light-dependent phase are suited for transferring chemical energy but not for long-term energy storage. For this reason, some of their energy is transferred to chemical bonds in carbohydrates, which can be produced in large quantities and stored for future use. Known as **carbon fixation,** or **CO_2 fixation,** these reactions "fix" carbon atoms from CO_2 to existing skeletons of organic molecules. Because the carbon fixation reactions have no direct requirement for light, they were formerly referred to as the "dark" reactions. However, they certainly do not require darkness; in fact, many of the enzymes involved in carbon fixation are much more active in the light than in the dark. Furthermore, carbon fixation reactions depend on the products of the light-dependent reactions. Carbon fixation reactions take place in the stroma of the chloroplast.

[1] Although the correct way to write the reduced form of $NADP^+$ is $NADPH + H^+$, for simplicity's sake we present the reduced form as NADPH throughout the chapter

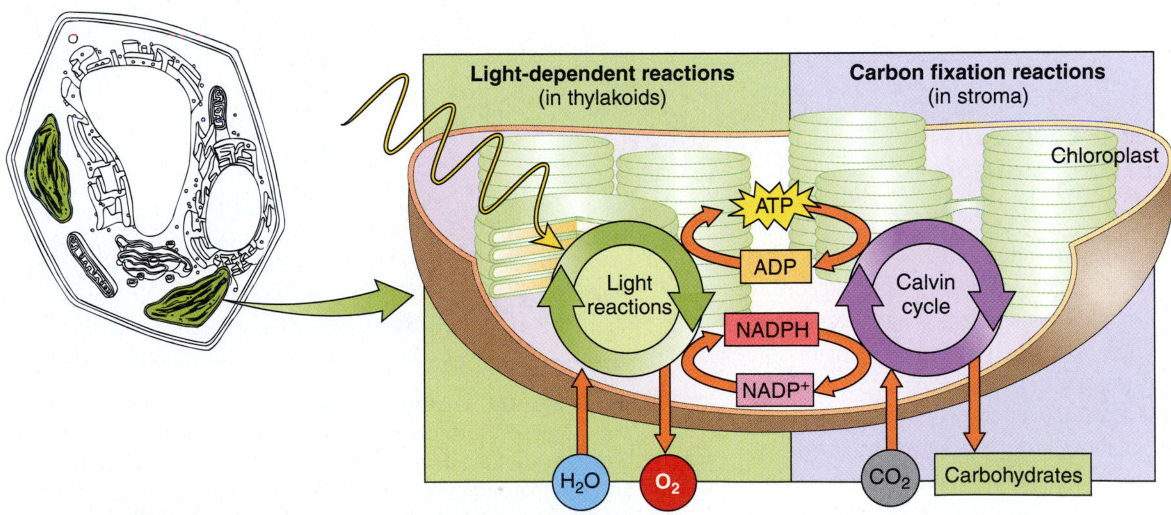

Figure 8–8 Overview of photosynthesis. Photosynthesis consists of light-dependent reactions, which occur in association with the thylakoids, and carbon fixation reactions, which occur in the stroma.

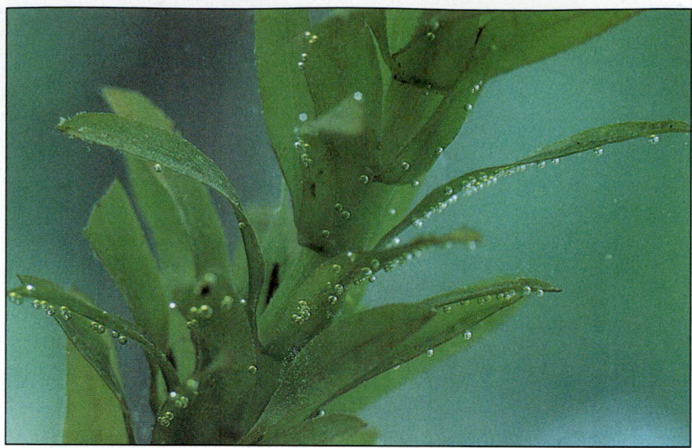

Figure 8–9 Oxygen produced by photosynthesis. On sunny days the oxygen released by aquatic plants may sometimes be visible as bubbles in the water. This plant (*Elodea* sp.) is actively carrying on photosynthesis. *(Bernd Wittich/Visuals Unlimited)*

Now that we have presented an overview of photosynthesis, let's examine the entire process more closely.

THE LIGHT-DEPENDENT REACTIONS CONVERT LIGHT ENERGY TO CHEMICAL ENERGY

In the light-dependent reactions, the radiant energy from sunlight is used to make ATP and to reduce $NADP^+$, forming NADPH. The light energy captured by chlorophyll is temporarily stored in these two compounds. The light-dependent reactions are summarized as follows:

$$12\ H_2O + 12\ NADP^+ + 18\ ADP + 18\ P_i \xrightarrow[\text{Chlorophyll}]{\text{Light}} 6\ O_2 + 12\ NADPH + 18\ ATP$$

Photosystems I and II each consist of a reaction center and multiple antenna complexes

The light-dependent reactions of photosynthesis begin when chlorophyll *a* and/or accessory pigments absorb light. According to the currently accepted model, chlorophylls *a* and *b* and accessory pigment molecules are organized with pigment-binding proteins in the thylakoid membrane into units called **antenna complexes.** The pigments and associated proteins are arranged as highly ordered groups of about 250 chlorophyll molecules associated with specific enzymes and other proteins. Each antenna complex absorbs light energy and transfers it to the **reaction center,** a complex of chlorophyll molecules and proteins, including electron transfer components, that participates directly in photosynthesis (Fig. 8–10). Light energy is converted to chemical energy in the reaction centers by a series of electron transfer reactions.

Two types of photosynthetic units, designated Photosystem I and Photosystem II, are involved in photosynthesis. Their reaction centers are distinguishable because they are associated with proteins in a way that causes a slight shift in their absorption spectra. Ordinary chlorophyll *a* has a strong absorption peak at about 660 nm. In contrast, the chlorophyll *a* molecule that makes up the reaction center associated with **Photosystem I** has an absorption peak at 700 nm and is referred to as **P700.** The reaction center of **Photosystem II (P680)** is made up of a chlorophyll *a* molecule with an absorption peak of about 680 nm.

When a pigment molecule absorbs light energy, that energy is passed from one pigment molecule to another until it reaches the reaction center. When the energy reaches a molecule of P700 (in a Photosystem I reaction center) or P680 (in a Photosystem II reaction center), an electron is then raised to a higher energy level. As we will see in the next section, this energized electron can be donated to an electron acceptor that becomes reduced in the process.

Noncyclic electron transport produces ATP and NADPH

We begin our discussion of noncyclic electron transport with the events associated with Photosystem I (Fig. 8–11). A pigment molecule in an antenna complex associated with Photosystem I absorbs a photon of light. The absorbed energy is transferred to the reaction center, where it excites an electron in a molecule of

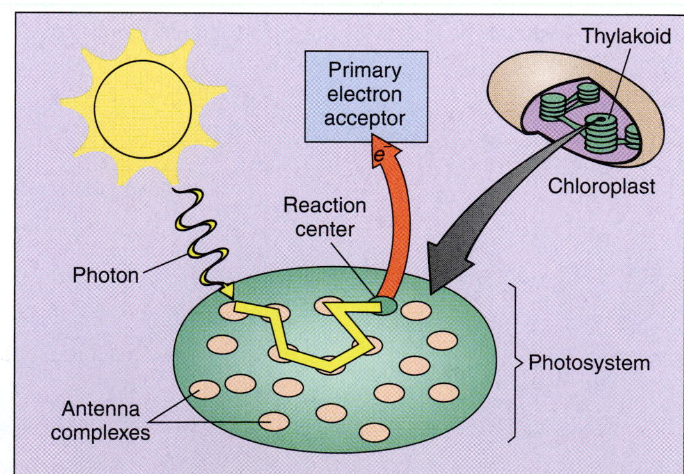

Figure 8–10 A photosystem. Chlorophyll molecules and accessory pigments are arranged in light-harvesting arrays known as antenna complexes. When a molecule in an antenna complex absorbs a photon, the energy of that photon is funneled into the reaction center. When this energy reaches the P700 (or P680) chlorophyll molecule in the reaction center, an electron becomes energized and is accepted by a primary electron acceptor.

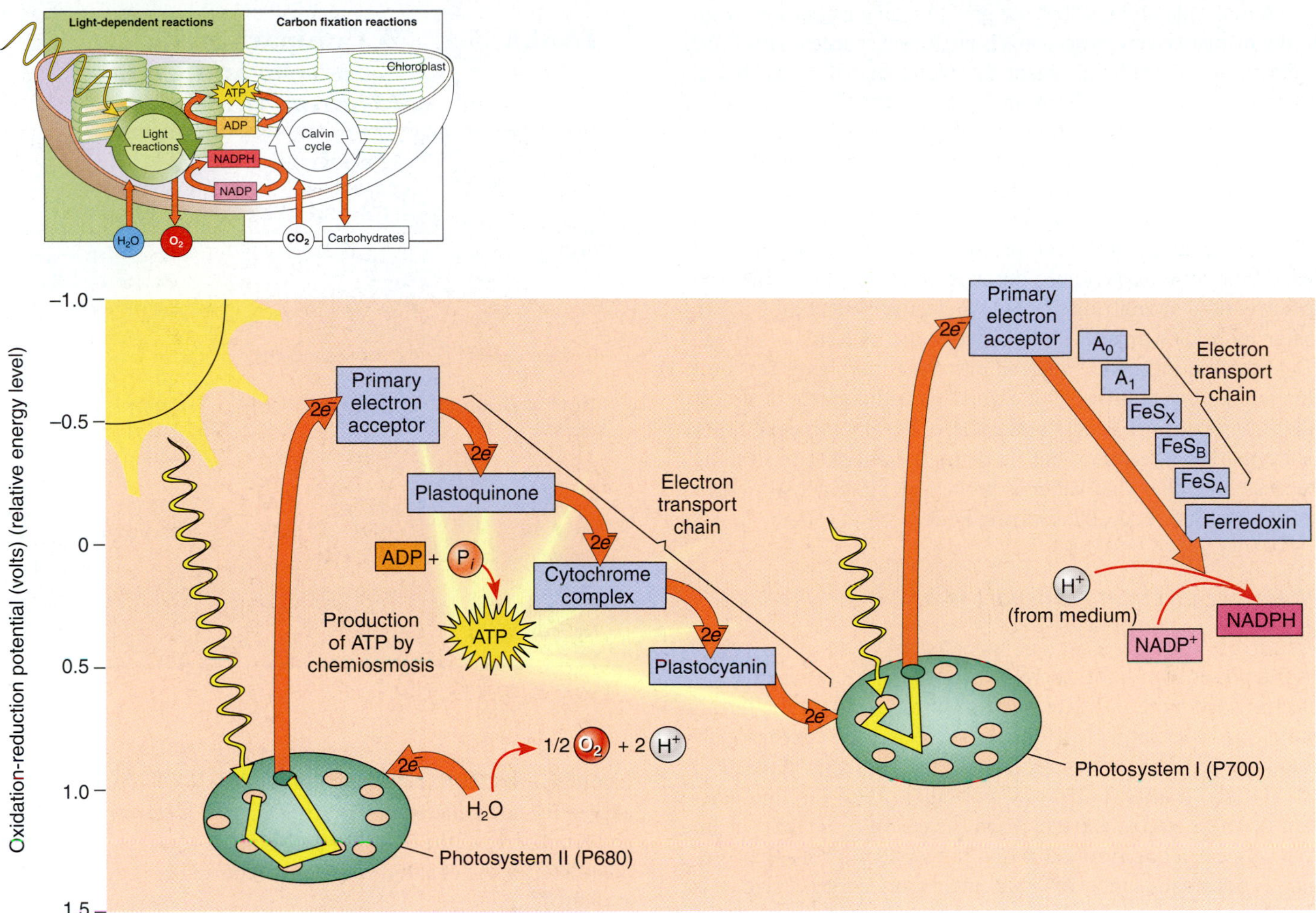

Figure 8-11 Noncyclic electron transport. In noncyclic electron transport, the formation of ATP is coupled to the one-way flow of energized electrons *(orange arrows)* from H_2O *(lower left)* to $NADP^+$ *(middle right).* Single electrons actually pass down the electron transport chain; two are shown in this figure because two are required to form one molecule of NADPH. Electrons are supplied to the system from the splitting of H_2O by Photosystem II, with the release of O_2 as a byproduct. When Photosystem II is activated by absorbing photons, electrons are passed along an electron transport chain and are eventually donated to Photosystem I. Electrons in Photosystem I are "re-energized" by the absorption of additional light energy and are passed to $NADP^+$, forming NADPH.

P700. This energized electron is transferred to a primary electron acceptor, which is the first of several electron acceptors in a series. (Uncertainty exists regarding the exact chemical nature of the primary electron acceptor for Photosystem I.) The energized electron is passed along an **electron transport chain** from one electron acceptor to another, until it is passed to **ferredoxin,** an iron-containing protein. Ferredoxin transfers the electron to $NADP^+$ in the presence of the enzyme *ferredoxin–NADP$^+$ reductase.* When $NADP^+$ accepts the two electrons, they unite with a proton (H^+); hence the reduced form of $NADP^+$ is NADPH, which is released into the stroma. P700 becomes positively charged when it gives up an electron to the primary electron acceptor; the missing electron is replaced by one donated by Photosystem II.

Like Photosystem I, Photosystem II is activated when a pigment molecule in an antenna complex absorbs a photon of light energy. The energy is transferred to the reaction center, where it causes an electron in a molecule of P680 to move to a higher energy level. This energized electron is accepted by a primary electron acceptor (a highly modified chlorophyll molecule known as pheophytin) and then passes along an electron transport chain until it is donated to P700 in Photosystem I.

A molecule of P680 that has given up an energized electron to the primary electron acceptor is positively charged. This P680 molecule is an oxidizing agent so strong that it is capable of pulling electrons away from an oxygen atom that is part of a H_2O molecule. In a reaction probably catalyzed by a unique, manganese-containing enzyme, the process of **photolysis** (literally "light-splitting") breaks water into its components: two electrons, two protons, and oxygen. Each electron is donated to a P680 molecule, and the protons are released into the thylakoid lumen. Because oxygen does not exist in atomic form, the oxygen produced by splitting one H_2O molecule is written $\frac{1}{2}O_2$. Two water molecules must be split to yield one molecule of oxygen (O_2), which is ultimately released into the atmosphere. This production of oxygen during photosynthesis is the source of almost all of the oxygen in Earth's atmosphere. The photolysis of water is a remarkable reaction, but its name is somewhat misleading because it implies that water is broken by light. Actually, light splits water indirectly, by oxidizing P680.

Noncyclic electron transport is a continuous linear process

In the presence of light, there is a continuous, one-way flow of electrons from the ultimate electron source, H_2O, to the terminal electron acceptor, $NADP^+$. Water undergoes enzymatically catalyzed photolysis to replace energized electrons donated to the electron transport chain by molecules of P680 in Photosystem II. These electrons travel down the electron transport chain that connects Photosystem II with Photosystem I. Thus they provide a continuous supply of replacements for energized electrons that have been given up by P700.

As electrons are transferred along the electron transport chain that connects Photosystem II with Photosystem I, they lose energy. Some of the energy released is used to pump protons across the thylakoid membrane, from the stroma to the thylakoid lumen, producing a proton gradient. The energy of this proton gradient is harnessed to produce ATP from ADP by chemiosmosis, which will be discussed shortly. ATP and NADPH, the products of the light-dependent reactions, are released into the stroma, where both are required in the carbon fixation reactions.

Cyclic electron transport produces ATP but no NADPH

Only Photosystem I is involved in cyclic electron transport, the simplest light-dependent reaction. The pathway is cyclic because energized electrons that originate from P700 at the reaction center eventually return to P700. In the presence of light, there is a continuous flow of electrons through an electron transport chain within the thylakoid membrane. As they are passed from one acceptor to another, the electrons lose energy, some of which is used to pump protons across the thylakoid membrane. An enzyme (ATP synthase, discussed shortly) in the thylakoid membrane uses the energy of the proton gradient to manufacture ATP. NADPH is not produced, H_2O is not split, and oxygen is not generated. By itself, cyclic electron transport could not serve as the basis of photosynthesis because, as we explain later in the chapter, NADPH is required to reduce CO_2 to carbohydrate.

The significance of cyclic electron transport to photosynthesis in plants is not yet certain. Cyclic electron transport may occur in plant cells when there is too little $NADP^+$ to accept electrons from ferredoxin. Biologists generally agree that this process was used by ancient bacteria to produce ATP from light energy (see *Focus On: The Evolution of Photosystems I and II*). A reaction pathway analogous to cyclic electron transport in plants is present in some modern photosynthetic bacteria. Noncyclic and cyclic electron transport are compared in Table 8–1.

TABLE 8–1 A Comparison of Noncyclic and Cyclic Electron Transport

	Noncyclic Electron Transport	Cyclic Electron Transport
Electron source	H_2O	None — electrons cycle through the system
Oxygen released?	Yes (from H_2O)	No
Terminal electron acceptor	$NADP^+$	None — electrons cycle through the system
Form in which energy is temporarily captured	ATP (by chemiosmosis); NADPH	ATP (by chemiosmosis)
Photosystem(s) required	PS I (P700) and PS II (P680)	PS I (P700) only

ATP synthesis occurs by chemiosmosis

Each member of the electron transport chain that links Photosystem II to Photosystem I can exist in an oxidized (lower energy) form and a reduced (higher energy) form. The electron accepted from P680 by the primary electron acceptor is highly energized; it is passed from one carrier to the next in a series of exergonic redox reactions, losing some of its energy at each step. Some of the energy given up by the electron is not lost by the system, however; it is used to drive the synthesis of ATP (an endergonic reaction). Because the synthesis of ATP (that is, the phosphorylation of ADP) is coupled to the transport of electrons that have been energized by photons of light, the process is called **photophosphorylation.**

FOCUS ON The Evolution of Photosystems I and II

Photosynthesis is an extremely ancient biological process that has apparently changed a great deal since it first appeared more than 3 billion years ago. The use of light energy to manufacture organic molecules first evolved in ancient bacteria that were similar to the green sulfur bacteria existing today.

Initially there was only one photosystem, Photosystem I, which used a green pigment called *bacteriochlorophyll* to gather light energy. Photosystem I operated alone, generating ATP from light energy by cyclic electron transport. However, cyclic electron transport does not provide the reducing power of NADPH, which is needed to manufacture carbohydrate molecules from CO_2. (Recall that in cyclic electron transport, the electrons from P700 are not passed to $NADP^+$ but are instead returned to P700.) Ancient photosynthetic bacteria, like some of their modern counterparts, used electron donors such as hydrogen sulfide (H_2S) rather than H_2O to generate the reducing power needed to manufacture carbohydrates in photosynthesis:

$$H_2S \rightarrow S + 2\,H^+ + 2\,e^-$$
$$2\,H^+ + 2\,e^- + NADP^+ \rightarrow NADPH + H^+$$

This process is not very efficient, however; bacteriochlorophyll has enough oxidative potential to extract electrons from H_2S but not sufficient oxidative potential to extract electrons from H_2O.

Around 3.5 to 3.1 billion years ago, a new group of prokaryotes called **cyanobacteria** evolved (see Chapter 20). These ancient cyanobacteria were probably similar to modern cyanobacteria, which have light-requiring reactions that are similar to those of photosynthetic eukaryotes, including plants. They possess chlorophyll *a* instead of bacteriochlorophyll and can carry out noncyclic electron transport because they have Photosystem II in addition to Photosystem I. Water provides the electrons required to generate NADPH, which in turn provides the reducing power required to manufacture carbohydrate molecules from CO_2.

The chemiosmotic model explains the coupling of ATP synthesis and electron transport

As discussed earlier, the pigments and electron acceptors of the light-dependent reactions are embedded in the thylakoid membrane. Energy released from electrons traveling through the chain of acceptors is used to pump protons from the stroma, across the thylakoid membrane, and into the thylakoid lumen (Fig. 8–12). Thus, the pumping of protons results in the formation of a proton gradient across the thylakoid membrane. Protons also accumulate in the thylakoid lumen as water is split during noncyclic electron transport. Because protons are actually hydrogen ions (H^+), the accumulation of protons causes the pH of the thylakoid interior to fall to a pH of about 5 in the thylakoid lumen, as compared with a pH of about 8 in the stroma. This difference of about 3 pH units across the thylakoid membrane means there is an approximately 1000-fold difference in hydrogen ion concentration.

The proton gradient has a great deal of free energy because of its state of low entropy. How does the chloroplast convert that energy to a more useful form? According to the general principles of diffusion, the concentrated protons inside the thylakoid might be expected to diffuse out readily. However, they are prevented from doing so because the thylakoid membrane is impermeable to H^+ except through certain channels formed by an enzyme called **ATP synthase.** ATP synthase, a transmembrane protein, forms complexes so large they can be seen in electron micrographs; these project into the stroma. As the protons diffuse through an ATP synthase complex, free energy decreases as a consequence of an increase in entropy. Each ATP synthase complex couples this exergonic process of diffusion down a

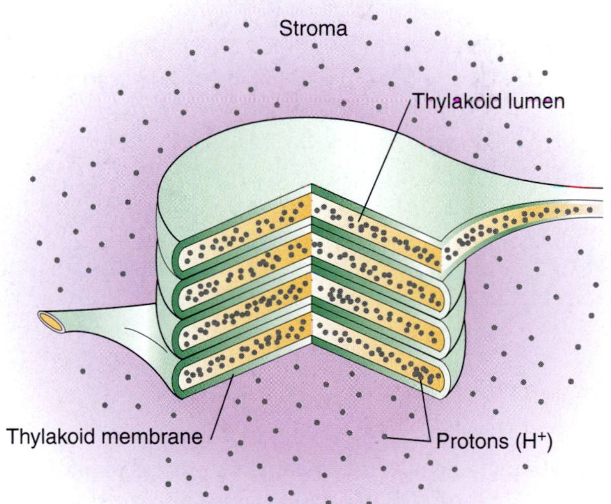

Figure 8–12 Accumulation of protons within the thylakoid lumen. As electrons move down the electron transport chain, protons (H^+) move from the stroma to the thylakoid lumen, creating a proton gradient. The greater concentration of H^+ in the thylakoid lumen lowers the pH.

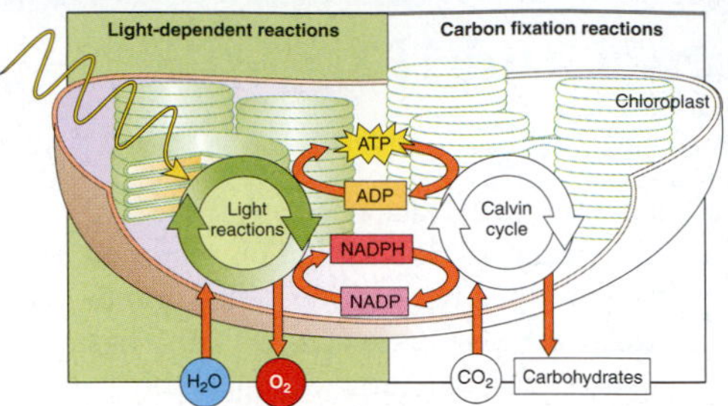

Figure 8–13 A detailed look at electron transport and chemiosmosis. The orange arrows indicate the pathway of electrons along the electron transport chain in the thylakoid membrane. The electron carriers within the membrane become alternately reduced and oxidized as they accept and donate electrons. The energy released during electron transport is used to transport H^+ from the stroma to the thylakoid lumen, where a high concentration of H^+ accumulates. The H^+ are prevented from diffusing back into the stroma except through special channels in ATP synthase in the thylakoid membrane. The flow of the H^+ through ATP synthase generates ATP.

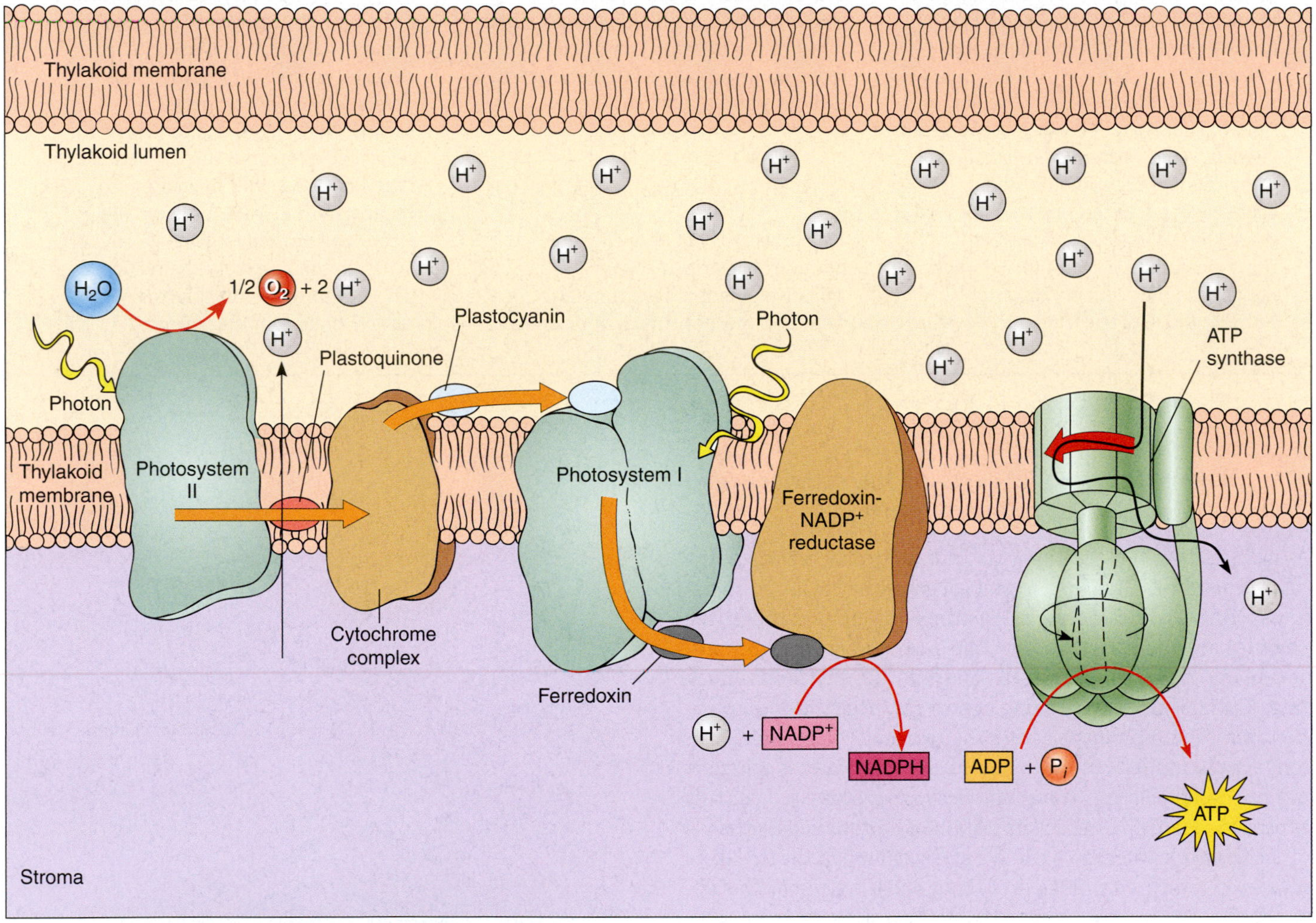

concentration gradient to the endergonic process of phosphorylation of ADP to form ATP, which is released into the stroma (Fig. 8–13). The movement of protons through ATP synthase is thought to induce changes in the conformation of the enzyme that are necessary for the synthesis of ATP. It is estimated that for every four protons that move through ATP synthase, one ATP molecule is synthesized.

The mechanism by which the phosphorylation of ADP is coupled to diffusion down a proton gradient is called **chemiosmosis.** As the essential connection between the electron transport chain and the phosphorylation of ADP, chemiosmosis is a basic mechanism of energy coupling in cells. You may recall from Chapter 7 that chemiosmosis also occurs in aerobic respiration (see Table 8–2).

THE CARBON FIXATION REACTIONS REQUIRE ATP AND NADPH

In carbon fixation, the energy of ATP and NADPH is used in the formation of organic molecules from CO_2. The carbon fixation reactions may be summarized as follows:

$$12\ \text{NADPH} + 18\ \text{ATP} + 6\ CO_2 \longrightarrow C_6H_{12}O_6 + 12\ \text{NADP}^+ + 18\ \text{ADP} + 18\ P_i + 6\ H_2O$$

TABLE 8–2 A Comparison of Photosynthesis and Aerobic Respiration

	Photosynthesis	Aerobic Respiration
Type of metabolic reaction	Anabolism	Catabolism
Raw materials	CO_2, H_2O	$C_6H_{12}O_6$, O_2
End products	$C_6H_{12}O_6$, O_2	CO_2, H_2O
Which cells have these processes?	Cells that contain chlorophyll (certain cells of plants, algae, and some bacteria)	Every actively metabolizing cell has aerobic respiration or some other energy-releasing pathway
Sites involved (in eukaryotic cells)	Chloroplasts	Cytosol (glycolysis); mitochondria
ATP production	By photophosphorylation (a chemiosmotic process)	By substrate-level phosphorylation and by oxidative phosphorylation (a chemiosmotic process)
Principal electron transfer compound	$NADP^+$ is reduced to form NADPH*	NAD^+ is reduced to form NADH*
Location of electron transport chain	Thylakoid membrane	Mitochondrial inner membrane (cristae)
Source of electrons for electron transport chain	In noncyclic electron transport: H_2O (undergoes photolysis to yield electrons, protons, and oxygen)	Immediate source: NADH, $FADH_2$ Ultimate source: glucose or other carbohydrate
Terminal electron acceptor for electron transport chain	In noncyclic electron transport: $NADP^+$ (becomes reduced to form NADPH)	O_2 (becomes reduced to form H_2O)

*NADPH and NADH are very similar hydrogen (i.e., electron) carriers, differing only in a single phosphate group. However, NADPH generally works with enzymes in anabolic pathways, such as photosynthesis. NADH is associated with catabolic pathways, such as cellular respiration.

Most plants use the Calvin (C_3) cycle to fix carbon

Carbon fixation occurs in the stroma through a sequence of 13 reactions known as the Calvin cycle. Melvin Calvin, Andrew Benson, and others at the University of California were able to elucidate the details of this cycle; for his work, Calvin was awarded a Nobel Prize in 1961 (Fig. 8–14).

The 13 reactions of the Calvin cycle are divided into three phases: CO_2 uptake, carbon reduction, and RuBp regeneration (Fig. 8–15). All 13 enzymes that catalyze steps in the Calvin cycle are located in the stroma of the chloroplast. Ten of the enzymes also participate in glycolysis (see Chapter 7). These enzymes are able to catalyze reversible reactions, degrading carbohydrate molecules in cellular respiration and synthesizing carbohydrate molecules in photosynthesis.

The *CO_2 uptake phase* of the Calvin cycle consists of a single reaction in which a molecule of CO_2 reacts with a phosphorylated five-carbon compound, **ribulose bisphosphate (RuBP).** This reaction is catalyzed by the enzyme **ribulose bisphosphate carboxylase/oxygenase,** also known as **Rubisco.** The Rubisco enzyme is present in the chloroplast in the largest amounts of any protein, and it may be one of the most abundant proteins in the biosphere. The product of this reaction is an unstable, six-carbon intermediate, which immediately breaks down into two molecules of **phosphoglycerate (PGA)** with three carbons each. The carbon that was originally part of a CO_2 molecule is now part of a carbon skeleton; the carbon has been "fixed." The Calvin cycle is also known as the **C_3 pathway** because the product of the initial carbon fixation reaction is a three-carbon compound.

The *carbon reduction phase* consists of two steps in which the energy and reducing power from ATP and NADPH (both produced in the light-dependent reactions) are used to convert the PGA molecules to **glyceraldehyde-3-phosphate (G3P).** As shown in Figure 8–15, for every six carbons that enter the cycle as CO_2, six carbons can leave the system as two molecules of G3P, to be used in carbohydrate synthesis. Each of these three carbon molecules of G3P is essentially half a hexose (six-carbon sugar) molecule. (In fact, you may recall that G3P is a key intermediate in the splitting of sugar in glycolysis; see Figs. 7–3 and 7–4.)

The reaction of two molecules of G3P is exergonic and can lead to the formation of glucose or fructose. In some plants, glucose and fructose are then joined to produce sucrose (common table sugar). Sucrose can be harvested from sugar cane, sugar beets, and maple sap. The plant cell also uses glucose to produce starch or cellulose.

Notice that although 2 G3P molecules are removed from the cycle, 10 G3P molecules remain; this represents 30 carbon atoms in all. Through a complex series of ten reactions that make up the *RuBP regeneration phase* of the Calvin cycle, these 30 carbons and their associated atoms become rearranged into 6 molecules of ribulose phosphate, each of which becomes phosphorylated to produce RuBP, the five-carbon compound with which the cycle started. These RuBP molecules can begin the process of CO_2 fixation and eventual G3P production once again.

In summary, the inputs required for the carbon fixation reactions are six molecules of CO_2, phosphates transferred from

Figure 8–14 Experimental apparatus of Calvin and Benson. The classic experiments carried out by Calvin, Benson, and others in the 1950s elucidated the steps in the carbon fixation reactions of photosynthesis. Calvin and his colleagues grew algae in the green "lollipop." CO_2 labeled with carbon-14 (^{14}C) was bubbled through the algae, which were periodically killed by dumping the "lollipop" contents into a beaker of boiling alcohol. By identifying which compounds in the algae contained the ^{14}C at different times, Calvin was able to determine the steps of carbon fixation in photosynthesis. *(Courtesy of Melvin Calvin, University of California, Berkeley)*

ATP, and electrons (as hydrogen) from NADPH. In the end, the six carbons from the CO_2 can be accounted for by the harvest of a hexose molecule. The remaining G3P molecules are used to synthesize the RuBP molecules with which more CO_2 molecules may combine. (Table 8–3 provides a detailed summary of photosynthesis.)

The initial carbon fixation step differs in C_4 plants and in CAM plants

Because CO_2 is not a very abundant gas (composing only about 0.03% of the atmosphere), it is not easy for plants to obtain the CO_2 they need. This problem is complicated by the fact that gas exchange can occur only across a moist surface. The surfaces of leaves and other exposed plant parts are covered with a waterproof layer that helps prevent excess loss of water vapor. Entry and exit of gases is therefore limited to microscopic pores, called **stomata** (sing., *stoma*), usually concentrated on the undersides of the leaves (see Fig. 8–4*a*). These openings lead to the interior of the leaf, which is made up of chloroplast-containing cells, known as the **mesophyll,** with many air spaces and a very high concentration of water vapor. The stomata open and close in response to such environmental factors as water content and light intensity. When conditions are hot and dry, the stomata close to reduce the loss of water vapor. As a result, the supply of CO_2 is greatly diminished. Ironically, CO_2 is potentially less available at the very times when maximum sunlight is available to power the light-dependent reactions.

Many plant species living in hot, dry environments have adaptations that allow them initially to fix CO_2 through one of two pathways that help minimize water loss. These pathways, known as the C_4 pathway and the CAM pathway, take place in the cytosol. Both the C_4 and CAM pathways merely precede the Calvin cycle (C_3 pathway); they do not replace it.

The C_4 pathway efficiently fixes CO_2 at low concentrations

Some plants, known as **C_4 plants,** first fix CO_2 into a four-carbon compound, **oxaloacetate,** prior to the C_3 pathway. The C_4 pathway not only occurs before the C_3 pathway, it also occurs in different cells.

Leaf anatomy is usually distinctive in C_4 plants. In addition to having mesophyll cells, C_4 leaves have prominent chloroplast-containing **bundle sheath cells** (Fig. 8–16). These cells tightly encircle the veins of the leaf. The mesophyll cells in C_4 plants are closely associated with the bundle sheath cells. The **C_4 pathway** (also called the **Hatch-Slack pathway,** after M.D. Hatch and C.R. Slack, who worked out many of its steps) occurs in the mesophyll cells, whereas the Calvin cycle takes place within the bundle sheath cells.

The key component of the C_4 pathway is a remarkable enzyme that has an extremely high affinity for CO_2, binding it effectively even at unusually low concentrations. This enzyme, **PEP carboxylase,** catalyzes the reaction by which CO_2 reacts with the three-carbon compound **phosphoenolpyruvate (PEP),** forming oxaloacetate (Fig. 8–17 on page 192).

In a step that requires NADPH, oxaloacetate is converted to some other four-carbon compound, usually malate. The malate then passes to chloroplasts within bundle sheath cells, where a

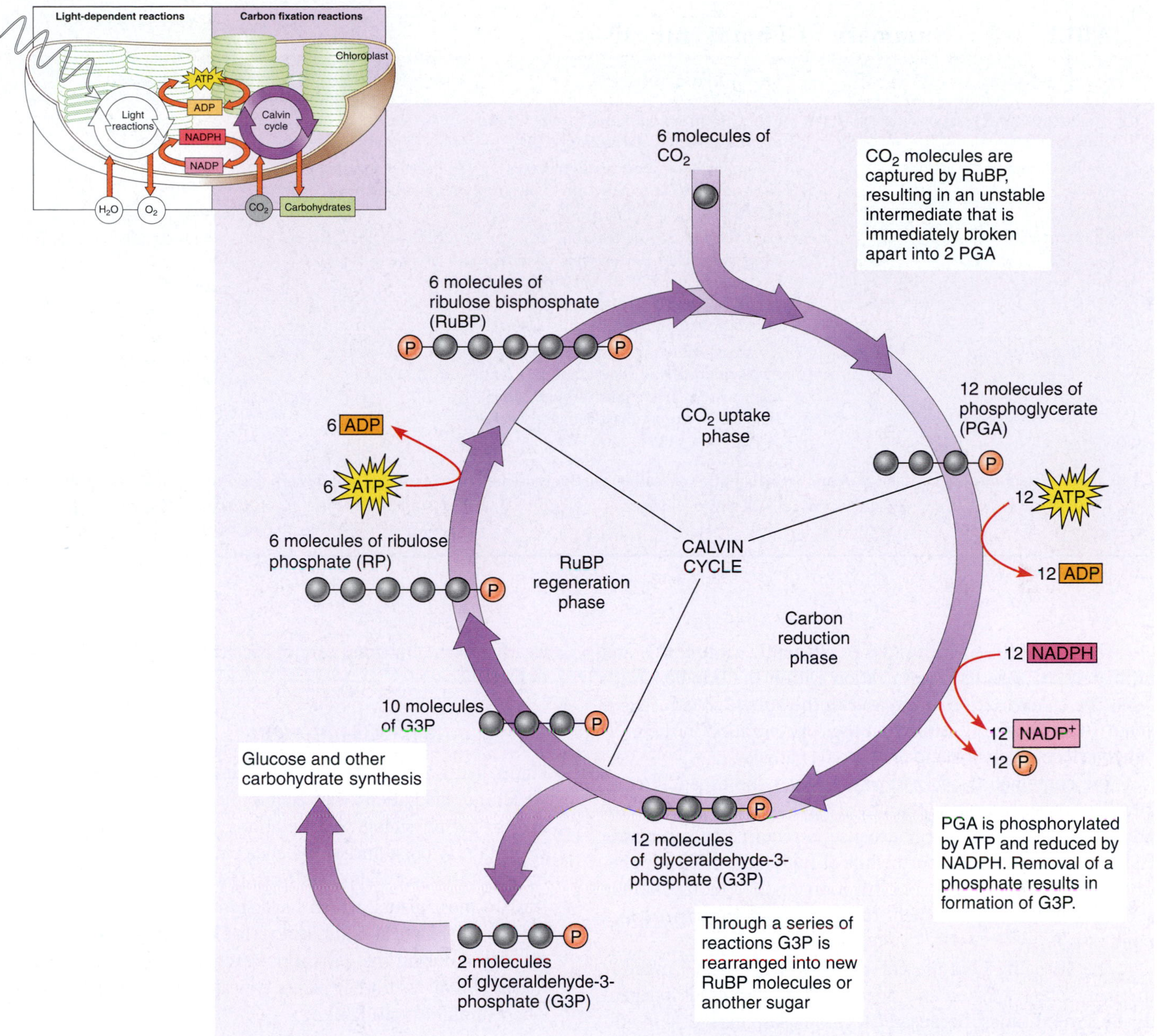

Figure 8–15 A detailed look at the Calvin cycle. This diagram, which shows carbon atoms as black balls, demonstrates that six molecules of CO_2 must be "fixed" (incorporated into preexisting carbon skeletons) to produce one molecule of a six-carbon sugar such as glucose. Two glyceraldehyde-3-phosphate (G3P) molecules "leave" the cycle for every glucose formed. Although these reactions do not require light directly, the energy that drives the Calvin cycle comes from ATP and NADPH, which are the products of the light-dependent reactions.

different enzyme catalyzes the decarboxylation of malate to yield pyruvate (which has three carbons) and CO_2. NADPH is formed, replacing the one used earlier.

$$\text{Malate} + \text{NADP}^+ \rightarrow \text{Pyruvate} + CO_2 + \text{NADPH}$$

The CO_2 released in the bundle sheath cell combines with ribulose bisphosphate and goes through the Calvin cycle in the usual manner. The pyruvate formed in the decarboxylation reaction returns to the mesophyll cell, where it reacts with ATP to regenerate phosphoenolpyruvate.

TABLE 8–3 Summary of Photosynthesis

Reaction Series	Summary of Process	Needed Materials	End Products
A. Light-dependent reactions (take place in thylakoid membranes)	Energy from sunlight used to split water, manufacture ATP, and reduce $NADP^+$		
1. Photochemical reactions	Chlorophyll-activated; reaction center gives up photoexcited electron to electron acceptor	Light energy; pigments (chlorophyll)	Electrons
2. Electron transport	Electrons are transported along chain of electron acceptors in thylakoid membranes; electrons reduce $NADP^+$; splitting of water provides some of H^+ that accumulates inside thylakoid space	Electrons, $NADP^+$, H_2O, electron acceptors	NADPH, O_2
3. Chemiosmosis	H^+ are permitted to move across the thylakoid membrane down their gradient; they cross the membrane through special channels in ATP synthase complex; energy released is used to produce ATP	Proton gradient, ADP + P_i	ATP
B. Carbon fixation reactions (take place in stroma)	Carbon fixation: carbon dioxide is used to make carbohydrate	Ribulose bisphosphate, CO_2, ATP, NADPH, necessary enzymes	Carbohydrates, ADP + P_i, $NADP^+$

The role of the C_4 pathway is to efficiently capture CO_2 and ultimately increase its concentration within the bundle sheath cells. The concentration of CO_2 within the bundle sheath cells is about 10 to 60 times greater than the concentration in the mesophyll cells of plants having only the C_3 pathway.

The combined C_3–C_4 pathway involves the expenditure of 30 ATPs per hexose, rather than the 18 ATPs used by the C_3 pathway alone. The extra energy expense is required to regenerate PEP from pyruvate. It is worthwhile at high light intensities because it ensures a high concentration of CO_2 in the bundle sheath cells and permits them to carry on photosynthesis at a rapid rate.

The C_4 pathway is present in addition to the C_3 pathway in many plant species and apparently has evolved independently several times. Because PEP carboxylase fixes CO_2 so efficiently, C_4 plants do not need to have their stomata open as much. C_4 plants therefore tolerate higher temperatures and greater light intensities, lose less water by transpiration (evaporation), and generally have higher rates of photosynthesis and growth than plants that use only the Calvin cycle. Among the many quick-growing and aggressive plants that use the C_4 pathway are sugar cane, corn, and crabgrass. When sunlight is abundant, the yields of C_4 crop plants can be two to three times greater than those of C_3 plants. If this pathway could be incorporated into more of our crop plants by genetic manipulation, we might be able to greatly increase food production in some parts of the world.

When light is abundant, the rate of photosynthesis is limited by the concentration of CO_2, so C_4 plants, with their higher levels of CO_2 in bundle sheath cells, have the advantage. At lower light intensities and temperatures, C_3 plants are favored. For example, winter rye, a C_3 plant, grows lavishly in cool weather when crabgrass cannot because it requires more energy to fix CO_2.

CAM plants fix CO_2 at night

Plants living in very dry, or *xeric,* conditions have a number of structural adaptations that enable them to survive. Many xeric plants have physiological adaptations as well. For example, their stomata may open during the cooler night and close during the hot day to reduce water loss from transpiration. This is in contrast to most plants, which have stomata that are open during the day and closed at night. But xeric plants that have their stomata closed during the day cannot exchange gases for photosynthesis. (Recall that other plants typically fix CO_2 during the day, when sunlight is available.)

Many xeric plants evolved a special carbon fixation pathway called **crassulacean acid metabolism (CAM)** that in effect solves this dilemma. The name comes from the stonecrop plant family (the Crassulaceae), which possesses the CAM pathway, although it has evolved independently in some members of more than 25 other plant families, including the cactus family (Cactaceae), the lily family (Liliaceae), and the orchid family (Orchidaceae) (Fig. 8–18).

CAM plants use the enzyme PEP carboxylase to fix CO_2 during the night when stomata are open, forming oxaloacetate, which is converted to malate and stored in cell vacuoles. During the day, when stomata are closed and gas exchange cannot occur between the plant and the atmosphere, CO_2 is removed from malate by a decarboxylation reaction. Now the CO_2 is available within the leaf tissue to be fixed into sugar by the Calvin cycle (C_3 pathway).

The CAM pathway is very similar to the C_4 pathway but with important differences. C_4 plants initially fix CO_2 into four-carbon organic acids in mesophyll cells. The acids are later decarboxy-

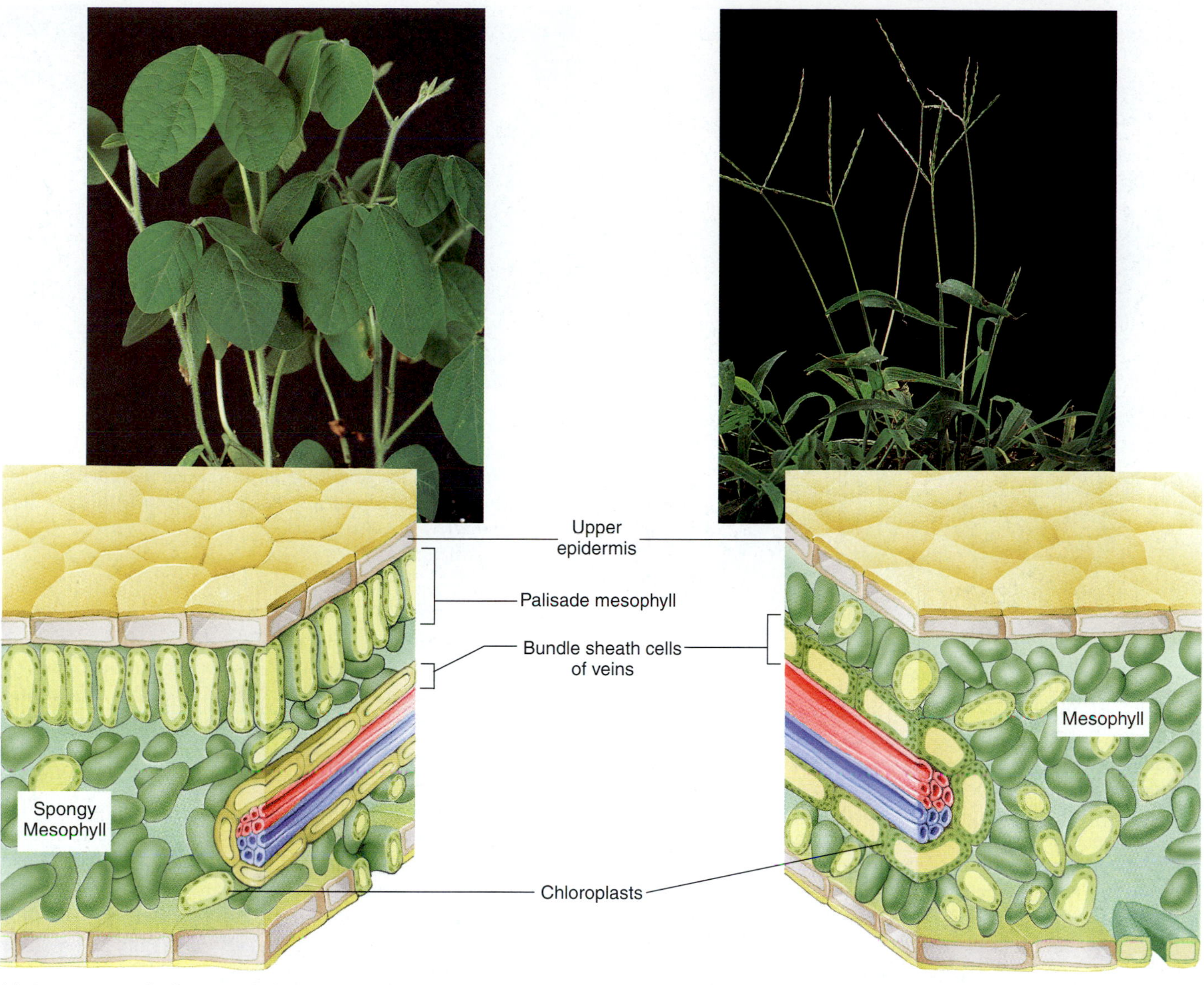

(a) Arrangement of cells in a C_3 leaf

(b) Arrangement of cells in a C_4 leaf

Figure 8–16 Comparison of C_3 and C_4 anatomy. **(a)** In C_3 plants, such as soybeans, the Calvin cycle takes place in the mesophyll cells, and the bundle sheath cells are nonphotosynthetic. **(b)** In C_4 plants, such as crabgrass, reactions that fix CO_2 into four-carbon compounds take place in the mesophyll cells. The four-carbon compounds are transferred from the mesophyll cells to the photosynthetic bundle sheath cells, where the Calvin cycle takes place. *(a and b, Dennis Drenner)*

lated to produce CO_2, which is fixed by the C_3 pathway in the bundle sheath cells. In other words, the C_4 and C_3 pathways occur in *different locations* within the leaf of a C_4 plant. In CAM plants, the initial fixation of CO_2 occurs at night. Decarboxylation of malate and subsequent production of sugar from CO_2 by the normal C_3 photosynthetic pathway occur during the day. In other words, the CAM and C_3 pathways occur at *different times* within the same cell of a CAM plant.

Although it does not promote rapid growth the way that the C_4 pathway does, the CAM pathway is a very successful adaptation to xeric conditions. CAM plants are able to exchange gases for photosynthesis and to reduce water loss significantly. Plants with CAM photosynthesis survive in deserts where neither C_3 nor C_4 plants can.

Photorespiration reduces photosynthetic efficiency

Many C_3 plants, including certain agriculturally important crops such as soybeans, wheat, and potatoes, do not yield as much carbohydrate from photosynthesis as might be expected. This reduction in yield is especially significant during very hot spells in summer. On hot, dry days, plants close their stomata to conserve water. Once the stomata close, photosynthesis rapidly uses up

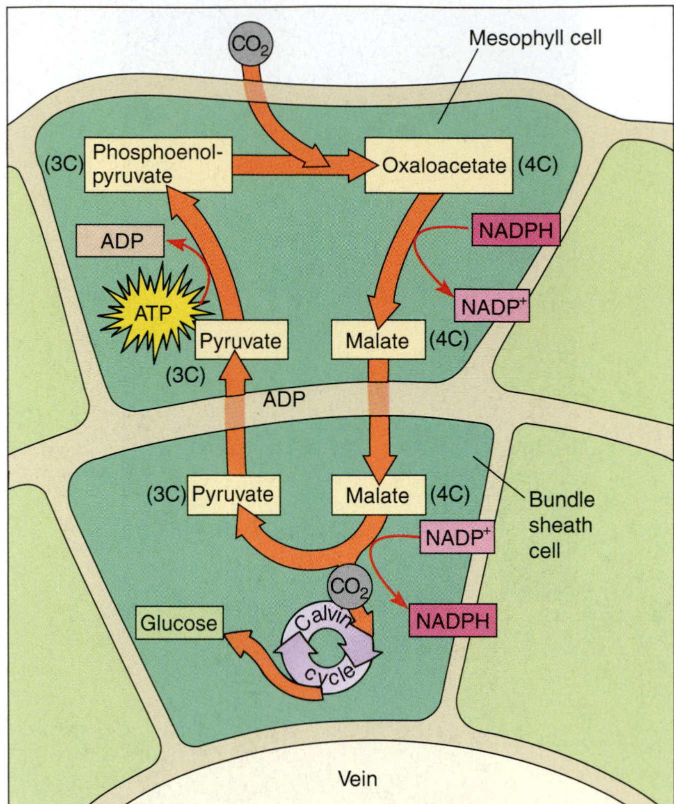

Figure 8–17 Summary of the C_4 pathway. CO_2 combines with phosphoenolpyruvate (PEP) in the chloroplasts of mesophyll cells, forming a four-carbon compound that is converted to malate. Malate goes to the chloroplasts of bundle sheath cells, where it is decarboxylated. The CO_2 thus released in the bundle sheath cell is used to make carbohydrate by way of the Calvin cycle.

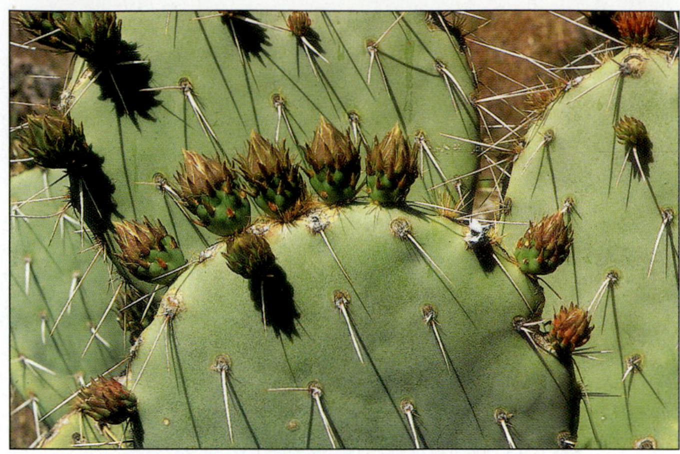

Figure 8–18 A typical CAM plant. Prickly pear cactus (*Opuntia* sp.) is a CAM plant. The more than 200 species of *Opuntia* living today originated in various xeric habitats in North and South America. *(Robert W. Domm/Visuals Unlimited)*

the CO_2 remaining in the leaf and produces O_2, which accumulates in the chloroplasts.

Recall that the enzyme RuBP carboxylase/oxygenase (Rubisco) is responsible for CO_2 fixation in the Calvin cycle by attaching CO_2 to RuBP. In addition to this carboxylase activity, Rubisco can also react with O_2, which competes with CO_2 for the active site of Rubisco. The activity of Rubisco toward the substrates CO_2 and O_2 depends on the relative amounts of these two gases in the chloroplast environment. When chloroplast oxygen levels are high and CO_2 levels are low, Rubisco is more likely to catalyze the reaction of RuBP with O_2 instead of with CO_2. When this occurs, some of the intermediates involved in the Calvin cycle are degraded to CO_2 and H_2O. This process is called **photorespiration** because (1) it occurs in the presence of light; (2) it requires oxygen, like aerobic respiration; and (3) it produces CO_2 and H_2O, like aerobic respiration. Unlike aerobic respiration, however, ATP is not produced during photorespiration. Photorespiration reduces photosynthetic efficiency because it removes some of the intermediates used in the Calvin cycle.

The reasons for photorespiration are incompletely understood, although it is thought to possibly reflect the origin of Rubisco at an ancient time when CO_2 levels were high and molecular oxygen levels were low.

Photorespiration is negligible in C_4 plants because the concentration of CO_2 in bundle sheath cells (where Rubisco is present) is always high. However, many important crop plants are C_3 plants that carry out photorespiration. This is yet another reason that some scientists are attempting to transfer genes for the C_4 pathway to C_3 crops such as soybeans and wheat. If this genetic transfer is accomplished, these plants should be able to produce much more carbohydrate during hot weather.

SUMMARY WITH KEY TERMS

I. Green plants, algae, and certain bacteria are **photoautotrophs** and use light as an energy source for manufacturing organic molecules from CO_2 and H_2O. All animals, fungi, and most bacteria are **chemoheterotrophs** that depend on photoautotrophs for organic molecules.

II. Light behaves as both a wave and a particle. Particles of light energy, called **photons,** can excite biological molecules. The resulting energized electrons may be accepted by electron acceptor compounds.

III. In plants, **photosynthesis** occurs in **chloroplasts,** which are located mainly within **mesophyll** cells inside the leaf. During photosynthesis, light energy is captured and converted to the chemical energy of carbohydrates; oxygen is released as a byproduct.

   A. Chloroplasts are organelles bounded by a double membrane; the inner membrane encloses the **stroma** in which membranous, saclike **thylakoids** are suspended. Thylakoids arranged in stacks are called **grana.**

   B. **Chlorophyll** and other photosynthetic pigments are components of the thylakoid membranes of chloroplasts. Each thylakoid encloses a **thylakoid lumen.**

C. The **absorption spectra** of chlorophylls *a* and *b* are similar to the **action spectrum** for photosynthesis.

D. **Photosystems I and II** are the two types of photosynthetic units involved in photosynthesis. Each photosystem includes **antenna complexes** and a **reaction center.**

1. Chlorophyll molecules and accessory pigments are organized with pigment-binding proteins into antenna complexes.
2. Only a special chlorophyll *a* in the **reaction center** actually gives up its energized electrons to a nearby electron acceptor.

IV. During the noncyclic **light-dependent reactions,** known as noncyclic electron transport, ATP and NADPH are formed.

A. The electrons in Photosystem I are energized by the absorption of light and passed through an **electron transport chain** to $NADP^+$, forming NADPH.

B. Electrons given up by P700 in Photosystem I are replaced by electrons from P680 in Photosystem II.

C. Electrons given up by P680 in Photosystem II are replaced by electrons made available by the **photolysis** of H_2O; oxygen is released in the process.

D. A series of redox reactions takes place as energized electrons are passed along the electron transport chain from Photosystem II to Photosystem I.

1. Some of the energy is used to pump protons across the thylakoid membrane, providing the energy to generate ATP by **chemiosmosis.**
2. **Photophosphorylation** is the synthesis of ATP coupled to the transport of electrons energized by photons of light.
3. As protons diffuse through **ATP synthase,** an enzyme complex in the thylakoid membrane, ADP is phosphorylated to form ATP.

V. During the cyclic light-dependent reactions, known as cyclic electron transport, electrons from Photosystem I are eventually returned to Photosystem I. ATP is produced by chemiosmosis, but no NADPH or oxygen is generated.

VI. During the **carbon fixation reactions,** the energy of ATP and NADPH is used to manufacture carbohydrate molecules from CO_2.

A. The carbon fixation reactions proceed by way of the **Calvin cycle,** also known as the **C_3 pathway.**

1. During the CO_2 uptake phase of the Calvin cycle, CO_2 is combined with **ribulose bisphosphate (RuBP),** a five-carbon sugar, by the enzyme **ribulose bisphosphate carboxylase/oxygenase (Rubisco).**
2. During the carbon reduction phase of the Calvin cycle, the energy and reducing power of ATP and NADPH are used to convert PGA molecules to **glyceraldehyde-3-phosphate (G3P).**
3. For every 6 CO_2 molecules fixed, 12 molecules of G3P are produced, and 2 molecules of G3P can leave the cycle.
4. The two G3P molecules are required to produce the equivalent of one molecule of glucose; the remaining G3P molecules are modified to regenerate RuBP during the RuBP regeneration phase of the Calvin cycle.

B. In the **C_4 pathway,** the enzyme **PEP carboxylase** binds CO_2 effectively, even when CO_2 is at a low concentration.

1. C_4 reactions take place within mesophyll cells. The CO_2 is fixed in **oxaloacetate,** which is then converted to **malate.**
2. The malate moves into a **bundle sheath cell,** and CO_2 is removed from it. The released CO_2 then enters the Calvin cycle.

C. The **crassulacean acid metabolism (CAM)** pathway is similar to the C_4 pathway. PEP carboxylase fixes carbon at night in the mesophyll cells, and the Calvin cycle occurs during the day in the same cells.

VII. In **photorespiration,** C_3 plants consume oxygen and generate CO_2 but do not produce ATP. Photorespiration is significant on bright, hot, dry days when plants close their stomata, conserving water but preventing the passage of CO_2 into the leaf.

Summary Reactions for Photosynthesis

The light-dependent reactions (noncyclic electron transport):

$$12\ H_2O + 12\ NADP^+ + 18\ ADP + 18\ P_i \xrightarrow[\text{Chlorophyll}]{\text{Light}} 6\ O_2 + 12\ NADPH + 18\ ATP$$

The carbon fixation reactions (Calvin cycle):

$$12\ NADPH + 18\ ATP + 6\ CO_2 \longrightarrow C_6H_{12}O_6 + 12\ NADP^+ + 18\ ADP + 18\ P_i + 6\ H_2O$$

By canceling out the common items on opposite sides of the arrows in these two coupled equations, we obtain the simplified overall equation for photosynthesis:

$$\underset{\text{Carbon dioxide}}{6\ CO_2} + \underset{\text{Water}}{12\ H_2O} \xrightarrow[\text{Chlorophyll}]{\text{Light energy}} \underset{\text{Glucose}}{C_6H_{12}O_6} + \underset{\text{Oxygen}}{6\ O_2} + \underset{\text{Water}}{6\ H_2O}$$

POST-TEST

1. Where is chlorophyll located in the chloroplast? (a) thylakoid membranes (b) stroma (c) mitochondrial matrix (d) thylakoid lumen (e) between the inner and outer membranes
2. In photolysis, some of the energy captured by chlorophyll is used to split (a) CO_2 (b) ATP (c) NADPH (d) H_2O (e) both b and c
3. Light is composed of particles of energy called (a) carotenoids (b) reaction centers (c) photons (d) antenna complexes (e) photosystems
4. The relative effectiveness of different wavelengths of light in photosynthesis is demonstrated by (a) an action spectrum (b) photolysis (c) carbon fixation reactions (d) photoheterotrophs (e) an absorption spectrum
5. In plants, the final electron acceptor in the light-dependent reactions is: (a) $NADP^+$ (b) CO_2 (c) H_2O (d) O_2 (e) G3P
6. In addition to chlorophyll, most plants contain accessory photosynthetic pigments such as (a) PEP (b) G3P (c) carotenoids (d) PGA (e) $NADP^+$
7. The part of a photosystem that absorbs light energy is its (a) antenna complexes (b) reaction center (c) terminal quinone electron acceptor (d) pigment-binding protein (e) thylakoid lumen
8. In ____________, electrons that have been energized by light contribute their energy to add phosphate to ADP, producing ATP. (a) crassulacean acid metabolism (b) the Calvin cycle (c) photorespiration (d) C_4 pathways (e) photophosphorylation
9. In ____________, there is a one-way flow of electrons to $NADP^+$, forming NADPH. (a) crassulacean acid metabolism (b) the Calvin cycle (c) photorespiration (d) cyclic electron transport (e) noncyclic electron transport
10. The mechanism by which electron transport is coupled to ATP production by means of a proton gradient is called (a) chemiosmosis (b) crassulacean acid metabolism (c) fluorescence (d) the C_3 pathway (e) the C_4 pathway
11. In photosynthesis in eukaryotes, the transfer of electrons through a sequence of electron acceptors provides energy to pump protons across the (a) chloroplast outer membrane (b) chloroplast inner membrane (c) thylakoid membrane (d) inner mitochondrial membrane (e) plasma membrane
12. The inputs for ____________ are CO_2, NADPH, and ATP. (a) cyclic electron transport (b) the carbon fixation reactions (c) noncyclic electron transport (d) Photosystems I and II (e) chemiosmosis
13. The Calvin cycle begins when CO_2 reacts with (a) phosphoenolpyruvate

(b) glyceraldehyde-3-phosphate (c) ribulose bisphosphate (d) oxaloacetate (e) phosphoglycerate

14. The enzyme directly responsible for almost all carbon fixation on Earth is (a) Rubisco (b) PEP carboxylase (c) ATP synthase (d) phosphofructokinase (e) ligase

15. In C_4 plants, C_4 and C_3 pathways occur at different __________, whereas in CAM plants, CAM and C_3 pathways occur at different __________. (a) times of day; locations within the leaf (b) seasons; locations (c) locations; times of day (d) locations; seasons (e) times of day; seasons

REVIEW QUESTIONS

1. Why does photosynthesis require light energy?
2. What is the role of chlorophyll in photosynthesis?
3. What is the significance of the fact that the combined absorption spectra of chlorophyll *a* and *b* roughly match the action spectrum of photosynthesis? Why do they not coincide exactly?
4. How is oxygen produced during photosynthesis?
5. In noncyclic electron transport, what molecule becomes phosphorylated? Why is the phosphorylation process referred to as *photo*phosphorylation? Why is it said to be noncyclic?
6. How are ATP and NADPH produced and used in the process of photosynthesis?
7. Describe the three phases of the Calvin cycle.
8. How do the C_4 and CAM pathways improve photosynthesis in hot, dry environments?
9. Label the figure to the right. Use Figure 8–8 to check your answers.

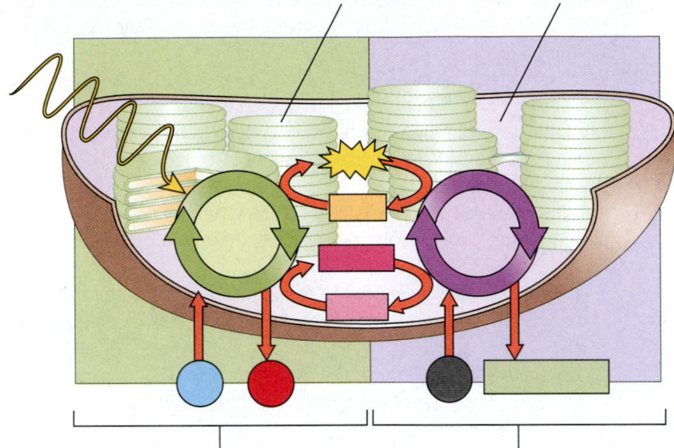

YOU MAKE THE CONNECTION

1. Only some plant cells have chloroplasts, but all actively metabolizing plant cells have mitochondria. Why?
2. Explain why the proton gradient formed during chemiosmosis represents a state of low entropy. (You may wish to refer to the discussion of entropy in Chapter 6.)
3. The electrons in glucose have relatively high free energies. How did they become so energetic?
4. What strategies might be employed in the future to increase world food supply? Base your answer on your knowledge of photosynthesis and related processes.

RECOMMENDED READINGS

Balzani, V. "Greener Way to Solar Power." *New Scientist,* Vol. 12, 12 Nov. 1994. Chemists are imitating the light-gathering photosystems of plants in an attempt to convert solar energy to electricity.

Cline, K. "Gateway to the Chloroplast." *Nature,* Vol. 403, 13 Jan. 2000. Examines recent insights into the process by which proteins are imported during chloroplast development.

Demmig-Adams, B., and W.W. Adams III. "Harvesting Sunlight Safely." *Nature,* Vol. 403, 27 Jan. 2000. A chlorophyll-binding protein within the light-harvesting complexes of the photosynthetic apparatus has a unique, energy-dissipating function, to protect against excess light.

Hendry, G. "Making, Breaking, and Remaking Chlorophyll." *Natural History,* May 1990. Examines the endless process by which plants make chlorophyll in the spring and break it down in the fall.

Moore, P.D. "Mixed Metabolism in Plant Pools." *Nature,* Vol. 399, 13 May 1999. Some aquatic plants use C_4 or CAM photosynthetic pathways, which allow them to coexist and optimize their use of environmental resources.

Riebesell, U. "Carbon Fix for a Diatom." *Nature,* Vol. 407, 21 Jan. 2000. This news article highlights the discovery of a marine diatom (a unicellular alga) that uses the same C_4 pathway as certain terrestrial plants.

Sharkey, T.D. "Some Like It Hot." *Science,* Vol. 287, 21 Jan. 2000. Many plants adapt to colder or warmer temperatures by adjusting the fatty acid content in chloroplast membranes.

Taiz, L., and E. Zeiger. *Plant Physiology,* 2nd edition. Sinauer Associates, Inc., Sunderland, MA, 1998. Chapters 7 through 9 contain an in-depth examination of photosynthesis.

- Visit our Web site at **http://www.info.brookscole.com/solomonbergmartin** for links to chapter-related resources on the World Wide Web. Additional on-line materials relating to this chapter can also be found on our Web site.

See chapter activity on BioActive Learner CD for additional help in mastering the chapter's material. Icon location in the chapter's margins shows which topics have tutorials or simulations in the CD.

CAREER VISIONS

Middle School Science Teacher

CHRISTIE GAYHEART

***Christie Gayheart** is a middle school science teacher in the Midland Public Schools, Midland, Michigan. Originally from Houghton, Michigan, Christie received her B.S. in biology, with an ecology option, at Michigan Technological University in 1996. After deciding in her senior year that she wanted to become a teacher, she pursued a B.S. in Secondary Education from Central Michigan University, with a major in biology and minor in physical science, completing her degree in 1999. She is currently working toward her M.S. in Educational Administration at Saginaw Valley State University. Christie divides her time between two schools, Central Middle and Northeast Middle, teaching eighth-grade science at both. She encourages her students to think like scientists, to conduct experiments and question their results. She tries to harness the natural enthusiasm of eighth graders—supplementing class work with hands-on experiments, guest speakers, and field trips—to keep science fun but challenging.*

Did you have a particular focus in your undergraduate biology coursework?

I started out in a biomedical engineering program, so I was mostly in general engineering classes at first. After taking some of the design courses, and being in the computer lab, the degree work seemed impersonal to me. I decided that was not me, that I needed to be talking to people. I stuck with biology. I was coaching high school golf and high school and college cheerleading at the time, and I decided I liked working with kids. I thought about teaching but I never really pursued it. Instead, I went with an ecology focus, thinking that I wanted to work for the Department of Natural Resources and do some kind of field interpretive work.

What influenced your decision to become a teacher?

The coaching, as well as some tutoring I did as an undergraduate, mostly in high school science. Then, when I started the teaching program, I began working as a substitute teacher, which really convinced me that teaching was what I wanted to do.

Was a second Bachelor of Science degree required for you to become a teacher, or could you have earned a teaching certificate instead?

I could have gotten a certificate, but I felt the bachelor's in education might be more helpful for my future plans—to obtain a master's degree and possibly go on for a doctorate in education.

The best thing about teaching is that it makes you feel good at the end of the day, that you're doing something for people and, hopefully, for humanity in general.

For your secondary education degree, did you have to take additional science courses?

For a secondary education degree, you have to declare a major and a minor, or a double minor, depending on the distinctions in your state. I wanted to be able to teach any secondary-level science course. It made sense to declare a biology major, because I had a lot of those courses. If I took a physical science minor, I could get a distinction called DX, which allows a person to teach any science discipline at the secondary level, 7th through 12th grade. I had to take astronomy and geology, but I had enough credits in the other sciences: chemistry, physics, and biology.

What were your education requirements?

I took education core classes and a two-semester interdisciplinary course in the biology program, "Teaching for Biology Majors," which included an internship in a classroom.

You currently teach eighth-grade science. Is the focus on biology, physical science, or some combination of disciplines?

It's general science. Our seventh- and eighth-grade program is a combination of the different disciplines, so students are exposed to a little bit of earth science, physical science, and life science.

How do you cope with students' differences in abilities?

You really try to vary your teaching style, working with three different approaches: hands-on, visual (giving them visual instruction), and auditory teaching. It is the same with assessment: Not all students are good at tests. You need to vary assessment strategies to meet everyone's needs, so they can all be successful.

Are eighth-grade students typically interested in science? How do you generate or direct their enthusiasm?

I find the students to be very interested. If you give them the right tools and make it meaningful to them, then they are right there with you. I use the word "experimenting," but they think it's play.

That's good, too; I think they should be excited about doing the hands-on activities, but you have to focus this excitement into experimenting, to get them thinking about science process skills. As a new teacher, I am still working on doing inquiry-based learning: giving them a thought question, having them use science process skills to achieve an answer to the problem, and showing them that science isn't always a cookie-cutter method. All of the activities have to be age-appropriate.

How does your biology background affect your approach to teaching?

Because I have a full degree in biology, I have gone through undergraduate research, and my focus is more on the discipline of science—thought processes and problem solving—and on experimentation. I am probably more disciplined and require more in the areas where you have to talk about processes. Also, because I have a varied science background, including coursework in astronomy and geology, I am able to easily connect all the different science disciplines.

What "tools of the trade" have you learned so far?

One is to always be prepared with lesson planning. If you're prepared, you can make it through a lesson. If you feel like you lacked creativity, you can work on that next time. Second, don't beat yourself up if your lesson goes wrong. Just reflect on the strength of the lesson and modify where needed. Finally—and this is important—obtain a mentor and utilize him or her; they are a great resource. Use them to find what works and what doesn't. Most teachers are very supportive of one another.

What are the greatest rewards of your profession? What are the biggest demands you face?

The best thing about teaching is that it makes you feel good at the end of the day, that you're doing something for people and, hopefully, for humanity in general. I like working with kids.

I probably have two big challenges. In the classroom, it's helping kids to feel that they fit in, teaching them that they can be successful in my class. Kids often have preconceived notions, like, "I'm not good in science." It's really hard for them to change that opinion of themselves; they first need to learn that they can be successful.

Another challenge lies with the teaching profession itself, that we are professionals, like any other area, but we aren't always perceived that way. I think that's mostly because salaries have not been in line with other professional areas or degrees, and it is a reason why you don't see a lot of people going into teaching. Young people can make more money elsewhere, but the jobs they choose may not be as rewarding. Many of my friends in engineering or business are making twice what I am, but they are unhappy because it's not rewarding. But when you work in teaching, people generally appreciate everything you do for kids, and the kids appreciate it.

What sort of person makes the best teacher?

Definitely someone who enjoys working with students. That's the number-one thing. You have to enjoy what you do. Second would be someone who's flexible, yet still firm, disciplined, and principled. Someone who is energetic, because teaching is trying on your body and mind. Definitely friendly. You work with the public and all different kinds of people. You have to be accepting of other people. The last thing I would say is that a teacher should be goal-oriented. Teachers are responsible for furthering their own professional development. You have to set goals and be working toward them all the time.

How would you advise undergraduate biology majors considering a career in teaching?

First, get experience working with students in the classroom. Even when you are working on your biology degree, do something with kids! Make sure you like working with them! Contact a high school; see if they need help with after-school tutoring. Ask to sit in on classes to see how they operate. You really need to know how it is, way before you even start education courses. Even working in after-school programs or church youth groups would be helpful—anything that lets you work with kids. Second, actively learn your subject area. I think it's very important to be sound in your discipline, to be able to explain it to someone else. The last suggestion I have is to network with your college professors and with local high school teachers that you knew in school or in the town where you go to college. Those are people who will be your contacts and your references. If there's a school district that you want to work in, get networking, start substitute teaching, do it all.

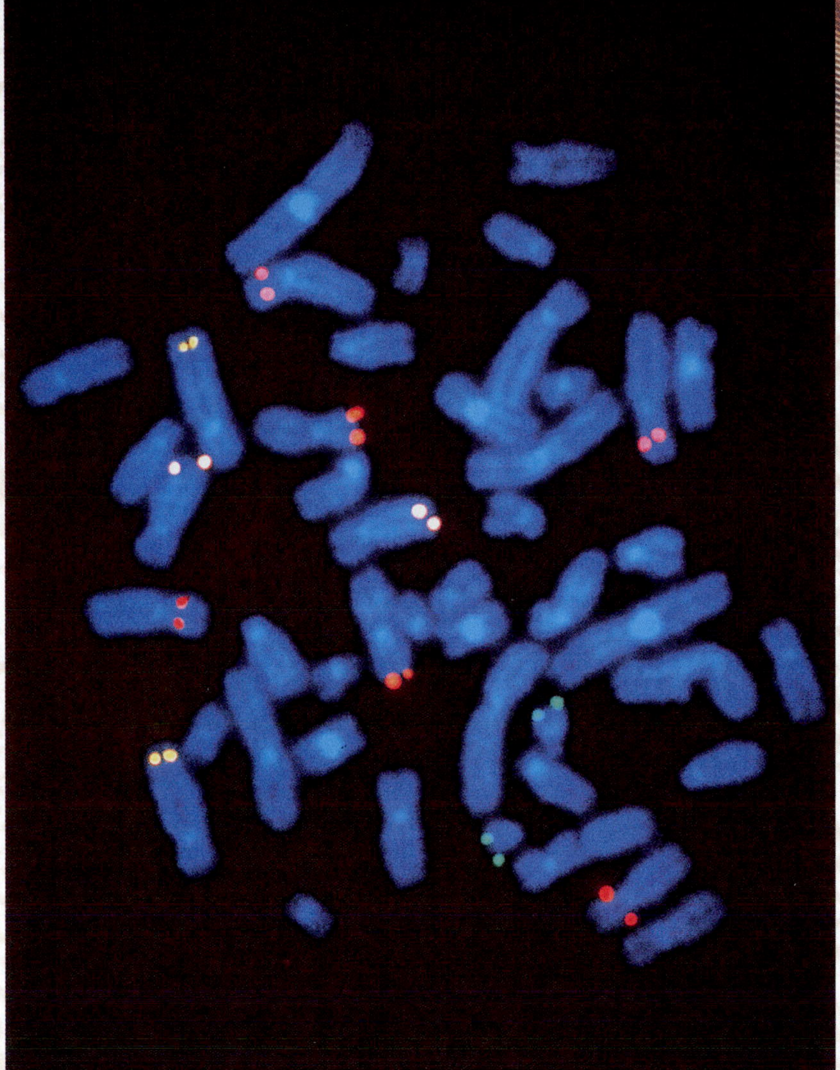

Researchers in human genetics have identified the genes responsible for many hereditary diseases. The colored dots in this fluorescence micrograph indicate the positions of specific genes on these human chromosomes. *(Drs. T. Ried and D. Ward/Peter Arnold, Inc.)*

Part 3

The Continuity of Life: Genetics

Chapters

9

Chromosomes, Mitosis, and Meiosis

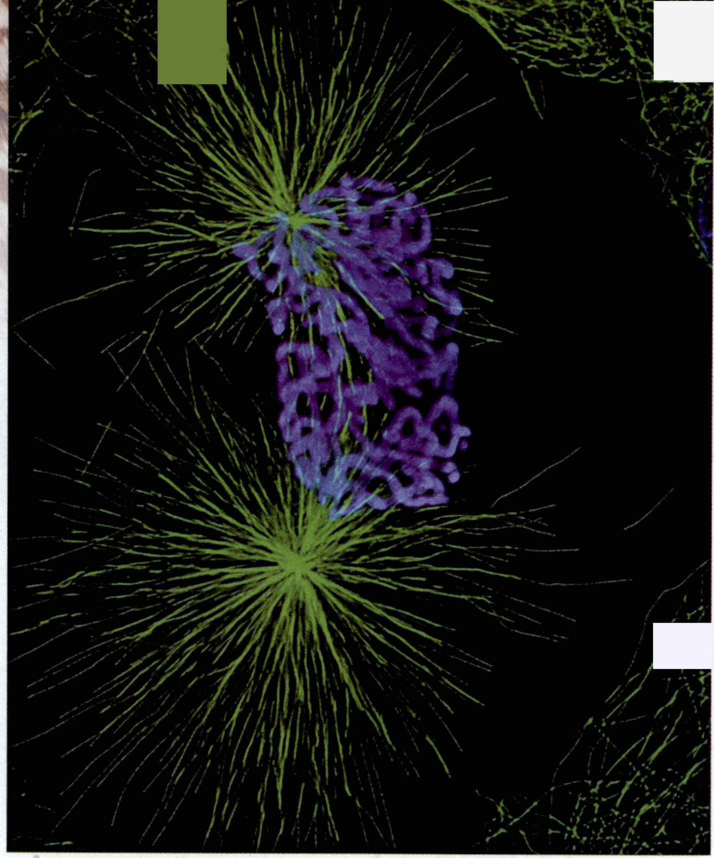

Fluorescence LM of a human lung cell in early mitosis. Computer processing is responsible for this unique three-dimensional view of the mitotic spindle *(green)* and chromosomes *(blue)*. *(Alexey Khodjakov, Wadsworth Center, Albany, NY)*

LEARNING OBJECTIVES

After you have studied this chapter you should be able to

1. Discuss the significance of chromosomes in terms of their information content.
2. Identify the stages in the eukaryotic cell cycle, describe the principal events characteristic of each, and point out some ways in which the cycle is controlled.
3. Illustrate the structure of a duplicated chromosome, labeling the sister chromatids, sister centromeres, and sister kinetochores.
4. Explain the significance of mitosis and diagram the process.
5. Discriminate between asexual and sexual reproduction.
6. Distinguish between haploid and diploid cells and define homologous chromosomes.
7. Explain the significance of meiosis and diagram the process.
8. Contrast mitosis and meiosis, emphasizing how differences in events lead to different outcomes.
9. Compare the roles of mitosis and meiosis and of haploidy and diploidy in various generalized life cycles.

Because all cells are formed by the division of preexisting cells, cells divide to enable the organism to grow, repair damaged parts, or reproduce. Cells can serve as the basic units of life and the essential link between generations because even the simplest contains a massive amount of coded information in the form of DNA, collectively referred to as the organism's **genome.** Today, through the Human Genome Project and related efforts introduced in Chapter 1, we are learning how genomes are organized into informational units called **genes,** which are used to control the activities of the cell and are passed on to its descendants.

When a cell divides, the information contained in the DNA first must be faithfully duplicated and the copies then transmitted to each daughter cell through a precisely choreographed series of steps. However, this presents a problem because DNA is a very long, thin molecule that could easily become tangled and broken. The potential difficulty is especially acute for a eukaryotic cell because its nucleus contains a very large amount of DNA. In this chapter we consider how eukaryotes solve this problem by packaging each DNA molecule with proteins and assembling the resulting complex into a structure called a **chromosome,** each of which contains hundreds or thousands of genes.

We then consider *mitosis,* the process that ensures that a parent cell transmits one copy of every chromosome to each of its two daughter cells. In this way the chromosome number is preserved through successive mitotic divisions. Most of the body cells of eukaryotes divide by mitosis.

Finally we discuss *meiosis,* a special process in which the chromosome number is reduced by half. Meiosis is required in sexual life cycles in eukaryotes. Sexual reproduction is characterized by the fusion of two sex cells, or *gametes,* to form a single cell called a *zygote.* Meiosis makes it possible for each gamete to contain only half the number of parental chromosomes, thereby preventing the zygotes from having twice as many chromosomes as the parents.

Bacterial reproduction is described in Chapter 23. Prokaryotic cells contain much less DNA than eukaryotic cells. Their DNA is usually in the form of a circle and is packaged with very few associated proteins. Although the distribution of genetic material in dividing prokaryotic cells is a simpler process than mitosis, it nevertheless must be very precise if the daughter cells are to be genetically identical to the parent cell.

EUKARYOTIC CHROMOSOMES CONTAIN DNA AND PROTEIN

The major carriers of genetic information in eukaryotes are the chromosomes contained within the cell nucleus. Although the term *chromosome* means "colored body," chromosomes are virtually colorless; the name refers to their ability to be stained by certain dyes.

Chromosomes are made up of **chromatin,** a complex material consisting of proteins and deoxyribonucleic acid (DNA). When a cell is not dividing, the chromosomes are present but in an extended, partially unraveled form. The chromatin consists of long, thin threads that are somewhat aggregated, which gives them a granular appearance when viewed with the electron microscope (see Fig. 4–12). During cell division, the chromatin fibers condense and the chromosomes become visible as distinct structures (Fig. 9–1). The structure of chromatin is described in more detail in Chapter 11.

DNA is organized into informational units called genes

Each chromosome may contain hundreds or even thousands of genes. For example, humans are thought to have about 35,000 to 45,000 genes that code for proteins. The precise number is not known, although scientists are engaged in a massive, coordinated effort to make a precise determination as part of the Human Genome Project (see Chapter 15). As will be evident in succeeding chapters, our concept of the gene has changed considerably since the beginnings of the science of genetics, but our definitions have always centered on the gene as an informational unit. By providing information needed to carry out one or more specific cellular functions, a gene ultimately affects some characteristic of the organism. For example, genes control eye color in humans, wing length in flies, and seed color in peas.

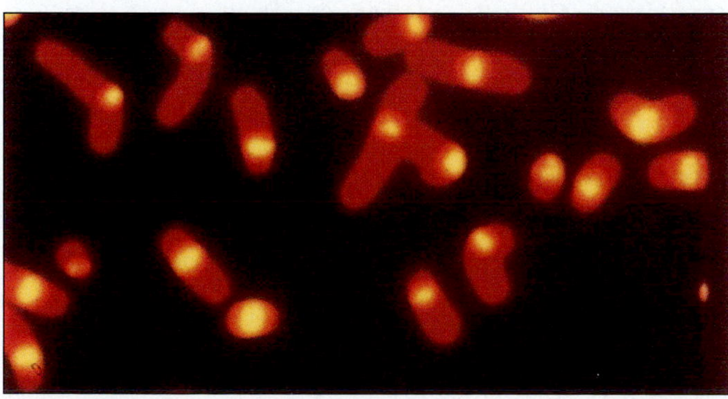

Figure 9–1 Chromosomes. The human chromosomes shown in this confocal fluorescence LM have been stained with a fluorescent antibody that binds to the centromere regions *(yellow)*. *(Courtesy of Oncor, Inc.)*

Chromosomes of different species differ in number and informational content

Every individual of a given species has a characteristic number of chromosomes in most nuclei of its body cells. However, it is not the *number* of chromosomes that makes each species unique but rather the *information* specified by the genes in the chromosomes.

Most human body cells have exactly 46 chromosomes, but humans are not humans merely because they have 46 chromosomes. In fact, some humans have abnormal karyotypes (chromosome assortments) with more or fewer than 46 (see Chapter 15). Humans are not unique in having 46 chromosomes; some other species of animals and plants also happen to have 46, whereas others have different chromosome numbers. For example, a certain species of roundworm has only 2 chromosomes in each cell, while some crabs have as many as 200 and some ferns have more than 1000. Most animal and plant species have between 10 and 50 chromosomes. Numbers above and below this are uncommon.

THE CELL CYCLE IS A SEQUENCE OF CELL GROWTH AND DIVISION

Usually when cells reach a certain size, they must either stop growing or divide. Not all cells divide; some, such as nerve, skeletal muscle, and red blood cells, do not normally divide once they are mature. Other cells undergo a sequence of activities required for growth and cell division referred to as the **cell cycle,** which is commonly represented in diagrams as a circle (Fig. 9–2). The **generation time,** the period from the beginning of one division

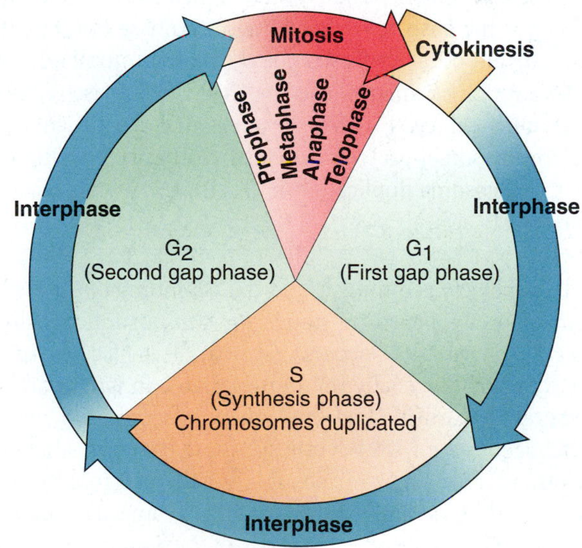

Figure 9–2 The eukaryotic cell cycle. The generation time, the time required to complete one cycle, includes cell division (mitosis and cytokinesis) and interphase. Proportionate amounts of time spent at each stage vary with the species, cell type, and growth conditions.

to the beginning of the next, can vary widely, but in actively growing plant and animal cells it is often about 8 to 20 hours.

Eukaryotic cell division involves two main processes, mitosis and cytokinesis. **Mitosis,** a complex process involving the nucleus, ensures that each new nucleus receives the same number and types of chromosomes as were present in the original nucleus. **Cytokinesis,** which generally begins before mitosis is complete, is the division of the cytoplasm of the cell to form two cells. Multinucleate cells are formed if mitosis is not followed by cytokinesis; this is a normal condition for some kinds of cells.

Chromosomes become duplicated during interphase

Most of the life of a cell is spent in **interphase,** the time when no cell division is occurring. Although the appearance of the nucleus is generally unremarkable (see Fig. 4–12), a cell that is capable of dividing is typically very active during this time, synthesizing needed materials and growing. Most proteins and other materials are synthesized throughout interphase.

Process of Science

In the early 1950s, researchers demonstrated that cells preparing to divide synthesize their chromosomes at a relatively restricted interval during interphase and not during early mitosis, as had been previously thought. These investigators used isotopes, such as ^{3}H, to synthesize radioactive thymidine, a nucleotide (see Chapter 3) that is incorporated specifically into DNA as it is synthesized. After the radioactive thymidine had been supplied for a brief period to actively growing cells, a fraction of the cells could be shown by autoradiography (see Chapter 2) to have silver grains over their chromosomes (Fig. 9–3). These labeled cells were those that had been engaged in DNA replication (known as the **synthesis phase,** or **S phase**) during the experiment. The S phase provides the major landmark that is the basis for subdividing interphase (see Fig. 9–2). The proportion of labeled cells out of the total number of cells provides a rough estimate of the length of the S phase relative to the rest of the cell cycle. Other chromosomal constituents, such as the chromosomal proteins, are also synthesized during the S phase. Chromosome duplication is a complex process discussed in Chapter 11.

The time between mitosis and the beginning of the S phase is termed the **G_1 phase** (*G* stands for *gap,* an interval during which no DNA synthesis occurs). Growth takes place during the G_1 phase, which is usually the most variable in length and also the longest. Toward the end of G_1 there is increased activity of enzymes required for DNA synthesis; these enzymes, along with many other factors, make it possible for the cell to enter the S phase. Cells that are not dividing usually become arrested in the part of the cell cycle prior to the onset of the S phase and are said to be in a state called G_0.

After it completes the S phase, the cell enters a second gap phase, the **G_2 phase.** At this time, increased protein synthesis occurs as the final steps in the cell's preparation for division take place. For many cells, the G_2 phase is short relative to the S and

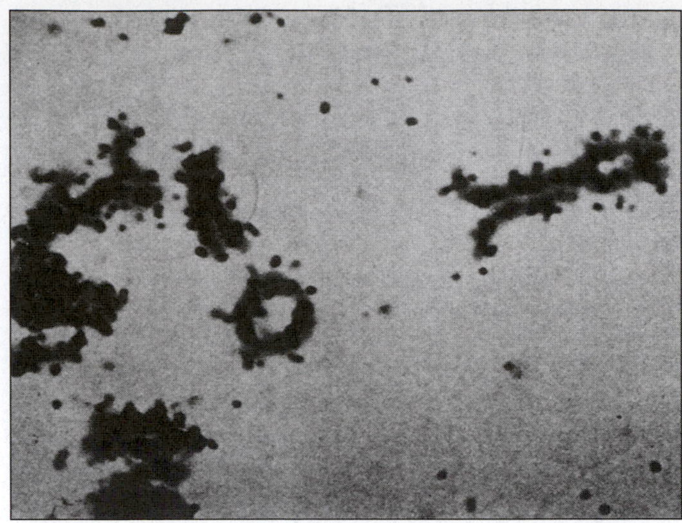

Figure 9–3 Detection of the time of DNA synthesis. The silver grains *(black dots)* lying on top of the bean chromosomes in this autoradiogram mark locations where the chromosomes incorporated ^{3}H-labeled thymidine, a DNA precursor, during the S phase of the interphase that preceded mitosis. *(Professor J.H. Taylor,* Proceedings of the National Academy of Science, *Vol. 43, 1957, courtesy of New York Academy of Medicine)*

G_1 phases. The completion of the G_2 phase is marked by the beginning of mitosis. The sequence of the substages of interphase is therefore:

G_1 phase → S phase → G_2 phase

Mitosis ensures orderly distribution of chromosomes

Each mitotic division is a continuous process. However, for descriptive purposes, mitosis has been divided into stages:

Prophase → Metaphase → Anaphase → Telophase

Refer to Figure 9–4 as you read the descriptions of these stages as they would occur in a typical plant or animal cell.

During prophase, duplicated chromosomes become visible with the microscope

The first stage of mitosis, **prophase,** begins when the long chromatin threads begin a coiling process in which chromosomes become simultaneously shorter and thicker. The chromatin can then be distributed to the daughter cells without tangling.

When stained with certain dyes and viewed through the light microscope, chromosomes become visible as darkly staining bodies as prophase progresses. Each chromosome has been duplicated during the preceding S phase and consists of a pair of units, termed **sister chromatids,** which contain identical DNA sequences. Each chromatid includes a constricted region called the **centromere.** Sister chromatids are tightly associated in the vicinity of their centromeres (Fig. 9–5). Although the chemical basis for this close association is not completely understood, evidence suggests that special DNA sequences and special proteins that bind to those DNA sequences are involved. Attached to each centromere is a **kinetochore,** a structure formed from proteins to which microtubules (see Chapter 4) can bind. These microtubules are instrumental to the process by which the chromosomes are distributed during mitosis.

A dividing cell is usually described as a globe, with an equator that determines the midplane, or equatorial plane, and two opposite poles. This terminology is used for all cells regardless of their actual shape. Microtubules radiate from each pole, and some of these protein fibers elongate toward the chromosomes, forming a complex structure known as the **mitotic spindle** (Fig. 9–6). The mitotic spindle is responsible for the separation of the chromosomes during anaphase.

Animal cells differ from the cells of complex plants in the details of mitotic spindle formation. In both types of dividing cells, each pole contains a region, referred to as a **microtubule-organizing center,** from which the microtubules grow outward. When viewed with the electron microscope, microtubule-organizing centers in plant cells consist of fibrils with little or no discernible structure.

In contrast, animal cells and some other cells have a pair of **centrioles** (see Chapter 4) in the middle of each microtubule-organizing center. The centrioles are surrounded by fibrils that make up the **pericentriolar material.** The ends of the spindle microtubules are found in the pericentriolar material, but they do not actually touch the centrioles themselves. Evidence that the pericentriolar material may be functionally similar to the material in the microtubule-organizing centers of plants derives from the fact that they are similar in appearance and in the specific types of proteins present.

Each of the two centrioles becomes duplicated during interphase, yielding two pairs of centrioles. Late in prophase, microtubules radiate from the pericentriolar material surrounding the centrioles, and one pair of centrioles migrates to each pole. The migration of the centrioles to the poles essentially marks the migration of the microtubule-organizing centers. Additional microtubules form clusters extending outward in many directions from the pericentriolar material at the poles; these structures are called **asters.** The function of the asters is not well understood, although there is evidence that these structures play a role in cytokinesis.

It is likely that both plant and animal spindles are organized by similar microtubule organizing centers. Although centrioles were long thought to be required for spindle formation in animal cells, their apparent involvement is probably coincidental. Because centrioles usually arise in association with preexisting centrioles, the localization of the centrioles in the microtubule-organizing center may have evolved to provide for the orderly distribution of these organelles to the daughter cells.

During prophase, the nucleolus (see Chapter 4) diminishes in size and usually disappears. Toward the end of prophase, the nuclear envelope breaks down, and each sister chromatid becomes attached to some of the spindle microtubules at its kinetochore. The chromosomes (each consisting of a pair of sister chromatids connected at their centromeres) undergo a series of movements and finally become aligned along the equatorial plane of the cell.

At metaphase duplicated chromosomes line up on the midplane

The period during which the duplicated chromosomes are lined up along the equatorial plane of the cell constitutes **metaphase.** The mitotic spindle is complete. It is composed of two types of microtubules: polar microtubules and kinetochore microtubules. **Polar microtubules** extend from each pole to the equatorial region, where they generally overlap. **Kinetochore microtubules** extend from each pole and attach to the kinetochores (see Fig. 9–6). At mitotic metaphase the sister chromatids that make up each duplicated chromosome are attached by kinetochore microtubules to *opposite* poles of the cell.

During metaphase each chromatid is completely condensed and appears quite thick and distinct. Because individual chromosomes can be seen more distinctly at metaphase than at any other time, they are usually photographed at this stage to be studied for certain chromosome abnormalities (Fig. 9–7 on page 205; also see Fig. 15–5*b*).

During anaphase chromosomes move toward the poles

Anaphase begins as the protein tethers holding the sister chromatids together in the vicinity of their centromeres are released. Each chromatid is now referred to as an *independent chromosome.* The now separate chromosomes slowly move to opposite poles. The kinetochores, still attached to kinetochore microtubules, lead the way, with the chromosome arms trailing behind. Anaphase ends when all the chromosomes have reached the poles.

The overall mechanism of chromosome movement in anaphase is still poorly understood, although significant progress is being made in this area. Microtubules lack elastic or contractile properties. So how do the chromosomes move apart? Are they pushed or pulled, or do other forces operate?

Process of Science Chromosome movements are studied in a variety of ways. Numbers of microtubules present at a particular stage or after certain treatments can be determined through careful analysis of electron micrographs. It is possible to physically perturb living cells that are dividing, using laser beams or mechanical devices known as *micromanipulators.* Skilled researchers can move chromosomes, break their connections to microtubules, and even remove them from the cell entirely.

Microtubules are constantly changing structures, with tubulin subunits being constantly removed from their ends and others being added, and in fact there is considerable evidence that kinetochore microtubules shorten during anaphase. Therefore, many current hypotheses to explain anaphase movement

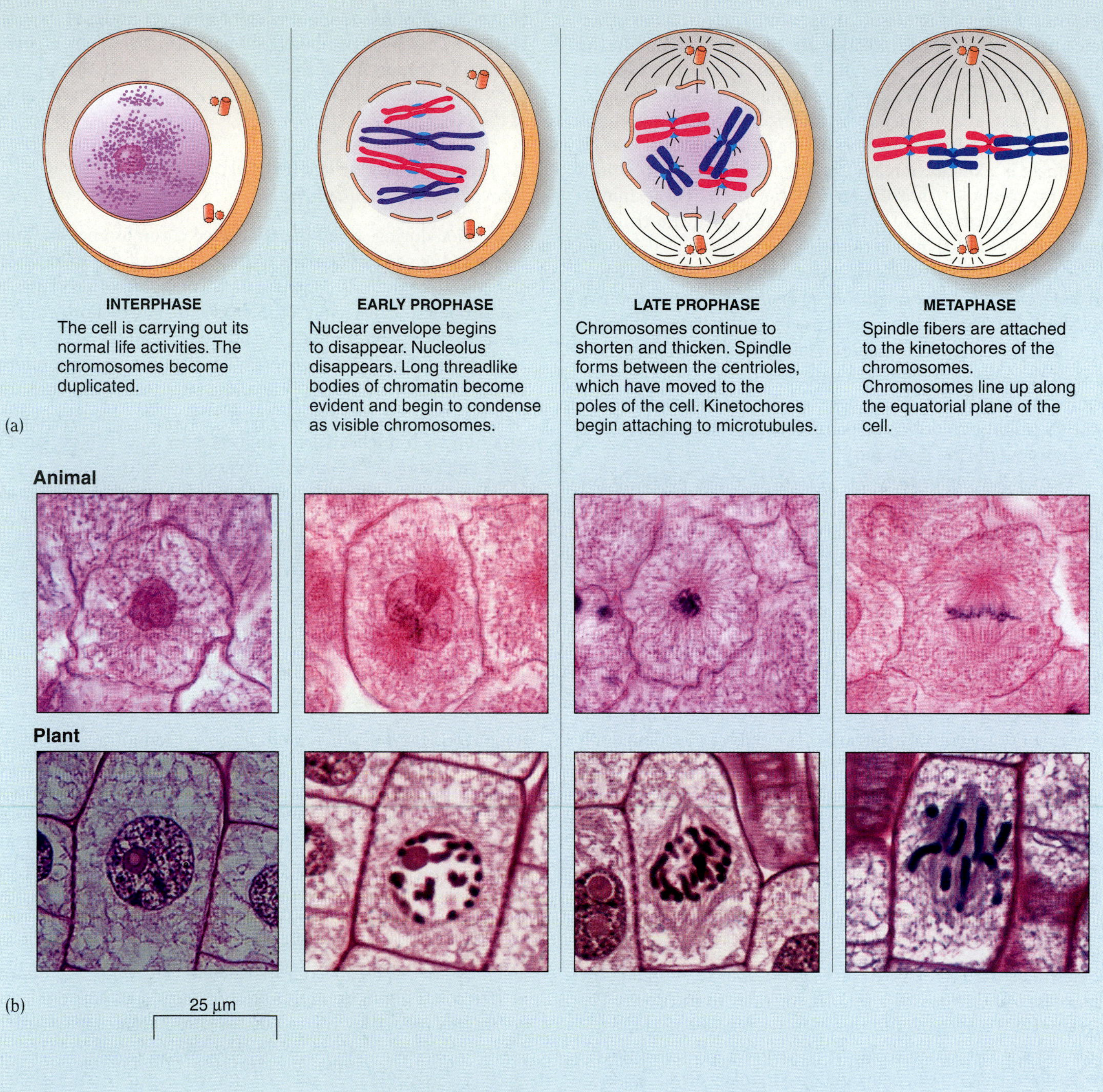

INTERPHASE
The cell is carrying out its normal life activities. The chromosomes become duplicated.
EARLY PROPHASE
Nuclear envelope begins to disappear. Nucleolus disappears. Long threadlike bodies of chromatin become evident and begin to condense as visible chromosomes.
LATE PROPHASE
Chromosomes continue to shorten and thicken. Spindle forms between the centrioles, which have moved to the poles of the cell. Kinetochores begin attaching to microtubules.
METAPHASE
Spindle fibers are attached to the kinetochores of the chromosomes. Chromosomes line up along the equatorial plane of the cell.
(a)
Animal
Plant
(b)
25 μm

Figure 9–4 Cell division. Interphase and the stages of mitosis are similar in plant and animal cells. Although both types have microtubule organizing centers, the plant cells lack centrioles. **(a)** The drawings depict generalized animal cells with a diploid chromosome number of four. The sizes of the nuclei and chromosomes are exaggerated to show the structures more clearly. **(b)** The upper row of LMs depicts cells of an animal (the whitefish, *Coregonus* sp.). The chromosomes have been stained and the cells flattened on microscope slides. Stained chromosomes in sectioned cells of the onion, *Allium cepa*, are shown in the lower row. *(Animal cells, Michael Abbey/Science Source/Photo Researchers, Inc.; Plant cells, first interphase through telophase, Ed Reschke; second interphase, Carolina Biological Supply Company/Phototake)*

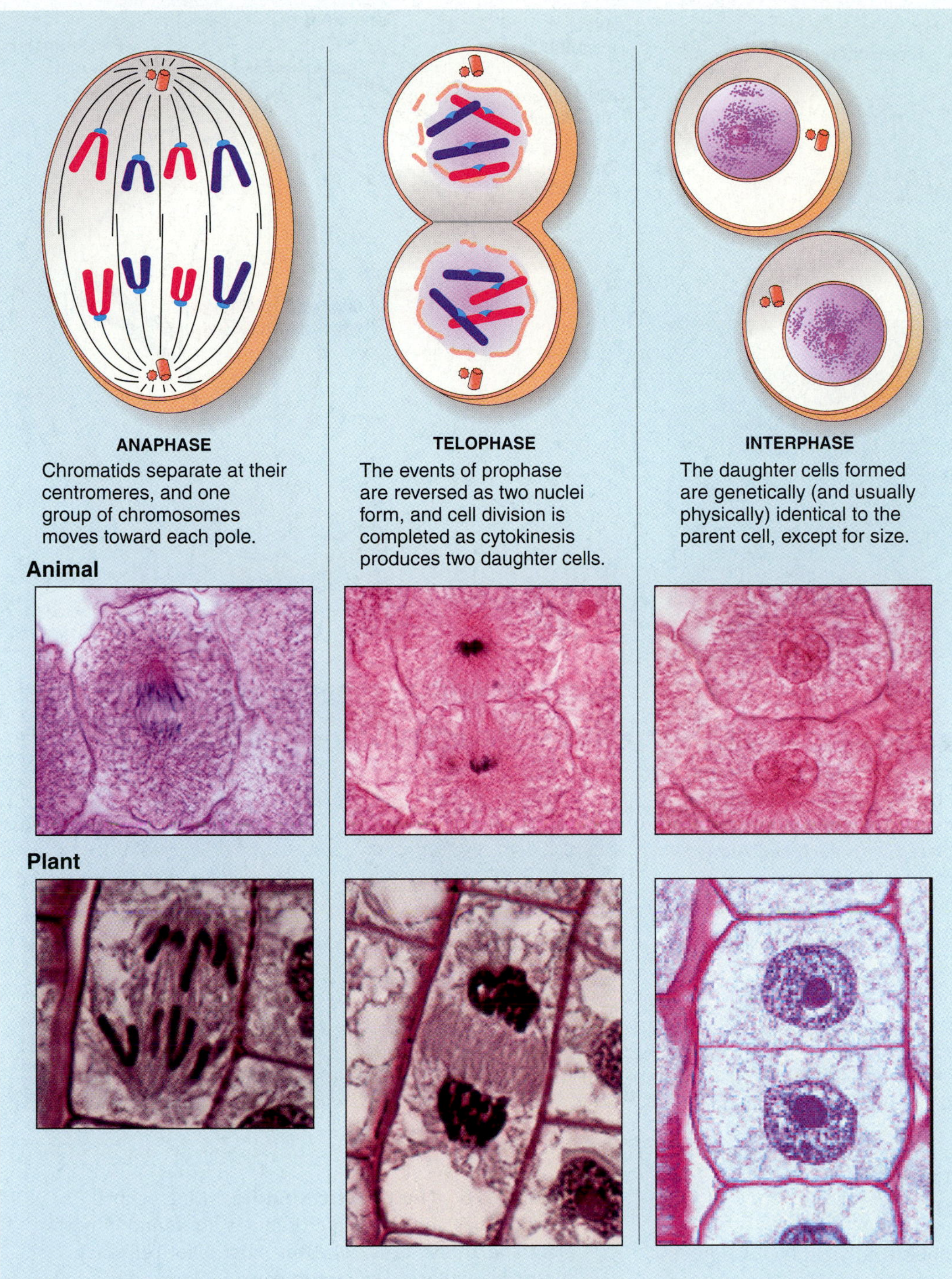

include the idea that chromosomes are able to move poleward because they are able to remain anchored to the kinetochore microtubules even as tubulin subunits are being removed. There is experimental evidence that multiple types of motor proteins, including forms of kinesin and dynein (see Chapter 4), play a role in this movement.

ANAPHASE
Chromatids separate at their centromeres, and one group of chromosomes moves toward each pole.
TELOPHASE
The events of prophase are reversed as two nuclei form, and cell division is completed as cytokinesis produces two daughter cells.
INTERPHASE
The daughter cells formed are genetically (and usually physically) identical to the parent cell, except for size.
Animal
Plant

A second phenomenon also plays a role in chromosome separation. During anaphase the spindle as a whole elongates, at least partly because polar microtubules originating at opposite poles are associated with motors that enable them to slide past one another at the equator, decreasing the degree to which they are overlapped and thereby "pushing" the poles apart. This mechanism indirectly causes the chromosomes to move apart because they are attached to the poles by kinetochore microtubules.

During telophase two separate nuclei are formed

The final stage of mitosis, **telophase,** is characterized by a return to interphase-like conditions. The chromosomes decondense by uncoiling. A new nuclear envelope forms around each set of

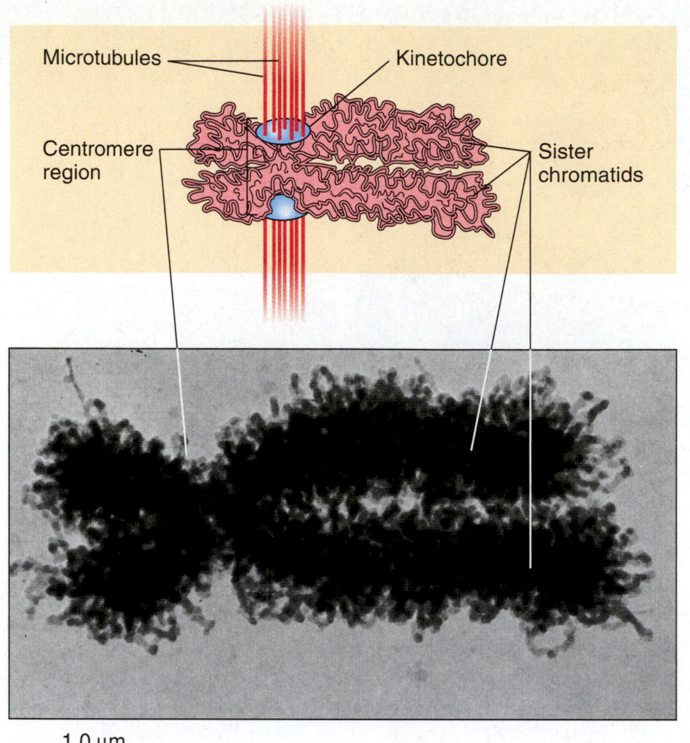

Figure 9–5 A metaphase chromosome. An SEM *(bottom)* is paired with an interpretive drawing. The sister chromatids, each consisting of tightly coiled chromatin fibers, are tightly associated at their centromere regions, indicated by the bracket. Associated with each centromere is a structure known as a kinetochore, which serves as a microtubule attachment site. The kinetochores and microtubules are not evident in the SEM. *(E.J. DuPraw)*

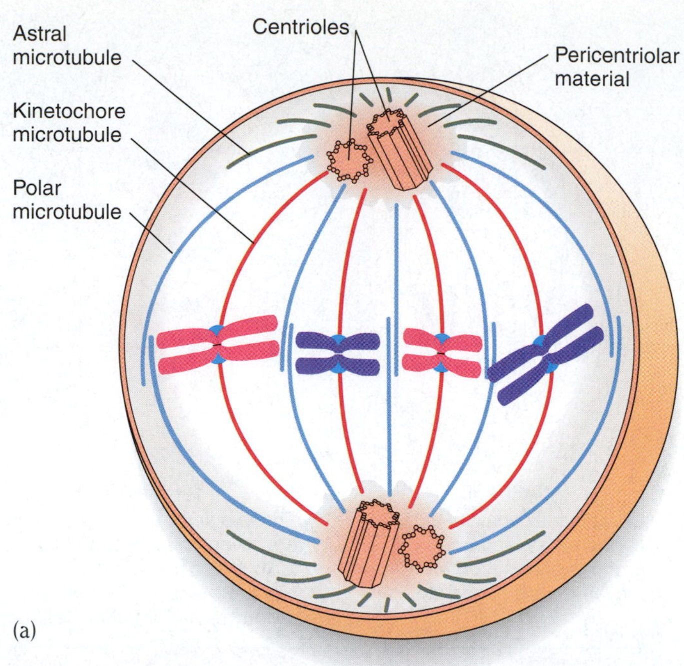

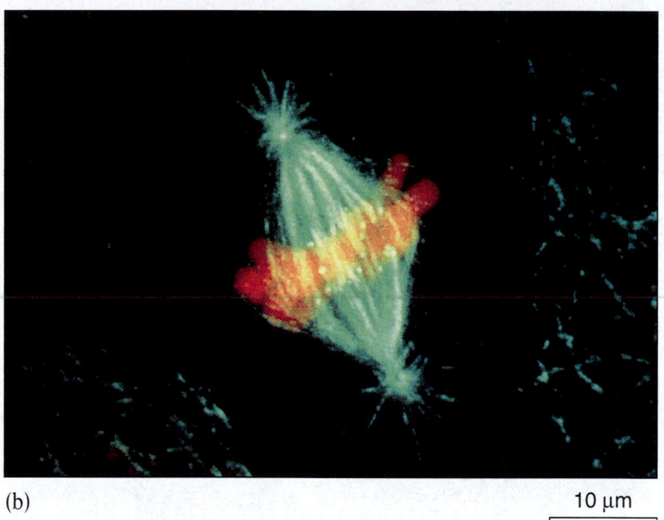

Figure 9–6 The mitotic spindle. Kinetochore microtubules and polar microtubules are found in both plant and animal cells, whereas most plant cells lack centrioles and astral microtubules. **(a)** One end of each microtubule of this animal cell is associated with one of the poles. Astral microtubules *(green)* radiate in all directions, forming the aster. Kinetochore microtubules *(red)* connect the kinetochores to the poles, and polar microtubules *(blue)* overlap at the midplane. **(b)** Fluorescent-stained LM of animal cell at metaphase with well-defined spindle and asters (chromosomes, *orange;* microtubules, *green*). *(b, CNRI/Phototake, NYC)*

chromosomes, made at least in part from small vesicles and other components derived from the old nuclear envelope. The spindle microtubules disappear, and the nucleoli reorganize.

Cytokinesis is the formation of two separate daughter cells

Cytokinesis, the division of the cytoplasm to yield two daughter cells, usually overlaps mitosis, generally beginning during telophase. Cytokinesis of an animal cell begins as a ring of actin microfilaments associated with the plasma membrane encircles the cell in the equatorial region, at right angles to the spindle (Fig. 9–8*a*). The ring contracts, producing a **cleavage furrow** that gradually deepens and separates the cytoplasm into two daughter cells, each with a complete nucleus.

In plant cells, cytokinesis occurs by the formation of a **cell plate** (Fig. 9–8*b*), a partition constructed in the equatorial region of the spindle and growing laterally toward the cell wall. The cell plate forms as a line of vesicles that originate in the Golgi complex (see Chapter 4). The vesicles contain materials to construct both a primary cell wall for each daughter cell and a middle

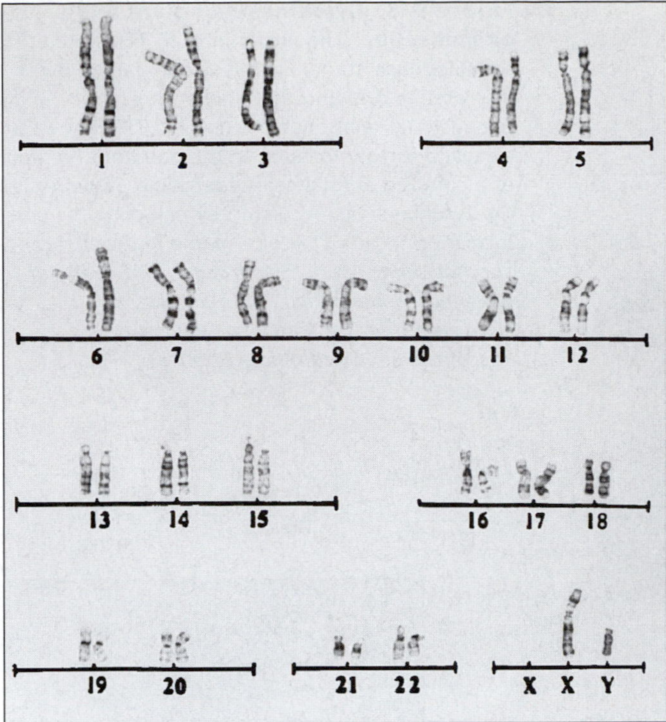

Figure 9–7 A normal human karyotype. Although these are duplicated chromosomes (late prophase), the sister chromatids are not clearly evident because they are closely aligned throughout their lengths. As discussed in Chapter 10, the X and Y chromosomes of this normal male are not strictly homologous; normal females have two X chromosomes and no Y chromosome. The process of karyotyping is discussed in Chapter 15. *(Courtesy of Dr. Leonard Sciorra)*

lamella that will cement the primary cell walls together. The vesicle membranes fuse to become the plasma membrane of each daughter cell.

Mitosis typically produces two cells genetically identical to the parent cell

The remarkable regularity of the process of cell division ensures that each of the daughter nuclei receives exactly the same number and kinds of chromosomes that the parent cell had. Thus, with a few exceptions, every cell of a multicellular organism has exactly the same genetic makeup. If a cell receives more or fewer than the characteristic number of chromosomes through some malfunction of the cell division process, the resulting cell may show marked abnormalities and is often unable to survive.

Most cytoplasmic organelles are distributed randomly to the daughter cells

Mitosis provides for the orderly distribution of chromosomes (and of centrioles, if present), but what about the various cytoplasmic organelles? For example, all eukaryotic cells, including plant cells, require mitochondria. Likewise, photosynthetic plant cells cannot carry out photosynthesis without chloroplasts. These organelles contain their own DNA and appear to form by the division of previously existing mitochondria or plastids or their precursors. This nonmitotic division process is similar to prokaryotic cell division (see Chapter 23) and generally occurs during interphase, not when the cell divides. Because many copies of each organelle are present in each cell, organelles are apportioned more or less equally between the daughter cells at cytokinesis.

The cell cycle is controlled by an internal genetic program interacting with external signals

When conditions are optimal, some prokaryotic cells can divide every 20 minutes. The generation times of eukaryotic cells are generally much longer, although the frequency of cell division varies widely among different species and among different tissues of the same species. Some cells in the central nervous system usually cease dividing after the first few months of life, whereas blood-forming cells, digestive tract cells, and skin cells divide frequently throughout the life of the organism. Under optimal conditions of nutrition, temperature, and pH, the eukaryotic cell cycle length is constant for any cell type. Under less favorable conditions, however, the generation time may be longer.

According to a considerable body of evidence that has accumulated in recent years, certain fundamental mechanisms of genetic control of the cell cycle are common to all eukaryotes. The cell cycle is controlled by regulatory molecules that trigger a specific sequence of events. Because a failure to carefully control cell division can have disastrous consequences, checkpoints are built into the program to ensure that all the events of a particular stage have been completed before the next stage begins. Among the key regulatory molecules are **protein kinases,** enzymes that activate or inactivate other proteins by adding phosphate groups (phosphorylation). The particular protein kinases involved in controlling the cell cycle are called **cyclin-dependent protein kinases (Cdk)** because they are only active when they form complexes with regulatory proteins called **cyclins.** The cyclins are so named because their levels fluctuate predictably during the cell cycle (i.e., they "cycle").

When a specific cyclin-dependent protein kinase forms a complex with a specific cyclin, it actively phosphorylates certain cellular proteins. Some of these proteins, including certain enzymes, become activated when they are phosphorylated, and others become inactivated. As some enzymes are activated and others are inactivated by phosphorylation, the activities of the cell change. First, G_1 Cdk complexes prepare the cell for the S phase, and then S-phase Cdk complexes allow the cell to enter the S phase. Mitotic Cdk complexes are responsible for causing chromosome condensation, nuclear envelope breakdown, and mitotic spindle formation. Mitotic Cdk complexes also activate the anaphase-promoting complex (APC) at the end of metaphase. APC initiates anaphase by allowing degradation of the proteins

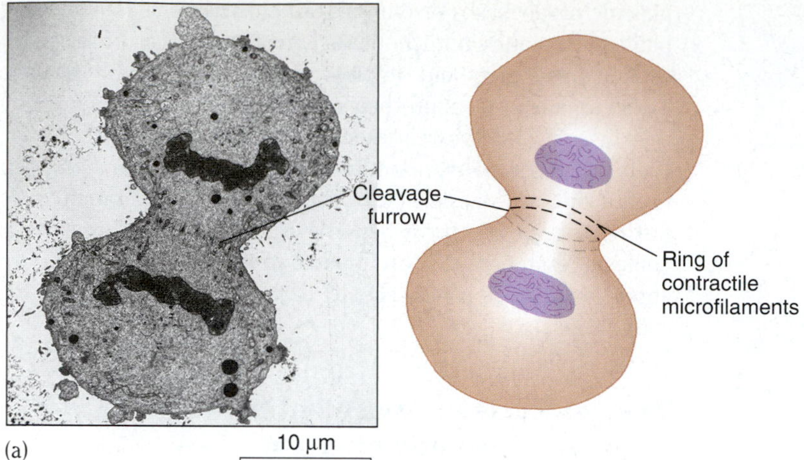

Figure 9–8 Cytokinesis in plant and animal cells. The nuclei in both TEMs are at the telophase stage. Each is accompanied by an interpretive drawing that shows three-dimensional relationships. **(a)** This TEM shows a cleavage furrow forming in the equatorial region of a cultured animal cell undergoing cytokinesis. **(b)** Cytokinesis is occurring by cell plate formation in this TEM of a maple leaf cell, *Acer saccharinum*. *(a, T.E. Schroeder, University of Washington/Biological Photo Service; b, E.H. Newcomb and B.A. Palevitz, University of Wisconsin/Biological Photo Service)*

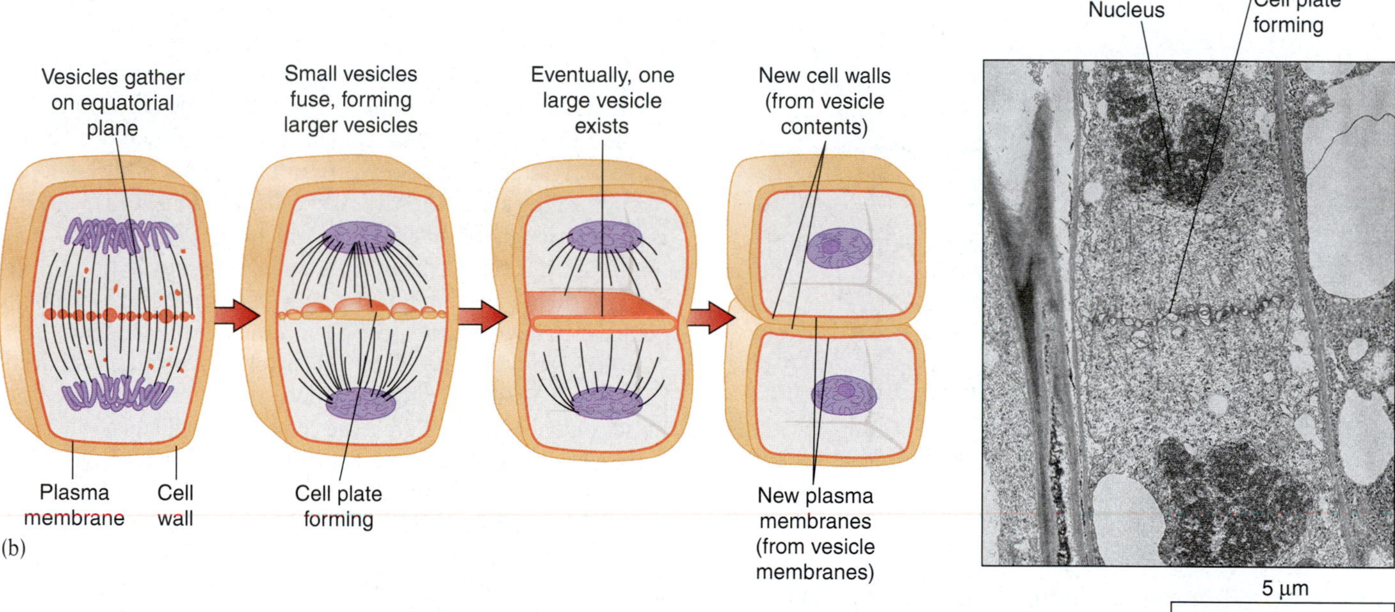

that hold the sister chromatids together. Figure 9–9 illustrates the activities of an important type of mitotic Cdk complex, known as mitosis-promoting factor (MPF). Although not all of the details are understood, these systems of regulating the cell cycle have been highly conserved during the evolution of eukaryotes; they are found in organisms as diverse as yeast (a unicellular fungus), clams, frogs, plants, and humans.

Certain drugs can stop the cell cycle at a specific checkpoint. Some of these prevent DNA synthesis, whereas others inhibit the synthesis of proteins that control the cycle or inhibit the synthesis of structural proteins that contribute to the mitotic spindle. Because one of the distinguishing features of most cancer cells is their high rate of cell division relative to most normal body cells, they can be most affected by these drugs. Many of the side effects of certain anticancer drugs (e.g., nausea, hair loss) are due to the drugs' effects on rapidly dividing normal cells in the digestive system and hair follicles.

Colchicine, a drug used to block cell division in eukaryotic cells, binds with unpolymerized tubulin subunits, preventing them from being added to the spindle microtubules. Under these conditions, the rate of microtubule breakdown far exceeds the rate of microtubule assembly, resulting in the disappearance of the spindle. Although the sister chromatids eventually become detached from one another and each becomes a chromosome, they cannot move to the poles in the absence of the spindle. As a result, a cell may end up with extra sets of chromosomes (a condition known as **polyploidy,** which is discussed in the next section).

In plant cells, certain hormones are known to stimulate mitosis. These include the **cytokinins,** a group of plant hormones that promote mitosis both in normal growth and in wound healing (see Chapter 36). Similarly, hormones such as certain steroid hormones can act as growth stimulators in animals (see Chapter 47).

Protein **growth factors,** which are active at extremely low concentrations, stimulate mitosis in certain animal cells by causing G_1 Cdk complexes to be formed. Of the approximately 50 protein growth factors known, some act only on specific cell classes,

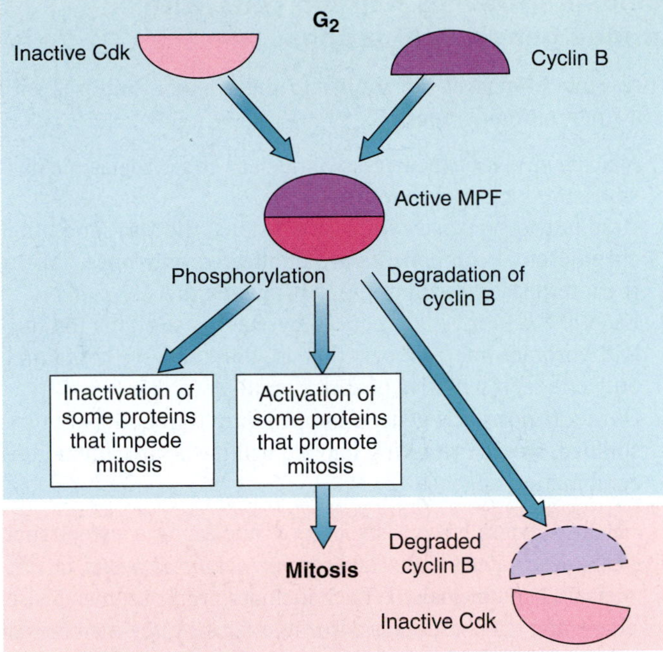

Figure 9–9 Genetic control of the cell cycle. In this example, mitosis-promoting factor (MPF) is required for the cell to pass through the G_2 checkpoint, and make the transition to mitosis. Active MPF is produced when an inactive cyclin-dependent protein kinase (Cdk) becomes complexed with cyclin B. MPF then phosphorylates many proteins, thereby activating those needed for mitosis and inactivating those that would impede mitosis. Later in mitosis, MPF becomes inactive as cyclin B is degraded. These events are repeated when cyclin B levels rise again in G_2 of the next cell cycle.

while others work over a broader range. For example, the effects of the growth factor erythropoietin are limited to cells that will develop into red blood cells, whereas epidermal growth factor stimulates many cell types to divide. Many types of cancer cells divide even in the absence of growth factors.

SEXUAL REPRODUCTION REQUIRES A MECHANISM TO REDUCE THE CHROMOSOME NUMBER

Although the details of the reproductive process vary greatly among different kinds of eukaryotes, we can distinguish two basic types of reproduction: asexual and sexual. In **asexual reproduction** a single parent usually splits, buds, or fragments to produce two or more individuals (see Figs. 1–5, 28–8*a*, and 35–19). In most forms of eukaryotic asexual reproduction, all the cells are the result of mitotic divisions, so their genes and inherited traits are like those of the parent. Such a group of genetically identical organisms is termed a **clone.** Asexual reproduction permits organisms well adapted to their environment to produce new generations of similarly adapted organisms. It can occur rapidly and efficiently, at least in part because time and energy do not need to be expended in finding a mate.

In contrast, **sexual reproduction** involves the union of two specialized sex cells, or **gametes,** to form a single cell called a **zygote.** Usually the gametes are contributed by two different parents, but in some cases a single parent furnishes both gametes. In the case of animals and plants, the egg and sperm cells are the gametes, and the fertilized egg is the zygote. Sexual reproduction results in genetic variation among the offspring. Because they are not genetically identical to their parents or to each other, some may be able to survive environmental changes or other stresses better than either parent. However, some others, with a different combination of traits, may be less likely to survive than their parents.

There is a potential for problems in eukaryotic sexual reproduction: If each gamete has the same number of chromosomes as did the parental cell that produced it, then the zygote would be expected to have twice as many chromosomes. This doubling would occur generation after generation. How do organisms avoid producing zygotes with ever-increasing chromosome numbers? To answer this question, we need more information about the types of chromosomes found in cells.

Each chromosome found in a somatic (body) cell of a higher plant or animal normally has a partner chromosome. The two partners, known as **homologous chromosomes,** are similar in size, shape, and the position of their centromeres. When stained by special techniques, the members of a pair generally share a characteristic pattern of bands. In most species, chromosomes vary enough in their morphological features that cytologists can distinguish the different homologous pairs and match up the partners. The 46 chromosomes in human cells constitute 23 different homologous pairs (see Fig. 9–7). The most important feature of homologous chromosomes is that they carry very similar, but not necessarily identical, genetic information. For example, members of a pair of homologous chromosomes might each carry a gene that specifies hemoglobin structure; however, one member might have the information for the normal hemoglobin β chain (see Fig. 3–23*a*), whereas the other might specify the abnormal form of hemoglobin associated with sickle cell anemia (see Chapter 15). Homologous chromosomes can therefore be contrasted with the two members of a pair of sister chromatids, which are precisely identical to each other.

A *set* of chromosomes has one of each kind of chromosome; in other words, it contains one member of each homologous pair. If a cell or nucleus contains two sets of chromosomes, it is said to have a **diploid** chromosome number. If it has only a single set of chromosomes, it has the **haploid** number. In humans the diploid chromosome number is 46 and the haploid number is 23. When a sperm and egg fuse at fertilization, each gamete is haploid, contributing one set of chromosomes; the diploid number is thereby restored in the fertilized egg (zygote). When the zygote divides by mitosis to form the first two cells of the embryo, each daughter cell receives the diploid number of chromosomes, and this is repeated in subsequent mitotic divisions. Thus, most human body cells are diploid.

If a cell or an individual has three or more sets of chromosomes, we say that it is **polyploid.** Polyploidy is relatively rare among animals but quite common among plants (see Chapter 19). In fact, polyploidy has been an important factor in plant evolution. As many as 80% of all flowering plants are polyploid. Polyploid plants are often larger and hardier than diploid members of the same group. Many commercially important plants, such as wheat and cotton, are polyploid.

The abbreviation for the chromosome number found in the gametes of a particular species is n, and the zygotic chromosome number is given as $2n$. If the organism is not polyploid, the haploid chromosome number is equal to n and the diploid number is equal to $2n$. For example, in humans, $n = 23$ and $2n = 46$. For simplicity, in the rest of this chapter we assume that the organisms used as examples are not polyploid. We therefore use the designations diploid and $2n$, and haploid and n, interchangeably, although these terms are not strictly synonymous.

DIPLOID CELLS UNDERGO MEIOSIS TO FORM HAPLOID CELLS

We have examined the process of mitosis, which ensures that each daughter cell receives exactly the same number and kinds of chromosomes that the parent cell had. A diploid cell that undergoes mitosis produces two diploid cells; similarly, a mitotic haploid cell produces two haploid cells. A division resulting in a reduction in chromosome number is called **meiosis.** The term *meiosis* means "to make smaller," referring to the fact that the chromosome number is reduced by one-half. In meiosis a diploid cell undergoes two cell divisions, potentially yielding four haploid cells.

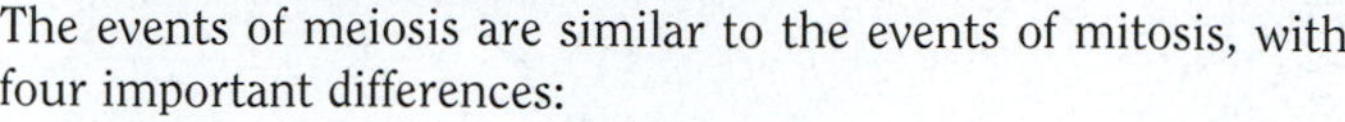

Meiosis produces haploid cells with unique gene combinations

The events of meiosis are similar to the events of mitosis, with four important differences:

1. Meiosis involves two successive nuclear and cytoplasmic divisions, producing up to four cells.
2. Despite two successive nuclear divisions, the DNA and other chromosomal components are duplicated only once, during the interphase preceding the first meiotic division.
3. Each of the four cells produced by meiosis contains the haploid chromosome number, that is, only one set containing only one representative of each homologous pair.
4. During meiosis, the genetic information from both parents is shuffled, so each resulting haploid cell has a virtually unique combination of genes.

Meiosis typically consists of two nuclear and cytoplasmic divisions, designated the *first* and *second meiotic divisions,* or simply **meiosis I** and **meiosis II.** Each includes prophase, metaphase, anaphase, and telophase stages. During meiosis I, the members of each homologous pair of chromosomes first join together and then separate and move into different nuclei. In meiosis II, the sister chromatids that make up each chromosome separate and are distributed to two different nuclei. The following discussion describes meiosis in an animal with a diploid chromosome number of four. Refer to Figures 9–10 and 9–11 as you read.

Prophase I includes synapsis and crossing-over

As in mitosis, the chromosomes are duplicated during the S phase of interphase, before meiosis actually begins. Each duplicated chromosome consists of two chromatids. During *prophase*

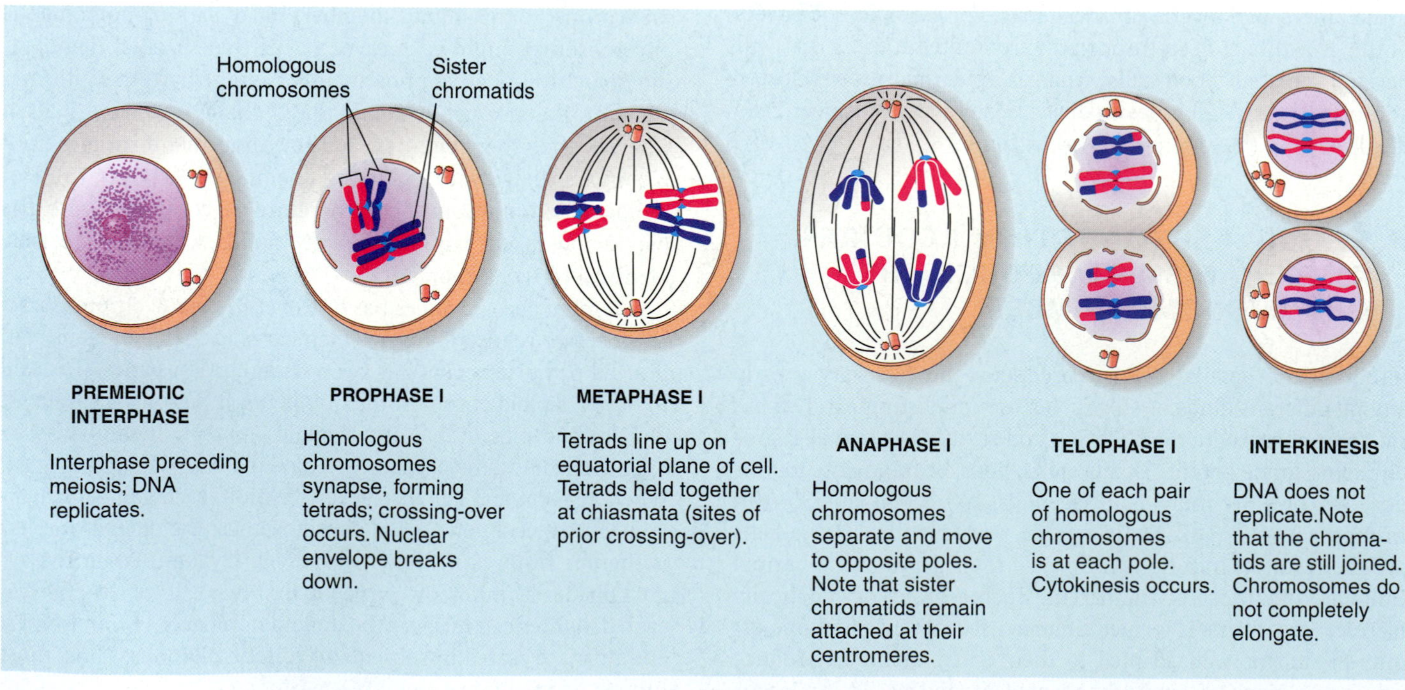

Homologous chromosomes
Sister chromatids
PREMEIOTIC INTERPHASE
Interphase preceding meiosis; DNA replicates.
PROPHASE I
Homologous chromosomes synapse, forming tetrads; crossing-over occurs. Nuclear envelope breaks down.
METAPHASE I
Tetrads line up on equatorial plane of cell. Tetrads held together at chiasmata (sites of prior crossing-over).
ANAPHASE I
Homologous chromosomes separate and move to opposite poles. Note that sister chromatids remain attached at their centromeres.
TELOPHASE I
One of each pair of homologous chromosomes is at each pole. Cytokinesis occurs.
INTERKINESIS
DNA does not replicate.Note that the chromatids are still joined. Chromosomes do not completely elongate.

I, while the chromatids are still elongated and thin, the homologous chromosomes come to lie lengthwise side by side. This process is called **synapsis,** which means "fastening together." In our example, because the diploid number is four, there are two homologous pairs.

It is customary when discussing higher organisms to refer to one member of each homologous pair as a **maternal homologue** because it was originally inherited from the female parent, and to the other as a **paternal homologue** because it was contributed by the male parent during the formation of the zygote. Because each chromosome was duplicated during the premeiotic interphase and now consists of two chromatids, synapsis results in the association of *four* chromatids. The resulting complex is known as a **bivalent** or a **tetrad.** The term bivalent, in which the prefix *bi-* refers to the two homologous chromosomes, is commonly used by cytogeneticists (scientists who study inheritance at the cellular level, particularly through the analysis of chromosomes). The term tetrad (*tetra* means "four") is preferred by some geneticists interested in following the fates of the four chromatids. We will use tetrad in further discussions.

The number of tetrads per prophase I cell is equal to the haploid chromosome number. In our example of an animal cell with a diploid number of four, there are two tetrads; in a human cell at prophase I, there are 23 tetrads (and a total of 92 chromatids).

Homologous chromosomes become closely associated during synapsis. Electron microscopic observations reveal that a characteristic structure, known as the **synaptonemal complex,** forms between the synapsed homologues (Fig. 9–12). This structure holds the synapsed homologues together and is thought to play a role in **crossing-over,** a process in which genetic material is exchanged between homologous (nonsister) chromatids. In crossing-over, enzymes break homologous chromatids and then join them to produce new combinations of genes. The resulting **genetic recombination** greatly enhances the amount of genetic variation among sexually produced offspring. This process is discussed in more detail in Chapter 10.

In many species, prophase I is a lengthy phase during which the cell grows and synthesizes nutrients. This is especially true during the formation of some egg cells because materials need to be made for the benefit of the future embryo. In many types of meiotic cells, the chromosomes assume unusual shapes during this phase. For example, lampbrush chromosomes, found in the female meiotic cells (oocytes) of some amphibians, are composed of hundreds of pairs of loops of chromatin projecting from the chromatid axis. They owe their name to their resemblance to the brushes used to clean old-fashioned oil lamps (Fig. 9–13). The loops are sites of intense synthesis of RNA, which is used to direct the synthesis of specific proteins.

In addition to the unique processes of synapsis and crossing-over, events similar to those seen during mitotic prophase also take place. A spindle composed of microtubules and other components forms. If centrioles are present (as in animal cells), one pair moves to each pole, and astral microtubules are formed. The

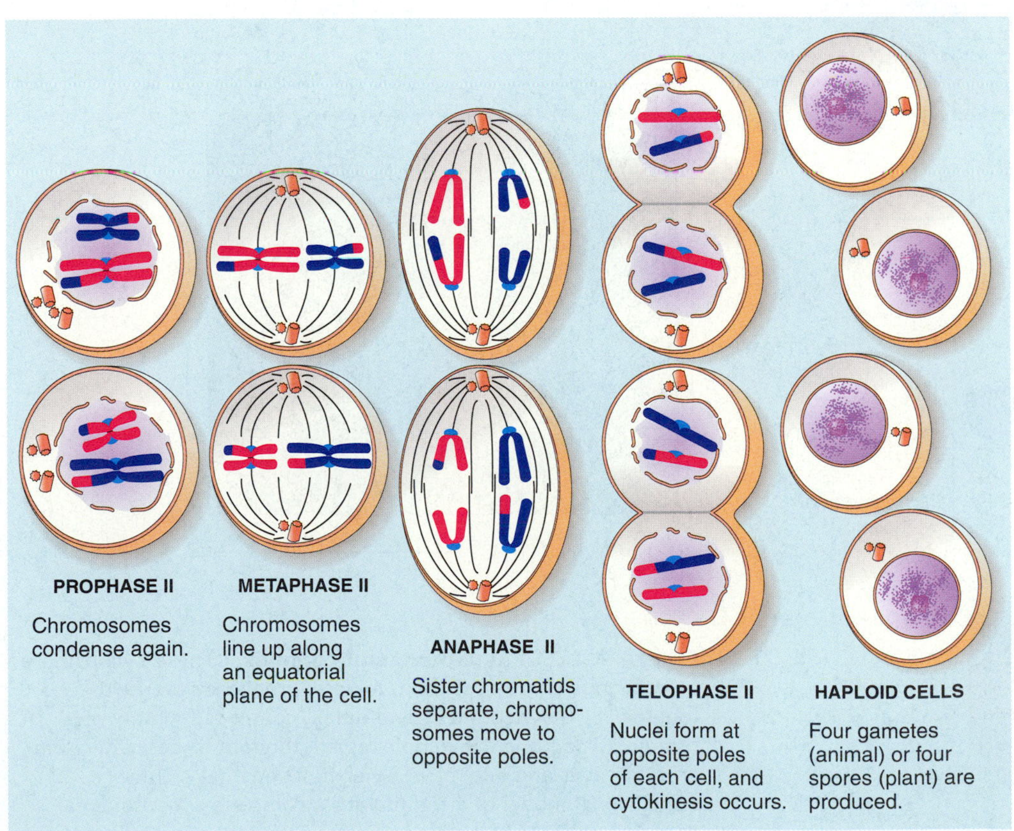
PROPHASE II
Chromosomes condense again.
METAPHASE II
Chromosomes line up along an equatorial plane of the cell.
ANAPHASE II
Sister chromatids separate, chromosomes move to opposite poles.
TELOPHASE II
Nuclei form at opposite poles of each cell, and cytokinesis occurs.
HAPLOID CELLS
Four gametes (animal) or four spores (plant) are produced.

Figure 9–10 Meiosis. Two nuclear divisions, meiosis I and meiosis II, are required. In this illustration, the process begins with a cell that has a diploid chromosome number of four and ends with the formation of four haploid cells with two chromosomes each. Centrioles are shown because the example is of an animal cell. The maternal chromosomes are shown in red; the paternal chromosomes are blue.

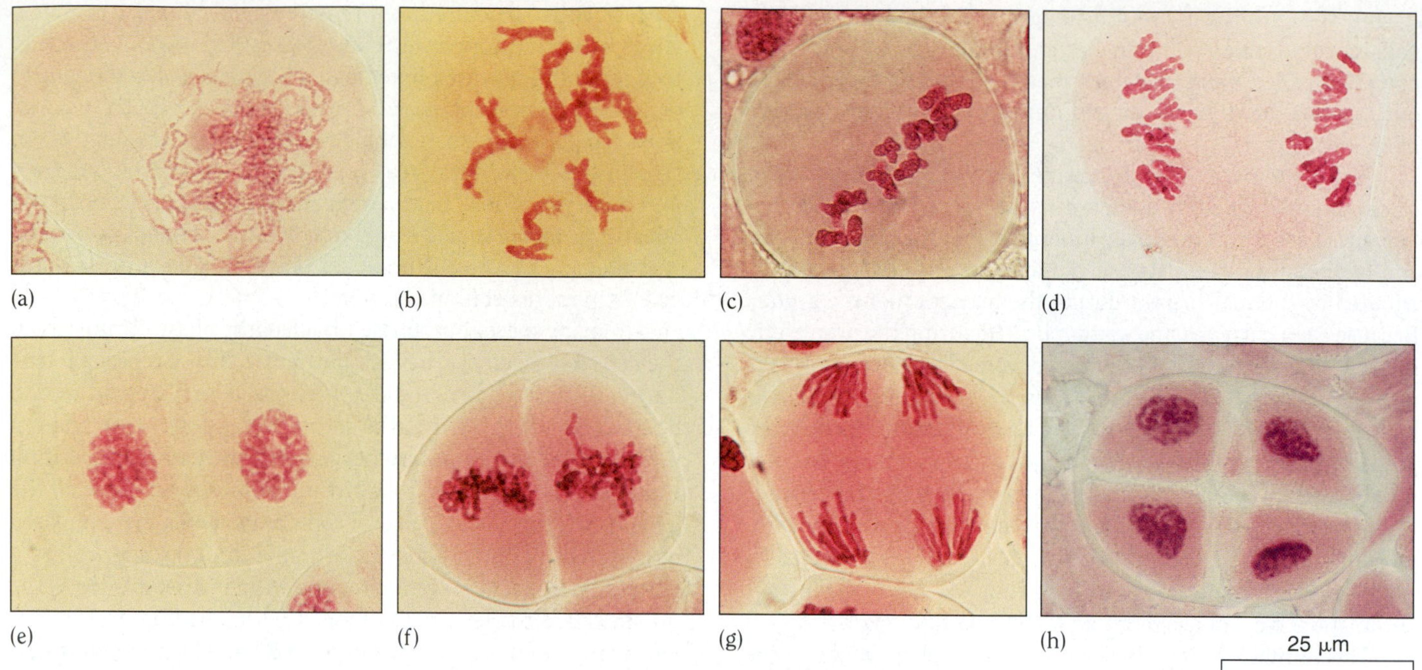

Figure 9–11 Meiosis in the trumpet lily, *Lilium longiflorum*. The chromosomes shown in these LM views have been stained and the cells flattened on microscope slides. **(a)** Mid-prophase I. **(b)** Late prophase I. **(c)** Metaphase I. **(d)** Anaphase I. **(e)** Prophase II. **(f)** Metaphase II. **(g)** Anaphase II. **(h)** Four daughter cells. *(Clare Hasenkampf/Biological Photo Service)*

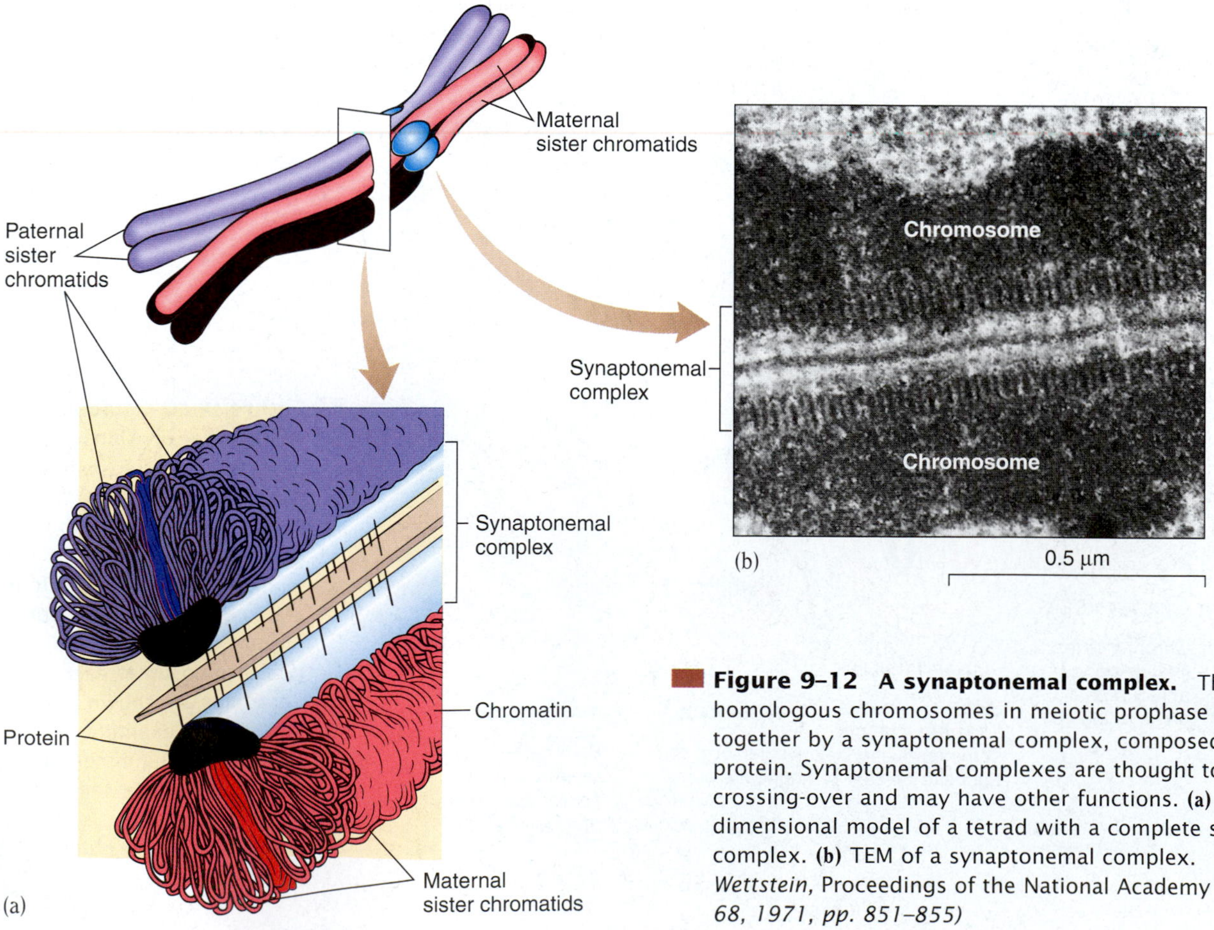

Figure 9–12 A synaptonemal complex. These synapsing homologous chromosomes in meiotic prophase I are held together by a synaptonemal complex, composed mainly of protein. Synaptonemal complexes are thought to be involved in crossing-over and may have other functions. **(a)** A three-dimensional model of a tetrad with a complete synaptonemal complex. **(b)** TEM of a synaptonemal complex. *(b, D. Von Wettstein,* Proceedings of the National Academy of Science, Vol. *68, 1971, pp. 851–855)*

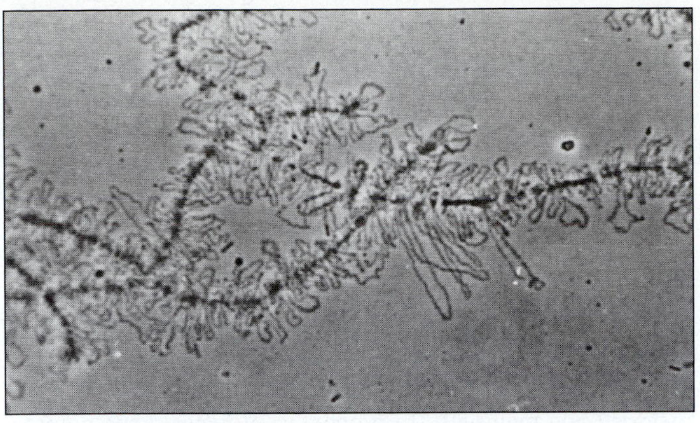

Figure 9–13 Lampbrush chromosomes. This LM shows parts of several tetrads from a female meiotic cell (oocyte) of a newt, *Triturus viridescens.* The loops, composed of chromatin, are sites of intense RNA synthesis. *(Dennis Gould)*

nuclear envelope disappears in late prophase I, and in cells with large and distinct chromosomes the structure of the tetrads can be seen clearly with the microscope (Fig. 9–14). The sister chromatids remain closely aligned along their lengths. However, the centromeres (and kinetochores) of the homologous chromosomes become separated from one another. In late prophase I, the homologous chromosomes are held together only at specialized regions, termed **chiasmata** (sing., *chiasma*). Each chiasma originates at a site of crossing-over, that is, a site at which homologous chromatids were previously broken by enzymes, exchanged genetic material, and rejoined, producing an X-shaped configuration. The genetic consequences of crossing-over will be discussed in Chapter 10.

During meiosis I homologous chromosomes separate

Prophase I ends when the tetrads become aligned on the equatorial plane; the cell is now said to be at **metaphase I.** Both sister kinetochores of one chromosome are attached by spindle fibers to the same pole, and both kinetochores of the homologous chromosome are attached to the opposite pole. (By contrast, in mitosis, sister kinetochores are attached to opposite poles.) During **anaphase I,** the paired homologous chromosomes separate, or disjoin, and move toward opposite poles. Each pole receives a random mixture of maternal and paternal chromosomes, but only one member of each homologous pair is present at each pole. The sister chromatids are united at their centromere regions. Again, this differs from mitotic anaphase, in which the sister chromatids pass to opposite poles.

During **telophase I,** the chromatids generally decondense somewhat, the nuclear envelope may reorganize, and cytokinesis may take place. Each telophase I nucleus contains the haploid number of chromosomes, but each chromosome is a duplicated chromosome (i.e., it consists of a pair of chromatids). In our example, there are two duplicated chromosomes at each pole, for a total of four chromatids; in humans, there are 23 duplicated chromosomes (46 chromatids) at each pole.

An interphase-like stage usually follows. Because it is not a true interphase (i.e., there is no S phase), it is given the name **interkinesis.** Interkinesis is very brief in most organisms and absent in some.

Chromatids separate in meiosis II

Because the chromosomes usually remain partially condensed between divisions, the prophase of the second meiotic division is brief. **Prophase II** is similar to mitotic prophase in many respects. There is no pairing of homologous chromosomes (indeed, only one member of each pair is present in each nucleus) and no crossing-over.

During **metaphase II,** the chromosomes line up on the equatorial planes of their cells. The first and second metaphases can be easily distinguished in diagrams; at metaphase I the chromatids are arranged in bundles of four (tetrads), and at metaphase II they are in groups of two (as in mitotic metaphase). This is not always so obvious in living cells.

During **anaphase II** the chromatids, attached to spindle fibers at their kinetochores, separate and move to opposite poles,

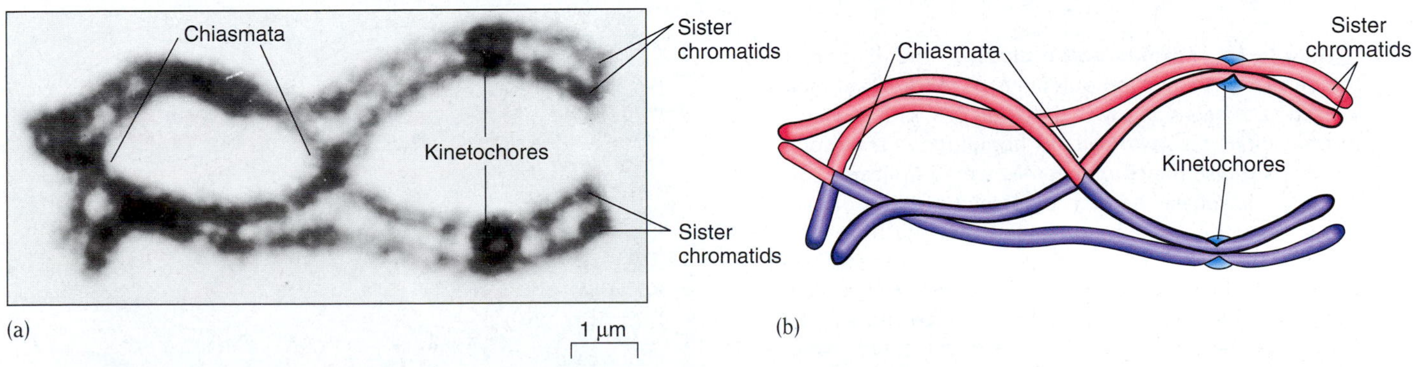

Figure 9–14 A meiotic tetrad. **(a)** This LM is of a tetrad during late prophase I of a male meiotic cell (spermatocyte) from a salamander. **(b)** Interpretive drawing indicating the structure of the tetrad. The paternal chromatids are purple, and the maternal chromatids are pink. *(a, Courtesy of J. Kezer)*

just as they would at mitotic anaphase. As in mitosis, each former chromatid is now referred to as a chromosome. Thus, at **telophase II** there is one representative for each homologous pair at each pole. Each is an unduplicated (i.e., single) chromosome. Nuclear envelopes then re-form, the chromosomes gradually elongate to form chromatin threads, and cytokinesis occurs.

The two successive divisions yield four haploid nuclei, each containing *one* of each kind of chromosome. Each resulting haploid cell has a different combination of genes. This genetic variation has two sources: (1) During meiosis the maternal and paternal chromosomes are "shuffled" so that one member of each pair becomes randomly distributed to the poles at anaphase I; (2) DNA segments are exchanged between maternal and paternal homologues during crossing-over.

The important genetic consequences of these events are discussed in more detail in Chapter 10.

THE EVENTS OF MITOSIS AND MEIOSIS LEAD TO CONTRASTING OUTCOMES

Although mitosis and meiosis share many similar features, specific distinctions between these processes result in the formation of different types of cells (Fig. 9–15).

Mitosis is a single division in which *sister chromatids* separate from each other. These are distributed to the two daughter cells, which are genetically identical to each other and to the original cell. Homologous chromosomes do not associate physically at any time in mitosis.

In meiosis, a diploid cell undergoes two successive divisions, meiosis I and meiosis II. In prophase I of meiosis, the homologous chromosomes undergo synapsis to form tetrads. If we ignore crossing-over, we can say that *homologous chromosomes* separate during meiosis I, and *sister chromatids* separate during

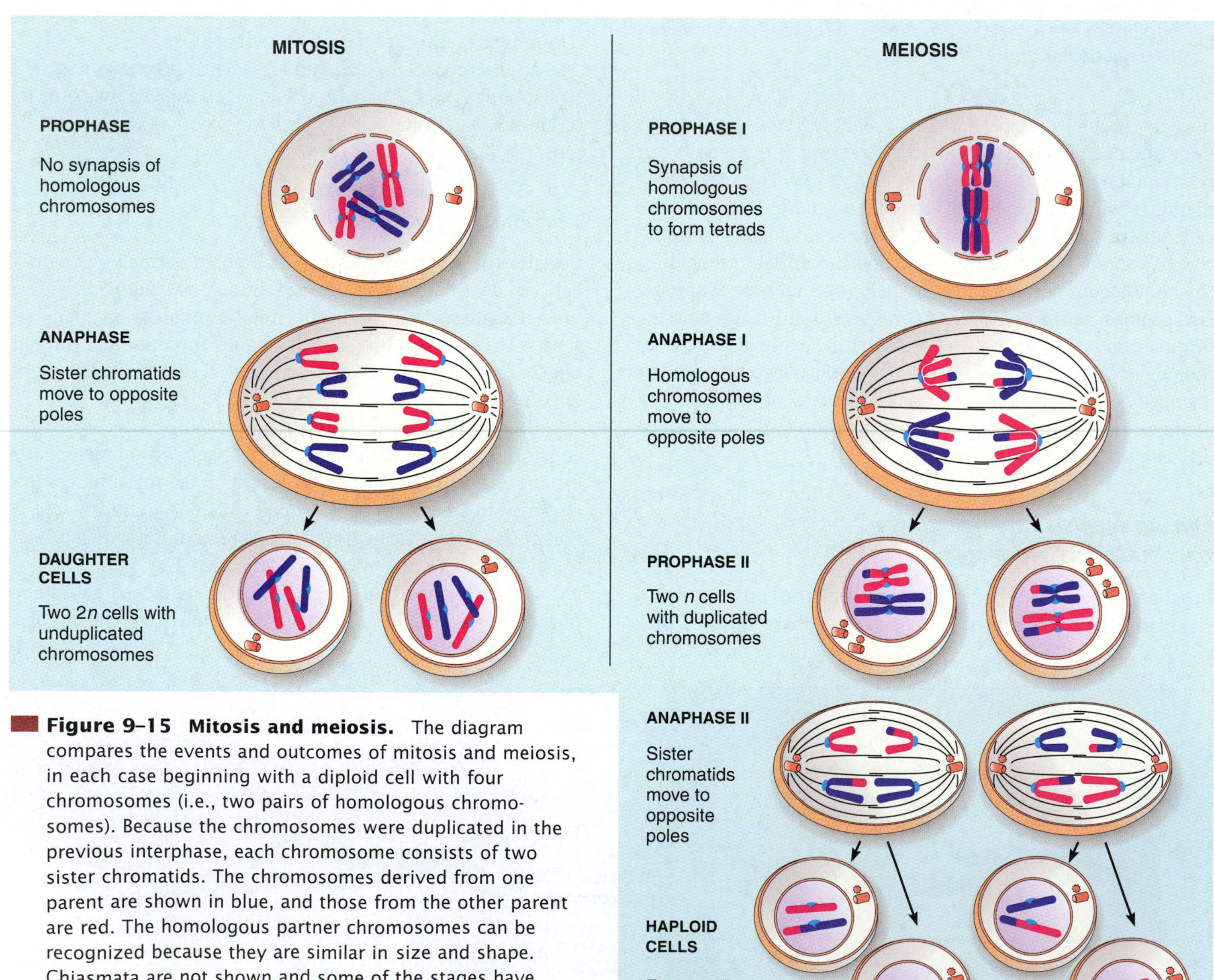

Figure 9–15 Mitosis and meiosis. The diagram compares the events and outcomes of mitosis and meiosis, in each case beginning with a diploid cell with four chromosomes (i.e., two pairs of homologous chromosomes). Because the chromosomes were duplicated in the previous interphase, each chromosome consists of two sister chromatids. The chromosomes derived from one parent are shown in blue, and those from the other parent are red. The homologous partner chromosomes can be recognized because they are similar in size and shape. Chiasmata are not shown and some of the stages have been omitted for simplicity.

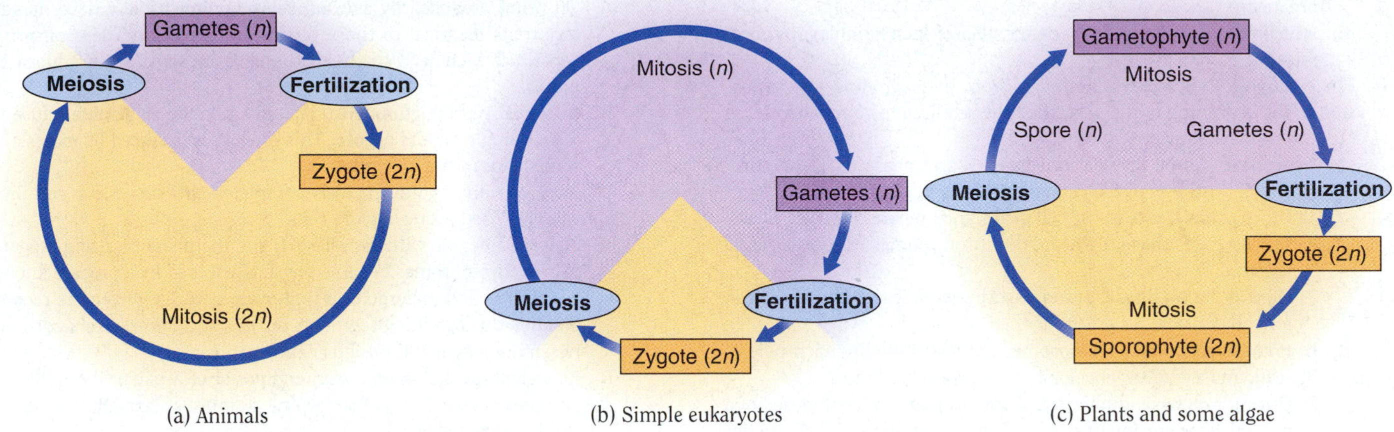

(a) Animals (b) Simple eukaryotes (c) Plants and some algae

Figure 9–16 Representative life cycles.

meiosis II. It is also correct to say that homologous centromeres (or kinetochores) separate during meiosis I, and sister centromeres (or kinetochores) disjoin during meiosis II. Meiosis ends with the formation of four, genetically different, haploid daughter cells. The fates of these cells depend on the type of life cycle; in animals they differentiate as gametes, whereas in plants they become spores.

THE POSITION OF MEIOSIS IN THE LIFE CYCLE VARIES AMONG SPECIES

Because sexual reproduction is characterized by the fusion of two haploid sex cells to form a diploid zygote, it follows that, in a sexual life cycle, meiosis must occur before gametes can be produced.

In animals and a few other organisms, meiosis leads directly to gamete formation (Fig. 9–16*a*). The **somatic** (body) cells of an individual organism multiply by mitosis and are diploid; the only haploid cells produced are the gametes. These are formed when certain cells, known as **germ line** cells, undergo meiosis. The formation of gametes is known as **gametogenesis.** Male gametogenesis, termed **spermatogenesis,** results in the formation of four haploid sperm cells for each cell that enters meiosis.

In contrast, female gametogenesis, termed **oogenesis,** results in the formation of a single egg cell, or **ovum,** for every cell that enters meiosis. This is accomplished by a process that apportions virtually all of the cytoplasm to only one of the two nuclei at each of the meiotic divisions. At the end of the first meiotic division, one nucleus is retained and the other, called the first **polar body,** is excluded from the cell and ultimately degenerates. Similarly, at the end of the second division, one nucleus becomes the second polar body and the other nucleus survives. In this way, one haploid nucleus becomes the recipient of most of the accumulated cytoplasm and nutrients from the original meiotic cell. (See Fig. 48–13 for a more detailed description.)

Although meiosis occurs at some point in a sexual life cycle, it does not always *immediately* precede gamete formation. Many simple eukaryotes (including some fungi and algae) remain haploid (their cells dividing mitotically) throughout most of their lives, with individuals being unicellular or multicellular. Two haploid gametes (produced by mitosis) fuse to produce a diploid zygote that undergoes meiosis to restore the haploid state (Fig. 9–16*b*). Examples of these types of life cycles can be found in Figures 24–15, 24–21, and 25–7.

The most complex life cycles are displayed by plants and some algae (Fig. 9–16*c*). These life cycles, characterized by an **alternation of generations,** consist of a multicellular diploid stage, termed the **sporophyte generation,** and a multicellular haploid stage, termed the **gametophyte generation.** Diploid sporophyte cells undergo meiosis to form haploid spores, each of which then divides mitotically to produce a multicellular haploid gametophyte. Gametophytes produce gametes by mitosis. The female and male gametes (eggs and sperm cells) then fuse to form a diploid zygote that divides mitotically to produce a multicellular, diploid sporophyte.

In higher plants, including flowering plants, the diploid sporophyte—which includes the roots, stems, and leaves of the plant body—is the dominant form. The gametophytes are small and inconspicuous. For example, a microscopic pollen grain contains a haploid male gametophyte that forms haploid sperm by mitosis. More detailed descriptions of alternation of generations can be found in Chapters 26 and 27.

SUMMARY WITH KEY TERMS

I. In the production of a new generation, cells transfer genetic information from parent to offspring.

II. **Genes** are made of DNA, which in eukaryotes is complexed with protein to form the **chromatin** fibers that make up **chromosomes.**

A. A **diploid** organism of a given species has a characteristic number of chromosome pairs per cell.

B. The two members of each chromosome pair, called **homologous chromosomes,** are similar in length, shape, and other structural features and carry genes affecting the same kinds of attributes of the organism.

C. A **haploid** cell contains only one member of each homologous chromosome pair.

III. The eukaryotic **cell cycle** is the period from the beginning of one division to the beginning of the next; the time required to complete one cycle is the **generation time.**

A. **Interphase** can be divided into the first gap phase (G_1), the chromosomal synthesis phase (S), and the second gap phase (G_2).

1. During the **G_1 phase,** the cell grows and prepares for the S phase.
2. During the **S phase,** DNA and the chromosomal proteins are synthesized.
3. During the **G_2 phase,** protein synthesis increases in preparation for cell division.

B. In **mitosis,** identical chromosomes are distributed to each pole of the cell, and a nuclear envelope forms around each set.

1. During **prophase,** duplicated chromosomes, each composed of a pair of **sister chromatids** associated with each other in the vicinity of their **centromeres,** become visible with the microscope. The nucleolus disappears, the nuclear envelope breaks down, and the **mitotic spindle** begins to form.
2. During **metaphase,** the chromosomes are aligned on the equatorial plane of the cell; the mitotic spindle is complete and the **kinetochores** of the sister chromatids are attached by microtubules to opposite poles of the cell.
3. During **anaphase,** the sister chromatids become separated and move to opposite poles. Each former chromatid is now referred to as a chromosome.
4. During **telophase,** a nuclear envelope re-forms around each set of chromosomes, nucleoli become apparent, the chromosomes uncoil, and the spindle disappears.

C. During **cytokinesis,** which generally begins in telophase and therefore overlaps mitosis, the cytoplasm divides to form two individual cells.

1. In animal cells, a ring of microfilaments contracts, producing a **cleavage furrow** that divides the cytoplasm.
2. In plant cells, the **cell plate** provides materials for new plasma membranes and cell walls.

IV. There are two major forms of reproduction: asexual and sexual.

A. Offspring produced by **asexual reproduction** usually have hereditary traits identical to those of the single parent. These offspring constitute a **clone.** Usually all the cells involved are produced by mitosis.

B. In **sexual reproduction,** two haploid sex cells, or **gametes,** fuse to form a single diploid **zygote.** This process is balanced by **meiosis** at some point in the life cycle.

V. A diploid cell undergoing meiosis completes two successive cell divisions to give rise to four haploid cells.

A. **Meiosis I** begins with **prophase I;** the members of a homologous pair of chromosomes become physically joined by a process known as **synapsis** and undergo **crossing-over,** a process of **genetic recombination** during which segments of DNA strands are exchanged between homologous (nonsister) chromatids.

B. At **metaphase I, tetrads,** each composed of a pair of homologous chromosomes held together by one or more **chiasmata,** line up on the equatorial plane.

C. The members of each pair of homologous chromosomes separate during meiotic **anaphase I** and are distributed to different nuclei. Each nucleus contains the haploid number of chromosomes; each chromosome consists of two chromatids.

D. During **meiosis II,** the two chromatids of each chromosome separate and one is distributed to each daughter cell. Each former chromatid is now referred to as a chromosome.

VI. The position of meiosis in the life cycle varies.

A. The **somatic** (body) cells of animals are diploid; the only haploid cells are the gametes (produced by **gametogenesis,** which in animals includes meiosis).

B. Simple eukaryotes may be regularly haploid; the only diploid stage is the zygote, which undergoes meiosis to restore the haploid state.

C. Plants and some algae have **alternation of generations.** A multicellular diploid **sporophyte** forms haploid spores by meiosis. Each spore divides mitotically to form a multicellular haploid **gametophyte,** which produces gametes by mitosis. Two haploid gametes then fuse to form a diploid zygote, which divides mitotically to produce a new diploid sporophyte.

POST-TEST

1. Chromatin fibers include (a) DNA and structural polysaccharides (b) RNA and phospholipids (c) protein and carbohydrate (d) DNA and protein (e) triacylglycerol and steroids
2. The term *S phase* refers to (a) DNA synthesis during interphase (b) synthesis of chromosomal proteins during prophase (c) active RNA synthesis in lampbrush chromosomes (d) synapsis of homologous chromosomes (e) fusion of gametes in sexual reproduction
3. At which of the following stages do human skin cell nuclei have the same DNA content? (a) early mitotic prophase; mitotic telophase (b) G_1; G_2 (c) G_1; early mitotic prophase (d) G_1; mitotic telophase (e) G_2; mitotic telophase
4. In a cell at ____________, each chromosome consists of a pair of attached chromatids. (a) mitotic prophase (b) meiotic prophase II (c) meiotic prophase I (d) meiotic anaphase I (e) all of the above
5. In an animal cell at mitotic metaphase, you would expect to find (a) two pairs of centrioles located on the metaphase plate (b) a pair of centrioles inside the nucleus (c) a pair of centrioles within each microtubule organizing center (d) a centriole within each centromere (e) no centrioles
6. Cell plate formation usually begins during (a) telophase in a plant cell (b) telophase in an animal cell (c) G_2 in a plant cell (d) G_2 in an animal cell (e) a and b are correct
7. The life cycle of a sexually reproducing organism includes (a) mitosis (b) meiosis (c) fusion of sex cells (d) b and c (e) a, b, and c
8. Which of the following are genetically identical? (a) two cells resulting from meiosis I (b) two cells resulting from meiosis II (c) four cells resulting from meiosis I followed by meiosis II (d) two cells resulting from a mitotic division (e) all of the above
9. You would expect to find a synaptonemal complex in a cell at (a) mitotic prophase (b) meiotic prophase I (c) meiotic prophase II (d) meiotic anaphase I (e) meiotic anaphase II
10. A particular plant species has a diploid chromosome number of 20. A haploid cell of that species at mitotic prophase contains a total of ____________ chromosomes and ____________ chromatids. (a) 20; 20 (b) 20; 40 (c) 10; 10 (d) 10; 20 (e) none of the above because haploid cells cannot undergo mitosis
11. A diploid nucleus at early mitotic prophase has ____________ set(s) of chromosomes; a diploid nucleus at mitotic telophase has ____________ set(s) of chromosomes. (a) 1; 1 (b) 1; 2 (c) 2; 2 (d) 2; 1 (e) not enough information has been given
12. A chiasma links a pair of (a) homologous chromosomes at meiotic metaphase II (b) homologous chromosomes at meiotic metaphase I (c) sister chromatids at meiotic metaphase II (d) sister chromatids at mitotic metaphase (e) sister chromatids at meiotic metaphase I

REVIEW QUESTIONS

1. Two species may have the same chromosome number and yet have very different attributes. Explain.
2. Sketch a duplicated chromosome and label the sister chromatids, the centromeres, and the kinetochores. What are the functions of centromeres and of kinetochores?
3. How does the DNA content of the cell change from the beginning of interphase to the end of interphase? Does the number of chromatids change? Does the number of chromosomes change?
4. Are homologous chromosomes present in a diploid cell? Are they present in a haploid cell?
5. How does meiosis differ from mitosis? Are there any points of similarity between these two processes? Explain.
6. What kinds of life cycles include a multicellular haploid stage? Can haploid cells divide by mitosis? By meiosis?
7. Assume that an animal has a diploid chromosome number of ten. (a) How many chromosomes would it have in a typical body cell, such as a skin cell? (b) How many chromosomes would be present in a cell at mitotic prophase? How many chromatids? (c) How many chromosomes would be present in each daughter cell produced by mitosis? Are these duplicated chromosomes? (d) How many tetrads would form in prophase I of meiosis? (e) How many chromosomes would be present in each gamete? Are these duplicated chromosomes?

YOU MAKE THE CONNECTION

1. Decide whether each of the following is an example of sexual or asexual reproduction and state why. (a) A diploid queen honeybee produces haploid eggs by meiosis. Some of these eggs are never fertilized and develop into haploid male honeybees (drones). (b) Haploid male honeybees produce haploid sperm by mitosis. These sperm fertilize haploid eggs produced by the queen, resulting in the development of diploid female worker bees. (c) Seeds develop after a flower has been pollinated with pollen from a different plant of the same species. (d) Seeds develop after a flower has been pollinated with pollen from the same plant. (e) A cutting from a plant develops roots after it has been placed in water. The plant survives and grows after it is transplanted to soil.

RECOMMENDED READINGS

Haber, J.E. "Searching for a Partner." *Science,* Vol. 279, 6 Feb. 1998. A discussion of experimental evidence that the order of events in meiotic prophase is not the same in all organisms.

Lodish, H., A. Berk, S.L. Zipursky, P. Matsudaira, D. Baltimore, and J. Darnell. *Molecular Cell Biology,* 4th ed. W.H. Freeman and Co., New York, 2000. An extensive, detailed, and well-written discussion of cell growth and division, covering the control of cell division, the cell cycle, and the events of mitosis and meiosis.

Nasmyth, K. "Viewpoint: Putting the Cell Cycle in Order." *Science,* Vol. 274, 6 Dec. 1996. This overview introduces a series of articles in the same issue that explore various aspects of the cell cycle.

Nicklas, R.B. "How Cells Get the Right Chromosomes." *Science,* Vol. 275, 31 Jan. 1997. A pioneer in the micromanipulation of chromosomes in living cells reviews the mechanisms that ensure proper distribution of chromosomes in mitosis and meiosis.

Sharp, D.J., G.C. Rogers, and J.M. Scholey. "Microtubule Motors in Mitosis." *Nature,* Vol. 407, 7 Sept. 2000. A review of the many types of motor proteins that interact with microtubules during mitosis.

- Visit our Web site at **http://www.info.brookscole.com/solomonbergmartin** for links to chapter-related resources on the World Wide Web. Additional on-line materials relating to this chapter can also be found on our Web site.

See chapter activity on BioActive Learner CD for additional help in mastering the chapter's material. Icon location in the chapter's margins shows which topics have tutorials or simulations in the CD.

10

The Basic Principles of Heredity

Gregor Mendel. A monk who bred pea plants, Mendel is depicted here in his monastery garden at Brünn, Austria (now Brüno, Czech Republic). *(The Bettmann Archive)*

LEARNING OBJECTIVES

After you have studied this chapter you should be able to

1. Define and use correctly the terms *allele, locus, genotype, phenotype, dominant, recessive, homozygous, heterozygous,* and *test cross.*
2. Apply Mendel's principles to solve genetics problems involving monohybrid and dihybrid crosses.
3. Apply the product rule and sum rule appropriately when predicting the outcomes of genetic crosses.
4. Solve genetics problems involving incomplete dominance, epistasis, polygenes, multiple alleles, and X-linked inheritance.
5. Explain some of the ways in which genes may interact to affect the phenotype; discuss how it is possible for a single gene to affect many features of the organism simultaneously.
6. Analyze data from a test cross involving alleles of two loci. Show how such data can be used to distinguish between independent assortment and linkage. Relate independent assortment and linkage to specific events in meiosis.
7. Discuss the genetic determination of sex and the role of the Y chromosome in determining male sex in humans; contrast the mechanism of sex determination in humans and other mammals with that in various other animals and some plants; compare dosage compensation of X-linked genes in mammals and fruit flies.
8. Assess the effects of inbreeding versus outbreeding on a population; discuss the genetic basis of hybrid vigor.

Heredity, the transmission of genetic information from parent to offspring, is generally a very regular process that follows predictable patterns. The basic rules of inheritance in eukaryotes were first discovered by Gregor Mendel (1822–1884), a monk who bred pea plants. Mendel was the first scientist to effectively apply quantitative methods to the study of inheritance. He did not merely describe his observations; he planned his experiments carefully, recorded the data, and subjected the results to mathematical analysis. Although his work was unappreciated in his lifetime, it was rediscovered in 1900. His major findings, including those now known as Mendel's principles of segregation and independent assortment, became the foundation of the science of **genetics.**

During the decades following the rediscovery of Mendel's findings, geneticists initially extended Mendel's principles by correlating the transmission of genetic information from generation to generation with the behavior of chromosomes during meiosis. They also refined his methods and, through their studies on a variety of organisms, both verified Mendel's findings and added to a growing list of so-called exceptions to his principles. These include such phenomena as linkage, sex linkage, and polygenic inheritance.

Some geneticists were very active in the development of the emerging science of statistical analysis (which had been in its infancy in Mendel's time), thereby providing scientists with increasingly sophisticated ways to analyze and interpret experimental data. These statistical methods were also essential to the study of the genetic makeup of natural populations of organisms. The results of these investigations on the genetics of populations were combined with Charles Darwin's theory of evolution by natural selection to develop a unified modern theory of evolution, firmly based on genetic principles (see Chapters 17 and 18).

Geneticists study not only the transmission of genes but also the expression of genetic information. As you will see in this and succeeding chapters, our understanding of the relationship between an organism's genes and its characteristics has become increasingly sophisticated as we have learned more about the flow of information in cells.

MENDEL FIRST DEMONSTRATED THE PRINCIPLES OF INHERITANCE

Process of Science

Gregor Mendel was not the first plant breeder; at the time he began his work, **hybrid** plants and animals (offspring of two genetically dissimilar parents) had been known for a long time. When Mendel began his breeding experiments in 1856, two main facts about inheritance were widely recognized: (1) All hybrid plants that are offspring of the same kinds of parents are similar in appearance. (2) When these hybrids are mated to each other they do not breed true; their offspring show a mixture of traits. Some look like their parents, and some have features like their grandparents.

Mendel's genius lay in his ability to recognize a pattern in the way the parental traits reappear in the offspring of hybrids. No one before had categorized and counted the offspring and analyzed these regular patterns over several generations.

Just as geneticists do today, Mendel chose the organism for his experiments very carefully. The garden pea, *Pisum sativum,* had several advantages. Pea plants are easy to grow, and many varieties were available through commercial sources. It is impossible to study inheritance without such genetic **variation.** (If every person in the population had blue eyes, it would be impossible to study the inheritance of eye color.) Another advantage of pea plants is that controlled pollinations are relatively easy to conduct. Pea flowers (Fig. 10–1) have both male and female parts, and are naturally self-pollinated. However, the anthers (the male parts of the flower that produce pollen) can be removed to prevent self-fertilization. Pollen from a different source can then be applied to the stigma (receptive surface of the female part). Pea flowers are easily protected from other sources of pollen because the reproductive structures are completely enclosed by the petals. Although Mendel did not mention having done this, plant hybridizers usually cover the flowers with small bags to provide additional protection from pollinating insects.

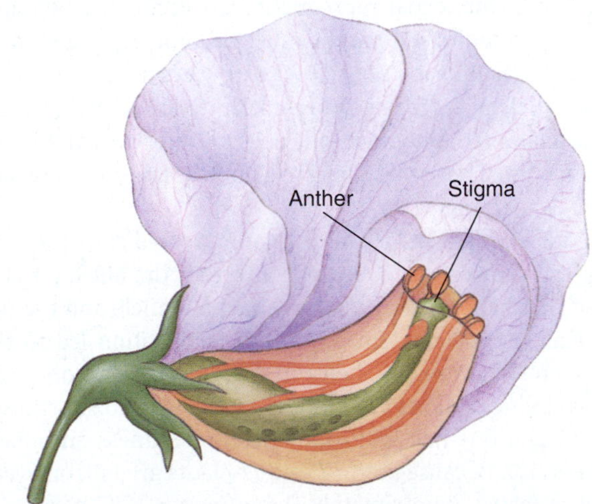

Figure 10–1 Reproductive structures of a pea flower. This cutaway view shows the pollen-producing anthers and the stigma, that portion of the female part of the flower that receives the pollen.

Although his original pea seeds were obtained from commercial sources, Mendel did some important preliminary work before he started his actual experiments. For several years he worked to develop genetically pure, or **true-breeding,** lines for various inherited features. Today we use the term **phenotype** to refer to the physical appearance of an organism. A true-breeding line produces only offspring expressing the same phenotype (e.g., round seeds or tall plants), generation after generation. During this time he apparently chose those characteristics of his pea strains that could be studied most easily and probably discarded or ignored others. He probably made the initial observations that would later form the basis of his theories.

Mendel eventually chose strains representing seven clearly contrasting pairs of phenotypes: yellow versus green seeds, round versus wrinkled seeds, green versus yellow pods, tall versus short plants, inflated versus constricted pods, white seed coats versus gray seed coats, and flowers borne on the ends of the stems versus flowers appearing all along the stems. Other plant breeders typically studied hybrids between parents that differed in many, often not clearly defined, ways. Mendel's results were much easier to analyze because he chose easily distinguishable phenotypes and limited the genetic variation studied in each experiment.

Mendel began his experiments by crossing plants from two different true-breeding lines with contrasting phenotypes; these genetically pure individuals constituted the **parental,** or **P, generation.** In every case, the members of the first generation of offspring all looked alike and resembled one of the two parents. For example, when he crossed tall plants with short plants, all the progeny were tall (Fig. 10–2). These offspring were the first filial (*filial* comes from Latin for "sons and daughters") generation, or **F_1 generation.** The second filial generation, or **F_2 generation,** was produced by a cross between F_1 individuals, or by self-pollination of F_1 individuals. Mendel's F_2 generation in this experiment included 787 tall plants and 277 short plants.

Most breeders of Mendel's time thought that inheritance was controlled by fluids that blended together when hybrids were formed. One implication of this idea is that a hybrid should be intermediate between the two parents and, in fact, plant breeders had obtained such hybrids. Although Mendel observed some types of hybrids that were intermediate, he chose for further study those F_1 hybrids in which hereditary factors from one of the parents apparently masked expression of the hereditary

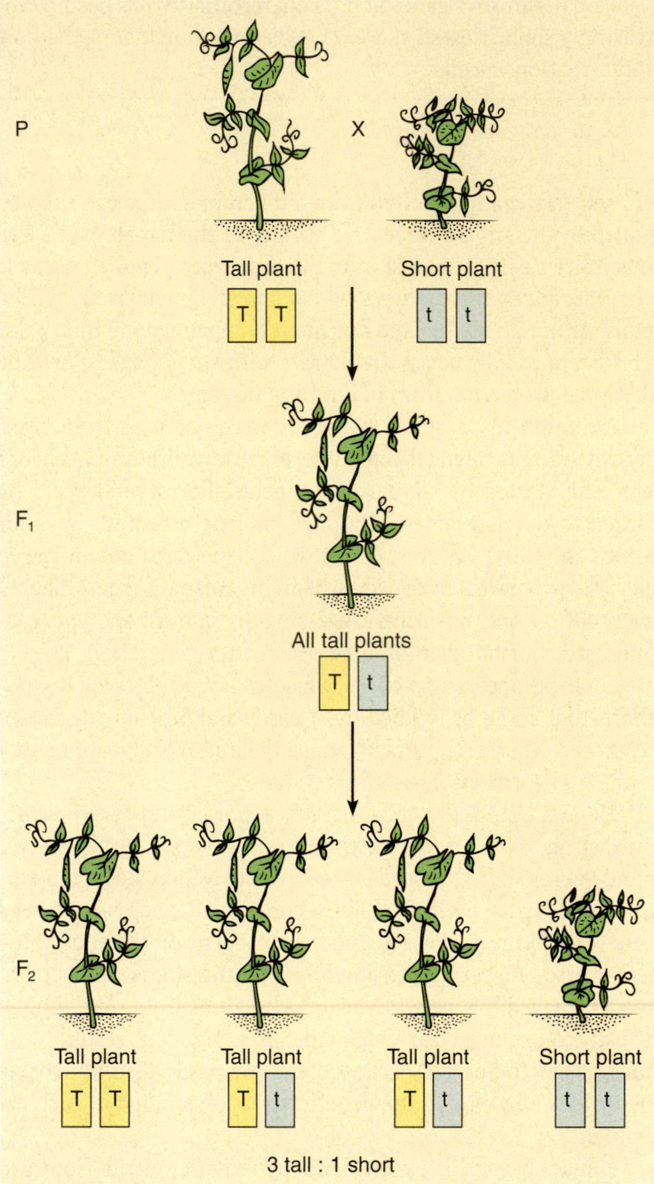

Figure 10–2 One of Gregor Mendel's many pea plant crosses. Crossing a true-breeding tall pea plant with a true-breeding short pea plant yielded only tall offspring in the F_1 generation. However, when these F_1 individuals self-pollinated, or when two F_1 individuals were crossed, the resulting F_2 generation included tall and short plants in a ratio of about 3:1.

factors from the other parent. These types of hybrids had also been observed by other breeders, but they explained them merely as cases in which the "fluids" from one parent were stronger than those from the other parent. Using modern terminology, the factor expressed in the F_1 generation (tallness in our example) is said to be **dominant;** the one hidden (shortness) is said to be **recessive.** Dominant traits mask recessive ones when both are present in the same individual. Although we know today that dominance is not always observed (exceptions are considered later in this chapter), the fact that dominance can occur was not entirely consistent with the notion of blending inheritance.

Mendel's results also argued against blending inheritance in a more compelling way. Once two fluids have blended, it is very difficult to imagine how they can be separated. However, in the example just discussed, in the F_1 generation the hereditary factor(s) that controlled shortness clearly were not lost or blended inseparably with the hereditary factor(s) that controlled tallness because shortness reappeared in the F_2 generation. Mendel was very comfortable with the theoretical side of biology because he was also a student of physics and mathematics. He therefore proposed that each kind of inherited feature of an organism is controlled by two factors that behave like particles and are present in every individual. To Mendel these "hereditary factors" were abstractions, because he knew nothing of chromosomes and DNA. They are essentially what we call **genes** today, so we will use that term in our discussion. Today we know that genes are not particles, but treating them as such allowed Mendel to develop precise mathematical models, testable by experiment, to predict the patterns by which genes are transmitted from generation to generation.

Mendel's experiments led to his discovery and explanation of the major principles of heredity, which we now know as the principles of segregation and independent assortment. We consider the first now and the second later in the chapter.

The principle of segregation states that alleles separate before gametes are formed

Today we use the term **alleles** to refer to the alternative forms of a gene. In the example in Figure 10–2, each F_1-generation tall plant had two different alleles that control plant height: one for tallness (which we designate *T*) and one for shortness (which we designate *t*), but because the tall gene was dominant, these plants were tall. To explain his experimental results, Mendel proposed an idea that we now refer to as the principle of segregation. Using modern terminology, the **principle of segregation** states that, in order for sexual reproduction to occur, the two alleles carried by an individual parent must become separated (segregated). As a result, each sex cell (egg or sperm) formed contains only one allele of each pair. An essential feature of the process is that the alleles remain intact (one does not mix with or eliminate the other); thus, recessive alleles are not lost and so can reappear in the F_2 generation.

In our example, before the F_1 plants formed gametes, the allele for tallness separated (segregated) from the allele for shortness, so that half the gametes contained a *T* allele and the other half a *t* allele. The random process of fertilization led to three possible combinations of alleles in the F_2 offspring: one-fourth with two tallness alleles *(TT),* one-fourth with two shortness alleles *(tt),* and one-half with one allele for tallness and one for shortness *(Tt).* Because both *TT* and *Tt* plants are tall, on average Mendel expected approximately three-fourths (787 of the 1064 plants he obtained) to express the phenotype of the dominant allele (tall) and about one-fourth (277/1064) the phenotype of the recessive allele (short). (The mathematical reasoning behind these predictions will be explained shortly.)

Process of Science Today we know that segregation of alleles is a direct result of the separation of homologous chromosomes during meiosis (Fig. 10–3). (Recall from Chapter 9 that in all sexual life cycles, meiosis must occur at some point prior to gamete formation.) Later, at the time of fertilization, each haploid gamete contributes one chromosome from each homologous pair and therefore one gene for each gene pair (either *T* or *t* in our example). Although gametes and fertilization were known at the time Mendel carried out his research, mitosis and meiosis had not yet been discovered. It is truly remarkable that Mendel was able to formulate his ideas mainly on the basis of mathematical abstractions. Today his principles are much easier to understand because we are able to think about them in concrete terms by relating the transmission of genes to the behavior of chromosomes.

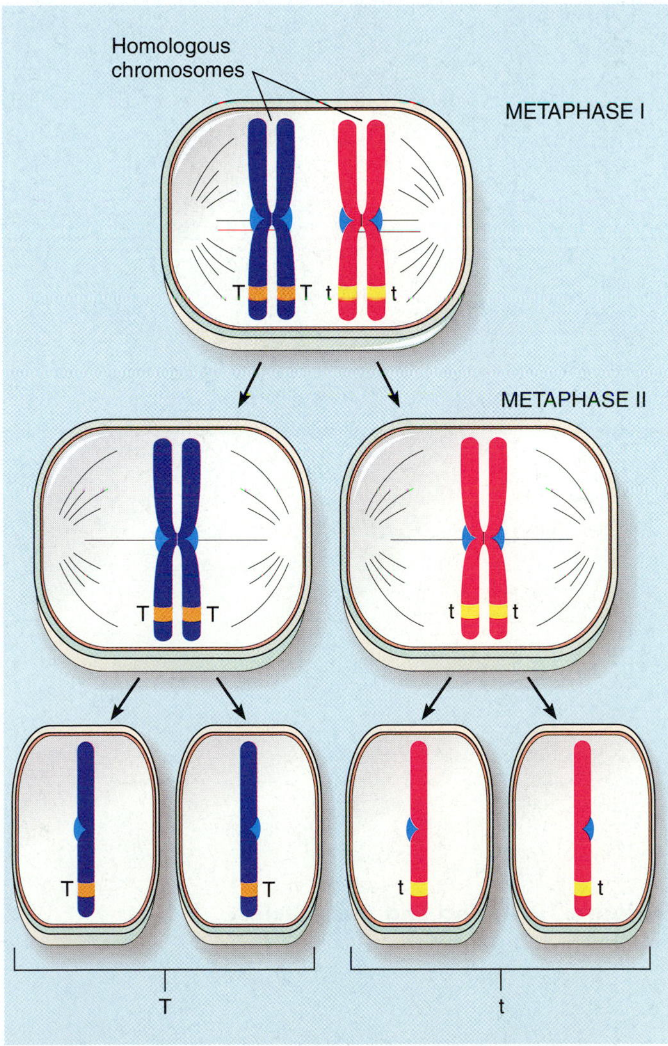

Figure 10–3 The chromosomal basis for segregation. The separation of homologous chromosomes during meiosis results in the segregation of alleles in a heterozygote. Note that half of the gametes will carry *T* and half will carry *t*. See also Figure 9–10.

Mendel reported these and other findings (discussed later in this chapter) at a meeting of the Brünn Society for the Study of Natural Science; he published his results in the transactions of that society in 1866. At that time biology was largely a descriptive science, and biologists had little interest in applying quantitative and experimental methods such as Mendel had used. The importance of his results and his interpretations of those results were not appreciated by other biologists of the time, and his findings were neglected for nearly 35 years.

In 1900 Hugo DeVries in Holland, Karl Correns in Germany, and Erich von Tschermak in Austria each rediscovered Mendel's paper and found that it provided explanations for their own research findings. They gave credit to Mendel by naming the basic laws of inheritance after him. By this time biologists had a much greater appreciation of the value of quantitative experimental methods. The details of mitosis, meiosis, and fertilization had been described, and in 1903 W.S. Sutton pointed out the connection between Mendel's segregation of genes and the separation of homologous chromosomes during meiosis. The time was right for wider acceptance and extension of these ideas and their implications.

Alleles occupy corresponding loci on homologous chromosomes

Today we know that each chromatid is made up of one long linear DNA molecule and that each gene is actually a segment of that DNA molecule. We also know that homologous chromosomes are not only similar in size and shape but usually have similar genes located in corresponding positions. The term **locus**[1] (pl., *loci*) was originally coined to designate the location of a particular gene on the chromosome (Fig. 10–4). Of course we are actually referring to a segment of the DNA that has the information required to control some aspect of the structure or function of the organism. One locus may be involved in determining seed color, another seed shape, still another the shape of the pods, and so on. The existence of a particular locus can be inferred (by traditional genetic methods at least) only if at least two allelic variants of that locus, producing contrasting phenotypes (e.g., yellow peas versus green peas) are available for study. In the simplest cases an individual can express one (yellow) or the other (green) but not both.

Thus alleles are genes that govern variations of the same feature (yellow versus green seed color) and occupy corresponding loci on homologous chromosomes. Each allele (variant) of a locus is assigned a single letter (or group of letters) as its symbol. Although more complicated forms of notation are often used by geneticists, it is customary when working simple genetics problems to indicate a dominant allele with a capital letter and a recessive allele with the same letter in lowercase. The choice of the letter is generally determined by the first allelic variant found for that locus. For example, the dominant allele that governs the

[1] In mathematics a locus is a dimensionless point; a genetic locus, being a segment of DNA, is obviously not dimensionless!

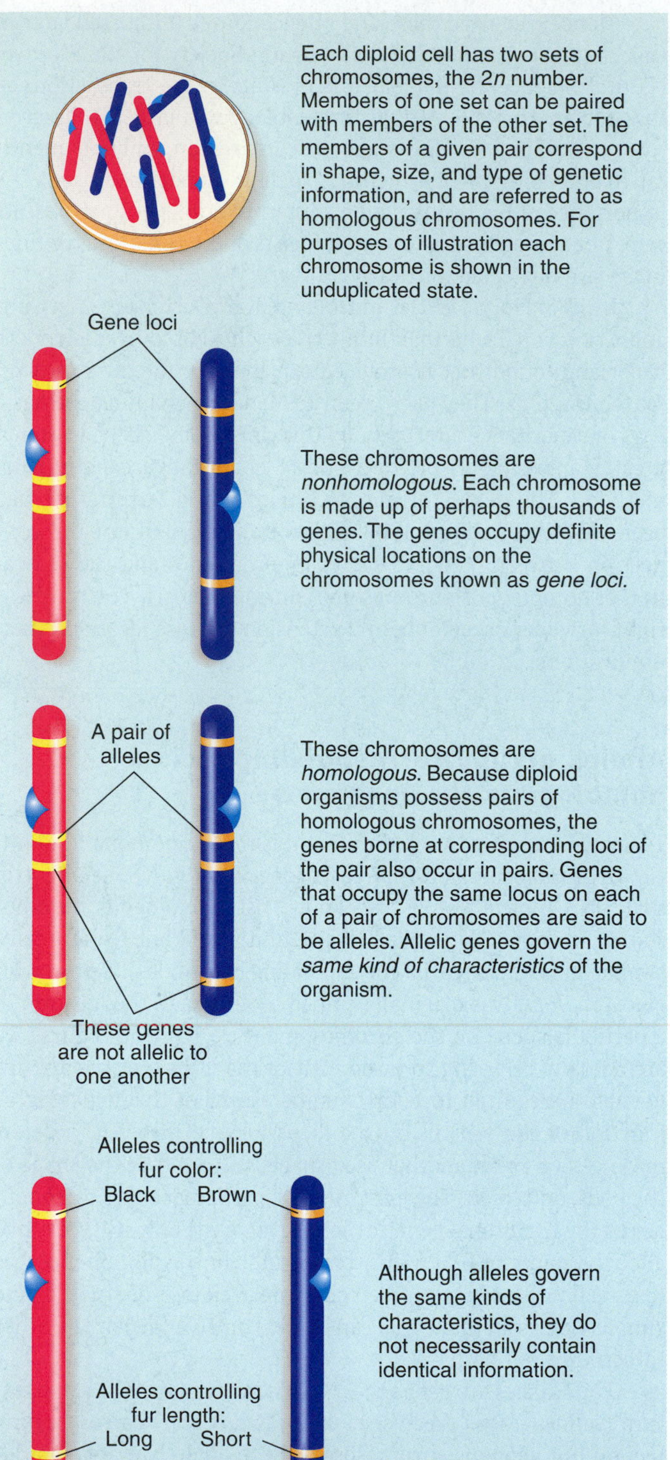

Each diploid cell has two sets of chromosomes, the 2n number. Members of one set can be paired with members of the other set. The members of a given pair correspond in shape, size, and type of genetic information, and are referred to as homologous chromosomes. For purposes of illustration each chromosome is shown in the unduplicated state.
Gene loci
These chromosomes are nonhomologous. Each chromosome is made up of perhaps thousands of genes. The genes occupy definite physical locations on the chromosomes known as gene loci.
A pair of alleles
These chromosomes are homologous. Because diploid organisms possess pairs of homologous chromosomes, the genes borne at corresponding loci of the pair also occur in pairs. Genes that occupy the same locus on each of a pair of chromosomes are said to be alleles. Allelic genes govern the same kind of characteristics of the organism.
These genes are not allelic to one another
Alleles controlling fur color:
Black
Brown
Although alleles govern the same kinds of characteristics, they do not necessarily contain identical information.
Alleles controlling fur length:
Long
Short

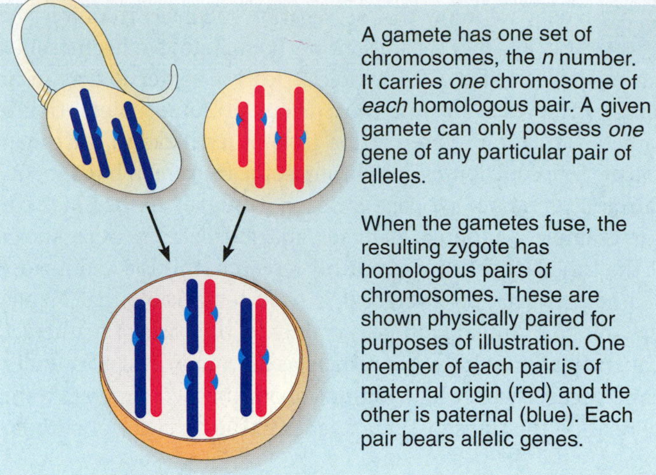

A gamete has one set of chromosomes, the n number. It carries one chromosome of each homologous pair. A given gamete can only possess one gene of any particular pair of alleles.
When the gametes fuse, the resulting zygote has homologous pairs of chromosomes. These are shown physically paired for purposes of illustration. One member of each pair is of maternal origin (red) and the other is paternal (blue). Each pair bears allelic genes.

Figure 10–4 Loci and their alleles.

yellow color of the seed might be designated *Y*, and the recessive allele responsible for the green color would then be designated *y*. Because discovery of the yellow allele made identification of this locus possible, we refer to the locus as the *yellow* locus, although pea seeds are most commonly green.

Remember that the term *locus* designates not only a position on a chromosome but also a type of gene controlling a particular kind of characteristic; thus, *Y* (yellow) and *y* (green) represent a specific pair of alleles of a locus involved in determining seed color in peas. *Although you may initially be uncomfortable with the fact that geneticists sometimes use the term gene to specify a locus and at other times to specify one of the alleles of that locus, the meaning is usually clear from the context.*

A MONOHYBRID CROSS INVOLVES INDIVIDUALS WITH DIFFERENT ALLELES OF A GIVEN LOCUS

The basic principles of genetics and the use of genetic terms are best illustrated by examples. In the simplest case, a **monohybrid cross,** the inheritance of two alleles of a single locus is studied. Our first example in this section deals with the expected ratios in the F_2 generation, as did our previous example of Mendel's work on tall and short pea plants.

Heterozygotes carry two different alleles of a locus; homozygotes carry identical alleles

Figure 10–5 illustrates a monohybrid cross featuring a locus that governs coat color in guinea pigs. The female comes from a true breeding line of black guinea pigs. We say that she is **homozygous** for black because the two alleles she carries for this locus are identical. The brown male is also from a true breeding line and is homozygous for brown. What color would you expect the F_1 offspring to be? Dark brown? Spotted? It is impossible to make such a prediction without more information.

In this particular case, the F_1 offspring are black, but they are **heterozygous,** meaning that they carry two different alleles for this locus. The brown allele influences coat color only in a homozygous brown individual; it is referred to as a recessive allele. The black allele influences coat color in both homozygous black and heterozygous individuals; it is a dominant allele. On the basis of this information, we can use standard notation to designate the dominant black allele as *B* and the recessive brown allele as *b*.

During meiosis in the female parent *(BB),* the two *B* alleles separate according to Mendel's principle of segregation so that each egg has only one *B* allele. In the male *(bb),* the two *b* alleles separate so that each sperm has only one *b* allele. The fertilization of each *B* egg by a *b* sperm results in heterozygous F_1 offspring, each with the alleles *Bb.* That is, each individual has one allele for brown coat and one for black coat. Because this is the only possible combination of alleles present in the eggs and sperm, all the F_1 offspring are *Bb.*

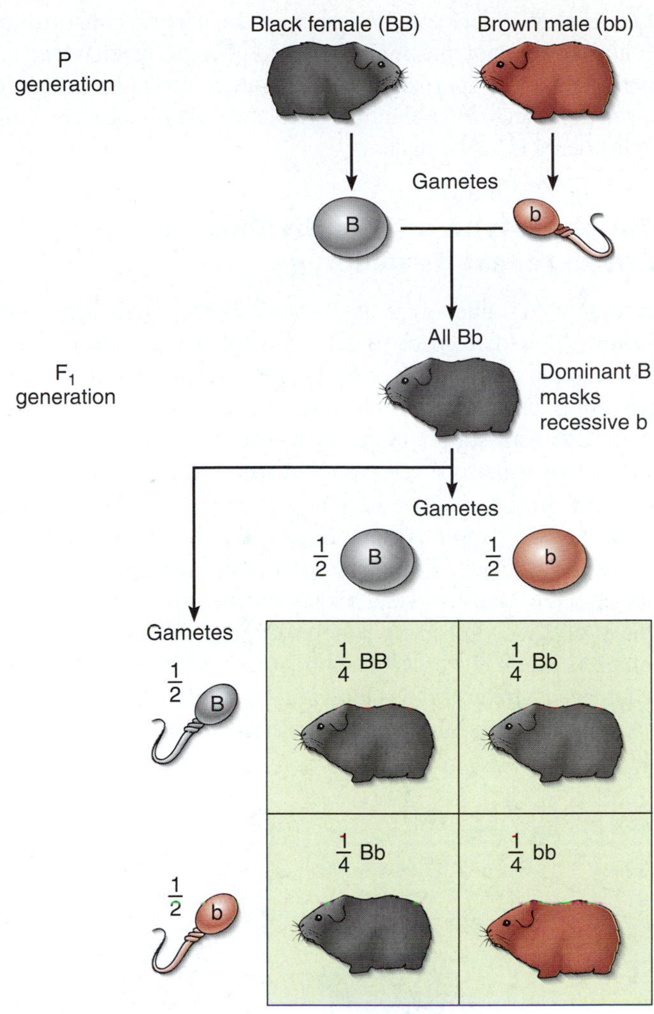

Figure 10–5 A monohybrid cross. In this example, a homozygous black guinea pig is mated with a homozygous brown guinea pig. The F_1 generation includes only black individuals. However, the mating of two of these offspring yields F_2 generation offspring in the expected ratio of 3 black to 1 brown, indicating that the F_1 individuals are heterozygous. The corresponding F_2 genotypic ratio is 1 *BB*:2 *Bb*:1 *bb.*

A Punnett square predicts the ratios of genotypes and phenotypes of the offspring of a cross

During meiosis in heterozygous black guinea pigs *(Bb),* the chromosome containing the *B* allele becomes separated from its homologue (the chromosome containing the *b* allele), so each normal sperm or egg contains *B* or *b* but never both. Gametes containing *B* alleles and those containing *b* alleles are formed in equal numbers by heterozygous *Bb* individuals. Because no special attraction or repulsion occurs between an egg and a sperm containing the same allele, fertilization is a random process.

As illustrated in Figure 10–5, the possible combinations of eggs and sperm at fertilization can be represented in the form of a "checkerboard" known as a **Punnett square,** devised by an early geneticist, Sir Reginald Punnett. The types of gametes (and their expected frequencies) from one parent are represented across the top, and those from the other parent are indicated along the left side. The squares are then filled in with the resulting F_2 zygote combinations. Three-fourths of all F_2 offspring have the genetic constitution *BB* or *Bb* and are phenotypically black; one-fourth have the genetic constitution *bb* and are phenotypically brown. The genetic mechanism responsible for the approximate 3:1 F_2 ratios (called *monohybrid F_2 phenotypic ratios*) obtained by Mendel in his pea-breeding experiments is again evident.

The phenotype of an individual does not always reveal its genotype

An organism's phenotype is its appearance (in a given environment) with respect to a certain inherited feature. However, because some alleles may be dominant and others recessive, we cannot always determine which alleles are carried by an organism simply by looking at it. The *genetic constitution* of that organism, most often expressed in symbols, is its **genotype.** In the cross we have been considering, the genotype of the female parent is homozygous dominant, *BB,* and her phenotype is black. The genotype of the male parent is homozygous recessive, *bb,* and his phenotype is brown. The genotype of all the F_1 offspring is heterozygous, *Bb,* and their phenotype is black. To prevent confusion we always indicate the genotype of a heterozygous individual by writing the symbol for the dominant allele first and the recessive allele second (always *Bb,* never *bB*).

The phenomenon of dominance partly explains why an individual may resemble one parent more than the other, even if the two parents make equal contributions to their offspring's genetic constitution. Dominance is not predictable and can be determined only by experiment. In one species of animal, black coat may be dominant to brown; in another species, brown may be dominant to black.

A test cross can detect heterozygosity

Guinea pigs with the genotypes *BB* and *Bb* are alike phenotypically; they both have black coats. How, then, can we know the genotype of a black guinea pig? Geneticists can accomplish this by performing a **test cross,** in which an individual of unknown genotype is crossed with a homozygous recessive individual (Fig. 10–6). In a test cross, the alleles carried by the gametes from the parent of unknown genotype are never "hidden" in the offspring by dominant alleles contributed by the other parent. Therefore, one can deduce the genotypes of all the classes of offspring directly from their phenotypes. If all the offspring were black, what inference would you make about the genotype of the black parent? If any of the offspring were brown, what conclusion would you draw regarding the genotype of the black parent? Would you be more certain about one of these inferences than the other?

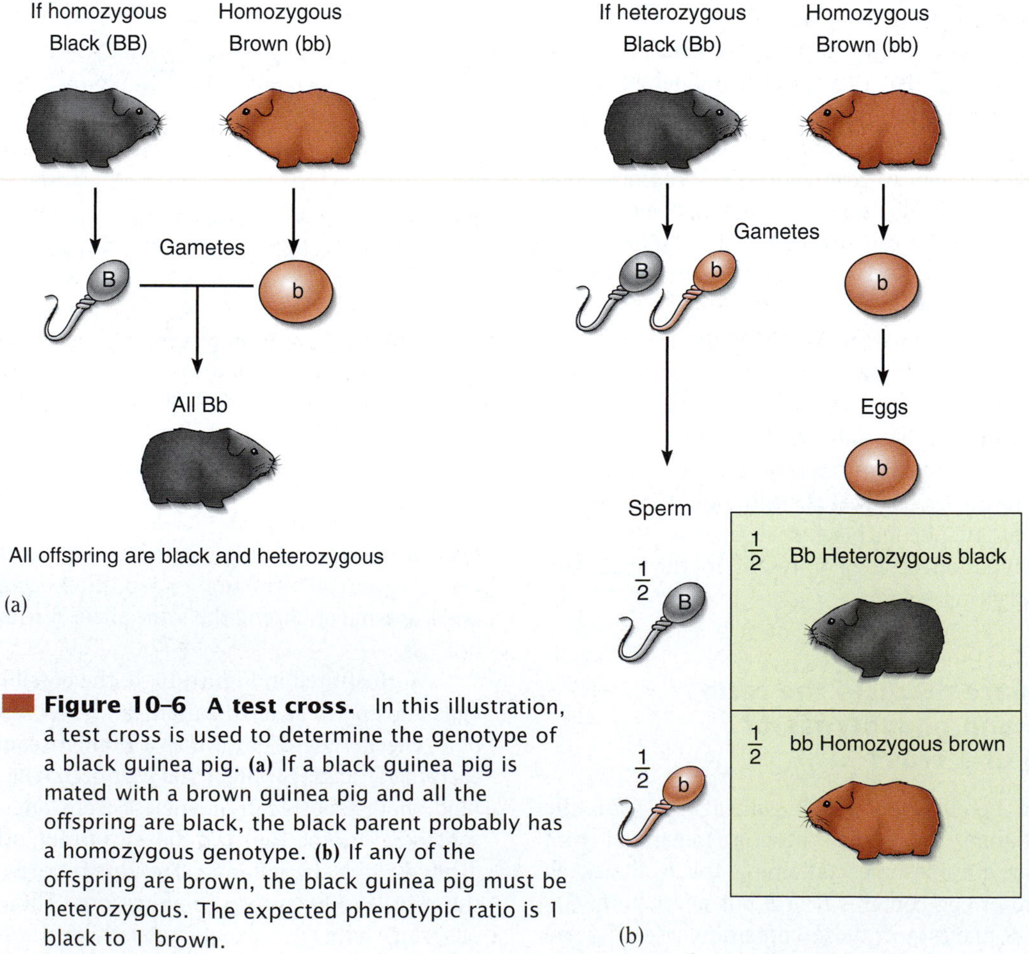

Figure 10–6 A test cross. In this illustration, a test cross is used to determine the genotype of a black guinea pig. **(a)** If a black guinea pig is mated with a brown guinea pig and all the offspring are black, the black parent probably has a homozygous genotype. **(b)** If any of the offspring are brown, the black guinea pig must be heterozygous. The expected phenotypic ratio is 1 black to 1 brown.

Mendel conducted numerous test crosses; for example, he bred F_1 (tall) pea plants with homozygous recessive *(tt)* short ones. He reasoned that the F_1 individuals were heterozygous *(Tt)* and would be expected to produce equal numbers of *T* and *t* gametes. Because the homozygous short parents *(tt)* were expected to produce only *t* gametes, Mendel predicted that he would obtain equal numbers of tall *(Tt)* and short *(tt)* offspring. His results agreed with his predictions, thereby providing additional evidence supporting the hypothesis that there is 1:1 segregation of the alleles of a heterozygous parent. Thus, Mendel's principle of segregation not only explained the known facts, such as the 3:1 monohybrid F_2 phenotypic ratio, but also enabled him to successfully anticipate the results of other experiments, in this case the 1:1 test cross phenotypic ratio.

THE RULES OF PROBABILITY PREDICT THE LIKELIHOOD OF GENETIC EVENTS

All genetic ratios are properly expressed in terms of probabilities. In the examples just discussed, among the offspring of two individuals heterozygous for the same gene pair, the expected ratio of the phenotypes of the dominant and recessive alleles is 3:1. The probability of an event is its expected frequency; therefore we can say that there are 3 chances in 4 (3/4) that any particular individual offspring of two heterozygous individuals will express the dominant allele phenotype and 1 chance in 4 (1/4) that it will express the recessive allele phenotype. Although we sometimes speak in terms of percentages, probabilities must always be calculated as fractions (e.g., 3/4) or decimal fractions (e.g., 0.75). If an event is certain to occur, its probability is 1; if it is certain not to occur, its probability is 0. A probability can be 0, 1, or some number between 0 and 1.

Often we wish to *combine* two or more probabilities. The Punnett square, which we use to predict the results of genetic crosses, is a device that allows us to combine probabilities. When we use a Punnett square we are intuitively following two important statistical principles known as the **product rule** and the **sum rule.**

The product rule predicts the combined probabilities of independent events

Events are independent if the occurrence of one does not affect the probability that the other will occur. For example, the probability of obtaining heads on the first toss of a coin is 1/2; the probability of obtaining heads on the second toss (an independent event) is also 1/2. If two or more events are *independent* of each other, the probability of their both occurring is the product of their individual probabilities. If this seems strange to you, keep in mind that when we multiply two numbers that are less than 1, the product is a smaller number. Therefore, the probability of obtaining heads two times in a row is $1/2 \times 1/2 = 1/4$, or 1 chance in 4 (Fig. 10–7).

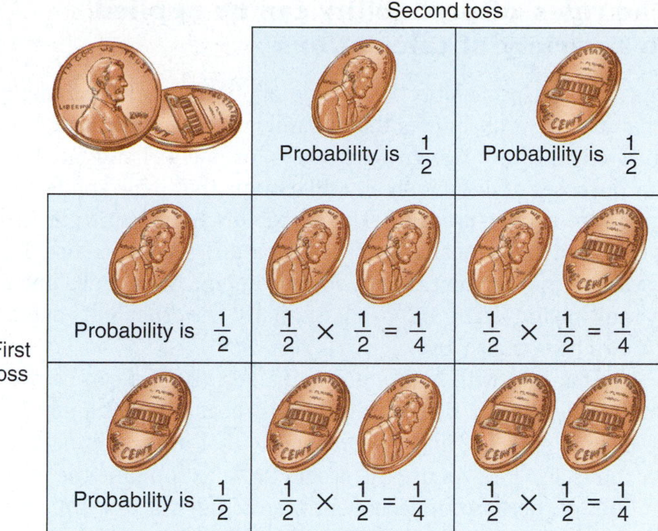

Figure 10–7 The rules of probability. For each coin toss the probability of heads is 1/2 and the probability of tails is also 1/2. Because the outcome of the first toss is independent of the outcome of the second, the combined probabilities of the outcomes of successive tosses are calculated by multiplying their individual probabilities (according to the product rule: $1/2 \times 1/2 = 1/4$). These same rules of probability are used to predict genetic events.

Similarly, we can apply the product rule to genetic events. If both parents are *Bb,* what is the probability that they will produce a child who is *bb?* For the child to be *bb,* he or she must receive a *b* gamete from each parent. The probability of a *b* egg is 1/2 and the probability of a *b* sperm is also 1/2. Like the outcomes of the coin tosses, these probabilities are independent, so we combine them by the product rule ($1/2 \times 1/2 = 1/4$). You may wish to check this result using a Punnett square.

The sum rule predicts the combined probabilities of mutually exclusive events

In some cases there is more than one way to obtain a specific outcome. These different ways are called *mutually exclusive* events because no more than one of them can happen; i.e., if one of them occurs, the other(s) cannot. For example, if both parents are *Bb,* what is the probability that their first child will also have the *Bb* genotype? There are two different ways these parents can have a *Bb* child: Either a *B* egg combines with a *b* sperm (probability 1/4), or a *b* egg combines with a *B* sperm (probability 1/4).

Naturally, if there is more than one way to obtain a result, the chances of its being obtained are improved; we therefore combine the probabilities of mutually exclusive events by summing (adding) their individual probabilities. The probability of obtaining a *Bb* child in our example is therefore $1/4 + 1/4 = 1/2$. (Because there is only one way these heterozygous parents can produce a homozygous recessive child, *bb,* that probability is only 1/4. The probability of a homozygous dominant child, *BB,* is likewise 1/4.)

The rules of probability can be applied to a variety of calculations

The rules of probability have wide applications. For example, what are the probabilities that a family with two (and only two) children will have two girls, two boys, or one girl and one boy? For purposes of discussion we will assume that male and female births are equally probable. The probability of having a girl first is 1/2, and the probability of having a girl second is also 1/2. These are independent events, so we combine their probabilities by multiplying: $1/2 \times 1/2 = 1/4$. Similarly, the probability of having two boys is also 1/4.

In families with both a girl and a boy, the girl can be born first or the boy can be born first. The probability that a girl will be born first is 1/2, and the probability that a boy will be born second is also 1/2. We use the product rule to combine the probabilities of these two independent events: $1/2 \times 1/2 = 1/4$. Similarly, the probability that a boy will be born first and a girl second is also 1/4. These two kinds of families represent mutually exclusive outcomes, that is, two different ways of obtaining a family with one boy and one girl. Having two different ways of obtaining the desired result improves our chances, so we use the sum rule to combine the probabilities: $1/4 + 1/4 = 1/2$. Notice that the probabilities of the three types of families—both boys (1/4), both girls (1/4), and one girl, one boy (1/2)—add up to 1. This serves as a useful check that the calculations have been done correctly. You may also wish to confirm these results by making a Punnett square.

In working with probabilities, it is important to keep in mind a point that many gamblers forget: Chance has no memory. This means that if events are truly random, past events have no influence on the probability of the occurrence of independent future events. For example, if two brown-eyed people have a child, what is the probability that it will have blue eyes? If their first child has blue eyes, what is the probability that their second child will also have blue eyes? The color of the iris of the human eye is controlled by alleles at several loci, but alleles at one locus are primarily responsible. The allele for brown eye color, *B,* is usually dominant to the allele for blue, *b.* If the two brown-eyed parents are heterozygous, there is 1 chance in 4 that any child of theirs will have blue eyes. Each fertilization is a separate, independent event; its result is not affected by the results of any previous fertilizations. If these two, heterozygous, brown-eyed parents already have three brown-eyed children and are expecting their fourth child, what is the probability that the child will have blue eyes? The uninformed might guess that this one *must* have blue eyes, but in fact there is still only 1 chance in 4 that the child will have blue eyes and 3 chances in 4 that the child will have brown eyes.

If two heterozygous people marry and *plan* to have four children, what is the probability that all four will have brown eyes? The probability of brown eyes for each child is 3/4, so we combine these independent events by the product rule: $3/4 \times 3/4 \times 3/4 \times 3/4 = 81/256$ or 0.32. Why is the answer to this question so different from the answer to the previous question? Remember that for the brown-eyed children that were already born, chance (3/4) is replaced by certainty (1), so the calculation becomes $1 \times 1 \times 1 \times 3/4 = 3/4$. The chance that the fourth (as yet unborn) child will have brown eyes is therefore 3/4 (and the chance of blue eyes is 1/4).

When working probability problems, common sense is more important than blindly memorizing rules. Examine your results to see whether they appear reasonable; if they do not, you should reevaluate your assumptions.

A DIHYBRID CROSS INVOLVES INDIVIDUALS THAT HAVE DIFFERENT ALLELES AT TWO LOCI

Simple monohybrid crosses each involve a pair of alleles of a single locus. Mendel also analyzed crosses involving alleles of two or more loci. A mating between individuals with different alleles at two loci is called a **dihybrid cross.** Consider the case when two pairs of alleles are located in nonhomologous chromosomes (i.e., one pair of alleles is located in one pair of homologous chromosomes, and the other pair of alleles is located in a *different* pair of homologous chromosomes). Each pair of alleles is inherited independently; that is, each pair segregates during meiosis independently of the other.

An example of a dihybrid cross carried through the F_2 generation is illustrated in Figure 10–8. When a homozygous, black, short-haired guinea pig (*BBSS,* because black is dominant to brown and short hair is dominant to long hair) and a homozygous, brown, long-haired guinea pig *(bbss)* are mated, the *BBSS* animal produces gametes that are all *BS,* and the *bbss* individual produces gametes that are all *bs.* Each gamete contains one and only one allele for each of the two loci. The union of the *BS* and *bs* gametes yields only individuals with the genotype *BbSs.* All these F_1 offspring are heterozygous for hair color and for hair length, and all are phenotypically black and short-haired.

The principle of independent assortment states that alleles on nonhomologous chromosomes are randomly distributed into gametes

Each F_1 guinea pig produces four kinds of gametes with equal probability: *BS, Bs, bS,* and *bs.* Hence, the Punnett square has 16 (4^2) squares representing the zygotes, some of which are genotypically or phenotypically alike. There are 9 chances in 16 of obtaining a black, short-haired individual; 3 chances in 16 of obtaining a black, long-haired individual; 3 chances in 16 of obtaining a brown, short-haired individual; and 1 chance in 16 of obtaining a brown, long-haired individual. This 9:3:3:1 phenotypic ratio is expected in a dihybrid F_2 if the hair color and hair length loci are on nonhomologous chromosomes.

On the basis of similar results, Mendel formulated the principle of inheritance, now called Mendel's **principle of independent assortment,** which states that members of any gene pair segregate from one another independently of the members of the other gene pairs. This occurs in a regular way that ensures that

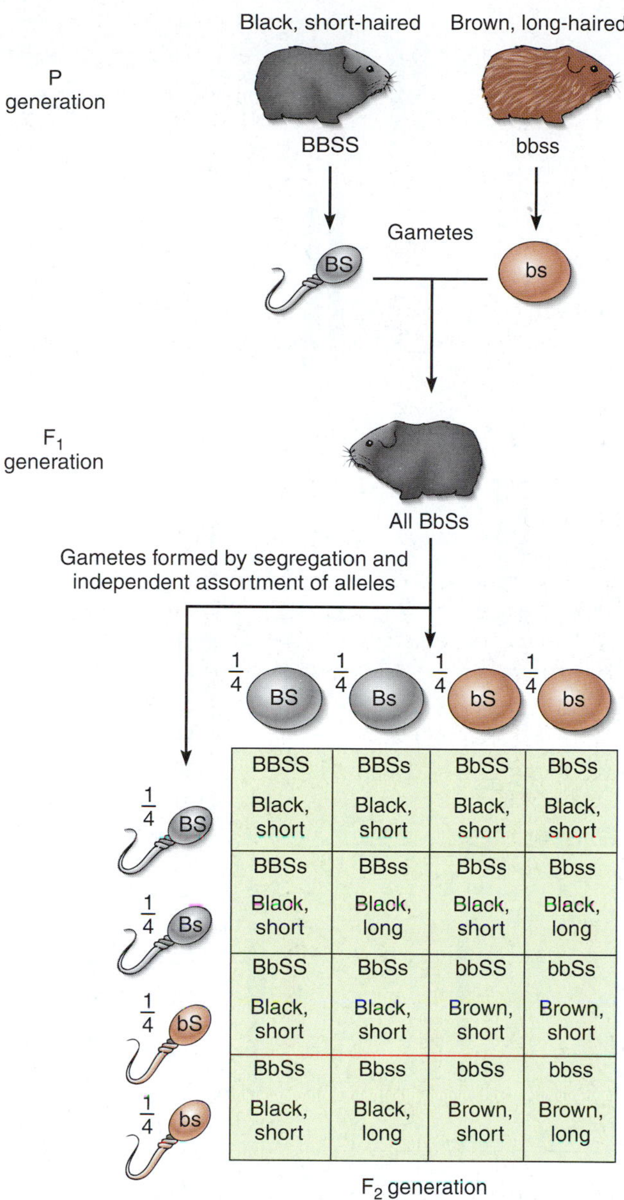

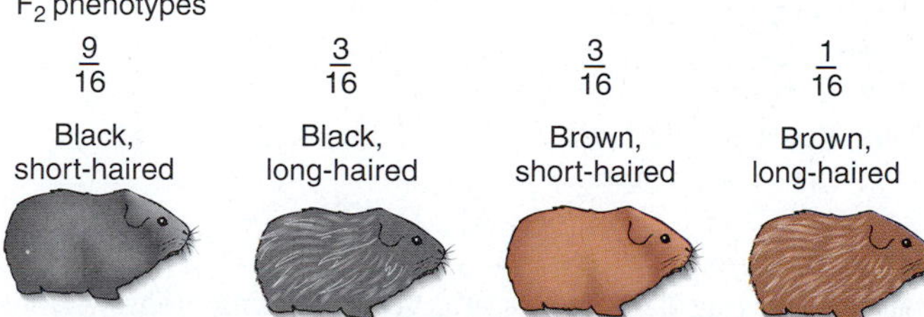

Figure 10-8 A dihybrid cross. When a black, short-haired guinea pig is crossed with a brown, long-haired one, all the offspring are black and have short hair. However, when two members of the F_1 generation are crossed, the ratio of phenotypes is 9:3:3:1. Note that the two pairs of alleles considered here assort independently.

each gamete contains one allele for each locus, but the alleles of different loci are assorted at random with respect to each other in the gametes. (Mendel reported on the results of crosses in which the genes assorted independently. As you will soon see, independent assortment does not always occur.)

Procedures used in solving genetics problems that illustrate Mendel's principles of segregation and independent assortment are summarized at the end of the chapter in *Focus On: Solving Genetics Problems*, and *Focus On: Deducing Genotypes.*

The mechanics of meiosis are the basis for independent assortment

Today we recognize that independent assortment is related to the events of meiosis. It occurs because there are two different ways in which two pairs of homologous chromosomes can be arranged at metaphase I of meiosis. These occur randomly, with approximately half the meiotic cells having one orientation, and the other half having the opposite orientation. The orientation of the homologous chromosomes on the metaphase plate then determines the way they subsequently separate and are distributed into the haploid cells (Fig. 10–9).

Linked genes do not assort independently

Independent assortment does not apply if the two loci are located in the same pair of homologous chromosomes. In fruit flies there is a locus controlling wing shape (the dominant allele *V* for normal wings and the recessive allele *v* for vestigial wings) and another locus controlling body color (the dominant allele *B* for gray and the recessive allele *b* for black). If a homozygous *BBVV* fly is crossed with a homozygous *bbvv* fly, the F_1 flies all have gray bodies and normal wings, and their genotype is *BbVv*.

Because these loci happen to be located in the *same pair* of homologous chromosomes, their alleles do not assort independently; instead they tend to be inherited together and are said to be **linked.** Linkage is most readily observed by analyzing the results of a test cross in which heterozygous F_1 flies *(BbVv)* are mated with homozygous recessive *(bbvv)* flies (Fig. 10–10). Because heterozygous individuals are mated to homozygous recessive individuals, this test cross is similar to the test cross described previously. However, it is called a **two-point test cross** because alleles of two loci are involved.

If the loci governing these characteristics were on different chromosomes (unlinked), the heterozygous parent in a test cross would produce four kinds of gametes (*BV, Bv, bV,* and *bv*) in equal numbers. As a result of this independent assortment, offspring with new gene combinations not present in the parental generation would be produced. Any process that leads to new gene combinations is called **recombination.** In our example, *Bv* and *bV* are both **recombinant gametes.** The other two kinds of gametes, *BV* and *bv*, are called **parental gametes** because they are identical to the gametes produced by the P generation. Of course, the homozygous recessive parent produces only one kind of gamete, *bv*. Thus, if independent assortment were to occur in the F_1 flies, approximately 25% of the test-cross offspring would be gray-bodied

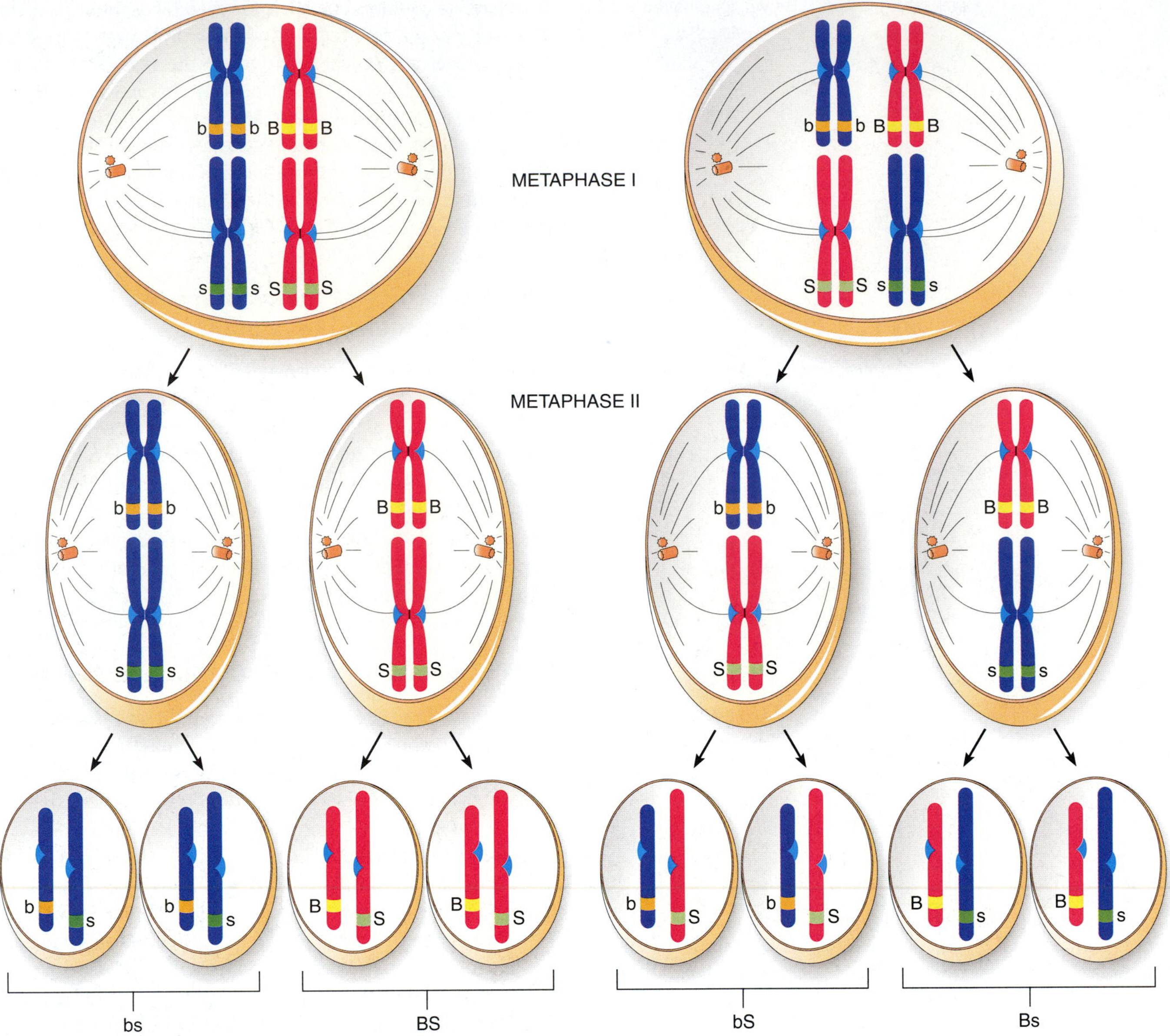

Figure 10–9 Meiosis and independent assortment. There are two equally likely ways that two different pairs of homologous chromosomes can line up at metaphase I and be subsequently distributed. A cell with the orientation shown at the left produces half *BS* and half *bs* gametes. Conversely, the cell at the right produces half *Bs* and half *bS* gametes. Because approximately half of the meiotic cells at metaphase I are of each type, the ratio of the four possible types of gametes is 1:1:1:1.

and normal-winged *(BbVv),* 25% black-bodied and normal-winged *(bbVv),* 25% gray-bodied and vestigial-winged *(Bbvv),* and 25% black-bodied and vestigial-winged *(bbvv).* Notice that the two-point test cross allows us to determine the genotypes of the offspring directly from their phenotypes.

By contrast, the alleles of the loci in our example do not undergo independent assortment because they are linked. Alleles at different loci on a given chromosome tend to be inherited together because chromosomes pair and separate during meiosis as units and therefore tend to be inherited as units. If linkage were complete, only parental type flies would be produced, with approximately 50% having gray bodies and normal wings *(BbVv),* and 50% having black bodies and vestigial wings *(bbvv).* However, in our example, the progeny also include some gray-bodied, vestigial-winged flies and some black-bodied, normal-winged flies. These are recombinant type flies, having received a recombinant gamete from the heterozygous F_1 parent. Each recombinant gamete arose by crossing-over between these loci in a meiotic cell of a heterozygous female[2] fly. Recall from Chapter 9 that

[2] Fruit flies are unusual in that crossing-over occurs only in females and not in males. It is far more common for crossing-over to occur in both sexes of a species.

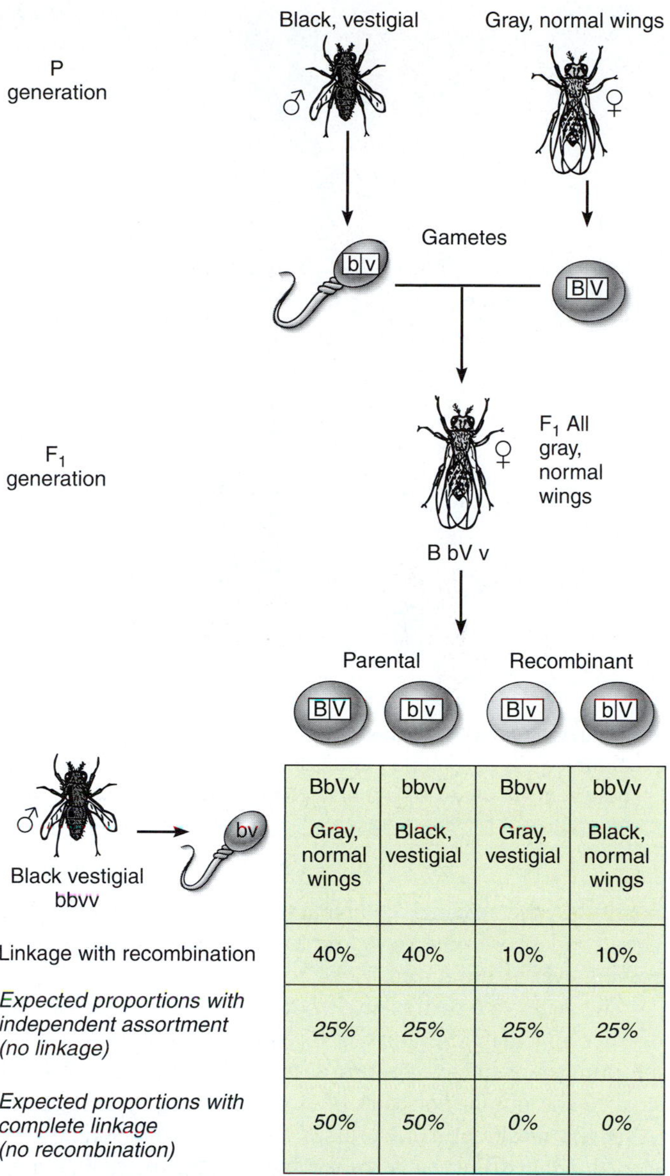

	BV	bv	Bv	bV
	BbVv Gray, normal wings	bbvv Black, vestigial	Bbvv Gray, vestigial	bbVv Black, normal wings
Linkage with recombination	40%	40%	10%	10%
Expected proportions with independent assortment (no linkage)	*25%*	*25%*	*25%*	*25%*
Expected proportions with complete linkage (no recombination)	*50%*	*50%*	*0%*	*0%*

Figure 10–10 A two-point test cross to detect linkage. Linkage can be recognized when an excess of parental type offspring and a deficiency of recombinant type offspring are produced in a two-point test cross. In this example, loci for wing length in fruit flies (vestigial versus normal wings) and for body color (black versus gray body) are linked; they are located on a homologous chromosome pair. This is evident in the proportions of offspring: about 40% of the offspring belong to each of the two parental classes (80% total), and 10% belong to each of the two recombinant classes (20% total). The numbers in italics allow us to contrast this outcome (linkage with recombination), with the expected proportions if either of two other alternatives were true: (1) independent assortment and (2) complete linkage with no recombination.

when chromosomes pair and undergo synapsis, **crossing-over** occurs as homologous (nonsister) chromatids exchange segments of chromosomal material by a process of breakage and rejoining (Fig. 10–11).

The linear order of linked genes on a chromosome is determined by calculating the frequency of crossing-over

In our example, about 20% of the offspring are recombinant types: gray flies with vestigial wings, *Bbvv* (approximately 10% of the total); and black flies with normal wings, *bbVv* (also about 10% of the total). The remaining 80% are parental types. These data can be used to calculate the percentage of crossing-over between the loci. This is done by adding the number of individuals in the two recombinant classes of offspring (10 + 10), dividing

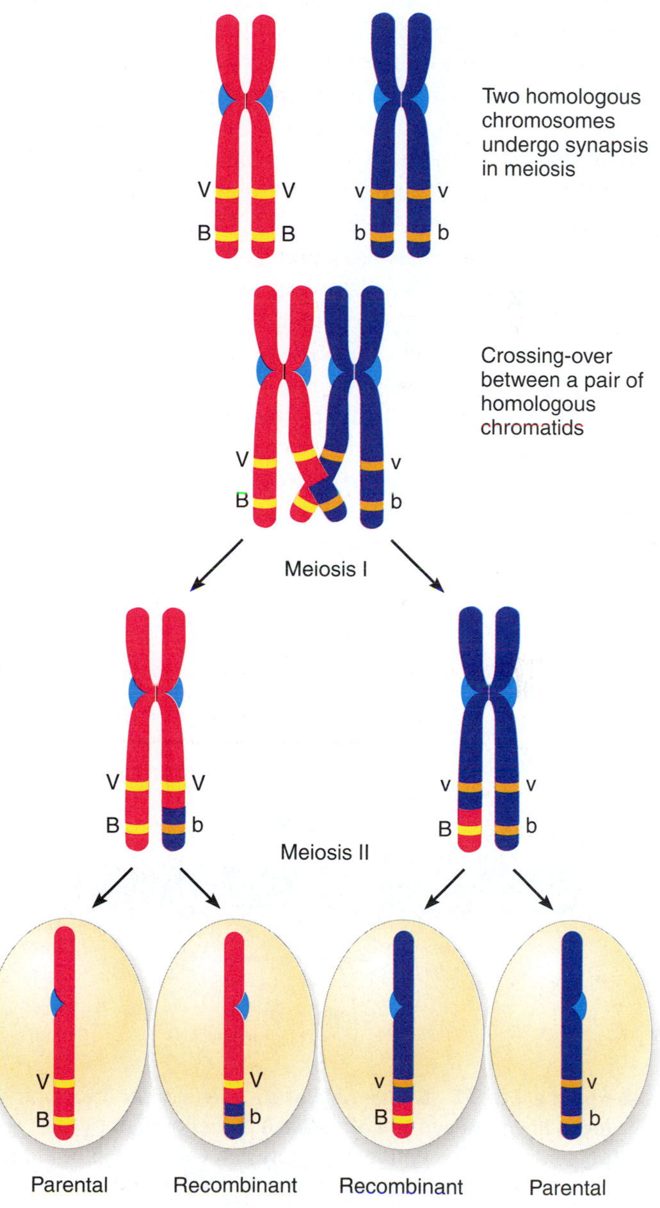

Figure 10–11 Crossing-over. The exchange of segments between chromatids of homologous chromosomes permits the recombination of linked genes. Genes located far apart on a chromosome have a greater probability of being separated by an exchange of segments than do genes that are closer together.

by the *total number of offspring* (40 + 40 + 10 + 10), and multiplying by 100. Thus, the *V* locus and the *B* locus have 20% recombination between them.

During a single meiotic division, crossing-over may occur at several different points along the length of each homologous chromosome pair. In general, a crossover is more likely to occur between two loci if they are far apart on the chromosome and less likely to occur if they are close together. Because this rough correlation exists between the frequency of recombination between two loci and the linear distance between them, a genetic map of the chromosome can be generated by converting the percentage of recombination to **map units.** By convention, 1% recombination between two loci equals a distance of 1 map unit, so the loci in our example are 20 map units apart.

The frequencies of recombination between specific linked loci have been measured in many species. All of the experimental results are consistent with the hypothesis that genes are present in a linear order in the chromosomes. Figure 10–12 illustrates the traditional method for determining the linear order of genes in a chromosome.

More than one crossover between two loci in a single tetrad can occur in a given cell undergoing meiosis. (Recall from Chapter 9 that a tetrad is a group of four chromatids that make up a pair of synapsed homologous chromosomes.) We can observe only the frequency of offspring receiving recombinant gametes from the heterozygous parent, not the actual number of crossovers. In fact, the actual frequency of crossing-over is slightly more than the observed frequency of recombinant gametes. This is because the simultaneous occurrence of two crossovers involving the same two homologous chromatids reconstitutes the original combination of genes (Fig. 10–13). When two loci are relatively close together, the effect of double crossing-over is minimized.

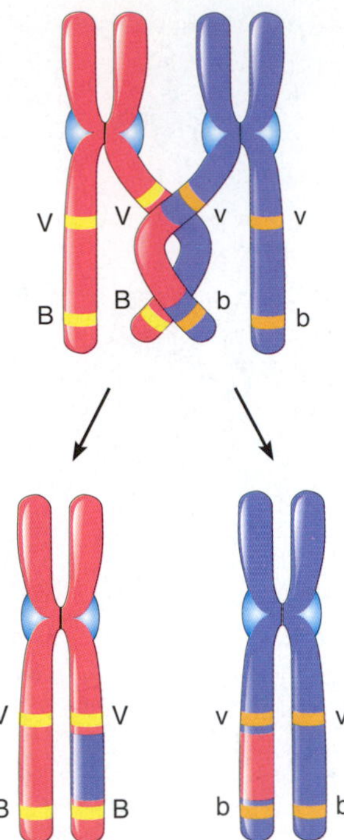

Figure 10–13 Double crossing-over. If the same homologous chromatids undergo double crossing-over between the genes of interest, the gametes formed are not recombinant for these genes.

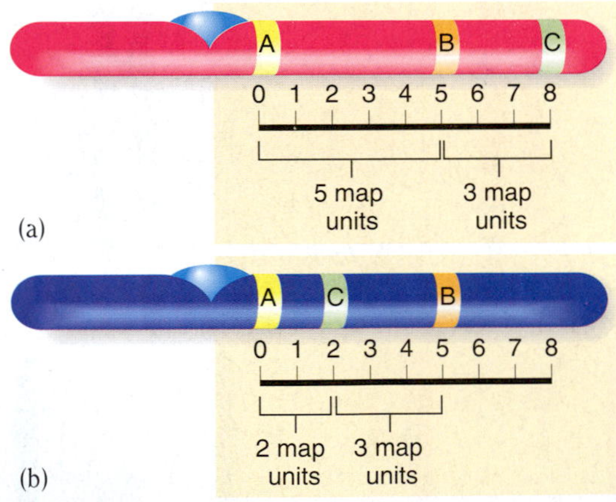

Figure 10–12 Gene mapping. Gene order (i.e., which locus lies between the other two) is determined by the percentage of recombination between each of the possible pairs. In this hypothetical example, the percentage of recombination between locus A and locus B is 5% (corresponding to 5 map units) and that between B and C is 3% (3 map units). There are two alternatives for the linear order of these alleles. **(a)** If the recombination between A and C is 8% (8 map units), B must be in the middle. **(b)** If the recombination between A and C is 2%, then C must be in the middle.

The genes in a particular chromosome tend to be inherited together and therefore are said to constitute a **linkage group.** The number of linkage groups determined by genetic tests is equal to the number of pairs of chromosomes. By putting together the results of many crosses, scientists painstakingly developed detailed linkage maps for many eukaryotes, including the fruit fly (which has four pairs of chromosomes), the mouse, yeast, *Neurospora* (a fungus), and many plants, especially those that are important crops. In addition, special genetic methods have made possible the development of a detailed map for *Escherichia coli,* a bacterium with a single, circular DNA molecule, and many other prokaryotes and viruses. Much more sophisticated maps of chromosomes have been made by means of recombinant DNA technology (see Chapter 14). These methods have been particularly useful in producing maps of human chromosomes through the Human Genome Project (see Chapter 15).

SEX IS COMMONLY DETERMINED BY SPECIAL SEX CHROMOSOMES

Although in some species sex is controlled mainly by the environment, genes are the most important sex determinants in most organisms. The major sex-determining genes of most animals are carried by sex chromosomes. Typically, members of one sex (the **homogametic** sex) have a pair of similar sex chromosomes and produce gametes that are all identical in sex chromosome constitution. The members of the other sex (the **heterogametic** sex) have two different sex chromosomes and produce two kinds of gametes, each bearing a single kind of sex chromosome.

The females of many animal species (including humans) are homogametic; their cells contain two **X chromosomes.** In contrast, the males are heterogametic, having a single X chromosome and a smaller **Y chromosome.** For example, human males have 22 pairs of **autosomes,** which are chromosomes other than the sex chromosomes, plus one X chromosome and one Y chromosome; females have 22 pairs of autosomes plus a pair of X chromosomes. Domestic cats have 19 pairs of autosomes, to which are added a pair of X chromosomes in females, or an X plus a Y in males.

The Y chromosome determines male sex in most species of mammals

Do male humans have a male phenotype because they have only one X chromosome or because they have a Y chromosome? Much of the traditional evidence bearing on this question comes from studies of persons with abnormal sex chromosome constitutions (see Chapter 15). A person with an XXY constitution is a nearly normal male in external appearance, although his testes are underdeveloped (Klinefelter syndrome). A person with one X but no Y chromosome has the appearance of an immature female (Turner syndrome). An embryo with a Y but no X does not survive. Hence all individuals require at least one X, and the Y is the male-determining chromosome. In fact, several genes on the Y chromosomes that are involved in male determination have been identified. The major male-determining gene on the Y acts as a "genetic switch" that causes the testes to develop in the fetus. The developing testes then secrete testosterone, which causes other features characteristic of males to develop. A small number of other genes on the Y also play a role in sex determination, as do many genes on the X and some on the autosomes, which explains why an XXY individual does not have a completely normal male phenotype.

The X and Y chromosomes are thought to have originated as a homologous pair. However, they are not truly homologous in their present forms because they are not similar in size, shape, or genetic constitution. Nevertheless, they have retained a short homologous "pairing region" that allows them to synapse and separate from one another during meiosis. Half the sperm contain an X chromosome and half contain a Y chromosome. All normal eggs bear a single X chromosome (Fig. 10–14). Fertilization of an X-bearing egg by an X-bearing sperm results in an XX (female) zygote; fertilization by a Y-bearing sperm results in an XY (male) zygote.

We would expect to have equal numbers of X- and Y-bearing sperm and a 1:1 ratio of females to males. In fact, however, in humans more males are conceived than females and more males die before birth. Even at birth the ratio is not 1:1; about 106 boys are born for every 100 girls. It is not known why this occurs, but Y-bearing sperm appear to have some competitive advantage.

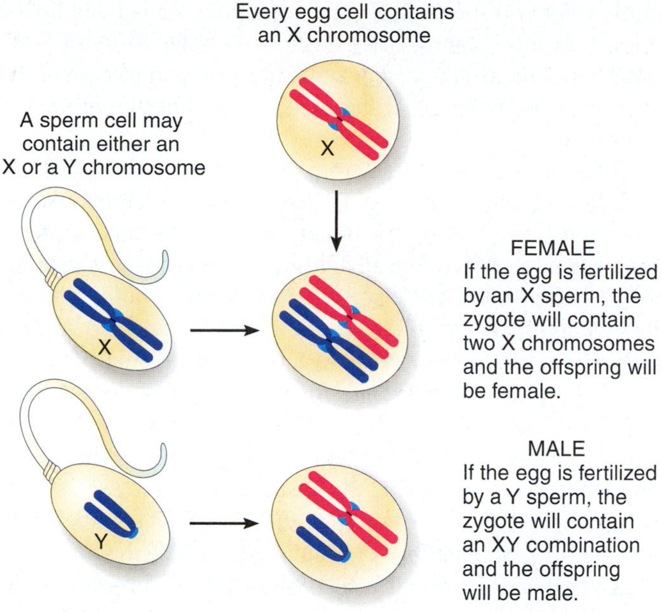

Figure 10–14 Sex is determined by the sperm in mammals. An X-bearing sperm produces a female; a Y-bearing sperm produces a male.

An XX/XY sex chromosome mechanism, similar to that of humans, operates in many species of animals. However, it is not universal, and many of the details may vary. For example, the fruit fly, *Drosophila,* has homogametic (XX) females and heterogametic (XY) males, but the Y is not male-determining; a fruit fly with an X chromosome and no Y chromosome has a male phenotype. In birds and butterflies the mechanism is reversed, with homogametic males (the equivalent of XX) and heterogametic females (the equivalent of XY).

Some organisms are hermaphroditic

In **hermaphroditic** organisms, organs of both sexes are found in the same individual. Hermaphroditic animals (see Chapters 28, 29, and 48) do not have sex chromosomes. Most flowering plants are hermaphrodites. The male and female sexual organs may be in the same flowers. If they are in separate flowers on the same plant, the plants are said to be **monoecious;** corn, walnuts, and oaks are examples. Far fewer flowering plants are not hermaphroditic; they are **dioecious,** having male and female floral organs on separate plants. A few dioecious plants, such as asparagus, apparently have sex chromosomes, although they are not necessarily comparable to those of animals.

X-linked genes have unusual inheritance patterns

The human X chromosome contains many loci that are required in both sexes, whereas the Y chromosome contains only a few genes, including one or more genes for maleness. Genes located in the X chromosome, such as colorblindness and the most common form of hemophilia, are sometimes called **sex-linked** genes. It is more appropriate, however, to refer to them as **X-linked** genes because they follow the transmission pattern of the X chromosome and, strictly speaking, are not linked to the sex of the organism per se.

A female receives one X from her mother and one X from her father. A male receives his Y chromosome, which makes him male, from his father. From his mother he inherits a single X chromosome and therefore all of his X-linked genes. In the male, every X chromosome allele present is expressed, whether that allele was dominant or recessive in the female parent. A male is neither homozygous nor heterozygous for his X-linked loci; instead he is always **hemizygous** for every X-linked locus (*hemi* means "half").

We will use a simple system of notation for problems involving X linkage, indicating the X and incorporating specific alleles as superscripts. For example, the symbol X^c signifies a recessive X-linked allele for colorblindness and X^C the dominant X-linked allele for normal color vision. The Y chromosome is written without superscripts because it does not carry the locus of interest. Two recessive X-linked alleles must be present in a female for the abnormal phenotype to be expressed (i.e., X^cX^c), whereas in the hemizygous male a single abnormal allele is expressed (X^cY). As a practical consequence, these abnormal alleles are much more frequently expressed in male offspring. A heterozygous female may be a *carrier,* an individual who possesses one copy of a mutant recessive allele but does not express it in the phenotype (i.e., X^CX^c).

To be expressed in a female, a recessive X-linked allele must be inherited from both parents. A colorblind female, for example, must have a colorblind father and a mother who is heterozygous or homozygous for colorblindness (Fig. 10–15). Such a combination is unusual because the frequency of the allele for colorblindness is relatively low. In contrast, a colorblind male need only have a mother who is heterozygous for colorblindness; his father can be normal. Hence, X-linked recessive traits are generally much more common in males than in females, a fact that may partially explain why human male embryos are more likely to die.

Dosage compensation equalizes the expression of X-linked genes in males and females

The X chromosome contains numerous genes required by both sexes, yet a normal female has two copies ("doses") for each locus, whereas a normal male has only one. **Dosage compensation** is a mechanism that makes the two doses in the female and the

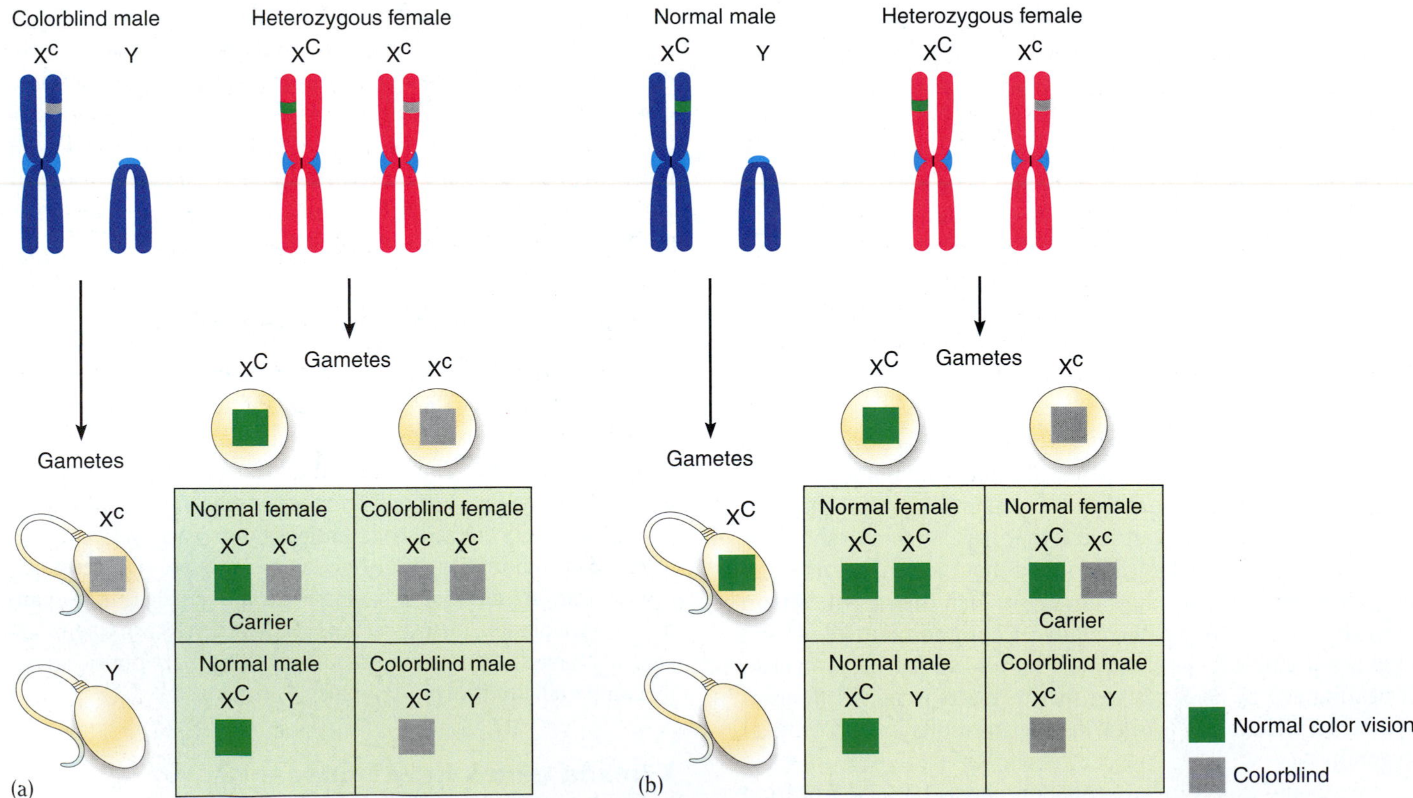

Figure 10–15 X-linked recessive colorblindness. Note that the Y chromosome does not carry a gene for color vision. **(a)** To be colorblind, a female must inherit alleles for colorblindness from both parents. **(b)** If a normal male mates with a carrier (heterozygous) female, half of their sons would be expected to be colorblind and half of their daughters would be expected to be carriers.

single dose in the male equivalent. Male fruit flies accomplish this by making their single X chromosome more active. In most tissues, the metabolic activity of a single male X chromosome is equal to the combined metabolic activity of the two X chromosomes present in the female.

Dosage compensation in mammals generally involves inactivation of one of the two X chromosomes in the female. During interphase a dark spot of chromatin, called a **Barr body,** is visible at the edge of the nucleus of each female mammalian cell (Fig. 10–16). The Barr body has been found to represent a dense, dark-staining, and metabolically inactive X chromosome. The other X chromosome resembles the metabolically active autosomes; during interphase it is a greatly extended thread that is not evident by light microscopy. From this and other evidence, the British geneticist Mary Lyon hypothesized in 1961 that in any cell of a female mammal, only one of the two X chromosomes is active; the other is inactive and is visible as a Barr body. (Actually, X chromosome inactivation is never complete; a small fraction of the genes are active.)

Because only one X chromosome is active in any one cell and because X chromosome inactivation is a random event, a female mammal that is heterozygous at an X-linked locus expresses one of the alleles in about half her cells and the other allele in the other half. This is sometimes (but not always) evident in the phenotype. Mice and cats have several X-linked genes for certain coat colors. Females that are heterozygous for such genes may show patches of one coat color in the midst of areas of the other coat color. This phenomenon, termed **variegation,** is evident in calico (Fig. 10–17) and tortoiseshell cats. Early in development, when relatively few cells are present, X chromosome inactivation occurs randomly in each cell. When any one of these cells divides by mitosis, the cells of the resulting clone (group of genetically identical cells) all have the same active X chromosome, and, therefore, a patch of cells that all express the same color develops.

Why, you might ask, is variegation not always apparent in females heterozygous at X-linked loci? The answer is that, although variegation usually occurs, we may need to use special techniques to observe it. For example, colorblindness is due to a defect involving the pigments in the cone cells in the retina of the eye (Chapter 41). In at least one type of red-green colorblindness, the retina of a heterozygous female actually contains patches of abnormal cones, but the patches of normal cones are sufficient to provide normal color vision. Variegation can be very hard to observe in the cases where cell products become mixed in bodily fluids. For instance, in females heterozygous for hemophilia, only half the cells responsible for producing a specific blood-clotting factor are capable of doing so, but they produce enough to ensure that the blood clots normally.

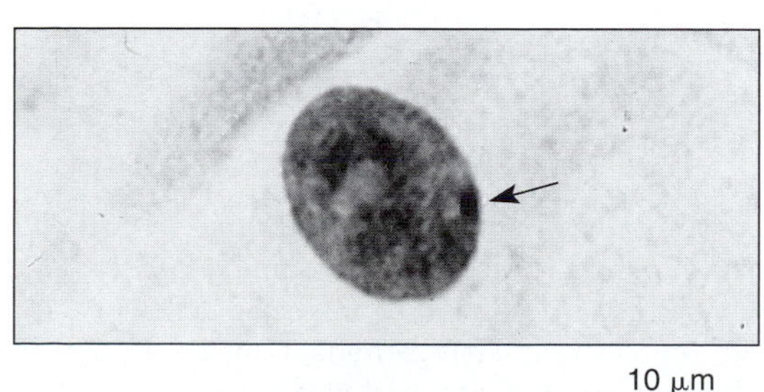

Figure 10–16 A Barr body. The darkly stained Barr body *(arrow)* at the edge of the nucleus in this LM is an inactivated X chromosome. The entire cell is not shown. *(Omikron/Photo Researchers, Inc.)*

Figure 10–17 A calico cat. This cat has X-linked genes for both black and yellow (or orange) pigmentation of the fur, but because of random X chromosome inactivation, black is expressed in some groups of cells and yellow (or orange) is expressed in others. Because other genes affecting fur color are also present, white patches are usually evident as well. *(Larime Photographic/Dembinsky Photo Associates)*

Sex-influenced genes are autosomal, but their expression is affected by the individual's sex

Not all characteristics that differ in the two sexes are X-linked. Certain **sex-influenced** traits are inherited through autosomal genes, but the *expression* of alleles at these loci can be altered or influenced by the sex of the animal. Therefore, males and females with the same genotype with respect to these loci may have different phenotypes.

Pattern baldness in humans, characterized by premature loss of hair on the front and top of the head, but not on the sides, is far more common among males than among females. It has been proposed that a single pair of autosomal alleles is involved, with the allele responsible for pattern baldness being dominant in males and recessive in females. Because of this unusual situation, we modify our notation, designating the pattern baldness allele as B_1 and the allele for normal hair growth as B_2. Individuals with the genotype B_1B_1 show pattern baldness, regardless of sex. Persons with a B_1B_2 genotype are bald if they are male but not bald if they are female. Individuals with the genotype B_2B_2 are not bald, regardless of sex.

Evidence suggests that the expression of most sex-influenced traits is strongly modified by sex hormones. For example, male hormones (see Chapter 48) are strongly implicated in the expression of pattern baldness.

THE RELATIONSHIP BETWEEN GENOTYPE AND PHENOTYPE IS OFTEN COMPLEX

The relationship between a given locus and the characteristic it controls may be simple: A single pair of alleles of a locus may regulate the appearance of a single characteristic of the organism (e.g., tall versus short in garden peas). Alternatively, the relationship may be more complex: A pair of alleles may participate in the control of several characteristics, or alleles of many loci may cooperate to regulate the appearance of a single characteristic. Not surprisingly, these more complex relationships are quite common.

We may assess the phenotype on one or many levels. It may be a morphological characteristic, such as shape, size, or color. It may be a physiological characteristic or even a biochemical trait, such as the presence or absence of a specific enzyme required for the metabolism of some specific molecule. In addition, the phenotypic expression of genes may be altered by changes in the environmental conditions under which the organism develops.

Dominance is not always complete

Studies of the inheritance of many traits in a wide variety of organisms have clearly shown that one member of a pair of alleles may not be completely dominant to the other. In such instances it is improper to use the terms *dominant* and *recessive.*

Process of Science

For example, the plants commonly known as four o'clocks *(Mirabilis jalapa)* may have red or white flowers. Each color breeds true when these plants are self-pollinated. What flower color might we expect in the offspring of a cross between a red-flowering plant and a white-flowering one? Without knowing which is dominant, we might predict that all would have red flowers or all would have white flowers. This cross was first made by the German botanist Karl Correns (one of the rediscoverers of Mendel's work), who found that all F_1 offspring have pink flowers! Does this result in any way prove that Mendel's assumptions about inheritance are wrong? Did the parental characteristics blend inseparably in the offspring? Quite the contrary, for when two of these pink-flowered plants are crossed, red-flowered, pink-flowered, and white-flowered offspring appear in a ratio of 1:2:1 (Fig. 10–18).

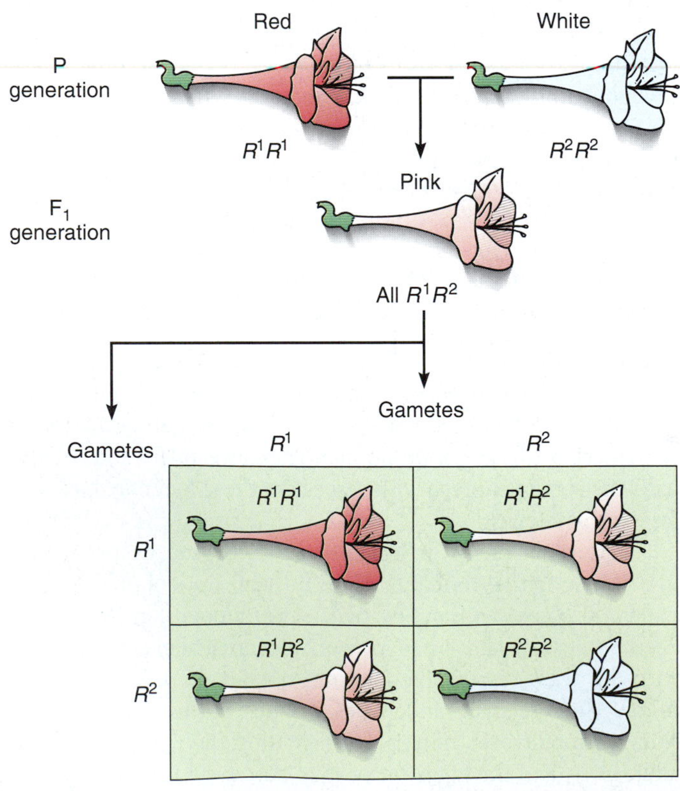

Figure 10–18 Incomplete dominance in four o'clocks. If a pair of alleles is incompletely dominant to each other, a heterozygote has a phenotype intermediate between its parents. Two incompletely dominant alleles, R^1 and R^2, are responsible for red, white, and pink flower colors. Red-flowered plants are R^1R^1; white-flowered plants are R^2R^2, and heterozygotes (R^1R^2) are pink. Note that uppercase notation is used for both alleles, because neither is recessive to the other.

In this instance, as in all other aspects of the scientific process, results that differ from those predicted prompt scientists to reexamine and modify their assumptions to account for the exceptional results. The pink-flowered plants are clearly the heterozygous individuals, and neither the red allele nor the white allele is completely dominant. When the heterozygote has a phenotype that is intermediate between those of its two parents, the genes are said to show **incomplete dominance.** In these crosses, the genotypic and phenotypic ratios are identical.

Incomplete dominance is not unique to four o'clocks. Numerous additional examples of incomplete dominance are known in both plants and animals. For example, white chickens and black chickens produce gray offspring when crossed.

In both cattle and horses, reddish coat color is not completely dominant to white coat color. Heterozygous individuals have a mixture of reddish hairs and white hairs, which is called roan. If you saw a white mare nursing a roan foal, what would you guess was the coat color of the foal's father? Because the reddish and white colors are expressed independently (hair by hair) in the roan heterozygote, we sometimes refer to this as a case of **codominance.** Strictly speaking, incomplete dominance refers to instances in which the heterozygote is intermediate in phenotype, and codominance refers to instances in which the heterozygote simultaneously expresses the phenotypes of both types of homozygotes.

The human ABO blood group provides a classic case of codominant alleles. Blood types O, A, B, and AB are controlled by three alleles representing a single locus. Allele I^A codes for the synthesis of a specific glycoprotein, antigen A, which is expressed on the surface of red blood cells. (Immunity is discussed in Chapter 43; for now we define antigens simply as substances capable of stimulating an immune response.) Allele I^B leads to the production of a different (but related) glycoprotein, antigen *B*. Allele i^O does not code for an antigen, although it is allelic to I^A and I^B. As shown in Table 10–1, persons with the genotype I^AI^A or I^Ai^O have blood type A; those with genotype I^BI^B or I^Bi^O have blood type B; those with genotype I^AI^B have blood type AB; and those with genotype i^Oi^O have blood type O. These results show that neither allele I^A nor allele I^B is dominant to the other; they are both expressed phenotypically in the heterozygote and are therefore codominant to each other, although each is dominant to allele i^O.

Determining blood types was one of the traditional ways of settling cases of disputed parentage. While blood type tests can exclude someone as a possible parent of a particular child, they can never prove that a certain person is the parent; they can only determine that he or she *could* be. Could a man with blood type AB be the father of a child with blood type O? Could a woman with blood type O be the mother of a child with blood type AB? Could a type-B child with a type-A mother have a type-A father or a type-O father?[3]

Multiple alleles for a locus may exist in a population

Most of the examples given so far have dealt with situations in which each locus was represented by a maximum of two allelic variants. It is true that a single diploid individual has a maximum of two different alleles for a particular locus and that a haploid gamete has only one allele for each locus. However, if we survey a population, we may find more than two alleles for a particular locus, as we saw with the ABO blood group. If three or more alleles for a given locus exist within the population, we say that locus has **multiple alleles.** A great many loci can be shown to have multiple alleles if the population is surveyed carefully. Some alleles can be identified by the activity of a certain enzyme or by some other biochemical feature but do not produce an obvious phenotype. Others produce a readily recognizable phenotype,

[3] The answer to all these questions is "no."

TABLE 10–1 ABO Blood Types*

Phenotype (blood type)	Genotypes	Antigen on RBC	Antibodies to A or B Antigens in Plasma	Frequency in U.S. Population (%)	
				Western European Descent	African Descent
A	I^AI^A, I^Ai^O	A	Anti-B	45	29
B	I^BI^B, I^Bi^O	B	Anti-A	8	17
AB	I^AI^B	A, B	None	4	4
O	i^Oi^O	None	Anti-A, anti-B	43	50

*This table and the discussion of the ABO system have been simplified somewhat. Note that persons produce antibodies against the antigens *lacking* on their own red blood cells (RBCs). Because of their specificity for the corresponding antigens, these antibodies are used in standard tests to determine blood types.

and certain patterns of dominance can be discerned when the alleles are combined in various ways.

In rabbits, for example, a *C* allele causes a fully colored coat. The homozygous recessive genotype, *cc,* causes albino coat color. There are two additional allelic variants of the same locus, c^h and c^{ch}. The genotype c^hc^h causes the "Himalayan" pattern, in which the body is white but the tips of the ears, nose, tail, and legs are colored (similar to the color pattern of a Siamese cat). An individual with the genotype $c^{ch}c^{ch}$ has the "chinchilla" pattern, in which the entire body has a light gray color. On the basis of the results of genetic crosses, these alleles can be arranged in the following series: $C > c^h > c^{ch} > c$. Each allele is dominant to those following it and recessive to those preceding it. For example, a c^hc^{ch} rabbit has the "Himalayan" pattern, whereas a $c^{ch}c$ rabbit has the "chinchilla" pattern. In some other series of multiple alleles, certain alleles may be codominant and others incompletely dominant; hence the heterozygotes commonly have phenotypes intermediate between those of their parents.

A single gene may affect multiple aspects of the phenotype

In the examples presented so far, the relationship between a gene and its phenotype has been direct, precise, and exact, and the loci considered have controlled the appearance of single traits. However, the relationship of gene to characteristic may be quite complex. Most genes probably have many different effects, a quality referred to as **pleiotropy.** Most cases of pleiotropy can be traced to a single fundamental cause. For example, a defective enzyme may affect the functioning of many types of cells. Pleiotropy is evident in many genetic diseases in which multiple symptoms are caused by a single pair of alleles. For example, individuals who are homozygous for the recessive allele that causes cystic fibrosis produce abnormally thick mucus in many parts of the body, including the respiratory, digestive, and reproductive systems (see Chapter 15).

Alleles of different loci may interact to produce a phenotype

Several pairs of alleles may interact to affect a single phenotype, or one pair may inhibit or reverse the effect of another pair. More than 12 pairs of alleles interact in various ways to produce coat color in rabbits, and more than 100 pairs are concerned with eye color and shape in fruit flies.

One type of gene interaction is illustrated by the inheritance of combs in poultry, where two genes may interact to produce a novel phenotype (Fig. 10–19). The allele for a rose comb, *R,* is dominant to that for a single comb, *r.* A second, unlinked pair of alleles governs the inheritance of a pea comb, *P,* versus a single comb, *p.* A single-combed fowl is homozygous for the recessive allele at both loci *(pprr).* A rose-combed fowl is either *ppRR* or *ppRr,* and a pea-combed fowl is either *PPrr* or *Pprr.* When an *R* and *P* occur in the same individual, the phenotype is neither a pea nor a rose comb but a completely different type, called a *walnut comb.* The walnut comb phenotype is produced whenever a fowl has one or two *R* alleles, plus one or two *P* alleles (i.e., *PPRR, PpRR, PPRr,* or *PpRr*). What would you predict about the types of combs among the offspring of two heterozygous walnut-combed fowl, *PpRr?* How does this form of gene interaction affect the ratio of phenotypes in the F_2 generation? Is it the typical Mendelian 9:3:3:1 ratio?

Epistasis is a common type of gene interaction in which the presence of certain alleles of one locus can prevent or mask the expression of alleles of a different locus. (Literally, epistasis means "standing upon.") We have already seen that coat color in guinea pigs can be determined by the *B* and *b* allelic pair, with the *B* allele for black coat dominant to the *b* allele for brown coat. The expression of either phenotype, however, depends on the presence of a dominant allele at yet another locus. This allele, *C,* codes for the enzyme tyrosinase, which converts a colorless precursor to the pigment melanin and hence is required for the production of any kind of pigment. The recessive allele *(c)* codes for an inactive form of the enzyme. Thus, an animal that is homozygous recessive for this allele lacks the enzyme and produces no melanin. It is therefore a white-coated, pink-eyed albino, regardless of the combination of *B* and *b* alleles. Albinism, or lack of melanin pigment, is not restricted to guinea pigs but is found in humans and a variety of other animals (Fig. 10–20).

When an albino guinea pig with the genotype *ccBB* is mated to a brown guinea pig with the genotype *CCbb,* the F_1 generation is black-coated, *CcBb.* When two such animals are mated, their offspring appear black-coated, brown-coated, and albino in a ratio of 9:3:4. (Make a Punnett square to verify this.)

You might wonder why heterozygous *Cc* individuals do not show at least some lightening of the coat color, given the fact

Figure 10–19 Gene interaction. Two gene pairs govern these types of genetically determined combs in roosters. Note that four different genotypes have walnut combs.

Figure 10–20 Epistasis. Homozygous alleles for albinism exhibit epistasis, masking the expression of alleles of other loci that govern production of melanin pigment. Albino individuals, such as this albino koala, occur occasionally in nature. *(Tom McHugh/Photo Researchers, Inc.)*

that they produce only about half the normal amount of tyrosinase enzyme. It turns out that half the normal amount of enzyme is usually adequate to produce normal amounts of pigment. This type of situation applies to many enzymes and explains many (although certainly not all) cases of dominance.

Polygenes act additively to produce a phenotype

The inherited components of many human characteristics, such as height, body form, and skin color, are not inherited through alleles at a single locus. The same holds true for many commercially important characteristics in domestic plants and animals, such as milk and egg production. Alleles at several, perhaps many, different loci affect each characteristic. The term **polygenic inheritance** is applied when multiple independent pairs of genes have similar and additive effects on the same characteristic.

Polygenes are responsible for the inheritance of skin color in humans. It is now thought that alleles representing three to four different loci are involved in determining skin color. We illustrate the principle of polygenic inheritance with pairs of alleles at three unlinked loci. These can be designated *A* and *a, B* and *b,* and *C* and *c* (Fig. 10–21). The capital letters represent incompletely dominant alleles producing dark skin. The more capital letters, the darker the skin, because the alleles affect skin color in an additive fashion. A person with the darkest skin would have the genotype *AABBCC,* and a person with the lightest skin would have the genotype *aabbcc.* The F_1 offspring of an *aabbcc* person and an *AABBCC* person are all *AaBbCc* and have an intermediate skin color. The F_2 offspring of two such triple heterozygotes would have skin colors ranging from very dark to very light.

Polygenic inheritance is therefore characterized by an F_1 generation that is intermediate between the two completely homozygous parents and by an F_2 generation that shows wide variation between the two parental types. Most of the F_2 generation individuals have one of the intermediate phenotypes; only a few show the extreme phenotypes of the grandparents (P generation). On average, only 1 of 64 is as dark as the very dark grandparent, and only 1 of 64 is as light as the very light grandparent (Fig. 10–22). The alleles *A, B,* and *C* each produce about the same amount of darkening of the skin; hence, the genotypes *AaBbCc, AABbcc, AAbbCc, AaBBcc, aaBBCc, AabbCC,* and *aaBbCC* all produce similar intermediate phenotypes.

The model used here for the inheritance of skin color in humans is a rather simple example of polygenic inheritance because only three major allelic pairs are used. The inheritance of height in humans involves alleles representing ten or more loci. Because many allelic pairs are involved and because height is modified by a variety of environmental conditions, the height of adults ranges from perhaps 125 to 215 cm. If we were to measure the heights of 1000 adult American men selected at random, we would find that only a few are as tall as 215 cm or as short as 125 cm. The heights of most would cluster around the mean, about 170 cm. When the number of men at each height is plotted against height (in centimeters) and the points are connected, the result is a bell-shaped curve, called a **normal distribution curve** (Fig. 10–23).

SELECTION, INBREEDING, AND OUTBREEDING ARE USED TO DEVELOP IMPROVED STRAINS

How do geneticists go about establishing a breed of cow that will give more milk, a strain of hens that will lay bigger eggs, or a variety of corn with more kernels per ear? By selecting organisms that manifest the desired phenotype and using these individuals in further matings, a true-breeding strain with the commercially advantageous phenotype is gradually developed. Such a strain is expected to be homozygous for all of the genes involved, whether they be dominant, recessive, or additive in their effects.

There is a limit to the effectiveness of breeding by selection. When a strain becomes homozygous for all of the genes involved, further selective breeding cannot increase the desired quality. Moreover, because of **inbreeding**—the mating of two closely related individuals—the strain may become homozygous for multiple undesirable traits as well. Human inbreeding, for example, increases the frequency with which otherwise very rare genetic disorders are observed to occur in the population, although the individual risk is relatively small.

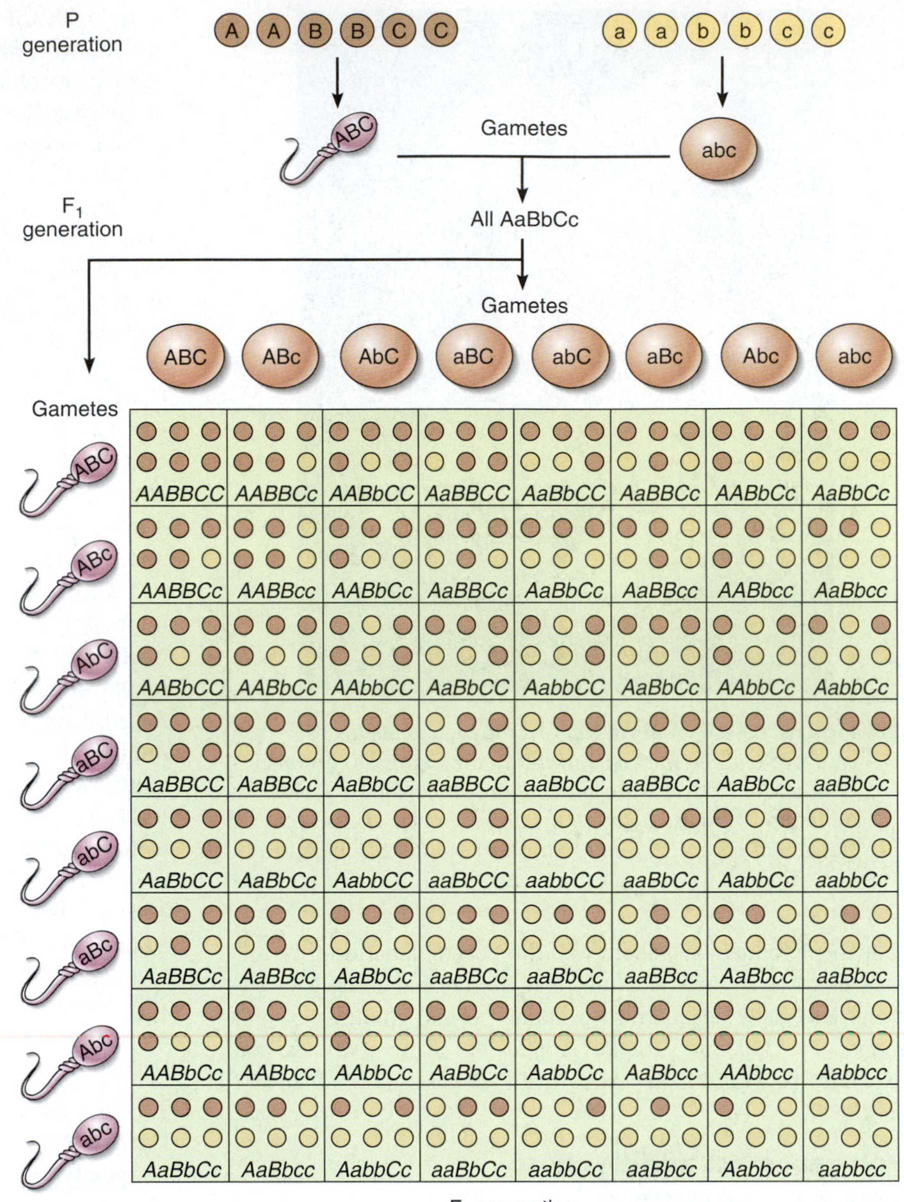

Figure 10–21 Polygenic inheritance. This model assumes that skin color in humans is governed by alleles of three unlinked loci. The alleles producing dark skin (*A, B,* and *C*) are represented by capital letters, but they are not dominant. Instead they have additive effects. If one parent is very dark and the other very light, their children (F_1) are intermediate in skin color. A wide range of skin colors are expected in the F_2. The number of dark dots (each signifying an allele producing dark skin) is counted to determine the phenotype. The results are summarized in Figure 10–22.

The mating of individuals of totally unrelated strains, termed **outbreeding,** frequently leads to offspring much better adapted for survival than either parent. Such improvement reflects a phenomenon called **hybrid vigor.** A large proportion of the corn, wheat, and other crops grown in the United States consists of hybrid strains. Each year the seed to grow these crops must be obtained by mating the original strains. The hybrids are heterozygous at a great many loci and give rise, even when self-fertilized, to a wide variety of forms, few of which are as good as the original hybrid. (The seeds produced by F_1 hybrid corn plants are not normally planted, but eaten instead!)

The reason for hybrid vigor has long been a matter of debate, and, in fact, there may be multiple causes. One explanation may be that each of the parental strains is homozygous for cer-

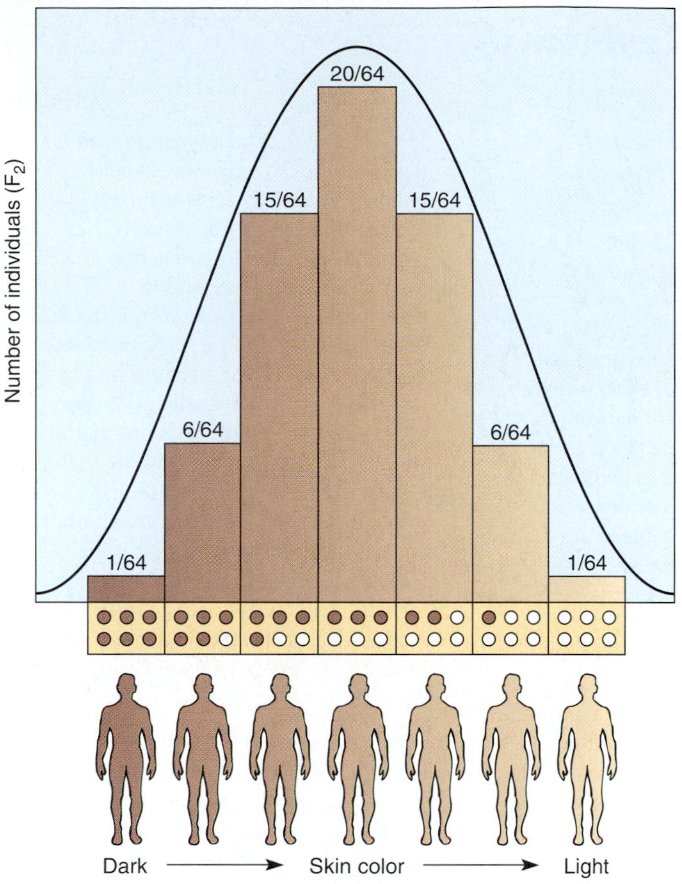

Figure 10–22 Distribution of phenotypes in polygenic inheritance. The bars indicate the expected phenotypic ratios in the F_2 generation shown in Figure 10–21. This expected distribution of phenotypes is consistent with the superimposed normal distribution curve.

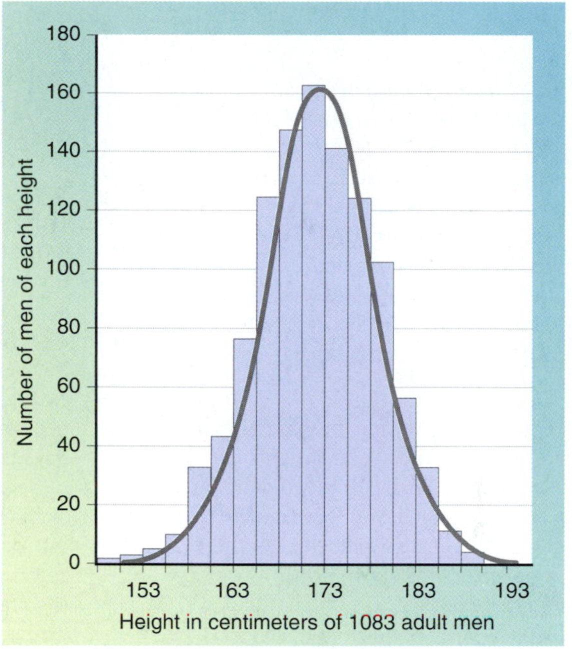

Figure 10–23 Polygenic inheritance in a population. The distribution of heights of 1083 adult males approximates a normal curve, which is consistent with continuous phenotypic variation in the population. The bars indicate the actual number of men whose heights were within the unit range. For example, there were 163 men whose heights were between 170.5 and 173.0 cm.

tain undesirable recessive genes, but any two strains are homozygous for different undesirable genes. Each strain contains dominant genes to make up for the recessive, undesirable genes of the other strain. The hybrid offspring would express all the desirable (dominant) traits and none of the undesirable (recessive) traits of the two parental strains.

Another explanation is that hybrid vigor is due to **heterozygote advantage,** that is, the superiority of the heterozygous genotype to either homozygous genotype. The key may be that a particular allele may have advantages under one set of conditions, but that a different allele may be favored when conditions change. Therefore an individual with two different alleles for a locus may be able to function over a wider range of conditions. This view is supported by experimental evidence that the metabolism of an individual heterozygous at many loci tends to be more stable and less affected by environmental changes.

A case of heterozygote advantage with a rather specific explanation is seen in humans who are heterozygous for the recessive sickle cell anemia allele *(s)* and the normal dominant allele *(S)* (see Chapter 18). These *Ss* individuals appear to have increased resistance to the parasite that lives inside red blood cells and causes malaria. Such resistance is a significant advantage in areas of the world where malaria is still uncontrolled. Homozygous normal individuals *(SS)* appear to be less resistant to malaria; homozygous sickle cell individuals *(ss)* are at a distinct disadvantage due to severe anemia and other serious effects of the sickle cell allele (see Chapter 15).

FOCUS ON Solving Genetics Problems

Simple Mendelian genetics problems are like puzzles. They can be fun and easy to work if you follow certain conventions and are methodical in your approach.

1. Always use standard designations for the generations. The generation with which a particular genetic experiment is begun is called the *P*, or *parental, generation.* Offspring of this generation (the "children") are called the F_1, or *first filial, generation.* The offspring resulting when two F_1 individuals are bred constitute the F_2, or *second filial, generation* (the "grandchildren").
2. Write down a key for the symbols you are using for the allelic variants of each locus. Use uppercase to designate a dominant allele and lowercase to designate a recessive allele. Use the same letter of the alphabet to designate both alleles of a particular locus. If you are not told which is dominant and which is recessive, the phenotype of the F_1 generation is a good clue.
3. Determine the genotypes of the parents of each cross by making use of the following types of evidence:
   a. Are they from true-breeding lines? If so, they should be homozygous.
   b. Can their genotypes be reliably deduced from their phenotypes? This is usually true if they express the recessive phenotype.
   c. Do the phenotypes of their offspring provide any information? *See Focus On: Deducing Genotypes* for an example of how these determinations can be made.
4. Indicate the possible kinds of gametes formed by each of the parents. It is helpful to draw a circle around the symbols for each kind of gamete.
   a. If it is a monohybrid cross, we must apply the principle of segregation; i.e., a heterozygote *Aa* forms two kinds of gametes: *A* and *a*. Of course a homozygote, such as *aa,* forms only one kind of gamete: *a.*
   b. If it is a dihybrid cross, we must apply both the principle of segregation *and* the principle of independent assortment. For example, an individual heterozygous for two loci would have the genotype *AaBb.* Allele *A* segregates from *a,* and *B* segregates from *b.* The assortment of *A* and *a* into gametes is independent of the assortment of *B* and *b.* Therefore *A* is equally likely to end up in a gamete with *B* or *b.* The same is true for *a.*

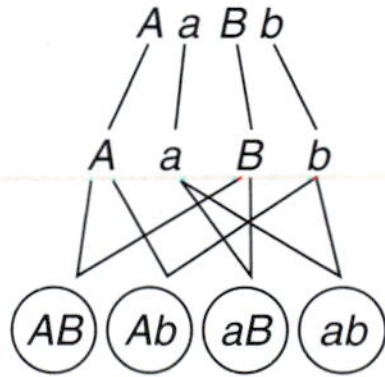

Segregation of the alleles of each locus

Independent assortment of alleles of different loci

5. Set up a Punnett square, placing the possible types of gametes from one parent down the left side and the possible types from the other parent across the top.
6. Fill in the Punnett square and read off (and sum up) the genotypic and phenotypic ratios of the offspring. Avoid confusion by consistently placing the dominant allele first and the recessive allele second in heterozygotes (*Aa,* never *aA*). If it is a dihybrid cross, it is very important to always write the two alleles of one locus first and the two alleles of the other locus second. It does not matter which locus you choose to write first, but once you have decided on the order, it is critical that you maintain it consistently. This means that if the individual is heterozygous for both loci, you will always use the form *AaBb.* Writing this particular genotype as *aBbA,* for example, would cause confusion.
7. If you do not need to know the frequencies of all the expected genotypes and phenotypes, you may use the rules of probability as a shortcut. For example, if both parents are *AaBb,* what is the probability of an *AABB* offspring? To be *AA,* the offspring must receive an *A* gamete from each parent. The probability that a given gamete is *A* is 1/2 and each gamete represents an independent event, so we combine their probabilities by multiplying ($1/2 \times 1/2 = 1/4$). The probability of *BB* is calculated similarly and is also 1/4. The probability of *AA* is independent of the probability of *BB,* so again we use the product rule to obtain their combined probabilities ($1/4 \times 1/4 = 1/16$).

FOCUS ON Deducing Genotypes

The science of genetics resembles mathematics in that it consists of a few basic principles, which, once grasped, enable the student to solve a wide variety of problems. Very often the genotypes of the parents can be deduced from the phenotypes of their offspring. In chickens, for example, the allele for rose comb *(R)* is dominant to the allele for single comb *(r)*. Suppose that a cock is mated to three different hens, as shown in the figure. The cock and hens A and C have rose combs; hen B has a single comb. Breeding the cock with hen A produces a rose-combed chick, with hen B a single-combed chick, and with hen C a single-combed chick. What types of offspring can be expected from further matings of the cock with these hens?

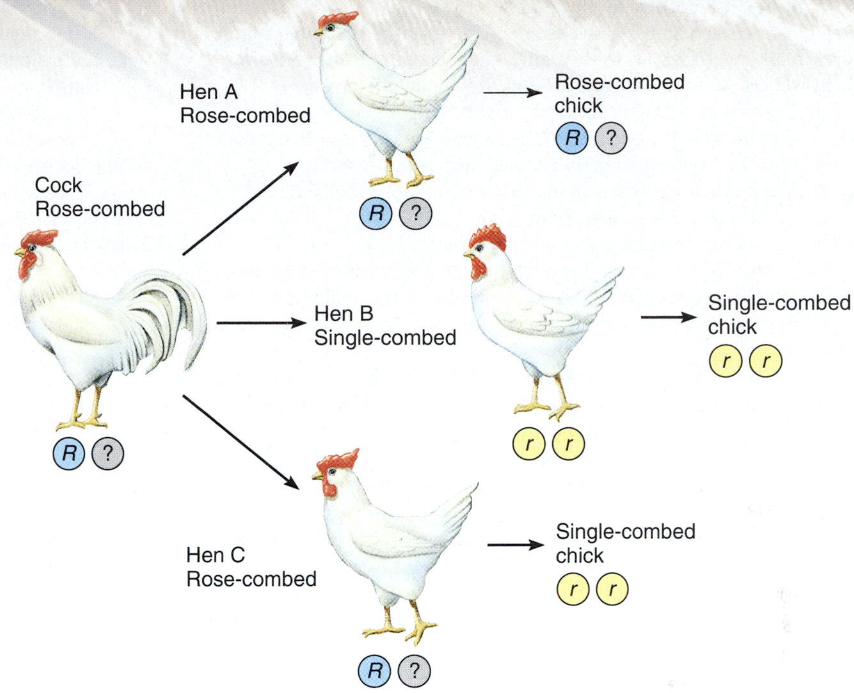

Because the allele for single comb, *r*, is recessive, all hens and chicks that are phenotypically single-combed must be *rr*. We can deduce that hen *B* and the offspring of hens *B* and *C* are genotypically *rr*.

All individuals that are phenotypically rose-combed must have at least one *R* allele. The fact that the offspring of the cock and hen B was single-combed proves that the cock is heterozygous *Rr*, because, although the single-combed chick received one *r* allele from its mother, it must have received the second one from its father.

The fact that the offspring of the cock and hen C had a single comb proves that hen C is heterozygous, *Rr*. It is impossible to determine from the data given whether hen A is homozygous *RR* or heterozygous *Rr;* further breeding would be necessary to determine this. (Can you suggest an appropriate mating?)

Additional matings of the cock with hen B should result in half rose-combed and half single-combed individuals; additional matings of the cock with hen C should produce three-fourths rose-combed and one-fourth single-combed chicks.

SUMMARY WITH KEY TERMS

I. **Genetics** is the study of the structure, expression, and transmission of genetic information. The transfer of genetic information from parent to offspring is called **heredity.**

II. Mendel's inferences about inheritance obtained from his garden pea-breeding experiments have been tested repeatedly in all kinds of diploid organisms and found to be generally true. These principles have been extended and now can be stated in a more modern form.

   A. Today we know that the **genes** are in chromosomes; the site a gene occupies in the chromosome is its **locus.**

   B. Different forms of a particular gene are **alleles;** they occupy corresponding loci on homologous chromosomes. Genes therefore exist as pairs of alleles in diploid individuals.

   C. An individual that carries two identical alleles is said to be **homozygous** for that locus. If the two alleles are different, that individual is said to be **heterozygous** for that locus.

   D. One allele (the **dominant allele**) may mask the expression of the other allele (the **recessive allele**) in a heterozygous individual. For this reason two individuals with the same appearance **(phenotype)** may differ from each other in genetic constitution **(genotype).**

   E. Dominance does not always apply, and alleles can be **incompletely dominant,** in which the heterozygote is intermediate in phenotype, or **codominant,** in which the heterozygote simultaneously expresses the phenotypes of both homozygotes.

   F. According to Mendel's **principle of segregation,** during meiosis the alleles for each locus separate, or segregate, from each other as the homologous chromosomes separate. When haploid gametes are formed, each contains only one allele for each locus.

   G. According to Mendel's **principle of independent assortment,** alleles of different loci are distributed randomly into the gametes. This can result in **recombination,** that is, production of new gene combinations that were not present in the **parental (P) generation.**

   H. Each chromosome behaves genetically as if it were composed of genes arranged in a linear order. Genes in the same chromosome are **linked.**

1. Recombination of linked genes can occur as a result of **crossing-over** (breaking and rejoining of homologous chromatids) in meiotic prophase I.
2. By measuring the frequency of recombination between linked genes, it is possible to construct a linkage map of a chromosome.

I. A cross between homozygous parents (P generation) that differ from each other with respect to their alleles at one locus is called a **monohybrid cross;** if they differ at two loci, it is a **dihybrid cross.**
   1. The first generation of offspring is heterozygous and is called the first filial, or **F_1, generation;** the generation produced by a cross of two F_1 individuals is the second filial, or **F_2, generation.**
   2. A **test cross** is between an individual of unknown genotype and a homozygous recessive individual.

III. Genetic ratios can be expressed in terms of probabilities.
   A. Any probability is expressed as a fraction or decimal fraction, calculated as the number of favorable events divided by the total number of events. This can range from 0 (an impossible event) to 1 (a certain event).
   B. According to the **product rule,** the probability of two independent events occurring together can be obtained by multiplying the probabilities of each occurring separately.
   C. According to the **sum rule,** the probability of an outcome that can be obtained in more than one way can be calculated by adding the separate probabilities.

IV. The sex of humans and many other animals is determined by the **X** and **Y sex chromosomes** or their equivalents. **Autosomes** are chromosomes other than sex chromosomes.
   A. Normal female mammals have two X chromosomes; normal males have one X and one Y.
   B. The fertilization of an X-bearing egg by an X-bearing sperm results in a female (XX) zygote. The fertilization of an X-bearing egg by a Y-bearing sperm results in a male (XY) zygote.
   C. The Y chromosome is responsible for determining male sex in mammals.
   D. The X chromosome contains many important genes unrelated to sex determination that are required by both males and females. A male receives all his **X-linked** genes from his mother. A female receives X-linked genes from both parents.
   E. A female mammal shows **dosage compensation** of X-linked genes. Only one of the two X chromosomes is expressed in each cell; the other is inactive and is seen as a dark-staining **Barr body** at the edge of the interphase nucleus.

V. **Multiple alleles** (three or more alleles that can potentially occupy a particular locus) may exist in a population. A diploid individual has any two of the alleles; a haploid individual or gamete has only one.

VI. The relationship between a gene and its phenotype may be quite complex.
   A. Most genes have many different effects; they are **pleiotropic.**
   B. Many types of gene interactions are known. In **epistasis,** an allele of one locus can mask the expression of alleles of a different locus.
   C. In **polygenic inheritance,** multiple independent pairs of genes may have similar and additive effects on the phenotype.
      1. Many human characteristics showing continuous variation, such as height and skin color, as well as many characteristics in other animals and plants, are inherited through polygenes.
      2. In polygenic inheritance, the F_1 generation is intermediate between the two parental types and shows little variation; the F_2 generation shows wide variation.

VII. **Inbreeding,** the mating of two closely related individuals, greatly increases the probability that an individual offspring will be homozygous for one or more recessive genes. **Outbreeding,** the mating of totally unrelated individuals, increases the probability that the offspring will be heterozygous at many loci and exhibit **hybrid vigor.**

POST-TEST

1. One reason why Mendel was able to discover the basic principles of inheritance is that he (a) understood the behavior of chromosomes in mitosis and meiosis (b) studied a wide variety of experimental organisms (c) began by establishing true-breeding lines (d) studied various types of linkage (e) studied hybrids between parents that differed in many, often not clearly defined, ways
2. One of the autosomal loci controlling eye color in fruit flies has two alleles, one for brown eyes and the other for red eyes. Fruit flies from a true-breeding line with brown eyes were crossed with flies from a true-breeding line with red eyes. The F_1 flies had red eyes. What conclusion can be drawn from this experiment? (a) these alleles underwent independent assortment (b) these alleles underwent segregation (c) these genes are X-linked (d) the allele for red eyes is dominant to the allele for brown eyes (e) all of the above are true
3. The F_1 flies described in question 2 were mated with brown-eyed flies from a true-breeding line. What phenotypes would you expect the offspring to have? (a) all red eyes (b) all brown eyes (c) half red eyes and half brown eyes (d) red-eyed females and brown-eyed males (e) brown-eyed females and red-eyed males
4. The type of cross described in question 3 is a(an) (a) F_2 cross (b) dihybrid cross (c) test cross (d) two-point test cross (e) none of the above

Use the following information to answer questions 5 through 8:

In peas, the allele for round seeds *(R)* is dominant to that for wrinkled seeds *(r);* the allele for yellow seeds *(Y)* is dominant to that for green seeds *(y).* These loci are unlinked. Plants from a true-breeding line with round, green seeds are crossed with plants from a true-breeding line with wrinkled, yellow seeds. These parents constitute the P generation.

5. The genotypes of the P generation are (a) *RRrr* and *Yyyy* (b) *RrYy* (c) *RRYY* and *rryy* (d) *RRyy* and *rrYY* (e) *RR* and *YY*
6. What are the expected genotypes of the F_1 hybrids produced by the described cross? (a) *RRrr* and *YYyy* (b) all *RrYy* (c) *RRYY* and *rryy* (d) *RRyy* and *rrYY* (e) *RR* and *YY*
7. What kinds of gametes can the F_1 individuals produce? (a) *RR* and *YY* (b) *Rr* and *Yy* (c) *RR, rr, YY,* and *yy* (d) *R, r, Y,* and *y* (e) *RY, Ry, rY,* and *ry*
8. What is the expected proportion of F_2 wrinkled, yellow seeds? (a) 9/16 (b) 1/16 (c) 3/16 (d) 1/4 (e) zero
9. Individuals of genotype *AaBb* were crossed with *aabb* individuals. Approximately equal numbers of the following classes of offspring were produced: *AaBb, Aabb, aaBb,* and *aabb.* These results illustrate Mendel's principle(s) of (a) linkage (b) independent assortment (c) segregation (d) a and c (e) b and c
10. Assume that the ratio of females to males is 1:1. A couple already has two daughters and no sons. If they plan to have a total of six children, what is the probability that they will have a family of all girls? (a) 1/4 (b) 1/8 (c) 1/16 (d) 1/32 (e) 1/64
11. Red-green colorblindness is an X-linked recessive disorder in humans. Your friend is the daughter of a colorblind father. Her mother had normal color vision, but her maternal grandfather was colorblind. What is the probability that your friend is colorblind? (a) 1 (b) 1/2 (c) 1/4 (d) 3/4 (e) zero
12. When homozygous, a particular allele of a locus in rats causes abnormalities of the cartilage throughout the body, an enlarged heart, slow development, and death. This is an example of (a) pleiotropy (b) polygenic inheritance (c) epistasis (d) codominance (e) dosage compensation

REVIEW QUESTIONS

1. In peas, yellow seed color is dominant to green. Predict the phenotypes (and their proportions) of the offspring of the following crosses: (a) homozygous yellow X green (b) heterozygous yellow X green (c) heterozygous yellow X homozygous yellow (d) heterozygous yellow X heterozygous yellow
2. If two animals heterozygous for a single pair of alleles are mated and have 200 offspring, about how many would be expected to have the phenotype of the dominant allele (i.e., to look like the parents)?
3. When two long-winged flies were mated, the offspring included 77 with long wings and 24 with short wings. Is the short-winged condition dominant or recessive? What are the genotypes of the parents?
4. A blue-eyed man, both of whose parents were brown-eyed, married a brown-eyed woman whose father was blue-eyed and whose mother was brown-eyed. If brown is dominant to blue, what are the genotypes of the individuals involved?
5. Outline a breeding procedure whereby a true-breeding strain of red cattle could be established from a roan bull and a white cow.
6. What is the probability of rolling a seven with a pair of dice? Which is a more likely outcome, rolling a six with a pair of dice or rolling an eight?
7. In rabbits, spotted coat *(S)* is dominant to solid color *(s)*, and black *(B)* is dominant to brown *(b)*. These loci are unlinked. A brown, spotted rabbit from a pure line is mated to a solid black one, also from a pure line. What are the genotypes of the parents? What would be the genotype and phenotype of an F_1 rabbit? What would be the expected genotypes and phenotypes of the F_2 generation?
8. The long hair of Persian cats is recessive to the short hair of Siamese cats, but the black coat color of Persians is dominant to the brown-and-tan coat color of Siamese. Make up appropriate symbols for the alleles of these two unlinked loci. If a pure black, long-haired Persian is mated to a pure brown-and-tan, short-haired Siamese, what will be the appearance of the F_1 offspring? If two of these F_1 cats are mated, what is the chance that a long-haired, brown-and-tan cat will be produced in the F_2 generation? (Use the shortcut probability method to obtain your answer; then check it with a Punnett square.)
9. Mr. and Mrs. Smith are concerned because their own blood types are A and B respectively, but their new son, Richard, is blood type O. Could Richard be the child of these parents?
10. The expression of an allele called *frizzle* in fowl causes abnormalities of the feathers. As a consequence, the animal's body temperature is lowered, adversely affecting the functions of many internal organs. When one gene affects many characteristics of the organism in this way, we say that gene is ____________.
11. A walnut-combed rooster is mated to three hens. Hen A, which is walnut-combed, has offspring in the ratio of 3 walnut : 1 rose. Hen B, which is pea-combed, has offspring in the ratio of 3 walnut:3 pea:1 rose:1 single. Hen C, which is walnut-combed, has only walnut-combed offspring. What are the genotypes of the rooster and the three hens?
12. What kinds of matings result in the following phenotypic ratios? (a) 3:1 (b) 1:1 (c) 9:3:3:1 (d) 1:1:1:1
13. The weight of the fruit in a certain variety of squash is determined by allelic pairs for two loci: *AABB* produces fruits that average 2 kg each, and *aabb* produces fruits that are 1 kg each. Each allele represented by a capital letter adds 0.25 kg. When a plant that produces 2-kg fruits is crossed with a plant that produces 1-kg fruits, all of the offspring produce fruits that are 1.5 kg each. What would be the weights of the fruits produced by the F_2 plants if two of these F_1 plants were crossed?
14. The X-linked *barred* locus in chickens controls the pattern of the feathers, with the alleles *B* for barred pattern and *b* for no bars. If a barred female (X^BY) is mated to a nonbarred male (X^bX^b), what will be the appearance of the male and female progeny? (Recall that in birds males are homogametic and females are heterogametic.) Do you see any commercial usefulness for this result? (*Hint:* It is notoriously difficult to determine the sex of newly hatched chicks.)
15. Individuals of genotype *AaBb* were mated to individuals of genotype *aabb*. One thousand offspring were counted, with the following results: 474 *Aabb*, 480 *aaBb*, 20 *AaBb*, and 26 *aabb*. What is this type of cross known as? Are these loci linked? What are the two parental classes and the two recombinant classes of offspring? What is the percentage of recombination between these two loci? How many map units apart are they?
16. Genes *A* and *B* are 6 map units apart, and *A* and *C* are 4 map units apart. Which gene is in the middle if *B* and *C* are 10 map units apart? Which is in the middle if *B* and *C* are 2 map units apart?

Answers to these Review Questions are included in Appendix A with the Post-Test answers.

YOU MAKE THE CONNECTION

1. Would the development of the science of genetics in the 20th century have been any different if Gregor Mendel had never lived?
2. Sketch a series of diagrams showing each of the following, making sure to end each series with haploid gametes:
   (a) how a pair of alleles for a single locus segregates in meiosis
   (b) how the alleles of two unlinked loci assort independently in meiosis
   (c) how the alleles of two linked loci undergo genetic recombination
3. Can you always ascertain an organism's genotype for a particular locus if you know its phenotype? Conversely, if you are given an organism's genotype for a locus, can you always reliably predict its phenotype? Explain.

RECOMMENDED READINGS

Corcos, A., and F. Monaghan. *Mendel's Experiments on Plant Hybrids: A Guided Study.* Rutgers University Press, New Brunswick, 1993. This interpretive study of Mendel's paper includes information on his life as a monk and a scientist.

Ganetzky, B. "Tracking Down a Cheating Gene." *American Scientist,* Vol. 88, Mar.-Apr. 2000. An analysis of the *Segregation-Distorter* system in fruit flies, which violates the Mendelian principle of 1:1 segregation of alleles.

Heim, W.G. "What Is A Recessive Allele?" *The American Biology Teacher,* Vol. 53, Feb. 1991. A discussion of the molecular basis for the traits Mendel studied in garden peas.

Mendel, G. "Experiments in plant hybridization." Reprinted in *Genetics: Readings from Scientific American.* W.H. Freeman and Co., San Francisco, 1990. Try reading this translation of Mendel's classic paper from the perspective of other scientists of his time who lacked knowledge of chromosomes, mitosis, and meiosis.

There are several well-written, college-level genetics texts that cover the principles of genetics in eukaryotes. The following are three representative examples:

Atherly, A.G., J.R. Girton, and J.F. McDonald. *The Science of Genetics,* 1st ed. Saunders College Publishing, Philadelphia, 1999.

Griffiths, A.J.F., J.H. Miller, D.T. Suzuki, R.C. Lewontin, and W.M. Gelbart. *An Introduction to Genetic Analysis,* 7th ed. W.H. Freeman, New York, 2000.

Russell, P. *Genetics,* 5th ed. Benjamin/Cummings, Menlo Park, CA, 1998.

- Visit our Web site at **http://www.info.brookscole.com/solomonbergmartin** for links to chapter-related resources on the World Wide Web. Additional on-line materials relating to this chapter can also be found on our Web site.

See chapter activity on BioActive Learner CD for additional help in mastering the chapter's material. Icon location in the chapter's margins shows which topics have tutorials or simulations in the CD.

11

DNA: The Carrier of Genetic Information

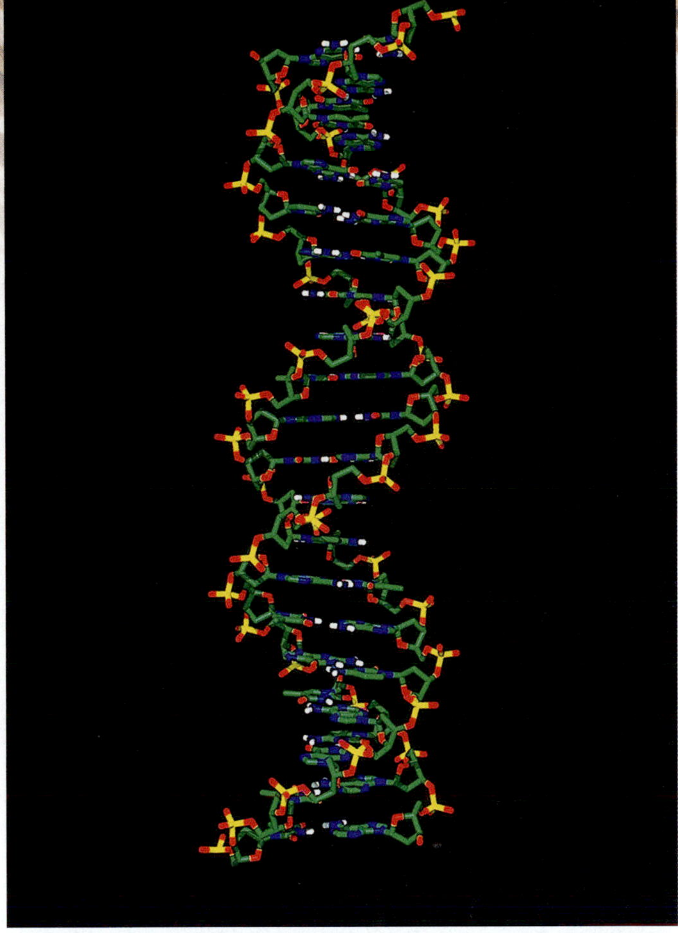

A model showing the structure of DNA (deoxyribonucleic acid). *(Prof. K. Seddon & Dr. T. Evans, Queen University, Belfast, Science Photo Library/Photo Researchers, Inc.)*

LEARNING OBJECTIVES

After you have studied this chapter you should be able to

1. Summarize the evidence that accumulated during the 1940s and early 1950s demonstrating that DNA is the genetic material.
2. Relate the chemical and physical features of DNA to the structure proposed by Watson and Crick.
3. Sketch how nucleotide subunits are linked together to form a single DNA strand.
4. Illustrate how the two strands of DNA are oriented with respect to each other.
5. State the base-pairing rules for DNA and describe how complementary bases bind to each other.
6. Cite experimental evidence that allowed scientists to differentiate between semiconservative replication of DNA and alternative models (conservative and dispersive replication).
7. Summarize how DNA replicates and identify some of the unique features of the process.
8. Explain the special constraints on DNA replication that cause it to be (1) bidirectional and (2) discontinuous in one strand and continuous in the other.
9. Compare the organization of DNA in prokaryotic and eukaryotic cells.

Following the rediscovery of Mendel's principles, geneticists conducted a variety of elegant experiments to learn how genes are arranged in chromosomes and how they are transmitted from generation to generation. However, the basic questions remained unanswered through most of the first half of the 20th century: What are genes made of? How do genes work? The studies of inheritance patterns described in Chapter 10 did not answer these questions, but they provided a foundation of knowledge that allowed scientists to make predictions about what the molecular (chemical) nature of genes might be and how genes might function.

Because scientists generally agreed that the function of genes must be to provide information, it followed that the molecules of which genes are made should have the ability to store information in a form that could be retrieved and used by the cell. But genes had other properties that had to be accounted for as well. For example, countless genetic experiments on a wide variety of organisms had demonstrated that genes are usually quite stable, being passed unchanged from generation to generation. However, occasionally a gene was observed to convert to a different form; such genetic changes, called **mutations,** were then transmitted unchanged to future generations.

As the science of genetics was developing, the science of biochemistry was flourishing as well. Not surprisingly, there was a growing effort to correlate the known properties of genes with the nature of the various biological molecules. What kind of molecule could store information? How could that information be retrieved and used to direct cellular functions? What kind of molecule could be relatively stable but have the capacity to change, resulting in a mutation, under some circumstances?

Some scientists thought that the problem could never be solved. They thought the information required by a cell to be so complex that no one type of molecule could function as the genetic material. However, as more was learned about the central role that proteins play in virtually every aspect of cellular structure and metabolism, other scientists considered them the prime candidates for the genetic material. However, protein did not turn out to be the molecule responsible for inheritance. In this chapter we discuss how **deoxyribonucleic acid (DNA),** a nucleic acid that was once thought to be dull and uninteresting, was found to be the molecular basis of inheritance, and we examine the unique features of DNA that allow it to carry out this role.

EVIDENCE THAT DNA IS THE HEREDITARY MATERIAL WAS FIRST FOUND IN MICROORGANISMS

During the 1930s and early 1940s, most geneticists and biochemists paid little attention to DNA and RNA, and were convinced that the genetic material must be protein. In light of the accumulating evidence that genes control the production of proteins, discussed in Chapter 12, it certainly seemed likely that genes themselves must also be proteins. Proteins were known to contain more than 20 different kinds of amino acids in many different combinations, allowing each type of protein to have unique properties. Given their complexity and diversity compared with other molecules, proteins seemed to be the "stuff" of which genes are made.

In contrast, it had been established that DNA and other nucleic acids were made of only four nucleotides, and what was known about their arrangement made them seem relatively uninteresting to most scientists. For this reason, several early clues to the role of DNA were not widely recognized.

One of these clues had its origin in 1928, when Frederick Griffith made a curious observation concerning two strains of pneumococcus bacteria (Fig. 11–1). A smooth (S) strain, named for its formation of smooth colonies on a solid growth medium, was known to be **virulent,** or lethal. When living cells of this strain were injected into mice, the animals contracted pneumonia and died. Not surprisingly, the injected animals survived if the cells were first killed with heat. A related rough (R) strain, which forms colonies with a rough surface, was known to be **avirulent,** or nonlethal; mice injected with either living or heat-killed cells of this strain survived. However, when Griffith injected mice with a mixture of *heat-killed,* virulent S-strain cells and live avirulent R-strain cells, a high proportion of the mice died. Griffith was then able to isolate living S-strain cells from the dead mice.

Because neither the heat-killed S strain nor the living R strain could be converted to the living virulent form when injected by itself in the control experiments, something in the heat-killed cells appeared to convert the avirulent cells to the lethal form. This type of permanent genetic change in which the properties of one strain of dead cells are conferred on a different strain of living cells became known as **transformation.** It was widely hypothesized that transformation was caused by some chemical substance (called the "transforming principle") that was transferred from the dead bacteria to the living cells.

In 1944, O.T. Avery, C.M. MacLeod, and M. McCarty of the Rockefeller Institute chemically identified Griffith's transform-

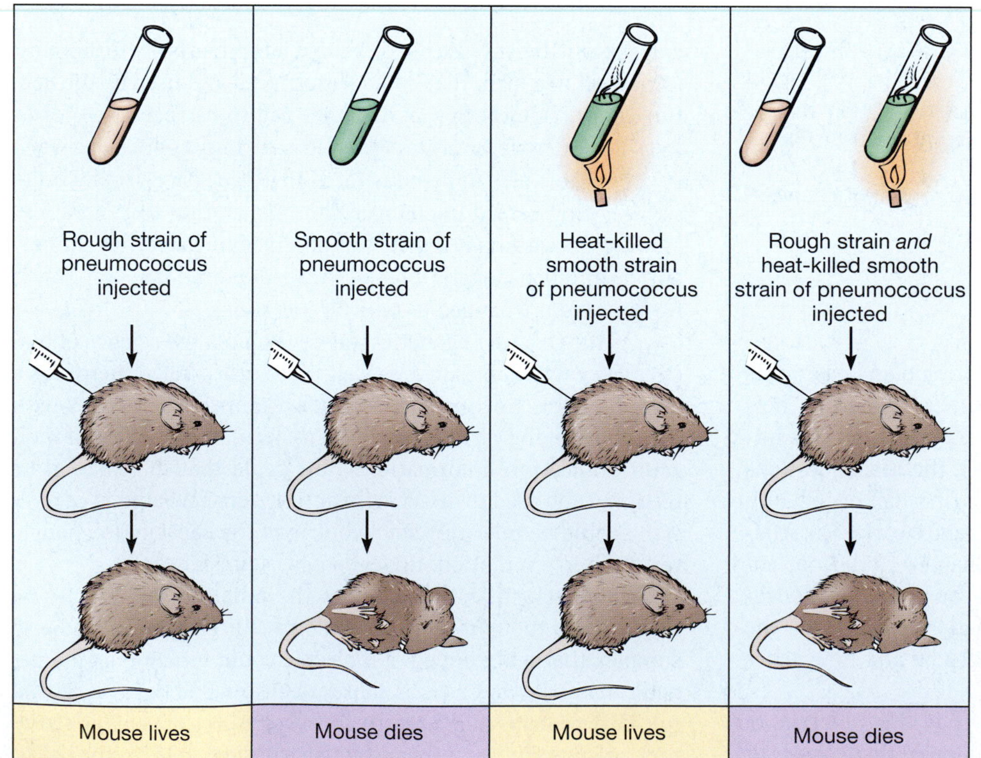

Figure 11–1 Griffith's transformation experiments. Although neither the rough (R) strain nor the heat-killed smooth (S) strain could kill a mouse, a combination of the two did. Autopsy of the dead mouse showed the presence of living, S-strain pneumococci. These results indicated that some substance in the heat-killed S strain was responsible for the transformation of the living R strain to a virulent form. Avery and his colleagues later showed that purified DNA isolated from the S strain confers virulence on the R-strain bacteria, establishing that DNA carries the necessary information for bacterial transformation.

ing principle as DNA. They did this through a series of very careful experiments in which they treated living R cells with highly purified DNA extracted from S cells, causing the R cells to become transformed into S cells. Although today we consider their findings to be the first demonstration that DNA is the genetic material, not all scientists of the time were convinced. One argument given was that the DNA preparations used might have been contaminated with a tiny amount of protein, which might have been responsible for the results. This was not a trivial objection, because it was well known that a very small amount of an enzyme could have significant biological effects. Furthermore, many scientists thought that their findings, even if true, might apply only to bacteria and not have any relevance for the genetics of eukaryotes.

During the next few years, new evidence accumulated that the haploid nuclei of pollen grains and gametes such as sperm contain only half the amount of DNA found in diploid somatic cells of the same species. Because the idea that genes are on chromosomes was generally accepted, these findings correlating DNA content with chromosome number provided strong circumstantial evidence of DNA's importance in inheritance in eukaryotes.

In 1952 Alfred Hershey and Martha Chase performed a series of elegant experiments (Fig. 11–2) on the reproduction of viruses that infect bacteria, known as **bacteriophages** (see Chapter 23). When they planned their experiments, they knew that bacteriophages reproduce inside a bacterial cell, eventually causing the cell to break open, releasing large numbers of new viruses. Because electron microscope studies had shown that only part of an infecting bacteriophage actually enters the cell, they reasoned that the genetic material should be included in that portion. They labeled the viral protein of one sample of bacteriophages with ^{35}S, a radioactive isotope of sulfur, and the viral DNA of a second sample with ^{32}P, a radioactive isotope of phosphorus. (Recall from Chapter 3 that proteins contain sulfur as part of the amino acid cysteine, and nucleic acids contain phosphate groups.) The bacteriophages in each sample were allowed to attach to bacteria and then shaken off by agitating them in a blender. The cells were then subjected to centrifugation (see Chapter 4). In the sample in which the proteins were labeled with ^{35}S, all of the label was subsequently found in the supernatant, indicating that the protein had not entered the cells. In the sample in which the DNA was labeled with ^{32}P, the label was found associated with the bacterial cells (in the pellet), indicating that DNA had actually entered the cells. Hershey and Chase

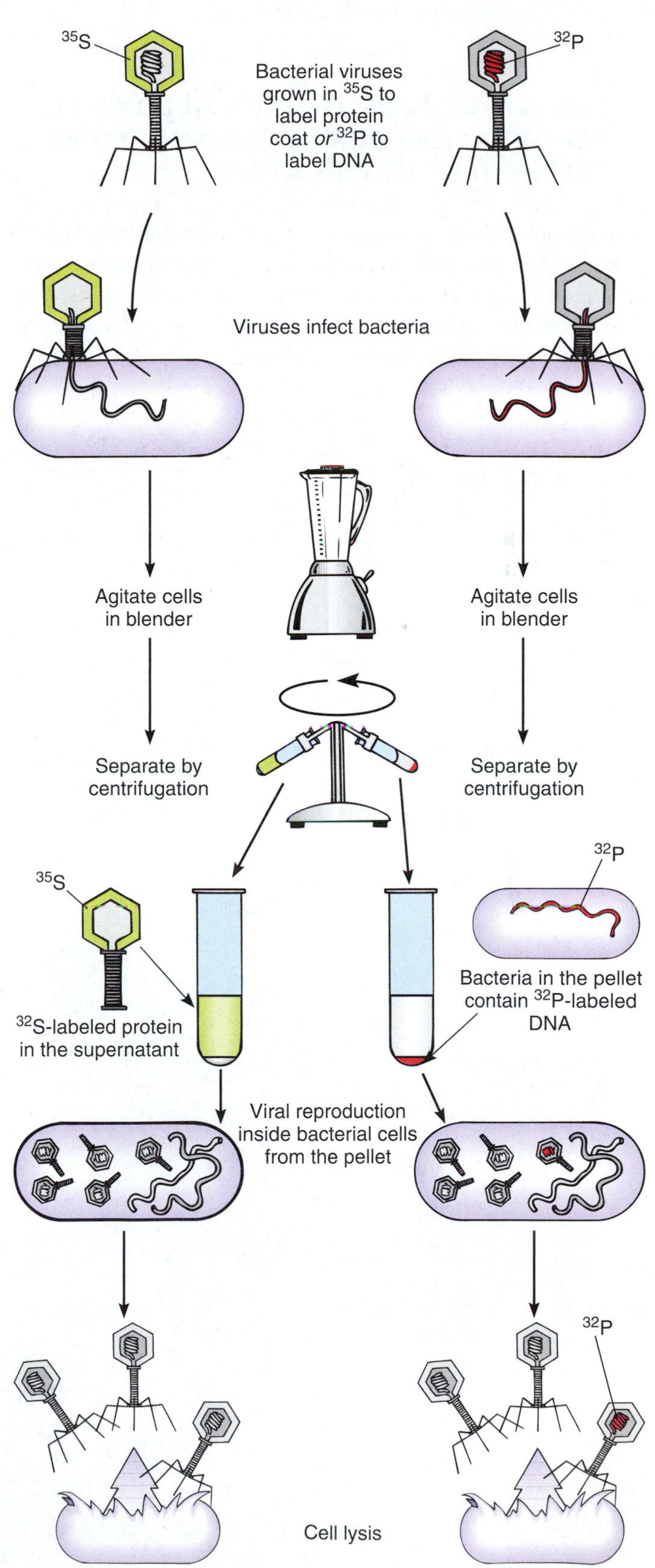

Figure 11–2 The Hershey-Chase experiments. Although bacteriophage protein coats labeled with the radioactive isotope ^{35}S *(left)* could be separated from infected bacterial cells without interfering with viral reproduction, viral DNA labeled with the radioactive isotope ^{32}P *(right)* could not, thus demonstrating that the DNA enters the cells and is required for synthesis of new protein coats and DNA.

concluded that bacteriophages inject their DNA into bacterial cells, leaving most of their protein on the outside. This finding emphasized the significance of DNA in viral reproduction and was seen by many scientists as another important demonstration of the role of DNA as the hereditary material.

THE STRUCTURE OF DNA ALLOWS IT TO CARRY INFORMATION AND TO BE FAITHFULLY DUPLICATED

DNA did not become widely accepted as the genetic material until 1953, when James Watson and Francis Crick proposed a model for its structure that had extraordinary explanatory power. The story of how the structure of DNA came to be determined is one of the most remarkable chapters in the history of modern biology.

As we will see in the subsequent discussion, a great deal was already known about the physical and chemical properties of DNA when Watson and Crick became interested in the problem, and in fact they did not conduct any experiments or gather any new data. Their all-important contribution was to integrate all the available information into a model that demonstrated how the molecule can both carry information for making proteins and serve as its own **template** (pattern or guide) for its duplication.

Nucleotides can be covalently linked in any order to form long polymers

As discussed in Chapter 3, each DNA building block is a **nucleotide** consisting of a pentose sugar **(deoxyribose)**, a phosphate, and one of four nitrogenous bases (Fig. 11–3*a*). It is conventional to number the atoms in a molecule using a system

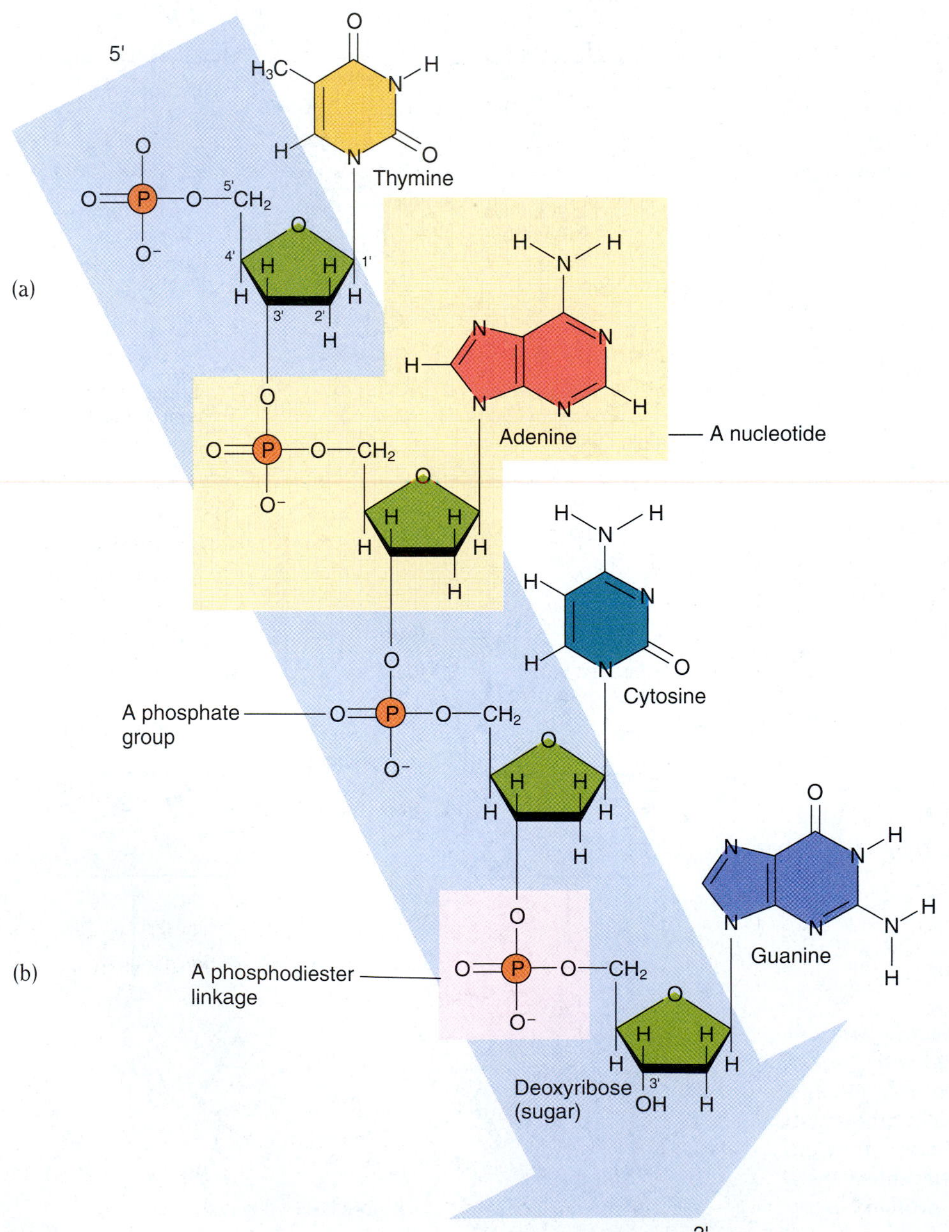

Figure 11–3 The nucleotide subunits of DNA. **(a)** A single strand of DNA consists of a backbone made of phosphate groups alternating with the sugar deoxyribose *(green)*. Linked to the 1' carbon of each sugar is one of four nitrogenous bases. The purine bases, adenine and guanine, have two-ring structures; the pyrimidine bases, thymine and cytosine, have one-ring structures. Note the polarity of the polynucleotide chain, with the 5' end at the top of the figure and the 3' end at the bottom. **(b)** Sugars of adjacent nucleotides are joined by phosphodiester linkages.

devised by organic chemists, and accordingly in nucleic acid chemistry the individual carbons in each sugar and each base are numbered. The carbons in a base are designated by ordinary numerals, but the carbons in a sugar are distinguished from those in the base by "prime" (') designations, such as 2'. The nitrogenous base is attached to the 1' carbon of the sugar, and the phosphate is attached to the 5' carbon. The bases include two **purines—adenine (A)** and **guanine (G)**—and two **pyrimidines—thymine (T)** and **cytosine (C).**

The nucleotides are linked by covalent bonds to form an alternating sugar-phosphate backbone. The 3' carbon of one sugar is bonded to the 5' phosphate of the adjacent sugar to form a 3',5' **phosphodiester linkage** (Fig. 11–3*b*). It is therefore possible to form a polymer of indefinite length, with the nucleotides linked in any order. We now know that most DNA molecules found in cells are millions of bases long. Figure 11–3*a* also illustrates that a single polynucleotide chain is directional. No matter how long the chain may be, one end (the **5' end**) has a 5' carbon attached to a phosphate and the other (the **3' end**) has a 3' carbon attached to a hydroxyl group.

DNA is made of two polynucleotide chains intertwined to form a double helix

Key information about the structure of DNA came from **x-ray diffraction** studies on crystals of purified DNA, carried out by Rosalind Franklin in the laboratory of M.H.F. Wilkins. X-ray diffraction is a powerful method for determining distances between atoms of molecules arranged in a regular, repeating crystalline structure (Fig. 11–4). X rays have such extremely short wavelengths that they can be scattered by the electrons surrounding the atoms in a molecule. Atoms with dense electron clouds (e.g., phosphorus, oxygen) tend to deflect electrons more strongly than do atoms with lower atomic numbers.

When a crystal is exposed to an intense beam of x rays, the regular arrangement of the atoms in the crystal causes the x rays to be diffracted, or scattered, in specific ways. The pattern of diffracted x rays is seen on film as dark spots. Mathematical analysis of the arrangement and distances between the spots can then be used to determine precise distances between atoms and their orientation within the molecules.

Franklin had already produced x-ray crystallographic films of DNA patterns when Watson and Crick began to pursue the problem of DNA structure. The pictures clearly showed that DNA has a type of helical structure, and three major types of regular, repeating patterns in the molecule (with the dimensions 0.34 nm, 3.4 nm, and 2.0 nm) were evident. Franklin and Wilkins had inferred from these patterns that the nucleotide bases (which are flat molecules) are stacked like rungs of a ladder. Using this information, Watson and Crick began to build scale models of the DNA components and then fit them together to agree with the experimental data.

After a number of trials, the two worked out a model that fit the existing data (Fig. 11–5). The nucleotide chains conformed to the dimensions of the x-ray data only if each DNA molecule consisted of *two* polynucleotide chains arranged in a coiled **double helix.** In their model, the sugar-phosphate backbones of the two chains form the outside of the helix. The bases belonging to the two chains associate as pairs along the central axis of the helix. The reasons for the repeating patterns of 0.34-nm and 3.4-nm measurements are readily apparent from the model: each pair of bases is exactly 0.34 nm from the adjacent pairs above and below. Because exactly ten base pairs are present in each full turn of the helix, each turn constitutes 3.4 nm of length. To fit the data, the two chains must run in opposite directions; therefore, each end of the double helix must have an exposed 5' phosphate on one strand and an exposed 3' hydroxyl group on the other (Fig. 11–6*a*). Because the two strands run in opposite directions, they are said to be **antiparallel** to each other.

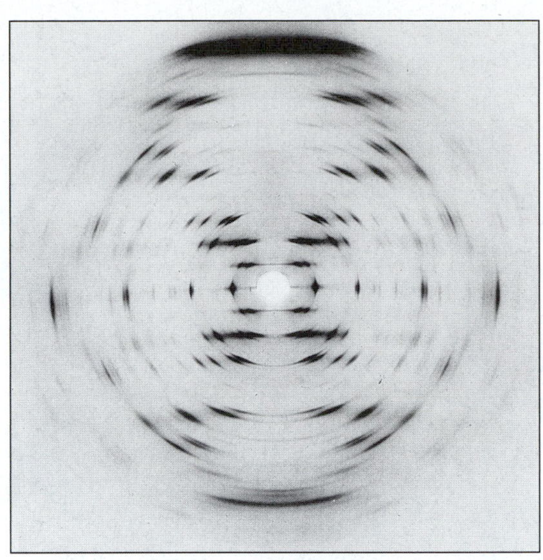

Figure 11–4 X-ray diffraction image of DNA. Important clues about DNA structure are provided by detailed mathematical analysis of measurements of x-ray diffraction images of the lithium salt of DNA. The diagonal pattern of spots stretching from 11 o'clock to 5 o'clock and from 1 o'clock to 7 o'clock provides evidence for the helical structure of DNA. The elongated horizontal patterns at the top and bottom indicate that the purine and pyrimidine bases are stacked 0.34 nm apart and are perpendicular to the axis of the DNA molecule. *(Dr. S.D. Dover, Division of Biomolecular Sciences, Kings College, London)*

In double-stranded DNA, hydrogen bonds form between adenine and thymine and between guanine and cytosine

Other features of the Watson and Crick model integrated critical information about the chemical composition of DNA with the x-ray diffraction data. By 1950, the base composition of DNA from a number of organisms and tissues had been determined by Erwin Chargaff and his coworkers at Columbia University. They found a simple relationship among the bases that turned out to be an important clue to the structure of DNA. Regardless of the

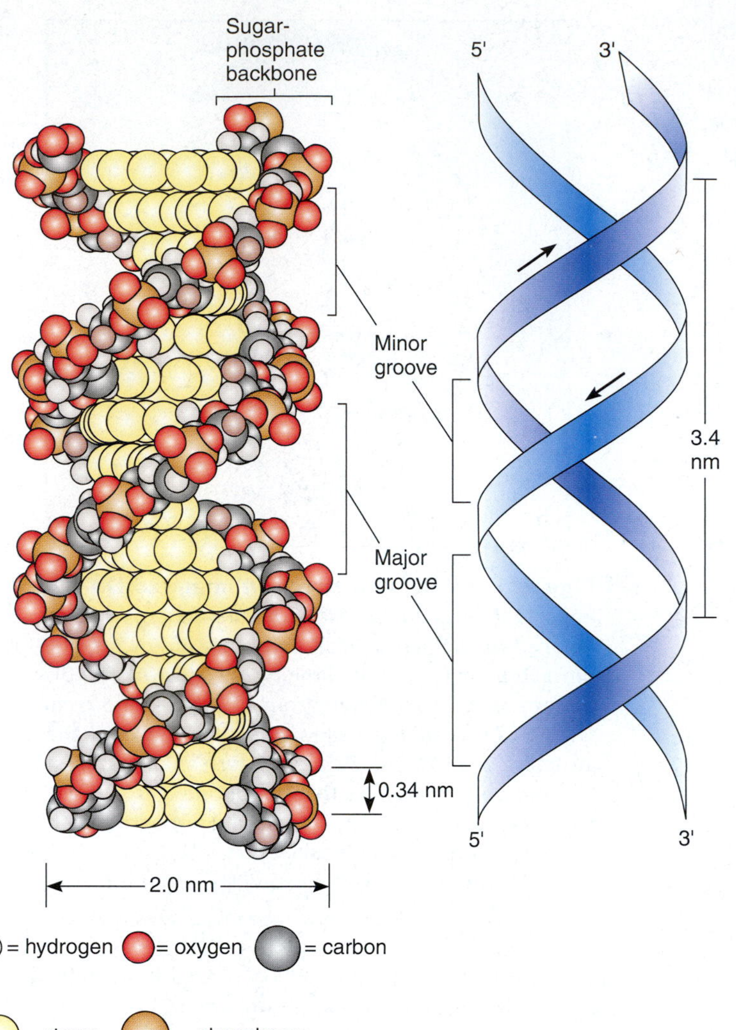

Figure 11–5 Structure of the DNA double helix. On the left is a space-filling model of the DNA double helix. The measurements on the diagrammatic model on the right match those derived from x-ray diffraction images. The ribbons represent the sugar-phosphate backbone of each strand; the arrows indicate that the two strands extend in opposite directions.

source of the DNA, in Chargaff's words, the "ratios of purines to pyrimidines and also of adenine to thymine and of guanine to cytosine were not far from 1." In other words, in double-stranded DNA molecules, the number of purines equals the number of pyrimidines, the number of adenines equals the number of thymines (A equals T), and the number of guanines equals the number of cytosines (G equals C).

The x-ray diffraction studies indicated that the double helix has a precise and constant width, as shown by the 2.0-nm measurements. This finding is actually connected to Chargaff's rules. Notice in Figure 11–3 that each pyrimidine (cytosine or thymine) contains only one ring of atoms, whereas each purine (guanine or adenine) contains two rings. Study of the models made it clear to Watson and Crick that if each cross-rung of the ladder were to contain one purine and one pyrimidine, the width of the helix at each base pair would be exactly 2.0 nm; by contrast, the combination of two purines (each of which is 1.2 nm wide) would be wider and that of two pyrimidines would be narrower, so the diameter would not be constant.

Further examination of the model showed that adenine can pair with thymine (and guanine with cytosine) in such a way that hydrogen bonds form between them; the opposite combinations, cytosine with adenine and guanine with thymine, do not lead to favorable hydrogen bonding.

The nature of the hydrogen bonding between adenine and thymine and between guanine and cytosine is shown in Figure 11–6*b*. Two hydrogen bonds can form between adenine and thymine, and three between guanine and cytosine. This concept of *specific base-pairing* neatly explains Chargaff's rules. The amount of cytosine has to equal the amount of guanine because every cytosine in one chain must have a paired guanine in the other chain. Similarly, every adenine in the first chain must have a thymine in the second chain. Thus, the sequences of bases in the two chains are **complementary** to each other. In other words, the sequence of nucleotides in one chain dictates the complementary sequence of nucleotides in the other. For example, if one strand has the sequence:

3' — AGCTAC —5'

then the other strand has the complementary sequence:

5' — TCGATG — 3'

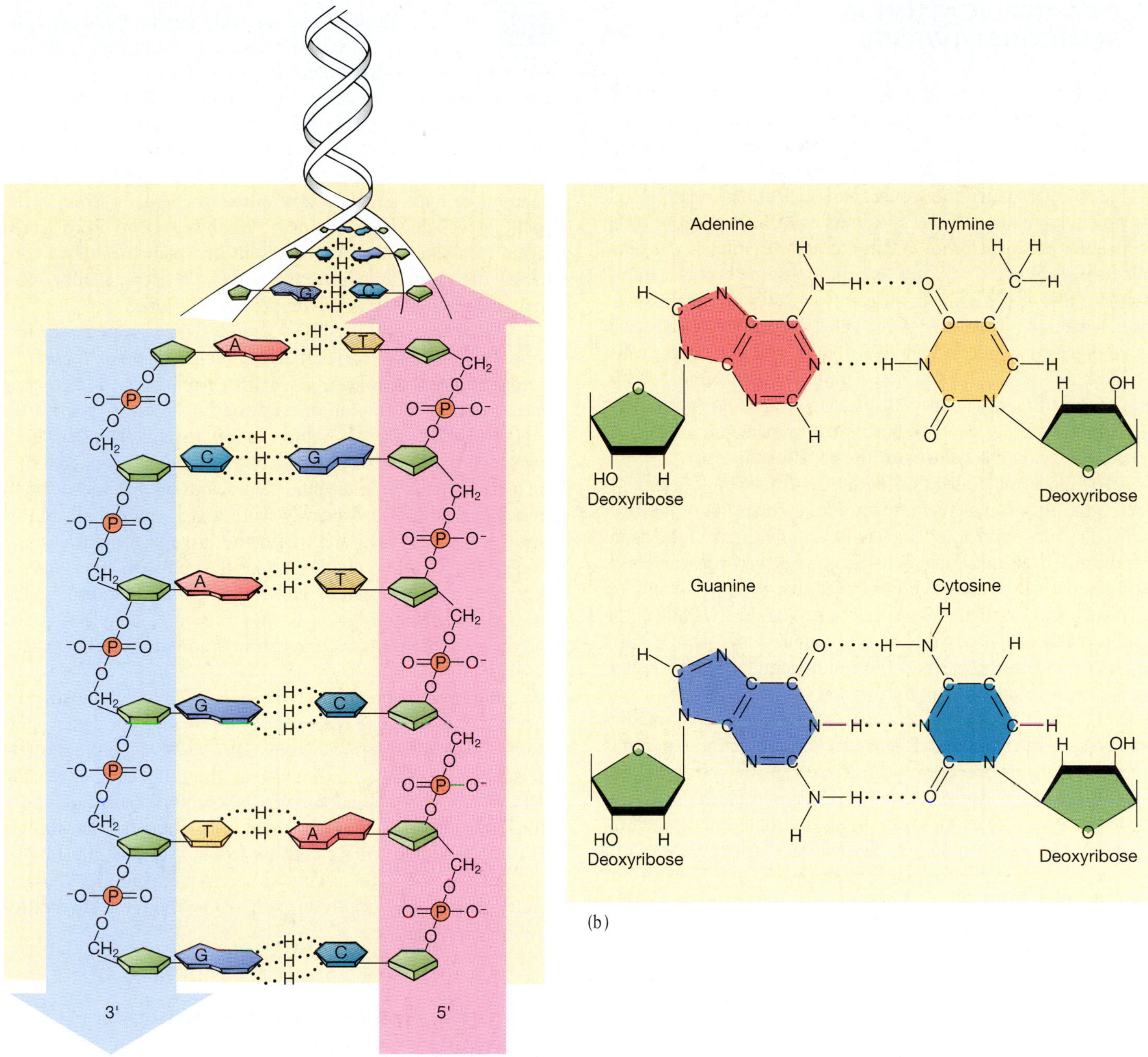

Figure 11-6 Hydrogen bonding between bases. The two strands of a DNA double helix are associated by hydrogen bonding between the bases. **(a)** The two sugar-phosphate chains run in opposite directions. This orientation permits the complementary bases to pair. **(b)** Diagram of the hydrogen bonding between base pairs adenine (A) and thymine (T) *(top)* and guanine (G) and cytosine (C) *(bottom)*. The AT pair has two hydrogen bonds; the GC pair has three.

The double-helix model strongly suggested that the sequence of bases in DNA can provide for the storage of genetic information and that this sequence could be ultimately related to the sequences of amino acids in proteins. Although there are restrictions on how the bases pair with each other, the number of possible sequences of bases in a strand is virtually unlimited. Because a DNA molecule in a cell can be millions of nucleotides long, it can store enormous amounts of information, usually comprising a large number of genes.

DNA REPLICATION IS SEMICONSERVATIVE

Two immediately apparent and distinctive features of the Watson-Crick model made it seem more plausible that DNA is the genetic material. We have already mentioned that the sequence of bases in DNA can carry coded information. The model also suggested a way in which the sequence of nucleotides in DNA could be precisely copied, a process known as **DNA replication.** The connection to the behavior of chromosomes in mitosis was obvious to Watson and Crick; just as a chromosome becomes duplicated so that it consists of two identical sister chromatids that later separate at anaphase, so the genetic material must be able to be precisely duplicated and distributed to the daughter cells. They noted in a classic and now famous understatement at the end of their first brief paper, "It has not escaped our notice that the specific pairing we have postulated immediately suggests a possible copying mechanism for the genetic material."

The model suggested that, because the nucleotides pair with each other in a complementary fashion, each strand of the DNA molecule could serve as a template for the synthesis of the opposite strand. It would simply be necessary for the hydrogen bonds between the two strands to break (recall that hydrogen bonds are relatively weak) and the two chains to separate. Each strand of the double helix could then pair with new complementary nucleotides to replace its missing partner. The result would be two DNA double helices, each identical to the original one and consisting of one original strand from the parent molecule and one newly synthesized complementary strand. This type of information copying is known as a **semiconservative replication** mechanism (Fig. 11–7*a*).

Although the semiconservative replication mechanism suggested by Watson and Crick was (and is) a simple and compelling model, experimental proof was needed to establish that DNA in fact duplicates in that manner. First it was necessary to rule out several other possibilities. For example, with a *conservative replication* mechanism, both parent (or old) strands would remain together, and the two newly synthesized strands would form a second double helix (Fig. 11–7*b*). As a third alternative, the parental and newly synthesized strands might become randomly mixed during the replication process; this possibility was known as *dispersive replication* (Fig. 11–7*c*). To discriminate among the semiconservative replication mechanism and the other possibilities, it was necessary to distinguish between old and newly synthesized strands of DNA.

One way to accomplish this is to use a heavy-nitrogen isotope, ^{15}N (ordinary nitrogen is ^{14}N), to label the bases of the DNA strands, making them more dense. Using the technique known as **density gradient centrifugation** (a technique that had been recently developed at that time), large molecules such as DNA can be separated on the basis of differences in their density. When DNA is mixed with a solution containing cesium chloride (CsCl) and centrifuged at high speed, the solution forms a density gradient in the centrifuge tube, ranging from a region of lowest density at the top to one of highest density at the bottom. During centrifugation, the DNA molecules migrate to the region of the gradient identical to their own density.

Process of Science

In 1957, Matthew Meselson and Franklin Stahl grew the bacterium *Escherichia coli* on a medium that contained ^{15}N in the form of ammonium chloride (NH_4Cl). The cells used the ^{15}N to synthesize bases, which then became incorporated into DNA (Fig. 11–7*d*). The resulting heavy nitrogen–containing DNA molecules were extracted from some of the cells. When they were subjected to density gradient centrifugation, they accumulated in the high-density region of the gradient. The rest of the bacteria (which also contained ^{15}N-labeled DNA) were transferred to a different growth medium in which the NH_4Cl contained the naturally abundant, lighter ^{14}N isotope; they were then allowed to undergo additional cell divisions.

The newly synthesized DNA strands were expected to be less dense because they incorporated bases containing the lighter ^{14}N isotope. Indeed, double-stranded DNA from cells isolated after one generation had an intermediate density, indicating that they contained half as many ^{15}N atoms as the "parent" DNA. This finding supported the semiconservative model, which predicted that each double helix should contain a previously synthesized strand (heavy in this case) and a newly synthesized strand (light in this case). It was also consistent with the dispersive model, which would also yield one class of molecules, all with intermediate density. It was inconsistent with the conservative model, which predicted that there should be two classes of double-stranded molecules, those with two heavy strands and those with two light strands.

After another cycle of cell division in the medium with the lighter ^{14}N isotope, two types of DNA appeared in the density gradient. One consisted of "hybrid" DNA helices (with one ^{15}N strand and one ^{14}N strand), whereas the other contained only DNA with the naturally occurring light isotope. This finding refuted the dispersive model, which predicted that all strands should have intermediate density. Instead, each strand of the parental double-helix molecule was conserved, but in a *different* daughter molecule, exactly as predicted by the semiconservative replication model.

Semiconservative replication explains the stability of mutations

The recognition that DNA could be copied by a semiconservative mechanism suggested how DNA could fulfill a third essential characteristic of genetic material—the ability to mutate.

It was long known that mutations, or genetic changes, could arise in genes and then be transmitted faithfully to succeeding generations. When the double-helix model was proposed, it seemed plausible that mutations could represent a change in the sequence of bases in the DNA. One could predict that if DNA is copied by a mechanism involving complementary base pairing, any change in the sequence of bases on one strand would result in a new sequence of complementary bases during the next replication cycle. The new base sequence would then be passed on to daughter molecules by the same mechanism used to copy the original genetic material, as if no change had occurred.

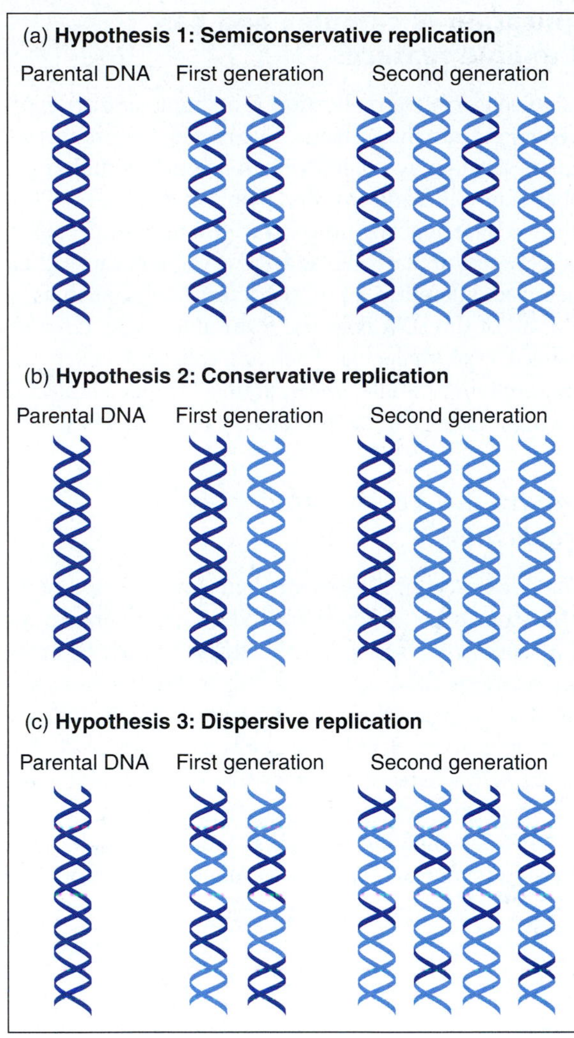
(a) Hypothesis 1: Semiconservative replication
Parental DNA
First generation
Second generation
(b) Hypothesis 2: Conservative replication
Parental DNA
First generation
Second generation

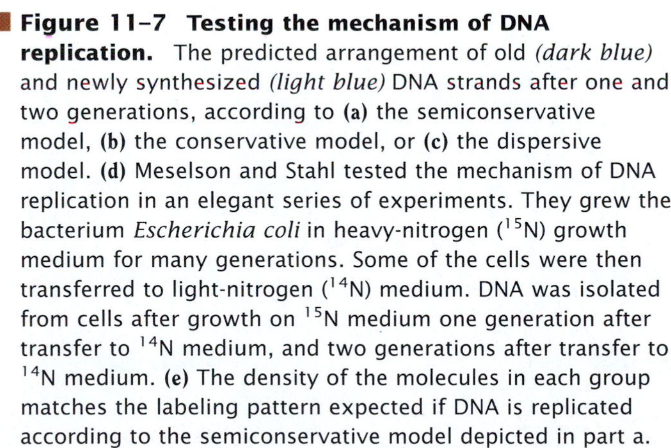
(c) Hypothesis 3: Dispersive replication
Parental DNA
First generation
Second generation

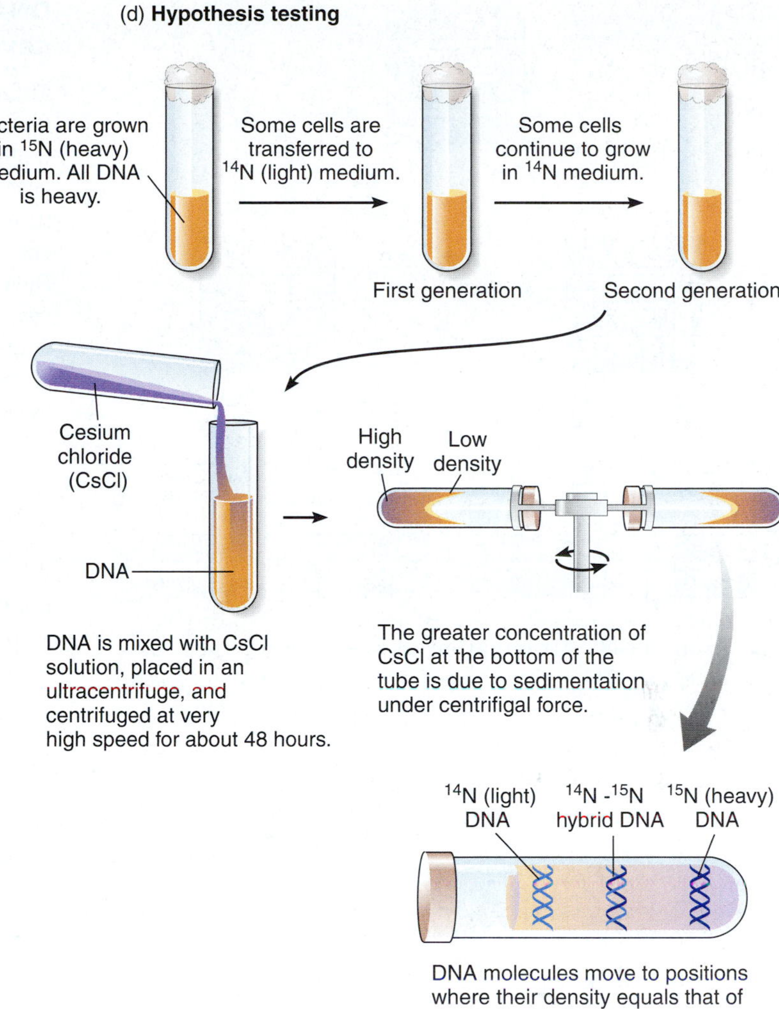
(d) Hypothesis testing
Bacteria are grown in ¹⁵N (heavy) medium. All DNA is heavy.
Some cells are transferred to ¹⁴N (light) medium.
Some cells continue to grow in ¹⁴N medium.
First generation
Second generation
Cesium chloride (CsCl)
DNA
High density
Low density
DNA is mixed with CsCl solution, placed in an ultracentrifuge, and centrifuged at very high speed for about 48 hours.
The greater concentration of CsCl at the bottom of the tube is due to sedimentation under centrifigal force.
¹⁴N (light) DNA
¹⁴N -¹⁵N hybrid DNA
¹⁵N (heavy) DNA
DNA molecules move to positions where their density equals that of the CsCl solution.

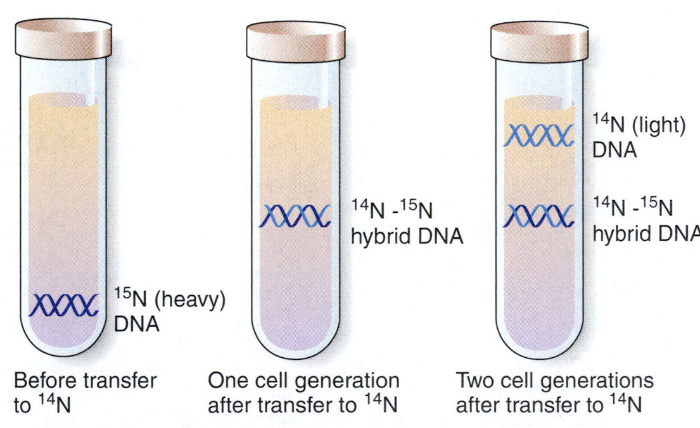
(e) Results
¹⁵N (heavy) DNA
¹⁴N -¹⁵N hybrid DNA
¹⁴N (light) DNA
¹⁴N -¹⁵N hybrid DNA
Before transfer to ¹⁴N
One cell generation after transfer to ¹⁴N
Two cell generations after transfer to ¹⁴N
The location of DNA molecules within the centrifuge tube can be determined by UV optics. DNA solutions absorb strongly at 260 nm.

Figure 11–7 Testing the mechanism of DNA replication. The predicted arrangement of old *(dark blue)* and newly synthesized *(light blue)* DNA strands after one and two generations, according to **(a)** the semiconservative model, **(b)** the conservative model, or **(c)** the dispersive model. **(d)** Meselson and Stahl tested the mechanism of DNA replication in an elegant series of experiments. They grew the bacterium *Escherichia coli* in heavy-nitrogen (^{15}N) growth medium for many generations. Some of the cells were then transferred to light-nitrogen (^{14}N) medium. DNA was isolated from cells after growth on ^{15}N medium one generation after transfer to ^{14}N medium, and two generations after transfer to ^{14}N medium. **(e)** The density of the molecules in each group matches the labeling pattern expected if DNA is replicated according to the semiconservative model depicted in part a.

In the example shown in Figure 11–8, an adenine base in one of the DNA strands has been changed to guanine. This could occur by a rare error in DNA replication or by one of several other known mechanisms. (There are systems of enzymes that repair errors when they occur, but not all mutations are corrected properly.) When the DNA molecule is replicated again, one of the strands gives rise to a molecule exactly like the parent strand; the other (mutated) strand gives rise to a molecule with a new combination of bases that will be stably transmitted generation after generation.

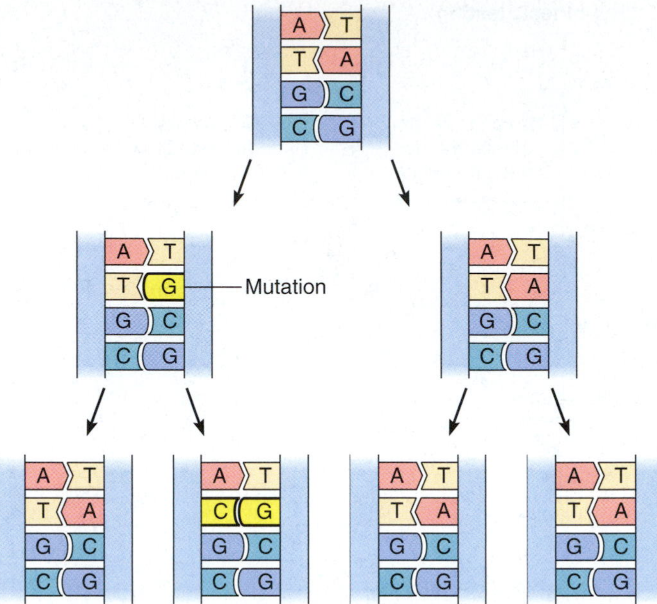

Figure 11-8 Perpetuation of a mutation. The process of DNA replication can stabilize a mutation *(bright yellow)* so that it will be passed to future generations.

DNA replication is complex and has several unique features

Although semiconservative replication is a simple and straightforward prediction from the Watson-Crick model, the process actually requires a complex "replication machine" containing a large number of proteins and enzymes. Many of the essential features of DNA replication are universal, although some differences exist between prokaryotes and eukaryotes because their DNA is organized differently. In most bacterial cells, such as *E. coli,* most or all of the DNA is in the form of a single, *circular,* double-stranded DNA molecule. Each unreplicated eukaryotic chromosome contains a single, *linear,* double-stranded molecule associated with at least as much protein (by mass) as DNA.

DNA strands must be unwound during replication

Watson and Crick recognized that in their double-helix model the two DNA strands are wrapped around one another like the strands of a rope. If we try to pull the strands apart, the rope must either rotate or twist into tighter coils. We would expect similar things to happen when complementary DNA strands are separated for replication. Separating the two strands of DNA is accomplished by **DNA helicase enzymes** that travel along the helix, opening the double helix as they move. Once the strands are separated, **helix-destabilizing proteins** bind to single DNA strands, preventing re-formation of the double helix until the

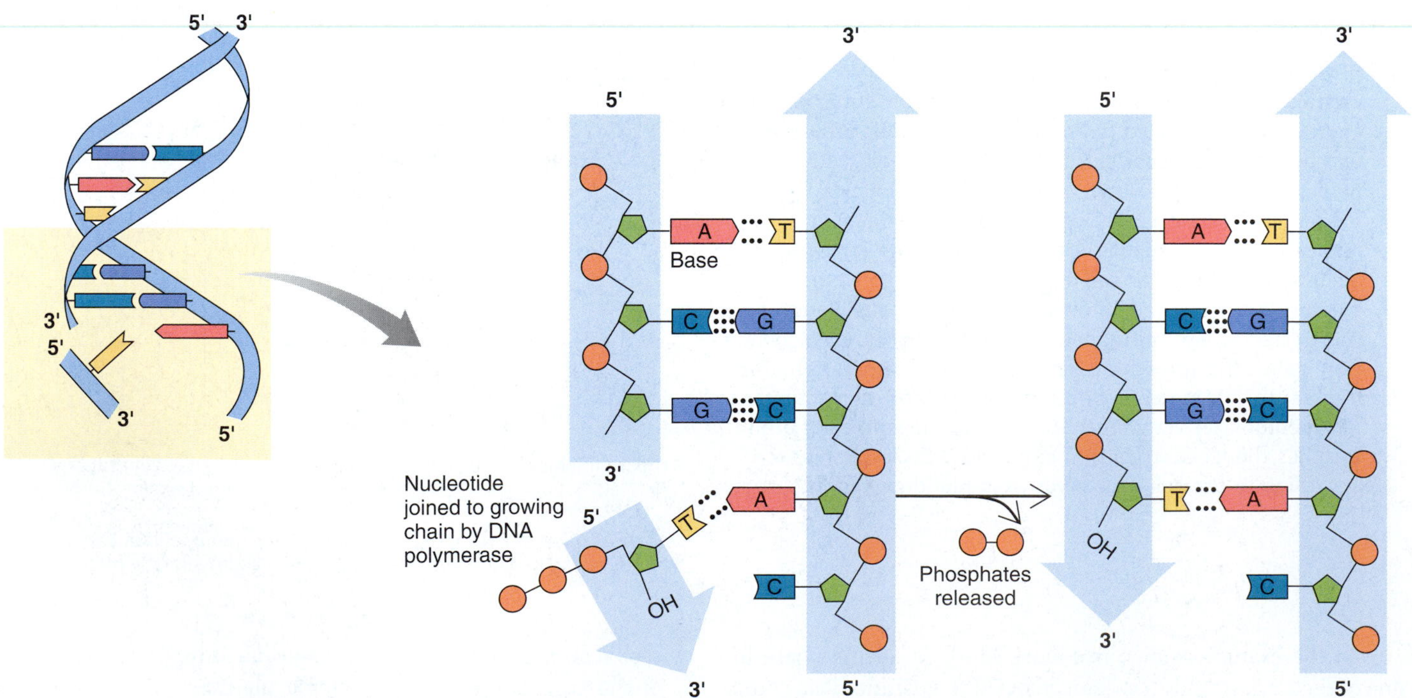

Figure 11-9 A simplified view of DNA replication. The polymerase enzymes that catalyze the polymerization reactions add one nucleotide at a time to the 3' end of a growing chain.

strands are copied. Enzymes called **topoisomerases** produce breaks in the DNA molecules and then rejoin the strands, relieving strain and effectively preventing the formation of knots during replication.

DNA synthesis always proceeds in a 5' → 3' direction

The enzymes that catalyze the linking together of the nucleotide subunits are called **DNA polymerases.** They have several limitations that contribute to the complexity of the replication process. They are able to add nucleotides only to the 3' end of a growing polynucleotide strand, and this strand must be paired with the strand being copied (Fig. 11–9). Nucleotides with three phosphate groups are used as substrates for the polymerization reaction. As the nucleotides are linked together, two of the phosphates are removed. Like the hydrolysis of ATP, these reactions are strongly exergonic (see Chapter 6) and do not require additional energy. Because the new polynucleotide chain is elongated by the linkage of the 5' phosphate group of the next nucleotide subunit to the 3' hydroxyl group of the sugar at the end of the preexisting strand, the new strand of DNA always grows in the 5' → 3' direction.

DNA synthesis requires an RNA primer

A second limitation of the DNA polymerases is that they can add nucleotides only to the 3' end of an *existing* polynucleotide strand. So how can DNA synthesis be initiated once the two strands are separated? The answer is that a short piece (usually about 5–14 nucleotides) of an **RNA primer** is first synthesized at the point of initiation of replication (Fig. 11–10).

RNA, or **ribonucleic acid** (see Chapters 3 and 12), is a nucleic acid polymer consisting of nucleotide subunits that can associate by complementary base-pairing with the single-stranded DNA template. The RNA primer is synthesized by **primase,** an enzyme that is able to start a new strand of RNA opposite a DNA strand. After a few nucleotides have been added, the primase is displaced by DNA polymerase, which can then add subunits to the 3' end of the short RNA primer. The primer is later degraded by specific enzymes, and the space is filled in with DNA.

DNA replication is discontinuous in one strand and continuous in the other

A major obstacle in understanding DNA replication was the fact that the complementary DNA strands are antiparallel. Because DNA synthesis proceeds only in the 5' → 3' direction, which

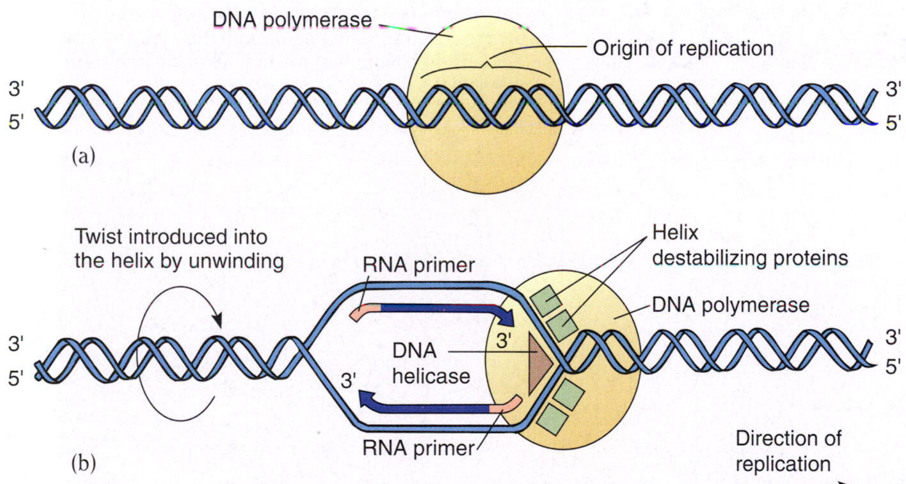

DNA synthesis begins at a specific base sequence, termed the *origin of replication.*

Strands are separated at the origin of replication and unwound by DNA helicase, which "walks" along the DNA molecule preceding the DNA-synthesizing enzymes. Single-stranded regions are prevented from re-forming into double strands by helix-destabilizing proteins, which bind to single-stranded DNA. The region of active DNA synthesis is associated with the "replication fork," formed at the junction of the single strands and the double-stranded region. Both strands are synthesized in the vicinity of the fork (each in a 5'→3' direction).

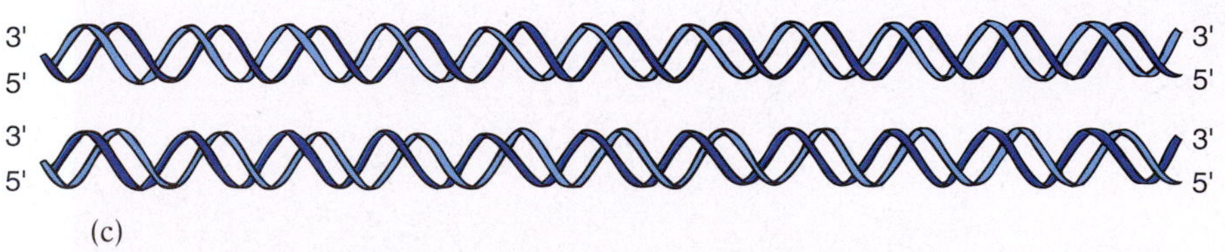

Completion of replication results in the formation of two daughter molecules, each containing one old and one newly synthesized strand. Each double helix is a chromatid of a duplicated eukaryotic chromosome.

Figure 11–10 Overview of DNA replication. This process requires a number of steps involving several enzymes and RNA primers.

means that the strand being copied is being read in a 3' ⟶ 5' direction, it would seem necessary to copy one of the strands starting at one end of the double helix and the other strand starting at the opposite end. Such a solution to this, the most difficult replication problem, would be extremely awkward at best, and probably completely unworkable; thus a very different mechanism has evolved.

DNA replication begins at specific sites on the DNA molecule, termed **origins of replication,** and both strands are replicated at the same time at a Y-shaped structure called the **replication fork** (Fig. 11–11). The position of the replication fork is constantly moving as replication proceeds. Two identical DNA polymerase molecules are responsible for replication. One of these adds nucleotides to the 3' end of the new strand that is always growing *toward* the replication fork. Because this strand can be formed smoothly and continuously, it is called the **leading strand.**

A separate (but identical) DNA polymerase molecule adds nucleotides to the 3' end of the other new strand, termed the **lag-**

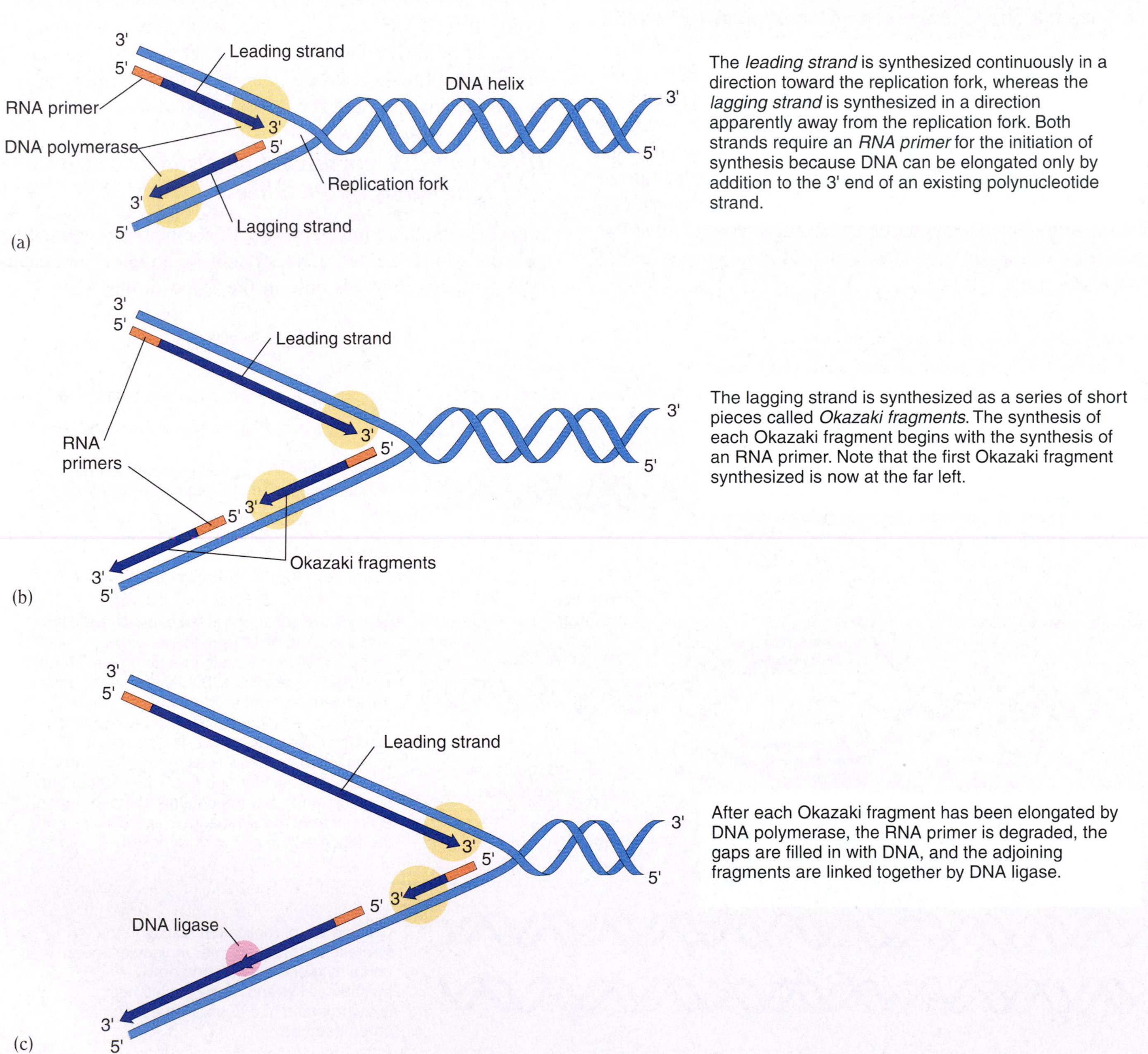

3'
5'
Leading strand
DNA helix
RNA primer
DNA polymerase
Replication fork
Lagging strand
(a)
The *leading strand* is synthesized continuously in a direction toward the replication fork, whereas the *lagging strand* is synthesized in a direction apparently away from the replication fork. Both strands require an *RNA primer* for the initiation of synthesis because DNA can be elongated only by addition to the 3' end of an existing polynucleotide strand.
Leading strand
RNA primers
Okazaki fragments
(b)
The lagging strand is synthesized as a series of short pieces called *Okazaki fragments*. The synthesis of each Okazaki fragment begins with the synthesis of an RNA primer. Note that the first Okazaki fragment synthesized is now at the far left.
Leading strand
DNA ligase
(c)
After each Okazaki fragment has been elongated by DNA polymerase, the RNA primer is degraded, the gaps are filled in with DNA, and the adjoining fragments are linked together by DNA ligase.

Figure 11–11 Leading and lagging DNA strands. Because elongation can proceed only in a 5' ⟶ 3' direction, the two strands at the replication fork are copied in different ways, each by a separate DNA polymerase molecule.

ging strand. The lagging strand is always growing *away* from the replication fork. Only short pieces can be synthesized because the DNA polymerase enzyme would need to move far away from the fork if it were to add continuously to the 3' end of that strand. These 100- to 2000-nucleotide pieces are called **Okazaki fragments** after their discoverer, Reijii Okazaki.

Each Okazaki fragment is initiated by a separate RNA primer and is then extended toward the 5' end of the previously synthesized fragment by DNA polymerase. When the RNA primer of the previously synthesized fragment is reached, the primer is degraded and replaced with DNA by DNA polymerase. The fragments are then joined together by **DNA ligase,** an enzyme that links the 3' OH of one DNA fragment to the 5' phosphate of another immediately adjacent, forming a phosphodiester linkage.

It has been suggested that simultaneous synthesis of both the leading and lagging strands is possible because the lagging strand (and the DNA strand from which it is copied) forms a loop. This loop allows DNA polymerase to synthesize the lagging strand while remaining close to the replication fork as it works.

Most DNA synthesis is bidirectional

When double-stranded DNA is separated, two replication forks are formed, allowing the molecule to be replicated in both directions from the origin of replication. Prokaryotic cells usually have only one origin of replication on each circular DNA molecule (Fig. 11–12*a*), so the two replication forks proceed around the circle and eventually meet at the other side to complete the formation of two new DNA molecules.

A eukaryotic chromosome is composed of one, extremely long, linear DNA molecule, so the process is speeded up by having multiple origins of replication (Fig. 11–12*b–d*). Synthesis continues at each replication fork until it meets one coming from the opposite direction, resulting in the formation of a chromosome containing two DNA double helices (each of which corresponds to a chromatid). The ends of eukaryotic chromosomes present special problems in replication. For this reason, they are capped by special regions known as **telomeres** (see *On the Cutting Edge: Telomerase, Cellular Aging, and Cancer*).

DNA IS PACKAGED IN A HIGHLY ORGANIZED WAY IN CHROMOSOMES

Prokaryotic and eukaryotic cells differ markedly in their DNA content as well as in the organization of DNA molecules. An *E. coli* cell normally contains about 4×10^6 base pairs (almost 1.35 mm) of DNA in its single circular DNA molecule. In fact, the total length of the DNA is about 1000 times greater than the length of the cell itself. Therefore, the DNA molecule must, with the help of special proteins, be twisted and folded compactly to fit inside the bacterial cell.

A typical eukaryotic cell contains much more DNA than a bacterium does, and it is organized in the nucleus as multiple chromosomes; these vary widely in size and number among different species. Although a human cell nucleus is about the size

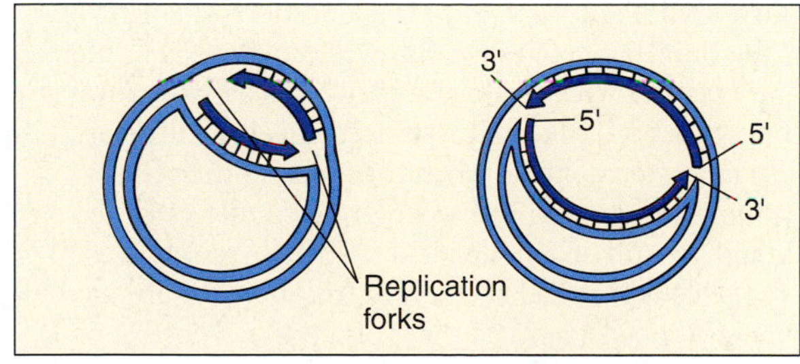

(a)

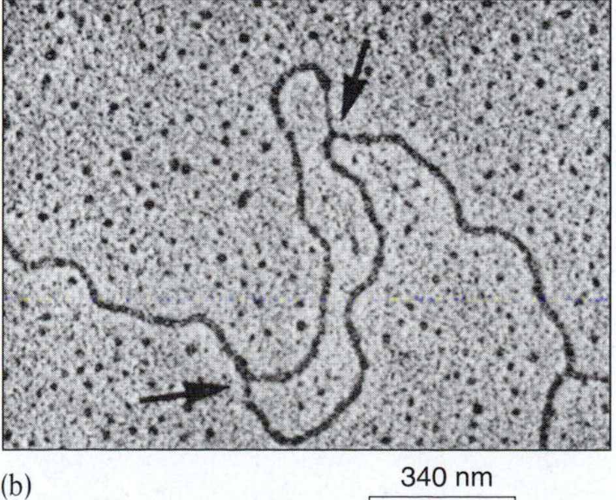

(b)

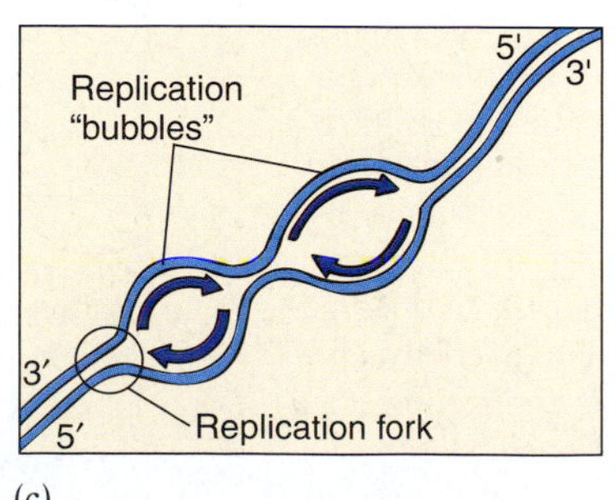

(c)

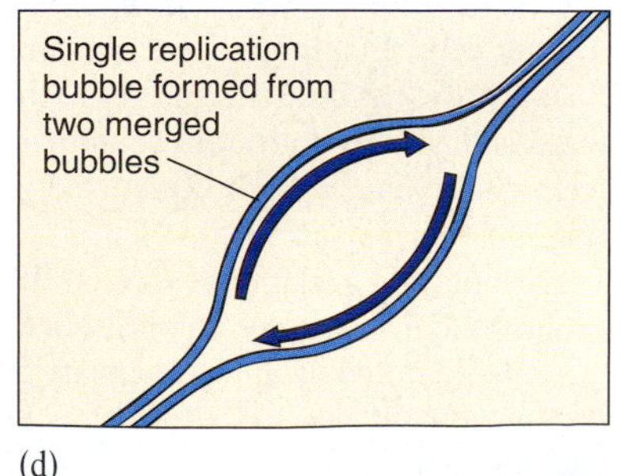

(d)

Figure 11–12 Bidirectional DNA replication. The leading strands and lagging strands are not represented in the illustrations. **(a)** The circular DNA in *E. coli* has only one origin of replication. Because DNA synthesis proceeds from that point in both directions, two replication forks are formed, travel around the circle, and eventually meet *(not shown)*. **(b)** This TEM shows two replication forks *(arrows)* in a segment of a eukaryotic chromosome that has been partly replicated. **(c)** Eukaryotic chromosomal DNA contains multiple origins of replication. DNA synthesis proceeds in both directions from each origin until adjacent "replication bubbles" eventually merge **(d)**. *(b, Kriegstein, H.J. and D.S. Hogness, 1974, Proc. Nat. Acad. Sci. USA, 71:135–139.)*

ON THE CUTTING EDGE

Telomerase, Cellular Aging, and Cancer

HYPOTHESIS: The number of cell divisions that normal human somatic cells can undergo when grown in culture is limited by the length of their telomeric DNA.

METHOD: To test the prediction that cells will be able to undergo more divisions if their telomeric DNA is lengthened, researchers used a virus to introduce the gene coding for the catalytic subunit of telomerase into cultured normal human cells, which normally lack telomerase activity.

RESULTS: The cells produced active telomerase, which elongated the telomeres. The cells continued to actively proliferate for many cell cycles beyond the point at which cell division would normally cease.

CONCLUSION: The life spans of cells can be extended by lengthening the telomeric DNA of their chromosomes.

Unlike prokaryotic DNA, which is circular, eukaryotic chromosomes have free ends. This simple fact has far-reaching consequences. Because DNA replication is discontinuous in the lagging strand, DNA polymerases are unable to complete replication of the strand neatly. When they reach the end of the DNA, they leave a small portion unreplicated, causing a small single-stranded segment of the DNA to be lost with each cell cycle. This situation is less dangerous than it sounds because chromosomes have end caps (telomeres) that do not contain protein-coding genes, but instead consist of short, simple, noncoding DNA sequences that are repeated many times. Therefore, although a small amount of telomeric DNA fails to replicate each time the cell divides, a cell can divide many times before it starts losing essential genetic information.

It has been suggested that the progressive loss of DNA at the telomeres could contribute to the phenomenon known as **cellular aging.** Cellular aging has been analyzed since the 1960s, following the pioneering studies of Leonard Hayflick, who showed that when normal cells of the human body, known as *somatic cells,* are grown in culture, they lose their ability to divide after a limited number of cell divisions. Furthermore, the number of cell divisions is determined by the age of the individual from whom the cells were taken. Cells from a 70-year-old can divide only 20 to 30 times, as compared with those from an infant, which can divide 80 to 90 times.

Telomeric DNA can be lengthened by **telomerase,** a special DNA replication enzyme. This enzyme, which was discovered in 1984 by Carol W. Greider and Elizabeth H. Blackburn, is typically present in cells that can divide an unlimited number of times, including unicellular organisms and many types of cancer cells. In animals such as humans, active telomerase is usually present in germ line cells (the cells that give rise to eggs and sperm) but not in somatic cells. It is these somatic cells that typically show evidence of cellular aging when placed in culture.

Although correlations between the ability of cells to undergo unlimited divisions and the presence of telomerase activity were repeatedly noted by different investigators, this connection remained controversial. Proof of a causal relationship was lacking until Andrea G. Bodnar and her colleagues at the Geron Corporation teamed up with researchers from the University of Texas Southwestern Medical Center to conduct a direct test.* Using the techniques of recombinant DNA technology (see Chapter 14), they infected cultured normal human cells with a virus that carried the genetic information coding for the catalytic subunit of telomerase. Not only did the infected cells produce active telomerase, which elongated the telomeres significantly, but the cells continued to divide long past the point at which cell divisions would normally cease.

These findings have revived interest in telomeres, both for theoretical and practical reasons. The ability to give cells the capacity to divide many times more than they would ordinarily has many potential therapeutic applications, especially if lost or injured cells need to be replaced. However, cancer cells also have the ability to divide many times in culture; in fact, they are virtually immortal. Would cells with active telomerase behave like cancer cells if transplanted into the human body?

To obtain a preliminary answer to this question, scientists at Geron and the University of Texas Southwestern Medical Center independently tested the cells to see if they behaved like cancer cells in other ways.[†,‡] Unlike cultured normal cells, cultured cancer cells fail to stop dividing when they touch other cells, can form tumor-like aggregations when grown under certain conditions, and are still able to divide if their DNA is damaged. By contrast, the cells that had been given the ability to produce active telomerase showed none of these properties. These results are reassuring, but of course they do not prove that there would be no risk for cancer from these cells if they were transplanted.

* Bodnar, A.G., M. Ouellette, M. Frolkis, S.E. Holt, C.P. Chiu, G.B. Morin, C.B. Harley, J.W. Shay, S. Lichsteiner, and W.E. Wright. "Extension of Life-Span by Introduction of Telomerase into Normal Human Cells." *Science,* Vol. 279,16 Jan. 1998.

† Jiang, X., G. Jimenez, E. Chang, M. Frolkis, B. Kusler, M. Sage, M. Beeche, A.G. Bodnar, G.M. Wahl, T.D. Tisty, and C. Chiu. "Telomerase Expression in Human Somatic Cells Does Not Induce Changes Associated With a Transformed Phenotype." *Nature Gen.,* Vol. 21, pp. 111–114, 1999.

‡ Morales, C.P., S.E. Holt, M. Ouellette, K.J. Kaur, Y. Yan, K.S. Wilson, M.A. White, W.E. Wright, and J.W. Shay. "Absence of Cancer-Associated Changes in Human Fibroblasts Immortalized With Telomerase." *Nature Gen.,* Vol. 21, pp. 115–118, 1999.

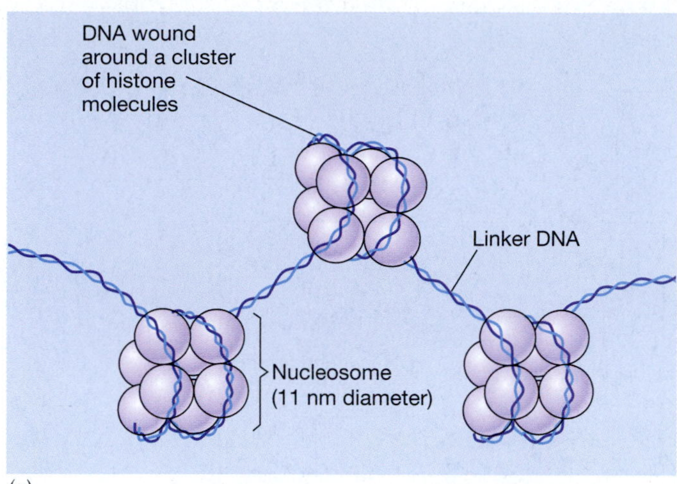

(a)

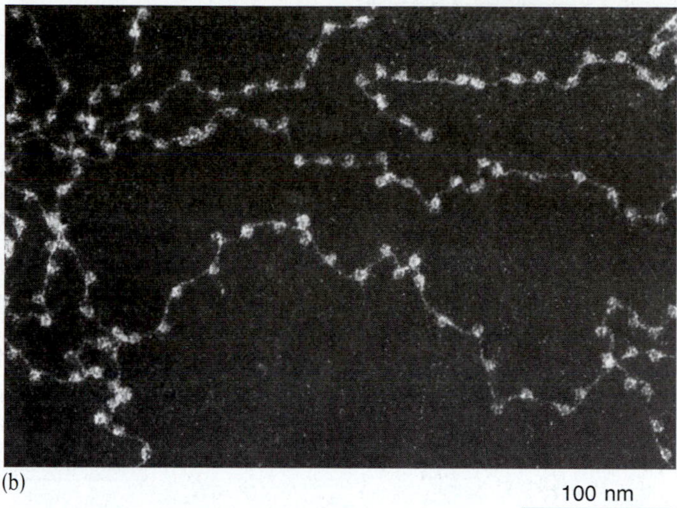

(b)

Figure 11–13 Nucleosomes. **(a)** A model for the structure of a nucleosome. Each nucleosome bead contains a set of eight histone molecules; these form a protein core around which the double-stranded DNA is wound. The DNA surrounding the histones consists of 146 nucleotide pairs; another segment of DNA, about 60 nucleotide pairs long, links nucleosome beads. **(b)** TEM of nucleosomes from the nucleus of a chicken red blood cell. Normally nucleosomes are packed more closely together, but the preparation procedure has spread them apart, revealing the DNA linkers. *(b, Courtesy of D.E. Olins and A.L. Olins)*

of a large bacterial cell, it contains almost 1000 times the amount of DNA found in *E. coli.* The haploid DNA content of a human cell is about 3×10^9 base pairs; if stretched end to end, it would be almost 1 m long.

So how does a eukaryotic cell pack its DNA into the chromosomes? It does so with the help of certain proteins known as **histones.**[1] Histones are positively charged because they have a high proportion of amino acids with basic side chains (see Chapter 3). The positively charged histones are able to associate with DNA, which is negatively charged due to its phosphate groups, to form structures called **nucleosomes.** The fundamental unit of each nucleosome complex consists of a beadlike structure with 146 base pairs of DNA wrapped around a disc-shaped core of eight histone molecules (two each of four different histone types) (Fig. 11–13). Although the nucleosome was originally defined as a bead plus a DNA segment that links it to an adjacent bead, today the term more commonly refers only to the bead itself (i.e., the eight histones and the DNA wrapped around them).

The nucleosomes are part of the **chromatin,** the complex of nucleic acids and protein that makes up the chromosomes. The higher order structures of chromatin leading to the formation of a highly condensed metaphase chromosome are illustrated in Figure 11–14. The nucleosomes themselves are 11 nm in diameter. The packed nucleosome state occurs when a fifth type of histone, known as histone H1, associates with the linker DNA, packing adjacent nucleosomes together to form a 30-nm-diameter thread. In extended chromatin, these 30-nm-diameter threads form large coiled loops held together by a set of nonhistone **scaffolding proteins.** The loops then interact in complex ways to form the condensed chromatin found in a metaphase chromosome.

Nucleosomes function like tiny spools, thereby preventing DNA strands from becoming tangled. The importance of this role is underscored by Figure 11–15, which illustrates the enormous number of DNA fibers that unravel from a mouse chromosome after the histones have been removed. However, their role is more than structural, for their arrangement also affects the activity of the DNA with which they are associated (Chapter 13).

[1] A few types of eukaryotic cells lack histones. Conversely, histones do occur in one group of prokaryotes, the Archaebacteria (Chapter 23).

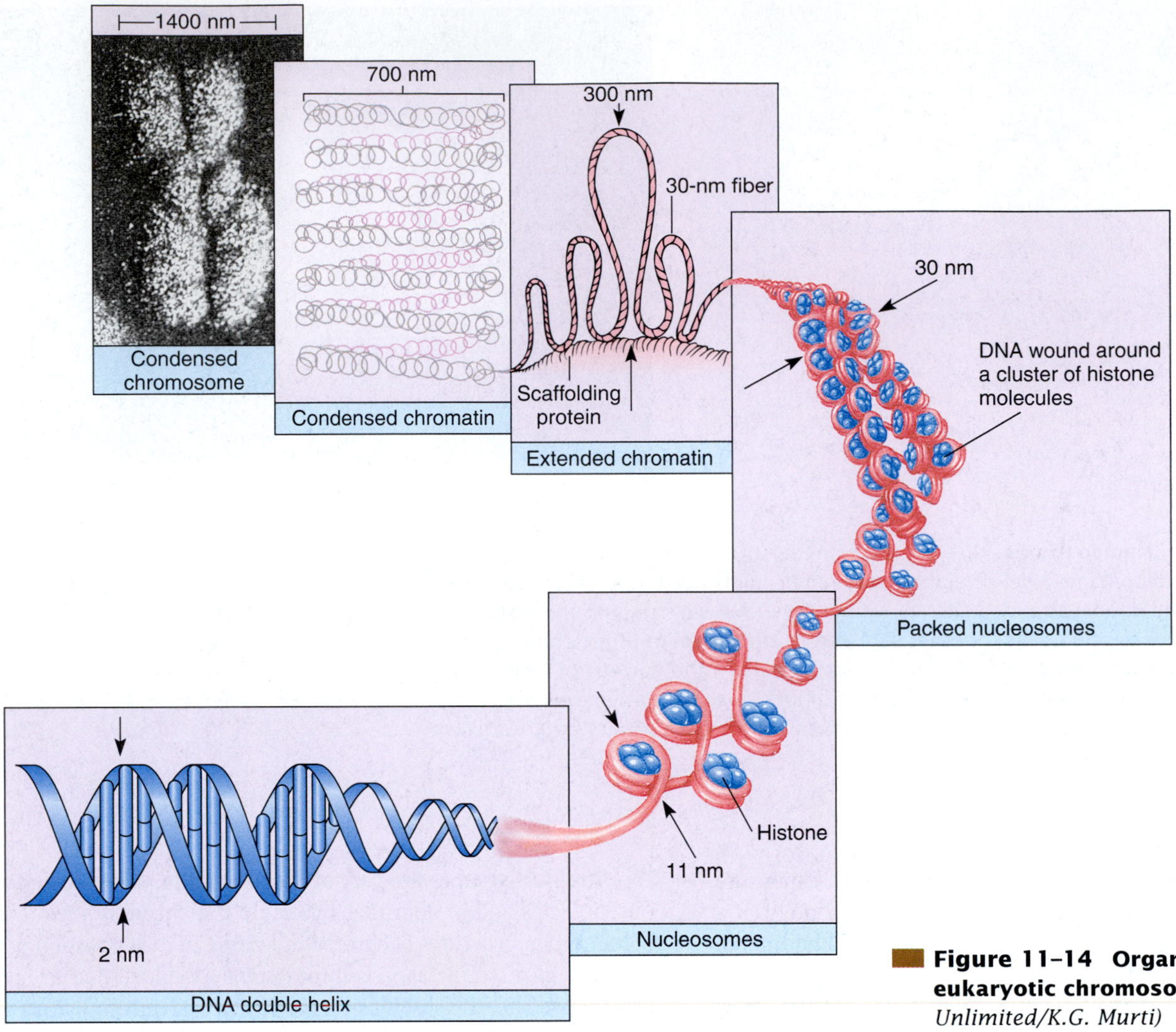

Figure 11–14 Organization of a eukaryotic chromosome. *(Visuals Unlimited/K.G. Murti)*

Figure 11–15 A chromosome depleted of histones. Note how densely packed the DNA fibrils are in this TEM of a mouse metaphase chromosome, even though they have been released from the histone proteins that organize them into tightly coiled structures. The dark structure extending from left to right across the bottom of the photograph is composed of scaffolding proteins. *(Courtesy of U. Laemmli, from* Cell, *Vol. 12, p. 817, 1988. Copyright by Cell Press)*

SUMMARY WITH KEY TERMS

I. Many early geneticists thought that genes were made of proteins. Proteins were known to be complex and variable, whereas nucleic acids were thought of as rather simple molecules with a limited ability to store information.

II. Several lines of evidence supported the idea that **DNA** is the genetic material.
   A. In **transformation** experiments, the DNA of one strain of bacteria can endow related bacteria with new genetic characteristics.
   B. When a bacterial cell becomes infected with a **bacteriophage** (virus), only the DNA from the virus enters the cell; this DNA is sufficient for the virus to reproduce and form new virus particles.
   C. Watson and Crick's studies on the structure of DNA demonstrated how information can be stored in the molecule's structure and how DNA molecules can serve as **templates** for their own duplication.

III. DNA is a very regular polymer of **nucleotides.**
   A. Each nucleotide subunit contains a nitrogenous base, which may be one of the **purines (adenine** or **guanine)** or one of the **pyrimidines (thymine** or **cytosine).** Each base is covalently linked to a five-carbon sugar, **deoxyribose,** which is covalently bonded to a phosphate group.
   B. The backbone of each single DNA chain is formed by alternating sugar and phosphate groups, joined by covalent **phosphodiester linkages.** Each phosphate group is attached to the 5' carbon of one deoxyribose and to the 3' carbon of the neighboring deoxyribose.
   C. Each DNA molecule is composed of two polynucleotide chains that associate as a **double helix.** The two chains are **antiparallel** (meaning they run in opposite directions); at each end of the DNA molecule one chain has a phosphate attached to a 5' deoxyribose carbon (the **5' end**), and the other has a hydroxyl group attached to a 3' deoxyribose carbon (the **3' end**).
   D. The two chains of the helix are held together by hydrogen bonding between specific base pairs. Adenine (A) forms two hydrogen bonds with thymine (T); guanine (G) forms three hydrogen bonds with cytosine (C).
      1. **Complementary** base-pairing between A and T and between G and C is the basis of Chargaff's rules, which state that A equals T and G equals C.
      2. Because the two strands of DNA are held together by complementary base-pairing, it is possible to predict the base sequence of one strand if one knows the base sequence of the other strand.

IV. During **DNA replication,** the two strands of the double helix unwind. Each strand serves as a template for the formation of a new, complementary strand.
   A. DNA replication is **semiconservative;** that is, each daughter double helix contains one strand from the parent molecule and one newly synthesized strand.
   B. DNA replication is a complex process requiring a number of different enzymes.
      1. The enzyme that adds new deoxyribonucleotides to a growing DNA strand is a **DNA polymerase.**
      2. Additional enzymes and other proteins are required to unwind and stabilize the separated DNA helix and to form **RNA primers. Topoisomerases** prevent tangling and knotting, and **DNA ligase** links together fragments of newly synthesized DNA.
   C. DNA synthesis always proceeds in a 5' ⟶ 3' direction. This requires that one DNA strand (the **lagging strand**) be synthesized discontinuously, as short **Okazaki fragments.** The opposite strand (the **leading strand**) is synthesized continuously.
   D. DNA replication is bidirectional, starting at the **origin of replication** and proceeding in both directions from that point. A eukaryotic chromosome may have multiple origins of replication and may be replicating at many points along its length at any one time.
   E. Eukaryotic chromosome ends, known as **telomeres,** shorten slightly with each cell cycle but can be extended by the enzyme **telomerase.** The absence of telomerase activity in certain cells may be a cause of **cellular aging.**

V. DNA molecules have to be organized in a cell because they are much longer than the nuclei or the cells that contain them.
   A. Prokaryotic cells usually have circular DNA molecules.
   B. The organization of eukaryotic DNA into chromosomes allows the DNA to be accurately replicated and segregated into daughter cells without tangling. Chromosomes have several levels of organization.
      1. The DNA is associated with **histones** (basic proteins) to form **nucleosomes,** each of which consists of a histone bead with DNA wrapped around it.
      2. The nucleosomes are organized into large coiled loops held together by nonhistone **scaffolding proteins.**

POST-TEST

1. Which of the following inspired Avery and his coworkers to do the experiments that demonstrated that the transforming principle in bacteria is DNA? (a) the fact that A is equal to T, and G is equal to C (b) Watson and Crick's model of DNA structure (c) Meselson and Stahl's studies on DNA replication in *E. coli* (d) Griffith's experiments on smooth and rough strains of pneumococci (e) Hershey and Chase's experiments on the reproduction of bacteriophages
2. In the Hershey-Chase experiment with bacteriophages (a) harmless bacterial cells became permanently transformed into virulent cells (b) DNA was demonstrated to be the transforming principle of earlier bacterial transformation experiments (c) the replication of DNA was conclusively shown to be semiconservative (d) viral DNA was shown to enter bacterial cells and be responsible for the production of new viruses within the bacteria (e) the viruses injected proteins, not DNA, into bacterial cells
3. The experiments in which Meselson and Stahl grew bacteria in heavy nitrogen conclusively demonstrated that DNA (a) is a double helix (b) replicates semiconservatively (c) consists of repeating nucleotide subunits (d) has complementary base pairing (e) is always synthesized in a 5' to 3' direction
4. The statement "DNA replicates by a semiconservative mechanism" means that (a) only one DNA strand is copied (b) first one DNA strand is copied, and then the other strand is copied (c) the two strands of a double helix have identical base sequences (d) some portions of a single DNA strand are old, and other portions are newly synthesized (e) each double helix consists of one old and one newly synthesized strand
5. Multiple origins of replication (a) speed up replication of eukaryotic chromosomes (b) allow the lagging strands and leading strands to be synthesized at different replication forks (c) help to relieve strain as the double helix is unwound (d) prevent mutations (e) are necessary for the replication of a circular DNA molecule in bacteria
6. Topoisomerases (a) synthesize DNA (b) synthesize RNA primers (c) join Okazaki fragments (d) break and rejoin DNA to resolve knots that have formed (e) prevent single DNA strands from joining to form a double helix
7. A phosphate in DNA is (a) hydrogen-bonded to a base (b) covalently linked to two bases (c) covalently linked to two deoxyriboses (d) hydrogen-bonded to two additional phosphates (e) covalently linked to a base, a deoxyribose, and another phosphate

8. Which of the following depicts the relative arrangement of the complementary strands of a DNA double helix?

   (a) 5′ — 5′ 3′ — 3′ (b) 3′ — 5′ 3′ — 5′ (c) 3′ — 3′ 3′ — 3′ (d) 5′ — 5′ 5′ — 5′ (e) 3′ — 5′ 5′ — 3′

9. A lagging strand is formed by (a) joining primers (b) joining Okazaki fragments (c) joining leading strands (d) breaking up a leading strand (e) joining primers, Okazaki fragments, and leading strands
10. The immediate source of energy for DNA replication is (a) the hydrolysis of the nucleotides, with the release of two phosphates (b) the oxidation of NADPH (c) the hydrolysis of ATP (d) electron transport (e) the breaking of hydrogen bonds
11. A nucleosome consists of (a) DNA and scaffolding proteins (b) scaffolding proteins and histones (c) DNA and histones (d) DNA, histones, and scaffolding proteins (e) histones only

REVIEW QUESTIONS

1. How did the experiments of Avery and coworkers point to DNA as the essential genetic material? Did the Hershey-Chase experiment establish that DNA is the genetic material in all organisms? Did either of these experiments demonstrate how DNA could function as the chemical basis of genes?
2. Sketch the structure of a single strand of DNA. What types of subunits make up the chain? How are they linked?
3. Describe the structure of double-stranded DNA as determined by Watson and Crick.
4. Does a single strand of DNA obey Chargaff's rules? How do Chargaff's rules relate to the structure of DNA?
5. What are some of the mechanical problems encountered in DNA replication? How are they dealt with by the cell?
6. Why is DNA replication continuous for one strand but discontinuous for the other?
7. Compare the structures of a bacterial DNA molecule and a eukaryotic chromosome. What effects do these differences have on replication?
8. Describe how both prokaryotic and eukaryotic cells cope with the large discrepancy between the length of their DNA molecules and the size of the cell or nucleus.

YOU MAKE THE CONNECTION

1. What characteristics must a molecule have if it is to serve as genetic material?
2. What important features of the structure of DNA are consistent with its role as the chemical basis of heredity?

RECOMMENDED READINGS

Greider, C.W., and E.H. Blackburn. "Telomeres, Telomerase, and Cancer." *Scientific American,* Vol. 274, No. 2, Feb. 1996. The discoverers of telomerase discuss the possible roles of telomere shortening and lengthening in cancer and aging.

Judson, H.F. *The Eighth Day of Creation: Makers of the Revolution in Biology.* Simon & Schuster, New York, 1979. A beautifully written and fascinating account of the early history of molecular biology.

Lodish, H., A. Berk, S.L. Zipursky, P. Matsudaira, D. Baltimore, and J. Darnell. *Molecular Cell Biology,* 4th ed. W.H. Freeman and Co., New York, 2000. An extensive, detailed, and well written discussion of DNA structure and replication.

Pool, R. "Dr. Tinkertoy." *Discover,* Feb. 1997. Chemist Ned Seeman uses short DNA molecules as construction components that join together by complementary base-pairing.

Rennie, J. "DNA's New Twists." *Scientific American,* Mar. 1993. A fascinating summary of novel studies in genetics that challenge previous ideas and open the way for a new understanding of the role of DNA in evolution and in genetic diseases.

Rebek, J., Jr. "Synthetic Self-Replicating Molecules." *Scientific American,* Jul. 1994. Experiments on certain molecules fabricated in the laboratory demonstrate that self-duplication is not unique to DNA.

Watson, J.D. *The Double Helix.* Atheneum, New York, 1968. Watson's view of the discovery of the structure of DNA. Somewhat controversial, but entertaining and insightful reading.

Watson, J.D., and F.H.C. Crick. "Molecular Structure of Nucleic Acids: A Structure for Deoxyribose Nucleic Acid." *Nature,* Vol. 171, 1953. Watson and Crick's original report—a simple, clearly written two-page paper that shook the scientific world.

- Visit our Web site at **http://www.info.brookscole.com/solomonbergmartin** for links to chapter-related resources on the World Wide Web. Additional on-line materials relating to this chapter can also be found on our Web site.

See chapter activity on BioActive Learner CD for additional help in mastering the chapter's material. Icon location in the chapter's margins shows which topics have tutorials or simulations in the CD.

12

RNA and Protein Synthesis: The Expression of Genetic Information

Visualization of transcription. In this TEM, RNA molecules *(lateral strands)* are being synthesized as complementary copies of a DNA template *(central axis)*. *(Professor Oscar Miller/Science Photo Library/Photo Researchers, Inc.)*

LEARNING OBJECTIVES

After you have studied this chapter you should be able to

1. Summarize the early evidence that most genes specify the structure of proteins.
2. Outline the flow of genetic information in cells, from DNA to protein.
3. Compare the structures of DNA and RNA and explain how the structure of each is related to its role in the cell.
4. Compare the processes of transcription and replication, identifying both similarities and differences.
5. Identify the features of tRNA that are important in decoding genetic information and converting it into "protein language."
6. Explain how the ribosome functions in protein synthesis.
7. Diagram the processes of initiation, chain elongation, and chain termination in protein synthesis.
8. Compare eukaryotic and prokaryotic mRNAs and explain the functional significance of their structural differences.
9. Analyze the differences in translation in prokaryotic and eukaryotic cells.
10. Explain why the genetic code is said to be redundant and virtually universal. Discuss how these features may reflect the evolutionary history of the code.
11. Give examples of the different classes of mutations that affect the base sequence of DNA and demonstrate the effects that each has on the protein produced.

In Chapter 11 we saw that the sequence of nucleotides in DNA is replicated extremely accurately by a cell so that it can be passed unaltered to its descendants. The basic features of the DNA double helix originally described by Watson and Crick are now known to be the same in all cells studied to date, from those of humans to bacteria.

By the mid-1950s it became evident that the sequence of bases in DNA contains the information required to specify all the proteins needed by the cell. However, more than a decade of intense investigation by many scientists was required before a fundamental understanding could be developed of how cells are able to convert DNA information into amino acid sequences of proteins. Much of that understanding came from studying the functions of bacterial genes. After the discovery of the structure of DNA, prokaryotic cells quickly became the organisms of choice for these investigations because they could be grown quickly and easily and because they seemed to contain only the minimal amount of DNA needed for growth and reproduction. The validity, as well as the utility, of this approach has been repeatedly confirmed, as researchers have learned that all organisms share fundamental genetic similarities.

In this chapter we first examine the evidence that accumulated in the first half of the 20th century that most genes specify the structure of proteins. We then consider at the molecular level how DNA is able to affect the phenotype of the organism through a process known as **gene expression.** Gene expression is accomplished by a complex series of events in which the information contained in the sequence of bases in DNA is used to specify the makeup of the proteins in the cell. The proteins produced then affect the phenotype in some way; these effects can range from readily observable visible physical traits to subtle changes detectable only at the biochemical level. The first major step of gene expression is **transcription,** the synthesis of RNA molecules complementary to the DNA. The second key event is **translation,** in which RNA is used as a coded template to direct protein synthesis. In Chapter 13 we consider some of the ways the entire process of gene expression is controlled.

We first focus our attention on gene expression in prokaryotes because these cells are best understood. We then extend our discussion to include eukaryotic cells. Our understanding of these cells is improving rapidly as a result of groundwork laid by study of the simpler bacterial systems.

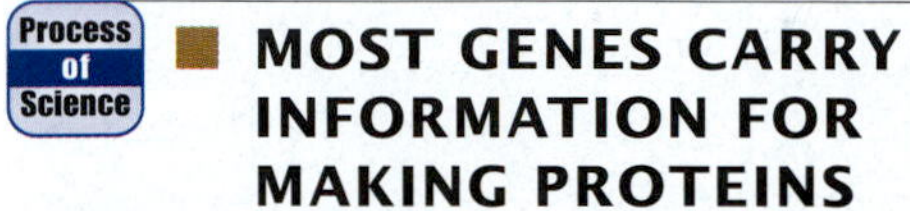

MOST GENES CARRY INFORMATION FOR MAKING PROTEINS

The idea that there is a connection between proteins and genes had its inception early in the 20th century, soon after the rediscovery of Mendel's principles. In the first edition of his book, *Inborn Errors of Metabolism* (1908), Archibald Garrod, an English physician, discussed a genetic disease called **alkaptonuria,** which was known to have a simple recessive inheritance pattern. The condition involves the metabolic pathway that breaks down the amino acids phenylalanine and tyrosine, ultimately converting them to carbon dioxide and water. The urine of affected individuals contains an intermediate in this pathway, homogentisic acid, which turns black when exposed to air (Fig. 12–1).

At that time, enzymes were known, but the protein nature of enzymes was yet to be established. Garrod hypothesized that persons with alkaptonuria lack the enzyme that normally oxidizes homogentisic acid and that this metabolic block causes homogentisic acid to accumulate in their tissues and blood, and to be excreted in their urine. Before the second edition of his book had been published in 1923, it was found that affected persons do indeed lack the enzyme that oxidizes homogentisic acid. Garrod's hypothesis was correct: A mutation in this specific gene is associated with the absence of a specific enzyme. The first clear identification of an enzyme as a protein came shortly thereafter, in 1926, when James Sumner purified a different enzyme, urease, and showed it to be a protein.

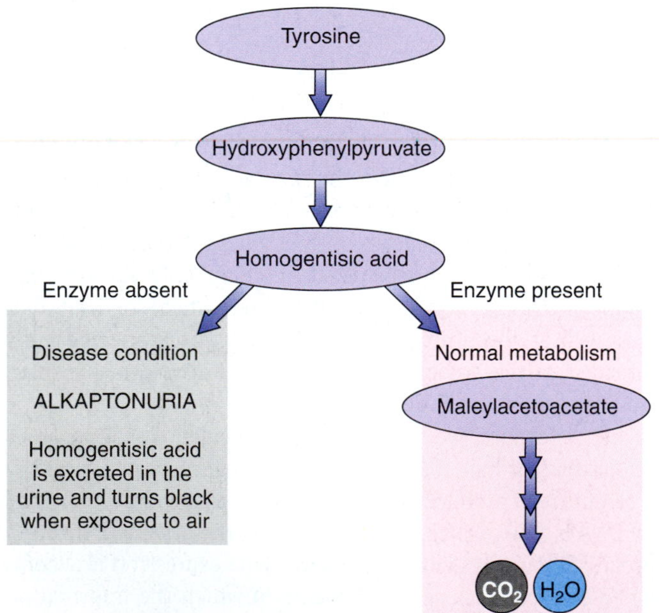

Figure 12–1 An "inborn error of metabolism." Garrod proposed that the alkaptonuria allele causes the absence of a specific enzyme, one that is part of the pathway by which the amino acid tyrosine is catabolized. That enzyme normally converts homogentisic acid to maleylacetoacetate. Homogentisic acid thus accumulates in the blood and is excreted through the urine. When the homogentisic acid in the urine comes in contact with air, it oxidizes and turns black.

Given the hindsight we have today, we might say that, taken together, these findings demonstrated that a mutation in a specific gene can cause a change in a specific protein. However, despite the implications, little work was done in this area, primarily because genetically transmitted errors in metabolism appeared to be rare. The lack of experimental subjects made genetic testing and statistical analysis very difficult.

A major advance in understanding the relationship between genes and enzymes came in the early 1940s, when George Beadle and Edward Tatum developed a new approach to the problem. Most efforts until that time had centered on studying known loci and attempting to determine what biochemical reactions they affected. Experimenters examined previously identified loci, such as those controlling eye color in *Drosophila* or pigments in plants. They found that specific phenotypes are controlled by a series of biosynthetic reactions, but it was not clear to the investigators whether the genes themselves were acting as enzymes or if they determined the workings of the enzymes in more complex ways.

Beadle and Tatum decided to take the opposite approach. Rather than try to identify the enzymes affected by single genes, they decided to look for mutations interfering with the known metabolic reactions that produce essential molecules such as amino acids and vitamins. They chose a fungus, *Neurospora,* which is a common bread mold, as an experimental organism for several important reasons. First, wild-type[1] *Neurospora* is easy to grow in culture. It can make all of its essential biological molecules when it is grown on a simple minimal growth medium containing only sugar, salts, and the vitamin biotin. However, a mutant *Neurospora* strain that cannot make a substance such as an amino acid can still grow if that substance is simply added to the growth medium.

Second, the life cycle of *Neurospora* includes both sexual and asexual reproduction, a fact that facilitates certain types of manipulations and genetic analysis. Third, *Neurospora* grows primarily as a haploid organism, allowing a recessive mutant al-

[1] *Wild type* refers to the genotypes and phenotypes most commonly found in natural populations of a particular species. Wild-type alleles are generally thought of as "normal," or nonmutant, alleles.

lele to be immediately identified because there is no homologous chromosome that could carry a dominant allele that would mask its expression. (For an illustration of the generalized life cycles of simple organisms such as fungi, see Figure 9–16*b;* a more detailed life cycle of organisms similar to *Neurospora* is given in Figure 25–9.)

Beadle and Tatum began by exposing thousands of haploid wild-type *Neurospora* asexual spores to x rays or ultraviolet radiation to induce mutant strains. Each irradiated strain was first grown on a complete growth medium, which contained all the amino acids and vitamins normally made by *Neurospora.* Each strain was then tested on the standard minimal growth medium described previously. About 1% to 2% of the strains that grew on the complete medium failed to grow after transfer to the minimal medium. Beadle and Tatum reasoned that such a strain carried a mutation that made it unable to produce one of the compounds essential for growth. Further testing of the mutant strain on media containing different combinations of amino acids, purines, vitamins, and so on enabled the investigators to determine the exact compound that was required (Fig. 12–2).

Their findings can be illustrated with a class of mutants that require the amino acid arginine. Beadle and Tatum found that some of the arginine-requiring mutants could grow on ornithine or citrulline, as well as arginine; others could grow on citrulline or arginine; and still others could grow only on arginine (Fig. 12–3*a*). This information was then used to deduce the order of these intermediates in the biochemical pathway leading to arginine (Fig. 12–3*b*).

Using this approach, Beadle and Tatum analyzed mutants affecting several metabolic pathways. Each mutant strain was verified by special genetic crossing experiments to have a mutation in only one gene locus. They found that for each individual gene locus identified, only one enzyme was affected. This one-to-one correspondence between genes and enzymes was succinctly stated as the *one gene, one enzyme hypothesis.*

Through the discoveries of Beadle and Tatum and others, the sciences of genetics and biochemistry became ever more closely allied, leading to an evolution of the definition of the gene and additional predictions regarding its chemical nature. The idea that a gene encodes the information required to produce a single enzyme held for almost a decade, until additional findings required a modification of this definition.

In the late 1940s it became evident that genes control not only enzymes, but other proteins as well. In 1949 Linus Pauling and his coworkers were able to demonstrate that the structure of hemoglobin can be altered by a mutation of a single locus. This particular mutant form of hemoglobin is associated with the genetic disease sickle cell anemia (Chapter 15). In addition, various studies showed that many proteins are constructed from two or more polypeptide chains, each of which may be controlled by a different locus. For example, hemoglobin was shown to contain two types of polypeptide chains, the α and β subunits (see Fig. 3–23*a*). Sickle cell anemia results from a mutation affecting the β subunits.

The definition of a gene was therefore extended to state that one gene is responsible for one polypeptide chain. Even this definition has proved to be only partially correct, although as we will see later in this chapter, we still define a gene in terms of its product.

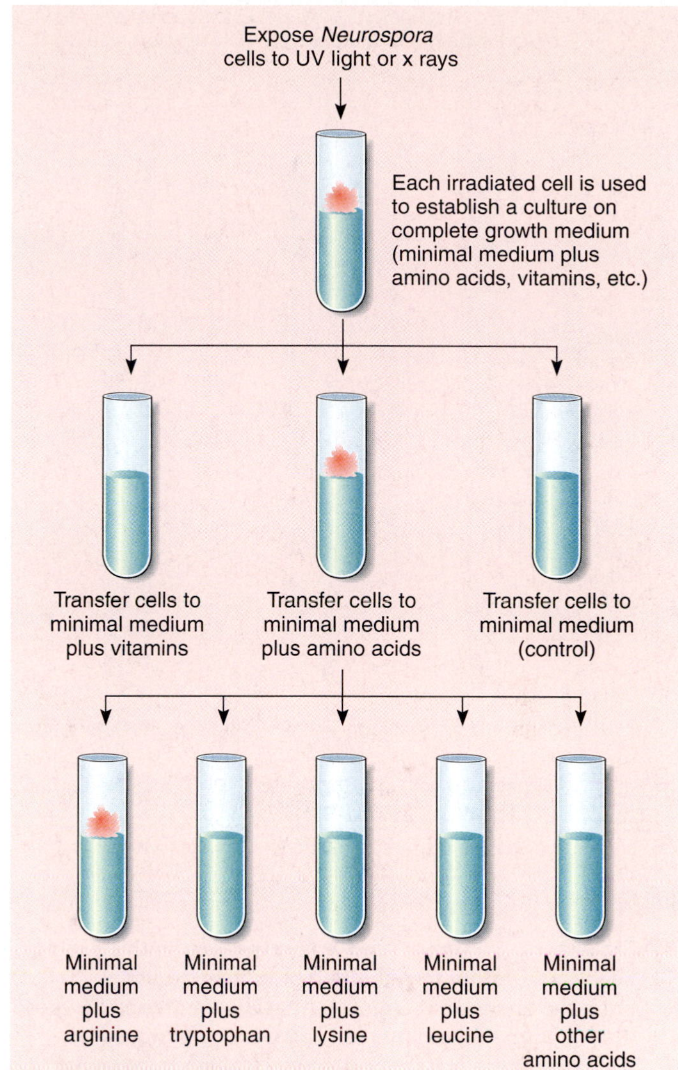

Figure 12–2 Mutations affecting biochemical pathways. Beadle and Tatum irradiated *Neurospora* cells in an effort to induce mutations. Cultures derived from these cells were first established on a complete growth medium, one that contained all the amino acids, vitamins, and so on that *Neurospora* normally makes for itself. Any strain that subsequently failed to grow when transferred to a simple minimal growth medium was thought to carry a mutation causing a block at a step in a biochemical pathway. The specific nutritional requirement in the mutant strain was determined by testing for growth on minimal media supplemented with individual vitamins or amino acids. In this example, only the medium containing the amino acid arginine supports growth, indicating that the mutation affects some part of the arginine biosynthetic pathway.

Although the elegant work of Beadle and Tatum and others demonstrated that genes code for proteins, the mechanism by

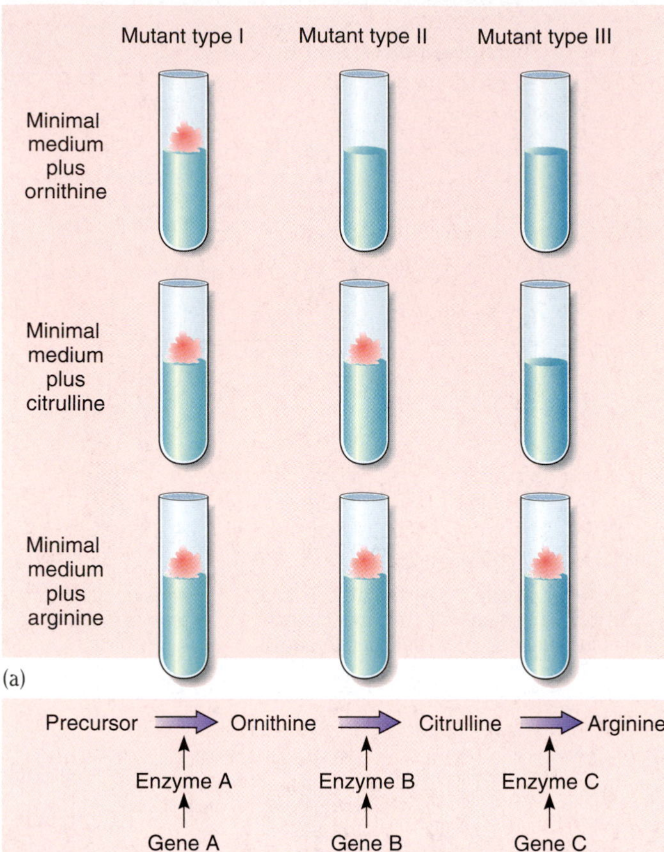

Figure 12–3 Genes and enzymes. **(a)** In this example, Beadle and Tatum tested different mutant strains that require the amino acid arginine. These were grouped (Types I, II, or III in the figure) by their response to various intermediates in the metabolic pathway leading to arginine, which were provided as supplements to the minimal growth medium. **(b)** Analysis of the experimental results led to this model for a portion of the arginine biosynthetic pathway (which is actually a cycle). Because type I mutant strains grew on minimal medium supplemented with ornithine, citrulline, or arginine, they were thought to be missing enzyme A, required for formation of all three compounds. Type II mutant strains were thought to be missing enzyme B because they were unable to grow on minimal medium supplemented with ornithine but allowed the conversion of citrulline to arginine. Type III mutant strains were thought to be missing enzyme C because they grew on minimal medium to which only arginine had been added. Special genetic crosses *(not shown)* were conducted to verify that there was a one-to-one correspondence between each specific enzyme and a specific gene locus.

which this could occur was completely unknown. After Watson and Crick's discovery of the structure of DNA, a great many scientists worked to understand gene expression in terms of the basic flow of information in the cell. We will begin with an overview of gene expression and then consider the various processes in more detail.

OVERVIEW: INFORMATION FLOWS FROM DNA TO RNA TO PROTEIN

Although the sequence of bases in DNA determines the sequence of amino acids in polypeptides, the information in DNA is not used directly. Instead, a related nucleic acid, **ribonucleic acid (RNA),** serves as an intermediary between DNA and protein. When a protein-coding gene is expressed, first an RNA copy is made of the information in the DNA. It is this RNA copy that provides the information to direct protein synthesis.

Like DNA, RNA is a polymer of nucleotides, but it has some important differences (Fig. 12–4). RNA is usually single-stranded, although internal regions of some RNA molecules may have complementary sequences that allow the strand to fold back and pair to form short, double-stranded segments. As shown in Figure 12–4, the sugar in RNA is **ribose,** which is similar to deoxyribose of DNA, but with an extra hydroxyl group at the 2' position. (Compare ribose with the deoxyribose of DNA, shown in Fig. 11–3, which has only a hydrogen at the 2' position.) The base **uracil** substitutes for thymine and, like thymine, is a pyrimidine that can form two hydrogen bonds with adenine. Hence, uracil and adenine are a complementary pair.

DNA is transcribed to form RNA

The process by which RNA is synthesized resembles DNA replication in that the sequence of bases in the RNA strand is determined by complementary base-pairing with one of the DNA strands, referred to as the *transcribed strand,* or *template strand* (Fig. 12–5). Because RNA synthesis involves taking the information in one kind of nucleic acid (DNA) and copying it in the form of another nucleic acid (RNA), we refer to this process as **transcription** ("copying"). The type of RNA that carries the specific information for making a protein is called **messenger RNA,** or **mRNA.**

RNA is translated to form a polypeptide

Figure 12–5 also shows the second stage of gene expression, in which the transcribed information in the mRNA is used to specify the amino acid sequence of a polypeptide. This process is called **translation** because it involves conversion of the "nucleic acid language" in the mRNA molecule into the "amino acid language" of protein.

A sequence of three consecutive bases in mRNA, called a **codon,** specifies one amino acid. For example, one codon that corresponds to the amino acid threonine is 5'— ACG — 3'. Because each codon requires three nucleotides, the code is referred to as a **triplet code.** Taken together, all of the assignments of codons to amino acids or to punctuation (stop and start signals) are referred to collectively as the **genetic code** (Table 12–1). Various aspects of the genetic code are discussed later in this chapter.

Transfer RNAs (tRNAs) are critical parts of the decoding machinery because they act as "adapters" that provide a connection between amino acids and nucleic acids. This is possible because each tRNA can (1) link with a specific amino acid and (2) recognize the appropriate mRNA codon for that particular amino acid. A particular tRNA can recognize a specific codon because it

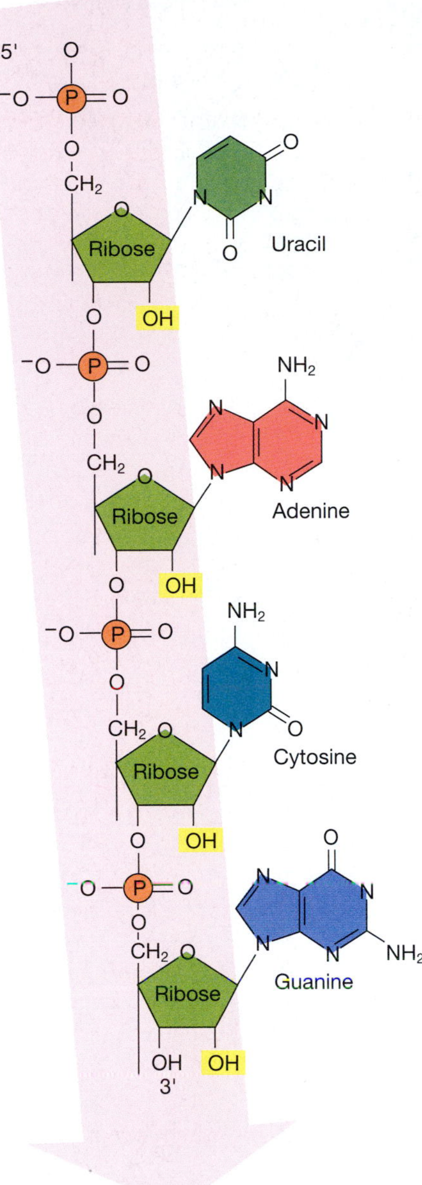

Figure 12–4 Nucleotide structure of RNA. The nucleotide subunits of RNA are joined by 5' ⟶ 3' phosphodiester linkages, like those found in DNA. Adenine, guanine, and cytosine are present, as in DNA, but the base uracil replaces thymine. All four nucleotide types contain the five-carbon sugar ribose, which has a hydroxyl group on its 2' carbon atom.

has a sequence of three bases, called the **anticodon,** that hydrogen bonds with the mRNA codon by complementary base-pairing. The exact anticodon that is complementary to the codon for threonine in our example is 3' — UGC — 5'.

Translation requires that (1) each tRNA anticodon be hydrogen bonded to the complementary mRNA codon and (2) the amino acids carried by the tRNAs be linked together in the order specified by the order of the codons in the mRNA. This is accomplished by **ribosomes** (see Chapter 4), complex organelles composed of two different subunits, each containing more than 50 proteins and **ribosomal RNA (rRNA).** Ribosomes attach to one end of the mRNA and travel along it, thereby allowing the tRNAs to attach sequentially to the codons of mRNA. In this way the amino acids carried by the tRNAs become properly positioned to be joined by peptide bonds in the correct sequence to form a polypeptide.

Now that we have presented an overview of information flow from DNA to proteins, let us examine the entire process more closely.

TRANSCRIPTION IS THE SYNTHESIS OF RNA FROM A DNA TEMPLATE

Three main kinds of RNA are transcribed from DNA: ribosomal RNA (rRNA) and transfer RNA (tRNA), as well as messenger RNA (mRNA). Most RNA is synthesized by **DNA-dependent RNA polymerases,** enzymes that are present in all cells. These enzymes require DNA as a template and have many similarities to the DNA polymerases discussed in Chapter 11. Like DNA polymerases, they carry out synthesis in a 5' ⟶ 3' direction; that is, they begin at the 5' end of the RNA molecule being synthesized and then continue to add nucleotides at the 3' end until the molecule is complete (Fig. 12–6). They use nucleotides with three phosphate groups as substrates, removing two of the phosphates as the nucleotides are covalently linked to the 3' end of the RNA. Like DNA replication and the hydrolysis of ATP, these reactions are strongly exergonic (see Chapter 6).

Whenever nucleic acid molecules associate by complementary base-pairing, the two strands are antiparallel. Just as the two paired strands of DNA are antiparallel (see Chapter 11), the transcribed strand of the DNA and the complementary RNA strand are also antiparallel. Therefore, as shown in Fig 12–6, when transcription takes place, as RNA is being synthesized in its 5' → 3' direction, the DNA template is being read in its 3' → 5' direction.

It is conventional to refer to a sequence of bases in a gene or the mRNA sequence transcribed from it as *upstream* or *downstream* of some reference point. **Upstream** means toward the 5' end of the mRNA sequence or the 3' end of the transcribed DNA strand. **Downstream** means toward the 3' end of the RNA or the 5' end of the transcribed DNA strand.

Upstream		Downstream
	5'—A—T—G—A—C—T—3'	(nontranscribed DNA strand)
	3'—T—A—C—T—G—A—5'	(transcribed DNA strand)
	Direction of transcription ⟶	
Triphosphate	5'—A—U—G—A—C—U—3' OH (RNA)	

Messenger RNA synthesis includes several steps

In both prokaryotes and eukaryotes, the DNA sequence to which RNA polymerase initially binds is called the **promoter.** Because

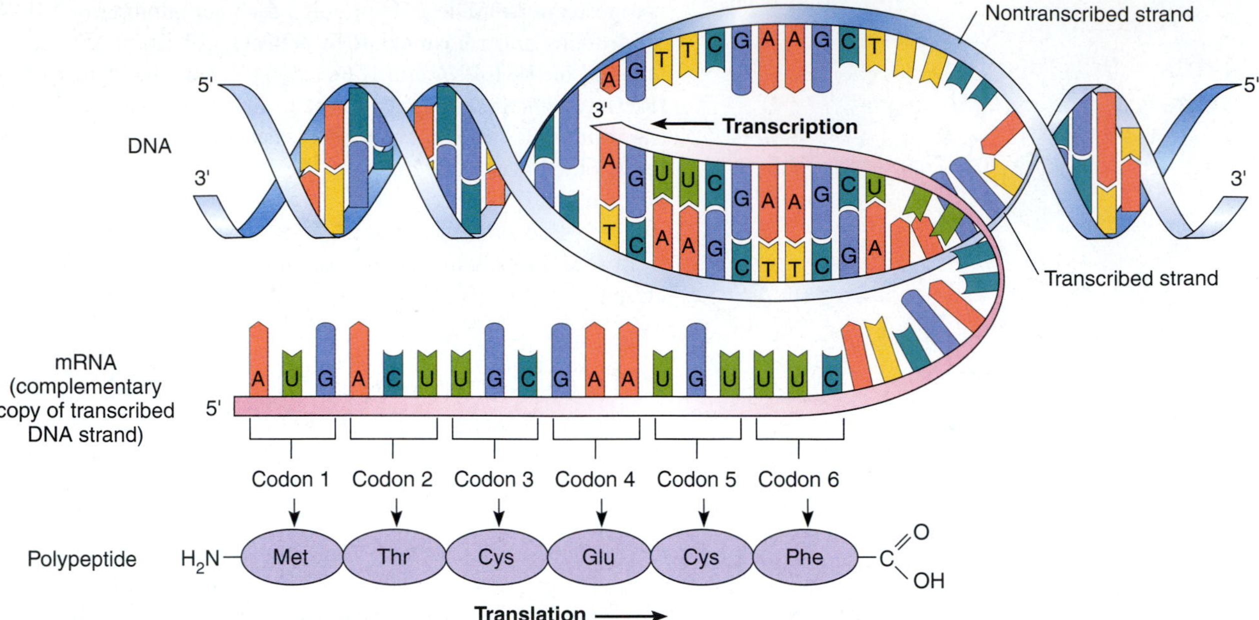

Figure 12–5 Overview of transcription and translation. In transcription, messenger RNA is synthesized as a complementary copy of one of the DNA strands, the transcribed strand. Messenger RNA carries genetic information in the form of sets of three bases called codons, each of which specifies one amino acid. Messenger RNA codons are translated consecutively, thus specifying the linear sequence of amino acids in the polypeptide chain. Translation requires transfer RNA and ribosomes *(not shown)*. The figure depicts transcription and translation in prokaryotes. In eukaryotes transcription takes place in the nucleus and translation occurs in the cytoplasm.

the promoter is not transcribed, the RNA polymerase must move past the promoter to begin transcription of the protein-coding sequence of DNA. Different genes may have slightly different promoter sequences, so the cell can direct which genes are transcribed at any one time (Chapter 13). Bacterial promoters are usually about 40 bases long and are positioned in the DNA just upstream of the point at which transcription will begin. Once the polymerase has recognized the correct promoter, it unwinds the helix and begins transcription.

Unlike DNA synthesis, RNA synthesis does not require a primer. However, transcription requires several proteins in addition to RNA polymerase; these are discussed in Chapter 13.

As illustrated in Figure 12–6, the first nucleotide at the 5' end of a new mRNA chain retains its triphosphate group, but as each additional nucleotide is incorporated at the 3' end of the growing RNA molecule, two of its phosphates are removed in an exergonic reaction that leaves the remaining phosphate to become part of the sugar-phosphate backbone (as in DNA replication). The last nucleotide to be incorporated has an exposed 3' hydroxyl group.

The termination of transcription, like its initiation, is controlled by a set of specific base sequences. These termination signals at the end of the gene cause transcription to stop.

Usually only one of the strands in a protein-coding region of DNA is transcribed (Fig. 12–7*a*). For example, consider a segment of DNA that contains the following DNA base sequence in the transcribed strand:

3' — TAACGGTCT — 5'

If the complementary DNA strand

5' — ATTGCCAGA — 3'

were to be transcribed, a message specifying an entirely different (and generally nonfunctional) protein would be produced. However, the fact that only one strand is transcribed does not mean that the same strand is always the template throughout the length of a chromosome-sized DNA molecule. Instead, a particular strand may serve as the transcribed strand for some genes and the nontranscribed strand for others (Fig. 12–7*b*).

Messenger RNA contains additional base sequences that do not directly code for protein

A completed messenger RNA contains more than the nucleotide sequence that codes for a protein. A typical bacterial messenger

TABLE 12–1 The Genetic Code: Codons of mRNA that Specify a Given Amino Acid

First Position (5' end)	Second Position	Third Position (3' end) U	C	A	G
U	U	UUU	UUC	UUA	UUG
		Phenylalanine		Leucine	
	C	UCU	UCC	UCA	UCG
		Serine			
	A	UAU	UAC	UAA	UAG
		Tyrosine		(Stop)	(Stop)
	G	UGU	UGC	UGA	UGG
		Cysteine		(Stop)	Tryptophan
C	U	CUU	CUC	CUA	CUG
		Leucine			
	C	CCU	CCC	CCA	CCG
		Proline			
	A	CAU	CAC	CAA	CAG
		Histidine		Glutamine	
	G	CGU	CGC	CGA	CGG
		Arginine			
A	U	AUU	AUC	AUA	AUG (start)
		Isoleucine			Methionine
	C	ACU	ACC	ACA	ACG
		Threonine			
	A	AAU	AAC	AAA	AAG
		Asparagine		Lysine	
	G	AGU	AGC	AGA	AGG
		Serine		Arginine	
G	U	GUU	GUC	GUA	GUG
		Valine			
	C	GCU	GCC	GCA	GCG
		Alanine			
	A	GAU	GAC	GAA	GAG
		Aspartic acid		Glutamic acid	
	G	GGU	GGC	GGA	GGG
		Glycine			

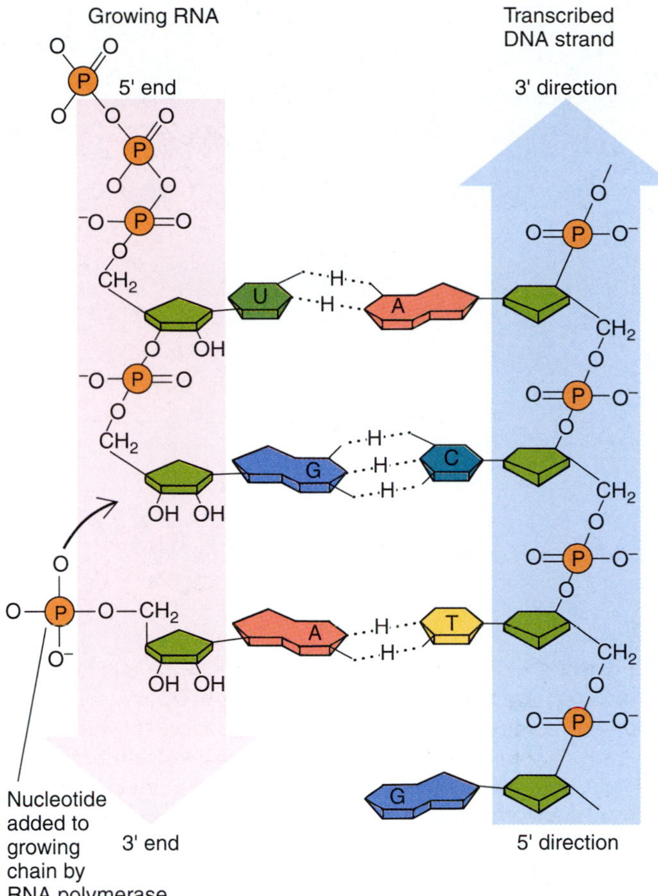

Figure 12–6 Transcription. Incoming nucleotides with three phosphates pair with complementary bases on the transcribed DNA strand *(right)*. RNA polymerase cleaves two phosphates *(not shown)* from each nucleotide and covalently links the remaining phosphate to the 3' end of the growing RNA chain. Thus, RNA, like DNA, is synthesized in a 5' ⟶ 3' direction.

RNA is shown in Figure 12–8. (The unique features of eukaryotic messenger RNA and transcription in eukaryotes are discussed later in the chapter.) In both prokaryotes and eukaryotes, RNA polymerase starts transcription of a gene well upstream of the protein-coding DNA sequence. As a result, the mRNA has a noncoding **leader sequence** at its 5' end. The leader contains recognition signals for ribosome binding, which allow the ribosomes to be properly positioned to translate the message. The leader sequence is followed by the **coding sequence,** which contains the actual messages for the proteins. Unlike eukaryotic cells, it is common for one or more polypeptides to be encoded by a single mRNA molecule in bacterial cells (see Chapter 13). At the end of each coding sequence is a special **termination,** or **stop, codon.** The stop codons—UAA, UGA, or UAG (see Table 12–1)— are present in both prokaryotic and eukaryotic messages. They do not code for amino acids but instead specify the end of the protein. These are followed by noncoding 3' trailing sequences, which can vary in length.

DURING TRANSLATION, THE NUCLEIC ACID MESSAGE IS DECODED

In eukaryotes, messenger RNA must move from the nucleus, the site of transcription, to the cytoplasm, the site of **translation,** or

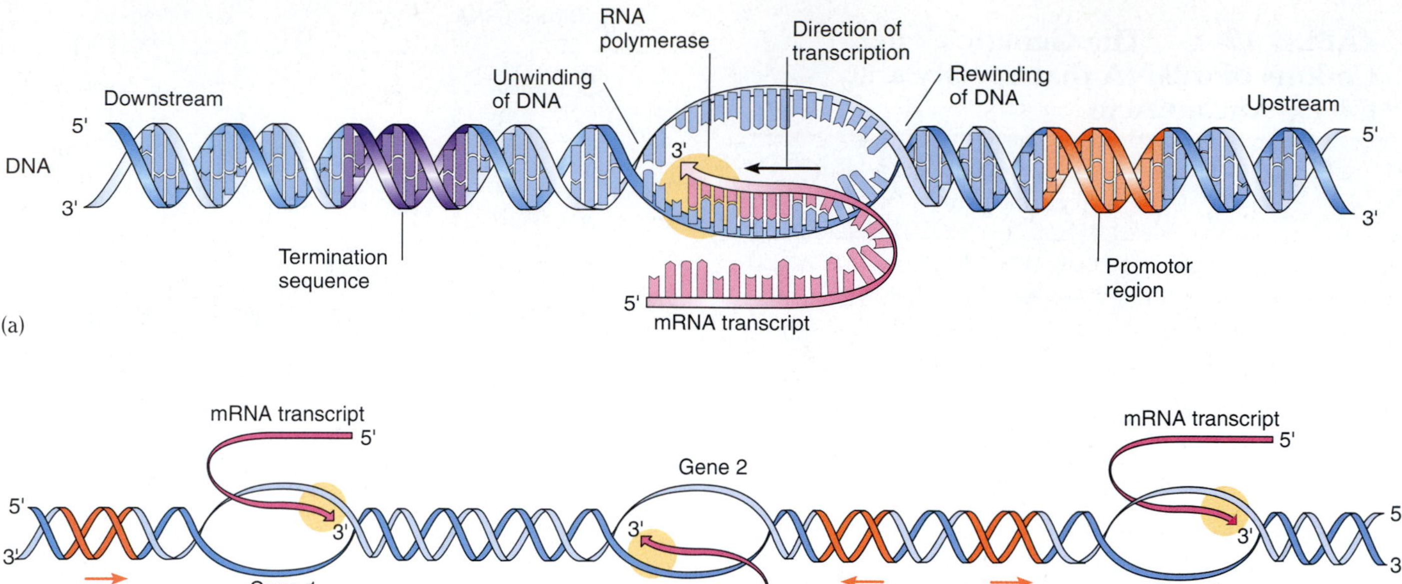

Figure 12-7 Synthesis of mRNA. (a) The mRNA is synthesized in a 5' ⟶ 3' direction from the transcribed strand of the DNA molecule. Transcription starts downstream from a DNA promoter sequence, to which the RNA polymerase initially binds. Termination sequences, found downstream from the protein-coding sequences, signal the RNA polymerase to stop transcription and be released from the DNA. (b) Usually only one of the two strands is transcribed for a given gene, but the opposite strand may be transcribed for a neighboring gene. Each transcript starts at its own promoter *(orange region)*.

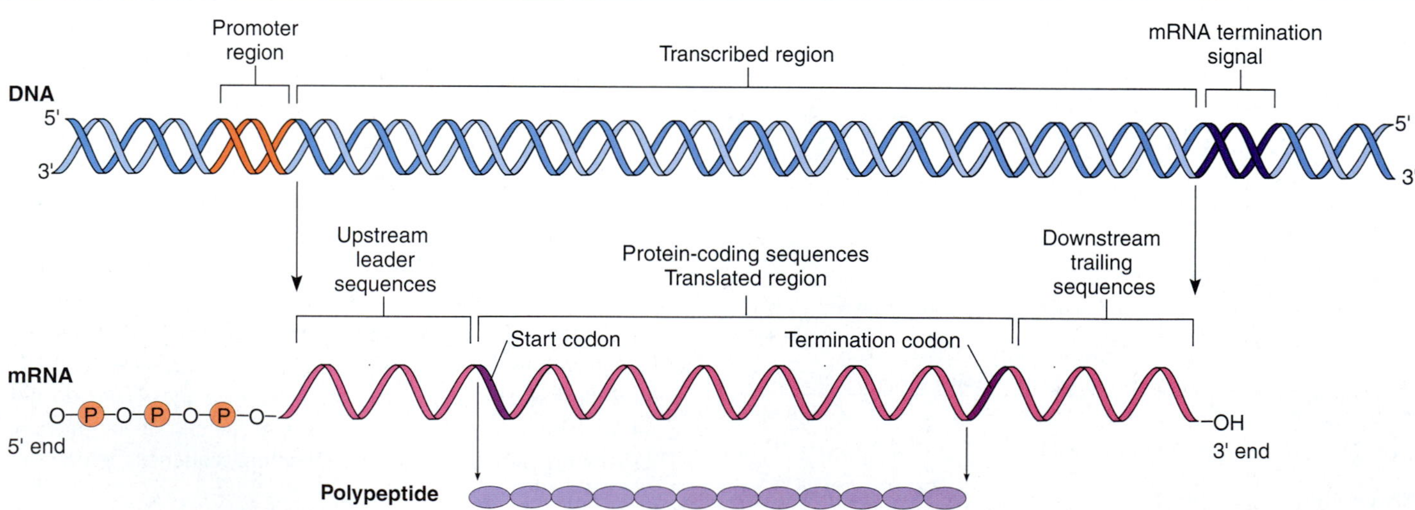

Figure 12-8 Bacterial mRNA. This figure compares a bacterial mRNA with the region of DNA from which it was transcribed. RNA polymerase recognizes, but does not transcribe, promoter sequences in the DNA. Initiation of RNA synthesis occurs five to eight bases downstream from the promoter. Ribosome-recognition sites are located in the 5' mRNA upstream leader sequences. Protein-coding sequences begin at a start codon, which follows the leader sequences, and end at a downstream termination codon near the 3' end of the molecule. Noncoding trailing sequences, which can vary in length, follow the protein-coding sequences.

protein synthesis. No such movement occurs in prokaryotes, which lack a nucleus. Translation adds another level of complexity to the process of information transfer because it involves the conversion of the triplet nucleic acid code to the 20-amino acid alphabet of proteins. The structural differences between a polynucleotide chain and a polypeptide chain are so great that no simple way exists for amino acids to interact directly with an mRNA molecule to make a protein. Translation therefore requires the coordinated functioning of more than 100 kinds of macromolecules, including the protein and RNA components of the ribosomes, mRNA, and amino acids linked to tRNAs.

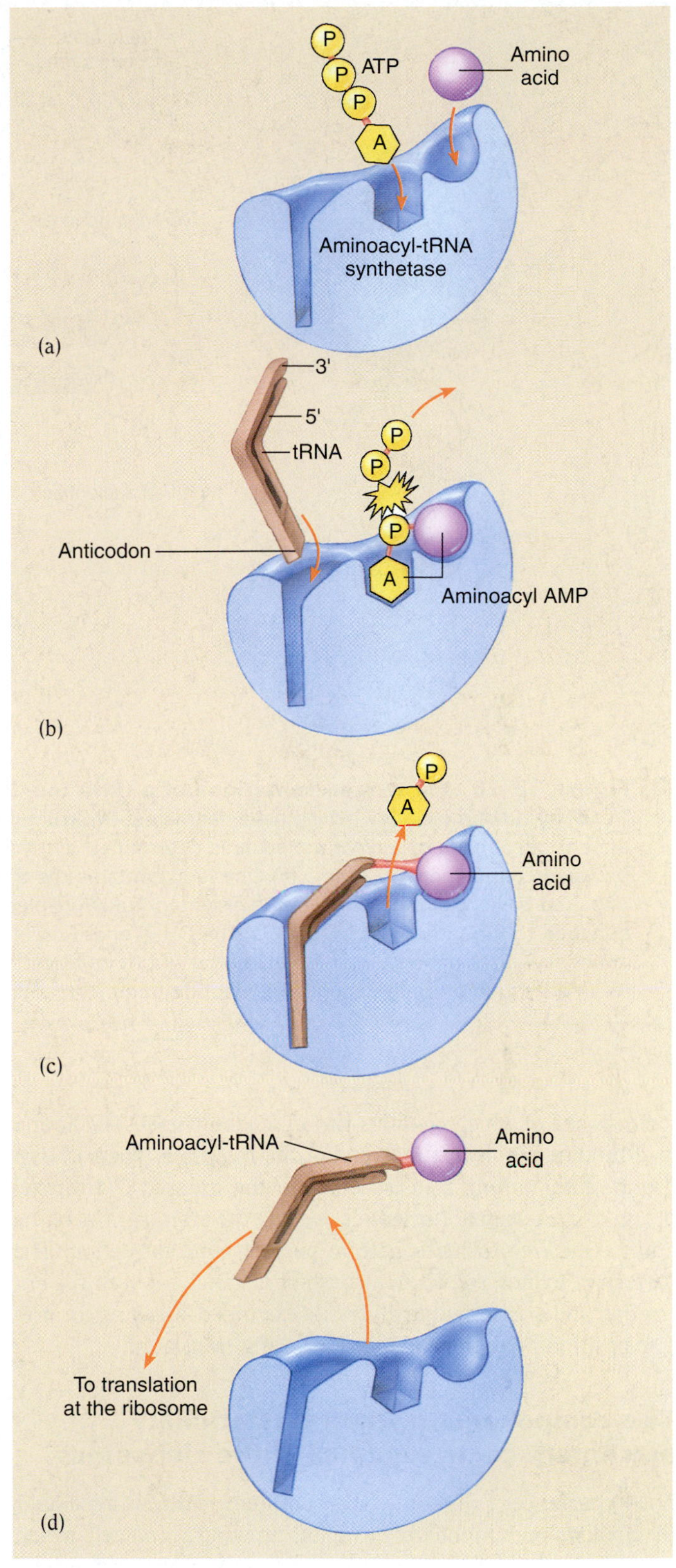

Figure 12–9 Formation of aminoacyl-tRNA. **(a)** A specific aminoacyl-tRNA synthetase is responsible for catalyzing the attachment of an amino acid to its correct tRNA. **(b,c)** In an ATP-requiring reaction, the carboxyl group of the amino acid becomes attached to the 3' end of the tRNA. **(d)** The resulting aminoacyl-tRNA complex is released.

An amino acid is attached to transfer RNA before becoming incorporated into a polypeptide

Amino acids are joined together by peptide bonds to form proteins (see Chapter 3). This joining involves linking the amino and carboxyl groups of adjacent amino acids. The translation process ensures that not only are peptide bonds formed but that the amino acids are linked in the correct sequence specified by the codons in the mRNA. How do the amino acids become aligned in the proper sequence so they can become linked?

Francis Crick, one of the codiscoverers of the structure of DNA (see Chapter 11), recognized this problem and proposed that a molecule was needed to serve as an "adapter" in protein synthesis and bridge the gap between mRNA and proteins. Crick's adapters turned out to be transfer RNA (tRNA) molecules. DNA contains special tRNA genes that are transcribed to form the tRNAs. Each kind of tRNA molecule binds to a specific amino acid. Amino acids are covalently linked to their respective tRNA molecules by specific enzymes called **aminoacyl-tRNA synthetases,** which use ATP as an energy source (Fig. 12–9). The resulting complexes, called **aminoacyl-tRNAs,** are able to bind to the mRNA coding sequence so as to align the amino acids in the correct order to form the polypeptide chain.

Transfer RNA molecules have specialized regions with specific functions

Although tRNA molecules are considerably smaller than mRNA or rRNA molecules, they have a complex structure. A tRNA molecule must have several properties:

1. It must have an **anticodon,** a specific complementary binding sequence for the correct mRNA codon.
2. It must be recognized by a specific aminoacyl-tRNA synthetase that adds the correct amino acid.
3. It must have a region that serves as the attachment site for the specific amino acid specified by the anticodon.
4. It must be recognized by ribosomes.

The tRNAs are polynucleotide chains 70 to 80 nucleotides long, each with several unique base sequences, as well as some sequences that are common to all (Fig. 12–10). Complementary base-pairing within each tRNA molecule causes it to be doubled back and folded. Three or more loops of unpaired nucleotides are

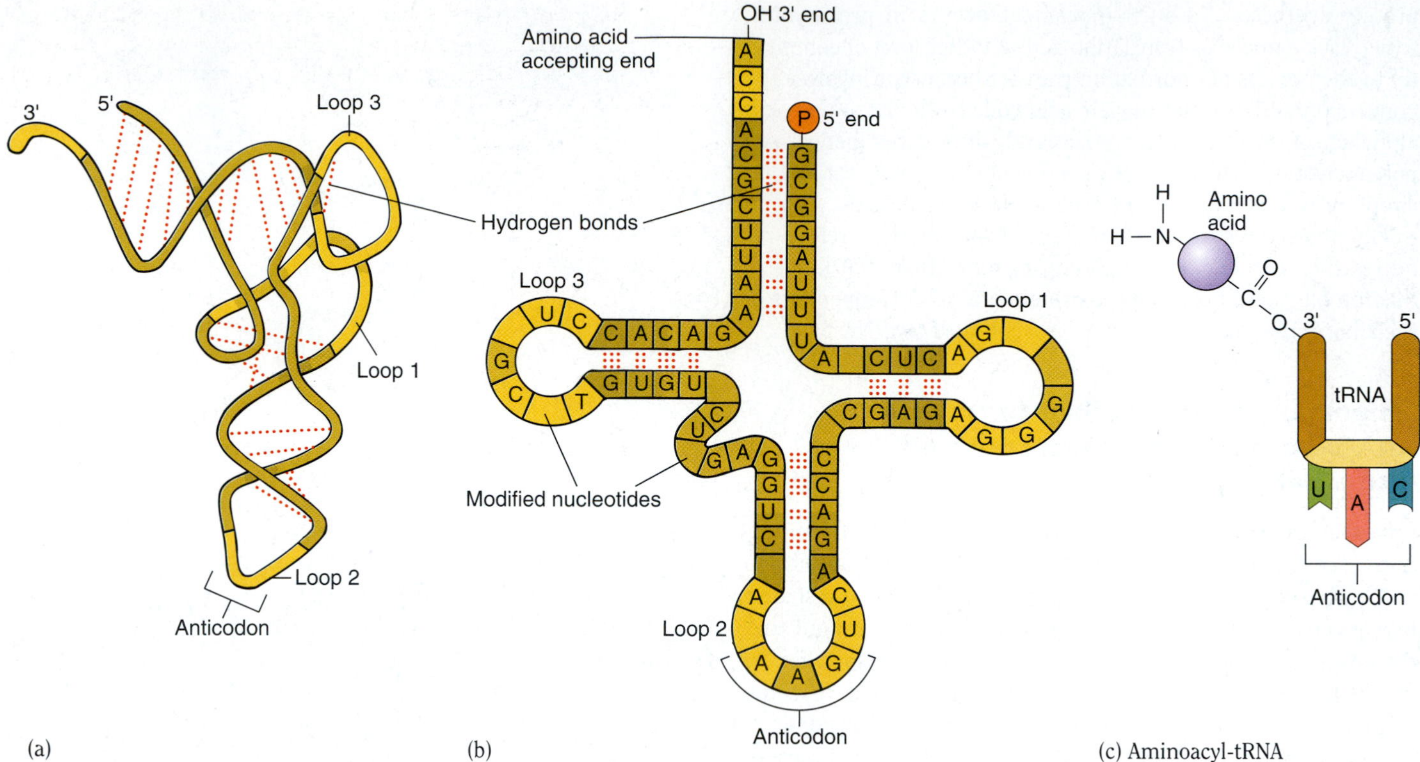

Figure 12–10 Three representations of a tRNA molecule. The genetic code is "read" by tRNA molecules, which have characteristic structures. **(a)** The three-dimensional shape of a tRNA molecule is determined by hydrogen bonds that form between complementary bases. **(b)** One loop contains the triplet anticodon; these unpaired bases can pair with a complementary mRNA triplet codon. The amino acid is attached to the terminal nucleotide at the OH 3' end. **(c)** This schematic diagram of an aminoacyl tRNA shows an amino acid attached to its tRNA by its carboxyl group, leaving its amino group exposed for peptide bond formation.

formed, one of which contains the anticodon triplet. The amino acid binding site is at the 3' end of the molecule. The carboxyl group of the amino acid is bound to the exposed 3' hydroxyl group of the terminal nucleotide, leaving the amino group of the amino acid free to participate in peptide bond formation. The pattern of folding results in a constant distance between the anticodon and amino acid in all tRNAs examined, allowing for precise positioning of the amino acids during translation.

The components of the translational machinery come together at the ribosomes

The importance of ribosomes and of protein synthesis in cellular metabolism is exemplified by a rapidly growing *E. coli* cell, which contains some 15,000 ribosomes—nearly one-third of the total mass of the cell. Although prokaryotic and eukaryotic ribosomes are not identical, ribosomes from all organisms share a great many fundamental features and basically work in the same way. All are composed of two subunits made up of protein and ribosomal RNA, which is transcribed from DNA. Unlike messenger RNA and transfer RNA, ribosomal RNA does not transfer specific information but instead has catalytic functions. The ribosomal proteins do not appear to be catalytic but instead contribute to the overall structure of the ribosome.

Each ribosomal subunit can be isolated intact in the laboratory and then separated into each of its RNA and protein constituents. For example, it has been found that in bacteria the smaller of these subunits contains 21 proteins and one ribosomal RNA molecule, and the larger contains 35 proteins and two ribosomal RNA molecules. Under certain conditions it is possible to reassemble each subunit into a functional form by adding each component in its correct order. Through this approach, together with sophisticated electron microscopic studies, it has been possible to determine the three-dimensional structure of the ribosome (Fig. 12–11*a*), as well as how it is assembled in the living cell. The large subunit contains a depression on one surface into which the small subunit fits. During translation, the mRNA fits in a groove formed between the contact surfaces of the two subunits.

The structure of the ribosome allows it to hold not only the mRNA template but also the aminoacyl-tRNA molecules and the growing peptide chain in the correct orientation so that the genetic code can be read and the next peptide bond formed. Trans-

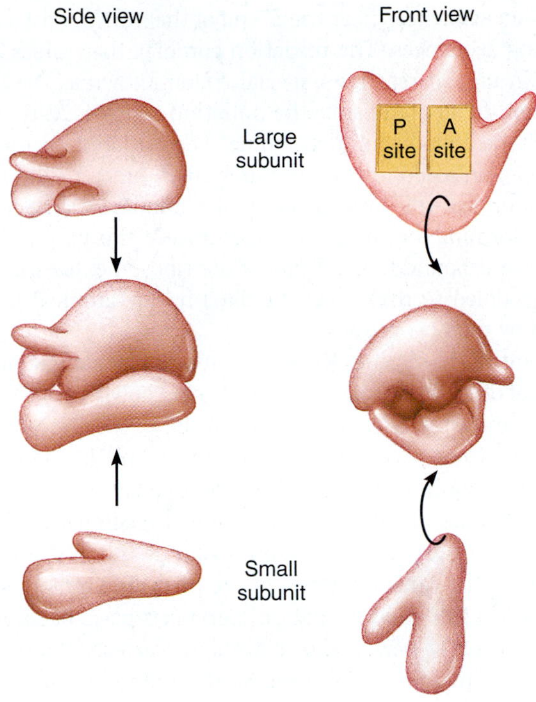

(a) This model of a ribosome is based on three-dimensional reconstruction of electron microscopic images.

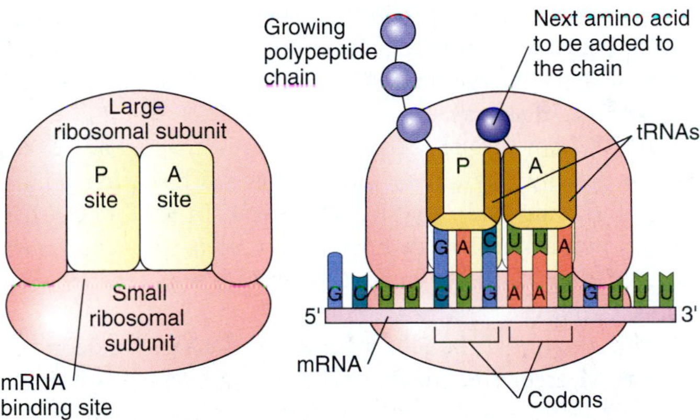

(b) The mRNA passes through a groove formed between the two ribosomal subunits. A ribosome contains two binding sites for tRNAs that recognize adjacent codons. The A site (aminoacyl-tRNA site) binds an aminoacyl-tRNA that will be used to add an amino acid to the growing chain. The P site (polypeptide-tRNA site) binds the tRNA that is linked to the growing polypeptide chain.

Figure 12–11 Ribosome structure. **(a)** A ribosome consists of two subunits, one larger and one smaller. **(b)** Each ribosome contains two binding sites for aminoacyl tRNA molecules.

fer RNA molecules attach to two depressions on the ribosome, the **A** and **P** binding sites (Fig. 12–11*b*). The tRNA holding the growing **p**olypeptide chain occupies the P site and is bound to an mRNA codon. The A site is so named because the **a**minoacyl-tRNA delivering the next amino acid in the sequence binds at this location, positioning it at the next codon to be read. Following peptide bond formation between the amino acid at the A site and the end of the growing polypeptide chain, the tRNA (now with the entire polypeptide chain attached) then moves to the P site of the ribosome, leaving the A site available for the next aminoacyl-tRNA molecule.

Translation includes initiation, elongation, and termination

For purposes of discussion, the process of protein synthesis is generally divided into three distinct stages: **initiation,** repeating cycles of **elongation,** and **termination.**

The initiation process (Fig. 12–12) consists of several steps and requires proteins called **initiation factors,** which become attached to the small ribosomal subunit, allowing it to bind to a special **initiator tRNA.**

Once the initiator tRNA is loaded on the small subunit, the resulting initiation complex binds to the special **ribosome-recognition sequences** near the 5' end of the mRNA, upstream of the coding sequences. The initiation complex then slides along the mRNA until it reaches a special codon known as the **initiation codon.** In all organisms, the initiation codon is AUG, which codes for the amino acid methionine. The initiation factors dissociate, the anticodon of the initiator tRNA binds to the initiation codon, and the large ribosomal subunit attaches to the complex, forming the completed ribosome. At this point, the initiator tRNA is bound to the P site of the ribosome, leaving the A site unoccupied so that it can be filled by the aminoacyl-tRNA specified by the next codon.

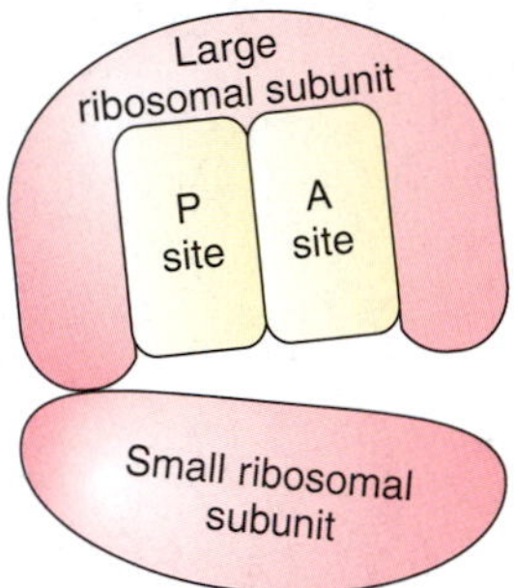

(a) Before translation begins, the ribosomes are dissociated into small and large subunits.

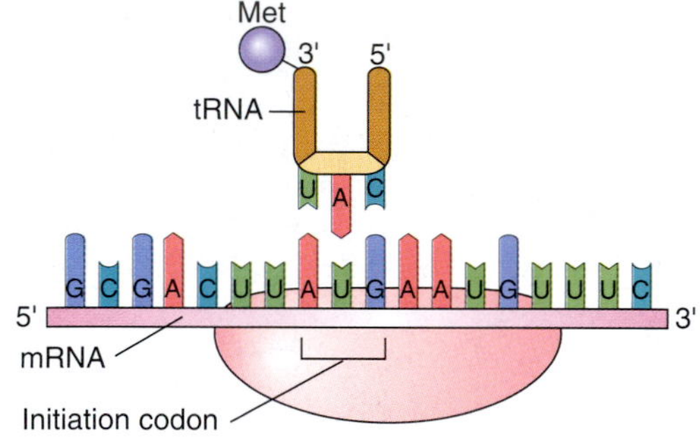

(b) An initiation complex forms, consisting of the small ribosomal subunit, mRNA, and formylated methionine (the initiator tRNA).

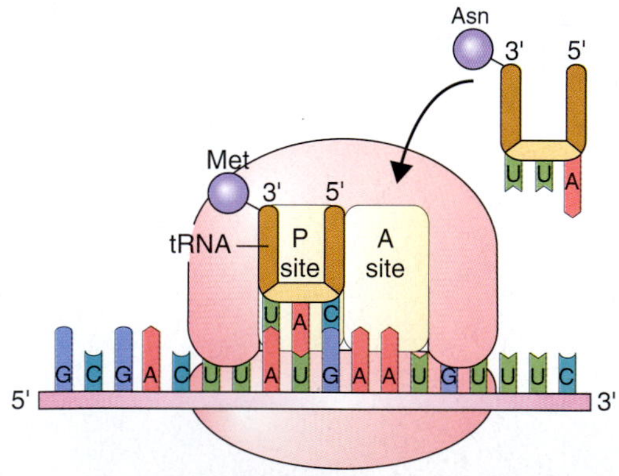

(c) The large ribosomal subunit binds to the initiation complex. The process of peptide elongation begins with the addition of the second tRNA with its amino acid.

Figure 12–12 Initiation of protein synthesis. The small ribosomal subunit participates in the formation of an initiation complex, which then associates with the large subunit. The initiation complex shown is found in *E. coli.* Initiation factors are not shown.

Figure 12–13 outlines the events involved in **elongation,** the addition of other amino acids to the growing polypeptide. The appropriate aminoacyl-tRNA binds to the A site by specific base-pairing of its anticodon with the complementary mRNA codon. This binding step requires energy, in this case supplied by GTP (guanosine triphosphate, an energy transfer molecule similar to ATP).

The amino group of the amino acid at the A site is now aligned with the carboxyl group of the preceding amino acid at the P site. Peptide bond formation then takes place between the amino group of the new amino acid and the carboxyl group of the preceding amino acid. In this process, the amino acid attached at the P site is released from its tRNA and becomes attached to the aminoacyl-tRNA at the A site. This reaction is spontaneous (i.e., it does not require additional energy) because energy was transferred from ATP during the formation of the aminoacyl-tRNA. It does, however, require an enzyme (known as *peptidyl transferase*). Remarkably, this enzyme is not a protein, but an rRNA component of the large ribosomal subunit. Such an RNA catalyst is known as a **ribozyme.**

Recall from Chapter 3 that polypeptide chains have direction, or polarity. The amino acid on one end has a free amino group (the amino end), and the amino acid at the other end has a free carboxyl group (the carboxyl end). Protein synthesis always proceeds from the amino end to the carboxyl end of the growing peptide chain.

After the peptide bond is formed, the tRNA molecule in the P site is released. The growing peptide chain, which is now attached to the tRNA in the A site, is then translocated to the P site, leaving the A site open for the next aminoacyl tRNA. This **translocation** process requires energy, which is again supplied by GTP.

During translocation, the ribosome and the mRNA move in relation to each other so that the mRNA codon specifying the next amino acid in the polypeptide chain becomes positioned in the unoccupied A site. This process involves movement of the ribosome in the 3' direction along the mRNA molecule; thus, translation of the mRNA always proceeds in a 5' to 3' direction. The end of the mRNA molecule that is synthesized first during transcription is also the first to be translated to form a polypeptide. Formation of each peptide bond requires a fraction of a second; by repeating the elongation cycle, an average-sized protein of about 360 amino acids can be assembled by a prokaryotic cell in about 18 seconds, and by a eukaryotic cell in a little over a minute.

The synthesis of the polypeptide chain is terminated by "release factors" that recognize the termination, or stop, codon at

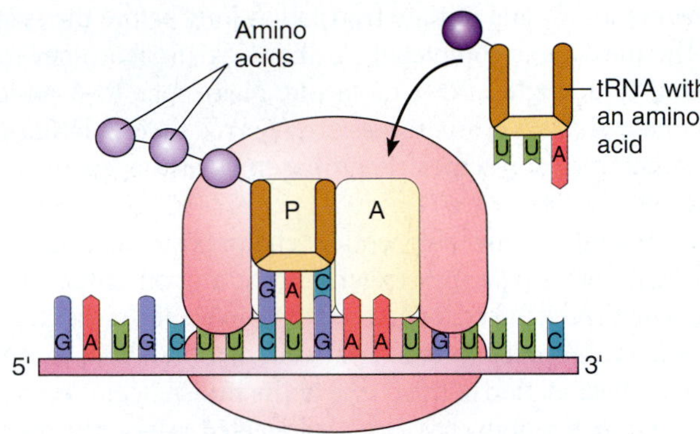

(a) The polypeptide chain is attached to the tRNA that carries the amino acid most recently added to the chain. This tRNA is in the P site of the ribosome.

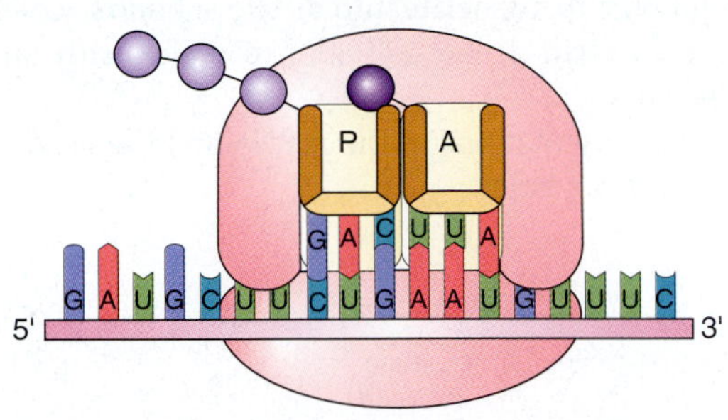

(b) A tRNA with its specific amino acid attached has bound to the A site. Base pairs have formed between the anticodon of tRNA and the codon of mRNA.

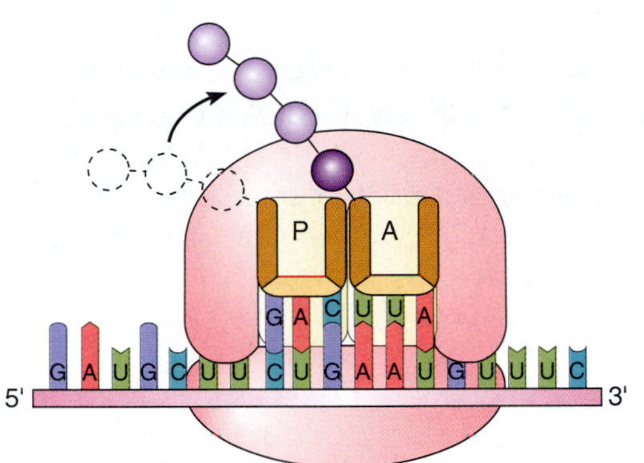

(c) The growing polypeptide chain is detached from the tRNA molecule in the P site and joined by a peptide bond to the amino acid linked to the tRNA at the A site.

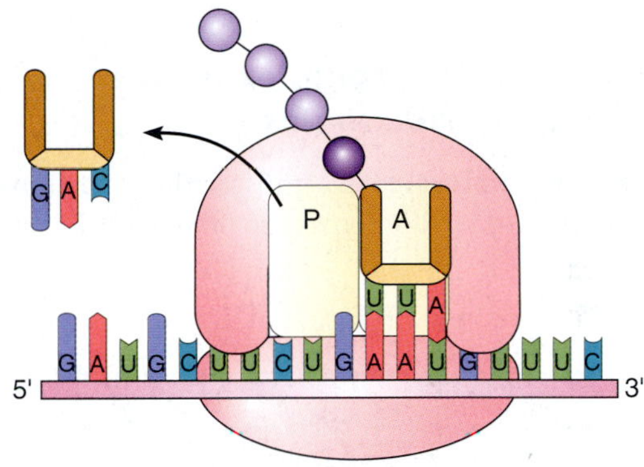

(d) The released tRNA joins the cytoplasmic pool of tRNA and can bind with another amino acid.

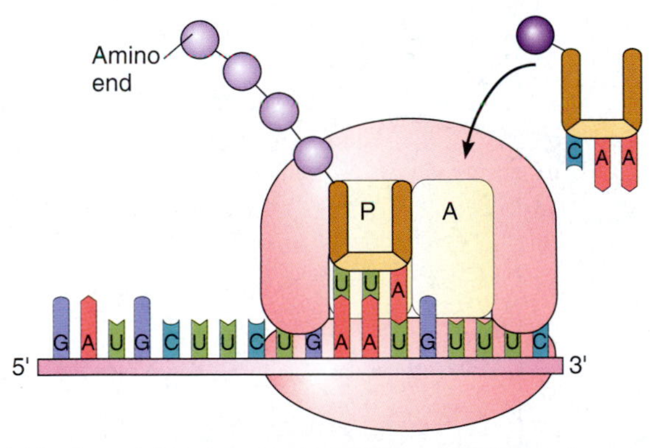

(e) In the translocation step the mRNA and tRNA move in one direction and the ribosome moves in the opposite direction. In this way the growing polypeptide chain is transferred to the P site.

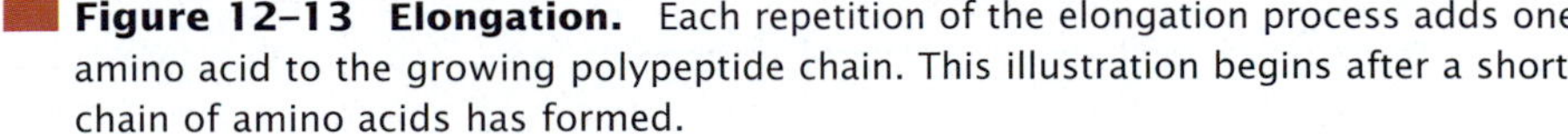

Figure 12–13 Elongation. Each repetition of the elongation process adds one amino acid to the growing polypeptide chain. This illustration begins after a short chain of amino acids has formed.

the end of the coding sequence. The release factors release the newly made protein, the mRNA, and the last tRNA used, and then cause the ribosome to dissociate into its two subunits, which can then be used to form a new initiation complex with another mRNA molecule.

The relative orientations of the mRNA and the synthesized polypeptide are given below:

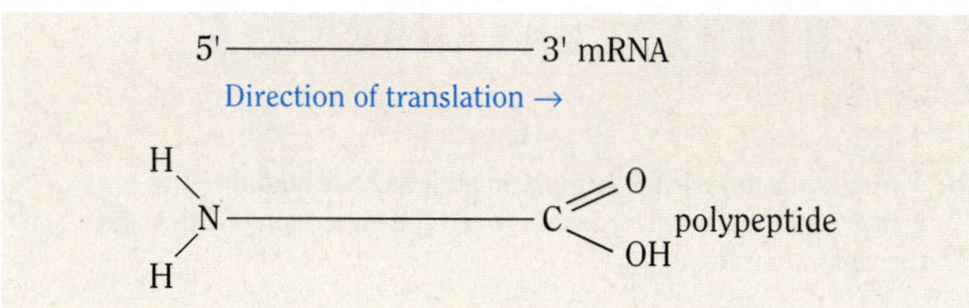

A polyribosome is a complex of one mRNA and many ribosomes

In *E. coli* and other prokaryotes, transcription and translation are *coupled* (Fig. 12–14). Ribosomes can bind to the 5' end of the growing mRNA and initiate translation long before the synthesis of the message is completed. As many as 15 ribosomes may be bound to a single mRNA molecule. Messenger RNA molecules bound to clusters of ribosomes are referred to as **polyribosomes,** or sometimes **polysomes.** Polyribosomes also occur in eukaryotic cells.

Although many polypeptide chains can be actively synthesized on a single messenger RNA at any one time, the half-life (the time it takes for half of the molecules to be degraded) of mRNA molecules in bacterial cells is only about 2 minutes. Usually, degradation of the 5' end of the mRNA begins even before the first polypeptide has been completed. Once the ribosome recognition sequences at the 5' end of the mRNA have been degraded, no more ribosomes can attach and initiate protein synthesis.

TRANSCRIPTION AND TRANSLATION ARE COMPLEX IN EUKARYOTES

Although the basic mechanisms of transcription and translation are quite similar in all organisms, some significant differences

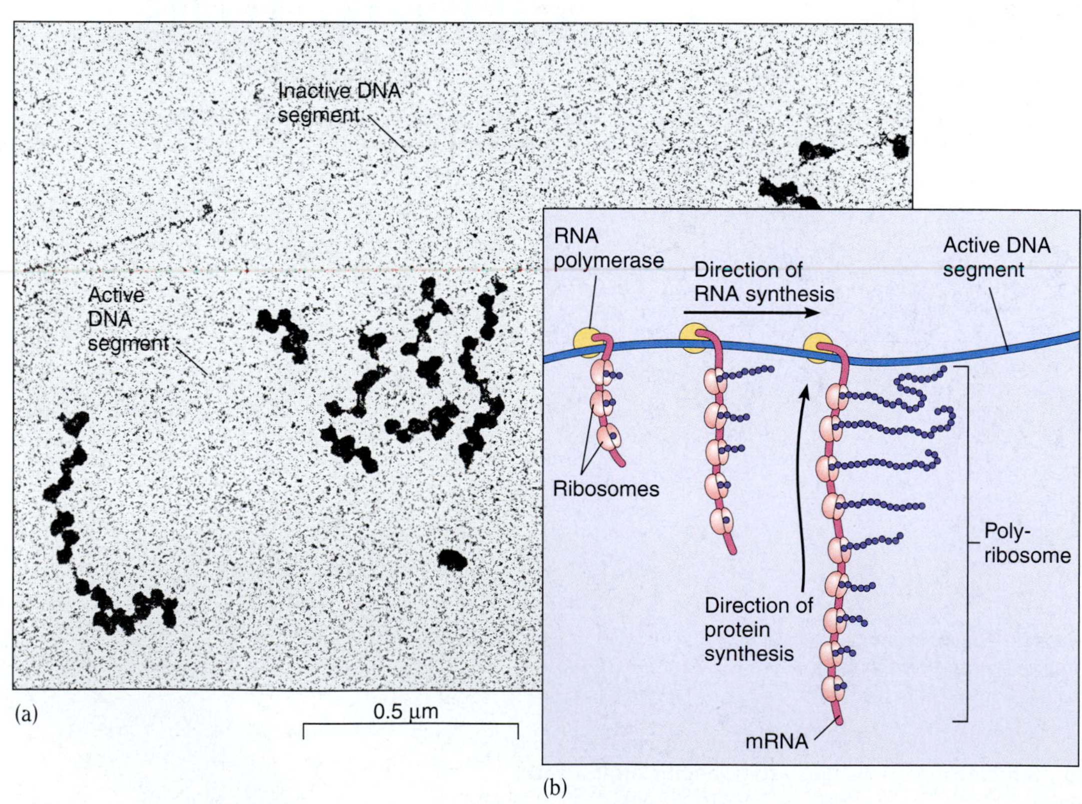

Figure 12–14 Coupled transcription and translation in bacteria. (a) TEM of two strands of *E. coli* DNA, one inactive and the other actively producing mRNA. Protein synthesis begins while the mRNA is being completed, as multiple ribosomes attach to the mRNA to form a polyribosome. **(b)** Diagram showing a sequence of coupled transcription and translation. *(Courtesy of Dr. Barbara Hamkalo, University of California, Irvine)*

exist between eukaryotes and prokaryotes. Some of these are a consequence of differences in cell structure. Bacterial mRNA is translated as it is being transcribed from the DNA. This cannot occur in eukaryotes because eukaryotic chromosomes are confined to the cell nucleus, and protein synthesis takes place in the cytoplasm. Therefore, eukaryotic mRNA must be transported through the nuclear envelope and into the cytoplasm before it can be translated.

There are many other differences between eukaryotic and prokaryotic mRNA. Bacterial mRNAs are used immediately after transcription without further processing. In eukaryotes, the original transcript, known as **precursor mRNA,** or **pre-mRNA,** must be modified in several ways while it is still in the nucleus. These **posttranscriptional modification and processing** activities are required to form mature mRNA that is competent for transport and translation.

Modification of the eukaryotic message begins when the growing RNA transcript is about 20 to 30 nucleotides long. At that point, enzymes add a **cap** to the 5' end of the mRNA chain. The cap is in the form of an unusual nucleotide, 7-methylguanylate, which is guanosine monophosphate with a methyl group added to one of the nitrogens in the base. Eukaryotic ribosomes cannot bind to an uncapped message.

Capping may also protect the RNA from certain types of degradation and may therefore be partially responsible for the fact that eukaryotic mRNAs are much more stable than prokaryotic mRNAs. Eukaryotic mRNAs have half-lives ranging from 30 minutes to as long as 24 hours; the average half-life of an mRNA molecule in a mammalian cell is about 10 hours (compared with 2 minutes in a bacterial cell).

A second modification of eukaryotic mRNA occurs at the 3' end of the molecule. Near the 3' end of a completed message there is usually a sequence of bases that serves as a signal for the addition of a "tail" with many adenines, known as a **polyadenylated** (or **poly-A**) **tail.** Within about 1 minute of completion of the transcript, enzymes in the nucleus recognize this **polyadenylation signal** and cut the mRNA molecule at that site. This is followed by the enzymatic addition of a string of 100 to 250 adenine nucleotides to the 3' end. The function of polyadenylation is not completely understood, although there is evidence that it helps in the export of the mRNA from the nucleus, stabilizes some mRNAs against degradation in the cytoplasm, and makes initiation of translation more efficient.

Both noncoding and coding sequences are transcribed from eukaryotic genes

One of the greatest surprises in the history of molecular biology was the finding that most eukaryotic genes have **interrupted coding sequences;** that is, there are long sequences of bases within the protein-coding sequences of the gene that do not code for amino acids in the final protein product! The noncoding regions within the gene are called **introns** (**in**tervening sequences), as opposed to **exons** (**ex**pressed sequences) which are parts of the protein-coding sequence.

A typical eukaryotic gene may have multiple exons and introns, although the number is quite variable. For example, the β-globin gene, which produces one component of hemoglobin, contains 2 introns; the ovalbumin gene of egg white contains 7; and the gene specifying another egg-white protein, conalbumin, contains 16. In many cases, the combined lengths of the introns are much greater than those of the exon sequences. For instance, the ovalbumin gene contains about 7700 base pairs, whereas the total of all the exon sequences is only 1859 base pairs.

When a gene that contains introns is transcribed, the entire gene is copied as a large RNA transcript, the precursor mRNA, or pre-mRNA (Fig. 12–15). A pre-mRNA molecule contains both exon and intron sequences. (Note that the terms *intron* and *exon* refer to corresponding nucleotide sequences in both DNA and RNA.) For the pre-mRNA to be made into a functional message, not only must it be capped and have a poly-A tail added, but also the introns must be removed and the exons spliced together to form a continuous protein-coding message. The splicing reactions are mediated by special base sequences within and to either side of the introns. Splicing itself can occur by several different mechanisms. In many instances it involves the association of **small nuclear ribonucleoprotein complexes (snRNPs)** which bind to the introns and catalyze the excision and splicing reactions. In some cases, the RNA within the intron acts as a ribozyme, splicing itself without the use of protein enzymes.

The evolution of eukaryotic gene structure is incompletely understood

The reason for the complex structure of eukaryotic genes is a matter of ongoing debate among molecular biologists. Why do introns occur in most eukaryotic nuclear genes but not in the genes of most prokaryotes (or of mitochondria and chloroplasts)? How did this remarkable genetic system involving interrupted coding sequences ("split genes") evolve, and why has it survived? It seems incredible that as much as 75% of the original transcript of a eukaryotic nuclear gene has to be removed to make a working message.

In the early 1980s, Walter Gilbert of Harvard University proposed that exons are nucleotide sequences that code for **protein domains,** regions of protein tertiary structure that may have specific functions. For example, the active site of an enzyme might comprise one domain. A different domain might enable that enzyme to bind to a particular cellular structure, and yet another might be a site involved in allosteric regulation (see Chapter 6). Analyses of the DNA and amino acid sequences of a great many eukaryotic genes have shown that most exons are too small to code for an entire protein domain, although a block of several exons can code for a domain.

Gilbert further postulated that new proteins with new functions can emerge rapidly when novel combinations of exons are produced by genetic recombination within intron regions of genes that code for different proteins. This hypothesis has become known as *evolution by "exon shuffling."* It has been

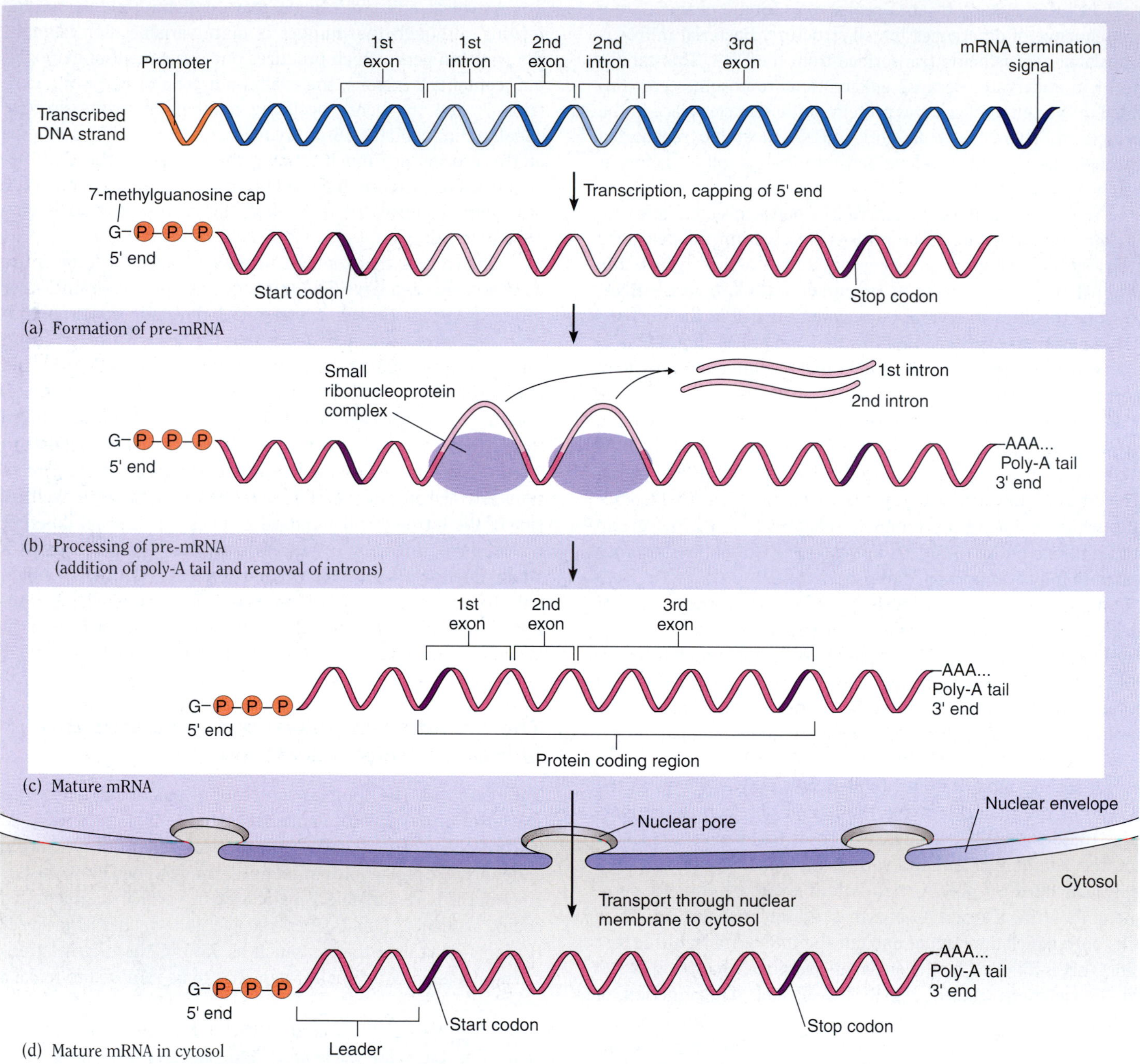

Figure 12-15 Posttranscriptional modification of eukaryotic RNA. **(a)** A DNA sequence containing both exons and introns is transcribed by RNA polymerase to make the primary transcript, or mRNA precursor. As it is synthesized, the pre-mRNA is "capped" by the addition of a modified base attached "backwards" (by a 5'–5' linkage) to its 5' end. **(b)** A poly-A tail (100–250 nucleotides long) is added to the 3' end; introns are removed, and the exons are spliced together. **(c)** The mature mRNA is transported through the nuclear envelope and into the cytosol **(d)** to be translated.

supported by examples such as the low-density lipoprotein (LDL) receptor protein, a protein found on the surface of human cells that binds to cholesterol transport molecules (see Chapter 5). The LDL receptor protein has several domains that are related to parts of several other proteins with totally different functions. However, many other genes and their corresponding proteins show no evidence of exon shuffling.

Some scientists think that introns first evolved in the nucleus of an early eukaryote and were propagated as mobile genetic elements, known as *transposons* (discussed later in this chapter). Regardless of how split genes originated, intron excision provides one of the many ways in which present-day eukaryotes regulate the expression of their genes (see Chapter 13). This opportunity for control, together with the fact that eukaryotic RNAs are far more stable than those of prokaryotes, may balance the energy cost of maintaining a large load of noncoding DNA.

THE GENETIC CODE HAS SPECIAL FEATURES

Process of Science Before the genetic code was deciphered, scientists had become interested in how a genetic code might work. The Watson and Crick model of DNA showed it to be a linear sequence of four nucleotides. If each nucleotide were to code for a single amino acid, it would be possible to specify only 4 amino acids, not the 20 found in the vast variety of proteins in the cell. Scientists saw that the DNA bases could serve as a four-letter "alphabet" and hypothesized that three-letter combinations of the four bases (4^3) would make it possible to form a total of 64 "words," more than sufficient to specify all the naturally occurring amino acids. In 1961, Francis Crick and his coworkers concluded from a mathematical analysis that the code was based on nonoverlapping triplets of bases. They predicted that the code is read, one triplet at a time, from a fixed starting point that establishes the **reading frame.** Because there are no "commas" separating the triplets, an alteration in the reading frame would result in the incorporation of incorrect amino acids.

Experimental evidence allowing the assignment of specific triplets to specific amino acids was first obtained by Marshall Nirenberg and Heinrich Matthaei. By constructing artificial mRNA molecules with known base sequences, they were able to determine which amino acids would be incorporated into protein in purified in vitro protein synthetic systems derived from *E. coli.* For example, when the synthetic mRNA polyuridylic acid (UUUUUUUUU . . .) was added to a mixture of purified ribosomes, aminoacyl tRNAs, and essential cofactors needed to synthesize protein, only phenylalanine was incorporated into the resulting polypeptide chain. The inference that UUU is the triplet that codes for phenylalanine was inescapable. Similar experiments showed that polyadenylic acid (AAAAAAAAA . . .) codes for a polypeptide of lysine, and polycytidylic acid (CCCCCCCCC . . .) codes for a polypeptide of proline.

By using mixed nucleotide polymers (e.g., a random polymer of A and C) as artificial messengers, it became possible to assign the other nucleotide triplet codons to specific amino acids. However, three of the codons, UAA, UGA, and UAG, were not found to specify any amino acid. These codons (the stop, or termination, codons mentioned earlier) are now known to be the signals that specify the end of the coding sequence for a polypeptide chain.

By 1967 the "cracking" of the genetic code was completed, and the coding assignments of all 64 possible codons listed in Table 12–1 were known. Investigators were also able to demonstrate conclusively that the code is a nonoverlapping triplet code.

Remember that the genetic code we define and use is an mRNA code. The tRNA anticodon sequences as well as the DNA sequence from which the message is transcribed are complementary to the sequences shown in Table 12–1. For example, the mRNA codon for the amino acid methionine is 5' — AUG — 3'. It is transcribed from the DNA base sequence 3' — TAC — 5', and the corresponding tRNA anticodon is 3' — UAC — 5'.

The genetic code is universal

This chapter has only one table to represent the genetic code. This is because of the single most remarkable feature of the code: *It is essentially universal!* Over the years, the genetic code has been examined in a diverse array of species and found to be the same in organisms as different as *E. coli,* redwood trees, and humans. These findings strongly suggest that the code is an ancient legacy that evolved very early in the history of life (see Chapter 20) and has been retained as a kind of "frozen accident" because all but the most minimal changes would be lethal.

Recently some very minor exceptions to the universality of the genetic code have been discovered. In several unicellular protozoa, two of the stop codons, UAA and UGA, code for the amino acid glutamine. The other exceptions are found in mitochondria, which contain their own DNA and protein-synthesis machinery for a small number of genes (see Chapters 4 and 20). These slight coding differences vary with the organism, but it is important to keep in mind that in each case all of the other coding assignments are identical to the standard genetic code.

The genetic code is redundant

Given that there are 64 possible codons and that there are only 20 common amino acids, it is not surprising that certain amino acids are specified by more than one codon. The genetic code is said to be degenerate because of this redundancy in the code, which has certain characteristic patterns. The codons CCU, CCC, CCA, and CCG are synonymous in that they all code for the amino acid proline. The only difference among the four codons involves the nucleotide at the 3' end of the triplet. Although the code may be read three nucleotides at a time, only the first two nucleotides appear to contain specific information for proline. A similar pattern can be seen for many other amino acids. Only methionine and tryptophan have single-triplet codes. All other amino acids are specified by two to six different codons.

Process of Science There are 61 codons that specify amino acids. Although most cells contain only about 40 different tRNA molecules, some of these tRNAs can pair with more than one codon, so all the codons can still be used. This apparent breach of the base-pairing rules was first proposed by Francis Crick as the **wobble hypothesis.** Crick reasoned that the third nucleotide of a tRNA anticodon (which is the 5' base of the anticodon sequence) may sometimes be capable of forming hydrogen bonds with more than one kind of third nucleotide (the 3' base) of a codon. Investigators later established this experimentally by determining the anticodon sequences of tRNA molecules and testing their specificities in artificial systems. Some tRNA molecules can recognize as many as three separate codons differing in their third nucleotide, but specifying the same amino acid.

A GENE IS DEFINED AS A FUNCTIONAL UNIT

At the beginning of this chapter, we traced the development of ideas regarding the nature of the gene. For a time it was useful to define a gene as a sequence of nucleotides that codes for one polypeptide chain. As we have continued to learn more about how genes work, we have revised our definition. We now know that some genes are transcribed to produce RNA molecules such as rRNA and tRNA, whereas others specify the RNA component of the small nuclear ribonucleoprotein complexes used to modify complex mRNA molecules. Studies have also shown that in eukaryotic cells a single gene may be capable of producing more than one polypeptide chain by modifications in the way the mRNA is processed (see Chapter 13).

It is perhaps most useful to define a gene in terms of its product. A **gene** can therefore be thought of as a nucleotide sequence that carries the information needed to produce a specific RNA or protein product. As we will see in Chapter 13, a gene includes noncoding regulatory sequences, as well as coding sequences.

THERE ARE EXCEPTIONS TO THE USUAL DIRECTION OF FLOW OF GENETIC INFORMATION

For several decades, one of the central premises of molecular biology was that genetic information always flows from DNA to RNA to protein. An important exception to this rule was discovered by Howard Temin in 1964 through his studies of certain viruses. Although viruses are noncellular, they contain a single type of nucleic acid and are capable of reproducing in a host cell. Temin was studying certain unusual, cancer-causing tumor viruses that have RNA, rather than DNA, as their genetic material. He found that infection of a host cell by one of these particular viruses is blocked by inhibitors of DNA synthesis and also by inhibitors of transcription. These findings suggested that DNA synthesis and transcription are required for the multiplication of RNA tumor viruses and that there must be a way for information to flow in the "reverse" direction (that is, from RNA to DNA).

Temin proposed that a **DNA provirus** is formed as an intermediary in the replication of RNA tumor viruses. This hypothesis required a new kind of enzyme that would synthesize DNA using RNA as a template. In 1970 Temin and David Baltimore discovered just such an enzyme, and in 1975 they shared the Nobel Prize for their discovery. This RNA-directed DNA polymerase, also known as **reverse transcriptase,** was found in all RNA tumor viruses. (Some RNA viruses that do not produce tumors, however, are replicated directly without using a DNA intermediate.) The steps of RNA tumor virus reproduction are shown in Figure 12–16. Because of their reversal of the usual direction of information flow, viruses that require reverse transcriptase have become known as **retroviruses.** The virus (HIV-1) that causes AIDS is the most widely known retrovirus. As we will see in Chapter 14, the reverse transcriptase enzyme has become an extremely important research tool for molecular biologists.

MUTATIONS ARE CHANGES IN DNA

One of the first major discoveries about genes was that they can undergo **mutations,** changes in the nucleotide sequence of the DNA. However, the overall observed mutation rate is much lower than the frequency of damage to DNA, because all organisms have special systems of enzymes that can repair certain kinds of alterations in the DNA.

As explained in Chapter 11, once the DNA sequence has been changed, DNA replication copies the altered sequence just as it would copy a normal sequence, making the mutation stable over an indefinite number of generations. In most cases, the mutant allele has no greater tendency than does the original allele to mutate again. Mutations provide the diversity of genetic material that makes it possible to study inheritance and the molecular nature of genes. As we will see in later chapters, mutations also provide the variation necessary for evolution to occur within a given species.

Genes can be altered by mutation in several ways (Fig. 12–17). It is now possible to determine where a specific mutation occurs in a gene by using recombinant DNA methods to isolate the gene and determine its sequence of bases (see Chapter 14).

Base substitution mutations result from the exchange of one base pair for another

The simplest type of mutation, called a **base substitution mutation,** involves a change in only one pair of nucleotides. Often these mutations result from errors in base pairing that occurred during the replication process. For example, an AT base pair might be replaced by a GC, CG, or TA pair. Such a mutation may cause the altered DNA to be transcribed as an altered mRNA. The altered mRNA may then be translated into a peptide chain with only one amino acid different from the normal sequence.

Base substitutions that result in the replacement of one amino acid by another are sometimes referred to as **missense mutations.** Missense mutations can have a wide range of effects. If the amino acid substitution occurs at or near the active site of an enzyme, the activity of the altered protein may be decreased or even destroyed. Some missense mutations involve a change in an amino acid that is not part of the active site. Others may result in the substitution of a closely related amino acid (one with very similar chemical characteristics). Such mutations may be *silent* (undetectable) if one simply examines its effects on the whole organism. Because silent mutations occur relatively frequently, the true number of mutations in an organism or a species is much greater than what is actually observed.

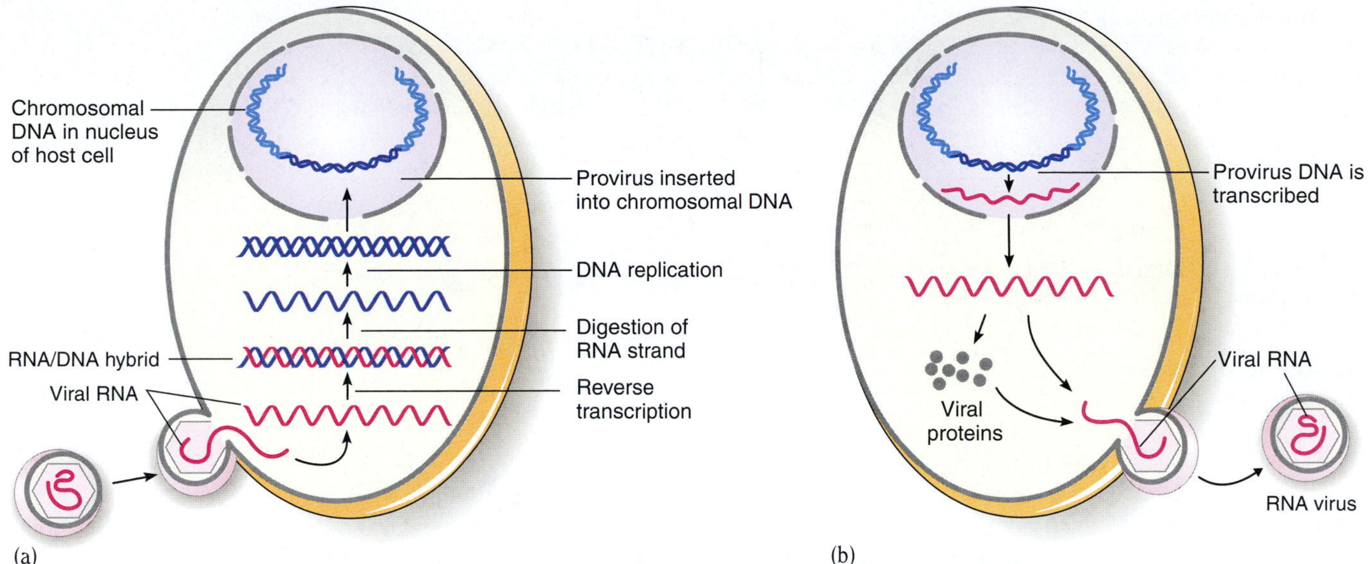

Figure 12-16 Infection cycle of an RNA tumor virus. (a) After an RNA tumor virus enters the host cell, the viral reverse transcriptase synthesizes a DNA strand that is complementary to the viral RNA. Next, the RNA strand is degraded and a complementary DNA strand is synthesized, thus completing the double-stranded DNA provirus, which is then integrated into the host cell's DNA. (b) The provirus DNA is transcribed, and the resulting viral mRNA is translated to form specific viral proteins. Additional viral RNA molecules are produced and then incorporated into mature virus particles enclosed by protein coats.

Nonsense mutations are base substitutions that convert an amino acid–specifying codon to a termination codon. A nonsense mutation usually destroys the function of the gene product; in the case of a protein-specifying gene, the part of the polypeptide chain that follows the termination codon is missing.

Frameshift mutations result from the addition or deletion of base pairs

In **frameshift mutations,** one or two nucleotide pairs are inserted into or deleted from the molecule, causing an alteration of the *reading frame.* As a result of this shift, codons downstream of the insertion or deletion site specify an entirely new sequence of amino acids. Depending on where the insertion or deletion occurs in the gene, different effects can be generated. In addition to producing an entirely new polypeptide sequence immediately after the change, frameshift mutations usually produce a stop or termination codon within a short distance of the mutation. This codon terminates the already altered polypeptide chain. A frame shift in a gene specifying an enzyme usually results in a loss of enzyme activity. If the enzyme is an essential one, the effect on the organism can be disastrous.

Process of Science One type of mutation is caused by DNA sequences that "jump" into the middle of a gene. These movable sequences of DNA not only disrupt the functions of some genes, but under some conditions also activate previously inactive genes. Jumping genes, now known as **mobile genetic elements, transposable elements,** or **transposons,** were discovered in maize (corn) by Barbara McClintock in the 1950s. She observed that certain genes appeared to be "turned off" and "turned on" spontaneously. She deduced that the mechanism involved a gene that moved from one region of a chromosome to another, where it would either activate or inactivate genes in that vicinity. It was not until the development of recombinant DNA methods (see Chapter 14) and the discovery of transposons in a wide variety of organisms that this phenomenon began to be understood. We now know that transposons are segments of DNA that range from a few hundred to several thousand bases. The elements themselves seem to require a special **transposase** enzyme in order to be incorporated into a new location within the chromosome. The longer transposable elements may contain other genes that "go along for the ride." In recognition of her insightful findings, McClintock was awarded a Nobel Prize in 1983.

Some mutations involve larger DNA segments

Some types of mutations are due to a change in chromosome structure (see Chapters 13, 15, and 16). These changes usually have a wide range of effects because they involve large numbers of genes.

Mutations have varied causes

Most types of mutations occur infrequently but spontaneously as a consequence either of mistakes in DNA replication or of defects in the mitotic or meiotic separation of chromosomes. Some regions of DNA, known as mutational **hot spots,** are much more

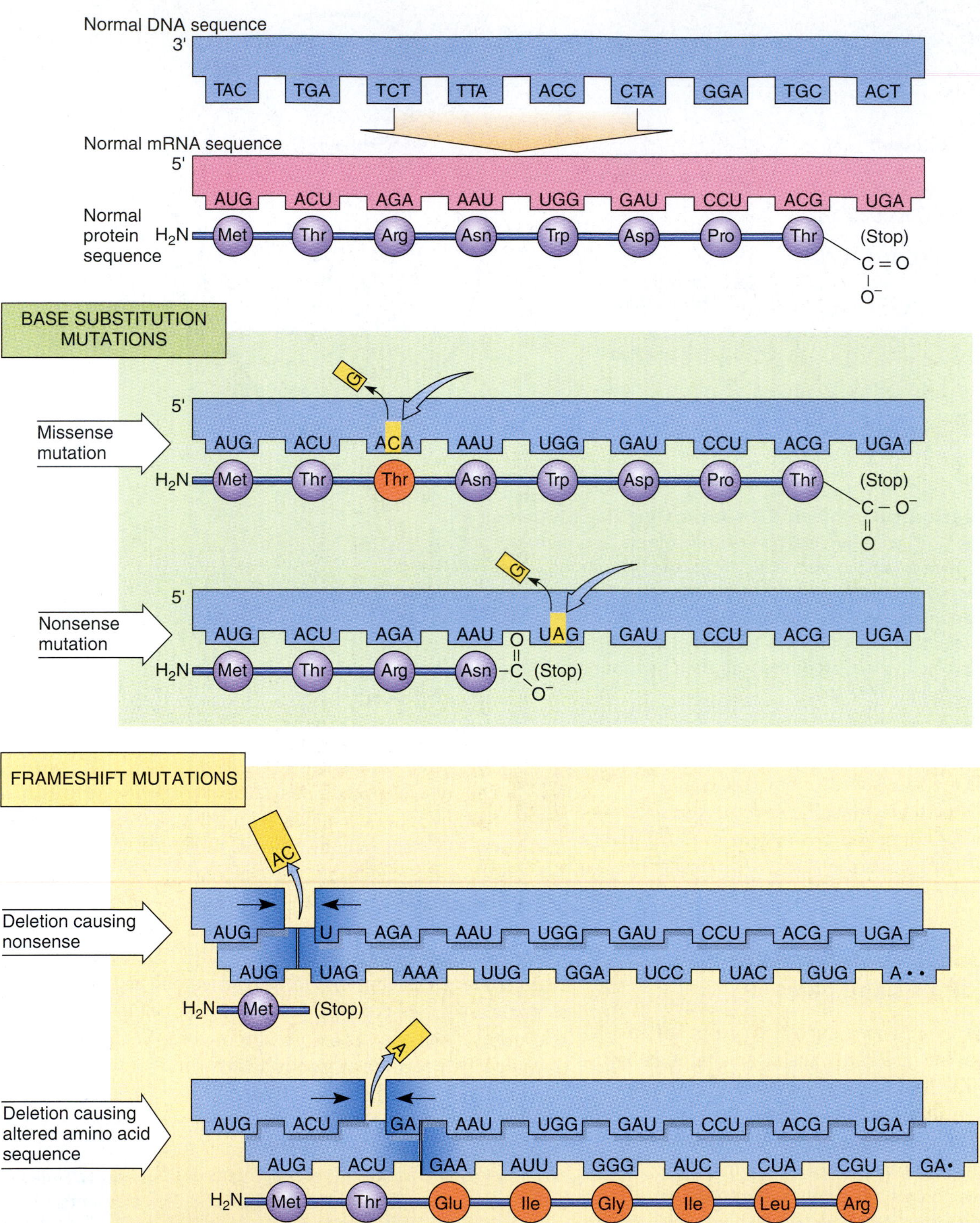

■ **Figure 12-17 Mutations.** Missense and nonsense mutations are types of base substitutions. A missense mutation results in a protein of normal length, but with an amino acid substitution. A nonsense mutation, caused by conversion of an amino acid–specifying codon to a termination (stop) codon, results in the production of a truncated (shortened) protein, which is usually not functional. A frameshift mutation, which results from the insertion *(not shown)* or deletion of one or two bases, causes the base sequence following the mutation to shift to a new reading frame. A frame shift may produce a termination codon downstream of the mutation (which would have the same effect as a nonsense mutation caused by base substitution), or it may produce an entirely new amino acid sequence.

likely than others to undergo mutation. An example would be a short stretch of repeated nucleotides, which can cause DNA polymerase to "slip" while reading the template during replication.

Mutations in certain genes can increase the overall mutation rate. For example, a mutation in a gene coding for DNA polymerase might make DNA replication less precise, or a mutation in a gene coding for a repair enzyme might cause more mutations to persist rather than to be repaired.

Not all mutations occur spontaneously; many of the types of mutations discussed previously can also be caused by agents known as **mutagens.** Among these are various types of radiation, including x rays, gamma rays, cosmic rays, and ultraviolet rays. Some chemical mutagens react with and modify specific bases in the DNA, leading to mistakes in complementary base pairing when the DNA molecule is replicated. Other mutagens are inserted into the DNA molecule and change the normal reading frame during replication.

Despite the presence of enzymes that repair mutations, some new mutations do persist. In fact, each of us is very likely to have some mutant allele that was not present in either of our parents. Although some of these mutations can produce an altered phenotype, most are not noticeable because they are recessive.

Mutations that occur in the cells of the body (somatic cells) are not passed on to the offspring. However, these mutations are of concern because there is a close relationship between somatic mutations and cancer. Many mutagens are also **carcinogens,** agents that produce cancer.

SUMMARY WITH KEY TERMS

I. Garrod's early work on inborn errors of metabolism and that of Beadle and Tatum in the 1940s with *Neurospora* mutants provided strong evidence that genes specify proteins.

II. The mechanism by which information encoded in DNA is used to specify the sequences of amino acids in proteins involves two processes: transcription and translation.
   - A. During **transcription,** an RNA molecule that is complementary to the transcribed or template DNA strand is synthesized. **Messenger RNA (mRNA)** molecules contain information that specifies the amino acid sequences of polypeptide chains.
   - B. During **translation,** a polypeptide chain specified by the mRNA is synthesized.
     1. Each **triplet** (three-base sequence) in the mRNA constitutes a **codon,** which specifies one amino acid in the polypeptide chain, or a stop or start signal.
     2. Translation requires tRNAs and complex cell machinery, including ribosomes.

III. Messenger RNA is synthesized by **DNA-dependent RNA polymerase** enzymes.
   - A. RNA is formed from nucleotide subunits, each of which contains the sugar ribose, a base (uracil, adenine, guanine, or cytosine), and three phosphates.
   - B. RNA polymerase initially binds to a special DNA sequence called the **promoter** region.
   - C. Like DNA, RNA subunits are covalently joined by a 5' — 3' linkage to form an alternating sugar-phosphate backbone. The same base-pairing rules are followed as in DNA replication, except that uracil is substituted for thymine.
   - D. RNA synthesis proceeds in a 5' → 3' direction, which means that the template DNA strand is "read" in a 3' → 5' direction.

IV. **Transfer RNAs (tRNAs)** are the "decoding" molecules in the translation process.
   - A. Each tRNA molecule is specific for only one amino acid.
     1. One part of the molecule contains a three-base **anticodon,** which is complementary to a codon of the mRNA.
     2. Attached to one end of the tRNA molecule is the amino acid specified by the complementary mRNA codon.
   - B. Amino acids are covalently bound to tRNA by **aminoacyl-tRNA synthetase** enzymes.

V. **Ribosomes** bring together all the mechanical machinery necessary for translation. They couple the tRNAs to their proper codons on the mRNA, facilitate the formation of peptide bonds between amino acids, and translocate the mRNA so that the next codon can be read.
   - A. Each ribosome is made of a large and a small subunit; each subunit contains **ribosomal RNA (rRNA)** and over 50 proteins.
   - B. **Initiation** is the first stage of translation. The small ribosomal subunit protein, plus **initiation factors** and the **initiator tRNA,** binds to AUG, the **initiation codon,** in the 5' region of the mRNA, followed by binding of the large ribosomal subunit.
   - C. **Elongation** is a cyclic process in which amino acids are added one by one to the growing polypeptide chain.
     1. Elongation proceeds in a 5' ⟶ 3' direction along the mRNA.
     2. The polypeptide chain grows from its amino end to its carboxyl end.
   - D. **Termination,** the final stage of translation, occurs when the ribosome reaches one of three special **termination,** or **stop, codons,** which triggers release of the completed polypeptide chain.

VI. In bacterial cells, transcription and translation are coupled. Translation of the mRNA molecule usually begins before the 3' end of the transcript is completed. A single mRNA molecule can be translated by groups of ribosomes called **polyribosomes.**

VII. The basic features of transcription and translation are the same in prokaryotic and eukaryotic cells, but eukaryotic genes and their mRNA molecules are more complex than those of bacteria.
   - A. After transcription, eukaryotic mRNA molecules are **capped** at the 5' end with a modified guanosine triphosphate. Many also have a **poly-A tail** composed of adenine-containing nucleotides added at the 3' end. These modifications appear to protect eukaryotic mRNA molecules from degradation, giving them longer lifetimes than bacterial mRNA.
   - B. In many eukaryotic genes the coding regions, called **exons,** are interrupted by noncoding regions, called **introns.** Both introns and exons are transcribed, but the introns are later removed from the **mRNA precursor,** and the exons are spliced together to produce a continuous protein-coding sequence.

VIII. The **genetic code** is defined at the mRNA level. There are 61 codons that code for amino acids, plus three codons that serve as stop signals.
   - A. The start signal for all proteins is the codon AUG, which also specifies the amino acid methionine.
   - B. The genetic code is read from mRNA as a series of nonoverlapping triplets that specify a single sequence of amino acids.
   - C. The genetic code is said to be degenerate because it is redundant; that is, some amino acids are specified by more than one codon.
   - D. The genetic code is virtually universal, strongly suggesting that all organisms are descended from a common ancestor. Only a few minor variations are exceptions to the standard code.

IX. A **gene** can be defined as a sequence of nucleotides (plus closely associated regulatory sequences) that can be transcribed to yield a specific protein or RNA product.

X. The flow of genetic information can be reversed by **reverse transcriptase,** an enzyme associated with RNA tumor viruses that can synthesize DNA from an RNA template.

XI. Types of **mutations** range from disruption of the structure of a chromosome to a change in only a single pair of nucleotide bases.

A. A **base substitution** can destroy the function of a protein if a codon is altered such that it specifies a different amino acid **(missense mutation)** or becomes a termination codon **(nonsense mutation).** A base substitution has minimal effects if the amino acid is not altered or if the codon is changed to specify a chemically similar amino acid.

B. Insertion or deletion of one or two base pairs in a gene invariably destroys the function of that protein because it results in a **frameshift mutation** that changes the codon sequences downstream from the mutation.

C. Mutations can be produced by errors in DNA replication, by physical agents such as x rays or ultraviolet rays, or by chemical **mutagens.** Mutations can also occur through **transposons** or **mobile genetic elements,** which move from one part of a chromosome to another, disrupting the function of a part of the DNA. Some damage to DNA can be repaired by special systems of enzymes.

POST-TEST

1. Beadle and Tatum (a) predicted that tRNA molecules would have anticodons (b) discovered the genetic disease alcaptonuria (c) showed that the genetic disease sickle cell anemia is due to a change in a single amino acid in a hemoglobin polypeptide chain (d) worked out the genetic code (e) studied the relationship between genes and enzymes in *Neurospora*
2. The genetic code is defined as a series of ____________ in ____________. (a) anticodons; tRNA (b) codons; DNA (c) anticodons; mRNA (d) codons; mRNA (e) codons and anticodons; rRNA
3. Transcription is the process by which ____________ is/are synthesized. (a) mRNA (b) mRNA and tRNA (c) mRNA, tRNA, and rRNA (d) protein (e) mRNA, tRNA, rRNA, and protein
4. RNA differs from DNA in that the base ____________ is substituted for ____________. (a) adenine; uracil (b) uracil; thymine (c) guanine; uracil (d) cytosine; guanine (e) guanine; adenine
5. RNA grows in the ____________ direction, as DNA-dependent RNA polymerase moves along the template DNA strand in the ____________ direction. (a) 5' ⟶ 3'; 3' ⟶ 5' (b) 3' ⟶ 5'; 3' ⟶ 5' (c) 5' ⟶ 3'; 5' ⟶ 3' (d) 3' ⟶ 3'; 5' ⟶ 5' (e) 5' ⟶ 5'; 3' ⟶ 3'
6. Which of the following is/are *not* found in a prokaryotic mRNA molecule? (a) protein termination codon (b) upstream leader sequences (c) downstream trailing sequences (d) start codon (e) promoter sequences
7. Which of the following is/are typically removed from pre-mRNA during nuclear processing in eukaryotes? (a) upstream leader sequences (b) poly-A tail (c) introns (d) exons (e) all of the above are removed
8. Which of the following is a spontaneous process, with no direct requirement for ATP or GTP? (a) formation of a peptide bond (b) translocation of the ribosome (c) formation of aminoacyl-tRNA (d) a and b (e) all of the above are spontaneous processes
9. Select the events of the elongation cycle of protein synthesis from the following list and place them in the proper sequence.
   1. peptide bond formation
   2. binding of the small ribosomal subunit to the 5' end of the mRNA
   3. binding of aminoacyl-tRNA to the A site
   4. translocation of the ribosome

   (a) 1 ⟶ 3 ⟶ 2 ⟶ 4 (b) 3 ⟶ 1 ⟶ 4 (c) 3 ⟶ 1 ⟶ 3 ⟶ 2 (d) 1 ⟶ 3 ⟶ 4 (e) 4 ⟶ 2 ⟶ 1 ⟶ 3
10. The statement "the genetic code is redundant" or "degenerate" means that (a) some codons specify punctuation (stop and start signals) rather amino acids (b) some codons specify more than one amino acid (c) certain amino acids can be specified by more than one codon (d) in some cases, the third nucleotide of an anticodon may be able to pair with more than one kind of base in a codon (e) all organisms have essentially the same genetic code
11. A nonsense mutation (a) causes one amino acid to be substituted for another in a polypeptide chain (b) results from the deletion of one or two bases, leading to a shift in the reading frame (c) results from the insertion of one or two bases, leading to a shift in the reading frame (d) results from the insertion of a transposon (e) usually results in the formation of an abnormally short polypeptide chain

REVIEW QUESTIONS

1. A certain transcribed DNA strand has the following nucleotide sequence:

   3' — TACTGCATAATGATT — 5'

   What would be the sequence of codons in the mRNA transcribed from this strand? What would be the nucleotide sequence of the complementary nontranscribed DNA strand? What would be the exact anticodon for each codon? Use Table 12–1 to determine the amino acid sequence of the polypeptide. Be sure to label the 5' and 3' ends of the nucleic acids and the carboxyl and amino ends of the polypeptide.
2. What are ribosomes made of? Do ribosomes themselves carry information to specify the amino acid sequence of proteins?
3. In what ways are DNA polymerase and RNA polymerase similar? How do they differ?
4. Describe initiation, elongation, and termination of protein synthesis.
5. Explain how the genetic code was deciphered. What experimental procedures needed to be developed before this could be accomplished?
6. What are the main types of mutations? What effects does each have on the protein product produced?
7. Why can't amino acids become incorporated into polypeptides without the aid of tRNA?

YOU MAKE THE CONNECTION

1. In Chapter 10 we discussed the fact that heritable variation is essential for the study of inheritance. What role did mutant strains play in Beadle and Tatum's development of the one gene, one enzyme hypothesis?
2. How many amino acids could be specified if two bases coded for one amino acid? Why is redundancy or degeneracy important to the idea of wobble? If you could "reinvent" the genetic code, would you make any changes? Why or why not?

3. Compare and contrast the formation of mRNA in prokaryotic and eukaryotic cells. How do the differences affect the way in which each type of mRNA is translated? Does one system have any obvious advantage in terms of energy cost? Which system offers greater opportunities for control of gene expression?

RECOMMENDED READINGS

Craig, P.P. "Jumping Genes: Barbara McClintock's Scientific Legacy." *Carnegie Institution of Washington Perspectives in Science,* No. 6, 1994. A fascinating illustrated essay on the life of Barbara McClintock and the far-reaching implications of her work. Copies may be obtained for $1.00 each by calling 202-939-1121 or by writing The Carnegie Institution of Washington, 1530 P Street NW, Washington, DC, 20005–1910.

Dahlberg, A.E. "The Ribosome in Action." *Science,* Vol. 292, 4 May 2001. This perspective introduces two articles that present new images of the bacterial ribosome, with an emphasis on interactions with tRNA and mRNA, and the mechanism by which certain antibiotics disrupt ribosome function.

Frank, J. "How the Ribosome Works." *American Scientist,* Vol. 86, No. 5, Sept.–Oct. 1998. Sophisticated methods are producing clear three-dimensional maps of ribosomes that permit function to be explained in terms of structure.

Hayes, B. "The Invention of the Genetic Code." *American Scientist,* Vol. 86, No. 1, Jan.–Feb. 1998. This article relates the history of early conjectures regarding the nature of the genetic code and how the genetic code was finally solved.

Lodish, H., A. Berk, S.L. Zipursky, P. Matsudaira, D. Baltimore, and J. Darnell. *Molecular Cell Biology,* 4th ed. W.H. Freeman and Co., New York, 2000. Contains an extensive, detailed, and well written discussion of transcription and translation.

Moffat, A.S. "Transposons Help Sculpt a Dynamic Genome." *Science,* Vol. 289, 1 Sept., 2000. Genomic restructuring caused by transposons is apparently common in both plants and animals and may have evolutionary significance.

Ochert, A. "Transposons." *Discover,* Dec. 1999. Transposons alter patterns of gene expression in ways that can cause genetic disease but also may foster evolutionary change.

Strasser, B.J. "Sickle Cell Anemia: A Molecular Disease." *Science,* Vol. 286, 19 Nov., 1999. The title of this perspective is taken from the famous 1949 *Science* paper by Linus Pauling et al., linking the genetic disease sickle cell anemia to abnormal hemoglobin (a protein).

- Visit our Web site at **http://www.info.brookscole.com/solomonbergmartin** for links to chapter-related resources on the World Wide Web. Additional on-line materials relating to this chapter can also be found on our Web site.

See chapter activity on BioActive Learner CD for additional help in mastering the chapter's material. Icon location in the chapter's margins shows which topics have tutorials or simulations in the CD.

13

Gene Regulation: The Control of Gene Expression

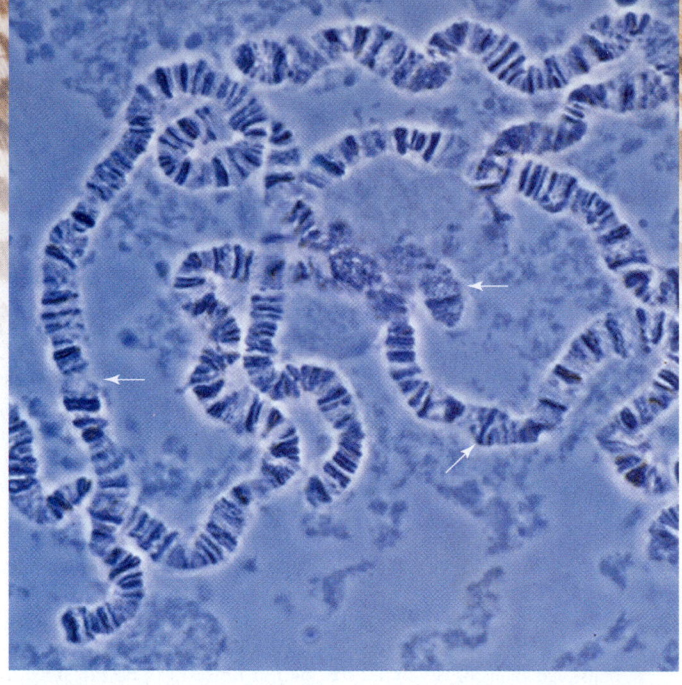

Regulated gene expression. LM of chromosomes from a cell in the salivary gland of the larva of the fruit fly, *Drosophila.* The expanded "puffed" chromosomal regions *(arrows)* contain genes that are actively transcribed in that cell, while the more compact banded regions include genes that are transcribed at very low rates, at different times, or not at all. *(Peter J. Bryant/Biological Photo Service)*

LEARNING OBJECTIVES

After you have studied this chapter you should be able to

1. Explain why the organization of genes into operons is advantageous to bacteria. Explain why some genes, such as those of the lactose operon, are inducible, while others, such as those of the tryptophan operon, are repressible.
2. Sketch the main elements of an inducible operon, such as the lactose operon, and explain the functions of the operator and promoter regions.
3. Diagram the major components of an mRNA molecule produced by the lactose operon, and relate that structure to the organization of the operon.
4. Differentiate between positive and negative control, and show how both types of control operate in the regulation of the lactose operon.
5. Sketch the structure of a typical eukaryotic gene and the DNA sequences involved in the regulation of that gene.
6. Give examples of some of the ways eukaryotic DNA-binding proteins bind to DNA.
7. Illustrate how a change in chromosome structure might affect the activity of a gene.
8. List two ways that a gene in a multicellular organism might be able to produce different products in different types of cells.
9. Identify some of the types of regulatory controls that can be exerted in eukaryotes after mature mRNA is formed.

Each type of cell in a multicellular organism has a characteristic shape, carries out very specific activities, and makes a distinct set of proteins. Yet, with few exceptions, they all contain the same genetic information. Why, then, are they not identical in structure and molecular composition? This is because gene expression is regulated; i.e., only certain subsets of the total genetic information are expressed in any given cell *(see photograph).*

What are the mechanisms controlling the expression of a gene? Let us consider a gene that codes for a protein that is an enzyme. Expression of that gene involves not only transcription to form messenger ribonucleic acid (mRNA); the mRNA must be translated into protein, and the protein must actively catalyze a specific reaction. Gene expression, then, is the result of a series of processes, each of which can be controlled in many different ways. The control mechanisms require information in the form of various signals (some originating within the cell and others coming from other cells or from the environment) that interact with deoxyribonucleic acid (DNA), RNA, or protein.

Some of the main strategies used to regulate gene expression include controlling (1) the amount of mRNA that is available, (2) the rate of translation of the mRNA, and (3) the activity of the protein product. This can be done in a variety of ways. For example, the amount of mRNA available can be controlled by regulating the rate of mRNA degradation as well as the rate of transcription.

Although bacteria are not multicellular, they still need to regulate the expression of their genes. Energy efficiency and economical use of resources are usually of primary importance to bacterial cells. As a result, most gene regulation in prokaryotes involves transcriptional level control of genes whose products are involved in resource utilization. In eukaryotes there is a much greater emphasis on fine-tuning the control systems, which is consistent with the greater complexity of these cells and the need to control development in multicellular organisms (see Chapter 16). Consequently, eukaryotic gene regulation occurs at many levels.

GENE REGULATION IN PROKARYOTES IS ECONOMICAL

Escherichia coli is a bacterium that is a common inhabitant of the intestine. It has 4288 genes that code for proteins, approximately 62% of which have known functions. Some of these genes encode proteins that are always needed (e.g., enzymes involved in glycolysis). These genes, which are constantly transcribed, are called **constitutive genes.** Other proteins are needed only when the bacterium is growing under special conditions.

For instance, the bacteria living in the colon of an adult cow are not normally exposed to the milk sugar lactose, a disaccharide (see Chapter 3). If those cells were to end up in the colon of a calf, however, they would have lactose available as a source of energy. This poses a dilemma. Should a bacterial cell invest energy and materials to produce lactose-metabolizing enzymes just in case it ends up in the digestive system of a calf? Given that the average lifetime of an actively growing *E. coli* cell is about 30 minutes, such a strategy appears wasteful. Yet if *E. coli* cells do not have the capacity to produce those enzymes, they might starve in the midst of an abundant food supply. *E. coli* handles this problem by regulating many of its enzymes to efficiently use available organic molecules.

Cells have two basic ways of controlling their metabolic activity. They can regulate the *activity* of certain enzymes (how effectively an enzyme molecule works), and/or they can control the *number* of enzyme molecules present in each cell. Some enzymes may be regulated in both ways in the same type of cell. An *E. coli* cell growing on glucose is estimated to need about 800 different enzymes. Some of these must be present in large numbers, whereas only a few molecules of others are required. For the cell to function properly, the quantity of each enzyme must be efficiently controlled.

Operons in prokaryotes permit coordinated control of functionally related genes

The French researchers François Jacob and Jacques Monod are credited with the first demonstration, in 1961, of how some genes are regulated at the biochemical level. They studied the genes that code for the enzymes that metabolize lactose. For *E. coli* to use lactose as an energy source, the sugar must be cleaved into the monosaccharides glucose and galactose by the enzyme β-galactosidase. Galactose is then converted to glucose by another enzyme, and the resulting two glucose molecules are further broken down by enzymes in the glycolysis pathway (see Chapter 7).

E. coli cells growing on glucose contain very little of the β-galactosidase enzyme, perhaps no more than one to three molecules per cell. However, cells grown on lactose have as many as 3000 β-galactosidase molecules per cell, accounting for about 3% of total cellular protein. Levels of two other enzymes, galactose permease and galactoside transacetylase, also increase when the cells are grown on lactose. The permease is needed to transport lactose efficiently across the bacterial plasma membrane; without it, only small amounts of lactose can enter the cell. The transacetylase may be involved in a minor aspect of lactose metabolism, although its function is not clear.

Jacob and Monod were able to identify mutant strains of *E. coli* in which a single genetic defect resulted in the loss of all three enzymes. This finding, along with other information, led them to the conclusion that the coding DNA sequences for all three enzymes are linked together as a unit on the bacterial DNA and are subject to a common control mechanism. Each protein coding sequence is known as a **structural gene.** Jacob and Monod coined the term **operon** to refer to a gene complex consisting of a group of structural genes with related functions, plus the closely linked DNA sequences responsible for controlling them. The structural genes of the lactose operon—*lac Z, lac Y,* and *lac A*—code for β-galactosidase, galactose permease, and galactoside transacetylase, respectively (Fig. 13–1).

Transcription of the lactose operon begins as RNA polymerase binds to a single **promoter** site upstream from the coding sequences. It then proceeds to transcribe the DNA, forming a single mRNA molecule that contains the coding information for all three enzymes. Between the enzyme-coding sequences this mRNA contains a translation termination codon for the previous sequence and a translation initiation codon for the sequence to follow; hence it is translated to form three separate protein molecules. Because all three enzymes are translated from the same mRNA molecule, their synthesis can be coordinated by turning a single molecular "switch" off or on.

The switch that controls mRNA synthesis is called the **operator;** it is a sequence of bases that overlaps part of the promoter region and is upstream from the first structural gene in the operon. In the absence of lactose, a **repressor protein** called the **lactose repressor** binds tightly to the operator region. Because the repressor protein is large enough to cover part of the promoter sequence, RNA polymerase is unable to bind to the lactose promoter site, and transcription of the lactose operon is effectively blocked.

The lactose repressor protein is encoded by a **regulatory gene,** which in this case is an adjacent structural gene located upstream from the promoter site. Unlike the lactose operon genes, the repressor gene is always "on," so small amounts of the repressor protein are produced continuously. Such a gene that is not regulated and is therefore constantly transcribed is said to be constitutive. The repressor protein binds specifically to the lactose operator sequence. When cells are grown in the absence of lactose, the operator site is nearly always occupied by a repressor molecule. A very small amount of mRNA can be synthesized when the operator site is briefly free of the repressor, but very few enzyme molecules are synthesized because *E. coli* mRNA is rapidly degraded (having a half-life of about 2 to 4 minutes).

Lactose is able to "turn on," or *induce,* the transcription of the lactose operon because the lactose repressor protein contains a second functional region separate from its DNA binding site; this is an allosteric binding site for allolactose, a structural isomer made from lactose. (Recall from Chapter 6 that an **allosteric regulator,** such as allolactose in this case, binds to a site in a protein other than its active site, changing its conformation and thereby altering its function.) If lactose is present in the growth medium, a few molecules are able to enter the cell and are converted to allolactose by the few β-galactosidase molecules

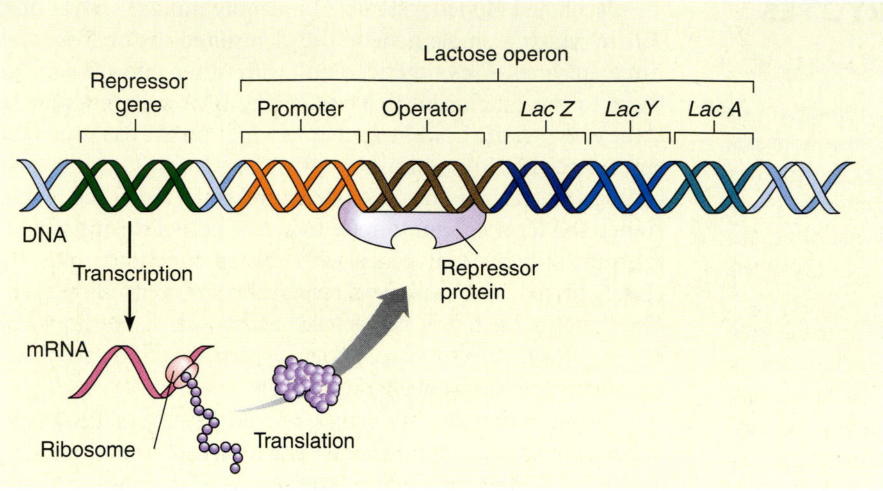

(a) In the absence of lactose, a repressor protein, encoded by an adjacent gene, binds a region known as the operator. By preventing RNA polymerase from binding to the promotor, the bound repressor protein blocks transcription of the structural genes.

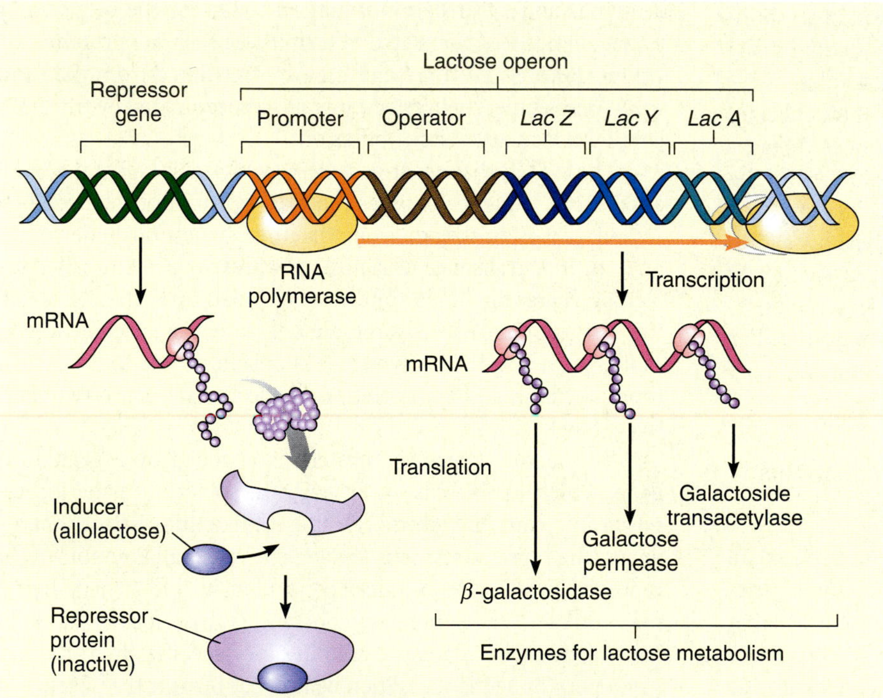

(b) When lactose is present, it is converted to allolactose, which binds to the repressor at an allosteric site, altering the structure of the protein so that it can no longer bind to the operator. This allows RNA polymerase to bind to the promoter and transcribe the structural genes.

Figure 13–1 Lactose operon. The structural genes coding for the three enzymes used by *E. coli* to metabolize the disaccharide lactose are transcribed as part of a single mRNA molecule.

present. When a molecule of allolactose binds to the repressor at the allosteric site, it alters the conformation of the protein so that its DNA-binding site can no longer recognize the operator. When all the repressor molecules have allolactose bound to them and are therefore inactivated, RNA polymerase binds to the unblocked promoter, and the operon is actively transcribed.

The *E. coli* cell continues to produce β-galactosidase and the other lactose operon proteins until virtually all of the lactose is used up. When intracellular levels of lactose drop, allolactose dissociates from the repressor proteins, which then assume a conformation that allows them to bind to the operator region and shut down transcription of the operon.

Jacob and Monod isolated genetic mutants to study the lactose operon

How were Jacob and Monod able to dissect the functioning of the lactose operon? Their approach centered around the use of mutant strains, which even today play an essential role in allowing researchers to unravel the components of a regulatory system. Mutant strains permit investigators to carry out genetic crosses to determine the map positions (linear order) of the genes on the DNA (see Chapter 10) and to infer normal gene functions by studying what happens when they are missing or altered. This information is usually combined with results of direct biochemical studies.

To understand the reasoning behind Jacob and Monod's experiments, follow the branching steps in Figure 13–2 as you read. They were able to divide their mutant strains into two groups based on whether a particular mutation affected only one enzyme or all three. In one group only one enzyme of the three—β-galactosidase, galactose permease, or galactoside transacetylase—was affected. Subsequent gene mapping studies showed that these were mutations in structural genes located next to each other in a linear sequence.

Jacob and Monod also studied strains they classified as regulatory mutants because a single mutation coordinately affected the expression of all three enzymes. Some of these regulatory mutants were constitutive; in these the structural genes of the lactose operon were always transcribed at a significant rate, even in the absence of lactose, causing the cell to waste energy producing unneeded enzymes. One group of constitutive mutations had map positions just outside the lactose operon itself. Using special genetic strains, it was possible to show that these particular mutations always caused constitutive expression, regardless of their location in the genome. On the basis of these findings, Jacob and Monod hypothesized the existence of a repressor gene that codes for a repressor molecule (later found to be a protein). Although the specific defect may vary, the members of this group of constitutive mutants do not produce active repressor proteins; hence, no binding to the lactose operator and promoter takes place, and the lactose operon is transcribed constitutively.

The genes responsible for the behavior of a second group of constitutive mutants had map positions within the lactose operon but did not directly involve any of the three structural genes. The members of this group were found to produce normal repressor molecules but to have abnormal operator sequences incapable of binding the repressor.

In contrast to the constitutive mutants, other mutants failed to transcribe the lactose operon even when lactose was present. Some of these abnormal genes had the same map position as the hypothesized regulatory gene. They were eventually found to have an altered binding site on the repressor protein that prevented allolactose from binding, although the ability of the repressor to bind to the operator was unaffected. Once bound to the operator, such a mutant repressor remains bound, keeping the operon "turned off."

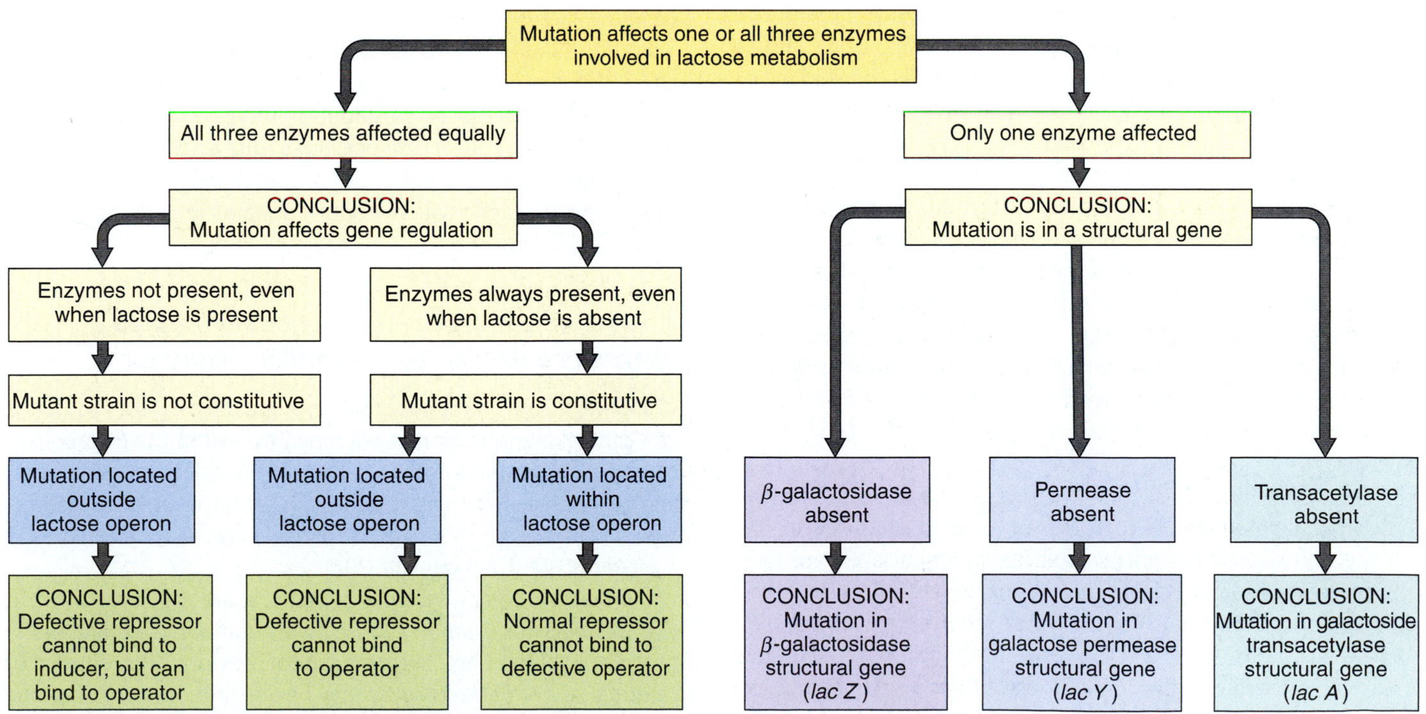

Figure 13–2 Genetic and biochemical characterization of the lactose operon. Jacob and Monod analyzed the properties of various mutant strains of *E. coli* to deduce the structure and function of the lactose operon.

An inducible gene is not transcribed unless a specific inducer inactivates its repressor

The lactose operon is called an **inducible** system. An inducible gene or operon is usually controlled by a repressor that keeps it in the "off" state. The presence of an **inducer molecule** (in this case allolactose) renders the repressor inactive so that the gene or operon is turned on. Inducible genes or operons usually code for enzymes that are part of catabolic pathways, which are responsible for breaking down molecules to provide both energy and components for anabolic reactions. This type of regulatory system enables the cell to save the energy costs of making enzymes when no substrates are available upon which they can act.

A repressible gene is transcribed unless a specific repressor-corepressor complex is bound to the DNA

Another type of gene regulation system in bacteria is associated mainly with anabolic pathways in which amino acids, nucleotides, and other essential molecules are synthesized from simpler precursors. Regulation of these pathways normally involves repressible enzymes, which are coded for by **repressible** genes.

Repressible genes and operons are usually on; they are turned off only under special conditions. In most cases the molecular signal used to regulate these genes is the end product of the anabolic pathway. When the supply of the end product (e.g., an amino acid) is low, all enzymes in the pathway are actively synthesized. When intracellular levels of the end product are high, enzyme synthesis is repressed. Because compounds such as amino acids are continuously needed by the growing cell, the most effective strategy is to keep the genes that control their production on except when a large supply of the amino acid is available. The ability to turn the genes off allows cells to avoid overproduction of amino acids and other molecules that are essential but energetically expensive to make.

The tryptophan operon is an example of a repressible system. In both *E. coli* and a related bacterium, *Salmonella,* the operon consists of five structural genes that code for the enzymes required for synthesis of the amino acid tryptophan; these are clustered together as a transcriptional unit with a single promoter and a single operator (Fig. 13–3). A distant repressor gene codes for a diffusible repressor protein, which differs from the lactose repressor in that it is synthesized in an inactive form that is unable to bind to the operator region of the tryptophan operon.

The DNA-binding site of the repressor becomes effective only when tryptophan, its **corepressor,** binds to an allosteric site on the repressor. When intracellular tryptophan levels are low, the repressor protein is inactive and unable to bind to the operator region of the DNA. The enzymes required for tryptophan synthesis are produced, and the concentration of tryptophan increases. As the concentration of intracellular tryptophan rises, some tryptophan binds to the allosteric site of the repressor, altering its conformation so that it binds tightly to the operator. This has the effect of switching the operon off, thereby blocking transcription.

Negative regulators repress transcription; positive regulators activate transcription

The features of the lactose and tryptophan operons described so far are examples of **negative control.** Systems under negative control are those in which the DNA-binding regulatory protein is a repressor that turns off transcription of the gene. Some regulatory systems involve **positive control,** that is, regulation by **activator proteins** that bind to DNA and thereby stimulate transcription of a gene. The lactose operon is controlled by both a negative regulator (the lactose repressor protein) and a positively acting activator protein (Fig. 13–4).

Positive control of the lactose operon requires that the cell be able to sense the absence of the sugar glucose, which is the initial substrate in the glycolysis pathway (see Chapter 7). Lactose, like glucose, can undergo stepwise breakdown to yield energy. However, because glucose is a product of the catabolic hydrolysis of lactose, it is most efficient for *E. coli* cells to use the available supply of glucose first, sparing the cell the considerable energy cost of making additional enzymes such as β-galactosidase.

The lactose operon has a very inefficient promoter sequence; that is, it has a low affinity for RNA polymerase even when the repressor protein is inactivated. However, a DNA sequence adjacent to the promoter site is a binding site for another regulatory protein, the **catabolite gene activator protein (CAP).**

In its active form CAP has **cyclic AMP**, or **cAMP,** an alternative form of adenosine monophosphate (see Fig. 3–27), bound to an allosteric site.[1] As the cells become depleted of glucose, cAMP levels increase. The cAMP molecules bind to CAP, and the resulting active complex then binds to the CAP-binding site near the lactose operon promoter. This binding of active CAP causes the double helix to bend (Fig. 13–5) and increases the affinity of the promoter region for RNA polymerase such that the rate of transcription initiation is increased in the presence of lactose. Thus, the lactose operon is fully active only if lactose is available and intracellular glucose levels are low. The properties of negative and positive control systems are summarized in Table 13–1.

A regulon is a group of functionally related operons controlled by a common regulator

CAP differs from the lactose and tryptophan repressors in that it can control transcription of operons involved in the metabolism of several catabolites, such as the sugars galactose, arabinose, and maltose, as well as of lactose. A group of operons coordinately controlled by a single regulator is generally referred to as a **regulon** (Fig. 13–6 on page 292).

Other multigene systems in bacteria are also controlled in this manner. For example, genes involved in nitrogen and phosphate metabolism are organized into regulons that consist of multiple sets of operons controlled by one or more combinations

[1] For this reason, CAP is also known as CRP, which stands for **c**AMP **r**eceptor **p**rotein.

(text continues on page 291)

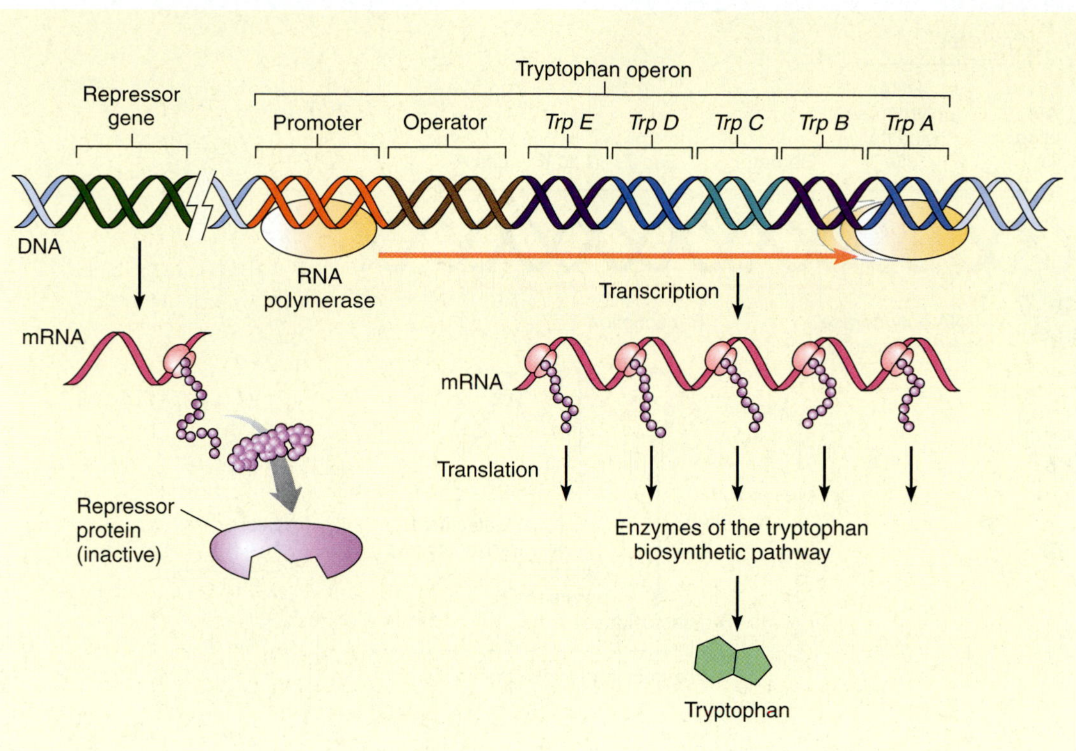

(a) **Intracellular tryptophan levels low.** Repressor protein is unable to prevent transcription because it cannot bind to the operator.

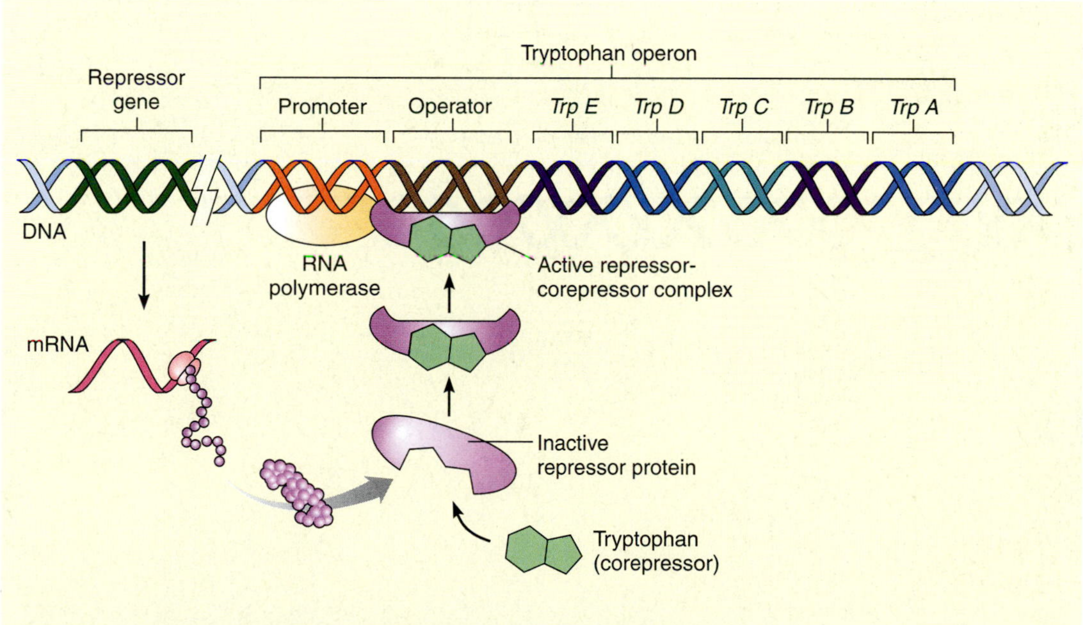

(b) **Intracellular tryptophan levels high.** The amino acid tryptophan binds to an allosteric site on the repressor protein, changing its conformation. The resulting active form of the repressor binds to the operator region, blocking transcription of the operon until tryptophan is again required by the cell.

Figure 13–3 Tryptophan operon. Genes coding for enzymes that synthesize the amino acid tryptophan are organized in a repressible operon.

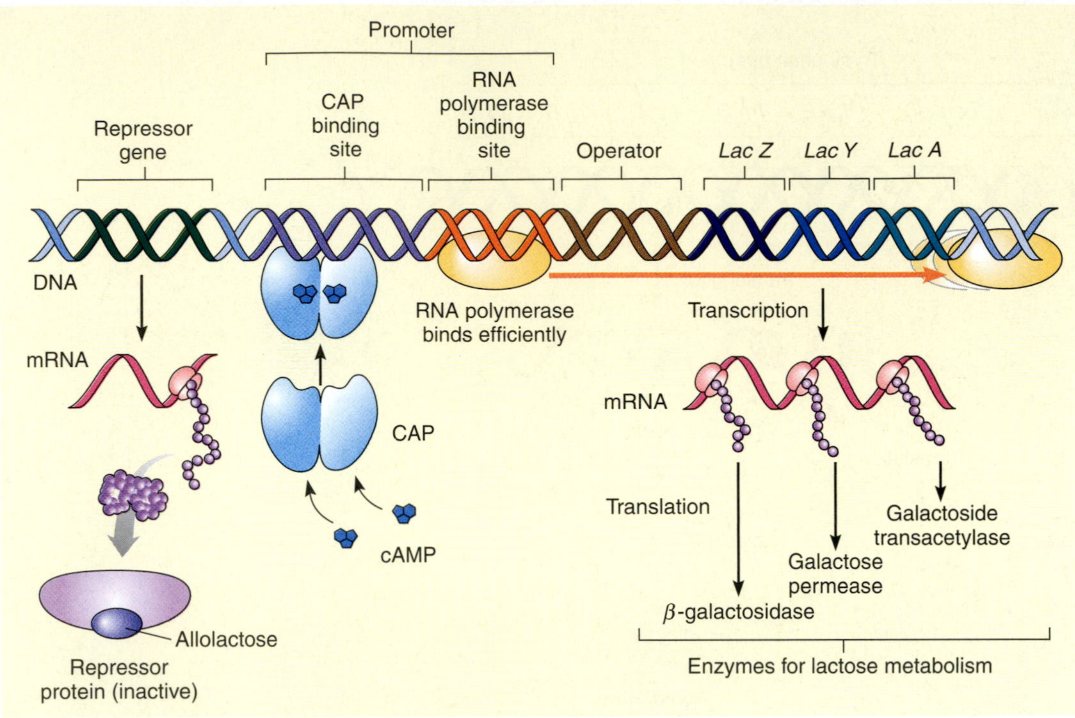

(a) **Lactose high, glucose low, cAMP high.** When glucose concentrations are low, each CAP polypeptide has cAMP bound to its allosteric site. CAP can bind to the DNA sequence and transcription becomes activated.

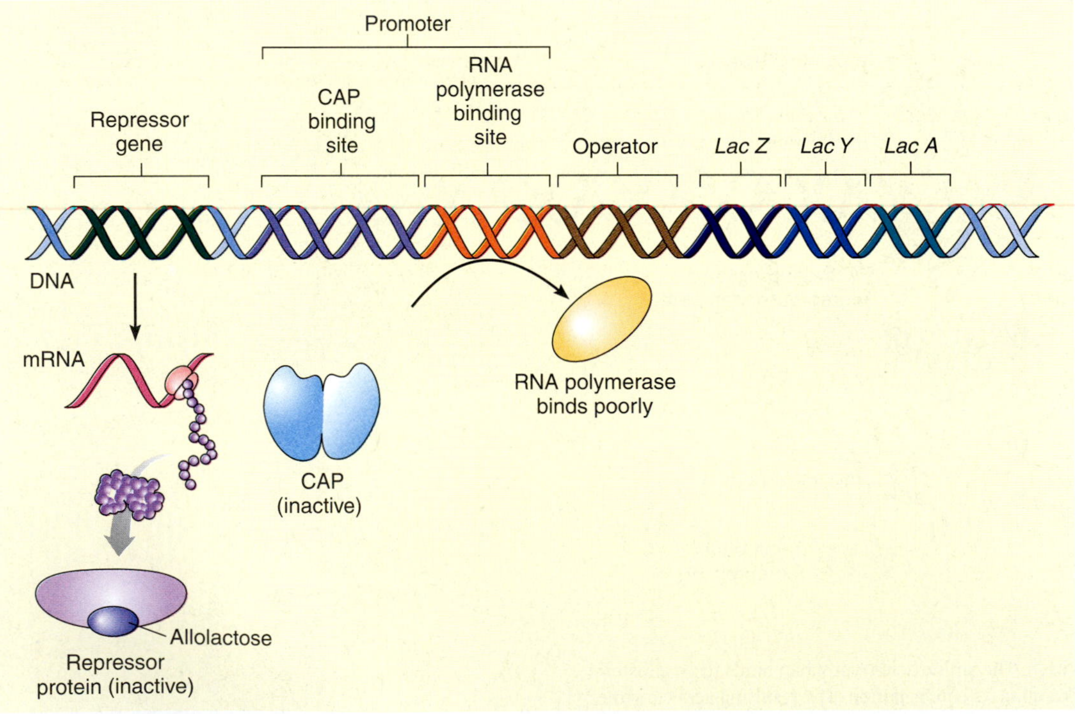

(b) **Lactose high, glucose high, cAMP** low. When glucose levels are high, cAMP is low. CAP is in an inactive form and cannot stimulate transcription. Transcription occurs at a low level or not at all.

Figure 13–4 Positive control of the lactose operon. The lactose promoter by itself is weak and binds RNA polymerase inefficiently even when the lactose repressor is inactive. The CAP is an allosteric regulator that can bind to a sequence of bases in the promoter, allowing RNA polymerase to bind efficiently, thereby stimulating transcription of the operon. The CAP molecule contains two polypeptides that can each bind to cAMP at allosteric sites. The cell's cAMP concentration is inversely proportional to the glucose concentration.

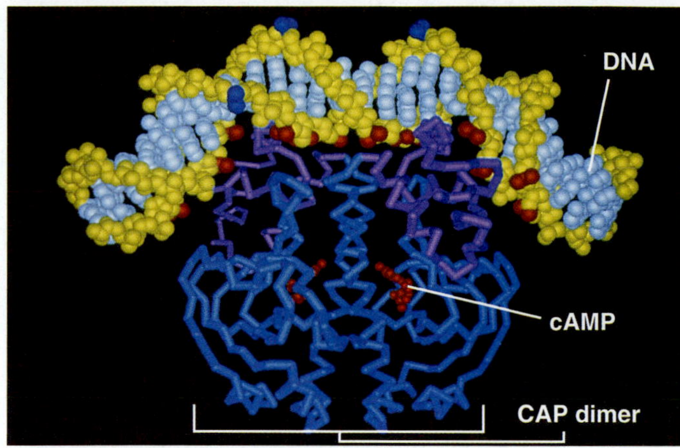

Figure 13–5 Binding of CAP to DNA. This computer-generated picture illustrates the bend formed in the DNA double helix when it binds to CAP. CAP is a dimer consisting of two identical polypeptide chains, each of which binds one molecule of cAMP. *(Courtesy of S.C. Schultz, G.C. Shields, and T.A. Steitz, Yale University)*

of regulatory genes. Other complex multigene systems respond to changes in environmental conditions, such as rapid shifts in temperature, exposure to radiation, changes in osmotic pressure, and changes in oxygen levels. Specific mutants often provide clues to the existence of a regulon system. A single mutation that destroys the activity of CAP, for example, prevents the cell from efficiently metabolizing not only lactose but many other sugars also regulated by CAP.

Not all constitutive genes are transcribed at the same rate

Many of the gene products encoded by the *E. coli* DNA are needed only under certain environmental or nutritional conditions. As we have seen, these genes are generally regulated at the level of transcription. They can be turned on and off as metabolic and environmental conditions change. By contrast, constitutive genes are continuously transcribed, but they are not necessarily transcribed (or their mRNAs translated) at the same rate. Some enzymes work more effectively or are more stable than others and consequently need to be present in smaller amounts. Constitutive genes that encode proteins required in large amounts are generally transcribed more rapidly than genes coding for proteins required at lower levels. The transcription rate of these genes is controlled by their promoter sequences. Constitutive genes with efficient ("strong") promoters bind RNA polymerase more frequently and consequently transcribe more mRNA molecules than those with inefficient ("weak") promoters.

Genes coding for repressor or activator proteins that regulate metabolic enzymes are usually constitutive and produce their protein products constantly. Because each cell usually needs relatively few molecules of any specific repressor or activator protein, promoters for those genes tend to be relatively weak.

TABLE 13–1 Types of Transcriptional Control in Prokaryotes*

Negative Control	Result
Inducible genes	
Repressor protein alone	**Active repressor "turns off" regulated gene(s)**
Lactose repressor alone	Lactose operon not transcribed
Repressor protein + inducer	**Inactive repressor-inducer complex fails to "turn off" regulated gene(s)**
Lactose repressor + allolactose	Lactose operon transcribed
Repressible genes	
Repressor protein alone	**Inactive repressor fails to "turn off" regulated gene(s)**
Tryptophan repressor alone	Tryptophan operon transcribed
Repressor protein + corepressor	**Active repressor-corepressor complex "turns off" regulated gene(s)**
Tryptophan repressor + tryptophan	Tryptophan operon not transcribed
Positive Control	**Result**
Activator protein alone	**Activator alone cannot stimulate transcription of regulated gene(s)**
CAP alone	Transcription of lactose operon not stimulated
Activator protein + coactivator	**Functional activator-coactivator complex stimulates transcription of regulated gene(s)**
CAP + cAMP	Transcription of lactose operon stimulated

*A general description of each type is followed by a specific example. A negative regulator is a repressor that "turns off" transcription of the regulated gene(s). Conversely, a positive regulator is an activator that stimulates transcription.

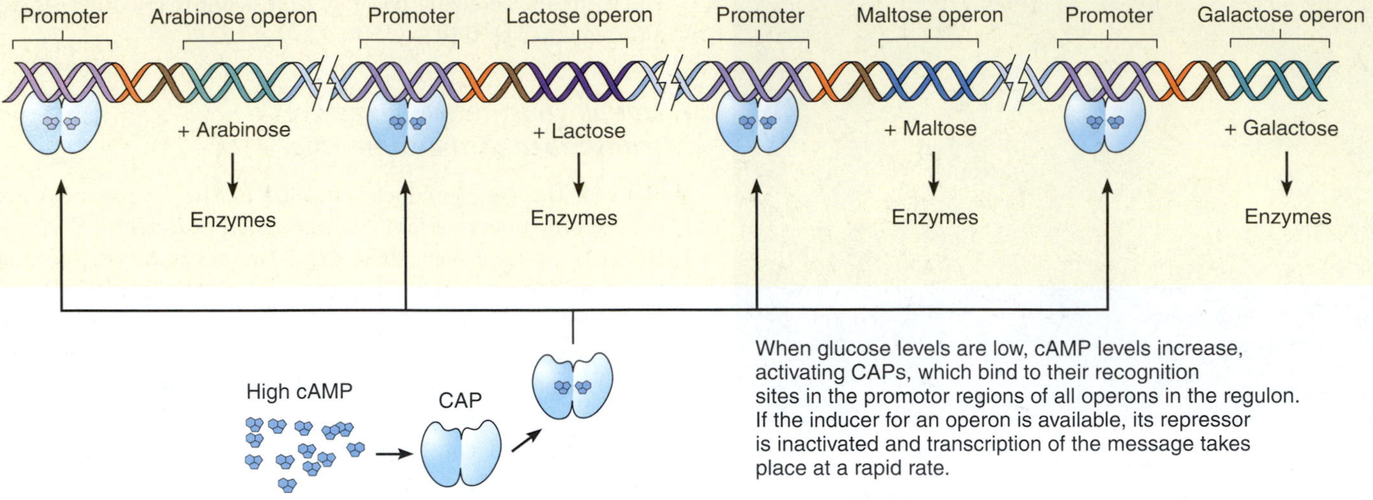
Promoter
Arabinose operon
Promoter
Lactose operon
Promoter
Maltose operon
Promoter
Galactose operon
+ Arabinose
Enzymes
+ Lactose
Enzymes
+ Maltose
Enzymes
+ Galactose
Enzymes
High cAMP
CAP
When glucose levels are low, cAMP levels increase, activating CAPs, which bind to their recognition sites in the promotor regions of all operons in the regulon. If the inducer for an operon is available, its repressor is inactivated and transcription of the message takes place at a rapid rate.

Figure 13–6 Regulon. Operons that convert several different sugars to glucose in *E. coli* make up a regulon that is under positive control by CAP.

Some posttranscriptional regulation occurs in prokaryotes

As we have seen, much of the variability in protein levels in *E. coli* is determined by **transcriptional level control.** However, regulatory mechanisms that operate after transcription, known as **posttranscriptional controls,** also occur, operating at various levels of gene expression.

Translational controls are posttranscriptional controls that regulate the rate at which a particular mRNA molecule is translated. Because the lifetime of an mRNA molecule in a bacterial cell is very short, an mRNA that is translated rapidly can produce more proteins than one that is translated slowly. Some mRNA molecules in *E. coli* are translated as much as 1000 times faster than others. Most of the differences appear to be due to the rate at which ribosomes can attach to the mRNA and begin translation.

Posttranslational controls generally act as switches that activate or inactivate one or more existing enzymes. These systems allow the cell to respond to changes in the intracellular concentrations of essential molecules, such as amino acids, by rapidly adjusting the activities of its enzymes. A common posttranslational control adjusts the rate of synthesis in a metabolic pathway through **feedback inhibition,** also known as **end product inhibition** (see Fig. 6–17). The end product binds to the first enzyme in the pathway at an allosteric site, temporarily inactivating the enzyme. When the first enzyme in the pathway does not function, all of the succeeding enzymes are deprived of substrates. Notice that feedback inhibition differs from the repression caused by tryptophan discussed previously. In that case, the end product of the pathway (tryptophan) prevented the formation of new enzymes. Feedback inhibition acts as a fine-tuning mechanism that regulates the activity of the existing enzymes in a metabolic pathway.

GENE REGULATION IN EUKARYOTES IS MULTIFACETED

Like bacteria, eukaryotic cells must respond to changes in their environment. In addition, multicellular eukaryotes require modes of regulation that permit individual cells to become committed to specialized roles and groups of cells to organize into tissues and organs (see *Focus On: Regulation in Prokaryotes and Eukaryotes*). This is mainly accomplished by transcriptional regulation, but posttranscriptional (translational and posttranslational) controls are also important. In previous chapters we observed that all aspects of information transfer, including replication, transcription, and translation, are far more complicated in eukaryotes. Not surprisingly, this complexity provides additional opportunities for control of gene expression.

Unlike many of the prokaryotic genes, most eukaryotic genes are not found in operon-like clusters. However, each eukaryotic gene has specific regulatory sequences that are essential in the control of transcription.

Many of the "housekeeping" enzymes (those needed by all cells) appear to be encoded by constitutive genes, which are expressed in all cells at all times. Some inducible genes have also been found; these respond to environmental threats or stimuli such as heavy metal ingestion, viral infection, and heat shock. For example, when a cell is exposed to high temperature, many proteins fail to fold properly. The presence of these unfolded proteins elicits a survival response in which genes known as heat-shock genes are transcribed, and heat-shock proteins are formed. Although the functions of most of the heat-shock proteins are not known, some are **molecular chaperones,** which are responsible for helping other proteins attain their proper conformation.

Some genes appear to be inducible only during certain periods in the life of the organism; they are thought to be controlled by **temporal regulation** mechanisms. Finally, a number of genes

FOCUS ON Regulation in Prokaryotes and Eukaryotes

Why do prokaryotic and eukaryotic cells have distinctly different strategies for regulating the activity of their genes? In large part these differences reflect the ways in which the organisms make their living. Bacterial cells exist independently, and each cell must be able to perform all its own essential functions. Because they grow rapidly and have relatively short lifetimes, bacterial cells carry little excess baggage.

The dominant theme of prokaryotic gene regulation is *economy,* and controlling transcription is usually the most cost-effective way to regulate gene expression. The organization of related genes into operons and regulons that can be rapidly turned on and off as units allows these cells to synthesize only the gene products needed at any particular time. This type of regulation requires rapid turnover of mRNA molecules to prevent messages from accumulating and continuing to be translated when they are not needed. Bacteria rarely regulate enzyme levels by degrading proteins. Once the synthesis of a protein ends, the previously synthesized protein molecules are diluted so rapidly in subsequent cell divisions that breaking them down is usually not necessary. Only when cells are starved or deprived of essential amino acids are protein-digesting enzymes used to recycle amino acids by breaking down proteins no longer needed for survival.

Eukaryotic cells have different regulatory requirements. In multicellular organisms, groups of cells cooperate with each other in a division of labor. Because a single gene may need to be regulated in different ways in different types of cells, eukaryotic gene regulation is complex. Although transcriptional level control predominates, especially in multicellular eukaryotes, control of other levels of gene expression is also very important. Eukaryotic cells usually have long lifetimes during which they may need to respond repeatedly to many different stimuli. Rather than synthesize new enzymes each time they respond to a stimulus, these cells make extensive use of preformed enzymes and other proteins that can be rapidly converted from an inactive to an active state. Some cells have a large store of inactive messenger RNA; for example, the mRNA of an egg cell becomes activated when it is fertilized.

Much of the emphasis of gene regulation in multicellular organisms is on *specificity* in the form and function of the cells in each tissue. Each type of cell has certain genes that are active and others that may never be used (see Chapter 16). Apparently the adaptive advantages of cellular cooperation in eukaryotes far outweigh the detrimental effects of carrying a load of inactive genes through many cell divisions. For example, developing red blood cells produce the oxygen transport protein hemoglobin, whereas muscle cells never produce hemoglobin but instead produce myoglobin, a related protein that stores oxygen in muscle tissues.

are under the control of **tissue-specific regulation.** For example, a gene involved in the production of a particular enzyme may be regulated by one stimulus (e.g., a hormone) in muscle tissue, by an entirely different stimulus in pancreatic cells, and by a third stimulus in liver cells. These types of regulation are explored in more detail in Chapter 16.

Eukaryotic transcription is controlled at many sites and by many different regulatory molecules

Most genes of multicellular eukaryotes are controlled at the transcriptional level. As we see in the following discussion, various base sequences in the DNA are important in transcriptional control. In addition, the rate of transcription is affected by regulatory proteins and by the way the DNA is organized in the chromosome.

Eukaryotic promoters vary in efficiency, depending on their upstream promoter elements

In eukaryotic as well as prokaryotic cells, the transcription of any gene requires a base pair where transcription begins, known as the **transcription initiation site,** plus a promoter to which RNA polymerase binds. A prokaryotic promoter (Fig. 13–7*a*) includes certain characteristic base sequences, known as a *Pribnow box* and a *-35 box,* located 35 base pairs upstream from the transcription initiation site.[2] In multicellular eukaryotes, RNA polymerase binds to a sequence of bases known as a **TATA box,** about 25 to 35 base pairs upstream from the transcription initiation site (Fig. 13–7*b*). A eukaryotic promoter generally contains one or more sequences of 8 to 12 bases within a short distance upstream of the RNA polymerase–binding site. These have been given various names; we will refer to them as **upstream promoter elements (UPEs).** Efficient initiation of transcription seems to be related to the number and type of UPEs. Thus, a constitutive gene containing only one UPE is generally weakly expressed, whereas one containing five or six UPEs is usually transcribed much more actively (Fig. 13–7*c* and *d*).

Enhancers are DNA sequences that increase the rate of transcription

Regulated eukaryotic genes commonly require not only the upstream promoter elements but also DNA sequences called **enhancers.** Whereas the promoter elements are required for

[2] By convention, the base pair that serves as the transcription initiation site is given the designation +1; upstream base pairs are given negative numbers, and downstream base pairs are given positive numbers.

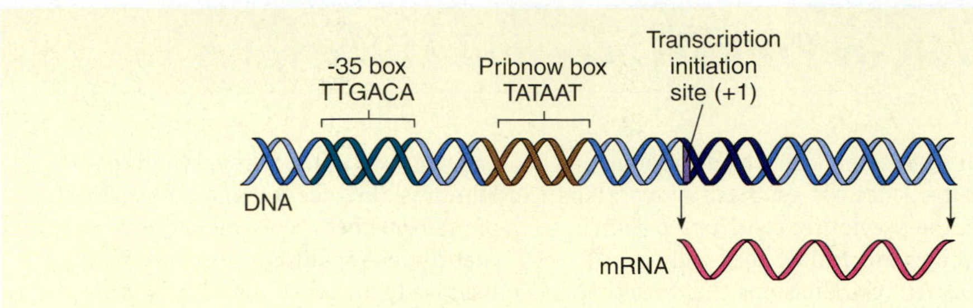

(a) **A prokaryotic promoter.** A typical prokaryotic promoter contains a "Pribnow box" and a "-35 box," usually found, respectively, 10 and 35 base pairs upstream from the transcription initiation site (the +1 base pair). The base sequences shown are those most commonly found.

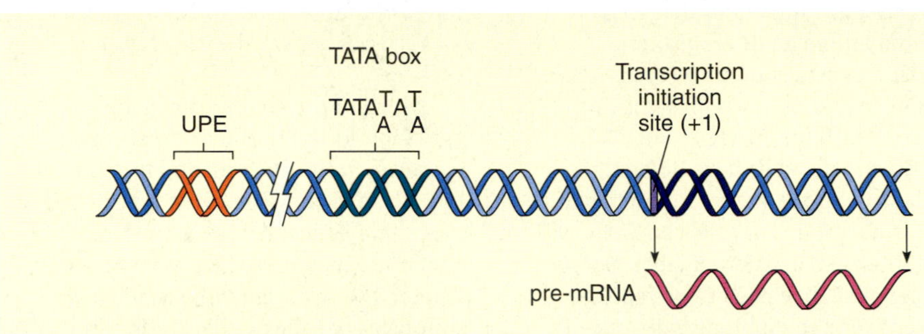

(b) **Eukaryotic promoter elements.** A eukaryotic promoter usually contains a "TATA box" located 25 to 35 base pairs upstream from the transcription initiation site. The most commonly found base sequence is shown (either T or A can be present at the positions where they appear together). One or more upstream promoter elements (UPEs) are usually present.

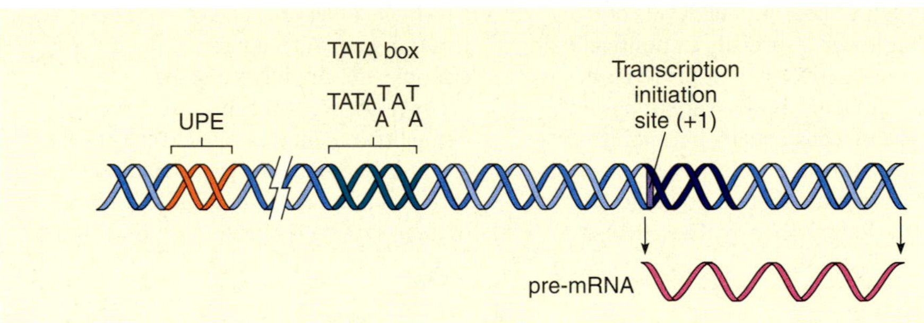

(c) **A weak eukaryotic promoter.** A weakly expressed gene contains only one UPE.

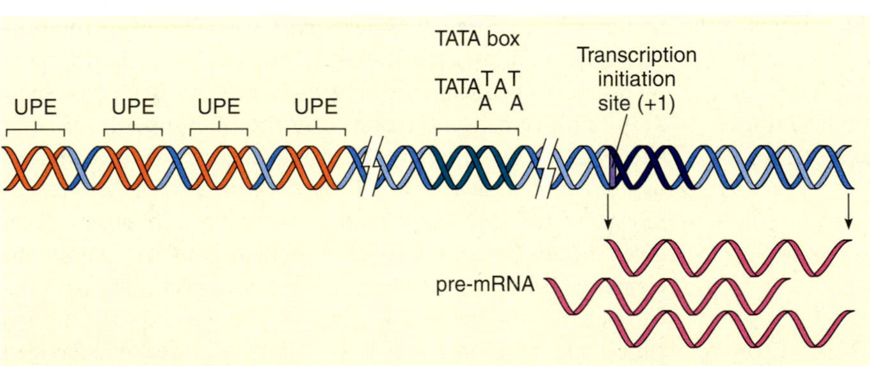

(d) **A strong eukaryotic promoter.** A strongly expressed gene is likely to contain several UPEs.

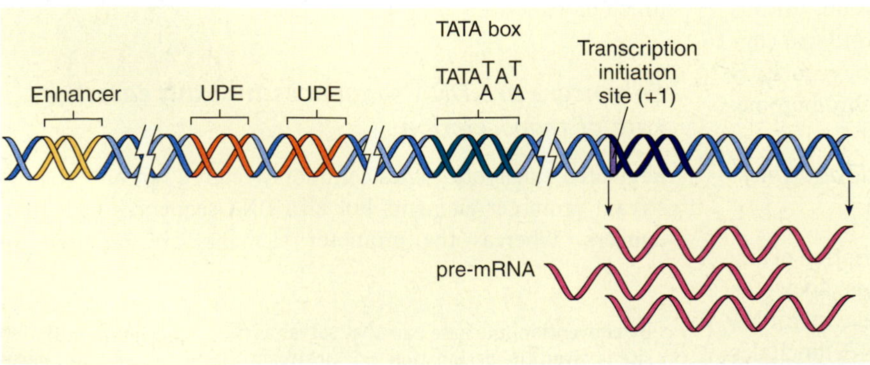

(e) **A strong eukaryotic promoter plus an enhancer.** Transcription of this eukaryotic gene is stimulated by an enhancer, located several thousand bases from the promoter.

Figure 13-7 Control of transcription. The DNA double helix and other elements are not drawn to scale.

accurate and efficient initiation of mRNA synthesis, enhancers increase the *rate* of RNA synthesis after initiation, often by several orders of magnitude (Fig. 13–7*e*).

Enhancer sequences are remarkable in many ways. Although present in all cells, a particular enhancer is functional only in certain types of cells. An enhancer can regulate a gene on the same DNA molecule from very long distances (up to thousands of base pairs away from the promoter) and can be either upstream or downstream of the promoters it controls. Furthermore, if an enhancer sequence is experimentally cut out of the DNA and inverted, it still regulates the gene it normally controls. As we shall see, evidence suggests that at least some enhancers work by interacting with proteins that regulate transcription.

Transcription factors are regulatory proteins that have several functional domains and may work in various combinations

We previously discussed some DNA-binding proteins that regulate transcription in prokaryotes. These include the lactose repressor, the tryptophan repressor, and the catabolite gene activator protein (CAP). Similarly, many regulators of transcription have been identified in eukaryotes; these eukaryotic proteins are known as **transcription factors.**

It is useful to compare transcriptional regulators in prokaryotes and eukaryotes. Many transcriptional regulators are modular molecules; that is, they have more than one **domain** (region with its own tertiary structure), and each domain has a different function. Each eukaryotic transcription factor, like the transcriptional regulators of prokaryotes, has a DNA-binding domain, plus at least one other domain, and may be either an activator or a repressor.

Many prokaryotic regulators and some eukaryotic transcription factors contain a *helix-turn-helix* motif, consisting of two α-helical segments. One of these, known as the *recognition helix,* is inserted into the major groove of the DNA without unwinding the double helix. The second helps to hold the first in place. (Fig. 13–8*a*). The "turn" is a sequence of amino acids that forms a sharp bend in the molecule.

Some other regulators have multiple "zinc fingers," loops of amino acids held together by zinc ions. Each loop includes an α-helix that fits into the major groove of the DNA (Fig. 13–8*b*). Certain amino acid functional groups exposed in each finger have been shown to recognize specific DNA sequences.

Many regulatory proteins are functional only as pairs, or *dimers,* and these have special domains required for dimer formation. Many of these transcription factors are known as **leucine zipper proteins** because they are held together by the side chains of leucine and other hydrophobic amino acids (Fig. 13–8*c*). In some cases the two polypeptides that make up the dimer may be identical and form a *homodimer.* In other instances they are different, and the resulting *heterodimer* may have very different regulatory properties. For a simple and speculative example, let us assume that three regulatory proteins—A, B, and C—are involved in controlling a particular set of genes. These three proteins might associate as dimers in six different ways: three kinds of homodimers (AA, BB, and CC) and three kinds of heterodimers (AB, AC, and BC). Such multiple combinations of regulatory

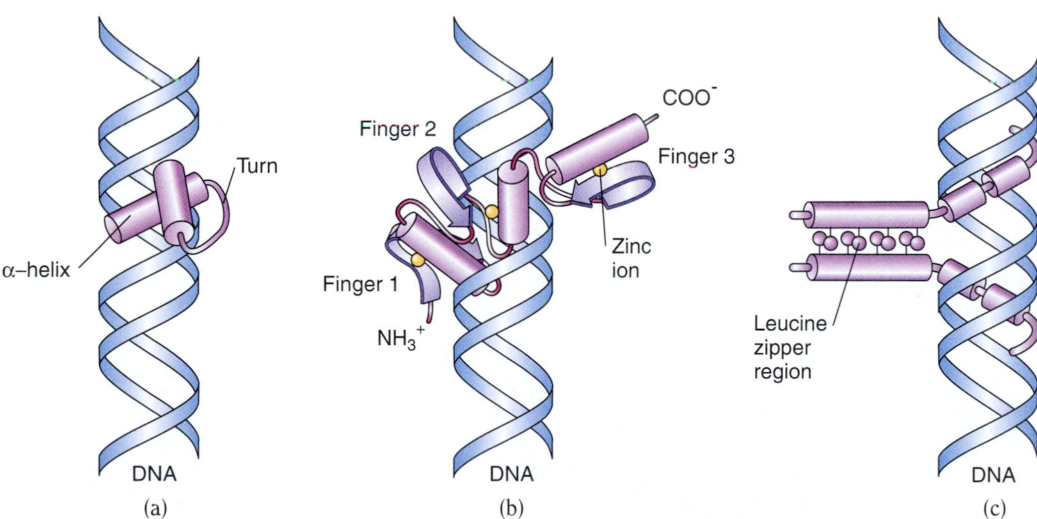

Figure 13–8 Regulatory proteins. In these illustrations, α-helical regions are shown as barrel shapes and β-pleated sheets are shown as ribbons. **(a)** A portion of a regulatory protein containing the helix-turn-helix arrangement. The recognition helix is inserted into the major groove of the DNA and is connected to a second helix that helps hold it in place by a sequence of amino acids that form a sharp bend. **(b)** Regions of certain transcription factors form projections known as "zinc fingers," which can insert into the grooves of the DNA and bind to specific base sequences. **(c)** This leucine zipper protein is a dimer, held together by hydrophobic interactions involving side chains of leucine and other amino acids.

proteins have the potential to greatly increase the number of possible ways that transcription can be controlled.

Transcription factors interact with the general transcription machinery

Transcription in eukaryotes requires multiple regulatory proteins that are bound to different parts of the promoter. The "general transcription machinery" is a protein complex that binds to the TATA region of the promoter near the transcription initiation site. That complex is required for RNA polymerase to bind and initiate transcription.

Both enhancers and UPEs apparently become functional when specific transcription factors are bound to them. Figure 13–9 illustrates interactions involving an enhancer and a transcription factor that acts as an activator. Each activator must have at least two functional domains: a DNA recognition site that usually binds to an enhancer or UPE, and a "gene activation site" that contacts the target in the general transcriptional machinery. The DNA between the enhancer and promoter sequences is thought to form a loop that allows an activator bound to an enhancer to come in contact with one or more target proteins associated with the general transcriptional machinery. When this occurs, the rate of transcription is increased.

The organization of the chromosome may affect the expression of some genes

A chromosome is not simply a bearer of genes. Various arrangements of its ordered components can result in increased or decreased expression of the genes it contains.

MULTIPLE COPIES OF GENES A single gene cannot always provide enough copies of its mRNA to meet the cell's needs. The requirement for high levels of certain products may be met if multiple copies of the genes that encode them are present in the chromosome. Genes of this type, whose products are essential for all cells, may be present as multiple copies arranged one after another along the chromosome. These are known as *tandemly repeated gene sequences.* Other genes, which may be required by only a small group of cells, may be selectively replicated in those cells in a process called **gene amplification** (see Chapter 16).

Within an array of repeated genes, each copy is almost identical to the others. Histone genes, which code for the proteins that associate with DNA to form nucleosomes (see Chapter 11), are usually found as multiple copies of 50 to 500 genes in cells of multicellular organisms. Similarly, multiple copies (150 to 450) of genes for rRNA and tRNA occur in all cells.

GENE INACTIVATION BY CHANGES IN CHROMATIN STRUCTURE In multicellular eukaryotes, only a subset of the genes present in a cell are active at any one time. The inactivated genes differ among cell types and in many cases seem to be irreversibly dormant.

Some of the inactive genes appear to be found in highly compacted chromatin, which can be seen as densely staining regions of chromosomes during cell division. These regions of chromatin remain tightly coiled throughout the cell cycle, and even during interphase are visible as darkly staining fibers called **heterochromatin** (Fig. 13–10). Evidence suggests that the DNA of heterochromatin is not transcribed. When one of the two X chromosomes is inactivated in female mammals, most of the inactive X chromosome

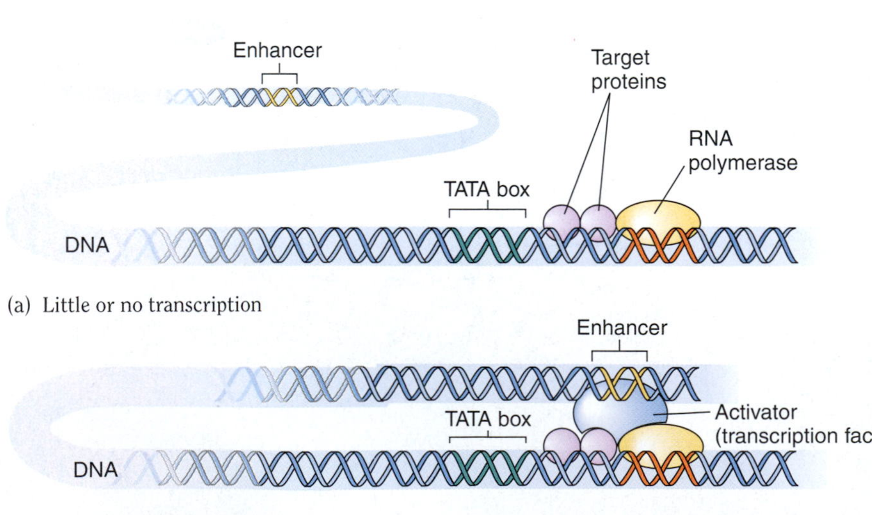

Figure 13–9 Stimulation of transcription by an enhancer. (a) This gene is transcribed at a very low rate or not at all, even though the general transcriptional machinery, including RNA polymerase, is bound to the promoter. (b) A regulatory protein that functions as a transcriptional activator becomes bound to an enhancer. The intervening DNA forms a loop, allowing the activator to contact one or more target proteins in the general transcriptional machinery, thereby increasing the rate of transcription.

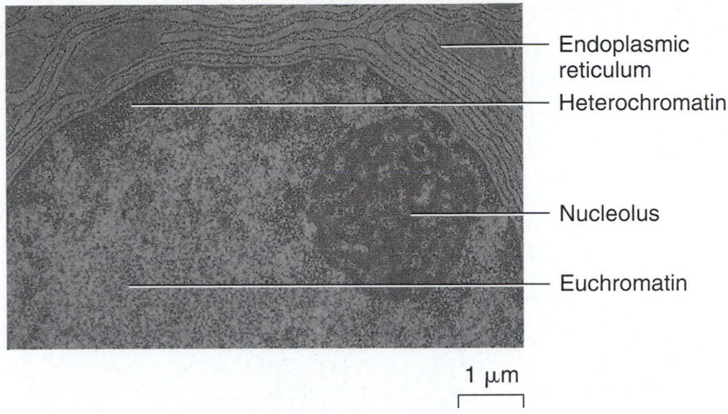

Figure 13–10 Heterochromatin. The heterochromatin in this TEM of a human pancreas cell is dense and darkly staining and tends to be associated with the nuclear envelope. The euchromatin consists of a much looser fibrillar structure. *(D. Fawcett)*

becomes heterochromatic and is seen as the Barr body (see Fig. 10–16). Active genes are associated with a more loosely packed chromatin structure called **euchromatin** (Fig. 13–11).

GENE INACTIVATION BY DNA METHYLATION Inactive genes of vertebrates and some other organisms typically exhibit a pattern of **DNA methylation** in which the DNA has been chemically altered by enzymes that add methyl groups to certain cytosines. (The resulting 5-methylcytosine is still able to base pair with guanine in the usual way.) There is evidence that certain proteins selectively bind to methylated DNA and make it inaccessible to the transcription machinery.

DNA methylation is thought to reinforce gene inactivation rather than to serve as the initial mechanism. It appears that once a gene has been turned off by some other means, DNA methylation ensures that it will remain inactive. For example, the DNA of the inactive X of a female mammal becomes methylated after the chromosome has become a condensed Barr body. Each time the DNA replicates, methylation enzymes perpetuate the preexisting methylation pattern; hence, the DNA continues to be transcriptionally inactive in both daughter cells.

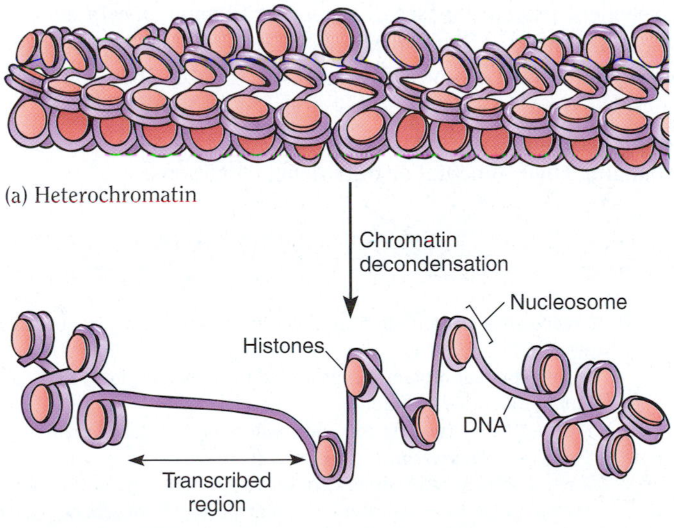

Figure 13–11 Effect of chromatin structure on transcription. **(a)** An inactive region of DNA (heterochromatin) is organized into tightly associated nucleosomes. **(b)** Active genes are found in decondensed chromatin (euchromatin). Chromatin decondensation is often a response to specific inducing signals. The loosely packed chromatin increases the accessibility to RNA polymerase required for transcription of the region. The histones are physically removed from the DNA in the region where transcription occurs.

The long-lived, highly processed mRNAs of eukaryotes provide many opportunities for posttranscriptional control

The half-life of prokaryotic mRNA is usually measured in minutes; eukaryotic mRNA, even when it turns over rapidly, is far more stable. Prokaryotic mRNA is transcribed in a form that can be translated immediately. In contrast, eukaryotic mRNA molecules require further modification and processing before they can be used in protein synthesis. The message is capped, polyadenylated, spliced, and then transported from the nucleus to the cytoplasm to initiate translation (see Chapter 12). These events represent potential control points at which translation of the message and production of its encoded protein can be regulated.

Some pre-mRNAs can be processed in more than one way

Several forms of regulation involving mRNA processing have been discovered. In some instances, the same gene is used to produce one type of protein in one tissue and a related but somewhat different type of protein in another tissue. This is possible because some genes produce pre-mRNA molecules that have multiple splicing patterns; that is, they can be spliced in more than one way depending on the tissue. Typically, such a gene includes at least one segment that can be either an intron or an exon. As an intron, the sequence would be removed, but as an exon it would be retained. Through **differential mRNA processing,** the cells in each tissue can produce their own version of mRNA corresponding to the particular gene (Fig. 13–12). For example, this mechanism allows different forms of troponin, a protein that regulates muscle contraction, to be produced in different muscle tissues.

The stability of mRNA molecules can vary

Controlling the lifetime of a particular kind of mRNA molecule makes it possible to control the number of protein molecules translated from it. In some cases messenger RNA stability is under hormonal control. This is true for mRNA that codes for vitellogenin, a protein synthesized in the livers of certain female animals such as frogs and chickens. Vitellogenin is transported to the oviduct, where it is used in the formation of egg yolk proteins.

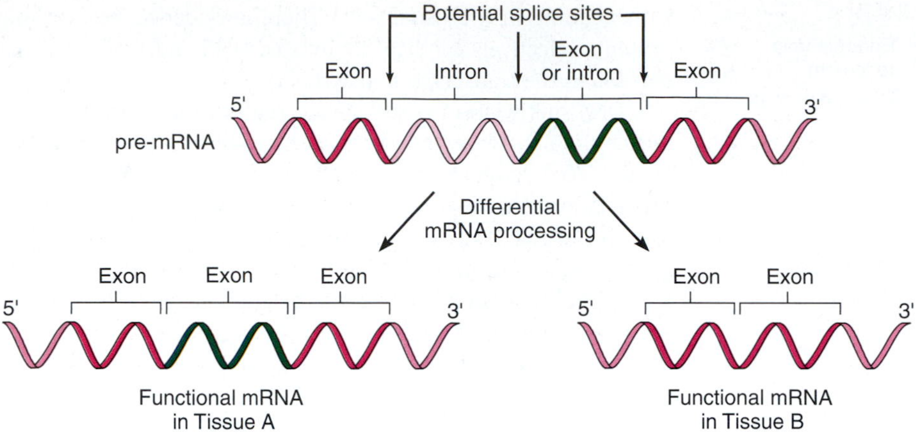

Figure 13–12 Differential mRNA processing. In some cases a complex transcriptional unit can be processed in more than one way to yield two or more mRNAs, each of which encodes a related, but different, protein. In this generalized example the gene contains a segment that can be an exon in tissue A *(left)*, but an intron in tissue B *(right)*.

Vitellogenin synthesis is regulated by the hormone estradiol. When estradiol levels are high, the half-life of vitellogenin mRNA in frog liver is about 500 hours. When cells are deprived of estradiol, the half-life of the mRNA drops rapidly to less than 165 hours. This leads to a rapid decrease in cellular vitellogenin mRNA levels and a decreased synthesis of the vitellogenin protein. In addition to affecting the stability of the mRNA, the hormone seems to control the rate at which the mRNA is synthesized.

The activity of eukaryotic proteins may be altered by posttranslational chemical modifications

The ultimate phenotypic expression of a gene may also be controlled by regulation of the activity of the gene product. As in bacteria, many metabolic pathways in eukaryotes contain allosteric enzymes that are regulated through feedback inhibition. In addition, many eukaryotic proteins are extensively modified after they are synthesized.

In *proteolytic processing* proteins are synthesized as inactive precursors, which are converted to an active form by removal of a portion of the polypeptide chain. For example, proinsulin contains 86 amino acids. The removal of 35 amino acids yields the hormone insulin, which consists of two polypeptide chains containing 30 and 21 amino acids respectively, linked by disulfide bridges. Other proteins may be regulated in part by a process of *selective degradation,* which keeps their numbers constant within the cell.

Chemical modification, through the addition or removal of functional groups, can reversibly alter the activity of an enzyme. One common way of modifying the activity of an enzyme or other protein is the addition or removal of phosphate groups. Enzymes that add phosphate groups are called **kinases;** those that remove them are **phosphatases.** For example, the cyclin-dependent protein kinases discussed in Chapter 9 help control the cell cycle by adding phosphate groups to certain key proteins, causing them to become activated or inactivated. Chemical modifications such as protein phosphorylation also allow the cell to respond rapidly to certain hormones (see Chapter 47), or to fast-changing environmental or nutritional conditions.

SUMMARY WITH KEY TERMS

I. Most regulated genes in bacteria are organized into **operons,** each of which may encode several proteins.
   A. Each operon has a single **promoter** region upstream from the protein-coding regions.
   B. The **operator** is a sequence of bases that overlaps the promoter and serves as the regulatory switch responsible for **transcriptional level control** of the operon.
      1. A **repressor protein** binds specifically to the operator sequence and blocks transcription by preventing RNA polymerase from binding to the promoter.
      2. When the repressor is not bound to the operator, RNA polymerase can bind to the promoter and transcription can proceed.
   C. An **inducible** operon such as the lactose operon is normally turned off.
      1. The repressor protein is synthesized in an active form that binds to the operator.
      2. If lactose is present, it is converted to allolactose (the **inducer**), which binds an allosteric site on the repressor protein, causing it to change shape.
      3. The altered repressor cannot bind to the operator, and the operon is transcribed.
   D. A **repressible** operon such as the tryptophan operon is normally turned on.
      1. The repressor protein is synthesized in an inactive form that cannot bind to the operator.
      2. A metabolite (usually the end product of a metabolic pathway) acts as a **corepressor.**
      3. When intracellular corepressor levels are high, one of the molecules binds to an allosteric site on the repressor, changing its shape so that it can bind to the operator and thereby turn off transcription of the operon.
   E. Repressible and inducible operons are under **negative control.** When the repressor protein binds to the operator, transcription of the operon is turned off.
   F. Some inducible operons are also under **positive control.** A separate protein can bind to the DNA and stimulate transcription of the gene.
      1. The lactose operon is activated by **CAP (catabolite gene activator protein),** which binds to the promoter region, stimulating transcription by binding RNA polymerase tightly.
      2. To bind to the lactose operon, CAP requires **cAMP (cyclic AMP).** Levels of cAMP increase as levels of glucose decrease.

G. A group of operons can be organized into a multigene system, known as a **regulon,** which is controlled by a single regulatory protein. CAP activates a regulon associated with the metabolism of carbohydrates.

II. **Constitutive genes** are neither inducible nor repressible; they are active at all times. Regulatory proteins such as CAP and the repressor proteins are produced constitutively. These proteins work by recognizing and binding to specific base sequences in the DNA. The activity of constitutive genes is controlled by how efficiently RNA polymerase binds to their promoter regions.

III. Some **posttranscriptional controls** operate in prokaryotes.
   A. A **translational control** regulates the rate of translation of a particular mRNA.
   B. **Posttranslational controls** include **feedback inhibition** of key enzymes in some metabolic pathways.

IV. Eukaryotic genes are generally not organized into operons. Regulation of eukaryotic genes can occur at the levels of transcription, mRNA processing, translation, and the protein product.
   A. The promoter of a regulated eukaryotic gene consists of an RNA polymerase-binding site and short DNA sequences known as **upstream promoter elements (UPEs).** The efficiency of the promoter is determined by the number and types of UPEs within the promoter region.
   B. Inducible eukaryotic genes are controlled by **enhancer** elements, which can operate thousands of bases away from the promoter. Proteins that bind to enhancers appear to facilitate the binding of RNA polymerase to the promoter.
   C. Eukaryotic genes are controlled by DNA-binding protein regulators known as **transcription factors.** Many of these are transcriptional activators; others are transcriptional repressors.
   D. The activity of eukaryotic genes is affected by chromosome structure.
      1. Some genes whose products are required in large amounts exist as multiple copies in the chromosome. Other genes may be selectively amplified (i.e., **gene amplification**) by DNA replication in some cells.
      2. Genes can be inactivated by changes in chromosome structure. Densely packed regions of chromosomes called **heterochromatin** contain inactive genes. Active genes are associated with a loosely packed chromatin structure called **euchromatin.**
      3. **DNA methylation** is a mechanism that perpetuates gene inactivation.
   E. Many eukaryotic genes are regulated after the RNA transcript is made.
      1. As a consequence of **differential mRNA processing,** a single gene can produce different forms of a protein in different tissues, depending on how the pre-mRNA is spliced.
      2. Certain regulatory mechanisms increase the stability of mRNA, allowing more protein molecules to be formed per mRNA molecule prior to degradation.
      3. Posttranslational control of eukaryotic genes can occur by feedback inhibition or by modification of the protein structure. The function of a protein can be changed by the addition of phosphate groups by **kinases,** or their removal by **phosphatases.**

POST-TEST

1. Regulation of most prokaryotic genes occurs at the level of (a) transcription (b) translation (c) replication (d) posttranslation (e) postreplication
2. The operator of an operon (a) encodes information for the repressor protein (b) is the binding site for the inducer (c) is the binding site for the repressor protein (d) is the binding site for RNA polymerase (e) encodes the information for the catabolite gene activator protein
3. A mutation that renders the regulatory gene of the lactose operon inactive would result in (a) the continuous transcription of the structural genes (b) no transcription of the structural genes (c) the binding of the repressor to the operator (d) no production of RNA polymerase (e) no difference in the rate of transcription
4. At a time when the lactose operon is actively transcribed (a) the operator is bound to the inducer (b) the lactose repressor is bound to the promoter (c) the operator is not bound to the promoter (d) the gene coding for the repressor is not expressed constitutively (e) the lactose repressor is bound to the inducer
5. A repressible operon codes for the enzymes of the following pathway. Which component of the pathway is most likely to be the corepressor for that operon?

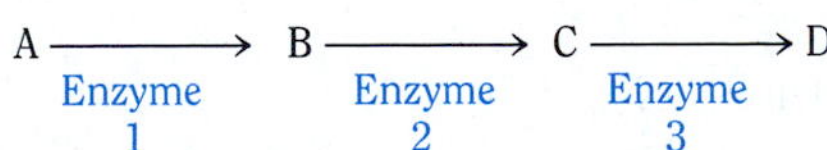

   (a) substance A (b) substance B or C (c) substance D (d) enzyme 1 (e) enzyme 3
6. An mRNA molecule transcribed from the lactose operon contains nucleotide sequences complementary to (a) structural genes coding for enzymes (b) the operator region (c) the promoter region (d) the repressor gene (e) introns
7. Feedback inhibition is an example of control at the __________ level. (a) transcriptional (b) translational (c) posttranslational (d) replicational (e) all of the above
8. Which of the following control mechanisms is generally the most economical in terms of conserving energy and resources? (a) control by means of operons and regulons (b) feedback inhibition (c) selective degradation of mRNA (d) selective degradation of enzymes (e) gene amplification
9. A repressible operon, such as the tryptophan operon, is "off" when (a) the gene that codes for the repressor is expressed constitutively (b) the repressor-corepressor complex binds to the operator (c) the repressor binds to the structural genes (d) the corepressor binds to RNA polymerase (e) CAP binds to the promoter
10. Which of the following is an example of positive control? (a) transcription can occur when a repressor binds to an inducer (b) transcription cannot occur when a repressor binds to a corepressor (c) transcription is stimulated when a transcription activator binds to DNA (d) a and b (e) a and c
11. Which of the following are typically absent in prokaryotes? (a) enhancers (b) proteins that regulate transcription (c) repressors (d) promoters (e) operators
12. The "zipper" of a leucine zipper protein attaches (a) specific amino acids to specific DNA base pairs (b) two polypeptide chains to each other (c) one DNA region to another DNA region (d) amino acids to zinc atoms (e) RNA polymerase to the operator
13. Inactive genes tend to be found in (a) highly condensed chromatin, known as euchromatin (b) decondensed chromatin, known as euchromatin (c) highly condensed chromatin, known as heterochromatin (d) decondensed chromatin, known as heterochromatin (e) chromatin that is not organized as nucleosomes

REVIEW QUESTIONS

1. Make a sketch of the lactose operon and briefly describe its function. Be sure to include the following elements: (a) structural genes, (b) promoter, (c) operator, (d) CAP-binding site.
2. What structural features does the tryptophan operon have in common with the lactose operon? What features are different?
3. Why do we define the tryptophan operon as repressible and the lactose operon as inducible?
4. How is glucose involved in the positive control of the lactose operon? How is CAP similar to the lactose repressor protein? How is it different?
5. Compare the structure of a prokaryotic promoter region with known eukaryotic promoter regions. How does the regulation of inducible eukaryotic genes differ from the regulation of inducible prokaryotic genes?
6. Explain why it is necessary for certain genes in eukaryotic cells to be present in multiple copies.
7. How can the activity of some eukaryotic genes be affected by the structure of the chromosome?
8. Make a sketch illustrating how differential mRNA processing can give rise to different forms of a eukaryotic protein.

YOU MAKE THE CONNECTION

1. Develop a simple hypothesis that would explain the behavior of each of the following types of mutants in *E. coli:*
   (a) *Mutant a:* The map position of this mutation is in the tryptophan operon. The mutant cells are constitutive; that is, they produce all of the enzymes coded for by the tryptophan operon, even if large amounts of tryptophan are present in the growth medium.
   (b) *Mutant b:* The map position of this mutation is in the tryptophan operon. The mutant cells do not produce any of the enzymes coded for by the tryptophan operon under any conditions.
   (c) *Mutant c:* The map position of this mutation is some distance from the tryptophan operon. The mutant cells are constitutive; that is, they produce all of the enzymes coded for by the tryptophan operon, even if the growth medium contains large amounts of tryptophan.
   (d) *Mutant d:* The map position of this mutation is some distance from the tryptophan operon. The mutant cells do not produce any of the enzymes coded for by the tryptophan operon under any conditions.
2. Compare the types of bacterial genes associated with inducible operons, those associated with repressible operons, and those that are constitutive. Predict the category into which each of the following would most likely fit: (a) a gene that codes for RNA polymerase; (b) a gene that codes for an enzyme required to break down maltose; (c) a gene that codes for an enzyme used in the synthesis of adenine.
3. The regulatory gene that codes for the tryptophan repressor is not tightly linked to the tryptophan operon. Would it be advantageous if it were? Explain your answer.

RECOMMENDED READINGS

Hagman, M. "How Chromatin Changes Its Shape." *Science,* Vol. 285, 20 Aug. 1999. This review examines how posttranslational alterations of histone proteins, including phosphorylation and methylation, can affect the structure of chromatin.

Hardison, R. "The Evolution of Hemoglobin" *American Scientist,* Vol. 87, Mar.-Apr. 1999. Various forms of hemoglobin differ little in their amino acid sequences but are regulated quite differently.

Pennisi, E. "Chemical Shackles for Genes?" *Science,* Vol. 273, 2 Aug. 1996. DNA methylation is involved in the regulation of gene activity in plants as well as vertebrates.

Stein, G.S., J.L. Stein, A.J. van Wijnen, and J.B. Lian. "The Maturation of a Cell." *American Scientist,* Vol. 84, Jan.-Feb. 1996. The authors describe the activation of genes in the maturation of a bone cell.

- Visit our Web site at **http://www.info.brookscole.com/solomonbergmartin** for links to chapter-related resources on the World Wide Web. Additional on-line materials relating to this chapter can also be found on our Web site.

See chapter activity on BioActive Learner CD for additional help in mastering the chapter's material. Icon location in the chapter's margins shows which topics have tutorials or simulations in the CD.

14

Genetic Engineering

Bioreactor. This complex apparatus is an automated bioreactor that creates an optimal environment for genetically engineered bacteria to produce useful proteins. *(Rosenfeld Images Ltd/Science Photo Library/Photo Researchers, Inc.)*

LEARNING OBJECTIVES

After you have studied this chapter you should be able to

1. Draw a sketch that demonstrates how a typical restriction enzyme cuts DNA molecules and give examples of the ways in which these enzymes are used in recombinant DNA technology.
2. Summarize the properties of plasmids that allow them to be used as DNA cloning vectors.
3. Distinguish between a genomic DNA library and a complementary DNA (cDNA) library.
4. Explain why one would clone the same eukaryotic gene from both a genomic library and a cDNA library.
5. Identify some of the uses of DNA hybridization probes.
6. Describe how specific primers can be used to amplify specific genes from a mixture of genomic DNA or cDNA.
7. Draw a diagram that illustrates the most widely used DNA sequencing technique.
8. List some important proteins and other products that can be produced by genetic engineering techniques.
9. List some of the difficulties encountered in using *Escherichia coli* to produce proteins coded by eukaryotic genes and explain the rationale behind using transgenic plants and animals to solve some of those problems.

Beginning in the mid-1970s, a revolution in the field of biology occurred as the development of new ways of studying deoxyribonucleic acid (DNA) led to radically new research approaches. These techniques have had a major impact not only on genetic studies but also in areas ranging from cell biology to evolution, as well as on society.

This chapter begins with a consideration of **recombinant DNA technology,** in which DNA from different organisms is spliced together in the laboratory. The primary goal of this technology is to allow scientists to obtain a great many copies of any specific DNA segment to study it biochemically. Recombinant DNA technology allows scientists to introduce foreign DNA into the cells of microorganisms. Under the right conditions, this DNA is replicated and transmitted to the daughter cells when a cell divides. In this way a particular DNA sequence can be amplified, or **cloned,** to provide millions of identical copies that can be isolated in pure form. Today these methods have been supplemented by extremely valuable techniques that permit the cloning of DNA **in vitro** (outside of a living organism). Because new recombinant DNA methods are continually emerging, we will not attempt to explore them all. (Several Internet sites, such as "Access Excellence" and the "National Center for Biotechnology Information," are good sources of updated information.) Instead, we will discuss some of the major approaches that have provided a foundation for the technology.

We also consider the ways in which studies of cloned **DNA sequences** have been of immense value in allowing scientists to understand the organization of genes and the relationship between genes and their products. In fact, most of our knowledge of the complex structure and control of eukaryotic genes (see Chapters 12 and 13) and of the roles of genes in development (see Chapter 16) is derived from the application of these methods.

This chapter also explores the many practical applications of recombinant DNA technology. One of the rapidly advancing areas of study is **genetic engineering**—the modification of the DNA of an organism to produce new genes with new characteristics. Genetic engineering can take many forms, ranging from basic research, to the production of strains of bacteria that manufacture useful protein products, to the development of plants and animals that express foreign genes. This wide range of applications of ongoing discoveries in molecular genetics is causing a transformation of our view of **biotechnology,** the use of organisms to benefit humanity. Traditional forms of biotechnology include such familiar examples as the selective breeding of plants and the use of yeast to make alcoholic beverages or cause bread to rise. However, the examples of biotechnology most frequently cited today are applications of genetic engineering in such diverse areas as medicine and the pharmaceutical industries, foods and agriculture, and others. Biotechnology does not stand alone; its advances are greatly facilitated by other kinds of technology, including powerful computer programs and automated systems *(see photograph).*

RECOMBINANT DNA METHODS GREW OUT OF RESEARCH IN MICROBIAL GENETICS

Recombinant DNA technology was not developed quickly. It actually had its roots in the 1940s with genetic studies of bacteria and **bacteriophages** (literally "bacteria eaters"), the viruses that infect them (see Chapters 11 and 23). In the mid-1970s, after decades of basic research and the accumulation of extensive knowledge, the technology became feasible and available to the many scientists who now use these methods.

In recombinant DNA technology, special enzymes from bacteria, known as **restriction enzymes,** are used to cut DNA molecules only in specific places. Restriction enzymes allow researchers to reproducibly cut DNA into manageable segments.

Each fragment is then incorporated into a suitable **vector** molecule, a carrier capable of transporting it into a cell. Either bacteriophages or special DNA molecules called plasmids are commonly used as vectors. Recall from Chapter 11 that bacterial DNA is in the form of a circle; a **plasmid** is a separate, much smaller, circular DNA molecule that may be present and able to replicate inside a bacterial cell, typically *E. coli.* Plasmids can be introduced into bacterial cells by a method called transformation (see Chapter 11). For **transformation,** the uptake of foreign DNA by cells, to be efficient, researchers must alter the bacterial cell walls to make them permeable to the plasmid DNA molecules. Once a plasmid enters a cell, it is replicated and distributed to the daughter cells during cell division. When a recombinant plasmid (one that has foreign DNA spliced into it) replicates in this way, many copies of the foreign DNA are made (i.e., the foreign DNA is cloned).

Restriction enzymes are "molecular scissors"

The discovery of restriction enzymes was a major breakthrough in the development of recombinant DNA technology. Today large numbers of different types of restriction enzymes, each with its own characteristics, are readily available to researchers. For example, a restriction enzyme known as Hind III recognizes and cuts a DNA molecule at the restriction site 5'—AAGCTT—3', whereas the sequence 5'—GAATTC—3' is cut by another, known as EcoRI.[1] Why do bacteria produce such enzymes? Recall from Chapter 11 that during infection a bacteriophage injects its DNA into a bacterial cell. Such a cell can defend itself if it possesses restriction enzymes capable of attacking the bacteriophage DNA. The bacteria protect their own DNA from breakdown by modifying it after replication. An enzyme adds a methyl group to one or more bases in each restriction site so that the restriction enzyme is unable to recognize and cut the bacterial DNA.

Restriction enzymes enable scientists to cut DNA from chromosomes into shorter fragments in a controlled way. Many of the restriction enzymes used for recombinant DNA studies cut **palindromic** sequences, which means that the base sequence of one strand reads the same as its complement, but in the opposite direction. (Thus, the complement of our example, 5'—AAGCTT—3', reads 3'—TTCGAA—5'.) By cutting both strands of the DNA, but in a staggered fashion, these enzymes produce fragments with identical, complementary, single-stranded ends.

[1] The names of restriction enzymes are generally derived from the names of the bacteria from which they were originally isolated. Hence Hind III and EcoRI are derived from *Hemophilus influenzae* and *Escherichia coli,* respectively.

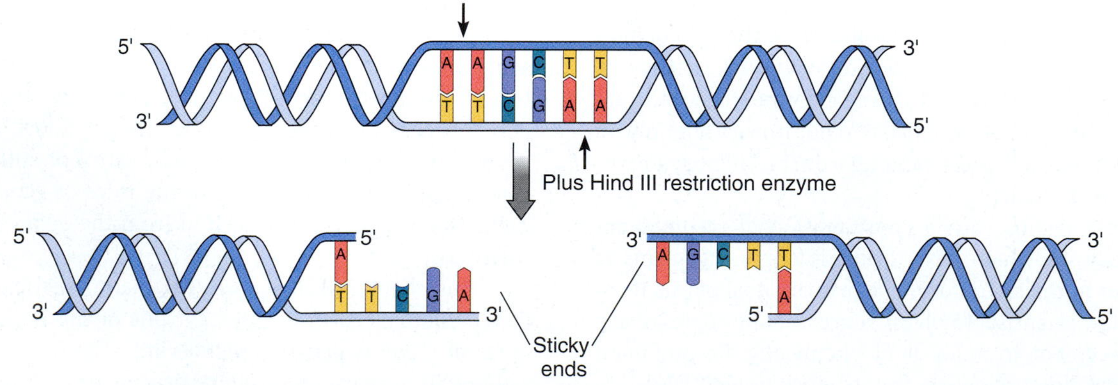

Figure 14–1 Cutting DNA with a restriction enzyme. Many restriction enzymes, like Hind III, cut DNA at sequences that are palindromic, thereby producing complementary sticky ends.

These ends are called "sticky ends" because they can pair (by hydrogen bonding) with the complementary, single-stranded ends of other DNA molecules that have been cut with the same enzyme (Fig. 14–1). Once two molecules have been joined together in this way, they can be treated with **DNA ligase** (see Chapter 11), an enzyme that covalently links the two DNA fragments to form a stable recombinant DNA molecule.

Restriction enzymes vary widely in the number of DNA bases that they recognize, ranging from as few as 4 to as many as 23 bases. If the restriction sites are randomly distributed in the DNA, we expect the restriction sequence of a "four-base cutter" to occur on the average of every 4^4, or 256, bases. A four-base cutter would therefore produce fragments with an average length of 256 bases, whereas a six-base cutter would produce fragments averaging 4^6, or 4096, bases.

Recombinant DNA is formed when DNA is spliced into a vector

In recombinant DNA technology, foreign DNA and plasmid DNA are both cut with the same restriction enzyme. The two types of DNA are then mixed together under conditions that facilitate hydrogen bonding between the complementary bases of the sticky ends, and the resulting recombinant DNA is stabilized by DNA ligase (Fig. 14–2).

The plasmids now used in recombinant DNA work have been extensively "engineered" in the laboratory to include features helpful in the isolation and analysis of cloned DNA (Fig. 14–3). Among these are (1) an origin of replication (see Chapter 11), (2) one or more restriction sites, and (3) genes that allow researchers to select cells that have been transformed by recombinant plasmids. These are genes that permit transformed cells to grow under specified conditions that do not allow growth of untransformed cells. In this way the researchers are making use of features that are also commonly found in naturally occurring plasmids. Typically, plasmids do not contain genes that are essential to the *E. coli* cells under normal conditions but often carry genes that are useful under specific environmental conditions, such as those that confer resistance to particular antibiotics or allow the cells to use a particular nutrient. For example, cells transformed with a plasmid that includes a gene for resistance to the antibiotic tetracycline can grow in a medium that contains tetracycline, but untransformed cells cannot.

A limiting property of any vector, however, is the size of the DNA fragment that it can effectively carry. The size of a DNA segment is often given in kilobases, with **1 kilobase (kb)** being equal to 1000 base pairs. Fragments smaller than 10 kb can usually be inserted into plasmids for use in *E. coli*. However, larger fragments require the use of bacteriophage vectors, which can handle up to 15 kb of DNA.

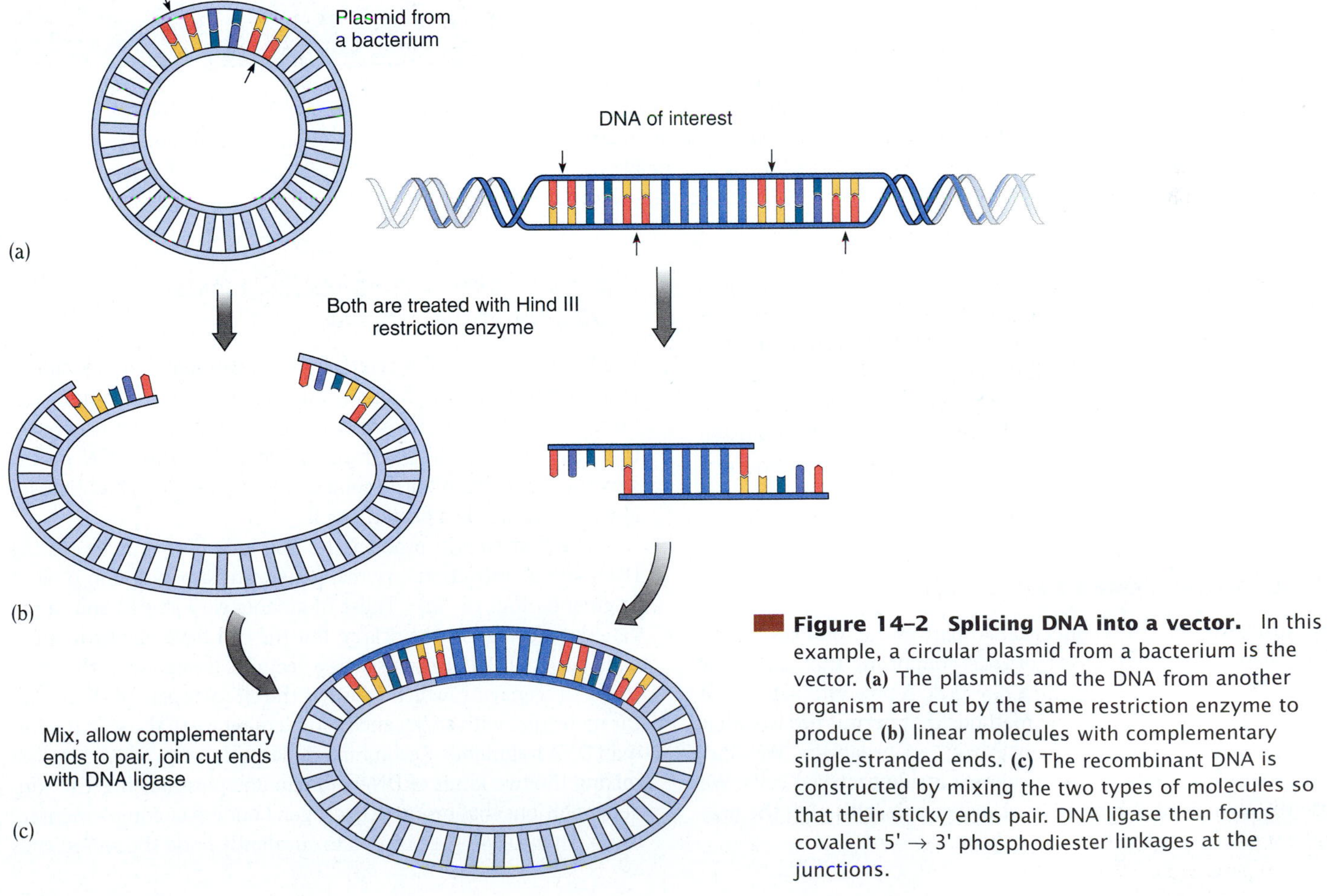
Plasmid from
a bacterium
DNA of interest
(a)
Both are treated with Hind III
restriction enzyme
(b)
Mix, allow complementary
ends to pair, join cut ends
with DNA ligase
(c)

Figure 14–2 Splicing DNA into a vector. In this example, a circular plasmid from a bacterium is the vector. **(a)** The plasmids and the DNA from another organism are cut by the same restriction enzyme to produce **(b)** linear molecules with complementary single-stranded ends. **(c)** The recombinant DNA is constructed by mixing the two types of molecules so that their sticky ends pair. DNA ligase then forms covalent 5' → 3' phosphodiester linkages at the junctions.

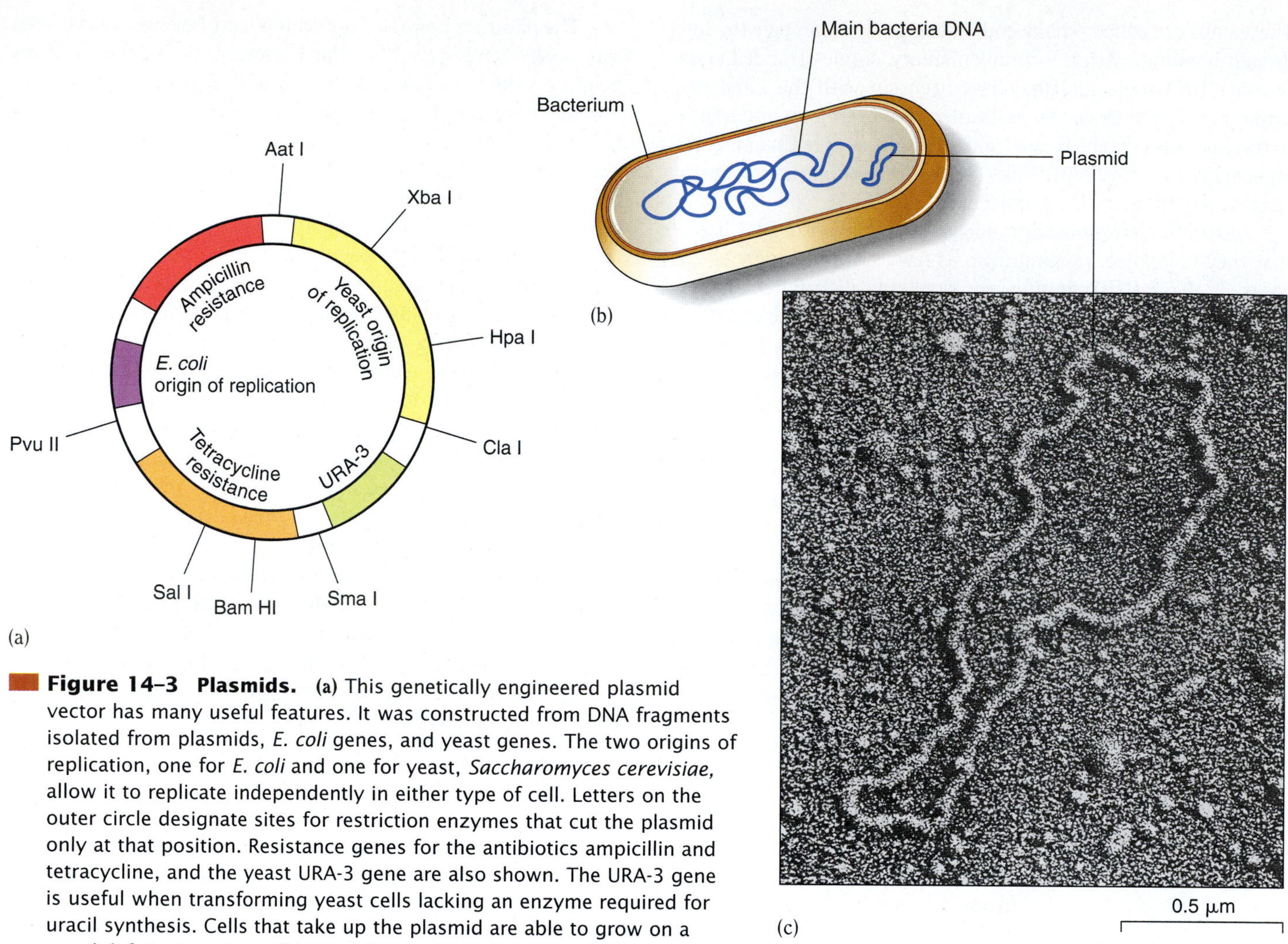

Figure 14–3 Plasmids. **(a)** This genetically engineered plasmid vector has many useful features. It was constructed from DNA fragments isolated from plasmids, *E. coli* genes, and yeast genes. The two origins of replication, one for *E. coli* and one for yeast, *Saccharomyces cerevisiae,* allow it to replicate independently in either type of cell. Letters on the outer circle designate sites for restriction enzymes that cut the plasmid only at that position. Resistance genes for the antibiotics ampicillin and tetracycline, and the yeast URA-3 gene are also shown. The URA-3 gene is useful when transforming yeast cells lacking an enzyme required for uracil synthesis. Cells that take up the plasmid are able to grow on a uracil-deficient medium. **(b)** The relative sizes of a plasmid and the main DNA of a bacterium. **(c)** TEM of a plasmid from *E. coli.* *(c, Dr. Stanley Cohen/Science Photo Library/Photo Researchers, Inc.)*

Recombinant DNA can also be introduced into cells of more complex organisms. For example, engineered viruses are used as vectors in mammalian cells. These viruses have been disabled in such a way that they do not kill the cells they infect; instead their DNA, and any foreign DNA they carry, becomes incorporated into the chromosomes of the cell following infection. As discussed later, other methods have been developed that do not require a biological vector.

DNA can be cloned inside cells

Because a single gene is only a small part of the total DNA in an organism, isolating the piece of DNA containing that particular gene is like finding a needle in a haystack. A powerful detector is needed. Today there are many methods that permit the isolation of a specific nucleotide sequence from an organism. We start with methods in which DNA is cloned inside bacterial cells. We use the cloning of human DNA as an example, although the procedure can be applied to any organism.

A genomic library contains fragments of all DNA in the genome

The total DNA per cell is referred to as a **genome.** For example, if DNA is extracted from human cells, we refer to it as human genomic DNA. A **genomic library** is a collection of DNA fragments that are more or less representative of all the DNA in the genome. Each fragment is spliced into a plasmid, which is usually inserted inside a bacterial cell.

The first step in producing a genomic library is to cut the DNA with a restriction enzyme, generating a population of DNA fragments (Fig. 14–4*a*) . These fragments vary in size and in the genetic information they carry, but they all have identical sticky ends. Plasmid DNA to be used as a vector is treated with the same restriction enzyme, which converts the circular plasmids into linear molecules with sticky ends complementary to those of the human DNA fragments. Recombinant plasmids are produced by first mixing the two kinds of DNA (human and plasmid) together under conditions that promote hydrogen bonding of complementary bases. Then DNA ligase is used to covalently bond the paired ends

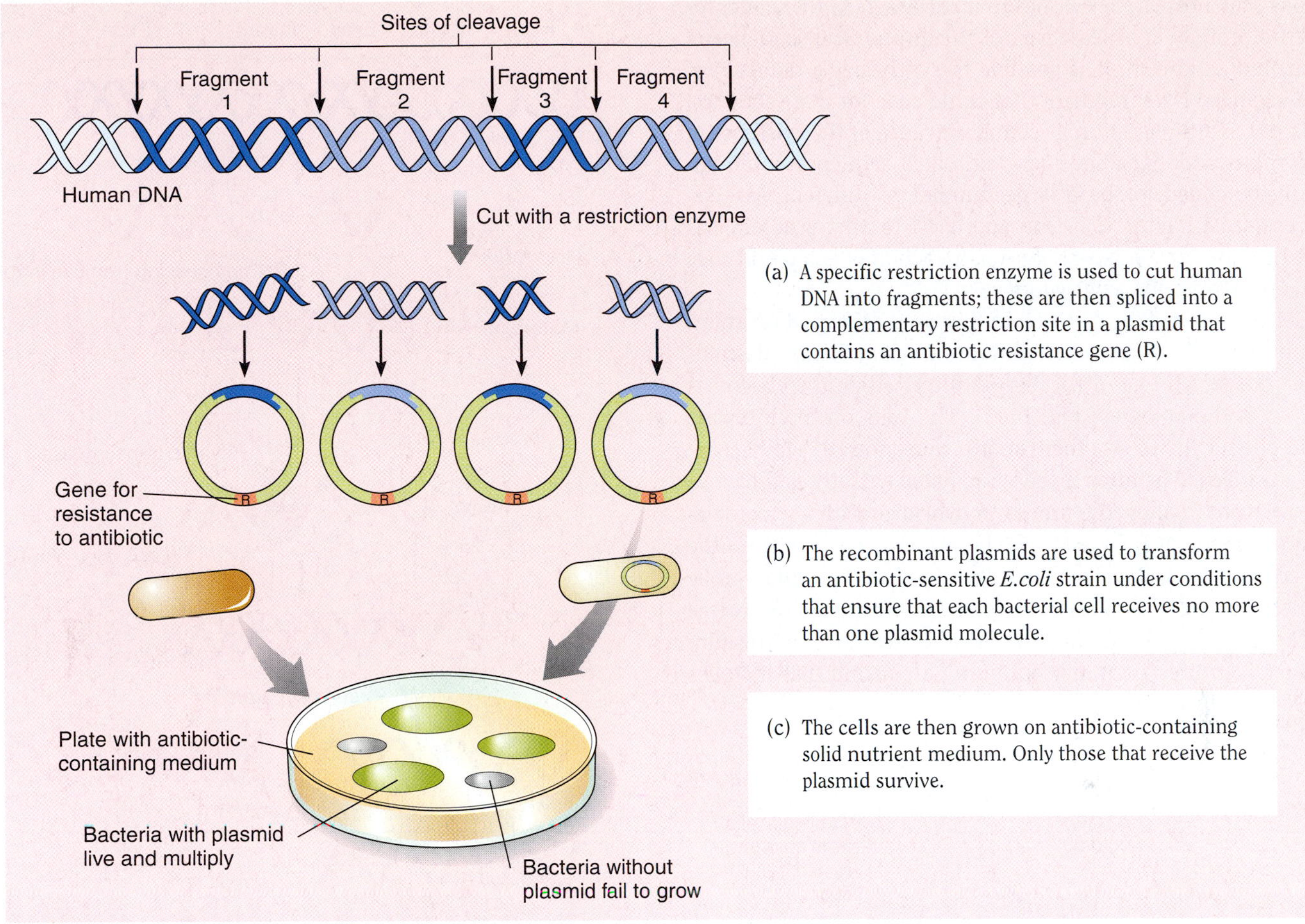

Sites of cleavage
Fragment 1
Fragment 2
Fragment 3
Fragment 4
Human DNA
Cut with a restriction enzyme
(a) A specific restriction enzyme is used to cut human DNA into fragments; these are then spliced into a complementary restriction site in a plasmid that contains an antibiotic resistance gene (R).
Gene for resistance to antibiotic
R
(b) The recombinant plasmids are used to transform an antibiotic-sensitive *E.coli* strain under conditions that ensure that each bacterial cell receives no more than one plasmid molecule.
Plate with antibiotic-containing medium
(c) The cells are then grown on antibiotic-containing solid nutrient medium. Only those that receive the plasmid survive.
Bacteria with plasmid live and multiply
Bacteria without plasmid fail to grow

Figure 14–4 Cloning genomic DNA. Only a small part of one chromosome is shown. A great many more DNA fragments would be produced from an entire genome.

of the plasmid and human DNA. Unavoidably, nonrecombinant plasmids are also formed because some plasmids revert to their original circular form without incorporating foreign DNA.

The plasmids are inserted into antibiotic-sensitive *E. coli* cells by transformation (Fig. 14–4*b*). Because the ratio of plasmids to cells is kept very low, it is rare for a cell to receive more than one plasmid molecule, and not all cells receive a plasmid. The normally antibiotic-sensitive cells are incubated on a nutrient medium that includes antibiotics, so only cells that have incorporated a plasmid (which contains a gene for antibiotic resistance) are able to grow (Fig. 14–4*c*). In addition, the plasmid has usually been engineered in ways that permit researchers to select only those cells containing *recombinant* plasmids.

A genomic library contains redundancies; that is, certain human DNA sequences have been inserted into plasmids more than once, purely by chance. However, each individual recombinant plasmid (analogous to a book in the library) contains only a single fragment of the total human genome. Each of these fragments is usually smaller than a gene; therefore, several fragments must be isolated to study the complete gene.

To allow identification of a plasmid containing a sequence of interest, each plasmid must be amplified, or cloned, until there are millions of copies to work with. This process occurs as the *E. coli* cells grow and divide. A dilute sample of the bacterial culture is spread on solid growth medium, so that the cells will be widely separated. Each cell divides many times, giving rise to a visible **colony,** which is a clone of genetically identical cells. All the cells of a particular colony contain the same recombinant plasmid, so during this process a specific sequence of human DNA is also cloned . The major task is to determine which colony (out of thousands) contains a cloned fragment of interest. There are many ways in which specific DNA sequences can be identified.

A specific DNA sequence can be detected by a complementary genetic probe

A common approach to the problem of detecting the DNA of interest involves the use of a **genetic probe,** which is usually a radioactively labeled segment of ribonucleic acid (RNA) or single-stranded DNA that can **hybridize** (become attached by base pairing) to complementary base sequences in the target gene.

Suppose that a researcher wishes to identify a gene that codes for a specific protein. If at least part of the amino acid sequence of that protein is known, it is possible to synthesize a radioactive, single-stranded DNA fragment that could code for that sequence. This is not as simple as it may sound: Because of the existence of synonymous codons, a specific amino acid sequence could potentially be coded for by a large number of different base sequences (see Chapter 12). One approach to this problem has been to synthesize a mixture of probes, each of which could code for the desired amino acid sequence.

Genetic probes can be used in a variety of ways. For example, cells from *E. coli* colonies containing recombinant plasmids can be transferred to a nitrocellulose filter, which then becomes a *replica* of the colonies (Fig. 14–5). The cells on the filter are treated chemically to lyse them and to cause the DNA to become single-stranded. The filter is then incubated with the radioactive probe mixture to allow the probes to hybridize with any complementary strands of DNA that may be present. Each spot on the filter containing DNA complementary to that particular probe becomes radioactive and can be detected by autoradiography (see Chapter 2), using a special x-ray film. Each spot on the film therefore identifies a colony containing a plasmid that includes the DNA of interest.

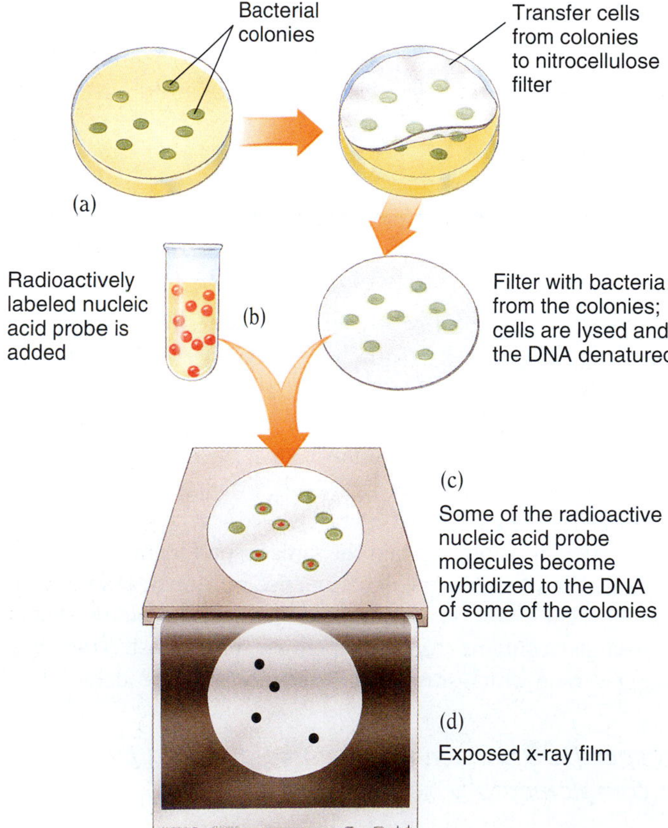

Figure 14–5 Use of a genetic probe. A radioactive nucleic acid probe (which can be either RNA or single-stranded DNA) reveals the presence of complementary sequences of DNA.

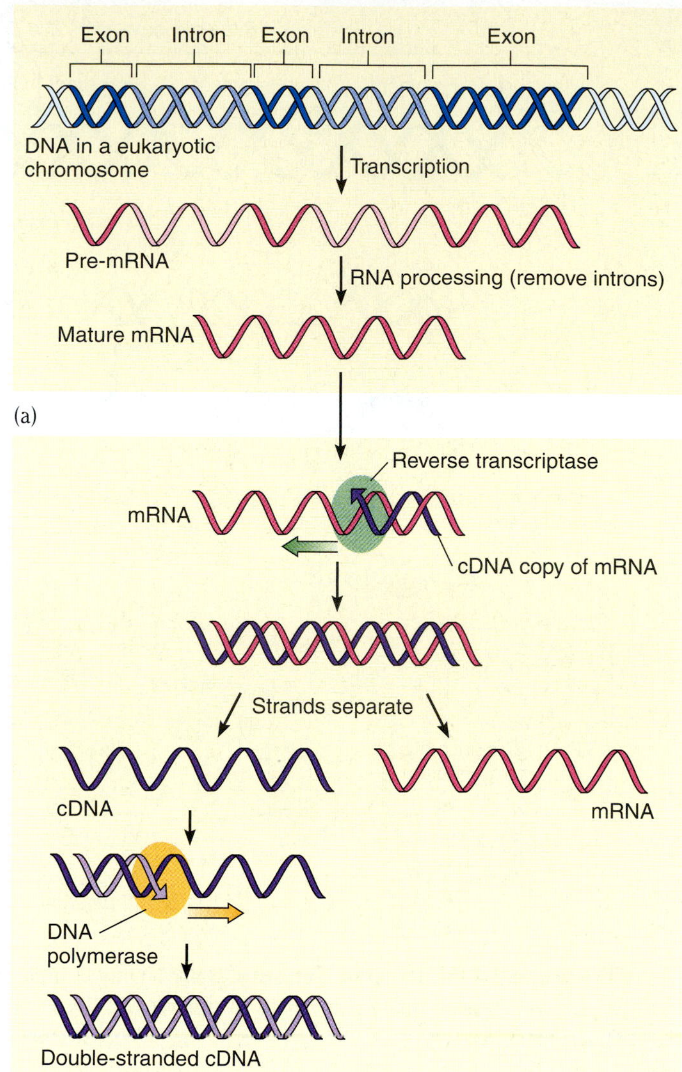

Figure 14–6 Formation of cDNA. **(a)** RNA processing occurs in the nucleus to form mature mRNA. **(b)** Researchers isolate mature mRNA and use the reverse transcriptase enzyme to produce single-stranded cDNA complementary to it. DNA polymerase is used to synthesize double-stranded DNA.

A cDNA library is complementary to mRNA and does not contain introns

For reasons that are discussed later, researchers frequently wish to avoid cloning introns and other parts of eukaryotic genes that do not directly code for proteins. They also may wish to clone only genes that are expressed in a particular cell type. In such cases they construct libraries consisting of DNA copies of mature messenger RNA (mRNA) from which introns have been removed. The copies, known as **complementary DNA (cDNA)** because they are complementary to RNA, also lack introns. **Reverse transcriptase** (see Chapter 12) is used to synthesize single-stranded cDNA, which is then separated from the mRNA and made double-stranded by DNA polymerase (Fig. 14–6).

A **cDNA library** is formed using mRNA from a single cell type as the starting material. The double-stranded cDNA molecules are inserted into plasmid or virus vectors, which then multiply in *E. coli* cells.

Cloning a gene from both a cDNA library and a genomic library has several advantages. Analysis of the genomic DNA clones gives useful information about the structure of the gene in the chromosome and the structure of the pre-mRNA transcript, as well as nontranscribed regulatory regions.

Analysis of cDNA clones allows investigators to determine certain characteristics of the protein encoded by the gene, including its exact amino acid sequence. The structure of the mature mRNA can also be studied. Furthermore, because the cDNA copy of the mRNA does not contain intron sequences, comparison of the cDNA and genomic DNA base sequences reveals the locations of intron and exon coding sequences in the gene.

Cloned cDNA sequences are also useful when it is desirable to produce a eukaryotic protein in *E. coli.* When an intron-containing human gene such as the gene for human growth hormone is introduced into *E. coli,* the bacterium is unable to remove the introns from the transcribed RNA to make a functional mRNA for the production of its protein product. If a cDNA clone of the gene is inserted into the bacterium, however, its transcript contains an uninterrupted coding region. A functional protein can be synthesized if the gene is inserted downstream of an appropriate bacterial promoter.

The polymerase chain reaction is a technique for amplifying DNA in vitro

The methods to amplify a specific DNA sequence described above all involve cloning DNA in cells, usually those of bacteria. These processes are time-consuming and require an adequate DNA sample as starting material. The **polymerase chain reaction (PCR)** technique allows researchers to amplify a tiny sample of DNA millions of times in a few hours (Fig. 14–7).

In PCR, DNA polymerase uses nucleotides and primers to replicate a DNA sequence in vitro, thereby producing two DNA molecules. The two strands of each molecule are then denatured (separated by heating) and replicated again, so then there are four double-stranded molecules. After the next cycle of heating and replication there are eight molecules, and so on, with the number of DNA molecules doubling in each cycle. After only 20 heating and cooling cycles (which are carried out using automated equipment) this exponential process yields 2^{20}, or more than 1 million, copies of the target sequence!

Because the reaction can only be carried out efficiently if the DNA polymerase can remain stable through many heating cycles, a special heat-resistant DNA polymerase, known as *Taq* polymerase, is used. The name of this enzyme reflects its source, *Thermus aquaticus,* a bacterium that lives in hot springs in Yellowstone Park. (Similar enzymes can be found in bacteria living in deep-sea thermal vents; see *Focus On: Life Without the Sun,* Chapter 53.)

The PCR technique is particularly valuable because only specific *target sequences* are replicated. Recall from Chapter 11 that DNA polymerase can add nucleotides only to a preexisting polynucleotide strand. In the cell, DNA synthesis begins with the formation of a short RNA primer, which is then extended by DNA polymerase. In the PCR technique, chemically synthesized short DNA single-stranded molecules with a specified nucleotide sequence are included in the reaction mixture. These attach to complementary target sequences of the single-stranded DNA and act as primers, thereby designating the starting point for replication by DNA polymerase. In this way, a specific sequence can be cloned from an unpurified mixture of DNA sequences.

The PCR technique has virtually limitless applications. It allows the amplification and analysis of tiny DNA samples from seemingly unlikely sources, ranging from crime scenes to archaeological remains. For example, in 1997 the first analysis of mitochondrial DNA obtained from the bones of Neandertals was reported (see Chapter 21).

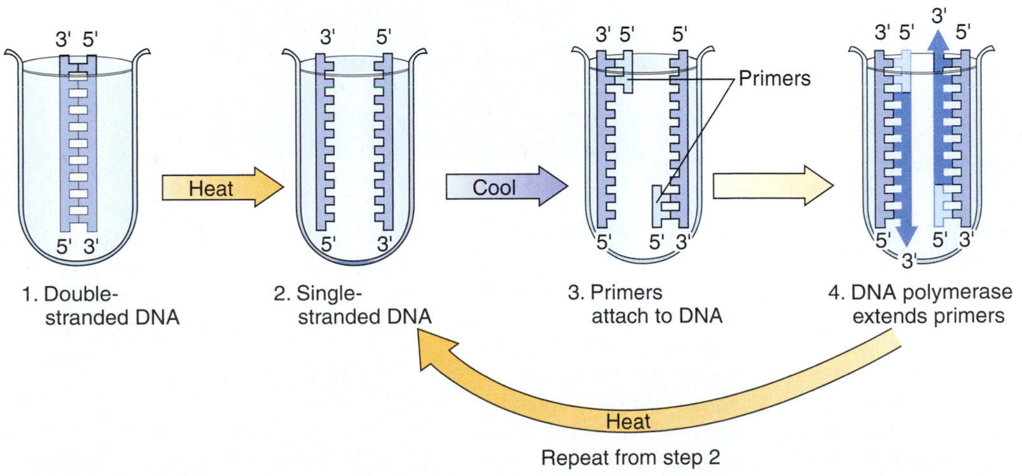

Figure 14–7 The polymerase chain reaction. (1) The initial reaction mixture includes a very small amount of double-stranded DNA *(shown)*, DNA precursors (deoxyribonucleotides), specific nucleic acid primers, and heat-resistant *Taq* DNA polymerase. (2) The DNA is denatured (separated into single strands) by heat. (3) Each DNA strand acts as a template for DNA synthesis catalyzed by the *Taq* DNA polymerase. (4) The number of double-stranded DNA molecules doubles each time the cycle of heating and cooling is repeated.

If the PCR technique has a flaw, it is the fact that it is almost too sensitive. Even a tiny amount of contaminant DNA in a sample could become amplified if it includes a DNA sequence complementary to the primer, potentially leading to an erroneous conclusion. Researchers are constantly improving their methods to avoid this and other technical pitfalls.

Gel electrophoresis is the most widely used technique to separate macromolecules

Mixtures of certain macromolecules such as polypeptides, DNA fragments, or RNA can be separated by **gel electrophoresis,** a method that exploits the fact that these molecules carry charged groups that cause them to migrate in an electrical field. Figure 14–8 illustrates gel electrophoresis of DNA molecules. Both DNA and RNA migrate through the gel toward the positive pole of the electrical field because they are negatively charged due to their phosphate groups (see Chapters 11 and 12). Because the gel retards the movement of the large molecules more than the small molecules, the rate at which they travel is inversely proportional to their length (molecular weight). Including DNA fragments of known size as standards allows accurate measurement of the molecular weights of the unknown fragments.

The DNA fragments can be identified if they hybridize with a complementary genetic probe. However, it can be very cumbersome to work with DNA fragments contained in a gel. For this reason the DNA is usually denatured and then transferred to a nitrocellulose filter, which picks up the DNA much as a blotter picks up ink. The resulting "blot," which is essentially a replica of the gel, is incubated with a radioactive genetic probe, which hybridizes with any complementary DNA fragments. It is then used for autoradiography (see Chapter 2). The resulting spots on the x-ray film correspond to the locations of the fragments in the gel that are complementary to the probe. This type of **blot hybridization** (called a **Southern blot** after its inventor, E.M. Southern) has widespread applications. It is often used to diagnose certain types of genetic disorders. For example, in some cases the DNA of a mutant gene can be detected because it migrates differently in the gel than the DNA of its normal counterpart.

Similar blotting techniques are used to study RNA and proteins. When RNA molecules separated by electrophoresis are transferred to a membrane, the result is, rather in jest, called a **Northern blot.** In the same spirit, the term **Western blot** is applied to a blot consisting of polypeptides previously separated by gel electrophoresis. (So far, no one has invented a type of blot that could be called an "Eastern blot.") In the case of Western blotting,

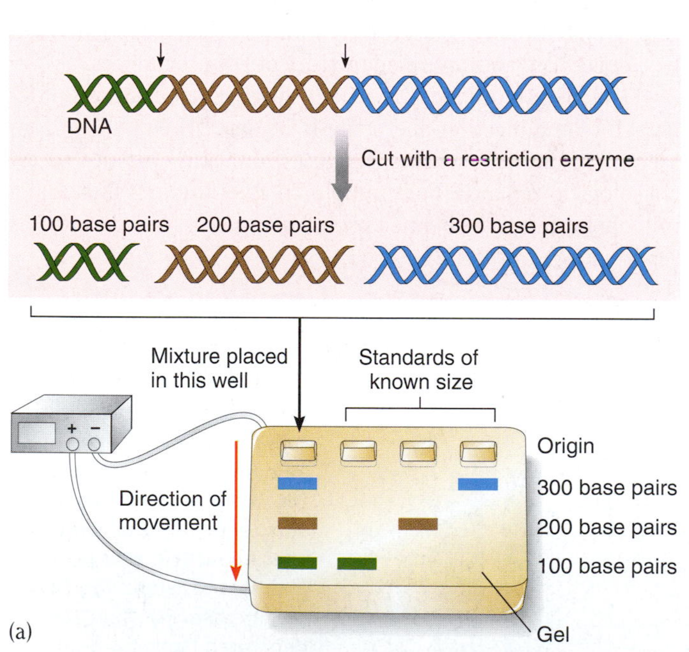

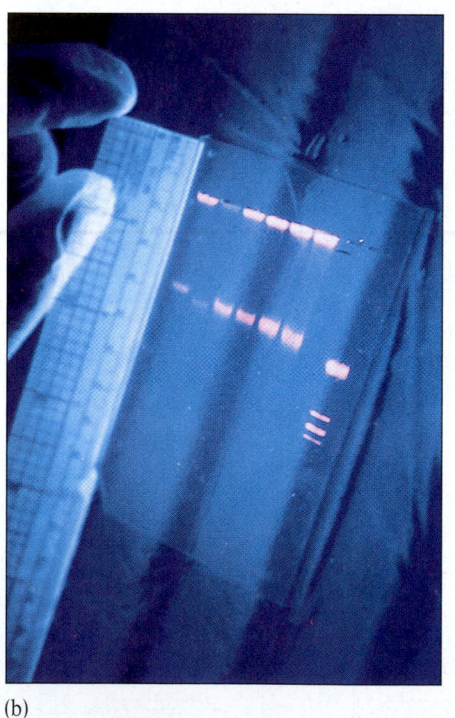

Figure 14–8 Gel electrophoresis. Charged molecules, such as DNA, RNA, or protein, can be separated based on the rate at which they migrate in an electrical field. **(a)** An electrical field is set up in a gel material, consisting of agarose or polyacrylamide, which is poured as a thin slab on a glass or Plexiglas holder. After the gel has solidified, samples containing a mixture of macromolecules of different sizes are loaded in wells formed at one end of the gel, and then an electrical current is applied. The smallest DNA fragments *(green)* travel the longest distance. **(b)** A gel containing separated DNA fragments. The gel is stained with ethidium bromide, a dye that binds to DNA and is fluorescent under UV light. *(b, Michael Gabridge/Visuals Unlimited)*

the polypeptides of interest are recognized by radioactive antibody molecules that bind to them specifically. For example, Western blotting is used diagnostically to detect the presence of proteins specific to human immunodeficiency virus-1 (HIV-1), the virus that causes acquired immunodeficiency syndrome (AIDS).

A great deal of information can be inferred from a DNA nucleotide sequence

A cloned piece of DNA can be used as a research tool for a wide variety of applications. Even if the purpose of cloning the gene is to obtain the encoded protein for some industrial or pharmaceutical process, a great deal must be known about the gene and how it functions before it can be engineered for a particular application. The usual first step is to determine the sequence of nucleotides.

The most commonly used method of DNA sequencing is based on the fact that a replicating DNA strand that has incorporated a modified synthetic nucleotide, known as a *dideoxynucleotide,* cannot elongate beyond that point. Unlike a "normal" deoxynucleotide (which lacks a hydroxyl group on its 2' carbon), a dideoxynucleotide also lacks a hydroxyl group on its 3' carbon (Fig. 14–9*a*). (Recall from Chapter 11 that a 3' hydroxyl group is needed to react each time a phosphodiester linkage is formed.) Thus, dideoxynucleotides terminate elongation during DNA replication.

Four different reaction mixtures are prepared. Each contains multiple single-stranded copies of the DNA to be sequenced, DNA polymerase, appropriate radioactively labeled primers, and all four deoxynucleotides needed to synthesize DNA: dATP, dCTP, dGTP, and dTTP. Each also includes a small amount of only one of the four dideoxynucleotides: ddATP, ddCTP, ddGTP, or ddTTP[2] (Fig. 14–9*b*).

For example, consider how the reaction proceeds in the mixture that includes ddATP. At each site where adenine is specified, occasionally a growing strand will incorporate a ddATP and will be unable to elongate further. Consequently, a mixture of DNA fragments of varying lengths is formed in the reaction mixture. Each fragment that contains a ddATP marks a specific location where adenine would be normally found in the newly synthesized strand. Similarly, in the reaction mixture that includes ddCTP, each fragment that contains ddCTP marks the position of a cytosine in the newly synthesized strand, and so on (Fig. 14–9*c*).

The radioactive fragments from each reaction are denatured and then separated by gel electrophoresis, with each reaction mixture (corresponding to A, T, G, or C) occupying its own lane in the gel. The positions of the newly synthesized fragments in the gel can then be determined by autoradiography (Fig. 14–9*d* and *e*). Because the high resolution of the gel makes it possible to distinguish between fragments that differ in length by only a single nucleotide, one can read off the sequence in the newly synthesized DNA one base at a time, beginning with the shortest fragment. To follow the example in Figure 14–9*d,* if the shortest fragment is in the "G" lane, then the first base is G; similarly, if the next shortest fragment is also in the G lane, then the next base is also G; if the third shortest fragment is in the "A" lane, then the third base is A, and so on. The entire sequence in the figure is 5'—GGAGCATAGCAT—3'. Of course, we are actually interested in the sequence of the original strand that served as the template, which is 3'—CCTCGTATCGTA—5'.

Knowing the DNA sequence of a cloned gene allows investigators to identify which parts of the DNA molecule contain the actual protein-coding sequences, as well as which parts may be regulatory regions involved in gene expression (see Chapter 13). Signals involved in mRNA processing and modification can be recognized, and the amino acid sequence of the encoded protein can be inferred directly from the base sequence. Prior to the development of DNA-sequencing methods, protein sequences were determined by laborious methods from highly purified protein samples. Although protein microsequencing technology has also advanced rapidly, in most cases cloning and sequencing a gene are easier than purifying and sequencing the encoded protein.

Advances in sequencing technology have made it possible for researchers to study the nucleotide sequences of a wide variety of organisms, both prokaryotic and eukaryotic. Much of this research received its initial impetus in conjunction with the Human Genome Project (see Chapter 15). The sequence of the 3 billion base pairs of the human genome was essentially completed in 2001. The genomes of many prokaryotes, both parasitic and free-living, have been completely sequenced. For example, the complete sequence of the 4.6 million base pairs (4288 genes) of *E. coli* was published in 1997. This was preceded by another major landmark, electronic publication of the complete sequence of the 12 million base pairs (6223 genes) of yeast, a unicellular eukaryote with an unusually small genome, in 1996. Sequencing projects encompass a variety of multicellular eukaryotes, with an emphasis on those that have been important research tools and/or are of agricultural and medical importance. For example, the genome of the fruit fly, *Drosophila melanogaster,* has been sequenced (more than 14,000 genes), as have those of two other widely used model organisms, *Caenorhabditis elegans* (a worm with almost 19,000 genes) and *Arabidopsis thaliana* (a plant with more than 26,000 genes) (see Chapter 16). We are in the middle of an extraordinary explosion of gene sequence data, largely because of the use of much more advanced automated sequencing methods.

DNA sequence information is now kept in large computer databases, many of which can be accessed through the Internet. Examples include databases maintained by the National Center for Biotechnology Information (a service of the U. S. National Library of Medicine and the National Institutes of Health) and by the Human Genome Organization (HUGO). These allow investigators to compare newly discovered sequences with those already known and to utilize many other kinds of information. By searching for DNA (and amino acid) sequences in a database, researchers can gain a great deal of insight into the function and structure of the gene product, as well as the evolutionary relationships among genes, and the variability among gene sequences in a population.

[2] The prefix "dd" is used for dideoxynucleotides, to distinguish them from deoxynucleotides, which are designated "d."

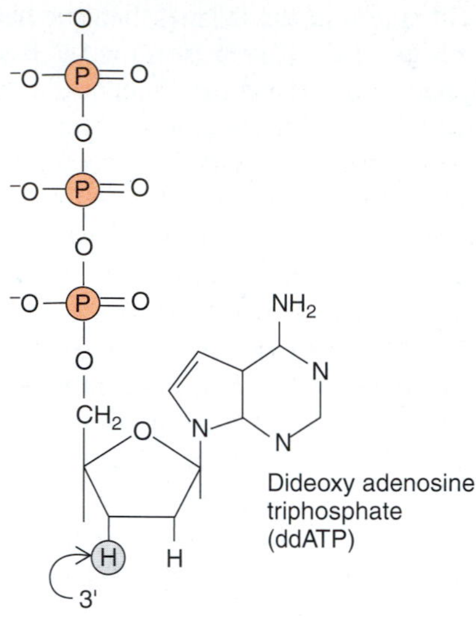

(a) Dideoxynucleotides are modified nucleotides that lack a 3' hydroxyl group and thus block further elongation of a new DNA chain.

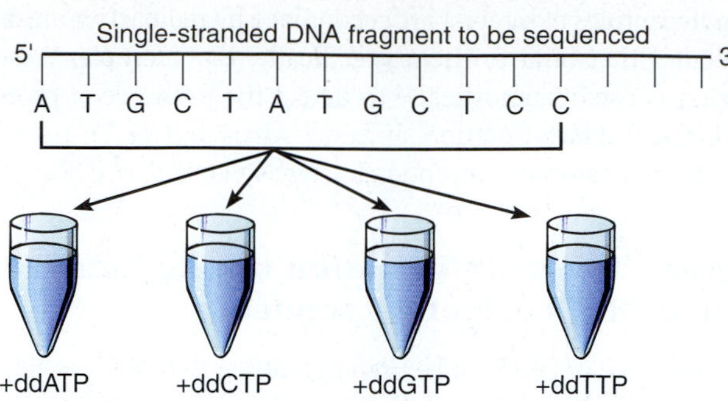

(b) Four different reaction mixtures are used to sequence a DNA fragment; each contains a small amount of a single dideoxynucleotide, such as ddATP. Larger amounts of the four normal deoxynucleotides (dATP, dCTP, dGTP, and dTTP) plus DNA polymerase and radioactively labeled primers are also included.

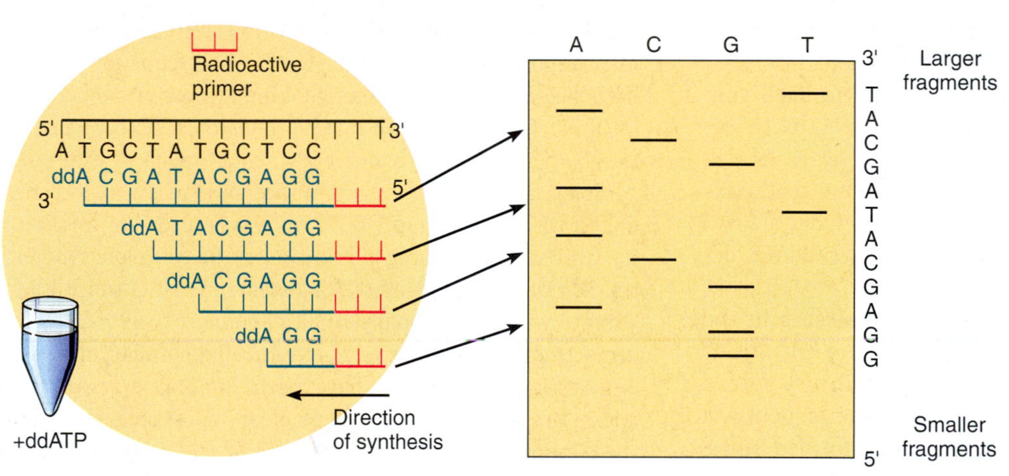

(c) The random incorporation of dideoxy ATP into the growing chain generates a series of smaller DNA fragments ending at all the possible positions where adenine is found in the newly synthesized fragments. These correspond to positions where thymine occurs in the original template strand.

(d) The radioactive products of each reaction mixture are separated by gel electrophoresis and located by exposing the gel to x-ray film. The nucleotide sequence of the newly synthesized DNA is read directly from the film (5' → 3'). The sequence in the original template strand is its complement (3' → 5').

A C G T

(e) An exposed x-ray film of a DNA sequencing gel. The four lanes represent A, C, G, and T dideoxy reaction mixes, respectively.

Figure 14–9 DNA sequencing. *(e, Courtesy of B. Slatko, New England Biolabs)*

Restriction fragment length polymorphisms are a measure of genetic relationships

The variability of genes within a population can be studied in several different ways. As mentioned in the previous section, a direct approach is to determine DNA sequences. Advances in DNA sequencing technology will make this the most widely used method in the future.

A more traditional procedure uses restriction enzymes. It is based on the fact that random DNA mutations and recombination may result in individuals differing in the number and location of sites where a particular restriction enzyme cuts the DNA.

Therefore, each individual differs in the lengths of the fragments produced by that enzyme. Such **restriction fragment length polymorphisms** (commonly known as **RFLPs,** or "riflips") can be used to determine how closely related different members of the population are. (The term *polymorphism* literally means "many forms." A **genetic polymorphism** is said to exist if individuals of two or more discrete genetic types, or "morphs," are found in a population or species; see Chapter 18.)

Restriction enzymes are used to cut the DNA from two or more individuals, and the fragments are separated by gel electrophoresis (with the DNA from each individual in a separate lane). A Southern blot is made of the DNA on the gel, which is then denatured and allowed to hybridize with a genetic probe. Usually the probe is complementary to a sequence that is repeated and interspersed throughout the genome. The resulting patterns of bands, commonly referred to as DNA fingerprints, can then be compared (Fig. 14–10).

RFLP analysis is an especially powerful tool in the fields of population and evolutionary biology because it can measure the degree of genetic relatedness between individuals. It has also been very useful in settling cases of disputed parentage.

The most controversial use of DNA fingerprinting is in the field of forensics. Traditional RFLP analysis can be used if there is a large enough amount of blood, semen, or other DNA-containing tissue left at the scene of a crime. However, if even a tiny amount of such physical evidence is available, one or more target DNA sequences can be amplified by the PCR technique, and subjected to electrophoresis, yielding a DNA fingerprint that can be analyzed.

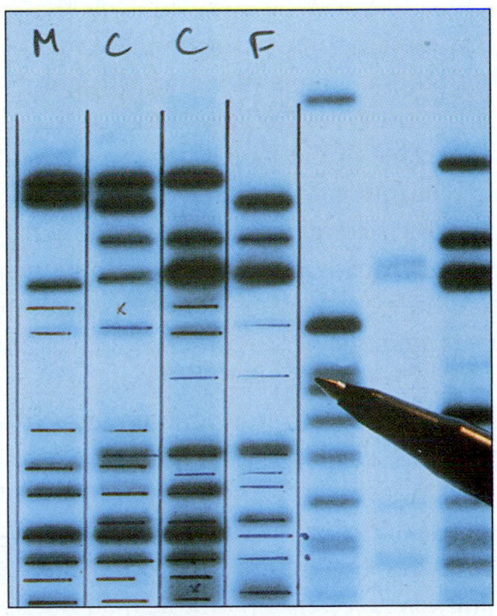

Figure 14–10 Gel electrophoresis showing a restriction fragment length polymorphism (RFLP). The lanes marked "M" and "F" contain DNA from a mother and father, respectively, and the two marked "C" contain DNA from their children. Note that every band present in one of the children is also found in at least one of the parents. *(David Parker/Science Photo Library/Photo Researchers, Inc.)*

If applied properly, DNA fingerprinting has the power to identify the guilty with a high degree of certainty. Conversely, it can exonerate the innocent. In fact, hundreds of convicted persons have won new trials and have been subsequently released from incarceration based on the application of DNA fingerprinting to physical evidence from the crime scene. Such evidence has been ruled admissible in many court cases, including certain trials that have received a great deal of attention in recent years. One limitation arises from the fact that the DNA samples are usually small and may have been degraded. Obviously, great care must be taken to prevent contamination of the samples. This is especially crucial if the PCR technique is to be used to amplify the DNA.

The development of DNA fingerprinting sparked a lively debate over how "unique" each individual pattern might be. For example, some scientists argued that a pattern that is quite rare in the general population might be more common in a particular ethnic group, and this might significantly affect the calculated probability of a match. As data on the frequency of particular patterns in various populations have accumulated, it has been learned that these concerns are of less practical importance than once thought. For example, the odds that two persons taken at random from the general population would have identical DNA fingerprints may be as low as one in several billion. If two persons are members of the same ethnic group, the odds of a match may increase but are usually still extremely low (perhaps one in several million).

GENETIC ENGINEERING HAS MANY APPLICATIONS

Recombinant DNA technology has provided not only a new and unique set of tools for examining fundamental questions about cells but also new approaches to problems of applied technology in many other fields. In some cases the production of genetically engineered proteins and organisms has begun to have considerable impact on our lives. The most striking of these have been in the fields of pharmacology and medicine.

In 1982 human insulin produced by *E. coli* became the first genetically engineered protein approved for use by humans. Prior to the use of recombinant DNA techniques to generate genetically altered bacteria capable of producing the human hormone, insulin was derived exclusively from other animals. Many diabetic persons become allergic to the insulin from animal sources because its amino acid sequence differs slightly from human insulin. The ability to produce the human hormone by recombinant DNA methods has resulted in significant medical benefits to diabetics.

Genetically engineered human growth hormone (see Chapter 47) is available to children who need it to overcome growth deficiencies. Human growth hormone could previously be obtained only from cadavers. Only small amounts were available, and evidence suggested that some of the preparations from cadavers were contaminated with infectious agents similar to those causing mad cow disease (see Chapter 23). The list of products that can be produced by genetic engineering is ever growing. These include treatments for multiple sclerosis, certain cancers, heart attacks, and certain forms of anemia. Recombinant DNA technology is also increasingly used to produce vaccines that provide safe and effective immunity against infectious diseases such as hepatitis B.

Additional engineering is required for a recombinant eukaryotic gene to be expressed in bacteria

Even if a gene has been isolated and successfully introduced into *E. coli,* the bacterium does not necessarily make the encoded protein in large quantities. Several obstacles stand in the way of producing gene products of eukaryotes in bacteria. One is that the gene has to be correctly associated with an appropriate set of regulatory and promoter sequences that the bacterial RNA polymerase can recognize. Recall from Chapters 12 and 13 that the regulatory regions of prokaryotic and eukaryotic genes are quite different. A usual approach to this problem is to combine the amino acid coding portion of a eukaryotic gene with a bacterial promoter sequence that can be strongly expressed. Some eukaryotic genes, for example, are fused to the lactose operon regulatory region (see Chapter 13); the protein product of the eukaryotic gene is synthesized when the bacterium is fed lactose in the growth medium.

We have already discussed the fact that bacterial cells cannot process RNA molecules containing eukaryotic intron sequences and that one solution to this problem is to introduce a cDNA copy of the gene. Other problems may arise in the expression of a recombinant protein in *E. coli* because of differences in the ways the proteins are expressed in prokaryotic and eukaryotic cells. Insulin, for example, is made in human cells from a large polypeptide that is folded in a specific way by the formation of three disulfide bonds, each joining two cysteines (sulfur-containing amino acids). After the polypeptide is folded, parts of the polypeptide are removed by proteolytic (protein-digesting) enzymes, leaving the insulin as two separate polypeptide chains held together by the disulfide bonds. *E. coli* lacks the specific enzymes necessary to cut the larger protein and is not able to fold the molecule properly. To overcome these problems, the gene was engineered to produce the two polypeptides separately. The recombinant proteins are then purified from the cells and allowed to associate in vitro. This procedure results in a relatively low yield of the active hormone, because the insulin can fold in several ways, only one of which results in a functional hormone.

It has been possible to circumvent some of these types of problems by introducing recombinant genes into eukaryotic cells such as yeast or other fungi, or cultured mammalian cells, that contain the protein-processing machinery required to produce fully functional proteins. Foreign proteins can also be produced by some types of genetically engineered plants or animals.

Transgenic organisms have incorporated foreign DNA into their cells

Plants and animals in which foreign genes have been incorporated are referred to as **transgenic** organisms. Varied approaches are used to insert foreign genes into plant or animal cells. Viruses are often used as vectors, although other methods, such as direct injection of DNA into cells, have also been applied.

Transgenic animals are valuable in research

Process of Science

Transgenic animals are usually produced by microinjecting the DNA of a particular gene into the nucleus of a recipient fertilized egg cell or embryonic **stem cells** (see Chapter 1). The eggs are then implanted into the uterus of a female and allowed to develop. Alternatively, genetically modified embryonic stem cells are injected into isolated blastocysts (see Chapter 49) and then implanted into a foster mother.

Transgenic offspring have already been shown to have numerous valuable research applications over a wide range of investigations, including regulation of gene expression, immune system function, genetic diseases, viral diseases, and genes responsible for the development of cancer. The laboratory mouse *(Mus)* has become a particularly important model organism for these studies.

In a classic pioneering study of the control of gene expression, reported by the laboratory of R.L. Brinster in 1983, transgenic mice carrying a gene for rat growth hormone were produced (Fig. 14–11; also see Fig. 16–18*a*). Brinster and his colleagues wanted to understand the controls that allow certain genes to be expressed in some tissues and not in others. A mouse normally produces small amounts of growth hormone in its pituitary gland, but these researchers reasoned that other tissues might also be capable of producing growth hormone. First the gene for growth hormone was isolated from a library of genomic rat DNA. It was then combined with the promoter region of a mouse gene that normally produces metallothionein, a protein that is active in the liver and whose synthesis is stimulated by the presence of toxic amounts of heavy metals such as zinc. The metallothionein regulatory sequences were used as a switch to turn the production of rat growth hormone on and off at will. After the engineered gene was injected with a microinjection pipette into mouse embryo cells, the embryos were implanted into the uterus of a mouse and allowed to develop. Because of the difficulty in manipulating the embryos without damaging them, the gene transplant was successful in only a small fraction of the animals. When exposed to small amounts of zinc, these transgenic mice produced large amounts of growth hormone because the liver is a much larger organ than the pituitary gland. The mice grew rapidly, and one mouse, which developed from an embryo that had received two copies of the growth hormone gene, grew to more than double the normal size. As might be expected, such mice are often able to transmit their increased growth capability to their offspring.

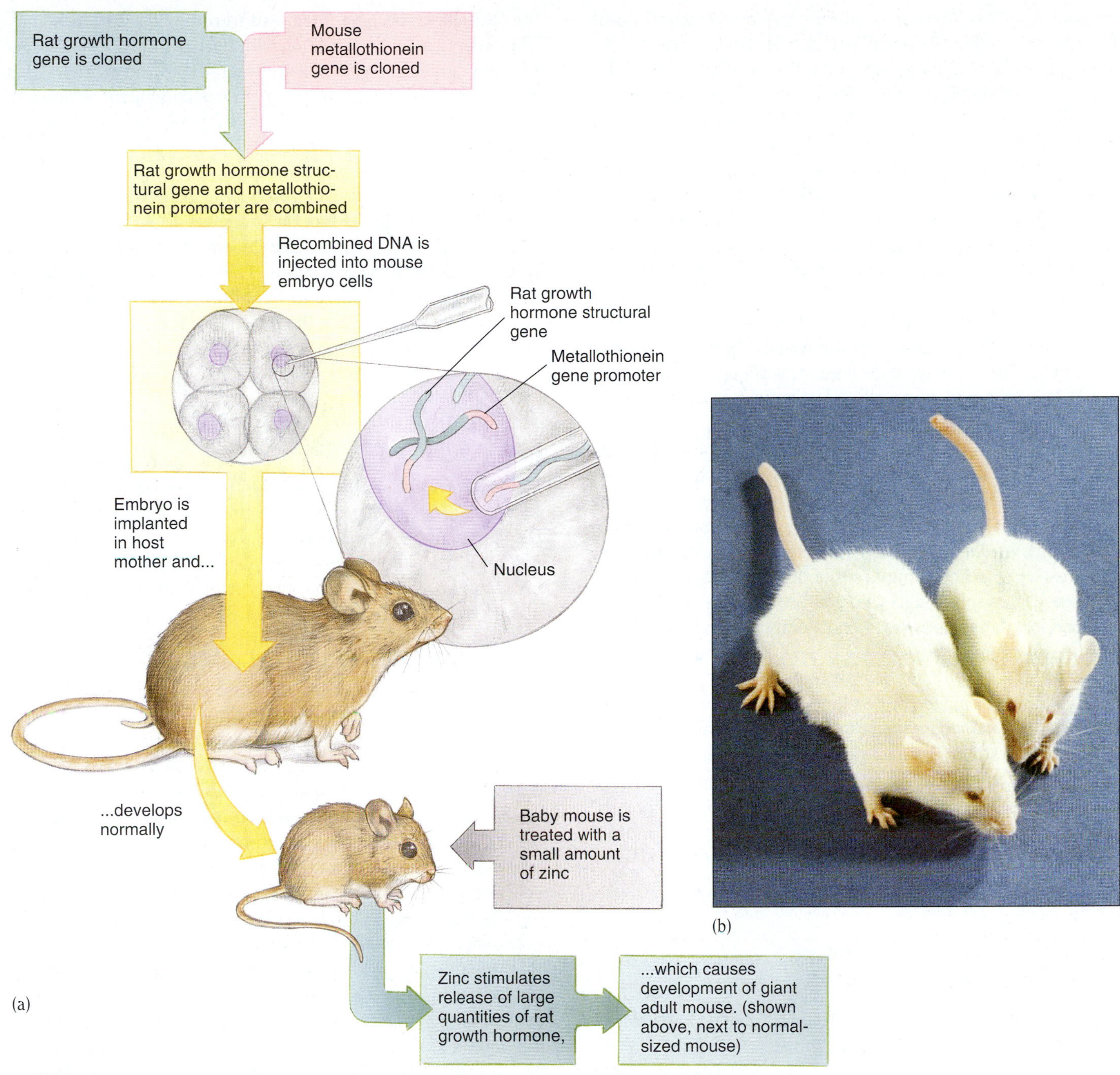
Rat growth hormone gene is cloned
Mouse metallothionein gene is cloned
Rat growth hormone structural gene and metallothionein promoter are combined
Recombined DNA is injected into mouse embryo cells
Rat growth hormone structural gene
Metallothionein gene promoter
Nucleus
Embryo is implanted in host mother and...
...develops normally
Baby mouse is treated with a small amount of zinc
Zinc stimulates release of large quantities of rat growth hormone,
...which causes development of giant adult mouse. (shown above, next to normal-sized mouse)
(a)
(b)

Figure 14–11 Transgenic mice. **(a)** How to make a giant mouse. **(b)** The mouse on the right is normal, while the mouse on the left is a transgenic animal that expresses rat growth hormone. *(Photo by R.L. Brinster, University of Pennsylvania Medical School)*

One extremely powerful research tool is **gene targeting,** a procedure in which a single gene is chosen and "knocked out" (inactivated) in a mouse. The roles of the inactivated gene can be determined by observing the phenotype of the mice bearing the knockout gene. For example, if the gene codes for a protein, the functions of that protein can be identified by studying individuals in which it is lacking. Because at least 99% of the loci of mice have human counterparts (although the specific alleles are usually different), information about *knockout genes* in mice provides details about human genes as well.

Gene targeting, pioneered by Mario Capecchi, a molecular geneticist at the University of Utah School of Medicine, is a rather complex and lengthy procedure; it takes about a year to develop a new strain of knockout mice. First, a nonfunctional (knockout) gene is introduced into mouse embryonic stem cells (ES cells). ES cells are particularly easy to handle because, like cancer cells, they can be grown in culture indefinitely. Most important, if they are placed into a mouse embryo, they are capable of dividing and producing all of the cell types normally found in the mouse. In a tiny fraction of these ES cells, the introduced gene will become physically associated with the corresponding gene in a chromosome. If this occurs, the chromosomal gene and the introduced gene will tend to exchange DNA segments in a poorly understood process known as *homologous recombination.* In this way, the normal allele in the mouse chromosome is replaced by the knockout allele.

Researchers inject ES cells they hope are carrying a knockout gene into early mouse embryos and allow the mice to develop to maturity. The mice are then bred for several generations, allowing researchers to eventually select any offspring that might be homozygous for the knockout gene.

If the gene is not lethal when inactivated, the researchers generally study homozygous animals that carry the knockout gene in every cell. However, because many genes are essential to life, researchers have modified the knockout technique to develop strains in which a specific gene is selectively inactivated in only one cell type. Today hundreds of different strains of knockout mice, each displaying its own characteristic phenotype, have been developed in various research laboratories, and the number continues to grow.

Gene targeting in mice is providing answers to basic biological questions relating to the development of embryos, the development of the nervous system, and the normal functioning of the immune system. This technique has great potential for revealing more about various human diseases, especially as we have learned that many thousands of diseases have a genetic component. Gene targeting is being used to study cancer, heart disease, respiratory diseases such as cystic fibrosis, sickle cell anemia, and other health problems.

Transgenic animals can produce genetically engineered proteins

Certain transgenic animal strains produce foreign proteins that are secreted into milk. For example, the gene for a human blood clotting factor has been introduced into sheep. These recombinant genes have been fused to the regulatory sequences of the milk protein genes and are therefore activated only in mammary tissues involved in milk production.

The advantage of obtaining the protein from milk is that potentially it can be produced in large quantities and can be harvested simply by milking the animal. The protein is then purified from the milk. The animals are not harmed by the introduction of the gene, and, because usually the progeny of the transgenic animal also produce the recombinant protein, transgenic strains can be established.

Transgenic animals are not only produced by microinjecting DNA into cells. Sometimes viruses are used as recombinant DNA vectors. RNA viruses called **retroviruses** make DNA copies of themselves by reverse transcription (see Chapter 12). Sometimes the DNA copies become integrated into the host chromosomes, where they are replicated along with host DNA. For example, genetically altered mouse leukemia viruses are retroviruses that can be used as vectors to incorporate recombinant genes into cultured cells. Under certain conditions genes carried by the engineered virus can be expressed in the animal cells to produce genetically engineered proteins.

A major disadvantage of introducing genes into cultured animal cells is that the yields of the proteins encoded by the foreign DNA carried by the viruses are generally low. However, these types of vectors show some promise as treatments for human genetic disorders. Procedures to treat human disorders by genetically modifying the cells of the patient are known as **gene transfer therapy** or, more simply, **gene therapy** (see Chapter 15).

Transgenic plants are increasingly important in agriculture

Plants have been selectively bred for thousands of years. The success of such efforts depends on the presence of desirable traits in the variety of plant being selected, or in closely related wild or domesticated plants whose traits can be transferred by crossbreeding. Local varieties or closely related species of cultivated plants often have traits, such as disease resistance, that could be advantageously introduced into varieties more suited to modern needs. If genes are introduced into plants from strains or species with which they do not ordinarily interbreed, the possibilities for improvement are greatly increased.

Unfortunately, a suitable vector for the introduction of recombinant genes into many types of plant cells has proved quite difficult to find. The most widely used vector system employs the crown gall bacterium, *Agrobacterium tumefaciens.* This bacterium normally produces plant tumors by introducing a special plasmid, called the *Ti* (for *tumor-inducing*) *plasmid,* into the cells of its host. The Ti plasmid induces abnormal growth by forcing the plant cells to produce elevated levels of a plant growth hormone called cytokinin (see Chapter 36).

It is possible to "disarm" the Ti plasmid so that it does not induce tumor formation and then to use it as a vector to insert genes into plant cells (Fig. 14–12). The cells into which the altered plasmid is introduced are essentially normal except for the genes that have been inserted. Genes placed in the plant genome in this fashion may be transmitted sexually, via seeds, to the next generation, but they can also be propagated asexually (e.g., by taking cuttings) if desired.

Unfortunately, not all plants take up DNA readily, and this is particularly true of the grain plants that are the major food source for humans. One useful approach has been the development of a genetic "shotgun." Microscopic gold fragments are coated with DNA and then shot into plant cells, penetrating the cell walls. Some of the cells retain the DNA and are transformed by it. Those cells can then be cultured and used to regenerate an entire plant (see Chapter 16). For example, such an approach has been successfully used to transfer a gene for resistance to a bacterial disease into cultivated rice from one of its wild relatives.

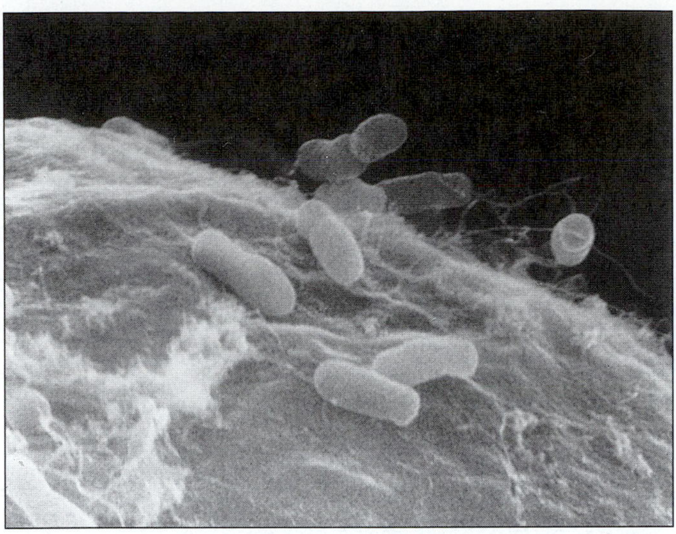

Figure 14–12 Transformation of plant cells. This SEM shows *Agrobacterium tumefaciens* infecting cultured plant cells. The close contact permits the transfer of plasmid DNA from the bacteria to the larger plant cells. *(Courtesy of Ann G. Matthysse)*

An additional complication of plant genetic engineering is that many important plant genes are located in the DNA of the chloroplasts (see Chapter 4). Chloroplasts are essential in photosynthesis, which is the basis for plant productivity. Obviously, it is useful to develop methods for changing the portion of the plant's DNA that resides within the chloroplast. Methods of chloroplast engineering are currently the focus of intense research.

The applications of transgenic plants are not limited to disease resistance and increased production of crops. Like some transgenic animals, certain transgenic plants can potentially be used to produce large quantities of medically important proteins, such as antibodies. The developers of this technology are conducting ongoing field trials to demonstrate the feasibility of these production methods.

Some people are concerned about the health effects of consuming foods derived from genetically modified plants, commonly known as GM foods (see Chapter 1). For this reason there is an ongoing controversy as to whether the use of such foods should be restricted or if they should be required to include special warning labels.

SAFETY GUIDELINES HAVE BEEN DEVELOPED FOR RECOMBINANT DNA TECHNOLOGY

When recombinant DNA technology was introduced in the early 1970s, many scientists considered the potential misuses to be at least as significant as the possible benefits. The possibility that an organism with undesirable environmental effects might be accidentally produced was a great concern, because it was anticipated that totally new strains of bacteria or other organisms, with which the world has no previous experience, might be difficult to control. This possibility was recognized by the scientists who developed the recombinant DNA methods and led them to insist on stringent guidelines for making the new technology safe.

Experiments in thousands of university and industrial laboratories over more than 25 years have seen recombinant DNA manipulations carried out safely. Laboratory strains of *E. coli* are poor competition for the wild strains in the outside world and quickly perish. Experiments thought to entail unusual risks are carried out in special facilities designed to contain disease-causing organisms and allow researchers to work with them safely. So far there is no evidence that hazardous genes have been accidentally cloned or that dangerous organisms have been released into the environment. However, malicious *intentional* manipulations of dangerous genes certainly remain a possibility.

Many of the restrictive guidelines for using recombinant DNA have been relaxed as the safety of the experiments has been established. Stringent restrictions still exist, however, in certain areas of recombinant DNA research where there are known dangers, or where questions about possible effects on the environment are still unanswered.

These restrictions are most evident in research that proposes to introduce recombinant organisms into the wild, such as agricultural strains of plants whose seeds or pollen might spread in an uncontrolled manner. A great deal of research activity is now concentrated on determining the effects of introducing transgenic organisms into a natural environment. Carefully conducted tests have shown that transgenic organisms are not dangerous to the environment simply because they are transgenic. However, it is important to assess the biology of each new recombinant organism. In this way scientists will be able to determine if it has characteristics that might cause it to present an environmental hazard under certain conditions. For example, if a transgenic crop plant has been engineered to resist an herbicide, might that gene be transferred, via pollen or by some other route, to that plant's weedy relatives, generating herbicide-resistant "superweeds?"

Other concerns relate to plants that have been engineered to produce pesticides, such as insecticides. For example, a transgenic plant may produce a relatively small amount of an insecticide, which may be sufficient to provide protection if the targeted insect population is not very numerous. However, the presence of low levels of the insecticide could potentially provide ideal conditions for selection for resistant individuals in the insect population. Another concern is that non-pest species could be harmed. For example, a great deal of attention has been paid to the finding that monarch butterfly larvae raised in the laboratory are harmed if they are fed pollen from corn plants genetically engineered to produce Bt toxin, a natural insecticide. Although more recent studies have suggested that monarch larvae living in a natural environment do not consume enough pollen to cause damage, such concerns persist and will have to be addressed on a case-by-case basis.

SUMMARY WITH KEY TERMS

I. **Recombinant DNA** technology is concerned with isolating and amplifying specific sequences of DNA by incorporating them into **vector** DNA molecules. The resulting recombinant DNA can then be propagated and amplified in organisms such as *E. coli.*
   A. **Restriction enzymes** are used to cut DNA into specific fragments.
      1. Each type of restriction enzyme recognizes and cuts DNA at a highly specific base sequence.
      2. Many restriction enzymes cleave DNA sequences to produce complementary, single-stranded cut ends (sticky ends).
   B. The most common recombinant DNA vectors are constructed from naturally occurring circular DNA molecules called **plasmids,** or from bacterial viruses called **bacteriophages;** both of these are found in some bacteria.
   C. Recombinant DNA molecules are often constructed by allowing the ends of a DNA fragment and a plasmid (which have both been cut with the same restriction enzyme) to associate by complementary base pairing. The DNA strands are then covalently linked by **DNA ligase** to form a stable recombinant molecule.
   D. Parts of genes are isolated from recombinant DNA libraries, which are mixtures of DNA fragments inserted into appropriate vectors.
      1. **Genomic libraries** are DNA fragments from the total DNA of an organism. Genes present in recombinant DNA genomic libraries from eukaryotes contain introns. Those genes can be amplified in *E. coli,* but the protein is not properly expressed.
      2. When a **cDNA library** is produced, **reverse transcriptase** is used to make DNA copies of mRNA isolated from eukaryotic cells; these copies, known as **complementary DNA (cDNA),** are then incorporated into recombinant DNA vectors. Because the introns have been removed from mRNA molecules, eukaryotic genes in cDNA libraries can sometimes be expressed in *E. coli* to make their protein products.
   E. The **polymerase chain reaction (PCR)** is a widely used, usually automated, **in vitro** technique in which a particular DNA sequence can be targeted by specific primers and then **cloned** by a special heat-resistant DNA polymerase.
   F. Analysis of a cloned sequence can yield useful information about the gene and its encoded protein and can enable investigators to identify and subclone DNA fragments for use as molecular probes.
      1. A **DNA sequence** gives information about the structure of the gene and the probable amino acid sequence of the encoded proteins. It can be compared with other sequences stored in massive databases.
      2. A radioactive DNA or RNA sequence can be used as a **genetic probe** to identify complementary nucleic acid sequences. In the **Southern blot technique,** DNA fragments are separated by gel electrophoresis, denatured, and then blotted onto a nitrocellulose filter. The radioactive probe is then hybridized by complementary base pairing to the DNA bound to the filter, and the radioactive band or bands of DNA can be identified by autoradiography.
   G. The degree of genetic relationship among the individuals in a population can be determined by comparing nucleotide sequences, or it can be estimated in other ways, including the study of **restriction fragment length polymorphisms (RFLPs).**

II. **Genetic engineering** is a technology that uses genetic and recombinant DNA methods to devise new combinations of genes to produce improved pharmaceutical and agricultural products.
   A. Genes isolated from one organism can be modified and expressed in other organisms ranging from *E. coli* to **transgenic** plants and animals.
      1. Expression of eukaryotic proteins in bacteria such as *E. coli* requires that the gene be linked to regulatory elements that the bacterium can recognize. In addition, bacterial cells do not contain many of the enzymes needed for the posttranslational processing of eukaryotic proteins.
      2. Expression of eukaryotic genes in eukaryotic host organisms shows great promise, because the processing and modification machinery for eukaryotic proteins is already present in these cells.
         a. Production of important pharmaceutical products can be engineered in transgenic animals and possibly in plants.
         b. Genetic engineering of plants and domestic animals holds the promise of increasing the availability of food, although some consumers are concerned about the safety of such foods.
   B. Recombinant DNA technology is carried out under certain safety guidelines.

POST-TEST

1. A plasmid (a) can be used as a DNA vector (b) is a type of bacteriophage (c) is a type of cDNA (d) is a retrovirus (e) b and c
2. DNA molecules with complementary "sticky ends" associate by (a) covalent bonds (b) hydrogen bonds (c) ionic bonds (d) disulfide bonds (e) phosphodiester linkages
3. Human DNA and a particular plasmid both have sites that can be cut by the restriction enzymes Hind III and EcoRI. To make recombinant DNA, one should (a) cut the plasmid with EcoRI and the human DNA with Hind III (b) use EcoRI to cut both the plasmid and the human DNA (c) use Hind III to cut both the plasmid and the human DNA (d) a or b (e) b or c
4. Which of the following sequences is *not* palindromic?

   (a) 5'—AAGCTT—3'
   3'—TTCGAA—5'

   (b) 5'—GATC—3'
   3'—CTAG—5'

   (c) 5'—GAATTC—3'
   3'—CTTAAG—5'

   (d) 5'—CTAA—3'
   3'—GATT—5'

   (e) b and d
5. The PCR technique uses (a) heat-resistant DNA polymerase (b) reverse transcriptase (c) DNA ligase (d) restriction enzymes (e) b and c
6. A cDNA clone contains (a) introns (b) exons (c) anticodons (d) a and b (e) b and c
7. The dideoxynucleotides ddATP, ddTTP, ddGTP, and ddCTP are important in DNA sequencing because they (a) cause premature termination of a growing DNA strand (b) are used as primers (c) cause the DNA fragments that contain them to migrate more slowly through a sequencing gel (d) are not affected by high temperatures (e) have more energy than deoxynucleotides
8. In restriction fragment length polymorphism (RFLP) analysis to determine parentage, (a) every band present in a child would be expected to be present in both of the true parents (b) every band present in a child would be expected to be present in at least one of the true parents (c) every band present in a true parent would be expected to be present in all of the children (d) a and b (e) b and c
9. In the Southern blot technique, ____________ is/are transferred from a gel to a special nitrocellulose filter. (a) protein (b) RNA (c) DNA (d) bacterial colonies (e) reverse transcriptase

10. Gel electrophoresis separates nucleic acids on the basis of differences in (a) length (molecular weight) (b) charge (c) nucleotide sequence (d) relative proportions of adenine and guanine (e) relative proportions of thymine and cytosine

11. The Ti plasmid, carried by *Agrobacterium tumefaciens,* is especially useful for introducing genes into (a) *E. coli* (b) plants (c) animals (d) yeast (e) all eukaryotes

REVIEW QUESTIONS

1. What is meant by the term *genetic engineering?*
2. What are restriction enzymes? How are they used in recombinant DNA research?
3. What characteristics should be engineered into a plasmid to make it a useful cloning vector?
4. Diagram the process by which recombinant DNA molecules are usually constructed.
5. How is a gene library constructed? What are the relative merits of genomic libraries and cDNA libraries?
6. Sketch an example illustrating how a restriction map of a gene is made.
7. Why is the PCR technique valuable?

YOU MAKE THE CONNECTION

1. What are some of the problems that might arise if you were trying to produce a eukaryotic protein in a bacterium? How might some of these problems be solved by using transgenic plants or animals?
2. Would genetic engineering be possible if we did not know a great deal about the genetics of bacteria? Explain.
3. What are some of the ecological concerns regarding transgenic organisms? What kinds of information are needed to determine if these concerns are valid?

RECOMMENDED READINGS

Brown, K. "Seeds of Concern." *Scientific American,* Vol. 284, No. 4, Apr. 2001. An explanation of the potential environmental risks and benefits of genetically modified crops.

Butler, D., T. Reichardt, et al. "Long-term Effect of GM Crops Serves Up Food for Thought." *Nature,* Vol. 398, 22 Apr. 1999. A thoughtful, balanced consideration of the use of genetically modified crops.

Hopkin, K. "Risks on the Table." *Scientific American,* Vol. 284, No. 4, Apr. 2001. A discussion of the safety of the genetically modified foods contained in about 60% of the processed foods sold in the United States, as well as those that might be produced in the future.

Langridge, W.H. "Edible Vaccines." *Scientific American,* Vol 282, No. 3, Sep. 2000. The feasibility of producing oral vaccines in plants is being actively studied. Such vaccines hold the promise of providing easily administered immunizations to the world population.

Marvier, M. "Ecology of Transgenic Crops." *American Scientist,* Vol. 89. Mar.–Apr. 2001. A consideration of the difficulties in assessing the significance of potential environmental risks posed by transgenic plants.

Miller, R.V. "Bacterial Gene Swapping in Nature." *Scientific American,* Vol. 278, No. 1, Jan. 1998. This article assesses the potential mechanisms and extent of transfer of genes to other microbes by genetically engineered bacteria released into natural environments.

Mullis, K.B. "The Unusual Origin of the Polymerase Chain Reaction." *Scientific American,* Vol. 262, No. 4, Apr. 1990. A highly personal, first-hand account of the development of the PCR technique and an excellent illustration of the nature of scientific insight.

Nemeck, S. "Does the World Need GM Foods?" *Scientific American,* Vol. 284, No. 4, Apr. 2001. The author interviews both an advocate for the development of GM foods, and an opponent.

Ronald, P.C. "Making Rice Disease-Resistant." *Scientific American,* Vol. 277, No. 5, Nov. 1997. The applications of genetic engineering to rice breeding are expected to increase the food supply for the 2 billion people worldwide for whom rice is the mainstay of the diet.

Somerville, C., and J. Dangl. "Plant Biology in 2010." *Science,* Vol. 290, 15 Dec. 2000. The completion of the sequencing of the genome of the model plant *Arabidopsis* will enable researchers to attain far-reaching goals.

Weiner, D.B., and R.C. Kennedy. "Genetic Vaccines." *Scientific American,* Vol. 281, No. 1, Jul. 1999. Genetic engineering techniques can be used to produce vaccines that are composed of DNA.

Wolfenbarger, L.L., and P.R. Phifer. "The Ecological Risks and Benefits of Genetically Engineered Plants." *Science,* Vol. 290, 15 Dec. 2000. The authors argue that the ecological effects of genetically modified plants are inherently difficult to assess because of the complexity of ecological systems.

- Visit our Web site at **http://www.info.brookscole.com/solomonbergmartin** for links to chapter-related resources on the World Wide Web. Additional on-line materials relating to this chapter can also be found on our Web site.

See chapter activity on BioActive Learner CD for additional help in mastering the chapter's material. Icon location in the chapter's margins shows which topics have tutorials or simulations in the CD.

15

The Human Genome

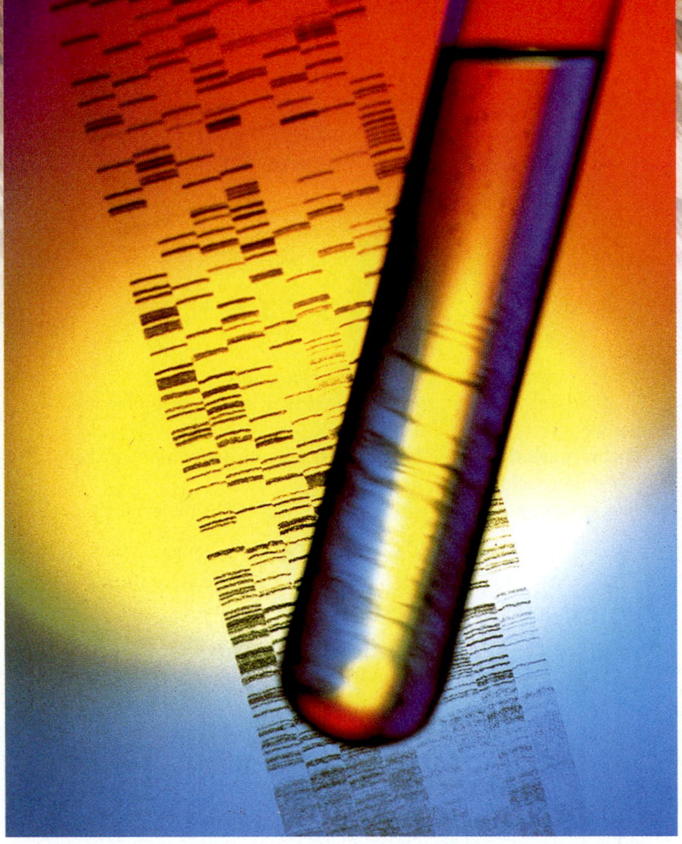

DNA autoradiogram. The autoradiogram shows the sequence of base pairs for a section of human DNA. *(TEK Image/Science Photo Library/Photo Researchers, Inc.)*

LEARNING OBJECTIVES

After you have studied this chapter you should be able to

1. Define human genetics, the human genome, bioinformatics, and pharmacogenomics.
2. Given a simple pedigree for a particular trait, determine the probable genotypes for all individuals in the pedigree.
3. Discuss the implications of the Human Genome Project, including the benefits and difficult ethical concerns.
4. Distinguish between chromosome abnormalities and single-gene defects.
5. Make a sketch illustrating how nondisjunction can occur in meiosis. Show how nondisjunction can be responsible for specific chromosome abnormalities such as Down syndrome, Klinefelter syndrome, and Turner syndrome.
6. State whether each of the following genetic defects is inherited as an autosomal recessive, autosomal dominant, or X-linked recessive: phenylketonuria (PKU), sickle cell anemia, cystic fibrosis, Tay-Sachs disease, Huntington disease, and hemophilia A.
7. Describe how amniocentesis is used in the prenatal diagnosis of human genetic abnormalities; state the relative advantages and disadvantages of amniocentesis and chorionic villus sampling.
8. Distinguish between genetic screening programs for newborns and adults.
9. Discuss the scope and implications of genetic counseling.

The principles of genetics apply to all organisms, including humans. There are, however, some important differences between genetic research on humans and genetic research on other organisms. To study aspects of inheritance in other species, geneticists ideally (1) have standard stocks of genetically identical individuals, that is, **isogenic strains**, that are homozygous at virtually all of their loci; (2) conduct **controlled matings** between members of different isogenic strains; and (3) raise the offspring under carefully controlled conditions.

Of course, experimental matings under controlled conditions are not feasible in the human population. In addition, human families are small, and 20 to 30 years or more elapse between generations. It is therefore virtually impossible to conduct genetic research in humans in the same way we do with other species.

Despite the inherent difficulties of studying inheritance in humans, knowledge in **human genetics**, the science of inherited variation in humans, is progressing very rapidly. Traditionally, human genetics was examined using such approaches as population studies of large extended families. More recently, the field of human genetics has been greatly facilitated by the medical attention given to genetic diseases in humans. The extensive medical records of diseases serve as a very useful data pool on which hypotheses may be based and against which they may be tested. Genetic studies of other organisms also have provided invaluable insights. Indeed, many phenomena in human inheritance that were initially puzzling have been explained by solving analogous problems in the inheritance of bacteria, yeasts, worms, fruit flies, or mice.

The **human genome**, the totality of genetic information in human cells, has been mapped and **sequenced** (the order of nucleotides in DNA has been identified) to determine how we are all alike—that is, what makes us human *(see figure)*. The human genome includes the DNA content of both the nucleus (which accounts for almost all genetic information in the human genome) and the mitochondria. Like genomes of other eukaryotic organisms, some of the human genome specifies the synthesis of polypeptides or RNA. However, much of the human genome consists of noncoding DNA, repetitive (multiple copies of) DNA, and gene segments for which there are currently no known functions. Relating specific genes to the proteins they code for and determining what these proteins do in the body are some of the avenues of human genetic research that will be pursued during the 21st century.

Recent advances in human genetics have enormous implications for the future. Already, the new knowledge has created several scientific fields, such as bioinformatics and pharmacogenomics. **Bioinformatics,** also called **biological computing,** includes the storage, retrieval, and comparison of DNA sequences within human DNA and between genomes of different species. The tools of bioinformatics are powerful computers and sophisticated software used to manage and analyze the data. For example, as new DNA sequences are determined, automated computer programs scan the sequences for patterns typically found in genes. Bioinformatics has already provided important insights into gene identification, gene function, and evolutionary relationships (by comparing databases of DNA sequences from different organisms).

Knowledge of the human genome promises a revolution in human health care, and some experts predict that in highly developed countries the average human life expectancy at birth will be 90 to 95 years by 2050. (As a comparison, the current average life expectancy in the United States is 77 years.) New health-related information about the human genome, from identifying genes associated with hypertension to identifying genes associated with specific cancers, is announced nearly every week.

Pharmacogenomics is a new field of gene-based medicine in which drugs are personalized to match a patient's genetic makeup. The subtle genetic differences among individuals are taken into account in pharmacogenomics. In as few as five to ten years, patients may take routine genetic screening tests before a doctor prescribes medicine for them. Many of these diagnostic tests will involve **DNA microarrays,** in which thousands of DNA molecules are placed on a glass slide or chip (see Fig. 16–5). DNA microarrays allow researchers to compare the activities of thousands of genes in normal and diseased cells. Since cancer and other diseases exhibit altered patterns of gene expression, the use of DNA microarrays has the potential to identify genes or the proteins they code for that can then be targeted by therapeutic drugs. Pharmacogenomics, like other fields in human genetics, poses difficult ethical questions. The genetic testing that will be an everyday part of pharmacogenomics raises issues of privacy, genetic bias, and potential discrimination.

In this chapter, we first examine how the human genome is studied, including the Human Genome Project. Then we discuss a variety of human genetic disorders and examine genetic testing, screening, and counseling, and how genetic diseases are treated. The chapter concludes with a consideration of ethical concerns that relate to our expanding knowledge of the human genome.

THE STUDY OF HUMAN GENETICS REQUIRES A VARIETY OF METHODS

Human geneticists use a variety of methods that allow them to make inferences about a trait's mode of inheritance. We consider three of these methods—the identification of chromosomes by karyotyping, the summary of family inheritance studies by pedigree construction, and DNA sequencing and mapping of genes by genome projects. Human inheritance is often studied most effectively by using a combination of these and other approaches.

Human chromosomes are studied by karyotyping

Cytogenetics is the study of chromosomes and their role in inheritance. Many of the basic principles of genetics were discovered by researchers working with simpler organisms, in which it was possible to relate genetic data to the number and structure of specific chromosomes. Some of the organisms used in genetics, such as the fruit fly *Drosophila,* have very few chromosomes (only four pairs in *Drosophila*). In *Drosophila* larval salivary glands and most other larval tissues, the chromosomes are large enough that their structural details are readily evident. This organism, therefore, has provided unique opportunities for correlating certain phenotypic changes with certain alterations in chromosome structure.

Process of Science

Recall from Chapter 10 that the normal number of chromosomes for the human species is 46: 44 **autosomes** (22 pairs) and 2 sex chromosomes (one pair). Until the mid-1950s, when modern methods of karyotyping began to be adopted, the accepted number of chromosomes for the human species was 48, based on a study of human chromosomes published in 1923. The reason researchers counted 48 human chromosomes was the difficulty in separating the chromosomes so they could be accurately counted. In 1951 T.C. Hsu treated cells with a hypotonic salt solution by mistake, which caused the chromosomes to spread apart beautifully. Other less serendipitous techniques were also developed, and in 1956, researchers Jo Hin Tjio and Albert Levan reported that humans have 46 chromosomes, not 48. Other researchers subsequently verified this report. The story of the human chromosome number is a valuable example of the self-correcting nature of science (although science sometimes takes a while to correct itself). The reevaluation of established facts and ideas, often by using improved techniques or new methods, is an essential part of the scientific process.

Human chromosomes are only visible in dividing cells (see Chapter 9), and it is difficult to obtain dividing cells directly from the human body. Blood is typically used because the white blood cells can be induced to divide in a culture medium by treating them with **lectins,** sugar-binding proteins extracted from plant seeds. Other sources of dividing cells include skin and, for prenatal chromosome studies, chorionic villi or fetal cells shed into the amniotic fluid (both discussed later in the chapter).

The term **karyotype** (from the Greek *kary,* meaning nucleus) refers both to the chromosome composition of an individual and to a photomicrograph showing that composition. In karyotyping, dividing human cells are cultured and then treated with the drug **colchicine,** which arrests the cells at mitotic metaphase or late prophase, when the chromosomes are most highly condensed. Next the cells are placed into a hypotonic solution that causes them to swell; this process spreads out the chromosomes so they can be readily observed. The cells are then flattened on microscope slides, and the chromosomes are stained to reveal the patterns of bands that are unique for each homologous pair. After the microscopic image has been scanned into a computer, the homologous pairs are electronically matched and placed together (Fig. 15–1*a;* also see Fig. 9–7 for a normal human karyotype).

Chromosomes are identified by length; position of the centromere; banding patterns, which are produced by staining chromosomes with dyes that produce dark and light cross-bands of varying widths; and other features such as *satellites,* which are tiny knobs of chromosome material at the tips of certain chromosomes (Fig. 15–1*b*). The largest human chromosome (chromosome 1) is about five times as long as the smallest one (chromosome 21), but there are only slight size differences among some of the intermediate-sized chromosomes. Differences from the normal karyotype—that is, deviations in chromosome number or structure—are associated with certain disorders.

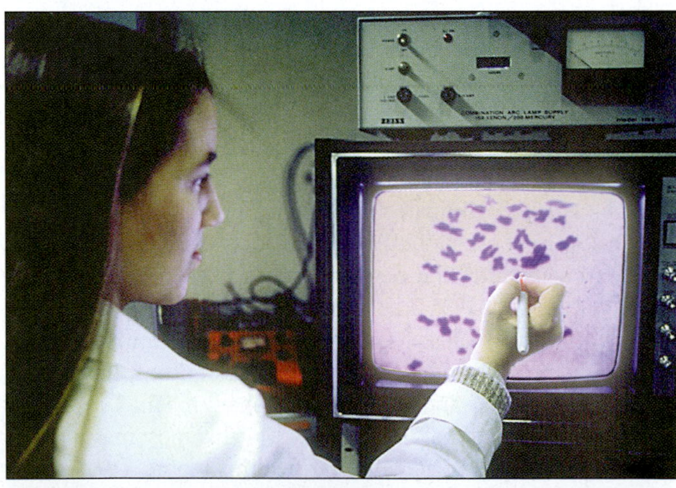

(a)

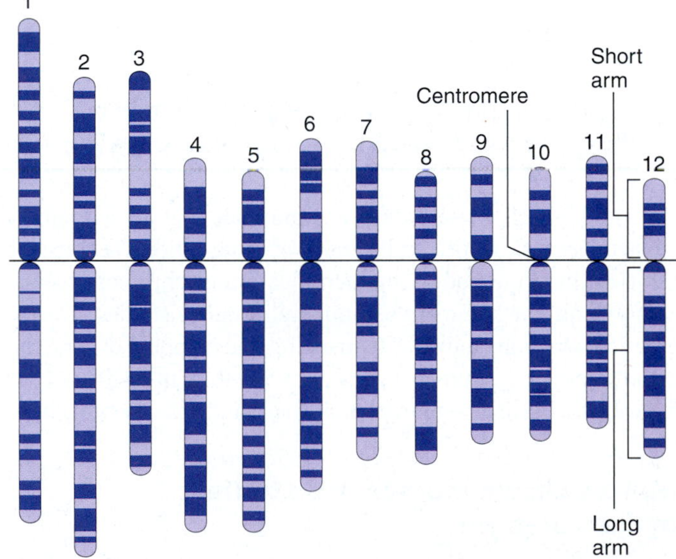

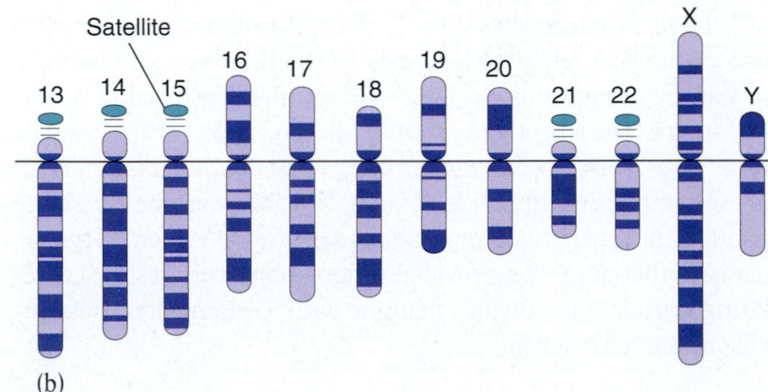

(b)

Figure 15–1 Karyotyping. (a) Using an image analysis computer to prepare a karyotype. The biologist is matching up homologous chromosomes and organizing them by size. Before computer use was widespread, researchers laboriously prepared karyotypes by cutting and pasting chromosomes from photographs. (b) Diagram of a human karyotype. The chromosomes are numbered in order of size, except that chromosome 21 is smaller than chromosome 22. Centromeres, satellites, and a compilation of banding patterns are shown. (a, *SIU Peter Arnold, Inc.;* b, *Adapted from Tobin, A.J., and R.E. Morel.* Asking About Cells. *Harcourt College Publishers, Philadelphia, 1997*)

Family pedigrees can help identify some inherited conditions

Early studies of human genetics usually dealt with readily identified pairs of contrasting traits and their distribution among members of a family. A chart that is constructed to show the transmission of genetic traits within a family over several generations is known as a **pedigree.** Pedigree analysis remains a useful technique, even in today's world of powerful molecular genetic techniques, because it helps molecular geneticists determine the exact interrelationships among the DNA molecules they analyze from related individuals. Pedigree analysis is also an important tool of genetic counselors and clinicians. However, because human families tend to be small and information on certain family members, particularly deceased relatives, may be lacking, pedigree analysis has certain limitations.

Pedigrees are produced using more or less standardized symbols. Examine Figure 15–2, which shows a hypothetical pedigree for **albinism,** a lack of pigmentation in the skin, hair, and eyes. Each horizontal row represents a separate generation, with the oldest generation (Roman numeral I) at the top and the most recent generation at the bottom. Within a given generation, the individuals are usually numbered consecutively, from left to right, using Arabic numerals. A horizontal line connects two parents, and a vertical line drops from the parents to their children.

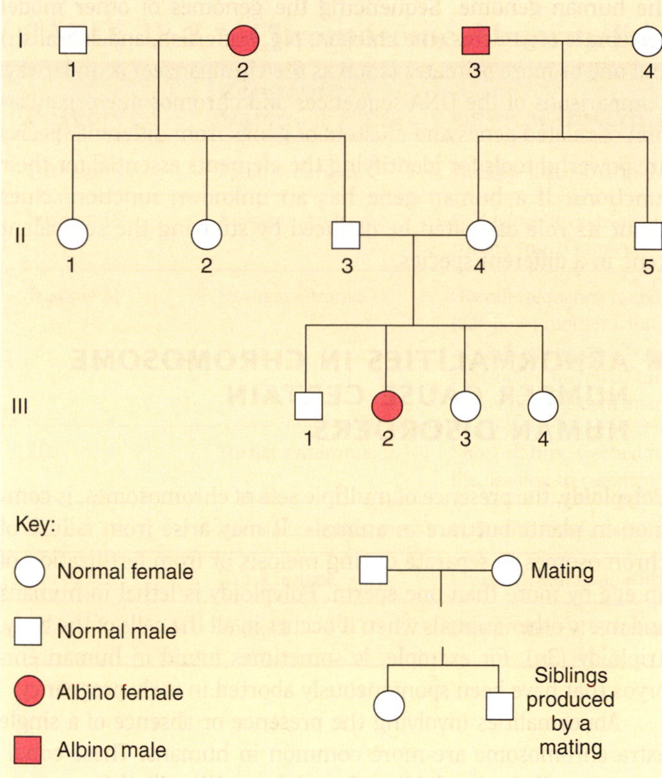

Figure 15–2 Pedigree analysis of albinism. By studying family histories, it is often possible to determine the genetic mechanism of the trait being studied. Consider III-2, an albino girl with two phenotypically normal parents, II-3 and II-4. The allele for albinism cannot be dominant because if it were, at least one of III-2's parents would have to be an albino. Also, albinism cannot be an X-linked recessive allele because if it were, her father would have to be an albino (and her mother would have to be a heterozygous carrier). This pedigree is easily explained if albinism is inherited as an autosomal (not carried on a sex chromosome) recessive allele. With an autosomal recessive allele, two phenotypically normal parents could produce an albino offspring (because they are heterozygotes and could each transmit a recessive allele).

For example, individuals II-3 and II-4 are parents of four offspring (III-1, III-2, III-3, and III-4). Note that it is possible for individuals in a given generation to be genetically unrelated. For example, II-1, II-2, and II-3 are unrelated to II-4 and II-5. Within a group of siblings, the oldest is on the left, and the youngest is on the right.

Studying pedigrees has enabled human geneticists to predict how phenotypic traits that are governed by the genotype at a single locus are inherited. Such traits are said to be *Mendelian.* About 10,000 Mendelian traits have been described in humans (see McKusick's on-line catalog, listed in the *Recommended Readings*). Pedigree analysis most often identifies three types of single-gene inheritance—autosomal dominant, autosomal recessive, and X-linked recessive; we define and discuss examples of these traits later in this chapter.

Traits that do not show a simple Mendelian inheritance pattern cannot be characterized as well using pedigree analysis. Such traits may be the result of interactions among genes at two, three, or more loci. In other cases, **imprinting**—the expression of a gene based on its parental origin—occurs. In some genes the paternally inherited allele is always repressed (not expressed), whereas in other genes the maternally inherited allele is always repressed. Environmental factors may also play a major role in non-Mendelian inheritance patterns. Common birth defects, such as cleft palate and congenital heart disease, are examples of traits that do not show a simple Mendelian inheritance pattern.

The Human Genome Project is mapping the genes on all human chromosomes

In 1999 a significant milestone was reached: Human chromosome 22 was the first human chromosome to have virtually all of its DNA sequence determined (Table 15–1). One of the smallest chromosomes, human chromosome 22 contains at least 545 genes encoded in its more than 33 million bases of adenine, guanine, cytosine, and thymine.

Sequencing human chromosome 22 is part of the **Human Genome Project,** which is mapping and sequencing all of the DNA in the nuclear human genome—about 2.9 billion base pairs (Fig. 15–3). (The human mitochondrial genome was sequenced in 1981.) This international undertaking, which is based on the DNA from six to ten anonymous individuals, was essentially completed and published in 2001 by two independent teams, the International Human Genome Sequencing Consortium and Celera Genomics. The project includes not only sequencing the entire human genome but also eventually identifying where all of the perhaps 35,000 to 45,000 genes[1] are located in the sequenced DNA.

[1] Biologists do not yet know how many genes are in the human genome. Estimates range from 30,000 to more than 150,000, but as more data have accumulated from the Human Genome Project, many researchers have revised their estimates downward. Our estimate of 35,000 to 45,000 is from *Science,* Vol. 291, 16 Feb. 2001. A more accurate estimate should be available in the next few years.

TABLE 15–1 Some Important Milestones in Genetics

Year	Scientific Advance
1866	Mendel proposed the existence of hereditary factors now known as genes
1871	Nucleic acids discovered
1953	Structure of DNA determined
1960s	Genetic code explained (how proteins are made from DNA)
1977	DNA sequencing began
1986	DNA sequencing automated
1999	Sequencing of first human chromosome completed
2001	Draft sequence of entire human genome published

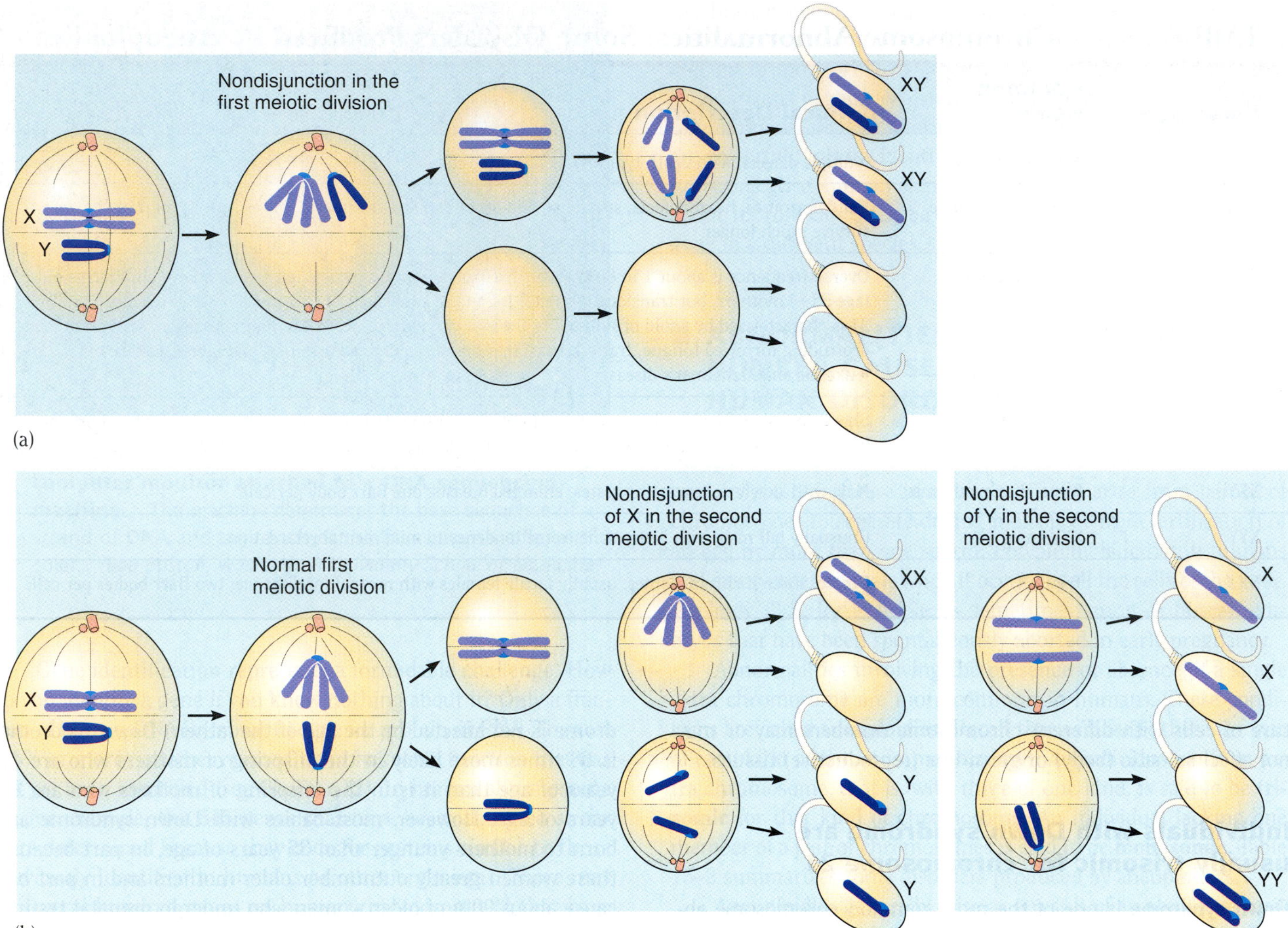

Figure 15–4 Meiotic nondisjunction. In these examples of nondisjunction of the sex chromosomes in the human male, only the X *(purple)* and Y *(blue)* chromosomes are shown. **(a)** Nondisjunction in the first meiotic division results in two XY sperm and two sperm with neither an X nor a Y. **(b)** Second-division nondisjunction of the X chromosome results in one sperm with two X chromosomes, two with one Y each, and one with no sex chromosomes. Nondisjunction of the Y results in one sperm with two Y chromosomes, two with one X each, and one with no sex chromosome.

According to the *single active X hypothesis,* mammals compensate for extra X chromosome material by rendering all but one X chromosome inactive. The inactive X is seen as a Barr body, a region of darkly staining, condensed chromatin next to the nuclear envelope of an interphase nucleus (see Fig. 10–16). The presence of the Barr body in the cells of normal females (but not of normal males) has been used as an initial screen to determine whether an individual is genetically female or male. As we will see in our discussion of sex chromosome aneuploidies, however, the Barr body test has serious limitations.

Individuals with **Klinefelter syndrome** have 47 chromosomes, including two Xs and one Y. They have small testes, produce few or no sperm and are therefore sterile. Evidence that the Y chromosome is the major determinant of the male phenotype has been substantiated by the fact that there is at least one gene on the Y chromosome that appears to act as a genetic switch, directing male development. Males with Klinefelter syndrome tend to be unusually tall and to have female-like breasts. About half show some degree of mental retardation, but many live relatively normal lives. However, when their cells are examined they are found to have one Barr body per cell. On the basis of such a test, they would be erroneously classified as females. About 1 in 600 to 1000 live-born males has Klinefelter syndrome.

We designate the sex chromosome constitution for **Turner syndrome,** in which an individual has only one sex chromosome, an X chromosome, as XO. The O refers to the absence of a second sex chromosome. Because of the absence of the male-determining effect of the Y chromosome, these individuals develop as females. However, both their internal and external genital structures are underdeveloped, and they are sterile. Apparently a second X chromosome is necessary for the normal development of the ovaries in a female embryo. Examination of

(a)

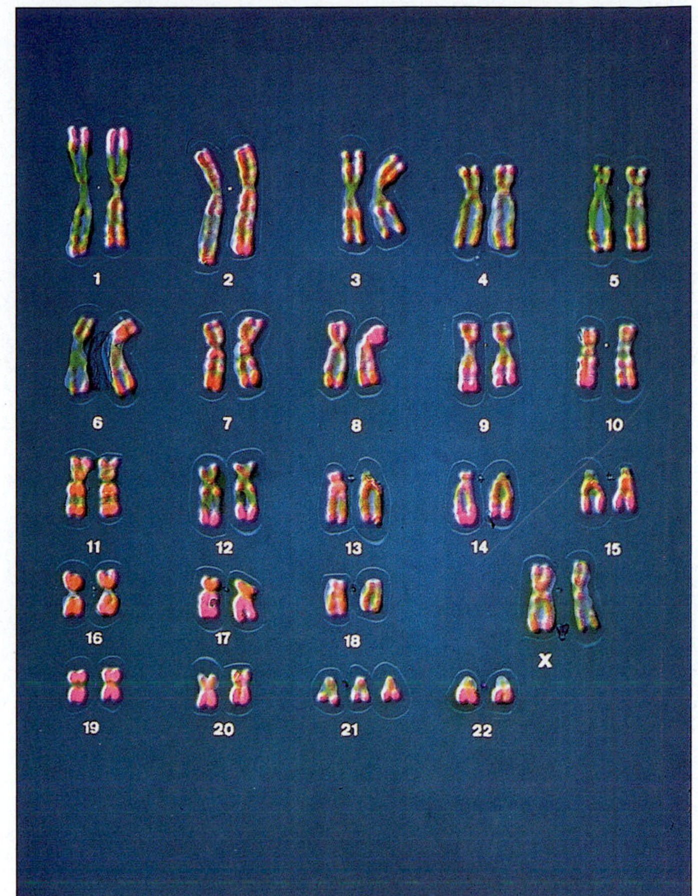

(b)

Figure 15–5 Down syndrome. **(a)** This male child with Down syndrome is working on a science experiment in his kindergarten class. Some individuals with Down syndrome learn to read and write. **(b)** Note the presence of an extra chromosome 21 in this color-enhanced karyotype of a female with Down syndrome. *(a, Richard Hutchings/Photo Researchers, Inc.; b, CNRI/Science Photo Library/Photo Researchers, Inc.)*

their cells reveals no Barr bodies because there is no extra X chromosome to be inactivated. Using the standards of the Barr body test, such an individual would be classified erroneously as a male. About 1 in 2500 live-born females has Turner syndrome.

People with an X chromosome plus two Y chromosomes are phenotypically males, and they are fertile. Other characteristics of these individuals (tall, with severe acne) hardly merit the term syndrome; hence the designation **XYY karyotype.** Some years ago there were several widely publicized studies that suggested that individuals with this condition are more likely to display criminal tendencies and to be imprisoned. However, these studies were flawed because they were based on small numbers of XYY men without adequate or well-matched control studies of XY males. The prevailing opinion in medical genetics today is that there are many undiagnosed XYY males in the general population who do not have overly aggressive or criminal behaviors and who are not incarcerated.

Aneuploidies usually result in prenatal death

Recognizable chromosome abnormalities are seen in less than one percent of all live births, but substantial evidence suggests that the rate at conception is much higher. At least 17% of pregnancies recognized at 8 weeks will end in spontaneous abortion (miscarriage). Approximately half of these spontaneously

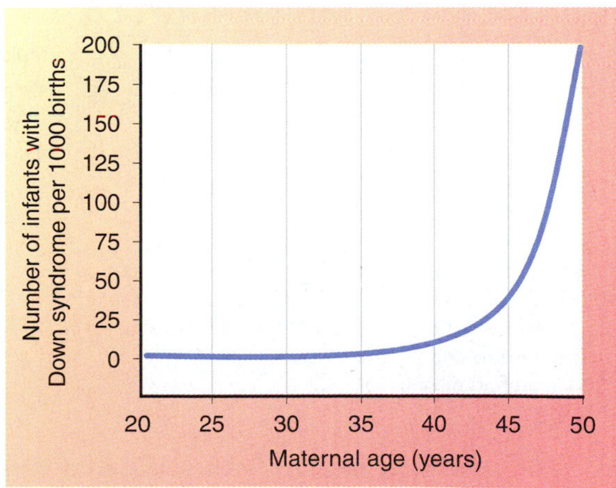

Figure 15–6 Down syndrome and maternal age. The estimated risk of a live birth of an infant with Down syndrome increases with maternal age. *(Adapted from Gardner, R.J.M. and G.R. Sutherland.* Chromosome Abnormalities and Genetic Counseling. *Oxford University Press, New York, 1996; and from Hecht, C.A. and E.B. Hook. "The Imprecision in Rates of Down Syndrome by One-Year Maternal Age Intervals: A Critical Analysis of Rates Used in Biochemical Screening."* Prenatal Diagnosis, *Vol. 14, 1994)*

aborted embryos have major chromosome abnormalities, including autosomal trisomies (e.g., trisomy 21), triploidy, tetraploidy, and Turner syndrome (XO). Autosomal monosomies are exceedingly rare, possibly because they induce a spontaneous abortion very early in the pregnancy, before a woman is even aware that she is pregnant. Some investigators place surprisingly high estimates (50% or more) on the rate of loss of very early embryos. It is widely assumed that chromosome abnormalities are responsible for a substantial fraction of these.

ABNORMALITIES IN CHROMOSOME STRUCTURE CAUSE CERTAIN DISORDERS

Chromosome abnormalities are not only caused by changes in chromosome *number* but also by distinct changes in the *structure* of one or more chromosomes. Here we consider three simple examples of structural abnormalities: translocations, deletions, and fragile sites.

Translocation is the attachment of part of a chromosome to a nonhomologous chromosome

In some cases part of one chromosome may break off and attach to a nonhomologous chromosome (a **translocation**), or two nonhomologous chromosomes may exchange parts (a **reciprocal translocation**). The consequences of translocations vary considerably but include situations in which some genes are missing **(deletions),** and extra copies of other genes are present **(duplications)** (Fig. 15–7).

In about 4% of individuals with Down syndrome, only 46 chromosomes are present, but one is abnormal. Extra genetic material from chromosome 21 has been translocated onto one of the larger chromosomes, such as chromosome 14. We refer to this kind of abnormal translocation chromosome as a *14/21 chromosome.* Affected individuals have one chromosome 14, one 14/21 chromosome, and two normal copies of chromosome 21. All or part of the genetic material from chromosome 21 is thus present in triplicate. When the karyotypes of such an individual's parents are studied, either the mother or the father is usually found to have only 45 chromosomes, although she or he is generally phenotypically normal. The parent with 45 chromosomes has one chromosome 14, one 14/21 chromosome, and one chromosome 21; although the karyotype is abnormal, there is no extra genetic material. In contrast to trisomy 21, this translocation form of Down syndrome can run in families, and its incidence is not related to maternal age.

A deletion is loss of part of a chromosome

Sometimes chromosomes break but fail to rejoin. Such breaks result in deletions of as little as a few base pairs to as much as an entire chromosome arm. As you might expect, large deletions are lethal, whereas small deletions cause several recognizable human disorders.

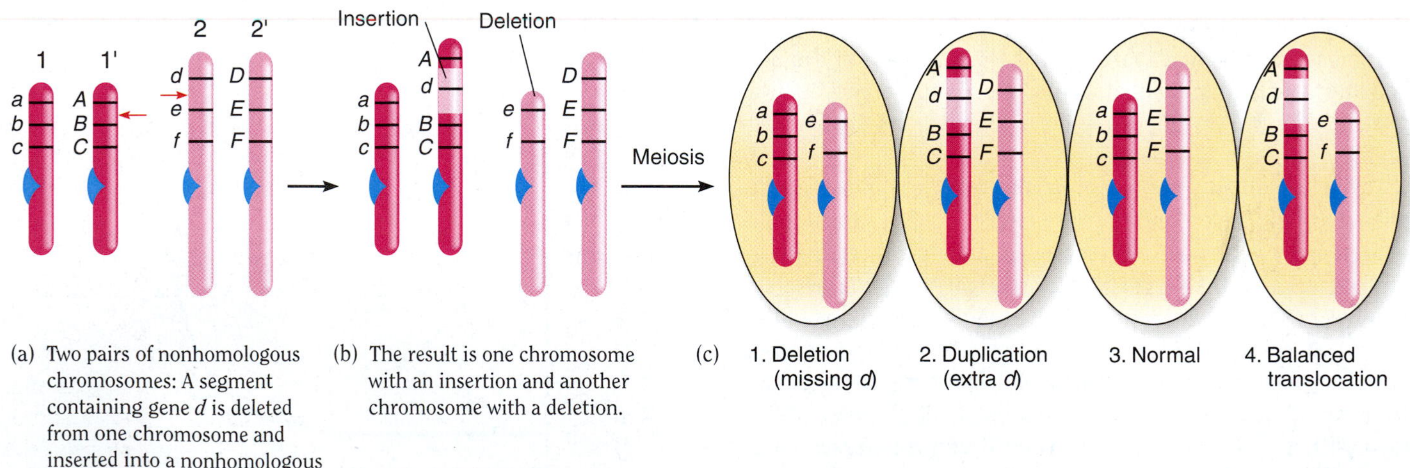

(a) Two pairs of nonhomologous chromosomes: A segment containing gene *d* is deleted from one chromosome and inserted into a nonhomologous chromosome.

(b) The result is one chromosome with an insertion and another chromosome with a deletion.

(c) 1. Deletion (missing *d*) 2. Duplication (extra *d*) 3. Normal 4. Balanced translocation

Figure 15–7 Translocation. **(a)** In this example, there are two pair of nonhomologous chromosomes, designated 1/1' and 2/2'. Three pairs of alleles, designated *A/a, B/b,* and *C/c* are on chromosomes 1/1', and three allelic pairs (*D/d, E/e,* and *F/f*) are on chromosomes 2/2'. Red arrows designate the breakage points in the chromosomes. **(b)** Allele *d* is deleted from chromosome 2 and inserted into chromosome 1'. **(c)** Following meiosis, four different combinations of chromosomes (and the genes on those chromosomes) are possible. (1) In one of the gametes, allele *d* is missing altogether. (2) In one gamete, there is an extra copy of *d.* (3) In one gamete, there is a complete set (one allele for each locus). (4) In one gamete, there is one allele for each locus, but allele *d* is not in its normal location on chromosome 2—that is, the gamete contains a balanced translocation.

One relatively common deletion disorder (1 in 50,000 live births) is **cri du chat syndrome,** in which part of the short arm of chromosome 5 is deleted. As in most deletions, the exact point of breakage in chromosome 5 varies from one individual to another; some cases of cri du chat involve a small loss, whereas others involve a more substantial deletion. Infants born with cri du chat syndrome typically have a small head with altered features described as a "moon face" and a distinctive cry that sounds like a kitten mewing. (The name *cri du chat* literally means "cry of the cat" in French.) Affected individuals usually survive beyond childhood but exhibit severe mental retardation.

Fragile sites are weak points at specific sites in chromatids

A **fragile site** occurs where part of a chromatid appears to be attached to the rest of the chromosome by a thin thread of DNA. Fragile sites may occur at a specific location on both chromatids of a particular chromosome. Such sites have been identified on the X chromosome as well as on certain autosomes. The location of a fragile site is exactly the same in all of an individual's cells, as well as in cells of other family members. Scientists report growing evidence that cancer cells may have breaks at these fragile sites. Whether cancer destabilizes the fragile sites, leading to breakage, or the fragile sites themselves contain genes that contribute to cancer is unknown at this time.

In **fragile X syndrome** the fragile site occurs near the tip of the X chromosome (Fig. 15–8). At the tip of an X chromosome, a nucleotide triplet (cytosine, guanine, guanine, or CGG) is normally repeated up to 50 times. The fragile X chromosome, however, repeats CGG from 200 to more than 1000 times. The effects of fragile X syndrome, which are more pronounced in males than in females, range from mild learning and attention disabilities to severe mental retardation and hyperactivity. According to the National Fragile X Foundation, about 80% of boys and 35% of girls with fragile X syndrome are at least mildly mentally retarded. Females with fragile X syndrome are usually heterozygous (because their other X chromosome is normal) and are therefore more likely to have normal intelligence. The discovery of the fragile X gene in 1991 and the development of the first fragile X mouse model in 1994 have provided researchers with ways to test potential treatments, including gene therapy.

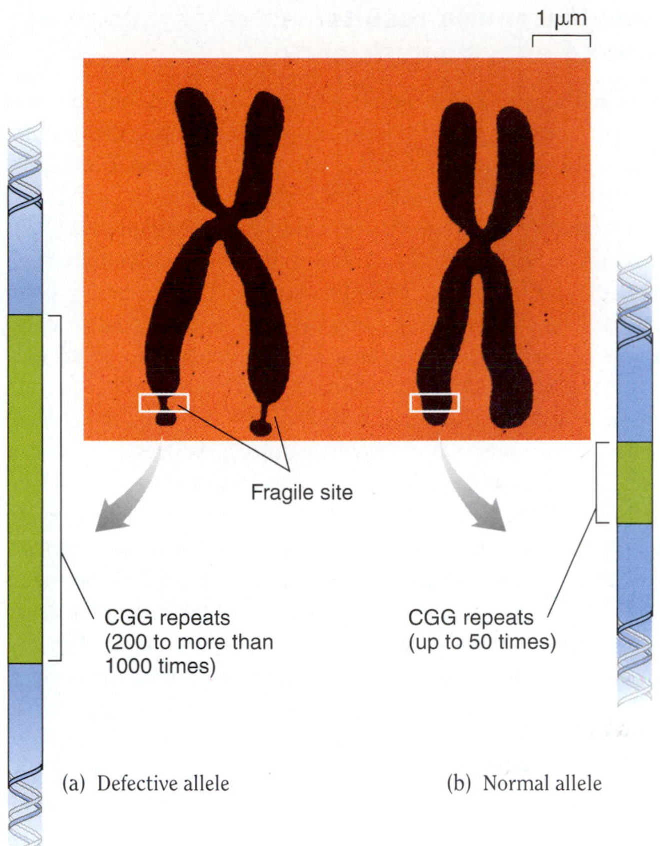

Figure 15–8 Fragile X syndrome. Color-enhanced SEM of *(left)* an X chromosome with a fragile site and *(right)* a normal X chromosome. **(a)** The defective allele at the tip of the X chromosome repeats the nucleotide triplet CGG from 200 to more than 1000 times. **(b)** The normal allele repeats CGG up to 50 times. *(Science VU)*

MOST GENETIC DISEASES ARE INHERITED AS AUTOSOMAL RECESSIVE TRAITS

We have seen that several human disorders involve chromosome abnormalities. Hundreds of human disorders, however, involve enzyme defects caused by mutations of single genes. Phenylketonuria (PKU) and alkaptonuria (see Chapter 12) are examples of these disorders, which are sometimes referred to as an **inborn error of metabolism,** a metabolic disorder caused by the mutation of a gene that codes for an enzyme needed in a biochemical pathway. Both PKU and alkaptonuria involve blocks in the metabolism of the amino acid phenylalanine. Only a small percentage of human genetic diseases have a simple inheritance pattern, but most of those that do are transmitted as autosomal recessive traits and so are expressed only in the homozygous state.

Why are these traits usually recessive? Most single-gene mutations that have an observable effect result in a gene that no longer works (it either does not produce enough gene product, or it produces a defective gene product). In most cases, the diploid cells are heterozygous and contain one normal copy of the gene and one mutated, nonfunctional copy. Since protein production is generally regulated by need, the cell will usually compensate by having the normal copy of the gene meet the demand. Thus, most genetic diseases are seen in homozygous recessive individuals—that is, they only exhibit the disorder when *both* copies of the gene are nonfunctional. In these relatively rare cases, in which both gene copies in the diploid cell contain mutations that make them nonfunctional, the cell's needs are not met, and the individual exhibits symptoms of the disorder.

Phenylketonuria results from an enzyme deficiency

Phenylketonuria (PKU), which is most common in individuals of western European descent, is an autosomal recessive disease that is a defect of amino acid metabolism. It affects about 1 in 10,000 live births in North America. Homozygous recessive individuals lack an enzyme that converts the amino acid phenylalanine to another amino acid, tyrosine. These individuals accumulate high levels of phenylalanine, phenylpyruvic acid, and similar compounds. The accumulating phenylalanine is converted to compounds known as phenylketones, which damage the central nervous system, including the brain. The ultimate result is severe mental retardation. An infant with PKU is usually healthy at birth because its mother, who is heterozygous, produces enough enzyme to prevent phenylalanine accumulation before birth. However, during infancy and early childhood, the toxic products eventually cause irreversible damage to the central nervous system.

In the 1950s it was found that if infants with PKU are identified and placed on a special low-phenylalanine diet early enough, the symptoms can be dramatically alleviated. (The diet is difficult to adhere to because it contains no meat, fish, dairy products, breads, or nuts. Also, the sugar substitute aspartame, found in many diet drinks and foods, should not be consumed by individuals with PKU because it contains phenylalanine.) Biochemical tests for PKU have been developed, and screening of newborns through a simple blood test is required in the United States. Because of these screening programs and the availability of effective treatment, thousands of PKU-diagnosed children have not developed severe mental retardation. Most such children must continue the diet through at least adolescence. Doctors recommend that the diet be continued throughout life because some adults who have discontinued the special diet show a decline in certain mental faculties, such as some loss of the ability to concentrate and some short-term memory loss.

Ironically, the success of PKU treatment in childhood presents a new problem today. If a homozygous female who has discontinued the special diet becomes pregnant, the high phenylalanine levels in her blood can result in damage to the brain of the fetus she is carrying, even though that fetus is heterozygous. Therefore, she must resume the diet, preferably before becoming pregnant. This procedure is usually (although not always) successful in preventing the effects of **maternal PKU.** Therefore it is especially important that females with PKU be aware of their condition so that they may obtain appropriate counseling and medical treatment during pregnancy.

Sickle cell anemia results from a hemoglobin defect

Sickle cell anemia is inherited as an autosomal recessive trait. The disease is most common in individuals of African descent (approximately 1 in 500 African-Americans), and about 1 in 12 African-Americans is heterozygous. The blood cells of an individual with sickle cell anemia are shaped like sickles, or half-moons, whereas normal red blood cells are biconcave discs.

The mutation that causes sickle cell anemia was first identified more than 50 years ago. The sickle cell contains abnormal hemoglobin molecules, which have the amino acid valine instead of glutamic acid at position 6 (the sixth amino acid from the amino terminal end) in the beta chain (see Chapter 3). The substitution of valine for glutamic acid makes the hemoglobin less soluble. As a result, it tends to form crystal-like structures that change the shape of the red blood cells (Fig. 15–9*a*). This sickling occurs in the veins after the oxygen has been released from the hemoglobin. The blood cells' abnormal sickled shape slows blood flow and blocks small blood vessels (Fig. 15–9*b*), with resulting tissue damage from lack of oxygen and essential nutrients, and painful episodes. Sickled red blood cells also have shorter life spans than normal red blood cells, leading to severe anemia in many affected individuals.

Available treatments for sickle cell anemia include pain relief measures, transfusions, and, more recently, medicines such as hydroxyurea, which activates the gene for the production of normal fetal hemoglobin (this gene is generally not expressed after birth). The presence of normal fetal hemoglobin in the blood dilutes the sickle cell hemoglobin so there are fewer painful episodes and less need for blood transfusions. The long-term effects of hydroxyurea are not known at this time, but there are concerns that it may induce tumor formation.

Ongoing research is directed toward eventually providing gene therapy for sickle cell anemia. Bone marrow transplants are also a promising future treatment for seriously ill individuals. The development of a mouse model of sickle cell anemia, which was reported in 1997, will allow researchers to test potential drug and genetic therapies.

The reason that the sickle cell allele occurs at a higher frequency in parts of Africa is well known. Individuals who are heterozygous (*Ss*) and carry alleles for both normal hemoglobin (*S*) and sickle cell hemoglobin (*s*) are more resistant to falciparum malaria, a severe form of malaria that is often fatal. The malarial parasite, which spends part of its life cycle inside red blood cells, does not thrive when sickle cell hemoglobin is present. (An individual heterozygous for sickle cell anemia produces both normal and sickle cell hemoglobin.) Areas in Africa where falciparum malaria occurs correlate well with areas in which the frequency of the sickle cell allele is more common in the human population. Thus, *Ss* individuals, who possess one copy of the mutant sickle cell allele, have a selective advantage over homozygous individuals, both *SS* (who may die of malaria) and *ss* (who may die of sickle cell anemia). This phenomenon, known as **heterozygote advantage,** is discussed further in Chapter 18 (see Fig. 18–7).

Cystic fibrosis results from defective ion transport

Cystic fibrosis is the most common autosomal recessive disorder in children of European descent (1 in 2500 births). About 1 in 25 individuals in the United States is a heterozygous carrier of the cystic fibrosis allele. Abnormal secretions in the body characterize this disorder. Its most severe effect is on the respiratory sys-

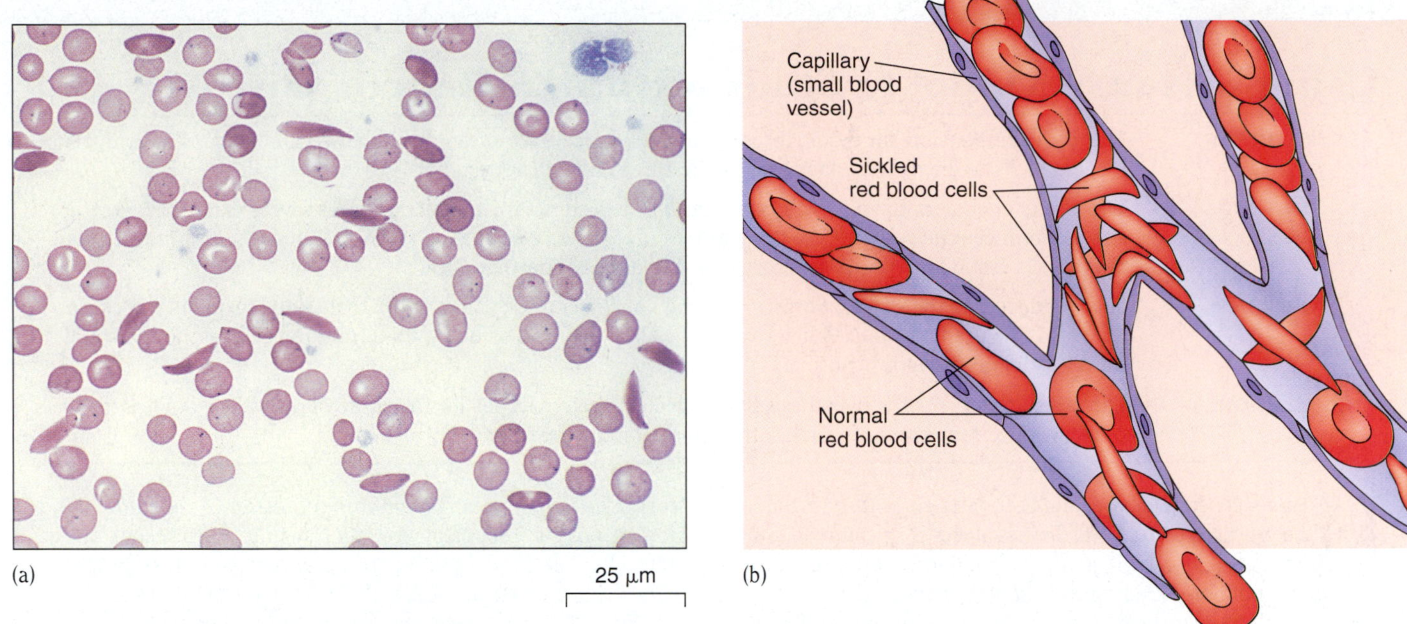

Figure 15–9 Sickle cell anemia. **(a)** LM of sickled red blood cells. Some of the red blood cells from this individual with sickle cell anemia show an abnormal sickle shape. **(b)** Sickled red blood cells do not pass through small blood vessels as easily as normal red blood cells. The sickled blood cells can cause blockages that prevent oxygen from being delivered to tissues. *(a, G.W. Willis/Biological Photo Service)*

tem, which produces abnormally viscous mucus. The cilia that line the bronchi (see Chapter 44) cannot easily remove the mucus, and it thus becomes a culture medium for dangerous bacteria. These bacteria or their toxins attack the surrounding tissues, leading to recurring pneumonia and other complications. The heavy mucus also occurs elsewhere in the body (e.g., in the ducts of the pancreas, liver, intestines, and sweat glands), causing digestive difficulties and other effects.

The gene responsible for cystic fibrosis codes for a protein that controls the transport of chloride and certain other ions across cell membranes. The mutant protein, found in plasma membranes of epithelial cells lining the ducts of the lungs, intestines, pancreas, liver, sweat glands, and reproductive organs, results in the production of an unusually thick mucus that eventually leads to tissue damage. Although many mutant forms exist and these vary somewhat in severity of symptoms, the disease is usually very serious.

Antibiotics are used to control bacterial infections, and daily physical therapy is required to clear mucus from the respiratory system (Fig. 15–10). Treatment with an enzyme (produced by recombinant DNA technology) that breaks down the mucus is also helpful. Without treatment, death would occur in infancy. With treatment, the average life expectancy for individuals with cystic fibrosis is now about 30 years. Because of the serious limitations of available treatments, gene therapy for cystic fibrosis is under development.

The most severe mutant allele for cystic fibrosis predominates in northern Europe, and another, somewhat less serious, mutant

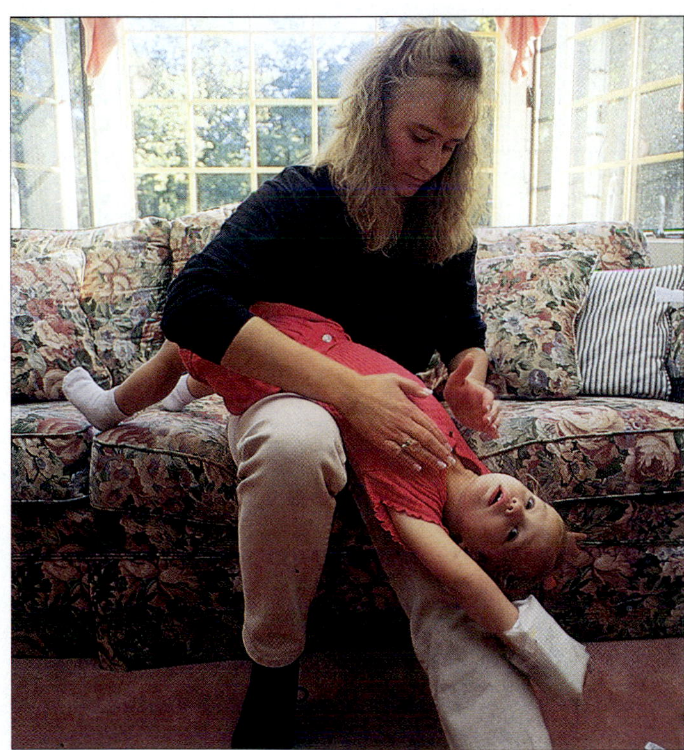

Figure 15–10 Treatment of cystic fibrosis. The traditional treatment for cystic fibrosis has been chest percussion, or gentle pounding on the chest, to clear mucus from clogged airways in the lungs. *(Abraham Menashe)*

Process of Science

ON THE CUTTING EDGE

Using a Mouse Model to Study a Human Genetic Disease

HYPOTHESIS: The alleles responsible for cystic fibrosis confer a heterozygote advantage because heterozygous individuals are less likely to die from certain types of life-threatening diarrhea.

METHOD: The effects of cholera toxin, which is produced by a bacterial infection and causes severe diarrhea, were studied in mice. The responses of three groups of mice were evaluated: (1) mice homozygous for an allele that causes cystic fibrosis, (2) heterozygous mice, and (3) normal mice.

RESULTS: Mice homozygous for a cystic fibrosis allele did not respond to cholera toxin; that is, the toxin did not cause diarrhea. Cholera toxin caused heterozygotes to lose only half as much fluid from the cells of the intestinal lining as did homozygous "normal" animals.

CONCLUSION: These results support the hypothesis that individuals heterozygous for the cystic fibrosis allele are less likely to die from certain kinds of diarrhea than normal individuals.

Human geneticists have long been puzzled by the fact that heterozygosity for cystic fibrosis alleles is so common among whites (1/20), particularly among Northern Europeans and those of Northern European descent. Recent advances in our understanding of the molecular biology of this genetic disease have led to a new explanation of why these alleles have persisted in the population. This new hypothesis is that heterozygous individuals are less likely to die from certain types of potentially fatal diarrhea, such as cholera.

In severe diarrhea, large amounts of water and electrolytes are lost from the intestine. If unchecked, particularly in infants and young children, this condition can result in death. It has been found that the allele that causes cystic fibrosis is a mutant form of a gene involved in controlling the body's water and electrolyte balance. This gene has been cloned and found to code for a protein, the *CFTR protein,* that serves as a chloride ion channel in the plasma membrane. (CFTR stands for *cystic fibrosis transmembrane conductance regulator.*) This ion channel is responsible for transporting chloride ions out of the cells lining the digestive tract and the respiratory system. When the chloride ions leave the cells, water follows by osmosis. Thus the normal secretions of these cells are relatively watery.

Because the cells of individuals with cystic fibrosis lack normal chloride ion channels, their secretions have a very low water content, and their sweat is very salty. Cells of heterozygous individuals have only half the usual number of functional CFTR ion channels, but these are sufficient to maintain normal fluidity of their secretions. However, this ion channel deficiency might be an advantage if an individual is infected by a pathogen that produces a toxin that causes the ion channels to remain constantly open, precipitating diarrhea. With only half as many normal ion channels, a heterozygote might lose only half as much chloride and water.

allele is more prevalent in southern Europe. Presumably these mutant alleles are independent mutations that have been maintained by natural selection. Some experimental evidence supports the hypothesis that heterozygous individuals are less likely to die from infectious diseases that produce severe diarrhea. (See *On the Cutting Edge: Using a Mouse Model to Study a Human Genetic Disease.*)

Tay-Sachs disease results from abnormal lipid metabolism in the brain

Tay-Sachs disease is an autosomal recessive disease that affects the central nervous system and results in blindness and severe mental retardation. The symptoms begin within the first year of life and result in death before the age of five years. Because of the absence of an enzyme, a normal membrane lipid in the brain cells fails to break down properly and accumulates in the lysosomes. Although research is ongoing, no effective treatment for Tay-Sachs disease is available at this time. However, an effective strategy for the treatment of Tay-Sachs in a mouse model was reported in 1997. This treatment, which involves oral administration of an inhibitor that reduces the synthesis of the lipid that accumulates in the lysosomes, offers the hope of an effective way to deal with Tay-Sachs disease in humans in the future.

The abnormal allele is especially common in the United States among Jews whose ancestors came from Eastern and Central Europe (Ashkenazi Jews). About 1 in 4000 live births in the North American Jewish population has the disease. By contrast, Jews whose ancestors came from the Mediterranean region (Sephardic Jews) have a very low frequency of the allele. Why the high frequency of such a harmful allele has been maintained in the Ashkenazi Jewish population is not known. It has been suggested that carriers of the Tay-Sachs allele may have some, as yet unknown, heterozygote advantage.

SOME GENETIC DISEASES ARE INHERITED AS AUTOSOMAL DOMINANT TRAITS

Huntington disease (HD), named after George Huntington, the American physician who first described it in 1872, is due to a rare autosomal dominant allele that affects the nervous system. The disease causes severe mental and physical deterioration, uncontrollable muscle spasms, personality changes, and ultimately death. No effective treatment has been found. Every child of an affected individual has a 50% chance of also being

Research on any disease is greatly facilitated if an animal model can be used for experimentation. Such a model became available when strains of mice homozygous for cystic fibrosis, as well as those that were heterozygous, were produced by gene targeting (see Chapter 14).

Sherif E. Gabriel and colleagues at the University of North Carolina used the cystic fibrosis mouse model to test the heterozygote advantage hypothesis (see Chapter 18).* They treated mice with cholera toxin produced by *Vibrio cholerae,* the bacterium that causes cholera. Cholera toxin is known to affect the functioning of the CFTR ion channels, causing the uncontrolled loss of chloride ions and water.

Their results neatly fit the predictions of the heterozygote advantage hypothesis. Animals homozygous for cystic fibrosis, that is, those with no functional CFTR channels, did not lose any fluid through their intestinal cells when exposed to cholera toxin. The toxin caused heterozygotes (with only half the number of normal CFTR channels) to lose only half as much fluid as mice with the normal number of channels.

Despite the success of this demonstration, it is not thought that cholera itself is the selective force responsible for the high incidence of cystic fibrosis alleles today. Cystic fibrosis is thought to have arisen more than 50,000 years ago, whereas European cholera epidemics were first recognized in the early 1800s. Rather, some other diarrhea-causing infection is probably the culprit.

* Gabriel, S.E., K.N. Brigman, B.H. Koller, R.C. Boucher, and M.J. Stutts. "Cystic Fibrosis Heterozygote Resistance to Cholera Toxin in the Cystic Fibrosis Mouse Model." *Science,* Vol. 266, 7 Oct. 1994.

Researchers have continued to test candidates likely to be responsible for a possible heterozygote advantage of cystic fibrosis alleles. In 1998 Gerald Pier and colleagues at the Harvard Medical School, University of Bristol, and University of Cambridge announced that the cystic fibrosis allele protects against typhoid fever, a diarrheal disease caused by the bacterium *Salmonella typhi.*† This bacterium enters human gastrointestinal cells with the CFTR channel protein but does not easily enter cells expressing a cystic fibrosis allele. These data are supported by studies in which no *S. typhi* entered cells of mice homozygous for the cystic fibrosis allele, and 86% fewer *S. typhi* entered heterozygous mouse cells compared with homozygous normal mouse cells.

Some researchers are now focusing their efforts on understanding the complex way in which the CFTR channel is activated or inactivated.‡ It is hoped that this information will allow researchers to design drugs that enhance or inhibit chloride transport through the CFTR channel. Such drugs have the potential to treat cystic fibrosis (by activating the mutant channel) and diarrhea (by inhibiting the normal channel).

† Pier, G.B., M. Grout, T. Zaidi, G. Meluleni, S.S. Mueschenborn, G. Banting, R. Ratcliff, M.J. Evans, and W.H. Colledge. "*Salmonella typhi* Uses CFTR to Enter Intestinal Epithelial Cells." *Nature,* Vol. 393, 7 May 1998.

‡ Naren, A.P., E. Cormet-Boyaka, J. Fu, M. Villain, J.E. Blalock, M.W. Quick, and K.L. Kirk. "CFTR Chloride Channel Regulation by an Interdomain Interaction." *Science,* Vol. 286, 15 Oct. 1999.

affected (and, if affected, of passing the abnormal allele to his or her offspring). Ordinarily we would expect a dominant allele with such devastating effects to occur only as a new mutation and not to be transmitted to future generations. This disease is characterized by onset of symptoms relatively late in life (most do not develop the disease until their 30s or 40s), so an individual may have children before the disease develops (Fig. 15–11). In North America HD occurs at a frequency of 1 in 20,000 live births.

The gene responsible for HD was cloned in 1993. The mutation is a nucleotide triplet (cytosine, adenine, guanine, or CAG) that is repeated many times; the normal allele repeats CAG from 6 to 35 times, whereas the mutant allele repeats CAG from 40 to more than 150 times. Because the triplet CAG codes for the amino acid glutamine, the resulting proteins have long strands of glutamine. Interestingly, the number of nucleotide triplet repeats seems to be important in determining the age of onset and the severity of the disease; larger numbers of repeats correlate with an earlier age of onset and greater severity.

Much research is now directed at studying how the mutation is linked to neurodegeneration in the brain. A mouse model of HD has been produced that is providing valuable clues about the development of the disease. Once HD's mechanism of

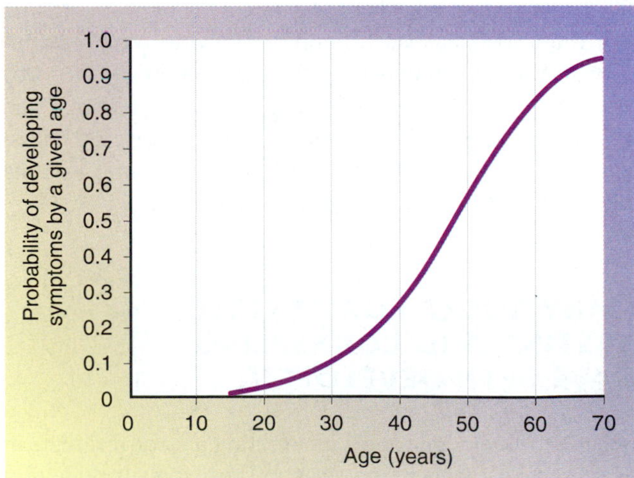

Figure 15–11 Age of onset of Huntington disease. The graph shows the probability that an individual affected with Huntington disease will have developed symptoms at a given age. The onset of Huntington disease occurs relatively late in life. *(Adapted from Harper, P.S.* Genetic Counseling, *5th edition. Butterworth-Heinemann, Oxford, 1998)*

action on nerve cells is better understood, it may be possible to develop effective treatments to slow the progression of the disease.

Cloning of the HD allele became the basis for tests that allowed those at risk to learn presymptomatically if they carry the allele. The decision to be tested for any genetic disease is understandably a highly personal one. Certainly, the information can be very useful for those who must make decisions such as whether or not to have children. However, someone who tests positive for the HD allele must then live with the virtual certainty of eventually developing this devastating and incurable disease. It is hoped that affected individuals who choose to be identified before the onset of symptoms may ultimately contribute to the development of effective treatments.

SOME GENETIC DISEASES ARE INHERITED AS X-LINKED RECESSIVE TRAITS

Hemophilia A was once referred to as a disease of royalty because of its high incidence among male descendants of Queen Victoria, but it is also found in many nonroyal pedigrees. Characterized by the lack of a blood clotting factor, Factor VIII, it causes severe internal bleeding in the head, joints, and other areas from even a slight wound. The mode of inheritance is X-linked recessive. Thus, affected individuals are almost exclusively male, having inherited the abnormal allele on the X chromosome from their heterozygous carrier mothers. (For a female to be affected by an X-linked trait, she would have to inherit the defective allele from both parents, whereas an affected male only needs to inherit one defective allele, from his mother.)

Treatments for hemophilia A consist of blood transfusions and administration of clotting factor VIII (the missing gene product) by injection. Unfortunately, these treatments are costly (as much as $40,000 a year for one individual). During the 1980s, many clotting factor VIII preparations made from human plasma were contaminated with human immunodeficiency virus-1 (HIV-1), the virus that causes AIDS, acquired immunodeficiency syndrome, and many men with hemophilia subsequently died. Since 1992, virus-free clotting factor VIII has been available from both human plasma and recombinant DNA technology (see Chapter 14).

MANY TOOLS FOR GENETIC TESTING AND COUNSELING HAVE BEEN DEVELOPED

There have been many advances in detecting genetic disorders in individuals, their families, and society. These tools include prenatal diagnosis and genetic screening. With these advances comes increased information for couples at risk of having children with genetic diseases. Helping couples understand and deal with genetic information now available has been the realm of the rapidly expanding field of genetic counseling.

Prenatal diagnosis can detect abnormalities in chromosome number and structure as well as many gene defects

Genetic abnormalities may become apparent during early prenatal development or not until late in adult life. Given that early detection increases the possibilities for prevention or alleviation of the effects of genetic abnormalities, efforts have been made over the years to detect such abnormalities before birth. In the past 20 years health care providers have become increasingly successful at prenatal diagnosis of many genetic diseases.

In one diagnostic technique known as **amniocentesis,** a sample of the fluid surrounding the fetus (the *amniotic fluid*) is obtained. A needle is inserted through the walls of the pregnant woman's abdomen, into the uterus, and then into the amniotic sac surrounding the fetus. Some of the amniotic fluid is withdrawn from the amniotic cavity into a syringe (Fig. 15–12). This procedure is normally safe from needle injuries because the positions of the fetus, placenta, and the needle can be determined through **ultrasound imaging** (see Fig. 49–21). There is a slight (about 0.5%, or 1 in 200) chance that amniocentesis will induce a miscarriage.

The amniotic fluid contains living cells sloughed off the body of the fetus and hence genetically identical to the cells of the fetus. These amniotic fluid cells can be cultured in the laboratory. After 10 to 14 days, dividing cells from the culture can be karyotyped to detect chromosome abnormalities. Amniocentesis, which has been performed since the 1960s, is routinely offered for pregnant women older than 35 years of age because their fetuses have a higher than normal risk of Down syndrome.

Other prenatal tests have been developed to detect many simply inherited genetic disorders, but these disorders are rare enough that the tests are usually done only if a particular problem is suspected. Enzyme deficiencies can often be detected through incubation of cells recovered from amniotic fluid with the appropriate substrate and measurement of the product; this technique has been useful in the prenatal diagnosis of disorders such as Tay-Sachs disease. The tests for several other diseases, including sickle cell anemia, HD, and cystic fibrosis, are carried out by directly testing the individual's DNA for the defective gene.

Amniocentesis is also useful in detecting a condition known as *spina bifida,* in which the spinal cord does not close properly during development. A relatively common (about 1 in 300 births) malformation, this birth defect is associated with abnormally high levels of a normally occurring protein, α-fetoprotein, in the amniotic fluid. Some of this protein crosses the placenta into the mother's blood, which can be tested for maternal serum α-fetoprotein (MSAFP) as a screen for spinal cord defects. If an elevated level of MSAFP is detected, diagnostic tests, such as ultrasound imaging and amniocentesis, are performed. (Interestingly, abnormally *low* levels of MSAFP are associated with Down syndrome and other trisomies.)

One problem with amniocentesis is that most of the conditions it detects are incurable, and the results are generally not obtained until well into the second trimester when terminating the pregnancy is both psychologically and medically more diffi-

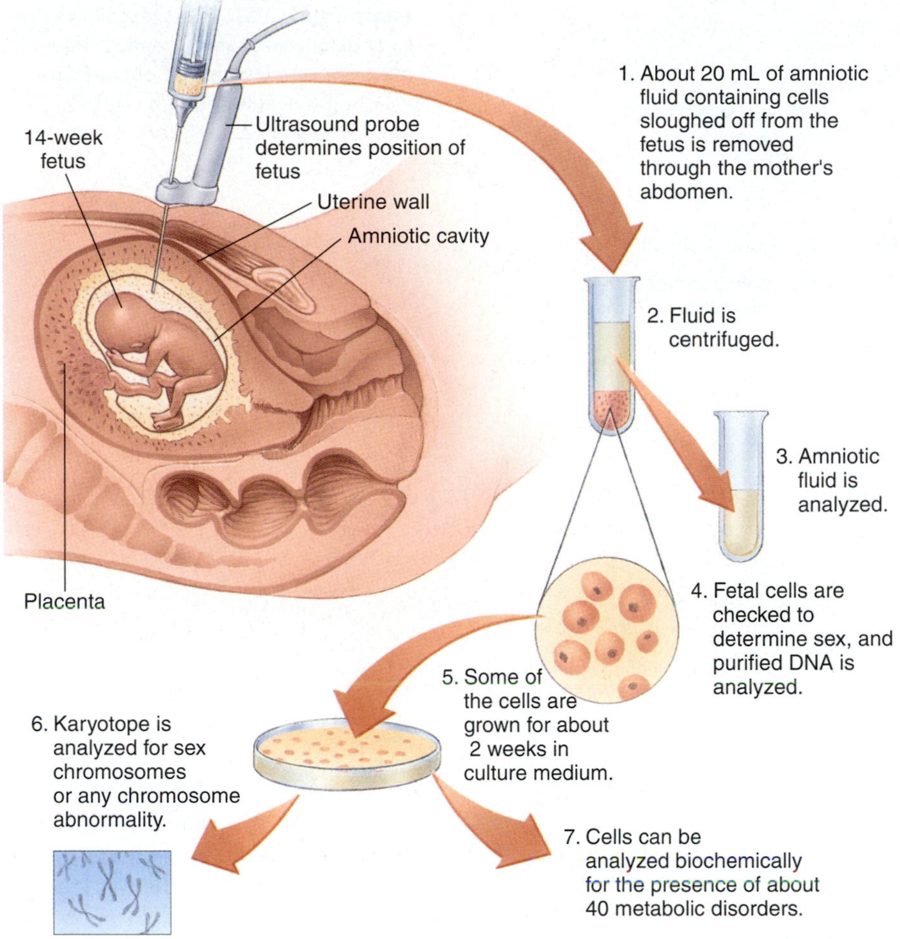

Figure 15–12 Amniocentesis. Certain genetic diseases and other abnormal conditions can be diagnosed prenatally by amniocentesis.

14-week fetus
Ultrasound probe determines position of fetus
Uterine wall
Amniotic cavity
Placenta
1. About 20 mL of amniotic fluid containing cells sloughed off from the fetus is removed through the mother's abdomen.
2. Fluid is centrifuged.
3. Amniotic fluid is analyzed.
4. Fetal cells are checked to determine sex, and purified DNA is analyzed.
5. Some of the cells are grown for about 2 weeks in culture medium.
6. Karyotope is analyzed for sex chromosomes or any chromosome abnormality.
7. Cells can be analyzed biochemically for the presence of about 40 metabolic disorders.

cult than earlier. Therefore, efforts have been made to develop tests that yield results earlier in the pregnancy. One such test, **chorionic villus sampling (CVS),** involves removing and studying cells that will form the fetal contribution to the placenta (Fig. 15–13). CVS, which has been performed in the United States since about 1983, is associated with a slightly greater risk of infection or miscarriage than amniocentesis, but its advantage is that results can be obtained earlier than in amniocentesis, usually within the first trimester.

Although both amniocentesis and CVS can diagnose certain genetic disorders with a high degree of accuracy, they are not foolproof, and many disorders cannot be diagnosed. Therefore, the lack of an abnormal finding is no guarantee of a normal pregnancy.

Genetic screening searches for a particular genotype or karyotype in infants or adults

Genetic screening is a systematic search through a population for individuals with a genotype or karyotype that might cause a serious genetic disease in them or their offspring. There are two main types of genetic screening, for newborns and for adults, and each serves a different purpose. Screening of newborns is done primarily as the first step in preventive medicine, and screening of adults is done to help them make informed reproductive decisions.

Newborns are screened to detect and treat one or more genetic diseases before the onset of serious symptoms. The routine screening of infants for PKU, discussed earlier in the chapter, began in 1962 in Massachusetts. Laws in all 50 states of the United States, as well as in several other countries, currently require PKU screening. Sickle cell anemia is another genetic disease that is more effectively treated if the diagnosis is made early. Screening newborns for sickle cell anemia enables doctors to reduce infant mortality about 15% because they can administer daily doses of antibiotics, thereby preventing deadly bacterial infections to which infants with sickle cell anemia are susceptible.

Genetic screening of adults is done to identify carriers (heterozygotes) of recessive genetic disorders. The carriers are then counseled about the risks involved in having children if both prospective parents are heterozygous. Since the 1970s, about 1 million young Jewish adults in the United States, Israel, and other countries have been screened voluntarily for Tay-Sachs disease, and about 1 in 30 has been identified as a carrier. Tay-Sachs screening programs have resulted in a reduction in the incidence of Tay-Sachs disease by more than 90%.

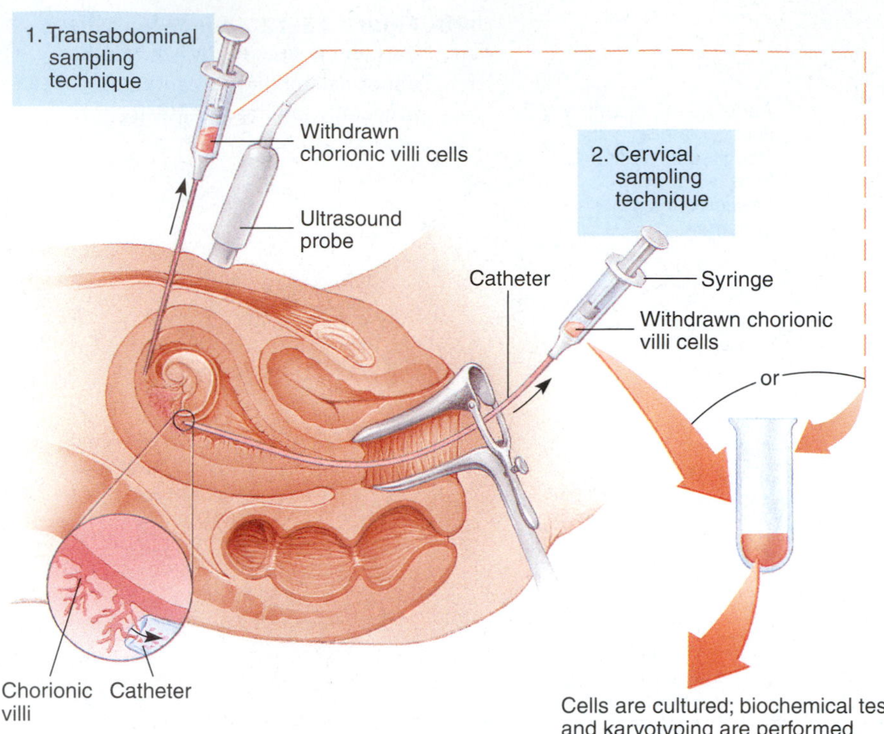

Figure 15–13 Chorionic villus sampling (CVS). This test allows the early diagnosis of some genetic abnormalities. Samples may be obtained by inserting a needle through (1) the uterine wall or (2) the cervical opening.

Genetic counselors educate people about genetic diseases

Couples who are concerned about the risk of abnormality in their children, either because they have had an abnormal child or have a relative affected by a hereditary disease, may seek **genetic counseling.** Genetic counselors provide medical and genetic information as well as support and guidance. Genetic clinics are available in most major metropolitan centers; these clinics are usually affiliated with medical schools.

Genetic counselors, who have received specialized training in counseling, medicine, and human genetics,[2] give people information needed to make reproductive decisions. Advice, tempered with respect and sensitivity, is given in terms of risk estimates—that is, the *probability* that any given offspring will inherit a particular condition. The counselor needs complete family histories of both the man and the woman and may screen for the detection of heterozygous carriers of certain conditions.

When a disease involves only a single gene locus, probabilities can usually be easily calculated. For example, if one prospective parent is affected with a trait that is inherited as an autosomal dominant disorder, such as Huntington disease, the probability that any given child will have the disease is 0.5, or 50%. The birth to phenotypically normal parents of a child affected with an autosomal recessive trait, such as albinism or PKU, establishes that both parents are heterozygous carriers, and the probability that any subsequent child will be affected is therefore 0.25, or 25%. For a disease inherited through a recessive allele on the X chromosome, such as hemophilia A, a normal woman and an affected man will have daughters who are carriers and sons who are normal. The probability that the son of a carrier mother and a normal father will be affected is 0.5, or 50%; the probability that their daughter will be a carrier is also 0.5.

It is important that identified carriers receive appropriate genetic counseling. A genetic counselor is trained not only to give information needed for reproductive decisions but also to help individuals understand their situation and avoid feeling stigmatized.

Genetic counselors often receive inquiries about mental retardation, epilepsy, deafness, congenital heart disease, and other conditions. It is possible that some environmental factor may have played a role in producing the abnormality in the affected child. Did the mother have an infectious disease such as rubella (German measles) during pregnancy? Was she receiving some kind of drug therapy, or was she subjected to ionizing radiation? Had the father been exposed to any potentially hazardous agents? By dissecting the environmental contributions, the geneticist can more accurately estimate the probability of the trait's recurrence in subsequent offspring.

GENE THERAPY IS BEING EXPLORED FOR SEVERAL GENETIC DISEASES

Because many difficulties are inherent in treating most serious genetic diseases, scientists have dreamed of developing actual cures. Today, genetic engineering may be bringing these dreams closer to reality. One strategy is to introduce the normal gene

[2] For information about obtaining a master's degree in genetic counseling, contact the National Society of Genetic Counselors, 233 Canterbury Drive, Wallingford, PA 19086; e-mail: nsgc@aol.com.

into certain body cells *(somatic cell gene therapy)*. The rationale is that, although a particular gene may be present in all cells, it is expressed only in some (see Chapter 16). Expression of the normal allele in only the cells that require it may be sufficient to give a normal phenotype.

This approach presents a number of technical obstacles. The solutions to these problems must be tailored to the nature of the gene itself, as well as to its product and the types of cells in which it must be expressed. First the gene must be cloned and the DNA introduced into the appropriate cells. One of the most successful techniques is to package the genetic material in a virus, creating a viral **vector.** Ideally the virus should infect a high percentage of the cells. Most important, the virus should do no harm, especially over the long term. For example, if the virus inserts the introduced gene into a part of the chromosome that codes for a **proto-oncogene,** a normal gene involved in growth and development, the proto-oncogene may change into an oncogene (see *Focus On: Oncogenes and Cancer* in Chapter 16).

Although many obstacles must be overcome, gene therapies for several genetic diseases are under development or are being tested on individuals in clinical trials. Scientists are currently addressing some of the unique problems presented by each disease.

Gene therapy programs are under increasing scrutiny

Until recently, major technical advances caused the number of clinical studies involving gene therapy to grow dramatically. However, in September 1999, a death in a gene therapy trial led to a shutdown of many trials, pending the outcome of investigations about health risks. The main safety concern in these investigations is the potential toxicity of viral vectors, the viruses that move the normal gene into target cells that currently have a defective gene. The vector that was used in the young patient who died was an adenovirus (see Fig. 23–1*b*), a virus required in large doses if enough genes are to be transferred to make the therapy effective. Unfortunately, the high viral doses triggered a fatally strong immune response in the patient's body.

Performing clinical studies on humans always has inherent risks. Patients must be carefully selected, and the potential benefits and risks, as far as they are known, must be explained thoroughly so the patient or, in the case of children, the parents can give an informed consent to the procedure.

BOTH HUMAN GENETICS AND BELIEFS ABOUT GENETICS AFFECT SOCIETY

Many misconceptions exist about genetic diseases and their effects on society. Some people erroneously think of certain individuals or populations as genetically unfit and responsible for many of society's ills. They argue, for example, that medical treatment of individuals affected with genetic diseases, especially those who are able to reproduce, increases the frequency of abnormal alleles in the population. This is true for autosomal dominant and X-linked diseases, but most genetic diseases that are simply inherited show an autosomal recessive inheritance pattern. Only homozygous individuals actually have the disease; heterozygous carriers, present in far greater numbers in the population, are phenotypically normal.

For example, if 1 in 25 individuals in the United States is heterozygous for cystic fibrosis, the chance that two parents will both be heterozygous is $1/25 \times 1/25 = 1/625$. On average, one-fourth of the children of such a couple would have cystic fibrosis, so the frequency of affected individuals in the population is about $1/625 \times 1/4 = 1/2500$. Because their numbers are usually very small compared with heterozygotes, reproduction by homozygotes does not greatly increase overall frequencies of abnormal alleles.

Abnormal alleles are present in *all* individuals and *all* ethnic groups; no one is exempt. According to one estimate, each of us is heterozygous for several (3 to 15) very harmful alleles, any of which could cause debilitating illness or death in the homozygous state. Why, then, are genetic diseases not more common? Each of us has many thousands of essential genes, any of which can be mutated. It is very unlikely that the abnormal alleles that one individual carries are also carried by that individual's mate. Of course, this possibility is more likely if the harmful allele is a relatively common one, such as the one responsible for cystic fibrosis.

Relatives are more likely than nonrelatives to carry the same harmful alleles, having inherited them from a common ancestor. In fact, a greater than normal frequency of a particular genetic disease among offspring of **consanguineous matings** (matings of close relatives) is often the first clue that the mode of inheritance is autosomal recessive. The offspring of consanguineous matings have a small but significantly increased risk of genetic disease. In fact, they can account for a disproportionately high percentage of those individuals in the population with autosomal recessive disorders. Because of this perceived social cost, first-cousin marriages are prohibited by most states in the United States. However, consanguineous marriages are still relatively common in many countries, particularly in developing countries, where other factors may mask the significance of genetic disease. A high incidence of infectious disease, for example, may result in so many deaths that the health effects of relatively uncommon genetic diseases are largely ignored in developing countries. The ability of a society to concern itself with genetic diseases is a luxury that comes with affluence.

Genetic discrimination has provoked heated debate

One of the fastest growing areas of medical diagnostics is genetic screening and testing, and the number of new genetic tests that screen for diseases such as cystic fibrosis, sickle cell anemia, HD, colon cancer, and breast cancer increases each year. However, genetic testing raises many social, ethical, and legal issues that society must address.

One of the most difficult issues is whether genetic information should be made available to health and life insurance companies. Many people think genetic information should not be given to insurance companies, but others, including employers, insurers, and many organizations representing people affected by genetic disorders, say such a view is unrealistic. If people use genetic tests to help them decide when to buy insurance and how much to buy, then insurers insist they should also have access to this information. Insurers say they need access to genetic data to help calculate equitable premiums (the whole idea of insurance is to average risk over a large population). However, there are concerns that insurers might use the results of genetic tests to discriminate against people with genetic diseases or to deny them coverage.

Doctors are concerned that people at risk for a particular genetic disease might delay being tested because they fear genetic discrimination from insurers and employers. The perception of genetic discrimination already exists in society. In a 1996 study, 25% of 332 people with family histories of one or more genetic disorders thought they had been refused life insurance, 22% thought they had been refused health insurance, and 13% thought they had been denied employment because of genetic discrimination. In a 1998 survey by the National Center for Genetic Resources, 63% of respondents said they probably or definitely would not take a genetic test if the results could be disclosed to either their employers or insurers.

Making the issue even more complicated is the fact that genetic tests are sometimes difficult to interpret, in part because there are many complex interactions between genes and the environment. If a woman tests positive for a gene that has been linked to breast cancer, for example, she is at significant risk, but testing positive does not necessarily mean that she will get breast cancer. Moreover, if she gets breast cancer, the age of onset and severity of the disease are not predicted by genetic tests. These uncertainties also make it difficult to decide what form of medical intervention, from frequent mammograms to surgical removal of healthy breasts, is appropriate.

The Ethical, Legal, and Social Implications (ELSI) Research Program of the National Human Genome Research Institute has developed principles designed to protect people against genetic discrimination. The Health Insurance Portability and Accountability Act of 1996 provides some safeguards against genetic discrimination, but ELSI has additional recommendations that have not yet been put into law. As this book goes to press, many bills that extend protection against workplace discrimination, health discrimination, and invasion of privacy based on genetic information are up for consideration by both federal and state legislatures. These issues will be debated for years to come.

SUMMARY WITH KEY TERMS

I. **Human genetics** is the science of inherited variation in humans. The **human genome** is the total genetic information in human cells.
   A. **Bioinformatics** includes the storage, retrieval, and comparison of DNA sequences within human DNA and between genomes of different species.
   B. **Pharmacogenomics** is a new field of gene-based medicine in which drugs are personalized to match a patient's genetic makeup.

II. Karyotyping, pedigree construction, and DNA sequencing and mapping of genes are some of the tools used in the study of human genetics.
   A. Studies of an individual's **karyotype** (the number and kinds of chromosomes present in the nucleus) permit detection of various chromosome abnormalities.
   B. A **pedigree** is a chart that shows the transmission of genetic traits within a family over several generations.
   C. The first phase of the Human Genome Project, an international effort to map and sequence all of the DNA in the human genome, was essentially completed in 2001.

III. **Aneuploidy,** in which there are either missing or extra copies of certain chromosomes, causes certain human disorders. Aneuploidies include **trisomy,** in which an individual possesses an extra chromosome, and **monosomy,** in which one member of a pair of chromosomes is lacking.
   A. **Trisomy 21,** the most common form of **Down syndrome,** and **Klinefelter syndrome** (XXY) are examples of trisomy.
   B. **Turner syndrome** (XO) is an example of monosomy.
   C. Trisomy and monosomy are caused by **nondisjunction,** in which sister chromatids or homologous chromosomes fail to disjoin (move apart) properly during meiosis or mitosis.

IV. Structural abnormalities in chromosomes cause certain human disorders.
   A. In a **translocation,** part of one chromosome becomes attached to another. About 4% of individuals with Down syndrome have a translocation in which part of chromosome 21 is attached onto one of the larger chromosomes, such as chromosome 14.
   B. A **deletion** can result in chromosome breaks that fail to rejoin. The deletion may range in size from a few base pairs to an entire chromosome arm. The most common deletion disorder in humans is **cri du chat syndrome,** in which part of the short arm of chromosome 5 is deleted.
   C. **Fragile sites** may occur at specific locations on both chromatids of a particular chromosome. In **fragile X syndrome** the fragile site occurs near the tip on the X chromosome; at that site the nucleotide triplet CGG is repeated many more times than is normal.

V. Most human genetic diseases that show a simple inheritance pattern are transmitted as autosomal recessive traits. An **inborn error of metabolism** is a metabolic disorder caused by the mutation of a gene that codes for an enzyme needed for a biochemical pathway.
   A. **Phenylketonuria (PKU)** is an autosomal recessive disorder in which toxic phenylketones damage the developing nervous system.
   B. **Sickle cell anemia** is an autosomal recessive disorder in which abnormal hemoglobin (the protein needed to carry oxygen in the blood) is produced.
   C. **Cystic fibrosis** is an autosomal recessive disorder in which abnormal secretions are produced in organs primarily of the respiratory and digestive systems.
   D. **Tay-Sachs disease** is an autosomal recessive disorder caused by abnormal lipid metabolism in the brain.

VI. **Huntington disease** has an autosomal dominant inheritance pattern. It results in mental and physical deterioration, usually beginning in middle age.

VII. **Hemophilia A** is an X-linked recessive disorder. It results in a defect in one of the components of blood required for clotting.

VIII. Prenatal diagnosis, genetic screening, and genetic counseling assist individuals and families in detecting and coping with genetic disorders.
   A. Prenatal diagnosis increases the possibilities for prevention or alleviation of the effects of genetic abnormalities.
      1. In **amniocentesis,** the amniotic fluid surrounding the fetus is sampled and the fetal cells suspended in the fluid are cultured and screened for genetic defects. Amniocentesis provides results in the second trimester of pregnancy.
      2. In **chorionic villus sampling (CVS),** some chorion cells are removed and studied. CVS provides results in the first trimester of pregnancy.
   B. **Genetic screening** is a systematic search through a population for individuals with a genotype or karyotype that might cause a serious genetic disease in them or their offspring. Screening of newborns is done primarily as the first step in preventive medicine, and screening of adults is done to help them make informed reproductive decisions.
   C. Couples who are concerned about the risk of abnormality in their children may seek **genetic counseling.** Genetic counselors provide medical and genetic information.
   D. Gene therapies for several genetic diseases are undergoing development or are being tested on individuals in clinical trials. Scientists are currently addressing some of the unique problems presented by each disease.

IX. The effect of human genetics on society is complex.
   A. The fact that a particular abnormal allele is especially common in a certain population does not mean that group has a higher frequency of abnormal alleles in general.
   B. Most abnormal alleles are recessive; therefore, they are manifested phenotypically only in homozygotes, who constitute a tiny fraction of the individuals with the allele. Virtually every individual in the population is a heterozygous carrier of several abnormal alleles.
   C. One of the most difficult issues is whether genetic information should be made available to employers and to health and life insurance companies. Doctors are concerned that people at risk for a particular genetic disease might delay being tested because they fear genetic discrimination from insurers and employers.

POST-TEST

1. The most important tool in bioinformatics is (a) controlled matings (b) karyotyping (c) pedigree analysis (d) a computer (e) chorionic villus sampling
2. A pedigree is a diagram showing (a) controlled matings between members of different isogenic strains (b) the total genetic information in human cells, currently being mapped and sequenced (c) a comparison of DNA sequences among genomes of humans and other species (d) the subtle genetic differences among people (e) the expression of genetic traits among the members of two or more generations of a family
3. The Human Genome Project is (a) mapping and sequencing all of the DNA in the nuclear human genome (b) exclusively concerned with the comparison of DNA sequences between human DNA and DNA of other species (c) personalizing drugs to match an individual's genetic makeup (d) a systematic search for individuals with a genotype that might cause a serious genetic disease in them or their offspring (e) providing risk estimates on human genetic diseases
4. An abnormality in which there is one more or one fewer than the normal number of chromosomes is called a(an) (a) karyotype (b) fragile site (c) aneuploidy (d) trisomy (e) translocation
5. An individual with one extra chromosome (three of one kind) is said to be (a) monosomic (b) triploid (c) trisomic (d) consanguineous (e) isogenic
6. An individual who is missing one chromosome, having only one member of a pair, is said to be (a) monosomic (b) haploid (c) trisomic (d) consanguineous (e) isogenic
7. The failure of chromosomes to separate normally during cell division is called (a) a fragile site (b) an inborn error of metabolism (c) a satellite knob (d) a translocation (e) nondisjunction
8. The transfer of a part of one chromosome to a nonhomologous chromosome is called (a) a karyotype (b) an inborn error of metabolism (c) a pedigree (d) a translocation (e) nondisjunction
9. A photomicrograph of the array of stained metaphase chromosomes present in a given cell is called a(an) (a) karyotype (b) nucleotide triplet repeat (c) pedigree (d) inborn error of metabolism (e) translocation
10. Individuals with trisomy 21, or ____________, are mentally and physically retarded and have abnormalities of the face, tongue, and eyelids. (a) Down syndrome (b) Klinefelter syndrome (c) Turner syndrome (d) Huntington disease (e) Tay-Sachs disease
11. An inherited disorder caused by a defective or absent enzyme is called a(an) (a) karyotype (b) trisomy (c) reciprocal translocation (d) inborn error of metabolism (e) aneuploidy
12. In ____________, a genetic mutation codes for an abnormal hemoglobin molecule that is less soluble than usual and more likely than normal to crystallize and deform the shape of the red blood cell. (a) Down syndrome (b) Tay-Sachs disease (c) sickle cell anemia (d) PKU (e) hemophilia A
13. In an individual with ____________, the mucus is abnormally viscous and tends to plug the ducts of the pancreas and liver and to accumulate in the lungs. (a) Down syndrome (b) Tay-Sachs disease (c) sickle cell anemia (d) PKU (e) cystic fibrosis
14. During this procedure, a sample of the fluid that surrounds the fetus is obtained by insertion of a needle through the walls of the abdomen and uterus. (a) DNA marker (b) chorionic villus sampling (c) ultrasound imaging (d) consanguineous mating (e) amniocentesis
15. For which of the following situations would a genetic counselor *not* recommend prenatal diagnosis involving amniocentesis or chorionic villus sampling? (a) there is an increased risk of a chromosomal abnormality (b) there is an increased risk of a single-locus (Mendelian) disease (c) there is an increased risk of a spinal cord defect (d) there is a desire to know the sex of the fetus (e) a pregnant woman is older than 35 years of age

REVIEW QUESTIONS

1. What means have been devised for overcoming some of the difficulties in studying human inheritance?
2. What is meant by nondisjunction? What are some human abnormalities that appear to be the result of nondisjunction?
3. What is meant by inborn errors of metabolism? Give an example.
4. To be expressed, an autosomal recessive genetic disease must be homozygous. What relationship does this fact have to consanguineous matings?
5. What are some of the ways that heterozygous carriers of certain genetic diseases can be identified?
6. What are the relative advantages and disadvantages of amniocentesis and chorionic villus sampling?
7. Examine the following pedigrees and decide whether each disorder is most likely inherited by an autosomal recessive, an autosomal dominant, or an X-linked recessive allele. Determine the probable genotypes for all individuals shown.

(a) 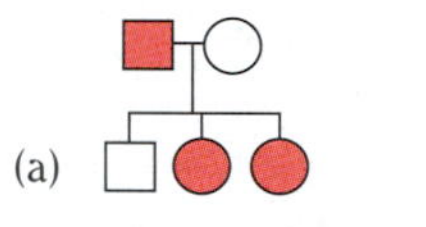

(b) 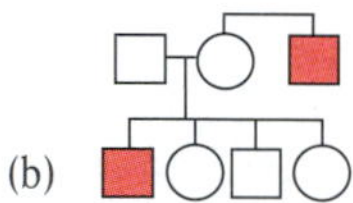

(c) 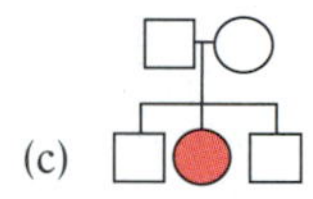

8. Complete the table at right by checking the correct box for each genetic disorder.

Disease	Chromosome Abnormality	Autosomal Recessive	Autosomal Dominant	X-Linked Recessive
Down syndrome				
Tay-Sachs disease				
Phenylketonuria				
Hemophilia A				
Sickle cell anemia				
Turner syndrome				
Huntington disease				
Klinefelter syndrome				
Cri du chat syndrome				
Fragile X syndrome				

YOU MAKE THE CONNECTION

1. Imagine that you are a genetic counselor. What advice or suggestions might you give in the following situations?
   a. A couple has come for advice because the woman had a sister who died of Tay-Sachs disease.
   b. A young man and woman who are not related are engaged to be married. However, they have learned that the man's parents are first cousins. They are worried about the possibility of increased risk of genetic defects in their own children.
   c. A young woman's paternal uncle (her father's brother) has hemophilia A. Her father is free of the disease, and there has never been a case of hemophilia A in her mother's family. Should she be concerned about the possibility of hemophilia A in her own children?
   d. A 20-year-old man is seeking counseling because his father recently was diagnosed with Huntington disease.
2. A deletion of part of an X chromosome may be lethal in a male but causes few problems in a female. Explain.
3. A common misconception about human genetics is that an individual's genes alone determine his or her destiny. Explain why this myth is incorrect. How is the perpetuation of this myth harmful to society?

RECOMMENDED READINGS

Brown, K. "The Human Genome Business Today." *Scientific American,* Vol. 283, No. 1, Jul. 2000. Considers the business potential of new fields such as bioinformatics and pharmacogenomics.

Ezzell, C. "Beyond the Human Genome." *Scientific American,* Vol. 283, No. 1, Jul. 2000. The push is on to understand the proteins that are encoded in the human genome.

Friedmann, T. "Principles for Human Gene Therapy Studies." *Science,* Vol. 287, 24 Mar. 2000. This paper examines principles of medical ethics that must be addressed for human clinical trials involving gene therapy.

Hesman, T. "The Meaning of Life." *Science News,* Vol. 157, 29 Apr. 2000. Computers and gene-hunting programs are important tools in the field of bioinformatics.

Howard, K. "The Bioinformatics Gold." *Scientific American,* Vol. 283, No. 1, Jul. 2000. The medical potential of bioinformatics is examined in detail in this article.

Mange, E.J., and A.P. Mange. *Basic Human Genetics,* 2nd ed. Sinauer Associates, Inc., Sunderland, Massachusetts, 1999. This text provides excellent coverage of all aspects of human genetics, from the diagnosis of diseases to ethical concerns.

Marshall, E. "Gene Therapy on Trial." *Science,* Vol. 288, 12 May 2000. This article provides a detailed analysis of the gene therapy trial at the University of Pennsylvania that led to the 1999 death of a young patient suffering from a defective gene that codes for an essential liver enzyme.

McKusick, V.A. *Mendelian Inheritance in Man: A Catalog of Human Genes and Genetic Disorders,* 12th ed. Johns Hopkins University Press, Baltimore, 1998. This is the print version of the on-line OMIM (Online Mendelian Inheritance in Man) database (http://www.ncbi.nlm.nih.gov/Omim/).

Nagel, R.L. "Molecule, Heal Thyself." *Natural History,* Sept. 2000. The genetic component of human illnesses and injuries (e.g., a broken leg) has been vastly underestimated.

Nature, Vol. 409, 15 Feb. 2001. This issue focuses on the landmark completion of sequencing the human genome.

Pennisi, E. "Finally, the Book of Life and Instructions for Navigating It." *Science,* Vol. 288, 30 Jun. 2000. Much of the human genome has been sequenced, a goal that has huge implications for both biology and medicine.

Science, Vol. 291, 16 Feb. 2001. This issue is devoted to the completion of the human genome sequence, which is a milestone in science.

Shreeve, J. "Secrets of the Gene." *National Geographic,* Vol. 196, No. 4, Oct. 1999. Discusses how medicine and other fields should benefit from recent advances in human genetics.

Strasser, B.J. "Sickle Cell Anemia, a Molecular Disease." *Science,* Vol. 286, 19 Nov. 1999. This article, published on the 50-year anniversary of a groundbreaking paper by Linus Pauling and colleagues on sickle cell anemia, considers both the advances and the lack of progress made in treating sickle cell anemia since that time.

Wheelwright, J. "Betting on Designer Genes." *Smithsonian,* Jan. 2001. Some of the challenges involved in using gene transfer to treat a disease.

- Visit our Web site at **http://www.info.brookscole.com/solomonbergmartin** for links to chapter-related resources on the World Wide Web. Additional on-line materials relating to this chapter can also be found on our Web site.

16

Genes and Development

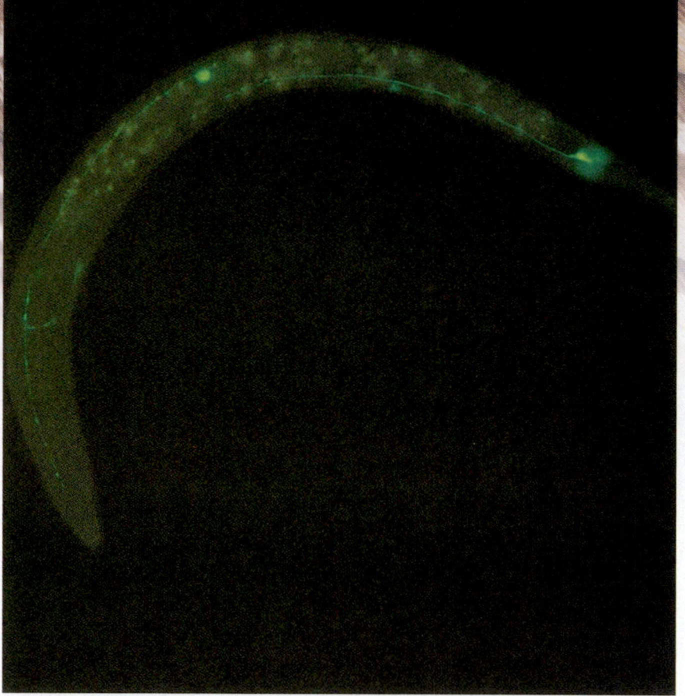

The model organism *Caenorhabditis elegans.* The small transparent body of this nematode worm allows researchers to locate cells in which a specific developmentally important gene is active. These cells show up as bright green spots in the photograph because they have been genetically engineered to produce a green fluorescent protein known as GFP. *(Courtesy of Dr. Martin Chalfie, Columbia University)*

LEARNING OBJECTIVES

After you have studied this chapter you should be able to

1. Distinguish between cell determination and cell differentiation.
2. Relate the process of pattern formation to morphogenesis.
3. Describe the kinds of experiments that indicate the totipotency of some differentiated plant cells and some animal nuclei. Discuss how these findings support the idea of nuclear equivalence.
4. Identify the attributes of an organism that would make it especially useful in studies on the genetic control of development. Discuss the value of transgenic organisms in research on the genetic control of development.
5. Indicate the features of the development and genetics of *Drosophila melanogaster, Caenorhabditis elegans,* the mouse *(Mus musculus),* and *Arabidopsis thaliana* that have made these organisms so valuable to researchers.
6. Distinguish among maternal effect genes, zygotic genes, and homeotic genes in *Drosophila.*
7. Explain the relationship between transcription factors and genes that control development. Provide some examples of genes that are known to function as genetic switches in development.
8. Define the phenomena of induction and programmed cell death, and give examples of the roles they play in development.
9. Describe the functions of some homeotic-like genes in plants.
10. Point out some of the known exceptions to the general phenomenon of nuclear equivalence.

The study of **development,** which is broadly defined as all the changes that occur in the life of an individual, encompasses some of the most fascinating and difficult problems in biology today. Of particular interest is the process by which cells specialize and organize into a complex organism. During the many cell divisions required for a single cell to develop into a multicellular organism, groups of cells become gradually committed to specific patterns of gene activity through a process called cell **determination.** The final step leading to cell specialization is cell **differentiation.** A differentiated cell can be recognized by its characteristic appearance and activities.

An even more intriguing part of the developmental puzzle is the building of the body, known as morphogenesis, or the development of form. In **morphogenesis,** cells in specific locations differentiate and become spatially organized into recognizable structures through a multistep process known as **pattern formation.**

Until the late 1970s, little was known about how certain genes interact with various signals from within the organism and from the environment to control development. Although certain genes affecting developmental pathways had been identified, their specific functions in the organism were not well understood. Because these networks are too complex to unravel using only traditional methods, it had been thought that it might be impossible to understand development. However, rapid progress in recombinant DNA technology led scientists to renew their search for developmental mutants and to apply the most sophisticated techniques to study them.

Today scientists interested in development study a variety of carefully chosen mutant organisms with altered developmental patterns. They use the tools of genetic engineering combined with more conventional descriptive and experimental approaches to derive fresh insights into the role of genetic information in the control of development. The organism in the photograph is *Caenorhabditis elegans,* a nematode worm with many attributes that make it unusually attractive for developmental studies. Work with such organisms has profound implications for our understanding of both normal human development and the kinds of malfunctions that can lead to birth defects and even "normal" aging. Although these worms may seem to have little in common with humans, scientists are learning that many of the genes important in development are quite similar in a wide range of organisms. These similarities have led to new ways to unravel evolutionary relationships, through the study of developmentally important genetic mechanisms that appear to be deeply rooted in the evolutionary history of multicellular organisms.

CELL DIFFERENTIATION USUALLY DOES NOT INVOLVE CHANGES IN DNA CONTENT

The human body, like those of other vertebrates, contains more than 200 recognizably different types of cells (Fig. 16–1). Combinations of these specialized cells, known as **differentiated cells,** are organized into remarkably diverse and complex structures such as the eye, the hand, and the brain, each capable of carrying out many sophisticated activities. Most remarkable of all, however, is the fact that all the structures of the body and the different cells within them are descended from a single fertilized egg.

All multicellular organisms undergo complex patterns of development. The root cells of plants, for example, have structures and functions very different from those of the various types of cells located in leaves. Remarkable diversity can also be found at the molecular level; most strikingly, each type of plant or animal cell makes a highly specific set of proteins (Fig. 16–2). In some cases, such as the protein hemoglobin in red blood cells, one cell-specific protein may make up more than 90% of the total mass of

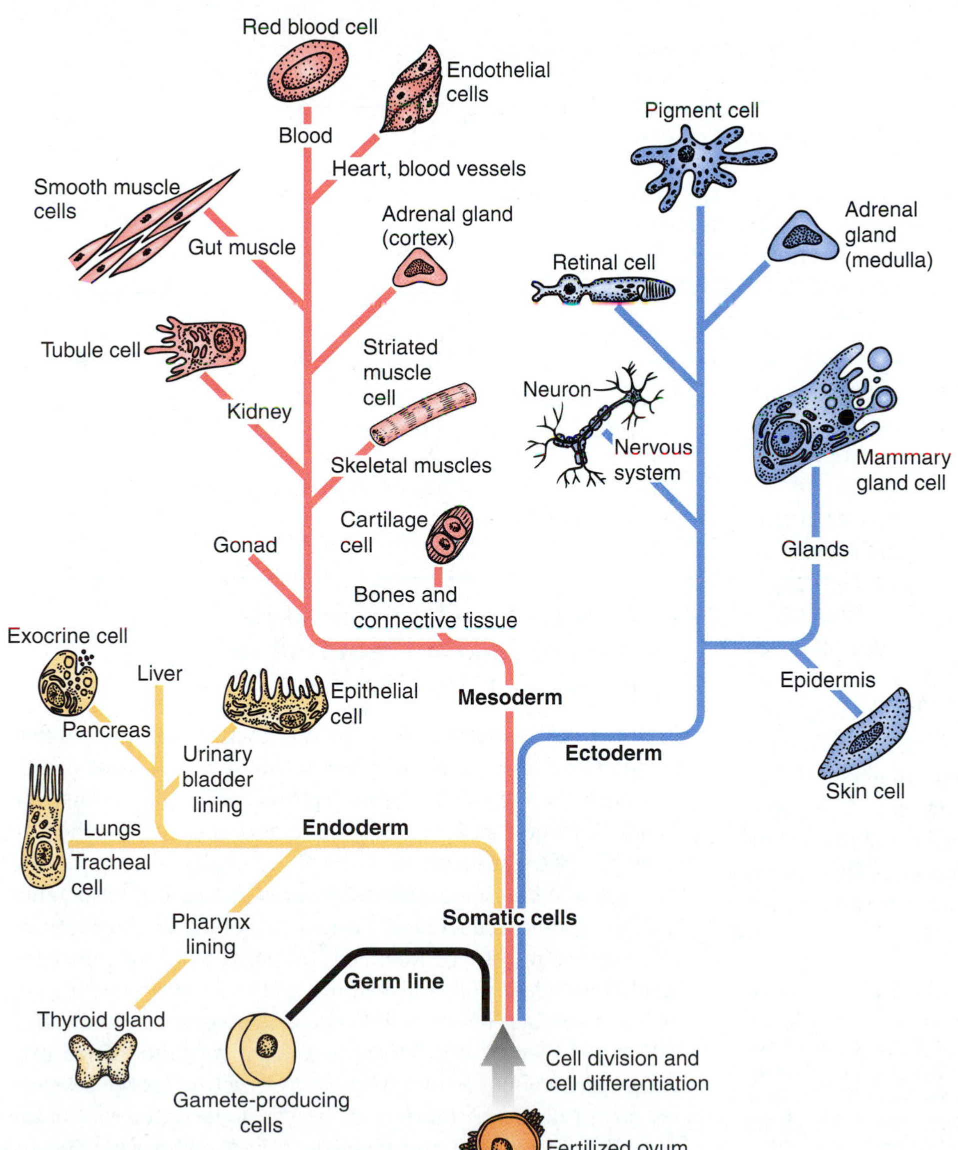

Figure 16–1 Vertebrate cell lineages. Repeated divisions of the fertilized egg *(bottom of the figure)* result in the establishment of tissues containing groups of specialized cells (see Chapter 49 for a discussion of how these are formed). Germ-line cells (cells that produce the gametes) are set aside early in development. Somatic cells progress along the developmental pathways, undergoing a series of commitments that progressively determine their fates.

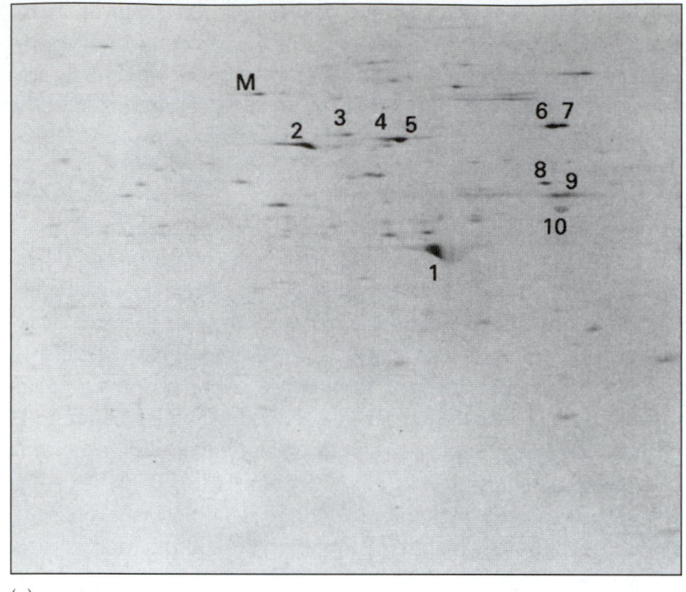

(a)

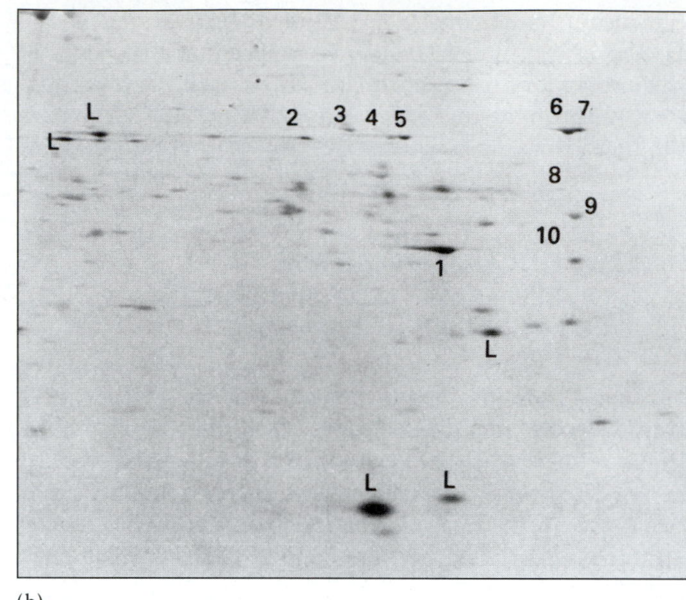

(b)

Figure 16–2 Cell-specific proteins. The spots in the photographs are proteins from **(a)** muscle and **(b)** liver cells of a mouse. The proteins were separated by two-dimensional gel electrophoresis, a method that separates the proteins in the horizontal direction by their electric charge, followed by a second separation in the vertical direction by molecular weight. Several hundred proteins can be distinguished in each panel. The spots that are labeled with numbers are present in all tissues, but notice that many of them are present in different amounts from one tissue to another. The proteins that are labeled with letters are found only in that specific tissue. *(Patrick O'Farrell, from Fig. 12–1 in Darnell, J., H. Lodish, and D. Baltimore,* Molecular Cell Biology. *Scientific American Books, New York, 1986)*

protein in the cell. Other cells may have a complement of cell-specific proteins that are each present in small amounts but still play essential roles. However, because certain proteins are required in every type of cell (all cells, for example, require certain enzymes for glycolysis), cell-specific proteins usually make up only a fraction of the total number of different kinds of proteins.

One explanation for the fact that each type of differentiated cell makes a unique set of proteins might be that during development each group of cells loses the genes it does not need and retains only those that are required. With just a few exceptions, however, this does not seem to be true. According to the concept of **nuclear equivalence,** the nuclei of essentially all differentiated adult cells of an individual are genetically (though not necessarily metabolically) identical to each other and to the nucleus of the fertilized egg cell from which they descended. This means that virtually all **somatic**[1] cells in an adult have the same genes, but different cells express different subsets of these genes.

[1] Somatic cells are cells of the body and are distinguished from *germ-line cells,* which ultimately give rise to a new generation. In animals, germ-line cells, whose descendants ultimately undergo meiosis and differentiate as gametes, are generally set aside early in development. In plants, the distinction between somatic cells and germ-line cells is not clear-cut, and the determination that certain cells will undergo meiosis is made much later in development.

The evidence for nuclear equivalence comes from cases in which differentiated cells or their nuclei have been found to be capable of supporting normal development. Such cells or nuclei are said to be **totipotent.**

A totipotent nucleus contains all the information required to direct normal development

In plants it is possible to demonstrate that at least some differentiated cells can be induced to become the equivalent of embryonic cells (Fig. 16–3). **Tissue culture** techniques are used to isolate individual cells from certain plants and to allow them to grow in a nutrient medium.

Some of the first experiments investigating cell totipotency in plants were conducted at Cornell University in the 1950s by F.C. Steward and others. Root cells from a carrot were induced to divide in a liquid nutrient medium and to form groups of cells called "embryoid" (embryo-like) bodies. These clumps of dividing cells could then be transferred to an agar medium, which provides nutrients plus a solid supporting structure for the developing plant cells. After transfer to the agar, some of the cells of the embryoid bodies gave rise to roots, stems, and leaves. The re-

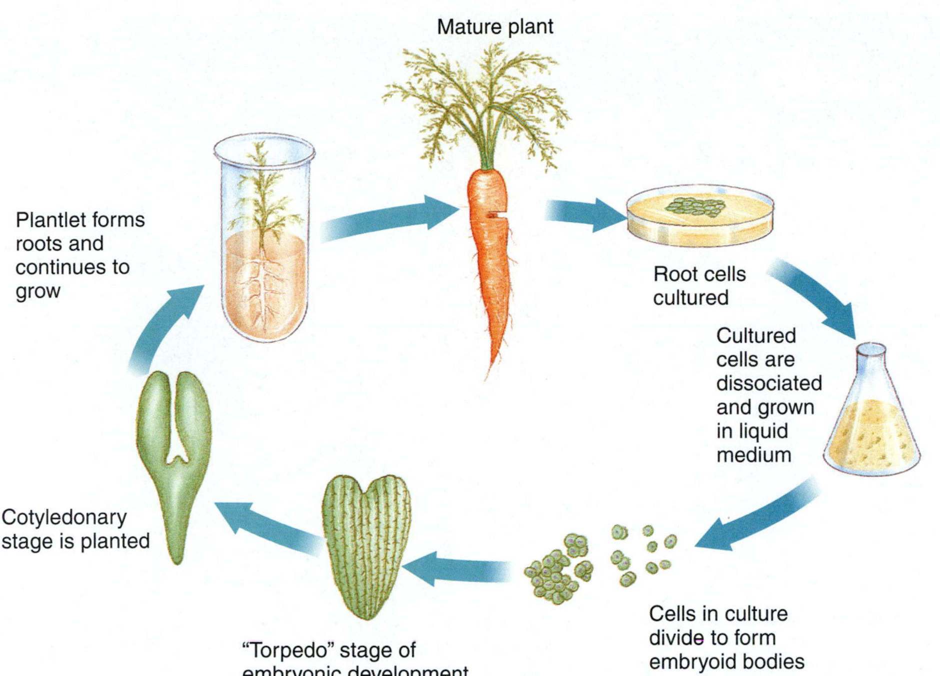

Figure 16–3 Cell totipotency. A complete carrot plant can develop from differentiated somatic cells. Carrot root tissues were cut into discs made up of phloem cells, which are specialized for nutrient transport. When these differentiated cells were cultured in a liquid nutrient medium, individual cells divided to form clumps of undifferentiated cells, known as embryoid bodies. The embryoid bodies, which closely resembled plant embryos in their early stages of development, then progressed to form embryonic shoots and roots. Transferring the embryonic tissue to a solid nutrient medium stimulated the tissues to form small plants, called plantlets, which then developed into mature plants.

sulting small plants, called *plantlets* to distinguish them from true seedlings, were then transplanted to soil, where they ultimately developed into adult plants capable of producing flowers and viable seeds. If these plants are all derived from the same parent plant, they are genetically alike and therefore constitute a **clone.** The methods of plant tissue culture are now extensively used to produce genetically engineered plants, for they allow the regeneration of whole plants from individual cells that have incorporated recombinant DNA molecules (see Chapter 36, *Focus On: Cell and Tissue Culture*).

Similar experiments have been attempted with animal cells, but so far it has not been possible to induce a fully differentiated somatic cell to behave like a zygote. Instead, it has been possible to test whether steps in the process of determination are reversible by transplanting the *nucleus* of a cell in a relatively late stage of development into an egg cell that has been enucleated (i.e., its own nucleus has been destroyed).

In the 1950s R. Briggs and T.J. King conducted experiments in which nuclei from amphibian cells at different stages of development were transplanted into egg cells. Some of the transplants proceeded normally through several developmental stages, and a few even developed into normal tadpoles. As a rule, the nuclei transplanted from cells at earlier stages were most likely to support development to the tadpole stage. As the fate of the cells became more and more determined, the probability that a transplanted nucleus could control normal development diminished rapidly (Fig. 16–4).

In experiments carried out in England in the 1960s by J. Gurdon, in a few cases nuclei isolated from the specialized intestinal cells of a tadpole were able to direct development up to the tadpole stage. This occurred infrequently, but in such experiments success counts more than failure, and we can safely conclude that at least some nuclei of differentiated animal cells are in fact totipotent.

For many years these successes with amphibians could not be repeated with mammalian embryos, leading many developmental biologists to conclude that some fundamental feature of mammalian reproductive biology might be an impenetrable barrier to mammalian cloning. This perception changed markedly in 1996 and 1997 with the first reports of the birth of cloned mammals. Since that time, successful cloning of several additional mammalian species has been accomplished (see *Focus On: Mammalian Cloning*).

It may not be particularly surprising that nuclei do not usually lose genetic material during development; after all, there should be no loss of chromosomes in the course of normal mitotic cell divisions. These demonstrations of nuclear totipotency also imply that apparently inactive genes are capable of being reactivated when cells or nuclei are placed in a suitable environment. Nevertheless, it is important to recognize that the nuclei of embryonic cells progressively undergo changes that make it more difficult to remain in a totipotent state. This is especially true of animal nuclei, although various kinds of animals may differ considerably in this regard.

Stem cells, undifferentiated cells that are not truly totipotent but nevertheless able to differentiate into more than one cell type, are of great scientific interest today. Studies on these cells not only help scientists understand the changes that occur when cells differentiate but also have far-reaching medical significance (see *Focus On: Stem Cells* on page 346).

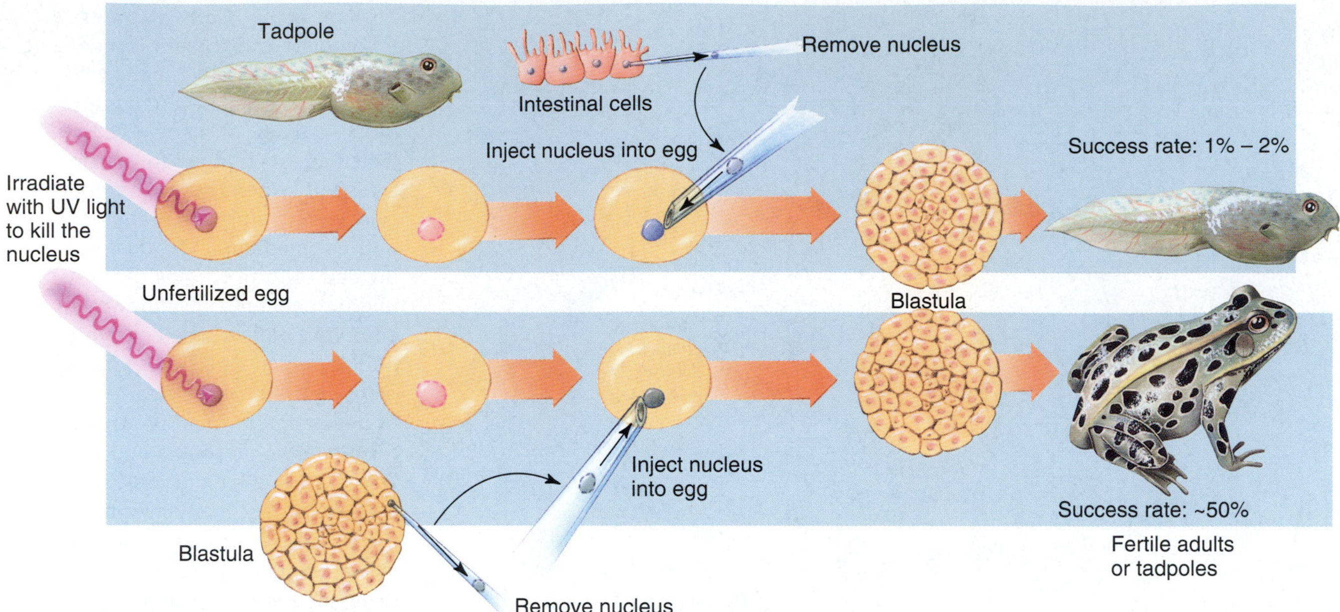

Figure 16–4 Nuclear totipotency. In nuclear transplantation experiments in amphibians, nuclei of differentiated cells at different stages of development were injected into eggs whose own nuclei had been destroyed by ultraviolet radiation. As seen in the upper panel, most trials using nuclei from tadpole intestinal cells (a relatively late developmental stage) resulted in no growth. In a small number of trials (about 1%–2% of the total) normal development did proceed to the tadpole stage, indicating that the genes necessary to program development up to that point were still present and able to be appropriately activated. However, the success rate improved dramatically if nuclei from earlier developmental stages were used. As shown in the lower panel, if a nucleus was taken from a cell at the blastula stage of development (when cell division has produced about 1000 cells formed in the shape of a ball), in about half the cases the transplanted nucleus could successfully program normal development resulting in a tadpole or fertile adult.

Most differences among cells are due to differential gene expression

Because genes do not appear to be lost regularly during development, the differences in the molecular composition of cells must occur by regulating the activities of different genes. This process of developmental gene regulation is often referred to as **differential gene expression.** As discussed in Chapter 13, the expression of eukaryotic genes can be regulated in many different ways and at many levels. For example, a particular enzyme may be produced in an inactive form and then be activated later. However, much of the regulation that is important in development occurs at the transcriptional level. The transcription of certain sets of genes is repressed, whereas other sets are activated. Even expression of genes that are **constitutive** (i.e., constantly transcribed) and active in all cells can be regulated during development so that the *quantity* of each product varies from one tissue type to another.

We can think of differentiation as a series of pathways leading from a single cell to cells in each of the different specialized tissues, arranged in an appropriate pattern. There are times when a cell makes genetic "commitments" to the developmental path its descendants will follow. These commitments gradually restrict the development of the descendants to a limited set of final tissue types. Determination, then, is a progressive fixation of the fate of a cell's descendants.

As the development of a cell becomes determined along a particular differentiation pathway, the physical appearance of the cell may not change significantly. Nevertheless, when a particular stage of determination is complete, the changes in the cell usually become self-perpetuating and are not easily reversed. Cell differentiation, then, is usually the last stage in the developmental process. At this stage, a precursor cell becomes structurally and functionally recognizable as a bone cell, for example, and its pattern of gene activity is different from that of a nerve cell, or any other cell type.

 DNA microarrays are a powerful approach for tracking gene expression

Biotechnology (see Chapter 14) is now making it possible for researchers to study differential gene expression in increasingly sophisticated ways. The science of determining the roles of genes

FOCUS ON Mammalian Cloning

In 1996 Ian Wilmut and coworkers at the Roslin Institute in Scotland reported that they had succeeded in cloning sheep, using nuclei from early sheep embryos (blastocyst stage; see Chapter 49). These scientists subsequently received worldwide attention in early 1997 when they announced the birth of a lamb, nicknamed "Dolly," derived from a cultured mammary gland cell (from an adult sheep) fused with an enucleated sheep's egg. The resulting cell divided and developed into an embryo that was then cultured in vitro until it reached a stage at which it could be transferred to a host mother. As might be expected, the overall success rate was low: Out of 277 fused cells, only 29 developed into embryos that could be transferred, and Dolly was the only live lamb produced. These researchers have also produced cloned transgenic lambs derived from fetal cells.

Why was Wilmut's team the first to succeed when so many other researchers had failed? Applying the basic principles of cell biology, they recognized that the cell cycles of the egg cytoplasm and the donor nucleus were incompatible; that is, the egg cell is a cell that is arrested at metaphase II of meiosis, whereas the actively growing donor cell is usually in the DNA synthesis phase (S), or in G_2. By withholding certain nutrients from the mammary gland cells used as donors, they were able to cause them to enter a nondividing state referred to as G_0 (see Chapter 9). This had the effect of synchronizing the donor nucleus to the cell cycle of the egg. They then used an electrical shock to cause the donor cell to fuse readily with the egg and initiate the development of the embryo.

Although an extremely high level of technical expertise is required, these and other researchers have been able to use modifications and extensions of these techniques to produce cloned calves, goats, pigs, and mice, and there is reason to think that the list of mammals that have been successfully cloned will continue to grow. However, the success rate for each set of trials continues to be extremely low, on the order of 1% to 2%, and the incidence of defects is high. The production of transgenic animals continues to be the main focus of this line of research (see Chapter 14). Researchers are very actively pursuing new techniques that they hope will dramatically improve the efficiency of the cloning process, because only then will it be possible to produce large numbers of cloned transgenic animals that can be used for a variety of purposes.

Results such as these continue to fuel an ongoing debate regarding the potential for human cloning and its ethical implications. In the United States, the National Bioethics Advisory Commission, whose recommendations are posted on the Internet, has been established to study this and other questions: http://www.nih.gov/nbac.htm.

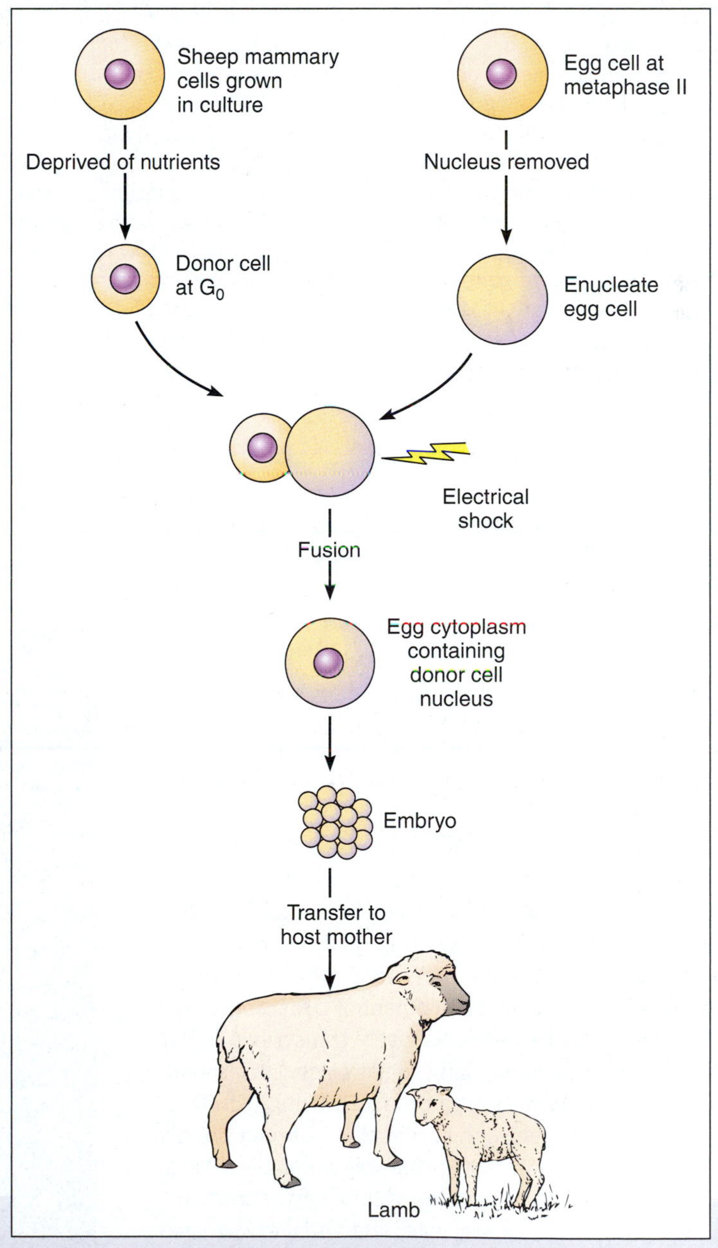

How to clone a sheep. An embryo produced by fusing a cultured adult sheep mammary cell with an enucleate sheep's egg is implanted into the uterus of a host mother and develops into a lamb.

FOCUS ON Stem Cells

Stem cells are undifferentiated cells that can divide to produce differentiated descendants, yet retain the ability to divide to reproduce themselves, thereby maintaining the stem cell population. The most versatile stem cells, those that have the ability to give rise to all tissues of the body, are known as **pluripotent** stem cells. Other stem cells appear to be more specialized; for example, neural stem cells can differentiate to form all types of brain cells, and stem cells in the bone marrow form various types of blood cells. However, recent studies have shown that even specialized stem cells may be more versatile than once thought. For example, neural stem cells have been shown to form blood cells when transplanted into bone marrow, and bone marrow stem cells can differentiate into muscle cells.

Stem cells are potential sources of cells to be transplanted into patients to treat serious conditions. For example, Parkinson's disease results from a progressive loss of cells that produce the neurotransmitter dopamine in a specific region of the brain. Transplantation of stem cells that have been induced to differentiate as dopamine-producing cells holds great promise as an effective long-term treatment. Similarly, stem cells may become a source of insulin-producing cells in the pancreas that are lacking in individuals with Type 1 diabetes (see Chapter 47) or of neurons in individuals who have suffered spinal cord injury or other types of neurological damage.

Researchers ultimately hope to establish lines of human pluripotent stem cells that can be grown indefinitely in culture, can be induced to differentiate under controlled conditions and stably maintain their differentiated state, and can be manipulated genetically. Although work on stem cells in mice and other mammals has been conducted for many years, similar studies in humans have progressed slowly despite the great promise of such research. Many of these studies have been funded by private companies. This is largely a consequence of governmental restrictions on research funding, due to ethical considerations that are related to the origins of stem cells. So far the only known sources of pluripotent stem cells are early embryos and specific types of cancer cells known as teratocarcinoma cells. Fetuses are also a source of some types of specialized stem cells. More recent findings that certain types of specialized stem cells also exist in tissues of adult mice, as well as human adults and children, may alleviate some ethical concerns. Unfortunately, these cells are rare and lack many of the advantages of embryonic cells.

in cells, known as **functional genomics,** includes the analysis of patterns of gene expression in different cell types. One approach is the use of DNA microarrays, which rely heavily on methods that originated in the computer industry (Fig. 16–5). Thousands of tiny spots of DNA, known as microdots, are "printed" on a "chip", which is usually a glass microscope slide. Each microdot contains many single-stranded copies of a fragment of DNA from a particular organism, and collectively all the microdots on a chip are a microarray that contains all or most of the DNA in that organism's genome.

RNA is then extracted from cells the researchers wish to study, and DNA complementary to it (cDNA) is synthesized using reverse transcriptase (see Chapter 14). The single-stranded cDNA molecules are tagged with a fluorescent dye and incubated with the DNA on the chip under conditions that promote complementary base pairing. If a particular segment of DNA on the chip corresponds to a gene that was very actively transcribed in the cells being studied, there will be many copies of the corresponding cDNA available to bind to that microdot, causing it to fluoresce very brightly. Conversely, a microdot that fluoresces dimly or not at all is one that contains DNA that was not actively transcribed in those cells. The chips are then scanned by instruments that detect the patterns of fluorescence, and the data are analyzed by powerful computers. These methods allow researchers to compare patterns of gene expression, as measured by RNA synthesis, in various cell types, or in the same cell type under different conditions.

Figure 16–5 Gene expression. Data from an experiment showing the expression of thousands of genes on a single GeneChip® probe array. *(Image courtesy of Affymetrix)*

MOLECULAR GENETICS HAS REVOLUTIONIZED THE STUDY OF DEVELOPMENT

Development has been an important area of research for many years, and considerable effort has been expended on studying the development of invertebrate and vertebrate animals. By identifying patterns of morphogenesis in different animals, researchers have been able to identify similarities, as well as differences, in the basic plan of development from a fertilized egg to an adult in organisms ranging from the sea urchin to mammals (see Chapter 49).

In addition to descriptive studies, many classic experiments have established important evidence concerning how groups of cells differentiate and undergo pattern formation. Researchers have developed elaborate screening programs to detect mutations that allow them to identify large numbers of developmental genes in both plants and animals. They then exploit a wide variety of molecular genetic techniques and other sophisticated methodologies to determine how those genes work and how they interact to coordinate developmental processes.

Certain organisms are well suited for studies of the genetic control of development

In studies of the genetic control of development, the choice of an organism to use as an experimental system has become increasingly important. One of the most powerful approaches involves the isolation of mutants with arrested or abnormal development at a particular stage. Not all organisms have useful characteristics that allow developmental mutants to be isolated and maintained for future study. The genetics of the fruit fly, *Drosophila melanogaster,* is so thoroughly understood that this organism has become one of the most important systems for such studies. Other organisms such as the nematode worm, *Caenorhabditis elegans;* the laboratory mouse, *Mus musculus;* certain plants, including *Arabidopsis thaliana,* a tiny weed with many convenient features; and some simple eukaryotes, such as the yeast *Saccharomyces cerevisiae,* have also become important models in developmental genetics. Each of these organisms has attributes that make it particularly useful for examining certain aspects of development.

DROSOPHILA MELANOGASTER PROVIDES RESEARCHERS WITH A WEALTH OF DEVELOPMENTAL MUTANTS

Undoubtedly the most extensive (and spectacular) examples of genes that control development have been identified in the fruit fly, *Drosophila melanogaster.* The *Drosophila* genome sequence became available in late 1999, and it has been determined that it includes about 13,600 protein-coding genes. One of the traditional advantages of using *Drosophila* as a research organism is the abundance of mutants (including developmental mutants) available for study and the relative ease with which a new mutation can be directly mapped on the chromosomes. The genetic analysis is greatly facilitated by special chromosomes found in certain tissues with large, metabolically active cells, including the salivary glands of the larvae. These **polytene** ("many-stranded") chromosomes are unusual interphase chromosomes formed when the DNA replicates many times but without mitosis and cytokinesis.

A typical polytene chromosome may consist of more than 1000 DNA double helices (along with associated histones and other proteins) aligned side by side (Fig. 16–6). Polytene chromosomes are therefore quite large and show a pattern of bands that is very useful in assigning a particular gene to a particular location on the chromosome. When a gene is active, the chromosome band in which it resides uncoils and forms a **puff,** which is a site of intense RNA synthesis. This evidence of gene activity is similar to that observed in lampbrush chromosomes of certain female meiotic cells (see Fig. 9–13).

Studies of *Drosophila* are also facilitated by the fact that foreign DNA can be injected into eggs and become incorporated into the fly's DNA in a process called **transformation** (by analogy with transformation in prokaryotes).

The *Drosophila* life cycle includes egg, larval, pupal, and adult stages

The life cycle of *Drosophila* consists of several distinct stages (Fig. 16–7). After the egg is fertilized, a period of embryogenesis occurs during which the zygote develops into a sexually immature form known as a **larva** (pl., *larvae*). After hatching from the egg, each larva undergoes several molts (shedding of the external covering or cuticle). Each molt allows an increase in size until the larva is ready to pupate. **Pupation** involves a molt and the hardening of the new external cuticle, so that the pupa is completely encased. The insect then undergoes a complete **metamorphosis** (change in form). During that time, most of the larval tissues degenerate and other tissues differentiate to form the body parts of the sexually mature adult fly.

The larvae are wormlike in appearance and look nothing like the adult flies. However, very early in embryogenesis of the developing larvae, precursor cells of many of the adult structures are organized as relatively undifferentiated paired structures called **imaginal discs.** This term comes from **imago,** the name given to the adult form of the insect. Each imaginal disc occupies a definite position in the larva and will form a specific structure, such as a wing or a leg, in the adult body (Fig. 16–8). The discs are formed by the time embryogenesis is complete and the larva is ready to begin feeding. In some respects the larva can be thought of as a complex developmental stage that is simply used to feed and nurture the precursor cells that give rise to the adult fly (which is the only form that can reproduce).

The organization of the precursors of the adult structures, including the imaginal discs, is under complex genetic control. So far more than 50 genes have been identified that specify the

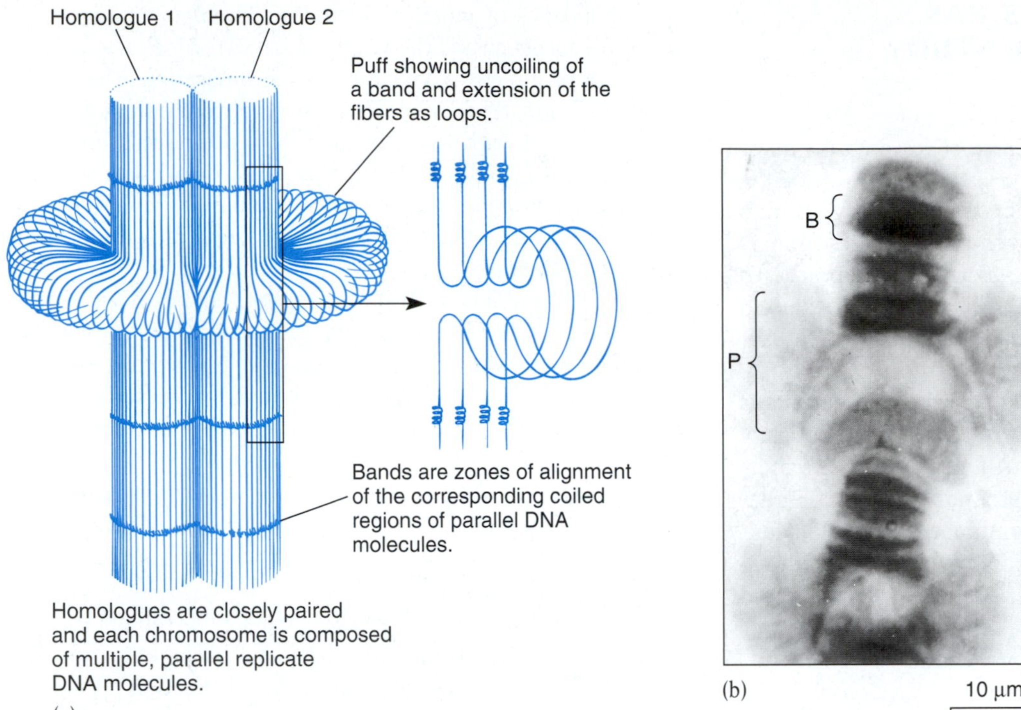
Homologue 1
Homologue 2
Puff showing uncoiling of a band and extension of the fibers as loops.
B
P
Bands are zones of alignment of the corresponding coiled regions of parallel DNA molecules.
Homologues are closely paired and each chromosome is composed of multiple, parallel replicate DNA molecules.
(a)
(b)
10 µm

Figure 16–6 ***Drosophila*** **polytene chromosomes.** These large chromosomes, found in cells of the salivary gland and some other tissues, aid in locating genes. **(a)** In contrast with the chromosomes of most somatic cells, homologous polytene chromosomes are paired, and each consists of more than 1000 parallel longitudinal chromosomes fibers, produced by repeated DNA synthesis without mitosis. **(b)** LM of a region of a polytene chromosome showing the pattern of stained bands of condensed chromatin (B) and decondensed puffed bands (P), which are the sites of intense RNA synthesis. Although the chromosome banding patterns vary from tissue to tissue, those in a particular tissue are constant and can be associated with the locations of mutant genes by genetic mapping and DNA hybridization methods. *(b, Courtesy of U. Clever)*

formation of the discs, their positions within the larva, and their ultimate functions within the adult fly. Those genes have been identified through mutations that either prevent certain discs from forming or alter their structure or ultimate fate.

Many *Drosophila* developmental mutants affect the body plan

Many developmental mutants of *Drosophila* have been identified. Their effects on development in various combinations have been examined and studied extensively at the molecular level. In our discussion we pay particular attention to those that affect the segmented body plan of the organism, both in the larva and in the adult.

Early *Drosophila* development occurs in the following way. The structure of the egg becomes organized as it develops in the ovary of the female, as stores of messenger RNA (mRNA), along with yolk proteins and other cytoplasmic molecules, are passed into it from the surrounding maternal cells. Immediately after fertilization, the zygote nucleus in the egg divides, beginning a remarkable series of 13 mitotic divisions. Each of these divisions takes only 5 or 10 minutes, which means that the DNA in the nuclei is replicated constantly at a very rapid rate. During that time the nuclei do not synthesize RNA. Cytokinesis does not take place, and the nuclei produced by the first seven divisions remain at the center of the embryo until the eighth division occurs. At that time, most of the nuclei start to migrate out from the center and to become localized at the periphery of the embryo. This is known as the *syncytial blastoderm* stage because the nuclei are not surrounded by individual plasma membranes. (A *syncytium* is a structure containing many nuclei.) Subsequently, cell membranes do form, and the embryo becomes known as a *cellular blastoderm.*

Maternal effect genes organize the egg cytoplasm

The genes that act to organize the structure of the egg cell are referred to as **maternal effect genes.** These are genes in the surrounding maternal tissues that are transcribed to produce mRNA

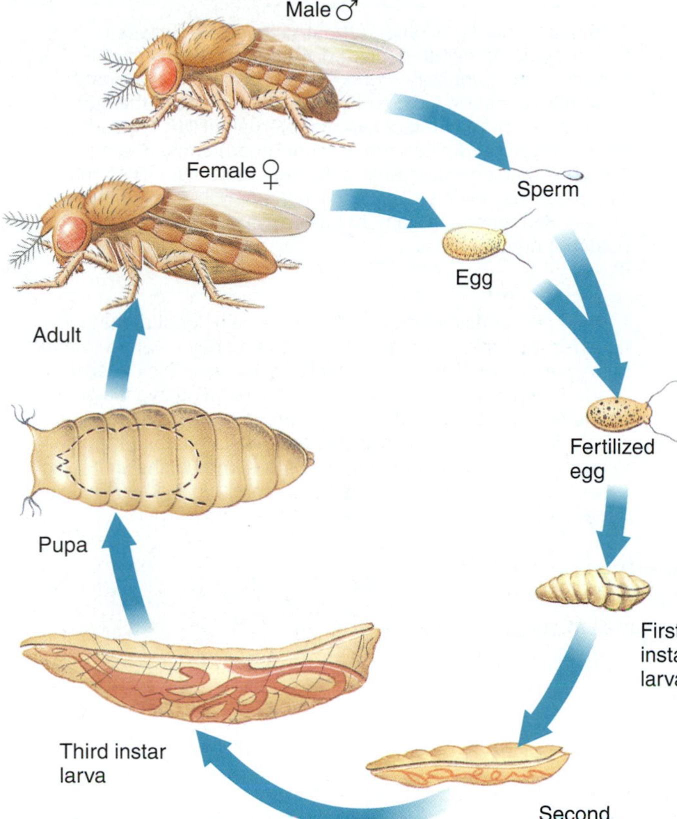

Figure 16–7 *Drosophila* development. A *Drosophila* passes through several stages as it develops from the egg to the sexually mature adult fly. About 12 days are required to complete the life cycle at 25° C. (The dotted lines within the pupa represent the animal undergoing metamorphosis.)

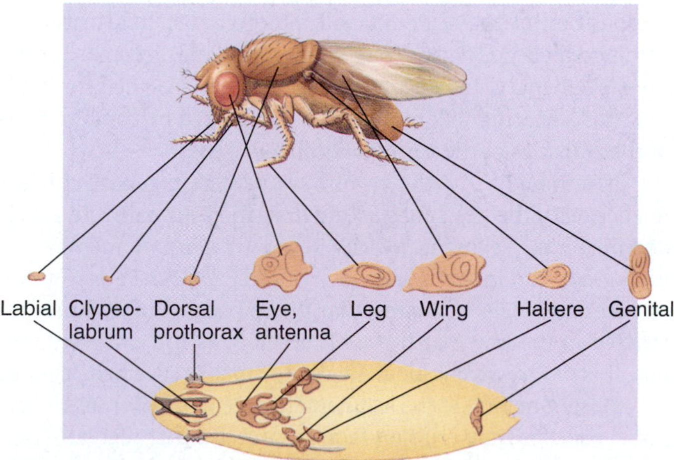

Figure 16–8 Imaginal discs. Each pair of discs in a *Drosophila* larva *(bottom)* develops into a specific pair of structures in the adult fly.

molecules to be transported into the developing egg. Analysis of mutants defective in these genes has revealed that many are involved in establishing the polarity of the embryo by designating which parts of the egg are dorsal or ventral and which are anterior or posterior (see Chapter 28); thus they are known as *egg polarity genes.*

Figure 16–9*a* illustrates concentration gradients for particular types of maternal mRNA in the very early embryo, at the beginning of nuclear migration. These mRNA transcripts of some of the maternal effect genes can be identified by their ability to hybridize with radioactive DNA probes derived from cloned genes. Alternatively, their protein products can be identified by antibodies that specifically bind to them. The protein produced by translation of the mRNA appears to be part of a system of determinants that organize the early pattern of development in the embryo. A combination of these protein gradients may provide positional information that specifies the fate of each nucleus or cell within the embryo. That information may then be interpreted by a cell as signals specifying the developmental path it should follow. For example, owing to the absence of specific signals in the egg, maternal effect mutations can produce an embryo with two heads or two posterior ends.

In many cases, the phenotype associated with a maternal effect mutation can be reversed by injecting normal maternal mRNA into the mutant embryo. When this is done, the fly develops normally, indicating that the gene product is needed only for a short time at the earliest stages of development.

Zygotic segmentation genes continue and extend the developmental program

As the nuclei start to migrate to the periphery of the embryo, **zygotic genes** begin to be expressed as production of some of the embryonic mRNA begins. (It is customary to refer to the genes of the embryo itself as zygotic genes, even though the embryo is no longer a zygote.) Zygotic genes that extend the developmental program beyond the pattern established by the maternal genome include the zygotic segmentation genes and the homeotic genes.

So far geneticists have identified at least 24 **zygotic segmentation genes** that are responsible for generating a repeating pattern of body segments within the embryo. The zygotic segmentation genes fall into three classes—gap genes, pair-rule genes, and segment polarity genes—representing a rough hierarchy of gene action.

The **gap genes** are apparently the first set of zygotic segmentation genes to act. These genes seem to interpret the maternal anterior-posterior information in the egg and begin organization of the body segments. A mutation in one of the gap genes usually causes the absence of one or more body segments in an embryo (Figs. 16–9*b* and 16–10*a*).

The other two classes of zygotic segmentation genes do not act on small groups of body segments but rather affect all segments. For example, mutations in the **pair-rule genes** delete every other segment (Fig. 16–10*b*), whereas mutations in **segment polarity genes** produce segments in which one part is

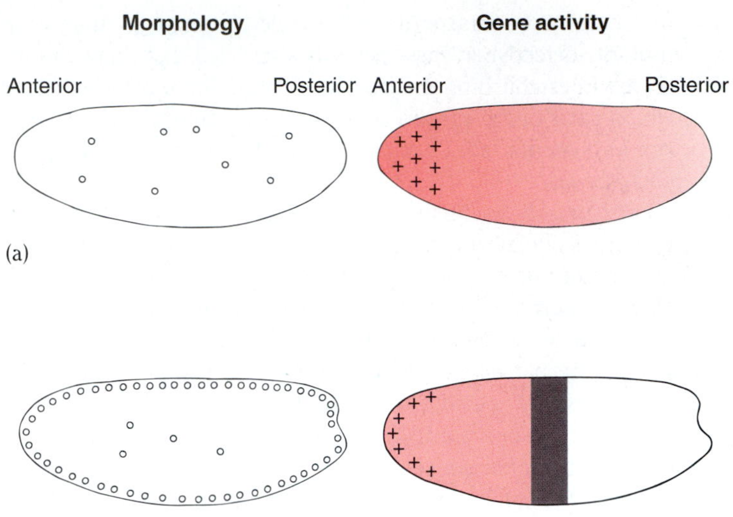

1.25 hours after fertilization (about 128 nuclei). Between the seventh and eighth nuclear divisions, the nuclei start to migrate to the periphery of the egg. The products of several maternal genes are localized in different regions of the egg. The crosses mark the location of maternal mRNA transcribed from a gene that defines the anterior (head) end of the egg. The pink shading represents a different maternal mRNA that was previously evenly distributed in the egg, but is now present in a concentration gradient that extends from the anterior to the posterior end.

2 hours after fertilization (about 1500 nuclei). Most of the nuclei have reached the perimeter of the egg and have started to make their own mRNA. The maternal mRNA shown in pink in part *(a)* is now being transcribed from the corresponding zygotic gene by the nuclei in the anterior part of the embryo *(crosses)*. The mRNA from a certain zygotic gap gene is transcribed from cells in only one segment in the middle of the embryo shown in black. If that gap gene is mutated, part of the embryo will be missing because that particular segment will fail to develop.

Figure 16–9 Early development of *Drosophila*. Longitudinal sections illustrating the morphology of the embryo at different times after fertilization *(left)* are matched with greatly simplified representations of the patterns of gene activity at each stage *(right)*. *(After Akam, M.E., "The Molecular Basis for Metameric Pattern in the* Drosophila *Embryo."* Development, *Vol. 101, 1987)*

missing and the remaining part is duplicated as a mirror image (Fig. 16–10*c*). The effects of the different classes of mutants are summarized in Table 16–1.

Each zygotic segmentation gene can be shown to have distinctive times and places in the embryo in which it is most active (Fig. 16–11). The observed pattern of expression of the maternal and zygotic genes controlling segmentation indicates that cells destined to form adult structures are determined by a progressive series of developmental decisions. First, the anterior-posterior (head-to-tail) axis and the dorsal and ventral regions of the embryo are determined by maternal segmentation genes thought to form gradients of **morphogens** in the egg. A morphogen is a chemical agent that affects the differentiation of cells and the development of form.

Zygotic segmentation genes then respond to the amounts of various morphogens at each location to control the production of a series of segments from the head to the posterior region. Then, within each segment, other genes are activated that read the position of the segment and interpret that information to specify which body part that segment should become. Within every segment, each cell's position is further specified so that it now has a specific "address" that is designated by combinations of the activities of the various regulatory genes.

It is thought that the zygotic segmentation genes act in sequence, with the gap genes acting first, then the pair-rule genes, and finally the segment polarity genes. In addition, members of each group can interact with each other. Each time a new group of genes acts, cells of a particular group become more finely restricted in the way that they will develop. As the embryo develops, it is progressively subdivided into smaller specified regions.

Most, if not all, of the segmentation genes (maternal and zygotic) code for **transcription factors,** proteins that regulate transcription in eukaryotic cells (see Chapter 13). For example, some of the segmentation genes code for a "zinc-finger" type of DNA-binding regulatory protein (see Fig. 13–8*b*). Others code for other types of transcription factors; these are discussed in the

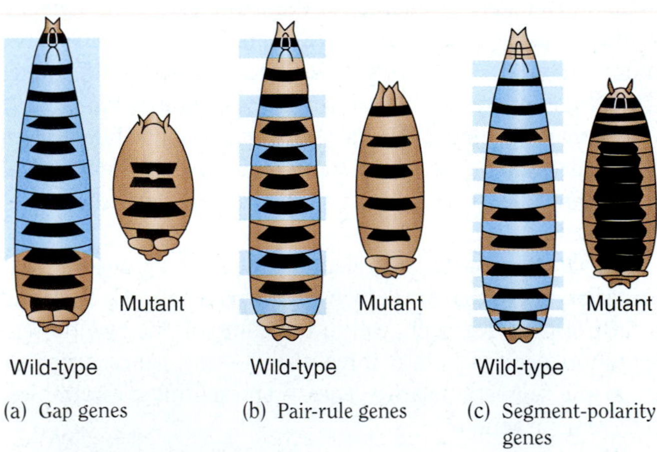

Figure 16–10 Zygotic segmentation genes in *Drosophila*. Gap genes **(a)**, pair-rule genes **(b)**, and segment polarity genes **(c)** control the pattern of body segments in a *Drosophila* embryo. The blue bands mark the regions in which the protein products of these genes are normally expressed in wild-type embryos. These same regions are absent in embryos in which the gene is mutated; the resulting phenotype is characteristic of the class to which the gene belongs. *(After Nüsslein-Volhard, C., and E. Wieschaus, "Mutations Affecting Segment Number and Polarity in* Drosophila.*"* Nature, *Vol. 287, 1980)*

TABLE 16–1 Classes of Genes Involved in Pattern Formation of Embryonic Segments in *Drosophila*

Type of Gene	Site of Gene Activity	Effects of Mutant Alleles and Proposed Function(s) of Genes
Maternal effect genes	Maternal tissues (ovary)	Many maternal effect mutations alter the polarity of the embryo; initiate pattern formation by activating zygotic regulatory genes in nuclei in certain locations in embryo
Zygotic segmentation genes		
Gap genes	Embryo	Mutant alleles cause one or more segments to be missing; some may influence activity of pair-rule genes, segment polarity genes, and homeotic genes
Pair-rule genes	Embryo	When mutated, cause alternate segments to be missing; some may influence activity of segment polarity genes and homeotic genes
Segment polarity genes	Embryo	Mutant alleles delete part of every segment and replace it with mirror image of remaining structure; may influence activity of homeotic genes
Homeotic genes	Embryo	Homeotic mutations cause parts of fly to form structures normally formed in other segments; control the identities of the segments

next section. The fact that many of the genes involved in the control of development code for transcription factors indicates that those proteins indeed act as genetic "switches" regulating the expression of other genes. Once proteins that function as transcription factors have been identified, it is possible to use the purified proteins to identify the DNA target sequences to which they bind. This approach has been increasingly useful in identifying additional parts of the regulatory pathway involved in different stages of development. Transcription factors also play a role in cancer (see *Focus On: Oncogenes and Cancer*).

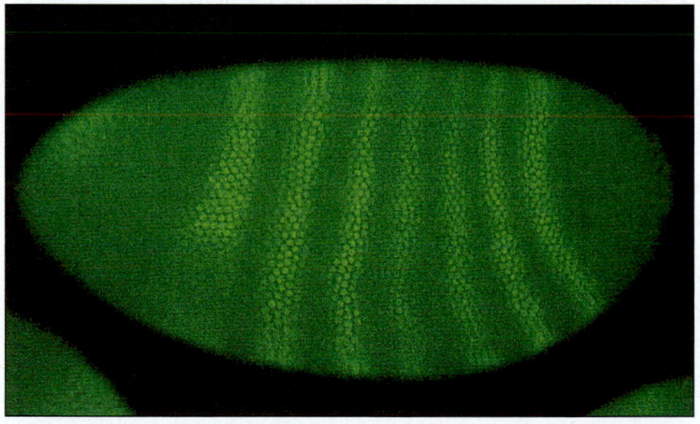

Figure 16–11 Activity of a zygotic segmentation gene. The bright bands in this fluorescence LM reveal the presence of mRNA transcribed from one of the zygotic pair-rule loci known as *fushi tarazu* (Japanese for "not enough segments"). The segments of the larva that are normally derived from these bands are absent when this locus is mutated. *(Courtesy of Steve Paddock, Jim Langeland, Sean Carroll, Howard Hughes Medical Institute, University of Wisconsin)*

Homeotic genes specify the identity of each body segment

One function of the zygotic segmentation genes is to regulate the expression of a separate set of genes that actually designates the final adult structure formed by each of the imaginal discs. These genes are called **homeotic genes.** Because of their involvement in segment identity, mutations in homeotic genes cause one body part to be substituted for another and therefore produce some peculiar changes in the adult. Among the most striking examples are the *Antennapedia* mutants, which have legs that grow from the head at a position where the antennae would normally be found (Fig. 16–12 on page 354).

Homeotic genes in *Drosophila* were originally identified by the altered phenotypes produced by mutant alleles. When geneticists analyzed the DNA sequences of several homeotic genes, they discovered a short DNA sequence of approximately 180 base pairs that is characteristic of many homeotic genes as well as some other genes that play a role in development. This DNA sequence has been termed the **homeobox.** Each homeobox codes for a protein functional region called a **homeodomain,** consisting of 60 amino acids that form four α-helices. One of these serves as a recognition helix, which can bind to specific DNA sequences and affect transcription. Thus the products of the homeotic genes, like those of the earlier acting zygotic segmentation genes, are transcription factors. In fact, some of the segmentation genes also contain homeoboxes.

MAKING THE CONNECTION How does the study of homeobox-containing genes provide insights into evolutionary relationships? Consider studies on *Hox* genes, clusters of homeobox-containing genes that specify the anterior-posterior axis during development. *Hox* genes were initially discovered in *Drosophila,* where they are arranged in two adjacent groups on

FOCUS ON Oncogenes and Cancer

A cancer cell lacks normal biological inhibitions. Normal cells are tightly regulated by control mechanisms that cause them to divide when necessary and prevent them from growing and dividing at inappropriate times. Cells of many tissues in the adult are normally prevented from dividing; they reproduce only to replace a neighboring cell that has died or become damaged. Cancer cells have escaped such controls and can divide continuously.

As a consequence of their abnormal growth pattern, some cancer cells eventually form a mass of tissue called a **tumor.** If the tumor remains at the spot where it originated, it can usually be removed by surgery. One of the major problems with certain forms of cancer is that the cells can escape from the controls that maintain them in their proper location. These cells can **metastasize,** or spread, to different parts of the body, invading other tissues and forming multiple tumors. Lung cancer, for example, is particularly deadly because its cells are highly metastatic and can enter the blood and spread to form tumors in other parts of the lungs, or in other organs such as the liver and the brain. Tumors with cells that can metastasize are referred to as **malignant tumors.**

We now know that cancer is a disease caused by altered gene expression. Using recombinant DNA methods, researchers identified many of the genes that, when they function abnormally, transform normal cells into cancer cells. Each kind of cancer cell apparently owes its traits to at least one, and possibly several, of a relatively small set of genes known as **oncogenes** (cancer-causing genes). Oncogenes arise from changes in the expression of certain genes called **proto-oncogenes,** which are *normal* genes found in all cells and involved in the control of growth and development.

Oncogenes were first discovered in viruses that can infect mammalian cells and transform them into cancer cells *(malignant transformation).* Such viruses

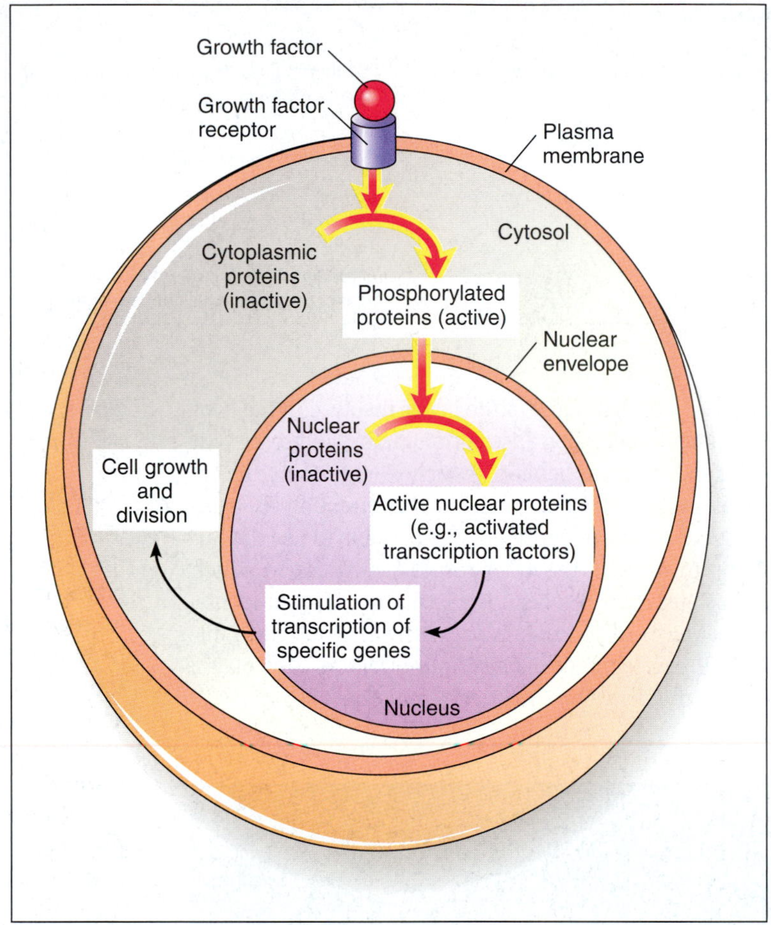

Simplified view of part of a growth control cascade. In this example, a growth factor stimulates cell growth. The growth factor receptor, as well as some of the other components of the system, are coded for by proto-oncogenes. When a proto-oncogene mutates, becoming an oncogene, the cell grows and divides even in the absence of the growth factor.

the chromosome: the *Antennapedia* complex and the *bithorax* complex. As homeobox-containing genes have been identified in other animals, including other arthropods, annelids (segmented worms), roundworms, and mammals, it has been found that these genes are also clustered and that their organization is remarkably similar to that seen in *Drosophila.* Figure 16–13 compares the organization of the *Hox* gene clusters of *Drosophila,* the roundworm *C. elegans* (to be discussed in the next section), and the laboratory mouse. These images are matched with the regions where they are expressed in the animals. Remarkably, the *Drosophila* and mouse *Hox* genes are located in the same order along the chromosome, although the correlation is less exact for *C. elegans.* Furthermore, the linear order of the genes on the chromosome reflects the order of the corresponding regions they control (from anterior to posterior) in the animal. This organization apparently reflects the need for these genes to be transcribed in a specific temporal sequence.

can incorporate DNA sequences of the proto-oncogenes into their own nucleic acid. In some cases, the viruses alter the expression of the proto-oncogenes, converting them into oncogenes. This may happen if the DNA sequences come under the control of viral regulatory elements, which cause the gene to be transcribed at much higher than normal levels, or if the captured gene mutates so that its protein product is more active than the product of the normal proto-oncogene.

A proto-oncogene in a cell that has not been infected by a virus can also mutate and become an oncogene. One of the first oncogenes identified was isolated from a bladder tumor. In the cell that gave rise to the tumor, a proto-oncogene had undergone a single base-pair mutation; the result was that the amino acid glycine was replaced by a valine in the protein product of the gene. This subtle change was apparently a critical factor in the conversion of the normal cell into a cancer cell.

By means of recombinant DNA technology and other techniques of molecular biology, it has been possible for researchers to identify more than 60 oncogenes and their corresponding proto-oncogenes. Because the fundamental controls of normal cell division and differentiation probably evolved very early in the evolutionary history of eukaryotes, it is not surprising that very similar proto-oncogenes have been found in a diverse array of organisms, ranging from yeasts to humans. For example, the proto-oncogene counterpart of the oncogene found in some human bladder tumors has also been found in yeast cells.

Some of these controls are illustrated in greatly simplified form in the accompanying figure. The growth and division of cells can be triggered by one or more external signal molecules (see Chapter 5). Some of these substances are **growth factors** that bind to specific **growth factor receptors** associated with the cell surface, initiating a cascade of events inside the cell. Often the growth factor receptor complex acts as a **protein kinase** (an enzyme that phosphorylates proteins), which then phosphorylates specific amino acids of several cytoplasmic proteins. This posttranslational modification usually results in the activation of previously inactive enzymes. These activated enzymes are then able to catalyze the activation of certain nuclear proteins, many of which are transcription factors. Activated transcription factors bind to their DNA targets and stimulate transcription of specific sets of genes that initiate growth and cell division.

Even in the simplified scenario presented in the figure, it is evident that multiple steps are required to control cell proliferation. Remarkably, the proto-oncogenes that encode the products responsible for many of these steps have been identified. The current list of known proto-oncogenes includes genes that code for various growth factors or growth factor receptors and genes that respond to stimulation by growth factors (including many transcription factors). When one of these proto-oncogenes is expressed inappropriately (i.e., becomes an oncogene), the cell may misinterpret the signal and respond by growing and dividing. For example, in some cases a proto-oncogene encoding a growth factor receptor becomes mutated in a way that causes the growth factor receptor to no longer be regulated. It is always switched "on," even in the absence of the growth factor that normally controls it.

Not all genes that cause cancer when mutated are proto-oncogenes. About one-half of all cancers are caused by a mutation in a **tumor suppressor gene.** These genes, also known as **anti-oncogenes,** normally interact with growth inhibiting factors to block cell division. When mutated they lose their ability to "put on the brakes," and uncontrolled growth ensues.

Certain oncogenes appear to be particularly common and are found in a variety of tumors. However, a change in a single proto-oncogene is usually insufficient to cause a cell to become malignant. The development of cancer is usually a multistep process involving both oncogenes and mutated tumor suppressor genes. Additional factors, such as the inappropriate activation of the enzyme responsible for the maintenance of telomeres, may also play a role (see Chapter 11, *On the Cutting Edge: Telomerase, Cellular Aging, and Cancer*). As more of these genes are discovered and their complex interactions are unraveled, we will gain a fuller understanding of the control of growth and development. This understanding is leading to improved diagnosis and treatment of various cancers.

Drosophila has only one *Antennapedia-bithorax* complex. However, humans and other vertebrates have four similar *Hox* gene clusters, each located in a different chromosome. These complexes probably arose through gene duplication. The fact that extra copies of these genes are present helps explain why mutations causing homeotic-like transformations are seldom seen in vertebrate animals. However, one particular type of *Hox* mutation that has been described in both mice and humans causes abnormalities in the limbs and genitalia. The involvement of the genitalia provides a further explanation for the rarity of these mutant alleles, because affected individuals are unlikely to reproduce.

The fact that very similar developmental controls are seen in organisms as diverse as insects, unsegmented roundworms, and vertebrates (including humans) indicates that the basic mechanism evolved early and has been highly conserved in all animals

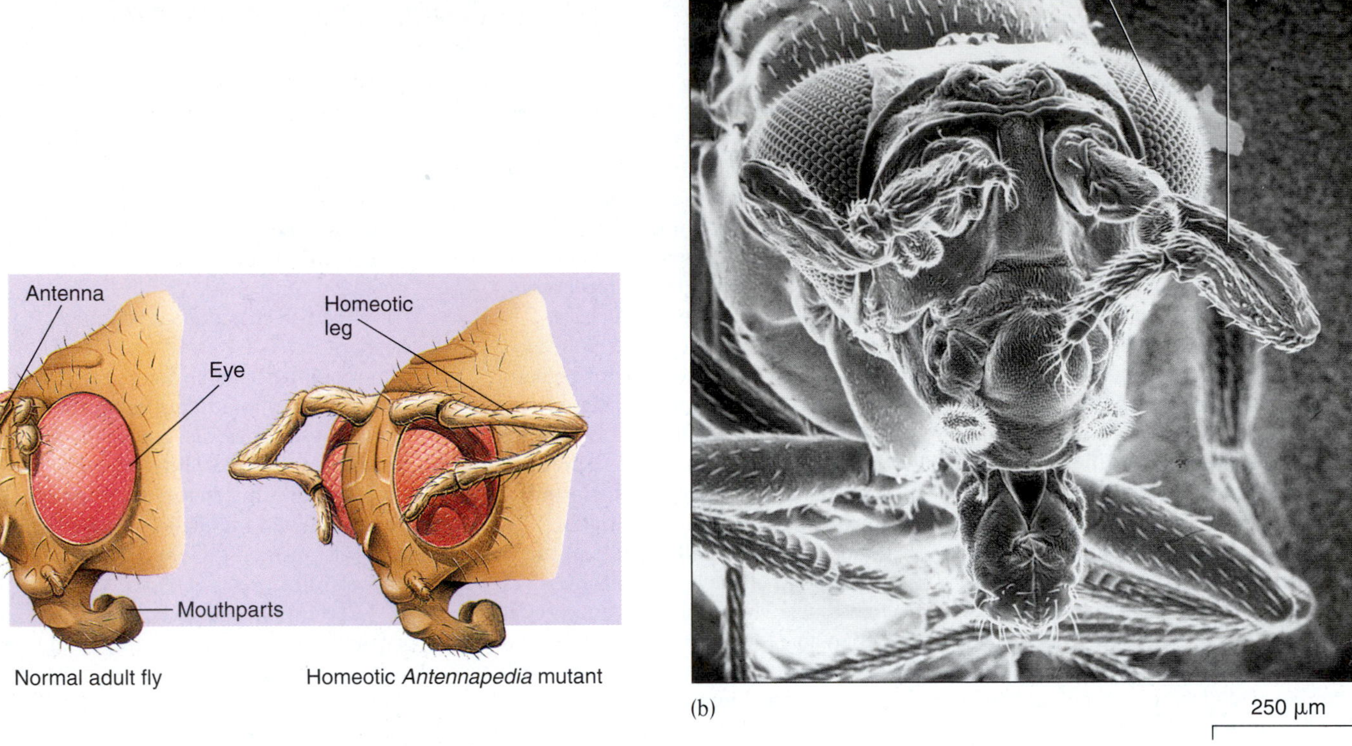

Figure 16–12 The *Antennapedia* locus. *Antennapedia* mutations of *Drosophila* cause homeotic transformations in which the antennae are replaced by legs or parts of legs. **(a)** Head of a normal fly and a fly with an *Antennapedia* mutation. **(b)** SEM of the head of a fly with a mutant allele of the *Antennapedia* locus that produces an extreme phenotype. Most of the mutant alleles of this locus produce only incomplete legs in place of the structures of the antennae. *(b, Dr. Thomas Kaufman)*

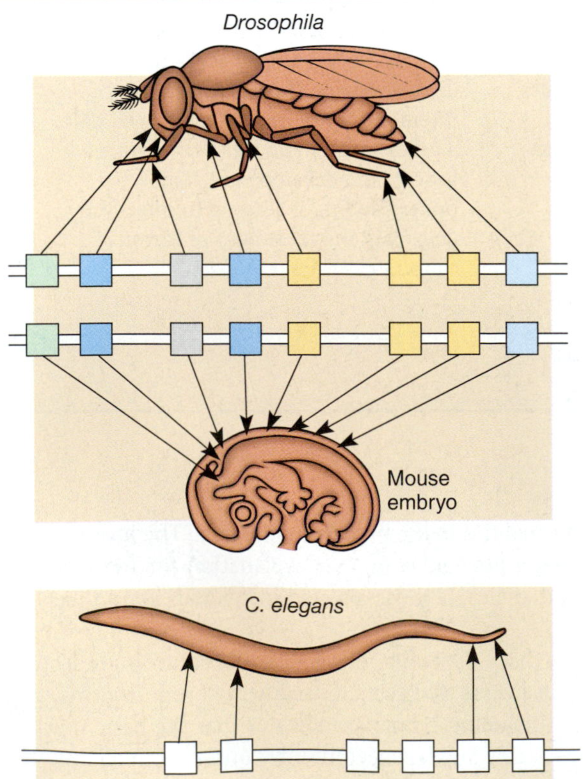

Figure 16–13 *Hox* gene clusters. The *Hox* gene clusters of *Drosophila* and the laboratory mouse are correlated with the parts of the body in which each gene is expressed. (Only one of the four mouse *Hox* gene clusters is shown.) Note that in each organism the order of the genes on the chromosome reflects their spatial order of expression in the embryo. Although not evident in the figure, some of these regions of expression overlap. The most anteriorly expressed genes are shown to the left, while those expressed most posteriorly are at the right. *C. elegans* also has similar clustered genes, although the gene order is not identical and there is evidence that not all are required for embryonic development. *(After Kenyon, C., and B. Wang. "A Cluster of* Antennapedia*—Class Homeobox Genes in a Nonsegmented Animal."* Science, *Vol. 253, 2 Aug, 1991; and Van Auken, K., et al. "*Caenorhabditis elegans *Embryonic Axial Patterning Requires Two Recently Discovered Posterior-Group* Hox *Genes."* Proceedings of the National Academy of Sciences, *Vol. 97, 25 Apr. 2000)*

that have an anterior-posterior axis, even those that are not segmented. It is becoming clear that once a successful way of controlling groups of genes and integrating their activities evolved, it was retained, although it has apparently been modified in various ways that provided for alterations of the body plan.

The finding of homeobox-like genes in plants suggests that these genes may be of ancient origin and may in fact be the genes that made multicellularity possible. Further investigations may allow researchers to develop an overall model of how the rudiments of morphogenesis are controlled in both plants and animals. These systems of master genes that control development are proving to be a rich source of "molecular fossils" that are illuminating evolutionary history in new and exciting ways.

CAENORHABDITIS ELEGANS HAS A VERY RIGID EARLY DEVELOPMENTAL PATTERN

C. elegans, a roundworm or nematode (see Chapter 29), has one of the simplest systems of genetic control of development. The study of this animal was begun in the 1960s by Sydney Brenner, a molecular biologist. Its genome of almost 19,000 genes was the first animal genome to be sequenced, and today it is an important tool for answering basic questions about the development of individual cells within a multicellular organism.

Even as an adult, *C. elegans* is only 1.5 mm long and contains only about 1000 somatic cells (the exact number depends on the sex) and about 2000 germ-line cells. Individuals can be either **hermaphrodites** (organisms with both sexes in the same individual) or males. Hermaphroditic individuals are self-fertilizing, which makes it easy to obtain offspring homozygous for newly induced recessive mutations. The availability of males that can mate with the hermaphrodites makes it possible to do genetic crosses as well.

Because the worm's body is transparent, researchers can follow the development of literally every one of its somatic cells (Fig. 16–14) using a Nomarski differential interference microscope (see Chapter 4). As a result of efforts by several laboratories, the lineage of each somatic cell in the adult has now been determined. Those studies have shown that the nematode has a very rigid developmental pattern. After fertilization, the egg undergoes repeated divisions to produce about 550 cells that make up the small, sexually immature larva. After the larva hatches from the egg case, further cell divisions give rise to the adult worm.

The lineage of each somatic cell in the adult can be traced to a single cell in a small group of **founder cells,** which are formed early in development (Fig. 16–15). If a particular founder cell is destroyed or removed, the structures that would normally develop from that cell are missing. A rigid developmental pattern, in which the fates of the cells are largely predetermined, is referred to as **mosaic development.** Each cell has a specific fate in the embryo, just as in art a specific tile forms a particular part of the pattern of a mosaic picture.

It was originally thought that each organ system in *C. elegans* might be derived from only one founder cell. Detailed analysis of cell lineages, however, reveals that many of the structures found in the adult, such as the nervous system and the musculature, are in fact derived from more than one founder cell (Fig. 16–15*h*). Conversely, a few lineages have been identified in

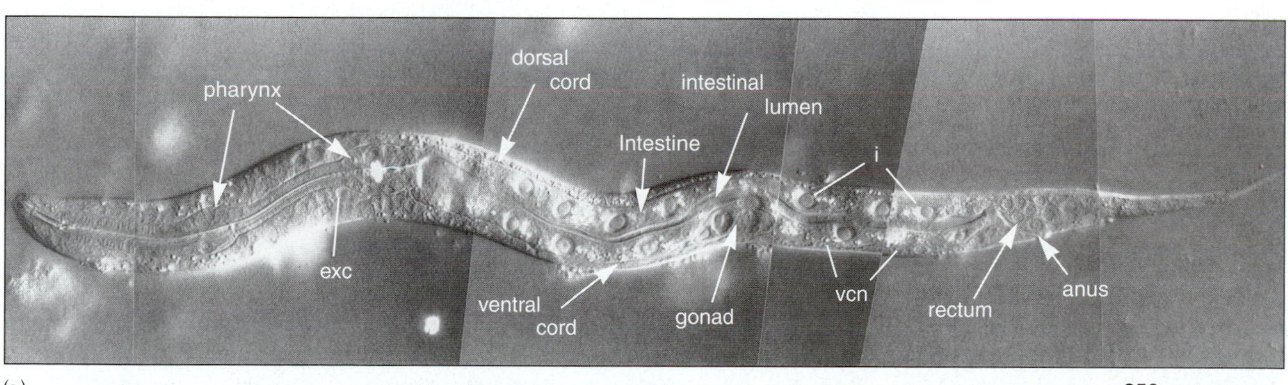

(a)

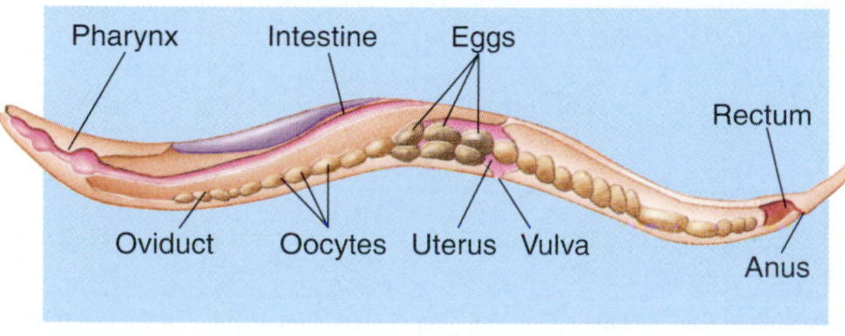

(b)

Figure 16–14 ***Caenorhabditis elegans.*** This transparent organism has a fixed number of somatic cells. **(a)** Nomarski interference LM of the adult hermaphrodite nematode. (The abbreviated labels "exc," "i," and "vcn" refer to certain cells of the excretory, digestive and nervous system, respectively.) **(b)** Diagram illustrating structures in the adult hermaphrodite. The sperm-producing structures are not shown. *(a, Courtesy of Dr. John Sulston, Medical Research Council; b, from Fig. 22–6a in Walbot, V., and N. Holder,* Developmental Biology. *Random House, New York, 1987)*

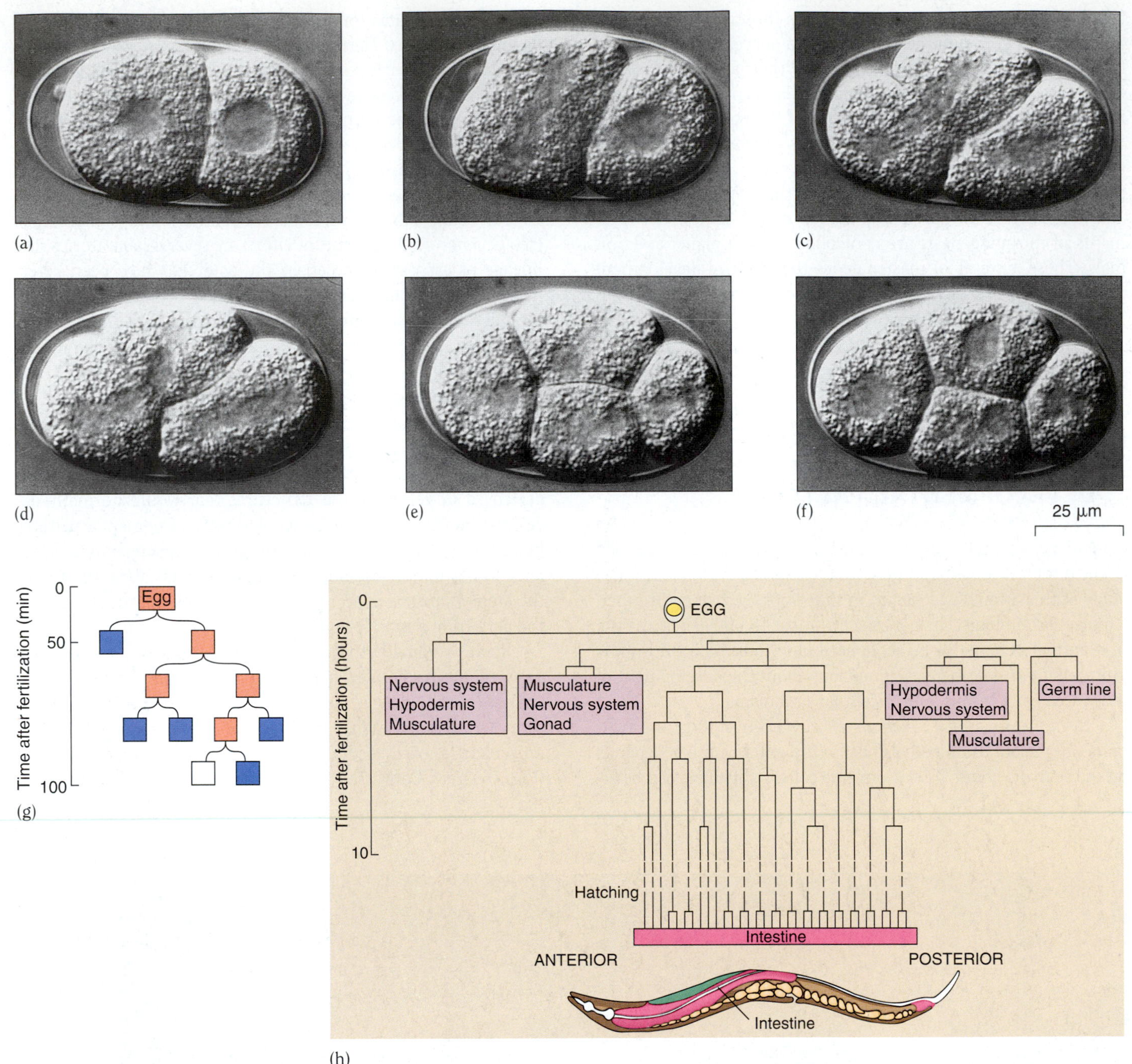

Figure 16–15 Cell lineages of *C. elegans*. All somatic cells of *C. elegans* are derived from five somatic founder cells produced during the early cell divisions of the embryo. **(a–f)** Nomarski interference LMs showing the early cell divisions of the embryo. **(g)** A lineage map showing the origins of the five somatic founder cells *(blue)*. The cell shown in white will give rise to the germ cells. **(h)** This lineage map traces the development of the cells that form the intestine. The dashed lines represent many cell divisions of a particular lineage. *(a–f, E. Schierenberg, from G. von Ehrenstein and E. Schierenberg, in* Nematodes as Biological Models, *Vol. 1, B. Zuckerman, ed. Academic Press, New York, 1980)*

which a nerve cell and a muscle cell are derived from the division of a single cell. Mutations affecting cell lineages have been isolated, and many of these appear to have properties that would be expected of genes involved in control of developmental decisions.

By using microscopic laser beams small enough to destroy individual cells, it is possible to determine what influence one cell may have on the development of a neighbor. Consistent with the rigid pattern of cell lineages, destruction of an individual cell in *C. elegans* results, in most cases, in the absence of all of the structures derived from that cell but with the normal differentiation of all of the neighboring somatic cells. This suggests that development in each cell is regulated through its own internal program.

However, the developmental pattern of *C. elegans* is not entirely mosaic. There are cases in which differentiation of a cell can be influenced by interactions with particular neighboring cells, a phenomenon known as **induction.** One example is the formation of the vulva (pl., *vulvae*), the structure through which the eggs are laid. A single nondividing cell, called the *anchor cell,* is a part of the ovary (the structure in which the germ-line cells undergo meiosis to produce the eggs). The anchor cell attaches to the ovary and to a point on the outer surface of the animal, triggering the formation of a passage through which the eggs pass to the outside. When the anchor cell is present, cells on the surface organize to form the vulva and its opening. If the anchor cell is destroyed by a laser beam, however, the vulva does not form and the cells that would normally form the vulva remain as surface cells (Fig. 16–16). The anchor cell therefore induces the surface cells to form a vulva.

Analysis of certain cell lineage mutations has been useful in understanding such inductive interactions. For example, several types of mutations cause more than one vulva to form. In such mutant animals, multiple vulvae form even if the anchor cell is destroyed. Thus, the mutant cells do not require an inductive signal from an anchor cell to form a vulva. Evidently in these mutants the gene or genes responsible for vulva formation are constitutive. Conversely, mutants lacking a vulva are also known. In some of these, the cells that would normally form the vulva appear unable to respond to the inducing signal from the anchor cell.

During development in *C. elegans,* there are instances in which cells die shortly after they are produced. This phenomenon, known as **apoptosis** or **programmed cell death** (see Chapter 4) has been observed in a wide variety of organisms, both plant and animal. For example, the human hand is formed as a webbed structure, but the fingers become individualized when the cells between them die. In *C. elegans,* these programmed cell deaths are under genetic control, and several mutants have been isolated that alter the pattern of these deaths. The loci identified by these mutations, in addition to mutations with similar effects in other organisms, are being analyzed at the molecular level and should shed considerable light on the general phenomena of cellular aging and programmed cell death.

Mutations are also known that appear to identify genes involved in developmental timing, known as **chronogenes.** One such locus has recessive alleles that cause certain cells to adopt fates that would ordinarily be seen later in development. Dominant alleles of the same locus cause certain cells to adopt fates that would usually be expressed earlier. Such genes appear to be good candidates for "switches" that control developmental timing.

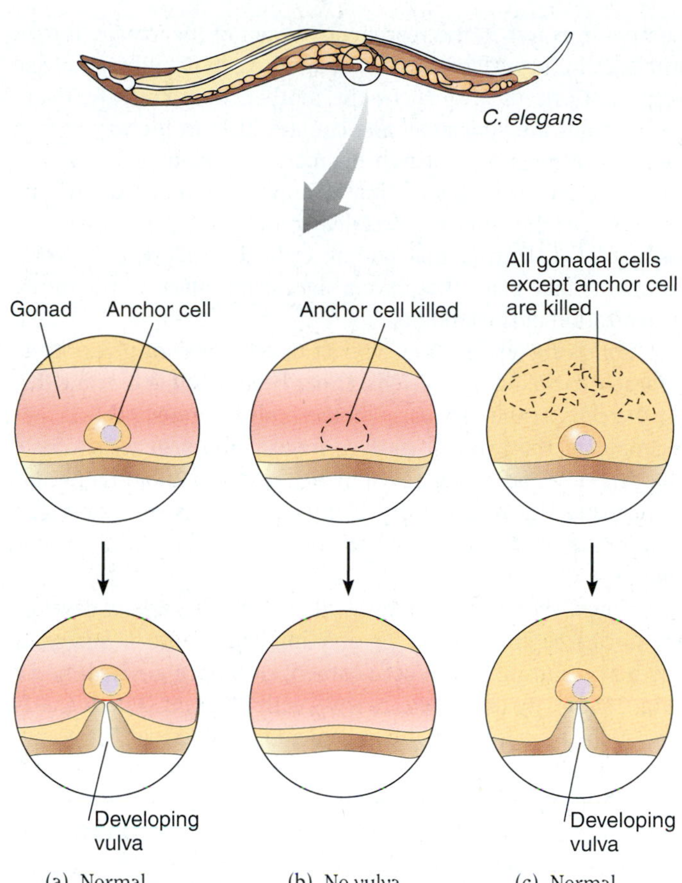

Figure 16–16 Induction. A single anchor cell induces neighboring cells to form the vulva in *C. elegans.* This schematic diagram shows how laser destruction of single cells or a group of cells can be used to demonstrate the influence of a cell on its neighbors.

THE MOUSE IS A MODEL FOR MAMMALIAN DEVELOPMENT

Mammalian embryos develop in markedly different ways from the embryos of *Drosophila* and *C. elegans.* The laboratory mouse, *Mus musculus,* is the best studied example of early mammalian development. Numerous genes affecting development have been identified in the mouse, and the mouse genome sequence was virtually completed in 2001.

Cells of very early mouse embryos are totipotent

The early development of the mouse and other mammals is similar in many ways to human development, which is described in detail in Chapter 49. During the early developmental period, the embryo lives free in the reproductive tract of the female. It then implants in the wall of the uterus, after which its nutritional and respiratory needs are met by the mother. Consequently, mammalian eggs are very small and contain little in the way of food reserves. Almost all research on mouse development has concentrated on the stages leading to implantation because during those stages the embryo is free-living and can be experimentally manipulated. During that period, critical developmental commitments take place that have a significant effect on the future organization of the embryo.

Following fertilization, a series of cell divisions gives rise to a loosely packed group of cells. It has been possible to show that all the cells in the very early mouse embryo are equivalent. For example, at the two-cell stage of mouse embryogenesis, one of the two cells can be destroyed by pricking it with a fine needle. Implanting the remaining cell into the uterus of a surrogate mother in most cases leads to the development of a normal mouse.

Conversely, two embryos at the eight-cell stage of development can be fused together and implanted into a surrogate mother, resulting in the development of a normal-sized mouse (Fig. 16–17). By using two embryos with different genetic markers (such as coat color), it can be demonstrated that the resulting mouse has four genetic parents. These mice have fur with patches of different colors derived from clusters of genetically different cells. Animals formed in this way are called **chimeras.** (The term *chimera,* derived from the name of a mythical beast that had the head of a lion, the body of a goat, and the tail of a snake, is used today to refer to any organism that contains two or more kinds of genetically dissimilar cells arising from different zygotes.) Chimeras have been important in allowing the use of genetically marked cells to trace the fates of certain cells during development.

The responses of mouse embryos to these kinds of manipulations are in marked contrast with the mosaic or predetermined nature of early C. *elegans* development, in which the destruction of one of the founder cells results in loss of a significant portion of the embryo. For this reason, we say that the mouse (and presumably other mammals) has highly **regulative development.** This means that the early embryo acts as a self-regulating whole that can accommodate missing or extra parts. On the other hand, it has not been possible so far to demonstrate totipotency of cells from slightly later stages of mouse development. However, as discussed in *Focus On: Mammalian Cloning* earlier in the chapter, successful nuclear transplantation experiments on mice leading to live births demonstrate that at least some nuclei of differentiated mouse cells are totipotent.

Transgenic mice are used in studies of developmental regulation

In transformation experiments similar to those done with *Drosophila,* foreign DNA injected into fertilized mouse eggs can be incorporated into the chromosomes and expressed (Fig. 16–18). The resulting **transgenic** mice have given researchers insights into how genes are activated during development. In addition, mouse genes can be inactivated ("knocked out") by the technique of **gene targeting** (see Chapter 14). For example, when a locus known as *Engrailed-1* is knocked out in mice, the resulting embryos exhibit lethal abnormalities in brain development.

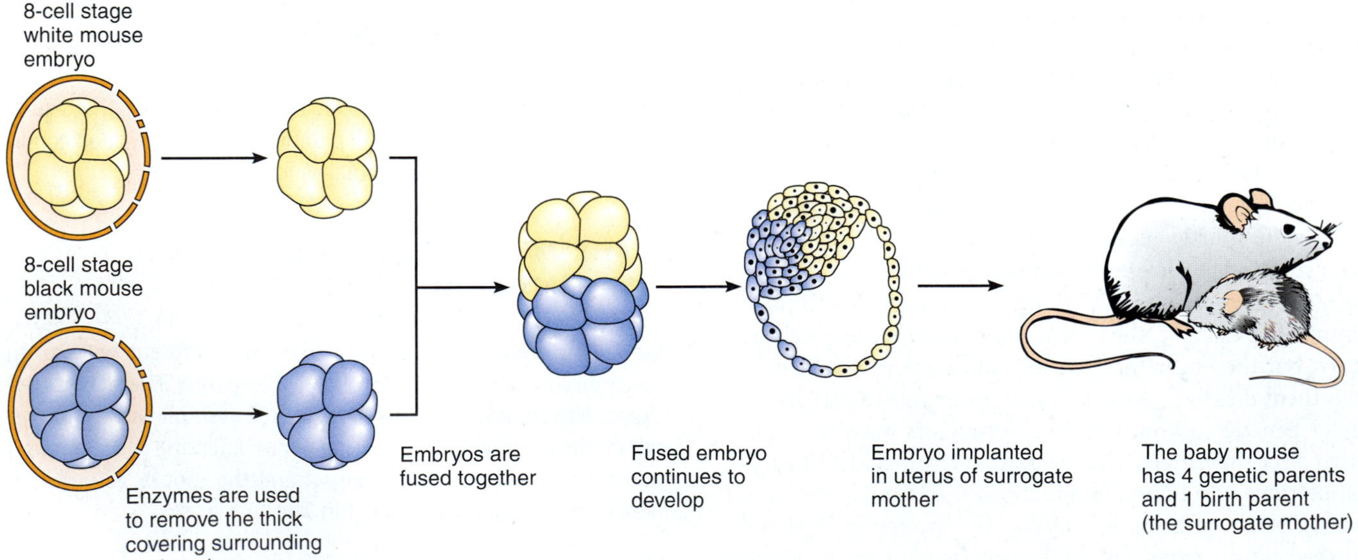
8-cell stage
white mouse
embryo
8-cell stage
black mouse
embryo
Enzymes are used
to remove the thick
covering surrounding
each embryo
Embryos are
fused together
Fused embryo
continues to
develop
Embryo implanted
in uterus of surrogate
mother
The baby mouse
has 4 genetic parents
and 1 birth parent
(the surrogate mother)

Figure 16–17 Chimeric mice. Embryos can be removed from females of two different strains, and the cells combined in vitro. The resulting aggregate embryo continues to develop and is implanted in the uterus of a surrogate mother. The offspring has four different genetic parents. Although the surrogate mother is the birth mother, she is not genetically related.

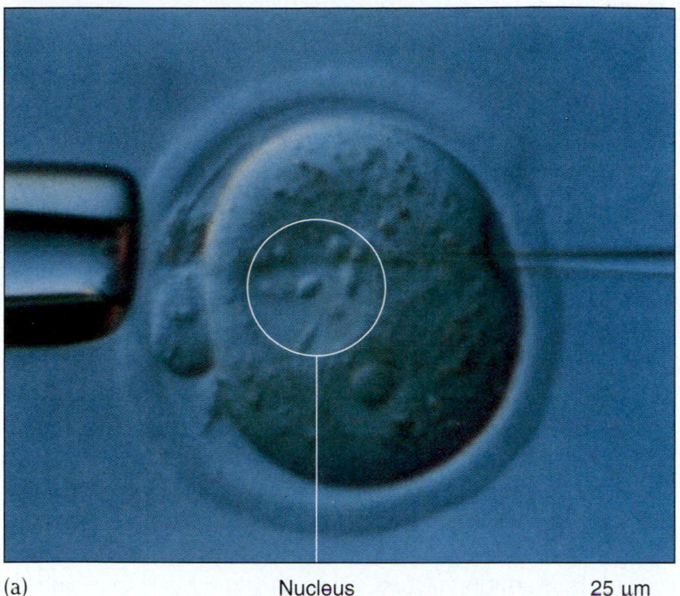

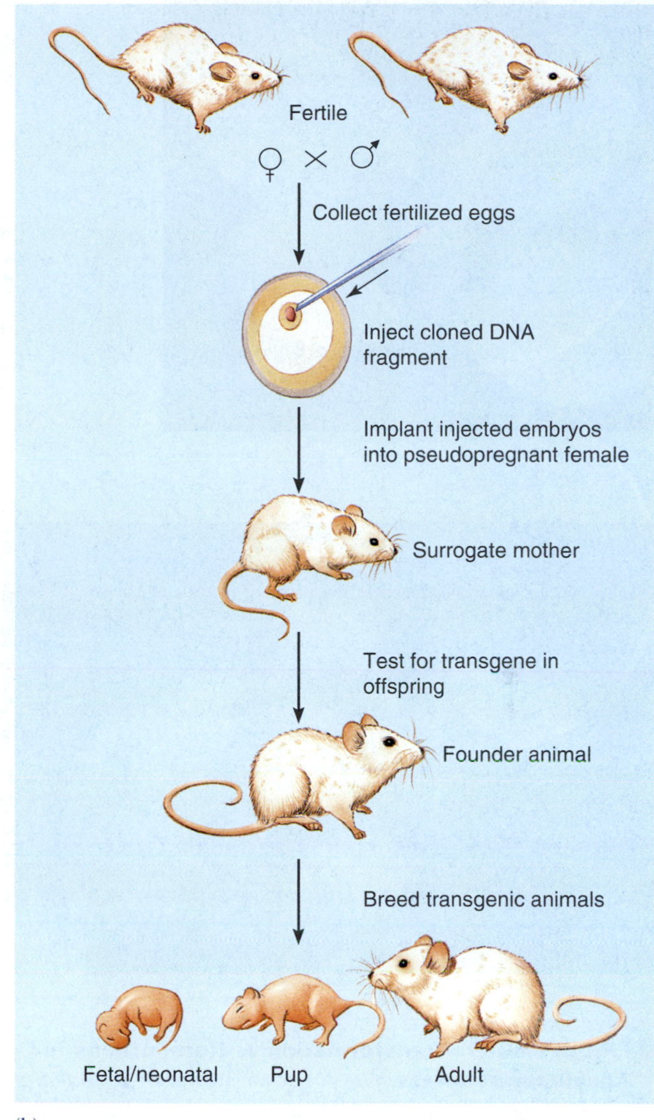

Figure 16–18 Producing a transgenic mouse. **(a)** Cloned DNA fragments are injected into the nucleus of a fertilized mouse egg by a glass needle, shown entering from the right, which is about 1 μm in diameter at the tip. The egg is held in place by suction on a holding pipet *(left).* **(b)** The injected eggs are then surgically transferred to a surrogate mother. The presence of the foreign gene can be examined in the transgenic animal, or the animal can be bred to establish a transgenic line of mice. *(a, R.L. Brinster, University of Pennsylvania School of Veterinary Medicine)*

Scientists can identify a transgene (foreign gene) that has been introduced into a mouse and determine whether it is active by marking the gene in several ways. Sometimes a similar gene from a different species is used; its protein can be distinguished from the mouse protein by specific antibodies. It is also possible to construct a hybrid gene that contains the regulatory elements of a mouse gene of interest together with part of another gene that codes for a "reporter" protein, such as an enzyme not normally found in the mouse. Such studies have been important in showing which DNA sequences of a mouse homeobox gene determine where the gene is expressed in the embryo.

Many developmentally controlled genes have been introduced into mice and have yielded important information about gene regulation. When developmentally controlled genes from other species such as humans or rats have been introduced into mice, they are regulated in the same way that they normally are in the donor animal. For example, when introduced into the mouse, human genes encoding insulin, globin, and crystallin—which are normally expressed in cells of the pancreas, blood, and eye lens, respectively—are expressed only in those same tissues in the mouse. The fact that these genes are correctly expressed in their appropriate tissues indicates that the signals for tissue-specific gene expression are highly conserved through evolution. This is an exciting finding because it means that information on the regulation of genes controlling development in one organism can have valuable applications to other organisms, including humans.

ARABIDOPSIS THALIANA IS A MODEL FOR PLANT DEVELOPMENT

Certain well-characterized plants are used in the study of the genetic control of development. Many of these are economically important crop plants such as corn, *Zea mays.* Genes with developmental effects are known in corn, including some that can be thought of as analogous to the homeotic genes of *Drosophila.*

Arabidopsis thaliana, the organism that is most widely used to study genetics and development in plants, is a member of the mustard family. Although *Arabidopsis* itself is a weed of no economic importance, it has several advantages for research. The plant completes its life cycle in just a few weeks and is small enough to be grown in a petri dish, permitting thousands of

(a)

(b)

Figure 16–19 Transformation of floral organs in *Arabidopsis* flowers. **(a)** A normal flower of *A. thaliana* has four outer leafy green sepals (hidden by the petals), four white petals, six stamens (the male reproductive structures), and a central pistil (the female reproductive structure). **(b)** This mutant has only sepals and petals because the C gene, which is responsible for the development of both the stamens and pistil, is nonfunctional and, therefore, not expressed. *(Dr. E. Meyerowitz, California Institute of Technology)*

individuals to be grown in limited space. Chemical mutagens can be used to produce mutant strains, and many developmental mutants have been isolated. Cloned foreign genes can be inserted into *Arabidopsis* cells, and these can be integrated into the chromosomes and expressed. These transformed cells can be induced to differentiate into transgenic plants.

In 2000 the *Arabidopsis* genome became the first plant genome to be sequenced. Although its genome is relatively small (the rice genome, for example, is about four times larger than that of *Arabidopsis*), it includes more than 25,000 protein-coding genes. In comparison, *Drosophila* has about 13,600 genes and *C. elegans* about 18,400. Many genes in *Arabidopsis* are functionally equivalent to genes in *Drosophila, C. elegans,* and other animal species.

Of particular importance to development are the more than 1500 genes that code for transcription factors (compared to 635 at last count for *Drosophila*). Not surprisingly, many of the genes that are known to specify the identities of the parts of the *Arabidopsis* flower code for transcription factors. During the development of *Arabidopsis* flowers, four distinct flower parts differentiate: sepals, petals, stamens, and carpels (Fig. 16–19*a*). Sepals cover and protect the flower when it is a bud, petals help attract animal pollinators to the flower, stamens produce pollen grains, and the pistil produces ovules, which develop into seeds following fertilization (see Chapter 27). The **ABC model** explains the molecular biology behind how these four organs develop. The A gene is needed to specify sepals, both A and B genes are needed to specify petals, both B and C genes are needed to specify stamens, and the C gene is needed to specify the pistil. Mutations in the A, B, or C organ-identity genes, all of which are homeotic and code for transcription factors, cause one flower part to be substituted for another. For example, class C homeotic mutants (which have an inactive C gene) have petals in place of stamens and sepals in place of the pistil. Therefore, the entire flower consists of only sepals and petals (Fig. 16–19*b*).

These findings in *Arabidopsis* vastly increase the number of molecular probes available from plants, which will allow many more genes that control development to be identified in various plant species and compared with genes from a wide range of organisms. The ongoing success of the *Arabidopsis* sequencing project has led to a bold new initiative, the 2010 Project, which has as its goal the understanding of the functions of all plant genes. This functional genomic information will lead to a far deeper understanding of plant development and evolutionary history.

SOME EXCEPTIONS TO THE PRINCIPLE OF NUCLEAR EQUIVALENCE HAVE BEEN FOUND

Although the concept of nuclear equivalence appears to apply to most cells in higher organisms, certain types of developmental regulation can involve physical changes in DNA. Such changes in the structure of the genome are not common.

Genomic rearrangements involve structural changes in DNA

The activity of some genes may be modified during development by different types of **genomic rearrangements** that lead to actual physical changes in the structure of the gene. In some cases, parts of genes are rearranged to make new coding sequences. This is an important mechanism for the development of the immune system (see Chapter 43).

One type of gene rearrangement is gene replacement. This takes place in the eukaryotic unicellular parasite *Trypanosoma,* which is a protozoon (see Fig. 24–5*b*). Trypanosomes carried by the tsetse fly in Africa cause sleeping sickness in humans and related diseases in other animals (Fig. 16–20). When the parasite infects humans, it is able to defeat the immune system by repeatedly changing the glycoprotein molecules that are exposed on the surface of its cell.

The trypanosome cell contains as many as 1000 different genes for cell surface molecules. The differences among their amino acid sequences are so great that an antibody that recognizes one of them would not recognize another. Only one or a few of those copies are expressed at any one time, depending on which copy is present at an **expression site,** which is usually located near the end of a chromosome. The genes in the expression site are exchanged in about one out of every 10^4 to 10^6 cells, causing trypanosomes with novel antigens to appear every 7 to 10 days. Thus the infection is maintained by a constant supply of new cells with antigens that are not being targeted by the immune system. Although gene replacement clearly offers a mechanism that could serve as a regulatory "genetic switch," it is not known at present whether these mechanisms are relevant to development in multicellular eukaryotes.

Gene amplification increases the number of copies of specific genes

Some gene products are required in such large quantities during certain stages of development that a single copy of a gene cannot

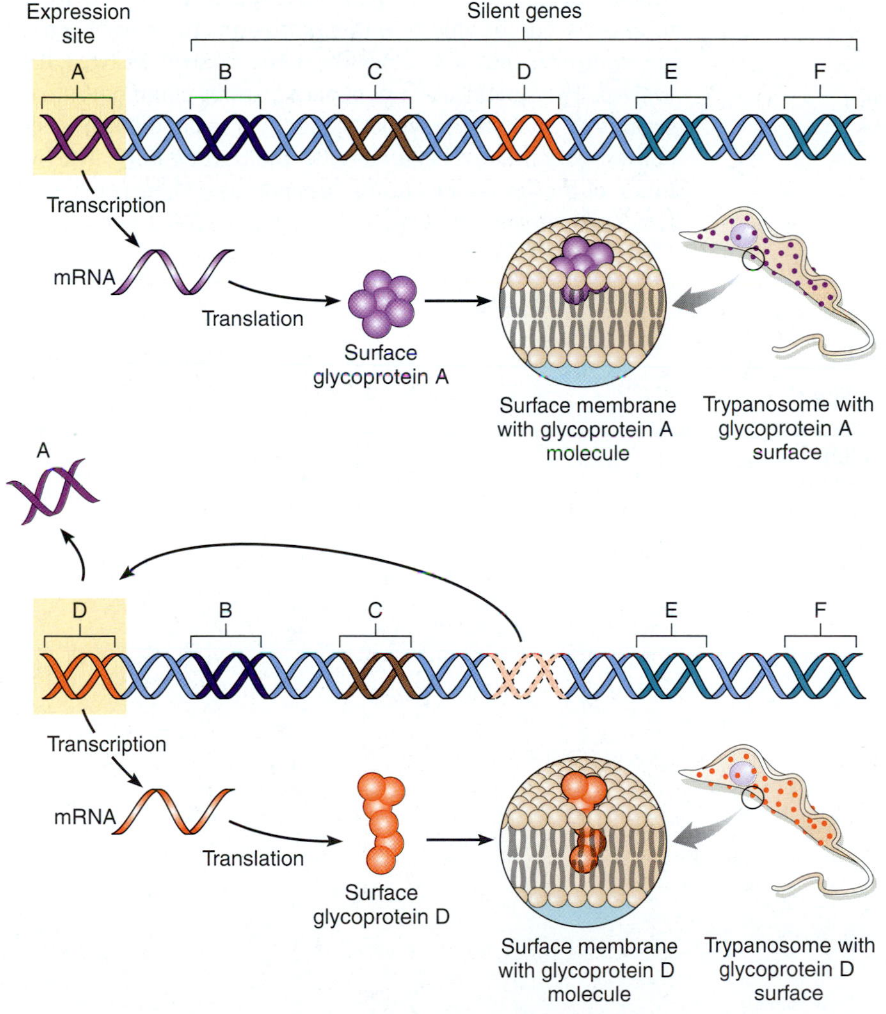

Figure 16–20 Gene rearrangement in trypanosomes. Trypanosomes change the molecules coating their surfaces frequently and are thus able to "outrun" the immune system of their human host. Each cell contains as many as a thousand silent genes, each coding for a different surface coat glycoprotein. Only one of those genes, which is located at a position near the end of a chromosome called the expression site, is active at any one time. As the trypanosomes multiply in the blood, occasionally a copy of one of the silent genes replaces the gene currently in the expression site, leading to the production of organisms with a new surface glycoprotein that is not recognized by the host's immune system.

be transcribed, nor can its mRNA be translated, rapidly enough to fill the needs of the developing cells. In certain cases, the number of gene copies may be increased, through a process known as **gene amplification,** to meet the demand. For example, the *Drosophila* chorion (eggshell) gene product is a protein made specifically in cells of the insect oviduct. These cells make massive amounts of the particular protein that envelops and protects the fertilized egg. The demand for chorion mRNA in those cells is met by specifically amplifying the chorion protein gene by DNA replication so that the DNA in that small region of the chromosome is copied many times (Fig. 16–21). In other cells of the insect body, however, the gene appears to exist as a single copy in the chromosome.

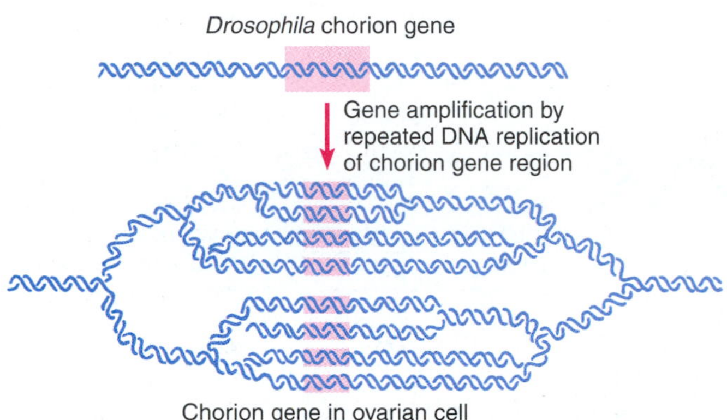

Figure 16–21 Gene amplification. In *Drosophila,* multiple replications of a small region of the chromosome result in the amplification of the chorion (eggshell) protein genes. Replication is initiated at a discrete chromosome origin of replication *(pink box)* for each copy of the gene that is produced; replication is randomly terminated, resulting in a series of forked structures in the chromosome.

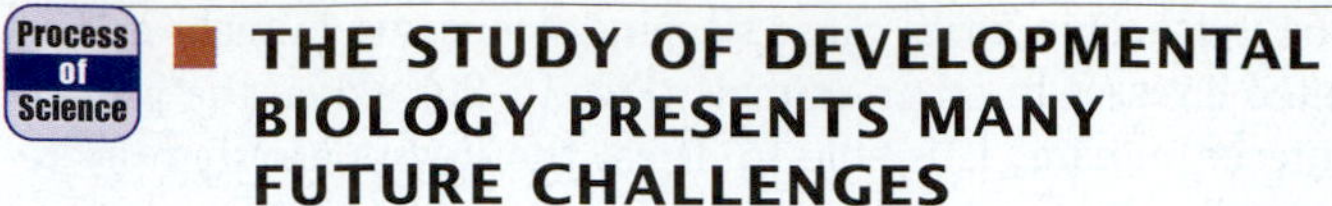

THE STUDY OF DEVELOPMENTAL BIOLOGY PRESENTS MANY FUTURE CHALLENGES

New model organisms, each with characteristics that uniquely suit it for developmental studies, are now being added to the list of well-characterized experimental systems. In a herculean effort spearheaded by Christiane Nüsslein-Volhard of the University of Tübingen, Germany, researchers have identified more than 2000 developmental mutants of the zebrafish *(Danio rerio),* and the sequencing of its genome is underway. Previously best known to tropical fish enthusiasts, the zebrafish was chosen as a research subject mainly for its transparent embryos.

Scientists are now learning how genes are activated, inactivated, and modified and how batteries of master regulatory genes interact to control development. Eventually researchers expect to understand not only how differentiation and morphogenesis are controlled but also how the basic control systems have evolved. The identification of certain features common to many organisms, such as homeobox genes, will make the task easier, but the work has just begun (see On the Cutting Edge: The Evolution of the Vertebrate Head in Chapter 49). Many complex interactions remain to be explored, and many revelations await us.

SUMMARY WITH KEY TERMS

I. **Development,** all the changes that occur in the life of an individual, includes the processes by which the descendants of a single cell specialize and organize into a complex organism.
   A. An organism contains many types of cells that are specialized both structurally and chemically to carry out specific functions. These cells are the product of a process of gradual commitment, called cell **determination,** which ultimately leads to the final step in cell specialization, called cell **differentiation.**
   B. **Morphogenesis,** the development of form, occurs through stages, referred to as **pattern formation,** in which cells become organized into structures.
   C. There is no evidence that genes are normally lost during most developmental processes.
      1. At least some nuclei from differentiated plant and animal cells are **totipotent** and contain all the genetic material that would be present in the nucleus of a zygote.
      2. **Nuclear equivalence** is the concept that, with a few exceptions, all of the nuclei of the differentiated **somatic** cells of an organism are identical to each other and to the nucleus of the single cell from which they descended.
      3. **Stem cells** can divide to produce differentiated descendants, yet retain the ability to divide to maintain the stem cell population. **Pluripotent** stem cells can give rise to all cell types.
      4. Differences among various cell types are apparently due to **differential gene expression. Functional genomics,** the study of the roles of specific genes, includes the analysis of patterns of RNA synthesis in cells.

II. Several organisms have characteristics that make them especially useful in studies of the genetic control of development.
   A. Many types of developmental mutants have been identified in the fruit fly, *Drosophila melanogaster.* Many of these mutations affect the segmented body plan of the organism.
      1. The earliest developmental program to operate in the egg is established by **maternal effect genes;** these are active prior to fertilization, and some produce gradients of **morphogens,** thereby affecting the segmentation pattern that is progressively established in the embryo.
      2. **Zygotic segmentation genes** do not become active until much later, when the embryo is no longer a zygote. They continue and extend the developmental program initiated by the maternal effect genes.
      3. The zygotic segmentation genes and their products interact with each other and with the products of the maternal effect genes according to a hierarchical pattern, with certain earlier acting genes controlling particular later acting genes.
      4. The later acting **homeotic genes** are responsible for specifying the identity of each segment.
      5. Many of the segmentation genes are known to code for **transcription factors.** Some of these contain a DNA sequence called a **homeobox,** which codes for a protein with a DNA-binding region called a **homeodomain.**
   B. *Caenorhabditis elegans* is a roundworm with **mosaic development,** a relatively rigid developmental pattern in which the fates of cells are largely predetermined.
      1. The lineage of every somatic cell in the adult is known, and each can be traced to a single **founder cell** in the early embryo.
      2. Mutations affecting cell lineages have been identified, and many of these appear to identify genes that control developmental processes such as **induction** (developmental interactions with neighboring cells), **programmed cell death (apoptosis),** and developmental timing (i.e., **chronogenes**).
   C. The laboratory mouse, *Mus musculus,* is extensively used in studies of mammalian development.
      1. In contrast to *C. elegans,* the mouse shows **regulative development,** which means that the very early embryo is a self-regulating whole and can develop normally even if it has extra or missing cells.
      2. **Transgenic** mice have been extremely useful in determining how genes are activated and regulated during development.
   D. Genes affecting the developmental pattern have also been identified in certain plants, including *Arabidopsis* and *Zea mays* (corn). Many of these code for transcription factors.
      1. The **ABC model** of interactions among three kinds of genes, designated the A gene, B gene, and C gene, explains how floral organs develop in *Arabidopsis.*
      2. All three genes are homeotic and code for **transcription factors.**
      3. Mutations in one or more of these genes cause one flower part to be substituted for another.
   E. Some homeobox genes are organized into complexes that appear to be systems of master genes specifying an organism's body plan. Remarkable parallels exist between the homeobox complex of *Drosophila* and those of other animals, including the laboratory mouse and *C. elegans.*

III. A few exceptions to the general rule of nuclear equivalence are known. Among these are **genomic rearrangement** (physical rearrangements of the DNA) and **gene amplification,** which provides more copies of certain genes for transcription.

POST-TEST

1. Morphogenesis occurs through a series of stages known as (a) differentiation (b) determination (c) pattern formation (d) totipotency (e) selection
2. The "cloning" experiments carried out on amphibians demonstrated that (a) all differentiated amphibian cells are totipotent (b) some differentiated amphibian cells are totipotent (c) all nuclei from differentiated amphibian cells are totipotent (d) some nuclei from differentiated amphibian cells are totipotent (e) the mechanism of cellular differentiation always requires the loss of certain genes
3. *Drosophila* is a particularly good model for developmental studies because (a) a large number of developmental mutants are available (b) it has a fixed number of somatic cells in the adult (c) its embryos are transparent (d) it is a vertebrate (e) all of the above
4. The anterior-posterior axis of a *Drosophila* embryo is first established by certain (a) homeotic genes (b) maternal effect genes (c) zygotic segmentation genes (d) chronogenes (e) pair-rule genes
5. You discover a new *Drosophila* mutant in which mouthparts appear where the antennae are normally found. You would predict that the mutated gene is most likely a (a) homeotic gene (b) gap gene (c) pair-rule gene (d) maternal effect gene (e) segment polarity gene
6. Most zygotic segmentation genes code for (a) special transfer RNAs (b) enzymes (c) transcription factors (d) histones (e) transport proteins
7. The developmental pattern of *C. elegans* is said to be relatively mosaic because (a) development is controlled by gradients of morphogens (b) part of the embryo fails to develop if a founder cell is destroyed (c) some individuals are self-fertilizing hermaphrodites (d) all development is controlled by maternal effect genes (e) programmed cell deaths never occur
8. The formation of the vulva, the structure through which eggs are laid, in *C. elegans* involves (a) maternal effect genes that organize the egg cytoplasm (b) gradients of morphogens in the eggs (c) groups of *Hox* genes that form the *Antennapedia* complex and *bithorax* complex (d) induction of surface cells by the anchor cell (e) mutations in chronogenes, which control developmental timing
9. Which of the following illustrates the regulative nature of early mouse development? (a) the mouse embryo is free-living prior to implantation in the uterus (b) it is possible to produce a transgenic mouse (c) it is possible to produce a mouse in which a specific gene has been "knocked out" (d) genes related to *Drosophila* homeotic genes have been identified in mice (e) a chimeric mouse can be produced by fusing two mouse embryos
10. When the human gene that codes for insulin is introduced into fertilized mouse eggs that are subsequently allowed to develop, the insulin gene is correctly expressed in the mouse's pancreatic cells. This indicates that (a) the gene that codes for insulin is analogous to the homeotic genes of *Drosophila* (b) the signals for tissue-specific gene expression are highly conserved through evolution (c) like humans, the mouse has polytene chromosomes (d) unlike the rigid developmental pattern of *C. elegans*, the development of mice and humans is highly regulative (e) genomic rearrangements have occurred in the mouse embryo
11. *Arabidopsis* is useful as a model organism for the study of plant development because (a) it is of great economic importance (b) it has large polytene chromosomes (c) many developmental mutants have been isolated (d) it contains a large amount of DNA per cell (e) it has a rigid developmental pattern
12. According to the ABC model of floral organ development in *Arabidopsis*, the A gene is needed to specify sepals, the A and B genes to specify petals, the B and C genes to specify stamens, and the C gene to specify the pistil. If a mutation occurs in one of the B genes, rendering it inactive, the resulting flowers will consist of (a) sepals, petals, stamens, and pistils, (b) sepals, stamens, and pistils (c) petals, stamens, and pistils (d) sepals and pistils (e) petals and stamens

REVIEW QUESTIONS

1. Development consists of four main processes: cell determination, cell differentiation, pattern formation, and morphogenesis. Define each process and describe how they relate to one another.
2. What lines of evidence support the concept of nuclear equivalence?
3. What are the relative merits of *Drosophila, C. elegans,* the mouse, and *Arabidopsis* as model organisms for the study of development?
4. What is the value of homeotic genes in developmental studies?
5. Describe how transgenic organisms are useful in the study of gene regulation in development.
6. Give some examples of genomic rearrangements that are known to occur as a part of some developmental processes.
7. Under what conditions are examples of gene amplification seen?
8. What are oncogenes, and what is their relationship to cellular genes involved in the control of normal growth and development?

YOU MAKE THE CONNECTION

1. Why is an understanding of gene regulation in eukaryotes crucial to an understanding of developmental processes?
2. Why do developmental biologists need to understand biological diversity? Why is it necessary for scientists to study development in more than one type of organism?

RECOMMENDED READINGS

Hamadel, H., and C.A. Afshari. "Gene Chips and Functional Genomics." *American Scientist,* Vol. 88, Nov.–Dec. 2000. A discussion of the current and potential uses of gene chip technology, with an emphasis on the ability to detect alterations in gene expression caused by environmental toxins.

Hines, P.J., B.A. Purnell, and J. Marx. "Stem Cells Branch Out." *Science,* Vol. 287, 25 Feb. 2000. The introduction to a special multiarticle section entitled "Stem Cell Research and Ethics."

Hodgkin, J., H.R. Horvitz, B.R. Jasny, and J. Kimble. "*C. elegans:* Sequence to Biology." *Science,* Vol. 282, 11 Dec. 1998. The introduction to multiple papers in a special section on the *C. elegans* genome, the first animal genome to be sequenced.

Jasny, B.R. "The Universe of *Drosophila* Genes." *Science,* Vol. 287, 24 Mar. 2000. The introduction to a special section including multiple papers on the *Drosophila* genome sequence.

Malakoff, D. "The Rise of the Mouse, Biomedicine's Model Mammal." *Science,* Vol. 288, 14 Apr. 2000. A history of the use of the laboratory mouse in research and a perspective on future developments.

McKinnell, R.G., and M.A. DiBernardino. "The Biology of Cloning: History and Rationale." *Bioscience,* Vol. 49, No. 11, Nov. 1999. This review emphasizes the insights into the fundamental controls of developmental processes that have been provided by cloning.

McLaren, A. "Cloning: Pathways to a Pluripotent Future." *Science,* Vol. 288, 9 Jun. 2000. Perspectives on both the history and future of cloning.

Nüsslein-Volhard, C. "Gradients that Organize Embryo Development." *Scientific American,* Vol. 275, No. 2, Aug. 1996. The author, who shared the 1995 Nobel Prize for Physiology or Medicine with Eric Wieschaus and Edward B. Lewis, describes gene interactions that control *Drosophila* development.

Riechmann, J.L., et al. "*Arabidopsis* Transcription Factors: Genome-Wide Comparative Analysis Among Eukaryotes." *Science,* Vol. 290, 15 Dec. 2000. A comparison of transcription factors in *Arabidopsis, Drosophila, C. elegans,* and *Saccharomyces cerevisiae* (yeast, a unicellular fungus).

Somerville, C., and J. Dangl. "Plant Biology in 2010." *Science,* Vol. 290, 15 Dec. 2000. This article lays out the goals of an ambitious project in plant functional genomics.

- Visit our Web site at **http://www.info.brookscole.com/solomonbergmartin** for links to chapter-related resources on the World Wide Web. Additional on-line materials relating to this chapter can also be found on our Web site.

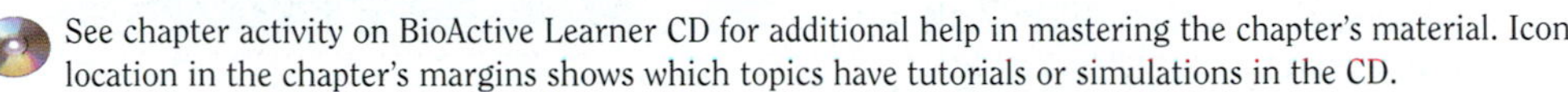

See chapter activity on BioActive Learner CD for additional help in mastering the chapter's material. Icon location in the chapter's margins shows which topics have tutorials or simulations in the CD.

CAREER VISIONS

Database Administrator

MARTIN SORICH

***Martin Sorich** is a database administrator at Genitope, a biotech "start-up" company that specializes in developing individualized therapies to boost the immune systems of cancer patients. A trained molecular biologist with seven years of experience in academic and industry labs, Martin has been at Genitope for two years. A Californian from the San Jose area, Martin started his higher education at De Anza College, a community college, where he earned an associate's degree in 1992. In 1994, he received a Bachelor of Science degree in Molecular Biology, with a Chemistry minor, at San Jose State University.*

You have been involved with researching patient-specific immunotherapy at Genitope. Can you describe the goal of this research?

Basically, the goal of this study was to reproduce the clinical research that Dr. Ron Levy at Stanford has done for the past 10 years, to provide cancer patients—specifically those with lymphoma—with a better chance of fighting their disease. Dr. Levy was the first to do this type of active-immunotherapy research. Dr. Levy's process for producing patient-specific immunotherapies is labor-intensive, but it has a success rate of 85%. The founder of Genitope, Dr. Dan Denny, developed our company's molecular approach, which allows Genitope to process many samples at a time, and our success rate has been 100% in producing the immunotherapies. We're still in clinical trials, and we have really good results.

Genitope uses a molecular approach to develop patient-specific therapies that are directed against an individual lymphoma patient's own cancerous lymphocytes (lymphocytes are white blood cells that function as part of the immune system). Each lymphoma patient has a unique population of cancerous lymphocytes, all comprised of the same genetic makeup, because, in theory, they originated from a single cell. These lymphocytes can be recognized by the presence of a lymphoma-specific protein on their surfaces. The goal is to create a vaccine that will induce the patient's immune system to attack cancerous lymphocytes but not affect normal ones.

To develop this vaccine, the messenger RNA (mRNA) that codes for the lymphoma-specific protein is extracted from the patient's lymphoma cells. This mRNA is used to synthesize complementary DNA (cDNA) copies, which are then put into an expression vector. The expression vector is used to generate a mammalian cell line that produces large amounts of the lymphoma-specific protein, which is purified and used as a vaccine to immunize the patient. The patient's immune system mounts an immune response that causes the death of the cancerous lymphocytes.

. . . after meeting him I couldn't stop smiling for the rest of the day, because I knew my work helped save this man's life.

What was your particular role in developing patient-specific immunotherapies?

I was primarily involved in the early stages of the process—characterizing the tumor-specific genes, subcloning them into expression vectors, and preparing them for introduction into a mammalian cell line system.

Have you met any of the patients you were helping, the ones involved in clinical trials?

Yes. The first patient sample I received belonged to the first person who was immunized in our first clinical trial, and he mounted an immune response, which was great—the first one worked just like it was supposed to. I had the privilege of meeting the patient when his family visited Genitope. His wife was thanking everybody and was very grateful. I remember after meeting him I couldn't stop smiling for the rest of the day, because I knew my work helped save this man's life.

Was it your work at community college that made you decide to go on for your bachelor's degree?

Yes. As a child I enjoyed watching nature programs, and I'm the type of person who likes to figure out how things work; studying biology is figuring out how living things work. There were some influential instructors at De Anza, my community college, who guided me in making my career choice in biology. One of the instructors, Doug Cheeseman, taught an introductory biology course. In the first part of that course, he discussed genetics and molecular biology, such as what DNA is and how genes work. It was fairly basic, but it really intrigued me, and I think that's where my interest in molecular biology began.

Do you have any suggestions for undergraduates interested in careers in molecular biology, genetics, or biochemistry?

If they are interested in pursuing a research career, I would encourage them to get involved early on in laboratory work, volunteering either for a biotech company or for a professor who is involved in research that they find intriguing. This experience will help them achieve the goal of becoming a research scientist or of going on to graduate school.

Part 4

These fossilized remains of a prehistoric fish *(Priscacara serata)* were discovered in Wyoming. *(Alfred Pasieka/Science Photo Library/Photo Researchers, Inc.)*

The Continuity of Life: Evolution

Chapters

17

Introduction to Darwinian Evolution

Charles Darwin. This portrait was made shortly after Darwin returned to England from his voyage around the world. *(The Granger Collection, New York)*

LEARNING OBJECTIVES

After you have studied this chapter you should be able to

1. Discuss the historical development of the theory of evolution.
2. Define evolution and explain the four premises of evolution by natural selection as proposed by Charles Darwin.
3. Compare the synthetic theory of evolution with Darwin's original theory of evolution.
4. Summarize the evidence for evolution obtained from the fossil record.
5. Summarize the evidence for evolution derived from comparative anatomy.
6. Define and give examples of homology, homoplasy, and vestigial structures.
7. Define biogeography and summarize how the distribution of organisms supports evolution.
8. Briefly explain how developmental biology provides insights into the evolutionary process.
9. Describe how scientists make inferences about evolutionary relationships from the sequence of amino acids in proteins and the sequence of nucleotides in DNA.
10. Give an example of how evolutionary hypotheses can be tested experimentally.

The biological diversity represented by the millions of species currently living on our planet is thought to have evolved from a single ancestor during Earth's long history. Thus, organisms that are radically different, such as slime molds and crocodiles, are in fact distantly related to one another and are linked through numerous intermediate ancestors to a single, common ancestor. The British naturalist Charles Darwin (1809–1882), shown here at age 31, developed a remarkably simple, scientifically testable mechanism to explain this. He argued persuasively that all the species that exist today, as well as the countless extinct species that existed in the past, arose from earlier ones by a process of *gradual divergence* (separation into separate evolutionary pathways), or evolution.

Evolution can be defined as the accumulation of inherited changes within populations over time. A **population** is a group of individuals of one species that live in the same geographical area at the same time. Evolution does not refer to changes that occur in an individual within its lifetime. Instead, evolution refers to changes in the characteristics of populations over the course of generations. These changes may be so small that they are difficult to detect or so great that the population differs markedly from its ancestral population. Eventually, two populations may diverge to such a degree that we refer to them as different species. (The concept of **species** is developed extensively in Chapter 19. For now, a simple working definition is that a species comprises a group of organisms with similar structure, function, and behavior that are capable of interbreeding with one another.) Thus, evolution has two main perspectives—the short-term adaptations of populations to changes in the environment (*microevolution,* see Chapter 18), and the long-term formation of different species from common ancestors (*macroevolution,* see Chapter 19).

The concept of evolution is the cornerstone of biology because it links all fields of the life sciences into a unified body of knowledge. As evolutionary geneticist Theodosius Dobzhansky put it, "Nothing in biology makes sense except in the light of evolution."[1] Biologists attempt to understand both the remarkable

[1] *American Biology Teacher,* Vol. 35, No. 125 (1973).

variety as well as the fundamental similarities of organisms within the context of evolution. Evolution allows biologists to compare common threads among organisms as seemingly different as bacteria, whales, lilies, and tapeworms. Behavioral evolution, evolutionary developmental biology, evolutionary genetics, evolutionary ecology, evolutionary systematics, and molecular evolution are examples of some of the biological disciplines that focus on evolution.

Evolution also has important practical applications. Agriculture must deal with the evolution of pesticide resistance in insects and other pests. Likewise, medicine must respond to the rapid evolutionary potential of disease-causing organisms such as bacteria. (Significant evolutionary change occurs in a very short time period in insects, bacteria, and other organisms with short lifespans.) The conservation management of rare and endangered species makes use of the evolutionary principles of population genetics (see Chapter 18). The rapid evolution of bacteria and fungi in polluted soils is used in the field of **bioremediation,** in which microorganisms are employed to clean up hazardous waste sites. Evolution even has applications beyond biology. For example, certain computer applications make use of algorithms that mimic natural selection in biological systems.

This chapter discusses Darwin and the scientific development of the theory of evolution by natural selection. It also presents several kinds of evidence that support evolution, including fossils, comparative anatomy, biogeography, developmental biology, molecular biology, and experimental studies of ongoing evolutionary change, both in the laboratory and in nature.

IDEAS ABOUT EVOLUTION ORIGINATED BEFORE DARWIN

Although Darwin is universally associated with evolution, ideas of evolution predate Darwin by centuries. Aristotle (384–322 B.C.) saw much evidence of natural affinities among organisms. This led him to arrange all of the organisms he knew in a "Scale of Nature" that extended from the exceedingly simple to the most complex. He visualized organisms as being imperfect but "moving toward a more perfect state." This idea has been interpreted by some scientific historians as the forerunner of evolutionary theory, but Aristotle was vague on the nature of this "movement toward perfection" and certainly did not propose that the process of evolution was driven by natural processes. Furthermore, modern evolutionary theory now recognizes that evolution does not move toward more "perfect" states, nor even necessarily toward greater complexity.

Long before Darwin, fossils had been discovered embedded in rocks. Some of these corresponded to parts of familiar species, but others were strangely unlike any known species. Fossils were often found in unexpected contexts. Marine invertebrates (sea animals without backbones), for example, were sometimes discovered in rocks high on mountains. Leonardo da Vinci (1452–1519) was among the first to correctly interpret these unusual finds as the remains of animals that had existed in previous ages but had become extinct.

The French naturalist Jean Baptiste de Lamarck (1744–1829) was the first scientist to propose that organisms undergo change over time as a result of some natural phenomenon rather than divine intervention. In his *Philosophie Zoologique,* published in 1809, Lamarck presented a possible explanation for how organisms evolved. Lamarck thought that all organisms were endowed with a vital force that drove them to change toward greater complexity over time. He also thought that organisms could pass traits acquired during their lifetimes to their offspring. For example, Lamarck suggested that the long neck of the giraffe developed when a short-necked ancestor stretched its neck to browse on the leaves of trees. Its offspring inherited the longer neck, which stretched still further as they ate. This process, repeated over many generations, supposedly resulted in the long necks of modern giraffes.

The proposed mechanism for Lamarckian evolution was discredited when the basis of heredity was later discovered. It remained for Darwin to discover the mechanism of evolution by natural selection.

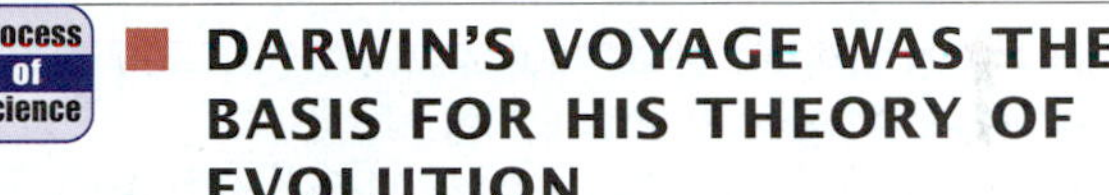

DARWIN'S VOYAGE WAS THE BASIS FOR HIS THEORY OF EVOLUTION

Darwin, the son of a prominent physician, was sent at the age of 15 to study medicine at the University of Edinburgh. Finding himself unsuited for medicine, he transferred to Cambridge University to study theology. During that time, he became the protégé of the Reverend John Henslow, who was a professor of botany. Henslow encouraged Darwin's interest in the natural world. Shortly after receiving his degree, Darwin embarked as a naturalist on the H.M.S. *Beagle,* which was taking a five-year exploratory cruise around the world to prepare navigation charts for the British Navy.

The *Beagle* left Plymouth, England, in 1831 and cruised along the east and west coasts of South America (Fig. 17–1). While other members of the crew mapped the coasts and harbors, Darwin spent many weeks ashore studying the animals, plants, fossils, and geological formations of both coastal and inland regions, areas that had not been extensively explored. He collected and catalogued thousands of plant and animal specimens and kept notes of his observations—information that would become essential in the development of his theory.

The *Beagle* spent almost two months at the Galapagos Islands, 965 km (600 mi) west of Ecuador, where Darwin continued his observations and collections. He compared the animals and plants of the Galapagos with those of the South American mainland. He was particularly impressed by their similarities and wondered why the organisms of the Galapagos should resemble those from South America more than those from other islands in different parts of the world. Moreover, although there were similarities between Galapagos and South American species, there were also distinct differences. There were even recognizable differences in the reptiles and birds from one island to the next (Fig. 17–2). After he returned home, Darwin pondered these

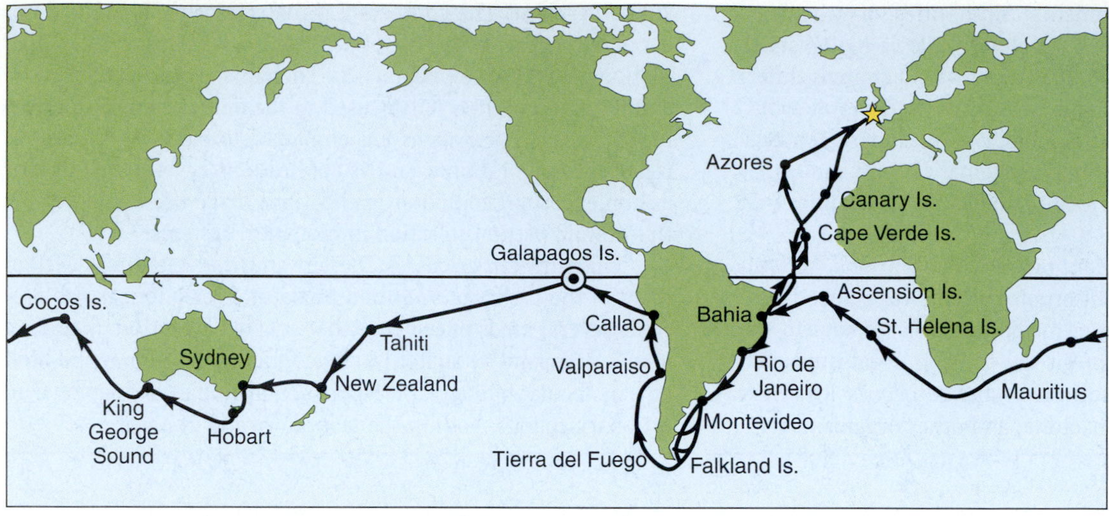

Figure 17-1 The voyage of H.M.S. *Beagle.* The five-year voyage began in Plymouth, England *(star)*, in 1831. Observations made in the Galapagos Islands *(bull's eye)* off the western coast of South America helped Darwin discover a satisfactory mechanism to explain how a population of organisms could change over time.

observations and attempted to develop a satisfactory explanation for the distribution of species among the islands.

Darwin drew on several lines of evidence when considering how species might have originated. Despite the work of Lamarck, the general notion in the mid-1800s was that the Earth was too young for organisms to have changed significantly since they had first appeared. During the early 19th century, however, geologists advanced the idea that mountains, valleys, and other physical features of Earth's surface did not originate in their present forms but developed slowly over long periods by the geological

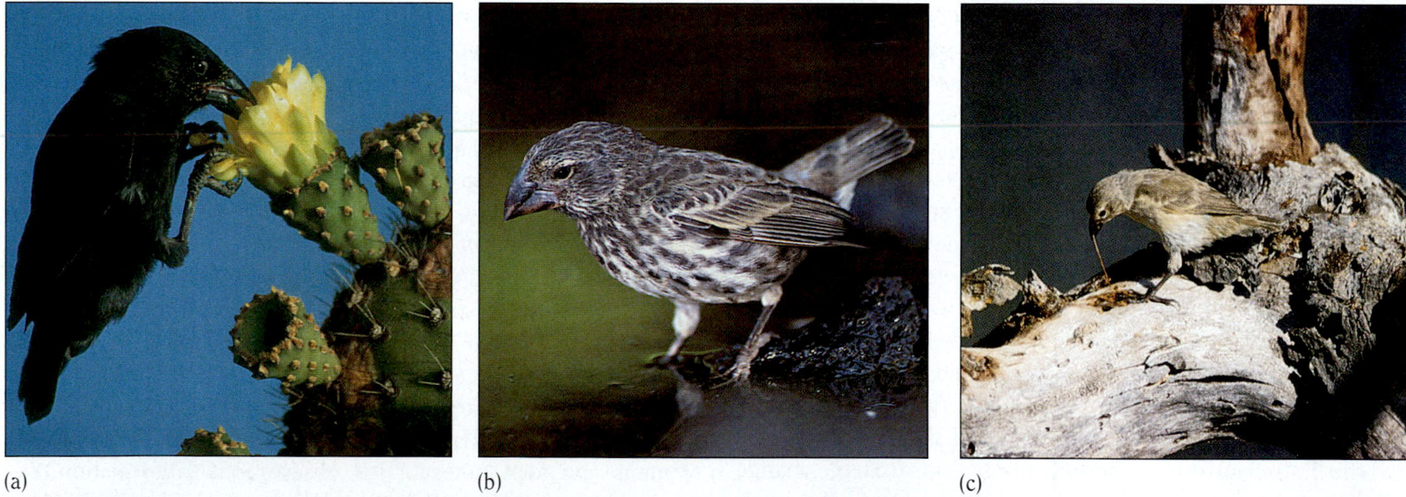

Figure 17-2 Galapagos finches. Darwin inferred that these birds were derived from a common ancestral population of seed-eating birds from South America. Although they are largely similar in body size and coloration (males are often darker than females), the Galapagos finches differ markedly in beak size and shape. Variation in their beaks is the result of adaptations to different kinds of food. **(a)** The cactus ground finch *(Geospiza scandens)* feeds on the fleshy parts of cacti. **(b)** The large ground finch *(Geospiza magnirostris)* has an extremely heavy, nutcracker-type bill adapted for eating thick, hard-walled seeds. **(c)** The woodpecker finch *(Camarhyncus pallidus)* has insectivorous habits similar to those of woodpeckers but lacks the complex beak and tongue adaptations that permit woodpeckers to reach their prey. The adaptations of the woodpecker finch to this lifestyle are almost entirely behavioral—it digs insects out of bark and crevices using cactus spines, twigs, or even dead leaves. *(a and b, Frans Lanting/Minden Pictures; c, Miguel Castro/Photo Researchers, Inc.)*

processes of volcanic activity, uplift, erosion, and glaciation. Darwin took *Principles of Geology,* published by geologist Charles Lyell in 1830, with him on his voyage and studied it carefully. Lyell provided an important concept for Darwin—that the slow pace of geological processes, which still occur today, indicated that Earth was extremely old.

Other important evidence that influenced Darwin was the fact that breeders and farmers could develop many varieties of domesticated animals in just a few generations (Fig. 17–3). This was accomplished by choosing certain traits and breeding only individuals that exhibited the desired traits, a procedure known as **artificial selection.** Breeders, for example, had produced numerous dog varieties—bloodhounds, Dalmatians, Airedales, border collies, and Pekinese, to name a few—by artificial selection.

Many plant varieties were also produced by artificial selection. For example, cabbage, broccoli, Brussels sprouts, cauliflower, collard greens, kale, and kohlrabi are distinct vegetable crops that are all members of the same species, *Brassica oleracea* (Fig. 17–4). All seven were produced by selective breeding of the colewort, or wild cabbage, a leafy plant native to Europe and Asia. Beginning more than 4000 years ago, some farmers artificially selected wild cabbage plants that formed overlapping leaves. Over time, these leaves became so prominent that the plants, which resembled modern-day cabbages, became recognized as separate and distinct from their wild cabbage ancestor. Other farmers selected different features of the wild cabbage, giving rise to the other modifications. For example, kohlrabi was produced by selection for an enlarged storage stem, and Brussels sprouts by selection for enlarged axillary buds. Thus, humans are responsible for the evolution of *B. oleracea* into seven distinct vegetable crops. Darwin was impressed by the changes induced by artificial selection and thought that a similar selective process occurred in nature. Darwin therefore used artificial selection as a model when he developed the concept of natural selection.

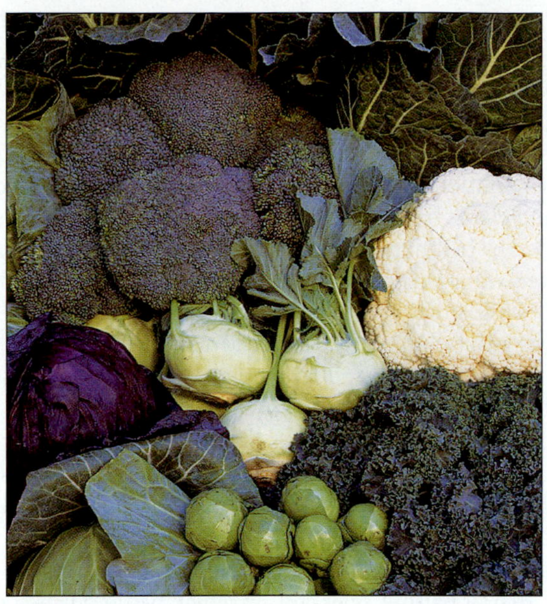

Figure 17–4 Artificial selection in *Brassica oleracea.* An enlarged terminal bud (the "head") was selected in cabbage *(lower left),* flower clusters in broccoli *(upper left)* and cauliflower *(middle right),* axillary buds in Brussels sprouts *(bottom middle),* leaves in collards *(upper right)* and kale *(lower right),* and stems in kohlrabi *(middle).* *(John Arnaldi)*

Figure 17–3 Artificial selection in chickens. Shown is a chicken that was deliberately bred to resemble Big Bird on *Sesame Street.* Many show breeds of chickens exist that exhibit a great deal of variation. Domestic chickens are not a recognizable breed but are hybrids bred for their meat or egg production. *(Eric Sander)*

The ideas of Thomas Malthus (1766–1834), a British clergyman and economist, were another important influence on Darwin. In *An Essay on the Principle of Population as It Affects the Future Improvement of Society,* published in 1798, Malthus noted that population growth is not always desirable—a view contrary to the beliefs of his day. He observed that populations have the capacity to increase geometrically ($1 \rightarrow 2 \rightarrow 4 \rightarrow 8 \rightarrow 16$) and thus outstrip the food supply, which has the capacity to increase only arithmetically ($1 \rightarrow 2 \rightarrow 3 \rightarrow 4 \rightarrow 5$). In the case of humans, Malthus suggested that the conflict between population and food supply generates famine, disease, and war, which serve as inevitable brakes on population growth.

DARWIN PROPOSED THAT EVOLUTION OCCURS BY NATURAL SELECTION

Malthus's idea that there is a strong and constant check on human population growth strongly influenced Darwin's explanation of evolution. Darwin's years of observing the habits of animals and plants had introduced him to the struggle for existence described by Malthus. It occurred to Darwin that in this struggle inherited variations favorable to survival would tend to be preserved, while unfavorable ones would be eliminated. The result would be **adaptation,** an evolutionary modification that improves the chances of survival and reproductive success in a given environment. Eventually, the accumulation of modifications might result in a new species. Time was the only thing required for new species to originate, and the geologists of the era, including Lyell, had supplied evidence that Earth was indeed old enough to provide an adequate period.

Darwin had at last developed a workable explanation of evolution, that of **natural selection,** in which better adapted organisms are more likely to survive and become the parents of the next generation. As a result of natural selection, the population changes over time; the frequency of favorable traits increases in successive generations while less favorable traits become scarce or disappear. Darwin spent the next 20 years formulating his arguments for natural selection, accumulating an immense body of evidence to support his theory, and corresponding with other scientists.

As Darwin was pondering his ideas, Alfred Russell Wallace (1823–1913), a British naturalist who studied the plants and animals of the Malay Archipelago for eight years, was similarly struck by the diversity of species and the peculiarities of their distribution. He wrote a brief essay on this subject and sent it to Darwin, by then a world-renowned biologist, asking his opinion. Darwin recognized his own theory and realized that Wallace had independently arrived at the same conclusion—that evolution occurs by natural selection. Darwin's colleagues persuaded him to present Wallace's manuscript along with an abstract of his own work, which he had prepared and circulated to a few friends several years earlier. Both papers were presented in July 1858 at a London meeting of the Linnaean Society. Darwin's monumental book, *The Origin of Species by Natural Selection; or, The Preservation of Favored Races in the Struggle for Life,* was published in 1859. Wallace's book, *Contributions to the Theory of Natural Selection,* was published in 1870, eight years after he returned from the Malay Archipelago.

Darwin's mechanism of evolution by natural selection consists of four observations about the natural world: variation, overproduction, limits on population growth, and differential reproductive success.

1. **Variation.** The individuals in a population exhibit variation. Each individual has a unique combination of traits, such as size, color, and ability to tolerate harsh environmental conditions. Some traits improve an individual's chances of survival and reproductive success, whereas others do not. It is important to remember that the variation necessary for evolution by natural selection must be inherited (Fig. 17–5). (Although Darwin recognized the importance to evolution of inherited variation, he did not know the mechanism of inheritance.)

Figure 17–5 Genetic variation in emerald tree boas. These snakes, all of the same species *(Corallus caninus)*, were caught in a small section of forest in French Guiana. Many snake species exhibit considerable variation in their coloration and patterns. *(BIOS/Peter Arnold, Inc.)*

2. **Overproduction.** The reproductive ability of each species causes its populations to geometrically increase in number over time. That is, in every generation, each species has the capacity to produce more offspring than can survive.
3. **Limits on population growth,** or a struggle for existence. There is only so much food, water, light, growing space, and other resources available to a population, and, consequently, organisms compete with one another for these limited resources. Because there are more individuals than the environment can support, not all will survive to reproduce. Other limits on population growth include predators, disease organisms, and unfavorable weather conditions.
4. **Differential reproductive success.** Those individuals that possess the most favorable combination of characteristics (those that make individuals better adapted to their environment) are more likely to survive and reproduce. Because offspring tend to resemble their parents, the next generation inherits the parents' genetically based traits. Successful reproduction is the key to natural selection: The best-adapted individuals are those that reproduce most successfully, whereas less fit individuals die prematurely or produce fewer or inferior offspring.

Over time, enough changes may accumulate in geographically separated populations (often with slightly different environments) to produce new species. Darwin noted that the Galapagos finches appeared to have evolved in this way. The 13 species are all descended from a single species that found its way from the South American mainland. (The blue-black grassquit, a bird that lives in western South America, may be a close relative of the Galapagos finches.) The different islands of the Galapagos kept the finches isolated from one another, thereby allowing them to diverge into 13 separate species. Six of the 13 species are seed eaters, six are insect eaters, and one is a woodpecker-type insect eater. The evolution of new species is considered in greater detail in Chapter 19.

THE SYNTHETIC THEORY OF EVOLUTION COMBINES DARWIN'S THEORY WITH GENETICS

Process of Science

One of the premises on which Darwin based his theory of evolution by natural selection is that individuals transmit traits to the next generation. However, Darwin was unable to explain *how* this occurs or *why* individuals vary within a population. Although he was a contemporary of Gregor Mendel (see Chapter 10), who elucidated the basic patterns of inheritance, Darwin was apparently not acquainted with Mendel's work, which was not recognized by the scientific community until the early part of the 20th century.

During the 1930s and 1940s, biologists combined the principles of genetics with Darwin's theory of natural selection to develop a unified explanation of evolution known as neo-Darwinism or, more commonly, the **synthetic theory of evolution.**[2] (*Synthesis* in this context refers to combining parts of several previous theories to form a unified whole.)

The synthetic theory of evolution is a conceptual breakthrough that explains Darwin's observation of variation among offspring in terms of **mutation,** or changes in DNA, such as nucleotide substitutions. That is, mutation provides the genetic variability on which natural selection acts during evolution. The synthetic theory of evolution, which emphasizes the genetics of populations as the central focus of evolution, has held up well since it was developed (see Chapter 18). It has dominated the thinking and research of biologists working in many areas and has resulted in an enormous accumulation of new discoveries that validate evolution by natural selection.

Most biologists not only accept the basic principles of the synthetic theory of evolution but also try to better understand the causal processes of evolution. For example, what is the role of chance in evolution? How rapidly do new species evolve? These and other questions have arisen in part from a reevaluation of the fossil record and in part from new discoveries in molecular aspects of inheritance. Such critical analyses are an integral part of the scientific process because they stimulate additional observation and experimentation, along with reexamination of previous evidence. Science is an ongoing process, and information obtained in the future may require modifications to certain parts of the synthetic theory of evolution.

We now consider one of the many evolutionary questions currently being addressed by biologists: the relative effects of chance and natural selection on evolution.

[2] Some of the founders of the synthetic theory of evolution were biologists Theodosius Dobzhansky, Ronald Fisher, J.B.S. Haldane, Julian Huxley, Ernst Mayr, George Gaylord Simpson, G. Ledyard Stebbins, Jr., and Sewell Wright.

Biologists debate the effect of chance on evolution

Biologists have wondered, if we were able to repeat evolution by starting with similar organisms exposed to similar environmental conditions, would we get the same results? That is, would the same kinds of changes evolve, as a result of natural selection? Or would the organisms be quite different, as a result of random chance? Several recently reported examples of evolution in action suggest that chance may not be as important as natural selection.

A fruit fly species *(Drosophila subobscura)* native to Europe inhabits areas from Denmark to Spain. Biologists had noted that the northern flies have larger wings than southern flies (Fig. 17–6). The same fly species was accidentally introduced to North and South America in the late 1970s in two separate introductions. Ten years after its introduction to the Americas, biologists determined that no statistically significant changes in wing size had occurred in the different regions of North America. However, 20 years after its introduction, the fruit flies in North America exhibited the same type of north-south wing changes as in Europe. (It is not known why larger wings evolve in northern areas and smaller wings in southern climates.)

A study of the evolution of fish known as sticklebacks in three coastal lakes of west Canada yielded intriguingly similar results to the fruit fly study. When the lakes first formed several thousand years ago, molecular evidence indicates they were populated with the same ancestral species. (Analysis of the mitochondrial DNA of sticklebacks in the three lakes supports the hypothesis of a common ancestor.) In each lake, the same two species have evolved from the common ancestral fish: One species is large and consumes invertebrates along the bottom of

Figure 17–6 Wing size in female fruit flies. In Europe, female fruit flies *(Drosophila subobscura)* in northern countries have larger wings than flies in southern countries. Shown are two flies: one from Denmark *(right)* and the other from Spain *(left)*. The same evolutionary pattern rapidly emerged in North America after the accidental introduction of *D. subobscura* to the Americas. *(George Gilchrist)*

the lake, whereas the other species is smaller and consumes plankton at the lake's surface. Members of the two species within a single lake do not interbreed with one another, but individuals of the larger species from one lake will interbreed in captivity with individuals of the larger species from the other lakes. Similarly, smaller individuals from one lake will interbreed with smaller individuals from the other lakes.

Thus, natural selection appears to be a more important agent of evolutionary change than chance. If chance were the most important factor influencing the direction of evolution, then fruit fly evolution would not have proceeded the same way on two different continents, and stickleback evolution would not have proceeded the same way in three different lakes. (For more information on stickleback evolution, see Chapter 52.)

MANY TYPES OF SCIENTIFIC EVIDENCE SUPPORT EVOLUTION

A vast body of scientific evidence supports evolution, including observations from the fossil record, comparative anatomy, biogeography, developmental biology, and molecular biology. In addition, evolutionary hypotheses are increasingly being tested experimentally.

The fossil record provides strong evidence for evolution

Perhaps the most direct evidence for evolution comes from the discovery, identification, and interpretation of **fossils,** which are the remains or traces typically left in sedimentary rock by previously existing organisms. (The term *fossil* comes from the Latin word *fossilis,* meaning "something dug up.") Sedimentary rock forms by the accumulation and solidification of particles produced by the weathering of older rocks, such as volcanic rocks. In an undisturbed rock sequence, the oldest layer is at the bottom, and upper layers are successively younger. The study of sedimentary rock layers, including their composition, arrangement, and correlation (similarity) from one location to another, enables geologists to place events recorded in rocks in their correct sequence.

The fossil record shows a progression from the earliest unicellular organisms to the many unicellular and multicellular organisms living today. The fossil record therefore demonstrates that life has evolved through time. To date, paleontologists (scientists who study extinct species) have described and named about 300,000 fossil species, and more are being discovered all the time.

Although most fossils are preserved in sedimentary rock, some more recent remains have been exceptionally well preserved in bogs, tar, amber (ancient tree resin), or ice (Fig. 17–7). For example, the remains of a woolly mammoth deep-frozen in Siberian ice for more than 25,000 years were so well preserved that part of its DNA could be analyzed.

The formation and preservation of a fossil require that an organism be buried under conditions that slow or prevent the decay process. This is most likely to occur if an organism's remains are covered quickly by a sediment of fine soil particles suspended in water. In this way remains of aquatic organisms may be trapped in bogs, mud flats, sandbars, or deltas. Remains of terrestrial organisms that lived on a flood plain may also be covered by water-borne sediments or, if the organism lived in an arid region, by wind-blown sand. Over time, the sediments harden to form sedimentary rock, and minerals usually replace the organism's remains so that many details of its structure, even cellular details, remain.

The fossil record is not a random sample of past life but instead is biased toward aquatic organisms and those living in the few terrestrial habitats conducive to fossil formation. Relatively few fossils of tropical rainforest organisms have been found, for example, because their remains decay extremely rapidly on the forest floor, before fossils can develop. Another reason for bias in the fossil record is that organisms with hard body parts such as bones and shells are more likely to form fossils than those with soft body parts.

Process of Science

Because of the nature of the scientific process, each fossil discovery represents a separate "test" of the theory of evolution. If any of the tests fail, the theory would have to be modified to fit the existing evidence. The verifiable discovery, for example, of fossil remains of modern humans *(Homo sapiens)* in Precambrian rocks, which are more than 570 million years old, would falsify the theory of evolution as currently proposed. However, Precambrian rocks examined to date contain only fossils of simple organisms, such as algae and small, soft-bodied animals, that are thought to have evolved early in the history of life. The earliest fossils of *H. sapiens* with anatomically modern features do not appear in the fossil record until approximately 100,000 years ago (see Chapter 21).

Fossils provide a record of ancient organisms and some understanding of where and when they lived. Using fossils of organisms from different geological ages, the lines of descent (evolutionary relationships) that gave rise to modern-day organisms can sometimes be inferred. In many instances, fossils provide direct evidence of the origin of new species from preexisting species, including many transitional forms.

Transitions found in the fossil record document whale evolution

One question that has excited biologists for about a century is how whales and other cetaceans (marine mammals) evolved from land-dwelling mammals. During the 1980s and 1990s, paleontologists discovered several fossil intermediates in whale evolution that document the whales' transition from land to water. Based on fossil evidence, one candidate for the ancestor of whales is a now-extinct group of four-legged, land-dwelling mammals called mesonychians (Fig. 17–8*a*). These animals had unusually large heads and teeth that were remarkably similar to those of the earliest whales. About 50 million to 60 million years ago (mya), some descendants of mesonychians had adapted to swimming in shallow seas.

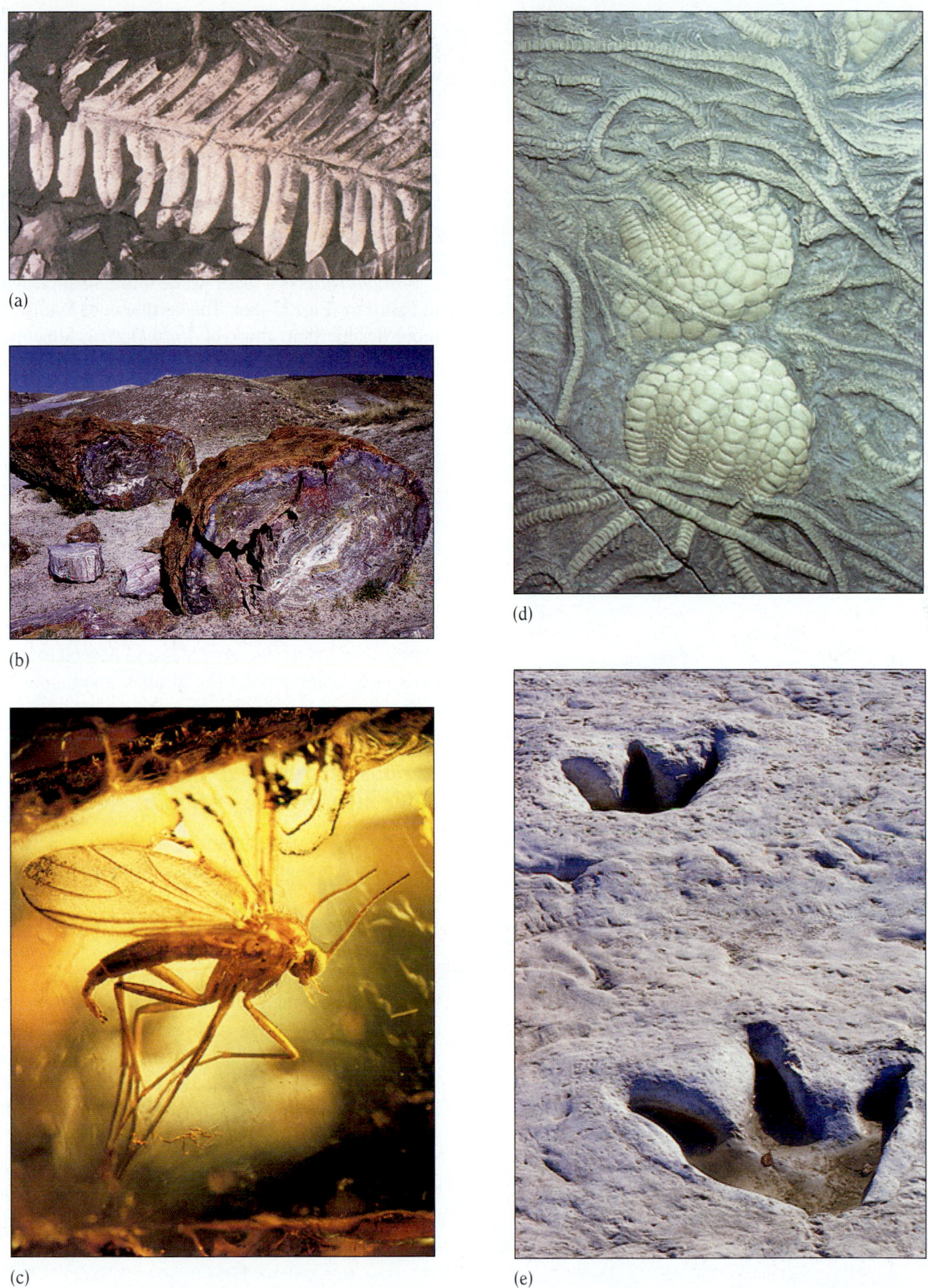

(a) (b) (c) (d) (e)

Figure 17–7 Fossils develop in different ways. **(a)** Although some fossils contain traces of organic matter, all that remains in this fossil of a seed fern leaf is an impression, or imprint, in the rock. **(b)** Petrified wood from the Petrified Forest National Park in Arizona consists of trees that were buried and infiltrated with minerals. **(c)** A 2-million-year-old insect fossil (a midge) was embedded in amber. **(d)** A cast fossil of ancient echinoderms called crinoids formed when the crinoids decomposed, leaving a mold that later filled with dissolved minerals that hardened. **(e)** Dinosaur footprints, each 75 to 90 cm (2.5 to 3 ft) in length, provide clues about the posture, gait, and behavior of these extinct animals. *(a, Carolina Biological Supply Company/Phototake, New York City; b, Kenneth Murray/Photo Researchers, Inc.; c, Alfred Pasieka/Science Photo Library/Photo Researchers, Inc.; d, A.J. Copley/Visuals Unlimited; e, Scott Berner/Visuals Unlimited)*

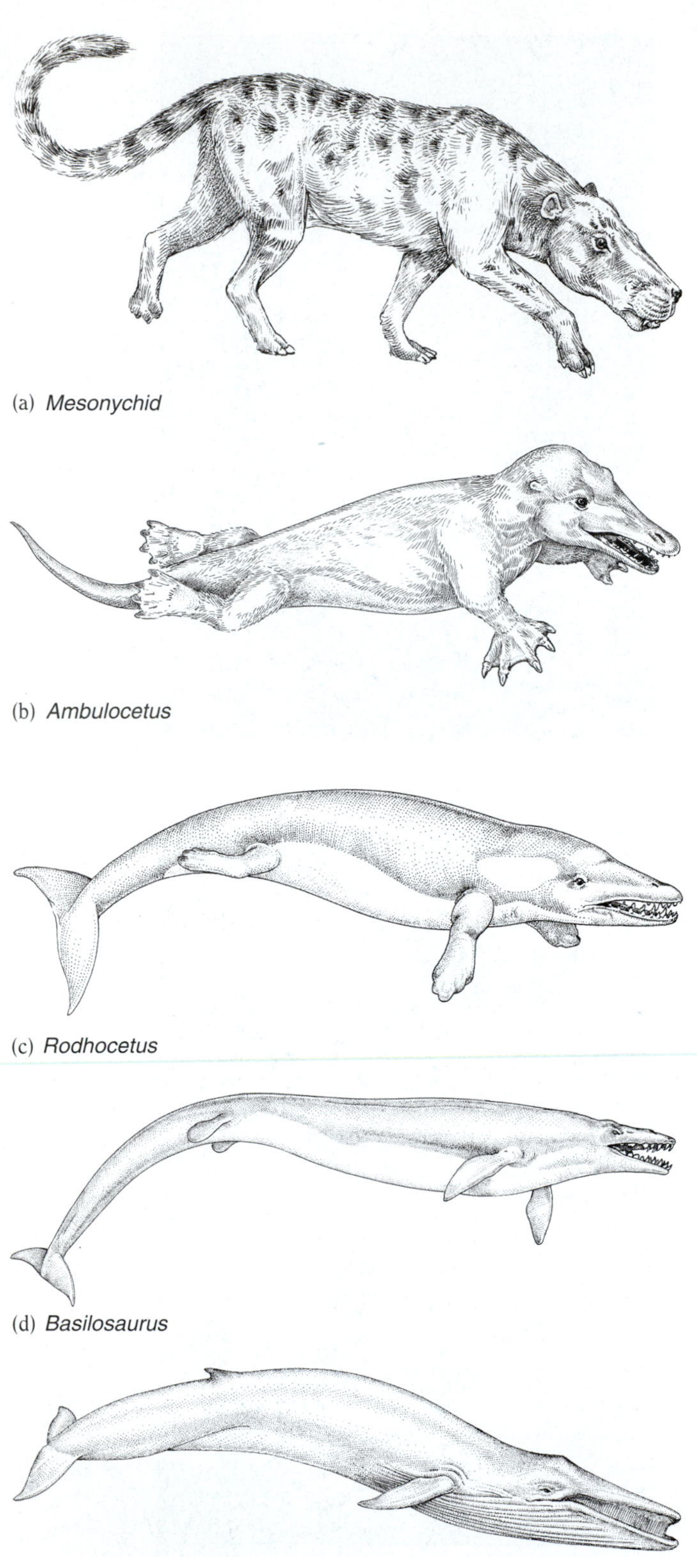

Fossils of *Ambulocetus natans,* a 50-million-year-old whale discovered in Pakistan, have many features of modern whales but also possess hind limbs and feet (Fig. 17–8*b*). (Modern whales do not have hind limbs, although vestigial pelvic and hind limb bones persist. Vestigial structures are discussed later in the chapter.) The vertebrae of *Ambulocetus'* lower back were very flexible, allowing its back to move dorsoventrally (up and down) during swimming and diving, like modern whales. In addition to swimming, this ancient whale moved about on land, perhaps as sea lions do today.

Rodhocetus is a fossil whale found in slightly younger rocks in Pakistan (Fig. 17–8*c*). The vertebrae of *Rodhocetus* were even more flexible than those of *Ambulocetus,* allowing a more powerful dorsoventral movement during swimming. *Rodhocetus* may have been totally aquatic.

By 40 mya, the whale transition from land to ocean was almost complete. Egyptian fossils of *Basilosaurus,* a whale that lived at that time, indicate a streamlined body and front flippers for steering, like modern-day whales (Fig. 17–8*d*). *Basilosaurus* retained vestiges of its land-dwelling ancestors—a pair of reduced hind limbs that were disjointed from the backbone and probably not used in locomotion. Reduction in the hind limbs continued to the present. The modern blue whale has vestigial pelvis and femur bones embedded in its body (Fig. 17–8*e*). (The closest living relatives of whales and porpoises are hoofed mammals such as antelopes, deer, giraffes, and hippos, discussed later in the chapter.)

Various methods can be used to determine the age of fossils

Because layers of sedimentary rock occur naturally in the sequence of their deposition, with the more recent layers on top of the older, earlier ones (Fig. 17–9), most fossils are dated by their relative position in sedimentary rock. However, geological events occurring after the rocks were initially formed have occasionally changed the relationships of some rock layers. Geologists identify specific sedimentary rocks not only by their positions in layers but also by features such as mineral content and by the fossilized remains of certain organisms, known as **index fossils,** that characterize a specific layer over large geographical areas. Index fossils are fossils of organisms that existed for a relatively short geological time but were preserved as fossils in large numbers. With this information, geologists can arrange rock layers and the fossils they contain in chronological order and identify comparable layers in widely separated locations.

Figure 17–8 Fossil intermediates in whale evolution. **(a)** *Mesonychid,* an extinct terrestrial mammal, may have been the ancestor of whales. **(b)** *Ambulocetus natans,* a transitional form between modern whale descendants and their terrestrial ancestors, possessed a number of recognizable whale features yet retained the hind limbs of its four-legged ancestors. **(c)** The more recent *Rodhocetus* had flexible vertebrae that permitted a powerful dorsoventral movement during swimming. **(d)** *Basilosaurus* was more streamlined and possessed tiny nonfunctional hindlimbs. **(e)** *Balaenoptera,* the modern blue whale, contains vestiges of pelvis and leg bones. *(a–d, adapted from Fig. 2 on pages 260–261 in Futuyma, D.J.* Science on Trial: The Case for Evolution. *Sinauer Associates, Inc., Sunderland, MA, 1995)*

Figure 17–9 Exposed layers of sedimentary rock. Shown are weathered rock layers near the Paria River in the Grand Staircase–Escalante National Monument, Utah. The younger layers overlie the older layers. Many layers date to the Mesozoic era, from 248 to 65 mya. Characteristic fossils are associated with each layer. *(Tom Till)*

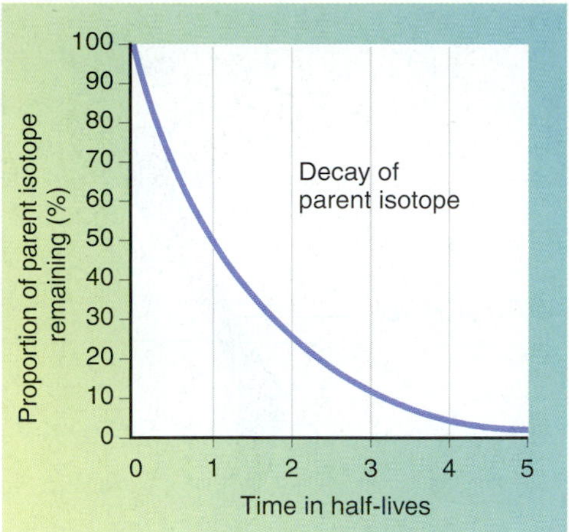

Figure 17–10 Radioisotope decay. At time zero, the sample is composed entirely of the radioisotope, and the radioactive clock begins ticking. After one half-life, only 50% of the original radioisotope remains. During each succeeding half-life, one-half of the remaining radioisotope is converted to decay product(s).

Radioactive isotopes, also called **radioisotopes,** present in a rock provide a means to accurately measure its age (see Chapter 2). Radioisotopes emit invisible radiations. As a radioisotope emits radiation, its nucleus changes into the nucleus of a different element in a process known as **radioactive decay.** For example, the radioactive nucleus of uranium-235 decays over time into lead-207.

Each radioisotope has its own characteristic rate of decay. The period of time required for one half of the atoms of a radioisotope to change into a different atom is known as its **half-life** (Fig. 17–10). Radioisotopes differ significantly in their half-lives. For example, the half-life of iodine-132 is only 2.4 hours, whereas the half-life of uranium-235 is 704 million years. The half-life of a particular radioisotope is constant and does not vary with temperature, pressure, or any other environmental factor.

The age of a fossil in sedimentary rock is usually estimated by measuring the relative proportions of the original radioisotope and its decay product in volcanic rock intrusions that penetrate the sediments. For example, the half-life of potassium-40 is 1.3 billion years, meaning that in 1.3 billion years half of the radioactive potassium will have decayed into its decay product, argon-40. The radioactive clock begins ticking when the magma solidifies into volcanic rock. The rock initially contains some potassium but no argon. Because argon is a gas, it escapes from hot rock as soon as it forms, but when potassium decays in rock that has cooled and solidified, the argon accumulates in the crystalline structure of the rock. If the ratio of potassium-40 to argon-40 in the rock being tested is 1:1, the rock is 1.3 billion years old.

Several radioisotopes are commonly used to date fossils. These include potassium-40 (half-life 1.3 billion years), uranium-235 (half-life 704 million years), and carbon-14 (half-life 5730 years). Potassium-40, with its long half-life, can be used to date fossils that are many hundreds of millions of years old. Radioisotopes other than carbon-14 are used to date the *rock* in which fossils are found, whereas carbon-14 is used to date the *carbon remains* of anything that was once living, such as wood, bones, and shells. Whenever possible, the age of a fossil is independently verified using two or more different radioisotopes.

Carbon-14, which is continuously produced in the atmosphere from nitrogen-14 (by cosmic radiation), subsequently decays back to nitrogen-14. Because the formation and the decay of carbon-14 occur at constant rates, the ratio of carbon-14 to carbon-12 (the more abundant, stable isotope of carbon) is constant in the atmosphere. Since each organism absorbs carbon from the atmosphere, its ratio of carbon-14 to carbon-12 is the same as the atmosphere.[3] When an organism dies, however, it no longer absorbs carbon, and the proportion of carbon-14 in its remains declines as carbon-14 decays to nitrogen-14. Because of its relatively short half-life, carbon-14 is useful for dating fossils that are 50,000 years old or less. It is particularly useful for dating archaeological sites.

[3]Organisms absorb carbon from the atmosphere either directly (by photosynthesis) or indirectly (by consuming photosynthetic organisms).

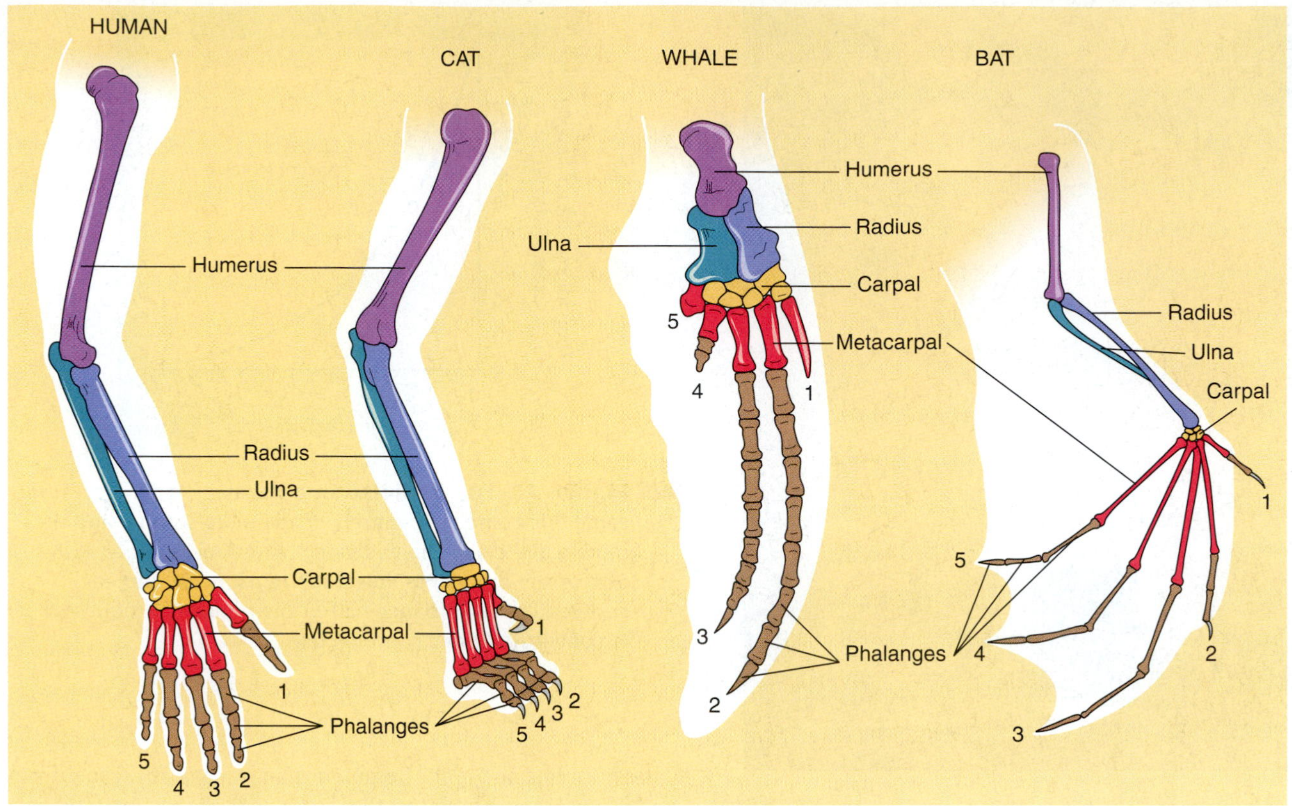

Figure 17–11 Homology in animals. The human arm, cat forelimb, whale flipper, and bat wing have a basic underlying similarity of structure because they are derived from a common ancestor. The five digits are numbered in each drawing.

Comparative anatomy of related species demonstrates similarities in their structures

Comparing the structural details of features found in different but related organisms reveals a basic similarity. Such features that are derived from the same structure in a common ancestor are termed **homologous features;** the condition is known as **homology.** For example, consider the limb bones of mammals. A human arm, a cat forelimb, a whale front flipper, and a bat wing, although quite different in appearance, have strikingly similar arrangements of bones, muscles, and nerves. Figure 17–11 shows a comparison of their skeletal structures. Each has a single bone (the humerus) in the part of the limb nearest the trunk of the body, followed by the two bones (radius and ulna) of the forearm, a group of bones (carpals) in the wrist, and a variable number of digits (metacarpals and phalanges). This similarity is particularly striking because arms, forelimbs, flippers, and wings are used for different types of locomotion, and there is no overriding mechanical reason for them to be so similar structurally. Similar arrangements of parts of the forelimb are evident in ancestral reptiles and amphibians and even in the first fishes that came out of water onto land hundreds of millions of years ago.

Leaves are an example of homology in plants. In many plant species, leaves have been modified for functions other than photosynthesis. A cactus spine and a pea tendril, although quite different in appearance, are homologous because both are modified leaves (Fig. 17–12). The spine protects the succulent stem tissue of the cactus, whereas the tendril, which winds around a small object once it makes contact, helps support the climbing stem of the pea plant. Such modifications in organs used in different ways are the expected outcome of a common evolutionary origin. The basic structure present in a common ancestor was modified in different ways for different functions as various descendants subsequently evolved.

Not all species with "similar" features have descended from a recent common ancestor, however. Structurally similar features that are not homologous but simply have similar functions in distantly related organisms are said to be **homoplastic features;** such a condition is called **homoplasy.**[4] For example, the wings of various distantly related flying animals, such as insects and birds, resemble each other superficially; they are homoplastic features that evolved over time to meet the common function of flight, though they are different in more fundamental aspects. Bird wings are modified forelimbs supported by bones, whereas insect wings may have evolved from gill-like appendages present in the aquatic ancestors of insects. Spines, which are modified

[4]An older, less precise term that some biologists still use for nonhomologous features with similar functions is *analogy*.

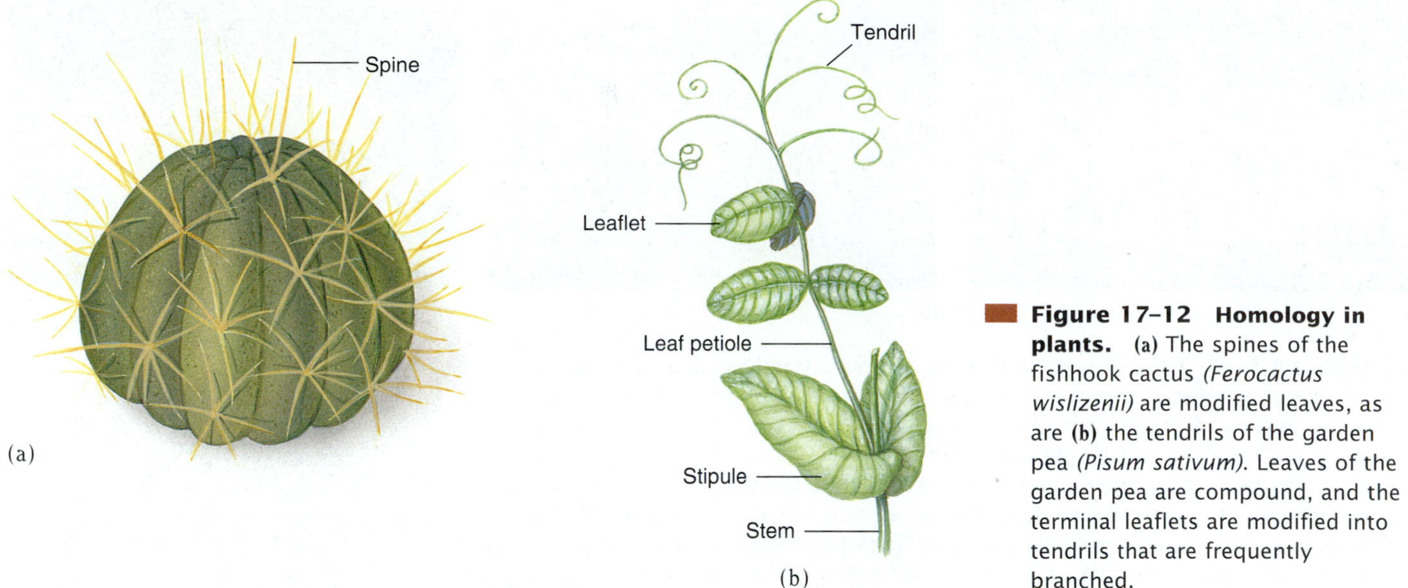

Figure 17–12 Homology in plants. **(a)** The spines of the fishhook cactus *(Ferocactus wislizenii)* are modified leaves, as are **(b)** the tendrils of the garden pea *(Pisum sativum)*. Leaves of the garden pea are compound, and the terminal leaflets are modified into tendrils that are frequently branched.

leaves, and thorns, which are modified stems, are an example of homoplasy in plants. Spines and thorns resemble one another superficially but are homoplastic features that evolved independently to solve the common need for protection from herbivores (Fig. 17–13).

Like homology, homoplasy offers crucial evidence of evolution. Homoplastic features are of evolutionary interest because they demonstrate that organisms with separate ancestries may adapt in similar ways to similar environmental demands. Such independent evolution of similar structures in distantly related organisms is known as **convergent evolution.** Aardvarks, anteaters, and pangolins are an excellent example of convergent evolution (Fig. 17–14). They resemble one another in lifestyle and certain structural features. All have strong, sharp claws to dig open ant and termite mounds and elongated snouts with long, sticky tongues to catch these insects. Yet aardvarks, anteaters, and pangolins evolved from three distantly related orders of mammals. (See Chapter 22 for further discussion of homology and homoplasy. Also, see Figure 30–27, which shows several examples of convergent evolution in placental and marsupial mammals.)

Comparative anatomy reveals the existence of **vestigial structures.** Many organisms contain organs or parts of organs that are seemingly nonfunctional and degenerate, often

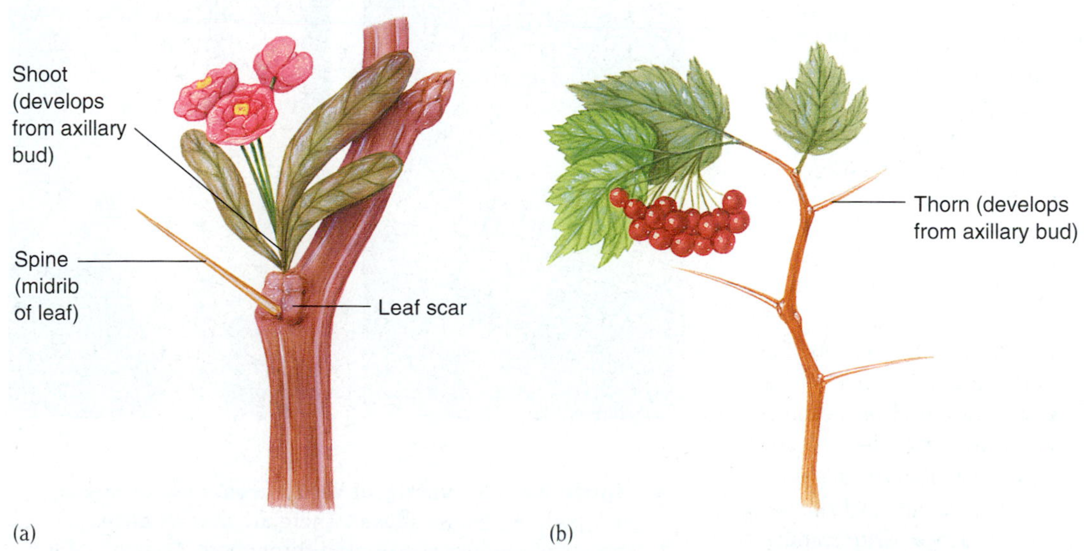

Figure 17–13 Homoplasy. **(a)** A spine of Japanese barberry *(Berberis thunbergii)* is a modified leaf. (In this example, the spine is actually the midrib of the original leaf, which has been shed.) **(b)** Thorns of downy hawthorn *(Crataegus mollis)* are modified stems that develop from axillary buds.

(a)

(b)

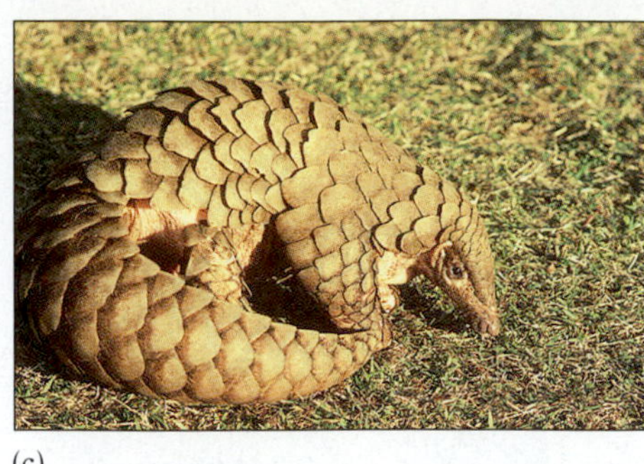
(c)

Figure 17–14 Convergent evolution. Three distantly related mammals exhibit similarities in lifestyle and certain structural features as a result of convergent evolution. These species adapted independently to eat ants and termites in similar grassland/forest environments in different parts of the world. **(a)** The aardvark *(Orycteropus afer)* is native to central, southern, and eastern Africa. **(b)** A giant anteater *(Myrmecophaga tridactyla)* at a termite mound. The anteater is native to Latin America, from southern Mexico to northern Argentina. **(c)** The pangolin *(Manis crassicaudata)* is native to Africa and southern and southeastern Asia. *(a, Kjell B. Sandved/Visuals Unlimited; b, Gunter Ziesler/Peter Arnold, Inc.; c, Mandal Ranjit/Photo Researchers, Inc.)*

undersized or lacking some essential part. Vestigial structures are remnants of more developed structures that were present and functional in ancestral organisms. In the human body, more than 100 structures are considered vestigial, including the appendix, coccyx (fused tailbones), third molars (wisdom teeth), and the muscles that move our ears. Whales and pythons have vestigial hindlimb bones (Fig. 17–15); pigs have vestigial toes that do not touch the ground; wingless birds such as the kiwi have vestigial wing bones; and many blind, burrowing or cave-dwelling animals have nonfunctioning, vestigial eyes.

The occasional presence of a vestigial structure is to be expected as a species adapts to a changing mode of life. Some structures become much less important for survival and may end up as vestiges. When a structure no longer confers a selective advantage, it usually becomes smaller and loses much or all of its function with the passage of time. Since the presence of the vestigial structure is usually not harmful to the organism, however, selective pressure for completely eliminating it is weak, and the vestigial structure can be found in many subsequent generations.

The distribution of plants and animals supports evolution

The study of the past and present geographical distribution of organisms is called **biogeography.** Darwin was interested in biogeography and considered why the species found on ocean islands tend to resemble species of the nearest mainland, even if the environment is different, but not to resemble species on islands with similar environments in other parts of the world. Darwin studied the plants and animals of two sets of arid islands—the Cape Verde Islands, some 650 km (about 400 mi) west of Dakar, Africa, and the Galapagos Islands, a comparable distance west of Ecuador, South America. On each group of islands, the plants and terrestrial animals were indigenous (native), but those of the Cape Verde Islands resembled African species and those of the Galapagos resembled South American species. The similarities of Galapagos species to South American species were particularly striking considering that the Galapagos Islands are dry and rocky and the nearest part of South America is humid

(a)

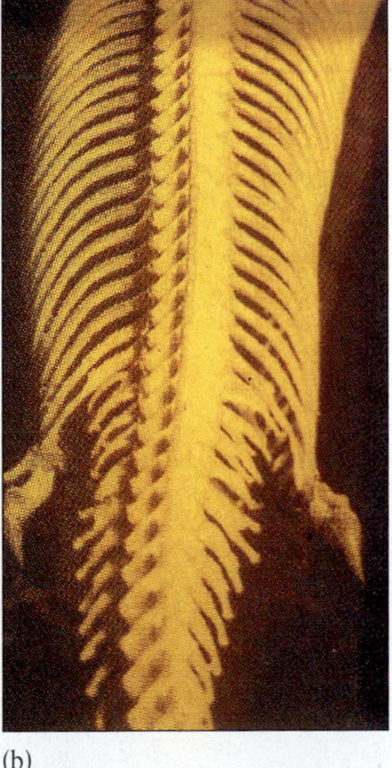
(b)

Figure 17–15 Vestigial structures. **(a)** An African rock python *(Python sebae)*. There are several dozen species of pythons distributed throughout Asia and Africa. **(b)** Close-up view of a python skeleton showing the hindlimb bones. All pythons have remnants of hindlimb bones embedded in their bodies. *(a, E.R. Degginger/Animals Animals; b, J.D. Cunningham/Visuals Unlimited)*

and has a lush tropical rain forest. Darwin concluded that species from the neighboring continent migrated or were carried to the islands, where they subsequently adapted to the new environment and, in the process, evolved into new species.

If evolution were not a factor in the distribution of species, we would expect to find a given species everywhere that it could survive. However, the geographical distribution of organisms that actually exists makes sense in the context of evolution. For example, Australia, which has been a separate land mass for millions of years, has distinctive organisms. Australia has populations of egg-laying mammals (monotremes) and pouched mammals (marsupials) not found anywhere else. Two hundred million years ago, Australia and the other continents were joined together in a major land mass. Over the course of millions of years, the Australian continent gradually separated from the others. The monotremes and marsupials in Australia continued to thrive and diversify. The isolation of Australia also prevented placental mammals, which arose elsewhere at a later time, from competing with its monotremes and marsupials. In other areas of the world where placental mammals occurred, most monotremes and marsupials became extinct.

We now consider how Earth's dynamic geology has affected biogeography and evolution.

Earth's geological history is related to biogeography and evolution

In 1915 the German scientist Alfred Wegener, who had noted a correspondence between the geographical shapes of South America and Africa, proposed that all the land masses had at one time been joined into one huge supercontinent, which he called Pangaea (Fig. 17–16*a*). He further suggested that Pangaea had subsequently broken apart and that the various land masses had separated in a process known as **continental drift.** Wegener did not know of any mechanism that could have caused continental drift, and so his idea, although debated initially, was largely ignored.

In the 1960s, scientific evidence accumulated that provided the explanation for continental drift. Earth's crust is composed of seven large plates (plus a few smaller ones) that float on the mantle, which is the mostly solid[5] layer of Earth lying beneath the crust and above the core. The land masses are situated on some

[5] Most of the rock in the upper portion of the mantle is solid, although 1% or 2% is melted. Because of its higher temperature, the solid rock of the mantle is more plastic than the solid rock of the lithosphere above it.

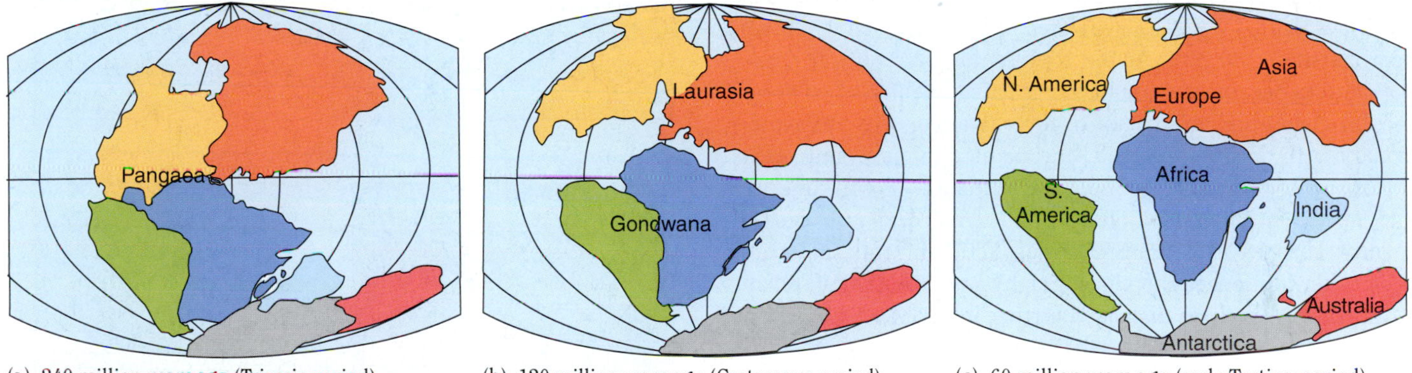

(a) 240 million years ago (Triassic period)
(b) 120 million years ago (Cretaceous period)
(c) 60 million years ago (early Tertiary period)

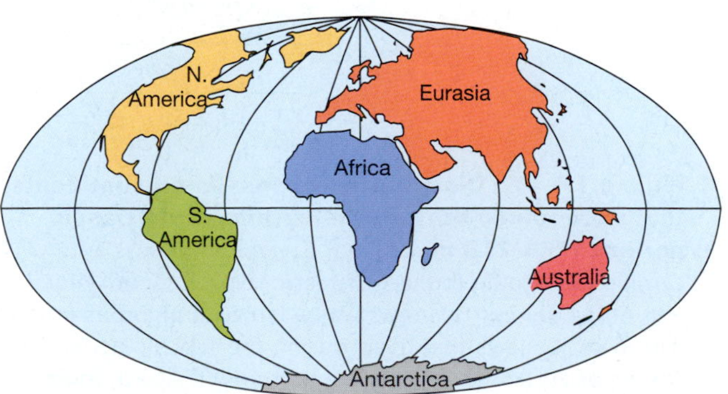

(d) Today

Figure 17–16 Continental drift. **(a)** The supercontinent Pangaea, about 240 mya. **(b)** Breakup of Pangaea into Laurasia (Northern Hemisphere) and Gondwana (Southern Hemisphere), 120 mya. **(c)** Further separation of land masses, 60 mya. Note that Europe and North America were still joined and that India was a separate land mass. **(d)** The continents today.

of these plates. As the plates move, the continents change their relative positions (Fig. 17–16*b, c,* and *d*). The movement of the crustal plates is termed **plate tectonics.**

Any area where two plates meet is a site of intense geological activity. Earthquakes and volcanoes are common in such a region. Both San Francisco, noted for its earthquakes, and the Mount Saint Helens volcano are situated where two plates meet. If land masses lie on the edges of two adjacent plates, mountains may form. The Himalayas formed when the plate carrying India rammed into the plate carrying Asia. When two plates grind together, one of them is sometimes buried under the other in a process known as subduction. When two plates move apart, a ridge of lava forms between them. The Atlantic Ocean is getting larger because of the expanding zone of lava along the Mid-Atlantic Ridge, where two plates are separating.

Knowledge that the continents were at one time connected and have since drifted apart is useful in explaining certain aspects of biogeography (Fig. 17–17). Likewise, continental drift has played a major role in the evolution of different organisms. When Pangaea originally formed during the late Permian period, it brought together terrestrial species that had evolved separately from one another, leading to competition and some extinctions. Marine life was adversely affected, in part because, with the continents joined as one large mass, less coastline existed. (Because coastal areas are shallower, they contain high concentrations of marine species.) Pangaea separated into several land masses approximately 180 mya. As the continents began to drift apart, populations became geographically isolated in different environmental conditions and began to diverge along separate evolutionary pathways. As a result, the plants, animals, and other organisms of previously connected continents—South America and Africa, for example—differ. Continental drift also caused gradual changes in ocean and atmospheric currents that have profoundly influenced the biogeography and evolution of organisms. (Biogeography is discussed further in Chapter 54.)

Developmental biology is increasingly being used to explain evolution

How snakes became elongated and lost their limbs has long intrigued evolutionary biologists. Comparative anatomy indicates, for example, that pythons have vestigial hind limb bones embedded in their bodies (see Fig. 17–15). Other snakes have lost their hind limbs entirely.

Increasingly, developmental biology, particularly at the molecular level, is providing answers to such questions. In many cases, evolutionary changes such as limblessness in snakes, occur as a result of changes in genes that affect the orderly sequence of events that occur during development. In pythons, for example, the loss of forelimbs and elongation of the body are linked to mutations in several *Hox* genes that affect the expression of body patterns and limb formation in a wide variety of animals (see Chapter 16 discussion of *Hox* gene complexes). Apparently the hind limbs do not develop because python embryonic tissue does not respond to internal signals that trigger leg elongation. Developmental biologists speculate that mutations in the *Hox* genes probably occurred first, resulting in an ancient snake with hind limbs but no forelimbs. The fossil record supports this hypothesis: A fossil of a primitive snake with such features has been identified *(Pachyrhachis problematicus).*

Scientific evidence overwhelmingly demonstrates that development in different animals is controlled by the same kinds of genes; these genetic similarities in a wide variety of organisms

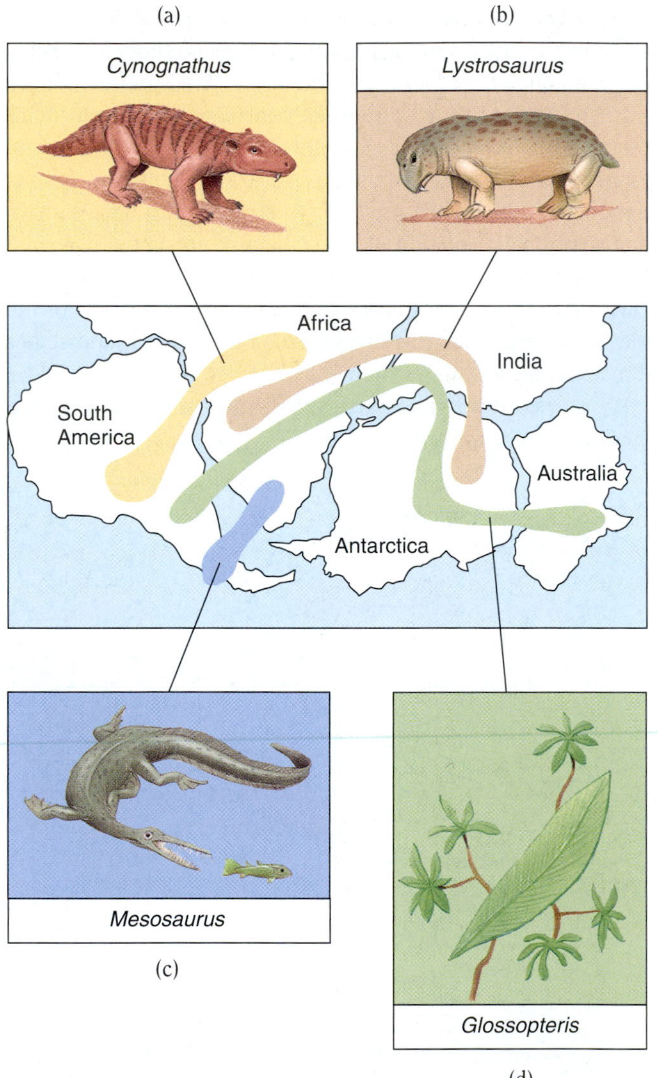

Figure 17–17 Distribution of fossils on continents that were joined during the Permian and Triassic periods (286–213 mya). **(a)** *Cynognathus* was a carnivorous reptile found in Triassic rocks in South America and Africa. **(b)** *Lystrosaurus* was a large herbivorous reptile with beaklike jaws that lived during the Triassic period. Fossils of *Lystrosaurus* have been found in Africa, India, and Antarctica. **(c)** *Mesosaurus* was a small freshwater reptile found in Permian rocks in South America and Africa. **(d)** *Glossopteris* was a seed-bearing tree dating from the Permian period. *Glossopteris* fossils have been found in South America, Africa, India, Antarctica, and Australia. *(Adapted from Colbert, E.H.* Wandering Lands and Animals. *Hutchinson, London, 1973)*

reflect a shared evolutionary history. For example, all vertebrates have similar patterns of embryological development that indicate they share a common ancestor. Segmented muscles, pharyngeal (gill) pouches, a tubular heart without left and right sides, a system of arteries known as aortic arches in the gill region, and many other features are found in all vertebrate embryos. All these structures are necessary and functional in the developing fish. The small, segmented muscles of the fish embryo give rise to the segmented muscles used by the adult fish in swimming. The gill pouches break through to the surface as gill slits. The adult fish heart remains undivided and pumps blood forward to the gills that develop in association with the aortic arches.

Since none of these embryonic features persists in the adults of reptiles, birds, or mammals, why are these fishlike structures present in their embryos? Evolution is a conservative process, and natural selection builds on what has come before rather than starting from scratch. The evolution of new features often does not require the evolution of new developmental genes but instead depends on a modification in developmental genes that already exist (see Chapter 19 discussion of preadaptations). Terrestrial vertebrates are thought to have evolved from fishlike ancestors; therefore, they share some of the early stages of development still found in fish today. The accumulation of genetic changes over time in these vertebrates has modified the basic body plan laid out in fish development.

Molecular comparisons among organisms provide evidence for evolution

Similarities and differences in the biochemistry and molecular biology of various organisms provide evidence for evolutionary relationships. Lines of descent based solely on biochemical and molecular characters often resemble lines of descent based on structural and fossil evidence. Molecular evidence for evolution includes the universal genetic code and the conserved sequences of amino acids in proteins and of nucleotides in DNA.

The genetic code is virtually universal

Organisms owe their characteristics to the types of proteins that they possess, which in turn are determined by the sequence of nucleotides in their messenger ribonucleic acid (mRNA), as specified by the order of nucleotides in their DNA. Evidence that all life is related comes from the fact that all organisms use a genetic code that is virtually identical.[6] Recall from Chapter 12 that the genetic code specifies a triplet (a sequence of three nucleotides in DNA) that codes for a particular codon (a sequence of three nucleotides in mRNA) that codes for a particular amino acid in a polypeptide chain. For example, "AAA" in DNA codes for "UUU" in mRNA, which codes for the amino acid phenylalanine in organisms as diverse as shrimp, humans, bacteria, and tulips. In fact, "AAA" codes for phenylalanine in all organisms examined to date.

[6]There is some minor variation in the genetic code. For example, mitochondria have some deviations from the standard code.

The universality of the genetic code—no other code has been found in any organism—is compelling evidence that all organisms arose from a common ancestor. The genetic code has been maintained and transmitted through all branches of the evolutionary tree since its origin in some extremely early (and successful) organism.

Proteins and DNA contain a record of evolutionary change

Thousands of comparisons of protein and DNA sequences from various species have been done during the past 25 years or so. In many cases, sequence-based relationships agree with earlier studies that based evolutionary relationships on similarities in structure among living organisms and in fossil data of extinct organisms.

Investigations of the sequence of amino acids in proteins that play the same roles in many species have revealed both great similarities and certain specific differences. Even organisms that are remotely related, such as humans, fruit flies, sunflowers, and yeasts, share some proteins, such as cytochrome *c,* which is part of the electron transport chain in aerobic respiration. To survive, all aerobic organisms need a respiratory protein with the same basic structure and function as the cytochrome *c* of their common ancestor. Consequently, not all amino acids that confer the structural and functional features of cytochrome *c* are free to change. Any mutations that changed the amino acid sequence at structurally important sites of the cytochrome *c* molecule would have been harmful, and natural selection would have prevented such mutations from being passed to future generations. However, in the course of the long, independent evolution of different organisms, mutations have resulted in the substitution of many amino acids at less important locations in the cytochrome *c* molecule. The greater the differences in the amino acid sequences of their cytochrome *c* molecules, the longer it is thought to have been since two species diverged.

Because a protein's amino acid sequences are coded in DNA,[7] the differences in amino acid sequences indirectly reflect the nature and number of underlying DNA base-pair changes that must have occurred during evolution. Such molecular information is determined directly by **DNA sequencing,** in which the order of nucleotide bases in DNA is determined. Generally, the more closely species are considered related on the basis of other scientific evidence, the greater the percentage of nucleotide sequences that their DNA molecules have in common. By using the DNA sequence data in Table 17–1, for example, you can conclude that the closest living relative of humans is the chimpanzee (because its DNA has the lowest percentage differences in the sequence examined). Which of the primates in Table 17–1 is the most distantly related to humans? (Primate evolution is discussed in Chapter 21.)

[7]Of course, not all DNA codes for proteins (witness introns and transfer RNA genes). DNA sequencing of non–protein-coding DNA is also useful in determining evolutionary relationships.

TABLE 17–1 Differences in Nucleotide Sequences in DNA as Evidence of Phylogenetic Relationships*

Species Pairs	Percent Divergence in a Selected DNA Sequence†
Human–chimpanzee	1.7
Human–gorilla	1.8
Human–orangutan	3.3
Human–gibbon	4.3
Human–rhesus monkey (Old World monkey)	7.0
Human–spider monkey (New World monkey)	10.8
Human–tarsier	24.6

*From Goodman, M., et al. "Primate Evolution at the DNA Level and a Classification of Hominoids." *Journal of Molecular Evolution,* Vol. 30, 1990.

†Noncoding sequences of β-globin genes.

Process of Science

In some cases, molecular evidence challenges traditional evolutionary ideas that were based on structural comparisons among living species and/or on studies of fossil skeletons. Consider artiodactyls, an order of even-toed hoofed mammals such as pigs, camels, deer, antelope, cattle, and hippopotamuses. Traditionally, whales, which do not possess toes, are not classified as artiodactyls (although early fossil whales possessed an even number of toes on their appendages). Figure 17–18 depicts a hypothetical phylogenetic tree for whales and selected artiodactyls based on molecular data. Such **phylogenetic trees**—diagrams showing lines of descent—can be derived from differences in a given DNA nucleotide sequence. This diagram suggests whales should be classified as artiodactyls and shows that hippopotamuses are more closely related to whales than any other artiodactyl. The branches representing whales and hippopotamuses are thought to have diverged relatively recently because of the close similarity of DNA sequences in these species. In contrast, camels, which have DNA sequences that are less similar to whales, diverged much earlier. The molecular evidence indicates that whales and hippopotamuses share a recent common ancestor, a hippo-like artiodactyl that split from the rest of the artiodactyl line some 55 mya. However, available fossil evidence does not currently provide support for the molecular hypothesis; a fossil ancestor common to both whales and hippos has not yet been discovered. (Recall from earlier in the chapter that most paleontologists currently suggest that the mesonychians, which are not ancient artiodactyls, may have been the ancestor of whales.) Paleontologists hope that future fossil discoveries will help clarify this discrepancy between molecular and fossil data.

DNA sequencing is used to estimate the time of divergence between two closely related species or taxonomic groups

Within a given taxonomic group, mutations are assumed to have occurred at a fairly steady rate over millions of years. Thus, if more differences occur in the same sequences of DNA of one species compared with another, more time must have elapsed since the two diverged from a common ancestor. (The *same* DNA sequence refers to the nucleotide sequence in homologous segments of DNA in the species being compared.)

From the number of alterations in homologous DNA sequences taken from different species, we can develop a **molecular clock** to estimate the time of divergence between two closely related species or higher taxonomic groups. A molecular clock makes use of the average rate at which a particular gene evolves. The clock is calibrated by comparing the number of nucleotide differences between two organisms with the dates of evolutionary branch points that are known from the fossil record. Once a molecular clock is calibrated, past evolutionary events whose

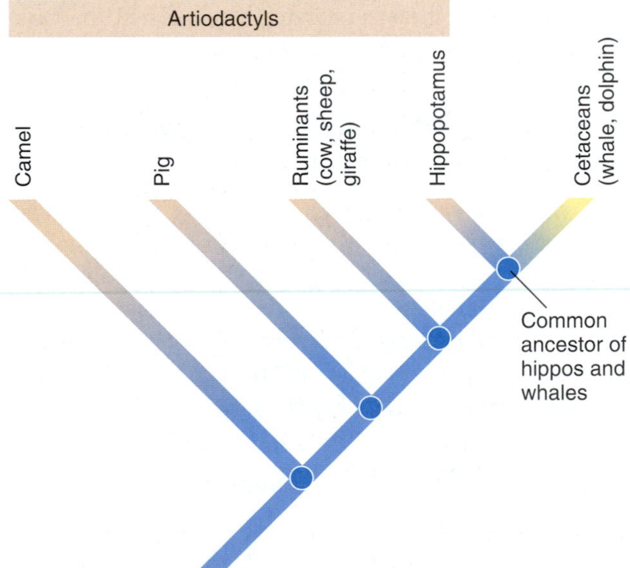

Figure 17–18 Phylogenetic tree of whales and their closest living relatives. This diagram, which is based on DNA sequence differences among selected mammals, shows hypothetical evolutionary relationships. It suggests that artiodactyls are the close relatives of whales, that the hippopotamus is the closest living relative of whales, and that artiodactyls and whales share a common ancestor in the distant past. However, this common ancestor has yet to be identified in the fossil record. (Ruminants are artiodactyls that have a multichambered stomach and that chew regurgitated plant material to improve its digestibility.) *(Adapted from Nikaido, M., et al. "Phylogenetic Relationships among Cetartiodactyls Based on Insertions of Short and Long Interspersed Elements: Hippopotamuses Are the Closest Extant Relatives of Whales."* Proceedings of the National Academy of Sciences, *Vol. 96, 31 Aug. 1999)*

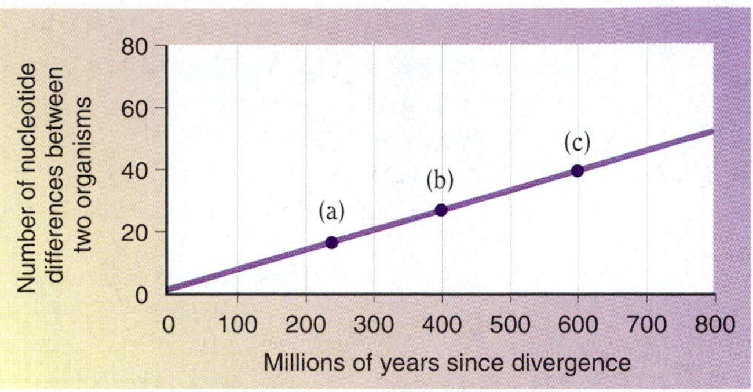

Figure 17–19 Calibration and use of a molecular clock. In this hypothetical example, the DNA of a specific gene is sequenced for birds, reptiles, fish, and insects. The number of nucleotide differences between birds and reptiles (a) is placed on a graph at the time at which birds and reptiles are thought to have diverged (based on fossil evidence). Likewise, the number of nucleotide differences between reptiles and fish (b) is placed on the graph at the time of branching indicated by the fossil record. A line is drawn through points (a) and (b) and extended, enabling scientists to estimate a much earlier time (c) at which the insect line diverged from the vertebrate line. Because a single gene may have large statistical errors when used as a molecular clock, molecular biologists typically obtain more reliable estimates of divergence times by constructing multiple molecular clocks from many different genes.

timing is not known with certainty can be estimated (Fig. 17–19).

Molecular clocks can be used to complement geological estimates of the divergence of species or to assign tentative dates to evolutionary events that lack fossil evidence. Where there is no fossil record of an evolutionary event, molecular clocks are the only way to estimate the timing of that event. Molecular clocks are also used, along with fossil evidence and structural data, to help reconstruct **phylogeny,** which is the evolutionary history of a group of related species (see Chapter 22). By assigning tentative dates to the divergence of species, molecular clocks show the relative order of branch points in phylogeny.

Molecular clocks must be developed and interpreted with care. Mutation rates vary among different genes and among distantly related taxonomic groups, causing molecular clocks to tick at different rates. Some genes, such as the gene for the respiratory protein cytochrome *c,* code for proteins that lose their function if the amino acid sequence changes slightly; these genes evolve slowly. Other genes, such as genes for blood-clotting proteins, code for proteins that are less constrained by changes in amino acid sequence; these genes evolve rapidly.

Although many dates estimated by molecular clocks are in agreement with fossil evidence, some discrepancies between molecular clocks and fossils exist. In most of these cases, the molecular clock's estimates of divergence times of particular organisms are much older than the dates at which the groups are first observed in the fossil record. Resolving these differences will require additional research in both molecular biology and paleontology.

Bacteria and other organisms that cause infectious disease are evolving resistance to drugs

Beginning in the late 1980s, an alarming increase in the incidence of tuberculosis (TB) has been documented worldwide. TB currently kills more than 2 million people each year. In the 30 or so years before the 1980s, the number of cases of TB had declined, at least in the developed world, largely as a result of treating TB with antibiotics, which are drugs intended to harm or kill bacteria and other microorganisms.

The evolution of drug-resistant strains in the bacterium that causes TB *(Mycobacterium tuberculosis)* is a disturbing trend. These strains are resistant to one or more antibiotics that traditionally were used to treat TB. Drug-resistant TB is deadly: As many as 80% of the people infected with multidrug-resistant TB (MDR-TB) die within two months of diagnosis—even with medical care. The problem with MDR-TB is particularly serious in five countries—Estonia, China, Latvia, Russia, and Iran—where 5% to 14% of all patients first diagnosed with TB are infected with multidrug-resistant strains.

Bacteria are continually evolving, even inside the bodies of human and animal hosts. Bacteria develop genetic resistance through mutations (see Chapters 11 and 12) and through acquiring new genes from plasmids (see Chapters 11 and 14, discussion of transformation) or viruses (see Chapter 23, discussion of transduction). When an antibiotic is used to treat a bacterial infection, a few bacteria may survive because they are genetically resistant to the antibiotic, and they pass these genes to future generations. As a result of selection, the bacterial population contains a larger percentage of antibiotic-resistant bacteria than before.

Poor prescribing practices by doctors and poor patient compliance with treatment are factors in the development of drug-resistant strains of the TB bacterium. A person infected with TB must take three to ten pills of antibiotics each day for at least six months. After the first month of treatment, the person usually feels better; many patients decide to quit taking their medication at this point. When this happens, the TB bacteria still lurking in their bodies—those with a resistance to the prescribed antibiotic—rally. The evolution of a strain of bacteria resistant to several drugs is a worst-case scenario.

Process of Science: Evolutionary hypotheses can be tested experimentally

Increasingly, biologists are designing imaginative experiments, often in natural settings, to test evolutionary hypotheses. David Reznick from the University of California at Santa Barbara and John Endler from James Cook University in Australia have studied evolution in guppy populations in Venezuela and Trinidad, a small island in the southern Caribbean.

Reznick and Endler observed that different streams have different kinds and numbers of fishes that prey on guppies. Predatory fish that prey on larger guppies are present at lower elevations in all streams; these areas of intense predation pressure are known as *high predation habitats*. Predators are often excluded from tributaries or upstream areas by rapids and waterfalls. The areas above such barriers are known as *low predation habitats* because they contain only one species of small predatory fish that occasionally eat smaller guppies.

Differences in predation are correlated with many differences in the guppies, such as male coloration, behavior, and attributes known as *life history traits,* including age and size at sexual maturity, the number of offspring per litter, the size of the offspring, and the frequency of reproduction. For example, guppy adults are larger in streams at higher elevations and smaller in streams at lower elevations. Reznick and colleagues have studied these life history traits and considered the role that predators in the two habitats might have played in the evolution of differences in life histories.

Reznick and his colleagues paid particular attention to the differences in guppy size, hypothesizing that these differences are related to predator preferences. However, they first had to rule out the possibility that some unknown environmental factor was responsible for the size differences. To determine this, they captured adult female guppies from high and low predation habitats in Trinidad and sent them to their laboratory. They exploited a convenient property of guppy reproductive biology—adult females are virtually always pregnant. Because these females store sperm, each female was able to produce a series of litters when she was isolated in the laboratory. The offspring of each female were then mated to produce a second generation. All the guppies were reared in identical predator free laboratory environments.

The second generation of guppies descended from females from low predation habitats grew into larger, later maturing adults. Conversely, the second generation from high predation habitats grew into smaller adults that matured at an earlier age. It is assumed that any differences among guppy populations that persist after two generations in identical laboratory environments have a genetic basis. This experiment therefore demonstrated that differences in size between the two populations were inherited; that is, they were not caused by some nongenetic factor, such as the availability of food.

Genetic differences between the two guppy populations are of particular interest because ecologists had predicted before this study that higher mortality rates found in high predation habitats would select for earlier maturity and smaller size at maturity. These predicted life history differences are consistent with the genetic differences observed among these populations of guppies.

Do predators actually cause these differences to evolve? Reznick and colleagues tested this evolutionary hypothesis by conducting field experiments in Trinidad. Taking advantage of waterfalls that prevent upstream movement of guppies, guppy predators, or both, they moved either guppies or guppy predators over such barriers. For example, guppies from a high predation habitat were introduced into a low predation habitat by moving them over a barrier waterfall into a section of stream that was free of guppies and large predators. The only fish species that lived in this section of stream prior to the introduction was the predator that occasionally preyed on small guppies.

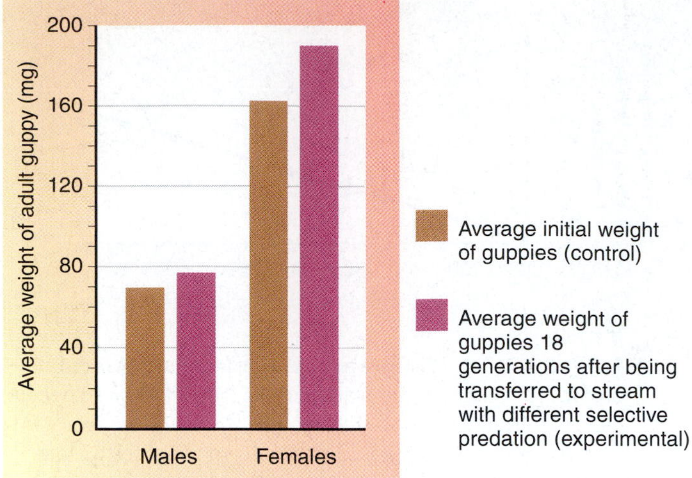

Figure 17–20 Experimental evidence of natural selection in guppies. Male and female guppies from a stream in which the predators preferred large adult guppies as prey *(brown bars)* were transferred to a stream in which the predators preferred juveniles and small adults. After 11 years, the descendants of these guppies *(pink bars)* were measurably larger in size, compared with their ancestors. *(Data from Reznick, D.N., et al. "Evaluation of the Rate of Evolution in Natural Populations of Guppies* (Poecilia reticulata)." Science, *Vol. 275, 28 Mar. 1997)*

Eleven years later, they captured adult females from the introduction site (low predation habitat) and the control site below the barrier waterfall (high predation habitat). They bred these females in their laboratory and compared the life history traits of succeeding generations, as in the earlier study. The descendants of guppies introduced into the low predation habitat matured at an older age and larger size than the descendants of guppies collected from the control site below the waterfall (Fig. 17–20). They also produced fewer, but larger, offspring. The life histories of the introduced fish had therefore evolved to be similar to those of fish that are typically found in such low predation habitats. These and other studies have also demonstrated that the predators have played an active role in the evolution of other traits, such as average number of offspring produced during the lifetime of an individual female (fecundity), male coloration, and behavior.

These and other experiments demonstrate that evolution is not only real but that it is occurring now, driven by selective environmental forces, such as predation, that can be experimentally manipulated. Darwin incorrectly assumed evolution is so gradual that humans cannot observe it. As Jonathan Weiner, author of *The Beak of the Finch: A Story of Evolution in Our Time,* puts it, "Darwin did not know the strength of his own theory. He vastly underestimated the power of natural selection. Its action is neither rare nor slow. It leads to evolution daily and hourly, all around us, and we can watch."

SUMMARY WITH KEY TERMS

I. **Evolution,** genetic change in a **population** of organisms over time, is the unifying concept of biology.
   A. The concept of evolution is the cornerstone of biology because it links all fields of the life sciences into a unified body of knowledge.
   B. Evolution has important practical applications in fields such as agriculture, medicine, conservation biology, **bioremediation,** and computers.

II. Charles Darwin's voyage on the H.M.S. *Beagle* was the basis for his theory of evolution.
   A. Darwin tried to explain his observations of the similarities between animals and plants of the arid Galapagos Islands and the humid South American mainland.
   B. Darwin was influenced by **artificial selection,** in which breeders develop many varieties of domesticated plants and animals in just a few generations.
   C. Darwin applied Thomas Malthus's ideas on the natural increase in human populations to natural populations.
   D. The idea that Earth was extremely old, which was promoted by Charles Lyell and other geologists, influenced Darwin.

III. Charles Darwin and Alfred Wallace independently proposed the theory of evolution by **natural selection,** which is based on four observations.
   A. *Variation:* Genetic variation exists among the individuals in a population.
   B. *Overproduction:* The reproductive ability of each species causes its populations to geometrically increase in number over time.
   C. *Limits on population growth:* Organisms compete with one another for the resources needed for life, such as food, living space, water, and light.
   D. *Differential reproductive success:* The offspring with the most favorable combination of characteristics are most likely to survive and reproduce, passing those genetic characteristics to the next generation. Thus, natural selection results in **adaptations,** which are evolutionary modifications that improve the chances of survival and reproductive success in a particular environment.
   E. Over time, enough changes may accumulate in geographically separated populations (often with slightly different environments) to produce new species.

IV. The **synthetic theory of evolution,** or **neo-Darwinism,** combines Darwin's theory of evolution by natural selection with modern genetics to explain how species adapt to their environment.
   A. **Mutation** provides the genetic variability that natural selection acts on during evolution.
   B. The synthetic theory of evolution emphasizes the genetics of populations rather than of individuals.

V. The concept that evolution has occurred and is occurring is now well documented.
   A. Direct evidence of evolution comes from **fossils,** the remains or traces of ancient organisms.
      1. Layers of sedimentary rock normally occur in their sequence of deposition, with the more recent layers on top of the older, earlier ones.
      2. **Index fossils** characterize a specific layer over large geographical areas.
      3. **Radioisotopes** present in a rock provide a way to accurately measure the rock's age.
   B. Evidence supporting evolution is derived from comparative anatomy.
      1. **Homologous features** have basic structural similarities, even though the structures may be used in different ways, because homologous features derive from the same structure in a common ancestor. Homologous features indicate evolutionary affinities among the organisms possessing them.
      2. **Homoplastic features** have similar functions in quite different, distantly related organisms. Homoplastic features demonstrate **convergent evolution,** in which organisms with separate ancestries adapt in similar ways to comparable environmental demands.
      3. **Vestigial structures** are nonfunctional or degenerate remnants of structures that were present and functional in ancestral organisms. Structures occasionally become vestigial as species adapt to different modes of life.
   C. **Biogeography,** the distribution of plants and animals, supports evolution.
      1. Areas that have been separated from the rest of the world for a long time have organisms that are unique to those areas.
      2. At one time the continents were joined to form a supercontinent. **Continental drift,** which caused the various land masses to break apart and separate, has played a major role in evolution.
   D. Developmental biology provides evidence of evolution.
      1. Evolutionary changes are often the result of mutations in genes that affect the orderly sequence of events during development.
      2. Scientific evidence demonstrates that development in different animals is controlled by the same kinds of genes, which indicates that these animals have a shared evolutionary history.
      3. The accumulation of genetic changes since organisms diverged, or took separate evolutionary pathways, has modified the pattern of development in more complex vertebrate embryos.
   E. Molecular biology provides evidence of evolution.
      1. The universality of the genetic code is compelling evidence that all life is related.
      2. The sequence of amino acids in common proteins reveals greater similarities in closely related species.
      3. **DNA sequencing** determines the order of nucleotide bases in DNA. A greater proportion of the nucleotide sequence in DNA is identical in closely related organisms.
      4. A **phylogenetic tree,** a diagram showing lines of descent, can be derived from differences in a given DNA nucleotide sequence in various organisms.
      5. A **molecular clock** estimates the time of divergence between two closely related species or higher taxonomic groups.
   F. Bacteria that cause infectious diseases such as tuberculosis are evolving resistance to antibiotics.
   G. Experimental data provide evidence of evolution.
      1. Reznick has studied the effects of predation intensity on the evolution of guppy populations in both the laboratory and nature.
      2. Such experiments are a powerful way for investigators to test the underlying processes of natural selection.

POST-TEST

1. Evolution is based on which of the following concepts? (a) organisms share a common origin (b) over time, organisms have diverged from a common ancestor (c) an animal's body parts can change over its lifetime, and these acquired changes can be passed to the next generation (d) a and b are correct (e) a, b, and c are correct
2. Evolution is the accumulation of genetic changes within ____________ over time. (a) individuals (b) populations (c) communities (d) a and b (e) a and c
3. Charles Darwin proposed that evolution could be explained by the differential reproductive success of organisms that resulted from their naturally occurring variation. Darwin called this process (a) coevolution (b) convergent evolution (c) natural selection (d) artificial selection (e) homoplasy
4. Which of the following statements is *false?* (a) Darwin was the first to supply convincing evidence for biological evolution (b) Darwin was the first to propose that organisms change over time (c) Wallace independently developed the same theory as Darwin (d) Darwin's theory is based on four observations about the natural world (e) Darwin's studies in the Galapagos strongly influenced his ideas about evolution
5. Which of the following is *not* part of Darwin's mechanism of evolution? (a) differential reproductive success (b) variation in a population (c) inheritance of acquired (nongenetic) traits (d) overproduction of offspring (e) struggle for existence
6. The synthetic theory of evolution (a) is based on the sequence of fossils in rock layers (b) uses genetics to explain the source of hereditary variation that is essential to natural selection (c) was first proposed by ancient Greek scholars (d) considers the influence of the geographical distribution of organisms on their evolution (e) is reinforced by homologies that are explained by common descent
7. Jewish and Muslim men have been circumcised for many generations, yet this practice has had no effect on the penile foreskin of their offspring. This observation disproves evolution as envisioned by (a) Lamarck (b) Darwin (c) Wallace (d) Lyell (e) Malthus
8. Which of the following is *least* likely to have occurred after a small population of finches reached the Galapagos Islands from the South American mainland? (a) after many generations, the finches became increasingly different from the original population (b) over time, the finches adapted to their new environment (c) after many generations, the finches were unchanged and unmodified in any way (d) the finches were unable to survive in their new home and died out
9. The fossil record (a) usually occurs in sedimentary rock (b) sometimes appears fragmentary (c) is relatively complete for tropical rainforest organisms but incomplete for aquatic organisms (d) a and b are correct (e) a, b, and c are correct
10. The molecular record found inside cells suggests that evolutionary changes are caused by an accumulation of (a) traits acquired through need (b) alterations in the order of nucleotides in DNA (c) characters acquired during an individual's lifetime (d) hormones (e) environmental changes
11. In ____________, the selecting agent is the environment, whereas in ____________, the selecting agent is humans. (a) natural selection; convergent evolution (b) mutation; artificial selection (c) homoplasy; homology (d) artificial selection; natural selection (e) natural selection; artificial selection
12. Structures that are similar in underlying form in different species due to a common evolutionary origin are called (a) homoplastic (b) homologous (c) vestigial (d) convergent (e) synthetic
13. Aardvarks, anteaters, and pangolins are only distantly related but are similar in structure and form as a result of (a) homology (b) convergent evolution (c) biogeography (d) vestigial structures (e) artificial selection
14. The species of the Galapagos Islands (a) are similar to those on the Cape Verde Islands (b) are similar to those on the South American mainland (c) are identical to those on the Cape Verde Islands (d) are identical to those on the South American mainland (e) are similar to those on both the African and South American mainlands
15. In the guppy experiments, Reznick determined that (a) different streams have the same kinds of fishes that prey on guppies (b) guppy adults are larger in streams at lower elevations (c) differences in guppy size are related to availability of food (d) guppies living in an environment with large predators tend to be smaller in size (e) fish that prey on guppies exert no influence on guppy evolution

REVIEW QUESTIONS

1. Explain briefly the concept of evolution by natural selection.
2. Why are only inherited variations important in the evolutionary process?
3. What part of Darwin's theory was he unable to explain? How does the synthetic theory of evolution fill this gap?
4. How do scientists date fossils? How do fossils provide evidence of evolution?
5. Distinguish among homologous features, homoplastic features, and vestigial structures. How does each provide evidence of evolution?
6. How does continental drift occur?
7. Explain why fossils of *Mesosaurus,* an extinct reptile that could not swim across open water, are found in the southern parts of both Africa and South America.
8. How does developmental biology provide evidence of a common ancestry for vertebrates as diverse as reptiles, birds, pigs, and humans?
9. Explain why there are many species of marsupials in Australia and only a few marsupials elsewhere.
10. What is indicated if the DNA from two species is found to be almost identical?
11. Explain how predator preference drives the evolution of size in guppies.

YOU MAKE THE CONNECTION

1. The use of model organisms, such as the laboratory mouse, for biomedical testing and research is based on the assumption that all organisms share a common ancestor. On what evidence is this assumption based?
2. What adaptations must an animal possess to swim in the ocean? Why are such genetically different organisms as porpoises, which are mammals, and sharks, which are fish, so similar in form?
3. The human fetus grows a coat of fine hair, the lanugo, that is shed before or shortly after birth. Fetuses of chimpanzee and other primates also grow coats of hair, but they are not shed. Explain these observations based on what you have learned in this chapter.
4. Write short paragraphs explaining each of the following statements:
   a. Natural selection chooses from among the individuals in a population those most suited to *current* environmental conditions. It does not guarantee survival under future conditions.
   b. Individuals do not evolve, but populations do.
   c. The organisms that exist today do so because their ancestors possessed traits that allowed them and their offspring to thrive.
   d. At the molecular level, evolution can take place by the replacement of one nucleotide by another.
   e. Evolution is said to have occurred within a population when measurable genetic changes are detected.
5. Although most salamanders have four legs, a few species that live in shallow water lack hind limbs and have extremely tiny forelimbs *(see photograph)*. Develop a hypothesis to explain how limbless salamanders came about according to Darwin's mechanism of evolution by natural selection. How could you test your hypothesis?

The narrow-striped dwarf siren *(Pseudobranchus striatus axanthus).* This aquatic salamander, which is native to Florida, resembles an eel. *(Suzanne L. Collins and Joseph T. Collins/Photo Researchers, Inc.)*

RECOMMENDED READINGS

Diamond, J. "Evolving Backward." *Discover,* Sep. 1998. An examination of vestigial structures that have been lost or reduced in the course of evolution.

Gould, S.J. "A Division of Worms." *Natural History,* Feb. 1999. This eminent paleontologist examines Jean-Baptiste Lamarck's considerable contributions to anatomy and taxonomy.

Grenard, S. "Is Rattlesnake Venom Evolving?" *Natural History,* Jul./Aug. 2000. Evidence suggests that the venom of North America's rattlesnakes is becoming more potent, possibly in response to prey that are evolving resistance to rattlesnake venom.

Huey, R.B., G.W. Gilchrist, M.L. Carlson, D. Berrigan, and L. Serra. "Rapid Evolution of a Geographic Cline in Size in an Introduced Fly." *Science,* Vol. 287, 14 Jan. 2000. This paper describes the study of rapid fruit fly evolution that was discussed in the chapter.

Mayr, E. "Darwin's Influence on Modern Thought." *Scientific American,* Vol. 283, No. 1, Jul. 2000. The author, one of the 20th century's most influential evolutionary biologists, considers the wide reach of Darwin's ideas, which permeate modern thinking on many subjects.

McComas, W.F. "The Discovery and Nature of Evolution by Natural Selection: Misconceptions and Lessons from the History of Science." *The American Biology Teacher,* Vol. 59, No. 8, Oct. 1997. The author provides some fascinating insights into Darwin's development of his theory of evolution by natural selection.

Monastersky, R. "The Whale's Tale." *Science News,* Vol. 156, 6 Nov. 1999. Although much is known about how whales evolved from life on land to life in the water, a few important questions remain to be answered.

Nesse, R.M., and G.C. Williams. "Evolution and the Origins of Disease." *Scientific American,* Vol. 279, No. 5, Nov. 1998. Increasingly, physicians are considering health, disease, and modern medical practices in an evolutionary context.

Pennisi, E., and W. Roush. "Developing a New View of Evolution." *Science,* Vol. 277, 4 Jul. 1997. How developmental biologists are illuminating some of the mysteries of evolution by studying the genes that control embryonic development.

Porter, D.M., and P.W. Graham. *The Portable Darwin.* Penguin Books, New York, 1993. This collection of Darwin's writings reveals the diverse interests of the man who profoundly changed the intellectual climate of the 19th and 20th centuries.

Science and Creationism: A View from the National Academy of Sciences, 2nd ed. National Academy Press, Washington, D.C., 1999. A great introduction to the evidence for evolution.

Weiner, J. *The Beak of the Finch: A Story of Evolution in Our Time.* Knopf, New York, 1994. This Pulitzer Prize-winning book focuses on the research of Peter and Rosemary Grant on evolution in the Galapagos finches.

Zimmer, C. "Hidden Unity." *Discover,* Jan. 1998. Discusses the similarities of developmental genes in a wide variety of organisms. Several very similar genes, for example, control the development of eyes in animals as diverse as humans and insects.

- Visit our Web site at **http://www.info.brookscole.com/solomonbergmartin** for links to chapter-related resources on the World Wide Web. Additional on-line materials relating to this chapter can also be found on our Web site.

See chapter activity on BioActive Learner CD for additional help in mastering the chapter's material. Icon location in the chapter's margins shows which topics have tutorials or simulations in the CD.

18

Evolutionary Change in Populations

LEARNING OBJECTIVES

After you have studied this chapter you should be able to

1. Define population, population genetics, genetic equilibrium, and microevolution.
2. Distinguish among genotype, phenotype, and allele frequencies.
3. Discuss the significance of the Hardy-Weinberg principle as it relates to evolution, and list the five conditions required for genetic equilibrium.
4. Use the Hardy-Weinberg principle to solve problems involving populations.
5. Discuss how each of the following alters allele frequencies in populations: nonrandom mating, mutation, genetic drift, gene flow, and natural selection.
6. Distinguish among stabilizing selection, directional selection, and disruptive selection, and give an example of each.
7. Describe the nature and extent of genetic variation, including genetic polymorphism, balanced polymorphism, neutral variation, and geographical variation.
8. Explain how the sickle cell allele illustrates heterozygote advantage.
9. Relate how frequency-dependent selection affects genetic variation.

Genetic variation in snail shells. Shown are the shell patterns and colors in a single snail species *(Cepaea nemoralis)*, native to Scotland. Variation in shell color may have adaptive value in these snails because some colors predominate in cooler environments, whereas other colors are more common in warmer habitats. *(G.I. Bernard/Animals Animals)*

As we saw in Chapter 17, evolution occurs in populations, not individuals. Although natural selection results from differential survival and reproduction of individuals, individuals do not evolve during their lifetimes. Evolutionary change, which includes modifications in structure, physiology, ecology, and behavior, is inherited from one generation to the next. Although Darwin recognized that evolution occurs in populations, he did not understand how traits are passed to successive generations. One of the most significant advances in biology since Darwin's time has been the demonstration of the genetic basis of evolution. As you will see in this chapter, Gregor Mendel's principles of inheritance (see Chapter 10) underlie Darwinian evolution.

Recall from Chapter 17 that a **population** consists of all the individuals of the same species that live in a particular place at the same time. Individuals within a population vary in many recognizable traits. A population of snails, for example, may vary in shell size, weight, or color *(see photograph)*. Some of this variation is due to the environment, and some is due to heredity.

Biologists study variation in a particular trait by taking measurements of that trait in a population. By comparing the trait in parents and offspring, it is possible to estimate the amount of observed variation that is genetic, as represented by the number, frequency, and kinds of alleles in a population. (Recall from Chapter 10 that an **allele** is one of two or more alternate forms of a gene. Alleles occupy corresponding positions, or **loci,** on homologous chromosomes.)

In this chapter we present some basic concepts of **population genetics,** the study of genetic variability within a population and of the forces that act on it. Population genetics represents an extension of Mendelian inheritance. We contrast genetic equilibrium with evolutionary change and discuss the five factors responsible for evolutionary change: nonrandom mating, mutation, genetic drift, gene flow, and natural selection. We then consider genetic variation as the raw material for evolution.

GENOTYPE, PHENOTYPE, AND ALLELE FREQUENCIES CAN BE CALCULATED

Each population possesses a **gene pool,** which includes all the alleles for all the genes present in the population. Because diploid organisms possess a maximum of two different alleles at each genetic locus, a single individual typically has only a small fraction of the alleles present in a population's gene pool. The genetic variation that is evident among individuals in a given population indicates that each individual has a different combination of the alleles in the gene pool.

The evolution of populations is best understood in terms of genotype, phenotype, and allele frequencies. Suppose, for example, that all 1000 individuals of a hypothetical population have their genotypes tested, with the following results:

Genotype	Number	Genotype Frequency
AA	490	0.49
Aa	420	0.42
aa	90	0.09
Total	1000	1.00

Each **genotype frequency** is the proportion of a particular genotype in the population. Genotype frequency is usually expressed as a decimal fraction, and the sum of all genotype frequencies is 1.0 (somewhat like probabilities, which were discussed in Chapter 10). For example, the genotype frequency for the *Aa* genotype is $420 \div 1000 = 0.42$.

A **phenotype frequency** is the proportion of a particular phenotype in the population. If each genotype corresponds to a specific phenotype, then the phenotype and genotype frequencies are the same. If allele *A* is dominant over allele *a,* however, the phenotype frequencies in our hypothetical population would be the following:

Phenotype	Number	Phenotype Frequency
Dominant	910	0.91
Recessive	90	0.09
Total	1000	1.00

(In this example, the dominant phenotype is the sum of two genotypes, *AA* and *Aa,* and so the number 910 is obtained by adding 490 + 420.)

An **allele frequency** is the proportion of a specific allele (that is, of *A* or *a*) in a particular population. As mentioned earlier, each individual, being diploid, has two alleles at each genetic locus. Since we started with a population of 1000 individuals, we must account for a total of 2000 alleles. The 490 *AA* individuals have 980 *A* alleles, whereas the 420 *Aa* individuals have 420 *A* alleles, making a total of 1400 *A* alleles in the population. The total number of *a* alleles in the population is $420 + 90 + 90 = 600$. Now it is easy to calculate allele frequencies:

Allele	Number	Allele Frequency
A	1400	0.7
a	600	0.3
Total	2000	1.0

THE HARDY-WEINBERG PRINCIPLE DESCRIBES GENETIC EQUILIBRIUM

In the example just discussed, we observe that only 90 of the 1000 individuals in the population exhibit the recessive phenotype characteristic of the genotype *aa.* The remaining 910 individuals exhibit the dominant phenotype and are either *AA* or *Aa.* You might assume that, after many generations, genetic recombination during sexual reproduction would cause the dominant allele to become more common in the population. You might also assume that the recessive allele would eventually disappear altogether. These were common assumptions of many biologists early in the 20th century. However, these assumptions were incorrect because the frequencies of alleles and genotypes do not change from generation to generation unless influenced by outside factors (discussed later).

A population whose allele and genotype frequencies do not change from generation to generation is said to be at **genetic equilibrium.** Such a population, with no net change in allele or genotype frequencies over time, is not undergoing evolutionary change. However, evolution is occurring in a population in which there are changes in allele frequencies over successive generations.

The explanation for the stability of successive generations in populations at genetic equilibrium was provided independently by Godfrey Hardy, an English mathematician, and Wilhelm Weinberg, a German physician, in 1908. They pointed out that the expected frequencies of various genotypes in a population can be described mathematically. The resulting **Hardy-Weinberg principle** shows that in large populations the process of inheritance does not by itself cause changes in allele frequencies. It also explains why dominant alleles are not necessarily more common than recessive ones. The Hardy-Weinberg principle represents an ideal situation that seldom occurs in the natural world. However, it is useful because it provides a model to help us understand the real world. Knowledge of the Hardy-Weinberg principle is essential to understanding the mechanisms of evolutionary change in sexually reproducing populations.

We now expand our original example to illustrate the Hardy-Weinberg principle. Keep in mind as we go through these calculations, that in most cases, we only know the phenotype frequencies. (When alleles are dominant and recessive, it is usually impossible to visually distinguish heterozygous individuals from homozygous dominant individuals.) The Hardy-Weinberg principle allows us to use phenotype frequencies to calculate the expected genotype frequencies and allele frequencies, assuming we have a clear understanding of the genetic basis for the trait under study.

As mentioned earlier, the frequency of either allele, *A* or *a*, is represented by a number that ranges from zero to one. An allele that is totally absent from the population has a frequency of zero. If all the alleles of a given locus are the same in the population, then the frequency of that allele is one.

Because only two alleles, *A* and *a*, exist at the locus in our example, the sum of their frequencies must equal one. If we let p represent the frequency of the dominant *(A)* allele in the population, and q the frequency of the recessive *(a)* allele, then we can summarize their relationship with a simple binomial equation, $p + q = 1$. When we know the value of either p or q, we can calculate the value of the other: $p = 1 - q$ and $q = 1 - p$.

Squaring both sides of $p + q = 1$ results in $(p + q)^2 = 1$. This equation can be expanded to describe the relationship of the allele frequencies to the genotypes in the population. When it is expanded, we obtain the frequency of the offspring genotypes:

p^2	+	$2pq$	+	q^2	=	1
Frequency of *AA*		Frequency of *Aa*		Frequency of *aa*		All the individuals in a population

We always begin Hardy-Weinberg calculations by determining the frequency of the homozygous recessive genotype. From the fact that we had 90 homozygous recessive individuals in our population of 1000, we infer that the frequency of the *aa* genotype, q^2, is 90/1000, or 0.09. Since q^2 equals 0.09, q (the frequency of the recessive *a* allele)is equal to the square root of 0.09, or 0.3. From the relationship between p and q, we conclude that the frequency of the dominant *A* allele, p, equals $1 - q = 1 - 0.3 = 0.7$.

Based on this information, we can calculate the frequency of homozygous dominant *(AA)* individuals: $p^2 = 0.7 \times 0.7 = 0.49$ (Fig. 18–1). The frequency of heterozygous individuals *(Aa)*, would be: $2pq = 2 \times 0.7 \times 0.3 = 0.42$. Thus, approximately 490 individuals are expected to be homozygous dominant, and 420 are expected to be heterozygous. Note that the sum of homozygous dominant and heterozygous individuals equals 910, the number of individuals with the dominant phenotype that we started with.

Any population in which the distribution of genotypes conforms to the relation $p^2 + 2pq + q^2 = 1$, whatever the absolute values for p and q may be, is at genetic equilibrium. The Hardy-Weinberg principle allows biologists to calculate allele frequencies in a given population if we know the genotype frequencies, and vice versa. These values can be used as a basis of comparison with a population's allele or genotype frequencies in succeeding generations. During that time, if the allele or genotype frequencies deviate from the values predicted by the Hardy-Weinberg principle, then the population is evolving. (You can test your understanding of the Hardy-Weinberg principle by solving the problems in the Review Questions at the end of this chapter.)

Genetic equilibrium occurs if certain conditions are met

The Hardy-Weinberg principle of genetic equilibrium tells us what to expect when a sexually reproducing population is not evolving. The relative proportions of alleles and genotypes in

Genotypes	*AA*	*Aa*	*aa*
Frequency of genotypes in population	0.49	0.42 (0.21 + 0.21)	0.09
Frequency of alleles in gametes	*A* = 0.49 + 0.21 = 0.7		*a* = 0.21 + 0.09 = 0.3

(a) Genotype and allele frequencies

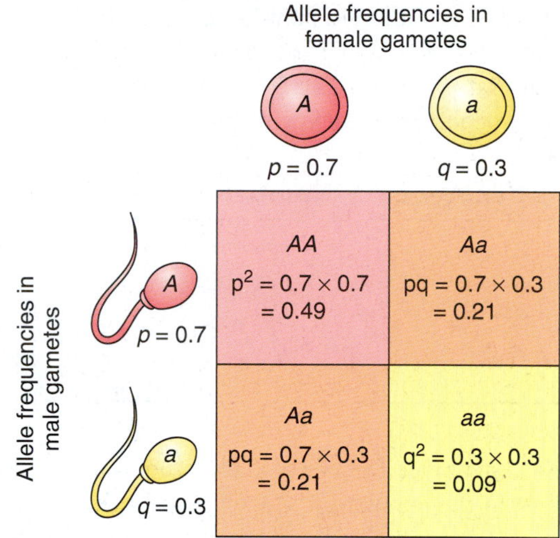

(b) Segregation of alleles and random fertilization

Figure 18–1 Hardy-Weinberg principle **(a)** How to calculate frequencies of the alleles *A* and *a* in the gametes. **(b)** When eggs and sperm containing *A* or *a* alleles unite randomly, the frequency of each of the possible genotypes *(AA, Aa, aa)* among the offspring is calculated by multiplying the frequencies of the alleles *A* and *a* in eggs and sperm.

successive generations will always be the same, provided the following five conditions are met:

1. **Random mating.** In unrestricted random mating, each individual in a population has an equal chance of mating with any individual of the opposite sex. In our example, the individuals represented by the genotypes *AA, Aa,* and *aa* must mate with one another at random and must not select their mates on the basis of genotype or any other factors that result in nonrandom mating.
2. **No net mutations.** There must be no mutations that convert *A* into *a* or vice versa. That is, the frequencies of *A* and *a* in the population must not change due to mutations.
3. **Large population size.** Allele frequencies in a small population are more likely to be changed by random fluctuations (i.e., by genetic drift, which is discussed later) than are allele frequencies in a large population.
4. **No migration.** There can be no exchange of genes with other populations that might have different allele frequencies. In

other words, there can be no migration of individuals into or out of a population.

5. **No natural selection.** If natural selection is occurring, certain phenotypes (and their corresponding genotypes) are favored over others. Consequently, the allele frequencies will change, and the population will evolve.

Human MN blood groups are a valuable illustration of the Hardy-Weinberg principle

Humans possess dozens of antigens on the surfaces of their blood cells. (An *antigen* is a molecule, usually a protein or carbohydrate, that can be recognized as foreign by cells of another organism's immune system.) One group of antigens, designated the MN blood group, stimulates the production of antibodies when injected into rabbits or guinea pigs. However, humans do not produce antibodies for M and N, so the MN blood group is not medically important, for example, when giving blood transfusions. (Recall the discussion of the medically important ABO alleles in Chapter 10.) The MN blood group is of interest to population geneticists, however, because the genotype frequencies can be observed directly (the alleles for the MN blood group, usually designated *M* and *N*, are codominant) and compared with calculated frequencies.

Genotype	Phenotype (Antigen on RBC)
MM	Antigen M only
MN	Antigens M and N
NN	Antigen N only

The following data are typical of the MN blood group in people in the United States:

Genotype	Observed
MM	320
MN	480
NN	200
Total	1000

Because there are 1000 diploid individuals in the sample, there are a total of 2000 alleles. The frequency of *M* alleles in the population $= p = (2 \times 320 + 480) \div 2000 = 0.56$. The frequency of *N* alleles in the population $= q = (2 \times 200 + 480) \div 2000 = 0.44$. As a quick check, the sum of the frequencies should equal one. Does it?

If this population is in genetic equilibrium, then the expected *MM* genotype frequency $= p^2 = (0.56)^2 = 0.31$. The expected *MN* genotype frequency $= 2pq = 2 \times 0.56 \times 0.44 = 0.49$. The expected *NN* genotype frequency $= q^2 = (0.44)^2 = 0.19$. As a quick check, the sum of the three genotype frequencies should equal 1. Does it?

You can use the calculated genotype frequencies to determine how many individuals in a population of 1000 should have the expected genotype frequencies. By comparing the expected numbers with the actual results observed, you can see how closely the population is to genetic equilibrium. Simply multiply each genotype frequency by 1000:

Genotype	Observed	Expected
MM	320	313.6
MN	480	492.8
NN	200	193.6
Total	1000	1000

The expected numbers closely match the observed numbers, indicating that the *MN* blood groups in the human population are almost at genetic equilibrium. This is not surprising, given that the *MN* characteristic has no medical significance and does not produce a visible trait that might affect random mating.

MICROEVOLUTION OCCURS WHEN A POPULATION'S ALLELE OR GENOTYPE FREQUENCIES CHANGE

Evolution represents a departure from the Hardy-Weinberg principle of genetic equilibrium. The degree of departure between the observed allele or genotype frequencies and those expected by the Hardy-Weinberg principle indicates the amount of evolutionary change. This type of evolution—generation-to-generation changes in allele or genotype frequencies *within* a population—is sometimes referred to as **microevolution** because it often involves relatively small or minor changes, usually over a few generations. Changes in the allele frequencies of a population result from five microevolutionary processes: nonrandom mating, mutation, genetic drift, gene flow, and natural selection. When one or more of these processes is acting on a population, allele or genotype frequencies will change from one generation to the next.

NONRANDOM MATING CHANGES GENOTYPE FREQUENCIES

When individuals select mates on the basis of phenotype (thereby selecting the corresponding genotype), they can bring about evolutionary change in the population. Two examples of nonrandom mating are inbreeding and assortative mating.

In many populations, individuals mate more often with close neighbors than with more distant members of the population. As a result, neighbors tend to be more closely related—that is, genetically similar—to one another. The mating of genetically similar individuals that are more closely related than if they had been chosen at random from the entire population is known as **inbreeding.** Although inbreeding does not change the overall allele frequency, the frequency of homozygous genotypes increases with each successive generation of inbreeding. The most extreme example of inbreeding is self-fertilization, which is particularly common in plants.

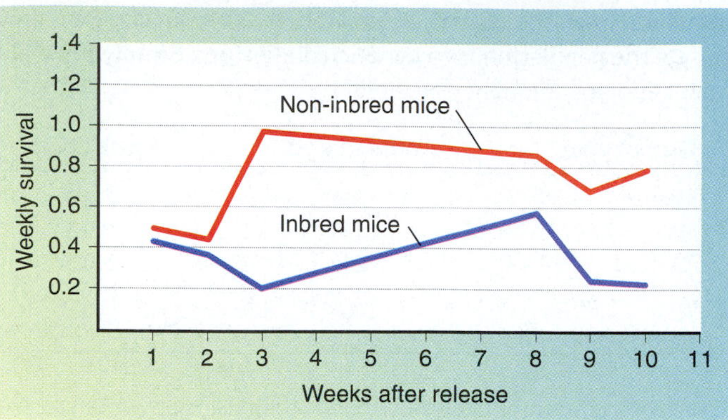

Figure 18–2 Survival of inbred and non-inbred mice. The mouse population was sampled six times (each for a three-day span) during a 10-week period. Non-inbred mice *(red)* had a higher survival rate than inbred mice *(blue)*. Values on the Y-axis are the estimated proportion of mice that survived from one week to the next. Hence, a value of 0.6 means that 60% of the mice alive at the beginning of the week survived through that week. *(Adapted from Jiménez, J.A., et al. "An Experimental Study of Inbreeding Depression in a Natural Habitat."* Science, *Vol. 266, 14 Oct. 1994)*

Inbreeding does not appear to be detrimental in some populations, but in others it causes **inbreeding depression,** in which inbred individuals have lower fitness than non-inbred individuals. **Fitness** is the relative ability of a given genotype to make a genetic contribution to subsequent generations; fitness is usually measured as the average number of surviving offspring of one genotype compared to the average number of surviving offspring of competing genotypes. Inbreeding depression, as evidenced by fertility declines and high juvenile mortality, is thought to be caused by the expression of harmful recessive alleles as homozygosity increases with inbreeding.

Several studies in the 1990s provided direct evidence of the deleterious consequences of inbreeding in nature. For example, white-footed mice *(Peromyscus leucopus)* were taken from a field and used to develop both inbred and non-inbred populations in the laboratory. When these laboratory-bred populations were returned to nature, their survivorship was estimated from release-recapture data. The non-inbred mice had a statistically significant higher rate of survival (Fig. 18–2). It is not known why the inbred mice had a lower survival rate. Some possibilities include higher disease susceptibility, poorer ability to evade predators, lesser ability to find food, and lesser ability to win fights with other white-footed mice.

Assortative mating, in which individuals select mates by their phenotypes, is another example of nonrandom mating. For example, biologists selected two phenotypes—high bristle number and low bristle number—in a fruit fly *(Drosophila melanogaster)* population. Although they made no effort to control mating, they observed that the flies preferentially mated with those of similar phenotypes. Females with high bristle number tended to mate with males with high bristle number, and females with low bristle number tended to mate with males with low bristle number. Such selection of mates with the same phenotype is known as *positive assortative mating* (as opposed to the less common phenomenon, *negative assortative mating,* in which mates with opposite phenotypes are selected). Positive assortative mating is practiced in many human societies, in which men and women tend to marry individuals like themselves in such characteristics as height or intelligence. Like inbreeding, assortative mating usually increases homozygosity at the expense of heterozygosity in the population and does not change the overall allele frequencies in the population. However, assortative mating changes genotype frequencies only at the loci involved in mate choice, whereas inbreeding affects genotype frequencies in the entire genome.

MUTATION INCREASES VARIATION WITHIN A POPULATION

Variation is introduced into a population through **mutation,** which is an unpredictable change in deoxyribonucleic acid (DNA) (Fig. 18–3). Mutations, which are the source of all new alleles, can result from (1) a change in the nucleotide base pairs of a gene, (2) a rearrangement of genes within chromosomes so that their interactions produce different effects, or (3) a change in chromosome structure. Mutations occur unpredictably and spontaneously. The rate of mutation appears to be relatively con-

(a)

(b)

Figure 18–3 Fruit fly *(Drosophila melanogaster)* mutation. **(a)** A normal fly. **(b)** A mutant with vestigial wings. Because mutations are random changes in genetic material, most mutations are neutral or harmful to the organism. Yet for island-dwelling insects, fully developed wings might be more of a disadvantage than an advantage, permitting the insect to be too easily blown away from land. Perhaps for this reason, flies and other insects that live on small islands frequently have reduced wings or are entirely wingless. *(a, b, Peter J. Bryant/Biological Photo Service)*

stant for a particular gene but may vary by several orders of magnitude among genes within a single species and among different species.

Not all mutations pass from one generation to the next. Those occurring in somatic (body) cells are not inherited. When an individual with such a mutation dies, the mutation is lost. Some mutations, however, occur in reproductive cells. These mutations may or may not overtly affect the offspring, because most of the DNA in a cell is "silent" and does not code for specific polypeptides or proteins that are responsible for physical characteristics. Even if a mutation occurs in the DNA that codes for a polypeptide, it may still have little effect on the structure or function of that polypeptide (we discuss such *neutral variation* later in the chapter). However, when a polypeptide is sufficiently altered to change its function, the mutation is usually harmful. By acting against seriously abnormal phenotypes, natural selection eliminates or reduces to low frequencies the most harmful mutations. Mutations with small phenotypic effects, even if slightly harmful, have a better chance of being incorporated into the population, where at some later time, under different environmental conditions, they may produce traits that are useful or adaptive for the population.

Mutations do not determine the *direction* of evolutionary change. Consider a population living in an increasingly dry environment. A mutation producing a new allele that helps an individual adapt to dry conditions is no more likely to occur than one for adapting to wet conditions or one with no relationship to the changing environment. The production of new mutations simply increases the genetic variability that can be acted on by natural selection and, therefore, the potential for new adaptations.

Mutation by itself causes small deviations in allele frequencies from those predicted by the Hardy-Weinberg principle. Although allele frequencies may be changed by mutation, these changes are typically several orders of magnitude smaller than changes caused by other evolutionary forces, such as genetic drift. As an evolutionary force, mutation is usually negligible, but it is important as the ultimate source of variation for evolution.

IN GENETIC DRIFT, RANDOM EVENTS CHANGE ALLELE FREQUENCIES

The size of a population has important effects on allele frequencies because random events, or chance, will tend to cause changes of relatively greater magnitude in a small population. If a population consists of only a few individuals, an allele present at a low frequency in the population could be completely lost purely by chance. Such an event would be most unlikely in a large population. For example, consider two populations, one with 10,000 individuals and one with 10 individuals. If an uncommon allele occurs at a frequency of 10%, or 0.1, in both populations, then 1900 individuals in the large population possess the allele.[1] That same frequency, 0.1, in the smaller population means that only about two individuals possess the allele.[2] From this exercise, it is easy to see that there is a greater likelihood of losing the rare allele from the smaller population than from the larger one. Predators, for example, might happen to kill one or two individuals possessing the uncommon allele in the smaller population purely by chance, so that these individuals would leave no offspring.

The production of random evolutionary changes in small breeding populations is known as **genetic drift.** Genetic drift results in changes in allele frequencies in a population from one generation to another. One allele may be eliminated from the population purely by chance, regardless of whether that allele is beneficial, harmful, or of no particular advantage or disadvantage. Thus, genetic drift can decrease genetic variation *within* a population, although it tends to increase the genetic differences *among* different populations.

When bottlenecks occur, genetic drift becomes a major evolutionary force

Because of fluctuations in the environment, such as depletion in food supply or an outbreak of disease, a population may periodically experience a rapid and marked decrease in the number of individuals. The population is said to go through a **bottleneck** during which genetic drift can occur in the small population of survivors. As the population again increases in size, many allele frequencies may be quite different from those in the population preceding the decline.

Scientists hypothesize that genetic variation in the cheetah (see Figs. 50–9 and 50–19*c*) was considerably reduced by a bottleneck that occurred at the end of the last Ice Age, some 10,000 years ago. At that time, cheetahs nearly became extinct, perhaps due to overhunting by humans. The few surviving cheetahs possessed greatly reduced genetic variability, and as a result, the cheetah population today is nearly genetically uniform or homogeneous.

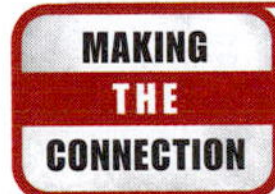

Are cheetahs endangered because of their extreme genetic uniformity or because of loss of habitat? The outlook for the cheetah's long-term survival is far from certain. There are currently about 12,000 cheetahs in sub-Saharan Africa and northern Iran.

For many years, lack of genetic diversity and inbreeding depression were considered the primary factors responsible for the cheetah's plight. Evidence of the extreme genetic uniformity in cheetahs was demonstrated in the 1980s when biologists discovered that unrelated cheetahs accepted skin grafts from one another. (Normally, only identical twins accept skin grafts so readily.) Geneticists hypothesized that inbreeding depression was responsible for problems with cheetahs in captive breeding programs. Compared with other large cats, cheetah males exhibit low sperm counts, many of their sperm have abnormalities, and

[1] $2pq + q^2 = 2(0.9)(0.1) + (0.1)^2 = 0.18 + 0.01 = 0.19; 0.19 \times 10{,}000 = 1900$

[2] $0.19 \times 10 = 1.9$

cheetah females do not bear as many offspring. Also, many cheetah offspring have health problems and are more susceptible to disease.

Notwithstanding the cheetah's genetic uniformity, ecologists who study cheetahs say that inbreeding depression is not impairing the cheetah's chance of breeding success. They have observed that the main cause of juvenile mortality is predation—lions and hyenas kill most young cheetahs—rather than the consequences of defective genes. Moreover, ecologists say that male cheetahs have no difficulty siring offspring in nature, despite their low sperm counts and high rates of sperm abnormalities.

Ecologists would like the focus on saving the cheetah to switch from genetic to environmental factors, that is, from captive breeding in zoos to what ecologists perceive as the real threat to the cheetah's survival—loss of habitat. Although geneticists agree that protection of habitat should be the main focus in attempts to save the cheetah (and all other endangered species) from extinction, they also emphasize that lack of genetic diversity should remain an important consideration in conservation strategies to save the cheetah.

Several studies reported in the late 1990s support the geneticists' view. For example, inbreeding and loss of genetic diversity were correlated with the extinction of local populations of butterflies (Glanville fritillaries) on Finnish islands. In addition, genetically uniform, declining populations of two different species—adders in Sweden and greater prairie chickens in Illinois (Fig. 18–4)—recovered when genetically diverse individuals from large populations were introduced into the genetically impoverished populations. Such experiments demonstrate that preserving genetic variability is an important way to increase the viability of wild populations.

Figure 18–4 Male greater prairie chicken *(Tympanuchus cupido)* in a courtship display. The greater prairie chicken's population plummeted in Illinois, from an estimated 100 million in 1900 to fewer than 50 in the mid-1990s. Studies comparing Illinois greater prairie chickens to greater prairie chickens living in nearby states showed the Illinois greater prairie chickens had lost much of their genetic diversity. This loss was substantiated by a comparison of DNA in Illinois greater prairie chickens with DNA in museum specimens collected in Illinois during the 1930s to 1960s. Fortunately, the Illinois population has recovered somewhat, after more than 500 greater prairie chickens from Minnesota, Kansas, and Nebraska (where greater prairie chicken populations are still thriving) were introduced during the 1990s and subsequently interbred with the Illinois birds. Photographed in Nebraska. *(Bob and Clara Calhoun/Bruce Coleman, Inc.)*

The founder effect occurs when a few "founders" establish a new colony

When one or a few individuals from a large population establish, or found, a colony (as when a few birds separate from the rest of the flock and fly to a new area), they bring with them only a small fraction of the genetic variation present in the original population. As a result, the only alleles represented among their descendants will be those few that the colonizers happened to possess. Typically, the allele frequencies in the newly founded population are quite different from those of the parent population. The genetic drift that results when a small number of individuals from a large population colonize a new area is called the **founder effect.**

The founder effect has been observed in populations of wild plants on islands off the Pacific coast of Canada. The Canadian mainland has several wild species of small, weedy annuals in the daisy family. These plants produce wind-dispersed seeds with fluffy parachutes similar to dandelion fruits. The seeds of mainland populations range in size from small to large, as do the fluffy parachutes. Sampling of these plants on 240 islands off the Canadian coast over a ten-year period revealed that the youngest island populations produced significantly smaller seeds than mainland populations. (Ages of island populations were easy to estimate because new colonizations occurred frequently.) This observation illustrates both the founder effect (only small seeds with large parachutes remain aloft to be blown by the wind to the nearby islands) and rapid evolutionary change by natural selection.

The Finnish people may also illustrate the founder effect. Geneticists who sampled DNA from Finns and from the European population at large found that Finns exhibit considerably less genetic variation than other Europeans. This evidence supports the hypothesis that Finns are descended from a small group of people who settled in that area that is now Finland about 4000 years ago and, because of the geography, remained separate from other European societies for centuries.

GENE FLOW GENERALLY INCREASES VARIATION WITHIN A POPULATION

Members of a species tend to be distributed in local populations that are genetically isolated to some degree from other populations. For example, the bullfrogs of one pond form a population separated from those in an adjacent pond. Some exchanges occur by migration between ponds, but the frogs in one pond are much more likely to mate with those in the same pond. Members of most species tend to be distributed in such local populations. Because each population is isolated to some extent from other populations of the species, they can have distinct genetic traits and gene pools.

The migration of breeding individuals between populations causes a corresponding movement of alleles, or **gene flow,** that can have significant evolutionary consequences. As alleles flow from one population to another, they usually increase the amount of genetic variability within the recipient population. If sufficient gene flow occurs between two populations, they become more similar genetically. Because gene flow has a tendency to reduce the amount of variation between two populations, it tends to counteract the effects of natural selection and genetic drift, both of which often cause populations to become increasingly distinct.

If migration by members of a population is considerable, and if populations differ in their allele frequencies, then significant genetic changes can result in local populations. For example, by 10,000 years ago modern humans occupied almost all of Earth's major land areas except a few islands. Because the population density was low in most locations, the small, isolated human populations underwent random genetic drift and natural selection. More recently (during the past 300 years or so), major migrations have caused an increase in gene flow, significantly altering allele frequencies within previously isolated human populations.

NATURAL SELECTION CHANGES ALLELE FREQUENCIES IN A WAY THAT INCREASES ADAPTATION

Natural selection is the mechanism of evolution first proposed by Darwin in which members of a population that possess more successful adaptations to the environment are more likely to survive and reproduce (see Chapter 17). Over successive generations, the proportion of favorable alleles increases in the population. In contrast with other microevolutionary processes (nonrandom mating, mutation, genetic drift, and gene flow), natural selection leads to adaptive evolutionary change. Natural selection not only explains why organisms are well adapted to the environments in which they live but also helps to account for the remarkable diversity of life. Natural selection enables populations to change, thereby adapting to different environments and different ways of life.

Natural selection is the differential reproduction of individuals with different traits, or phenotypes (and therefore different genotypes), in response to the environment. Natural selection results in the preservation of individuals with favorable phenotypes and elimination of those with unfavorable phenotypes. Individuals that are able to survive and produce fertile offspring have a selective advantage.

The mechanism of natural selection does not cause the development of a "perfect" organism. Rather, it weeds out those individuals whose phenotypes are less adapted to environmental challenges, while allowing better adapted individuals to survive and pass their alleles to their offspring. By reducing the frequency of alleles that result in the expression of less favorable traits, the probability is increased that favorable alleles responsible for an adaptation will come together in the offspring.

Natural selection operates on an organism's phenotype

Natural selection does not act directly on an organism's genotype. Instead, it acts on the phenotype, which is, at least in part, an expression of the genotype. The phenotype represents an interaction of all the alleles in the organism's genotype with the environment. It is rare that a single gene has complete control over a single phenotypic trait, such as Mendel originally observed in garden peas. Much more common is the interaction of alleles of several loci for the expression of a single trait (see Chapter 10). Many plant and animal characteristics are under this type of polygenic control.

When traits (e.g., human height) are under polygenic control, a range of phenotypes occurs, with most of the population located in the median range and fewer at either extreme. This is a normal distribution or standard bell curve (Fig. 18–5*a;* see also Fig. 10–22). Three kinds of selection occur that cause changes in the normal distribution of phenotypes in a population: stabilizing, directional, and disruptive selection. Although we consider each process separately, their influences generally overlap in nature.

Stabilizing selection favors intermediate phenotypes

The process of natural selection associated with a population that is well adapted to its environment is known as **stabilizing selection.** Most populations are probably under the influence of stabilizing forces most of the time. Stabilizing selection selects against phenotypic extremes. In other words, individuals with an average, or intermediate, phenotype are favored.

Because stabilizing selection tends to decrease variation by favoring individuals near the mean of the normal distribution at the expense of those at either extreme, the bell curve narrows (Fig. 18–5*b*). Although stabilizing selection decreases the amount of variation in a population, variation is rarely eliminated by this process because other microevolutionary processes act against a decrease in variation. For example, mutation is slowly but continually adding to the genetic variation within a population.

One of the most widely studied cases of stabilizing selection involves human birth weight, which is under polygenic control and is also influenced by environmental factors. Extensive data from hospitals have shown that infants born with intermediate weights are most likely to survive (Fig. 18–6). Infants at either extreme (too small or too large) have higher rates of mortality. When newborn infants are too small, their body systems are immature, and when they are too large, they have difficult deliveries because they cannot pass as easily through the cervix and vagina. Stabilizing selection operates to reduce the variability in birth weight so that it is close to the weight with the minimum mortality rate.

Directional selection favors one phenotype over another

If an environment changes over time, **directional selection** may favor phenotypes at one of the extremes of the normal distribution (Fig. 18–5*c*). Over successive generations, one phenotype gradually replaces another. So, for example, if greater size is

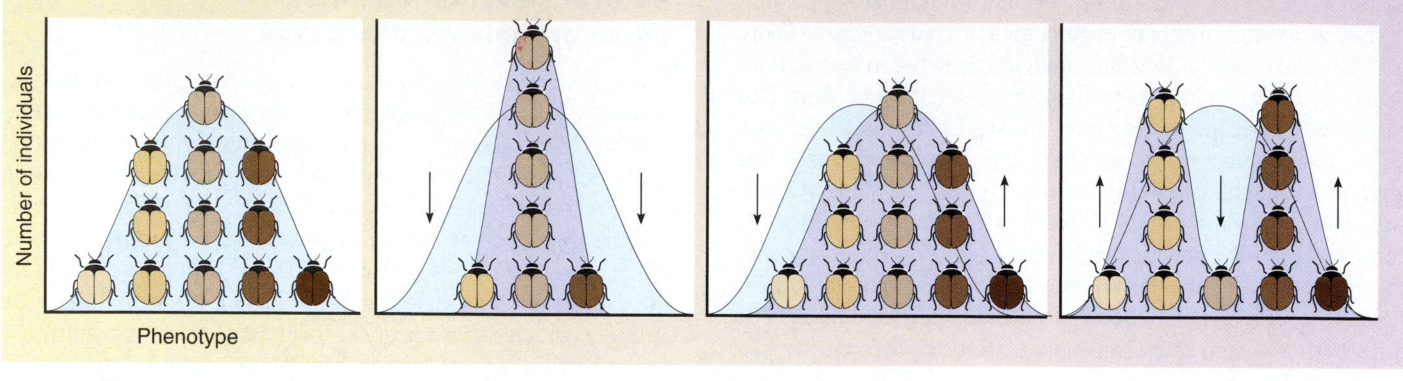

(a) No selection (b) Stabilizing selection (c) Directional selection (d) Disruptive selection

Figure 18–5 Modes of selection. The blue screen represents the distribution of individuals by phenotype (in this example, color variation) in the original population. The purple screen represents the distribution by phenotype in the evolved population. The arrows represent the pressure of natural selection on the phenotypes. **(a)** A trait that is under polygenic control (in this example, wing colors in a hypothetical population of beetles) exhibits a normal distribution of phenotypes in the absence of selection. **(b)** As a result of stabilizing selection, which trims off extreme phenotypes, variation about the mean is reduced. **(c)** Directional selection shifts the curve in one direction, changing the average value of the trait. **(d)** Disruptive selection, which trims off intermediate phenotypes, results in two or more peaks.

advantageous in a new environment, larger individuals will become increasingly common in the population. Directional selection can only occur, however, if alleles favored under the new circumstances are already present in the population.

Darwin's Galapagos finches provide an excellent example of directional selection. Since 1973, Peter and Rosemary Grant of Princeton University have studied the Galapagos finches. The Grants did a meticulous analysis of finch eating habits and beak sizes on Isla Daphne Major during three extended droughts (1977–1978, 1980, and 1982), one of which was followed by an extremely wet El Niño event (1983). During the droughts, the number of insects and small seeds declined, and large, heavy seeds became the finches' primary food source. Many finches died during this time, and most of the survivors were larger birds whose beaks were larger and deeper. In a few generations, these larger birds became more common in the population (Table 18–1). Following the wet season, however, smaller seeds became the primary food source, and smaller finches with average-sized beaks were favored. In this example, natural selection is directional: During the drought, natural selection operated in favor of the larger phenotype, whereas following the wet period, selection occurred in the opposite direction, favoring the smaller phenotype. The guppy populations studied in Venezuela and Trinidad (see Chapter 17) are another example of directional selection.

Disruptive selection favors phenotypic extremes

Sometimes extreme changes in the environment may favor two or more different phenotypes at the expense of the mean. That is, more than one phenotype may be favored in the new environment. **Disruptive selection** is a special type of directional selection in which there is a trend in several directions rather than just one (Fig. 18–5*d*). It results in a divergence, or splitting apart, of distinct groups of individuals within a population. Disruptive selection, which is relatively rare, selects against the average, or intermediate, phenotype.

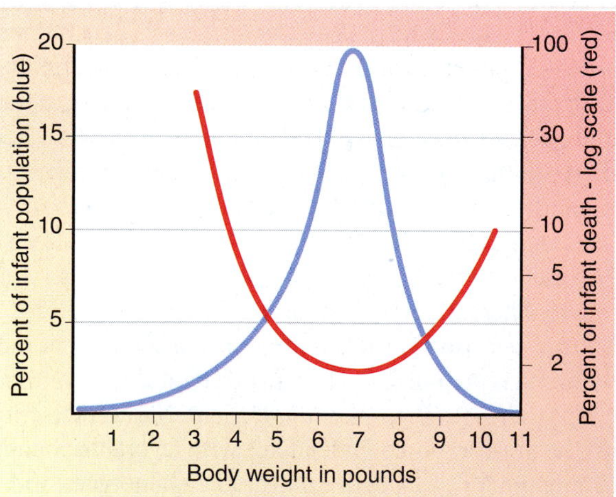

Figure 18–6 Stabilizing selection. The blue curve shows the distribution of birth weights in a sample of 13,730 infants. The red curve shows mortality (death) at each birth weight. This figure shows that infants with very low or very high birth weights have higher death rates than infants of average weight. The optimum birth weight, that is, the one with the lowest mortality, is close to the average birth weight (about 7 pounds). *(Adapted from Cavalli-Sforza, L.L., and W.F. Bodmer.* The Genetics of Human Populations. *W.H. Freeman and Company, San Francisco, 1971)*

TABLE 18–1 Population Changes in *Geospiza fortis* Before and After the 1976–1977 Drought

Trait	Average Before Drought (634)*	Average After Drought (135)*	Difference
Weight (g)	16.06	17.13	+1.07
Wing length (mm)	67.88	68.87	+0.99
Tarsus (leg, just above the foot) length (mm)	19.08	19.29	+0.21
Bill length (mm)	10.63	10.95	+0.32
Bill depth (mm)	9.21	9.70	+0.49
Bill width (mm)	8.58	8.83	+0.25

*Number of birds in sample.

From Grant, P.R. and B.R. Grant. "Predicting Microevolutionary Responses to Directional Selection on Heritable Variation." *Evolution,* Vol. 49, 1995.

Limited food supply during a severe drought caused a population of finches on another island in the Galapagos to experience disruptive selection. The finch population initially exhibited a variety of beak sizes and shapes. Because the only foods available on this island during the drought were wood-boring insects and seeds from cactus fruits, natural selection favored birds with beaks suitable for obtaining these types of food. Finches with longer beaks survived because they could open cactus fruits, and those with wider beaks survived because they could strip off tree bark to expose insects. However, finches with intermediate beaks were unable to use either food source efficiently and consequently had a lower survival rate.

Natural selection can induce change in the types and frequencies of alleles in populations only if there is preexisting inherited variation. Genetic variation is the raw material for evolutionary change, as it provides the diversity on which natural selection can act. Without genetic variation, evolution cannot occur. We now examine the genetic basis for variation that is acted on by natural selection.

GENETIC VARIATION IS NECESSARY FOR NATURAL SELECTION

Populations contain abundant genetic variation that was originally introduced by mutation. Sexual reproduction, with its associated crossing-over, independent assortment of chromosomes during meiosis, and random union of gametes, also contributes to genetic variation. The sexual process allows the variability introduced by mutation to be combined in new ways, which may be expressed as new phenotypes.

Genetic polymorphism exists among genes and the proteins for which they code

One way of evaluating genetic variation in a population is to examine **genetic polymorphism,** which is the presence in a population of two or more alleles for a given locus. Genetic polymorphism is extensive in populations, although many of the alleles are present at low frequencies. Much of genetic polymorphism is not evident because it does not produce distinct phenotypes.

One way that biologists estimate the total amount of genetic polymorphism in populations is by comparing the different forms of a particular protein. Each form consists of a slightly different amino acid sequence that is coded for by a different allele. For example, tissue extracts containing a particular enzyme may be analyzed by gel electrophoresis for different individuals. In gel electrophoresis, the enzymes are placed in slots on an agarose gel, and an electric current is applied that causes each enzyme to migrate across the gel (see Fig. 14–8). Slight variations in amino acid sequences in the different forms of a particular enzyme cause each to migrate at a different rate, which can be detected using special stains or radioactive labels. Table 18–2 shows the degree of polymorphism in selected plant and animal groups based on gel electrophoresis of several enzymes. Note that genetic polymorphism tends to be greater in plants than in animals.

Determining the sequence of nucleotides in DNA from individuals in a population provides a *direct* estimate of genetic polymorphism. One method of DNA sequencing is described in Figure 14–9. DNA sequencing of specific genes in an increasing number of organisms, including humans, indicates that genetic polymorphism is extensive in most populations.

Balanced polymorphism can exist for long periods of time

Balanced polymorphism is a special type of genetic polymorphism in which two or more alleles persist in a population over many generations as a result of natural selection. Heterozygote advantage and frequency-dependent selection are mechanisms that preserve balanced polymorphism.

TABLE 18–2 Genetic Polymorphism of Selected Enzymes Within Plant and Animal Species

Organism	Number of Species Examined	Percentage of Enzymes Studied That Are Polymorphic
Plants		
Gymnosperms	55	70.9
Flowering plants (monocots)	111	59.2
Flowering plants (dicots)	329	44.8
Invertebrates		
Marine snails	5	17.5
Land snails	5	43.7
Insects	23	32.9
Vertebrates		
Fishes	51	15.2
Amphibians	13	26.9
Reptiles	17	21.9
Birds	7	15.0
Mammals	46	14.7

*Plant data adapted from Hamrick, J.L., and M.J. Godt. "Allozyme Diversity in Plant Species." In Brown, A.H.D., M.T. Clegg, A.L. Kahler, and B.J. Weir (eds.). *Plant Population Genetics, Breeding, and Genetic Resources,* Sunderland, MA, Sinauer Associates, 1990. Animal data adapted from Hartl, D. *Principles of Population Genetics,* Sunderland, MA, Sinauer Associates, 1980, and Hedrick, P.W. *Genetics of Populations,* Boston, Science Books International, 1983.

Genetic variation may be maintained by heterozygote advantage

We have seen that natural selection often causes unfavorable alleles to be eliminated from a population while favorable alleles are retained. However, natural selection sometimes helps to maintain genetic diversity, including alleles that are unfavorable in the homozygous state, in a population. This happens, for example, when the heterozygote, *Aa,* has a higher degree of fitness than either homozygote, *AA* or *aa.* This phenomenon, known as **heterozygote advantage,** is demonstrated in humans by the selective advantage of heterozygous carriers of the sickle cell allele.

The mutant allele *(s)* for sickle cell anemia produces an altered hemoglobin that deforms or sickles the red blood cells, making them more likely to form dangerous blockages in capillaries and to be destroyed in the liver, spleen, or bone marrow (see Chapter 15). Individuals who are homozygous for the sickle cell allele *(ss)* usually die at an early age if medical treatment is not available.

Heterozygous individuals carry alleles for both normal *(S)* and sickle cell hemoglobin. The heterozygous condition *(Ss)* causes an individual to be more resistant to a type of severe malaria (caused by the parasite *Plasmodium falciparum*) than those individuals who are homozygous for the normal hemoglobin allele *(SS).* In a heterozygous individual, each allele produces its own specific kind of hemoglobin, and the red blood cells contain the two kinds in roughly equivalent amounts. Such cells do not ordinarily sickle as readily as cells containing only the *s* allele. In addition, the red blood cells containing the abnormal hemoglobin are more resistant to infection by the malaria-causing parasite, which lives in red blood cells, than are the red blood cells containing only normal hemoglobin.

Where malaria is a problem, each of the two types of homozygous individuals is at a disadvantage. Those homozygous for the sickle cell allele are likely to die of sickle cell anemia, whereas those homozygous for the normal allele may die of malaria. The heterozygote is therefore more fit than either homozygote. In parts of Africa, the Middle East, and southern Asia where *P. falciparum* malaria is prevalent, heterozygous individuals survive in greater numbers than either homozygote (Fig. 18–7). The *s* allele is maintained at a high frequency in the population even though the homozygous recessive condition is almost always lethal.

What happens to the frequency of *s* alleles in Africans and others who possess it when they migrate to the United States and other nonmalarial countries? As might be expected, the frequency of the *s* allele gradually declines in such populations, possibly because it confers a selective disadvantage by causing sickle cell anemia in homozygous individuals but no longer confers a selective advantage by preventing malaria in heterozygous individuals. The *s* allele never disappears from the population, however, because it is "hidden" from selection in heterozygous individuals and because it is reintroduced into the American population by gene flow from the African population.

Genetic variation may be maintained by frequency-dependent selection

Thus far in our discussion of natural selection, we have assumed that the fitness of particular phenotypes (and their corresponding genotypes) is independent of their frequency in the population. There are, however, cases of **frequency-dependent selection,** in which the fitness of a particular phenotype depends on how frequently it appears in the population. Often, a phenotype has a greater selective value when it is rare than when it is common in the population. Such phenotypes lose their selective advantage as they become more common.

Frequency-dependent selection often acts to maintain genetic variation in populations of prey species. In this case, the predator catches and consumes the more common phenotype but may ignore the rarer phenotypes. Consequently, the less common phenotype produces more offspring and therefore makes a greater relative contribution to the next generation. Frequency-dependent selection has been demonstrated with aquatic insects called water boatmen (Fig. 18–8*a*), which have three distinct color phenotypes. When all three phenotypes are present at equal frequencies, fish are more likely to consume the most obvious (i.e., least camouflaged) form. However, in popula-

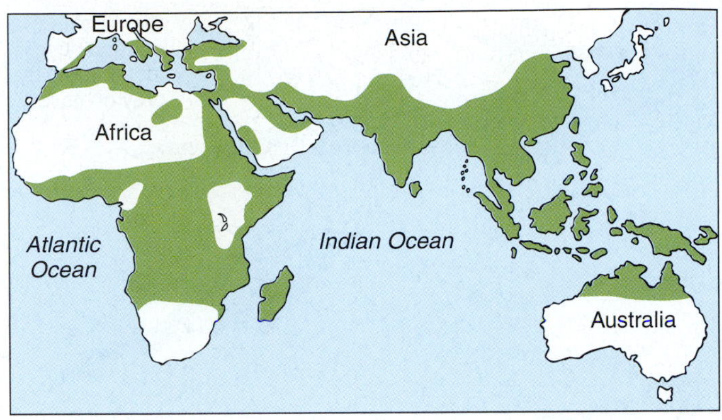

(a) *P. falciparum* malaria

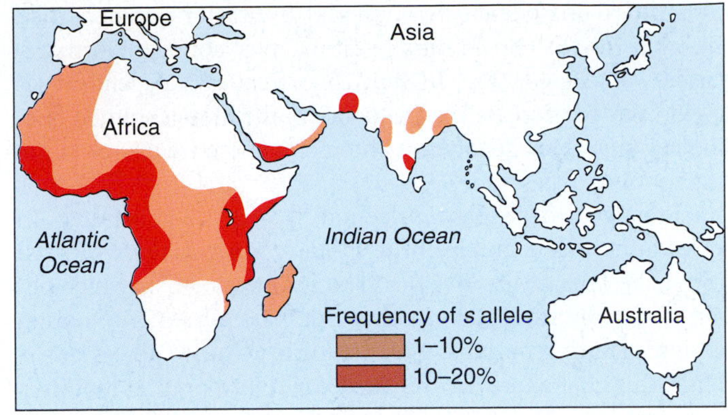

(b) Sickle cell anemia

Figure 18-7 Heterozygote advantage. The geographical distribution of **(a)** *P. falciparum* malaria *(green)* is compared with that of **(b)** sickle cell anemia *(red and orange)*. The greater fitness of heterozygous individuals in malarial regions supports the hypothesis of heterozygote advantage, which can be seen by the large area of codistribution. *(Adapted from Allison, A.C. "Protection Afforded by Sickle-Cell Traits Against Subtertian Malarial Infection."* Brit. Med. J., *Vol. 1, 1954)*

(a)

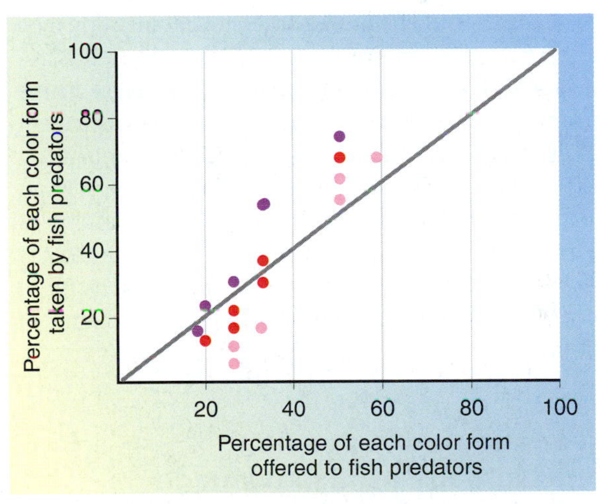

(b)

Figure 18-8 Frequency-dependent selection in water boatmen. **(a)** Water boatmen are aquatic insects that swim in ponds and streams by using their middle and hind legs as oars. These insects, which are the preferred food of many fishes, occur in three color forms that range from light to dark brown (only one color is shown). **(b)** In this experiment, different proportions of the three color phenotypes of water boatmen *(Sigara distincta)* were offered to predatory fish in aquaria. The solid line represents the situation that would occur if the percentage of each color phenotype eaten by predators were the same as the percentage offered to the predators. The results indicate that the predators preferentially ate the more common form, regardless of color. (For each of the color phenotypes, the data points at higher percentages offered are above the solid line, whereas the data points at lower percentages offered tend to be below the solid line.) Thus, frequency-dependent selection operates to maintain all three phenotypes within a given population. *(a, Stephen Dalton/ Photo Researchers, Inc.; b, Adapted from Clark, B. "Balanced Polymorphism and the Diversity of Sympatric Species."* Syst. Assoc. Publ., *Vol. 4, 1962)*

tions where one phenotype is present in greater numbers than the other two, fish preferentially eat the most abundant form, regardless of its color (Fig. 18–8*b*). Thus, frequency-dependent selection acts to decrease the frequency of the more common phenotypes (and their genotypes) and increase the frequency of the less common types.

Frequency-dependent selection is also demonstrated in scale-eating fish (cichlids of the species *Perissodus microlepsis*) from Lake Tanganyika in Africa. The scale-eating fish, which obtain food by biting scales off other fish, have either left-pointing or right-pointing mouths. A single locus with two alleles determines this characteristic; the allele for right-pointing mouth is dominant over the allele for left-pointing mouth. These fish attack their prey from behind, and from a single direction, depending on mouth morphology. Those with left-pointing mouths always attack the right flanks of their prey, whereas those with right-pointing mouths always attack the left flanks (Fig. 18–9*a*).

The prey species are more successful at evading attacks from the more common form of scale-eating fish. For example, if the cichlids with right-pointing mouths are more common than those with left-pointing mouths, the prey are attacked more often on their left flanks. They therefore become more wary against such attacks, conferring a selective advantage to the less common cichlids with left-pointing mouths. The cichlids with left-pointing mouths would be more successful at obtaining food and would therefore have more offspring. Over time, the frequency of fish with left-pointing mouths would increase in the population, until their abundance causes frequency-dependent selection to work against them and confer an advantage on the now less-common fish with right-pointing mouths. Thus, frequency-dependent selection maintains both populations of fish at approximately equal numbers (Fig. 18–9*b*). (See *On the Cutting Edge: Explaining the Rapid Loss of Cichlid Diversity in Lake Victoria* in Chapter 19 for another fascinating example of cichlid evolution.)

Neutral variation may give no selective advantage or disadvantage

Some of the genetic variation observed in a population may confer no apparent selective advantage or disadvantage in a particular environment. For example, random changes in DNA that do not alter protein structure usually do not affect the phenotype. Variation that does not alter the ability of an individual to survive and reproduce and is, therefore, not adaptive is called **neutral variation.**

The extent of neutral variation in organisms is difficult to determine. It is relatively easy to demonstrate that an allele is beneficial or harmful, provided that its effect is observable. But the variation in alleles that involves only slight differences in the proteins they code for may or may not be neutral. These alleles may be influencing the organism in subtle ways that are difficult to measure or assess. Also, an allele that is neutral in one environment may be beneficial or harmful in another.

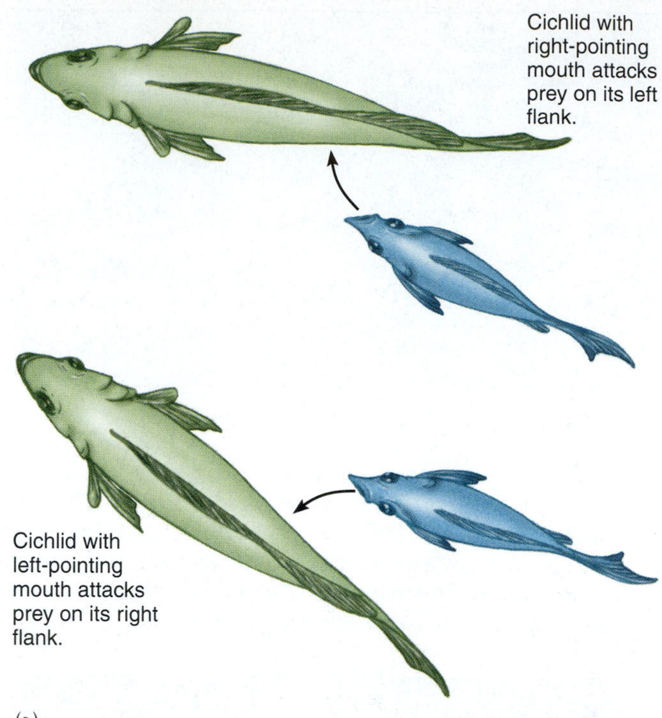

(a)

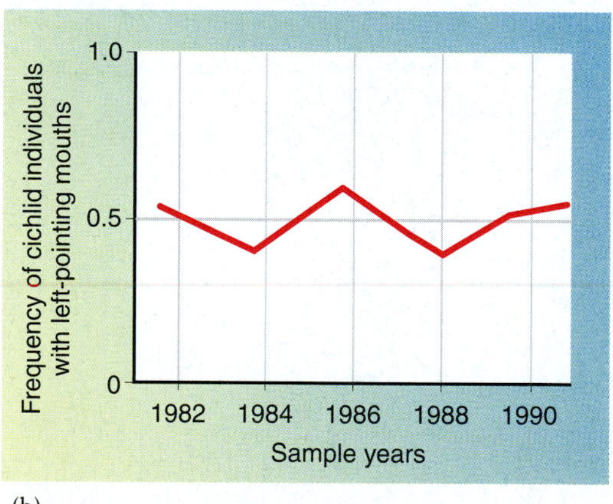

(b)

Figure 18–9 Frequency-dependent selection in scale-eating cichlids. **(a)** Scale-eating cichlids have two forms, right-pointing mouths and left-pointing mouths. Individuals with right-pointing mouths attack prey on the left flank, whereas individuals with left-pointing mouths attack prey on the right flank. **(b)** The frequency of fish with left-pointing mouths over a ten-year period. Frequency-dependent selection maintains the frequencies of left-pointing and right-pointing fish in approximately equal numbers, that is, at about 0.5. *(b, Adapted from Hori, M. "Frequency-Dependent Natural Selection in the Handedness of Scale-Eating Cichlid Fish."* Science, *Vol. 260, 9 Apr. 1993)*

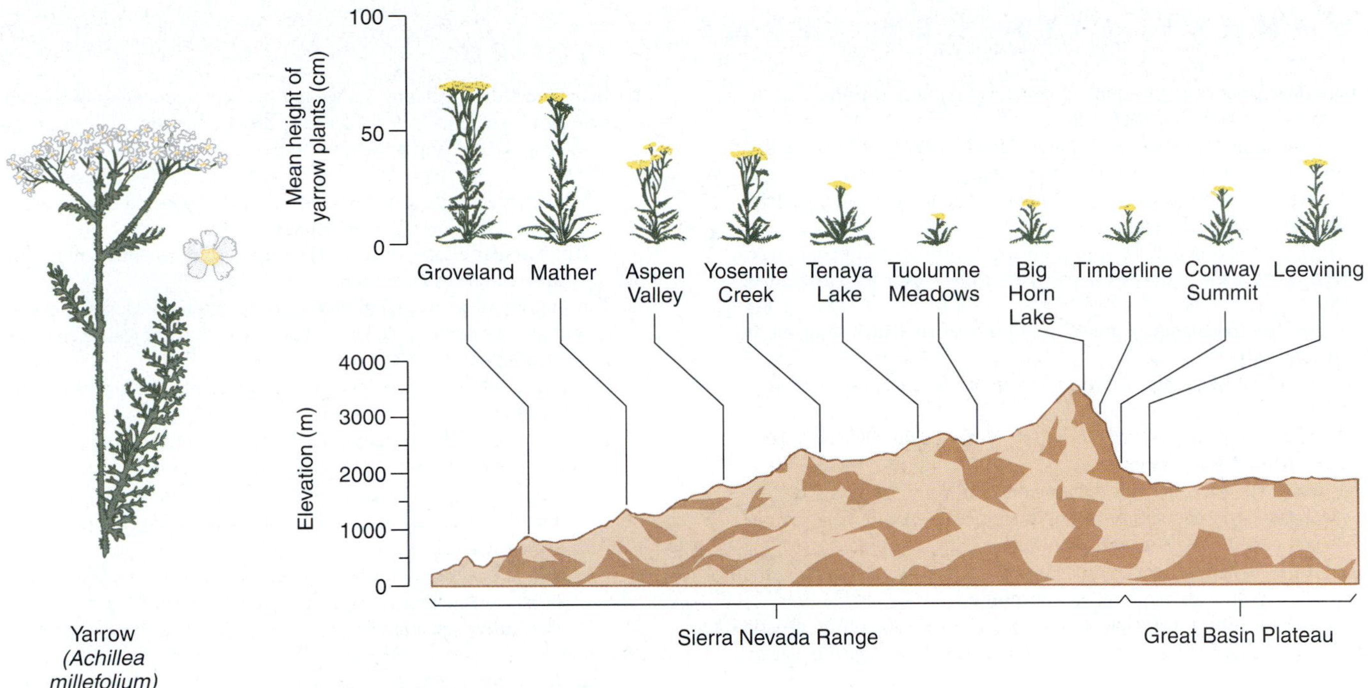

Figure 18–10 Clinal variation in yarrow *(Achillea millefolium).* Seeds from widely dispersed populations in the Sierra Nevada of California and Nevada were collected and grown for several generations under identical conditions in the same test garden at Stanford, California. The plants retained their distinctive heights, revealing genetic differences that were related to the elevation where the plants were collected. Plants descended from individuals adapted to higher elevations were shorter than those from lower elevations. *(After Clausen, J., D.D. Keck, and W.M. Hiesey. "Experimental Studies on the Nature of Species: III. Environmental Responses of Climatic Races of* Achillea.*"* Carnegie Institute Washington Publication, *Vol. 58, 1948)*

Populations in different geographical areas often exhibit genetic adaptations to local environments

In addition to the genetic variation among individuals within a population, genetic differences often exist among different populations within the same species, a phenomenon known as *geographical variation.* One type of geographical variation is a **cline,** which is a gradual change in a species' phenotype and genotype frequencies through a series of geographically separate populations as a result of an environmental gradient. A cline exhibits variation in the expression of such traits as color, size, shape, physiology, or behavior. Clines are common among species with continuous ranges over large geographical areas. For example, the body sizes of many widely distributed birds and mammals increase gradually as the latitude increases, presumably because larger animals are better able to withstand the colder temperatures of winter.

The common yarrow *(Achillea millefolium),* a wildflower that grows in a variety of North American habitats from lowlands to mountain highlands, exhibits clinal variation in height in response to different climates at different altitudes. Although substantial variation exists among individuals within each population, individuals in populations at higher altitudes are, on average, shorter than those at lower altitudes. The genetic basis of these clinal differences can be experimentally demonstrated by growing a series of populations from different geographical areas in the same environment (Fig. 18–10). Despite being exposed to identical environmental conditions, each experimental population exhibits the traits characteristic of the altitude from which it was collected.

SUMMARY WITH KEY TERMS

I. **Population genetics** is the study of genetic variability within a population and of the forces that act on it.
   A. A **population** consists of all the individuals of the same species that live in a particular place at the same time.
   B. Each population possesses a **gene pool,** which includes all the **alleles** for all the genes present in the population.

II. The evolution of populations is best understood in terms of phenotype, genotype, and allele frequencies, which are usually expressed as decimal fractions.
   A. A **genotype frequency** is the proportion of a particular genotype in the population.
   B. A **phenotype frequency** is the proportion of a particular phenotype in the population.
   C. An **allele frequency** is the proportion of a specific allele of a given gene locus in the population.
      1. If the allele frequencies remain constant from generation to generation, the population is not undergoing evolutionary change and is said to be at **genetic equilibrium.**
      2. If the allele frequencies change over successive generations, the population is undergoing **microevolution.**

III. The **Hardy-Weinberg principle** states that in a population at genetic equilibrium, allele and genotype frequencies do not change from generation to generation.

IV. Allele and/or genotype frequencies may be changed by nonrandom mating, mutation, genetic drift, gene flow, and natural selection.
   A. In nonrandom mating, individuals select mates on the basis of phenotype and indirectly on its corresponding genotype(s).
      1. **Inbreeding** is the mating of genetically similar individuals that are more closely related than if they had been chosen at random from the entire population. Inbreeding in some populations causes **inbreeding depression,** in which inbred individuals have lower **fitness** than non-inbred individuals.
      2. In **assortative mating** individuals select mates by their phenotypes.
      3. Both inbreeding and assortative mating increase the frequency of homozygous genotypes.
   B. New alleles originate as **mutations,** unpredictable changes in DNA. Mutations increase the genetic variability that can be acted on by natural selection.
   C. **Genetic drift** is a random change in the allele frequencies of a small population. Genetic drift can decrease genetic variation within a population, and the changes caused by genetic drift are usually not adaptive.
      1. A sudden decrease in population size due to adverse environmental factors is known as a **bottleneck.**
      2. The **founder effect** is genetic drift that occurs when a small population colonizes a new area.
   D. The migration of individuals between populations causes a corresponding movement of alleles, or **gene flow,** that can cause changes in allele frequencies.
   E. Changes in allele frequencies that lead to adaptation are caused by **natural selection.**
      1. Natural selection operates on an organism's phenotype.
      2. Natural selection can change the genetic composition of a population in a favorable direction for a particular environment.
         a. **Stabilizing selection** favors the mean at the expense of phenotypic extremes.
         b. **Directional selection** favors one phenotypic extreme over another, causing a shift in the phenotypic mean.
         c. **Disruptive selection** favors two or more phenotypic extremes.

V. Most populations have abundant genetic variability.
   A. **Genetic polymorphism** is the presence in a population of two or more alleles for a given locus.
   B. **Balanced polymorphism** is a special type of genetic polymorphism in which two or more alleles persist in a population over many generations as a result of natural selection.
      1. **Heterozygote advantage** occurs when the heterozygote exhibits greater fitness than either homozygote. Genetic variation is maintained in the population.
      2. In **frequency-dependent selection,** a genotype's selective value varies with its frequency of occurrence.
   C. Genetic variation that confers no detectable selective advantage is called **neutral variation.**
   D. Geographical variation is genetic variation that exists among different populations within the same species. A **cline** is a gradual change in a species' phenotype and genotype frequencies through a series of geographically separate populations.

POST-TEST

1. The genetic description of an individual is its genotype, whereas the genetic description of a population is its (a) phenotype (b) gene pool (c) genetic drift (d) founder effect (e) changes in allele frequencies
2. In a diploid species, each individual possesses (a) one allele for each locus (b) two alleles for each locus (c) three or more alleles for each locus (d) all the alleles found in the gene pool (e) half of the alleles found in the gene pool
3. The MN blood group is of interest to population geneticists because (a) people with genotype *MN* cannot receive blood transfusions from either *MM* or *NN* people (b) the *MM, MN, and NN* genotype frequencies can be observed directly and compared with calculated expected frequencies (c) the *M* allele is dominant to the N allele (d) people with the *MN* genotype exhibit frequency-dependent selection (e) people with the *MN* genotype exhibit heterozygote advantage
4. If all copies of a given locus have the same allele, then the allele frequency is (a) 0 (b) 0.1 (c) 0.5 (d) 1.0 (e) 10.0
5. If a population's allele and genotype frequencies remain constant from generation to generation (a) the population is undergoing evolutionary change (b) the population is said to be at genetic equilibrium (c) microevolution has taken place (d) directional selection is occurring, but only for a few generations (e) genetic drift is a significant evolutionary force
6. Comparing the different forms of a particular protein in a population provides biologists with an estimate of (a) genetic drift (b) genetic polymorphism (c) gene flow (d) heterozygote advantage (e) frequency-dependent selection
7. The continued presence of the allele that causes sickle cell anemia in areas where *P. falciparum* malaria is prevalent demonstrates which of the following phenomena? (a) inbreeding depression (b) frequency-dependent selection (c) heterozygote advantage (d) genetic drift (e) bottleneck
8. Frequency-dependent selection often acts to maintain ____________ in a population. (a) assortative mating (b) genetic drift (c) gene flow (d) genetic variation (e) stabilizing selection
9. According to the Hardy-Weinberg principle (a) allele frequencies are not dependent on dominance or recessiveness but remain essentially un-

changed from generation to generation (b) the sum of allele frequencies for a given locus is always greater than 1 (c) if a locus has a single allele, its frequency must be zero (d) allele frequencies change from generation to generation (e) the process of inheritance, by itself, causes changes in allele frequencies

10. What is the correct equation for the Hardy-Weinberg principle?
    (a) $p^2 + pq + 2q^2 = 1$ (b) $p^2 + 2pq + 2q^2 = 1$
    (c) $2p^2 + 2pq + 2q^2 + 1$ (d) $p^2 + pq + q^2 = 1$
    (e) $p^2 + 2pq + q^2 = 1$
11. The Hardy-Weinberg principle is applicable provided (a) population size is small (b) migration only occurs at the beginning of the breeding season (c) mutations occur at a constant rate (d) matings occur exclusively between individuals of the same genotype (e) natural selection does not occur
12. Which of the following is *not* an evolutionary agent that causes change in allele frequencies? (a) mutation (b) natural selection (c) genetic drift (d) random mating (e) gene flow from migration
13. Mutation (a) leads to adaptive evolutionary change (b) adds to the genetic variation of a population (c) is the result of genetic drift (d) almost always benefits the organism (e) a and b are correct
14. Which of the following is *not* true of natural selection? (a) Natural selection acts to preserve favorable traits and eliminate unfavorable traits. (b) The offspring of individuals that are better adapted to the environment will make up a larger proportion of the next generation. (c) Natural selection directs the course of evolution by preserving the traits acquired during an individual's lifetime. (d) Natural selection acts on a population's genetic variability, which arises through mutation. (e) Natural selection may result in changes in allele frequencies in a population.
15. In ____________ individuals with a phenotype near the phenotypic mean of the population are favored over those with phenotypic extremes. (a) microevolution (b) stabilizing selection (c) directional selection (d) disruptive selection (e) genetic equilibrium

REVIEW QUESTIONS

1. In a human population of 1000, 840 are tongue rollers (*TT* or *Tt*), and 160 are non–tongue rollers (*tt*). What is the frequency of the dominant allele *(T)* in the population?
2. In a population at genetic equilibrium, the frequency of allele *A* is 0.5. What is the frequency of the homozygous dominant genotype *(AA)*? What is the frequency of the heterozygous genotype *(Aa)*?
3. If 96% of the garden peas in a population at genetic equilibrium are tall (*TT* or *Tt*), what is the frequency of the dominant allele *(T)*?
4. If 16% of the individuals in a population at genetic equilibrium are homozygous recessive *(aa)*, what is the frequency of the recessive allele in the population? What is the frequency of the dominant allele?
5. The genotype frequencies of a population are determined to be 0.6 *AA*, 0.0 *Aa* (there are no heterozygotes), and 0.4 *aa*. Is this population at genetic equilibrium? Why or why not?
6. The chemical phenylthiocarbamide (PTC) tastes bitter to most people but is tasteless to others. About 30% of Americans are PTC nontasters (*tt*), and 70% are PTC tasters (*TT* or *Tt*). Assuming the population is at genetic equilibrium, estimate the frequency of the nontaster *(t)* and taster *(T)* alleles.
7. If a population of 2000 is at genetic equilibrium but contains only 180 individuals with the recessive phenotype *(rr)*, what is the expected frequency of the recessive allele *(r)* nine generations later if equilibrium is maintained?
8. If the genotype frequencies in a population at genetic equilibrium are 0.36 *TT*, 0.48 *Tt*, and 0.16 *tt*, what are the allele frequencies of *T* and *t*?
9. If the genotype frequencies in a population at genetic equilibrium are 0.64*TT*, 0.32*Tt*, and 0.04*tt*, what are the allele frequencies of *T* and *t*?
10. The frequency of the allele *A* in a population at genetic equilibrium is 0.7. What is the expected frequency of the *Aa* genotype?

Answers to these Review Questions are included in the Appendix with the Post-Test answers.

YOU MAKE THE CONNECTION

1. Why are mutations almost always neutral or harmful?
2. Explain this apparent paradox: We discuss evolution in terms of *genotype* fitness (the selective advantage that a particular genotype confers on an individual), yet natural selection acts on an organism's *phenotype.*
3. How are the cichlids with right-pointing and left-pointing mouths an example of balanced polymorphism?

RECOMMENDED READINGS

Case, T.J. "Natural Selection Out on a Limb." *Nature,* Vol. 387, 1 May 1997. This "News and Views" article discusses the significance of a long-term experiment (also in this issue) that documents microevolution in lizards introduced on several small Caribbean islands.

Christensen, D. "Weight Matters, Even in the Womb." *Science News,* Vol. 158, 9 Dec. 2000. An interesting extension of research on the possible relationship between birth weight and chronic diseases later in life.

Grant, P.R. "Natural Selection and Darwin's Finches." *Scientific American,* Vol. 265, No. 4, Oct. 1991. A study of the finches of the Galapagos Islands reveals natural selection in action during a drought.

King, R.B., and R. Lawson. "Microevolution in Island Water Snakes." *BioScience,* Vol. 47, No. 5, May 1997. The authors demonstrated the interaction between natural selection and gene flow in determining color pattern differences in populations of Lake Erie water snakes.

Madsen, T., R. Shine, M. Olsson, and H. Wittzell. "Restoration of an Inbred Adder Population." *Nature,* Vol. 402, 4 Nov. 1999. A genetically uniform population of adders was declining in number and in danger of extinction until biologists added genetically diverse adders from other populations. The introduced adders bred with local adders, resulting in greater genetic diversity, as demonstrated by gel electrophoresis, and allowing the adder population to make a dramatic recovery.

Mange, E.J., and A.P. Mange. *Basic Human Genetics,* 2nd ed. Sinauer Associates, Inc., Sunderland, MA, 1999. This text has an excellent chapter on population genetics that includes the *MN* blood groups as well as many other human examples. Much of our chapter's section on calculating frequencies was modeled after the presentation in this text.

Mayr, E. *Population, Species, and Evolution.* Harvard University Press, Cambridge, MA, 1970. A classic discussion of evolution.

Milius, S. "Dull Birds and Bright Ones Beat So-So Guys." *Science News,* Vol. 158, 28 Oct. 2000. An example of disruptive selection in male lazuli buntings.

Milius, S. "Wild Inbred Butterflies Risk Extinction." *Science News,* Vol. 153, 4 Apr. 1999. The study summarized here provides convincing evidence that inbreeding figures in extinction.

Pennisi, E. "Tracking a Sparrow's Fall." *Science,* Vol. 285, 9 Jul. 1999. This "New Focus" reports on a documented example in nature of the effects of inbreeding in a population of song sparrows.

Soulé, M.E., and L.S. Mills. "No Need to Isolate Genetics." *Science,* Vol. 282, 27 Nov. 1998. This perspective introduces a study (page 1695 of the same issue) of the decline and subsequent recovery of the greater prairie chicken population in Illinois.

Weiner, J. *The Beak of the Finch: A Story of Evolution in Our Time.* Alfred A. Knopf, New York, 1994. This Pulitzer Prize–winning book focuses on the research of Peter and Rosemary Grant on evolution in the Galapagos finches.

- Visit our Web site at **http://www.info.brookscole.com/solomonbergmartin** for links to chapter-related resources on the World Wide Web. Additional on-line materials relating to this chapter can also be found on our Web site.

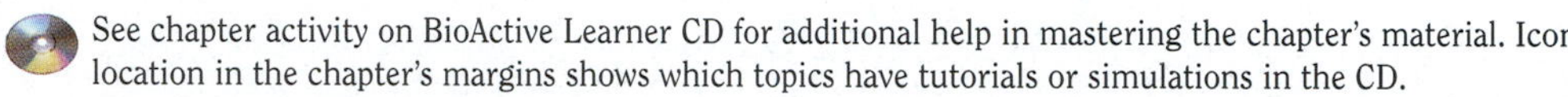

See chapter activity on BioActive Learner CD for additional help in mastering the chapter's material. Icon location in the chapter's margins shows which topics have tutorials or simulations in the CD.

19

Speciation and Macroevolution

Interbreeding between different species. Shown is a "zebrass," a sterile hybrid formed by a cross between a zebra and a donkey that retains features of both parental species. Although such matings may occur under artificial conditions, such as the wildlife ranch in Texas where this cross took place, zebras and donkeys do not interbreed in the wild. *(Gary Retherford/Photo Researchers, Inc.)*

LEARNING OBJECTIVES

After you have studied this chapter you should be able to

1. Define a species and explain the limitations of the biological species concept.
2. Explain the significance of reproductive isolating mechanisms and distinguish among the different prezygotic and postzygotic barriers.
3. Explain the mechanism of allopatric specification and give an example.
4. Explain the mechanisms of sympatric specification and give both plant and animal examples.
5. Take either side in a debate on the pace of evolution by representing the opposing views of gradualism and punctuated equilibrium.
6. Define macroevolution and distinguish among microevolution, speciation, and macroevolution.
7. Discuss macroevolution in the context of novel features, including preadaptations, allometric growth, and paedomorphosis.
8. Discuss the macroevolutionary significance of adaptive radiation and extinction.
9. Explain how the course of evolution was affected by continental drift.

In Chapters 17 and 18, we examined natural selection and how populations evolve. We now focus our attention on how species and higher taxa (for example, new classes)[1] evolve. We do not know exactly how many species exist, but biologists estimate there may be something on the order of 10 million to 100 million different species.[2] About 1.75 million species have been scientifically named and described. These include 250,000 plant species, 42,000 vertebrate animals, and some 750,000 insects.

The concept of distinct kinds of organisms, known as **species** (from Latin, meaning "kind") is not new. However, every definition of species has some sort of limitation. Linnaeus, the 18th century biologist who is considered the founder of modern taxonomy, classified plants and other organisms into separate species based on structural differences (see Chapter 22). This method, known as the *morphological species concept,* is still used to help characterize species, but structure alone is not adequate to explain what constitutes a species.

Population genetics did much to clarify the concept of species. According to the **biological species concept,** first expressed by Ernst Mayr in 1940, a species consists of one or more populations whose members are capable of interbreeding in nature to produce fertile offspring and do not interbreed with—that is, are reproductively isolated from—members of different species. In other words, each species has a gene pool that is separate from that of other species, and reproductive barriers restrict each species from interbreeding with other species.

One of the problems with the biological species concept is that it applies only to sexually reproducing organisms. Organisms that reproduce asexually do not interbreed, so we cannot think of them in terms of reproductive isolation. These organisms and extinct organisms are classified on the basis of structural and biochemical characteristics. Another potential problem with the biological species concept is that organisms assigned to different

[1] The taxonomic categories (taxa) above the level of species are artificial constructs used by humans to indicate degrees of relatedness among organisms. As described in Chapter 1, closely related species are grouped into the same genus (pl., *genera*), similar genera into the same family, similar families into the same order, similar orders into the same class, and similar classes into the same phylum.

[2] Reaka-Kudla, M.L., D.E. Wilson, and E.O. Wilson, eds. *Biodiversity II.* Joseph Henry Press, Washington, D.C., 1997.

species may successfully interbreed if brought into the artificial environment and caging of a wildlife ranch *(see photograph),* circus, zoo, greenhouse, aquarium, or laboratory. For that reason, natural matings alone can be used to test biological species identity; captive matings are not valid.

This chapter discusses reproductive barriers that isolate species from one another, the possible evolutionary mechanisms that explain how the millions of species that live today or lived in the past originated from ancestral species, and the rates of evolutionary change. We then examine *macroevolution,* which is large-scale phenotypic changes (such as the appearance of wings with feathers during the evolution of birds from reptiles) that permit the evolution of taxonomic groups at the species level and higher—new genera, families, orders, classes, and even phyla.

SPECIES ARE REPRODUCTIVELY ISOLATED IN VARIOUS WAYS

A number of **reproductive isolating mechanisms** prevent interbreeding between two different species whose ranges overlap. These mechanisms preserve the genetic integrity of each species because gene flow between the two species is prevented. Most species have two or more mechanisms that block a chance occurrence of individuals from two closely related species overcoming a single reproductive isolating mechanism. Most occur before mating or fertilization occurs *(prezygotic),* whereas others work after fertilization has taken place *(postzygotic).*

Prezygotic barriers interfere with fertilization

Prezygotic barriers are reproductive isolating mechanisms that prevent fertilization from taking place. Because male and female gametes never come into contact, an interspecific zygote (fertilized egg formed by the union of an egg from one species and a sperm from another species) is never produced. Prezygotic barriers include temporal isolation, habitat isolation, behavioral isolation, mechanical isolation, and gametic isolation.

Sometimes genetic exchange between two groups is prevented because they reproduce at different times of the day, season, or year. Such examples demonstrate **temporal isolation.** For example, two very similar species of fruit flies, *Drosophila pseudoobscura* and *Drosophila persimilis,* have ranges that overlap to a great extent, but they do not interbreed. *D. pseudoobscura* is sexually active only in the afternoon and *D. persimilis* only in the morning. Similarly, two frog species have overlapping ranges in eastern Canada and the United States. The wood frog *(Rana sylvatica)* usually mates in late March or early April, when the water temperature is about 7.2° C (45° F), whereas the northern leopard frog *(Rana pipiens)* usually mates in mid-April, when the water temperature is 12.8° C (55° F) (Fig. 19–1).

Although two closely related species may be found in the same geographical area, they usually live and breed in different habitats in that area. This causes **habitat isolation** between the two species. For example, the five species of small *Empidonax* flycatchers in the eastern part of North America are nearly identical in appearance and have overlapping ranges (Fig. 19–2). They exhibit habitat isolation because, during the breeding season, each species is found in a particular habitat within its range, so potential mates from different species do not meet. The least flycatcher *(Empidonax minimus)* frequents open woods, farms, and orchards; the acadian flycatcher *(Empidonax virescens)* is found in deciduous forests, particularly in beech trees, and swampy woods; the alder flycatcher *(Empidonax alnorum)* prefers wet thickets of alders; the yellow-bellied flycatcher *(Empidonax flaviventris)*

(a)

(b)

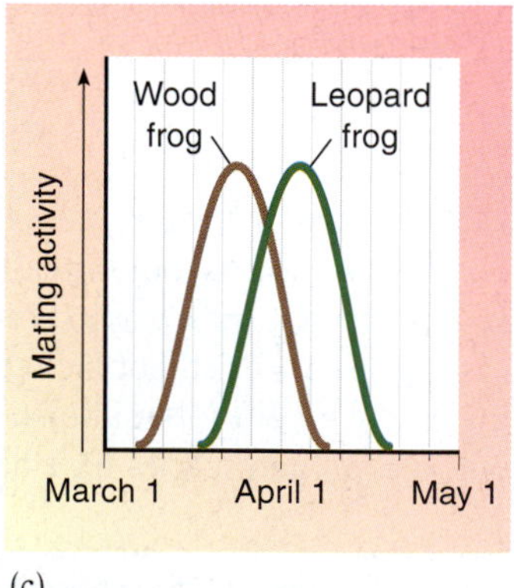

(c)

Figure 19–1 Temporal isolation in wood and leopard frogs. (a) The wood frog *(Rana sylvatica)* mates in early spring, often before the ice has completely melted in the ponds. **(b)** The leopard frog *(R. pipiens)* typically mates a few weeks later. **(c)** Graph of peak mating activity in wood and leopard frogs. In nature, wood and leopard frogs do not interbreed, although they have done so in the laboratory. *(a, L. & D. Klein/Photo Researchers, Inc.; b, Rod Planck/Photo Researchers, Inc.)*

Figure 19–2 Habitat isolation in flycatchers. Each flycatcher species is found in a particular habitat during mating. The acadian flycatcher *(Empidonax virescens)* frequents deciduous forests and swampy woods during its breeding season. Shown is a male at his nest in West Virginia. *(Bill Beatty/Visuals Unlimited, Inc.)*

nests in conifer woods; and the willow flycatcher *(Empidonax traillii)* frequents brushy pastures and willow thickets.

Many animal species exchange a distinctive series of signals before mating. Such courtship behaviors illustrate **behavioral isolation** (also known as **sexual isolation**). Bowerbirds, for example, exhibit species-specific courtship patterns. The male satin bowerbird of Australia constructs an elaborate bower of twigs, adding decorative blue parrot feathers and white flowers at the entrance (Fig. 19–3). When a female approaches the bower, the male dances about her, holding a particularly eye-catching decoration in his beak. While dancing, he sings a courtship song that consists of a variety of sounds, including buzzes and laughlike hoots. These specific courtship behaviors keep similar bird species reproductively isolated from the satin bowerbird. If a male and female of two different species begin courtship, it stops when one member does not recognize or respond to the signals of the other. Another example of behavioral isolation involves the wood frogs and northern leopard frogs (just discussed as an example of temporal isolation). Males of these two species have very specific vocalizations to attract females of their species for breeding. These vocalizations reinforce the reproductive isolation of these species.

Sometimes members of different species court and even attempt copulation, but the incompatible structures of their genital organs prevent successful mating. Structural differences that inhibit mating between species produce **mechanical isolation.** For example, many flowering plant species have physical differences in their flower parts that help them maintain their reproductive isolation from one another. In such plants, the flower parts are adapted for specific insect pollinators. Two species of sage, for example, have overlapping ranges in southern California. Black sage *(Salvia mellifera),* which is pollinated by small bees, has a floral structure different from that of white sage *(Salvia apiana),* which is pollinated by large carpenter bees (Fig. 19–4). Interestingly, black sage and white sage are also prevented from mating by a temporal barrier: black sage flowers in early spring, and white sage flowers in late spring and early summer. Presumably, mechanical isolation prevents insects from cross-pollinating the two species should they happen to flower at the same time.

If mating has taken place between two species, their gametes may still not combine. Molecular and chemical differences between species cause **gametic isolation,** in which the egg and sperm of different species are incompatible. In aquatic animals that release their eggs and sperm into the surrounding water simultaneously, interspecific fertilization is extremely rare (Fig. 19–5). The surface of the egg contains specific proteins that bind only to complementary molecules on the surface of sperm cells of the same species (see Chapter 49). A similar type of molecular recognition often occurs between pollen grains and the stigma (receptive surface of the female part of the flower) so that pollen does not germinate on the stigma of a different plant species.

Figure 19–3 Behavioral isolation in bowerbirds. Each bowerbird species has highly specialized courtship patterns that prevent its mating with another species. The male satin bowerbird *(Ptilonorhynchus violacens)* constructs an enclosed place, or bower, of twigs to attract a female. (The bower is the dark "tunnel" on the left.) Note the flowers and blue decorations, including human-made objects such as bottle caps, that he has arranged at the entrance to his bower. *(R. Brown/VIREO)*

Figure 19–4 Mechanical isolation in black sage and white sage. Differences in floral structures between black sage and white sage allow them to be pollinated by different insects. Because the two species exploit different pollinators, they cannot interbreed. **(a)** The petal of the black sage functions as a landing platform for small bees. Larger bees cannot fit on this platform. **(b)** The larger landing platform and longer stamens of white sage allow pollination by larger California carpenter bees (a different species). If smaller bees land on white sage, their bodies do not brush against the stamens. (The upper part of the white sage flower has been removed.)

Figure 19–5 Gametic isolation in sponges. A sponge *(Ectyoplasia ferox)* releases masses of unfertilized eggs into the water. If one of these eggs encounters a sperm of a different sponge species, biochemical differences will prevent their union. Photographed in the Atlantic Ocean in the Bahamas. *(Doug Perrine/Seapics.com)*

Postzygotic barriers prevent gene flow when fertilization occurs

Fertilization sometimes occurs between gametes of two closely related species despite the existence of **prezygotic barriers.** When this happens, postzygotic barriers that increase the likelihood of reproductive failure come into play. Generally, the embryo of an interspecific hybrid spontaneously aborts. Embryonic development is a complex process requiring the precise interaction and coordination of many genes. Apparently, the genes from parents belonging to different species do not interact properly in regulating the mechanisms for normal development. In this case, reproductive isolation occurs by **hybrid inviability.** For example, nearly all the hybrids die in the embryonic stage when the eggs of a bullfrog are fertilized artificially with sperm from a leopard frog. Similarly, in crosses between different species of irises, the embryos die before reaching maturity.

If an interspecific hybrid does live, it may not be able to reproduce. There are several reasons for this. Hybrid animals may exhibit courtship behaviors incompatible with those of either parental species and, as a result, they will not mate. More often, **hybrid sterility** occurs when problems during meiosis cause the gametes of an interspecific hybrid to be abnormal. Hybrid sterility is particularly common if the two parental species have different chromosome numbers. For example, a mule is the offspring of a female horse ($2n = 64$) and a male donkey ($2n = 62$) (Fig. 19–6). This type of union almost always results in sterile offspring ($2n = 63$) because chromosomal synapsis (the pairing of homologous chromosomes during meiosis) and segregation cannot occur properly. The zebrass shown in the chapter introduction also exhibits hybrid sterility due to different chromosome numbers. Donkeys have 62 chromosomes, and zebras, depending on the species, have 32, 44, or 46 chromosomes.

Occasionally, a mating between two F_1 hybrids produces a second hybrid generation (F_2). The F_2 hybrid may exhibit **hybrid breakdown,** the inability of a hybrid to reproduce due to some defect. For example, hybrid breakdown in the F_2 generation of a

Figure 19–6 Hybrid sterility in mules. Mules are interspecific hybrids formed by mating a female horse with a male donkey. Although the mule *(center)* exhibits valuable characteristics of each of its parents, it is sterile. *(John Eastcott/Yva Momatiuk/Animals Animals)*

TABLE 19–1 Reproductive Isolating Mechanisms

Mechanism	How It Works
Prezygotic Barriers	Prevent fertilization
Temporal isolation	Similar species reproduce at different times
Habitat isolation	Similar species reproduce in different habitats
Behavioral isolation	Similar species have distinctive courtship behaviors
Mechanical isolation	Similar species have structural differences in their reproductive organs
Gametic isolation	Gametes of similar species are chemically incompatible
Postzygotic Barriers	Reduce viability or fertility of hybrid
Hybrid inviability	Interspecific hybrid dies at early stage of embryonic development
Hybrid sterility	Interspecific hybrid survives to adulthood but is unable to reproduce successfully
Hybrid breakdown	Offspring of interspecific hybrid are unable to reproduce successfully

cross between two sunflower species was 80%. In other words, 80% of the F_2 generation were defective in some way and could not reproduce successfully. Hybrid breakdown can also occur in the F_3 and later generations.

Table 19–1 summarizes the various prezygotic and postzygotic barriers that prevent interbreeding between two species.

The genetic basis of isolating mechanisms are being elucidated

Recent progress has been made in identifying some of the genes involved in reproductive isolation. For example, scientists have determined the genetic basis for prezygotic isolation in species of abalone, large mollusks found along the Pacific coast of North America. In abalone, the fertilization of eggs by sperm requires lysin, a sperm protein that attaches to a lysin receptor protein located on the egg envelope. After attachment, the lysin produces a hole in the egg envelope that permits the sperm to penetrate the egg. Scientists cloned the lysin receptor gene and demonstrated that this gene varies among abalone species. Differences in the lysin receptor protein in various abalone species determine sperm compatibility with the egg. Sperm of one abalone species will not attach to a lysin receptor protein of an egg of a different abalone species.

A gene involved in postzygotic isolating mechanisms has been identified and cloned in fruit flies (*Drosophila* sp.). The gene causes sterility in male hybrids formed by a cross between *Drosophila mauritiana* and *Drosophila simulans.* The actual function of the gene is currently unknown, but scientists hypothesize it may control the transcription of a different gene that is involved in spermatogenesis, the production of sperm.

REPRODUCTIVE ISOLATION IS THE KEY TO SPECIATION

We are now ready to consider how entirely new species may arise from previously existing ones. The evolution of a new species is called **speciation.** The fossil record suggests that there are two patterns of speciation: Some speciation events are anagenic, whereas others are cladogenic. **Anagenesis,** or **phyletic evolution,** refers to the relatively small, progressive evolutionary changes in a single lineage over long periods. Given enough time, anagenesis results in the conversion of one species into a different species. Thus, a *sequence* of species occurs over time without an increase in the *number* of species. Beginning with Charles Darwin, many evolutionary biologists have focused on anagenesis and tried to determine how a single species changes over time, by means of natural selection.

Other evolutionary biologists concentrate on species multiplication. In **cladogenesis,** or **branching evolution,** two or more populations of an ancestral species split and diverge, eventually forming two or more new species. Such a cluster of species (or other taxa) that are all derived from a single common ancestor is known as a **clade** (see Chapter 22). Over time, the process of cladogenesis increases *species richness,* which is the number of species (see Chapter 52).

The formation of two species from a single species occurs when a population becomes reproductively isolated from other members of the species, and the gene pools of the two separated populations begin to diverge in genetic composition. When a population is sufficiently different from its ancestral species that no genetic exchange can occur between them, we say that speciation has occurred. Speciation is thought to occur in two ways: allopatric speciation and sympatric speciation.

Long physical isolation and different selective pressures result in allopatric speciation

Speciation that occurs when one population becomes geographically separated from the rest of the species and subsequently evolves by natural selection and/or genetic drift is known as **allopatric speciation** (from the Greek *allo,* "different," and *patri,* "native land"). Allopatric speciation is thought to be the most common method of speciation, and the evolution of new animal species has been almost exclusively by allopatric speciation.

The geographical isolation required for allopatric speciation may occur in several ways. Earth's surface is in a constant state of change. Such change includes rivers shifting their courses; glaciers migrating; mountain ranges forming; land bridges developing that separate previously united aquatic populations; and large lakes diminishing into several smaller, geographically separated pools.

What might be an imposing geographical barrier to one species may be of no consequence to another. Birds and cattails, for example, do not become isolated when a lake subsides into smaller pools; birds can easily fly from one pool to another, and cattails disperse their pollen and fruits by air currents. Fish, on the other hand, are usually unable to cross the land barriers between the pools and so become reproductively isolated. In the Death Valley region of California and Nevada, large interconnected lakes formed during wetter climates of the last Ice Age. These lakes were populated by one or several species of pupfish. Over time, the climate became drier, and the large lakes dried up, leaving isolated pools. Presumably, each pool contained a small population of pupfish that gradually diverged from the common ancestral species by genetic drift and natural selection in response to the high temperatures, high salt concentrations, and low oxygen levels characteristic of desert springs. Today, there are more than 20 species of pupfish, and many, such as the Devil's Hole pupfish *(Cyprinodon diabolis)* and the Owens pupfish *(Cyprinodon radiosus),* are restricted to one or two isolated springs (Fig. 19–7).

Allopatric speciation also occurs when a small population migrates or is dispersed (e.g., by a chance storm) and colonizes a new area away from the range of the original species. This colony is geographically isolated from its parental species, and the small microevolutionary changes that accumulate in the isolated gene pool over many generations may eventually be sufficient to form a new species. Because islands provide the geographical isolation required for allopatric speciation, they offer excellent opportunities to study this evolutionary mechanism. A few individuals of a few species probably colonized the Galapagos Islands and the Hawaiian Islands, for example. The hundreds of unique species presently found on each island presumably descended from these original colonizers (Fig. 19–8; also see Chapter 17).

Figure 19–8 Allopatric speciation of the Hawaiian goose (the nene). Nene (pronounced "nay–nay"; *Branta sandvicensis*) are geese originally found only on volcanic mountains on the geographically isolated islands of Hawaii and Maui, which are some 4200 km (2600 mi) from the nearest continent. Compared with other geese, the feet of Hawaiian geese are not completely webbed, their toenails are longer and stronger, and their foot pads are thicker; these adaptations enable Hawaiian geese to walk easily on lava flows. Nene are thought to have evolved from a small population of geese that originated in North America. Photographed in Hawaii Volcanoes National Park, Hawaii. *(Victoria McCormick/Animals Animals)*

Figure 19–7 Allopatric speciation of pupfish *(Cyprinodon).* Shown is one of the more than 20 pupfish species that apparently evolved when larger lakes in southern Nevada dried up about 10,000 years ago, leaving behind small, isolated desert pools fed by springs. The pupfish's short, stubby body is characteristic of fish that live in springs; fish that live in larger bodies of water are more streamlined. *(Steinhart Aquarium, Tom McHugh/Photo Researchers, Inc.)*

Speciation is more likely to occur if the original isolated population is small. Recall that genetic drift, including the founder effect, is more consequential in small populations (see Chapter 18). Genetic drift tends to result in rapid changes in allele frequencies in the small, isolated population. The divergence caused by genetic drift is further accentuated by the different selective pressures of the new environment to which the population is exposed.

The Kaibab squirrel is an example of allopatric speciation in progress

About 10,000 years ago, when the American Southwest was less arid, the forests in the area supported a tree squirrel with conspicuous tufts of hair sprouting from its ears. A small tree squirrel population living on the Kaibab Plateau of the Grand Canyon became geographically isolated when the climate changed, causing areas to the north, west, and east to become desert. Just a few miles to the south lived the rest of the squirrels, known as Abert squirrels, but the two groups were separated by the Grand Canyon. With changes over time in both its appearance and its ecology, the Kaibab squirrel is on its way to becoming a new species.

During its many years of geographical isolation, the small population of Kaibab squirrels has diverged from the widely distributed Abert squirrels in several ways. Perhaps most evident are changes in fur color. The Kaibab squirrel now has a white tail and a black belly, in contrast to the gray tail and white belly of the Abert squirrel (Fig. 19–9). It is not clear why these striking changes arose in Kaibab squirrels.

Some scientists consider the Kaibab squirrel and the Abert squirrel as distinct populations of the same species *(Sciurus aberti)*. Because the Kaibab and Abert squirrels are reproductively isolated from each other, however, some scientists have classified the Kaibab squirrel as a different species *(Sciurus kaibabensis)*.

Porto Santo rabbits may be an example of extremely rapid allopatric speciation

Allopatric speciation has the potential to occur quite rapidly. Early in the 15th century, a small population of rabbits was released on Porto Santo, a small island off the coast of Portugal. Because there were no other rabbits or competitors and no predators on the island, the rabbits thrived. By the 19th century, these rabbits were markedly different from their European ancestors. They were only half as large (weighing slightly more than 500 g, or 1.1 lb), with a different color pattern and a more nocturnal lifestyle. Most significantly, attempts to mate Porto Santo rabbits with mainland European rabbits failed. Many biologists concluded that, within 400 years, an extremely brief period in evolutionary history, a new species of rabbit had evolved.

Not all biologists agree that the Porto Santo rabbit is a new species. The objection stems from a more recent breeding experiment and is based on biologists' lack of a consensus about the definition of a species. In the experiment, foster mothers of the wild Mediterranean rabbit raised newborn Porto Santo rabbits. When they reached adulthood, these Porto Santo rabbits mated successfully with Mediterranean rabbits to produce healthy, fertile offspring. To some biologists, this experiment clearly demonstrated that Porto Santo rabbits are not a separate species but instead are an example of speciation in progress, much like the Kaibab squirrels just discussed. Other biologists think the Porto Santo rabbit is a separate species because it does not interbreed with other rabbits under natural conditions. They point out that the breeding experiment was successful only after the baby Porto Santo rabbits were raised under artificial conditions that probably modified their natural behavior.

Two populations diverge in the same physical location by sympatric speciation

Although geographical isolation is an important factor in many cases of evolution, it is not an absolute requirement. In **sympatric speciation** (from the Greek *sym,* "together," and *patri,* "native land"), a new species evolves within the same geographical region as the parental species. The divergence of two gene pools in the same geographical range occurs when reproductive isolating mechanisms evolve at the start of the speciation process. Sympatric speciation is especially common in plants. The role of sympatric speciation in animal evolution is probably much less important than allopatric speciation; until recently, sympatric speciation in animals has been difficult to demonstrate in nature.

There are at least two ways in which sympatric speciation can occur: a change in **ploidy** (the number of chromosome sets making up an organism's genome) and a change in ecology. We now examine each of these mechanisms.

Allopolyploidy is an important mechanism of sympatric speciation in plants

As discussed earlier, the union of two gametes from different species rarely forms viable offspring; if offspring are produced, they are usually sterile. Before gametes form, meiosis occurs to reduce the chromosome number (see Chapter 9). For the chromosomes to be parceled correctly into the gametes, homologous chromosome pairs must come together (a process called *synapsis*) during prophase I. This cannot usually occur in interspecific hybrid offspring because not all the chromosomes are homologous. However, if the chromosome number doubles *before*

(a)

(b)

Figure 19–9 Allopatric speciation in progress. **(a)** The Kaibab squirrel, with its white tail, is found north of the Grand Canyon. **(b)** The Abert squirrel, with its gray tail and white belly, is found south of the Grand Canyon. *(a, Tom and Pat Leeson; b, Kent and Donna Dannen)*

meiosis, then pairing of homologous chromosomes can take place. Although not a common occurrence, this spontaneous doubling of chromosomes has been documented in a variety of plants and a few animals. It produces nuclei with multiple sets of chromosomes.

Polyploidy, the possession of more than two sets of chromosomes, is a major factor in plant evolution. Reproductive isolation occurs in a single generation when a polyploid species with multiple sets of chromosomes arises from diploid parents. There are two kinds of polyploidy: autopolyploidy and allopolyploidy. An **autopolyploid** contains multiple sets of chromosomes from a single species, and an **allopolyploid** contains multiple sets of chromosomes from two or more species. We restrict our discussion to allopolyploidy because its occurrence in nature is much more common.

Allopolyploidy occurs in conjunction with **hybridization,** which is sexual reproduction between individuals from closely related species. Allopolyploidy can produce a fertile interspecific hybrid because the polyploid condition provides the homologous chromosome pairs necessary for synapsis during meiosis. As a result, gametes may be viable (Fig. 19–10). An allopolyploid, that is, an interspecific hybrid produced by allopolyploidy, can reproduce with itself (self-fertilize) or with a similar individual. However, allopolyploids are reproductively isolated from both parents because their gametes have a different number of chromosomes than those of either parent.

If a population of allopolyploids (i.e., a new species) becomes established, selective pressures cause one of three outcomes. First, the new species may be unable to compete successfully against species that are already established, and so it becomes extinct. Second, the allopolyploid individuals may assume a new role in the environment and so coexist with both parental species. Third, the new species may successfully compete with either or both of its parental species. If it has a combination of traits that confers greater fitness than one or both parental species for all or part of the original range of the parent(s), the hybrid species may replace the parent(s).

Although allopolyploidy is extremely rare in animals, it is considered a significant factor in the evolution of flowering plant species. As many as 80% of all flowering plant species are thought to be polyploids, and most of these are allopolyploids. Moreover, allopolyploidy provides a mechanism for extremely rapid speciation. A single generation is all that is needed to form a new, reproductively isolated species. Allopolyploidy may explain the rapid appearance of many flowering plant species in the fossil record and their remarkable diversity (about 235,000 species) today.

The kew primrose *(Primula kewensis)* is an example of sympatric speciation that was documented at the Royal Botanic Gardens at Kew, England, in 1898 (Fig. 19–11). The interspecific hybrid of two primrose species, *Primula floribunda* ($2n = 18$) and *Primula verticillata* ($2n = 18$), *P. kewensis* had a chromosome number of 18 but was sterile. Then, at three different times, it was reported to have spontaneously formed a fertile branch, which was an allopolyploid ($2n = 36$) that produced viable seeds of *P. kewensis.*

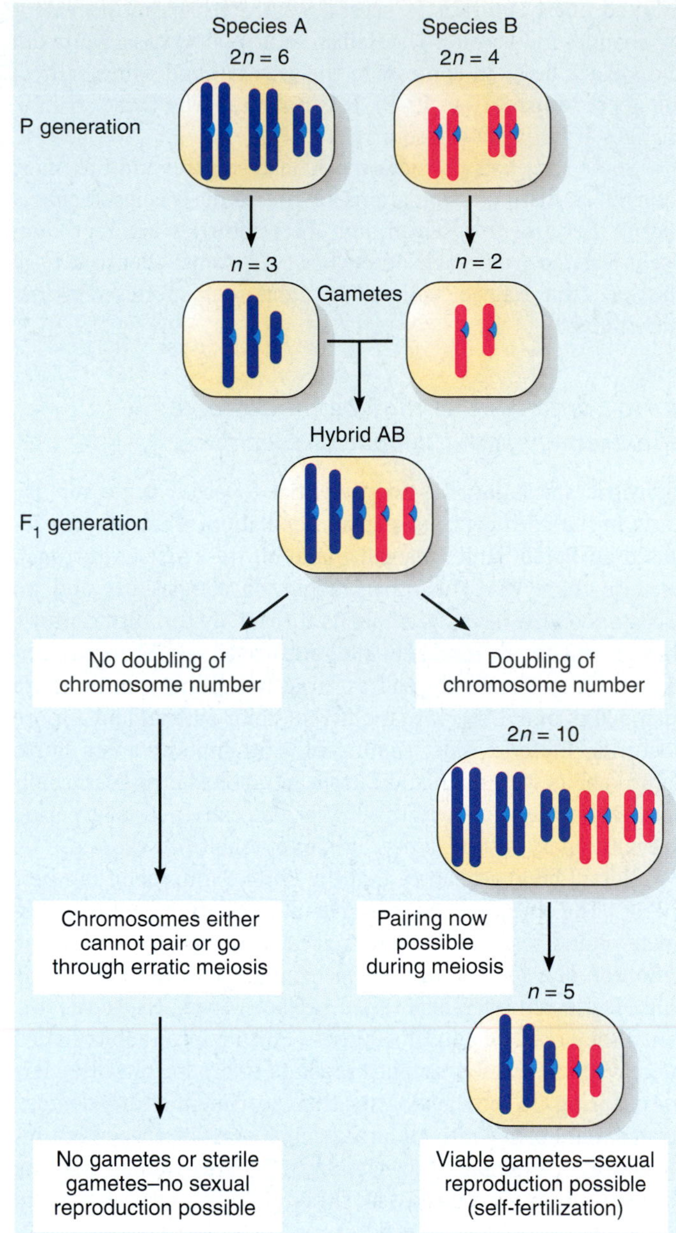

Figure 19–10 Sympatric speciation by allopolyploidy in plants. When two species (designated the P generation) successfully interbreed, the interspecific hybrid offspring (the F_1 generation) are almost always sterile *(bottom left).* If the chromosomes double, proper synapsis and segregation of the chromosomes can occur, and viable gametes can be produced *(bottom right).* (Unduplicated chromosomes are shown for clarity.)

The mechanism of sympatric speciation has been experimentally verified for many plant species. One example is a group of species, collectively called hemp nettles, that occurs in temperate parts of Europe and Asia. One hemp nettle, *Galeopsis tetrahit* ($2n = 32$), is a naturally occurring allopolyploid thought to have formed by the hybridization of two species, *Galeopsis pu-*

Figure 19–11 Sympatric speciation of a primrose. An allopolyploid primrose, *Primula kewensis,* arose in 1898 as an allopolyploid derived from the interspecific hybridization of *P. floribunda* and *P. verticillata.* Today *P. kewensis* is a popular houseplant.

bescens ($2n = 16$) and *Galeopsis speciosa* ($2n = 16$). This process occurred in nature but was experimentally reproduced. *Galeopsis pubescens* and *G. speciosa* were crossed to produce F_1 hybrids, most of which were sterile. Nevertheless, both F_2 and F_3 generations were produced. The F_3 generation included a polyploid plant with $2n = 32$ that self-fertilized to yield fertile F_4 offspring that could not mate with either of the parental species. These allopolyploid plants had the same appearance and chromosome number as the naturally occurring *G. tetrahit.* When the experimentally produced plants were crossed with the naturally occurring *G. tetrahit,* a fertile F_1 generation was formed. Thus, the experiment duplicated the speciation process that occurred in nature.

Changing ecology can cause sympatric speciation in animals

Biologists have observed the occurrence of sympatric speciation in animals, but its significance—how often it occurs and under what conditions—is still actively debated. Many examples involve parasitic insects and rely on genetic mechanisms other than polyploidy. For example, in the 1860s in the Hudson River Valley of New York, a population of fruit maggot flies *(Rhagoletis pomonella)* parasitic on the small red fruits of native hawthorn trees was documented to have switched to a new host, domestic apples, which had been introduced from Europe. Although the sister populations (hawthorn maggot flies and apple maggot flies) continue to occupy the same geographical area, no gene flow occurs between them because they eat, mate, and lay their eggs on different hosts (Fig. 19–12). In other words, because the hawthorn and apple maggot flies have diverged and are reproductively isolated from each other, they have effectively become separate species. Most entomologists, however, still recognize hawthorn and apple maggot flies as a single species because their appearance is virtually identical.

In situations like the fruit maggot flies, it is thought that a mutation arises in an individual and spreads through a small group of insects by sexual reproduction. The mutation causes disruptive selection (see Chapter 18), which favors a distinct phenotype by allowing the mutants to have a different ecological opportunity—in this case, to parasitize a different host species—than the original phenotype. Additional mutations may occur that cause the sister populations to diverge even further.

In 1988 biologists reported that hawthorn and apple maggot flies have genetic differences in three chromosomal regions and identified genetic markers associated with changes in the timing of fly development. Both hawthorn and apple larvae (maggots) tunnel out of the fruit before winter, drop to the ground, and burrow into the soil. However, hawthorn maggot flies emerge from the ground as adults later in the summer than apple

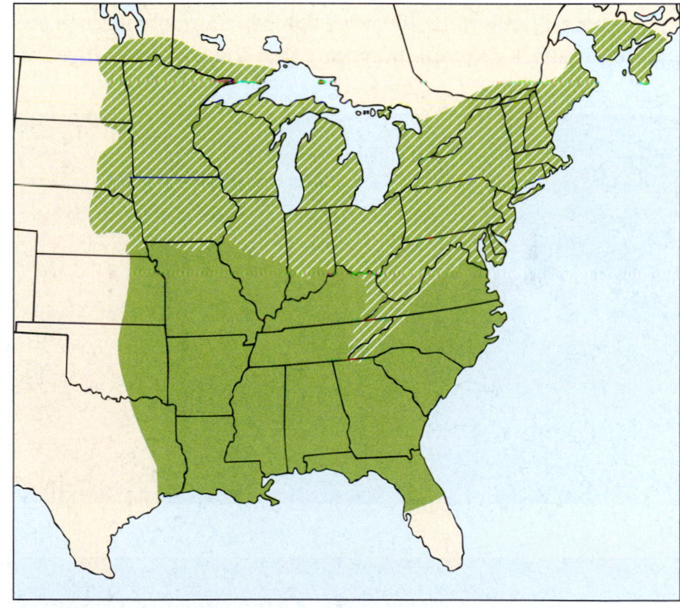

Hawthorn maggot fly range

Apple maggot fly and hawthorn maggot fly range

Figure 19–12 Ranges of apple and hawthorn maggot flies. Apple and hawthorn maggot flies are sympatric throughout the northern half of the hawthorn maggot fly's range. *(Adapted from Bush, G.L. "Sympatric Host Race Formation and Speciation in Frugivorous Flies of the Genus* Rhagoletis *[Diptera, Tephritidae]." Evolution, Vol. 23, No. 2, Jun. 1969)*

maggot flies, a difference that further contributes to their reproductive isolation.

Biologists have studied the speciation of colorful fishes known as cichlids (pronounced "sik-lids") in several East African lakes (Fig. 19–13). The different species of cichlids in a given lake have remarkably different eating habits. Some graze on algae; some consume dead organic material at the bottom of the lake; and others are predatory and eat plankton (microscopic aquatic organisms), insect larvae, the scales off fish, or even other cichlid species. In some cichlids food preferences are related to size (e.g., smaller cichlids consume plankton), which in turn is related to mating preference (small, plankton-eating cichlids mate only with other small, plankton-eating cichlids). In other cichlids related species do not differ in size but only in color, and females mate with males of specific colors. Recent DNA sequence data indicate that the cichlid species within a single lake are more closely related to one another than they are to fishes in nearby lakes or rivers. These molecular data suggest that cichlid species evolved sympatrically, or at least within the confines of a lake, rather than by repeated colonizations by fish populations in nearby rivers.

How rapidly did sympatric speciation occur in cichlids? In 1996 scientists published seismic and drill core data that suggest that the more than 500 endemic (i.e., found nowhere else) cichlid species in Lake Victoria evolved in a remarkably short time—less than 12,400 years. This inference is based on evidence that Lake Victoria dried up completely during the late Pleistocene (from about 17,000 to 12,400 years ago) when much of north and equatorial Africa was arid. It appears that the cichlids evolved after the climate became wetter and Lake Victoria refilled. If future data substantiate this conclusion, it means that the evolution of Lake Victoria's cichlids is the fastest known for such a large number of vertebrate species. (See *On the Cutting Edge: Explaining the Rapid Loss of Cichlid Diversity in Lake Victoria* for a discussion of how human activities have altered the evolutionary mechanism that maintains cichlid diversity.)

Reproductive isolation breaks down in hybrid zones

When two populations have significantly diverged as a result of geographical separation, there is no easy way to determine if the speciation process is complete (recall the disagreement about whether Porto Santo rabbits are a separate species or a *subspecies,* which is a taxonomic subdivision of a species). If such populations, subspecies, or species come into contact, they may hybridize where they meet, forming a **hybrid zone,** or area of overlap in which interbreeding occurs. Hybrid zones are typically narrow, presumably because the hybrids are not well adapted for either parental environment, and the hybrid population is typically very small compared with the parental populations.

On the Great Plains of North America, red-shafted and yellow-shafted flickers (types of woodpeckers) meet and interbreed. The red-shafted flicker, named for the male's red underwings and tail, is found in the western part of North America, from the Great Plains to the Pacific Ocean. Yellow-shafted flicker males, which have yellow underwings and tails, range east of the Rock-

(a)

(b)

(c)

Figure 19–13 Color variation in Lake Victoria cichlids. **(a)** *Pundamilia pundamilia* males have bluish-silver bodies. **(b)** *Pundamilia nyererei* males have red backs. **(c)** *Pundamilia* "red head" males have a red chest. (The "red head" species has not yet been scientifically named.) Some evidence suggests that changes in male coloration may be the first step in speciation of Lake Victoria cichlids. Later, other traits, including ecological characteristics, diverge. (Female cichlids generally have cryptic coloration; their drab colors help them blend into their surroundings.) *(Courtesy of Ole Seehausen/Leiden University, The Netherlands, and Hull University, United Kingdom)*

ies. Hybrid flickers, which form a stable hybrid zone from Texas to southern Alaska, are varied in appearance, although many have orange-colored underwings and tails.

Process of Science Biologists are not in agreement about whether the red-shafted and yellow-shafted flickers are separate species or geographical subspecies. According to the biological species concept, if red-shafted and yellow-shafted flickers are two species, they should maintain their reproductive isolation. On the other hand, the flicker hybrid zone has not expanded, that is, the two types of flickers have maintained their distinctiveness and have not rejoined into a single, freely interbreeding population.

The study of hybrid zones has made important contributions to what is known about speciation. As in other fields of science, disagreements and differences of opinion are an important part of the scientific process because they stimulate new ideas, hypotheses, and experimental tests that expand our base of scientific knowledge.

EVOLUTIONARY CHANGE CAN OCCUR RAPIDLY OR GRADUALLY

We have seen that speciation is hard for us to directly observe as it occurs. Does the fossil record provide clues about how rapidly new species arise? Biologists have long recognized that the fossil record lacks many transitional forms; the starting points (ancestral species) and the end points (new species) are present, but the intermediate stages in the evolution from one species to another are often lacking. This observation has traditionally been explained by the incompleteness of the fossil record. Biologists have attempted to fill in the missing parts with new fossil discoveries, much as a writer might fill in the middle of a novel when the beginning and end are already there.

Two different models—punctuated equilibrium and gradualism—have been developed to explain evolution as observed in the fossil record (Fig. 19–14). The **punctuated equilibrium** model was proposed by paleontologists who question whether the fossil record really is as incomplete as it initially appeared. First advanced by Stephen Jay Gould and Niles Eldredge in 1972, the punctuated equilibrium model suggests that the fossil record accurately reflects evolution as it actually occurs. That is, in the history of a species, long periods of **stasis** (little or no evolutionary change) are punctuated, or interrupted, by short periods of rapid speciation that are perhaps triggered by changes in the environment, that is, periods of great evolutionary stress. Thus, speciation normally proceeds in "spurts." These relatively short periods of active evolution (e.g., 100,000 years) are followed by long periods (e.g., 2 million years) of stability.

With punctuated equilibrium, speciation can occur in a relatively short period. Keep in mind, however, that a "short" amount of time for speciation may be thousands of years. Such a span is short when compared with the several million years of a species'

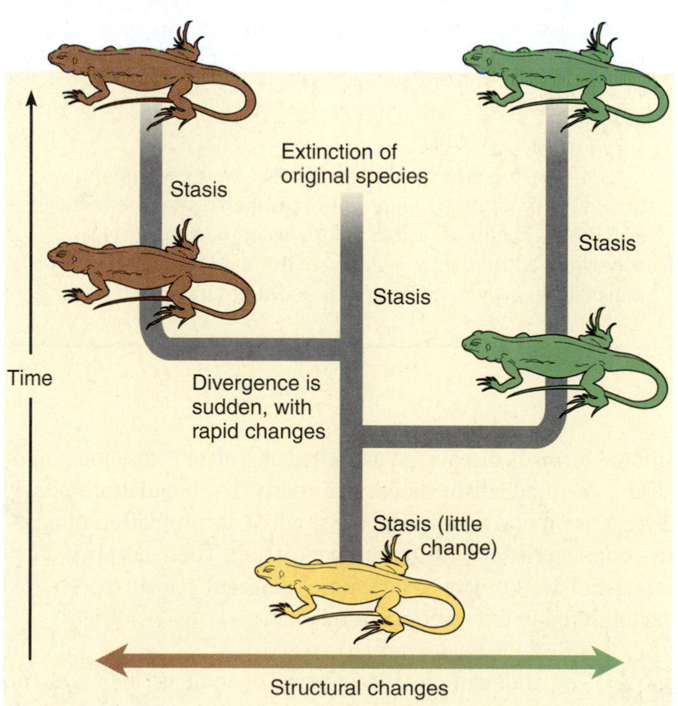

(a) Punctuated equilibrium

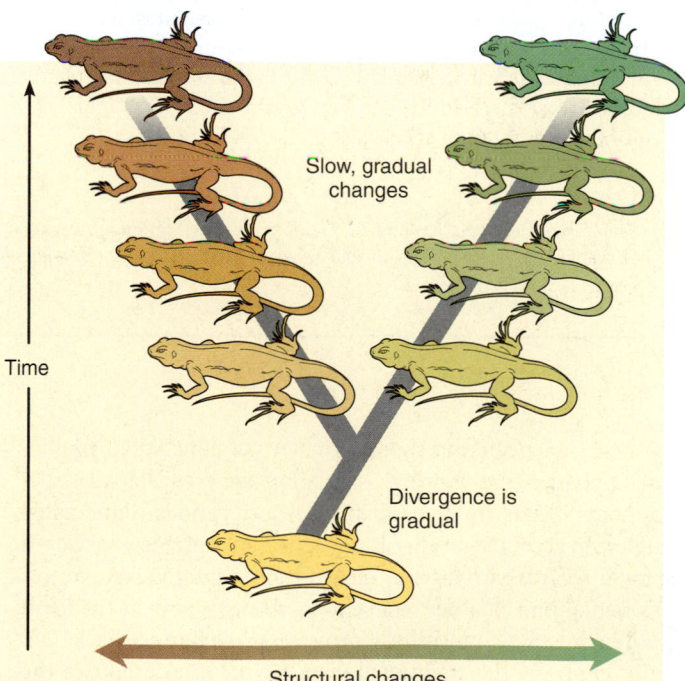

(b) Gradualism

Figure 19–14 Punctuated equilibrium and gradualism. In this figure, structural changes in the lizards are represented by changes in skin color. **(a)** In punctuated equilibrium, long periods of stasis are interrupted by short periods of rapid speciation. **(b)** In gradualism, a slow, steady change in species occurs over time.

Process of Science

ON THE CUTTING EDGE

Explaining the Rapid Loss of Cichlid Diversity in Lake Victoria

HYPOTHESIS: The number of cichlid species in Lake Victoria is declining because of changes in their habitat that affect sexual isolation between species.

METHOD: Take field measurements in 22 different locations in Lake Victoria. Verify these field observations with laboratory experiments.

RESULTS: Increasing turbidity of Lake Victoria's water is preventing females of cichlid species from choosing mates of their own species, which they do based on color variations among the males of different species. As a result, females are interbreeding with males of other species, and the number of cichlid species declines as the distinctive gene pools of different species are merging.

CONCLUSION: Cichlid diversity has declined and will continue to decline unless effective measures are taken to reduce pollution in Lake Victoria immediately.

The hundreds of small, colorful cichlid species that diversified in Lake Victoria have always intrigued evolutionary biologists. However, many of these species are rapidly disappearing, and about half of the 500 species that biologists estimate lived in Lake Victoria in 1978 are now extinct. The Nile perch, a voracious predator that was deliberately introduced into the lake in 1960 to stimulate the local fishing economy and whose population exploded during the 1980s, has been blamed for the extinction of most of the cichlid species that have disappeared.

Biologists from Leiden University in the Netherlands observed, however, that species were also disappearing in lake habitats where the Nile perch is known to have little or no impact. Ole Seehausen and colleagues measured environmental features at 22 localities with such habitats and found strong correlations between bright and diverse male colors, the number of species, and clear, well-lit water.*

The better the light in a given part of the lake, the more colorful the males and the greater the number of species present at a specific site. Each species has its own distinct male coloration, which matches the preferences of females of that species in choosing mates, and has evolved under sexual selection. **Sexual selection,** which is choosing a mate based on its color or some other characteristic, is discussed in Chapter 50. In clear water, members of each species do not interbreed with members of other species.

As nearby forests have been cut, soil erosion has made the water of Lake Victoria more turbid (cloudy). Increased agricultural activity in the area has also contributed fertilizer as well as sediment pollution. Water transparency in open (deep) lake water has decreased from an average of 6.8 m in the 1920s to 2.2 m in the 1990s, whereas water transparency in shallow areas along the shoreline has decreased from 3 m to 1.5 m in the past decade.

Seehausen demonstrated that as the water became more turbid from pollution, light could not penetrate as effectively, and the spectrum of visible colors became narrower. He hypothesized that, as a result, females could not distinguish males of their own species from males of closely related

* Seehausen, O., J.J.M. Van Alphen, and F. Witte, "Cichlid Fish Diversity Threatened by Eutrophication that Curbs Sexual Selection." *Science,* Vol. 277, 19 Sept. 1997.

existence. Biologists who support the idea of punctuated equilibrium emphasize that sympatric speciation and even allopatric speciation can occur in such relatively short periods. Punctuated equilibrium accounts for the abrupt appearance of a new species in the fossil record, with little or no evidence of intermediate forms. Proponents think that few transitional forms appear in the fossil record because few transitional forms occurred during speciation.

In contrast, the traditional view of evolution espouses the **gradualism** model, in which evolution proceeds continuously over long periods. Gradualism is rarely observed in the fossil record because the record is incomplete. (Recall from Chapter 17 that the conditions required for fossil formation are quite precise. Most organisms decompose when they die, leaving no trace of their existence.) Occasionally, a complete fossil record of transitional forms is discovered and cited as a strong case for gradualism. The gradualism model maintains that populations slowly diverge from one another by the gradual accumulation of adaptive characteristics within each population. These adaptive characteristics accumulate as a result of different selective pressures encountered in different environments.

Process of Science The abundant fossil evidence of long periods with no change in a species has been used to argue against the gradualism model of evolution. Gradualists, however, maintain that any periods of stasis evident in the fossil record are the result of stabilizing selection (see Chapter 18). They also emphasize that stasis in fossils is deceptive because fossils do not reveal all aspects of evolution. While fossils display changes in external

species. As discussed in the chapter, the cichlids of Lake Victoria are young species, having evolved during the past 12,000 years. Their relative youth means that these species may not have yet evolved reproductive isolating mechanisms other than mate preference based on color. It also means that closely related species are able to interbreed without loss of fertility. With increasing turbidity, males lost their bright colors because producing them was no longer rewarded, and females have mated with males of other species, producing fertile, healthy offspring. Thus, males are losing their bright colors, and many original species are being replaced by fewer hybrids.

Further support for this hypothesis comes from laboratory experiments. When aquaria were well lit, females of two closely related cichlid species (one with blue males, the other one with red) consistently chose mates of their own species over those of the other species. When biologists blocked the light to simulate the turbid conditions in parts of Lake Victoria, however, females often chose males of the other species.

More recently, Seehausen has evaluated possible explanations for the extraordinary diversity in cichlids. Based on his observations and the observations of others, Seehausen hypothesized that cichlids speciate by disruptive sexual selection on polymorphic coloration.† **Polymorphic coloration** is the presence in a population of two or more colors in either males or females. If cichlid mating preferences are based on color, and if color preferences vary between individuals (and it appears they often do), this mating behavior tends to cause disruptive sexual selection. Recall from Chapter 18 that disruptive selection results in a divergence, or splitting apart, of distinct groups of individuals within a population. **Disruptive sexual selection** is a special type of disruptive selection that is based on mate preference.

Seehausen tested predictions of his hypothesis by studying the geographical distribution, coloring, and relatedness of 41 rock-dwelling cichlid species endemic to Lake Victoria. In some cases, he found that speciation was "in progress," with members of the same species in the process of diverging into two or more groups based solely on differences in male or female coloration. These groups (known as *color morphs*) were not completely reproductively isolated and so were considered incipient species (species just beginning to form).‡ Seehausen also observed that closely related species that were completely reproductively isolated exhibited the same color variations he had found associated with disruptive sexual selection in incipient species, and these closely related species usually had sympatric distributions.

Seehausen's research, like all good science, seeks to reduce the apparent complexity of the world—in this case, the extraordinary number of cichlid species—to general principles that can provide new insights—in this case, into evolutionary processes. Although Seehausen's hypothesis on disruptive sexual selection is well supported for rock-dwelling Lake Victoria cichlids, it must be further tested with cichlid species that inhabit other environments and other lakes before it can be widely accepted.

† Seehausen, O. and J.J.M. Van Alphen. "Can Sympatric Speciation by Disruptive Sexual Selection Explain Rapid Evolution of Cichlid Diversity in Lake Victoria?" *Ecology Letters,* Vol. 2, 1999.

‡ Seehausen, O., J.J.M. Van Alphen, and R. Lande. "Color Polymorphism and Sex Ratio Distortion in a Cichlid Fish as an Incipient Stage in Sympatric Speciation by Sexual Selection." *Ecology Letters,* Vol. 2, 1999.

structure and skeletal structure, genetic changes in physiology, internal structure, and behavior—all of which also represent evolution—are not evident. Gradualists recognize rapid evolution only when strong directional selection occurs.

Punctuated equilibrium and gradualism are related to cladogenesis and anagenesis, the two patterns of evolution discussed earlier in this chapter. The key prediction of punctuated equilibrium is that structural changes should be concentrated at the point of speciation through cladogenesis, as shown in Figure 19–14*a*. Gradualism predicts that structural changes need not be associated with cladogenesis and may take place at any point in a species' history. Note that anagenic change is shown in the right-hand and left-hand branches of Figure 19–14*b*. This difference between punctuated equilibrium and gradualism is important to the process of science since what each model predicts can be tested against future discoveries in the fossil record.

Many biologists embrace both models to explain the fossil record; they also contend that the pace of evolution may be abrupt in certain instances and gradual in others and that neither punctuated equilibrium nor gradualism exclusively characterizes the complex changes associated with evolution. Other biologists do not view the distinction between punctuated equilibrium and gradualism as real. They suggest that genetic changes occur gradually and at a roughly constant pace and that the majority of these mutations do not cause speciation. When the mutations that do cause speciation occur, they are dramatic and produce a pattern consistent with the punctuated equilibrium model.

MACROEVOLUTION INVOLVES MAJOR EVOLUTIONARY EVENTS

Macroevolution includes dramatic changes that occur over long time spans in evolution. One concern of macroevolution is to explain evolutionary novelties, which are large phenotypic changes such as the appearance of wings with feathers during the evolution of birds from reptiles. These phenotypic changes are so great that the new species possessing them are assigned to different genera or higher taxonomic categories. Studies of macroevolution also attempt to discover and explain major changes in species diversity through time, such as occur during adaptive radiation, when many species appear, and mass extinction, when many species disappear. Thus, evolutionary novelties, adaptive radiation, and mass extinction are important aspects of macroevolution.

Evolutionary novelties originate through modifications of preexisting structures

New designs arise from structures already in existence. A change in the basic pattern of an organism can produce something unique, such as wings on insects, flowers on plants, and feathers on birds. Usually these evolutionary novelties are variations of some preexisting structures, called **preadaptations,** that originally fulfilled one role but were subsequently modified in a way that was adaptive for a different role. Feathers, which evolved from reptilian scales and may have originally provided thermal insulation in primitive birds and some dinosaurs (see Chapter 20), represent a preadaptation for flight. That is, with gradual modification, feathers evolved to function in flight as well as to fulfill their original thermoregulatory role. (Interestingly, a few feather-footed birds exist; this phenotype is the result of a change in gene regulation that alters scales, normally found on bird feet, into feathers.)

Similarly, the mammalian middle ear bones originated from modified jawbones of reptiles. Consider the malleus, a hammer-shaped bone that is in contact with the eardrum in the mammalian middle ear (see Chapter 41). During the evolution of mammals, it is hypothesized that the articular bone of the reptile lower jaw became smaller and lighter and was gradually transformed into the malleus bone. Supporting evidence includes embryological studies, which indicate that the early stages of articular bone development in reptiles are the same as developmental stages in the malleus bone in mammals. In addition, fossils of mammal-like reptiles show some of the intermediate steps between articular and malleus bones.

How do such evolutionary novelties originate? Many are probably due to changes during **development,** which is the orderly sequence of events that occur as an organism grows and matures. Regulatory genes may exert control over hundreds of other genes during development, and very slight genetic changes in regulatory genes could ultimately cause major structural changes in the organism (see section on developmental biology in Chapter 17).

For example, during development, most organisms exhibit varied rates of growth for different parts of the body, known as **allometric growth** (from the Greek *allo,* "different," and *metr,* "measure"). The size of the head in human newborns is large in proportion to the rest of the body. As a human grows and matures, its torso, hands, and legs grow more rapidly than the head. Allometric growth is found in many organisms, including the male fiddler crab with its single, oversized claw, and the ocean sunfish with its enlarged tail (Fig. 19–15). If growth rates are al-

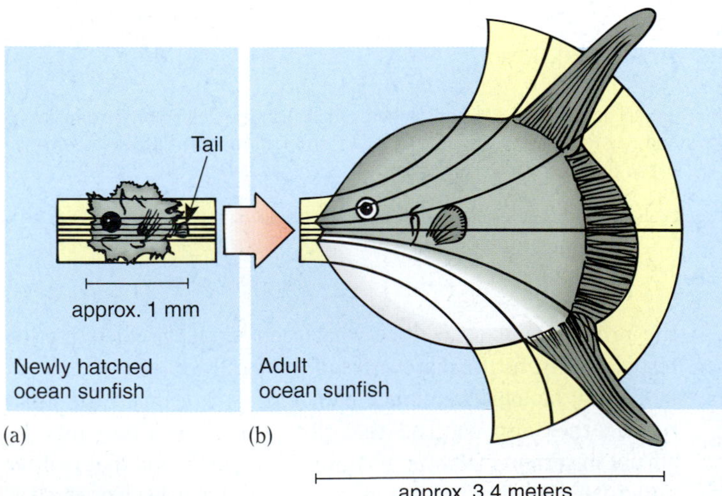

(c)

Figure 19–15 Allometric growth in the ocean sunfish. The tail end of an ocean sunfish *(Mola mola)* grows faster than the head end, resulting in the unique shape of the adult. **(a)** A newly hatched ocean sunfish, only 1 mm long, has an extremely small tail. **(b)** This allometric transformation can be visualized by drawing rectangular coordinate lines through a picture of the juvenile fish and then changing the coordinate lines mathematically. **(c)** An ocean sunfish swims off the coast of southern California. The adult ocean sunfish may reach 4 m (13 ft) and weigh about 1500 kg (3300 lb). *(Richard Herrmann)*

tered even slightly, drastic changes in the shape of an organism may result, changes that may or may not be adaptive. For example, allometric growth may help explain the extremely small and relatively useless forelegs of the dinosaur *Tyrannosaurus rex,* as compared with its ancestors.

Sometimes novel evolutionary changes occur when a species undergoes changes in the *timing* of development in comparison with its ancestor. Consider, for example, the changes that would occur if juvenile characteristics were retained in the adult stage, a phenomenon known as **paedomorphosis** (from the Greek *paed,* "child," and *morph,* "form"). Adults of some salamander species have external gills and tail fins, features found only in the larval (immature) stages of other salamanders. Retention of external gills and tail fins throughout life obviously alters the salamander's behavioral and ecological characteristics (Fig. 19–16). Perhaps such salamanders succeeded because they had a selective advantage over "normal" adult salamanders, that is, by remaining aquatic, they did not have to compete for food with the terrestrial adult forms of related species. The paedomorphic forms also escaped the typical predators of terrestrial salamanders (although they had other predators in their aquatic environment). The biological basis of paedomorphosis in salamanders has been studied and is probably the result of mutations in genes that block the production of hormones that stimulate metamorphic changes. When paedomorphic salamanders receive hormone injections, they develop into normal adults.

Figure 19–16 Paedomorphosis in a salamander. An adult axolotl salamander *(Ambystoma mexicanum)* retains the juvenile characteristics of external gills (feathery structures protruding from the neck) and a tail fin *(not visible).* Paedomorphosis allows the axolotl to remain permanently aquatic and to reproduce without developing typical adult characteristics. *(Jane Burton/Bruce Coleman, Inc.)*

Adaptive radiation is the diversification of an ancestral species into many species

Adaptive radiation is the evolutionary diversification of many related species from one or a few ancestral species in a relatively short period. The concept of adaptive zones was developed to help explain why adaptive radiations take place. **Adaptive zones** are new ecological opportunities that were not exploited by an ancestral organism. At the species level, an adaptive zone is essentially identical to one or more similar *ecological niches* (the functional roles of species within a community; see Chapter 52). Examples of adaptive zones include nocturnal flying to catch small insects; grazing on grass while migrating across a savanna; and swimming at the ocean's surface to filter out plankton. When many adaptive zones are empty, as was the case of Lake Victoria when it refilled some 12,000 years ago (discussed earlier in the chapter), colonizing species such as the cichlids are able to rapidly diversify and exploit them.

Because islands have fewer species than do mainland areas of similar size, latitude, and topography, vacant adaptive zones are more common on islands than on continents. Consider the Hawaiian honeycreepers, a group of related birds found on the Hawaiian Islands. When the honeycreeper ancestors reached Hawaii, few other birds were present. The succeeding generations of honeycreepers quickly diversified into many new species and, in the process, occupied the many available adaptive zones that are normally occupied by finches, honeyeaters, treecreepers, and woodpeckers on the mainland. The diversity of their bills, like those of Galapagos finches (see Chapter 17), is a particularly good illustration of adaptive radiation (Fig. 19–17). Some honeycreeper bills are curved to extract nectar out of tubular flowers, for example, whereas others are short and thickened for ripping away bark in search of insects.

Another example of vacant adaptive zones involves the Hawaiian silverswords, 28 species of closely related plants found only on the Hawaiian Islands. When the silversword ancestor, a California plant related to daisies, reached the Hawaiian Islands, many diverse environments, such as exposed lava flows, dry woodlands, moist forests, and bogs, were present and more or less unoccupied. The succeeding generations of silverswords quickly diversified, occupying the many adaptive zones available to them. The diversity in their leaves, which changed during the course of natural selection to enable different populations to adapt to various levels of light and moisture, is a particularly good illustration of adaptive radiation (Fig. 19–18). Leaves of silverswords that are adapted to shady moist forests are large, for example, whereas those of silverswords living in exposed dry areas are small. The leaves of silverswords living on exposed volcanic slopes are covered with dense silvery hairs that may reflect some of the intense ultraviolet radiation off the plant.

Adaptive radiation appears to be more common during periods of major environmental change, but it is difficult to determine if these changes actually induce adaptive radiation. It is possible that major environmental change indirectly affects

Figure 19–17 Adaptive radiation in Hawaiian honeycreepers. Compare the various beak shapes and methods of obtaining food. Many honeycreeper species are now extinct or nearing extinction as a result of human activities, including the destruction of habitat and the introduction of predators such as rats, dogs, and pigs. By late 2000, for example, the poo-uli population consisted of three known individuals.

adaptive radiation by increasing the rate of extinction. Extinction produces empty adaptive zones, which provide new opportunities for those species that remain. Mammals, for example, had existed as small nocturnal insectivores (insect eaters) for millions of years before undergoing adaptive radiation leading to the modern mammalian orders. This radiation was presumably triggered by the extinction of the dinosaurs. Mammals diversified and exploited a variety of adaptive zones relatively soon after the dinosaurs' demise. Flying bats, running gazelles, burrowing moles, and swimming whales all originated from the small, insect-eating, ancestral mammals.

The appearance of novel features is usually associated with major periods of adaptive radiation. For example, shells and skeletons may have been the evolutionary novelties responsible for a period of adaptive radiation at the beginning of the Paleozoic era (see Chapter 20), in which most animal phyla, living and extinct, appeared. Care must be taken in interpreting a cause-and-effect relationship between the appearance of a novel feature and adaptive radiation, however. It is tempting to take a simplistic approach and assume, for example, that the evolution of the flower facilitated the adaptive radiation of thousands of species of flowering plants. It is true that the flowering plants diversified after the evolution of the flower, which may have presented a more competitive method of sexual reproduction because it permitted pollination by insects and other animals. However, adaptive radiation in the flowering plants may instead be a consequence of other adaptations that also evolved. (Chapter 27 discusses other flowering plant adaptations as well as their highly successful mode of reproduction.)

Extinction is an important aspect of evolution

Extinction, the end of a lineage, occurs when the last individual of a species dies. The loss is permanent, for once a species is extinct it can never reappear. Extinctions have occurred continually since the origin of life; by one estimate, only 1 species is alive today for every 2000 that have become extinct. Extinction is the eventual fate of all species, in the same way that death is the eventual fate of all individual organisms.

Although extinction has a negative short-term impact on species richness, it can facilitate evolution over a period of thousands to millions of years. As mentioned previously, when species become extinct, their adaptive zones become vacant. Consequently, those organisms still living are presented with new opportunities for speciation and may diverge, filling in some of the unoccupied zones. In other words, the extinct species may eventually be replaced by new species.

(a) (b) (c)

Figure 19–18 Adaptive radiation in Hawaiian silverswords. The ancestor of the Hawaiian silverswords was a California plant similar to the daisy. The 28 silversword species are found in three closely related genera. They live in a variety of habitats, from rain forests to exposed mountain slopes. **(a)** The Haleakala silversword (*Argyroxyphium sandwicense* ssp. *macrocephalum*) is found only in the cinders on the upper slope of Haleakala Crater on the island of Maui. This plant is adapted to low precipitation and high levels of ultraviolet radiation. **(b)** This silversword species (*Wilkesia gymnoxiphium*), which superficially resembles a yucca, is found only along the slopes of Waimea Canyon on the island of Kauai. **(c)** *Daubautia scabra* is a small, herbaceous silversword found in moist to wet environments on several Hawaiian islands. (The fern fronds in the background give an idea of the small size of *D. scabra*.) *(a–c, Jack Jeffrey Photography)*

During the long history of life, extinction appears to have occurred at two different rates. The continuous, low-level extinction of species is sometimes called **background extinction.** In contrast, five or possibly six times during Earth's history, **mass extinctions** of numerous species and higher taxa have taken place in both terrestrial and marine environments. The most recent mass extinction, which occurred about 65 million years ago (mya), killed off many marine organisms, terrestrial plants, and vertebrates, including the last of the dinosaurs (Fig. 19–19). The time span over which a mass extinction occurred may have been several million years, but that is relatively short compared with the 3.5 billion years or so of Earth's history of life. Each period of mass extinction has been followed by a period of adaptive radiation of some of the surviving groups.

The causes of past episodes of mass extinction are not well understood. Both environmental and biological factors seem to have been involved. Major changes in climate could have adversely affected those plants and animals that lacked the genetic flexibility to adapt. Marine organisms, in particular, are adapted to a steady, unchanging climate. If Earth's temperature were to increase or decrease by just a few degrees overall, many marine species would probably perish.

It is also possible that past mass extinctions were due to changes in the environment induced by catastrophes. If a large comet or small asteroid collided with Earth, for example, the dust ejected into the atmosphere on impact could have blocked much of the sunlight. In addition to disrupting the food chain by killing many plants (and therefore terrestrial animals), this event would have lowered Earth's temperature, leading to the death of many marine organisms. Evidence that the extinction of dinosaurs was caused by an extraterrestrial object's collision with Earth continues to accumulate (see Chapter 20).

Biological factors also trigger extinction. Competition among species may lead to the extinction of those species that cannot compete effectively. The human species, in particular, has had a profound impact on the rate of extinction. The habitats of many animal and plant species have been altered or destroyed by humans, and habitat destruction can result in a species' extinction. Some biologists think that we have entered the greatest mass extinction episode in Earth's history. (Extinction is discussed further in Chapter 55.)

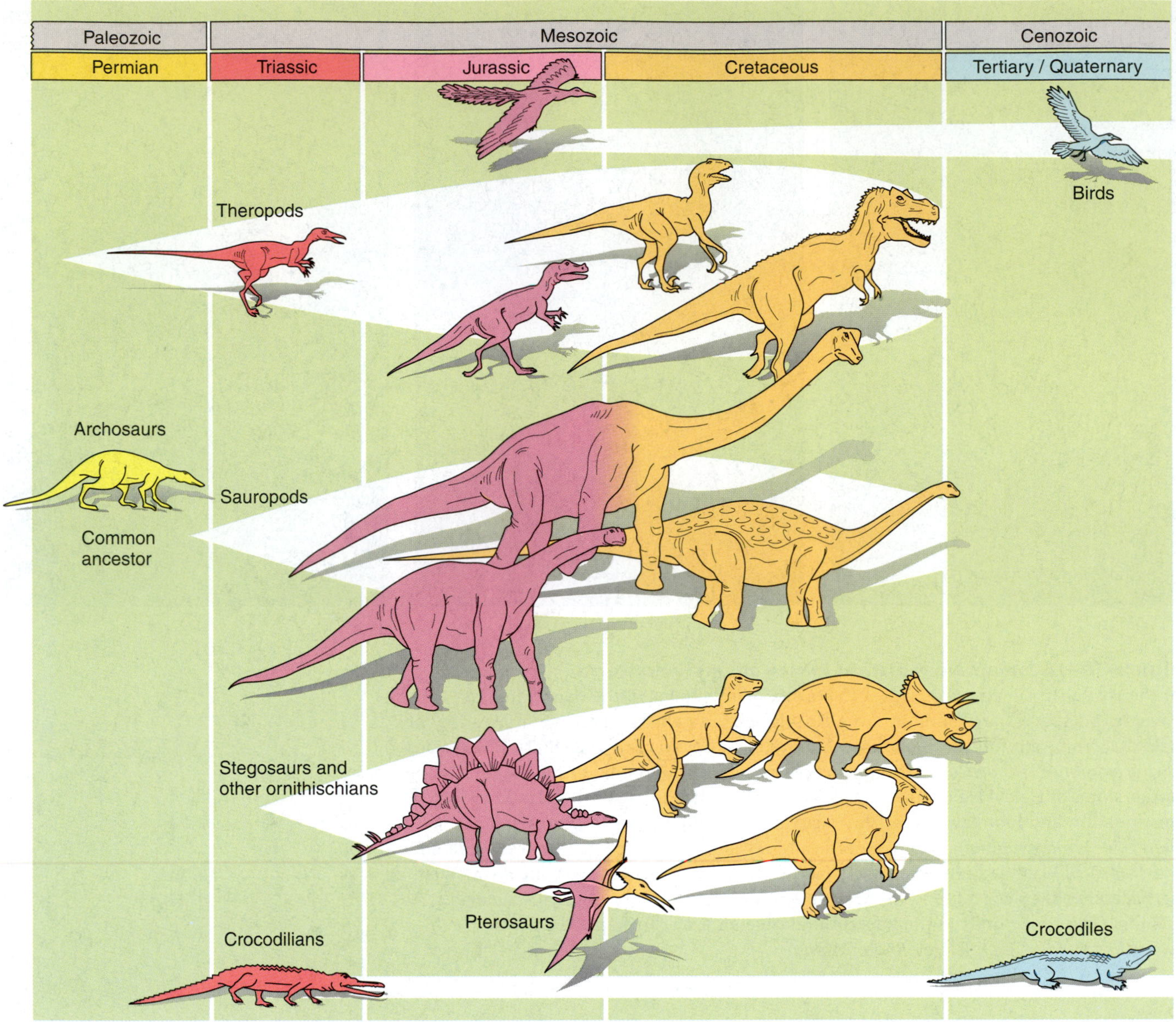

Figure 19–19 Mass extinction of the archosaurs. At the end of the Cretaceous period, approximately 65 mya, a mass extinction of many organisms, including the remaining dinosaurs, occurred. (Dinosaurs had already been declining in diversity throughout the latter part of the Cretaceous period.) Most of the archosaurs (one of five main groups of reptiles) became extinct. The only lines to survive were crocodiles and birds, both of which are archosauran descendants.

IS MICROEVOLUTION RELATED TO SPECIATION AND MACROEVOLUTION?

The concepts presented in Chapters 17 and 18 represent the **synthetic theory of evolution,** in which mutation provides the genetic variation on which natural selection acts. The synthetic theory of evolution combines Darwin's theory with important aspects of genetics. Many aspects of the synthetic theory of evolution have been tested and verified at the population and subspecies levels. Many biologists contend that microevolutionary processes (natural selection, mutation, genetic drift, and gene flow) account for the genetic variation within species and for the origin of new species. These biologists also think that macroevolution can be explained by microevolutionary processes.

A considerable body of data from many fields supports the synthetic theory of evolution as it relates to speciation and macroevolution. Consider, for example, the evolution of amphibians from fish, which was a major macroevolutionary event in the history of vertebrates. Study of the few known fossil intermediates has demonstrated that the transition from aquatic fish to terrestrial amphibian occurred as evolutionary novelties, such as changes in the limbs and skull roof, were added. These novelties accumulated as a succession of small changes over a period of 9 million to 14 million years. This time scale is sufficient to have allowed natural selection and other microevolutionary processes to have produced the novel characters.

Although few biologists doubt the role of natural selection and microevolution in generating specific adaptations, some question the *extent* of microevolution's role in the overall pattern of life's history. These biologists ask whether speciation and macroevolution have been dominated by microevolutionary processes or by external, chance events (e.g., an impact by an asteroid). Chance events do not "care" about adaptive superiority but instead lead to the random extinction or survival of species. In the case of an asteroid impact, for example, those species that survive may do so because they were "lucky" enough to be in a protected environment at the time of impact. If chance events have been the overriding factor during life's history, then microevolution cannot be the exclusive explanation for the biological diversity we have today (see additional discussion of the role of chance in evolution in Chapter 17).

SUMMARY WITH KEY TERMS

I. According to the **biological species concept,** a **species** consists of one or more populations whose members freely interbreed in nature to produce fertile offspring and do not freely interbreed with members of different species.

II. **Reproductive isolating mechanisms** restrict gene flow between species.
   A. **Prezygotic barriers** are reproductive isolating mechanisms that prevent fertilization from taking place.
      1. **Temporal isolation** occurs when two species reproduce at different times of the day, season, or year.
      2. In **habitat isolation,** two closely related species live and breed in different habitats in the same geographical area.
      3. In **behavioral isolation,** distinctive courtship behaviors prevent mating between species.
      4. **Mechanical isolation** is due to incompatible structural differences in the reproductive organs of similar species.
      5. In **gametic isolation,** gametes from different species are incompatible owing to molecular and chemical differences.
   B. **Postzygotic barriers** are reproductive isolating mechanisms that prevent gene flow after fertilization has taken place.
      1. **Hybrid inviability** is the death of interspecific embryos during development.
      2. **Hybrid sterility** prevents interspecific hybrids that survive to adulthood from reproducing successfully.
      3. **Hybrid breakdown** prevents the offspring of hybrids that survive to adulthood and successfully reproduce from reproducing beyond one or a few generations.
   C. Some of the genes involved in reproductive isolation in certain species have been identified.

III. **Speciation** is the evolution of a new species from an ancestral population. **Anagenesis** refers to relatively small, progressive evolutionary changes in a single lineage over long periods. **Cladogenesis** occurs when two or more populations of an ancestral species split and diverge, eventually forming two or more new species.
   A. **Allopatric speciation** occurs when one population becomes geographically isolated from the rest of the species and subsequently diverges.
      1. Speciation is more likely to occur if the original isolated population is small, because genetic drift is more significant in small populations.
      2. Examples of allopatric speciation include Death Valley pupfish, Kaibab squirrels, and Porto Santo rabbits.
   B. **Sympatric speciation** does not require geographical isolation.
      1. Sympatric speciation in plants results almost exclusively from **allopolyploidy,** in which a **polyploid** individual (one with more than two sets of chromosomes) is a hybrid derived from two species. Two examples of sympatric speciation by allopolyploidy are the kew primroses and hemp nettles.
      2. Sympatric speciation occurs in animals, such as fruit maggot flies and cichlids, but how often it occurs and under what conditions remain to be determined.
   C. A **hybrid zone** is an area in which two related populations, subspecies, or species meet and interbreed. Hybrid zones are typically narrow, presumably because the hybrids are not well adapted for either parental environment. Red-shafted and yellow-shafted flickers have a hybrid zone in the western part of North America.

IV. The interpretation of evolution, as observed in the fossil record, is currently being debated.
   A. According to the **punctuated equilibrium** model, evolution of species proceeds in spurts. Short periods of active speciation intersperse long periods of **stasis.**
   B. According to the **gradualism** model, populations slowly diverge from one another by the accumulation of adaptive characteristics within a population.

V. **Macroevolution** refers to dramatic evolutionary changes that occur over long time spans.
   A. Macroevolution includes the appearance of evolutionary novelties, which are phenotypic changes so great that the new species possessing them are assigned to different genera or higher taxonomic categories.
      1. Evolutionary novelties may be due to changes during **development.** Slight genetic changes in regulatory genes, for example, could ultimately cause major structural changes in the organism.
      2. Evolutionary novelties may originate from **preadaptations,** structures that originally fulfilled one role but changed in a way that was adaptive for a different role. Feathers and mammalian middle ear bones are examples of preadaptations.
      3. Changes in **allometric growth,** varied rates of growth for different parts of the body, result in overall changes in the shape of an organism. Examples include the ocean sunfish and the male fiddler crab.
      4. **Paedomorphosis,** the retention of juvenile characteristics in the adult, can occur owing to changes in the timing of development. Adult axolotl salamanders, with external gills and tail fins, are an example of paedomorphosis.
   B. **Adaptive radiation** is the process of diversification of an ancestral species into many new species.
      1. **Adaptive zones** are new ecological opportunities that were not exploited by an ancestral organism. When many adaptive zones are empty, colonizing species are able to rapidly diversify and exploit them.
      2. Hawaiian honeycreepers and silverswords both underwent adaptive radiation after their ancestors colonized the Hawaiian Islands.
   C. **Extinction** is the death of a species. When species become extinct, the adaptive zones that they occupied become vacant, allowing other species to evolve and fill them.
      1. **Background extinction** is the continuous, low-level extinction of species.
      2. **Mass extinction** is the extinction of numerous species and higher taxa in both terrestrial and marine environments.

POST-TEST

1. Two populations belong to the same species if (a) their members freely interbreed in nature (b) individuals from the two populations produce fertile offspring (c) their members do not interbreed with individuals of different species (d) a and c are correct (e) a, b, and c are correct
2. The zebrass is an example of (a) a fertile hybrid (b) a sterile hybrid (c) prezygotic barriers (d) a biological species (e) allopolyploidy
3. A prezygotic barrier prevents (a) the union of egg and sperm (b) reproductive success by an interspecific hybrid (c) the development of the zygote into an embryo (d) allopolyploidy from occurring (e) changes in allometric growth
4. The reproductive isolating mechanism in which two closely related species live in the same geographical area but reproduce at different times is (a) temporal isolation (b) behavioral isolation (c) mechanical isolation (d) gametic isolation (e) hybrid inviability
5. Interspecific hybrids, if they survive, are (a) always sterile (b) always fertile (c) usually sterile (d) usually fertile (e) never sterile
6. Which of the following evolutionary patterns is/are responsible for an increase in species diversity? (a) anagenesis (b) cladogenesis (c) prezygotic barriers (d) postzygotic barriers (e) hybrid zones
7. The first step leading to allopatric speciation is (a) hybrid inviability (b) hybrid breakdown (c) adaptive radiation (d) geographical isolation (e) paedomorphosis
8. The pupfish in the Death Valley region are an example of which evolutionary process? (a) background extinction (b) allopatric speciation (c) sympatric speciation (d) allopolyploidy (e) paedomorphosis
9. Sympatric speciation (a) is most common in animals (b) does not require geographical isolation (c) accounts for the evolution of the Hawaiian nene (d) involves the accumulation of gradual genetic changes (e) usually takes millions of years
10. Which of the following evolutionary processes is associated with allopolyploidy? (a) gradualism (b) allometric growth (c) sympatric speciation (d) mass extinction (e) preadaptation
11. According to the punctuated equilibrium model (a) populations slowly diverge from one another (b) the evolution of species occurs in spurts interspersed with long periods of stasis (c) evolutionary novelties originate from preadaptations (d) reproductive isolating mechanisms restrict gene flow between species (e) the fossil record, being incomplete, does not accurately reflect evolution as it actually occurred
12. The evolutionary conversion of reptilian scales into a bird's feathers is an example of (a) allometric growth (b) paedomorphosis (c) gradualism (d) hybrid breakdown (e) preadaptation
13. Adaptive radiation is common following a period of mass extinction, probably because (a) the survivors of a mass extinction are remarkably well adapted to their environment (b) the unchanging environment following a mass extinction drives the evolutionary process (c) many adaptive zones are empty (d) many ecological niches are filled (e) the environment induces changes in the timing of development for many species
14. Adaptive radiations do not appear to have ever occurred (a) on isolated islands (b) in birds such as honeycreepers (c) in environments colonized by few species (d) in plants such as silverswords (e) in environments with many existing species
15. The Hawaiian silverswords are an excellent example of which evolutionary process? (a) allometry (b) anagenesis (c) microevolution (d) adaptive radiation (e) extinction

REVIEW QUESTIONS

1. Compare the biological species concept with the morphological species concept.
2. Give an example of each of the following: (a) temporal isolation; (b) habitat isolation; (c) behavioral isolation; (d) mechanical isolation; (e) gametic isolation.
3. Describe the three types of postzygotic barriers and give an example of each.
4. Identify at least five geographical barriers that might lead to allopatric speciation.
5. Explain how hybridization and polyploidy can cause a new plant species to form in as little as one generation.
6. If you were in a debate and had to support the gradualism model, what would you say? How would you support the punctuated equilibrium model? Are these two ideas mutually exclusive?
7. Why are evolutionary novelties a concern of scientists studying macroevolution?
8. Give an example of each of the following: (a) preadaptation; (b) allometric growth; (c) paedomorphosis.
9. What roles do extinction and adaptive radiation play in macroevolution?

YOU MAKE THE CONNECTION

1. Why is allopatric speciation more likely to occur if the original isolated population is small?
2. Using the definition of the biological species concept given in the chapter introduction, is the Porto Santo rabbit an example of a speciation event that occurred in historical times, or is it an example of speciation in progress? Explain your answer.
3. Based on what you have learned about prezygotic and postzygotic isolating mechanisms, which reproductive isolating mechanism(s) would you say is/are probably at work between the Porto Santo rabbit and its mainland relative?
4. Based on what you have learned in the chapter, hypothesize what the common ancestor of the more than 20 species of desert pupfish may have looked like. (*Hint:* The ancestral species lived in one or more large lakes.) How could you test your hypothesis?
5. Could hawthorn and apple maggot flies be considered an example of assortative mating (discussed in Chapter 18)? Explain your answer.
6. Since mass extinction is a natural process that may facilitate evolution during the period of thousands to millions of years that follow it, should humans be concerned about the current mass extinctions that we are causing? Why or why not?

RECOMMENDED READINGS

Barlow, G.W. *The Cichlid Fishes: Nature's Grand Experiment in Evolution.* Perseus Publishing, Cambridge, MA, 2000. Examines the numerous specialized adaptations of cichlids and the many unanswered questions about cichlid behavior and evolution.

Boake, C.R.B. "Flying Apart: Mating Behavior and Speciation." *BioScience,* Vol. 50, No. 6, Jun. 2000. Discusses research on speciation as it relates to behavioral isolation in Hawaiian fruit flies. The author focuses on four fruit fly species, each endemic to a single island, although all four species live and mate in comparable habitats on their respective islands.

Eldredge, N. *Reinventing Darwin: The Great Debate at the High Table of Evolutionary Theory.* Wiley, New York, 1995. A lively, if partisan, survey of the intense debate in evolutionary theory stimulated by punctuated equilibrium and other ideas. (Recall that Niles Eldredge was one of the first to advance the punctuated equilibrium model.)

Hoffman, A.A., and M.J. Hercus. "Environmental Stress as an Evolutionary Force." *BioScience,* Vol. 50, No. 3, Mar. 2000. Examines what is known about how stressful environmental conditions affect evolutionary change.

Mayr, E. *Animal Species and Evolution.* Harvard University Press, Cambridge, 1963. This classic on animal evolution was written by the biologist who introduced the biological species concept and is a strong proponent of the allopatric speciation model.

Milius, S. "Superstud Grass Menaces San Francisco Bay." *Science News,* Vol. 154, 14 Nov. 1998. An introduced species of cordgrass threatens a native species with hybridization.

Milner, R. "Ernst Mayr at 93." *Natural History,* May 1997. An interview with one of the 20th century's greatest biologists.

Morell, V. "Earth's Unbounded Beetlemania Explained." *Science,* Vol. 281, 24 Jul. 1998. Describes research that seeks to explain why beetles had such an explosive adaptive radiation. (There are about 330,000 species of beetles.)

Stiassny, M.L.J., and A. Meyer. "Cichlids of the Rift Lakes." *Scientific American,* Vol. 280, No. 2, Feb. 1999. Beautiful illustrations accompany this fascinating account of the diversity in cichlids.

Turner, G. "Small Fry Go Big Time." *New Scientist,* 2 Aug. 1997. An overview of the evolution of cichlid species in Lake Victoria and other African great lakes.

Wright, K. "Pupfish in Peril." *Discover,* Jul. 1999. Pupfish species, a wonderful example of allopatric speciation, are increasingly threatened by human activities such as agriculture and mining. These activities pump groundwater from the aquifer that supplies the desert pools in which pupfish live.

- Visit our Web site at **http://www.info.brookscole.com/solomonbergmartin** for links to chapter-related resources on the World Wide Web. Additional on-line materials relating to this chapter can also be found on our Web site.

See chapter activity on BioActive Learner CD for additional help in mastering the chapter's material. Icon location in the chapter's margins shows which topics have tutorials or simulations in the CD.

20

The Origin and Evolutionary History of Life

Fossil trilobites *(Phacops rana).* These extinct arthropods, which were about 3 cm (1.2 in) long, flourished in the ocean during the Paleozoic era. They ranged from 1 mm to 1 m in length, depending on the species. Note the large, well-developed eyes (visible on either side of the head region). *(William E. Ferguson)*

LEARNING OBJECTIVES

After you have studied this chapter you should be able to

1. Describe the conditions thought to have existed on early Earth.
2. Contrast the prebiotic broth hypothesis and the iron-sulfur world hypothesis.
3. Outline the major steps hypothesized to have occurred in the origin of cells.
4. Explain how the evolution of photosynthetic autotrophs affected both the atmosphere and other organisms.
5. Describe the endosymbiont theory and summarize the evidence supporting it.
6. List the geological eras in chronological order and give approximate dates for each.
7. Briefly describe the distinguishing organisms and major biological events of Precambrian time and of the Paleozoic, Mesozoic, and Cenozoic eras.

The preceding three chapters were concerned with the evolution of organisms, but we have not yet dealt with what many regard as a fundamental question of biological evolution: How did life begin? Although biologists generally accept the hypothesis that life developed from nonliving matter, exactly how this process, called **chemical evolution,** occurred is not certain. Chemical evolution probably involved several stages. Current models suggest that small organic molecules first formed spontaneously and accumulated over time. These molecules may have been able to accumulate rather than being broken down (as occurs today) because conditions were different. The two factors presently responsible for breaking down organic molecules—free oxygen and living organisms—were absent from early Earth.

Large organic macromolecules such as proteins and nucleic acids could have then assembled from the smaller molecules. The macromolecules interacted with one another, combining into more complicated structures that could eventually metabolize and replicate. Natural selection favored macromolecular assemblages with cell-like structures. Their descendants eventually became the first true cells. After the first cells originated, they diverged over several billion years into the rich biological diversity that characterizes our planet today. Photosynthesis, aerobic respiration, and eukaryotic cell structure represent several major advances that evolved during the history of life.

Geological evidence, in particular the fossil record, provides us with much of what is known about the history of life, such as what kinds of organisms existed and where and when they lived. Certain organisms appear in the fossil record, then disappear and are replaced by others. Initially, unicellular prokaryotes predominated, followed by unicellular eukaryotes. The first multicellular eukaryotes—soft-bodied animals that did not leave many fossils—appeared in the ocean approximately 630 million years ago (mya). Shelled animals and many other marine invertebrates (animals without backbones) appeared next, as exemplified by trilobites *(see photograph),* members of a large group of primitive aquatic arthropods. Marine invertebrates were followed by the first vertebrates. The first fishes with jaws appeared and diversified; some of these gave rise to amphibians, which also spread and diversified. About 300 mya, amphibians gave rise to reptiles, which diversified and populated the land. Reptiles in turn gave rise independently to birds and to mammals. Plants underwent a comparable evolutionary history and diversification.

In this chapter we survey life over a vast span of time, starting some 3.8 billion years ago (bya) when our planet was relatively young. We examine proposed models about how life began and trace life's long evolutionary history from its beginnings to the present.

EARLY EARTH PROVIDED THE CONDITIONS FOR CHEMICAL EVOLUTION

Many biologists speculate that life originated only once and that life's beginnings occurred under environmental conditions quite different from those of today. We must therefore examine the conditions of early Earth to understand the origin of life. Although we will never be certain about the exact conditions that existed when life arose, scientific evidence from a number of sources provides us with valuable clues that help us formulate plausible scenarios. Study of the origin of life is an active area of scientific research today, and many important contributions are adding to our understanding of how life began.

Astrophysicists and geologists have determined that Earth is approximately 4.6 billion years old. The atmosphere of early Earth apparently included carbon dioxide (CO_2), water vapor (H_2O), carbon monoxide (CO), hydrogen (H_2), and nitrogen (N_2). It may also have contained some ammonia (NH_3), hydrogen sulfide (H_2S), and methane (CH_4), although these reduced molecules may have been rapidly broken down by ultraviolet radiation from the sun. The early atmosphere probably contained little or no free oxygen (O_2).

Four requirements must have existed for the chemical evolution of life: little or no free oxygen, a source of energy, the availability of chemical building blocks, and time. First, life could have begun only in the absence of free oxygen. Oxygen is quite reactive and would have oxidized the organic molecules that are necessary building blocks in the origin of life. Earth's early atmosphere was probably strongly reducing, which means that any free oxygen would have reacted with other elements to form oxides. Thus, oxygen would have been tied up in compounds.

The origin of life would also have required energy to do the work of building biological molecules from simple inorganic chemicals. Early Earth was a place of high energy with violent thunderstorms; widespread volcanic activity; bombardment from meteorites and other extraterrestrial objects; and intense radiation, including ultraviolet radiation from the sun (Fig. 20–1). The young sun probably produced more ultraviolet radiation than it does today, and ancient Earth had no protective ozone layer to filter it.

A third requirement would have been the presence of the chemical building blocks needed for chemical evolution. These included water, dissolved inorganic minerals (present as ions), and the gases present in the early atmosphere. A final requirement for the origin of life was time for molecules to accumulate and react with one another. Earth is approximately 4.6 billion years old, and the earliest traces of life are approximately 3.8 billion years old; therefore, life had a maximum of 800 million years to get started.

Process of Science — Organic molecules formed on primitive Earth

Because organic molecules are the building materials for organisms, it is reasonable to first consider how they might have originated. There are two main models that try to explain how the organic precursors of life originated: The **prebiotic broth hypothesis** proposes that these molecules formed near Earth's surface, whereas the **iron-sulfur world hypothesis** proposes that organic precursors formed at cracks in the ocean's floor.

Figure 20–1 An artist's interpretation of conditions on early Earth. The strongly reducing atmosphere lacked oxygen; volcanoes erupted, spewing gases that contributed to the atmosphere; and violent thunderstorms produced torrential rainfall that eroded the land. Meteorites and other extraterrestrial objects continually bombarded Earth, causing cataclysmic changes in the crust, ocean, and atmosphere. *(Courtesy of Reader's Digest Books. Drawing by H.K. Wimmer)*

The prebiotic broth hypothesis suggests that organic molecules were produced at Earth's surface

The concept that simple organic molecules such as sugars, nucleotide bases, and amino acids could form spontaneously from simpler raw materials was first advanced in the 1920s by two scientists working independently: A.I. Oparin, a Russian biochemist, and J.B.S. Haldane, a Scottish physiologist and geneticist.

Their hypothesis was tested in the 1950s by American biochemists Stanley Miller and Harold Urey, who designed a closed apparatus that simulated conditions that presumably existed on early Earth (Fig. 20–2). They exposed an atmosphere rich in H_2, CH_4, H_2O, and NH_3 to an electrical discharge that simulated lightning. Their analysis of the chemicals produced in a week revealed that amino acids and other organic molecules had formed. Although more recent data suggest that Earth's early atmosphere was not rich in methane or ammonia, similar experiments using different combinations of gases have produced a wide variety of organic molecules that are important in contemporary organisms. These include all 20 amino acids, several sugars, lipids, the nucleotide bases of RNA and DNA, and ATP (when phosphate is present). Thus, before life began, its chemical building blocks may have been accumulating as a necessary step in chemical evolution.

Oparin envisioned that the organic molecules would, over vast spans of time, accumulate in the shallow seas to form a "sea of organic soup." Under such conditions, he envisioned smaller organic molecules (monomers) combining to form larger ones (polymers). Evidence gathered since Oparin's time indicates that organic polymers may have formed and accumulated on rock or clay surfaces rather than in the primordial seas. Clay, which consists of microscopic particles of weathered rock, is particularly intriguing as a possible site for early polymerizations because it binds organic monomers and contains zinc and iron ions that might have served as catalysts. Laboratory experiments have confirmed that organic polymers form spontaneously from monomers on hot rock or clay surfaces.

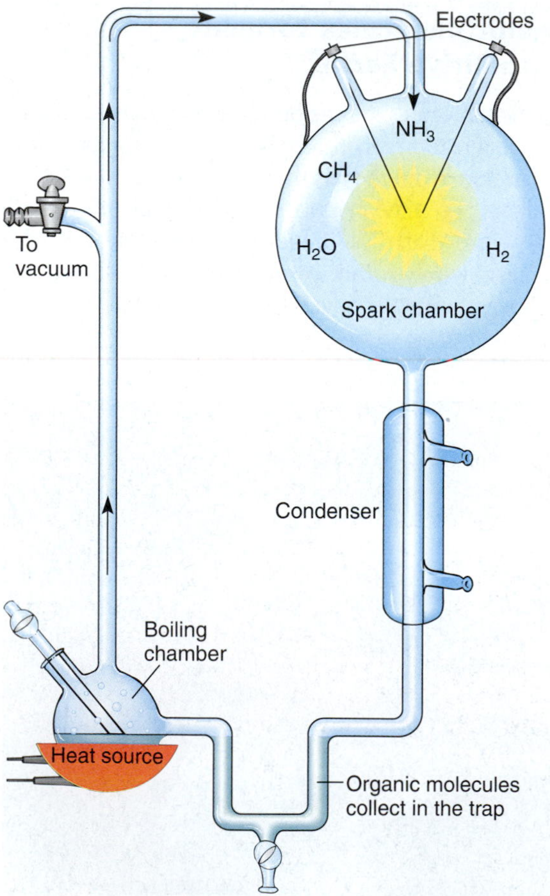

Figure 20–2 Testing the prebiotic broth hypothesis. Diagram of the apparatus that Miller and Urey used to simulate the reducing atmosphere of early Earth. An electrical spark was produced in the upper right flask to simulate lightning. The gases present in the flask reacted together, forming a variety of simple organic compounds that accumulated in the trap at the bottom.

The iron-sulfur world hypothesis suggests that organic molecules were produced at hydrothermal vents

In a different scenario of chemical evolution, some biologists have hypothesized that early polymerizations leading to the origin of life may have occurred in cracks in the deep ocean floor where hot water, carbon monoxide, and minerals such as iron and nickel sulfides spew forth. Such **hydrothermal vents** would have been better protected than Earth's surface from the catastrophic effects of meteorite bombardment.

Today these hot springs produce precursors of biological molecules and of energy-rich "food," including hydrogen sulfide and methane. These chemicals support a diverse community of microorganisms, clams, crabs, tube worms, and other animals (see *Focus On: Life Without the Sun* in Chapter 53).

Testing the iron-sulfur world hypothesis at hydrothermal vents is difficult, but laboratory experiments simulating the high pressures and temperatures at the vents have yielded intriguing results. For example, experiments have demonstrated that ammonia, one of the precursors of proteins and nucleic acids, is produced in abundance, suggesting that vents may have been ammonia-rich environments in the prebiotic world.

THE FIRST CELLS PROBABLY ASSEMBLED FROM ORGANIC MOLECULES

After the first polymers formed, could they have assembled spontaneously into more complex structures? Scientists have synthesized several different **protobionts,** which are assemblages of abiotically produced (i.e., not produced by organisms) organic polymers. They have been able to recover protobionts that resemble living cells in several ways, thus providing clues as to how aggregations of complex nonliving molecules took that "giant leap" and became living cells. These protobionts exhibit many functional and structural attributes of living cells. They often divide in half (binary fission) after they have sufficiently "grown." Protobionts maintain an internal chemical environment that is different from the external environment (homeostasis), and some of them show the beginnings of metabolism (catalytic activity). They are highly organized, considering their relatively simple composition.

Microspheres are a type of protobiont formed by adding water to abiotically formed polypeptides (Fig. 20–3). Some microspheres demonstrate excitability: They produce an electrical potential across their surfaces, reminiscent of electrochemical gradients in cells. Microspheres can also absorb materials from their surroundings (selective permeability) and respond to changes in osmotic pressure as though they were enveloped by membranes, even though they contain no lipid.

The study of protobionts allows us to appreciate that relatively simple "pre-cells" can exhibit some of the properties of contemporary life. However, it is a major step (or several steps) to go from simple molecular aggregates such as protobionts to living cells. Although much has been learned about how organic molecules may have formed on primitive Earth, the problem of how pre-cells evolved into living cells remains to be solved.

It is not known exactly when life first appeared on Earth. **Microfossils** (ancient remains of microscopic life) indicate that cells were thriving 3.5 bya, and nonfossil evidence—isotopic "fingerprints" of carbon from living organisms—indicates that life existed even earlier, perhaps 3.8 bya.

The earliest cells were prokaryotic. Australian and South African rocks have yielded microscopic fossils of prokaryotic cells 3.1 to 3.5 billion years old. **Stromatolites,** another type of fossil evidence of the earliest cells, are rocklike columns composed of many minute layers of prokaryotic cells, usually cyanobacteria (Fig. 20–4). Over time, sediment collects around the cells and mineralizes. Meanwhile, a new layer of living cells grows over the older, dead cells. Fossil stromatolite reefs are found in a number of places in the world, including Great Slave Lake in Canada and the Gunflint Iron Formations along Lake Superior in the United States. Some fossil stromatolites are extremely ancient. One group in Western Australia, for example, is several billion years old. Living stromatolite reefs are still found in hot springs and in warm, shallow pools of fresh and salt water.

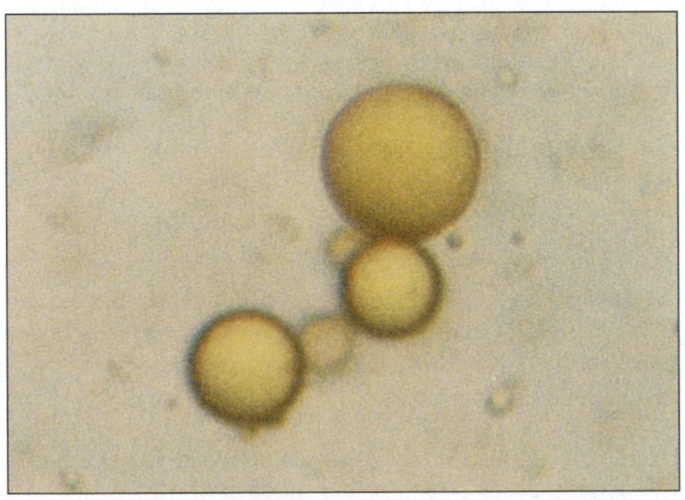

Figure 20–3 Microspheres. These tiny protobionts exhibit some of the properties of life. *(Steven Brooke and Richard LeDuc)*

(a)

(b)

Figure 20–4 Stromatolites. **(a)** Living stromatolites at Hamlin Pool in Western Australia, which are composed of mats of cyanobacteria and minerals such as calcium carbonate, are several thousand years old. **(b)** Cutaway view of a fossil stromatolite showing the layers of cyanobacteria and sediments that accumulated over time. This stromatolite, also from Western Australia, is about 3.5 billion years old. Such ancient stromatolites were much more common than living stromatolites today. *(a, Fred Bavendam/Peter Arnold, Inc.; b, Biological Photo Service)*

We have said that the origin of cells from macromolecular assemblages was a major step in the origin of life. Actually, the evolution of cells probably occurred in a series of small steps. One of the most significant parts of that process would have been the evolution of molecular reproduction.

Molecular reproduction was a crucial step in the origin of cells

In living cells, genetic information is stored in the nucleic acid DNA, which is transcribed into messenger RNA (mRNA), which in turn is translated into the proper amino acid sequence in proteins. All three macromolecules in the DNA → RNA → protein sequence contain precise information, but only DNA and RNA are capable of self-replication, although only in the presence of the proper enzymes. Because both RNA and DNA can form spontaneously on clay in much the same way that other organic polymers do, the question becomes which molecule, DNA or RNA, first appeared in the prebiotic world.

Some scientists have suggested that RNA was the first informational molecule to evolve in the progression toward a self-sustaining, self-reproducing cell and that proteins and DNA came along later. According to a model known as the **RNA world,** the chemistry of prebiotic Earth gave rise to self-replicating RNA molecules that functioned as both enzymes and substrates for their own replication. We represent the replication of RNA in the RNA world scenario as a circular arrow:

One of the features of RNA is that it often has catalytic properties; such enzymatic RNAs are called **ribozymes.** In contemporary cells, ribozymes help catalyze the synthesis of RNA and process precursors into rRNA, tRNA, and mRNA (see Chapter 12). Before the evolution of true cells, ribozymes may have catalyzed their own replication in the clays, shallow rock pools, or hydrothermal vents where life originated. When RNA strands are added to a test tube containing RNA nucleotides but no enzymes, the nucleotides combine to form short RNA molecules. The rate of this reaction is increased if zinc is added as a catalyst. (Recall that zinc is bound to clay.)

The occurrence of an RNA world early in the history of life can never be proven, but experiments with **in vitro evolution,** also called **directed evolution,** have shown that it is feasible. These experiments address an important question about the RNA world, namely, could RNA molecules have catalyzed the many different chemical reactions needed for life? In directed evolution, a large pool of RNA molecules with different sequences are mixed together and selected for their ability to catalyze a single biologically important reaction (Fig. 20–5). Those molecules that have at least some catalytic ability are then amplified and mutated before being exposed to another round of selection. After this cycle is repeated several times, the RNA molecules at the end of the selection process are able to function efficiently as catalysts for the reaction. In vitro evolution studies have shown that RNA has a large functional repertoire—that is, RNA can catalyze a variety of biologically important reactions.

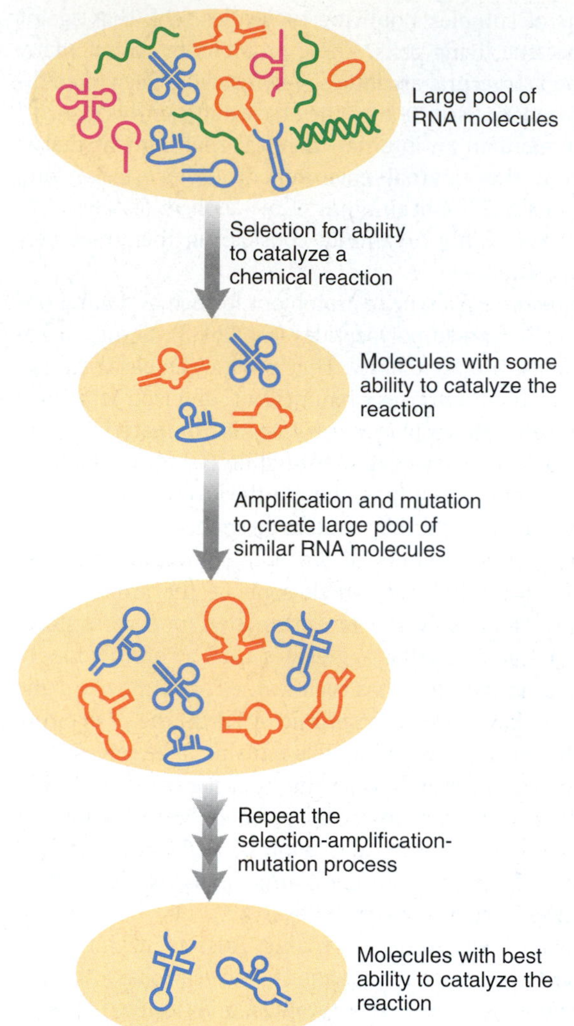

Figure 20–5 In vitro evolution of RNA molecules. RNA molecules are selected from a large pool based on their ability to catalyze a specific reaction, then amplified and mutated before undergoing 7 to 20 additional repetitions of the same process. The final group of RNA molecules is the most efficient at catalyzing the chemical reaction that was selected for. More than two dozen synthetic RNA catalysts have been developed by in vitro evolution.

In the RNA world, ribozymes (catalytic RNAs) initially catalyzed protein synthesis and other important biological reactions; only later did protein enzymes catalyze these reactions.

RNA → protein

Interestingly, RNA can direct protein synthesis by catalyzing peptide bond formation. Some single-stranded RNA molecules fold back on themselves as a result of interactions among the nucleotides composing the RNA strand. Sometimes the conformation (shape) of the folded RNA molecule is such that it weakly binds to an amino acid. If amino acids are held close together by RNA molecules, they may bond together, forming a polypeptide.

We have considered how the evolution of informational molecules may have given rise to RNA and later to proteins. If a self-replicating RNA capable of coding for proteins appeared before DNA, how did DNA, the universal molecule of heredity in cells, become involved? Perhaps RNA made double-stranded copies of itself that eventually evolved into DNA.

$$\text{DNA} \leftarrow \text{RNA} \rightarrow \text{protein}$$

The incorporation of DNA into the information transfer system would have been advantageous because the double-helix conformation of DNA is more stable (i.e., less reactive) than the single-stranded conformation of RNA. Such stability in a molecule that stores genetic information would have provided a decided advantage in the prebiotic world (as it does today).

In the DNA/RNA/protein world, then, DNA became the information storage molecule, RNA remained involved in protein synthesis, and protein enzymes catalyzed most cellular reactions, including DNA replication, RNA synthesis, and protein synthesis.

$$\circlearrowright\text{DNA} \rightarrow \text{RNA} \rightarrow \text{protein}$$

RNA is still a necessary component of the information transfer system because DNA is not catalytic. Thus natural selection at the molecular level favored the DNA $\rightarrow$ RNA $\rightarrow$ protein information sequence. Once DNA was incorporated into this sequence, RNA molecules assumed their present role as an intermediary in the transfer of genetic information.

Several additional steps had to occur before a true living cell could evolve from macromolecular aggregations. For example, the genetic code must have arisen extremely early in the prebiotic world because all organisms possess it, but how did it originate? Also, how did a plasma membrane of lipid and protein come to envelop the pre-cell assemblages, thereby permitting the accumulation of some molecules and the exclusion of others?

The first cells were probably heterotrophs, not autotrophs

The earliest cells probably obtained the organic molecules they needed from the environment, rather than synthesizing them. These primitive **heterotrophs** probably consumed many types of organic molecules that had spontaneously formed—sugars, nucleotides, and amino acids, to name a few. By fermenting these organic compounds, they obtained the energy needed to support life. Fermentation is, of course, an anaerobic process (i.e., performed in the absence of oxygen), and the first cells were almost certainly **anaerobes.**

When the supply of spontaneously generated organic molecules was gradually depleted, only certain organisms could survive. Mutations had probably already occurred that permitted some cells to obtain energy directly from sunlight, perhaps by using sunlight to make ATP. These cells, which did not require the energy-rich organic compounds that were now in short supply in the environment, had a distinct selective advantage.

Photosynthesis requires not only light energy but also a source of electrons, which are used to reduce CO_2 when organic molecules such as glucose are synthesized (see Chapter 8). Most likely, the first photosynthetic **autotrophs** (organisms that produce their own food from simple raw materials) used the energy of sunlight to split hydrogen-rich molecules such as H_2S, releasing elemental sulfur (not oxygen) in the process. Indeed, the green sulfur bacteria and the purple sulfur bacteria still use H_2S as a hydrogen source for photosynthesis.[1]

The first photosynthetic autotrophs to obtain hydrogen by splitting water were the cyanobacteria. Water was quite abundant on early Earth, as it is today, and the selective advantage that splitting water bestowed on them allowed cyanobacteria to thrive. In the process of splitting water, oxygen was released as a gas (O_2). Initially, the oxygen released from photosynthesis oxidized minerals in the ocean and in Earth's crust, and oxygen did not begin to accumulate in the atmosphere for a long time. Eventually, however, oxygen levels increased in the ocean and the atmosphere.

The timing of the events just described can be estimated on the basis of geological and fossil evidence. Fossils from that period, which include rocks that contain traces of chlorophyll, as well as the fossil stromatolites discussed previously, indicate that the first photosynthetic organisms appeared approximately 3.1 to 3.5 bya. This evidence suggests that heterotrophic forms existed even earlier.

Aerobes appeared after oxygen increased in the atmosphere

By 2 bya, cyanobacteria had produced sufficient oxygen to begin significantly changing the composition of the atmosphere. The increase in atmospheric oxygen had a profound effect on life. Obligate anaerobes (those organisms that cannot use oxygen for cellular respiration) were poisoned by the oxygen, and many species undoubtedly perished. Some anaerobes, however, survived in environments where oxygen does not penetrate; others evolved adaptations to neutralize the oxygen so that it could not harm them. Some organisms, called **aerobes,** evolved a respiratory pathway that used the oxygen to extract more energy from food and convert it to ATP energy. Aerobic respiration was joined to the existing anaerobic process of glycolysis.

The evolution of organisms that could use oxygen in their metabolism had several consequences. Organisms that respire aerobically gain much more energy from a single molecule of glucose than anaerobes gain by fermentation. As a result, the newly evolved aerobic organisms were more efficient and more competitive than anaerobes. Coupled with the poisonous nature of oxygen to anaerobes, the efficiency of aerobes forced anaerobes into relatively minor roles. Today the vast majority of organisms, including plants, animals, and most fungi, protists, and prokaryotes, use aerobic respiration, whereas only a few bacteria and even fewer protists and fungi are anaerobic.

The evolution of aerobic respiration had a stabilizing effect on both oxygen and carbon dioxide levels in the biosphere.

[1]Members of a third group of bacteria, the purple nonsulfur bacteria, use other organic molecules or hydrogen gas as a hydrogen source.

Photosynthetic organisms used carbon dioxide as a source of carbon for the synthesis of organic compounds. This raw material would have been depleted from the atmosphere in a relatively brief period without the advent of aerobic respiration, which releases carbon dioxide as a waste product from the complete breakdown of organic molecules. Carbon thus started cycling in the biosphere, moving from the nonliving physical environment, to photosynthetic organisms, to heterotrophs that ate the photosynthetic organisms (see Chapter 53). Carbon was released back into the physical environment as carbon dioxide by aerobic respiration, and the carbon cycle continued. In a similar manner, molecular oxygen was produced by photosynthesis and used during aerobic respiration.

Another significant consequence of photosynthesis occurred in the upper atmosphere, where molecular oxygen reacted to form **ozone,** O_3 (Fig. 20–6). A layer of ozone eventually blanketed Earth, preventing much of the sun's ultraviolet radiation from penetrating to the surface. With the ozone layer's protection from the mutagenic effect of ultraviolet radiation, organisms were able to live closer to the surface in aquatic environments and eventually to move onto land. Because the energy in ultraviolet radiation may have been necessary to form organic molecules, however, their abiotic synthesis decreased.

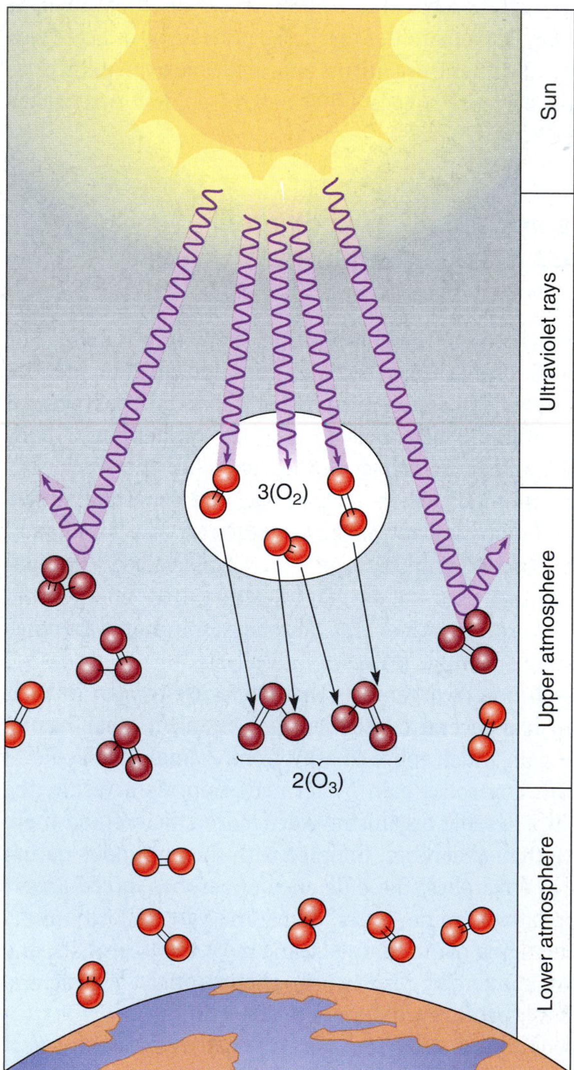

Figure 20–6 How ozone forms. Ozone (O_3) forms in the upper atmosphere when ultraviolet radiation from the sun breaks the double bonds of oxygen molecules.

Eukaryotic cells descended from prokaryotic cells

Eukaryotes may have appeared in the fossil record as early as 1.5 to 1.6 bya, and geochemical evidence suggests that eukaryotes were present much earlier. In 1999 Australian geochemists reported steranes, molecules derived from steroids, in Australian rocks dated at 2.7 billion years. Because bacteria are not known to produce steroids, the steranes are thought to be biomarkers for eukaryotes. (These ancient rocks lack fossil traces of ancient organisms because they have since been exposed to heat and pressure that would have destroyed any fossilized cells. Steranes, however, are very stable in the presence of heat and pressure.)

Eukaryotes arose from prokaryotes. Recall from Chapter 4 that prokaryotic cells lack nuclear envelopes as well as other membranous organelles such as mitochondria and chloroplasts. The **endosymbiont theory,** advanced by Lynn Margulis, declares that organelles such as mitochondria and chloroplasts may each have originated from mutually advantageous symbiotic relationships between two prokaryotic organisms (Fig. 20–7). Chloroplasts apparently evolved from photosynthetic bacteria (cyanobacteria) that lived inside larger heterotrophic cells, while mitochondria presumably evolved from aerobic bacteria (perhaps purple bacteria) that lived inside larger anaerobic cells. Thus, early eukaryotic cells were composed of assemblages of formerly free-living prokaryotes.

How did these bacteria come to be **endosymbionts,** which are organisms that live symbiotically inside a host cell? They may have originally been ingested, but not digested, by a host cell. Once incorporated, they could have survived and reproduced along with the host cell so that future generations of the host also contained endosymbionts. The two organisms developed a mutualistic relationship in which each contributed something to the other. Eventually the endosymbiont lost the ability to exist outside its host, and the host cell lost the ability to survive without its endosymbionts. This theory stipulates that each of these partners brought to the relationship something the other lacked. For example, mitochondria provided the ability to carry out the aerobic respiration lacking in the original anaerobic host cell. Chloroplasts provided the ability to use a simple carbon source (carbon dioxide) to produce needed organic molecules. The host cell provided a safe habitat and raw materials or nutrients.

The principal evidence in favor of the endosymbiont theory is that mitochondria and chloroplasts possess some (although not all) of their own genetic material and translational components. They have their own DNA (as a circular molecule similar to that of prokaryotes; see Chapter 23) and their own ribosomes (which resemble prokaryotic rather than eukaryotic ribosomes). Mitochondria and chloroplasts also possess some of the machin-

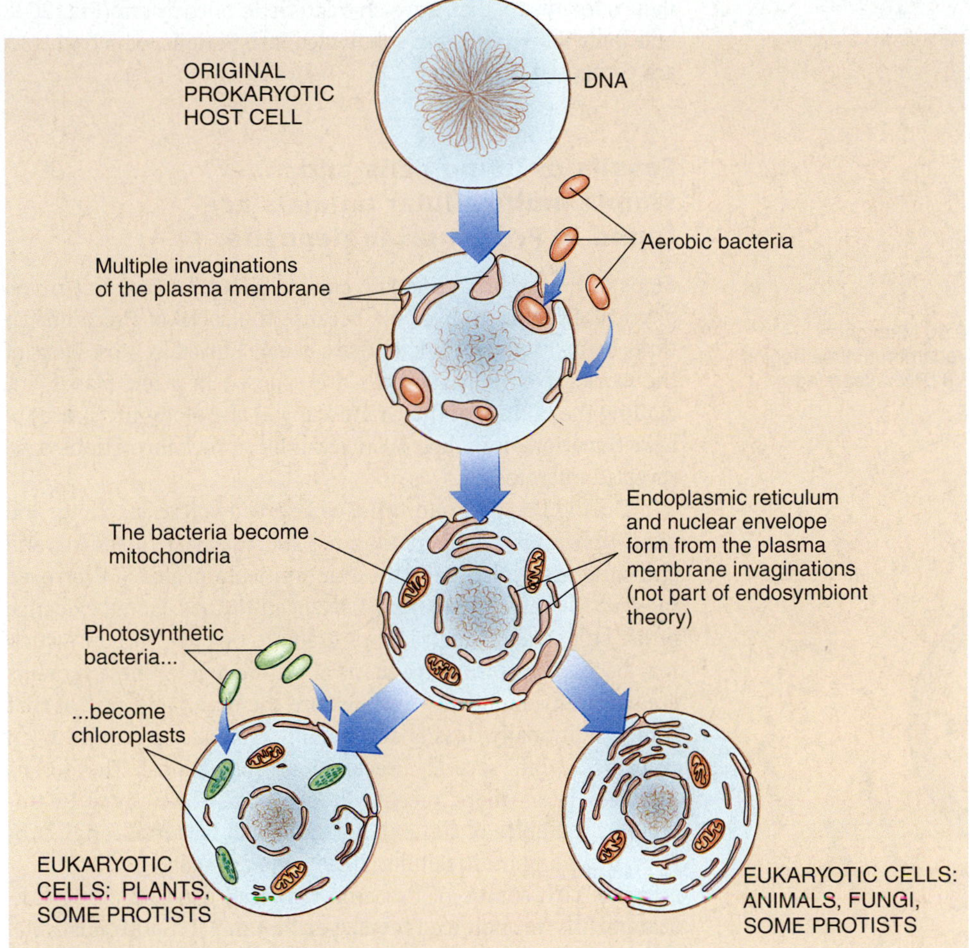

Figure 20–7 The endosymbiont theory. Chloroplasts and mitochondria of eukaryotic cells are thought to have originated from various bacteria that lived as endosymbionts inside other cells.

ery for protein synthesis, including tRNA molecules, and are able to conduct protein synthesis on a limited scale independent of the nucleus. Furthermore, it is possible to poison mitochondria and chloroplasts with an antibiotic that affects prokaryotic but not eukaryotic cells. As discussed in Chapter 4, mitochondria and chloroplasts are enveloped by double membranes. The outer membrane apparently developed from the invagination of the host cell's plasma membrane, whereas the inner membrane is derived from the endosymbiont's plasma membrane. (The endosymbiont theory is discussed in greater detail in Chapter 24.)

Many endosymbiotic relationships exist today. For example, many corals have algae living as endosymbionts within their cells (see Fig. 52–12). In the gut of the termite lives a protozoon *(Myxotricha paradoxa)* that in turn has several different endosymbionts, including spirochete bacteria that are attached to the protozoon and function as whiplike flagella, allowing it to move.

The endosymbiont theory does not completely explain the evolution of eukaryotic cells from prokaryotes. It does not explain, for example, how the genetic material in the nucleus came to be surrounded by a double membranous envelope. A few biologists reject the endosymbiont theory and subscribe to the **autogenous model,** in which eukaryotes arose from prokaryotes by the proliferation of internal membranes to form cellular compartments; these internal membranes were derived from the prokaryotic plasma membrane. Regardless of how eukaryotic cells evolved, their advent set the stage for further evolutionary developments.

THE FOSSIL RECORD PROVIDES CLUES TO THE HISTORY OF LIFE

The sequence of biological, climate, and geological events that make up the history of life is recorded in rocks and fossils. The sediments of Earth's crust consist of five major rock strata (layers), each subdivided into minor strata, lying one on top of the other. Very few places on Earth are covered by all layers, but the strata that are present typically occur in the correct order, with younger rocks on top of older ones. These sheets of rock were formed by the accumulation of mud and sand at the bottoms of the ocean, seas, and lakes. Each layer contains certain

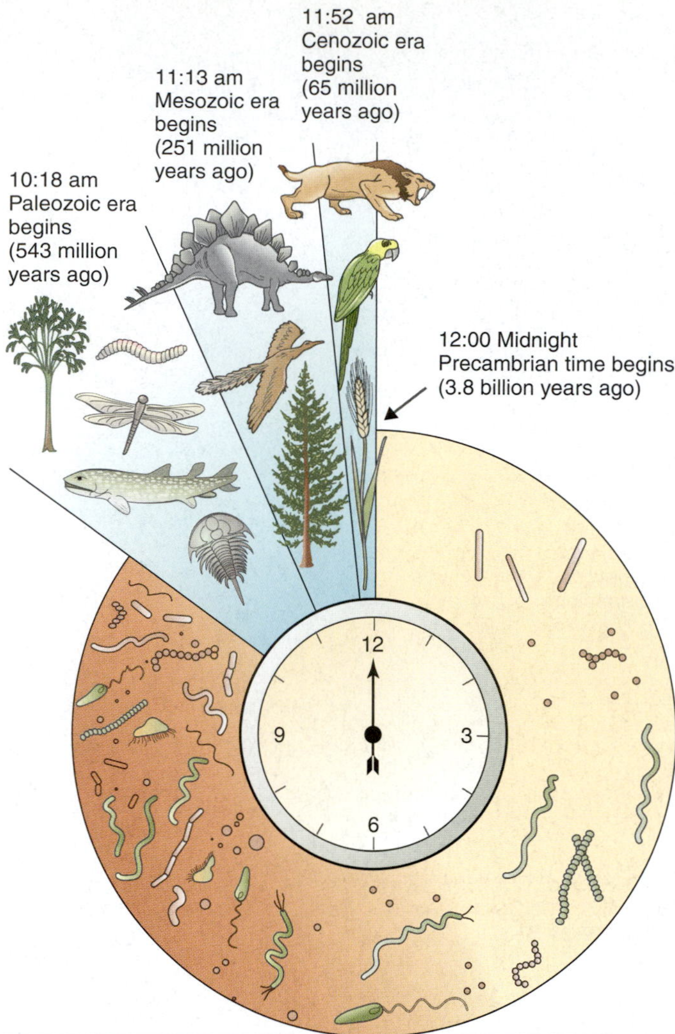

Figure 20–8 An interpretive scale of biological time. Because it is difficult to interpret time in millions or billions of years, using a clock may help represent such vast spans of time. Life began 3.8 bya, at 12:00 midnight. More than 10 hours later, at 10:18 A.M., the Paleozoic era began. The beginning of the Mesozoic era, 251 mya, would be at 11:13 A.M. The Cenozoic era, which began 65 mya, would start at 11:52 A.M. The last epoch of the Cenozoic era, the Holocene epoch, began 10,000 years ago, which would be represented by the last 0.1 second before 12:00 noon. Representative life forms for each era are included.

characteristic fossils, known as **index fossils,** that serve to identify deposits made at approximately the same time in different parts of the world (see Chapter 17).

Geologists divide Earth's 4.6-billion-year history into units of time based on major geological, climate, and biological events. Relatively little is known about Earth from its beginnings approximately 4.6 bya up to 543 mya, a period known informally as **Precambrian time.** The fossil record of ancient organisms is abundant beginning about 543 mya. This most recent time, from 543 mya to the present, is divided into three **eras** based primarily on organisms that were characteristic of each era (Fig. 20–8 and Table 20–1). Eras are subdivided into **periods,** which in turn are composed of **epochs.**

Fossils of living cells and simple multicellular animals are found in Precambrian deposits

Signs of Precambrian life date back to about 3.8 bya. Not much physical evidence is available because the rocks of Precambrian time, being extremely ancient, are deeply buried in most parts of the world. Precambrian rocks are exposed in a few places, including the bottom of the Grand Canyon and along the shores of Lake Superior. More than 400 Precambrian rock formations have revealed microfossils.

During Precambrian time, widespread volcanic activity and giant upheavals raised mountains, and the heat, pressure, and churning associated with these movements probably destroyed most of whatever fossils may have been formed. Some evidence of life still remains as traces of graphite or pure carbon, which may be the transformed remains of primitive life. These remains are especially abundant in what were the ocean and seas of that time. Additionally, fossils resembling cyanobacteria have been recovered from several Precambrian formations. The fossils found in later (more recent) Precambrian rocks show unambiguous examples of some major groups of bacteria, fungi, protists (including multicellular algae), and animals.

One rich source of Precambrian fossil deposits is the Ediacaran Hills (pronounced "ee-dee-ack'a-ran") in South Australia. **Ediacaran fossils,** the oldest known fossils of multicellular animals, are from very late in Precambrian time—from 600 to 543 mya. Biologists have not yet resolved the phylogenetic affinities of the simple, soft-bodied animals found there and at other late Precambrian sites around the world. The oldest, simplest Ediacaran fossils are from 600-million-year-old rocks in the Mackenzie Mountains of northwest Canada.

Some Ediacaran animals appear to be early examples of jellyfish, soft corals, segmented worms, mollusks, and soft-bodied arthropods, whereas others show no resemblance to any other known fossil or living organism (Fig. 20–9). If this interpretation is correct, then at least some of the Ediacaran animals were ancestral to animals that followed. Other biologists who have studied the fossils, however, think that the Ediacaran animals have a body plan that is different from all known animal phyla. If this interpretation is correct, then these animals probably went extinct by the end of the Precambrian and would not be directly related to modern animals.

A considerable diversity of organisms evolved during the Paleozoic era

The **Paleozoic era** began approximately 543 mya and lasted approximately 192 million years. It is divided into six periods: Cambrian, Ordovician, Silurian, Devonian, Carboniferous, and Permian.

TABLE 20–1 Some Important Biological Events in Geological Time

Time*	Era	Period	Epoch	Geological/ Climatic Conditions	Plants and Microorganisms	Animals
0.01	Cenozoic	Quaternary	Holocene	End of last Ice Age; warmer climate; higher sea levels as glaciers melt	Decline of some woody plants; rise of herbaceous plants	Age of *Homo sapiens*
2			Pleistocene	Multiple ice ages; glaciers in Northern Hemisphere	Extinction of some plant species	Extinction of many large mammals at end
5		Tertiary	Pliocene	Uplift and mountain-building; volcanoes; climate much cooler; North and South America join at Isthmus of Panama	Expansion of extensive grasslands and deserts; decline of forests	Many grazing mammals; large carnivorous mammals; first known human-like primates
24			Miocene	Mountains form; climate drier and cooler	Flowering plants continue to diversify	Great diversity of grazing mammals and songbirds
33			Oligocene	Rise of Alps and Himalayas; most land low; volcanic activity in Rockies; climate cool and dry	Spread of forests; flowering plant communities expand	Apes appear; present mammalian families are represented
55			Eocene	Climate warmer	Flowering plants dominant	Modern mammalian orders appear and diversify; modern bird orders appear
65			Paleocene	Continental seas disappear; climate mild to cool and wet	Semitropical vegetation (flowering plants and conifers) widespread	Primitive mammals diversify rapidly
144	Mesozoic	Cretaceous		Continents separate; most continents low; large inland seas and swamps; climate warm	Rise of flowering plants	Dinosaurs reach peak, then become extinct at end; toothed birds become extinct; primitive mammals
206		Jurassic		Continents low; inland seas; mountains form; continental drift begins; climate mild	Gymnosperms common	Large, specialized dinosaurs; first toothed birds; primitive insectivorous mammals diversify
251		Triassic		Many mountains form; widespread deserts; climate warm and dry	Gymnosperms dominant; ferns common	First dinosaurs; first mammals
290	Paleozoic	Permian		Glaciers; continents rise and merge as Pangaea; climate variable	Conifers diversify; cycads appear	Modern insects appear; mammal-like reptiles; extinction of many Paleozoic invertebrates and vertebrates at end of Permian
354		Carboniferous		Lands low and swampy; climate warm and humid, becoming cooler later	Forests of ferns, club mosses, horsetails, and gymnosperms; mosses and liverworts	First reptiles; spread of ancient amphibians; many insect forms; ancient sharks abundant
408		Devonian		Glaciers; inland seas	Vascular plants diversify and become well established; first forests; gymnosperms appear; bryophytes appear	Many trilobites; fishes with jaws appear and diversify; amphibians appear; wingless insects appear
439		Silurian		Most continents remain covered by seas; climate warm	Algae dominant in aquatic environments; vascular plants appear	Jawless fishes diversify; coral reefs common; terrestrial arthropods
495		Ordovician		Sea covers most continents	Marine algae dominant; fossil spores of terrestrial plants (bryophytes?)	Invertebrates dominant; coral reefs appear; first fishes appear
543		Cambrian		Oldest rocks with abundant fossils; lands low; climate mild and wet	Algae; bacteria and cyanobacteria; fungi	Age of marine invertebrates; modern and extinct animal phyla represented; first chordates

*Time from beginning of period to present (millions of years).

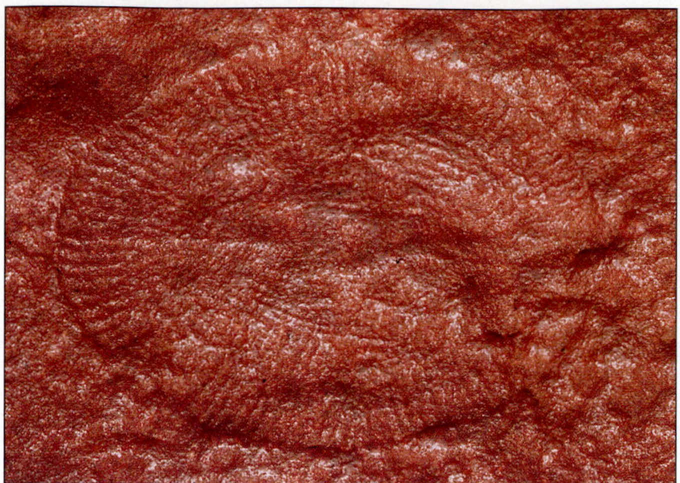

Figure 20–9 A Precambrian fossil *(Dickinsonia costata).* The organism, from the Ediacaran Hills of South Australia, lived in shallow marine waters and is unlike any known modern organism. The fossil is about 5 cm (2 in) long. *(William E. Ferguson)*

Rocks rich in fossils represent the oldest subdivision of the Paleozoic era, the Cambrian period. From about 565 mya to 525 mya, evolution was in such high gear, with the sudden appearance of many new animal groups, that this period has been nicknamed the **Cambrian explosion.** Fossils of all contemporary animal phyla are present, along with many bizarre, extinct phyla, in marine sediments. The sea floor was covered with sponges, corals, sea lilies, sea stars, snails, clamlike bivalves, primitive squidlike cephalopods, lamp shells (brachiopods), trilobites (see chapter opening photograph; also see Fig. 29–16). In addition, small vertebrates—cartilaginous fishes, first reported in 1999—became established in the marine environment. Scientists have not determined the factor or factors responsible for the Cambrian explosion, which has been unmatched in the evolutionary history of life. There is some evidence that oxygen concentrations, which had continued to gradually increase in the atmosphere, passed some critical environmental threshold (10% of present-day oxygen, or higher) late in Precambrian time. Scientists who advocate the *oxygen enrichment hypothesis* note that until late in Precambrian time, Earth possessed insufficient oxygen to support larger animals. The most important fossil sites that document the Cambrian explosion are the **Chenjiang site** in China (for early Cambrian fossils) and the **Burgess Shale** in British Columbia (for middle Cambrian fossils; Fig. 20–10).

The major animal body plans, which are discussed in Chapter 28, were established so early in the history of the eukaryotes that no major change in body plan or basic structure has occurred since. This probably indicates that, early in the Cambrian period, each animal phylum had reached a degree of adaptation that allowed it to exploit its environment and accommodate changes in its surroundings with relatively limited modifications in its body plan.

According to geologists, the continents were gradually flooded during the Cambrian period. In the Ordovician period, much land was covered by shallow seas, in which there was another burst of evolutionary diversification, although not as dramatic as the Cambrian explosion. The Ordovician seas were inhabited by giant cephalopods, squidlike animals with straight shells 5 to 7 m (16 to 23 ft) long and 30 cm (12 in) in diameter. Coral reefs first appeared during this period, as did small, jawless, bony-armored fishes called *ostracoderms* (Fig. 20–11). Lacking jaws, these fishes typically had round or slitlike mouth openings that may have sucked in small food particles from the water or scooped up bottom organic debris. Ordovician deposits also contain fossil spores of terrestrial (land-dwelling) plants, which suggests that the colonization of land had begun.

During the Silurian period, jawless fishes diversified considerably, and jawed fishes first appeared. Definitive evidence of two life forms of great biological significance appeared in the Silurian period: terrestrial plants and air-breathing animals. The first

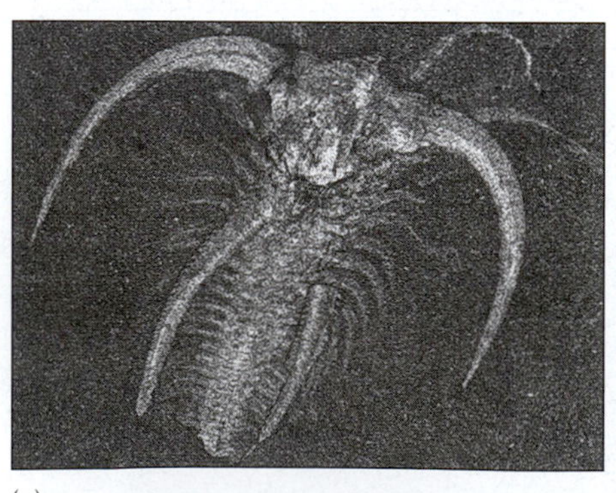

(a)

(b)

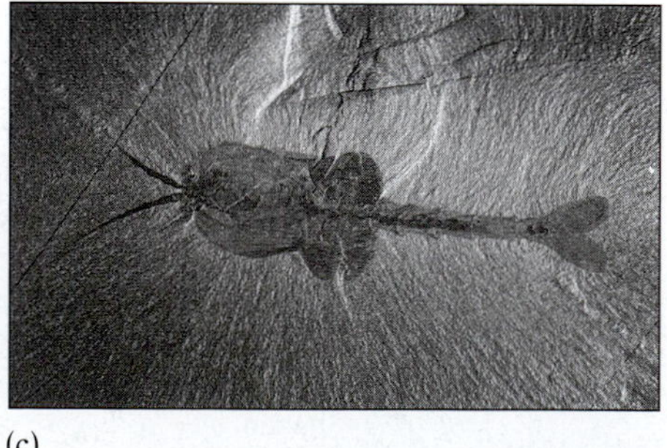

(c)

Figure 20–10 Fossils from the Cambrian explosion. **(a)** *Marrella splendens* was a small arthropod. **(b)** *Wiwaxia* was a bristle-covered marine worm that was distantly related to earthworms. It had scaly armor and needle-like spines for protection. **(c)** *Waptia fieldensis* was a crustacean that may have been an ancestor of modern crustaceans such as shrimp. All three of these fossils were discovered in the Burgess Shale in the Canadian Rockies of British Columbia. *(a–c, Chip Clark)*

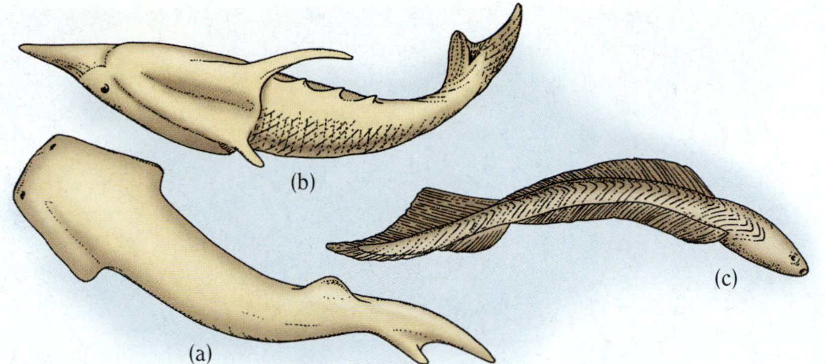

Figure 20–11 Representative ostracoderms. (a) *Thelodus,* (b) *Pterapsis,* and (c) *Jamoytius* are fossil ostracoderms, primitive jawless fishes that appeared in the Devonian period. Ostracoderms ranged from 10 to 50 cm (4 to 20 in) in length.

known plants resembled ferns in that they possessed vascular (conducting) tissue and reproduced by spores (see Chapter 26). The evolution of plants allowed animals to colonize the land because plants provided the first terrestrial animals with food and shelter. All air-breathing land animals discovered in Silurian rocks were arthropods—millipedes, spider-like arthropods, and possibly centipedes. From an ecological perspective, the energy flow from plants to animals probably occurred via detritus, which is organic debris from decomposing organisms, rather than directly from living plant material. Millipedes eat plant detritus today, and spiders and centipedes prey on millipedes and other animals.

A great variety of fishes appeared in the Devonian period. In fact, the Devonian period is frequently called the Age of Fishes. Jawless ostracoderms persisted into the Devonian, but this period also witnessed the explosive radiation of fishes with jaws, an adaptation that enables a vertebrate to chew and bite. Armored *placoderms,* an extinct group of jawed fishes, diversified to exploit varied lifestyles, from bottom-dwelling filter-feeders to the most voracious predators of the time (see Fig. 30–11*b*). Appearing in Devonian deposits are sharks and the two predominant types of bony fish: lobe-finned fishes (including the coelacanths and lungfishes) and ray-finned fishes, which gave rise to the major orders of modern fishes. Coelacanths, primitive bony fish with lobed fins, were originally thought to have become extinct; in 1938, however, the first living coelacanth was discovered in the deep waters off the coast of Madagascar (see Fig. 30–16). This discovery was of great scientific significance because it gave paleontologists an opportunity to test their hypotheses about fossil coelacanths by comparing them with the living species. Lungfishes, an ancient group of air-breathing fish, were most common during the Devonian period; only about six species persist today, in South America, Africa, and Australia.

Upper (more recent) Devonian sediments contain fossil remains of salamander-like amphibians (labyrinthodonts) that were often quite large, with short necks and heavy, muscular tails (see Fig. 30–17). These animals, whose skulls were encased in bony armor, were in many respects quite similar to the lobe-finned fishes, one of which may have been their immediate ancestor; for example, early amphibians possessed fishlike tail fins and scaly body coverings. Early amphibians probably spent most of their time in and around water. Wingless insects also originated in the late Devonian period.

The early vascular plants diversified during the Devonian period in a burst of evolution that rivaled that of animals during the Cambrian explosion. With the exception of flowering plants, all major plant groups appeared during the Devonian. Forests of ferns, club mosses, horsetails, and seed ferns (an extinct group of ancient plants that had fernlike foliage but reproduced by forming seeds) flourished.

The Carboniferous period is named for the great swamp forests whose remains persist today as major coal deposits. Much of the land during this time was covered with low swamps filled with horsetails, club mosses, ferns, seed ferns, and gymnosperms (seed-bearing plants such as conifers) (Fig. 20–12).

Amphibians, which underwent an **adaptive radiation** and exploited both aquatic and terrestrial ecosystems, were the dominant terrestrial carnivores of the Carboniferous period. Reptiles first appeared and diverged to form two major lines at this time. One line (sauropsids) consisted of mostly small and mid-sized, insectivorous (insect-eating) lizards; this line would later lead to lizards, snakes, crocodiles, dinosaurs, and birds. The other reptilian line (synapsids) led to a diverse group of Permian and early Mesozoic mammal-like reptiles. Two groups of winged insects, cockroaches and dragonflies, appeared in the Carboniferous period. The dragonflies ranged in size from those smaller than today's dragonflies to some with wingspans of 75 cm (2.5 ft).

Amphibians continued in importance during the Permian period, but they were no longer the dominant carnivores in terrestrial ecosystems. During the Permian period, the mammal-like reptiles diversified explosively and dominated both carnivorous and herbivorous terrestrial lifestyles. One important group of mammal-like reptiles, originating in the Permian and extending into the Mesozoic era, were the *therapsids,* a group that included the ancestor of mammals (discussed shortly; also see Fig. 30–23).

During the Permian period, seed plants diversified and dominated most plant communities. Cone-bearing conifers were widespread, and cycads (plants resembling palms, with crowns of fernlike leaves and large, seed-containing cones) and ginkgoes (trees with broad, fan-shaped leaves and exposed, fleshy seeds) appeared.

The greatest mass extinction of all time occurred at the end of the Paleozoic era, between the Permian and Triassic periods, 251 mya. More than 90% of all existing marine species became

KEY

1–13	club mosses
14–16	seed ferns
17–19	ferns
20–21	horsetails
22	early gymnosperm
23	primitive insect
24	early dragonfly
25, 26	early roaches

Figure 20–12 Reconstruction of a Carboniferous forest. Plants of this period included giant ferns, horsetails, and club mosses as well as seed ferns and early gymnosperms. *(No. GEO85638c, Field Museum of Natural History, Chicago)*

extinct at this time, as did more than 70% of the vertebrate genera living on land. There is also evidence of a major extinction of plants at this time. The Permian period was characterized by great changes in climate and topography. By the late Permian, the sea level had dropped, and the distribution of shallow seas on continental shelves shrank to less than one-third of their distribution during the early Permian. In the early Triassic, the sea level rose and shallow seas expanded again.

The cause of the late Permian mass extinction is controversial. Changes in sea level may account for the massive extinctions of marine invertebrates. The reduction of shallow seas also would have caused climate instability on land, perhaps triggering the extinction of terrestrial organisms observed at that time as well. Another hypothesis for the Permian-Triassic extinction episode is the occurrence of widespread oxygen depletion in the ocean, an event that is supported by geochemical evidence. Cataclysmic volcanic eruptions that occurred in Siberia over a period of 1 million years have also been linked to the late Permian mass extinction; these eruptions may have caused global cooling. Regardless of the cause of this mass extinction episode, evidence reported during the late 1990s and early 2000s suggests that the extinction occurred globally in a very compressed period, within a few hundred thousand years. This is extremely short in the geological time scale and suggests that some sort of catastrophic event caused the extinctions.

Dinosaurs and other reptiles dominated the Mesozoic era

The **Mesozoic era** began about 251 mya and lasted some 186 million years. It is divided into the Triassic, Jurassic, and Cretaceous periods. Fossil deposits from the Mesozoic era occur worldwide. Notable sites include the **Yixian formation** in northeast China, the Solnhofen Limestone in Germany, northwestern Patagonia in Argentina, the Sahara Desert in central Niger, the badlands in South Dakota, and other sites in western North America.

The outstanding feature of the Mesozoic era was the origin, differentiation, and ultimately the extinction of a large variety of reptiles. For this reason, the Mesozoic era is commonly called the Age of Reptiles. Most of the modern orders of insects appeared during the Mesozoic era. Snails and bivalves (clams and their relatives) increased in number and diversity, and sea urchins reached their peak diversity. From a botanical viewpoint, the Mesozoic era was dominated by gymnosperms until the mid-Cretaceous period, when the flowering plants first diversified.

During the Triassic period, reptiles underwent an adaptive radiation leading to many groups. On land, the dominant Triassic groups were the mammal-like therapsids, which ranged from small-sized insectivores to moderately large herbivores, and a diverse group of *thecodonts,* early "ruling reptiles," that were primarily carnivores (Fig. 20–13*a*). The thecodont group was ancestral to dinosaurs, flying reptiles, and birds.

In the ocean, several important marine reptile groups, the plesiosaurs and ichthyosaurs, appeared in the Triassic and persisted into the Cretaceous. *Plesiosaurs* were aquatic reptiles with bodies up to 15 m (about 49 ft) long and paddle-like fins (Fig. 20–13*b*). *Ichthyosaurs,* also aquatic reptiles, had body forms superficially resembling those of sharks or porpoises, with short necks, large dorsal fins, and shark-type tails (Fig. 20–13*c*). Ichthyosaurs had very large eyes, which may have enabled them to see when diving to depths of 500 m or more.

During the late Triassic period, many new reptiles and their descendants appeared. Turtles appeared more than 210 mya. Both marine and land turtles have survived to the present with few skeletal changes. The first mammals to appear in the Triassic period were small insectivores that evolved from the mammal-like therapsids of the Triassic. Mammals diversified into a variety of mostly small, nocturnal insectivores during the remainder of the Mesozoic, with marsupial and placental mammals appearing in the Cretaceous period. *Pterosaurs,* the first flying reptiles, appeared and underwent considerable diversification during the Mesozoic era (Fig. 20–13*d*). This group produced some quite spectacular forms, most notably the giant *Quetzalcoatlus,* known from fragmentary Cretaceous fossils in Texas to have a wingspan of 11 to 15 m (36 to 49 ft). In addition, the two main dinosaur lines (discussed later) were established by the end of the Triassic period.

During the Jurassic and Cretaceous periods, other important groups—such as crocodiles, lizards, snakes, and birds—appeared, and the dinosaurs diversified dramatically to "inherit the Earth." Crocodiles originated in the early Jurassic, probably from a thecodont ancestor. Lizards and snakes appeared in the late Jurassic and early Cretaceous periods, respectively. Most of the Mesozoic snakes and lizards were similar to their present-day descendants. One group of lizards, the *mosasaurs,* were large, voracious marine predators during the late Cretaceous period. The mosasaurs, which attained lengths of 10 m (33 ft) or more, did not survive to the present (Fig. 20–13*e*).

Dinosaurs underwent an impressive radiation throughout the Jurassic and Cretaceous periods. There were two main groups of dinosaurs: the *saurischians,* with pelvic bones similar to those of modern-day lizards, and the *ornithischians,* with pelvic bones similar to those of birds (Fig. 20–14). Some saurischians were fast, bipedal forms ranging from those the size of a dog to the ultimate representatives of this group, the gigantic carnivores of the Cretaceous period, *Tyrannosaurus, Giganotosaurus,* and *Carcharodontosaurus* (Figs. 20–13*f* and 20–15). Other saurischians were huge, quadrupedal dinosaurs that ate plants. Some of these were the largest terrestrial animals that have ever lived, including *Argentinosaurus,* with an estimated length of 30 m (98 ft) and an estimated weight of 72 to 90 metric tons (80 to 100 tons) (Fig. 20–13*g*). It is thought that *Argentinosaurus* and other plant-eating saurischians ate huge quantities of vegetation such as needles (leaves) from tall conifers.

The other group of dinosaurs, the ornithischians, was entirely herbivorous. Although some ornithischians were bipedal, most were quadrupedal. Some had no front teeth and possessed stout, horny, birdlike beaks. In some species these beaks were broad and ducklike, hence the common name, duck-billed dinosaurs (Fig. 20–13*h*). Other ornithischians had great armor plates, possibly as protection against carnivorous saurischians. *Ankylosaurus,* for example, had a broad, flat body covered with armor plates (actually bony scales embedded in the skin) and large, laterally projecting spikes (Fig. 20–13*i*).

Many traditional ideas about dinosaurs—that they were cold-blooded, slow-moving monsters living in swamps, for example—have been reconsidered over the last 30 years. Recent evidence suggests that at least some dinosaurs may have been warm-blooded, agile, and capable of moving extremely fast. Many dinosaurs appear to have had complex social behaviors, including courtship rituals and parental nurturing of their young. Some species lived in social groups and hunted in packs.

Birds appeared by the late Jurassic period, and most paleontologists think they evolved directly from theropod dinosaurs. A few paleontologists dispute this idea and think that birds evolved from more primitive reptiles. Excellent bird fossils, many showing the outlines of feathers, have been preserved from the Juras-

Figure 20-13 Representative animals of the Mesozoic era. Figures are not drawn to scale. **(a)** This Triassic thecodont, *Euparkia,* was about 150 cm (5 ft) long. Because its forelimbs are shorter than its hind limbs, *Euparkia* was probably bipedal. **(b)** *Elasmosaurus* was a long-necked plesiosaur. Other plesiosaurs had short necks and superficially resembled seals. Some plesiosaurs attained a length of 15 m (about 49 ft). **(c)** *Opthalmosaurus* was an ichthyosaur that superficially resembled a shark or porpoise. It was about 3.6 m (12 ft) long. Other ichthyosaurs were much larger; one ichthyosaur reported in 1996 may have been 45 m (146 ft) long. **(d)** *Pteranodon* was a pterosaur from the Cretaceous period with a wingspan of 7 to 9.2 m (23–30 ft), depending on the species. Pterosaur wings were membranes of skin that were supported by an elongated fourth finger bone. Some pterosaurs had long tails, whereas others lacked tails. **(e)** *Tylosaurus* was a large (about 10 m [33 ft] long) marine lizard (a mosasaur). **(f)** *Giganotosaurus,* whose fossil remains were discovered in Argentina, was the largest (more than 12 m [39 ft] in length) predatory saurischian. **(g)** *Argentinosaurus,* a herbivorous saurischian from Argentina, is the largest known animal to have ever walked on land. It was 35 to 40 m (115–130 ft) long. **(h)** *Hadrosaurus* was a duck-billed, plant-eating ornithischian. It was 7 to 10 m (23–33 ft) long and had hundreds of cheek teeth (its bill was toothless). **(i)** *Ankylosaurus* was a heavily armored ornithischian. Ankylosaurs ranged from 2 to 6 m (7–20 ft) in length.

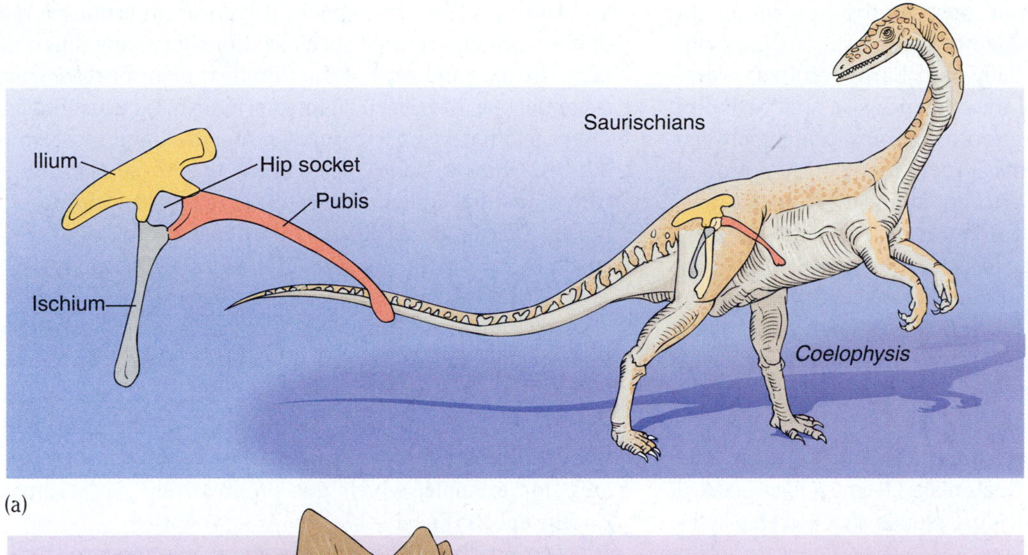

(a)

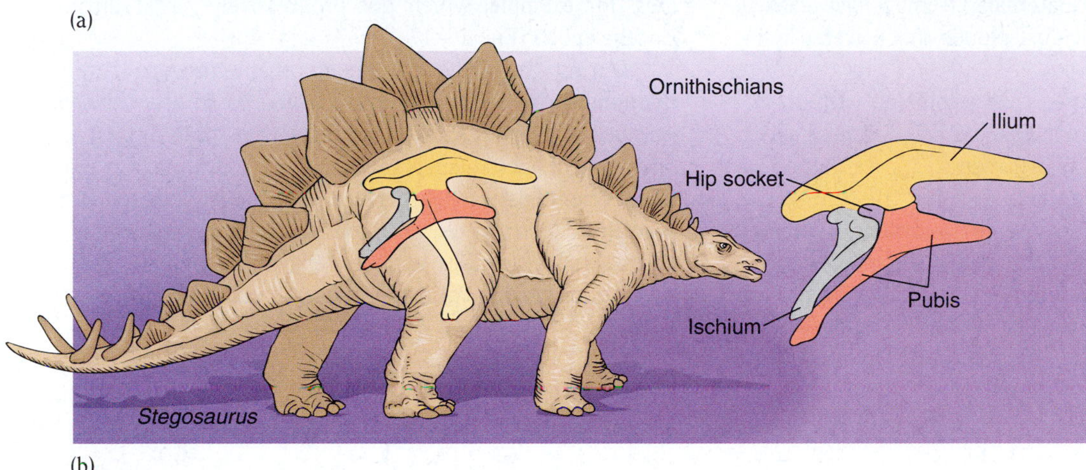

(b)

Figure 20–14 Saurischian and ornithischian dinosaurs. The two orders of dinosaurs are distinguished primarily by differences in their pelvic bones. (In each dinosaur figure, the pale yellow femur is shown relative to the pelvic bone.) **(a)** The saurischian pelvis. Note the opening (hip socket), a trait possessed by no quadrupedal vertebrates other than dinosaurs. **(b)** The ornithischian pelvis also has the hole in the hip socket but differs from the saurischian pelvis in that it has a backward-directed extension of the pubis.

sic period. *Archaeopteryx,* one of the oldest known birds, lived about 150 mya (see Fig. 30–21*b*). It was about the size of a pigeon and had rather feeble wings that it used to glide rather than actively fly. Although *Archaeopteryx* is considered a bird (witness the feathers), it had many reptilian features, including a mouthful of teeth and a long, bony tail.

Thousands of well-preserved bird fossils have been found in early Cretaceous deposits in China. These include *Sinornis,* a 135-million-year-old sparrow-sized bird capable of perching, and the magpie-sized *Confuciusornis,* the earliest known bird with a toothless beak. *Confuciusornis* may date back as far as 142 million years. The Chinese fossils from the Yixian formation document a variety of very primitive birds that possessed many reptilian features yet were clearly able to fly.

At the end of the Cretaceous period, 65 mya, dinosaurs, pterosaurs, and many other animals abruptly became extinct. Many gymnosperms, with the exception of conifers, also perished. Although several explanations for the mass extinction at

Figure 20–15 Reconstruction of a skull of *Carcharodontosaurus.* The fossil remains of this fearsome predator were discovered in North Africa. *(Paul Sereno. Reprinted with permission of* Discover*)*

the end of the Cretaceous period have been proposed, an increasing amount of evidence suggests that a catastrophic collision of a large extraterrestrial body with Earth resulted in dramatic climate changes. Part of the evidence is a small band of dark clay with a high concentration of iridium located between Mesozoic and Cenozoic sediments at more than 200 sites around the world. Iridium is rare on Earth but abundant in meteorites, leading many to conclude that Earth was hit by a large extraterrestrial object at that time. (The force of the impact would have driven the iridium into the atmosphere to be deposited later on the land by precipitation.) The Chicxulub crater, which is buried under the Yucatán Peninsula in Mexico, is the apparent site of the collision at the close of the Cretaceous period. The impact produced giant tsunamis (tidal waves) that deposited materials from the extraterrestrial body around the perimeter of the Gulf of Mexico, from Alabama to Guatemala. It may have caused worldwide forest fires and giant dust clouds that lowered temperatures for many years.

Although it is widely accepted that a collision with an extraterrestrial body occurred 65 mya, there is no consensus about the effects of such an impact on organisms. The extinction of many marine organisms at or immediately after the time of the impact was probably the result of the environmental upheaval produced by the collision. However, a number of clam species associated with the mass extinction at the end of the Cretaceous period appear to have become extinct *before* the impact, suggesting that some of the massive extinctions occurring then were caused by other factors.

The Cenozoic era is known as the Age of Mammals

With equal justice the **Cenozoic era** could be called the Age of Mammals, the Age of Birds, the Age of Insects, or the Age of Flowering Plants. This era is marked by the appearance of all these forms in great variety and numbers of species. The Cenozoic era extends from 65 mya to the present. It is subdivided into two periods: the Tertiary period, encompassing some 63 million years, and the Quaternary period, which covers the last 2 million years. The Tertiary period is subdivided into five epochs, named from earliest to latest: Paleocene, Eocene, Oligocene, Miocene, and Pliocene. The Quaternary period is subdivided into the Pleistocene and Holocene epochs.

Flowering plants, which arose during the Cretaceous period, continued to diversify during the Cenozoic era. During the Paleocene and Eocene epochs, fossils indicate that tropical to semitropical plant communities extended to relatively high latitudes. Palms, for example, are found in Eocene deposits in Wyoming. Later in the Cenozoic era, there is evidence of more open habitats. Grasslands and savannas occurred throughout much of North America during the Miocene epoch, with deserts developing later in the Pliocene and Pleistocene epochs. During the Pleistocene epoch, plant communities changed dynamically in response to fluctuating climates associated with the multiple advances and retreats of continental glaciers.

During the Eocene epoch, there was an explosive radiation of birds, which acquired specializations for many different habitats. The jaws and beak of the flightless giant bird *Diatryma,* for example, may have been adapted primarily for crushing and slicing vegetation in Eocene forests, marshes, and grasslands (Fig. 20–16). Other paleontologists argue that these giant birds were carnivores that killed or scavenged mammals and other vertebrates. The songbirds diversified extensively during the Miocene epoch to become the most diverse order of living birds.

During the Paleocene epoch, an explosive radiation of primitive mammals occurred. Most of these were small forest dwellers that are not closely related to modern mammals. During the Eocene epoch, mammals continued to diverge, and all the modern orders first appeared. Again, many of the mammals were small, but there were also some larger herbivores—the *titanotheres,* for example, which got progressively larger during the Eocene epoch (Fig. 20–17*a*).

During the Oligocene epoch, many modern families of mammals appeared, including the first fossil apes in Africa. A number of lineages showed specializations that suggest a more open type of habitat, such as grassland or savanna. For example, many mammals were larger and had longer legs for running, specialized teeth for chewing coarse vegetation or for preying on animals, and increases in their relative brain sizes. These specializations continued in the Miocene and Pliocene epochs. Es-

Figure 20–16 A representative bird from the Eocene epoch. *Diatryma,* a giant, flightless bird, stood 2.1 m (7 ft) tall and weighed about 175 kg (385 lb).

Figure 20–17 Representative North and South American mammals of the Cenozoic era. Figures are not drawn to scale. **(a)** *Brontotherium* was a titanothere from the late Eocene epoch. It was 2.5 m (8 ft) tall at the shoulder. **(b–f)** Mammals of the Pliocene and Pleistocene epochs. **(b)** The elephant-like mastodon *(Mammut)* was at home in forests, lakes, and rivers. *Mammut* was 2.5 m (8 ft) tall at the shoulder. **(c)** The saber-toothed cat *(Smilodon)*, which had short, powerful legs and was about the size of the modern African lion, was found in both North and South America. Its enlarged canine teeth were approximately 18 cm (7 in) long and were curved like sabers. **(d)** This camel-like mammal *(Macrauchenia)* from South America probably browsed in forest clearings. *Macrauchenia,* which was about 3 m (10 ft) tall, is usually depicted with a short trunk because, like elephants, its nasal openings are on the skull roof rather than in the front of the skull. **(e)** *Megatherium* was a South American giant ground sloth. As long as 6 m (20 ft), *Megatherium* had 18-cm (7-in) claws that may have been used to strip bark from trees. Other paleontologists think *Megatherium* used its claws to stab prey. **(f)** The giant armadillo *(Glyptodon)* lived on the pampas of South America. Encased in bony armor, *Glyptodon* was about 2 m (6.5 ft) long. Later glyptodont species reached 4 m (13 ft).

pecially diverse during these epochs were hoofed mammals such as horses, which underwent an adaptive radiation to include both browsing and grazing lifestyles. Carnivores specialized for long-distance running after prey also appeared in the Pliocene epoch. (Before that, carnivores were primarily ambush predators.) Human ancestors are found during the Pliocene epoch, about 4.4 mya, in Africa; the genus *Homo* appeared approximately 2.3 mya. (Primate evolution, including human evolution, is discussed in Chapter 21.)

The Pliocene and Pleistocene epochs witnessed a spectacular North and South American large-mammal fauna, including mastodons, saber-toothed cats, camels, giant ground sloths, giant armadillos, and numerous other species (Fig. 20–17*b–f*). However, many of the large mammals became extinct at the end of the Pleistocene. This extinction may have been due to climate change—the Pleistocene epoch was marked by several ice ages—and/or the influence of humans, which had spread from Africa to Europe and Asia, and later to North and South America by crossing a land bridge between Siberia and Alaska. Strong archaeological evidence exists that this mass extinction event was concurrent with the appearance of human hunters that possessed Clovis spear-point technology.

SUMMARY WITH KEY TERMS

I. Biologists generally agree that life originated from nonliving matter by **chemical evolution.** Although chemical evolution is very difficult to test experimentally, hypotheses about the origin of life are testable.
  A. Four requirements for chemical evolution are
    1. The absence of oxygen, which would have reacted with and oxidized abiotically produced organic molecules.
    2. Energy to form organic molecules.
    3. Chemical building blocks, including water, minerals, and gases present in the atmosphere, to form organic molecules.
    4. Sufficient time for molecules to accumulate and react.
  B. Four steps are hypothesized in chemical evolution.
    1. Small organic molecules formed spontaneously and accumulated.
      a. The **prebiotic broth hypothesis** proposes that organic molecules formed near Earth's surface in a "sea of organic soup" or on rock or clay surfaces.
      b. The **iron-sulfur world hypothesis** suggests that organic molecules were produced at **hydrothermal vents,** cracks in the deep ocean floor.
    2. Macromolecules assembled from the small organic molecules.
    3. Macromolecular assemblages called **protobionts** formed from macromolecules.
      a. According to a model known as the **RNA world,** RNA was the first informational molecule to evolve in the progression toward a self-sustaining, self-reproducing cell.
      b. Natural selection at the molecular level resulted in the DNA → RNA → protein information sequence.
      c. Experiments with **in vitro evolution,** also called **directed evolution,** have demonstrated how RNA could have evolved to catalyze a variety of biologically important reactions.
    4. Cells arose from the macromolecular assemblages.

II. The oldest cells in the fossil record are 3.1 to 3.5 billion years old. Nonfossil evidence places earliest life at 3.8 billion years old.
  A. The first cells were prokaryotic **heterotrophs** that obtained organic molecules from the environment. They were almost certainly **anaerobes.**
  B. Later, **autotrophs,** organisms that produce their own organic molecules by photosynthesis, arose.
  C. The evolution of photosynthesis ultimately changed early life because it generated oxygen, which accumulated in the atmosphere, and permitted the evolution of **aerobes,** organisms that could use oxygen for a more efficient type of cellular respiration.
  D. Eukaryotic cells arose from prokaryotic cells. According to the **endosymbiont theory,** certain eukaryotic organelles (mitochondria and chloroplasts) evolved from prokaryotic **endosymbionts** within larger prokaryotic hosts.

III. Earth's history is divided into three **eras.** Each era is divided into **periods,** which are divided into **epochs.**
  A. During **Precambrian time,** which extended from approximately 4.6 bya up to 543 mya, life began and diverged into different groups of bacteria, protists (including algae), fungi, and simple multicellular animals.
    1. Signs of Precambrian life date back to about 3.8 bya.
    2. **Microfossils,** ancient remains of microscopic life, and **stromatolites,** rocklike columns containing many minute layers of prokaryotic cells, date back about 3.5 billion years.
    3. One rich source of Precambrian fossil deposits is the Ediacaran Hills in South Australia. **Ediacaran fossils** of multicellular animals are from very late in Precambrian time—from 590 to 543 mya.
  B. During the **Paleozoic era,** which began approximately 543 mya and lasted approximately 192 million years, all major groups of plants, except for flowering plants, and all animal phyla appeared.
    1. During the Cambrian period, the pace of evolution was so rapid that this period has been nicknamed the **Cambrian explosion.** The most important fossil sites that document the Cambrian explosion are the **Chenjiang site** in China (for early Cambrian fossils) and the **Burgess Shale** in British Columbia (for middle Cambrian fossils).
    2. Fish and amphibians flourished, and reptiles appeared and diversified during the Paleozoic era.
    3. The greatest mass extinction of all time occurred at the end of the Paleozoic era, 251 mya. More than 90% of all existing marine species and more than 70% of land-dwelling vertebrate genera became extinct at this time, as well as many plant species.
  C. The **Mesozoic era** began about 251 mya and lasted some 186 million years.
    1. The Mesozoic era was characterized by the appearance of flowering plants and the evolutionary diversification of reptiles. Dinosaurs, which descended from early reptiles, dominated Earth during the Mesozoic era. Insects flourished, and birds and early mammals evolved.
    2. The two main groups of dinosaurs were the saurischians, with pelvic bones similar to modern-day lizards, and the ornithischians, with pelvic bones similar to those of birds.
    3. At the end of the Cretaceous period, 65 mya, a great many animals abruptly became extinct. A collision of Earth with a large extraterrestrial body may have resulted in dramatic climate changes that played a role in this mass extinction episode.
  D. In the **Cenozoic era,** which extends from 65 mya to the present, flowering plants, birds, insects, and mammals diversified greatly.
    1. All the modern orders of mammals appeared and diversified during the Eocene epoch. During the Oligocene epoch, many modern families of mammals appeared, including the first apes in Africa. A spectacular North and South American large-mammal fauna evolved during the Pliocene and Pleistocene epochs.
    2. Birds diversified during the Eocene epoch, adapting to various lifestyles and habitats. The songbirds diversified extensively during the Miocene epoch.

POST-TEST

1. Energy, the absence of oxygen, chemical building blocks, and time were the requirements for (a) chemical evolution (b) biological evolution (c) the Cambrian explosion (d) the mass extinction episode at the end of the Cretaceous period (e) directed evolution
2. Protobionts (a) form spontaneously in hydrothermal vents in the ocean floor (b) are heterotrophs that obtain the organic molecules they need from the environment (c) are assemblages of abiotically produced organic polymers that resemble living cells in several ways (d) are autotrophs that use sunlight to split hydrogen sulfide (e) are fossilized mats of cyanobacteria
3. Many scientists think that ____________ was the first information molecule to evolve. (a) DNA (b) RNA (c) a protein (d) an amino acid (e) a lipid
4. The first cells were probably (a) heterotrophs (b) autotrophs (c) anaerobes (d) both a and c (e) both b and c
5. According to the endosymbiont theory (a) life originated from nonliving matter (b) the pace of evolution quickened at the start of the Cambrian period (c) chloroplasts, mitochondria, and possibly other organelles originated from intimate relationships among prokaryotic organisms (d) banded iron formations reflect the buildup of sufficient oxygen in the atmosphere to oxidize iron at Earth's surface (e) the first photosynthetic organisms appeared 3.1 to 3.5 bya
6. All geological time prior to the beginning of the Paleozoic era some 543 mya is informally known as (a) the Cenozoic era (b) the Paleozoic era (c) the Mesozoic era (d) Precambrian time (e) the Cambrian period
7. Geologists divide Earth's history, from Precambrian time to the present, into (a) three periods (b) three epochs (c) three eras (d) five periods (e) five eras
8. Ediacaran fossils (a) are the oldest known fossils of multicellular animals (b) come from the Burgess Shale in British Columbia (c) contain remains of large salamander-like organisms (d) are the oldest fossils of early vascular plants (e) contain a high concentration of iridium
9. The correct chronological order of geological eras, starting with the oldest, is (a) Paleozoic, Cenozoic, and Mesozoic (b) Mesozoic, Cenozoic, and Paleozoic (c) Mesozoic, Paleozoic, and Cenozoic (d) Paleozoic, Mesozoic, and Cenozoic (e) Cenozoic, Paleozoic, and Mesozoic
10. The time of greatest evolutionary diversification in the history of life occurred during the (a) Cambrian period (b) Ordovician period (c) Silurian period (d) Carboniferous period (e) Permian period
11. The greatest mass extinction episode in the history of life occurred at what boundary? (a) Pliocene-Pleistocene (b) Permian-Triassic (c) Mesozoic-Cenozoic (d) Cambrian-Ordovician (e) Triassic-Jurassic
12. The Mesozoic era is divided into three periods, which are (a) Cambrian, Ordovician, and Silurian (b) Devonian, Carboniferous, and Permian (c) Triassic, Jurassic, and Cretaceous (d) Cretaceous, Tertiary, and Quaternary (e) Pliocene, Pleistocene, and Holocene
13. The Age of Reptiles corresponds to the (a) Paleozoic era (b) Mesozoic era (c) Cenozoic era (d) Pleistocene epoch (e) Permian period
14. Evidence exists that a catastrophic collision between Earth and a large extraterrestrial body occurred 65 mya, resulting in the extinction of (a) Precambrian worms, mollusks, and soft-bodied arthropods (b) jawless ostracoderms and jawed placoderms (c) dinosaurs, pterosaurs, and many gymnosperm species (d) mastodons, saber-toothed cats, and giant ground sloths
15. Flowering plants and mammals diversified and became dominant during the (a) Paleozoic era (b) Mesozoic era (c) Cenozoic era (d) Devonian period (e) Cambrian period

REVIEW QUESTIONS

1. What are the four requirements for chemical evolution, and why is each essential?
2. How did the presence of molecular oxygen in the atmosphere affect early life?
3. Give at least two types of evidence that support the endosymbiont theory.
4. Arrange the following sets of organisms in order of appearance in the fossil record, starting with the earliest: (a) eukaryotic cells, multicellular organisms, prokaryotic cells; (b) reptiles, mammals, amphibians, fish; (c) flowering plants, ferns, gymnosperms.

YOU MAKE THE CONNECTION

1. If you were experimenting on how protobionts evolved into cells and you developed a protobiont that was capable of self-replication, would you consider it a living cell? Why or why not?
2. If living cells were created in a test tube from nonbiological components by chemical processes, would this accomplishment prove that life evolved in a similar manner billions of years ago? Why or why not?
3. Why did the evolution of complex multicellular organisms such as plants and animals have to be preceded by the evolution of oxygen-producing photosynthesis?
4. How might studying outer space help us reconstruct the evolutionary history of life on Earth?

RECOMMENDED READINGS

Doebler, S.A. "The Dawn of the Protein Era." *BioScience,* Vol. 50, No. 1, Jan. 2000. Discusses various hypotheses about the origin of proteins or nucleic acids as the first molecules of life.

Doolittle, W.F. "Uprooting the Tree of Life." *Scientific American,* Vol. 282, No. 2, Feb. 2000. This article, which describes what is currently known about the origin of prokaryotes and eukaryotes, is an excellent example of how scientific knowledge progresses from general controversy to consensus.

Erickson, G.M. "Breathing Life into *Tyrannosaurus rex.*" *Scientific American,* Vol. 281, No. 3, Sep. 1999. Paleontologists are beginning to unravel secrets regarding the feeding behavior of the tyrannosaurs.

Fortey, R. "Crystal Eyes." *Natural History,* Oct. 2000. Examines the eyes of trilobites, marine animals that have been extinct for 250 million years.

Gould, S.J. "Tales of a Feathered Tail." *Natural History,* Nov. 2000. This article is one of a long series of essays written by the popular evolutionary biologist at Harvard University.

Hoffmann, H.J. "Messel: Window on an Ancient World." *National Geographic,* Vol. 197, No. 2, Feb. 2000. Some of the world's most remarkable fossils of the Eocene epoch in the Cenozoic era come from an abandoned mine pit in Messel, Germany.

Hoffmann, H.J. "When Life Nearly Came to an End: The Permian Extinction." *National Geographic,* Vol. 198, No. 3, Sep. 2000. The largest mass extinction episode in Earth's history occurred 250 mya, at the end of Earth's Permian period.

Landweber, L.F., P.J. Simon, and T.A. Wagner. "Ribozyme Engineering and Early Evolution." *BioScience,* Vol. 48, No. 2, Feb. 1998. This article explains in vitro evolution of RNA molecules and how it may mimic early RNA evolution.

Monastersky, R. "Life Grows Up." *National Geographic,* Vol. 193, No. 4, Apr. 1998. Highlights Ediacaran fossils of ancient, soft-bodied animals.

Motani, R. "Rulers of the Jurassic Seas." *Scientific American,* Vol. 283, No. 6, Dec. 2000. Presents some of the fascinating information known about the reptilian ichthyosaurs that lived in the ocean when dinosaurs roamed the land.

Nisbet, E.G., and N.H. Sleep. "The Habitat and Nature of Early Life." *Nature,* Vol. 409, 22 Feb. 2001. This insightful review article examines the environment of early Earth, when life was first evolving.

Simpson, S. "Life's First Scalding Steps." *Science News,* Vol. 155, 9 Jan. 1999. Reviews the hypothesis that chemicals produced at hydrothermal vents could have led to the evolution of living organisms.

Webster, D. "A Dinosaur Named Sue." *National Geographic,* Vol. 195, No. 6, Jun. 1999. Highlights what has been learned from studying the most complete *Tyrannosaurus rex* fossil ever discovered.

Westenberg, K. "From Fins to Feet." *National Geographic,* Vol. 195, No. 5, May 1999. Highlights the evolution of tetrapods—animals with backbones and four limbs—during the Devonian period.

- Visit our Web site at **http://www.info.brookscole.com/solomonbergmartin** for links to chapter-related resources on the World Wide Web. Additional on-line materials relating to this chapter can also be found on our Web site.

See chapter activity on BioActive Learner CD for additional help in mastering the chapter's material. Icon location in the chapter's margins shows which topics have tutorials or simulations in the CD.

21

The Evolution of Primates

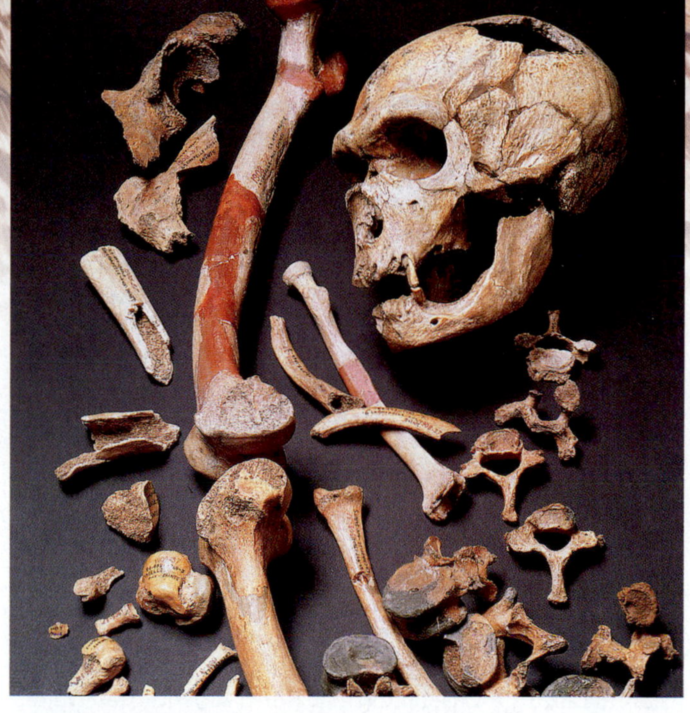

Fossilized remains of a Neandertal man. Each of these fossils, which were found in La Chapelle-aux-Saints, France, has been carefully catalogued and studied. Some of the bones are deformed due to osteoarthritis. *(John Reader/Science Photo Library/Photo Researchers, Inc.)*

LEARNING OBJECTIVES

After you have studied this chapter you should be able to

1. Describe the structural adaptations that primates possess for life in treetops and explain why even primates that live on the ground have these adaptations.
2. List the three suborders of primates and give representative examples of each.
3. Explain the significance of each of the following fossil primates: *Eosimias, Aegyptopithecus, Proconsul,* and the dryopithecines.
4. Distinguish among mammals, primates, anthropoids, hominoids, and hominids.
5. Describe skeletal and skull differences between apes and hominids.
6. Compare the following early hominids: *Ardipithecus ramidus, Australopithecus anamensis, Australopithecus afarensis,* and *Australopithecus africanus.*
7. Distinguish among the following members of genus *Homo: Homo habilis, Homo ergaster, Homo erectus, Homo heidelbergensis, Homo neanderthalensis,* and *Homo sapiens.*
8. Discuss the current debate over the origin of modern humans and briefly describe the opposing out of Africa and multiregional hypotheses.
9. Describe cultural evolution and its impact on the biosphere.

Twelve years after Darwin wrote *The Origin of Species by Natural Selection,* he published another controversial book, *The Descent of Man,* which addressed human evolution. In it, Darwin hypothesized that humans and apes share a common ancestry. For nearly a century after Darwin, fossil evidence of human ancestry remained fairly incomplete. However, research over the last few decades, especially in Africa, has yielded fossils that provide an increasingly clear answer to the question, "Where did we come from?"

Humans and other primates, such as lemurs, tarsiers, monkeys, and apes, are mammals. Mammals (class Mammalia) arose from mammal-like reptiles known as therapsids more than 200 million years ago (mya), during the Mesozoic era (see Chapter 20). These early mammals remained a minor component of life on Earth for almost 150 million years before rapidly diversifying during the Cenozoic era (the last 65 million years). Mammals are **endothermic** (they use metabolic energy to maintain a constant body temperature); produce body hair for such functions as insulation, protective coloration, and waterproofing; and feed their young with milk from mammary glands. Most mammals are **viviparous,** which means that their eggs develop into young offspring within the female body.

It is currently thought that the three groups of living mammals—the monotremes, marsupials, and placental mammals—all are descended from the same common ancestor. The **monotremes** are mammals, such as the duck-billed platypus, that lay eggs (see Chapter 30). The **marsupials,** such as kangaroos and opossums, carry their young in an abdominal pouch after giving birth to them in a very underdeveloped condition. **Placental mammals,** the largest and most successful group, possess a **placenta,** an organ that exchanges materials between the mother and the embryo/fetus developing in the uterus. Placental mammals give birth to their young in a more developed condition than marsupials.

The first primates appeared about 55 mya, apparently descendants of small shrewlike placental mammals that lived in trees and ate insects. Many traits of the 233 living primate species are related to their **arboreal** (tree-dwelling) past. Humans and their ancestors *(see photograph),* who differ from most other primates because they did not remain in the trees but instead adapted to a terrestrial way of life, are the main focus of this chapter.

Fossil evidence has allowed **paleoanthropologists,** scientists who study human evolution, to infer not only the structure but also the habits of early humans and other primates. Teeth and bones are the main fossil evidence studied by paleoanthropologists. Much information can be obtained, for example, by studying teeth, which have changed dramatically during the course of primate and human evolution. Because tooth enamel is more mineralized (harder) than bone, teeth are more likely to be fossilized. The teeth of each primate species, living or extinct, are distinctive enough to identify the species, approximate age, diet, and even sex of the individual. Consider *Australopithecus robustus,* which lived in southern Africa about 2 mya, at a time that the climate was becoming more arid and the forests were giving way to grasslands. Its large jaws and molars (broad-ridged teeth in the back of the mouths of adult mammals) indicate that its diet included tough foods such as roots, tubers, and seeds.

EARLY PRIMATE EVOLUTION REFLECTED AN ARBOREAL EXISTENCE

Fossil evidence indicates that the first true primates appeared by the early Eocene epoch about 55 mya. These early primates had digits with nails, and their eyes were directed more forward on the head. The climate was milder then, and early primates were widely distributed over much of North America, Europe, and Asia. (Recall from Chapter 17 that North America was still attached to Europe at that time.) As the climate became cooler and drier toward the end of the Eocene epoch, many of these early primates became extinct.

Several novel adaptations evolved in early primates that allowed them to live in trees. One of the most significant features of primates is that they have five grasping digits: four lateral digits (fingers) plus a partially or fully opposable thumb or big toe (Fig. 21–1). The opposable first digit enables primates to grasp objects such as branches. Nails (instead of claws) provide a protective covering for the tips of the digits, and the fleshy pads at the ends of the digits are sensitive to touch. Another arboreal feature is long, slender limbs that rotate freely at the hips and shoulders, giving primates full mobility to climb and search for food in the treetops. The location of eyes in front of the head provides stereoscopic, or three-dimensional, vision. Stereoscopic vision is vital in an arboreal environment, especially for species that leap from branch to branch, because an error in depth perception might cause a fatal fall. In addition to sharp sight, hearing is acute in primates.

Primates share several other characteristics, including a relatively large brain size. It has been suggested that increased sensory input associated with their sharp vision and greater agility favored the evolution of larger brains. Primates are generally very social and intelligent animals that reach sexual maturity relatively late in life. They typically have long life spans. Females usually bear one offspring at a time; the baby is helpless and requires a long period of nurturing and protection.

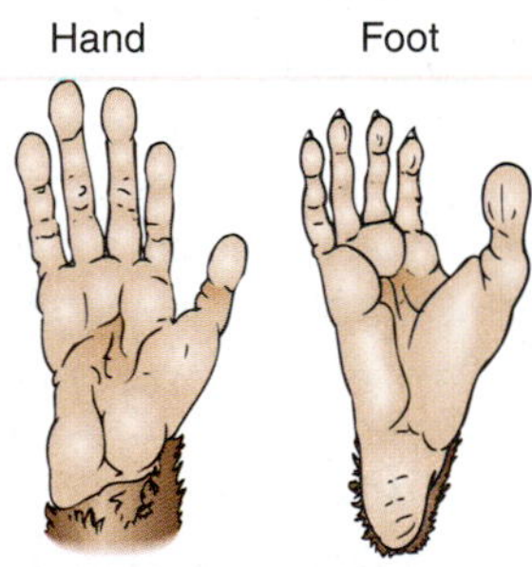

(a) Lemur *(Eulemur mongoz)*

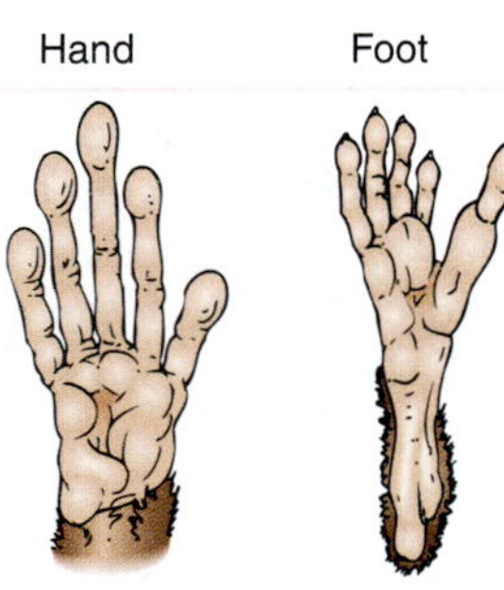

(b) Tarsier *(Tarsius spectrum)*

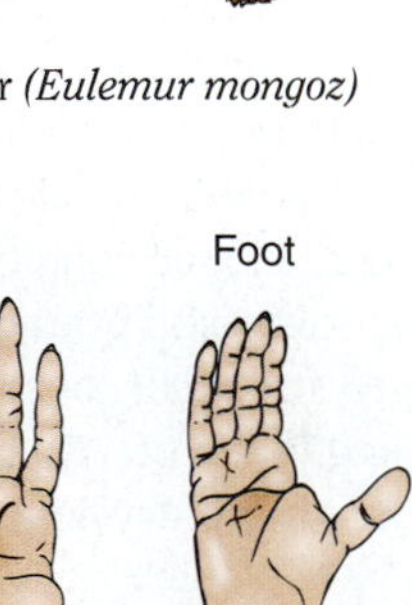

(c) Woolly spider monkey *(Brachyteles arachnoides)*

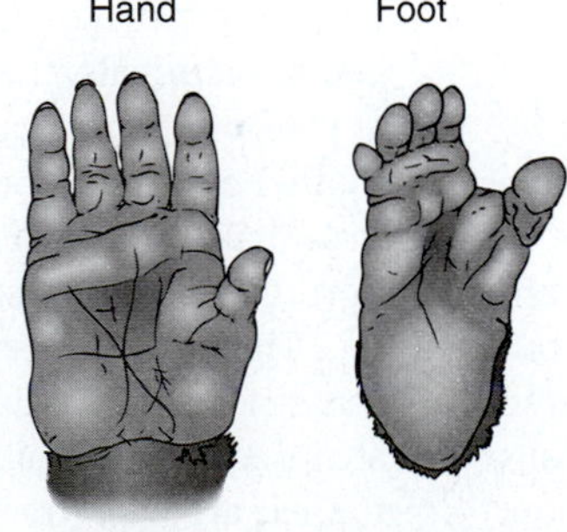

(d) Gorilla *(Gorilla gorilla)*

Figure 21–1 Right hands and feet of selected primates. Primates have five grasping digits, and the thumb or big toe is often partially or fully opposable. **(a)** Lemur. **(b)** Tarsier. **(c)** Wooly spider monkey. **(d)** Gorilla. *(Figures not drawn to scale.) (Adapted from Schultz, A.H.* The Life of Primates. *Weidenfeld & Nicholson, London, 1969)*

LIVING PRIMATES MAY BE CLASSIFIED INTO THREE SUBORDERS

Many biologists currently divide the order Primates into three suborders (Table 21–1 and Fig. 21–2). The suborder Prosimii includes lemurs, galagos, and lorises, the suborder Tarsiiformes includes tarsiers, and the suborder Anthropoidea includes **anthropoids** (monkeys, apes, and humans).

All lemurs are restricted to the island of Madagascar off the coast of Africa (Fig. 21–3). Because of extensive habitat destruction and hunting, they are highly endangered. Lorises, which are found in tropical areas of Southeast Asia and Africa, resemble lemurs in many respects, as do galagos, which live in sub-Saharan Africa. Lemurs, lorises, and galagos have retained several early mammalian features, such as elongated, pointed faces.

Tarsiers are found in rain forests of Indonesia and the Philippines (Fig. 21–4). They are small primates (about the size of a small rat) and are very adept leapers. These nocturnal primates resemble anthropoids in a number of ways, including their shortened snouts and forward-pointing eyes.

TABLE 21–1 A Classification of Living Groups in the Order Primates*

Suborder Prosimii
Infraorder Lemuriformes (lemurs)
Infraorder Lorisoformes (galagos, lorises)
Suborder Tarsiiformes (tarsiers)
Suborder Anthropoidea
Infraorder Platyrrhini (New World primates)
Infraorder Catarrhini (Old World primates)
Superfamily Cercopithecoidea (Old World monkeys)
Superfamily Hominoidea (Old World apes and humans)
Family Hylobatidae (gibbons)
Family Pongidae (orangutans, gorillas, chimpanzees)
Family Hominidae (humans)

* There are many alternative classifications of primates. For example, classifications based on molecular data place humans and the great apes (orangutans, gorillas, and chimpanzees) in a single family.

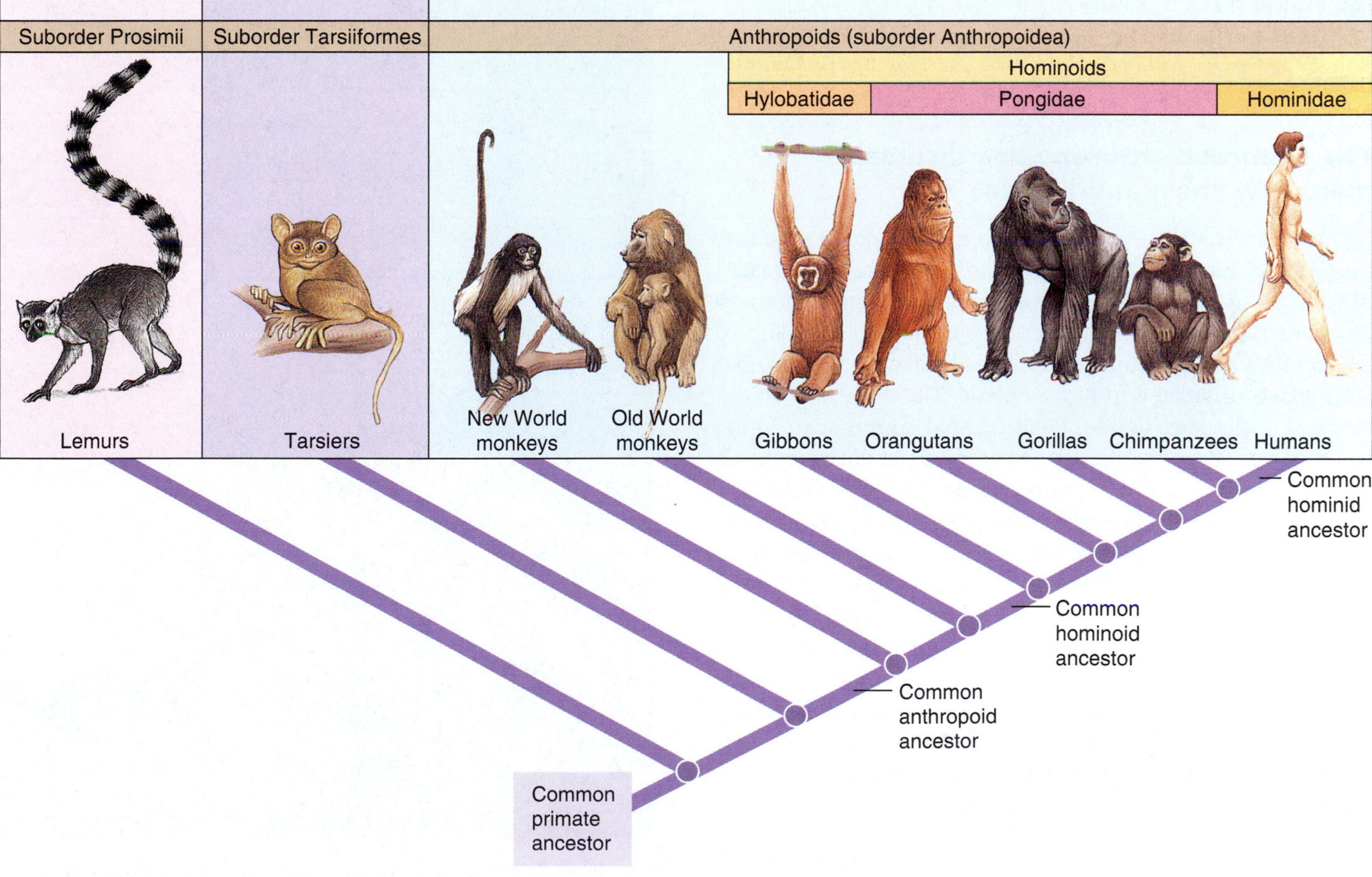

Figure 21–2 Primate evolution. This diagram shows hypothetical phylogenetic relationships among living primates, based on current scientific evidence. The three families of hominoids recognized by many biologists are shown. *(Figures not drawn to scale.)*

Figure 21–3 Lemurs. A mother ring-tailed lemur *(Lemus catta)* and her baby share a piece of fruit. Lemurs are native to Madagascar. *(Frans Lanting/Minden Pictures)*

The suborder Anthropoidea includes monkeys, apes, and humans

Anthropoid primates arose during the middle Eocene epoch, at least 45 mya. Several different fossil anthropoids have been identified from Asia and North Africa, and there is a growing scientific consensus about the relationships of these fossil groups to one another and to living anthropoids. Evidence indicates that anthropoids originated in Africa or Asia. The oldest known anthropoid fossils, such as 42-million-year-old *Eosimias,* are found in China and Myanmar. Based on details about their dentition and the few bones that have been discovered, scientists infer that *Eosimias* and other ancestral anthropoids were small, insect-eating arboreal primates that were active during the day. Once they arose, anthropoids quickly spread throughout Europe, Asia, and Africa and arrived in South America much later. (The oldest known South American primate, *Branisella,* which is from Bolivia, is dated at 26 million years.)

One significant difference between anthropoids and other primates is in the size of their brains. The cerebrum, in particular, is more developed in monkeys, apes, and humans, where it functions as the center for learning, voluntary movement, and interpretation of sensation.

Monkeys are generally diurnal (active during the day) tree dwellers. They tend to eat fruit and leaves, with nuts, seeds, buds, insects, spiders, birds' eggs, and even small vertebrates playing a smaller part in their diets. The two main groups of monkeys, New World monkeys and Old World monkeys, are named for the hemispheres where they diversified. Monkeys in South and Central America are called New World monkeys, whereas monkeys in Africa, Asia, and Europe are called Old World monkeys. New and Old World monkeys have been evolving separately for tens of millions of years.

One of the most important unanswered questions in anthropoid evolution concerns *how* monkeys arrived in South America. Africa and South America had already drifted apart (see Chapter 17), so the ancestors of New World monkeys may have rafted from Africa to South America on floating masses of vegetation. (The South Atlantic Ocean would have been about half as wide as it is today, and any islands that may have been present could have provided "stepping stones.") Alternatively, the ancestors of New World monkeys may have dispersed from Asia to North America to South America. Once established in the New World, these monkeys rapidly diversified.

New World monkeys are restricted to Central and South America and include marmosets, capuchins, howler monkeys, squirrel monkeys, and spider monkeys. New World monkeys are arboreal, and some possess long, slender limbs that permit easy movement in the trees (Fig. 21–5*a*). A few have **prehensile** tails

Figure 21–4 Tarsiers. The huge eyes of the tarsier *(Tarsius bancanus)* help it find insects, lizards, and other prey when it hunts at night. When a tarsier sees an insect, it pounces on it and grasps the prey with its hands. Tarsiers live in the rain forests of Indonesia and the Philippines. *(Frans Lanting/Minden Pictures)*

(a)

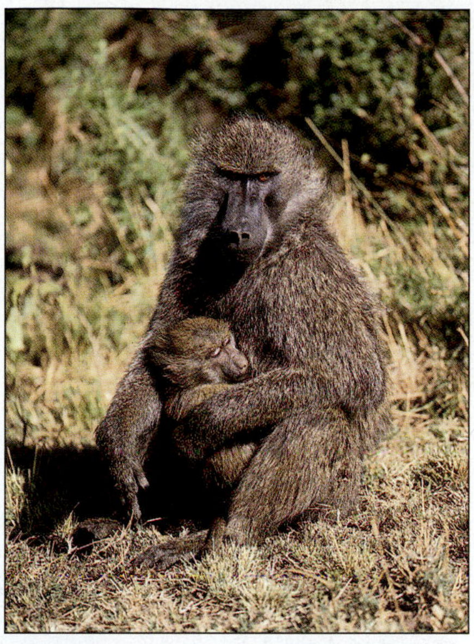

(b)

Figure 21–5 New and Old World monkeys. **(a)** The white-faced monkey *(Cebus capucinus)* is a New World monkey that has a prehensile tail and a flattened nose with nostrils directed to the side. **(b)** The Anubis baboon *(Papio anubis)* is an Old World monkey native to Africa. Note that its nostrils are directed downward *(a, C.C. Lockwood/DRK Photo; b, S. Meyers/Okapia/Photo Researchers, Inc.)*

capable of wrapping around branches and serving as fifth limbs. Some New World monkeys have shorter thumbs, and in certain cases the thumbs are totally absent. Their facial anatomy is different from that of the Old World monkeys; they have flattened noses with the nostrils opening to the side. They live in groups and exhibit complex social behaviors.

Old World monkeys are distributed in tropical parts of Africa and Asia. In addition to baboons and macaques (pronounced muh-kacks′), the Old World group includes guenons, mangabeys, langurs, and colobus monkeys. Most Old World monkeys are arboreal, although some, such as baboons and macaques, spend much of their time on the ground (Fig. 21–5*b*). The ground dwellers, which are **quadrupedal** ("four-footed"; they walk on all fours), arose from arboreal monkeys. None of the Old World monkeys has a prehensile tail, and some have extremely short tails. They have a fully opposable thumb, and unlike the New World monkeys, their nostrils are closer together and directed downward. Old World monkeys are intensely social animals.

Many classification schemes place apes and humans in three families

Old World monkeys shared a common ancestor with the **hominoids,** a group composed of apes and **hominids** (humans and their ancestors). A fairly primitive anthropoid was discovered in Egypt and named *Aegyptopithecus* (Fig. 21–6*a*). *Aegyptopithecus,* a cat-sized, forest-dwelling arboreal monkey with a few apelike characteristics, lived during the Oligocene epoch, approximately 34 mya.

During the Miocene epoch, approximately 20 mya, apes and Old World monkeys diverged. The oldest fossils with hominoid features were discovered in East Africa, mostly in Kenya. *Proconsul,* for example, appears early in the Miocene epoch, about 20 mya. It has a larger brain than monkeys, apelike teeth and diet (fruits), but a monkey-like body. At least 30 other early hominoid species lived during the Miocene epoch, but most of them became extinct and were not the common ancestor of modern apes and humans.

Miocene fossils of forest-dwelling, chimpanzee-sized apes called *dryopithecines,* which lived about 15 mya, are of special

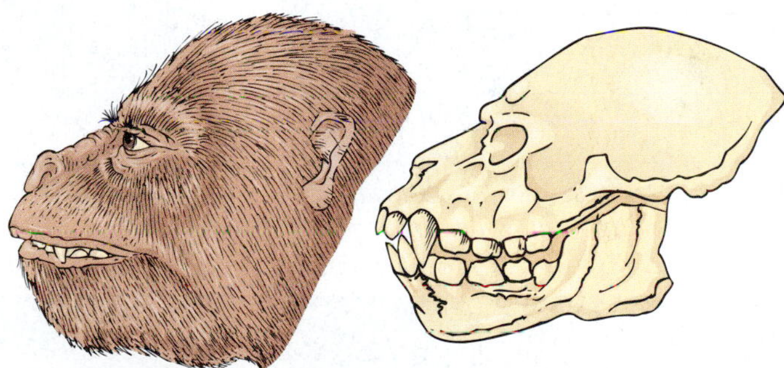

(a) Oligocene anthropoid, *Aegyptopithecus*

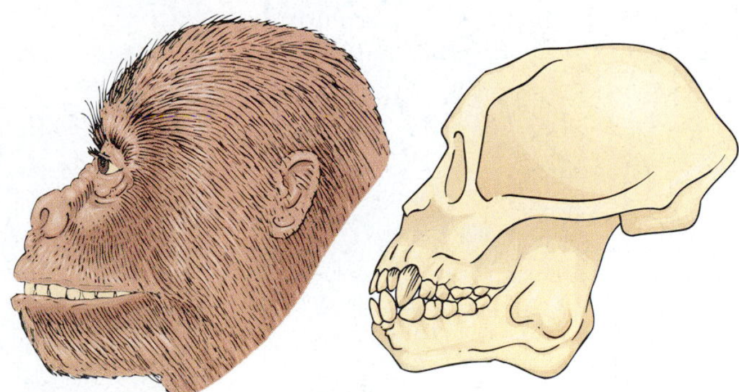

(b) Miocene ape, *Dryopithecus*

Figure 21–6 *Aegyptopithecus* and *Dryopithecus*. **(a)** Fossils of *Aegyptopithecus*, a fairly primitive anthropoid, were discovered in Egypt. **(b)** *Dryopithecus*, a more advanced ape, may have given rise to modern hominoids. *(Figures not drawn to scale.)*

interest because this hominoid lineage may have given rise to modern apes as well as to the human line (Fig. 21–6*b*). The dryopithecines, such as *Dryopithecus, Kenyapithecus,* and *Morotopithecus,* were distributed widely across Europe, Africa, and Asia. As the climate gradually cooled and became drier, their range became more limited. These apes had highly modified bodies for swinging through the branches of trees, although there is also evidence that some of them may have left the treetops for the ground as dense forest gradually changed into open woodland. Many questions about the relationships among the various early apes have been generated by the discovery of these and other Miocene hominoids. As future fossil finds are evaluated, they may lead to a rearrangement of ancestors in the hominoid family tree.

Many biologists classify the five genera of hominoids alive today into three families: Gibbons *(Hylobates)* are known as lesser apes and are placed in the family Hylobatidae. The family Pongidae includes orangutans *(Pongo),* gorillas *(Gorilla),* and chimpanzees *(Pan),* and the family Hominidae includes humans *(Homo)*. Re-

(a) (b) (c) (d)

Figure 21–7 Apes. **(a)** White-handed gibbons *(Hylobates lar)* are extremely acrobatic and often move through the trees by brachiation. **(b)** An orangutan *(Pongo pygmaeus)* mother and baby. **(c)** A young lowland gorilla *(Gorilla gorilla)* in knuckle-walking stance. **(d)** A bonobo chimpanzee *(Pan paniscus)* grooms another member of the group. Bonobos are endemic to a single country, the Democratic Republic of Congo (formerly Zaire). *(a, Joe McDonald/Visuals Unlimited; b, BIOS/Peter Arnold, Inc.; c, Nancy Adams/Tom Stack & Associates; d, K. & K. Ammann/Bruce Coleman, Inc.)*

cent molecular evidence indicates a closer relationship between humans and the greater apes, particularly chimpanzees, and some scientists now classify them in the same family.

Gibbons are natural acrobats that can **brachiate,** or arm-swing, with their weight supported by one arm at a time (Fig. 21–7*a*). Orangutans are also tree dwellers, but chimpanzees and especially gorillas have adapted to life on the ground (Fig. 21–7*b–d*). Gorillas and chimpanzees have retained long arms typical of brachiating primates but use these to assist in quadrupedal walking, sometimes known as **knuckle-walking** because of the way they fold (flex) their digits when moving. Apes, like humans, lack tails. They are generally much larger than monkeys, although gibbons are a notable exception.

Evidence of the close relatedness of orangutans, gorillas, chimps, and humans is abundant at the molecular level. The amino acid sequence of the chimpanzee's hemoglobin is identical to that of the human; hemoglobin molecules of the gorilla and rhesus monkey differ from the human's by 2 and 15 amino acids, respectively. DNA sequence analyses indicate that chimpanzees are likely to be our nearest living relatives among the apes (see Table 17–1). Molecular evidence suggests that gorillas may have diverged from the chimpanzee and hominid lines some 8 to 10 mya, whereas chimpanzee and hominid lines probably separated about 6 mya.

THE FOSSIL RECORD PROVIDES CLUES TO HOMINID EVOLUTION

Scientists have a growing storehouse of hundreds of hominid fossils, which provide useful information about general trends in the body design, appearance, and behavior of ancestral humans. It is evident, for example, that early hominids adopted a **bipedal** (two-footed) posture before their brains enlarged. Despite the wealth of fossil evidence, interpretations of hominid characteristics, taxonomy, and phylogeny continue to be vigorously debated, and new discoveries raise new questions. Furthermore, hominid evolution, like other scientific fields, is influenced by the different perspectives of the various workers studying it. The lack of a scientific consensus regarding certain aspects of hominid evolution is, therefore, an expected part of the scientific process.

Evolutionary changes from the earliest hominids to modern humans are evident in some of the characteristics of the skeleton and skull. Compared with the ape skeleton, the human skeleton possesses distinct differences that reflect our ability to stand erect and walk on two feet (Fig. 21–8). These differences also reflect the habitat change for early hominids, from an arboreal existence in the forest to a life spent at least partly on the ground.

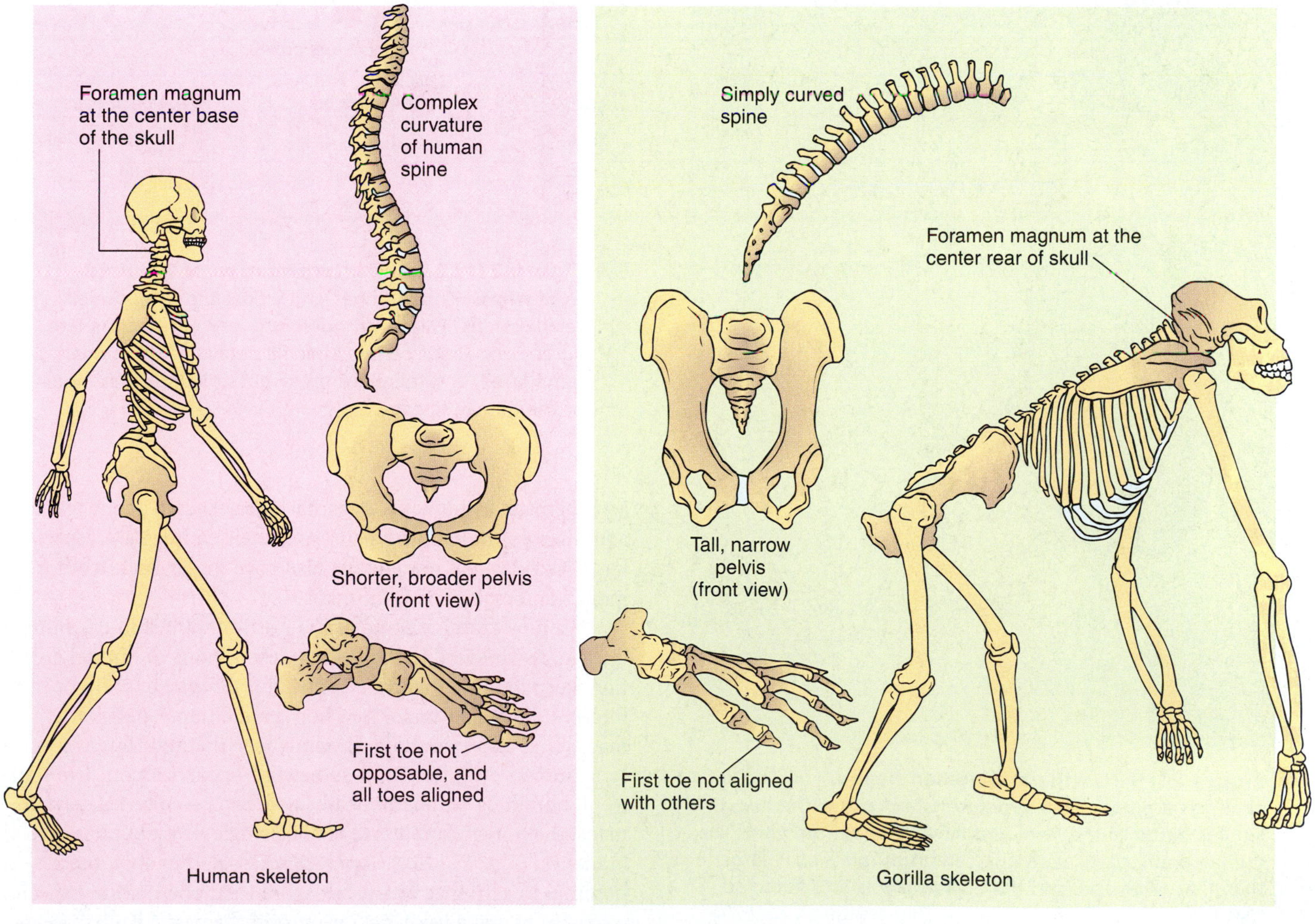

Figure 21–8 Gorilla and human skeletons. When gorilla and human skeletons are compared, the skeletal adaptations for bipedalism in humans become apparent.

The curvature of the human spine provides better balance and weight distribution for bipedal locomotion. The human pelvis is shorter and broader than the ape pelvis, providing a better attachment of muscles used for upright walking. The hole in the base of the skull for the spinal cord, called the **foramen magnum,** is located in the middle of the rear of the skull in apes. In contrast, the human foramen magnum is centered in the skull base, positioning the head for erect walking. An increase in the length of the legs relative to the arms, and alignment of the big toe with the rest of the toes, further adapted the early hominids for bipedalism.

Another major trend in hominid evolution was an increase in the size of the brain relative to the size of the body (Fig. 21–9). The ape skull possesses prominent bony ridges above the eye sockets, whereas these **supraorbital ridges** are lacking in modern human skulls. Human faces are flatter than those of apes, and the jaws are different. The arrangement of teeth in the ape jaw is somewhat rectangular, compared with a rounded, or U-shaped, arrangement in humans. Apes have larger front teeth (canines and incisors) than do humans, and their canines are especially large. Gorillas and orangutans also have larger back teeth (premolars and molars) than humans.

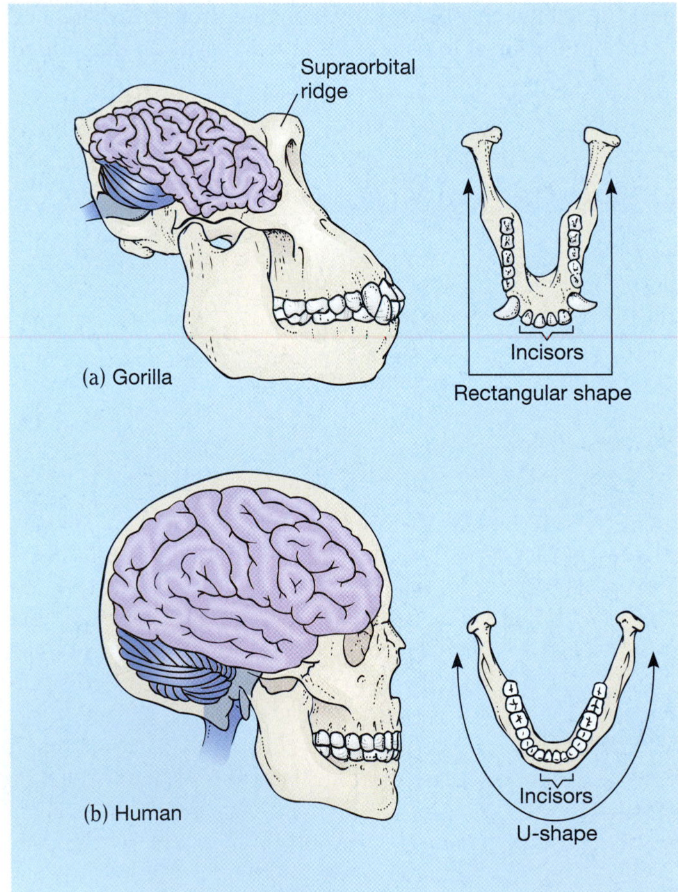

Figure 21–9 Gorilla and human heads. **(a)** The ape skull has a pronounced supraorbital ridge. **(b)** The human skull is flatter in the front and has a pronounced chin. The human brain, particularly the cerebrum *(purple),* is larger than that of an ape, and the human jaw is structured so that the teeth are arranged in a U shape. Human canines and incisors are also smaller than those of apes.

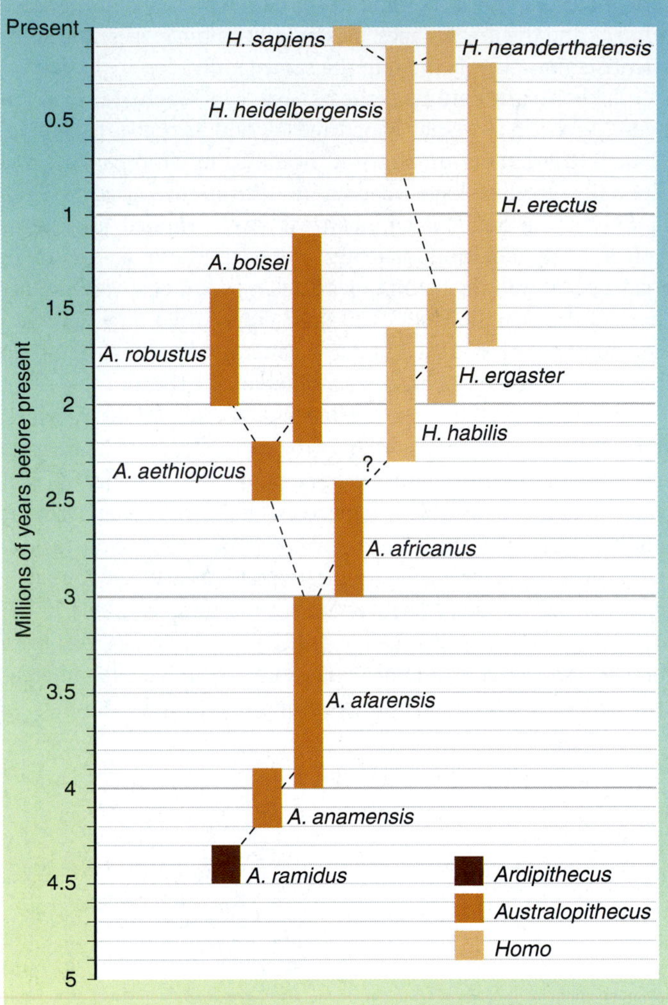

Figure 21–10 One interpretation of hominid evolution. Dashed lines show possible evolutionary relationships. Paleoanthropologists are not in complete agreement about certain specific details of the human family tree, and there are many possible interpretations of the human lineage.

We now examine some of the fossil hominids in the human lineage. As you read the following descriptions of human evolution, keep in mind that much of what is discussed is still open to interpretation and major revision as additional discoveries are made. It is also important to remember that, although we present human evolution in a somewhat linear fashion, from ancient hominids to anatomically and behaviorally modern humans, the human family tree is not a single trunk but has several branches (Fig. 21–10). *Homo sapiens* is the only species of hominid in existence today, but more than one hominid species coexisted at any given time for most of the past 4 million years.

The earliest hominid belongs to the genus *Ardipithecus*

Hominid evolution began in Africa. The earliest hominid, which belongs to the genus *Ardipithecus,* appeared at least 4.4 mya. *Ardipithecus* gave rise to *Australopithecus,* a genus that includes several species that lived between 4 and 1 mya. These two genera of early hominids, often referred to as **australopithecines,** or "southern man apes," had longer arms, shorter legs, and smaller brains relative to modern humans.[1] The actual number of australopithecine species for which fossil evidence has been found is under debate, because in some cases, differences in the relatively few skeletal fragments could indicate either variation among individuals within a species or evidence of separate species. Most paleoanthropologists recognize at least six species of australopithecines.

The first fossils of the earliest hominids were discovered by Tim White, Berhane Asfaw, and Gen Suwa in 1992 and assigned to ***Ardipithecus ramidus*** in 1995. The specific epithet *ramidus* is derived from a word meaning "root" in the Afar language, spoken in the region of Ethiopia where the fossils were found. This hominid, which is more primitive than any other known hominid, is quite close to the "root" of the human family tree, that is, to the last common ancestor of modern hominids and chimpanzees. Because no leg bones were found in the initial discovery, it has not yet been determined if *A. ramidus* was bipedal. Future discoveries may clarify this important point.

The genus *Australopithecus* contains the immediate ancestors of the genus *Homo*

Hominids that existed between 3.9 and 4.2 mya are assigned to the species ***Australopithecus anamensis,*** first named in 1995 by Meave Leakey and her coworkers from fossils discovered in East Africa. This hominid species, which has a mixture of apelike and human-like features, presumably arose from *Ardipithecus ramidus.* A comparison of male and female *A. anamensis* body sizes and canine teeth reveals **sexual dimorphism,** marked phenotypic differences between the two sexes of the same species. (The modern-day gorilla exhibits sexual dimorphism.) The back teeth and jaws of *A. anamensis* are larger than those of modern chimpanzees, whereas the front teeth are smaller and more like those of later hominids. A fossil leg bone, the tibia, indicates that *A. anamensis* had an upright posture and was bipedal, although it also may have foraged in the trees. Thus, bipedalism occurred early in human evolution and may have been the first human adaptation.

Australopithecus afarensis, another primitive hominid, appears to have arisen directly from *A. anamensis.* Many fossils of *A. afarensis* skeletal remains have been discovered in Africa, including a remarkably complete skeleton nicknamed Lucy found in Ethiopia in 1974 by a team led by Donald Johanson. Lucy, a small hominid approximately 1.04 m (3 ft, 5 in) is thought to be about 3.2 million years old. In 1978, Mary Leakey and coworkers discovered beautifully preserved fossil footprints of three *A. afarensis* individuals who walked more than 3.6 mya. In 1994 the first adult skull of *A. afarensis* was found. The skull, characterized by a relatively small brain, pronounced supraorbital ridges, a jutting jaw, and large canine teeth, is an estimated 3 million years old. It is probable that *A. afarensis* did not construct tools or make fires, since no evidence of tools or fire has been found at fossil sites.

Many paleoanthropologists think *A. afarensis* gave rise to several australopithecine species, including ***Australopithecus africanus,*** which may have appeared as early as 3 mya. The first *A. africanus* fossil was discovered in South Africa in 1924, and since then hundreds have been found. This hominid walked erect and possessed hands and teeth that were distinctly human-like. Based on characteristics of the teeth, it is thought that *A. africanus* ate both plants and animals. Like *A. afarensis,* it had a small brain, more like that of its primate ancestors than of present-day humans.

Three australopithecine species (*A. robustus* from South Africa, and *A. aethiopicus* and *A. boisei,* both from East Africa) are larger than *A. africanus* and have extremely large molars, very powerful jaws, relatively small brains, and bony skull crests. Most females lacked the skull crests and had substantially smaller jaws, another example of sexual dimorphism in early hominids. The teeth and jaws suggest a diet, perhaps of tough roots and tubers, that would require powerful grinding. These so-called *robust australopithecines* may or may not be closely related but are generally thought to represent evolutionary offshoots, or side branches, of human evolution. The first robust australopithecine, *A. aethiopicus,* appeared about 2.5 mya. Some researchers classify robust australopithecines in a separate genus *(Paranthropus).*

Homo habilis is the oldest member of the genus *Homo*

The first hominid to have enough uniquely human features to be placed in the same genus as modern humans is ***Homo habilis.*** It was first discovered in the early 1960s at Olduvai Gorge in Tanzania. Since then other fossils of *H. habilis* have been discovered in East and South Africa. *Homo habilis* was a small hominid with a larger brain and smaller premolars and molars than the australopithecines. This hominid appeared approximately 2.3 mya and persisted for about 0.75 million years. Fossils of *H. habilis* have been found in numerous areas in Africa. These sites contain primitive tools, stones that had been chipped, cracked, or hammered to make sharp edges for cutting or scraping.[2] *Oldowan* pebble choppers and flakes, for example, were probably used to cut through animal hides to obtain meat and to break bones for their nutritious marrow.

[1] Do not make the mistake of thinking that our smaller-brained ancestors were inferior to ourselves. Ancestral hominids were evolutionarily successful in that they were well adapted to their environment and survived for millions of years.

[2] The oldest known stone tools, discovered in the mid-1990s by Sileshi Semaw and colleagues, were found in Gona, Ethiopia. These ancient tools were made some 2.6 mya, but because no hominid remains have been found at the site yet, it is not known who made them.

The relationship between the australopithecines and *H. habilis* is not clear. Using physical characteristics of their fossilized skeletons as evidence, many paleoanthropologists have inferred that the australopithecines were ancestors of *H. habilis*. Some researchers do not think *H. habilis* belongs in the genus *Homo*, and they suggest it should be reclassified as *Australopithecus habilis*. Discoveries of additional fossils may help clarify these relationships.

Homo erectus apparently evolved from *Homo habilis*

Numerous fossils of ***Homo erectus*** have been found throughout Africa and Asia. (The first fossil evidence of *H. erectus* was found in Indonesia in the 1890s.) *Homo erectus* is thought to have originated in Africa about 1.7 mya and then to have spread quickly to Europe and Asia. The oldest fossils of *H. erectus* that have been found in Southeast Asia, for example, may be as old as 1.8 million years, although the most widely accepted date is about 1 million years. Peking man and Java man, discovered in Asia, were later examples of *H. erectus*, which existed until at least 200,000 years ago; some populations of *H. erectus* may have persisted more recently.

Homo erectus was taller than *H. habilis*. Its brain, which was larger than that of *H. habilis*, got progressively larger during the course of its evolution. Its skull, although larger, did not possess totally modern features, retaining the heavy supraorbital ridge and projecting face that are more characteristic of its ape ancestors (Fig. 21–11). *Homo erectus* is the first hominid to have fewer differences between the sexes.

The increased mental faculties associated with an increased brain size enabled these early humans to make more advanced stone tools, known as *Acheulean* tools, including hand axes and other implements that have been interpreted as choppers, borers, and scrapers. Their intelligence also allowed them to survive in cold areas. *Homo erectus* obtained food by hunting or scavenging and may have worn clothing, built fires, and lived in caves or shelters. Evidence of weapons (spears) has been unearthed at *Homo erectus* sites in Europe.

Ideas regarding *Homo erectus*, like many other aspects of human evolution, are changing with each new fossil discovery. Many scientists now hypothesize that the fossils classified as *H. erectus* really represent two species, ***Homo ergaster***, an earlier African species, and *H. erectus*, a later East Asian offshoot. The best known fossils of *H. ergaster* come from the Lake Turkana region in Kenya. Researchers who support this split speculate that *Homo ergaster* may be the direct ancestor of later humans, whereas *Homo erectus* may be an evolutionary dead end. It is hoped that future fossil discoveries will help clarify the status of *Homo erectus*.

Archaic *Homo sapiens* appeared about 800,000 years ago

Archaic *Homo sapiens* are regionally diverse descendants of *Homo erectus* or *Homo ergaster* that lived in Africa, Asia, and Europe from about 800,000 to 100,000 years ago. They thus overlapped both *H. erectus* populations and the later appearing Neandertals (discussed later). Some researchers classify archaic *Homo sapiens* as a separate species, ***Homo heidelbergensis***.

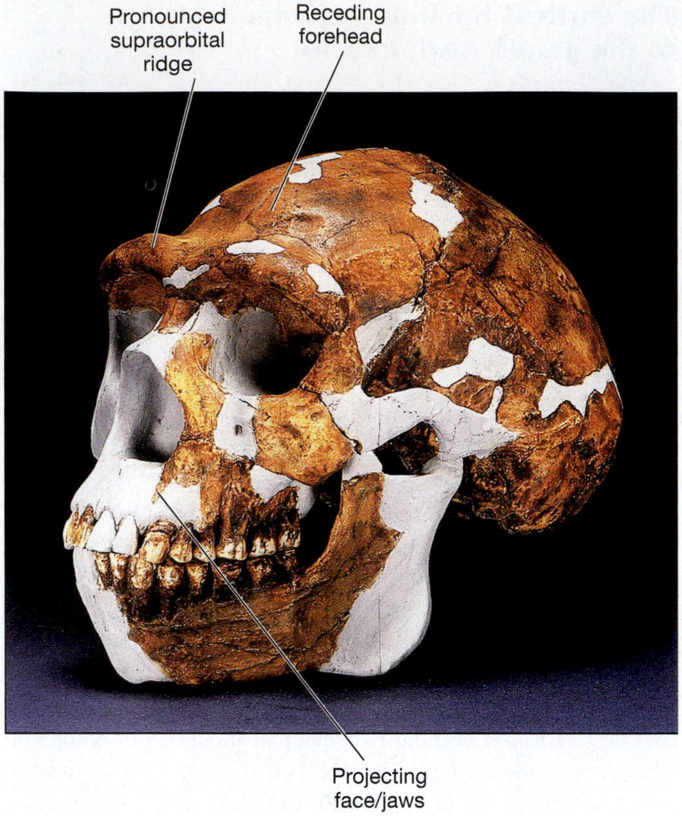

Pronounced supraorbital ridge

Receding forehead

Projecting face/jaws

Figure 21–11 *Homo erectus* skull from China. The reconstructed portions are white. Note the receding forehead, pronounced supraorbital ridge, and projecting face and jaws. *(Ken Mowbray)*

Neandertals appeared approximately 230,000 years ago

Neandertals[3] were first discovered in the Neander Valley in Germany. They lived throughout Europe and western Asia from about 230,000 to 30,000 years ago. These early humans had short, sturdy builds. Their faces projected slightly, their chins and foreheads receded, they had heavy supraorbital ridges and jawbones, and their brains and front teeth were larger than those of modern humans. Their nasal cavities were large, and their cheekbones were receding. Scientists have suggested that the large noses provided larger surface areas in Neandertal sinuses, enabling them to better warm the cold air of Ice Age Eurasia as air traveled through the head to the lungs.

Scientists have not reached a consensus about whether the Neandertals are a separate species from modern humans. Many think the anatomical differences between Neandertals and mod-

[3] Neandertal was formerly spelled "Neanderthal." The silent "h" has been dropped in modern German but not in the scientific name.

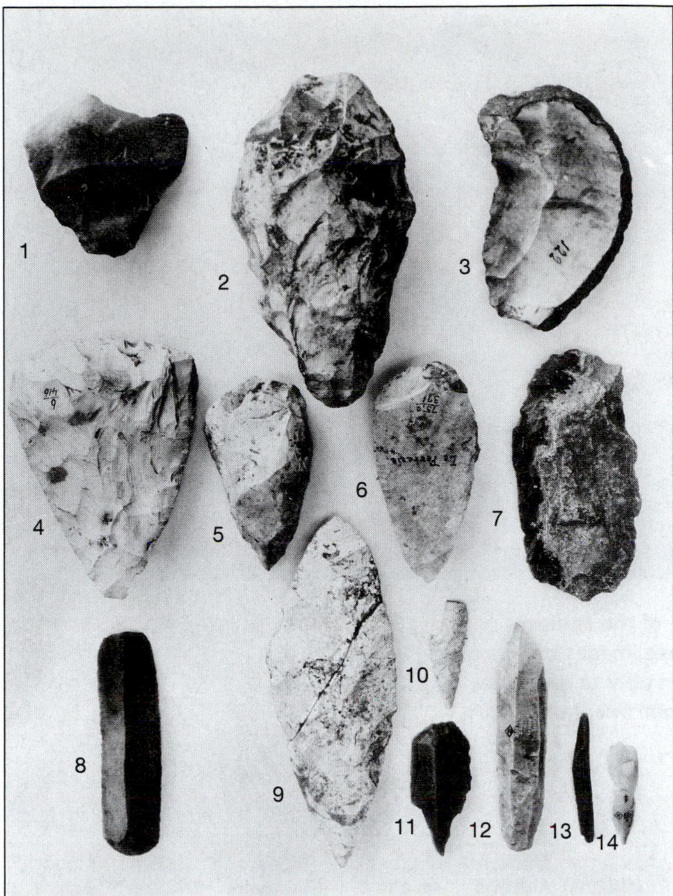

Figure 21–12 Mousterian tools. Mousterian tools are named after a Neandertal site in Le Moustier, France. Mousterian tools included a variety of skillfully made stone tools, such as hand axes, flakes, scrapers, borers, and spear points. (*1* to *4* are earlier tools, and *8* to *14* are later tools.) *(Photo Researchers, Inc.)*

ern humans mean that they were separate species, ***Homo neanderthalensis*** and ***Homo sapiens***. Other scientists disagree and think that Neandertals were a group of *Homo sapiens*.

Neandertal tools, known as *Mousterian* tools, include the oldest known spear points (Fig. 21–12). Neandertal tools were more sophisticated than those of *H. erectus*. Studies of Neandertal sites indicate that they hunted large animals. The existence of skeletons of elderly Neandertals and of Neandertals with healed fractures may demonstrate that they cared for the aged and the sick, an indication of advanced social cooperation. They apparently had rituals, possibly of religious significance, and sometimes buried their dead.

Process of Science

The disappearance of the Neandertals some 30,000 years ago is a mystery that has sparked debate among paleoanthropologists. Other groups of *H. sapiens* with more modern features coexisted for tens of thousands of years with the Neandertals. Perhaps the other humans out-competed or exterminated the Neandertals, leading to their extinction. It is also possible that the Neandertals interbred with these humans, diluting their features beyond recognition.

Analysis of **mitochondrial DNA (mtDNA)** contributes useful data to such controversies. Each of the several hundred mitochondria within a cell has about ten copies of a small loop of DNA that codes for transfer RNAs, ribosomal RNAs, and certain respiratory enzymes. Mitochondrial DNA mutates more rapidly than nuclear DNA, so mtDNA is a sensitive indicator of evolution. In 1997 Svante Pääbo and colleagues analyzed mtDNA extracted from a Neandertal bone and reported that it differs significantly from all modern human mtDNA sequences, although it is more similar to human than to chimpanzee mtDNA. This finding suggests that Neandertals are an evolutionary dead end that did not interbreed with more modern humans.

To be considered authentic, the results of scientific research must be independently reproducible. Hence, a molecular analysis of mtDNA from a different Neandertal bone, reported by Igor V. Ovchinnikov in 2000, is significant. Ovchinnikov's findings corroborate those of Pääbo, that is, that Neandertals did not contribute to the modern human mtDNA.

The question of the relationship between Neandertals and anatomically modern humans remains controversial, however. In 1999 researchers reported the discovery of a 4-year-old child's remains in Portugal. The skeleton, dated at 24,500 years of age, has traits of both modern humans and Neandertals (short lower limb bones). The child lived several thousand years after Neandertals are thought to have disappeared, and the researchers who discovered it view it as an example of mixed ancestry, meaning that Neandertals and anatomically modern humans are members of the same species who interbred freely. Other scientists disagree with their interpretation and think the so-called Neandertal features of the child may reflect normal variation inherent in the human species.

The origin of modern *Homo sapiens* is hotly debated

Homo sapiens with anatomically modern features existed in Africa and the Middle East at least 100,000 years ago. The *H. sapiens* skull lacked a heavy brow ridge and possessed a distinct chin. By about 30,000 years ago, anatomically modern humans were the only members of genus *Homo* remaining. European remains of these ancient people are referred to as **Cro-Magnons.** Their weapons and tools were complex and often made of materials other than stone, including bone, ivory, and wood. They made stone blades that were extremely sharp. Cro-Magnons developed art, including cave paintings, engravings, and sculpture, possibly for ritualistic purposes (Fig. 21–13). Their sophisticated tools and art indicate that they may have possessed language, which would have been used to transmit their culture to younger generations.

Process of Science

Two opposing hypotheses currently exist about the origin of these modern humans: the *out of Africa* hypothesis and the *multiregional* hypothesis (Fig. 21–14). The out of Africa hypothesis holds that modern *H. sapiens* arose in Africa

Figure 21–13 Cro-Magnon cave paintings. These are some of the earliest known examples of human art. Discovered in Lascaux, France, these images of reindeer have been interpreted as having religious significance, possibly to guarantee a successful hunt. *(Photo by J. Beckett/D. Stipkovich, courtesy Department of Library Services, American Museum of Natural History)*

between 200,000 and 100,000 years ago and then migrated to Europe and Asia, displacing the Neandertals and other more primitive humans living there. According to the multiregional hypothesis, modern humans evolved from *H. erectus*. They originated, beginning as early as 2 mya, as separately evolving populations living in several parts of Africa, Asia, and Europe. Each of these populations evolved in its own distinctive way but occasionally met and interbred with other populations, thereby preventing complete reproductive isolation. The variation found today in different geographical populations therefore represents a continuation of this multiregional process.

Data from *Homo* fossils, as well as molecular biology and population genetics studies of modern humans, have been cited in support of both hypotheses, and both have vigorous defenders and strong detractors. Such disagreement is an important part of the scientific process because it stimulates research that may ultimately resolve the issue.

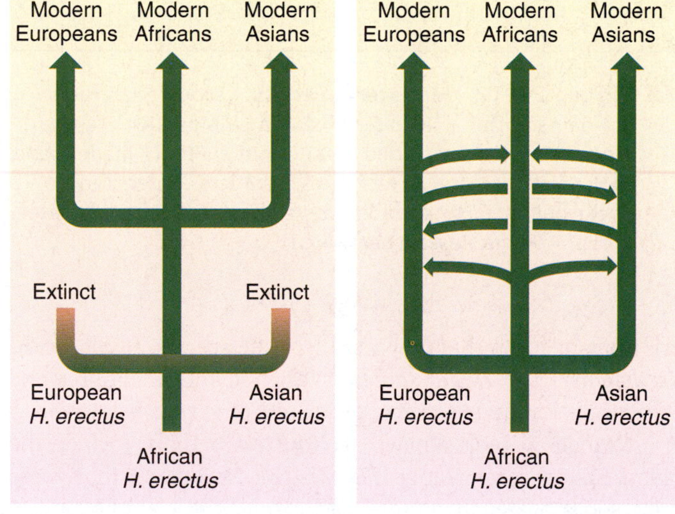

(a) Out of Africa hypothesis (b) Multiregional hypothesis

Figure 21–14 Competing hypotheses on the origin of modern humans. Scientists agree that *Homo erectus* arose in Africa and migrated to other continents. What happened then is the controversy. **(a)** According to the out of Africa hypothesis, all but the African line of *H. erectus* went extinct. In Africa, *H. erectus* evolved into modern humans, which migrated to other continents. **(b)** According to the multiregional hypothesis, modern humans evolved from *H. erectus* populations in different regions of the world. The smaller horizontal arrows in **(b)** represent gene flow (migration and interbreeding among the different populations).

Analysis of DNA provides evidence on the origin of modern humans

Molecular anthropology, the comparison of biological molecules from present-day individuals of regional human populations, provides clues that help scientists unravel the origin of modern humans and trace human migrations. The out of Africa hypothesis was originally supported by studies in the late 1980s of mtDNA from various human populations. In 1992 the statistical assumptions used in one analysis of mtDNA were found to be erroneous, leading to questions about the validity of this purported test of the out of Africa hypothesis. Multiple subsequent molecular studies comparing DNA sequences from different human groups, however, all have produced essentially the same answer—that modern humans are descended from an early human population that lived in southern Africa.

For example, a 1996 study examined the genetic variation in two stretches of DNA on human chromosome 12 from 1600 people living in 42 different populations around the world (13 African, 2 Middle Eastern, 7 European, 9 Asian, 3 Pacific, and 8 Amerindian populations). The DNA segments varied depending on where the populations lived. Based on the results, scientists divided the present-day human population into three groups: sub-Saharan Africans, northeastern Africans, and non-Africans. (Sub-Saharan Africa refers to all countries located south of the Sahara Desert.) The sub-Saharan Africans exhibited the greatest genetic diversity, whereas the non-African populations were the least diverse. These findings are consistent with the predictions of the out of Africa hypothesis for two reasons. First, according to the hypothesis, the sub-Saharan populations are expected to be more diverse because they are older and have had a longer time to accumulate that diversity. Second, the small populations thought to have emigrated from Africa could not have been representative of the total diversity present in the larger African population (recall the discussion in Chapter 18 on the founder effect and genetic drift).

Such research has not disproved the multiregional hypothesis, but it indicates the direction for additional research on other segments of human DNA. A series of recent genetic studies of both mtDNA and nuclear DNA has strengthened the case for Africa as the birthplace of modern humans, and in 1999 scientists announced that the ancestors common to all humans were probably ancient Khoisans, an indigenous group in southern Africa. This conclusion was based on separate analyses of DNA from both mitochondria and the Y chromosome. Both studies indicated that the Khoisan people are the most ancient of all human groups.

Some molecular research has suggested that the out of Africa hypothesis may not be as simple as originally envisioned—that is, humans from Africa may not have completely replaced the archaic humans on other continents but may have interbred to some extent with them as they evolved into modern humans. Although many of the human genes that were analyzed have demonstrated an African ancestry, a few appear to have arisen in Asia and to have been introduced at a later time into African populations, probably by migration from Asia to Africa. Thus, certain ancestral human populations in both Africa and Asia may have contributed to the gene pool of modern humans.

HUMANS UNDERGO CULTURAL EVOLUTION

Genetically speaking, humans are not very different from other primates. At the level of our DNA sequences, we are roughly 98% identical to gorillas and 99% identical to chimpanzees. Our relatively few genetic differences, however, give rise to several important distinguishing features, such as greater intelligence and the ability to capitalize on it through **cultural evolution,** which is the transmission of knowledge from one generation to the next. (It should be noted that researchers are increasingly in agreement that humans are not the only animals to possess culture. Chimpanzees have primitive cultures that include tool-using techniques, hunting methods, and social behaviors, all of which vary from one population to another. These cultural traditions are passed from one generation to the next by teaching and imitation; see *On the Cutting Edge: Cultural Variation among Chimpanzee Groups* in Chapter 50.)

Human culture is dynamic; it is modified as we obtain new knowledge. Human cultural evolution is generally divided into three stages: (1) the development of hunter-gatherer societies; (2) the development of agriculture; and (3) the Industrial Revolution.

Early humans were hunters and gatherers who relied on what was available in their immediate environment. They were nomadic, and as the resources in a given area were exhausted or as the population increased, they migrated to a different area. These societies required a division of labor and the ability to make tools and weapons, which were needed not only to kill game but also to scrape hides, dig up roots and tubers, and cook food. Although we are not certain when hunting was incorporated into human society, we do know that it declined in importance approximately 15,000 years ago. This may have been due to a decrease in the abundance of large animals, triggered in part by overhunting. A few isolated groups of hunter-gatherer societies, including the Inuit of northern polar regions and the Mbuti of Africa, have survived into the 21st century.

Development of agriculture resulted in a more dependable food supply

Evidence that humans had begun to cultivate crops approximately 10,000 years ago includes the presence of agricultural tools and plant material at archaeological sites. Agriculture, which involves keeping animals as well as cultivating plants, resulted in a more dependable food supply. Recent archaeological evidence suggests that agriculture arose in several steps. Although there is variation from one site to another, plant cultivation, in combination with hunting, usually occurred first. Animal domestication followed later. Agriculture, in turn, often led to more permanent dwellings because considerable time was invested in growing crops in one area. Villages and cities often grew up around the farmlands, but the connection between agriculture and the establishment of villages and towns is complicated by recent discoveries. For example, Abu Hureyra in Syria was a village founded *before* agriculture arose. The villagers subsisted on the rich plant life of the area and the migrating herds of gazelle. Once people turned to agriculture, however, they seldom went back to hunting and gathering to obtain food.

Archaeological evidence indicates that agriculture developed independently in several different regions. There were three main centers of agriculture and several minor ones. Each of the main centers was associated with cultivation of a cereal crop, although other foods were grown as well. Cereals are grasses, which are members of the monocot group of flowering plants (see Chapter 27). The cereals associated with the three main centers of agriculture are wheat, corn, and rice.

Wheat was cultivated in the semiarid regions along the eastern edge of the Mediterranean. Other crops that originated there include peas, lentils, grapes, and olives. Central and South America were the sites of the maize (corn) culture. Squash, chili peppers, beans, and potatoes were also cultivated there. In southern China, evidence exists of the early cultivation of rice and other crops such as soybeans.

Corn, wheat, and rice all are propagated by seed, which requires fairly sophisticated agricultural practices. Growing plants that could be propagated vegetatively may have occurred earlier. Plants raised in this manner, such as bananas, yams, potatoes, and manioc, do not preserve as well as grains because of their high water content. That may be the reason we have little archaeological evidence of their cultivation.

Other advances in agriculture include the domestication of animals, which were kept to supply food, milk, and hides. Archaeological evidence indicates that wild goats and sheep were probably the first animals to be domesticated in southwest Turkey, northern Iraq, and Iran. In the Old World, animals were also used to prepare fields for planting. Another major advance in agriculture was irrigation, which began more than 5000 years ago in Egypt.

Producing food agriculturally was more time-consuming than hunting and gathering, but it was also more productive. In hunter-gatherer societies, everyone shares the responsibility for obtaining food. In agricultural societies, fewer people are needed to provide food for everyone. Thus agriculture freed some people to pursue other endeavors, including religion, art, and various crafts.

Cultural evolution has had a profound impact on the biosphere

Cultural evolution has had far-reaching effects on both human society and on other organisms. The Industrial Revolution, which began in the 18th century, caused populations to concentrate in urban areas near centers of manufacturing. Advances in agriculture encouraged urbanization because fewer and fewer people were needed in rural areas to produce food for everyone. The spread of industrialization increased the demand for natural resources to supply the raw materials for industry.

Cultural evolution has permitted the human population, which reached 6.14 billion in 2001, to expand so dramatically that there are serious questions about Earth's ability to support so many people indefinitely (see Chapter 51). According to the U.N. Food and Agricultural Organization, about 828 million people lack access to the food needed to be healthy and lead productive lives. To further compound the problem, the United Nations projects that 3 billion *additional* people will be added to the world population by the year 2050.

Cultural evolution has resulted in large-scale disruption and degradation of the environment. Tropical rain forests and other natural environments are rapidly being eliminated. Soil, water, and air pollution occur in many places. Since World War II, soil degradation due to poor agricultural practices, overgrazing, and deforestation has occurred in an area equal to 17% of the Earth's total vegetated surface area. Many species cannot adapt to the rapid environmental changes caused by humans and thus are becoming extinct. The decrease in biological diversity due to extinction is alarming (see Chapter 55).

On a positive note, we are aware of the damage we are causing, and we have the intelligence to modify our behavior to improve these conditions. Education, including the study of biology, may help future generations develop environmental sensitivity, making cultural evolution our salvation rather than our destruction.

SUMMARY WITH KEY TERMS

I. Primates are **placental mammals** that arose from small, **arboreal** (tree dwelling), shrewlike mammals.
   A. Primates are adapted for an arboreal existence by the presence of five grasping digits, including an opposable thumb or toe; long, slender limbs that move freely at the hips and shoulders; and eyes located in front of the head.
   B. Primates are divided into three suborders.
      1. The suborder Prosimii includes lemurs, galagos, and lorises.
      2. The suborder Tarsiiformes includes tarsiers.
      3. The suborder Anthropoidea includes **anthropoids** (monkeys, apes, and humans).

II. Anthropoids arose from early primate ancestors.
   A. The early anthropoids branched into two groups: the New and Old World primates.
   B. **Hominoids** (apes and humans) arose from the Old World monkey lineage.
   C. There are four modern genera of apes: gibbons, orangutans, gorillas, and chimpanzees.

III. The **hominid** line consists of humans and their ancestors.
   A. Hominid evolution began in Africa.
      1. The earliest known hominids belong to ***Ardipithecus ramidus,*** which appeared about 4.4 mya.
      2. *Ardipithecus ramidus* presumably gave rise to ***Australopithecus anamensis.*** The genus *Australopithecus,* which includes at least six species that lived between about 4 and 1.25 mya, contains the immediate ancestors of the genus *Homo*.
      3. *Ardipithecus* and *Australopithecus* species are often referred to as **australopithecines.** *Australopithecus* species were **bipedal** (walked on two feet), a hominid feature. It is not yet known if *Ardipithecus* was bipedal.

B. ***Homo habilis*** was the earliest known hominid with some of the human features lacking in the australopithecines, including a slightly larger brain. *H. habilis* fashioned crude tools from stone.

C. ***Homo erectus*** had a larger brain than *H. habilis;* made more sophisticated tools; and may have worn clothing, built fires, and lived in caves or shelters. Some scientists now think that fossils identified as *Homo erectus* represent two different species, ***Homo ergaster,*** an earlier African species that gave rise to archaic *Homo sapiens,* and *Homo erectus,* a later Asian offshoot that may be an evolutionary dead end.

D. **Archaic *Homo sapiens*** lived in Africa, Asia, and Europe from about 800,000 to 100,000 years ago. Some researchers classify archaic *Homo sapiens* as a separate species, ***Homo heidelbergensis.***

E. **Neandertals** existed from about 230,000 to 30,000 years ago.
   1. Neandertals had short, sturdy builds; receding chins and foreheads; heavy supraorbital ridges and jawbones; larger front teeth; and nasal cavities with unusual triangular bony projections.
   2. Many scientists think that Neandertals were a separate species, ***Homo neanderthalensis,*** whereas some scientists think Neandertals were a type of modern human.
   3. The disappearance of Neandertals is a mystery.

F. Anatomically modern humans ***(Homo sapiens)*** existed 100,000 years ago.
   1. By about 30,000 years ago, anatomically modern humans were the only members of genus *Homo* remaining. European remains of these ancient people are referred to as **Cro-Magnons.**
   2. The origin of modern humans is controversial. Two hypotheses, the out of Africa and the multiregional hypotheses, purport to explain the origin of modern humans. **Molecular anthropology,** the comparison of biological molecules from individuals of regional human populations, generally favors the African origin of modern humans.

IV. **Cultural evolution** is the transmission of knowledge from one generation to the next.
   A. Large human brain size makes cultural evolution possible.
   B. Two significant advances in cultural evolution were the development of agriculture and the Industrial Revolution.

POST-TEST

1. The first primates evolved about 55 mya from (a) shrewlike monotremes (b) therapsids (c) shrewlike placental mammals (d) tarsiers (e) shrewlike marsupials
2. The anthropoids are more closely related to ___________ than to ___________. (a) tarsiers; lemurs (b) lemurs; monkeys (c) tree shrews; tarsiers (d) lemurs; tarsiers (e) tree shrews; monkeys
3. Unlike Old World monkeys, some New World monkeys possess (a) body hair (b) five grasping digits (c) a well-developed cerebrum (d) a bipedal walk (e) a prehensile tail
4. Apes and humans are collectively called (a) mammals (b) primates (c) anthropoids (d) hominoids (e) hominids
5. With what group do hominoids share the most recent common ancestor? (a) Old World monkeys (b) New World monkeys (c) tarsiers (d) lemurs (e) lorises and galigos
6. The ___________ in humans is centered at the base of the skull, positioning the head for erect walking. (a) supraorbital ridge (b) foramen magnum (c) pelvis (d) bony skull crest (e) femur
7. The oldest evidence of bipedalism is found in the tibia of (a) *Australopithecus afarensis* (b) *Australopithecus anamensis* (c) *Australopithecus africanus* (d) *Homo erectus* (e) *Homo habilis*
8. Humans and their *immediate* ancestors are collectively called (a) mammals (b) primates (c) anthropoids (d) hominoids (e) hominids
9. The earliest hominid belongs to the genus (a) *Aegyptopithecus* (b) *Dryopithecus* (c) *Ardipithecus* (d) *Australopithecus* (e) *Homo*
10. The earliest hominid to be placed in the genus *Homo* is (a) *H. habilis* (b) *H. ergaster* (c) *H. erectus* (d) *H. heidelbergensis* (e) *H. neanderthalensis*
11. Some scientists now think that fossils identified as *Homo erectus* represent which two different species? (a) *H. habilis* and *H. erectus* (b) *H. ergaster* and *H. erectus* (c) *H. heidelbergensis* and *H. ergaster* (d) *H. neanderthalensis* and *H. erectus* (e) *H. neanderthalensis* and *H. sapiens*
12. Archaic *Homo sapiens* appeared about ___________ years ago. (a) 5 million (b) 800,000 (c) 230,000 (d) 100,000 (e) 5000
13. ___________were an early group of humans with short, sturdy builds and heavy supraorbital ridges that lived throughout Europe and western Asia from about 230,000 to 30,000 years ago. (a) Australopithecines (b) Dryopithecines (c) Archaic *Homo sapiens* (d) Neandertals (e) Cro-Magnons
14. The modern human skull *lacks* (a) small canines (b) a foramen magnum centered in the skull (c) pronounced supraorbital ridges (d) a U-shaped arrangement of teeth on the jaw (e) a large cranium (brain case)
15. The comparison of genetic material from individuals of regional populations of humans, used to help unravel the origin and migration of modern humans, is known as (a) paleoarchaeology (b) cultural anthropology (c) molecular anthropology (d) cytogenetics (e) genetic dimorphism

REVIEW QUESTIONS

1. Distinguish between each of the following pairs: (a) mammals and primates (b) anthropoids and hominoids (c) hominoids and hominids (d) australopithecines and the genus Homo
2. Describe three different ways primates are adapted to an arboreal existence.
3. Identify at least three differences between the skulls of apes and humans.
4. Explain at least three ways in which an ape skeleton differs from a human skeleton.
5. Distinguish between each of the following pairs: (a) *Homo habilis* and *Homo erectus* (b) *Homo ergaster* and *Homo erectus* (c) *Homo erectus* and *Homo heidelbergensis* (d) *Homo neanderthalensis* and *Homo sapiens*
6. Describe the two currently proposed hypotheses that explain where modern humans originated.
7. What is cultural evolution, and how has it affected Earth?
8. Add the following labels to the diagram: suborder Anthropoidea, suborder Prosimii, suborder Tarsiformes, Hominids, Hominoids, Hylobatidae, and Pongidae. Use Figure 21–2 to check your answers.

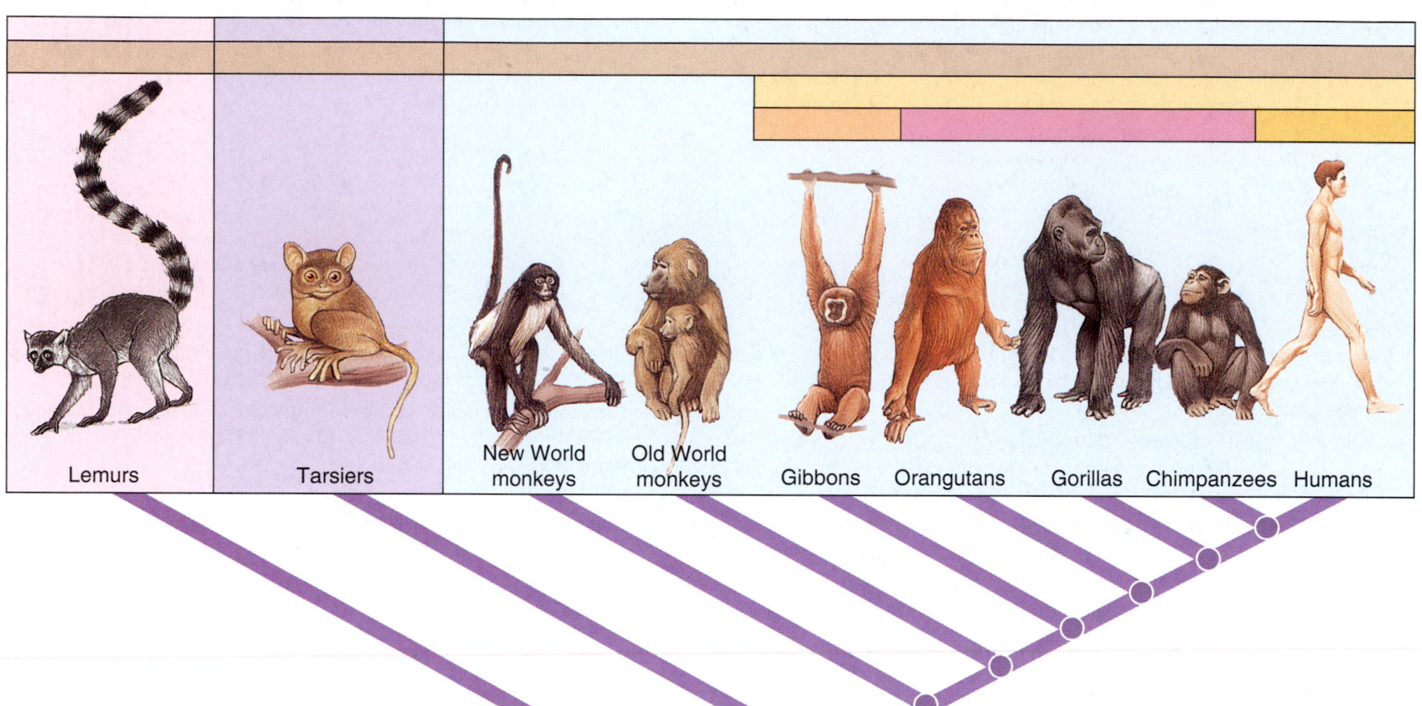

YOU MAKE THE CONNECTION

1. What types of as-yet-undiscovered scientific evidence would help explain how monkeys got to South America from the Old World?
2. Why are classifying fossil hominid species and determining their evolutionary relationships to one another such controversial endeavors?
3. Which hypothesis of the origin of modern humans (out of Africa or multiregional) seems to more closely match our understanding of evolutionary processes in general? Explain your answer.
4. If you were evaluating whether other early humans exterminated the Neandertals, what kinds of archaeological evidence might you look for?
5. The remains of Cro-Magnons have been found in southern Europe alongside reindeer bones, but reindeer currently exist only in northern Europe and Asia. Can you explain the apparent discrepancy?

RECOMMENDED READINGS

Bower, B. "Out on a Limb." *Science News,* Vol. 158, 25 Nov. 2000. The fact that relatively minor alterations in certain developmental genes can lead to major changes in body form may affect the validity of cladistic analysis of human evolution.

Bower, B. "Salvaged DNA Adds to Neandertals' Mystique." *Science News,* Vol. 157, 1 Apr. 2000. A second retrieval of ancient DNA from a Neandertal supports the hypothesis that Neandertals did not interbreed with anatomically modern humans.

Brainard, J. "Giving Neandertals Their Due." *Science News,* Vol. 154, 1 Aug. 1998. Recent studies suggest that the Neandertals produced relatively sophisticated ornaments and tools.

Gore, R. "People Like Us." *National Geographic,* Vol. 198, No. 1, Jul. 2000. This article examines what is known about human culture and behavior from 100,000 years ago to 10,000 years ago.

Kahn, P., and A. Gibbons. "DNA from an Extinct Human." *Science,* Vol. 277, 11 Jul. 1997. This research news article highlights the first successful attempt to extract and analyze DNA from a fossil Neandertal bone.

Keyser, A.W. "The Dawn of Humans: Finds in South Africa." *National Geographic,* Vol. 197, No. 5, May 2000. Discusses recent fossil discoveries of *Australopithecus robustus,* an evolutionary side branch of the human tree.

Larick, R., R.L. Ciochon, and Y. Zaim. "Fossil Farming in Java." *Natural History,* Jul./Aug. 1999. Indonesian farmers have found many hominid fossils while they cultivate their crops.

Leakey, M., and A. Walker. "Early Hominid Fossils from Africa." *Scientific American,* Vol. 276, No. 6, Jun. 1997. An overview of exciting discoveries of early hominid fossils (*Ardipithecus* and *Australopithecus*) and their significance to human evolution.

Morell, V. "Forming the Robust Australopithecine Face." *Science,* Vol. 284, 9 Apr. 1999. This news article reports on a research paper in the same issue. It emphasizes that many of the unique facial characteristics of the robust australopithecines are the result of a change in dental proportions (i.e., the extremely large premolars and molars characteristic of robust australopithecines).

Tattersall, I. "A Hundred Years of Missing Links." *Natural History,* Dec. 2000/Jan. 2001. The author reviews the 20th century's significant discoveries in human evolution in this centennial issue of *Natural History.*

Tattersall, I. "Once We Were Not Alone." *Scientific American,* Vols. 282, No. 1, Jan. 2000. Although *Homo sapiens* are the only hominids that exist today, during the course of human evolution, many hominid species coexisted.

Vogel, G. "Chimps in the Wild Show Stirrings of Culture." *Science,* Vol. 284, 25 Jun. 1999. A growing number of researchers say that chimpanzees have a primitive form of culture, including variation in tool use and social customs from one population to another.

Wong, K. "Who Were the Neandertals?" *Scientific American,* Vol. 282, No. 4, Apr. 2000. New (and controversial) evidence suggests that Neandertals interbred with anatomically modern humans.

- Visit our Web site at **http://www.info.brookscole.com/solomonbergmartin** for links to chapter-related resources on the World Wide Web. Additional on-line materials relating to this chapter can also be found on our Web site.

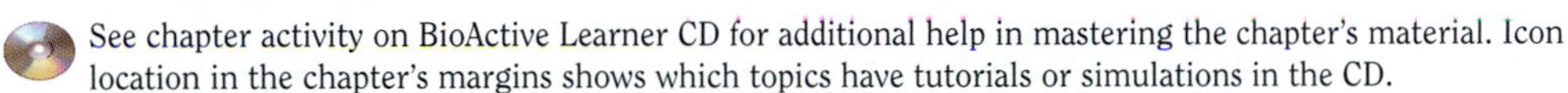

See chapter activity on BioActive Learner CD for additional help in mastering the chapter's material. Icon location in the chapter's margins shows which topics have tutorials or simulations in the CD.

CAREER VISIONS

Wildlife Forensic Specialist

COOKIE SIMS

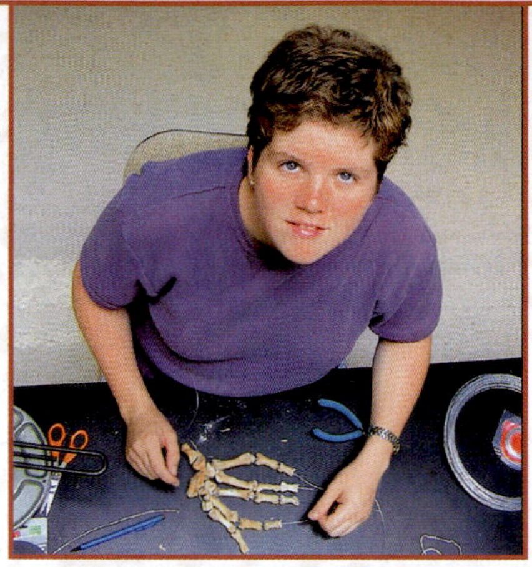

***Cookie Sims** carries out wildlife forensics work at the National Fish and Wildlife Forensic Laboratory in Ashland, Oregon, which is part of the U.S. Fish and Wildlife Service Office of Law Enforcement. She earned a B.S. in biology at Southern Oregon University and is currently enrolled in the biology department's master's program. Cookie began her work in wildlife forensics as a college student, when she became intrigued by the lab's function and volunteered her time there.*

What is the mission of the wildlife forensics lab?

The mission of the wildlife forensics lab is the same as that of the U.S. Fish and Wildlife Service, which is to work with others to conserve, protect, and enhance fish, wildlife, plants, and their habitats, for the continuing benefit of the American people. The specific duty of the forensics lab is to provide forensic identification services to wildlife law enforcement officers at field stations and ports of entry in the United States.

What are the responsibilities and duties of a "forensic specialist"? Specifically, what does your job involve?

My title is forensic specialist in mammals, birds, and reptiles. The responsibilities of a forensic specialist can vary. Specialty areas in the lab include morphology, genetics, pathology, photography and videography, firearms examination, computer technology, and criminalistics (such as trace evidence and fingerprint analysis). I work in the morphology section. In our section, we use visual and microscopic techniques to identify a wide range of wildlife parts and products that come to the lab as evidence in wildlife law enforcement cases. Another important duty is providing expert witness testimony in court cases.

Give examples of specific crimes against wildlife that the lab pursues.

Some involve poaching, such as illegally hunting deer out of season. There are also crimes that violate particular wildlife laws, such as the Migratory Bird Treaty Act or the Marine Mammal Protection Act.

I feel like any advancement in your education is another open door down the road for you.

What is the most memorable case you've been involved in?

That's a hard question, because so many of the cases are memorable. Much of the evidence the lab receives results from violations of importation and exportation laws in the United States. Animal products are seized by wildlife inspectors at ports of entry. We get the weirdest items, such as zebra foot bookends, alligator purses, and feather crafts. One of the most memorable things we received was a smoked, dried animal nose. We had never gotten a nose before to identify. We found out that the nose originated in Southeast Asia, so we had an idea of the range of animals it could be from. We searched through our collection and through the literature available, and our mammalogist identified it as a serow nose. A serow is an animal in the same order as goats and antelopes that lives in Southeast Asia. We happened to have a taxidermied head of this animal, and the nose matched perfectly. It was amazing. That was definitely a memorable case.

You don't deal with the importation of live animals?

No, live animals are taken care of at the ports by wildlife inspectors. We only deal with parts or products. We sometimes receive whole dead animals to determine cause of death or answer some other question.

How is wildlife forensics similar or different to forensics pursued by a typical city's law enforcement agencies?

They are very similar. All crime labs do two things: identify evidence, and attempt to link the suspect, victim, and crime scene. Our lab is much like a human crime lab, except the victims are animals. Also, typical city law enforcement agencies work with one species—humans. But in wildlife forensics we have to consider virtually all the animals in the world.

Do you specialize in the protection of any particular animal species?

We basically deal with endangered or threatened animals that are protected by certain laws. The Endangered Species Act, a federal law, and CITES, an international treaty, both protect certain animals.

You are currently working on your master's degree; why did you decide to enter a graduate program?

I feel like any advancement in your education is another open door down the road for you. In biology, and I would think in any science field, a higher degree will allow you to go more places and do more things.

How would you advise an undergraduate interested in a career in wildlife forensics?

I would really emphasize keeping your options open by taking general biology courses, so that you are well rounded and have more opportunities when job searching. For government job availability, the Web site www.usajobs.opm.gov is a great resource for summer jobs. One of the things I would really advise is to volunteer your time, either at your school or with a local agency.

A tortoise *(Geochelone elephantopus)* from Isla Española in the Galapagos Islands, Ecuador. *(Joe McDonald/Tom Stack & Associates)*

Part 5

The Diversity of Life

Chapters

22

Understanding Diversity: Systematics

The dwarf monkey *(Callithrix humilis).* This new primate species, only about 10 cm (3.9 in) long and weighing about 190 g (7 oz), was identified in 1997 in Brazil. Ten new monkey species have been described in Brazil alone since 1990. *(AP/Wide World Photos)*

LEARNING OBJECTIVES

After you have studied this chapter you should be able to

1. Offer at least two justifications for the use of scientific names and classifications of organisms.
2. Arrange the Linnaean categories in hierarchical fashion, from most inclusive to least inclusive.
3. List the three domains and six kingdoms of organisms introduced in this chapter, give the rationales for and against this system of classification, and describe the distinguishing characters of the organisms assigned to each.
4. Critically summarize the difficulties encountered in choosing taxonomic criteria.
5. Apply the concept of shared derived characteristics to the classification of organisms.
6. Describe methods of molecular biology now used by taxonomists, and summarize their advantages and disadvantages.
7. Contrast monophyletic, paraphyletic, and polyphyletic taxa.
8. Compare and contrast two approaches to systematics: evolutionary systematics and phylogenetic systematics (cladistics).

About 1.75 million species[1] of living organisms have been described, and new species are being discovered almost daily. The dwarf monkey shown here represents a primate species first identified in 1997 in the Amazon jungle of Brazil. Biologists speculate that at least several million additional species remain to be identified. The total number of living species is estimated to be between 10 million and 100 million.

The variety of living organisms and the variety of ecosystems that they form are referred to as **biological diversity,** or **biodiversity.** The totality of life on Earth represents our biological heritage, and the quality of life for all organisms depends on the health and balance of this worldwide web of organisms. For example, we depend on organisms to maintain the life-sustaining composition of gases in our atmosphere, to form soil, break down wastes, recycle nutrients, and provide food for one another. We humans exploit many species for economic benefit. One very anthropocentric (human-centered) example of the practical importance of biodiversity is that more than 40% of prescriptions dispensed by pharmacists in the United States are derived from living organisms. Investigators are just learning how to effectively screen organisms for potential drugs. Unfortunately, human activity is resulting in a serious reduction in biodiversity, and species are becoming extinct more rapidly than researchers can study them (see Chapter 55).

Biologists are alarmed and have begun to mobilize to study and conserve biodiversity. Biologist Edward O. Wilson has proclaimed that describing and classifying all the surviving species of the world should be one of the great scientific goals of the 21st

[1]Recall that a **species** is a group of similar organisms defined by their common gene pool and ability to interbreed in their natural environment; members of a species are reproductively isolated from other organisms.

century. An important step toward that goal is the work of a collaborative international research team that is setting up a global database available on the Internet.

To study the diverse life forms that share our planet and to effectively communicate their findings, biologists must organize their knowledge. The scientific study of the diversity of organisms and their evolutionary relationships is called **systematics.** An important aspect of systematics is **taxonomy,** the science of naming, describing, and classifying organisms. The term **classification** means arranging organisms into groups based on their similarities, which reflect historical relationships among lineages. Because biologists often debate the best methods for inferring the history of life from the data we have, classifying organisms is a complex and often controversial endeavor.

Systematics is a dynamic science that proceeds by the constant reevaluation of data, hypotheses, and theoretical constructs. As new data are discovered and old data are subjected to reinterpretation, the ideas of systematists change. As a result, our understanding of how organisms are related and the way we classify organisms are continuously changing. In this chapter, we explore some of the approaches and methods used by systematists to infer the evolutionary history of life on Earth. In Chapters 23 through 30 we introduce the major groups of organisms that share our planet. We describe how systematists have classified these organisms in domains and kingdoms based on current data from morphology, development, the fossil record, behavior, and molecular biology.

ORGANISMS ARE NAMED USING THE BINOMIAL SYSTEM OF NOMENCLATURE

With millions of kinds of organisms, we need a system for accurately identifying them. Imagine that you were going to develop a classification system. How would you use what you already know about living things to assign them to categories? Would you place insects, bats, and birds in one category because they all have wings and fly? And would you, perhaps, place squid, whales, fish, penguins, and Olympic backstroke champions in another category because they all swim? Or would you classify organisms according to a culinary scheme, placing lobsters and tuna in the same part of the menu, perhaps identifying them as "seafood"? Any of these schemes might be valid, depending on your purpose. Similar methods have been used throughout history. Animals, for example, were classified by St. Augustine in the fourth century as useful, harmful, or superfluous—to humans. During the Renaissance, scholars began to develop categories based on the characteristics of the organisms themselves.

Of the many classification systems that were developed, the one designed by Carolus Linnaeus in the mid-18th century (described briefly in Chapter 1) has survived with some modification to the present day. As we discuss in the next section, Linnaeus grouped organisms according to their similarities, mainly structural similarities.

Before the mid-18th century, each species had a lengthy descriptive name, sometimes composed of ten or more Latin words! Linnaeus simplified scientific classification, developing a **binomial system of nomenclature** in which each species is assigned a unique two-part name.The first part of a scientific name designates the **genus** (pl., *genera*), and the second part is called the **specific epithet.** Note that the specific epithet alone is *not* the species' name. In fact, the same specific epithet can be used as the second name of species in different genera. For example, *Quercus alba* is the scientific name for the white oak, and *Salix alba* is the name for the white willow. (*Alba* comes from a Latin word meaning "white.") Thus, both parts of the name must be used to accurately identify the species.

The genus name is always capitalized, whereas the specific epithet is usually not. Both names must be underlined or italicized. The genus, or generic, name can be used alone to designate all species in the genus (e.g., *Quercus* is the genus that includes all oak species). However, the specific epithet is never used alone; it must always be preceded by the full or abbreviated genus name (e.g., *Quercus alba* or *Q. alba*).

Scientific names are generally derived from Greek or Latin roots or from Latinized versions of the names of persons, places, or characteristics. For example, the generic name for the bacterium *Escherichia coli* is based on the name of the scientist, Theodor Escherich, who first described it. The specific epithet *coli* reminds us that *E. coli* lives in the colon (large intestine).

Most areas of biology depend on scientific names and classification. For example, to study the effects of pollution on an aquatic community, biologists must be able to accurately record the relative numbers of each type of organism present and changes in their populations over time. This requires that all investigators identify each species accurately by name.

Scientific names permit biology to be a truly international science. Even though the common names of an organism may vary in different locations and languages, an organism can be universally identified by its scientific name. A researcher in Puerto Rico can know exactly which organisms were used in a study published by a Russian scientist and therefore can repeat or extend the Russian's experiments using the same species.

EACH TAXONOMIC LEVEL IS MORE GENERAL THAN THE ONE BELOW IT

Linnaeus devised a system for assigning species to a hierarchy of groups. As you move up the hierarchy, each group is more inclusive. Linnaeus did not have a theory of evolution in mind when he set up his system. Nor did he have any idea of the vast number of extant (living) and extinct organisms that would later be discovered. Yet his system has proved to be remarkably flexible and adaptable to new biological knowledge and theory. Very

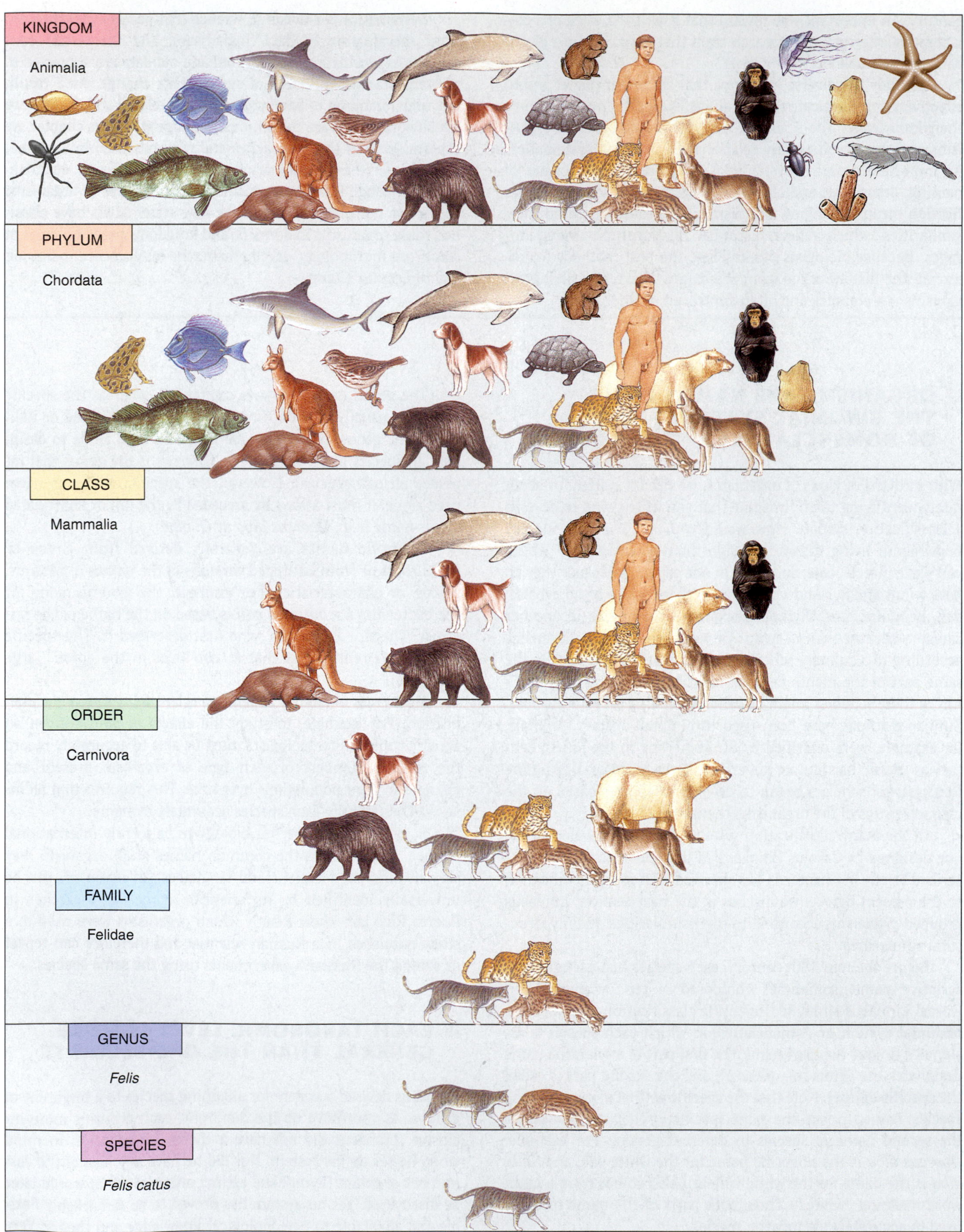
KINGDOM
Animalia
PHYLUM
Chordata
CLASS
Mammalia
ORDER
Carnivora
FAMILY
Felidae
GENUS
Felis
SPECIES
Felis catus

◀ **Figure 22–1 The principal categories used in classification.** The domestic cat *(Felis catus)* is classified here to illustrate the hierarchical organization of our taxonomic system. Each level is more inclusive than the one below it.

few other 18th-century inventions survive today in a form that would still be recognizable to their originators.

The range of taxonomic categories from species to kingdom forms a hierarchy (Fig. 22–1; Table 22–1). Closely related species are assigned to the same genus, and closely related genera may be grouped together in a single **family.** Families are grouped into **orders,** orders into **classes,** classes into **phyla,**[2] and phyla into **kingdoms** and/or **domains.** These groupings can also be separated into subgroupings, for example, subphyla or superclasses.

A **taxon** (pl., *taxa*) is a formal grouping of organisms at any given level, such as the species, genus, or phylum. For example, class Mammalia is a taxon that includes many different orders. Similarly, subphylum Vertebrata is a taxon that contains several classes, including Amphibia and Mammalia.

Some taxonomists ignore minor variations and group organisms into already existing taxa. This practice is referred to as **"lumping."** Other taxonomists subdivide taxa on the basis of minor differences, establishing separate categories for forms that do not fall naturally into one of the existing classifications. This practice is called **"splitting."** Lumpers acknowledge very few animal and plant phyla, whereas splitters may recognize numerous phyla.

Systematists have experienced some difficulty renaming and reclassifying organisms as quickly as new species are identified

[2] Plants and fungi have been traditionally classified in divisions rather than phyla. However, the International Botanical Congress has approved the phylum designation for plants and fungi.

TABLE 22–1 Classification of Corn

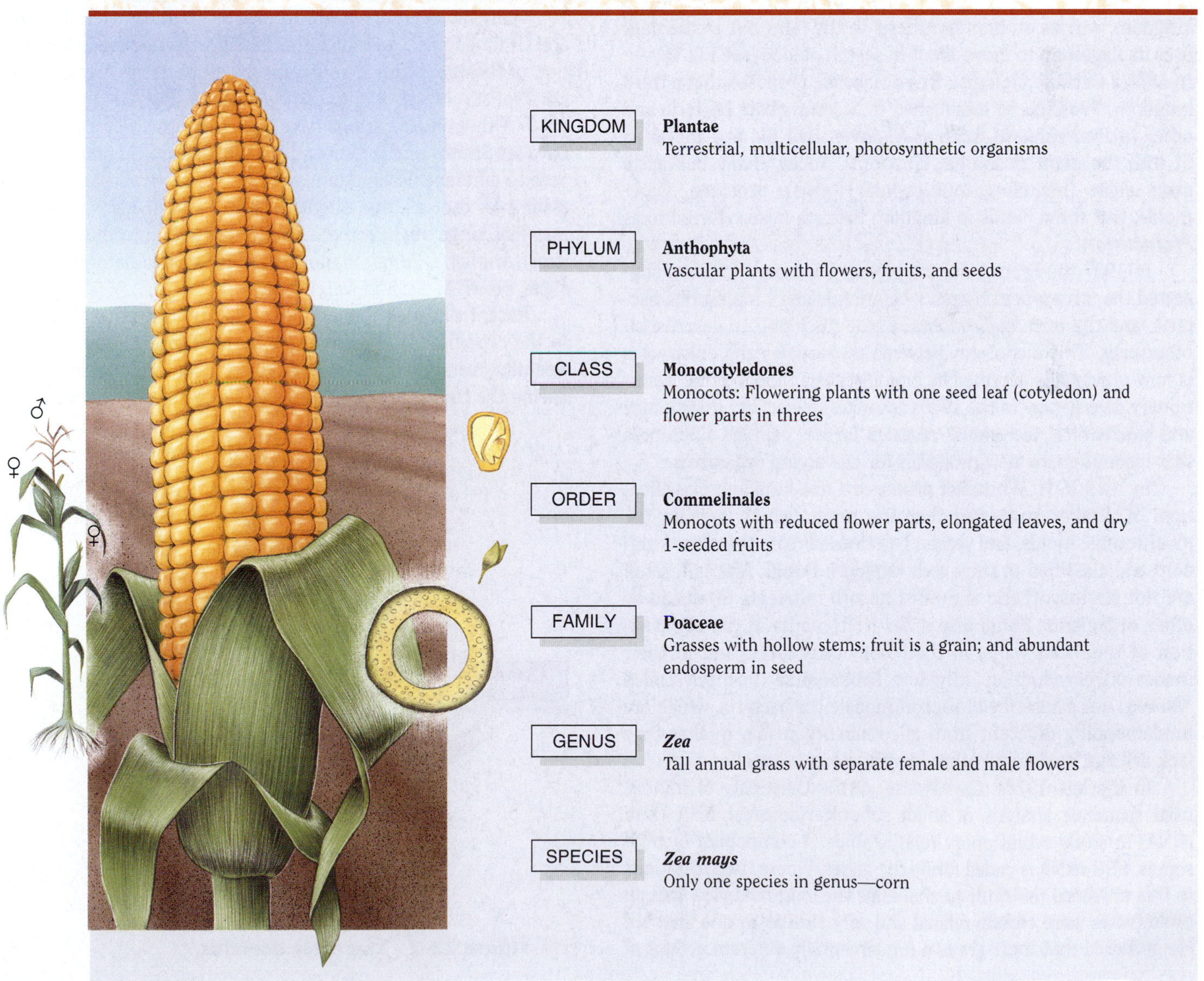

KINGDOM	**Plantae** Terrestrial, multicellular, photosynthetic organisms
PHYLUM	**Anthophyta** Vascular plants with flowers, fruits, and seeds
CLASS	**Monocotyledones** Monocots: Flowering plants with one seed leaf (cotyledon) and flower parts in threes
ORDER	**Commelinales** Monocots with reduced flower parts, elongated leaves, and dry 1-seeded fruits
FAMILY	**Poaceae** Grasses with hollow stems; fruit is a grain; and abundant endosperm in seed
GENUS	***Zea*** Tall annual grass with separate female and male flowers
SPECIES	***Zea mays*** Only one species in genus—corn

and relationships among organisms are adjusted. One group of biologists has proposed a classification system that does away with Linnaeus' hierarchy. Known as PhyloCode, this system groups organisms into clades; a clade is defined as a group of organisms with a common ancestor. Whether biologists will give up conventional nomenclature and the hierarchical system of classification remains to be seen.

NEW DATA INFLUENCE CLASSIFICATION IN KINGDOMS

The history of taxonomy at the kingdom level is a good example of the process of science. From the time of Aristotle to the mid-19th century, biologists divided organisms into two kingdoms: **Plantae** and **Animalia.** After the development of microscopes, it became increasingly obvious that many organisms could not be easily assigned to either the plant or the animal kingdom. For example, the unicellular organism *Euglena,* which has been classified at various times in the plant kingdom and in the animal kingdom, carries on photosynthesis in the light but in the dark uses its flagellum to move about in search of food (see Fig. 24–9). In 1866 a German biologist, Ernst Haeckel, proposed that a third kingdom, **Protista,** be established to accommodate bacteria and other microorganisms, such as *Euglena,* that did not appear to fit into the plant or animal kingdoms. Today, many biologists place algae (including multicellular forms), protozoa, water molds, and slime molds in kingdom Protista (also referred to as *Protoctista*).

In 1937 the French marine biologist Edouard Chatton suggested the term *procariotique* ("before nucleus") to describe bacteria, and the term *eucariotique* ("true nucleus") to describe all other cells. This dichotomy between prokaryotes and eukaryotes is now universally accepted by biologists as a fundamental evolutionary divergence. In the 1960s advances in electron microscopy and biochemical techniques revealed further cellular differences that inspired many new proposals for classifying organisms.

In 1969 R. H. Whittaker proposed a five-kingdom classification. Whittaker suggested that the fungi (which include the mushrooms, molds, and yeasts) be removed from the plant kingdom and classified in their own kingdom **Fungi.** After all, fungi are not photosynthetic and must absorb nutrients produced by other organisms. Fungi also differ from plants in the composition of their cell walls, in their body structures, and in their modes of reproduction. Kingdom **Prokaryotae** (formerly called *Monera*) was established to accommodate the bacteria, which are fundamentally different from all other organisms in that they lack distinct nuclei and other membranous organelles.

In the late 1970s, Carl Woese, of the University of Illinois, used sequence analysis of small subunit ribosomal RNA (SSU rRNA) to study evolutionary relationships. A component of ribosomes, SSU rRNA is coded for by the genes. Woese used variations in this universal molecule to challenge the long-held view that all prokaryotes were closely related and very similar to one another. He proposed that there are two fundamentally different groups of bacteria, Archaebacteria and Eubacteria, and that the prokaryotes account for two of three of the major branches of organisms. This concept gained support in 1996 when Carol J. Bult of the Institute for Genomic Research in Rockville, Maryland reported in the journal *Science* that she and her colleagues had sequenced the complete genome of a methane-producing archaebacterium *Methanococcus jannaschii.* When these researchers compared gene sequences with those of two previously sequenced eubacteria, they found that fewer than half of the genes matched. Based on this molecular evidence, many biologists now divide the prokaryotes into kingdom **Eubacteria** and kingdom **Archaebacteria.**

Based on fundamental molecular differences among the Eubacteria, Archaebacteria, and eukaryotes, many systematists now use a level of classification above the kingdom, called a **domain.** They classify organisms in three domains: **Archaea** (which corresponds to kingdom Archaebacteria), **Eubacteria** (also called Bacteria), and **Eukarya** (eukaryotes; Fig. 22–2). Gene sequencing indicates that the Archaea have a combination of bacteria-like and eukaryote-like genes and appear to be more closely related to the Eukarya than to the Eubacteria.

Many biologists now group organisms into three domains and six kingdoms, and that is the classification we use in this edition of *Biology*. The six kingdoms are Eubacteria, Archaebacteria, Protista, Fungi, Plantae, and Animalia (Fig. 22–3 and Table 22–2). Other classifications have been proposed. For example, T. Cavalier-Smith, of the University of British Columbia, has proposed a different six-kingdom system (Fig. 22–4). Cavalier-Smith assigns all bacteria to a single kingdom and divides eukaryotes into five kingdoms: Protozoa (heterotrophic unicellular organisms), Animalia, Fungi, Plantae, and Chromista (diatoms, golden algae, brown algae, and water molds).

Recent evidence suggests that evolution was not linear and so the classification and trees we use to depict the process are woefully oversimplified. Biologists are now hypothesizing that during the course of evolution genes were not only passed down

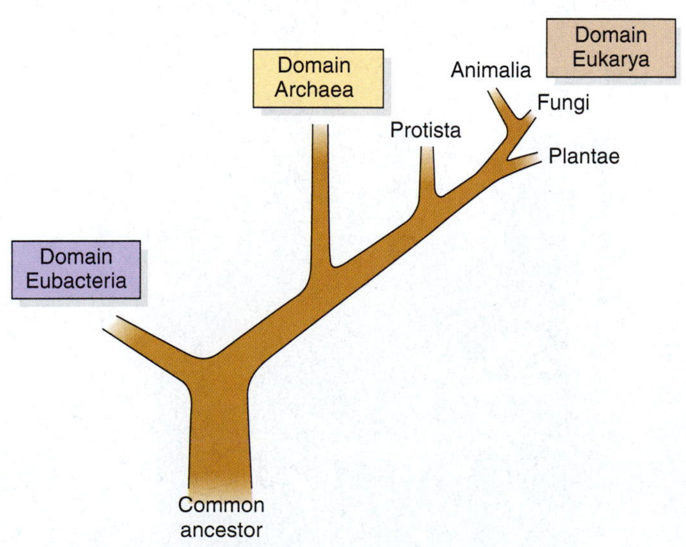

Figure 22–2 The three domains.

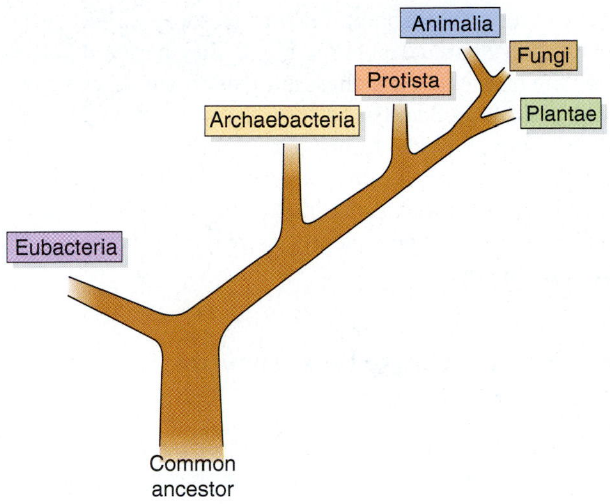

Figure 22–3 The six-kingdom system of classification.

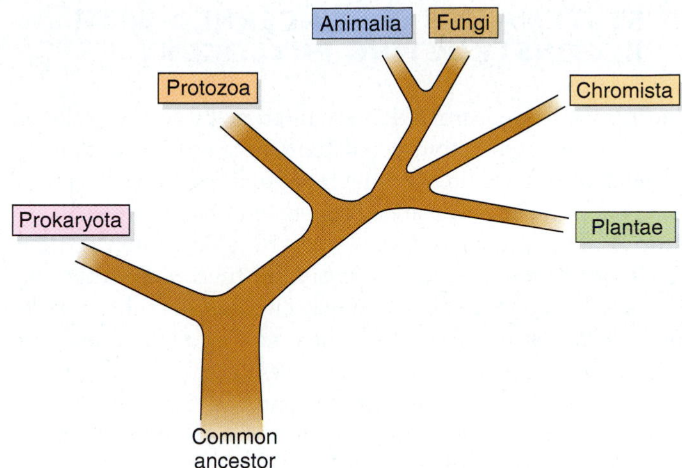

Figure 22–4 An alternative six-kingdom system. This classification is based on the work of T. Cavalier Smith. Kingdom Chromista includes certain algal groups and the water molds.

vertically from one generation to the next but were also transferred laterally (horizontally), for example, by occasional interbreeding between closely related species.

The evolution of systematics reflects the creative and dynamic process of science. Systematists are very responsive to new data, and consequently classification of organisms at all levels is a challenging and continuously changing process. An important goal is to base classification on evolutionary relationships, making it useful in reconstructing the history of life.

TABLE 22–2 Six Kingdoms: Eubacteria, Archaebacteria, Protista, Fungi, Plantae, and Animalia

Kingdom	Characteristics	Ecological Role and Comments
Eubacteria	Prokaryotes (lack distinct nuclei and other membranous organelles); unicellular; microscopic; cell walls generally composed of peptidoglycan.	Most are decomposers; some parasitic (and pathogenic); some chemosynthetic autotrophs; some photosynthetic; important in recycling nitrogen and other elements; some used in industrial processes.
Archaebacteria	Prokaryotes; unicellular; microscopic; peptidoglycan absent in cell walls; differ biochemically from eubacteria.	Methanogens are anaerobes that inhabit sewage, swamps, and animal digestive tracts; extreme halophiles inhabit salty environments; extreme thermophiles inhabit hot, sometimes acidic environments.
Protista	Eukaryotes; mainly unicellular or simple multicellular. Three informal groups (not taxa) include protozoa; algae; and slime molds and water molds.	Protozoa are an important part of zooplankton. Algae are important producers, especially in marine and freshwater ecosystems; important oxygen source.
Fungi	Eukaryotes; heterotrophic; absorb nutrients; do not photosynthesize; body composed of threadlike hyphae that form tangled masses that infiltrate food or habitat; cell walls of chitin.	Decomposers; some parasitic (and pathogenic); some used as food; yeast used in making bread and alcoholic beverages; some used to make industrial chemicals or antibiotics; responsible for much spoilage and crop loss.
Plantae	Eukaryotes; multicellular; photosynthetic; possess multicellular reproductive organs; alternation of generations; cell walls of cellulose.	Terrestrial biosphere depends on plants in their role as primary producers; important source of oxygen in Earth's atmosphere.
Animalia	Eukaryotes; multicellular heterotrophs; many exhibit tissue differentiation and complex organ systems; most able to move about by muscular contraction; specialized nervous tissue coordinates responses to stimuli.	Consumers; some specialized as herbivores, carnivores, or detritus feeders.

SYSTEMATICS IS CONCERNED WITH RECONSTRUCTING PHYLOGENY

In Chapter 17 we learned that evolution is the accumulation of inherited changes within populations over time. Recall that a **population** is made up of all the individuals of the same species that live in a particular area. A population has a dimension in space—its geographical range—and also a dimension in time. Each population extends backward in time. Somewhat like branches of a tree, a population may diverge from other populations sufficiently to become a new species (i.e., a new tree branch; see Chapters 18 and 19). Species have various degrees of evolutionary relationships with one another, depending on the degree of genetic divergence since their populations branched from a common ancestor.

Process of Science

The goal of systematics is to determine the evolutionary relationships, or **phylogeny** (literally, "production of phyla"), among and between species and higher taxa. Once these relationships are established, systematists can use these evolutionary relationships to build classifications based on common ancestry. Once phylogenies are determined, they can be used to help answer other questions in biology. For example, phylogenies can help us understand evolutionary patterns that might provide clues to the origin and spread of HIV and other pathogens. Phylogenies are also useful in predicting characteristics of newly discovered species. Systematists may classify a new species based on specific characters that it shares with other organisms in a particular grouping. They might then infer that the new species shares other characters with organisms in that group.

As you read the following sections, it is important to remember that phylogenies are testable hypotheses. They are supported or falsified by the available data. Thus, systematics is a dynamic science that changes as biologists discover new species and use ever more sophisticated techniques to investigate the evolutionary relationships among organisms.

Homologous structures are important in determining evolutionary relationships

Just how to determine phylogeny and how to group species into higher taxa—genera, families, orders, classes, or phyla—can be difficult decisions. Traditionally, many biologists have based their judgments about the degree of relationship on the extent of similarity among living species and, when available, on the fossil record.

When they evaluate similarities, systematists now consider structural, physiological, developmental, behavioral, and molecular traits. When comparing these similarities, they look for homology in different organisms. Recall from Chapter 17 that **homology** refers to the presence in two or more species of a structure derived from a recent common ancestor. For example, the bones in the wings of a bird and a bat are homologous. In contrast, the possession by two or more species of similar structures not derived from a recent common ancestor sometimes occurs when unrelated or distantly related species become adapted to similar environmental conditions. For example, although the wings of a butterfly are adapted for flight, they have a different structure than bird wings, and these animals do not share a common winged ancestor. Sharks and dolphins have similar, but independently derived, body forms because they have become adapted to similar environments (aquatic) and lifestyles (predatory). A characteristic that superficially appears to be homologous, but is actually independently acquired, is described as exhibiting **homoplasy.**

Shared derived characters provide clues about phylogeny

Organisms sharing many homologous structures are considered to be closely related, while organisms sharing few homologous characters are presumed less closely related. However, distinguishing between homology and homoplasy is not always straightforward. Therefore, carefully selecting similarities that are indicative of evolutionary relationships is extremely important.

How does a systematist interpret the significance of these similarities? In making decisions about taxonomic relationships, the systematist first examines the characteristics found in the largest group (e.g., phylum, class) of organisms being studied and interprets them as indicating the most remote common ancestry. These **shared ancestral characters,** or **plesiomorphic characters,** are traits that were present in an ancestral species and remain present in all the groups that descended from that ancestor. For example, the vertebral column, present in all vertebrates, is an ancestral character for study of classes within the subphylum Vertebrata. Studying the presence or absence of the vertebral column would not help us discriminate between various classes of vertebrates. For example, we could not distinguish between amphibians and mammals because all individuals in these classes possess a vertebral column.

Shared derived characters, or **synapomorphic characters,** are traits found in two or more taxa that first appeared in their most recent common ancestor. Birds and reptiles, for example, have a common ancestor, and both share the ancestral (plesiomorphic) character of laying eggs. The feathers and beaks of birds, however, are not shared by reptiles. They are derived characters that evolved in birds.

A feature viewed as a *derived character* in a more inclusive (broader) taxon may also be considered an *ancestral character* in a less inclusive (narrower) taxon. More recent common ancestry is indicated by classification into less and less inclusive taxonomic groups with more and more specific shared derived characters. For example, the three small bones in the middle ear are useful in identifying a branch point between reptiles and mammals. The evolution of this derived character was a unique event, and only mammals have these bones. However, if we compare mammals with one another, the three ear bones are an ancestral character because all mammals have them. Consequently, they have no value in distinguishing among mammalian taxa. Other characters must be used to establish branch points among the mammals.

If we compare dogs, goats, and dolphins (all of which are mammals), we find that dogs and goats have abundant hair whereas dolphins do not. Hair is an ancestral trait in mammals and therefore cannot be used as evidence that dogs and goats share a more recent common ancestor. In contrast, the virtual absence of hair in dolphins is a derived character within mammals. When we compare dogs, dolphins, and whales, we find that dolphins and whales share this derived character, providing evidence that these animals evolved from a common ancestor not shared by dogs.

Biologists carefully choose taxonomic criteria

Both fishes and dolphins have streamlined body forms, but this characteristic is homoplastic and does not indicate close evolutionary relationships. The dolphin shares important homologous derived characters (synapomorphies) with mammals such as humans: mammary glands, which produce milk for the young, three small bones in the middle ear, and a muscular diaphragm that helps move air into and out of the lungs. Thus, the dolphin is classified as a mammal.

Although dolphins have more shared derived characters in common with humans than with fishes, some characters are shared by all three of these animals. Among these ancestral characters are a dorsal tubular nerve cord and, during embryonic development, a notochord (skeletal rod) and rudimentary gill slits. These shared ancestral characters (plesiomorphies) indicate a common ancestry and serve as a basis for classification. The ancestry is more remote between the dolphin and the fish than between the dolphin and the human. Therefore, although fishes, humans, and dolphins are grouped together in a more inclusive taxon, the phylum Chordata, humans and dolphins are also classified together in class Mammalia, a less inclusive taxon within phylum Chordata, indicating that they are more closely related.

Process of Science Determining which traits best illustrate evolutionary relationships can be challenging. What, for example, are the most important taxonomic characteristics of a bird? We might list feathers, beak, wings, absence of teeth, egg-laying, and the fact that they are endotherms (animals that use metabolic heat to maintain a constant body temperature despite variations in surrounding temperature). Some mammals, for example, monotremes such as the duck-billed platypus, have many of these same characteristics: beaks, endothermy, absence of teeth, and egg-laying. Yet we do not classify them as birds (Fig. 22–5). No mammal, however, has feathers. Is this trait absolutely diagnostic of birds? According to the conventional taxonomic wisdom, the presence or absence of feathers determines what is and is not a bird. This applies only to modern birds, however. A number of dinosaur fossils suggest that some extinct reptiles had feathers.

Usually, organisms are classified on the basis of a combination of traits rather than on any single trait. The significance of these combinations is determined inductively, that is, by an integration and interpretation of the data about traits. Such induction is a necessary part of the process of science. Taxonomists may hypothesize, for example, that all birds should have beaks, feathers, no teeth, and so forth. Then, they reexamine the living world and observe whether there are organisms that might reasonably be called birds that do not fit the current definition of "birdness." If not, the definition is permitted to stand. If too many exceptions emerge, the definition may be modified or abandoned. Sometimes, the taxonomist determines that an apparent exception—the bat, for instance—resembles a bird only superficially and should not be considered one. The bat possesses all of the basic characteristics of a mammal, such as hair and mammary glands that produce milk for the young.

Figure 22–5 Is this animal a bird? A few mammals, such as the duck-billed platypus, lay eggs, have beaks, and lack teeth. However, the platypus does not have feathers, and it nourishes its young with milk secreted from mammary glands. *(Jean Philippe Varin/Jacana/Photo Researchers, Inc.)*

Molecular biology provides additional characters

When a new species evolves, it does not always exhibit obvious phenotypic differences relative to closely related species. For example, two distinct species of fruit flies may appear identical. Some of their DNA, proteins, and other molecules, however, are different. Such variations in the structure of specific macromolecules among species, just like differences in anatomical structure, result from mutations. Macromolecules that are functionally similar in two different types of organisms are considered homologous if their subunit sequence is similar.

The science of **molecular systematics** focuses on the use of molecular structure to elucidate evolutionary relationships. Advances in molecular biology have provided the tools for biologists to compare the macromolecules of various organisms. Amino acid sequencing techniques, immunological methods, and DNA and RNA sequencing are used to compare macromolecules.

Comparisons of amino acid sequences of proteins and nucleotide sequences of nucleic acids provide systematists with valuable information about the degree of relatedness of organisms. The more subunit sequences of two species correspond, the more closely they are considered to be related. The number of differences in DNA or RNA nucleotide sequences or in amino acid sequences in two groups of organisms may reflect how much time has passed since the groups branched from a common ancestor. (This can be true only if the rates of change have remained constant.) Thus, specific genes and specific proteins can sometimes be used as **molecular clocks** (see Chapter 17).

In Chapter 17 we discussed the respiratory protein cytochrome *c*. Although the structure and function of cytochrome *c* are similar in all aerobic organisms, some differences in amino acid sequences exist among species. Comparisons of these differences among about 100 species provide a good example of how data gained through amino acid sequencing contribute to our understanding of phylogeny. Interestingly, chimpanzee cytochrome *c* has the same amino acid sequences as human cytochrome *c*, but in a more distantly related primate, the rhesus monkey, 1 of the 104 amino acids in the sequence is different. In the dog, a nonprimate, 13 amino acids are different. Systematists use this type of information to help make decisions about evolutionary relationships. For example, these data support the hypothesis that humans and chimps are more closely related to each other than to any other taxa. Confirming data derived from studies of several different proteins strengthen such hypotheses.

We have learned that among related species the DNA sequences for the same structural genes are very similar. Detailed restriction maps within large homologous regions of chromosomes of related organisms are also very similar (see Chapter 14). For example, the DNA region that codes for the equivalent of the human hemoglobin beta chain has been mapped in several primates. Even though the gorilla may have diverged from the human line some 8 million to 10 million years ago (mya), 65 of the 70 sites attacked by specific restriction enzymes are identical.

Researchers have determined the nucleotide sequence of a portion of DNA from each of three species of primates (humans, gorillas, and chimpanzees). From their analysis of this 7000 nucleotide sequence, the investigators inferred a common ancestral gene. The simplest branching pattern that would account for the results suggests that the line leading to the gorilla split off first from the common ancestor it shared with the chimpanzee and human. The lines leading to the chimpanzees and humans diverged later (probably about 6 mya).

Systematists are currently comparing ribosomal RNA structure to help determine phylogenies (evolutionary histories). All known organisms have ribosomes that function in protein synthesis, and certain ribosomal RNA nucleotide sequences have been highly conserved in evolution. The division of organisms into three domains was based, in large part, on the comparison of ribosomal RNA by Carl Woese and his research team at the University of Illinois. Prokaryotic ribosomes all contain three types of RNA named in order of increasing size: 5S, 16S, and 23S. The 5S and 16S RNAs have been extensively used to determine evolutionary relationships among bacteria.

Comparison of ribosomal RNA sequences has also been used to challenge the once widely accepted idea that fungi are closely related to plants. Based on ribosomal RNA analysis, fungi are more closely related to animals than to plants; that is, animals and fungi share a more recent common ancestor, perhaps a flagellated protist.

In 1997 molecular taxonomist Robert Wayne of the University of California, Los Angeles, and his international team of researchers compared mitochondrial DNA sequences from 162 wolves at 27 different geographical locations with those from domestic dogs of 67 different breeds (Fig. 22–6). Their results supported the hypothesis that dogs evolved more than 100,000 years ago from a common ancestor that was a wolf. Dog and wolf nucleotide sequences were found to be similar, differing by no more than 12 substitutions. In contrast, dog sequences differ by at least 20 substitutions from jackal and coyote DNA. Using DNA sequences as a molecular clock, these investigators proposed that dogs evolved from wolves about 135,000 years ago, much earlier than the 14,000-year date formerly estimated.

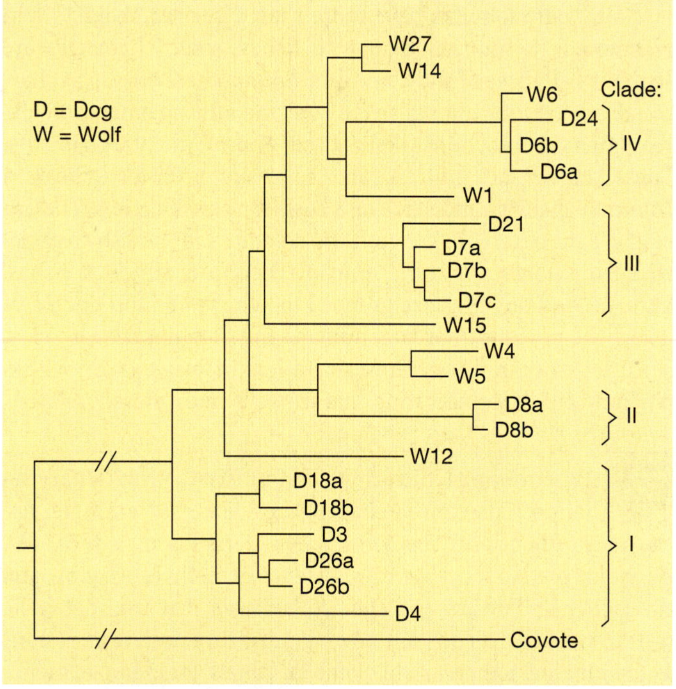

Figure 22–6 Molecular taxonomy. Diagram of the relationships of 8 wolves and 15 dogs based on differences in mitochondrial DNA. The researchers compared 1030 base pairs of a segment of mitochondrial DNA. Wolf and dog sequences were very similar, differing by no more than 12 substitutions. Dog and coyote sequences differed by at least 20 substitutions. Four separate clades (lineages) of dogs were identified among various wolf lineages, suggesting multiple origins. The numbers refer to specific mitochondrial genotypes. For example, clade IV contained three dog genotypes (D6a, D6b, and D24) that were identical or very similar to a wolf genotype (W6) (D, dog; W, wolf). *(From Carles Vila et al.* Science, *Vol. 276, 13 June 1997)*

Taxa should reflect evolutionary relationships

Based on molecular data and other taxonomic criteria, systematists recognize three kinds of taxonomic groupings: monophyletic, paraphyletic, and polyphyletic. A **monophyletic taxon** includes an ancestral species and all its descendants (Fig. 22–7*a*). Mammals, for example, form a monophyletic taxon because all mammals are thought to have evolved from a common ancestral mammal, and all descendants of this ancestor are mammals. Monophyletic taxa are, therefore, natural groupings because they represent true evolutionary relationships and include all close relatives.

A **paraphyletic taxon** is a group that contains a common ancestor and some, but not all, of its descendants. As discussed later in this chapter, the class Reptilia is paraphyletic because it does not include all descendants of the most recent common ancestor of reptiles. Birds are thought to share a recent common ancestor with reptiles, but evolutionary systematists assign birds to their own class.

Some currently recognized taxa are **polyphyletic,** consisting of several evolutionary lines and not including the most recent common ancestor of the included lineages (Fig. 22–7*b*). The members of such a taxon might have been mistakenly grouped together because they share similar features arising from convergent evolution (see Chapter 17). Systematists attempt to avoid constructing polyphyletic taxa because they are unnatural and may misrepresent evolutionary relationships. Cladistic systematists (discussed later in this chapter) also avoid constructing paraphyletic taxa.

HOW DATA ARE ANALYZED AND INTERPRETED DEPENDS ON THE SYSTEMATIST'S APPROACH

In determining the relationships among organisms, systematists use a variety of data and methods. **Phenetics,** or **numerical taxonomy,** was an early approach to the quantitative analysis of characters. First gaining attention of systematists in the 1960s, phenetics is based on as many phenotypic similarities (i.e., similarities in *appearance*) as possible. In the phenetic approach, computer programs are used to analyze data and group organisms according to the number of shared characteristics. No attempt is made to determine whether their similarities arose from a common ancestor or from convergent evolution. Pheneticists argue that it is not important to try to differentiate between homology and homoplasy because many more similarities are due to homology than to homoplasy. Thus, the number of similarities that two organisms have in common reflects the degree of homology.

A taxonomist who follows the phenetic system might explain that dolphins and porpoises are classified as mammals

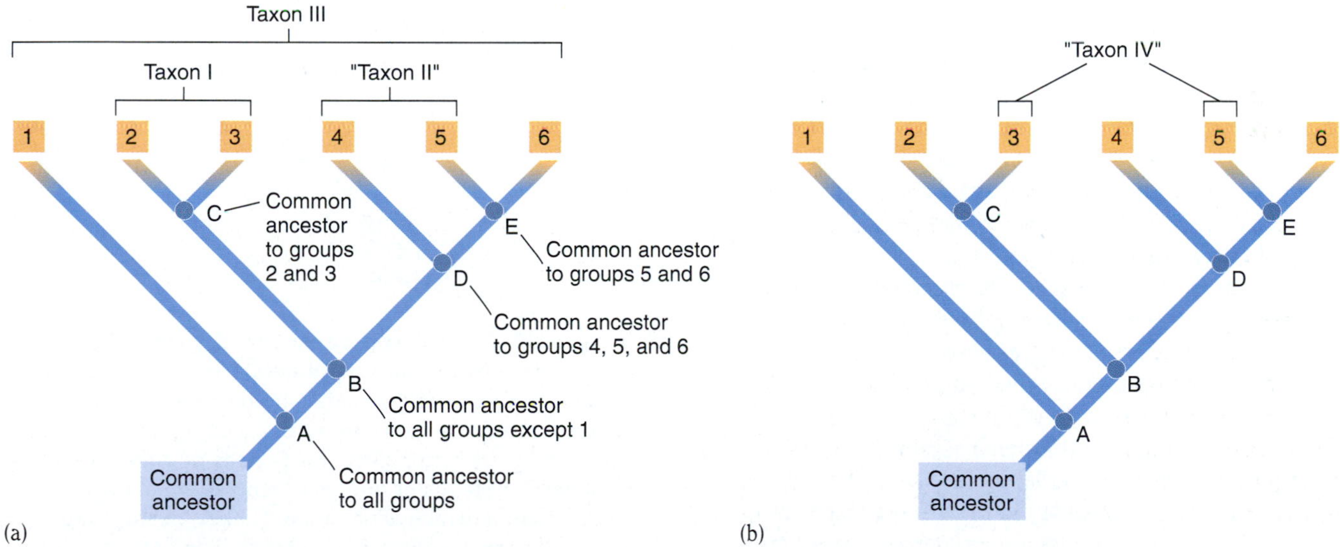

Figure 22–7 Evolutionary relationships. In both figures, node A represents the common ancestor of all groups shown. Node C represents the common ancestor for groups 2 and 3. Node D represents the common ancestor for groups 4, 5, and 6. **(a)** Monophyletic and paraphyletic groups. Taxon I and taxon III are monophyletic. Each includes a common ancestor and all of its descendants. "Taxon II" is paraphyletic; it does not include all of the descendants of a common ancestor. **(b)** "Taxon IV" is polyphyletic; members of this group do not share a most recent common ancestor. (Paraphyletic and polyphyletic "taxa" are indicated with quotation marks because they are not natural evolutionary groupings.)

rather than as fish because they share more similarities with mammals. Phenetics is not widely used by taxonomists today because the inclusion of homoplastic similarities can lead to inaccurate conclusions about evolutionary relationships. However, pheneticists made an important contribution to systematics by introducing the use of computers for making quantitative comparisons among characters.

In addition, some phenetic techniques are currently used in molecular taxonomy. For example, if each amino acid in a protein is considered as a trait, amino acid sequences of various animals can be determined in the laboratory and compared by computer. The information about differences in amino acid sequences can be used to construct phylogenetic diagrams. Species are placed at relative distances from each other, reflecting the extent of difference in amino acid sequence.

The two major approaches to systematics that are widely used today are **evolutionary systematics** and **phylogenetic systematics,** also known as **cladistics.** Although we discuss them separately here, many modern systematists use aspects of both approaches in their work. Although evolutionary and cladistic systematists evaluate phenotypic traits as pheneticists would, they also consider whether similarities arise through common ancestry or convergent evolution. In addition, they consider dissimilar traits that may share a common origin but have diverged over evolutionary time.

Evolutionary systematics allows paraphyletic groups

Evolutionary systematics, sometimes called classical evolutionary systematics, uses a system of phylogenetic classification and presents evolutionary relationships in phylogenetic trees. Evolutionary systematists consider both evolutionary branching (as do cladists) and the extent of divergence (structural and other changes) that has occurred in a lineage since it branched from a stem group. They use a combination of shared ancestral characters and shared derived characters to establish evolutionary relationships and build classifications.

Many, perhaps most, of the taxa currently recognized by evolutionary systematists are monophyletic, based on the possession of shared derived characters. These taxa are also recognized by cladists. For example, proponents of both approaches agree that birds are a monophyletic taxon based on shared derived characters such as feathers. However, evolutionary systematics also recognizes paraphyletic taxa, groups that include some, but not all, subgroups of organisms that share the same most recent common ancestor. Paraphyletic groups are recognized based on a combination of shared ancestral and shared derived characters.

Evolutionary systematists recognize class Reptilia as a valid group that does not include birds because (1) it includes the common ancestor of all reptiles and (2) it lacks the derived features of birds (e.g., feathers). Class Reptilia does not include all subgroups, such as birds, that evolved from the ancestral reptile (Fig. 22–8*a*); they assign birds to a separate class because they have diverged markedly from the reptiles. In contrast, cladists do not recognize reptiles as a natural grouping containing snakes, lizards, crocodiles, dinosaurs, and turtles because they do not form a monophyletic group, or **clade.** Instead, they recognize a class Reptilia that includes birds.

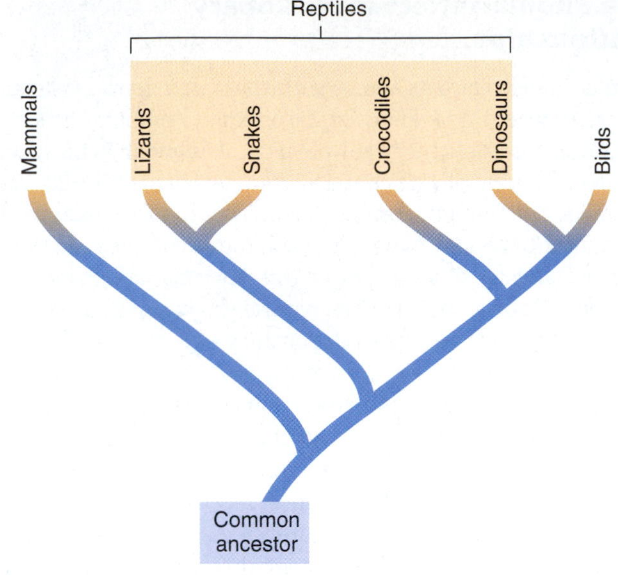

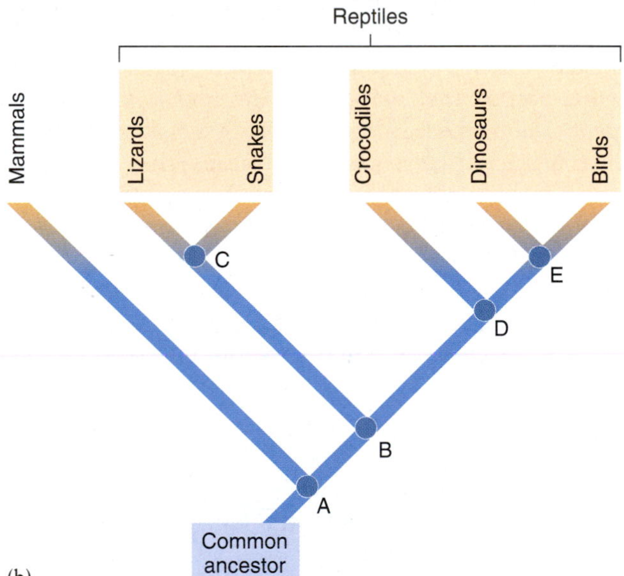

Figure 22–8 Two approaches to the classification of reptiles, birds, and mammals. **(a)** The evolutionary systematics approach considers both common ancestry and extent of divergence that has occurred since two taxa split. The branching points and degrees of difference in the evolution of the major groups of reptiles are shown. Reptiles are a paraphyletic group. Snakes, lizards, and crocodiles are phenotypically most similar, but crocodiles, dinosaurs, and birds are most closely related because they evolved most recently from a common ancestor. **(b)** Cladists classify birds and some reptiles together because they have a common ancestor. Node D represents the common ancestor of crocodiles, dinosaurs, and birds.

Phylogenetic systematics (cladistics) emphasizes phylogeny

The cladistic approach to systematics emphasizes common ancestry, rather than phenotypic similarity, as the basis for classification. Cladists base their assessment on shared derived characters that can be structural, behavioral, physiological, or molecular. The characters must be homologous. According to cladistics, dolphins are classified with mammals rather than with fishes because dolphins and mammals share derived characters not present in fishes, indicating a more recent common ancestor. Cladistics uses shared derived characters to reconstruct phylogenies.

A crucial step in most cladistic analysis is **outgroup analysis,** a method for estimating which attributes are shared derived characters in a given group of taxa. An **outgroup** is a taxon that is considered to have diverged earlier than any of the taxa under investigation (ingroup) and thus to represent an approximation of the ancestral condition. An ideal outgroup is the closest relative of the group being studied (its sister group) and has not been highly modified since its origin. Systematists argue that such a group is likely to retain the ancestral state for characters being used in the analysis, allowing them to identify the evolutionary changes leading to derived characters. A specific example of outgroup analysis is given in the next section.

Cladists determine the evolutionary relationships of organisms and express them in branching diagrams called **cladograms.** Each branch on a cladogram represents the divergence, or splitting, of two or more new groups from a common ancestor. Consider the evolutionary grouping of mammals, lizards, snakes, crocodiles, dinosaurs, and birds (Fig. 22–8*b*). Birds, along with dinosaurs, are thought to share a common ancestor with modern crocodiles and alligators (node D on Fig. 22–8*b*). Crocodiles, dinosaurs, and birds, then, constitute a monophyletic group, and cladists would classify them in the same clade. Similarly, snakes and lizards form a clade that is the closest group to birds, dinosaurs, and crocodiles. Mammals form an additional clade.

When cladists consider the evolutionary relationships of organisms, they must choose between multiple, competing cladograms. How do they make this choice? The most common criterion is the principle of **parsimony**—they choose the simplest explanation to interpret the data. Parsimony, a guiding principle in many areas of research, is based on the experience that the simplest explanation is most probably the correct one. When applied to choice of cladograms, parsimony requires that the cladogram with the fewest changes in characters (i.e., the one with the fewest homoplasies) be accepted as most probable (Fig. 22–9). In actual practice it is often possible to generate several cladograms that are equally parsimonious.

Phylogenies can be determined by building and interpreting cladograms[3]

The first step in constructing a cladogram is to select the taxa, which may consist of individuals, species, genera, or other taxonomic levels. Here we use a representative group of eight chordates (Fig. 22–10). The next step is to select the homologous characters to be analyzed. In our example, we use seven characters. For each character, we must define all the different conditions, or states, as they exist in our taxa. For simplicity,

[3] This discussion of building and interpreting cladograms is based on an essay contributed to *Biology,* 4th ed., by Dr. John Beneski, Department of Biology, West Chester University, West Chester, Pennsylvania.

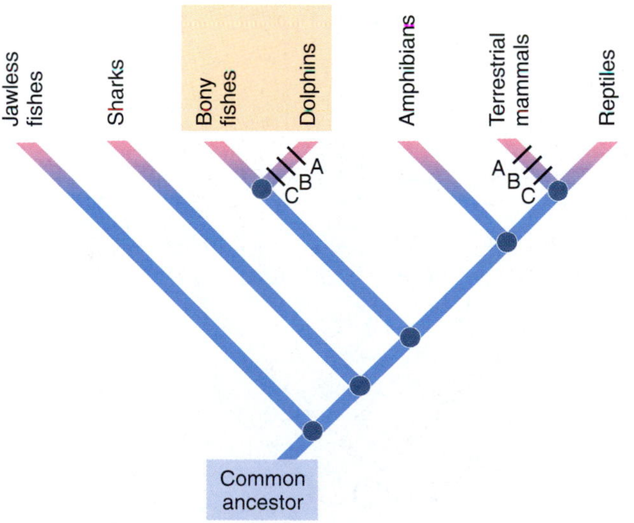

(a) Hypothesis 1: Dolphins and bony fishes are close relatives

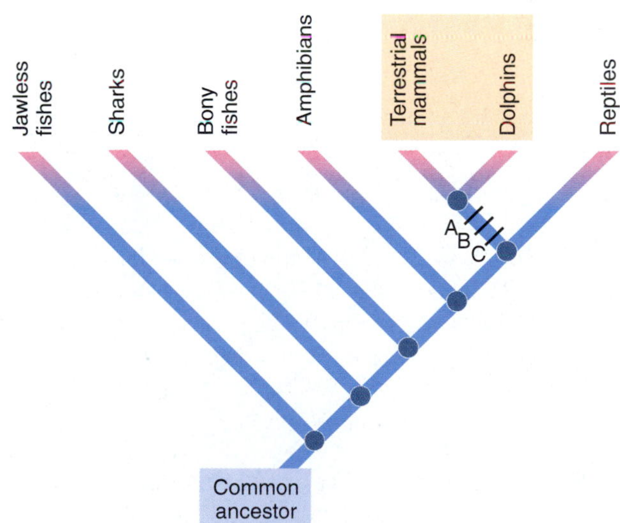

(b) Hypothesis 2: Dolphins and terrestrial mammals are close relatives

Figure 22–9 Using parsimony to choose alternative hypotheses of relationships. Three characters are examined: Character A = hair; B = mammary glands; C = middle ear ossicles. In hypothesis 1, all three characters examined had to evolve independently twice. Hypothesis 2 is chosen because it requires that each character evolve only once.

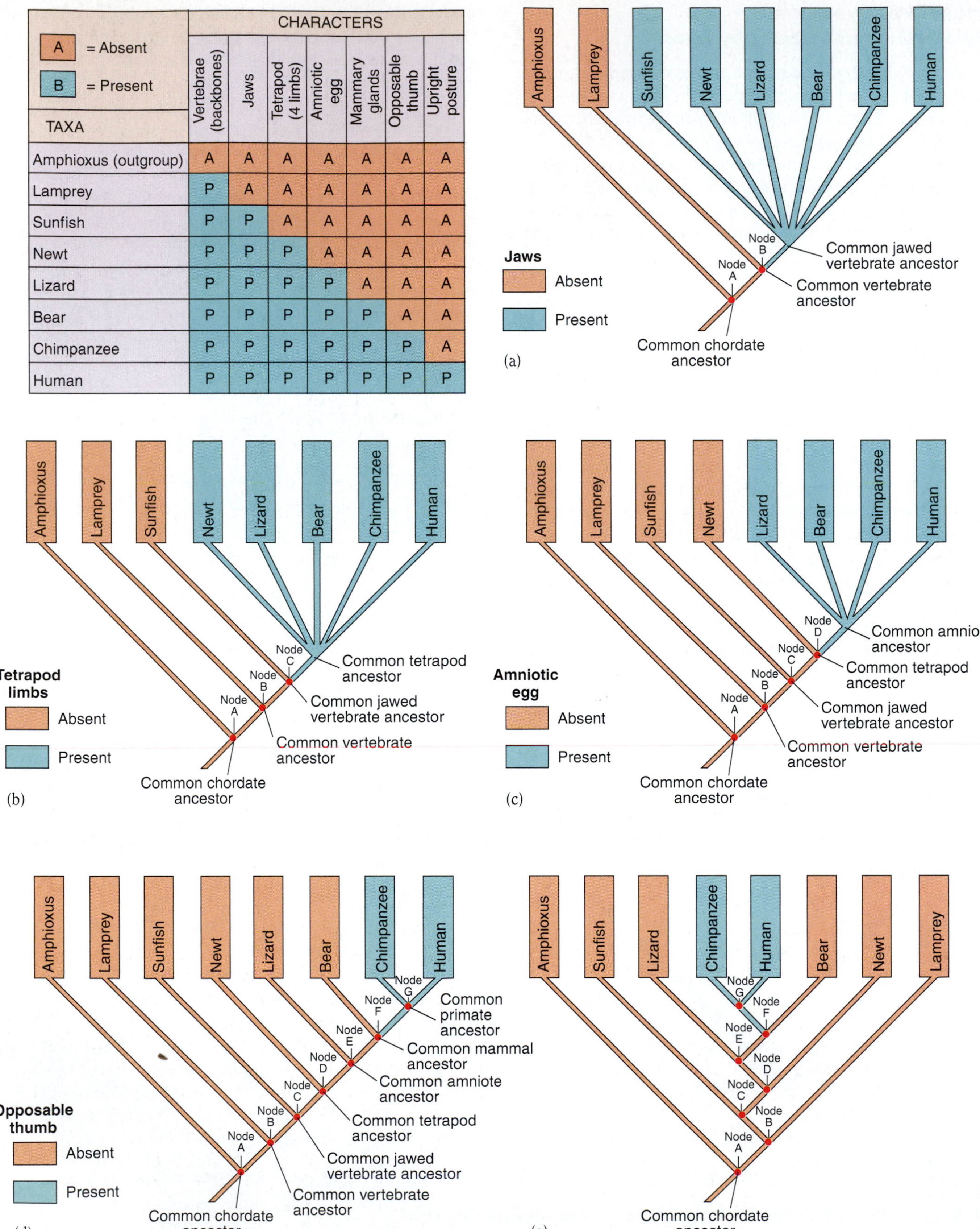

TAXA	Vertebrae (backbones)	Jaws	Tetrapod (4 limbs)	Amniotic egg	Mammary glands	Opposable thumb	Upright posture
Amphioxus (outgroup)	A	A	A	A	A	A	A
Lamprey	P	A	A	A	A	A	A
Sunfish	P	P	A	A	A	A	A
Newt	P	P	P	A	A	A	A
Lizard	P	P	P	P	A	A	A
Bear	P	P	P	P	P	A	A
Chimpanzee	P	P	P	P	P	P	A
Human	P	P	P	P	P	P	P

Figure 22–10 Building a cladogram. In this example, amphioxus is the chosen outgroup that represents an approximation of the ancestral condition.

we consider our characters to have only two different states: present or absent. Keep in mind that many characters used in cladistics have more than two states. For example, black, brown, yellow, and red may be only a few of the many possible states for the character of hair color.

The last, and often the most difficult, step in preparing the data is to organize the character states into their correct evolutionary order. Outgroup analysis is used. In our example, amphioxus, a small fishlike chordate, is the chosen outgroup. It belongs to a taxon that is considered to have diverged earlier than any of the other taxa under investigation but to be closely related to vertebrates; thus amphioxus represents an approximation of the ancestral condition for vertebrates. Therefore, the character state "absent" for a particular character state such as vertebrae is the ancestral condition, and the character state "present" is the derived condition for the characters in Figure 22–10.

A cladogram can be constructed by considering shared derived characters

Our objective is to construct a cladogram that requires the fewest number of evolutionary changes in the characters. Recall that, in cladistics, taxa are grouped by the presence of shared derived characters. To form a valid monophyletic group, all members must share at least one derived character. Membership in a clade cannot be established by shared ancestral characters.

In our example, notice that all taxa except the outgroup amphioxus possess vertebrae. We may therefore conclude that these seven vertebrate taxa form a valid clade. Next, among the seven vertebrate taxa, notice that jaws are present in all groups except for lampreys. Using these data, we may construct a preliminary cladogram (Fig. 22–10*a*).

The base of the cladogram represents the common ancestor for all taxa being analyzed. Each branch point (referred to as a *node*) represents the immediate common ancestor of the next monophyletic group depicted in the cladogram. In Figure 22–10*a,* node A represents the common chordate ancestor from which the outgroup (amphioxus) and the seven vertebrate taxa evolved. Similarly, node B represents the common ancestor of the vertebrates and node C, the common ancestor of the jawed vertebrates. Continuing with this procedure, notice that among the six jawed taxa, all but sunfish are tetrapods (animals with four limbs) (Fig. 22–10*b*). Among the five tetrapods, all but newts have amniotic eggs (Fig. 22–10*c*). (In an amniotic egg, the embryo is surrounded by a fluid-filled sac, called an amnion.) The branching process is continued using the data in the table in the upper left-hand corner of the figure until all clades are established (Fig. 22–10*d*).

In a cladogram each branch point represents a major evolutionary step

In Figure 22–10*d,* notice that humans and chimpanzees share a recent common ancestor at node G. This hypothesis is supported by the derived characters (synapomorphies) that they share. In the same way, bears are more closely related to the human-chimpanzee clade than to any other clade considered in our example, as indicated by the common ancestor at node F. In comparing the nodes, the order of divergence (branching) is indicated by distance from the base of the diagram. The further a node is located up the cladogram, the more recently the group diverged. In our example, node G represents the most recent divergence, and node A represents the most ancient divergence. Thus, in our example, humans are closely related to chimpanzees (through node G) but more distantly related to bears (through node F). Therefore, humans and chimpanzees are assigned to a less inclusive taxon (order Primates) whereas humans, chimpanzees, and bears are assigned to a broader, more inclusive taxon (class Mammalia). In addition, the cladogram reveals that lizards are more closely related to the mammal clade than to newts, sunfish, or any other clade. Can you explain why?

Two important concepts guide our interpretation of cladograms. First, the relationships among taxa can be determined only by tracing along the branches back to the most recent common ancestor (i.e., node) and not by the relative placement of the branches along the horizontal axis. It is possible to represent the same relationships with many different types of branching diagrams. For example, the cladogram in Figure 22–10*e* is equivalent to the one in Figure 22–10*d*. (You should verify this by comparing the numbered nodes and by checking the relationships described earlier.)

A second key concept is that the cladogram tells us which taxa shared a common ancestor and how recently they shared a common ancestor. The ancestor itself remains unspecified. The cladogram does not establish direct ancestor-descendant relationships among taxa. In other words, a cladogram does not suggest that a taxon gave rise to any other taxon.

SYSTEMATICS WILL CONTINUE TO CHALLENGE BIOLOGISTS

Many evolutionary questions remain. New hypotheses about how organisms are related suggest new taxonomic schemes. Systematics offers tools to help us test these hypotheses. In Chapters 23 through 30 we explore many evolutionary puzzles and discuss current hypotheses about relationships and classifications of organisms.

SUMMARY WITH KEY TERMS

I. **Systematics** is the scientific study of the diversity of organisms and their evolutionary relationships. **Biological diversity,** or **biodiversity,** refers to the variety of living organisms and the ecosystems they form.
   A. **Taxonomy** is the branch of systematics devoted to naming, describing, and classifying organisms.
   B. The process of assigning organisms into groups based on their similarities or relationships is **classification.**

II. Biologists name organisms using the **binomial system of nomenclature** developed by Linnaeus in the mid-18th century.
   A. In this system the basic unit of classification is the **species.**
   B. The name of each species has two parts: the **genus** name followed by the **specific epithet.** For example, the scientific name of the human is *Homo sapiens,* and that of the white oak is *Quercus alba.*

III. The hierarchical system of classification currently used includes **domain, kingdom, phylum, class, order, family, genus,** and **species.** Each formal grouping at any given level is a **taxon.**

IV. The three-domain classification system assigns organisms to domain **Archaea,** domain **Eubacteria,** or domain **Eukarya.** The six-kingdom classification recognizes the kingdoms **Archaebacteria, Eubacteria, Protista, Fungi, Plantae,** and **Animalia.**

V. The goal of systematics is to determine evolutionary relationships, or **phylogeny,** based on shared characteristics.
   A. **Homology** implies evolution from a common ancestor. **Homoplasy** refers to superficially similar characters that are not homologous because they evolved independently.
   B. **Shared ancestral characters,** or **plesiomorphic characters,** suggest a distant common ancestor. **Shared derived characters,** or **synapomorphic characters,** indicate a more recent common ancestor.
   C. **Molecular systematics** provides methods for comparing macromolecules such as nucleic acids and proteins for assessing evolutionary relationships. Comparison of nucleotide sequences in ribosomal RNA has led to important taxonomic decisions regarding domains, kingdoms, and species.
   D. A **monophyletic group** includes all of the descendants of the most recent common ancestor. A **paraphyletic group** consists of a common ancestor and some, but not all, of its descendants. The organisms in a **polyphyletic group** evolved from different ancestors.

VI. Three approaches to systematics are phenetics, evolutionary systematics, and phylogenetic systematics, also known as cladistics.
   A. **Phenetics,** or **numerical taxonomy,** is based on similarities of many characters; this was an early approach in which organisms were classified according to the number of characteristics they shared without trying to determine whether their similarities were homologous or homoplastic.
   B. **Evolutionary systematics** considers both evolutionary branching and the extent of divergence. Evolutionary systematics is based on shared ancestral characters as well as shared derived characters.
   C. **Cladistics,** also called **phylogenetic systematics,** insists that taxa be monophyletic. Each monophyletic taxon, or **clade,** consists of a common ancestor and all its descendants. Shared derived characters are used to determine these relationships. Diagrams called **cladograms** are used to illustrate phylogeny.
      1. Cladists use **outgroup analysis** to determine which characters are ancestral and which are derived in a given group of taxa. An **outgroup** is a taxon that represents the ancestral condition because it diverged earlier than any of the other taxa being investigated.
      2. Cladists use the principle of **parsimony:** They choose the simplest explanation to interpret the data.
      3. In interpreting cladograms, the relationships among taxa can only be determined by tracing along the branches back to the most recent common ancestor (represented by a node on the cladogram).
      4. The cladogram indicates which taxa shared a common ancestor and how recently they shared that ancestor.

POST-TEST

1. The science of describing, naming, and classifying organisms is (a) systematics (b) taxonomy (c) cladistics (phylogenetic systematics) (d) phenetics (e) evolutionary systematics
2. Using the binomial system of nomenclature, the scientific name of each species consists of two parts (a) class, specific epithet (b) family, genus (c) genus, specific epithet (d) family, species (e) genus, species
3. The mold that produces penicillin is *Penicillium notatum. Penicillium* is the name of its (a) genus (b) order (c) family (d) species (e) specific epithet
4. Closely related genera may be grouped together in a single (a) phylum (b) domain (c) species (d) family (e) kingdom
5. Related classes are grouped together in the same (a) genus (b) phylum (c) order (d) paraphyletic taxon (e) family
6. In the six-kingdom system, the kingdom that includes the protozoa is (a) Plantae (b) Protista (c) Archaea (d) Eukarya (e) Fungi
7. Decomposers such as molds and mushrooms belong to kingdom (a) Plantae (b) Protista (c) Archaebacteria (d) Eukarya (e) Fungi
8. A taxon that contains a recent common ancestor and all of its descendants is (a) polyphyletic (b) paraphyletic (c) monophyletic (d) phyletic (e) plesiomorphic
9. The presence of homologous structures in different organisms suggests that (a) the organisms evolved from a common ancestor (b) convergent evolution has occurred (c) they belong to a polyphyletic group (d) homoplasy has occurred (e) independently acquired characters may evolve when organisms inhabit similar environments
10. The dolphin and the human both have the ability to nurse their young, whereas the less closely related fish does not. The ability to nurse their young is a (a) shared derived character of mammals (b) shared ancestral character of all vertebrates (c) plesiomorphy (d) homologous behavior (e) homoplasy
11. Relative constancy in the rates of DNA and protein evolution permits biologists to use these macromolecules as (a) molecular clocks (b) polymerase chains (c) clades (d) paraphyletic clues (e) outgroups
12. Phenetics (a) is a numerical taxonomy based on phenotypic similarities (b) emphasizes common ancestry (c) emphasizes polyphyletic groups (d) focuses on ancestral characters (e) strives to differentiate between homology and homoplasy.
13. Systematists who classify crocodiles and birds in the same taxon because they are monophyletic follow which approach? (a) phyletic (b) cladistic (c) evolutionary systematic (d) polyphyletic (e) either cladistic or evolutionary systematic
14. In cladistic analysis (a) ancestral characters are used to reconstruct phylogenies (b) characters must be homoplastic (c) polyphyletic groups are preferred (d) ancestral character analysis is commonly used (e) outgroup analysis is used
15. When cladists use the principle of parsimony, they (a) choose the simplest explanation to interpret the data (b) select multiple hypotheses to explain each relationship (c) do not use outgroups (d) typically use a polyphyletic approach (e) are hypothesizing that the most complex explanation is most probably the correct one

REVIEW QUESTIONS

1. Briefly describe the binomial system of nomenclature. Define the term *species*.
2. Compare and contrast the three domains.
3. What are the advantages of a "six-kingdom" system over a "two-kingdom" one? What types of organisms are especially difficult to assign to a kingdom?
4. In which kingdom would you classify each of the following? (a) oak tree (b) amoeba (c) *Escherichia coli* (a bacterium) (d) tapeworm (e) black bread mold
5. Compare the cladistic and evolutionary approaches to systematics. Are their methods similar? Are their conclusions ever the same?
6. Of what use to a taxonomist is knowledge of the amino acid sequences of the proteins of various organisms?
7. Shared ancestral traits cannot provide evidence for relationships between organisms within a taxon that has those traits. Give an example of this principle and explain why it is true.
8. In the illustration, what kind of grouping is represented by the bracketed area?

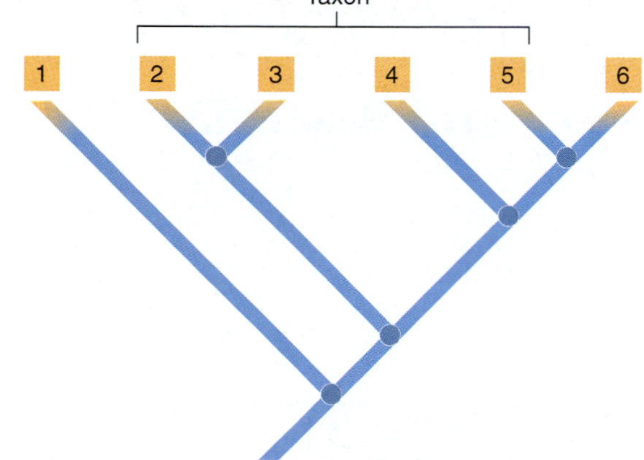

YOU MAKE THE CONNECTION

1. What difficulties do we encounter when we attempt to use the concept of a species?
2. Are members of a genus similar because they share a common ancestor, or do they belong to the same genus because they are similar? How might your answer vary depending on which approach to systematics you are following?
3. After many years of being considered old-fashioned, the science of systematics has reemerged on the cutting edge of biological research. Why do you think this shift has occurred?
4. The TATA-binding protein (TBP) is thought to be necessary for transcription in all eukaryotic cell nuclei. Studies show that archaebacteria, but not eubacteria, have a protein structurally and functionally similar to TBP. What does this similarity suggest regarding the evolution of archaebacteria and eukaryotes? How might knowledge of this similarity affect how systematists classify these organisms?

RECOMMENDED READINGS

Davis, J.I. "Phylogenetics, Molecular Variation, and Species Concepts." *BioScience,* Vol., 46 No.7, Jul./Aug. 1996. The author discusses definitions of species and how they relate to phylogenetics.

Doolittle, W. F. "Uprooting the Tree of Life." *Scientific American,* Vol. 282, No. 2, Feb. 2000. Recent discoveries challenge our current view of phylogeny and call for new hypotheses.

Gould, S.J. "Linnaeus's Luck?" *Natural History,* Sep. 2000. An interesting discussion of the historical and current role of Linnaeus's classification.

Olsen, G. J., and C.R. Woese. "Archaeal Genomics: An Overview." *Cell,* Vol. 89, 27 Jun. 1997. Examines the genome of archaebacterium *M. jannaschii* and its evolutionary relationships to eubacteria and eukarya.

Pennisi, E. "Genome Data Shake Tree of Life." Research News. *Science,* Vol. 280, 1 May 1998. Researchers are finding unexpected connections between domains.

Vila, C., et al. "Multiple and Ancient Origins of the Domestic Dog." *Science,* Vol. 276, 13 Jun. 1997. Mitochondrial DNA sequences were analyzed and compared in wolves, dogs, and jackals.

Wainright, P.O., G. Hinkle, M.L. Sogin, and S.K. Stickel. "Monophyletic Origins of the Metazoa: An Evolutionary Link with Fungi." *Science,* Vol. 260, 16 Apr. 1993. The authors use comparisons of ribosomal RNA sequences to hypothesize evolutionary relationships of animals and fungi.

Whittaker, R.H. "New Concepts of Kingdoms of Organisms." *Science,* Vol. 163, 1969. Whittaker's original proposal for classifying organisms according to a five-kingdom system.

Wilson, E.O. "A Global Biodiversity Map." *Science,* Vol. 289, 29 Sep. 2000. The author urges scientists to make identification and classification of the world's species an important goal for this century.

Withgott, J. "Is It 'So Long, Linnaeus?'" *Bioscience,* Vol. 50, Aug. 2000. Proponents of PhyloCode challenge the taxonomic rankings of the Linnaean system.

- Visit our Web site at **http://www.info.brookscole.com/solomonbergmartin** for links to chapter-related resources on the World Wide Web. Additional on-line materials relating to this chapter can also be found on our Web site.

See chapter activity on BioActive Learner CD for additional help in mastering the chapter's material. Icon location in the chapter's margins shows which topics have tutorials or simulations in the CD.

23

Viruses and Bacteria

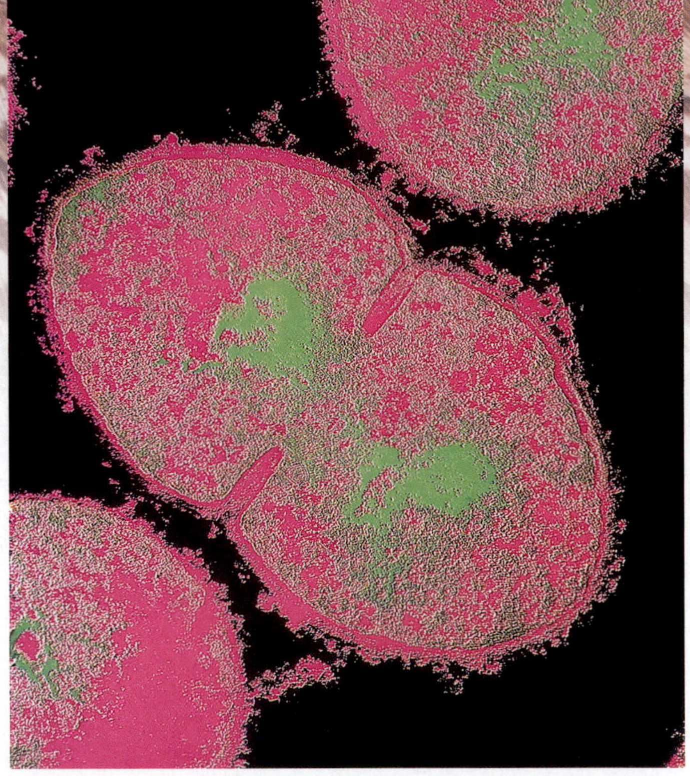

Color-enhanced TEM of a bacterium *(Streptococcus pyogenes)* dividing by binary fission. *(Dr. Kari Lounatmaa/Science Photo Library/Photo Researchers, Inc.)*

LEARNING OBJECTIVES

After you have studied this chapter you should be able to

1. Describe the structure of a virus and compare a virus with a free-living cell.
2. Trace the evolutionary origin of viruses according to current hypotheses.
3. Characterize bacteriophages and contrast a lytic cycle with a lysogenic cycle.
4. Explain how viruses infect animal and plant cells.
5. Describe the reproductive cycle of a retrovirus such as human immunodeficiency virus (HIV).
6. Compare and contrast viroids and prions.
7. Compare the three domains: Eubacteria, Archaea, and Eukarya.
8. Describe asexual reproduction in bacteria and summarize three mechanisms (transformation, conjugation, and transduction) that may lead to genetic recombination.
9. Characterize the metabolic diversity of autotrophic and heterotrophic bacteria, including aerobes, facultative anaerobes, and obligate anaerobes.
10. Distinguish among the three main groups of archaea and among several groups of eubacteria as described in Table 23–3. Give examples of each group.
11. Discuss the ecological roles of bacteria, their importance as pathogens, and their commercial importance.

Bacteria have inhabited our planet for more than 3.5 billion years—much longer than eukaryotes, which evolved about 1.5 to 1.6 billion years ago. Although bacteria are microscopic, they are so numerous that they, together with the fungi, account for approximately one-half of Earth's biomass, the mass of living material. In contrast, plants account for about 35% and animals for about 15% of the biomass of our planet.

Anton van Leeuwenhoek discovered bacteria and other microorganisms in 1674 when he looked at a drop of lake water through a glass lens. During the late 1800s, many microorganisms, including bacteria, fungi, and protozoa, were identified as **pathogens,** agents that cause disease. The bacterium *(Streptococcus pyogenes)* shown dividing in the TEM is a pathogen that inhabits the human nose and throat. This bacterium can cause scarlet fever and inflammation of the heart tissue. The strain shown here is resistant to antibiotics, and infection can be fatal.

Although bacteria cause many diseases, including respiratory infections and food poisoning in humans, only a small minority of bacterial species are pathogens. In fact, bacteria play an essential role in the biosphere as decomposers, breaking down organic molecules into their components. Along with fungi, bacteria are nature's recyclers. Without these microorganisms, the available carbon, nitrogen, phosphorus, and sulfur would eventually be tied up in the wastes and dead bodies of plants and animals. Life as we know it would cease to exist because of the lack of raw materials for the synthesis of new cell components.

Some bacteria are producers that carry on photosynthesis. Other bacteria perform a key role in agriculture by converting atmospheric nitrogen to ammonia and then to nitrates, a form that can be used by plants (see Fig. 53–7). This conversion of nitrogen enables plants and animals (because they eat plants) to manufacture essential nitrogen-containing compounds such as proteins and nucleic acids.

In contrast to bacteria, viruses are not cellular. Biologists long considered them to be nonliving particles, but some biologists now view them as life forms. They contain the nucleic acid necessary to make copies of themselves, and they reproduce by invading living cells and commandeering their metabolic machinery. Viruses are intimately involved with living organisms. These tiny, but potent, particles infect other organisms and are responsible for a wide variety of diseases in plants and animals. Viruses influence many ecological processes including the recycling of nutrients and the biodiversity of bacteria and algae.

This chapter examines the diversity and characteristics of viruses, bacteria, and the smaller viroids and prions. They are not a natural group of closely related organisms, and we discuss them in a single chapter only for convenience, not to reflect shared ancestry.

VIRUSES ARE NONCELLULAR INFECTIOUS AGENTS

Process of Science During the late 1800s, botanists searched for the cause of tobacco mosaic disease, which stunts the growth of tobacco plants and gives the infected tobacco leaves a spotted, mosaic appearance. They found that the disease could be transmitted to healthy plants by daubing their leaves with the sap of diseased plants. In 1892 Dmitrii Ivanowsky, a Russian botanist, showed that the sap was infective even after it had been passed through filters fine enough to remove particles the size of all known bacteria. A few years later his work was expanded by Martinus Beijerinck, who provided evidence that the agent that caused tobacco mosaic disease had many characteristics of a living organism. He hypothesized that the agent could reproduce only within a living cell.

Early in the 20th century, scientists discovered infectious agents that could cause disease in animals or kill bacteria. Like the agents that cause tobacco mosaic disease, these pathogens passed through filters that removed known bacteria and were so small that they could not be seen with the light microscope (LM). Curiously, they could not be grown in laboratory cultures unless living cells were present. These pathogens that infected plants or animals came to be known as viruses. Those that killed bacteria were called **bacteriophages** ("bacteria eaters"), or **phages**.

A virus particle consists of nucleic acid surrounded by a protein coat

A **virus,** or **virion,** is a tiny, infectious particle consisting of a nucleic acid core (its genetic material) surrounded by a protein coat called a **capsid.** Some viruses are also surrounded by an outer membranous envelope containing proteins, lipids, carbohydrates, and traces of metals. A typical small virus, like the poliovirus, is about 20 nm in diameter (about the size of a ribosome), whereas a larger virus, such as the poxvirus that causes smallpox, might be 400 nm long and 200 nm wide.

Viruses are not cellular and cannot independently perform metabolic activities. They do not have the components necessary to carry on cellular respiration or to synthesize proteins and other molecules. Other living organisms contain both deoxyribonucleic acid (DNA) and ribonucleic acid (RNA), but a virus contains either DNA or RNA, not both. Viruses can reproduce, but only within the complex environment of the living cells they infect. In a sense, viruses come "alive" only when they infect a cell. Viruses have genetic information that can force the host cell to replicate the viral nucleic acid and can take over the translational and transcriptional mechanisms of the host cell. The host then synthesizes the capsid and envelope components of the virus.

The shape of a virus is determined by the organization of protein subunits that make up the capsid. Viral capsids are generally either helical or polyhedral, or a complex combination of both shapes (Fig. 23–1). Helical viruses, such as the tobacco mosaic virus, appear as long rods or threads. The capsid is a hollow cylinder made up of proteins that form a groove into which the RNA fits. Polyhedral viruses, such as the adenovirus (which causes a number of human illnesses, including some respiratory infections), appear to be somewhat spherical. The T4 phage that infects *Escherichia coli* consists of a polyhedral "head" attached to a helical "tail."

Viruses may have "escaped" from cells

Where did viruses come from? The most widely held hypothesis is that viruses are bits of nucleic acid that "escaped" from cellular organisms. Viruses may have originated as mobile genetic elements such as transposons (see Chapter 12) or plasmids (small, circular DNA fragments discussed later in this chapter). Such fragments could have moved from one cell and entered another through damaged cell membranes.

According to this *escaped gene hypothesis,* some viruses may trace their origin to animal cells, others to plant cells, and still others to bacterial cells. Their multiple origins might explain why viruses are species-specific; perhaps they infect only those species that are closely related to the organisms from which they originated. This hypothesis is supported by the genetic similarity between some viruses and their host cells—a closer similarity than exists between one type of virus and another.

Another hypothesis, proposed in 2000, suggests that viruses arose early in the history of life, even before the three domains diverged. Evidence for this hypothesis comes from similarities found recently in the protein structures of some viral capsids and also in genetic similarities between some viruses that infect Archaea and some that infect Eubacteria. Molecular biologists studying this issue have suggested that it is improbable that these similarities evolved independently. This suggests that viruses diverged very early.

Phages are viruses that attack bacteria

Much of our knowledge of viruses has come from studying phages because they can be cultured easily within living bacteria in the laboratory. Phages are among the most complex viruses (Fig. 23–1*c*). Their most common structure consists of a long nucleic acid molecule (usually DNA) coiled within a polyhedral head. Many phages have a tail attached to the head. The phage may use fibers extending from the tail to attach to a bacterium.

Before the age of sulfa drugs and antibiotics, phages were used clinically to treat infection. Then in the 1940s, they were

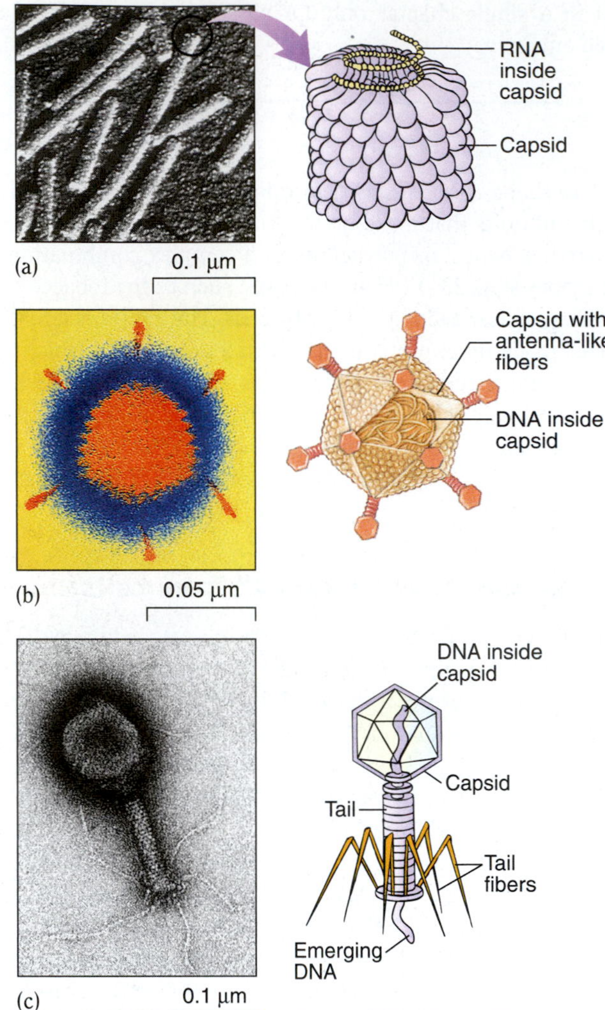

Figure 23–1 Virus structure. Viral capsids typically have a helical or polyhedral structure, or a combination of both. **(a)** TEM of tobacco mosaic virus, which is a rod-shaped virus with a helical arrangement of capsid proteins. **(b)** Color-enhanced TEM of an adenovirus, which has a capsid composed of 252 subunits (visible as tiny ovals) arranged into a 20-sided polyhedron. Twelve of the subunits have projecting protein spikes that permit the virus to recognize host cells. The adenovirus infects the upper respiratory tract, causing symptoms that resemble those of a common cold. **(c)** TEM of the bacteriophage known as T4, which has a polyhedral head and a helical tail. *(a, Omikron/Science/Photo Researchers, Inc.; b, Biozentrum/Science Photo Library/Photo Researchers, Inc.; c, Lee D. Simon/Science Source/Photo Researchers, Inc.)*

abandoned (at least in Western countries) in favor of antibiotics, which were more dependable and easier to use. Now, with the widespread and growing problem of bacterial resistance to antibiotics, these bacteria-killing viruses are once again the focus of research. With new knowledge of phages and sophisticated technology, several research groups are investigating phages to determine which ones kill which species of bacteria. Some groups are genetically engineering phages so that bacteria will be slower to evolve resistance to them. Clinical trials are under way.

A lytic reproductive cycle destroys the host cell

Two types of viral reproduction are lytic and lysogenic. In a **lytic** reproductive cycle, the virus lyses (destroys) the host cell. When the virus infects a susceptible host cell, it forces the host to use its metabolic machinery to replicate viral particles. Viruses that have only a lytic cycle are described as **virulent.**

Five steps are typical in lytic viral reproduction (Fig. 23–2):

1. **Attachment (or absorption).** The virus attaches to receptors on the host cell wall.
2. **Penetration.** The nucleic acid of the virus moves through the plasma membrane and into the cytoplasm of the host cell. The capsid of a phage remains on the outside. In contrast, many viruses that infect animal cells enter the host cell intact.
3. **Replication.** The viral genome contains all the information necessary to produce new viruses. Once inside, the virus induces the host cell to synthesize the necessary components for its replication.
4. **Assembly.** The newly synthesized viral components are assembled into new viruses.
5. **Release.** Assembled viruses are released from the cell. Generally, lytic enzymes produced by the phage destroy the host cell.

The new viruses infect other cells, and the process begins anew. The time required for viral reproduction, from attachment to the bacterium to the release of new viruses, varies from less than 20 minutes to more than an hour.

Temperate viruses can integrate their DNA into the host DNA

Temperate viruses do not always destroy their hosts. In a **lysogenic cycle** the viral genome usually becomes integrated into the host bacterial DNA and is then referred to as a **prophage.** When the bacterial DNA replicates, the prophage also replicates (Fig. 23–3). The viral genes that code for viral structural proteins may be repressed indefinitely. Bacterial cells carrying prophages are called **lysogenic cells.** Certain external conditions (such as ultraviolet light and x rays) can cause temperate viruses to revert to a lytic cycle and then destroy their host. Sometimes temperate viruses become lytic spontaneously.

Bacterial cells containing certain temperate viruses may exhibit new properties. This is called **lysogenic conversion.** An interesting example involves the bacterium *Corynebacterium diphtheriae,* which causes diphtheria. Two strains of this species exist, one that produces a toxin (and causes diphtheria) and one that does not. The only difference between these two strains is that the toxin-producing bacteria contain a specific temperate phage. The phage DNA codes for the powerful toxin that causes the symptoms of diphtheria. Similarly, the bacterium *Clostridium botulinum,* which causes botulism, a serious form of food poisoning, is harmless unless it contains certain prophage DNA that induces synthesis of the toxin.

Some viruses infect animal cells

Hundreds of different viruses infect humans and other animals (Fig. 23–4*a*). Most viruses cannot survive very long outside a liv-

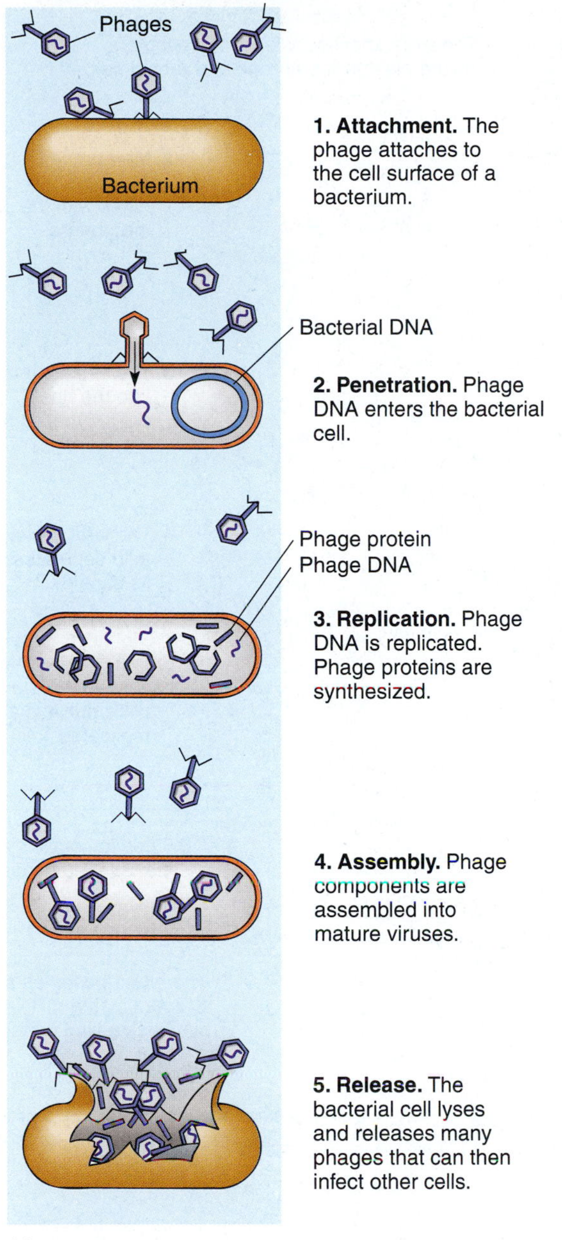

Phages
Bacterium
1. Attachment. The phage attaches to the cell surface of a bacterium.
Bacterial DNA
2. Penetration. Phage DNA enters the bacterial cell.
Phage protein
Phage DNA
3. Replication. Phage DNA is replicated. Phage proteins are synthesized.
4. Assembly. Phage components are assembled into mature viruses.
5. Release. The bacterial cell lyses and releases many phages that can then infect other cells.
(a)

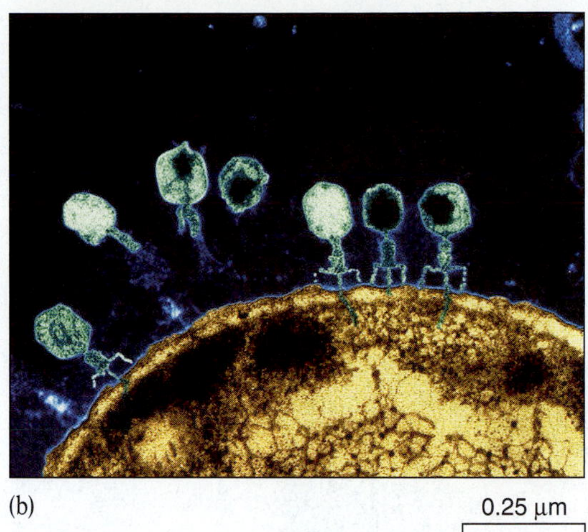

Figure 23–2 Lytic cycle. The host cell is destroyed in a lytic infection. **(a)** The sequence of events in a lytic infection. **(b)** Color-enhanced TEM of phages infecting a bacterium, *Escherichia coli.* *(b, Oliver Meckes/MPL-Tübingen/Photo Researchers, Inc.)*

ing host cell, so their survival depends on their being transmitted from animal to animal. The type of attachment proteins on the surface of a virus determines what type of cell it can infect. Some viruses, such as the adenoviruses, have fibers that project from the capsid and are thought to help the virus adhere to complementary receptor sites on the host cell. Other viruses, such as those that cause herpes, influenza, and rabies, are surrounded by a lipoprotein envelope with projecting glycoprotein spikes that aid in attachment to a host cell.

Receptor sites typically vary with each species and sometimes with each type of tissue. Thus, most human viruses can infect only humans because their attachment proteins combine only with receptor sites found on human cell surfaces. The measles virus and pox viruses can infect many types of human tissue because their attachment proteins combine with receptor sites on a variety of cells. In contrast, polioviruses attach to specific types of cells, such as those that line the digestive tract and motor neurons of the brain and spinal cord.

Viruses have several ways to penetrate animal cells (Fig. 23–4*b,c*). After attachment to a host-cell receptor, some enveloped viruses fuse with the animal cell's plasma membrane. The viral capsid and nucleic acid are both released into the animal cell. Other viruses enter the host cell by endocytosis. In this process, the plasma membrane of the animal cell invaginates to form a membrane-bounded vesicle that contains the virus.

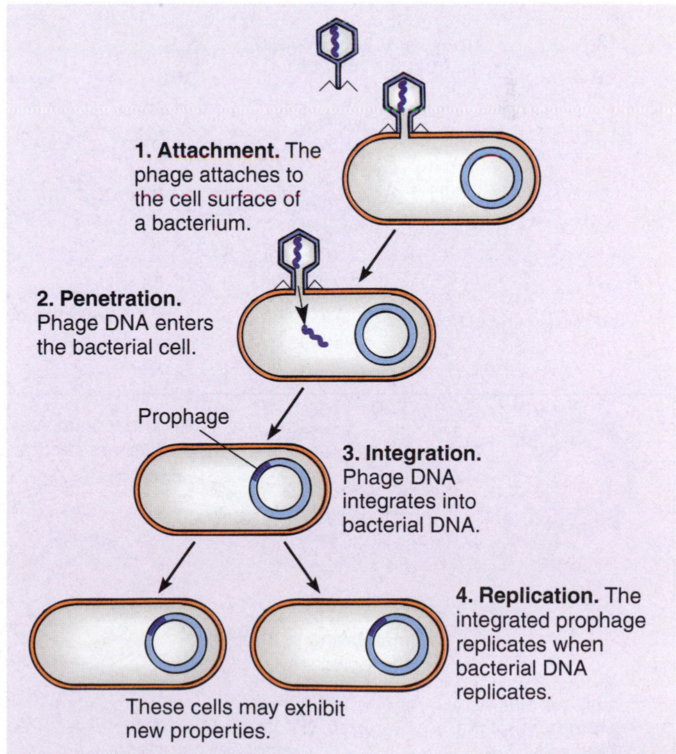

Figure 23–3 Lysogenic cycle. Temperate phages can integrate their nucleic acid into the host cell DNA, making it a lysogenic cell.

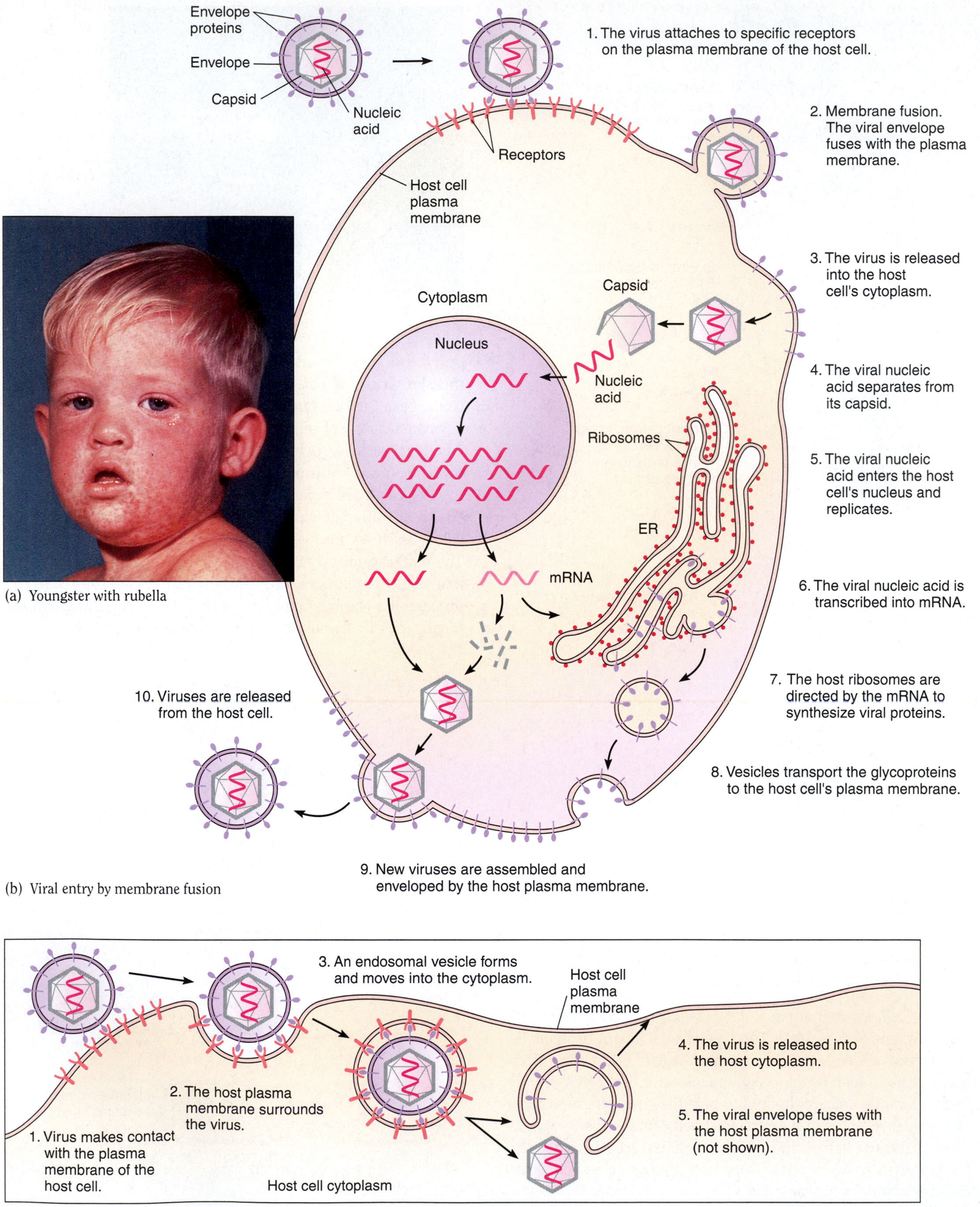

Envelope proteins
Envelope
Capsid
Nucleic acid

1. The virus attaches to specific receptors on the plasma membrane of the host cell.

Receptors

Host cell plasma membrane

2. Membrane fusion. The viral envelope fuses with the plasma membrane.

3. The virus is released into the host cell's cytoplasm.

Capsid

Cytoplasm

Nucleus

Nucleic acid

4. The viral nucleic acid separates from its capsid.

Ribosomes

5. The viral nucleic acid enters the host cell's nucleus and replicates.

ER

mRNA

6. The viral nucleic acid is transcribed into mRNA.

(a) Youngster with rubella

10. Viruses are released from the host cell.

7. The host ribosomes are directed by the mRNA to synthesize viral proteins.

8. Vesicles transport the glycoproteins to the host cell's plasma membrane.

9. New viruses are assembled and enveloped by the host plasma membrane.

(b) Viral entry by membrane fusion

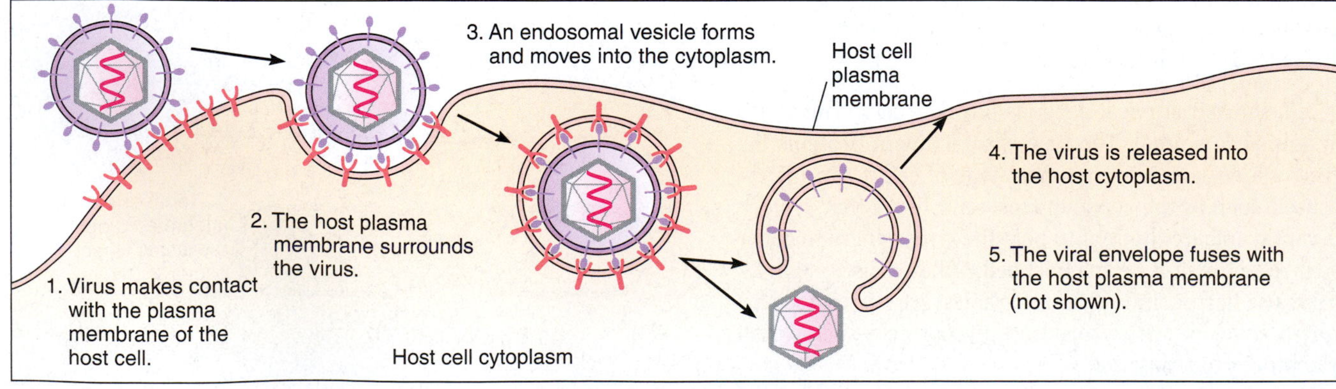

3. An endosomal vesicle forms and moves into the cytoplasm.
Host cell plasma membrane
4. The virus is released into the host cytoplasm.
2. The host plasma membrane surrounds the virus.
5. The viral envelope fuses with the host plasma membrane (not shown).
1. Virus makes contact with the plasma membrane of the host cell.
Host cell cytoplasm

(c) Viral entry by endocytosis

Figure 23–4 Viruses cause diseases in humans and other animals. (a) This child's rash is a symptom of rubella (German measles), which is caused by an RNA virus spread by close contact. When contracted during pregnancy, rubella can cause birth defects. Vaccination has greatly decreased the incidence of this disease. (b) Some viruses enter animal cells by membrane fusion. The process of viral replication in animal cells is summarized in the diagram. (c) Some viruses enter the host cell by endocytosis. *(a, Centers for Disease Control and Prevention, U.S. Department of Health and Human Services)*

Like other viruses, those that infect animal cells multiply and produce new virus particles. During the time that viral nucleic acid is being replicated and viral proteins are being synthesized, the synthesis of host DNA, RNA, and protein may be inhibited.

Animal viruses can contain either DNA or RNA. In DNA viruses, the synthesis of viral DNA and protein is similar to the processes by which the host cell would normally carry out its own DNA and protein synthesis. In most RNA viruses, RNA replication and transcription take place with the help of an RNA-dependent RNA polymerase. However, **retroviruses** are RNA viruses that have a DNA polymerase called **reverse transcriptase** used to transcribe the RNA genome into a DNA intermediate (Fig. 23–5).

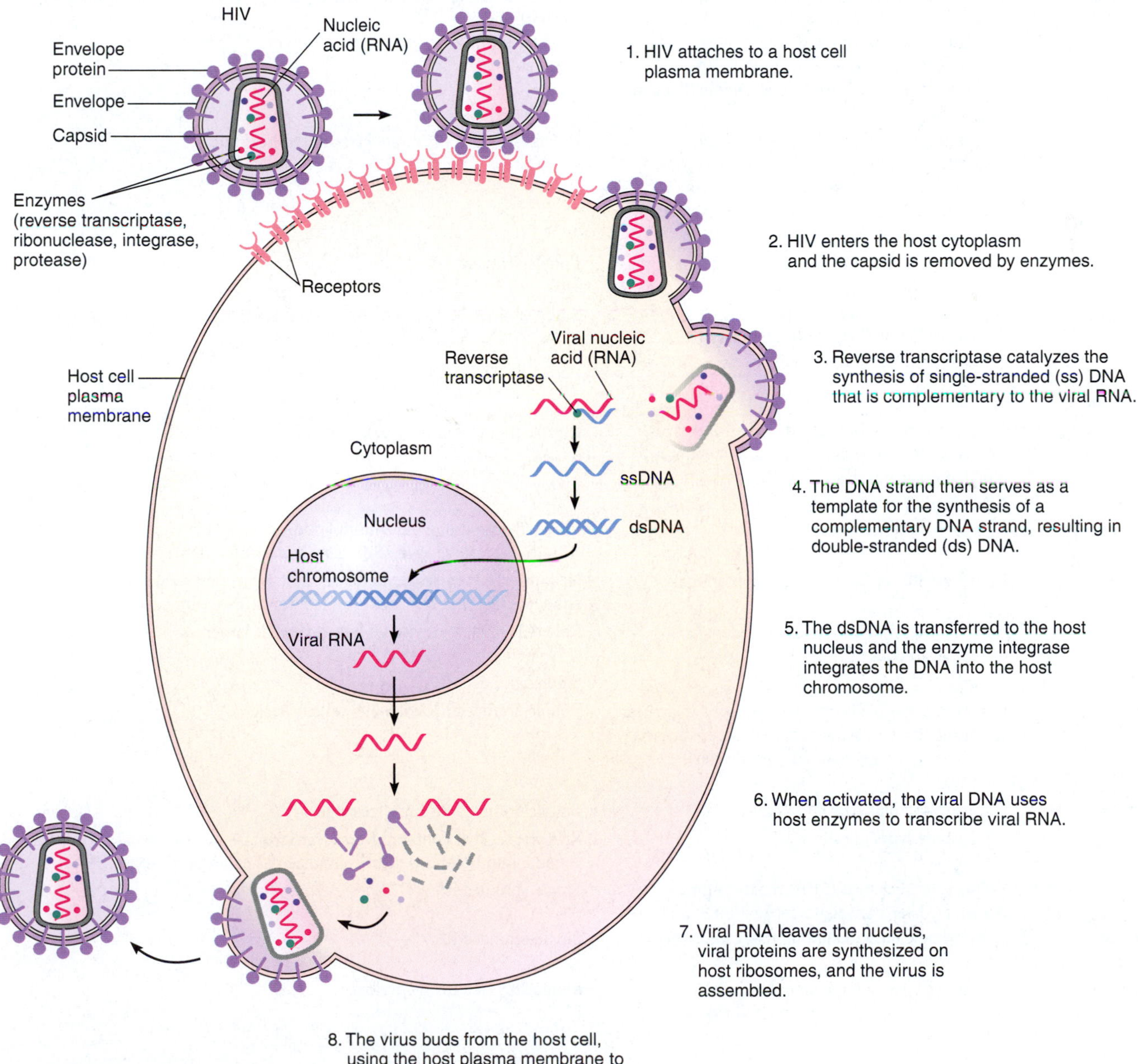
HIV
Envelope protein
Envelope
Capsid
Nucleic acid (RNA)
Enzymes (reverse transcriptase, ribonuclease, integrase, protease)
Receptors
1. HIV attaches to a host cell plasma membrane.
2. HIV enters the host cytoplasm and the capsid is removed by enzymes.
Viral nucleic acid (RNA)
Reverse transcriptase
Host cell plasma membrane
3. Reverse transcriptase catalyzes the synthesis of single-stranded (ss) DNA that is complementary to the viral RNA.
Cytoplasm
ssDNA
dsDNA
4. The DNA strand then serves as a template for the synthesis of a complementary DNA strand, resulting in double-stranded (ds) DNA.
Nucleus
Host chromosome
Viral RNA
5. The dsDNA is transferred to the host nucleus and the enzyme integrase integrates the DNA into the host chromosome.
6. When activated, the viral DNA uses host enzymes to transcribe viral RNA.
7. Viral RNA leaves the nucleus, viral proteins are synthesized on host ribosomes, and the virus is assembled.
8. The virus buds from the host cell, using the host plasma membrane to make the viral envelope.

Figure 23–5 Life cycle of HIV, the retrovirus that causes AIDS. HIV infects certain cells of the immune system, including helper T cells. The virus attaches to protein receptors, known as CD4, on the host plasma membrane. HIV has two identical single-stranded RNA molecules.

This DNA becomes integrated into the host DNA by an enzyme also carried by the virus. Copies of the viral RNA are synthesized as the incorporated DNA is transcribed by host RNA polymerases. The human immunodeficiency virus (HIV) that causes acquired immunodeficiency syndrome (AIDS) is a retrovirus. Certain cancer-causing viruses are also retroviruses.

After the viral genes are transcribed, the viral structural proteins are synthesized. The capsid is produced, and new virus par-

TABLE 23–1 Animal Viruses

Group	Diseases Caused	Characteristics
DNA Viruses		
Poxviruses	Smallpox, cowpox, and economically important diseases of domestic fowl	Large, complex viruses; double-stranded DNA; replicates in the cytoplasm of the host cell*
Herpesviruses	Cold sores (herpes simplex virus type 1); genital herpes, a sexually transmitted disease (herpes simplex virus type 2); chickenpox and shingles (herpes varicella-zoster virus); infectious mononucleosis and Burkitt's lymphoma (Epstein-Barr virus)	Medium to large, enveloped viruses; double-stranded DNA; replicates in the host nucleus†
Adenoviruses	Respiratory tract disorders (e.g., sore throat, tonsillitis), conjunctivitis, and gastrointestinal disorders are caused by more than 40 types of adenoviruses in humans; other varieties infect other animals	Double-stranded DNA; replicates in the host nucleus
Papovaviruses	Human warts and some degenerative brain diseases; some cancers	Double-stranded DNA‡
Parvoviruses	Infections in dogs, swine, arthropods, rodents; gastroenteritis in humans (transmitted by consumption of infected shellfish)	Single-stranded DNA; some require a helper virus in order to multiply
RNA Viruses		
Picornaviruses	Polio (poliovirus); hepatitis A (hepatitis A virus); intestinal disorders (enterorviruses); common cold (rhinoviruses); aseptic meningitis (coxsackievirus, echovirus)	Diverse group of small viruses; single-stranded RNA that can serve as mRNA
Togaviruses	Rubella (German measles)	Large, diverse group of medium-sized, enveloped viruses; single-stranded RNA that can serve as mRNA; many transmitted by arthropods
Orthomyxoviruses	Influenza (flu) in humans and other animals	Medium-sized viruses that often exhibit projecting spikes; single-stranded RNA that serves as template for mRNA synthesis
Paramyxoviruses	Rubeola (measles) and mumps in humans; distemper in dogs	Resemble orthomyxoviruses but somewhat larger
Rhabdoviruses	Rabies	Single-stranded RNA
Rotavirus	Common cause of severe gastroenteritis (diarrhea, vomiting); most common cause of severe diarrhea in children; causes about 25% of traveler's diarrhea	Double-stranded RNA; double-walled capsid
Reoviruses	Vomiting and diarrhea; encephalitis	Double-stranded RNA; no envelope
Retroviruses	AIDS; some types of cancer	RNA viruses that contain reverse transcriptase for transcribing the RNA genome into DNA; two identical molecules of single-stranded RNA
Flaviviruses	Yellow fever; hepatitis C (the most common reason for liver transplants in the United States)	Single-stranded RNA
Filoviruses	Hemorrhagic fever, including that caused by the Ebola virus	Single-stranded RNA
Bunyaviruses (hanta viruses belong to this group)	St. Louis encephalitis; hantavirus pulmonary syndrome (caused by Sin Nombre Virus, a hantavirus)	Single-stranded RNA; envelope

*The vaccinia (cowpox) virus is used to produce genetically engineered vaccines.
†These viruses frequently cause latent infections; some cause tumors.
‡The virus SV40 has been used as a vector to transport genes into cells.

FOCUS ON Emerging Viruses

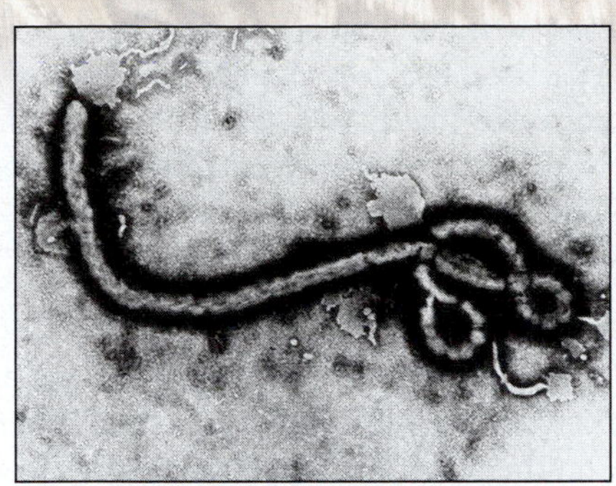

TEM of Ebola virus. *(Courtesy of Frederick A. Murphy)*

The 2000–2001 Ebola virus outbreak in Uganda serves as a grim reminder that pathogens can strike quickly and fatally. Emerging diseases are those that either are new to the human population or unpredictably cause an epidemic, sometimes in a new geographical area. According to the U.S. Centers for Disease Control and Prevention, more than 200 new, continual, or reemerging pathogens have the potential to strike globally. Historically, new viral strains have claimed many human lives. For example, in 1918 an influenza epidemic killed more than 20 million people. Even at the level of our current knowledge about viruses and epidemiology, just how well could we contain a particularly virulent virus? How well are we containing HIV (the virus that causes AIDS), which claims more than 3 million lives each year?

Let us consider the deadly Ebola virus as an example of an emerging virus. Ebola outbreaks have occurred several times during the past few decades. The virus was first identified in 1976 in Zaire and Sudan (outbreaks that killed more than 400 people). During the 2000–2001 outbreak in Uganda several hundred people were infected , with a fatality rate of about 50%.

Ebola is an elongated, single-stranded RNA, enveloped virus (a filovirus; see Table 23–1). Early symptoms of Ebola infection resemble those of influenza or dysentery. Within about three days of infection, victims develop a fever and weakness, followed by a rash and vomiting. The victim hemorrhages internally, and bleeds from the mouth, eyes, ears, and other body openings. Internal organs shut down, and 50% to 90% of victims die within one to two weeks after infection. The disease is spread by contact with infected body fluids. Like many RNA viruses, Ebola makes frequent mistakes when it duplicates its RNA. The resulting high mutation rate leads to the rapid evolution of new strains, making it difficult for researchers to develop an effective vaccine.

An important aspect of understanding an infectious agent is identifying *patient zero,* the first patient who contracts the virus during an outbreak. If the origin of the epidemic can be found, investigators might be able to trace where the virus came from and how it infects humans. Between outbreaks, Ebola remains hidden, apparently surviving in an animal host that has not yet been identified. Ebola infects chimpanzees, but because they die quickly, they are not thought to be the reservoir species. However, some other primate host might be the reservoir that would allow the virus to maintain itself indefinitely.

Human activity, including social factors such as urbanization and global travel, contributes to epidemics of infectious disease. For example, as human populations concentrate in cities, large numbers of people are in close contact, permitting the rapid spread of viruses. Living conditions, including sanitation, nutrition, physical stress, level of health care, and sexual practices, are important factors in the spread of disease. In the United States and other highly developed countries, infectious disease accounts for about 4% to 8% of deaths compared with death rates of 30% to 50% in developing regions.

Changing natural habitats can create the conditions necessary for new pathogens to emerge. For example, cutting down forests can bring disease-carrying insects into contact with humans. Sometimes, even naturally occurring ecological changes can spawn outbreaks of disease. The 1993 outbreak of hantavirus in the Southwestern United States killed more than 50 persons. A mild winter coupled with heavy rainfall resulted in a large crop of seeds, which supported a population explosion of field mice. The mice carry the virus. Lessons learned from dealing with emerging viruses and the resurgence of old ones will help us contain future epidemics, but much more research is needed.

ticles are assembled. Viruses that do not have an outer envelope exit by cell lysis. The plasma membrane ruptures, releasing many new viral particles. Enveloped viruses receive their lipoprotein envelopes by picking up a fragment of the host plasma membrane as they leave the infected cell (Figs. 23–4*b* and 23–5).

Viral proteins damage the host cell in a variety of ways. These proteins may alter the permeability of the plasma membrane or may inhibit synthesis of host nucleic acids or proteins. Viruses sometimes damage or kill their host cells by their sheer numbers. A poliovirus may produce 100,000 new viruses within a single host cell!

Animal viruses cause hog cholera, foot-and-mouth disease, canine distemper, swine influenza, and certain types of cancer (e.g., feline leukemia). Most humans suffer from two to six viral infections each year, including common colds. Other human diseases caused by viruses include chickenpox, herpes simplex (one type causes genital herpes), mumps, rubella (German measles), rubeola (measles), rabies, warts, infectious mononucleosis, influenza, hepatitis, and AIDS (Table 23–1 and *Focus On: Emerging Viruses*). Hepatitis C virus (HCV) has infected an estimated 170 million people throughout the world—more than four times as many as HIV, the virus that causes AIDS.

Viruses can infect plant cells

Viral diseases can be spread among plants by insects such as aphids and leafhoppers as they feed on plant tissues. Because of the thick cell wall, plant cells cannot be penetrated by viruses unless the cells are damaged. Plant viruses can be inherited by way of infected seeds or by asexual propagation. Once a plant is infected, the virus spreads through the plant body by passing through plasmodesmata (cytoplasmic connections) that penetrate the walls between adjacent cells (see Fig. 5–25). The genome of most plant viruses consists of RNA.

Symptoms of viral infection include reduced plant size, and spots, streaks, or mottled patterns on leaves, flowers, or fruits (Fig. 23–6). Infected crops almost always produce lower yields. Cures are not known for most viral diseases of plants, so it is common to burn plants that have been infected. Some agricultural scientists are focusing their efforts on prevention of viral disease by developing virus-resistant strains of important crop plants.

VIROIDS AND PRIONS ARE SMALLER THAN VIRUSES

The discovery of infective agents even smaller than viruses has challenged old ideas. Until **viroids** were studied, most biologists assumed that protein was necessary for an infectious agent to duplicate itself. However, no proteins are associated with viroids, and evidence suggests that the viroid genome does not code for any proteins. Each viroid consists of a very short strand of RNA (only 250–400 nucleotides) with no protective protein coat. Viroids are copied by host RNA polymerases; the viroid is used as a template. Viroids cause a variety of plant diseases and may also infect animals. These infective agents are generally found within the host cell nucleus and may interfere with gene regulation.

Even more heretical is the idea that a pathogen could exist and transfer information without nucleic acids. The **prion,** a protein-like infectious particle, has been linked to a group of fatal degenerative brain diseases that have been identified in birds and mammals. These diseases are called **transmissible spongiform encephalopathies (TSEs)** because the infected brain appears to develop holes, becoming somewhat spongelike.

Process of Science

In 1997 Stanley Prusiner, professor of neurology and biochemistry at the University of California School of Medicine, San Francisco, won the Nobel Prize in Physiology or Medicine "for his discovery of prions—a new biological principle of infection." Prusiner began his studies of prions in the early 1970s, motivated by the death of a patient from Creutzfeldt-Jakob disease (CJD), one of the human TSEs. Prusiner found that the infective agent was not affected by radiation (which typically mutates nucleic acids), and he could not find DNA or RNA in the particles. In 1982 he named the infective agent *prion* for "proteinaceous infectious particle." Researchers have determined the structure of the prion protein, which consists of 208 amino acids. Despite extensive searching by many investigators, no nucleic acid component has been found.

Because nucleic acids are the molecules replicated during cell division and reproduction, how prions reproduce has been of great interest. Prusiner and other researchers have shown that mammals have a gene that encodes the prion protein. Although its function is not known, this protein is normally harmless. However, the prion protein sometimes converts to a different

(a)

(b)

Figure 23–6 Plant viruses. **(a)** Virus-streaked tulips. The virus that causes this relatively harmless disease affects pigment formation in the petals. **(b)** Pepper leaves infected with tobacco mosaic virus. The leaf is characteristically mottled with light green areas. *(a, Kenneth M. Corbett; b, Jack M. Bostrack/Visuals Unlimited)*

shape, an insoluble variant that accumulates in the brains of patients with TSE. Genetically engineered mice that lack the prion protein gene are immune to TSE infection.

According to Prusiner, mutations in the prion protein gene increase the risk that the protein will change shape. The prion protein changes shape by a posttranslational process that involves refolding of the molecule. Twenty different mutations in the prion gene have been identified that have been linked with inherited prion diseases. Prusiner has genetically engineered mice that produce the mutant prion protein found in humans with inherited TSE. These mice develop TSE.

Many biologists have remained skeptical about prions, believing that eventually a viral component would be discovered. These biologists hold that proving the prion theory would require production of insoluble prion protein in a nonbiological system to ensure no viral contamination. Researchers would then need to demonstrate that the prion protein causes TSE. In 2000 researchers reported that they made an artificial prion by fusing part of a yeast prion to a normal rat protein. They demonstrated that this chimeric protein changed the property of yeast cells and that these changes could be inherited. These experiments provide convincing support for the protein-only hypothesis.

One of the most studied prion diseases is scrapie in sheep and goats. When infected, animals lose coordination, become irritable, and itch so severely that they scrape off their wool or hair. Bovine spongiform encephalopathy (BSE), popularly referred to as "mad cow disease," became epidemic in cattle in the United Kingdom in the 1990s. Scientists hypothesized that cattle became infected when they ate feed that had been mixed with sheep offal—brains and other organs—containing prions. The epidemic was slowly brought under control after sheep offal was banned as feed. However, approximately 200,000 cattle died as a result of the disease, and nearly 4.5 million cattle suspected of exposure were destroyed. This disease led to widespread bans on the export and consumption of British cattle.

By 2001, more than 90 people in the United Kingdom had died of what appeared to be a human variety of BSE, providing evidence that the disease is transmissible from cow to human. The human disease thought to be transmitted from infected cows is similar to CJD, which was known before the BSE epidemic. The infectious agent, which has been recovered from infected human neural tissue, appears to be similar to the prion that causes scrapie in sheep. Human-to-human transmission has been associated with tissue and organ transplants and transfusion with contaminated blood.

BACTERIA ARE PROKARYOTES

Thousands of species of bacteria have been described. In contrast with viruses, viroids, and prions, which consist only of nucleic acid and/or protein, **bacteria** are cellular organisms. Bacteria are prokaryotes, and their cell structure is fundamentally different from the cells of other living organisms. For this reason some biologists have assigned them to their own kingdoms: **Archaebacteria** and **Eubacteria.** Other biologists, including most microbiologists, prefer to assign the bacteria to two domains: **Archaea** and **Eubacteria.**

Most bacterial cells are very small. Typically, their diameter ranges from 0.5 to 1.0 μm. Their cell volume is only about one-thousandth that of small eukaryotic cells, and their length is only about one-tenth. Most prokaryotes are unicellular, but some form colonies or filaments containing specialized cells.

Epulopiscium fishelsoni is a spectacular exception to the generalization that all bacteria are microscopic. This bacterium lives as a symbiont in the intestine of surgeonfish (Fig. 23–7).

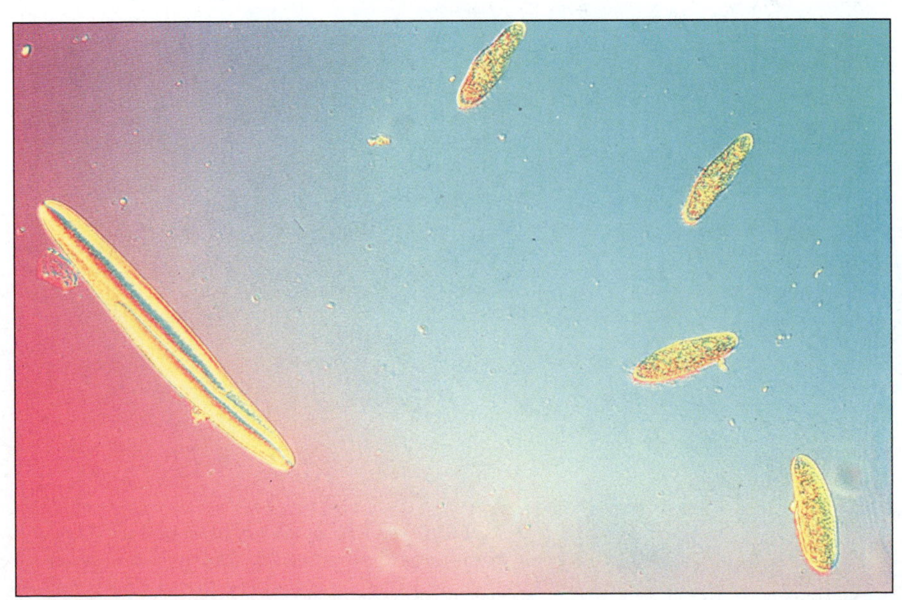

Figure 23–7 Color-enhanced LM of the giant bacterium. This bacterium *(Epulopiscium fishelsoni),* about a million times larger than a typical bacterium, is shown here *(on left)* with four paramecia (which are eukaryotes). *(Esther R. Angert and Norman R. Pace, Indiana University)*

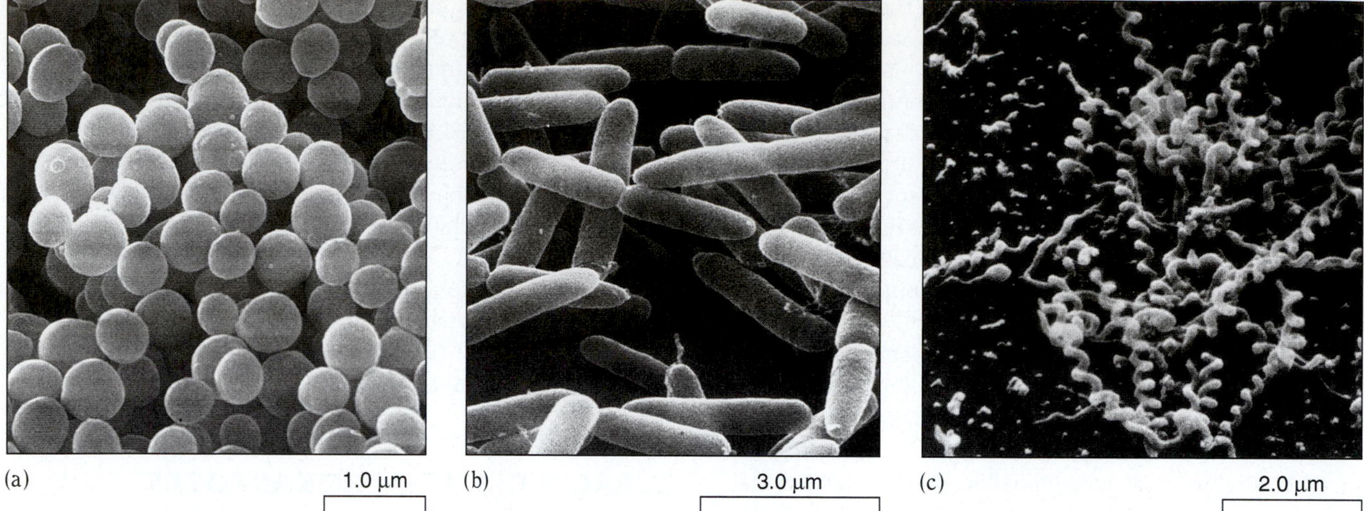

Figure 23–8 Common shapes of bacteria. **(a)** SEM of *Micrococcus*, cocci bacteria. **(b)** SEM of *Salmonella*, bacilli bacteria. **(c)** SEM of *Spiroplasma*, spirilla bacteria. *(David M. Phillips/Visuals Unlimited)*

About 600 μm long and 80 μm wide, its volume is about a million times larger than that of a typical bacterium.

Bacteria have several common shapes

Although many species have irregular shapes, shape is an important criterion in identifying bacterial species. Three common shapes are spherical, rod-shaped, and spiral (Fig. 23–8). Spherical bacteria, known as **cocci** (sing., *coccus*), occur singly in some species and in groups of independent cells in others. Cells may be grouped in twos *(diplococci)*, in long chains *(streptococci)*, or in irregular clumps that look like bunches of grapes *(staphylococci)*. Rod-shaped bacteria, called **bacilli** (sing., *bacillus*), may occur as single rods or as long chains of rods. Some bacteria are helical. A bacterium shaped like a very short helix is called a **vibrio.** One that is a longer helix is known as a **spirillum** (pl., *spirilla*) if rigid, and as a **spirochete** if flexible.

Prokaryotic cells lack membrane-bounded organelles

Prokaryotic cells do not have membrane-bounded organelles typical of eukaryotic cells. Thus, bacterial cells do not have nuclei, mitochondria, chloroplasts, endoplasmic reticula, Golgi complexes, or lysosomes (Fig. 23–9). The dense cytoplasm of the bacterial cell contains ribosomes and storage granules that hold glycogen, lipid, or phosphate compounds. Enzymes needed for metabolic activities may be located in the cytoplasm. Although the membranous organelles of eukaryotic cells are absent, in some bacterial cells the plasma membrane is extensively folded inward. Enzymes needed for cellular respiration and photosynthesis may be associated with the plasma membrane or its folds.

A cell wall typically covers the cell surface

Most prokaryotic cells have a cell wall surrounding the plasma membrane. The cell wall provides a rigid framework that supports the cell, maintains its shape, and keeps it from bursting under hypotonic conditions (see Chapter 5). Most bacteria seem to be adapted to hypotonic surroundings. When wall-less forms of

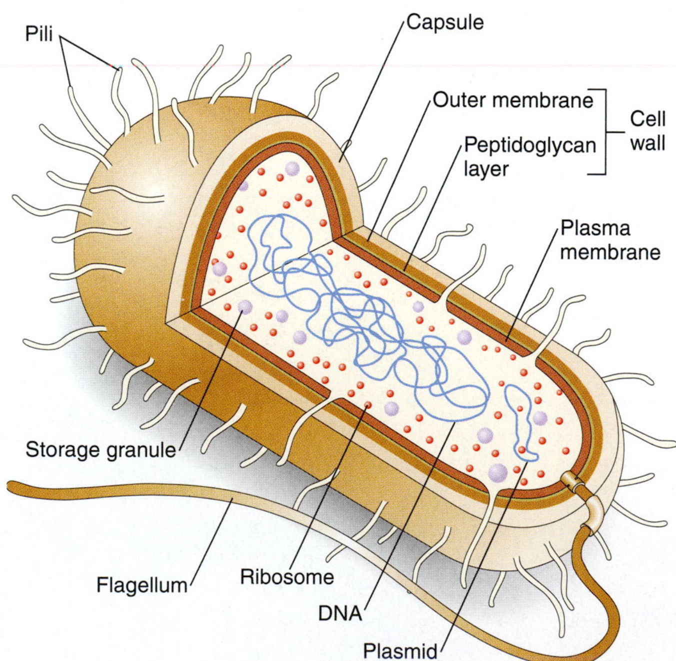

Figure 23–9 Bacteria are prokaryotic cells. This bacillus is a gram-negative bacterium (discussed in text). Note the absence of a nuclear envelope surrounding the bacterial DNA.

bacteria are produced experimentally, they must be maintained in isotonic solutions to keep them from bursting. However, cell walls are of little help when a bacterium is in a hypertonic environment, as found in food preserved by means of a high sugar or salt content. That is why most bacteria grow poorly in jellies, jams, salted fish, and other foods preserved in these ways.

Bacteria were once classified with plants because they have cell walls. However, eukaryotic cell walls are homoplastic, not homologous, to bacterial cell walls; their structure and composition are different. Unlike plant cell walls, the bacterial cell wall is not made of cellulose. The eubacterial cell wall includes **peptidoglycan,** a complex polymer that consists of two unusual types of sugars (amino sugars) linked with short polypeptides. The sugars and polypeptides are linked to form a single macromolecule that surrounds the entire plasma membrane. The cell wall structure varies among species.

Differences in bacterial cell wall composition are of great interest to microbiologists and are important clinically. In 1888 the Danish physician Christian Gram developed the Gram staining procedure. Bacteria that absorb and retain crystal violet stain in the laboratory are referred to as **gram-positive,** whereas those that do not retain the stain when rinsed with alcohol are **gram-negative.** The cell walls of gram-positive bacteria are very thick and consist primarily of peptidoglycan. The cell walls of gram-negative bacteria consist of two layers: a thin peptidoglycan wall and a thick outer membrane. The outer membrane resembles the plasma membrane but contains polysaccharides bonded to lipids (Fig. 23–10).

Distinguishing between gram-positive and gram-negative bacteria is important in treating certain diseases. For example, the antibiotic penicillin interferes with peptidoglycan synthesis, ultimately resulting in a fragile cell wall that cannot protect the cell (see section on drugs that are enzyme inhibitors in Chapter 6). Predictably, penicillin works most effectively against gram-positive bacteria.

Some species of bacteria produce a **capsule** or slime layer that surrounds the cell wall. In free-living species, the capsule may provide the cell with added protection against phagocytosis (engulfment; see Chapter 5) by other microorganisms. In disease-causing bacteria, the capsule may protect against phagocytosis by the host's white blood cells. For example, the ability of *Streptococcus pneumoniae* to cause bacterial pneumonia depends on its capsule. A strain of *S. pneumoniae* that lacks a capsule does not cause the disease. Bacteria also use their capsules to attach to surfaces such as rocks, plant roots, or human teeth (where they cause dental plaque).

Some bacteria have hundreds of hairlike appendages known as **pili** (sing., *pilus*). These protein structures are organelles of attachment that help bacteria adhere to one another or to certain surfaces, such as the cells they infect. Some pili are involved in the transmission of DNA between bacteria.

Many types of bacteria are motile

Can you imagine trying to swim through molasses? Water has the same relative viscosity to bacteria that molasses has to humans. Most motile bacteria move by means of rotating **flagella.** The number and location of flagella are important in classification of some bacterial species.

Prokaryotic flagella are quite different from eukaryotic flagella (see Chapter 4). The bacterial flagellum is not composed of microtubules. It consists of three parts: a basal body, a hook, and

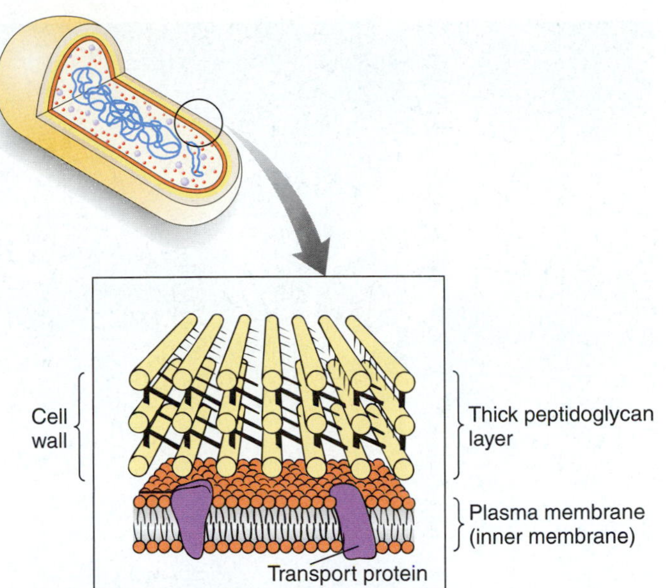

(a) Gram-positive cell wall

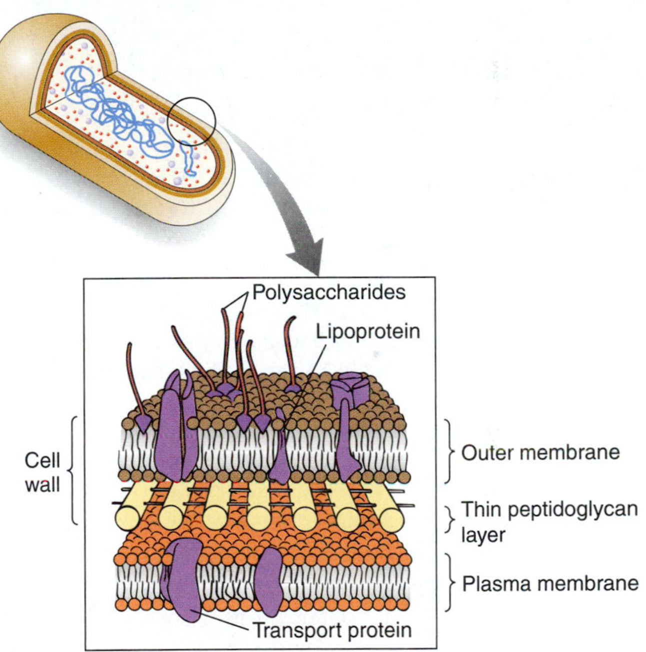

(b) Gram-negative cell wall

Figure 23–10 Bacterial cell walls. **(a)** In the gram-positive cell wall, a thick layer of peptidoglycan molecules is held together by amino acids. **(b)** In the gram-negative cell wall, a thin peptidoglycan layer is covered by a thick outer membrane.

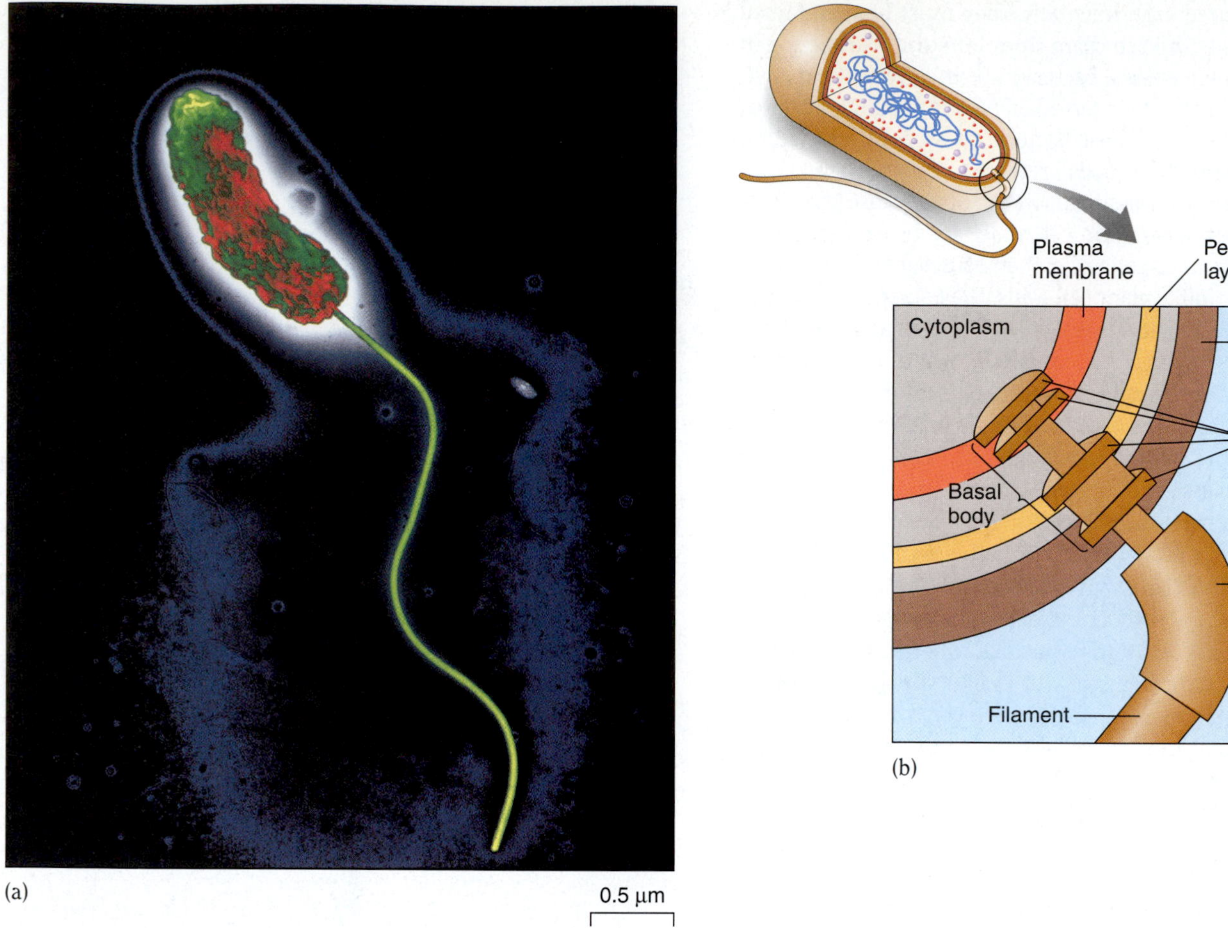

Figure 23–11 Bacterial flagella. (a) Color-enhanced TEM of *Vibrio cholerae,* the flagellated bacterium that causes cholera. (b) Structure of a bacterial flagellum. The motor is the basal body, which consists of a series of disc-shaped plates that (1) anchor the flagellum to the cell wall and plasma membrane and (2) spin the hook and filament of the flagellum. *(a, Oliver Meckes/Gelderblom/Photo Researchers, Inc.)*

a single filament (Fig. 23–11). The basal body is a complex structure that anchors the flagellum into the cell wall by disc-shaped plates. The curved hook connects the basal body to the long, hollow filament that extends into the outside environment. The basal body is a motor. The bacterium uses energy from ATP to pump protons out of the cell. Diffusion of these protons back into the cell powers the motor that spins the flagellum like a propeller. Thus, the flagellum produces a rotary motion that pushes the cell much as a propeller pushes a ship through the water.

Bacteria have a circular DNA molecule

The genetic material of a bacterium lies in the cytoplasm and is not surrounded by a nuclear envelope. In most species, it is contained in a single, circular DNA molecule. If stretched out to its full length, this molecule would be about 1000 times longer than the cell itself. Unlike eukaryotic chromosomes, the bacterial DNA has little protein associated with it.

In addition to the genomic DNA, most bacteria have a small amount of genetic information present as one or more **plasmids,** smaller circular fragments of DNA. Plasmids can replicate independently of the genomic DNA (see Chapter 14) or become integrated into it. Bacterial plasmids often have genes that code for catabolic enzymes, for genetic exchange, or for resistance to antibiotics.

Most bacteria reproduce by binary fission

Bacteria reproduce asexually, generally by **binary fission,** a process in which one cell divides into two similar cells *(see chapter introduction photograph).* First the circular bacterial DNA replicates, and then a transverse wall is formed by an ingrowth of both the plasma membrane and the cell wall.

Binary fission occurs with remarkable speed. Under ideal conditions some bacterial species divide in less than 20 minutes. At this rate, if nothing interfered, one bacterium could give rise to more than 1 billion bacteria within 10 hours! However, bacteria cannot reproduce at this rate for very long before their population expansion is affected by lack of food or by the accumulation of waste products.

A less common form of asexual reproduction among bacteria is **budding.** In budding a cell develops a bulge, or bud, that enlarges, matures, and eventually separates from the mother cell. A few species of bacteria (actinomycetes) divide by **fragmentation.** Walls develop within the cell, which then separates into several new cells.

Although sexual reproduction involving the fusion of gametes does not occur in bacteria, genetic material is sometimes exchanged between individuals. This exchange takes place by three different mechanisms: transformation, transduction, and conjugation.

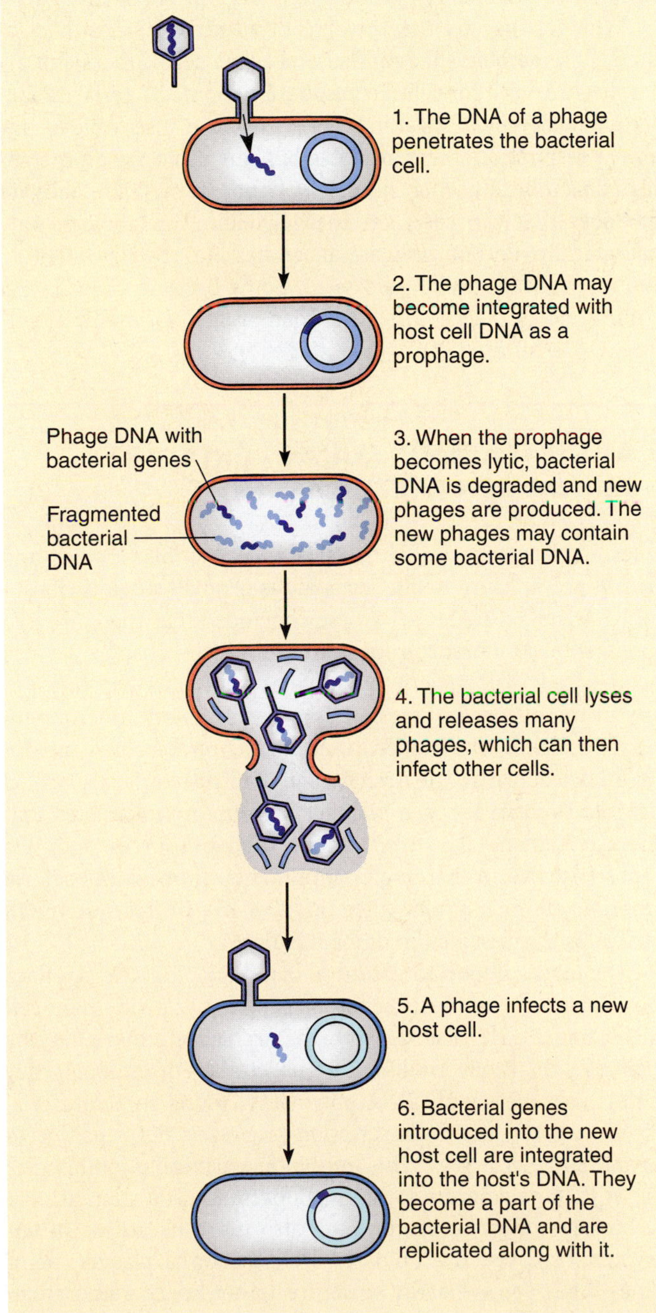

1. The DNA of a phage penetrates the bacterial cell.
2. The phage DNA may become integrated with host cell DNA as a prophage.
Phage DNA with bacterial genes
Fragmented bacterial DNA
3. When the prophage becomes lytic, bacterial DNA is degraded and new phages are produced. The new phages may contain some bacterial DNA.
4. The bacterial cell lyses and releases many phages, which can then infect other cells.
5. A phage infects a new host cell.
6. Bacterial genes introduced into the new host cell are integrated into the host's DNA. They become a part of the bacterial DNA and are replicated along with it.

Figure 23–12 Transduction. In this process, a phage transfers bacterial DNA from one bacterium to another.

1. In **transformation,** fragments of DNA released by a cell are taken in by another bacterial cell. Recall that this mechanism was used experimentally to demonstrate that genes can be transferred from one bacterium to another and that DNA is the chemical basis of heredity (see Figure 11–1).
2. In a different process of gene transfer, **transduction,** a phage carries bacterial genes from one bacterial cell into another (Fig. 23–12).
3. In **conjugation,** two cells of different mating types come together, and genetic material is transferred from one to the other (Fig. 23–13). In contrast with transformation and transduction, conjugation involves contact between two cells.

Conjugation has been most extensively studied in the bacterium *Escherichia coli.* In the *E. coli* population there are donor cells (sometimes referred to as male cells) that have plasmids that can be transmitted to recipient (female) cells. A pilus on the donor cell recognizes the recipient cell and makes the first contact. A cytoplasmic bridge forms between the two cells, and DNA is transferred from donor to recipient cell.

Some bacteria form endospores

When the environment becomes unfavorable, such as when it gets very dry, the cells of some bacteria become dormant. Cells lose water, shrink slightly, and remain inactive until water is again available. Some other bacteria form dormant, extremely durable cells called **endospores.** After the endospore forms, the

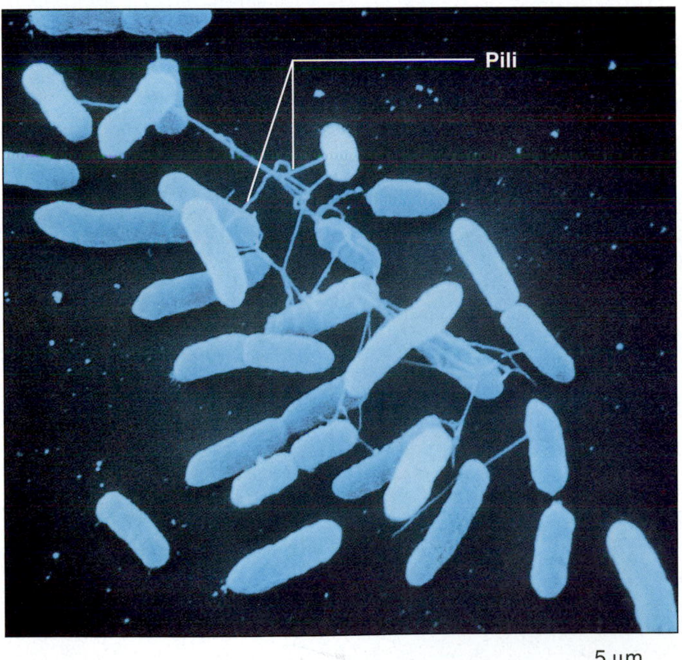

Figure 23–13 Conjugation. Color-enhanced SEM of *Serratia marcescens* bacteria, which are connected by pili. Stimulated by the contact, the cells pull close together and form cytoplasmic bridges between donor and recipient cells *(not shown).* *(Manfred Kage/Peter Arnold, Inc.)*

cell wall of the original cell lyses, releasing the endospore. Formation of endospores is not a type of reproduction in bacteria; endospores are not comparable to the reproductive spores of fungi and plants. Only one endospore is formed per original cell, so the total number of individuals does not increase.

Endospores can survive in very dry, hot, or frozen environments, or when food is scarce (Fig. 23–14). Some endospores are so resistant that they can survive an hour or more of boiling or centuries of freezing. When environmental conditions are again suitable for growth, the endospore germinates, forming an active, growing bacterial cell. Several types of bacteria that form endospores are medically important. For example, when surgical instruments are not effectively sterilized, endospores of pathogenic bacteria may survive. Patients may then be exposed to serious diseases such as tetanus (caused by *Clostridium tetani*) and gas gangrene (caused by several species of *Clostridium*). The endospore of *Bacillus anthracis,* the bacterium that causes anthrax, is so hardy that this pathogen has become a concern as an agent of biological warfare.

Metabolic diversity is evident among bacteria

A bacterial cell contains about 5000 different kinds of molecules. The functions of these molecules, how they interact, and how the bacterium synthesizes them from the nutrients it takes in are complex biochemical problems that have interested researchers for years. Much of the knowledge gained from studying these mechanisms in bacterial cells has been successfully applied to cells of humans and other organisms. The basic biochemical processes of all organisms are impressively similar.

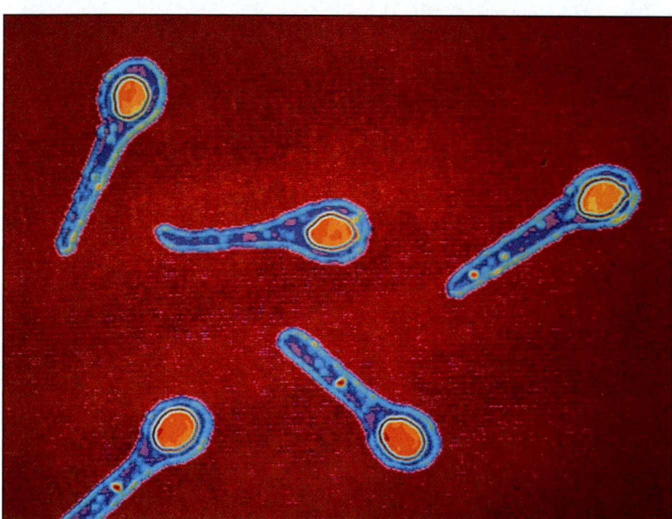

Figure 23–14 Endospores. Color-enhanced TEM of *Clostridium tetani*, the bacterium that causes tetanus. Each bacterial cell *(blue)* contains one endospore *(orange)*, a resistant, dehydrated cell that develops within the original cell. *(Alfred Pasieka/Peter Arnold, Inc.)*

Most bacteria are **heterotrophs** that must obtain organic compounds from other organisms. The majority of these heterotrophic bacteria are free-living **saprotrophs** (also called **saprobes**), organisms that get their nourishment from dead organic matter. Some heterotrophic bacteria obtain their nourishment from living organisms, in some cases harming them by causing diseases and in other cases benefiting them, for example by producing vitamins needed by the host.

Some bacteria are **autotrophs;** they are able to manufacture their own organic molecules from simple raw materials. **Photosynthetic autotrophs,** or simply **photoautotrophs,** obtain their energy from light. **Chemosynthetic autotrophs,** or **chemoautotrophs,** obtain energy by oxidizing inorganic chemicals.

Whether they are heterotrophs or autotrophs, most bacterial cells are **aerobic** (like animal and plant cells), requiring atmospheric oxygen for cellular respiration. Some bacteria are **facultative anaerobes,** meaning that they can use oxygen for cellular respiration if it is available but can also carry on metabolism anaerobically when necessary. Other bacteria are **obligate anaerobes** that can carry on energy-yielding metabolism only anaerobically. Certain obligate anaerobes are actually killed by even low concentrations of oxygen. Some bacteria respire with terminal electron acceptors other than oxygen, for example, sulfate (SO_4^{2-}), nitrate (NO_3^-), or iron (Fe^{2+}).

ARCHAEA AND EUBACTERIA ARE FUNDAMENTALLY DISTINCT

Under a microscope, most bacteria appear similar in size and form. However, using sequence analysis of small subunit ribosomal RNA (SSU rRNA), Carl Woese concluded that there are two fundamentally different groups of bacteria (see Chapter 22). He hypothesized that ancient prokaryotes split into two lineages early in the history of life. Based on Woese's work, microbiologists now classify the modern descendants of these two ancient lines in two domains: the **Archaea** and the **Eubacteria,** also called **Bacteria.** Domain Archaea includes a group of prokaryotes that produce methane gas from simple carbon sources and two groups that live in extreme environments. Domain Eubacteria comprises all other prokaryotes (Fig. 23–15). In this system, all eukaryotes are classified in domain **Eukarya.**

The most important feature distinguishing the Archaea from Eubacteria is the absence of peptidoglycan in their cell walls (Table 23–2). In some ways, the Archaea are more like the eukaryotes than they are like the Eubacteria. For example, they do not have the simple RNA polymerase found in Eubacteria. Like eukaryotes, their transcription process is not sensitive to the antibiotic rifamycin. This antibiotic specifically inhibits the type of RNA polymerase found in Eubacteria. The translational mechanisms of archaebacteria are also more like those of eukaryotes. However, the lipids in archaebacterial plasma membranes differ somewhat in structure (ether-linked rather than ester-linked) from those of both Eubacteria and Eukarya.

More than 1220 genera and 4000 species of prokaryotes have been classified. The editors of *Bergey's Manual of Systematic Bac-*

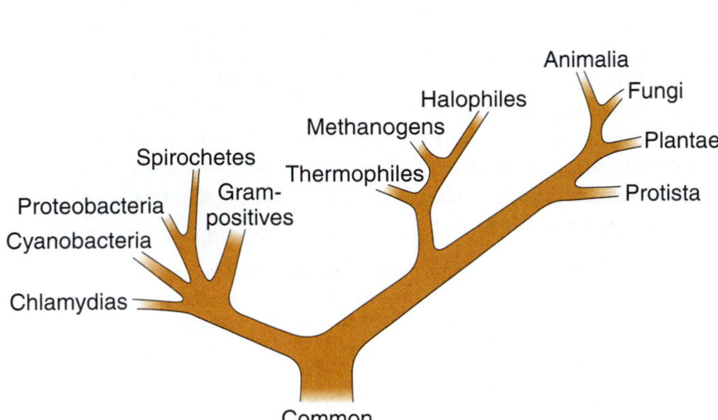

Figure 23–15 Three domains. Representative groups of each domain are shown. Gram-positive bacteria include actinomycetes, lactic acid bacteria, mycobacteria, streptococci, staphylococci, and clostridia. Spirochetes, proteobacteria, cyanobacteria, and chlamydias are gram-negative bacteria.

teriology,[1] considered the definitive reference text by microbiologists, have divided prokaryotes into several groups based on their ribosomal RNA. Each group has distinctive nucleotide sequences known as **signature sequences** in their ribosomal RNA. Several representative groups of prokaryotes are described in Table 23–3.

The Archaea include methanogens, extreme halophiles, and extreme thermophiles

Many extant members of the Archaea survive in environments thought to be similar to conditions on early Earth. The **methanogens** are a large, diverse group that inhabit oxygen-free environments in sewage and swamps and are common in the digestive tracts of humans and other animals. They are strict anaerobes that produce methane gas from simple carbon compounds. The methanogens are important in recycling components of organic products of organisms that inhabit swamps.

Extreme halophiles are heterotrophs that live in saturated brine solutions such as salt ponds (Fig. 23–16). Some extreme halophiles capture the energy of sunlight with a purple pigment (bacteriorhodopsin) that is very similar to the pigment rhodopsin involved in animal vision. The process is different from photosynthesis carried out by plants, algae, or cyanobacteria.

Extreme thermophiles normally grow in hot (45°–110° C), sometimes acidic, environments. One species is found in the hot sulfur springs of Yellowstone Park at temperatures near 60° C and pH values of 1 to 2 (the pH of concentrated sulfuric acid). Others inhabit volcanic areas under the sea. One species, found in hot deep-sea vents on the sea floor, lives at temperatures from 80° to 110° C.

Members of the Archaea also inhabit less extreme conditions. For example, they are abundant in the soil and in the cold ocean surface waters near Antarctica. Microbiologists have suggested that Archaea may be important in biogeochemical cycles and in marine food chains. The three main groups of Archaea are summarized in Table 23–3.

Eubacteria are the most familiar prokaryotes

The Eubacteria are widely distributed in the environment and are better known to microbiologists than the Archaea. Several groups of eubacteria are described in Table 23–3.

Bacteria are of great ecological importance

Eubacteria are vital members of the biosphere. Bacterial photosynthesis accounts for up to half of primary production in aquatic habitats. Like algae and plant cells, cyanobacteria carry out photosynthesis, using water as the electron source and generating oxygen. Other photosynthetic bacteria use compounds other than water, for example, hydrogen sulfide, as the electron source and do not produce oxygen.

[1] *Bergey's Manual of Systematic Bacteriology,* 2nd ed., Springer-Verlag, New York, 2000.

TABLE 23–2 Comparison of the Three Domains

Characteristic	Eubacteria	Archaea	Eukarya
Peptidoglycan in cell wall	Present	Absent	Absent
Nuclear envelope	Absent	Absent	Present
Membrane-bounded organelles	Absent	Absent	Present
Simple RNA polymerase	Present	Absent	Absent
70S ribosomes*	Present	Present	Absent
Flagellum (if present) has single filament	Yes	Yes	No
Transcription is sensitive to the antibiotic rifamycin	Yes	No	No

*The number 70S refers to the sedimentation coefficient (a measure of relative size) when centrifuged.

Figure 23–16 Sea water evaporating ponds. These salt ponds near San Francisco Bay are colored pink, orange, and yellow from the large number of extreme halophiles that inhabit them. The colors are a result of pigments (carotenoids) in the cell wall. These bacteria are harmless, and the ponds are used to produce salt commercially. *(Helen E. Carr, Biological Photo Service)*

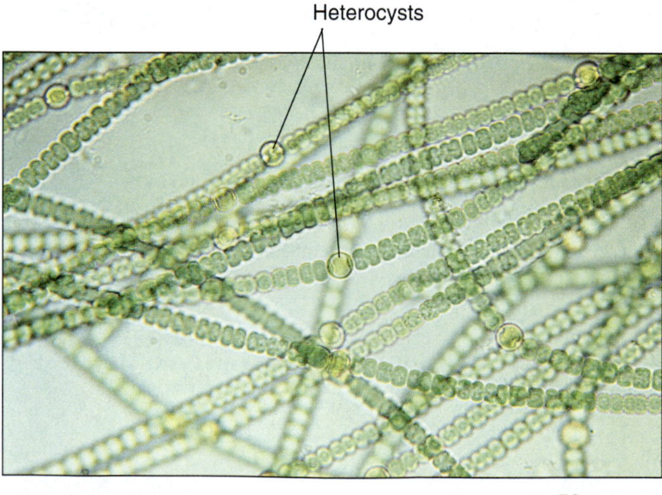

Figure 23–17 LM of *Anabaena*, a filamentous cyanobacterium that fixes nitrogen. Nitrogen fixation takes place in the rounded cells, called heterocysts. *(Dennis Drenner)*

Bacteria, especially actinomycetes and myxobacteria, are the most numerous inhabitants of soil. As described in the chapter introduction, bacteria are essential as decomposers and in recycling nutrients, including nitrogen, oxygen, carbon, phosphorus, sulfur, and many trace elements. (A discussion of the roles of bacteria in biogeochemical cycles, particularly the nitrogen cycle, is found in Chapter 53.)

Because nitrogen is constantly removed from the soil by plants and other natural processes, as well as by human activity, it must be continually added to the soil. Plant growth depends on the availability of usable nitrogen. Several types of bacteria transform atmospheric nitrogen to forms that can be used by plants (Fig. 23–17).

MAKING THE CONNECTION How does the mutualistic relationship between certain bacteria and plants benefit agriculture? Rhizobial bacteria (bacteria in the genus *Rhizobium*) form symbiotic associations with the roots of legumes, a large family of herbs, shrubs, and trees. Legumes include such important crops as peas, beans, lentils, soybeans, and peanuts, as well as clover and alfalfa, which are grown for livestock feed and to fertilize the soil.

Rhizobial bacteria are motile bacilli that inhabit the soil. After they infect the roots of a legume, the roots develop nodules (see Fig. 53–8*a*). The nodules consist of plant tissue in which the bacteria reside and fix nitrogen. Rhizobial bacteria and the roots of legumes form a mutualistic type of symbiotic relationship. In a mutualistic relationship both partners benefit (see Chapter 52). The bacteria living in nodules supply the plant with all the nitrogen it requires, and the plant provides the bacteria with organic compounds, including sugar needed for cellular respiration.

Because legumes, like other plants, produce sugar by photosynthesis, a correlation exists between photosynthesis and nitrogen fixation. When a legume is photosynthesizing at a higher rate, it provides more sugar for its bacterial partners. The bacteria are then able to fix larger amounts of nitrogen. Plants without nodules must obtain nitrogen from the soil (see Chapter 34). Because many soils are deficient in nitrogen, legumes that have formed mutualistic associations with rhizobial bacteria have a decided advantage over other plants.

Some Eubacteria cause disease

Some bacterial species have coevolved with eukaryotes and are interdependent with them. All plants and animals harbor a population of microorganisms that are considered normal microbiota—harmless symbiotic bacteria. In fact, the number of bacteria that normally inhabit the human body (approximately 700 trillion cells) exceeds the number of its own human cells (about 70 trillion cells). Some of these bacteria provide a beneficial service for their host. For example, the presence of certain bacterial populations prevents harmful microorganisms (including other bacteria) from flourishing. Some bacteria that inhabit the human intestine produce vitamin K and some of the B vitamins, which are then absorbed and utilized.

A small percentage of bacterial species are important pathogens of plants and animals. Some of the normal bacterial (and viral) inhabitants are opportunistic pathogens that cause disease only under certain conditions. For example, if one's immune system is compromised, opportunistic bacteria may increase in number and cause disease.

Process of Science The idea that some unknown agent causes disease was debated long before Leeuwenhoek discovered microorganisms with his microscope in the late 1600s. However, scientists did not develop the tools or the methods needed to accurately understand the relationships between bacteria and disease until much later. During the late 19th century several physicians, microbiologists, and chemists working independently laid the foundations for the science of microbiology. Chemist Louis Pasteur disproved the prevailing views of spontaneous generation by demonstrating that sterilization of a sugar and protein culture prevented bacterial growth. Pasteur also developed a rabies vaccine, showing that people can be stimulated to develop immunity to disease.

Physician Robert Koch was the first to clearly demonstrate that bacteria cause infectious disease. In 1876 he showed that *Bacillus anthracis* was the cause of anthrax, a serious disease of humans, sheep, and some other animals. Using a microscope, Koch observed the bacteria in the blood and spleens of dead sheep. When he inoculated mice with the infected sheep blood, he was able to identify *B. anthracis* in the blood of the mice. He also cultured *B. anthracis* and showed that when he injected the bacteria into healthy mice, they developed anthrax. Koch proposed a set of guidelines, now known as **Koch's postulates,** that are still used to demonstrate that a specific pathogen causes specific disease symptoms: (1) the pathogen must be present in every individual with the disease; (2) a sample of the microorganism taken from the diseased host can be grown in pure culture; (3) when a sample of the pure culture is injected into a healthy host, it causes the same disease; and (4) the microorganism can be recovered from the experimentally infected host.

Pathogenic microorganisms can enter the body in food, dust, droplets, or through wounds. Many diseases are transmitted by insect or animal bites. To cause disease, a pathogen must adhere to a specific cell type, multiply, and produce toxic substances. Adherence and multiplication can occur only when the pathogen competes successfully with the normal microbiota and counteracts the host's defenses against invasion.

Pathogens produce a variety of substances that increase their success. Some bacteria produce **exotoxins,** strong poisons that either are secreted from the cell or leak out when the bacterial cell is destroyed. The toxin, not the presence of the bacteria themselves, is responsible for the disease. As mentioned previously, diphtheria is caused by a gram-positive bacillus *(C. diphtheriae)* that is lysogenized by a phage. The diphtheria toxin kills cells and causes inflammation.

Botulism, a type of food poisoning that can lead to paralysis and sometimes death, results from ingestion of improperly canned food. Botulism is caused by an exotoxin released by the gram-positive, endospore-forming *Clostridium botulinum.* During the canning process, food must be heated sufficiently to kill any highly heat-resistant endospores that might be present. If not, the endospores can germinate. The resulting bacterial population grows and releases an exotoxin so powerful that 1 g could kill a million humans! Like many exotoxins, the one that causes botulism can be inactivated by heating. (Food must be heated to 80° C for 10 minutes, or boiled for 3 to 4 minutes.)

Endotoxins are not secreted by pathogens but are components of the cell walls of most gram-negative bacteria. These compounds affect the host only when they are released from dead bacteria. Endotoxins bind to macrophages and stimulate them to release substances that cause fever and other symptoms of infection. Unlike exotoxins, which cause specific symptoms, endotoxins appear to cause systemic symptoms such as fever. Endotoxins are not destroyed by heating.

Bacteria are used in many commercial processes

Many foods and beverages are produced with the help of microbial fermentation. Lactic acid bacteria are used in the production of acidophilus milk, yogurt, pickles, olives, and sauerkraut (Fig. 23–18). Several types of bacteria are used to produce cheese. Bacteria are involved in making fermented meats such as salami and in the production of vinegar, soy sauce, chocolate, and certain B vitamins (B_{12} and riboflavin). Bacteria are also used in the production of citric acid, a compound added to candy and to most soft drinks.

Some types of microorganisms produce antibiotics, compounds that limit competition for nutrients by inhibiting or destroying other microorganisms. By the 1950s antibiotics had become important clinical tools that transformed the treatment of infectious disease. Today about 100 clinically useful antibiotics

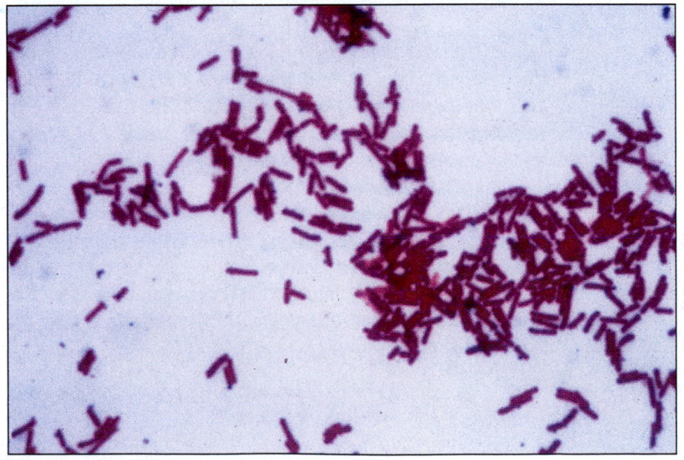

Figure 23–18 LM of lactic acid bacteria *(Lactobacillus acidophilus). (George J. Wilder/Visuals Unlimited)*

are available, and literally tons of antibiotics are produced annually. Pharmaceutical companies obtain most antibiotics from three groups of microorganisms: a large group of soil bacteria known as actinomycetes, gram-positive bacteria of the genus *Bacillus,* and molds (eukaryotes belonging to kingdom Fungi).

Because of their prolific reproduction rates, bacteria are ideal "factories" for the production of biomolecules. They have been genetically engineered (see Chapter 14) to produce certain vaccines, human growth hormone, insulin, and many other clinically important compounds. Most of the insulin used to treat diabetics is now derived from genetically engineered bacteria. Researchers are in the process of developing genetically engineered bacteria for production of many other medically and agriculturally useful products. For example, genetic engineers have inserted a bacterial gene (the *Bt* gene) into certain crops such as cotton. Plants with this gene produce a toxin that is harmful to some insect pests. These genetically engineered plants require significantly fewer pesticide applications than ordinary plants.

Bacteria are used in sewage treatment and to break down solid wastes in landfills. They are also used in **bioremediation,** a process in which a contaminated site is exposed to microorganisms that break down the toxins, leaving behind harmless metabolic byproducts such as carbon dioxide and chlorides. More than 1000 different species of bacteria and fungi have been used to clean up various forms of pollution.

SUMMARY WITH KEY TERMS

I. A **virus,** or **virion,** is a tiny particle consisting of a DNA or RNA genome surrounded by a capsid (protein coat).
   A. Viruses are subcellular particles that cannot metabolize on their own. In the past, biologists considered them to be nonliving particles, but some now view them as life forms.
   B. Viruses may be bits of nucleic acid that originally "escaped" from other organisms. They infect all types of organisms.
      1. **Phages (bacteriophages)** are viruses that infect bacteria.
      2. Many different viruses infect humans and other animals. Examples of human viral diseases include chickenpox, mumps, warts, influenza, hepatitis, AIDS, as well as certain types of cancer. **Retroviruses,** such as HIV, use **reverse transcriptase** to transcribe their RNA genome to synthesize a DNA intermediate.
      3. Plant viruses cause serious agricultural losses. Viral diseases can be spread among plants by insect vectors.
   C. A viral reproductive cycle can be lytic or lysogenic.
      1. In a **lytic** cycle, the virus destroys the host cell. The five steps in a lytic cycle are: **attachment** to the host cell; **penetration** of viral nucleic acid into the host cell; **replication** of the viral nucleic acid; **assembly** of newly synthesized components into new viruses; and **release** from the host cell.
      2. **Temperate** viruses do not always destroy their hosts. In a **lysogenic** cycle the viral genome is replicated along with the host DNA.
         a. In some phages, the phage nucleic acid becomes integrated into the bacterial DNA; it is then called a **prophage.** Bacterial cells that carry prophages are **lysogenic** cells.
         b. In **lysogenic conversion,** bacterial cells containing certain temperate viruses exhibit new properties.

II. **Viroids** and **prions** are smaller than viruses.
   A. A viroid consists of a short strand of RNA with no protein coat.
   B. The prion consists only of protein. Prions cause **transmissible spongiform encephalopathies (TSEs).**

III. Prokaryotes are assigned to kingdom **Archaebacteria** and kingdom **Eubacteria.** Microbiologists prefer classifying the bacteria into two domains: **Archaea** and **Eubacteria.**
   A. Prokaryotic cells do not have membrane-bounded organelles such as nuclei and mitochondria. Common shapes of bacterial cells include **coccus,** or spherical; **bacillus,** or rod-shaped; and spiral. Spiral bacteria include the **vibrio,** which is a short helix; **spirillum,** a longer, rigid helix; and **spirochete,** a longer, flexible helix.
   B. Most eubacteria have cell walls composed of **peptidoglycan.** The walls of **gram-positive** bacteria are very thick and consist mainly of peptidoglycan. The cell walls of **gram-negative** bacteria consist of a thin peptidoglycan layer and a thick outer membrane resembling the plasma membrane. Some species produce a **capsule** that surrounds the cell wall.
   C. Some bacteria have **pili,** protein organelles that extend out from the cell and help bacteria adhere to one another or to certain other surfaces.
   D. Bacterial **flagella** are structurally different from eukaryotic flagella; each flagellum consists of a basal body, hook, and filament. Unlike eukaryotic flagella, they produce a rotary motion.
   E. The genetic material of a prokaryote typically consists of a circular DNA molecule, and most bacteria also have one or more **plasmids,** smaller circular fragments of DNA.
   F. Bacteria reproduce asexually by **binary fission, budding,** or **fragmentation.**
   G. Genetic material may be exchanged by **transformation, transduction,** or **conjugation.**
   H. Some bacteria form dormant, durable cells called **endospores.**
   I. Bacteria are metabolically diverse. Most are **heterotrophs;** some are **autotrophs.**
      1. The majority of heterotrophic bacteria are free-living **saprotrophs,** organisms that obtain nourishment from dead organic matter.
      2. Autotrophs may be **photoautotrophs (photosynthetic autotrophs)** that obtain energy from light or **chemoautotrophs (chemosynthetic autotrophs)** that obtain energy by oxidizing inorganic chemicals.
      3. Most bacteria are aerobic; some are **facultative anaerobes** that can metabolize anaerobically when necessary; others are **obligate anaerobes** that can carry on metabolism only anaerobically.

IV. The cell walls of **Archaea** do not have peptidoglycan, and their translational mechanisms more closely resemble eukaryotic mechanisms than those of other prokaryotes. The three main groups of Archaea are **methanogens, extreme halophiles,** and **extreme thermophiles.**
   A. Methanogens produce methane gas from simple carbon compounds. They inhabit anaerobic environments such as marshes, marine sediments, and the digestive tracts of animals.
   B. Extreme halophiles inhabit saturated salt solutions.
   C. Extreme thermophiles can inhabit environments at temperatures higher than 100° C.

V. See Table 23–3 to review some of the major groups of bacteria.
   A. Bacteria play essential ecological roles as decomposers and are important in recycling nutrients.
   B. Many bacteria are symbiotic with other organisms; some are important **pathogens** of plants and animals.
      1. Anton van Leeuwenhoek, Louis Pasteur, and Robert Koch were important pioneers in microbiology. **Koch's postulates** are a set of guidelines used to demonstrate that a specific pathogen causes specific disease symptoms.
      2. Some pathogenic bacteria produce **exotoxins;** others produce **endotoxins.**

POST-TEST

1. Pathogens (a) are agents that cause disease (b) include most bacteria (c) include most myxobacteria (d) include many autotrophs (e) are cyanobacteria that cause disease
2. The genome of a virus consists of (a) DNA (b) RNA (c) prions (d) DNA and RNA (e) DNA or RNA
3. The capsid of a virus consists of (a) protein (b) nucleic acid (c) helical proteins (d) a simple carbohydrate (e) RNA and lipid
4. Viruses that kill host cells are (a) lysogenic (b) lytic (c) viroids (d) prophages (e) temperate
5. The correct sequence in viral reproduction is (a) attachment, penetration, assembly, replication, release (b) penetration, absorption, assembly, replication, release (c) attachment, penetration, replication, assembly, release (d) release, assembly, penetration, replication, attachment (e) attachment, penetration, replication, release, assembly
6. In lysogenic conversion (a) bacterial cells may exhibit new properties (b) the host cell dies (c) prions sometimes convert to viroids (d) reverse transcriptase transcribes DNA into RNA (e) prion proteins convert to infective agents
7. Peptidoglycan is a chemical compound found in the cell walls of (a) most viroids (b) most Archaea (c) all prokaryotes (d) most Eubacteria (e) most Eukarya
8. Retroviruses (a) are DNA viruses (b) are RNA viruses (c) use reverse transcriptase to transcribe DNA into RNA (d) use prion protein to transcribe RNA (e) are DNA viruses that cause AIDS
9. Bacterial flagella (a) are homologous with eukaryotic flagella (b) are used in classifying some bacterial species (c) consist of a basal body and nine pairs of microtubules (d) are important in transduction (e) are characteristic of gram-positive bacteria
10. In conjugation (a) two bacterial cells of different mating types come together, and genetic material is transferred from one to another (b) a bacterial cell develops a bulge that enlarges and eventually separates from the mother cell (c) fragments of DNA released by a broken cell are taken in by another bacterial cell (d) a phage carries bacterial genes from one bacterial cell into another (e) walls develop in the cell, which then divides into six new cells
11. Endospores (a) are formed by some viruses (b) are extremely durable cells (c) are comparable to the reproductive spores of fungi and plants (d) cause fever and other symptoms in the host (e) are extremely vulnerable to infection by phages
12. The majority of heterotrophic bacteria are (a) free-living saprotrophs (b) photoautotrophs (c) chemoautotrophs (d) facultative anaerobes (e) obligate anaerobes
13. Which of the following do *not* belong to domain Archaea ? (a) bacteria that produce methane from carbon dioxide and hydrogen (b) thermophiles (c) halophiles (d) gram-positive bacteria (e) bacteria with cell walls that lack peptidoglycan
14. Which of the following scientists proposed a set of guidelines to demonstrate that a specific pathogen causes specific disease symptoms? (a) Leeuwenhoek (b) Pasteur (c) Koch (d) Prusiner (e) Woese
15. Bacteria that thrive in puncture wounds are likely to be (a) saprotrophs (b) photoautotrophs (c) chemoautotrophs (d) endospores (e) obligate anaerobes
16. Prions (a) consist of a short strand of RNA with no protein coat (b) are deadly viruses that have caused several hundred deaths in Uganda and other parts of Africa (c) appear to consist only of protein; no nucleic acid component has been found (d) are responsible for causing hepatitis C (e) are not transmissible from human to human
17. Bacteria that are autotrophs (a) do not require atmospheric oxygen for cellular respiration (b) manufacture their own organic molecules from simple raw materials (c) must obtain organic compounds from other organisms (d) get their nourishment from dead organisms (e) produce endospores when oxygen levels are too low for active growth
18. Rhizobial bacteria (a) lack cell walls (b) inhabit saturated salt solutions (c) inhabit the mouth and cause dental caries (d) are opportunistic pathogens that cause diseases such as boils and skin infections (e) live symbiotically in root nodules of leguminous plants
19. Lactic acid bacteria (a) have gram-negative cell walls (b) fix nitrogen under anaerobic conditions (c) were formerly known as blue-green algae (d) contribute to the unique taste of yogurt, pickles, and sauerkraut (e) secrete slime on which they glide
20. Which group of bacteria contains the gram-positive, anaerobic bacterium that causes botulism? (a) clostridia (b) actinomycetes (c) enterobacteria (d) spirochetes (e) streptococci
21. Actinomycetes (a) inhabit the mouth and digestive tract (b) are important decomposers in the soil (c) are photosynthetic bacteria (d) form endospores that produce a potent toxin (e) thrive at temperatures greater than 100° C
22. Eubacteria (a) lack peptidoglycan (b) possess a nuclear envelope (c) possess membrane-bounded organelles (d) are sensitive to the antibiotic rifamycin (e) lack simple RNA polymerase

REVIEW QUESTIONS

1. What characteristics does a virus share with a living cell? What characteristics of life are absent in a virus?
2. List the steps in a lytic cycle. Draw diagrams to illustrate your answer.
3. Contrast the sequence of events in a lysogenic cycle with that of a lytic cycle.
4. Contrast and compare viroids, prions, viruses, and prokaryotes.
5. Compare and contrast members of the Archaea with members of Eubacteria.
6. Contrast the cell wall of a gram-positive bacterium with that of a gram-negative bacterium.
7. Give the distinguishing characteristics of each of the groups of bacteria described in Table 23–3, and give examples of each group.
8. Label the diagram. Use Figure 23–9 to check your answers.
9. Fill in the table. Use Table 23–2 to check your answers.

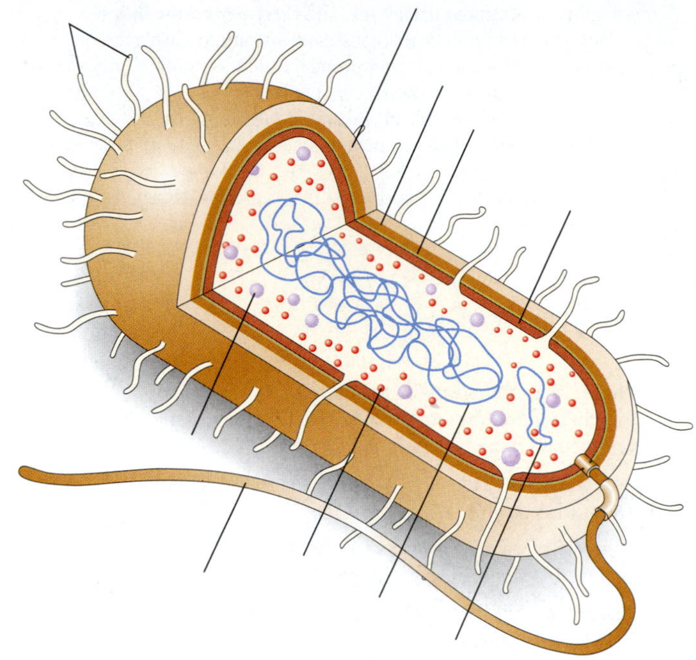

Comparison of the Three Domains

Characteristic	Eubacteria	Archaea	Eukarya
Peptidoglycan in cell wall			
Nuclear envelope			
Membrane-bounded organelles			
Simple RNA polymerase			
70S ribosomes*			
Flagellum (if present) has single filament			
Transcription is sensitive to the antibiotic rifamycin			
*The number 70S refers to the sedimentation coefficient (a measure of relative size) when centrifuged.			

YOU MAKE THE CONNECTION

1. Historically, biologists thought that viruses, because of their simple structure, evolved before cellular organisms. Later, the hypothesis that viruses escaped from cells gained favor. Recently, microbiologists have again suggested that viruses evolved early in the history of life. Based on what you have learned about viruses, present an argument for one of these hypotheses.
2. Imagine that you discover a new microorganism. After careful study you determine that it should be classified in the domain Archaea. What characteristics might lead you to this classification?
3. How might life on Earth be different if bacteria had never evolved?

RECOMMENDED READINGS

Bessen, R.A. "Neurodegenerative Prion Diseases." *Science & Medicine,* Vol. 3, No. 5, Sept./Oct. 1996. A discussion of prions and their interspecies transmission.

Carmichael, W.W. "The Toxins of Cyanobacteria." *Scientific American,* Vol. 270, No. 1, Jan. 1994. Many cyanobacteria produce unusual toxins that are deadly but potentially valuable.

Finlay, B.B. "Bacterial Virulence Factors." *Scientific American,* Vol. 272, No. 5, May/Jun. 1995. A fascinating discussion of the mechanisms that make pathogenic bacteria successful.

Jarrel, K.F., D.P. Bayley, J.D. Correia, and N.A. Thomas. "Recent Excitement about the Archaea." *Bioscience,* Vol. 49, No. 7, Jul. 1999 . A discussion of the importance of the Archaea in both basic biological research and in biotechnology.

Krieg, A.M. "Human Endogenous Retroviruses." *Science & Medicine,* Vol. 4, No. 2, Mar./Apr. 1997. The genomes of humans and other vertebrates include retroviral sequences, some of which can express viral proteins.

Losick, R., and D. Kaiser. "Why and How Bacteria Communicate." *Scientific American,* Vol. 276, No. 2, Feb. 1997. An overview of the chemical signals bacteria send to one another and to their hosts.

Perry, J.J., and J.T. Staley. *Microbiology: Dynamics and Diversity.* Saunders College Publishing, Philadelphia, 1997. An excellent introduction to microbiology.

Prusiner, S.B. "Prion Diseases and the BSE Crisis." *Science,* Vol. 278, 10 Oct. 1997. A discussion of the mechanisms by which prion proteins may become infective agents; includes a discussion of prevention.

Wilhelm, S.W., and C.A. Suttle. "Viruses and Nutrient Cycles in the Sea." *Bioscience,* Vol. 49, No.10, Oct. 1999. Viruses destroy the cells of their prokaryotic and eukaryotic hosts, releasing their contents into the pool of organic matter in the ocean. In this way, viruses affect the flow of carbon and other nutrients.

Woese, C.R., O. Kandler, and M.L. Wheelis. "Towards a Natural System of Organisms: Proposal for the Domains Archaea, Bacteria, and Eucarya." *Proceedings of the National Academy of Science,* Vol. 87, Jun. 1990. Proposal of a classification system consisting of three domains.

- Visit our Web site at **http://www.info.brookscole.com/solomonbergmartin** for links to chapter-related resources on the World Wide Web. Additional on-line materials relating to this chapter can also be found on our Web site.

See chapter activity on BioActive Learner CD for additional help in mastering the chapter's material. Icon location in the chapter's margins shows which topics have tutorials or simulations in the CD.

24

The Protists

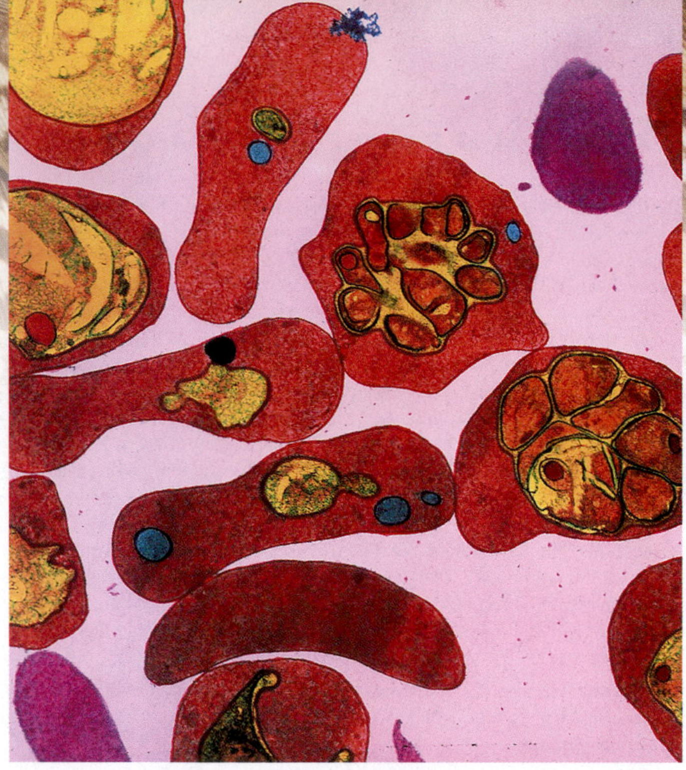

Color-enhanced TEM of human red blood cells filled with malarial parasites. Merozoites, a stage of *Plasmodium,* the parasite that causes malaria, multiply inside human red blood cells, safe from the body's immune system. *Plasmodium* is a type of protist classified as an apicomplexan. *(Gopal Murti/Science Photo Library/Photo Researchers, Inc.)*

LEARNING OBJECTIVES

After you have studied this chapter you should be able to

1. Characterize the features common to the members of kingdom Protista and explain why there are several different classification schemes for protists currently in use.
2. Discuss in general terms the diversity inherent in the protist kingdom, including modes of nutrition, body forms, motility, and reproduction.
3. Briefly describe and compare these representative protozoa: amoebas, foraminiferans, actinopods, zooflagellates, ciliates, and apicomplexans.
4. Briefly describe and compare these representative algae: euglenoids, dinoflagellates, diatoms, golden algae, brown algae, green algae, and red algae.
5. Briefly describe and compare these representative fungus-like protists: plasmodial slime molds, cellular slime molds, and water molds.
6. Discuss the uncertainties surrounding the evolutionary relationships both among the protists and between the protists and other eukaryotic kingdoms.
7. Describe the endosymbiont theory and briefly explain some of the evidence that supports it.

In Chapter 22 you learned that the kingdom **Protista** was first proposed in 1866 to include bacteria and other microorganisms that differ from plants and animals. When Robert Whittaker recommended the five-kingdom system of classification in 1969, only unicellular eukaryotic organisms were placed in kingdom Protista. The boundaries of this kingdom have expanded since then to include any eukaryote that is not a plant, animal, or fungus. More recently, some biologists have proposed that the kingdom Protista be split into several kingdoms based largely on molecular data that have clarified certain evolutionary relationships within the protists. Many relationships among protists remain uncertain, however.

Kingdom Protista currently consists of a vast assortment of primarily aquatic eukaryotic organisms whose diverse body forms, types of reproduction, modes of nutrition, and lifestyles make them difficult to characterize. **Protists** are unicellular, colonial, or simple multicellular organisms that possess a eukaryotic cell organization. The word *protist,* from Greek, meaning "the very first," reflects the idea that protists were the first eukaryotes to evolve (discussed later in the chapter).

Eukaryotic cells, the unifying feature of protists, are also characteristic of complex multicellular organisms belonging to the kingdoms Fungi, Animalia, and Plantae. However, possessing a eukaryotic cell structure clearly differentiates protists from members of the prokaryotic kingdoms Eubacteria and Archaebacteria. Recall from Chapter 4 that, unlike prokaryotic cells, eukaryotic cells possess nuclei and other membrane-bounded organelles such as mitochondria and plastids, 9 + 2 flagella, and multiple chromosomes in which DNA is complexed with proteins. Sexual reproduction, meiosis, and mitosis are also characteristic of eukaryotes.

Size varies considerably within the protist kingdom, from microscopic protozoa to giant kelps, which are brown algae that can reach 75 m (about 250 ft) in length. Although most protists are unicellular, some form **colonies** (loosely connected groups of cells), some are **coenocytic** (consisting of a multinucleate mass of cytoplasm), and some are **multicellular** (composed of many cells). Unlike animals, plants, and many fungi, most multicellular protists have relatively simple body forms without specialized tissues.

Because of their huge numbers, members of kingdom Protista are important to the natural balance of the living world. Protists are an important source of food for other organisms, and photosynthetic protists supply oxygen to aquatic and terrestrial ecosystems. Certain protists are economically important, while others cause diseases *(see photograph).*

Biologists currently recognize as many as 60 phyla of protists. Consideration of all protist phyla is beyond the scope of this text, but we will discuss 16 representative phyla. As an aid to learning, we divide the protists into three main groups based on overall characteristics: protozoa, algae, and fungus-like protists. These three groups exist strictly for convenience, and this presentation in no way indicates close evolutionary relationships within a given group; in many cases, similarities in characteristics within a group are the result of **convergent evolution** (see Chapter 17). For example, the brown algae are thought to be more closely related to water molds than to red algae or green algae. For purposes of discussion, however, we consider the brown algae with other algae, and the water molds with other fungus-like protists.

PROTISTS HAVE EVOLVED DIVERSE CELL STRUCTURES, ECOLOGICAL ROLES, AND LIFE HISTORIES

Size and structural complexity are not the only variable features of protists. During the course of evolutionary history, organisms in the kingdom Protista have evolved diversity in their (1) means of locomotion, (2) ways of obtaining nutrients, (3) interactions with other organisms, (4) habitat, and (5) modes of reproduction.

Protists, most of which are motile at some point in their life cycle, have various means of locomotion. They may move by pushing out cytoplasmic extensions **(pseudopodia)** as an amoeba does, by flexing individual cells, by gliding over surfaces, by waving **cilia** (short, hairlike organelles), or by lashing **flagella** (long, whiplike organelles). Some protists use a combination of two or more means of locomotion, for example, both flagella and pseudopodia.

Methods of obtaining nutrients differ widely in kingdom Protista. Most of the algae are autotrophic and photosynthesize as plants do. Some heterotrophic protists obtain their nutrients by absorption, as fungi do, whereas others resemble animals in that they ingest food. Some protists switch their modes of nutrition and are autotrophic at certain times and heterotrophic at others.

Although many protists are free-living, others form stable symbiotic associations with unrelated organisms. These intimate associations range from *mutualism,* a more or less equal partnership where both partners benefit, to *commensalism,* where one partner benefits and the other is unaffected, to *parasitism,* where one partner (the parasite) lives on or in another (the host) and is metabolically dependent on it (see Chapter 52). Some parasitic protists are important pathogens (disease-causing agents) of plants or animals. Specific examples of symbiotic associations involving protists are described throughout this chapter.

Most protists are aquatic and live in the ocean or in freshwater ponds, lakes, and streams. They make up most of the **plankton,** the floating, often microscopic organisms that inhabit surface waters and are the base of the food web in aquatic ecosystems. Other aquatic protists attach to rocks or other surfaces in the water. Even parasitic protists are aquatic because they live in the watery environments of other organisms' body fluids. Terrestrial protists are restricted to damp places such as soil, cracks in bark, and leaf litter.

Reproduction is varied in the kingdom Protista. Almost all protists reproduce asexually, and many also reproduce sexually, often by **syngamy,** the union of gametes. However, most protists do not develop multicellular reproductive organs, nor do they form embryos the way more complex organisms do.

Based on recent molecular and ultrastructure research, systematists propose major changes in protist classification

Two types of modern research, molecular and ultrastructure, have contributed substantially to our understanding of the phylogenetic relationships among protists. Comparative molecular data were initially obtained for the small subunit ribosomal RNA (SSU rRNA; see Chapter 23). More recently, other nuclear genes, many of which code for proteins, have been analyzed and compared for different groups. **Ultrastructure** is the use of electron microscopy to study cell structure. In many cases, ultrastructure data complement molecular data. That is, electron microscopy reveals similar structural patterns among protists that comparative molecular evidence suggests are **monophyletic.**

Based on the diversity in their structures and molecules, most biologists regard the protist kingdom as a **paraphyletic** group of organisms—that is, protists contain some, but not all, of the descendants of a common eukaryote ancestor. Molecular and ultrastructural analyses continue to help systematists clarify relationships among the various protist phyla and among protists and the other eukaryotic kingdoms (Fig. 24–1). Systematists also use these data to develop various proposals that split organisms in kingdom Protista into several kingdoms. In one proposal, for example, red algae and green algae are removed from the protists and classified within the plant kingdom. Figure 24–1 supports such a classification change, because it suggests they share a common ancestor; the node at "A" in Figure 24–1 represents the common ancestor shared by red algae, green algae, and plants.

We now examine the remarkable diversity found in 16 representative phyla of protists.

PROTOZOA ARE A DIVERSE GROUP OF UNICELLULAR HETEROTROPHIC PROTISTS

The name **protozoa** (from the Latin, meaning "first animals;" sing., *protozoon*) was originally given to animal-like organisms that are not multicellular. The term *protozoa* is used today as an

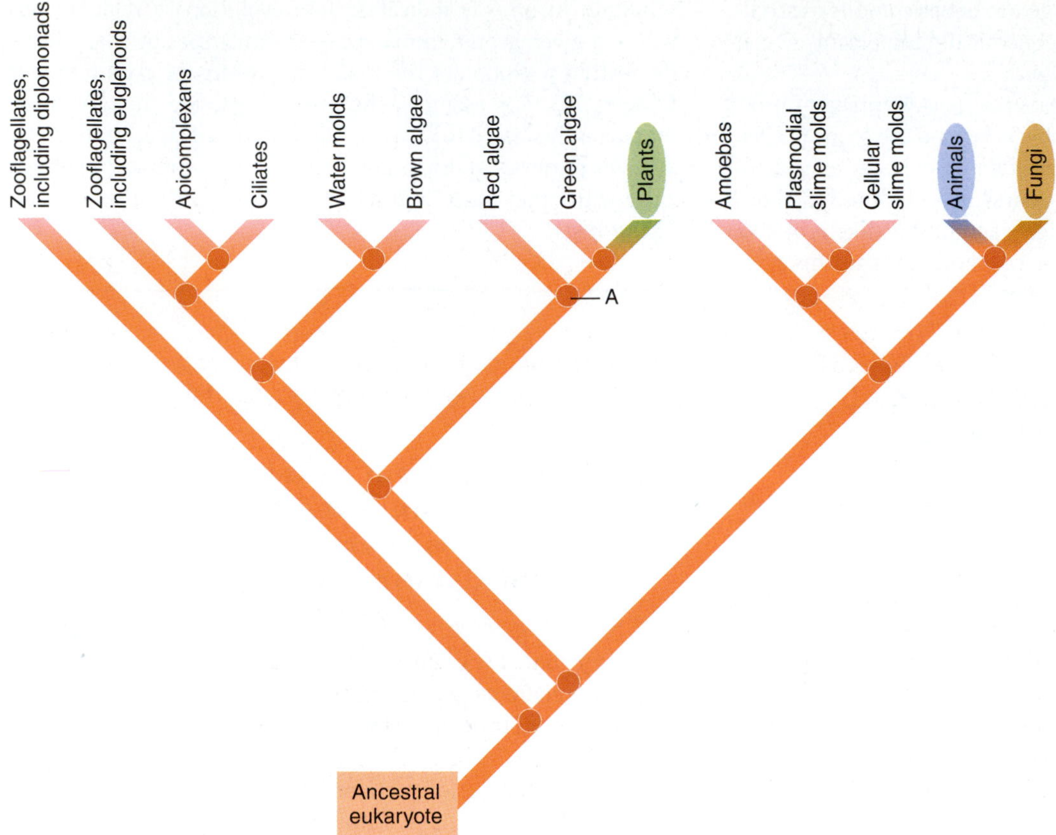

Figure 24–1 One hypothesis of evolutionary relationships among eukaryotes. Eukaryotes are a diverse group, with many relationships still unclear, and there are several interpretations of the eukaryote lineage other than the cladogram depicted here. This evolutionary tree is based on a comparison of sequences of four protein molecules among many eukaryote phyla. Foraminiferans, actinopods, dinoflagellates, diatoms, and golden algae are not shown. The encircled eukaryotes represent separate kingdoms in the six-kingdom system adopted in this text. All other eukaryotes are currently placed in the protist kingdom, which this diagram clearly shows is a paraphyletic taxon. (Node "A" is discussed in the text.) *(Based on Baldauf, S.L., A.J. Roger, I. Wenk-Siefert, and W.F. Doolittle. "A Kingdom-Level Phylogeny of Eukaryotes Based on Combined Protein Data."* Science, *Vol. 290, 3 Nov. 2000)*

informal designation for those protists that are heterotrophic and ingest food (as animals do). Protozoa are a **polyphyletic** group (they lack a common ancestor), and their evolutionary relationships are continually being evaluated as additional evidence becomes available. In this chapter we consider six phyla of protozoa: amoebas, foraminiferans, actinopods, zooflagellates, ciliates, and apicomplexans (Table 24–1).

Amoebas move by forming pseudopodia

Amoebas (phylum Rhizopoda) are unicellular organisms found in soil, fresh water, the ocean, and other organisms (as parasites). Because of the extreme flexibility of their outer plasma membrane, many members of this group have an asymmetrical body form and continually change shape as they move. (The word *amoeba* is derived from a Greek word meaning "change.") An amoeba moves by pushing out temporary cytoplasmic projections called pseudopodia (sing., *pseudopodium,* meaning "false foot") from the surface of the cell. More cytoplasm flows into the pseudopodia, enlarging them until all the cytoplasm has entered and the organism as a whole has moved. Pseudopodia are also used to capture and engulf food by surrounding and forming a vacuole around it (Fig. 24–2). This process is called **phagocytosis** (see Chapter 5). Food particles are digested when the food vacuole fuses with a lysosome containing digestive enzymes. Digested materials are absorbed from the food vacuole, which gradually shrinks as it empties. Amoebas reproduce asexually by **binary fission,** splitting into two equal parts, after mitotic division of the nucleus; sexual reproduction has not been observed.

Parasitic amoebas include *Entamoeba histolytica,* which causes amoebic dysentery, a serious human intestinal disease characterized by severe diarrhea with blood in the stools and the formation of large ulcers in the intestinal wall. In especially severe cases, the organism spreads from the large intestine and

TABLE 24–1 Representative Protist Phyla

Phylum and Common Name	Representative Genera	Morphology	Locomotion	Photosynthetic Pigments	Energy Reserves	Special Features
Rhizopoda (Amoebas)	*Acanthamoeba, Chaos, Entamoeba*	Unicellular, no definite shape	Pseudopodia	—	—	Some have shells (tests)
Foraminifera (Foraminiferans or forams)	*Globigerina*	Unicellular	Cytoplasmic projections	—	—	Pore-studded shells (tests)
Actinopoda (Actinopods)	*Actinophrys*	Unicellular	Some produce flagellated reproductive cells	—	—	Axopods protrude through pores in shell
Zoomastigina (Zooflagellates)	*Giardia, Trichonympha, Trypanosoma*	Unicellular; some colonial	One to many flagella; some amoeboid	—	—	Symbiotic forms often highly specialized
Ciliophora (Ciliates)	*Euplotes, Paramecium, Stentor*	Unicellular	Cilia	—	—	Macronuclei and micronuclei
Apicomplexa (Apicomplexans or Sporozoa)	*Plasmodium*	Unicellular	Move by flexing	—	—	Parasitic; develop resistant spores
Euglenophyta (Euglenoids)	*Euglena, Peranema*	Unicellular	Two flagella (one very short)	Chlorophylls *a* and *b*: carotenoids	Paramylon	Flexible outer covering (pellicle)
Dinoflagellata (Dinoflagellates)	*Ceratium, Gonyaulax, Gymnodinium, Pfiesteria, Protoperidinium*	Unicellular; some colonial	Two flagella	Chlorophylls *a* and *c*; carotenoids, including fucoxanthin	Oils or polysaccharides	Many covered with cellulose plates
Bacillariophyta (Diatoms)	*Pseudo-nitzschia*	Unicellular; some colonial	Most nonmotile; some glide over secreted slime	Chlorophylls *a* and *c*; carotenoids, including fucoxanthin	Oils or carbohydrates	Silica in shell
Chrysophyta (Golden algae)	*Dinobryon, Emiliania*	Unicellular; some colonial	Two flagella	Chlorophylls *a* and *c*; carotenoids, including fucoxanthin	Oils or chrysolaminarin	Covered with tiny scales of silica or calcium carbonate
Phaeophyta (Brown algae)	*Laminaria, Macrocystis, Sargassum*	Multicellular	Usually two flagella on reproductive cells	Chlorophylls *a* and *c*; carotenoids, including fucoxanthin	Laminarin	Kelps have bodies with blades, stipes, and holdfasts
Chlorophyta (Green algae)	*Chara, Chlamydomonas, Codium, Micrasterias, Spirogyra, Ulva, Volvox*	Unicellular; coenocytic; colonial; multicellular	Most flagellated at some stage in life; some entirely nonmotile	Chlorophylls *a* and *b*; carotenoids	Starch	Reproduction highly variable
Rhodophyta (Red algae)	*Bossiella, Polysiphonia, Rhodymenia*	Most multicellular; some unicellular	No flagellated stages	Chlorophylls *a* and *d*; carotenoids; phycocyanin; phycoerythrin	Floridean starch	Some reef builders
Myxomycota (Plasmodial slime molds)	*Physarum*	Multinucleate plasmodium	Cytoplasmic streaming; flagellated or amoeboid reproductive cells	—	—	Reproduce by spores formed in sporangia
Acrasiomycota (Cellular slime molds)	*Dictyostelium*	Unicellular feeding stage; multicellular (slug and fruiting body)	Pseudopodia (for single cells); cytoplasmic streaming (for multicellular forms)	—	—	Aggregation of cells signaled by cyclic AMP
Oomycota (Water molds)	*Phytophthora, Saprolegnia*	Coenocytic mycelium	Biflagellate zoospores	—	—	Cellulose and/or chitin in cell walls

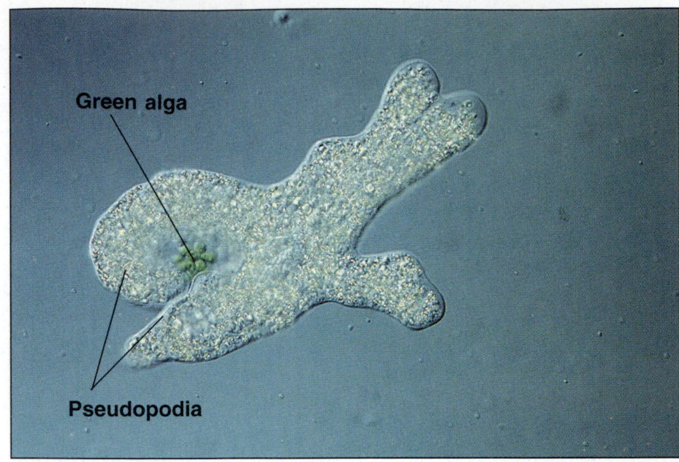

Figure 24–2 Amoebas. LM of a giant amoeba *(Chaos carolinense),* a unicellular protist that moves and feeds by means of pseudopodia, which are shown surrounding and ingesting a colonial green alga. *Chaos* amoebas are generally scavengers that feed on debris in freshwater habitats, but they ingest living organisms when the opportunity arises. *(Michael Abbey/Photo Researchers, Inc.)*

causes abscesses in the liver, lungs, or brain. According to the World Health Organization, about 48 million people are newly infected with this amoeba each year, and 70,000 people die. *Entamoeba histolytica* is transmitted as **cysts** in contaminated drinking water. (A cyst is a thick-walled, resistant, resting stage in the life history of some protists.) Other amoebas, like *Acanthamoeba,* are usually free-living but can produce opportunistic infections such as eye infections in contact lens users.

Foraminiferans extend cytoplasmic projections through tests

Almost all **foraminiferans,** informally called **forams** (phylum Foraminifera), are marine organisms that produce shells, or **tests** (Fig. 24–3*a*). The ocean contains enormous numbers of forams, which secrete chalky, many-chambered tests with pores through which cytoplasmic projections can be extended. The group gets its phylum name from this characteristic, as *foraminifera* is derived from Latin words that mean "bearing openings." The cytoplasmic projections form a sticky, interconnected net that entangles prey. Many forams contain unicellular algal **endosymbionts** (green algae, red algae, or diatoms) that provide food by photosynthesis. (An endosymbiont lives inside the body of the organism with which it has formed a close relationship.) Many foram species live on the ocean floor, but others are part of the plankton. *Globigerina,* a common organism in plankton, is a representative genus.

Dead forams settle on the bottom of the ocean, where their tests form a gray mud that is gradually transformed into chalk. With geological uplifting, these chalk formations can become part of the land, like the White Cliffs of Dover in England (Fig. 24–3*b*). (The White Cliffs of Dover are composed of the remains of a variety of carbonate organisms, not only forams.) Because foram tests are often found in rock layers covering oil deposits, geologists involved in oil exploration look for foram tests in rock strata. Forams are well preserved in the fossil record, and some are used as **index fossils,** markers to help identify sedimentary rock layers (see Chapter 17).

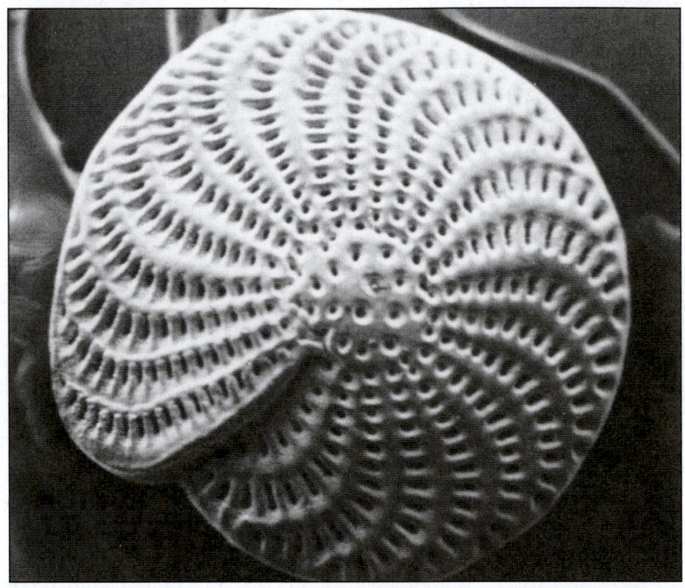

(a)

(b)

Figure 24–3 Foraminiferans. These protists with shells, or tests, live primarily in marine environments. **(a)** SEM of a foram test. Note the pores through which cytoplasm can extrude. Most forams have multichambered shells like this one. As the foram grows, the cytoplasm secretes another compartment; these compartments are secreted in a symmetrical pattern. The entire test, comprising all of its compartments, is filled with cytoplasm. **(b)** The White Cliffs of Dover, England, consist largely of the tests of forams. *(a, Biophoto Associates; b, Lynn McLaren/Photo Researchers, Inc.)*

Actinopods project slender axopods

Actinopods (phylum Actinopoda) are mostly marine plankton organisms with long, filamentous cytoplasmic projections called **axopods** that protrude through pores in their shells (Fig. 24–4*a*). A cluster of microtubules strengthens each axopod. Unicellular algae and other prey become entangled in these axopods and are engulfed outside the main body of the actinopod; cytoplasmic streaming carries the prey back within the shell. Many actinopods contain algal endosymbionts that provide them with the products of photosynthesis. *Actinophrys* is a representative genus.

Some actinopods secrete elaborate and beautiful glassy shells made of silica (Fig. 24–4*b*). Known as **radiolarians,** these actinopods are an important constituent of marine plankton. When radiolarians and other actinopods die, their shells settle and become an ooze (sediment) several meters thick on the ocean floor.

Zooflagellates move by means of flagella

Zooflagellates (phylum Zoomastigina) are mostly unicellular (a few are colonial) organisms with spherical or elongated bodies, a single central nucleus, and from one to many long, whiplike flagella that enable them to move. Zooflagellates move rapidly, pulling themselves forward by lashing flexible flagella that are usually located at the anterior (front) end. Some zooflagellates are also amoeboid and engulf food by forming pseudopodia. Other zooflagellates ingest food by means of a definite "mouth," or *oral groove,* and "throat," or *cytopharynx*.

Zooflagellates are heterotrophic and obtain their food either by ingesting living or dead organisms or by absorbing nutrients from dead or decomposing organic matter. They may be free-living or endosymbionts. For example, trichonymphs, complex, specialized zooflagellates with many flagella, live in the guts of termites and wood-eating cockroaches (Fig. 24–5*a*). These zooflagellates possess the enzymes to digest cellulose in the wood that termites or roaches eat, and both the insect and its zooflagellates obtain their nutrients from this source. (Recall from Chapter 3 that cellulose joins glucose molecules by β 1—4 linkages. These bonds are not split by the enzymes that hydrolyze starch.) Termites and wood roaches would starve to death without their flagellated endosymbionts, as would the trichonymphs, which have the ability to ingest but not to digest wood.[1] This is an excellent example of mutualism.

Some parasitic zooflagellates cause disease. For example, the zooflagellate *Trypanosoma* is a human parasite that causes African sleeping sickness (Fig. 24–5*b*). It is transmitted by the bite of infected tsetse flies. Early symptoms include recurring attacks of fever. Later, when the trypanosomes have invaded the central nervous system, infected individuals have difficulty speaking or walking; if untreated, African sleeping sickness can cause death.

Giardia is a primitive parasitic zooflagellate known as a **diplomonad.** Diplomonads are zooflagellates with one or two nuclei, no mitochondria, and one to four flagella. *Giardia intestinalis* causes backpackers' diarrhea, a common infection among campers and hikers, particularly in the mountains of the western United States. *Giardia* is eliminated as a resistant cyst in the feces of many vertebrate animals. These cysts are a common contaminant in mountain streams. Campers and hikers become

[1] The zooflagellate endosymbionts of termites and wood roaches in turn possess endosymbionts, bacteria that reside within the zooflagellates. These bacteria, rather than the zooflagellates, may produce the enzymes that digest cellulose.

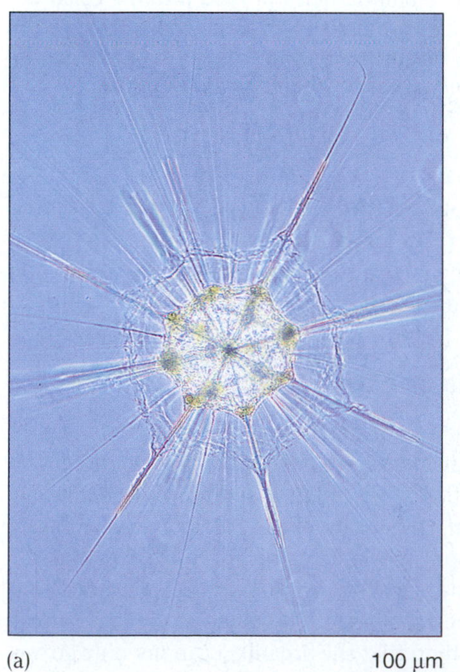
(a) 100 µm

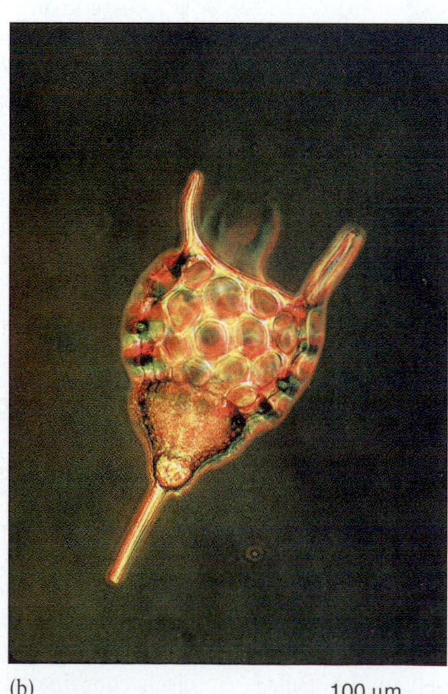
(b) 100 µm

Figure 24–4 Actinopods. **(a)** LM of an unidentified living actinopod from the Red Sea. Note the many slender axopods that project from the cell. The shell is not visible because it is covered on all sides by cytoplasm (the shell is an endoskeleton). **(b)** LM of an unidentified radiolarian, an actinopod with a glassy shell. Radiolarian axopods are generally not visible by light microscopy. *(a, Robert Brons/Biological Photo Service; b, Eric Grave, Photo Researchers, Inc.)*

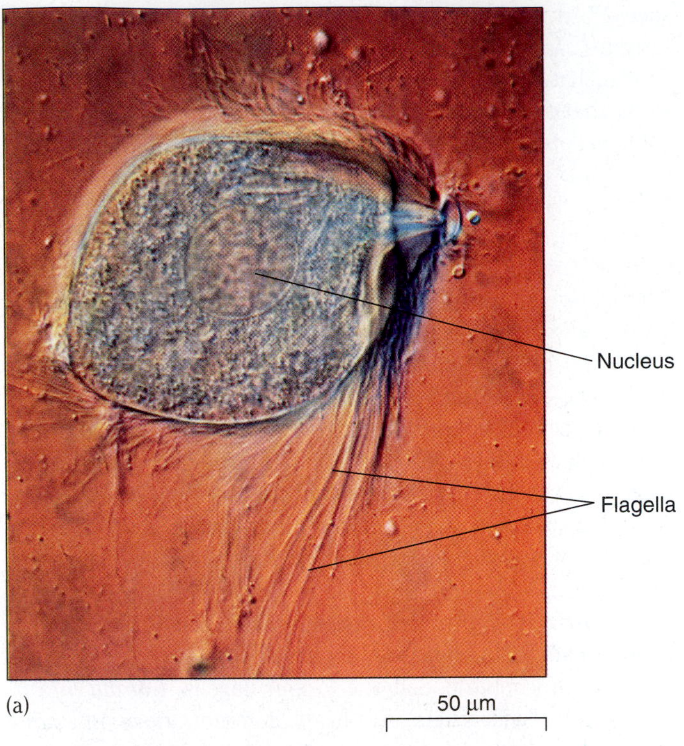

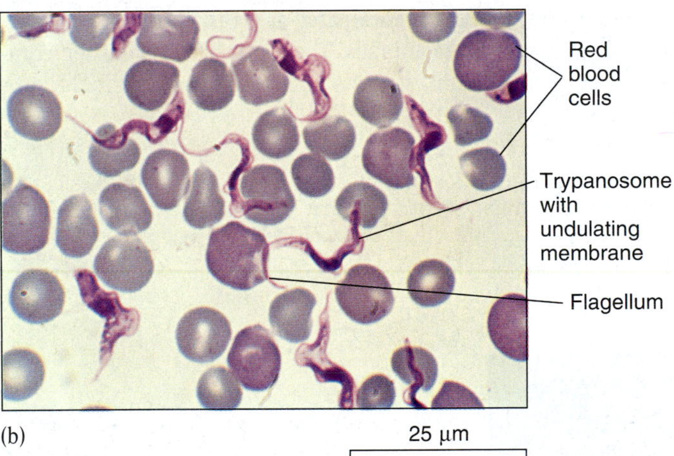

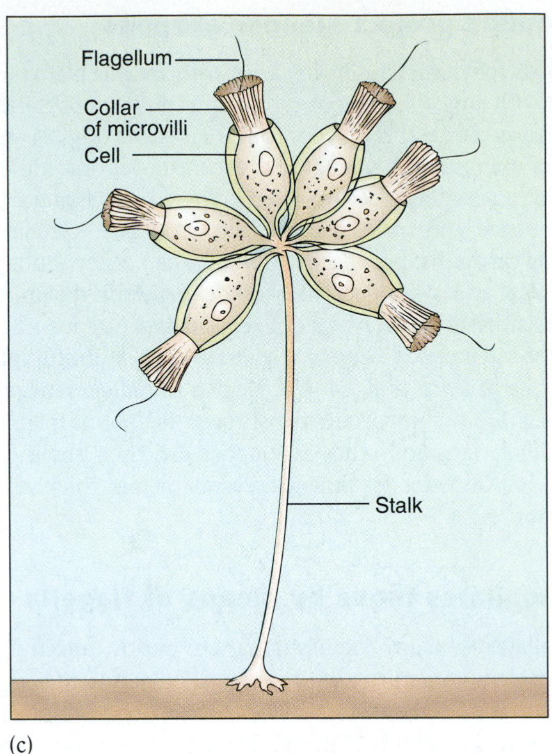

Figure 24–5 Zooflagellates. **(a)** LM of a complex zooflagellate (*Trichonympha* sp.) that lives in the gut of wood-eating termites and digests the cellulose of the wood particles eaten by its host. In this way, it obtains sugar for itself and its host. *Trichonympha* has hundreds of flagella. **(b)** LM of the zooflagellate *Trypanosoma gambiense* that causes sleeping sickness, among red blood cells in a human blood smear. **(c)** Choanoflagellates obtain food by waving their flagella, causing water currents to carry bacteria and other small particles of food into the collar of microvilli. A colonial form is shown. Each cell is 5 to 10 μm in length, not including the flagellum. *(a, M.A. Abbey/Visuals Unlimited; b, Biophoto Associates/Photo Researchers, Inc.)*

infected when they drink or rinse dishes in the "clean" mountain water. In a heavy infection, much of the wall of the small intestine is coated with these zooflagellates, which interfere with the absorption of digested nutrients and cause weight loss, abdominal cramps, and diarrhea. Unlike most protozoa, *Giardia* lacks mitochondria (discussed later in the chapter).

Choanoflagellates (collared zooflagellates) are one of the classes of zooflagellates in the phylum Zoomastigina. These marine and freshwater zooflagellates include both free-swimming and **sessile** species that are permanently attached by a thin stalk to bacteria-rich debris. Their single flagellum is surrounded at the base by a delicate collar of microvilli (Fig. 24–5*c*). They are of special interest because of their striking resemblance to collar cells in sponges (see Fig. 28–6). Other animal phyla, such as flatworms and rotifers, also contain choanoflagellate-like cells. Based on structural similarities and comparative ribosomal RNA and DNA sequence data, many biologists hypothesize that choanoflagellates are related to the ancestor of animals—that is, that living choanoflagellates and animals may share a common choanoflagellate-like ancestor.

Ciliates use cilia for locomotion

Ciliates (phylum Ciliophora) are among the most complex of cells. These unicellular organisms possess a flexible outer covering called a **pellicle** that gives them a definite but changeable shape. In *Paramecium,* the surface of the cell is covered with several thousand fine, short, hairlike organelles, called cilia, that extend through pores in the outer covering and permit movement (Fig. 24–6*a,b;* also see Fig. 1–1*a*). The cilia beat in such a precisely coordinated fashion that the organism can not only go forward but can also back up and turn around. Just under the pelli-

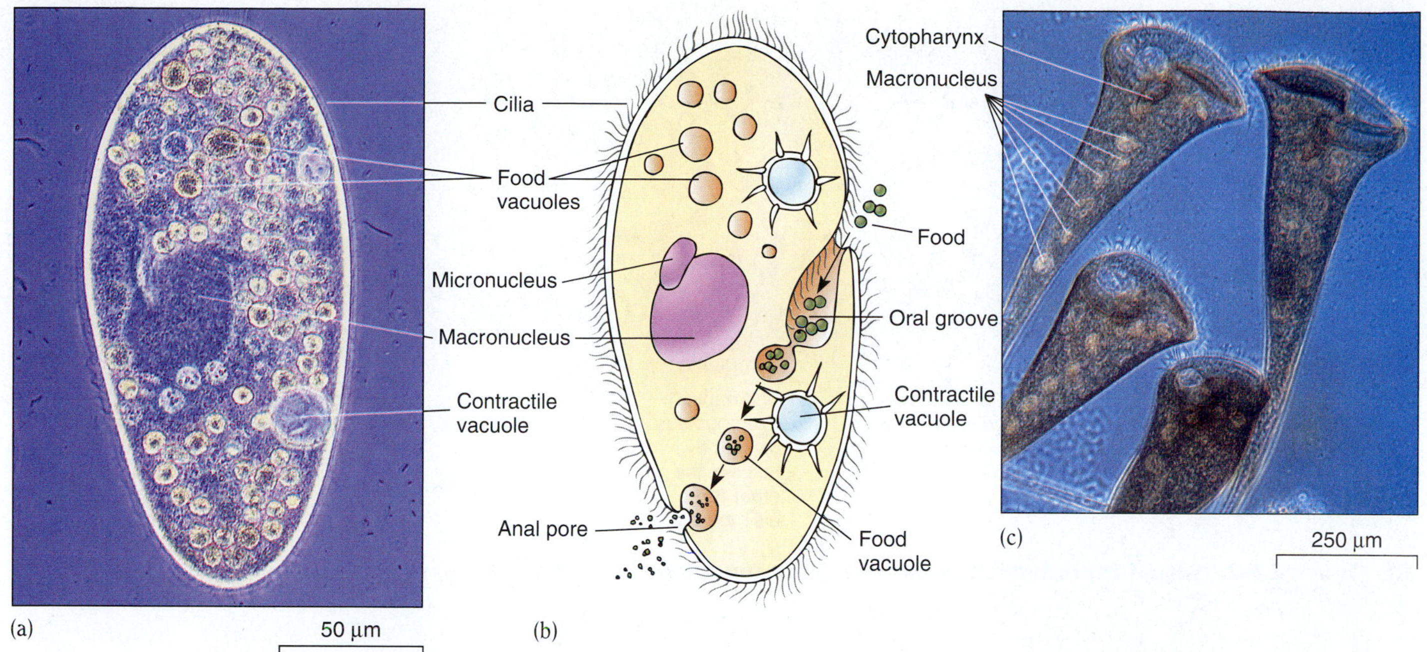

Figure 24-6 Ciliates. **(a)** Note the complex cell structure in this phase contrast LM of *Paramecium* sp., a freshwater ciliate. Like many ciliates, *Paramecium* has multiple nuclei: a macronucleus and one or more smaller micronuclei. **(b)** Food particles are swept into *Paramecium*'s ciliated oral groove and incorporated into food vacuoles. Lysosomes fuse with the food vacuoles, and the food is digested and absorbed; undigested wastes are eliminated through the anal pore. *Paramecium* absorbs water by osmosis from its freshwater surroundings, but it does not swell because contractile vacuoles fill with excess water and then contract to discharge it into the environment. **(c)** LM of *Stentor* sp., a sessile ciliate. Note the numerous cilia that direct food particles into its funnel-like cytopharynx. The macronucleus is elongated and resembles a string of beads. **(d)** LM of *Euplotes*, a hypotrich ciliate. Note the stiff ciliary tufts, called cirri. *(a, Robert Brons/Biological Photo Service; c, Eric Grave/Photo Researchers, Inc.; d, Cabisco/Visuals Unlimited)*

cle, many ciliates possess numerous small **trichocysts,** organelles that discharge filaments thought to aid in trapping and holding prey.

Not all ciliates are motile. Some sessile forms are stalked, and others, although capable of some swimming, are more likely to remain attached to a rock or other surface at one spot (Fig. 24–6*c*). Their cilia set up water currents that draw food toward them.

One group of ciliates, the **hypotrichs,** have greatly modified body cilia and exhibit an unusual creeping-darting locomotion. In hypotrichs, the cilia are absent over much of the body except on the ventral (lowermost) surface, where they occur in stiff tufts called **cirri** (sing., *cirrus,* from the Latin word meaning "a curl of hair") (Fig. 24–6*d*). Hypotrichs crawl about by means of these cirri, which beat in a coordinated manner.

Ciliates have a wide range of feeding habits and diets; most ingest bacteria or other tiny protists. Their cilia draw the food into a simple opening in some species and into a funnel-like oral groove in others. A vacuole forms around the food at the end of the opening, and the food is digested. Water regulation in freshwater ciliates is controlled by special organelles called **contractile vacuoles.** Being hypertonic to their environment, freshwater ciliates continually take in water by osmosis; the contractile vacuole continually expels excess water from the ciliate back to the environment.

Ciliates differ from other protozoa in having two kinds of nuclei: one or more small, diploid **micronuclei** that function in reproduction, and a larger, polyploid **macronucleus** that controls cell metabolism and growth (through protein synthesis). Most ciliates are capable of a sexual process called **conjugation,** in which two individuals come together and exchange genetic material. Each ciliate species is divided into several different mating types. Individuals with different mating types are identical in appearance but genetically different in terms of sexual compatibility. Because there are no known physical differences between the mating types, it is not appropriate to refer to them as "male" and "female."

During conjugation in *Paramecium,* two individuals of compatible mating types press their oral surfaces together (Fig. 24–7*a*). Within each individual the macronucleus disintegrates and the micronucleus undergoes meiosis, forming four haploid nuclei (Fig. 24–7*b,c*). Three of these degenerate, leaving one,

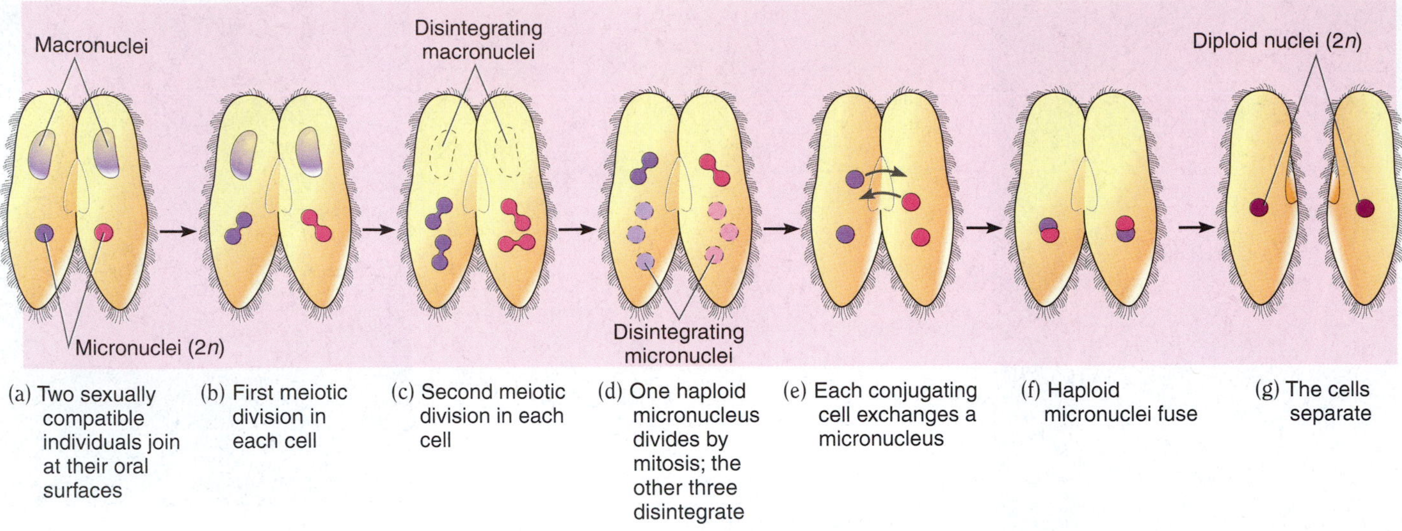
Macronuclei
Disintegrating macronuclei
Diploid nuclei (2n)
Micronuclei (2n)
Disintegrating micronuclei
(a) Two sexually compatible individuals join at their oral surfaces
(b) First meiotic division in each cell
(c) Second meiotic division in each cell
(d) One haploid micronucleus divides by mitosis; the other three disintegrate
(e) Each conjugating cell exchanges a micronucleus
(f) Haploid micronuclei fuse
(g) The cells separate

Figure 24-7 Sexual reproduction in *Paramecium caudatum*. See text for details.

which divides mitotically to form two identical haploid nuclei (Fig. 24–7*d*). One of these remains within the cell, and the other nucleus crosses through the oral region into the other individual (Fig. 24–7*e*). A type of syngamy occurs; that is, the migrant haploid micronucleus fuses with the stationary haploid nucleus in that cell (Fig. 24–7*f*), and the two ciliates separate (Fig. 24–7*g*). The new diploid micronucleus in each cell divides a varying number of times, with one micronucleus developing into a new macronucleus *(not shown)*. Thus, the normal nuclear condition of that *Paramecium* species is restored.

A single act of conjugation yields two cross-fertilizations as each cell fertilizes the other. Conjugation results in two "new" cells that are genetically identical to each other but different from what they were before conjugation. Actual cell division need not follow immediately after conjugation. Cell division is a complex process involving more than simply splitting in half, because complex organelles must be duplicated.

Ciliates also reproduce asexually, by binary fission, in which a unicellular organism undergoes mitosis and divides equally into two parts. Ciliates usually divide perpendicular to their longitudinal axis.

Apicomplexans are spore-forming parasites of animals

Apicomplexans (phylum Apicomplexa) are a large group of parasitic, spore-forming protozoa, some of which cause serious diseases such as malaria in humans. They may have evolved from parasitic dinoflagellates living in the intestines of marine invertebrates (dinoflagellates are discussed later in this chapter). Apicomplexans lack specific structures, such as cilia, flagella, or pseudopodia, for locomotion, but move by flexing. Apicomplexans possess an *apical complex* of microtubules that attaches the parasite to its host cell; the apical complex is only visible using electron microscopy. At some stage in their life cycles, apicomplexans produce **sporozoites,** small infective agents transmitted to the next host; for this reason, many biologists call these organisms **sporozoa.** Many apicomplexans spend part of their complex life cycle in one host species and part in a different host species.

Malaria, which is caused by an apicomplexan, is the world's most serious infectious disease. According to the World Health Organization, approximately 300 million to 500 million people currently have malaria, and as many as 3 million people, many of them children in developing countries, die from the disease each year. Although the disease was described for centuries in Chinese, Greek, Arabic, and Roman writings, its causal agent and mode of transmission were not identified until the end of the 19th century. Ronald Ross, an Englishman, received the Nobel Prize in 1902 for his role in elucidating the life cycle. Four species of *Plasmodium* cause malaria in humans. *Plasmodium* sporozoites enter human blood through the bite of an infected female *Anopheles* mosquito (Fig. 24–8). *Plasmodium* first enters liver cells, where it multiplies, and then red blood cells, where it continues to proliferate. When each infected red blood cell bursts, many new parasites are released. The released parasites infect new red blood cells, and the process is repeated. The simultaneous bursting of millions of red cells causes the symptoms of malaria: a chill, followed by high fever caused by toxic substances that are released and affect other organs of the body.

Malaria is currently recurring in many countries where it had been under control for decades. Formerly effective control methods—antimalarial drugs and pesticides—have lost much of their effectiveness. Chloroquine and several other antimalarial drugs are taken prophylactically to prevent malaria, but *Plasmodium* has evolved resistance to several of these in many areas, necessitating the use of a combination of drugs. Pesticides are used to control *Plasmodium's* vector, the mosquito, but mosquitoes have evolved resistance to many pesticides. New antimalarial drugs and several possible vaccines against malaria are currently

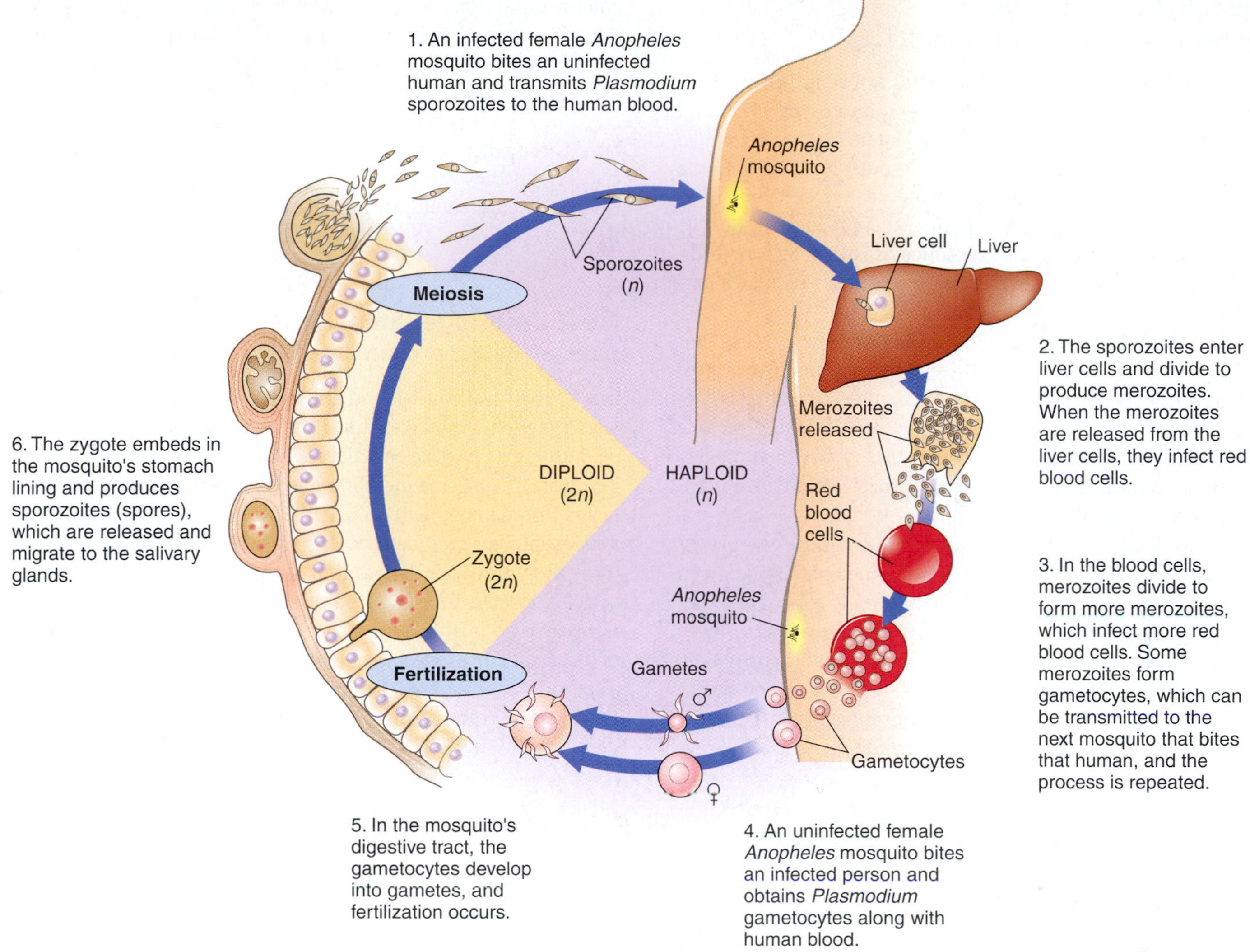
1. An infected female Anopheles mosquito bites an uninfected human and transmits Plasmodium sporozoites to the human blood.
Anopheles mosquito
Liver cell
Liver
Sporozoites (n)
Meiosis
2. The sporozoites enter liver cells and divide to produce merozoites. When the merozoites are released from the liver cells, they infect red blood cells.
Merozoites released
6. The zygote embeds in the mosquito's stomach lining and produces sporozoites (spores), which are released and migrate to the salivary glands.
DIPLOID (2n)
HAPLOID (n)
Red blood cells
Zygote (2n)
3. In the blood cells, merozoites divide to form more merozoites, which infect more red blood cells. Some merozoites form gametocytes, which can be transmitted to the next mosquito that bites that human, and the process is repeated.
Anopheles mosquito
Fertilization
Gametes
♂
♀
Gametocytes
5. In the mosquito's digestive tract, the gametocytes develop into gametes, and fertilization occurs.
4. An uninfected female Anopheles mosquito bites an infected person and obtains Plasmodium gametocytes along with human blood.

Figure 24–8 The life cycle of *Plasmodium* sp., the causal agent of malaria.

being tested. In addition, the Malaria Genome Project, which is currently sequencing the genome of *Plasmodium falciparum,* has the potential to provide new diagnostics, drugs, and vaccines.

ALGAE ARE A DIVERSE GROUP OF PHOTOSYNTHETIC PROTISTS

Algae (sing., *alga*) are an informal group of mostly photosynthetic protists that range in size from unicellular, microscopic forms to large, multicellular seaweeds. (The word *alga* is derived from a Latin word for "seaweed.")

Although most algae are autotrophic, they are not considered plants because they lack many plant structures, such as roots, stems, and leaves. Algae lack a cuticle, which is a waxy covering over the aerial parts of plants that reduces water loss. When actively growing, algae are restricted to damp or wet environments, such as the ocean; freshwater ponds, lakes, and streams; hot springs; polar ice; alpine snow; moist soil, trees, and rocks; and the bodies of certain animals, including sloths, sea anemones, corals, and worms. Also, most algae do not have multicellular **gametangia** (sing. *gametangium;* reproductive structures in which gametes are produced). Algal gametangia generally are formed from single cells.

In addition to green chlorophyll *a* and yellow and orange **carotenoids,** which are photosynthetic pigments found in all algae (and plants), different algal phyla possess a variety of other pigments that are also important in photosynthesis. Classification into phyla is largely based on their pigment composition and what kinds of materials they produce to store energy reserves (fuel molecules). Other characteristics used to classify algae include their cell wall composition, the number and placement of flagella, and chloroplast structure. In the following sections, we consider seven phyla of algae: euglenoids, dinoflagellates, diatoms, golden algae, brown algae, green algae, and red algae (see Table 24–1).

Euglenoids are unique freshwater unicellular flagellates

Most **euglenoids** (phylum Euglenophyta) are unicellular flagellates, and about one-third of them are photosynthetic (Fig. 24–9). They generally possess two flagella: one long and whiplike and one so short that it does not extend outside the cell. Some euglenoids, such as *Euglena* sp., change shape continually as they move through the water because their pellicle is flexible rather than rigid. Euglenoids reproduce asexually by longitudinal cell division; none has ever been observed to reproduce sexually.

Euglenoids have at various times been classified in the plant kingdom (with the algae) and in the animal kingdom (when protozoa were considered animals). Based on molecular data, euglenoids are probably closely related to zooflagellates. We include euglenoids in our discussion of algal protists rather than the discussion of zooflagellates because many species contain chloroplasts and photosynthesize. They have chlorophylls *a* and *b* and carotenoids, the same pigments found in green algae and plants. (Although the euglenoids have the same pigments as the green algae and plants, they do not appear to be closely related to either group, as shown in Fig. 24–1.) Their energy reserves are stored as *paramylon,* a polysaccharide. Some photosynthetic euglenoids lose their chlorophyll when grown in the dark and obtain their nutrients heterotrophically, by ingesting organic matter. Other species of euglenoids, such as *Peranema* sp., are always colorless and heterotrophic. Some heterotrophic species absorb organic compounds from the surrounding water, whereas others engulf bacteria and protists by phagocytosis; the prey are digested within food vacuoles following digestion.

Euglenoids inhabit freshwater ponds and puddles, particularly those with large concentrations of organic material. For that reason they are used as indicator species of organic pollution. Some euglenoids are also found in marine waters and mud flats.

Most dinoflagellates are a part of marine plankton

One of the most unusual protist phyla is that of the **dinoflagellates** (phylum Dinoflagellata). Most dinoflagellates are unicellular, although a few are colonial. Their cells often have intracellular (inside the plasma membrane) shells of interlocking cellulose plates impregnated with silicates (Fig. 24–10*a*). The typical dinoflagellate has two flagella: One flagellum is wrapped around a transverse groove in the center of the cell like a belt, and the other is located in a longitudinal groove (perpendicular to the transverse groove) and projects behind the cell. The undulation of these flagella propels the dinoflagellate through the water like a spinning top. Indeed, the dinoflagellates' name is derived from the Greek *dinos,* meaning "whirling." Many marine dinoflagellates are bioluminescent.

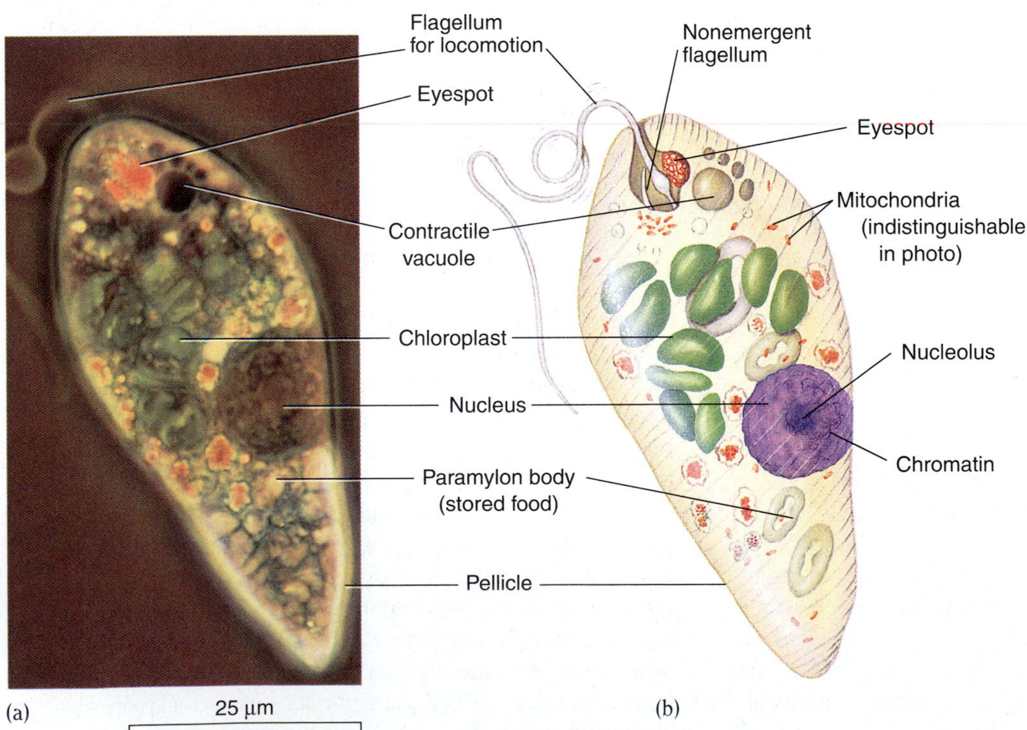

Figure 24–9 Euglenoid. **(a)** Note the complex cell structure shown in this LM of a unicellular, flagellated euglenoid *(Euglena gracilis).* **(b)** The eyespot is thought to shield a light detector at the base of the long flagellum, thereby helping *Euglena* move to light of an appropriate intensity. *Euglena*'s pellicle is flexible and enables it to change shape easily. *(Biophoto Associates)*

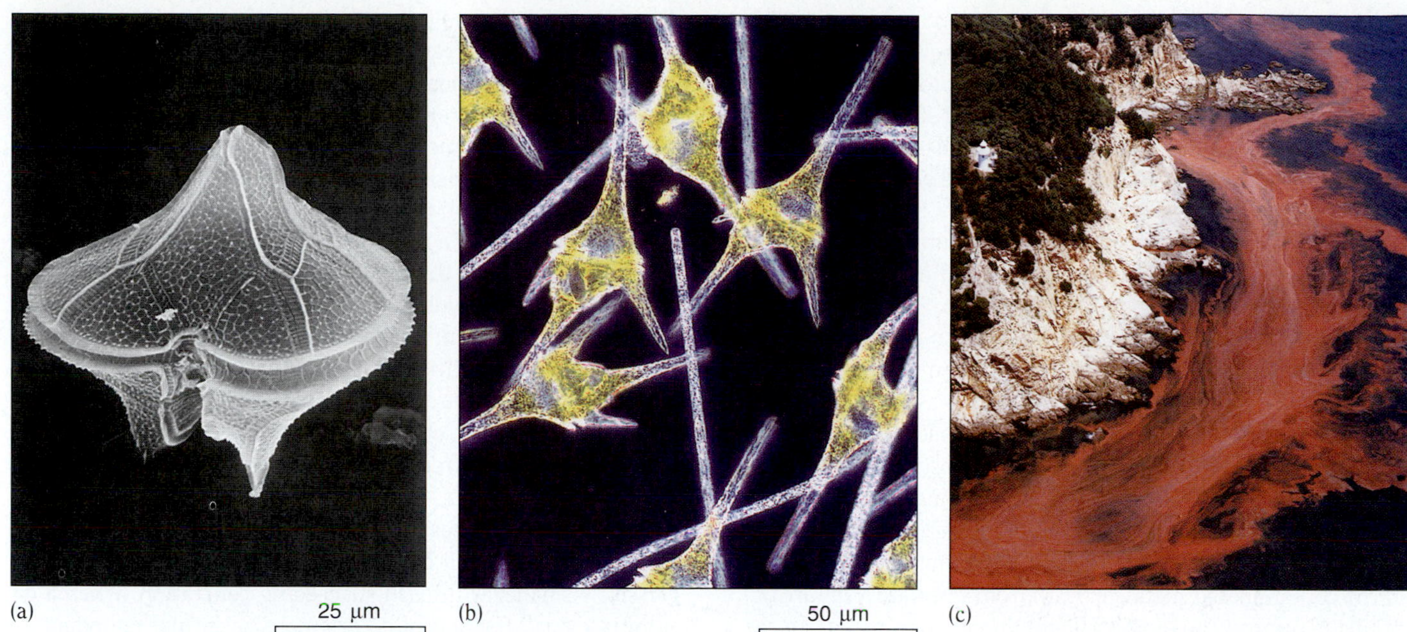

Figure 24–10 Dinoflagellates. **(a)** SEM of *Protoperidinium* sp., a unicellular, photosynthetic dinoflagellate. Note the cellulose plates that encase the unicellular body and the sutures, or junctions, between adjacent plates. The two flagella *(not visible)* are located in grooves. **(b)** LM of *Ceratium* sp. Many species of this unicellular dinoflagellate are photosynthetic (notice the green chlorophyll). The cellulose plates are difficult to see in living cells because of the pigments. A few species of *Ceratium* supplement photosynthesis by also engulfing particulate food. **(c)** Aerial view of a red tide around Matushima Island, Japan. The water is colored by the presence of billions of dinoflagellates. *(a, Courtesy of T.K. Maugel, University of Maryland; b, from* Freshwater Algae: Their Microscopic World Explored. *Bristol, United Kingdom, Biopress, 1995. Photo by Hilda Canter-Lund; c, Suisan Aviastion Company)*

Most dinoflagellates are photosynthetic and possess the photosynthetic pigments chlorophylls *a* and *c* and carotenoids, including **fucoxanthin,** a special yellow-brown carotenoid (Fig. 24–10*b*). However, a number are colorless; some of these ingest other microorganisms for food. Dinoflagellates usually store energy reserves as oils or polysaccharides.

Many dinoflagellates are endosymbionts that reside in the bodies of marine invertebrates such as mollusks, jellyfish, and corals (see Fig. 52–12). These symbiotic dinoflagellates lack cellulose plates and flagella and are called **zooxanthellae.** Zooxanthellae photosynthesize and provide carbohydrates for their invertebrate partners. The contribution of zooxanthellae to the productivity of coral reefs is substantial. Other dinoflagellates that are endosymbionts lack pigments and do not photosynthesize; these heterotrophs are parasites that live off their hosts.

Reproduction in the dinoflagellates is primarily asexual, by longitudinal cell division, although a few species have been reported to reproduce sexually. The dinoflagellate nucleus is distinctive because the chromosomes are permanently condensed and always evident. Meiosis and mitosis are unusual because the nuclear envelope remains intact throughout cell division and the spindle is located *outside* the nucleus. (The chromosomes do not make direct contact with the spindle microtubules; instead, the chromosomes appear to be attached to the nuclear envelope, and the spindle separates the new nuclei from each other.)

Ecologically, dinoflagellates are one of the most important groups of producers in marine ecosystems. A few dinoflagellates are known to have occasional population explosions, or blooms. These blooms frequently color coastal waters orange, red, or brown and are known as **red tides** (Fig. 24–10*c*). It is not known what environmental conditions initiate dinoflagellate blooms, but they are more common in the warm waters of late summer. Many experts think that human-produced coastal pollution triggers red tides, presumably by providing nutrients to the dinoflagellates. Some of the dinoflagellate species that form red tides produce a toxin that attacks the nervous systems of fishes, leading to massive fish kills. Birds such as cormorants suffer and sometimes die when exposed to the toxin from eating the dead fish. Scientists determined that a red tide of the dinoflagellate *Gymnodinium breve* was responsible for the deaths of 158 manatees on Florida's gulf coast in 1996.

Since 1991, thousands to millions of fish with open, bleeding sores have died each year in estuaries and tributaries of North Carolina and Maryland waters. Water skiers, fishermen, and scientists that came into contact with the contaminated water developed breathing difficulty, skin rashes, and temporary memory losses. These events, which have received national attention, are probably caused by a population explosion of toxic dinoflagellates that are relatively new to science. *Pfiesteria piscicida* was the first dinoflagellate of this type to be scientifically named and identified, in 1991. *Pfiesteria* and similar dinoflagellates have complex life cycles with about 24 stages, including dormant cysts, nontoxic amoeboid forms, and both nontoxic and toxic flagellated forms. The algal blooms have been circumstantially linked to nitrogen and phosphorus pollution in shallow estuaries near hog and chicken operations and municipal sewage treatment plants.

Humans sometimes get paralytic shellfish poisoning by eating filter-feeding mollusks, such as oysters, mussels, or clams, that have fed on certain dinoflagellates such as *Gonyaulax*. Recent data suggest that the toxin is not produced directly by the dinoflagellates but instead by a bacterial endosymbiont. The poison is a neurotoxin that blocks sodium channels in cells of the nervous system, preventing action potentials (see Chapter 39). Paralytic shellfish poisoning impairs breathing in humans who eat the contaminated shellfish; death from respiratory failure occasionally occurs. The dinoflagellates do not appear to harm the shellfish.

Diatoms have shells composed of two parts

Most **diatoms** (phylum Bacillariophyta) are unicellular, although a few exist as colonies. The cell wall of each diatom consists of two shells that overlap where they fit together, much like a petri dish. Silica is deposited in the shell, and this glasslike material is laid down in intricate patterns (Fig. 24–11*a*). There are two basic groups of diatoms: those with radial symmetry (wheel-shaped) and those with bilateral symmetry (boat-shaped or needle-shaped). Although some diatoms are part of the floating plankton, others live on rocks and sediments, where they move by gliding. This gliding movement is facilitated by the secretion of a slimy material from a small groove along the shell.

Diatoms contain the photosynthetic pigments chlorophylls *a* and *c* and carotenoids, including fucoxanthin; their pigment composition gives them a yellow or brown color. Energy reserves are stored as oils or the water-soluble carbohydrate *chrysolaminarin,* which is similar to the laminarin stored in brown algae (discussed later in the chapter).

Diatoms most often reproduce asexually by cell division. When a diatom divides, the two halves of its shell separate, and each becomes the larger half of a new diatom shell (Fig. 24–11*b*). Because the glass shell cannot grow, some diatom cells get progressively smaller with each succeeding generation. When a diatom is a small fraction of its original size, sexual reproduction is triggered, with the production of shell-less gametes. Sexual reproduction restores the diatom to its original size because the resulting **zygote,** a $2n$ cell that results from the fusion of n gametes, grows substantially before producing a new shell.

Diatoms are common in fresh water, but they are especially abundant in relatively cool ocean water. They are major producers in aquatic ecosystems because of their extremely large num-

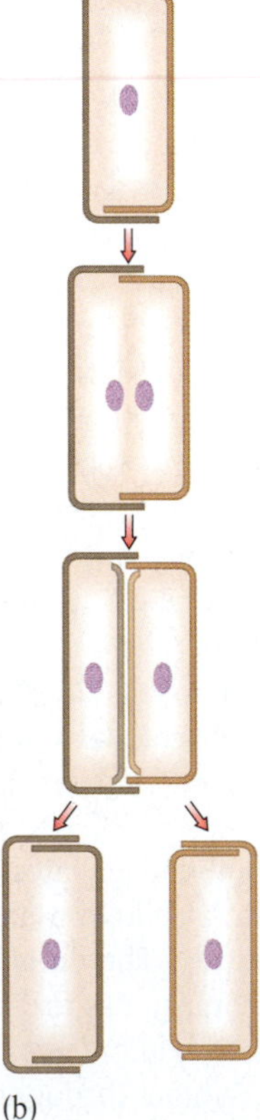

Figure 24–11 Diatoms. **(a)** LM of diatoms, unicellular algae with shells that contain silica. These algae have strikingly beautiful patterns on their symmetrical shells. **(b)** Asexual reproduction in diatoms. As cell division occurs, each new cell retains half of the original shell. The newly synthesized half of the shell always fits *inside* the original half. As a result, one of the new cells is smaller than the other. *(a, Philip Harrington/The Stock Market)*

bers. At least one species *(Pseudo-nitzchia australis)* is toxic and has been linked to shellfish poisonings, marine mammal strandings, and the deaths of more than 400 sea lions along the central California coast. When diatoms die, their shells trickle down to the ocean floor and accumulate in layers that eventually become sedimentary rock. After millions of years, some of these deposits have been exposed on land by geological upheaval. Called diatomaceous earth, these deposits are mined and used as filtering, insulating, and soundproofing material. As a filtering agent, diatomaceous earth is used to refine raw sugar and to process vegetable oils. Because of its abrasive properties, diatomaceous earth is a common ingredient in scouring powders and metal polishes; it is no longer added to most toothpastes because it is too abrasive for tooth enamel. The intricately detailed diatom shells are often used to test microscope resolution down to 1 μm.

Most golden algae are flagellated, unicellular organisms

Golden algae (phylum Chrysophyta) are a complex group found in both freshwater and marine environments. Most species are biflagellated, unicellular organisms, although some are colonial (Fig. 24–12*a*). A few lack flagella and are similar to amoebas in appearance except that they contain chloroplasts. Tiny scales of either silica or calcium carbonate may cover the cells. Reproduction in golden algae is primarily asexual and involves the production of flagellated, motile spores called **zoospores.**

Most golden algae are photosynthetic and produce the same pigments as diatoms: chlorophylls *a* and *c* and carotenoids, including fucoxanthin. The pigment composition of golden algae gives them a golden or golden brown color. As in diatoms, energy reserves are stored as oils or carbohydrates. A few species ingest bacteria and other particles of food. Ecologically, golden algae are an important group of producers in marine environments. They comprise a significant portion of the ocean's **nanoplankton,** extremely minute algae that are major producers because of their great abundance.

Classification of golden algae is controversial. Some biologists lump diatoms and golden algae in a single phylum, whereas others think they should be classified as brown algae. At the other extreme, some biologists divide the golden algae into two phyla by placing many of the marine species, such as **coccolithophorids** (Fig. 24–12*b*), in a separate phylum.

Brown algae are multicellular seaweeds

Brown algae (phylum Phaeophyta) include the giants of the protist kingdom. They are the largest and most complex of all algae commonly called seaweeds. All brown algae are multicellular and range in size from a few centimeters (an inch or so) to approximately 75 m (about 250 ft) in length. Their body forms may be branched filaments, tufts, fleshy "ropes," or thick, flattened branches. The largest brown algae, called kelps, are tough and leathery in appearance; many kelps possess leaflike **blades** in which most photosynthesis occurs, stemlike **stipes,** and rootlike anchoring **holdfasts** (Fig. 24–13*a*). They often have gas-filled floats *(bladders)* that provide buoyancy. (The blades, stipes, and holdfasts of brown algae are not homologous to the leaves, stems, and roots of plants. Brown algae and plants arose from different unicellular ancestors, as shown in Fig. 24–1.)

Brown algae are photosynthetic and possess chlorophylls *a* and *c* and carotenoids, including fucoxanthin, in their chloroplasts. The main energy storage reserve in brown algae is a carbohydrate called *laminarin.*

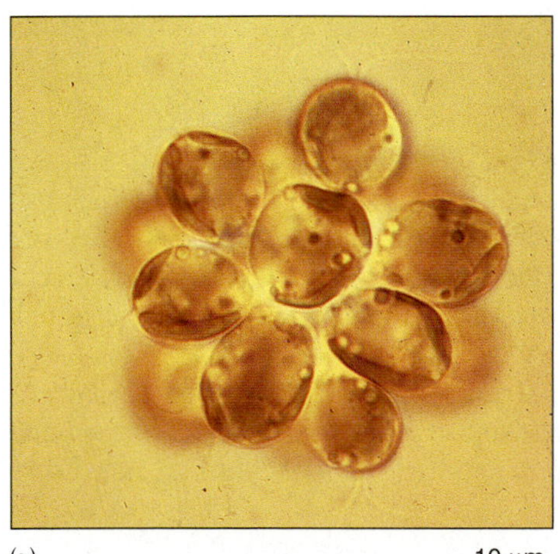

(a) 10 µm

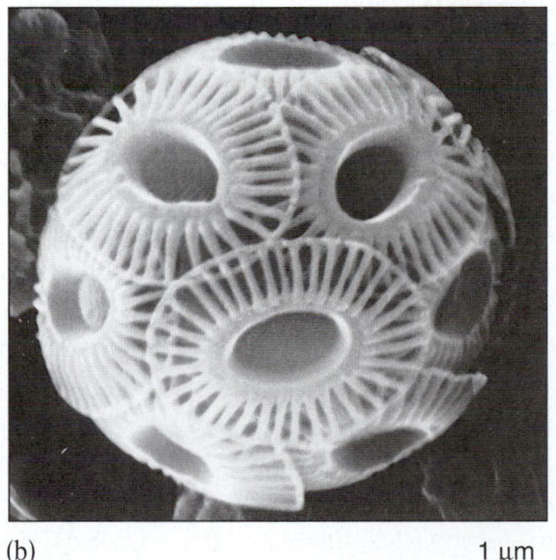

(b) 1 µm

Figure 24–12 Golden algae. **(a)** LM of a colonial golden alga (*Synura* sp.). The free-swimming colonies are composed of a few to many cells. Each cell has two flagella *(not shown). Synura* is commonly found in freshwater lakes and ponds and muddy ditches. **(b)** SEM of a coccolithophorid *(Emiliania huxleyi),* a golden alga that is an important component of nanoplankton. Tiny overlapping scales of calcium carbonate cover coccolithophorids. *(a, Philip Sze/Visuals Unlimited; b, Dr. Elizabeth Venrick/Scripps Institution of Oceanography)*

(a)

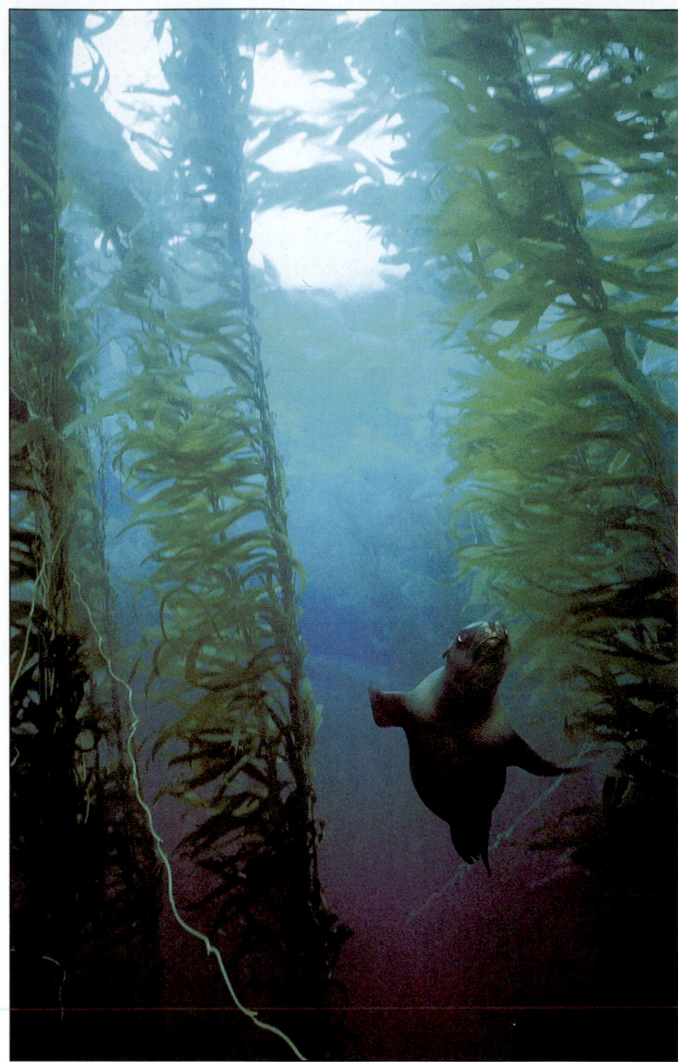

(b)

Figure 24–13 Brown algae. **(a)** *Laminaria* sp. is a representative brown alga. Note its blade, stipe, and holdfast, all of which are visible in this photograph because this individual was removed from the water. Species of *Laminaria,* which grow to 2 m (6.5 ft) in length, are widely distributed on rocky coastlines of temperate and polar seas. **(b)** A kelp *(Macrocystis pyrifera)* bed off the coast of California. These underwater forests are ecologically important, supporting many aquatic organisms, including the sea lion *(shown).* *(a, J.R. Waaland/Biological Photo Service; b, Gregory Ochocki/Photo Researchers, Inc.)*

Reproduction is varied and complex in the brown algae. Their reproductive cells, both asexual zoospores and sexual gametes, are usually flagellated. Most have life cycles that exhibit **alternation of generations,** in which they spend a portion of their life cycles as haploid organisms and a portion as diploid organisms (see Fig. 9–16*c*).

Brown algae are commercially important for several reasons. Their cell walls contain a polysaccharide called algin that possibly helps cement the cell walls of adjacent cells together. Algin is harvested from kelps such as *Macrocystis* and used as a thickening and stabilizing agent in ice creams, toothpastes, shaving creams, hair sprays, and hand lotions. Brown algae are an important human food, particularly in East Asian countries, and they are rich sources of certain vitamins and of minerals such as iodine.

Brown algae are common in cooler marine waters, especially along rocky coastlines, where they can be found mainly in the intertidal zone or relatively shallow offshore waters. Kelps form extensive underwater "forests" called kelp beds (Fig. 24–13*b*). They are essential in that ecosystem for two reasons: They are the primary food producers and they provide habitats for many marine invertebrates, fish, and mammals. The diversity of life supported by kelp beds rivals that found in coral reefs. There is also an extensive population of floating brown algae in a central area of the North Atlantic Ocean called the Sargasso Sea, named for the brown alga *Sargassum*. The Sargasso Sea is not greatly affected by the surface ocean currents rotating around the margins of the North Atlantic, and so the floating *Sargassum* remains there.

Green algae share many similarities with plants

Green algae (phylum Chlorophyta) have pigments, energy reserve products, and cell walls that are chemically identical to those of plants. Green algae are photosynthetic, with chlorophylls *a* and *b* and carotenoids present in chloroplasts of a wide variety of shapes. Their main energy reserves are stored as starch. Most green algae possess cell walls with cellulose, although some lack walls. Because of these and other similarities, it is generally accepted that plants arose from ancestral green algae (see Fig. 24–1). Using recent molecular and ultrastructure data, many systematists suggest that this extremely diverse group should be reclassified from the protists to the plant kingdom.

Green algae exhibit a variety of body types, from single cells to colonial forms, to coenocytic algae (multinucleate), to multicellular filaments and sheets (Fig. 24–14; also see Fig. 8–7*a*). The multicellular forms do not have cells differentiated into tissues, a characteristic that separates them from plants. Most green algae are flagellated or produce flagellated cells during their life history, although a few are totally nonmotile.

Reproduction in the green algae is as varied as their body forms, with both sexual and asexual reproduction, and many green algae have life cycles that exhibit alternation of generations. Asexual reproduction may be by binary fission in single cells or by fragmentation in multicellular forms. Many green algae produce spores asexually by mitosis; if these spores are flagellated and motile, they are called zoospores (Fig. 24–15).

Sexual reproduction in the green algae involves the formation of gametes in unicellular gametangia. Three types of sexual reproduction—isogamous, anisogamous, and oogamous—are

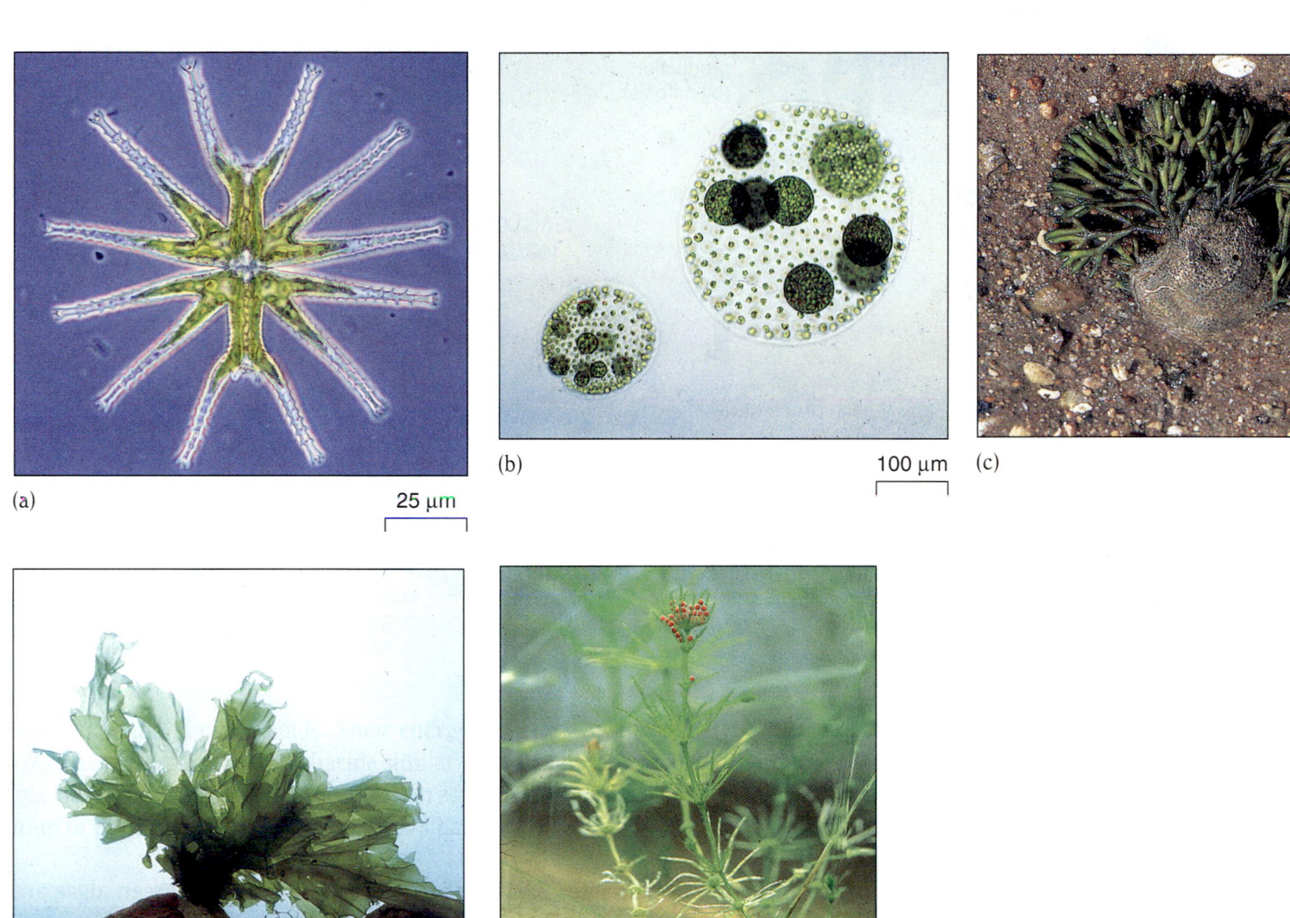

Figure 24–14 Green algae. **(a)** LM of a widely distributed desmid (*Micrasterias* sp.), a unicellular green alga with mirror-image halves. **(b)** LM of two *Volvox* colonies, each composed of up to 50,000 cells. New colonies can be observed inside the parental colonies, which eventually break apart to release them. **(c)** Dead man's fingers *(Codium fragile)*, a coenocytic green alga, exposed during low tide in Megansett Bay, Cape Cod, Massachusetts. When covered with water, the branches ("fingers") float. *Codium* is 10 to 30 cm (3.9–11.7 in) tall. **(d)** Some multicellular green algae are sheetlike. *Ulva*'s thin, leaflike form suggests its common name: sea lettuce. *Ulva* typically grows to 30 cm (11.7 in) in length, but it can grow to 1 m (39 in) in calm waters. **(e)** *Chara* sp., a green alga commonly called a stonewort, is closely related to plants. *Chara* is widely distributed in fresh water, where it grows to 30 cm (11.7 in) or longer. *(a, Biophoto Associates; b, J. Robert Waaland/Biological Photo Service; c, Andrew J. Martinez/Science Source/Photo Researchers, Inc.; d, J.M. Kingsbury; e, James W. Perry)*

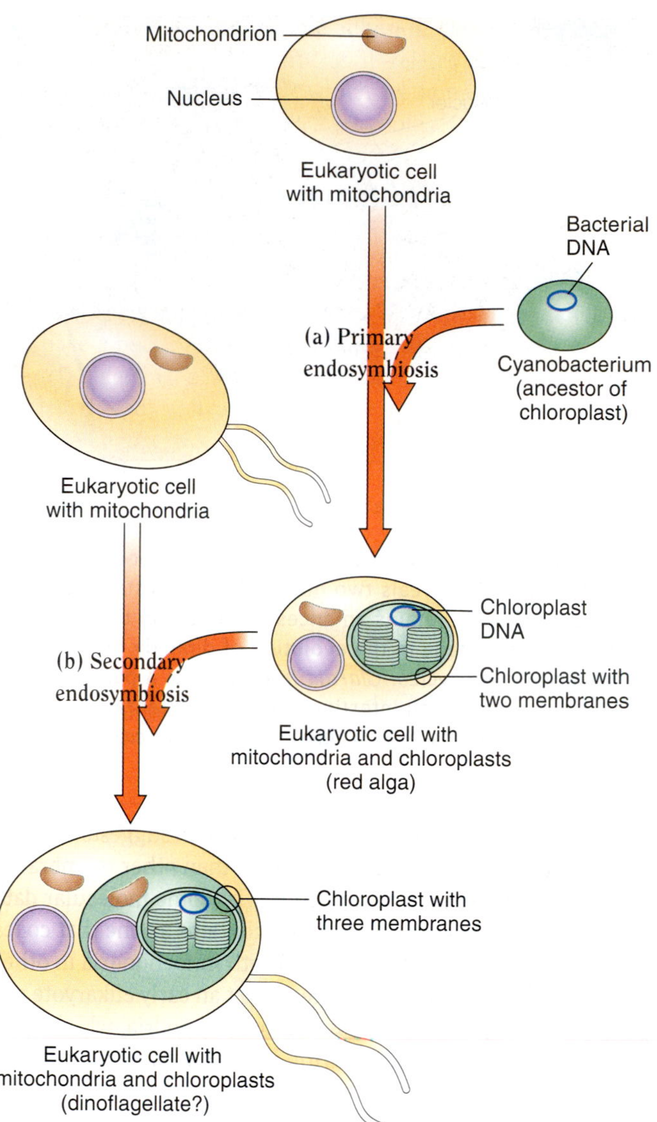

Figure 24-23 Chloroplast evolution by primary and secondary endosymbiosis. **(a)** In primary endosymbiosis, an ancient eukaryotic cell engulfed a cyanobacterium, which survived and over a vast period evolved into a chloroplast. The ancient cell is depicted as a eukaryotic cell because it is thought that mitochondria evolved before chloroplasts. **(b)** In secondary endosymbiosis, a heterotrophic eukaryotic cell (with mitochondria) engulfed a eukaryotic cell with chloroplasts (a red alga is depicted). The red alga survived and over a vast period evolved into a chloroplast surrounded by three membranes (a dinoflagellate chloroplast is depicted). Other secondary endosymbiotic events resulted in more complex chloroplast membrane structures.

nonphotosynthetic eukaryotes with their chloroplasts during *secondary endosymbiotic* events (Fig. 24–23).

Secondary endosymbiosis occurred frequently in eukaryote evolution, as evidenced by the presence of additional chloroplast membranes. For example, *three* membranes envelop the chloroplasts of euglenoids and dinoflagellates, and *four* membranes surround the chloroplasts of diatoms, golden algae, and brown algae. Understanding how these membranes originated is an essential aspect of the endosymbiont theory, and many researchers are addressing the origin of chloroplasts in different organisms.

Even nonphotosynthetic protists may contain chloroplast relics from secondary endosymbiotic events. Apicomplexans such as *Plasmodium* possess a nonfunctional chloroplast surrounded by four membranes. This chloroplast, which is probably derived from either a red alga or a green alga, has great medical potential in treating malaria because some researchers hypothesize that drugs that inhibit chloroplasts also may kill *Plasmodium*.

SUMMARY WITH KEY TERMS

I. **Protists** are "simple" eukaryotic organisms, most of which are unicellular and live in aquatic environments.
   A. Protists range in size from microscopic unicellular organisms to **colonies** (loosely connected groups of cells) to **coenocytes** (multinucleate masses of cytoplasm) to **multicellular** organisms (composed of many cells).
   B. Protists obtain their nutrients autotrophically or heterotrophically.
   C. Protists may be free-living or symbiotic, with symbiotic relationships ranging from mutualism to parasitism. An **endosymbiont** lives inside the body of the organism with which it has formed a close relationship.
   D. Many protists reproduce both sexually, often by **syngamy** (union of gametes), and asexually; others reproduce only asexually.
   E. Protists have various means of locomotion, including pseudopodia, flagella, and cilia. A few are nonmotile.
   F. Phylogenetic relationships among protists are determined largely by **ultrastructure** (the use of electron microscopy to study cell structure) and comparative molecular data. Most systematists consider the protist kingdom as a **paraphyletic** group of organisms that contains some, but not all, descendants of a common eukaryotic ancestor.

II. **Protozoa** are unicellular heterotrophic protists.
   A. **Amoebas** move and obtain food by **phagocytosis** using cytoplasmic extensions called **pseudopodia.** The parasitic amoeba *Entamoeba histolytica* causes amoebic dysentery.
   B. **Foraminiferans (forams)** secrete many-chambered **tests** (shells) with pores through which cytoplasmic projections extend to move and obtain food. Many fossils of foram tests are **index fossils,** markers that help identify sedimentary rock layers.

C. **Actinopods** are mostly marine plankton that obtain food by means of **axopods,** slender cytoplasmic projections that extend through pores in their skeletons. Actinopods with glassy shells are known as **radiolarians.**

D. **Zooflagellates** are mostly unicellular heterotrophs that move by means of whiplike **flagella.** The zooflagellate *Trypanosoma* causes African sleeping sickness. The zooflagellate *Giardia* is a primitive **diplomonad. Choanoflagellates** are zooflagellates that may be related to the ancestor of animals.

E. **Ciliates** move by hairlike **cilia,** have **micronuclei** (for sexual reproduction) and **macronuclei** (for controlling cell metabolism and growth), and undergo a complex sexual reproduction called **conjugation.** The **hypotrichs** are ciliates with body cilia arranged in stiff tufts called **cirri.**

F. **Apicomplexans** are parasites that produce **spores** and are nonmotile. They possess an apical complex of microtubules that attach the apicomplexan to its host cell. An apicomplexan *(Plasmodium)* causes malaria.

III. **Algae** are mostly autotrophic protists with **gametangia** (gamete-containing reproductive structures) formed from single cells.

A. **Euglenoids** are unicellular, flagellated algae. Many, for example, *Peranema,* are not photosynthetic.

B. **Dinoflagellates** are mostly unicellular, biflagellated, photosynthetic organisms of great ecological importance as major producers in marine ecosystems. Their cells often have intracellular shells of interlocking cellulose plants impregnated with silicates. Many dinoflagellates are endosymbionts, for example, **zooxanthellae** in corals and other marine invertebrates. Some dinoflagellates produce toxic blooms known as **red tides.**

C. **Diatoms,** which are major producers in aquatic ecosystems, are mostly unicellular, with shells containing silica. Some diatoms are part of floating **plankton,** and others live on rocks and sediments where they move by gliding.

D. **Golden algae** are mostly unicellular, biflagellated freshwater and marine algae that are of ecological importance as a major component of the ocean's extremely minute **nanoplankton. Coccolithophorids** are golden algae covered by tiny, overlapping scales of calcium carbonate.

E. **Brown algae** are multicellular seaweeds that are ecologically important in cooler ocean waters. The largest brown algae (kelps) possess leaflike **blades,** stemlike **stipes,** anchoring **holdfasts** and gas-filled bladders for buoyancy. They produce flagellated cells during their complex reproductive cycles, which involve **alternation of generations.**

F. **Green algae** exhibit a wide diversity in size, structural complexity, and reproduction. They share many similarities with plants, and it is thought that ancestral green algae gave rise to plants. Many systematists think green algae should be reclassified in the plant kingdom.

G. **Red algae,** which are mostly multicellular seaweeds, are ecologically important in warm tropical ocean waters. Red algae that incorporate calcium carbonate in their cell walls are important in reef building. Red algae lack motile cells and have complex reproductive cycles.

IV. Fungus-like protists were originally classified with the fungi but have several protist features.

A. The feeding stage of **plasmodial slime molds** is a multinucleate **plasmodium.** Reproduction is by haploid spores produced within **sporangia.** *Physarum* is a model organism used to study many biological processes.

B. **Cellular slime molds** feed as individual amoeboid cells. They reproduce by aggregating into a **pseudoplasmodium** (slug), then forming asexual spores. *Dictyostelium* is a model organism used to study many biological processes such as **cell signaling.**

C. **Water molds** have a coenocytic **mycelium.** They reproduce asexually by forming biflagellate **zoospores** and sexually by forming **oospores.** The water mold *Phytophthora* causes serious plant diseases, such as late blight of potato.

V. Protists originated about 1.5 billion to 1.6 billion years ago and were the first eukaryotes.

A. Based on RNA sequencing data and cell ultrastructural studies, it is thought that diplomonads, a group of zooflagellates, may represent one of the more ancient lineages of extant protists.

B. According to the **endosymbiont theory,** mitochondria and chloroplasts arose from symbiotic relationships between larger cells and smaller prokaryotes that were incorporated and lived within them.

1. Mitochondria probably originated from aerobic eubacteria. Rickettsias are the closest living relatives of mitochondria.
2. Chloroplasts of red algae, green algae, and plants probably arose in a single primary endosymbiotic event in which a cyanobacterium was incorporated into a cell. Multiple secondary endosymbioses led to chloroplasts in euglenoids, dinoflagellates, diatoms, golden algae, and brown algae, and to the nonfunctional chloroplasts in apicomplexans.

POST-TEST

1. Which of the following is *not* true of the protists? (a) they are unicellular, colonial, coenocytic, or simple multicellular organisms (b) their cilia and flagella have a 9 + 2 arrangement of microtubules (c) they are prokaryotic, like the eubacteria and archaebacteria (d) some are free-living, and some are endosymbionts (e) most are aquatic and live in the ocean or in freshwater ponds
2. Amoebas move and obtain food by means of (a) pseudopodia (b) flagella (c) cilia (d) gametangia (e) trichocysts
3. Foraminiferans (a) are endosymbionts in many marine invertebrates (b) were responsible for the Irish potato famine in the 19th century (c) secrete many-chambered tests with pores through which cytoplasmic extensions project (d) possess numerous trichocysts that may aid in trapping and holding prey (e) contain phycocyanin and phycoerythrin, pigments found in no other protist group
4. Unicellular protozoa that are free-living or parasitic, move by means of flagella, and do not photosynthesize are called (a) euglenoids (b) dinoflagellates (c) myxamoebas (d) zooflagellates (e) apicomplexans
5. *Paramecium* and other ciliates often display a sexual phenomenon called (a) oogamy (b) conjugation (c) anisogamy (d) red tide (e) macronucleus
6. Parasitic protozoa that form spores at some stage in their life belong to which group? (a) actinopods (b) ciliates (c) coccolithophorids (d) apicomplexans (e) dinoflagellates
7. Malaria (a) is transmitted by the bite of a female tsetse fly (b) is caused by a parasitic zooflagellate, *Giardia intestinalis* (c) is a more serious form of amebic dysentery, caused by *Entamoeba histolytica* (d) is caused by an apicomplexan that spends part of its life cycle in the *Anopheles* mosquito and part in humans (e) is transmitted when people drink water tainted by a red tide
8. Algae characterized by two flagella, one wrapped around the center of the cell like a belt and the other projecting behind the cell, are known as (a) actinopods (b) ciliates (c) coccolithophorids (d) apicomplexans (e) dinoflagellates
9. Photosynthetic protists with shells composed of two halves that fit together like a petri dish are known as (a) golden algae (b) diatoms (c) euglenoids (d) brown algae (e) foraminiferans

10. Chloro gae, r (c) bro and go
11. Which cyanob gae (e)
12. The m blades, (c) eug
13. The fee modiur (e) myc

R E V

1. What ar classify
2. Why ar mans?
3. What ar three e
4. Disting (b) amo
5. Disting noids, a

Y O U

1. Based on place ther the dinofl
2. Indicate w brown alg The comn

25

Kingdom Fungi

Fungal spores. The rounded earthstar *(Geastrum saccatum)* releases a puff of microscopic spores after the sac, which is about 1.3 cm (0.5 in) wide, is hit by a raindrop. This fungus is common in leaf litter under trees throughout North America. *(Jeff Lepore/Photo Researchers, Inc.)*

LEARNING OBJECTIVES

After you have studied this chapter you should be able to

1. Describe the distinguishing characteristics of the kingdom Fungi.
2. Describe the body plan of a typical fungus.
3. Trace the fate of a fungal spore that lands in a favorable location, such as an overripe peach, and describe conditions that permit fungal growth.
4. Explain why chytridiomycetes may have been the earliest fungal group to have evolved from the flagellated protist hypothesized as the common ancestor of fungi.
5. List distinguishing characteristics, describe a typical life cycle, and give examples of each of the following fungal groups: chytridiomycetes, zygomycetes, ascomycetes, basidiomycetes, and imperfect fungi.
6. Characterize the unique nature of a lichen.
7. Explain the ecological significance of fungi as decomposers.
8. Describe the special ecological role of mycorrhizae.
9. Summarize some of the ways humans use fungi.
10. Identify at least three fungal diseases of plants and three of animals.

Mushrooms, morels, and truffles, delights of the gourmet, have much in common with the black mold that forms on stale bread and the mildew that collects on damp shower curtains. These life forms belong to the kingdom Fungi, a diverse group of more than 81,000 known species, most of which are terrestrial. Although they vary strikingly in size and shape, all **fungi** (sing. *fungus*) are eukaryotes; their cells contain membrane-bounded nuclei, mitochondria, and other organelles. Like the cells of bacteria, certain protists, and plants, fungal cells are enclosed by cell walls during at least some stage in their life cycle. Fungal cell walls, however, have a different chemical composition than cell walls of other organisms. In most fungi, the cell wall is composed of complex carbohydrates, including **chitin,** a polymer that consists of subunits of a nitrogen-containing sugar (see Fig. 3–11). Chitin, which is coincidentally a component of the external skeletons of insects and other arthropods, is resistant to breakdown by most microorganisms.

Fungi lack chlorophyll and are nonphotosynthetic. As heterotrophs, they cannot produce their own organic materials from a simple carbon source (carbon dioxide) but instead obtain preformed carbon molecules produced by other organisms. Fungi do not ingest food as animals do, but they secrete digestive enzymes and then absorb the predigested food (as small organic molecules) through their cell walls and plasma membranes. They obtain their nutrients from dead organic matter (as decomposers) or from other living organisms (as parasites). As decomposers, fungi, along with the bacteria (see Chapter 23), play an important ecological role.

Fungi grow best in moist habitats, but they are found universally wherever organic material is available. They require moisture to grow, and they can obtain water from a humid atmosphere as well as from the medium on which they live. When the environment becomes dry, fungi survive by going into a resting stage or by producing spores *(see photograph)* that are resistant to desiccation (drying out).

Although the optimum pH for most species is about 5.6, various fungi can tolerate and grow in environments where the pH ranges from 2 to 9. Many fungi are less sensitive to high osmotic pressures than are bacteria. As a result, they can grow in concentrated salt solutions, or in sugar solutions such as jelly, that discourage or prevent bacterial growth. Fungi also thrive over a wide temperature range. Even refrigerated food is not immune to fungal invasion.

Fungi were classified originally in the plant kingdom, but biologists today recognize that they are not plants, and recent molecular studies suggest that fungi are related more closely to animals than to plants (see Fig. 24–1). Because fungi are distinct from plants, animals, and protists in many ways, they are assigned to a separate kingdom.

MOST FUNGI HAVE A FILAMENTOUS BODY PLAN

The vegetative (nonreproductive) body plan of most fungi consists of long, branched, threadlike filaments called **hyphae** (sing., *hypha*) (Fig. 25–1*a, b*). As hyphae elongate, the fungus expands into new food sources and absorbs nutrients from its large surface area. Hyphae form a tangled mass or tissue-like aggregation known as a **mycelium** (pl., *mycelia*). The cobweb-like mold sometimes seen on bread is the mycelium of a fungus. What is not seen is the extensive mycelium that grows into the bread.

Some hyphae are **coenocytic;** that is, they are not divided into individual compartments or cells and are instead an elongated, multinucleated giant cell (Fig. 25–1*c*). Other hyphae are divided by cross walls called **septa** (sing., *septum*) into individual cells containing one or more nuclei (Fig. 25–1*d,e*). The septa of many septate fungi are perforated by a pore that may be large enough to permit organelles to flow from cell to cell.

MOST FUNGI REPRODUCE BY SPORES

Most fungi reproduce by means of microscopic **spores,** which are nonmotile (or, less commonly, motile) reproductive cells dispersed by wind, water, or animals. Spores are usually produced on specialized aerial hyphae or in fruiting structures. This arrangement permits the spores to be more easily dispersed. The aerial hyphae of some fungi form large, complex reproductive structures, called **fruiting bodies,** in which spores are produced. The familiar part of a mushroom or toadstool is a large fruiting body. We do not normally see the bulk of the fungus, a nearly invisible mycelium buried out of sight in the rotting material or soil on which it grows.

Fungi may produce spores either sexually or asexually. Unlike most animal and plant cells, fungal cells usually contain haploid nuclei. In sexual reproduction, the hyphae of two genetically compatible mating types come together, and their cytoplasm and nuclei fuse, forming a diploid cell known as a **zygote.**[1] In two fungal groups—the ascomycetes and basidiomycetes—the hyphae fuse, but the two different nuclei do not fuse immediately; rather, they remain separate within the fungal cytoplasm. Hyphae that contain two genetically distinct, sexually compatible nuclei within each cell are **dikaryotic,** a condition that is described as $n + n$ rather than $2n$ because there are two

[1] In certain fungi (e.g., *Allomyces*), zygotes are typically formed by the union of two haploid gametes rather than the union of two haploid nuclei in compatible hyphae.

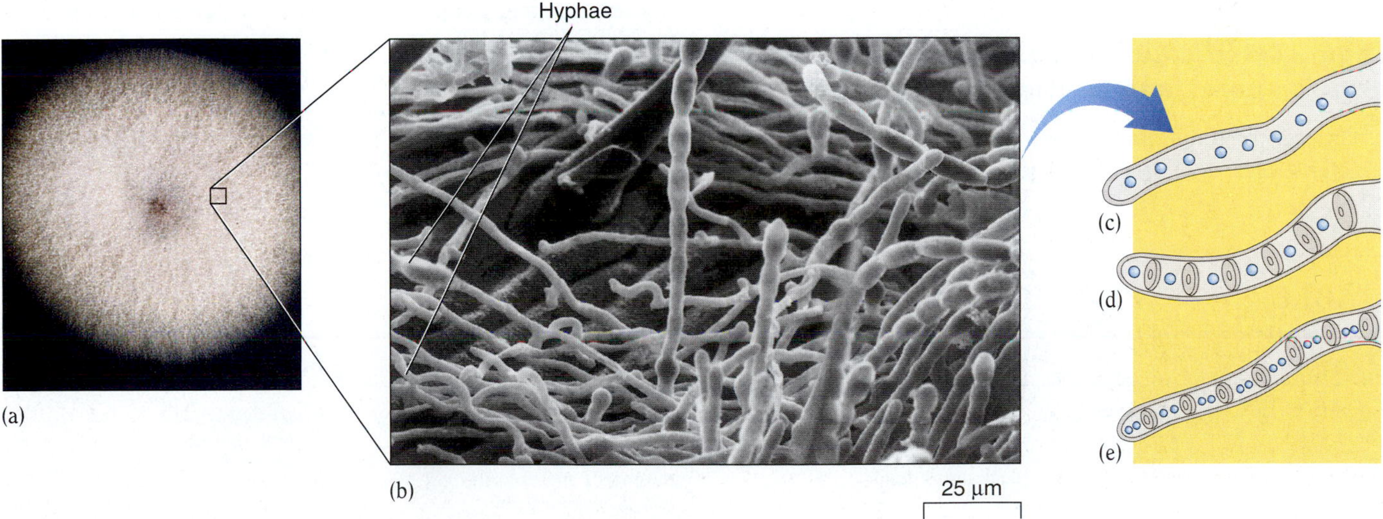

Figure 25–1 Fungus body plan. **(a)** A fungal mycelium, a mass of threadlike hyphae (10 cm [4 in] wide), growing on agar in a culture dish. In nature, fungal mycelia are rarely so symmetrical. **(b)** SEM of a mycelium of *Blumeria graminis*, growing on a leaf *(darker area underneath the mycelium)*. **(c)** A coenocytic hypha. **(d)** A hypha divided into cells by septa; each cell is monokaryotic (has one nucleus). In some taxa the septa are perforated *(as shown)*, permitting cytoplasm to stream from one cell to another. **(e)** A septate hypha in which each cell is dikaryotic (has two genetically distinct nuclei). *(a, Dennis Drenner; b, G.T. Cole/University of Texas/BPS)*

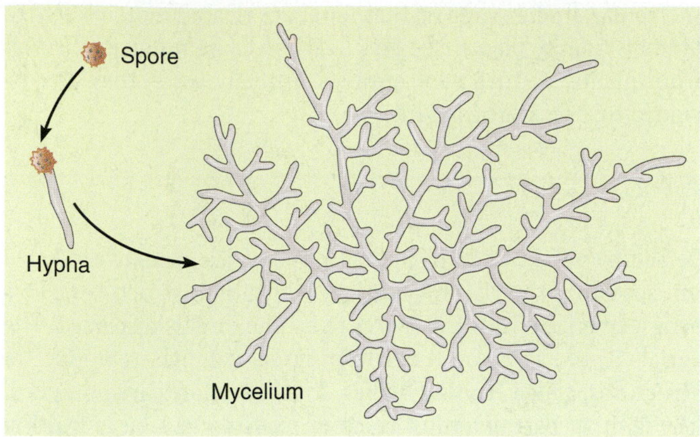

Figure 25–2 Germination of a spore to form a mycelium. Because the filamentous mycelium spreads throughout its substrate (such as overripe fruit, wood, or soil), it is difficult to estimate the actual size of an individual fungus.

separate haploid nuclei. Hyphae that contain only one nucleus per cell are said to be **monokaryotic.**

When a fungal spore comes into contact with an appropriate food source, perhaps an overripe peach that has fallen to the ground, it germinates and begins to grow (Fig. 25–2). A threadlike hypha emerges from the tiny spore and grows at its tip, branching frequently. Soon a mycelium infiltrates the peach, degrading its complex organic compounds to small organic molecules that the fungus can absorb. Fungi are very efficient at converting nutrients into new cell material. If excessive amounts of nutrients are available, fungi are able to store them, usually as lipid droplets or glycogen, in the mycelium.

Fungal classification is based in part on their spores and fruiting bodies

Classification of fungi traditionally was based mainly on the characteristics of their sexual spores and fruiting bodies, but molecular data, such as comparative DNA and RNA sequences, are playing an increasingly important role in determining relationships among fungi. Most biologists assign these diverse organisms to five phyla: Chytridiomycota, Zygomycota, Ascomycota, Basidiomycota, and Deuteromycota (Fig. 25–3). Table 25–1 summarizes fungal classification. Slime molds and water molds were originally classified as fungi but are now generally considered protists (see Chapter 24).

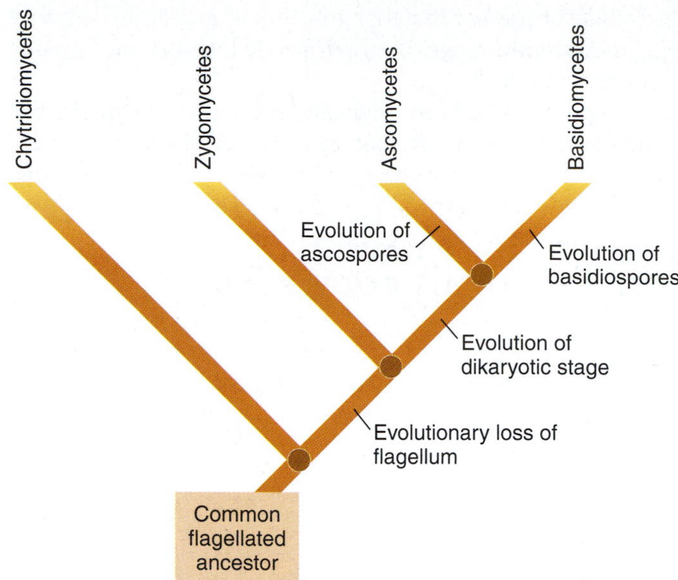

Figure 25–3 Fungal evolution. There are few examples of fossilized fungi, and fossil evidence has not helped biologists unravel evolutionary relationships among different fungal groups. Instead, molecular evidence has been used to clarify aspects of fungal evolution. This cladogram shows phylogenetic relationships among living fungi, based on comparisons of ribosomal and nuclear gene sequence data of many species. The chytridiomycetes were the first fungi to split from other fungal phyla. The deuteromycetes are not shown on this figure because, as discussed later in the chapter, the group is polyphyletic.

TABLE 25–1 Phyla of Kingdom Fungi

Phylum	Common Types	Asexual Reproduction	Sexual Reproduction
Chytridiomycota (chytridiomycetes or chytrids)	*Allomyces*	Zoospores	Flagellated gametes in some chytrids
Zygomycota (zygomycetes)	Black bread mold	Nonmotile spores form in a sporangium	Zygospores
Ascomycota (ascomycetes or sac fungi)	Yeasts, powdery mildews, molds, morels, truffles	Conidia pinch off from conidiophores	Ascospores
Basidiomycota (basidiomycetes or club fungi)	Mushrooms, bracket fungi, puffballs, rusts, smuts	Conidia (uncommon)	Basidiospores
Deuteromycota (deuteromycetes or imperfect fungi)	Molds	Conidia	Sexual stage not observed

CHYTRIDIOMYCETES ARE THE MOST PRIMITIVE FUNGI

Until recently, **chytridiomycetes** (phylum Chytridiomycota), also called *chytrids,* were considered by most biologists to be fungus-like protists similar in many respects to the water molds. However, molecular comparisons, particularly of DNA and RNA sequences, have provided compelling evidence that the approximately 750 species of chytridiomycetes are closely allied to fungi and not closely related to water molds or other fungus-like protists. The molecular evidence also indicates that chytridiomycetes, which are small, relatively simple aquatic fungi, are the most primitive group of fungi living today. *Primitive* in this context implies that chytridiomycetes were probably the earliest fungal group to evolve from the ancient flagellated protist hypothesized as the common ancestor of all fungi. Chytridiomycetes, which produce flagellated cells at some stage in their life history, have presumably retained this feature of their protist ancestor. All other fungal phyla are nonflagellated; they apparently lost the ability to produce motile cells at some point in their evolutionary past, perhaps during the transition from aquatic to terrestrial habitats.

Chytridiomycetes are parasites or decomposers that are found principally in fresh water (Fig. 25–4), although a few species occur on land (in moist environments) and in marine water. Like other fungi, chytridiomycetes have cell walls containing chitin and store their food reserves as glycogen. Their motile cells possess a single, posterior flagellum. They reproduce both sexually and asexually.

One common chytridiomycete is *Allomyces,* which has an unusual life cycle compared to most fungi because it has an **alternation of generations,** spending part of its life as a haploid *(n)* **thallus** and part as a diploid ($2n$) thallus (Fig. 25–5). (The term *thallus* is used to describe the simple body plan of certain algae, fungi, or plants.) The haploid and diploid thalli are similar in appearance and consist of a stout trunklike part with slender branches. The haploid thallus bears male and female **gametangia,** which are structures in which gametes are formed by mitosis, at the tips of the branches. When a flagellated male gamete fuses with a slightly larger, flagellated female gamete, the resulting zygote develops into a diploid thallus. The diploid thallus bears two kinds of spore cases, zoosporangia and resting sporangia. Zoosporangia produce flagellated diploid **zoospores** that may settle down and develop into new diploid thalli. Meiosis occurs within resting sporangia to form haploid zoospores, each of which has the potential to settle down and develop into a haploid thallus.

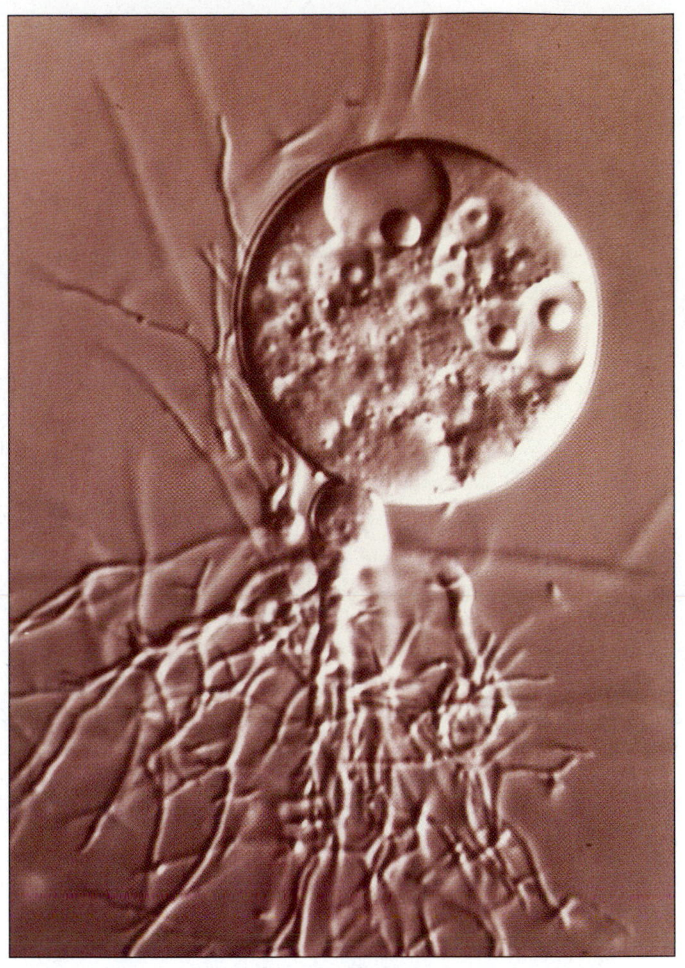

Figure 25–4 Chytrid. LM (Nomarski differential interference micrograph) of a common chytrid *(Chytridium convervae).* Many chytrids have a microscopic body form consisting of a coenocytic globose (rounded) thallus and branched rhizoids that superficially resemble roots. The rhizoids may anchor the chytrid thallus and absorb predigested food. Chytrids are widely distributed in nature and common in many water and soil environments. *(John Taylor, University of California–Berkeley)*

ZYGOMYCETES REPRODUCE SEXUALLY BY FORMING ZYGOSPORES

All of the approximately 800 species of **zygomycetes** (phylum Zygomycota) produce sexual spores, called **zygospores.** Their hyphae are coenocytic; that is, they lack regularly spaced septa. Septa do form, however, to separate the hyphae from reproductive structures. Most zygomycetes are decomposers that live in the soil on decaying plant or animal matter (Fig. 25–6), although some are parasites of plants and animals and others are mycorrhizal fungi associated with plant roots (discussed shortly).

One common zygomycete is the black bread mold, *Rhizopus stolonifer,* a decomposer that breaks down bread and other foods (Fig. 25–7). If preservatives are not added, bread left at room temperature often becomes covered with a black, fuzzy growth in a few days. Bread becomes moldy when a spore falls on it and then germinates and grows into a mycelium. Hyphae penetrate the bread and absorb nutrients. Eventually, certain hyphae grow upward and develop spore sacs called **sporangia** (sing., *sporangium*) at their tips. Clusters of black asexual spores develop within each sporangium and are released when the delicate

(text continues on page 539)

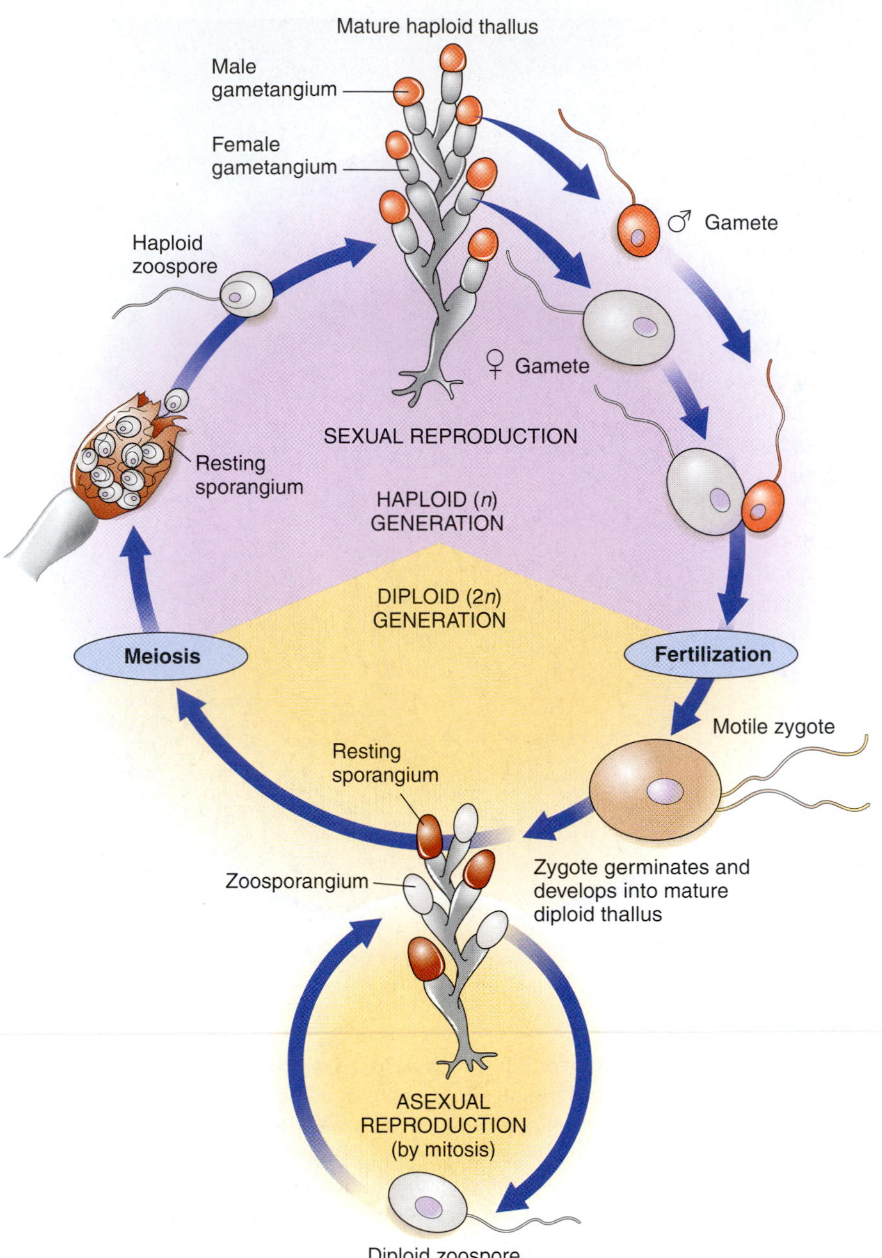

Figure 25-5 Life cycle of *Allomyces arbuscula*, a chytridiomycete. *Allomyces* alternates between haploid and diploid stages, which are similar in appearance. (Alternation of generations is common in plants but rare in fungi.) The haploid thallus gives rise to male and female gametes that fuse and subsequently develop into the diploid thallus. Meiosis occurs in the diploid resting sporangia, forming haploid zoospores that develop into haploid thalli, and the cycle continues. *Allomyces* also reproduces asexually by forming diploid zoospores.

Figure 25-6 *Pilobolus*, a zygomycete that grows in animal dung. Shown are the stalked sporangia protruding from a pile of dung, which contains an extensive mycelium of the fungus. The stalked sporangia, which are 5 to 10 mm tall, act like shotguns and forcefully discharge sporangia (the black tips) away from the dung, onto nearby grass. When animals such as cattle or horses eat the grass, the spores pass unharmed through the animal's digestive tract to be deposited in a fresh pile of dung. *(Cabisco/Visuals Unlimited)*

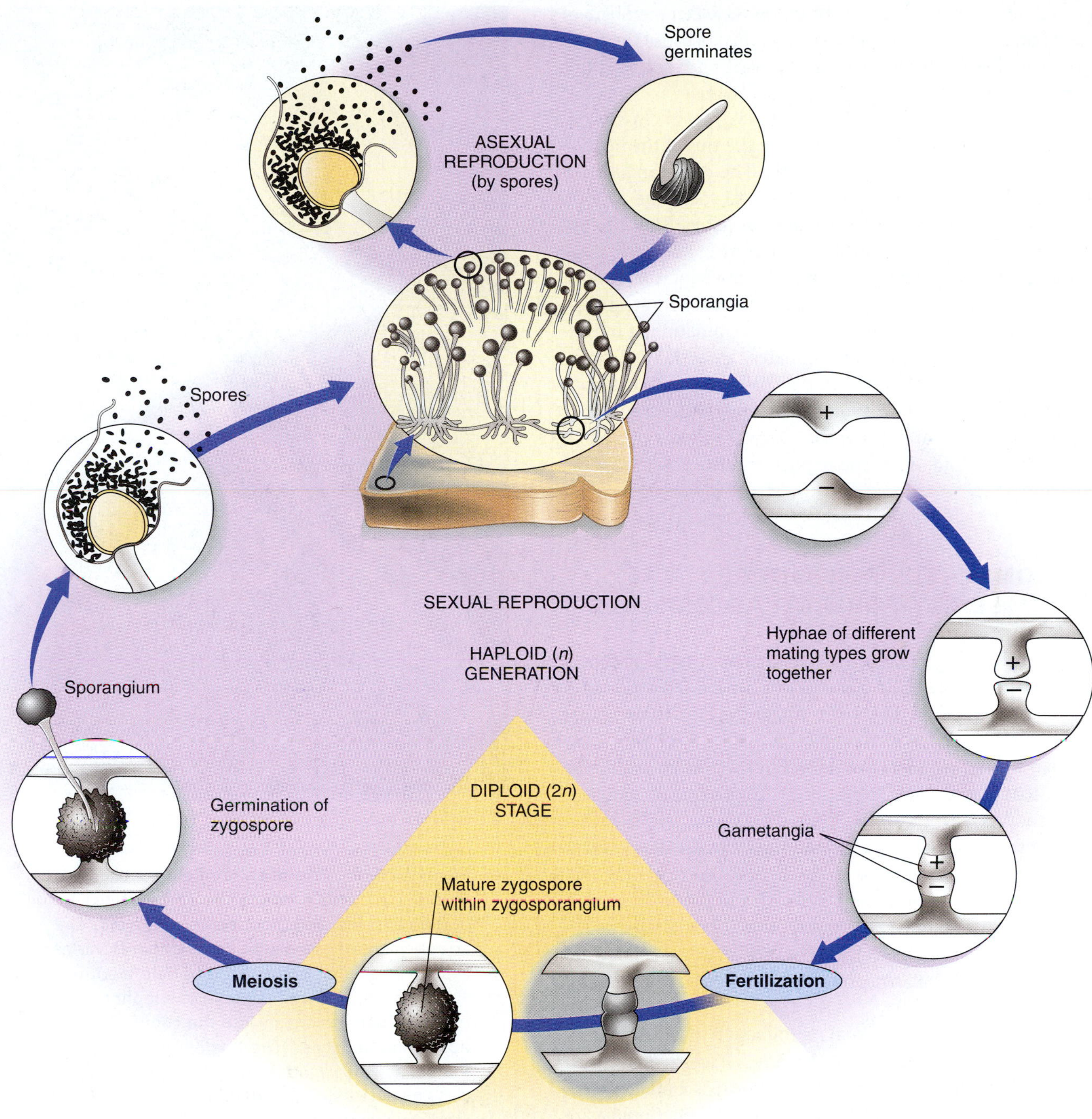

Figure 25-7 Life cycle of the black bread mold *(Rhizopus stolonifer)*, a zygomycete. Asexual reproduction involves the formation of haploid spores. Sexual reproduction takes place only between different mating types, designated (+) and (−). After their hyphae meet and the nuclei fuse, a single zygospore develops within a thick-walled, black zygosporangium. Meiosis occurs, the zygospore germinates, and the emerging hypha develops a sporangium at its tip. The sporangium may contain all (+) spores, all (−) spores, or a mixture of (+) and (−) spores.

sporangium ruptures. The spores give the black bread mold its characteristic color.

Sexual reproduction in the black bread mold occurs when the hyphae of two different mating types, designated as plus (+) and minus (−), grow into contact with one another. The bread mold is **heterothallic,** meaning that an individual fungal hypha is self-sterile and mates only with a hypha of a different mating type. That is, sexual reproduction occurs only between a member of a (+) strain and one of a (−) strain, not between members of two (+) strains or members of two (−) strains. Because there are no physical differences between the two mating types, it is not appropriate to refer to them as "male" and "female."

When hyphae of opposite mating types grow in close proximity, hormones are produced that cause the tips of the hyphae to come together and form gametangia. The gametangia then unite, and the (+) and (−) nuclei fuse to form the zygote, which is a diploid nucleus. A zygospore develops from the zygote, surrounded by a thick protective covering known as a **zygosporangium.** The zygospore may lie dormant for several months and can survive desiccation and extreme temperatures. Meiosis probably occurs at or just before germination of the zygospore. When the zygospore germinates, an aerial hypha develops with a sporangium at the tip. Mitosis within the sporangium produces haploid spores that are released and that may germinate to form new hyphae. Only the zygote of a black bread mold is diploid; all of the hyphae and the asexual spores are haploid.

ASCOMYCETES REPRODUCE SEXUALLY BY FORMING ASCOSPORES

Ascomycetes (phylum Ascomycota) comprise a large group of fungi consisting of about 30,000 described species. Ascomycetes are sometimes referred to as *sac fungi* because their sexual spores are produced in microscopic sacs called **asci** (sing., *ascus*). Their hyphae usually have septa, but these cross walls have pores so that cytoplasm is continuous from one cell compartment to another.

The diverse ascomycetes include most yeasts; the powdery mildews; most of the blue-green, pink, and brown molds that cause food to spoil; decomposer cup fungi; and the edible morels and truffles. Some ascomycetes cause serious plant diseases such as Dutch elm disease, ergot disease on rye, powdery mildew on fruits and ornamental plants, and chestnut blight (see *Focus On: Fighting Chestnut Blight*).

In most ascomycetes, asexual reproduction involves production of spores called **conidia** (sing., *conidium*), which are formed at the tips of certain specialized hyphae known as **conidiophores** (from the Greek, meaning "dust-bearers"; Fig. 25–8). Conidia are a means of rapidly propagating new mycelia when environmental conditions are favorable. Conidia occur in various shapes, sizes, and colors in different species. The color of the conidia is responsible for the characteristic brown, blue-green, pink, or other tints of many of these molds.

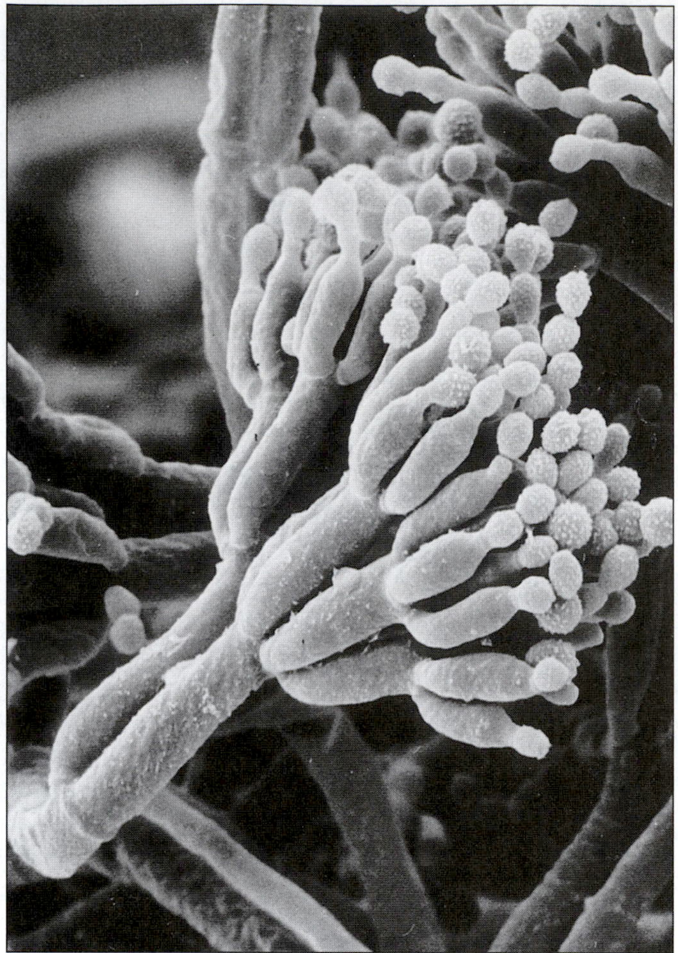

Figure 25–8 Conidia. Conidia are asexual reproductive cells produced by ascomycetes, some basidiomycetes, and most deuteromycetes. The arrangement of conidia on conidiophores varies from species to species and is used to help identify these fungi. Shown is a SEM of *Penicillium* conidiophores, which resemble paintbrushes. Note the conidia pinching off the tips of the "brushes." The taxonomy of fungi classified in the genus *Penicillium* is somewhat confusing, because certain species are classified as ascomycetes (because they produce a sexual stage) and other species are classified as deuteromycetes (because they have no known sexual stage and only reproduce asexually). DNA sequence analyses and structural evidence suggest that the genus *Penicillium* is not monophyletic. *(Biophoto Associates/Photo Researchers, Inc.)*

Some species of ascomycetes are heterothallic and have different mating strains; others are **homothallic,** which means that they are self-fertile and have the ability to mate with themselves. In both heterothallic and homothallic ascomycetes, sexual reproduction takes place after two gametangia come together and their cytoplasm mingles (Fig. 25–9). Within this fused structure, pairs of haploid nuclei, one from each parent hypha, associate but do not fuse. New hyphae, the cells of which are dikaryotic, develop from the fused structure and branch repeatedly until the hyphal tips reach the site where asci will be produced. As the many sac-shaped asci develop, each containing two dissimilar nuclei (one from each parent), they are surrounded by intertwining haploid (monokaryotic) hyphae that develop into a fruiting body known as an **ascocarp** (Fig. 25–10*a* on page 543).

FOCUS ON Fighting Chestnut Blight

American chestnuts. During the first part of the 20th century, the chestnut blight fungus killed almost all of these massive trees, which grew to 30.5 m (100 ft). The historical photograph showing several live chestnut trees was taken in the late 1800s in the Great Smoky Mountains. *(Inset)* Chestnuts that sprout from the roots of old stumps do not grow very large before they are killed back by the chestnut blight. *(Photo courtesy of the American Chestnut Foundation; inset, Marion Lobstein)*

Prior to the 20th century, the American chestnut *(Castanea dentata)* was a very important tree, both ecologically and economically, in eastern forests. Throughout its natural range, which extended from Maine to Georgia and west to Illinois, the chestnut tree provided food and shelter for forest animals, including squirrels, birds, and insects. The tree grew rapidly, forming a straight, tall trunk that provided valuable, decay-resistant wood for house foundations and interior trim, furniture, railroad ties, and fence posts. The wood was also an important source of tannin, which is a chemical used to convert animal hides into leather. The edible nuts of the American chestnut were harvested, roasted, and eaten directly or ground into flour.

Early in the 20th century the chestnut blight fungus *(Cryphonectria parasitica),* an ascomycete, was accidentally imported on diseased oriental chestnuts and quickly attacked native chestnuts, which had no resistance to the fungus. First identified in New York in 1904, the blight had killed or damaged several billion mature trees—almost every North American chestnut tree throughout its entire natural range—by the late 1940s *(see figure).*

A spore of the fungus enters a chestnut tree through a wound or crack in the bark, and the hyphae grow rapidly, attacking the inner bark. When the trunk is girdled (completely encircled) by the fungus, the portion of the plant above the injury dies. The fungus does not kill the roots, however, and they periodically produce new shoots, only to be killed back by the blight once again. Wind, water, and animals carry the spores from tree to tree.

Although the American chestnut is on the brink of extinction, there are hopes for its eventual recovery. Some biologists have bred the American chestnut with the Chinese chestnut, a related species that is resistant to chestnut blight. With careful selective breeding over many years, it may be possible to develop a disease-resistant variety of American chestnut.

Some biologists are approaching the problem in a different way, by trying to isolate strains of the chestnut blight fungus that are less virulent (less deadly). A less virulent strain of the chestnut blight fungus was first identified in Europe in the 1950s; some of the trees there have been able to recover from the disease. More recently, a less virulent strain has appeared in the United States, and, even more encouraging, it appears that the less virulent strain is slowly spreading, replacing the more deadly strain.

Biologists determined that the reduced virulence of the chestnut blight fungus is caused by the presence of a virus. They genetically engineered a synthetic form of the virus and introduced it into the virulent strain of the chestnut blight fungus, which subsequently lost its virulence. They plan to treat infected chestnut trees with the benign fungal strain, which they hope will spread rapidly and eventually replace the virulent strain.

Within a cell that develops into an ascus, the two nuclei fuse and form a diploid nucleus, the zygote, that then undergoes meiosis to form four haploid nuclei. One mitotic division of each of the four nuclei usually follows, resulting in eight haploid nuclei. Each haploid nucleus becomes incorporated into a heavy-walled **ascospore,** so there are usually eight haploid ascospores within the ascus (Fig. 25–10*b*). The ascospores are usually released through a pore, slit, or hinged lid at the tip of the ascus. Air currents carry individual ascospores, often for long distances. If one lands in a suitable location, it germinates and forms a new mycelium.

Phylum Ascomycota includes more than 300 species of unicellular **yeasts** (see Fig. 7–15*a*). Asexual reproduction of yeasts is

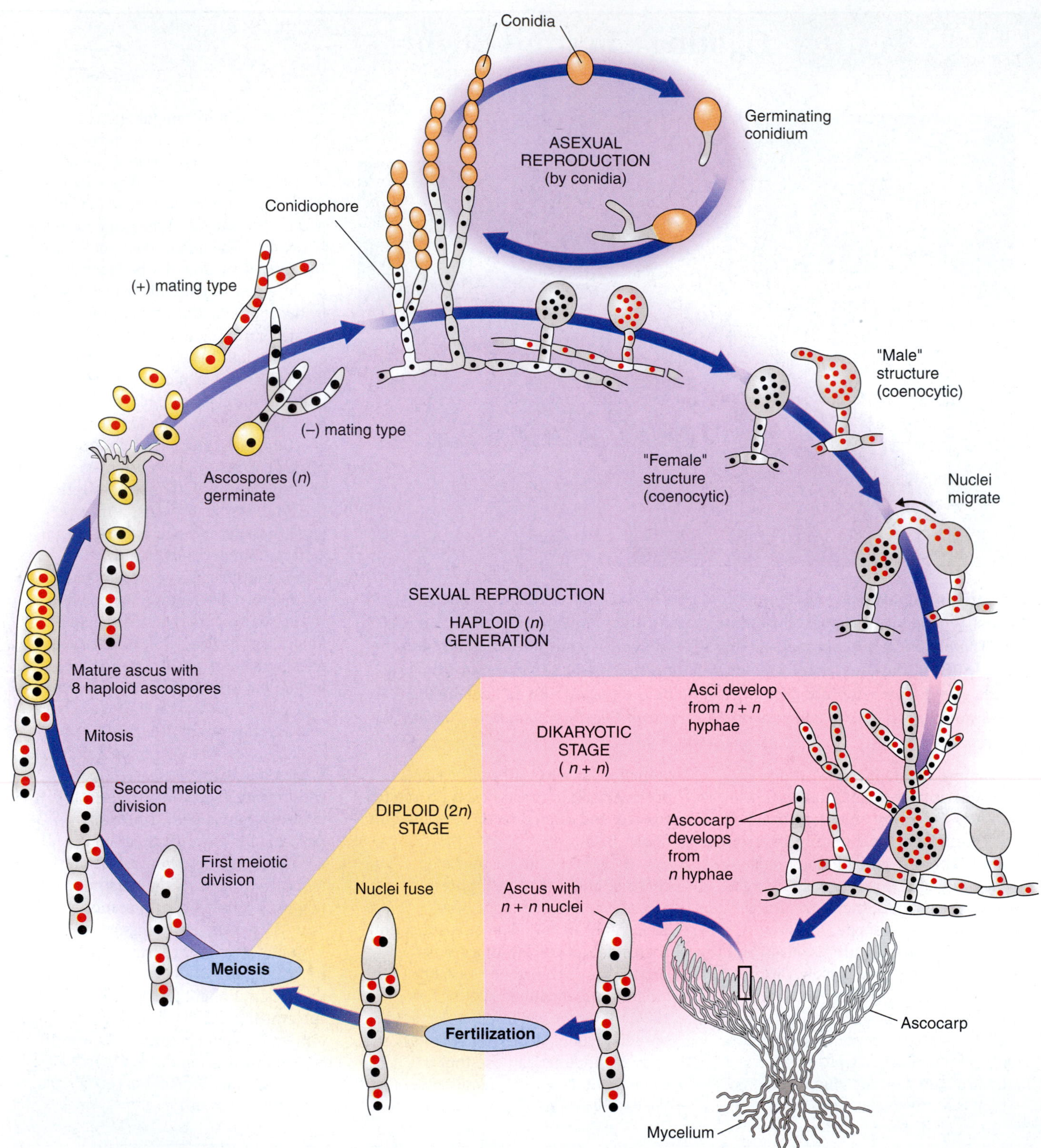

Figure 25-9 Life cycle of a typical ascomycete. Asexual reproduction involves the formation of haploid conidia. Sexual reproduction involves the fusion of two haploid gametangia of different mating types to form dikaryotic ($n + n$) hyphae from which asci develop. In each ascus the $n + n$ nuclei fuse, followed by meiosis and mitosis to produce eight ascospores. The asci form the inner layer of a fruiting body called the ascocarp.

(a)

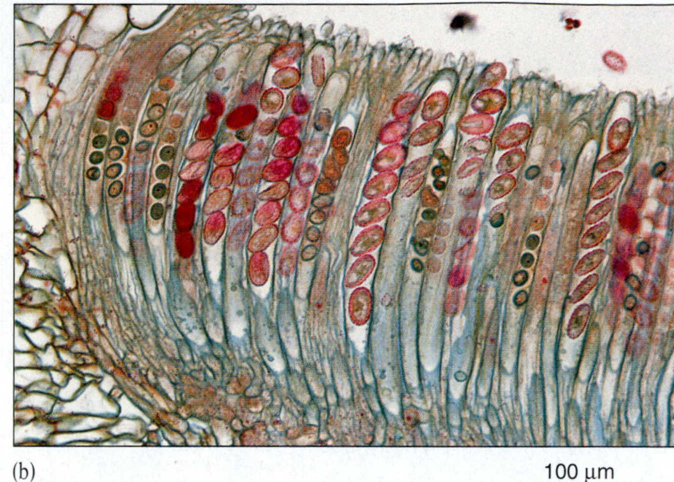

(b)

100 μm

Figure 25–10 Sexual reproduction in the ascomycetes. **(a)** The ascocarp of the common brown cup *(Peziza badio-confusa)* is saucer- or bowl-shaped and 3 to 10 cm (1 to 4 in) wide. It is found on damp soil in woods throughout North America. Photographed in Muskegon, Michigan. **(b)** Asci, each containing eight ascospores, line the inner portion of the saucer. *(a, Ed Reschke; b, Robert and Linda Mitchell)*

mainly by **budding;** in this process a small protuberance (bud) grows and eventually separates from the parent cell. Each bud can grow into a new yeast cell. Some yeasts also reproduce asexually by **binary fission,** in which they undergo mitosis and then divide in half, and sexually by forming ascospores. During sexual reproduction, two haploid yeasts fuse, forming a diploid zygote. The zygote undergoes meiosis, and the resulting haploid nuclei are incorporated into ascospores. These spores remain enclosed for a time within the original cell wall, which corresponds to an ascus. Yeasts are essential in making bread and fermenting alcoholic beverages (discussed later in this chapter). The yeast *Saccharomyces cerevisiae* is an important model eukaryotic organism. Its genome has been sequenced, and scientists are now trying to determine the functions of the 6000 newly identified proteins coded for by its genes.

BASIDIOMYCETES REPRODUCE SEXUALLY BY FORMING BASIDIOSPORES

The 25,000 or more species of **basidiomycetes** (phylum Basidiomycota) include the most familiar of the fungi: the mushrooms, bracket fungi, and puffballs (Fig. 25–11). Some destructive plant parasites of important crops, such as wheat rust and corn smut, are also basidiomycetes. Basidiomycetes, sometimes called *club fungi,* derive their name from the fact that they develop microscopic club-shaped **basidia** (sing., *basidium*), structures comparable in function to the asci of ascomycetes. Each basidium is an enlarged hyphal cell, on the tip of which develop four **basidiospores** (Fig. 25–12). Note that basidiospores develop on the *outside* of a basidium, whereas ascospores develop *within* an ascus.

Each individual fungus produces millions of basidiospores, and each basidiospore has the potential to give rise to a new **primary mycelium** (Fig. 25–13). Hyphae of a primary mycelium are composed of monokaryotic cells. The mycelium of a basidiomycete such as the meadow mushroom, *Agaricus brunnescens,*[2] which is commonly cultivated, consists of a mass of white, branching, threadlike hyphae that live mostly below ground. The hyphae are divided into cells by septa, but, as in ascomycetes, the septa are perforated and allow cytoplasmic streaming between cells.

When in the course of its growth a hypha of a primary mycelium encounters another monokaryotic hypha of a different mating type, the two hyphae fuse. As in the ascomycetes, the two haploid nuclei remain separate within each cell. In this way a dikaryotic **secondary mycelium** with dikaryotic hyphae is produced, in which each cell contains two haploid nuclei.

The $n + n$ hyphae of the secondary mycelium grow extensively and, when environmental conditions are favorable, form compact masses, called buttons, along the mycelium. Each button grows into a fruiting body that we call a mushroom. A mushroom, which consists of a stalk and a cap, is more formally referred to as a **basidiocarp.** Each basidiocarp consists of intertwined hyphae that are matted together. The lower surface of the cap usually consists of many thin perpendicular plates called **gills** that radiate from the stalk to the edge of the cap. On the gills of the mushroom, haploid nuclei fuse in the dikaryotic cells within the young basidia to form diploid zygotes. These are the only diploid cells that form during a basidiomycete's life history. Meiosis then takes place, forming four haploid nuclei that move to the outer edge of the basidium. Finger-like extensions of the basidium develop, into which the nuclei and some cytoplasm

(text continues on page 546)

[2] Also called *Agaricus bisporus.*

(a)

(b)

(c)

(d)

Figure 25–11 Basidiomycete fruiting bodies. **(a)** Basidia line the gills of the Jack-o'lantern mushroom *(Omphalotus olearius)*, a poisonous species whose gills produce a greenish glow in the dark. Each cap is about 15 cm (6 in) wide. Photographed at the base of an oak tree in Maryland, the Jack-o'lantern occurs throughout eastern North America and California. **(b)** A giant puffball *(Calvatia gigantea)* in White Plains, New York. Giant puffballs grow to 51cm (20 in). At maturity, a dried-out puffball often has a pore through which the basidiospores are discharged as a puff of dust. **(c)** The elegant stinkhorn *(Phallus ravenelii)* has a foul smell that attracts flies, which help disperse the slimy mass of basidiospores. Fruiting bodies of elegant stinkhorns grow to 18 cm (7 in) tall. Photographed in Pennsylvania. **(d)** Turkey-tail *(Trametes versicolor)* is a common bracket fungus. Bracket fungi grow on both dead and living trees, producing shelflike fruiting bodies. Basidiospores are produced in pores located underneath each shelf. Photographed in Pennsylvania. *(a, Dennis Drenner; b, Ed Kanze/Dembinsky Photo Associates; c, Richard D. Poe/Visuals Unlimited; d, J.L. Lepore/Photo Researchers, Inc.)*

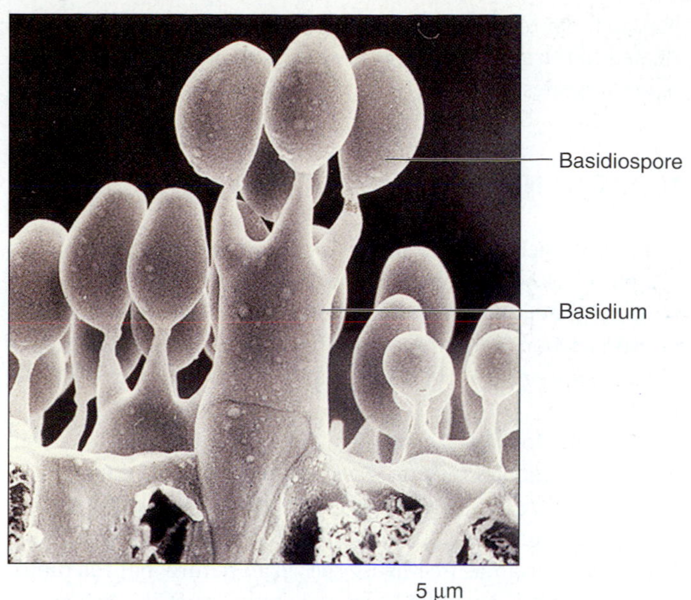

Figure 25–12 SEM of a basidium. Each basidium produces four basidiospores. *(Biophoto Associates/Photo Researchers, Inc.)*

Figure 25–13 Life cycle of a typical basidiomycete. Sexual reproduction involves the fusion of two haploid hyphae of different mating types to form a secondary mycelium composed of dikaryotic ($n + n$) hyphae. Fruiting bodies (basidiocarps) periodically develop from the secondary mycelium. Basidia form along the gills of the basidiocarps. In each basidium the $n + n$ nuclei fuse, followed by meiosis to produce four basidiospores. In basidiomycetes in which the fruiting body is a mushroom, the basidia line the outer surfaces of the gills.

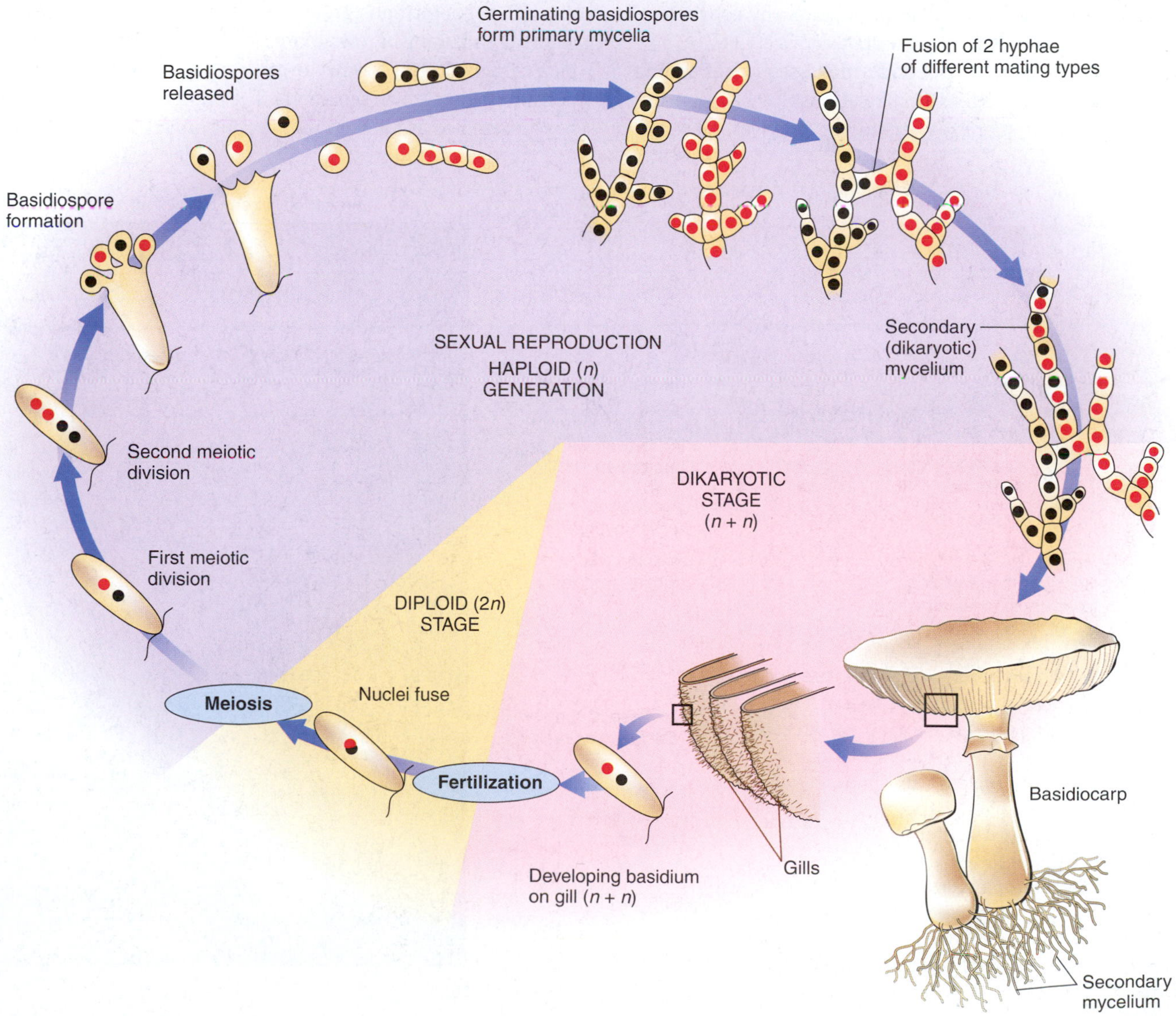

move; each of these extensions becomes a basidiospore. A septum forms that separates the basidiospore from the rest of the basidium by a delicate stalk that breaks when the basidiospore is forcibly discharged.

IMPERFECT FUNGI HAVE NO KNOWN SEXUAL STAGE

About 25,000 species of fungi have been assigned to a *form phylum* referred to as the **deuteromycetes** (form phylum Deuteromycota). Members of a form phylum are similar to one another in certain respects but do not necessarily share a common ancestry; that is, the group is probably **polyphyletic,** which means that individuals classified as deuteromycetes lack a common ancestor. Deuteromycetes, also known as **imperfect fungi** because in most of them no sexual stage has been observed at any point during their life cycle, are classified together simply as a matter of convenience. Some of these fungi have lost the ability to reproduce sexually, whereas others may reproduce sexually only rarely. Should further study reveal a sexual stage, the sexual state of these species will be reassigned to a different phylum. (By convention, the asexual stage is still identified as a deuteromycete.)

Increasingly, comparing DNA and/or RNA sequences among various species helps biologists find relationships between imperfect fungi and their sexually reproducing relatives. Most imperfect fungi reproduce only by means of conidia and are closely related to the ascomycetes. A few are more closely related to the basidiomycetes.

LICHENS ARE DUAL "ORGANISMS"

Although a **lichen** looks like a single organism, it is actually a symbiotic association between two organisms: a *phototroph* (a photosynthetic organism) and a fungus (Fig. 25–14*a*). (A symbiotic association is an intimate relationship between organisms of different species.) About 13,500 kinds of lichens have been described.

The phototrophic component of a lichen is either a green alga or a cyanobacterium. The fungus is most often an ascomycete, although in some tropical lichens the fungal partner is a basidiomycete. Most of the phototrophic organisms found in lichens also occur as free-living species in nature, but the fungal components are generally found only as a part of the lichen.

In the laboratory the fungal and phototrophic components can be isolated and grown separately in appropriate culture media. The phototroph grows more rapidly when separated, whereas the fungus grows more slowly and requires many complex carbohydrates. Neither organism resembles a lichen in appearance when grown separately. The phototroph and fungus can be reassembled as a lichen thallus, but only if they are placed in

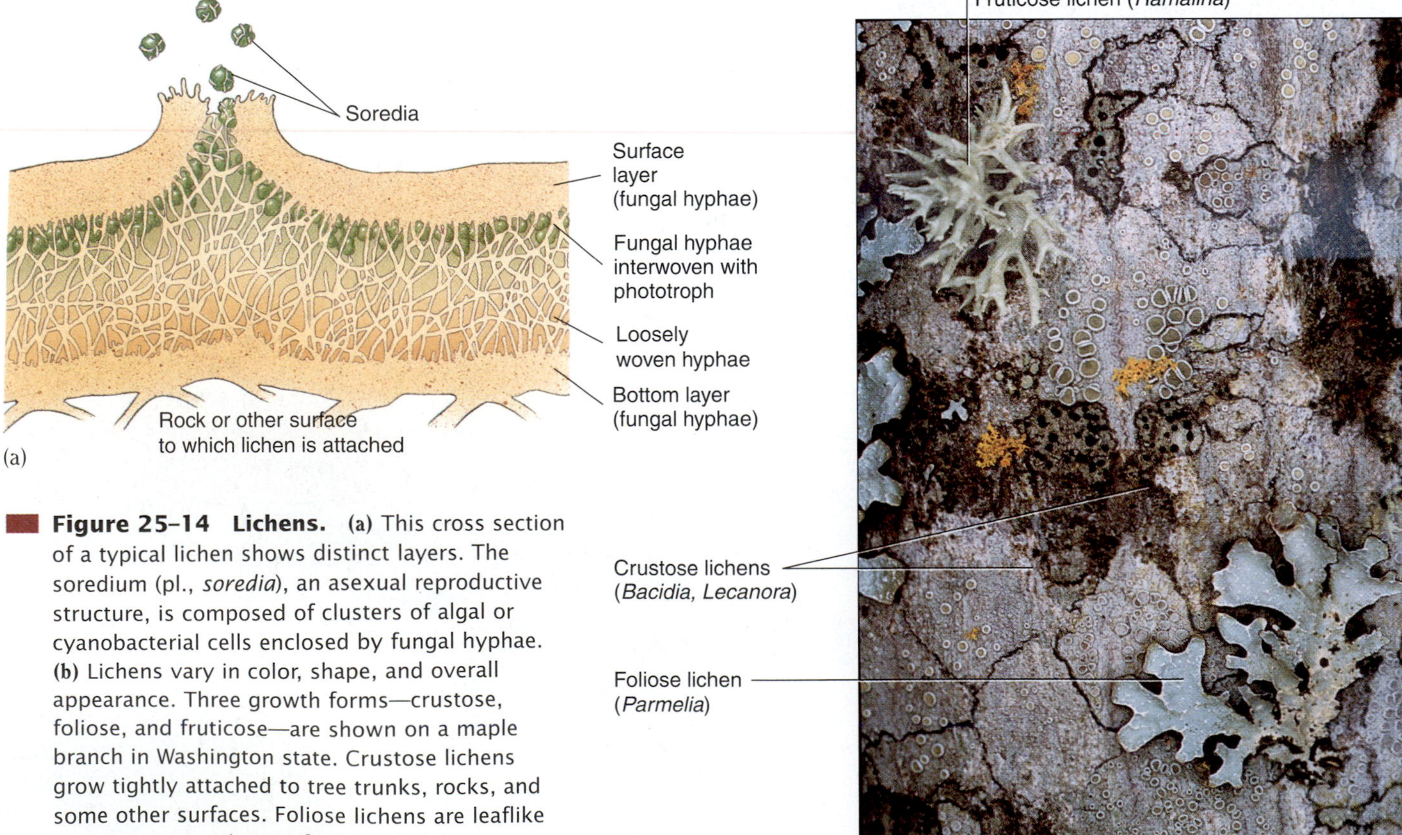

Figure 25–14 Lichens. **(a)** This cross section of a typical lichen shows distinct layers. The soredium (pl., *soredia*), an asexual reproductive structure, is composed of clusters of algal or cyanobacterial cells enclosed by fungal hyphae. **(b)** Lichens vary in color, shape, and overall appearance. Three growth forms—crustose, foliose, and fruticose—are shown on a maple branch in Washington state. Crustose lichens grow tightly attached to tree trunks, rocks, and some other surfaces. Foliose lichens are leaflike in appearance, whereas fruticose lichens are branching and shrublike. *(Fred M. Rhoades)*

a culture medium under conditions that cannot support either of them independently.

What is the nature of this partnership? The lichen was originally considered a definitive example of mutualism, a symbiotic relationship that is beneficial to both species. The phototroph carries on photosynthesis, producing food for both members of the lichen, but it is unclear how the phototroph benefits from the relationship. It has been suggested that the phototroph obtains water and nutrient minerals from the fungus as well as protection against desiccation. More recently some biologists have suggested that the lichen partnership is not really a case of mutualism but one of controlled parasitism of the phototroph by the fungus.

Lichens typically possess one of three different growth forms (Fig. 25–14*b*): *Crustose lichens* are flat and grow tightly against their substrate (the surface they are growing on); *foliose lichens* are also flat, but they have leaflike lobes and are not so tightly pressed to the substrate; and *fruticose lichens* grow erect and have many branches.

Able to tolerate extremes of temperature and moisture, lichens grow in almost all terrestrial environments except polluted cities. They exist farther north than any plants of the arctic region and are equally at home in the steaming equatorial rain forest. They grow on tree bark, leaves, and exposed rock surfaces, from solidified lava to tombstones. In fact, lichens are often the first organisms to inhabit rocky areas. Lichen growth in these areas plays an important role in the formation of soil from rock because they gradually etch tiny cracks in rock, which eventually breaks into particles of soil. This process sets the stage for further disintegration of the rock by wind and rain.

Reindeer mosses of the arctic region, which serve as the main source of food for migrating herds of caribou, are not mosses but lichens. Some lichens produce colored pigments. One of them, orchil, is used to dye woolens, and another, litmus, is widely used in chemistry laboratories as an acid-base (pH) indicator.

Lichens vary greatly in size. Some are almost invisible, whereas others, like the reindeer mosses, may cover many square kilometers of land with an ankle-deep growth. Growth proceeds slowly; the radius of a lichen may increase by less than 1 mm each year. Some mature lichens are thought to be thousands of years old.

Lichens absorb minerals from the air, rainwater, and the surface on which they grow. They cannot excrete the elements they absorb, and perhaps for this reason they are extremely sensitive to toxic compounds. A reduction in lichen growth has been used as a sensitive indicator of air pollution, particularly sulfur dioxide. In 1997, for example, Italian researchers were able to establish a relationship between lung cancer and air pollution by comparing the locations of low lichen biodiversity (and therefore of air pollution) with the locations of lung cancer deaths in young male residents. The return of lichens to an area indicates an improvement in air quality.

Lichens reproduce mainly by asexual means, usually by fragmentation, a process in which special dispersal units of the lichen, called **soredia,** break off and, if they land on a suitable surface, establish themselves as new lichens. Soredia contain cells of both partners. In some lichens, the fungus produces ascospores, which may be dispersed by wind and find an appropriate algal partner only by chance.

FUNGI ARE ECOLOGICALLY IMPORTANT

Fungi make important contributions to the ecological balance of our world. Like bacteria, most fungi are **saprotrophs** (also called *saprobes*), decomposers that absorb nutrients from organic wastes and dead organisms. For example, many fungal saprotrophs are able to degrade cellulose and lignin, the main components of plant cell walls. When fungi degrade wastes and dead organisms, water, carbon (as CO_2), and mineral components of organic compounds are released, and these elements are recycled (see biogeochemical cycles in Chapter 53). Without this continuous decomposition, essential nutrients would soon become locked up in huge mounds of animal carcasses, feces, branches, logs, and leaves. The nutrients would be unavailable for use by new generations of organisms, and life would eventually cease.

Although most fungi are saprotrophs, others form symbiotic relationships of various kinds. More than 200 ant species in Central and South America and in the southern United States gather leaves and other organic material to form an underground mulch in which they cultivate fungi. These fungal crops provide food for the ant colony.

Some fungi are **parasites,** organisms that live in or on other organisms and are harmful to their hosts. Parasitic fungi absorb food from the living bodies of their hosts. Since the 1970s, many of the world's amphibian populations have declined, and a parasitic chytrid has been implicated in at least some of the die-offs (see *Focus On: Declining Amphibian Populations,* in Chapter 55).

Some types of fungi form mutualistic relationships with other organisms. **Mycorrhizae** (from Greek words meaning "fungus-roots") are mutualistic relationships between fungi and the roots of plants (see Fig. 34–10). Such relationships occur in more than 90% of all plant families. The mycorrhizal fungus benefits the plant by decomposing organic material in the soil and providing water and nutrient minerals such as phosphorus to the plant (Fig. 25–15). (The fungus in effect increases the absorptive surface area of the plant, enabling it to take in more water and nutrient minerals.) At the same time, the roots supply the fungus with sugars, amino acids, and other organic substances. Scientists have also measured the movement of organic materials from one tree species to another through shared mycorrhizal connections (see *On the Cutting Edge: Mycorrhizae and Carbon Exchange Between Plant Species in a Forest Ecosystem* in Chapter 34).

The importance of mycorrhizae first became evident when horticulturalists observed that orchids do not grow unless an appropriate fungus lives with them. Similarly, many forest trees such as pines decline and eventually die from mineral deficiencies when transplanted to mineral-rich grassland soils that lack the appropriate mycorrhizal fungi. When forest soil containing the appropriate fungi or their spores is added to the soil around

(a)

(b)

Figure 25–15 Western red cedar *(Thuja plicata)* seedlings respond to mycorrhizae. **(a)** Control plants grown in low phosphorus in the absence of the fungus. **(b)** These seedlings are the same age as the control plants and were grown under conditions identical to the control, except that their roots have formed mycorrhizal associations. *(a, b, Courtesy of Randy Molina, U.S. Forest Service)*

TABLE 25–2 Total Plant Dry Mass of Two Grass Species Grown in Prairie Soil with and without Mycorrhizal Fungi*

	Total Dry Mass (g) with Mycorrhizae Present†	Total Dry Mass (g) with Mycorrhizae Suppressed†,‡
Andropogon gerardii	88.84	65.60
Sorghastrum nutans	120.88	61.61

*From Wilson, G.W.T., and D.C. Hartnett. "Effects of Mycorrhizae on Plant Growth and Dynamics in Experimental Tallgrass Prairie Microcosms." *American Journal of Botany,* Vol. 84, No. 4, 1997.
†Per plastic container (40 × 52 × 32 cm).
‡Soil was treated with a fungicide to suppress mycorrhizal fungi.

these trees, they quickly resume normal growth. Similar results were reported in 1997 for tallgrass prairie plant species with and without mycorrhizal fungi (Table 25–2).

The discovery of 460 million-year-old fossilized fungal hyphae and spores that resemble modern mycorrhizal fungi is intriguing. It suggests that these fungi may have had a significant role in the colonization of land by plants.

FUNGI ARE ECONOMICALLY IMPORTANT

The same powerful digestive enzymes that enable fungi to decompose wastes and dead organisms also permit fungi to reduce wood, fiber, and food to their basic components with great efficiency. Thus, various fungi cause incalculable damage to stored goods and building materials each year. Bracket fungi, for example, cause enormous losses by decaying wood, both in living trees and in stored lumber.

Fungi cause economic gains as well as losses. People eat them and grow them to make various chemicals. At the other extreme, some fungi are harmful from a human perspective because they cause diseases in humans and other animals and are the most destructive disease-causing organisms of plants. Their activities cost billions of dollars in agricultural damage yearly.

Fungi provide beverages and food

The capability of yeasts to produce ethyl alcohol and carbon dioxide from sugars such as glucose by fermentation (see Chapter 7) is exploited to make wine, beer, and other fermented beverages, and also to make bread. Wine is produced when yeasts ferment fruit sugars, and beer results when yeasts ferment sugars derived from starch in grains (usually barley). During the process of making bread, carbon dioxide produced by yeast becomes trapped in dough as bubbles, causing the dough to rise; this is what gives leavened bread its light texture. Both the carbon dioxide and the alcohol produced by the yeast escape during baking.

The unique flavor of cheeses such as Roquefort, Brie, Gorgonzola, and Camembert is produced by the action of species of *Penicillium. Penicillium roquefortii,* for example, is found in caves near the French village of Roquefort; only cheeses produced in this area can be called Roquefort cheese. In Roquefort and certain other cheeses, the blue mycelium of the fungus is visible in the cheese.

Aspergillus tamarii and other imperfect fungi are used in Asia to produce soy sauce by fermenting soybeans with the fungi

(a)

(b)

Figure 25–16 Edible ascomycetes. Expensive gourmet treats, both morels and truffles are ascocarps that produce ascospores. **(a)** The yellow morel *(Morchella esculenta),* which grows 6 to 10 cm (2.5 to 4 in) tall, is found throughout North America. It is often associated with old apple orchards and dead elm trees. Photographed in Michigan. **(b)** The Oregon white truffle *(Tuber gibbosum),* which is found underground near Douglas firs and possibly oak trees in British Columbia and Northern California, is 1 to 5 cm (0.4 to 2 in) wide. People find these subterranean ascocarps with the help of trained dogs or pigs. Here truffles are shown whole and sectioned to show the conspicuous white marbled tissue. *(a, Richard Shiell/Dembinsky Photo Associates; b, John D. Cunningham/Visuals Unlimited)*

for three or more months.[3] Soy sauce enriches other foods with more than just its special flavor. It also adds vital amino acids from both the soybeans and the fungi themselves to supplement the low-protein rice diet.

Among the basidiomycetes, there are some 200 kinds of edible mushrooms and about 70 species of poisonous ones, sometimes called toadstools. Some edible mushrooms are cultivated commercially. According to the U.S. Department of Agriculture, the 1999–2000 U.S. mushroom crop was more than 389,000 metric tons (about 858 million lb). The meadow mushroom *(Agaricus brunnescens)* is the only fungal species grown extensively for food, although about 30 other mushroom species, such as oyster, shiitake, portobello, and straw mushrooms, are increasingly available in supermarkets. Morels, which superficially resemble mushrooms, and truffles, which produce underground fruiting bodies, are ascomycetes (Fig. 25–16). These gourmet delights are now being cultivated as mycorrhizal fungi on the roots of tree seedlings.

Edible and poisonous mushrooms can look very much alike and may even belong to the same genus. There is no simple way to tell them apart; an expert must identify them. Some of the most poisonous mushrooms belong to the genus *Amanita* (Fig. 25–17). Toxic species of this genus have been appropriately called such names as "destroying angel" *(Amanita virosa)* and "death cap" *(Amanita phalloides).* Eating a single mushroom of either species can be fatal.

[3] In the United States soy sauce is often made by adding flavoring to salt water rather than soaking fermented soybeans.

Ingestion of certain species of mushrooms causes intoxication and hallucinations. The sacred mushrooms of the Aztecs—*Conocybe* and *Psilocybe*—are still used in religious ceremonies by Native Americans of Central America for their hallucinogenic properties. The chemical ingredient *psilocybin,* related to lysergic acid diethylamide (LSD), is responsible for the trances and visions experienced by those who eat these mushrooms. Ingestion of psychoactive mushrooms is dangerous because negative reactions vary considerably, from mild indigestion, sweating, and heart palpitations, to death. In addition, the possession and use of such mushrooms are illegal in the United States and some other countries.

Fungi produce useful drugs and chemicals

In 1928 Alexander Fleming noticed that one of his petri dishes containing a bacterial culture was contaminated by mold. The bacteria were not growing in the vicinity of the mold, leading Fleming to the conclusion that the mold was releasing some substance harmful to them. Within a decade or so of Fleming's discovery, *penicillin* produced by *Penicillium notatum* was purified and used in treating bacterial infections. Penicillin is still among the most widely used and effective antibiotics. Other drugs derived from fungi include the antibiotic griseofulvin (used clinically to inhibit the growth of fungi), lovastatin (used to lower blood cholesterol levels), and cyclosporine (used to suppress immune responses in patients who receive organ transplants). Fumagillin is a fungal chemical that inhibits the formation of new blood vessels; because solid tumors need a rich blood supply, fumagillin shows promise as an anticancer agent.

Figure 25–17 Poisonous mushrooms. The destroying angel *(Amanita virosa)* is an extremely poisonous mushroom that is distinguished, as are other amanitas, by the ring of tissue around its stalk and by the underground cup from which the stalk protrudes. About 50 g (2 oz) of this mushroom could kill an adult man. Initial symptoms include vomiting, diarrhea, and cramps. If the poisoning goes untreated, liver and kidney failure occur, resulting in death in five to ten days. Survival rate is low, even with treatment, and most survivors require liver transplants. The destroying angel, which is 7.5 to 20 cm (3 to 8 in) tall is found in grass or near trees throughout North America. *(James W. Richardson/CBR Images)*

The ascomycete *Claviceps purpurea* infects the flowers of rye plants and other cereals. It produces a structure called an *ergot* where a seed would normally form in the grain head. When livestock eat this grain or when humans eat bread made from ergot-contaminated rye flour, they may be poisoned by the extremely toxic substances in the ergot. These substances may result in nerve spasms, convulsions, psychotic delusions, and even gangrene. This condition, called ergotism, was known as St. Anthony's fire during the Middle Ages, when it occurred often. In the year 994, for example, an epidemic of St. Anthony's fire caused more than 40,000 deaths. In 1722, the cavalry of Czar Peter the Great was felled by ergotism on the eve of the battle for the conquest of Turkey. This was one of several recorded times that a fungus changed the course of human history. Lysergic acid, one of the constituents of ergot, is an intermediate in the synthesis of LSD. Some of the compounds produced by ergot are now used clinically in small quantities as drugs to induce labor, to stop uterine bleeding, to treat high blood pressure, and to relieve one type of migraine headache.

Some fungi are grown commercially to produce citric acid and other industrial chemicals. Also, biologists are using recombinant DNA techniques to manipulate yeasts and certain filamentous fungi to produce important biological molecules such as hormones.

Fungi cause many important plant diseases

Fungi are responsible for many serious plant diseases, including epidemic diseases that spread rapidly and often result in complete crop failure. All plants are apparently susceptible to some fungal infection. Damage may be localized in certain tissues or structures of the plant, or the disease may be systemic and spread throughout the entire plant. Fungal infections may cause stunting of plant parts or of the entire plant, they may cause growths like warts, or they may kill the plant.

A plant often becomes infected after hyphae enter through stomata (pores) in the leaf (Fig. 25–18) or stem or through wounds in the plant body. Alternatively, the fungus may produce an enzyme called cutinase that dissolves the waxy cuticle that covers the surface of leaves and stems. After dissolving the cuticle, the fungus easily invades the plant tissues. As the mycelium grows, it may remain mainly between the plant cells or it may penetrate the cells. Parasitic fungi often produce special hyphal branches called **haustoria** (sing., *haustorium*) that penetrate the host cells and obtain nourishment from the cytoplasm.

Some important plant diseases caused by ascomycetes are powdery mildew; chestnut blight; Dutch elm disease; apple scab; and brown rot, which attacks cherries, peaches, plums, and apricots (Fig. 25–19*a*). Diseases caused by basidiomycetes include smuts and rusts that attack various plants, for example, corn, wheat, oats, and other grains (Fig. 25–19*b*). Certain imperfect fungi also cause plant diseases. Examples include verticillium wilt on potatoes, which is caused by the deuteromycete *Verticillium* sp., and bean anthracnose, which is caused by the deuteromycete *Colletotrichum lindemuthianum*.

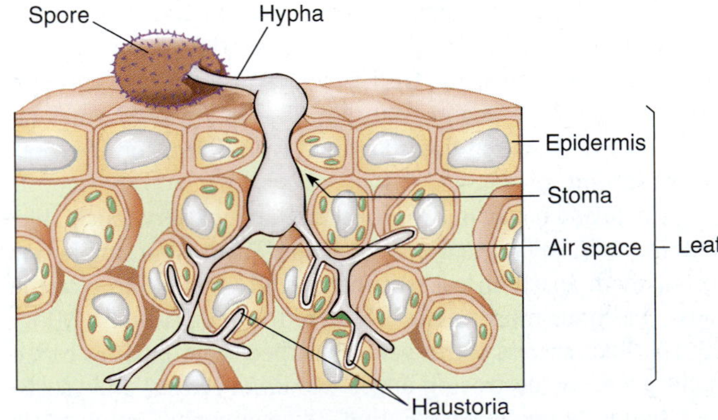

Figure 25–18 How a fungus parasitizes a plant. In this example, the hypha enters the leaf through a stoma and then grows, branching extensively through the internal air spaces and penetrating plant cells with specialized hyphal extensions called haustoria.

(a)

(b)

Figure 25–19 Fungi that cause plant diseases. **(a)** Brown rot of peaches is caused by *Monilinia fruticola,* an ascomycete. Photographed in Oregon. **(b)** Corn smut on an ear of sweet corn is caused by *Ustilago maydis,* a basidiomycete. Photographed in Pennsylvania. *(a, Kathy Merrifield/ Photo Researchers, Inc.; b, Runk/Schoenberger, from Grant Heilman)*

Some fungal parasites, such as the stem rust of wheat, have complex life cycles that involve two or more different host plants and the production of several kinds of spores. For example, wheat rust must infect a barberry plant at one stage in its life cycle. Since this fact was discovered, the eradication of barberry plants in wheat-growing regions has reduced infection with wheat rust. Wheat rust has not been eliminated by eradication of the barberry, however, because the fungus overwinters on wheat at the southern end of the Grain Belt and forms asexual spores. During the spring, wind blows these spores for hundreds of kilometers, reinfecting northern areas of the United States and Canada.

Fungi cause certain animal diseases

Some fungi cause superficial infections in which only the skin, hair, or nails are infected. Ringworm and athlete's foot are examples of superficial fungal infections; both are caused by imperfect fungi. Candidiasis, a yeast infection of mucous membranes of the mouth, throat, or vagina, is among the most common fungal infections; it too is caused by an imperfect fungus.

Other fungi cause systemic infections that infect internal tissues and organs and may spread through many regions of the body. Histoplasmosis, for example, is a serious infection of the lungs caused by inhaling spores of a fungus common in soil contaminated with bird feces. Most people in the eastern and midwestern parts of the United States have been exposed to this fungus at some time, and an estimated 40 million Americans have had mild infections by this pathogen (disease-causing organism). Fortunately, the infection is usually confined to the lungs and is of short duration, but if the infection spreads into the blood and from there into the heart, brain, eyes, kidneys, or other parts of the body, it can be serious and sometimes fatal.

Most pathogenic fungi are opportunists that cause infections only when the body's immunity is lowered. For example, the deuteromycete *Aspergillus fumigatus* does not usually cause disease in humans but does cause aspergillosis in people with defective immune systems, such as patients with AIDS. During the course of this disease, the fungus can invade the lungs, heart, brain, kidneys, and other vital organs and can cause death. Other patients at high risk of acquiring life-threatening fungal infections include those with cancer or organ transplants.

Some fungi produce poisonous compounds collectively called **mycotoxins.** A few species of *Aspergillus,* for example, produce potent mycotoxins called *aflatoxins* that harm the liver and are known carcinogens. Foods on which aflatoxin-producing fungi commonly grow include peanuts, pecans, corn, and other grains. Other foods that may contain traces of aflatoxins include animal products such as milk, eggs, and meat (from animals that consumed feed contaminated by aflatoxin). Avoiding aflatoxin in the diet is impossible, but exposure should be minimized as much as possible. Any human food or animal forage product that has become moldy should be suspected of aflatoxin contamination and discarded.

SUMMARY WITH KEY TERMS

I. **Fungi** are eukaryotes with cell walls containing **chitin.**
   A. Fungi lack chlorophyll and are heterotrophic; they absorb predigested food molecules.
   B. A fungus may be unicellular **(yeast),** coenocytic (chytrids, molds), or multicellular (mold).
      1. The structure of a multicellular fungus consists of long, branched **hyphae** that form a **mycelium.**
      2. In the **zygomycetes,** the hyphae are **coenocytic** (undivided by septa).
      3. In other fungi, perforated **septa** are present that divide the hyphae into individual cells.
   C. Most fungi reproduce both sexually and asexually by means of **spores.** When a fungal spore lands in a suitable spot, it germinates.
      1. Some hyphae infiltrate the substrate and digest its organic compounds.
      2. Spores are produced on aerial hyphae.

II. Fungi are classified into phyla based largely on their modes of sexual reproduction.
   A. **Chytridiomycetes** are simple aquatic fungi that produce motile cells with single, posterior flagella for both sexual (gametes) and asexual reproduction **(zoospores).**
      1. *Chytridium* and *Allomyces* are representatives of this group.
      2. Chytridiomycetes may have been the earliest fungal group to evolve from the ancient flagellated protist hypothesized as the common ancestor of all fungi.
   B. **Zygomycetes** produce both asexual spores and sexual spores **(zygospores).** *Pilobolus* and the black bread mold *Rhizopus* are representatives of this group.
   C. **Ascomycetes** produce asexual spores called **conidia;** sexual spores called **ascospores** are produced in **asci.** Ascomycetes include yeasts, cup fungi, morels, truffles, and pink, brown, and blue-green molds.
   D. **Basidiomycetes** produce sexual spores called **basidiospores** on the outside of a **basidium;** basidia develop on the surface of **gills** in mushrooms, a type of **basidiocarp.** Basidiomycetes include mushrooms, puffballs, bracket fungi, rusts, and smuts.
   E. **Imperfect fungi (deuteromycetes)** lack a sexual stage. Most reproduce asexually by forming conidia. Members of this group include *Aspergillus tamarii* (used to produce soy sauce), some species of *Penicillium,* and fungi that cause certain human infections.
   F. A **lichen** is a symbiotic combination of a fungus and a phototroph (an alga or cyanobacterium); the association is one of controlled parasitism. Lichens have three main growth forms: crustose, foliose, and fruticose.

III. Fungi are ecologically significant.
   A. Many fungi are **saprotrophs** that break down organic compounds.
   B. **Mycorrhizae** are mutualistic relationships between fungi and the roots of plants. The fungus supplies water and nutrient minerals to the plant, and the plant secretes organic compounds needed by the fungus.
   C. Lichens play an important role in soil formation.

IV. Fungi have both positive and negative economic importance.
   A. Mushrooms, morels, and truffles are used as food; yeasts are vital in the production of beer, wine, and bread; certain fungi are used to produce cheeses and soy sauce.
   B. Fungi are used to make penicillin and other antibiotics; ergot is used to produce certain drugs; other fungi make citric acid and many other industrial chemicals.
   C. Fungi cause many plant diseases, including wheat rust, Dutch elm disease, and chestnut blight; they cause human diseases such as ringworm, athlete's foot, candidiasis, and histoplasmosis.

POST-TEST

1. Most fungi produce cell walls containing the polymer ____________. (a) chitin (b) penicillin (c) cellulose (d) aflatoxin (e) lysergic acid
2. Which of the following fungi does *not* have a mycelium? (a) black bread mold (b) yeast (c) decomposer cup fungus (d) cultivated mushroom (e) *Penicillium*
3. The condition described as $n + n$ is said to be (a) monokaryotic (b) diploid (c) a primary mycelium (d) coenocytic (e) dikaryotic
4. With the exception of chytridiomycetes, fungi are generally disseminated by (a) water currents (b) fragmentation of hyphae (c) soredia (d) airborne spores (e) flagellated zoospores
5. Which statement is *not* true of the chytridiomycetes? (a) they are simple aquatic fungi (b) they produce motile cells with single, posterior flagella (c) they have both sexual and asexual reproduction (d) the black bread mold is a representative of this group (e) they are the most primitive group of fungi
6. Which statement is *not* true of the zygomycetes? (a) they are simple, aquatic fungi (b) their sexual spores are called zygospores (c) they have both sexual and asexual reproduction (d) the black bread mold is a representative of this group (e) they have coenocytic hyphae
7. Which statement is *not* true of the ascomycetes? (a) their sexual spores are produced in asci (b) they produce motile cells with single, posterior flagella (c) they have both sexual and asexual reproduction (d) their asexual spores are called conidia (e) some species cause serious plant diseases
8. The ascomycete life cycle includes (a) an alternation of generations between haploid and diploid thalli (b) the formation of a thick zygosporangium (c) the production of eight haploid ascospores within an ascus (d) intertwined hyphae to form a basidiocarp (e) the production of ascospores, zoospores, and conidia at different stages
9. Which statement is *not* true of the basidiomycetes? (a) they have a diploid thallus that produces zoospores (b) their sexual spores are called basidiospores (c) they produce a secondary mycelium with $n + n$ hyphae (d) mushrooms, bracket fungi, and puffballs are examples of this group (e) basidiomycetes include both edible and poisonous species
10. The familiar portion of a mushroom is actually a large fruiting body called a(an) ____________. (a) ascocarp (b) basidium (c) basidiocarp (d) gametangium (e) ascus
11. Which statement is *not* true of deuteromycetes? (a) many are ascomycetes that lost the ability to reproduce by forming ascospores (b) they are also known as imperfect fungi (c) they have both sexual and asexual reproduction (d) their asexual spores are called conidia (e) some are important plant pathogens
12. A ____________ is a symbiotic association between a phototroph and a fungus. (a) soredium (b) basidiomycete (c) lichen (d) haustorium (e) saprotroph
13. Which characteristic is true of *all* fungi? (a) saprotrophic (b) parasitic (c) nonflagellated (d) pathogenic (e) heterotrophic
14. Mutualistic relationships between fungi and the roots of plants are called (a) lichens (b) mycorrhizae (c) deuteromycetes (d) haustoria (e) conidiophores
15. *Agaricus brunnescens* (a) is the meadow mushroom commonly cultivated for food (b) is the yeast that ferments wine and beer (c) produces the unique flavor of many cheeses (d) is a highly toxic mushroom (e) has been ingested for its hallucinogenic properties

REVIEW QUESTIONS

1. What characteristics distinguish fungi from other organisms?
2. How does the body of a typical yeast differ from that of a mold?
3. What is the ecological importance of saprotrophic fungi? Of lichens? Of mycorrhizae?
4. Which fungal phylum is thought to be the most primitive? What evidence supports your answer?
5. Diagram the life cycle of the black bread mold.
6. Distinguish among each of the following: (a) ascocarp, ascus, and ascospore (b) basidiocarp, basidium, and basidiospore (c) conidium and ascospore (d) ascus and basidium (e) sporangium and conidium.
7. Some dictionaries erroneously define a morel as a type of mushroom. Why isn't a morel a mushroom?
8. Briefly describe three important fungal diseases of plants and three fungal diseases of humans.

YOU MAKE THE CONNECTION

1. How are the life cycles of the marine alga *Ulva* (see Fig. 24–16) and *Allomyces* similar?
2. What measures can you suggest to prevent bread from becoming moldy?
3. If you do not see any mushrooms in your lawn, can you conclude that no fungi live there? Why or why not?
4. Explain the following statement: Mushrooms are like the tips of icebergs.
5. Biologists have discovered that many mycorrhizal fungi are sensitive to a low pH. What human-caused environmental problem might prove catastrophic for these fungi? How might this problem affect their plant partners?

RECOMMENDED READINGS

Alexopoulos, C.J., C.W. Mims, and M. Blackwell. *Introductory Mycology,* 4th ed. John Wiley & Sons,, New York, 1996. A mycology textbook that provides comprehensive information on the biology of the fungi.

Ariniello, L. "Protecting Paradise." *BioScience,* Vol. 49, No. 10, Oct. 1999. Ants that farm fungi have an intricate, highly sophisticated agricultural system.

Cislaghi, C., and P.L. Nimis. "Lichens, Air Pollution, and Lung Cancer." *Nature,* Vol. 387, 29 May 1997. The authors demonstrated a relationship between lung cancer and air pollution by comparing biodiversity maps of pollution-sensitive lichens with mortality maps in northeastern Italy.

Conover, A. "Hunting Slime Molds." *Smithsonian,* Mar. 2001. Examines the unusual world of slime molds through the eyes of biologists who study them.

Lincoff, G.H. *National Audubon Society Field Guide to North American Mushrooms,* 12th Printing. Alfred A. Knopf, New York, 1997. A guide to common fungi with color photographs of each species discussed.

Milius, S. "Yikes! The Lichens Went Flying." *Science News,* Vol. 158, 26 Aug. 2000. Examines lichen biology and conservation issues.

Murawski, D.A. "Fungi." *National Geographic,* Vol. 192, No. 2, Aug. 2000. A beautiful photoessay about fungi.

Sharnoff, S.D. "Lichens: More Than Meets the Eye." *National Geographic,* Vol. 191, No. 2, Feb. 1997. Beautiful photographs accompany this essay on the ecological and economic value of lichens.

Stutz, B. "Midsummer's Madness." *Natural History,* Jul./Aug. 1999. Examines the fascinating hobby of collecting edible mushrooms.

Walsh, R. "Seeking the Truffle." *Natural History,* Jan. 1996. A fascinating account of the history of truffles in France.

- Visit our Web site at **http://www.info.brookscole.com/solomonbergmartin** for links to chapter-related resources on the World Wide Web. Additional on-line materials relating to this chapter can also be found on our Web site.

See chapter activity on BioActive Learner CD for additional help in mastering the chapter's material. Icon location in the chapter's margins shows which topics have tutorials or simulations in the CD.

26

The Plant Kingdom: Seedless Plants

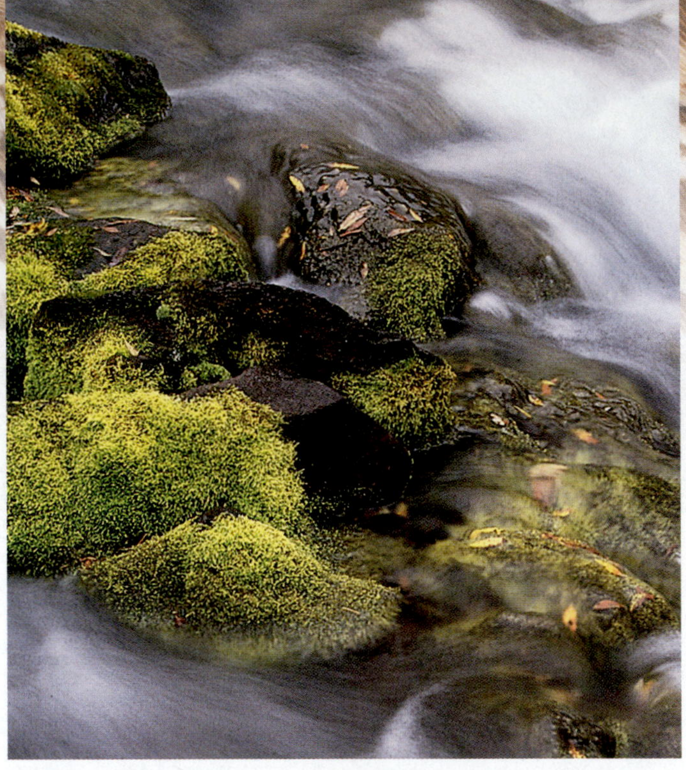

Moss-covered rocks. Shown is Marron Creek in Snowmass Wilderness, Colorado. *(Sydney Karp/Photo/Nats, Inc.)*

LEARNING OBJECTIVES

After you have studied this chapter you should be able to

1. Discuss some of the environmental challenges of living on land, and describe several adaptations that plants possess to meet these challenges.
2. Name the protist group from which plants are thought to have descended and describe supporting evidence.
3. Summarize the features that distinguish bryophytes from green algae and from other plants.
4. Name and briefly describe the three phyla of bryophytes.
5. Diagram the life cycle of mosses and compare their gametophyte and sporophyte generations.
6. Discuss the features that distinguish ferns and other seedless vascular plants from algae and from other plants.
7. Name and briefly describe the four phyla of seedless vascular plants.
8. Diagram the life cycle of ferns and compare their sporophyte and gametophyte generations.
9. Compare the generalized life cycles of homosporous and heterosporous plants.

About 440 million years ago (mya) planet Earth would have seemed a most inhospitable place because, although life abounded in the oceans, it did not yet exist in abundance on land. The oceans were filled with vast numbers of fish, mollusks, and crustaceans as well as countless microscopic algae, and the water along rocky coastlines was home to large seaweeds. Occasionally, perhaps, an animal would crawl out of the water onto land, but it never stayed there permanently because there was little to eat on land—not a single blade of grass, no fruit, and no seeds.

During the next 30 million years, a time corresponding roughly to the Silurian period of the Paleozoic era in geological time (see Table 20–1), plants appeared in abundance and colonized the land. Where did they come from? Although plants living today exhibit great diversity in size, form, and habitat, they all are thought to have evolved from a common ancestor: an ancient green alga. Biologists infer this relationship because modern green algae share a number of biochemical and metabolic traits with modern plants. Both green algae and plants contain the same photosynthetic pigments: chlorophylls *a* and *b* and accessory pigments, the yellow and orange carotenoids, including xanthophylls (yellow pigments) and carotenes (orange pigments). Also, both store their excess carbohydrates as starch and possess cellulose as a major component of their cell walls (see Chapter 31). In addition, plants and some green algae share certain details of cell division, including the formation of a cell plate during cytokinesis (see Chapter 9).

Recent ultrastructural and molecular data indicate that plants probably descended from a group of green algae called **charophytes** or **stoneworts** (see Fig. 24–14*e*). Molecular comparisons, particularly of DNA and RNA sequences, have provided compelling evidence that charophytes are closely allied to plants. These data include comparisons among plants and various green algae, including charophytes, of chloroplast DNA sequences, of certain nuclear DNA sequences, and of ribosomal RNA sequences. In each case, the closest match occurs between charophytes and plants, indicating that modern charophytes and plants probably share a common charophyte ancestor.

Today the plant kingdom comprises hundreds of thousands of species that live in varied habitats, from frozen Arctic tundra to lush tropical rain forests to harsh deserts to moist stream banks, as shown in the photograph. Plants are complex multicellular organisms that range in size from minute, almost microscopic duckweeds to massive giant sequoias, some of the largest organisms that have ever lived.

PLANTS HAVE ADAPTED TO LIFE ON LAND

What are some of the features of plants that have permitted them to colonize so many different environments? One important difference between plants and algae is that the aerial portion of a plant is covered by a waxy **cuticle.** A cuticle is essential for existence on land because it helps prevent the desiccation, or drying out, of plant tissues by evaporation. Most plants that are adapted to moister habitats may have a very thin layer of wax, whereas those adapted to drier environments often possess a thick, crusty cuticle. (Many desert plants also possess a reduced surface area, particularly of leaves, that minimizes water loss.)

Plants obtain the carbon they need for photosynthesis from the atmosphere as carbon dioxide (CO_2). For CO_2 to be fixed into organic molecules such as sucrose, it must first diffuse into the chloroplasts that are inside green plant cells. Because the external surfaces of leaves and stems are covered by a waxy cuticle, however, gas exchange through the cuticle between the atmosphere and the insides of cells is negligible. To facilitate gas exchange, tiny pores called **stomata** (sing., *stoma*) dot the surfaces of leaves and stems of almost all plants; algae lack stomata.

Most plants possess multicellular sex organs called **gametangia** (sing., *gametangium*), whereas the gametangia of algae are unicellular (Fig. 26–1). Each plant gametangium has a layer of sterile (nonreproductive) cells that surrounds and protects the delicate gametes (eggs and sperm cells). In plants, the fertilized egg develops into a multicellular **embryo** (young plant) within the female gametangium. Thus, the embryo is protected during its development. In algae, the fertilized egg develops away from its gametangium; in some algae, the gametes are released before fertilization, whereas in others the fertilized egg is released.

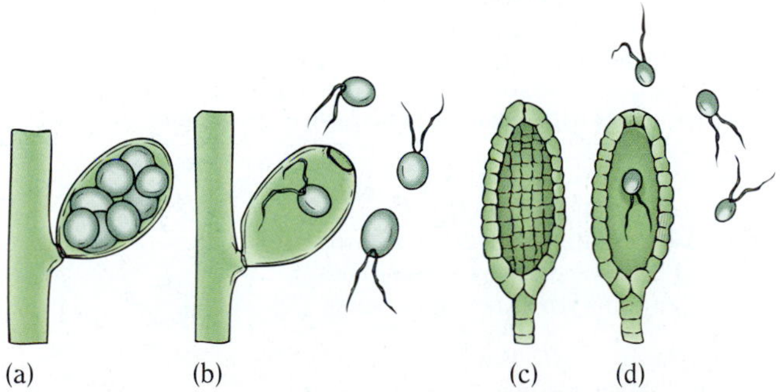

Figure 26–1 Generalized reproductive structures of algae and plants. **(a, b)** In algae, gametangia are generally unicellular. When the gametes are released, only the wall of the original cell remains. **(c, d)** In plants, the gametangia are multicellular, but only the inner cells become gametes. The gametes are surrounded by a protective layer of sterile cells.

The plant life cycle alternates haploid and diploid generations

Plants have a clearly defined **alternation of generations** in which they spend part of their lives in a multicellular haploid stage and part in a multicellular diploid stage[1] (Fig. 26–2). The haploid portion of the life cycle is called the **gametophyte generation** because it gives rise to haploid gametes by mitosis. When two gametes fuse, the diploid portion of the life cycle, called the **sporophyte generation,** begins. The sporophyte generation produces haploid spores by the process of meiosis; these spores represent the first stage in the gametophyte generation.

Let us examine alternation of generations more closely. The haploid gametophytes produce **antheridia** (male gametangia), in which sperm cells form, and/or **archegonia** (female gametangia), each bearing a single egg (Fig. 26–3). Sperm cells reach the female gametangium in a variety of ways, and one sperm cell fertilizes the egg to form a **zygote,** or fertilized egg.

The diploid zygote is the first stage in the sporophyte generation. The zygote divides by mitosis and develops into a multicellular embryo, the young sporophyte plant. Embryo development takes place within the archegonium; thus, during its development, the embryo is protected. Eventually the embryo

[1] For convenience we limit our discussion to plants that are not polyploid, although polyploidy is very common in the plant kingdom. We therefore use the terms diploid and $2n$ (and haploid and n) interchangeably, although these terms are not actually synonymous.

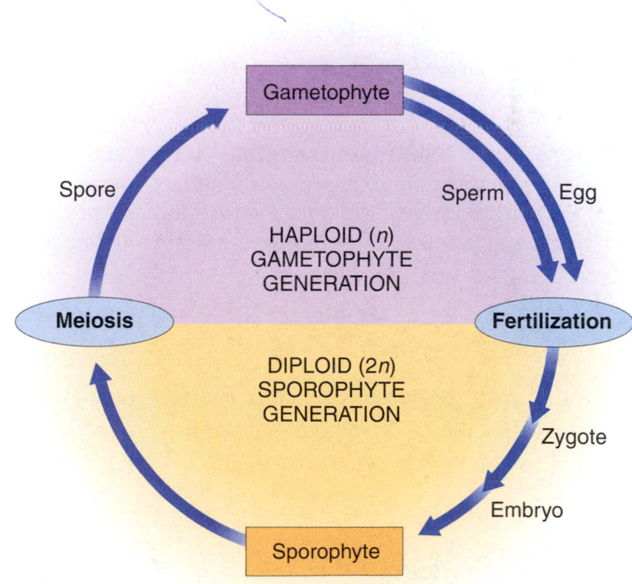

Figure 26–2 The basic plant life cycle. Plants alternate generations, spending part of the cycle in a haploid gametophyte stage and part in a diploid sporophyte stage. All plants have modifications of this cycle. Depending on the plant group, the haploid or the diploid stage may be greatly reduced.

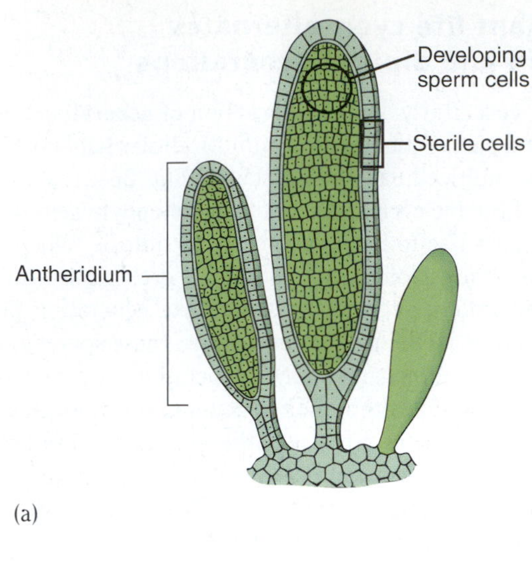

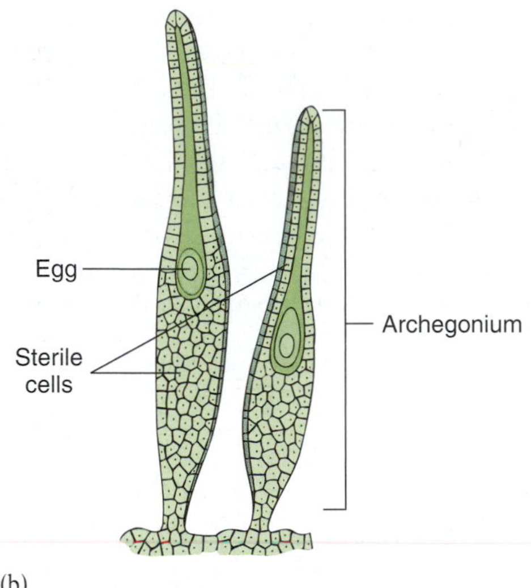

Figure 26–3 Plant gametangia. **(a)** Each antheridium, the male gametangium, produces numerous sperm cells. **(b)** Each archegonium, the female gametangium, produces a single egg. Shown are generalized moss gametangia.

grows into a mature sporophyte plant. The mature sporophyte has special cells called *sporogenous cells* (spore-producing cells, also called *spore mother cells*) that divide by meiosis to form haploid **spores.**

Fertilization of egg by sperm cell ⟶ zygote ⟶ embryo ⟶ mature sporophyte plant ⟶ sporogenous cells

All plant spores are produced by meiosis, in contrast with algae and fungi, which may produce spores by meiosis or mitosis. The spores represent the first stage in the gametophyte generation. Each spore divides by mitosis to produce a multicellular gametophyte, and the cycle continues. Plants therefore have an alternation of generations, alternating between a haploid gametophyte generation and a diploid sporophyte generation.

Meiosis of sporogenous cells ⟶ spores ⟶
mature gametophyte plants → archegonia ⟶ eggs
mature gametophyte plants → antheridia ⟶ sperm cells

THERE ARE FOUR MAJOR GROUPS OF PLANTS

The plant kingdom consists of four major groups of plants: bryophytes, seedless vascular plants, gymnosperms, and flowering plants (Fig. 26–4; see also Table 26–1, which is an overview of living plant phyla discussed in Chapters 26 and 27). The mosses and other bryophytes are small nonvascular plants that lack a specialized vascular, or conducting, system to transport dissolved nutrients, water, and essential minerals throughout the plant body. In the absence of such a system, bryophytes rely on diffusion and osmosis to obtain needed materials. This reliance means that bryophytes are restricted in size; if they were much larger, some of their cells could not obtain the necessary materials in sufficient quantities. Bryophytes are known as seedless plants because they do not form seeds, which are reproductive structures discussed in Chapter 27. Bryophytes reproduce and disperse primarily via haploid spores. Recent molecular and fossil evidence, discussed later in the chapter, indicates that bryophytes may have been some of the earliest plants to colonize land.

The other three groups of plants—ferns, gymnosperms, and flowering plants—possess vascular tissues and are thus known as vascular plants. The two vascular tissues are **xylem,** for water and mineral conduction, and **phloem,** for conduction of dissolved organic molecules such as sugar. A key step in the evolution of vas-

TABLE 26–1 The Plant Kingdom

Nonvascular plants with a dominant gametophyte generation (bryophytes)
Phylum Bryophyta (mosses)
Phylum Hepaticophyta (liverworts)
Phylum Anthocerotophyta (hornworts)
Vascular plants with a dominant sporophyte generation
Seedless plants
Phylum Pterophyta (ferns)
Phylum Psilotophyta (whisk ferns)
Phylum Sphenophyta (horsetails)
Phylum Lycophyta (club mosses)
Seed plants
Plants with naked seeds (gymnosperms)
Phylum Coniferophyta (conifers)
Phylum Cycadophyta (cycads)
Phylum Ginkgophyta (ginkgoes)
Phylum Gnetophyta (gnetophytes)
Seeds enclosed within a fruit
Phylum Anthophyta (angiosperms or flowering plants)
Class Dicotyledones (dicots)
Class Monocotyledones (monocots)

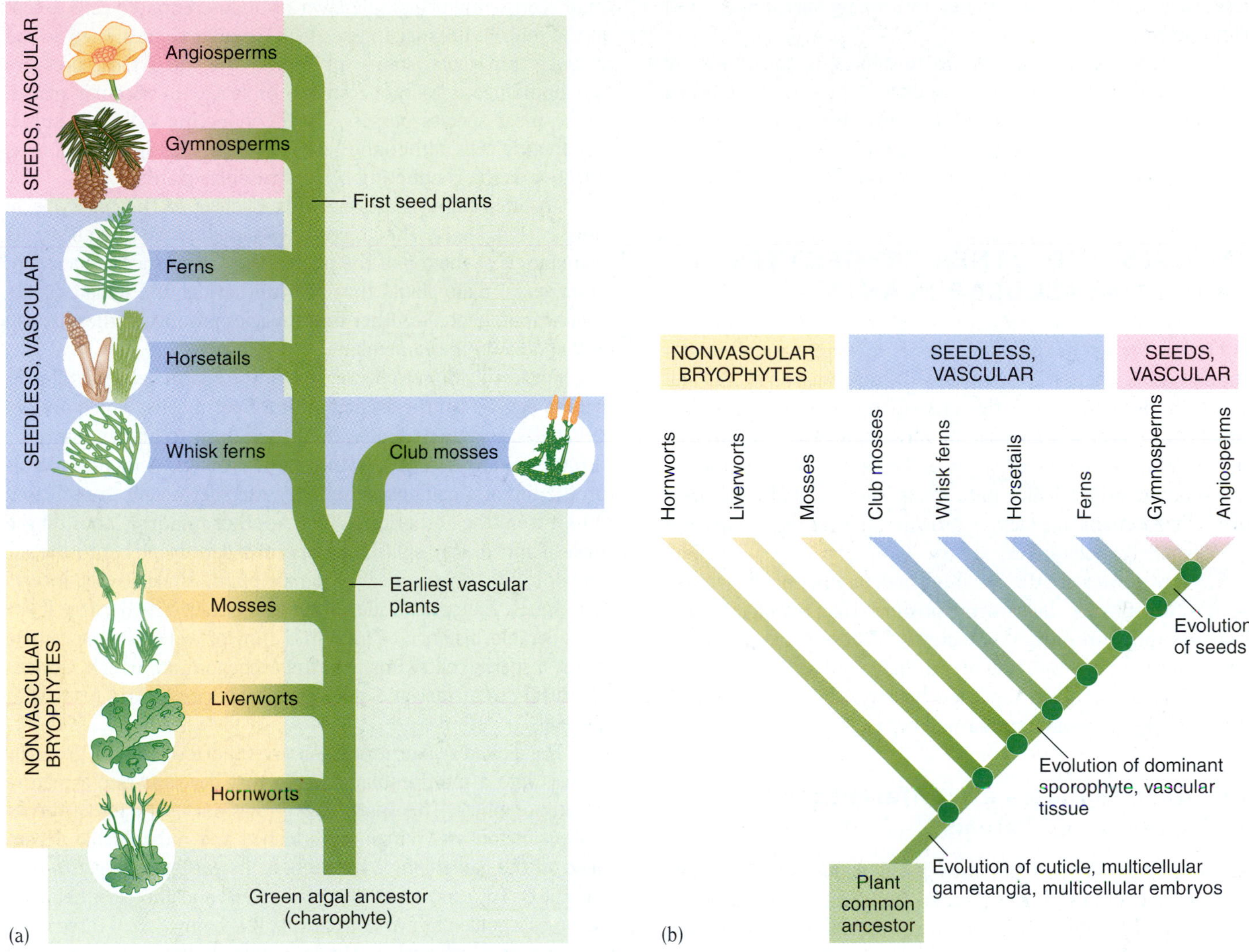

Figure 26–4 Plant evolution. These diagrams illustrate two different ways to show hypothetical evolutionary relationships among living plants, based on current scientific evidence. Although the arrangement of nonvascular, seedless vascular, and seed plant groupings is widely recognized, the exact positions of various phyla remain uncertain. The order in which the hornworts, liverworts, and mosses evolved is not yet resolved, and the exact position of the whisk ferns is still uncertain. **(a)** A classical evolutionary tree of plant groups. The club mosses are placed in a separate line from other seedless vascular plants because their ancestry has been traced to a different group of extinct seedless vascular plants. **(b)** A cladogram of plant phylogeny.

cular plants was the ability to produce **lignin,** a strengthening polymer found in the walls of cells that function for support and conduction (see Chapter 31 for a discussion of plant cell wall chemistry, including lignin). The stiffening property of lignin enabled plants to grow tall (to maximize light interception) and dominate the landscape. The successful occupation of the land by plants in turn made the evolution of terrestrial animals possible by providing them with both habitat and food.

Ferns and their allies (whisk ferns, horsetails, and club mosses) are seedless vascular plants that, like the bryophytes, reproduce and disperse primarily via spores. Seedless vascular plants arose and diversified during the Silurian and Devonian periods of the Paleozoic era, between 420 and 360 mya (discussed later in the chapter). Ferns and fern allies extend back more than 420 million years and were of considerable importance as Earth's dominant plants in past ages. Fossil evidence indicates that many species of these plants were the size of immense trees. Many ferns and most fern allies are extinct today; a few small representatives of the ancient groups survive.

The gymnosperms are vascular plants that reproduce by forming seeds (see Chapter 27). Gymnosperms produce seeds borne exposed (unprotected) on a stem or in a cone. Plants with seeds as their primary means of reproduction and dispersal first appeared about 360 mya, at the end of the Devonian period.

These early seed plants diversified into many varied species of gymnosperms.

The most recent plant group to appear is the flowering plants, or angiosperms, which arose during the early Cretaceous period of the Mesozoic era about 130 mya. Like gymnosperms, flowering plants reproduce by forming seeds. Flowering plants, however, produce seeds enclosed within a fruit.

MOSSES AND OTHER BRYOPHYTES ARE NONVASCULAR PLANTS

The **bryophytes** (from the Greek words meaning "moss plant") comprise more than 15,000 species of mosses, liverworts, and hornworts; bryophytes are the only living nonvascular plants (Table 26–2). Because they have no means for extensive internal transport of water, dissolved sugar, and essential nutrient minerals, bryophytes are typically quite small. They generally require a moist environment for active growth and reproduction, but some bryophytes tolerate dry areas.

The bryophytes are divided into three distinct phyla: mosses (phylum Bryophyta), liverworts (phylum Hepaticophyta), and hornworts (phylum Anthocerotophyta).[2] These three groups of plants differ in many ways and may or may not be closely related. They are usually studied together, however, because they lack vascular tissues and have similar life cycles.

Moss gametophytes are differentiated into "leaves" and "stems"

Mosses (phylum Bryophyta), with about 9000 species, usually live in dense colonies or beds (Fig. 26–5*a*). Each individual plant has tiny, hairlike absorptive structures called *rhizoids,* and an upright, stemlike structure that bears leaflike blades, each normally consisting of a single layer of undifferentiated cells except at the midrib. Because mosses lack vascular tissues, they do not possess true roots, stems, or leaves; the moss structures are not homologous to roots, stems, or leaves in vascular plants. Some moss species possess water-conducting cells and sugar-conducting cells, although these cells are not as specialized or as effective as the conducting cells of vascular plants.

An alternation of generations is clear in the life cycle of mosses (Fig. 26–6). The green moss gametophyte often bears its gametangia at the top of the plant. Many moss species have separate sexes: male plants that bear antheridia and female plants that bear archegonia. Other moss species produce antheridia and archegonia on the same plant.

For fertilization to occur, one of the sperm cells must fertilize the egg within the archegonium. Sperm cells, which are flagellated, are transported from antheridium to archegonium by flowing water, such as splashing rain droplets. A raindrop lands on the top of a male gametophyte, and sperm cells are released into it from the antheridia. When another raindrop lands on the male plant, it may splash the sperm-laden droplet into the air and onto the top of a nearby female plant. Alternatively, insects may touch the sperm-laden fluid and inadvertently carry it for considerable distances. Once in a film of water on the female moss, a sperm cell swims into the archegonium, which secretes chemicals to attract and guide the sperm cells, and fuses with the egg.

The diploid zygote, formed as a result of fertilization, grows by mitosis into a multicellular embryo that develops into a mature moss sporophyte. This sporophyte grows out of the top of the female gametophyte, remaining attached and nutritionally dependent on the gametophyte throughout its existence (Fig. 26–7 on page 561). The sporophyte is initially green and photosynthetic but becomes a golden brown at maturity. It is composed of three main parts: a *foot,* which anchors the sporophyte to the gametophyte and absorbs minerals and nutrients from it; a *seta,* or stalk; and a *capsule,* which contains sporogenous cells (spore mother cells). The capsule of some species is covered by a caplike structure, known as the *calyptra,* which is derived from the archegonium.

[2] Until 1993 plants were classified into divisions rather than phyla. The International Botanical Congress, however, has approved the use of both designations for plants.

TABLE 26–2 A Comparison of Major Groups of Seedless Plants

Plant Group	Dominant Stage of Life Cycle	Representative Genera
Nonvascular; reproduce by spores (bryophytes)		
Mosses (phylum Bryophyta)	Gametophyte: leafy plant	*Polytrichum, Sphagnum, Physcomitrella*
Liverworts (phylum Hepaticophyta)	Gametophyte: thalloid or leafy plant	*Marchantia*
Hornworts (phylum Anthocerotophyta)	Gametophyte: thalloid plant	*Anthoceros*
Vascular; reproduce by spores		
Ferns (phylum Pterophyta)	Sporophyte: roots, rhizomes, and leaves (megaphylls)	*Pteridium, Polystichum, Azolla, Platycerium*
Whisk ferns (phylum Psilotophyta)	Sporophyte: rhizomes and erect stems; no true roots or leaves	*Psilotum*
Horsetails (phylum Sphenophyta)	Sporophyte: roots, rhizomes, erect stems, and leaves (reduced megaphylls)	*Equisetum*
Club mosses (phylum Lycophyta)	Sporophyte: roots, rhizomes, erect stems, and leaves (microphylls)	*Lycopodium, Selaginella*

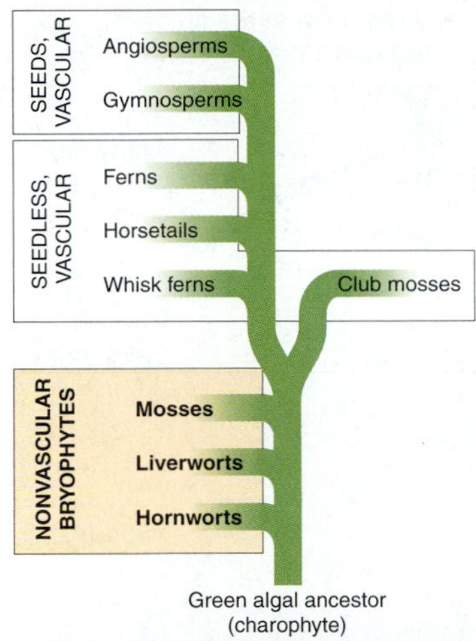

(a)

(b)

(c)

Figure 26–5 Mosses, liverworts, and hornworts. **(a)** A closeup of haircap moss *(Polytrichum commune)* gametophytes, which grow in dense clusters. The haircap moss is a popular ground cover in rock gardens, particularly in Japan. **(b)** Flattened, ribbon-like lobes characterize the gametophyte of the common liverwort *(Marchantia polymorpha). Marchantia* liverworts grow on moist soil, from the tropics to arctic regions. **(c)** The gametophyte with mature sporophytes (the "horns" projecting out of the gametophyte) of the common hornwort *(Anthoceros natans).* *(a and b, Rod Planck/Dembinsky Photo Associates; c, Robert A. Ross)*

The sporogenous cells undergo meiosis to form haploid spores. When the spores are mature, the capsule opens to release the spores. These microscopic cells are transported by wind or rain. If a moss spore lands in a suitable spot, it germinates and grows into a filament of cells called a **protonema.** The protonema, which superficially resembles a filamentous green alga, forms buds, each of which grows into a green gametophyte, and the life cycle continues.

The haploid gametophyte generation is considered the dominant generation in mosses because it is capable of living independently of the diploid sporophyte. In contrast, the moss sporophyte is attached to and nutritionally dependent on the gametophyte.

Mosses make up an inconspicuous but significant part of their environment. They play an important role in forming soil. Because they grow tightly packed together in dense colonies, mosses hold the soil in place and help prevent erosion.

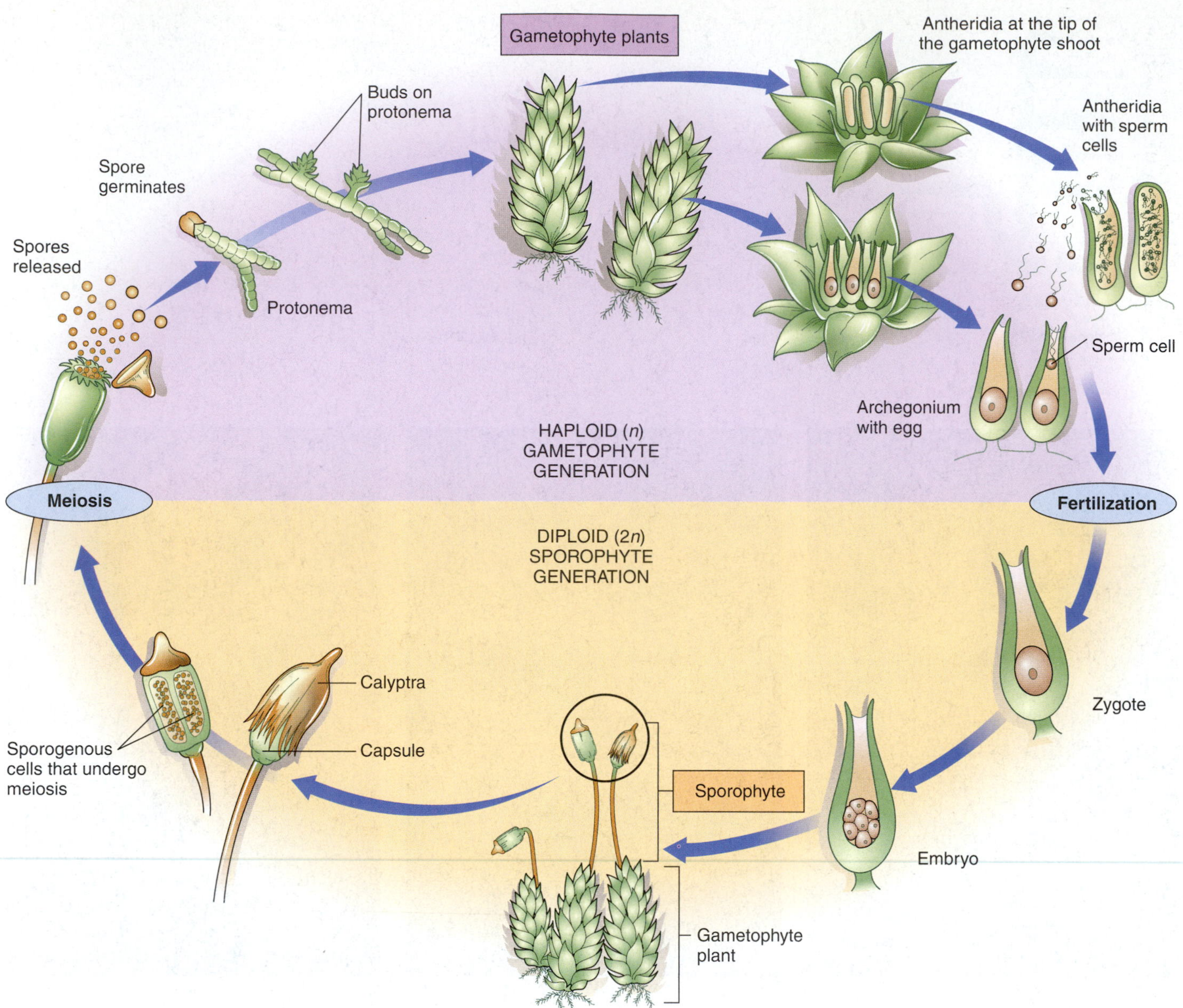

Figure 26-6 The life cycle of mosses. The gametophyte generation is dominant in the moss life cycle. After sexual reproduction takes place in mosses, the sporophyte grows out of the gametophyte. Mosses require water as a transport medium for sperm cells during fertilization.

Commercially, the most important mosses are the peat mosses in the genus *Sphagnum*. One of the distinctive features of *Sphagnum* "leaves" is the presence of many large dead cells that are able to absorb and hold water. This feature makes peat moss particularly beneficial as a soil conditioner. When added to sandy soils, for example, peat moss helps to absorb and retain moisture. In some countries such as Ireland and Scotland, layers of dead peat moss that have accumulated for hundreds of years are extracted from peat bogs, dried, and burned for fuel.

The name "moss" is often commonly misused to refer to plants that are not truly mosses. For example, reindeer moss is a lichen that is a dominant form of vegetation in the Arctic tundra, Spanish moss is a flowering plant, and club moss (discussed later in this chapter) is a relative of ferns.

Liverwort gametophytes are either thalloid or leafy

Liverworts (phylum Hepaticophyta) comprise about 6000 species of nonvascular plants with a dominant gametophyte generation, but the gametophytes of some liverworts are quite different from those of mosses. Their body form is often a flattened, lobed struc-

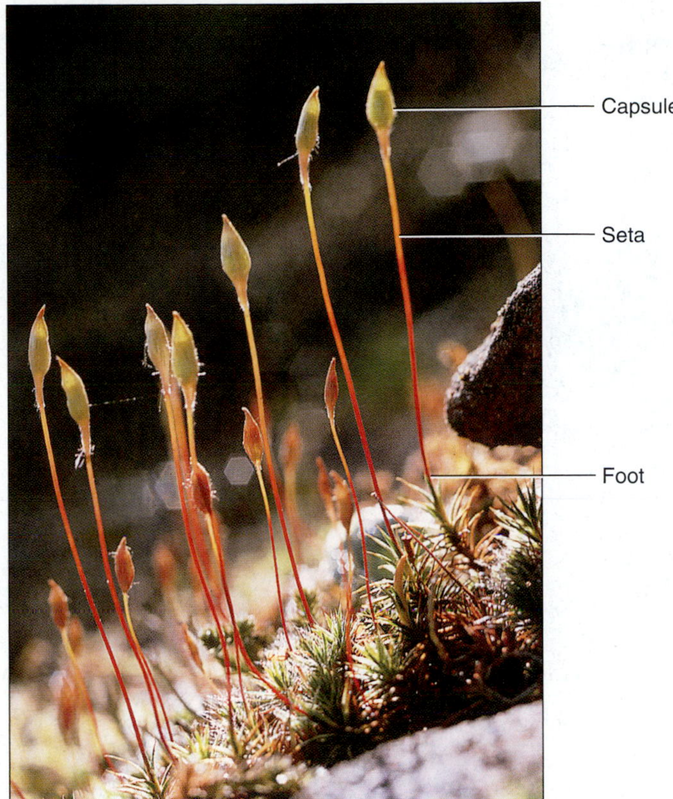

Figure 26–7 Moss sporophytes. The moss sporophytes, each consisting of a foot, seta, and capsule, grow out of the top of the gametophytes. Spores are produced within the capsule at the tip of each sporophyte. Shown is the haircap moss *(Polytrichum commune)*, which prefers light shade and moist soil. *(David Cavagnaro)*

ture called a **thallus** that is not differentiated into leaves, stems, or roots. The common liverwort, *Marchantia polymorpha,* is thalloid (see Fig. 26–5*b*). Liverworts are so named because the lobes of their thalli superficially resemble the lobes of the human liver; *wort* is derived from the Old English word *wyrt,* meaning "plant." On the underside of the liverwort thallus are hairlike rhizoids that anchor the plant to the soil. Other liverworts, known as leafy liverworts, superficially resemble mosses, with leaflike blades, "stems," and rhizoids rather than a lobed thallus. As in the mosses, leafy liverwort "leaves" consist of a single layer of undifferentiated cells. Like other bryophytes, liverworts are small, generally inconspicuous plants that are largely restricted to damp environments. Unlike other plants, including mosses and hornworts, liverworts lack stomata (although some liverworts have surface pores thought to be analogous to stomata).

Liverworts reproduce both sexually and asexually (Fig. 26–8). Their sexual reproduction involves the production of archegonia and antheridia on the haploid gametophyte. In some liverworts, these gametangia are borne on stalked structures called archegoniophores and antheridiophores; *archegoniophores* bear archegonia, and *antheridiophores* bear antheridia. Their life cycle is basically the same as that of mosses, although some of the structures look quite different. The liverwort sporophyte, which is usually somewhat spherical, is attached to the gametophyte, as in mosses.

Some liverworts reproduce asexually by forming tiny balls of tissue called **gemmae** (sing., *gemma*), which are borne in a saucer-shaped structure, the gemmae cup, directly on the liverwort thallus. Splashing raindrops and small animals aid in the dispersal of gemmae. When a gemma lands in a suitable place, it grows into a new liverwort thallus. Liverworts may also reproduce asexually by thallus branching and growth. The individual thallus lobes elongate, and each becomes a separate plant when the older part of the thallus that originally connected the individual lobes dies. Both of these mechanisms of reproduction, gemmae and thallus branching, are asexual because they do not involve fusion of gametes.

Hornwort gametophytes are inconspicuous thalloid plants

Hornworts (phylum Anthocerotophyta) are a small group of about 100 species of bryophytes whose gametophytes superficially resemble those of the thalloid liverworts. Hornworts are found in disturbed habitats such as fallow fields and roadsides.

Hornworts may or may not be closely related to other bryophytes. For example, their cell structure, particularly the presence of a single large chloroplast in each cell, is more like certain algal cells than it is like plant cells. In contrast, mosses, liverworts, and other plants have many disk-shaped chloroplasts per cell.

In hornworts, as for example, the common hornwort *(Anthoceros natans),* archegonia and antheridia are embedded in the gametophyte thallus rather than on archegoniophores and antheridiophores. After fertilization and development, the needle-like sporophyte projects out of the gametophyte thallus, forming a spike or "horn"—hence the name *hornwort* (see Fig. 26–5*c*). A single gametophyte often produces multiple sporophytes. Meiosis occurs within each **sporangium** (pl., *sporangia*), or spore case, and spores are formed. The sporangium splits open from the top to release the spores; each spore has the potential to give rise to a new gametophyte thallus. One unique feature of hornworts is that the sporophyte, unlike those of mosses and liverworts, continues to grow from its base for the remainder of the gametophyte's life.

Bryophytes are used for experimental studies

Scientists use certain bryophytes as experimental models to study many fundamental aspects of plant biology, including genetics, growth and development, plant ecology, plant hormones, and **photoperiodism** (plant responses to varying periods of night and day length; see Chapter 36). As experimental organisms, bryophytes are easy to grow on artificial media and do not require much space because they are so small (Fig. 26–9).

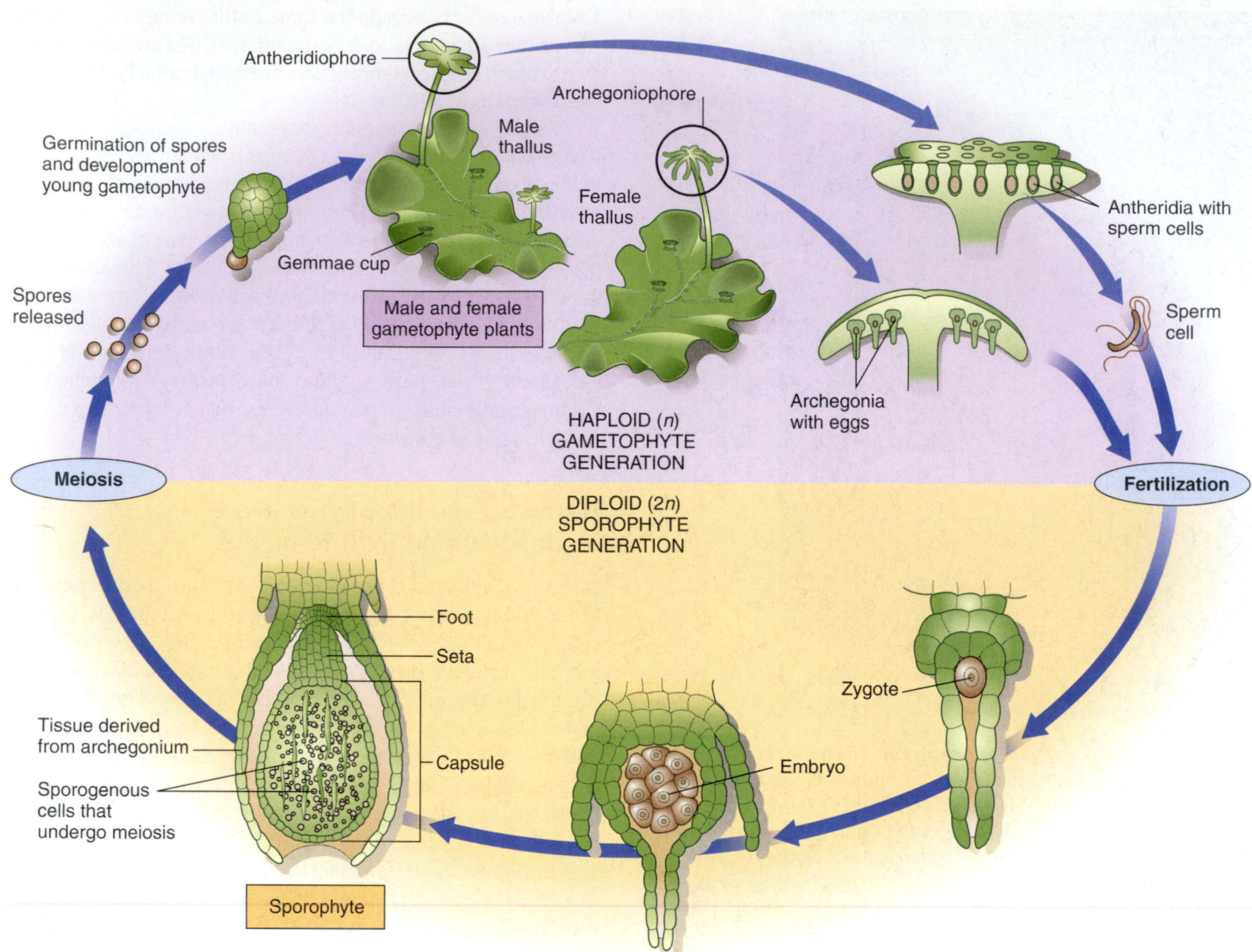

Figure 26-8 The life cycle of the common liverwort *(Marchantia polymorpha)*. The dominant generation is the gametophyte, represented by separate male and female thalli. The stalked, umbrella-shaped structures are the antheridiophores, with antheridia that produce sperm cells, and the archegoniophores (only one is shown) that each bear an egg cell.

Process of Science: The evolution of bryophytes is based on fossil evidence of ancient plants and on structural and molecular evidence of living plants

Plants are a **monophyletic group**—that is, all plants are thought to have evolved from a common ancestral green alga. Fossil evidence indicates that the bryophytes are ancient plants, probably the first group of plants to arise from the common plant ancestor. The fossil record of ancient bryophytes is incomplete, consisting mostly of spores and small tissue fragments, and can be interpreted in different ways. As a result, it does not provide a definite answer on bryophyte evolution.

The oldest known complete fossil plant, dated at about 400 million years old, was reported in 1995. This fossil, obtained from a coal deposit in the United Kingdom, resembles modern-day liverworts in many respects, but its spores are virtually identical to those found in 460-million-year-old rocks. This discovery suggests that liverwort-like plants may have been the earliest plants to colonize land.

Recent structural and molecular evidence, however, supports the hypothesis that *hornworts* may be the most ancient group of plants alive today. A 1998 cladistic analysis using structural data was the first to strongly support this hypothesis, which was reinforced in 2000 by molecular (DNA) evidence. The DNA study of nuclear, chloroplast, and mitochondrial genes from numerous plant species, from bryophytes to flowering plants, suggests that an early plant ancestor gave rise to two lineages, one from which the hornworts descended and the other from which all other plants descended.

Figure 26–9 Mosses as research organisms. This petri dish has been inoculated with offspring from a genetic cross in the moss *Physcomitrella patens*. The cultures that result will be tested for their phenotypes with respect to vitamin requirements. (The culture medium in which the mosses are currently growing is supplemented with all the vitamins required by the parental strains.) *(Courtesy of David J. Cove, University of Leeds)*

SEEDLESS VASCULAR PLANTS INCLUDE FERNS AND THEIR ALLIES

About 11,000 species of ferns exist today. Ferns are especially common in temperate woodlands and tropical rain forests, where they are found in the greatest variety. Three groups of vascular plants—whisk ferns (about 12 species), club mosses (about 1000 species), and horsetails (15 species)—are considered fern allies because their life cycles are similar to those of ferns (Table 26–2).

The most important adaptation found in ferns and their allies, though absent in algae and bryophytes, is the presence of specialized vascular tissues—xylem and phloem—for support and conduction. This system of conduction enables vascular plants to achieve larger sizes than the bryophytes do because water, dissolved minerals, and dissolved sugar can be transported over great distances to all parts of the plant. Although ferns in temperate environments are relatively small, tree ferns in the tropics may grow to heights of 18 m (60 ft). The ferns and fern allies all have true stems with vascular tissues, and most also have true roots and leaves.

The evolution of the leaf as the main organ of photosynthesis has been studied extensively. There are two basic types of leaves: microphylls and megaphylls (Fig. 26–10). The **microphyll,** which is usually small and possesses a single vascular strand, is thought to have evolved from small, projecting extensions of stem tissue. Only one group of living plants, the club mosses, possesses microphylls. In contrast, **megaphylls** are thought to have evolved from stem branches that gradually filled in with additional tissue to form most leaves as we know them today. Megaphylls possess more than one vascular strand, as would be expected if they evolved from branch systems. Ferns, horsetails, gymnosperms, and flowering plants have megaphylls.

Ferns have a dominant sporophyte generation

Most of the 11,000 species of **ferns** (phylum Pterophyta) are terrestrial, although a few have adapted to aquatic habitats (Fig. 26–11*a, b*). Ferns range from the tropics to the Arctic Circle, with most species living in tropical rain forests where they perch high in the branches of trees (Fig. 26–11*c*). In temperate regions, ferns commonly inhabit swamps, marshes, moist woodlands, and stream banks, although some species can be found in fields, rocky crevices on cliffs or mountains, or even deserts. The most common fern species throughout the world is bracken fern *(Pteridium aquilinum)*, a rugged, coarse, weedy plant that grows well on poor soil and is uncommon in the moist habitats favored by other ferns (Fig. 26–11*d*).

The life cycle of ferns involves a clearly defined alternation of generations (Fig. 26–12). The ferns grown as houseplants (such as the Boston fern, maidenhair fern, and staghorn fern) represent the larger, more conspicuous sporophyte generation. The fern sporophyte is composed of a horizontal underground stem, or *rhizome,* that bears leaves, called *fronds,* and true roots. As each young frond first emerges from the ground, it is tightly coiled and resembles the top of a violin, resulting in the name *fiddlehead* (Fig. 26–13 on page 567). As fiddleheads grow, they unroll and expand to form fronds. Fern fronds are usually compound (i.e., the blade is divided into several leaflets), with the leaflets forming beautifully complex leaves. Fronds, roots, and rhizomes all contain vascular tissues.

Spore production usually occurs in certain areas on the fronds, which develop sporangia in which sporogenous cells (spore mother cells) undergo meiosis to form haploid spores. In many species, the sporangia are borne in clusters, called **sori** (sing., *sorus*). When the sporangia burst open and the spores are discharged, they may germinate and grow by mitosis into mature gametophytes.

The mature fern gametophyte, which bears no resemblance to the sporophyte, is a tiny (less than half the size of one of your fingernails), green, often heart-shaped structure that grows flat against the ground. Called a **prothallus** (pl., *prothalli*), the fern gametophyte lacks vascular tissues and has tiny, hairlike absorptive rhizoids to anchor it (Fig. 26–14 on page 567). The prothallus usually produces both archegonia and antheridia on its underside. Each archegonium contains a single egg, whereas numerous sperm cells are produced in each antheridium.

Although ferns are considered more advanced than mosses because they possess vascular tissues, they have retained a primitive fertilization technique: the use of water as a transport medium. The flagellated sperm cells swim to the neck of an

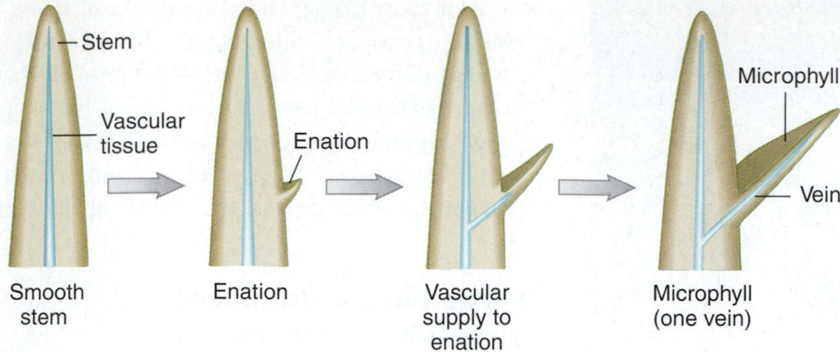

(a) Microphyll evolution

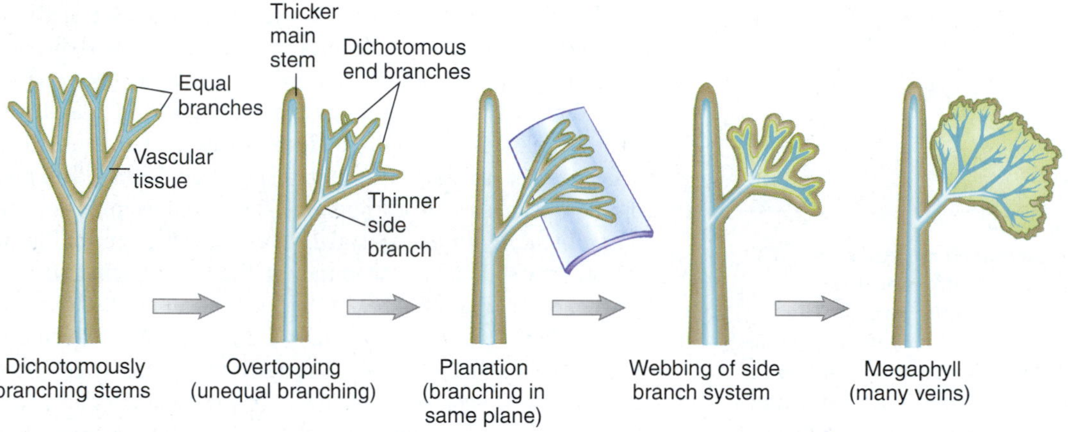

(b) Megaphyll evolution

Figure 26–10 Evolution of microphylls and megaphylls. **(a)** Microphylls probably originated as outgrowths (enations) of stem tissue that later developed a single vascular strand. **(b)** Megaphylls, which are more complex and have multiple veins, probably evolved from the evolutionary modification of side branches. Webbing is the evolutionary process in which the spaces between close branches become filled with chlorophyll-containing cells.

archegonium through a thin film of water on the ground underneath the prothallus. After one of the sperm cells fertilizes the egg, a diploid zygote grows by mitosis into a multicellular embryo. At this stage in its life, the sporophyte embryo is attached to and dependent on the gametophyte, but as the embryo matures, the prothallus withers and dies, and the sporophyte becomes free-living.

The fern life cycle alternates between the dominant, diploid sporophyte with its rhizome, roots, and fronds, and the haploid gametophyte (prothallus). The sporophyte generation is dominant not only because it is larger than the gametophyte but also because it persists for an extended period (most fern sporophytes are perennials), whereas the gametophyte dies soon after reproducing.

Whisk ferns are the simplest vascular plants

Only about 12 species of **whisk ferns** (phylum Psilotophyta) exist today, and the fossil record contains several extinct species. All are relatively simple in structure and lack true roots and leaves but possess vascularized stems. *Psilotum nudum,* a representative whisk fern, has both a horizontal underground rhizome and vertical aerial stems (Fig. 26–15*a* on page 568). Whenever the stem forks or branches, it always divides into two equal halves. Botanists consider this **dichotomous branching** a primitive characteristic. In contrast, when most plant stems branch, one stem is more vigorous and becomes the main trunk.

The upright stems of *Psilotum* are green and are the main organs of photosynthesis. Tiny, round sporangia, borne directly on the erect, aerial stems, contain sporogenous cells that undergo meiosis to form haploid spores. After being dispersed, the spores germinate to form haploid prothalli. The prothalli of whisk ferns are difficult to study because they grow underground. They are nonphotosynthetic, owing to their subterranean location, and they apparently have a symbiotic relationship with mycorrhizal fungi (see Chapter 25) that provides them with sugar and essential minerals.

Most species of whisk ferns are extinct, and the few surviving species are found mainly in the tropics and subtropics. Although whisk ferns do not closely resemble ferns in appearance,

Figure 26–11 Ferns. **(a)** The Christmas fern *(Polystichum acrostichoides)* is green at Christmas, making it a popular holiday decoration. This fern, photographed in the Great Smoky Mountains in Tennessee, has fronds that grow to 0.6 m (2 ft) in length. **(b)** The tiny mosquito fern *(Azolla caroliniana)* is a free-floating aquatic fern that does not resemble "typical" ferns. Although each individual plant grows to only about 1.3 cm (0.5 in) in length, *Azolla* populations sometimes grow so densely across ponds that they reportedly smother mosquito larvae. Photographed in Massachusetts. **(c)** The staghorn fern *(Platycerium bifurcatum)* is native to Australian rain forests and is widely cultivated elsewhere. In nature the staghorn fern is an epiphyte, a plant that grows attached to another organism (in this case, a tree trunk) but derives no nourishment from it. The individual leaves grow to 0.9 m (3 ft) in length. **(d)** Bracken fern *(Pteridium aquilinum)* is one of the few ferns considered weedy in many parts of the world. The fronds, which are as tall as 1.2 m (4 ft), are poisonous to livestock. The fiddleheads, which once were widely consumed by humans, are now known to contain carcinogens. Photographed in Colorado. *(a, Ed Reschke; b, W. Ormerod/Visuals Unlimited; c, Carlyn Iverson; d, Doug Sokell/Visuals Unlimited)*

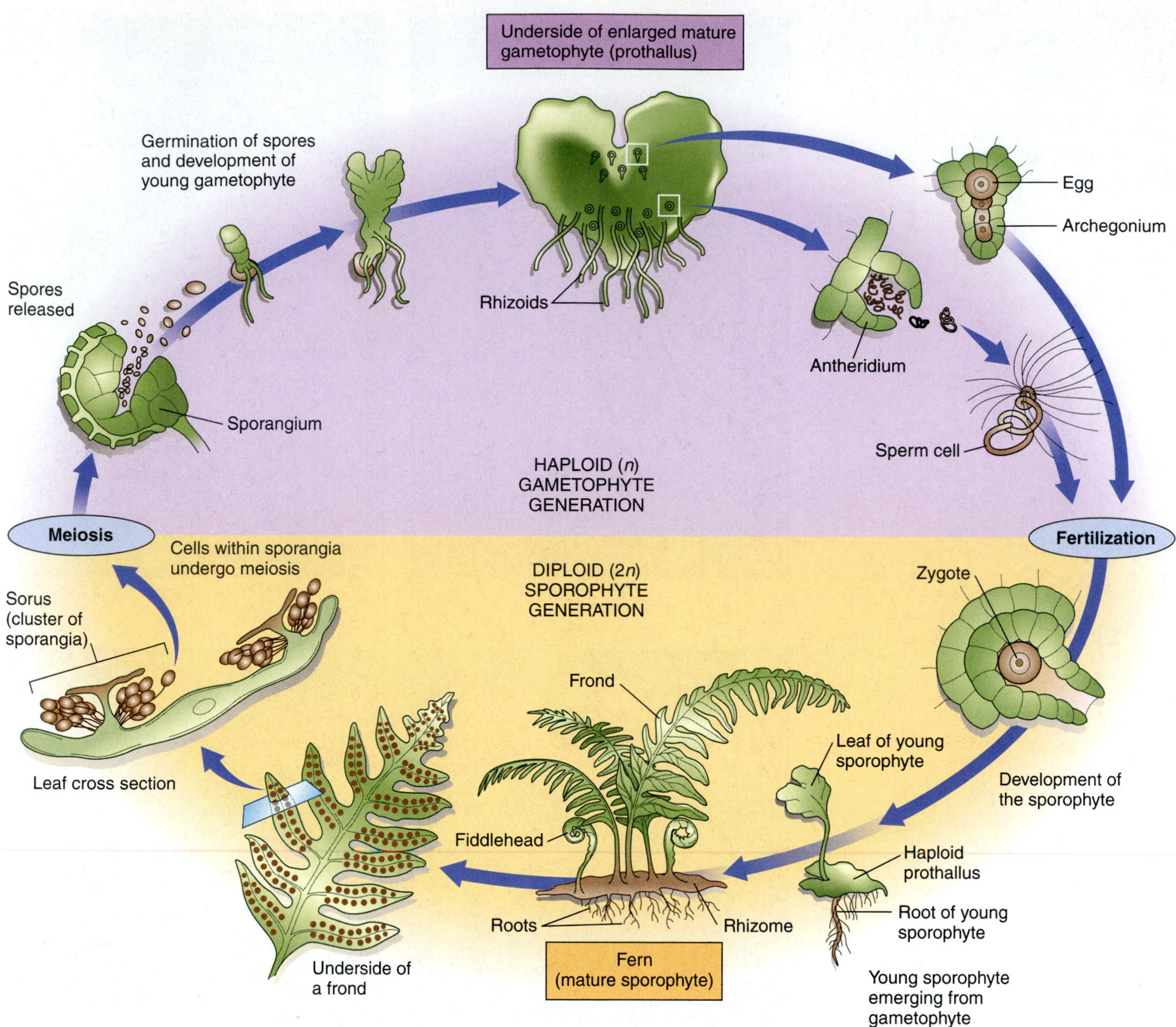

Figure 26–12 The life cycle of ferns. Note the clearly defined alternation of generations between the gametophyte (prothallus) and sporophyte (leafy plant) generations. Fertilization in ferns requires water as a transport medium for sperm cells.

they are considered fern allies because of similarities in their life cycles.

Whisk ferns have been carefully studied in recent years, but botanists disagree about how to interpret their structures. Some botanists consider whisk ferns to be surviving representatives of very primitive vascular plants (see discussion of rhyniophytes later in the chapter). Other botanists think whisk ferns are highly modified relatives of ferns. Recent molecular data, including comparisons of nucleotide sequences of ribosomal RNA, chloroplast DNA, and mitochondrial DNA in living species, support the hypothesis that the whisk ferns are more closely related to ferns than to other seedless vascular plants.

Horsetails have hollow, jointed stems

About 300 mya the **horsetails** (phylum Sphenophyta) were among the dominant plants and grew as large as modern trees (Fig 26–16 on page 569). These ancient horsetails are still significant today because they contributed to Earth's vast coal deposits (see *Focus On: Ancient Plants and Coal Formation*). The few

(text continues on page 569)

FOCUS ON Ancient Plants and Coal Formation

Our industrial society depends on energy from fossil fuels that formed from the remains of ancient organisms. One of our most important fossil fuels is coal, which is burned to produce electricity and to manufacture items made of steel and iron. Although coal is mined as a mineral, it is not an inorganic mineral like gold or aluminum, but an organic material formed from the remains of ancient vascular plants, particularly those of the Carboniferous period (approximately 300 mya). Five main groups of plants contributed to coal formation. Three were seedless vascular plants: the club mosses, horsetails, and ferns. The other two important groups were seed plants: seed ferns (now extinct) and primitive gymnosperms.

It is hard to imagine that relatives of the small, relatively inconspicuous club mosses, ferns, and horsetails of today could have been so significant in forming vast beds of coal. However, many of the members of these groups that existed during the Carboniferous period were giants compared with their modern counterparts and formed immense forests (see Fig. 20–12).

The climate during the Carboniferous period was warm, moist, and mild. Plants in most locations could grow year-round because of the favorable conditions. Forests of these plants often occurred in low-lying, swampy areas that were periodically flooded when the sea level rose. When the sea level receded, these plants would become reestablished.

When these large plants died or were blown over during storms, they decomposed incompletely because they were covered by swamp water. (The anaerobic conditions of the water prevented wood-rotting fungi from decomposing the plants, and anaerobic bacteria do not decompose wood rapidly.) Thus, over time the partially decomposed plant material accumulated and consolidated.

Layers of sediment formed over the plant material each time the water level rose and flooded the low-lying swamps. With time, heat and pressure built up in these accumulated layers and converted the plant material to coal and the sediment layers to sedimentary rock. Much later, geological upheavals raised the layers of coal and sedimentary rock. Coal is usually found in seams, underground layers that vary in thickness from 2.5 cm (1 in) to more than 30 m (100 ft).

The various grades of coal (lignite, bituminous, and anthracite) were formed as a result of the different temperatures and pressures to which the layers were exposed. Coal that was exposed to high heat and pressure during its formation is drier, is more compact (and therefore harder), and has a higher heating value (that is, a higher energy content).

Figure 26–13 Fiddleheads. The tightly coiled young fronds, or fiddleheads, are characteristic of ferns. *(Marion Lobstein)*

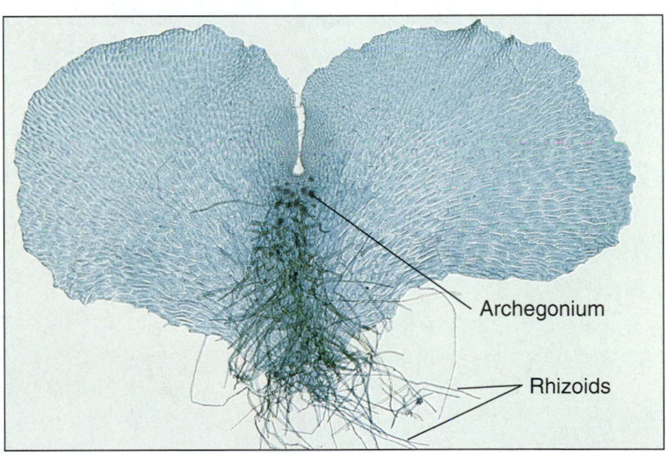

Figure 26–14 Fern prothallus. The dark spots near the notch of the "heart" are archegonia; no antheridia are visible. (In many prothalli, the archegonia and antheridia mature at different times.) Shown is Virginia chain fern *(Woodwardia virginica)*. *(Carolina Biological Supply Company/Phototake NYC)*

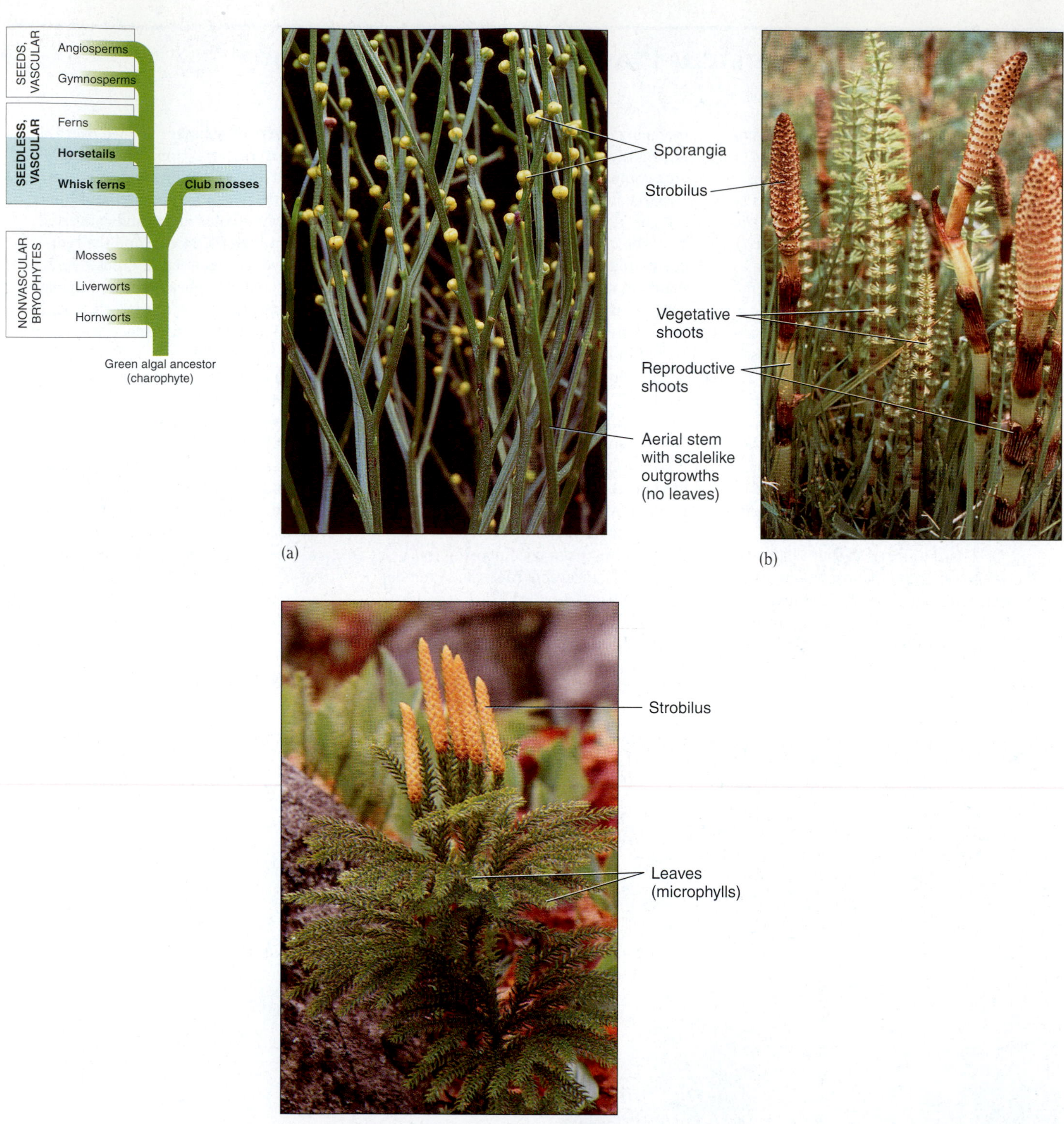

Figure 26–15 Fern allies. **(a)** The sporophyte of *Psilotum nudum,* a whisk fern. The stem is the main organ of photosynthesis in this rootless, leafless, vascular plant. Sporangia, which are initially green but turn yellow as they mature, are borne on short lateral branches directly on the stems. **(b)** *Equisetum telematia,* a horsetail with a wide distribution in Eurasia, Africa, and North America, has unbranched reproductive shoots bearing conelike strobili and separate, highly branched vegetative (nonreproductive) shoots. In some horsetail species, both reproductive and vegetative shoots are unbranched. **(c)** The sporophyte of *Lycopodium* sp., a club moss, has small, scalelike leaves (microphylls) that are evergreen. Spores are produced in sporangia on reproductive leaves clustered in a conelike strobilus *(as shown)* or, in other species, scattered along the stem. *(a, John Arnaldi; b, J. Robert Waaland/Biological Photo Service; c, Ed Reschke)*

both photosynthetic and nutritionally independent at maturity. Like ferns, horsetails require water as a medium for flagellated sperm cells to swim to the egg.

Figure 26–16 Reconstruction of *Calamites* sp. This ancient horsetail was as tall as many modern-day trees—to 20 m (about 65 ft). *Calamites* had an underground rhizome where roots and aerial shoots originated. *(Redrawn from Emberger, L.* Les Plantes Fossiles. *Masson et Cie, Paris, 1968)*

surviving horsetails, about 15 species in the genus *Equisetum,* grow mostly in wet, marshy habitats and are less than 1.3 m (4 ft) tall but extremely distinctive (see Fig. 26–15*b*). They are widely distributed on every continent except Australia.

Horsetails have true roots, stems (both rhizomes and erect aerial stems), and small leaves. The hollow, jointed stems are impregnated with silica, which gives them a gritty texture. Small leaves, interpreted as reduced megaphylls, are fused in whorls at each node (the area on the stem where leaves attach). The green stem is the main organ of photosynthesis. Horsetails are so-named because certain vegetative (nonreproductive) stems have whorls of branches that give the appearance of a bushy horse's tail. In pioneer days horsetails were called "scouring rushes" and were used to scrub out pots and pans along stream banks.

Each reproductive branch of a horsetail bears a terminal conelike **strobilus** (pl., *strobili*). The strobilus is composed of several stalked, umbrella-like structures, each of which bears five to ten sporangia in a circle around a common axis.

The horsetail life cycle is similar in many respects to the fern life cycle. In horsetails, as in ferns, the sporophyte is the conspicuous plant, whereas the gametophyte is a minute, lobed thallus ranging in width from the size of a pinhead to about 1 cm (less than 0.5 in) across. The sporophyte and gametophyte are

Club mosses are small plants with rhizomes and short, erect branches

Like horsetails, **club mosses** (phylum Lycophyta) were important plants millions of years ago, when species that are now extinct often attained great size (Fig. 26–17). These large, treelike plants, like the ancient horsetails, were major contributors to our present-day coal deposits. The 1000 or so species of club mosses living today, such as *Lycopodium* sp. (see Fig. 26–15*c*), are small (less than 25 cm [10 in] tall), attractive plants commonly found in temperate woodlands. They possess true roots; both rhizomes and erect aerial stems; and small, scale-like leaves

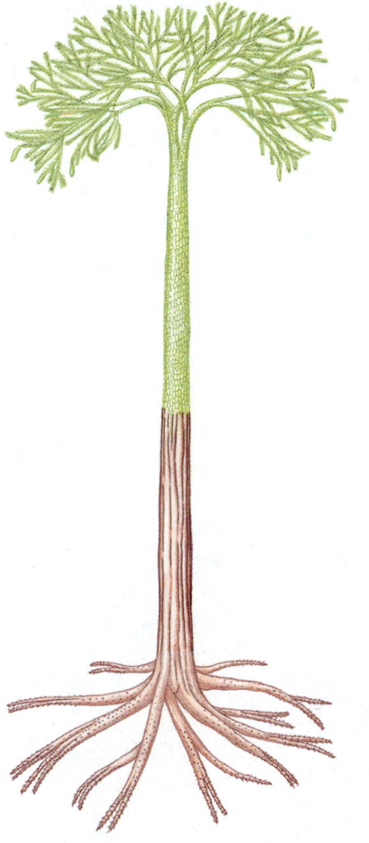

(a)

(b)

Figure 26–17 *Lepidodendron.* **(a)** Reconstruction of *Lepidodendron* sp. This ancient club moss was the size of a large tree—to 40 m (about 130 ft). Numerous fossils of *Lepidodendron* were preserved in coal deposits, particularly in Great Britain and the central United States. **(b)** Closeup of fossil *Lepidodendron* bark. The diamond-shaped structures are scars where leaves were once attached. Each leaf scar is approximately 1.9 cm (0.75 in) wide. *(a, Redrawn from Hirmer, M.* Handbuch der Paläobotanik. *R. Olderbourg, Munich, 1927; b, Ted Clutter/The National Audubon Society Collection/Photo Researchers, Inc.)*

(microphylls). Sporangia are borne on reproductive leaves that are either clustered in conelike strobili at the tips of stems or scattered in reproductive areas along the stem. Club mosses are evergreen and often fashioned into Christmas wreaths and other decorations. In some areas they are endangered from overharvesting.

The fact that common names can be misleading in biology is vividly evident in this group of plants. The most common names for the phylum Lycophyta are "club mosses" and "ground pines," yet these plants are neither mosses nor pines and are most closely allied to the ferns.

Some ferns and club mosses are heterosporous

In the life cycles examined thus far, plants produce only one type of spore as a result of meiosis. This condition, known as **homospory,** is characteristic of bryophytes, horsetails, whisk ferns, and most ferns and club mosses. However, certain ferns and club mosses exhibit **heterospory,** in which they produce two different types of spores: microspores and megaspores. Figure 26–18 illustrates the generalized life cycle of a heterosporous plant.

Spike moss (*Selaginella* sp.), a small, delicate club moss, is an example of a heterosporous plant (Fig. 26–19). Each strobilus usually bears two kinds of sporangia: microsporangia and megasporangia. *Microsporangia* are sporangia that produce *microsporocytes* (also called *microspore mother cells*), which undergo meiosis to form microscopic, haploid **microspores.** Each microspore can develop into a male gametophyte that produces sperm cells within antheridia. *Megasporangia* in the *Selaginella* strobilus produce *megasporocytes* (also called *megaspore mother cells*). When megasporocytes undergo meiosis, they form haploid **megaspores,** each of which can develop into a female gametophyte that produces eggs in archegonia. In *Selaginella,* the development of male gametophytes from microspores and of female gametophytes from megaspores occurs within their respective spore walls, using stored food provided by the sporophyte. As a result, the male and female gametophytes are not truly free-living, unlike the gametophytes of other seedless vascular plants.

Heterospory was a significant development in plant evolution because it was the forerunner of the evolution of seeds. Heterospory is found in the two most successful groups of plants existing today, the gymnosperms and the flowering plants, both of which produce seeds (see Chapter 27).

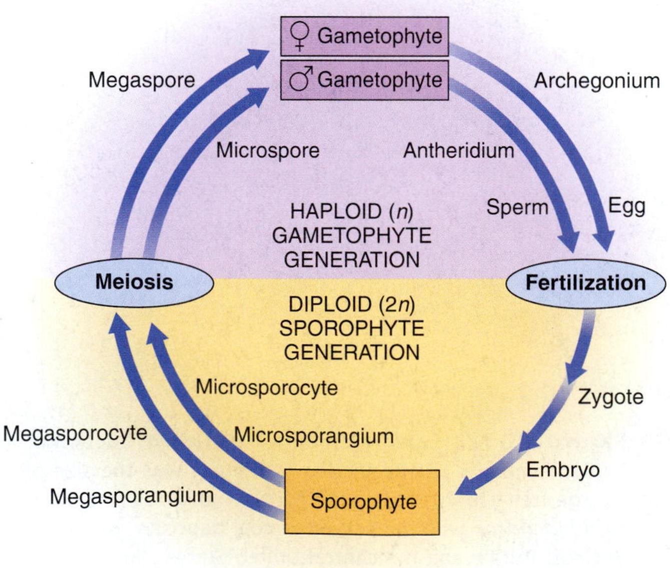

Figure 26–18 The basic life cycle of heterosporous plants. Two types of spores, microspores and megaspores, are produced during the life cycle of heterosporous plants.

Seedless vascular plants are used for experimental studies

Many seedless vascular plants are used as experimental models to study certain aspects of plant biology, such as physiology, growth, and development. Some ferns, for example, produce haploid sporophytes and diploid gametophytes. (Recall that the sporophyte is normally diploid, and the gametophyte is normally haploid.) Research involving alternation of generations implicates environmental conditions, rather than chromosome number, in the expression of the characteristic sporophyte and gametophyte morphologies.

Ferns and other seedless vascular plants have been useful in studying how apical meristems give rise to plant tissues. (As discussed in Chapter 31, an **apical meristem** is the area at the tip of a root or shoot where growth—cell division, elongation, and differentiation—occurs.) Ferns and other seedless vascular plants have a single large *apical cell* located at the center tip of the apical meristem. This apical cell is the source, by mitosis, of all the cells that eventually make up the root or shoot. The apical cell divides in an orderly fashion, and the smaller daughter cells produced by the apical cell in turn divide, giving rise to different parts of the root or shoot. It is possible to trace mature cells in the root or shoot back to their origin from the single apical cell.

Seedless vascular plants arose more than 420 mya

Currently, the oldest known megafossils of early vascular plants are from mid-Silurian (420 mya) deposits in Europe. (Plant *megafossils* are fossilized roots, stems, leaves, and reproductive structures.) Megafossils of several kinds of small, seedless vascular plants have also been discovered from Silurian deposits in Bolivia, Australia, and northwestern China. Microscopic spores of early vascular plants appear in the fossil record earlier than megafossils, suggesting that even older megafossils of simple vascular plants remain to be discovered.

The oldest known vascular plants are assigned to phylum Rhyniophyta which, according to the fossil record, arose some

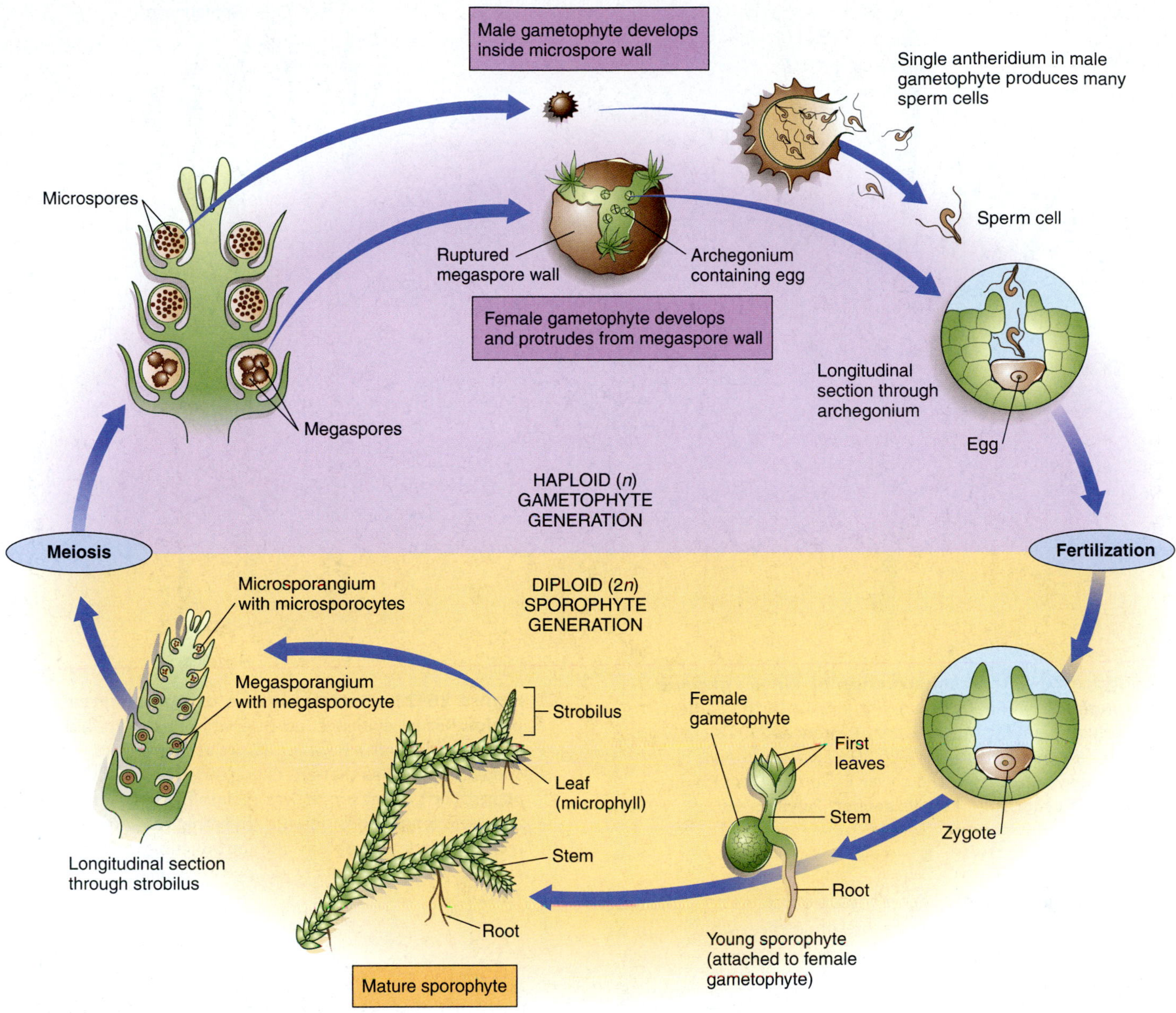

Figure 26–19 The life cycle of spike moss (*Selaginella* sp.). Spike moss is heterosporous, producing two types of spores in one strobilus. The megaspores develop into female gametophytes, and the microspores become male gametophytes.

420 mya and became extinct about 380 mya. The rhyniophytes are so-named because many of the fossils of these primitive plants were found in fossil beds near the village of Rhynie, Scotland. *Rhynia gwynne-vaughanii* is an example of an early vascular plant that superficially resembled whisk ferns in that it consisted of leafless upright stems that branched dichotomously from an underground rhizome (Fig. 26–20). *Rhynia* lacked roots, although it possessed absorptive rhizoids. Sporangia formed at the ends of short branches. The internal structure of its rhizome contained a central core of xylem cells for conducting water and dissolved nutrient minerals.

Process of Science

For more than 60 years, *Rhynia major,* a plant that grew about 50 cm (20 in) tall and probably lived in marshes, was considered a classic example of a rhyniophyte. Fossils indicate that this plant had rhizoids, dichotomously branching rhizomes and upright stems that terminated in sporangia (Fig. 26–21). However, recent microscopic studies of fossil rhizomes indicate that the central core of tissue lacked the xylem cells characteristic of vascular plants. For that reason, *R. major* was reclassified into a new genus, *Aglaophyton,* and is no longer considered a rhyniophyte. Science is an ongoing enterprise, and over time, existing scientific knowledge is reevaluated in light of

Figure 26–20 Reconstruction of *Rhynia gwynne-vaughanii.* This leafless plant, one of Earth's earliest vascular plants, is now extinct. It grew about 18 cm (7 in) tall. *(Redrawn from Edwards, D. "Evidence for the Sporophytic Status of the Lower Devonian Plant* Rhynia gwynne-vaughanii." Rev. Palaeobot. Palynol., *Vol. 29, 1980)*

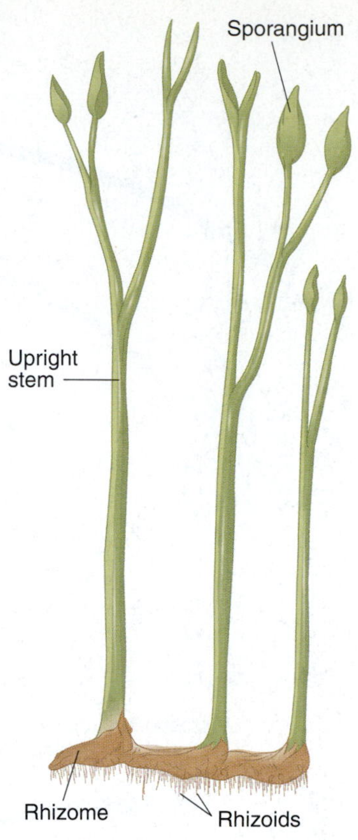

Figure 26–21 Reconstruction of *Aglaophyton major.* Recent evidence indicates that this plant, although superficially similar to other early vascular plants, lacked conducting tissues that are characteristic of vascular plants. For that reason, it was reclassified into a new genus and is no longer considered a rhyniophyte. *(Redrawn from Mauseth, J.D.* Botany: An Introduction to Plant Biology, *2nd ed. Saunders College Publishing, Philadelphia, 1995)*

newly discovered evidence. *Aglaophyton major* is an excellent example of the self-correcting nature of science, which is in a perpetually dynamic state and changes in response to newly available techniques and data.

SUMMARY WITH KEY TERMS

I. Plants are complex multicellular organisms that obtain energy by photosynthesis. Recent ultrastructural and molecular data indicate that plants probably arose from a group of green algae called **charophytes.**
   A. Plants and green algae have similar biochemical characteristics: the same photosynthetic pigments, cell wall components, and carbohydrate storage material.
   B. Plants and green algae share similarities in certain fundamental processes such as cell division.

II. The colonization of land by plants required the evolution of a number of anatomical, physiological, and reproductive adaptations.
   A. Plants possess a waxy **cuticle** to protect against water loss and **stomata** for gas exchange needed for photosynthesis.
   B. An evolutionary trend in plants has been toward a larger, more dominant **sporophyte generation** and a smaller, less dominant **gametophyte generation.**
   C. Most plants produce multicellular **gametangia** with a protective jacket of sterile cells surrounding the gametes. **Antheridia** are gametangia that produce sperm cells, and **archegonia** are gametangia that produce eggs.
   D. Mosses and ferns, although adapted to life on land, have motile sperm cells and require water as a transport medium for fertilization.
   E. Vascular plants possess **xylem** to conduct water and dissolved nutrient minerals, and **phloem** to conduct dissolved sugar.

III. Plant life cycles have an **alternation of generations** in which they spend part of their life cycle as a multicellular haploid gametophyte and part as a multicellular diploid sporophyte.
   A. The mature gametophyte plant produces haploid gametes by mitosis.
   B. These gametes fuse to form a diploid **zygote** during fertilization.
   C. The first stage in the sporophyte generation is the zygote, which develops into a multicellular **embryo** that is protected and nourished by the gametophyte.

D. The mature sporophyte plant produces sporogenous cells (spore mother cells). These undergo meiosis to form haploid **spores,** which are the first stage in the gametophyte generation.

IV. **Mosses** and other **bryophytes** have several adaptations that green algae lack, including a cuticle, stomata, and multicellular gametangia. They are nonvascular (lacking xylem and phloem).
   A. Bryophytes are the only plants with a dominant gametophyte generation. Their sporophytes remain permanently attached and nutritionally dependent on the gametophyte.
   B. Moss gametophytes are green plants that grow from a filamentous **protonema.**
   C. **Liverwort** gametophytes are either leafy or **thalloid.**
   D. **Hornworts** have thalloid gametophytes.

V. Fossil evidence of ancient plants and structural and molecular evidence of living plants are used to infer bryophyte evolution.
   A. The bryophytes may be the first group of plants to arise from the common plant ancestor.
   B. Molecular evidence supports the hypothesis that hornworts may be the most ancient group of plants alive today.

VI. Two types of leaves evolved in the plant kingdom.
   A. **Microphylls,** small leaves thought to have evolved from lateral projections of stem tissue, are characteristic of club mosses.
   B. **Megaphylls,** leaves that probably evolved from branch systems, are characteristic of all vascular plants other than whisk ferns, which lack leaves, and club mosses.

VII. **Ferns** and fern allies have several adaptations that algae and bryophytes lack, including vascular tissues and a dominant sporophyte generation.
   A. Ferns are the largest and most diverse group of seedless vascular plants.
      1. Fern sporophytes have roots, rhizomes, and leaves that are megaphylls. Their leaves, called fronds, bear sporangia in clusters called **sori.** Meiosis in sporangia produces haploid spores.
      2. The fern gametophyte, called a **prothallus,** develops from a haploid spore and bears both archegonia and antheridia.
      3. Reproduction in ferns depends on water as a transport medium for their motile sperm cells.
   B. Sporophytes of **whisk ferns** consist of **dichotomously branching** rhizomes and erect stems; they lack true roots and leaves. *Psilotum* is a representative whisk fern.
   C. **Horsetail** sporophytes have roots, rhizomes, aerial stems that are hollow and jointed, and leaves that are reduced megaphylls. *Equisetum* is a representative horsetail.
   D. Sporophytes of **club mosses** consist of roots, rhizomes, erect branches, and leaves that are microphylls. *Lycopodium* and *Selaginella* are representative club mosses.

VIII. Vascular plants are either homosporous or heterosporous.
   A. **Homospory,** the production of one kind of spore, is characteristic of bryophytes, whisk ferns, horsetails, most club mosses, and most ferns.
   B. **Heterospory,** the production of two kinds of spores (microspores and megaspores), occurs in certain club mosses and ferns and in all seed plants. The evolution of heterospory was an essential step in the evolution of seeds.
      1. **Microspores** give rise to male gametophytes that produce sperm cells.
      2. **Megaspores** give rise to female gametophytes that produce eggs.

IX. The fossil record indicates that several lines of vascular plants had evolved by the mid-Silurian period, about 420 mya.
   A. *Rhynia gwynne-vaughannii* is an example of one of the earliest vascular plants. It superficially resembles whisk ferns.
   B. The ferns and fern allies were Earth's dominant plants in past ages.
   C. Coal formed largely from the prehistoric remains of ancient ferns, club mosses, and horsetails. These plants existed during the Carboniferous period, approximately 300 mya.

POST-TEST

1. The bryophytes (a) include mosses, liverworts, and hornworts (b) include whisk ferns, horsetails, and club mosses (c) are small plants that lack a vascular system (d) both a and c (e) both b and c
2. The waxy layer that covers aerial parts of plants is the (a) cuticle (b) archegonium (c) protonema (d) stoma (e) thallus
3. A strengthening compound found in cell walls of vascular plants is (a) xanthophyll (b) lignin (c) cutin (d) cellulose (e) carotenoid
4. Stomata (a) help prevent desiccation of plant tissues (b) transport water and minerals through plant tissues (c) allow gas exchange for photosynthesis (d) strengthen cell walls (e) produce male gametes
5. The female gametangium, or ____________, produces an egg; the male gametangium, or ____________, produces sperm cells. (a) antheridium; archegonium (b) archegonium; megaphyll (c) megasporangium; antheridium (d) archegonium; antheridium (e) megasporangium; megaphyll
6. Liverworts and hornworts share life cycle similarities with (a) ferns (b) mosses (c) horsetails (d) club mosses (e) whisk ferns
7. The green, gametangia-bearing moss plant (a) is the gametophyte generation (b) is the sporophyte generation (c) is called a protonema (d) contains cells with single large chloroplasts (e) both b and c
8. Seedless vascular plants possess ____________ to conduct water and dissolved minerals and ____________ to conduct dissolved sugar. (a) cuticle; xylem (b) phloem; stoma (c) phloem; xylem (d) stoma; cuticle (e) xylem; phloem
9. Whisk ferns, horsetails, and club mosses share life cycle similarities with (a) ferns (b) mosses (c) hornworts (d) liverworts (e) b, c, and d
10. A(an) ____________ is a leaf that arose from a branch system. (a) antheridium (b) microphyll (c) megaphyll (d) sorus (e) microspore
11. These plants have vascularized stems but lack true roots and leaves. (a) mosses (b) club mosses (c) horsetails (d) whisk ferns (e) hornworts
12. These plants have hollow, jointed stems that are impregnated with silica. (a) mosses (b) club mosses (c) horsetails (d) whisk ferns (e) hornworts
13. Spike moss *(Selaginella)* is a (a) homosporous fern (b) homosporous horsetail (c) heterosporous fern (d) heterosporous horsetail (e) heterosporous club moss
14. Which of the following statements about ferns is *not* true? (a) Ferns have motile sperm cells that swim through water to the egg-containing archegonium. (b) Bracken is the world's most common type of fern. (c) Ferns are the most economically important group of bryophytes. (d) The fern sporophyte consists of a rhizome, roots, and fronds. (e) The diversity of ferns is greatest in the tropics.
15. Plants are thought to have descended from a group of green algae known as (a) rhyniophytes (b) *Calamites* (c) epiphytes (d) charophytes (e) club mosses
16. Which of the following is *not* a characteristic of plants? (a) cuticle (b) unicellular gametangia (c) stomata (d) multicellular embryo (e) alternation of generations
17. In plant life cycles (a) the first products of meiosis are gametes (b) spores are part of the diploid sporophyte generation (c) the embryo gives rise to a zygote (d) the first stage in the diploid sporophyte generation is the zygote (e) the first stage in the haploid gametophyte generation is the prothallus
18. Which of these is a nonvascular plant with a dominant gametophyte generation? (a) ferns (b) whisk ferns (c) club mosses (d) horsetails (e) hornworts
19. Ferns are (a) seedless, vascular plants (b) vascular plants with seeds (c) seedless, nonvascular plants (d) nonvascular plants with seeds (e) seedless plants with a dominant gametophyte generation

20. The sporophyte generation in mosses consists of (a) protonema, buds, and archegonia (b) foot, seta, and capsule (c) thallus, archegoniophore, and antheridiophore (d) frond, rhizome, and roots (e) strobilus, microphyll, and root
21. A body form consisting of a flattened, lobed structure called a thallus is characteristic of many (a) mosses (b) liverworts (c) whisk ferns (d) horsetails (e) club mosses
22. Microphylls (a) probably evolved from stem branches that gradually filled in with additional tissues (b) are usually small and possess a single vascular strand (c) are characteristics of club mosses (d) both a and c (e) both b and c
23. Ancient plants that contributed significantly to coal formation included (a) mosses, liverworts, and hornworts (b) mosses, club mosses, and whisk ferns (c) hornworts, horsetails, and whisk ferns (d) club mosses, horsetails, and ferns (e) liverworts, club mosses, and hornworts
24. Which of the following is a genus of club mosses, with small, scalelike leaves and conelike strobili? (a) *Equisetum* (b) *Psilotum* (c) *Lycopodium* (d) *Marchantia* (e) *Azolla*
25. In the life cycle of the heterosporus club moss *Selaginella* (a) meiosis occurs in the capsule of the sporophyte, producing haploid spores (b) eggs are contained in archegonia that are borne on archegoniophores (c) the gametophyte stage is a free-living prothallus (d) both egg and sperm cells are produced within conelike strobili (e) the male gametophyte develops inside the microscopic wall, and the female gametophyte develops inside and then protrudes from the megaspore wall

REVIEW QUESTIONS

1. What are the most important environmental challenges that plants face living on land, and what adaptations do they possess to meet these challenges?
2. Plants are thought to have descended from which group of protists? What kinds of evidence support this idea?
3. Define alternation of generations, and distinguish between gametophyte and sporophyte generations.
4. Sketch the generalized life cycle of a moss and label the following: gametophyte, antheridia, archegonia, sperm cells, egg cell, fertilization, zygote, embryo, sporophyte, capsule, spore mother cells, meiosis, spores, and protonema.
5. How are mosses, liverworts, and hornworts similar? How is each group distinctive?
6. State the adaptations of bryophytes that algae lack. What adaptations do ferns have that both algae and bryophytes lack?
7. Distinguish between megaphylls and microphylls.
8. Sketch the generalized life cycle of a fern and label the following: sporophyte, frond, rhizome, roots, sorus, sporangium, meiosis, spores, gametophyte (prothallus), antheridium, archegonium, sperm cells, egg cell, fertilization, and zygote.
9. Name the three groups of plants known as fern allies. What features do these plants share with ferns? How is each group distinctive?
10. How does heterospory modify the life cycle?
11. Label the diagram. Use Figure 26–2 to check your answers.

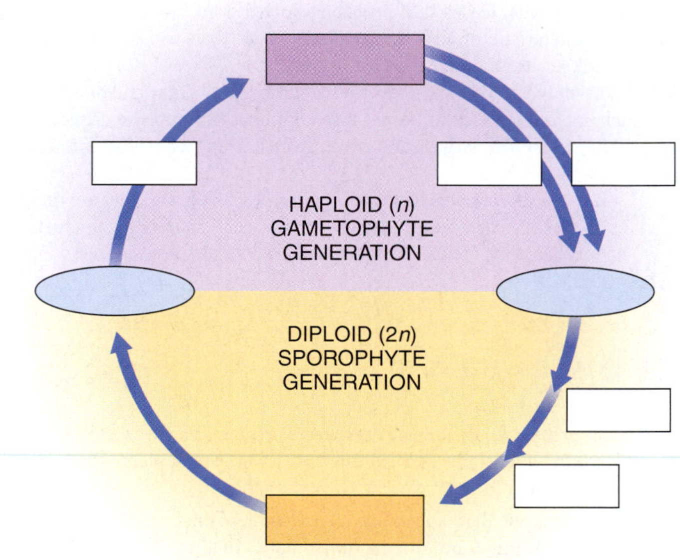

YOU MAKE THE CONNECTION

1. Which group probably colonized the land first, plants or animals? Explain.
2. How might the following trends in plant evolution be adaptive to living on land?
   a. Dependence on water for fertilization ⟶ no need for water as a transport medium
   b. Dominant gametophyte generation ⟶ dominant sporophyte generation
   c. Homospory ⟶ heterospory

RECOMMENDED READINGS

Berg, L.R. *Introductory Botany: Plants, People, and the Environment.* Saunders College Publishing, Philadelphia, 1997. An introduction to general botany with an environmental emphasis.

Camus, J.M., A.C. Jermy, and B.A. Thomas. *A World of Ferns*. Natural History Museum Publications, London, 1991. An outstanding book on ferns.

Graham, L.E. *Origin of Land Plants.* Wiley, New York, 1993. The algal relatives of plants provide insights into plant evolution in a terrestrial environment.

Kenrick, P., and P.R. Crane. "The Origin and Early Evolution of Plants on Land." *Nature,* Vol. 389, 4 Sept. 1997. This review article discusses how recent fossil discoveries and advances in plant systematics have added to our understanding of plant evolution.

Levanthes, L.E. "Mysteries of the Bog." *National Geographic,* Vol. 171, No. 3, Mar. 1987. Discusses peat bogs and the mosses that form them.

Niklas, K.J. "One Giant Step for Life." *Natural History,* Jun. 1994. An overview of the transition from water to land during the evolution of plants.

Pearce, F. "Peat Bogs Hold Bulk of Britain's Carbon." *New Scientist,* Vol. 144, 19 Nov. 1994. A strong case for the preservation of Scottish peat bogs, which contain far more carbon than would be stored in equivalent areas of forest. Because it is not in the atmosphere as CO_2, stored carbon helps prevent global warming.

Raven, P.H., R.F. Evert, and S.E. Eichhorn. *Biology of Plants,* 6th ed. W.H. Freeman & Company, New York, 1999. An excellent general botany textbook, with detailed descriptions of plant evolution.

- Visit our Web site at **http://www.info.brookscole.com/solomonbergmartin** for links to chapter-related resources on the World Wide Web. Additional on-line materials relating to this chapter can also be found on our Web site.

See chapter activity on BioActive Learner CD for additional help in mastering the chapter's material. Icon location in the chapter's margins shows which topics have tutorials or simulations in the CD.

27

The Plant Kingdom: Seed Plants

Papaya seeds. Papaya *(Carica papaya)* is a flowering plant, and its seeds are enclosed within a fruit. *(Scott Camazine/The National Audubon Society Collection/Photo Researchers, Inc.)*

LEARNING OBJECTIVES

After you have studied this chapter you should be able to

1. Compare the features of seeds with those of spores, and discuss the advantages of plants that reproduce primarily by seeds rather than by spores.
2. Trace the steps in the life cycle of a pine and compare its sporophyte and gametophyte generations.
3. Summarize the features that distinguish gymnosperms from other plants.
4. Name and briefly describe the four phyla of gymnosperms.
5. Summarize the features that distinguish flowering plants from other plants.
6. Diagram a generalized life cycle of a flowering plant and describe double fertilization.
7. Contrast dicots and monocots, the two classes of flowering plants.
8. Discuss the evolutionary adaptations of flowering plants.
9. Trace the evolution of gymnosperms from seedless vascular plants and that of flowering plants from gymnosperms.

Our discussion of plants in Chapter 26 focused on bryophytes and on ferns and their allies, all of which reproduce by means of *spores,* which are haploid reproductive units that give rise to gametophytes. Although the most successful and widespread plants also produce spores, their primary means of reproduction and dispersal is by **seeds,** each of which consists of an embryonic sporophyte and nutritive tissue surrounded by a protective coat. Seeds develop from the fertilized egg cell, the female gametophyte, and its associated tissues. The two groups of the seed plants, gymnosperms and angiosperms (flowering plants), exhibit the greatest evolutionary complexity in the plant kingdom and are the dominant plants in most terrestrial environments.

Seeds, such as the papaya seeds shown in the photograph, are reproductively superior to spores for three main reasons. First, development is further advanced in seeds than in spores. A seed contains a multicellular young plant with embryonic root, stem, and one or more leaves already formed, whereas a spore is a single cell. Second, a seed contains an abundant food supply. After germination, the plant embryo is nourished by food stored in the seed until it becomes self-sufficient. Because a spore is a single cell, few food reserves exist for the plant that develops from a spore. Third, a seed is protected by a multicellular **seed coat** that is very thick and hard in some plants, as for example, in seeds of lima beans. Like spores, seeds can live for extended periods at reduced rates of metabolism, germinating when conditions become favorable.

Seeds and seed plants have been intimately connected with the development of human civilization. From prehistoric times, early humans collected and used seeds for food. The food stored in seeds is a concentrated source of proteins, oils, carbohydrates, and vitamins, which are nourishing for humans as well as for germinating plants. Also, seeds are easy to store (provided they are kept dry); this has allowed humans to collect them during times of plenty and save them for times of need. Few other foods can be stored as conveniently or for as long. Although most seeds that humans consume are produced by flowering plants, seeds of certain gymnosperms, the piñon pine *(Pinus edulis),* for example, are edible. They are usually sold as "pine nuts."

GYMNOSPERMS AND FLOWERING PLANTS BEAR SEEDS

In seed plants, each seed develops from an **ovule,** which is a megasporangium and its enclosed structures, following fertilization of an egg cell within. Seed plants also have **integuments,** layers of sporophyte tissue surrounding and enclosing the megasporangium (see Fig. 35–2). The seed coat develops from the integuments after fertilization takes place.

Seed plants are divided into two groups based on whether or not their ovules (unfertilized seeds) are surrounded by an ovary wall (an *ovary* is a structure that contains one or more ovules). The two groups of seed plants are the **gymnosperms** and the **angiosperms** (Table 27–1). The word *gymnosperm* is adapted from the Greek for "naked seed." Gymnosperms produce seeds that are totally exposed or borne on the scales of cones. In other words, the ovules of gymnosperms are not surrounded by an ovary wall. Pine, spruce, fir, hemlock, and *Ginkgo* are examples of gymnosperms.

The Greek expression from which the term *angiosperm* is derived translates as "seed enclosed in a vessel or case." Angiosperms are flowering plants that produce their seeds within a fruit (a mature ovary). Thus the ovules of angiosperms are protected. Flowering plants are extremely diverse and include corn, oaks, water lilies, cacti, apples, grasses, palms, and buttercups.

Both gymnosperms and flowering plants possess vascular tissues: **xylem** for the conduction of water and dissolved nutrient minerals, and **phloem** for the conduction of dissolved sugar. Both have life cycles with an **alternation of generations,** that is, they spend a portion of their lives in the diploid sporophyte stage and a portion in the haploid gametophyte stage. The sporophyte generation is the dominant stage in each group, and the gametophyte generation is significantly reduced in size and entirely dependent on the sporophyte generation. Unlike the plants we have considered so far (bryophytes, ferns, and most of their allies), gymnosperms and flowering plants do not have free-living gametophytes (see Chapter 26). Instead, the female gametophyte is attached to and nutritionally dependent on the sporophyte generation. All gymnosperms and flowering plants are *heterosporous* and produce two types of spores: microspores and megaspores.

GYMNOSPERMS ARE THE "NAKED SEED" PLANTS

The gymnosperms include some of the most interesting members of the plant kingdom, including a number of record holders. For example, a giant sequoia *(Sequoiadendron giganteum)* known as the General Sherman Tree, located in Sequoia National Park in California, is one of the world's most massive organisms. It is 82 m (267 ft) tall and has a girth of 23.7 m (77 ft) measured 1.5 m (5 ft) above ground level. Another gymnosperm, a coast redwood *(Sequoia sempervirens)* known as the Mendocino tree, is among the world's tallest trees, measuring 112 m (364 ft) in the year 2000. One of the oldest living trees, a bristlecone pine *(Pinus aristata)* in the White Mountains of California, has been determined by tree ring analysis to be 4900 years old!

Gymnosperms are usually classified into four phyla, which represent four different evolutionary lines (Fig. 27–1). Numbering 550 species, the largest phylum of gymnosperms is the phylum Coniferophyta, commonly called conifers. Two phyla of gymnosperms, the ginkgo (phylum Ginkgophyta) and the cycads (phylum Cycadophyta), represent evolutionary remnants of groups that were more significant in the past. The fourth phylum of gymnosperms, the gnetophytes (phylum Gnetophyta), is a collection of some unusual plants that share certain traits not found in other gymnosperms.

Conifers are woody plants that produce seeds in cones

The **conifers** (phylum Coniferophyta), which include pines, spruces, hemlocks, and firs, are the most familiar group of gymnosperms (Fig. 27–2*a*). They are woody trees or shrubs that produce annual additions of secondary tissues (wood and bark; see

TABLE 27–1 A Comparison of Gymnosperms and Angiosperms

Characteristic	Gymnosperms	Angiosperms
Growth habit	Woody trees and shrubs	Woody or herbaceous
Conducting cells in xylem	Tracheids*	Vessel elements and tracheids*
Reproductive structures	Cones (usually)	Flowers
Pollen grain transfer	Wind (usually)	Animals or wind
Fertilization	Egg and sperm → zygote; double fertilization in gnetophytes	Double fertilization: egg and sperm → zygote; 2 polar nuclei and sperm → endosperm
Seeds	Exposed or borne on scales of cones	Enclosed within fruit
Number of species	About 760	More than 235,000
Geographical distribution	Worldwide	Worldwide

*See Chapter 31.

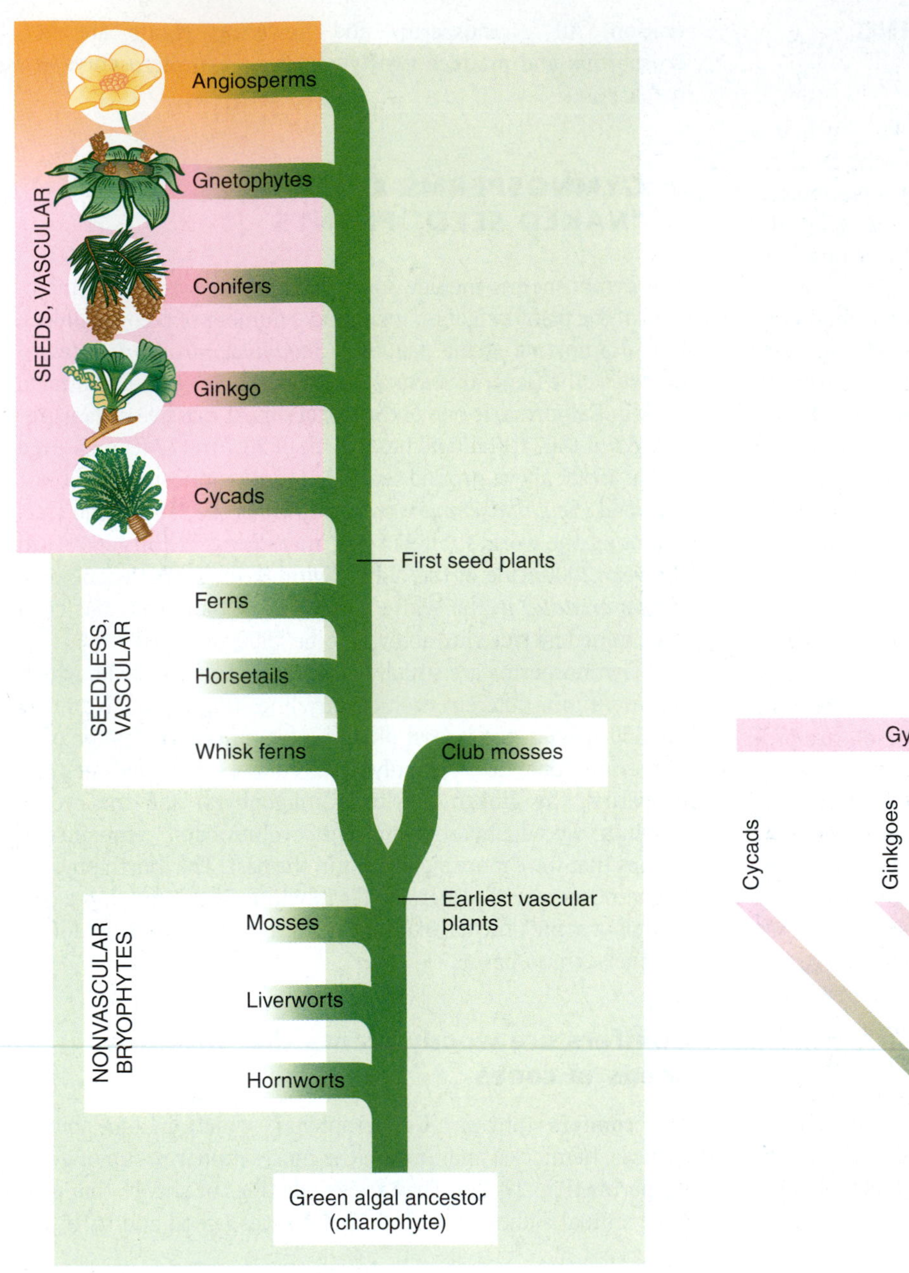

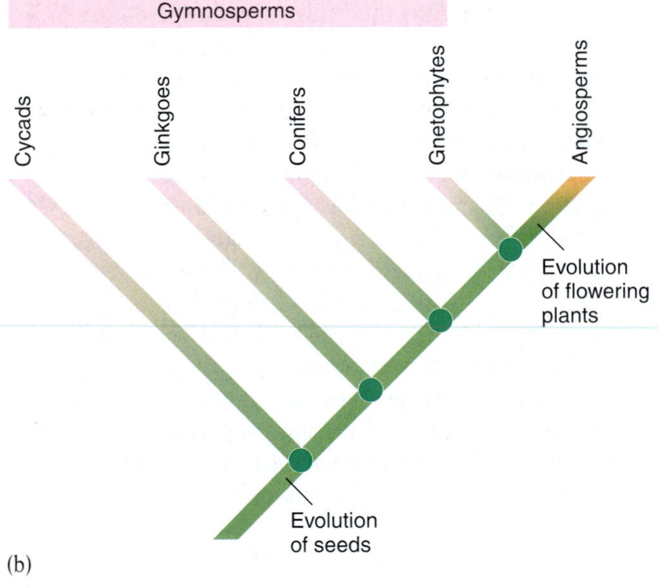

Figure 27-1 Gymnosperm and angiosperm evolution. These diagrams illustrate two different ways to show one hypothesis of phylogenetic relationships among living seed plants. The diagrams are based on several lines of scientific evidence, including structural and some molecular comparisons. In this scheme, the gnetophytes are closely related to the angiosperms. There is much uncertainty about the exact order of relationships, however. Analyses of certain DNA sequences of gymnosperm and angiosperm species suggest that the gnetophytes are more closely related to other gymnosperms than to the angiosperms, unlike the depiction shown here. The arrangement of the phyla shown here is likely to change as future analyses help us to clarify relationships. **(a)** A classical evolutionary tree. The pink branches are gymnosperms; the angiosperm branch is orange. **(b)** A cladogram.

Chapter 33); there are no herbaceous (nonwoody) conifers. The wood (secondary xylem) is composed of **tracheids,** which are long, tapering cells with pits through which water and dissolved nutrient minerals move from one cell to another.

Many conifers produce **resin,** a viscous, clear or translucent substance consisting of several organic compounds that may protect the plant from attack by fungi or insects. The resin collects in resin ducts, tubelike cavities that extend throughout the

Figure 27–2 Conifers. **(a)** The sacred fir *(Abies religiosa)* is a conifer native to Mexico, where it grows to 30.5 m (100 ft) or more. **(b, c)** Leaf variation in conifers. **(b)** The leaves of white pine *(Pinus strobus)* are long, slender needles that occur in clusters of five. **(c)** In American arborvitae *(Thuja occidentalis)*, leaves are small and scalelike *(see inset)*. *(a, Geoff Bryant/The National Audubon Society Collection/Photo Researchers, Inc.)*

roots, stems, and leaves. Resin is produced and secreted by cells lining the resin ducts.

Most conifers have leaves (*megaphylls;* see Chapter 26) called **needles** that are commonly long and narrow, tough, and leathery (Fig. 27–2*b*). Pines bear clusters of two to five needles, depending on the species. In a few conifers such as American arborvitae *(Thuja occidentalis),* the leaves are scalelike and cover the stem (Fig. 27–2*c*). Most conifers are evergreen and bear their leaves throughout the year. Only a few, such as the dawn redwood, larch, and bald cypress, are deciduous and shed their needles at the end of each growing season.

Most conifers are **monoecious,** which means that they have separate male and female reproductive parts in different locations on the same plant. These reproductive parts are generally borne in *strobili* that are commonly called cones, hence the name *conifer,* which means "bears cones."

The approximately 550 species of conifers occupy extensive areas, ranging from the Arctic to the tropics, and are the dominant vegetation in the forested regions of Alaska, Canada, northern Europe, and Siberia. In addition, they are important in the Southern Hemisphere, particularly in wet, mountainous areas of temperate and tropical regions in South America, Australia, New Zealand, and Malaysia. Southwestern China, with more than 60 species of conifers, has the greatest regional diversity of conifer species on Earth. California, New Caledonia (an island east of Australia), southeastern China, and Japan also have considerable diversity of conifer species.

Ecologically, conifers contribute food and shelter to animals and other organisms, and their roots hold the soil in place and help prevent soil erosion. Humans use conifers for their wood (for building materials as well as paper products), medicine (e.g., the anticancer drug taxol from the Pacific yew), turpentine, and resins. Because of their attractive appearance, conifers such as firs, spruces, pines, and cedars are grown for landscape design and decorative holiday trees and wreaths.

Pines represent a typical conifer life cycle

The genus *Pinus,* by far the largest genus in the conifers, consists of about 100 species. A pine tree is a mature sporophyte. Pine is heterosporous and therefore produces microspores and megaspores in separate cones.[1] Male cones, usually 1 cm or less in length, are smaller than female cones and are generally produced on the lower branches each spring (Fig. 27–3). The more familiar, woody female cones, which are on the tree year-round, are usually found on the upper branches of the tree and bear seeds after reproduction. Female cones vary considerably in size. The sugar pine *(Pinus lambertiana)* of California produces the world's longest female cones, which reach lengths of 60 cm (2 ft).

Each male cone, also called a pollen cone, is composed of **sporophylls,** leaflike structures that bear sporangia on the underside. At the base of each sporophyll are two **microsporangia,** which contain numerous **microsporocytes,** also called *microspore mother cells* (Fig. 27–4). Each microsporocyte undergoes meiosis to form four haploid microspores. **Microspores** then develop into extremely reduced male gametophytes. Each immature male gametophyte, also called a **pollen grain,** consists of four cells, two of which—a *generative cell* and a *tube cell*—are involved in reproduction. The other two cells soon degenerate. Two large air sacs on each pollen grain provide buoyancy for wind dissemination. Pollen grains are shed from male cones in great numbers, and wind currents carry some to the immature female cones.

[1] It might be helpful to review alternation of generations in Chapter 26, including Figure 26–18, which depicts a heterosporous life cycle, before studying the pine life cycle.

Figure 27–3 Male and female cones in lodgepole pine *(Pinus contorta).* *(Top)* Mature, dark brown female cones have opened to shed their seeds. *(Bottom)* Clusters of golden male cones produce copious amounts of pollen grains in the spring. The location of female cones above male cones facilitates cross-pollination between different trees, because it is unlikely that wind will blow pollen grains directly upward to female cones on the same tree. *(Walt Anderson/Visuals Unlimited)*

The female cones (also called seed cones) are considered by many botanists to be modified branch systems. Each cone scale bears two **megasporangia** on its upper surface. Within each megasporangium, meiosis of a **megasporocyte,** or *megaspore mother cell,* produces four haploid **megaspores.** One of these divides mitotically, developing into the female gametophyte, which produces an egg within each of several archegonia. The other three megaspores are nonfunctional and soon degenerate.

When the ovule is ready to receive pollen, it produces a sticky droplet at the opening where the pollen grains land. **Pollination,** the transfer of pollen to the female cones, occurs in the spring during a period of a week or ten days, after which the pollen cones wither and drop off the tree. One of the many pollen grains that adhere to the sticky female cone grows a **pollen tube,** an outgrowth that digests its way through the megasporangium to the egg within the archegonium. The germinated pollen grain with its pollen tube is the mature male gametophyte. The tube cell, which is involved in pollen tube growth, and the generative cell enter the pollen tube. The generative cell divides and forms a *stalk cell* and a *body cell;* the body cell later divides and forms two nonmotile sperm cells. When the pollen tube reaches the female gametophyte, it discharges the two sperm cells near the egg. One of these sperm cells fuses with the egg, in a process known as **fertilization,** to form a zygote, or fertilized egg, which subsequently grows into a young pine embryo in the seed. The other sperm cell degenerates.

The developing embryo, which consists of an embryonic root and an embryonic shoot with several cotyledons (embryonic leaves), is embedded in haploid female gametophyte tissue that becomes the nutritive tissue in the mature pine seed. The embryo and nutritive tissue are surrounded by a tough, protective seed coat derived from the integuments. The seed coat extends out to form a thin papery wing at one end of the seed. This wing enables the seed to be dispersed by air currents.

A long time elapses between the appearance of female cones on a tree and the maturation of seeds in those cones. When pollination occurs in the spring, the female cone is still immature, and meiosis of the megasporocytes (megaspore mother cells) has not yet occurred. During the following year, the female reproductive tissue gradually matures, and eggs form within archegonia. Meanwhile, the pollen tube grows slowly through the megasporangium to the archegonia. Fertilization occurs during the spring of the following year, and the embryo begins to develop. Seed maturation takes several additional months, although some seeds remain within the female cones for a number of years before being shed.

In the pine life cycle, the sporophyte generation is dominant, and the gametophyte generation is restricted in size to microscopic structures in the cones. Although the female gameto-

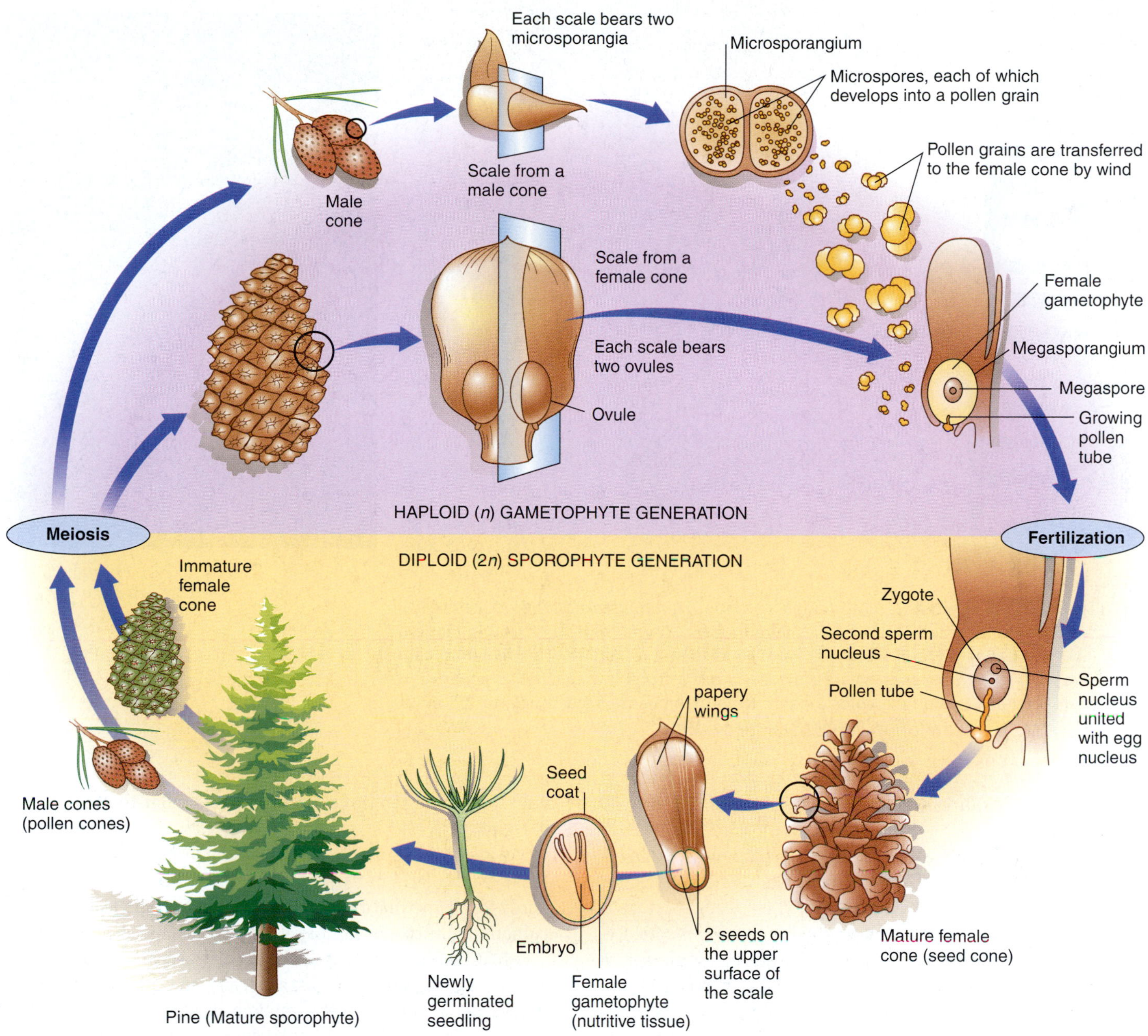

Figure 27–4 Life cycle of pine. One major advantage of gymnosperms over the seedless vascular plants is the production of wind-borne pollen grains. Pines and other gymnosperms do not depend on water as a transport medium for sperm cells.

phyte produces archegonia, the male gametophyte is so reduced that it does not produce antheridia. The gametophyte generation in pines, as in all seed plants, depends totally on the sporophyte generation for nourishment.

A major adaptation in the pine life cycle is elimination of the need for external water as a sperm transport medium. Instead, pine pollen grains are carried to female cones by air currents, and nonmotile sperm cells accomplish fertilization by moving through a pollen tube to the egg. Pine and other conifers are plants whose reproduction is totally adapted for life on land.

Cycads have seed cones and compound leaves

Cycads (phylum Cycadophyta) were very important during the Triassic period, which began approximately 248 million years ago (mya) and is sometimes referred to as the "Age of Cycads." Most species are now extinct, and the few surviving cycads, about 140 species, are tropical and subtropical plants with stout, trunklike stems and compound leaves that resemble those of palms or tree ferns (Fig. 27–5). Many cycads are endangered, primarily because

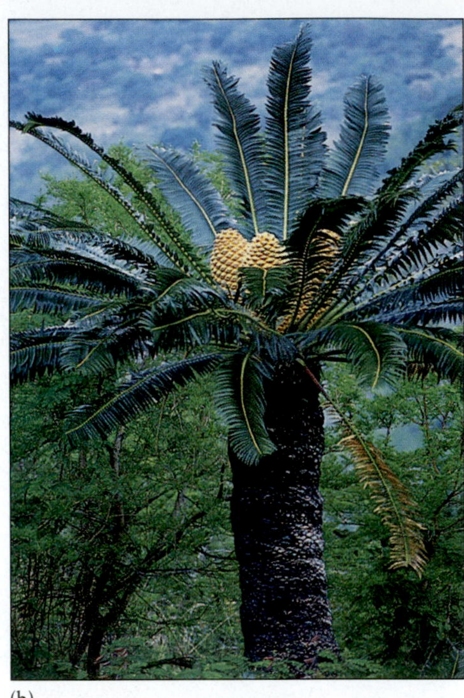

Figure 27–5 Cycads. **(a)** A female comptie *(Zamia integrifolia)* produces seed cones. The only cycad native to the United States, *Z. integrifolia* is found in Florida. Like *Zamia,* most cycads are short plants (less than 2 m tall). **(b)** This female cycad *(Encephalartos transvenosus)* in South Africa has a trunk that reaches a height of about 9 m (30 ft) and resembles a palm. Note its immense seed cones, to 0.8 m (30 in) in length. *(b, Walter H. Hodge/Peter Arnold, Inc.)*

they are popular as ornamentals and are gathered from the wild and sold to collectors.

Cycad reproduction is similar to that in pines except that cycads are **dioecious** and therefore have seed cones on female plants and pollen cones on male plants. Their seed structure is most like that of the earliest seeds found in the fossil record. Cycads have also retained the primitive feature of motile sperm cells, each of which possesses many hairlike flagella. These motile sperm cells are a vestige retained from the ancestors of cycads, in which sperm cells swam from antheridia to archegonia. Cycad pollen grains, however, are carried by air or insects to the female plants and their cones; there the pollen grain germinates and grows a pollen tube. The sperm cells are released at the top of this tube and swim to get to the egg. In other words, motile sperm cells occur in cycads.

Ginkgo biloba is the only living species in its phylum

Ginkgo (phylum Ginkgophyta) is represented by a single living species, the maidenhair tree *(Ginkgo biloba)* (Fig. 27–6). A native of eastern China, it has been found growing in the wild in only two locations, although it has been cultivated in China and Japan for centuries for its edible seeds. *Ginkgo* is the oldest genus (and species) of living trees. Fossil ginkgoes 200 million years old have been discovered that are similar to the modern-day ginkgo.

Ginkgo is commonly planted in North America and Europe today, particularly in parks and along city streets, because it is hardy and somewhat resistant to air pollution. Its leaves are deciduous and turn a beautiful yellow before being shed in the fall.

Like cycads, ginkgo is dioecious, with separate male and female trees. It has flagellated sperm cells, an evolutionary vestige that is not required because ginkgo produces airborne pollen grains. Ginkgo seeds are completely exposed rather than occurring within cones. Male trees are typically planted because the female trees bear seeds whose fleshy outer coverings give off a foul odor that smells like rancid butter. In China and Japan, where people eat the seeds, the female trees are more common.

Ginkgo has been an important medicinal plant for centuries. Extracts from the leaves may enhance neurological functioning by increasing blood flow to the brain. Several studies suggest that ginkgo may improve memory in elderly people suffering from Alzheimer's disease; no data suggest that it improves memory in young people, however.

Gnetophytes include three unusual genera

The **gnetophytes** (phylum Gnetophyta) consist of about 70 species in three diverse genera *(Gnetum, Ephedra,* and *Wel-*

Figure 27–6 Ginkgo, or maidenhair tree *(Gingko biloba).* **(a)** A young male ginkgo tree. **(b)** Close-up of a branch from a female ginkgo, showing the exposed seeds and the distinctive, fan-shaped leaves. *(a, Carlyn Iverson; b, Marion Lobstein)*

witschia). Gnetophytes share certain features that make them clearly more advanced than the rest of the gymnosperms. For example, gnetophytes have more efficient water-conducting cells, called **vessel elements,** in their xylem (see Chapter 31). Flowering plants also have vessel elements in their xylem, but no gymnosperms except the gnetophytes do. Also, the cone clusters produced by some of the gnetophytes resemble flower clusters, and certain details in their life cycles resemble those of flowering plants.

The genus *Gnetum* contains tropical vines, shrubs, and trees with broad leaves that resemble those of flowering plants (Fig. 27–7*a*). Species in the genus *Ephedra* include many shrubs and vines found in deserts and other dry temperate and tropical regions. Some *Ephedra* species resemble horsetails in that they possess jointed green stems with tiny leaves (Fig.27–7*b*). Commonly called joint fir, *Ephedra* has been used medicinally for centuries. An Asiatic *Ephedra* is the source of ephedrine, which stimulates the heart and raises blood pressure. Ephedrine is sold over the counter in weight-control medications and herbal energy-boosters; several deaths have been reported from chronic use or overdose of products containing ephedrine.

The third gnetophyte genus, *Welwitschia,* contains a single species found in deserts of southwestern Africa (Fig. 27–7*c*). Most of *Welwitschia*'s body—a long taproot—grows underground. Its short, wide stem forms a shallow disk, up to 0.9 m (3 ft) in diameter, from which two ribbon-like leaves extend. These two leaves continue to grow from the stem throughout the plant's life, but their ends are usually broken and torn by the wind, giving the appearance of numerous leaves. Each leaf grows to about 2 m (6.5 ft) in length. When *Welwitschia* reproduces, it forms cones around the edge of its disclike stem.

FLOWERING PLANTS PRODUCE FLOWERS, FRUITS, AND SEEDS

Flowering plants or **angiosperms** (phylum Anthophyta)[2] are the most successful plants today, surpassing even the gymnosperms in importance. They have adapted to almost every habitat and, with at least 235,000 species, are Earth's dominant plants. Flowering plants come in a wide variety of sizes and forms, from herbaceous violets to massive eucalyptus trees. Some flowering plants—tulips and roses, for example—have large, conspicuous flowers, whereas others, such as grasses and oaks, produce flowers that are small and inconspicuous.

Flowering plants are vascular plants that reproduce sexually by forming flowers, and, after a unique double fertilization process that is discussed later, seeds within fruits. The fruit protects the developing seeds and often aids in their dispersal (see Chapter 35). Flowering plants possess efficient water-conducting cells called vessel elements in their xylem, and efficient sugar-

[2] Some botanists prefer to classify flowering plants in phylum Magnoliophyta instead of phylum Anthophyta. The International Botanical Congress has not yet standardized phylum names, although the matter is currently under study.

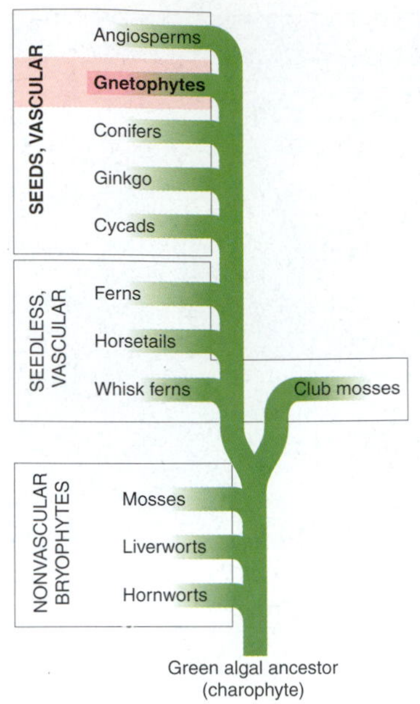

(a)

(b)

(c)

Figure 27–7 Gnetophytes. **(a)** The leaves of *Gnetum gnemon* resemble those of flowering plants. The exposed seeds are red to yellow when ripe. **(b)** A male joint fir (*Ephedra* sp.) has pollen cones clustered at the nodes. Most species of *Ephedra* are dioecious and have separate male and female plants. Nineteenth-century European pioneers used species native to desert areas in the American Southwest to make a beverage called Mormon tea. **(c)** *Welwitschia mirabilis* is a gnetophyte that is native to deserts in southwestern Africa. It survives on moisture-laden fogs that drift inland from the ocean. (This region receives less than 2.5 cm of annual precipitation.) The two leaves of this extremely unusual plant grow from the edges of a short, wide stem and become tattered and torn over time. This particular specimen, known as the Giant Hasab *Welwitschia*, is estimated to be at least 1500 years old. Photographed in the Namib Desert, Namibia. *(a, John D. Cunningham/Visuals Unlimited; b, David Cavagnaro; c, Robert and Linda Mitchell)*

conducting cells called **sieve tube members** in their phloem (see Chapter 31).

Flowering plants are extremely important to humans because our survival as a species literally depends on them. All of our major food crops are flowering plants, including cereal crops such as rice, wheat, corn, and barley. Woody flowering plants such as oak, cherry, and walnut provide valuable lumber. Flowering plants give us fibers like cotton and linen and medicines like digitalis and codeine. Products as diverse as rubber, tobacco, coffee, chocolate, and aromatic oils for perfumes come from flowering plants. Economic botany is the subdiscipline of botany that deals with plants of economic importance, most of which are flowering plants.

Monocots and dicots are the two classes of flowering plants

Traditionally, phylum Anthophyta is divided into two classes: the monocots (class Monocotyledones) and the dicots (class Dicotyledones)[3] (Fig. 27–8). Monocots include palms, grasses, orchids, irises, onions, and lilies. The dicot class includes oaks, roses, mustards, cacti, blueberries, and sunflowers. Dicots are more diverse and include many more species (at least 170,000) than the monocots (at least 65,000). Table 27–2 provides a comparison of some of the general features of the two classes.

Monocots are mostly herbaceous plants with long, narrow leaves that have parallel veins (the main leaf veins run parallel to one another). The flower parts of monocot flowers usually occur in threes or multiples of three. For example, a flower might have three sepals, three petals, six stamens, and a compound pistil consisting of three fused carpels (these flower parts are discussed shortly). Monocot seeds have a single **cotyledon,** or embryonic seed leaf, and **endosperm,** a nutritive tissue, is usually present in the mature seed.

Dicots may be herbaceous (e.g., a tomato plant) or woody (e.g., a hickory tree). Their leaves vary in shape but usually are broader than monocot leaves, with netted veins (branched veins resembling a net). Flower parts usually occur in fours or fives or multiples thereof. Two cotyledons are present in dicot seeds, and endosperm is usually absent in the mature seed, having been absorbed by the two cotyledons.

[3] Some botanists prefer to classify monocots in class Liliopsida and dicots in class Magnoliopsida. The International Botanical Congress has not yet standardized class names, although the matter is currently under study.

Flowers are involved in sexual reproduction

Flowers are reproductive shoots usually composed of four parts—sepals, petals, stamens, and carpels—arranged in whorls (circles) on the end of a flower stalk, or **peduncle** (Fig. 27–9). The peduncle may terminate in a single flower or a cluster of flowers known as an **inflorescence.** The tip of the peduncle enlarges to form a **receptacle** that bears some or all of the flower parts.

All four floral parts are important in the reproductive process, but only the stamens (the "male" organs) and carpels (the "female" organs) produce gametes. A flower that has all four parts is said to be **complete,** whereas an **incomplete** flower lacks one or more of these four parts. A flower with both stamens and carpels is said to be **perfect,** whereas an **imperfect** flower has stamens or carpels, but not both.

Sepals, which make up the lowermost and outermost whorl on a floral shoot, are leaflike in appearance and often green (Fig. 27–10*a*). Sepals cover and protect the other flower parts when the flower is a bud. As the blossom opens, the sepals fold back to reveal the more conspicuous petals. The collective term for all the sepals of a flower is the **calyx.**

The whorl just above the sepals consists of **petals,** which are broad, flat, and thin (like sepals and leaves) but vary in shape and

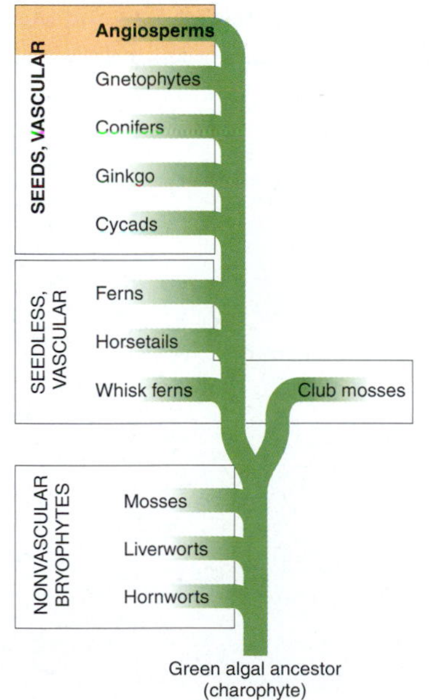

(a)

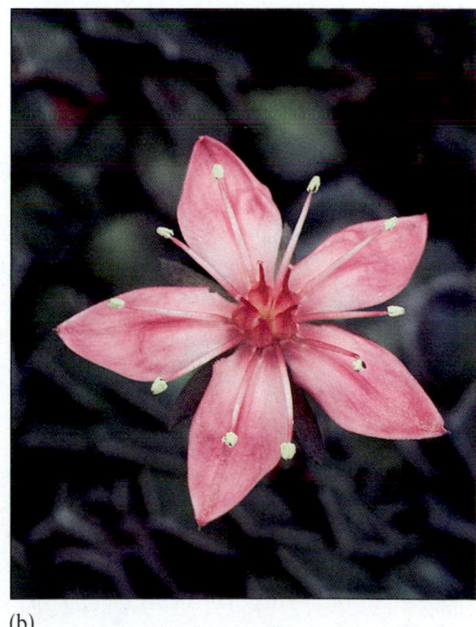
(b)

Figure 27–8 Monocots and dicots. **(a)** Most monocots such as this *Trillium erectum* have their floral parts in threes. Note the three green sepals, three red petals, six stamens, and three stigmas (the compound pistil is composed of three fused carpels). **(b)** Most dicots such as this *Tacitus* sp. have their floral parts in fours or fives. Note the five petals, ten stamens, and five separate pistils. Five sepals are also present, but they are barely visible against the green background. *(a, John Gerlach/Tom Stack & Associates; b, Richard H. Gross)*

TABLE 27–2 Distinguishing Features of Dicots and Monocots

Feature	Dicots	Monocots
Flower parts	Usually in fours or fives	Usually in threes
Pollen grains	Three furrows or pores	One furrow or pore
Leaf venation	Usually netted	Usually parallel
Vascular bundles in stem cross section	Arranged in a circle (ring)	Usually scattered or more complex arrangement
Roots	Taproot system	Fibrous root system
Seeds	Embryo with two cotyledons	Embryo with one cotyledon
Secondary growth (wood and bark)	Often present	Absent

are frequently brightly colored. Petals play an important role in attracting animal pollinators to the flower (see Chapter 35). Sometimes petals are fused to form a tube (e.g., trumpet honeysuckle flowers) (Fig. 27–10*b*) or other floral shape (e.g., snapdragons). The petals of a flower are referred to collectively as the **corolla.**

Just inside the petals is a whorl of **stamens** (Fig. 27–10*c*). Each stamen is composed of a thin stalk, called a **filament,** and a saclike **anther,** where meiosis occurs to form microspores that develop into pollen grains. Each pollen grain produces two cells surrounded by a tough outer wall. One cell eventually divides to form two male gametes, or sperm cells, and the other produces a pollen tube through which the sperm cells travel to reach the ovule.

In the center of most flowers is one or more closed **carpels,** the "female" reproductive organs (Fig. 27–10*c*). Carpels bear ovules, which, as you may recall, are structures with the potential to develop into seeds. The carpels of a flower may be separate or fused into a single structure. The female part of the flower is also referred to as a **pistil.** A pistil may consist of a single carpel (a simple pistil) or a group of fused carpels (a compound pistil) (Fig. 27–11). Each pistil generally has three sections: a **stigma,** on which the pollen grain lands; a **style,** a necklike structure through which the pollen tube grows; and an **ovary,** an enlarged structure that contains one or more ovules. Each young ovule contains a female gametophyte that forms one female gamete (an egg), two polar nuclei, and several other haploid cells. Following fertilization, the ovule develops into a seed and the ovary into a fruit.

The life cycle of flowering plants includes double fertilization

Flowering plants have an alternation of generations in which the sporophyte generation is larger and nutritionally independent and the gametophyte generation is microscopic in size and nutritionally dependent on the sporophyte (Fig. 27–12). Flowering plants, like gymnosperms and certain other vascular plants, are

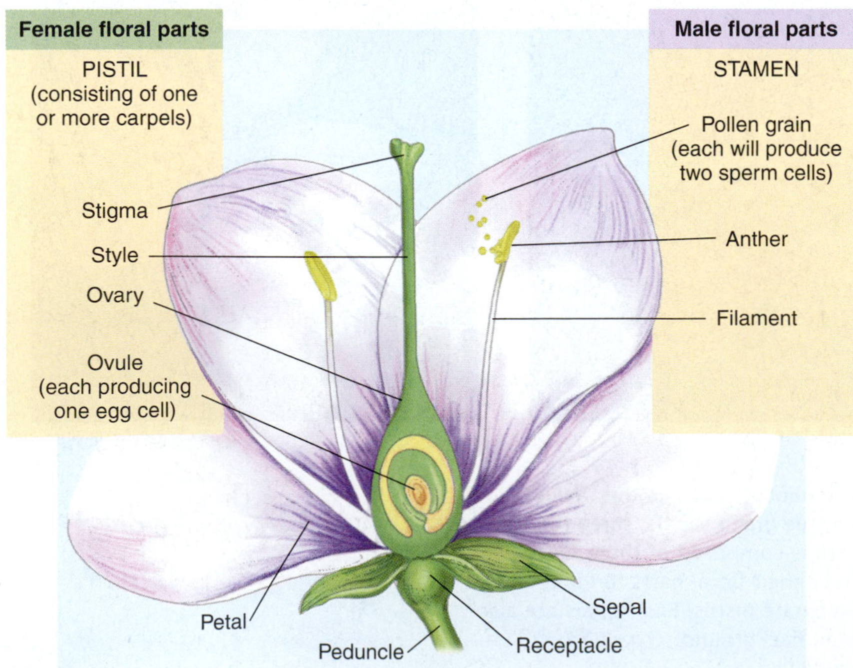

Figure 27–9 Floral structure. This cutaway view of a "typical" flower shows the details of basic floral structure. This flower is both a complete and a perfect flower. Not all flowers possess all these structures.

(a)

(b)

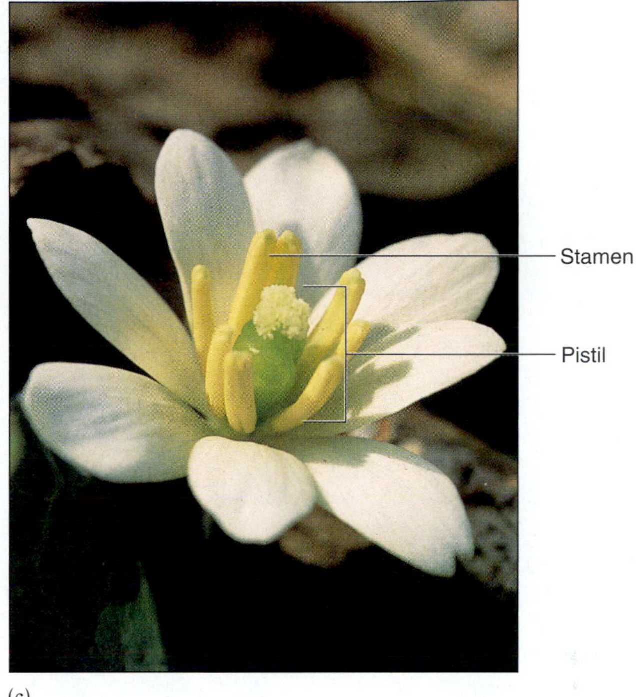

(c)

Figure 27–10 Parts of a flower. **(a)** The leaflike sepals of a rose (*Rosa* sp.) bud enclose and protect the inner flower parts. **(b)** The five petals of each trumpet honeysuckle *(Lonicera sempervirens)* flower are fused to form a tubular corolla. **(c)** A twinleaf *(Jeffersonia diphylla)* flower has eight yellow stamens. Note the simple pistil with its green ovary in the center of the flower. *(a, James Mauseth, University of Texas; b, James L. Castner; c, Marion Lobstein)*

heterosporous and produce two kinds of spores: microspores and megaspores. Sexual reproduction occurs in the flower.

Each young ovule within an ovary contains a megasporocyte (megaspore mother cell) that undergoes meiosis to produce four haploid megaspores. Three of these usually disintegrate, and one divides mitotically and develops into a mature female gametophyte, also called an **embryo sac.** The most widely studied type of embryo sac contains seven cells with eight haploid nuclei. Six of these cells, including the egg cell, contain a single nucleus each, and a central cell has two nuclei, called **polar nuclei.** The egg and the central cell with two polar nuclei are directly involved in fertilization; the other five cells in the embryo sac apparently have no direct role in the fertilization process and disintegrate. It has been hypothesized, however, that as the *synergids* (the two cells closely associated with the egg) disintegrate, they release chemicals that may affect the direction of pollen tube growth.

Each pollen sac, or microsporangium, of the anther contains numerous microsporocytes (microspore mother cells), each of which undergoes meiosis to form four haploid microspores. Every microspore develops into an immature male gametophyte, also called a pollen grain. Pollen grains are extremely small; each consists of two cells: the *tube cell* and the *generative cell.*

The anther sacs split open and begin to shed pollen. Pollen grains are transferred to the stigma by a variety of agents, including wind, water, insects, and other animal pollinators (see Chapter 35). If compatible with the stigma, the pollen grain germinates; that is, the tube cell forms a pollen tube that grows down the style and into the ovary. The germinated pollen grain with its pollen tube is the mature male gametophyte. Next, the generative cell divides to form two nonmotile male gametes called sperm cells. The sperm cells move down the pollen tube and are discharged into the embryo sac. Both sperm cells are involved in fertilization.

Something happens during sexual reproduction in flowering plants that does not occur anywhere else in the living world.

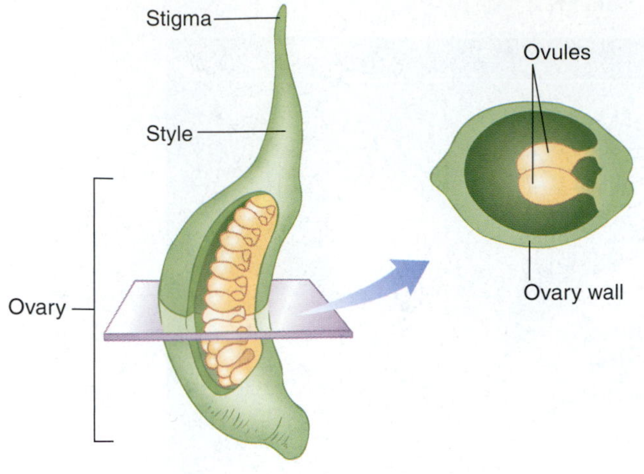

(a) Simple pistil

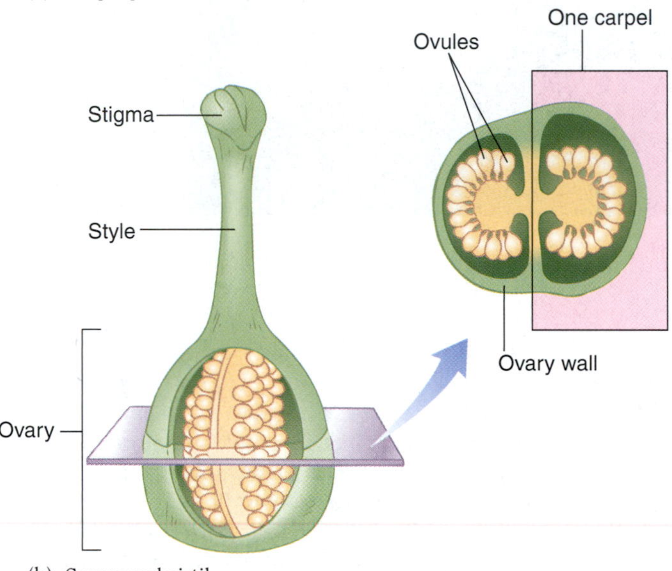

(b) Compound pistil

Figure 27-11 Simple and compound pistils. **(a)** This simple pistil consists of a single carpel. **(b)** This compound pistil has two united carpels. In most flowers with single pistils, the pistils are compound, consisting of two or more fused carpels.

When the two sperm cells enter the embryo sac, *both* participate in fertilization. One sperm cell fuses with the egg, forming a zygote that grows by mitosis and develops into a multicellular embryo in the seed. The second sperm cell fuses with the two haploid polar nuclei of the central cell to form a triploid ($3n$) cell that grows by mitosis and develops into **endosperm,** a nutrient tissue rich in lipids, proteins, and carbohydrates that nourishes the growing embryo. This fertilization process, which involves two separate nuclear fusions, is called **double fertilization** and is, with two exceptions, unique to flowering plants.[4]

[4] Double fertilization was reported in the gymnosperms *Ephedra nevadensis* (in 1990) and *Gnetum gnemon* (in 1996). This process differs from double fertilization in flowering plants in that an additional zygote, rather than endosperm, is produced. The second zygote later disintegrates.

Seeds and fruits develop after fertilization

As a result of double fertilization and subsequent growth and development, each seed contains a young plant embryo and nutritive tissue (the endosperm), both of which are surrounded by a protective seed coat. In monocots the endosperm persists and is the main source of food in the mature seed. In most dicots the endosperm nourishes the developing embryo, which subsequently stores food in its cotyledons.

As a seed develops from an ovule following fertilization, the ovary wall surrounding it enlarges dramatically and develops into a **fruit.** In some instances, other tissues associated with the ovary also enlarge to form the fruit. Fruits serve two purposes: to protect the developing seeds from desiccation as they grow and mature and to aid in the dispersal of seeds (see Chapter 35). For example, dandelion fruits have feathery plumes that enable the entire fruit to be carried by air currents. Once a seed lands in a suitable place, it may germinate and develop into a mature sporophyte that produces flowers, and the life cycle continues as described.

Flowering plants have many adaptations that account for their success

The evolutionary adaptations of flowering plants account for their success in terms of their ecological dominance and their great number of species. Seed production as the primary means of reproduction and dispersal, an adaptation shared with the gymnosperms, is clearly significant and provides a definite advantage over seedless vascular plants. Closed carpels, which give rise to fruits surrounding the seeds, and the process of double fertilization increase the likelihood of reproductive success. The evolution of a variety of interdependencies with many types of insects, birds, and bats, which disperse pollen from one flower to another of the same species, is another reason for angiosperm success (see Chapter 35). Pollen transfer results in cross fertilization, which mixes the genetic material and promotes genetic variation among the offspring.

In addition to their highly successful reproduction involving flowers, fruits, and seeds, flowering plants possess a number of distinctive features that have contributed to their success. Recall that most flowering plants have very efficient water-conducting cells, called vessel elements, in their xylem, in addition to tracheids. In contrast, the xylem of almost all seedless vascular plants and gymnosperms consists exclusively of tracheids. Most flowering plants also have efficient carbohydrate-conducting cells, called sieve tube members, in their phloem. With the exception of the flowering plants and gnetophytes, vascular plants generally lack vessel elements and sieve tube members.

The leaves of flowering plants, with their broad, expanded blades, are very efficient at absorbing light for photosynthesis. Abscission (shedding) of these leaves during cold or dry spells reduces water loss and thus has enabled some flowering plants to expand into habitats that would otherwise be too harsh for survival. The stems and roots of flowering plants are often modified for food or water storage, another feature that helps flowering plants to survive in severe environments.

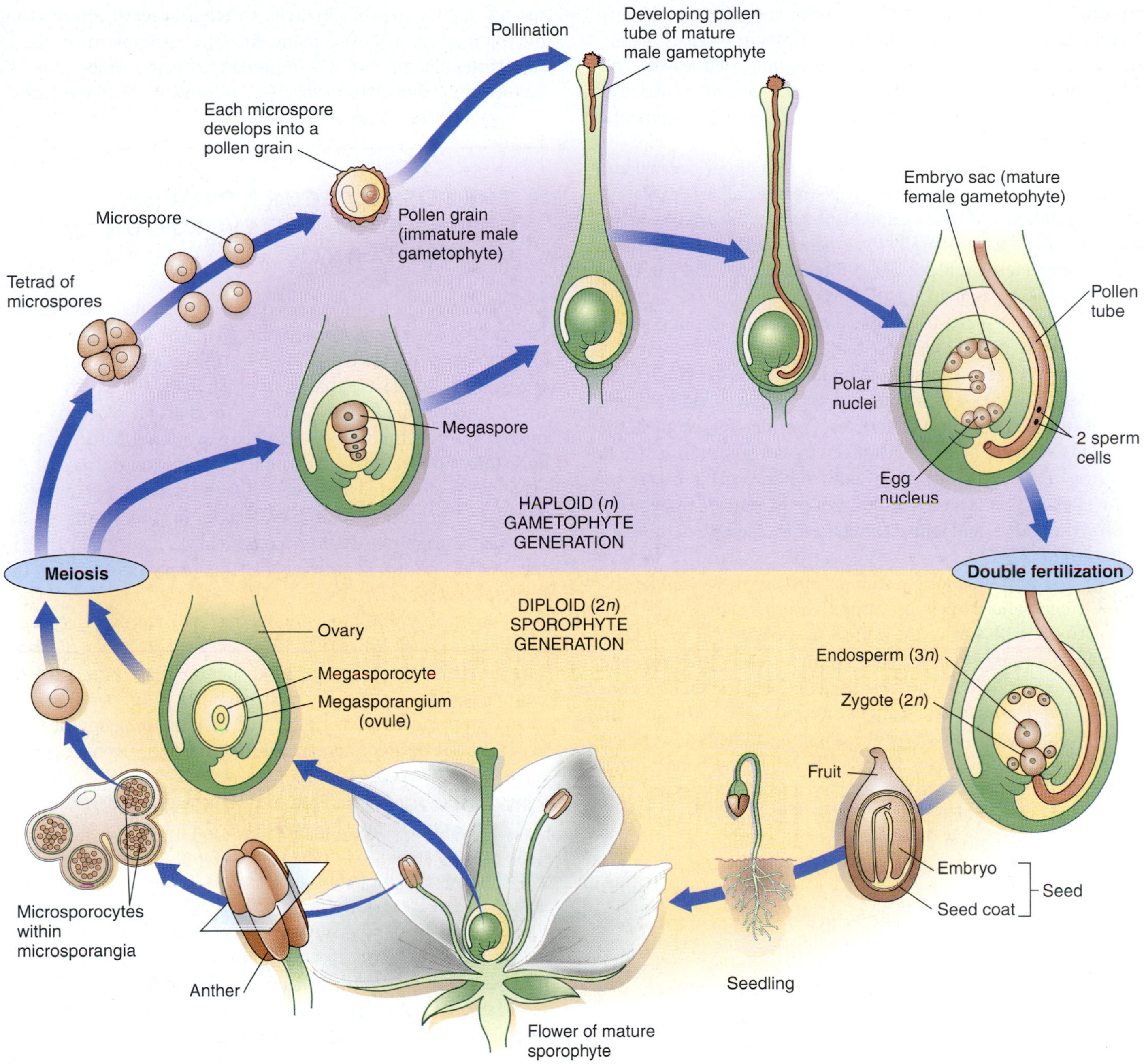

Figure 27-12 Life cycle of flowering plants. A significant feature of the flowering plant life cycle is double fertilization, in which one sperm cell unites with the egg, forming a zygote, and the other sperm cell unites with the two polar nuclei, forming a triploid cell that gives rise to endosperm.

Probably most crucial to the evolutionary success of flowering plants, however, is the overall adaptability of the sporophyte generation. As a group, flowering plants readily adapt to new habitats and changing environments. This adaptability is evident in the great diversity exhibited by the various species of flowering plants. For example, the cactus is remarkably well adapted to desert environments. Its stem stores water; its leaves (spines) have a reduced surface area available for transpiration (loss of water vapor; see Chapter 32) and may also protect against thirsty herbivorous animals; and its thick, waxy cuticle reduces water loss. On the other hand, the water lily is well adapted for wet environments, in part because it has air channels that provide adequate oxygen to stems and roots living in oxygen-deficient water and mud.

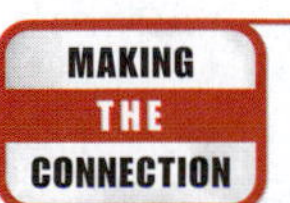

How did flowers evolve? Recall from Chapter 19 that, in evolution, new structures or organs often originate by modification of previously existing structures or organs. Scientific evidence supports the classical interpretation that the four organs of a flower—sepals, petals, stamens, and carpels—arose from highly modified leaves. This evidence includes comparisons of the arrangement of vascular tissues in both flowers and leafy stems and of the developmental stages of floral parts and leaves.

Sepals are the most leaflike of the four floral organs, and botanists generally agree that sepals are specialized leaves. Although petals of many flowering plant species are leaflike in appearance, botanists generally view petals as modified stamens that became sterile and leaflike. Cultivated roses and camellias provide evidence supporting this hypothesis; in some varieties the stamens have been transformed into petals, forming showy flowers with large numbers of petals.

The remarkably leaflike stamens and carpels of certain tropical trees and other species support the origin of stamens and carpels from leaves or leaflike organs. Consider, for example, the carpel of *Drimys,* a genus of evergreen trees and shrubs native to Southeast Asia, Australia, and South America. This carpel resembles a leaf that is folded inward along the midrib, thereby enclosing the ovules, and joined along the entire length of the leaf's margin (Fig. 27–13).

Process of Science The fundamental question is whether these leaflike stamens and carpels are primitive organs that were conserved (retained) during the course of evolution or are highly specialized organs that do not resemble early stamens and carpels. Many botanists who have studied this question have concluded that stamens and carpels are probably derived from leaves. Not all botanists accept the origin of stamens and carpels from highly modified leaves, however. Recall from Chapter 1 that uncertainty and debate are part of the scientific process and that scientists can never claim to know a final answer.

During the course of more than 130 million years of angiosperm evolution, flower structure diversified as floral organs fused together or became reduced in size or number. These changes led to greater complexity in floral structure in some species and to greater simplicity in other species. Interpreting the floral structures of so many different angiosperm species is sometimes difficult, but it is important because correct interpretations are essential to devising a classification scheme that is phylogenetic (see Chapter 22).

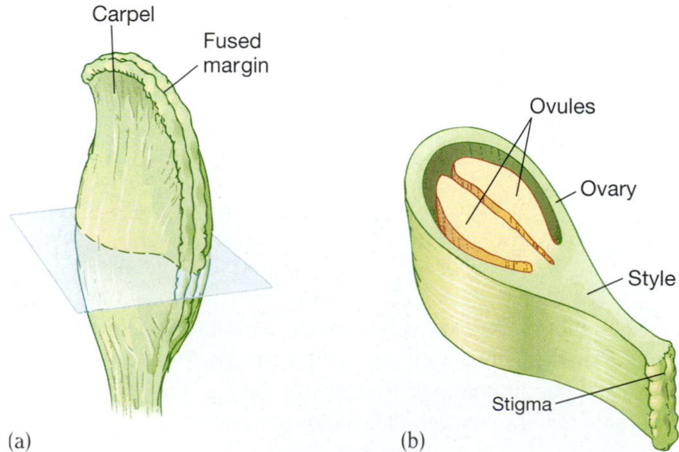

Figure 27–13 Carpel of *Drimys piperita*. **(a)** The carpel resembles a folded leaf in which the ovules borne on its upper surface are enclosed. **(b)** A cross section of the carpel, cut along the dashed line in part **(a)**.

THE FOSSIL RECORD PROVIDES CLUES ABOUT THE EVOLUTION OF SEED PLANTS

One of the groups that descended from ancestral seedless vascular plants was the **progymnosperms,** all of which are now extinct. Progymnosperms had two derived features that their immediate ancestors lacked: leaves that were megaphylls, and woody tissue, that is, secondary xylem, similar to that of modern gymnosperms. Progymnosperms retained a primitive feature of their ancestors, however: They reproduced by spores, not seeds. *Archaeopteris* was a progymnosperm that lived about 370 mya (Fig. 27–14*a*). The discovery and evaluation of 150 fossils of *Archaeopteris* in southeast Morocco, which was reported in 1999 in the journal *Nature,* confirmed that this progymnosperm is the earliest known modern tree.

Fossils of several progymnosperms with reproductive structures intermediate between those of spore plants and seed plants have been discovered. For example, the evolution of microspores into pollen grains and of megasporangia into ovules (seed-producing structures) can be traced in fossil progymnosperms. Plants producing seeds appeared during the late Devonian period, more than 360 mya. The fossil record indicates that different groups of seed plants apparently arose independently several times.

As mentioned previously, fossilized remains of ginkgo are found in 200-million-year-old rocks, and other groups of gymnosperms were well established by 160 to 100 mya. Although the gymnosperms are an ancient group, some questions persist about the exact pathways of gymnosperm evolution. The fossil record indicates that progymnosperms probably gave rise to conifers and to another group of extinct plants called **seed ferns,** which were seed-bearing woody plants with fernlike leaves (Fig. 27–14*b*). The seed ferns in turn probably gave rise to cycads and possibly ginkgo, as well as to several gymnosperm groups that are now extinct. The origin of gnetophytes remains unclear, although molecular data indicate they are closely related to conifers.

Flowering plants are the most recent group of plants to evolve. The fossil record, although incomplete, indicates that flowering plants probably descended from gymnosperms. By the middle of the Jurassic period, approximately 180 mya, several gymnosperm lines existed with features reminiscent of flowering plants. Among other traits, these derived gymnosperms possessed leaves with broad, expanded blades and the first modified seed-bearing leaves, which nearly enclose the ovules. It is also evident that beetles were visiting these plants, and biologists have suggested that perhaps this relationship was the beginning of **coevolution** (mutual adaptation) between plants and their animal pollinators.

One important task facing paleobotanists (biologists who study fossil plants) is determining which of the ancient gym-

Figure 27–14 Evolution of seed plants. **(a)** *Archaeopteris,* a progymnosperm that existed about 370 mya, had a number of structural features in common with modern seed plants but did not produce seeds. **(b)** *Emplectopteris,* a seed fern, had seeds that were borne on fernlike leaves. Seed ferns existed from about 360 to 250 mya. *(a, Redrawn from Beck, C.B. "Reconstructions of* Archaeopteris *and Further Consideration of Its Phylogenetic Position."* American Journal of Botany, *Vol. 49, 1962; b, Redrawn from Andrews, H.N.* Ancient Plants and the World They Lived In. *Comstock, New York, 1947)*

nosperms are in the direct line of evolution leading to the flowering plants. Based on structural data, most botanists think that flowering plants arose only once, that is, that there is only one line of evolution from the gymnosperms to the flowering plants. The gnetophytes are the gymnosperm group that some botanists consider the closest living relatives of flowering plants; both structural similarities and certain comparative molecular data support this conclusion. Gnetophytes are similar to angiosperms in that they possess vessels, lack archegonia, possess flower-like compound strobili, and undergo double fertilization. However, other molecular data refute a close link between the gnetophytes and flowering plants. Additional studies will hopefully clarify the relationships among the various gymnosperm phyla and the flowering plants.

The oldest definitive trace of flowering plants to appear in the fossil record is of ovules enclosed in tiny podlike fruits interpreted as carpels in Jurassic and Lower Cretaceous rocks some 125 to 145 million years old (Fig. 27–15). The fossil fruits were discovered in Northeast China in 1998 by Ge Sun of Academia Sinica, China, and David Dilcher of the University of Florida. The oldest fossilized flowers are about 118 to 120 million years old.

Based on the fossil record as well as both structural and molecular data of living angiosperms, the first flowering plants

Figure 27–15 The oldest known fossil angiosperm. This fossil of *Archaefructus* shows a carpel-bearing stem. It was discovered in northeast China in 1998. *(David Dilcher and Ge Sun)*

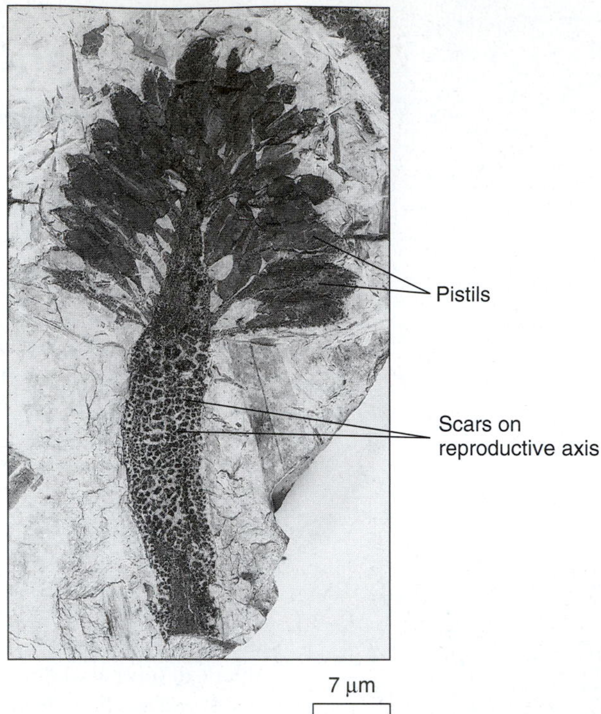

Figure 27–16 Fossil flower. The fossilized flower of the extinct plant *Archaeanthus linnenbergeri,* which lived during the Cretaceous period, about 98 to 100 mya. The scars on the reproductive axis (receptacle) may have been where stamens, petals, and sepals were originally attached but abscised (fell off). Many spirally arranged pistils were still attached at the time this flower was fossilized. *(Courtesy of David Dilcher)*

are thought to have been small, weedy shrubs or herbaceous plants that were adapted to disturbed habitats. They may have been fragile plants that were not easily preserved. If this is the case, it explains why there are few fossils of early angiosperms: The environment in which they evolved was repeatedly disturbed and not favorable for their preservation as fossils.

Botanists hypothesize that the rapid diversification of angiosperms did not occur until early flowering plants had invaded lowland regions. By 90 mya, during the late Cretaceous period, flowering plants had diversified and had begun to replace gymnosperms as Earth's dominant plants. Fossils of flowering plant leaves, stems, flowers, fruits, and seeds are numerous and diverse. They outnumber fossils of gymnosperms and ferns in late Cretaceous deposits, indicating the rapid success of flowering plants once they appeared (Fig. 27–16). Many angiosperm species apparently arose as a result of changes in chromosome number (see discussion of sympatric evolution in Chapter 19).

Within the angiosperms, current evidence suggests that the dicots evolved before the monocots and are therefore an older group. The monocots are thought to have evolved from certain ancient dicots (Fig. 27–17). The cladistic analysis of recent molecular evidence suggests that the monocots are a monophyletic group but the dicots are not. Instead, the dicots are paraphyletic. Recall from Chapter 22 that a **paraphyletic** group contains a common ancestor and some, but not all, of its descendants. According to molecular evidence such as DNA sequence comparisons, some dicots—the magnolias and laurels—are more closely related to the monocots than to other dicots. Despite these data, many botanists currently recognize the classes Monocotyledones and Dicotyledones for convenience.

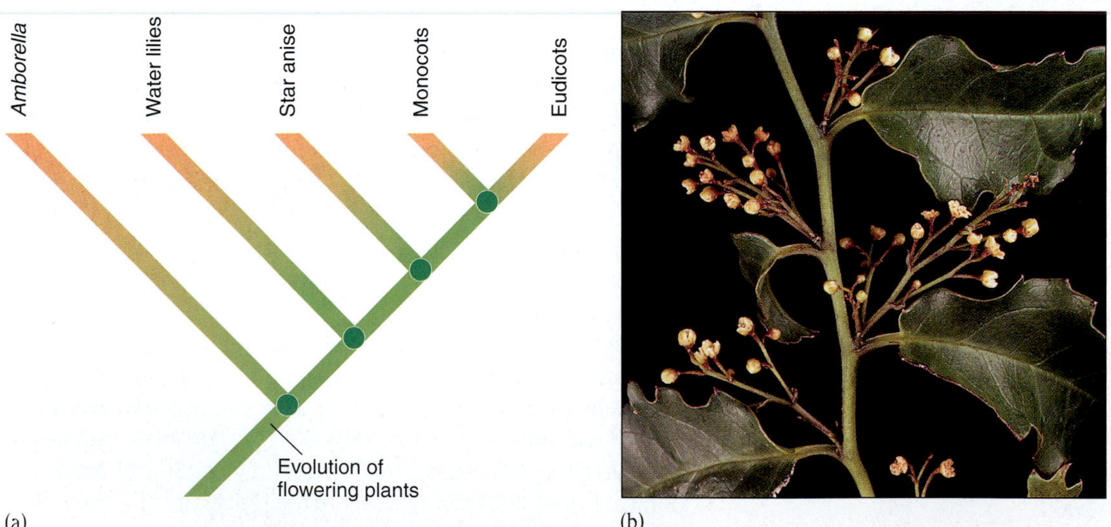

Figure 27–17 Evolution of flowering plants. **(a)** One hypothesis of relationships among the major groups of flowering plants, based on certain molecular evidence. *Amborella,* water lilies (*Nymphaea* sp.), and star anise *(Illicium verum)* are living dicots whose ancestors apparently branched off the angiosperm family tree early. These early dicot taxa were followed by the monocot branch and then by all other dicots (the Eudicots). **(b)** *Amborella trichopoda,* which is native to New Caledonia, an island in the South Pacific, may be the nearest living relative to the ancestor of all flowering plants. Photographed in the National Tropical Botanical Garden World Collection Greenhouse in Kauai, Hawaii. *(b, David Liittschwager and Susan Middleton with Environmental Defense)*

SUMMARY WITH KEY TERMS

I. **Seeds** are the primary means of reproduction and dispersal of gymnosperms and angiosperms (flowering plants). Each seed contains a well-developed embryonic sporophyte and a food supply, surrounded by a protective **seed coat.**

II. **Gymnosperms** are vascular plants with seeds that are totally exposed or borne on the scales of cones.
   A. Gymnosperms produce wind-borne **pollen grains,** a feature that seedless vascular plants lack.
   B. There are four phyla of gymnosperms.
      1. **Conifers,** which are the largest phylum of gymnosperms, are woody plants that bear **needles** (leaves that are usually evergreen) and produce seeds in cones. Most conifers are **monoecious** and have male and female reproductive parts in separate cones on the same plant.
         a. Male cones produce **pollen grains** (immature male gametophytes) that are carried by air currents to female cones.
         b. Female cones produce female gametophytes within **ovules (megasporangia).**
         c. Following fertilization, the zygote develops into an embryo that is encased inside a seed adapted for wind dispersal.
      2. **Cycads** are palmlike or fernlike in appearance. They are **dioecious**—that is, they have male and female reproductive structures on separate plants—but reproduce with pollen and seeds in conelike structures. There are relatively few living members of this once large phylum.
      3. *Ginkgo biloba,* the only surviving species in its phylum, is a deciduous, dioecious tree. The female **ginkgo** produces fleshy seeds directly on branches.
      4. **Gnetophytes** share a number of traits with angiosperms, such as efficient water-conducting cells **(vessel elements)** in their xylem.

III. **Flowering plants (angiosperms)** constitute the phylum of vascular plants that produce seeds enclosed within a **fruit.** They are the most diverse and most successful group of plants.
   A. Flowering plants have several specialized features that account for their success.
      1. The flower, which may contain **sepals, petals, stamens,** and **carpels,** functions in sexual reproduction.
      2. **Double fertilization,** which results in the formation of a diploid zygote and triploid **endosperm,** is characteristic of flowering plants.
      3. Unlike those of gymnosperms, the ovules of flowering plants are enclosed within an **ovary.** After fertilization, the ovules become seeds, and the ovary develops into a fruit.
      4. Flowering plants possess efficient water-conducting cells, called vessel elements, in their xylem and efficient carbohydrate-conducting cells, called **sieve tube members,** in their phloem.
      5. Wind, water, insects, or other animals transfer pollen grains in various flowering plants.
   B. There are two classes of flowering plants.
      1. Most **monocots** have floral parts in multiples of three, and their seeds each contain one **cotyledon.** The nutritive tissue in their mature seeds is endosperm.
      2. **Dicots** usually have floral parts in multiples of four or five, and their seeds each contain two cotyledons. The nutritive organs in their mature seeds are usually the cotyledons, which have absorbed the nutrients in the endosperm.

IV. Seed plants arose from seedless vascular plants.
   A. **Progymnosperms** were seedless vascular plants that had megaphylls and "modern" woody tissue. *Archaeopteris* is a representative progymnosperm.
      1. Progymnosperms probably gave rise to conifers.
      2. Progymnosperms probably gave rise to **seed ferns** as well, which in turn probably gave rise to cycads and possibly ginkgo.
   B. The evolution of the gnetophytes, particularly their relationship to flowering plants, is unclear.
   C. Flowering plants probably descended from ancient gymnosperms that had specialized features, such as leaves with broad, expanded blades and closed carpels.
      1. Flowering plants probably arose only once.
      2. The first flowering plants were probably dicots that were weedy shrubs or small herbaceous plants. *Amborella* is a dicot that may be the nearest living relative to the ancestor of all flowering plants.

POST-TEST

1. Seed plants do *not* possess which of the following structures? (a) ovules surrounded by integuments (b) microspores and megaspores (c) vascular tissues (d) a large, nutritionally independent sporophyte (e) a large, nutritionally independent gametophyte
2. Conifers, cycads, ginkgo, and gnetophytes are collectively called (a) fern allies (b) gymnosperms (c) angiosperms (d) dicots (e) seedless vascular plants
3. Most conifers are ____________, having male and female reproductive parts at different locations on the same plant. (a) incomplete (b) imperfect (c) monoecious (d) dioecious (e) perfect
4. The immature male gametophytes of pine are called (a) ovules (b) stamens (c) seed cones (d) pollen grains (e) polar nuclei
5. The transfer of pollen grains from the male to the female reproductive structure is known as (a) pollination (b) fertilization (c) embryo sac development (d) seed development (e) fruit development
6. Motile sperm cells are found as vestiges in these two gymnosperm groups: (a) monocots, dicots (b) gnetophytes, conifers (c) gnetophytes, flowering plants (d) cycads, conifers (e) cycads, ginkgo
7. More than ____________ species of flowering plants have been identified. (a) 235 (b) 2350 (c) 23,500 (d) 235,000 (e) 2,350,000
8. This class of flowering plants includes the palms, grasses, and orchids. (a) dicots (b) gnetophytes (c) cycads (d) monocots (e) conifers
9. The pistil has three sections: (a) stigma, style, and anther (b) anther, filament, and ovule (c) stigma, style, and ovary (d) ovary, ovule, and sepal (e) corolla, stamen, and sepal
10. A simple pistil consists of a single: (a) calyx (b) carpel (c) ovule (d) filament (e) petal
11. A flower that lacks stamens is said to be both ____________ and ____________. (a) complete; imperfect (b) incomplete; perfect (c) complete; perfect (d) incomplete; imperfect
12. After fertilization, the ____________ develops into a fruit and the ____________ develops into a seed. (a) ovary; ovule (b) polar nuclei; ovule (c) ovary; endosperm (d) ovule; ovary (e) ovule; polar nuclei
13. The female gametophyte in flowering plants is also called the (a) polar nuclei (b) anther (c) embryo sac (d) endosperm (e) sporophyll
14. This living dicot may be the nearest living relative to the ancestor of all flowering plants (a) *Amborella* (b) *Archaeopteris* (c) *Gnetum* (d) water lily (e) *Archaeanthus*
15.  This cross section through an ovary reveals that the pistil is (a) simple, with one carpel (b) compound, with two fused carpels (c) compound, with three fused carpels (d) compound, with six separate carpels (e) compound, with six fused carpels

REVIEW QUESTIONS

1. Why are seeds such a significant evolutionary innovation?
2. What distinguishing features do cycads, ginkgo, and gnetophytes share with conifers?
3. How are the vegetative adaptations of flowering plants different from those of gymnosperms?
4. How does the gymnosperm life cycle differ from that of flowering plants?
5. What are the two classes of flowering plants, and how can one distinguish between them?
6. How does pollination occur in the gymnosperms? In flowering plants?
7. How does fertilization differ in gymnosperms and flowering plants?

YOU MAKE THE CONNECTION

1. How are cones and flowers alike? How are they different? (*Hint:* Your answer should consider microspores/megaspores and seeds.)
2. How do the life cycles of seedless plants (see Chapter 26) and seed plants differ? In what fundamental way are they alike?
3. Describe the evolutionary changes that occurred as ancient seedless vascular plants evolved into gymnosperms and as ancient gymnosperms evolved into flowering plants.
4. In contrast with the cones of gymnosperms, which are either male or female, most flowers contain both male and female reproductive structures. Explain how bisexual flowers might be advantageous to flowering plants.
5. Contrast the algae, mosses, ferns, gymnosperms, and angiosperms with respect to their dependence on water as a transport medium for reproductive cells. Suggest a hypothesis to explain the differences.

RECOMMENDED READINGS

Berg, L.R. *Introductory Botany: Plants, People, and the Environment.* Saunders College Publishing, Philadelphia, 1997. An introduction to general botany that stresses environmental links between plants and humans.

Bernhardt, P. *The Rose's Kiss: A Natural History of Flowers.* Island Press, Washington, D.C., 1999. This eloquent, immensely readable account of floral biology will interest and entertain students and professors alike.

Crepet, W. "Early Bloomers." *Natural History,* May 1999. Describes some of the 90-million-year-old fossil flowering plants that have been unearthed from a site in New Jersey.

Hershey, D. "The Truth Behind Some Great Plant Stories." *The American Biology Teacher,* Vol. 62, No. 6, June 2000. The myth about ancient Chinese monasteries saving *Ginkgo* from extinction is debunked.

Heywood, V.H. *Flowering Plants of the World.* Oxford University Press, New York, 1993. This beautifully illustrated guide describes more than 300 families of angiosperms, including their economic uses.

Holden, C. "Ancient Trees Down Under." *Science,* 20 Jan. 1995. A conifer species thought to be extinct for the past 50 million years was recently discovered in Australia.

Mlot, C. "Early Flowering Tree Rediscovered." *Science News,* Vol. 152, 2 Aug. 1997. A rare flowering tree with many primitive features, such as lack of vessel elements in the xylem, has been found in a rain forest in Madagascar.

Niklas, K. "What's So Special about Flowers?" *Natural History,* May 1999. The evolutionary achievement represented by flowers is examined in this article.

Palmer, D. "First Flowers Emerge from Triassic Mud." *New Scientist,* 29 Jan. 1994. Two hundred-million-year-old rocks may contain the oldest traces of flowering plant pollen.

Poinar, G.O. "Still Life in Amber." *The Sciences,* Vol. 34, No. 2, 1993. The fossilized remains of many insects and other small animals have been preserved in amber, which is the fossilized resin of prehistoric gymnosperms.

Raven, P.H., R.F. Evert, and S.E. Eichhorn. *Biology of Plants,* 6th ed. W.H. Freeman & Company, New York, 1999. An excellent general botany textbook, with detailed descriptions of plant evolution.

- Visit our Web site at **http://www.info.brookscole.com/solomonbergmartin** for links to chapter-related resources on the World Wide Web. Additional on-line materials relating to this chapter can also be found on our Web site.

See chapter activity on BioActive Learner CD for additional help in mastering the chapter's material. Icon location in the chapter's margins shows which topics have tutorials or simulations in the CD.

28

The Animal Kingdom: An Introduction to Animal Diversity

LEARNING OBJECTIVES

After you have studied this chapter you should be able to

1. List several characteristics common to most animals.
2. Describe the ecological roles and distribution of animals and compare the advantages and disadvantages of life in the ocean, in fresh water, and on land.
3. Justify the classification and relationships of animal phyla on the basis of symmetry, type of body cavity, pattern of development, and molecular data.
4. Identify distinguishing characteristics of the two major animal clades—protostomes and deuterostomes—and describe the hypothesized split of the protostomes into two large clades: Lophotrochozoa and Ecdysozoa.
5. Identify distinguishing characteristics of phyla Porifera, Cnidaria, Ctenophora, and Platyhelminthes, and classify a given animal into the appropriate phylum.
6. Summarize the life cycles of the blood fluke and the beef tapeworm, and identify several adaptations that these animals possess for their parasitic lifestyles.
7. Describe the adaptive advantages of the following characteristics: bilateral symmetry, cephalization, multiple tissue layers, and a motile life stage.

The tube sponge *(Callyspongia vaginalis).* This animal, which ranges in color from purple to blue to gray, is common on coral reefs in the Caribbean, from Florida to Mexico. *(Charles V. Angelo/Photo Researchers, Inc.)*

Biologists have described and named more than a million species of animals, and several million more remain to be discovered and classified. Although most animal species are readily recognizable as animals, the identity of some others is less obvious. Early naturalists thought that sponges were plants because they did not move from place to place. Some people still mistake certain marine animals, such as sponges and corals, for plants. Locomotion is not a requirement for being classified as an animal.

Animals are so diverse that for almost any definition, we can find exceptions. Still, the following characteristics describe most animals:

1. Animals are multicellular eukaryotes.
2. Cells that make up the animal body are specialized to perform specific functions. In all but the simplest animals, cells are organized to form tissues, and tissues are organized to form organs. In many animals with simple body plans, life processes such as gas exchange, circulation of materials, and waste disposal take place by diffusion directly to and from the environment. In large, complex animals, specialized organ systems and mechanisms have evolved that perform these life processes.
3. Animals are **heterotrophs.** As consumers, they depend on producers for their raw materials and energy. Most animals ingest their food first and then digest it inside the body, usually within a digestive system.
4. Most animals are capable of locomotion at some time during their life cycle. Some animals (e.g., sponges and corals) move about as larvae (immature forms) but are **sessile** (firmly attached to the ground or some other surface) as adults *(see photograph)*.
5. Most animals have nervous systems and muscle systems that enable them to respond rapidly to stimuli in their environment.
6. Most animals are diploid organisms that reproduce sexually, with large, nonmotile eggs and small, flagellated sperm. A haploid sperm unites with a haploid egg, forming a **zygote** (fertilized egg) that undergoes **cleavage,** a series of mitotic cell divisions. During cleavage the zygote is converted to a hollow ball of cells called a **blastula.** Cells of the blastula undergo **gastrulation,** a process of forming and segregating specific layers of tissue, called germ layers. Some animals develop directly into adults. Others develop into a **larva,** a sexually immature form that may look very different from the adult. Larvae then undergo **metamorphosis,** a developmental process that converts the immature animal into a juvenile form that can then grow into an adult.

Most biologists divide **kingdom Animalia** into about 35 phyla. The animals most familiar to us, fishes, amphibians (e.g., frogs and salamanders), reptiles (e.g., snakes and alligators), birds, and mammals (lions, humans) are **vertebrates,** a subphylum of phylum Chordata. A vertebrate is an animal with a backbone (vertebral column). You may be surprised to learn that vertebrates account for fewer than 5% of the species in the animal kingdom. The majority of animals are the less familiar **invertebrates,** animals without backbones. The invertebrates include diverse animals such as sponges, jellyfish, worms, mollusks, insects, crustaceans, and sea stars.

Biologists have long debated the evolutionary relationships among animal groups. Recently developed molecular methods are providing new tools for understanding these relationships. Animal phylogeny has become an exciting field as biologists use new data to redraw the tree of animal life. For example, historically biologists viewed round worms and ribbon worms as members of simple groups that appeared early in animal evolution. After studying recent molecular data, many biologists hypothesize that some of these groups are closely related to structurally complex animals. These worms may have evolved from more complex animals and then become structurally simpler over time.

In this chapter we discuss animal phylogeny and describe four animal phyla that have simpler body plans than other animal groups. The term **body plan** refers to the basic structure of the body. In Chapters 29 and 30 we discuss two major clades of animals: the protostomes and the deuterostomes.

ANIMALS INHABIT MOST ENVIRONMENTS

Fossil evidence suggests that animals evolved during the Precambrian in shallow, marine environments (see Chapter 20). Although animals are now distributed in virtually every environment, members of most animal phyla still inhabit marine environments.

Marine environments offer many advantages. The buoyancy of sea water helps support its inhabitants, and the temperature is relatively stable owing to the large volume of water. The body fluids of most invertebrates have about the same osmotic concentration as sea water, so fluid and salt balance can be more easily maintained than in fresh water. **Plankton,** which consists of the mainly microscopic animals and protists that are suspended in water and float with its movement, provides a ready source of food.

Life in the ocean also presents some challenges. Although the continuous motion of water brings nutrients to animals and washes their wastes away, they must be able to cope with the water's movements and the currents that could sweep them away. Squids, fishes, and marine mammals are strong swimmers and can usually direct their movements and maintain their location. However, most invertebrates and young vertebrates are unable to swim strongly, and they have adapted in many different ways to the tides and currents. Some sessile animals attach permanently to a stable structure such as a rock. Others burrow in the sand and silt that cover the sea bottom. Many invertebrates have adapted by maintaining a small body size and becoming part of the plankton. As they are tossed about, their food supply continues to surround them.

Far fewer kinds of animals make their homes in fresh water than in the ocean. Fresh water is hypotonic to the tissue fluids of animals, so water tends to move into the animal by osmosis. Freshwater species must have mechanisms for removing excess water while retaining salts. This *osmoregulation* requires an expenditure of energy. Fresh water offers a much less constant environment than sea water and generally contains less food. Oxygen content and temperature vary, and turbidity (owing to sediments suspended in the water) and even water volumes fluctuate.

Living on land is even more difficult than living in fresh water and requires many adaptations. Based on the fossil record, many biologists hypothesize that the first air-breathing land animals were scorpion-like arthropods that came ashore in the Silurian period about 450 million years ago (mya). The first vertebrates to inhabit terrestrial environments, the amphibians, did not appear until the latter part of the Devonian period, about 30 million years later.

The chief problem facing all terrestrial organisms is that of desiccation (drying out). Water is constantly lost by evaporation and is often difficult to replace. A body covering adapted to minimize fluid loss helps solve this problem in many land animals. Location of the respiratory surface deep within the animal also helps prevent fluid loss. Thus, gills are typically located externally, and lungs and tracheal tubes (found in insects and some other arthropods) are internal.

Reproduction on land also poses challenges to protect gametes and the developing offspring from desiccation. Aquatic animals typically shed their gametes in the water, where fertilization occurs. The surrounding water also serves as an effective shock absorber, protecting the delicate embryos as they develop. Some land animals, including most amphibians, return to the water for reproduction, and their larval forms develop in the water.

The evolution of internal fertilization has permitted many terrestrial animals, including earthworms, land snails, insects, reptiles, birds, and mammals, to meet the desiccation challenge. Because they transfer sperm from the body of the male directly into the body of the female by copulation, the sperm are continuously surrounded by a watery medium. Another important adaptation to reproduction on land is the tough, protective shell that surrounds the eggs of many species. Secreted by the female, this shell protects the developing embryo from drying out. An alternative adaptation for terrestrial reproduction is the development of the embryo within the moist body of the mother.

The temperature extremes of terrestrial habitats also present challenges. In later chapters we will learn about behavioral and physiological adaptations for maintaining body temperature, that have evolved in animals.

RECONSTRUCTING PHYLOGENY IS A MAJOR CHALLENGE

Historically, biologists have depended on fossils, on similarities in body plan, and on patterns of development to determine evolutionary relationships among various groups of animals. Now, molecular methods are providing additional data that help answer questions about phylogeny. A basic question still being debated is whether animals evolved once or multiple times. Recent molecular analysis indicates that the structure of RNA, genes that control development, and many other molecules are very similar among all animal groups that have been studied. The *principle of parsimony* supports the hypothesis that animals evolved only once (see Chapter 22). Such complex molecules are unlikely to have evolved multiple times.

Biologists generally agree that animals evolved from protists, probably from colonial flagellates. In some colonial flagellates, cells became specialized to perform specific functions such as movement, feeding, or reproduction. As this division of labor evolved, a colony of flagellates reached the level of cooperation and coordination that qualified it to be considered a single organism—the first animal.

Biologists divide animals into **parazoa** (*para,* "alongside" and *zoa,* "animals") and **eumetazoa** (*eu,* "true"; *meta,* "later"; *zoa,* "animals"). The only extant animals classified as parazoa are sponges (phylum Porifera). Sponges diverged early from the lineage leading to other animals. Because they have been evolving independently for so long, they are very different from other animals. Although they are multicellular and can be large, sponges function much like colonial, unicellular protozoa. Most cells of sponges are extremely versatile and can change form and function; they are not organized into tissues. Other animals have true tissues and are classified as eumetazoa.

Phylogenetic relationships among various animal groups is an issue of continuing controversy. For example, the simplest extant animals, the **Placozoans,** were probably not the earliest to evolve. These elusive marine animals are very small (less than 3 mm). Their flat, rounded bodies consist of only a few thousand cells of four types. Biologists formerly classified them along with sponges as parazoa, but recent data suggest that they evolved later and are more closely related to certain complex invertebrates. According to this view, placozoans became simplified later in the history of animal evolution.

Animals can be classified according to body symmetry

Symmetry refers to the arrangement of body structures in relation to the axis of the body. Most sponges are not symmetrical, so that when cut in half the two halves are not similar to one another. Most other animals exhibit either radial or bilateral body symmetry. Members of phylum Cnidaria (jellyfish, sea anemones, and their relatives) and adult echinoderms (sea stars and their relatives) have **radial symmetry.** In radial symmetry, the body has the general form of a wheel or cylinder, and similar structures are regularly arranged as spokes from a central axis (Fig. 28–1*a, b*). Multiple planes can be drawn through the central axis, each dividing the organism into two mirror images. An animal with radial symmetry receives stimuli equally from all directions in the environment. Many radially symmetrical animals actually have modified radial symmetry. For example, sea anemones and ctenophores (comb jellies) have **biradial symmetry** in which parts of the body have become specialized so that only two planes can divide the body into similar halves.

Most animals are **bilaterally symmetrical,** at least in their larval stages. A bilaterally symmetrical animal can be divided through only one plane (which passes through the midline of the body) to produce roughly equivalent right and left halves that are mirror images (Fig. 28–1*c, d*). As bilateral symmetry evolved, there was a corresponding evolutionary trend toward **cephalization,** the development of a head where sensory structures are concentrated. In many animal groups, concentrations of nerve cells in the head form a brain, and a nerve cord extends from the brain toward the rear end of the animal. Bilateral symmetry and cephalization are considered adaptations to locomotion. The head end of the animal meets its environment first and is best equipped to capture food or respond to danger.

To locate body structures in bilaterally symmetrical animals, it is helpful to define some basic terms and directions. The back surface of an animal is its **dorsal** surface; the underside (belly) is its **ventral** surface. **Anterior** (or **cephalic**) means toward the head end of the animal; **posterior,** or **caudal,** means toward the tail end. A structure is said to be **medial** if it is located toward the midline of the body and **lateral** if it is toward one side of the body; for example, the human ear is lateral to the nose. In human anatomy, the term **superior** refers to a structure located above some point of reference, or toward the head end of the body. The term **inferior** is used in human anatomy to mean located below some point of reference, or toward the feet.

A bilaterally symmetrical animal has three axes, each at right angles to the other two: an anterior-posterior axis extending from head to tail, a dorsoventral axis extending from back to belly, and a left-right axis extending from side to side. We can distinguish three planes or sections that divide the body into specific parts. A **sagittal plane** divides the body into right and left parts. A sagittal plane passes from anterior to posterior and from dorsal to ventral. A **frontal** plane divides a bilateral body into dorsal and ventral parts. A **transverse section,** or **cross section,** cuts at right angles to the body axis and separates anterior and posterior parts.

Animals can be grouped according to type of body cavity

The structures of most animals develop from three embryonic tissue layers, called **germ layers.** The outer germ layer, called the **ectoderm,** gives rise to the outer covering of the body and to the nervous system (if the animal has one). The inner layer, or **endoderm,** forms the lining of the digestive tube and other digestive organs. Cnidarians and ctenophores develop only these two germ layers and are referred to as **diploblastic.** All of the other eumetazoa are **triploblastic.** They develop a middle layer, called

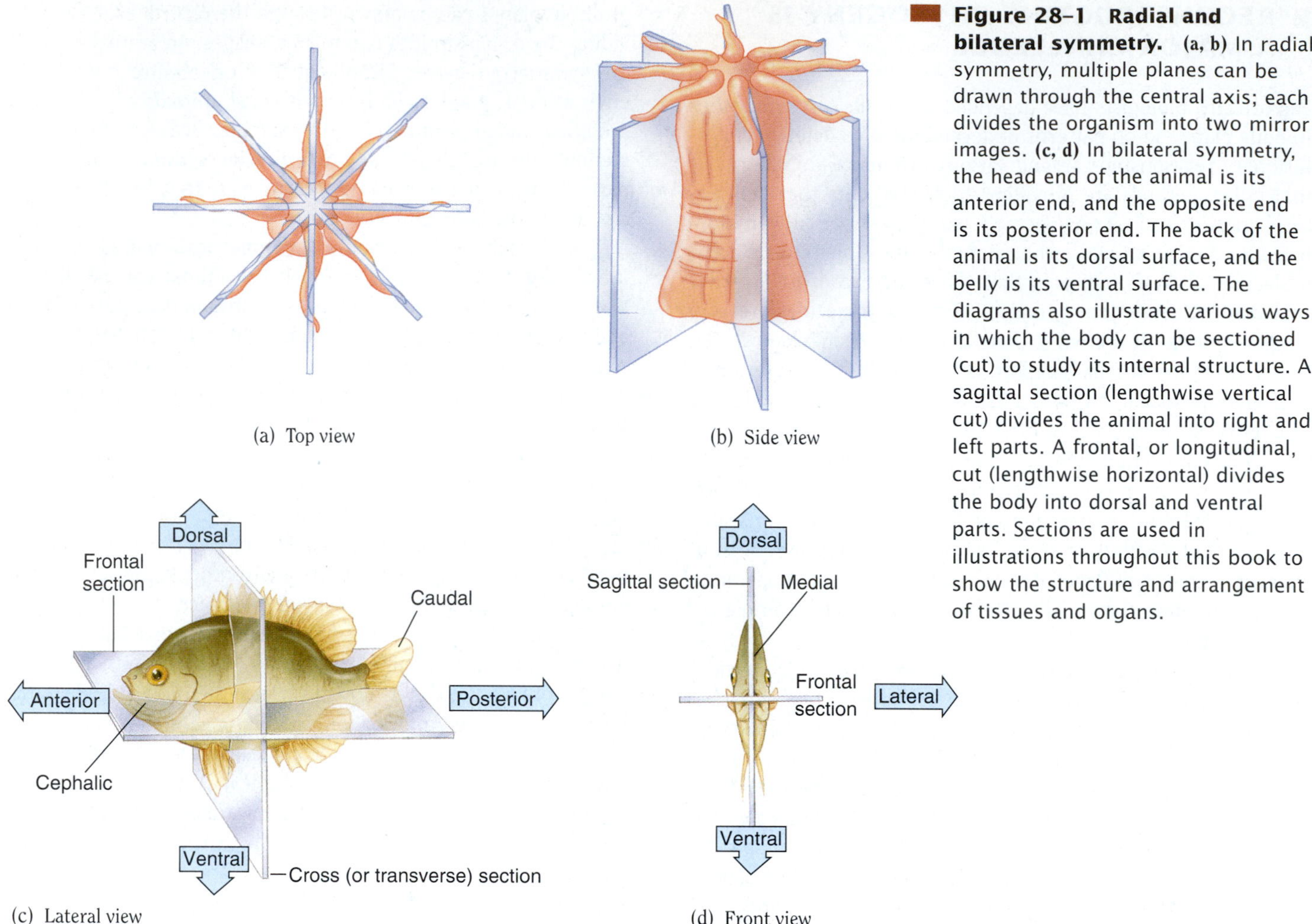

(a) Top view

(b) Side view

(c) Lateral view

(d) Front view

Figure 28–1 Radial and bilateral symmetry. **(a, b)** In radial symmetry, multiple planes can be drawn through the central axis; each divides the organism into two mirror images. **(c, d)** In bilateral symmetry, the head end of the animal is its anterior end, and the opposite end is its posterior end. The back of the animal is its dorsal surface, and the belly is its ventral surface. The diagrams also illustrate various ways in which the body can be sectioned (cut) to study its internal structure. A sagittal section (lengthwise vertical cut) divides the animal into right and left parts. A frontal, or longitudinal, cut (lengthwise horizontal) divides the body into dorsal and ventral parts. Sections are used in illustrations throughout this book to show the structure and arrangement of tissues and organs.

mesoderm, that gives rise to most other body structures, including muscles, skeletal structures, and circulatory system (when they are present).

A widely held traditional system for grouping triploblastic animals is based on the presence and type of **body cavity,** or **coelom** (see′ lum), a fluid-filled space between the body wall and the digestive tube. The flatworms (phylum Platyhelminthes) and proboscis worms (phylum Nemertea) are triploblastic but have a solid body; that is, they have no body cavity. They are referred to as **acoelomates** (*a,* "without" ; *coelom,* "cavity") (Fig. 28–2).

Generally animals with a body cavity have a **tube-within-a-tube body plan.** The body wall, which forms the outer tube, is covered with tissue that develops from ectoderm. Tissue derived from endoderm lines the inner tube—the digestive tube, or gut—which has an opening at each end: the mouth and the anus. The space that exists between the two tubes is the body cavity. If the body cavity is not completely lined with mesoderm it is called a **pseudocoelom** ("false coelom"). Animals with a pseudocoelom, like nematodes (roundworms) and rotifers, are referred to as **pseudocoelomates.** Traditionally, biologists have classified them as a separate group.

In most animals, the body cavity is completely lined with mesoderm. Such a body cavity is a true coelom. Animals with true coeloms are referred to as **coelomates.** Recent evidence suggests that pseudocoelomates are not a monophyletic group. Rather, they probably evolved through a process of simplification from more than one group of animals with a true coelom. Thus, the system of grouping animals based on type of body cavity is no longer considered valid from an evolutionary standpoint.

Coelomate animals form two main groups based on differences in development

Animals with a coelom form two main evolutionary lines: **protostomes,** which include mollusks, annelids, and arthropods, and **deuterostomes,** which include the echinoderms (e.g., sea stars and sea urchins) and chordates (the phylum that includes the vertebrates). Protostomes and deuterostomes are distinguished by basic differences in their pattern of early development.

An important difference in the development of protostomes and deuterostomes is the pattern of cleavage, the first several cell divisions of the embryo. In many protostomes, the early cell divisions are diagonal to the polar axis (the long axis of the egg), resulting in a somewhat spiral arrangement of cells; any one cell is located between the two cells above or below it (Fig. 28–3). This pattern of division is known as **spiral cleavage.** In **radial cleavage,**

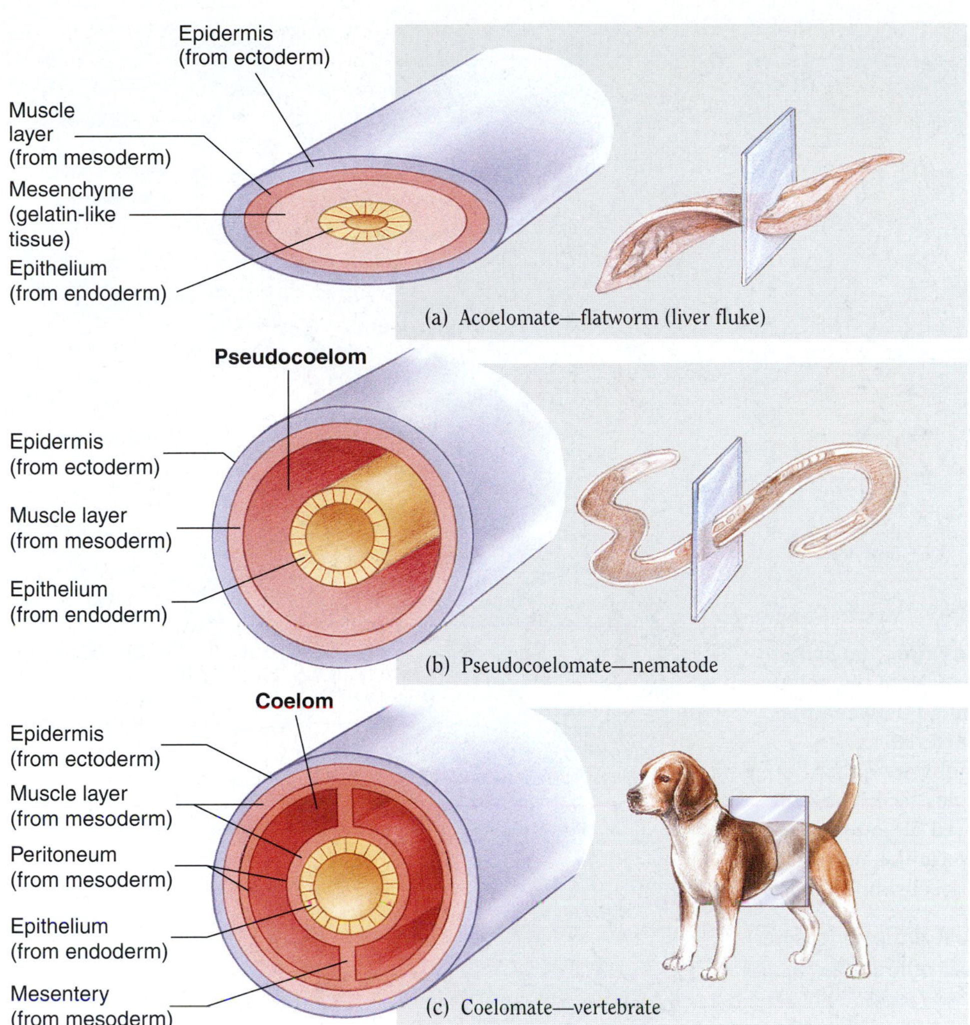

Figure 28–2 Three basic body plans in triploblastic animals. The germ layer from which each tissue was derived is indicated in parentheses. **(a)** An acoelomate animal has no body cavity. **(b)** A pseudocoelomate animal has a body cavity that is not completely lined with mesoderm. **(c)** In a coelomate animal the body cavity, or coelom, is completely lined with tissue that develops from mesoderm.

characteristic of the deuterostomes, the early divisions are either parallel or at right angles to the polar axis. The resulting cells are located directly above or below one another.

In the protostomes, the developmental fate of each embryonic cell is typically fixed very early. For example, if the first four cells of an annelid embryo are separated, each cell develops into only a fixed quarter of the larva; this is referred to as **determinate cleavage.** In deuterostomes, cleavage is usually **indeterminate.** If the first four cells of a sea star embryo, for instance, are separated, each cell is capable of forming a complete, though small, larva. If a few cells are removed from the blastula of an embryo undergoing determinate cleavage, some structure such as a limb, does not develop. In contrast, if a few cells are removed from a blastula undergoing indeterminate cleavage, other cells are able to compensate, and the embryo develops normally.

During gastrulation a group of cells moves inward, forming a sac that becomes the embryonic gut. The opening to the outside is called the **blastopore.** In most protostomes, the blastopore develops into the mouth. The word *protostome* comes from Greek words meaning "first, the mouth." In deuterostomes the blastopore does not give rise to the mouth. Instead it generally develops into the anus. A second opening that forms later in development gives rise to the mouth. The word *deuterostome* means "second, the mouth."

Another, though less reliable, difference between protostome and deuterostome development is the manner in which the coelom is formed. In most protostomes, the mesoderm splits, and the split widens into a cavity that becomes the coelom (Fig. 28–4). This method of coelom formation is known as **schizocoely.** In deuterostomes, the mesoderm forms as "outpocketings" of the developing gut. These outpocketings eventually pinch off and form pouches; the cavity within the pouches becomes the coelom. This type of coelom formation is called **enterocoely.**

Process of Science — Biologists are rethinking animal relationships based on molecular data

Many biologists hypothesize that most of the approximately 35 extant animal phyla evolved during the Cambrian explosion, a geologically brief 40-million-year span during the Precambrian and the early Cambrian Period. They suggest that basic body plans of all extant and an amazing variety of extinct animals appeared during this time. Other biologists have challenged this view, tracing at least the beginnings of animal diversity to a much earlier time. According to this latter view, a reorganization of the metazoan genome took place during the Cambrian explosion, leading to new body plans. Whichever hypothesis is correct, animal body plans diversified quickly, and soft-bodied forms left

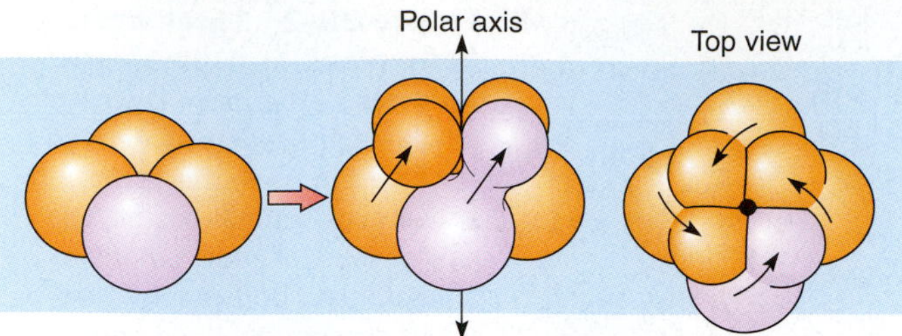

(a) Protostomes are characterized by spiral cleavage.

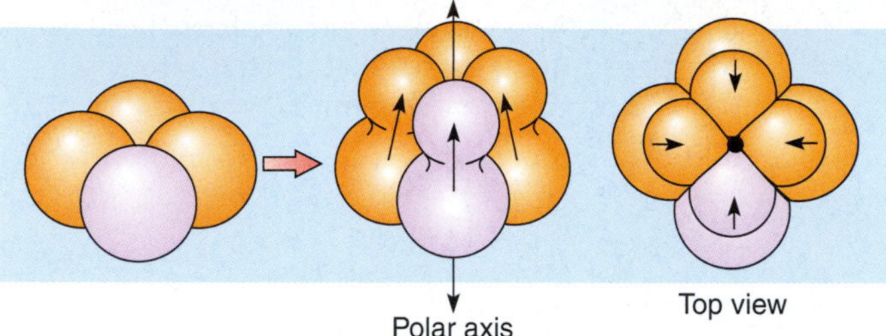

(b) Deuterostomes are characterized by radial cleavage.

Figure 28–3 Spiral and radial cleavage. (a) Spiral cleavage is characteristic of protostomes. Note the spiral arrangement, with the upper cells centered between the lower cells. **(b)** In radial cleavage, characteristic of deuterostome embryos, the early divisions are either parallel to the polar axis or at right angles to it. The cells are stacked, with the upper cells centered directly above the lower cells. The pattern of cleavage can be appreciated by comparing the position of the purple cells in *(a)* and *(b)*.

few fossils. The scarcity of fossils has made it difficult to determine the age, rate of divergence, and number of branches of animal groups.

With the emergence of new molecular techniques, biologists have begun to explore biochemical similarities such as similarities in ribosomal RNA sequences and *Hox* genes. Recall from Chapter 16 that *Hox* genes are a group of regulatory genes that help control early development. They are found in all major animal groups except sponges. Investigators suggest that all the *Hox* gene groups had evolved by the beginning of the Cambrian.

Although molecular data have thus far validated much of the phylogeny that systematists have based on morphology and development, these data suggest changes in the placement of several animal groups. Biologists are responding by redrawing parts of the phylogenetic tree for animals. Some of the more widely held current hypotheses are presented in this chapter.

As explained in the last section, biologists classify coelomate animals into protostome and deuterostome groups. Based on an evolving hypothesis of phylogeny, many systematists now subdivide the protostomes into two branches: the **Lophotrochozoa** and the **Ecdysozoa** (Fig. 28–5; Table 28–1). The Lophotrochozoa include the nemerteans (proboscis worms), mollusks, annelids, and the lophophorate phyla (groups that have a ciliated ring of tentacles surrounding the mouth; described in Chapter 29). The Ecdysozoa, animals that molt, include the nematodes and arthropods. Thus, in this phylogeny there are three major clades of coelomate animals.

Molecular data suggest that the evolution of animals was not an orderly progression from simple to complex. The acoelomate

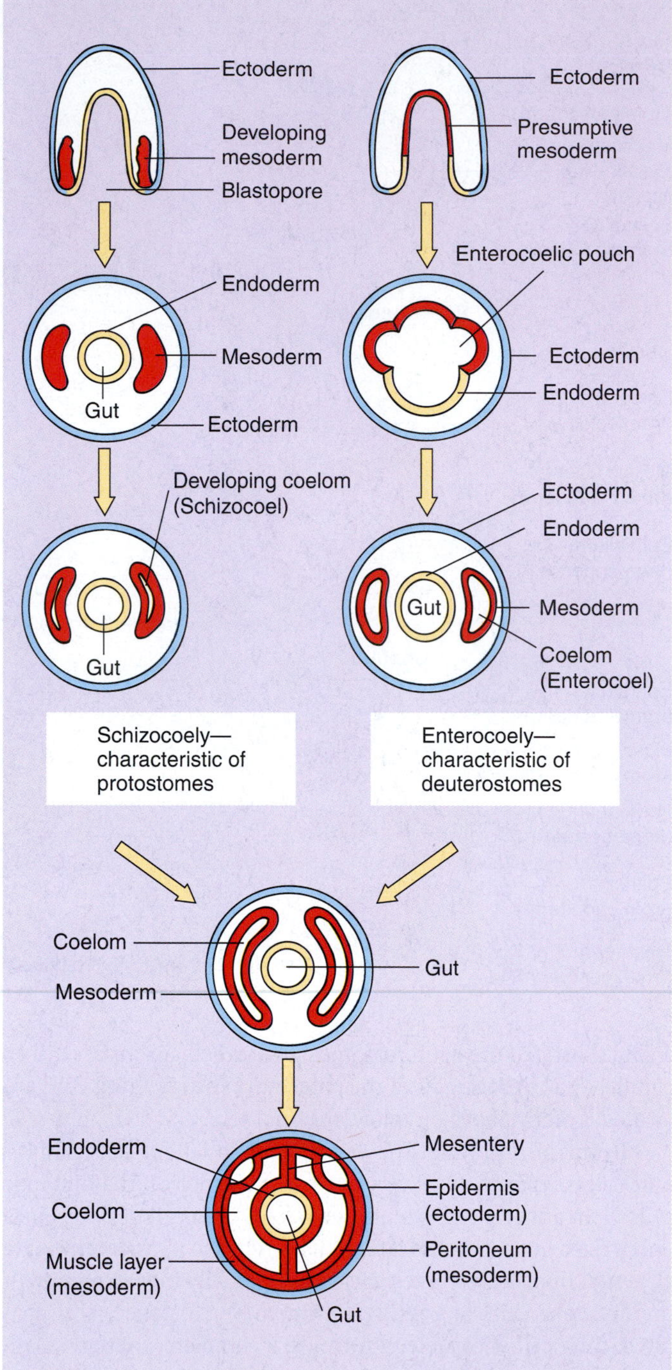

Figure 28–4 Coelom formation. Two types of coelom formation are shown here, using longitudinal (top two) and cross-section diagrams of developing embryos. The coelom originates in the embryo from blocks of mesoderm that split off from each side of the embryonic gut. In protostomes the coelom typically forms by a process called schizocoely in which the mesoderm *(red)* splits. The split widens, forming a cavity that becomes the coelom. In enterocoely, characteristic of deuterostomes, the mesoderm outpockets from the gut, forming pouches. The cavity within these pouches becomes the coelom. Ectoderm is shown in blue, endoderm in yellow.

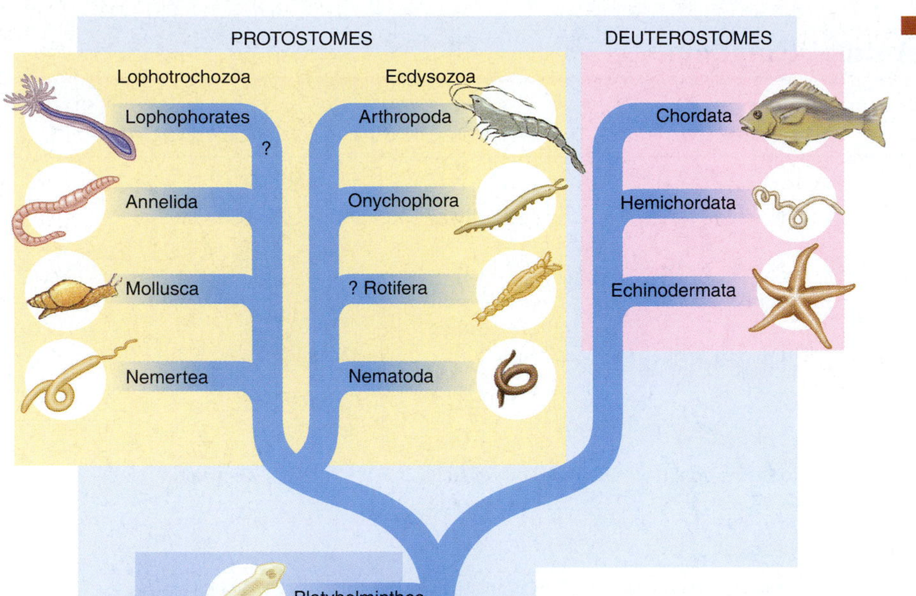

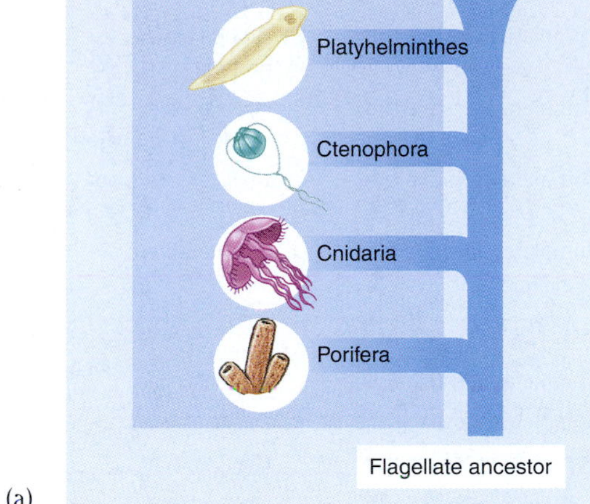

(a)

PROTOSTOMES
DEUTEROSTOMES
Lophotrochozoa
Ecdysozoa
Porifera
Cnidaria
Ctenophora
Platyhelminthes
Nemertea
Mollusca
Annelida
Lophophorate phyla
Nematoda
Rotifera
Onychophora
Arthropoda
Echinodermata
Hemichordata
Chordata
?
?
Radial symmetry
Protostome pattern of development
Deuterostome pattern of development
Development of a body cavity
Three tissue layers (ectoderm, endoderm, and mesoderm), bilateral symmetry
Differentiation of tissues (ectoderm and endoderm)
Multicellularity
Flagellate ancestor

(b)

Figure 28–5 Evolutionary relationships of major animal phyla. The relationships among animal groups shown here reflect current hypotheses. As new data are collected and considered, our understanding of these relationships is likely to change. Two main evolutionary branches of coelomates are protostomes and deuterostomes (echinoderms, hemichordates, and chordates). Two major clades make up the protostomes: the Lophotrochozoa (which includes nemerteans, mollusks, annelids, and lophophorates) and Ecdysozoa (which includes nematodes, rotifers, onychophorans, and arthropods). **(a)** A phylogenetic tree of animal groups. **(b)** A cladogram of animal phylogeny.

groups were not necessarily among the earliest to evolve. In the future, they may no longer be shown at the base of the phylogenetic tree. Some biologists now hypothesize that the acoelomate nemerteans are more closely related to coelomate phyla in the Lophotrochozoa. Similarly, as we have discussed, the pseudocoelomates do not form a monophyletic group but are related to various coelomate groups. In the discussion of various animal groups, we indicate some of the phylogenetic relationships that are vigorously debated at this time. The phylogenetic trees shown in Figure 28–5 depict one current view of the relationships among the major phyla of animals.

Now that we have briefly discussed animal body plans and some of the criteria for determining phylogenetic relationships, we survey representative animal phyla. In the remainder of this chapter, we discuss animals without a coelom: sponges, cnidarians, comb jellies, and flatworms.

SPONGES HAVE UNIQUE FLAGELLATED CELLS

About 10,000 species of sponges have been identified and assigned to phylum **Porifera.** The name *Porifera,* meaning "to have pores," aptly describes the sponges, whose bodies are perforated by tiny holes. Sponges are aquatic, mainly marine, animals that range in size from 1 to 200 cm (0.4–79 in) in height. Many are asymmetrical, but they vary in shape from flat, encrusting growths to balls, cups, fans, or vases. Living sponges may be brightly colored—green, orange, red, yellow, blue, or purple—or they may be white or drab (Fig. 28–6). Some species are inhabited by symbiotic bacteria or algae that give them color.

Sponges are thought to have evolved from choanoflagellates. These protozoa have a single flagellum surrounded by a collar of microvilli (see Fig. 24–5*c*). Sponges are the only animals with **choanocytes,** or **collar cells,** flagellated cells that are strikingly similar to the choanoflagellates.

Sponge larvae are flagellated and able to swim about. Adult sponges attach to some solid object and have long been described as sessile. Recently, several species have been observed to move slowly (about 4 mm per day) by extending footlike appendages.

TABLE 28–1 Overview of the Animal Kingdom

Phylum or group	Symmetry	Level of Organization	Digestion	Circulation
Parazoa				
Porifera (sponges)	None	Cells loosely arranged	Intracellular	Diffusion
Animals with Radial Symmetry				
Cnidarians (hydras, jellyfish, corals)	Radial	Diploblastic; tissues	Gastrovascular cavity with one opening	Diffusion
Ctenophores (comb jellies)	Biradial	Diploblastic; tissues	Gastrovascular cavity with mouth and anal pores	Diffusion
Acoelomates*				
Platyhelminthes (flatworms)	Bilateral	Triploblastic; organs	Gastrovascular cavity with one opening	Diffusion
Protostome Coelomates of the Lophotrochozoan Branch				
Nemertea (proboscis worms)	Bilateral	Triploblastic; organ systems	Complete digestive tube†	Blood vessels; no heart
Mollusca (clams, snails, squids)	Bilateral	Triploblastic; organ systems	Complete digestive tube	Open system†† (closed in cephalopods)
Annelida (some marine worms, earthworms, leeches)	Bilateral	Triploblastic; organ systems	Complete digestive tube	Closed system
Lophophorates (brachiopods, phoronids, bryozoans)	Bilateral	Triploblastic; organ systems	Complete digestive tube	Closed system
Protostome Coelomates of the Ecdysozoan Branch				
Rotifera (wheel animals)	Bilateral	Triploblastic; organ systems	Complete digestive tube	Closed system
Nematoda (roundworms)	Bilateral	Triploblastic; organ systems	Complete digestive tube	Diffusion
Onychophora	Bilateral	Triploblastic; organ systems	Complete digestive tube	Open system
Arthropoda (crustaceans, insects, arachnids)	Bilateral	Triploblastic; organ systems	Complete digestive tube	Open system
Deuterostome Coelomates				
Echinodermata (sea stars, sea urchins)	Embryo bilateral; adult pentaradial	Triploblastic; organ systems	Complete digestive tube	Open system; reduced
Hemichordata	Bilateral	Triploblastic; organ systems	Complete digestive tube	Open system
Chordata (tunicates, lancelets, vertebrates)	Bilateral	Triploblastic; organ systems	Complete digestive tube	Closed system

*Triploblastic, but have solid body (i.e., no body cavity).

†A complete digestive tube has a mouth for food intake and an anus for elimination of wastes.

††A type of circulatory system in which the heart pumps blood through vessels into tissues that have open ends, and the blood bathes the tissues directly.

Biologists divide sponges into three main classes based on the type of skeleton they secrete. Members of class **Calcarea** secrete a chalky skeleton composed of small calcium carbonate spikes, or **spicules.** Members of class **Hexactinellida,** the glass sponges, have a skeleton made of six-rayed spicules containing silica. Most sponges belong to class **Demospongiae,** characterized by variable skeletons: Some are made of a fibrous protein material known as *spongin,* others contain spicules of silica, and most have a combination of both.

In a simple sponge, water enters through hundreds of tiny pores *(ostia)*, passes into the central cavity, or **spongocoel** (not a digestive cavity), and flows out through the sponge's open end, the **osculum.** In most types of sponges, the body wall is extensively folded, and there are complicated systems of canals that provide increased surface area for food capture.

Although sponges are multicellular, their cells are loosely associated and do not have the cell junctions necessary to form definite tissues. However, a division of labor exists among the several types of cells that make up the sponge, with certain cells specializing in nutrition, support, contraction, or reproduction. Epidermal cells form the outer layer of the sponge and line the canals. Specialized tubelike cells, called *porocytes,* form the pores of a simple sponge. These cells regulate the diameter of the pores by contracting.

The collar cells, which make up the inner layer of certain sponges, create the water current that brings food and oxygen to

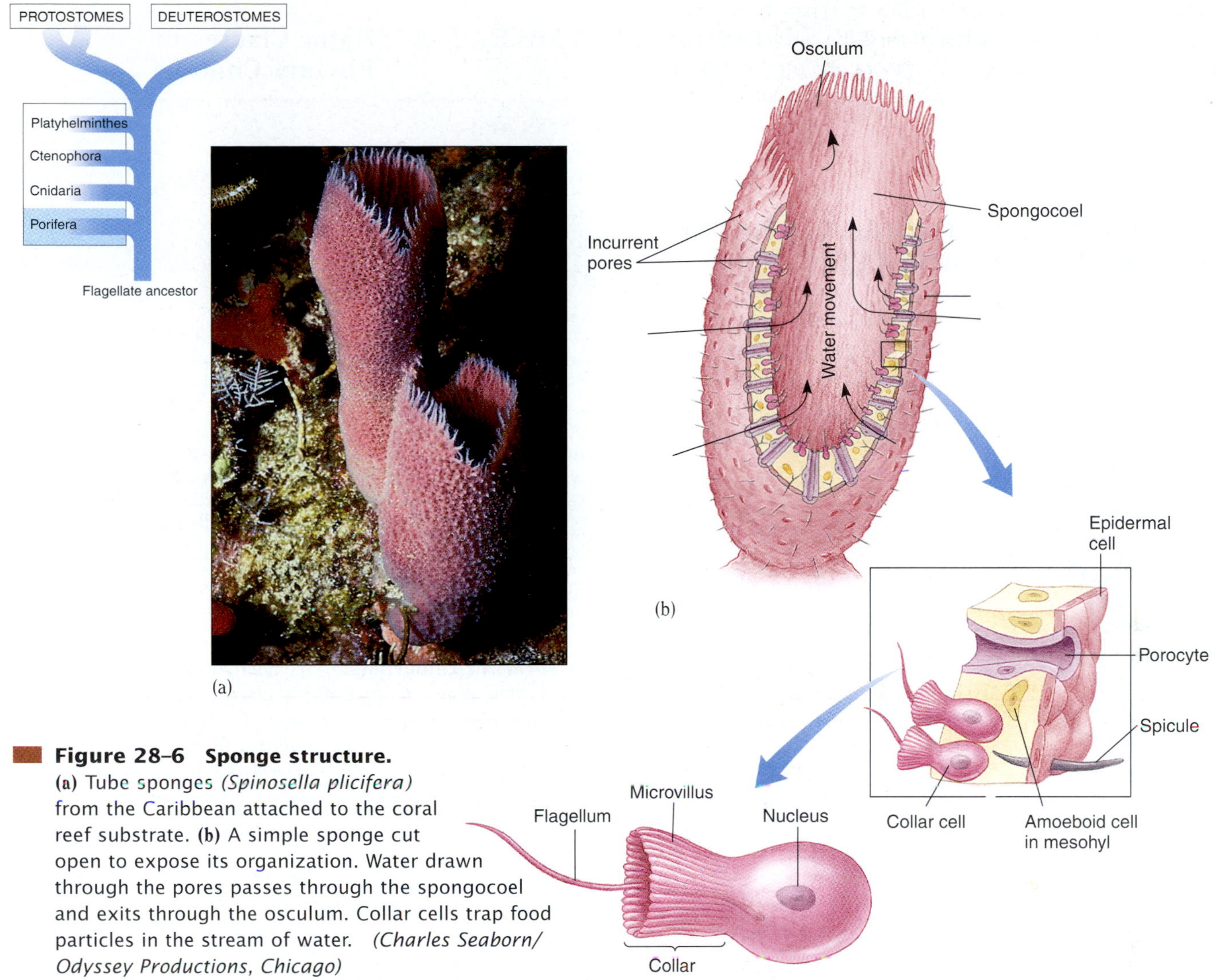

Figure 28–6 Sponge structure. **(a)** Tube sponges *(Spinosella plicifera)* from the Caribbean attached to the coral reef substrate. **(b)** A simple sponge cut open to expose its organization. Water drawn through the pores passes through the spongocoel and exits through the osculum. Collar cells trap food particles in the stream of water. *(Charles Seaborn/ Odyssey Productions, Chicago)*

the cells and carries away carbon dioxide and other wastes. These cells also trap and phagocytize food particles. Each of these cells is equipped with a tiny collar that surrounds the base of the flagellum. The collar is an extension of the plasma membrane and consists of microvilli. Together, the collar cells of some complex sponges can pump a volume of water equal to the volume of the sponge each minute!

Between the outer and inner cell layers of the sponge body is a gelatin-like layer, the *mesohyl,* supported by skeletal spicules. Amoeboid cells, which wander about in this layer, are important in digestion and food transport. Other amoeboid cells in the mesohyl secrete the spicules.

Sponges are *suspension feeders,* adapted for trapping and eating whatever food the water brings to them. As water circulates through the body, food is trapped along the sticky collars of the choanocytes. Food particles are either digested within the collar cell or transferred to an amoeboid cell for digestion and transport of nutrients to epidermal cells. Undigested food passes out through the osculum and is simply eliminated into the water.

Gas exchange and excretion of wastes depend on diffusion into and out of individual cells. Although cells of the sponge are irritable and can react to stimuli, sponges do not have specialized nerve cells and so cannot react as a whole. Behavior appears limited to basic metabolic necessities such as capturing food and regulating the flow of water through the body.

Sponges can reproduce asexually. A small fragment or bud may break free from the parent sponge and give rise to a new sponge. Such fragments may attach to the parent sponge, forming a colony. Sponges also reproduce sexually. Most sponges are **hermaphroditic,** meaning that the same individual can produce both eggs and sperm. Some of the amoeboid cells develop into sperm cells, others into egg cells. However, hermaphroditic sponges usually produce eggs and sperm at different times, and they cross-fertilize with other sponges. Mature sperm are released into the water and are taken in by other sponges of the same species.

Fertilization and early development take place within the jelly-like mesohyl. Embryos eventually move into the spongocoel and leave the parent along with the stream of outflowing water. After swimming about for awhile, a larva finds a solid object, attaches to it, and settles down to a sessile life.

Sponges possess a remarkable ability to repair themselves when injured and to regenerate lost parts. If the cells of a sponge are separated from one another in the laboratory, they recognize one another and their place in the whole and reaggregate, forming a complete sponge again.

CNIDARIANS HAVE RADIAL SYMMETRY AND UNIQUE STINGING CELLS

Most of the 10,000 or so species of phylum **Cnidaria** (pronounced "ni-dah′-ree-ah") are marine. These animals get their name from specialized cells, called **cnidocytes** (from a Greek word meaning "sea nettles"), that contain stinging organelles. Biologists assign cnidarians to three main classes (Table 28–2). Class **Hydrozoa** in-

TABLE 28–2 Major Classes of Phylum Cnidaria

Class and Representative Animals	Characteristics
Hydrozoa *Hydra, Obelia,* Portuguese man-of-war	Mainly marine, but some freshwater species; alternation of polyp and medusa stages in most species (polyp form only in *Hydra*); some form colonies.
Scyphozoa Jellyfish	Mainly marine; typically inhabit coastal water, free-swimming medusa most prominent form; polyp stage often reduced.
Anthozoa Sea anemones, corals, sea fans	Marine; solitary or colonial polyps; in most no medusa stage; gastrovascular cavity divided by partitions into chambers, increasing area for digestion; sessile.

PROTOSTOMES
DEUTEROSTOMES
Platyhelminthes
Ctenophora
Cnidaria
Porifera
Flagellate ancestor

Figure 28–7 Polyp and medusa body forms of cnidarians. **(a)** This hydrozoan, *(Gonothyraea loveni)* forms a colony of polyps. The drawing illustrates a single polyp. **(b)** The sea nettle *(Chrysaora fuscescens)*, like other jellyfish, uses its tentacles equipped with cnidocytes to capture small animals (zooplankton) suspended in the water and to carry this food to the mouth. **(c)** Coral polyps *(Montastrea cavernosa)* extended for feeding. *(a, Robert Brons/Biological Photo Service; b, Brian Parker/Tom Stack & Associates; c, Mike Bacon/Tom Stack & Associates)*

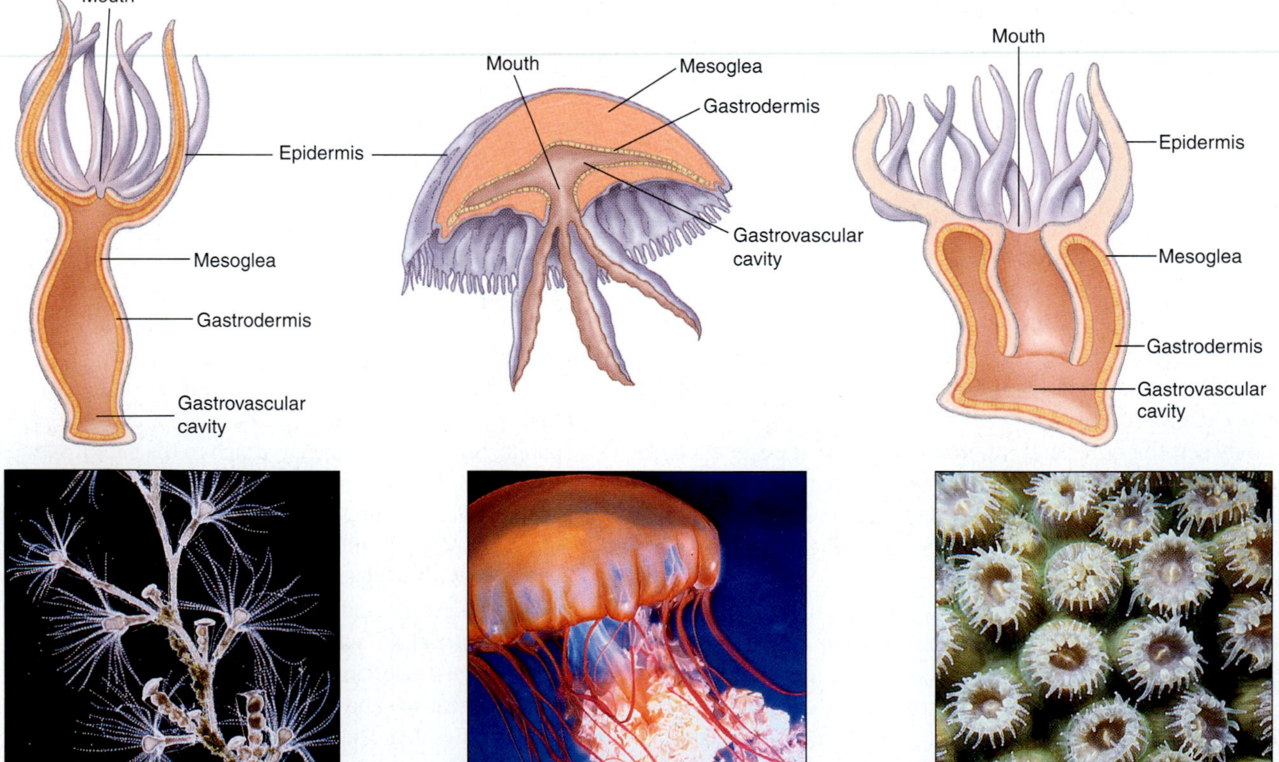

(a) Class Hydrozoa (polyp)

(b) Class Scyphozoa (medusa)

(c) Class Anthozoa (polyp)

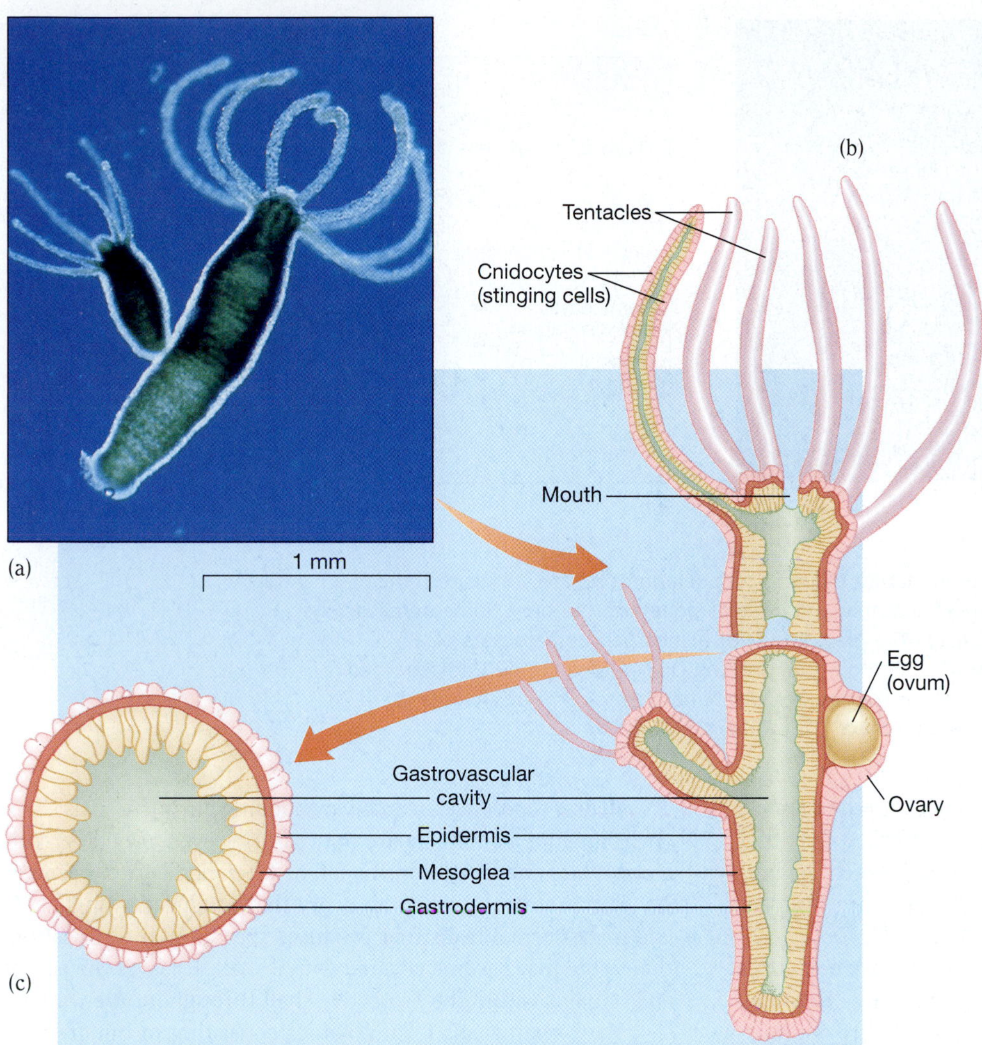

Figure 28–8 *Hydra*, a freshwater hydrozoan. **(a)** *Hydra viridis* with a large bud. When the bud separates from its parent, it becomes an independent individual. **(b)** Hydra cut longitudinally to show its internal structure. Asexual reproduction by budding is represented on the left; sexual reproduction is represented by the ovary on the right. Male hydras develop testes that produce sperm. **(c)** Cross section through the body of a hydra. *(Biophoto Associates/Photo Researchers, Inc.)*

cludes hydras, and *hydroids,* such as *Obelia* and the Portuguese man-of-war; class **Scyphozoa** includes jellyfish; and class **Anthozoa** includes sea anemones and corals (Fig. 28–7). Some cnidarians live a solitary existence, whereas many others, such as corals, form colonies.

Basically, the cnidarian body is radially symmetrical and is organized as a hollow sac with the mouth and surrounding tentacles located at one end. The mouth leads into the digestive cavity, called the **gastrovascular cavity.** The mouth is the only opening into the gastrovascular cavity and so must serve for both ingestion of food and expulsion of wastes.

Much more highly organized than the sponge, the cnidarian is diploblastic; that is, it has two definite tissue layers. The ectoderm gives rise to the outer **epidermis,** a protective layer covering the body. The endoderm gives rise to the inner **gastrodermis,** which lines the gut and functions in digestion. These thin layers are separated by a gelatin-like, mainly acellular, **mesoglea.**

Cnidarians have two body shapes: the polyp and the medusa. The **polyp** form, represented by *Hydra,* typically has a dorsal mouth surrounded by tentacles. Some cnidarians have the polyp shape during one stage of their life cycle and the **medusa** (pl., *medusae*), or jellyfish, form during another stage. In the medusa, the mouth is located in the lower concave, or *oral,* surface; the convex upper surface is the *aboral* surface. The Portuguese man-of-war and some other cnidarians actually consist of many individuals, some of which are polyps and others medusae.

Cnidarians have nerve cells that form **nerve nets** connecting sensory cells in the body wall to contractile and gland cells. Sense organs, for example photoreceptors that detect light, are positioned around the edge of the body. An impulse set up by one of them passes in all directions more or less equally. Nerve cells are not organized to form a brain or nerve cord.

Class Hydrozoa includes solitary and colonial forms

Although not really typical, the cnidarian most often studied by beginning biology students is the tiny, solitary *Hydra* found in freshwater ponds. To the naked eye *Hydra* looks like a bit of frayed string (Fig. 28–8). Because it has a remarkable ability to regenerate, *Hydra* is named after the multiheaded monster of Greek mythology that was able to grow two new heads for each head cut off. When *Hydra* is cut into several pieces, each piece may regrow all the missing parts and become a whole animal.

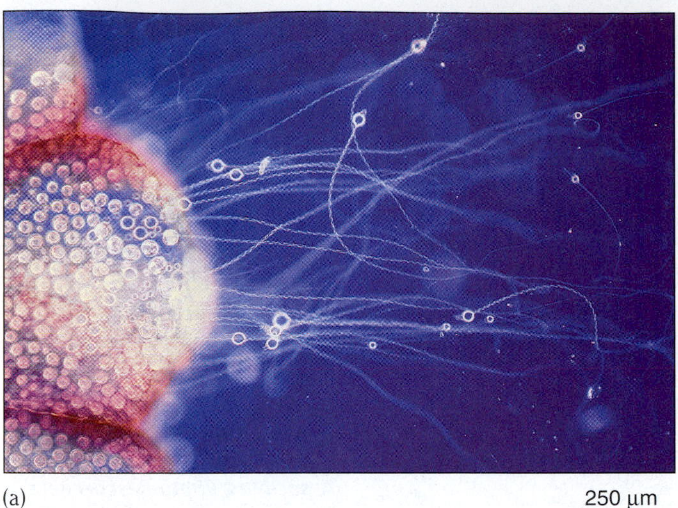

(a)

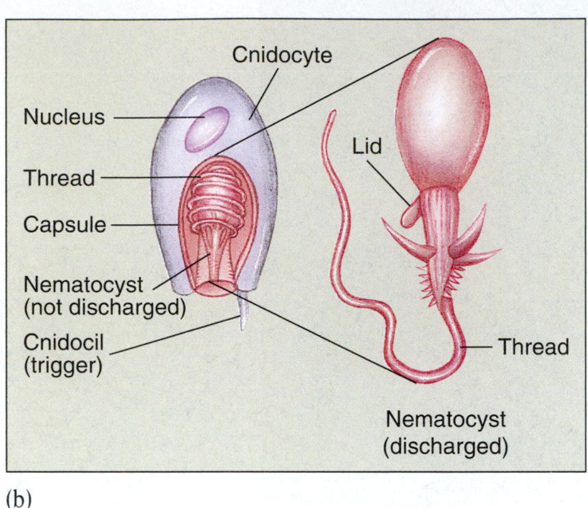

(b)

Figure 28–9 Nematocysts. When cnidarian stinging cells (cnidocytes) are stimulated, the nematocyst discharges, ejecting a thread that may entangle or penetrate the prey. Some nematocysts secrete a toxic substance that immobilizes the prey. **(a)** LM of discharged nematocysts of a Portuguese man-of-war *(Physalia physalis)*. Photographed in the Gulf of Mexico. **(b)** Undischarged and discharged nematocyst. The cnidocil, or trigger, is a mechanoreceptor that discharges the nematocyst when it senses contact with an object. *(a, Robert & Linda Mitchell)*

Hydra lives in fresh water and is typically attached to a rock, aquatic plant, or detritus by a disc of cells at its base. At the other end, the mouth connects the gastrovascular cavity with the outside. The mouth is surrounded by tentacles that are used for feeding.

The hydra's body consists of an outer, protective epidermis and an inner gastrodermis that functions in digestion. Both layers have cells specialized to contract (however, they are not muscle cells). Contractile fibers in the epidermal cells run lengthwise, and those in the gastrodermis run circularly. These two sets of contractile cells act on the water-filled gastrovascular cavity, which forms a **hydrostatic skeleton** that supports the body and allows movement. By the contraction of one set of contractile cells or the other, the hydra can shorten, lengthen, or bend its body.

Cnidocytes are located mainly in the epidermis, especially on the tentacles. The cnidocytes contain stinging "thread capsules," or **nematocysts** (Fig. 28–9). When stimulated, a nematocyst releases a coiled, hollow thread. Some types of nematocyst threads are sticky. Others are long and coil around prey. A third type bears barbs or spines and can inject a protein toxin that paralyzes prey animals such as small crustaceans. Each cnidocyte has a small, projecting trigger *(cnidocil)* on its outer surface. Stimuli such as touch or chemicals dissolved in the water cause the nematocyst to fire its thread.

Captured prey are pushed into the mouth by the tentacles. Digestion begins in the gastrovascular cavity. Partially digested fragments are taken up by the gastrodermal cells, and digestion is completed within food vacuoles in these cells. Gas exchange and excretion occur by diffusion. The body wall of a hydra is thin enough that no cell is far from the surface. The motion of the body as it stretches and shortens helps circulate the contents of the gastrovascular cavity.

Hydras reproduce asexually by budding during periods when environmental conditions are optimal. However, they differentiate as males and females in the fall or when pond water becomes stagnant. Females develop an ovary that produces a single egg, and males form a testis that produces sperm. After fertilization, the zygote may become covered with a shell. It leaves the parent and remains within the protective shell throughout the winter.

Many hydrozoans form colonies consisting of hundreds or thousands of individuals. A colony begins with a single polyp that reproduces asexually by budding. However, instead of separating from the parent, the bud remains attached and eventually forms additional buds. Several types of individuals may arise in the same colony, some specialized for feeding, some for reproduction, and others for defense.

Some marine cnidarians are remarkable for an alternation of sexual and asexual stages. This alternation of stages differs from the alternation of generations in plants in that both sexual and asexual forms are diploid; only sperm and eggs are haploid. The life cycle of the colonial marine hydrozoan *Obelia* illustrates alternation of sexual and asexual stages (Fig. 28–10). The asexual stage consists of a polyp colony composed of two types of polyps: those specialized for feeding and those for reproduction. Free-swimming male and female medusae bud off from the reproductive polyps. These medusae eventually release sperm and eggs, and fertilization takes place externally. The zygote develops into a ciliated swimming larva called a **planula.** The larva attaches to some solid object and begins to form a new generation of polyps by asexual reproduction (budding).

The Portuguese man-of-war, *Physalia,* superficially resembles a jellyfish but is actually a hydrozoan colony of polyps and medusae. An iridescent purple, gas-filled sac (a modified medusa) helps maintain the animal's position in the water. *Physalia's* long tentacles may hang down for several meters below the float. Its

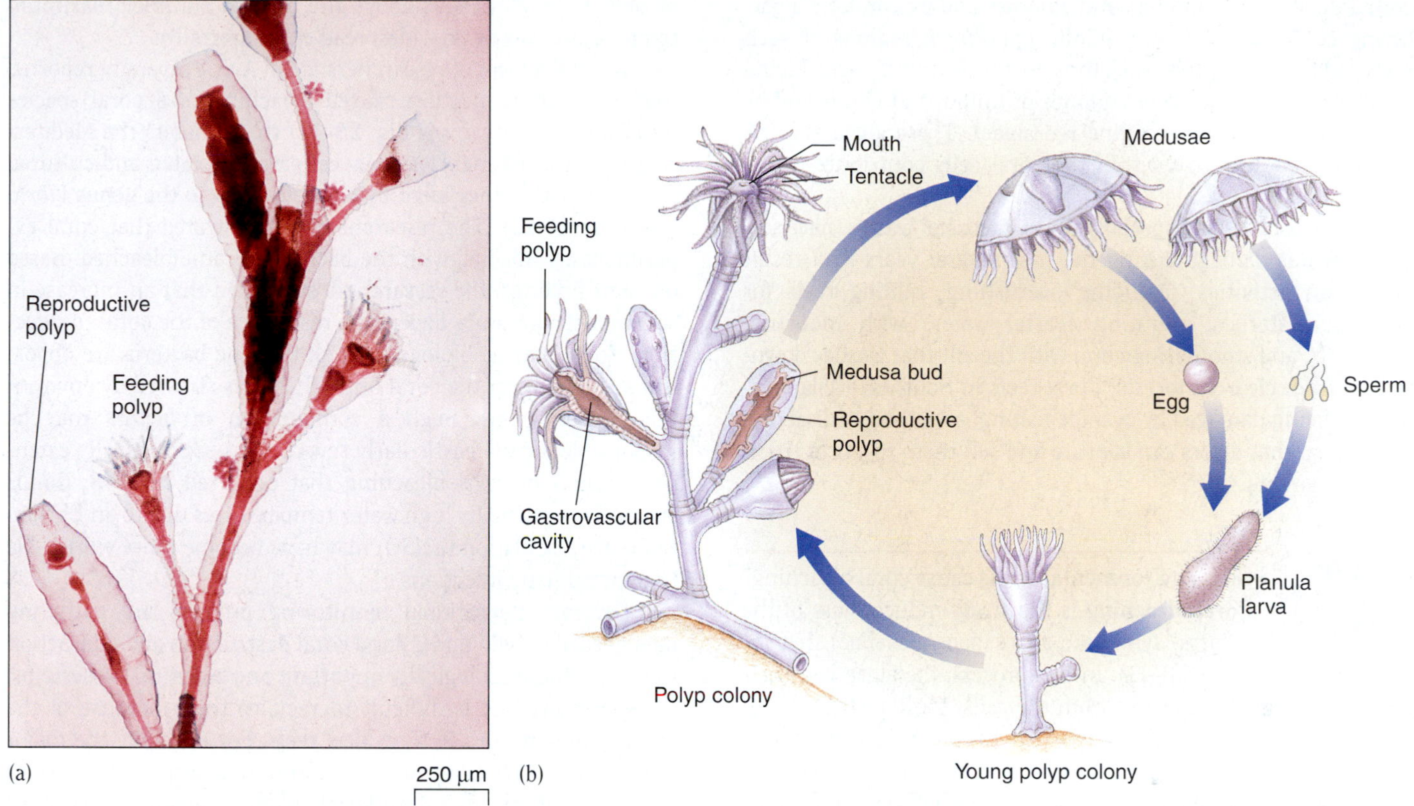

Figure 28–10 ***Obelia*, a marine colonial hydrozoan.** **(a)** LM of *Obelia*. Some polyps have tentacles and are specialized for feeding, whereas others are specialized for reproduction. **(b)** Life cycle of *Obelia*. Reproductive polyps give rise asexually to medusae. The free-swimming medusae reproduce sexually, and the zygote develops into a planula larva. The larva develops into a polyp that forms a new colony, which grows as new polyps bud and remain attached. *(David M. Phillips/Visuals Unlimited)*

cnidocytes are capable of paralyzing a large fish and can severely wound a human swimmer.

The medusa stage is dominant among the jellyfish

Among the jellyfish, members of class Scyphozoa, the medusa is the predominant body form. Scyphozoan medusae are generally larger than hydrozoan medusae, and they possess a thick, viscous mesoglea that gives firmness to the body. In scyphozoans, the polyp stage is small and inconspicuous or may even be absent. The largest jellyfish, *Cyanea,* may be more than 2 m (6.5 ft) in diameter and have tentacles 30 m (98 ft) long. These orange and blue "monsters," among the largest invertebrates, are dangerous to swimmers in the North Atlantic Ocean.

Anthozoans have no medusa stage

Sea anemones and corals are anthozoans, members of class Anthozoa. These animals have either individual or colonial polyps but no free-swimming medusa stage. The polyp produces eggs and sperm, and the fertilized egg develops into a small ciliated **planula.** This larval form may swim to a new location before attaching to develop into a polyp.

Anthozoans differ from hydrozoans in that the gastrovascular cavity is partially divided into a number of connected chambers by a series of vertical partitions. The partitions increase the surface area for digestion, enabling an anemone to digest an animal as large as a crab. Although corals can capture prey, many tropical species depend for nutrition on photosynthetic algae *(zooxanthellae)* that live within cells lining the coral's digestive cavity (see Chapter 24 and Fig. 52–12). The relationship between coral and zooxanthellae is symbiotic and mutually beneficial. The algae provide the coral with oxygen and with carbon and nitrogen compounds. In exchange, the coral supplies the algae with waste products such as ammonia, from which the algae make nitrogenous compounds for both partners.

In warm, shallow seas, much of the bottom is covered with coral or anemones, most of them brightly colored. Coral communities (reefs) are among the most productive of all ecosystems, rivaling the tropical rain forests in species diversity. A single reef can serve as home for more than 3000 species of fishes and other marine organisms, and an estimated one-fourth of all marine species depend on coral reefs. Reef organisms form complex food webs. Some organisms attach to the reef, whereas others find shelter within its crevices. Many terrestrial organisms also benefit from coral reefs, which form and maintain the foundation of thousands of islands. By providing a barrier against waves, reefs also protect shorelines against storms and erosion (see Chapter 54).

The reefs and atolls of the South Pacific are the remains of billions of microscopic, cup-shaped calcareous skeletons, secreted during past ages by coral colonies and by coralline algae. Living colonies occur only in the uppermost regions of such reefs, adding their own skeletons to the forming rock. Living coral reefs are made up of colonies of millions of corals and by certain algae (mainly coralline red algae). These algae and the zooxanthellae that live within the coral cells contribute to the reef's brilliant colors.

Many coral reefs, especially those in coastal waters, have suffered serious damage during the past several years as a result of human activities, including overfishing, mining reefs for building materials, polluting coastal waters with industrial chemicals, and smothering coral with the silt that washes downstream from clearcut forests. Coral reefs in Southeast Asian waters are being damaged by cyanide fishing, a method of disabling reef fish so that divers can capture and sell them to world tropical fish markets.

MAKING THE CONNECTION

Does environmental stress cause coral bleaching? **Coral bleaching** is the stress-induced loss of the colorful symbiotic algae (zooxanthellae) that inhabit coral cells (Fig. 28–11). In this process, the algae lose their pigmentation or are expelled from the cells. Without their algae, coral become malnourished and die. Although biologists have known about coral bleaching for more than 75 years, bleaching has become widespread and has become an important research focus.

The causes of coral bleaching are not well understood. Suspected environmental factors include pollution, changes in salinity, disease, high doses of ultraviolet radiation (associated with the destruction of the ozone layer), and unusually high or low temperatures. Some scientists think that one of the most important factors in recent years has been abnormally high water temperatures, possibly caused by global warming (see Chapter 55). Healthy coral thrives in a narrow temperature range. An increase of only 1° or 2° Celsius above the normal summer maximum temperature can cause widespread coral mortality.

In 1996 a team of researchers at Tel Aviv University reported that a bacterial infection caused bleaching of a coral species (*Oculina patagonica;* see Fig. 28–11) that inhabits the Mediterranean coast of Israel. These investigators isolated and cultured the bacteria that they identified as belonging to the genus *Vibrio* (see Chapter 23). The researchers demonstrated that coral experimentally infected with the bacteria became bleached. Based on their findings, the researchers suggested that an increase in seawater temperature lowers the resistance of the coral to infection. Some marine biologists think that the bacteria are opportunistic and infect the coral only when it is stressed by environmental conditions. Such a combination of factors may be responsible for the particularly severe and geographically extensive episode of coral bleaching that occurred in 1998. Corals stressed by unusually high water temperatures due to an El Niño Southern Oscillation (ENSO) may have become more vulnerable to opportunistic infections.

Several international monitoring projects are gathering data needed to help understand coral destruction and take action to protect these ecologically important and beautiful ecosystems. Once destroyed, it is difficult to reclaim them because of the lengthy time needed to form new reefs. For example, the major reef builder in Florida and Caribbean waters, a species known as star coral *(Montastrea annularis),* requires about 100 years to form a reef just 1 m high. (The ecological significance of coral reefs is further discussed in Chapter 54.)

Figure 28–11 Bleached coral *(Oculina patagonica)* from the Mediterranean coast of Israel. These polyps are 2 to 4 mm in diameter. *(Courtesy of A. Kushmaro, Y. Loya, M. Fine, and E. Rosenberg. "Bacterial Infection and Coral Bleaching."* Nature, *Vol. 380, Apr. 1996. Photo by A. Shoob)*

COMB JELLIES MOVE BY CILIA

The 100 or so species of comb jellies, members of phylum **Ctenophora,** all are marine. They are fragile, luminescent animals that may be as small as a pea or larger than a tomato. The outer surface of a ctenophore bears eight rows of cilia, resembling combs (Fig. 28–12). The coordinated beating of the cilia in these combs moves the animal through the water. A sense organ functions in balance and helps the animal orient itself. Some ctenophores have two tentacles. They do not have the stinging nematocysts characteristic of the cnidarians. However, their tentacles are equipped with adhesive glue cells that trap prey.

Ctenophores are biradially symmetrical, meaning that you could obtain equal halves by cutting through the body (oral-aboral) axis in two different ways. Because of their radial symmetry and feeding tentacles, the ctenophore body plan seems somewhat similar to that of a cnidarian medusa. Ctenophores are also like cnidarians in that they consist of two cell layers separated by a thick jelly-like mesoglea. Because of these similarities, biologists have historically classified ctenophores near the cnidarians. However, their development is different, and unlike cnidarians, their digestive system has a mouth for food intake at one end and two anal pores for the egestion of wastes at the other end. The similarities between ctenophores and cnidarians might be a result of convergence from living in a similar environment,

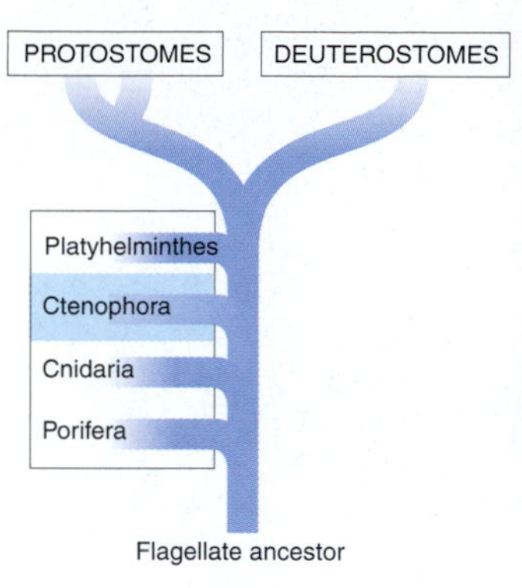

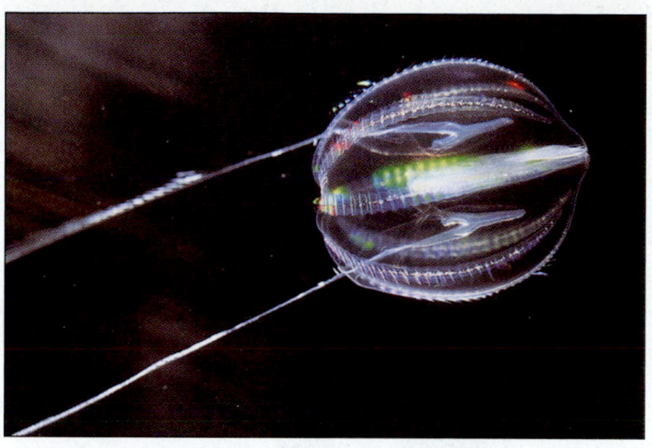

Figure 28–12 Ctenophore (comb jelly). Ctenophores are free-swimming, bioluminescent hermaphrodites capable of self-fertilization. The sea gooseberry (*Pleurobrachia* sp.) has two long aboral tentacles with adhesive cells used to capture prey. *(David Wrobel/Biological Photo Service)*

the ocean. Where, then, do the ctenophores belong in the phylogenetic tree? Biologists will need to decide as they gather new data.

FLATWORMS ARE BILATERAL ACOELOMATES

Members of phylum **Platyhelminthes,** the **flatworms,** are flat, elongated, acoelomate animals that exhibit bilateral symmetry. They are soft-bodied animals that left few fossils. Biologists have historically assigned the 20,000 species to four classes. Class **Turbellaria** comprises the free-living flatworms, including planarians and their relatives. Classes **Trematoda** and **Monogenea** include the flukes, which are either internal or external parasites. Class **Cestoda** includes the tapeworms, which as adults are intestinal parasites of vertebrates (Table 28–3).

TABLE 28–3 Classes of Phylum Platyhelminthes

Class and Representative Animals	Characteristics
Turbellaria Planarians	Mainly free-living; mainly marine; body covered by ciliated epidermis; typically carnivorous; prey on tiny invertebrates.
Trematoda and Monogenea Flukes	All parasites with a wide range of vertebrate and invertebrate hosts; may require intermediate hosts; adults have suckers for attachment to host.
Cestoda Tapeworms	Parasites of vertebrates; complex life cycle usually with one or two intermediate hosts; larval host may be invertebrate; typically have suckers and sometimes hooks for attachment to host; eggs produced within proglottids, which are shed; no digestive or nervous systems.

Process of Science

At the time of this writing the phylogenetic relationships of flatworms are being hotly debated. Molecular data, specifically 18S ribosomal RNA sequences, indicate that Platyhelminthes is not a monophyletic group. Some researchers are proposing that the flatworms known as acoels, currently classified in one order of class Turbellaria, differ significantly from other flatworms. The acoels may be the only group of flatworms that are closely related to the common ancestor that linked radially symmetrical animals with bilateral animals. Other flatworms may be more closely related to coelomates. If so, they became simplified later in evolutionary history. In this edition of *Biology* we use the traditional classification of Platyhelminthes. If further research supports an alternative hypothesis, our next edition will likely separate flatworm groups, and only the acoels will remain in this phylogenetic position. Other flatworms will likely be classified with the lophotrochozoa.

Flatworms exhibit *bilateral symmetry* and *cephalization.* Along with their symmetry, flatworms have definite anterior and posterior ends. The beginnings of cephalization are evident in flatworms. A simple brain and paired sense organs are concentrated at the anterior end. An animal with a front end ("head") generally moves forward. With a concentration of sense organs in the part of the body that first meets its environment, the animal can find food or detect an enemy quickly.

Flatworms are triploblastic (have all three germ layers) and have *three definite tissue layers.* In addition to an outer epidermis derived from ectoderm, and an inner endodermis derived from endoderm, the flatworm has a middle tissue layer that develops from mesoderm. The mesoderm provides tissue for developing **organs,** functional structures made of two or more kinds of tissue. Flatworms typically have a muscular pharynx that takes in food; a simple brain, eyespots and other sensory organs in the head; **protonephridia,** structures that function in osmoregulation and metabolic waste disposal (excretion); and complex reproductive organs. Flatworms have no organs for circulation or gas exchange. These functions depend largely on diffusion through the body wall.

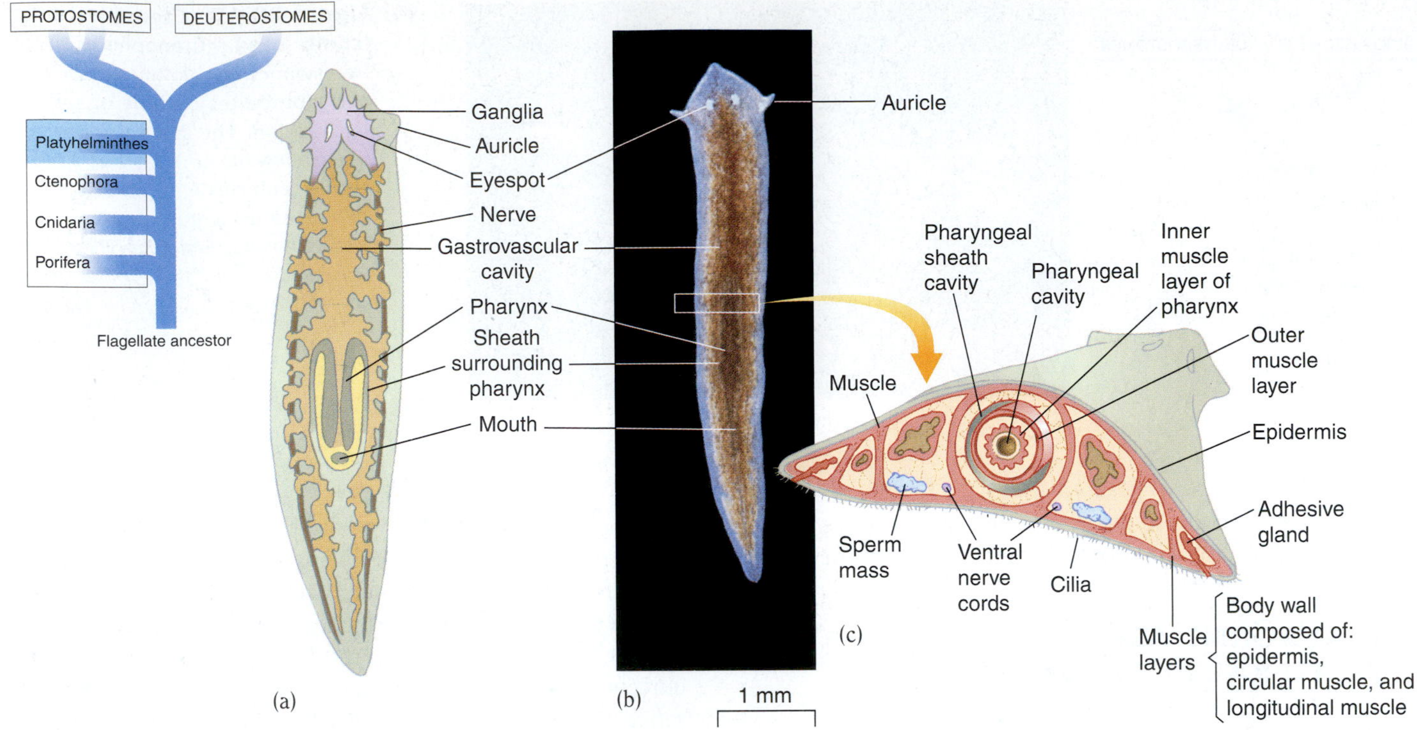

Figure 28–13 The common planarian, *Dugesia.* **(a)** Internal structure. **(b)** LM of a living planarian *(Dugesia dorotocephala).* Note the prominent auricles, used to locate food. **(c)** Cross section through a planarian. *(b, T.E. Adams/Peter Arnold, Inc.)*

The activities of some organs are coordinated, and form simple organ systems, for example, the digestive and nervous systems. As in the cnidarians, the digestive system has only one opening, a mouth. The *gastrovascular cavity* is often extensively branched.

Flatworms typically have a simple **nervous system.** The brain consists of two masses of nervous tissue, called **ganglia,** in the head region. In many species, the ganglia are connected to two nerve cords that extend the length of the body. This nervous system is sometimes referred to as a ladder-type nervous system because a series of nerves connects the cords like the rungs of a ladder.

Parasitic flatworms—flukes and tapeworms—are highly adapted to and modified for their parasitic lifestyle. They have suckers or hooks for holding onto their hosts. The bodies of those that live in digestive tracts are resistant to the digestive enzymes secreted by their hosts. Many have complicated life cycles and produce large numbers of eggs. Other adaptations include the loss of certain structures such as sense organs. Tapeworms have also lost the digestive system.

Class Turbellaria includes planarians

Most members of class Turbellaria are free-living flatworms. Most are marine, but many inhabit freshwater habitats, and some are terrestrial. **Planarians** are turbellarian flatworms found in ponds and quiet streams throughout the world. The common American planarian *Dugesia* is about 15 mm (0.6 in) long, with what appear to be crossed eyes and flapping "ears" called **auricles** (Fig. 28–13). The auricles actually serve as organs of chemoreception, important in locating food.

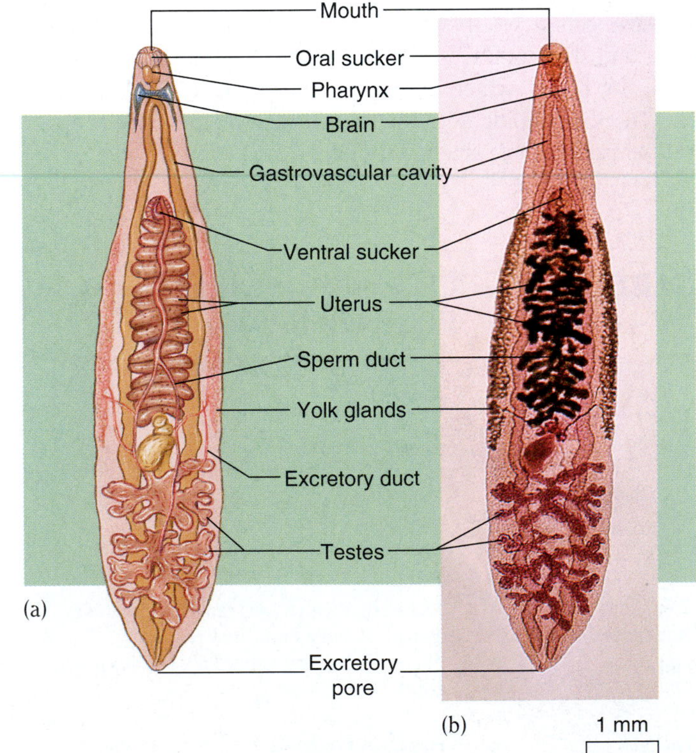

Figure 28–14 Structure of a fluke. **(a)** The liver fluke's oral sucker and well-developed reproductive system are adaptations to its parasitic life style. **(b)** LM of a human liver fluke *(Clonorchis sinensis).* *(b, Carolina Biological Supply Company/Phototake)*

Planarians are carnivorous, trapping small animals in a mucous secretion. The digestive system consists of a single opening (the mouth), a tubelike, muscular **pharynx** (the first portion of the digestive tube), and a branched gastrovascular cavity. A planarian can project its pharynx outward through its mouth, using it to suck in prey. Extracellular digestion takes place in the gastrovascular cavity by enzymes secreted by gland cells. Digestion is completed after the nutrients have been absorbed into individual cells. Undigested food is eliminated through the mouth. The long, highly branched gastrovascular cavity helps distribute food to all parts of the body, so that each cell can receive nutrients by diffusion.

A planarian's flattened body ensures that gases can reach all of its cells by diffusion. Excretion also takes place mainly by diffusion, but some metabolic wastes are excreted by the *protonephridia.* These structures function primarily in fluid balance, or *osmoregulation.* Protonephridia are blind tubules that end in **flame bulbs,** collecting cells equipped with cilia. The beating of the cilia channels waste through the system of tubules and eventually out of the body through excretory pores. Osmoregulation and protonephridia are discussed further in Chapter 46.

Planarians are capable of learning. Memory is not localized within the ganglia but appears to be retained throughout the nervous system. Planarians can reproduce either asexually or sexually. In asexual reproduction, an individual constricts in the middle and divides into two planarians. Each regenerates its missing parts. Sexually, these animals are hermaphroditic. During the warm months of the year, each is equipped with a complete set of male and female organs. Two planarians come together in copulation and exchange sperm cells so that their eggs are cross-fertilized.

Flukes parasitize other animals

Although their body plan resembles that of the free-living flatworms, specialized adaptations make the **flukes,** members of classes **Trematoda** and **Monogenea**, successful parasites. For example, most adult flukes have structures, such as hooks and suckers, for attachment to the host. Flukes also have extremely complex and prolific reproductive organs (Fig. 28–14).

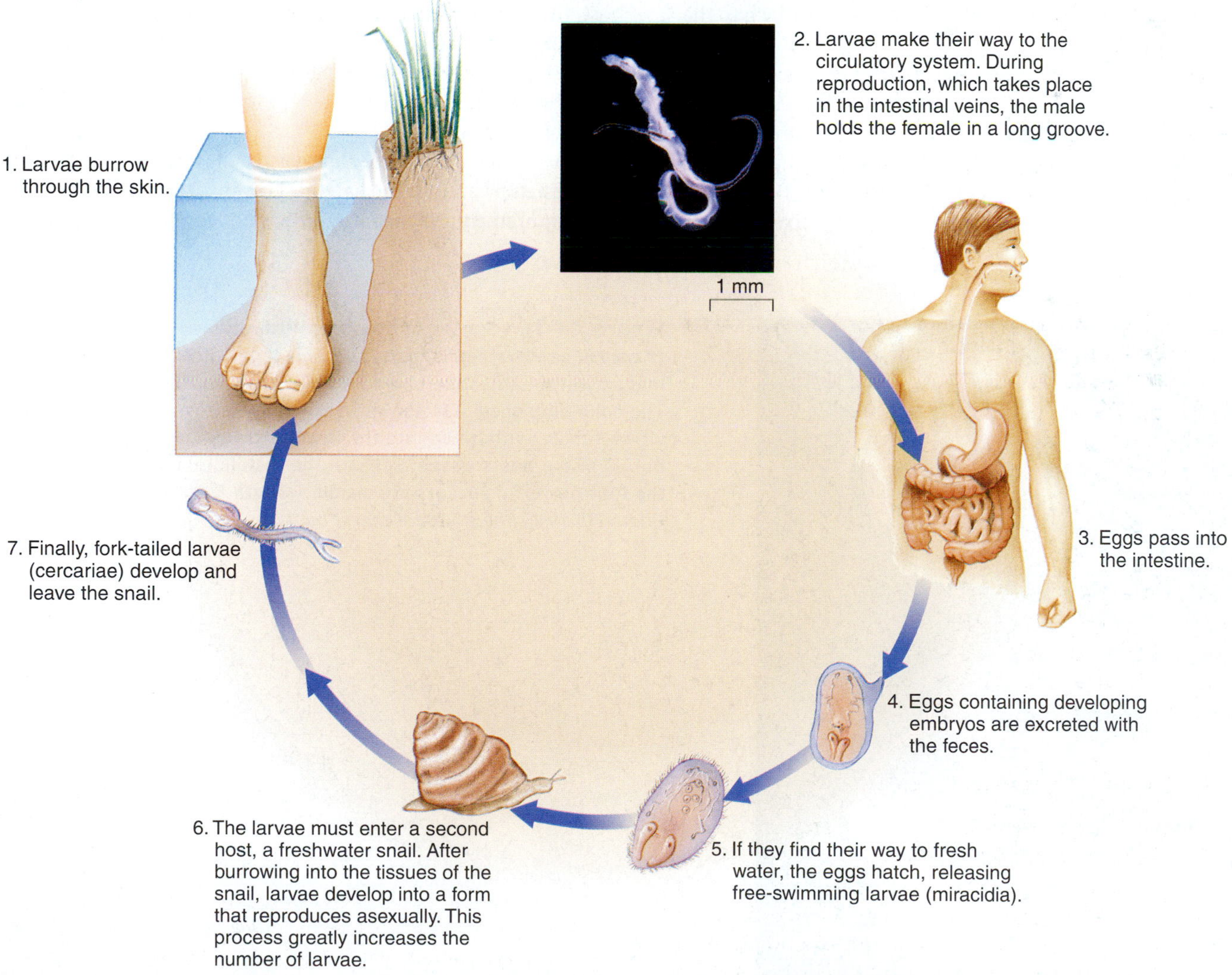

1. Larvae burrow through the skin.
2. Larvae make their way to the circulatory system. During reproduction, which takes place in the intestinal veins, the male holds the female in a long groove.
1 mm
3. Eggs pass into the intestine.
4. Eggs containing developing embryos are excreted with the feces.
5. If they find their way to fresh water, the eggs hatch, releasing free-swimming larvae (miracidia).
6. The larvae must enter a second host, a freshwater snail. After burrowing into the tissues of the snail, larvae develop into a form that reproduces asexually. This process greatly increases the number of larvae.
7. Finally, fork-tailed larvae (cercariae) develop and leave the snail.

Figure 28–15 Life cycle of the blood fluke (*Schistosoma* sp.). The LM shows an adult male enfolding a smaller female. *(U.S. Centers for Disease Control and Prevention, Atlanta, GA/Biological Photo Service)*

Flukes that are parasitic in humans include blood flukes, widespread in tropical areas of the world, and liver flukes, common in Asia, particularly in areas where human feces are used for fertilizing crops. Blood flukes of the genus *Schistosoma* infect about 200 million people who live in tropical areas. Both blood flukes and liver flukes go through complicated life cycles involving a number of different forms, alternation of sexual and asexual stages, and parasitism on one or more intermediate hosts (e.g., snails and fishes) (Fig. 28–15 on page 611). The aquatic snails that serve as intermediate hosts thrive in ponds, rice paddies, and marshy areas that form when dams are built.

Tapeworms inhabit the intestine of vertebrates

Adult members of the more than 5000 different species of class Cestoda live as parasites in the intestine of probably every kind of vertebrate, including humans. Tapeworms are long, flat, ribbonlike animals strikingly specialized for their parasitic mode of life. Among their many adaptations are suckers and sometimes hooks on the "head," or **scolex,** that enable the parasite to attach to the host's intestine (Fig. 28–16).

The reproductive adaptations and abilities of tapeworms are extraordinary. The body of the tapeworm consists of a long chain of segments called **proglottids.** Each proglottid is an entire reproductive machine equipped with both male and female reproductive organs and containing up to 100,000 eggs. Because an adult tapeworm may have as many as 2000 segments, its reproductive potential is staggering. A single tapeworm can produce 600 million eggs in a year. Proglottids farthest from the tapeworm's head contain the ripest eggs; these segments are shed from the host's body along with the feces.

The tapeworm has no mouth or digestive system. Digested food from the host is absorbed across the worm's body wall. The tapeworm also lacks well-developed sense organs. Some tapeworms have complex life cycles, spending their larval stage within the body of an intermediate host and their adult life within the body of a different, final host. Let us consider the life cycle of the beef tapeworm, so named because humans can become infected when they eat undercooked beef containing the larvae (Fig. 28–17).

The microscopic tapeworm larva spends part of its life cycle encysted within the muscle tissue of cattle. When a human ingests raw or rare infected beef, digestive juices break down the cyst, releasing the larva. Soon the larva attaches itself to the intestinal lining and within a few weeks matures into an adult tapeworm, which may grow to a length of about 15 m (50 ft). The parasite reproduces sexually within the human intestine and sheds proglottids filled with zygotes. Once established within a human host, the tapeworm makes itself very much at home and may remain for the rest of its life, as long as ten years. A person infected with a tapeworm may suffer pain or discomfort, increased appetite, weight loss, and other symptoms, or may be totally unaware of its presence.

For the life cycle of the tapeworm to continue, its eggs must be ingested by an intermediate host, in this case, a cow or steer. This requirement explains why we are not completely overrun by

Figure 28–16 Scolex of the small tapeworm *(Acanthrocirrus retrisrostris).* This tapeworm reaches maturity in the intestine of wading birds that eat barnacles. The color-enhanced SEM shows the piston-like rostellum, which can be withdrawn into the head or thrust out and buried in the host's tissue. Beneath the rostellum, two of the four powerful suckers are visible. *(Cath Ellis/Science Photo Library/ Photo Researchers, Inc.)*

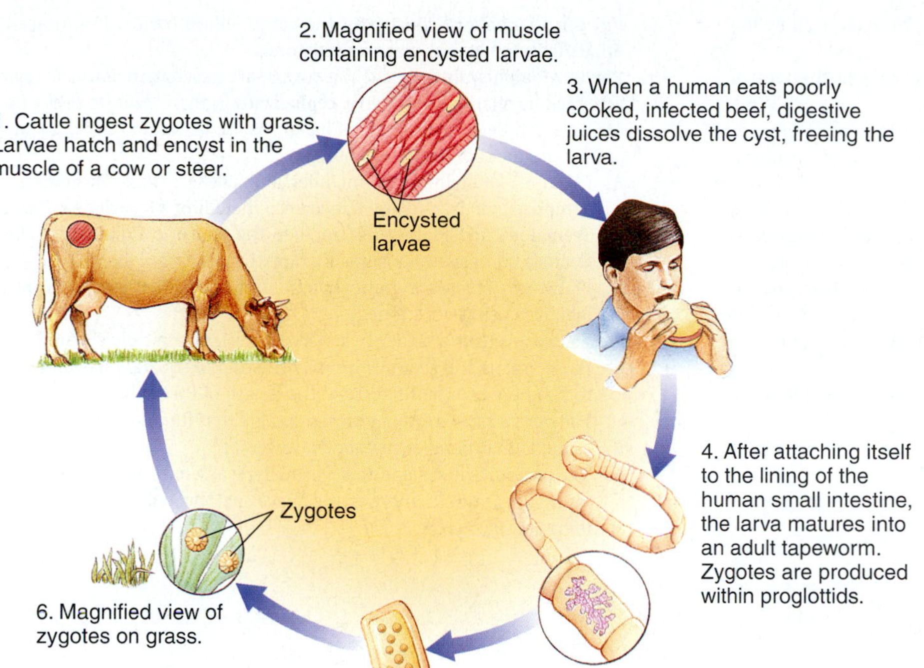
2. Magnified view of muscle containing encysted larvae.
1. Cattle ingest zygotes with grass. Larvae hatch and encyst in the muscle of a cow or steer.
3. When a human eats poorly cooked, infected beef, digestive juices dissolve the cyst, freeing the larva.
Encysted larvae
4. After attaching itself to the lining of the human small intestine, the larva matures into an adult tapeworm. Zygotes are produced within proglottids.
Zygotes
6. Magnified view of zygotes on grass.
5. Mature proglottids leave the body with the feces and burst open, releasing zygotes.

Figure 28–17 Life cycle of the beef tapeworm.

tapeworms and why a tapeworm must produce millions of eggs to ensure that at least a few survive. When cattle eat grass or other foods contaminated with human feces, eggs may be ingested. The eggs hatch in the cattle's intestines, and the larvae make their way into muscle. There they encyst and remain until released by a final host, perhaps a human eating very rare steak.

Two other tapeworms that infect humans are the pork tapeworm, found in undercooked, infected pork, and the fish tapeworm, found in raw or undercooked, infected fish. Like most parasites, tapeworms tend to be species-specific; that is each can infect only certain specific species. For example, the beef tapeworm can spend its adult life only in a human host.

SUMMARY WITH KEY TERMS

I. Members of **kingdom Animalia** are eukaryotic, multicellular, **heterotrophic** organisms with cells specialized to perform specific functions. Typically, they are capable of locomotion at some time during their life cycle, can reproduce sexually, and can respond adaptively to external stimuli.
   A. In animal sexual reproduction, sperm and egg unite to form a **zygote** that undergoes **cleavage.** Multiple cell divisions result in the development of a hollow ball of cells, a **blastula.** The blastula undergoes **gastrulation,** forming embryonic tissues.
   B. Most animals are **invertebrates,** animals without backbones. **Vertebrates** are a subphylum of phylum Chordata.

II. Animals inhabit marine environments, fresh water, and land. Only a few groups have successfully adapted to terrestrial environments.

III. Body structure, developmental pattern, and molecular similarities provide clues to animal phylogeny.
   A. Cnidarians, ctenophores, and adult echinoderms have **radial symmetry;** most other animals are **bilaterally symmetrical,** at least in their larval stages. In bilaterally symmetrical animals, the location of body structures can be defined by relative directional terms such as **dorsal** or **ventral; anterior (cephalic)** or **posterior (caudal); medial** or **lateral.**
   B. **Triploblastic** animals have traditionally been classified as **acoelomate** (no body cavity), **pseudocoelomate** (body cavity not completely lined with mesoderm), or **coelomate** (body cavity completely lined with mesoderm). Recent data indicate that pseudocoelomate animals do not form a natural group and probably evolved from coelomate ancestors.
      1. Most structures develop from embryonic tissues called **germ layers.** The germ layers include the outer **ectoderm** that gives rise to the body covering and the nervous system; the inner **endoderm** that lines the gut and other digestive organs; and a middle **mesoderm** that gives rise to most other body structures.
      2. A true **coelom** is a body cavity that is completely lined with mesoderm.
      3. Two major evolutionary branches of coelomates are **protostomes** (mollusks, annelids, arthropods) and **deuterostomes** (echinoderms, hemichordates, and chordates). In protostomes the blastopore develops into the mouth; in deuterostomes the blastopore typically becomes the anus.
      4. Based on molecular data, some biologists have subdivided the protostomes into two branches, the **Lophotrochozoa** and the **Ecdysozoa.** The Lophotrochozoa include the nemerteans (proboscis worms), mollusks, annelids, and the lophophorate phyla (groups that have a ciliated ring of tentacles surrounding the mouth). The Ecdysozoa, animals that molt, include the nematodes and arthropods. In this phylogeny, there are three major clades.

IV. Phylum **Porifera** consists of the sponges, animals characterized by flagellated **collar cells (choanocytes).**
   A. Sponges are divided into three classes on the basis of the type of skeleton they secrete.
   B. The sponge body is a sac with tiny openings through which water enters; a central cavity **(spongocoel);** and an open end, or **osculum,** through which water exits.

V. Phylum **Cnidaria,** which includes the hydras, jellyfish, and corals, is characterized by radial symmetry, two tissue layers, and **cnidocytes,** cells that contain stinging organelles **(nematocysts).**
   A. Phylum Cnidaria includes three main classes. Class **Hydrozoa** includes hydras, hydroids, and the Portuguese man-of-war; class **Scyphozoa** comprises the jellyfish; and class **Anthozoa** includes sea anemones and corals.
   B. In many types of cnidarians, the life cycle includes a sessile **polyp** stage and a free-swimming **medusa** stage.
   C. The **gastrovascular cavity** has a single opening that serves as both mouth and anus.
   D. Nerve cells form irregular, nondirectional **nerve nets** that connect sensory cells with contractile and gland cells.

VI. Phylum **Ctenophora** consists of the comb jellies: fragile, luminescent, biradially symmetrical marine predators.

VII. Phylum **Platyhelminthes,** the flatworms, are acoelomate animals characterized by bilateral symmetry, **cephalization,** three definite tissue layers, and well-developed organs. Flatworms are **hermaphroditic;** a single animal produces both sperm and eggs.
   A. Phylum Platyhelminthes includes four classes: Class **Turbellaria** is comprised of free-living flatworms, including **planarians.** Classes **Trematoda** and **Monogenea** include the parasitic **flukes,** and class **Cestoda** comprises the parasitic tapeworms.
   B. Flatworms are not a monophyletic group, and the traditional phylogeny is being questioned.
   C. Flatworms have a ladder-type nervous system, typically consisting of sense organs and a simple brain consisting of two **ganglia** connected to two nerve cords that extend the length of the body.
   D. Flatworms have **protonephridia,** organs that function in osmoregulation and disposal of metabolic wastes.
   E. The parasitic flukes and tapeworms typically have suckers or hooks for holding onto their hosts; they have complicated reproductive systems and life cycles.

POST-TEST

1. Invertebrates are (a) animals with vertebral columns (b) acoelomates (c) deuterostomes (d) animals without backbones (e) animals that lack *Hox* genes
2. Radial symmetry is characteristic of (a) protostomes (b) acoelomates (c) deuterostomes (d) cnidarians (e) nematodes
3. The auricles of a planarian are ____________ to its cerebral ganglia (a) medial (b) cephalic (c) caudal (d) anterior (e) lateral
4. The germ layer that gives rise to the outer covering of the body and the nervous system is the (a) gastrodermis (b) ectoderm (c) contractile layer (d) endoderm (e) mesoderm
5. A true coelom is completely lined with (a) flagella (b) ectoderm (c) a contractile layer (d) endoderm (e) mesoderm
6. Collar cells are characteristic of phylum (a) Porifera (b) Cnidaria (c) Platyhelminthes (d) Ctenophora (e) Eumetazoa
7. Rudimentary cephalization, three tissue layers, and protonephridia characterize phylum (a) Porifera (b) Cnidaria (c) Platyhelminthes (d) Ctenophora (e) Parazoa
8. Tapeworms are classified in phylum (a) Porifera (b) Cnidaria (c) Platyhelminthes (d) Ctenophora (e) Coelomata
9. Corals (a) are coelomates (b) have cnidocytes (c) lack a polyp stage (d) have ganglia (e) have protonephridia
10. Which of the following is *not* an adaptation of parasitic life? (a) production of a few well-protected eggs (b) hooks (c) suckers (d) reduced digestive system (e) intermediate host
11. Protostomes are characterized by (a) spiral cleavage (b) indeterminate cleavage (c) enterocoely (d) radial symmetry (e) a distinctive body plan that includes a pseudocoelom
12. The evolution of animals (a) followed an orderly progression from simple to complex (b) can be better understood as biologists continue to collect molecular and other types of data (c) has been determined by studying classification (d) began with the placozoans (e) began with their common ancestor, which was a sponge
13. Cephalization (a) evolved along with bilateral symmetry (b) refers to the development of a digestive system (c) is characteristic of cnidarians (d) involves a concentration of excretory organs (e) first evolved in deuterostomes
14. During cleavage an animal (a) undergoes metamorphosis (b) becomes a larva (c) undergoes a series of mitotic divisions and becomes a blastula (d) becomes diploid (e) reproduces sexually
15. Marine cnidarians (a) lack cnidocytes (b) may alternate sexual and asexual stages (c) exhibit bilateral symmetry (d) are the common ancestor of protostomes and deuterostomes (e) are all free-swimming medusae

REVIEW QUESTIONS

1. For centuries sponges were classified as plants. Justify their current classification as animals.
2. Why is the sea a more hospitable environment for many animals than the land or fresh water?
3. How does the body plan of flatworms differ from that of cnidarians? In what ways are these animals alike?
4. What are the advantages of bilateral symmetry and cephalization?
5. Identify the phyla (from among those studied in this chapter) that have the following characteristics: (a) radial symmetry (b) protonephridia (c) acoelomate (d) alternation of sexual and asexual stages (e) cnidocytes (f) nerve net (g) a body bearing eight rows of cilia (h) collar cells
6. Describe the alternation of stages exhibited by *Obelia*.
7. How do flatworms survive without specialized structures for gas exchange and internal transport of materials?
8. What special adaptations do tapeworms have for their parasitic mode of life?
9. Describe the life cycles of the following animals: (a) beef tapeworm (b) liver fluke

10. Label the branches of the diagram. Use Figure 28–5*a* to check your answers.

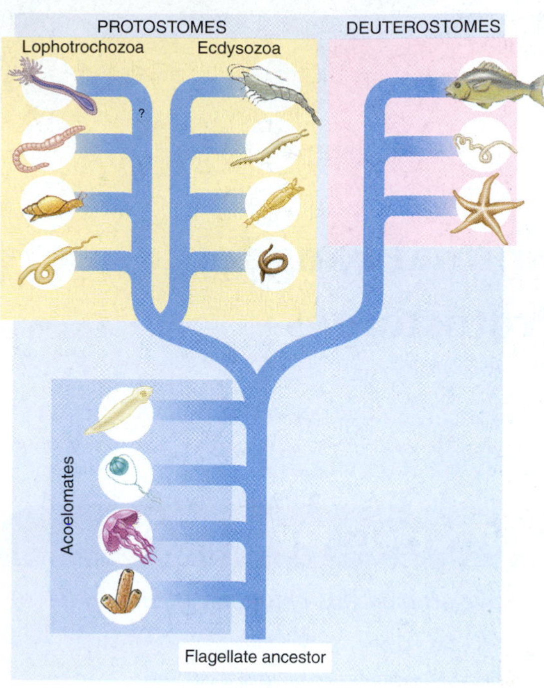

YOU MAKE THE CONNECTION

1. Every evolutionary adaptation has both benefits and costs. Explain this concept in terms of each of the following: (a) cephalization (b) hermaphroditism (c) living on land
2. Several international monitoring projects are gathering data needed to help us understand coral reef destruction. Why is it important to take action to protect coral reefs?
3. Imagine that you discover a new animal in a rain forest. How would you decide to which phylum it belongs? What are some characteristics that might contribute to your decision?

RECOMMENDED READINGS

Fabricius, K.E., Y. Benayahu, and A. Genin. "Herbivory in Asymbiotic Soft Corals." *Science,* Vol. 268, 7 Apr. 1995. Unlike most corals, a group of soft corals feeds on phytoplankton.

Fagoonee I., H.B. Wilson, M.P. Hassell, and J.R. Turner. "The Dynamics of Zooxanthellae Populations: A Long-Term Study in the Field." *Science,* Vol. 283, 5 Feb. 1999. A field study of algal density and coral bleaching.

Freeman, S., and J.C. Herron. *Evolutionary Analysis.* Prentice-Hall, Upper Saddle River, New Jersey, 1998. A readable textbook on evolution that includes a discussion of animal phylogeny.

Genthe, H. "The Incredible Sponge." *Smithsonian,* Aug. 1998. An interesting and beautifully illustrated discussion of sponges.

Johnsen, S. "Transparent Animals." *Scientific American,* Vol. 282, No. 2., Feb. 2000. Transparency is an adaptation for survival in the open ocean.

Knoll, A.H., and S. B. Carroll. "Early Animal Evolution: Emerging Views from Comparative Biology and Geology." *Science,* Vol. 284, 25 Jun. 1999. A discussion of body plans, genetic changes, and environmental events that may have contributed to the evolution of animals.

Morse, A.N.C., and D.E. Morse. "Flypapers for Coral and Other Planktonic Larvae." *BioScience,* Vol. 46, No. 4, Apr. 1996. Coral and other larvae recognize specific chemical signals that induce attachment and metamorphosis. Some of the chemical compounds involved may prove important to commercial aquaculture or medicine.

Raff, R.A. *The Shape of Life: Genes, Development, and the Evolution of Animal Form.* University of Chicago, Chicago, 1996. This readable book elaborates on many themes discussed in this chapter.

Simpson, S. "Fishy Business." *Scientific American,* Vol. 285, No. 1, Jul. 2001. Cyanide fishing is threatening coral reefs in Southeast Asian waters.

Travis, J. "Bleaching Power: Marine Bacteria Rout Coral's Colorful Algae." *Science News,* Vol. 149, 15 Jun. 1996. An interesting account of a research project suggesting that bacteria cause coral bleaching.

Weiss, P. "Soaking Up Rays." *Science News,* Vol. 160, 4 Aug. 2001. Materials scientists look to sponges to understand the composition of their natural optical devices.

- Visit our Web site at **http://www.info.brookscole.com/solomonbergmartin** for links to chapter-related resources on the World Wide Web. Additional on-line materials relating to this chapter can also be found on our Web site.

See chapter activity on BioActive Learner CD for additional help in mastering the chapter's material. Icon location in the chapter's margins shows which topics have tutorials or simulations in the CD.

29

The Animal Kingdom: The Protostomes

A bearded fireworm *(Hermodice carunculata).* The fireworm is a segmented worm (an annelid) that is related to earthworms. It lives in shallow marine waters—in coral reefs, beds of turtle grass, or under rocks. Fireworm bristles are toxic and can break off in the skin, causing a severe burning sensation. *(Marty Snyderman/Visuals Unlimited)*

LEARNING OBJECTIVES

After you have studied this chapter you should be able to

1. Identify several advantages of having a coelom.
2. Describe the distinguishing characteristics of the phyla discussed in this chapter, including nemerteans, mollusks, annelids, lophophorates, rotifers, nematodes, and arthropods; and properly classify an animal that belongs to any of these phyla.
3. Describe the classes of mollusks discussed and give examples of animals that belong to each.
4. Describe and give examples of each of the three classes of annelids discussed.
5. Distinguish among the subphyla and classes of arthropods, and give examples of animals that belong to each group.
6. Discuss factors that have contributed to the great biological success of insects.

The **coelom** was one of the most important early animal adaptations. As explained in Chapter 28, the coelom is a fluid-filled space completely lined by mesoderm that lies between the digestive tube and the outer body wall. Evolution of the coelom produced a new body design, a **tube-within-a-tube plan.** The inner tube, the digestive tube and its associated organs, no longer had to be attached to the body wall (the outer tube) except at its ends. Because the coelom separates the muscles of the body wall from those in the wall of the digestive tract, the digestive tube can move food along independently of body movements. Typically, the digestive tube has a mouth at one end and an anus for elimination of wastes at the other end.

Animals with a coelom, and to a lesser extent those with a pseudocoelom, have another major advantage over acoelomate animals. As an enclosed compartment (or series of compartments) of fluid under pressure, the coelom can serve as a **hydrostatic skeleton** in which contracting muscles push against a tube of fluid. In many coelomates, like the bristleworm *(Hermodice)* shown here, a hydrostatic skeleton permits a greater range of movement than in acoelomate animals. The evolution of various shapes and divisions of the coelom provided the opportunity for animals to become specialized in swimming, crawling, or walking. Animals that can move quickly to capture food or avoid predators are more likely to survive. The hydrostatic skeleton also confers shape to the body of soft animals.

Evolution of the coelom provided a space for the development of internal organs. Most coelomate animals have well-developed circulatory, excretory, and nervous systems. Many internal organs are suspended within folds of the tissue lining the coelom and can move independently of the outer wall of the body. For example, the pumping action of the heart is possible because of the surrounding space provided by the coelom. The fluid-filled coelom also protects internal organs by cushioning them.

In some animals, fluid within the coelom helps transport materials such as food, oxygen, and wastes. Cells bathed by the coelomic fluid can exchange materials with it. The cells receive nutrients and oxygen from the coelomic fluid and excrete wastes into it. Some coelomates have excretory structures that remove wastes directly from the coelomic fluid.

The coelom provides space for the gonads to develop. During the breeding season of many animals such as birds, the gonads enlarge within the coelom as they fill with ripe gametes.

Historically, the **protostome coelomates** included the annelids, mollusks, and arthropods, as well as several smaller, related phyla. Based on molecular data, biologists now classify several other groups as coelomate protostomes. Recall from Chapter 28 that the coelomate protostomes appear to have evolved along two major branches: the **Lophotrochozoa** and the **Ecdysozoa** (see Fig. 28–5). The Lophotrochozoa include the nemerteans (ribbon worms), mollusks, annelids, and the lophophorate phyla, characterized by a ciliated ring of tentacles surrounding the mouth. The Ecdysozoa include the rotifers, nematodes (roundworms), and arthropods. Ecdysozoans are characterized by a **cuticle,** a noncellular body covering secreted by the epidermis. The name Ecdysozoa refers to the process of **ecdysis,** or molting, characteristic of animals in this group. During ecdysis, an animal sheds its cuticle. The phylogeny presented here is supported by RNA data. In this chapter we discuss the evolution, relationships, body plans, and life history of several groups of the Lophotrochozoa and Ecdysozoa.

PHYLUM NEMERTEA IS CHARACTERIZED BY THE PROBOSCIS

Phylum **Nemertea,** known as ribbon worms or proboscis worms, is a relatively small group (about 900 species) of free-living animals (Fig. 29–1). Almost all are marine, although a few inhabit fresh water or damp soil. They are carnivorous, feeding on crustaceans and other worms. Nemerteans have long narrow bodies, either cylindrical or flattened, generally ranging in length from 5 cm (2 in) to about 2 m (6.5 ft), although some are much longer. Some are a vivid orange, red, or green, with black or colored stripes.

Their most remarkable organ, the **proboscis,** from which they get one of their common names, *proboscis worms,* is a long, hollow, muscular tube that can be rapidly everted (turned inside-out) from the anterior end of the body. The proboscis can be wrapped around prey. In some species it is sharp and in various species it secretes sticky or toxic fluids that aid in trapping or immobilizing prey. Although the nemerteans are functionally acoelomate, the chamber surrounding the proboscis is a true coelomic space, known as a **rhynchocoel,** derived like the coelom in coelomate protostomes (see Chapter 28). The proboscis is a novel derived character that distinguishes nemerteans from all other invertebrate groups.

Historically, biologists thought the nemerteans were closely related to the flatworms. This group was of special evolutionary interest because its members have a circulatory system and a tube-within-a-tube body plan. Based on the presence of the rhyncocoel and on recent molecular evidence, nemerteans are now thought to be more closely related to the lophotrochozoan coelomates. The rhyncocoel is viewed as a remnant of the coelom of a coelomate ancestor.

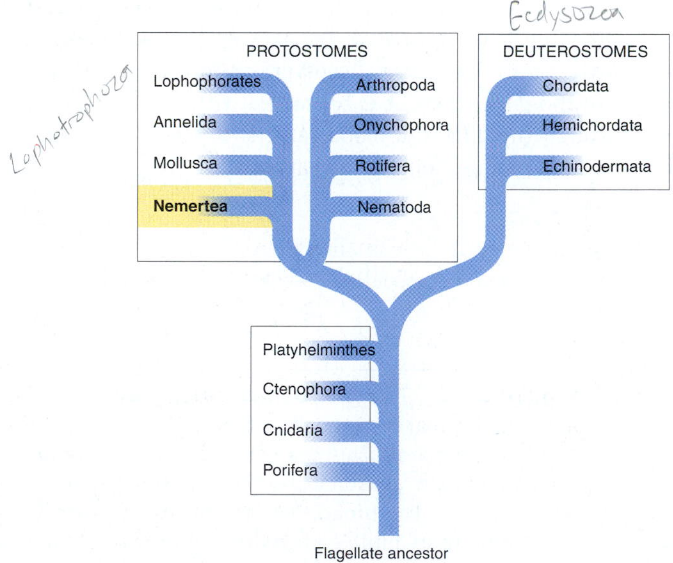

(a)

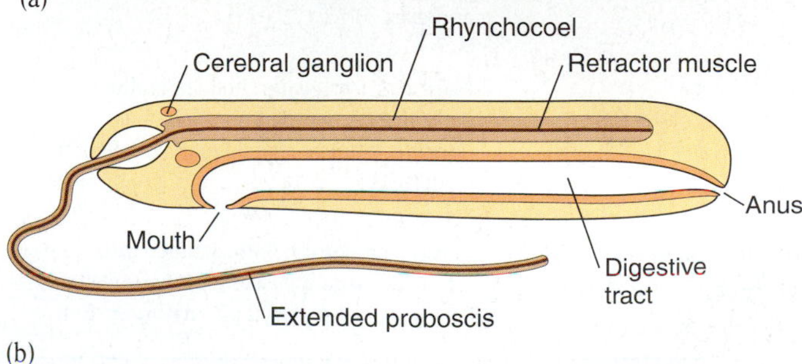

(b)

Figure 29–1 Nemerteans (proboscis worms). **(a)** Proboscis worm (*Lineus* sp.) from the Pacific coast of Panama. **(b)** Lateral view of a typical nemertean. Note the complete digestive tract that extends from mouth to anus, giving this animal a tube-within-a-tube body plan. *(Kjell B. Sandved)*

In nemerteans the digestive and circulatory functions are separated. The primitive circulatory system consists of blood vessels, muscular tubes extending the length of the body and connected by transverse vessels. Nemerteans have no heart to pump the blood. The blood transports materials as it is circulated through the vessels by movements of the body and contractions of the muscular blood vessels.

MOLLUSKS HAVE A FOOT, VISCERAL MASS, AND MANTLE

Mollusks are among the best known of the invertebrates. Most of us have walked along the shore of an ocean or lake collecting their shells. Phylum **Mollusca** includes clams, oysters, octopods, snails, slugs, and the largest of all the invertebrates, the giant squid, which averages 9 to 16 m (about 30–53 ft) in length, including its tentacles. More than 50,000 living species and 35,000 fossil species (second only to the arthropods in number) have been described. Representative mollusks are illustrated in Figure 29–2. Four of the eight recognized classes are discussed in this chapter and are listed in Table 29–1.

Although most mollusks are marine, many snails and clams live in fresh water, and some species of snails and slugs inhabit the land. Mollusks probably evolved early in the history of the protostome clade, soon after the evolution of the coelom but before the origin of the segmented body that is characteristic of annelids and arthropods. Although mollusks vary widely in outward appearance, most share certain basic characteristics:

TABLE 29–1 Major Classes of Phylum Mollusca

Class and Representative Animals	Characteristics
Polyplacophora Chitons	Primitive marine animals with shell consisting of eight separate transverse plates; head reduced; broad foot used for locomotion.
Gastropoda Snails, slugs, nudibranchs	Marine, freshwater, or terrestrial; coiled shell in many species; torsion of visceral mass; well-developed head with tentacles and eyes.
Bivalvia Clams, oysters, mussels	Marine or freshwater; body laterally compressed; two-part shell hinged dorsally; hatchet-shaped foot; suspension feeders.
Cephalopoda Squids, octopods	Marine; predatory; foot modified into tentacles, usually bearing suckers; well-developed eyes; closed circulatory system.

1. A soft body, usually covered by a dorsal shell composed mainly of calcium carbonate.
2. A broad, flat, muscular **foot,** located ventrally, which is used for locomotion.
3. The body organs (viscera) concentrated as a **visceral mass** located above the foot.
4. A **mantle,** a thin sheet of tissue that covers the visceral mass and usually contains glands that secrete a shell. The mantle generally overhangs the visceral mass, forming a mantle cavity that contains gills and other structures.
5. A rasplike structure called a **radula,** which is a belt of teeth in the mouth region. (The radula is not present in clams or their relatives, or in other suspension feeders.)
6. The coelom is generally reduced to small compartments around certain organs, including the heart and excretory organs *(metanephridia)*. The main body cavity is typically a **hemocoel,** a space containing blood (see discussion of open circulatory system that follows).

All organ systems typical of complex animals are present in mollusks. The digestive system is a tube, often coiled, consisting of a mouth, buccal cavity (mouth cavity), esophagus, stomach, intestine, and anus. The radula, located within the buccal cavity, can be projected out of the mouth and used to scrape particles of food from the surface of rocks or the ocean floor. Sometimes the radula is used to drill a hole in the shell of a prey animal or to tear off pieces of a plant.

Most mollusks have an **open circulatory system** in which the blood, called **hemolymph,** bathes the tissues directly. In mollusks, the heart pumps blood into a single blood vessel, the aorta, which may branch into other vessels. Eventually blood flows into a network of large spaces called sinuses, where the tissues are bathed directly; this network makes up the **hemocoel,** or blood cavity. From the sinuses, blood drains into vessels that conduct it to the gills, where it is recharged with oxygen. After passing through the gills, the blood returns to the heart. Thus, blood flow in a mollusk follows the pattern:

heart ⟶ aorta ⟶ smaller blood vessels ⟶
blood sinuses (hemocoel) ⟶ blood vessels to gills ⟶ heart

Figure 29–2 The molluskan body plan. **(a)** Chitons are sluggish marine animals with shells composed of eight overlapping plates. This sea cradle chiton *(Tonicella lineata)* inhabits coastal waters off the U.S. Pacific Northwest. **(b)** The broad, flat foot of the gastropod is an adaptation to its mobile lifestyle. The terrestrial banded garden snail *(Cepaea nemoralis)* is widely distributed. **(c)** The compressed body of the horseneck clam *(Tresus capax)* is adapted for burrowing in the mud. **(d)** The squid body is streamlined for swimming. To avoid being seen by potential predators, the squid can change color to blend with its background. This squid *(Sepioteuthis lessoniana)* is native to Hawaii. *(a, Kjell B. Sandved/Visuals Unlimited; b, William E. Ferguson; c, Tom McHugh/Photo Researchers, Inc.; d, Mike Severns/Tom Stack & Associates)*

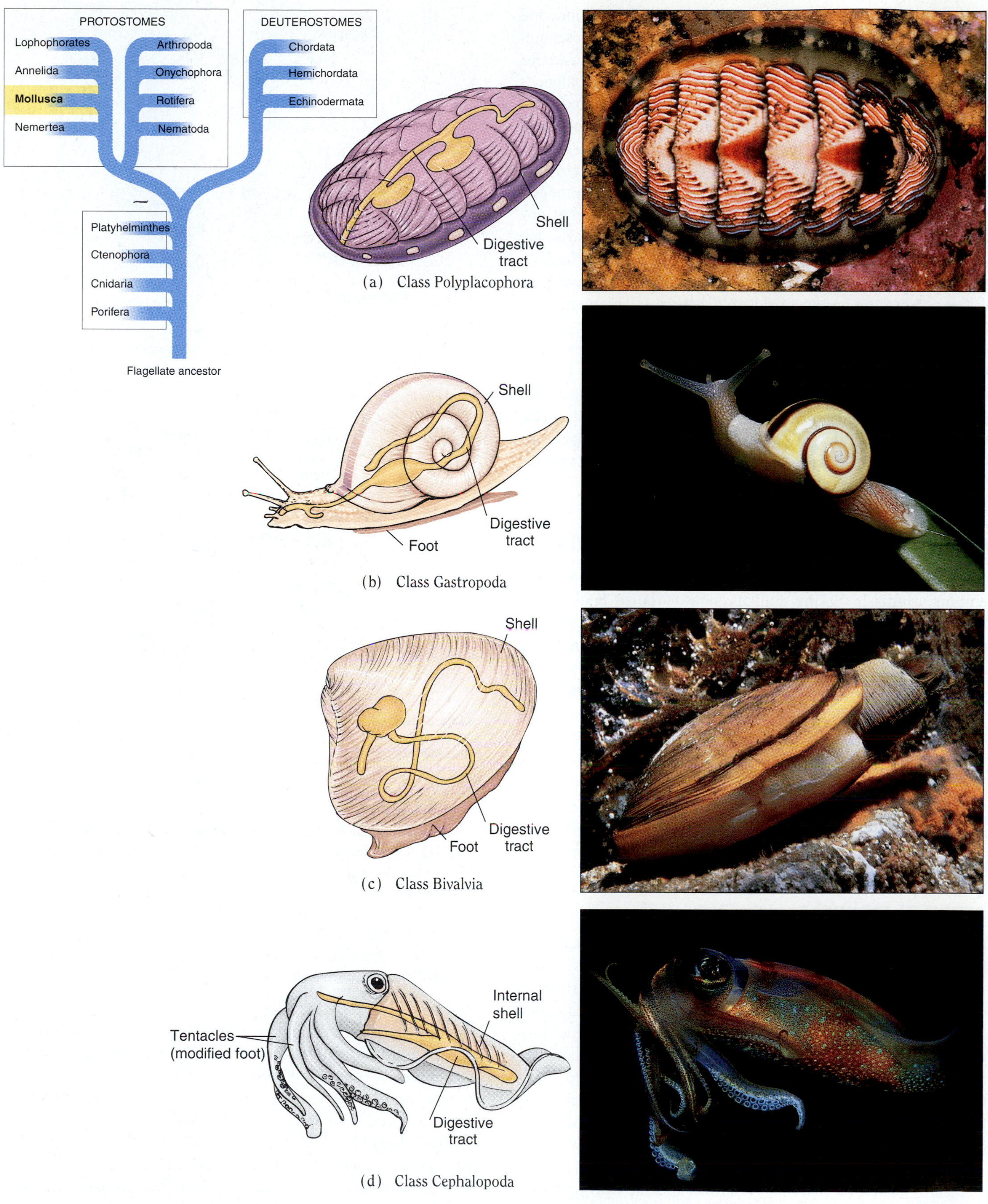

(a) Class Polyplacophora

(b) Class Gastropoda

(c) Class Bivalvia

(d) Class Cephalopoda

In open circulatory systems, blood pressure tends to be low, and tissues are not very efficiently oxygenated. However, most mollusks are slow-moving animals with low metabolic rates, and this type of circulatory system is adequate. In the active cephalopods (the class that includes the squids and octopods), the circulatory system is closed. In a **closed circulatory system,** blood flows through a complete circuit of blood vessels.

Large, paired **metanephridia** (sing., *metanephridium*) are excretory structures characteristic of mollusks and annelids. Each metanephridium is a ciliated tubule with a funnel at one end that removes wastes from the fluid in the coelom. Wastes are discharged through the other end of the tubule, which leads to an excretory pore. Open at both ends, metanephridia are quite different from the protonephridia of flatworms, which are blind tubules that open only to the exterior (see Chapter 28).

Most marine mollusks pass through one or more larval stages. The first larval stage is typically a **trochophore larva,** a free-swimming, ciliated, top-shaped larva characteristic of mollusks and annelids (Fig. 29–3). In many mollusks (gastropods, e.g., snails, and bivalves, e.g., clams), the trochophore larva develops into a **veliger larva,** which has a shell, foot, and mantle. The veliger larva is unique to the mollusks.

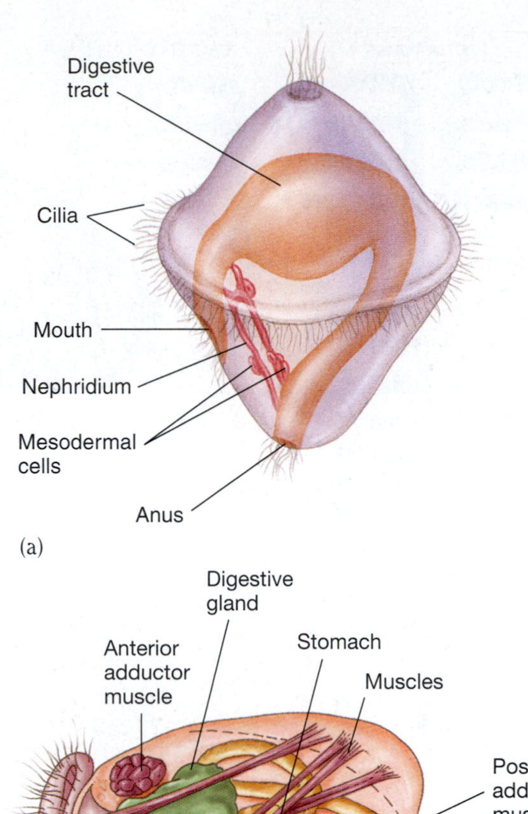

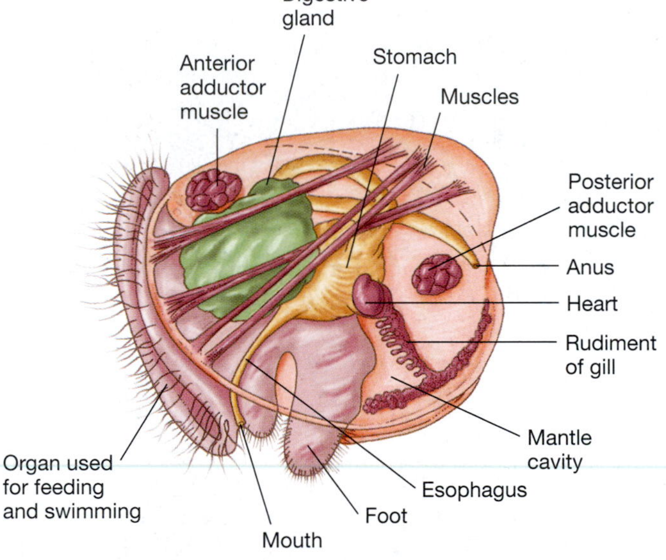

Figure 29–3 Larval stages of marine mollusks. **(a)** A trochophore larva, the first larval stage of a marine mollusk, is also characteristic of annelids. Just above the mouth there is a characteristic band of ciliated cells that functions as a swimming organ and, in some species, collects suspended food particles. **(b)** Typically, the trochophore larva develops into a bilateral veliger larva with foot, shell, and mantle.

Chitons may be similar to ancestral mollusks

Class **Polyplacophora** (meaning "many plates") comprises the **chitons,** sluggish marine animals with flattened bodies (Fig. 29–2*a*). Their most distinctive feature is a shell composed of eight separate but overlapping dorsal plates. The head is reduced, and there are no eyes or tentacles.

The chiton inhabits rocky intertidal zones, using its broad, flat foot to move and to hold firmly onto rocks. By pressing its mantle against the substratum and lifting the inner edge of the mantle, the chiton can produce a partial vacuum. The resulting suction enables the animal to adhere powerfully to its perch. Using its radula for grazing, the chiton scrapes algae and other small organisms off rocks and shells.

Gastropods are the largest group of mollusks

Class **Gastropoda,** which includes snails, slugs, and their relatives, is the largest and most diverse group of mollusks (Fig. 29–4). In fact, with more than 40,000 extant species, gastropods comprise the second largest class in the animal kingdom, second only to insects. Most gastropods inhabit marine waters, but others make their homes in brackish or fresh water, or on land. We think of snails as having a single, spirally coiled shell into which they can withdraw the body, and many do. However, other gastropods, such as limpets, have shells like flattened dunce caps. Still others, such as garden slugs and the beautiful marine slugs known as **nudibranchs,** have no shell at all.

Many gastropods have a well-developed head with tentacles. Two simple eyes may be located on stalks that extend from the head. The broad, flat foot is used for creeping. Most land snails do not have gills. Instead, the mantle is highly vascularized and functions as a lung. These garden snails and slugs are described as **pulmonate** (meaning "having a lung").

A unique feature of gastropods is **torsion,** a twisting of the visceral mass. (This twisting is unrelated to the coiling of the shell.) As the bilateral larva develops, one side of the visceral mass grows more rapidly than the other side. This uneven growth results in rotation of the visceral mass. The visceral mass and mantle twist permanently up to 180 degrees relative to the head. As a result, the digestive tract becomes somewhat U-shaped, and the

(a)

(b)

Figure 29–4 Gastropods. **(a)** A rough keyhole limpet *(Diodora aspera)* escapes predators by exuding slippery mucus from its mantle. **(b)** Porter's nudibranch *(Chromodoris porterae)* feeding on a lightbulb tunicate *(Clavelina huntsmani)*. *(a, Norbert Wu/Peter Arnold, Inc.; b, Bruce Watkins/Animals Animals)*

anus comes to lie above the head and gill (Fig. 29–5). Subsequent growth is dorsal and usually in a spiral coil. Torsion limits space in the body, and typically the gill, metanephridium (excretory tubules), and gonad are absent on one side. Some biologists hypothesize that torsion is an adaptation that protects the head by allowing it to enter the shell first during withdrawal from potential predators. Without torsion, the foot would be withdrawn first.

Bivalves typically burrow in the mud

Class **Bivalvia** includes the clams, oysters, mussels, scallops and their relatives. Typically, the soft body of members of class Bivalvia is laterally compressed and completely enclosed by a two-part shell that hinges dorsally and opens ventrally (Fig. 29–6). This arrangement allows the hatchet-shaped foot to protrude ventrally for locomotion and for burrowing in mud. The two parts, or **valves,** of the shell are connected by an elastic ligament. Stretching of the ligament opens the shell. Large, strong adductor muscles attached to the shell permit the animal to close its shell.

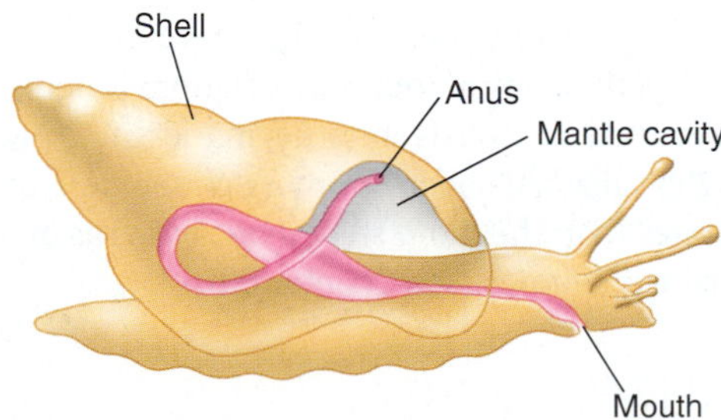

Figure 29–5 Embryonic torsion in a gastropod. As the bilateral larva develops, the visceral mass twists 180 degrees relative to the head. The digestive tube coils, and the anus becomes relocated so that it is near the mouth.

The inner pearly layer of the bivalve shell is made of calcium carbonate secreted in thin sheets by the epithelial cells of the mantle. Known as *mother-of-pearl,* this material is valued for making jewelry and buttons. Should a bit of foreign matter lodge between the shell and the mantle, the epithelial cells covering the mantle may be stimulated to secrete concentric layers of calcium carbonate around the intruding particle. Many species of oysters and clams form pearls in this way.

The bivalve nervous system has three pairs of ganglia, two pairs of nerve cords, and a variety of sense organs. Pigmented eyespots *(ocelli)* may be present along the edge of the mantle, enabling the animal to detect changes in light intensity that could be caused by the shadow of a predator. In scallops and in some oysters and clams, the ocelli are well developed, permitting formation of images.

Some bivalves, such as oysters, attach permanently to the substrate. Others burrow slowly through rock or wood, seeking protected dwellings. The shipworm, *Teredo,* a clam that damages dock pilings and other marine installations, is just looking for a home. A few bivalves, such as scallops, swim rapidly by clapping their two shells together with the contraction of a large adductor muscle (the part of the scallop that is eaten by humans).

Clams and oysters are suspension feeders that trap food particles in sea water. They take water in through an extension of the mantle called the **incurrent siphon.** Water leaves by way of an **excurrent siphon.** As the water passes over the gills, food particles in the water are trapped in mucus secreted by the gills. Cilia

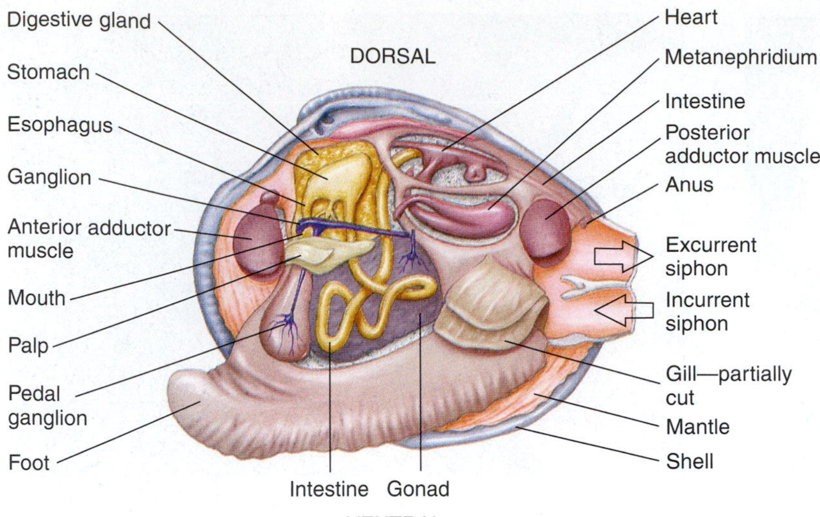

Figure 29–6 Internal anatomy of a clam. The two shells of a bivalve hinge dorsally and open ventrally.

move the food to the mouth. An oyster can filter more than 30 L (about 32 quarts) of water per hour. As suspension feeders, bivalves have no radula, and indeed they are the only group of mollusks that lack this structure.

Most bivalves have two distinct sexes. In some marine and nearly all freshwater bivalves, sperm are shed into the water and fertilize the eggs within the mantle cavity of the female. In these species, the female also carries her young within the mantle cavity, and development of the young takes place among the female's gill filaments. A marine bivalve typically develops as a trochophore larva, which later becomes a veliger larva. Some freshwater species have a modified larval stage, called a *glochidium*. In some species, glochidia spend several weeks as parasites on the gills or fins of fishes.

Cephalopods are active, predatory animals

In contrast with most other mollusks, members of the class **Cephalopoda** (meaning "head-foot") are fast-swimming predatory animals (see Fig. 29–2*d*). The cephalopod mouth is surrounded by tentacles, or arms: 10 in squids, 8 in octopods, and as many as 90 in the chambered nautilus. The large head has well-developed eyes that form images. Although they develop differently, the eyes are structurally similar to vertebrate eyes and function in much the same way. The octopus has no shell, and the shell of the squid is greatly reduced and is located inside the body.

Nautilus has a coiled shell consisting of many chambers built up over time (see Fig. 6–3). Each year the animal lives in the newest and largest chamber of the series. *Nautilus* secretes a mixture of gases similar to air into the other chambers. By regulating the amount of gas in the chambers, the animal can control its depth in the ocean.

The tentacles of squids, octopods, and cuttlefish are covered with suckers for seizing and holding prey. In addition to a radula, the mouth is equipped with two strong, horny beaks used to kill prey and tear it to bits. The mantle is thick and muscular and fitted with a funnel-like **siphon.** By filling the cavity with water and ejecting it through the siphon, the cephalopod achieves forceful jet propulsion. Squids and cuttlefish have streamlined bodies adapted for efficient swimming.

Besides speed, the cephalopod has two other important adaptations that enable it to escape from its predators, which include certain whales and moray eels. One is its ability to confuse the enemy by rapidly changing colors. By expanding and contracting *chromatophores,* cells in the skin that contain pigment granules, the cephalopod can display an impressive variety of mottled colors. Another defense mechanism is its **ink sac,** which produces a thick, black liquid that is released in a dark cloud when the animal is alarmed. While its enemy pauses, temporarily blinded and confused, the cephalopod escapes. The ink has been shown to inactivate the chemical receptors of some predators, rendering them incapable of detecting their prey.

The octopus feeds on crabs and other arthropods, catching and killing them with a poisonous secretion of its salivary glands. During the day the octopus usually hides among the rocks; in the evening it emerges to hunt for food. Its motion is incredibly fluid, giving little hint of the considerable strength in its eight arms.

Process of Science Small octopods survive well in aquaria and have been studied extensively. Because they have a relatively high degree of intelligence and can make associations among stimuli, they have been used as models for studying learning and memory. Their highly adaptable behavior more closely resembles that of the vertebrates than the more stereotypic patterns of behavior seen in other invertebrates (see Chapter 50).

ANNELIDS ARE SEGMENTED WORMS

Members of phylum **Annelida,** the segmented worms, have bilateral symmetry and a tubular body that may be partitioned into more than 100 ringlike segments. This phylum, composed of about 15,000 species, includes three main classes: the polychaetes, a group of marine and freshwater worms; the earthworms; and the leeches (Table 29–2 and Fig. 29–7).

The term *Annelida* (meaning "ringed") refers to the series of rings, or segments, that make up the annelid body. Both the body wall and many of the internal organs are segmented. Some structures, such as the digestive tract and certain nerves, extend the length of the body, passing through successive segments. Other structures, such as excretory organs, are repeated in each segment. In polychaetes and earthworms, segments are separated from one another internally by transverse partitions called **septa.**

One advantage of **segmentation** is that it facilitates locomotion. The coelom is divided into segments, and each segment has its own muscles. This arrangement allows the animal to elongate one part of its body while shortening another part. The annelid's hydrostatic skeleton is important in movement. In polychaetes and earthworms, bristle-like structures called **setae** (sing., *seta*), located on each segment, provide traction as the worm moves by contracting its muscles (alternating contraction of longitudinal and circular muscles).

The annelid nervous system generally consists of a simple brain composed of a pair of ganglia, and ventral nerve cord. A pair of ganglia and lateral nerves are repeated in each segment. Annelids have a large, well-developed coelom, a closed circulatory system, and a complete digestive tract extending from mouth to anus. Respiration is *cutaneous,* that is, through the skin, or by gills. Typically, a pair of metanephridia (excretory tubules) is found in each segment (described in Chapter 46).

TABLE 29–2 Major Classes of Phylum Annelida

Class and Representative Animals	Characteristics
Polychaeta Sandworms, tubeworms	Mainly marine; each segment bears a pair of parapodia with many setae; well-developed head; separate sexes; trochophore larva.
Oligochaeta Earthworms	Terrestrial and freshwater worms; few setae per segment; lack well-developed head; hermaphroditic.
Hirudinea Leeches	Most are blood-sucking parasites that inhabit fresh water; appendages and setae absent; prominent muscular suckers.

Most polychaetes are marine worms with parapodia

Class **Polychaeta** includes marine worms that swim freely in the sea, burrow in the mud near the shore, or live in tubes secreted by the animal or made by cementing bits of shell and sand together with mucus. Each body segment typically bears a pair of paddle-shaped appendages called **parapodia** (sing., *parapodium*) that function in locomotion and in gas exchange. These fleshy structures bear many stiff setae (the name *Polychaeta* means "many bristles"; see the fireworm in the chapter introduction). Most polychaetes have a well-developed head bearing eyes and antennae. The head may also be equipped with tentacles and palps (feelers). Polychaetes develop from free-swimming trochophore larvae similar to those of mollusks.

Many polychaete species have evolved behavioral patterns that ensure fertilization. By responding to certain rhythmic variations, or cycles, in the environment, nearly all of the females and males of a given species release their gametes into the water

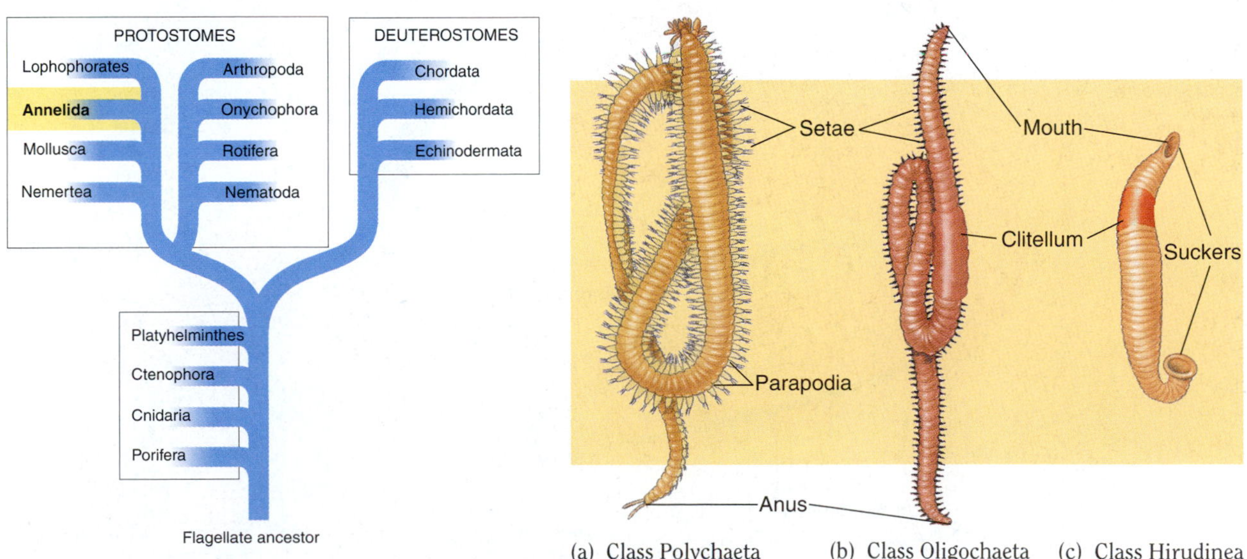

Figure 29–7 Three major classes of annelids. **(a)** Polychaetes are marine worms with paddle-shaped parapodia. **(b)** Oligochaetes, which include the earthworms, inhabit fresh water and moist terrestrial areas. **(c)** Many leeches, members of class Hirudinea, are parasites that feed on blood.

at the same time. For example, more than 90% of reef-dwelling *Palolo* worms of the South Pacific shed their eggs and sperm within a 2-hour period on one night of the year. In this animal the seasonal rhythm limits the reproductive period, which is related to the lunar cycle and tides, to October or November. The posterior portion of the *Palolo* worm, loaded with eggs or sperm, breaks off from the rest of the body. It comes to the surface and bursts, releasing its gametes. Local islanders eagerly await this annual event when they can gather great numbers of the swarming polychaetes, which they bake or eat raw.

Earthworms help maintain fertile soil

The 3000 or so species of the class **Oligochaeta** are found almost exclusively in fresh water and in moist terrestrial habitats. These worms lack parapodia, have few bristles per segment (the name *Oligochaeta* means "few bristles"), and lack a well-developed head. All oligochaetes are hermaphroditic.

Lumbricus terrestris, the common earthworm, is about 20 cm (8 in) long. Its body is divided into more than 100 segments, separated externally by grooves that indicate the internal position of the septa (Fig. 29–8). The earthworm's body is somewhat protected from drying by a thin, transparent cuticle, secreted by the cells of the epidermis. Mucus secreted by gland cells of the epidermis forms an additional protective layer over the body surface. The body wall has an outer layer of circular muscles and an inner layer of longitudinal muscles.

An earthworm literally eats its way through the soil, ingesting its own weight in soil and decaying vegetation every 24 hours. As the earthworm moves about, the soil is turned, aerated, and enriched by its nitrogenous wastes. This is how earthworms enhance the formation and maintenance of fertile soil. The earthworm's meal of soil, containing nutritious decaying vegetation, is processed in its complex digestive system.

Food is swallowed through the muscular **pharynx** and passes through the **esophagus,** which may be modified to form a thin-walled **crop** where food is stored and a thick-walled, muscular **gizzard.** In the gizzard, food is ground to bits as it is moved against sand grains consumed along with the meal. The rest of the digestive system is a long, straight **intestine** where food is chemically digested and absorbed. The absorptive area of the intestine is increased by a dorsal, longitudinal fold, called the *typhlosole*. Wastes pass out of the intestine to the exterior through the **anus.**

The efficient, closed circulatory system consists of two main blood vessels that extend longitudinally. The dorsal blood vessel, just above the digestive tract, collects blood from vessels in the segments, then contracts to pump the blood anteriorly. In the region of the esophagus, five pairs of blood vessels propel blood from the dorsal to the ventral blood vessel. Located just below the digestive tract, the ventral blood vessel carries blood posteriorly. Small vessels branch from it and deliver blood to the various structures in each segment, as well as to the body wall. Within these structures blood flows through tiny capillaries before returning to the dorsal blood vessel.

Gas exchange takes place through the moist skin. Oxygen is transported by the respiratory pigment hemoglobin in the blood plasma (the liquid part of the blood). The excretory system consists of paired metanephridia, repeated in almost every segment

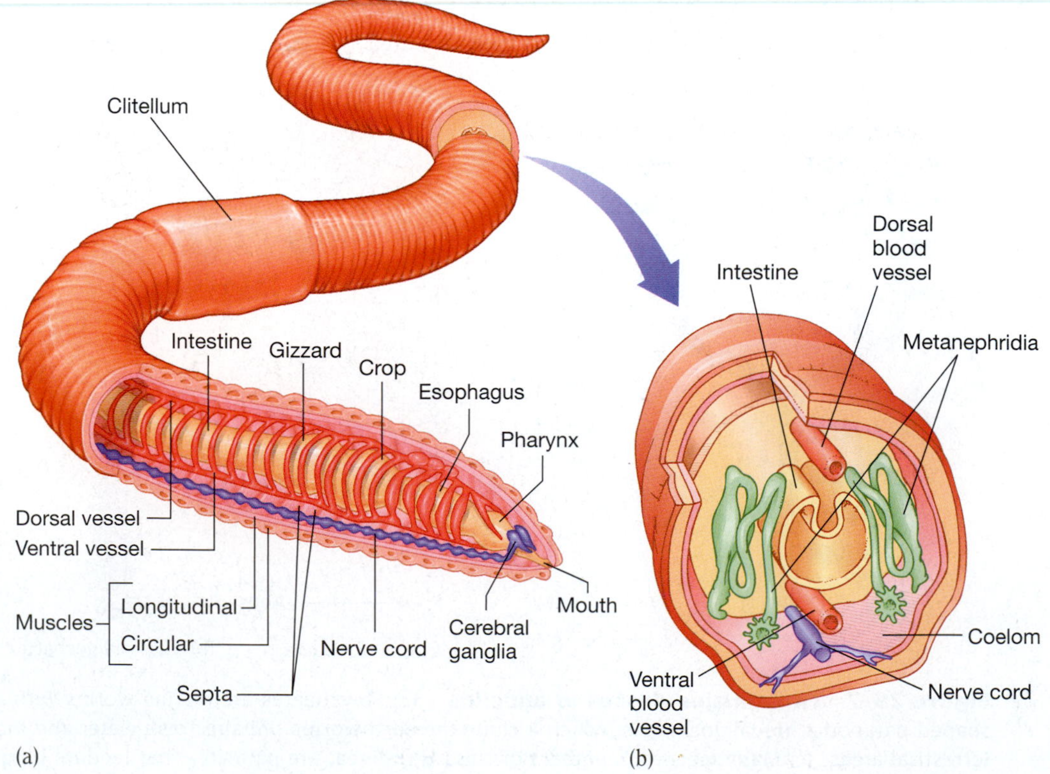

Figure 29–8 Body plan of an earthworm. **(a)** The internal structure has been exposed at the anterior end of an earthworm. **(b)** Cross section of an earthworm.

of the body. The ciliated funnel of each metanephridium opens into the next anterior coelomic cavity.

The nervous system consists of a pair of **cerebral ganglia** (a simple brain) just above the pharynx, and a subpharyngeal ganglion, just below the pharynx. A ring of nerve fibers connects these ganglia. From the lower ganglion a ventral nerve cord extends beneath the digestive tract to the posterior end of the body. In each segment along the nerve cord, there is a pair of fused segmental ganglia. Nerves extend laterally from the segmental ganglia to the muscles and other structures of that segment. The segmental ganglia coordinate the contraction of the muscles of the body wall, allowing the worm to creep along.

Like other oligochaetes, earthworms are hermaphroditic. During copulation, the worms exchange sperm. Two worms, headed in opposite directions, press their ventral surfaces together. These surfaces become glued together by the thick mucous secretions of each worm's *clitellum,* a thickened ring of epidermis. Sperm are then transferred and stored in the *seminal receptacles* (small sacs) of the other worm. The worms then separate. A few days later each clitellum secretes a membranous cocoon containing a sticky fluid. As the cocoon is slipped forward, eggs are laid into it. Sperm are added as the cocoon passes over the openings of the seminal receptacles. As the cocoon slips free over the worm's head, its openings constrict so that a spindle-shaped capsule is formed. The fertilized eggs develop into tiny worms within this capsule. This complex reproductive pattern is an adaptation to terrestrial life; the cocoon protects the delicate gametes and young worms from drying out.

Many leeches are blood-sucking parasites

Most leeches, members of class **Hirudinea,** inhabit fresh water, but some live in the sea or in moist areas on land. About 75% of the known species of leeches are blood-sucking parasites. Some leeches are nonparasitic predators that capture small invertebrates such as earthworms and snails. Leeches differ from other annelids in having neither setae nor parapodia.

Leeches have muscular suckers at both anterior and posterior ends. Most parasitic leeches attach themselves to a vertebrate host, bite through the skin, and suck out a quantity of blood, which is stored in pouches in the digestive tract. *Hirudin,* an anticoagulant secreted by glands in the crop, ensures leeches a full meal of blood. Some leeches feed only about twice each year because they digest their food slowly over several months.

Leeches have been used since ancient times for drawing blood from areas swollen by poisonous stings and bites. During the 19th century they were widely used to remove "bad blood," thought to be the cause of many diseases. Leeches have been used in modern medicine to remove excess fluid and blood that accumulates within body tissues as a result of injury, disease, or surgery (Fig. 29–9). The leech attaches its sucker near the site of injury, makes an incision, and secretes hirudin. The hirudin prevents the blood from clotting and dissolves already existing clots. In 30 minutes, a leech can suck out as much as ten times its own weight in blood.

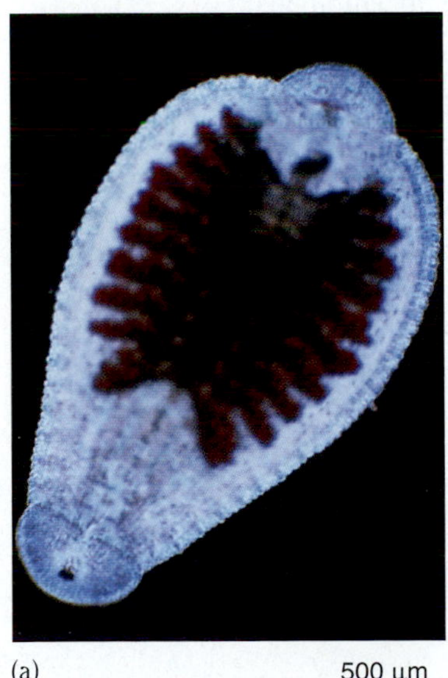

(a) 500 μm

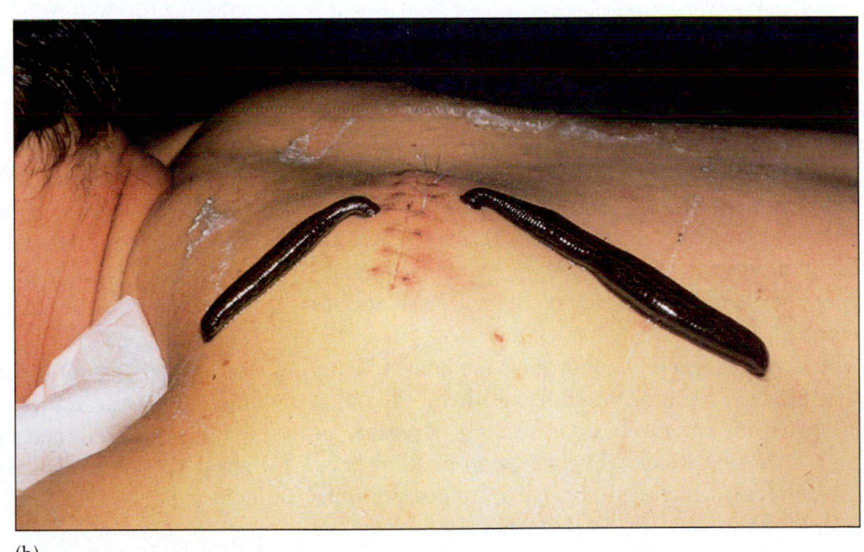

(b)

Figure 29–9 Leeches. **(a)** LM of a blood-sucking leech *(Helobdella stagnalis)* that feeds on mammals. The digestive tract *(dark area)* of its swollen body is filled with ingested blood. **(b)** The medicinal leech, *Hirudo medicinalis,* is used to treat hematoma, an accumulation of blood within tissues that results from injury or disease. The leech releases hirudin, an anticoagulant that prevents blood from clotting and dissolves preexisting clots. *(a, T.E. Adams/ Visuals Unlimited; b, St. Bartholomew's Hospital/Science Photo Library/Photo Researchers, Inc.)*

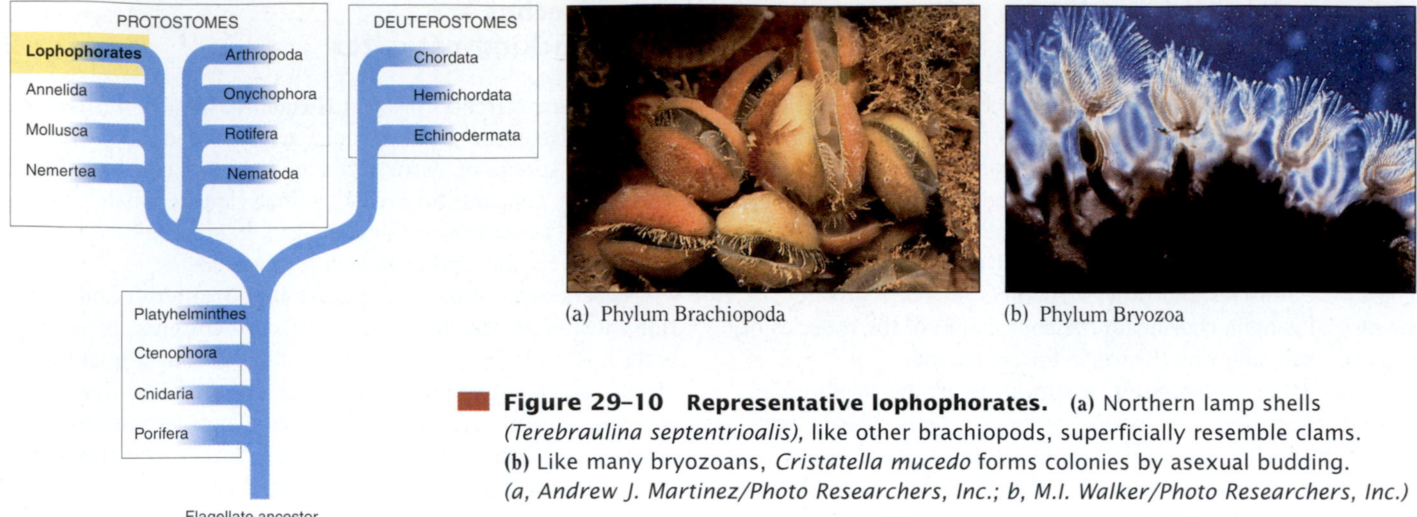

(a) Phylum Brachiopoda

(b) Phylum Bryozoa

Figure 29–10 Representative lophophorates. **(a)** Northern lamp shells *(Terebraulina septentrioalis),* like other brachiopods, superficially resemble clams. **(b)** Like many bryozoans, *Cristatella mucedo* forms colonies by asexual budding. *(a, Andrew J. Martinez/Photo Researchers, Inc.; b, M.I. Walker/Photo Researchers, Inc.)*

THE LOPHOPHORATE PHYLA ARE DISTINGUISHED BY A CILIATED RING OF TENTACLES

Although their evolutionary place is controversial, the **lophophorate phyla**—Brachiopoda, Phoronida, and Bryozoa—are now classified with the Lophotrochozoa by some biologists. With a few exceptions, lophophorates are marine animals that are adapted for life on the ocean floor. The **lophophore,** a ciliated ring of tentacles that surrounds the mouth, is specialized for capturing suspended particles in the water.

Brachiopods, or lampshells, are solitary marine animals that inhabit cold water. They superficially resemble clams and other bivalve mollusks because the body is enclosed between two shells (Fig. 29–10*a*). However, they differ from bivalve mollusks in that the shells are dorsal and ventral, rather than lateral; each shell is symmetrical about the midline, and the two shell valves are typically of unequal size. Brachiopods attach to the substrate by a long stalk. Cilia on the lophophore beat, bringing water laden with food into the slightly opened shell. Although there are only about 350 living species, brachiopods were abundant and diversified during the Paleozoic era. About 30,000 fossil species have been described.

Only about 12 extant species of **phoronids** are known. Some phoronids secrete tubes of sediment and live in them, extending their lophophores from their tubes for feeding. **Bryozoa,** or "moss animals," form sessile colonies by asexual budding (Fig. 29–10*b*). Their colonies, which can consist of millions of individuals, may appear plantlike. Other bryozoan colonies have the appearance of coral. About 5000 extant species are known.

ROTIFERS HAVE A CROWN OF CILIA

Among the less familiar invertebrates are the "wheel animals" of phylum **Rotifera.** Although no larger than many protozoa, these aquatic, microscopic animals are multicellular. More than 1800 species have been described. Most inhabit fresh water, but some live in marine environments or damp soil. Rotifers were formerly classified as pseudocoelomates. Many biologists now think that rotifers evolved from animals with a coelom. Because they have a cuticle, they classify them as ecdysozoans related to nematodes and arthropods. However, other biologists classify rotifers with the lophotrochozoans.

Rotifers have a characteristic crown of cilia on their anterior end. The cilia beat rapidly during swimming and feeding, giving the appearance of a spinning wheel (Fig. 29–11). Rotifers feed on tiny organisms suspended in the stream of water drawn into the mouth by the action of the cilia. A muscular organ posterior to the mouth grinds the food. The complete digestive tract ends in an anus through which wastes are eliminated.

Rotifers have a nervous system with a "brain" and sense organs, including eyespots. Protonephridia with flame cells remove excess water from the body and may also excrete metabolic wastes.

Process of Science Like some other animals with a pseudocoelom, rotifers are "cell constant." Each member of a given species is composed of exactly the same number of cells. Indeed, each part of the body is made of a precisely fixed number of cells arranged in a characteristic pattern. Cell division does not take place after embryonic development, and mitosis cannot be induced; growth and repair are not possible. One of the challenging problems of biological research is discovering the difference between such nondividing cells and the dividing cells of other animals. Do you think rotifers could develop cancer?

ROUNDWORMS ARE OF GREAT ECOLOGICAL IMPORTANCE

Members of phylum **Nematoda,** the **roundworms,** play key roles in decomposition and nutrient recycling. Nematodes are numerous and widely distributed in soil and in marine and freshwater sediments. More than 15,000 species have been identified, and many remain to be discovered. A spadeful of soil may contain

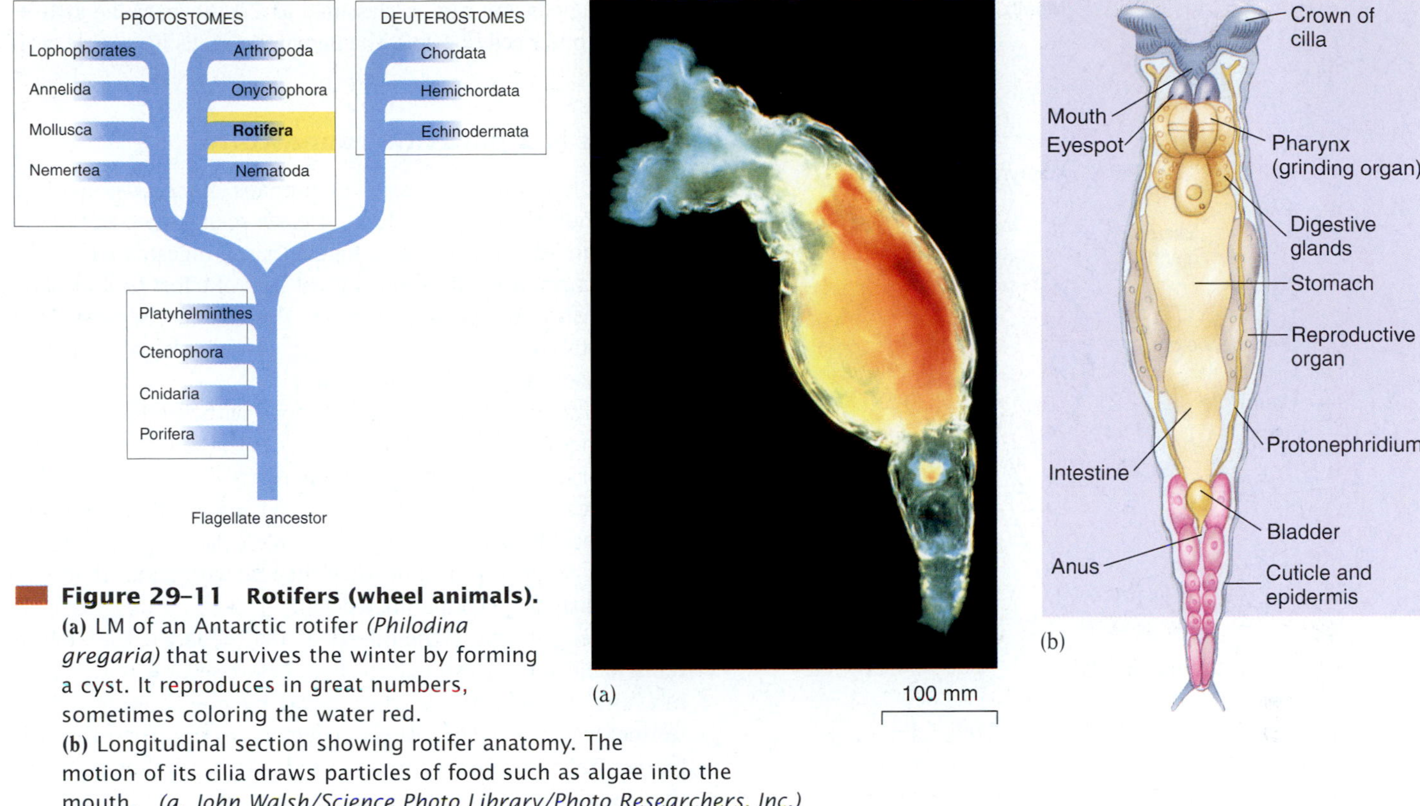

Figure 29–11 Rotifers (wheel animals). **(a)** LM of an Antarctic rotifer *(Philodina gregaria)* that survives the winter by forming a cyst. It reproduces in great numbers, sometimes coloring the water red. **(b)** Longitudinal section showing rotifer anatomy. The motion of its cilia draws particles of food such as algae into the mouth. *(a, John Walsh/Science Photo Library/Photo Researchers, Inc.)*

more than a million of these mainly microscopic worms, which thrash around, coiling and uncoiling. These movements are produced by contraction of longitudinal muscles in the body.

Although most nematodes are free-living (Fig. 29–12), others are important parasites in plants and animals. More than 30 roundworms, including hookworms, pinworms, trichina worms, filarial worms, and the intestinal roundworm *Ascaris,* are human parasites. *Caenorhabditis elegans,* a free-living nematode, is an important research organism for biologists studying the genetic control of development (see Chapter 16).

The elongated, cylindrical, threadlike nematode body is pointed at both ends and covered with a tough, flexible *cuticle* (Fig. 29–13). Secreted by the underlying epidermis, the thick cuticle gives the nematode body shape and offers some protection.

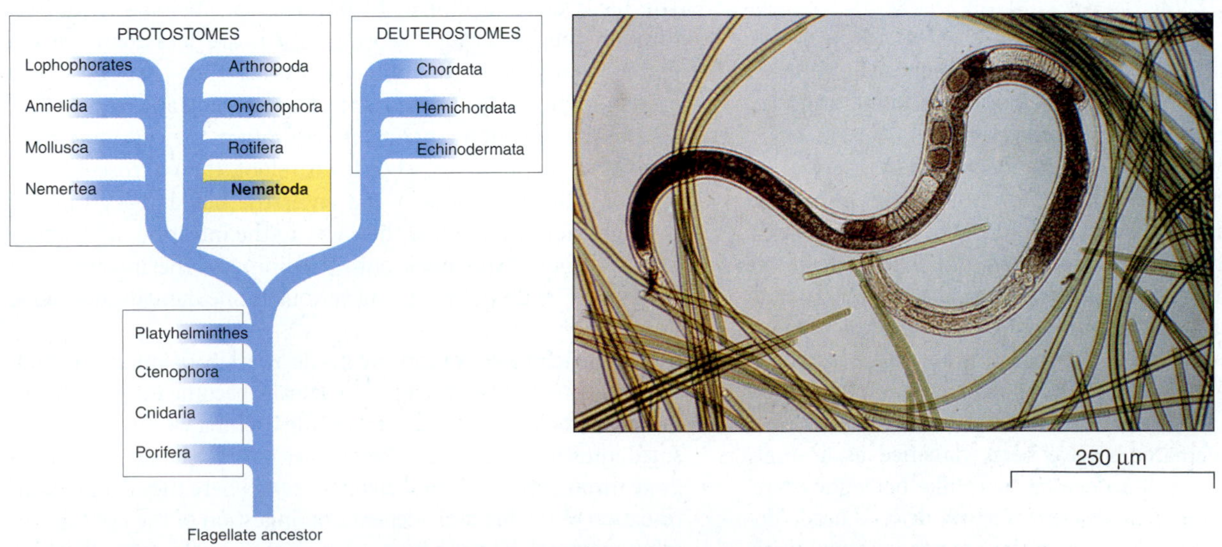

Figure 29–12 LM of a free-living nematode. This unidentified aquatic nematode is shown among the cyanobacterium *Oscillatoria,* which it eats. *(T.E. Adams/Visuals Unlimited)*

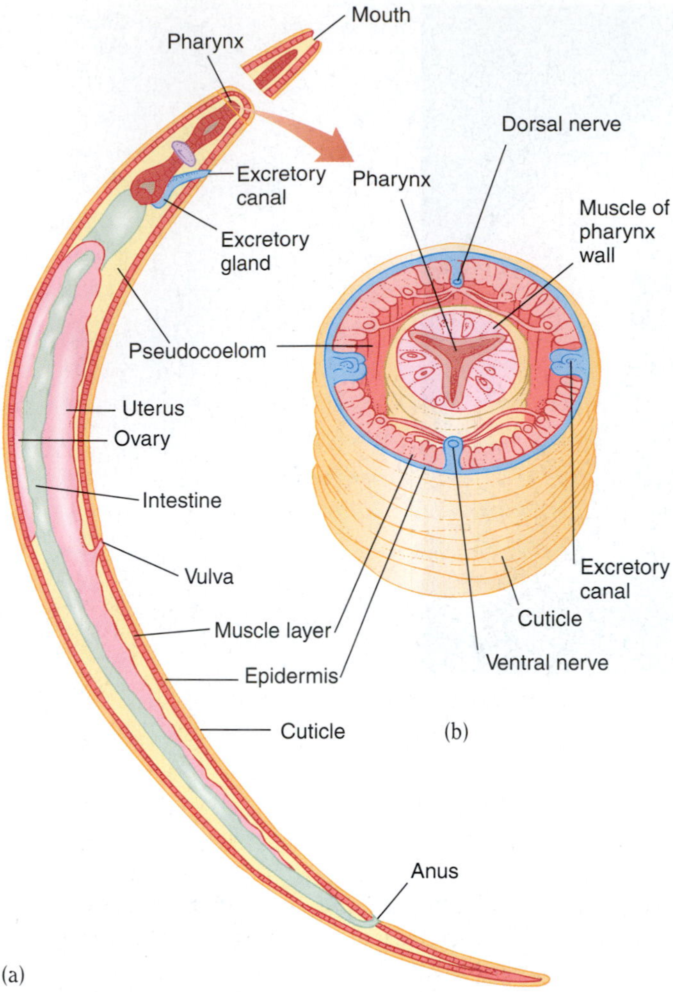

Figure 29–13 The roundworm *Ascaris*. **(a)** Longitudinal section. Note the complete digestive tract that extends from mouth to anus. **(b)** Cross section through *Ascaris* shows the tube-within-a-tube body plan. The protective cuticle that covers the body helps the animal resist drying.

The epidermis is unusual in that it is not composed of distinct cells. Beneath the epidermis is a layer of longitudinal muscles. No circular muscles are present in the body wall.

Nematodes have a fluid-filled pseudocoelom that serves as a hydrostatic skeleton. It transmits the force of muscle contraction to the enclosed fluid. Movement of fluid in the pseudocoelom is also important in nutrient transport and distribution. Like the ribbon worms, the nematodes exhibit bilateral symmetry, a complete digestive tract, three definite tissue layers, and definite organ systems; however, they lack specific circulatory structures. The sexes are usually separate, and the male is generally smaller than the female.

Traditionally, nematodes have been classified as a single phylum based on the pseudocoelom, wormlike body structure, and other characters such as absence of a well-defined head. Molecular evidence suggests that nematodes are not a monophyletic group and that they evolved from more complex animals by a simplification in body plan. Nematodes are probably related to the arthropods and can be classified as Ecdysozoa. Like arthropods and other ecdysozoans, the nematode sheds its cuticle periodically.

Ascaris is a parasitic roundworm

A common intestinal parasite of humans, ***Ascaris*** is a white worm about 25 cm (10 in) long. *Ascaris* spends its adult life in the human intestine, where it ingests partly digested food. Like the tapeworm, it must devote a great deal of effort to reproduction to ensure survival of its species. The sexes are separate, and copulation takes place within the host. A mature female may produce as many as 200,000 eggs a day.

Ascaris eggs leave the human body with the feces. Where sanitation is poor, they find their way onto the soil. In many parts of the world, human wastes are used as fertilizer, a practice that encourages the survival of *Ascaris* and many other parasites. People become infected when they ingest *Ascaris* eggs on unwashed vegetables or fruit or when they eat with hands that have become dirty from contaminated soil.

The eggs hatch in the intestine, and the larvae then take a remarkable journey through the body before settling in the small intestine. To begin the journey, larvae burrow through the intestinal wall into blood vessels or lymph vessels. Then they are carried in blood vessels through the heart to the lungs. There, they break through into the air sacs and move up the air passageways to the throat, where they are swallowed. Finally, the larvae pass through the stomach and into the intestine, where they settle, feed, and mature. During their migration, the larvae damage the lungs and other tissues. When worms settle in the small intestine, they may cause abdominal pain, allergic reactions, or malnutrition. Sometimes a tangled mass of these worms blocks the intestine.

Several other parasitic roundworms infect humans

The life cycle of a human **hookworm** also includes a journey through the body. Adult worms, which are less than 1.5 cm (0.6 in) long, live in the human intestine. They lay eggs, which pass out of the body with the feces. Larvae hatch and feed on bacteria in the soil. After a period of maturation, they become infective. When a potential host walks barefoot, the microscopic larvae bore through the skin and enter the blood. They migrate through the body and find their way to the intestine, where they mature. Hookworms hook onto the lining of the intestine and suck blood, potentially causing serious tissue damage and blood loss.

The **trichina worm** can live inside a variety of animals including pigs, rats, and bears. Humans typically become infected by eating undercooked, infected meat. Adult trichina worms live in the small intestine of the host. The females produce larvae, which migrate through the body to skeletal muscle, where they encyst. Continuation of the life cycle depends on ingestion of the cysts by another mammal. Because humans are not normally eaten, trichina larvae are not liberated from their cysts and eventually die. The cysts become calcified and remain in the muscles, causing stiffness

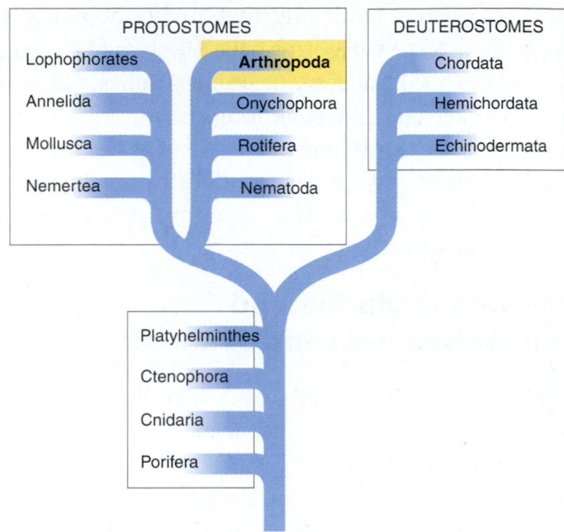

(a)

(b)

(c)

Figure 29–14 Arthropods. **(a)** The emperor scorpion *(Pandinis imperator)* is an example of an arachnid. Probably among the first terrestrial arthropods, scorpions have an elongated, segmented abdomen ending with a poisonous stinger. **(b)** This sponge crab *(Criptodromia octodenta)*, photographed in Australia, is a crustacean. It wears a living sponge on its back as camouflage. **(c)** The green lacewing (*Chrysopoda* sp.) is an insect. Lacewings are seldom attacked by predators, possibly because they have an unpleasant odor. *(a, Carlyn Iverson; b, Fred Baverdam/Peter Arnold, Inc.; c, Runk/Schoenberger/Grant Heilman)*

and discomfort. Migrating larvae and adults cause other symptoms. No cure has been found for trichina infection.

Pinworms are the most common worms found in children. The tiny pinworm eggs are often ingested by eating with hands contaminated with them. Adult worms, less than 1.3 cm (0.5 in) long, live in the large intestine. Female pinworms often migrate to the anal region at night to deposit their eggs. Irritation and itching caused by this practice induce scratching. Eggs may be distributed in the air and are in this way scattered throughout the house. Mild infestations may go unnoticed, but those more serious may result in discomfort, irritation, and injury to the intestinal wall.

ARTHROPODS ARE CHARACTERIZED BY JOINTED APPENDAGES AND AN EXOSKELETON OF CHITIN

Phylum **Arthropoda** includes more than 1 million species, and millions more remain to be identified. Biologists consider arthropods the most biologically successful group of animals because they are the most diverse and live in a greater range of habitats than do the members of any other animal phylum (Fig. 29–14). The following adaptations are major contributions to their success:

1. The arthropod body, like that of the annelid, is segmented. An important advantage of **segmentation** is that groups of segments can be specialized (by virtue of their shape, muscles, or the appendages they bear) to perform particular functions.
2. A hard **exoskeleton,** composed of chitin and protein, covers the entire body and appendages. The exoskeleton serves as a

coat of armor protecting against predators and helps prevent excessive loss of moisture. It also gives support to the underlying soft tissues. Distinct muscles attach to the inner surface of the exoskeleton and operate the joints of the body and appendages. The exoskeleton has an important disadvantage, however. As the arthropod grows, it periodically outgrows this nonliving shell. The process of shedding an old exoskeleton and growing a larger one is known as **molting.** The shed exoskeleton represents a net metabolic loss, and molting also leaves the arthropod temporarily vulnerable to predators.

3. **Paired, jointed appendages,** from which this phylum gets its name (*arthropod* means "jointed foot"), are modified for many different functions. They serve as swimming paddles, walking legs, mouth parts for capturing and manipulating food, sensory structures, or organs for transferring sperm.
4. The nervous system, which resembles that of the annelids, consists of a "brain" (cerebral ganglia) and ventral nerve cord with ganglia. In some arthropods, successive ganglia may fuse together. Arthropods have a variety of very effective sense organs. Many have organs of hearing, and **antennae** that sense taste and touch. Most insects and many crustaceans have **compound eyes** composed of many light-sensitive units called **ommatidia** (see Chapter 41). The compound eye can form an image and is especially adapted for detecting movement.
5. Arthropods have an *open circulatory system*. A dorsal, tubular heart pumps hemolymph into a dorsal artery, which may branch into smaller arteries. From the arteries, hemolymph flows into large spaces that collectively make up the hemocoel. Hemolymph in the hemocoel bathes the tissues directly. Eventually hemolymph reenters the heart through openings, called *ostia,* in its walls.

The exoskeleton presents a barrier to diffusion of oxygen and carbon dioxide through the body wall, necessitating the evolution of specialized respiratory systems for gas exchange. Most aquatic arthropods have gills that function in gas exchange, whereas many terrestrial forms have a system of internal branching air tubes called **tracheae,** or **tracheal tubes.** Other terrestrial arthropods have platelike **book lungs.**

The coelom is small and filled chiefly by the organs of the reproductive system. The arthropod digestive system is a tube similar to that of the earthworm. Excretory structures vary somewhat from class to class.

Arthropod evolution and classification are controversial

Process of Science Arthropod fossils have been identified that date back to the Precambrian era. Arthropods evolved rapidly during the Cambrian explosion, as evidenced by many diverse fossils in the Burgess shale and other ancient sites.

How arthropods are related to other protostomes continues to be a topic of lively debate. Some biologists have argued that arthropods share a common ancestor with annelids. The relationship between arthropods and annelids is evident in their basic body plans. Like annelids, arthropods are segmented, at least as embryos. The segments develop in the same way in both phyla. However, in contrast with most annelids, each arthropod has a fixed number of segments that remains the same throughout life.

Segmentation is important from an evolutionary perspective because it provides the opportunity for specialization of body regions. In most annelids the individual segments are almost all alike, but in many segmented animals (arthropods and chordates), different segments and groups of segments are specialized to perform different functions. In some groups, the specialization is so pronounced that the basic segmentation of the body plan may not be apparent. For example, in humans, although segmentation of the body is not obvious, muscles and nerves are segmentally organized. Some biologists now hypothesize that

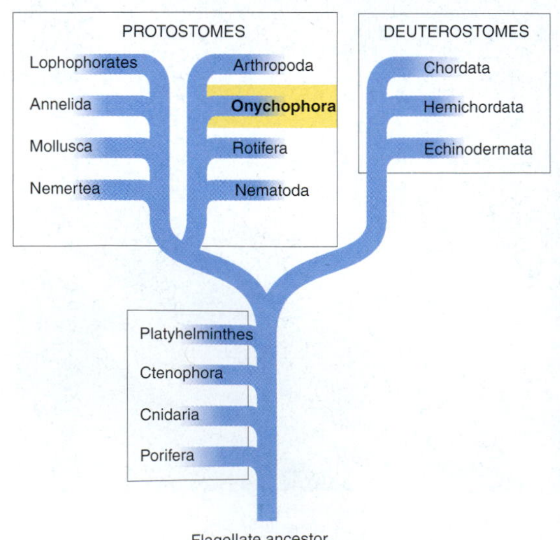

Figure 29–15 Onychophorans (velvet worms). *Peripatus* sp. with young. Note the soft, sluglike body and the series of nonjointed legs. Onychophorans have features of both arthropods and annelids. *(Thomas C. Boydean/Visuals Unlimited)*

segmentation is an ancestral character of protostomes and that some protostome groups have lost this trait.

The basic plan of the nervous system is similar in both annelids and arthropods. A ventral nerve cord extends from a dorsal, anterior brain, and ganglia are present in each segment.

Some biologists have viewed the wormlike **onychophorans** as the "missing link" between annelids and some groups of arthropods because they have some annelid and some arthropod characters (Fig. 29–15). Members of phylum **Onychophora,** also known as velvet worms, inhabit humid tropical rain forests. Like annelids, onychophorans are internally segmented, and many organs are duplicated serially. However, like arthropods, onychophorans have an open circulatory system and a respiratory system consisting of tracheal tubes. Although they are saclike, these animals have unbranched legs. Their jaws are derived from appendages, as in arthropods.

Recent molecular data suggest that despite their similarities, annelids and arthropods are not closely related and that onychophorans are not closely related to annelids. Some taxonomists now place the onychophorans, along with the **tardigrades,** close to the arthropods. Also known as water bears, members of phylum Tardigrada are tiny animals with unbranched, clawed legs (see Fig. 2–12). They inhabit soil or live on vegetation.

Some biologists debate the majority view that the arthropods are a monophyletic group, asserting that they are polyphyletic. Some propose that the chelicerates (horseshoe crabs, spiders, ticks) and crustaceans (lobsters, crabs, and their relatives) evolved from a marine ancestor and that the uniramians (insects) evolved from a terrestrial ancestor. According to this polyphyletic view, basic arthropod features such as jointed appendages and an exoskeleton of chitin evolved independently at least twice.

TABLE 29–3 Arthropod Subphyla

Subphylum and Selected Classes	Characteristics
Subphylum Chelicerata Class Merostomata (horseshoe crabs) Class Arachnida (spiders, scorpions, ticks, mites)	First pair of appendages are the chelicerae used to manipulate food; body consists of cephalothorax and abdomen; no mandibles; no antennae.
Subphylum Crustacea Class Malacostraca (lobsters, crabs, shrimp, isopods) Class Copepoda (copepods) Class Cirripedia (barnacles)	Biramous appendages; mandibles; two pairs of antennae.
Subphylum Uniramia Class Insecta (grasshoppers, roaches, ants, bees, butterflies, flies, beetles) Class Chilopoda (centipedes) Class Diplopoda (millipedes)	Uniramous appendages; mandibles; single pair of antennae.

In this book we present arthropod classification based on the more widely accepted monophyletic view and accordingly assign the living arthropods to three subphyla: Chelicerata, Crustacea, and Uniramia (Tables 29–3 and 29–4). Subphylum **Chelicerata** includes the horseshoe crabs and arachnids (spiders, scorpions, ticks, mites). Chelicerates, the only arthropods without antennae, have fanglike feeding appendages called **chelicerae** (sing., *chelicera*). Members of subphylum **Crustacea** (lobsters, crabs,

TABLE 29–4 General Characteristics of the Principal Arthropod Groups

	Arachnida (about 65,000 species)	*Crustacea* (about 32,000 species)	*Insecta* (more than 1,000,000 species)	*Chilopoda* (about 3000 species)	*Diplopoda* (about 7500 species)
Main Habitat	Mainly terrestrial	Marine or freshwater; a few are terrestrial	Mainly terrestrial	Terrestrial	Terrestrial
Body Divisions	Cephalothorax and abdomen	Head, thorax, and abdomen	Head, thorax, and abdomen	Head with segmented body	Head with segmented body
Gas Exchange	Book lungs or tracheae	Gills	Tracheae	Tracheae	Tracheae
Appendages	Uniramous	Biramous	Uniramous	Uniramous	Uniramous
Antennae	None	2 pairs	1 pair	1 pair	1 pair
Mouthparts	Chelicerae, pedipalps	Mandibles, 2 pairs of maxillae (for food handling)	Mandibles, maxillae	Mandibles, maxillae	Mandibles, maxillae
Legs	4 pairs on cephalothorax	Typically 1 pair per segment	3 pairs on thorax	1 pair per segment	Usually 2 pairs per segment
Development	Direct, except mites and ticks	Usually larval stages (nauplius larva)	Typically, larval stages; most with complete metamorphosis	Direct	Direct

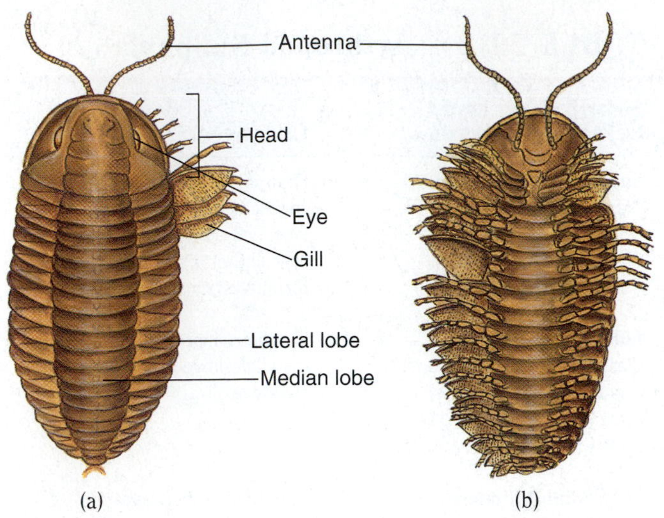

Figure 29–16 Trilobites. These extinct marine arthropods are considered the most primitive members of the phylum. **(a)** Dorsal view of a trilobite. **(b)** Ventral view.

shrimp, and barnacles) and members of subphylum **Uniramia** (insects, centipedes, and millipedes) have jawlike **mandibles** instead of chelicerae. Other than the extinct trilobites (discussed next), the crustaceans are the only group to have **biramous appendages,** that is, appendages with two jointed branches arising from their base. They are also the only group to have two pairs of antennae. Members of subphylum **Uniramia** have **uniramous** (unbranched) **appendages** and a single pair of antennae.

Trilobites were early arthropods

Among the earliest arthropods to evolve, **trilobites** inhabited shallow Paleozoic seas more than 500 million years ago. These arthropods, which have been extinct for about 250 million years, lived on the sea bottom and dug into the mud. Most ranged from 3 to 10 cm, but a few reached almost 1 m in length.

Covered by a hard, segmented exoskeleton, the trilobite body was a flattened oval divided into three parts: an anterior head bearing a pair of antennae and a pair of compound eyes; a thorax; and a posterior abdomen (Fig. 29–16). At right angles to these divisions, two dorsal grooves extended the length of the animal, dividing the body into a median lobe and two lateral lobes. (The name *trilobite* derives from this division of the body into these three longitudinal parts.) Each segment had a pair of segmented biramous appendages, consisting of an inner walking leg and an outer branch with gills.

The trilobites may have given rise to the chelicerates, which also lived in Paleozoic waters. Only a few species of marine chelicerates, most notably the horseshoe crabs, exist today. Fossil evidence suggests that some extinct marine chelicerates invaded fresh water and may have given rise to the arachnids. The early arachnids were aquatic. The first terrestrial arachnids appeared more than 360 million years ago (in the Devonian period), and most living arachnids are terrestrial.

Chelicerates have no antennae

Subphylum Chelicerata includes the merostomes (horseshoe crabs) and the arachnids. The chelicerate body consists of a cephalothorax (fused head and thorax) and an abdomen. These animals have no antennae and no chewing mandibles. Instead, the first pair of appendages, located immediately anterior to the mouth, are the chelicerae. The second pair of appendages are the **pedipalps.** The chelicerae and pedipalps are modified to perform different functions in various groups, including manipulation of food, locomotion, defense, and copulation. The four pairs of legs on the cephalothorax are specialized for walking.

Horseshoe crabs are the only living merostomes

Almost all of the merostomes are extinct. Only the horseshoe crabs have survived, essentially unchanged for 350 million years or more. *Limulus polyphemus,* the species common along the shore of North America, is horseshoe-shaped (Fig. 29–17*a*). Its long, spikelike tail is used in locomotion, not for defense or offense. Horseshoe crabs feed on mollusks, worms, and other invertebrates that they find on the sandy ocean floor.

Most arachnids are carnivores

Arachnids include spiders, scorpions, ticks, harvestmen (daddy long-legs), and mites (Fig. 29–17*b, c*). Most of the 65,000 or so named species are carnivorous and prey on insects and other small arthropods. The arachnid body consists of a cephalothorax and abdomen, and there are six pairs of jointed appendages. In spiders, the first pair, the chelicerae, are fanglike structures used to penetrate prey. In some spider species, the chelicerae are used to inject poison into the prey. The second pair of appendages, the pedipalps, are used by spiders to hold and chew food. In many arachnid species the pedipalps are modified as sense organs for tasting food or for reproduction (for sperm transfer and courtship displays). In scorpions, the very large pedipalps are used as pincers for capturing prey. The other four pairs of appendages are used for walking.

Typically, spiders have eight eyes arranged in two rows of four along the anterior dorsal edge of the cephalothorax. The eyes detect movement and locate objects. Some spiders, notably jumping spiders, are capable of forming a relatively sharp image.

Gas exchange in arachnids takes place by either tracheal tubes, book lungs, or both. A book lung consists of 15 to 20 plates, like pages of a book, that contain tiny blood vessels. Air enters the body through abdominal slits and circulates between the plates. As air passes over the blood vessels in the plates, oxygen diffuses into the blood and carbon dioxide diffuses out of the blood into the air. As many as four pairs of book lungs may be present, providing an extensive surface area for gas exchange.

The spider has unique glands in its abdomen that secrete silk, an elastic protein that is spun into fibers by organs called spinnerets. The silk is liquid as it emerges from the spinnerets but hardens after it leaves the body. Spiders use silk to build nests, to encase their eggs in a cocoon, and, in some species, to trap prey in a web. Many spiders lay down a silken dragline as they venture forth. The dragline serves as a safety line and also as a means of

(a)

(b)

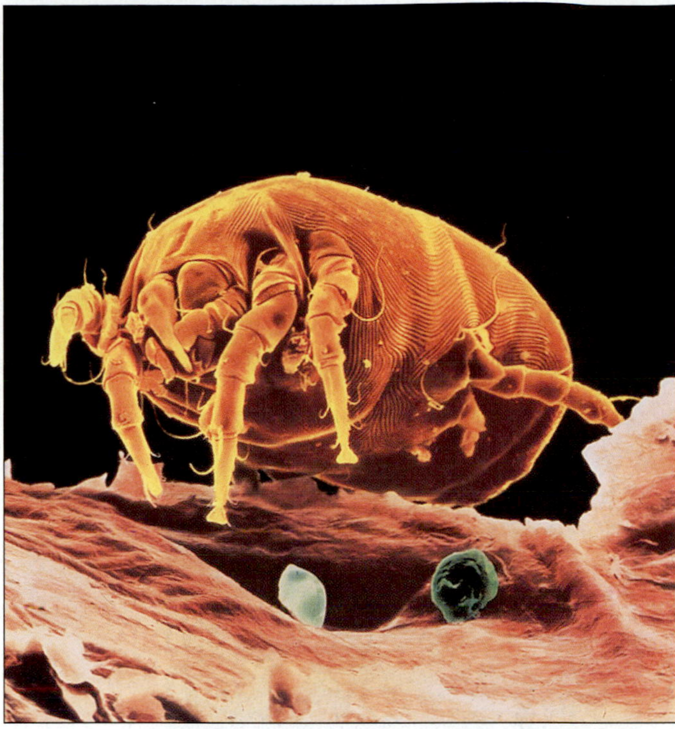

(c) 100 µm

Figure 29–17 Subphylum Chelicerata. **(a)** The only living merostomes are a few closely related species of horseshoe crabs. Seasonally, horseshoe crabs *(Limulus polyphemus)* return to beaches for mating. In this photograph, males are competing for a female. **(b)** A female black widow spider *(Latrodectus mactans)* rests on her web. The venom of the black widow spider is a neurotoxin. **(c)** Color-enhanced SEM of a house-dust mite (*Dermatophagoides* sp.), a common inhabitant of homes. This mite has been implicated in house-dust allergies. *(a, Milton H. Tierney, Jr./Visuals Unlimited; b, Steve Maslowski/Visuals Unlimited; c, K. H. Kjeldsen/Science Photo Library/Photo Researchers, Inc.)*

communication between members of a species. From a dragline another spider can determine the sex and maturity level of the spinner.

Although all spiders have poison glands useful in capturing prey, only a few produce poison that is toxic to humans. The most widely distributed poisonous spider in the United States is the black widow (*Latrodectus mactans;* Fig. 29–17*b*). Its poison is a neurotoxin that interferes with transmission of messages from nerves to muscles. The brown recluse spider *(Loxosceles reclusa)* is more slender than the black widow and has a violin-shaped dorsal stripe on its cephalothorax. Its venom destroys the tissues surrounding the bite and, like the venom of the black widow, can occasionally be fatal. Although painful, spider bites cause fewer than five fatalities per year in the United States. Poison glands are also present in scorpions.

Although some mites contribute to soil fertility by breaking down organic matter, many mites and ticks are serious nuisances. They eat crops, infest livestock and pets, and inhabit our own bodies. Many live unnoticed, owing to their small size, but others transmit disease. Certain mites cause mange in dogs and other domestic animals. Chiggers (red bugs), the larval form of red mites, attach themselves to the skin and secrete an irritating digestive fluid that may cause itchy red welts.

Larger than mites, ticks are parasites on dogs, deer, and many other animals. They can transmit diseases such as Rocky Mountain spotted fever, Texas cattle fever, relapsing fever, and Lyme disease (see Fig. 52–1).

Crustaceans are vital members of marine food webs

Subphylum Crustacea includes the lobsters, crabs, shrimp, and their relatives (Fig. 29–18). Many crustaceans are primary consumers of algae and detritus. Countless billions of microscopic crustaceans contribute to marine zooplankton, the free-floating, mainly microscopic organisms found in the upper layers of the ocean. These crustaceans are food for many fishes and other marine animals, such as certain baleen whales. Crustacean fossils have been dated back to the Cambrian period (more than 505 million years ago), but neither their origin nor their relationship to other arthropod subphyla is clear.

A distinctive feature of crustaceans is the *nauplius larva,* which is often the first stage after hatching. This larva has only the most anterior three pairs of appendages. Crustaceans are also characterized by mandibles, biramous appendages, and two pairs of antennae. Their antennae serve as sensory organs for touch and taste. The hard **mandibles,** which are the third pair of appendages, are located on each side of the ventral mouth and are used for biting and grinding food. Posterior to the mandibles are two pairs of appendages, the first and second **maxillae,** used for manipulating and holding food. Several other pairs of appendages are present. Some may be modified for walking. Others may be specialized for swimming, sperm transmission, carrying eggs and young, or sensation.

As the only group of arthropods that are primarily aquatic, crustaceans usually have gills for gas exchange. Two large **antennal glands** located in the head remove metabolic wastes from the blood and body fluids. Wastes are excreted through ducts opening at the base of each antenna. Most adult crustaceans have compound eyes, and many crustaceans have **statocysts,** sense organs that detect gravity.

Crustaceans characteristically have separate sexes. During copulation, the male uses specialized appendages to transfer sperm into the female. The fertilized eggs are usually carried on some part of the body. The newly hatched animals may resemble adults, or they may pass by successive molts through a series of larval stages before they develop the adult body. The lobster, for example, molts seven times during its first summer; after each molt it gets larger and more closely resembles the adult. After it becomes a small adult, additional molts allow for further growth.

Barnacles, the only sessile crustaceans, differ markedly in their external anatomy from other members of the subphylum. They are marine suspension feeders that secrete complex limestone cups within which they live (Fig. 29–18*a*). The larvae of barnacles are free-swimming forms that go through several

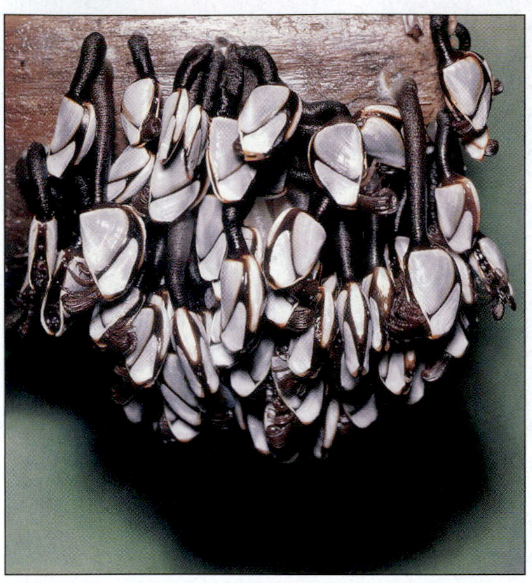

(a)

(b)

(c)

Figure 29–18 Subphylum Crustacea. **(a)** Goose barnacles *(Lepas fascicularis)* hanging from driftwood. These stalked barnacles occur in large numbers on intertidal rocks along the West Coast of the United States. **(b)** Antarctic krill *(Euphausia superba)* are shrimplike crustaceans that are an important component of marine plankton. **(c)** Broken-back shrimp (*Heptocarpus* sp.) from Monterey Bay. *(a, Hal Harrison/Grant Heilman; b, Flip Nicklin/Minden Pictures; c, Frans Lanting/Minden Pictures)*

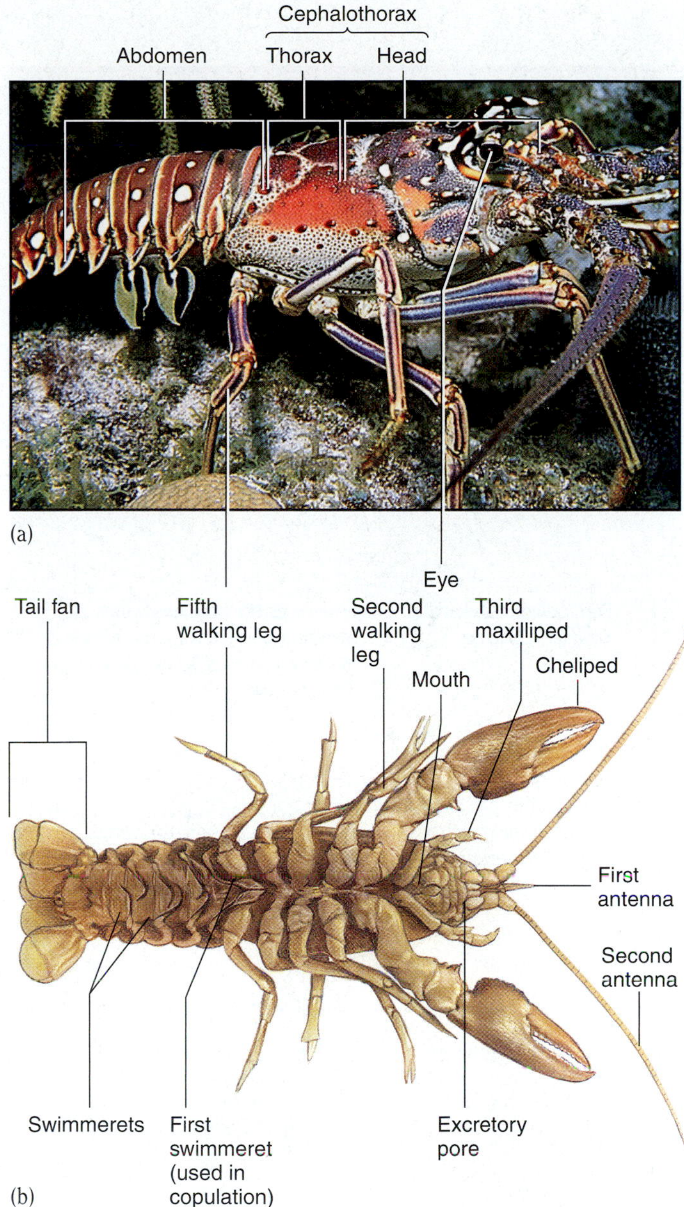

Figure 29–19 Anatomy of the lobster. **(a)** Like other decapods, the spiny lobster *(Panulirus argus)* has five pairs of walking legs. The first pair of walking legs is modified as chelipeds *(large claws)*. **(b)** Ventral view of a lobster. Note the variety of specialized appendages. *(a, Larry Lipsky/Tom Stack & Associates)*

molts. They eventually become sessile and develop into the adult form. The barnacle was described by the 19th century naturalist Louis Agassiz as "nothing more than a little shrimplike animal standing on its head in a limestone house and kicking food into its mouth." The bane of marine boaters, barnacles can proliferate on ship bottoms in great numbers. Their presence may reduce the speed of a ship by more than 30%.

Isopods are mainly tiny (5–15 mm in length) marine crustaceans that inhabit the ocean floor. However, this group includes some familiar terrestrial animals: pillbugs and sowbugs. The mainly microscopic *copepods* are the largest group of crustaceans. Marine copepods are the most numerous component of zooplankton.

The largest order of crustaceans, *Decapoda,* contains more than 10,000 species of lobsters, crayfish, crabs, and shrimp. Most decapods are marine, but a few, such as the crayfish, certain shrimp, and a few crabs, live in fresh water. The crustaceans in general and the decapods in particular show striking specialization and differentiation of parts in the various regions of the animal. In the lobster, the appendages in the different parts of the body differ markedly in form and function (Fig. 29–19).

The five segments of the lobster's head and the eight segments of its thorax are fused into a cephalothorax, which is covered on the top and sides by a shield, the **carapace,** composed of chitin impregnated with calcium salts. The two pairs of antennae serve as chemoreceptors and tactile sense organs. The mandibles are short and heavy, with opposing surfaces used in grinding and biting food. Behind the mandibles are two pairs of accessory feeding appendages: the first and second maxillae. The appendages of the first three segments of the thorax are the maxillipeds, which aid in chopping up food and passing it to the mouth. The fourth segment of the thorax has a pair of large **chelipeds,** or pinching claws. The last four thoracic segments bear walking legs.

The appendages of the first abdominal segment are part of the reproductive system and function in the male as sperm-transferring structures. The following four abdominal segments bear paired **swimmerets,** small paddle-like structures used by some decapods for swimming and by the females of all species for holding eggs. Each branch of the sixth abdominal appendages (called *uropods*) consists of a large flattened structure. Together with the flattened posterior end of the abdomen (the *telson*), they form a fan-shaped structure used for swimming backward.

Subphylum Uniramia includes the insects, centipedes, and millipedes

The insects, centipedes, and millipedes are grouped together in subphylum Uniramia because they all have uniramous (unbranched) appendages, similar mouthparts, and a single pair of antennae (in contrast to two pairs in crustaceans).

Class Insecta is the most successful group of animals

With more than 1 million described species (and perhaps millions more not yet identified), class **Insecta** is the most successful group of animals on our planet in terms of diversity, geographic distribution, number of species, and number of individuals[1] (Table 29–5). More species of insects have been identified than of all other classes of animals combined. What they lack in size, insects make up in sheer numbers. If we could weigh all the insects in the world, their weight would exceed that of all the remaining terrestrial animals. Although primarily terrestrial, some insect species live in fresh water, a few are truly marine, and others inhabit the shore between the tides.

The earliest fossil insects—primitive, wingless species—date back to the Devonian period more than 360 million years ago (mya). Insect fossils from the Carboniferous period more

(text continues on page 638)

[1] The nematodes rival the insects in number of individuals.

TABLE 29–5 Some Orders of Insects

Order, Number of Species, and Examples	Representative Adult		Some Characteristics
Insects with No Metamorphosis (egg → immature form → adult)			
Thysanura (320) Silverfish		Silverfish *Lepisma saccharina*	No wings; biting-chewing mouthparts; 2–3 "tails" extend from posterior tip of abdomen; inhabit dead leaves; eat starch in books
Insects with Incomplete Metamorphosis (egg → larva → adult)			
Odonata (5000) Dragonflies Damselflies		Damselfly *Ischnura* sp.	Two pairs of long, membranous wings; chewing mouthparts; large, compound eyes; active predators; larvae very different from adult
Orthoptera (20,000) Grasshoppers Crickets		Fork-tailed bush katydid *Scudderia furcata*	Forewings leathery, hindwings membranous; chewing mouthparts; most herbivorous, some cause crop damage; some predatory
Blattaria (3700) Cockroaches		American cockroach *Periplaneta americana*	When wings present, forewings leathery, hindwings membranous; chewing mouthparts; legs adapted for running
Isoptera (2000) Termites		Eastern subterranean termite *Reticulitermes flavipes*	2 pairs of wings, or none; wings shed by sexual forms after mating; chewing mouthparts; social insects, form large colonies; eat wood
Anoplura (500) Sucking lice		Human head and body louse *Pediculus humanus*	No wings; piercing-sucking mouthparts; ectoparasites of mammals; head and body louse and crab louse are human parasites; vectors of typhus fever
Hemiptera (true bugs) (35,000) Chinch bugs Bed bugs Water striders		Chinch bug *Blissus leucopterus*	Hindwings membranous; forewings smaller; piercing-sucking mouthparts form beak; most herbivorous; some parasitic

TABLE 29–5 Continued

Order, Number of Species, and Examples	Representative Adult		Some Characteristics
Homoptera (33,000) Aphids Leafhoppers Cicadas Scale insects		Buffalo treehopper *Stictocephala bubalus*	2 pairs membranous wings; piercing-sucking mouthparts form beak; parasites of plants; vectors of plant diseases
Insects with Complete Metamorphosis (egg → larva → pupa → adult)			
Lepidoptera (120,000) Moths Butterflies		Luna moth *Aetias luna*	Usually 2 pairs of membranous, colorful, scaled wings; sucking mouthparts; larvae are wormlike caterpillars that eat plants; adults suck flower nectar; important pollinators
Diptera (150,000) Houseflies Mosquitos		Deerfly *Chrysops vittatus*	Only forewings functional in flying; hindwings small, knoblike halteres; mouthparts usually adapted for sucking (and piercing in some); larvae are maggots or wrigglers and may damage domestic animals or food; adults may transmit disease such as sleeping sickness, yellow fever, or malaria
Siphonaptera (1750) Fleas		Dog flea *Ctenocephalides canis*	No wings; piercing-sucking mouthparts; legs adapted for clinging and jumping; parasites on birds and mammals; vectors of bubonic plague and typhus
Coleoptera (300,000) Beetles Weevils		Colorado potato beetle *Leptinotarsa decemlineata*	Forewings modified as protective coverings for membranous hindwings (which are sometimes absent); chewing mouthparts; largest order of insects; most herbivorous; some aquatic
Hymenoptera (130,000) Ants Bees Wasps		Bald-faced hornet *Vespula maculata*	Usually 2 pairs of membranous wings; mouthparts may be modified for sucking or lapping nectar; many are social insects; some sting; important pollinators

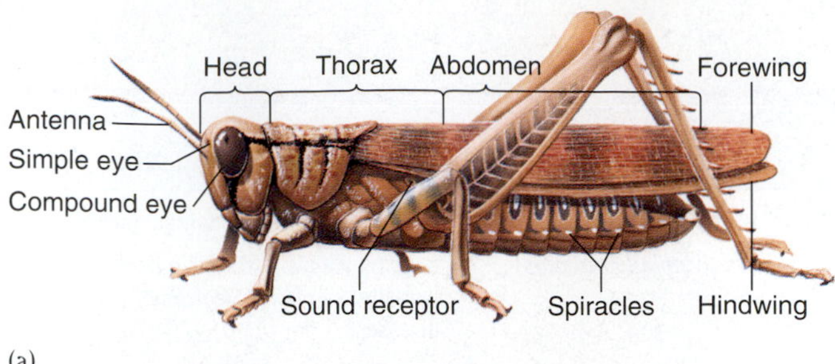

(a)

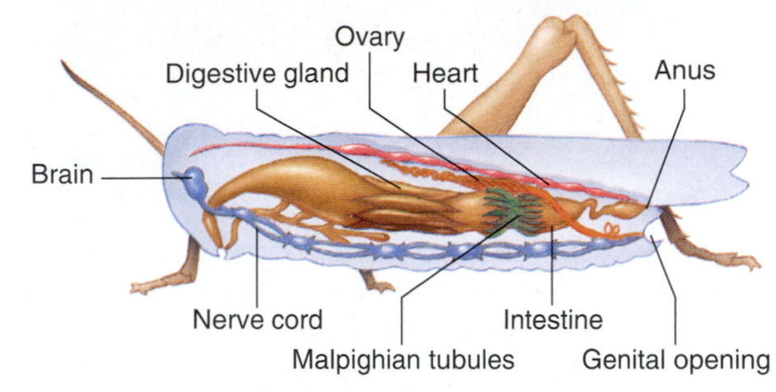

(b)

Figure 29–20 Anatomy of the grasshopper. **(a)** External structure. Note the three pairs of segmented legs. **(b)** Internal anatomy.

than 286 mya include both wingless and primitive winged species. Cockroaches, mayflies, and cicadas are among the insects that have survived relatively unchanged from the Carboniferous period to the present day.

We can describe an insect as an **articulated** (jointed), **tracheated** (having tracheal tubes for gas exchange) **hexapod** (having six feet). The insect body consists of three distinct parts: the head, thorax, and abdomen (Fig. 29–20). Three pairs of legs emerge from the adult thorax, and insects in many orders have one or two pairs of wings. One pair of antennae protrudes from the head, and the sense organs include both simple and compound eyes. Their complex mouthparts may be adapted for piercing, chewing, sucking, or lapping.

The tracheal system of insects has made an important contribution to their diversity. Air enters the tracheal tubes (also called tracheae) through **spiracles,** tiny openings in the body wall. Oxygen passes directly to the internal organs. This effective oxygen delivery system permits insects to have the high metabolic rate necessary for activities such as flight, even though they have an open circulatory system.

Excretion is accomplished by two or more **Malpighian tubules,** which receive metabolic wastes from the blood, concentrate the wastes, and discharge them into the intestine. Unique to terrestrial arthropods, Malpighian tubules also perform the extremely important function of conserving water (discussed further in Chapter 46).

The sexes are separate, and fertilization takes place internally. Several molts occur during development. The immature stages between molts are called **instars.** Primitive insects with no wings, such as silverfish, have direct, or simple, development: The young hatch as juveniles that resemble the adult form. Other insects, such as grasshoppers and cockroaches, undergo **incomplete metamorphosis** in which the egg gives rise to a larva that resembles the adult in many ways but lacks functional wings and reproductive structures. The larva goes through a series of molts during which it becomes more like the adult.

Most insects, including bees, butterflies, and fleas, undergo **complete metamorphosis** with four distinct stages in the life cycle: **egg, larva, pupa,** and **adult.** The wormlike larva does not look at all like the adult. For example, caterpillars have an entirely different body form from butterflies. Typically, an insect spends most of its life as a larva. Eventually the larva stops feeding, molts, and enters a pupal stage, usually within a protective cocoon or underground burrow. The pupa does not feed and typically cannot defend itself. Energy reserves stored during the larval stage are spent remodeling its body. When it emerges as an adult, it is equipped with functional wings and reproductive organs.

Certain species of bees, ants, and termites exist as colonies or societies made up of several different types of individuals, each adapted for some particular function. The members of some insect societies communicate with each other by "dances" and by means of **pheromones,** substances secreted to the external environment. Social insects and their communication are discussed in Chapter 50.

Many adaptations contribute to the biological success of insects

What are the secrets of insect success? One important feature is the tough exoskeleton, which protects insects from predators and helps prevent water loss. At the same time, the segmented exoskeleton permits mobility and flexibility. Segmentation permits regional specialization of the body. For example, cephalization and highly developed sense organs are important in both offense and defense. The insect body plan has been modified and specialized in so many ways that insects are adapted to a remarkable number of lifestyles. The jointed appendages are specialized for various types of feeding, sensory functions, and different types of locomotion (walking, jumping, swimming). For example, grasshopper mouthparts are adapted for biting and chewing leaves. Moth and butterfly mouthparts are adapted for sucking nectar from flowers, and mosquito mouthparts are adapted for sucking blood. Aphids and leafhoppers have mouthparts specialized to pierce plants and feed on plant juices.

Another reason for insect success is the ability to fly. Unlike other invertebrates, which creep slowly along on (or under) the ground, many insects fly rapidly through the air. Their wings and small size facilitate their wide distribution and, in many instances, their immediate survival. For example, when a pond dries up, adult aquatic insects can fly to another habitat.

The reproductive capacity of insects is amazing. Under ideal conditions, the fruit fly *Drosophila* can produce 25 generations in one year! Insect eggs are protected by a thick membrane, and several eggs may, in addition, be enclosed in a protective egg case. Typically, 50 to 100 eggs are laid at a time. Metamorphosis divides the insect life cycle into different stages. This strategy reduces intraspecific competition by placing larval forms in different lifestyles so that they do not compete with adults for food or habitats.

Insects have many interesting adaptations for offense and defense. We all are familiar with the stingers of bees and wasps, which are specialized egg-laying structures (ovipositors). Many insects have **cryptic coloration,** blending beautifully with the background of their habitat; some look like dead twigs or leaves. Others mimic poisonous insects, deriving protection from this resemblance. Still others "play dead" by remaining motionless.

Insects communicate by tactile, auditory, visual, or chemical signals. For example, certain ant species use pheromones to mark trails and to warn of danger. Other insects use pheromones to attract a mate.

Insects have a great impact on humans

Not all insects compete with us for food or cause us to scratch, swell up, or recoil from their presence. Bees, wasps, beetles, and many other insects pollinate flowers of crops and fruit trees. Some insects destroy other insects that are harmful. For example, dragonflies eat mosquitos. And some organic farmers and home gardeners purchase ladybird beetles, adept at ridding plants of aphids and other insect pests. Insects are important members of many food webs. Many birds, mammals, amphibians, reptiles, and some fishes depend on insects for food. Many beetles and the larvae (maggots) of flies are detritus feeders—they break down dead plants and animals and their wastes, permitting nutrients to be recycled.

Many insect products are useful to us. Bees produce honey as well as beeswax, which is used in making candles, lubricants, and other products. Shellac is made from lac, a substance given off by certain scale insects that feed on the sap of trees. And the labor of silkworms provides us with beautiful fabric.

On the negative side, billions of dollars worth of crops are destroyed each year by insect pests. Whole buildings may be destroyed by termites, and clothing can be damaged by moths. Fire ants not only inflict painful stings but cause farmers serious economic loss because of their large mounds, which damage mowers and other farm equipment. Ant mounds also reduce grazing land because livestock quickly learn to avoid them.

Blood-sucking flies, screw worms, lice, fleas, and other insects annoy and transmit disease in both humans and domestic animals. Mosquitos are vectors of malaria (see Fig. 24–8) and yellow fever. Body lice can transmit the typhus rickettsia, and houseflies sometimes transmit typhoid fever and dysentery. Tsetse flies transmit African sleeping sickness, and fleas may be vectors of bubonic plague.

Classes Chilopoda and Diplopoda include centipedes and millipedes

Members of class **Chilopoda** are called *centipedes* ("hundred-legged"), and members of class **Diplopoda** are known as *millipedes* ("thousand-legged"). These animals are all terrestrial and are typically found beneath stones or wood in the soil in both temperate and tropical regions.

(a)

(b)

Figure 29–21 Chilopods and diplopods.
(a) Centipede *(Lithobius sp.)*, a member of class Chilopoda. Centipedes have one pair of uniramous appendages per segment. **(b)** Millipede *(Diplopoda pachydesmus)*, a member of the class Diplopoda. Millipedes have two pairs of uniramous appendages per segment. *(a, Dwight Kuhn; b, John R. MacGregor/Peter Arnold, Inc.)*

Centipedes and millipedes are similar in having a head and an elongated trunk with many segments, each bearing uniramous legs (Fig. 29–21). The centipedes have one pair of legs on each segment behind the head. Most centipedes do not have enough legs to merit their name, the most common number is 30 or so, although a few species have 100 or more. The legs of centipedes are long, enabling them to run rapidly. Centipedes are carnivorous and feed on other animals, mostly insects. Larger centipedes will eat snakes, mice, and frogs. The prey is captured and killed with poison claws located just behind the head on the first trunk segment.

Millipedes have two pairs of legs on most body segments. Diplopods are not as agile as chilopods, and most species can crawl only slowly over the ground, although they can powerfully force their way through earth and rotting wood. Millipedes are generally herbivorous and feed on both living and dead vegetation.

The main phyla discussed in this chapter are compared in Table 29–6.

TABLE 29–6 Comparison of Characteristics of Several Protostome Coelomate Phyla*

	Nemertea (ribbon worms) (900 species)	Mollusca (50,000 species)	Annelida (segmented worms) (15,000 species)	Nematoda (roundworms) (>15,000 species)	Arthropoda (joint-footed animals) (>1 million species)
Representative Animals	Proboscis worms	Clams, snails, squids	Earthworms, leeches, marine worms	Ascarids, hookworms, nematodes	Crustaceans, insects, arachnids
Circulation	Pulsating longitudinal blood vessels; no heart	Open system (closed in cephalopods)	Closed system	Diffusion	Open system
Gas Exchange	Diffusion	Gills and diffusion through mantle	Diffusion through moist skin; oxygen circulated by blood	Diffusion	Tracheae in insects; gills in crustaceans; book lungs or tracheae in arachnids
Waste Disposal	Lateral excretory canals with flame cells	Metanephridia	Pair of metanephridia in each segment	Excretory canals; most with unique excretory cells	Malphighian tubules in insects; antennal (green) glands in crustaceans
Nervous System	Simple brain; nerve cords; cross nerves; simple sense organs	Simple brain (cerebral ganglia); simple sense organs	Simple brain; ventral nerve cord; simple sense organs	Simple brain; dorsal and ventral nerve cords; simple sense organs	Brain; ventral nerve cord; well-developed sense organs
Reproduction	Asexual, by fragmentation; sexual, sexes separate	Sexual	Sexual	Sexual, sexes separate	Sexual, sexes separate
Support and Movement	Support by its tissues; locomotion by muscles or cilia	Most have hydrostatic skelton; most have shell and ventral foot for locomotion	Fluid-filled coelom serves as hydrostatic skeleton; well-developed muscles in body wall	Support by tough cuticle; fluid in pseudocoelom serves as hydrostatic skeleton; longitudinal muscles in body wall	Tough exoskeleton; jointed appendages (some have wings); well-developed muscles
Environment and Lifestyle	Mainly marine; mainly carnivores; use proboscis for capturing food and defense	Mainly marine, some inhabit fresh water or are terrestrial; herbivores, carnivores, scavengers, or suspension feeders	Marine, freshwater, terrestrial; herbivores, carnivores, scavengers, suspension feeders	Widely distributed in the soil, salt water, and fresh water; carnivores, scavengers, parasites	Most diverse group in habitat and lifestyle; marine, freshwater, and terrestrial; herbivores, carnivores, scavengers

*Members of these phyla are at the organ system level of organization, have bilateral symmetry, and have a complete digestive tract.

SUMMARY WITH KEY TERMS

I. The evolution of the **coelom** permitted many innovations, including the **tube-within-a-tube body plan** and the **hydrostatic skeleton.** The coelom provides space for internal organs and for gonads to develop. It can also help transport materials, and it protects internal organs.

II. The **protostome coelomates** include two main branches: the **Lophotrochozoa** and the **Ecdysozoa.** The Lophotrochozoa include the nemerteans, mollusks, annelids, and the lophophorate phyla, groups that have a ciliated ring of tentacles surrounding the mouth. The Ecdysozoa include the rotifers, nematodes (roundworms), onychophorans, and arthropods.

III. Members of phylum **Nemertea** (ribbon worms) have a tube-within-a-tube body plan, a complete digestive tract with mouth and anus, and a separate circulatory system. The **proboscis** is a muscular tube used in capturing food and in defense. The **rhynchocoel** is a coelomic space surrounding the proboscis.

IV. Members of phylum **Mollusca** are soft-bodied animals usually covered by a shell.
  A. **Mollusks** have a ventral **foot** for locomotion and a **mantle** that covers the **visceral mass,** a concentration of body organs.
  B. Mollusks have an **open circulatory system** with the exception of cephalopods, which have a **closed circulatory system.** Mollusks have paired excretory tubules called **metanephridia.** A rasplike **radula** functions as a scraper in feeding in all groups except the bivalves, which are suspension feeders.
  C. Typically, marine mollusks have a **trochophore larva** stage; in many marine gastropods and bivalves, the trochophore larva develops into a **veliger larva.**
  D. Class **Polyplacophora** includes the sluggish marine **chitons,** which have shells that consist of eight overlapping plates.
  E. Class **Gastropoda,** the largest and most successful group of mollusks, includes the snails, slugs, and their relatives. In gastropods, the body undergoes **torsion,** a twisting of the visceral mass. The shell (when present) is coiled.
  F. Class **Bivalvia** includes the aquatic clams, scallops, and oysters; the bodies of these suspension feeders are enclosed by a two-part shell that is hinged dorsally.
  G. Class **Cephalopoda** includes the squids, octopods, and *Nautilus*. Cephalopods are active, predatory swimmers. Tentacles surround the mouth, located in the large head.

V. Phylum **Annelida,** the segmented worms, includes many aquatic worms, earthworms, and leeches. Annelids have conspicuously long bodies with **segmentation** both internally and externally; their large compartmentalized coelom serves as a hydrostatic skeleton.
  A. Class **Polychaeta** consists of marine worms characterized by bristled **parapodia,** used for locomotion. The parapodia bear many **setae.**
  B. Class **Oligochaeta,** the earthworms, is characterized by a few short setae per segment. The body is divided into more than 100 segments separated internally by **septa.**
  C. Class **Hirudinea,** the leeches, is characterized by the absence of setae and appendages. Parasitic leeches are equipped with suckers for holding onto their host.

VI. The **lophophorate phyla** include the **brachiopods, phoronids,** and **bryozoa.** Lophophorates are marine animals that have a lophophore, a ciliated ring of tentacles surrounding the mouth. The **lophophore** is specialized for capturing suspended particles in the water.

VII. Members of phylum **Rotifera** (wheel animals) are characterized by a crown of cilia. Rotifers are aquatic, microscopic, pseudocoelomate animals that exhibit cell constancy.

VIII. Phylum **Nematoda,** the **roundworms,** includes species of great ecological importance. Some species are parasitic in plants or animals.
  A. Nematodes are pseudocoelomates with bilateral symmetry, three tissue layers, and a complete digestive tract.
  B. The nematode body is covered by a tough **cuticle** that helps prevent desiccation.
  C. Parasitic nematodes in humans include ***Ascaris,*** **hookworms, trichina worms,** and **pinworms.**

IX. Phylum **Arthropoda** is composed of segmented animals with **paired, jointed appendages** and an armor-like **exoskeleton** of chitin. **Molting** is necessary for the arthropod to grow.
  A. Arthropods have an open circulatory system with a dorsal heart that pumps **hemolymph.**
  B. Aquatic forms have gills; terrestrial forms have either **tracheae** or **book lungs.**
  C. Many biologists consider the arthropods a monophyletic group, but some argue that they are polyphyletic. Some biologists have suggested that phylum **Onychophora** may be a link between annelids and arthropods, but molecular data do not support that hypothesis. The onychophorans, along with the **tardigrades,** appear to be closely related to the arthropods.
  D. The **trilobites** are extinct marine arthropods covered by a hard, segmented shell.
  E. Subphylum **Chelicerata** includes the merostomes (horseshoe crabs) and the **arachnids** (spiders, mites, and their relatives).
    1. The chelicerate body consists of a cephalothorax and abdomen; there are six pairs of jointed appendages, of which four pairs serve as legs.
    2. The first pair of appendages are **chelicerae,** the second pair are **pedipalps.** These appendages may be adapted for manipulation of food, locomotion, defense, or copulation. Chelicerates have no antennae and no mandibles.
  F. Subphylum **Crustacea** includes lobsters, crabs, shrimp, pillbugs, and barnacles.
    1. The crustacean body consists of a cephalothorax and abdomen; typically, five pairs of walking legs are present.
    2. Crustaceans have two pairs of **antennae** that sense taste and touch. The third appendages are **mandibles** used for chewing. Two pairs of **maxillae,** posterior to the mandibles, are used for manipulating and holding food.
  G. Subphylum **Uniramia** includes class **Insecta,** class **Chilopoda,** and class **Diplopoda;** members of this subphylum have unbranched appendages and a single pair of antennae.
    1. An insect (class Insecta) is an **articulated, tracheated hexapod;** its body consists of head, thorax, and abdomen.
    2. Insects have a system of tracheae for gas exchange and **Malpighian tubules** for excretion.
    3. The biological success of the insects results from many adaptations, including the versatile exoskeleton, segmentation, specialized jointed appendages, ability to fly, highly developed sense organs, **metamorphosis** (which reduces intraspecific competition), effective reproductive strategies, effective mechanisms for defense and offense, and the ability to communicate.
    4. The centipedes (class Chilopoda) have one pair of legs per body segment, whereas the millipedes (class Diplopoda) have two pairs of legs per body segment. Centipedes are carnivorous, whereas millipedes are typically herbivorous.

POST-TEST

1. Which of the following is associated with the evolution of the coelom? (a) radial symmetry (b) tube-within-a-tube body plan (c) incomplete metamorphosis (d) bilateral symmetry (e) development of ganglia
2. The hydrostatic skeleton (a) results in torsion (b) permits contracting muscles to recover quickly (c) permits a greater range of movement than possible in acoelomate animals (d) is a unique characteristic of annelids (e) permits ecdysis
3. Which of the following is *not* characteristic of nemerteans (proboscis worms)? (a) large coelom (b) tube-within-a-tube body plan (c) complete digestive tube (d) muscular tube for capturing food (e) circulatory system
4. Which of the following characteristics is associated with phylum Mollusca? (a) mandibles (b) mantle (c) pedipalps (d) chelipeds (e) setae
5. Which of the following belong to phylum Mollusca? (a) gastropods and crustaceans (b) polychaetes (c) chelicerates and bryozoans (d) crustaceans (e) gastropods and cephalopods
6. Which of the following belong to phylum Annelida? (a) gastropods (b) polychaetes (c) uniramians (d) nudibranchs (e) cephalopods

7. Which of the following is *not* a nematode? (a) hookworm (b) trichina worm (c) *Ascaris* (d) leech (e) pinworm
8. Which of the following is *not* true of the lophophorates? (a) this group includes the bryozoa (b) their evolutionary position is controversial (c) characterized by a ciliated ring of tentacles at anterior end (d) mainly marine animals (e) articulated, tracheated hexapods
9. In mollusks the first larval stage is typically a (a) pupa (b) veliger larva (c) trochophore larva (d) chiton (e) a miniature copy of the adult
10. Torsion in mollusks (a) is characteristic of bivalves (b) is a twisting of the visceral mass (c) involves coiling of the molluscan shell (d) begins in the adult stage (e) depends on action of parapodia
11. Which of the following is characteristic of members of phylum Annelida? (a) open circulatory system (b) hemocoel (c) segmentation (d) mantle (e) mandibles
12. Trilobites (a) were early mollusks (b) are members of phylum Onychophora (c) are characterized by parapodia and setae (d) were early arthropods (e) are an evolutionary link between annelids and arthropods
13. Which of the following belong to subphylum Uniramia? (a) insects (b) polychaetes (c) crustaceans (d) nudibranchs (e) chelicerates
14. The correct sequence in insect complete metamorphosis is (a) egg → immature form → adult (b) egg → trochophore larva → veliger larva → adult (c) egg → pupa → larva → adult (d) egg → larva → pupa → adult (e) adult → larva → egg → pupa
15. Which of the following is characteristic of insects? (a) gills (b) six legs (c) metanephridia (d) eight legs (e) two pairs of antennae
16. Which of the following is *not* characteristic of arthropods? (a) exoskeleton (b) veliger larva (c) paired, jointed appendages (d) chitin in exoskeleton (e) segmentation
17. Biramous appendages are characteristic of (a) crustaceans (b) insects (c) centipedes (d) millipedes (e) spiders
18. Which of the following is *not* associated with insect success? (a) exoskeleton (b) metamorphosis (c) ability to fly (d) sense organs (e) radula
19. Spiders are likely to have (a) mandibles and maxillae (b) six pairs of legs on the abdomen (c) one pair of antennae (d) biramous appendages (e) chelicerae and pedipalps

REVIEW QUESTIONS

1. Identify characteristics that mollusks and annelids share. In what ways are these animals different?
2. Give two distinguishing characteristics for each of the following: (a) mollusks (b) annelids (c) roundworms (d) arthropods
3. What are the advantages of each of the following? (a) presence of a coelom (b) the arthropod exoskeleton (c) metamorphosis
4. Contrast the lifestyles of a gastropod and a cephalopod. Identify adaptations that have evolved in each for its particular lifestyle.
5. Describe (a) a trochophore larva (b) a veliger larva (c) a larva of an insect that undergoes complete metamorphosis
6. Describe five adaptations that have contributed to insect success.
7. Distinguish between insects and arachnids.
8. What are the distinguishing features of each of the arthropod subphyla? Identify animals that belong to each group.
9. Compare complete and incomplete metamorphosis in insects.

YOU MAKE THE CONNECTION

1. Discuss the idea that every evolutionary adaptation has both advantages and disadvantages, using each of the following characteristics as an example: (a) the arthropod exoskeleton (b) segmentation with specialization (c) closed circulatory system
2. Hypothesize why oysters secrete calcium carbonate layers around foreign particles.
3. Insects that undergo complete metamorphosis outnumber those that do not by more than ten to one. Hypothesize an explanation.

RECOMMENDED READINGS

Berenbaum, M.R. *Bugs in the System: Insects and Their Impact on Human Affairs.* Addison-Wesley, Reading, MA, 1995. A readable book on insects that describes their life histories, physiology, behavior, and relationships with humans.

Chadwick, D.H. "Planet of the Beetles." *National Geographic,* Vol. 193, No. 3, Mar. 1998. This article highlights the remarkable diversity of beetles, which make up about 25% of all known animal species.

Milius, S. "Cicada Subtleties." *Science News,* Vol. 157, 24 Jun. 2000. An account of current research on these interesting insects.

Milius, S. "Why Fly into a Forest Fire?" *Science News,* Vol. 159, 3 Mar 2001. More than a dozen species of insects have been identified that rush into a fire, mate, and lay eggs in smoking trees.

Morell, V. "Life on a Grain of Sand." *Discover,* Apr. 1995. An interesting look at some unusual marine animals.

Peterson, I. "Calculating Swarms." *Science News,* Vol. 158, 11 Nov. 2000. The foraging behavior of ants suggests models for sophisticated robots and for finding traffic routes over congested telephone lines.

Natural History, Vol. 104, No. 3, Mar. 1995. This issue features an interesting collection of articles devoted to spiders.

- Visit our Web site at **http://www.info.brookscole.com/solomonbergmartin** for links to chapter-related resources on the World Wide Web. Additional on-line materials relating to this chapter can also be found on our Web site.

See chapter activity on BioActive Learner CD for additional help in mastering the chapter's material. Icon location in the chapter's margins shows which topics have tutorials or simulations in the CD.

30

The Animal Kingdom: The Deuterostomes

Representative deuterostomes This marine habitat photographed in Hawaii includes an echinoderm, known as a pencil urchin *(Heterocentrus mammillatus)*, and chordates, which are represented by fishes, the clown wrasse *(Coris gaimard)* with its yellow tail and the Moorish idol *(Zanchus cornutus)*. *(Ed Robinson/Tom Stack & Associates)*

LEARNING OBJECTIVES

After you have studied this chapter you should be able to

1. Discuss the relationship of the echinoderms, hemichordates, and chordates, describing shared derived characters that support grouping them together as deuterostomes.
2. Describe shared derived characters of the echinoderms.
3. Describe and give examples of each of the five main classes of echinoderms.
4. Trace the evolution of chordates and describe their shared derived characters.
5. Distinguish among the subphyla of phylum Chordata.
6. Trace the evolution of vertebrates according to current hypotheses, describe the shared derived characters of vertebrates, and distinguish among major taxa.
7. Identify adaptations that terrestrial vertebrates have made to life on land.
8. Contrast monotremes, Metatheria (marsupials), and Eutheria (placental mammals), and give examples of animals that belong to each group.
9. Identify the major orders of eutherian mammals, and give examples of animals that belong to each order.

What does a sea star have in common with a fish, frog, hawk, or human? It may seem strange to group the **echinoderms**—the sea stars, sea urchins, and sand dollars—with the **chordates,** the phylum to which humans and other **vertebrates** (animals with a backbone) belong. However, even though these animals look and behave very differently, evidence suggests that echinoderms and chordates share a common ancestor and are closely related. They are both **deuterostomes,** animals that make up the second main evolutionary branch of the animal kingdom. Biologists also include the **hemichordates,** a small group of wormlike marine animals, with the deuterostomes.

Deuterostomes share several derived characters (traits found in two or more taxa that first appeared in their most recent common ancestor; see Chapter 22) involving similarities in their patterns of development (see Chapter 28). They are characterized by radial, rather than spiral, cleavage. Their cleavage is indeterminate, which means that the fate of their cells is fixed later in development than is the case in protostomes. In less complex deuterostomes, the mesoderm develops from paired pouches that pocket out from the embryonic gut. The coelom forms from cavities within the mesodermal out-pocketings. The blastopore becomes the anus (or is located near the future site of the anus), and the mouth develops from a second opening at the anterior end of the embryo.

In this chapter we focus on echinoderms, such as the pencil urchin, and chordates, represented in the photograph by the fishes. The largest chordate subphylum is Vertebrata, animals with backbones (vertebral columns). The vertebrates are the animals with which we are most familiar—fishes, amphibians, reptiles, birds, and mammals. Fossil evidence of the first chordates, and perhaps also the first vertebrates, has been found in rocks from the Cambrian period more than 550 million years old.

Just how the **protochordate** (invertebrate chordates) and vertebrate groups are related is the focus of active research. Many investigators are comparing *Hox* genes in various groups. Recall that *Hox* genes are important in determining pattern development in embryos (see Chapter 16). Specifically, they specify the fate of cells in the anterior-posterior axis. These genes occur in clusters on specific chromosomes. Chordates that evolved earlier have one *Hox* cluster, whereas amphibians, reptiles, birds, and mammals have four clusters. Investigators have suggested that duplication of *Hox* clusters may have contributed to the evolution of vertebrate body plans. Interestingly, the zebrafish and its close relatives have seven *Hox* clusters. We discuss the evolutionary implications of this fact later in the chapter.

ECHINODERMS ARE CHARACTERIZED BY THEIR WATER VASCULAR SYSTEM

All of the members of phylum **Echinodermata** inhabit marine environments. They are found in the ocean at all depths. About 7000 living and more than 13,000 extinct species have been identified. The echinoderms probably evolved during the early Cambrian period and achieved maximum diversity by the middle of the Paleozoic era, about 400 million years ago (mya). By the beginning of the Mesozoic era 248 mya, they had declined, leaving six principal groups that have survived to the present day (Fig. 30–1): class Crinoidea, sea lilies and feather stars; class Asteroidea, sea stars; class Ophiuroidea, basket stars and brittle stars; class Echinoidea, sea urchins and sand dollars; class Holothuroidea, sea cucumbers; and class Concentricycloidea, sea daisies.

The echinoderms are in many ways unique in the animal kingdom. Echinoderm larvae are bilaterally symmetrical, ciliated, and free-swimming. During development the body form reorganizes, and the adult exhibits **pentaradial symmetry,** in which the body is arranged in five parts around a central axis. Echinoderms have a skeleton that consists of $CaCO_3$ plates and spines. This skeleton is an **endoskeleton** (internal skeleton) because it is covered by a thin, ciliated epidermis. This group of animals gets its name *Echinodermata,* meaning "spiny-skinned," from the spines that project outward from the endoskeleton. Some groups have pincer-like, modified spines called **pedicellariae** on the body surface (Fig. 30–2). These structures, unique to the echinoderms, keep the surface of the animal free of debris.

A unique derived character of echinoderms is the hydraulic **water vascular system,** a network of fluid-filled canals that functions in locomotion, feeding, and gas exchange. Branches of the water vascular system lead to numerous tiny **tube feet** that extend when filled with fluid. Each tube foot receives fluid from the main system of canals. A rounded muscular sac, or **ampulla,** at the base of the foot, stores fluid and is used to operate the tube foot. The tube foot is separated from other parts of the system by a valve. When the valve shuts, the ampulla contracts, forcing fluid into the tube foot so that it extends. At the bottom of the foot is a suction-type structure that presses against and adheres to whatever surface the tube foot is on.

Echinoderms have a well-developed coelom, and the coelomic fluid transports materials. Although its structure varies in different groups, the complete digestive system is the most prominent body system. A variety of respiratory structures are found in the various classes. No excretory organs are present. The nervous system is simple, generally consisting of nerve rings with radiating nerves about the mouth. Echinoderms have no brain. The sexes are usually separate, and eggs and sperm are generally released into the water, where fertilization takes place.

Members of class Crinoidea are suspension feeders

Class **Crinoidea,** the oldest class of living echinoderms, includes the feather stars and the sea lilies (see Fig. 30–1*a*). Although a great many extinct crinoids are known, there are relatively few living species. The feather stars are motile crinoids, although they often remain in the same location for long periods. Sea lilies are sessile and remain attached to the ocean floor by a stalk.

Crinoids remove suspended food from the water. In all other echinoderms, the mouth is located on the underside of the disk toward the substratum, but in crinoids the **oral** (mouth) **surface** is on the upper side of the disk. A number of branched, feathery arms also extend upward. Along the feathery arms are numerous tube feet that are shaped like small tentacles and coated with mucus that traps microscopic organisms.

Many members of class Asteroidea capture prey

Sea stars, or starfish, are members of class **Asteroidea** (see Fig. 30–1*b*). Their bodies consist of a central disk from which radiate from 5 to more than 20 arms, or rays (see Fig. 30–2). The undersurface of each arm is equipped with hundreds of pairs of tube feet. The mouth lies in the center of the underside of the disk. The endoskeleton consists of a series of calcareous plates that permit some movement in the arms. Delicate dermal gills, small extensions of the body wall, carry on gas exchange.

Most sea stars are carnivorous predators and scavengers that feed on crustaceans, mollusks, annelids, and even other echinoderms. Occasionally they catch small fish. The sea star's water vascular system does not permit rapid movement, so its prey are usually slow-moving or stationary animals such as clams.

To attack a clam or other bivalve mollusk, the sea star mounts it and assumes a humped position as it straddles the edge opposite the hinge (Fig. 30–3). Then, holding itself in position with its tube feet, the sea star slides its thin, flexible stomach out through its mouth and between the closed, or slightly gaping, valves (shell parts) of the clam. The sea star secretes enzymes that digest the soft parts of the clam to the consistency of a thick soup while the clam is still in its own shell. The partly digested meal passes into the sea star body, where it is further digested by enzymes secreted from glands located in each arm.

The circulatory system in sea stars is poorly developed and probably of little help in circulating materials. Instead, this function is assumed by the coelomic fluid, which fills the large coelom and bathes the internal tissues. Metabolic wastes pass to the outside by diffusion across the tube feet and dermal gills. The nervous system consists of a ring of nervous tissue encircling the mouth and a nerve cord extending from this ring into each arm.

Class Ophiuroidea is the largest class of echinoderms

Basket stars and brittle stars (serpent stars), members of class **Ophiuroidea,** are the largest group of echinoderms, both in number of species and in number of individuals (see Fig. 30–1*c*). These animals resemble sea stars in that their bodies consist of a central disk with arms, but the arms are long and slender and more sharply set off from the central disk. Ophiuroids can move more rapidly than sea stars, using their arms to perform rowing

PROTOSOMES
Lophophorates
Annelida
Mollusca
Nemertea
Arthropoda
Onychophora
Rotifera
Nematoda

DEUTEROSTOMES
Chordata
Hemichordata
Echinodermata

Platyhelminthes
Ctenophora
Cnidaria
Porifera

Flagellate ancestor

Figure 30–1 Echinoderms. **(a)** Feather stars use their slender, jointed appendages to cling to the surface of a rock or coral reef. They are able to creep and often move away to escape predators. **(b)** Orange and red sea star *(Fromia monilis)* on bubble coral. **(c)** Daisy brittle star *(Ophiopholis aculeata)*, photographed in Muscongus Bay, Maine. **(d)** With its flattened, circular body, the sand dollar *(Derdraser excentricus)* is adapted for burrowing on the ocean floor. **(e)** A sea cucumber (*Thelonota* sp.), raises its body to spawn. *(a,d, D. J. Wrobel, Monterey Bay Aquarium/ Biological Photo Service; b, Marc Chamberlain/ Tony Stone Images; c, Robert Dunne/Photo Researchers, Inc.; e, Peter Scoones/Seaphot, Ltd.)*

(a) Crinoidea

(b) Asteroidea

(c) Ophiuroidea

(d) Echinoidea

(e) Holothuroidea

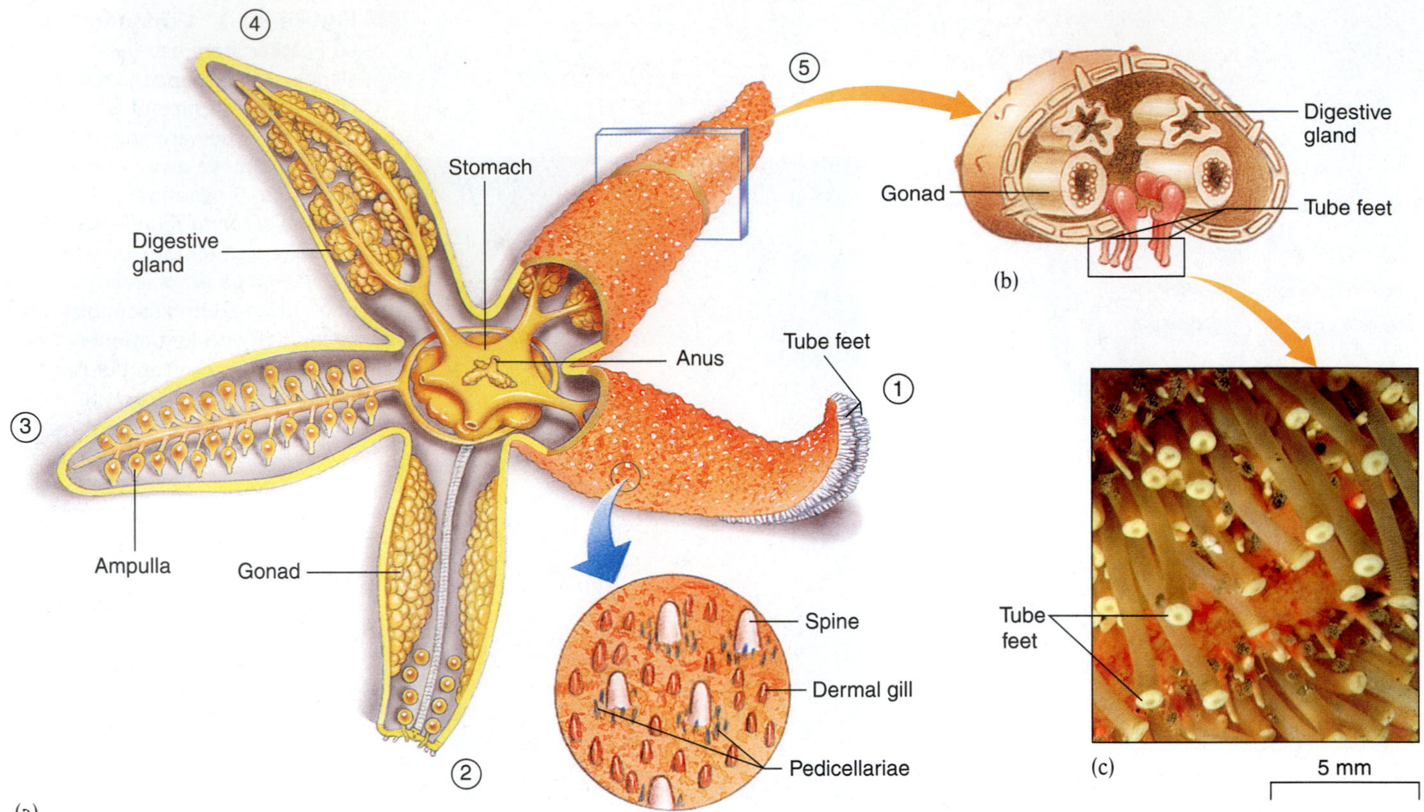

Figure 30–2 Body plan of a sea star. **(a)** A sea star viewed from above, with its arms in various stages of dissection. Similar structures are present in each arm. The two-part stomach is in the central disc with the anus on the aboral *(upper)* surface and the mouth beneath on the oral surface. (1) Upper surface with magnified detail. The end is turned up to show the tube feet on the lower surface. (2) The arm is dissected to show well-developed gonads. (3) Upper body and digestive glands have been removed, exposing the ampullae of some of the hundreds of tube feet *(magnified view)*. (4) Other organs have been removed to show the digestive glands. (5) Upper surface. **(b)** Cross section through arm and tube feet. **(c)** LM of tube feet of a sea star. *(c, Charles Seaborn/Odyssey Productions, Chicago)*

Figure 30–3 Painted sea star *(Orthiasterias koehleri)* attacking a clam. The sea star inserts its thin, everted stomach between the valves of the clam and begins to digest the clam while it is still in its shell. *(Richard Chesher/Seaphot, Ltd.)*

or even swimming movements. Their tube feet lack suckers and are not used in locomotion; they are used to collect and handle food and may also serve a sensory function, perhaps that of taste.

Members of class Echinoidea have moveable spines

Sea urchins and sand dollars, the animals that comprise class **Echinoidea,** have no arms (see Fig. 30–*d* and chapter opening photograph). Their skeletal plates are flattened and fused to form a solid shell called a **test.** The flattened body of the sand dollar is adapted for burrowing in the sand, where it feeds on tiny organic particles. Sand dollars have smaller spines than do sea urchins.

The sea urchin body is covered with spines that in some species can penetrate flesh and are difficult to remove. So threatening are these spines that swimmers on tropical beaches are often cautioned to wear shoes when venturing offshore, where these living pincushions may be found in abundance. Sea urchins use their tube feet for locomotion. They also push themselves along with their moveable spines. Many sea urchins graze on algae, scraping the sea floor with their calcareous teeth.

Members of class Holothuroidea are elongated, sluggish animals

Sea cucumbers, members of class **Holothuroidea,** are appropriately named, for some species are about the size and shape of a cucumber. The elongated sea cucumber body is a flexible, muscular sac (see Fig. 30–1*e*). The mouth is usually surrounded by a circle of tentacles that are modified tube feet. Another characteristic of sea cucumbers is the reduction of the endoskeleton to microscopic plates embedded in the body wall. The circulatory system is more highly developed than that of other echinoderms and functions to transport oxygen and perhaps nutrients as well.

Sea cucumbers are sluggish animals that usually live on the bottom of the sea, sometimes burrowing in the mud. Some graze with their tentacles, whereas others stretch their branched tentacles out in the water and wait for dinner to float by. Algae and other morsels are trapped in mucus along the tentacles.

An interesting habit of some sea cucumbers is evisceration, in which the digestive tract, respiratory structures, and gonads are ejected from the body, usually when environmental conditions are unfavorable. When conditions improve, the lost parts are regenerated. Even more curious is the fact that when certain sea cucumbers are irritated or attacked, they direct their rear end toward the enemy and shoot red tubules out of their anus! These unusual weapons are sticky, and the attacking animal may become hopelessly entangled. Some of these tubules release a toxic substance.

Members of class Concentricycloidea have a unique water vascular system

First discovered in 1983, sea daisies inhabit bacteria-rich wood sunk in deep water. These small (less than 1 cm in diameter), disk-shaped echinoderms appear to have two ring canals with the tube feet projecting from the outer one. Class Concentricycloidea was established to accommodate these interesting echinoderms.

HEMICHORDATES ARE MARINE DEUTEROSTOMES

Animals assigned to phylum **Hemichordata** are bilateral with a characteristic ring of cilia surrounding the mouth. A distinguishing derived character is the three-part body consisting of a proboscis, collar, and trunk. Each part contains a coelomic cavity.

The name "hemichordate" means "half chordate," and these animals do share several, but not all, chordate characteristics. Like chordates, they have pharyngeal gill slits and a dorsal nerve cord. However, molecular data suggest that hemichordates are actually more closely related to the echinoderms than to the chordates. This hypothesis is supported by the similarities in the larvae of some hemichordates and echinoderms.

The most familiar of the two clades of extant hemichordates are the acorn worms, animals that live buried in mud or sand (Fig. 30–4). Although most are a few centimeters in length, members of some species measure more than 2 m. Food particles that are trapped in mucus on the proboscis (feeding tube) are transported to the mouth by cilia.

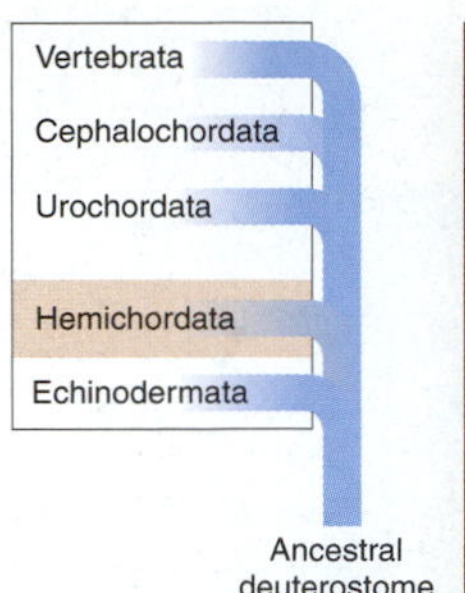

Figure 30–4 Acorn worm *(Saccoglossus kowalevskii).* This small worm (about 15 cm, or 6 in) lives in shallow intertidal mudflats and sand flats along the European and eastern North American coasts. *(C.R. Wyttenbach/Biological Photo Service)*

CHORDATE CHARACTERS INCLUDE A NOTOCHORD, DORSAL TUBULAR NERVE CORD, AND PHARYNGEAL SLITS

Chordata, the phylum of animals to which humans belong, is divided into three subphyla: subphylum **Urochordata,** which consists of marine animals called tunicates; subphylum **Cephalochordata,** composed of marine animals called lancelets; and subphylum **Vertebrata,** animals with backbones (Fig. 30–5).

Chordates are deuterostome coelomates with bilateral symmetry, a tube-within-a-tube body plan, and three well-developed germ layers (Fig. 30–6). Typically, they have an endoskeleton and a closed circulatory system with a ventral heart. Chordates are compared to echinoderms in Table 30–1.

Four derived characters that distinguish the chordates follow:

1. All chordates have a **notochord** during some time in their life cycle. The notochord is a dorsal longitudinal rod that is firm, but flexible, and supports the body.
2. At some time in their life cycle, chordates have a **dorsal tubular nerve cord.** The chordate nerve cord is different from the nerve cord of most other animals in that it is located dorsally rather than ventrally, is hollow rather than solid, and is single rather than double.
3. Chordates have **pharyngeal slits** (also called *pharyngeal gill slits*) during some time in their life cycle. In the embryo, a series of alternating branchial (gill) arches and grooves develop in the body wall in the pharyngeal (throat) region. Pharyngeal pouches extend laterally from the anterior portion of the digestive tract toward the grooves (described in more detail in Chapter 49).

   Early chordates, like many living chordates, were probably suspension feeders. The arrangement of pharyngeal pouches and slits permitted them to take water in through the mouth, concentrate small particles of food in the gut, and let the water escape from the body through the slits. The

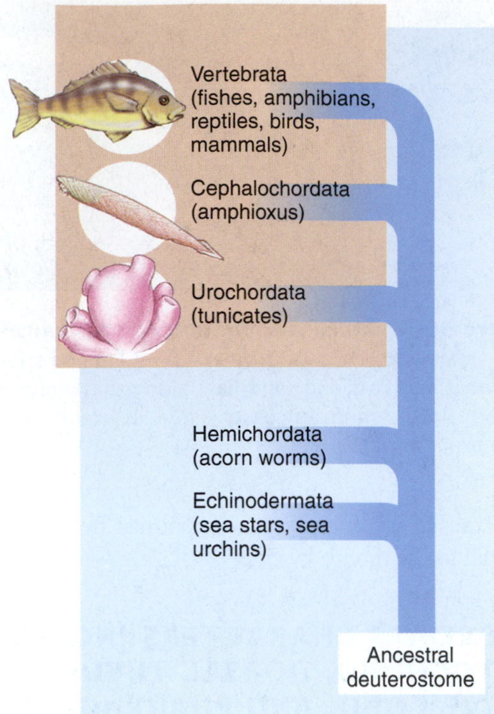

Figure 30–5 Chordate evolution. This diagram shows possible phylogenetic relationships among deuterostomes, including chordates (tan region), based on current data from studies of morphology and DNA.

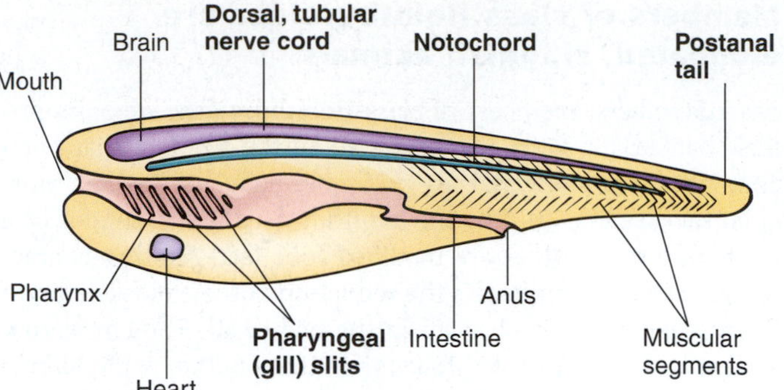

Figure 30–6 Generalized chordate body plan. Note the notochord; dorsal, tubular nerve cord; pharyngeal gill slits; and postanal tail.

pharyngeal slits and the arches supporting them became modified for gas exchange in fishes and some other aquatic vertebrates. Early in vertebrate evolution, anterior pharyngeal arches evolved into jaws.

4. Most chordate species have a larva with a **muscular postanal tail,** an appendage that extends posterior to the anus; other chordate species have embryos with postanal tails.

No clear fossil record of the chordate ancestors exists, but fossil evidence suggests that they were small, soft-bodied animals. A fossil of what is thought to be an early cephalochordate, a lancelet-like animal named ***Pikaia,*** has been found in the Burgess Shale of British Columbia, Canada. These rocks, which date back

TABLE 30–1 Comparison of Echinoderms and Chordates*

	Echinodermata (7000 species)	Chordata (48,000 species)
Representative animals	Sea stars, sea urchins, sand dollars	Tunicates, lancelets, vertebrates
Body symmetry	Embryo: bilateral Adult: pentaradial	Bilateral
Gas exchange	Skin; gills; various other structures	Gills or lungs; skin in some
Circulation	Open system; reduced	Closed system; ventral heart
Fluid balance and waste disposal	Diffusion	Kidneys; also lungs, skin, gills
Nervous system	Nerve rings; no brain	Dorsal tubular nerve cord with brain (when present) at anterior end
Reproduction	Sexual; sexes almost always separate	Sexual; sexes typically separate
Support and movement	Endoskeleton bearing spines; muscles; tube feet	Notochord; endoskeleton of cartilage, bone, or both; well-developed muscles
Environment and lifestyle	Marine; mainly carnivores	Diverse habitats and lifestyles; herbivores, carnivores, omnivores, scavengers, suspension feeders
Other characteristics	Water vascular system; pedicellariae	Pharyngeal (gill) slits; postanal tail

*Members of these phyla are deuterostomes, so they are at the organ system level of organization and have a complete digestive tract.

to the Cambrian period, have been a rich source of fossils. In 1995 scientists reported finding an earlier chordate, which they named ***Yunnanozoon,*** in an early Cambrian fossil site in China. This site, known as **Chengjiang,** has fine-grained rocks about 530 million years old that have preserved soft-bodied animals in great detail. In 1999 a team of researchers working at the same site found more than 300 fossils of an animal they named ***Haikouella.*** These fossils suggest that *Haikouella* was about 3 cm long and had a nerve cord and notochord. Because these specimens also indicate that *Haikouella* had a relatively large brain, eyes, and segmented muscles, some biologists hypothesize that this animal may represent a transition between protochordates and vertebrates.

Tunicates are common marine animals

The **tunicates,** which comprise subphylum Urochordata, include the sea squirts, or *ascidians,* and their relatives. Larval tunicates have typical chordate characteristics and superficially resemble tadpoles. The expanded body has a pharynx with slits, and the long muscular tail contains a notochord and a dorsal, tubular nerve cord. Some tunicates *(appendicularians)* are common members of the zooplankton that retain their chordate features and ability to swim about.

Most tunicates are sea squirts (class **Ascidiacea**). A sea squirt larva swims for a time, then attaches itself to the sea bottom and loses its tail, notochord, and much of its nervous system. Adult sea squirts are barrel-shaped, sessile marine animals unlike other chordates. Indeed, they are often mistaken for sponges or cnidarians (Fig. 30–7). Only the pharyngeal slits and the structure of its larva suggest that the sea squirt is a chordate.

Adult tunicates develop a protective covering, or **tunic,** that may be soft and transparent or quite leathery. Curiously, the tunic is composed of a carbohydrate much like cellulose. The tunic has two openings: the incurrent siphon, through which water and food enter, and the excurrent siphon, through which water, waste products, and gametes pass to the outside. Sea squirts get their name from their practice of forcefully expelling a stream of water from the excurrent siphon when irritated.

Tunicates are suspension feeders that remove plankton from the stream of water passing through the pharynx. Food particles

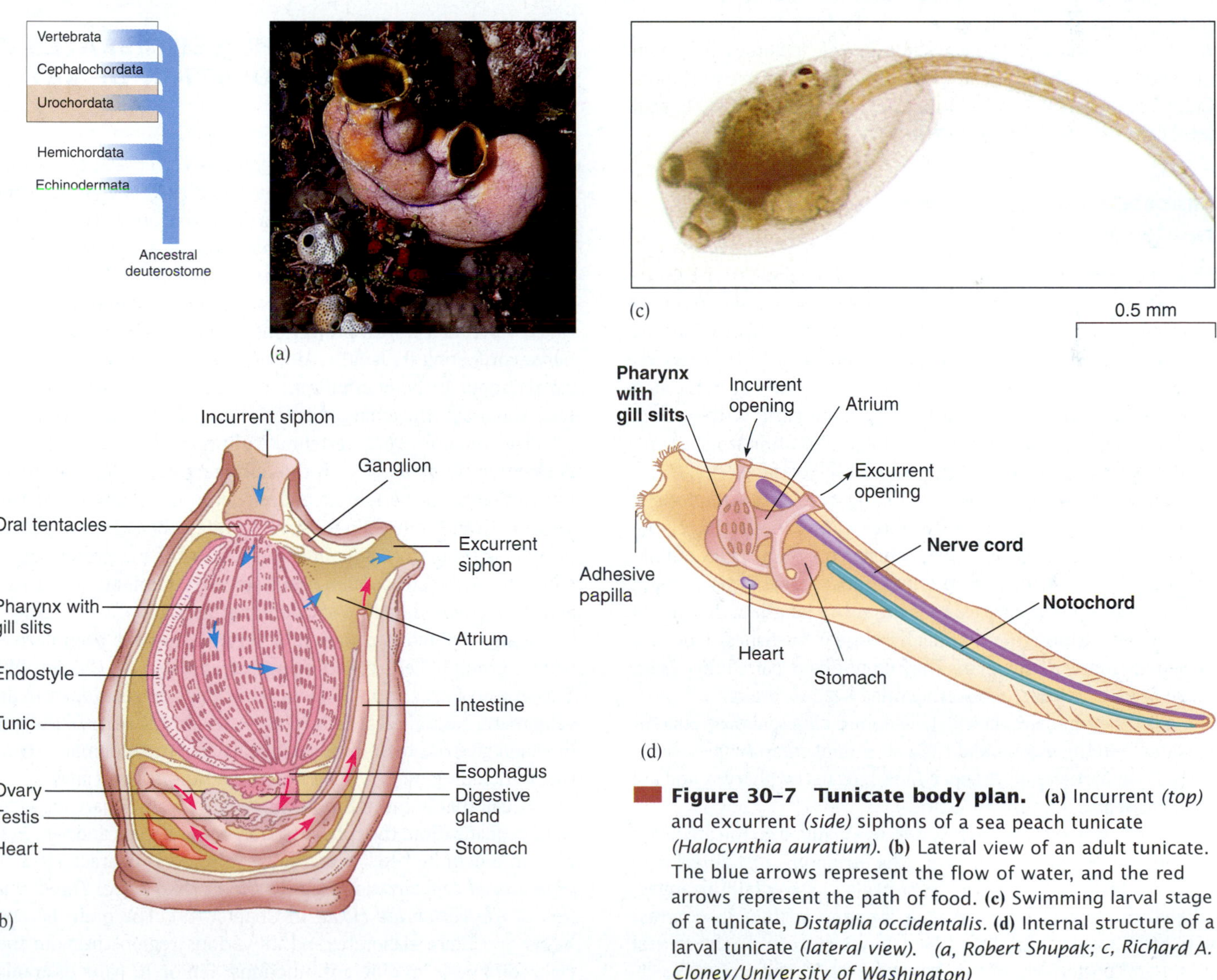

Figure 30–7 Tunicate body plan. **(a)** Incurrent *(top)* and excurrent *(side)* siphons of a sea peach tunicate *(Halocynthia auratium).* **(b)** Lateral view of an adult tunicate. The blue arrows represent the flow of water, and the red arrows represent the path of food. **(c)** Swimming larval stage of a tunicate, *Distaplia occidentalis.* **(d)** Internal structure of a larval tunicate *(lateral view).* *(a, Robert Shupak; c, Richard A. Cloney/University of Washington)*

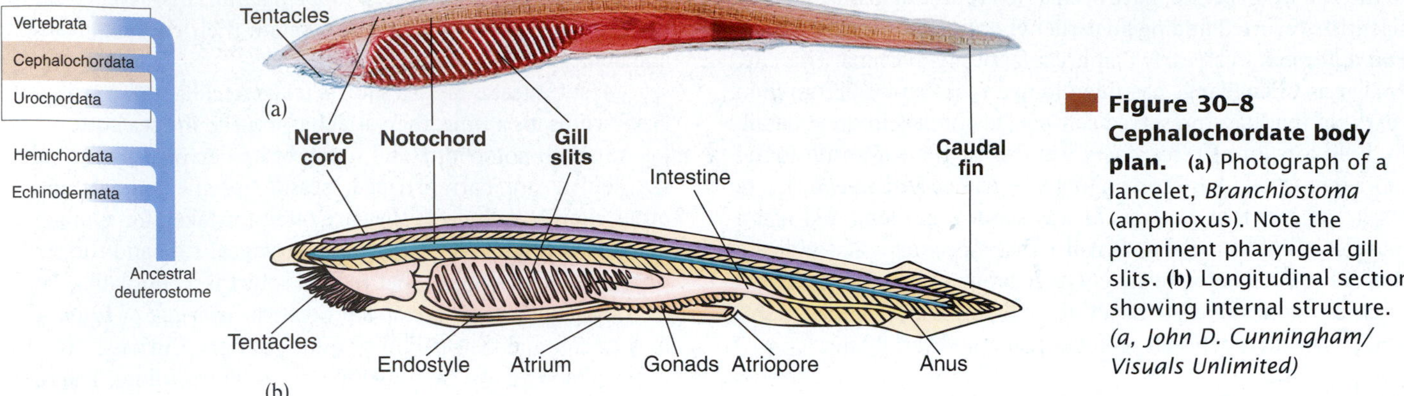

Figure 30–8 Cephalochordate body plan. (a) Photograph of a lancelet, *Branchiostoma* (amphioxus). Note the prominent pharyngeal gill slits. (b) Longitudinal section showing internal structure. *(a, John D. Cunningham/Visuals Unlimited)*

are trapped in mucus secreted by cells of the *endostyle,* a groove that extends the length of the pharynx. Ciliated cells of the pharynx move the stream of food-laden mucus into the esophagus. Much of the water entering the pharynx passes out through the pharyngeal slits into an **atrium** (chamber) and is discharged through the excurrent siphon.

Some species of tunicates form large colonies in which members may share a common tunic and excurrent siphon. Colonial forms often reproduce asexually by budding. Sexual forms are usually hermaphroditic.

Lancelets may be closely related to vertebrates

Most species of subphylum Cephalochordata belong to the genus *Branchiostoma,* which consists of animals commonly known as **lancelets,** or **amphioxus.** Lancelets are translucent, fish-shaped animals, 5 to 10 cm long and pointed at both ends. They are widely distributed in shallow seas, either swimming freely or burrowing in the sand near the low-tide line. In some parts of the world, lancelets are an important source of food. One Chinese fishery reports an annual catch of 35 tons (about 1 billion lancelets).

Chordate characteristics are highly developed in lancelets. The notochord extends from the anterior tip ("head"; hence the name *Cephalochordata*) to the posterior tip. Many pairs of pharyngeal slits are evident in the large pharyngeal region, and a tubular, dorsal nerve cord extends the entire length of the animal (Fig. 30–8). Although superficially similar to fishes, lancelets have a far simpler body plan. They do not have paired fins, jaws, sense organs, a heart, or a well-defined head or brain.

Like the tunicates, lancelets use their cilia to draw a current of water into the mouth and then strain out microscopic organisms. Food particles are trapped in mucus in the pharynx and are then carried back to the intestine.

Water passes through the pharyngeal slits into the atrium, a chamber with a ventral opening (the *atriopore*) just anterior to the anus. Metabolic wastes are excreted by segmentally arranged, ciliated protonephridia that open into the atrium. In contrast with other invertebrates, the blood flows anteriorly in the ventral vessel and posteriorly in the dorsal vessel. This circulatory pattern is similar to that of fishes.

Recent molecular and morphological research suggests that lancelets may well be the vertebrates' closest living relative. One similarity is the lancelet nerve cord, which has specialized regions that correspond to the vertebrate forebrain and midbrain.

THE SUCCESS OF THE VERTEBRATES IS LINKED TO THE EVOLUTION OF KEY ADAPTATIONS

The vertebrates, members of subphylum Vertebrata, are distinguished from other chordates in having a backbone, or **vertebral column,** that forms the skeletal axis of the body. This flexible support develops around the notochord, and in most species it largely replaces the notochord during embryonic development. The vertebral column consists of cartilaginous or bony segments called **vertebrae.** Dorsal projections of the vertebrae enclose the nerve cord along its length. Anterior to the vertebral column, a cartilaginous or bony **cranium,** or braincase, encloses and protects the brain, the enlarged anterior end of the nerve cord.

The cranium and vertebral column are part of the **endoskeleton.** In contrast with the nonliving exoskeleton of many invertebrates, the vertebrate endoskeleton is a living tissue that grows with the animal. In most vertebrates, the skeleton is made of bone, a tissue that contains fibers made of the protein collagen. The hard matrix of bone consists of the compound hydroxyapatite, composed mainly of calcium phosphate.

Many characters common to vertebrates have been derived from a group of cells called **neural crest cells** (see Chapter 49). This group of cells, found only in vertebrates, appears early in development. Neural crest cells migrate to various parts of the embryo and give rise to or influence the development of many structures, including nerves, head muscles, cranium, and jaws.

Recall that in invertebrates there is an evolutionary trend toward **cephalization,** the concentration of nerve cells and sense organs in a definite head. Vertebrate evolution is characterized by *pronounced* cephalization (see *On the Cutting Edge: The Evolution of the Vertebrate Head,* in Chapter 49). The brain became larger and more elaborate, and its various regions became specialized to perform different functions. Ten or 12 pairs of **cranial nerves** emerge from the brain and extend to various organs of the

TABLE 30–2 Extant Vertebrate Classes

Class	Examples	Characteristics
Myxiniformes	Hagfishes	Jawless, marine fishes with skeleton of cartilage; notochord persists throughout life; partial cranium; no vertebrae (so not true vertebrates); secrete sticky slime; development direct, no larval stage known
Petromyzontiformes	Lampreys	Jawless, freshwater and marine fishes with skeleton of cartilage; complete cranium and rudimentary vertebrae; hatch as small larvae
Chondrichthyes	Sharks, rays, skates, chimeras	Jawed marine and freshwater fishes with skeleton of cartilage; notochord replaced by vertebrae in adult; gills; placoid scales; two pairs of fins; oviparous, ovoviparous, or viviparous (a few species); well-developed sense organs (including lateral line system)
Actinopterygii (ray-finned fishes)	Perch, salmon, tuna, trout	Bony, marine and freshwater fishes; gills; swim bladder; generally oviparous
Actinistia (lobe-finned fishes)	Coelacanths	Bony fishes; marine, nocturnal predators on fish; seven-lobed fins
Dipnoi (lungfishes)	Lungfishes	Bony freshwater fishes; four similarly sized limbs, which are similar in structure and position to those of tetrapods
Amphibia	Salamanders, frogs and toads, caecilians	Aquatic larva typically undergoes metamorphosis into terrestrial adult; gas exchange through lungs and/or moist skin; heart consists of two atria and single ventricle; systemic and pulmonary circulation
Reptilia	Turtles, lizards, snakes, alligators	Tetrapods with hard scales; mainly terrestrial; adapted for reproduction on land (internal fertilization, leathery shell, amnion); lungs; ventricle of heart partially divided
Aves	Robins, pelicans, eagles, ducks, penguins, ostriches	Tetrapods with feathers; anterior limbs modified as wings; compact streamlined body; lungs; four-chambered heart; complete separation of oxygen-rich and oxygen-poor blood; endotherms; vocal calls and complex songs
Mammalia	Egg-laying mammals (platypus), marsupials (kangaroos), placentals (whales, mice, humans)	Tetrapods with hair; females nourish young with mammary glands; diaphragm moves air in and out of lungs; four-chambered heart; highly developed nervous system; endotherms; most are viviparous

body. Vertebrates have well-developed sense organs concentrated in the head: eyes; ears that serve as organs of balance and, in some vertebrates, for hearing as well; and organs of smell and taste.

Two pairs of appendages are present in most vertebrates. The fins of fishes are appendages that stabilize them in the water. Paired pectoral and pelvic fins are also used in steering. Biologists hypothesize that jointed appendages that facilitated locomotion on land evolved from the lobed (divided) fins of lungfishes.

Other shared derived characters of vertebrates include a closed circulatory system with a ventral heart and blood containing hemoglobin. Vertebrates have a complete digestive tract with specialized regions and large digestive glands (the liver and pancreas). They have several complex **endocrine glands,** ductless glands that secrete hormones. Vertebrates have paired kidneys that regulate fluid balance. The sexes are typically separate.

Vertebrate phylogeny is controversial

With about 48,000 species, the vertebrates are less diverse and much less numerous than the insects but rival them in their adaptation to an enormous variety of lifestyles. The study of evolutionary relationships of vertebrates is an important focus of research.

Process of Science Historically, biologists assigned extant vertebrates to three classes of fishes, referred to as superclass **Pisces,** and four classes of land vertebrates, superclass **Tetrapoda.** In this classification scheme, the classes of fishes include class **Agnatha** (*a,* "without"; *gnathos,* "jaw"), the jawless fishes, for example, lampreys; class **Chondrichthyes,** the sharks and rays with cartilaginous skeletons; and class **Osteichthyes,** the bony fishes. The four-limbed land vertebrates, or **tetrapods,** comprise class **Amphibia,** the frogs, toads, and salamanders; class **Reptilia,** the lizards, snakes, turtles, and alligators; class **Aves,** the birds; and class **Mammalia,** the mammals. Although not all the tetrapods have four limbs (e.g., the snakes), all evolved from four-limbed ancestors. Aquatic tetrapods (e.g., sea turtles, penguins, whales, seals) evolved from terrestrial ancestors.

During the past several years, as investigators have analyzed new data, different classification schemes have been proposed. For example, Philip C.J. Donoghue of the University of Birmingham in the United Kingdom, and his research team analyzed more than 100 different traits in 17 groups of fossil and living chordates. Based on their work and that of other investigators, many biologists currently recognize ten extant vertebrate classes: six classes of fishes and four classes of tetrapods (Table 30–2 and Fig. 30–9). According to this view, the vertebrate lineage split from the lancelets. The hagfishes branched off first, followed by the lampreys. Hagfishes are assigned to class **Myxiniformes** and lampreys are assigned to class **Petromyzontiformes.**

Some biologists consider hagfishes to be vertebrates because they exhibit many vertebrate characters, including a cranium. Others argue that hagfishes are not vertebrates because they have no trace of vertebrae. The notochord is their only axial support. Based on their cranium, many biologists now use the term **Craniata** to indicate the clade that includes the vertebrates

Figure 30–9 Evolutionary relationships of vertebrates. The relationships among animal groups shown here reflect current data from studies of morphology and DNA. As new data are collected and considered, our understanding of these relationships is likely to change. Note that the vertebrates plus the Myxiniformes make up the craniate group. **(a)** Vertebrate relationships as might be depicted by traditional evolutionary taxonomists. **(b)** This cladogram (on facing page) represents one phylogenetic interpretation of the relationships shown in **(a).** The evolution of some major changes in body plan is indicated.

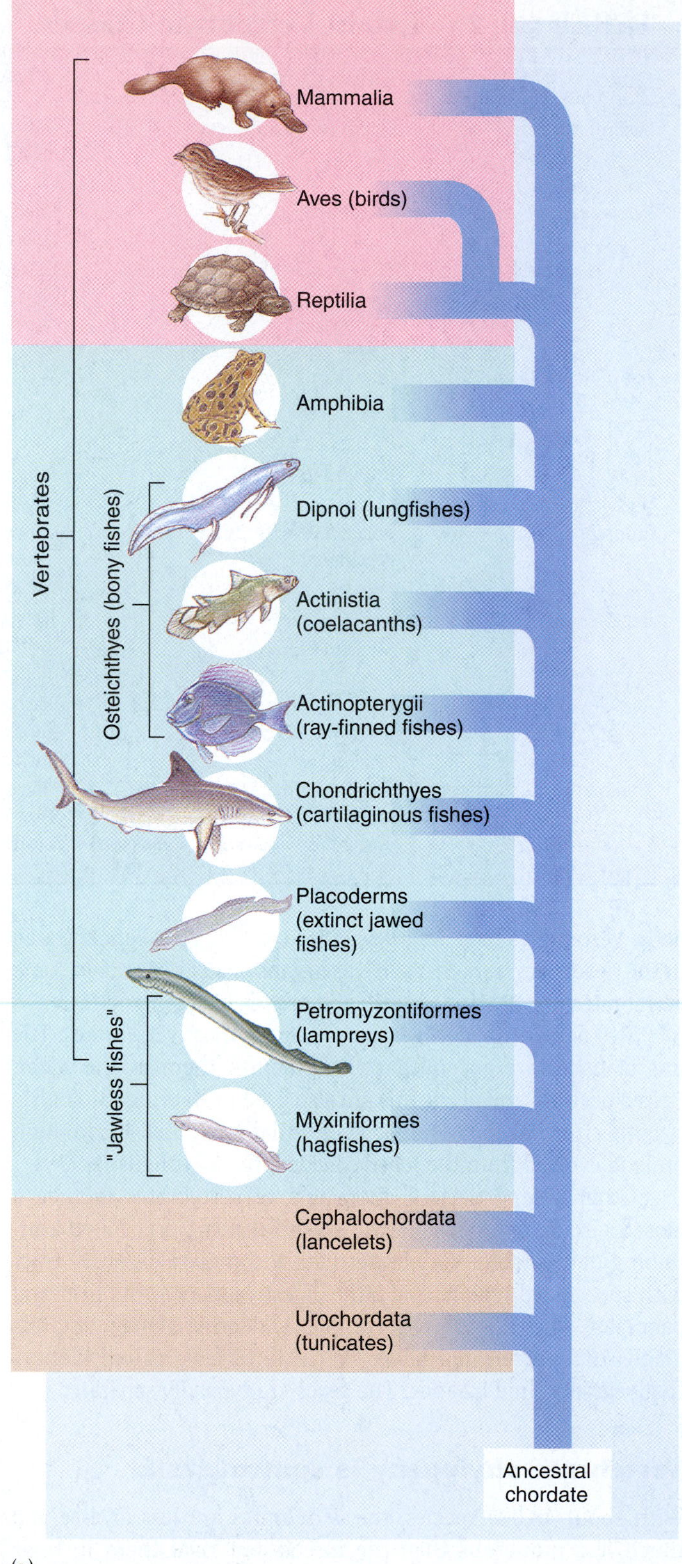

(a)

plus the hagfishes. They view the Myxiniformes as the sister group of all other craniates.

Class Chondrichthyes (sharks and rays) is considered monophyletic. Most systematists now assign the extant bony fishes to three classes: class **Actinopterygii,** which includes the ray-finned fishes; class **Actinistia,** the coelacanths; and class **Dipnoi,** the lungfishes. Many biologists recognize four classes of tetrapods: Amphibia, Reptilia, Aves (birds), and Mammalia. However, this classification is also being reevaluated. For example, as we discuss later, many cladists do not assign birds to a separate class; instead, they place them with reptiles. As new research results are reported, vertebrate classification will continue to change.

The jawless fishes represent the earliest vertebrates

Some of the earliest known vertebrates, collectively called **ostracoderms,** consisted of several groups of small, armored, jawless fishes that lived on the bottom and strained their food from the water (see Fig. 20–11). Their heads were protected from predators by thick bony plates, and their trunks and tails were covered with thick scales. Most ostracoderms lacked fins. Fragments of ostracoderm scales have been found in rocks from the Cambrian period, but most ostracoderm fossils date back to the Ordovician and Silurian periods. By the Silurian and Devonian periods, ostracoderms had radiated extensively. They became extinct by the end of the Devonian period.

Like ostracoderms, present-day hagfishes and lampreys have neither jaws nor paired fins. They are eel-shaped animals, up to 1 m long. Their smooth skin lacks scales, and they are supported by a cartilaginous skeleton and well-developed notochord.

Hagfishes, now assigned to class Myxiniformes, are marine scavengers that burrow for worms and other invertebrates or prey on dead and disabled fishes. As a defense mechanism, hagfishes secrete large amounts of sticky slime.

Hagfishes are of interest to medical researchers because, in addition to a heart, they have several other contractile structures that help circulate blood. These additional "hearts," which help maintain blood pressure, are regulated by a compound called *eptatretin*. Researchers are studying eptatretin in the hope that it might be used to treat certain types of abnormal heart rhythms in humans.

Lampreys, now assigned to class Petromyzontiformes, live in fresh water. Some spend their adult lives in the ocean and return to fresh water to reproduce. Many species of adult lampreys are parasites on other fishes (Fig. 30–10). Adult parasitic lampreys have a circular sucking disk around the mouth, which is located on the ventral side of the anterior end of the

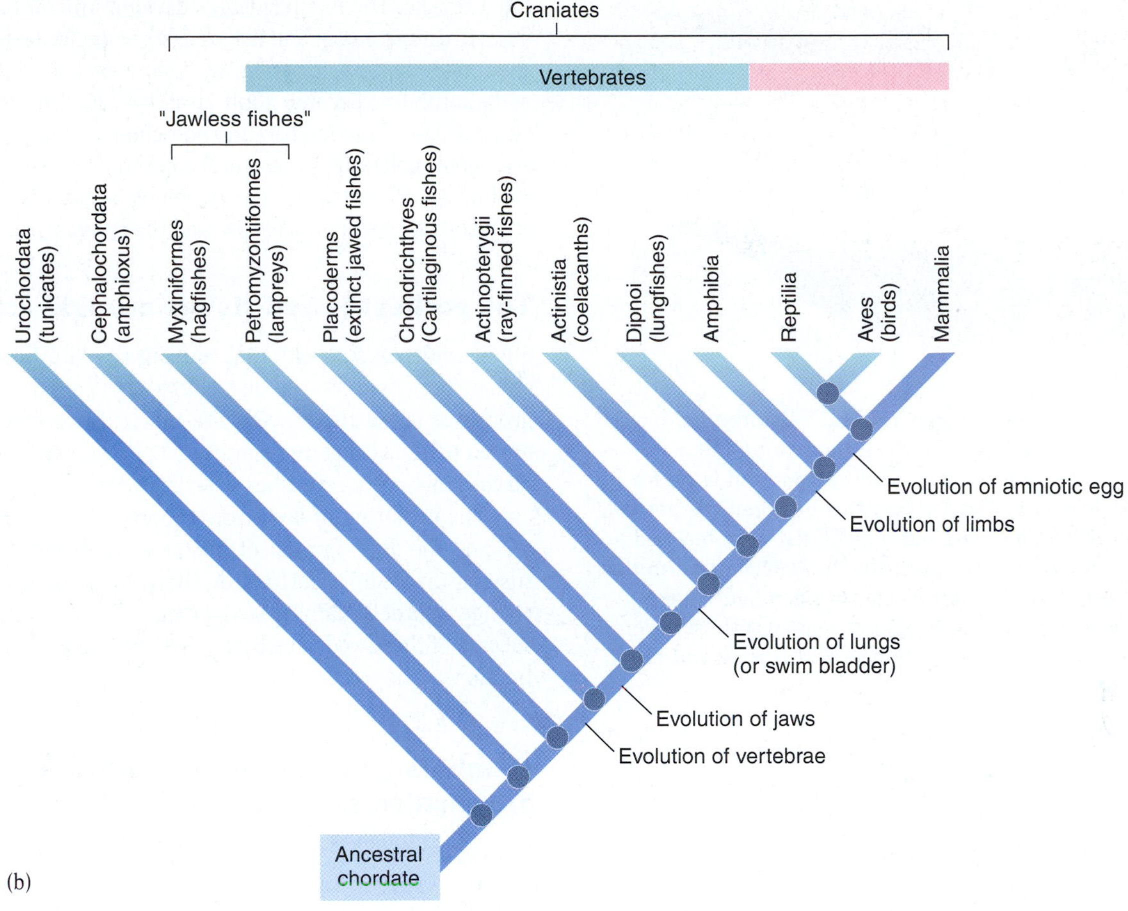

Figure 30–10 Lampreys. (a) Three lampreys are attached to a carp by their suction-cup mouths. Note the absence of jaws and paired fins. (b) Suction-cup mouth of adult lamprey *(Estosphenus japonicus)*. Note the rasplike teeth. *(a, Tom Stack/Tom Stack &Associates; b, courtesy of Dr. Kiyoko Uehara)*

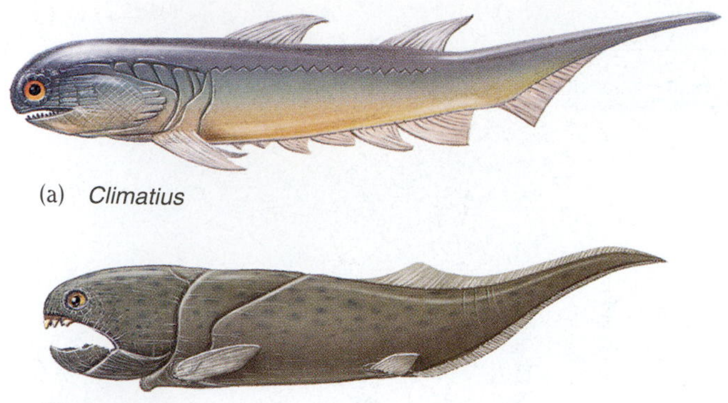

Figure 30–11 Early jawed fishes. Acanthodians and placoderms flourished in the Devonian period. **(a)** *Climatius,* a spiny-skinned acanthodian with large fin spines and five pairs of accessory fins between the pectoral and pelvic pairs. *Climatius* was a small fish that reached a length of 8 cm (3 in). **(b)** *Dinichthys,* a giant placoderm that grew to a length of 8 m (26 ft). (Most placoderms were only about 20 cm, or 8 in, long.) Its head and thorax were covered by bony armor, but the rest of the body and tail were naked.

body. Using this disk to attach to a fish, the lamprey bores through the skin of its host with horny (made of keratin rather than bone) teeth on the disk and tongue. Then the lamprey injects an anticoagulant into its host and sucks blood and soft tissues.

Adult lampreys leave the ocean or lake and swim upstream to spawn. They build a nest, a shallow depression in the gravelly stream bed, into which they shed eggs and sperm. After spawning, they die. The fertilized eggs develop into larvae, which drift downstream to a pool and live as suspension feeders in burrows in the muddy bottom for three to seven years. They then undergo metamorphosis, become adult lampreys, and migrate back to the ocean or lake. As in hagfish, the notochord persists throughout life and is not replaced by a vertebral column. However, lampreys have rudiments of vertebrae, cartilaginous segments, called neural arches, that extend dorsally around the spinal cord.

The earliest jawed fishes are now extinct

Fossil evidence suggests that, during the late Silurian and Devonian periods, fishes evolved jaws and paired fins. Two early groups of jawed fishes were the now extinct **acanthodians,** armored fishes with paired spines and pectoral and pelvic fins, and **placoderms,** armored fishes with paired fins (Fig. 30–11).

The evolution of jaws from a portion of the gill arch skeleton and the development of fins enabled fishes to change from suspension-feeding bottom dwellers to active predators. This change afforded many new opportunities for finding food. The success of the jawed vertebrates may have contributed to the extinction of the ostracoderms.

Members of class Chondrichthyes are cartilaginous fishes

Members of class Chondrichthyes, the cartilaginous fishes, evolved as successful marine forms in the Devonian period. This class includes the sharks, rays, and skates (Fig. 30–12). Most species are ocean dwellers, but a few have invaded fresh water.

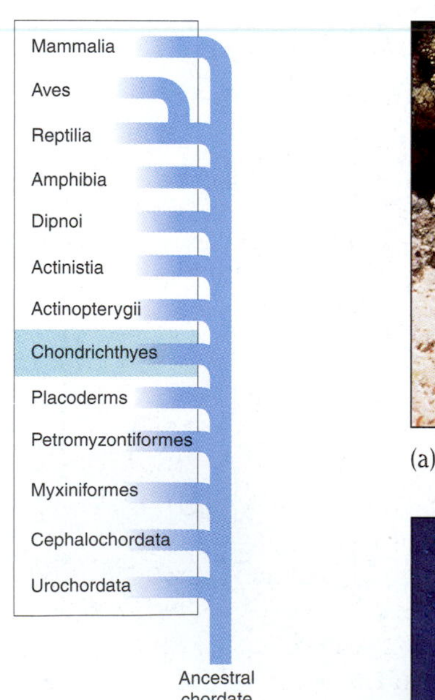

(a)

(b)

Figure 30–12 Cartilaginous fishes. **(a)** Blue-spotted sting ray *(Taeniura lymma).* Sting rays typically feed on shellfish and bottom-dwelling fishes. **(b)** The great white shark *(Carcharodon carcharias),* photographed in Australia, is considered the most dangerous shark to humans. This shark is actually white only on its ventral aspect; the rest of the body is brownish-gray or bluish-gray. *(a, Jeffrey L. Rotman/Peter Arnold, Inc.; b, Kelvin Aitken/Peter Arnold, Inc.)*

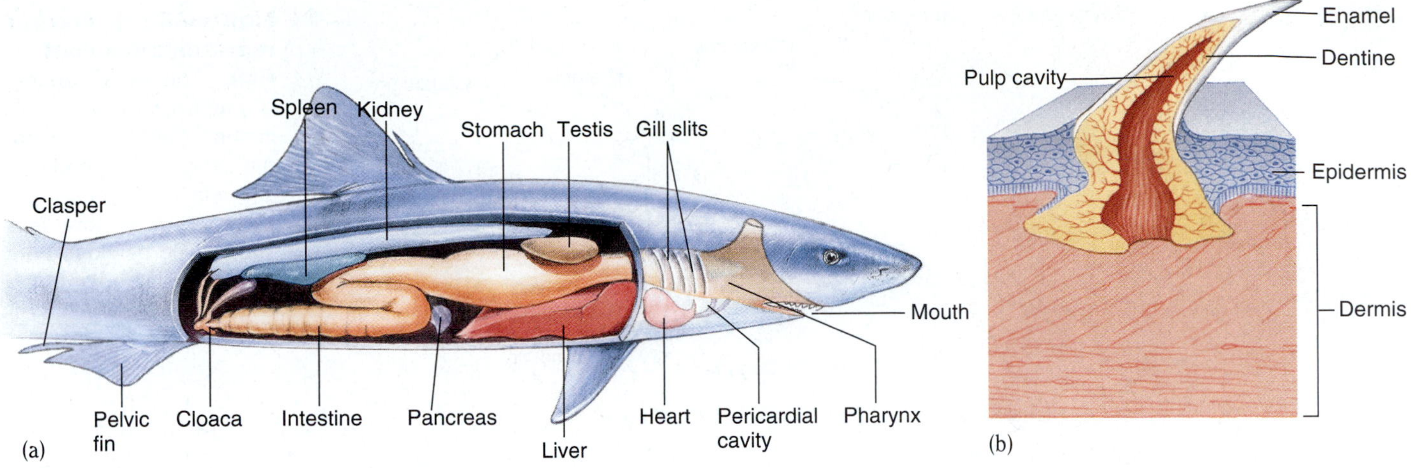

Figure 30–13 Sharks. **(a)** Internal structure of a shark. **(b)** Structure of a placoid scale.

Except for the whales, the sharks are the largest living vertebrates. The whale shark *(Rhincodon)* may exceed 15 m (49 ft) in length, making it the largest fish.

Most rays and skates are sluggish, flattened creatures that live partly buried in the sand. Their enormous pectoral fins propel them along the bottom, where they feed on mussels and clams. The sting ray has a whiplike tail with a barbed spine at its base that can inflict a painful wound. The electric ray has electric organs on either side of the head. These modified muscles can discharge enough electric current (up to 2500 watts) to stun fairly large fishes, as well as human swimmers.

The chondrichthyes retain their cartilaginous embryonic skeleton. Although this skeleton is not replaced by bone, it may be strengthened by a deposit of calcium salts. All chondrichthyes have paired jaws and two pairs of fins. The skin contains **placoid scales.** Each scale is a toothlike structure composed of an outer layer of enamel and an inner layer of dentine (Fig. 30–13). The lining of the mouth contains larger, but essentially similar, scales that serve as teeth. The teeth of other vertebrates are homologous with these scales. Shark teeth are embedded in the flesh and not attached to the jawbones; new teeth develop continuously in rows behind the functional teeth and migrate forward to replace any that are lost.

The shark body is adapted for swimming. Lift is provided by body shape and fins and by swimming swiftly. The shark stores a great deal of oil in its large liver. In some sharks the liver accounts for up to 30% of the body weight, and most of that weight is due to stored oil. Fats and oils decrease the overall density of fishes and contribute to buoyancy. Even so, the shark body is denser than water, so sharks tend to sink unless they are actively swimming.

Most sharks are streamlined predators that swim actively and catch and eat other fishes as well as crustaceans and mollusks. The largest sharks and rays, like the largest whales, are suspension feeders that strain plankton from the water. They gulp water through the mouth. Then, as the water passes through the pharynx and out the gill slits, food particles are trapped in a sievelike structure.

Predatory sharks are attracted to blood, so a wounded swimmer or a skin diver towing speared fish is a target. However, although books and films portray sharks as monstrous enemies, most do not go out of their way to attack humans. In fact, of the approximately 350 known shark species, fewer than 30 have been known to attack humans.

The shark has a complex brain, and a spinal cord protected by vertebrae. Their well-developed sense organs very effectively locate prey in the water. Sharks may detect other animals electrically before sensing them by sight or smell. **Electroreceptors** on the shark's head can sense weak electric currents generated by the muscular activity of animals. The **lateral line organ,** found in all fishes, is a groove along each side of the body with many tiny openings to the outside. Sensory cells in the lateral line organ are sensitive to waves and other motion in the water, alerting the shark to the presence of predator or prey (see Fig. 41–4).

Cartilaginous fishes have no lungs. They have five to seven pairs of gills. A current of water enters the mouth and passes over the gills and out the gill slits, constantly providing the fish with a fresh supply of dissolved oxygen. Sharks that actively swim depend on their motion to enhance gas exchange. Sharks that spend time on the ocean floor, and rays and skates, use muscles of the jaw and pharynx to pump water over their gills.

The digestive tract of sharks consists of the mouth cavity; a long pharynx leading to the stomach; a short, straight intestine; and a **cloaca,** which opens on the underside of the body and is characteristic of many vertebrates. The liver and pancreas discharge digestive juices into the intestine. The cloaca receives digestive wastes, as well as metabolic wastes from the urinary system. In females, the cloaca also serves as a reproductive organ.

The sexes are separate, and fertilization is internal. In the mature male, each pelvic fin has a slender, grooved section, known as a **clasper,** used to transfer sperm into the female's cloaca. The eggs are fertilized in the upper part of the female's oviducts. Part of the oviduct is modified as a shell gland, which secretes a protective coat around the egg.

Skates and some species of sharks are **oviparous;** that is, they lay eggs. Many species of sharks, however, are **ovoviviparous,** meaning that their young are enclosed in eggs that are incubated within the mother's body. During development, the young depend on stored yolk for their nourishment rather than

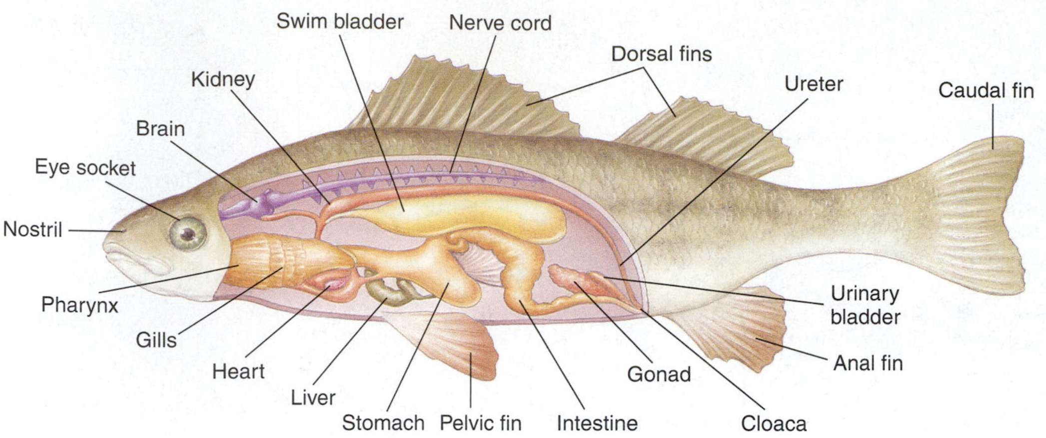

Figure 30–14 Perch, a representative bony fish. The swim bladder is a hydrostatic organ that enables the fish to change the density of its body and remain stationary at a given depth. Pectoral *(not shown)* and pelvic fins are paired.

on transfer of materials from the mother. The young are born after hatching from the eggs.

A few species of sharks are **viviparous.** Not only do the embryos develop within the uterus, but much of their nourishment is delivered to them by the mother's blood. This is accomplished by transfer of nutrients between the blood vessels in the lining of the uterus and the yolk sac surrounding each embryo.

The ray-finned fishes gave rise to modern bony fishes

Although bony fishes appear earlier in the fossil record than cartilaginous fishes, both groups may have evolved about the same time, during the Devonian period. The two groups share many characteristics (such as continuous tooth replacement), but they also differ in important ways.

Most bony fishes are characterized by a bony skeleton with many vertebrae. Bone has advantages over cartilage: It provides excellent support and serves as a very effective storage site for calcium. The body is covered with overlapping, bony dermal scales. Most species have flexible median and paired fins, supported by long rays made of cartilage or bone. A lateral bony flap, the **operculum,** extends posteriorly from the head and protects the gills.

Unlike most sharks, bony fishes are generally oviparous. They lay an impressive number of eggs that they fertilize externally. The ocean sunfish, for example, lays more than 300 million eggs! Of course, most of the eggs and young become food for other animals. The probability of survival is increased by certain behavioral adaptations. For example, many species of fishes build nests for their eggs and protect them.

During the Devonian period, the bony fishes diverged into two major groups: the **Sarcopterygii** and the **ray-finned fishes,** class **Actinopterygii.** Fossils of the earliest sarcopterygians date back to the Devonian about 400 mya. These fishes are characterized by lungs and fleshy, **lobed fins.** These flexible fins had a central appendage with many bones and muscles and could support the body.

Early sarcopterygians evolved along two separate lines: the lungfishes (class **Dipnoi**) and lobe-finned fishes (class **Actinistia**). Three genera of lungfishes have survived to the present day in the rivers of tropical Africa, Australia, and South America.

The ray-finned fishes underwent two important adaptive radiations. The first gave rise during the late Paleozoic era to a group of fishes that are now mostly extinct. The second radiation began during the early Mesozoic era and gave rise to the modern bony fishes. The ancestors of the modern bony fishes had lungs, which were retained by the lobe-finned fishes and lungfishes. In the ray-finned fishes, however, the lungs became modified as a **swim bladder,** an air sac that helps regulate buoyancy (Fig. 30–14). Bones and muscles are heavier than water, and without the swim bladder the fish would sink. By regulating gas exchange between the blood and the swim bladder, a fish can control the amount of gas in the swim bladder, changing the overall density of its body. This ability allows a bony fish, in contrast to a shark, to hover at a given depth of water without much muscular effort.

There are more species of bony fishes than any other group of vertebrates. Biologists have identified about 49,000 living species of freshwater and saltwater bony fishes, of many shapes and colors (Fig. 30–15). Bony fishes range in size from the Philippine goby, which is only about 1 cm (0.4 in) long, to the ocean sunfish (or *Mola;* see Fig. 19–15*c*), which may reach 4 m and weigh about 1500 kg (about 3300 lb).

The diversity of bony fishes may have resulted from the action of *Hox* genes. Early chordates have only one *Hox* cluster. Tetrapods have four clusters. The zebrafish has seven *Hox* clusters. The researchers who made this interesting discovery suggest that during the radiation of ray-finned fishes (once before and once after the lobe-finned fishes had split off), *Hox* clusters duplicated by the duplication of entire chromosomes. These *Hox* genes could have provided the genetic material for the evolution of the diverse species of ray-finned fish.

Did descendants of the lobe-finned fishes or lungfishes move onto the land?

Biologists had thought that the lobe-finned fishes were extinct by the end of the Paleozoic era, so the scientific community was very excited in 1938 when a commercial fisherman caught a lobe-finned fish off the coast of South Africa. After a long search, a second specimen was captured in 1952. Since that time more than 200 specimens of these giant "living fossils" have been

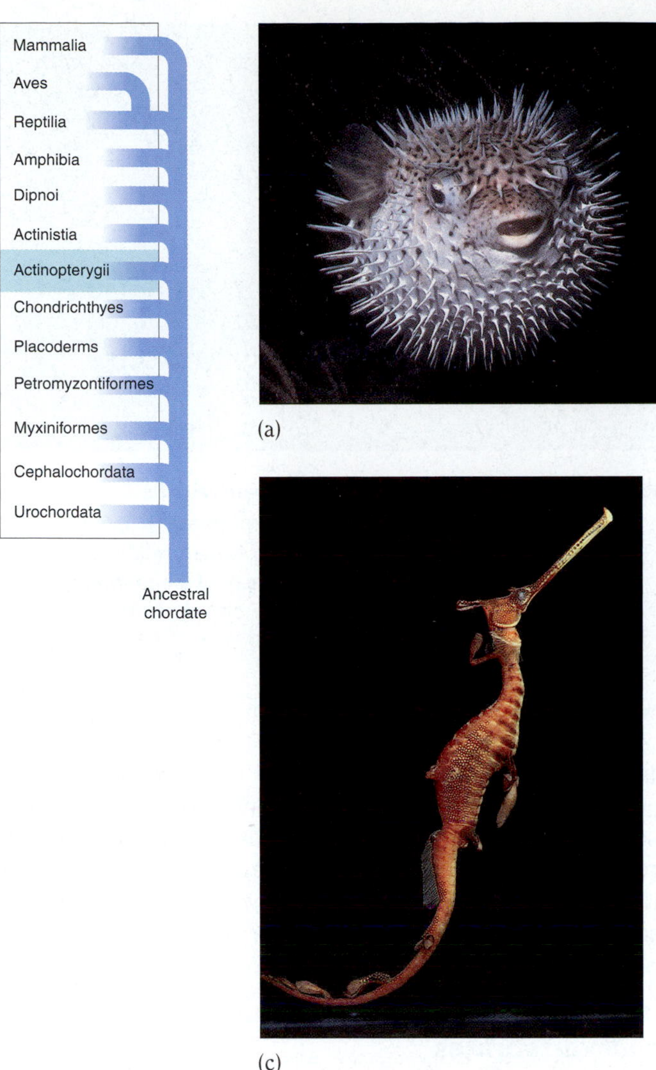

Figure 30–15 Modern bony fishes. **(a)** The porcupinefish *(Diodon hystrix)* can swallow air or water and inflate its body, a strategy to discourage potential predators. Photographed in the Virgin Islands. **(b)** The parrotfish *(Scarus gibbus)* feeds on coral, grinds it in its digestive tract, and extracts the coralline algae. The fish eliminates a fine white sand. These fishes contribute to white sand beaches in many parts of the world. Parrotfish begin life as females and later become males. **(c)** The leaflike extensions of the body wall of this Australian leafy sea dragon *(Phyllapteryx taeniolatus)* help camouflage it in surrounding kelp. *(a, David Hall/Photo Researchers, Inc.; b, Jeffrey L. Rotman/Peter Arnold, Inc.;* c, *Norbert Wu/Peter Arnold, Inc.)*

found in the deep waters off the east coast of Africa near the Comoro Islands (Fig. 30–16). These fishes, known as **coelacanths** and given the species name *Latimeria chaluminae,* measure nearly 2 m (about 6 ft) in length. They are nocturnal predators on other fishes. In 1998 biologists were again surprised to discover a second population of coelacanths off the coast of Indonesia. This second population appears to be a distinct species.

The pieces seemed to fit together so perfectly. The coelacanths were surely the living fossil representatives of the group that gave rise to the land vertebrates. This was the prevailing hypothesis until the 1980s, when a few investigators began to question it based on morphological data. The external nasal openings of fossil and living lungfish suggested that they, rather than the coelacanths, might be the ancestors of the land vertebrates. Other morphological similarities between lungfishes and land vertebrates are tooth enamel and the presence of four similarly sized limbs that have a similar structure and position. (Recall that the fins of lungfishes are also lobed.)

In 1990 the lungfish hypothesis gained support from comparisons of mitochondrial DNA. Many biologists were still not convinced until in 1998 researchers presented more extensive comparisons of mitochondrial DNA. Based on current data, lungfishes appear to be likely tetrapod ancestors. Future research on nuclear DNA may support or refute this hypothesis.

The following is one hypothesis that explains the origin of tetrapod limbs. During the Devonian period, frequent seasonal droughts caused swamps to become stagnant or even to dry up completely. Fishes with lobed fins would have had a tremendous advantage for survival under those conditions. They were strikingly preadapted for moving onto the land. They had lungs for breathing air, and their sturdy, fleshy fins may have allowed them to "walk" along in shallow water. These fins could support the fish's weight, enabling it to emerge onto dry land and make its way to another pond or stream.

Some biologists question this account. A competing view holds that limbs may have evolved in a fully aquatic environment. Animals with more developed limbs could move along in shallow water or creep through dense aquatic vegetation more efficiently than their lobe-finned ancestors. Whichever view of the evolution of tetrapods is correct, natural selection favored those individuals adapted for making their way on land.

The ability to move about, however awkwardly, on dry land gave animals access to new food sources. Terrestrial plants were already established, and terrestrial insects and arachnids were

Figure 30–16 Diver swimming with coelacanth. For many years, biologists thought that ancestors of this lobe-finned fish *(Latimeria)* gave rise to the amphibians. Living coelacanths are difficult to observe because they inhabit deep ocean waters, and when brought to the surface, they do not survive the change in pressure. *(Mark Erdmann)*

Figure 30–17 An artist's conception of labyrinthodonts. These early amphibians existed about 150 mya, from the late Paleozoic to the early Mesozoic eras. *(Photograph by Logan, courtesy of Department of Library Services, American Museum of Natural History)*

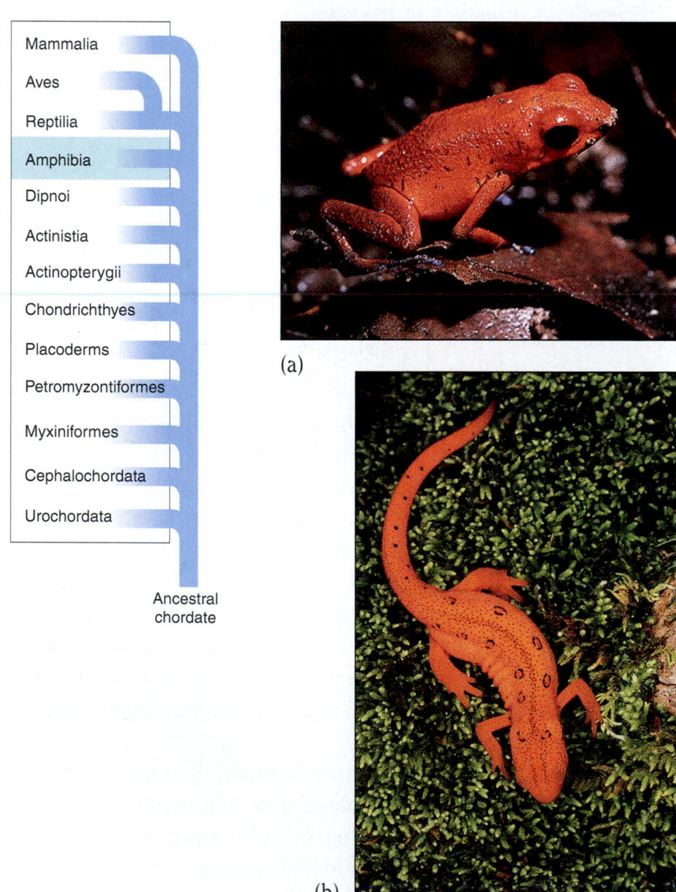

Figure 30–18 Modern amphibians. **(a)** This red dart frog *(Dendrobates pumioio)* is a poison arrow frog. **(b)** The red eft newt *(Notophthalmus viridescens)* is mainly aquatic but spends one to three years as a pre-adult in a moist terrestrial environment. *(a, Gerald and Buff Corsi/Tom Stack; b, Roy Morsch/The Stock Market)*

rapidly evolving. A vertebrate that could survive on land had less competition for food. Laying eggs on land, away from the many ocean predators, also increased their chances for successful reproduction. The early vertebrate experience on land was so successful that their lineage gave rise to the amphibians and, later, the reptiles, birds, and mammals.

Amphibians were the first successful land vertebrates

The first successful **tetrapods,** or land vertebrates, were the **labyrinthodonts** (Fig. 30–17), clumsy animals with short necks and heavy, muscular tails. They had limbs strong enough to support the weight of their bodies on land. The largest labyrinthodonts were the size of crocodiles. They flourished during the late Paleozoic and early Mesozoic eras, then became extinct. Biologists hypothesize that the labyrinthodonts gave rise to frogs and salamanders.

Modern amphibians are classified in three orders: Order **Urodela** ("visible tail") includes salamanders, mudpuppies, and newts, all animals with long tails; order **Anura** ("no tail") is made up of frogs and toads, with legs adapted for hopping; and order **Apoda** ("no feet") contains the wormlike caecilians (Fig. 30–18). Although some adult amphibians are quite successful as land animals and can live in dry environments, most return to the water to reproduce. Eggs and sperm are generally released in the water.

Amphibians undergo a single, although complex, change, or **metamorphosis,** from larva to adult. The embryos of most frogs and toads develop into larvae called **tadpoles.** These larvae have tails and gills, and most feed on aquatic plants. After a time, the tadpole undergoes metamorphosis, a process regulated by hormones secreted by the thyroid gland. During metamorphosis, gills and gill slits disappear, the tail is resorbed, and limbs emerge. The digestive tract shortens, and food preference shifts from plant material to a carnivorous diet; the mouth widens; a tongue develops; the tympanic membrane (ear drum) and eyelids appear; and the eye lens changes shape. Many biochemical changes also accompany the transformation from a completely aquatic life to an amphibious one.

Several salamanders, such as the mudpuppy *Necturus,* do not undergo complete metamorphosis; they retain many larval characteristics even when sexually mature adults. Recall from Chapter 19 that this is an example of **paedomorphosis** (see Fig. 19–16). This type of development permits these salamanders to remain aquatic rather than having to compete on land.

The coloration of amphibians may conceal them in their habitat or may be very bright and striking. Many of the brightly colored species are poisonous (Fig. 30–18*a*). Their distinctive colors warn predators that they are not encountering an ordinary amphibian. Some frogs can camouflage themselves by changing color.

Adult amphibians do not depend solely on their primitive lungs for the exchange of respiratory gases. Their moist, glandular skin, which lacks scales and is plentifully supplied with blood vessels, also serves as a respiratory surface. The numerous mucous glands within the skin help to keep the body surface moist, which is important in gas exchange. The mucus also makes the animal slippery, facilitating its escape from predators. Some amphibians have glands in their skin that secrete poisonous substances harmful to predators.

The amphibian heart is divided into three chambers: Two **atria** receive blood, and a single **ventricle** pumps it into the arteries. A double circuit of blood vessels keeps oxygen-rich and oxygen-poor blood partially separate. Blood passes through the **systemic circulation** to the various tissues and organs of the body. Then, after returning to the heart, it is directed through the **pulmonary circulation** to the lungs and skin, where it is recharged with oxygen. The oxygen-rich blood returns to the heart to be pumped out into the systemic circulation again. The comparative anatomy of the heart and circulation of various vertebrate classes is discussed in Chapter 42.

Amniotes are fully adapted to life on land

The evolution of reptiles from ancestral amphibians required many adaptations that allowed them to be completely terrestrial. Evolution of the **amniotic egg** was an extremely important event because it allowed terrestrial vertebrates to complete their life cycles on land. The egg contains an **amnion,** a membrane that forms a fluid-filled sac around the embryo. The amnion provides the embryo with its own private "pond," permitting independence from a watery external environment. In addition to keeping the embryo moist, the amniotic fluid serves as a shock absorber, cushioning the developing embryo.

The evolution of the amniotic egg is so important to the success of terrestrial vertebrates—reptiles, birds, and mammals—that biologists refer to these animals as **amniotes.** Systematists view amniotes as a monophyletic group because these animals have a common ancestor that was itself an amniote (Fig. 30–19).

In addition to the amnion, the amniotic egg has three other extraembryonic (not part of the developing body itself) membranes that protect the developing embryo, store nutrients *(yolk sac),* carry on gas exchange *(chorion and allantois),* and store wastes *(allantois).* Development of the extraembryonic membranes is discussed in Chapter 49. Although most extant mammals do not lay eggs, their embryos have an amnion and other extraembryonic membranes.

Another important adaptation to terrestrial life is a body covering that minimizes water loss. Such a covering severely decreases gas exchange across the body surface. Amniotes depend on efficient lungs and circulatory systems for exchange of oxygen and carbon dioxide.

Amniotes also have physiological mechanisms for conserving water. For example, to avoid water loss during excretion of metabolic wastes, much of the fluid filtered from the blood by the kidneys is reabsorbed in the kidney tubules and urinary bladder. In aquatic animals, nitrogenous wastes from protein and nucleic acid metabolism are excreted as ammonia. Because it is quite toxic, large amounts of water are needed to dilute it. Reptiles and birds (as well as most terrestrial arthropods) convert ammonia to uric acid, which is much less toxic and can be excreted as relatively insoluble crystals. These solid crystals do not require much water for their excretion, and so water is conserved. Mammals convert ammonia to urea (see Chapter 46).

Our understanding of amniote phylogeny is changing

Until recently, biologists divided amniotes into three distinct classes: Reptilia, Aves, and Mammalia. This classification was based on shared primitive characters. For example, reptiles have an amnion; dry, scaly skin; a three-chambered heart; and internal fertilization. Although birds share many characters with reptiles, they also have distinguishing characteristics, such as feathers, and so they were assigned to a separate group, Class Aves.

Cladistic analysis indicates that reptiles are not a monophyletic group. Rather, class Reptilia is paraphyletic because it includes some, but not all, of its descendants (see Fig. 22–8). Reflecting the current change in scientific views, we discuss the amniotes both in classical terms of three separate classes and also according to the cladistic view, which is rapidly gaining favor.

Amniotes underwent two major adaptive radiations

The earliest amniotes are thought to have somewhat resembled lizards. By the late Carboniferous period, about 290 mya, amniotes had undergone an impressive adaptive radiation and diverged into two branches. One of these branches, the **synapsids,** gave rise to the mammals. The second branch, the **sauropsids,** gave rise to all of the other reptiles and to the birds. The sauropsids divided to form two sub-branches: **anapsids** and **diapsids.** The turtles may be the only extant anapsids. The diapsids include the snakes, lizards, crocodilians, and many extinct reptiles including the dinosaurs. Based on cladistic analysis, many biologists also classify the birds as diapsids.

A second great amniote adaptive radiation occurred during the Mesozoic era that ended about 65 mya. In fact, the Mesozoic era is known as the Age of Reptiles. During that time reptiles were the dominant terrestrial animals (see Chapter 20). They had radiated into an impressive variety of ecological lifestyles (see

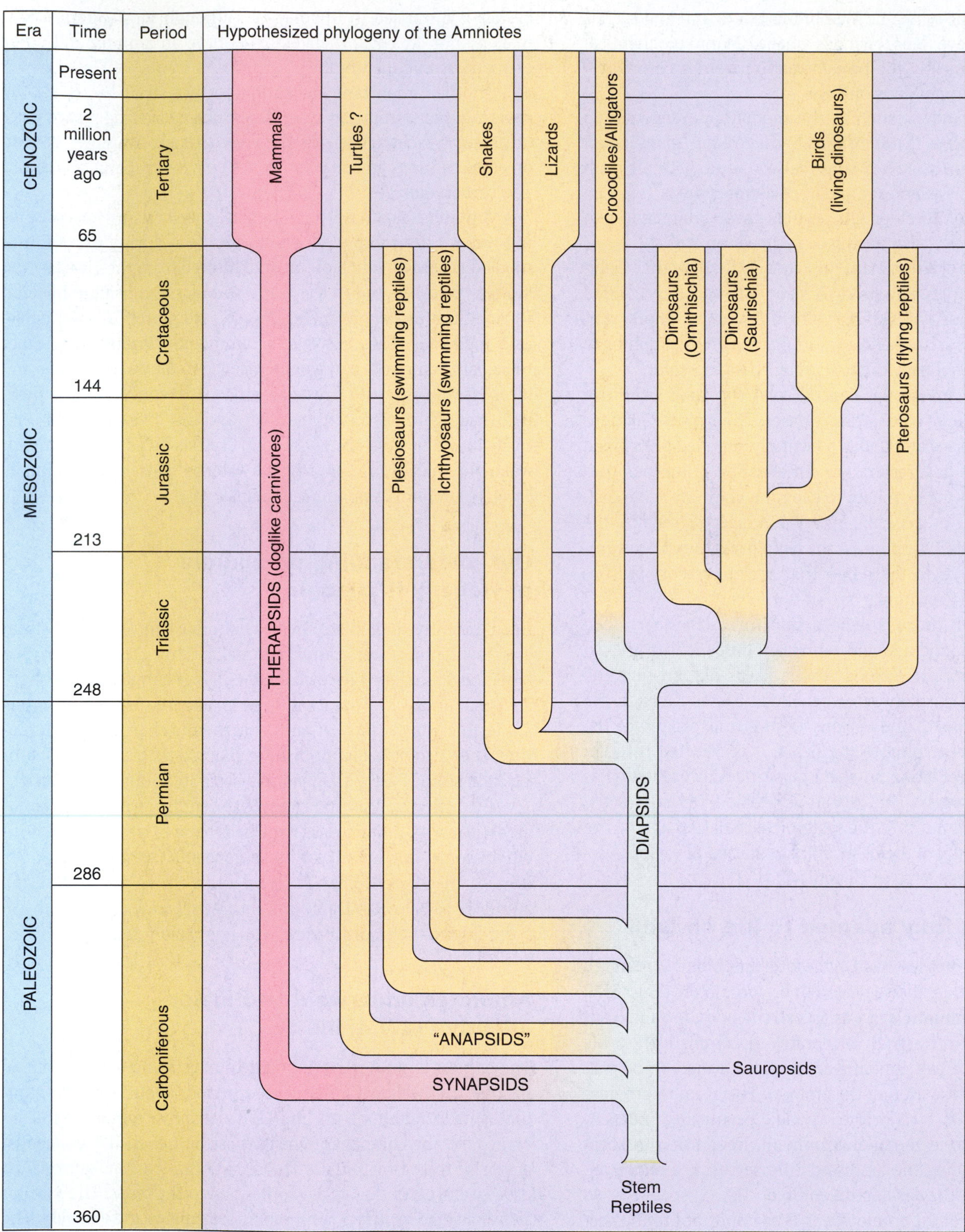

Figure 30–19 Phylogeny of the amniotes. Some proposed evolutionary relationships among extinct and present-day reptiles, birds, and mammals. Some of these relationships, for example, classification of the turtle as an anapsid, are currently being questioned and may be reinterpreted.

Fig. 20–13). Some were able to fly, others became marine, and many filled terrestrial habitats. Two major groups were the **dinosaurs** and the **pterosaurs,** the flying reptiles.

Some of the dinosaurs were among the largest land animals to have ever lived (see Fig. 20–15). Some dinosaurs apparently traveled in social groups and took care of their young. In 1998 a fossil of a four-chambered dinosaur heart was discovered. This finding supports the hypothesis that at least some dinosaurs were **endothermic,** meaning that they used metabolic energy to maintain a constant body temperature despite changes in the temperature of the environment. An advantage of endothermy is that it permits animals to be more active.

The reptiles were the dominant land animals for almost 200 million years. Then, toward the end of the Mesozoic era, many, including all the dinosaurs and pterosaurs, disappeared from the fossil record. In fact, more than half of all animal species became extinct at that time (see Chapter 20). As a result of this mass extinction, there are many more extinct reptiles than living species.

Reptiles possess many terrestrial adaptations

Many reptilian characters are adaptations to terrestrial life. The female reptile secretes a protective leathery shell around the egg, which helps prevent the developing embryo from drying out. Because sperm cannot penetrate this shell, internal fertilization must take place within the body of the female before the shell is added. In this process, the male uses a copulatory organ (penis) to transfer sperm into the female reproductive tract. As the embryo develops within the protective shell, an amnion forms and surrounds the embryo. The hard, dry, horny scales that protect the reptilian body from drying are another adaptation to life on land. This scaly protective armor, which also helps protect the reptile from predators, is shed periodically.

The dry reptilian skin is a poor organ for gas exchange. Reptilian lungs are better developed than the saclike lungs of amphibians. Divided into many chambers, the reptilian lung provides a greatly increased surface area for gas exchange. Most reptiles have a three-chambered heart that is more efficient than the amphibian heart. The single ventricle has a partition, though incomplete, that separates oxygen-rich and oxygen-poor blood, facilitating oxygenation of body tissues. In crocodiles the partition is complete, and the heart has four chambers.

Like fishes and amphibians, reptiles generally lack metabolic mechanisms for regulating body temperature. They are **ectothermic,** meaning that their body temperature fluctuates with the temperature of the surrounding environment. Some reptiles have behavioral adaptations that enable them to maintain a body temperature higher than that of their environment. You may have observed a lizard basking in the sun, which raises its body temperature and so increases its metabolic rate. This permits the lizard to hunt actively for food. When the body of a reptile is cold, its metabolic rate is low and it tends to be sluggish. Ectothermia may explain why reptiles are more successful in warm than in cold climates.

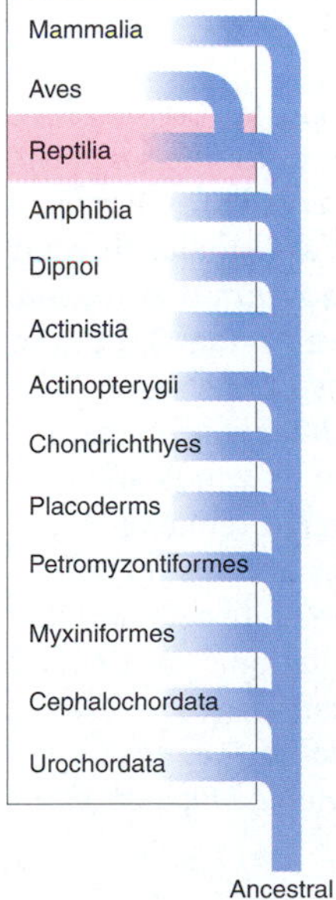

(a) Order Chelonia

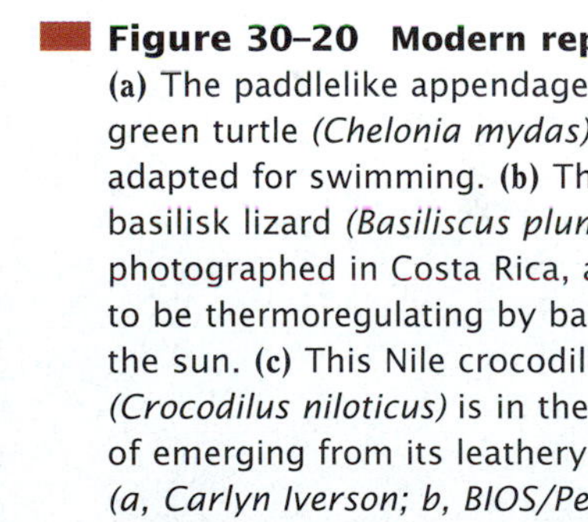

Figure 30–20 Modern reptiles. **(a)** The paddlelike appendages of this green turtle *(Chelonia mydas)* are adapted for swimming. **(b)** This basilisk lizard *(Basiliscus plumifrons),* photographed in Costa Rica, appears to be thermoregulating by basking in the sun. **(c)** This Nile crocodile *(Crocodilus niloticus)* is in the process of emerging from its leathery egg. *(a, Carlyn Iverson; b, BIOS/Peter Arnold, Inc.; c, Frans Lanting/Minden Pictures)*

(b) Order Squamata

(c) Order Crocodilia

Many reptiles are carnivores (animals that eat meat). Their paired limbs, usually with five toes, are well adapted for running and climbing, and their well-developed sense organs enable them to locate prey.

Ignoring birds for the moment, the extant reptiles are assigned to three main orders: Order **Chelonia** includes turtles, terrapins, and tortoises; order **Squamata** includes lizards and snakes; and order **Crocodilia** contains crocodiles, alligators, caimans, and gavials (Fig. 30–20 on page 661).

Members of order Chelonia are enclosed in a protective shell made of bony plates overlaid by horny scales. Some terrestrial species can withdraw their heads and legs completely into their shells. The size of adult turtles ranges from about 8 cm (3 in) long to that of the great marine species, which measure more than 2 m (6.5 ft) long and may weigh 450 kg (almost 1000 lb). The land species are usually referred to as tortoises, whereas the aquatic forms are called turtles; freshwater types are sometimes called terrapins.

Lizards and snakes are the most common of the modern reptiles. These animals have rows of scales that overlap like shingles on a roof, forming a continuous, flexible armor that may be shed periodically. Lizards range in size from certain geckos, which may weigh as little as 1 g, to the Komodo dragon of Indonesia, which may weigh 100 kg (220 lb). Their body sizes and shapes vary greatly. Some, like the glass snake, which is really a lizard, are legless.

Snakes are characterized by a flexible, loosely jointed jaw structure that permits them to swallow animals larger than the diameter of their own jaws. Snakes lack legs, and their bodies are elongated. Their eyes, covered by a transparent scale, do not have moveable eyelids. Also absent are an external ear opening, a tympanic membrane (ear drum), and a middle ear cavity.

The forked tongue of a snake, which often darts quickly from its mouth, is used as an accessory sensory organ for touch and smell. Chemicals from the ground or air adhere to the tongue. The tip is then rubbed across the opening to a sense organ located in the roof of the mouth that detects odors. Pit vipers and some boas also have a prominent **pit organ** on each side of the head that enables them to detect heat from endothermic prey (see Fig. 41–11). These sense organs permit them to locate and capture small nocturnal mammals.

Some snakes, for example, king snakes, pythons, and boa constrictors, kill their prey by rapidly wrapping themselves around the animal and squeezing it so that it cannot breathe. Others have fangs, which are hollow teeth connected to venom glands. When the snake bites, the venom is pumped through the fangs into the prey's body. Some snake venoms cause the breakdown of red blood cells; others, such as that of the coral snake, are neurotoxins that interfere with nerve function. Venomous snakes of the United States include rattlesnakes, copperheads, cottonmouths, and coral snakes. All except the coral snakes are pit vipers.

Three groups of crocodilians are (1) the crocodiles of Africa, Asia, and America; (2) the alligators of the southern United States and China, plus the caimans of Central America; and (3) the gavials of South Asia. Most species live in swamps, in

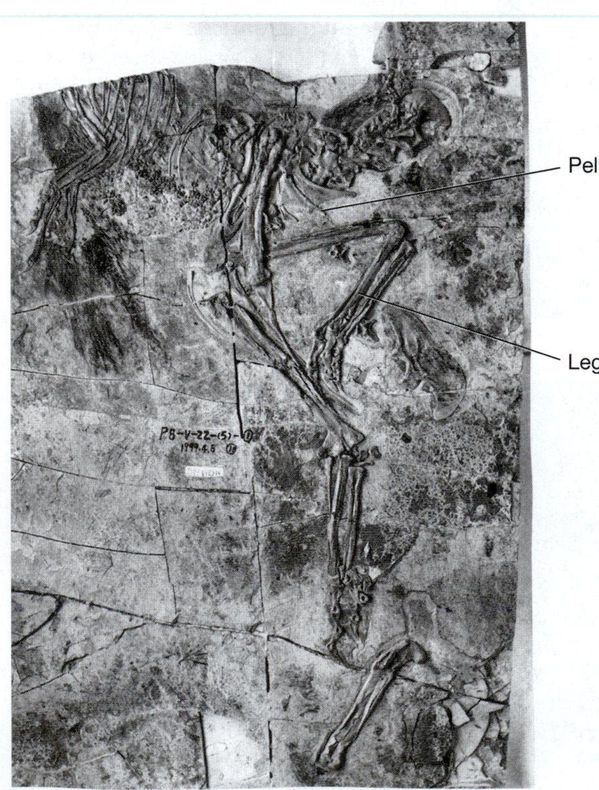

(a)

(b)

Figure 30–21 ***Caudipteryx* and *Archaeopteryx*.** **(a)** *Caudipteryx* exhibits both dinosaur and bird characteristics. This *Caudipteryx* fossil lacks a head, but the rest of its skeleton is well preserved. The bones in the foot and the shape and orientation of the pelvis are similar to what is found in the theropods (a group of long-tailed dinosaurs that moved about on two feet and had forelimbs with three clawed fingers), but the fossil also has bird characters, including impressions of feathers. **(b)** This reconstruction represents the view that *Archaeopteryx* was a climbing animal that had at least some ability to use its wings and feathers for gliding. Another view is that *Archaeopteryx* remained mainly on the ground, using its wings to trap small insects and its feathers for insulation. *(a, Chinese Academy of Sciences; b, From a painting by Rudolph Freund, courtesy of Carnegie Museum of Natural History)*

rivers, or along sea coasts, feeding on various kinds of animals. Crocodiles are the largest living reptiles; some exceed 7 m (23 ft) in length. The crocodile can be distinguished from the alligator or caiman by its long, slender snout and by the large fourth tooth on the bottom jaw that is visible when the mouth is closed.

Are birds really dinosaurs?

An increasing number of biologists object to treating birds as a separate class (Aves) of vertebrates. Cladistic analyses indicate that birds evolved from saurischian dinosaurs (see Fig. 20–14*a*), specifically carnivorous theropods. Many theropods were long-tailed animals that moved about on two feet and had forelimbs with three clawed fingers. Some biologists view birds as living dinosaurs!

Birds are the only extant animals with feathers. Feathers may have first evolved as an adaptation that conserved body heat. Bird ancestors became endothermic, which permitted them to be more active. According to one hypothesis, bird ancestors ran along the ground, using their feathered forelimbs to swat insects, which they then ate. Another hypothesis holds that bird ancestors climbed trees and used their feathered forelimbs for gliding.

Chinese investigators have described fossils of *Caudipteryx* as a possible transition form between theropod dinosaurs and birds (Fig. 30–21*a*). *Caudipteryx,* which has many dinosaur characters, also has imprints of feathers, and its bones show many bird characteristics not found in dinosaurs.

Although the bones of birds are fragile and disintegrate quickly, a few fossils of early birds have been found. The first birds looked very much like reptiles. They had teeth (which modern birds lack), a long tail, and bones with thick walls. Unlike reptiles, their jaws were elongated into beaks, and they had feathers and wings.

One of the earliest known birds, *Archaeopteryx* (meaning "ancient wing"), was about the size of a pigeon and had rather feeble wings (Fig. 30–21*b*). Unlike extant birds, its jawbones were armed with reptilian-type teeth, and it had a long reptilian tail covered with feathers. (Its feathers are more complex than those in *Caudipteryx.*) Each *Archaeopteryx* wing was equipped with three claw-bearing digits. Like modern birds, *Archaeopteryx* had wings, feathers, reduced digits, and a **furcula** ("wishbone"). Several specimens of this genus have been found in the Jurassic limestone of Bavaria, which was laid down about 145 mya. *Archaeopteryx* did not itself give rise to modern birds but is thought to have evolved from a common ancestor with other birds.

Cretaceous rocks have yielded fossils of other early birds. *Hesperornis,* which lived in North America, was a toothed, aquatic diving bird with rudimentary wings and broad, webbed feet for swimming. *Ichthyornis* was a toothed, flying bird about the size of a sea gull. From the Tertiary period onward, the fossil record of birds shows an absence of teeth and progressive changes leading to the modern birds.

Modern birds are adapted for flight

About 9000 species of extant birds have been described, and they have been classified into about 23 monophyletic groups. Birds inhabit a wide variety of habitats and can be found on all continents, most islands, and even the open sea. Birds are a diverse group (Fig. 30–22). The largest living birds are the ostriches

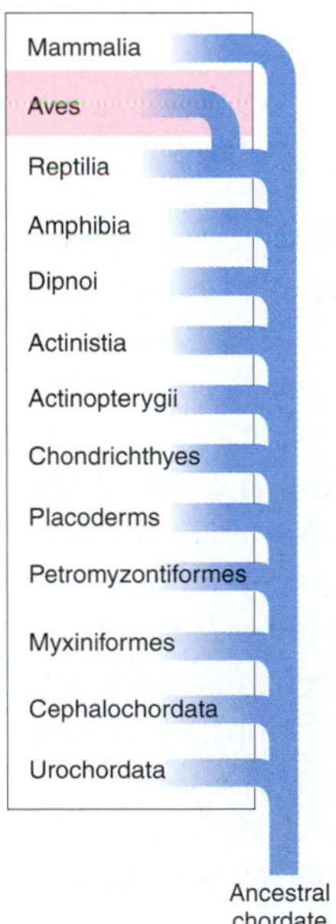

(a)

(b)

(c)

Figure 30–22 Modern birds. Although they are diverse, birds are highly adapted for flight, and their basic structure is similar. **(a)** A male peacock *(Pavo cristatus)* impressively displaying his feathers. **(b)** This Atlantic puffin *(Fratercula arctica)* has caught several fish. **(c)** Two fledgling wood storks *(Mycteria americana)*, about 8 to 9 weeks old, preen their feathers atop their nest in South Carolina's White Hall rookery. Young wood storks have yellow bills and dark feathers on their heads and necks. By adulthood they lose these feathers and the blackish skin on their heads and necks is exposed. *(a, Frans Lanting/Minden Pictures; b, Gary Meszaros/Dembinsky Photo Associates; c,* Smithsonian, *Vol. 31, Feb. 2001/Tom Blagden)*

of Africa, which may be up to 2 m (6.5 ft) tall and weigh 136 kg (300 lb), and the great condors of the Americas, with wingspans of up to 3 m (10 ft). The smallest known bird is Helena's hummingbird of Cuba, with a length of less than 6 cm (2 in) and a weight of less than 4 g (0.14 oz).

Like their reptilian ancestors, modern birds lay eggs and have reptilian-type scales on their legs. Birds have evolved remarkable specializations for flight. Feathers, which evolved from reptilian scales, aid in flight by presenting a flat surface to the air. They are very light, yet flexible and strong. Feathers also protect the body and decrease water and heat loss.

The anterior limbs of birds are wings, usually adapted for flight. The posterior limbs are modified for walking, swimming, or perching. Not all birds fly. Some, such as penguins, have small, flipper-like wings used in swimming. Ratites are a group of mainly large, flightless birds, including the ostriches, cassowarys, and kiwis. These birds have only vestigial wings, but they have well-developed legs, which they use for running.

In addition to feathers and wings, birds have many other adaptations for flight. Their bodies are compact and streamlined, and the fusion of many bones gives them the rigidity needed for flying. Their bones are strong but very light. Many are hollow, containing large air spaces. The avian jaw is light, and, instead of teeth, there is a light, horny beak. The breastbone is broad and keeled for the attachment of the large flight muscles.

Birds have efficient lungs with thin-walled extensions, called **air sacs,** that occupy spaces between the internal organs and within certain bones. They have a unique "one-way" flow of air through their respiratory system (discussed in Chapter 44). Birds have a four-chambered heart and a double circuit of blood flow. Blood delivers oxygen to the tissues and then is recharged with oxygen in the lungs before being pumped out into the systemic circulation again. The very effective respiratory and circulatory systems provide the cells with enough oxygen to permit a high metabolic rate, which is necessary for the tremendous muscular activity that flying requires. Some of the heat generated by metabolic activities is used to maintain a constant body temperature. Birds are *endothermic,* which permits them to remain active in cold climates.

Birds excrete nitrogenous wastes mainly as semisolid uric acid. Because birds typically do not have a urinary bladder, these solid wastes are delivered into the cloaca. They leave the body with the feces, which are dropped frequently. This adaptive mechanism helps to maintain a light body weight.

Birds have become adapted to a variety of environments, and various species have very different types of beaks, feet, wings, tails, and behavioral patterns. Bills are specifically adapted for the type of food the bird eats. Although all birds must eat frequently (because they have a high metabolic rate and typically do not store much fat), the choice of food varies widely among species. Most birds eat energy-rich foods such as seeds, fruits, worms, mollusks, or arthropods. Warblers and some other species eat mainly insects. Owls and hawks eat rodents, rabbits, and other small mammals. Vultures feed on dead animals. Pelicans, gulls, terns, and kingfishers catch fish. Some hawks catch snakes and lizards.

An interesting feature of the bird digestive system is the **crop,** an expanded, saclike portion of the digestive tract below the esophagus, in which food is temporarily stored. The stomach is divided into a **proventriculus,** which secretes gastric juices, and a thick, muscular **gizzard,** which grinds food. The bird swallows small bits of gravel that act as "teeth" in the gizzard, mechanically breaking down food.

Birds have a well-developed nervous system with a brain that is proportionately larger than that of reptiles. Birds rely heavily on vision, and their eyes are relatively larger than those of other vertebrates. Hearing is also well developed.

In striking contrast with the relatively silent reptiles, birds are very vocal. Most have short, simple *calls* that signal danger or influence feeding, flocking, or interaction between parent and young. *Songs* are usually more complex than calls and are performed mainly by males. Songs are related to reproduction, attracting and keeping a mate, and claiming and defending a territory.

One of the most fascinating aspects of bird behavior is the annual migration that many species make. Some birds, such as the golden plover and Arctic tern, fly from Alaska to Patagonia, South America, and back each year, covering perhaps 40,250 km (25,000 mi) en route. Migration and navigation are discussed in Chapter 50.

Beautiful and striking colors are found among birds. Color is due partly to pigments deposited during the development of the feathers and partly to reflection and refraction of light of certain wavelengths. Many birds, especially females, are protectively colored by their plumage. Brighter colors are often assumed by the male during the breeding season to help in attracting a mate.

Mammals are characterized by hair and mammary glands

Derived characters that distinguish mammals (class Mammalia) include hair, which insulates and protects the body; **mammary glands,** which produce milk for the young; the differentiation of teeth into incisors, canines, premolars, and molars; and three middle ear bones (malleus, incus, and stapes) that conduct vibrations.

A muscular **diaphragm** helps move air into and out of the lungs. Like birds, mammals are *endotherms,* but mammals are thought to have evolved endothermy independently of birds. The process of maintaining a constant body temperature is enhanced by the covering of insulating hair, by the four-chambered heart, and by separate pulmonary and systemic circulations. Red blood cells without nuclei serve as efficient oxygen transporters.

Contributing significantly to the success of the mammals, the complex nervous system is more highly developed than in any other group of animals. The cerebrum is especially large and complex, with an outer gray region called the **cerebral cortex.**

Fertilization is always internal, and, except for the primitive monotremes that lay eggs, mammals are viviparous. Most mammals develop a **placenta,** an organ of exchange between developing embryo and mother, through which the embryo receives its nourishment and oxygen and rids its blood of wastes. By carrying their developing young internally, mammals avoid the hazards of having their eggs consumed by predators. By nourishing the young and caring for them, the parents offer both protection and an "education" on how to obtain food and avoid being eaten.

The limbs of mammals are variously adapted for walking, running, climbing, swimming, burrowing, or flying. In four-legged mammals, the limbs are more directly under the body than they are in extant reptiles, which contributes to speed and agility. Life processes of mammals are discussed in detail in Part 7.

Early mammals were small, endothermic animals

Mammals descended from a group of synapsid reptiles known as **therapsids** during the Triassic period some 200 mya. The therapsids were doglike carnivores with differentiated teeth and legs adapted for running (Fig. 30–23). The fossil record indicates that early mammals were small, about the size of a mouse or shrew. Some may have been endothermic, and some may have had fur.

How did the mammals manage to coexist with the reptiles during the 160 million years or so that reptiles ruled Earth? Many adaptations permitted the early mammals to compete for a place on our planet. Perhaps one of the most important was their skill at being inconspicuous. They were **arboreal** (tree-dwelling) and **nocturnal** (active at night), searching for food (mainly insects and plant material, and perhaps reptile eggs) at night while the reptiles were inactive. Although their eyes may have been small, early mammals had well-developed sense organs for hearing and smell.

Figure 30–23 Therapsid *(Lycaenops).* The therapsids were synapsid reptiles. *Lycaenops*, a carnivore about the size of a coyote, lived in South Africa during the late Permian period. *(Painting by John C. Germann, Department of Library Services, American Museum of Natural History)*

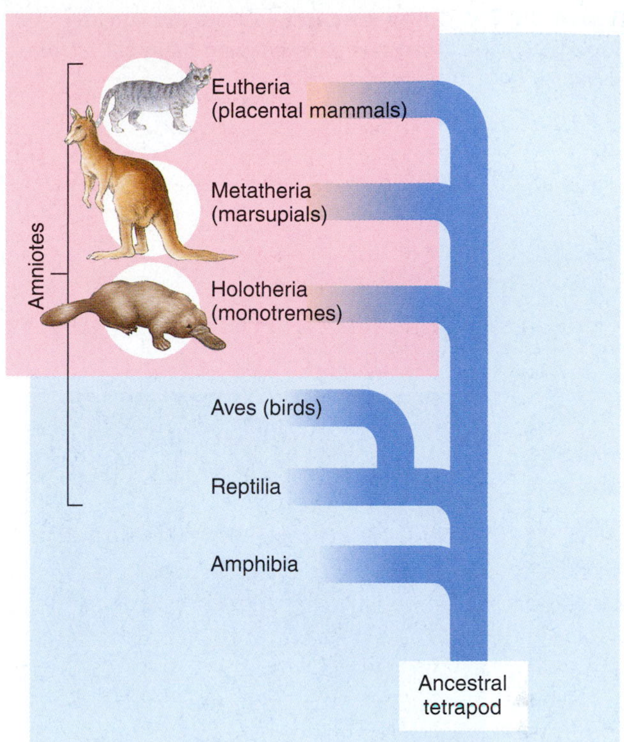

Figure 30–24 Evolution of mammals. One current interpretation of the evolution of mammals. Reptiles, birds, and mammals are amniotes.

As many reptiles died out, the mammals adapted to their abandoned niches (lifestyles). During this time, the flowering plants, including many trees, underwent adaptive radiation, providing new habitats, sources of food, and protection from predators. Larger forms and numerous varieties of mammals evolved. During the early Cenozoic era (more than 55 mya), the mammals underwent adaptive radiation, becoming widely distributed and adapted to an impressive variety of ecological lifestyles (see Fig. 20–16).

Modern mammals are assigned to three subclasses

By the end of the Cretaceous period, three main groups of mammals had evolved (Fig. 30–24). Today, mammals inhabit virtually every corner of Earth; they are found on land, in fresh and salt water, and in the air. They range in size from the tiny pigmy shrew, weighing about 2.5 g (less than 0.1 oz), to the blue whale, which may weigh up to 136,000 kg (150 tons) and is thought to be one of the largest animals that has ever lived.

Modern mammals are classified in three subclasses: **Holotheria** includes the egg-laying mammals, also called monotremes; **Metatheria** includes the marsupials, or pouched mammals; and **Eutheria** includes the placental mammals.

The **monotremes** are the only living order of subclass Holotheria. One genus includes the duck-billed platypus *(Ornithorhynchus)* and a second, the spiny anteaters or echidnas *(Tachyglossus)* (Fig. 30–25; see also Figs. 1–17 and 22–5). These animals are found in Australia and Tasmania; two of the three

Figure 30–25 Spiny anteater. The spiny anteater *(Tachyglossus aculeatus)* is an egg-laying mammal. *(Tom McHugh/Photo Researchers, Inc.)*

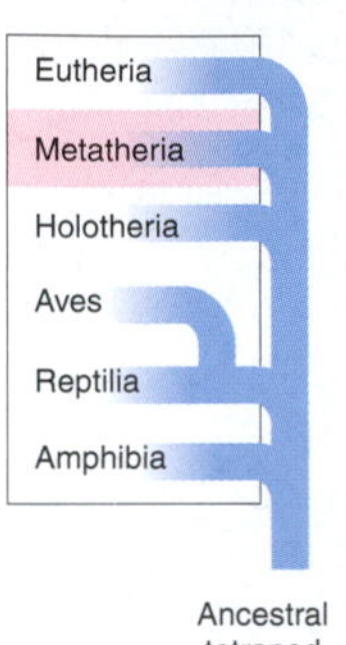

(a)

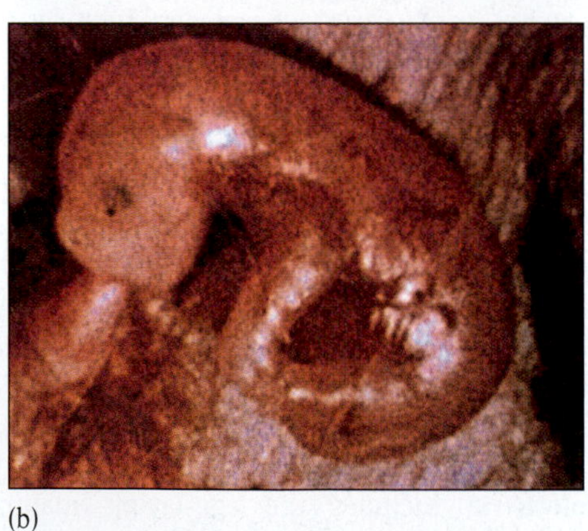

(b)

Figure 30–26 Marsupials. **(a)** Eastern gray kangaroo *(Macropus giganteus)* with young, known as a joey. The kangaroo is native to Australia. **(b)** A kangaroo soon after birth. Marsupials are born in an embryonic state and continue to develop in the safety of the marsupium (pouch). *(a, John Cancalosi/Peter Arnold, Inc.; b, Robert Anderson, reprinted with permission of Hubbard Scientific Company)*

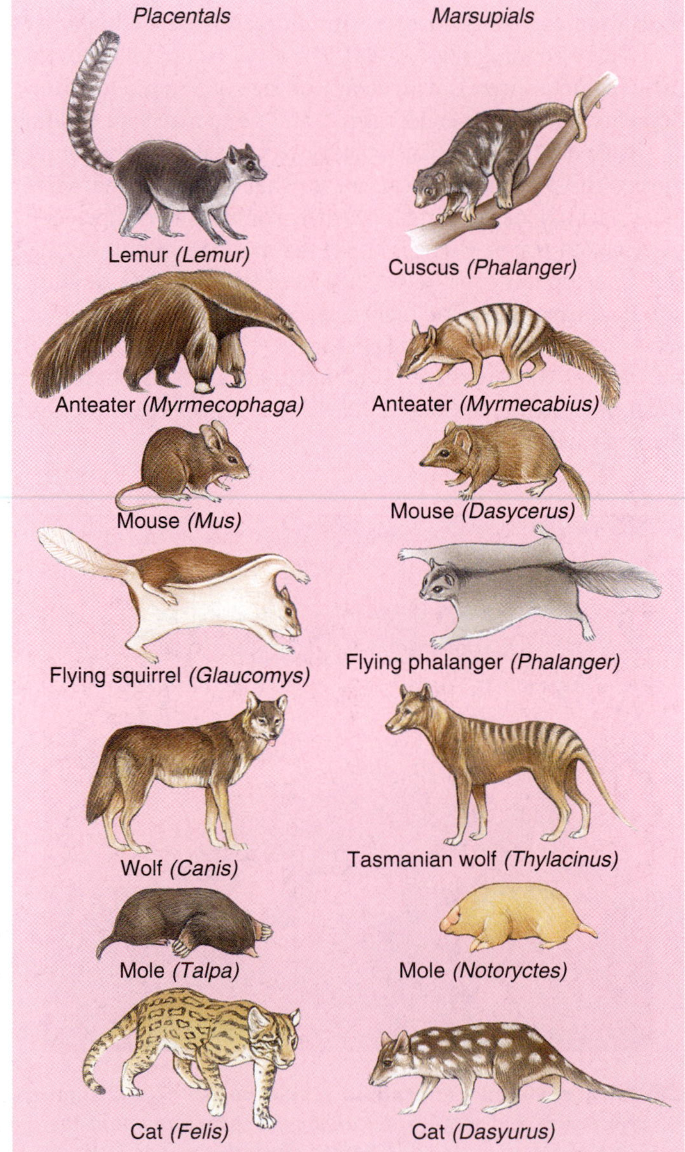

Figure 30–27 Convergent evolution in placental and marsupial mammals. For each mammal in one group, there is a counterpart in the other group. Their similarities include both lifestyles and structural features.

Figure 30–28 Placental mammals. (a) Wildebeest *(Connochaetes taurinus)*, photographed in Tanzania. The wildebeest, the dominant plains antelope in many areas of eastern and southern Africa, migrates long distances during the dry season in search of food and water. (b) Humpback whales *(Megaptera novaeangliae)* exhibiting a rare double breach. (c) Polar bears *(Ursus maritimus)*, the largest living land carnivores, feed mainly on seals. *(a, McMurray Photography; b, James D. Watt/Animals Animals; c, Michio Hoshino/Minden Pictures)*

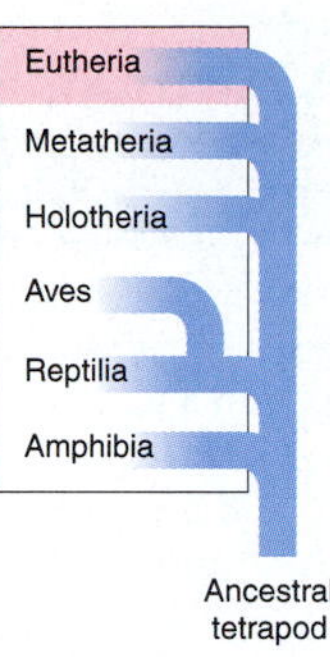

(b)

(a)

(c)

spiny anteater species are also found in New Guinea. The females lay eggs that may be carried in a pouch on the abdomen or kept warm in a nest. When the young hatch, they lap up milk secreted by mammary glands. Unlike other mammals, nipples are absent.

Spiny anteaters capture ants and termites with the long beak and long, sticky tongue. The duck-billed platypus, which lives in burrows along river banks, preys on freshwater invertebrates. It has webbed feet and a flat, beaver-like tail, which aids in swimming.

Marsupials include pouched mammals such as kangaroos and opossums (Fig. 30–26*a*). Embryos begin their development in the mother's uterus, where they are nourished by fluid and yolk. After a few weeks, still in a very undeveloped stage, the young are born. In many species, the young crawl to the **marsupium** (pouch), where they complete their development. The young marsupial attaches its mouth to a mammary gland nipple and is nourished by its mother's milk (Fig. 30–26*b*).

At one time, marsupials probably inhabited much of the world, but they were largely out-competed by placental mammals. Now marsupials are found mainly in Australia and in Central and South America. The only marsupial species that ranges into the United States is the house cat–sized Virginia opossum *(Didelphis virginiana)*. Australia became geographically isolated from the rest of the world before placental mammals reached it, and there the marsupials remained the dominant mammals (see discussion of continental drift in Chapter 17). Marsupials underwent adaptive radiation, paralleling the evolution of placental mammals elsewhere. Thus, in Australia and adjacent islands we find marsupials that correspond to North American placental wolves, bears, rats, moles, flying squirrels, and even cats (Fig. 30–27).

Most familiar to us are **placental mammals,** characterized by development of a placenta, an organ of exchange between the developing embryo and its mother (Fig. 30–28). The placenta forms from both embryonic membranes and the maternal uterine wall. In it the blood vessels of the embryo come very close to the blood vessels of the mother, so materials can be exchanged by diffusion. (The two circulations do not normally mix.) The placenta enables the young to remain within the mother's body until embryonic development is complete.

Placental mammals are born at a more mature stage than are marsupials. Indeed, among some species the young can walk around and begin to interact with other members of the group within a few minutes after birth. Extant placental mammals are classified into about 16 orders. A brief summary of some of these orders is given in Table 30–3.

TABLE 30–3 Some Orders of Living Placental Mammals

Order and Representative Members			Some Characteristics
Insectivora Moles, hedgehogs, and shrews		African hedgehog	Nocturnal; eat insects; considered most primitive placental mammals. Shrew is the smallest living mammal; some weigh less than 5 g.
Chiroptera Bats		Bat	Adapted for flying; a fold of skin extends from the elongated fingers to the body and legs, forming a wing. Guided in flight by a type of biological sonar; they emit high-frequency squeaks and are guided by the echoes from obstructions. Eat insects and fruit, or suck blood of other animals.
Carnivora Cats, dogs, wolves, foxes, bears, otters, mink, weasels, skunks		Wolf	Carnivores with sharp, pointed canine teeth and molars for shearing. Keen sense of smell; complex social interactions. Among fastest, strongest, and smartest animals.
Xenarthra Sloths, anteaters, armadillos		Nine-banded armadillo	Teeth reduced or no teeth. Sloths are sluggish animals that hang upside down from branches; often protectively colored by green algae that grow on their hair. Armadillos are protected by bony plates; eat insects and small invertebrates.
Rodentia Squirrels, beavers, rats, mice, hamsters, porcupines, guinea pigs		Flying squirrel	Gnawing animals with chisel-like incisors. As they gnaw, teeth are worn down, and so must grow continually.
Lagomorpha Rabbits, hares, pikas		Pika	Like rodents, have chisel-like incisors. Typically have long hind legs adapted for jumping. Many have long ears.
Primates Lemurs, monkeys, apes, humans		Ring-tailed lemur	Highly developed brains and eyes. Nails instead of claws. Opposable thumb. Eyes directed forward. Omnivores. Most species arboreal. (Primate evolution is discussed in Chapter 21.)

TABLE 30–3 Continued

Order and Representative Members			Some Characteristics
Perissodactyla Horses, zebras, tapirs, rhinoceroses		Tapir	Herbivores. Hoofed with an odd number of digits per foot; one or three toes. Teeth adapted for chewing. Usually large animals with long legs.
Artiodactyla Cattle, sheep, pigs, deer, giraffes		American elk	Hoofed with even numbers of digits per foot; most have two toes, some have four. Most have antlers or horns. Herbivores; most are ruminants that chew a cud and have a series of stomachs in which bacteria that digest cellulose are incubated; this contributes to their success as herbivores.
Proboscidea Elephants		African elephant	Largest land animals; weigh up to 7 tons; large head; broad ears; long, muscular trunk (proboscis) that is very flexible. Thick, loose skin is characteristic. The two upper incisors are elongated as tusks. This order includes the extinct mastodons and wooly mammoths.
Sirenia Sea cows, manatees		Manatee	Herbivorous, aquatic mammals with finlike forelimbs and no hind limbs. They are probably the basis for most tales about mermaids.
Cetacea Whales, dolphins, porpoises		Humpback whale	Adapted for aquatic life with fish-shaped body and broad, paddle-like forelimbs (flippers). Posterior limbs absent. Many have thick layer of fat (blubber) under the skin. Some are suspension feeders. Mate and bear their young in the water. Very intelligent.
Pinnipedia Seals, sea lions, walruses		Sea lions	Marine; limbs adapted as flippers for swimming. Carnivores; eat fish.

SUMMARY WITH KEY TERMS

I. **Deuterostomes** include echinoderms, hemichordates, and chordates.

II. Phylum **Echinodermata** includes marine animals with a spiny "skin," **water vascular system, tube feet,** and **endoskeleton.** The larvae have bilateral symmetry; most of the adults exhibit **pentaradial symmetry.**

   A. Class **Crinoidea** includes the sea lilies and feather stars. In these echinoderms, the **oral surface** is turned upward; some crinoids are sessile.

   B. Class **Asteroidea** consists of the sea stars, echinoderms with a central disk from which radiate five or more arms. Tube feet are used for locomotion.

   C. Class **Ophiuroidea** includes the brittle stars, which resemble sea stars but have longer, more slender arms that are set off more distinctly from the central disk. The arms are used for locomotion. Their tube feet lack suckers and are not used in locomotion.

   D. Class **Echinoidea** includes the sea urchins and sand dollars, animals that lack arms; they have a solid shell and are covered with spines.

   E. Class **Holothuroidea** consists of sea cucumbers, animals with elongated flexible bodies; the mouth is surrounded by a circle of modified tube feet that serve as tentacles.

III. **Hemichordates** are marine deuterostomes with a three-part body, including proboscis, collar, and trunk. The acorn worms are the more familiar of the two extant classes.

IV. Phylum **Chordata** consists of three subphyla: Urochordata, Cephalochordata, and Vertebrata. At some time in its life cycle, a chordate has a flexible, supporting **notochord,** a **dorsal tubular nerve cord,** and **pharyngeal (gill) slits.** Chordates are also characterized by a **postanal tail.**

   A. The **tunicates,** which belong to subphylum **Urochordata,** are suspension-feeding, marine animals with **tunics.** Larvae have typical chordate characteristics and are free-swimming. Adults of most groups are sessile suspension feeders, but some species can move about and are part of the plankton.

   B. Subphylum **Cephalochordata** consists of the **lancelets,** small, segmented, fishlike animals that exhibit chordate characteristics.

   C. Subphylum **Vertebrata** includes animals with a **vertebral column** forming the chief skeletal axis of the body. Vertebrates also have a braincase called a **cranium,** pronounced cephalization, a complex brain, muscles attached to the **endoskeleton** for movement, and **neural crest cells.**

      1. Some of the earliest known vertebrates, called **ostracoderms,** were jawless fishes that are now extinct.

      2. Biologists formerly assigned the jawless fishes, the lampreys and hagfishes, to class **Agnatha.** Most biologists now divide the jawless fishes into two classes: **Myxiniformes** (hagfishes) and **Petromyzontiformes** (lampreys). Some systematists no longer consider hagfishes to be vertebrates because they have no trace of vertebrae. They classify the vertebrates plus the hagfishes as **craniates.**

      3. Class **Chondrichthyes,** the cartilaginous fishes, includes the sharks, rays, and skates.

         a. These fishes have jaws, two pairs of fins, and **placoid scales.**

         b. Skates and some species of sharks are **oviparous,** meaning that they lay eggs. Many species of sharks are **ovoviparous;** their young are enclosed by eggs that are incubated in the mother's body. A few shark species are **viviparous;** the young develop in the mother's uterus and are nourished by transfer of nutrients from the mother's blood.

      4. The bony fishes were formerly assigned to Class **Osteichthyes.** Biologists now divide them into three classes: **Actinopterygii,** ray-finned fishes; **Actinistia,** coelancanths; and **Dipnoi,** lungfishes.

         a. The bony fishes and cartilaginous fishes are thought to have evolved at about the same time. During the Devonian, bony fishes gave rise to two evolutionary lines: the Actinopterygii, or **ray-finned fishes,** and the **Sarcopterygii,** or **lobe-finned fishes.**

         b. The ray-finned fishes gave rise to the modern bony fishes. In these fishes, the lungs have been modified as a **swim bladder,** an air sac for regulating buoyancy.

         c. The sarcopterygian fishes are thought to have given rise to the **lungfishes** (class **Dipnoi**) and **coelacanths** (class **Actinistia**).

         d. Both coelancanths and lungfishes were apparently preadapted for life on land. Biologists now think that the lungfish gave rise to the **tetrapods,** the land vertebrates.

      5. The first successful tetrapods were the now-extinct **labyrinthodonts.** These animals are thought to have been the ancestors of frogs and salamanders.

      6. Class **Amphibia** includes salamanders, frogs and toads, and wormlike caecilians.

         a. Most amphibians return to the water to reproduce. Frog embryos develop into **tadpoles,** which undergo **metamorphosis** to become adults.

         b. Amphibians use their moist skin as well as lungs for gas exchange. They have a three-chambered heart and **systemic** and **pulmonary circulations.**

      7. Terrestrial vertebrates, or **amniotes,** include reptiles, birds, and mammals. The development of the **amniotic** egg with its shell and **amnion** was an important adaptation for life on land.

      8. Class **Reptilia,** as traditionally defined, is paraphyletic. It includes dinosaurs, turtles, lizards, snakes, and alligators. Many biologists now include the birds in the dinosaur clade.

         a. In reptiles, fertilization is internal; most reptiles secrete a leathery protective shell around the egg; the embryo develops protective membranes, including an amnion that keeps it moist.

         b. A reptile has a dry skin with horny scales, lungs with many chambers, and a three-chambered heart with some separation of oxygen-rich and oxygen-poor blood.

         c. Reptiles dominated Earth during the Mesozoic era; then, toward the end of the Cretaceous period, many reptiles, including the dinosaurs, became extinct.

      9. Birds (traditionally assigned to class **Aves**) have many adaptations for flight, including feathers, wings, and light, hollow bones containing air spaces.

         a. Birds have a four-chambered heart, very efficient lungs, a high metabolic rate, and a constant body temperature; they excrete solid metabolic wastes (uric acid).

         b. Birds have a well-developed nervous system and excellent vision and hearing.

         c. Birds communicate with simple calls and complex songs, as well as with color and behavior.

      10. **Mammals** have hair, **mammary glands,** differentiated teeth, and three middle ear bones. They maintain a constant body temperature and have a highly developed nervous system and a muscular **diaphragm.**

         a. **Monotremes,** mammals that lay eggs, include the duck-billed platypus and spiny anteaters.

         b. **Marsupials** include pouched mammals, such as kangaroos and opossums. The young are born in an embryonic stage and complete their development in the **marsupium,** where they are nourished with milk from the mammary glands.

         c. **Placental mammals** are characterized by an organ of exchange, the **placenta,** that develops between the embryo and the mother. Both oxygen and nutrients diffuse across the placenta from mother to embryo, permitting development to take place within the uterus. Living placental mammals are classified in about 16 orders.

POST-TEST

1. Which of the following is *not* a deuterostome? (a) echinoderm (b) chordate (c) hemichordate (d) arthropod (e) hagfish
2. Which of the following belongs to subphylum Vertebrata? (a) lophophorate (b) lamprey (c) lancelet (d) tunicate (e) crinoid
3. Which of the following is (are) found in tunicates? (a) notochord (b) tube feet (c) anal gill slits (d) two pairs of appendages (e) vertebral column
4. Which of the following was an early jawed fish? (a) placoderm (b) crinoid (c) therapsid (d) labyrinthodont (e) ostracoderm.
5. An evolutionary sequence accepted by many biologists is (a) lancelet → ancestral chordate → tunicate (b) lungfish → labyrinthodont → modern amphibian (c) ancestral chordate → turtle→ ostracoderm → amniote (d) therapsid → placental mammal → monotreme (e) dinosaur → *Archaeopteryx* → therapsid → marsupial.
6. Which of the following characteristics is *not* associated with sea stars? (a) tube feet (b) water vascular system (c) central disk with five or more arms (d) spiny skeleton (e) notochord
7. Which of the following characteristics is associated with amphibians? (a) mantle (b) placoid scales (c) three-chambered heart (d) swim bladder (e) amnion
8. Which of the following is *not* characteristic of birds? (a) feathers (b) ectothermic (c) amnion (d) high metabolic rate (e) reptilian-like scales on legs
9. Which of the following is *not* true of the duck-billed platypus? (a) classified as a marsupial (b) lays eggs (c) embryo has a notochord (d) embryo has pharyngeal slits (e) is a predator of freshwater invertebrates
10. Which of the following is *not* a placental mammal? (a) shrew (b) human (c) dolphin (d) spiny anteater (e) bat
11. Which of the following is true of mammals? (a) all have placentas (b) evolved from saurischian dinosaurs (c) three middle ear bones (d) most are ovoviviparous (e) evolved during the Devonian period
12. The vertebral column (a) is characteristic of all chordates (b) forms the skeletal axis of the chordate body (c) may consist of cartilaginous or bony segments (d) is well developed in lancelets (e) three of the preceding answers are correct
13. A shark is characterized by (a) placoid scales (b) bony skeleton (c) water vascular system (d) amnion (e) muscular diaphragm
14. Reptiles (a) have dry, scaly skin (b) are amniotes (c) gave rise to the birds and mammals (d) answers a, b, and c are correct (e) answers a and c only are correct
15. Amniotes (a) include amphibians, reptiles, birds, and mammals (b) are especially well adapted for fresh water (c) typically have physiological mechanisms for conserving water (d) typically excrete ammonia (e) gave rise to the amphibians

REVIEW QUESTIONS

1. Echinoderms and chordates are both deuterostomes. What are some characteristics that they have in common?
2. What are four derived characters of a chordate? How are these evident in a tunicate larva? In an adult tunicate? In a lancelet? In a human?
3. What derived characters distinguish the vertebrates from the rest of the chordates?
4. How do hagfishes differ from other fishes?
5. Compare the skins of sharks, frogs, snakes, and mammals.
6. Identify organisms that possess each of the following, and give the location and function of each structure: (a) swim bladder (b) placenta (c) operculum (d) amnion (e) marsupium
7. Classify each of the following animals, giving the phylum, subphylum, class (and order if you can): (a) lizard (b) perch (c) lamprey (d) *Branchiostoma* (amphioxus) (e) shark (f) whale (g) frog (h) pelican (i) bat
8. Which vertebrate groups are endothermic? How do they accomplish this? Why is it advantageous?
9. From an evolutionary perspective, what is the significance of each of the following: (a) lungfishes (b) placoderms (c) labyrinthodonts (d) therapsids (e) *Archaeopteryx*
10. What is the significance of *Hox* gene clusters?

YOU MAKE THE CONNECTION

1. What is the function of gills? In general terms, how do they work? Why do you suppose aquatic mammals do not have them?
2. Some paleontologists consider monotremes to be therapsid reptiles rather than mammals. Give arguments for and against this position.
3. Which are more specialized, birds or mammals? Explain your answer.
4. Imagine that you discover an interesting new animal. You find that it has a dorsal, tubular nerve cord, a cranium, moist skin, and a heart with two atria and a ventricle. How would you classify the animal? Explain each step in your decision.

RECOMMENDED READINGS

Diamond, J. "Eat Dirt." *Discover,* Vol. 19, No. 2, Feb. 1998. A discussion of adaptations of parrots, one of the most successful groups of birds.

Houlahan J.E., et al. "Quantitative Evidence for Global Amphibian Population Declines." *Nature,* Vol. 404, 13 Apr. 2000. This paper examines the results of more than 900 studies of amphibian declines that were conducted between 1950 and 1997.

Martini, F.C. "Secrets of the Slime Hag." *Scientific American,* Vol. 279, No. 4, Oct. 1998. Hagfishes are important predators and scavengers; they produce large amounts of slime when attacked.

McClanahan, L.L., Ruibal, R., and V.H. Shoemaker. "Frogs and Toads in the Desert." *Scientific American.* Vol. 270, No. 3, Mar. 1994. Certain amphibians have adaptations that permit them to inhabit desert areas.

Padian, K., and L.M. Chiappe. "The Origin of Birds and Their Flight." *Scientific American,* Vol. 279, No. 2, Feb. 1998. An account of the evolution of birds from dinosaurs.

Rismiller, P.D., and R.S. Seymour. "The Echidna." *Scientific American,* Vol. 264, No. 2, Feb. 1991. Considers the natural history and reproductive behavior of the spiny anteater, a mammal that lays eggs.

Wray, G.A., and R.A. Raff. "Body Builders of the Sea." *Natural History,* 12/98–1/99. An interesting account of the evolution of the echinoderm body plan.

Zimmer, C. "In Search of Vertebrate Origins: Beyond Brain and Bone." *Science,* Vol. 287, 3 Mar. 2000. The evolution of neural crest cells, a group of embryonic cells that gives rise to nerves and many other structures, was an important step in developing a new body plan.

• Visit our Web site at **http://www.info.brookscole.com/solomonbergmartin** for links to chapter-related resources on the World Wide Web. Additional on-line materials relating to this chapter can also be found on our Web site.

CAREER VISIONS

Virologist

DR. JONATHAN TOWNER

***Dr. Jonathan Towner** is a virologist in the Special Pathogens Branch of the Centers for Disease Control and Prevention (CDC) in Atlanta. He utilizes this group's high containment BioSafety Level 4 laboratory to investigate the deadly and incurable Ebola virus. A native of Palos Verde, California, Jon earned a B.A. in Microbiology and Immunology in 1989 at the University of California, Berkeley, and a Ph.D. in Biology in 1997 at the University of California, Irvine.*

When did you decide to become a biologist?

Probably when I was an undergraduate at Berkeley. I started college as a premed student, like it seems everybody does. A senior student I knew told me how interesting and challenging it was to be a microbiology major. I then got involved in microbiology and started performing undergraduate research in an immunology lab, which I really liked. That was the point at which I started to take a different path, and I felt less that I wanted to be a medical doctor, and more that I liked basic science. The prospect of scientific discovery in the laboratory started to intrigue me.

What led to your interest in viruses?

I started getting interested in viruses my first year in graduate school, when I did rotations in different research labs. One faculty member there, Dr. Bert Semler, was excited about his work, which really made others enthusiastic about it. As a mentor, he got me excited about working with viruses.

In your doctoral work, you studied poliovirus. What was the goal of your research?

The broad goal of that work for me was really to learn how to do research. There's a certain kind of mental discipline or mindset you have to get into to learn how to think critically. With respect to poliovirus, I was interested in studying the replication of the virus, how some of the viral proteins interact to form the RNA replication complex, which makes multiple copies of the viral genome, the genetic material.

What is meant by the term "emerging viruses"?*

To me, an emerging virus is one that is becoming a new human health problem that we are beginning to recognize. Some viruses, like the Sin Nombre virus (a hantavirus) of the Four Corners area of the Southwestern United States, which was considered a new, emerging virus, have been around for quite awhile. Sin Nombre virus just wasn't recognized until 1994, when a lot of people came down with the same kind of disease. The same is true for Ebola virus. It's been around for a long time, too, but again, it's newly recognized. Human contact with the natural reservoirs of Ebola virus doesn't happen very often, but when it does, it can have devastating results.

Describe the type of work accomplished by the Special Pathogens Branch.

At Special Pathogens, we have the BioSafety Level 4 Laboratory pretty much under our jurisdiction. Our branch deals with the nastier, highly pathogenic viral diseases for which there's absolutely no cure, and with new viruses that need to be contained because their mechanism of spread is unknown. The primary goal of the Special Pathogens Branch is disease surveillance. When there's an outbreak going on, Special Pathogens sends a team of epidemiologists into the field to examine patients and determine what they are coming down with. They send specimens back to us to be identified. Based on what the virus is, we recommend precautions to take and offer suggestions on how to curtail the outbreak.

What is the focus of your work on Ebola?

I'm not a clinician; I'm trying to develop a genetic system for the virus, to figure out how it works. A good way to find out how an organism works is to manipulate it genetically. It's kind of a means to an end. I can screw something up genetically with the virus and say, "Okay, what did I screw up?" That's the power of a genetic system— you can figure out how the organism works, based on how the it behaves differently when you manipulate its genome.

What safety precautions and procedures do you carry out in the lab?

Any work that's done with actual, viable viruses is done in the BioSafety Level 4 lab. To enter the lab, I go through a series of sealed doors and sterile chambers. I dress in surgical scrubs; a thick, rubber, positive-pressure suit; and thick gloves. I test the suit before entering the lab, to make sure there aren't any tears or rips. To breathe, I hook up the suit to air lines hanging from the ceiling all over the lab. All the air within the lab itself, which I don't breathe, is double- or triple-filtered. Anything I put down the sink, all the water that comes off if I wash anything, goes to giant cooking tanks, where it all gets sterilized for many hours before it's released. When I leave the lab, I take a Lysol shower with my suit on, go back through the series of doors and chambers, and then take a personal shower.

Are there opportunities at CDC for those with bachelor's degrees in biology?

Yes. CDC offers Emerging Infectious Disease (EID) laboratory fellowships, which are short-term programs. Because CDC is a government agency, there are also federal jobs available here, laboratory technicians and many other positions.

*Also see Chapter 23, *Focus On: Emerging Viruses.*

Part

6

Texas bluebonnet *(Lupinus texensis).* These beautiful wildflowers, which carpet Texas prairies in spring, grow to 60 cm (2 ft). Photographed near Llano, Texas. *(Rod Planck/Photo Researchers, Inc.)*

Structure and Life Processes in Plants

Chapters

31

Plant Structure, Growth, and Differentiation

A massive oak tree may grow from this young oak seedling. *(Runk/Schoenberger from Grant Heilman)*

LEARNING OBJECTIVES

After you have studied this chapter you should be able to

1. Discuss the functions of various parts of the vascular plant body, including the nutrient- and water-absorbing root system and the photosynthesizing shoot system.
2. Describe the ground tissue system (parenchyma tissue, collenchyma tissue, and sclerenchyma tissue).
3. Describe the structure and function of the vascular tissue system (xylem and phloem).
4. Describe the dermal tissue system (epidermis and periderm).
5. Discuss what is meant by growth in plants, and relate how it differs from growth in animals.
6. Distinguish between primary and secondary growth.

About 90% of the approximately 262,000 species of plants are flowering plants. These vascular plants are characterized by flowers, double fertilization, endosperm, and seeds enclosed within fruits (see Chapter 27). Because flowering plants comprise the largest, most successful group of plants, they are the focus of much of this chapter and of Chapters 32 to 36. Let us begin our study of the structure and functions of plants by examining a familiar species, the oak, and its fruit, the acorn.

Squirrels often temporarily store acorns in holes in the ground to provide winter food. Many of these are never retrieved, and so a new oak often begins its life after a squirrel has buried an acorn in the soil. The seed within the acorn first absorbs water from the surrounding soil. Then germination occurs as a root emerges and works its way down into the soil, absorbing additional water and anchoring the young plant. A miniature shoot begins to grow upward and breaks through the soil. At this point the young basket oak *(Quercus prinus)* shown in the photograph is called a seedling because it has just emerged from the seed and because it is still dependent on the food supply stored within the seed.

The stem continues to elongate, and small leaves develop, expand, and begin to photosynthesize. The young plant is now independently established: It is anchored in the ground; it absorbs water and dissolved nutrient minerals from the soil; and it uses energy from organic molecules, such as glucose, that it has produced by photosynthesis.

Smaller roots branch off the original root, and the young tree grows taller, always by growth at the tips of its branches. The roots likewise elongate at their tips. As it ages, the tree also increases in girth (thickness) by forming additional wood and bark. This growth occurs along the sides of the more mature stems and roots. Thus, the oak progressively accumulates more wood and bark, more stem and root tissues, and more leaves.

Growth and expansion of both the root and shoot systems continue throughout the life of the oak. As in all plants, growth is flexible and dynamic, enabling the oak to respond to its environment. For example, the young oak may grow very slowly for several years, particularly if it is shaded by mature trees surrounding it. When conditions change, perhaps when an older tree nearby dies and falls to the ground, thereby permitting direct sunlight to reach the young tree, it is poised for rapid growth.

After several years of growth, the oak tree becomes reproductively mature. Oaks reproduce by forming separate male and female flowers in spring. The minute male flowers are more conspicuous than the female flowers because they are found together on slender, drooping, caterpillar-shaped catkins (clusters); in contrast, the tiny female flowers are often solitary. After pollen from the male flower is blown by the wind to the female flower, a sperm cell fertilizes the egg, and an acorn develops, partially enclosed by a scaly cup. First green in color, the acorn fruit turns a deep brown as it matures. Within the acorn is a seed containing an embryo of a miniature oak along with a supply of stored food. Thus the cycle of life repeats itself, as squirrels busily collect these acorns and cache them for later retrieval.

In this chapter we examine the external structure of the flowering plant body; the organization of its cells, tissues, and tissue systems; and its basic growth patterns. Information of this type, which is based on observation and description, rarely receives as much attention as experimental science. However, observation and description are crucial aspects of the scientific process.

PLANTS VARY IN STRUCTURE AND LIFESPAN

Plants range in size from the minute floating water-meal *(Wolffia microscopica),* the smallest flowering plant known, to Australian gum trees (*Eucalyptus* spp.), some of Earth's tallest trees (Fig. 31–1). The 235,000 or so species of flowering plants that live in and are adapted to the many environments offered by our planet represent remarkable variety. Yet all of these—from desert cacti with enormously swollen stems, to cattails partly submerged in marshes, to orchids growing in the uppermost branches of lush tropical rainforest trees—are recognizable as plants. Almost all plants have the same basic body plan, which consists of roots, stems, and leaves.

Plants are either herbaceous or woody. *Herbaceous* plants are nonwoody plants. In temperate climates, the aerial parts (stems and leaves) of herbaceous plants die back to the ground at the end of the growing season. In contrast, the aerial parts of *woody* plants (trees and shrubs) persist. Botanically speaking, woody plants produce hard, lignified secondary tissues (i.e., cell walls contain lignin), and herbaceous plants do not. (Lignin and the production of secondary tissues are discussed later in the chapter.)

Annuals are herbaceous plants (such as corn, geranium, and marigold) that grow, reproduce, and die in one year or less. Other herbaceous plants (such as carrot, Queen Anne's lace, cabbage, and foxglove) are **biennials** and take two years to complete their life cycles before dying. During their first season biennials produce extra carbohydrates, which they store and use during their second year when they typically form flowers and reproduce. **Perennials** are herbaceous and woody plants that have the potential to live for more than two years. In temperate climates, the aerial (aboveground) stems of herbaceous perennials such as iris, rhubarb, onion, and asparagus die back each winter. Their underground parts (roots and underground stems) become

(a)

(b)

Figure 31–1 Size variation in flowering plants. (a) Duckweeds (*Spirodela* sp.), the larger green plants in the photograph, are tiny floating aquatic plants about 1 cm (3/8 in) across. If you look closely at the frog's body, you will see minute green dots. Each is a water-meal (*Wolffia* sp.) plant. These tiny floating herbs, about 1.5 mm (1/16 in) wide, are the smallest known flowering plants. (b) This giant gum tree *(Eucalyptus regnans)* was photographed in Bushy Park, Australia. Although most giant gums grow to 75m (246 ft), some specimens have been measured at heights of 100m (328 ft) and are the world's tallest flowering plants. *(a, Carlyn Iverson; b, Joyce Photographers/ Photo Researchers, Inc.)*

dormant during the winter and send out new growth each spring. (In **dormancy,** an organism reduces its metabolic state to a minimum level to survive unfavorable conditions.) Likewise, in certain tropical climates with pronounced wet and dry seasons, the aerial parts of herbaceous perennials die back and the underground parts become dormant during the unfavorable dry season. Other tropical plants, such as orchids, are herbaceous perennials that grow year-round.

All woody plants are perennials, and some of them live for hundreds or even thousands of years. In temperate climates, the aboveground stems of woody plants become dormant during the winter. Many temperate woody perennials are **deciduous;** that is, they shed their leaves before winter and produce new stems with new leaves the following spring. Other woody perennials are **evergreen** and shed their leaves over a long period, so that some leaves are always present. Because they have permanent woody stems that are the starting points for new growth the following season, many trees attain massive sizes.

MAKING THE CONNECTION Under what conditions is it more favorable for individuals of a species to be long-lived or short-lived? Woody perennials often live for hundreds of years, whereas some herbaceous annuals may live for only a few weeks or months. Such characteristic features of an organism's life cycle, particularly as they relate to its survival and successful reproduction, are known as life history strategies. Biologists try to understand the relative advantages of each **life history strategy,** including any tradeoffs involved in the allocation of various resources. It appears that in some environments a longer lifespan is advantageous, whereas in others a shorter lifespan increases a species' chances for reproductive success.

When an environment is relatively favorable, it is filled with plants competing for available space. Because such an environment is so crowded, it has few open spots in which new plants can become established. When a plant dies, the empty area is quickly filled by another plant, but not necessarily by the same species as before. Thus, an adult perennial survives well, but young plants, whether perennials or annuals, do not. A plant with a long lifespan thrives in this type of environment because it can occupy a piece of soil and continue to produce seeds for many years. In a tropical rain forest, for example, competition prevents most young plants from becoming established, and woody perennials predominate.

In a relatively unfavorable environment, many possible sites are usually available. This type of environment is not crowded, and young plants usually do not have to compete against large, fully established plants. Here, smaller, short-lived plants have the reproductive advantage. These plants are opportunists: They grow and mature quickly during the brief periods when environmental conditions are most favorable. As a result, all their resources are directed into producing as many seeds as possible before dying. In deserts following a rainy spell, for example, annuals are more prevalent than woody perennials.

Thus, each species has its own characteristic life history strategy, with some plants adapted to variable environments and others adapted to stable environments. The longer lifespan characteristic of woody perennials is just one of several successful life history plans. We return to life history traits in our discussion of population ecology (see Chapter 51).

ROOTS, STEMS, LEAVES, FLOWERS, AND FRUITS MAKE UP THE PLANT BODY

The plant body of flowering plants (and other vascular plants) is usually organized into a root system and a shoot system (Fig. 31–2). The **root system** is generally underground. The aboveground portion, the **shoot system,** generally consists of a vertical stem that bears leaves, and, in flowering plants, flowers and fruits that contain seeds.

Each plant typically grows in two different environments: the dark, moist soil and the illuminated, relatively dry air. Plants usually have both roots and shoots because they require resources from both environments. Thus, roots branch extensively through the soil, forming a network that anchors the plant firmly in place and absorbs water and dissolved nutrient minerals from the soil. Leaves, the flattened organs of photosynthesis, are attached more or less regularly on the stem, where they absorb the sun's light and atmospheric CO_2 used in photosynthesis.

THE PLANT BODY IS COMPOSED OF CELLS AND TISSUES

As in other organisms, the basic structural and functional unit of plants is the cell. During the course of evolution, plants have developed a diversity of cell types, each specialized for particular functions.

Like animal cells, plant cells are organized into tissues. A **tissue** is a group of cells that form a structural and functional unit. Some plant tissues, known as *simple tissues,* are composed of only one kind of cell, whereas other plant tissues, *complex tissues,* have two or more kinds of cells.

In vascular plants, tissues are organized into three tissue systems, each of which extends throughout the plant body (Fig. 31–3). Each tissue system contains two or more kinds of tissues (Table 31–1). Most of the plant body is composed of the **ground tissue system,** which has a variety of functions, including photosynthesis, storage, and support. The **vascular tissue system,** an intricate conducting system that extends throughout the plant body, is responsible for conduction of various substances, including water, dissolved nutrient minerals, and food (dissolved sugar). The vascular tissue system also functions in strengthening and supporting the plant. The **dermal tissue system** provides a covering for the plant body.

Roots, stems, leaves, flower parts, and fruits are referred to as organs because each is composed of all three tissue systems. The tissue systems of different plant organs form an interconnected network throughout the plant. For example, the vascular

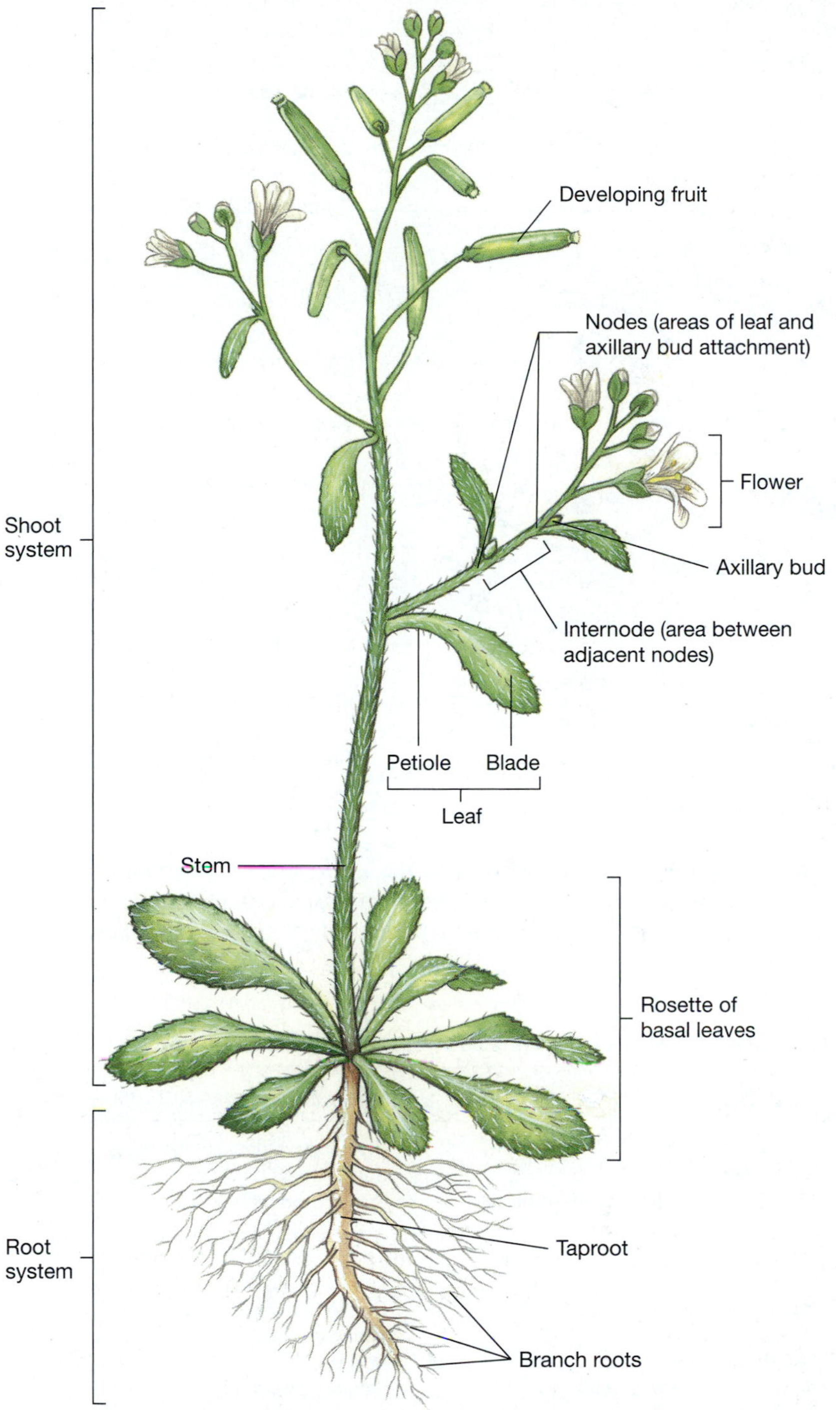

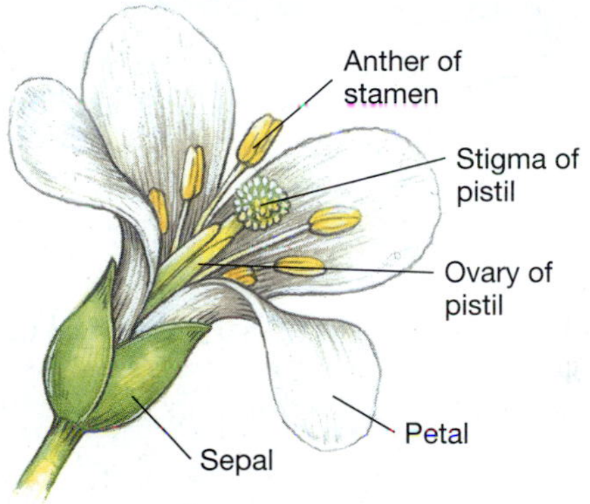

Figure 31-2 The plant body. A plant body consists of a root system, usually underground, and a shoot system, usually aboveground. The shoot system bears flowers and fruits and performs photosynthesis, whereas the root system provides anchorage and water and nutrient mineral uptake. Shown is mouse-ear cress *(Arabidopsis thaliana),* a small plant in the mustard family that is a model plant for biological research. *Arabidopsis* is native to North Africa and Eurasia and has been naturalized (i.e., it was introduced and is now growing in the wild) in California and the eastern half of the United States.

tissue of a leaf is continuous with the vascular tissue of the stem to which it is attached, and the vascular tissue of the stem is continuous with the vascular tissue of the root.

The ground tissue system is composed of three simple tissues

The bulk of an herbaceous plant is its ground tissue system, which is composed of three tissues: parenchyma, collenchyma, and sclerenchyma (Table 31-2 on page 680). These tissues can be distinguished by their cell wall structures. Recall that plant cells are surrounded by a cell wall that provides structural support (see Chapter 4). A growing plant cell secretes a thin *primary cell wall,* which stretches and expands as the cell increases in size. After the cell stops growing, it sometimes secretes a thick, strong *secondary cell wall,* which is deposited *inside* the primary cell wall—that is, between the primary cell wall and the plasma membrane.

Plant cell walls have several important roles in plants. Cell walls are involved in growth; the extensibility of primary cell

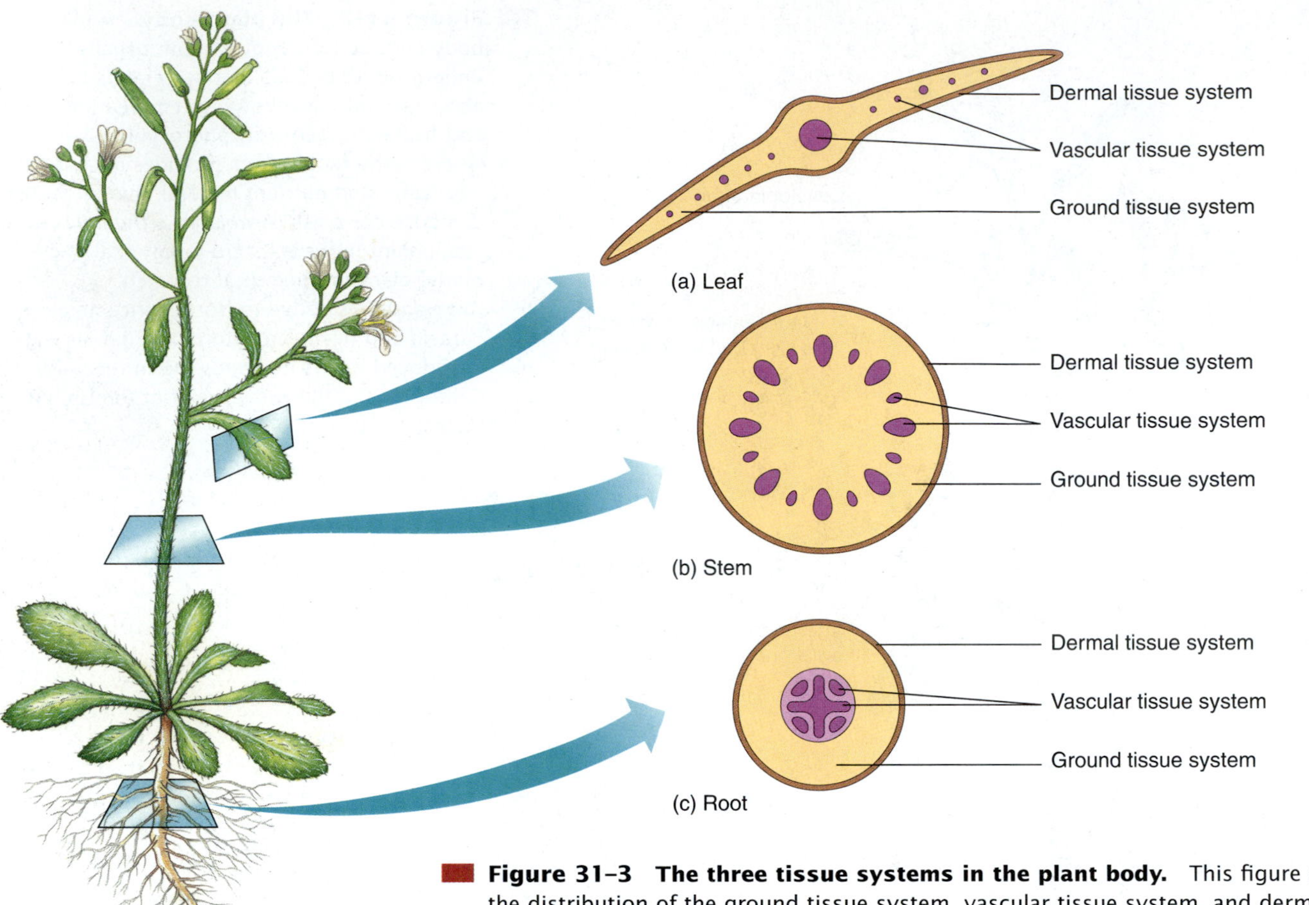

Figure 31–3 The three tissue systems in the plant body. This figure shows the distribution of the ground tissue system, vascular tissue system, and dermal tissue system in the leaves, stems, and roots of a herbaceous dicot such as *Arabidopsis*.

TABLE 31–1 Tissue Systems, Tissues, and Cell Types of Flowering Plants

Tissue System	Tissue	Cell Types	Main Functions of Tissue
Ground tissue system	Parenchyma tissue	Parenchyma cells	Storage, secretion, photosynthesis
	Collenchyma tissue	Collenchyma cells	Support
	Sclerenchyma tissue	Sclerenchyma cells (sclereids or fibers)	Support, strength
Vascular tissue system	Xylem	Tracheids	Conduction of water and nutrient minerals, support
		Vessel elements	Conduction of water and nutrient minerals, support
		Xylem parenchyma cells	Storage
		Fibers (sclerenchyma cells)	Support, strength
	Phloem	Sieve tube members	Conduction of sugar in solution, support
		Companion cells	May control functioning of sieve tube members
		Phloem parenchyma cells	Storage, sugar loading into sieve tube member
		Fibers (sclerenchyma cells)	Support, strength
Dermal tissue system	Epidermis	Epidermal cells	Protective covering over surface of plant body
		Guard cells	Regulate stomata
		Trichomes	Variable functions
	Periderm	Cork cells	Protective covering over surface of plant body
		Cork cambium cells	Meristematic
		Cork parenchyma cells	Storage

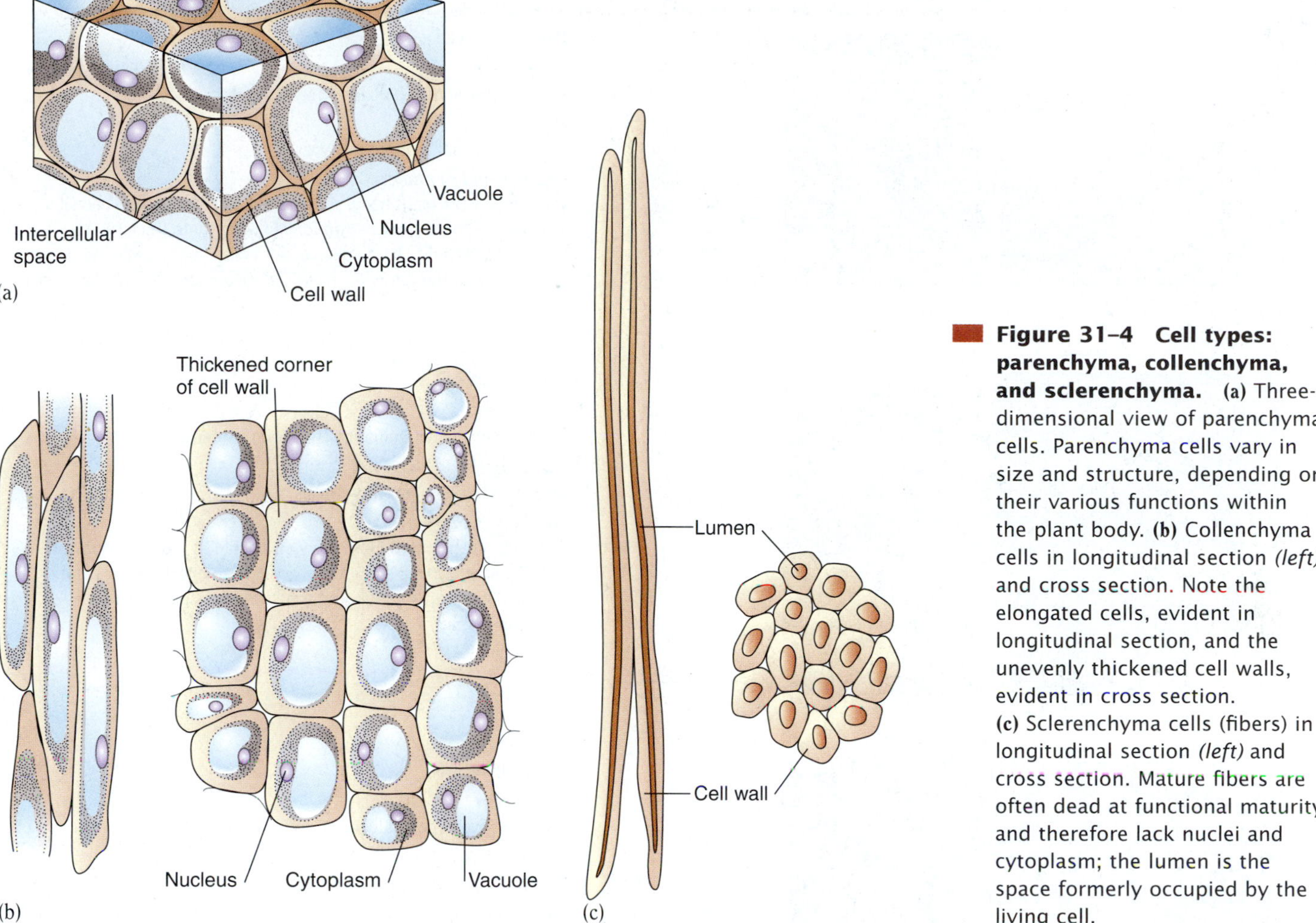

Figure 31–4 Cell types: parenchyma, collenchyma, and sclerenchyma. **(a)** Three-dimensional view of parenchyma cells. Parenchyma cells vary in size and structure, depending on their various functions within the plant body. **(b)** Collenchyma cells in longitudinal section *(left)* and cross section. Note the elongated cells, evident in longitudinal section, and the unevenly thickened cell walls, evident in cross section. **(c)** Sclerenchyma cells (fibers) in longitudinal section *(left)* and cross section. Mature fibers are often dead at functional maturity and therefore lack nuclei and cytoplasm; the lumen is the space formerly occupied by the living cell.

walls allows cells to increase in size. Scientists are increasingly aware that **signal transduction** is taking place in plant cell walls, because many carbohydrate and protein molecules in plant cell walls communicate with other molecules both inside and outside the cell. The cell wall is also the first line of defense against disease-causing organisms.

Parenchyma cells have thin primary walls

Parenchyma tissue, a simple tissue composed of parenchyma cells, is found throughout the plant body and is the most common type of cell and tissue (Fig. 31–4*a*). The soft parts of a plant, such as the edible part of an apple or a potato, consist largely of parenchyma.

Parenchyma cells perform a number of important functions, such as photosynthesis, storage, and secretion. Parenchyma cells that function in photosynthesis contain green chloroplasts, whereas nonphotosynthetic parenchyma cells lack chloroplasts and are often colorless. Materials stored in parenchyma cells include starch grains, oil droplets, water, and salts (sometimes visible as crystals). Resins, tannins, hormones, enzymes, and sugary nectar are examples of substances that may be secreted by parenchyma cells. The various functions of parenchyma require that they be living, metabolizing cells.

Parenchyma cells have the ability to differentiate into other kinds of cells, particularly when a plant has been injured. If xylem (water-conducting) cells are severed, for example, adjacent parenchyma cells may divide and differentiate into new xylem cells within a few days. (Recall from Chapter 16 that it is possible to induce certain plant cells to become the equivalent of embryonic cells that can then differentiate into specialized cells.)

Collenchyma cells have unevenly thickened primary walls

Collenchyma tissue, a simple plant tissue composed of collenchyma cells, is an extremely flexible structural tissue that provides much of the support in soft, nonwoody plant organs (Fig. 31–4*b*). Support is a crucial function in plants, in part because it allows them to grow upward, thus enabling them to compete with other plants for available sunlight in a plant-crowded area. Plants lack the bony skeletal system that is typical of many animals; instead, individual cells, including collenchyma cells, support the plant body.

Collenchyma cells, which are usually elongated, are alive at maturity. Their primary cell walls are unevenly thickened and are especially thick in the corners. Collenchyma is not found

TABLE 31–2 Cell Types in the Ground Tissue System

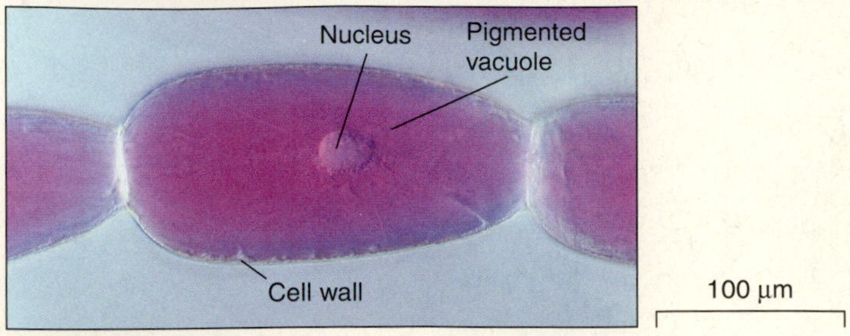

Parenchyma cell

Description
Living, actively metabolizing; thin primary cell walls

Functions
Storage; secretion; photosynthesis

Location and comments
Throughout the plant body; shown is an LM of a stamen hair of a spiderwort *(Tradescantia virginiana)* flower; note the large pigmented vacuole; the nucleus is not inside the vacuole but is lying on *top* of the vacuole *(Phil Gates/Biological Photo Service)*

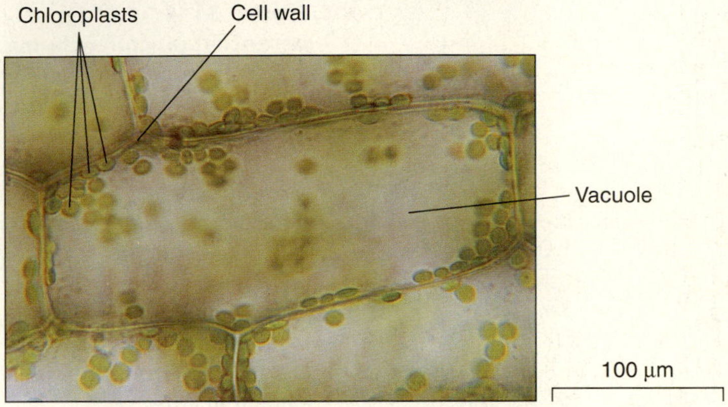

Parenchyma cell

Description
Living, actively metabolizing; thin primary cell walls

Functions
Storage; secretion; photosynthesis

Location and comments
Throughout the plant body; shown is an LM of leaf cells from an aquatic plant (*Elodea* sp.); note the many chloroplasts in the thin layer of cytoplasm surrounding the large, transparent vacuole *(Dennis Drenner)*

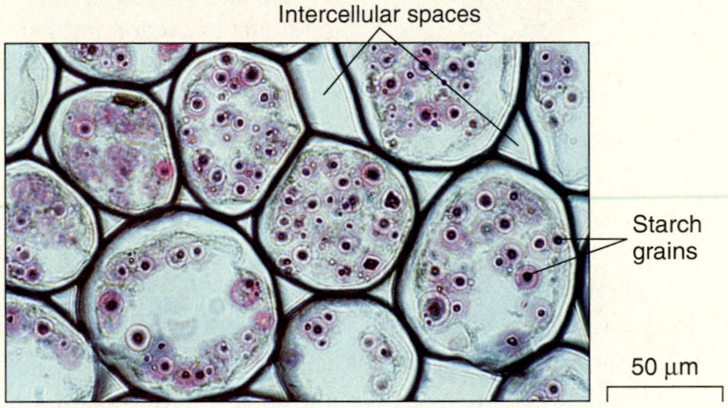

Parenchyma cell

Description
Living, actively metabolizing; thin primary cell walls

Functions
Storage; secretion; photosynthesis

Location and comments
Throughout the plant body; shown is an LM of a cross section of part of a buttercup (*Ranunculus* sp.) root; note the starch grains filling the cells *(Dennis Drenner)*

uniformly throughout the plant and often occurs as long strands near stem surfaces and along leaf veins. The "strings" in a celery stalk, for example, consist of collenchyma tissue.

Sclerenchyma cells have both primary walls and thick secondary walls

Another simple plant tissue specialized for structural support is **sclerenchyma** tissue, whose cells have both primary and secondary cell walls. The root of the word *sclerenchyma* is derived from a Greek word *(sclero)* meaning "hard." The secondary cell walls of sclerenchyma cells become strong and hard due to extreme thickening. Thus, sclerenchyma cells are unable to stretch or elongate. At functional maturity, when sclerenchyma tissue is providing support for the plant body, its cells are often dead.

Sclerenchyma tissue may be located in several areas of the plant body. Two types of sclerenchyma cells are sclereids and fibers. **Sclereids** are cells of variable shape common in the shells of nuts and in the pits of stone fruits such as cherries and peaches. Pears owe their slightly gritty texture to the presence of clusters of sclereids. **Fibers,** which are long, tapered cells that often occur in patches or clumps, are particularly abundant in the wood, inner bark, and leaf ribs (veins) of flowering plants (Fig. 31–4*c*).

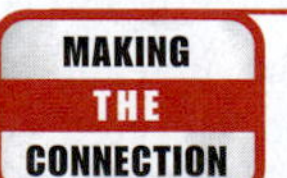

How is cell wall chemistry related to different cell types? Parenchyma, collenchyma, and sclerenchyma cells can be distinguished by the chemistry of their cell walls. We must first examine the main

TABLE 31–2 Continued

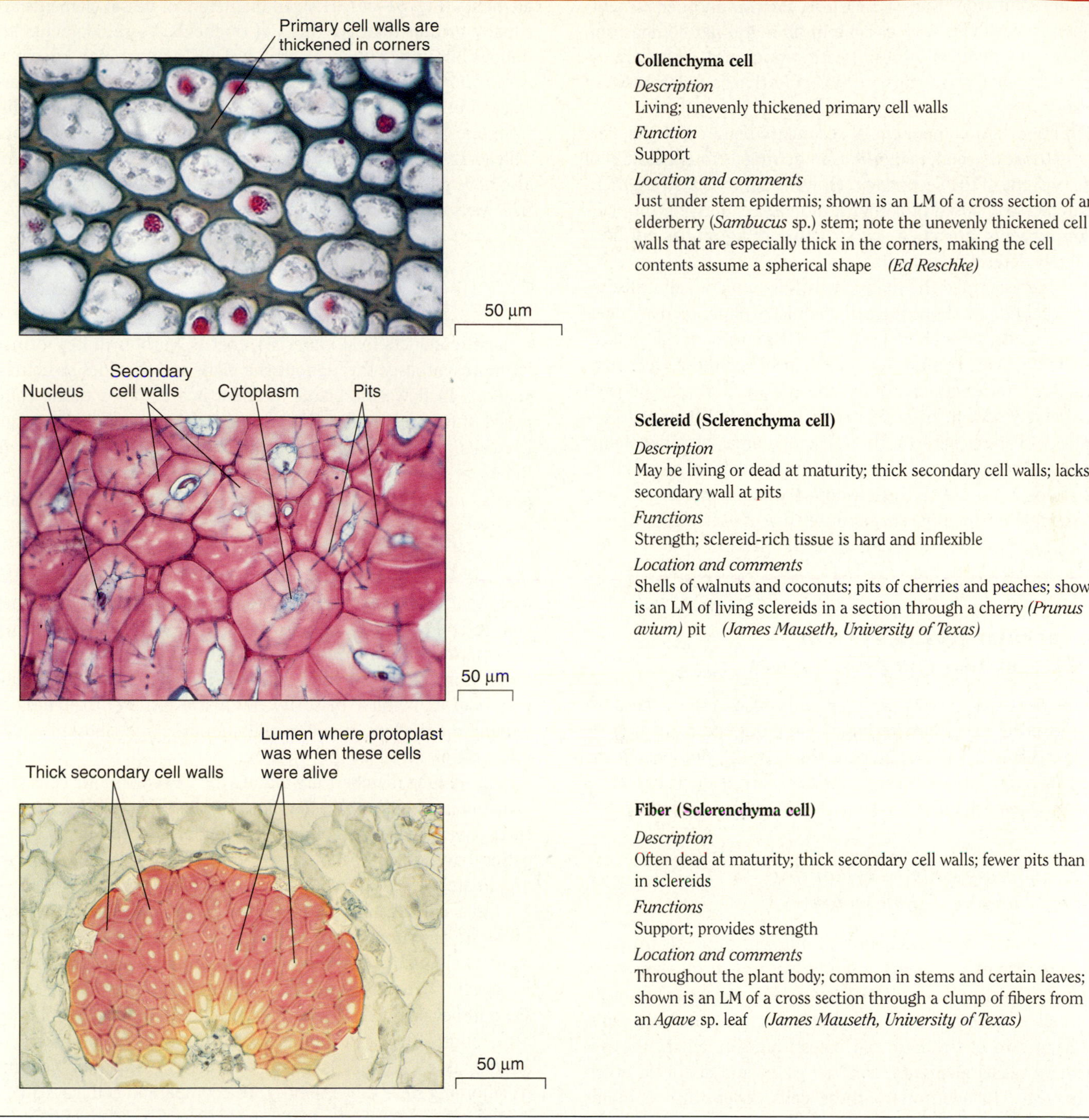

Collenchyma cell

Description
Living; unevenly thickened primary cell walls

Function
Support

Location and comments
Just under stem epidermis; shown is an LM of a cross section of an elderberry (*Sambucus* sp.) stem; note the unevenly thickened cell walls that are especially thick in the corners, making the cell contents assume a spherical shape *(Ed Reschke)*

Sclereid (Sclerenchyma cell)

Description
May be living or dead at maturity; thick secondary cell walls; lacks secondary wall at pits

Functions
Strength; sclereid-rich tissue is hard and inflexible

Location and comments
Shells of walnuts and coconuts; pits of cherries and peaches; shown is an LM of living sclereids in a section through a cherry *(Prunus avium)* pit *(James Mauseth, University of Texas)*

Fiber (Sclerenchyma cell)

Description
Often dead at maturity; thick secondary cell walls; fewer pits than in sclereids

Functions
Support; provides strength

Location and comments
Throughout the plant body; common in stems and certain leaves; shown is an LM of a cross section through a clump of fibers from an *Agave* sp. leaf *(James Mauseth, University of Texas)*

chemical components of plant cell walls. Cell walls may contain cellulose, hemicelluloses, pectin, and lignin. **Cellulose,** the most abundant polymer in the world, accounts for about 40% to 60% of the dry weight of plant cell walls. As discussed in Chapter 3, cellulose is a polymer of glucose units joined by β-1,4 bonds. Each cellulose molecule consists of thousands of glucose subunits joined to form a flat, ribbon-like chain. From 40 to 70 of these chains lie parallel to one another and connect by hydrogen bonding to form a **cellulose microfibril,** a strong, tiny strand visible under the electron microscope (see Fig. 3–10*a*).

Cellulose microfibrils are cemented together by a matrix of hemicelluloses and pectins. **Hemicelluloses** are a group of polysaccharides that are more soluble than cellulose. Hemicelluloses vary in their chemical composition from one species to another. Some hemicelluloses are composed of *xyloglucan,* which consists of a backbone of β-1,4-glucose molecules to which side chains of *xylose,* a five-carbon sugar, are attached. Despite their name, the chemical structure of the hemicelluloses is significantly different from that of cellulose. **Pectin,** another cementing polysaccharide, is less variable in its monomer composition than are the hemicelluloses. Pectin monomer units are *α-galacturonic acid,* a six-carbon molecule that is a derivative of glucose.

An important component of secondary plant cell walls, particularly those of wood, is **lignin.** Comprising as much as 35% of the dry weight of the secondary cell wall, lignin is a strengthening polymer composed of complex monomers derived from certain amino acids. The chemical structure of lignin has not been completely determined at this time.

Having examined the main chemicals in plant cell walls, we can now generalize about the cell chemistry of parenchyma, collenchyma, and sclerenchyma cells. The thin primary cell walls of parenchyma cells contain predominantly cellulose, although they also contain hemicelluloses and pectin. Both parenchyma and collenchyma cells have primary cell walls, but their walls are chemically distinct because the thickened areas of collenchyma walls contain large quantities of pectin in addition to cellulose and hemicelluloses. The thick secondary walls of sclerenchyma cells are chemically different because they are rich in lignin, in addition to cellulose, hemicelluloses, and pectin.

The vascular tissue system consists of two complex tissues

The vascular tissue system, which is embedded in the ground tissue, transports needed materials throughout the plant via two complex tissues: xylem and phloem (Table 31–3). Both xylem and phloem are continuous throughout the plant body. (Chapter 33 discusses the mechanisms of transport in xylem and phloem.)

The conducting cells in xylem are tracheids and vessel elements

Xylem conducts water and dissolved nutrient minerals from the roots to the stems and leaves and provides structural support. In flowering plants, xylem is a complex tissue composed of four different cell types: tracheids, vessel elements, parenchyma cells, and fibers. Two of the four cell types found in xylem, the **tracheids** and **vessel elements,** conduct water and dissolved nutrient minerals. In addition to these cells, xylem also contains parenchyma cells, known as *xylem parenchyma,* that perform storage functions, and fibers that provide support.

Tracheids and vessel elements are highly specialized for conduction. When mature, both cell types are dead and therefore hollow; only their cell walls remain. Tracheids, the chief water-conducting cells in gymnosperms (such as pine) and seedless vascular plants (such as ferns), are long, tapering cells located in patches or clumps (Fig. 31–5*a*). Water is conducted upward, from roots to shoots, passing from one tracheid into another through *pits,* thin areas in the tracheids' cell walls where a secondary wall did not form.

In addition to a relatively few tracheids, flowering plants possess extremely efficient water-conducting cells called vessel elements (Fig. 31–5*b*). The cell diameters of vessel elements are usually greater than are those of tracheids. Vessel elements are hollow, but unlike tracheids, the end walls have holes, known as *perforations,* or are entirely dissolved away. Vessel elements are stacked one on top of the other, and water is conducted readily from one vessel element into the next. A stack of vessel elements, called a *vessel,* resembles a miniature water pipe. Vessel elements also have pits in their side walls that permit the lateral transport (sideways movement) of water from one vessel to another.

Sieve tube members are the conducting cells of phloem

Phloem conducts food materials, that is, carbohydrates formed in photosynthesis, throughout the plant and provides structural support. In flowering plants, phloem is a complex tissue composed of four different cell types: sieve tube members, companion cells, fibers, and phloem parenchyma cells (Fig. 31–5*c, d*). Fibers are frequently extensive in the phloem of herbaceous plants, providing additional structural support for the plant body.

Food materials are conducted in *solution,* that is, dissolved in water, through the **sieve tube members,** which are among the most specialized cells in nature. Sieve tube members are joined end-on-end to form long sieve tubes. The cells' end walls, called *sieve plates,* have a series of holes through which cytoplasm extends from one sieve tube member into the next. Sieve tube members are living at maturity, but many of their organelles, including the nucleus, vacuole, mitochondria, and ribosomes, disintegrate or shrink as they mature.

Sieve tube members are among the few eukaryotic cells that can function without nuclei. These cells typically live for less than a year. There are, however, notable exceptions: Certain palms have sieve tube members that have remained alive approximately 100 years!

Adjacent to each sieve tube member is a **companion cell** that assists in the functioning of the sieve tube member. The companion cell is a living cell, complete with a nucleus. This nucleus is thought to direct the activities of both the companion cell and sieve tube member. Numerous **plasmodesmata**—cytoplasmic connections through which cytoplasm extends from one cell to another (see Chapter 5)—occur between a companion cell and its adjoining sieve tube member. The companion cell plays an essential role in moving sugar into the sieve tube members for transport to other parts of the plant.

The dermal tissue system consists of two complex tissues

The dermal tissue system, the epidermis and periderm, provides a protective covering over plant parts (Table 31–4 on page 685). In herbaceous plants, the dermal tissue system is a single layer of cells called the epidermis. Woody plants initially produce an epidermis, but it splits apart as the plant increases in girth owing to

TABLE 31–3 Selected Cell Types in the Vascular Tissue System

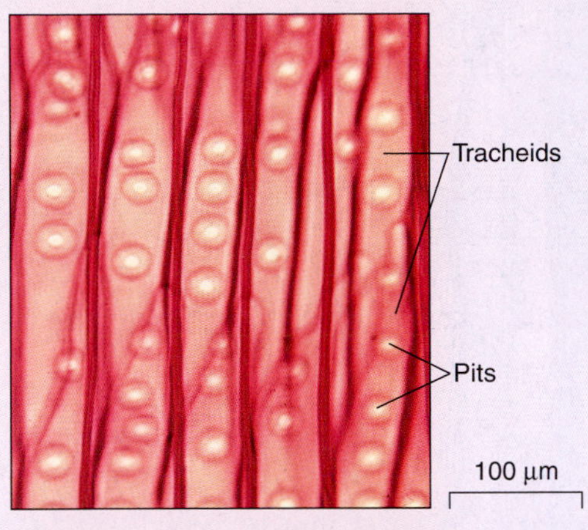

Tracheid

Description
Dead at maturity; lacks secondary wall at pits

Functions
Conduction of water and nutrient minerals, support

Location and comments
Occurs in clumps in xylem throughout plant body; shown is an LM of a longitudinal section of tracheids from white pine *(Pinus strobus)* wood *(John D. Cunningham/Visuals Unlimited)*

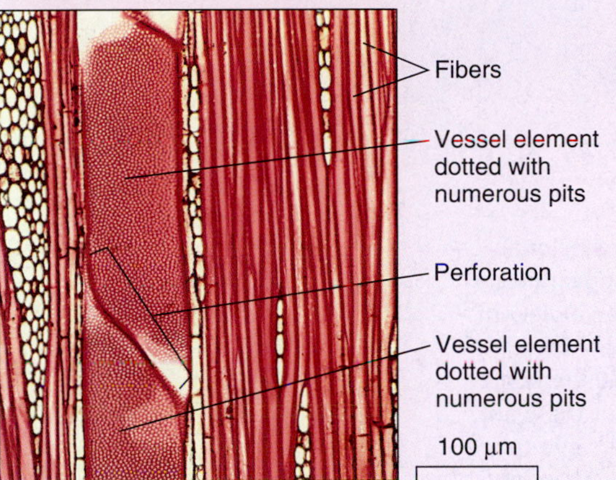

Vessel element

Description
Dead at maturity; end walls have perforations; lacks secondary wall at pits

Functions
Conduction of water and nutrient minerals, support

Location and comments
Xylem throughout plant body; vessel elements are more efficient than tracheids in conduction; shown is an LM of a longitudinal section of two vessel elements from an unknown woody dicot *(James Mauseth, University of Texas)*

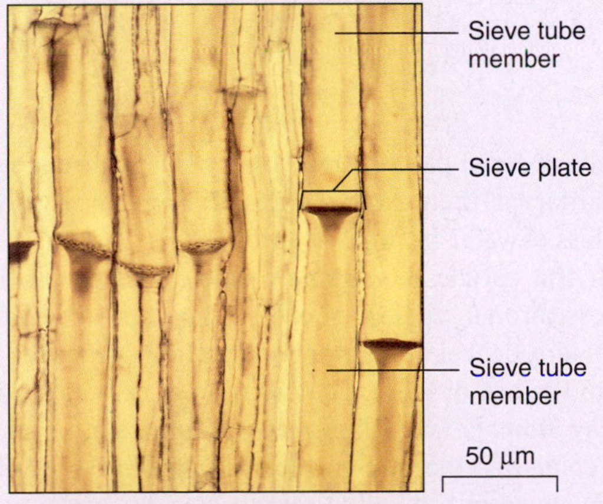

Sieve tube member

Description
Living but lacks nucleus and other organelles at maturity; end walls are sieve plates

Function
Conduction of sugar in solution

Location and comments
Phloem throughout plant body; shown is an LM of a longitudinal section through a clump of sieve tube members in a squash (*Cucurbita* sp.) petiole (leaf stalk) *(James Mauseth, University of Texas)*

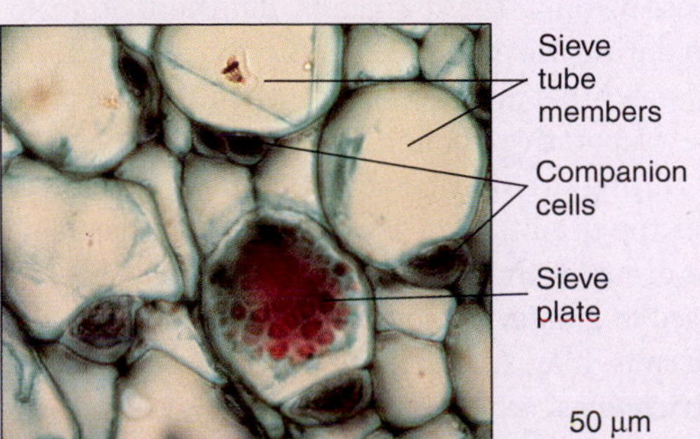

Companion cell

Description
Living; has cytoplasmic connections with sieve tube member

Function
Assists in moving sugars into and out of sieve tube member

Location and comments
Phloem throughout plant body; shown is an LM of phloem from a squash (*Cucurbita* sp.) petiole (leaf stalk) in cross section *(J. Robert Waaland/Biological Photo Service)*

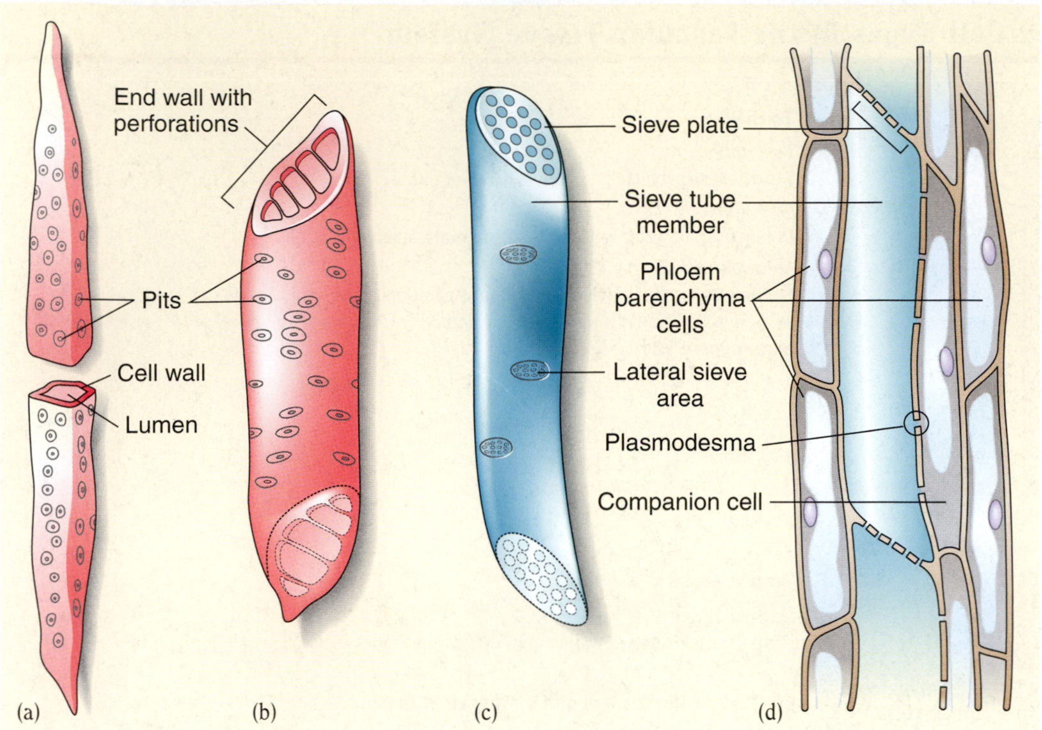

Figure 31–5 Cell types in xylem and phloem tissues. **(a)** Three-dimensional view of a tracheid (xylem cell). The tracheid is opened to reveal the cell's appearance in cross section. At maturity, tracheids are usually dead. **(b)** Three-dimensional view of a vessel element (xylem cell). The end walls of vessel elements have a variety of openings, or perforations, depending on the species being examined. Vessel elements are joined end-to-end to produce minute "water pipes" that run the length of the plant, from roots to leaves and other shoot parts. **(c)** Three-dimensional view of a sieve tube member, showing the sieve plate. **(d)** Longitudinal section of phloem tissue, showing sieve tube members, companion cells, and phloem parenchyma cells. Fiber cells are not shown.

the production of additional woody tissues inside the epidermis. Periderm, a tissue several to many cell layers thick, forms under the epidermis to provide a new protective covering as the epidermis is destroyed. Periderm, which replaces the epidermis in the stems and roots of older woody plants, composes the outer bark.

Epidermis is the outermost layer of cells of a herbaceous plant

The **epidermis** is a complex tissue composed of relatively unspecialized living cells. Dispersed among these cells are more specialized guard cells and outgrowths called trichomes (discussed shortly). In most plants, the epidermis consists of a single layer of flattened cells (Fig. 31–6). Epidermal cells generally contain no chloroplasts and are therefore transparent, allowing light to penetrate into the interior tissues of stems and leaves. In both stems and leaves, photosynthetic tissues lie *beneath* the epidermis.

An important requirement of the aerial parts (stems, leaves, flowers, and fruits) of a plant is the ability to control water loss. Epidermal cells of aerial parts secrete a waxy layer called a **cuticle** over the surface of their exterior walls; this wax layer greatly restricts the loss of water from plant surfaces.

Although the cuticle is extremely efficient at preventing most water loss through epidermal cells, it also slows the diffusion of carbon dioxide (required for photosynthesis) from the atmosphere into the leaf or stem. The diffusion of carbon dioxide is facilitated by **stomata** (singular, *stoma*). Stomata are minute pores in the epidermis that are surrounded by two cells called **guard cells.** Many gases, including carbon dioxide, oxygen, and water vapor, pass through the stomata by diffusion. Stomata are generally open during the day when photosynthesis is occurring, and the loss of water that also takes place when stomata are open provides some evaporative cooling. During the night, stomata usually close. During drought conditions, the need to conserve water overrides the need to cool the leaves and exchange gases. Thus, during a drought, the stomata close in the daytime. Stomata are discussed in greater detail in Chapter 32.

The epidermis may also contain special outgrowths, or hairs, called **trichomes,** which occur in many sizes and shapes

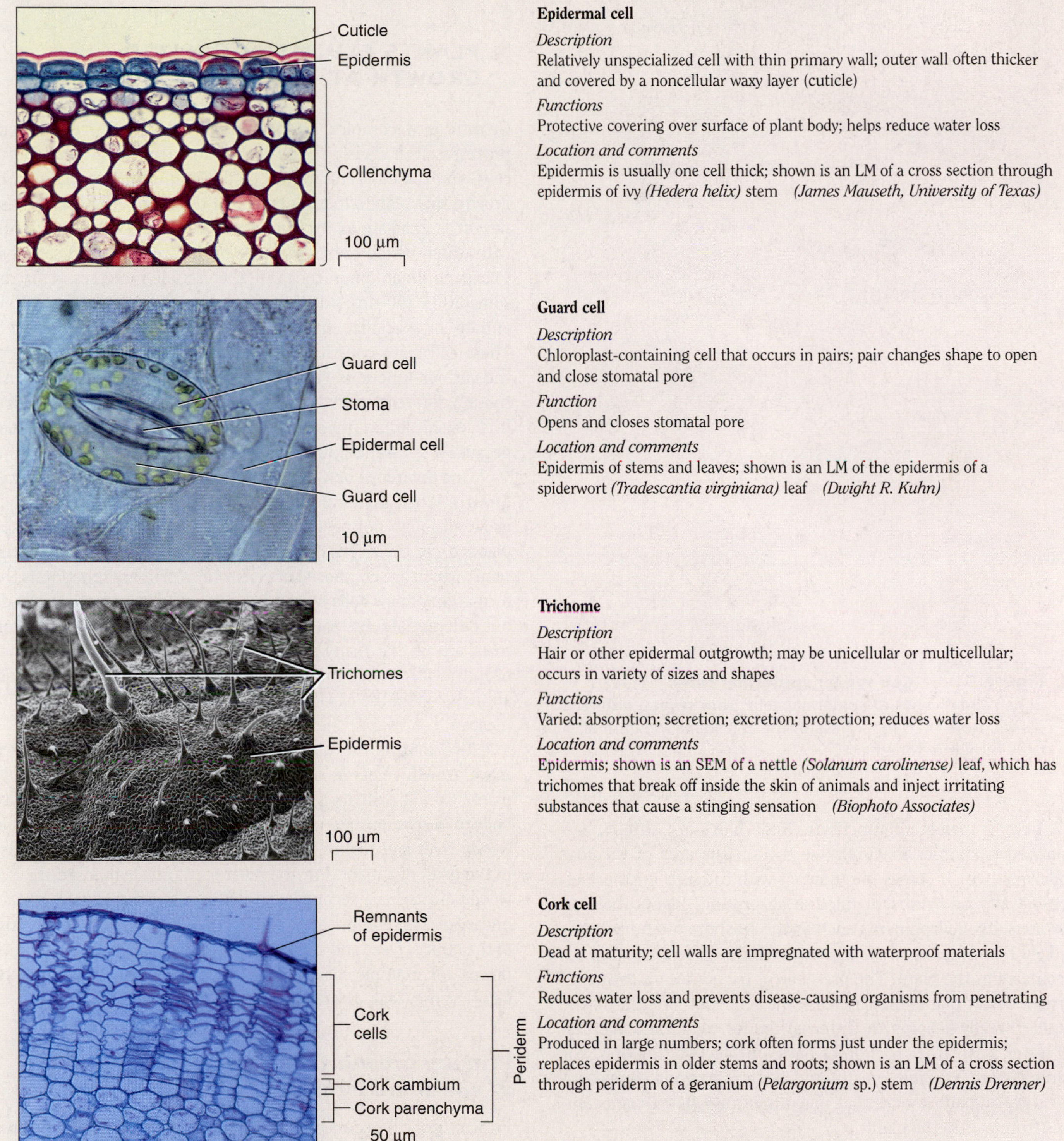

TABLE 31–4 Selected Cell Types in the Dermal Tissue System

Epidermal cell

Description
Relatively unspecialized cell with thin primary wall; outer wall often thicker and covered by a noncellular waxy layer (cuticle)

Functions
Protective covering over surface of plant body; helps reduce water loss

Location and comments
Epidermis is usually one cell thick; shown is an LM of a cross section through epidermis of ivy *(Hedera helix)* stem *(James Mauseth, University of Texas)*

Guard cell

Description
Chloroplast-containing cell that occurs in pairs; pair changes shape to open and close stomatal pore

Function
Opens and closes stomatal pore

Location and comments
Epidermis of stems and leaves; shown is an LM of the epidermis of a spiderwort *(Tradescantia virginiana)* leaf *(Dwight R. Kuhn)*

Trichome

Description
Hair or other epidermal outgrowth; may be unicellular or multicellular; occurs in variety of sizes and shapes

Functions
Varied: absorption; secretion; excretion; protection; reduces water loss

Location and comments
Epidermis; shown is an SEM of a nettle *(Solanum carolinense)* leaf, which has trichomes that break off inside the skin of animals and inject irritating substances that cause a stinging sensation *(Biophoto Associates)*

Cork cell

Description
Dead at maturity; cell walls are impregnated with waterproof materials

Functions
Reduces water loss and prevents disease-causing organisms from penetrating

Location and comments
Produced in large numbers; cork often forms just under the epidermis; replaces epidermis in older stems and roots; shown is an LM of a cross section through periderm of a geranium (*Pelargonium* sp.) stem *(Dennis Drenner)*

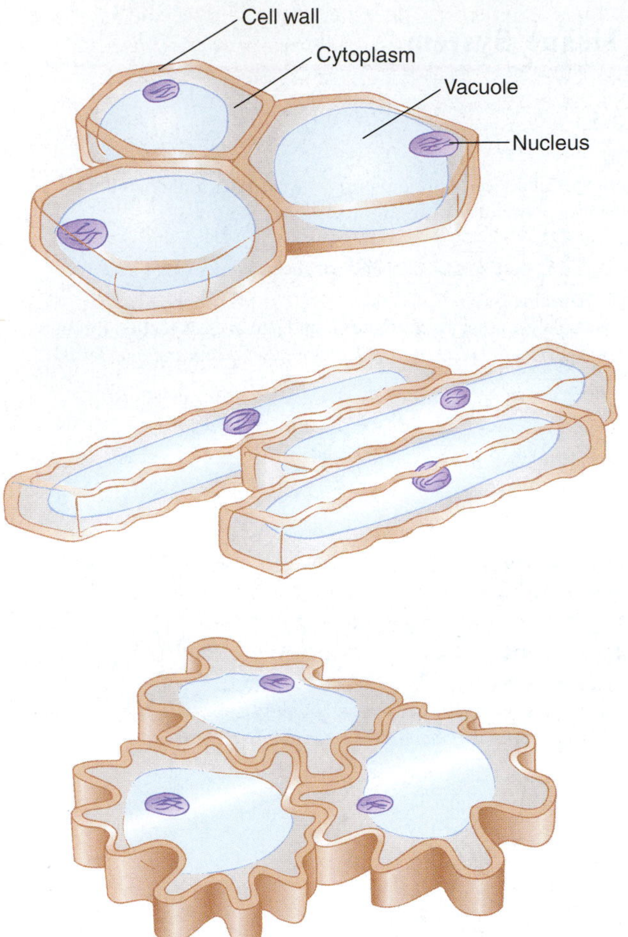

Figure 31–6 Cell types: epidermal cells. Three-dimensional views of epidermal cells from several different plants. Note that, regardless of the cell shape, epidermal cells fit tightly together.

and have a variety of functions. Root hairs are simple, unbranched trichomes that increase the surface area of the root epidermis (which comes into contact with the soil) for more effective water and nutrient mineral absorption. Plants that can tolerate salty environments such as the seashore often have specialized trichomes on their leaves to remove excess salt that accumulates in the plant. The presence of trichomes on the aerial parts of desert plants may increase the reflection of light off the plants, thereby keeping the internal tissues cooler and decreasing water loss. Other trichomes have a protective function. For example, the trichomes on stinging nettle leaves and stems contain irritating substances that may discourage herbivorous animals from eating the plant.

Epidermis is replaced by periderm in woody plants

As a woody plant begins to increase in girth, its epidermis is sloughed off and replaced by **periderm.** Periderm forms the outer bark of older stems and roots. It is a complex tissue composed mainly of cork cells and cork parenchyma cells. **Cork cells** are dead at maturity, and their walls are heavily coated with a waterproof substance called *suberin,* which helps reduce water loss. **Cork parenchyma** cells function primarily in storage.

PLANTS EXHIBIT LOCALIZED GROWTH AT MERISTEMS

Growth is a complex phenomenon involving three different processes: cell division, cell elongation (the lengthening of a cell), and cell differentiation. Cell division is an essential part of growth that results in an increase in the number of cells. These new cells elongate as the cytoplasm grows and the vacuole fills with water, which exerts pressure on the cell wall and causes it to expand. In an onion root cell, the vacuole increases in size by some 30 to 150 times during elongation. Plant cells also **differentiate,** or specialize, into the various cell types just discussed. These cell types constitute the mature plant body and perform the various functions required in a multicellular organism. Although differentiation does not contribute to an increase in size, it is considered an important aspect of growth and development because it is essential for tissue formation.

One difference between plants and animals is the *location* of growth. When a young animal is growing, all parts of its body grow, although not necessarily at the same rate. However, when plants grow, their cells divide only in specific areas, called **meristems,** which are composed of cells whose primary function is the formation of new cells by mitotic division. Meristematic cells do not differentiate. Instead, they retain the ability to divide by mitosis, a trait that many differentiated cells lose. The persistence of mitotically active meristems means that plants, unlike most animals, retain the capability for growth throughout their entire lifespan.

Two kinds of meristematic growth may occur in plants. **Primary growth** refers to an increase in stem and root length. All plants exhibit primary growth, which produces the entire plant body in herbaceous plants and the young, soft shoots and roots in woody trees and shrubs. **Secondary growth** refers to an increase in the girth of a plant. For the most part, only gymnosperms and woody dicots have secondary growth.[1] Tissues produced by secondary growth comprise the wood and bark, which make up most of the bulk of trees and shrubs. A few annuals, geranium and sunflower, for example, have limited secondary growth despite the fact that they lack obvious wood and bark tissues.

Primary growth takes place at apical meristems

Primary growth occurs as a result of the activity of **apical meristems,** areas located at the tips of roots and shoots, including within

[1] Recall from Chapter 27 that, on the basis of structural features, flowering plants are divided into two groups, informally called dicots and monocots. Oak, sycamore, ash, cherry, apple, and maple are examples of woody dicots, whereas bean, daisy, and snapdragon are examples of herbaceous dicots. Palm, corn, bluegrass, lily, and tulip are examples of monocots.

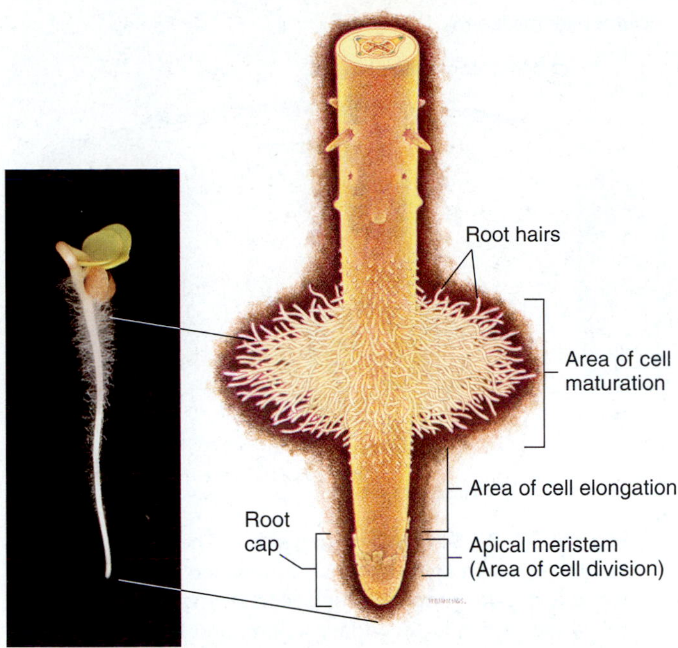

Figure 31–7 A root tip. The root apical meristem (where cells divide and thus increase in number) is protected by a root cap. Farther from the tip is an area of cell elongation, where cells enlarge and begin to differentiate. The area of cell maturation has fully mature, differentiated cells. Note the young root hairs in this area and on the root of a young radish *(Raphanus sativus)* seedling, which is approximately 5 cm (2 in) long *(left)*. *(Dennis Drenner)*

the buds of stems. (**Buds** are dormant embryonic shoots that develop into branches the following spring.) Such growth is evident when a root tip (Fig. 31–7) is examined. A protective layer of cells called a root cap covers the root tip. Directly behind the root cap is the root apical meristem, which consists of meristematic cells. Meristematic cells, which are quite small and cube-shaped, remain small because they are continually dividing. (In meristems, as daughter cells begin to enlarge, one or both will divide again.)

Farther from the root tip, just behind the area of cell division, is an area of cell elongation where the cells have been displaced from the meristem. Here the cells are no longer dividing but instead are growing longer, pushing the root tip ahead of them, deeper into the soil. Some differentiation also begins to occur in the area of cell elongation, and immature tissues such as differentiating xylem and phloem, become evident. The tissue systems continue to develop and differentiate into primary tissues (epidermis, xylem, phloem, and ground tissues) of the adult plant. Farther from the tip, following the area of cell elongation, most cells have completely differentiated and are fully mature. Root hairs are evident in this area.

A shoot apex (e.g., the terminal bud) is quite different in appearance from a root tip (Fig. 31–8). A dome of minute, regularly arranged meristematic cells—the shoot apical meristem—is located in each shoot apex. *Leaf primordia* (developing leaves) and *bud primordia* (developing buds) arise from the shoot apical meristem. The tiny leaf primordia tend to cover and protect the shoot apical meristem. As the cells formed by the shoot apical meristem elongate, the shoot apical meristem is pushed upward. Subsequent cell divisions produce additional stem tissue and cause new leaf and bud primordia to appear. Further from the stem tip, the immature cells differentiate into the three tissue systems of the mature plant body.

Secondary growth takes place at lateral meristems

In addition to primary growth, trees and shrubs have secondary growth. These plants increase in length by primary growth and increase in girth by secondary growth. This increase in girth, which occurs in areas that are no longer elongating, is due to cell divisions that take place in **lateral meristems,** areas extending

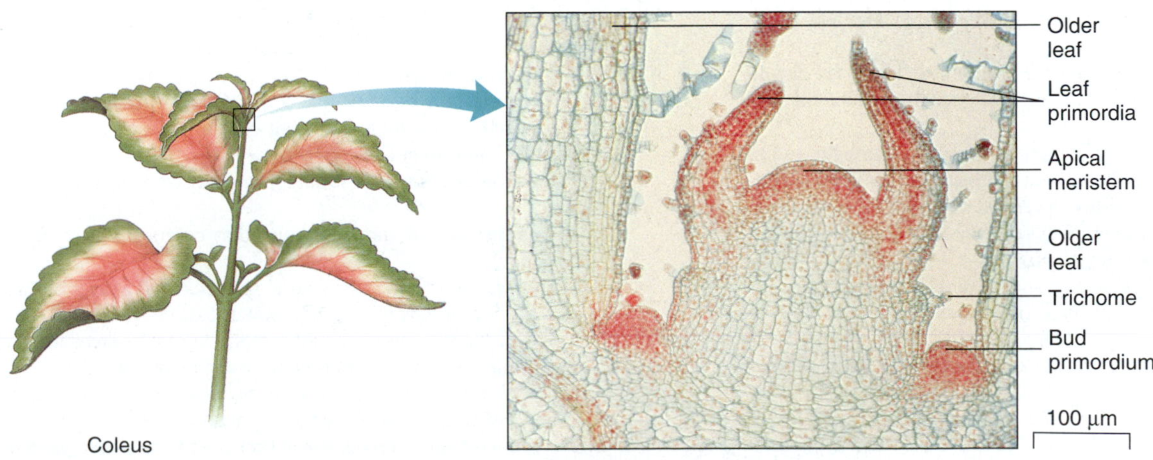

Figure 31–8 A shoot apex. LM of a longitudinal section through a shoot apex of coleus (*Coleus* sp.) showing the shoot apical meristem, leaf primordia, and bud primordia. *(James Mauseth, University of Texas)*

along the entire length of the stems and roots, except at the tips. Two lateral meristems, the vascular cambium and the cork cambium, are responsible for secondary growth, which is the formation of secondary tissues: secondary xylem, secondary phloem, and periderm (Fig. 31–9).

The **vascular cambium** is a layer of meristematic cells that forms a long, thin, continuous cylinder within the stem and root. It is located between the wood and bark of a woody plant. Cells of the vascular cambium divide, adding more cells to the wood (secondary xylem) and inner bark (secondary phloem).

The **cork cambium** is composed of a thin cylinder or irregular arrangement of meristematic cells and is located in the outer bark. Cells of the cork cambium divide and form cork cells toward the outside and one or more underlying layers of cork parenchyma cells that function in storage. Collectively, cork cells, cork cambium, and cork parenchyma make up the periderm (discussed previously).

We are now ready to give a more precise definition of bark. **Bark,** the outermost covering over woody stems and roots, consists of all plant tissues located outside the vascular cambium. Bark has two regions, a living inner bark composed of secondary phloem and a mostly dead outer bark composed of periderm. A more comprehensive discussion of secondary growth is given in Chapters 33 and 34.

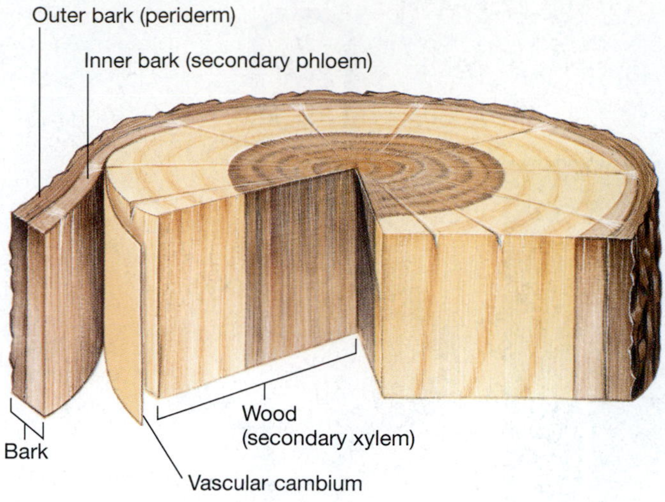

Figure 31–9 Secondary growth. The vascular cambium, a thin layer of cells sandwiched between the wood and bark, produces the secondary vascular tissues: the wood, which is secondary xylem, and the inner bark, which is secondary phloem. The cork cambium produces the periderm, the outer bark tissue that replaces the epidermis in a plant with secondary growth.

SUMMARY WITH KEY TERMS

I. Plants are either herbaceous (nonwoody) or woody.
   - A. **Annuals** are herbaceous plants that grow, reproduce, and die in one year or less.
   - B. **Biennials** take two years to complete their life cycles before dying.
   - C. **Perennials** are herbaceous and woody plants that have the potential to live for more than two years.

II. The vascular plant body typically consists of a root system and a shoot system.
   - A. The **root system** is generally underground and obtains water and dissolved nutrient minerals for the plant. Roots also anchor the plant firmly in place.
   - B. The **shoot system** is generally aerial and obtains sunlight and exchanges gases such as carbon dioxide, oxygen, and water vapor.
     1. The shoot system consists of a vertical stem that bears leaves (the main organs of photosynthesis) and reproductive structures (flowers and fruits in flowering plants).
     2. **Buds,** undeveloped embryonic shoots, develop on stems.

III. The plant body is composed of three tissue systems: ground, vascular, and dermal.
   - A. The **ground tissue system** consists of three tissues with a variety of functions.
     1. **Parenchyma** tissue is composed of living parenchyma cells that possess thin primary cell walls. Functions of parenchyma tissue include photosynthesis, storage, and secretion.
     2. **Collenchyma** tissue is composed of collenchyma cells with unevenly thickened primary cell walls. This tissue provides flexible structural support.
     3. **Sclerenchyma** tissue is composed of sclerenchyma cells (**sclereids** or **fibers**) that have both primary and secondary cell walls. Sclerenchyma cells are often dead at maturity, but they provide structural support.
   - B. The **vascular tissue system** conducts materials throughout the plant body and provides strength and support.
     1. **Xylem** is a complex tissue that conducts water and dissolved nutrient minerals. The actual conducting cells of xylem are **tracheids** and **vessel elements.**
     2. **Phloem** is a complex tissue that conducts sugar in solution. **Sieve tube members** are the conducting cells of phloem; they are assisted by **companion cells.**
   - C. The **dermal tissue system** is the outer protective covering of the plant body.
     1. The **epidermis** is a complex tissue that covers the herbaceous plant body.
        - a. The epidermis that covers aerial parts secretes a layer of wax, called the **cuticle,** that reduces water loss.
        - b. Gas exchange between the interior of the shoot system and the surrounding atmosphere occurs through **stomata.**
        - c. **Trichomes** are outgrowths, or hairs, that occur in many sizes and shapes and have a variety of functions.
     2. The **periderm** is a complex tissue that covers the woody parts of the plant body in woody plants.
   - D. Although separate organs (roots, stems, leaves, flower parts, and fruits) exist in the plant, tissues are integrated throughout the plant body, providing continuity from organ to organ.

IV. Growth in plants is localized in specific regions, called **meristems,** and involves cell division, cell elongation, and cell differentiation.
   A. **Primary growth** is an increase in stem and root length.
      1. Primary growth occurs in all plants.
      2. Primary growth results from the activity of **apical meristems** that are localized at the tips of roots and shoots and within the buds of stems.
   B. **Secondary growth** is an increase in stem and root girth (thickness).
      1. In addition to primary growth, woody plants also have secondary growth.
      2. Secondary growth is localized, typically occurring in long cylinders of meristematic cells throughout the length of older stems and roots.
      3. The two **lateral meristems** responsible for secondary growth are the **vascular cambium** and the **cork cambium.**

POST-TEST

1. Plants that complete their life cycles in one year are called ____________; those that complete them in two years are ____________; and those that live year after year are ____________. (a) annuals; perennials; biennials (b) biennials; annuals; perennials (c) annuals; biennials; perennials (d) perennials; annuals; biennials (e) perennials; biennials; annuals
2. Most of the plant body consists of the ____________ tissue system. (a) ground (b) vascular (c) periderm (d) dermal (e) cortex
3. Which tissue system provides a covering for the plant body? (a) ground (b) vascular (c) periderm (d) dermal (e) cortex
4. Storage, secretion, and photosynthesis are the functions of (a) collenchyma (b) vessel elements (c) lateral meristems (d) sclerenchyma (e) parenchyma
5. The two simple tissues that are specialized for support are (a) parenchyma and collenchyma (b) collenchyma and sclerenchyma (c) sclerenchyma and parenchyma (d) parenchyma and xylem (e) xylem and phloem
6. Sclereids and fibers are examples of which plant tissue? (a) parenchyma (b) collenchyma (c) sclerenchyma (d) xylem (e) epidermis
7. Which of the following statements about the vascular tissue system is *not* true? (a) Xylem and phloem are continuous throughout the plant body. (b) Xylem not only conducts water and dissolved nutrient minerals, but also provides support. (c) Four different cell types occur in phloem: sieve tube members, companion cells, tracheids, and vessel elements. (d) Sieve tube members lack nuclei. (e) Vessel elements are hollow, and their end walls have perforations or are entirely dissolved away.
8. Conduction of water and nutrient minerals in the xylem occurs in vessel elements and (a) sieve tube members (b) tracheids (c) collenchyma (d) cork cells (e) phloem
9. Conduction of sugar in solution in the sieve tube members is aided by (a) cork cells (b) sclerenchyma (c) parenchyma (d) guard cells (e) companion cells
10. The outer covering of plants with primary growth is ____________, whereas plants with secondary growth are covered by ____________. (a) cuticle; cork parenchyma (b) periderm; phloem (c) epidermis; periderm (d) epidermis; collenchyma (e) cellulose; lignin
11. The noncellular layer of wax secreted by the epidermis over its surface is called (a) lignin (b) cuticle (c) periderm (d) cellulose (e) trichome
12. Minute pores known as ____________ dot the surface of the epidermis of leaves and stems; each pore is bordered by two ____________. (a) stomata; guard cells (b) stomata; fibers (c) sieve tube members; companion cells (d) sclereids; guard cells (e) cuticle; guard cells
13. Localized areas within the plant body where cell divisions occur are known as (a) organs (b) fibers (c) meristems (d) cork parenchyma (e) stomata
14. Primary growth, an increase in the length of a plant, occurs at the (a) cork cambium (b) apical meristem (c) vascular cambium (d) lateral meristem (e) periderm
15. The two lateral meristems responsible for secondary growth are the (a) cork cambium and apical meristem (b) apical meristem and cork parenchyma (c) vascular cambium and apical meristem (d) vascular cambium and cork cambium (e) cork cambium and cork parenchyma

REVIEW QUESTIONS

1. What are some of the functions of roots? Of shoots?
2. What are the three tissue systems in plants? Describe the functions of each.
3. Compare the cellular structures and functions of parenchyma, collenchyma, and sclerenchyma tissues.
4. What are the functions of xylem and phloem?
5. Compare and contrast epidermis and periderm.
6. What is the role of plant meristems?
7. Distinguish between primary and secondary growth. Between apical and lateral meristems.

8. Identify each type of tissue in the figures. Use Tables 31–2, 31–3, and 31–4 to check your answers.

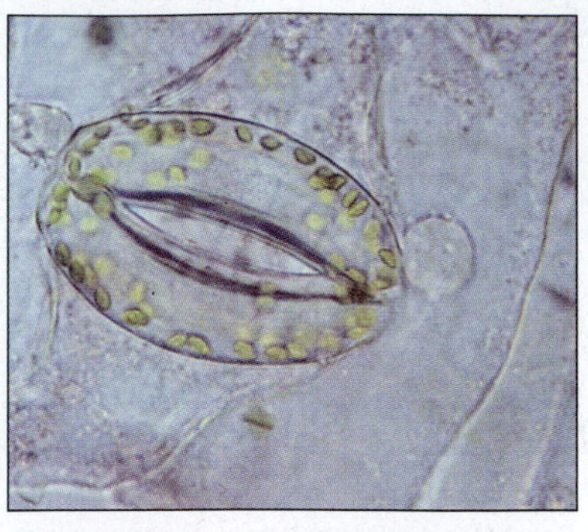
(a)

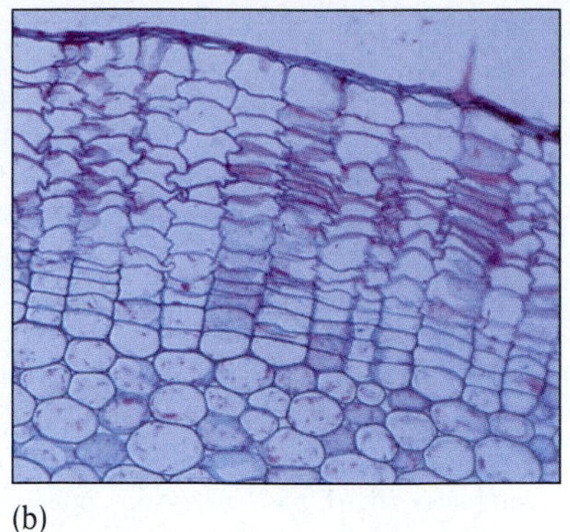
(b)

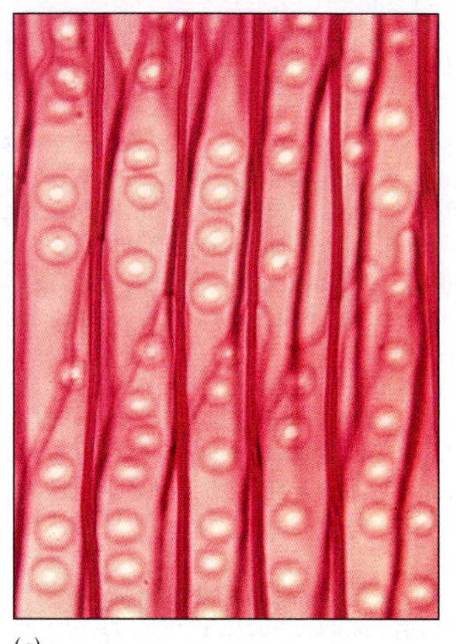
(c)

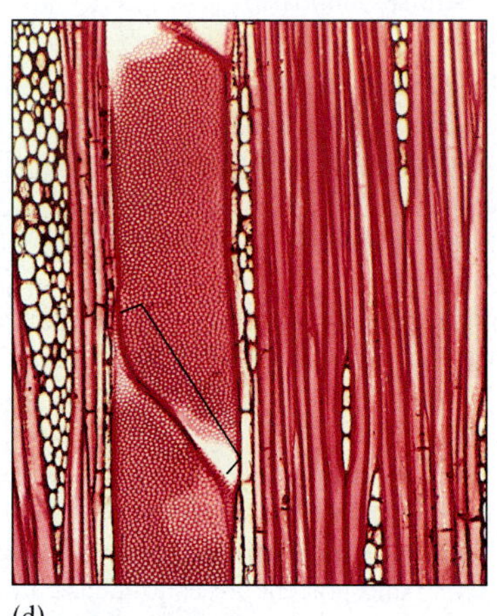
(d)

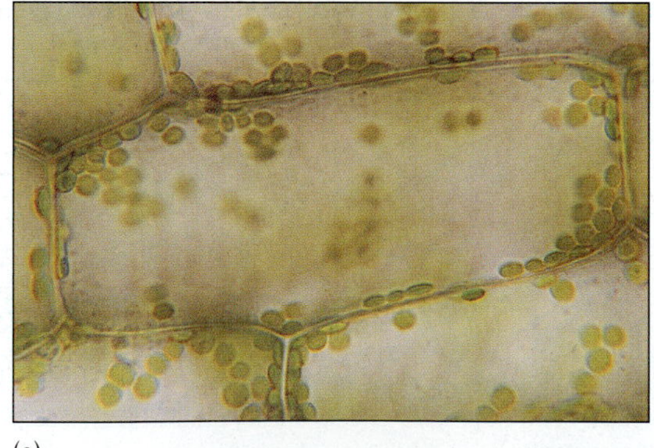
(e)

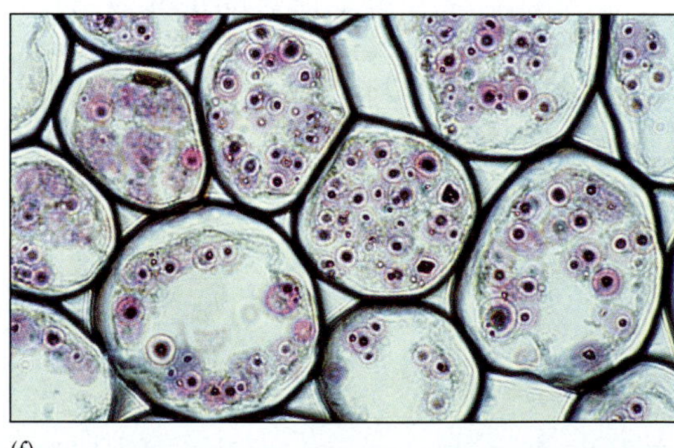
(f)

YOU MAKE THE CONNECTION

1. Grasses have a special meristem situated at the base of the leaves. Relate this information to what you know about the growth of grass after you mow the lawn.
2. A couple carved a heart with their initials into a tree trunk four feet above ground level; the tree was 25 feet tall at the time. Twenty years later the tree was 50 feet tall. How far above the ground were the initials? Explain your answer.
3. Sclerenchyma in plants is the functional equivalent of bone in humans (i.e., both sclerenchyma and bone provide support). However, sclerenchyma is dead, whereas bone is living tissue. What are some of the advantages of a plant having dead support cells? Can you think of any disadvantages?

RECOMMENDED READINGS

Bell, A.D. *Plant Form: An Illustrated Guide to Flowering Plant Morphology.* Oxford University Press, Oxford, England, 1991. The external features of plants are examined in this beautifully illustrated book.

Berg, L.R. *Introductory Botany: Plants, People, and the Environment.* Saunders College Publishing, Philadelphia, 1997. A general botany text with an environmental emphasis.

Carpita, N., and C. Vergara. "A Recipe for Cellulose." *Science,* Vol. 279, 30 Jan. 1998. Biologists have identified the first genes involved in making cellulose for plant cell walls.

Fullick, A. "Roots of History." *New Scientist,* Inside Science, Issue 125, 13 Nov. 1999. A brief overview of how plants are important to humans.

Pimm, S.L. "In Search of Perennial Solutions." *Nature,* Vol. 389, 11 Sep. 1997. This short article highlights the environmental problems caused by traditional agriculture, which makes use of annual crops. The possibility of developing and farming perennial crops is explored.

Raven, P.H., R.F. Evert, and S.E. Eichhorn. *Biology of Plants,* 6th ed. W.H. Freeman & Company, New York, 1999. This is an excellent general botany textbook, with detailed descriptions of plant structure and life processes.

Sivitz, L. "First Plant Genome Thrills Biologists." *Science News,* Vol. 158, 16 Dec. 2000. Biologists have published the genome of the model research plant *Arabidopsis thaliana*.

Strauss, E. "When Walls Can Talk, Plant Biologists Listen." *Science,* Vol. 282, 2 Oct. 1998. Plant cell walls play an active role in signal transduction that determines the fate of cells during development.

- Visit our Web site at **http://www.info.brookscole.com/solomonbergmartin** for links to chapter-related resources on the World Wide Web. Additional on-line materials relating to this chapter can also be found on our Web site.

See chapter activity on BioActive Learner CD for additional help in mastering the chapter's material. Icon location in the chapter's margins shows which topics have tutorials or simulations in the CD.

32

Leaf Structure and Function

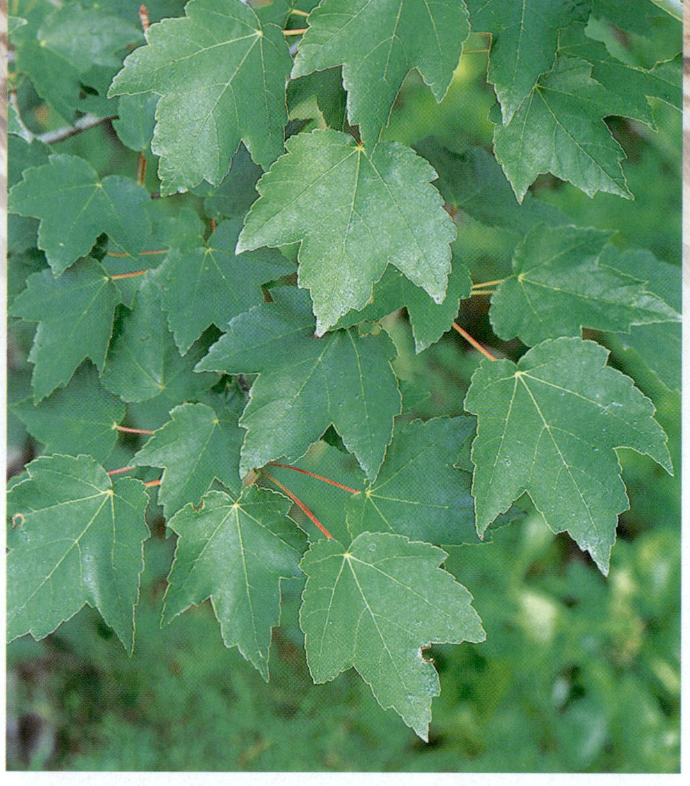

Leaves. Note how the red maple *(Acer rubrum)* leaves are arranged to efficiently capture light. *(David Sieren/Visuals Unlimited)*

LEARNING OBJECTIVES

After you have studied this chapter you should be able to

1. Discuss leaf structure, including simple versus compound leaves, leaf arrangement on the stem, and venation patterns.
2. Describe the major tissues of the leaf (epidermis, photosynthetic ground tissue, xylem, and phloem), and label them on a diagram of a leaf cross section.
3. Compare leaf anatomy in dicots and monocots.
4. Relate leaf structure to its function of photosynthesis.
5. Outline the physiological changes that accompany stomatal opening and closing.
6. Discuss transpiration and its effects on plants.
7. Define leaf abscission, explain why it occurs, and describe the physiological and anatomical changes that precede it.
8. List five examples of modified leaves and give the function of each.

Suppose for a moment that you are taking a course in engineering and have been asked to design an efficient solar collector capable of converting the radiant energy it collects into chemical energy of organic molecules. Where would you start? It might be helpful to check library books and periodicals to see how solar collectors/energy converters have been designed in the past. In this instance, it would also be wise to ask a biologist, or even a biology student like yourself, whether anything comparable exists in nature. The answer, of course, is yes: Plants have organs that are effective solar collectors and energy converters. They are called *leaves*.

Plants allocate many resources into the production of leaves. According to John E. Dale, a University of Edinburgh botanist who specializes in leaf development, a large maple tree annually produces 46.5 m^2 (500 ft^2) of leaves, which weigh more than 113 kg (250 lb). The metabolic cost of producing so many leaves is high, but leaves are essential to the survival of the tree.

Leaves gather the sunlight necessary for **photosynthesis,** the biological process that converts radiant energy into the chemical energy of carbohydrate molecules. Plants then use these molecules as starting materials to synthesize all other organic compounds and as fuel to provide energy for their metabolic processes. During a single summer, the leaves of the red maple shown here will "fix" about 454 kg (1000 lb) of carbon dioxide into sugars and other organic compounds.

Leaves are the main photosynthetic organs of most plants, and the structure of a leaf is superbly adapted for its primary function of photosynthesis. Most leaves are thin and flat, a shape that allows maximum absorption of light energy and the efficient internal diffusion of gases such as CO_2 and O_2. As a result of their ordered arrangement on the stem, leaves efficiently catch the sun's rays. The leaves of plants form an intricate green mosaic, bathed in sunlight and atmospheric gases.

The same structural adaptations that allow leaves to meet the requirements of photosynthesis also allow the loss of considerable water vapor. However, leaves possess several features that help reduce or control water loss, the most important of which is a thin, transparent layer of wax that covers the leaf surface. Structural adaptations are compromises between competing needs, and some features that optimize photosynthesis actually *promote* water loss. Plants have minute pores that allow an exchange of gases for photosynthesis, but these tiny openings also allow water to escape into the atmosphere as water vapor. Thus, leaf structure represents a trade-off between photosynthesis and water conservation.

FOLIAGE LEAVES VARY GREATLY IN EXTERNAL FORM

Leaves are the most variable of plant organs, so much so that plant biologists had to develop specific terminology to describe their shapes, margins (edges), vein patterns, and the way they attach to stems. Because each leaf is characteristic of the plant species on which it grows, many plants can be identified by their leaves alone. Leaves may be round, needle-like, scalelike, cylindrical, heart-shaped, fan-shaped, or thin and narrow. They vary in size from those of the raffia palm *(Raphia ruffia),* whose leaves often grow more than 20 m (65 ft) long, to those of water-meal (*Wolffia* sp.), whose leaves are so small that 16 of them laid end-to-end measure 2.5 cm (1 in) (see Fig. 31–1*a*).

The broad, flat portion of a leaf is the **blade;** the stalk that attaches the blade to the stem is the **petiole.** Some leaves also have **stipules,** which are leaflike outgrowths usually present in pairs at the base of the petiole (Fig. 32–1). Some leaves do not have petioles or stipules.

Leaves may be *simple* (having a single blade) or *compound* (having a blade divided into two or more leaflets) (Fig. 32–2*a*). Sometimes it is difficult to tell whether a plant has formed one compound leaf or a small stem bearing several simple leaves. One easy way to determine if a plant has simple or compound leaves is to look for axillary buds, so called because each develops in a leaf *axil* (the angle between the stem and petiole). Axillary buds form at the base of a leaf, whether it is simple or compound. However, axillary buds never develop at the base of leaflets. Also, the leaflets of a compound leaf lie in a single plane (you can lay a compound leaf flat on a table), whereas simple leaves usually are not arranged in one plane on a stem.

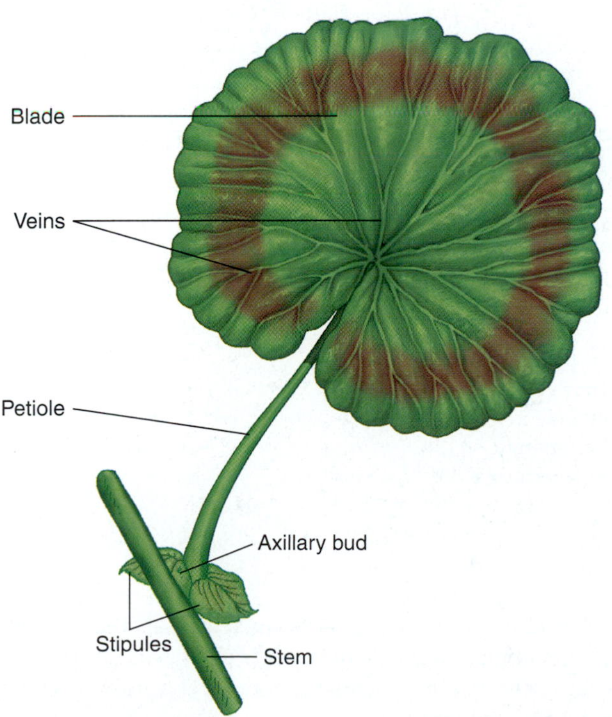

Figure 32–1 Parts of a leaf. A geranium leaf consists of a blade, a petiole, and two stipules at the base of the leaf. Note the axillary bud in the leaf axil.

Leaves are arranged on a stem in one of three possible ways (Fig. 32–2*b*). Plants such as beeches and walnuts have an *alternate leaf arrangement,* with one leaf at each node. (A *node* is the area of the stem where one or more leaves are attached.) In an *opposite leaf arrangement,* as occurs in maples and ashes, two leaves grow at each node. In a *whorled leaf arrangement,* as occurs in catalpa, three or more leaves grow at each node.

Leaf blades may possess *parallel venation,* in which the primary **veins** (strands of vascular tissue) run approximately parallel to one another (generally characteristic of monocots), or *netted venation,* in which veins are branched in such a way that they resemble a net (generally characteristic of dicots; Fig. 32–2*c*).[1] Netted veins can be *palmately netted,* in which several major veins radiate out from one point, or *pinnately netted,* in which major veins branch off along the entire length of the **midvein** (main or central vein of a leaf).

THE LEAF CONSISTS OF AN EPIDERMIS, GROUND TISSUE THAT IS PHOTOSYNTHETIC, AND VASCULAR TISSUE

The leaf is a complex organ composed of several tissues, all of which are organized to optimize photosynthesis (Fig. 32–3). The leaf blade has upper and lower surfaces that are covered by an epidermal layer. The **upper epidermis** covers the upper surface, and the **lower epidermis** covers the lower surface. Most cells in these layers lack chloroplasts and are relatively transparent. One interesting feature of leaf epidermal cells is that the cell wall facing toward the outside of the leaf is somewhat thicker than the cell wall facing inward. This extra thickness may provide the plant with additional protection against injury or water loss.

Because leaves have such a large surface area exposed to the atmosphere, water loss by evaporation from the leaf's surface is unavoidable. However, epidermal cells secrete a waxy layer, the **cuticle,** that reduces water loss from their exterior walls (see Table 31–4). The cuticle, which consists primarily of a waxy substance called **cutin,** varies in thickness in different plants, in part owing to environmental conditions. As one might expect, the leaves of plants adapted to hot, dry climates have extremely thick cuticles. Furthermore, a leaf's exposed (and warmer) upper epidermis generally has a thicker cuticle than its shaded (and cooler) lower epidermis.

The epidermis of many leaves is covered with various hairlike structures called **trichomes** (see Table 31–4). Some leaves,

[1] Recall that flowering plants, the focus of this chapter, are divided into two groups, informally called dicots and monocots, based on a number of structural features (see Chapter 27). Dicots include well-known plants such as beans, petunias, oaks, cherry trees, roses, and snapdragons; examples of monocots include corn, lilies, grasses, palms, tulips, orchids, and bananas.

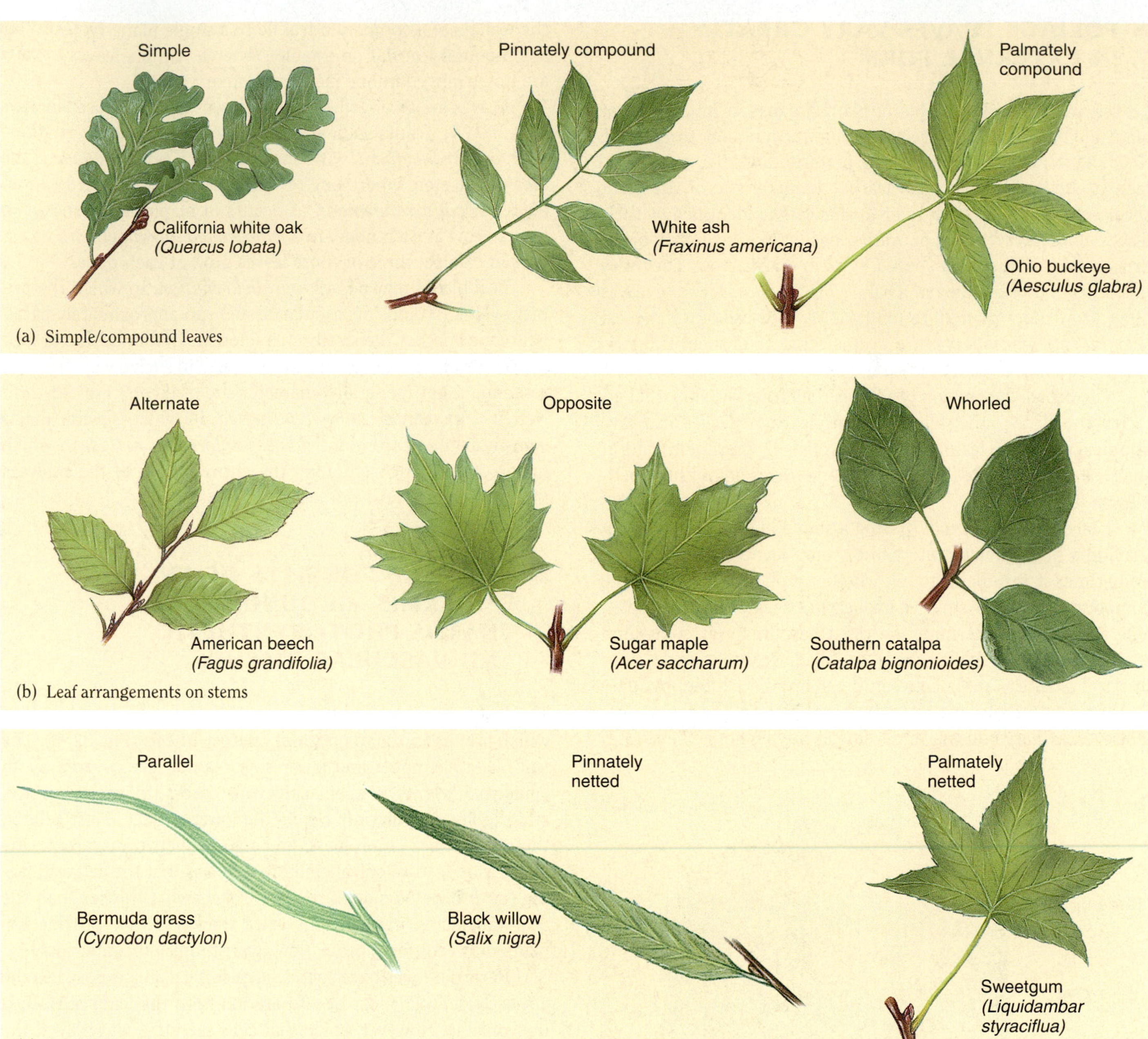

Figure 32–2 Leaf morphology. **(a)** Simple, pinnately compound, and palmately compound leaves. **(b)** Leaf arrangement may be alternate, opposite, or whorled, depending on the number of leaves at each node. **(c)** Venation patterns include parallel, which is characteristic of monocot leaves, pinnately netted, and palmately netted. All leaves shown are woody dicot trees from North America, except bermuda grass, which is a herbaceous monocot native to Europe and Asia.

such as those of the popular cultivated plant called lamb's ear, have so many trichomes that they feel fuzzy (Fig. 32–4). Trichomes may have several functions. Trichomes of some plants help reduce water loss from the leaf surface by retaining a layer of moist air next to the leaf and by reflecting sunlight, thereby protecting the plant from overheating. Some trichomes secrete stinging irritants for deterring **herbivores,** animals that feed on plants. In addition, a leaf covered with trichomes is difficult for an insect to walk over or eat. Other trichomes excrete excess salts absorbed from a salty soil.

The leaf epidermis contains minute openings, or **stomata** (sing., *stoma*), for gas exchange. Each stoma is flanked by two

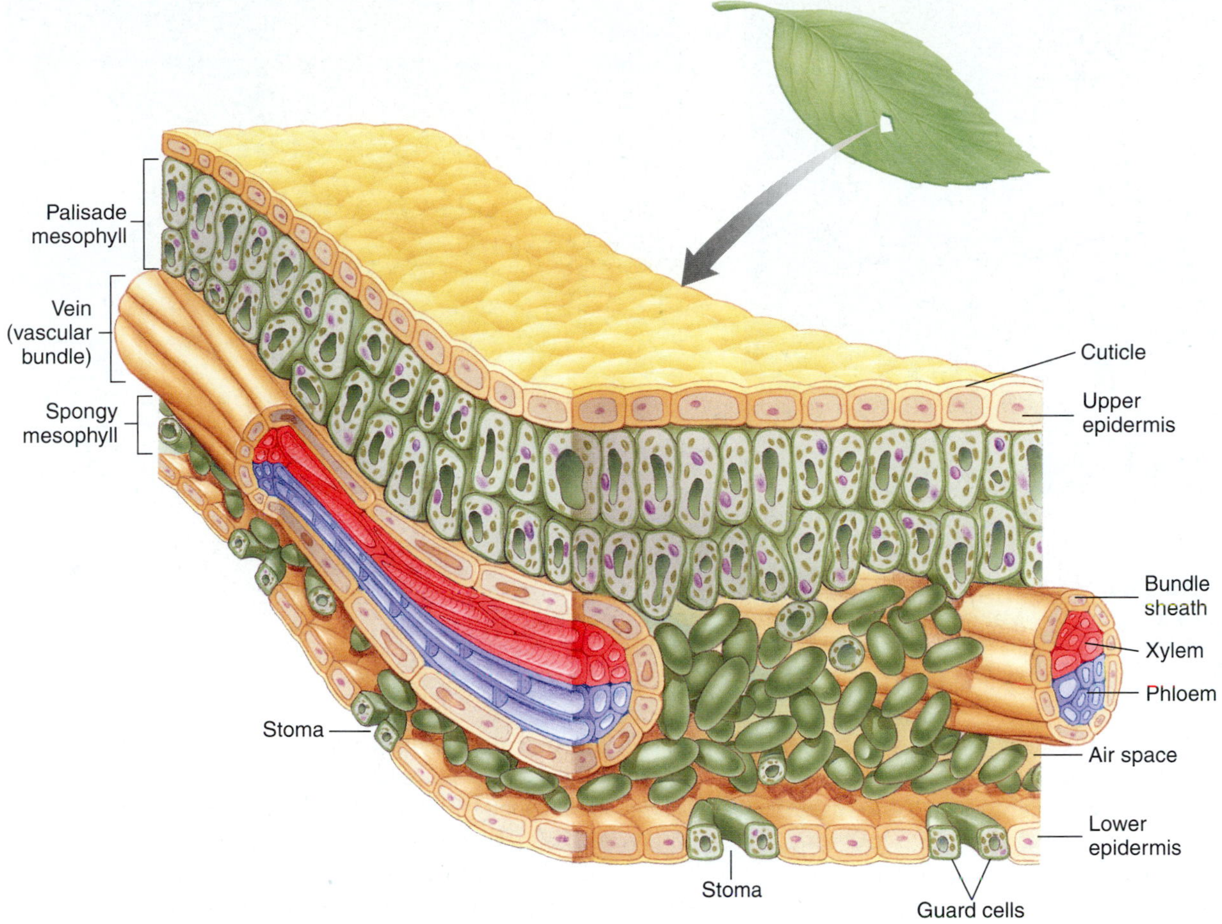

Figure 32–3 Tissues in a typical leaf blade. An upper epidermis and lower epidermis cover the blade. The photosynthetic ground tissue, called mesophyll, is often arranged into palisade and spongy layers. Veins branch throughout the mesophyll.

Figure 32–4 Trichomes on leaves of lamb's ear *(Stachys byzantina).* This popular garden plant has so many trichomes (the tiny white hairs covering the leaf's surface) that the leaves look and feel like velvet. Each leaf grows to 10 cm (4 in) long. *(Alan L. Detrick/Photo Researchers, Inc.)*

specialized epidermal cells called **guard cells** (Fig. 32–5). The guard cells are responsible for opening and closing the stoma. Guard cells are usually the only epidermal cells with chloroplasts.

Stomata are especially numerous on the lower epidermis of horizontally oriented leaves (an average of about 100 stomata per mm^2), and in many species are located *only* on the lower surface. The lower epidermis of apple *(Malus sylvestris)* leaves, for example, has almost 400 stomata per mm^2, whereas the upper epidermis has none. This adaptation reduces water loss because stomata on the lower epidermis are shielded from direct sunlight and are therefore cooler than those on the upper epidermis. On the other hand, leaves of aquatic plants such as water lilies have stomata only on the upper epidermis.

Guard cells are associated with special epidermal cells called **subsidiary cells** that are often structurally different from all other epidermal cells. Subsidiary cells provide a reservoir of water and ions that move into and out of the guard cells as they change shape during stomatal opening and closing (discussed later in this chapter).

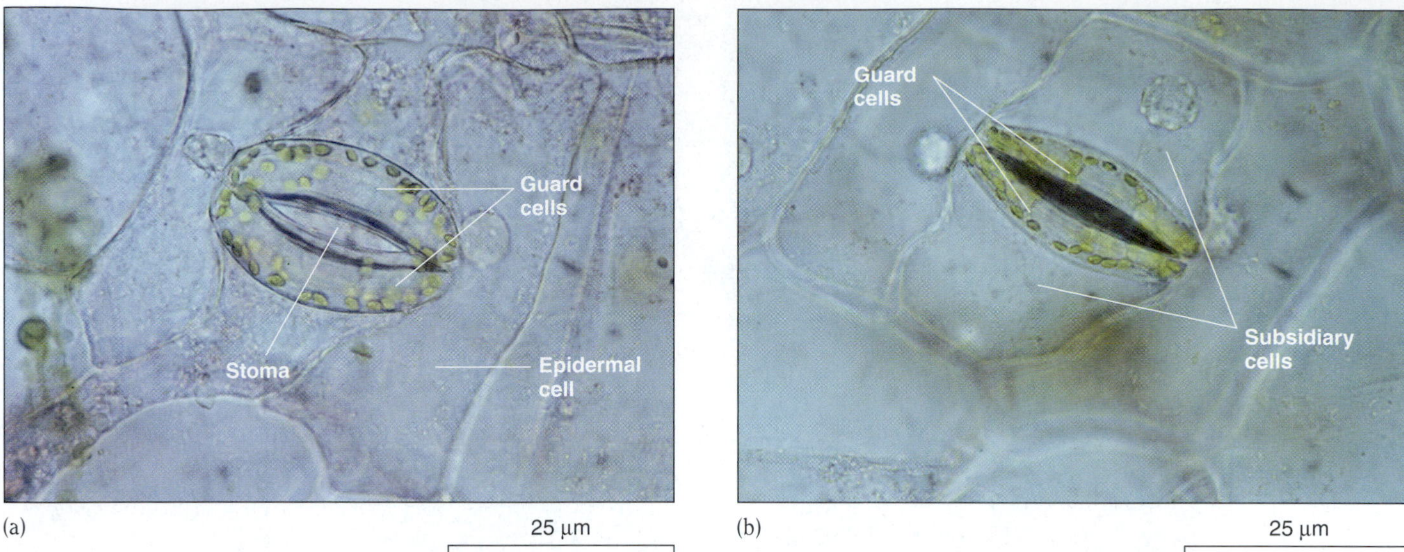

Figure 32–5 Stomata. LMs of **(a)** an open stoma and **(b)** a closed stoma from the leaf epidermis of wandering jew *(Zebrina pendula).* Note the chloroplasts present in the guard cells and the thicker inner wall of each guard cell. *(a, b, Dwight R. Kuhn)*

The photosynthetic ground tissue of the leaf, called the **mesophyll** (from the Greek *meso,* "the middle of," and *phyll,* "leaf"), is sandwiched between the upper epidermis and the lower epidermis. Mesophyll cells, which are parenchyma cells (see Chapter 31) packed with chloroplasts, are loosely arranged with many air spaces between them that facilitate gas exchange. These intercellular air spaces account for as much as 70% of the leaf's volume.

In many plants, the mesophyll is divided into two sublayers. Toward the upper epidermis, the columnar cells are stacked closely together in a layer called **palisade mesophyll.** In the lower portion, the cells are more loosely and irregularly arranged in a layer called **spongy mesophyll.** The two layers have different functions. Palisade mesophyll is the main site of photosynthesis in the leaf. Photosynthesis also occurs in the spongy mesophyll, but the primary function of the spongy mesophyll is to allow for diffusion of gases (particularly CO_2) within the leaf.

Palisade mesophyll may be further organized into one, two, three, or even more layers of cells. The presence of additional lay-

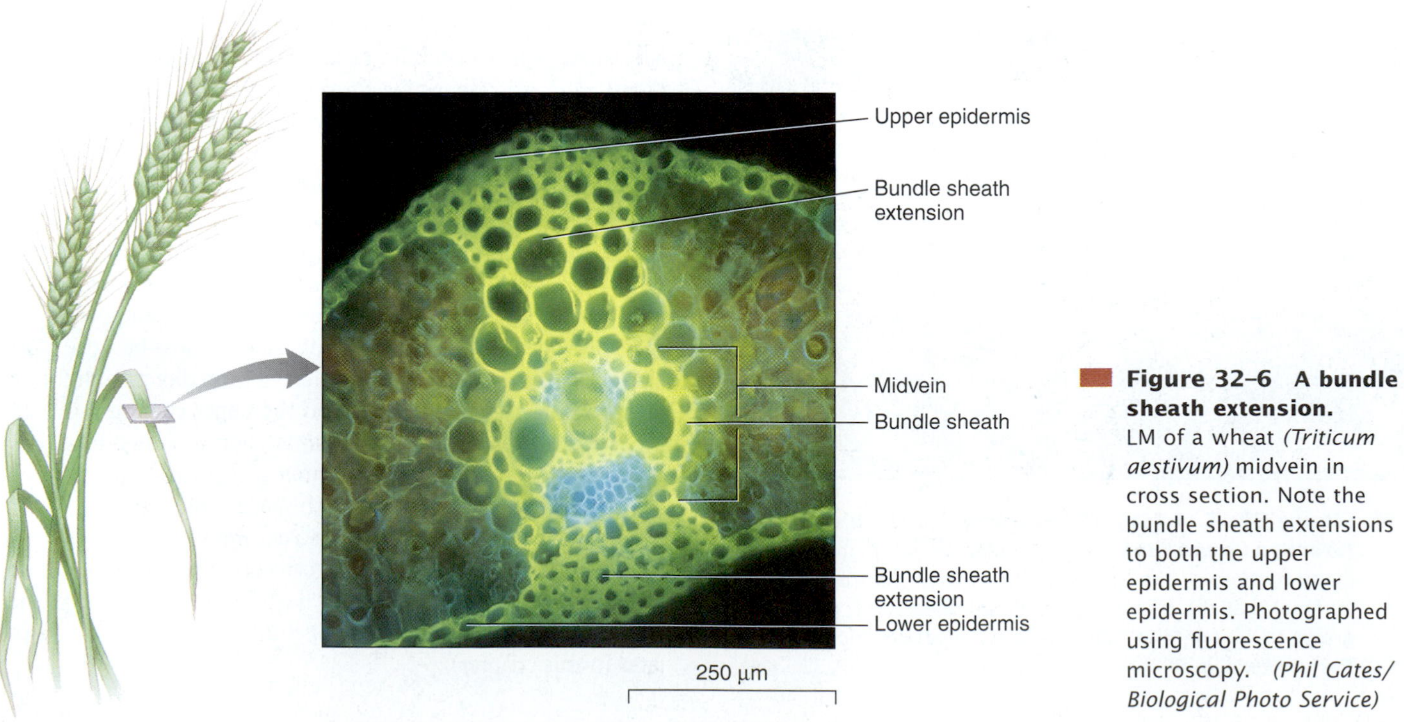

Figure 32–6 A bundle sheath extension. LM of a wheat *(Triticum aestivum)* midvein in cross section. Note the bundle sheath extensions to both the upper epidermis and lower epidermis. Photographed using fluorescence microscopy. *(Phil Gates/ Biological Photo Service)*

ers of palisade mesophyll is at least partly an adaptation to environmental conditions. Leaves exposed to direct sunlight contain more layers of palisade mesophyll than do shaded leaves on the same plant. In direct sunlight, the light is strong enough to effectively penetrate multiple layers of palisade mesophyll, allowing all layers to photosynthesize efficiently.

The veins, or **vascular bundles,** of a leaf extend through the mesophyll. Branching is extensive, and no mesophyll cell is more than two or three cells away from a vein. Therefore, the slow process of diffusion does not limit the movement of needed resources between mesophyll cells and veins. Each vein contains two types of vascular tissue: xylem and phloem (see Chapter 31). **Xylem,** which conducts water and dissolved nutrient minerals, is usually located in the upper part of a vein, toward the upper epidermis, whereas **phloem,** which conducts dissolved sugars, is usually confined to the lower part of a vein.

One or more layers of nonvascular cells surround the larger veins and make up the **bundle sheath.** Bundle sheaths are composed of parenchyma or sclerenchyma cells (see Chapter 31). Frequently, the bundle sheath has support columns, called **bundle sheath extensions,** that extend through the mesophyll from the upper epidermis to the lower epidermis (Fig. 32–6). Bundle sheath extensions may be composed of parenchyma, collenchyma, or sclerenchyma cells.

Leaf structure differs in dicots and monocots

A dicot leaf is usually composed of a broad, flattened blade and a petiole. As mentioned previously, dicot leaves typically have netted venation. In contrast, monocot leaves often lack a petiole; they are narrow and the base of the leaf often wraps around the stem, forming a sheath. Parallel venation is characteristic of monocot leaves.

Dicots and certain monocots also differ in internal leaf anatomy (Fig. 32–7). Although most dicots and monocots have both palisade and spongy layers, some monocots (corn and other

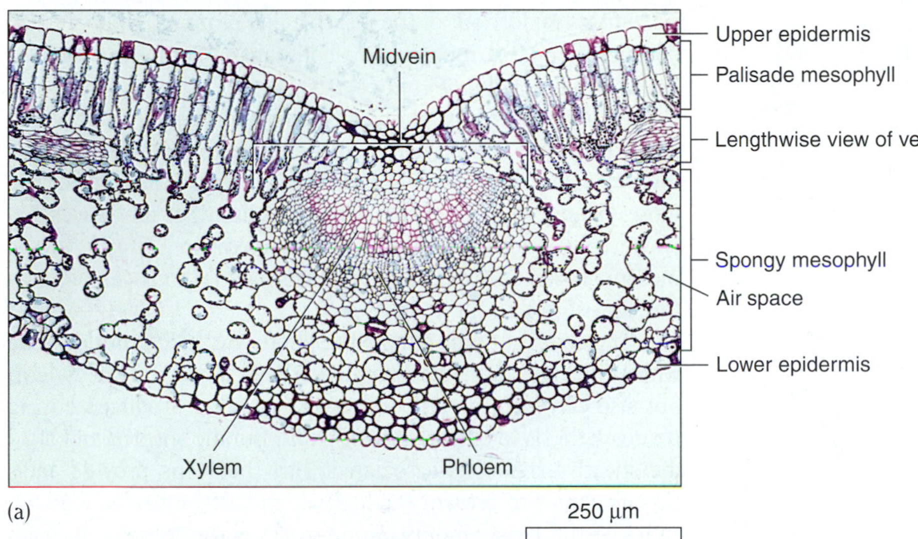

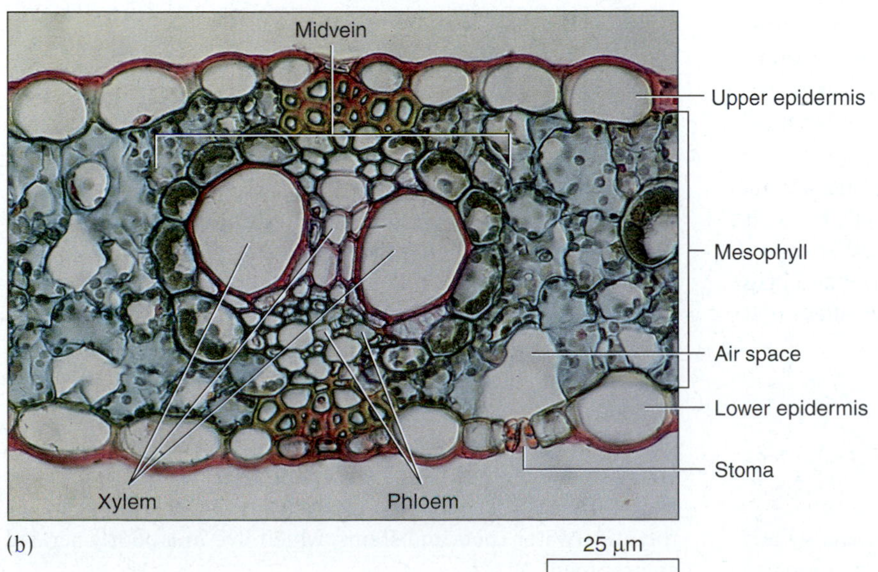

Figure 32–7 Dicot and monocot leaf cross sections. **(a)** LM of part of a leaf cross section of privet *(Ligustrum vulgare)*, a dicot. The mesophyll has distinct palisade and spongy sections. **(b)** LM of part of a leaf cross section of corn *(Zea mays)*, a monocot. Corn leaves lack distinct regions of palisade and spongy mesophyll. *(a, Ed Reschke; b, Dwight R. Kuhn)*

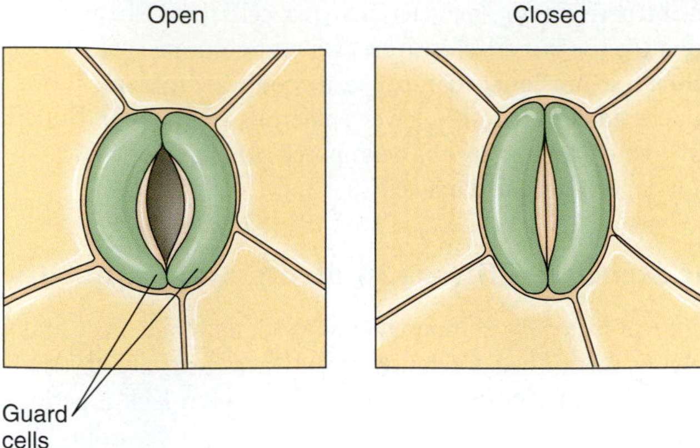

(a) Dicots and some monocots

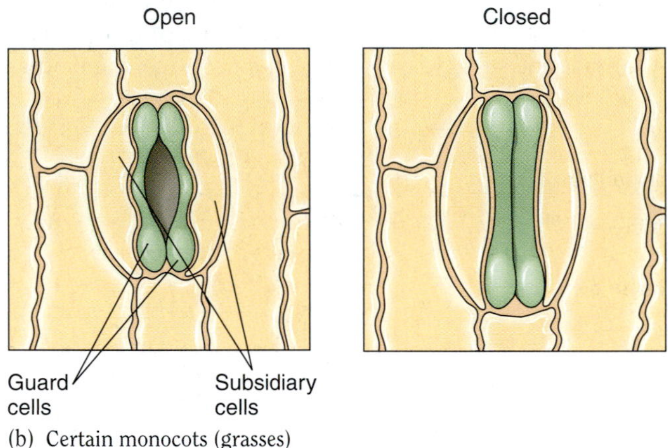

(b) Certain monocots (grasses)

Figure 32–8 Variation in guard cells. **(a)** Guard cells of dicots and many monocots are bean-shaped. **(b)** Other monocot guard cells (those of grasses, reeds, and sedges) are narrow in the center and thicker at each end. Guard cells are associated with special epidermal cells called subsidiary cells.

grasses) do not have mesophyll differentiated into distinct palisade and spongy layers. Because dicots have netted veins, a cross section of a dicot blade often shows veins in both cross-sectional and lengthwise views. In a cross section of a monocot leaf, on the other hand, the parallel venation pattern produces evenly spaced veins, all of which appear in cross section.

Differences between the guard cells in dicot and certain monocot leaves also occur (Fig. 32–8). The guard cells of dicots and many monocots are shaped like kidney beans. Other monocot leaves (those of grasses, reeds, and sedges) have guard cells shaped like dumbbells. These structural differences affect how the cells swell or shrink to open or close the stoma.

LEAF STRUCTURE IS RELATED TO FUNCTION

The primary function of leaves is to collect radiant energy and convert it to the chemical energy stored in the bonds of organic molecules such as glucose. This process, called photosynthesis, was examined in detail in Chapter 8. During photosynthesis, plants take relatively simple molecules (carbon dioxide and water) and convert them into sugars. Oxygen is given off as a byproduct. The plant uses the sugar formed during photosynthesis in two ways. First, it is broken down by aerobic respiration to release the chemical energy stored in its bonds for other cellular activities (see Chapter 7). Second, sugar molecules provide the cell with basic building materials. The cell modifies sugars, converting them into all the organic compounds used by the plant (starch and cellulose, amino acids, lipids, nucleic acids, and so forth).

How is leaf structure related to its primary function of photosynthesis? The epidermis of a leaf is relatively transparent and allows light to penetrate to the interior of the leaf where the photosynthetic ground tissue, the mesophyll, is located. Stomata, which dot the leaf surfaces, permit the exchange of gases between the atmosphere and the leaf's internal tissues. Carbon dioxide, a raw material of photosynthesis, diffuses into the leaf through stomata, and the oxygen produced during photosynthesis diffuses rapidly out of the leaf through stomata. Stomata also permit other gases, including air pollutants, to enter the leaf (see *Focus On: Air Pollution and Leaves*).

Water required for photosynthesis is obtained from the soil and transported in the xylem to the leaf, where it diffuses into the mesophyll and moistens the surfaces of mesophyll cells. The loose arrangement of the mesophyll cells, with air spaces between cells, allows for rapid diffusion of carbon dioxide to the mesophyll cell surfaces; there it dissolves in a film of water before diffusing into the cells.

The veins not only supply the photosynthetic ground tissue with water and minerals (from the roots, by way of the xylem) but also carry (in the phloem) dissolved sugar produced during photosynthesis to all parts of the plant. Bundle sheaths and bundle sheath extensions associated with the veins provide additional support to prevent the leaf, which is structurally weak because of the large amount of air space in the mesophyll, from collapsing under its own weight.

The leaves of each type of plant help it survive in the environment to which it is adapted

The environment to which a particular plant is adapted is reflected in its leaf structure. (Leaf evolution is discussed in Chapter 26.) Although both aquatic plants and those adapted to dry conditions perform photosynthesis and have the same basic leaf anatomy, their leaves are modified to enable them to survive different environmental conditions. The leaves of water lilies, for example, have petioles long enough to allow the blade to float on the water's surface (Fig. 32–9). Large air spaces in the mesophyll make the floating blade buoyant. The petioles and other submerged parts have an internal system of air ducts; oxygen moves through these ducts from the floating leaves to the underwater roots and stems, which live in a poorly aerated environment.

FOCUS ON Air Pollution and Leaves

The air we breathe is often dirty and contaminated with many pollutants, particularly in urban areas. Air pollution consists of gases, liquids, or solids present in the atmosphere in levels high enough to harm humans and other organisms, as well as nonliving materials. Although air pollutants can come from natural sources, as, for example, a lightning-caused fire or a volcanic eruption, human activities make a major contribution to global air pollution. Motor vehicles and industry are the two main human sources of air pollution.

All parts of a plant can be damaged by air pollution, but leaves are particularly susceptible because of their structure and function. The thin blade provides a large surface area that comes into contact with the surrounding air. The thousands of tiny stomatal pores that dot the epidermis and allow gas exchange with the atmosphere also permit pollutants to diffuse into the leaf. Just as lungs, the organs of gas exchange in humans, often are affected by air pollution, so too the leaves of a plant are most affected.

Trees provide a dramatic demonstration of the effect of air pollution on biological longevity. According to American Forests (formerly the American Forestry Association), the average lifespan for a tree living in a rural setting, where air pollution is generally low, is 150 years. Even in the most ideal urban setting, a tree typically lives only 60 years, and in a normal city setting the average life span is 32 years. Trees living in a downtown setting, where air quality is lowest, live only 7 years on average. It should be noted, however, that air pollution is not the only cause of short lifespan in urban trees. Other factors include inadequate room for root growth, lack of water, and polluted runoff from streets and sidewalks.

Many studies have shown that high levels of most forms of air pollution reduce the overall productivity of crop plants. The worst pollutant in terms of yield loss is ozone, a toxic gas produced when sunlight catalyzes a reaction between pollutants emitted by motor vehicles and industries. Ozone has been observed to reduce soybean yields by as much as 35%, and the average yield reduction from exposure to ozone is 15%. In plants, ozone inhibits photosynthesis because it damages the mesophyll cells, probably by altering the permeability of their cell membranes. Exposure to low levels of air pollution often causes a decline in photosynthesis without any other symptoms of injury. Lesions on leaves and other obvious symptoms appear at much higher levels of air pollution.

When air pollution is combined with other environmental stressors (such as low winter temperatures; prolonged droughts; insects; and bacterial, fungal, and viral diseases), it can cause plants to decline and die. More than half of the red spruce trees in the mountains of the northeastern United States have died since the mid-1970s, and sugar maples in eastern Canada and the United States are also dying. Many still-living trees are exhibiting symptoms of **forest decline,** characterized by gradual deterioration and often death of trees. The general symptoms of forest decline are reduced vigor and growth, but some plants exhibit specific symptoms, such as yellowing of needles in conifers. Forest decline is more pronounced at higher elevations, possibly because most trees growing at high elevations are at the limit of their normal range and are therefore more susceptible to wind and low temperatures.

Many factors can interact to decrease the health of trees, and no single factor accounts for the recent instances of forest decline. Several human-induced air pollutants have been implicated, including acid rain (see Fig. 2–18); ozone; and toxic heavy metals such as lead, cadmium, and copper. Power plants, ore smelters, refineries, and motor vehicles produce these pollutants. Insects and weather factors such as drought and severe winters may also be important. To complicate matters further, the actual causes of forest decline may vary from one tree species to another and from one location to another. Thus, forest decline appears to result from the combination of multiple stressors. When one or more stressors weaken a tree, then an additional stressor may be enough to cause its death.

The leaves of conifers, an important group of woody trees and shrubs that includes pine, spruce, fir, redwood, and cedar, are waxy needles.[2] Most conifers are evergreen, which means they produce and lose leaves throughout the year rather than during certain seasons. Conifers dominate a large portion of Earth's land area, particularly in northern forests and mountains. Their needles have structural adaptations that help them survive winter, the driest part of the year. (Winter is arid even in areas of heavy snows because roots cannot absorb water from soil when the soil temperature is low.) Indeed, many of the structural features of needles are also found in many desert plants.

Figure 32–10 shows a cross section of a pine needle. Note that the needle is somewhat thickened rather than thin and bladelike. The needle's relative thickness, which results in less surface area exposed to the air, reduces water loss. Other features that help conserve water include the thick, waxy cuticle and sunken stomata; these permit gas exchange while minimizing water loss. Thus, needles help conifers tolerate the dry (low relative humidity) winds that occur during winter. With the warming of spring, soil water again becomes available, and the needles quickly resume photosynthesis.

[2] As discussed in Chapter 27, conifers are gymnosperms, one of the two groups of seed plants (the other group is the flowering plants, or angiosperms). Unlike flowering plants, whose seeds are enclosed in fruits, conifers bear "naked" seeds on the scales of female cones.

Figure 32-9 Water lily (*Nymphaea* sp.) leaves. This underwater view of water lily leaves shows their petioles as well as their blades, which float on the water's surface. One of the water lily's adaptations for an aquatic lifestyle is that the length of its petioles depends on the depth of the water. *(Frans Lanting/Minden Pictures)*

STOMATAL OPENING AND CLOSING ARE DUE TO CHANGES IN GUARD CELL ION TRANSPORT AND TURGIDITY

Stomata are adjustable pores that are usually open during the day when carbon dioxide is required for photosynthesis and closed at night when photosynthesis is shut down (see the section on CAM photosynthesis in Chapter 8 for an interesting exception). The opening and closing of stomata are controlled by changes in the shape of the two guard cells that surround each pore. When water moves into guard cells from surrounding cells, they become turgid (swollen) and bend, producing a pore. When water leaves the guard cells, they become flaccid (limp) and collapse against one another, closing the pore.

Although light or darkness triggers the opening and closing of stomata, other environmental factors are also involved, including carbon dioxide concentration. A low concentration of CO_2 in the leaf induces stomata to open even in the dark. The effects of light and CO_2 concentration on stomatal opening are interrelated. Photosynthesis, which occurs in the presence of light, reduces the internal concentration of CO_2 in the leaf, triggering stomatal opening. Another environmental factor that affects stomatal opening and closing is dehydration (water stress). During a prolonged drought, stomata will remain closed even during the day. Stomatal opening and closing are under hormonal control, particularly by the plant hormone *abscisic acid* (see Chapter 36).

The opening and closing of stomata also appear to be under the control of an internal biological clock that in some way measures time. For example, after plants are placed in continual darkness, their stomata continue to open and close at more or less the same time each day. Such biological rhythms that follow an approximate 24-hour cycle are known as **circadian rhythms.** Other examples of circadian rhythms are provided in Chapter 36.

Stomata open and close in response to the movement of ions across the guard cells' plasma membranes

Data from numerous experiments and observations suggest that the movement of H^+ and K^+ across the plasma membranes of guard cells explains the opening and closing of stomata (Fig. 32–11). Light, particularly blue light, triggers the movement of

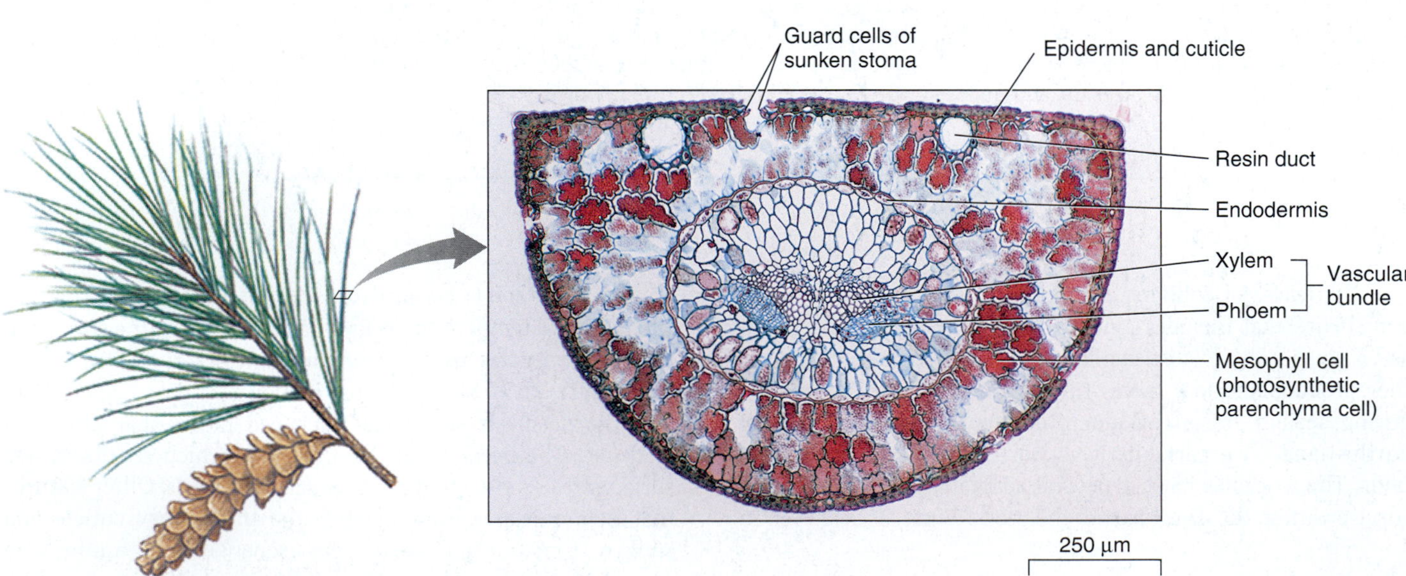

Figure 32-10 LM of a pine (*Pinus* sp.) needle in cross section. The thick, waxy cuticle and sunken stomata are two structural adaptations that enable the pine tree to retain its needles throughout the winter. *(John D. Cunningham/Visuals Unlimited)*

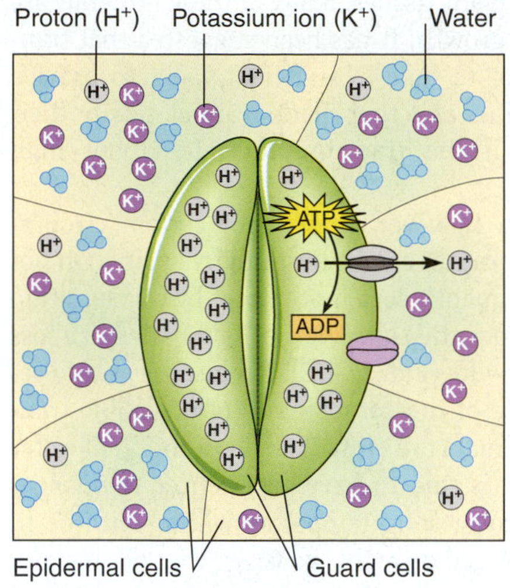

(a) Protons from malic acid are actively pumped out of guard cells. Potassium (K^+) ions are randomly distributed throughout the epidermis.

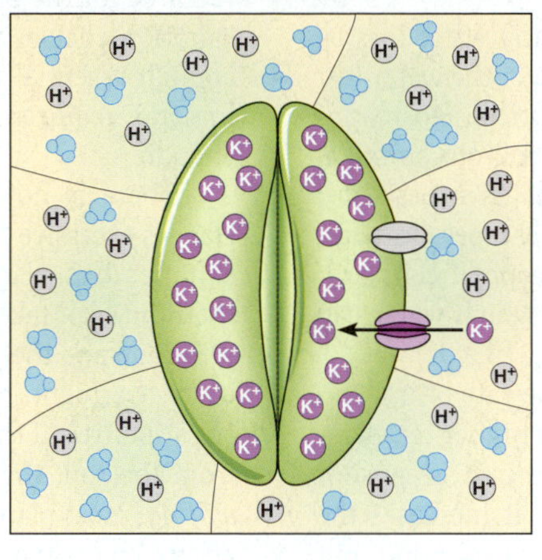

(b) Potassium (K^+) ions are actively transported into guard cells from surrounding epidermal cells.

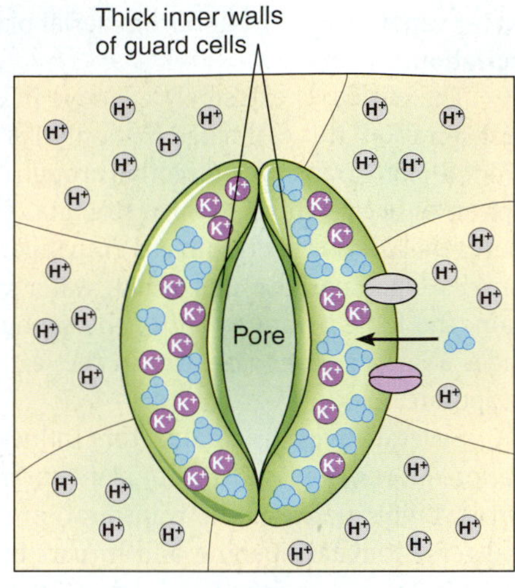

(c) Water moves into the guard cells by osmosis. This changes the shape of the guard cells, and the pore appears.

Figure 32–11 The mechanism of stomatal opening. Evidence suggests that stomatal opening is caused by **(a)** the movement of H^+ out of the guard cells, followed by **(b)** the movement of K^+ and then **(c)** water into the guard cells. For simplicity, a single H^+ and K^+ linked cotransport system is shown only in the right guard cell of each stoma.

potassium ions (K^+) into the guard cells from subsidiary cells or ordinary epidermal cells. This movement of K^+, which has been experimentally measured by patch clamp techniques (see Chapter 5), occurs by active transport through specific ion channels in the guard cells' plasma membranes and requires ATP. The ATP supplies energy to pump protons (H^+) out of the guard cells; the H^+ is formed when malic acid produced in the chloroplasts of the guard cells ionizes. The removal of H^+ from the guard cells produces an electrochemical gradient that can drive the uptake of K^+ through specific K^+ channels. This ion movement is an example of a linked cotransport system (see Chapter 5). The K^+ accumulates in the vacuoles of the guard cells.

The active uptake of K^+ in the guard cells increases the solute concentration in the vacuoles. As a result, water enters the guard cells from surrounding epidermal cells by osmosis. (Recall from the discussion of osmosis in Chapter 5 that when a cell has a solute concentration greater than that of surrounding cells, water flows into the cell.) The increased turgidity of the guard cells changes their shape, because the thickened inner cell walls do not expand as much as the outer walls. Thus, the stoma opens:

Light ⟶ proton pump moves H^+ out of guard cells ⟶ K^+ actively transported into guard cells ⟶ water diffuses into guard cells ⟶ guard cells change shape and stoma opens

In the late afternoon or early evening, stomata close, possibly by a reversal of the opening process. The K^+ is pumped out of the guard cells into the surrounding epidermal cells. Water leaves the guard cells by osmosis, the cells lose their turgidity, and the pore closes as the guard cells collapse.

Stomatal closure, however, may not be an exact reversal of stomatal opening. There is evidence that increases in the level of Ca^{2+} in guard cells trigger stomatal closure. In 2000, researchers from the University of California at San Diego reported in the journal *Science* about a mutant *Arabidopsis* plant that cannot produce oscillations in Ca^{2+} concentrations in guard cells under certain conditions. In the absence of the Ca^{2+} oscillations, the stomata of these mutant plants are unable to close. The possible role of Ca^{2+} in stomatal closure in plants is not surprising because oscillating Ca^{2+} levels are known to be an important part of **cell signaling,** communication between cells, in many animal cells.

As in other plant responses to light, stomata vary in their sensitivity to different colors, that is, different wavelengths, of light. Stomatal opening is most pronounced in blue and, to a lesser extent, in red light. Also, dim blue light induces stomatal opening, whereas dim red light does not. Any plant response to light must involve a pigment, a molecule that absorbs the light prior to the induction of a particular biological response. Data such as the responses of stomata to different colors of light suggest that the pigment involved in stomatal opening and closing is yellow (yellow pigments strongly absorb blue light and weakly absorb red light). The yellow pigment is thought to be located in the guard cells, probably in their plasma membranes.

LEAVES LOSE WATER BY TRANSPIRATION AND GUTTATION

Despite leaf adaptations such as the cuticle, approximately 99% of the water that a plant absorbs from the soil is lost by evaporation from the leaves and, to a lesser extent, the stems. Loss of water vapor by evaporation from aerial plant parts is called **transpiration.**

The cuticle is extremely effective in reducing water loss by transpiration. It is estimated that only 1% to 3% of the water lost from a plant transpires directly through the cuticle. Most transpiration occurs through open stomata. The numerous stomatal pores that are so effective in gas exchange for photosynthesis also provide openings through which water vapor escapes. In addition, the loose arrangement of the spongy mesophyll cells provides a large surface area within the leaf from which water can evaporate.

Several environmental factors influence the rate of transpiration. More water is lost from plant surfaces at higher temperatures. Light increases the transpiration rate, in part because it triggers stomatal opening and in part because it increases the leaf's temperature. Wind and dry air increase transpiration, but humid air *decreases* transpiration because the air is already saturated, or nearly so, with water vapor.

Although transpiration may seem wasteful, particularly to farmers in arid lands, it is an essential process that has adaptive value. Transpiration is responsible for water movement in plants, and without it water would not reach the leaves from the soil (see Chapter 33 discussion of the tension-cohesion model). The large amount of water plants lose by transpiration may provide some additional benefits. Transpiration, like sweating in humans, cools the leaves and stems. When water passes from a liquid state to a vapor, it absorbs a great deal of heat. When the water molecules leave the plant as water vapor, they carry this heat with them. Thus, the cooling effect of transpiration may prevent the plant from overheating, particularly in direct sunlight.

A second benefit of transpiration is that it distributes essential minerals throughout the plant. The water a plant transpires is initially absorbed from the soil, where it is not present as pure water but rather as a dilute solution of dissolved mineral salts. The water and dissolved nutrient minerals are then transported in the xylem throughout the plant, including its leaves. Water moves from the plant to the atmosphere during transpiration, but minerals remain in plant tissues. Many of these minerals are required for the plant's growth. It has been suggested that transpiration enables a plant to take in sufficient water to provide enough essential minerals and that plants cannot satisfy their mineral requirements if the transpiration rate is not high enough.

There is no doubt, however, that under certain circumstances excessive transpiration can be harmful to a plant. On hot summer days, plants frequently lose more water by transpiration than they can take in from the soil. Their cells experience a loss of turgor, and the plant wilts (Fig. 32–12). If a plant is able to recover overnight, because of the combination of negligible transpiration (recall that stomata are closed) and absorption of water from the soil, the plant is said to have experienced *temporary wilting*. Most plants recover from temporary wilting with no ill effects. In cases of prolonged drought, however, the soil may not contain sufficient moisture to permit recovery from wilting. A plant that cannot recover is said to be *permanently wilted* and will die.

MAKING THE CONNECTION Does transpiration influence an area's climate, that is, the average weather conditions that occur in that area over a period of years? Environmental factors that determine climate include temperature and precipitation, both of which are influenced by transpiration. If you have ever walked into a forest on a hot summer day, you have probably noticed that the air is cooler and moister there than outside the forest. Transpired water has a cooling effect not only for the plant but also for the local area.

Transpiration is an important part of the **hydrological cycle** (see Chapter 53), in which water cycles from the ocean and land to the atmosphere, and then back to the ocean and land. As a result of transpiration, water evaporates from leaves and stems to form clouds in the atmosphere. Thus, transpiration eventually results in precipitation. As you might expect, forest trees release

(a)

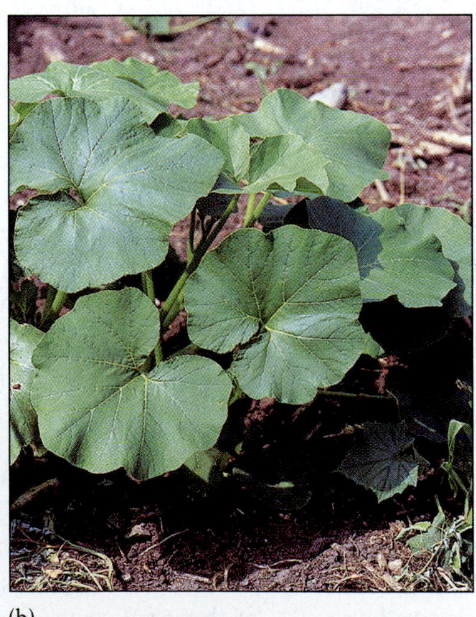

(b)

Figure 32–12 Temporary wilting. Shown are the same squash *(Cucurbita pepo)* leaves **(a)** in the late afternoon of a hot day, after the leaves have wilted owing to water loss, and **(b)** the following morning, after water in the leaves has been replenished from the roots. Note that wilting helps reduce the surface area from which transpiration occurs. During the night while transpiration is negligible, the plants recover by absorbing water from the soil. *(Carlyn Iverson)*

substantial amounts of moisture into the air by transpiration. Researchers have determined that at least half the rain that falls in the Amazon rainforest basin is recycled again and again by transpiration and precipitation.

Climate modelers who simulate deforestation in the Amazon rain forest also have determined that less moisture is carried into the deforested basin by trade winds; therefore, they predicted that rainfall would decline and droughts become more common in that region. This prediction has been observed and verified where large tracts of Amazon rain forest have been burned to grow crops; often people have observed droughts in areas that formerly had rainfall almost every day of the year. The explanation for less rainfall in a deforested region is that the heat carried into the atmosphere by transpiring trees accelerates the moist air movement upward, producing a low pressure area below it that pulls air in from over the Atlantic Ocean. (Air is not pulled in from over the Pacific Ocean because the Andes Mountains act as a barrier.) Similar declines in precipitation have also been observed in arid regions such as Australia, where extensive clearing of eucalyptus woodlands between 1950 and 1980 was followed by a 20% decline in winter rainfall.

Although the trees/rain connection holds in many areas, the relationship between trees and climate is not as simple as this brief discussion indicates. For example, climate modelers have determined that deforestation in India could *increase* precipitation there. The reason for the difference between the Amazon basin and India is related to geography. India is surrounded on three sides by ocean. During the day, the land heats up more than the surrounding ocean water. The heated land surface causes warm air masses to rise over the land, which pulls in moist air masses from the ocean. It is thought that deforestation will heat the land more, thereby magnifying this effect. Thus, complex interactions among ocean, atmosphere, land, and plants determine regional climate.

Figure 32–13 Guttation. Shown is a compound strawberry (*Fragaria* sp.) leaf. Many people mistake guttation for early morning dew. *(Ed Reschke/Peter Arnold, Inc.)*

Some plants exude liquid water

Many leaves have special structures through which liquid water is literally forced out. This loss of liquid water, known as **guttation,** occurs when transpiration is negligible and available soil moisture is high. Guttation typically occurs at night because the stomata are closed, but water continues to move into the roots by osmosis. People sometimes think erroneously that the early morning water droplets on leaf margins are dew rather than guttation (Fig. 32–13). Unlike dew, which condenses from cool night air, guttation droplets come from within the plant. (The mechanism for guttation is discussed in Chapter 33.)

LEAF ABSCISSION ALLOWS PLANTS TO SURVIVE AN UNFAVORABLE SEASON

All trees shed leaves; many conifers, for example, shed their needles in small numbers year-round. The leaves of deciduous plants, however, turn color and **abscise,** or fall off, once a year—as winter approaches in temperate climates or at the beginning of the dry period in tropical climates with pronounced wet and dry seasons. In temperate forests, most woody plants with broad leaves shed their leaves to survive the low temperatures of winter. During winter, the plant's metabolism, including its photosynthetic machinery, slows down or halts temporarily.

Another reason for abscission is related to a plant's water requirements, which become critical during the physiological drought of winter. As mentioned previously, as the ground chills, absorption of water by the roots is inhibited. When the ground freezes, *no* absorption occurs. If the broad leaves were to stay on the plant during the winter, the plant would continue to lose water by transpiration but would be unable to replace it with water absorbed from the soil.

Leaf abscission is a complex process that involves many physiological changes, all of which are initiated and orchestrated by changing levels of plant hormones, particularly *ethylene* (see Chapter 36). Briefly, the process is this: As autumn approaches, sugars, amino acids, and many essential minerals (such as nitrogen, phosphorus, and possibly potassium) are mobilized and transported from the leaves other plant parts. Chlorophyll breaks down, allowing the orange carotenes and yellow xanthophylls, some of the accessory pigments in the chloroplasts of leaf cells, to become evident. Recall from Chapter 8 that accessory pigments are always present in the leaf but are masked by the green of the chlorophyll. In addition, red water-soluble pigments called

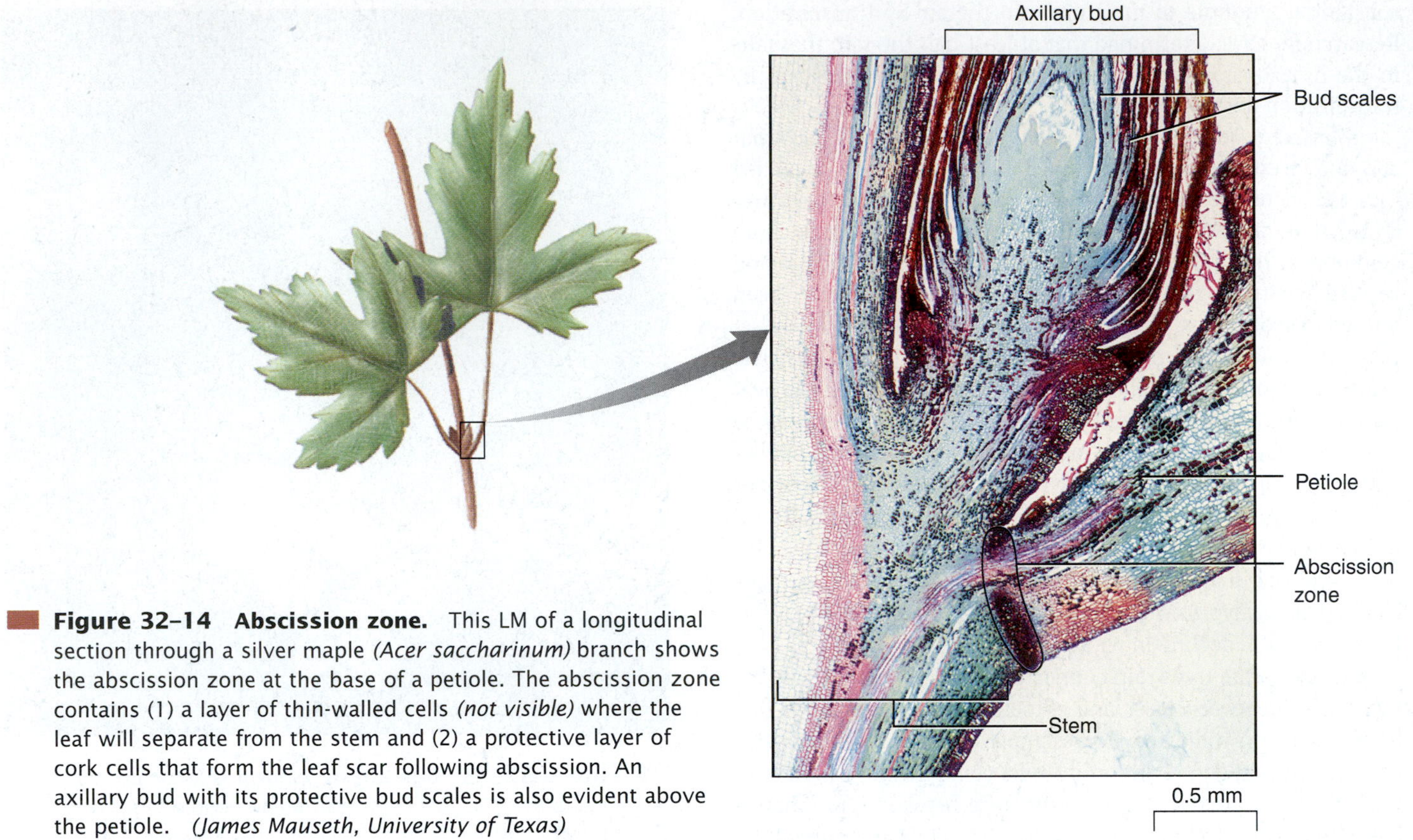

Figure 32–14 Abscission zone. This LM of a longitudinal section through a silver maple *(Acer saccharinum)* branch shows the abscission zone at the base of a petiole. The abscission zone contains (1) a layer of thin-walled cells *(not visible)* where the leaf will separate from the stem and (2) a protective layer of cork cells that form the leaf scar following abscission. An axillary bud with its protective bud scales is also evident above the petiole. *(James Mauseth, University of Texas)*

anthocyanins may be synthesized and stored in the vacuoles of leaf cells in some species; their function is unknown. The various combinations of these pigments are responsible for the brilliant colors found in autumn landscapes in temperate climates.

In many leaves, abscission occurs at an abscission zone near the base of the petiole

The area where a petiole detaches from the stem is structurally different from surrounding tissues. This structurally distinct area, called the **abscission zone,** is composed primarily of thin-walled parenchyma cells and, because it contains few fibers, is anatomically weak (Fig. 32–14). A protective layer of cork cells develops on the stem side of the abscission zone. These cells have a waxy, waterproof material impregnated in their walls. Enzymes then dissolve the *middle lamella* (the "cement" that holds the primary cell walls of adjacent cells together) in the abscission zone. Once this process is completed, nothing holds the leaf to the stem but a few xylem cells. A sudden breeze is enough to make the final break, and the leaf detaches. The protective layer of cork remains, sealing off the area and forming a leaf scar.

Process of Science Although the structure of abscission zones and the physiological changes associated with abscission are well known, the genes responsible for the development of an abscission zone have been investigated only recently. In 2000, researchers from Clemson University and the University of Iowa reported in the journal *Nature* that they had isolated a gene involved in the development of an abscission zone on tomato flowers and fruits. (Like leaves, flowers and fruits have abscission zones and are also shed.) Tomato plants with a mutated version of the gene did not develop normal abscission zones on their flower stalks.

MODIFIED LEAVES MAY HAVE VARIED FUNCTIONS

Although photosynthesis is the main function of leaves, certain leaves possess special modifications for other functions. Some plants have leaves specialized for deterring herbivores. **Spines,** modified leaves that are hard and pointed, may be found on many desert plants such as cacti (Fig. 32–15*a*). In the cactus, the main organ of photosynthesis is the stem rather than the leaf. Spines discourage animals from eating the succulent stem tissue.

Vines are climbing plants whose stems cannot support their own weight, and so they often possess **tendrils** that help keep the vine attached to the structure on which it is growing. The tendrils of many vines, such as peas, cucumbers, and squash, are specialized leaves (Fig. 32–15*b*). However, some tendrils, such as those of ivy, Virginia creeper, and grape, are specialized stems.

The winter buds of a dormant woody plant are covered by **bud scales,** modified leaves that protect the delicate meristematic tissue of the bud from being injured, frozen, or drying out (Fig. 32–15*c*).

Figure 32–15 Leaf modifications. (a) The leaves of cacti such as this pincushion cactus (*Mammilaria* sp.) are modified as spines for deterring herbivores. Photographed in Arizona. (b) Tendrils, which wind around objects and aid vines in climbing, may be modified leaves or stems. Shown are tendrils of bur cucumber (*Echinocystis lobata*), which are modified leaves. (c) Overlapping bud scales are leaves modified to protect buds. Shown here are a terminal bud and two axillary buds of a maple (*Acer* sp.) twig. (d) The leaves of bulbs such as the onion (*Allium cepa*) are fleshy for storage of food materials and water. (e) Some plants have thick, succulent leaves modified for water storage as well as photosynthesis. The succulent leaves of the string-of-beads plant (*Senecio rowleyanus*) are spherical to minimize surface area, thereby conserving water. *(a, Barbara Gerlach/Dembinsky Photo Associates; b, Stephen P. Parker/Photo Researchers, Inc.; c, d, Dennis Drenner;* e, *James Mauseth, University of Texas)*

Leaves may also be modified for storage of water or food. For example, a **bulb** is a short underground stem to which large, fleshy leaves are attached (Fig. 32–15*d*). Onions and tulips form bulbs. Many plants adapted to arid conditions, such as jade plant (*Crassula arborescens*), medicinal aloe (*Aloe barbadensis*), and string-of-beads (*Senecio rowleyanus*), have succulent leaves for water storage (Fig. 32–15*e*). These leaves are usually green and function in photosynthesis as well.

Figure 32–16 A common pitcher plant *(Sarracenia purpurea).* This species, whose pitchers grow to 30.5 cm (12 in), is widely distributed in acidic bogs and marshes in eastern North America. Young pitchers are green but turn red as they age. The insectivorous pitcher plant has leaves modified to form a water-collecting pitcher that drowns its prey. Note the dead beetle in the "pitcher." *(Bill Lea/Dembinsky Photo Associates)*

Some leaves are modified for sexual reproduction (see discussion of flower evolution in Chapter 27, including Fig. 27–13) or asexual reproduction (see Fig. 35–19).

Modified leaves of insectivorous plants capture insects

Insectivorous plants are plants that capture insects. Most insectivorous plants grow in poor soil that is deficient in certain essential minerals, particularly nitrogen. These plants meet some of their mineral requirements by digesting insects and other small animals. The leaves of insectivorous plants are adapted to attract, capture, and digest their animal prey.

Some insectivorous plants have passive traps. The leaves of a pitcher plant, for example, are shaped so that rainwater collects and forms a reservoir that also contains acid secreted by the plant (Fig. 32–16). Some pitchers are quite large; in the tropics, for example, pitcher plants may be large enough to hold 1 L (approximately 1 qt) or more of liquid. An insect attracted by the odor or nectar of the pitcher may lean over the edge and fall in. Although it may make repeated attempts to escape, the insect is prevented from crawling out by the slippery sides and the rows of stiff hairs that point downward around the lip of the pitcher. The insect eventually drowns, and part of its body eventually disintegrates and is absorbed. Although most insects are killed in pitcher plants, the larvae of several insects (certain fly, midge, and mosquito species), as well as a large community of microorganisms, actually live inside the pitchers. These insect species obtain their food from the insect carcasses, and the pitcher plant digests what remains. It is not known how these insects survive the acidic environment inside the pitcher.

The Venus flytrap is an insectivorous plant with active traps. Its leaf blades resemble tiny bear traps (see Fig. 1–4). Each side of the leaf blade contains three small, stiff hairs. If an insect alights and brushes against two of the hairs, or against the same hair twice in quick succession, the trap springs shut with amazing rapidity. The spines along the margins of the blades fit closely together to prevent the insect from escaping. Digestive glands on the surface of the trap secrete enzymes in response to the insect pressing against them. After the insect has died and been digested, the trap reopens and the indigestible remains fall out.

SUMMARY WITH KEY TERMS

I. Leaves exhibit variation in shape and form.
   A. Leaves typically consist of a broad, flat **blade** and a stalklike **petiole.** Some leaves also possess small, leaflike outgrowths from the base called **stipules.**
   B. Leaves may be simple or compound.
   C. Leaf arrangement on a stem may be alternate, opposite, or whorled.
   D. Leaves may have parallel or netted (either pinnately netted or palmately netted) venation.

II. Leaf structure is adapted for its primary function of **photosynthesis.**
   A. Most leaves have a broad, flattened blade that is quite efficient in collecting the sun's radiant energy.
   B. The transparent **epidermis** allows light to penetrate into the mesophyll, where photosynthesis occurs.
      1. **Stomata** are small pores in the epidermis that permit gas exchange needed for photosynthesis.
         a. Each pore is surrounded by two **guard cells** that are often associated with special epidermal cells called **subsidiary cells.**
         b. Subsidiary cells provide a reservoir of water and ions that move into and out of the guard cells as they change shape during stomatal opening and closing.
      2. The waxy **cuticle** coats the epidermis, enabling the plant to survive the dry conditions of a terrestrial existence.
   C. **Mesophyll** consists of photosynthetic parenchyma cells.
      1. Mesophyll contains air spaces that permit the rapid diffusion of carbon dioxide and water into, and oxygen out of, mesophyll cells.
      2. Mesophyll is divided into **palisade mesophyll,** which functions primarily for photosynthesis, and **spongy mesophyll,** which functions primarily for gas exchange.

D. Leaf **veins** have **xylem** to conduct water and essential minerals to the leaf and **phloem** to conduct sugar produced by photosynthesis to the rest of the plant.

III. Monocot and dicot leaves can be distinguished based on their external and internal structures.
  A. Monocot leaves have parallel venation, whereas dicot leaves have netted venation.
  B. Some monocots (corn and other grasses) do not have mesophyll differentiated into distinct palisade and spongy layers.
  C. Some monocots (grasses, reeds, and sedges) have guard cells shaped like dumbbells.

IV. Stomata generally open during the day and close at night.
  A. The movement of ions across the plasma membranes of guard cells explains the opening and closing process.
    1. Light, particularly blue light, initiates a proton pump in the guard cells' plasma membranes that pumps H^+ out of the guard cells. The H^+ is produced by the ionization of malic acid, which is produced in the chloroplasts of guard cells.
    2. The electrochemical gradient that was established by the proton pump drives the active transport of potassium ions into the guard cells.
    3. The resulting osmotic movement of water into the guard cells causes them to become turgid, forming a pore.
    4. Stomata close when potassium ions leave the guard cells, causing water to flow out by osmosis. As the guard cells become less turgid and collapse, the pore closes.
    5. Stomatal closure may also involve oscillating Ca^{2+} levels in guard cells. Oscillating Ca^{2+} levels are known to be an important part of **cell signaling** in animal cells.
  B. A number of factors affect stomatal opening and closing, including light or darkness, CO_2 concentration, water stress, and the plant's circadian rhythm.

V. **Transpiration** is the loss of water vapor from aerial parts of plants.
  A. Transpiration occurs primarily through the stomata.
  B. The rate of transpiration is affected by environmental factors such as temperature, wind, and relative humidity.
  C. Transpiration appears to be both beneficial and harmful to the plant (the trade-off between the CO_2 requirement for photosynthesis and the need for water conservation).

VI. **Guttation,** the release of liquid water from leaves of some plants, occurs through special structures when transpiration is negligible and available soil moisture is high.

VII. Leaf **abscission** is a complex process involving physiological and anatomical changes prior to leaf fall.
  A. Sugars, amino acids, and many essential minerals are transported from the leaves to other plant parts.
  B. Chlorophyll breaks down, and accessory pigments become evident.
  C. An **abscission zone** develops where the petiole detaches from the stem.

VIII. Leaves may be modified for functions other than photosynthesis.
  A. **Spines** are leaves adapted to deter herbivores.
  B. Some **tendrils** are leaves modified for grasping and holding onto other structures (to support weak stems).
  C. **Bud scales** are leaves modified to protect delicate meristematic tissue or dormant buds.
  D. **Bulbs** are short underground stems with fleshy leaves specialized for storage.
  E. Insectivorous plants have leaves modified to trap insects.

POST-TEST

1. Plants with an alternate leaf arrangement have (a) blades divided into two or more leaflets (b) major veins that radiate out from one point (c) one leaf at each node (d) major veins branching off along the entire length of the midvein (e) two leaves at each node
2. The photosynthetic ground tissue in the middle of the leaf is called (a) cutin (b) mesophyll (c) the abscission zone (d) subsidiary cells (e) palisade and spongy stomata
3. The *primary* function of the spongy mesophyll is (a) reduction of water loss from the leaf surface (b) changing the shape of the guard cells (c) support to prevent the leaf from collapsing under its own weight (d) diffusion of gases within the leaf (e) deterrence of herbivores
4. Gas exchange occurs through microscopic pores formed by two (a) subsidiary cells (b) abscission cells (c) mesophyll cells (d) guard cells (e) stipules
5. Most stomata are usually located in the ____________ of the leaf. (a) upper epidermis (b) lower epidermis (c) cuticle (d) spongy mesophyll (e) palisade mesophyll
6. The thin, noncellular layer of wax secreted by the epidermis of leaves is the (a) stoma (b) subsidiary cell (c) trichome (d) bundle sheath (e) cuticle
7. The ____________ encircles a vein. (a) palisade mesophyll (b) guard cell (c) bundle sheath (d) blade (e) cuticle
8. The ____________ of a leaf vein transports water and dissolved nutrient minerals, whereas the ____________ transports sugars produced by the leaf during photosynthesis. (a) xylem; phloem (b) xylem; bundle sheath (c) phloem; xylem (d) phloem; vein (e) vascular bundle; bundle sheath
9. Which of the following is *not* an adaptation of pine needles to conserve water? (a) less surface area exposed to the air than thin-bladed leaves (b) a relatively thick cuticle (c) sunken stomata (d) netted veins instead of parallel veins (e) both c and d are not adaptations of pine needles
10. Most of the water that a plant absorbs from the soil is lost by the process of (a) guttation (b) circadian rhythm (c) abscission (d) transpiration (e) photosynthesis
11. When transpiration is negligible, plants such as grasses exude excess water by (a) guttation (b) circadian rhythm (c) abscission (d) pumping H^+ out of and K^+ into guard cells (e) photosynthesis
12. Place the following events of stomatal opening in correct order.
    (1) proton pump moves H^+ out of guard cells
    (2) guard cells change shape and pore appears
    (3) leaf is exposed to light
    (4) water diffuses into guard cells
    (5) K^+ actively transported into guard cells
    (a) 1-5-4-2-3 (b) 5-3-4-2-1 (c) 1-3-4-2-5 (d) 3-5-1-2-4 (e) 3-1-5-4-2
13. The seasonal detachment of leaves is known as (a) forest decline (b) transpiration (c) abscission (d) guttation (e) dormancy
14. Modified leaves that enable a stem to climb are called ____________ whereas modified leaves that cover the winter buds of a dormant woody plant are called ____________ (a) spines; bud scales (b) bud scales; tendrils (c) tendrils; bud scales (d) tendrils; spines (e) insectivorous leaves; spines
15. Leaves have a trade-off, or compromise, between photosynthesis and transpiration, which results from (a) the numerous stomatal pores that provide both gas exchange for photosynthesis and openings through which water vapor escapes (b) the secretion of a waxy layer, the cuticle, that reduces water loss (c) blue light triggering an influx of potassium ions (K^+) into the guard cells (d) the abscission of leaves of deciduous plants as winter approaches in temperate climates (e) the stomata being closed at night, although water continues to move into the roots by osmosis

REVIEW QUESTIONS

1. Give the general equation for photosynthesis (see Chapter 8), and discuss how the leaf is organized to deliver the raw materials and remove the products of photosynthesis.
2. How is leaf structure related to photosynthesis and transpiration?
3. Relate the series of physiological changes that occur in guard cells during stomatal opening and closing.
4. Discuss at least two ways that leaves are adapted to conserve water.
5. How do environmental factors (sunlight, temperature, humidity, and wind) influence the rate of transpiration? How does the environment influence stomatal opening and closing?
6. Why do many temperate woody plants lose their leaves in autumn?
7. Discuss the specialized features of the leaves of insectivorous plants.
8. Label the diagram. Use Figure 32–3 to check your answers.

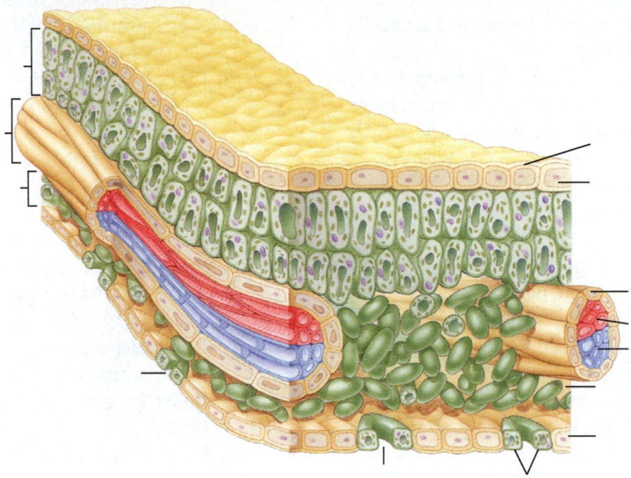

9. For each of the following plants specify if it has simple or compound leaves; alternate, opposite, or whorled leaf arrangement; and parallel, pinnately netted, or palmately netted venation. Use Figure 32–2 to check your answers.

(a)

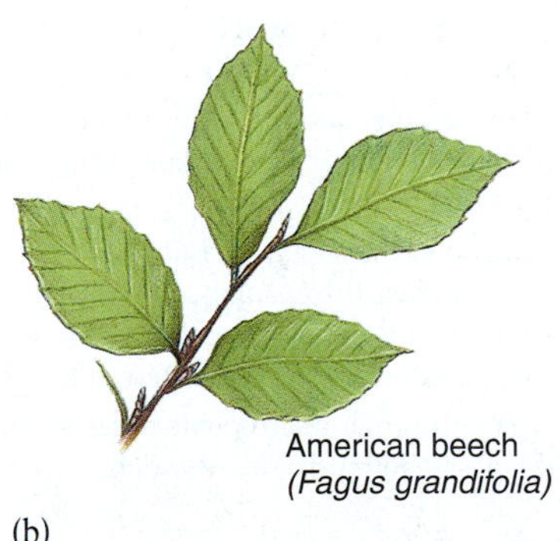

(b)

YOU MAKE THE CONNECTION

1. Suppose you are asked to observe a micrograph of a leaf cross section and distinguish between the upper and lower epidermis. How would you make this decision?
2. Given that (1) xylem is located toward the upper epidermis in leaf veins while phloem is toward the lower epidermis, and (2) the vascular tissue of a leaf is continuous with that of the stem, suggest one possible arrangement of vascular tissues in the stem that might account for the arrangement of vascular tissue in the leaf.
3. What might be some of the advantages of a plant having a few very large leaves? Disadvantages? What might be some advantages of having many very small leaves? Disadvantages? How would your answer differ along a moisture gradient, from a humid environment to a desert?
4. Briefly explain how additional research on the molecular mechanism of stomatal closure might be of future use in agriculture.

RECOMMENDED READINGS

Baskin, Y. *The Work of Nature: How the Diversity of Life Sustains Us.* Island Press, Washington, D.C., 1997. Chapter 8, "Climate and Atmosphere," examines some of the connections between regional climate and trees.

Beerling, D.J., and C.K. Kelly. "Stomatal Density Responses of Temperate Woodland Plants over the Past Seven Decades of CO_2 Increase: A Comparison of Salisbury (1927) with Contemporary Data." *American Journal of Botany,* Vol. 84, No. 11, 1997. Some plants have responded to human-induced changes in the concentration of atmospheric CO_2 by adjusting their stomatal density.

Berg, L.R. *Introductory Botany: Plants, People, and the Environment.* Saunders College Publishing, Philadelphia, 1997. A general botany text with an environmental emphasis.

Raven, P.H., R.F. Evert, and S.E. Eichhorn. *Biology of Plants,* 6th ed. W.H. Freeman & Company, New York, 1999. This is an excellent general botany textbook, with detailed descriptions of leaf structure and adaptations.

Smith, W.K., T.C. Vogelmann, E.H. DeLucia, D.T. Bell, and K.A. Shepherd. "Leaf Form and Photosynthesis." *BioScience,* Vol. 47, No. 11, Dec. 1997. Leaf structure is strongly linked to internal light and CO_2 levels.

Weiss, P. "The Puzzle of Flutter and Tumble." *Science News,* Vol. 154, 31 Oct. 1998. Some scientists are studying the physics behind leaf movements when leaves fall to the ground.

Zimmer, C. "The Processing Plant." *Discover,* Sep. 1995. Discusses the unique ecosystem and food chain inside a pitcher plant, which involves three insect species as well as the plant.

- Visit our Web site at **http://www.info.brookscole.com/solomonbergmartin** for links to chapter-related resources on the World Wide Web. Additional on-line materials relating to this chapter can also be found on our Web site.

See chapter activity on BioActive Learner CD for additional help in mastering the chapter's material. Icon location in the chapter's margins shows which topics have tutorials or simulations in the CD.

33

Stems and Plant Transport

Baobab tree. Baobabs *(Adansonia digitata)*, which are native to Africa, Madagascar, and Australia, store large volumes of water and starch in their massive trunks. Baobab trees are relatively short (growing to 18 m, or 60 ft), but their trunks can be as much as 9 m, or 30 ft, in diameter. Photographed in Namibia, with children around the tree as a size reference. *(Michael Fairchild/Peter Arnold, Inc.)*

LEARNING OBJECTIVES

After you have studied this chapter you should be able to

1. List three functions of stems.
2. Describe the external features of a woody twig.
3. Label cross sections of herbaceous dicot and monocot stems and describe the functions of each tissue.
4. Outline the transition from primary growth to secondary growth in a woody stem, name the two lateral meristems, and describe the tissues that arise from each.
5. Describe the pathway of water movement in plants.
6. Explain tension-cohesion and root pressure as mechanisms to explain the rise of water and dissolved nutrient minerals in xylem.
7. Describe the pathway of sugar translocation in plants.
8. Discuss the pressure-flow hypothesis of sugar translocation in phloem.

A vegetative (not sexually reproductive) vascular plant has three parts: roots, leaves, and stems. As discussed in Chapter 31, roots serve to anchor the plant and absorb materials from the soil, whereas leaves are primarily for photosynthesis, converting radiant energy into the chemical energy of carbohydrate molecules. Stems, the focus of this chapter, link a plant's roots to its leaves and are usually located above ground, although many plants have underground stems. Stems exhibit varied forms, ranging from ropelike vines to massive tree trunks. They can be either herbaceous (consisting of soft, nonwoody tissues) or woody (with extensive hard tissues of wood and bark).

Stems perform three main functions in plants. First, stems of most species support leaves and reproductive structures. The upright position of most stems and the arrangement of the leaves on them allow each leaf to absorb light for use in photosynthesis. Reproductive structures (flowers and fruits) are located on stems in areas accessible to insects, birds, and air currents, which transfer pollen from flower to flower and help disperse seeds and fruits.

Second, stems provide internal transport. They conduct water and dissolved nutrient minerals from the roots, where these materials are absorbed from the soil, to leaves and other plant parts. Stems also conduct the sugar produced in leaves by photosynthesis to roots and other parts of the plant. Remember, however, that stems are not the only plant organs that conduct materials. The vascular system is continuous throughout all parts of a plant, and conduction occurs in roots, stems, leaves, and reproductive structures.

Third, stems produce new living tissue. They continue to grow throughout a plant's life, producing buds that develop into stems with new leaves and/or reproductive structures. In addition to the main functions of support, conduction, and production of new stem tissues, stems of some species are modified for asexual reproduction (see Chapter 35) or, if green, to manufacture sugar by photosynthesis. Also, some stems are specialized to store starch *(see photograph)*.

A WOODY TWIG DEMONSTRATES EXTERNAL STEM STRUCTURE

Although stems exhibit great variation in structure and growth, they all have **buds,** which are embryonic shoots. A **terminal bud** is the embryonic shoot located at the tip of a stem. The dormant (i.e., not actively growing) apical meristem of a terminal bud is covered and protected by an outer layer of **bud scales,** which are modified leaves (see Fig. 32–15*c*). **Axillary buds,** also called **lateral buds,** are located in the axils of a plant's leaves (see Fig. 32–1). An axil is the upper angle between a leaf and the stem to which it is attached. When terminal and axillary buds grow, they form branches that bear leaves and/or flowers. The area on a stem where each leaf is attached is called a **node,** and the region between two successive nodes is an **internode.**

A woody twig of a deciduous tree that has shed its leaves can be used to demonstrate certain structural features of the stem (Fig. 33–1). Bud scales cover the terminal bud and protect its delicate apical meristem during dormancy. When the bud resumes growth, the bud scales covering the terminal bud fall off, leaving **bud scale scars** on the stem where the bud scales were attached. Because temperate-zone woody plants form terminal buds at the end of each year's growing season, the number of sets of bud scale scars on a twig indicates its age. A **leaf scar** shows where each leaf was attached on the stem; the pattern of leaf scars can be used to determine leaf arrangement on a stem—alternate, opposite, or whorled (see Fig. 32–2*b*). The vascular (conducting) tissue that extends from the stem out into the leaf forms **bundle scars** within a leaf scar. Axillary buds may be found above the leaf scars. Also, the bark of a woody twig has **lenticels,** sites of loosely arranged cells that allow oxygen to diffuse into the interior of the woody stem. Lenticels look like tiny specks on the bark of a twig.

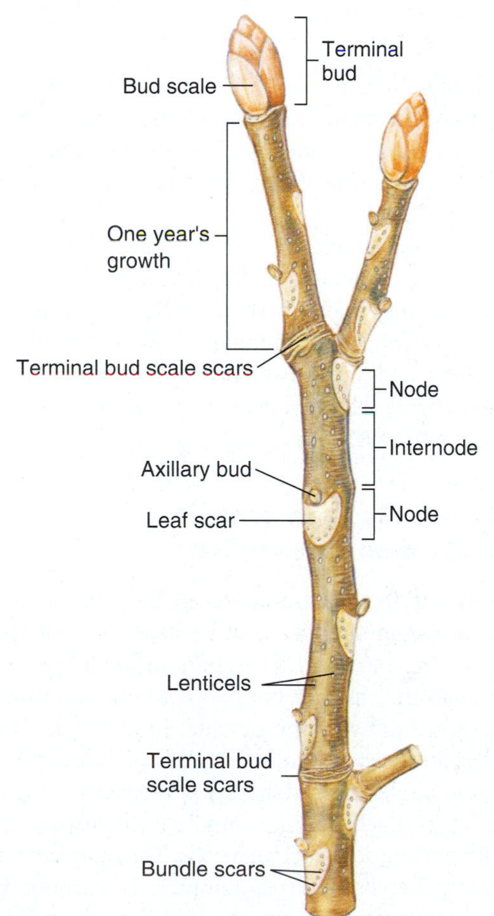

Figure 33–1 External structure of a woody twig in its winter condition. The age of a woody twig can be determined by the number of sets of bud scale scars (do not count side branches). How old is this twig?

STEMS ORIGINATE AND DEVELOP AT MERISTEMS

Recall from Chapter 31 that plants have two different types of growth. Primary growth is an increase in the length of a plant and occurs at **apical meristems** located at the tips of roots and shoots and also within the buds of stems. Secondary growth is an increase in the girth (thickness) of a plant owing to the activity of **lateral meristems** located within stems and roots. The new tissues formed by the lateral meristems are called *secondary tissues* to distinguish them from *primary tissues* produced by apical meristems.

All plants have primary growth; some plants have both primary and secondary growth. Recall that stems with only primary growth are herbaceous, whereas those with both primary and secondary growth are woody.[1] A woody plant increases in length by primary growth at the tips of its stems and roots, while its older stems and roots farther back from the tips increase in girth by secondary growth. In other words, at the same time that secondary growth is adding wood and bark (thereby causing the stem to thicken), primary growth (which increases the length of the stem) continues.

HERBACEOUS DICOT AND MONOCOT STEMS DIFFER IN INTERNAL STRUCTURE

Although considerable structural variation exists in stems, they all possess an outer protective covering (epidermis or periderm), one or more types of ground tissue, and vascular tissues (xylem and phloem). Let us first consider the structure of herbaceous dicot stems and then of monocot stems.

Vascular bundles of herbaceous dicot stems are arranged in a circle in cross section

A young sunflower stem is a representative herbaceous dicot stem that exhibits primary growth (Fig. 33–2). Its outer covering, the **epidermis,** provides protection in herbaceous stems, as it does in leaves and herbaceous roots (see Table 31–4 and Fig. 31–6). The cuticle, a waxy layer of *cutin,* covers the stem epidermis

[1] Many herbaceous stems (e.g., those of geranium and sunflower) also have a limited amount of secondary growth.

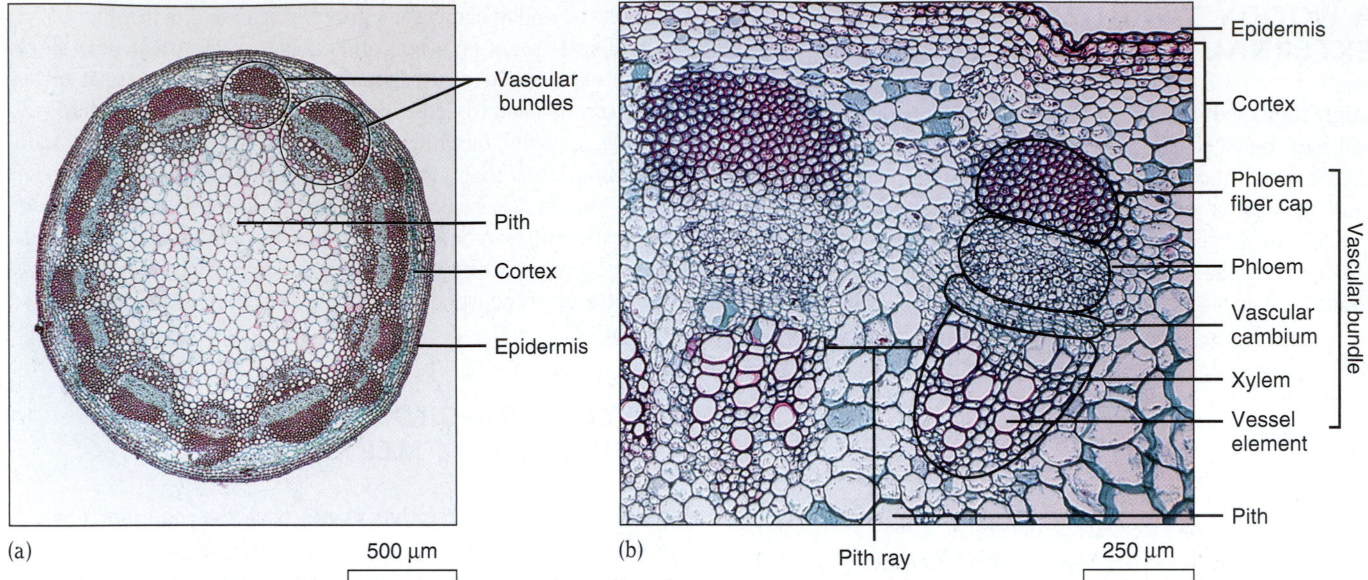

Figure 33–2 LMs of a herbaceous dicot stem. **(a)** Cross section of a sunflower *(Helianthus annuus)* stem. Note the vascular bundles arranged in a circle around a central core of pith. **(b)** Close-up of two vascular bundles. Xylem is located toward a stem's interior, and phloem is toward the exterior. Each vascular bundle is "capped" by a batch of fibers for additional support. *(a, b, Ed Reschke)*

and reduces water loss from the stem surface. (Recall from Chapter 32 that a cuticle also covers the leaf epidermis.)

Inside the epidermis is the **cortex,** a cylinder of ground tissue that may contain parenchyma, collenchyma, and sclerenchyma cells (see Table 31–2 and Fig. 31–4). As might be expected from the various types of cells that it contains, the cortex in herbaceous dicot stems can have several functions, such as photosynthesis, storage, and support. If a stem is green, photosynthesis occurs in chloroplasts of cortical parenchyma cells. Parenchyma in the cortex also stores starch (in amyloplasts) and crystals (in vacuoles). Collenchyma and sclerenchyma in the cortex confer strength and structural support for the stem.

The vascular tissues provide conduction and support. In herbaceous dicot stems, the vascular tissues are located in bundles that, when viewed in cross section, are arranged in a circle. However, viewed lengthwise, these bundles extend as long strands throughout the length of a stem and are continuous with vascular tissues of both roots and leaves.

Each vascular bundle contains both **xylem,** which transports water and dissolved nutrient minerals from roots to leaves, and **phloem,** which transports dissolved sugar (see Table 31–3 and Fig. 31–5). Xylem is located on the inner side of the vascular bundle, and phloem is found toward the outside. Sandwiched between xylem and phloem in some herbaceous stems is a single layer of cells called the *vascular cambium,* a lateral meristem responsible for secondary growth (discussed later).

Because most stems support the aerial plant body, they are much stronger than roots. The thick walls of tracheids and vessel elements in xylem help support the plant. Fibers also occur in both xylem and phloem, although they are usually more extensive in phloem. These fibers add considerable strength to the herbaceous stem. In sunflowers and certain other herbaceous dicot stems, phloem contains a cluster of fibers toward the outside of the vascular bundle, called a **phloem fiber cap,** that helps strengthen the stem. The phloem fiber cap is not present in all herbaceous dicot stems.

The **pith** is a ground tissue at the center of the herbaceous dicot stem that is composed of large, thin-walled parenchyma cells that function primarily in storage. Because of the arrangement of the vascular tissues in bundles, there is no distinct separation of cortex and pith between the vascular bundles. The areas of parenchyma between the vascular bundles are often referred to as **pith rays.**

Vascular bundles are scattered throughout monocot stems

An epidermis with its waxy cuticle covers monocot stems such as the herbaceous stem of corn. As in herbaceous dicot stems, the vascular tissues run in strands throughout the length of a stem. In cross section the vascular bundles contain xylem toward the inside and phloem toward the outside. In contrast with herbaceous dicots, however, vascular bundles of monocots are not arranged in a circle but instead are scattered throughout the stem (Fig. 33–3). Each vascular bundle is enclosed in a bundle sheath of supporting sclerenchyma cells. The monocot stem does not have distinct areas of cortex and pith. The ground tissue in which the vascular tissues are embedded performs the same functions as cortex and pith in herbaceous dicot stems.

Monocot stems do not possess the lateral meristems (vascular cambium and cork cambium), that give rise to secondary growth. Monocots have primary growth only and do not produce

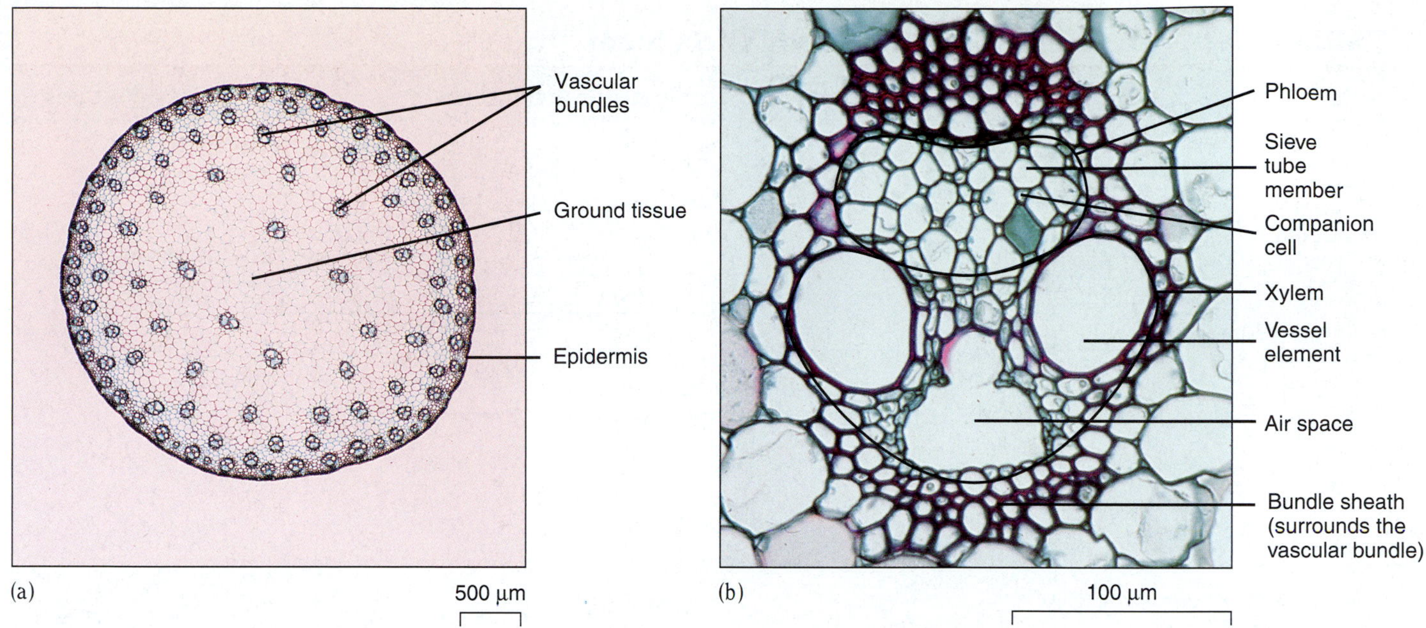

Figure 33–3 LMs of a monocot stem. **(a)** Cross section of a corn *(Zea mays)* stem shows vascular bundles scattered throughout ground tissue. **(b)** Close-up of a vascular bundle. The air space is the site where the first xylem elements were formed and later disintegrated. The entire bundle is enclosed in a bundle sheath of sclerenchyma for additional support. *(a, b, Ed Reschke)*

wood and bark. Although some treelike monocots such as palms attain considerable size, they do so by a modified form of primary growth in which parenchyma cells divide and enlarge. Stems of some monocots such as bamboo and palm contain a great deal of sclerenchyma tissue, which makes them hard and woodlike in appearance.

WOODY PLANTS HAVE STEMS WITH SECONDARY GROWTH

Woody plants undergo secondary growth, an increase in the girth of stems and roots. Secondary growth occurs as a result of the activity of two lateral meristems: vascular cambium and cork cambium. Among flowering plants, only woody dicots (such as apple, hickory, and maple) have secondary growth. Cone-bearing gymnosperms (such as pine, juniper, and spruce) also have secondary growth. Table 33–1 summarizes the relationships of meristems and tissues in a woody dicot stem.

Cells in the **vascular cambium** divide and produce two conducting and supporting tissues: secondary xylem (wood) to replace primary xylem, and secondary phloem (inner bark) to replace primary phloem. Primary xylem and primary phloem are not able to transport materials indefinitely and so are replaced in plants that have extended lifespans.

Cells of the outer lateral meristem, called **cork cambium,** divide and produce cork cells and cork parenchyma. Cork cambium and the tissues it produces are collectively referred to as **periderm** (outer bark), which functions as a replacement for the epidermis (see the last LM in Table 31–4).

Vascular cambium gives rise to secondary xylem and secondary phloem

Primary tissues in woody dicot stems are organized similarly to those in herbaceous dicot stems, with the vascular cambium a thin layer of cells sandwiched between xylem and phloem in the vascular bundles. Once secondary growth begins, the internal structure of a stem changes considerably (Fig. 33–4). Although vascular cambium is not initially a continuous cylinder of cells (because the vascular bundles are separated by pith rays), it becomes continuous when production of secondary tissues begins. This continuity develops because certain parenchyma cells in each pith ray retain the ability to divide. These cells connect to vascular cambium cells in each vascular bundle, thus forming a complete ring of vascular cambium.

Cells in the vascular cambium divide and produce daughter cells in two directions. The cells formed from the dividing vascular cambium are located either *inside* the ring of vascular cambium (to become secondary xylem, or wood), or *outside* it (to become secondary phloem, or inner bark) (Fig. 33–5). When a cell in the vascular cambium divides tangentially (inward or outward), one of the daughter cells remains meristematic; that is, it remains as a part of the vascular cambium. The other cell may divide again several times, but eventually it stops dividing and develops into mature secondary tissue. Thus, vascular cambium is a thin layer of cells sandwiched between the wood and inner bark, the two tissues that it produces (Fig. 33–6 on page 716).

As the stem increases in circumference, the number of cells in the vascular cambium also increases. This occurs by an occasional radial division of a vascular cambium cell, at right angles

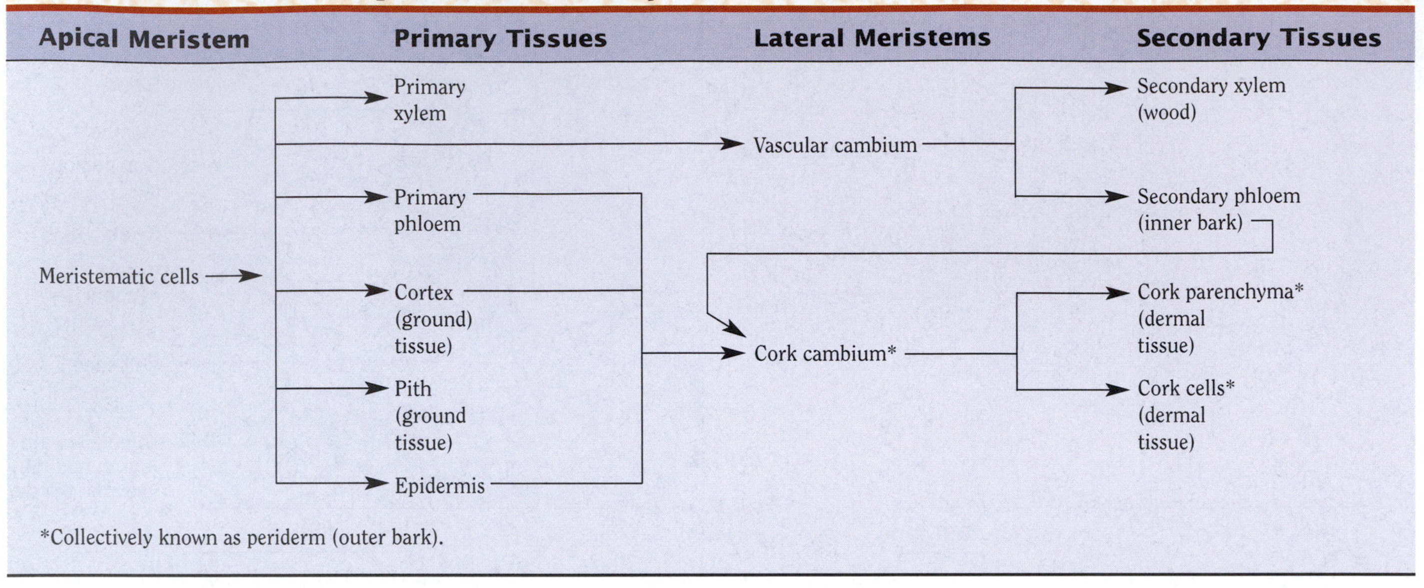

TABLE 33–1 Development in a Woody Dicot Stem

Apical Meristem	Primary Tissues	Lateral Meristems	Secondary Tissues
Meristematic cells	Primary xylem		
		Vascular cambium	Secondary xylem (wood)
			Secondary phloem (inner bark)
	Primary phloem		
	Cortex (ground) tissue	Cork cambium*	Cork parenchyma* (dermal tissue)
	Pith (ground tissue)		Cork cells* (dermal tissue)
	Epidermis		

*Collectively known as periderm (outer bark).

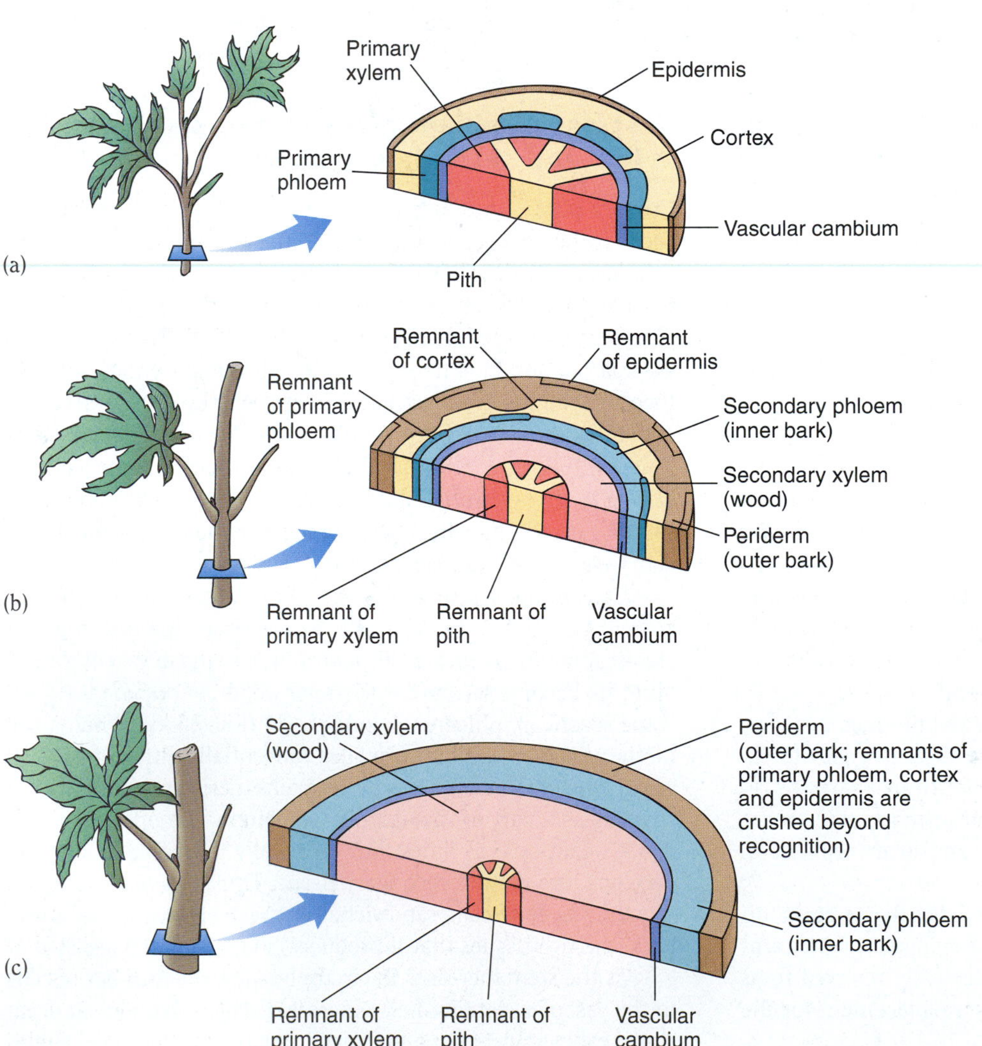

Figure 33–4 Development of secondary growth in a dicot stem. Vascular cambium and the tissues it produces are shown in cross section; cork cambium is not depicted. **(a)** A herbaceous dicot stem. Vascular cambium is sandwiched between primary xylem and primary phloem in each vascular bundle. At the onset of secondary growth, vascular cambium arises in the parenchyma between the vascular bundles (i.e., in the pith rays), forming a cylinder of meristematic tissue that appears as a blue circle in cross section. **(b)** Vascular cambium begins to divide, forming secondary xylem on the inside and secondary phloem on the outside. Note that the primary xylem and primary phloem in the original vascular bundles become separated during secondary growth. **(c)** A young woody stem. Vascular cambium produces significantly more secondary xylem than secondary phloem. (The figures change in scale owing to space limitations; the pith and primary xylem are actually the same size in all three diagrams, but the change in scale makes those tissues appear to shrink from part *a* to part *b* to part *c*.)

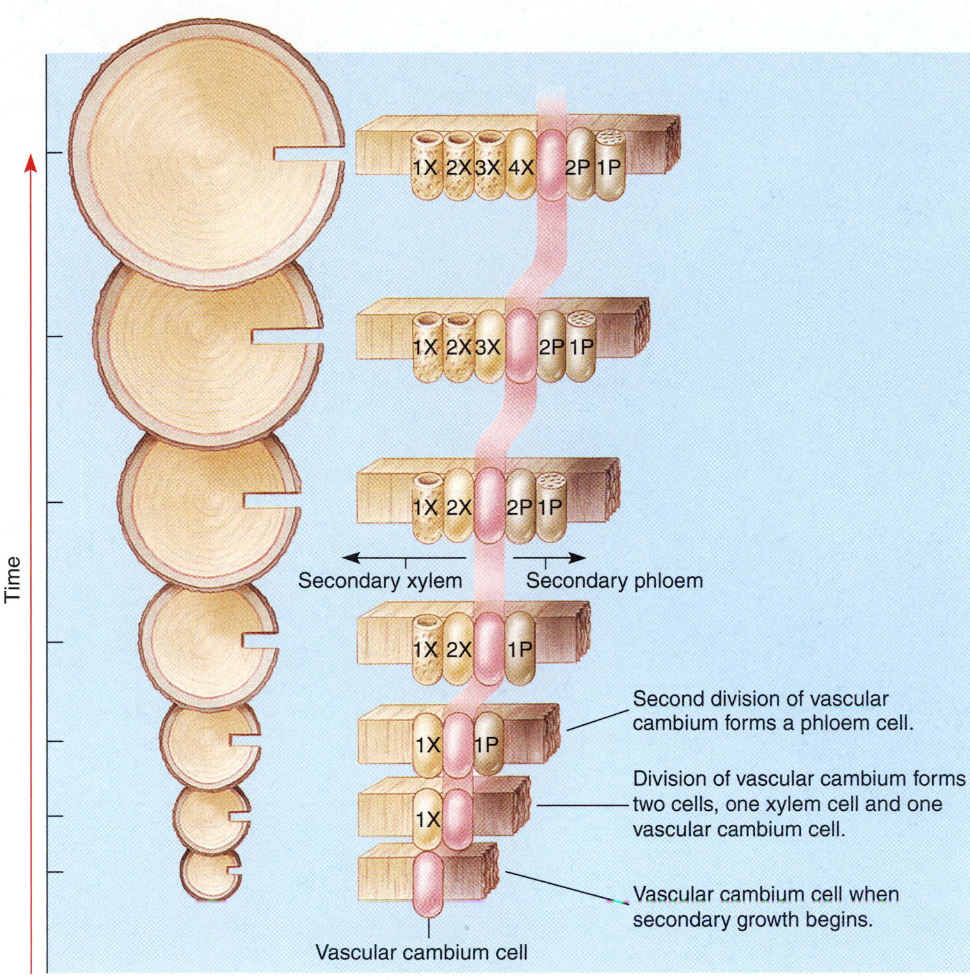

Figure 33–5 Development of secondary xylem and secondary phloem. To study the figure, which shows a radial view of a dividing vascular cambium cell, start at the bottom and move up. Note that vascular cambium (the pink cell) divides in two directions, forming secondary xylem (X) to the inside and secondary phloem (P) to the outside. These cells, which are numbered in the order in which they are produced, differentiate to form the mature cell types associated with xylem and phloem. As secondary xylem accumulates, vascular cambium "moves" outward, and the woody stem increases in diameter.

to its normal direction of division. In this case, both daughter cells remain meristematic.

What happens to the original primary tissues of a stem once secondary growth develops? As a stem increases in thickness, the orientation of the original primary tissues changes. For example, secondary xylem and secondary phloem are laid down between the primary xylem and primary phloem within each vascular bundle. Therefore, as vascular cambium forms secondary tissues, the primary xylem and primary phloem in each vascular bundle become separated from one another (Fig. 33–7). The primary tissues located outside the cylinder of secondary growth (i.e., primary phloem, cortex, and epidermis) are subjected to the mechanical pressures produced by secondary growth and are gradually crushed or torn apart and sloughed off.

Secondary tissues replace the primary tissues in function. Secondary xylem conducts water and dissolved nutrient minerals from roots to leaves in the woody plant. It contains the same types of cells found in primary xylem: water-conducting tracheids and vessel elements (see Chapter 31), in addition to parenchyma cells and fibers. The arrangement of the different cell types in secondary xylem produces the distinctive wood characteristics of each species.

Secondary phloem conducts dissolved sugar from its place of manufacture (leaves) to a place of use and storage (such as roots). The same types of cells found in primary phloem (sieve tube members, companion cells, parenchyma cells, and fibers) are also found in secondary phloem, although there are usually more fibers in secondary phloem than in primary phloem.

Whereas secondary xylem and secondary phloem transport water, minerals, and sugar vertically throughout the woody plant body, materials must also move horizontally, that is, laterally. Lateral movement occurs through **rays,** which are chains of parenchyma cells that radiate out from the center of the woody stem or root (see Fig. 33–6). Rays, which are often continuous from the secondary xylem to the secondary phloem, are formed by the vascular cambium. Water and dissolved nutrient minerals are transported laterally through rays, from the secondary xylem to the secondary phloem. Likewise, rays form pathways for the lateral transport of dissolved sugar, from the secondary phloem to the secondary xylem, and of waste products to the center, or heart, of the tree (discussed later).

Cork cambium produces periderm

Cork cambium, which usually arises from parenchyma cells in the outer cortex, produces **periderm,** the functional replacement for the epidermis. Cells of the cork cambium retain their ability to divide. Cork cambium is either a continuous cylinder of dividing

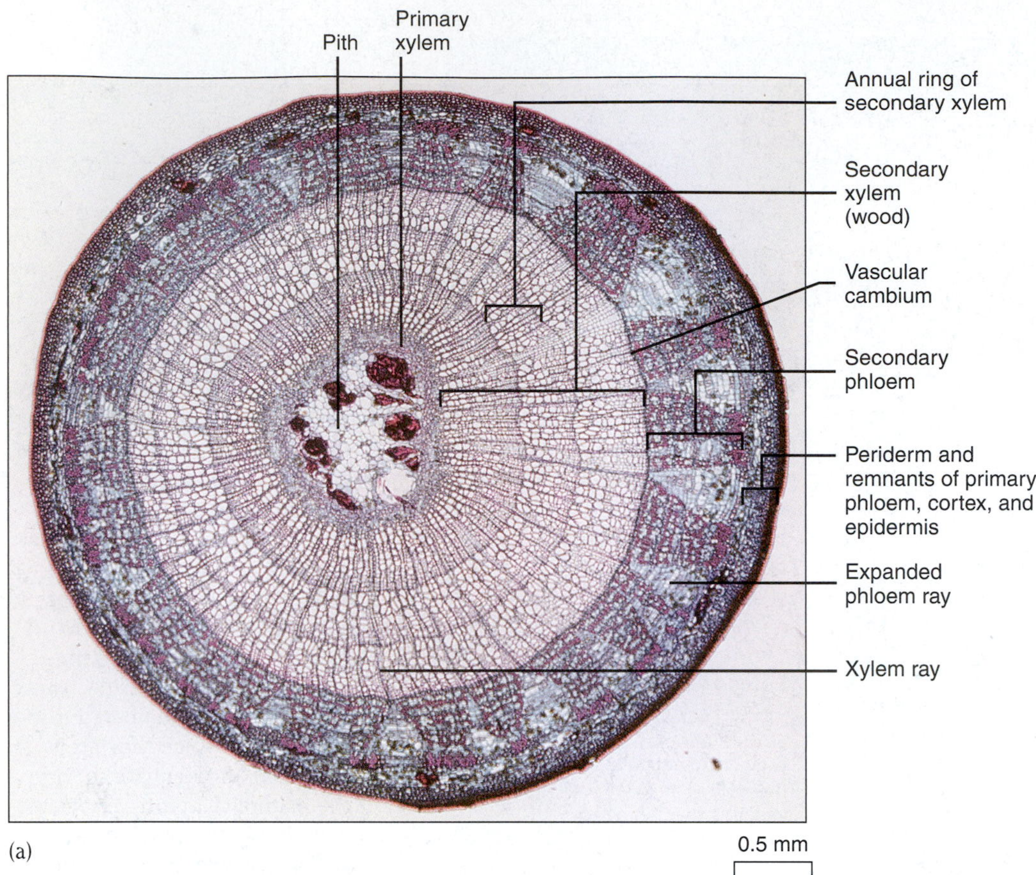

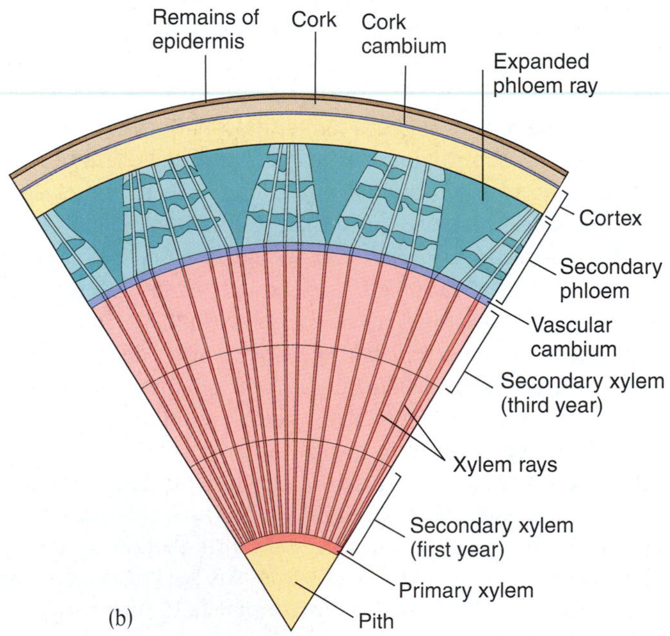

Figure 33–6 Three-year-old basswood *(Tilia americana)* stem in cross section. **(a)** LM of entire cross section **(b)** Sketch of a pie-shaped segment of the cross section. It may be easier to study the parts labeled on the sketch and then locate them on the micrograph. Note the location of the vascular cambium between the secondary xylem (wood) and secondary phloem (inner bark). Primary phloem is not labeled because it is crushed beyond recognition. *(a, Carolina Biological Supply Company/Phototake-NYC)*

cells (similar to vascular cambium) or a series of overlapping arcs of meristematic cells that form from parenchyma cells in successively deeper layers of the cortex and, eventually, secondary phloem. Variation in cork cambia and their rates of division explain why the outer bark of some tree species is fissured (e.g., bur oak), rough and shaggy (shagbark hickory), scaly (Norway pine), or smooth and peeling (paper birch) (Fig. 33–8).

As is true of vascular cambium, cork cambium divides to form new tissues in two directions—to its inside and its outside. Cork cells, formed to the outside of cork cambium, are dead at

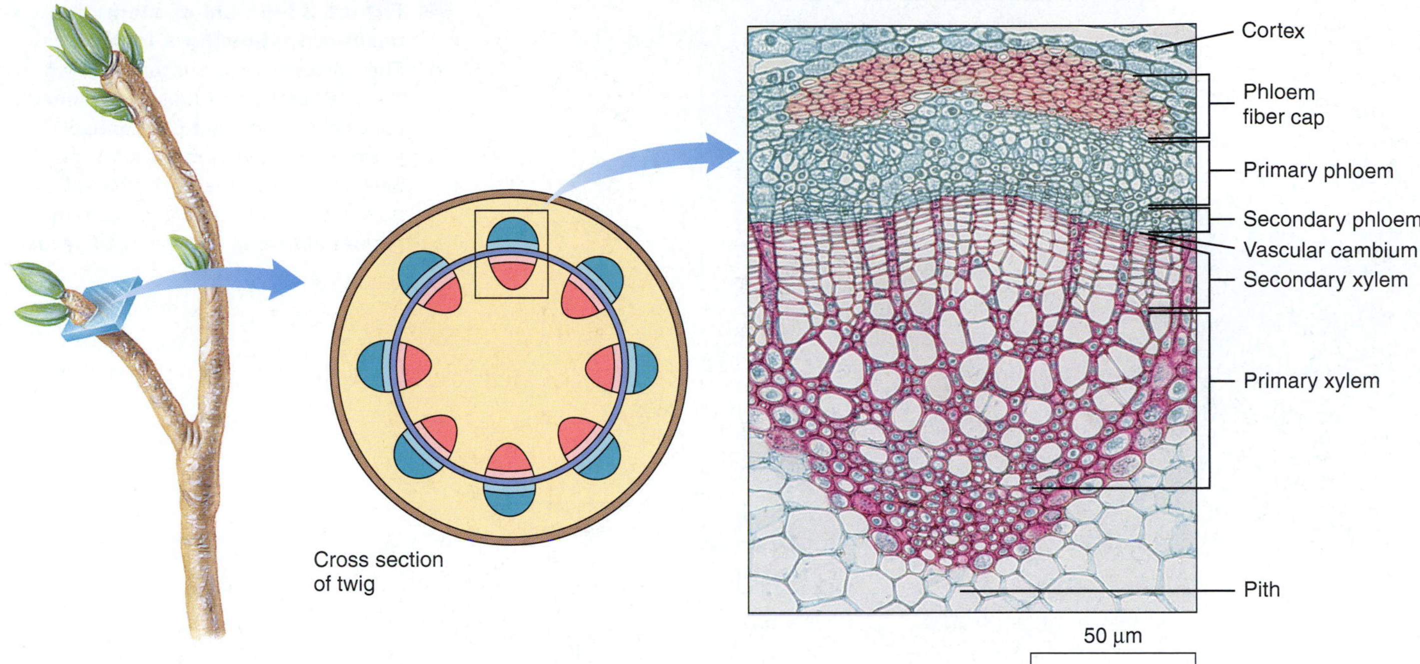

Figure 33–7 LM of part of a magnolia *(Magnolia* sp.*)* stem in cross section. Note that the vascular bundle has been split apart by secondary growth. Compare this vascular bundle with the one in Figure 33–2*b*, which has primary growth only. *(Dennis Drenner)*

maturity and have walls that contain layers of *suberin* and waxes, making them waterproof. These cork cells protect the woody stem against mechanical injury, mild fires, attacks by insects and fungi, temperature extremes, and water loss. To its inside, cork cambium sometimes forms cork parenchyma cells that store water and starch granules. Cork parenchyma is only one to several cells thick, much thinner than the cork cell layer.

Cork cells are impermeable to water and gases, yet the living internal cells of the woody stem require oxygen and must be able to exchange gases with the surrounding atmosphere. As a stem thickens from secondary growth, the epidermis, including stomata that allowed gas exchange for the herbaceous stem, dies. Stomata are replaced by lenticels, which permit gas exchange (Fig. 33–9).

Figure 33–8 Variation in bark. **(a)** Bur oak *(Quercus macrocarpa)* bark is deeply fissured. **(b)** Shagbark hickory *(Carya ovata)* has a rough, "shaggy" bark. **(c)** Bark from Norway pine *(Pinus resinosa)* is scaly. **(d)** Paper birch *(Betula papyrifera)* has a smooth, peeling bark. *(a,b,d, Carlyn Iverson; c, R.C. Bonner)*

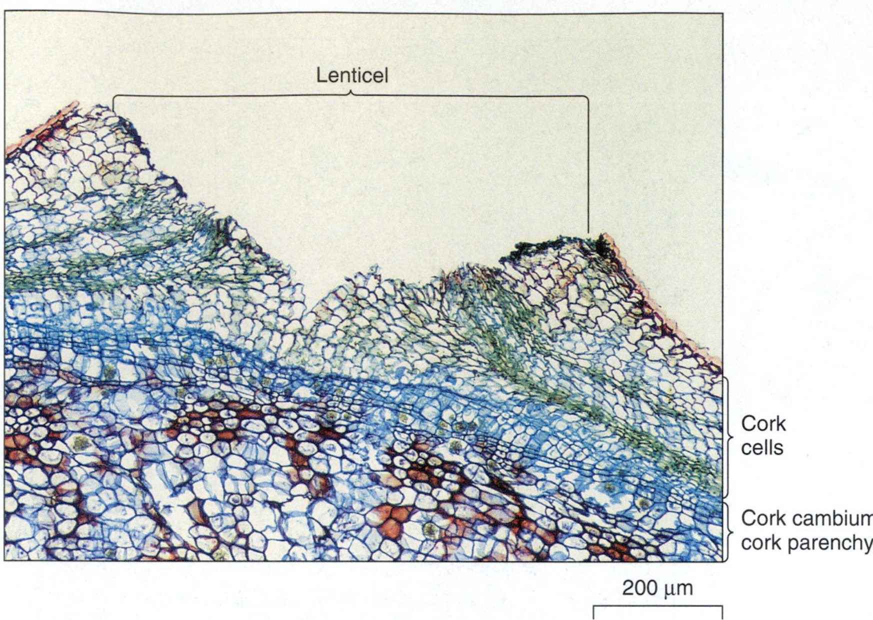

Figure 33–9 LM of stem periderm, showing a lenticel. The epidermis has ruptured owing to the proliferation of loosely arranged cork cells in the lenticel. Lenticels permit gas exchange through the periderm. From the bark of a calico flower *(Aristolochia elegans)* stem. *(James Mauseth, University of Texas)*

Common terms associated with wood are based on plant structure

If you have ever examined different types of lumber, you may have noticed that some trees have wood with two different colors (Fig. 33–10). The functional secondary xylem, that is, the part that conducts water and dissolved nutrient minerals, is the *sapwood,* a thin layer of younger, lighter colored wood that is closest to the bark. *Heartwood,* the older wood in the center of the tree, is typically a brownish-red color. A microscopic examination of heartwood reveals that its vessels and tracheids are plugged with pigments, tannins, gums, resins, and other materials. Therefore, heartwood no longer functions in conduction but instead functions as a storage site for waste products. Heartwood is denser than sapwood and therefore provides structural support for trees. Some evidence suggests that heartwood is also more resistant to decay.

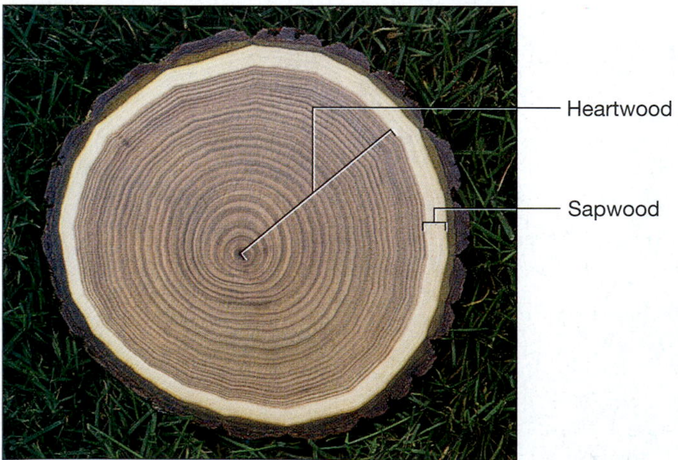

Figure 33–10 Heartwood and sapwood. The wood of older trees consists of a dense, central heartwood and an outer layer of sapwood. The sapwood is the functioning xylem that conducts water and dissolved nutrient minerals. The annual rings in the heartwood are very conspicuous. *(Carlyn Iverson)*

Almost everyone has heard of hardwood and softwood. Botanically speaking, *hardwood* is the wood of flowering plants and *softwood* is the wood of conifers (cone-bearing gymnosperms). The wood of pine and other conifers typically lacks fibers (with their thick secondary cell walls) and vessel elements; the conducting cells in gymnosperms are tracheids. These cell differences generally make conifer wood softer than the wood of flowering plants, although there is a substantial variation from one species to another. The balsa tree, for example, is a "hardwood" whose extremely light, soft wood is used to fashion airplane models.

Woody plants that grow in temperate climates where there is a growing period (during spring and summer) and a dormant period (during winter) exhibit *annual rings,* concentric circles found in cross sections of wood. To determine the age of a woody stem in the temperate zone, simply count the annual rings. In the tropics, environmental conditions, particularly seasonal or year-round precipitation patterns, determine the presence or absence of rings, and rings are not a reliable method of determining the ages of most tropical trees.

Examination of annual rings with a magnifying lens reveals no actual "ring," or line, separating one year's growth from the next. The appearance of a ring in cross section is due to differences in cell size and cell wall thickness between secondary xylem formed at the end of the preceding year's growth and that formed at the beginning of the following year's growth. In the spring, when water is plentiful, wood formed by vascular cambium has large-diameter conducting cells (tracheids and vessel elements) and few fibers and is appropriately called *springwood* or *early wood.* As summer progresses and water becomes less plentiful, the wood formed, known as *summerwood* or *late wood,* has nar-

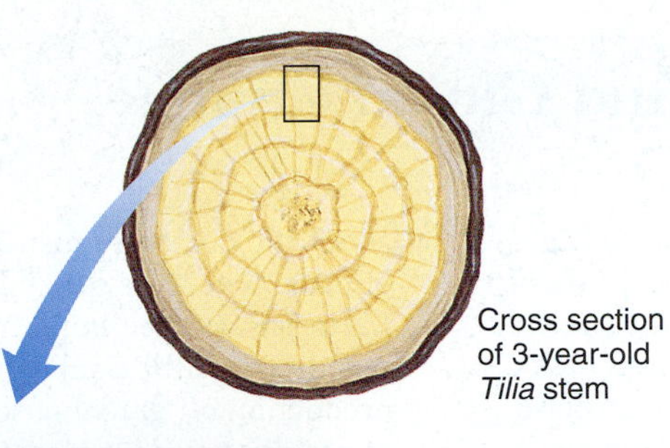

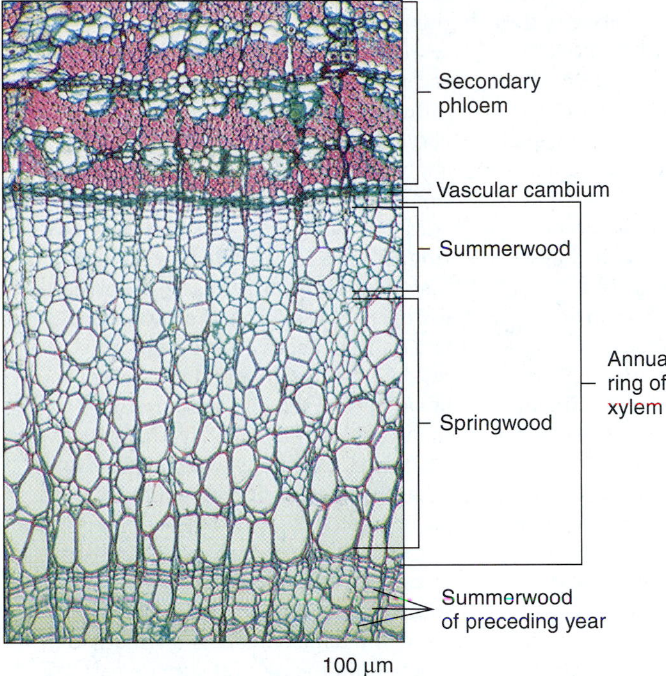

Figure 33–11 LM of a portion of a basswood *(Tilia americana)* stem cross section. One annual ring, or growth increment, is shown. Note the differences in cell size between the vessel elements of springwood and summerwood. The pink cells in the secondary phloem are fibers. *(Dennis Drenner)*

rower conducting cells and many fibers. It is this difference in cell size between the summerwood of one year and the springwood of the following year that gives the appearance of rings (Fig. 33–11). A great deal of information about climate in past times can be learned from the study of annual rings of ancient trees (see *Focus On: Tree Ring Analysis and Global Climate*).

As a woody stem increases in girth over the years, the branches that it bears grow along with it as long as they are alive. If a branch dies, it no longer continues to grow with the stem. In time, as the stem increases in girth, it surrounds the base of the dead branch. The basal portion of an embedded dead branch is called a *knot*. It is possible for a knot to contain bark as well as wood. The presence of knots in wood reduces its commercial value, except for ornamental purposes.

Now that we have discussed stem structure and primary and secondary growth, we examine internal transport in the vascular system of the plant body.

TRANSPORT IN PLANTS OCCURS IN XYLEM AND PHLOEM

Roots obtain water and dissolved nutrient minerals from the soil. Once inside roots, these materials are transported upward to stems, leaves, flowers, fruits, and seeds. Furthermore, sugar molecules manufactured in leaves by photosynthesis are transported in solution (i.e., dissolved in water) throughout the plant, including into the subterranean roots. Water and dissolved nutrient minerals are transported from roots to other parts of the plant in xylem, whereas dissolved sugar is **translocated** in phloem.

Xylem transport and phloem translocation do not resemble the movement of materials in animals, because in plants nothing *circulates* in a system of vessels. Water and minerals, transported in xylem, travel in one direction only (upward), whereas translocation of dissolved sugar may occur upward or downward in separate phloem cells. In addition, xylem transport and phloem translocation differ from internal circulation in animals because movement in both xylem and phloem is driven largely by natural physical processes rather than by a pumping organ, or heart.

How, exactly, do materials travel in the continuous system of the plant's vascular tissues? We first examine water and its movement through the plant, and later we discuss the translocation of dissolved sugar.

WATER AND MINERALS ARE TRANSPORTED IN XYLEM

Water initially moves horizontally into roots from the soil, passing through several tissues until it reaches xylem. Once the water moves into the tracheids and vessel elements of root xylem, it travels upward through a continuous network of these hollow, dead cells from root to stem to leaf. Dissolved nutrient minerals are carried along passively in the water. The plant does not expend any energy of its own to transport water, which moves as a result of natural physical processes. The transport of xylem sap is the most rapid of any movement of materials in plants (Table 33–2 on page 722).

How does water move to the tops of plants? It is either pushed up from the bottom of the plant or pulled up to the top of the plant. Although both mechanisms exist, current evidence indicates that most water is transported through xylem by being *pulled* to the top of the plant.

Water movement can be explained by a difference in water potential

To understand how water moves, it is helpful to introduce **water potential,** which is defined as the free energy (see Chapter 6) of water. Water potential is important in plant physiology because it is a measure of a cell's ability to absorb water by osmosis (see Chapter 5). Water potential also provides a measure of water's tendency to evaporate from cells.

The water potential of pure water is conventionally set at 0 megapascals (MPa) because it cannot be measured directly. (A megapascal is a unit of pressure equal to about 10 atmospheres,

FOCUS ON Tree Ring Analysis and Global Climate

In temperate climates, the age of a tree can be determined by counting the number of annual rings. Other useful information can be determined by analyzing tree rings as well. For example, the size of each ring varies depending on local weather conditions, including precipitation and temperature. Sometimes the variation in tree rings can be attributed to a single environmental factor, and similar patterns appear in the rings of different tree species over a large geographical area. For example, trees in the southwestern United States have similar ring patterns due to variations in the amount of annual precipitation. Years with adequate precipitation produce wider rings of growth, whereas years of drought produce much narrower rings.

It is possible to study ring sequences going back several thousand years. First, a *master chronology,* a complete sample of rings dating back as far as possible, is developed *(see figure).* A small core of wood is bored out of the trunk of an old living tree to obtain a sample of rings. The oldest rings (those toward the center of the tree) are matched with the youngest rings (those toward the outside) of an older tree or even an old piece of wood from a house. A master chronology of the area is obtained by using successively older and older sections of wood, even those found in prehistoric dwellings, and overlapping their matching ring sequences. The longest master chronology is of bristlecone pines in the western United States; it goes back almost 9000 years.

Dendrochronology, the study of both visible and microscopic details of tree rings, has been used extensively in several fields. Tree ring analysis has been extremely useful in dating prehistoric sites of Native Americans in the Southwest. For example, the Cliff Palace in the Mesa Verde National Park dates back to AD 1073. Tree ring analysis indicates that an extended drought forced the original inhabitants to abandon their homes. Tree ring analysis is also useful in other disciplines, including ecology (to study changes in a forest community over time), environmental science (to study the effects of air pollution on tree growth), and geology (to date earthquakes and volcanic eruptions).

Climatologists are increasingly using tree ring data to study past climatic patterns. Annual ring widths of certain tree species that grow at high elevations are sensitive to yearly temperature variations; the rings of these trees are wider in warm years and narrower in cool years. By studying tree rings across long time sequences, the natural pattern of global temperature fluctuations can be determined. This information is particularly important because of concerns about the human influence on global climate. Scientists generally agree that Earth has warmed in recent decades, and there is little doubt that human production of "greenhouse gases" such as carbon dioxide has contributed to this warming (see Chapter 55). Researchers are not sure how much of the recent warming is the result of human influence, as opposed to natural climate variability. It is thought that tree ring analysis will help answer this vital question. Scientists in nine European countries are cooperating in a massive tree ring analysis to construct an annual history of temperatures across northern Europe and Asia since the end of the last Ice Age, about 10,000 years ago. Local diving clubs are obtaining samples of well-preserved logs that are thousands of years old from river and lake sediments. Scientists will use these and other samples, including living trees, to piece together a 10,000-year master chronology. When that is accomplished (sometime before the year 2005), it should be possible to determine conclusively if today's climate is distinctly different from the natural climate patterns of the past.

Tree ring dating. A master chronology is developed using progressively older pieces of wood from the same geographical area. By matching the rings of a wood sample of unknown age to the master chronology, the age of the sample can be accurately determined. Ring matching, once a painstakingly tedious job, is usually performed by computers.

or 145.1 pounds per square inch.) It is possible to measure differences in the free energy of water molecules in different situations, however. When solutes are dissolved in water, the free energy of water decreases.[2] This means that dissolved solutes lower the water potential to a negative number. *Water moves from a region of higher (less negative) water potential to a region of lower (more negative) water potential.*

The water potential of the soil varies, depending on how much water it contains. When a soil is extremely dry, its water potential is very low (very negative). When a soil is moister, its water potential is higher, although it is still a negative number because dissolved nutrient minerals are present in dilute concentrations.

The water potential in root cells is also negative owing to the presence of dissolved solutes. Roots contain more dissolved materials than does soil water, unless the soil is extremely dry. This

[2] Solutes induce hydration, in which water molecules surround polar molecules and ions, keeping them in solution by preventing them from coming together (see Chapter 2). The association of water molecules with hydrated molecules and ions reduces the motion of water molecules, thereby decreasing their free energy.

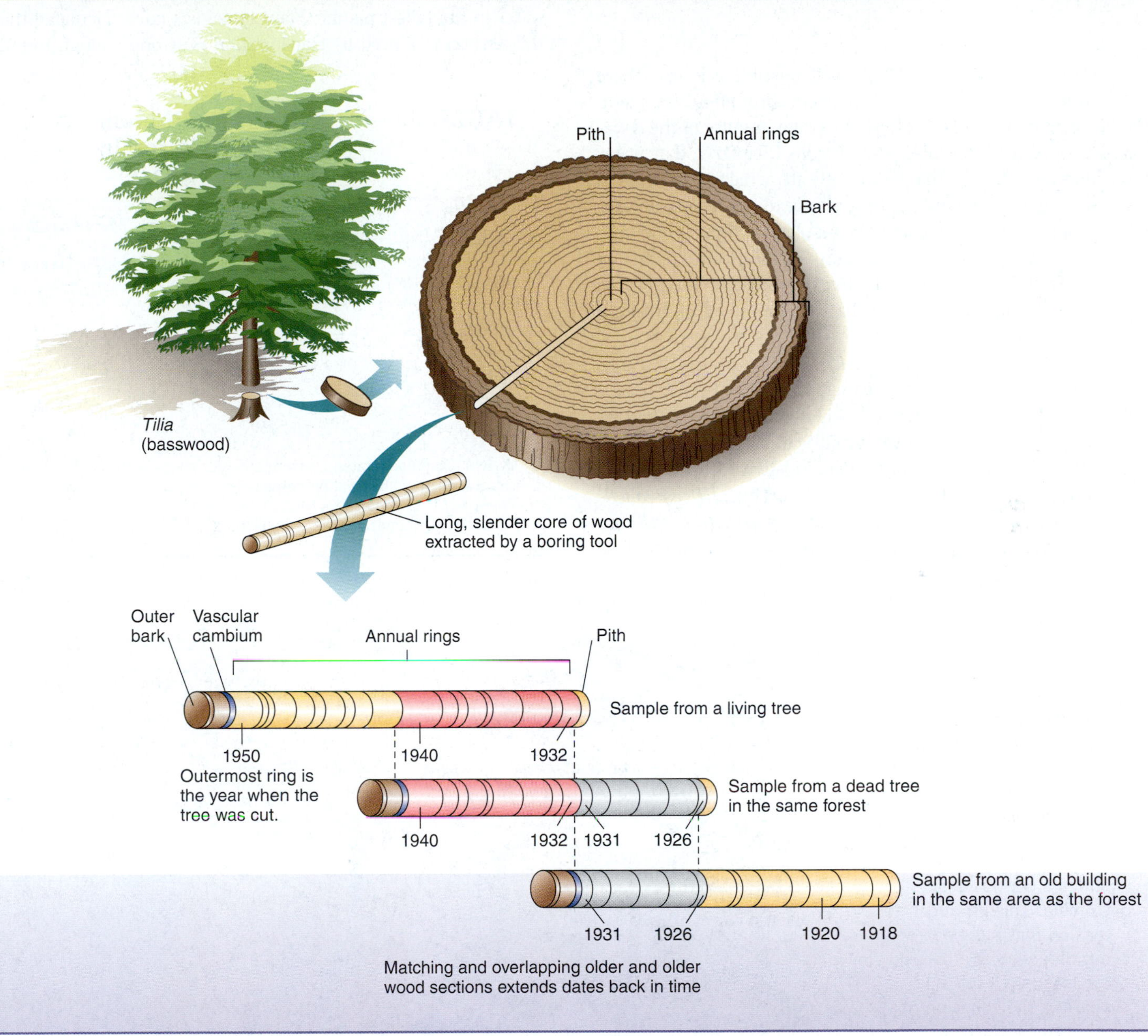

means that *under normal conditions the water potential of the root is more negative than the water potential of the soil.* Thus, water moves by osmosis from the soil into the root.

Tension-cohesion pulls water up a stem

According to the **tension-cohesion model,** also known as the **transpiration-cohesion model,** water is pulled up the plant as a result of a *tension* produced at the top of the plant (Fig. 33–12). This tension, which resembles that produced when drinking a liquid through a straw, is caused by the evaporative pull of transpiration. Recall from Chapter 32 that **transpiration** is the evaporation of water vapor from plants. Most water loss from transpiration takes place through the numerous microscopic pores (stomata) present on leaf and stem surfaces. The tension extends from leaves, where most transpiration occurs, down the stems and into the roots. It draws water up stem xylem to leaf cells that have lost water as a result of transpiration and pulls water from root xylem into stem xylem. As water is pulled upward, additional water from the soil is drawn into the roots. Thus, the pathway of water movement is as follows:

Soil ⟶ root tissues (epidermis, cortex, and so forth) ⟶ root xylem ⟶ stem xylem ⟶ leaf xylem ⟶ leaf mesophyll ⟶ atmosphere

This upward pulling of water is only possible as long as there is an unbroken column of water in xylem throughout the plant. Water forms an unbroken column in xylem because of the *cohesiveness* of water molecules. Recall from Chapter 2 that water molecules are cohesive, that is, strongly attracted to one another, because of *hydrogen bonding*. In addition, the *adhesion* of water to the walls of xylem cells, also the result of hydrogen bonding, is an important factor in maintaining an unbroken column of water. Thus, the cohesive and adhesive properties of water enable it to form an unbroken column that can be pulled up through the xylem.

The movement of water in xylem due to the tension-cohesion mechanism can be explained in terms of water potential. The atmosphere has an extremely negative water potential. For example, air with a relative humidity of 50% has a water potential of −100 MPa; even moist air at a relative humidity of 90% has a negative water potential of −13 MPa. Thus, *there is a water potential gradient from the least negative (the soil) up through the plant to the most negative (the atmosphere).* This gradient literally pulls the water from the soil up through the plant.

Is tension-cohesion powerful enough to explain the rise of water in the tallest plants? Plant biologists have calculated that the tension produced by transpiration is strong enough to pull

TABLE 33–2 Xylem and Phloem Transport Rates in Selected Plants*

Plant	Maximum Rate in Xylem (cm/min)	Maximum Rate in Phloem (cm/min)
Conifer	2	0.8
Woody dicot	73	2
Herbaceous dicot/monocot	100	2.8–11
Herbaceous vine	250	1.2

*Xylem and phloem rates are from different plants within each general group and should be used for comparative purposes only. Adapted from Mauseth, J.D. *Botany: An Introduction to Plant Biology,* 2nd ed. Philadelphia, Saunders College Publishing, 1995.

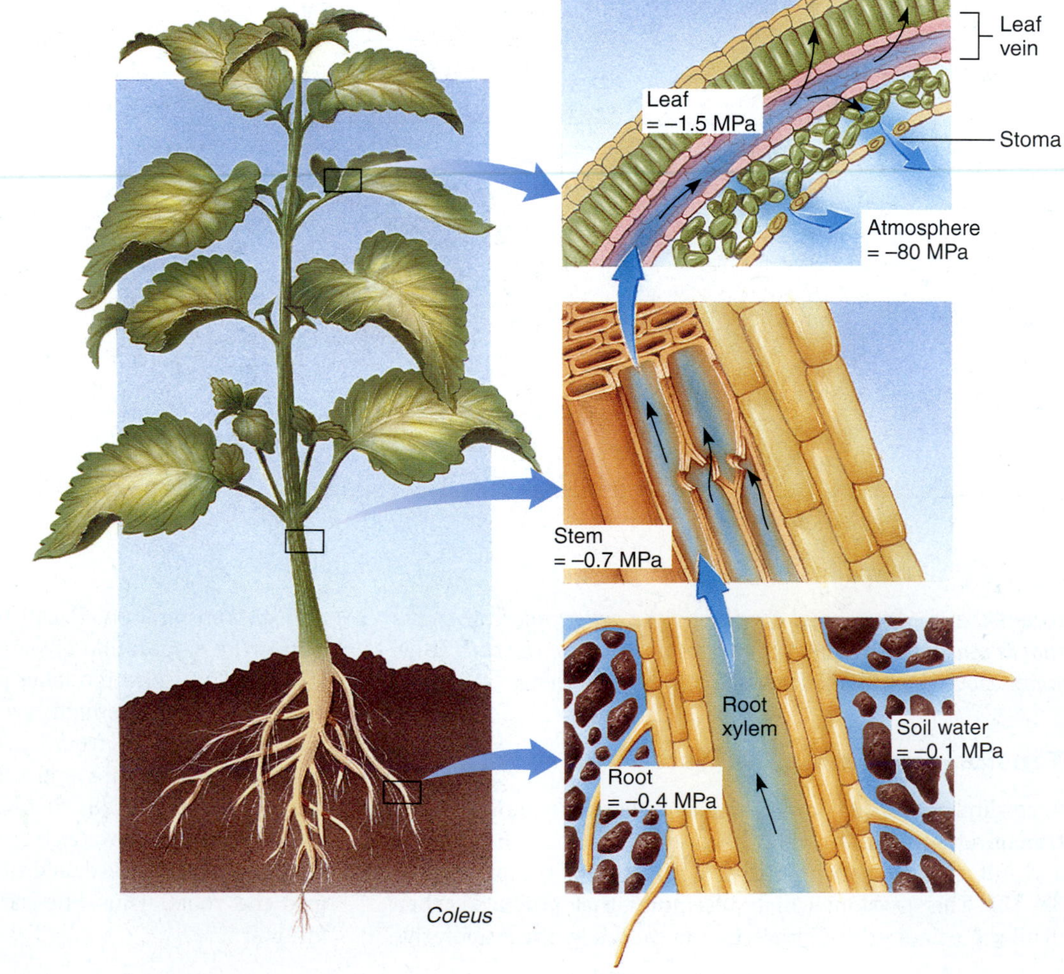

Figure 33–12 The tension-cohesion model. *(Top)* This model hypothesizes that water vapor diffuses from the surfaces of leaf mesophyll cells to the drier atmosphere through stomata. This produces a tension that pulls water out of leaf xylem toward the mesophyll cells. *(Middle)* The cohesion of the water molecules, caused by hydrogen bonding, allows unbroken columns of water to be pulled up the narrow vessels and tracheids of stem xylem. *(Bottom)* This in turn pulls water up root xylem, forming a continuous column of water from root xylem to stem xylem to leaf xylem. As water moves upward in the root, it produces a pull that causes soil water to diffuse into the root.

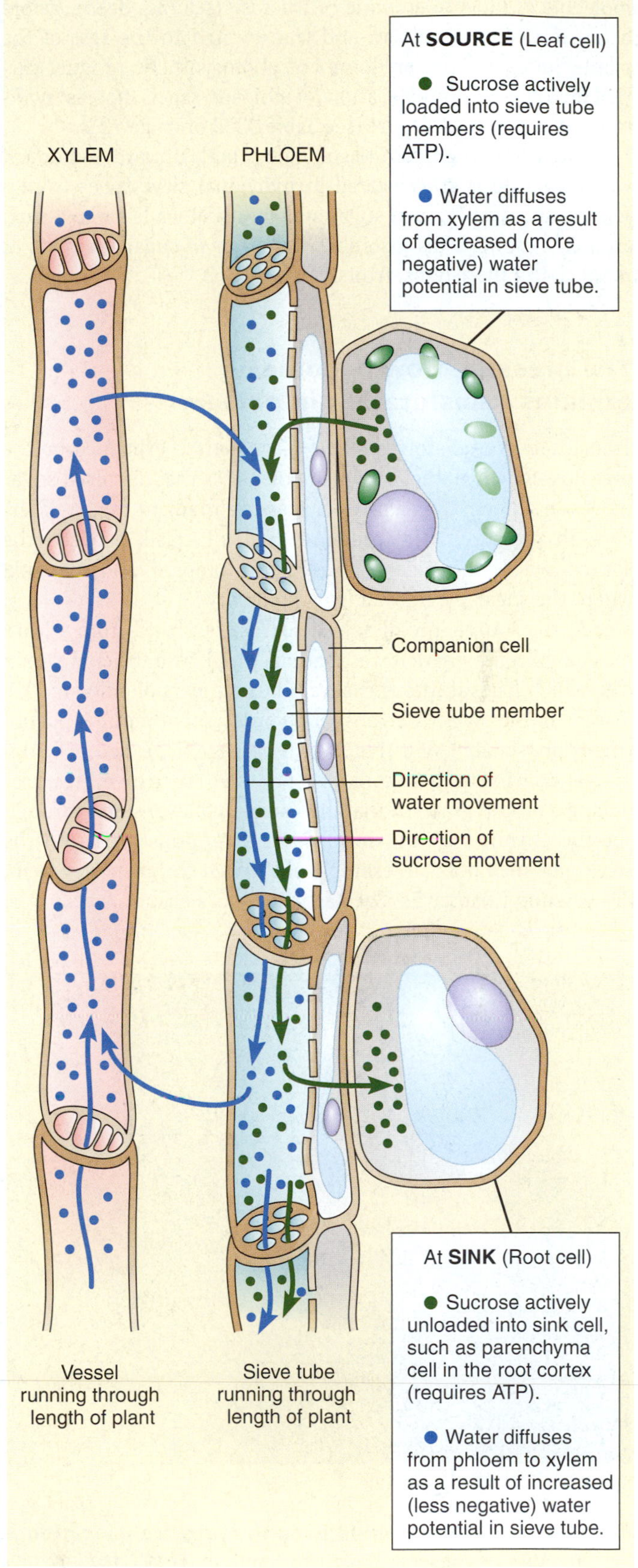

At SOURCE (Leaf cell)
● Sucrose actively loaded into sieve tube members (requires ATP).
● Water diffuses from xylem as a result of decreased (more negative) water potential in sieve tube.
XYLEM
PHLOEM
Companion cell
Sieve tube member
Direction of water movement
Direction of sucrose movement
At SINK (Root cell)
● Sucrose actively unloaded into sink cell, such as parenchyma cell in the root cortex (requires ATP).
● Water diffuses from phloem to xylem as a result of increased (less negative) water potential in sieve tube.
Vessel running through length of plant
Sieve tube running through length of plant

Figure 33–13 The pressure-flow hypothesis. Sugar is actively loaded into the sieve tube member at the source. As a result, water diffuses from the xylem into the sieve tube member. At the sink, the sugar is actively or passively unloaded, and water diffuses from the sieve tube member into the xylem. The pressure gradient within the sieve tube, from source to sink, causes translocation from the area of higher hydrostatic pressure (the source) to the area of lower hydrostatic pressure (the sink).

water upward 150 m (500 ft) in tubes the diameter of xylem vessels. Because the tallest trees are no more than about 117 m (375 ft) high, the tension-cohesion model easily accounts for the transport of water. Currently, most botanists consider the tension-cohesion model to be the dominant mechanism of xylem transport in most plants.

Although the tension-cohesion model was proposed more than 100 years ago to explain water transport in xylem, conclusive experimental evidence to support this mechanism was first obtained in 1995 by two research groups working independently. Both groups demonstrated that large negative pressures exist in xylem and that the water potential gradients in root, stem, and leaf xylem are adequate to explain the observed movement of water.

Root pressure pushes water from the root up a stem

In the less important mechanism for water transport, known as **root pressure,** water that moves into a plant's roots from the soil is *pushed* up through xylem toward the top of the plant. Root pressure occurs because nutrient mineral ions that are actively absorbed from the soil are pumped into the xylem, decreasing its water potential. Water then moves into xylem cells from surrounding root cells. In turn, water moves into roots by osmosis because of the difference in water potential between the soil and root cells. The accumulation of water in root tissues produces a positive pressure (as high as +0.2 MPa) that forces the water up through the xylem.

Guttation, a phenomenon in which liquid water is forced out through special openings in the leaves (see Chapter 32), is a manifestation of root pressure. However, root pressure is not strong enough to explain the rise of water to the tops of coastal redwoods and other tall trees. Root pressure exerts an influence in smaller plants, particularly in the spring when the soil is quite wet, but it clearly does not cause water to rise 100 m (330 ft) or more in the tallest plants. Furthermore, root pressure does not occur to any appreciable extent in summer (when water is often not plentiful in soil), yet the movement of water is greatest during hot summer days.

SUGAR IN SOLUTION IS TRANSLOCATED IN PHLOEM

The sugar produced during photosynthesis is converted into sucrose (common table sugar), a disaccharide composed of one molecule of glucose and one of fructose (see Fig. 3–8*b*), before being loaded into phloem and translocated to the rest of the plant. Sucrose is the predominant photosynthetic product carried in phloem. Translocation of phloem sap is not as swift-moving as xylem transport (see Table 33–2 on page 722).

Fluid within phloem tissue moves both upward and downward. Sucrose is translocated in individual sieve tubes from a *source,* an area of excess sugar supply (usually a leaf), to a *sink,* an area of storage (as insoluble starch) or of sugar use such as roots, apical meristems, fruits, and seeds.

The pressure-flow hypothesis explains translocation in phloem

Translocation of dissolved sugar in phloem is explained by the **pressure-flow hypothesis,** which postulates that dissolved sugar moves in phloem by means of a pressure gradient (i.e., a difference in pressure). The pressure gradient exists between the source, where the sugar is loaded into phloem, and the sink, where the sugar is removed from phloem.

At the source, the dissolved sucrose is moved from a leaf's mesophyll cells, where it was manufactured, into the companion cell, which loads it into the sieve tube members of phloem. This loading occurs by active transport, a process that requires adenosine triphosphate (ATP) (Fig. 33–13 on page 723). The ATP supplies energy to pump protons out of the sieve tube members, producing a proton gradient that drives the uptake of sugar through specific channels by the cotransport of protons back into the sieve tube members (an example of a linked cotransport system, discussed in Chapter 5). The sugar therefore accumulates in the sieve tube member. The increase in dissolved sugars in the sieve tube member at the source—a concentration that is two to three times greater than in surrounding cells—decreases (makes more negative) the water potential of that cell. As a result, water moves by osmosis from the xylem cells into the sieve tubes, increasing the hydrostatic pressure inside them. Thus, phloem loading at the source is as follows:

> Proton pump moves H^+ out of sieve tube member ⟶ sugar is actively transported into sieve tube member ⟶ water diffuses from xylem into sieve tube member ⟶ hydrostatic pressure increases within sieve tube

At its destination (the sink), sugar is unloaded by various mechanisms, both active and passive, from the sieve tube members. With the loss of sugar, the water potential in the sieve tube members at the sink increases (becomes less negative). Therefore, water moves out of the sieve tubes by osmosis and into surrounding cells; most of this water diffuses back to the xylem to be transported upward. This water movement decreases the hydrostatic pressure inside the sieve tubes at the sink. Thus, phloem unloading at the sink is as follows:

> Sugar is transported out of sieve tube member ⟶ water diffuses out of sieve tube member and into xylem ⟶ hydrostatic pressure decreases within sieve tube

Thus, the pressure-flow hypothesis explains the movement of dissolved sugar in phloem by means of a pressure gradient. It is the difference in sugar concentrations between the source and

(a)

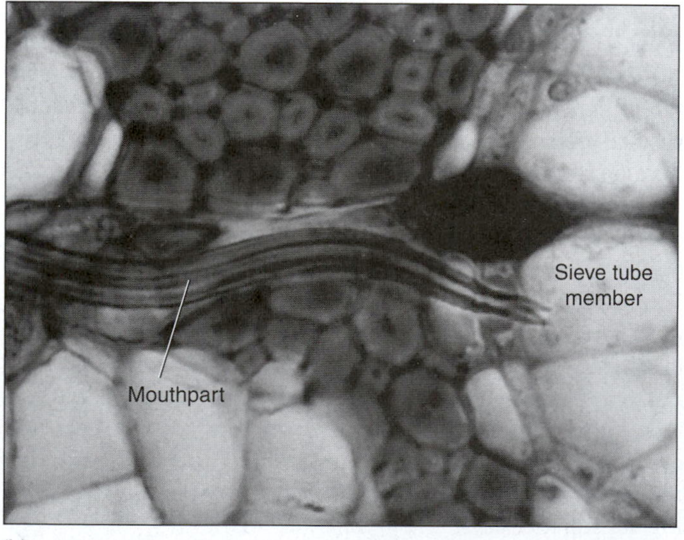

(b)

Figure 33–14 Aphids used to study translocation in phloem. **(a)** Mature aphid, a tiny insect about 3 to 6 mm in length, feeding on a stem. **(b)** LM of phloem cells, showing a sieve tube member that has been penetrated by the aphid mouthpart. *(a, Dwight Kuhn; b, M.H. Zimmerman,* Science, *Vol. 133, pp. 73–79 [Fig. 4], 13 Jan. 1961. Copyright 2002 by the American Association for the Advancement of Science)*

the sink that causes translocation in phloem, as water and dissolved sugar flow along the pressure gradient. This pressure gradient pushes the sugar solution through phloem much as water is forced through a hose.

The actual translocation of dissolved sugar in phloem does not require metabolic energy. However, the loading of sugar at the source and the active unloading of sugar at the sink require energy derived from ATP to move the sugar across cell membranes by active transport.

Process of Science Although the pressure-flow hypothesis adequately explains current data on phloem translocation, much remains to be learned about this complex process. Phloem translocation is difficult to study in plants. Because phloem cells are under pressure, cutting into phloem to observe it releases the pressure and causes the contents of the sieve tube members (the phloem sap) to exude and mix with the contents of other severed cells that are also unavoidably cut. In the 1950s botanists developed a unique research tool to avoid contaminating the phloem sap: aphids, which are small insects that insert their mouthparts into phloem sieve tubes for feeding (Fig. 33–14 on facing page). The pressure in the punctured phloem drives the sugar solution through the aphid's mouthpart and into its digestive system. When the aphid's mouthpart is severed from its body by a laser beam, the sugar solution continues to flow through the mouthpart at a rate proportional to the pressure in phloem. This rate can be measured, and the effects on phloem transport of different environmental conditions—varying light intensities, darkness, and mineral deficiencies, for example—can be ascertained.

The identity and proportions of translocated substances can also be determined using severed aphid mouthparts. This technique has verified that in most plant species the sugar sucrose is the main or only carbohydrate transported in phloem; however, some species transport other sugars, such as raffinose, or sugar alcohols, such as sorbitol. Additional substances transported in phloem include certain plant hormones, ATP, amino acids, K^+ and other inorganic ions, and disease-causing plant viruses.

SUMMARY WITH KEY TERMS

I. The main functions of stems are support, conduction, and production of new living tissues.

II. Woody twigs demonstrate the external structure of stems.
   A. **Buds** are undeveloped embryonic shoots. A **terminal bud** is located at the tip of a stem, whereas **axillary buds (lateral buds)** are located in leaf axils.
   B. A dormant bud is covered and protected by **bud scales.** When the bud resumes growth, bud scales covering the bud fall off, leaving **bud scale scars.**
   C. The area on a stem where each leaf is attached is called a **node,** and the region of a stem between two successive nodes is an **internode.**
   D. A **leaf scar** shows where each leaf was attached to the stem. The areas within a leaf scar where the vascular tissue extended from the stem to the leaf are called **bundle scars.**
   E. **Lenticels** are sites of loosely arranged cells that allow oxygen to diffuse into the interior of a woody stem.

III. Herbaceous stems possess an epidermis, vascular tissue, and either ground tissue or cortex and pith.
   A. The **epidermis** is a protective layer covered by a water-conserving cuticle. Stomata permit gas exchange.
   B. **Xylem** conducts water and dissolved nutrient minerals, and **phloem** conducts dissolved sugar and small quantities of other substances, such as hormones, ATP, amino acids, inorganic ions such as K^+, and viruses.
   C. The **cortex, pith,** and **ground tissue** function primarily for storage.

IV. Although herbaceous stems all have the same basic tissues, their arrangement varies considerably.
   A. Herbaceous dicot stems have the vascular bundles arranged in a circle (in cross section) and have a distinct cortex and pith.
   B. Monocot stems have vascular bundles scattered in ground tissue.

V. Secondary growth occurs in some flowering plants (woody dicots) and in all cone-bearing gymnosperms.
   A. **Vascular cambium** develops between the primary xylem and the primary phloem. Vascular cambium produces secondary xylem (wood) to the inside and secondary phloem (inner bark) to the outside.
   B. **Cork cambium** arises near a stem's surface.
      1. Cork cambium produces **periderm,** which consists of cork parenchyma to the inside and cork cells to the outside.
      2. Cork cells are the functional replacement for epidermis in a woody stem. Cork parenchyma functions primarily for storage in a woody stem.
      3. **Rays** are chains of parenchyma cells that radiate out from the center of a stem and form pathways for the lateral movement of materials between the secondary xylem and the secondary phloem.

VI. Water and dissolved nutrient minerals move upward in xylem from roots to stems and leaves.
   A. **Water potential** is a measure of the free energy of water.
      1. Pure water has a water potential of 0 megapascals, whereas water with dissolved solutes has a negative water potential.
      2. Water moves from an area of higher (less negative) water potential to an area of lower (more negative) water potential.
   B. The **tension-cohesion model** explains the rise of water in even the largest plants.
      1. The evaporative pull of **transpiration** causes tension at the top of the plant. This tension is the result of a water potential gradient that ranges from the slightly negative water potentials in the soil and roots to the very negative water potentials in the atmosphere.
      2. As a result of the gradient in water potentials, water moves along the following pathway: soil → root xylem → stem xylem → leaf xylem → mesophyll cells → the atmosphere.
      3. As a consequence of the cohesive and adhesive properties of water, the column of water pulled up through the plant remains unbroken.
   C. **Root pressure,** caused by the movement of water into roots from the soil as a result of the active absorption of nutrient mineral ions from the soil, helps explain the rise of water in smaller plants, particularly when the soil is wet. Root pressure pushes water up through xylem.

VII. Dissolved sugar is **translocated** upward or downward in phloem.
   A. Sucrose is the predominant sugar translocated in phloem.

B. The movement of materials in phloem is explained by the **pressure-flow hypothesis.**
   1. Companion cells actively load sugar into the sieve tubes at the source; ATP is required for this process.
      a. The ATP supplies energy to pump protons out of the sieve tube members.
      b. The proton gradient drives the uptake of sugar by the cotransport of protons back into the sieve tube members.
      c. Sugar therefore accumulates in the sieve tube member, causing the movement of water into the sieve tubes by osmosis.
   2. Companion cells actively (requiring ATP) and passively (not requiring ATP) unload sugar from the sieve tubes at the sink. As a result, water leaves the sieve tubes by osmosis, decreasing the hydrostatic pressure inside the sieve tubes.
   3. The flow of materials between source and sink is driven by the hydrostatic pressure gradient produced by water entering phloem at the source and water leaving phloem at the sink.

POST-TEST

1. The three main functions of stems are (a) support, conduction, and photosynthesis (b) support, anchorage in soil, and production of new living tissues (c) conduction, production of new living tissues, and sexual reproduction (d) conduction, asexual reproduction, and sexual reproduction (e) support, conduction, and production of new living tissues
2. All stems have undeveloped embryonic shoots called (a) lenticels (b) buds (c) lianas (d) phloem fiber caps (e) periderm
3. Axillary buds are located (a) at the tips of stems (b) in unusual places, such as on roots (c) in the region between two successive nodes (d) in the upper angle between a leaf and the stem to which it is attached (e) within the loosely arranged cells of lenticels
4. The tissue in monocot stems in which the vascular tissues are embedded is (a) cork cambium (b) cortex (c) ground tissue (d) pith (e) phloem
5. The protective outer layer of cells covering herbaceous stems is the (a) periderm (b) cork cambium (c) lateral meristem (d) epidermis (e) bud scale
6. Ground tissue in monocot stems performs the same functions as ____________ and ____________ in herbaceous dicot stems (a) phloem; xylem (b) cork cambium; vascular cambium (c) epidermis; periderm (d) primary xylem; secondary xylem (e) cortex; pith
7. The two lateral meristems responsible for secondary growth are (a) phloem and xylem (b) cork cambium and vascular cambium (c) epidermis and periderm (d) primary xylem and secondary xylem (e) cortex and pith
8. Cork cambium and the tissues it produces are collectively called (a) periderm (b) lenticels (c) cortex (d) epidermis (e) wood
9. Horizontal movement of materials in woody plants occurs in (a) bud scales (b) cortex (c) rays (d) lenticels (e) pith rays
10. The older wood in the center of a tree trunk is commonly called (a) hardwood (b) softwood (c) sapwood (d) heartwood (e) cork
11. Water potential is (a) the formation of a proton gradient across a cell membrane (b) the transport of a watery solution of sugar in phloem (c) the transport of water in both xylem and phloem (d) the removal of sucrose at the sink, causing water to move out of the sieve tubes (e) the free energy of water in a particular situation
12. Which of the following is a mechanism of water movement in xylem that does *not* generate sufficient force to explain the rise of water to the tops of the tallest trees? (a) pressure-flow hypothesis (b) tension-cohesion (c) root pressure (d) active transport of potassium into guard cells (e) transpiration
13. Which of the following is a mechanism of water movement in xylem in which a tension is produced at the top of the plant by the evaporative pull of transpiration and the cohesive and adhesive properties of water? (a) pressure-flow (b) tension-cohesion (c) root pressure (d) active transport of potassium into guard cells (e) guttation-transpiration hypothesis
14. Which of the following is a mechanism of phloem transport in which dissolved sugar is moved by means of a pressure gradient that exists between the source and the sink? (a) pressure-flow (b) tension-cohesion (c) root pressure (d) active transport of potassium into guard cells (e) guttation-transpiration hypothesis
15. How does increasing solute concentration affect water potential? (a) water potential becomes more positive (b) water potential becomes more negative (c) water potential becomes more positive under certain conditions and more negative under other conditions (d) water potential is not affected by solute concentration (e) water potential is always zero when solutes are dissolved in water

REVIEW QUESTIONS

1. List several functions of stems and describe the tissue(s) responsible for each.
2. If you are examining a cross section of a herbaceous stem of a flowering plant, how would you determine whether it is a dicot or a monocot?
3. What happens to the primary tissues of a stem when secondary growth occurs?
4. When a strip of bark is peeled off a tree branch, what tissues are usually removed?
5. Distinguish between vascular cambium and cork cambium, and describe the tissues that arise from each.
6. Name and briefly describe the model used to explain the rise of water in the tallest trees.
7. Describe root pressure. What are its limitations?
8. Describe the pressure-flow hypothesis of sugar movement in phloem, including the activities at source and sink.

9. Label the various tissues of this herbaceous dicot stem. Give at least one function for each of the tissues. Use Figure 33–2 to check your answers.

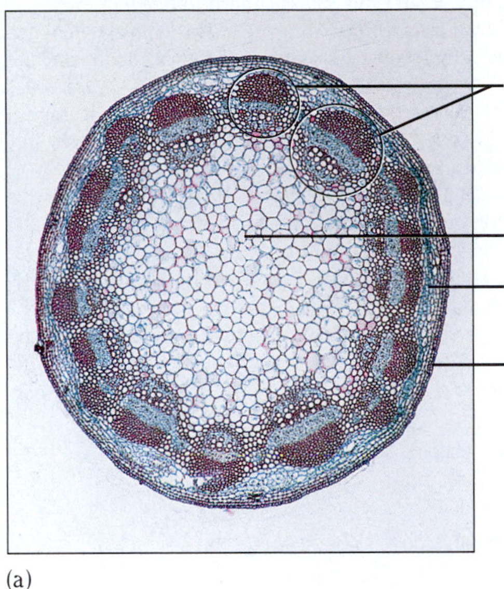

(a)

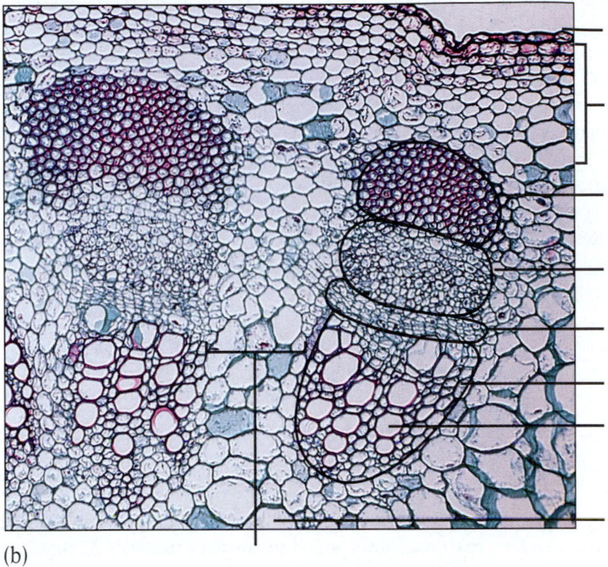

(b)

10. Label the various tissues of this three-year-old basswood stem. Use Figure 33–6 to check your answers.

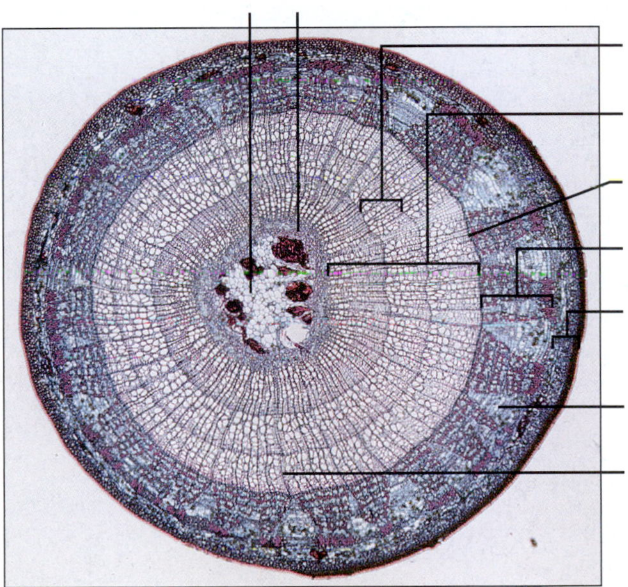

YOU MAKE THE CONNECTION

1. When secondary growth is initiated, certain cells become meristematic and begin to divide. Could a mature tracheid ever do this? A sieve tube member? Why or why not?
2. Why does the wood of many tropical trees lack annual rings? Why does the wood of other tropical trees possess annual rings?
3. Why is hardwood more desirable than softwood for making furniture? Explain your answer based on the structural differences between hardwood and softwood.
4. Why should you cut off a few inches from the stem ends of cut flowers before placing them in water? Base your answer on what you have learned about the tension-cohesion model.

RECOMMENDED READINGS

Berg, L.R. *Introductory Botany: Plants, People, and the Environment.* Saunders College Publishing, Philadelphia, 1997. A general botany text with an environmental emphasis.

Menon, S. "Grimness of Mythic Proportions." *Discover,* Jan. 1999. Tree ring analysis has revealed a serious drought that affected the early history of English and Spanish colonization of North America.

Niklas, K.J. "How to Build a Tree." *Natural History,* Feb. 1996. The ability to make woody tissues has evolved independently many times during the history of plants.

Norris, S. "Reading Between the Lines." *BioScience,* Vol. 50, No. 5, May 2000. This article provides an excellent overview of the science of dendrochronology.

Pearce, F. "Lure of the Rings." *New Scientist,* Vol. 152, 14 Dec. 1996. The analysis of tree rings produced over thousands of years may determine the extent of human influence on global climate.

Raven, P.H., R.F. Evert, and S.E. Eichhorn. *Biology of Plants,* 6th ed. W.H. Freeman & Company, New York, 1999. This is an excellent general botany textbook, with detailed descriptions of stem structure and adaptations.

Renfrew, C. "Kings, Tree Rings, and the Old World." *Nature,* Vol. 381, 27 Jun. 1996. This perspective of a research paper in the same issue explains how tree rings have been used to date archaeological sites in the eastern Mediterranean region, from 2220 to 718 BC.

Zimmer, C. "High and Dry." *Natural History,* Oct. 2000. Examines the biology of tree size, including why trees do not grow taller than they actually do.

Zürcher, E., and M.G. Cantiani. "Tree Stem Diameters Fluctuate with Tide." *Nature,* Vol. 392, 16 Apr. 1998. Scientists have determined that the diameter of tree stems fluctuates depending on the timing and strength of tides; this correlation suggests that the Moon affects the flow of water within trees.

- Visit our Web site at **http://www.info.brookscole.com/solomonbergmartin** for links to chapter-related resources on the World Wide Web. Additional on-line materials relating to this chapter can also be found on our Web site.

See chapter activity on BioActive Learner CD for additional help in mastering the chapter's material. Icon location in the chapter's margins shows which topics have tutorials or simulations in the CD.

34

Roots and Mineral Nutrition

LEARNING OBJECTIVES

After you have studied this chapter you should be able to

1. Describe the functions of roots and how their structure correlates with these functions.
2. Distinguish between the structures of roots and stems.
3. Label cross sections of a primary dicot root and a monocot root and describe the functions of each tissue.
4. Trace the pathway of water from the soil through the various root tissues.
5. Discuss the structure of roots with secondary growth and describe how secondary tissues form.
6. Describe several roots that are modified to perform uncommon functions.
7. List the four components of soil and give the ecological significance of each.
8. Describe how roots absorb positively charged nutrient mineral ions by the process of cation exchange.
9. Describe the factors involved in soil formation.
10. Distinguish between plant macronutrients and micronutrients and give a physiological role for each essential element.

Storage roots. Carrots *(Daucus carota)* are biennials and live for two years. During the first year's growth, food is stored in the fleshy root system; during the second year, the shoot elongates and produces flowers. Carrots and other root crops are important sources of human food. *(R. Calentine/Visuals Unlimited)*

In Chapters 32 and 33 we discussed the aerial vegetative structures of a plant: the leaves and stems. In this chapter we turn to the third major vegetative organ: the roots. Branching underground root systems are often more extensive than a plant's aerial parts. The roots of a corn plant, for example, may grow to a depth of 2.5 m (about 8 ft) and spread outward 1.2 m (4 ft) from the stem. Desert-dwelling tamarisk (*Tamarix* sp.) trees reportedly have roots that grow to a depth of 50 m (163 ft) to tap underground water. The total root length, not counting root hairs, of a four-month-old rye *(Secale cereale)* plant was found to exceed 500 km (310 mi)! The extent of a plant's root depth and spread varies considerably among different species and even among different individuals in the same species. Soil conditions, discussed in this chapter, greatly affect the extent of root growth.

Because roots are usually underground and out of sight, people do not always appreciate the important functions that they perform. First, as anyone who has ever pulled weeds can attest, roots anchor a plant securely in the soil. A plant needs a solid foundation from which to grow. Firm anchorage is also essential to a plant's survival so that the stem remains upright, enabling leaves to absorb sunlight effectively.

Second, roots absorb water and dissolved nutrient mineral salts such as nitrates, phosphates, and sulfates, which are necessary for the synthesis of important organic molecules. These dissolved nutrient minerals are then transported throughout the plant in the xylem.

Storage is the third main function performed by many roots. Carrots *(see photograph),* sweet potatoes, cassava, and other root crops are important sources of human food. Surplus sugars produced in the leaves by photosynthesis are transported in the phloem to the roots for storage (usually as starch or sucrose) until needed. Carrot roots have extensive phloem for this purpose. Although roots use some photosynthetic products for their own respiratory needs, most are stored and later transported out of the roots for use by the plant. Both *taproots* (carrots, beets, radishes, and turnips) and *fibrous roots* (sweet potatoes and yams) may be modified for storage. Plants with storage taproots are often *biennials* (see Chapter 31) that, as part of the strategy to survive winter, store their food reserves in the root during the first year's growth and use these reserves to reproduce during the second year's growth. Other plants, particularly those living in arid regions, possess storage roots adapted to store water.

In certain species, roots are modified for functions other than anchorage, absorption, conduction, and storage. Roots specialized to perform uncommon functions are discussed later in this chapter.

THERE ARE TWO BASIC TYPES OF ROOT SYSTEMS

Two types of root systems, a taproot system and a fibrous root system, may develop (Fig. 34–1). A **taproot** system consists of one main root (formed from the seedling's enlarging *radicle,* or embryonic root) with many lateral roots of various sizes coming out of it. Lateral roots often initially occur in regular rows along the length of the main root. Taproots are characteristic of many dicots and gymnosperms. A dandelion is a good example of a common herbaceous plant with a taproot system. A few trees, hickory, for example, retain their taproots, which become quite massive as the plants age. Most mature trees, however, do not retain their taproots and have root systems that consist of large, shallow lateral roots from which other roots branch off and grow downward.

A **fibrous root** system has several to many roots of the same size developing from the end of the stem, with lateral roots of various sizes branching off these roots. Fibrous root systems form in plants that have a short-lived embryonic root. The roots first originate from the base of the embryonic root and later from stem tissue. Because the main roots of a fibrous root system do not arise from preexisting roots, but rather from the stem, they are said to be **adventitious.** Adventitious organs occur in an unusual location, such as roots that develop on a stem, or buds that develop on roots. Onions, crabgrass, and other monocots have fibrous root systems.

Taproot and fibrous root systems are adapted to obtain water in different sections of the soil. Taproot systems often extend down into the soil to obtain water located deep underground, whereas fibrous root systems, which are located relatively close to the soil surface, are adapted to obtain rainwater from a larger area as it drains into the soil.

Figure 34–1 Root systems. **(a)** A taproot system develops from the embryonic root in the seed. **(b)** The roots of a fibrous root system are adventitious and develop from stem tissue.

ROOTS POSSESS ROOT CAPS AND ROOT HAIRS

Because of the need to adapt to the soil environment instead of the atmospheric environment, roots have several structures, such as root caps and root hairs, that stems lack. Although stems and leaves have various types of hairs, they are distinct from root hairs in structure and function.

Each root tip is covered by a **root cap,** a protective, thimble-like layer many cells thick that covers the delicate root apical meristem (Fig. 34–2*a;* also see Fig. 31–7). As the root grows, pushing its way through the soil, parenchyma cells of the root cap are sloughed off by the frictional resistance of the soil particles and replaced by new cells formed by the root apical meristem. The root cap cells secrete lubricating polysaccharides that reduce friction as the root passes through the soil. The root cap also appears to be involved in orienting the root so that it grows downward (see discussion of gravitropism in Chapter 36). When a root cap is removed, the root apical meristem grows a new cap. However, until the root cap has regenerated, the root grows randomly rather than in the direction of gravity.

Root hairs are short-lived tubular extensions of single epidermal cells located just behind the growing root tip. Root hairs continually form in the area of cell maturation closest to the root tip to replace those that are dying off at the more mature end of the root hair zone (Fig. 34–2*b;* also see Fig. 31–7). Each root hair is short (typically less than 1 cm, or 0.4 in, in length), but they are quite numerous. Root hairs greatly increase the absorptive capacity of roots by increasing their surface area in contact with moist soil. Soil particles are coated with a microscopically thin layer of water in which nutrient minerals are dissolved. The root hairs establish an intimate contact with soil particles, which allows absorption of much of the water and nutrient minerals.

Unlike stems, roots lack nodes and internodes and do not usually produce leaves or buds. Although herbaceous roots have certain primary tissues (such as epidermis, xylem, phloem, and parenchyma of the cortex and pith) found in herbaceous stems, these tissues are arranged quite differently. Table 34–1 summarizes the major differences between roots and stems in herbaceous dicots.

THE ARRANGEMENT OF VASCULAR TISSUES DISTINGUISHES HERBACEOUS DICOT ROOTS AND MONOCOT ROOTS

Although considerable variation exists in herbaceous dicot and monocot roots, they all possess an outer protective covering (epidermis), a cortex for storage of starch and other organic molecules, and vascular tissues for conduction. Let us first consider the structure of herbaceous dicot roots.

In most herbaceous dicot roots, the central core of vascular tissue lacks pith

The buttercup root is a representative dicot root with primary growth (Fig. 34–3). Like other parts of this herbaceous dicot, a single layer of protective tissue, the epidermis, covers its roots. The root hairs are a modification of the root epidermis that enables it to absorb more water from the soil. The root epidermis

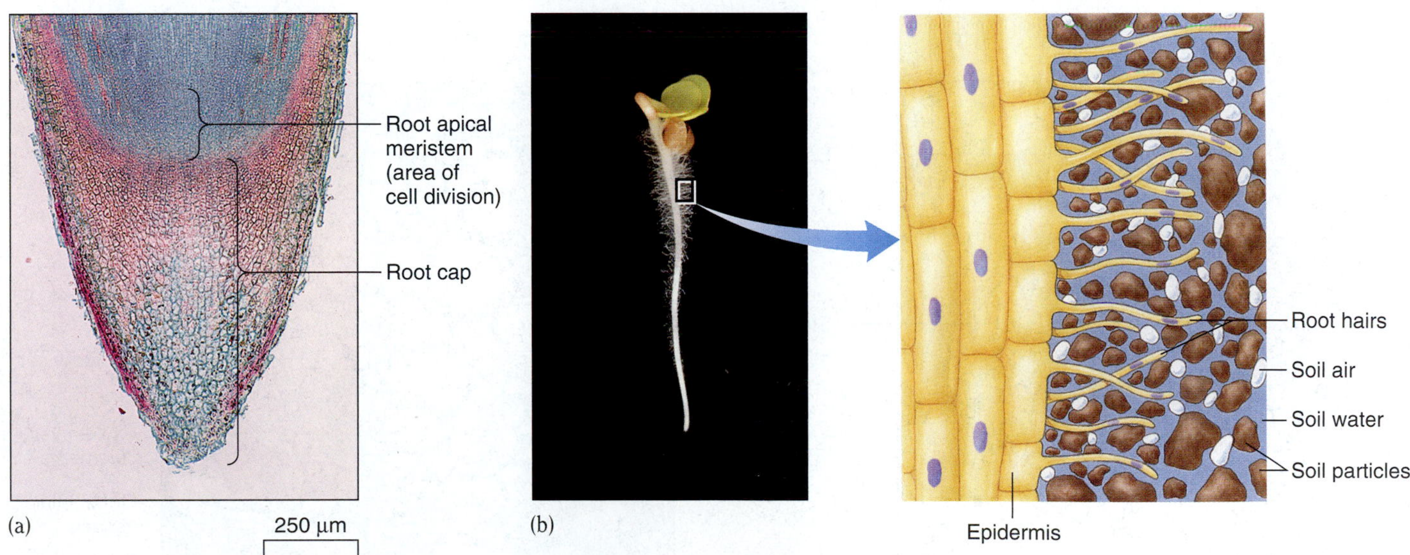

Figure 34–2 Structures unique to roots. **(a)** LM of an oak (*Quercus* sp.) root tip showing its root cap. The root apical meristem is protected by the root cap. **(b)** Root hairs on a radish seedling. Each delicate hair is an extension of a single cell of the root epidermis. Root hairs increase the surface area of the root in contact with the soil. The seedling is approximately 5 cm (2 in) long. *(a, Runk/Rannels/Grant Heilman Photography; b, Dennis Drenner)*

TABLE 34–1 Differences Between Roots and Stems of Herbaceous Dicots*

Roots	Stems
No nodes and internodes	Nodes and internodes
No leaves or buds	Leaves and buds
Nonphotosynthetic	Photosynthetic
No pith	Pith
No cuticle	Cuticle
Root cap	No cap
Root hairs	Trichomes
Pericycle	No pericycle
Endodermis	No endodermis
Lateral roots form internally from the pericycle	Branches form externally from lateral buds

*Some exceptions to these general differences exist.

does not secrete a thick, waxy cuticle in the region of root hairs because this layer would impede the absorption of water from the soil. Both the lack of a cuticle and the presence of root hairs increase absorption. (*Why* water moves from the soil into the root was explained in Chapter 33.)

Most of the water that enters the root moves along the cell walls rather than entering the cells. One of the major components of cell walls is cellulose, which absorbs water like a sponge. An example of the absorptive properties of cellulose is found in cotton balls, which are almost pure cellulose.

The **cortex,** which is primarily composed of loosely packed parenchyma cells, comprises the bulk of a herbaceous dicot root. Roots usually lack supporting collenchyma cells, probably because the soil supports the root, although roots may develop some sclerenchyma (another supporting tissue; see Chapter 31) as they age. The primary function of the root cortex is storage. A microscopic examination of the parenchyma cells that form the cortex often reveals numerous amyloplasts (see Fig. 34–3*b*), which store starch. Starch, an insoluble carbohydrate composed of glucose subunits, is the most common form of stored energy in plants. When used at a later time, these reserves provide energy for such activities as growth following winter or cell replacement following an injury.

The large intercellular (between-cell) spaces, a common feature of the root cortex, provide a pathway for water uptake and allow for aeration of the root. The oxygen that root cells need for aerobic respiration (see Chapter 7) diffuses from air spaces in the soil into the intercellular spaces of the cortex and from there into the cells of the root.

The inner layer of the cortex, the **endodermis,** regulates the movement of nutrient minerals that enter the xylem in the root's interior. Structurally, the endodermis differs from the rest of the cortex. Endodermal cells fit snugly against each other, and each has a special bandlike region, called a **Casparian strip** (Fig. 34–4), on its radial (side) and transverse (upper and lower) walls. If you compare the endodermis to a cylinder constructed of bricks, endodermal cells correspond to the bricks, and the Casparian strips correspond to the mortar between them. Casparian strips contain *suberin,* a fatty material that is waterproof. (Recall from Chapter 33 that suberin is also the waterproof material in cork cell walls.)

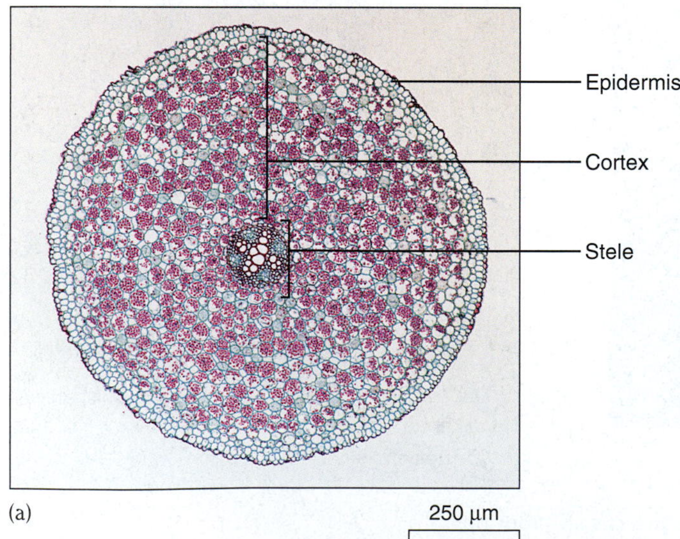

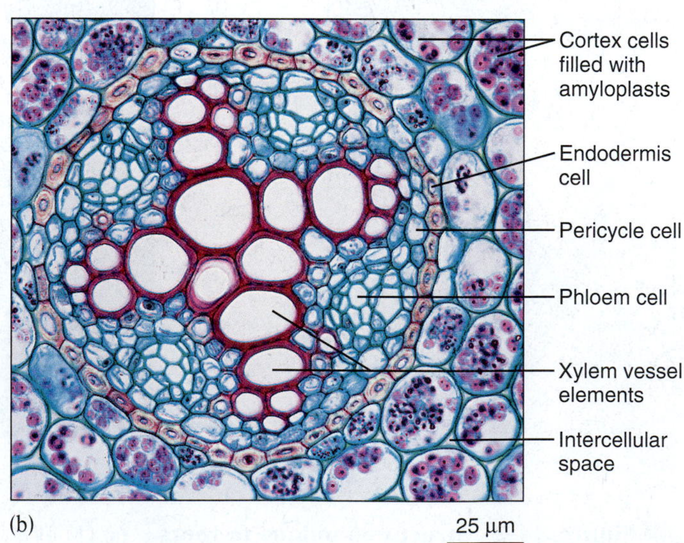

Figure 34–3 LMs of cross sections of a herbaceous dicot root. Shown is a buttercup (*Ranunculus* sp.) root. **(a)** Cortex comprises the bulk of herbaceous dicot roots. Note the X-shaped xylem in the center of the root. **(b)** Close-up of the root's stele. Surrounding the solid core of vascular tissues is the layer of pericycle, which is meristematic in growing roots. *(a, b, Ed Reschke)*

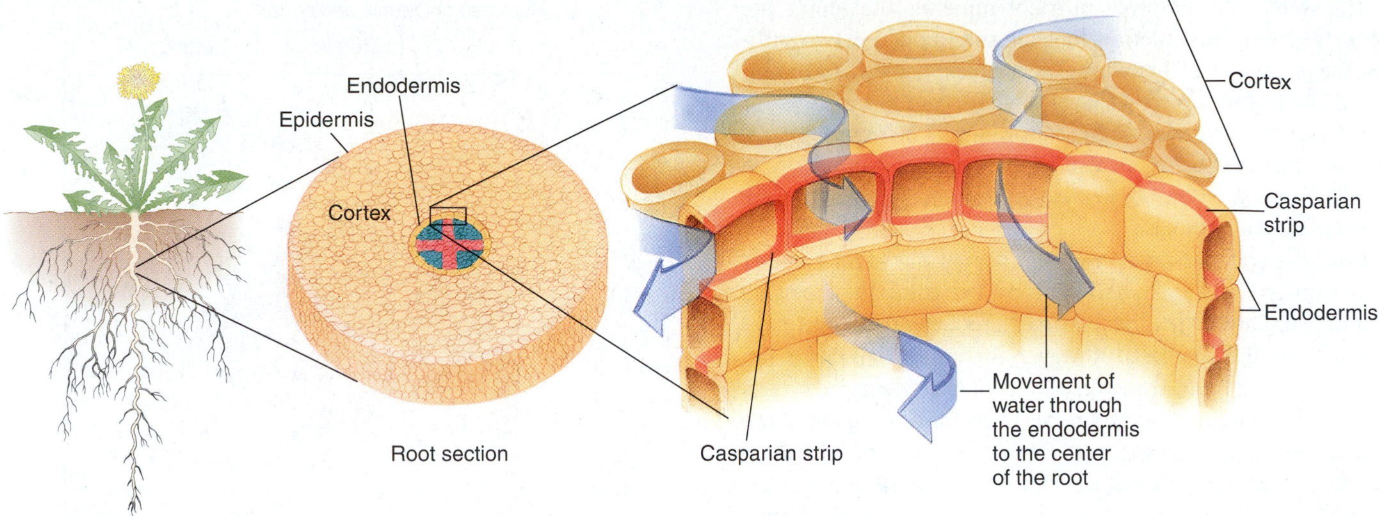

Figure 34–4 Endodermis and nutrient mineral uptake. Note the Casparian strip around the radial and transverse walls that prevents water from passing into the stele along endodermis cell walls.

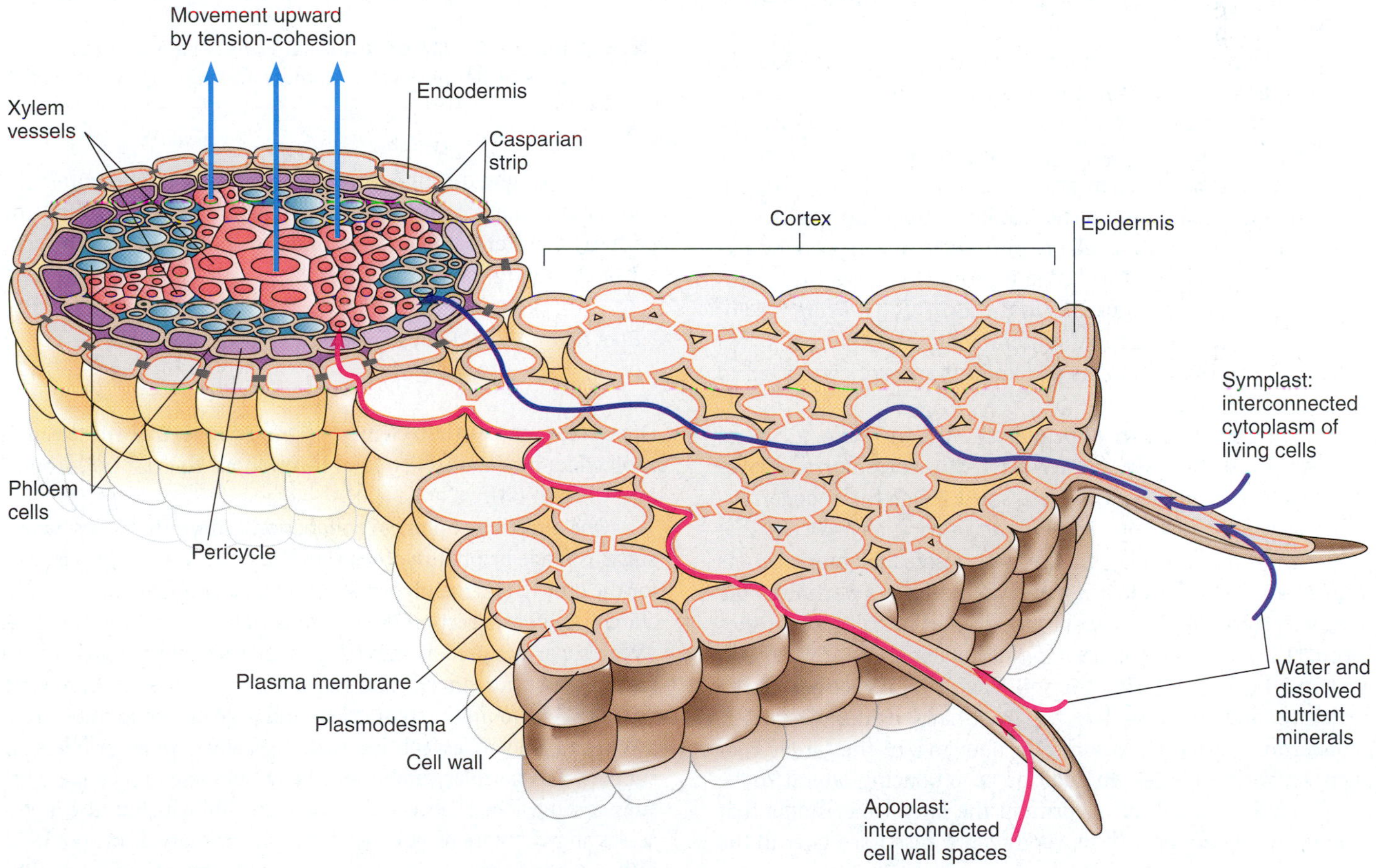

Figure 34–5 Pathways of water and dissolved nutrient minerals in the root. Water and dissolved nutrient minerals travel from cell to cell along the interconnected porous cell walls (the apoplast) or from one cell's cytoplasm to another through plasmodesmata (the symplast). On reaching the endodermis, water and nutrient minerals can only continue to move into the root's center if they pass through a plasma membrane and enter the cytoplasm of an endodermal cell. The Casparian strip blocks the passage of water and nutrient minerals along the cell walls between adjoining endodermal cells.

The water and dissolved nutrient minerals that enter the root cortex from the epidermis move in solution along two pathways: the apoplast and symplast. The **apoplast** consists of the interconnected porous cell walls of a plant, along which water moves freely. The **symplast** is the continuum of living cytoplasm, which is connected from one cell to the next by plasmodesmata (Fig. 34–5 on page 733). Some dissolved minerals move from the epidermis via the symplast.

Until the endodermis is reached, most of the water and dissolved nutrient minerals have not passed through a plasma membrane or entered the cytoplasm of a root cell (i.e., most have traveled along the apoplast). However, the waterproof Casparian strip on the radial and transverse walls of the endodermal cells prevents water and nutrient minerals from continuing along the cell walls. For substances to pass further into the root interior, they must move from the cell walls into the cytoplasm of the endodermal cells. Water enters by osmosis, whereas nutrient minerals enter the endodermal cells by passing through carrier proteins in their plasma membranes. Thus, the endodermis is responsible for controlling the movement of nutrient minerals into the root, even though it is an internal cell layer, and nutrient minerals must pass through the epidermis and cortex to reach it.

Dissolved nutrient mineral ions are actively transported through carrier proteins in the plasma membranes of endodermal cells (see Chapter 5). In active transport, the nutrient mineral ions move *against* their concentration gradient—that is, from an area of *low* concentration of that mineral in the soil solution to an area of *high* concentration in the plant's cells. One of many reasons that root cells require sugar and oxygen for aerobic respiration is that this active transport requires the expenditure of cellular energy, usually in the form of ATP. From the endodermis, water and nutrient mineral ions enter the root xylem (precisely how this is done is not known) and are conducted to the rest of the plant.

At the center of a dicot primary root is the **stele,** a central cylinder of vascular tissues (see Fig. 34–3*b*). The outermost layer of the stele is the **pericycle,** which is just inside the endodermis. The pericycle consists of a single layer of parenchyma cells that gives rise to multicellular lateral roots, also called branch roots (Fig. 34–6). Lateral roots originate when cells in a portion of the pericycle start dividing. As it grows, the lateral root pushes through several layers of root tissue (endodermis, cortex, and epidermis) before entering the soil. Each lateral root has all the structures and features—root cap, root hairs, epidermis, cortex, endodermis, pericycle, xylem, and phloem—of the larger root from which it emerges. In addition to producing lateral roots, the pericycle is involved in forming the lateral meristems that produce secondary growth in woody roots (discussed later in the chapter).

Xylem, the center-most tissue of the stele, often has two, three, four, or more extensions, or "xylem arms." **Phloem** is located in patches between the xylem arms. The xylem and phloem of the root have the same functions as in the rest of the plant: Water and dissolved nutrient minerals are conducted in xylem, and dissolved sugar (sucrose) is conducted in phloem.

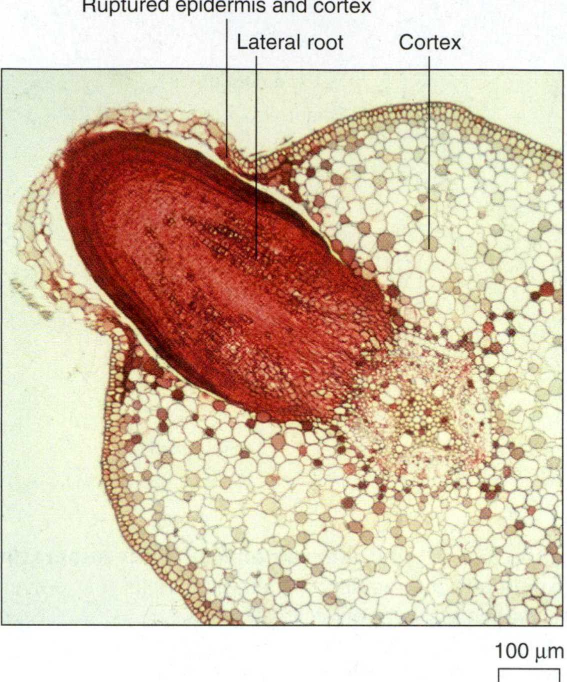

Figure 34–6 LM of a lateral root. Lateral roots originate at the pericycle. *(James Mauseth, University of Texas)*

After passing through the endodermal cells, water enters the root xylem, often at one of the xylem arms. Up to this point the pathway of water has been horizontal from the soil into the center of the root:

Root hair ⟶ epidermis ⟶ cortex ⟶ endodermis ⟶ pericycle ⟶ root xylem

Once water enters the xylem, it is transported upward through root xylem into stem xylem and throughout the rest of the plant (see Chapter 33).

One direction of phloem conduction is from the leaves, where sugar is made by photosynthesis, to the root, where sugar is used for the growth and maintenance of root tissues or stored, usually as starch. Another direction of phloem conduction is from the root, where sugar is stored as starch, to other parts of the plant, where sugar is used for growth and maintenance of tissues. The same kinds of cells found in stem xylem and phloem are found in root xylem and phloem, except that roots typically have fewer fibers for support. The **vascular cambium,** which gives rise to secondary tissues, is sandwiched between the xylem and phloem. Because it possesses an inner core of vascular tissue, the primary dicot root lacks **pith,** a ground tissue found in the centers of many stems and roots.

Xylem does not form the central tissue in some monocot roots

Monocot roots exhibit considerable variation in internal structure when compared with dicot roots. Starting at the outside of

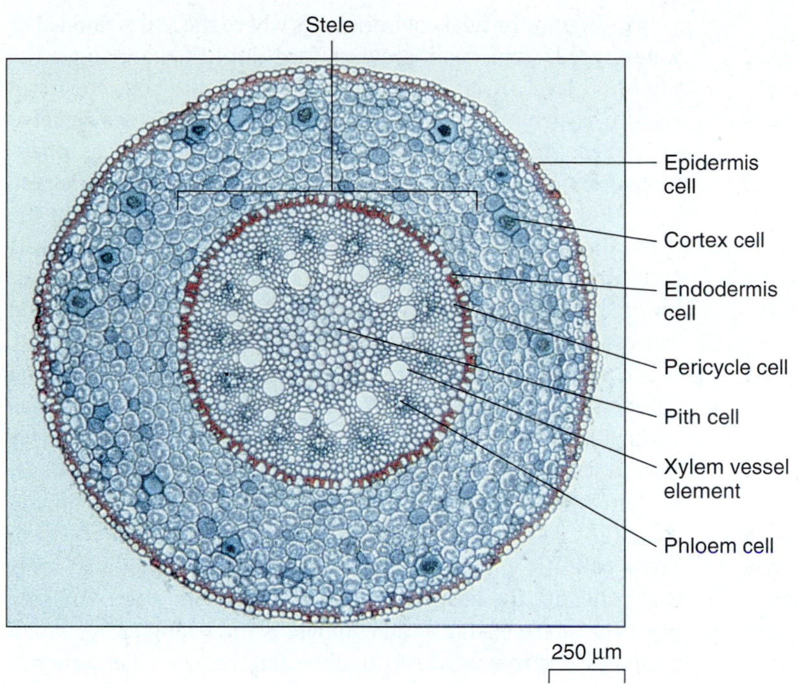

Figure 34–7 LM of a cross section of a monocot root. Shown is a greenbriar *(Smilax)* root. As in herbaceous dicot roots, the cortex of a monocot root is extensive. *(Dennis Drenner)*

some monocot roots, there is epidermis, then cortex, endodermis, and pericycle (Fig. 34–7). Unlike herbaceous dicot roots, the xylem in a monocot root does not form a solid cylinder in the center. Instead, the phloem and xylem are located in separate alternating bundles arranged around the central pith, which is composed of parenchyma cells.

Because virtually all monocots do not have secondary growth, no vascular cambium exists in monocot roots. Despite their lack of secondary growth, long-lived monocots, such as palms, may have thickened roots produced by a modified form of primary growth in which parenchyma cells in the cortex divide and enlarge.

WOODY PLANTS HAVE ROOTS WITH SECONDARY GROWTH

Plants that produce stems with secondary growth also produce roots with secondary growth. Recall from Chapter 33 that these plants, gymnosperms and woody dicots, have primary growth at apical meristems, while secondary growth occurs at lateral meristems. The production of secondary tissues occurs some distance back from the root tips and is the result of the activity of the same two lateral meristems found in woody stems: the vascular cambium and the cork cambium. Major roots of trees are often massive and possess both wood and bark. In temperate climates, the wood of both roots and stems exhibits annual rings in cross section.

Before secondary growth starts in a root, the vascular cambium is sandwiched between the primary xylem and the primary phloem (Fig. 34–8*a*). At the onset of secondary growth, the vascular cambium extends out to the pericycle, which develops into vascular cambium opposite the xylem arms. As a result, the pericycle links the separate sections of vascular cambium so that the vascular cambium becomes a continuous, noncircular loop of cells in cross section (Fig. 34–8*b*). As the vascular cambium divides to produce secondary tissues, it eventually forms a cylinder of vascular cambium that continues to divide, producing secondary xylem (wood) to the inside and secondary phloem (inner bark) to the outside (Fig. 34–8*c,d*). The root increases in girth (thickness), and the vascular cambium continues to move outward.

The epidermis, cortex, endodermis, and primary phloem are gradually torn apart as the root increases in girth. The root epidermis is replaced by periderm, composed of cork cells and cork parenchyma, both produced by the cork cambium (the last figure in Table 31–4 shows a LM of periderm). The cork cambium in the root initially arises from regions in the pericycle.

SOME ROOTS ARE SPECIALIZED FOR UNUSUAL FUNCTIONS

Adventitious roots often arise from the nodes of stems. Many aerial adventitious roots are adapted for functions other than anchorage, absorption, conduction, or storage. **Prop roots** are adventitious roots that develop from branches or a vertical stem and that grow downward into the soil to help support the plant in an upright position (Fig. 34–9*a*). Prop roots are more common in monocots than in dicots. Corn and sorghum, both monocots, are herbaceous plants that produce prop roots. Many tropical and subtropical dicot trees, such as red mangrove, banyan, and screw pine (*Pandanus* sp.), also produce prop roots.

The roots of many tropical rainforest trees are shallow and concentrated near the surface in a mat only a few centimeters (an inch or so) thick. The root mat catches and absorbs almost

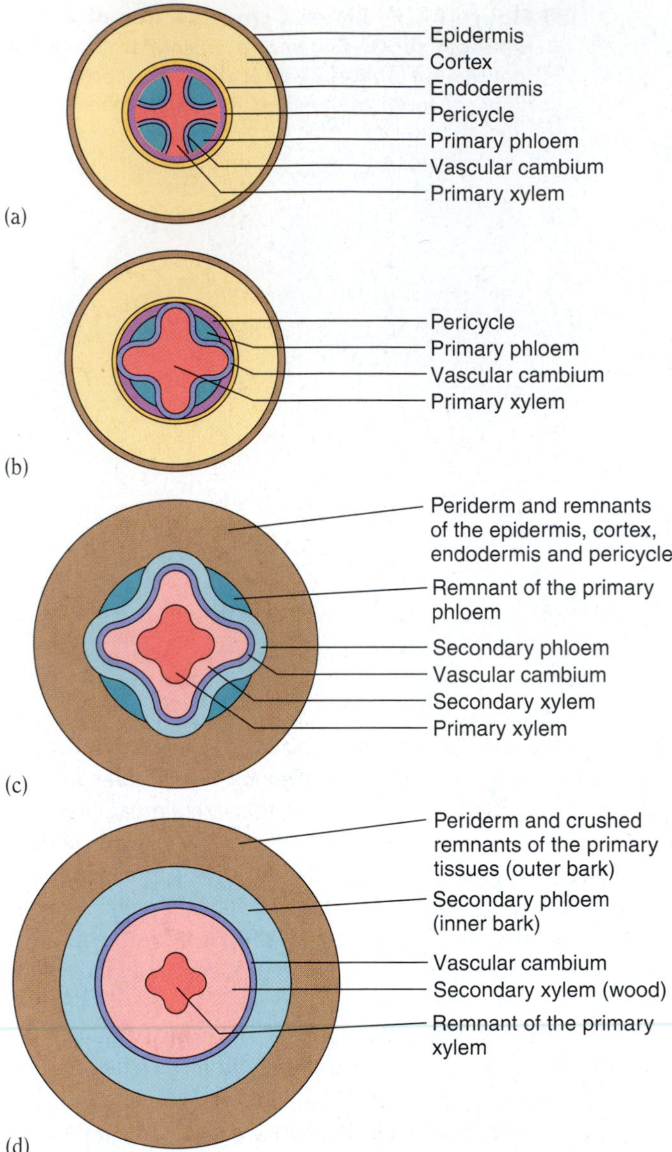

Figure 34–8 Development of secondary vascular tissues in a primary root. **(a)** The tissues in a primary root. **(b)** At the onset of secondary growth, the vascular cambium extends out to the pericycle, forming a continuous, noncircular loop. **(c)** The vascular cambium produces secondary xylem to its inside and secondary phloem to its outside. Note that the primary phloem is pushed outward. **(d)** Over time, the ring of vascular cambium gradually becomes circular. As the vascular cambium continues to divide, the epidermis, cortex, and primary phloem located in the outer bark are torn apart. (The figures are not drawn to scale because of space limitations; the primary xylem is actually the same size in all four diagrams, but differences in scale make the xylem different sizes.)

all nutrient minerals released from leaves by decomposition. Swollen bases or braces called **buttress roots** hold the trees upright and aid in the extensive distribution of the shallow roots (Fig. 34–9*b*).

In swampy or tidal environments where the soil is flooded or waterlogged, some roots grow upward until they are above the high-tide level. Even though roots live in the soil, they still require oxygen for aerobic respiration. Flooded soils are depleted of oxygen, so these aerial "breathing" roots, known as **pneumatophores,** may assist in getting oxygen to the submerged roots (Fig. 34–9*c*). Pneumatophores, which also help anchor the plant, have a well-developed system of internal air spaces that is continuous with the submerged parts of the root, presumably allowing gas exchange. Black mangrove, white mangrove, and bald cypress are examples of plants with pneumatophores.

Epiphytes (plants that grow attached to other plants) and climbing plants have aerial roots that anchor the plant to the bark, branch, or other surface on which it grows. Some epiphytes have aerial roots specialized for functions other than anchorage. Certain epiphytic orchids, for example, have photosynthetic roots (Fig. 34–9*d*). Epiphytic roots may absorb moisture as well. Some parasitic epiphytes such as mistletoe have modified roots that penetrate the host plant tissues and absorb water. Another plant that starts its life as an epiphyte is the strangler fig, which produces long roots that eventually reach the ground and anchor the plant (now a tree rather than an epiphyte) in the soil. The tree that the strangler fig originally grew on is often killed as the strangler fig grows around it, competing with it for light and other resources and crushing its secondary phloem.

Plants that produce corms or bulbs (underground stems or buds specialized for asexual reproduction; see Chapter 35) often have wiry **contractile roots** in addition to their "normal" roots. The contractile roots grow into the soil and then contract (the cortical cells shorten or totally collapse), thus pulling the corm or bulb deeper into the soil (Fig. 34–9*e*). Contractile roots are necessary for corms because each succeeding year's growth is *on top of* the preceding year's growth. As a result, corms tend to move upward in the soil over time. Without contractile roots they would eventually be exposed at the soil's surface. Contractile roots are more common in monocots, but certain dicots and ferns also possess them.

ROOTS FORM RELATIONSHIPS WITH OTHER SPECIES

As roots of certain trees grow through the soil, they sometimes encounter roots of other trees of the same or different species. When this occurs, they may grow together by secondary growth to form a natural **graft.** Because their vascular tissues are connected in the graft, dissolved sugars and other materials such as hormones pass between the two trees; disease organisms can also be transmitted in this way. Root grafts have been observed in more than 160 tree species.

The roots of most plant species form *mutualistic* (mutually beneficial) relationships with certain soil fungi (see Chapters 25 and 52). These subterranean associations, known as **mycorrhizae,** permit the transfer of materials (such as sugars) from roots to fungus. At the same time, essential nutrient minerals such as phosphorus move from fungus to roots. The threadlike

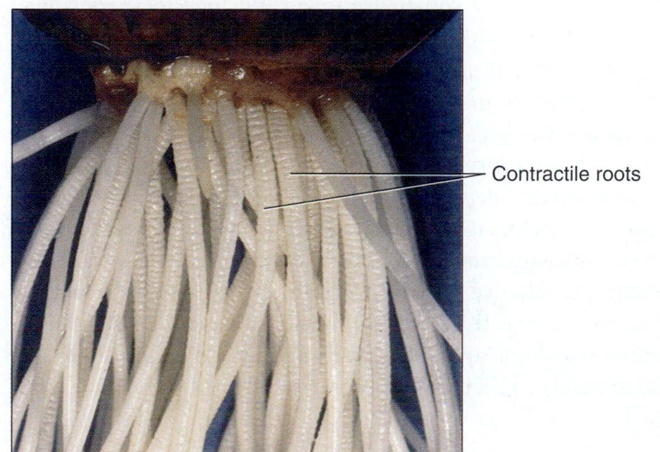

Figure 34–9 Specialized roots. **(a)** Prop roots are adventitious roots that arise near the base of the stem and provide additional support. The screw pine (*Pandanus* sp.) has an elaborate set of aerial prop roots. Photographed in Kauai, Hawaii. **(b)** Tropical rain-forest trees typically possess buttress roots that support them in the shallow, often wet soil. Shown are buttress roots on Australian banyan *(Ficus macrophylla)*. Photographed at Selby Gardens in Sarasota, Florida. **(c)** White mangrove *(Laguncularia racemosa)* produces pneumatophores, shown protruding from the wet mud in the foreground. Pneumatophores may provide oxygen for roots buried in anaerobic (oxygen-deficient) soil. Photographed in Isla del Carmen, Mexico. **(d)** The moth orchid (*Phalaenopsis* hybrid) has photosynthetic aerial roots. **(e)** Plants that produce corms or bulbs often have contractile roots. During successive seasons, contractile roots pull the corm or bulb deeper into the soil. *(a, Dr. Linda R. Berg; b and d, John Arnaldi; c, Robert and Linda Mitchell; e, Courtesy of Judith Jernstedt, University of California, Davis)*

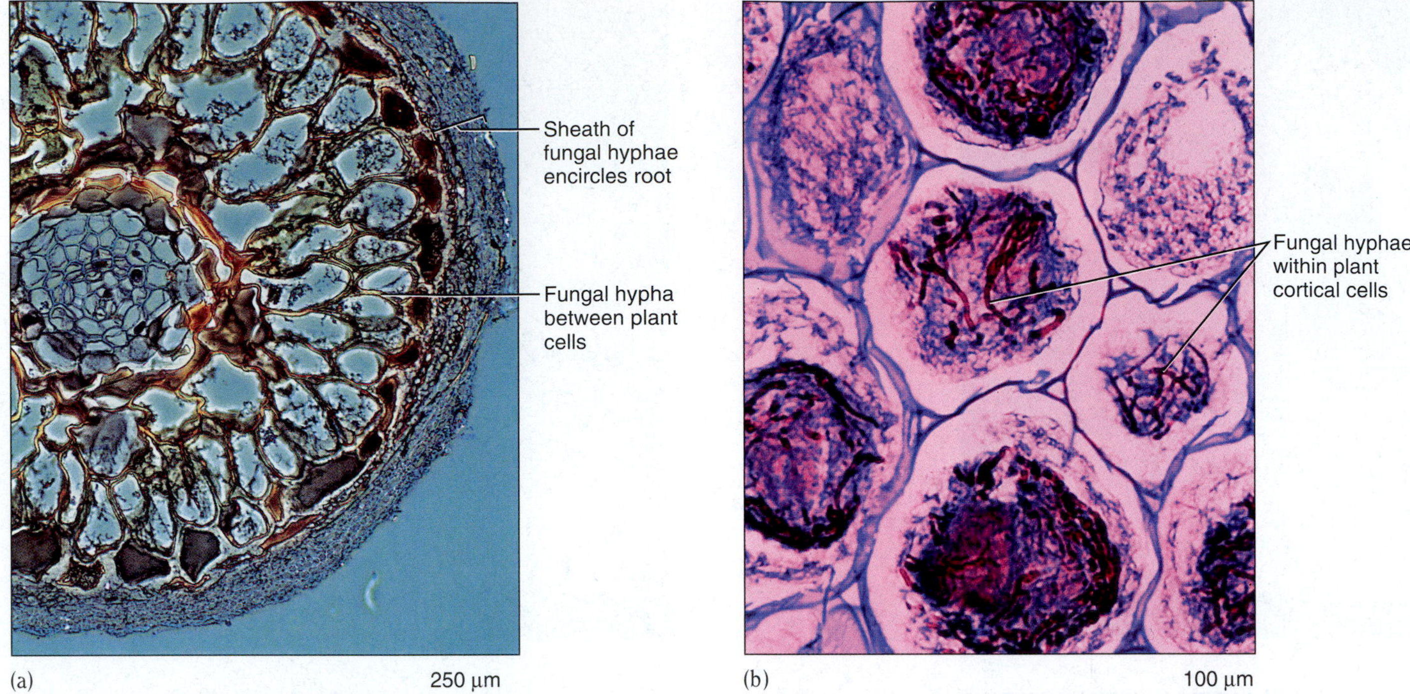

Figure 34–10 Mycorrhizae. Mycorrhizae enhance plant growth by providing soil nutrients to the roots. **(a)** LM of ectomycorrhizae, fungal associations that form a sheath around the root. The fungal hyphae penetrate the root between cortical cells but do not enter the cells. **(b)** LM of endomycorrhizae, fungal associations in which the fungal hyphae penetrate root cortical cells and form branched haustoria (absorbing organs) within the cells to aid in delivering and receiving nutrients. Roots of the majority of vascular plant species are colonized by endomycorrhizae. *(a, Robert Knauft/Biology Media/Photo Researchers, Inc.; b, Cabisco/Visuals Unlimited)*

body of the fungal partner extends into the soil, extracting nutrient minerals well beyond the reach of the plant's roots. In some mycorrhizae, the fungal mycelium encircles the root like a sheath, whereas in others, the fungus penetrates root cells (Fig. 34–10). The relationship is considered mutually beneficial because when mycorrhizae are not present, neither the fungus nor the plant grows as well (see Fig. 25–15). Recent evidence indicates that the hyphal network of mycorrhizae simultaneously interconnects different plants in the community (see *On the Cutting Edge: Mycorrhizae and Carbon Exchange Between Plant Species in a Forest Ecosystem*).

Certain nitrogen-fixing bacteria, collectively called *rhizobia,* form associations with the roots of leguminous plants—clover, peas, and soybeans, for example. **Nodules** (swellings) that house millions of the rhizobia develop on the roots (see Fig. 53–8*a*). Like mycorrhizae, the association between nitrogen-fixing bacteria and the roots of plants is mutually beneficial. Bacteria receive the photosynthetic products from plants while helping them to meet their nitrogen requirements by producing excess ammonia (NH_3) from atmospheric nitrogen.

Biologists have studied the molecular basis of associations between plants and rhizobia for several decades. The initial contact between the two partners in this association involves **cell signaling,** that is, an exchange of signal molecules. Some of the signal molecules produced by the bacteria are nodulation factors that cause the roots of leguminous plants to produce nodules.

SOIL ANCHORS PLANTS AND PROVIDES WATER AND NUTRIENT MINERALS

We now examine the soil environment in which most roots live. Soil is a relatively thin layer of Earth's crust that has been modified by the natural actions of weather, wind, water, and organisms. It is easy to take soil for granted. We walk on and over it throughout our lives but rarely stop to think about how important it is to our survival. Vast numbers and kinds of organisms colonize soil and depend on it for shelter and food. Most plants anchor themselves in soil, and from it they receive water and essential nutrient minerals. Most elements essential for plant growth are obtained directly from the soil (these elements are discussed later in this chapter). Most plants cannot survive on their own without soil, and because we depend on plants for our food, humans could not exist without soil either.

Process of Science

ON THE CUTTING EDGE

Mycorrhizae and Carbon Exchange Between Plant Species in a Forest Ecosystem

HYPOTHESIS: The exchange of carbon compounds between tree species connected by mycorrhizal fungi can be measured and is influenced by field conditions such as relative light intensity.

METHOD: Isotope tracers were used to measure the transfer of photosynthetically produced carbon compounds between same-age trees of different species (Douglas fir and paper birch) connected by mycorrhizal fungi. The net transfer between healthy trees exposed to full sunlight and less healthy trees growing in deep shade was compared.

RESULTS: Carbon transfer between Douglas fir and paper birch is bidirectional, but net transfer is always from paper birch to Douglas fir. Transfer of carbon compounds is greatest when paper birches are grown in full sun and Douglas firs are shaded.

CONCLUSION: Douglas firs receive a net gain of carbon compounds from paper birches by way of mycorrhizal fungi connections between the two trees. The magnitude of carbon transfer is affected by relative light intensity.

Biologists have studied mycorrhizae for many years, but most have used a reductionist approach (see Chapter 1) in which the fungus and plant are isolated from the natural environment and grown together in a pot. Many important aspects of the fungus-plant relationship have been determined using this approach, but other questions about the ecological significance of mycorrhizae in natural communities remained unanswered.

In 1997 Suzanne Simard of the British Columbia Ministry of Forests and her colleagues from Oregon State University, Okanagan University College in British Columbia, and the U. S. Department of Agriculture reported one of the first field studies involving mycorrhizae in a forest community.* This experiment demonstrated that in natural ecosystems the subterranean relationship between plants and fungi is far more complex than originally supposed. Simard and colleagues studied ectomycorrhizae, whose fungal hyphae ensheathe the root and then extend into the soil. These fungi are not host-specific; many ectomycorrhizal fungus species can colonize a single plant species, and each fungus species can form mycorrhizal relationships with many plant species. For example, as many as ten fungal species form mycorrhizal relationships with both paper birch *(Betula papyrifera)* and Douglas fir *(Pseudotsuga menziesii)*. The mycelia of these fungi often connect the different plant species, and materials such as phosphorus, nitrate, and carbon compounds pass from one plant to another by way of the mycorrhizal fungus.

The work of Simard and colleagues is unique because they were able to quantify the amount of exchange between two plant species. They exposed young (two-year-old seedlings in one experiment and three-year-old seedlings in another) birch and Douglas fir trees growing close together in the forest to *reciprocal isotope labeling.* One plant received ^{14}C-labeled CO_2 and the other received ^{13}C-labeled CO_2. The appropriate isotope was injected into a sealed bag that enclosed the aerial parts of each plant. After a short time to allow for the accumulation of carbon compounds produced by photosynthesis, the plants were harvested. It was hypothesized that if carbon compounds were not being transferred from one plant to the other by their fungal connection, then each plant would contain carbon compounds labeled with either ^{14}C or ^{13}C, but not both. Each plant was evaluated for the presence of ^{14}C- and ^{13}C-labeled carbon-containing molecules, and each contained both isotopes. These data indicate that the fungal connections linking the two plant species had transferred materials between them in both directions.

Although the movement of carbon compounds was bidirectional, Douglas fir received more carbon compounds from paper birch than paper birch received from Douglas fir. Perhaps even more interesting, the *net gain* of carbon compounds from paper birch to Douglas fir was influenced by relative light intensity. When Douglas firs were grown in deep shade by covering them with tents four weeks before the labeling experiment, the amount of carbon transferred from the paper birch to the shaded Douglas fir increased dramatically. On average, the paper birches provided 6% of their total photosynthetic production of carbon compounds to Douglas firs growing in deep shade.

Why does a shaded Douglas fir receive more carbon compounds from paper birch than does a Douglas fir grown in full sunlight? Simard and colleagues suggest that having mycorrhizal relationships with two strong, vigorously growing trees benefits the fungus more than such connections with a strong tree and a weaker one. (The weaker tree is the one growing in shade because it cannot produce enough energy-storing carbon compounds to grow vigorously.) Thus, the fungus may control the net transfer of carbon compounds between plants to optimize the overall health of their mutualistic partners. An alternative explanation is that there may be a pressure-flow (see Chapter 33) type of transfer from source (paper birch) to sink (Douglas fir). This transfer would be caused by a physical mechanism and would not be regulated by the mycorrhizal fungus. Additional research will have to be done to determine the cause of the differential transfer of carbon compounds.

The results of Simard and colleagues have raised several ecological questions. To what extent do mycorrhizal relationships promote forest diversity? Do mycorrhizae influence the diversity of plant communities other than forests? How do Simard and colleagues's results affect our understanding of the importance of competition among plant species? (Chapter 52 examines the role of competition, including whether competition is the main ecological determinant of community structure.)

* Simard, S. W., D. A. Perry, M. D. Jones, D. D. Myrold, D. M. Durall, and R. Molina. "Net Transfer of Carbon Between Ectomycorrhizal Tree Species in the Field." *Nature,* Vol. 388, 7 Aug. 1997.

Most soils are formed from rock (called parent material) that is gradually broken down, or fragmented, into smaller and smaller particles by biological, chemical, and physical **weathering processes.** Two important factors that work together in the weathering of rock are climate and organisms. When plant roots and other organisms living in the soil respire, they produce carbon dioxide (CO_2), which diffuses into the soil and reacts with soil water to form carbonic acid (H_2CO_3). Soil organisms such as lichens[1] also produce other kinds of acids. These acids etch tiny cracks, or fissures, in the rock surface; water then seeps into these cracks. If the parent material is located in a temperate climate, the alternate freezing and thawing of the water during winter cause the cracks to enlarge, breaking off small pieces of rock. Small plants can then become established and send their roots into the larger cracks, fracturing the rock further.

Topography, a region's surface features, such as the presence or absence of mountains and valleys, is also involved in soil formation. Steep slopes often have little or no soil on them because the soil and rock are continually transported down the slopes by gravity; runoff from precipitation tends to amplify erosion on steep slopes. Moderate slopes, on the other hand, may encourage the formation of deep soils.

The disintegration of solid rock into finer and finer mineral particles and the accumulation of organic material (discussed in the next section) in the soil take an extremely long time, sometimes thousands of years. Soil forms continually as the weathering of parent material beneath already-formed soil continues to add new soil. Soil thickness varies from a thin film to more than 3 m (10 ft) deep.

Soil is composed of inorganic minerals, organic matter, air, and water

Four distinct components comprise soil—inorganic mineral particles (which make up about 45% of a "typical soil"), organic matter (about 5%), water (about 25%), and air (about 25%). Soil occurs in layers, each of which has a certain composition and special properties. The plants, animals, fungi, and microorganisms that inhabit soil interact with it, and nutrient minerals are continually cycled from the soil to organisms, which use them in their biological processes. When the organisms die, bacteria and other soil organisms decompose the remains, returning the nutrient minerals to the soil.

The inorganic mineral particles, which come from weathered rock, constitute most of what we call soil. It provides anchorage and essential nutrient minerals for plants, as well as pore space for water and air. Because different rocks are composed of different minerals, soils vary in mineral composition and chemical properties. Rocks rich in aluminum form acidic soils, for example, whereas rocks that contain silicates of magnesium and iron form soils that may be deficient in calcium, nitrogen, and phosphorus. Also, soils formed from the same kind of parent material may not develop in the same way because other factors, such as weather, topography, and organisms, differ.

The texture, or structural characteristic, of a soil is determined by the percentages (by weight) of the different-sized inorganic mineral particles—sand, silt, and clay—in it. The size assignments for sand, silt, and clay give soil scientists a way to classify soil texture. Particles larger than 2 mm in diameter, called gravel or stones, are not considered soil particles because they do not have any direct value to plants. The largest soil particles are called *sand* (0.02 to 2 mm in diameter), the medium-sized particles are called *silt* (0.002 to 0.02 mm in diameter), and the smallest particles are called *clay* (less than 0.002 mm in diameter). Sand particles are large enough to be seen easily with the eye, and silt particles (about the size of flour particles) are barely visible with the eye. Most individual clay particles are too small to be seen with an ordinary light microscope; they can be seen only under an electron microscope.

A soil's texture affects many of that soil's properties, in turn influencing plant growth. The clay component of a soil is particularly important in determining many of its characteristics because clay particles have the greatest surface area of all soil particles. If the surface areas of about 450 g (1 lb) of clay particles were laid out side by side, they would occupy 1 hectare (2.5 acres).

Each clay particle has predominantly negative electrical charges on its outer surface that attract and reversibly bind cations (positively charged mineral ions). Clay particles retain many cations, such as potassium (K^+) and magnesium (Mg^{2+}), that are essential for plant growth. Roots secrete protons (H^+), which are exchanged for other positively charged mineral ions in a process known as **cation exchange.** These "freed" ions and the water that forms a film around the soil particles are absorbed by the plant's roots (Fig. 34–11). In contrast, anions (negatively

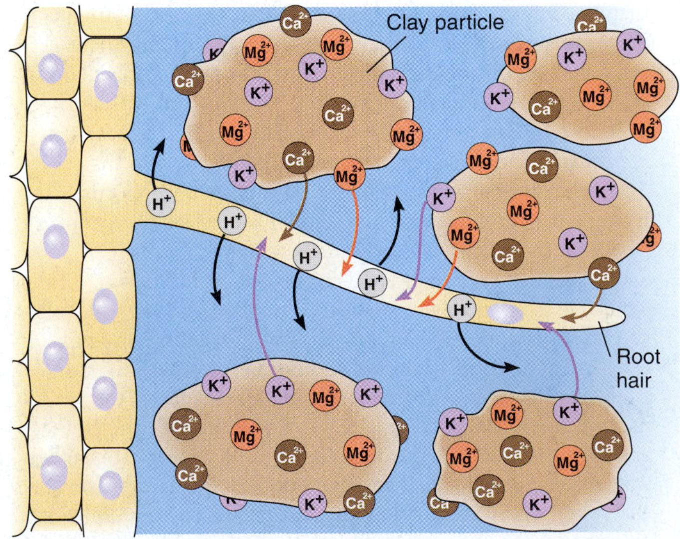

Figure 34–11 Cation exchange. Negatively charged clay particles bind to positively charged nutrient mineral cations, holding them in the soil. Roots pump out protons (H^+), which are exchanged for the cations, facilitating their absorption.

[1]A lichen is a dual organism composed of a fungus and a phototroph (photosynthetic organism). See Chapter 25 for further discussion of lichens.

charged mineral ions) are usually not held as tightly in the soil and are often washed out of the root zone.

Soil always contains a mixture of different-sized particles, but the proportions vary from one soil to another. A *loam,* which is an ideal agricultural soil, has an optimum combination of different soil particle sizes: It contains approximately 40% each of sand and silt and about 20% of clay. Generally, the larger particles provide structural support, aeration, and permeability to the soil, whereas the smaller particles bind together into aggregates, or clumps, and hold nutrient minerals and water. Soils with larger proportions of sand are not as desirable for most plants because they do not hold water and nutrient mineral ions well. Plants grown in such soils are more susceptible to drought and nutrient mineral deficiencies. Soils with larger proportions of clay are also not desirable for most plants because they provide poor drainage and often do not provide enough oxygen. Clay soils used in agriculture tend to get compacted, which reduces the number of soil spaces that can be filled by water and air.

A soil's organic matter consists of the wastes and remains of soil organisms

The organic matter in soil is composed of litter (dead leaves and branches on the soil's surface); droppings (animal dung); and the dead remains of plants, animals, and microorganisms in various stages of decomposition. Organic matter is decomposed by microorganisms, particularly bacteria and fungi, that inhabit the soil. During decomposition, essential nutrient mineral ions are released into the soil, where they may be bound by soil particles or absorbed by plant roots. Organic matter increases the soil's water-holding capacity by acting much like a sponge. For this reason gardeners often add organic matter to soils, especially sandy soils, which are naturally low in organic matter.

The partly decayed organic portion of the soil is referred to as **humus** (Fig. 34–12). Humus, which is not a single chemical compound but a mix of many organic compounds, binds nutrient mineral ions and holds water. On average, humus persists in agricultural soil for about 20 years. Certain components of humus may persist in the soil for hundreds of years. Although humus is somewhat resistant to decay, a succession of microorganisms gradually reduces it to carbon dioxide, water, and nutrient minerals.

About 50% of soil volume is composed of pore spaces

Soil has numerous pore spaces of different sizes around and among the soil particles. Pore spaces occupy roughly 50% of a soil's volume and are filled with varying proportions of air and water (Fig. 34–13). Both soil air and soil water are necessary to produce a moist but aerated soil that sustains plants and other soil-dwelling organisms. Water is usually held in the smaller pores, whereas air is found in the larger pores. After a prolonged rain, almost all of the pore spaces may be filled with water, but water drains rapidly from the larger pore spaces, drawing air from the atmosphere into those spaces.

Figure 34–12 Humus. Humus is brownish-black and consists of partially decomposed organic material that comes primarily from plant and animal remains. Soil rich in humus has a loose, somewhat spongy structure with several properties, such as increased water-holding capacity, that are beneficial for plants and other organisms living in it. *(USDA/Natural Resources Conservation Service)*

Soil air contains the same gases as atmospheric air, although they are usually present in different proportions. As a result of respiration by soil organisms, there is less oxygen and more carbon dioxide in soil air than in atmospheric air. (Recall from Chapter 7 that aerobic respiration uses oxygen and produces carbon dioxide.) Among the important gases in soil are oxygen (O_2), required by soil organisms for aerobic respiration; nitrogen (N_2), used by nitrogen-fixing bacteria; and carbon dioxide (CO_2), involved in soil weathering. As mentioned earlier in the chapter, carbon dioxide, a product of aerobic respiration, dissolves in water to form carbonic acid, a weak acid that accelerates the weathering process during soil formation.

Soil water originates as precipitation, which drains downward, or as groundwater (water stored in porous underground rock), which rises upward from the water table (the uppermost level of groundwater). Soil water contains low concentrations of dissolved nutrient minerals that enter the roots of plants when they absorb water. Water not bound to soil particles or absorbed by roots percolates (moves down) through soil, carrying dissolved nutrient minerals with it. The removal of dissolved materials from soil by percolating water is called **leaching.** The deposition of leached material in the lower layers of soil is known as **illuviation.** Iron and aluminum compounds, humus, and clay are some illuvial materials that can gather in the subsurface portion of the soil. Some substances completely leach out of the soil because they are so soluble that they migrate down into the groundwater. It is also possible for water that is moving *upward* through the soil to carry dissolved materials with it.

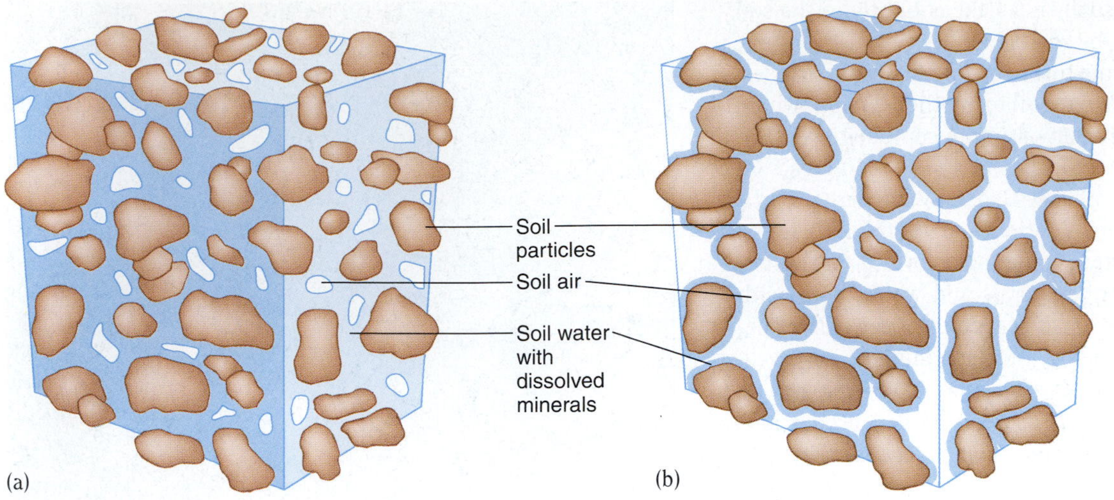

Figure 34–13 Pore space, soil air, and water. (a) In a wet soil, most of the pore space is filled with water. (b) In a dry soil, a thin film of water is tightly bound to soil particles, and most of the pore space is occupied by air.

The organisms living in the soil form a complex ecosystem

A single teaspoon of fertile agricultural soil may contain millions of microorganisms such as bacteria, fungi, algae, protozoa, and microscopic worms. Many other organisms also colonize the soil ecosystem, including plant roots, earthworms, insects, moles, snakes, and groundhogs (Fig. 34–14). Most numerous in soil are bacteria, which number in the hundreds of millions per gram of soil. Scientists have identified about 170,000 species of soil organisms, but thousands remain to be identified. Little is known about the roles of soil organisms, in part because it is hard to study their activities under natural conditions.

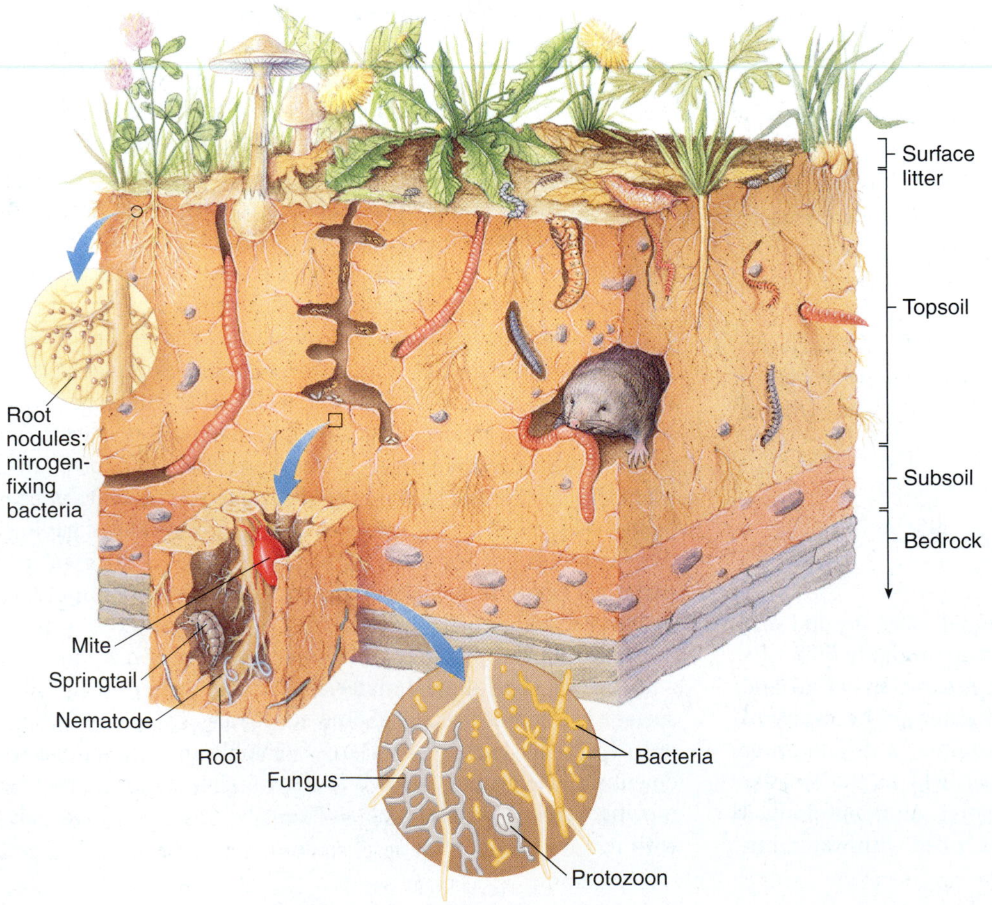

Figure 34–14 Diversity of life in fertile soil. Plants, algae, protozoa, fungi, bacteria, earthworms, flatworms, roundworms, insects, spiders and mites, and burrowing animals such as moles and groundhogs live in soil.

Worms are some of the most important organisms living in soil. Earthworms, probably one of the most familiar soil inhabitants, ingest soil and obtain energy and raw materials by digesting humus. *Castings,* bits of soil that have passed through the gut of an earthworm, are deposited on the soil surface. In this way, nutrient minerals from deeper layers are brought to upper layers. Earthworm tunnels serve to aerate the soil, and the worms' waste products and corpses add organic material to the soil.

Ants live in the soil in enormous numbers, constructing tunnels and chambers that help aerate it. Members of soil-dwelling ant colonies forage on the surface for bits of food, which they carry back to their nests. Not all of this food is eaten, however, and its eventual decomposition helps increase the organic matter in the soil. Many ants are also indispensable in plant reproduction because they bury seeds in the soil.

Soil pH affects soil characteristics and plant growth

As discussed in Chapter 2, soil acidity is measured using the pH scale, which extends from 0 (extremely acidic) through 7 (neutral) to 14 (extremely alkaline). The pH of most soils ranges from 4 to 8, but some soils are outside this range. The soil of the Pygmy Forest in Mendocino County, California, is extremely acidic (pH 2.8–3.9). At the other extreme, certain soils in Death Valley, California, have a pH of 10.5.

Plants are affected by soil pH, partly because the solubility of certain nutrient minerals varies with differences in pH. Soluble nutrient minerals can be absorbed by the plant, whereas insoluble forms cannot. At a low pH, for example, aluminum and manganese in soil water are more soluble and are sometimes absorbed by the roots in toxic concentrations. Certain nutrient mineral salts essential for plant growth, such as calcium phosphate, become less soluble and thus less available to plants at a higher pH.

Soil pH also affects the leaching of nutrient minerals. An acidic soil has less ability to bind positively charged ions to it because the soil particles also bind the abundant H^+ (Fig. 34–15). As a result, certain nutrient minerals essential for plant growth, such as potassium (K^+), are leached more readily from acidic soil. The optimum soil pH for most plant growth is 6.0 to 7.0, because most essential nutrient minerals are available to plants in that pH range.

Soil pH affects plants, and plants and other organisms in turn influence soil pH. The decomposition of humus and the cellular respiration of soil organisms, including roots, decrease the pH of the soil. When roots and other soil organisms respire, they release carbon dioxide, which dissolves in soil water to form carbonic acid (H_2CO_3).

Acid precipitation, a type of air pollution in which sulfuric and nitric acids produced by human activities fall to the ground as acid rain, sleet, snow, or fog, can seriously decrease soil pH. Acid precipitation is one of several factors implicated in **forest decline,** the gradual deterioration and, often, death of trees that has been observed in many European and North American forests. Forest decline may be partly the result of soil changes

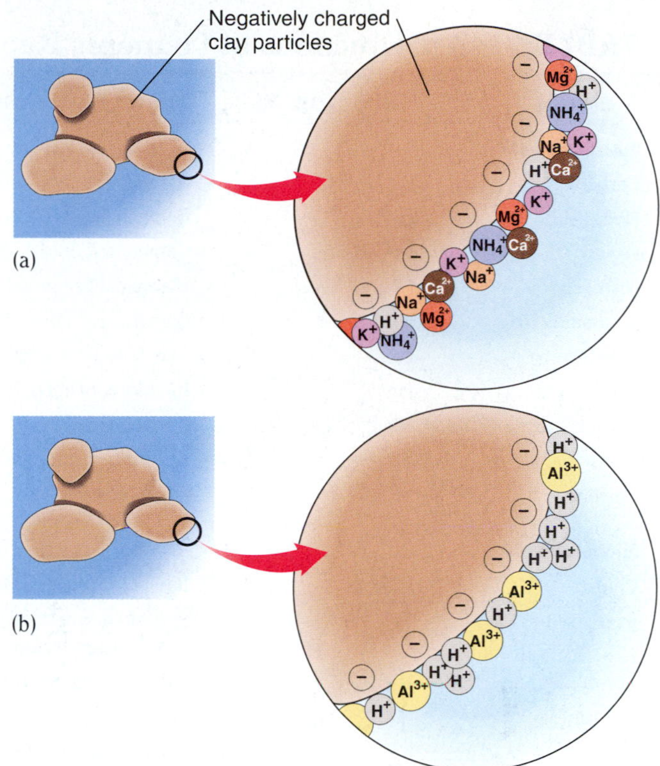

Figure 34–15 How acid alters soil chemistry. (a) In normal soil, positively charged nutrient mineral ions are attracted to the negatively charged soil particles. (b) In acidified soil, hydrogen ions displace the cations. Aluminum ions released from inorganic mineral particles when the soil becomes acidified also adhere to soil particles.

caused by acid precipitation. A 1989 study of Central European forests that have experienced forest decline, for example, found a strong correlation between forest damage and soil chemistry altered by acid precipitation (see Fig. 2–18).

SOIL PROVIDES MOST OF THE MINERALS FOUND IN PLANTS

More than 90 naturally occurring elements exist on Earth, and more than 60 of these, including elements as common as carbon and as rare as gold, have been found in plant tissues. Not all of these elements are considered essential for plant growth, however.

Nineteen elements have been found essential for plant growth (Table 34–2). Ten of these are required in fairly large quantities (greater than 0.05% dry weight) and are therefore known as **macronutrients.** These include carbon, hydrogen, oxygen, nitrogen, potassium, calcium, magnesium, phosphorus, sulfur, and silicon. The remaining nine **micronutrients** are needed in trace amounts (less than 0.05% dry weight) for normal plant growth and development. These include chlorine, iron, boron, manganese, sodium, zinc, copper, nickel, and molybdenum.

TABLE 34–2 Functions of Elements Required by Most Plants

Element	Taken Up As	Major Functions
Macronutrients		
Carbon	CO_2	Component of carbohydrate, lipid, protein, and nucleic acid molecules
Hydrogen	H_2O	Component of carbohydrate, lipid, protein, and nucleic acid molecules
Oxygen	CO_2, H_2O	Component of carbohydrate, lipid, protein, and nucleic acid molecules
Nitrogen	NO_3^-, NH_4^+	Component of proteins, nucleic acids, chlorophyll, certain coenzymes
Potassium	K^+	Osmotic and ionic balance; opening and closing of stomata; enzyme activator (for 40+ enzymes)
Calcium	Ca^{2+}	In cell walls; involved in membrane permeability; enzyme activator; second messenger
Magnesium	Mg^{2+}	In chlorophyll; enzyme activator in carbohydrate metabolism
Phosphorus	HPO_4^{2-}, $H_2PO_4^-$	In nucleic acids, phospholipids, ATP (energy transfer compound)
Sulfur	SO_4^{2-}	In certain amino acids and vitamins
Silicon	SiO_3^{2-}	In cell walls
Micronutrients		
Chlorine	Cl^-	Ionic balance; involved in photosynthesis
Iron	Fe^{2+}, Fe^{3+}	Part of enzymes and electron transport molecules involved in photosynthesis, respiration, and nitrogen fixation
Boron	$H_2BO_3^-$	Exact role unclear; involved in membrane transport and calcium utilization
Manganese	Mn^{2+}	Part of enzymes involved in respiration and nitrogen metabolism; required for photosynthesis
Sodium	Na^+	Involved in photosynthesis; substitutes for potassium in osmotic and ionic balance
Zinc	Zn^{2+}	Part of enzymes involved in respiration and nitrogen metabolism
Copper	Cu^+, Cu^{2+}	Part of enzymes involved in photosynthesis
Nickel	Ni^{2+}	Part of enzymes (urease) involved in nitrogen metabolism
Molybdenum	MoO_4^{2-}	Part of enzymes involved in nitrogen metabolism

Four of the 19 elements—carbon, oxygen, hydrogen, and nitrogen—come directly or indirectly from soil water or from gases in the atmosphere. Carbon is obtained from carbon dioxide in the atmosphere during photosynthesis. Oxygen is obtained from atmospheric oxygen (O_2) and water (H_2O). Water also supplies hydrogen to the plant. Plants absorb their nitrogen from the soil as ions of nitrogen salts, but the nitrogen in nitrogen salts ultimately comes from atmospheric nitrogen.

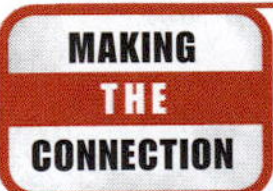

How does the nitrogen requirement by organisms show the interdependence of life? Because nitrogen is an essential part of biologically important molecules such as proteins and nucleic acids, all organisms must have nitrogen to survive. Animals, including humans, cannot use nitrogen as atmospheric N_2 or as inorganic ions (nitrate, NO_3^-, and ammonium, NH_4^+), and so they get their nitrogen from proteins and other nitrogen-containing compounds in the foods they eat. Ultimately, this nitrogen is traced back to plant sources.

Plants obtain their nitrogen as nitrate or ammonium ions from the soil. But where do the NO_3^- and NH_4^+ come from? Nitrogen is converted into those forms from atmospheric N_2 by billions upon billions of individual bacteria of just a few species. Thus, the nitrogen supply of the entire biosphere, including humans, comes from a few, at first glance insignificant, species. Without these bacteria, the supply of nitrogen in a form suitable for plants would be rapidly depleted, and plants would die. With the demise of plants, animals, including humans, would be unable to survive. We say more about the crucial role of microorganisms in the nitrogen cycle in Chapter 53.

The remaining 15 essential elements are obtained from the soil as dissolved nutrient mineral ions. Their ultimate source is the parent material from which the soil was formed.

Carbon, hydrogen, and oxygen are found as part of the structure of all biologically important molecules, including lipids, carbohydrates, nucleic acids, and proteins. Nitrogen is part of proteins, nucleic acids, and chlorophyll.

Potassium, which plants use in fairly substantial amounts, is not found in a specific organic compound in plant cells. Instead, it remains as free K^+ and plays a key physiological role in maintaining the turgidity of cells because it is osmotically active. The presence of K^+ in cytoplasm causes the cell to have a greater solute concentration than surrounding cells. As a result, water passes through the plasma membrane into the cell by osmosis. Potassium is also involved in the opening and closing of stomata (see Chapter 32).

Calcium plays a key structural role as a component of the middle lamella (the cementing layer between cell walls of adjacent plant cells). Calcium ions (Ca^{2+}) have also been implicated in a number of physiological roles in plants, such as altering membrane permeability and various cell signaling responses as second messengers (see Chapter 5).

Magnesium is part of the chlorophyll molecule. Phosphorus is critical for plants because it is found (as phosphate) in nucleic acids, phospholipids (an essential part of cell membranes), and energy transfer molecules such as ATP. Sulfur is essential because it is found in certain amino acids and vitamins. Many plants accumulate silicon in their cell walls and intercellular spaces. Silicon enhances the growth and fertility of some species and may help reinforce cell walls.

Chlorine and sodium are micronutrients that have a role in maintaining turgidity of cells. In addition to their osmotic role, chloride (Cl^-) and sodium ions are essential for photosynthesis. Six of the micronutrients (iron, manganese, zinc, copper, nickel, and molybdenum) are associated with various plant enzymes, often as cofactors, and are involved in certain enzymatic reactions. Although the role of boron in plants is unclear, some data suggest that it is involved in both the transport of carbohydrates across cell membranes and calcium utilization.

How do biologists determine whether an element is essential?

It is impossible to conduct mineral nutrition experiments by growing plants in soil because soil is too complex and contains too many elements. Thus, one of the most useful methods to test whether an element is essential is **hydroponics,** which is the growing of plants in aerated water to which nutrient mineral salts have been added. Hydroponics also has commercial applications in addition to its scientific use (Fig. 34–16).

Figure 34–16 Hydroponically grown lettuce. Lettuce is one of several hydroponic crops now grown commercially. The white boards support the plants, which are not anchored in soil. A chemically defined liquid solution of nutrient minerals trickles over their roots, which also have access to atmospheric oxygen. *(Hank Morgan/Photo Researchers, Inc.)*

If biologists suspect that a particular element is essential for plant growth, they grow plants in a nutrient solution that contains all known essential elements *except* the one in question. If plants grown in the absence of that element are unable to develop normally or to complete their life cycle, the element may be essential. Additional criteria are used to confirm whether an element is essential. For example, it must be demonstrated that the element has a direct effect on the plant's metabolism and that the element is essential for a wide variety of plants.

SOIL CAN BE DAMAGED BY HUMAN MISMANAGEMENT

Soil is a valuable natural resource on which humans depend for food. Many human activities generate or aggravate soil problems, including nutrient mineral depletion, soil erosion, and accumulation of salt.

Nutrient mineral depletion may occur in soils that are farmed

In a natural ecosystem, the essential nutrient minerals removed from the soil by plants are returned when the plants or the animals that eat them die and are decomposed. An agricultural system disrupts this pattern of nutrient cycling when the crops are harvested. Much of plant material, containing nutrient minerals, is removed from the cycle, so it fails to decay and release its nutrients back to the soil. Thus, over time, soil that is farmed inevitably loses its fertility, that is, its ability to produce abundant crops. In similar manner, homeowners often mow their lawns and remove the clippings, preventing decomposition and cycling of nutrient minerals that were in the grass blades.

The essential material (water, sunlight, or some essential element) that is in shortest supply usually limits plant growth. This phenomenon is sometimes called the concept of **limiting resources.** The three elements that are most often limiting resources for plants are nitrogen, phosphorus, and potassium. To sustain the productivity of agricultural soils, fertilizers are periodically added to depleted soils to replace the nutrient minerals that limit plant growth.

The two main types of fertilizers are organic and commercial inorganic. *Organic fertilizers* include such natural materials as cow manure, crop residues, bone meal, and compost. Green manure, a special type of organic fertilizer, is actually a crop that is planted and deliberately plowed into the soil to decompose instead of being harvested. Frequently the plants grown as green manure have nitrogen-fixing bacteria living in root nodules, thereby increasing the amount of nitrogen in the soil. Organic fertilizers are complex, and their exact compositions vary. The nutrient minerals in organic fertilizers become available to plants only as the organic material decomposes. For that reason, organic fertilizers are slow-acting and long-lasting.

Commercial inorganic fertilizers are manufactured from chemical compounds, and their exact compositions are known. Because they are soluble, they are immediately available to plants. Commercial inorganic fertilizers are available in the soil for only a short period (relative to organic fertilizers), because they quickly leach away, often into groundwater or surface runoff, polluting the water. Most commercial inorganic fertilizers contain the three elements (nitrogen, phosphorus, and potassium) that are usually the limiting resources in plant growth. The numbers on fertilizer bags (e.g., 10–20–20) tell the percentage concentrations of each of these three elements (N, P, and K).

Soil erosion is the loss of soil from the land

Water, wind, ice, and other agents cause **soil erosion,** the wearing away or removal of soil from the land. Water and wind are particularly effective in moving soil from one place to another. Rainfall loosens soil particles that can then be transported away by moving water (Fig. 34–17). Wind loosens soil and blows it away, particularly if the soil is barren and dry. Soil erosion is a natural process that can be greatly accelerated by human activities.

Soil erosion is a national and international problem that does not make the headlines very often. To get a feeling for how serious the problem is, consider that the U.S. Department of Agriculture (USDA) estimates that approximately 4.2 billion metric tons (4.6 billion tons) of topsoil are lost each year from U.S. croplands and rangelands as a result of soil erosion. Approximately one-fifth of U.S. cropland is vulnerable to soil erosion damage. Erosion causes a loss in soil fertility because essential nutrient minerals and organic matter that are a part of the soil are removed. As a result of these losses, the productivity of eroded agricultural soils declines, and more fertilizer must be used to replace the nutrients lost to erosion.

Figure 34–17 Soil erosion in an open field. Removal of plant cover exposes bare soil to erosion in heavy rains. Precipitation can cause gullies to enlarge rapidly because they provide channels for the runoff of water. *(USDA/Natural Resources Conservation Service)*

Humans often accelerate soil erosion through poor soil management practices. Agriculture is not the only culprit, since the removal of natural plant communities during surface mining, unsound logging practices (such as clearcutting large areas of forest), and the construction of roads and buildings also accelerate erosion.

Soil erosion has an impact on other natural resources. Sediment that enters streams, rivers, and lakes degrades water quality and fish habitats. Pesticides and fertilizer residues may be present in sediment, adding additional pollutants to the water. Also, when forests are removed within the watershed of a hydroelectric power facility, accelerated soil erosion can cause the reservoir behind the dam to fill in with sediment much faster than usual. This process results in a loss of electricity production at that facility.

Sufficient plant cover reduces the amount of soil erosion: Leaves and stems cushion the impact of rainfall, and roots help hold the soil in place. Although soil erosion is a natural process, abundant plant cover makes it negligible in many natural ecosystems.

Salt accumulates in soil that is improperly irrigated

Although irrigation improves the agricultural productivity of arid and semiarid lands, it sometimes causes salt to accumulate in the soil, a process called **salinization.** In a natural scenario, as a result of precipitation runoff, rivers carry dissolved salts away. Irrigation water, however, normally soaks into the soil and does not run off the land into rivers, so when it evaporates, the salt remains behind and accumulates in the soil. Salty soil results in a decline in productivity and, in extreme cases, renders the soil completely unfit for crop production.

Most plants cannot obtain all the water they need from salty soil because a water balance problem exists: Water moves by osmosis *out* of plant roots and into the salty soil. Obviously, most plants cannot survive under these conditions (see Fig. 5–13). Plant species that thrive in saline soils have special adaptations that enable them to tolerate the high amount of salt. Black mangroves, for example, excrete excess salt through their leaves.

Most crops, unless they have been genetically selected to tolerate high salt, are not productive in saline soil. Research on the molecular aspects of salt tolerance suggests that mutations in membrane transport proteins confer salt tolerance. A salt-tolerant variety of *Arabidopsis* was genetically engineered to overexpress a single gene that codes for a sodium transport protein in the vacuolar membrane. These plants are able to store large quantities of Na^+ in their vacuoles, thereby tolerating saline soils. Other salt-resistant plants exclude Na^+ from entering the cell through the plasma membrane.

SUMMARY WITH KEY TERMS

I. Anchorage, absorption, conduction, and storage are the main functions of roots.

II. Two types of root systems exist in plants.
   A. A **taproot** system has one main root (formed from the radicle), from which many lateral roots extend.
   B. A **fibrous root** system has several to many **adventitious roots** of the same size developing from the end of the stem. Lateral roots branch from these adventitious roots.

III. Roots differ structurally from stems in several ways.
   A. Each root tip is covered by a **root cap,** a protective layer that covers the delicate root apical meristem and may orient the root so that it grows downward.
   B. **Root hairs,** short-lived extensions of epidermal cells, increase the surface area of the root in contact with the moist soil.
   C. Roots lack nodes and internodes and do not usually produce leaves or buds.

IV. Primary roots possess an epidermis, ground tissues (cortex and in certain roots, pith), and vascular tissues (xylem and phloem).
   A. **Epidermis** protects the root, and its root hairs aid in absorption of water and dissolved nutrient minerals.
   B. **Cortex** consists of parenchyma cells that often store starch. The water and dissolved nutrient minerals move through the cortex along one of two pathways, the **apoplast** (along the interconnected porous cell walls) or the **symplast** (from one cell's cytoplasm to the next through plasmodesmata).
   C. **Endodermis,** the innermost layer of the cortex, regulates the movement of nutrient minerals into the root xylem.
      1. Cells of the endodermis possess a **Casparian strip** around their radial and transverse walls that is impermeable to water and dissolved nutrient minerals.
      2. Nutrient minerals are actively transported through membrane carrier proteins into the cytoplasm of the endodermal cells.
   D. **Pericycle** gives rise to lateral roots and lateral meristems.
   E. **Xylem** conducts water and dissolved nutrient minerals; **phloem** conducts dissolved sugar.

V. Roots vary in internal structure.
   A. Xylem of a herbaceous dicot root forms a solid core in the center of the root. In contrast, the center of a monocot root often consists of **pith** surrounded by a ring of alternating bundles of xylem and phloem.
   B. Monocot roots lack a **vascular cambium** and therefore do not have secondary growth.
   C. Roots of gymnosperms and woody dicots develop secondary tissues (wood and bark).

VI. Some roots are modified for specialized functions.
   A. **Prop roots** develop from branches or from a vertical stem and grow downward into the soil to help support the plant in an upright position.
   B. **Pneumatophores** are aerial "breathing" roots that may assist in getting oxygen to the submerged roots.
   C. Certain epiphytes have roots modified to photosynthesize, absorb moisture, or, if parasitic, penetrate host tissues and absorb nutrients.
   D. Corms and bulbs often have **contractile roots** that grow into the soil and then contract, thereby pulling the corm or bulb deeper into the soil.

VII. Roots often form associations with other species.
   A. A root **graft** is a natural union between the roots of trees belonging to the same or different species.
   B. **Mycorrhizae** are mutually beneficial associations between roots and soil fungi.
   C. Root **nodules** are swellings that develop on roots of leguminous plants and house millions of rhizobia (nitrogen-fixing bacteria).

VIII. Soil is the complex material in which roots grow.
   A. Factors influencing soil formation include parent material, climate, organisms, the passage of time, and topography.
   B. Soil is composed of inorganic minerals, organic matter, air, and water.
   C. Soil organisms such as plants, algae, fungi, worms, insects, spiders, and bacteria are important not only in forming soil, but also in cycling nutrient minerals.

IX. Plants require 19 essential elements for normal growth.
   A. Ten elements are **macronutrients:** carbon, hydrogen, oxygen, nitrogen, potassium, calcium, magnesium, phosphorus, sulfur, and silicon.
   B. Nine elements are **micronutrients:** chlorine, iron, boron, manganese, sodium, zinc, copper, nickel, and molybdenum.
   C. Plants obtain positively charged nutrient mineral ions from clay particles in the soil by **cation exchange,** in which roots secrete protons (H^+), which are exchanged for other positively charged mineral ions, freeing them to be absorbed by roots.

X. Soil is a valuable natural resource.
   A. Mineral depletion may occur in soils that are farmed.
   B. **Soil erosion** is the removal of soil from the land by the actions of agents such as water and wind.
   C. **Salinization,** the accumulation of excess salt, is a problem in some irrigated soils.

POST-TEST

1. One main root, formed from the enlarging embryonic root, with many smaller lateral roots coming out of it is a(an) (a) fibrous root system (b) adventitious root system (c) taproot system (d) contractile root system (e) prop root system
2. Roots produced at unusual places on the plant are (a) fibrous (b) adventitious (c) taproots (d) contractile (e) mycorrhizae
3. Root hairs (a) cover and protect the delicate root apical meristem (b) increase the absorptive capacity of roots (c) secrete a waxy cuticle (d) orient the root so it grows downward (e) store excess sugars produced in the leaves
4. Plants with corms often have __________ roots that pull the bulb deeper into the ground. (a) fibrous (b) adventitious (c) tap (d) contractile (e) prop
5. Certain plants adapted to flooded soil produce aerial "breathing" roots known as (a) fibrous roots (b) pneumatophores (c) mycorrhizae (d) contractile roots (e) prop roots
6. Unlike stems, roots produce (a) nodes and internodes (b) root caps and internodes (c) axillary buds and root hairs (d) terminal buds and axillary buds (e) root caps and root hairs
7. The waterproof region around the radial and transverse walls of endodermal cells is the (a) Casparian strip (b) pericycle (c) apoplast (d) symplast (e) pneumatophore
8. The apoplast is (a) a layer of cells that surrounds the vascular region in roots (b) the layer of cells just inside the endodermis (c) a system of interconnected plant cell walls through which water moves (d) the central cylinder of the root that comprises the vascular tissues (e) a continuum of cytoplasm of many cells, all connected by plasmodesmata

9. Plants obtain positively charged nutrient mineral ions from clay particles in the soil by cation exchange, in which (a) roots passively absorb the positively charged mineral ions they require (b) mineral ions flow freely along porous cell walls (c) roots secrete protons (H^+), which free other positively charged mineral ions to be absorbed by roots (d) the Casparian strip effectively blocks the passage of water and nutrient mineral ions along the endodermal cell wall (e) a well-developed system of internal air spaces in the root allows both gas exchange and cation exchange
10. The cell layer from which lateral roots originate is (a) epidermis (b) cortex (c) endodermis (d) pericycle (e) vascular cambium
11. The center of a herbaceous dicot root is composed of __________, whereas the center of a monocot root is composed of __________ (a) pith; cortex (b) xylem; phloem (c) phloem; xylem (d) xylem; pith (e) pith; xylem
12. Mutually beneficial associations between certain soil fungi and the roots of most plant species are called (a) mycorrhizae (b) pneumatophores (c) nodules (d) rhizobia (e) humus
13. Which of the following statements about a soil is *true?* (a) pore spaces are always filled with about 50% air and 50% water (b) a single teaspoon of fertile agricultural soil may contain as many as several hundred living microorganisms (c) the texture of a soil is determined by the soil's pH (d) a soil's organic matter includes litter, droppings, and the dead remains of plants, animals, and microorganisms (e) soil formation is unaffected by a region's climate or topography
14. The technique of growing plants in aerated water with dissolved nutrient mineral salts is known as (a) hydration (b) hydroponics (c) hydrophilic (d) hydrostatic (e) hydrolysis
15. Carbon, hydrogen, oxygen, nitrogen, potassium, calcium, magnesium, phosphorus, sulfur, and silicon are collectively known as (a) micronutrients (b) microvilli (c) micronuclei (d) macronuclei (e) macronutrients

REVIEW QUESTIONS

1. List several functions of roots and describe the tissue(s) or cell layer(s) responsible for each function.
2. How would you distinguish between a root hair and a small lateral root?
3. If you were examining a cross section of a primary root of a flowering plant, how would you determine whether it was a dicot or a monocot?
4. Trace the pathway of water from the soil through the various tissues in a herbaceous dicot root.
5. How does a herbaceous dicot root develop secondary tissues?
6. Distinguish among the following specialized roots: (a) storage root (b) prop root (c) aerial "breathing" root (d) photosynthetic root (e) contractile root.
7. List the four components of soil and describe how each is important to plants.
8. Explain how weathering processes convert rock into soil.
9. Distinguish between macronutrients and micronutrients.
10. Label the various tissues of this herbaceous dicot root. Use Figure 34–3 to check your answers.

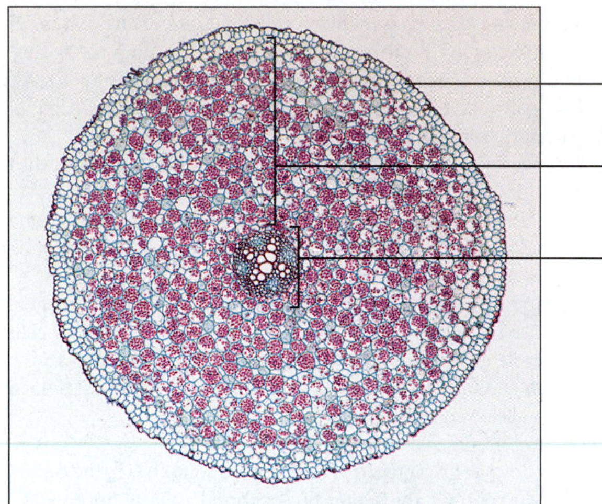

YOU MAKE THE CONNECTION

1. A mesquite root is found penetrating a mine shaft about 46 m (150 ft) below the surface of the soil. How might you determine when the root first grew into the shaft? (*Hint:* Mesquite is a woody plant.)
2. A barrel cactus that is 2 ft tall and 1 ft in diameter has roots more than 10 ft long. However, all of the plant's roots are found in the soil at a depth of 2 to 6 in. What possible adaptive value does such a shallow root system confer on a desert plant?
3. You are given a plant structure that was found growing in the soil and are asked to determine whether it is a root or an underground stem. How would you identify the plant part without a microscope? With a microscope?
4. How would you design an experiment to determine whether gold is essential for plant growth? What would you use for an experimental control?
5. Why does overwatering a plant often kill it?
6. Explain why, once secondary growth has occurred, that portion of the root is no longer involved in absorption. Where does absorption of water and dissolved nutrient minerals occur in plants that have roots with secondary growth?

RECOMMENDED READINGS

Baskin, Y. *The Work of Nature: How the Diversity of Life Sustains Us.* Island Press, Washington, D.C., 1997. This excellent book, written by a distinguished science writer, examines threats posed by the loss of biodiversity, including how changes in soil-dwelling species have affected soil fertility.

Berg, L.R. *Introductory Botany: Plants, People, and the Environment.* Saunders College Publishing, Philadelphia, 1997. A general botany text with an environmental emphasis.

Copley, J. "Ecology Goes Underground." *Nature,* Vol. 406, 3 Aug. 2000. The biological diversity of soils appears to influence the functioning of terrestrial ecosystems.

Raven, P.H., R.F. Evert, and S.E. Eichhorn. *Biology of Plants,* 6th ed. W.H. Freeman & Company, New York, 1999. This is an excellent general botany textbook, with detailed descriptions of root structure and adaptations.

Wolkomir, R. "Unearthing Secrets Locked Deep Inside Each Fistful of Soil." *Smithsonian,* Mar. 1997. Highlights the research of scientists at the National Soil Tilth Laboratory in Ames, Iowa.

Zimmer, C. "The Web Below." *Discover,* Nov. 1997. This article highlights the research of Simard and colleagues, which was featured in this chapter's *On the Cutting Edge* box.

- Visit our Web site at **http://www.info.brookscole.com/solomonbergmartin** for links to chapter-related resources on the World Wide Web. Additional on-line materials relating to this chapter can also be found on our Web site.

See chapter activity on BioActive Learner CD for additional help in mastering the chapter's material. Icon location in the chapter's margins shows which topics have tutorials or simulations in the CD.

35

Reproduction in Flowering Plants

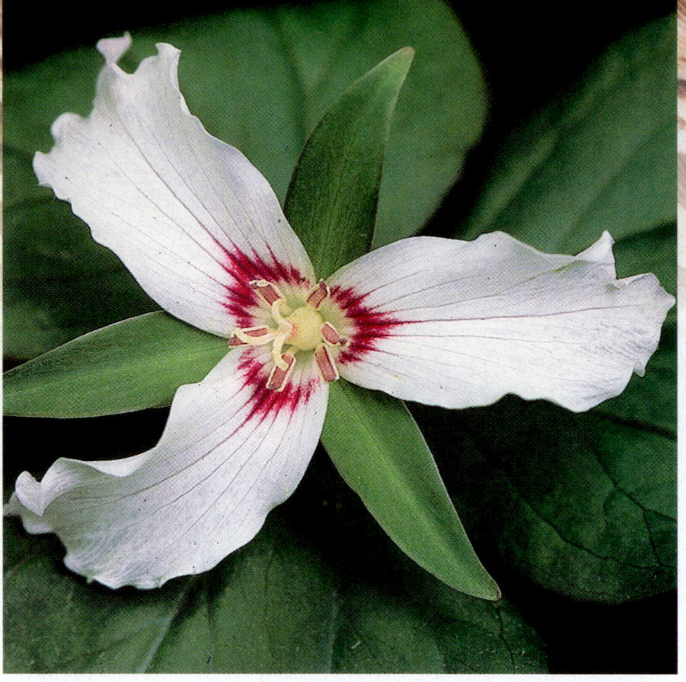

Painted trillium *(Trillium undulatum).* This plant, which grows to 50 cm (20 in), inhabits bogs and moist, acid woodlands in southern Canada and the northeastern United States, as well as the mountains south to Georgia. *(Skip Moody/Dembinsky Photo Associates)*

LEARNING OBJECTIVES

After you have studied this chapter you should be able to

1. State the differences between sexual and asexual reproduction, and discuss the evolutionary advantages and disadvantages of each.
2. Label the parts of a flower on a diagram, and describe the functions of each part.
3. Relate where eggs and pollen grains are formed within the flower and distinguish between pollination and fertilization.
4. Compare the evolutionary adaptations that characterize flowers pollinated in different ways (by insects, birds, bats, and wind).
5. Define coevolution and give examples of ways that plants and their animal pollinators have affected one another's evolution.
6. Trace the stages of embryo development in flowering plants, and list and define the main parts of seeds.
7. Explain the relationships among the following terms: ovules, ovaries, seeds, and fruits.
8. Distinguish among simple, aggregate, multiple, and accessory fruits; give examples of each type; and cite several different methods of seed and fruit dispersal.
9. Explain how the following structures may be used to propagate plants asexually: rhizomes, tubers, stolons, corms, bulbs, plantlets, and suckers. Explain how apomixis can occur.

In Chapter 31 you learned that flowering plants, which include at least 235,000 species, are the largest, most successful group of plants. One reason for the success of flowering plants, or **angiosperms,** is their ability to reproduce both sexually and asexually. You may have admired flowers for their often pleasing fragrances as well as their various colors and shapes *(see photograph).* The biological function of flowers, however, is sexual reproduction. Their colors, shapes, and fragrances are adaptations that increase the likelihood that pollen grains, which produce sperm cells, are carried from plant to plant.

As in most organisms, sexual reproduction in plants includes meiosis and the fusion of reproductive cells—egg and sperm cells, collectively called gametes. The union of gametes, which is called *fertilization,* occurs within the flower's ovary. After fertilization, flowering plants produce seeds inside fruits.

The offspring of parents that reproduce sexually exhibit considerable genetic variation. They may resemble one of the parent plants, both of the parents, or neither of the parents. Sexual reproduction offers the advantage of new gene combinations, not found in either parent, that might make an individual plant better suited to its environment. These new gene combinations largely result from the independent assortment of chromosomes that occurs during meiosis before the production of both egg and sperm cells. (How this variation occurs, that is, the details of independent assortment and genetic recombination, is discussed in Chapters 9 and 10.)

Many flowering plants also reproduce asexually. Asexual reproduction does not usually involve the formation of flowers, seeds, and fruits. Instead, offspring generally form when a vegetative part of an existing plant expands, grows, and then becomes separated from the rest of the plant, often by the death of tissues. This part subsequently grows to form a complete, independent plant.

Flowering plants have many methods of asexual reproduction, most of which involve modified vegetative organs such as stems, roots, and leaves. In particular, many modified stems, such as rhizomes, tubers, corms, and stolons, reproduce asexually. Roots that form suckers are an important means of asexual reproduction in other plants. Because asexual reproduction requires only one parent, and no meiosis or fusion of gametes occurs, the offspring of asexual reproduction are virtually genetically identical to each other and to the parent plant from which they came.[1]

In this chapter we examine various aspects of both sexual and asexual reproduction in flowering plants, including floral adaptations that are important in pollination, seed and fruit structure and dispersal, and several kinds of asexual reproduction. We conclude with a discussion of the evolutionary advantages and disadvantages of sexual and asexual reproduction.

THE FLOWERING PLANT LIFE CYCLE ALTERNATES BETWEEN A CONSPICUOUS SPOROPHYTE AND A REDUCED GAMETOPHYTE

In Chapters 26 and 27 you learned that angiosperms and other plants have a cyclic **alternation of generations** in which they spend a portion of their lives in a multicellular haploid stage and a portion in a multicellular diploid stage. The haploid portion, called the **gametophyte generation,** gives rise to gametes by mitosis. When two gametes fuse during fertilization, the diploid portion of the life cycle, called the **sporophyte generation,** begins. The sporophyte generation produces haploid spores by meiosis. Each spore has the potential to give rise to a gametophyte plant, and the cycle continues.

In flowering plants the diploid sporophyte generation is larger and nutritionally independent. The haploid gametophyte generation, which is located in the flower, is microscopic and nutritionally dependent on the sporophyte. We revisit alternation of generations in flowering plants after our discussion of the specialized reproductive structures called flowers. It may be helpful for you to review Figure 27–12, which shows the main stages in the flowering plant life cycle.

FLOWERS ARE INVOLVED IN SEXUAL REPRODUCTION

Flowers are reproductive shoots usually composed of four kinds of organs—sepals, petals, stamens, and carpels—arranged in whorls (circles) on the end of a flower stalk (Fig. 35–1; also see Figs. 27–9 and 27–10). In flowers with all four organs, the normal order of whorls from the flower's periphery to the center (or from the flower's base upward) is:

Sepals → petals → stamens → carpels

The tip of the stalk enlarges to form a **receptacle** on which some or all of the flower parts are borne. All four floral parts are important in the reproductive process, but only the stamens (the "male" organs) and carpels (the "female" organs) participate directly in sexual reproduction; sepals and petals are sterile.

Sepals, which constitute the outermost and lowest whorl on a floral shoot, cover and protect the flower parts when the flower is a bud. Sepals are leaflike in shape and form and are often green. Some sepals, such as those in lily flowers, resemble petals. As the blossom opens from a bud, the sepals fold back to reveal the more conspicuous petals. The collective term for all the sepals of a flower is **calyx.**

The whorl just inside and above the sepals consists of **petals,** which are broad, flat, and thin (like sepals and leaves) but tremendously varied in shape and frequently brightly colored, attracting pollinators. As discussed later, petals play an important role in ensuring that sexual reproduction will occur. Sometimes petals are fused to form a tube or other floral shape. The collective term for all the petals of a flower is **corolla.**

Just inside and above the petals are the **stamens,** the male reproductive organs. Each stamen is composed of a thin stalk, called a **filament,** at the top of which is an **anther,** a saclike structure in which **pollen grains** form. For sexual reproduction to occur, pollen grains must be transferred from the anther to the female reproductive structure (the carpel), usually of another flower of the same species. At first, each pollen grain consists of two cells surrounded by a tough outer wall. One cell (the *generative cell*) divides mitotically to form two nonflagellated male gametes, known as sperm cells, and the other (the *tube cell*) produces a **pollen tube,** through which the sperm cells travel to reach the ovule.

In the center or top of most flowers are one or more **carpels,** the female reproductive organs. Carpels bear **ovules,** which are structures with the potential to develop into seeds. The carpels of a flower may be separate or fused together into a single structure. The female part of the flower, often referred to as a **pistil,** may consist of a single carpel (a *simple pistil*) or a group of fused carpels (a *compound pistil*) (see Fig. 27–11). Each pistil has three sections: a **stigma,** on which the pollen grains land; a **style,** a necklike structure through which the pollen tube grows; and an **ovary,** a juglike structure that contains one or more ovules and that has the potential to develop into a fruit.

Female gametophytes are produced in the ovary, male gametophytes in the anther

Before we proceed further, it might be helpful to relate the stages in alternation of generations to floral structure. As discussed in

[1]Although offspring of asexual reproduction are generally considered to be genetically uniform, somatic mutations can result in some variability among asexually derived offspring.

(a)

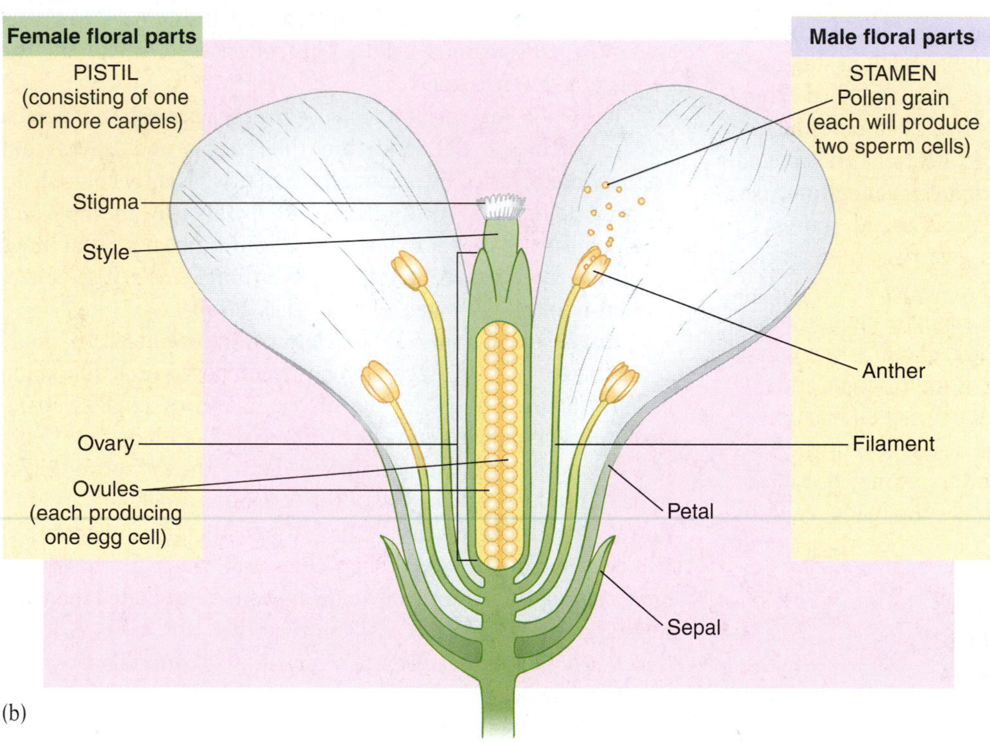

(b)

Figure 35–1 Structure of a flower of *Arabidopsis thaliana*, the model research plant. **(a)** A normal *Arabidopsis* flower. **(b)** Cutaway view of an *Arabidopsis* flower. Each flower has four sepals (two are shown), four petals (two are shown), six stamens, and one pistil. Four of the stamens are long, and two are short (two long and two short are shown). Pollen grains develop within sacs in the anthers. The compound pistil consists of two carpels that each contain numerous ovules. *(a, Dr. Elliot Meyerowitz/ California Institute of Technology)*

Chapters 26 and 27, angiosperms and certain other plants are heterosporous and produce two kinds of spores: megaspores and microspores (Fig. 35–2).

Each young ovule within an ovary contains a diploid cell, the *megasporocyte,* which undergoes meiosis to produce four haploid **megaspores.** Three of these usually disintegrate, and the fourth, the functional megaspore, divides mitotically to produce a multicellular female gametophyte, also called an **embryo sac.** The embryo sac, which is embedded in the ovule, typically contains seven cells with eight haploid nuclei. Six of these cells, including the egg cell (i.e., the female gamete), contain a single nucleus each; a large central cell has two nuclei, called **polar nuclei.** The egg and both polar nuclei participate directly in fertilization.

Pollen sacs within the anther contain numerous diploid cells called *microsporocytes,* each of which undergoes meiosis to produce four haploid cells called **microspores.** Each microspore divides mitotically to produce an immature male gametophyte, also called a pollen grain, that consists of two cells, the **tube cell** and the **generative cell.** The pollen grain becomes mature when its generative cell divides to form two nonmotile sperm cells.

POLLINATION IS THE FIRST STEP TOWARD FERTILIZATION

Before fertilization can occur, pollen grains must travel from the anther (where they form) to the stigma. The transfer of pollen grains from anther to stigma is known as **pollination.** Plants are *self-pollinated* if pollination occurs within the same

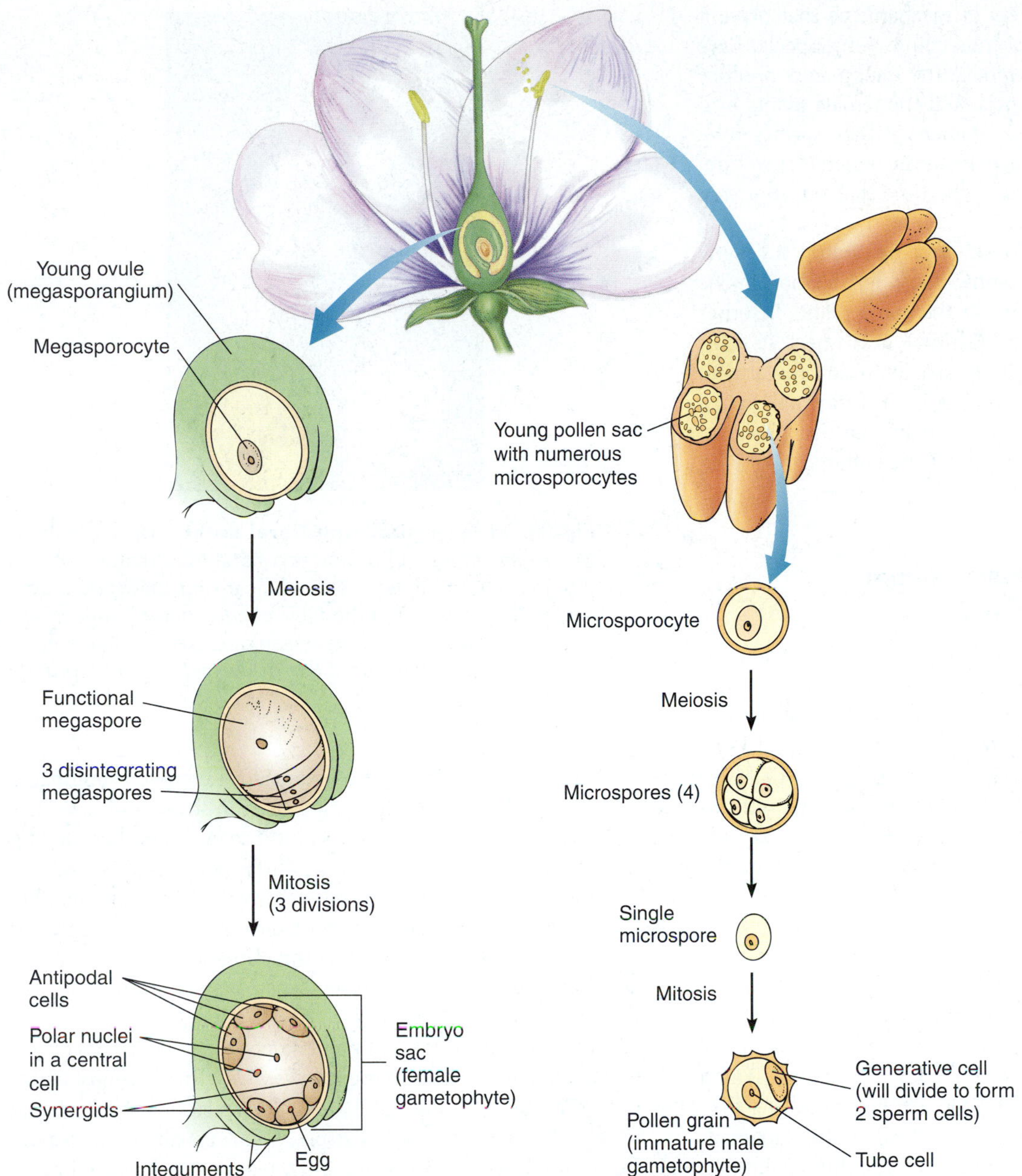

Figure 35–2 Development of female and male gametophytes. *(Left side)* The female gametophyte, or embryo sac, develops within the ovule. *(Right side)* The immature male gametophytes, or pollen grains, develop within pollen sacs. Each male gametophyte becomes mature when its generative cell divides mitotically to produce two sperm cells.

flower or a different flower on the same individual plant. When pollen grains are transferred to a flower on another individual of the same species, we say the plant is *cross-pollinated*. Flowering plants accomplish pollination in a variety of ways. Beetles, bees, flies, butterflies, moths, wasps, and other insects pollinate many flowers. Other animals such as birds, bats, snails, and small rodents also pollinate plants. Wind is the agent of pollination for certain flowers, whereas a few aquatic flowers are pollinated by water.

Many plant species have mechanisms to prevent self-pollination

In sexual reproduction in plants, the two gametes that unite to form a zygote may be from the same parent or from two different parents. The combination of gametes from two different parents increases the variation in offspring, and this variation may confer a selective advantage. Some offspring, for example, may be able to survive environmental changes better than either parent.

Plants have evolved a variety of mechanisms that prevent self-pollination and thus prevent inbreeding. Some species have separate male and female individuals; the male plants produce staminate flowers that lack carpels, and the female plants produce pistillate flowers that lack stamens. Other species have flowers with both stamens and pistils, but the pollen is shed from a given flower either before or after the time that the stigma of that flower is receptive to pollen.

Many species have genes for **self-incompatibility,** a genetic condition in which the pollen cannot cause fertilization in the same flower or in other flowers on the same plant. In other words, an individual plant can "identify" and reject its own pollen. Genes for self-incompatibility usually inhibit the growth of the pollen tube in the stigma and style. Self-incompatibility, which is more common in wild species than in cultivated plants, ensures that reproduction occurs only if the pollen comes from a genetically different individual.

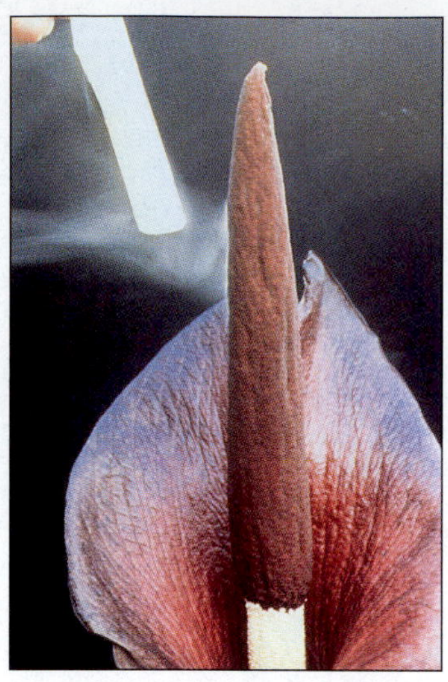

Figure 35–3 Visualizing floral scent. Devil's tongue (*Amorphophallus* sp.) produces a floral odor that is disagreeable to humans. The odor-causing chemicals react with HCl released from the filter paper *(shown in upper left corner)*, producing a visible precipitate. Devil's tongue is pollinated by beetles and flies. *(Courtesy of J.M. Patt and B.J.D. Meeuse)*

Flowering plants and their animal pollinators have coevolved

Animal pollinators and the plants they pollinate have had such close, interdependent relationships over time that they have affected the evolution of certain physical and behavioral features in each other. The term **coevolution** describes such reciprocal adaptation, in which two different species interact so closely that they become increasingly adapted to one another as they each undergo evolutionary change by natural selection. We now examine some of the features of flowers and their pollinators that are the products of coevolution.

Flowers pollinated by animals have various features to attract their pollinators, including showy petals (a visual attractant) and scent (an olfactory attractant) (Fig. 35–3). One of the rewards for the animal pollinator is food. Nectar is a sugary solution produced by some flowers in special floral glands called nectaries and used as an energy-rich food by pollinators. Pollen grains are also a protein-rich food for many animals. As they move from flower to flower searching for food, pollinators inadvertently carry pollen grains on their body parts, thus facilitating sexual reproduction in plants.

Biologists estimate that insects pollinate about 70% of all flowering plant species. Bees are particularly important as pollinators of crop plants. Crops pollinated by bees provide about 30% of human food. Plants pollinated by insects often have blue or yellow petals. The insect eye does not perceive color in the same way that the human eye does. Most insects see well in the violet, blue, and yellow range of visible light but do not perceive red as a distinct color. Consequently, flowers pollinated by insects are not usually red. Insects can also see in the ultraviolet range, wavelengths that are invisible to the human eye. Insects see ultraviolet radiation as a color called *bee's purple.* Many flowers have dramatic markings that may or may not be visible to humans but that direct insects to the center of the flower where the pollen grains and nectar are located (Fig. 35–4).

Insects have a well-developed sense of smell, and many insect-pollinated flowers have a strong scent that may be pleasant or foul to humans. The carrion plant, for example, is pollinated by flies and smells like the rotting flesh in which flies lay their eggs. As flies move from one reeking flower to another looking for a place to lay their eggs, they transfer pollen grains.

Birds such as hummingbirds are important pollinators (Fig. 35–5*a*). Flowers pollinated by birds are usually red, orange, or yellow because birds see well in this range of visible light. Because birds do not have a strong sense of smell, bird-pollinated flowers usually lack a scent.

Bats, which feed at night and do not see well, are important pollinators, particularly in the tropics where they are most abundant (Fig. 35–5*b*). Bat-pollinated flowers bloom at night and have dull white petals and a strong scent, usually of fermented fruit. Nectar-feeding bats are attracted to the flowers by the scent; they lap up the nectar with their long, extendible tongues. As they move from flower to flower, they transfer pollen grains. At least one bat-pollinated flower (the tropical vine *Mucana holtonii*) has evolved an unusual adaptation to encourage pollination by bats. When the pollen in a given flower is mature, a concave petal lifts up. The petal acts as an acoustic guide that reflects energy from the bat's echolocating calls back to the bat, thereby helping it to find the flower.

During the time that plants were coevolving specialized features such as petals, scent, and nectar to attract pollinators, animal pollinators also coevolved. Specialized body parts and behaviors adapted animals to aid pollination as they obtain nectar and pollen grains as a reward. For example, coevolution has selected for the bumblebees' hairy bodies, which catch and hold the sticky pollen grains for transport from one flower to another.

Coevolution may also have led to the long, curved beaks of the 'i'iwi, one of the Hawaiian honeycreepers, that inserts its

(a)

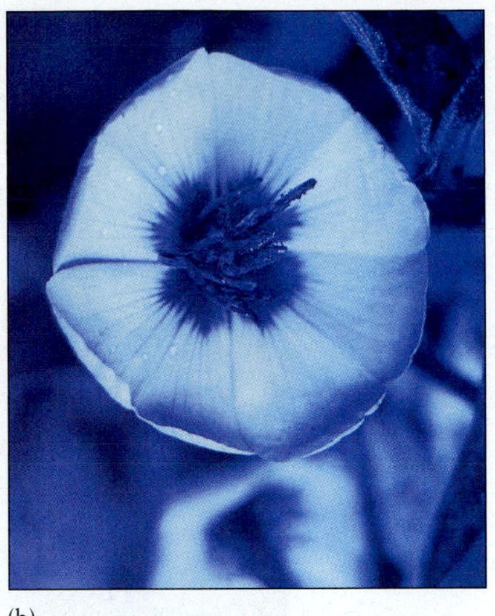

(b)

Figure 35–4 Ultraviolet markings on insect-pollinated flowers. **(a)** A flower as seen by the human eye is solid yellow. **(b)** The same flower viewed under ultraviolet radiation provides clues about how the insect eye perceives it. The light blue portions of the petals appear purple to a bee, whereas the dark blue inner parts appear yellow. These differences in color draw attention to the center of the flower, where the pollen grains and nectar are located. *(a, b, Thomas Eisner, Cornell University)*

beak into tubular flowers to obtain nectar (see Figs. 1–12*b* and 19–17). The long, tubular corolla of the flowers that 'i'iwis visit also came about through coevolution. During the 20th century, some of the tubular flower species (e.g., lobelias) became rare, largely as a result of grazing by non-native cows and feral goats, and about 25% of lobelioid species have become extinct. The 'i'iwi now feeds largely on the petalless flowers of the 'ohi'a tree, and the 'i'iwi bill appears to be slowly adapting to this change in feeding preference. A comparison of the bills of 'i'iwi museum specimens collected in 1902 with the bills of live birds captured in the 1990s show that 'i'iwi bills are now about 3% shorter than they were in 1902.

Animal behavior has also coevolved, sometimes in bizarre and complex ways. The flowers of certain orchids (*Ophrys* sp.), for example, resemble female bees in coloring and shape (Fig. 35–6). These flowers also secrete a scent similar to that produced by female bees, and the males are irresistibly attracted to it. The resemblance between *Ophrys* flowers and female bees is so strong that male bees mount the flowers and attempt to copulate with them. During this misdirected activity, known as *pseudo-copulation,* a pollen sac usually attaches to the back of the bee. When the frustrated bee departs and attempts to copulate with another orchid flower, pollen grains are transferred to that flower.

Some flowering plants depend on wind to disperse pollen

Some flowering plants, such as grasses, ragweed, maples, and oaks, are pollinated by wind and produce many small, inconspic-

(a)

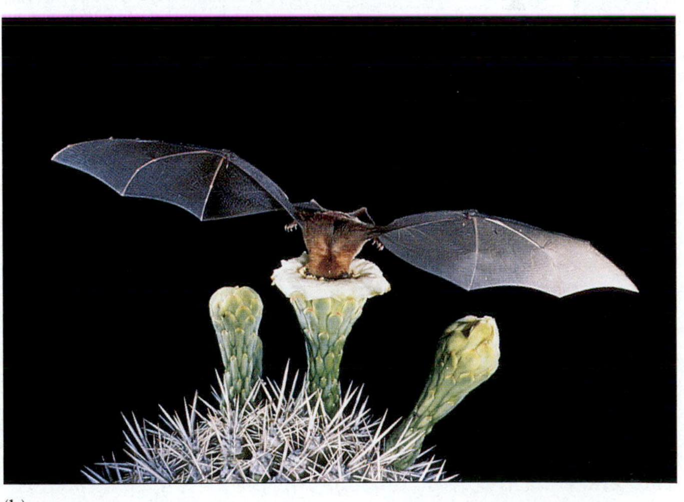

(b)

Figure 35–5 Animal pollinators. **(a)** A ruby-throated hummingbird obtains nectar from a trumpet vine flower. The pollen grains on the bird's feathers will be carried to the next plant. **(b)** A lesser long-nosed bat obtains nectar from a cardon cactus flower and transfers pollen as it moves from flower to flower. Bats pollinate several hundred species of plants. *(a, Dan Dempster/Dembinsky Photo Associates; b, Merlin Tuttle/Bat Conservation International/Photo Researchers, Inc.)*

Figure 35–6 Beelike flowers. The flowers of this orchid *(Ophrys scolopax)* resemble a female bee *(Eucera longicormis).* Male bees are attracted to the flowers and transfer pollen during misdirected attempts to copulate with the flowers. (The yellow sac attached near the bee's antennae is a pollen sac that became attached during pseudocopulation.) Photographed in Sardinia. *(Nuridsany & Perennou/Photo Researchers, Inc.)*

Figure 35–7 Wind pollination. Each cluster of male oak flowers dangles from a tree branch and sheds a shower of pollen when the wind blows. These flowers lack petals. *(Dr. Jeremy Burgess/Science Photo Library/Photo Researchers, Inc.)*

uous flowers (Fig. 35–7). Wind-pollinated plants do not produce large, colorful petals, scent, or nectar. Some have large, feathery stigmas, presumably to trap wind-borne pollen grains. Because wind pollination is a hit-or-miss affair, the likelihood of a particular pollen grain landing on a stigma of the same species of flower is slim. Wind-pollinated plants produce large quantities of pollen grains, which increase the likelihood that some pollen grains will land on the appropriate stigma.

FERTILIZATION IS FOLLOWED BY SEED AND FRUIT DEVELOPMENT

Once pollen grains have been transferred from anther to stigma, the tube cell, one of the two cells in the pollen grain, grows a thin pollen tube down through the style and into an ovule in the ovary. The second cell within the pollen grain divides to form two male gametes (the sperm cells), which move down the pollen tube and enter the ovule (Fig. 35–8).

A unique double fertilization process occurs in flowering plants

The egg within the ovule unites with one of the sperm cells, forming a zygote (fertilized egg) that will develop into an embryonic plant contained in a seed. The two polar nuclei in the central cell of the ovule fuse with the second sperm cell to form the first cell of the triploid ($3n$) **endosperm,** the tissue with nutritive and hormonal functions that surrounds the developing embryonic plant in a seed. This process, in which two separate cell fusions occur, is called **double fertilization.** It is, with two exceptions, unique to flowering plants. (A type of double fertilization has been reported in two gymnosperm species, *Ephedra nevadensis* and *Gnetum gnemon.*)

After double fertilization has occurred, the ovule develops into a seed, and the ovary surrounding it develops into a fruit. (Double fertilization and the flowering plant life cycle are discussed in greater detail in Chapter 27.)

Embryonic development in seeds is orderly and predictable

Flowering plants produce a young plant embryo complete with stored nutrients in a compact package, the **seed,** that develops from the ovule after fertilization. The nutrients in seeds are not only used by germinating plant embryos but also consumed by animals, including humans (see *Focus On: Seed Banks*). Development of the embryo and endosperm following fertilization is possible because of the constant flow of nutrients into the developing seed from the parent plant.

Cell divisions of the fertilized egg to form a multicellular embryo proceed in a variety of ways in flowering plants. The following description is of dicot embryonic development; monocot embryonic development is similar in the early stages.

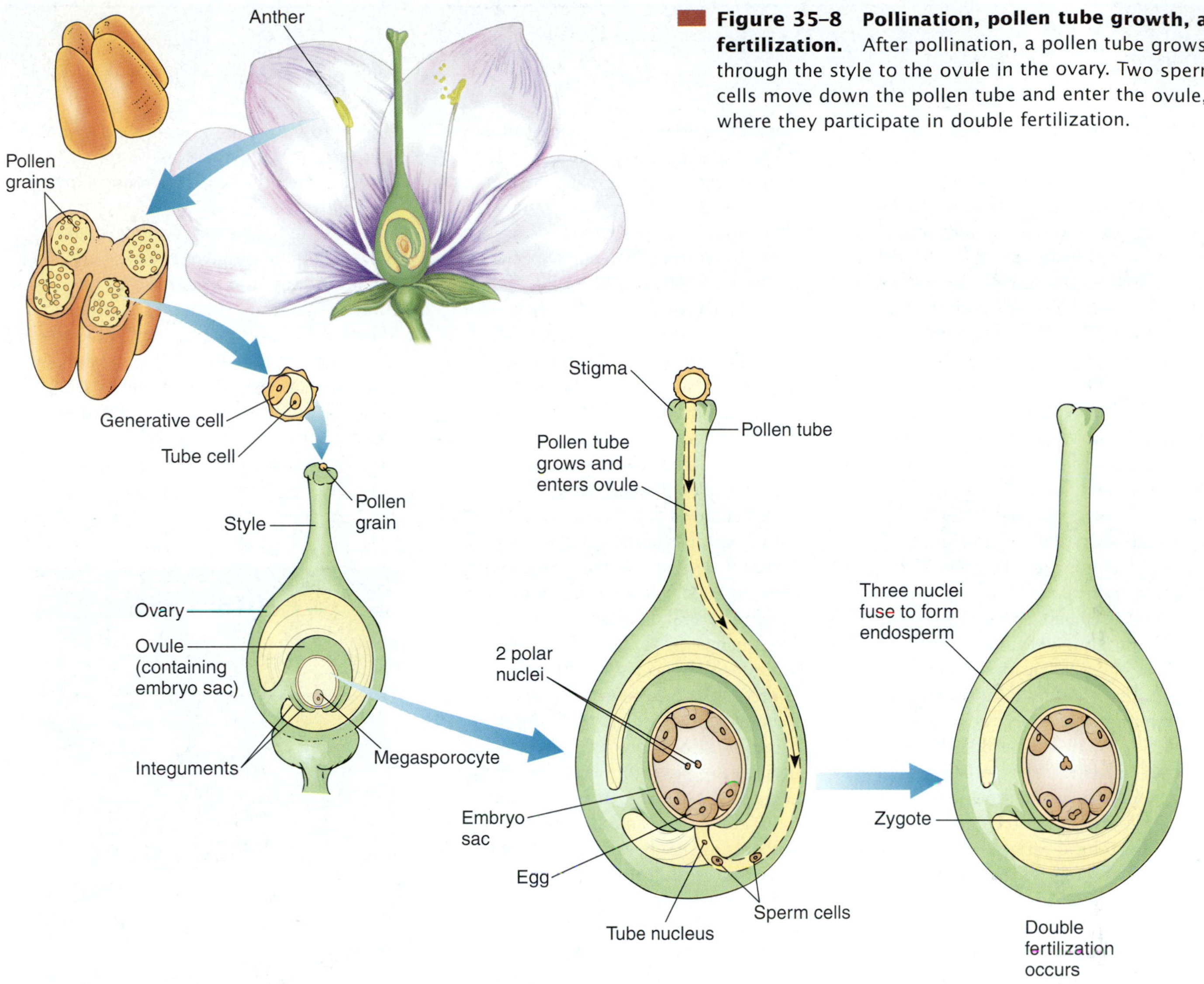

Figure 35–8 Pollination, pollen tube growth, and fertilization. After pollination, a pollen tube grows through the style to the ovule in the ovary. Two sperm cells move down the pollen tube and enter the ovule, where they participate in double fertilization.

The two cells (basal cell and apical cell) that are formed as a result of the first division of the fertilized egg establish polarity, or direction, in the embryo. The large basal cell (toward the outside of the ovule) typically develops into a **suspensor,** which is a multicellular structure that anchors the embryo and aids in nutrient uptake from the endosperm. The apical cell (toward the inside of the ovule) develops into the actual embryo.

Initially, the top cell divides to form a short cluster of cells, called a *proembryo* (Fig. 35–9). As cell division continues, a small sphere of cells, often called a *globular embryo,* develops. Cells begin to develop into specialized tissues during this stage. When the dicot embryo starts to develop its two cotyledons (seed leaves), it has two lobes and resembles a heart; this is often called the *heart stage*. During the *torpedo stage,* the embryo continues to grow as the cotyledons elongate. As the embryo enlarges, it often curves back on itself and crushes the suspensor beyond recognition.

The mature seed contains an embryonic plant and storage materials

A mature seed contains an embryonic plant and food (stored in the cotyledons or endosperm), surrounded by a tough, protective **seed coat** derived from the **integuments** (the outermost layers of an ovule). The seed in turn is enclosed within a fruit.

The mature embryo within the seed consists of a short embryonic root, or **radicle;** an embryonic shoot; and one or two seed leaves, or **cotyledons.** Monocots have a single cotyledon, whereas dicots have two (Fig. 35–10). The short portion of the embryonic shoot connecting the radicle to one or two cotyledons is known as the **hypocotyl.** The shoot apex, or terminal bud, located above the point of attachment of the cotyledon(s), is the **plumule.** After the radicle, hypocotyl, cotyledon(s), and plumule have formed, the young plant's development is arrested (by desiccation or dormancy). When conditions are right for continuation of the developmental program, the seed germinates.

FOCUS ON Seed Banks

To preserve older, more diverse varieties of plants, many countries are collecting plant **germplasm,** which is any plant material that may be used in breeding. Germplasm includes seeds, plants, and plant tissues of traditional crop varieties. The International Plant Genetics Resource Institute in Rome, Italy, is the scientific organization that oversees plant germplasm collections worldwide. More than 100 seed collections, called *seed banks,* exist around the world and collectively hold more than 3 million samples of thousands of different kinds of plants *(see figure).* The U.S. National Plant Germplasm System in Fort Collins, Colorado, stores seeds of about 250,000 different species and varieties.

Most seed banks help to preserve the genetic variation within different varieties of crops. Farmers typically discontinue planting local varieties when newer, improved varieties become available. The newer varieties have desirable genetic characteristics, such as a greater yield, for example, but the local, discarded varieties also contain valuable genes. Each local variety's characteristic combination of genes gives it distinctive nutritional value, size, color, flavor, resistance to disease, and adaptability to different climates and soil types. Maintaining the genetic diversity present in local crop varieties will help preserve genes that we may need in the future. The gene combinations of local varieties are potentially valuable to agricultural breeders because they can be transferred to other varieties, either by traditional breeding methods or by genetic engineering.

Seed banks offer the advantage of storing a large amount of live plant genetic material in a very small space. Seeds stored in seed banks are safe from habitat destruction, climate changes, and general neglect. There have even been some instances of using seeds from seed banks to reintroduce a plant species that has become extinct in the wild.

There are some disadvantages to seed banks. Seeds do not remain alive indefinitely and must be germinated periodically so that new ones can be collected. Growing, harvesting, and returning seeds to storage is the most expensive aspect of storing plant material in seed banks. Because accidents such as fires and power failures can result in the permanent loss of the genetic diversity represented by the seeds, biologists typically subdivide seed samples and store them in several different seed banks. Also, many types of plants, such as avocados and coconuts, cannot be stored as seeds. The seeds of these plants do not tolerate being dried out, which is a necessary step before they can be sealed in moisture-proof containers for storage at −18° C (−0.4° F). (If 20% or more moisture remains in a frozen seed, it will probably die.) Some seeds cannot be stored successfully because they only remain viable for a short time—a few months or even just a few days. (Cryopreservation in liquid nitrogen at −160° C, or −256° F, is a new method being developed for certain kinds of seeds. Seeds stored at this temperature are able to survive for longer periods than seeds stored at warmer temperatures.)

Perhaps the most important disadvantage of seed banks is that plants stored in this manner remain stagnant in an evolutionary sense. Removed from their natural habitats, they are no longer subject to the forces of natural selection. As a result, they may be less fit for survival when they are reintroduced into the wild.

Despite their shortcomings, seed banks are increasingly viewed as an important way to safeguard seeds for future generations. Other international efforts to preserve plant genetic diversity are also being implemented. For example, some farmers may soon be paid to set aside some of their land for cultivating local varieties of crops, thereby preserving the genetic diversity in agriculturally important plants.

Seed storage. These moisture-proof bottles, vials, and packets of seeds are stored at very low temperatures in the seed bank in Svalbard, Norway. *(Courtesy of Nordiska Genbanken, Alnarp, Svenge)*

Because the embryonic plant is nonphotosynthetic, it must have nutrients during germination until it becomes photosynthetic and, therefore, self-sufficient. The cotyledons of many plants function as storage organs and become large, thick, and fleshy as they absorb the food reserves (starches, oils, and proteins) initially produced as endosperm. Seeds that store nutrients in cotyledons have little or no endosperm at maturity. Examples of such seeds are peas, beans, squashes, sunflowers, and peanuts. Other plants, wheat and corn, for example, have thin cotyledons that function primarily to help the young plant digest and absorb food stored in the endosperm. (See Chapter 36 for a discussion of seed germination and early growth.)

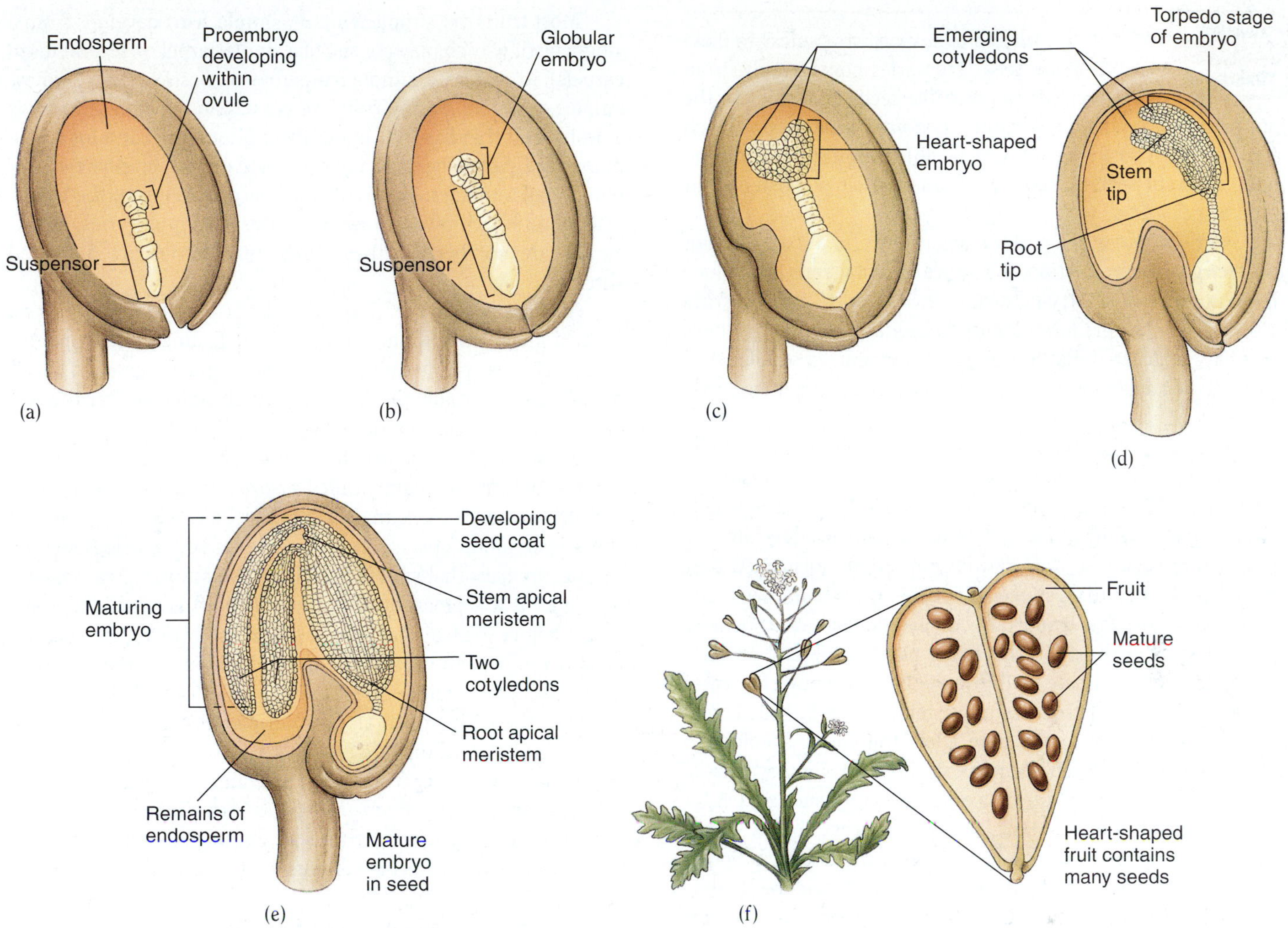

Figure 35–9 Embryonic development in shepherd's purse *(Capsella bursa-pastoris).* **(a)** The proembryo is the earliest multicellular stage of the embryo (the ovule is shown outside the ovary). **(b)** As cell division continues, the embryo becomes a ball of cells, called the globular stage. **(c)** As the two cotyledons begin to emerge, the embryo is shaped like a heart. **(d)** The cotyledons continue to elongate, forming the torpedo stage. **(e)** A maturing embryo within the seed. The food originally stored in the endosperm has been almost completely depleted during embryonic growth and development. Most of the food for the embryonic plant is in its cotyledons. **(f)** A longitudinal section through a heart-shaped fruit of shepherd's purse reveals numerous tiny seeds, each containing a mature embryo. Each seed developed from an ovule.

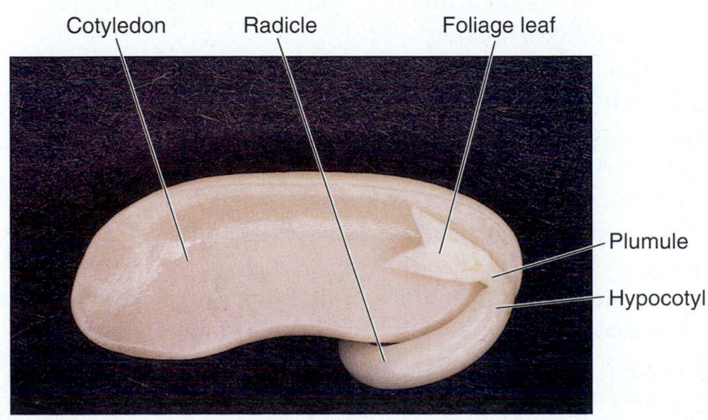

Figure 35–10 Seed structure. A bean seed has been dissected—its seed coat and one of its two cotyledons were removed—to show the radicle, hypocotyl, remaining cotyledon, and plumule with its foliage leaf at the shoot apex. This particular seed had begun germinating, so its radicle is larger than in an ungerminated seed. *(James Mauseth, University of Texas)*

Is seed size related to survival strategies? In flowering plants seed size varies considerably, from the microscopic, dustlike seeds of orchids to the giant seeds of the double coconut *(Lodoicea maldivica),* which weigh as much as 27 kg (almost 60 lb). Despite this variation among different species, seed size is a remarkably constant trait within a species.

Assuming that a given plant species invests a fixed amount of its energy in reproduction, is it more advantageous to produce a large number of small seeds, or a few big ones? By observing the seed sizes that predominate in different environments, biologists have suggested that in some environments, a smaller seed appears to be advantageous, whereas in others, a larger seed may be better.

Seed size for each plant species probably represents a trade-off between the requirements for dispersal and for successful establishment of seedlings. For example, plants that grow in widely scattered open sites (such as old fields) usually produce smaller seeds, perhaps because they can be more easily dispersed over large areas than can larger seeds. On the other hand, wide dispersal is probably less important for plants adapted to densely vegetated areas such as forests. These plants generally produce larger seeds with an ample food reserve that may confer a greater likelihood of becoming successfully established in a shaded environment. The stored energy may allow the young seedling to grow tall enough to reach adequate sunlight for photosynthesis.

Other ecological factors are associated with seed size. Larger seeds are typical of many plants that live in arid habitats, possibly because the food stored in a large seed allows a young seedling to establish an extensive root system quickly, thereby enabling it to survive the dry climate. Island plants also produce larger seeds than similar species on the nearby mainland. In this case, biologists hypothesize that large seeds are less likely to be widely dispersed and therefore less likely to fall into the ocean, where chances of survival are slim. On the other hand, seeds that remain dormant in the soil for a long period tend to be smaller; the exact reason for this observation is not known.

Despite many associations of seed size with specific environments, the general principles concerning the adaptive advantages of large seeds versus small seeds remain to be determined.

Fruits are mature, ripened ovaries

After double fertilization takes place within the ovule, the ovule develops into a seed (just described), and the ovary surrounding it develops into a **fruit.** For example, a pea pod is a fruit, and the peas within it are seeds. A fruit may contain one or more seeds; some orchid fruits contain several thousand to a few million seeds! Fruits provide protection for the enclosed seeds and sometimes aid in their dispersal.

There are several types of fruits; their differences result from variations in the structure or arrangement of the flowers from which they were formed. The four basic types of fruits—simple fruits, aggregate fruits, multiple fruits, and accessory fruits—are summarized in Figure 35–11.

Most fruits are simple fruits. A **simple fruit** develops from a single pistil (which may consist of a single carpel or several fused carpels). At maturity, simple fruits may be fleshy or dry. Two examples of simple, fleshy fruits are berries and drupes. A **berry** is a fleshy fruit that has soft tissues throughout and contains few to many seeds; a blueberry is a berry, as are grapes, cranberries, bananas, and tomatoes. Many so-called berries do not fit the botanical definition. Strawberries, raspberries, and mulberries, for example, are not berries; these three non-berries are discussed shortly.

A **drupe** is a simple, fleshy (or fibrous) fruit that contains a hard, stony pit surrounding a single seed. Examples of drupes include peaches, plums, olives, avocados, and almonds. The almond shell is actually the stony pit, which remains after the rest of the fruit has been removed.

Many simple fruits are dry at maturity; some of these split open, usually along seams, called *sutures,* to release their seeds. A milkweed pod is an example of a **follicle,** a simple, dry fruit that splits open along one suture to release its seeds. A **legume** is a simple, dry fruit that splits open along two sutures. Pea pods are legumes, as are green beans, although both are generally harvested before the fruit has dried out and split open. Pea seeds are usually removed from the fruit and consumed, whereas in green beans the entire fruit and seeds are eaten. A **capsule** is a simple, dry fruit that splits open along multiple sutures or pores. Iris, poppy, and cotton fruits are capsules.

Other simple, dry fruits, such as **grains,** for example, do not split open at maturity. Each grain contains a single seed. Because the seed coat is fused to the fruit wall, a grain appears to be a seed rather than a fruit. Kernels of corn and wheat are fruits of this type.

An **achene** is similar to a grain in that it is simple and dry, does not split open at maturity, and contains a single seed. However, the seed coat of an achene is not fused to the fruit wall. Instead, the single seed is attached to the fruit wall at one point only, permitting an achene to be separated from its seed. The sunflower fruit is an example of an achene. One can peel off the fruit wall (the shell) to reveal the sunflower seed within.

Nuts are simple, dry fruits that have a stony wall and do not split open at maturity. Unlike achenes, nuts are usually large and are often derived from a compound pistil. Examples of nuts include chestnuts, acorns, and hazelnuts. Many so-called nuts do not fit the botanical definition. Peanuts and Brazil nuts, for example, are seeds, not nuts.

Aggregate fruits are a second main type of fruit. An **aggregate fruit** is formed from a single flower that contains several to many separate (free) carpels (Fig. 35–12). After fertilization, each ovary from each individual carpel enlarges. As they enlarge, the ovaries may fuse together to form a single fruit. Raspberries, blackberries, and magnolia fruits are examples of aggregate fruits.

A third type is the **multiple fruit,** formed from the ovaries of many flowers that grow in proximity on a common floral stalk. The ovary from each flower fuses with nearby ovaries as it develops and enlarges after fertilization. Pineapples, figs, and mulberries are multiple fruits (Fig. 35–13).

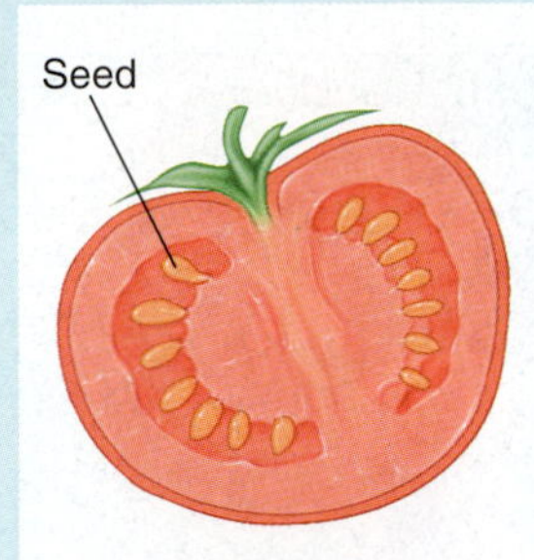

Berry (simple fruit)
A simple, fleshy fruit in which the fruit wall is soft throughout.

Tomato *(Lycopersicon lycopersicum)*

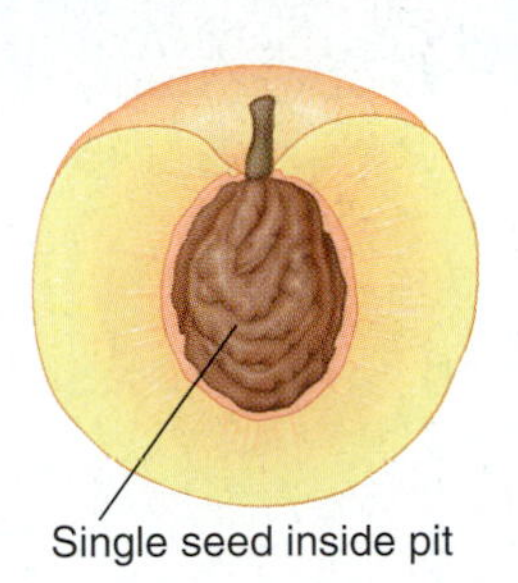

Drupe (simple fruit)
A simple, fleshy fruit in which the inner wall of the fruit is hard and stony (the pit).

Peach *(Prunus persica)*

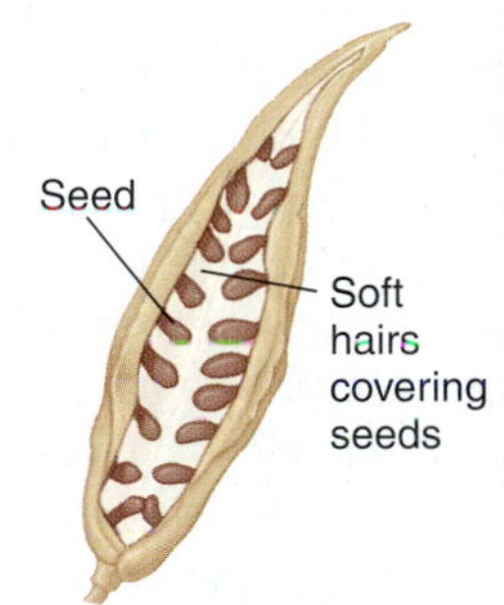

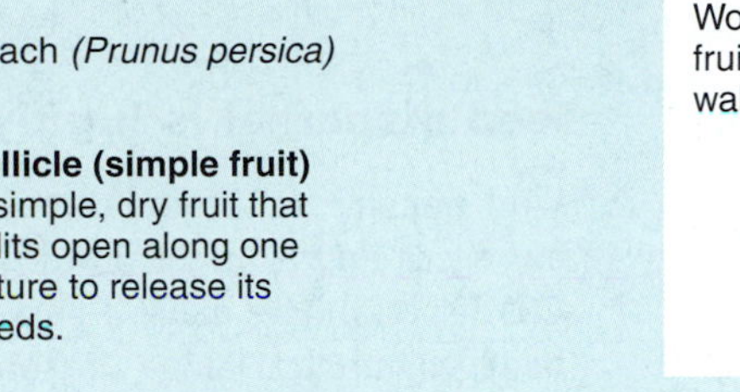

Follicle (simple fruit)
A simple, dry fruit that splits open along one suture to release its seeds.

Milkweed *(Asclepias syriaca)*

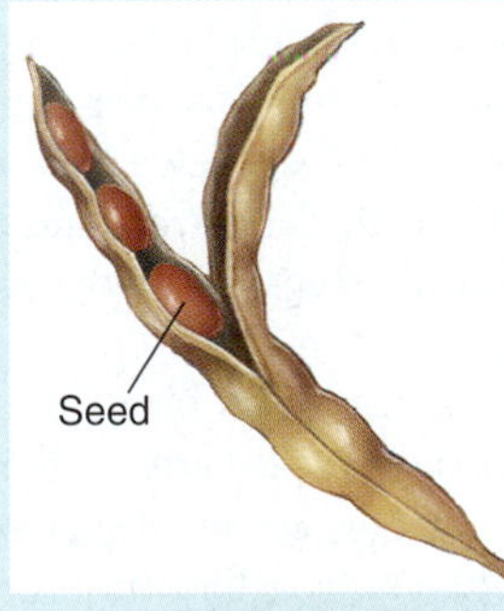

Legume (simple fruit)
A simple, dry fruit that splits open along two sutures to release its seeds.

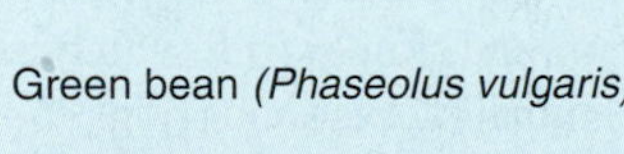

Green bean *(Phaseolus vulgaris)*

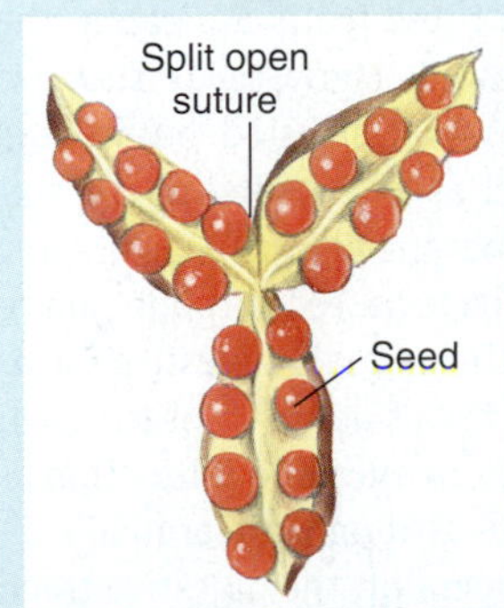

Capsule (simple fruit)
A simple, dry fruit that splits open along three or more sutures or pores to release its seeds.

Iris *(Iris* sp.)

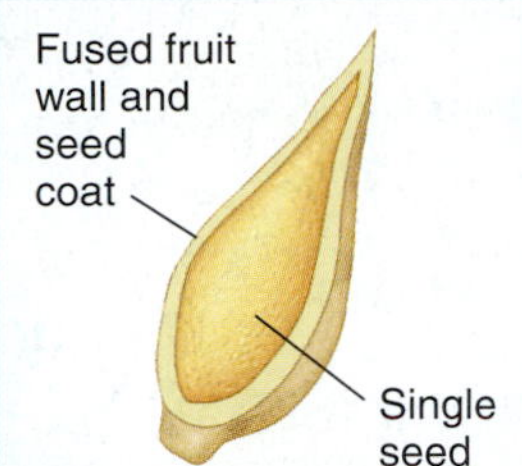

Grain (simple fruit)
A simple, dry fruit in which the fruit wall is fused to the seed coat.

Wheat *(Triticum* sp.)

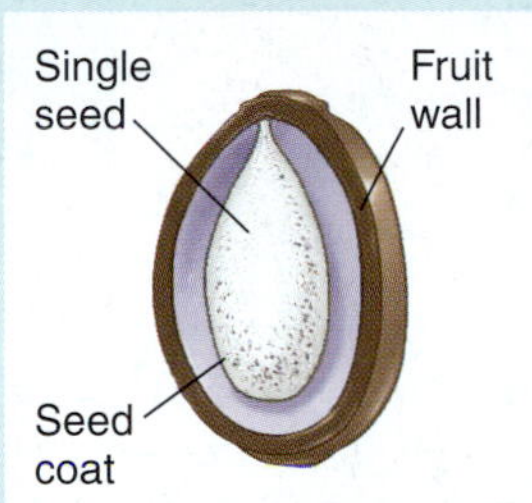

Achene (simple fruit)
A simple, dry fruit in which the fruit wall is separate from the seed coat.

Sunflower *(Helianthus annuus)*

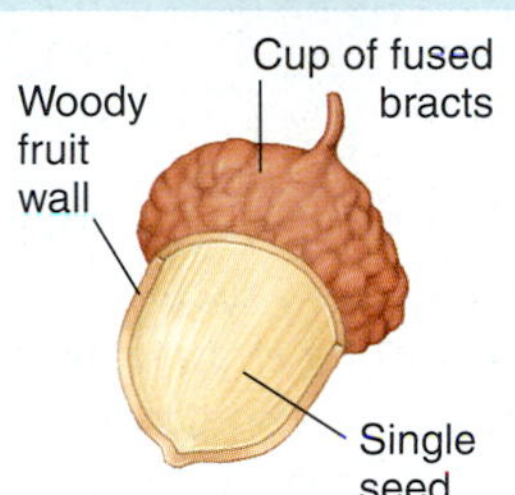

Nut (simple fruit)
A simple, dry fruit that has a stony wall, is usually large, and does not split open at maturity.

Oak *(Quercus* sp.)

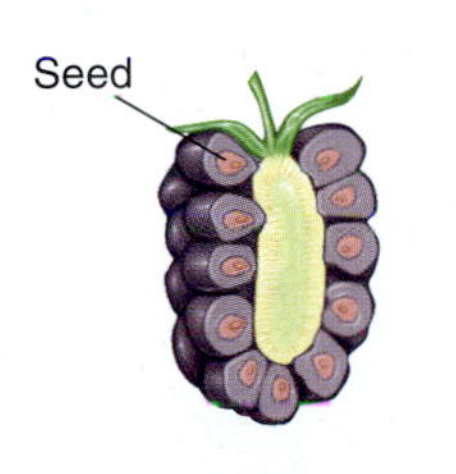

Aggregate fruit
A fruit that develops from a single flower with several to many pistils (i.e., carpels are not fused into a single pistil).

Blackberry *(Rubus* sp.)

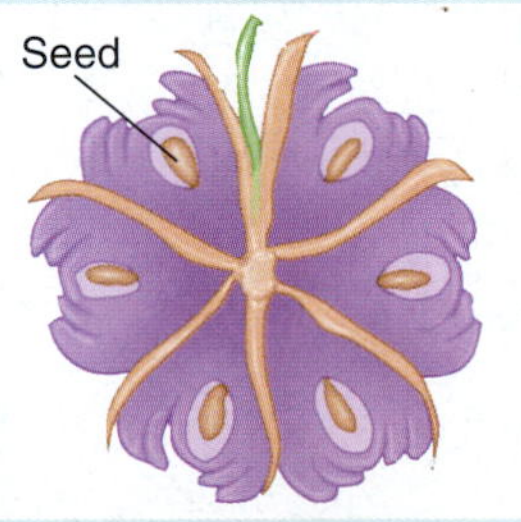

Multiple fruit
A fruit that develops from the ovaries of a group of flowers.

Mulberry *(Morus* sp.)

Accessory fruit
A fruit composed primarily of tissue (such as the receptacle) other than ovary tissue.

Apple *(Malus sylvestris)*

Figure 35-11 Fruit types. Fruits are botanically classified into four groups—simple, aggregate, multiple, and accessory fruits—based on their structure and mechanism of seed dispersal.

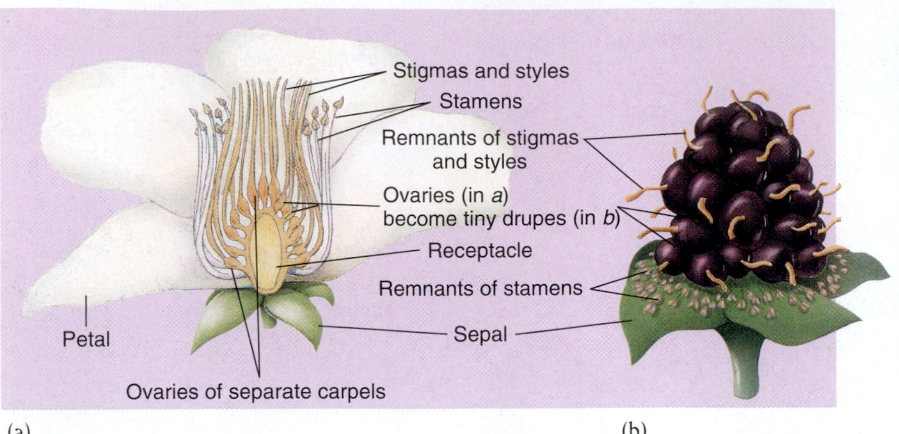

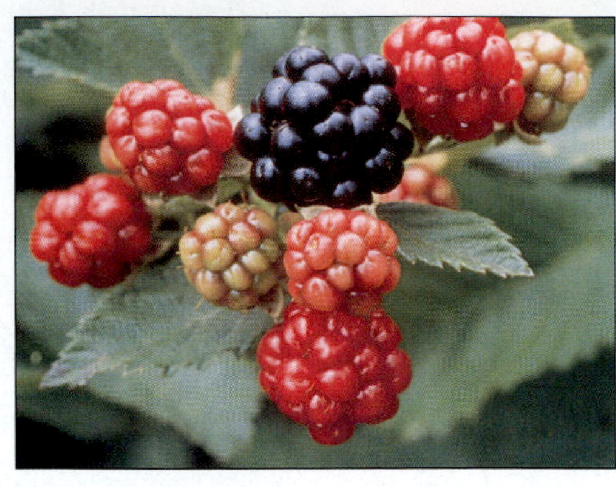

Figure 35–12 An aggregate fruit. **(a)** Cutaway view of a blackberry (*Rubus* sp.) flower, showing the many separate carpels in the center of the flower. **(b)** A developing blackberry fruit is an aggregate of tiny drupes. The little "hairs" on the blackberry are remnants of stigmas and styles. **(c)** Developing blackberry fruits at various stages of maturity. *(c, Dennis Drenner)*

Accessory fruits are the fourth type. They differ from other fruits in that other plant tissues in addition to ovary tissue make up the fruit. For example, the edible portion of a strawberry is the red, fleshy receptacle. Apples and pears are also accessory fruits; the outer part of each of these fruits is an enlarged *floral tube,* consisting of receptacle tissue along with portions of the calyx, that surrounds the ovary (Fig. 35–14).

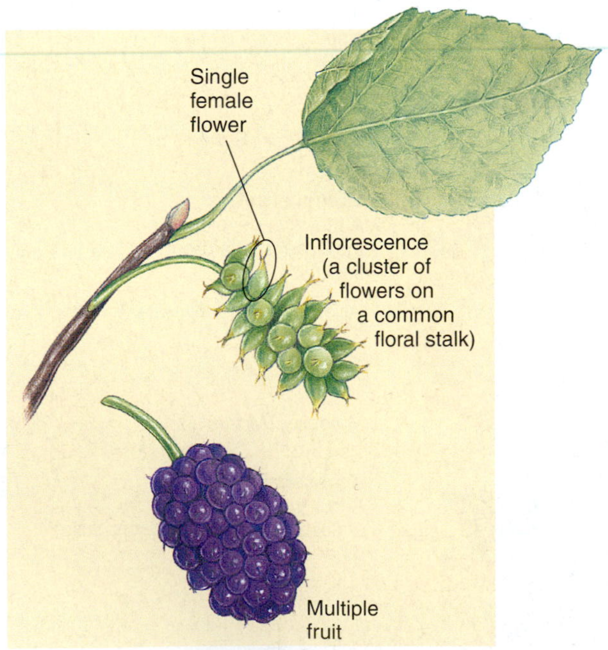

Figure 35–13 A multiple fruit. Mulberry (*Morus* sp.) is formed from the ovaries of many flowers that fused to become a multiple fruit. Mulberry flowers are imperfect and contain either stamens or pistils. The inflorescence of female flowers from which the mulberry fruit develops is also shown.

Seed dispersal is highly varied

Wind, animals, water, and explosive dehiscence disperse the various seeds and fruits of flowering plants. Effective methods of seed dispersal have made it possible for certain plants to expand their geographical range. In some cases, the seed is the actual agent of dispersal, whereas in others, the fruit performs this role. In tumbleweeds, such as Russian thistle, the entire plant is the agent of dispersal because it detaches and blows across the ground, scattering seeds as it bumps along. Tumbleweeds are lightweight but have a great deal of wind resistance and are sometimes blown many kilometers by the wind.

Wind is responsible for seed dispersal in many plants. Plants such as maple have winged fruits adapted for wind dispersal. Light, feathery plumes are other structures that allow seeds or fruits to be transported by wind, often for considerable distances. Both dandelion fruits and milkweed seeds (Fig. 35–15*a*) have this type of adaptation.

Some plants have special structures that aid in dispersal of their seeds and fruits by animals. The spines and barbs of burdock burs and similar fruits catch in animal fur and are dispersed as the animal moves about (Fig. 35–15*b*). Fleshy, edible fruits are also adapted for animal dispersal. As these fruits are eaten, the seeds are either discarded or swallowed. Many seeds that are swallowed have thick seed coats and are not digested, but instead are passed through the digestive tract and deposited in the animal's feces some distance from the parent plant. In fact, some seeds will not germinate unless they have passed through an animal's digestive tract; it is likely that the animal's digestive juices facilitate germination by helping to break down the seed coat. Some edible fruits apparently contain chemicals that function as laxatives to *speed* the passage of seeds through the animal's digestive tract; the longer these seeds spend in the digestive tract, the less likely they are to germinate.

Animals such as squirrels and many bird species also help disperse acorns and other fruits and seeds by burying them for

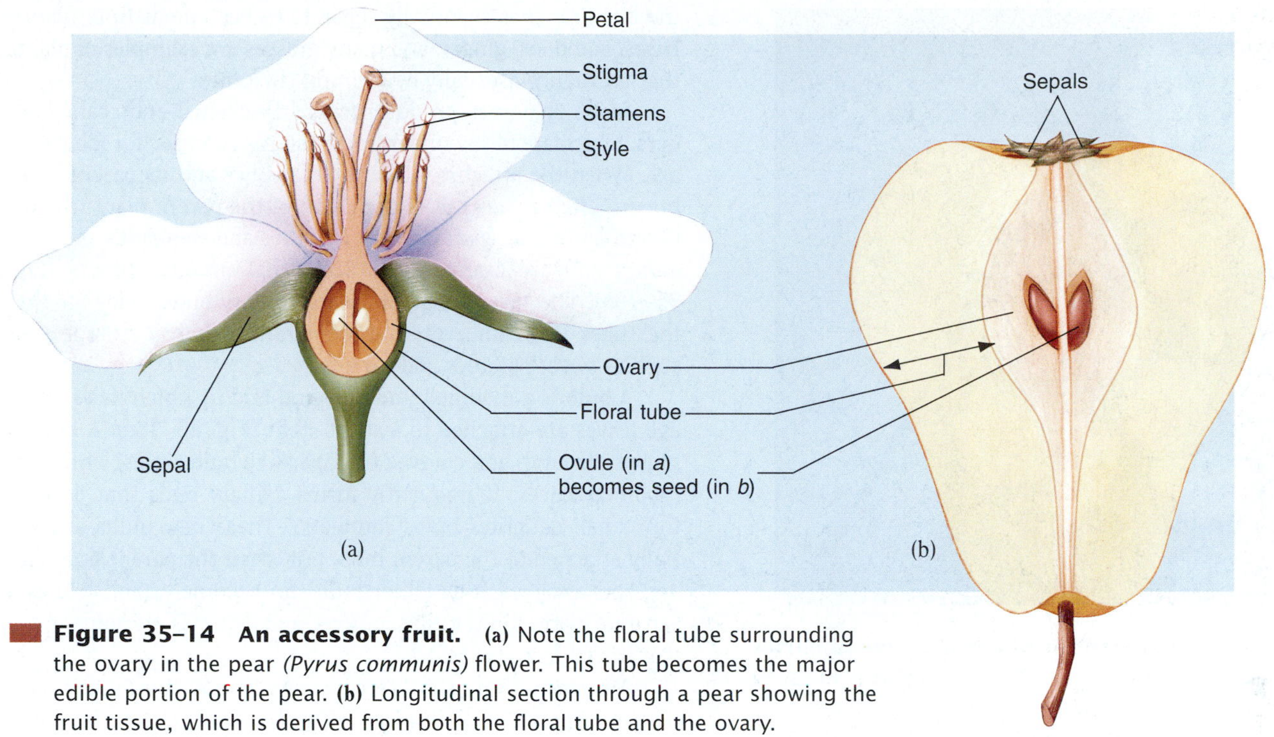

Figure 35–14 An accessory fruit. **(a)** Note the floral tube surrounding the ovary in the pear *(Pyrus communis)* flower. This tube becomes the major edible portion of the pear. **(b)** Longitudinal section through a pear showing the fruit tissue, which is derived from both the floral tube and the ovary.

winter use. Many buried seeds are never used by the animal and germinate the following spring. The role of burying seeds is performed for many plants by ants, which collect the seeds and take them underground to their nests. Ants disperse and bury seeds for hundreds of different plant species in almost every terrestrial environment, from northern coniferous forests to tropical rain forests to deserts.

Both ants and flowering plants benefit from their association. The ants ensure the reproductive success of the plants whose seeds they bury, and the plants supply food to the ants. A seed that is collected and taken underground by ants often contains a special structure called an *elaiosome,* or *oil body,* that protrudes from the seed (Fig. 35–16). Elaiosomes are a nutritious food for ants, which carry seeds underground before removing the elaiosome. Once an elaiosome is removed from a seed, the ants discard the undamaged seed in an underground refuse pile, which happens to be rich in organic material (such as ant droppings and dead ants) and contains the minerals required

(a)

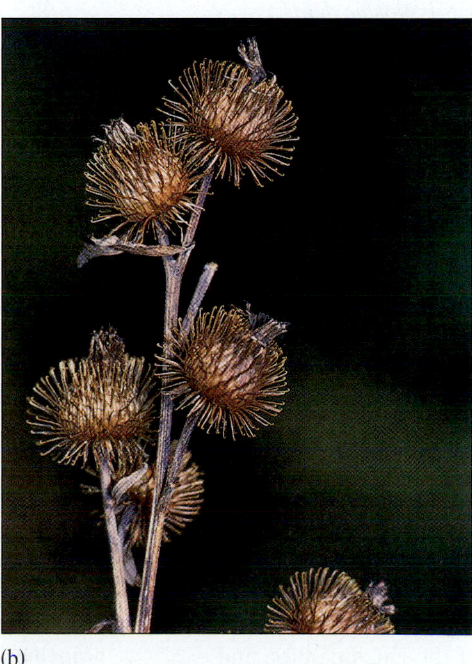

(b)

Figure 35–15 Seed (and fruit) dispersal. **(a)** The feathery plumes of milkweed *(Asclepias syriaca)* seeds make them buoyant for dispersal by wind. **(b)** Burdock *(Arctium minus)* burs (the hooked fruits) are carried from the parent plant after they become matted in bird feathers, mammal fur, or human clothing. *(a, John Serrao/Visuals Unlimited; b, DPA/Dembinsky Photo Associates)*

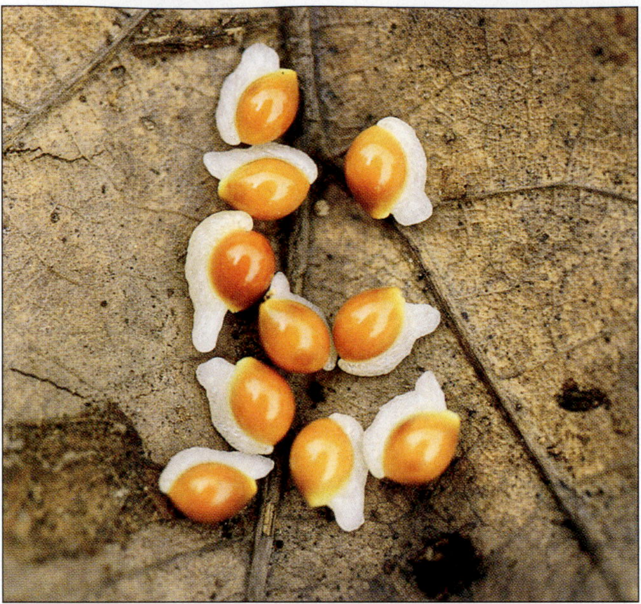

Figure 35–16 Dispersal of bloodroot *(Sanguinaria canadensis)* seeds by ants. The brown part is the seed proper, and the white part is the elaiosome, or oil body. The seeds have been placed along the midvein of an oak leaf to indicate scale. *(Marion Lobstein)*

by young seedlings. Thus, ants not only bury the seeds away from animals that might eat them but also place the seeds in rich soil that is ideal for germination and seedling growth.

The coconut is an example of a fruit adapted for dispersal by water. The coconut has air spaces and corky floats that make it buoyant and capable of being carried by ocean currents for thousands of kilometers. When it washes ashore, the seed within it may germinate and grow into a coconut palm tree.

Some seeds are dispersed neither by wind, animals, nor water. Such seeds are found in fruits that use *explosive dehiscence,* in which the fruit bursts open suddenly and quite often violently, forcibly discharging its seeds (Fig. 35–17). Pressures due to differences in turgor or to drying out cause these fruits to burst open. The fruits of plants such as touch-me-not and bitter cress split open so explosively that seeds are scattered a meter or more.

ASEXUAL REPRODUCTION IN FLOWERING PLANTS MAY INVOLVE MODIFIED STEMS, LEAVES, OR ROOTS

Flowering plants have many kinds of asexual reproduction, a number of which involve modified stems (rhizomes, tubers, bulbs, corms, and stolons). A **rhizome** is a horizontal underground stem that may or may not be fleshy. Fleshiness indicates that the rhizome is used for storing food materials such as starch (Fig. 35–18*a*). Although rhizomes resemble roots, they are really stems, as indicated by the presence of scalelike leaves, buds, nodes, and internodes. Rhizomes frequently branch in different directions. Over time, the old portion of the rhizome dies, and the two branches eventually separate to become distinct plants. Irises, bamboos, ginger, and many grasses are examples of plants that reproduce asexually by forming rhizomes.

Some rhizomes produce greatly thickened ends called **tubers,** which are fleshy underground stems enlarged for food storage. When the attachment between a tuber and its parent plant breaks, often as a result of the death of the parent plant, the tuber grows into a separate plant. Potatoes and elephant's ear (*Caladium* sp.) are examples of plants that produce tubers (Fig. 35–18*b*). The "eyes" of a potato are axillary buds, evidence that the tuber is an underground stem rather than a storage root such as sweet potatoes and carrots.

A **bulb** is a modified underground bud in which fleshy storage leaves are attached to a short stem (Fig. 35–18*c*). A bulb is globose (round) and covered by paper-like bulb scales, which are modified leaves. It frequently forms axillary buds that develop into small daughter bulbs (bulblets). These new bulbs are initially attached to the parent bulb, but when the parent bulb dies and rots away, each daughter bulb can become established as a separate plant. Lilies, tulips, onions, and daffodils are some of the plants that form bulbs.

A **corm** is a very short, erect underground stem that superficially resembles a bulb (Fig. 35–18*d*). Unlike the bulb, whose food is stored in underground leaves, the corm's storage organ is a thickened underground stem covered by papery scales (modified leaves). Axillary buds that give rise to new corms frequently arise; the death of the parent corm separates these daughter corms, which then become established as separate plants. Familiar garden plants that produce corms include crocus, gladiolus, and cyclamen.

Stolons, or **runners,** are horizontal above-ground stems that grow along the surface and are characterized by long internodes (Fig. 35–18*e*). Buds develop along the stolon, and each

Figure 35–17 Explosive dehiscence in bitter cress *(Cardamine pratensis).* **(a)** An intact fruit before it has opened. **(b)** The fruit splits open with explosive force, propelling the seeds some distance from the plant.

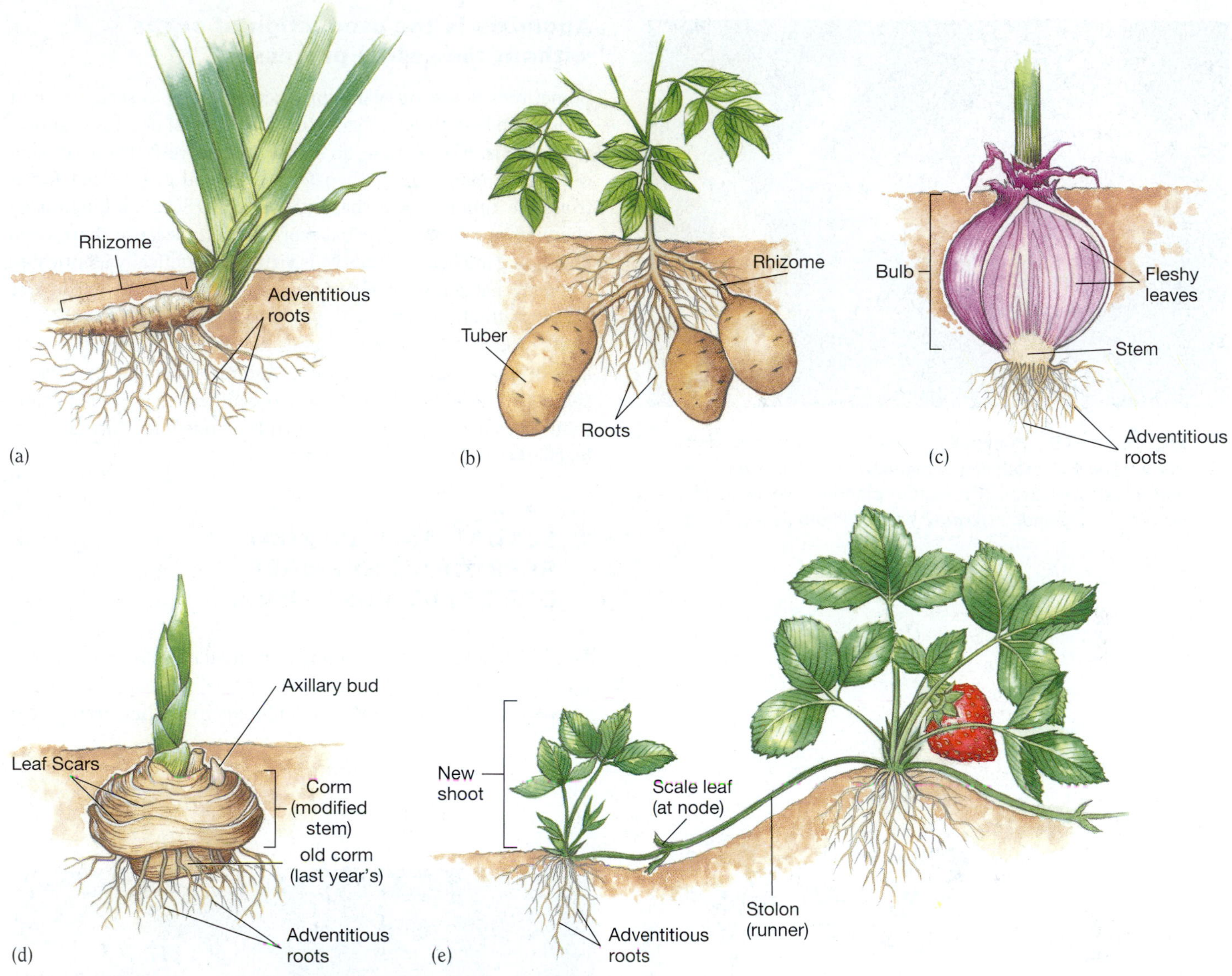

Figure 35–18 Modified stems. **(a)** Irises have horizontal underground stems called rhizomes. New aerial shoots arise from buds that develop on the rhizome. **(b)** Potato plants form rhizomes, which enlarge into tubers (the potatoes) at the ends. **(c)** A bulb is a short underground stem to which overlapping, fleshy leaves are attached; most of the bulb consists of leaves. **(d)** A corm is an underground stem that is almost entirely stem tissue surrounded by a few papery scales. **(e)** Strawberries reproduce asexually by forming stolons, or runners. New plants (shoots and roots) are produced at every other node.

bud gives rise to a new shoot that roots in the ground. When the stolon dies, the daughter plants live separately. The strawberry plant produces stolons.

Some plants are capable of forming detachable **plantlets** (small plants) in notches along their leaf margins. *Kalanchoe* sp., commonly called "mother of thousands," has meristematic tissue that gives rise to an individual plantlet at each notch in the leaf (Fig. 35–19). When these plantlets attain a certain size, they drop to the ground, root, and grow.

Some plants reproduce asexually by producing **suckers,** above-ground shoots that develop from adventitious buds on the roots. Each sucker grows additional roots and becomes an independent plant when the parent plant dies. Examples of plants that form suckers include black locust, pear, apple, cherry, blackberry, and aspen (Fig. 35–20). A quaking aspen *(Populus tremuloides)* colony in the Wasatch Mountains of Utah contains at least 47,000 tree trunks formed from suckers that can be traced back to a single individual; this massive "organism" occupies almost 43 hectares (106 acres). Some weeds, such as field bindweed, for example, can produce many suckers. These plants are difficult to control, because pulling the plant out of the soil seldom removes all of the roots, which can grow as deep as 3 m (10 ft). In fact, in response to wounding, the roots produce additional suckers, which can be a considerable nuisance.

Figure 35–19 Plantlets. The "mother of thousands" (*Kalanchoe* sp.) produces detachable plantlets along the margins of its leaves. The young plantlets will drop off and root in the ground. *(Jerome Wexler/Photo Researchers, Inc.)*

Apomixis is the production of seeds without the sexual process

Sometimes flowering plants produce embryos in seeds without meiosis and the fusion of gametes. This asexual process is known as **apomixis.** For example, an embryo may develop from a diploid cell in the ovule rather than from a diploid zygote that forms from the union of two haploid gametes. Seed production by apomixis is a form of asexual reproduction; because there is no fusion of gametes, the embryo is virtually genetically identical to the maternal genotype. However, the advantage of apomixis over other methods of asexual reproduction is that the seeds and fruits produced by apomixis can be dispersed by methods associated with sexual reproduction. Apomixis occurs in various species of more than 40 angiosperm families. Examples of plants that reproduce by apomixis include dandelions, citrus trees, blackberries, garlic, and certain grasses.

SEXUAL AND ASEXUAL REPRODUCTION HAVE DIFFERENT FUNCTIONS

Sexual and asexual reproduction are suited for different environmental circumstances. As you know, sexual reproduction results in offspring that are genetically different from the parents; that

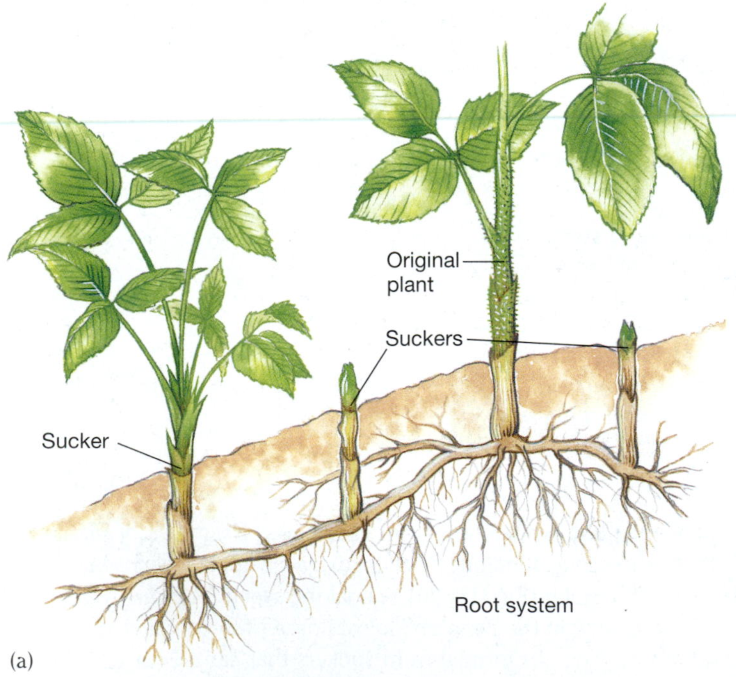

Figure 35–20 Suckers. **(a)** Suckers in various stages of development. All of the suckers are connected to one another through the root system. **(b)** A grove of quaking aspen trees *(Populus tremuloides)* is often descended from a single tree that reproduced asexually by forming suckers. Because all the trees in the grove are genetically identical, their responses to the environment are uniform. In spring they break dormancy simultaneously, and in the fall their leaves turn color at the same time. *(b, Sharon Cummings/ Dembinsky Photo Associates)*

is, the parental genotypes are not preserved in the offspring. This genetic diversity of offspring may be selectively advantageous, particularly in an unstable, or changing, environment. If a plant species that reproduces sexually (and is therefore genetically diverse) is exposed to increasing annual temperatures as a result of global warming, for example, some of the individuals may be more fit than either the parents or other offspring. The genetic diversity that results from sexual reproduction may also permit the individuals of a species to exploit new environments, thereby expanding their range.

You have seen that asexual reproduction results in offspring that are virtually genetically identical to the parent; that is, the parental genotype is preserved. Assuming that the parent is well adapted to its environment (i.e., has a favorable combination of alleles), this genetic similarity may be selectively advantageous if the environment remains stable (unchanging) for several generations. None of the offspring of asexual reproduction are more fit than the parent, but neither are any of them less fit.

Despite the apparent advantages of asexual reproduction, most plant species whose reproduction is primarily asexual occasionally reproduce sexually. Even in what appears to be a stable environment, plants are exposed to changing selective pressures, such as changes in the number and kinds of predators and parasites, the availability of food, competition from other species, and climate. Sexual reproduction permits species whose reproduction is primarily asexual to increase their genetic variability so at least some individuals are adapted to the changing selective pressures in a stable environment.

Sexual reproduction has some disadvantages

Although the genetic diversity produced by sexual reproduction is advantageous to a species' survival, sexual reproduction is a "costly" form of reproduction. In sexual reproduction, both male and female gametes are required, and these gametes have to meet for reproduction to occur. The many adaptations of flowers for different modes of pollination represent one of the costs of sexual reproduction.

Sexual reproduction produces some individuals with genotypes that are well adapted to the environment, but it also produces some individuals that are less well adapted. Therefore, sexual reproduction is usually accompanied by high death rates among offspring, particularly when selective pressures are strong. As discussed in Chapter 17, however, this aspect of sexual reproduction is an important part of evolution by natural selection.

Every biological process involves trade-offs, and sexual reproduction is no exception. Although sexual reproduction has its costs, the adaptive advantages of sexual reproduction clearly outweigh any disadvantages.

SUMMARY WITH KEY TERMS

I. The offspring produced by sexual reproduction are genetically variable.
   A. Sexual reproduction occurs in the flower. A flower may contain sepals, petals, stamens, and carpels (pistils).
      1. **Sepals** cover and protect the flower parts when the flower is a bud.
      2. **Petals** play an important role in attracting animal pollinators to the flower.
      3. **Stamens** produce pollen grains. Each stamen consist of a thin stalk (the **filament**) with a saclike structure (the **anther**).
      4. The **carpel** is the female reproductive unit. A **pistil** may consist of a single carpel or a group of fused carpels. Each pistil has three sections: a **stigma,** on which the pollen grains land; a **style,** through which the pollen tube grows; and an **ovary** that contains one or more **ovules.**
   B. Each pollen grain contains two cells. One generates two sperm cells, and the other produces a **pollen tube** through which the sperm cells will reach the ovule.
   C. An egg and two **polar nuclei,** along with several other nuclei, are formed in the ovule. Both egg and polar nuclei participate directly in fertilization.

II. **Pollination** is the transfer of pollen grains from anther to stigma.
   A. Some plants rely on animals to transfer pollen grains.
      1. **Coevolution** is reciprocal adaptation caused by two different species (such as flowering plants and their animal pollinators) forming an interdependent relationship and affecting the course of one another's evolution.
      2. Flowers pollinated by insects are often yellow or blue and possess a scent.
      3. Bird-pollinated flowers are often yellow, orange, or red and do not have a strong scent.
      4. Bat-pollinated flowers often have dusky white petals and are scented.
   B. Plants pollinated by wind often have smaller petals or lack petals altogether and do not produce a scent or nectar; wind-pollinated flowers make copious amounts of pollen.
   C. Plants have a variety of mechanisms that prevent self-pollination. Many species, for example, are **self-incompatible;** reproduction occurs only if the pollen comes from a genetically different individual.

III. After pollination, fertilization (fusion of gametes) occurs.
   A. A pollen tube grows down the style into the ovary, and the two sperm cells travel down the pollen tube into the ovule.
   B. Flowering plants have **double fertilization.**
      1. In the ovule, the egg fuses with one sperm cell, forming a zygote (fertilized egg) that eventually develops into a multicellular embryo in the seed.
      2. The two polar nuclei fuse with the second sperm cell, forming a nutritive tissue called **endosperm.**

IV. The seed and fruit develop as a result of successful fertilization.
   A. A dicot embryo develops in the seed in an orderly fashion, from proembryo to globular embryo to the heart stage to the torpedo stage. The mature flowering plant embryo consists of a **radicle,** a **hypocotyl, cotyledons** (one in monocots or two in dicots), and a **plumule.**
   B. A mature **seed** contains both a young embryo and nutritive tissue (stored in the endosperm or cotyledons) for use during germination.
   C. Seeds are enclosed within **fruits,** which are mature, ripened ovaries.
      1. **Simple fruits** develop from a single pistil that consists of one carpel or several fused carpels. Some simple fruits **(berries, drupes)** are fleshy at maturity, whereas others **(follicles, legumes, capsules, grains, achenes, nuts)** are dry.

2. **Aggregate fruits** develop from a single flower with many separate ovaries.
3. **Multiple fruits** develop from the ovaries of many flowers growing in close proximity on a common axis.
4. In **accessory fruits,** the major part of the fruit consists of tissue other than ovary tissue.

D. Seeds and fruits are adapted for various means of dispersal, including animals, wind, water, and explosive dehiscence.

V. Asexual reproduction involves the formation of offspring without the fusion of gametes. The offspring are virtually genetically identical to the single parent plant.

A. **Rhizomes, tubers, bulbs, corms,** and **stolons** are stems specialized for asexual reproduction.

B. Some leaves have meristematic tissue along their margins and give rise to detachable **plantlets.**

C. Roots may develop adventitious buds that develop into suckers. Suckers produce additional roots and may give rise to new plants.

D. **Apomixis** is the production of seeds and fruits without sexual reproduction.

VI. Sexual and asexual reproduction have different functions.

A. The parental genotypes are not preserved in the offspring of sexual reproduction.

1. Genetic diversity among offspring produced by sexual reproduction may be selectively advantageous, particularly in an unstable, or changing, environment.
2. Genetic diversity may also permit individuals to exploit new environments.
3. However, sexual reproduction is costly because both male and female gametes have to meet.

B. The parental genotype is preserved in asexual reproduction.

1. Genetic similarity may be selectively advantageous if the environment remains stable (unchanging) for several generations.
2. In asexual reproduction all individuals have the potential to produce offspring.
3. Despite the apparent advantages of asexual reproduction, most plant species whose reproduction is primarily asexual occasionally reproduce sexually, thereby increasing their genetic variability.

POST-TEST

1. In flowering plants, the ____________ is/are large (multicellular) and nutritionally independent. (a) gametes (b) microspores (c) megaspores (d) mature gametophyte (e) mature sporophyte
2. The normal order of whorls from the flower's periphery to the center is (a) sepals → petals → carpels → stamens (b) stamens → carpels → sepals → petals (c) sepals → petals → stamens → carpels (d) petals → carpels → stamens → sepals (e) carpels → stamens → petals → sepals
3. The pistil is composed of (a) stigma, style, and stamen (b) anther and filament (c) sepal and petal (d) stigma, style, and ovary (e) radicle, hypocotyl, and plumule
4. The petals of a flower are collectively called a(an) (a) calyx (b) capsule (c) carpel (d) cotyledon (e) corolla
5. The transfer of pollen grains from anther to stigma is known as (a) fertilization (b) double fertilization (c) pollination (d) germination (e) apomixis
6. The observation that long tubular flowers are pollinated by insects with long mouthparts, whereas short flowers are pollinated by insects with short mouthparts, is explained by (a) coevolution (b) germination (c) double fertilization (d) apomixis (e) explosive dehiscence
7. The process of ____________ in flowering plants involves one sperm cell fusing with an egg cell and one sperm cell fusing with two polar nuclei. (a) coevolution (b) germination (c) double fertilization (d) apomixis (e) pollination
8. The nutritive tissue in the seeds of flowering plants that is formed from the union of a sperm cell and two polar nuclei is called the (a) plumule (b) endosperm (c) cotyledon (d) hypocotyl (e) radicle
9. The ____________ is a multicellular structure that anchors the embryo and aids in nutrient uptake from the endosperm (a) proembryo (b) ovule (c) suspensor (d) cotyledon (e) pollen tube
10. After fertilization the ovule develops into a ____________, and the ovary into a ____________ (a) fruit; seed (b) seed; fruit (c) calyx; corolla (d) corolla; calyx (e) follicle; legume
11. In plants that lack endosperm in their mature seeds, the cotyledons function (a) to enclose and protect the seed (b) to aid in seed dispersal (c) as an absorptive embryonic root (d) to store food reserves (e) to attach the embryo within the ovule
12. ____________ fruits develop from many ovaries of a single flower, whereas ____________ fruits develop from the ovaries of many separate flowers (a) multiple; accessory (b) simple; accessory (c) aggregate; multiple (d) accessory; aggregate (e) simple; multiple
13. Apples, strawberries, and pears are examples of what kind of fruit? (a) accessory (b) simple (c) multiple (d) aggregate (e) legume
14. A horizontal, underground stem that may or may not be fleshy and that is often specialized for asexual reproduction is called a (a) stolon (b) bulb (c) corm (d) rhizome (e) tuber
15. Place the following events in the correct order: (1) pollen tube grows into ovule (2) insect lands on flower to drink nectar (3) embryo develops within the seed (4) fertilization occurs (5) pollen carried by insect contacts stigma (a) 2-5-1-4-3 (b) 1-4-2-5-3 (c) 3-2-5-1-4 (d) 5-1-3-4-2 (e) 2-5-4-3-1

REVIEW QUESTIONS

1. Distinguish between sexual and asexual reproduction, including the advantages and disadvantages of each.
2. What is the difference between pollination and fertilization? Which process occurs first?
3. Describe a "typical" insect-pollinated flower.
4. What is coevolution?
5. Put the following stages of dicot embryonic development in order and briefly describe each: torpedo stage, globular stage, proembryo, and heart stage.
6. Distinguish among simple, aggregate, multiple, and accessory fruits, and give examples of each.
7. Explain some of the features possessed by seeds and fruits that are dispersed by animals.
8. List and describe four modified stems that are involved in asexual reproduction.

9. Identify the fruit types below. Use Figure 35–11 to check your answers.

YOU MAKE THE CONNECTION

1. Draw pictures to show the kinds of flowers that might form simple, aggregate, multiple, and accessory fruits.
2. Could seed dispersal by ants be considered an example of coevolution? Why or why not?
3. Based on what you have learned in this chapter, speculate whether it is more likely that offspring of asexual reproduction develop in close proximity to or widely dispersed from the parent plant. Explain your reasoning. How could you design an experiment to test your hypothesis?
4. Which type of reproduction, sexual or asexual, might be more beneficial in the following circumstances, and why? (a) a perennial (plant that lives more than two years) in a stable environment; (b) an annual (plant that lives one year) in a rapidly changing environment; (c) a plant adapted to an extremely narrow climate range.
5. Based on what you have learned in this chapter, offer an explanation of why telephone poles and wires strung across grassy fields or plains often have tree seedlings growing under them.

RECOMMENDED READINGS

Berg, L.R. *Introductory Botany: Plants, People, and the Environment.* Saunders College Publishing, Philadelphia, 1997. A general botany text with an environmental emphasis.

Bernhardt, P. *The Rose's Kiss: A Natural History of Flowers.* Island Press, Washington, D.C., 1999. This eloquent, immensely readable account of floral biology will interest and entertain students and professors alike.

Hansen, E. "Bee Bop." *Natural History,* Mar. 1999. An engaging account of the bucket orchid's intriguing interaction with its bee pollinator.

Kearns, C.A., and D.W. Inouye. "Pollinators, Flowering Plants, and Conservation Biology." *BioScience,* Vol. 47, No. 5, May 1997. Threats to animal pollinators, such as habitat alteration and pesticides, also affect the well-being of the plants that require pollinators.

Marchand, P.J. "Seeds of Fortune." *Natural History,* Oct. 2000. Considers animal transport of seeds to suitable new territories.

Nicholson, R. "The Blackest Flower in the World." *Natural History,* May 1999. The first natural flower that is truly black has been documented in Mexico. Its animal pollinator is unknown.

Primack, R. "Life in Bloom." *Natural History,* May 1999. This article examines why some flowers are short-lived whereas others are long-lived (that is, last weeks or even months).

Raven, P.H., R.F. Evert, and S.E. Eichhorn. *Biology of Plants,* 6th ed. W.H. Freeman & Company, New York, 1999. This is an excellent general botany textbook, with detailed descriptions of reproduction in angiosperms.

Schwartz, D.M. "Birds, Bees, and Even Nectar-Feeding Bats Do It." *Smithsonian,* April 2000. Many animal pollinators are facing a variety of threats that may affect their effectiveness at pollinating flowering plants.

Tanksley, S.D., and S.R. McCouch. "Seed Banks and Molecular Maps: Unlocking Genetic Potential from the Wild." *Nature,* Vol. 277, 22 Aug. 1997. This article discusses how we may use the genetic potential of seed banks to benefit society.

Temeles, E.J., and P.W. Ewald. "Fitting the Bill?" *Natural History,* May 1999. This article examines some of the fascinating coevolution between birds and the flowers they pollinate.

- Visit our Web site at **http://www.info.brookscole.com/solomonbergmartin** for links to chapter-related resources on the World Wide Web. Additional on-line materials relating to this chapter can also be found on our Web site.

See chapter activity on BioActive Learner CD for additional help in mastering the chapter's material. Icon location in the chapter's margins shows which topics have tutorials or simulations in the CD.

36

Growth Responses and Regulation of Growth

Black-eyed Susan *(Rudbeckia hirta).* This plant, which grows to 0.9 m (3 ft), produces flowers in response to the shortening nights of spring and early summer. *(Dwight Kuhn)*

LEARNING OBJECTIVES

After you have studied this chapter you should be able to

1. Summarize the influence of environmental factors on the germination of seeds.
2. Discuss genetic and environmental factors that affect plant growth and development.
3. Explain how varying amounts of light and darkness induce flowering, and describe the role of phytochrome in flowering, including a brief discussion of phytochrome signal transduction.
4. Distinguish between a nastic movement and a tropism and describe phototropism, gravitropism, thigmotropism, and heliotropism.
5. Define circadian rhythm and give an example.
6. Diagram a general mechanism of action for plant hormones.
7. List several different ways each of the following hormones affects plant growth and development: auxins, gibberellins, cytokinins, ethylene, and abscisic acid.
8. Summarize what is known about the following plant hormones and hormone-like signaling molecules: brassinolides, salicylic acid, systemin, oligosaccharins, and jasmonates.

As you learned in Chapter 16, the ultimate control of plant growth and **development,** which includes all the changes that take place during the entire life of an individual, is genetic. If the genes required for development of a particular trait, such as the shape of the leaf, the color of the flower, or the type of root system, are not present, that characteristic does not develop. When a particular gene is present, its *expression,* that is, how it exhibits itself as an observable feature of an organism, is determined by a variety of factors, including signals from other genes and from the environment. The location of a cell in the young plant body, for example, has a profound effect on gene expression during development. Experiments suggest that chemical signals from adjacent cells may help the cell "perceive" its location within the plant body. Each cell's spatial environment, therefore, helps to determine what that cell ultimately becomes.

Environmental cues, such as changing day length and variations in precipitation and temperature, also exert an important influence on gene expression, as they do on all aspects of plant growth and development. The initiation of sexual reproduction is often under environmental control, particularly in temperate latitudes, and plants switch from vegetative growth (i.e., their leaves, stems, and roots grow) to reproductive growth after receiving the appropriate signals from the environment *(see photograph).* Such control is important for the plant's survival because the timing of sexual reproduction is critical for reproductive success: All flowering plants in temperate climates must flower and form seeds before the onset of winter induces **dormancy,** which is a temporary state of reduced physiological activity. Many flowering plants are sensitive to changes in the relative amounts of daylight and darkness that accompany the changing seasons, and these plants flower in response to those changes. Other plants have temperature requirements that induce sexual reproduction.

Plants, then, continually perceive information from the environment and use this information to help regulate normal growth and development. Growth and development are under the control of plant **hormones,** organic compounds that are present in very low concentrations in plant tissues and that act as highly specific chemical signals between cells. **Germination** (when a seed sprouts) and the growth of seedlings (young plants that develop from germinating seeds) into mature plants are aspects of growth and development. A plant's responses to changes in various conditions in its environment, including temperature, light, gravity, and touch, are other aspects of growth and development.

EXTERNAL AND INTERNAL FACTORS AFFECT GERMINATION AND EARLY GROWTH

In Chapter 35 you learned how pollination and fertilization are followed by seed and fruit development in flowering plants. Each seed develops from an ovule and contains an embryonic plant and food to provide nourishment for the embryo during germination. A mature seed, that is, a seed in which the embryo is fully developed, is often dormant and may not germinate immediately even if growing conditions are ideal.

Seed germination requires favorable environmental conditions

Within a given species, the precise requirements for seed germination represent evolutionary adaptations that protect the young seedlings from adverse environmental conditions. Environmental cues, such as the presence of water and oxygen, proper temperature, and sometimes the presence of light penetrating the soil surface, influence whether or not a seed germinates.

No seed germinates unless it has absorbed water. The embryo in a mature seed is dehydrated, and a watery environment is necessary for active metabolism. When a seed germinates, its metabolic machinery is turned on, and numerous materials are synthesized and others degraded. Therefore, water is an absolute requirement for germination. The absorption of water by a dry seed is known as **imbibition.** As a seed imbibes water, it often swells to several times its original, dry size (Fig. 36–1). Cells imbibe water by adhesion of water onto and into materials such as cellulose, pectin, and starches within the seed. Water molecules are attracted to and bound to these materials by adhesion, the attraction between unlike materials (see Chapter 2).

Figure 36–1 Imbibition. Dry seeds must imbibe water before they germinate. Pinto bean *(Phaseolus vulgaris)* seeds before imbibition *(left)* and after *(right)*. *(Marion Lobstein)*

Germination and subsequent growth require a great deal of energy. Because young plants obtain this energy by converting the energy of food molecules stored in the endosperm or cotyledons to ATP during aerobic respiration, much oxygen is usually needed during germination. Some plants, such as rice, can respire anaerobically during the early stages of germination and seedling growth. This enables rice plants to grow and become established in flooded soil, an environment that would suffocate most young plants.

Temperature is another environmental factor that affects germination. Each species has an optimal, or ideal, temperature at which the germination percentage is highest. For most plants, the optimal germination temperature is between 25° and 30° C (77° and 86° F). Some seeds, such as those of apples, require prolonged exposure to low temperatures before their seeds are able to break dormancy and germinate at any temperature. Some of the environmental factors that are needed for germination help ensure the survival of the young plant. The requirement of a prolonged low-temperature period ensures that seeds adapted to temperate climates germinate in the spring rather than in the fall.

Some plants, especially those with tiny seeds, such as lettuce, require light for germination. A light requirement ensures that a tiny seed germinates only if it is close to the surface of the soil. If such a seed were to germinate several centimeters below the soil surface, it might not have enough food reserves to grow to the surface. On the other hand, if this light-dependent seed remains dormant until the soil is disturbed and it is brought to the surface, it has a much greater likelihood of survival.

Some seeds do not germinate immediately, even when environmental requirements are met

In certain seeds, internal factors, which are under genetic control, prevent germination even when all external conditions are favorable. Many seeds are dormant because certain chemicals are present or absent or because the seed coat restricts germination. The presence of chemical inhibitors or the absence of chemicals that promote growth helps ensure the survival of the plant. For example, the seeds of many desert plants contain high levels of *abscisic acid* (discussed later in this chapter), which inhibits germination under unfavorable conditions. Abscisic acid is washed out only when rainfall is sufficient to support the plant's growth after the seed germinates.

Some seeds, such as certain legumes, have extremely hard, thick seed coats that prevent water and oxygen from entering, thereby inducing dormancy. *Scarification,* the process of scratching or nicking the seed coat (physically with a knife or chemically with an acid) before sowing it, induces germination in these plants. Two examples of how scarification occurs in nature are when these seeds pass through the digestive tracts of animals or when the seed coats are partially digested by soil bacteria.

Dicots and monocots exhibit characteristic patterns of early growth

Once conditions are right for germination, the first part of the plant to emerge from the seed is the **radicle,** or embryonic root. As the root grows and forces its way through the soil, it encounters considerable friction from soil particles. The delicate apical meristem of the root tip is protected by a root cap (see Chapter 34). The shoot is next to emerge from the seed. (Recall from Chapter 31 that a *shoot* is a collective term for a vertical stem with its leaves and reproductive structures.) Stem tips are not protected by a structure comparable to a root cap, but plants have ways to protect the delicate stem tip as it grows through the soil to the surface. The stem of a bean seedling (a dicot), for instance, curves over to form a hook so that the stem tip and cotyledons are actually *pulled up* through the soil (Fig. 36–2*a*). Corn and other grasses (monocots) have a special sheath of cells called a **coleoptile** that surrounds and protects the young shoot (Fig. 36–2*b*). First, the coleoptile pushes up through the soil, and then the leaves and stem grow through the tip of the coleoptile.

Certain parts of a plant grow throughout its life. This **indeterminate growth,** the ability to grow indefinitely, is characteristic of stems and roots, both of which arise from apical meristems. Hypothetically, stems and roots could continue to grow forever. On the other hand, many leaves and flowers have **determinate growth;** that is, they stop growing after reaching a certain size. The size of leaves and flowers with determinate growth varies from species to species and from individual to individual, depending on the plant's genetic programming and on environmental conditions, such as availability of sunlight, water, and essential nutrient minerals.

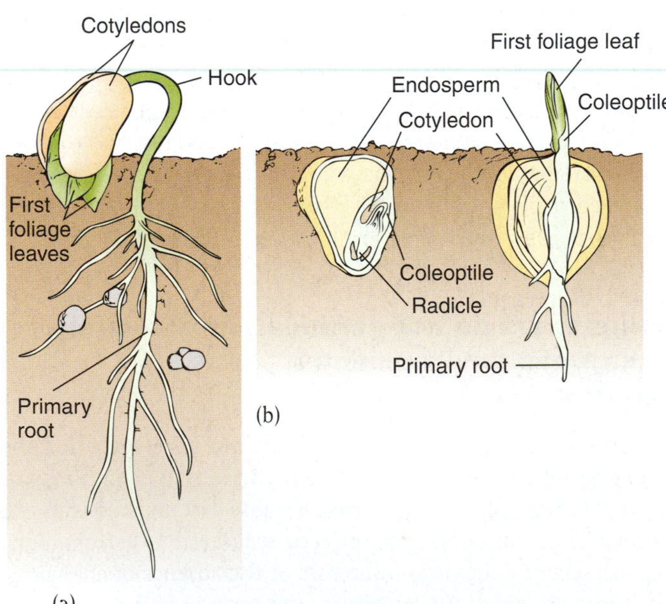

Figure 36–2 Germination and seedling growth. **(a)** Kidney bean, a dicot. Note the hook in the stem, which protects the delicate stem tip as it moves up through the soil. Once the shoot has emerged from the soil, the hook straightens. **(b)** Corn, a monocot. Note the coleoptile, a sheath of cells that emerges first from the soil. The shoot and leaves grow up through the middle of the coleoptile.

LIGHT CUES AFFECT FLOWERING AND OTHER PLANT RESPONSES

Photoperiodism is any response of a plant to the relative lengths of daylight and darkness. Initiation of flowering at the shoot apical meristem is one of several physiological activities that are photoperiodic in many plants. Plants are classified into four main groups—short-day, long-day, intermediate-day, and day-neutral—on the basis of how photoperiodism affects their transition from vegetative growth to flowering.

Short-day plants (also called **long-night plants**) flower when the night length is equal to or greater than some critical period (Fig. 36–3*a*). There are two kinds of short-day plants: qualitative and quantitative. In *qualitative short-day plants,* flowering occurs only in short days, whereas in *quantitative short-day plants,* flowering is accelerated by short days. The initiation of flowering in short-day plants is not due to the shorter period of daylight but to the long, uninterrupted period of darkness. The minimum critical night length varies considerably from one plant species to another but falls between 12 and 14 hours for many. Examples of short-day plants are florist's chrysanthemum, cocklebur, and poinsettia, which typically flower in later summer or fall. Poinsettias, for example, typically initiate flower buds in early October in the Northern Hemisphere and flower about eight to ten weeks later; hence, their traditional association with Christmas. Short-day plants can detect the lengthening nights of late summer or fall, and they flower at that time.

Long-day plants (also called **short-night plants**) flower when the night length is equal to or less than some critical period (Fig. 36–3*b*). In *qualitative long-day plants,* flowering occurs only in long days, whereas in *quantitative long-day plants,* flowering is accelerated by long days. Plants such as spinach, black-eyed Susan (see chapter introduction photo), and the model research organism *Arabidopsis thaliana* flower in late spring or summer and are long-day plants. These plants can detect the shortening nights of spring and early summer, and they flower at that time.

Intermediate-day plants flower when they are exposed to days and nights of intermediate length (Fig. 36–3*c*). Sugarcane and coleus are intermediate-day plants. These plants do not flower when night length is either too long or too short.

Some plants, called **day-neutral plants,** do not initiate flowering in response to seasonal changes in the period of daylight and darkness but instead respond to some other type of stimulus, external or internal. Cucumber, sunflower, corn, and onion are

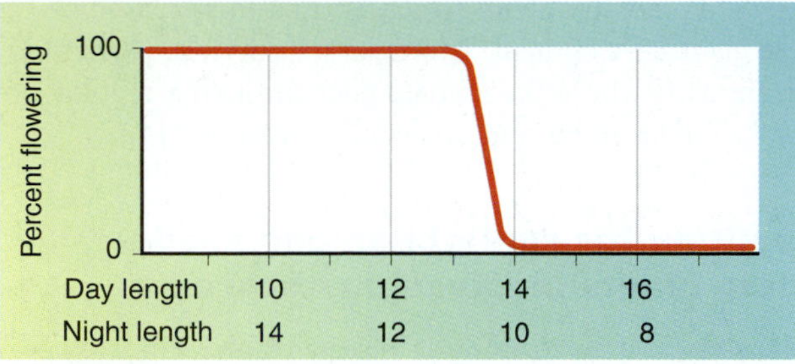

(a) Short-day plant

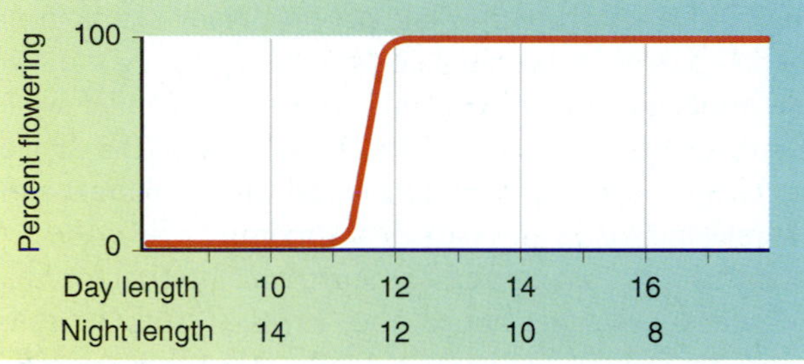

(b) Long-day plant

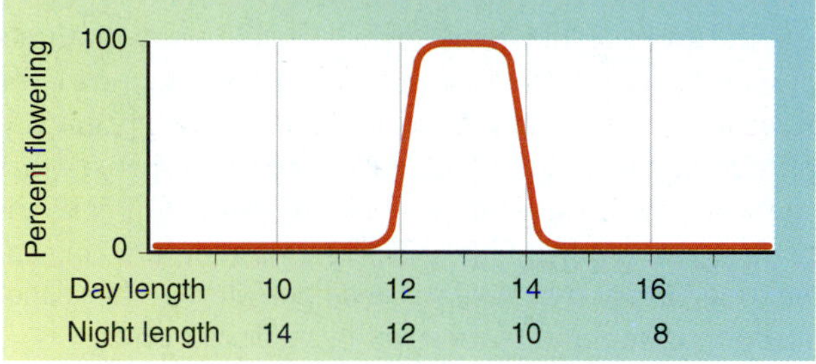

(c) Intermediate-day plant

Figure 36–3 Generalized photoperiodic responses in short-day, long-day, and intermediate-day plants. **(a)** Short-day plants, such as rice, florist's chrysanthemum, and cocklebur, flower when the night length is equal to or exceeds a certain critical length in a 24-hour period. **(b)** Long-day plants, such as spinach and black-eyed Susan, flower when the night length is equal to or less than a certain critical length in a 24-hour period. The critical length varies among different species; note that short-day plants may require a longer period of day length (i.e., a shorter night length) than long-day plants, as shown. **(c)** Intermediate-day plants, such as sugarcane and coleus, have a narrow night length requirement. If nights are longer or shorter than the period required, flowering does not occur.

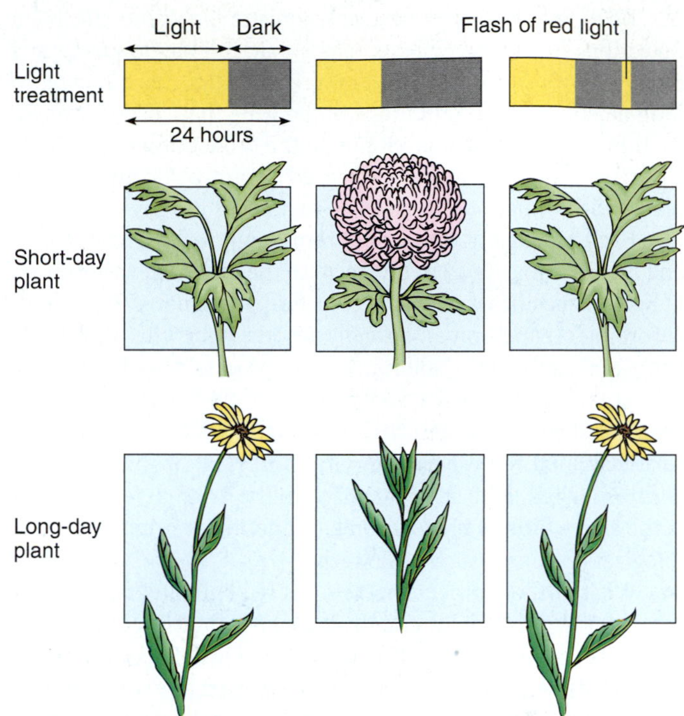

Figure 36–4 Photoperiodic responses of short-day and long-day plants. The short-day plant flowers when it is grown under long-night conditions *(top middle)*, but it does not flower when exposed to a long night interrupted with a brief flash of red light *(top right)*. The long-day plant does not flower when grown under long-night conditions *(bottom middle)* unless the long night is interrupted with a brief flash of red light *(bottom right)*.

examples of day-neutral plants. Many of these plants originated in the tropics where day length does not vary appreciably during the year. (In contrast, short-day, long-day, and intermediate-day plants are temperate, or mid-latitude, species.)

Plant biologists have experimented with the effects of various light regimens on flowering. Figure 36–4 shows how flowering in long-day and short-day plants is affected by different light treatments, including a short-day/long-night regimen with a *night break,* a short burst of light in the middle of the night.

Phytochrome detects day length

For a plant or any organism to have a biological response to light, it must contain a light-sensitive substance, called a *photoreceptor,* to absorb the light. The main photoreceptor for photoperiodism and many other light-initiated plant responses (such as germination, seedling establishment, and the structure, or architecture, of the mature plant) is **phytochrome,** a family of five or so blue-green pigment proteins, each of which is coded for by a different gene. A mixture of phytochrome proteins is present in cells of all vascular plants examined so far. For example, five members of the phytochrome family, designated phyA, phyB, phyC, phyD, and phyE, have been identified in the model research organism *Arabidopsis thaliana.*

Much of our current knowledge of phytochrome in *Arabidopsis* is based on various mutant plants that do not express a specific phytochrome gene, such as the gene that codes for phyA. By studying the physiological response of plants that do not produce an individual phytochrome, biologists have concluded that the individual forms of phytochrome have both unique and overlapping functions. PhyB appears to exert its influence at all stages of the plant life cycle, whereas the other forms of phytochrome have narrower functions at specific stages in the life cycle.

PhyA and phyB may have antagonistic (opposite) effects on flowering. In long-day plants, phyB may inhibit flowering and phyA may induce flowering. Flowering occurs more rapidly in long-day plants with mutations in the gene that codes for phyB; such mutations reduce or eliminate the production of phyB. On the other hand, flowering is delayed or prevented in long-day plants with mutations in the gene that codes for phyA.

Each member of the phytochrome family exists in two forms and readily converts from one form to the other after absorption of light of specific wavelengths. One form, designated Pr (for *r*ed-absorbing *p*hytochrome), strongly absorbs light with a relatively short red wavelength (660 nm). In the process, the shape of the molecule changes to the second form of phytochrome, Pfr, so designated because it absorbs *f*ar-*r*ed light, which is light with a relatively long red wavelength (730 nm) (Fig. 36–5). When Pfr absorbs far-red light, it reverts back to the original form, Pr. Pfr is the active form of phytochrome, triggering or inhibiting physiological responses such as flowering.

What does a pigment that absorbs red light and far-red light have to do with daylight and darkness? Sunlight is composed of various amounts of the entire spectrum of visible light, in addition to ultraviolet and infrared radiation. Because sunlight contains more red than far-red light, however, when a plant is exposed to sunlight, its level of Pfr increases. During the night, its level of Pfr slowly decreases as Pfr is degraded.

The importance of phytochrome to plants cannot be overemphasized. Timing of day length and darkness is the most reliable way for plants to measure the change from one season to the next. This measurement, which is used to synchronize the stages of plant development, is crucial for survival, particularly in environments where the climate goes through a regular, annual pattern of favorable and unfavorable seasons.

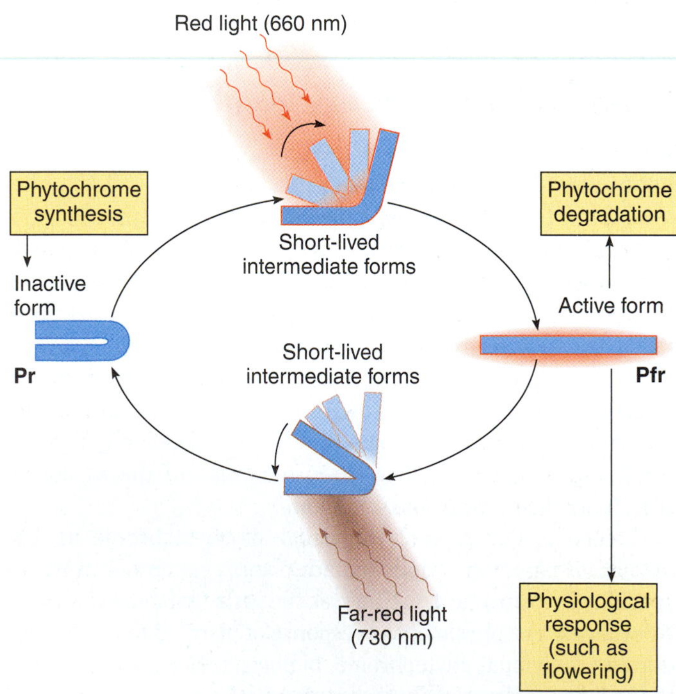

Figure 36–5 Phytochrome. This pigment occurs in two forms, designated Pr and Pfr, and readily converts from one form to the other. Red light (660 nm) converts Pr to Pfr, and far-red light (730 nm) converts Pfr to Pr.

Competition for sunlight among shade-avoiding plants involves phytochrome

Can plants sense the proximity of nearby plants? The answer is yes. Plants not only detect the presence of their neighbors but they also react to these plants, which are potential competitors, by changing the way they grow and develop. Many plants, from small herbs to large trees, compete for light, a response known as **shade avoidance,** in which plants tend to grow taller when closely surrounded by other plants. If successful, the shade-avoiding plant is able to project its new growth into direct sunlight, thereby increasing its chances of survival.

Botanists have recognized the environmental factor that triggers shade avoidance since the 1970s: Plants perceive changes in the ratio of red to far-red light that result from the presence of nearby plants. The leaves of neighboring plants absorb much more red light than far-red light. (Recall from Chapter 8 that the green pigment chlorophyll used in photosynthesis strongly absorbs red light.) In a densely plant-populated area, the ratio of red light to far-red light (r/fr) decreases. The greater relative amount of far-red light triggers a series of responses that cause the shade-avoiding plant, which is adapted to full light environments, to grow taller. Figure 36–6 shows an interesting application of additional (reflected) far-red light without the reduction in incoming sunlight that occurs in shade avoidance.

When a plant is using many of its resources for stem elongation, it has fewer resources to allocate for new leaves and branches, storage tissues, or reproductive tissues. However, for a shade-avoiding plant that is shaded by its neighbors, a rapid increase in stem length is advantageous because once this plant is taller than its neighbors, it obtains a larger share of unfiltered sunlight.

Phytochrome is involved in many other responses to light, including a light requirement for germination

Phytochrome is involved in the light requirement that some seeds have for germination. Seeds with a light requirement must be exposed to light containing the red wavelengths. Exposure to red light converts Pr to Pfr, and germination occurs. Many temperate forest species with small seeds require light for germination. (Larger seeds generally do not have a light requirement.) This requirement is an adaptation that enables the seeds to germinate at the optimal time to ensure their survival. During early spring, sunlight, including red light, penetrates the bare branches of overlying deciduous trees and reaches the soil between the trees. As spring temperatures warm, the seeds on the soil absorb red light and germinate. During their early growth, the newly germinated seedlings do not have to compete with the taller trees for sunlight.

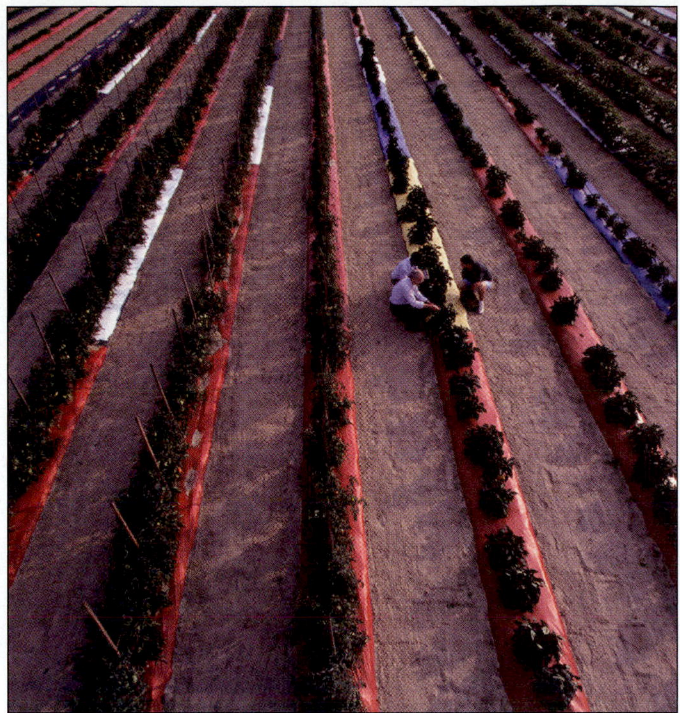

Figure 36-6 Using colored mulches to affect plant growth processes. The red mulch reflects far-red light from the ground to the tomato plants. The plants, absorbing the extra far-red light, react as if nearby plants are present. Their aboveground growth is greater, and their fruits ripen earlier. Manipulating mulch colors can modify yield, flavors, and nutrient content of plants such as strawberries. *(Cary Wolinsky)*

Other physiological functions under the influence of phytochrome include sleep movements in leaves (discussed later); shoot dormancy; leaf abscission (see Chapter 32); and pigment formation in flowers, fruits, and leaves.

Process of Science: Progress is being made in phytochrome signal transduction

Some phytochrome-induced responses are very rapid and short-term, whereas others are slower and long-term. The rapid responses probably involve reversible changes in the properties of membranes, thereby altering cellular ionic balances. Red light, for example, causes potassium ion (K^+) channels to open in plant cell membranes that are involved in plant movements caused by changes in turgor (discussed later). Exposure to far-red light causes the K^+ channels to close. Slower phytochrome-induced responses involve the selective regulation of transcription of numerous genes. For example, phytochrome activates transcription of the gene for the small subunit of **Rubisco,** an enzyme involved in photosynthesis (see Chapter 8).

Each phytochrome molecule consists of a protein attached to a light-absorbing photoreceptor. It is thought that the absorption of light by the photoreceptor portion of phytochrome elicits a change in the shape of the larger protein component. This change in turn triggers one or more **signal transduction** pathways that amplify the original signal and ultimately result in a physiological or developmental response, such as promotion or inhibition of flowering. (It may be helpful to review the general discussion of signal transduction in Chapter 5.)

Elucidating phytochome signal transduction has been the focus of intense research interest. Investigators have analyzed mutant plants that respond as if they have been exposed to a particular light stimulus even when they have not. One of the initial goals was to determine the molecule that interacts directly with phytochrome. Research indicated that phytochrome affects gene expression by activating a **transcription factor** in the nucleus (see Chapter 13). When activated, the transcription factor, which is bound to the promoter of a gene, either turns on or represses transcription that leads to protein synthesis. (Recall from Chapters 12 and 13 that the **promoter** is the nucleotide sequence in DNA to which RNA polymerase attaches to begin transcription.) Because there are many transcription factors, biologists have tried and finally succeeded in identifying a transcription factor involved with phytochrome signaling (see *On the Cutting Edge: How Light Regulates Gene Expression*).

TEMPERATURE MAY AFFECT REPRODUCTION

Certain plants have a temperature requirement that must be met if they are to flower. The promotion of flowering by exposure to low temperature for a period of time is known as **vernalization.** Although the exact temperature and amount of time required vary among species, vernalization typically occurs between 0° and 10° C (32° to 50° F) for a three- to eight-week period. The shoot apical meristem appears to be the site of the low-temperature perception; flowering is promoted when the shoot apex is chilled but the rest of the plant is not.

Different plants have different sensitivities to vernalization. In some plants the requirement of a low-temperature period is absolute, meaning that they will not flower unless they have been vernalized. Other plants flower sooner if exposed to low temperatures but will still flower later if they are not exposed to low temperatures.

Examples of plants with a low-temperature requirement include annuals such as winter wheat, which grow, reproduce, and die in one year, and biennials such as carrots, which take two years to complete their life cycles. Winter wheat is planted in the fall and germinates at that time. The shoot apical meristems of the young seedlings are exposed to low temperatures during the winter and subsequently flower after resuming growth the following spring. Winter wheat is a particularly suitable crop in the Midwest and Great Plains states because it can be harvested early, before the summer drought.

Carrots and other biennials grow vegetatively the first year and store surplus food in their roots. If the roots are not harvested, the plants flower and reproduce sexually during the second year, after their apical meristems are exposed to the low temperatures of winter. Carrots growing in a warm environment and

(text continues on page 778)

ON THE CUTTING EDGE

How Light Regulates Gene Expression

HYPOTHESIS: The transcription factor PIF3 is involved in the mechanism of phytochrome action.

METHOD: To test the hypothesis, researchers used molecular techniques to determine if PIF3 binds to DNA at a specific promoter sequence and, if so, whether phytochrome interacts with the DNA-bound PIF3.

RESULTS: PIF3 binds to a DNA sequence that is present in several light-regulated gene promoters. Phytochrome binds reversibly to the PIF3-promoter complex.

CONCLUSION: PIF3 plays a crucial role in phytochrome signal transduction.

For several years, Dr. Peter H. Quail and colleagues at the University of California, Berkeley, and the U.S.Department of Agriculture Western Regional Research Center in Albany, California, have actively worked on the mechanism that links phytochrome action to light-responsive genes. They identified and synthesized in vitro* a nuclear protein, designated **phytochrome-interacting factor-3 (PIF3),** a transcription factor that binds directly with phytochrome.

These investigators then convincingly demonstrated that phytochrome binds to PIF3 only when it is in its active form, Pfr, by using a series of pulse irradiations in vitro. For example, a 5-minute pulse of red light (which would convert Pr to Pfr) induced radioactively labeled phytochrome to bind to PIF3, which was immobilized on beads. However, no binding was detected when the 5-minute pulse of red light was followed by a 5-minute pulse of far-red light (which would convert Pfr to Pr).†

Additional studies indicated that PIF3 binds to DNA at a *palindromic sequence* (-CACGTG-) that reads the same as its complement, but in the opposite direction (see Chapter 14). A molecular technique known as the *electrophoretic mobility shift assay* (*EMSA*) verified the binding of PIF3 to a DNA fragment containing -CACGTG-. The EMSA technique is based on the observation that the binding of a protein to a fragment of DNA slows its movement during gel electrophoresis (see Fig. 14–8). By mixing PIF3 with the DNA fragment containing -CACGTG-, and then running it on a gel along with control samples in which PIF3 and the DNA fragments were not mixed, the scientists were able to identify a low-mobility band containing the PIF3-DNA complex.‡

The researchers also used EMSA to determine whether phytochrome would bind to the PIF3-DNA complex. Phytochrome in the Pr form did not bind to the PIF3 bound to DNA. They found, however, that an additional low-mobility band corresponding to a phytochrome/PIF3/DNA complex, was present only in the samples that had been irradiated with red light prior to electrophoresis. These data indicate that phytochrome in the Pfr form binds to the DNA-bound PIF3.

To determine if the binding of phytochrome to the PIF3-DNA complex is reversible, researchers tried several combinations of red and far-red treatments. They incubated phytochrome with the DNA-bound PIF3 in the dark after they exposed it to one of the following light treatments: red light only; far-red light only; red light followed by far-red light; and far-red light followed by red light. When red light was the final treatment before dark incubation, gel electrophoresis (EMSA) revealed a low-mobility band corresponding to a complex of phytochrome, PIF3, and DNA (Fig. A). When far-red light was the final treatment, the band was not present.

These and similar data suggest the main steps in phytochrome signal transduction with PIF3. The pathway is elegantly simple (Fig. B). Inactive phytochrome (Pr) in the cytoplasm absorbs red light and is converted into the active form, Pfr, which moves into the nucleus. There, phytochrome binds to the transcription factor PIF3 (which is already bound to the promoter) and stimulates the transcription of light-responsive genes. The signal transduction pathway, as currently understood, may be summarized as follows:

Red light → conversion of Pr to Pfr → movement of Pfr to nucleus → formation of Pfr-PIF3 complex that is bound to promoter region → light-responsive gene is switched on (or off)

The signal transduction pathway is shut down by far-red light, which is absorbed by the Pfr in the Pfr-PIF3 complex in the nucleus. When Pfr is converted to Pr, it dissociates from PIF3.

The signal transduction pathway just described is not the end of the story. There are probably other non-PIF3 pathways by which phytochrome regulates light-responsive gene expression. Some of the molecules that may be involved in other phytochrome-mediated signaling pathways include Ca^{2+}, calmodulin, G proteins, and cyclic guanosine monophosphate (cGMP); these molecules have also been implicated in signal transduction pathways in animals.

***In vitro* refers to a biological process occurring under experimental conditions outside an intact cell or organism; *in vivo* refers to a biological process that takes place in an intact cell or organism.

†Ni, M., J.M. Tepperman, and P.H. Quail. "Binding of Phytochrome B to Its Nuclear Signaling Partner PIF3 Is Reversibly Induced by Light." *Nature*, Vol. 400, 19 Aug. 1999.

‡Martinez-Garcia, J.F., E. Huq, and P.H. Quail. "Direct Targeting of Light Signals to a Promoter Element-Bound Transcription Factor." *Science*, Vol. 288, 5 May 2000.

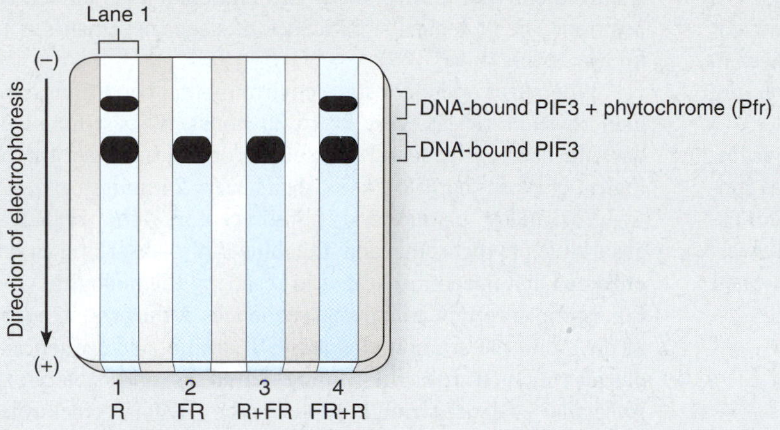

Figure A Reversibility of the phytochrome-PIF3-DNA complex. The electrophoretic mobility shift assay (EMSA) was performed after a mixture of phytochrome and DNA-bound PIF3 was exposed to a variety of light treatments: (1) red light only; (2) far-red light only; (3) red light followed by far-red light; and (4) far-red light followed by red light. When red light was the final treatment before dark incubation (lanes 1 and 4), gel electrophoresis revealed a low-mobility band corresponding to a complex of phytochrome (as Pfr), PIF3, and DNA. When far-red light was the final treatment (lanes 2 and 3), the band was absent because phytochrome as Pr did not bind to the DNA-bound PIF3. *(Adapted from Martinez-Garcia et al., 2000)*

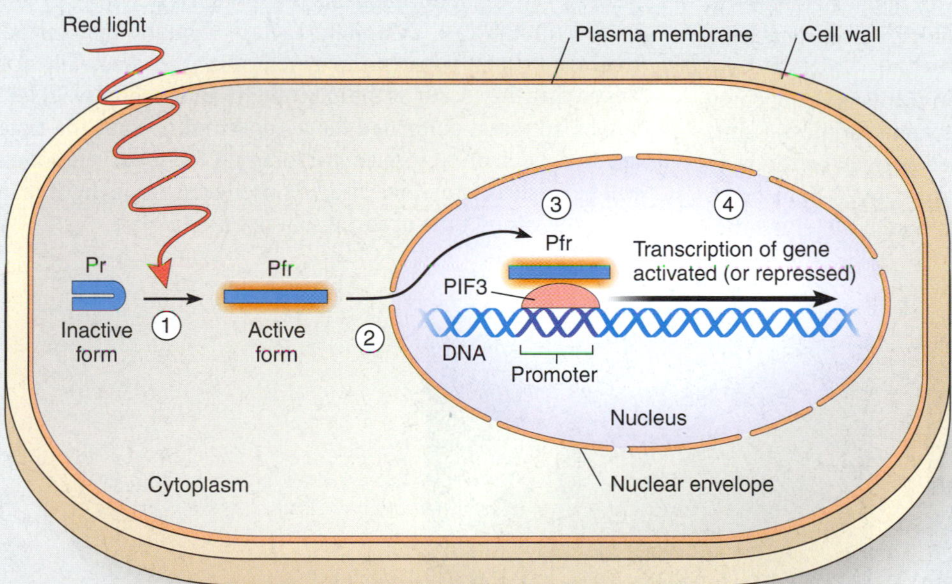

Figure B Phytochrome signal transduction. (1) Pr in the cytoplasm absorbs red light, converting it to the active form, Pfr. (2) Pfr moves into the nucleus and (3) binds to the transcription factor PIF3, which has already formed a complex with the promoter. (4) The binding of phytochrome to the PIF3-DNA complex activates (or represses) the transcription of light-regulated gene(s).

not exposed to low temperatures continue vegetative growth and do not initiate sexual reproduction.

Many plants that must be vernalized are also under the influence of photoperiodism. The most common combination is a low-temperature requirement followed by a requirement for long days. Both *Arabidopsis* and winter wheat, which flower in early summer, are examples of long-day plants that also require a cold treatment.

An external stimulus to which a plant responds, such as low temperature, may be moderated and influenced by internal conditions, such as hormone levels in the plant. It is possible, for example, to eliminate the low-temperature requirement for flowering in biennials by treating the plants with *gibberellin,* a plant hormone discussed later in this chapter.

A BIOLOGICAL CLOCK INFLUENCES MANY PLANT RESPONSES

Almost all organisms, including plants, animals, fungi, eukaryotic microorganisms, and many prokaryotes, appear to have an internal timer, or biological clock, that approximates a 24-hour cycle, the time it takes for Earth to rotate around its own axis. These internal cycles, known as **circadian rhythms** (from the Latin *circum,* "around," and *diurn,* "daily"), help the organism detect the time of day. In contrast, photoperiodism enables a plant to detect the time of year.

One example of a circadian rhythm in plants is the opening and closing of stomata, independent of light and darkness. Plants placed in continual darkness for extended periods continue to open and close their stomata on an approximate 24-hour cycle. **Sleep movements** observed in the common bean and other plants are another example of a circadian rhythm in plants (Fig. 36–7). During the day, bean leaves are horizontal, possibly for optimal light absorption, but at night the leaves fold down or up, a movement that orients them perpendicular to their daytime position. The biological significance of sleep movements is unknown at this time.

When kept under constant environmental conditions, circadian rhythms repeat every 20 to 30 hours, at least for several days. In nature, the rising and setting of the sun reset the biological clock so that the cycle repeats every 24 hours.

For many plants, two photoreceptors—the red light–absorbing phytochrome and the blue light–absorbing **cryptochrome**—have been implicated in resetting the biological clock. For example, certain amino acid sequences of the protein portion of phytochrome are homologous with amino acid sequences of clock proteins in fruit flies, fungi, mammals, and bacteria; this molecular evidence strongly supports the circadian clock role of phytochrome. The evidence for the universality of cryptochrome is also convincing. First discovered in plants, cryptochrome counterparts have also been reported in the fruit fly and mouse biological clocks.

Why do plants and other organisms exhibit circadian rhythms? Predictable environmental changes, such as sunrise and sunset, occur during the course of each 24-hour period. These predictable changes may be important to an individual organism, causing it to change its physiological activities or its behavior (in the case of animals). It is thought that circadian rhythms help an organism to synchronize repeated daily activities so that they occur at the appropriate time each day. If, for example, an insect-pollinated flower does not open at the time of day that pollinating insects are foraging for food, reproduction will be unsuccessful. Likewise, if a firefly flashes its light on and off at the wrong time of day, it will not find a mate.

(a)

(b)

Figure 36–7 Sleep movements in a bean *(Phaseolus vulgaris)* seedling. **(a)** Leaf position at noon. **(b)** Leaf position at midnight. It is not known why some plants exhibit sleep movements. *(a, b, Dennis Drenner)*

CHANGES IN TURGOR CAN INDUCE NASTIC MOVEMENTS

The sensitive plant *(Mimosa pudica)* dramatically folds its leaves and droops in response to touch (or to an electrical, chemical, or thermal stimulus) (Fig. 36–8). The response, which typically occurs in a few seconds, spreads throughout the plant even if only one leaflet is initially stimulated. When a sensitive plant is touched, an electrical impulse moves down the leaf to special cells housed in structures called **pulvini** (sing., *pulvinus*) at the base of each leaflet, each cluster of leaflets, and each petiole. Each pulvinus is a somewhat swollen joint that acts as a hinge. When the electrical signal reaches cells in the pulvinus, it induces a chemical signal that increases membrane permeability to certain ions. A loss of turgor occurs in certain pulvinus cells as potassium ions (K^+) exit through the now permeable plasma membrane, causing water to leave the cells by osmosis (see Chapter 5 for a discussion of turgor pressure). The sudden change in turgor causes the leaf movement. Such **nastic movements** occur in response to external stimuli, but the direction of movement is predetermined and is independent of the direction of the stimulus. Nastic movements are temporary and reversible. The movement of K^+ and water back into the pulvinus cells causes the plant part to return to its original position, although recovery takes several to many minutes longer than the original movement.

A TROPISM IS DIRECTIONAL GROWTH IN RESPONSE TO AN EXTERNAL STIMULUS

A plant may respond to an external stimulus, such as light, gravity, or touch, by directional growth (i.e., the direction of growth depends on the direction of the stimulus). Such a directional growth response, called a **tropism,** results in a change in the position of a plant part. Tropisms are irreversible and may be positive or negative, depending on whether the plant grows toward the stimulus (a positive tropism) or away from it (a negative tropism). Tropisms are under hormonal control, which is discussed later in this chapter.

Phototropism is the directional growth of a plant caused by light (Fig. 36–9). Most growing shoot tips exhibit positive phototropism by bending (growing) toward light, something you may have observed if you place houseplants near a sunny window. This growth response increases the likelihood that stems and leaves will receive adequate light for photosynthesis. The bending response of phototropism is triggered by blue light with wavelengths less than 500 nm.

The photoreceptor that absorbs blue light and triggers the phototropic response is thought to be a yellow pigment. Traditionally, two families of yellow pigments—flavins and carotenoids—have been leading candidates for the blue-light photoreceptor. In 1998, researchers convincingly demonstrated

Figure 36–8 Nastic movements. The sensitive plant *(Mimosa pudica)* is a compact shrub that grows to 0.9 m (3 ft). Leaves on the plant were photographed before **(a)** and several seconds after **(b)** being touched. Note how the compound leaves have folded and drooped. **(c)** Pulvini are located in three areas—at the base of each leaflet, at the base of each cluster of leaflets, and at the base of each leaf. Only changes in the pulvini at the base of leaflets are shown. *(Top right)* Section through two leaflets, showing their pulvini when the leaf is undisturbed. *(Bottom right)* Section through the two leaflets, showing how a loss of turgor produces the folding of the leaves. *(a, b, Dennis Drenner)*

Figure 36-9 Phototropism. Stems of corn *(Zea mays)* seedlings grow in the direction of light and therefore exhibit positive phototropism. The bending toward light is caused by greater elongation on the shaded side of a stem than on the lighted side. *(Runk/Schoenberger/Grant Heilman)*

in *Arabidopsis* that a flavoprotein (a flavin attached to a protein molecule) designated NPH1 is the photoreceptor for phototropism. Knowledge of the signal transduction pathway by which blue light triggers phototropism and other responses in plants is presently incomplete, but there is evidence that NPH1 becomes phosphorylated—that is, a phosphate group is added—in response to blue light. Thus, phosphorylation appears to be the first step in the blue light–signaling pathway.

Growth in response to the direction of gravity is called gravitropism. Most stem tips exhibit negative **gravitropism** by growing away from the center of Earth, whereas most root tips exhibit positive gravitropism (Fig. 36–10). The root cap is the site of gravity perception in roots; when the root cap is removed, the root continues to grow, but it loses any ability to perceive gravity. Special cells in the root cap possess starch-containing **amyloplasts** that collect toward the bottoms of the cells in response to gravity, and it is thought that these amyloplasts may initiate at least some of the gravitropic response. If the root is moved, as when a potted plant is laid on its side, the amyloplasts tumble to a new position, always settling in the direction of gravity. The gravitropic response (bending) occurs shortly thereafter and involves the hormone *auxin* (discussed later in the chapter). Despite the movement of amyloplasts in response to gravity, their role in gravitropism has been questioned. A mutant *Arabidopsis* plant that lacks amyloplasts in its root cap cells still responds gravitropically when placed on its side, indicating that roots do not necessarily need amyloplasts to respond to gravity. Ongoing research may help clarify how roots perceive gravity.

Thigmotropism is growth in response to a mechanical stimulus, such as contact with a solid object. The twining or curling growth of tendrils or stems, which helps attach a climbing plant such as a vine to some type of support, is an example of thigmotropism (see Fig. 32–15*b*).

Heliotropism, also called **solar tracking,** is the ability of leaves or flowers of certain plants, such as sunflower, soybean, and cotton, to follow the sun's movement across the sky (Fig. 36–11). Frequently, the leaves of such plants arrange themselves so that they are perpendicular to the sun's rays, regardless of the

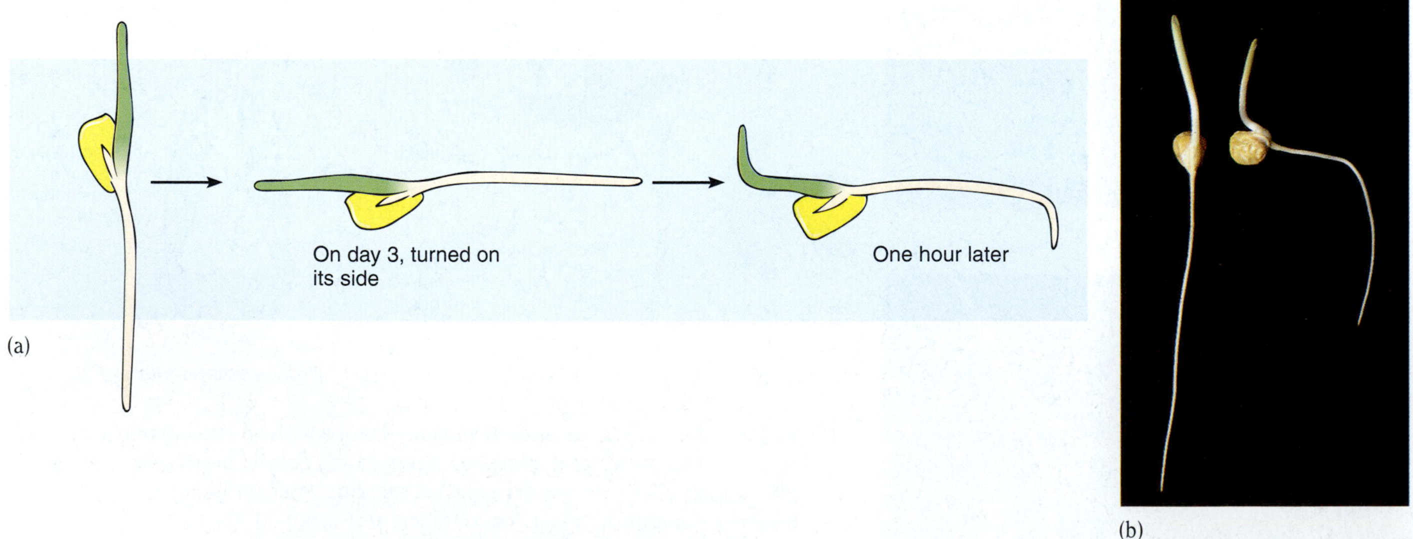

Figure 36-10 Gravitropism. Two corn *(Zea mays)* seeds were germinated at the same time. **(a)** One seedling was turned on its side on day 3. In 1 hour the root and shoot tips had curved. **(b)** In 24 hours *(seedling on the right)*, the new root growth was downward (positive gravitropism), and the new shoot growth was upward (negative gravitropism). (Control seedling is on the left.) *(Dennis Drenner)*

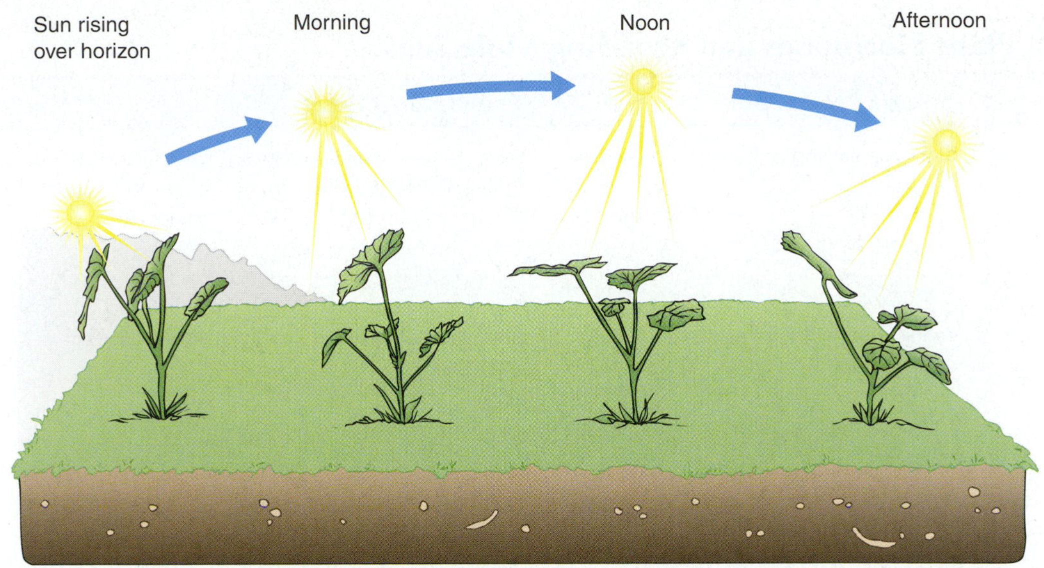

Figure 36–11 Heliotropism. Note that the leaves are oriented so that they are perpendicular to the sun's rays throughout the day, thereby maximizing the amount of light absorbed.

time of day or the sun's position in the sky. This positioning allows for maximal light absorption. Many solar trackers have pulvini at the bases of their petioles. Changes in turgor in the cells of the pulvinus help position the leaf in its optimum orientation relative to the sun. Like phototropism, heliotropism is triggered by blue light and is a growth response. During solar tracking, the shaded side of a heliotropic flower stem grows (i.e., elongates) more rapidly than the sunlit side.

HORMONES ARE CHEMICAL MESSENGERS THAT REGULATE GROWTH AND DEVELOPMENT

As mentioned earlier in the chapter, a *hormone* is an organic compound that acts as a highly specific chemical signal between cells. The study of plant hormones and their effects is challenging because hormones are effective in extremely small amounts (less than 10^{-6} *M*), and each hormone elicits many different responses. In addition, the effects of different plant hormones overlap, and it is difficult to determine which hormone, if any, is the primary cause of a particular response. Moreover, plant hormones may stimulate a certain response at one concentration and inhibit that same response at a different concentration.

For many years, biologists studied five major classes of plant hormones: auxins, gibberellins, cytokinins, ethylene, and abscisic acid. More recently, biologists have uncovered compelling evidence that a variety of signaling molecules, such as brassinolide, salicylic acid, systemin, oligosaccharins, and jasmonic acid, exist. Together, these hormones and signaling molecules regulate the growth and development of a plant. (Table 36–1 summarizes the ten plant hormones and hormone-like signaling molecules discussed in this chapter.)

Biologists have made dramatic advances in understanding the mechanisms of hormone action

Process of Science

Recent progress in the biology of plant hormones has been made using molecular genetic techniques, particularly with mutants of the model organism *Arabidopsis thaliana.* Various mutants have been identified that have defects in hormone synthesis, hormone transport, signal reception, or signal transduction. These mutants have allowed plant biologists to identify and clone genes involved in these aspects of hormone biology. Studying the mutant phenotypes has allowed plant biologists to establish connections between the mutant genes and specific physiological activities involved in growth and development.

Plant and animal hormones are similar in their basic mechanism of action. Like animal hormones, plant hormones bind to specific receptor proteins on the plasma membrane or in the target cells (Fig. 36–12). Each receptor has a three-dimensional shape that binds only with one kind of hormone molecule. Sometimes this binding triggers the production of a **second messenger,** an intracellular signaling molecule that affects the function of the cell. Ions such as Ca^{2+} serve as second messengers in many plant cells. Cyclic AMP (cAMP), an important second messenger in a variety of prokaryotes and eukaryotes, may also be a signaling molecule (for the plant hormone auxin). Once the concentration of Ca^{2+} or some other second messenger increases in the cell, it may bind to proteins and activate or inactivate certain enzymes. This activation or inactivation may lead to an altered membrane permeability and/or altered gene expression, that is, transcription and/or translation. (Chapter 47 discusses second messengers as they relate to animal hormones.)

TABLE 36–1 Selected Plant Hormones and Signaling Molecules

Hormone	Site of Production	Principal Actions
Auxin (e.g., IAA)	Shoot apical meristem, young leaves, seeds	Stem elongation, apical dominance, root initiation, fruit development
Gibberellins (e.g., GA_3)	Young leaves and shoot apical meristems, embryo in seed	Seed germination, stem elongation, flowering, fruit development
Cytokinins (e.g., Zeatin)	Roots	Cell division, delay of leaf senescence, inhibition of apical dominance, flower development, embryo development, seed germination
Ethylene	Stem nodes, ripening fruit, damaged or senescing tissue	Fruit ripening, responses to environmental stressors, seed germination, maintenance of apical hook on seedings, root initiation, senescence and abscission in leaves and flowers
Abscisic acid	Almost all cells that contain plastids (leaves, stems, roots)	Seed dormancy, responses to water stress

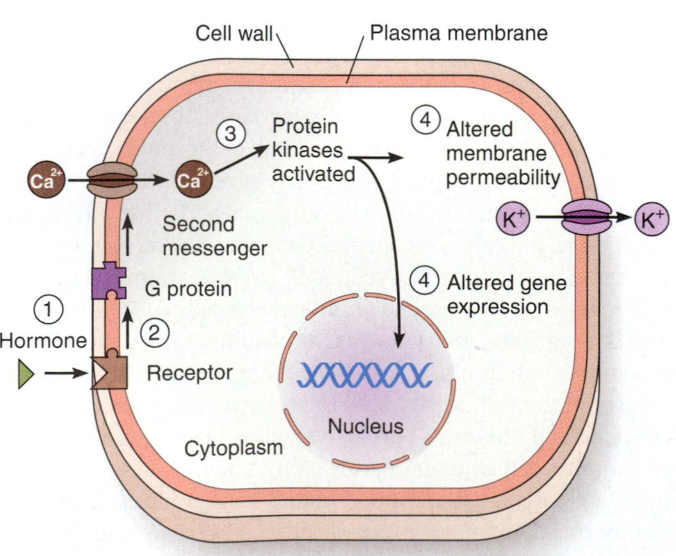

Figure 36–12 General mechanism of action of plant hormones. This diagram represents a hypothetical model that includes aspects of signal reception, signal transduction, and altered cell activity (responses) that have been demonstrated for various plant hormones. (1) The hormone binds with a receptor on the plasma membrane of a target cell. (2) The receptor is a signal transducer that converts the hormone signal to an intracellular signal. (3) A second messenger such as Ca^{2+} may relay the signal, and typically, a sequence of several signaling molecules relays the message. (4) The phosphorylated proteins produced by protein kinases alter the activity of the cell in some way, such as altering membrane permeability or gene expression.

Hormone	Site of Production	Principal Actions
Brassinosteroids (e.g., Brassinolide) 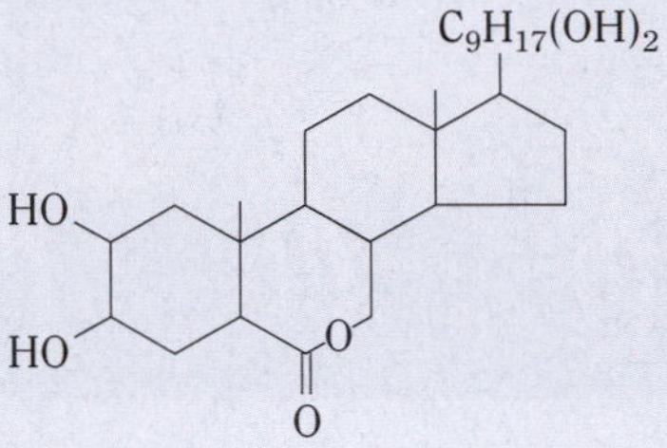	Unknown	Light-mediated gene expression, cell division, stem elongation, flower development, leaf senescence
Salicylic acid 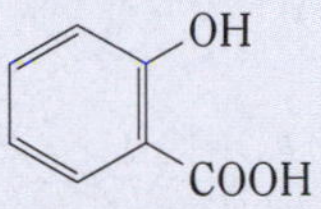	Wound (site of infection)	Resistance to disease organisms
Systemin* 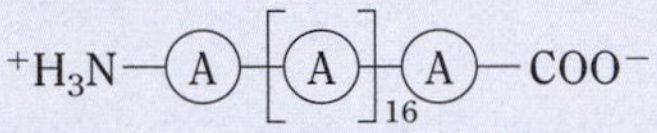 $^{+}H_3N-(A)-[(A)]_{16}-(A)-COO^{-}$	Wound (site of herbivore or pathogen attack)	Initiation of defenses against predators (herbivores) or disease organisms
Oligosaccharins*  $-[Sug]-[[Sug]]_{8-13}-[Sug]-$	Unknown	May function in normal cell growth and development; defense responses to disease organisms
Jasmonates (e.g., Jasmonic acid) 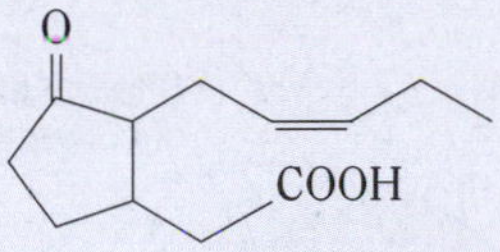	Leaves? Probably many tissues	Initiation of defenses against predators or disease organisms

*Key: Ⓐ = amino acid; [Sug] = sugar.

AUXIN PROMOTES CELL ELONGATION

Process of Science

Charles Darwin, best known for developing the theory of natural selection to explain evolution, also provided the first evidence for the existence of auxin. The experiments that Darwin and his son Francis performed in the 1870s involved positive phototropism, the directional growth of plants toward light. The plants they used were newly germinated canary grass seedlings. As in all grasses, the first part of a canary grass seedling to emerge from the soil is the coleoptile, a protective sheath that encircles the stem. When coleoptiles are exposed to light from only one direction, they bend toward the light. The bending occurs below the tip of the coleoptile.

The Darwins tried to influence this bending in several ways (Fig. 36–13). For example, they covered the tip of the coleoptile as soon as it emerged from the soil. When they covered that part of the coleoptile above where the bend would be expected to occur, the plants did not bend. On other plants, they removed the coleoptile tip and found that bending did not occur. When the bottom of the coleoptile where the curvature would occur was shielded from the light, the coleoptile bent toward light. From these experiments, the Darwins concluded that "some influence is transmitted from the upper to the lower part, causing it to bend."

In the 1920s Frits Went, a young Dutch scientist, isolated the phototropic hormone from oat coleoptiles. He removed the coleoptile tips and placed them on tiny blocks of agar for a period of time. When he put one of these agar blocks squarely on a decapitated coleoptile, normal growth resumed. When he placed one of these agar blocks to one side of the tip of a decapitated coleoptile in the dark, bending occurred (Fig. 36–14). This indicated that the substance had diffused from the coleoptile tip into the agar and, later, from the agar into the decapitated coleoptile. Went named this substance **auxin** (from the Greek *aux,* "enlarge" or "increase"). The purification and elucidation of auxin's chemical structure were accomplished in the mid-1930s by a research team led by Kenneth Thimann at the California Institute of Technology.

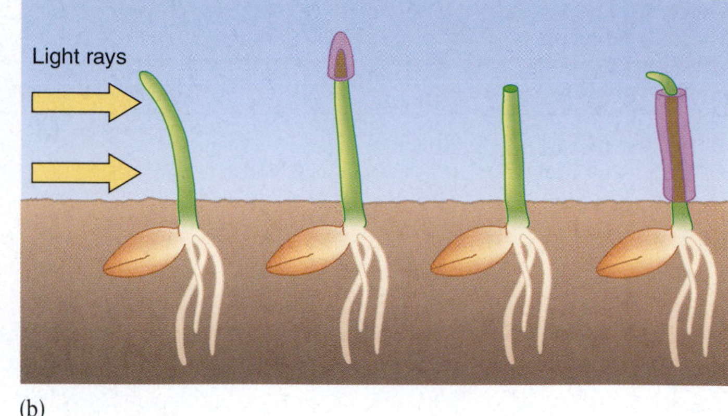

Figure 36-13 The Darwins' phototropism experiments with coleoptiles of canary grass seedlings. **(a)** Some plants were uncovered, some were covered only at the tip, some had the tip removed, and some were covered everywhere but at the tip. The covers were impervious to light. **(b)** After exposure to light coming from one direction, the uncovered plants and the plants with uncovered tips grew toward the light. The plants with tips covered or removed did not bend toward light.

Auxin is a group of natural and artificial plant hormones, the most important of which is **indoleacetic acid (IAA).** The movement of auxin in the plant is said to be *polar*, or unidirectional. Auxin moves downward along the shoot-root axis from its site of production, usually the shoot apical meristem. Young leaves and seeds are also sites of auxin production.

Auxin's most characteristic action is promotion of cell elongation in stems and coleoptiles. This effect, apparently exerted by acidification of cell walls, increases their plasticity, allowing them to expand under the force of the cell's internal turgor pressure. Auxin's effect on cell elongation also provides an explanation for phototropism. When a plant is exposed to a light from only one direction, some of the auxin migrates laterally to the shaded side of the stem before moving down the stem by polar

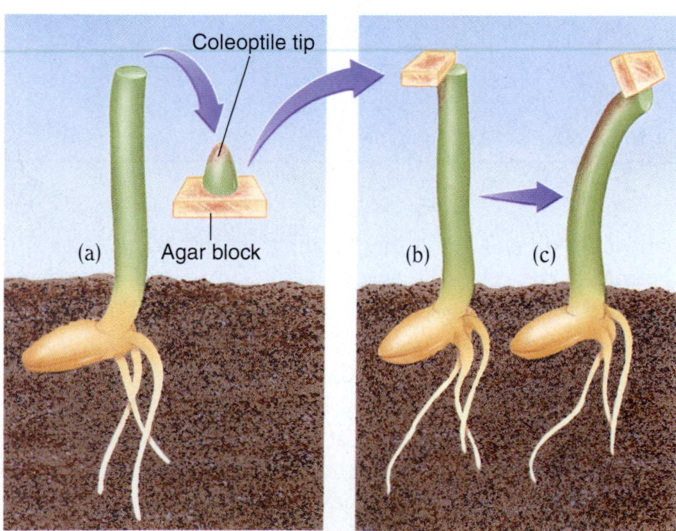

Figure 36-14 Isolating auxin from coleoptiles. **(a)** Coleoptile tips were placed on agar blocks for a period of time. **(b)** The agar block was transferred to a decapitated coleoptile. It was placed off-center, and the coleoptile was left in continual darkness. **(c)** The coleoptile bent. This indicated that a chemical had been transferred from the original coleoptile tip to the agar block and from there to one side of the decapitated coleoptile, causing that side to elongate.

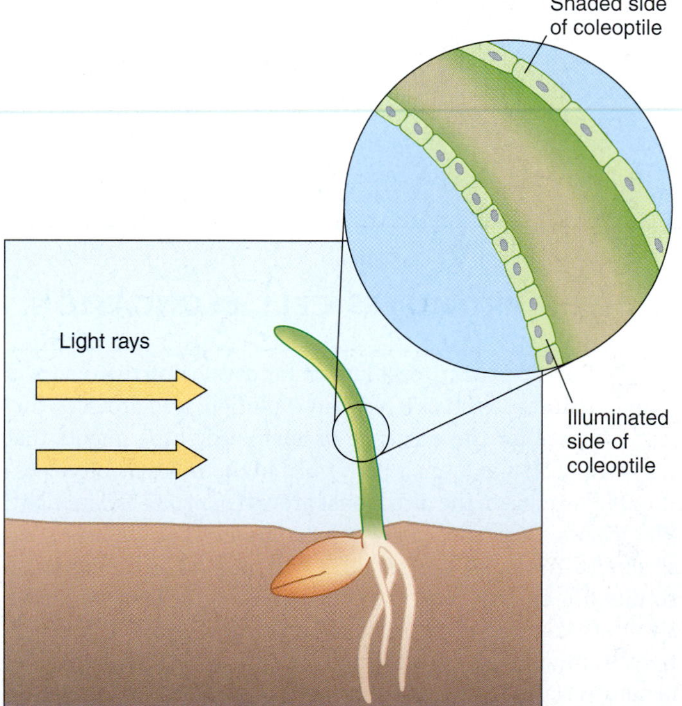

Figure 36-15 Phototropism and the unequal distribution of auxin. Auxin travels down the side of the stem or coleoptile *away* from the light, causing cells on the shaded side to elongate. Therefore, the stem or coleoptile bends toward light.

transport. Because of the greater auxin concentration on the shaded side of the stem, the cells there elongate more than the cells on the light side, and the stem bends toward the light (Fig. 36–15 on facing page). Auxin is involved in gravitropism, thigmotropism, and possibly heliotropism as well.

Auxin exerts other effects on plants. For example, some plants tend to branch out very little when they grow. Growth in these plants occurs almost exclusively from the apical meristem rather than from axillary buds, which do not develop as long as the terminal bud is present. Such plants are said to exhibit **apical dominance,** the inhibition of axillary bud growth by the apical meristem. In plants with strong apical dominance, it appears that auxin produced in the apical meristem inhibits axillary buds near the apical meristem from developing into actively growing shoots. When the apical meristem is pinched off, the auxin source is removed, and axillary buds grow to form branches. Apical dominance is often quickly reestablished, however, as one branch begins to inhibit the growth of others. Recent evidence indicates that other hormones (ethylene and cytokinin, both discussed later) are also involved in apical dominance. As with other physiological activities, the changing ratios of these hormones may be the factor responsible for apical dominance.

Auxin produced by developing seeds stimulates the development of the fruit. When auxin is applied to certain flowers in which fertilization has not occurred (and, therefore, in which seeds are not developing), the ovary enlarges and develops into a seedless fruit. Seedless tomatoes have been produced in this manner.[1] Auxin is not the only hormone involved in fruit development, however.

Some manufactured, or synthetic, auxins have been made that have structures similar to IAA. The synthetic auxin naphthalene acetic acid is used to stimulate root development on stem cuttings for asexual propagation, particularly of woody plants with horticultural importance (Fig. 36–16). The synthetic auxins 2,4-D and 2,4,5-T are used as selective herbicides (weed killers). The compounds 2,4-D and 2,4,5-T kill plants with broad leaves but, for reasons not completely understood at this time, do not kill grasses. Both herbicides are similar in structure to IAA and disrupt the plants' normal growth processes. Because many of the world's most important crops are grasses (e.g., wheat, corn, and rice), both 2,4-D and 2,4,5-T can be used to kill broadleaf weeds that compete with these crops. The use of 2,4,5-T is no longer allowed in the United States, however, because of its association with dioxins, a group of mildly to very toxic compounds formed as byproducts during the manufacture of 2,4,5-T.

Figure 36–16 Auxin and root development on stem cuttings. *(Left)* Many adventitious roots developed on a honeysuckle *(Lonicera fragrantissima)* cutting placed in a solution with a high concentration of synthetic auxin. *(Middle)* Fewer roots developed in a lower auxin concentration. *(Right)* The cutting placed in water (no auxin) served as a control and did not form roots in the same time period. *(Joe Eakes, Color Advantage/Visuals Unlimited)*

GIBBERELLINS PROMOTE STEM ELONGATION

In the 1920s, a Japanese biologist was studying a disease of rice in which the young rice seedlings grew extremely tall and spindly, fell over, and died. The cause of the disease, dubbed the "foolish seedling" disease, was a fungus *(Gibberella fujikuroi)* that produces a chemical substance named **gibberellin.** Not until after World War II did scientists in Europe and North America learn of the exciting work done by the Japanese. During the 1950s and 1960s, studies in the United States and Great Britain showed that gibberellins are produced by healthy plants as well as by the fungus that causes foolish seedling disease. Gibberellins were found to be hormones involved in many normal plant functions. The symptoms of foolish seedling disease were caused by an abnormally high gibberellin concentration in the plant tissue (because both plant and fungus were producing gibberellin). Currently, more than 110 naturally occurring gibberellins are known, although it is thought that many of these are inactive precursors; there are no synthetic gibberellins.

As in foolish seedling disease, gibberellins promote stem elongation in many plants. When gibberellin is applied to a plant, particularly to certain dwarf varieties, this elongation may be spectacular. Some corn and pea plants that are dwarf as a result of one or more mutations grow to a normal height when treated with gibberellin (Fig. 36–17*a*). Short-stemmed, high-yielding varieties of wheat are short-stemmed because they have a reduced response to gibberellin. These varieties put less of their resources into stem height and more resources into grain production. Gibberellins are also involved in **bolting,** the rapid elongation of a floral stalk that occurs naturally in many plants when they initiate flowering (Fig. 36–17*b*).

Gibberellins cause stem elongation by stimulating cells to divide as well as elongate. The actual mechanism of cell elongation appears to be different from that caused by auxin, however. Recall that IAA-induced cell elongation involves the acidification of the cell wall; in gibberellin-induced cell elongation, cell wall acidification does not occur.

Gibberellins are involved in several reproductive processes in plants. They stimulate flowering, particularly in long-day

[1]Not all seedless fruits are produced by treatment with auxin. In Thompson seedless grapes, for example, fertilization occurs but the embryos abort, and therefore the seeds fail to develop.

(a)

(b)

Figure 36–17 Gibberellin and stem elongation. **(a)** An experiment testing the effects of gibberellin on normal and dwarf corn *(Zea mays)* plants shows that dwarf plants respond to gibberellin much more dramatically than normal plants. (This dwarf variety is a mutant with a single recessive gene that impairs gibberellin biosynthesis.) From left to right: dwarf, untreated; dwarf, treated with gibberellin; normal, treated with gibberellin; normal, untreated. Normal corn grows to 4.5 m (15 ft). **(b)** Like many biennials, Indian blanket *(Gaillardia pulchella)* grows as a rosette, which is a circular cluster of leaves close to the ground, during its first year *(left)* and then bolts when it initiates flowering in the second year *(right)*. The plant in bloom grows to 0.6 m (2 ft). *(a, Courtesy of B.O. Phinney, University of California, Los Angeles; b, Robert E. Lyons)*

plants. In addition, they can substitute for the low temperature that biennials require before the initiation of flowering. If gibberellins are applied to biennials during their first year of growth, flowering occurs without exposure to a period of low temperature. Gibberellins, like auxin, affect the development of fruits. Gibberellins are applied to several varieties of grapes to produce larger fruits.

Gibberellins are also involved in the germination of seeds in many plants. In a classic experiment involving the germination of barley seeds, the release of gibberellin from the embryo was shown to trigger the synthesis of α-amylase, an enzyme that digests starch in the endosperm. As a result, glucose is available for absorption by the embryo. Although the formation of enzymes to mobilize starch reserves occurs in many types of seeds, control of enzyme formation in seeds by gibberellin appears to be restricted to cereals and other grasses. In addition to mobilizing food reserves in newly germinated grass seeds, the application of gibberellins can substitute for low-temperature or light requirements for germination in seeds such as lettuce, oats, and tobacco.

CYTOKININS PROMOTE CELL DIVISION

During the 1940s and 1950s, researchers were trying to find substances that might induce plant cells to divide in **tissue culture,** a technique in which cells are isolated from plants and grown in a nutrient medium (see *Focus On: Cell and Tissue Culture*). It was discovered that cells would not divide without a substance found in coconut milk. Because coconut milk has a complex chemical composition, the division-inducing substance was not chemically identified for some time. Finally, an active substance was isolated from a different source, aged DNA from herring sperm. It was called **cytokinin** because it induces cytokinesis, or cytoplasmic division. In 1963 the first naturally occurring plant cytokinin, zeatin, was identified from corn, and since that time similar molecules have been identified from other plants; several synthetic cytokinins have also been synthesized. Cytokinin is structurally similar to adenine, a purine base that is a part of DNA and RNA molecules.

FOCUS ON Cell and Tissue Culture

Cells can be isolated from certain plants and grown in a chemically defined, sterile nutrient medium. In initial experiments with such cultures, plant cells could be kept alive, but they did not divide. It was later discovered that the addition of certain natural materials such as the liquid endosperm of coconut, also known as coconut milk, induced cells to divide in culture. By the late 1950s, plant cells from a variety of sources could be cultured successfully, dividing to produce a mass of disorganized, relatively undifferentiated cells, or **callus.**

In 1958, F.C. Steward, a plant physiologist at Cornell University, succeeded in generating an entire carrot plant from a single callus cell derived from a carrot root (see Fig. 16–3). This demonstrated conclusively that each plant cell contains a genetic blueprint for all features of an entire organism. His work also showed that an entire plant can be grown from a single cell, provided the proper genes are expressed at the appropriate times.

Since Steward's pioneering work, many plants have been successfully cultured using a variety of cell sources. Plants have been regenerated from different tissues, organ explants (excised organs or parts such as root apical meristems or young embryos), and single cells. Under certain conditions, the genes that control embryonic development are expressed in plant tissue culture, and all of the plant's embryonic stages in the seed can be observed in their normal progression *(see figure).*

Cell and tissue culture techniques are used to help answer many fundamental questions involving growth and development in plants. These techniques also have great practical potential. Using tissue culture, it is possible to regenerate large numbers of genetically identical plants from cells of a single, genetically superior plant; this has been done for many different kinds of plants, from orchids and African violets to coastal redwoods.

It is also possible to alter the genetic composition of a cell while it is in culture and then have these changes expressed in the whole plant during regeneration. This provides a valuable tool to genetic engineers who wish to introduce desirable new traits, such as better nutritional properties, into crop species such as rice (see Chapter 14). About half the world's people eat rice as their staple food, and rice is a poor source of many vitamins, including vitamin A. According to the World Health Organization, 250 million of the world's children are at risk of vitamin A deficiency. Lack of vitamin A can cause poor vision, protein deficiency (vitamin A helps the body absorb and use amino acids), and an impaired immune system. In 2000, an international team of scientists reported that they had successfully engineered rice grains to produce beta-carotene, which the body uses to produce vitamin A; the new rice is commonly called golden rice. This feat, which was accomplished in part by using cell and tissue culture methods, has great potential for improved world health.

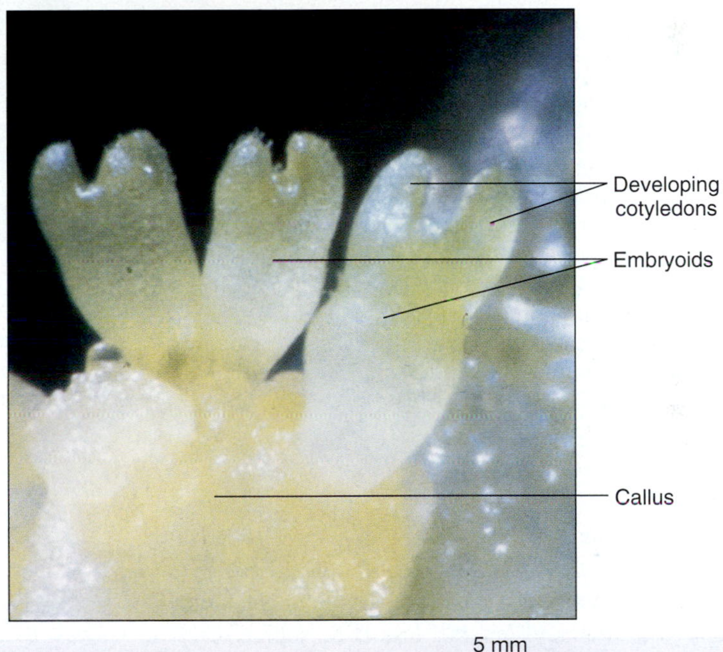

Tiny embryoids bud from callus. Compare these embryoids with the embryo depicted in Figure 35–9*d*. *(Courtesy of Dennis Gray, University of Florida)*

Cytokinins promote cell division and differentiation of young, relatively unspecialized cells into mature, more specialized cells in intact plants. They are a required ingredient in any plant tissue culture medium and must be present for cells to divide. In tissue culture, cytokinins interact with auxin during the formation of plant organs such as roots and stems (Fig. 36–18). For example, in tobacco tissue culture, a high ratio of auxin to cytokinin induces root formation, whereas a low ratio of auxin to cytokinin induces shoot formation.

Cytokinins and auxin also interact in the control of apical dominance. Here their relationship is antagonistic: Auxin inhibits the growth of axillary buds, and cytokinin promotes their growth.

One effect of cytokinins on plant cells is to delay the aging process, known as **senescence**. Plant cells, like all living cells, go

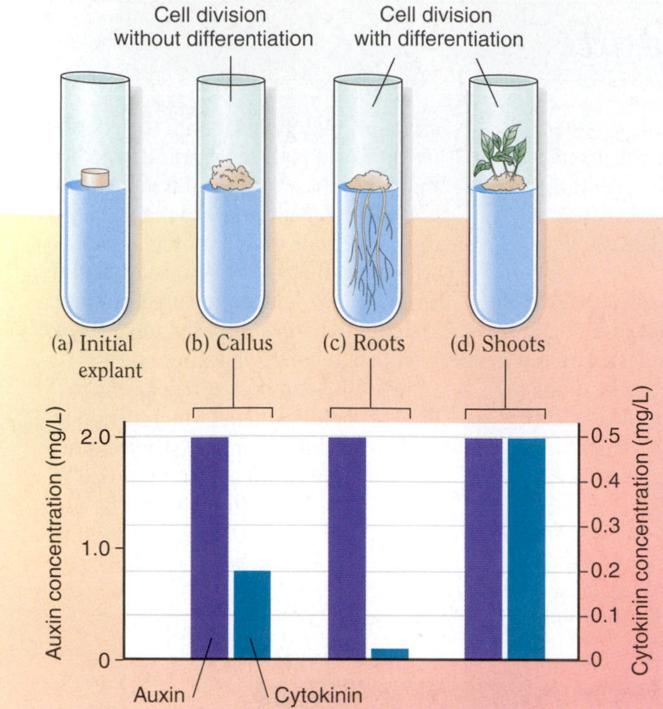

Figure 36–18 Auxin-cytokinin interactions in tobacco tissue culture. Varying amounts of auxin and cytokinin in the culture media produce different growth responses. **(a)** The initial explant is a small piece of sterile tissue from the pith of a tobacco stem, which is placed on nutrient agar media. **(b)** Nutrient agar containing 2.0 mg/L of auxin and 0.2 mg/L of cytokinin causes cells to divide and form a clump of undifferentiated tissue called a callus. **(c)** Medium with a high ratio of auxin (2.0 mg/L) to cytokinin (0.02 mg/L) stimulates root growth. **(d)** Medium containing a lower ratio of auxin (2.0 mg/L) to cytokinin (0.5 mg/L) stimulates shoot growth.

Figure 36–19 Cytokinin synthesis and delay of senescence. The tobacco *(Nicotiana tabacum)* plant on the left was genetically engineered to produce additional cytokinin as it aged, whereas the tobacco plant of the same age on the right was not genetically engineered and served as a control. Note the extensive senescence and death of older leaves on the control plant. Depending on the variety, tobacco grows 0.9 to 3 m (3 to 10 ft) tall. *(Courtesy of Dr. Richard M. Amasino, University of Wisconsin)*

through a natural aging process. Senescence is accelerated in cells of plant parts that are cut, such as flower stems. It is thought that plants must have a continual supply of cytokinins from the roots. Cut stems, of course, lose their source of cytokinins and therefore age rapidly. When cytokinins are sprayed on leaves of a cut stem of many species, they remain green, whereas leaves that are not sprayed turn yellow and die.

Despite their involvement in many aspects of plant growth and development, cytokinins currently have few commercial applications other than plant tissue culture. However, in 1995, molecular biologists at the University of Wisconsin combined a promoter from a gene that is activated during normal senescence with a gene that encodes an enzyme involved in cytokinin synthesis. The leaves of transgenic tobacco plants that contained this recombinant DNA produced more cytokinin and therefore lived longer and continued to photosynthesize (Fig. 36–19). Some biologists think this system has the potential to increase the longevity and productivity of certain crops into which it is engineered.

ETHYLENE STIMULATES ABSCISSION AND FRUIT RIPENING

During the early 20th century, scientists had observed that the gas **ethylene** (C_2H_4) has several effects on plant growth, but it was not until 1934 that the production of ethylene by plants was demonstrated. Many diverse plant processes are influenced by this natural plant hormone. Ethylene inhibits cell elongation, promotes the germination of seeds, promotes apical dominance, and is involved in plant responses to wounding or invasion by disease-causing microorganisms.

Ethylene also has a major role in many aspects of senescence, including fruit ripening. As a fruit ripens, it produces ethylene, which triggers an acceleration of the ripening process. This induces the fruit to produce more ethylene, which further accelerates ripening. The expression "one rotten apple spoils the barrel" is true. A rotten apple is one that is overripe and produces large amounts of ethylene, which diffuses and triggers the ripening process in nearby apples. Ethylene is used commercially to promote the uniform ripening of bananas. Bananas are picked while green and shipped to their destination, where they are exposed to ethylene before they are delivered to grocery stores.

Plants growing in a natural environment are subjected to rain, hail, wind, and contact with passing animals. All of these mechanical stressors alter the growth and development of plants, making them shorter and stockier than plants grown in a greenhouse. Such developmental responses to mechanical stimuli, known as **thigmomorphogenesis,** are regulated by ethylene. Plants that are mechanically disturbed produce additional ethylene, which in turn inhibits stem elongation and enhances cell wall thickening in supporting cells such as collenchyma and sclerenchyma. These changes are adaptive because shorter, thicker stems are less likely to be damaged by mechanical stressors.

Ethylene has been implicated as the hormone that induces leaf abscission. However, abscission is actually influenced by two antagonistic plant hormones, ethylene and auxin. As a leaf ages (when autumn approaches, for deciduous trees in temperate climates), the level of auxin in the leaf decreases. Concurrently, cells in the abscission layer at the base of the petiole (where the leaf will break away from the stem) begin producing ethylene.

Figure 36–20 Abscisic acid and seed germination. In a corn *(Zea mays)* mutant unable to produce abscisic acid, some of the kernels have germinated while still on the ear, producing roots *(arrows)*. *(Courtesy of M.G. Neuffer)*

ABSCISIC ACID PROMOTES SEED DORMANCY

Abscisic acid was discovered in 1963 by two different research teams. Despite its name, abscisic acid is involved in seed dormancy and in a plant's response to stress; abscisic acid does not induce abscission in most plants. As an environmental stress hormone, abscisic acid particularly promotes changes in plant tissues that are water-stressed, or exposed to droughts. (Recall that ethylene also affects plant responses to certain stressors, such as mechanical stressors and wounding.)

The effect of abscisic acid on plants suffering from water stress is understood best. The level of abscisic acid increases dramatically in the leaves of plants that are exposed to severe drought conditions. The high level of abscisic acid in the leaves activates a signal transduction process in which abscisic acid triggers an increase in the concentration of Ca^{2+} in guard cells, which leads to the closing of stomata. The closing of stomata saves the water that would normally be lost by transpiration, thereby increasing the plant's likelihood of survival.

The onset of winter is also a type of stress on plants. A winter adaptation that involves abscisic acid is dormancy in seeds. Many seeds have high levels of abscisic acid in their tissues and are unable to germinate until the abscisic acid washes out. In a corn mutant unable to synthesize abscisic acid, the seeds germinate as soon as the embryos are mature, even while attached to the ear (Fig. 36–20).

Abscisic acid is not the only hormone involved in seed dormancy. For example, addition of gibberellin reverses the effects of dormancy. In seeds, the level of abscisic acid decreases during the winter, and the level of gibberellin increases. Cytokinins have also been implicated in breaking dormancy. Once again we see that a single physiological activity such as seed dormancy may be controlled in plants by the interaction of several hormones. The plant's actual response may result from changing ratios of hormones rather than the effect of each individual hormone.

ADDITIONAL CLASSES OF SIGNALING MOLECULES ARE INVOLVED IN PLANT GROWTH AND DEVELOPMENT

The number and kinds of hormones and hormone-like signaling molecules that are known to function in plants continues to increase. Here we briefly consider five groups—the brassinolides, salicylic acid, systemin, oligosaccharins, and jasmonates.

Brassinolides are steroids

Although steroid hormones are known to have crucial roles in animals, their roles in plants have been unclear. The **brassinolides** are a group of steroid hormones that have long been suspected of being plant hormones. They appear to be involved in several aspects of plant growth and development. *Arabidopsis* mutants that cannot synthesize brassinolides are dwarf plants. This defect can be reversed by the application of brassinolides.

Studies of these mutants suggest that the brassinolides are involved in developmental processes such as cell division and cell elongation.

Salicylic acid is a phenolic compound

For centuries native Americans chewed willow (*Salix* sp.) bark to treat headaches and other types of pain. **Salicylic acid** was first extracted from willow bark and is chemically related to aspirin (acetylsalicylic acid). More recently, salicylic acid has been shown to help plants defend against insect pests and pathogens such as viruses. When a plant is under attack, the concentration of salicylic acid increases, and it spreads systemically throughout the plant. It is thought that salicylic acid binds to a cell receptor, thereby switching on genes that code for proteins that fight infection and promote wound healing.

Plants also use salicylic acid to signal nearby plants. Tobacco plants infected with tobacco mosaic virus release into the air a volatile form of salicylic acid, known as methyl salicylate, or oil of wintergreen. When nearby healthy plants receive the airborne chemical signal, they begin synthesizing antiviral proteins that enhance their resistance to the virus.

Systemin is a small polypeptide

Although many animal hormones are known to be polypeptides, the first plant polypeptide with hormonal properties was not isolated until 1991. Known as **systemin,** this polypeptide is transported systemically throughout the plant in response to wounding by insects. Systemin appears to stimulate natural defense mechanisms at extremely low concentrations, as low as one part per trillion. Systemin may trigger the plant to produce protease inhibitors, molecules that disrupt insect digestion, thereby curbing leaf damage done by caterpillars and other herbivorous (plant-eating) insects. The discovery of systemin caused a flurry of research in search of polypeptide regulators in plants, and several additional ones have been reported.

Oligosaccharins are composed of sugar residues

Oligosaccharins are cell wall carbohydrate fragments consisting of short, branched chains of sugar molecules. They are present in extremely small quantities in cells and active at much lower concentrations (100 to 1000 times lower) than hormones such as auxin. Different oligosaccharins appear to have distinct functions. Some trigger the production of **phytoalexins** (from the Greek *phyto,* "plant," and *alexi,* "to ward off"), antimicrobial compounds that limit the spread of plant pathogens such as fungi. Other oligosaccharins inhibit flowering and induce vegetative growth.

Jasmonates are derivatives of fatty acids

Jasmonates affect several plant processes, such as pollen development, root growth, fruit ripening, and senescence. These plant hormones are also produced in response to the presence of insect pests and disease-causing organisms. Jasmonates trigger the production of enzymes that confer an increased resistance against herbivorous insects. For example, caterpillar-infested tomato plants release volatile chemicals into the air that attract natural caterpillar enemies, such as parasitic wasps that lay their eggs on the caterpillar's bodies. When the eggs hatch, the larvae consume, and ultimately kill, the host insects. In one study, treating plants with jasmonic acid increased parasitism on caterpillar pests twofold over control fields in which the plants did not have jasmonic acid applications. Jasmonates may be of practical value in controlling certain insect pests without the use of chemical pesticides.

Process of Science ■ OTHER AS YET UNIDENTIFIED PLANT HORMONES REMAIN TO BE DISCOVERED

Experiments in which different tobacco species are grafted together indicate that both flower-promoting and flower-inhibiting substances may exist. *Nicotiana silvestris* is a long-day tobacco plant, and a variety of *Nicotiana tabacum* is a day-neutral tobacco plant. When a long-day tobacco is grafted to a day-neutral tobacco and exposed to short nights, both plants flower (Fig. 36–21). The day-neutral tobacco plant flowers sooner than it normally would. It has been suggested that a flower-promoting substance may be induced in the long-day tobacco and transported to the day-neutral tobacco through the graft union, causing the day-neutral tobacco to flower sooner than expected. The

Figure 36–21 Evidence for the existence of a flower-promoting substance. When a long-day tobacco *(Nicotiana silvestris)* is grafted to a day-neutral tobacco *(N. tabacum)* and both plants are exposed to a long-day/short-night regimen, they both flower. The day-neutral plant flowers sooner than it normally would, presumably because a flower-promoting substance passes from the long-day plant to the day-neutral one through the graft.

hypothetical flower-promoting substance is known as **florigen.** In an intact plant, florigen may be produced in the leaves and transported in the phloem to the shoot apical meristem. There, it induces a transition from vegetative to reproductive development—that is, to a meristem that produces flowers.

When a long-day tobacco is grafted to a day-neutral tobacco and exposed to long nights, neither plant flowers. As long as these conditions continue, the day-neutral plants do not flower even when they would normally do so. In this case, it has been suggested that the long-day tobacco may produce a flower-inhibiting substance that is transported to the day-neutral tobacco through the graft union. This substance prevents the day-neutral tobacco from flowering. The many attempts that have been made to isolate and chemically characterize substances that are clearly identifiable as flower promoters (florigen) or flower inhibitors have been unsuccessful.

SUMMARY WITH KEY TERMS

I. Plant growth and **development** are controlled by internal factors, such as a cell's location in the plant body, and by environmental factors, such as changing day length. The location of a cell in the young plant body, that is, a cell's spatial environment, affects gene expression during development by causing some genes in that cell to be turned off and others to be turned on.

II. **Germination** (when a seed sprouts) is affected by both internal and external factors.
   - A. External environmental factors that may affect germination include requirements for oxygen, water, temperature, and light. For example, dry seeds must **imbibe** water before germinating.
   - B. Internal factors affecting whether a seed germinates include the maturity of the embryo, the presence or absence of chemical inhibitors, and the presence or absence of hard, thick seed coats.

III. Many plants initiate flowering at the shoot apical meristem in response to specific cues from the environment.
   - A. **Photoperiodism** is any response of plants to the duration and timing of light and dark.
     1. In many plants, flowering is a photoperiodic response; some are **short-day plants,** some are **long-day plants,** and others are **intermediate-day plants.** In **day-neutral plants,** flowering is not affected by photoperiod.
     2. The photoreceptor in photoperiodism is **phytochrome,** a blue-green pigment with two forms, Pr and Pfr, named by the wavelength of light they absorb.
        - a. Pfr is the active form of phytochrome, triggering or inhibiting physiological responses such as flowering, **shade avoidance,** and a light requirement for germination.
        - b. Vascular plants contain a mixture of five or so phytochrome proteins. Each of the several types of phytochrome may have both unique and overlapping physiological functions. Phytochrome B appears to exert its influence at all stages of the plant life cycle.
        - c. Some phytochrome-induced responses are rapid and short-term (such as changes in membrane properties); others are slower and long-term (i.e., the regulation of gene transcription).
        - d. The absorption of light by phytochrome triggers one or more **signal transduction** pathways that amplify the original signal and ultimately result in a physiological or developmental response, such as promotion or inhibition of flowering.
   - B. **Vernalization** is the promotion of flowering by exposure to low temperatures for a period of time; the shoot apical meristem appears to be the site of the low-temperature requirement.

IV. **Circadian rhythms** are regular periodicities in growth or activities of a plant or other organism that approximate the 24-hour day and are reset by the rising and setting of the sun.
   - A. The red-light photoreceptor phytochrome and the blue-light photoreceptor **cryptochrome** have been implicated in resetting the biological clock in plants.
   - B. Two examples of circadian rhythms in plants are the opening and closing of stomata and **sleep movements.**

V. Nastic movements and tropisms are the two kinds of plant movements that occur in response to external stimuli.
   - A. **Nastic movements,** which are temporary and reversible, occur in response to external stimuli, but the direction of movement is independent of the direction of the stimulus.
   - B. **Tropisms** are directional growth responses (i.e., the direction of growth is dependent on the direction of the stimulus).
     1. **Phototropism** is plant growth in response to the direction of light.
     2. **Gravitropism** is plant growth in response to the influence of gravity.
     3. **Thigmotropism** is plant growth in response to contact with a solid object.
     4. **Heliotropism** is the ability of leaves or flowers to track the sun across the sky.

VI. Plants produce and respond to **hormones,** organic compounds that act as highly specific chemical signals between cells.
   - A. Hormones are effective in extremely small concentrations.
   - B. The functions of some plant hormones overlap.
   - C. Many physiological activities of plants may be due to the interactions of several hormones rather than to the effect of a single hormone.
   - D. *Arabidopsis* mutants with defects in hormone synthesis, hormone transport, signal reception, or signal transduction have been identified and studied.
     1. Plant hormones bind to specific receptor proteins in target cells.
     2. Sometimes this binding triggers the production of a **second messenger** such as Ca^{2+}.
     3. The second messenger may, in turn, bind to and activate or inactivate certain enzymes.
     4. This activation or inactivation may lead to an altered membrane permeability and/or altered gene expression.
   - E. For many years, biologists studied five major classes of plant hormones.
     1. **Auxin** is involved in cell elongation, tropisms, **apical dominance** (the inhibition of axillary buds by the apical meristem), and fruit development. Auxin also stimulates root development on stem cuttings. Some synthetic auxins (2,4-D and 2,4,5-T) are selective herbicides.
     2. **Gibberellins** are involved in stem elongation, flowering, and germination. Gibberellins were first recognized in the 1920s by a Japanese biologist studying "foolish seedling" disease of rice, which is caused by a fungus.
     3. **Cytokinins** promote cell division and differentiation, delay **senescence** (the natural aging process), and interact with auxin and ethylene in apical dominance. Cytokinins induce cell division in **tissue culture,** a technique in which cells are isolated from plants and grown in a nutrient medium.
     4. **Ethylene** has a role in the ripening of fruits, apical dominance, leaf abscission, wound response, **thigmomorphogenesis** (a developmental response to mechanical stressors such as wind), and senescence.
     5. **Abscisic acid** is an environmental stress hormone involved in seed dormancy and in stomatal closure due to water stress.

F. Several other hormones or hormone-like signaling molecules are implicated in specific aspects of plant growth and development.
   1. The plant steroids known as **brassinolides** appear to be involved in several aspects of plant growth and development, such as cell division and cell elongation.
   2. **Salicylic acid** helps defend plants against pathogens and insect pests. It may bind to a cell receptor, thereby switching on genes that fight infection and promote wound healing. A volatile form of salicylic acid, known as methyl salicylate, may serve as an airborne chemical signal from virus-infected plants to healthy ones.
   3. **Systemin,** a plant polypeptide with hormonal properties, stimulates a natural defense mechanism in which the plant produces protease inhibitors, molecules that disrupt insect digestion.
   4. **Oligosaccharins,** short, branched chains of sugar molecules that are fragments of cell wall carbohydrates, have a variety of functions, such as inhibition of flowering and stimulation of vegetative growth, and exert their effects by binding to membrane receptors and altering gene expression.
   5. **Jasmonates** affect several plant processes, such as pollen development, root growth, fruit ripening, and senescence. They are also produced in response to the presence of insect pests and disease-causing organisms.
   6. Grafting experiments indicate the possible existence of both flower-promoting **(florigen)** and flower-inhibiting substances, none of which has been successfully isolated.

POST-TEST

1. A plant's response to the relative amounts of daylight and darkness is known as (a) apical dominance (b) bolting (c) gravitropism (d) photoperiodism (e) phototropism
2. Pfr, the active form of __________, is formed when red light is absorbed. (a) photosystem I (b) phytochrome (c) far-red light (d) PIF3 (e) pyrimidines
3. Which of the following represents the correct order in the phytochrome signal transduction pathway? (1) red light (2) light-responsive gene is switched on (or off) (3) movement of Pfr to nucleus (4) conversion of Pr to Pfr (5) formation of PFr-PIF3 complex that is bound to promoter region (a) 1-3-5-4-2 (b) 1-5-3-2-4 (c) 1-2-3-4-5 (d) 1-4-3-2-5 (e) 1-4-3-5-2
4. Which of the following statements about phytochrome is *false*? (a) Phytochrome is the main photoreceptor for photoperiodism. (b) Phytochrome is a family of about five blue-green pigment proteins, each coded by a different gene. (c) Much of our knowledge of phytochrome function is based on mutant corn *(Zea mays)* plants. (d) Each member of the phytochrome family exists in two forms: Pr and Pfr. (e) Phytochrome helps plants sense the presence of nearby plants with which they must compete for light.
5. The promotion of flowering by the exposure of shoot apical meristems to a low-temperature treatment is known as (a) thigmomorphogenesis (b) bolting (c) gravitropism (d) photoperiodism (e) vernalization
6. In __________, leaves or other plant organs track the sun's movement across the sky. (a) heliotropism (b) thigmotropism (c) phototropism (d) bolting (e) photoperiodism
7. The orientation of the growth of a plant due to the direction of light is called __________, whereas the twining of tendrils is an example of __________. (a) heliotropism; gravitropism (b) photoperiodism; nastic movements (c) phototropism; gravitropism (d) phototropism; thigmotropism (e) photoperiodism; thigmotropism
8. When you prune shrubs to make them "bushier"—that is, to prevent apical dominance—you are affecting the distribution and action of which plant hormone? (a) auxin (b) jasmonic acid (c) systemin (d) ethylene (e) abscisic acid
9. A synthetic __________ known as 2,4-D is used as a selective herbicide. (a) auxin (b) gibberellin (c) cytokinin (d) ethylene (e) abscisic acid
10. Research on a fungal disease of rice provided the first clues about this plant hormone. (a) auxin (b) gibberellin (c) cytokinin (d) ethylene (e) abscisic acid
11. This plant hormone interacts with auxin during the formation of plant organs in tissue culture. (a) florigen (b) gibberellin (c) cytokinin (d) ethylene (e) abscisic acid
12. The plant hormone __________ delays senescence, whereas the plant hormone __________ promotes it. (a) cytokinin; auxin (b) auxin; cytokinin (c) cytokinin; ethylene (d) abscisic acid; ethylene (e) gibberellin; auxin
13. The stress hormone that helps plants respond to drought is (a) auxin (b) gibberellin (c) cytokinin (d) ethylene (e) abscisic acid
14. This hormone promotes seed dormancy. (a) auxin (b) gibberellin (c) cytokinin (d) ethylene (e) abscisic acid
15. Which signaling molecule triggers the release of volatile substances that attract parasitic wasps to plant-eating caterpillars? (a) brassinolide (b) jasmonic acid (c) systemin (d) salicylic acid (e) oligosaccharin

REVIEW QUESTIONS

1. What factors influence the germination of seeds? Explain the role that each of these factors plays in germination and discuss why the germinating seed responds the way it does.
2. Why are plant growth and development so sensitive to environmental cues?
3. What is phytochrome? Describe several roles of phytochrome.
4. Define signal transduction. Explain the interaction between phytochrome and the transcription factor PIF3.
5. Distinguish between nastic movements and tropisms.
6. Describe a general mechanism of action for plant hormones, including signal reception, signal transduction, and altered cell activity.
7. How is auxin involved in phototropism?
8. Discuss the interactions of various hormones involved in each of the following physiological processes: (a) seed germination, (b) stem elongation, (c) fruit ripening, (d) leaf abscission, (e) seed dormancy.

YOU MAKE THE CONNECTION

1. Predict whether flowering in a short-day plant with a minimum critical night length of 14 hours would be expected to occur in the following situations. Explain each answer. (a) The plant is exposed to 15 hours of daylight and 9 hours of darkness. (b) The plant is exposed to 9 hours of daylight and 15 hours of darkness. (c) The plant is exposed to 9 hours of daylight and 15 hours of darkness, with a 10-minute exposure to red light in the middle of the night.
2. If you transplanted the short-day plant discussed in Question No. 1 to the tropics, would it flower? Explain your answer.
3. Many solar-tracking flowers live in cool arctic or alpine environments. The temperature of certain heliotropic flowers has been measured to be up to 14° F warmer than that of the surrounding air. Insects are drawn to these warm flowers. Based on these observations, suggest a possible adaptive explanation for heliotropism in such environments.
4. What biological advantages are conferred on a plant whose stems are positively phototropic and whose roots are positively gravitropic?

RECOMMENDED READINGS

Berg, L.R. *Introductory Botany: Plants, People, and the Environment*. Saunders College Publishing, Philadelphia, 1997. A general botany text with an environmental emphasis.

Galen, C. "Sun Stalkers." *Natural History*, May 1999. Examines heliotropism in sunflowers and other solar trackers.

Jones, M. "Keeping in Time." *New Scientist*, Vol. 149, 16 Mar. 1996. An excellent overview of circadian rhythms and internal biological clocks in diverse organisms.

Leyser, H.M.O. "Plant Hormones." *Current Biology*, Vol. 8, No. 1, 1998. Discusses plant hormones and the recent advances biologists have made in understanding how they regulate growth and development.

Mestel, R. "Rhythm and Blooms." *Natural History*, Jun. 2000. Summarizes some of the factors that affect the time at which plants initiate flowering.

Milius, S. "Boosting Tomato's SOS Gets Pests Killed." *Science News*, Vol. 155, 19 Jun. 1999. Spraying jasmonic acid on tomato plants infested with beet armyworms resulted in a greater number of parasitic wasps responding to the chemical distress signal.

Raloff, J. "When Tomatoes See Red." *Science News*, Vol. 152, 13 Dec. 1997. Scientists have discovered that using different colors of plastic mulches to cover the soil results in different plant growth responses.

Raven, P.H., R.F. Evert, and S.E. Eichhorn. *Biology of Plants*, 6th ed. W.H. Freeman & Company, New York, 1999. This is an excellent general botany textbook, with detailed descriptions of growth responses and regulation of growth.

Smith, H. "Phytochromes and Light Signal Perception by Plants—An Emerging Synthesis." *Nature*, Vol. 407, 5 Oct. 2000. This review article discusses what is known about the phytochromes.

Taiz, L., and E. Zeiger. *Plant Physiology*, 2nd ed. Sinauer Associates, Sunderland, MA, 1998. This text, which contains contributions from many leading plant physiologists, contains both an historical perspective of plant physiology and current research knowledge.

Whitelam, G.C., and K.J. Halliday. "Photomorphogenesis: Phytochrome Takes a Partner!" *Current Biology*, Vol. 9, 1999. Reports on the identification of PIF3, the transcription factor that is involved in the phytochrome signal transduction pathway.

- Visit our Web site at **http://www.info.brookscole.com/solomonbergmartin** for links to chapter-related resources on the World Wide Web. Additional on-line materials relating to this chapter can also be found on our Web site.

See chapter activity on BioActive Learner CD for additional help in mastering the chapter's material. Icon location in the chapter's margins shows which topics have tutorials or simulations in the CD.

CAREER VISIONS

Assistant Professor of Biology

Dr. REGINA McCLINTON

***Dr. Regina McClinton** is an assistant professor of biology at the University of Louisiana at Lafayette (ULL). A native and long-term resident of Los Angeles, she became fascinated with biology early in life. She earned her B.A. in Biology at the University of California, Riverside, in 1989, and her Ph.D. in Molecular and Physiological Plant Biology at the University of California, Berkeley, in 1997. She joined the ULL biology faculty in 1998. Regina's job combines two professional passions: teaching and research. Her research interests center on plant development and cell biology.*

At what point in your academic career or personal life did you realize you were interested in biology?

It really happened for me in junior high school. I went to a science magnet junior high school, and we started doing karyotyping, where you basically can take a cell when the chromosomes are condensed and you can read them. I just thought that was really cool. I thought, "Would they pay me to do this?" And I started looking into it. When I got my hand in biology, I knew, "Oh, this is fabulous; I've got to do this."

What inspired you to pursue a career in research and academics?

In high school, I wanted to play with cells. I thought, if I want to do this, I'll have to do research. I spent the day with a geneticist in her lab and saw in a way how mundane it was, to do the same thing day after day, but also how it felt to be inspired and driven by those questions. As an undergraduate, I worked for a researcher who studied yeasts and became her lab tech for awhile. I was in charge of supervising her undergraduates' research, troubleshooting their projects, managing lab funds. But I hated being under somebody. I had initially wanted to pursue a Ph.D., and my desire to work independently really cemented that decision.

What made you decide to work with plants?

When I started graduate school, plant biology was really coming into its own. In animal biology we had all these tools; we were just beginning to figure out how to use them in plant biology. There's so much about the way plants develop that is regular and predictable, yet it's also flexible. Plants can't run away to deal with their environment. Whatever their environment is, they have to deal with it right then and there. I'm just fascinated by that.

How did you come to the position that you have now?

I actually knew a couple of the faculty members here. I was familiar with their work, which was similar to mine, so I hunted them down at a meeting when I was a graduate student. When they found out about this job opening, they contacted me. Those two faculty members knew I would fit in because I was using the same kinds of techniques they were using and asking similar questions, but our work did not overlap directly. Therefore, there was lots of potential for synergy in our work.

How important do you feel getting published is to a research biology career?

It's the core. You can sit in the lab and try to answer questions all you want, but unless you publish, your newfound knowledge is not disseminated. And there is something extremely valid, and I think extremely necessary, in having colleagues evaluate your work. The system is not perfect, but getting published in peer-reviewed journals is pertinent, and it's the way research scientists get evaluated.

What do you see as some of the more rewarding aspects of your profession? What are some of the drawbacks?

I'm doing what I love. When I wake up in the morning, I think, what am I going to be doing in the lab? I love being at the bench. I am decently paid, and I like being at a place with equipment I can use and colleagues I can interact with, a place where I can afford to live and not be afraid to bring up a family. I think this particular job has a great ratio of teaching and research for me. Drawbacks? This job shapes your life. I don't really mind the long hours, I come in because I want to most of the time. But, it's hard to relate to friends outside of academia. I have to work at keeping a foot in the real world and a foot in the academic world, which causes complications.

What advice would you give a college student considering a career in academic biology?

Several things: First, think about some of the research areas you might like to study, and what questions interest you, then find out what your options are. Second, find out what requirements you need before you go to a particular graduate school. Take care of those requirements as soon as you can. Lastly, find out if you are going to like being in a lab, and decide what you like about research. Can you do the same thing over and over and over again? Because that's part of research, and it's a part that you have to be okay with so that you can keep your questions focused.

This grasshopper *(Tropidacris dux)* represents one of more than 1 million species of insects that have been described—more than all other animal species combined! Grasshoppers have long, powerful hind legs adapted for jumping. They have large compound eyes, antennae, and mouthparts adapted for chewing the vegetation on which they feed. Animals are extremely diverse, with a wide variety of size, form, and function. Each animal is adapted to its habitat and to the role it plays in the biosphere. *(Stephen Dalton)*

Part 7

Structure and Life Processes in Animals

Chapters

37

The Animal Body: Introduction to Structure and Function

Larger body size does not mean bigger cells. The cells of the Yacare caiman *(Caiman yacare)* and the flambeau butterfly *(Dryas julia)* on its head are all about the same size. The caiman is larger because its genes specify that its body should consist of a larger number of cells. *(Fritz Polking/Deminsky Photo Associates)*

LEARNING OBJECTIVES

After you have studied this chapter you should be able to

1. Discuss advantages of multicellularity.
2. Define tissue, organ, and organ system.
3. Compare the general structure and function of the four principal kinds of animal tissues: epithelial, connective, muscle, and nervous tissues.
4. Describe the main types of epithelial tissue and give their functions.
5. Compare the main types of connective tissue and summarize their functions.
6. Contrast the three types of muscle tissue and their functions.
7. Identify the cells that make up nervous tissue and give their functions.
8. List and briefly describe the organ systems of a complex animal.
9. Define homeostasis and describe how blood sugar is regulated.
10. Describe strategies used by ectotherms and endotherms to maintain body temperature and to survive in extreme temperature conditions.
11. Contrast the cells of a malignant neoplasm with those of normal tissue and discuss current thinking about the mechanisms, causes, and treatment of cancer.

Why are most animals larger than bacteria, protists, and fungi? Many biologists think the answer to this question is related to *ecological niches,* which are the functional roles of a species within a community. By the time animals evolved, bacteria, protists, and fungi already occupied most available ecological niches. For new species to be successful they had to displace an organism from the niche it successfully occupied or adapt to a new one. Success in a new niche required a new body plan, and new body plans often involved larger size. Because predators are typically larger than their prey, large size afforded more opportunity for capturing food.

To grow to a larger size than their bacterial and protist competitors, animals had to be multicellular. Recall that the size of a single cell is limited by the ratio of its surface area (plasma membrane) to its volume (see Chapter 4). The plasma membrane needs to be large enough relative to the cell's volume to permit passage of materials into and out of the cell, so that the conditions necessary for life can be maintained. In a multicellular animal, each cell has a large enough surface-to-volume ratio to effectively regulate its internal environment. Individual cells can live and die and be replaced while the organism continues to maintain itself and thrive. The number of cells, not their individual sizes, is responsible for the size of an animal.

In unicellular organisms, such as bacteria and protists, the single cell must carry on all the activities necessary for life. Recall that unicellular and small, flat organisms depend on diffusion for many life processes, including gas exchange and disposal of metabolic wastes. One reason they can be small is that they do not have complex organ systems. They can move about without muscles, coordinate their activities without a nervous system, and process food without a digestive system.

Consider how different the caiman and the butterfly are, not only in size but also in body form and lifestyle *(see photograph)*. Multicellularity permits the specialization of cells. In a multicellular organism, cells specialize to perform specific tasks. A **tissue** consists of a group of closely associated, similar cells that carry out specific functions. Animal tissues are classified as epithelial, connective, muscle, or nervous tissue. Classification of tissues is based on their origin as well as on their structure. Each kind of tissue (and subtissue) is composed of cells with characteristic sizes, shapes, and arrangements. For example, some tissues are specialized to transport materials, whereas others contract, enabling the animal to move. Still others secrete hormones that regulate metabolic processes.

Tissues associate to form **organs** such as the heart or stomach. Groups of tissues and organs form the **organ systems** of the body. In this unit we discuss how cells associate with one another and how tissues, organs, and organ systems perform specialized functions such as circulating blood, hearing, or digestion.

Cells, tissues, organs, and organ systems work together to maintain appropriate conditions in the body. The tendency to maintain a relatively constant internal environment is called **homeostasis,** and the processes that accomplish the task are **homeostatic mechanisms.** For example, in mammals nervous, endocrine, and circulatory systems work together to regulate body temperature.

In this chapter we focus on the types and functions of tissues characteristic of animals. We discuss homeostasis using the regulation of blood sugar and of body temperature as examples. In the next several chapters we discuss how animals carry out life processes and how the organ systems of a complex animal work together to maintain homeostasis of the organism as a whole.

EPITHELIAL TISSUES COVER THE BODY AND LINE ITS CAVITIES

Epithelial tissue (also called **epithelium**) covers body surfaces and lines cavities. It forms the outer layer of the skin and the linings of the digestive, respiratory, excretory, and reproductive tracts. Epithelial tissue consists of cells fitted tightly together to form a continuous layer, or sheet, of cells. One surface of the sheet is typically exposed because it lines a cavity, such as the lumen of the intestine, or covers the body (outer layer of the skin). The other surface of an epithelial layer is attached to the underlying tissue by a noncellular **basement membrane** composed of tiny fibers and nonliving polysaccharide material produced by the epithelial cells. Table 37–1 illustrates the main types of epithelial tissue, indicates their locations in the body, and describes their functions. Each type of tissue is specialized to perform a specific function or group of functions. As we discuss each tissue type, notice the relationship between its form and its function.

Epithelial tissues perform a wide variety of functions, including protection, absorption, secretion, and sensation. The epithelial layer of the skin, the **epidermis,** covers the entire body and protects it from mechanical injury, chemicals, bacteria, and fluid loss. The epithelial tissue lining the digestive tract absorbs nutrients and water into the body. Some epithelial cells are organized into **glands** that secrete cell products such as hormones, enzymes, or sweat. Other epithelial cells are specialized as sensory receptors that receive information from the environment. For example, epithelial cells in taste buds and in the nose are specialized as chemical receptors.

Everything that enters or leaves the body must cross at least one layer of epithelium. Food taken into the mouth and swallowed is not really "inside" the body until it is absorbed through the epithelium of the gut and enters the blood. To a large extent, the permeabilities of the various epithelia regulate the exchange of substances between the different parts of the body, as well as between the animal and the external environment.

Many epithelial membranes, such as the epidermis of the skin, are subjected to continuous wear and tear. As outer cells are sloughed off, they must be replaced by new ones from below. Such epithelial tissues generally have a rapid rate of cell division that continuously produces new cells to replace those lost.

Three types of epithelial cells can be distinguished on the basis of shape (see Table 37–1). **Squamous** epithelial cells are thin, flattened cells shaped like flagstones. **Cuboidal** epithelial cells are short cylinders that from the side appear cube-shaped, resembling dice. Actually, each cell is typically hexagonal in cross section, making it an eight-sided polyhedron. **Columnar** epithelial cells look like columns or cylinders when viewed from the side. The nucleus is usually located near the base of the cell. Viewed from above or in cross section, these cells often appear hexagonal. On its free surface, a columnar epithelial cell may have cilia that beat in a coordinated way, moving materials over the tissue surface. Most of the upper respiratory tract is lined with ciliated columnar epithelium that moves particles of dust and other foreign material away from the lungs.

Epithelial tissue may also be classified by the number of layers. **Simple epithelium** is composed of one layer of cells. It is usually located where substances are secreted, excreted, or absorbed, or where materials diffuse between compartments. For example, simple squamous epithelium lines the air sacs in the lungs. The structure of this thin tissue is wonderfully suited to permit the diffusion of gases in and out of the air sacs.

Stratified epithelium, composed of two or more layers, is found where protection is required. Stratified squamous epithelium, which makes up the outer layer of the skin, regenerates itself as quickly as it is sloughed off during normal wear and tear.

The cells of **pseudostratified epithelium** falsely appear to be layered. Although all its cells rest on a basement membrane, not every cell extends to the exposed surface of the tissue. This arrangement gives the impression of two or more cell layers. Some of the respiratory passageways are lined with pseudostratified epithelium equipped with cilia.

The cells lining blood and lymph vessels have a different embryonic origin than "true" epithelium, and are referred to as **endothelium.** However, these cells are structurally similar to squamous epithelial cells, and can be included in that category.

A gland consists of one or more epithelial cells specialized to produce and secrete a product such as sweat, milk, mucus, wax, saliva, hormones, or enzymes (Fig. 37–1). Epithelial tissue lining the cavities and passageways of the body typically contains some specialized mucus-secreting cells called **goblet cells.** The mucus

TABLE 37–1 Epithelial Tissues

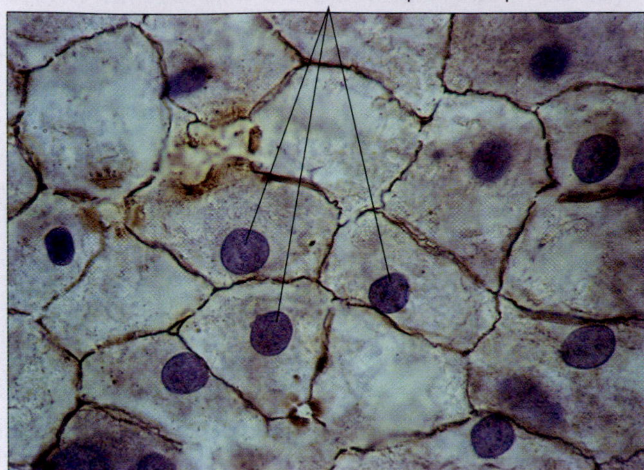

LM of simple squamous epithelium. *(Ed Reschke)*

Simple Squamous Epithelium

Main Locations

Air sacs of lungs; lining of blood vessels

Functions

Passage of materials where little or no protection is needed and where diffusion is major form of transport

Description and Comments

Cells are flat and arranged as single layer

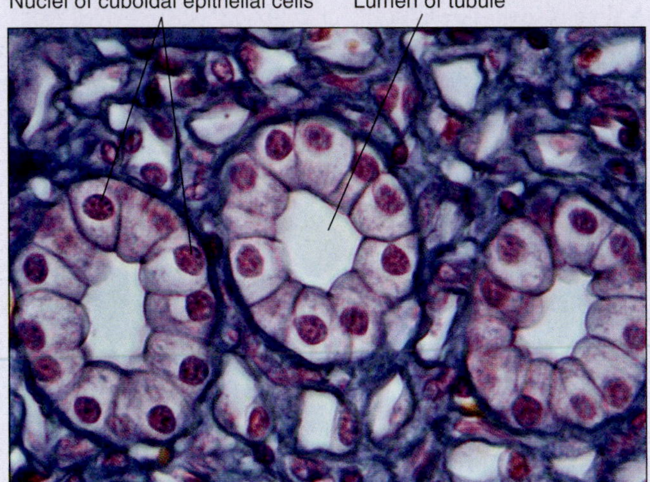

LM of simple cuboidal epithelium. *(Ed Reschke)*

Simple Cuboidal Epithelium

Main Locations

Linings of kidney tubules; gland ducts

Functions

Secretion and absorption

Description and Comments

Single layer of cells; LM shows cross section through tubules; from the side each cell looks like a short cylinder; some have microvilli for absorption

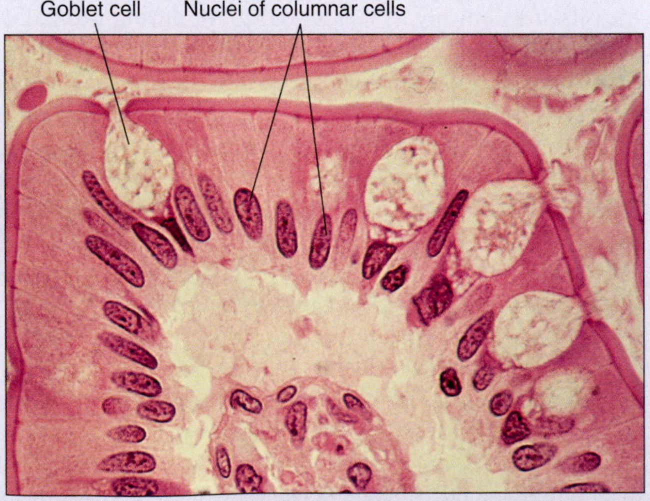

LM of simple columnar epithelium. *(Ed Reschke)*

Simple Columnar Epithelium

Main Locations

Linings of much of digestive tract and upper part of respiratory tract

Functions

Secretion, especially of mucus; absorption; protection; movement of layer of mucus

Description and Comments

Single layer of columnar cells; sometimes with enclosed secretory vesicles (in goblet cells); highly developed Golgi complex; often ciliated

TABLE 37–1 Epithelial Tissues *continued*

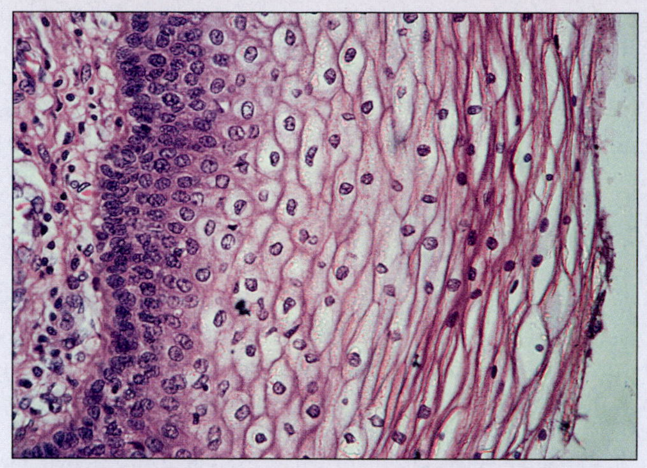

LM of stratified squamous epithelium. *(Ed Reschke)*

Stratified Squamous Epithelium

Main Locations

Skin; mouth lining; vaginal lining

Functions

Protection only; little or no absorption or transit of materials; outer layer continuously sloughed off and replaced from below

Description and Comments

Several layers of cells, with only the lower ones columnar and metabolically active; division of lower cells causes older ones to be pushed upward toward surface, becoming flatter as they move

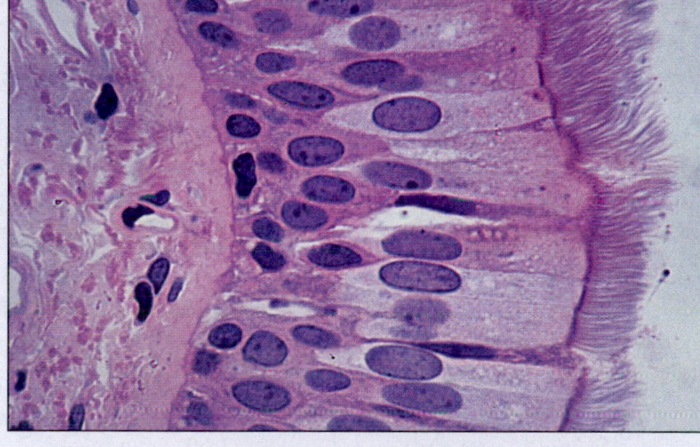

LM of pseudostratified columnar epithelium, ciliated. *(Ed Reschke)*

Pseudostratified Epithelium

Main Locations

Some respiratory passages; ducts of many glands

Functions

Secretion; protection; movement of mucus

Description and Comments

Ciliated, mucus-secreting, or with microvilli; comparable in many ways to columnar epithelium except that not all cells are the same height. Thus, though all cells contact the same basement membrane, the tissue appears stratified

lubricates these surfaces, offers protection, and facilitates the movement of materials.

Glands can be classified as exocrine or endocrine. **Exocrine glands,** like goblet cells and sweat glands, secrete their products onto a free epithelial surface, typically through a duct (tube). **Endocrine glands** lack ducts. These glands release their products, called **hormones,** into the **interstitial fluid** (tissue fluid) or blood; hormones may be transported to other parts of the body by the circulatory system. (Endocrine glands are discussed in Chapter 47.)

An **epithelial membrane** consists of a sheet of epithelial tissue and a layer of underlying connective tissue. Types of epithelial membranes include mucous membranes and serous membranes. A **mucous membrane,** or mucosa, lines a body cavity that opens to the outside of the body, such as the digestive or respiratory tract. The epithelial layer secretes mucus that lubricates the tissue and protects it from drying.

A **serous membrane** lines a body cavity that does not open to the outside of the body. It consists of simple squamous epithelium over a thin layer of loose connective tissue. This type of membrane secretes fluid into the cavity it lines. Some serous membranes with which you may be familiar are the pleural membranes lining the pleural cavities around the lungs, the pericardial membranes lining the pericardial cavity around the heart, and the peritoneum lining the abdominal cavity.

CONNECTIVE TISSUES JOIN AND SUPPORT OTHER BODY STRUCTURES

Almost every organ in the body has a framework of **connective tissue** that supports and cushions it. Cartilage and bone are examples of connective tissues that support the vertebrate body

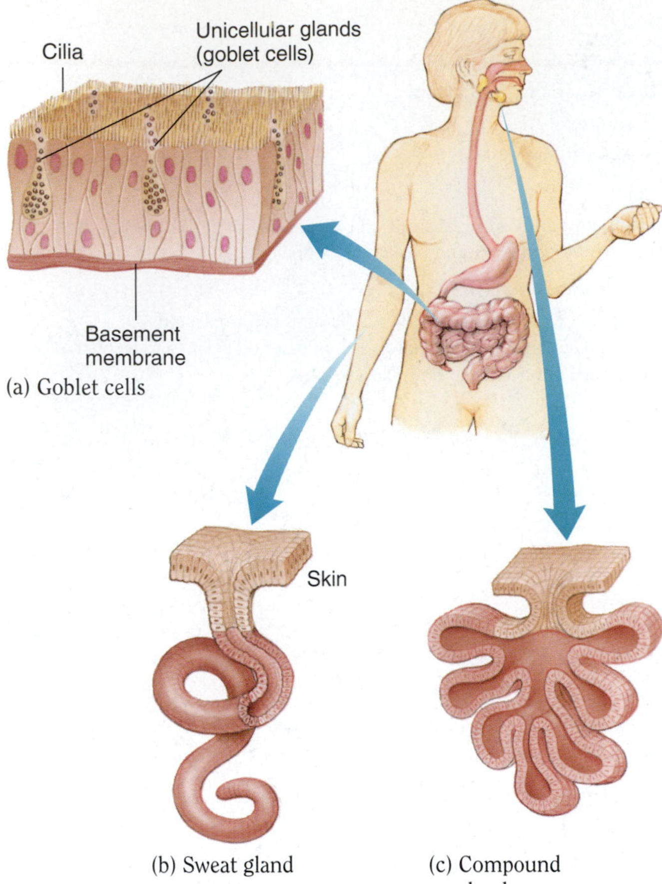

Figure 37–1 Glands. A gland consists of one or more epithelial cells. **(a)** Goblet cells are unicellular glands that secrete mucus. **(b)** Sweat glands are simple tubular glands with coiled tubes. Their walls are constructed of simple cuboidal epithelium. **(c)** Compound glands, such as the parotid salivary glands, have branched ducts.

and protect organs such as the heart and lungs. Blood is a connective tissue that transports materials, thereby connecting distant parts of the body. Adipose tissue provides a protective cushion and stores fat.

There are many systems for classifying connective tissues. Some of the main types are (1) loose and dense connective tissues; (2) elastic connective tissue; (3) reticular connective tissue; (4) adipose tissue; (5) cartilage; (6) bone; and (7) blood, lymph, and tissues that produce blood cells. These tissues vary widely in their structural details and in the functions they perform (Table 37–2).

Typically, connective tissues contain relatively few cells; these are embedded in an extensive **intercellular substance** consisting of threadlike, microscopic **fibers** scattered throughout a **matrix,** a thin gel composed of polysaccharides secreted by the cells. The cells of different kinds of connective tissues differ in their shapes and structures and in the kinds of fibers and matrices they secrete. The nature and function of each kind of connective tissue are determined in part by the structure and properties of the intercellular substance.

Connective tissue contains collagen, elastic, and reticular fibers

Connective tissue typically contains three types of fibers: collagen, elastic, and reticular. **Collagen fibers,** the most numerous type, are composed of collagens, the most abundant proteins in the body. Collagen is a very tough material. (Meat is tough because of its collagen content.) The tensile strength (ability to be stretched without tearing) of collagen fibers is comparable to that of steel. Collagen fibers are wavy and flexible, allowing them to remain intact when tissue is stretched. When treated with hot water, collagen is converted into gelatin, a soluble protein.

Elastic fibers branch and fuse to form networks. They can be stretched by a force and then (like a stretched rubber band) return to their original size and shape when the force is removed. Elastic fibers, composed of the protein elastin, are an important component of structures that must stretch.

Reticular fibers are very small branched fibers that form delicate networks not visible in ordinary stained slides; however, they become apparent when tissue is stained with silver. Reticular fibers are composed of collagen and some glycoprotein. The framework of many organs such as the liver and lymph nodes consists of reticular fibers.

Connective tissue contains specialized cells

Fibroblasts are connective tissue cells that produce the fibers, as well as the protein and carbohydrate complexes, of the matrix. Fibroblasts release protein components that become arranged to form the characteristic fibers. These cells are especially active in developing tissue and are important in healing wounds. As tissues mature, the number of fibroblasts decreases and they become less active. **Macrophages,** the scavenger cells of the body, commonly wander through connective tissues, cleaning up cellular debris and phagocytizing foreign matter, including bacteria.

Loose connective tissue is widely distributed

Loose connective tissue is the most widely distributed connective tissue in the body. Found as a thin filling between body parts, it serves as a reservoir for fluid and salts. Nerves, blood vessels, and muscles are wrapped in this tissue. Together with adipose tissue, loose connective tissue forms the subcutaneous (below the skin) layer that attaches skin to the muscles and other structures beneath. Loose connective tissue consists of fibers running in all directions though a semifluid matrix. Its flexibility permits the parts it connects to move.

Dense connective tissue consists mainly of fibers

Dense connective tissue, found in the lower layer (dermis) of the skin, is very strong, though somewhat less flexible than loose connective tissue. Collagen fibers predominate. In **irregular dense connective tissue,** the collagen fibers are arranged in bundles distributed in all directions through the tissue.

In **regular dense connective tissue,** collagen bundles are arranged in a definite pattern, making the tissue greatly resistant to stress. *Tendons,* the cords that connect muscles to bones, and *ligaments,* the cables that connect bones to one another, consist of regular dense connective tissue.

Elastic tissue is found in structures that must expand

Elastic connective tissue consists mainly of bundles of parallel elastic fibers. This tissue type is found in structures that must expand and then return to their original size, such as lung tissue and the walls of large arteries.

Reticular connective tissue provides support

Reticular connective tissue is composed mainly of interlacing reticular fibers. It forms a supporting internal framework in many organs, including the liver, spleen, and lymph nodes.

Adipose tissue stores energy

Adipose tissue is rich with fat cells that store fat and release it when fuel is needed for cellular respiration. Adipose tissue is found in the subcutaneous layer and in tissue that cushions internal organs. An immature fat cell is somewhat star-shaped. As fat droplets accumulate within the cytoplasm, the cell assumes a more rounded appearance (Fig. 37–2). Fat droplets merge with one another, eventually forming a single large drop of fat that occupies most of the volume of the mature fat-storing cell. The cytoplasm and its organelles are pushed to the cell edges, where a bulge is typically formed by the nucleus. In cross section a fat cell looks like a ring (the cytoplasm) with a single gemstone (the nucleus).

When you study a section of adipose tissue through a microscope, it may remind you of chicken wire. The rings of cytoplasm are the "wire," and the large spaces indicate where fat drops existed before they were dissolved by chemicals used to prepare the tissue. These spaces may cause the cells to collapse, giving the tissue a wrinkled appearance.

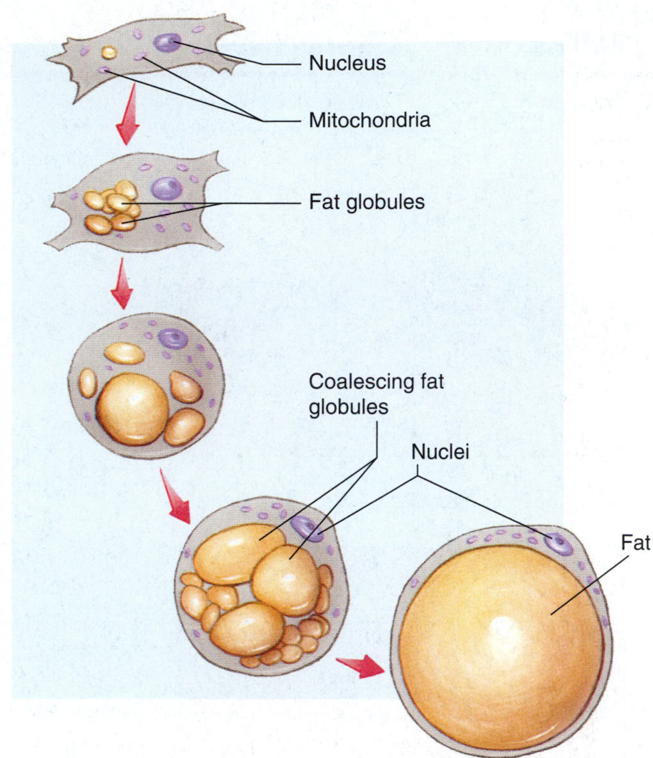

Figure 37–2 Development of a fat cell. As fat droplets accumulate in the cytoplasm, they coalesce to form a very large globule of fat. Such a fat globule may occupy most of the cell, pushing the cytoplasm and the organelles to the periphery. Fat cells make up adipose tissue. See Table 37–2 for a LM of fat cells filled with fat.

Cartilage and bone provide support

The supporting skeleton of a vertebrate is composed of cartilage, or of both cartilage and bone. Recall that **cartilage** is the supporting skeleton in the embryonic stages of all vertebrates. In most vertebrates, cartilage is largely replaced by bone during development. In humans, cartilage forms the supporting structure of the external ear, the supporting rings in the walls of the respiratory passageways, the tip of the nose, the ends of some bones, and the discs that serve as cushions between our vertebrae.

Cartilage is firm yet elastic. Its cells, called **chondrocytes,** secrete a hard, rubbery matrix around themselves and also secrete collagen fibers, which become embedded in the matrix and strengthen it. Chondrocytes eventually come to lie, singly or in groups of two or four, in small cavities in the matrix called **lacunae** (see Table 37–2). These cells remain alive and are nourished by nutrients and oxygen that diffuse through the matrix. Cartilage tissue lacks nerves, lymph vessels, and blood vessels.

Bone, the main vertebrate skeletal tissue, is similar to cartilage in that it consists mostly of matrix material containing lacunae. The bone cells, called **osteocytes,** secrete and maintain the matrix (Fig. 37–3). Unlike cartilage, however, bone is a highly vascular tissue with a substantial blood supply. Osteocytes communicate with one another and with capillaries by tiny channels (canaliculi) that contain long cytoplasmic extensions of the osteocytes.

Diffusion alone would not provide sufficient nourishment for the osteocytes. This is because bone matrix consists not only of collagen, mucopolysaccharides, and other organic materials but also of hydroxyapatite crystals, composed mainly of calcium phosphate. Diffusion through this substance is very slow.

A typical bone has an outer layer of *compact bone* surrounding a filling of *spongy bone.* Compact bone consists of spindle-shaped units called **osteons.** Within each osteon, osteocytes are arranged in concentric layers called *lamellae,* which are formed by the matrix. In turn, the lamellae surround central microscopic channels known as **Haversian canals,** through which capillaries and nerves pass (Fig. 37–3*c*). Thus, each osteon consists of a central blood vessel, surrounded by lamellae, and osteocytes.

TABLE 37–2 Connective Tissues

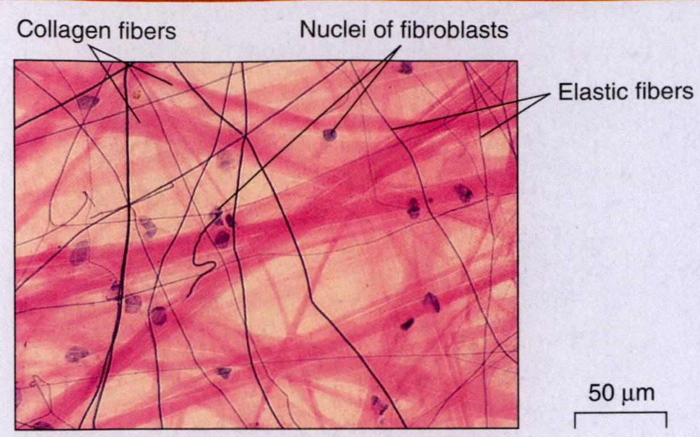

LM of loose connective tissue. *(Ed Reschke)*

Loose (Areolar) Connective Tissue

Main Locations

Everywhere that support must be combined with elasticity, such as subcutaneous tissue (the layer of tissue beneath the dermis of the skin)

Functions

Support; reservoir for fluid and salts

Description and Comments

Fibers produced by fibroblast cells embedded in semifluid matrix and mixed with miscellaneous other cells

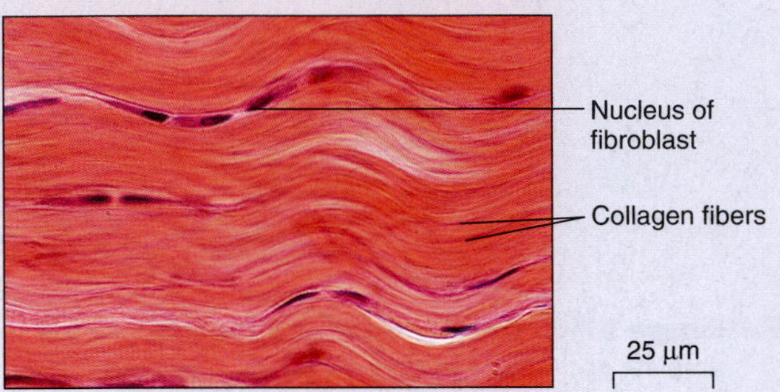

LM of dense connective tissue. *(Ed Reschke)*

Dense Connective Tissue

Main Locations

Tendons; many ligaments; dermis of skin

Functions

Support; transmission of mechanical forces

Description and Comments

Collagen fibers may be regularly or irregularly arranged

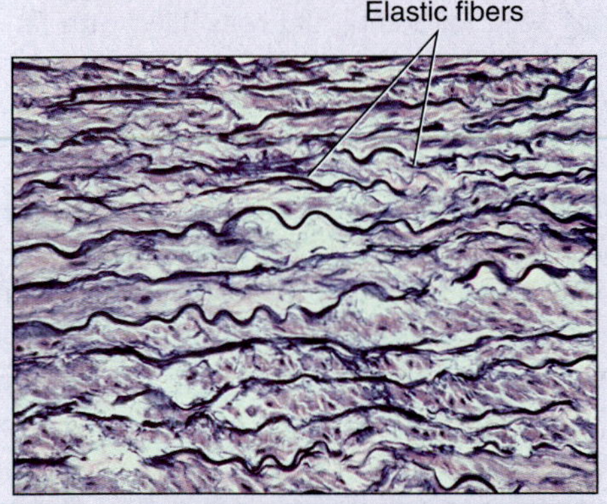

LM of elastic connective tissue. *(Ed Reschke)*

Elastic Connective Tissue

Main Locations

Structures that must both expand and return to their original size, such as lung tissue and large arteries

Function

Confers elasticity

Description and Comments

Branching elastic fibers interspersed with fibroblasts

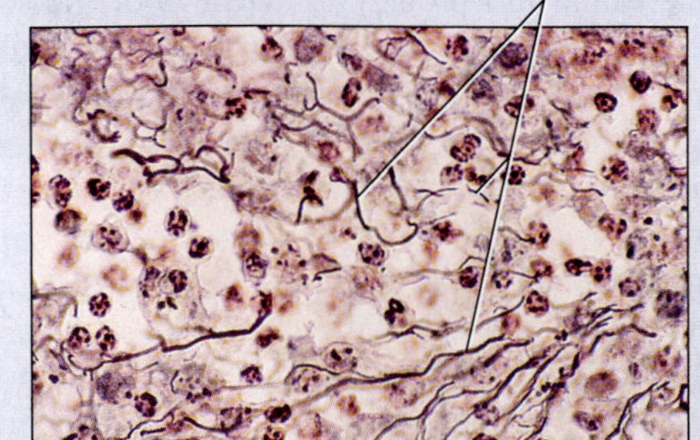

LM of reticular connective tissue. *(Ed Reschke)*

Reticular Connective Tissue

Main Locations

Framework of liver; lymph nodes; spleen

Function

Support

Description and Comments

Consists of interlacing reticular fibers

TABLE 37–2 Connective Tissues *continued*

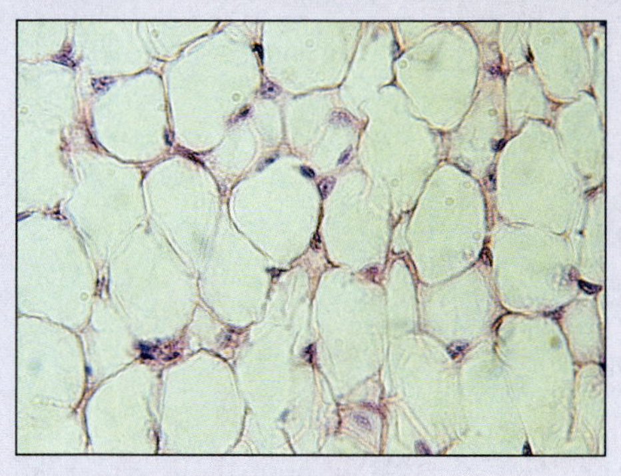

LM of adipose tissue. *(Dennis Drenner)*

Adipose Tissue

Main Locations

Subcutaneous layer; pads around certain internal organs

Functions

Food storage; insulation; support of such organs as mammary glands, kidneys

Description and Comments

Fat cells are star-shaped at first; fat droplets accumulate until typical ring-shaped cells are produced

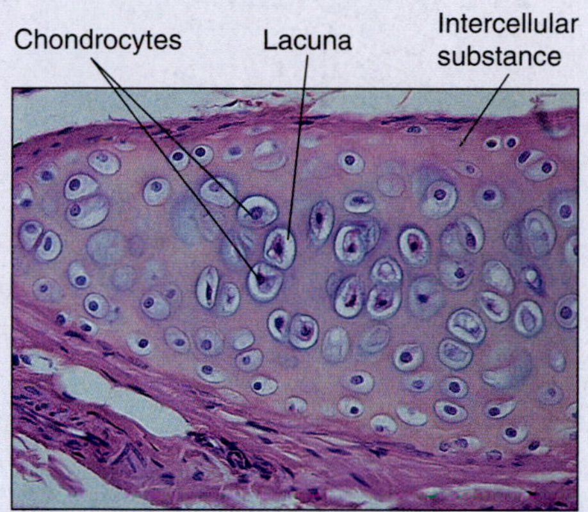

LM of cartilage. *(Ed Reschke)*

Cartilage

Main Locations

Supporting skeletons in sharks and rays; ends of bones in mammals and some other vertebrates; supporting rings in walls of some respiratory tubes; tip of nose; external ear

Function

Flexible support

Description and Comments

Cells (chondrocytes) separated from one another by intercellular substance; cells occupy lacunae

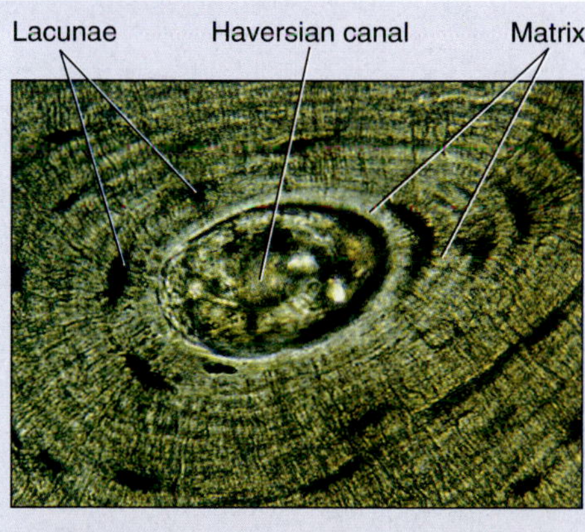

LM of bone. *(Dennis Drenner)*

Bone

Main Locations

Forms skeletal structure in most vertebrates

Functions

Support and protection of internal organs; calcium reservoir; skeletal muscles attach to bones

Description and Comments

Osteocytes in lacunae; in compact bone, lacunae arranged in concentric circles surrounding Haversian canals

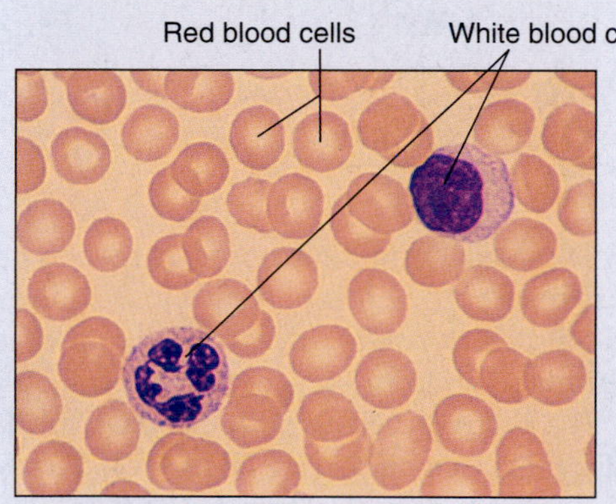

LM of blood. *(Ed Reschke)*

Blood

Main Locations

Within heart and blood vessels of circulatory system

Functions

Transports oxygen, nutrients, wastes, and other materials

Description and Comments

Consists of cells dispersed in fluid intercellular substance

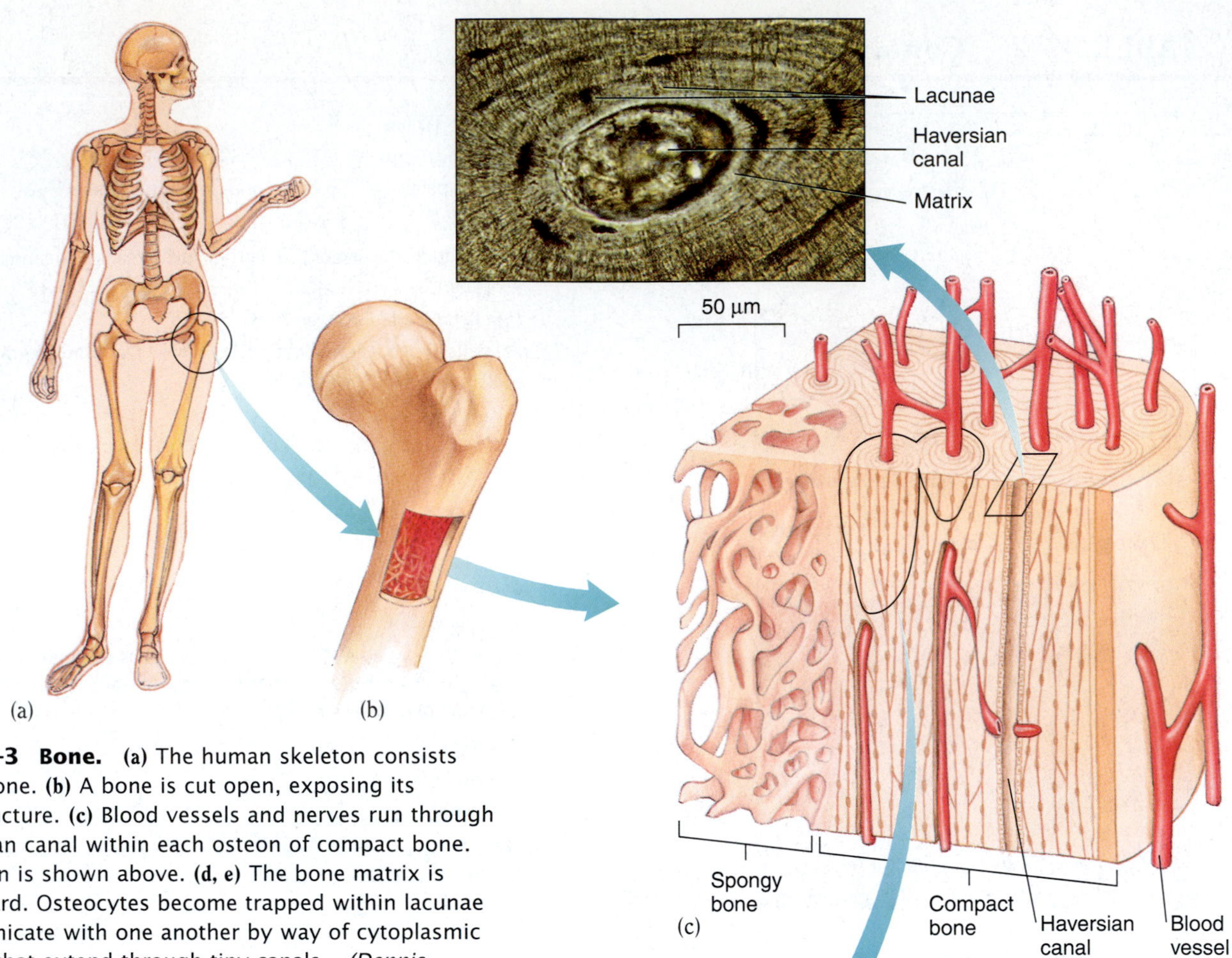

Figure 37–3 Bone. **(a)** The human skeleton consists mainly of bone. **(b)** A bone is cut open, exposing its internal structure. **(c)** Blood vessels and nerves run through the Haversian canal within each osteon of compact bone. LM of osteon is shown above. **(d, e)** The bone matrix is rigid and hard. Osteocytes become trapped within lacunae but communicate with one another by way of cytoplasmic extensions that extend through tiny canals. *(Dennis Drenner)*

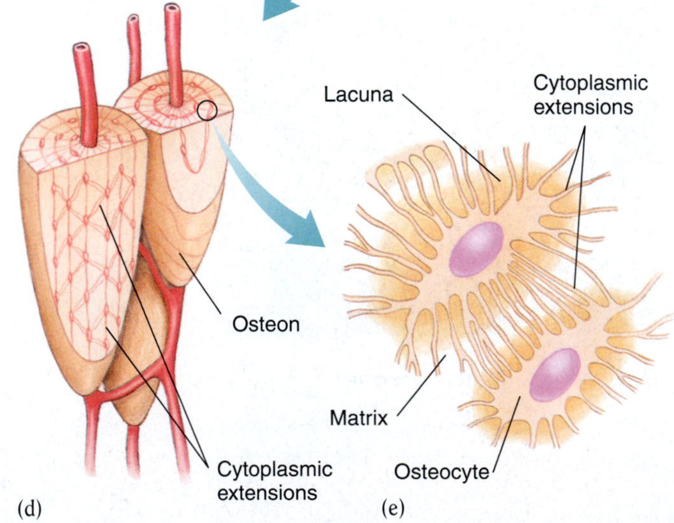

Bones are amazingly light and strong. Calcium salts of bone render the matrix very hard, and collagen prevents the bony matrix from being overly brittle. Most bones have a large central **marrow cavity** that may contain a spongy tissue called marrow. Yellow marrow consists mainly of fat. Red marrow is the connective tissue in which blood cells are produced.

Bone is a dynamic, living tissue that gradually changes its shape and internal architecture in response to normal growth processes and physical stress. In growth and remodeling, large multinucleated cells, called **osteoclasts,** help sculpt bone by dissolving and removing parts of the bony substance. Bone is discussed in more detail in Chapter 38.

Blood and lymph are circulating tissues

Blood and **lymph** are circulating tissues that help other parts of the body communicate and interact. Like other connective tissues, they consist of specialized cells dispersed in an intercellular substance. Blood and lymph are discussed in Chapter 42.

In mammals, blood is composed of red blood cells, white blood cells, and platelets all suspended within **plasma,** the liquid, noncellular part of the blood. Plasma consists of water, proteins, salts, and a variety of soluble chemical messengers such as hormones that it transports from one part of the body to another.

The **red blood cells** *(erythrocytes)* of humans and other vertebrates contain the red respiratory pigment *hemoglobin,* which transports oxygen. The red blood cells of most mammals are flattened, biconcave discs that lack nuclei (see Table 37–2). Those of other vertebrates are oval and have nuclei. Human blood contains five main types of **white blood cells** *(leukocytes),* each with distinct size, shape, structure, and functions. The white blood cells defend the body against disease-causing microorganisms (discussed in Chapter 43). **Platelets** are small fragments broken off from large cells in the bone marrow. In complex vertebrates they play a key role in blood clotting.

MUSCLE TISSUE IS SPECIALIZED TO CONTRACT

Movements of most animals result from the contraction of the elongated, cylindrical, or spindle-shaped cells of **muscle tissue.** Each muscle cell is referred to as a **muscle fiber** because of its length. A muscle fiber contains many thin, longitudinal, parallel contractile units called **myofibrils.** Two proteins, **myosin** and **actin,** are the chief components of myofibrils and play a key role in contraction of muscle fibers.

Some invertebrates have skeletal and smooth muscle. Vertebrates have three types of muscle tissue: skeletal, cardiac, and smooth (Table 37–3). **Skeletal muscle** makes up the large muscle masses attached to the bones of the body. Skeletal muscle fibers are very long—up to 2 or 3 cm (about 1 in). Each skeletal muscle fiber has many nuclei, a consequence of its formation from the fusion of several embryonic cells. The nuclei of skeletal muscle fibers are also unusual in their position. They lie just under the plasma membrane, which frees the entire central part of the skeletal muscle fiber for the myofibrils. This arrangement appears to be an adaptation that increases the efficiency of contraction. (Muscle contraction is discussed in Chapter 38.) Whereas skeletal muscle fibers are generally under voluntary control, cardiac and smooth muscle fibers are normally not contracted at will.

Light microscopy reveals that both skeletal and cardiac fibers have alternating light and dark transverse stripes, or **striations,** that change their relative sizes during contraction. Striated muscle fibers can contract rapidly but cannot remain contracted for a long period. They must relax and rest momentarily before contracting again.

Cardiac muscle is the main tissue of the heart. The fibers of cardiac muscle are joined end-to-end, and they branch and rejoin, forming complex networks. One or two nuclei are found within each fiber. A characteristic feature of cardiac muscle tissue is the presence of *intercalated discs,* specialized junctions where the fibers join.

Smooth muscle occurs in the walls of the digestive tract, uterus, blood vessels, and many other internal organs. Each spindle-shaped fiber contains a single, central nucleus.

NERVOUS TISSUE CONTROLS MUSCLES AND GLANDS

Nervous tissue is composed of **neurons,** cells specialized for transmitting electrochemical nerve impulses, and **glial cells,** which support and nourish the neurons (Fig. 37–4). Certain neurons receive signals from the external or internal environment and transmit them to the spinal cord and brain. Other neurons relay, process, or store information. Still others transmit signals from the brain and spinal cord to the muscles and glands. Neurons communicate at junctions called synapses. A nerve consists of a great many neurons bound together by connective tissue.

A typical neuron has a **cell body** containing the nucleus, and two types of cytoplasmic extensions (see Chapter 39). **Dendrites** are cytoplasmic extensions specialized for receiving impulses and transmitting them to the cell body. The single **axon** transmits

TABLE 37–3 Muscle Tissues

	Skeletal	Cardiac	Smooth
Location	Attached to skeleton	Walls of heart	Walls of stomach, intestines, etc.
Type of Control	Voluntary	Involuntary	Involuntary
Shape of Fibers	Elongated, cylindrical, blunt ends	Elongated, cylindrical, fibers that branch and fuse	Elongated, spindle-shaped, pointed ends
Striations	Present	Present	Absent
Number of Nuclei per Fiber	Many	One or two	One
Position of Nuclei	Peripheral	Central	Central
Speed of Contraction	Most rapid	Intermediate (varies)	Slowest
Resistance to Fatigue (with repetitive contraction)	Least	Intermediate	Greatest

(a) Skeletal muscle fibers

(b) Cardiac muscle fibers

(c) Smooth muscle fibers

FOCUS ON Unwelcome Tissues: Cancers

In this chapter we have focused on normal tissues. A **neoplasm** ("new growth"), or **tumor,** is an abnormal mass of cells. A neoplasm may be benign or malignant (cancerous). A **benign** ("kind") tumor tends to grow slowly, and its cells stay together. Because benign tumors form masses with distinct borders, they can usually be removed surgically.

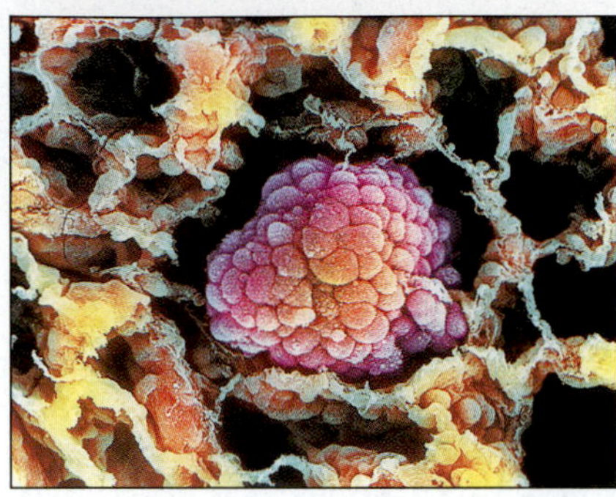

Color-Enhanced SEM of lung cancer. Cancer cells multiply rapidly and invade normal tissues, interfering with normal function. Cancer cells *(pink and tan)* have invaded the air sacs (alveoli) in this tissue from human lung. A mass of malignant cells occupies the air sac in the center of the SEM, and cancer cells that have separated from the main tumor can be seen in other air sacs. Microvilli on the surface of the cancer cells give them a fuzzy appearance. *(Moredum Animal Health LTD/Photo Researchers, Inc.)*

A **malignant** ("wicked") **neoplasm,** or **cancer,** usually grows much more rapidly and invasively than a benign tumor. In fact, two basic defects in behavior that characterize most cancer cells are rapid multiplication and abnormal relations with neighboring cells. Unlike normal cells, which respect one another's boundaries and form tissues in an orderly, organized manner, cancer cells grow helter-skelter on one another and infiltrate normal tissues. They are apparently no longer able to receive or respond appropriately to signals from surrounding cells; communication is lacking *(see figure)*.

When a cell that has transformed into a cancer cell multiplies, all the cells derived from it are also abnormal. Unlike the cells of benign tumors, cancer cells do not retain normal structural features. Cancers that develop from connective tissues or muscle are referred to as **sarcomas.** Those that originate in epithelial tissue are called **carcinomas.** Most human cancers are epithelial.

Death from cancer almost always results from **metastasis,** a migration of cancer cells through blood or lymph channels to distant parts of the body. Once there, they multiply, forming new malignant neoplasms that can interfere with the normal functions of the tissues being invaded. Cancer often spreads so rapidly and extensively that surgeons are unable to locate or remove all the malignant masses.

Studies suggest that many neoplasms grow to several millimeters in diameter and then enter a dormant stage, which may last for months or even years. Some researchers hypothesize that eventually cells of the neoplasm release a chemical substance that stimulates nearby blood vessels to develop new capillaries that grow into the abnormal mass of cells. Nourished by its new blood supply, the neoplasm may begin to grow rapidly. Newly formed blood vessels have leaky walls and so are an important route for metastasis. Malignant cells enter the blood through these walls and are transported to new sites.

Cancers appear to arise when a cell accumulates multiple mutations that act together to prevent normal control of cell division. **Proto-oncogenes** are normal genes found in all cells that are important in controlling growth and development (see *Focus On: Oncogenes and Cancer,* in Chapter 16). Mutations in proto-oncogenes can convert them to **oncogenes,** which can transform normal cells into cancer cells. More than 60 oncogenes have been identified.

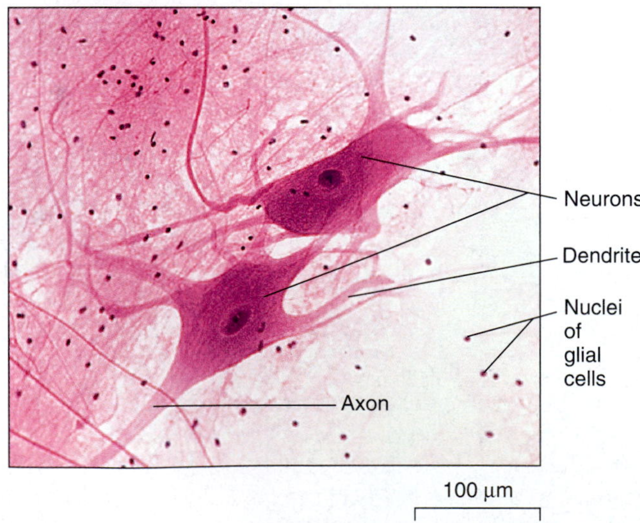

Figure 37–4 LM of nervous tissue. Neurons transmit impulses. Glial cells support and nourish neurons. *(Ed Reschke)*

impulses away from the cell body. Axons are usually long and smooth but may give off an occasional branch. They typically end in a group of fine branches. Axons range in length from 1 or 2 mm to more than a meter. Those extending from the spinal cord down the arm or leg in a human, for example, may be a meter or more in length.

In this chapter, we have focused on normal tissues. For a discussion of some abnormal tissues, see *Focus On: Unwelcome Tissues: Cancers.*

COMPLEX ANIMALS HAVE ORGANS AND ORGAN SYSTEMS

Although an animal organ may be composed predominantly of one type of tissue, other types are needed to provide support, protection, and a blood supply and to allow transmission of nerve impulses. For example, the heart consists mainly of cardiac mus-

Cancer can result from the expression of active oncogenes and also from the inactivation of **tumor suppressor genes** (anti-oncogenes) that normally inhibit proliferation of transformed cells. A tumor suppressor gene that has been extensively studied is known as ***p53.*** Since its discovery in 1979, thousands of studies have been published about this gene, which appears to prevent the cell from copying damaged DNA. The *p53* gene appears to be a key sensor that detects damage and then activates a gene circuit that controls *apoptosis* (self-destruction). Inactivation of the *p53* gene (by mutation or indirectly by changes in other molecules with which it interacts) has been found in at least 50% of all human cancer cells. Cancers associated with *p53* malfunction appear to be very aggressive, and investigators have suggested that the mutated gene may stimulate cell division instead of inhibiting it.

Cancer is the second greatest cause of death in the United States. One in three persons in the United States gets cancer at some time in his or her life, and two in three cancer patients die within five years of diagnosis. Currently, the key to survival is early diagnosis and treatment with some combination of surgery, hormonal treatment, radiation therapy, and drugs that suppress mitosis, such as chemotherapy. In patients with a functional *p53* gene, radiation treatment and some forms of chemotherapy (e.g., tamoxifen) appear to work by activating the gene. Investigators are developing several approaches to *p53*-based cancer treatment. For example, viruses can be used to deliver functional *p53* genes into malignant neoplasms. Another treatment approach that appears promising involves the use of immune therapy in which customized cancer vaccines destroy neoplasms.

Because cancer is actually an entire family of closely related diseases (there are hundreds of distinct varieties), it is probable that no single cure exists. The pace of cancer research has greatly increased with the recent application of a powerful new technique: **microarrays** (see Fig. 16–5). A microarray is a chip dotted with DNA from thousands of genes that can be used as probes to determine which genes are active in various types of cells including different types of cancer cells. This technique is being used to distinguish among cancer types and subtypes based on gene expression patterns. Different treatments might be effective for each subtype. For example, investigators might be able to determine which genes might be effective targets for treatment with a particular drug in each subtype of breast cancer.

Alleles of some genes appear to affect an individual's level of tolerance to carcinogens (cancer-producing agents). More than 80% of cancer cases are thought to be triggered by carcinogens in the environment. Risk of developing cancer can be decreased by following these recommendations:

1. Do not smoke or use tobacco. Smoking is responsible for more than 80% of lung cancer cases, and increases the risk for many other cancers.
2. Avoid prolonged exposure to the sun. When in the sun, use sunscreen or sun block. Exposure to the sun is responsible for almost all of the 400,000 cases of skin cancer reported each year in the United States.
3. Eat a healthy diet including fruit and vegetables. Although research results are not clear, it appears prudent to increase the fiber content of your diet and avoid high-fat, smoked, salt-cured, and nitrite-cured foods.
4. Avoid unnecessary exposure to x rays.
5. Women should examine their breasts each month, have regular mammograms, and obtain annual Papanicolaou's (Pap) tests. Men should regularly examine their testes and should have prostate examinations yearly after age 50.

cle tissue, but it is lined with endothelium, regulated by nervous tissue, and contains blood vessels composed of endothelium, smooth muscle, and connective tissue.

Several tissues and organs may work together, performing a specialized set of functions. Such an organized group of structures makes up an organ system. We can identify ten major organ systems that work together to make up a complex animal: **integumentary, skeletal, muscle, nervous, circulatory, digestive, respiratory, urinary, endocrine,** and **reproductive systems.** Table 37–4 and Figure 37–5 summarize their principal organs and functions.

Consider the digestive system as an example of an organ system. Its organs include the mouth, esophagus, stomach, small and large intestines, liver, pancreas, and salivary glands. The digestive system processes food, reducing it to small molecular components. The products of digestion are absorbed into the blood, which transports them to all cells of the body.

ORGAN SYSTEMS WORK TOGETHER TO MAINTAIN HOMEOSTASIS

In a large, complex animal, billions of cells may be organized to form tissues, organs, and organ systems. The organism functions effectively, in large part because very precise control mechanisms maintain homeostasis. If the organism is to survive and function, the composition of the fluids that bathe its cells must be carefully regulated. An appropriate concentration of nutrients, oxygen and other gases, ions, and compounds needed for metabolism must be available at all times. In addition, internal temperature and pressure must be maintained within relatively narrow limits.

Homeostasis is a basic concept in physiology. First coined by the physiologist Walter Cannon, the word *homeostasis* is derived from the Greek *homoios,* meaning "same," and *stasis,* "standing." Actually, the internal environment never really stays the same. Homeostasis is always being challenged by **stressors,** changes in the internal or external environment that affect

(text continues on page 810)

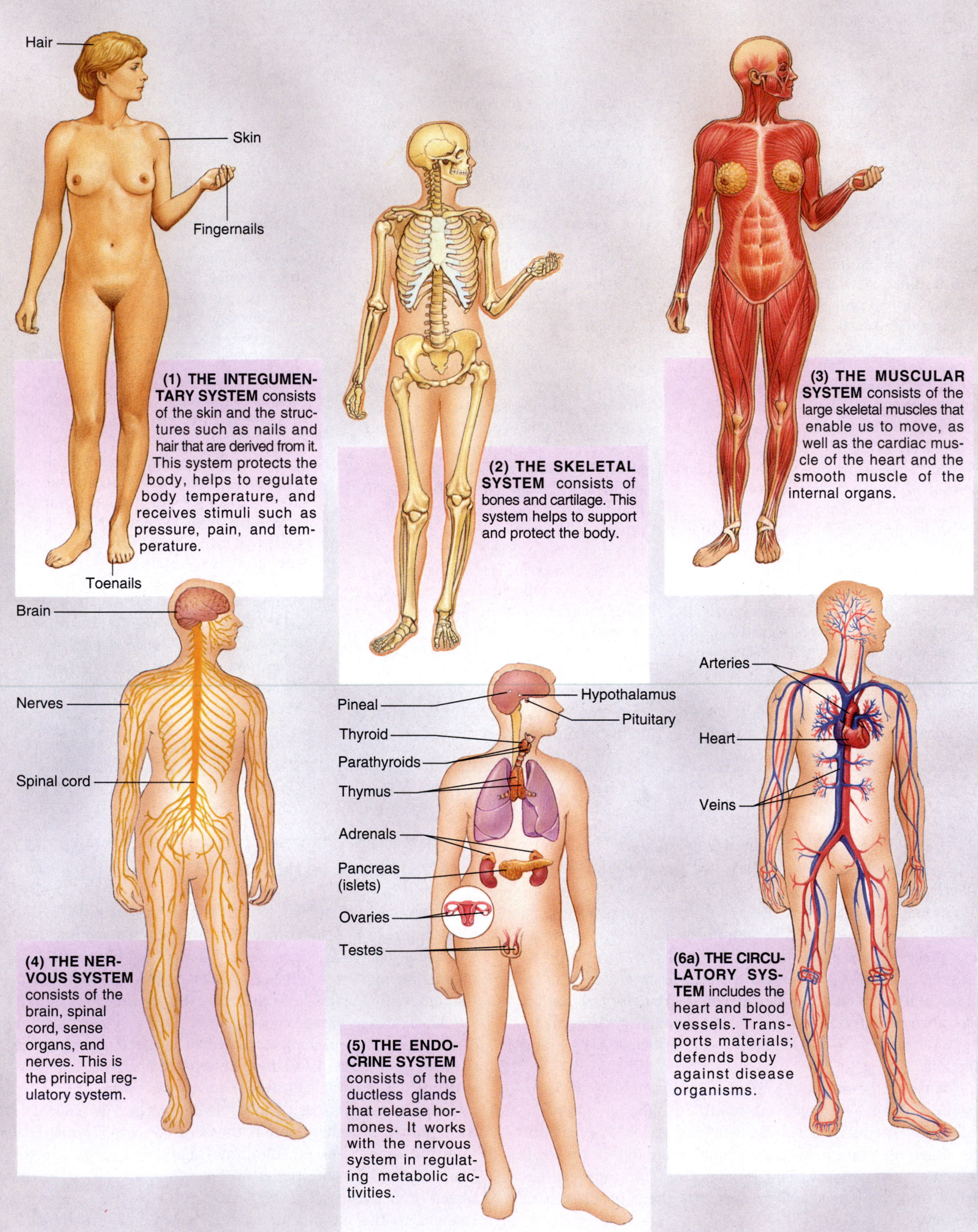

Hair
Skin
Fingernails
Toenails
(1) THE INTEGUMENTARY SYSTEM consists of the skin and the structures such as nails and hair that are derived from it. This system protects the body, helps to regulate body temperature, and receives stimuli such as pressure, pain, and temperature.
(2) THE SKELETAL SYSTEM consists of bones and cartilage. This system helps to support and protect the body.
(3) THE MUSCULAR SYSTEM consists of the large skeletal muscles that enable us to move, as well as the cardiac muscle of the heart and the smooth muscle of the internal organs.
Brain
Nerves
Spinal cord
(4) THE NERVOUS SYSTEM consists of the brain, spinal cord, sense organs, and nerves. This is the principal regulatory system.
Pineal
Hypothalamus
Pituitary
Thyroid
Parathyroids
Thymus
Adrenals
Pancreas (islets)
Ovaries
Testes
(5) THE ENDOCRINE SYSTEM consists of the ductless glands that release hormones. It works with the nervous system in regulating metabolic activities.
Arteries
Heart
Veins
(6a) THE CIRCULATORY SYSTEM includes the heart and blood vessels. Transports materials; defends body against disease organisms.

Figure 37–5 The ten principal organ systems of the human body.

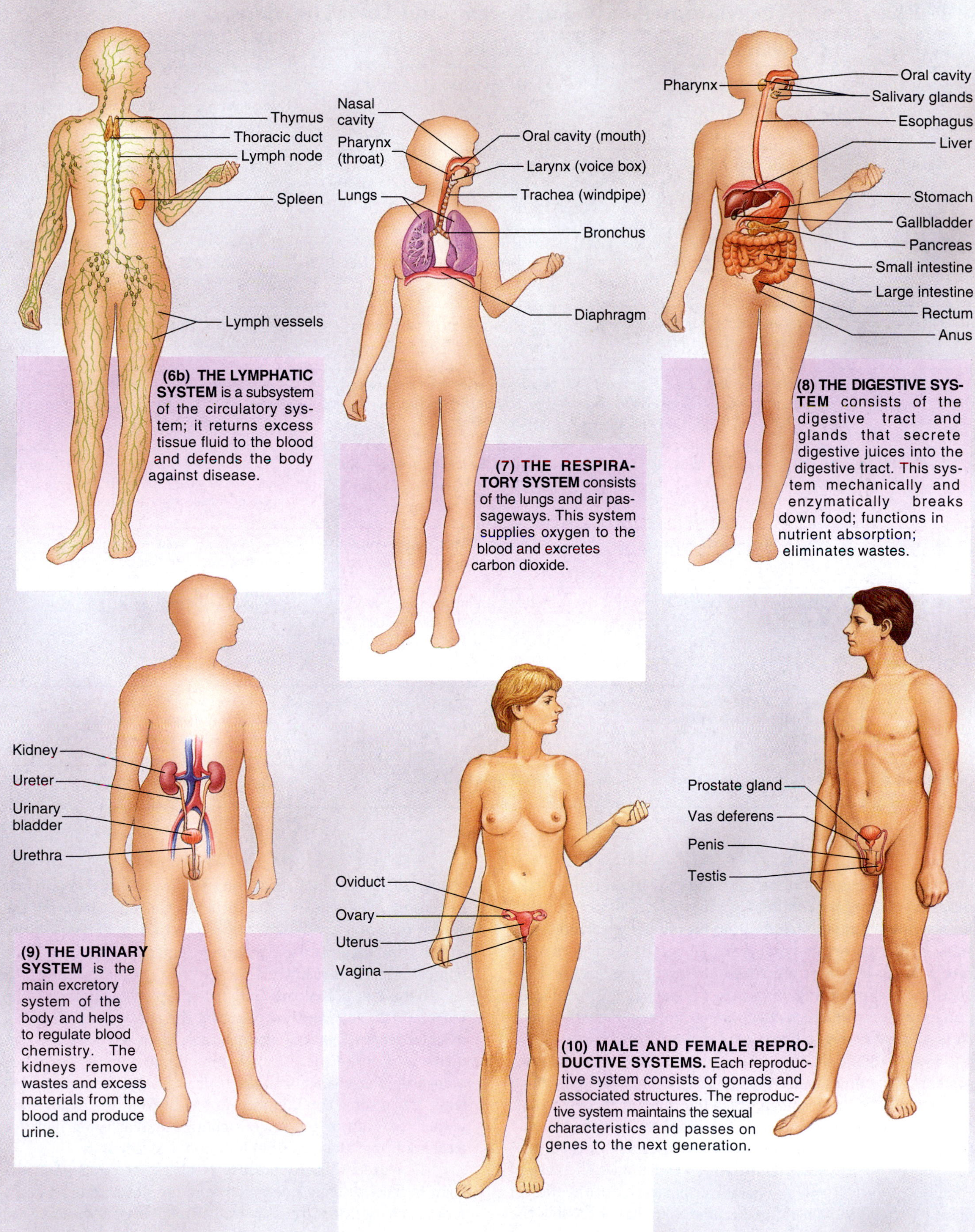

(6b) THE LYMPHATIC SYSTEM is a subsystem of the circulatory system; it returns excess tissue fluid to the blood and defends the body against disease.

(7) THE RESPIRATORY SYSTEM consists of the lungs and air passageways. This system supplies oxygen to the blood and excretes carbon dioxide.

(8) THE DIGESTIVE SYSTEM consists of the digestive tract and glands that secrete digestive juices into the digestive tract. This system mechanically and enzymatically breaks down food; functions in nutrient absorption; eliminates wastes.

(9) THE URINARY SYSTEM is the main excretory system of the body and helps to regulate blood chemistry. The kidneys remove wastes and excess materials from the blood and produce urine.

(10) MALE AND FEMALE REPRODUCTIVE SYSTEMS. Each reproductive system consists of gonads and associated structures. The reproductive system maintains the sexual characteristics and passes on genes to the next generation.

TABLE 37–4 The Mammalian Organ Systems and Their Functions

System	Components	Functions	Homeostatic Ability
Integumentary	Skin, hair, nails, sweat glands	Covers and protects body	Sweat glands help control body temperature; as a barrier, the skin helps maintain a steady state
Skeletal	Bones, cartilage, ligaments	Supports and protects body; provides for movement and locomotion; stores calcium	Helps maintain constant calcium level in blood
Muscular	Skeletal muscle; cardiac muscle; smooth muscle	Moves parts of skeleton; provides locomotion; moves internal materials	Ensures vital functions requiring movement, e.g., cardiac muscle circulates the blood
Digestive	Mouth, esophagus, stomach, intestine, liver, pancreas, salivary glands	Ingests and digests food; absorbs nutrients into blood	Maintains adequate supplies of fuel molecules and building materials
Circulatory	Heart, blood vessels, blood; lymph and lymph structures (lymphatic system is a subsystem of the circulatory system)	Transports materials from one part of body to another; defends body against disease organisms	Transports oxygen, nutrients, hormones, wastes; maintains water and ionic balance of tissues
Respiratory	Lungs, trachea, and other air passageways	Exchanges gases between blood and external environment	Maintains adequate blood oxygen content and helps regulate blood pH; removes carbon dioxide from the blood
Urinary	Kidney, bladder, and associated ducts	Excretes metabolic wastes; removes excessive substances from blood	Helps regulate volume and composition of blood and body fluids
Nervous	Nerves and sense organs; brain and spinal cord	Receives stimuli from external and internal environment; conducts impulses; integrates activities of other systems	Principal regulatory system
Endocrine	Ductless glands (e.g., pituitary, adrenal, thyroid) and tissues that secrete hormones	Regulates blood chemistry and many body functions	In conjunction with nervous system, regulates metabolic activities and blood levels of various substances
Reproductive	Testes, ovaries, and associated structures	Sexual reproduction	Maintains sexual characteristics

normal conditions within the body, causing **stress.** Homeostatic mechanisms interact continuously to manage stress, maintaining the internal environment within the physiological limits that support life. All organ systems participate in these regulatory mechanisms, but most are controlled by the nervous and endocrine systems. During our study of organ systems, we discuss numerous ways in which they interact to maintain the *steady state* (normal homeostatic state) of the organism.

How do homeostatic mechanisms work? Many are **feedback systems,** sometimes called biofeedback systems. Such a system consists of a cycle of events in which information about a change (e.g., a change in temperature) is fed back into the system so that the regulator (the temperature-regulating center in the brain) can control the process (temperature regulation). The desired condition (normal body temperature) is referred to as the **set point.** When body temperature becomes too high or too low, the change serves as input, triggering the regulator to counteract the change. The regulator activates mechanisms that bring the system back to the set point. The return to normal temperature signals the temperature-regulating center to "shut off" the homeostatic mechanisms.

In this type of feedback system, the response counteracts the inappropriate change, thus restoring the steady state. This is a **negative feedback mechanism,** because the response of the regulator is opposite (negative) to the output (Fig. 37–6*a*). Most homeostatic mechanisms in the body are negative feedback systems. When some condition varies too far from the steady state (either too high or too low), a control system using negative feedback brings the condition back to the steady state.

There are also a few **positive feedback mechanisms** in the body. In these systems a deviation from the steady state sets off a series of changes that intensify (rather than reverse) the changes.

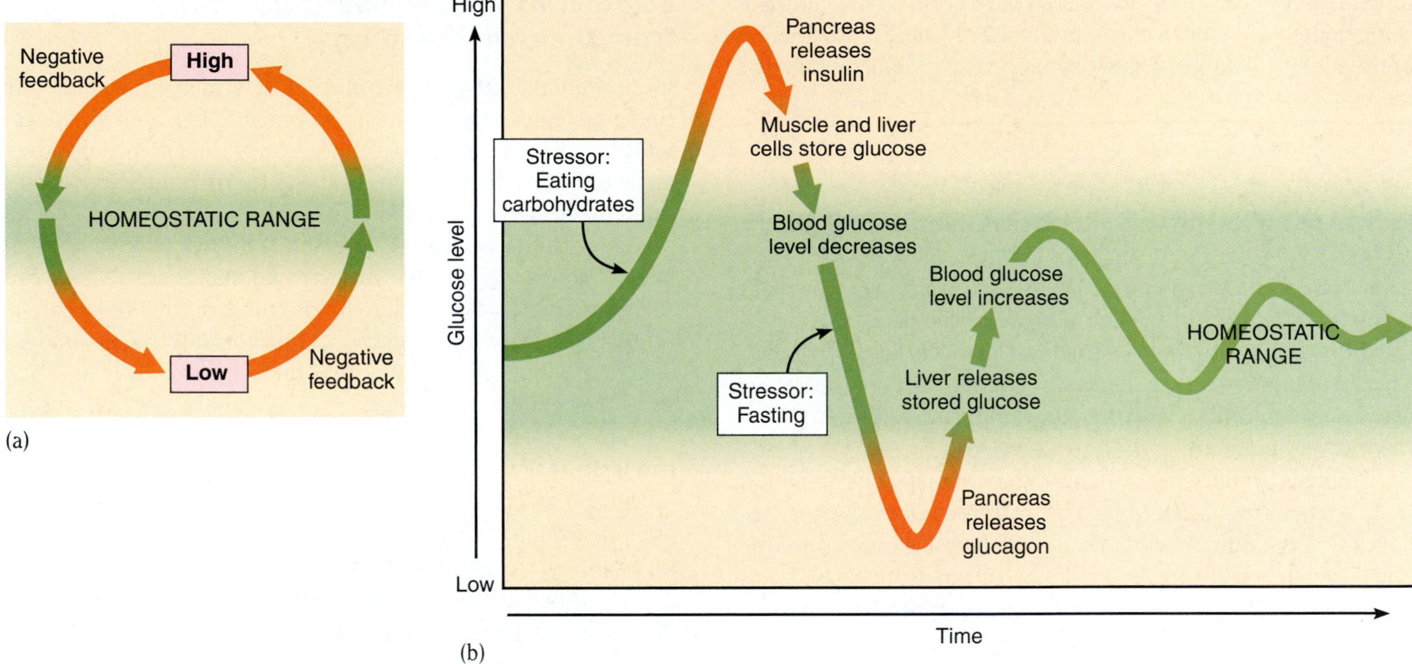

Figure 37–6 Negative feedback mechanisms maintain homeostasis. (a) In negative feedback, the response of the regulator is opposite to the output. Regulatory mechanisms bring conditions back to a homeostatic range. (b) Regulation of blood sugar concentration by negative feedback.

Therefore, although many positive feedback mechanisms are beneficial, they do not maintain homeostasis. A positive feedback cycle operates during the birth of a baby. As the baby's head pushes against the opening of the uterus (cervix), a reflex action causes the uterus to contract. The contraction forces the head against the cervix again, resulting in another contraction, and the positive feedback cycle is repeated again and again until the baby is born. Some positive feedback sequences, such as those that deepen circulatory shock following severe hemorrhage, can lead to disruption of steady states and even to death.

Regulation of blood sugar level provides a good example of homeostatic mechanisms at work (Fig. 37–6*b*). When you wake up in the morning your blood sugar (glucose) level is about 90 mg of glucose per 100 mL of blood. Perhaps you eat a big breakfast that includes pastry or a doughnut. Many of the starches and sugars in your breakfast are digested to glucose. The glucose is then absorbed into the circulatory system, causing the blood sugar level to rise.

An increase in blood sugar level stimulates the pancreas to release *insulin.* This hormone causes the body cells to take up glucose from the blood and stimulates the liver and muscle cells to store glucose (as glycogen). As a result, the glucose level in the blood decreases and returns to the normal fasting level of 90 mg/100 mL.

After several hours, when the glucose level of the blood begins to fall below the normal level, the pancreas releases *glucagon.* This hormone raises the blood sugar level by stimulating the liver cells to slowly convert glycogen to glucose and thereby release their stored glucose. In this way, insulin and glucagon act in seesaw fashion to maintain a steady state of glucose in the blood.

How does stress affect the body? Stressors are changes in the internal or external environment that disturb homeostasis, causing stress. Examples of external stressors are heat, cold, noise, abnormal atmospheric pressure, and lack of oxygen. Internal stressors include changes in blood pressure, pH, and salt concentration, as well as high and low blood sugar levels. Many stressors occur routinely. The body responds automatically by activating homeostatic mechanisms that expertly manage the stress. Other stressors are more severe and may cause serious disruption of homeostasis.

Under conditions of stress, cells produce **stress proteins.** These compounds belong to a group of proteins known as **molecular chaperones** that bind to polypeptide segments and stabilize their correct conformation. **Heat shock proteins** are stress proteins that are produced when cells are stressed by elevated temperature (42° C or higher). Recall that heat denatures proteins. Heat shock proteins have been conserved in evolution. In one group of these proteins, 50% of the amino acid sequence is the same in both *Escherichia coli* (a bacterium) and humans.

When homeostatic mechanisms are unable to restore the steady (normal) state, stress may cause malfunction that leads to disease or even death. In fact, we can view death as the failure of some homeostatic mechanism. In Chapter 47 we discuss the role of the adrenal glands in regulating homeostatic adjustments that help the body effectively cope with stress.

MANY ANIMALS THERMOREGULATE

Many animals have elaborate homeostatic mechanisms for regulating body temperature. Some of these mechanisms are physiological, whereas others are structural or behavioral strategies. **Thermoregulation** is the ability to maintain body temperature within certain limits, even when the temperature of the environment is very different.

Animals produce heat as a byproduct of metabolic activities. Body temperature is determined by the rate at which heat is produced and the rate at which heat is lost to, or gained from, the outside environment.

The strategies for maintaining body temperature that are available to an animal may restrict the type of environment it can inhabit. Each animal species has an optimal environmental temperature range. Some animals, for example, snowshoe hares, snowy owls, and weasels, are able to survive in cold arctic regions. Others, such as the Cape ground squirrel, are adapted to hot tropical climates. Although some animals can survive at temperature extremes, most can survive only within moderate temperature ranges.

Ectotherms absorb heat from their surroundings

In most animals body temperature is determined mainly by the changing temperature of the environment. Such animals are **ectotherms.** An advantage of ectothermy is the absence of direct metabolic cost. Most of the heat for thermoregulation comes from the sun. The metabolic rate of an ectotherm tends to change with the weather. As a result, ectotherms have a much lower daily energy expenditure than do endotherms. They can survive on less food, and more of the energy in their food can be converted to growth and reproduction. One of the disadvantages of ectothermy is that activity may be limited by daily and seasonal temperature conditions.

Many ectotherms use behavioral strategies to adjust body temperature. For example, lizards may keep warm by burrowing in the soil at night. During the day, lizards take in heat from their environment by basking in the sun, orienting their bodies to expose the maximum surface area to the sun's rays (Fig. 37–7*a*). Other behavioral strategies for regulating temperature include hibernation and migration.

Endotherms derive heat from metabolic processes

Birds and mammals, as well as some species of fish (e.g., tuna) and insects, are **endotherms,** which means that they have homeostatic mechanisms that maintain body temperature despite

(a)

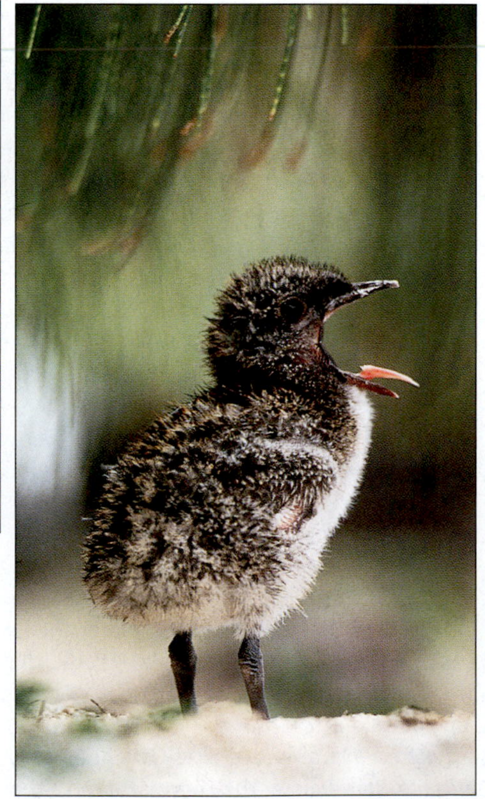

(b)

Figure 37–7 Behavioral adaptations for thermoregulation. **(a)** This marine iguana *(Amblyrhynchus cristatus)* is an ectotherm that increases its body temperature by sunning itself. **(b)** The body temperature of this baby sooty tern *(Sterna fuscata)* of Hawaii, an endotherm, is reduced as heat leaves the body through its open mouth. *(a, Breck P. Kent/Animals Animals; b, Frans Lanting/Minden Pictures)*

changes in the external temperature. Important advantages of endothermy include increased rate of enzyme activity and constant body temperature. Endotherms can carry out their activities even in low winter temperatures. However, endotherms must pay the high energy cost of thermoregulation during times when they are inactive. We must maintain our body temperature even when we are asleep.

Endotherms have structural adaptations for maintaining body temperature. For example, the insulating feathers of birds, hair of mammals, insulating layers of fat in birds and mammals, and hairs or fur of chitin on some insects are structures that help reduce heat loss from the body. Birds and mammals also have behavioral adaptations for maintaining body temperature. For example, the Cape ground squirrel uses its tail to shade its body from the direct rays of the sun. Elephants spray themselves with cool water. Humans put on warm coats and install furnaces in their homes.

Some insects use a combination of behavioral and physiological mechanisms to regulate body temperature. The "furry" body of the moth helps conserve body heat. When a moth prepares for flight, it contracts its flight muscles with little movement of its wings. The metabolic heat generated enables the moth to sustain the intense metabolic activity needed for flight.

Endotherms have a variety of physiological mechanisms for maintaining temperature homeostasis. They regulate heat production and regulate heat exchange with the environment. Most of their body heat comes from their own metabolic processes (see *Focus On: Electron Transport and Heat,* Chapter 7).

In mammals, receptors located in the hypothalamus of the brain and in the spinal cord regulate temperature. Heat from metabolic activities can be increased either directly, or indirectly by the action of hormones (e.g., thyroid hormones) that increase metabolic rate. Heat production can be increased by contracting muscles, and in cold weather many animals shiver.

When body temperature rises, birds and many mammals pant and some mammals sweat (Fig. 37–7*b*). These processes provide fluid for **evaporation,** the conversion of a liquid, such as sweat, to water vapor. Recall from Chapter 2 that when molecules enter the vapor phase, they take their heat energy with them. Heat is transferred from the body to the surroundings, resulting in evaporative cooling.

In humans information about body temperature is sent to the temperature-regulating center in the hypothalamus (Fig. 37–8). When body temperature rises above normal, the hypothalamus sends messages by way of nerves to the sweat glands, increasing sweat secretion. At the same time the hypothalamus

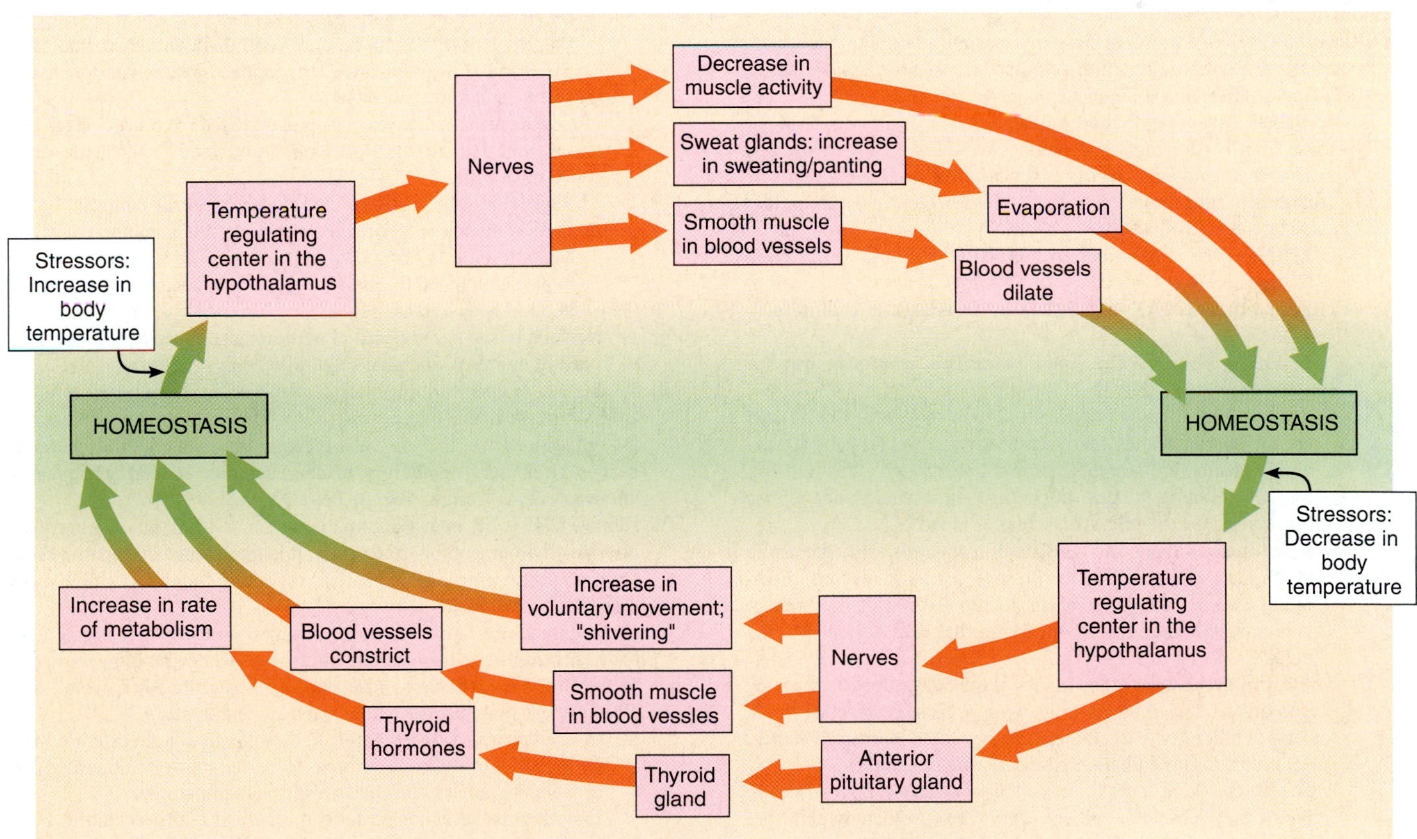

Figure 37–8 Regulation of temperature in the human body. When stressors increase body temperature above homeostatic levels, receptors send messages to the temperature-regulating center in the hypothalamus. Homeostatic mechanisms are activated that restore normal body temperature. When stressors decrease body temperature, the temperature-regulating center activates mechanisms that increase temperature, again restoring homeostasis.

sends messages to smooth muscle in the walls of capillaries in the skin, causing them to dilate. More blood circulates through the skin, bringing heat to the body surface. The skin acts as a heat radiator, allowing heat to radiate from the body surface into the environment. These homeostatic mechanisms help return body temperature to normal.

When body temperature decreases below normal, capillaries in the skin constrict so that less heat is brought to the body surface. Hormonal messages increase heat production by body tissues. Nerves signal muscles to shiver, or allow us to move muscles voluntarily to increase body temperature.

Many animals respond physiologically to changes in environmental temperature

Animals adjust to seasonal changes, a process called **acclimatization.** A familiar example is the thickening of a dog's coat in the winter. As water temperature decreases during fall and winter, a trout's enzyme systems decrease their level of activity, allowing the trout to remain active at the lowest metabolic cost.

When stressed by cold, many animals induce **torpor,** a decrease in body temperature below normal levels (an adaptive hypothermia). The decrease in body temperature is brought about by a decrease in the temperature set point in the hypothalamus. When in a state of torpor, metabolism may be dramatically depressed. Heart and respiratory rates decrease, and animals are less responsive to external stimulation. **Hibernation** is long-term torpor in response to winter cold and scarcity of food.

Estivation is a state of torpor caused by lack of food or water during periods of high temperature. During estivation some mammals retreat to their burrows. The cactus mouse *(Peromyscus eremicus)* enters a state of torpor during the winter in response to cold and scarcity of food. In summer, it estivates in response to lack of either food or water.

SUMMARY WITH KEY TERMS

I. Multicellular organisms can be much larger and more diverse than unicellular ones, and their cells can specialize, performing specific functions.

II. A **tissue** consists of a group of similarly specialized cells that associate to perform one or more functions. Animal tissues are classified as epithelial, connective, muscular, or nervous.
  A. **Epithelial tissue (epithelium)** may form a continuous layer, or sheet, of cells covering a body surface or lining a body cavity. One surface of an epithelial layer is attached to the underlying tissue by a **basement membrane,** consisting of fibers and polysaccharides made by the epithelial cells.
    1. Epithelial tissue functions in protection, absorption, secretion, or sensation.
    2. Epithelial cells may be **squamous, cuboidal,** or **columnar** in shape.
    3. Epithelial tissue may be **simple, stratified,** or **pseudostratified** (summarized in Table 37–1).
    4. Some epithelial tissue is specialized to form **glands. Goblet cells** are unicellular glands that secrete mucus. Goblet cells are **exocrine glands** that secrete their product through a duct onto an exposed epithelial surface. In contrast, **endocrine glands** release hormones into the **interstitial fluid** or blood.
    5. An **epithelial membrane** consists of a sheet of epithelial tissue and a layer of underlying connective tissue. A **mucous membrane** lines a cavity that opens to the outside of the body. A **serous membrane** lines a body cavity that does not open to the outside.
  B. **Connective tissue** joins other tissues of the body, supports the body and its organs, and protects underlying organs. Connective tissue consists of relatively few cells separated by **intercellular substance,** which is composed of **fibers** scattered through a **matrix.**
    1. The intercellular substance contains **collagen fibers, elastic fibers,** and/or **reticular fibers.** Connective tissue contains specialized cells such as **fibroblasts** and **macrophages.**
    2. Some main types of connective tissue are **loose connective tissue, dense connective tissue, elastic connective tissue, reticular connective tissue, adipose tissue, cartilage, bone,** and **blood** (summarized in Table 37–2).
    3. Cartilage cells, called **chondrocytes,** are located in **lacunae,** small cavities in the cartilage matrix.
    4. **Osteocytes** secrete and maintain the matrix of bone. As bone is remodeled, **osteoclasts** dissolve and remove parts of the bony substance. Compact bone consists of **osteons,** which are spindle-shaped units that consist of a central blood vessel that runs through a **Haversian canal,** surrounded by lamellae, concentric layers containing osteocytes.
  C. **Muscle tissue** is composed of cells specialized to contract. Each cell is an elongated muscle fiber containing many contractile units called **myofibrils.**
    1. **Skeletal muscle** is **striated** and under voluntary control.
    2. **Cardiac muscle** is striated; its contraction is involuntary.
    3. **Smooth muscle** contracts involuntarily. It is responsible for movement of food through the digestive tract and for movement of other body organs.
  D. **Nervous tissu**e is composed of **neurons,** cells specialized for transmitting impulses, and **glial cells,** which are supporting cells.

III. Tissues and **organs** work together, forming **organ systems.** In complex animals, ten principal organ systems work together, making up the living organism. These include the **integumentary, skeletal, muscular, digestive, circulatory, respiratory, urinary, nervous, endocrine** and **reproductive systems** (summarized in Table 37–4).

IV. **Homeostasi**s is the body's automatic tendency to maintain a constant internal environment, or steady state. It is maintained by **negative feedback** systems in which the regulators respond to counteract the changes caused by **stressors.**

V. Many animals are capable of **thermoregulation,** the ability to maintain body temperature within certain limits even when the temperature of the environment changes. Animals have structural, behavioral, and physiological strategies for regulating body temperature.
  A. In **ectotherms,** body temperature depends to a large extent on the temperature of the environment. However, many ectotherms have behavioral strategies for regulating body temperature.
  B. **Endotherms** have homeostatic mechanisms for regulating body temperature within a narrow range.
  C. **Acclimitization** is the process of adjustment to seasonal changes.
    1. **Torpor** involves an adaptive hypothermia; body temperature is maintained below normal levels. **Hibernation** is long-term torpor in response to winter cold.
    2. **Estivation** is torpor caused by lack of food or water during summer heat.

POST-TEST

1. A group of closely associated cells that carry out specific functions is a(an) (a) colony (b) tissue (c) organ system (d) organelle (e) homeostatic mechanism
2. Epithelial cells that produce and secrete a product into a duct form a(an) (a) endocrine gland (b) exocrine gland (c) pseudostratified membrane (d) endocrine or exocrine gland (e) serous membrane
3. A serous membrane (a) lines a body cavity that does not open to the outside of the body (b) consists of loose connective tissue (c) has a framework of reticular connective tissue (d) covers the body (e) lines the digestive and respiratory systems
4. The most numerous fibers in connective tissue are (a) reticular (b) myofibrils (c) collagen (d) elastic (e) glial
5. Dense connective tissue is likely to be found in (a) the outer layer of skin (b) tendons (c) bone (d) the air sacs of lungs (e) the framework of the lymph nodes
6. Chondrocytes are most likely to be found in (a) the outer layer of skin (b) bone (c) cartilage (d) lungs (e) blood
7. Tissue that contains fibroblasts and a great deal of intercellular substance is (a) connective tissue (b) muscle tissue (c) nervous tissue (d) pseudostratified epithelium (e) adipose tissue
8. Muscle that is striated and involuntary is (a) connective (b) smooth (c) skeletal (d) pseudostratified (e) cardiac
9. Glial cells are most characteristic of (a) muscle that is striated and voluntary (b) cardiac muscle tissue (c) nervous tissue (d) stratified epithelium (e) loose connective tissue
10. Tissue that has a hard, rubbery matrix and lacks blood vessels is (a) integumentary tissue (b) bone (c) nervous tissue (d) stratified epithelium (e) cartilage
11. The contractile elements in muscle tissue are (a) myofibrils (b) elastic (c) collagen (d) lacunae (e) stress proteins
12. Which system consists of glands that secrete hormones? (a) integumentary (b) skeletal (c) nervous (d) endocrine (e) exocrine
13. Which system has the homeostatic function of helping to maintain constant calcium level in the blood? (a) integumentary (b) skeletal (c) nervous (d) reproductive (e) exocrine
14. Which system has the homeostatic function of helping to regulate volume and composition of blood and body fluids? (a) integumentary (b) muscular (c) reproductive (d) urinary (e) exocrine
15. Many homeostatic functions are maintained by (a) negative feedback systems (b) positive feedback systems (c) set points (d) stressors (e) exocrine glands
16. An ectotherm (a) may use behavioral strategies to help adjust body temperature (b) has a variety of homeostatic mechanisms to precisely regulate body temperature (c) depends on sensors in the hypothalamus to regulate temperature (d) has a higher rate of enzyme activity than a typical endotherm (e) must expend more energy on thermoregulation than an endotherm
17. The state of torpor caused in some animals by lack of food or water during periods of high temperature is known as (a) positive feedback (b) hibernation (c) endothermy (d) ectothermy (e) estivation

REVIEW QUESTIONS

1. What advantages do multicellular organisms have over unicellular organisms? Can you think of any disadvantages?
2. What are the functions of epithelial tissues? How are these tissues adapted to carry out these functions?
3. Describe the structure of bone. Of adipose tissue. Of loose connective tissue. How is each adapted to carry out its special functions?
4. Compare the properties of the three types of muscle.
5. How is the neuron uniquely adapted for its function?
6. What kinds of tissues would you expect to find in the following organs: (a) lung (b) heart c) intestine (d) salivary glands?
7. How do the cells of a malignant neoplasm differ from those of a normal tissue? What is metastasis?
8. How do each of the following organ systems help maintain homeostasis? (a) respiratory (b) urinary (c) endocrine? (You can consult Table 37–4 for help.)
9. What is acclimatization? Give an example. Compare hibernation with estivation.
10. Identify each type of tissue in the figures below and on page 816. Refer to Tables 37–1 and 37–2 to check your answers.

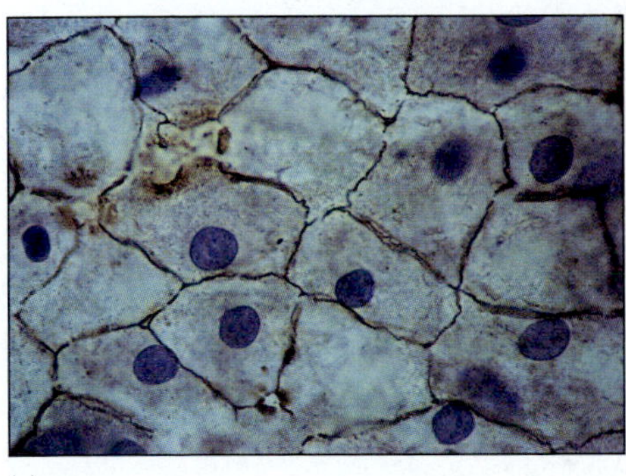
(a)

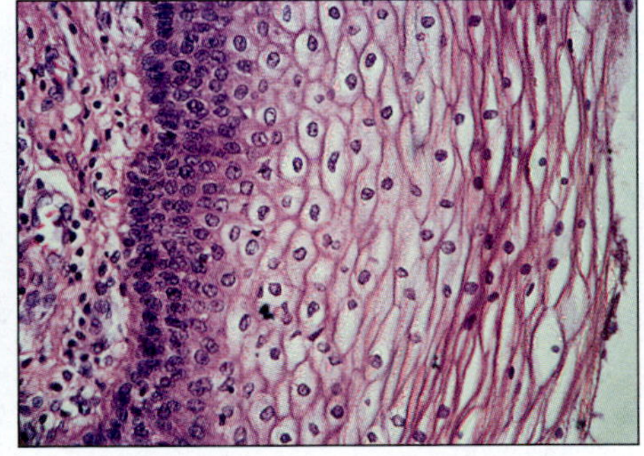
(b)

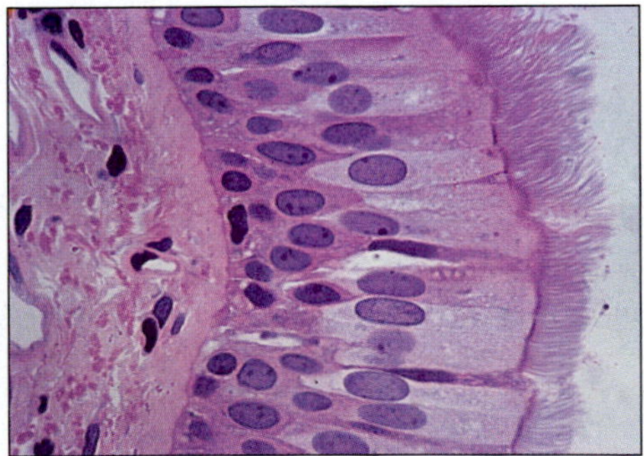
(c)

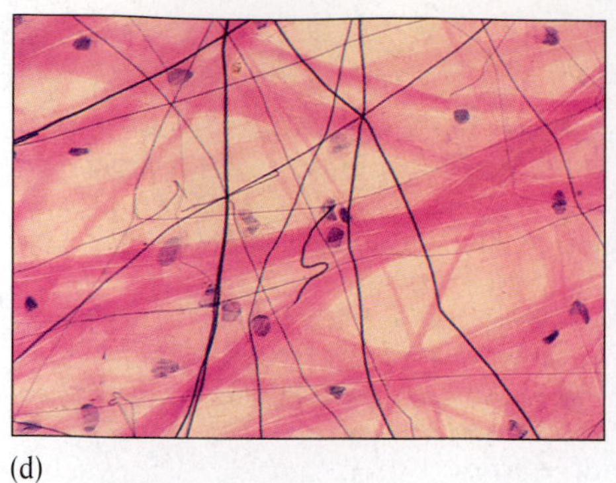

(d)

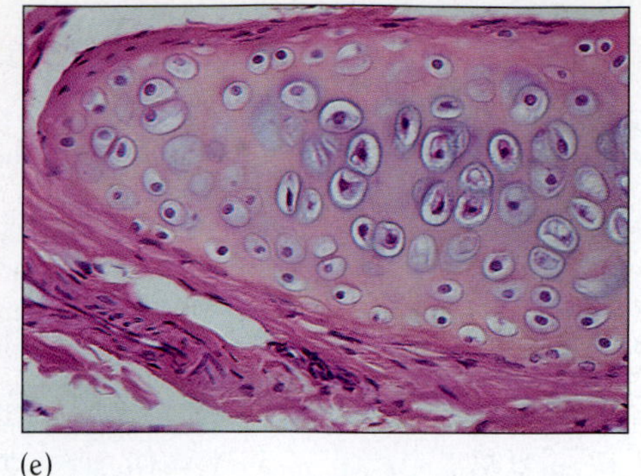

(e)

(Ed Reschke)

YOU MAKE THE CONNECTION

1. Imagine that all the epithelium in a complex animal, such as a human, suddenly disappeared. What effects might this have on the body and its ability to function?
2. What would connective tissue be like if it had no intercellular substance? What effect would the absence of intercellular substance have on the body?
3. A high concentration of carbon dioxide in the blood and interstitial fluid results in more rapid breathing. Explain this observation in terms of homeostasis.

RECOMMENDED READINGS

Engel, J. "Biochemistry: Versatile Collagens in Invertebrates." *Science,* Vol. 277, 19 Sep. 1997. An interesting discussion of the structure and versatility of collagen.

Fackelman, K. "Immune Attack on Cancer." *Science News,* Vol. 153, 13 Jun. 1998. An overview of the development of immune therapy in treating cancer.

Marshall, E. "Cancer: The Roadblocks to Angiogenesis Blockers." *Science,* Vol. 280, 15 May 1998. A brief summary of the status of a group of antitumor agents that block the development of new blood vessels.

Marx, J. "Medicine: DNA Arrays Reveal Cancer in Its Many Forms." *Science,* Vol. 289, 28 Sep. 2000. Investigators are using microarrays to determine patterns of gene expression, a strategy that is providing new insights into cancer and potential treatments.

Rhoades, R., and R. Pflanzer. *Human Physiology,* 3rd ed. Saunders College Publishing, Philadelphia, 1996. A readable presentation of how the human body works.

Ruben, J.A., et al. "The Metabolic Status of Some Late Cretaceous Dinosaurs." *Science,* Vol. 273, 30 Aug. 1996. The cross-sectional area of nasal passages was used to assess the metabolic rates of three theropod dinosaurs.

Vogelstein, B., D. Lane, and A.J. Levine. "Surfing the *p53* Network." *Nature,* Vol. 408, 16 Nov. 2000. The *p53* tumor suppressor gene integrates many cell signals that regulate life and death of the cell. Inactivation of this gene can lead to cancer.

- Visit our Web site at **http://www.info.brookscole.com/solomonbergmartin** for links to chapter-related resources on the World Wide Web. Additional on-line materials relating to this chapter can also be found on our Web site.

See chapter activity on BioActive Learner CD for additional help in mastering the chapter's material. Icon location in the chapter's margins shows which topics have tutorials or simulations in the CD.

38

Protection, Support, and Movement: Skin, Skeleton, and Muscle

Some epithelial coverings are specialized to secrete products. This striped land snail *(Helicella candicans)* glides through a slime track *(visible at lower left)* consisting of mucus secreted by the epithelial covering of its foot. *(Y. Momatiuk/Photo Researchers, Inc.)*

LEARNING OBJECTIVES

After you have studied this chapter you should be able to

1. Compare the structure and function of the external epithelium of invertebrates and vertebrates, and identify the principal derivatives of vertebrate skin.
2. Compare the advantages and disadvantages of different types of skeletal systems, including the hydrostatic skeleton, exoskeleton, and endoskeleton.
3. Identify the main divisions of the vertebrate skeleton and the bones that make up each division.
4. Describe the structure of a typical long bone and differentiate between endochondral and intramembranous bone development.
5. Describe the macroscopic and microscopic structures of skeletal muscle.
6. List, in sequence, the events that take place during muscle contraction.
7. Compare the roles of glycogen, creatine phosphate, and ATP in providing energy for muscle contraction.
8. Describe the antagonistic action of muscles.
9. Summarize the functional relationship between skeletal and muscle tissues.
10. Compare slow and fast muscle fibers.

In our survey of the animal kingdom (see Chapters 28 to 30), we made many references to protective coverings, movement, and skeletal systems in terms of adaptation of various animal groups. In Chapter 37 we described animal tissues, organs, and systems, laying the foundation for the more detailed discussions in this and later chapters. Here we focus on epithelial coverings, skeleton, and muscle—systems that are closely interrelated in function and significance. We compare these systems in several animal groups and then focus on their structures and functions in the human body.

In addition to protecting underlying tissues, the **epithelial coverings** of animals may be specialized to exchange gases, excrete wastes, regulate temperature, or secrete substances such as poison or mucus *(see photograph)*. Many animals are also protected by a **skeletal system.** Whether it is a fluid-filled compartment, a shell, or a system of bones, the skeletal system provides support for the body and typically protects internal organs. In vertebrates, bones store calcium and are important in maintaining homeostatic levels of calcium in the blood.

The skeletal system and **muscular system** work together. Muscles responsible for locomotion are anchored to the skeleton, which gives them something firm on which to act. In humans and most other vertebrates, bones serve as levers that transmit the force necessary to move various parts of the body.

Some animals walk, run, or jump; others swim; some fly. Although a few animals remain rooted to one spot, sweeping their surroundings with tentacles, locomotion (movement from place to place in the environment) is characteristic of most animals. In fact, dependable, rapid, and responsive movement is key to survival for most animals, and animal body plans have evolved along with types of locomotion.

Internal function also depends on movement. Many animals have internal circulating fluids, pumped by hearts and contained by hollow vessels that maintain their pressure with gentle squeezing. Most have digestive systems that push food along with peristaltic contractions. Both locomotion and the physiological actions necessary to maintain homeostasis are powered by muscle, a tissue specialized to contract. In complex animals, effective movement depends on the cooperation and interactions of several body systems, including muscular, skeletal, nervous, circulatory, respiratory, and endocrine.

EPITHELIAL COVERINGS PROTECT THE BODY

Epithelial tissue covers all external and internal surfaces of the animal body. The epithelial covering forms a protective shield around the body.

Invertebrate epithelium may function in secretion or gas exchange

In invertebrates, the external epithelium protects the body and may also be specialized for secretion, absorption, or gas exchange. Epithelial cells may be modified as sensory cells that are selectively sensitive to light, chemical stimuli, or mechanical stimuli such as contact or pressure.

In many species, the epithelium contains secretory cells that produce a protective cuticle or secrete lubricants or adhesives. In some species, secretory cells release odorous secretions used for communication among members of the species or for marking trails. In other species, the cells produce poisonous secretions used for offense or defense. In earthworms, a lubricating mucous secretion serves as a moist slime that promotes efficient diffusion of gases across the body wall and also reduces friction during movement through the soil.

An epithelial secretion may be limited to a particular region of the body surface. The foot of the gastropod mollusk, for example, releases a mucous secretion, producing a slime track along which the snail glides. Certain insects such as weaver ants release epithelial secretions as fine, very strong threads that are used to construct nests. The spinning glands of spiders develop from epithelial cells. Lepidopteran insects such as butterflies and moths synthesize silk protein in silk-forming glands.

Vertebrate skin functions in protection and temperature regulation

The **integumentary system** of vertebrates includes the skin and structures that develop from it. In many fishes, in the African ant-eating pangolin (a mammal; see Fig. 17–14*c*), and in some reptiles, skin has developed into a set of scales formidable enough to be considered armor. Even human skin has considerable strength.

Derivatives of skin differ considerably among vertebrates. Fish have bony or toothlike scales. Amphibians have naked skin covered with mucus, and some species are equipped with poison glands. Reptiles have epidermal scales, mammals have hair, and birds have feathers that provide even more effective insulation than fur. Skin and its derivatives are often brilliantly colored in connection with courtship rituals, territorial displays, and other kinds of communication.

In mammals, structures derived from skin include fingernails and toenails, hair, sweat glands, oil (sebaceous) glands, and several types of sensory receptors that give us the ability to feel pressure, temperature, and pain. The skin of mammals also contains mammary glands, specialized in females for secretion of milk.

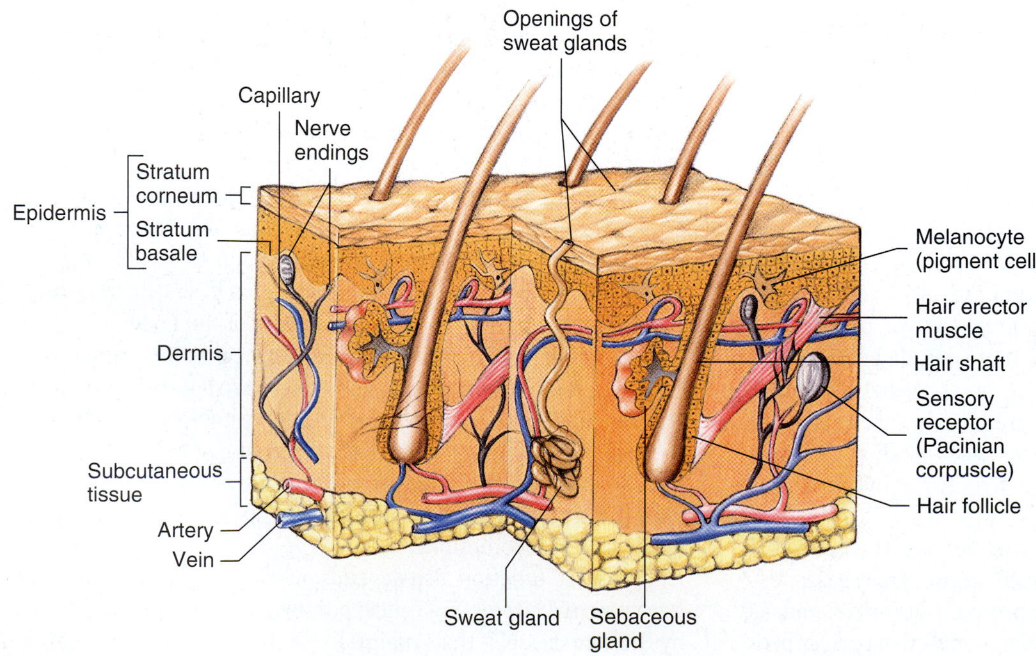

Figure 38–1 Human skin. The integumentary system of humans and other mammals is a complex organ system consisting of skin and structures such as hair and nails that are derived from the skin. Stratum basale consists of a single layer of epithelial cells.

Oil glands in human skin empty via short ducts into hair follicles. A **hair follicle** is the part of a hair below the skin surface, together with the epithelial tissue that surrounds it. Oil glands secrete **sebum,** a complex mixture of fats and waxes. In humans these glands are especially numerous on the face and scalp. The oil secreted keeps the hair moist and pliable and prevents the skin from drying and cracking. Sebum also contains substances that inhibit the growth of harmful bacteria. (At puberty, excessive sebum, produced in response to increased levels of sex hormones, may fill the glands and follicles, producing acne.)

In humans (and some other mammals) the skin helps maintain body temperature (see Chapter 37). About 2.5 million sweat glands secrete sweat, and its evaporation from the surface of the skin lowers the body temperature. Constriction and dilation of capillaries in the skin are also part of homeostatic mechanisms for regulating body temperature.

The epidermis is a waterproof protective barrier

The outer layer of skin, the **epidermis,** is the interface between the delicate tissues within the vertebrate body and the hostile outside environment. The epidermis consists of several strata, or sublayers. The deepest is **stratum basale,** and the most superficial is **stratum corneum** (Fig. 38–1). Pigment cells in stratum basale and in the dermis produce **melanin,** a pigment that contributes to the color of the skin. In stratum basale, cells divide and are pushed outward as other cells are produced below them. The epidermal cells mature as they move toward the skin surface. Because most vertebrates have no capillaries in the epidermis, the maturing cells receive less and less nourishment, which diminishes their metabolic activity.

As they move toward the body surface, epidermal cells manufacture **keratin,** an elaborately coiled protein that gives the skin considerable mechanical strength and flexibility. Keratin is quite insoluble and serves as a diffusion barrier for the body surface. As epidermal cells move through stratum corneum, they die. When they reach the outer surface of the skin, they wear off and must be continuously replaced.

The dermis contains blood vessels and other structures

Beneath the epidermis lies the **dermis** (see Fig. 38–1), which is composed of a dense, fibrous connective tissue made up mainly of collagen fibers. Collagen imparts strength and flexibility to the skin. Sweat glands and hair follicles are embedded in the dermis (Fig. 38–2). The dermis also contains blood vessels that nourish the skin and sensory receptors for touch, pain, and temperature. Mammalian skin rests on a layer of **subcutaneous tissue** composed mainly of adipose tissue that insulates the body from outside temperature extremes.

How does ultraviolet (UV) radiation, the short, invisible rays from the sun, affect the skin? UV radiation is a high-energy form of radiation; its wavelengths are just shorter than visible light. Although exposure to UV is necessary for the body to make vitamin D, too much can cause sunburn and skin damage, and can eventually result in wrinkling of the skin, cataracts, and skin cancer. Based on their wavelength, three bands of UV radiation have been identified: UVA, UVB, and UVC. UVA contributes to wrinkling, sunburn, and skin cancer. UVB has a shorter wavelength than UVA, making it more energetic and more dangerous. UVB is the most common cause of sunburn. It contributes to cataracts and skin cancer and damages the immune system. With the shortest wavelength, UVC is the most dangerous type of UV radiation.

The amount of UV radiation that reaches Earth's surface depends on the time of year, elevation, distance from the equator, and cloud cover. UV radiation is most intense near the equator, at high elevations, and around noon on a clear summer day. The amount of UV radiation that reaches us is also affected by the ozone layer, a layer of gas in the stratosphere that absorbs incoming UVB and UVC radiation. Scientists warn that the ozone

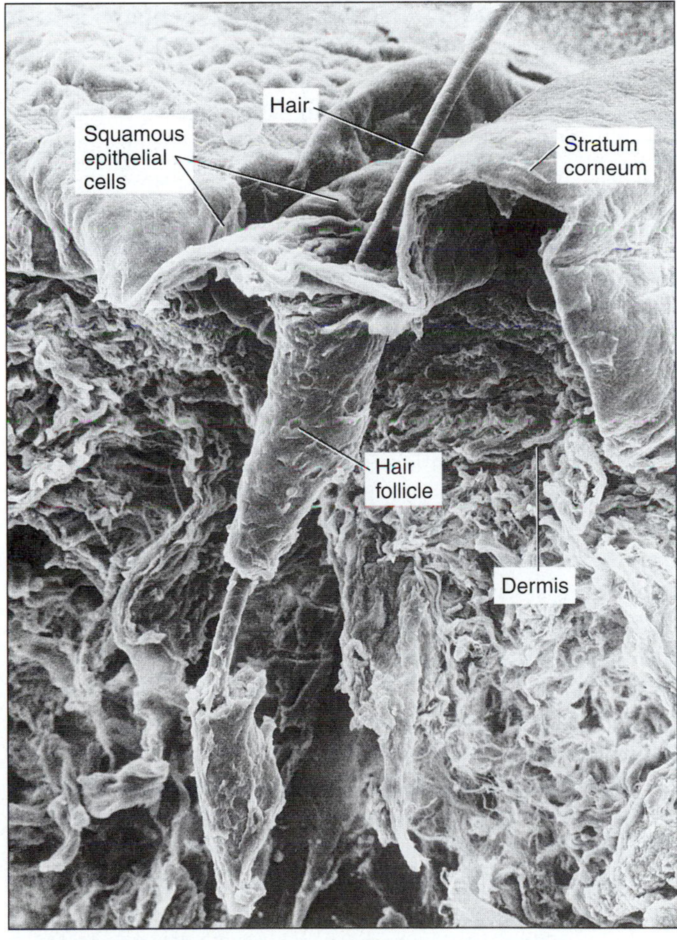

Figure 38–2 SEM of human skin showing a hair follicle. Part of the hair follicle was torn during preparation of the tissue. *(Courtesy of Dr. Karen A. Holbrook)*

layer is being destroyed by pollutants such as chlorofluorocarbons, called CFCs. These compounds are used as propellants for aerosol cans, as coolants in air conditioners and refrigerants (e.g., Freon), and in the production of foam for insulation and packaging (e.g., Styrofoam).

Stratospheric ozone levels have been decreasing worldwide for several decades. In 1985 scientists discovered a thinning in the ozone layer over a large part of Antarctica, allowing more UV radiation to reach Earth's surface. Thinning has also been found over a smaller area of the Arctic. Despite international agreements to reduce CFC production, depletion of the ozone layer continues. CFCs are stable compounds that will continue to damage the ozone layer for many years after they are no longer used. (Stratospheric ozone depletion is discussed in greater detail in Chapter 55.)

Exposure to UV rays causes the epidermis to thicken and stimulates pigment cells in the skin to produce melanin at an increased rate. An increase in melanin causes the skin to become darker. Melanin is an important protective screen against the sun because it absorbs some of the harmful UV rays. The suntan so prized by sun worshipers is actually a sign that the skin has been exposed to too much UV radiation. When the melanin is not able to absorb all the UV rays, the skin becomes inflamed, or sunburned. Because dark-skinned people have more melanin, they suffer less sunburn, wrinkling, and skin cancer, although they are still at risk. UV radiation damages DNA, causing mutations that lead to malignant transformation (see Chapter 16, *Focus On: Oncogenes and Cancer*).

Skin cancer is on the rise. Most cases are caused by excessive, chronic exposure to UV radiation. The incidence of malignant melanoma, the most lethal type of skin cancer, is increasing faster than any other type of cancer. Some forms of malignant melanoma spread rapidly through the body and may cause death within a few months after diagnosis. Most skin cancer can be prevented by protecting the body from sunlight and other forms of UV radiation. Tanning machines expose the body to UV rays and so also pose risks. Those who choose to expose themselves to sunlight can protect their skin by limiting the time they spend in the sun and by applying effective sun screens. Dermatologists suggest that everyone wear sun screen on exposed skin every day. Consumer organizations have persuaded the National Weather Service to include a UV Index as part of the daily weather forecast. The index indicates the time it would take for a fair-skinned North American to burn when exposed to the sun at noon.

New Zealand, southern Australia, parts of the United States, and certain other geographical areas are reporting significant increases in harmful radiation. Evidence suggests that continuing depletion of the ozone shield around our planet will cause 1 million extra cases of skin cancer each year and that about 30,000 of these victims will die. The highest incidences of skin cancer have been reported in Australia, New Zealand, Hawaii, and certain areas in the United States. About 75% of Australians of European ancestry will have at least one skin cancer removed in their lifetime.

SKELETONS FUNCTION IN LOCOMOTION, PROTECTION, AND SUPPORT

Some animals do not have hard skeletons. In these animals, muscle may act on a thickened epidermis or on a fluid-filled body cavity. In more complex animals, muscles act on hard structures such as chitin or bone that transmit and transform muscle contraction into the variety of motions animals use. In addition to its function in locomotion, the skeleton supports the body and protects the internal organs.

In echinoderms and chordates the skeleton is internal, an **endoskeleton** composed of plates or shafts of calcium-impregnated tissue (such as cartilage or bone). But in most animals, the skeleton is not a living tissue but a lifeless shell or **exoskeleton** deposited atop the epidermis.

In hydrostatic skeletons, body fluids are used to transmit force

Imagine an elongated balloon full of water. If you were to pull on it, it would lengthen and become thinner. It would do the same if you squeezed it. Conversely, if you pushed the ends toward the center, it would shorten and thicken. Many animals, including cnidarians and annelids, have a **hydrostatic skeleton** that works something like a balloon filled with water. Fluid in a closed compartment of the body is held under pressure. When muscles in the wall of the compartment contract, they push against the tube of fluid. Because fluids cannot be compressed, the force is transmitted through the fluid, causing change in shape and movement of the body.

In *Hydra* and other cnidarians, cells of the two body layers are capable of contraction. The contractile cells in the outer epidermal layer are arranged longitudinally, whereas the contractile cells of the inner layer (the gastrodermis) are arranged circularly around the central body axis (Fig. 38–3). The two groups of cells work in **antagonistic** fashion: What one can do, the other can undo. When the epidermal (longitudinal) layer contracts, the hydra shortens. Because of the fluid in the gastrovascular cavity, force is transmitted so that the hydra thickens as well. On the other hand, when the inner (circular) layer contracts, the hydra thins, and its fluid contents force it to lengthen.

Mechanically, the hydra is a bag of fluid. The fluid acts as a hydrostatic skeleton because it transmits force when the contractile cells contract against it. (Although technically not a closed compartment, the gastrovascular cavity can function as a hydrostatic skeleton because its opening is small.) Hydrostatic skeletons permit only crude mass movements of the body or its appendages. Delicate movements are difficult because force tends to be transmitted equally in all directions throughout the entire fluid-filled body of the animal. For example, it is not easy for the hydra to thicken one part of its body while thinning another.

The more sophisticated hydrostatic skeleton of the annelid worm enables it to be more versatile in movement. An earthworm's body consists of a series of segments divided by trans-

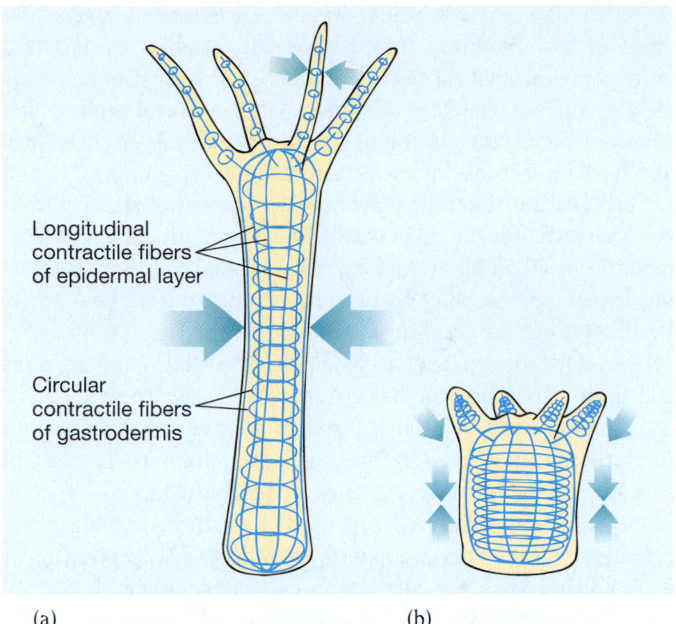

Figure 38–3 Hydrostatic skeleton of *Hydra*. The longitudinally arranged contractile cells are antagonistic to the cells arranged in circles around the body axis. **(a)** Contraction of the circular contractile fibers elongates the body. **(b)** Contraction of the longitudinal fibers shortens the body. **(c)** *Hydra* in motion. The body of this *Hydra* sp. appears to be shortening as it moves in on its prey, a water flea (*Daphnia* sp.). *(c, Larry Stepanowicz/Visuals Unlimited)*

verse partitions, or septa (see Fig. 29–8). The septa isolate portions of the body cavity and its contained fluid, permitting the hydrostatic skeletons of each segment to be largely independent of one another. Thus, contraction of the circular muscle in the elongating anterior end need not interfere with the action of the longitudinal muscle in the segments of the posterior end.

Some examples of hydrostatic skeletons occur even in complex invertebrates equipped with shells or endoskeletons, and in vertebrates with endoskeletons of cartilage or bone. Among mollusks, for example, the clam extends and anchors its foot by a hydrostatic blood pressure mechanism similar to that used by earthworms. Sea stars and sea urchins move their tube feet by an ingenious version of the hydrostatic skeleton (see Chapter 30). And even the human penis becomes erect and stiff because of the turgidity of pressurized blood in its internal spaces.

Mollusks and arthropods have nonliving exoskeletons

In both mollusks and arthropods, the exoskeleton is a nonliving product of the epidermal cells. In mollusks, the exoskeleton provides protection, a retreat used in emergencies, with the bulk of the naked, tasty body exposed at other times.

Exoskeletons of arthropods serve not only to protect but also to transmit forces. In this respect they are comparable to the skeletons of vertebrates. Although the arthropod exoskeleton is a continuous, one-piece sheath covering the entire body, it varies greatly in thickness and flexibility. Large, thick, inflexible plates are separated from one another by thin, flexible joints arranged segmentally. Enough joints are provided to make the arthropod's body as flexible as those of many vertebrates. The exoskeleton is also extensively modified to form specialized tools or weapons or is otherwise adapted to a vast variety of lifestyles.

A disadvantage of the rigid arthropod exoskeleton is that it interferes with growth. To accommodate growth, an arthropod must **molt,** that is, shed its exoskeleton and replace it with a new, larger one (Fig. 38–4). During molting the animal is weak and vulnerable to predators.

Internal skeletons are capable of growth

Endoskeletons, or internal skeletons, are well developed in echinoderms and chordates. Composed of living tissue, the endoskeleton grows along with the animal as a whole. The echinoderm endoskeleton consists of spicules and plates of calcium salts embedded in the tissue of the body walls, that is, beneath an epidermis that covers the body. This endoskeleton forms what amounts to an internal shell that provides support and protection (Fig. 38–5). Many echinoderm endoskeletons bear spines that project to the outer surface.

The internal skeletons of vertebrates provide support and protection, and transmit forces. Members of class Chondrichthyes (sharks and rays) have skeletons composed of cartilage, but in most vertebrates the skeleton consists mainly of bone. Although the skeletons of adult vertebrates vary, their bones have been shown to be homologous. For example, the

Figure 38–4 Molting. A greengrocer cicada (*Magicicada* sp.) requires 13 years to mature. It then emerges from the soil, climbs a tree, and molts prior to reproducing. *(Judy Davidson/Science Photo Library/Photo Researchers, Inc.)*

bones of the human arm, cat forelimb, whale flipper, and bat wing have a similar structure because they are derived from a common ancestor (see Fig. 17–11).

The vertebrate skeleton has two main divisions: the axial and appendicular skeletons. The **axial skeleton,** located along the central axis of the body, consists of the skull, vertebral column, ribs, and sternum (breastbone). The **appendicular skeleton** consists of the bones of the limbs (arms and legs) plus the bones making up the shoulder (pectoral) girdle and most of the hip (pelvic) girdle; these girdles connect the limbs to the axial skeleton (Fig. 38–6).

The **skull,** the bony framework of the head, consists of the cranial and facial bones. In the human, 8 cranial bones enclose the brain, and 14 bones make up the facial portion of the skull. Several cranial bones that are single in the adult human result from the fusion of 2 or more bones that are separate in the embryo or even in the newborn.

The vertebrate spine, or **vertebral column,** supports the body and bears its weight. In humans it consists of 24 **vertebrae** and 2 bones composed of fused vertebrae, the **sacrum** and **coccyx.** The vertebral column consists of the **cervical** (neck) region, with 7 vertebrae; the **thoracic** (chest) region, with 12 vertebrae; the **lumbar** (back) region, with 5 vertebrae; the **sacral** (pelvic) region, with 5 fused vertebrae; and the **coccygeal** region, also composed of fused vertebrae.

Although vertebrae differ in size and shape in various regions of the vertebral column, a typical vertebra consists of a bony **centrum** (central portion), which bears most of the body weight, and a dorsal ring of bone called the **neural arch,** which surrounds and protects the delicate spinal cord. Vertebrae may also have projections for the attachment of ribs and muscles and for articulating (joining) with neighboring vertebrae. The first cervical vertebra, the **atlas** (named for the mythical Greek who held the world on his shoulders), has rounded depressions on its upper surface into which fit two projections from the base of the skull, allowing the head to nod up and down. The second cervical vertebra, the **axis,** serves as a pivot for rotation of the atlas and skull, permitting the head to move from side to side.

The **rib cage** is a bony "basket" formed by the sternum (breastbone), thoracic vertebrae, and, in mammals, 12 pairs of ribs. (Males and females have the same number of ribs.) The rib cage protects the internal organs of the chest, including the heart and lungs. It also supports the chest wall, preventing it from collapsing as the diaphragm contracts with each breath. Each pair of ribs is attached dorsally to a separate vertebra. Of the 12 pairs of ribs in the human, the first 7 are attached ventrally to the sternum; the next 3 are attached indirectly by cartilages; and the last 2, the "floating ribs," have no attachments to the sternum.

The **pectoral girdle** consists of two collarbones, or **clavicles,** and two shoulder blades, or **scapulas.** The **pelvic girdle** consists of a pair of large bones, each composed of three fused hipbones. Whereas the pelvic girdle is securely fused to the vertebral column, the pectoral girdle is loosely and flexibly attached to it by muscles.

Each human limb consists of 30 bones and terminates in five **digits,** the fingers and toes. The more specialized appendages of other tetrapods may be characterized by four digits (as in the

Figure 38–5 The echinoderm endoskeleton. The endoskeleton provides support and protection. As in other echinoderms, the endoskeleton of the red sea star *(Milleporella lulamben)* is composed of spicules and plates of nonliving calcium salts embedded in tissues of the body wall. Photographed in Indonesia. *(Patricia Jordan/Peter Arnold, Inc.)*

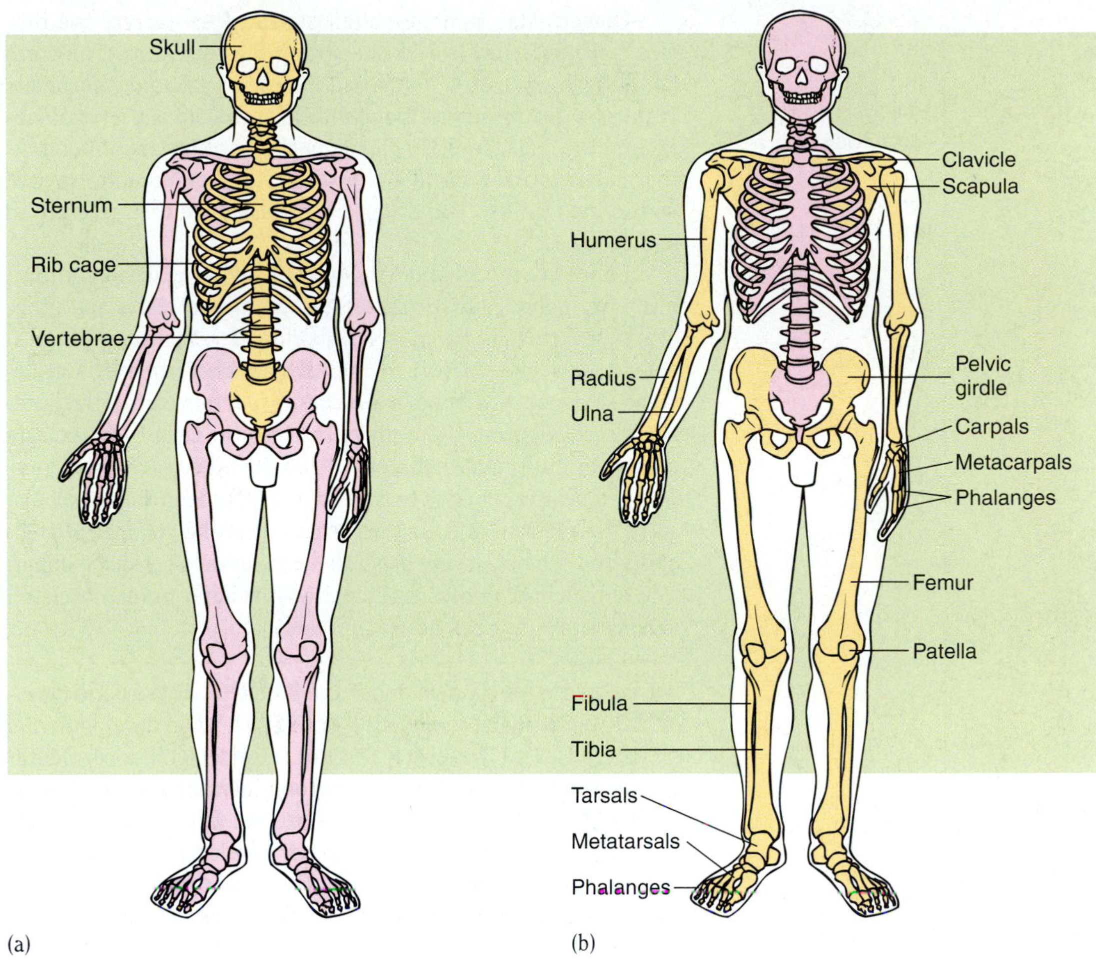

Figure 38–6 The human skeletal system. **(a)** Bones of the axial skeleton *(tan)*, anterior view. **(b)** Bones of the appendicular skeleton *(tan)*, anterior view.

pig), three (as in the rhinoceros), two (as in the camel), or one (as in the horse).

Great apes and humans have a highly specialized feature: the opposable thumb. (Great apes also have an opposable big toe. Although the human big toe is not opposable, it is similar enough in structure to the thumb that it can be surgically substituted.) The opposable thumb can be readily wrapped around objects, such as a tree limb, in climbing, and it is especially useful in grasping and manipulating objects.

In humans, the upper limbs are not generally used for locomotion as they are in other mammals, including the great apes. The combination of opposable thumbs and upright posture enables us to use our hands to write, shape, build, and use weapons. These abilities have permitted the human species to change its environment to a greater extent than any other organism in the history of Earth.

A typical long bone consists of compact and spongy bone

The radius, one of the two bones of the forearm, is a typical long bone (Fig. 38–7). Its numerous muscle attachments are arranged in such a way that the bone rotates about its long axis and also operates as a lever, amplifying the motion generated by the muscles. By themselves, muscles cannot shorten enough to produce large movements of the body parts to which they are attached.

Like other bones, the radius is covered by a connective tissue membrane, the **periosteum,** to which muscle tendons and ligaments attach. The periosteum is capable of producing fresh layers of bone and thus increasing the bone's diameter. The main shaft of a long bone is its **diaphysis;** each expanded end is called an **epiphysis.** In children, a disk of cartilage, the **metaphysis,** is found between the epiphyses and the diaphysis. Metaphyses are growth centers that disappear at maturity, becoming vague **epiphyseal lines.** Long bones contain a central cavity that contains **bone marrow.** Yellow marrow consists mainly of a fatty connective tissue; the red marrow found in certain bones produces blood cells.

The radius has a thin outer shell of **compact bone,** which is very dense and hard. Compact bone is found primarily near the surfaces of a bone, where it provides great strength. Compact bone consists of interlocking spindle-shaped units called **osteons,** or *Haversian systems* (see Fig. 37–3). Within an osteon, **osteocytes** (bone cells) are found in small cavities called **lacunae**

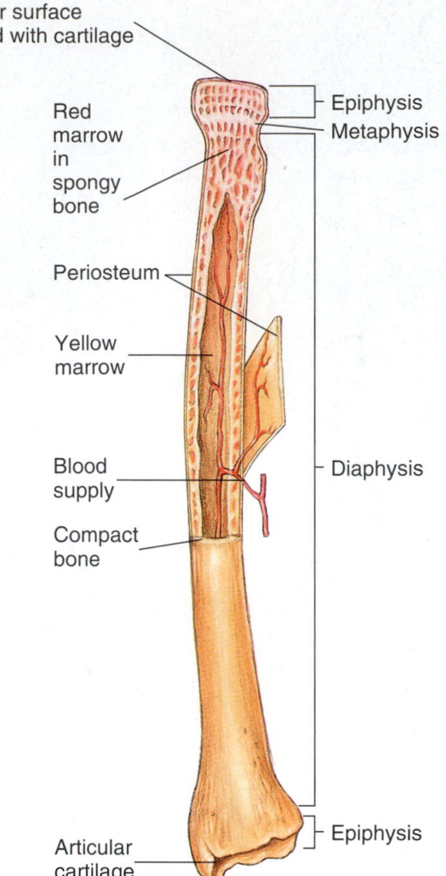

Figure 38–7 A typical long bone. A long bone has a thin shell of compact bone and a filling of spongy bone that contains marrow.

(sing., *lacuna*). The lacunae are arranged in concentric circles around central **Haversian canals.** Blood vessels that nourish the bone tissue pass through the Haversian canals. Osteocytes are connected by threadlike extensions of their cytoplasm that extend through narrow channels called *canaliculi.*

Interior to the thin shell of compact bone is a filling of **spongy bone** (also called *cancellous bone*). Spongy bone, which consists of a network of thin strands of bone, provides mechanical strength. The spaces within spongy bone are filled with bone marrow.

Bones are remodeled throughout life

During fetal development, bones form in two ways. Long bones, such as the radius, develop from cartilage templates in a process called **endochondral** bone development. A bone begins to ossify in its diaphysis, and secondary sites of bone production develop in the epiphyses. The part of the bone between the ossified regions can grow. Eventually the ossified regions fuse. In contrast, other bones, including the flat outer bones of the skull, develop from a noncartilage, connective tissue scaffold. This process is known as **intramembranous bone development.**

Osteoblasts are bone-building cells. They secrete the protein collagen, which forms the strong fibers of bone. The compound hydroxyapatite, composed mainly of calcium phosphate, is present in the interstitial fluid. It automatically crystallizes around the collagen fibers, forming the hard matrix of bone. As the matrix forms around the osteoblasts, they become isolated within the lacunae. The trapped osteoblasts are then referred to as osteocytes.

Bones are modeled during growth and remodeled continuously throughout life in response to physical stresses and other changing demands. As muscles develop in response to physical activity, the bones to which they are attached thicken and become stronger. As a bone grows, tissue is removed from its interior, especially from the walls of the marrow cavity. **Osteoclasts** are large, multinucleated cells that break down *(resorb)* bone. The osteoclasts move about, secreting hydrogen ions that dissolve the crystals, and enzymes that digest the collagen. Osteoclasts and osteoblasts are synergistic; together they shape bones. The remodeling process is extensive—the adult human skeleton is completely replaced every ten years!

Process of Science Little was known about the biochemistry of bone development and remodeling until 1994, when developmental geneticist Gerard Karsentry, at the M.D. Anderson Cancer Center of the University of Texas, Houston, identified genes in the mouse that code for osteocalcin, a protein that has been shown to inhibit osteoblast function. Another developmental geneticist, Patricia Ducy, working with Karsentry, then identified *Cbfa1,* the first transcription factor specific to osteoblasts. Ducy found that *Cbfa1* regulates the expression of the osteocalcin gene in differentiated osteoblasts. In 2000, Ducy reported that mice deficient in the hormone **leptin** or its receptor had two to three times higher bone mass than normal mice. Leptin, which inhibits osteoblasts, appears to be an important regulator of bone formation. Other hormones that help regulate bone formation will likely be identified.

In **osteoporosis,** the most common progressive degenerative bone disease, bone resorption takes place more rapidly than bone formation. Patients experience a decrease in bone mass that makes the bones fragile and greatly increases their risk for fracture. Drugs have been developed that inhibit osteoclast-mediated bone resorption. Researchers are testing drugs that promote osteoblast activity with the goal of promoting bone formation.

Joints are junctions between bones

Joints, or articulations, are junction sites of two or more bones of the skeleton. Joints facilitate flexibility and movement. At the joint, the outer surface of each bone consists of articular cartilages. One way to classify joints is according to the degree of movement they allow. The *sutures* found between bones of the human skull are **immovable joints.** In a suture, bones are held together by a thin layer of dense fibrous connective tissue, which may be replaced by bone in the adult. **Slightly movable joints,** found between vertebrae, are made of cartilage and help absorb shock.

Most joints are **freely movable joints.** Each is enclosed by a joint capsule composed of connective tissue and lined with a membrane that secretes a lubricant called **synovial fluid.** This viscous fluid reduces friction during movement and also absorbs shock. The joint capsule is typically reinforced by **ligaments,** bands of fibrous connective tissue that connect bones and limit movement at the joint.

Joints wear down with time and use. In osteoarthritis, a common joint disorder, cartilage repair does not keep up with degeneration, and the articular cartilage wears out. In rheumatoid arthritis, an autoimmune disease, the synovial membrane thickens and becomes inflamed. Synovial fluid accumulates, causing pressure, pain, stiffness, and progressive deformity, leading to loss of function.

MUSCLE TISSUE GENERATES MOVEMENT IN MOST ANIMALS

All eukaryotic cells contain the contractile protein **actin,** the major component of microfilaments (see Fig. 4–26*b*). Actin is important in many cell processes, including amoeboid movement and attachment of cells to surfaces. In most cells, actin is functionally associated with the contractile protein **myosin.** Actin and myosin are most highly organized in **muscle** cells.

Earlier in the chapter, we discussed the contractile cells of the hydrostatic skeleton of *Hydra* and other cnidarians. In flatworms and most other animal groups, muscle occurs as a specialized tissue organized into definite layers, or straplike bands. Skeletal and smooth muscle are found among some invertebrate phyla. The three types of vertebrate muscle—skeletal, smooth, and cardiac—are compared in Table 37–3.

Bivalve mollusks, such as clams, have two sets of muscles for opening and closing the shell. Smooth muscle, which is capable of slow, sustained contraction, can be used to keep the two shells tightly closed for long periods, even days or weeks at a time. Striated muscle, which contracts rapidly, is used for swimming and to shut the shell quickly when the mollusk is threatened.

Arthropod muscles are striated, even in the walls of the digestive tract. Insect flight muscles contract more rapidly than any other known muscle, up to 1000 contractions per second. Throughout the animal kingdom, muscle tissue generates the mechanical forces and motion necessary for locomotion, manipulation of objects, circulation of blood, movement of food through the digestive tract, and many other life processes.

A vertebrate muscle may consist of thousands of muscle fibers

In vertebrates, skeletal muscle is the most abundant tissue in the body. Its elongated cells, referred to as **muscle fibers,** are organized in bundles that are wrapped by connective tissue. The biceps in your arm, for example, consists of thousands of individual muscle fibers and their connective tissue coverings.

Each striated muscle fiber is a long, cylindrical cell with many nuclei (Fig. 38–8). The plasma membrane, known as the **sarcolemma** in a muscle fiber, has multiple inward extensions that form a set of **T tubules** (transverse tubules). The cytoplasm of a muscle fiber is referred to as **sarcoplasm,** and the endoplasmic reticulum as the **sarcoplasmic reticulum.**

Threadlike structures called **myofibrils** run lengthwise through the muscle fiber. They are composed of even smaller structures, the **myofilaments** or simply **filaments.** There are two types of myofilaments: myosin and actin filaments. **Myosin filaments** are thick myofilaments consisting mainly of the protein myosin. The thin **actin filaments** consist mostly of the protein actin; they also contain the proteins **tropomyosin** and the **troponin complex** that regulate the actin filament's interaction with myosin filaments.

Myosin and actin filaments are organized into repeating units called **sarcomeres,** the basic units of muscle contraction. Hundreds of sarcomeres connected end-to-end make up a myofibril. Sarcomeres are joined at their ends by an interweaving of filaments called the **Z line.** Each sarcomere consists of overlapping myosin and actin filaments. The filaments overlap lengthwise in the muscle fibers, producing the pattern of transverse bands or striations, characteristics of striated muscle (Figs. 38–8 and 38–9). The bands are designated by the letters A, H, and I.

The **I bands,** which consist of actin filaments, are located at both ends of the sarcomere, immediately adjacent to the Z line. The **A band** is the wide, dark region made up of overlapping myosin and actin filaments. Within the A band, there is a narrow, light area, the **H zone,** made up exclusively of myosin filaments; the actin filaments do not extend into this region.

Contraction occurs when actin and myosin filaments slide past each other

Muscle contraction occurs when the sarcomeres, and thus the muscle fibers, shorten. The I band and H zone decrease in length, but neither the actin nor myosin filaments themselves change in length. This theory of muscle contraction, known as the **sliding filament model,** was developed in the 1950s by two British biologists, Hugh Huxley and Andrew Huxley. We now know that the length of the muscle shortens as the actin and myosin filaments slide past each other, increasing their overlap. You might think of an extension ladder. The overall ladder length changes as the ends get closer or farther apart, but the length of each ladder section stays the same.

In a **motor unit,** a motor neuron (a neuron that stimulates muscle) is functionally connected with an average of about 150 muscle fibers (Fig. 38–10). Each junction of a motor neuron with a muscle fiber is called a **neuromuscular junction.** When a motor neuron transmits a message, it releases the neurotransmitter **acetylcholine** into the synaptic cleft (a small space) between the motor neuron and each muscle fiber. Acetylcholine binds with receptors on each muscle fiber, causing **depolarization,** a change in the distribution of electrical charge across its sarcolemma. Depolarization may cause an electrical impulse, or **action potential,** to be generated in the muscle fiber.

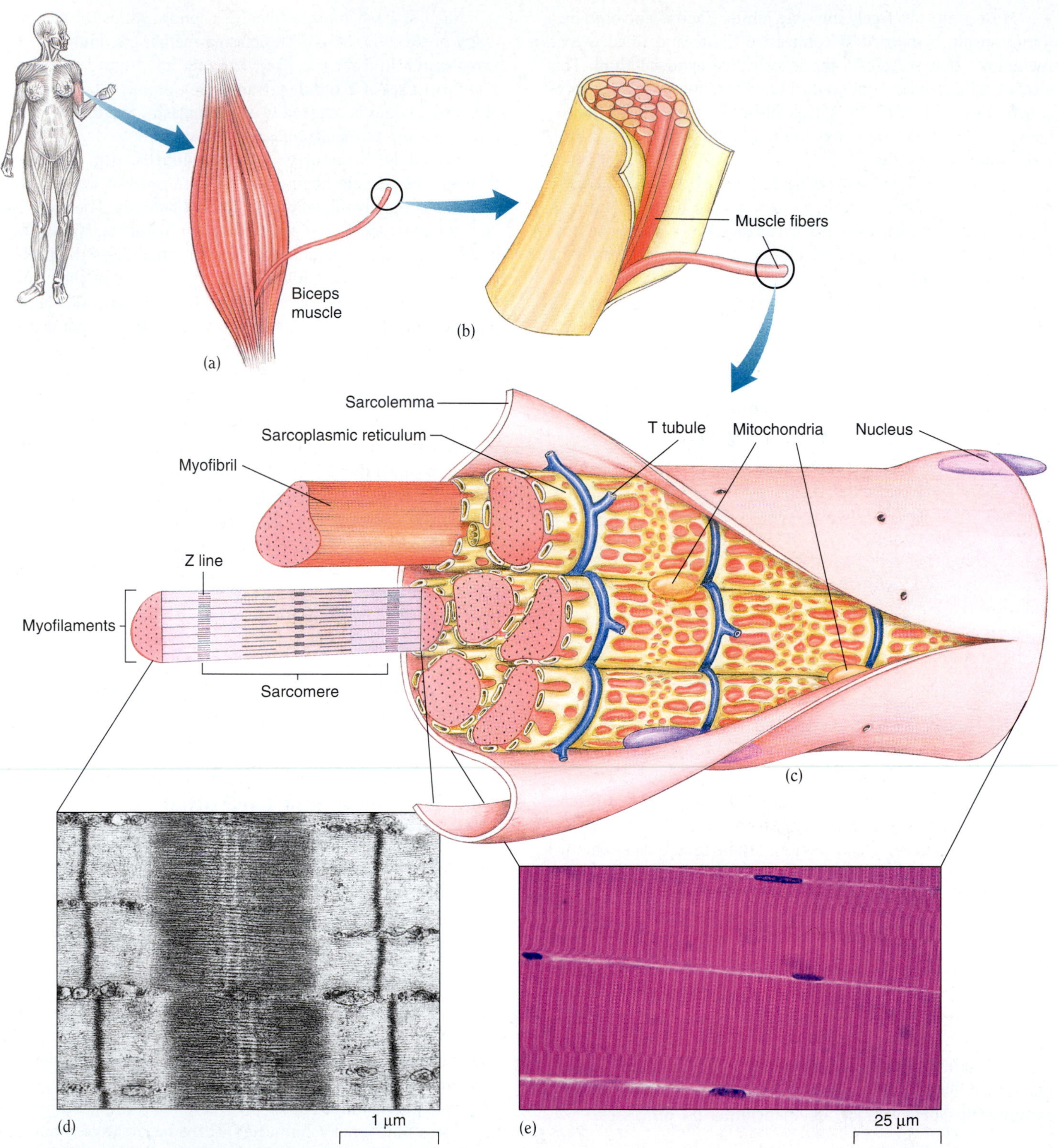

Figure 38–8 Muscle structure. **(a)** A muscle such as the biceps in the arm consists of many fascicles (bundles) of muscle fibers. **(b)** A fascicle wrapped in a connective tissue covering. **(c)** Part of a muscle fiber showing the structure of myofibrils. The Z lines mark the ends of the sarcomeres. **(d)** TEM of striated muscle. **(e)** LM showing striations. *(d, D.W. Fawcett; e, Ed Reschke)*

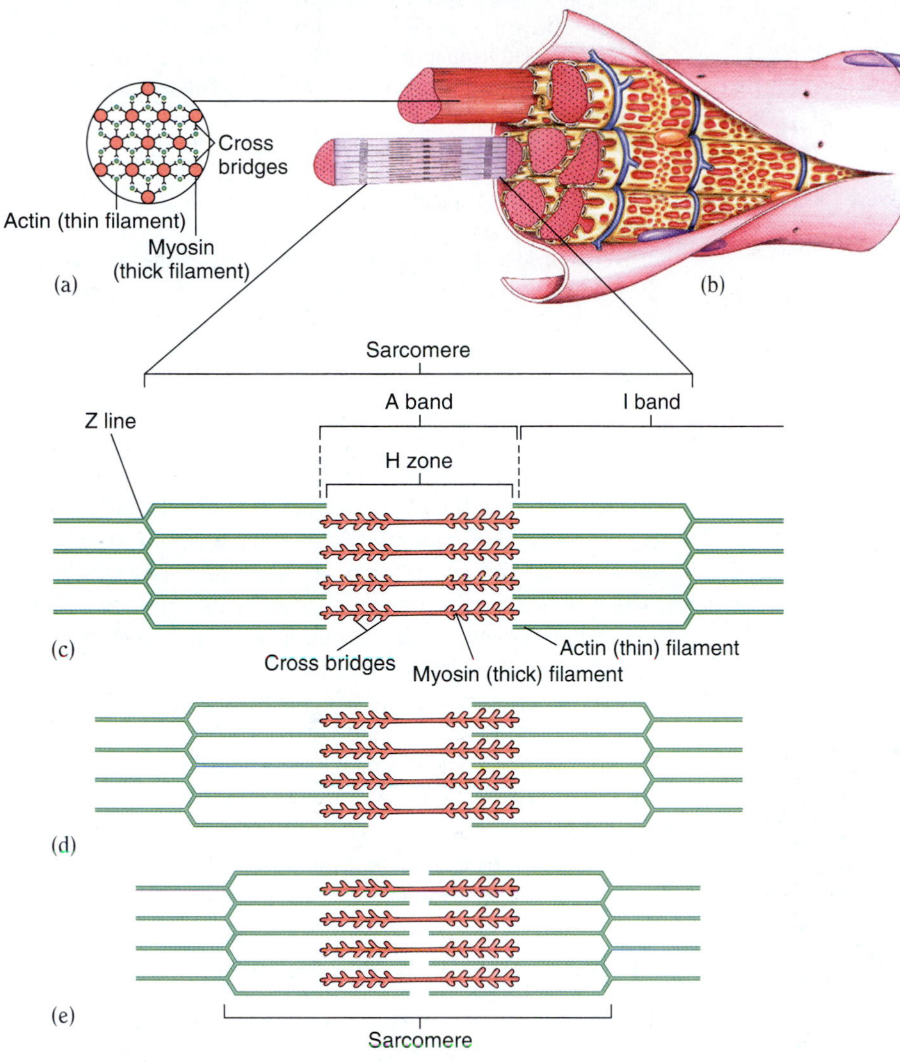

Figure 38–9 Muscle contraction. Myofibrils are threadlike structures that contain actin and myosin filaments. **(a)** Cross section of a myofibril shows the arrangement of actin and myosin filaments. **(b)** Part of a muscle fiber showing the location of the filaments. **(c)** The regular pattern of overlapping filaments gives skeletal and cardiac muscle their striated appearance. **(d)** During contraction actin filaments slide toward each other, increasing the amount of overlap between actin and myosin filaments. **(e)** As the sarcomeres become shorter, the muscle fiber becomes shorter. Note that the I band and H zone decrease as the filaments slide past each other. The filaments themselves do not become shorter. (The I band consists of actin filaments [only] of two adjacent sarcomeres.)

In a muscle fiber, an action potential is a wave of depolarization that travels along the sarcolemma and into the system of T-tubule membranes. Depolarization of the T tubules opens calcium channels in the sarcoplasmic reticulum, causing the release of stored calcium ions into the myofibrils. Calcium ions bind to a protein in the troponin complex on the actin filaments, which causes the troponin to change shape. This change results in the troponin pushing the tropomyosin away from the active sites on the actin filament (Fig. 38–11). These active sites, also referred to as myosin-binding sites, are now exposed.

One end of each myosin molecule is folded into two globular structures called heads. The rounded heads of the myosin molecules extend away from the body of the myosin filament. Each myosin molecule also has a long tail that joins other myosin tails to form the body of the thick filament. ATP is bound to the myosin when the muscle fiber is at rest (not contracting). Myosin is an adenosine triphosphatase (ATPase), an enzyme that splits ATP, forming ADP and inorganic phosphate (P_i). The ADP and P_i initially remain attached to the myosin head. Myosin heads (with ADP and P_i still bound to them) bind to exposed active sites on the actin filament, forming **cross bridges** that link the myosin and actin filaments.

According to the currently accepted model of muscle contraction, the ADP and P_i are then released, causing the myosin head to undergo a conformational change. The myosin head bends about 45 degrees in a flexing motion (the *power stroke*) that pulls the actin filament closer to the center of the sarcomere. A new ATP binds to the myosin head, allowing the myosin to detach from the actin, thereby breaking the cross bridge and beginning a new cycle.

Energized once again, the myosin heads contact a second set of active sites on the actin filament, closer to the end of the sarcomere. The process is repeated with a third set, and so on (see Fig. 38–11). This series of stepping motions pulls the actin filaments toward the center of the sarcomere. Thus, each time the myosin heads attach, move 45 degrees, detach, and then reattach farther along the actin filament, the muscle shortens. One way to visualize this process is to imagine the myosin heads engaging "hand-over-hand" with the actin filaments. When many sarcomeres contract simultaneously, they produce the

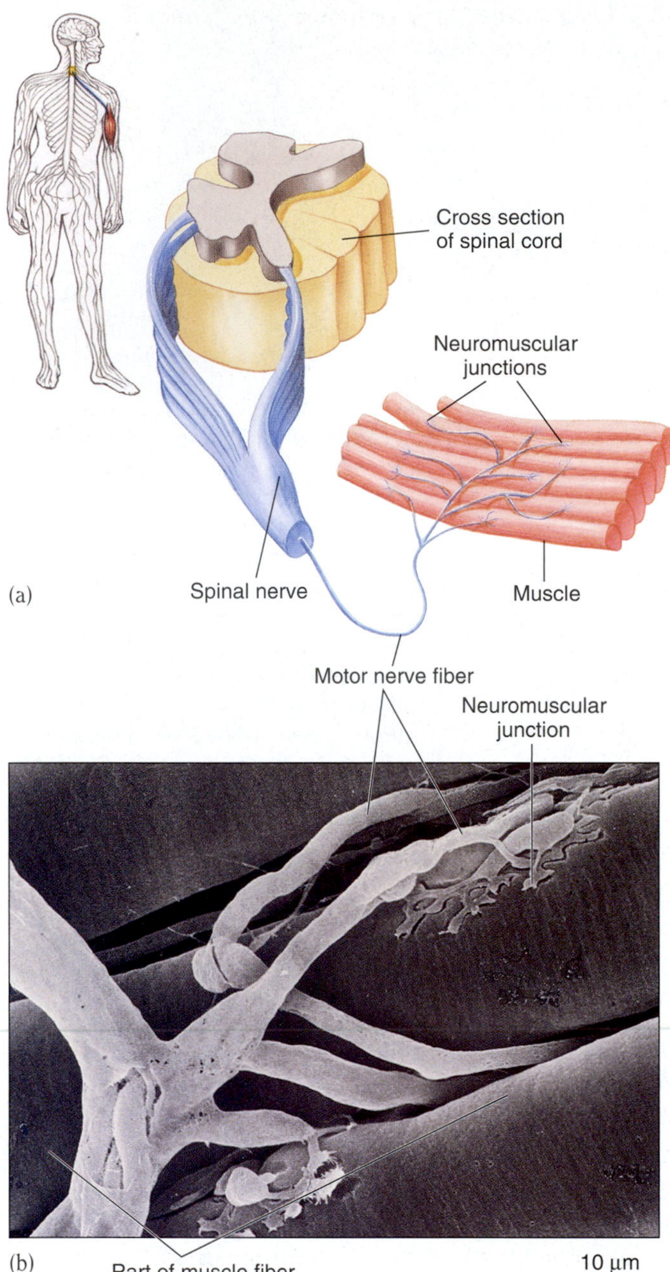

Figure 38–10 A motor unit. A motor unit consists of a motor nerve fiber and the muscle fibers that it innervates. A motor unit typically includes about 150 muscle fibers, but some units have less than a dozen fibers, while others have several hundred. **(a)** The motor unit illustrated here shows only a single motor nerve fiber. **(b)** SEM of some of the fibers in a motor unit. Note how the neuron branches to innervate all cells in the motor unit. *(b, Don Fawcett/Science Source/Photo Researchers, Inc.)*

contraction of the muscle as a whole. The sequence of events in muscle contraction can be summarized as follows:

(1) Motor neuron releases acetylcholine →
(2) acetylcholine combines with receptors on muscle fiber →
(3) depolarization of sarcolemma →
(4) action potential spreads through T tubules →
(5) Ca^{2+} released from sarcoplasmic reticulum →
(6) Ca^{2+} binds to troponin, causing conformational change →
(7) troponin pushes tropomyosin away, exposing active sites on actin filaments →
(8) ATP (attached to myosin) is split →
(9) myosin head binds to exposed active site on actin filament, forming cross bridge →
(10) P_i and ADP released from myosin head →
(11) cross bridge flexes and actin filament pulled toward center of sarcomere →
(12) myosin head binds ATP and detaches from actin →
(13) if sufficient Ca^{2+}, sequence repeats from step (8)

When impulses from the motor neuron cease, muscle fibers return to their resting state. Acetylcholine in the synaptic cleft is inactivated by the enzyme acetylcholinesterase. Calcium ions are pumped back into the sarcoplasmic reticulum by active transport, a process that requires ATP. Without calcium ions, the active sites on the actin filaments are again covered by the tropomyosin-troponin complex. The actin filaments slide back to their original position, and the muscle relaxes. This entire series of events happens in milliseconds.

Even when we are not moving, our muscles are in a state of partial contraction known as **muscle tone.** At any given moment some muscle fibers are contracted, stimulated by messages from motor neurons. Muscle tone is an unconscious process that helps keep muscles prepared for action. When the motor nerve to a muscle is cut, the muscle becomes limp (completely relaxed, or flaccid) and eventually atrophies (decreases in size).

ATP powers muscle contraction

Muscle cells are often called on to perform strenuously and must be provided with large amounts of energy. The immediate source of energy necessary for muscle contraction is ATP, needed not only for the pull exerted by the cross bridges but also for their release from each active site. Rigor mortis, the temporary but very marked muscular rigidity that appears after death, results from ATP depletion following the cessation of cellular respiration after death.[1] Without ATP, the cross bridges that have been formed are not released.

Sufficient energy can be held in ATP molecules for only a few seconds of strenuous activity. Fortunately, muscle cells have a backup energy storage compound called **creatine phosphate** that can be stockpiled. The energy stored in creatine phosphate is transferred to ATP as needed. But during vigorous exercise the supply of creatine phosphate is also short-lived. As ATP and cre-

[1] Rigor mortis does not persist indefinitely, however, for the entire contractile apparatus of the muscles eventually decomposes, restoring pliability. The phenomenon is temperature-dependent, so, given the prevailing temperature, a medical examiner can estimate the time of death of a cadaver from its degree of rigor mortis. Perhaps it should be said that rigor mortis by itself is not muscular contraction; it only tends to freeze the corpse in its position at the time of death. Thus, tales of corpses sitting up and pointing to their murderers, and otherwise carrying on posthumously may be entertaining but have no factual basis.

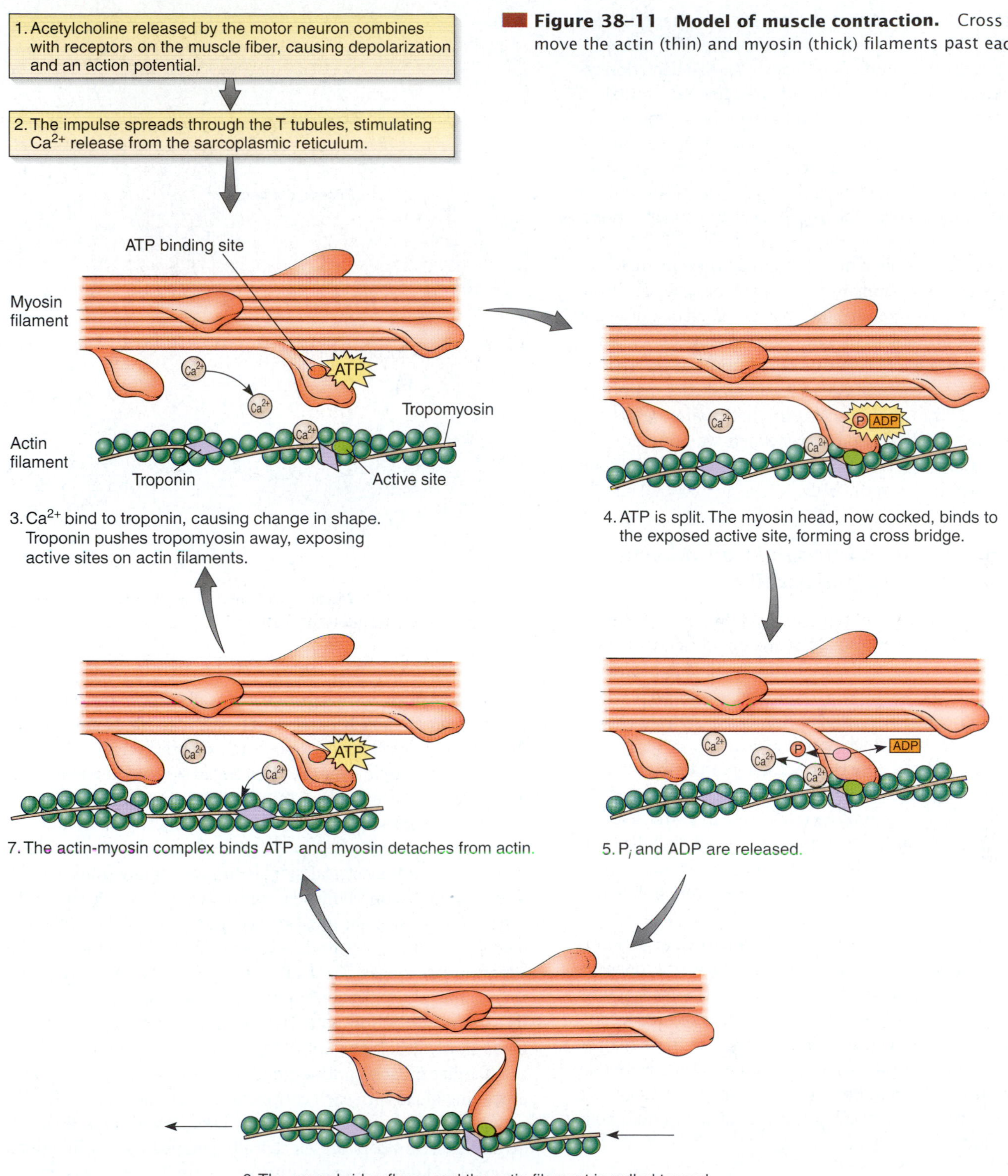

1. Acetylcholine released by the motor neuron combines with receptors on the muscle fiber, causing depolarization and an action potential.
2. The impulse spreads through the T tubules, stimulating Ca^{2+} release from the sarcoplasmic reticulum.
ATP binding site
Myosin filament
ATP
Tropomyosin
Actin filament
Troponin
Active site
3. Ca^{2+} bind to troponin, causing change in shape. Troponin pushes tropomyosin away, exposing active sites on actin filaments.
P ADP
4. ATP is split. The myosin head, now cocked, binds to the exposed active site, forming a cross bridge.
ADP
P
5. P_i and ADP are released.
6. The cross bridge flexes and the actin filament is pulled toward the center of the sarcomere.
ATP
7. The actin-myosin complex binds ATP and myosin detaches from actin.

Figure 38–11 Model of muscle contraction. Cross bridges move the actin (thin) and myosin (thick) filaments past each other.

atine phosphate stores are depleted, muscle cells must replenish their supplies of these energy-rich compounds.

Fuel is stored in muscle fibers in the form of **glycogen,** a large polysaccharide formed from hundreds of glucose molecules. Stored glycogen is degraded, yielding glucose, which is then broken down in cellular respiration. When sufficient oxygen is available, enough energy is captured from the glucose to produce needed quantities of ATP and creatine phosphate.

During a burst of strenuous exercise, the circulatory system cannot deliver sufficient oxygen to keep up with the demand of the rapidly metabolizing muscle cells. This results in an **oxygen debt.** Under these conditions, muscle cells can break down fuel molecules anaerobically (without oxygen) for short periods. Lactic acid fermentation is a method of generating ATP anaerobically, but not in great quantity (see Chapter 7). ATP depletion results in weaker contractions and muscle fatigue. Accumulation of the waste product lactic acid also contributes to muscle fatigue. Well-conditioned athletes develop the ability to tolerate the high levels of lactic acid generated during high-performance activity. The period of rapid breathing that generally follows strenuous exercise pays back the oxygen debt by consuming lactic acid.

The energy conversion of muscle contraction is relatively efficient compared with human-made machines. Recall from Chapter 7 that a steam power plant has an efficiency of about 35%. In muscle cells, about 40% of the chemical energy of the glucose fuel is actually converted to mechanical work. The remaining energy is accounted for as heat, produced mainly by frictional forces within the muscle cell. That is why we get hot during hard physical labor and shiver when we are cold. The muscle contractions involved in shivering are one way that animals produce heat to warm the body.

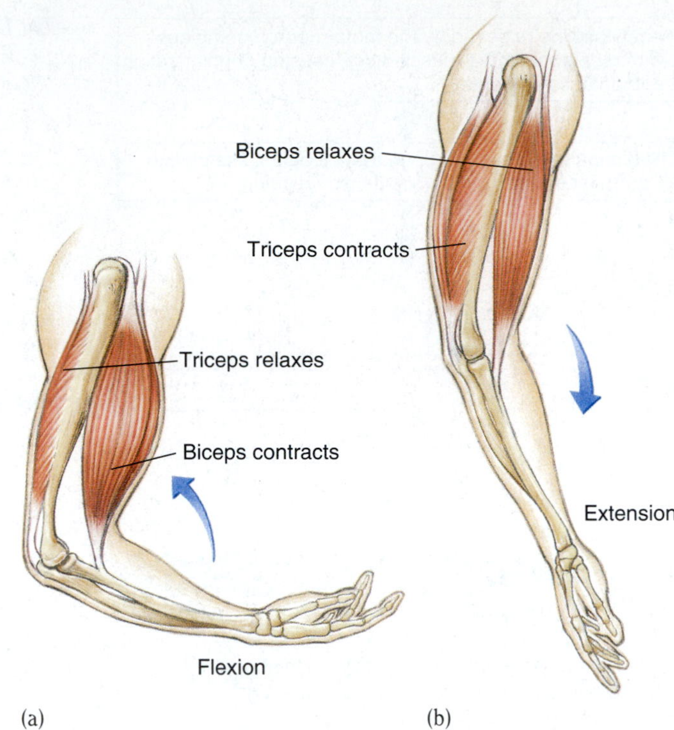

Figure 38–12 Muscle action. The biceps and triceps muscles function antagonistically.

Skeletal muscle action depends on muscle pairs that work antagonistically

Skeletal muscles produce movements by pulling on **tendons,** tough cords of connective tissue that anchor muscles to bone. Tendons then pull on bones. Skeletal muscles, or their tendons, pass across a joint and are attached to the bones on each side of the joint. When the muscle contracts, it pulls one bone toward or away from the bone with which it articulates.

Because muscles can only contract, they can only pull; they cannot push. Muscles act **antagonistically** to one another, which means that the movement produced by one can be reversed by another. The biceps muscle, for example, flexes (bends) your arm, whereas contraction of the triceps muscle extends it once again (Fig. 38–12). Thus, the biceps and triceps work antagonistically.

The muscle that contracts to produce a particular action is known as the **agonist.** The muscle that produces the opposite movement is the **antagonist.** When the agonist is contracting, the antagonist is relaxed. Generally, movements are accomplished by groups of muscles working together, so there may be several agonists and several antagonists in any action. Note that muscles that are agonists in one movement may be antagonists in another. The superficial skeletal muscles of the human body are shown in Figures 38–13 and 38–14.

Smooth, cardiac, and skeletal muscle respond in specific ways

Each type of muscle is specialized for particular types of responses. **Smooth muscle** is not attached to bones but forms tubes that squeeze like the muscle tissue in the body wall of the earthworm. Smooth muscle often contracts in response to simple stretching, and its contraction tends to be sustained. It is well adapted to performing such tasks as the regulation of blood pressure by sustained contraction of the walls of the arterioles. Although smooth muscle contracts slowly, it shortens much more than striated muscle does; it squeezes superlatively.

Smooth muscle is not striated because its actin and myosin filaments are not organized into myofibrils or into sarcomeres. The fibers of smooth muscle tissue can function as a unit because they are connected by *gap junctions* (Chapter 5). These junctions permit electrical signals to pass rapidly from fiber to fiber. Although smooth muscle contraction is basically similar to contraction of skeletal muscle (occurring by a sliding filament mechanism), the cross bridges in smooth muscle remain in the attached state longer. This translates into less ATP being required to maintain a high level of force. Some types of smooth muscle fibers generate action potentials without neural signals. These fibers slowly depolarize until they produce an action potential.

Cardiac muscle contracts and relaxes in alternating rhythm, propelling blood with each contraction. Sustained contraction of cardiac muscle would be disastrous! Like smooth muscle fibers, cardiac muscle fibers are electrically coupled by gap junctions (intercalated discs). Cardiac muscle can produce its own signals for contraction (discussed in Chapter 42).

Skeletal muscle, when stimulated by a single brief electrical stimulus, contracts with a single quick contraction called a **simple twitch.** Typically, skeletal muscle receives a series of separate stimuli timed very close together. These do not produce a series of simple twitches, however, but a single, smooth, sustained contraction called **tetanus.** Depending on the identity and number of our muscle cells tetanically contracting, we might thread a needle, rock a baby, or run a mile.

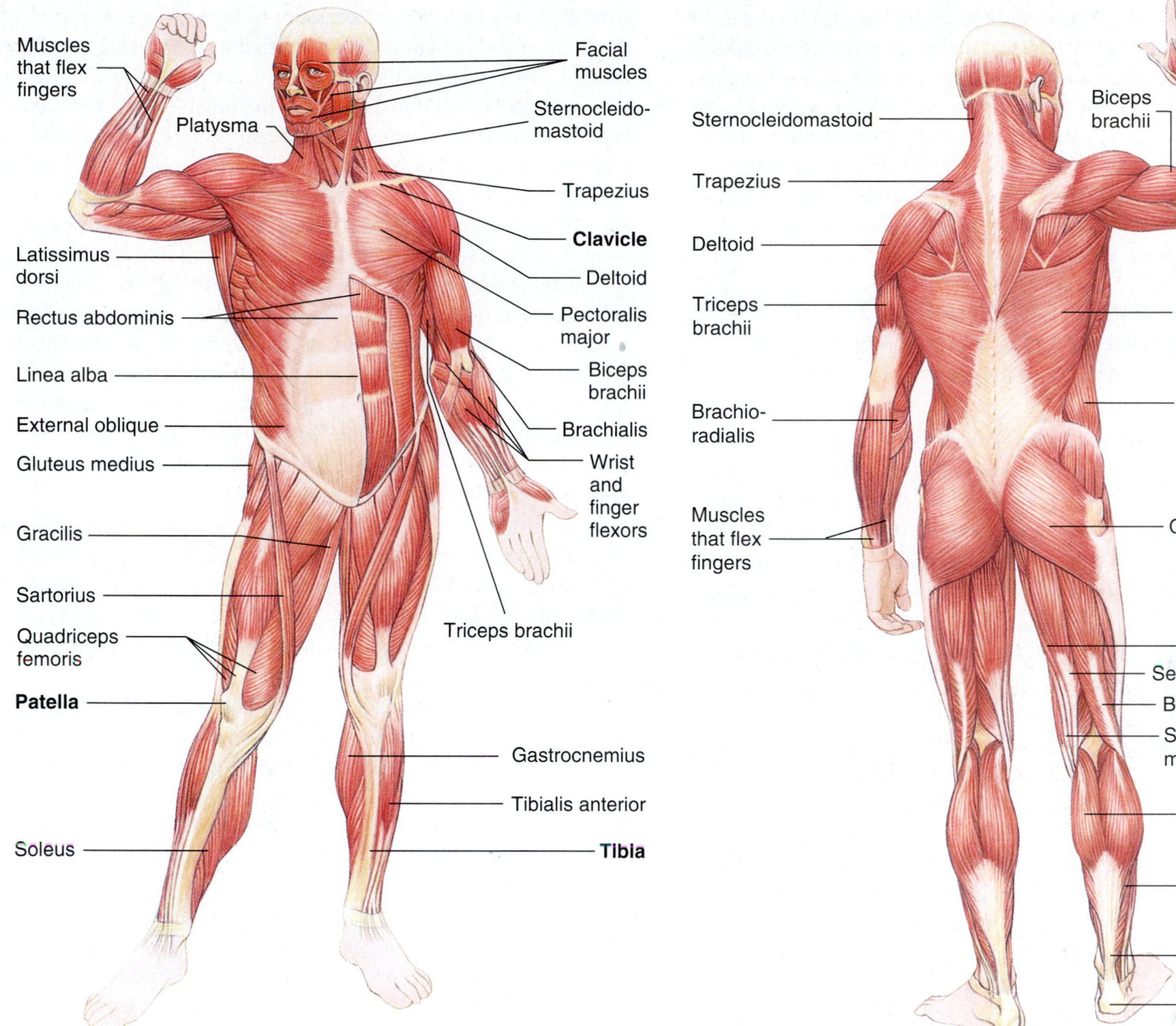

Figure 38–13 Some superficial muscles of the human body. Anterior view.

Figure 38–14 Some superficial muscles of the human body. Posterior view.

Muscle fibers may be specialized for slow or quick responses. One component of myosin, the heavy chain, exists in three different types: Types I, IIa, and IIx. The three types of myosin apparently evolved early and have been conserved throughout the animal kingdom. Muscle fibers that contain mainly Type I myosin are called Type I fibers, or **slow fibers.** These fibers are specialized for endurance activities such as swimming, long distance running, or maintaining posture. Slow fibers require a steady supply of oxygen. They derive most of their energy from aerobic metabolism and are rich in mitochondria and capillaries. Slow fibers are also referred to as **red fibers** because they are rich in **myoglobin,** a red pigment similar to hemoglobin in red blood cells. Myoglobin, like hemoglobin, stores oxygen. Myoglobin enhances the rapid diffusion of oxygen from the blood into the muscles during strenuous muscle exertion.

Muscle fibers that contain Type IIa and IIx myosin are known as **fast fibers,** or **white fibers.** These fibers can generate a great deal of power and carry out rapid movements but can sustain that activity for only a brief period. They are important in activities such as sprinting and weight lifting. (Type IIx fibers are the fastest.) White fibers obtain most of their energy from glycolysis; they have few mitochondria. When their glycogen supply is depleted, they fatigue rapidly. People who are sedentary have higher numbers of Type IIx fibers than physically fit individuals. With physical training, Type IIx fibers change to Type IIa fibers.

Entire muscles may be specialized for quick or slow responses. In chickens, for instance, the white breast muscles are efficient for quick responses, perhaps because a short flight is an escape mechanism for chickens. On the other hand, they walk about on the ground all day; the dark (red) meat of the leg and thigh is composed of muscle specialized for more sustained activity. Birds that fly have red breast muscles specialized to support sustained activity.

The proportions of slow and fast fibers vary from person to person and from muscle to muscle in the same person. Although the relative proportions of the two types of fibers appear to be genetically influenced, the proportions can be changed by appropriate training. Someone whose leg and thigh muscles contain a high proportion of fast fibers could, with proper training, become a good sprinter. An athlete with a greater proportion of slow fibers may be better suited to marathon activities.

Process of Science ■ WE ARE BEGINNING TO UNDERSTAND THE DYNAMICS OF INSECT FLIGHT

Insects were the first animals to evolve flight, an adaptation that has contributed to their impressive biological success (see Chapter 29). Just how insects fly has been an aerodynamic mystery, and scientists are working to understand their remarkable ability to maneuver. A central question has been how their flapping wings can generate sufficient force to keep them airborne. Insect flight involves much more than just flapping the wings up and down. The flapping motion of the insect wing changes direction and speed, and upstrokes alternate with downstrokes at very high rates. At each shift of stroke, the wing rotates about its long axis and tilts to just the right angle for the new direction of motion.

In many flying insects, the striated flight muscles are attached not directly to the wings but to the flexible portions of the exoskeleton that articulate with the wings. Each contraction of the muscles produces a dimpling of the exoskeleton in association with a downstroke and sometimes, depending on the exact arrangement of the muscles, on the upstroke as well. When the dimple springs back into its resting position, the muscles attached to it are stretched. The stretching immediately initiates another contraction, and the cycle is repeated.

The deformation of the exoskeleton is transmitted as a force to the wings, which beat so fast that we may perceive the sound as a musical tone. In the common blowfly, for instance, the wings may beat at a frequency of 120 cycles *per second*. Yet, in the same blowfly, the neurons that innervate those furiously contracting flight muscles are delivering impulses to them at the astonishingly low frequency of 3 per second. The mechanical properties of the musculoskeletal arrangement described provide the stimuli for contraction, by stretching the muscle fibers at a high frequency, but the nerve impulses are needed to maintain it.

Insect flight muscles in action have the highest known metabolic rate. (In fact cytochromes, the iron-containing proteins of electron transport systems, were first discovered in insect flight muscle.) Accordingly, these muscles contain more mitochondria than any known variety of muscle, and these muscles are elaborately infiltrated with tiny air-filled tubes called *tracheae* that carry oxygen directly to each fiber.

Flight muscles must be kept at appropriate operating temperature if they are to function. You have probably noticed the constant twitching of the wings of such insects as wasps even when they are crawling instead of flying. Biologists have hypothesized that this behavior is necessary to keep the temperature of the flight muscles high for instant readiness. You may also have noticed that the bodies of many moths are quite furry. Moth fur serves the same function as fur in a mammal—to conserve body heat. When the moth awakens and prepares for flight, it shivers its flight muscles at a low frequency to warm them up, constricting its abdominal blood vessel to keep the heat in its thorax. Gradually the frequency of the shivering increases until, at a critical moment, the moth spreads its wings and hums off into the darkness.

Many insects have special adaptations to rid the body of the excess heat produced by the flight muscles. The rapidly flying sphinx moth, for example, has in its abdomen what amounts to a radiator—a great blood vessel that carries heat from the thorax, where it is generated, and emits it into the cool of the night.

Somehow insects are able to create lift that is 20 or more times their body weight. In 1996 Charles P. Ellington and colleagues at the University of Cambridge in England reported that as insects flap their wings downward, some of the air flowing over the wings rolls up along the front edge, forming a vortex (like a whirlpool) that becomes larger as it moves along toward the tip of the wing. The vortex generates a low-pressure region above the wing that sucks the wing upward, generating an extra lifting force.

Researchers are using a variety of approaches to unravel the mysteries of insect flight. Some groups of investigators are using computer simulations, and one group built a fly robot. Christina M. McGraw, a graduate student at the University of Washington, and her advisors James B. Xallis and Martin Couterman, are painting the wings of honeybees in an effort to better understand the flight dynamics in living insects (Fig. 38–15). The paint contains a dye that responds to changes in air pressure. Under UV light, the dye phosphoresces bright red. Oxygen in the air diminishes the red glow, so regions on the wings that are impacted by the highest air pressure phosphoresce the least (more oxygen molecules are present in denser air). The researchers can map out the forces acting on the wings by recording the intensity of the red glow. Because a bee's wing-beat cycle takes place in 5 milliseconds, mapping these forces will be a major challenge.

Figure 38–15 Honeybee *(Apis mellifera)* in flight. *(Stephen Dalton/Animals Animals)*

SUMMARY WITH KEY TERMS

I. **Epithelial coverings** protect underlying tissues. They may be specialized for sensory or respiratory functions, to produce a protective cuticle, to secrete lubricants or adhesives, to produce odorous or poisonous secretions, or to produce threads for nests or webs.

II. The **integumentary system** of vertebrates includes the skin and structures that develop from it. In humans this system includes the skin, nails, hair, sweat glands, oil glands, and sensory receptors. The vertebrate skin protects, may help prevent dehydration, may be specialized for secretion or reception of stimuli, and may help regulate body temperature.
   A. In human skin, cells in **stratum basale** of the **epidermis** divide and are pushed upward toward the skin surface. These cells mature, flatten, produce **keratin,** and eventually die. **Stratum corneum,** the most superficial layer of the epidermis, consists of dead cells filled with keratin.
   B. The **dermis,** which consists of dense, fibrous connective tissue, rests on a layer of **subcutaneous tissue** composed largely of fat.

III. The **skeletal system** transmits mechanical forces generated by muscle and also supports and protects the body.
   A. Many invertebrates have a **hydrostatic skeleton** in which fluid in a closed body compartment is used to transmit forces generated by contractile cells or muscle.
   B. **Exoskeletons** are characteristic of mollusks and arthropods. The arthropod skeleton, composed mainly of chitin, is jointed for flexibility. This nonliving skeleton does not grow, making it necessary for arthropods to **molt** periodically.

IV. The **endoskeletons** of echinoderms and chordates are composed of living tissue and therefore are capable of growth.
   A. The vertebrate skeleton consists of an **axial skeleton** and an **appendicular skeleton.**
      1. The axial skeleton consists of **skull, vertebral column,** ribs, and sternum.
      2. The appendicular skeleton consists of bones of the limbs, **pectoral girdle,** and **pelvic girdle.**
   B. A typical long bone consists of a thin outer shell of **compact bone** surrounding the inner **spongy bone** and a central cavity that contains **bone marrow.**
   C. Long bones develop from cartilage templates during **endochondral bone formation.** Other bones, such as the flat bones of the skull, develop from a noncartilage connective tissue model by **intramembranous bone development.**
   D. **Osteoblasts,** cells that produce bone, and **osteoclasts,** cells that break down bone, work together to shape and remodel bone.
   E. **Joints** are junctions of two or more bones.
      1. The sutures of the skull are **immovable joints;** joints between vertebrae are **slightly movable joints.** A **freely movable joint** is enclosed by a joint capsule lined with a membrane that secretes **synovial fluid.**
      2. **Ligaments** are connective tissue bands that connect bones and limit movement at the joint.

V. All animals have the ability to move; most can move from place to place in the environment (locomotion). A **muscular system** is found in most invertebrate phyla and in vertebrates. As muscle tissue contracts (shortens), it moves body parts by pulling on them. Three types of muscle are **skeletal, smooth,** and **cardiac muscle.**
   A. A muscle such as the biceps is an organ made up of hundreds of **muscle fibers.** Each fiber is made up of threadlike **myofibrils** composed of smaller **myofilaments,** or simply **filaments.**
   B. The striations of skeletal muscle fibers reflect the overlapping of their **actin filaments** and **myosin filaments.** A **sarcomere** is a contractile unit of actin (thin) and myosin (thick) filaments.
   C. During muscle contraction, the actin filaments are pulled by the myosin filaments toward the center of the myofibril.
      1. **Acetylcholine** released by a motor neuron combines with receptors on the surface of a muscle fiber. This may cause **depolarization** of the sarcolemma and transmission of an **action potential.**
      2. The action potential spreads through the **T tubules,** resulting in the release of calcium ions from the **sarcoplasmic reticulum.**
      3. Calcium ions bind to **troponin** in the actin filaments, causing the troponin to change shape. Troponin pushes **tropomyosin** away from the active sites on the actin filaments.
      4. ATP binds to myosin; ATP is split, putting the myosin head in a high energy state (it is "cocked"). Energized myosin heads bind to the exposed active sites on the actin filaments, forming **cross bridges** that link the myosin and actin filaments.
      5. After myosin attaches to the actin filament, phosphate and ADP are released.
      6. It is thought that the release of ADP and P_i causes the cross bridge to flex, producing the power stroke that pulls the actin filament toward the center of the sarcomere.
      7. The myosin head binds a new ATP, which allows the myosin head to detach from the actin. As long as the calcium ion concentration remains elevated, the new ATP is split, and the sequence of steps repeats. The myosin reattaches to new active sites so that the filaments are pulled past one another, and the muscle continues to shorten.
   D. ATP is the immediate source of energy for muscle contraction. The energy from ATP hydrolysis provides the energy to "cock" the myosin. The binding of a new ATP detaches the myosin from the actin after movement has taken place.
      1. Muscle tissue has an intermediate energy storage compound, **creatine phosphate. Glycogen** is the fuel stored in muscle fibers.
      2. An **oxygen debt** occurs when the circulatory system cannot deliver enough oxygen to keep up with the demand of muscle cells during strenuous exercise.
   E. **Muscle tone** is the state of partial contraction characteristic of muscles.
      1. Muscles act by pulling on **tendons,** connective tissue cords that attach muscles to bones.
      2. Muscles act **antagonistically** to one another. The muscle that produces a particular action is the **agonist;** the **antagonist** produces the opposite movement.
   F. When activated by a brief electrical stimulus, skeletal muscle responds with a **simple twitch.** Typically skeletal muscle is stimulated with a series of separate stimuli timed close together and responds with a smooth, sustained contraction called **tetanus.**
   G. **Slow fibers,** also called **red fibers,** are rich in mitochondria and **myoglobin.** They are specialized for endurance activities. **Fast fibers,** also called **white fibers,** are specialized for rapid response. They can generate a great deal of power for a brief period using ATP from glycolysis.

POST-TEST

1. The vertebrate skin consists of (a) outer epidermis, inner hypodermis (b) outer epidermis, inner endoskeleton (c) outer endodermis, inner epidermis (d) outer epidermis, inner dermis (e) outer subcutaneous layer, inner dermis
2. Cells actively divide in (a) stratum basale (b) stratum corneum (c) stratum dermis (d) the layer with cells that contain keratin (e) two of the preceding answers are correct
3. An endoskeleton (a) is typically composed of dead tissue (b) is characterized by fluid in a closed compartment (c) is typical of echinoderms (d) is typical of arthropods (e) requires the animal to molt
4. Which of the following is *not* part of the axial skeleton? (a) skull (b) vertebral column (c) pelvic girdle (d) atlas (e) sacrum
5. The thin outer shell of a long bone is composed of (a) compact bone (b) spongy bone (c) epiphyses (d) cancellous bone (e) mainly chondrocytes
6. Which of the following connects bones to one another? (a) tendons (b) ligaments (c) osteoclasts (d) synovial membranes (e) smooth fibers
7. In endochondral bone formation (a) osteoclasts produce bone (b) joints connect fibers (c) the skeleton consists of cartilage (d) bones develop from cartilage templates (e) bones form in noncartilage connective tissue
8. All animals have (a) muscles (b) actin (c) bones (d) endoskeleton or exoskeleton (e) dermis
9. An energy storage compound that can be stockpiled in muscle cells for short-term use is (a) creatine phosphate (b) ATP (c) troponin (d) myosin (e) myoglobin
10. Myosin binds to actin, forming a cross bridge. What happens next? (a) acetylcholine is released (b) calcium ions stimulate process that leads to exposure of active sites (c) filaments slide past each other, and the muscle fiber shortens (d) myosin is activated (e) ADP and P_i are released, and the cross bridge flexes
11. When skeletal muscle is stimulated by a series of closely timed separate stimuli (a) it responds with a smooth, sustained contraction called tetanus (b) a simple twitch occurs (c) white fibers respond (d) red fibers respond (e) muscle tone occurs
12. Glycogen is (a) produced by actin (b) depleted within 1 second of strenuous activity (c) a form of long-term energy storage (d) synthesized when cross bridges form (e) depleted before creatine phosphate reserves are used
13. Slow fibers (a) are also called white fibers (b) do not depend on ATP (c) have few mitochondria (d) obtain most of their energy from glycolysis (e) are rich in myoglobin
14. Insect flight muscles (a) work best at low temperature (b) have a very high metabolic rate (c) do not create much lift (d) evolved after bird wings (e) are always attached directly to the wing and make up the bulk of the wing
15. Which of the following pairs of systems function together most closely? (a) integumentary/skeletal (b) integumentary/digestive (c) muscular/epithelial (d) skeletal/muscular (e) epithelial/muscular

REVIEW QUESTIONS

1. Compare the external epithelium of invertebrates and vertebrates.
2. What properties does keratin give human skin?
3. Describe a hydrostatic skeleton. What functions does it perform? How do the septa in the annelid worm contribute to the versatility of its hydrostatic skeleton?
4. What are some disadvantages of an exoskeleton? Some advantages?
5. Describe the divisions of the human skeleton.
6. Draw a typical long bone, such as the radius, and label its parts.
7. What are the functions of osteoblasts and osteoclasts? Why is it important that bones be continuously remodeled?
8. Describe a skeletal muscle fiber and compare its two types of filaments.
9. Outline the sequence of events thought to occur when a muscle fiber contracts, beginning with the release of acetylcholine and including cross-bridge action.
10. What is the role of ATP in muscle contraction? What are the functions of creatine phosphate and glycogen?
11. Label the diagram. Use Figure 38–8 to check your answers.

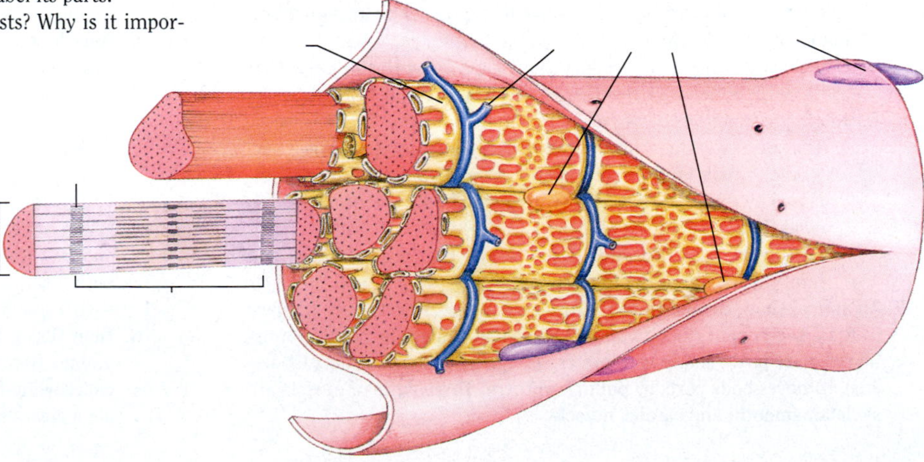

YOU MAKE THE CONNECTION

1. Compare the arthropod exoskeleton with the vertebrate endoskeleton. What are some benefits and some disadvantages of each type of skeleton?
2. What are some examples of hydrostatic support in plants?
3. Why is it important that a muscle be able to shift functions, sometimes acting as an agonist and at other times acting as an antagonist?

RECOMMENDED READINGS

Anderson, J.L., and B. Saltin. "Muscles, Genes, and Athletic Performance." *Scientific American,* Vol. 283, Sep. 2000. An interesting account of types of muscle fibers and muscle response to exercise.

Dickinson, M.H., et al. "How Animals Move: An Integrative View." *Science,* Vol. 288, 7 Apr. 2000. Several general principles of animal locomotion can be applied to walking, running, swimming, and flying.

Dickman, S. "No Bones about a Genetic Switch for Bone Growth." *Science,* Vol. 276, 6 Jun. 1997. A brief account of research on the genetics of bone development.

Ducy, P., T. Schinke, and G. Karsenty. "The Osteoblast: A Sophisticated Fibroblast under Central Surveillance." *Science,* Vol. 289, 1 Sep. 2000. A review of the research on bone-forming cells.

Fackelman, K. "Melanoma Madness: The Scientific Flap over Sunscreens and Skin Cancer." *Science News,* Vol. 153, 6 Jun. 1998. A reasoned discussion of the costs and benefits of using sunscreen.

Kilberstis, P., O. Smith, and C. Norman. "Bone Health in the Balance." *Science,* Vol. 289, 1 Sep. 2000. Introduction to a special issue on bone remodeling.

Rodan, G.A., and T.J. Martin. "Therapeutic Approaches to Bone Diseases." *Science,* Vol. 289, 1 Sep. 2000. A review of recent research on the development of bone diseases and progress on treatment strategies.

Simpson, S. "For the Bees." *Scientific American,* Vol. 282, May 2000. A brief summary of ongoing research on insect flight.

Teitelbaum, S.L. "Bone Resorption by Osteoclasts." *Science,* Vol. 289, 1 Sep. 2000. A review of the research on bone resorption.

- Visit our Web site at **http://www.info.brookscole.com/solomonbergmartin** for links to chapter-related resources on the World Wide Web. Additional on-line materials relating to this chapter can also be found on our Web site.

See chapter activity on BioActive Learner CD for additional help in mastering the chapter's material. Icon location in the chapter's margins shows which topics have tutorials or simulations in the CD.

39

Neural Signaling

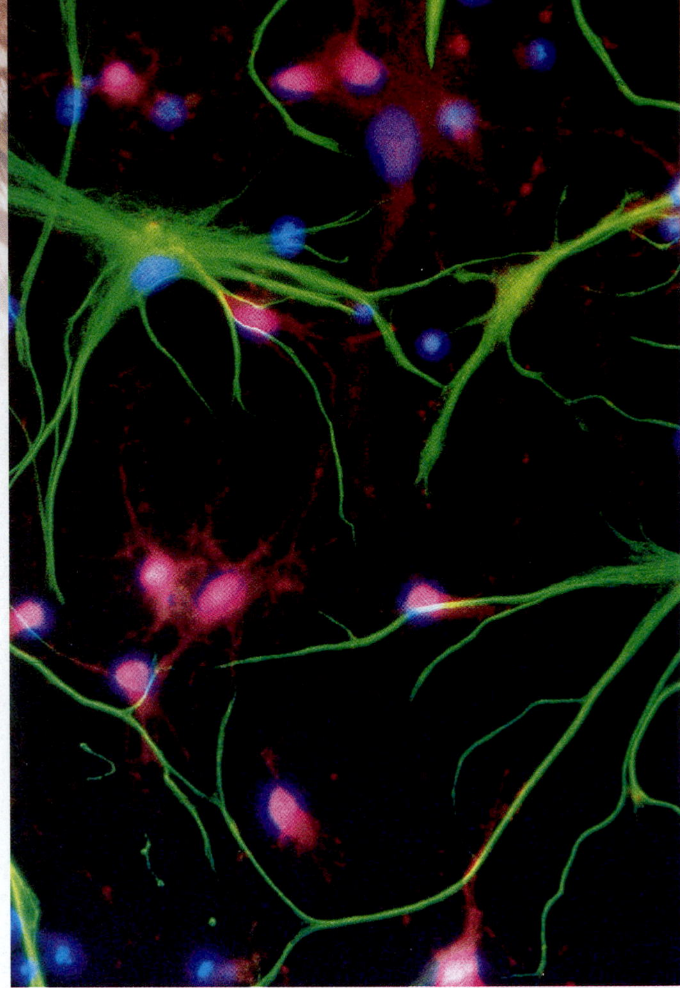

Immunofluorescent LM of several astrocytes *(green)* from mammalian spinal cord tissue. The smaller cells *(red)* are neurons. Their cell bodies appear pink. The blue dots are the nuclei of other glial cells. *(Nancy Kedersha, UCLA/ Science Photo Library/Photo Researchers, Inc.)*

LEARNING OBJECTIVES

After you have studied this chapter you should be able to

1. Trace the flow of information through the nervous system. Describe the four processes involved in neural signaling: reception, transmission, integration, and action by effectors.
2. Draw a typical neuron, label its parts (including myelin and cellular sheaths), and give the function of each.
3. Explain how the neuron develops and maintains a resting potential.
4. Compare a graded potential with an action potential, describing the production and transmission of each.
5. Trace the events that take place in synaptic transmission and draw diagrams to support your description.
6. Describe the actions of the neurotransmitters identified in the chapter and describe mechanisms for removing them from the synapse.
7. Compare excitatory and inhibitory signals and their effects.
8. Describe how a postsynaptic neuron integrates incoming stimuli and "decides" whether to fire.
9. Draw diagrams of neural circuits showing convergence, divergence, and reverberation, and explain why each is important.

An organism's ability to survive and to maintain homeostasis depends largely on how effectively it responds to internal signals such as hunger or lowered blood pressure and to external signals such as changes in temperature or the presence of predators. Changes within the body or in the outside world that can be detected by an organism are called **stimuli.** In all animals except the sponges, responses to stimuli depend on the activities of networks of nerve cells, or **neurons.** These cells are specialized for transmitting **neural signals,** which are electrical signals and chemical messages.

In most animals, neurons and supporting cells are organized as a **nervous system** that, like a computer, takes in information, integrates it, and responds. Just how animals respond to stimuli depends on how their neurons are organized and connected to one another. A single neuron in the vertebrate brain may be functionally connected to thousands of other neurons. The endocrine system works with the nervous system to regulate many behaviors and physiological processes. The endocrine system generally provides relatively slow and long-lasting regulation, whereas the nervous system typically permits more rapid, but brief, responses.

Neurobiology is one of the most exciting areas of biological research. Recently, much effort has been focused on **neurotransmitters,** the chemical messengers used by neurons to signal other neurons, and on the receptors that bind with the neurotransmitters. Mood states are influenced by the concentration of certain neurotransmitters, and disorders such as schizophrenia and depression are related to abnormal amounts. Many antidepressant medications act by selectively inhibiting the reuptake of the neurotransmitter serotonin by neurons in the brain. This action results in a greater concentration of serotonin, leading to an elevation in mood.

Another active area of research is the role of **glial cells** in the nervous system. These cells support and protect the neurons and have many regulatory functions. The glial cells shown in the photograph are *astrocytes,* cells that provide glucose for neurons and also help regulate the composition of the extracellular fluid in the brain and spinal cord. Because astrocytes can generate weak electrical signals and have many neuron characteristics (e.g., they have receptors for neurotransmitters), some researchers speculate that they may participate in information signaling in the brain.

Investigators have demonstrated that certain populations of astrocytes function as stem cells in the brain and spinal cord. These cells can give rise to neurons, additional astrocytes, and certain other glial cells. When taken out of their normal environment in the adult mouse brain, astrocytes are capable of giving rise to cells of all germ layers (i.e., the embryonic tissue layers: ectoderm, mesoderm, and endoderm). These neural stem cells may some day be used to produce specific types of cells needed for treating various medical conditions.

INFORMATION FLOW OCCURS THROUGH NEURAL SIGNALING

Thousands of stimuli constantly bombard an animal, and its survival depends on identifying and either ignoring or responding appropriately to these stimuli. Appropriate response to a stimulus depends on **neural signaling,** which is communication among neurons. In most animals, neural signaling involves four processes: reception, transmission, integration, and action by **effectors** (muscles or glands). **Reception,** the process of detecting a stimulus, is the job of the neurons and of specialized sense organs such as eyes and ears. **Transmission** is the process of sending messages along a neuron, from one neuron to another or from a neuron to a muscle or gland. In vertebrates, a neural message is transmitted from a receptor to the **central nervous system (CNS),** which consists of the brain and spinal cord. Neurons that transmit information to the CNS are called **sensory neurons,** also known as **afferent** (meaning "to carry toward") **neurons.**

Afferent neurons generally transmit information to **interneurons,** also referred to as **association neurons,** that integrate input and output. Most neurons, perhaps 99%, are interneurons. The cell body and axon of an interneuron are located within the CNS. **Integration** involves sorting and interpreting incoming sensory information and determining the appropriate response. Neural messages are transmitted from the CNS by **efferent** (meaning "to carry away") **neurons,** or **motor neurons,** to effectors (muscles and glands). The **action by effectors** is the actual response to the stimulus (Fig. 39–1). Sensory receptors and afferent and efferent neurons are part of the **peripheral nervous system (PNS).**

NEURONS AND GLIAL CELLS ARE THE CELLS OF THE NERVOUS SYSTEM

Two types of cells unique to the nervous system are neurons and glial cells. Neurons are specialized to receive and send information. Glial cells provide support and protection for the neurons.

Glial cells provide metabolic and structural support

Sometimes glial cells are referred to collectively as the **neuroglia,** which literally means "nerve glue." Vertebrates have three main types of glial cells in the CNS: microglia, astrocytes, and oligodendrocytes. **Microglia** are phagocytic cells that remove cellular debris. They are found near blood vessels in the nervous system and migrate into the CNS. Microglia are not neural in origin and may originate in the bone marrow.

As described in the chapter opener, **astrocytes** are star-shaped glial cells that provide neurons with glucose and perform several other functions. Astrocytes help regulate the composition of the extracellular fluid in the CNS by removing potassium ions and excess neurotransmitters. Some astrocytes position the ends of their long **processes** (cytoplasmic extensions) on blood vessels in the brain. In response, endothelial cells lining the blood vessels form *tight junctions* (Chapter 5) that prevent many substances in the blood from entering the brain tissue. This barrier is the *blood-brain barrier.* Astrocytes are also thought to play a role in development by guiding neurons to appropriate locations in the body.

Oligodendrocytes are glial cells that envelop neurons in the CNS, forming insulating sheaths around them. This covering consists of myelin, a white, fatty substance found in the plasma membrane of the glial cell. Because it is an excellent electrical insulator, its presence speeds the transmission of neural impulses. In **multiple sclerosis,** a neurological disease that affects about 300,000 people in the United States, patches of myelin deteriorate at irregular intervals along axons and are replaced by scar tissue. This damage interferes with conduction of neural impulses, and the victim suffers loss of coordination, tremor, and partial or complete paralysis of parts of the body. Evidence suggests that multiple sclerosis is an *autoimmune disease,* in which the body attacks its own tissue, in this case glial cells (autoimmune diseases are discussed in Chapter 43).

In vertebrates, **Schwann cells,** another type of glial cell, are located outside the CNS. Schwann cells form sheaths around some axons.

A typical neuron consists of a cell body, dendrites, and an axon

The neuron is highly specialized to receive stimuli and to produce and transmit electrical signals called **nerve impulses,** or **action potentials.** This cell also synthesizes and releases neurotransmitters. The neuron is distinguished from all other cells by its long processes. Examine the structure of a common type of neuron, the multipolar neuron in Figure 39–2.

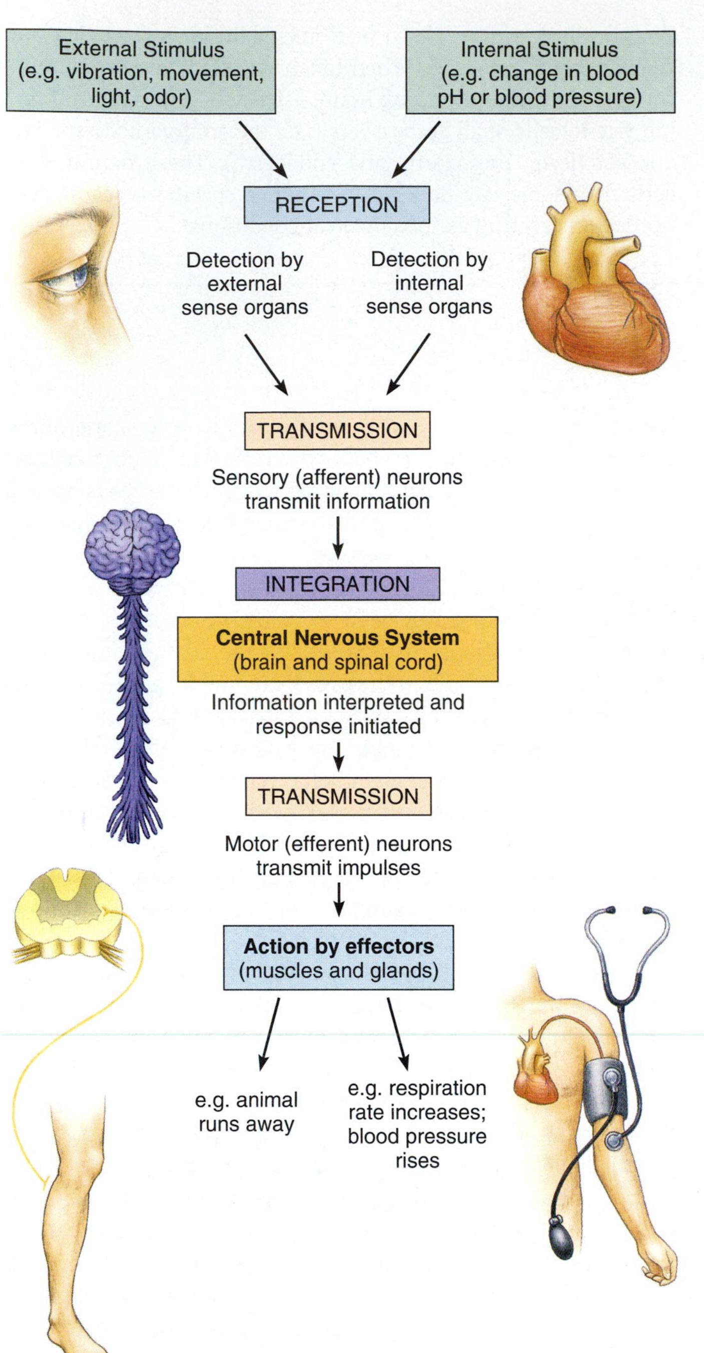

Figure 39–1 Response to a stimulus. Whether a stimulus originates in the outside world or inside the body, information must be received, transmitted to the central nervous system, integrated, then transmitted to effectors, muscles or glands that carry out some action.

The largest portion of the neuron, the **cell body,** contains the nucleus, the bulk of the cytoplasm, and most of the organelles. Typically, two types of cytoplasmic extensions project from the cell body of a multipolar neuron: Numerous dendrites extend from one end, and a long, single axon projects from the opposite end. **Dendrites** are typically short, highly branched processes specialized to receive stimuli and send signals to the cell body. The cell body integrates incoming signals.

Although microscopic in diameter, an **axon** may be 1 m (over 1 yd) or more in length and may divide, forming branches called **axon collaterals.** The axon conducts nerve impulses away from the cell body to another neuron, or to a muscle or gland. At its end the axon divides, forming many **terminal branches** that end in **synaptic terminals.** The synaptic terminals release **neurotransmitters,** chemicals that transmit signals from one neuron to another, or from neuron to effector. The junction between a synaptic terminal and another neuron (or effector) is called a **synapse.** Typically, a small space exists between the membranes of these two cells.

In vertebrates, the axons of many neurons outside the CNS are surrounded by a series of Schwann cells that form an insulating covering, the **myelin sheath.** Gaps in the myelin sheath, called **nodes of Ranvier,** occur between successive Schwann cells. At these points the axon is not insulated with myelin. Axons more than 2 μm in diameter have myelin sheaths and are described as *myelinated.* Those of smaller diameter are generally unmyelinated.

A **nerve** consists of hundreds or even thousands of axons wrapped together in connective tissue (Fig. 39–3). We can compare a nerve to a telephone cable. The individual axons correspond to the wires that run through the cable, and the sheaths

Figure 39–2 Structure of a multipolar neuron. The cell body contains most of the organelles. Many dendrites and a single axon extend from the cell body. Schwann cells form a myelin sheath around the axon.

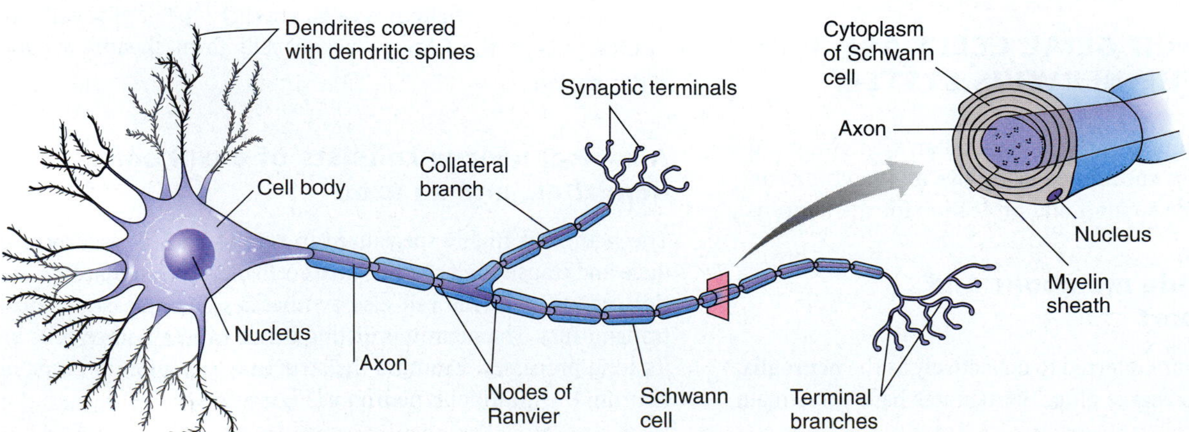

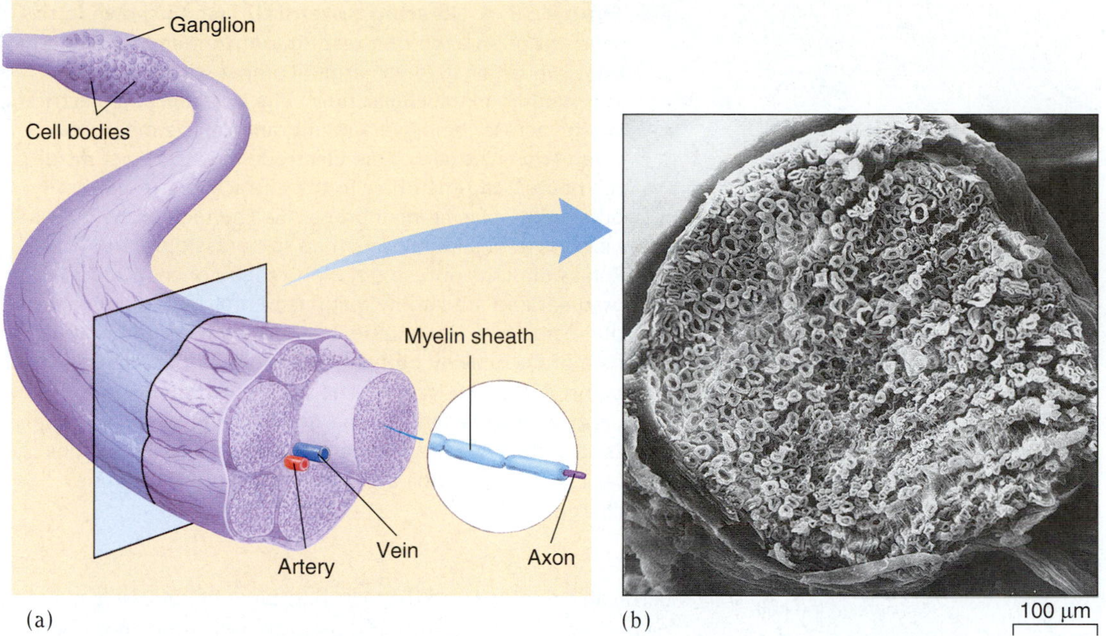

Figure 39–3 Structure of a nerve. **(a)** A nerve consists of bundles of axons held together by connective tissue. The cell bodies belonging to the axons of a nerve are grouped together in a ganglion. **(b)** SEM showing a cross section through a myelinated afferent nerve of a bullfrog. *(b, E.R. Lewis/Biological Photo Service)*

and connective tissue coverings correspond to the insulation. Within the CNS, bundles of axons are referred to as **tracts** or **pathways** rather than nerves. Outside the CNS, the cell bodies of neurons are usually grouped together in masses called **ganglia** (sing., *ganglion*). Inside the CNS, collections of cell bodies are generally referred to as *nuclei* rather than ganglia.

NEURONS USE ELECTRICAL SIGNALS TO TRANSMIT INFORMATION

Most animal cells have a difference in electrical charge across the plasma membrane—a more negative electrical charge inside the cell compared with the electrical charge of the extracellular fluid outside. The plasma membrane is said to be electrically **polarized,** meaning that one side, or pole, has a different charge from the other side. When electrical charges are separated in this way, a potential energy difference exists across the membrane. This difference in electrical charge across the plasma membrane gives rise to an electrical voltage gradient. The voltage measured across the plasma membrane is referred to as the **membrane potential.** Should the charges be permitted to come together, they have the ability to do work. Thus, the cell can be thought of as a biological battery. In excitable cells, such as neurons and muscle cells, the membrane potential can be rapidly changed, and such changes can transmit signals to other cells.

The neuron membrane has a resting potential

The membrane potential in a resting (not excited) neuron or muscle cell is called its **resting potential.** The resting potential is generally expressed in units called *millivolts* (mV). (A millivolt equals one-thousandth of a volt.) Voltage is the force that causes charged particles to flow between two points. Like other cells that can produce electrical signals, the neuron has a resting potential of about 70 mV. By convention this is expressed as −70 mV because the cytosol close to the plasma membrane is negatively charged relative to the extracellular fluid (Fig. 39–4).

We can measure the potential across the membrane by placing one electrode inside the cell and a second electrode outside the cell, and connecting them through a very sensitive voltmeter or oscilloscope. If we place both electrodes on the outside surface of the neuron, no potential difference between them is registered. All points on the same side of the membrane are at the same potential. However, once one of the electrodes penetrates the cell, the voltage changes from zero to approximately −70 mV.

Two main factors determine the magnitude of the membrane potential: (1) differences in the concentrations of specific ions inside the cell compared with the extracellular fluid, and (2) selective permeability of the plasma membrane to these ions. The distribution of ions inside neurons and in the extracellular fluid surrounding them is similar to that of most other cells in the body. The K^+ concentration is about ten times greater inside than outside the cell. In contrast, the Na^+ concentration is about ten times greater outside than inside.

This asymmetric distribution of ions across the plasma membrane at rest is brought about by the action of selective ion channels and ion pumps. In vertebrate neurons (and skeletal muscle fibers), the resting membrane potential depends mainly on the diffusion of ions down their concentration gradients.

Ions cross the plasma membrane by diffusion through ion channels formed by membrane proteins. Net movement of ions

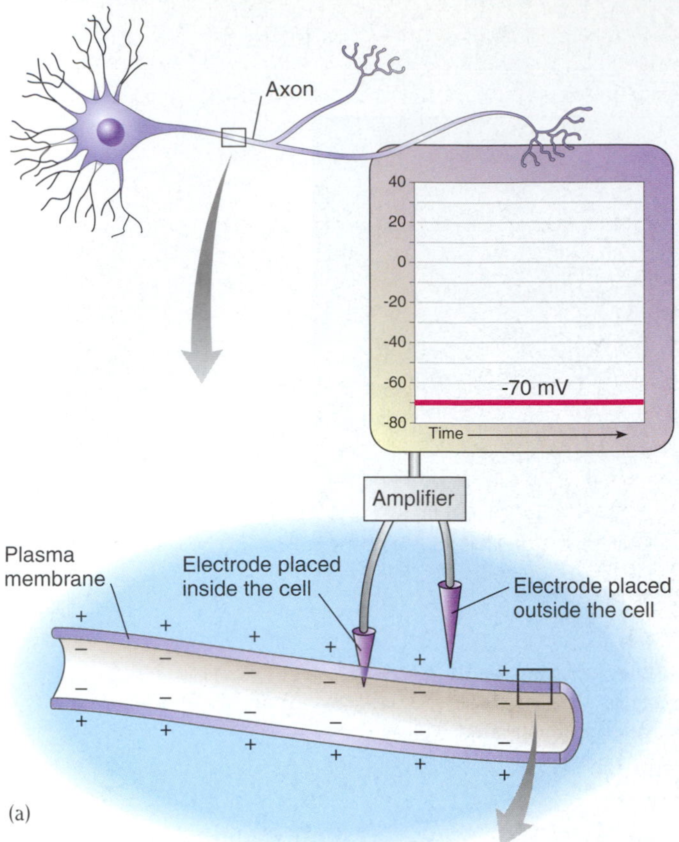

(a)

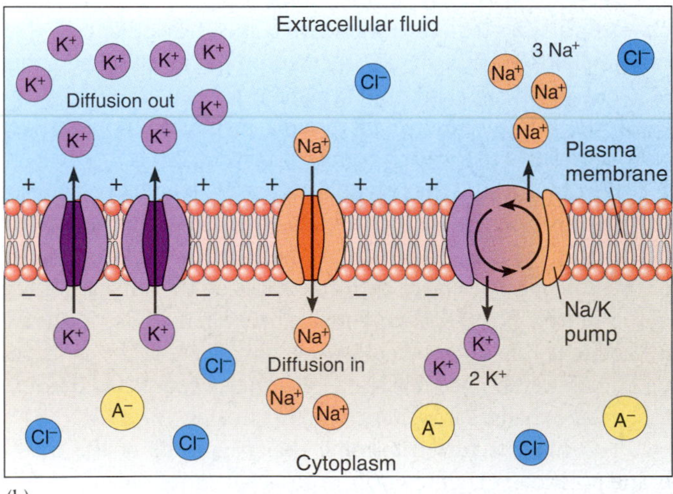

(b)

Figure 39–4 Resting potential. **(a)** As shown in this segment of an axon of a resting (nonconducting) neuron, the axon is negatively charged compared with the surrounding extracellular fluid. The difference in electrical charge across the plasma membrane can be measured using two electrodes. One electrode is placed just inside the neuron and the other in the extracellular fluid just outside the plasma membrane. **(b)** The membrane is less permeable to sodium ions than to potassium ions. Potassium ions diffuse out along their concentration gradient and are largely responsible for the voltage across the membrane. Sodium-potassium pumps in the plasma membrane actively pump sodium ions out of the cell and pump potassium ions in. The plasma membrane is permeable to negatively charged chloride ions, and they contribute slightly to the negative charge. Large anions (A^-) such as proteins also contribute to the negative charge.

occurs from an area of higher concentration to one of lower concentration (see discussion of diffusion in Chapter 5). Typically, these channels allow only specific types of ions to pass. Neurons have three types of ion channels: *passive ion channels, voltage-activated channels,* and *chemically activated ion channels* (discussed in a later section). **Passive ion channels** permit the passage of specific ions such as Na^+, K^+, Cl^-, and Ca^{2+} (see Fig. 39–4). (Unlike voltage-activated and chemically activated ion channels, passive ion channels are not controlled by gates.)

Potassium channels are the most common type of passive ion channel in the plasma membrane, and neurons are more permeable to potassium than to other types of ions. In fact, in the resting neuron the membrane is up to 100 times more permeable to K^+ than to Na^+. Sodium ions pumped out of the neuron cannot easily pass back into the cell, but K^+ pumped into the neuron can diffuse out.

Potassium ions leak out through passive ion channels following their concentration gradient. As these positively charged ions diffuse out of the neuron, they increase the positive charge in the extracellular fluid outside the cell relative to the charge inside the cell. The resultant change in the electrical gradient across the membrane influences the flow of ions. This electrical gradient forces positively charged potassium ions back into the cell.

The membrane potential at which the flux of K^+ inward (due to the electrical gradient) equals the flux of K^+ outward (due to the concentration gradient) is the **equilibrium potential** for K^+. The equilibrium potential for any particular ion is a steady state in which opposing chemical and electrical fluxes are equal and there is no net movement of the ion. The equilibrium potential for K^+ in the typical neuron is −80 mV. The equilibrium potential for Na^+ is +40 mV; it differs from the equilibrium potential of K^+ because of the difference in concentration of Na^+ across the membrane (concentration is high on the outside and low inside the cell). Because the membrane is much more permeable to K^+ than to Na^+, the resting potential of the neuron is closer to the potassium equilibrium potential than to the equilibrium potential of sodium. (The resting potential of the neuron is about −70 mV.)

Although the resting potential is primarily established by K^+, and less so by Na^+, chloride ions also contribute slightly because the plasma membrane is permeable to negatively charged chloride ions. These ions accumulate in the cytosol close to the plasma membrane. Also contributing to the negative charge in the cytosol are negative charges on large molecules such as proteins. These large anions cannot cross the plasma membrane.

The gradients that determine the resting potential are maintained by ion pumping. The neuron plasma membrane has very efficient **sodium-potassium pumps** that actively transport sodium ions out of the cell and potassium ions into the cell (see Fig. 5–15). Both sodium and potassium ions are pumped against their concentration and electrical gradients, and these pumps require ATP. For every three sodium ions pumped out of the cell, two potassium ions are pumped in. Thus, more positive ions are pumped out than in. The sodium-potassium pumps maintain a higher concentration of K^+ inside the cell than outside, and a higher concentration of Na^+ outside than inside.

In summary, the development of the resting potential depends mainly on the diffusion of K^+ out of the cell. However, once the neuron reaches a steady state, Na^+ diffusion into the cell is greater than K^+ diffusion out of the cell. This situation is offset by the sodium-potassium pumps that pump three Na^+ out of the cell for every two K^+ pumped in.

Graded local signals vary in magnitude

Neurons are excitable cells. They have the ability to respond to stimuli and to convert stimuli into nerve impulses. An electrical, chemical, or mechanical stimulus may alter the resting potential by increasing the membrane's permeability to sodium ions.

When a stimulus causes the membrane potential to become less negative (closer to zero) than the resting level, the membrane is described as **depolarized.** Because depolarization brings a neuron closer to transmitting a neural impulse, it is described as *excitatory.* In contrast, when the membrane potential becomes more negative than the resting potential, the membrane is **hyperpolarized.** Hyperpolarization decreases the ability of the neuron to generate a neural impulse and is described as *inhibitory.*

A stimulus may result in a change in potential in a relatively small region of the plasma membrane. Such a **graded potential** is a local response that can function as a signal only over a very short distance, because it fades out within a few millimeters of its point of origin. A graded potential varies in magnitude; i.e., the potential charge varies depending on the strength of the stimulus applied.

An action potential is generated by an influx of Na^+ and an efflux of K^+

If a stimulus is strong enough, a neuron fires a nerve impulse, or **action potential.** An action potential is an electrical excitation that travels rapidly down the axon into the synaptic terminals. All cells can generate graded potentials, but only neurons, muscle cells, and a few other cell types (certain cells of the endocrine and immune systems) can generate action potentials.

In a series of experiments in the 1940s, pioneering researchers Alan Hodgkin and Andrew Huxley inserted electrodes into large axons found in squids. By measuring voltage changes as they varied ion concentrations, they demonstrated that passage of Na^+ ions into the neuron and K^+ ions out of the neuron resulted in an action potential.

We now know that specific **voltage-activated ion channels** (also called **voltage-gated channels**) in the plasma membrane of the axon and cell body are regulated by changes in voltage. These channels have been studied using the *patch clamp technique,* also known as single-channel recording (see Fig. 5–16). Using this technique, biologists are able to measure currents across a very small segment of the plasma membrane rather than across the entire membrane. The patch of membrane measured is so small that it may contain a single channel.

In their investigations of membrane channels and how they function, biologists have made use of certain toxins that affect the nervous system. For example, the poison tetrodotoxin (TTX) binds to voltage-activated sodium channels, blocking the passage of Na^+. Using TTX and other toxins, researchers were able to identify the channel protein. TTX was first isolated from the Japanese pufferfish *(Fugu rubripes),* which is eaten as a delicacy (Fig. 39–5). In small amounts TTX tingles the taste buds, but in a larger dose it can prevent breathing. (The Japanese government operates a certification program for pufferfish chefs!)

Investigators have established that voltage-activated ion channels have charged regions that act as gates. Voltage-activated Na^+ channels have two gates—an *activation gate* and an *inactivation gate* (see Fig. 39–6*a*). Voltage-activated K^+ channels have a single gate (see Fig. 39–6*b*).

An action potential is generated when the voltage reaches a certain critical point known as the **threshold level.** The membrane of most neurons can depolarize by about 15 mV, that is, to a potential of about −55 mV, without initiating an action potential. However, when the extent of depolarization is greater than −55 mV, the threshold level is reached, and an action potential is generated. The neuron membrane quickly reaches zero potential and even **overshoots** to +35 mV or more as a momentary reversal in polarity takes place. The sharp rise and fall of the action potential is referred to as a spike. Figure 39–7 illustrates an action potential that has been recorded by placing one electrode inside an axon and one just outside.

At resting potentials, Na^+ activation gates are closed, and Na^+ cannot pass through them into the neuron (Fig. 39–7*a*). When the voltage reaches the threshold level, the protein making up the channel changes shape, opening the activation gate.

Figure 39–5 Japanese pufferfish *(Fugu rubripes).* A toxin, tetrodotoxin, isolated from this fish binds to voltage-activated sodium ion channels, allowing neurobiologists to study them. *(Jeffrey L. Rotman/Peter Arnold, Inc.)*

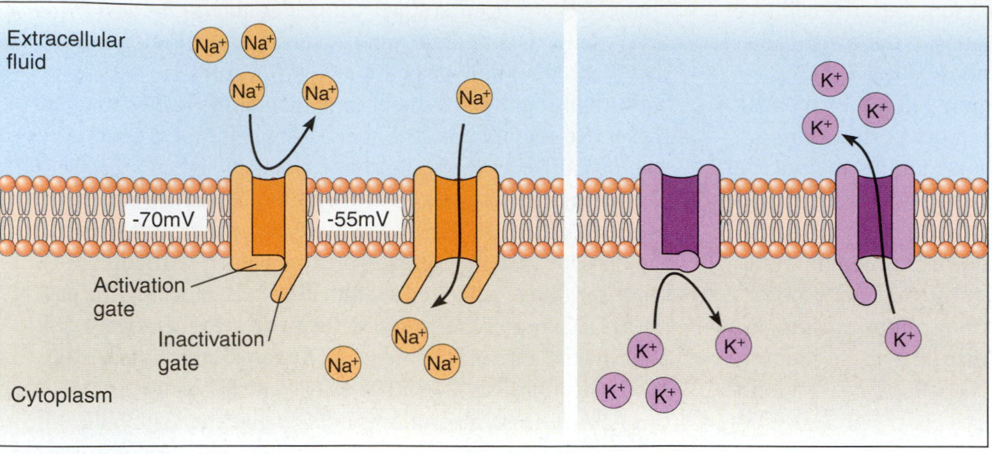

(a) Sodium channels (b) Potassium channels

Figure 39–6 Voltage-activated ion channels. **(a)** In the resting state, voltage-activated Na^+ channels are closed. When the voltage reaches threshold level, the voltage-activated gates open, allowing Na^+ to pass into the cell. After a certain amount of time elapses, these gates close, blocking the channels. Inactivation gates are open when a neuron is in the resting state. These gates close slowly in response to depolarization. The ion channel is open only during the brief period when both gates are open. **(b)** Potassium channels have activation gates that open slowly in response to depolarization. These channels have no deactivation gates.

When the gate opens, Na^+ flow through the channel and the inside of the neuron becomes positively charged relative to the extracellular fluid outside the plasma membrane (Fig. 39–7*b*). Inactivation gates that are open in the resting neuron close slowly in response to depolarization. The Na^+ ion channel is open only during the brief period when both activation and inactivation gates are open.

The generation of an action potential depends on a *positive feedback mechanism.* As just discussed, when a neuron is depolarized, voltage-activated sodium channels open, increasing the membrane permeability to Na^+. Sodium ions diffuse into the cell, moving from an area of higher concentration to an area of lower concentration. As Na^+ flows into the cell, the cytosol becomes positively charged. This charge further depolarizes the neuron so that more voltage-gated sodium channels open, further increasing membrane permeability to sodium. Note that not all voltage-activated channels in a neuron have precisely the same threshold. When some channels open, depolarization increases, leading to the opening of more channels by a positive feedback mechanism. After a certain period, inactivation gates close the channels.

When, after a certain period, inactivation sodium gates close, the membrane again becomes impermeable to sodium. This inactivation begins the process of **repolarization.** Voltage-gated potassium ion channels slowly open in response to depolarization. When potassium ion channels open, potassium leaks out of the neuron (Fig. 39–7*c*). This decrease in intracellular K^+ returns the interior of the membrane to its relatively negative state, repolarizing the membrane. Potassium channels remain open until the resting potential has been restored (Fig. 39–7*d*). As the wave of depolarization moves down the membrane of the neuron, the normal polarized state is quickly reestablished behind it. The entire mechanism—depolarization and then repolarization—can take place in less than 1 millisecond.

During the millisecond or so in which it is depolarized, the axon membrane is in an **absolute refractory period:** It cannot transmit another action potential no matter how great a stimulus is applied. This is because the voltage-activated Na^+ channels are inactivated. Until their gates are reset, they cannot be reopened. When enough Na^+ channel gates have been reset, the neuron enters a **relative refractory period** that lasts for a few additional milliseconds. During this period, the axon can transmit impulses, but the threshold is higher (i.e., more negative). Even with the limits imposed by their refractory periods, most neurons can transmit several hundred impulses per second.

Local anesthetics such as procaine (novocaine), lidocaine (Xylocaine), and also cocaine bind to voltage-activated sodium ion channels and block them. The channels are not able to open in response to depolarization, and the neuron cannot transmit an impulse through the anesthetized region. Signals do not reach the brain so pain is not experienced.

The action potential is an all-or-none response

Any stimulus too weak to depolarize the plasma membrane to threshold level cannot fire the neuron. It merely sets up a local signal that fades and dies within a few millimeters from the point of stimulus. Only a stimulus strong enough to depolarize the membrane to its critical threshold level results in transmission of an impulse along the axon. An action potential is an **all-or-**

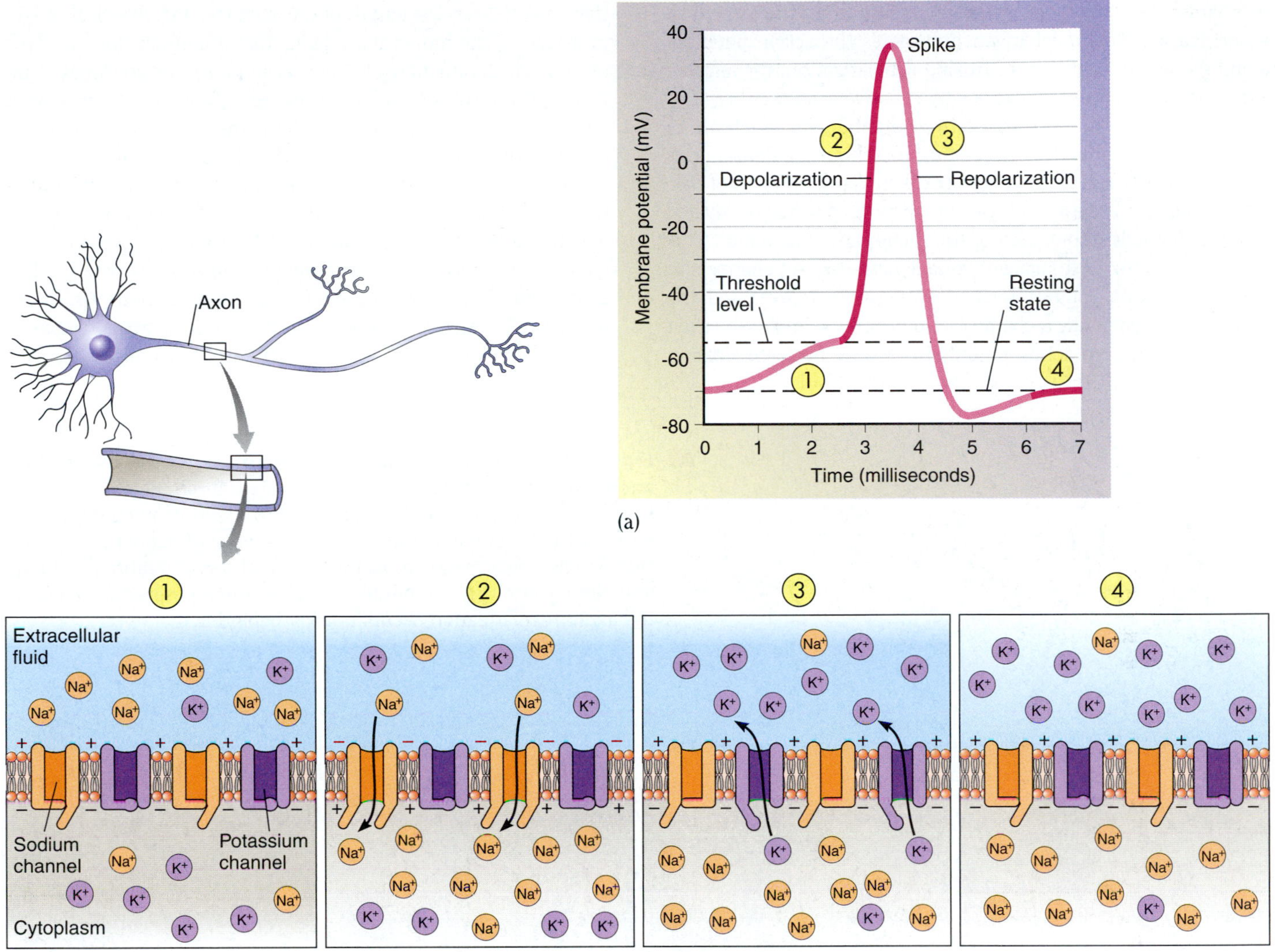

(b) Resting state.
Voltage-activated Na^+ and K^+ channels are closed.

(c) Depolarization.
Voltage-activated Na^+ channels open. Na^+ ions enter cell; inside of neuron becomes positive relative to outside.

(d) Repolarization.
Voltage-activated Na^+ channels close; K^+ channels are open; K^+ moves out of cell, restoring negative charge to inside of cell.

(e) Return to resting state.
Voltage-activated Na^+ and K^+ channels close.

Figure 39–7 Action of voltage-activated ion channels. **(a)** Action potential. When the axon depolarizes to about −55 mV, an action potential is generated. (The numerical values vary for different nerve cells.) **(b)** In the resting state, voltage-activated ion channels are closed. **(c)** At threshold, voltage-activated Na^+ channels open and Na^+ enters the neuron, causing further depolarization. K^+ channels slowly begin to open. **(d)** Repolarization begins when the Na^+ channels close and voltage-activated K^+ channels open, permitting K^+ to leave the cell. **(e)** Resting conditions are restored.

none response because either it occurs or it does not. No variation exists in the strength of a single impulse.

If the all-or-none law is valid, how can we explain differences in the levels of intensity of sensations? After all, we have no difficulty distinguishing between the pain of a severe toothache and that of a minor cut on the arm. This apparent inconsistency is explained by the fact that intensity of sensation depends on the *number of neurons stimulated and on their frequency of discharge.* For example, suppose you burn your hand. If a large area is burned, more pain receptors are stimulated and more neurons become depolarized. Also, each neuron transmits more action potentials per unit of time.

An action potential is self-propagating

Once initiated, the action potential is self-propagating. Recall that when the inside of the membrane becomes more positive relative to the outside during depolarization, a positive feedback mechanism intensifies the changing condition. In the same way, a positive feedback mechanism propagates the action potential along the length of the axon. During depolarization, the affected area of the membrane is more positive relative to adjacent regions where the membrane is still at resting potential. The difference in potentials between active and resting membrane regions causes ions to flow between them (an electrical current). The flow of Na^+ into the adjacent region induces the opening of voltage-activated sodium channels in one area of the membrane (Fig. 39–8). This action causes the next adjacent voltage-activated ion channels to open, permitting sodium ions to enter in that area, and the process is repeated like a chain reaction until the end of the axon is reached. Thus, an action potential is a **wave of depolarization** that travels down the length of the axon (see Fig. 39–8). It is self-propagating and moves at a velocity that is constant for each type of neuron. Conduction of a neural impulse is sometimes compared with burning a trail of gunpowder. Once the gunpowder is ignited at one end of the trail, the flame moves steadily to the other end by igniting the powder particles ahead of it.

Compared with the flow of electrons in electrical wiring or the speed of light, a nerve impulse travels rather slowly. Most axons transmit impulses at about 1 to 10 m per second. The smooth, progressive impulse transmission just described, called **continuous conduction,** occurs in unmyelinated neurons. In unmyelinated axons, the speed of transmission is proportional to the diameter of the axon. Axons with larger diameters transmit

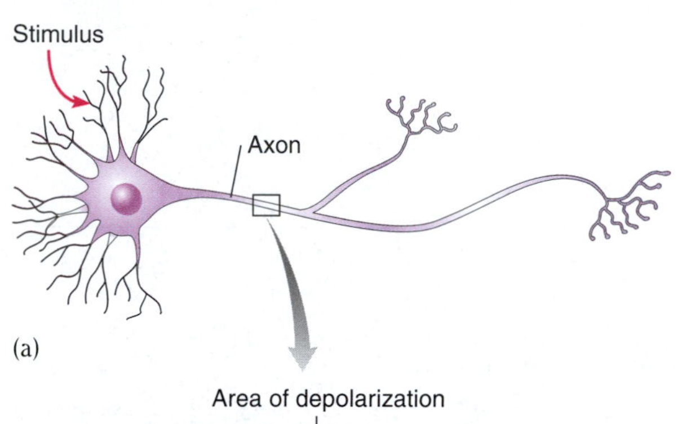

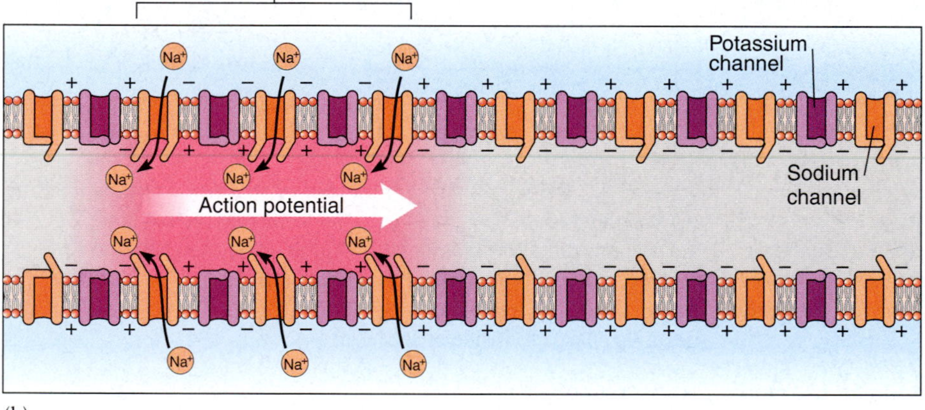

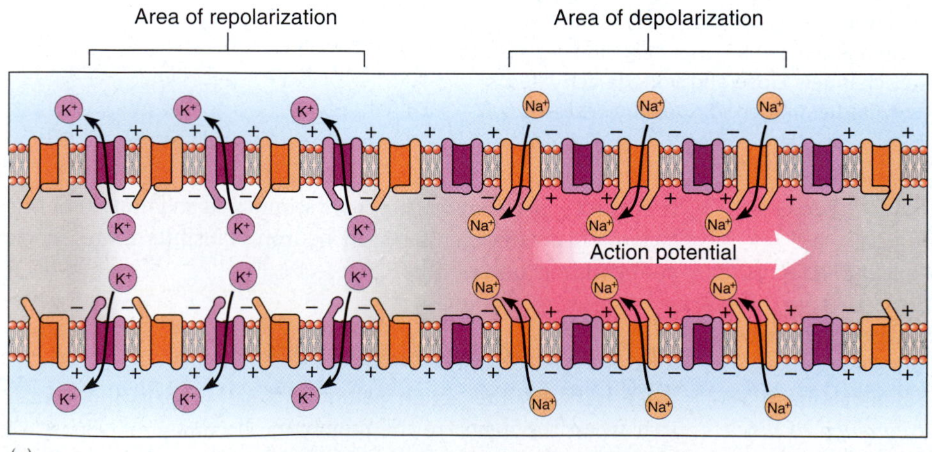

Figure 39–8 Transmission of an action potential along the axon. **(a)** The dendrites (or cell body) of a neuron are stimulated sufficiently to depolarize the membrane to its firing level. **(b)** and **(c)** An action potential is transmitted as a wave of depolarization that travels down the axon. At the region of depolarization, sodium ions diffuse into the cell. As the action potential progresses along the axon, repolarization occurs quickly behind it.

more rapidly than those with smaller diameters. This is because an axon with a large diameter presents less internal resistance to the ions flowing along its length. Squids and certain other invertebrates have giant axons, up to 1 mm in diameter, that permit them to respond rapidly when escaping from predators.

In vertebrates, another strategy has evolved that speeds transmission—myelinated neurons. Myelin acts as an effective electrical insulator around the axon except at the nodes of Ranvier, which are not myelinated. The axon plasma membrane makes direct contact with the surrounding extracellular fluid only at the nodes, and voltage-activated sodium and potassium ion channels are concentrated there. Ion movement across the membrane occurs only at the nodes. Because the ion activity at the active node results in diffusion of ions that depolarize the next node along the axon, the action potential jumps from one node of Ranvier to the next (Fig. 39–9). This type of impulse transmission is known as **saltatory conduction** (from the Latin word *saltus,* which means "to leap").

In myelinated neurons, the distance between successive nodes of Ranvier affects speed of transmission: When nodes are farther apart, less of the axon must depolarize, and the axon conducts the impulse faster. Using saltatory conduction, a myelinated axon can conduct an impulse up to 50 times faster than the fastest unmyelinated axon. Saltatory conduction has another advantage over continuous conduction: It requires less energy. Ions move across the plasma membrane only at the nodes, so fewer sodium and potassium ions are displaced. As a result, the sodium-potassium pumps do not have to expend as much ATP to reestablish resting conditions each time an impulse is conducted.

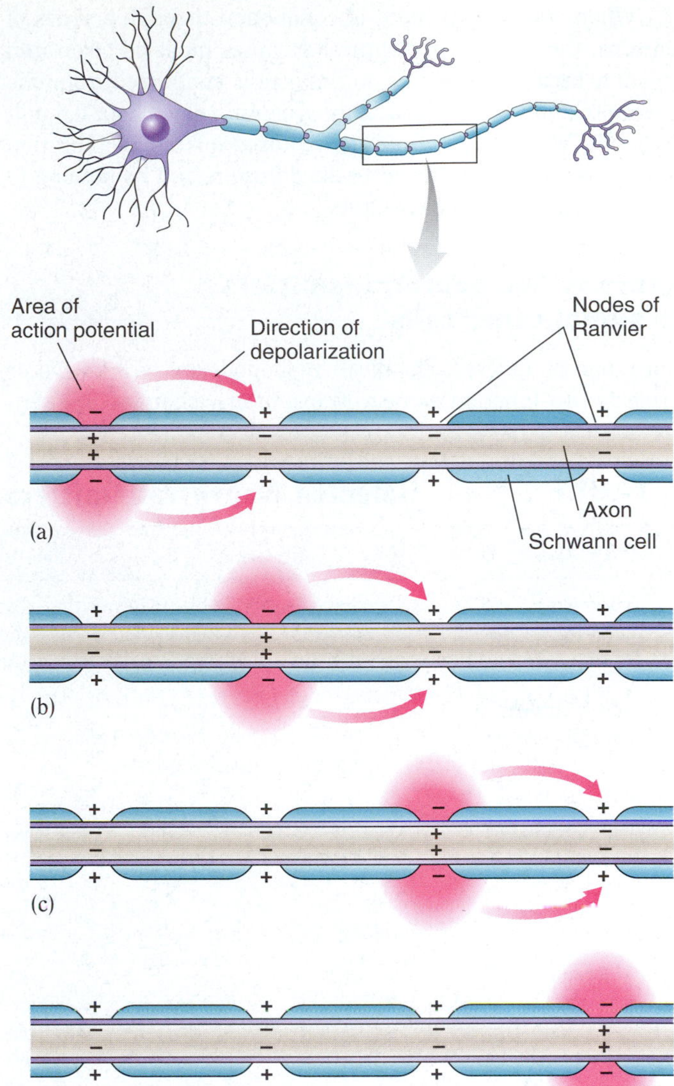

Figure 39–9 Saltatory conduction. In a myelinated axon, action potentials leap along from one node of Ranvier to the next.

NEURONS SIGNAL OTHER CELLS ACROSS SYNAPSES

A **synapse** is a junction between two neurons or between a neuron and an effector. You may recall from Chapter 38 that the synapse between a neuron and a muscle cell is referred to as a **neuromuscular junction.** A neuron that *terminates* at a specific synapse is referred to as a **presynaptic neuron,** whereas a neuron that *begins* at that synapse is known as a **postsynaptic neuron.** Note that these terms are relative to a specific synapse. A neuron that is postsynaptic with respect to one synapse may be presynaptic to the next synapse in the sequence.

Signals across synapses can be electrical or chemical

Based on how presynaptic and postsynaptic neurons communicate, two types of synapses have been identified: **electrical synapses** and **chemical synapses.** In electrical synapses, the presynaptic and postsynaptic neurons occur very close together (within 2 nm of one another) and form *gap junctions* (see Chapter 5). The interiors of the two cells are physically connected by a protein channel. Electrical synapses allow the passage of ions from one cell to another, permitting an impulse to be directly and rapidly transmitted from a presynaptic to a postsynaptic neuron. The escape responses of many invertebrates and vertebrates involve electrical synapses. For example, the "tail-flick" escape response of the crayfish involves giant neurons in the nerve cord that form electrical synapses with large motor neurons that then signal muscles in the abdomen, causing them to contract.

The majority of synapses are chemical synapses. Presynaptic and postsynaptic neurons are separated by a space, the **synaptic cleft,** about 20 nm wide (less than one-millionth of an inch). Because depolarization is a property of the plasma membrane, when an action potential reaches the end of the axon it is unable to jump the gap. The electrical signal must be converted into a chemical one. Neurotransmitters are the chemical messengers that conduct the neural signal across the synapse and bind to chemically activated ion channels in the membrane of the postsynaptic neuron. When a postsynaptic neuron reaches threshold depolarization, it transmits an action potential.

When considering speed of conduction through a series of neurons, the number of chemical synapses must be taken into account, because each time an impulse is conducted from one neuron to another there is a slight synaptic delay (about 0.5 millisecond). This delay is the time required for the mobilization and release of neurotransmitter, its diffusion, and its binding to postsynaptic membrane receptors.

Neurons use neurotransmitters to signal other cells

More than 60 different chemical compounds are now known or suspected to function as neurotransmitters, chemical messengers used by neurons to signal other neurons or to signal muscle or gland cells (Table 39–1). Other compounds function as **neuromodulators,** messengers that modify the effects of specific neurotransmitters. For example, some neuromodulators amplify or dampen the response by the postsynaptic cell.

In Chapter 38, we discussed **acetylcholine,** a low-molecular-weight neurotransmitter that is released from motor neurons and triggers muscle contraction. Acetylcholine is also released by some neurons in the brain and in the autonomic nervous system (see Chapter 40). Cells that release this neurotransmitter are referred to as **cholinergic neurons.**

Neurons that release **norepinephrine** are called **adrenergic neurons.** Norepinephrine and the neurotransmitters **serotonin**

TABLE 39–1 Selected Neurotransmitters

Substance	Source	Actions	Comments
Acetylcholine	Some neurons in the brain and certain neurons in the autonomic nervous system; motor neurons	Excitatory effect on skeletal muscle; inhibitory effect on cardiac muscle	Inactivated by acetylcholinesterase
Biogenic Amines			
Norepinephrine	Neurons in both the central and autonomic nervous systems	Excitatory or inhibitory effect; may be involved in dreaming during REM sleep	Mainly inactivated by reuptake into the axon terminal; also inactivated by enzymes such as monoamine oxidase (MAO); level in brain affects mood
Dopamine	Central nervous system (CNS)	Helps maintain balance between inhibition and excitation of neurons; important in motor functions	Level in brain affects mood; amount reduced in Parkinson's disease; abnormal amounts in schizophrenia and attention deficit disorder
Serotonin (5-hydroxytryptamine, 5-HT)	CNS	Excitatory effect on pathways that control muscle action; inhibitory effect on pathways involved in sensations	Highest level of secretion during alert states, lowest level during sleep; serotonin pathways also function in regulation of food intake and affect mood
Amino Acids			
Glutamate	Brain (major excitatory neurotransmitter in the brain)	*In vertebrates:* excitatory effect on pathways in the brain *In invertebrates:* excitatory effect at neuromuscular junctions	Functions in memory and learning
Aspartate	CNS	Excitatory effect on pathways in the CNS	Functions in memory and learning
Glycine	Inhibitory interneurons in the CNS	Inhibitory effect on pathways in the CNS	
Gamma-aminobutyric acid (GABA)	Inhibitory interneurons in CNS	*In vertebrates:* inhibitory effect on pathways in the CNS *In invertebrates:* inhibitory effect at neuromuscular junctions	Actions of GABA enhanced by barbiturates and benzodiazepines
Neuropeptides			
Endorphins (opioids)	Brain	Pain regulation	Morphine-like properties
Enkephalins (opioids)	Brain	Pain regulation	
Substance P	Sensory neurons	Pain regulation	
Gaseous Neurotransmitter			
Nitric oxide (NO)	Most neurons?	Retrograde messenger; transmits signals in opposite direction from other neurotransmitters, i.e., from postsynaptic to presynaptic neuron	Plays role in learning; modulates sensory and motor responses

and **dopamine** belong to a class of compounds called **biogenic amines,** or **catecholamines.** Biogenic amines affect mood, and their imbalance has been linked to several mental disorders including major depression, attention deficit disorder (ADD), and schizophrenia. Antidepressants and many other mood-affecting drugs work by altering the levels of biogenic amines in the brain.

The major excitatory neurotransmitter in the brain is the amino acid **glutamate.** One type of glutamate receptor is the target of several mind-altering drugs, including phencyclidine ("angel dust").

The amino acid **glycine** and **gamma-aminobutyric acid (GABA),** a modified amino acid, are neurotransmitters that inhibit neurons in the brain and spinal cord. Drugs that reduce anxiety such as the benzodiazepines (e.g., diazepam [Valium] and alprazolam [Xanax]) and barbiturates (e.g., phenobarbital) enhance the actions of GABA.

Opiate drugs, for example, morphine and codeine, are powerful analgesics, drugs that relieve pain without causing loss of consciousness. The body has its own endogenous opioids **enkephalin** and **beta-endorphin,** neuropeptides that bind to opioid receptors (the same receptors to which opiate drugs bind) and block pain signals. Opioids modulate the effect of other neurotransmitters. (Pain perception is discussed in Chapter 41.)

Nitric oxide (NO), a signaling molecule that has been the focus of much recent research, is a gas. NO may act as a retrograde messenger which means that it may transmit information from the postsynaptic neuron to the presynaptic neuron, the opposite direction of other neurotransmitters.

Many neurotransmitters, including acetylcholine, the biogenic amines, and amino acids, are small molecules that act rapidly. These low-molecular-weight neurotransmitters are synthesized within synaptic terminals. Mitochondria provide the ATP required for this synthesis. Enzymes needed for neurotransmitter synthesis are produced in the cell body and transported down the axon to the synaptic terminals. Higher-molecular-weight neuropeptide neurotransmitters, such as the opioids, are synthesized in the cell body.

Neurotransmitters bind with receptors on postsynaptic cells

Neurotransmitters are stored in the synaptic terminals within small membrane-bounded sacs called **synaptic vesicles.** Each time an action potential reaches a synaptic terminal, voltage-gated calcium channels open. Calcium ions from the extracellular fluid then flow into the synaptic terminal. The Ca^{2+} ions cause synaptic vesicles to fuse with the presynaptic membrane and release neurotransmitter molecules into the synaptic cleft by exocytosis (Fig. 39–10).

Neurotransmitter molecules diffuse across the synaptic cleft and combine with specific receptors on the dendrites or cell bodies of postsynaptic neurons (or on the plasma membranes of effector cells). Identification of these receptors is an area of intense biomedical research. Many neurotransmitter receptors are chemically activated ion channels known as **ligand-gated ion channels.** When the neurotransmitter, the **ligand,** binds with the receptor, the ion channel opens. The acetylcholine receptor, for example, is an ion channel that permits the passage of Na^+ and K^+.

Some neurotransmitters, such as serotonin, operate via a different mechanism. They work indirectly through the production or intracellular release of a **second messenger.** Binding of the neurotransmitter with a receptor activates a **G protein.** The G protein then activates an enzyme, such as adenylyl cyclase, in the postsynaptic membrane. Adenylyl cyclase converts ATP to **cyclic AMP (cAMP),** which acts as a second messenger (see Chapters 5 and 47). Cyclic AMP activates a protein kinase that phosphorylates a protein that closes K^+ channels.

If repolarization of a postsynaptic neuron is to occur quickly, any excess neurotransmitter in the synaptic cleft must be removed. Some neurotransmitters are inactivated by enzymes. For example, excess acetylcholine is degraded into its chemical components, choline and acetate, by the enzyme acetylcholinesterase. Other neurotransmitters, for example, the biogenic amines, are actively transported back into the synaptic terminals, a process known as **reuptake.** These neurotransmitters are repackaged in vesicles and recycled.

Many drugs inhibit the reuptake of neurotransmitters. For example, some antidepressants (selective serotonin reuptake inhibitors, or SSRIs, such as fluoxetine [Prozac]) work by inhibiting the reuptake of serotonin, thus increasing its concentration in the synaptic cleft. Cocaine inhibits the reuptake of dopamine.

Because NO diffuses through plasma membranes, it is not packaged in vesicles like most other neurotransmitters. Nitric oxide is also different in that it affects receptors inside the neuron rather than on the membrane.

Neurotransmitters and their receptors can send excitatory or inhibitory signals

Depending on the type of postsynaptic receptor with which it combines, the same neurotransmitter can have different effects. For example, acetylcholine has an excitatory effect on skeletal muscle. It opens sodium ion channels, which increases the permeability of the muscle fiber membrane to Na^+. The influx of Na^+ depolarizes the membrane, leading to contraction. In contrast, acetylcholine has an inhibitory effect on cardiac muscle, resulting in a decreased heart rate. A postsynaptic neuron may have receptors for several types of neurotransmitters. Some of its receptors may be excitatory and others may be inhibitory.

When neurotransmitter molecules bind to receptors, they directly or indirectly affect ion channels. The resulting redistribution of ions changes the electrical potential of the membrane, and the membrane may become depolarized. If sufficiently intense, a local depolarization can set off an action potential. Alternatively, it may become hyperpolarized, in which case the membrane potential becomes more negative.

For example, when neurotransmitter molecules combine with receptors that open sodium channels, the resulting influx of sodium ions partially depolarizes the membrane. A change in membrane potential that brings the neuron closer to firing is called an **excitatory postsynaptic potential (EPSP)** (Fig. 39–11*a*). Let us say that sufficient sodium ions enter to change

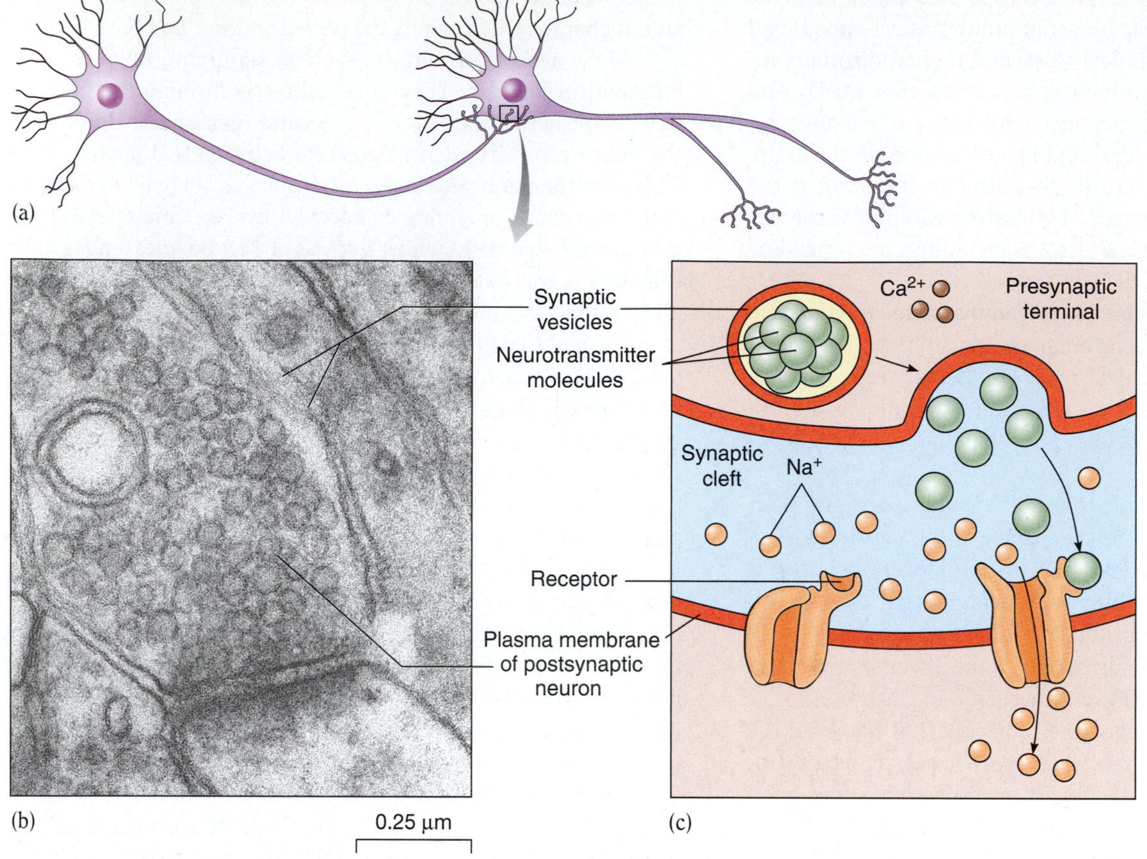

Figure 39–10 Synaptic transmission. Impulses are transmitted between neurons, or from a neuron to an effector by neurotransmitters. **(a)** At most synapses the action potential is unable to jump across the synaptic cleft between the two neurons. The problem is solved by chemical signaling across synapses. **(b)** The TEM shows synaptic terminals filled with synaptic vesicles. **(c)** When an action potential reaches the synaptic terminals at the end of a presynaptic neuron, calcium ions enter the synaptic terminal from the extracellular fluid. The calcium ions cause the synaptic vesicles to fuse with the plasma membrane and release a neurotransmitter into the synaptic cleft. The neurotransmitter diffuses across the synaptic cleft and combines with receptors in the membrane of the postsynaptic neuron. This changes the permeability of the postsynaptic membrane to certain ions, resulting in either depolarization or hyperpolarization. Sufficient depolarization can trigger an action potential in the postsynaptic neuron. *(b, J.F. Gennaro/Photo Researchers, Inc.)*

the membrane potential, from −70 mV to −60 mV. The membrane would be only −5 mV away from threshold. Under such conditions, an additional relatively weak stimulus, causing a small influx of Na^+, can cause the neuron to fire.

Some neurotransmitters act indirectly to close K^+ channels. When these channels are closed, K^+ cannot diffuse out of the cell. As they accumulate, the inner surface of the plasma membrane becomes more positive, leading to depolarization.

Some neurotransmitter-receptor combinations hyperpolarize the postsynaptic membrane. Because such an action takes the neuron farther away from the firing level, a potential change in this direction is called an **inhibitory postsynaptic potential (IPSP)** (Fig. 39–11*b*). For example, if the membrane potential changes from −70 mV to −80 mV, the membrane is farther away from threshold, and a stronger stimulus will be required to fire the neuron.

Like an EPSP, an IPSP can be produced in several ways. Two types of GABA receptors that produce IPSPs are known. Binding of GABA to a $GABA_B$ receptor opens K^+ channels. As K^+ diffuses out, the neuron becomes more negative, hyperpolarizing the membrane. Activated $GABA_A$ receptors produce IPSPs by opening Cl^- channels. In this case, the influx of negative ions hyperpolarizes the membrane. Although the mechanisms are different, the inhibiting effect of GABA is the same in both cases.

When certain types of receptors on the postsynaptic neuron are activated, the reactivity of the neuron can be affected for a long time, even years. Such long-term effects result from the actions of certain neuromodulators. They appear to work by affect-

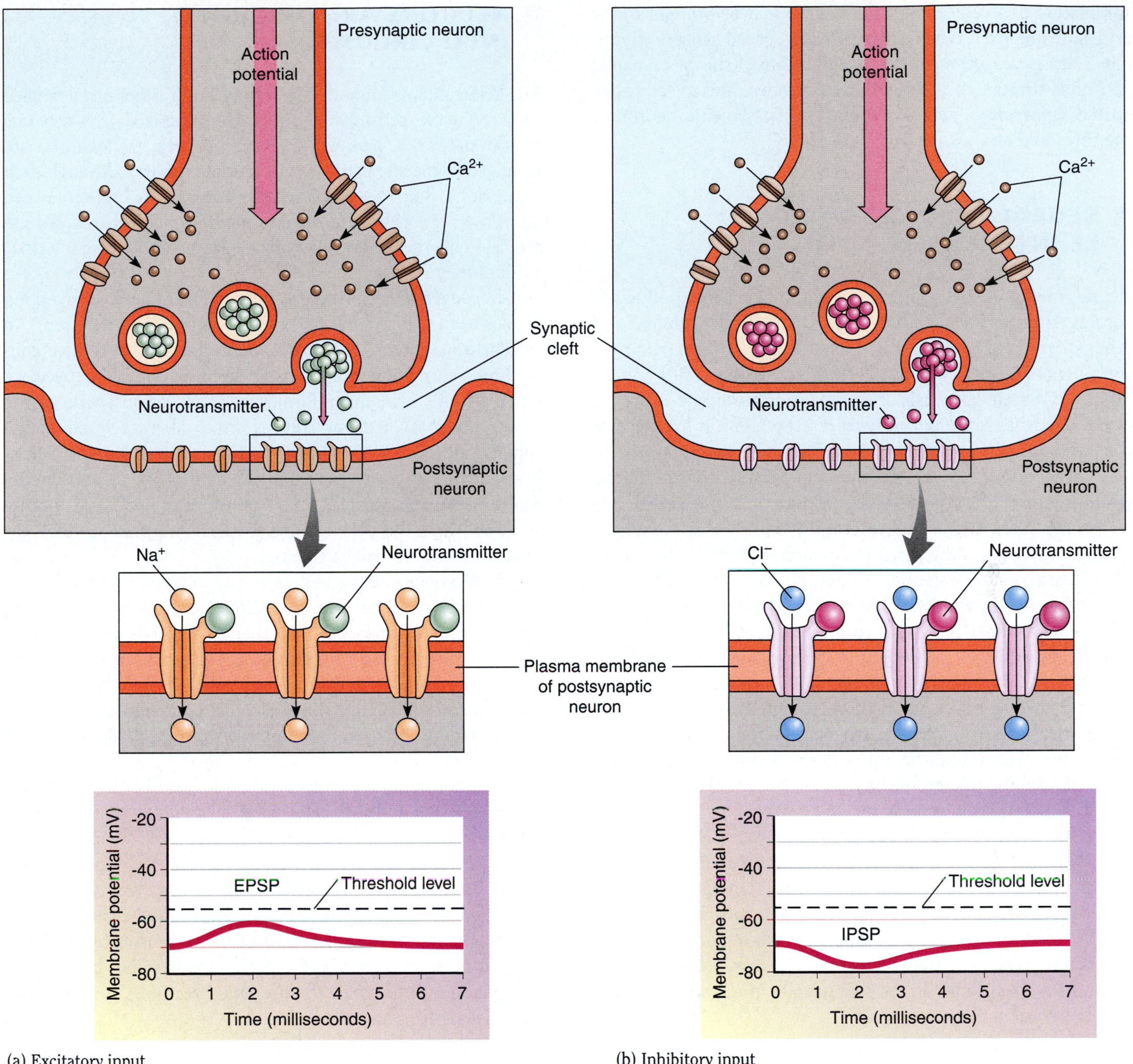

(a) Excitatory input

(b) Inhibitory input

Figure 39–11 Comparison of (a) an excitatory postsynaptic potential (EPSP) and (b) an inhibitory postsynaptic potential (IPSP). Note that a neurotransmitter that opens Na^+ channels generates EPSPs that bring the neuron closer to firing. A neurotransmitter that opens Cl^- channels generates IPSPs, thus hyperpolarizing the membrane.

ing genes or activating enzymes that increase or decrease the number of receptors. These mechanisms are discussed further in Chapter 40 in connection with the process of learning.

Each EPSP or IPSP is a graded potential that varies in magnitude depending on the strength of the stimulus applied. One EPSP is usually too weak to trigger an action potential by itself. Its effect is below threshold level. Even though subthreshold EPSPs do not produce an action potential, they do affect the membrane potential. EPSPs may be added together in a process known as **summation.**

Temporal summation occurs when repeated stimuli cause new EPSPs to develop before previous EPSPs have decayed. By summation of several EPSPs, the neuron may be brought to the critical firing level. When several closely spaced synaptic terminals release neurotransmitter simultaneously, the postsynaptic neuron is stimulated at several places at once. This effect, called **spatial summation,** can also bring the postsynaptic neuron to the threshold level.

NEURAL IMPULSES MUST BE INTEGRATED

Neural integration is the process of summing, or integrating, incoming signals. Each neuron may synapse with hundreds of other neurons. Indeed, as much as 40% of a postsynaptic neuron's dendritic surface and cell body may be covered by thousands of synaptic terminals of presynaptic neurons. It is the job of the dendrites and the cell body of every neuron to integrate the hundreds of messages that continually bombard them.

EPSPs and IPSPs are produced continually in postsynaptic neurons, and IPSPs cancel the effects of some of the EPSPs. It is important to remember that each EPSP and IPSP is not an all-or-none response. Rather, each is a graded response that does not travel like an action potential but may be added to or subtracted from other EPSPs and IPSPs. As the postsynaptic neuron membrane continuously updates its molecular tabulations, the neuron may be inhibited or brought to threshold level. This mechanism provides for integration of hundreds of signals (EPSPs and IPSPs) before an all-or-none action potential is actually transmitted along the axon of a postsynaptic neuron. Local responses permit the neuron and the entire nervous system a far greater range of response than would be the case if every EPSP generated an action potential.

Where does neural integration take place? Every neuron sorts through (on a molecular level) the hundreds and thousands of bits of information simultaneously bombarding it. In vertebrates more than 90% of the neurons in the body are located in the CNS, so most neural integration takes place there, within the brain and spinal cord. These neurons are responsible for making most of the "decisions." In the next chapter, the brain and spinal cord are examined in some detail.

NEURONS ARE ORGANIZED INTO CIRCUITS

The CNS contains millions of neurons, but it is not just a tangled mass of nerve cells. Its neurons are organized into separate **neural networks,** and within each network the neurons are arranged in specific pathways, or **neural circuits.** Although each network has some special characteristics, the neural circuits in all of the networks share many organizational features. For example, convergence and divergence are probably characteristic of all of them.

In **convergence,** a single neuron is controlled by converging signals from two or more presynaptic neurons (Fig. 39–12*a*). An interneuron in the spinal cord, for instance, may receive converging information from sensory neurons entering the cord, from neurons bringing information from various parts of the brain, and from neurons coming from different levels of the spinal cord. Information from all these sources has to be integrated before an action potential can be sent and an appropriate motor neuron stimulated. Convergence is an important mechanism by which the CNS can integrate the information that impinges on it from various sources.

In **divergence,** a single presynaptic neuron stimulates many postsynaptic neurons (Fig. 39–12*b*). Each presynaptic neuron may branch and synapse with as many as 5000 or more different postsynaptic neurons. For example, a single neuron transmitting an impulse from the motor area of the brain may synapse with hundreds of interneurons in the spinal cord, and each of these may diverge in turn, so that hundreds of muscle fibers may be stimulated.

Another important type of neural circuit is the **reverberating circuit,** a neural pathway arranged so that an axon collateral synapses with an interneuron (Fig. 39–13). The interneuron synapses with a neuron in a sequence that can send new impulses through the circuit. This is an example of positive feedback. New impulses can be generated again and again until the synapses fatigue (from depletion of neurotransmitter) or until stopped by some sort of inhibition. Reverberating circuits are thought to be important in rhythmic breathing, mental alertness, and perhaps short-term memory.

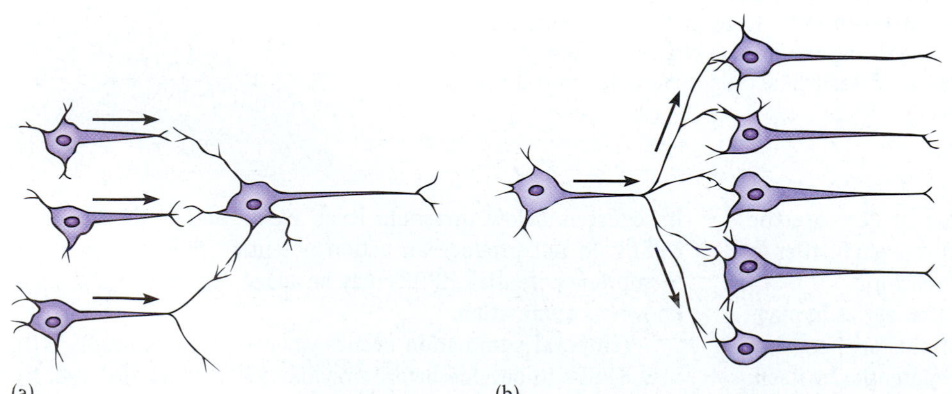

Figure 39–12 Neural circuits. **(a)** Convergence of neural input. Several presynaptic neurons synapse with one postsynaptic neuron. This organization permits one neuron to receive signals from many sources. **(b)** Divergence of neural output. A single presynaptic neuron synapses with several postsynaptic neurons. This organization allows one neuron to communicate with many others.

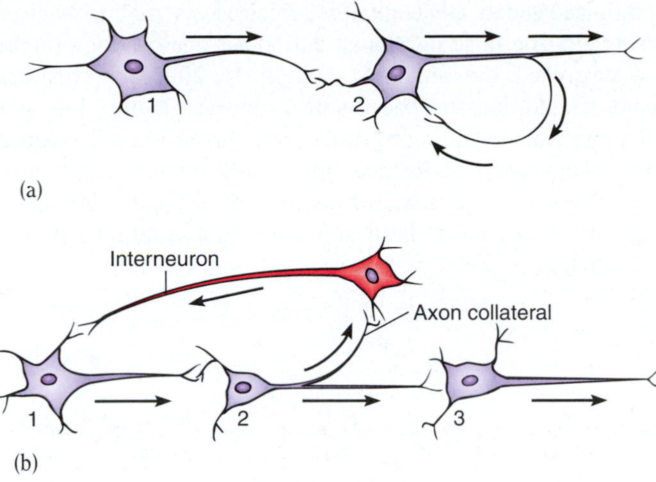

Figure 39–13 A reverberating circuit. A neuron may fire multiple times. **(a)** A simple reverberating circuit in which an axon collateral of the second neuron turns back on its own dendrites, so the neuron continues to stimulate itself. **(b)** In this neural circuit an axon collateral of the second neuron synapses with an interneuron. The interneuron synapses with the first neuron in the sequence. New impulses are triggered again and again in the first neuron, causing reverberation.

UNDERSTANDING NEURAL FUNCTION IS A KEY TO TREATING ALZHEIMER'S DISEASE

We have examined neurons and how they communicate with one another. In **Alzheimer's disease (AD),** a progressive, degenerative brain disorder, neural signaling is disrupted, and neurons are damaged and die prematurely. Alzheimer's disease is the leading cause of senile dementia, which is the loss of memory, judgment, and the ability to reason that may be associated with aging. This disease afflicts more than 4 million people in the United States alone.

The development of AD may be a lifelong process. One startling study demonstrated that a person's writing early in life can predict the disease. Young adults whose writings had a lower density of ideas were more likely to develop the disease. One approach to detecting development of AD is the use of positron emission tomography (PET) scans to study glucose uptake in the brain. Low glucose uptake in certain areas of the brain indicates that neurons in these areas may be damaged.

Researchers must integrate their knowledge of neural signaling, gene function, and proteins to understand AD and develop strategies to prevent and treat it. In AD, neurons are damaged in certain parts of the brain including the cerebral cortex and hippocampus, areas that are important in thinking and remembering. Neurons that secrete the neurotransmitter acetylcholine are especially affected. Investigators have demonstrated that two of the abnormalities that develop in brain tissue as we age, amyloid plaques and neurofibrillary tangles, are especially characteristic of AD. Understanding the biochemistry and genetic basis of both plaques and tangles may provide clues to the causes and cures of AD. As is typical of the scientific process, various groups of researchers are approaching the AD problem in different ways. Some investigators are working on understanding amyloid plaques, while others are focusing on neurofibrillary tangles.

Amyloid plaques were first identified in autopsied brains in 1906 by the German physician Alois Alzheimer. These plaques are extracellular deposits of a peptide called β-amyloid. Neurons near these plaques appear swollen and misshapen. Microglia are typically present, suggesting an attempt to remove the damaged cells or the plaques themselves.

Researchers have found that β-amyloid is cut from a protein called β-amyloid precursor protein, or β-APP, a large transmembrane protein coded by a gene located on chromosome 21. β-Amyloid is cut from the protein in a two-step process. An enzyme called β-secretase cleaves the protein. Then, a second enzyme, gamma-secretase, cleaves the fragment, producing β-amyloid. Normal β-amyloid molecules have 40 amino acids. However, about 10% of the peptides produced have two additional amino acids. This longer peptide appears to form the plaques. Investigators have identified mutations in specific genes that result in the production of abnormal β-amyloid.

Just how the longer β-amyloid leads to AD is not known. The abnormal peptide appears to disrupt calcium regulation, which can lead to the death of neurons in the brain. The peptide may also damage mitochondria, resulting in release of harmful free radicals.

In 2000 Dennis Selkoe, of Harvard Medical School, and his colleagues reported that insulin-degrading enzyme (IDE), a protein found in neurons and microglia, degrades β-amyloid. Selkoe's team is knocking out this gene in mice to determine the effect on β-amyloid. In 2001, Takaomi Saido of the RIKEN Brain Institute in Saitama, Japan, reported that neprilysin, a protease in mice, appears to be a natural β-amyloid–degrading enzyme. To demonstrate that malfunction of either IDE or neprilysin leads to AD, researchers will have to demonstrate that deficiency of the compound increases plaque formation in the AD mouse model.

Another important piece of the AD puzzle appears to be **neurofibrillary tangles,** which form in the neuron cytoplasm. A cytoskeletal protein called **tau** normally associates with the protein tubulin that forms microtubules. Mutations in the *tau* gene may interfere with the way the tau protein binds to microtubules. Neurons that have abnormal tau proteins have abnormal microtubules that cannot effectively transport materials through their axons. Abnormal tau proteins accumulate in the cytoplasm, forming fibrous deposits that make up the neurofibrillary tangles. These tangles interfere with neural signaling.

In 2001 Mel Feany of Harvard Medical School and her colleagues reported that they had used genetic engineering techniques to introduce either a normal or mutant *tau* gene into fruit flies. Presence of the mutant gene resulted in damaged neurons in the brain, and neurons that use acetylcholine to signal other neurons were most affected. This fruit fly model may help researchers in their search for genes and proteins that enhance or block the effects of the tau protein.

Recent findings indicate that the mechanisms that cause amyloid plaques and neurofibrillary tangles are connected. In 2001 two groups of investigators working with transgenic mice independently demonstrated that β-amyloid deposits in the brain influence the formation of neurofibrillary tangles in areas of the brain known to be affected in Alzheimer's disease.

Several drugs designed to slow the progress of Alzheimer's disease are in clinical trials. One of the challenges of developing a drug has been getting the medication through the blood-brain barrier (see Chapter 40). In 2000 Dale Schenk and his colleagues at Elan Pharmaceuticals demonstrated that when they immunized genetically engineered mice with a vaccine made of β-amyloid, the mice developed antibodies against the protein and destroyed the amyloid plaques. In 2001 investigators demonstrated that this vaccine also reduces memory loss and enhances learning in aging mice. Vaccinated mice developed fewer plaques and performed significantly better on memory tests. The vaccine is now in human clinical trials. Future AD drugs will likely target both amyloid plaques and neurofibrillary tangles.

SUMMARY WITH KEY TERMS

I. Neural signaling through the nervous system typically follows this sequence: **reception** of information; **transmission** to **interneurons** in the **central nervous system (CNS)** via **afferent (sensory) neurons; integration** by interneurons in the CNS; transmission from the CNS via **efferent neurons;** and **action by effectors,** the muscles and glands.

II. **Glial cells** and **neurons** are the two types of cells characteristic of neural tissue.
   - A. Glial cells are supporting cells. Vertebrates have three main types of glial cells in the CNS.
     1. **Microglia** are phagocytic cells.
     2. Some **astrocytes** are phagocytic; others help regulate the composition of the extracellular fluid in the CNS.
     3. **Oligodendrocytes** are glial cells that form myelin sheaths around neurons in the CNS.
   - B. Neurons are specialized to receive stimuli and transmit electrical signals.
     1. A typical neuron consists of a **cell body** with many branched **dendrites** and a single long **axon.** Branches at the end of the axon give rise to **synaptic terminals.**
     2. Many axons are surrounded by a **myelin sheath.** Outside the CNS, the myelin sheath is produced by **Schwann cells,** another type of glial cell.
     3. A **nerve** consists of several hundred axons wrapped in connective tissue; a **ganglion** is a mass of neuron cell bodies.

III. Neurons use electrical signals to transmit information.
   - A. In a resting neuron, one that is not transmitting an impulse, the inner surface of the plasma membrane is negatively charged compared with the outside; the membrane is **polarized.** A potential difference of about −70 mV exists across the membrane. This is the **resting potential.**
     1. The magnitude of the resting potential is determined by (1) differences in concentrations of specific ions (mainly Na^+ and K^+) inside the cell relative to the extracellular fluid, and (2) selective permeability of the plasma membrane to these ions.
     2. Ions pass through specific **passive ion channels.** K^+ leaks out more readily than Na^+ can leak in. Cl^- ions accumulate along the inner surface of the plasma membrane. Large anions such as proteins that cannot cross the plasma membrane contribute negative charges.
     3. The membrane potential at which the flow of K^+ into the cell (due to the electrical gradient) equals the flow of K^+ out of the cell (due to the concentration gradient) is the **equilibrium potential** for K^+.
     4. The gradients that determine the resting potential are maintained by **sodium-potassium pumps** that continuously transport three sodium ions out of the neuron for every two potassium ions transported in.
   - B. If a stimulus causes the membrane potential to become less negative, the membrane becomes **depolarized.** If the membrane potential becomes more negative than the resting potential, the membrane is **hyperpolarized.**
   - C. A **graded potential** is a local response that can vary in magnitude depending on the strength of the applied stimulus.
   - D. If the voltage across the membrane is decreased to a critical point, called the **threshold level, voltage-activated ion channels** open, allowing Na^+ to flow into the neuron, and an **action potential** is generated.
     1. The action potential is a **wave of depolarization** that moves down the axon.
     2. As the action potential moves down the axon, **repolarization** occurs behind it.
        - a. The axon enters an **absolute refractory period** during depolarization.
        - b. When enough gates controlling Na^+ channels have been reset, the neuron enters a **relative refractory period.**
     3. The action potential conforms to an **all-or-none law;** no variation exists in the strength of a single impulse. The membrane potential either exceeds threshold level, leading to transmission of an action potential, or it does not.
     4. An action potential is self-propagating.
     5. **Saltatory conduction** takes place in myelinated neurons. In this type of transmission, depolarization skips along the axon from one **node of Ranvier** to the next—sites where the axon is not covered by myelin and where Na^+ channels are concentrated.

IV. The junction between two neurons or between a neuron and effector is a **synapse.** Although there are electrical synapses, most synapses are chemical. Synaptic transmission generally depends on release of a **neurotransmitter** from **synaptic vesicles** in the synaptic terminals of a **presynaptic neuron.**
   - A. Some of the better studied neurotransmitters include **acetylcholine;** the **biogenic amines,** including **norepinephrine, serotonin,** and **dopamine; GABA,** a widespread inhibitory neurotransmitter; and the neuropeptides, which include **beta-endorphin,** and **enkephalin.**
   - B. A neurotransmitter diffuses across the **synaptic cleft** and combines with specific receptors on a **postsynaptic neuron.**
   - C. Many neurotransmitter receptors are proteins that form **ligand-gated ion channels.** Others work through a **second messenger** such as **cAMP.**
   - D. Receptor activation can cause either an **excitatory postsynaptic potential (EPSP)** or an **inhibitory postsynaptic potential (IPSP),** depending on the type of receptor.

V. Each EPSP or IPSP is a graded potential that varies in magnitude depending on the strength of the stimulus applied. The mechanism of neural integration is **summation,** the process of adding and subtracting incoming signals.

VI. Complex **neural circuits** are possible because of **reverberating circuits** and neural associations such as **convergence** and **divergence.**

POST-TEST

1. Summing incoming neural signals is part of (a) reception (b) transmission (c) integration (d) action by effectors (e) afferent neuron transmission
2. Which of the following are phagocytic cells that remove debris from tissue in the CNS? (a) Schwann cells (b) axons (c) oligodendrocytes (d) effectors (e) microglia
3. Most of the neuron cytoplasm is located in the (a) cell body (b) axon (c) dendrites (d) synaptic terminals (e) nodes
4. Action potentials are transmitted to synaptic terminals by the (a) ganglia (b) axon (c) dendrites (d) cell body (e) nodes
5. In responding to a stimulus, a motor neuron signals a(an) (a) efferent neuron (b) receptor (c) afferent neuron (d) interneuron (e) effector
6. The myelin sheath is produced around axons outside the CNS by the (a) axon (b) neuron cell body (c) dendrites (d) Schwann cells (e) synaptic terminals
7. Neurotransmitters are released by the (a) axon (b) neuron cell body (c) dendrites (d) astrocytes (e) synaptic terminals
8. Which of the following does *not* contribute to the resting potential of a neuron? (a) sodium-potassium pumps (b) ion channels (c) differences in concentration of ions across the membrane (d) differences in permeability to certain ions (e) graded potentials
9. Which of the following occurs first when voltage reaches the threshold level? (a) gates of certain voltage-activated ion channels open (b) K^+ channels close (c) the membrane hyperpolarizes (d) neurotransmitter is released (e) neurotransmitter reuptake takes place at the synapse
10. Saltatory conduction (a) requires more energy than continuous conduction (b) occurs in unmyelinated neurons (c) occurs when the action potential jumps from one node of Ranvier to the next (d) slows transmission of an impulse (e) depends on the presence of GABA
11. Acetylcholine (a) is a biogenic amine (b) is recycled (c) is released by motor neurons and by some neurons in the brain (d) is deactivated by G proteins (e) binds to opioid receptors
12. Neurotransmitter receptors often (a) are voltage-activated ion channels (b) permit influx of chloride ions, leading to depolarization of the membrane (c) work through a second messenger (d) inhibit reuptake of the neurotransmitter (e) are passive ion channels
13. IPSPs (a) excite presynaptic neurons (b) excite postsynaptic neurons (c) cancel the effects of some EPSPs (d) release large amounts of neurotransmitters (e) are released by postsynaptic cell bodies
14. A presynaptic neuron in the cerebrum synapses with hundreds of other neurons. This is an example of (a) convergence (b) divergence (c) summation (d) a reverberating circuit (e) a graded potential
15. During a relative refractory period (a) IPSPs are generated (b) opioid neurotransmitters are released (c) voltage-activated sodium channels are inactivated (d) an axon can transmit impulses but the threshold is higher (more negative) (e) an axon cannot transmit an action potential

REVIEW QUESTIONS

1. Distinguish between a neuron and a nerve.
2. Imagine that you are swimming and you suddenly spot a shark fin moving in your direction. What processes must take place within your nervous system before you can make your escape?
3. What is meant by the resting potential of a neuron? How do ion channels and sodium-potassium pumps contribute to the resting potential?
4. Describe the sequence of events that occurs when a neuron exceeds threshold level.
5. Describe the functions of the following: (a) myelin (b) voltage-activated ion channels (c) glial cells (d) dendrites (e) synaptic terminals
6. What is an action potential? How does it differ from a graded potential?
7. Contrast saltatory conduction with conduction in an unmyelinated neuron.
8. Describe the functions and mechanisms of action of the following substances: (a) acetylcholine (b) cholinesterase; (c) norepinephrine (d) GABA
9. Contrast convergence and divergence.
10. What is summation?
11. Label the diagram. Use Figure 39–2 to check your answers.

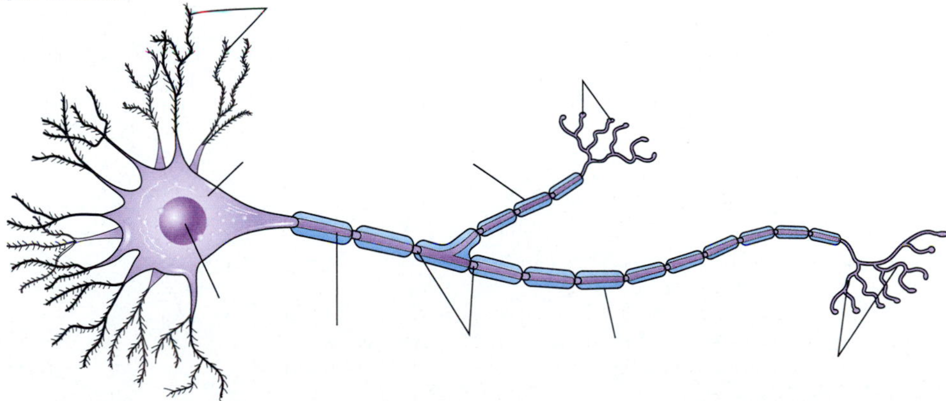

YOU MAKE THE CONNECTION

1. Discuss several specific ways that the nervous system helps maintain homeostasis.
2. Stimulant drugs such as amphetamines increase the activity of the nervous system. Propose two mechanisms involving synaptic transmission that could explain the action of such drugs.
3. Develop a hypothesis to explain the fact that acetylcholine has an excitatory effect on skeletal muscle but an inhibitory effect on cardiac muscle.
4. Investigators have genetically engineered cells to produce acetylcholine, dopamine, GABA, and other neurotransmitters. In what ways might these cells be useful in neurobiological research? How might they be useful in clinical medicine?

RECOMMENDED READINGS

Kandel, E.R., J.H. Schwartz, and T.M. Jessell. *Essentials of Neural Science and Behavior,* 4th ed. McGraw-Hill, New York, 2000. A well-written, comprehensive exploration of neurobiology.

Lee V. M.-Y. "Tauists and Baptists United—Well Almost!" *Science,* Vol. 293, 24 Aug. 2001. A summary of the findings of two independent research teams linking amyloid plaques and neurofibrillary tangles. Their reports are published in the same issue.

Marx, J. "Neurobiology: New Clue to the Cause of Alzheimer's." *Science,* Vol. 292, 25 May 2001. A protease called neprilysin may degrade β-amyloid and thus become a target for new drugs developed to treat Alzheimer's disease.

Marx, J. "Neuroscience: Elusive Protein Auditions for Several New Roles." *Science,* Vol. 293, 6 Jul. 2001. Researchers are working to understand the actions of amyloid precursor protein (APP) which is involved in Alzheimer's disease.

Matthews, G.G. *Neurobiology: Molecules, Cells, and Systems.* Blackwell Science, Malden, MA, 1998. An excellent and very readable discussion of selected topics in neurobiology.

McGuire, S.E., T. Le Phuong, and R.L. Davis. "The Role of *Drosophila* Mushroom Body Signaling in Olfactory Memory." *Science,* Vol. 293, 17 Aug. 2001. Insect brain stuctures, called mushroom bodies, are important during memory retrieval, but not during acquisition or consolidation of memories. Also see Perspective article "Learning How a Fruit Fly Forgets" by S. Waddell and W.G. Quinn in the same issue.

Neher, E., and B. Sakmann. "The Patch Clamp Technique." *Scientific American,* Vol. 266, No. 3, Mar. 1992. The 1991 Nobel Prize–winning authors discuss their technique for isolating ion channels in cell membranes.

Netting, J. "Gray Matters." *Science News,* Vol. 159, 7 Apr. 2001. Scientists debate the role of astrocytes.

St. George-Hyslop, P.H. "Piecing Together Alzheimer's." *Scientific American,* Vol. 283, No. 6, Dec. 2000. A discussion of the molecules, genes, and mechanisms linked to Alzheimer's disease.

Tobin, A.J., and R.E. Morel. *Asking About Cells.* Saunders College Publishing, Philadelphia, 1997. Chapter 23 provides an excellent introduction to neurons.

Vander, A., J. Sherman, and D. Luciano. *Human Physiology,* 8th ed. McGraw-Hill, New York, 2001. Chapter 8 provides a clear introduction to neural control mechanisms.

- Visit our Web site at **http://www.info.brookscole.com/solomonbergmartin** for links to chapter-related resources on the World Wide Web. Additional on-line materials relating to this chapter can also be found on our Web site.

See chapter activity on BioActive Learner CD for additional help in mastering the chapter's material. Icon location in the chapter's margins shows which topics have tutorials or simulations in the CD.

40

Neural Regulation

LEARNING OBJECTIVES

After you have studied this chapter you should be able to

1. Contrast nerve nets and radial nervous systems with bilateral nervous systems.
2. Compare the vertebrate nervous system with a bilateral invertebrate nervous system.
3. Trace the development of the principal vertebrate brain regions from the forebrain, midbrain, and hindbrain, and compare the brains of fishes, amphibians, reptiles, birds, and mammals.
4. Describe the functions and structure of the vertebrate spinal cord.
5. Locate the following parts of the human brain and give the functions of each: medulla, pons, midbrain, thalamus, hypothalamus, cerebellum, and cerebrum. Include a description of the functional areas of the cerebrum.
6. Describe the sleep-wake cycle and contrast rapid eye movement (REM) and non-REM sleep.
7. Describe the actions of the limbic system.
8. Summarize neurophysiological changes that take place during various types of learning, citing experimental evidence to support your descriptions.
9. Describe experimental evidence linking environmental stimuli with changes in the brain and with learning and motor abilities.
10. Describe the organization of the vertebrate nervous system, comparing the central nervous system with the peripheral nervous system and comparing the somatic system with the autonomic system.
11. Contrast the sympathetic and parasympathetic divisions of the autonomic system, giving examples of the effects of these systems on specific organs such as the heart and intestine.
12. Discuss the biological actions and effects on mood of the following types of drugs: alcohol, antidepressants, barbiturates, antianxiety drugs, antipsychotic drugs, opiates, stimulants, hallucinogens, and marijuana.

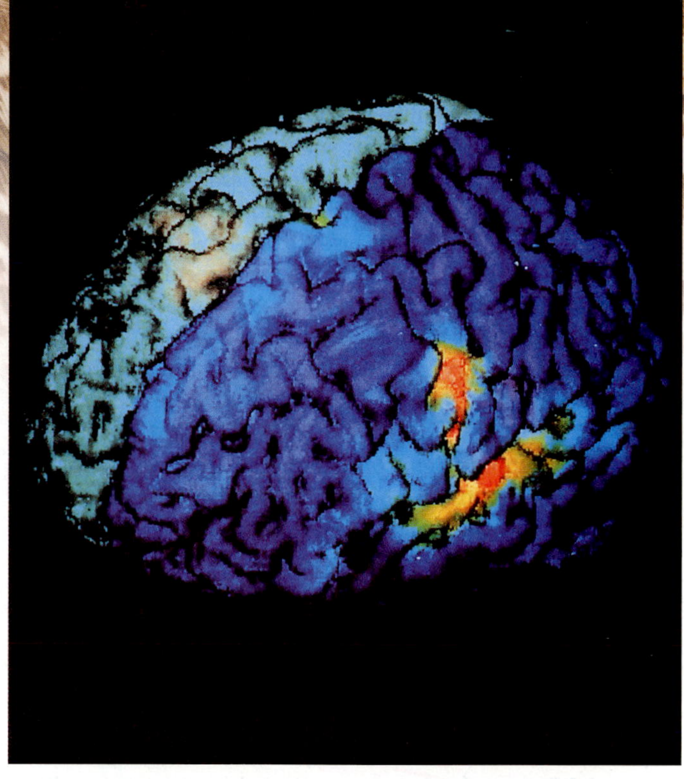

Functional magnetic resonance imaging (fMRI) of the brain. Functional MRI is sensitive to the increased blood flow that occurs with neural activity. In the image shown here, red indicates areas of greatest activation during speech; yellow indicates medium activation. *(Volker Steger/Peter Arnold, Inc.)*

A frog ejects its tongue with lightning speed to capture a fly, a rabbit escapes from a predator, and a human learns biology. In a complex nervous system millions of neurons work together to produce effective responses to stimuli in the external environment. The nervous system also regulates heart rate, breathing, and hundreds of other internal activities. Just how neurons interact has been the focus of extensive research, and we are only beginning to understand the mechanisms that permit the complex functions of the nervous system.

Neurobiologists have focused on two major aspects of neural function: "hard-wiring" of the nervous system and neural plasticity. **Hard-wiring** refers to how neurons signal one another, how they connect, and how they carry out basic functions such as regulating heart rate, blood pressure, respiration, and sleep-wake cycles. Hard-wiring was the major focus of neurobiologists for decades.

Eventually, neurobiologists began to focus on their observation that even animals with very simple nervous systems can learn to repeat behaviors associated with reward and avoid behaviors that cause pain. Such changes in behavior are possible as a result of **neural plasticity,** the ability of the nervous system to change in response to experience. Such change involves modification of structure and function. Many areas of the brain that were once thought to be hard-wired are now known to be flexible and capable of change. Familiar examples of neural plasticity involving human motor skills include learning to walk, ride a bicycle, or hit a baseball. At first you were probably clumsy, but with practice, your performance became smoother and more precise. Similarly, the ability to learn languages, solve problems, and perform scientific experiments depends on neural plasticity.

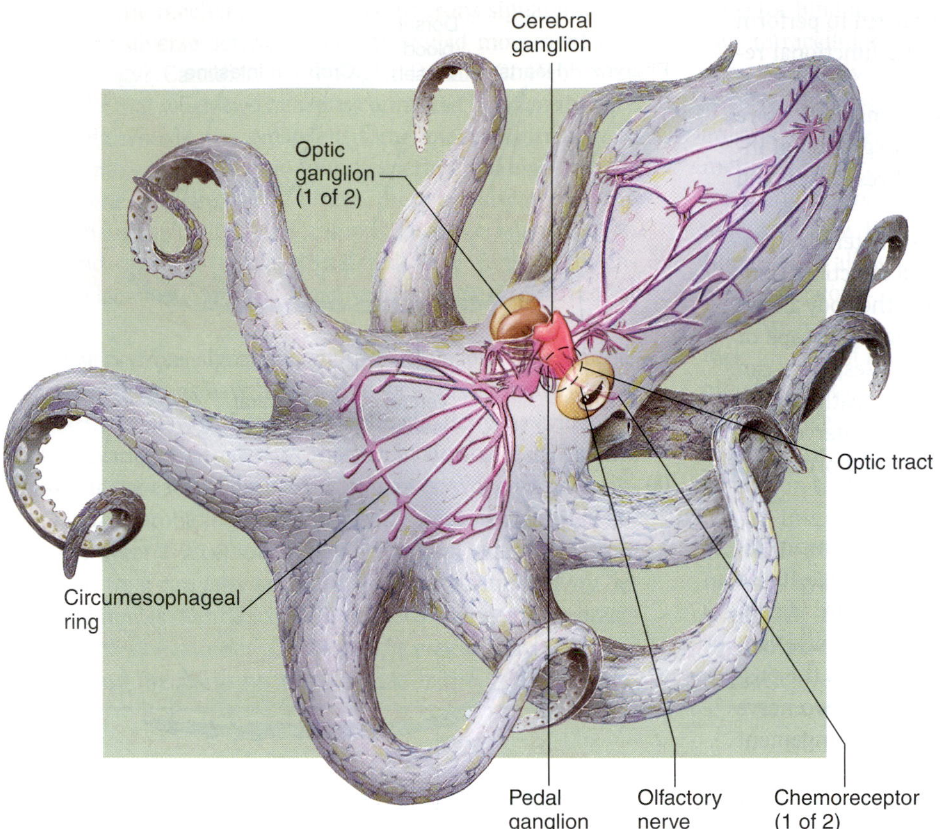

Figure 40–4 The cephalopod nervous system. Millions of neurons are concentrated in ganglia. Nerves extend from the ganglia into the head and other parts of the body.

contains about 168 million nerve cells. This complex nervous system, which includes well-developed sense organs, is correlated with the active, predatory lifestyle of these animals. With its highly developed nervous system, the octopus is capable of considerable learning and can be taught quite complex tasks. In fact, cephalopods are considered the most intelligent invertebrates.

THE VERTEBRATE NERVOUS SYSTEM HAS TWO MAIN DIVISIONS: CNS AND PNS

An animal's range of possible responses depends in large part on the number of its neurons and how they are organized in the nervous system. As animal groups evolved, nervous systems became increasingly complex. The vertebrate nervous system has two main divisions: the **central nervous system (CNS)** and the **peripheral nervous system (PNS)** (Table 40–1). The CNS consists of a complex brain that is continuous with the dorsal tubular spinal cord. Serving as central control, these organs integrate incoming information and determine appropriate responses.

The PNS is made up of the sensory receptors (e.g., tactile, auditory, and visual receptors) and the nerves, which are the communication lines. Various parts of the body are linked to the brain by cranial nerves and to the spinal cord by spinal nerves.

TABLE 40–1 Divisions of the Vertebrate Nervous System

Central Nervous System (CNS)

Brain

Spinal cord

Peripheral Nervous System (PNS)

Somatic division

1. Receptors
2. Afferent (sensory) nerves—transmit information from receptors to CNS
3. Efferent (motor) nerves—transmit information from CNS to skeletal muscles

Autonomic division

1. Receptors
2. Afferent (sensory) nerves—transmit information from receptors in internal organs to CNS
3. Efferent nerves—transmit information from CNS to glands and involuntary muscle in organs
   a. Sympathetic nerves—generally stimulate activity that results in mobilization of energy (e.g., speeds heartbeat)
   b. Parasympathetic nerves—action results in energy conservation or restoration (e.g., slows heart but speeds digestion)

Afferent neurons in these nerves continuously inform the CNS of changing conditions. Then efferent neurons transmit the "decisions" of the CNS to appropriate muscles and glands, which make the adjustments needed to maintain homeostasis.

For convenience the PNS can be subdivided into somatic and autonomic divisions. Most of the receptors and nerves concerned with changes in the external environment are somatic. Those that regulate the internal environment are autonomic. Both divisions have afferent nerves, which transmit messages from receptors to the CNS, and efferent nerves, which transmit information back from the CNS to the structures that must respond. In the autonomic division there are two kinds of efferent pathways: sympathetic and parasympathetic nerves. Sympathetic nerves generally stimulate activities that mobilize energy; for example, they stimulate the heart to contract more rapidly. Parasympathetic nerves stimulate activities that conserve or restore energy; for example, they cause the heart to decrease its rate of contraction.

THE EVOLUTION OF THE VERTEBRATE BRAIN IS MARKED BY INCREASING COMPLEXITY

All vertebrates, from fishes to mammals, have the same basic brain structure. Different parts of the brain are specialized in the various vertebrate classes, and the evolutionary trend is toward increasing complexity, especially of the cerebrum and cerebellum.

In the early vertebrate embryo, the brain and spinal cord differentiate from a single tube of tissue, the **neural tube.** Anteriorly, the tube expands and develops into the brain. Posteriorly, the tube becomes the spinal cord. Brain and spinal cord remain continuous, and their cavities communicate. As the brain begins to differentiate, three bulges become visible: the hindbrain, midbrain, and forebrain (Fig. 40–5).

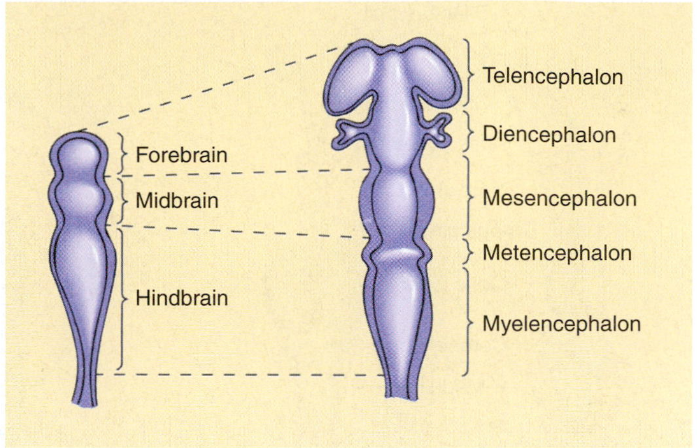

Figure 40–5 Early development of the vertebrate nervous system. Early in the development of the vertebrate embryo, the anterior end of the neural tube differentiates into the forebrain, midbrain, and hindbrain. These primary divisions subdivide and eventually give rise to specific structures of the adult brain (see Table 40–2).

The hindbrain develops into the medulla, pons, and cerebellum

The **hindbrain** subdivides to form the **metencephalon,** which gives rise to the **cerebellum** and **pons,** and the **myelencephalon,** which gives rise to the **medulla.** The medulla, pons, and midbrain make up the **brain stem,** the elongated portion of the brain that looks like a stalk holding up the cerebrum.

The medulla, the most posterior part of the brain, is continuous with the spinal cord. Its cavity, the **fourth ventricle,** is continuous with the central canal of the spinal cord and with a channel that runs through the midbrain. The walls of the medulla are thick and made up largely of *nerve tracts* (bundles of axons) that connect the spinal cord with various parts of the brain. In complex vertebrates, the medulla contains centers that regulate respiration, heartbeat, and blood pressure. Other reflex centers in the medulla regulate activities such as swallowing, coughing, and vomiting.

The size and shape of the cerebellum vary among the vertebrate classes (Fig. 40–6). Development of the cerebellum in different animals is roughly correlated with the extent and complexity of muscular activity, reflecting the principle that the relative size of a brain part correlates with its importance to the behavior of the species. In some fishes, birds, and mammals, the cerebellum is highly developed, whereas it tends to be small in jawless fishes, amphibians, and reptiles. The cerebellum coordinates muscle activity and is responsible for muscle tone, posture, and equilibrium.

Injury or removal of the cerebellum results in impaired muscle coordination. A bird without a cerebellum cannot fly, and its wings thrash about jerkily. When the human cerebellum is injured by a blow or by disease, muscular movements are uncoordinated. Any activity requiring delicate coordination, such as threading a needle, is very difficult, if not impossible.

In mammals, a large mass of fibers known as the pons forms a bulge on the anterior surface of the brain stem. The pons is a bridge connecting the spinal cord and medulla with upper parts of the brain. It contains centers that help regulate respiration and *nuclei* that relay impulses from the cerebrum to the cerebellum. (A **nucleus** is a group of cell bodies within the CNS.)

The midbrain is prominent in fishes and amphibians

In fishes and amphibians, the **midbrain,** or **mesencephalon,** is the most prominent part of the brain, serving as the main association area. It receives incoming sensory information, integrates it, and sends decisions to appropriate motor nerves. The dorsal portion of the midbrain is differentiated to some extent. For example, the optic lobes are specialized for visual interpretations.

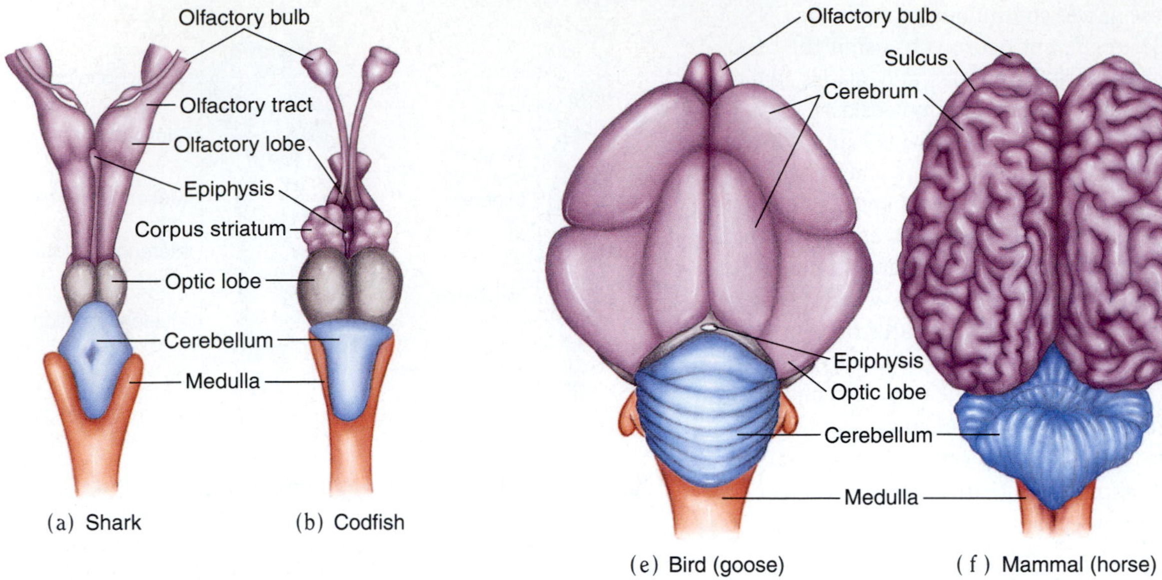

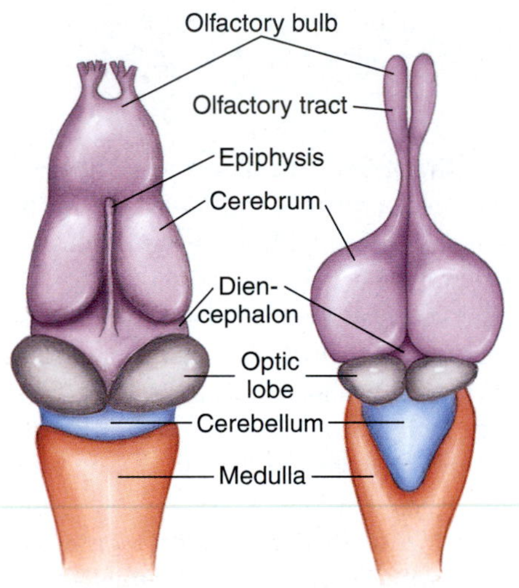

Figure 40–6 Evolution of the vertebrate brain. Comparison of the brains of members of six vertebrate classes reveals basic similarities and evolutionary trends. (Brains are not drawn to scale.) Note that different parts of the brain are specialized in the various groups. For example, the large olfactory lobes in the shark brain **(a)** are essential to this predator's highly developed sense of smell. **(b** through **f)** During the course of evolution, the cerebrum and cerebellum have become larger and more complex. In mammals **(f)**, the cerebrum is the most prominent part of the brain; the cerebral cortex, the thin outer layer of the cerebrum, is highly convoluted (folded), which greatly increases its surface area.

In reptiles, birds, and mammals, many of the functions of the optic lobes are assumed by the cerebrum, which develops from the forebrain. In mammals, the midbrain consists of the **superior colliculi,** centers for visual reflexes such as pupil constriction, and the **inferior colliculi,** centers for certain auditory reflexes. The mammalian midbrain also contains a center (the *red nucleus*) that helps maintain muscle tone and posture.

The forebrain gives rise to the thalamus, hypothalamus, and cerebrum

As indicated in Table 40–2, the **forebrain** subdivides to form the **telencephalon** and **diencephalon.** The telencephalon gives rise to the **cerebrum,** and the diencephalon to the **thalamus** and **hypothalamus.** The lateral ventricles (also called the first and second ventricles) are located within the cerebrum. Each lateral ventricle is connected with the third ventricle (within the diencephalon) by way of a channel.

In all vertebrate classes, the thalamus is a relay center for motor and sensory messages. In mammals, all sensory messages (except those from the olfactory receptors) are delivered to the thalamus before they are relayed to the sensory areas of the cerebrum.

The hypothalamus, which lies below the thalamus, forms the floor of the third ventricle. The hypothalamus contains olfactory centers and is the principal integration center for the regulation of the viscera (internal organs). It provides input to centers in the medulla and spinal cord that regulate activities such as heart rate, respiration, and digestive system function. In reptiles, birds, and mammals, the hypothalamus controls body temperature. It also regulates appetite and water balance and is involved in emotional and sexual responses. As discussed in Chapter 47, the hypothalamus links the nervous and endocrine systems and produces certain hormones.

The telencephalon gives rise to the cerebrum and, in most vertebrate groups, the olfactory bulbs. These structures are im-

TABLE 40–2 Differentiation of CNS Structures

Early Embryonic Divisions	Subdivisions	Derivatives in Adult	Cavity
Brain			
Forebrain	Telencephalon	Cerebrum	Lateral ventricles (first and second ventricles)
	Diencephalon	Thalamus, hypothalamus, epiphysis (pineal body)	Third ventricle
Midbrain	Mesencephalon	Optic lobes in fish and amphibians; superior and inferior colliculi	Cerebral aqueduct
Hindbrain	Metencephalon	Cerebellum, pons	
	Myelencephalon	Medulla	Fourth ventricle
Spinal Cord		Spinal cord	Central canal

portant in the chemical sense of smell—the dominant sense in most aquatic and terrestrial vertebrates. In fact, much of brain development in vertebrates appears to be focused on the integration of olfactory information. In fish and amphibians, the cerebrum is almost entirely devoted to this function.

Birds are an exception among the vertebrates in that their sense of smell is generally poorly developed. A part of their cerebrum, the *corpus striatum,* however, is highly developed. This structure is thought to control the innate, stereotypical, yet complex action patterns characteristic of birds. Just above the corpus striatum is a region thought to govern learning.

In most vertebrates, the cerebrum is divided into right and left **hemispheres.** Most of the cerebrum is made of **white matter,** which consists mainly of myelinated axons that connect various parts of the brain. In mammals and most reptiles, a layer of **gray matter,** the **cerebral cortex,** makes up the outer portion of the cerebrum. Gray matter contains cell bodies and dendrites.

Certain reptiles and all mammals have a type of cerebral cortex, the **neocortex,** not found in less complex vertebrates. The neocortex serves as an association area—a region that links sensory and motor functions and is responsible for higher functions such as learning. The neocortex is very extensive in mammals, making up the bulk of the cerebrum. In humans, about 90% of the cerebral cortex is neocortex, consisting of six distinct cell layers.

In mammals, the cerebrum is the most prominent part of the brain. During embryonic development, it expands and grows backward, covering many other brain structures. The cerebrum is responsible for many of the functions performed by other parts of the brain in other vertebrates. In addition, it has many complex association functions that are lacking in reptiles, amphibians, and fishes.

In small or simple mammals, the cerebral cortex may be smooth. However, in large complex mammals, the surface area is greatly expanded by numerous folds called **convolutions.** The furrows between them are called **sulci** (sing., *sulcus*) if shallow and **fissures** if deep. The number of folds (not the size of the brain) has been associated with complexity of brain function.

THE HUMAN CENTRAL NERVOUS SYSTEM IS REMARKABLY COMPLEX

The soft, fragile human brain and spinal cord are well protected. Encased within bone, they are covered by three layers of connective tissue, the **meninges.** The three meningeal layers are the tough, outer **dura mater,** the middle **arachnoid,** and the thin, vascular **pia mater,** which adheres closely to the tissue of the brain and spinal cord (Fig. 40–7). Meningitis is a disease in which these coverings become infected and inflamed.

Between the arachnoid and the pia mater is the subarachnoid space, which contains **cerebrospinal fluid (CSF).** The fluid is produced by special networks of capillaries, collectively called the **choroid plexus,** that extend from the pia mater into the ventricles. The choroid plexus extracts nutrients from the blood and adds them to the CSF. Together the choroid plexus and the arachnoid serve as a barrier between blood and CSF. They prevent harmful substances in the blood from entering the brain.

CSF is a shock-absorbing fluid that cushions the brain and spinal cord against mechanical injury. This fluid also serves as a medium for exchange of nutrients and waste products between the blood and brain. CSF circulates down through the ventricles and passes into the subarachnoid space surrounding the brain and spinal cord. It is then reabsorbed into large blood sinuses within the dura mater.

The spinal cord transmits impulses to and from the brain

The tubular **spinal cord** extends from the base of the brain to the level of the second lumbar vertebra. A cross section through the spinal cord reveals a small **central canal** surrounded by an area of gray matter shaped somewhat like the letter H (Fig. 40–8). The gray matter is composed of large masses of cell bodies, dendrites, unmyelinated axons, and glial cells.

The white matter, found outside the gray matter, consists of myelinated axons arranged in bundles called **tracts** or **pathways.**

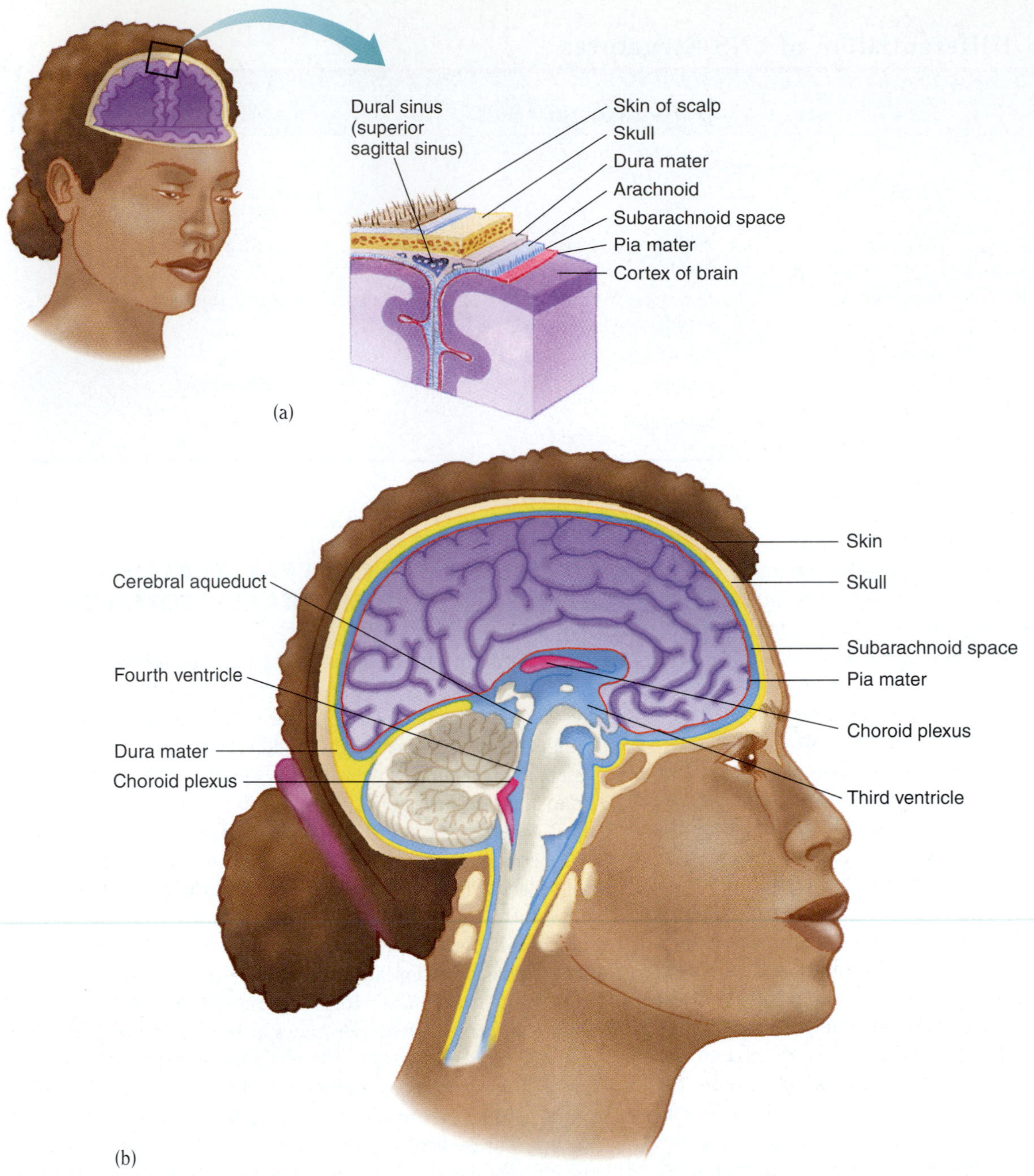

Figure 40–7 Protection of the brain and spinal cord. The CNS is well protected by the skull and meninges and by cerebrospinal fluid (CSF). **(a)** Frontal section through the superior part of the brain. Note the large blood sinus shown between two layers of the dura mater. Blood leaving the brain flows into such sinuses and then circulates to the large jugular veins in the neck. **(b)** Sagittal section. The CSF cushions the brain and spinal cord. Produced by the choroid plexi in the walls of the ventricles, this fluid circulates through the ventricles and subarachnoid space. It is continuously produced and continuously reabsorbed into the blood of the dural sinuses.

Long ascending tracts conduct impulses up the cord to the brain. For example, the spinothalamic tracts in the anterior and lateral columns of the white matter conduct pain and temperature information from sensory neurons in the skin. The pyramidal tracts are descending tracts that convey impulses from the cerebrum to motor nerves at various levels in the cord. The spinal nerves are described in a later section of this chapter.

In addition to transmitting impulses to and from the brain, the spinal cord controls many reflex activities. A **reflex action** is a relatively fixed response pattern to a simple stimulus. The re-

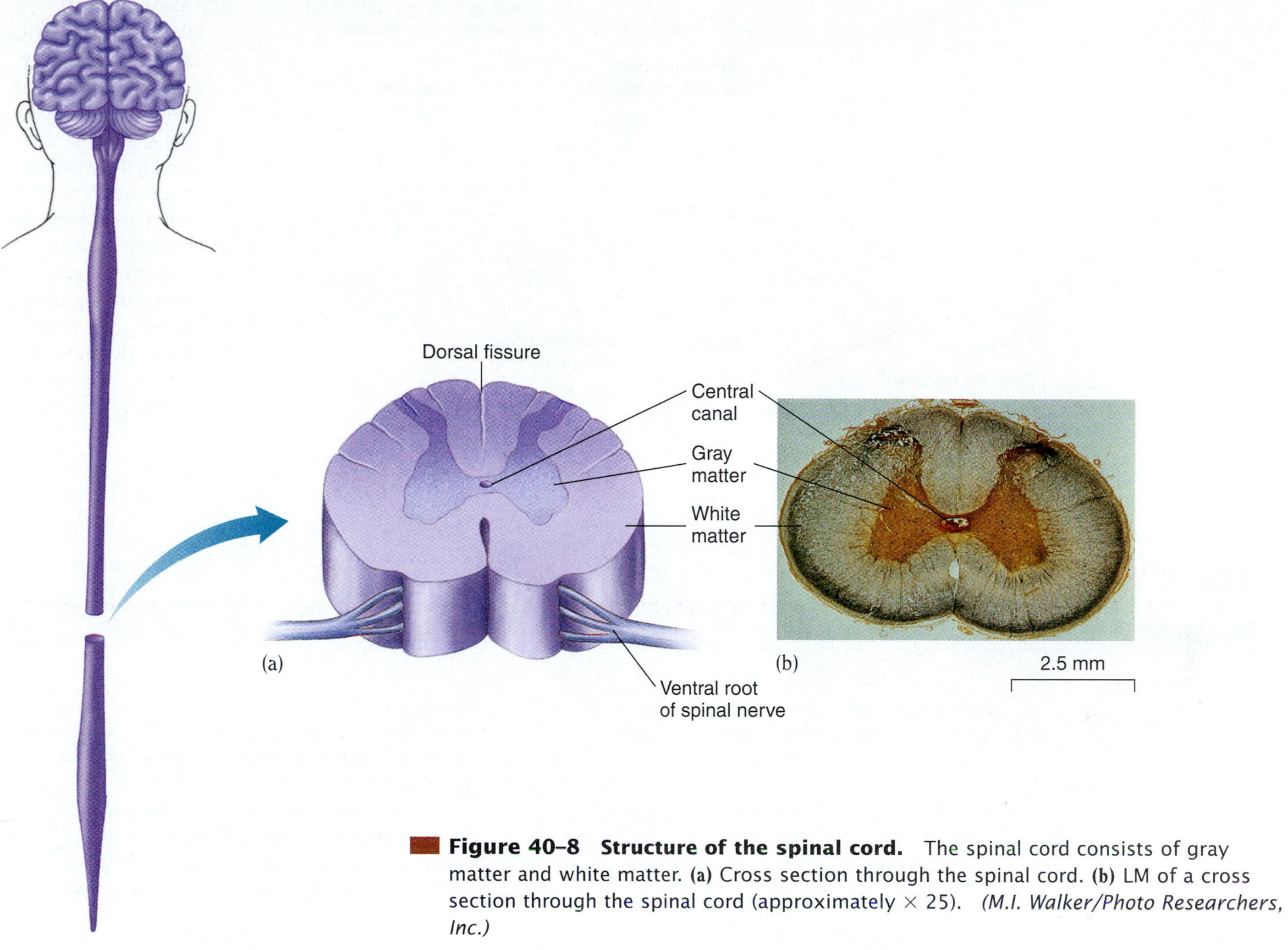

Figure 40–8 Structure of the spinal cord. The spinal cord consists of gray matter and white matter. **(a)** Cross section through the spinal cord. **(b)** LM of a cross section through the spinal cord (approximately × 25). *(M.I. Walker/Photo Researchers, Inc.)*

sponse is predictable and automatic, not requiring conscious thought. Many of the activities of the body, such as breathing, are regulated by reflex actions.

Although most reflex actions are much more complex, let us consider a **withdrawal reflex,** in which a neural circuit consisting of only three types of neurons is needed to carry out a response to a stimulus (Fig. 40–9). Suppose you touch a hot stove. Almost instantly, and even before you are consciously aware of what has happened, you jerk your hand away. In this brief instant a message has been carried by a sensory neuron from pain receptors in the skin to the spinal cord. In the tissue of the spinal cord, the message is transmitted from the sensory neuron to an association neuron. Finally, the message is transmitted to an appropriate motor neuron, which conducts it to groups of muscles that respond by contracting and moving the limb toward the midline of the body, moving the hand away from the stove. Actually, many neurons in sensory, association, and motor nerves participate in such a reaction, and complicated switching is involved. Generally, we are not even consciously aware that all these responding muscles exist.

At the same time that the reflex pathway is activated, a message may be also sent up the spinal cord to the conscious areas of the brain. As you withdraw your hand from the hot stove, you become aware of what has happened and feel the pain. This awareness, however, is not part of the reflex response.

Recent research suggests that contrary to long accepted dogma, the spinal cord has plasticity and is capable of training. For example, sensory feedback from walking exercise may increase the strength of neural connections in the spinal cord. Such findings raise hope that, providing the descending connections with the brain are intact, some victims of spinal cord injury (some 200,000 patients in the United States alone) could potentially develop at least some limited ability to walk again.

The most prominent part of the human brain is the cerebrum

The structure and functions of the main parts of the human brain are summarized in Table 40–3, and the brain is illustrated in Figures 40–10 and 40–11. As in other mammals, the human cerebral cortex consists of right and left **cerebral hemispheres** and is functionally divided into three areas: (1) **sensory areas** that receive incoming signals from the sense organs; (2) **motor areas** that control voluntary movement; and (3) **association areas** that link the

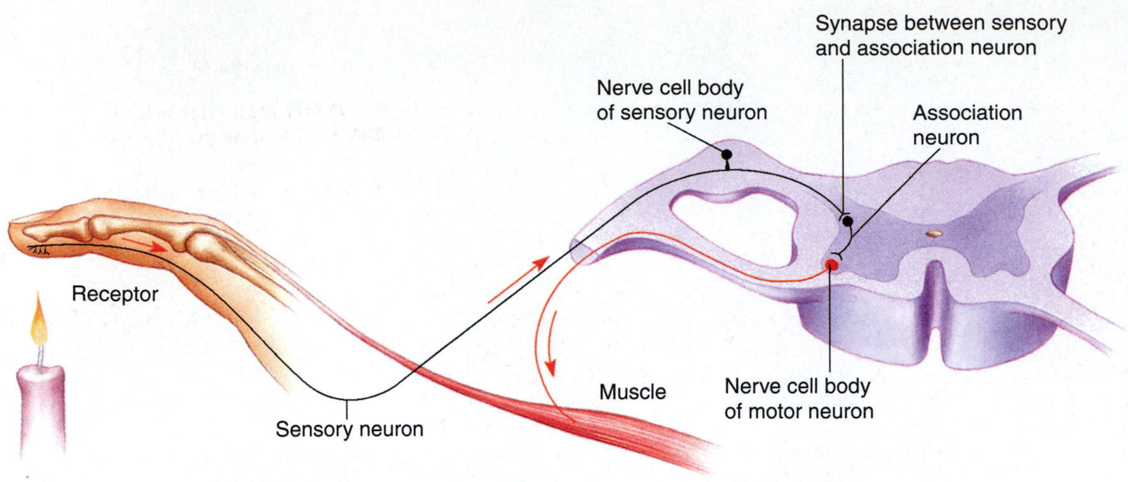

Figure 40–9 Withdrawal reflex. A sensory neuron transmits impulses from the receptor to the CNS, where it synapses with an association neuron. Then an appropriate motor neuron *(shown in red)* transmits impulses to the muscles that move the hand away from the flame (the actual response).

TABLE 40–3 The Human Brain

Structure	Description	Function
Brain stem		
Medulla	Continuous with spinal cord; primarily made up of nerves passing from spinal cord to rest of brain	Contains vital centers (clusters of neuron cell bodies) that control heartbeat, respiration, and blood pressure; contains centers that control swallowing, coughing, vomiting
Pons	Forms bulge on anterior surface of brain stem	Connects various parts of brain with one another; contains respiratory and sleep centers
Midbrain	Just above pons	Center for visual and auditory reflexes (e.g., pupil reflex, blinking, adjusting ear to volume of sound)
Thalamus	At top of brain stem	Main sensory relay center for conducting information between spinal cord and cerebrum; neurons in thalamus sort and interpret all incoming sensory information (except olfaction) before relaying messages to appropriate neurons in cerebrum
Hypothalamus	Just below thalamus; pituitary gland is connected to hypothalamus by stalk of neural tissue	Contains centers for control of body temperature, appetite, fat metabolism, and certain emotions; regulates pituitary gland
Cerebellum	Second largest division of brain	Reflex center for muscular coordination and refinement of movements
Cerebrum	Largest, most prominent part of human brain; longitudinal fissure divides cerebrum into right and left hemispheres, each divided into lobes: frontal, parietal, temporal, and occipital lobes	Center of intellect, memory, consciousness, and language; also controls sensation and motor functions
Cerebral cortex (outer gray matter)	Arranged into convolutions (folds) that increase surface area; functionally, cerebral cortex is divided into	
	1. Motor cortex	Controls movement of voluntary muscles
	2. Sensory cortex	Receives incoming information from eyes, ears, pressure and touch receptors, etc.
	3. Association cortex	Site of intellect, memory, language, and emotion; interprets incoming sensory information
White matter	Consists of myelinated axons of neurons that connect various regions of brain; these axons are arranged into bundles (tracts)	Connects the following: 1. Neurons within same hemisphere 2. Right and left hemispheres 3. Cerebrum with other parts of brain and spinal cord

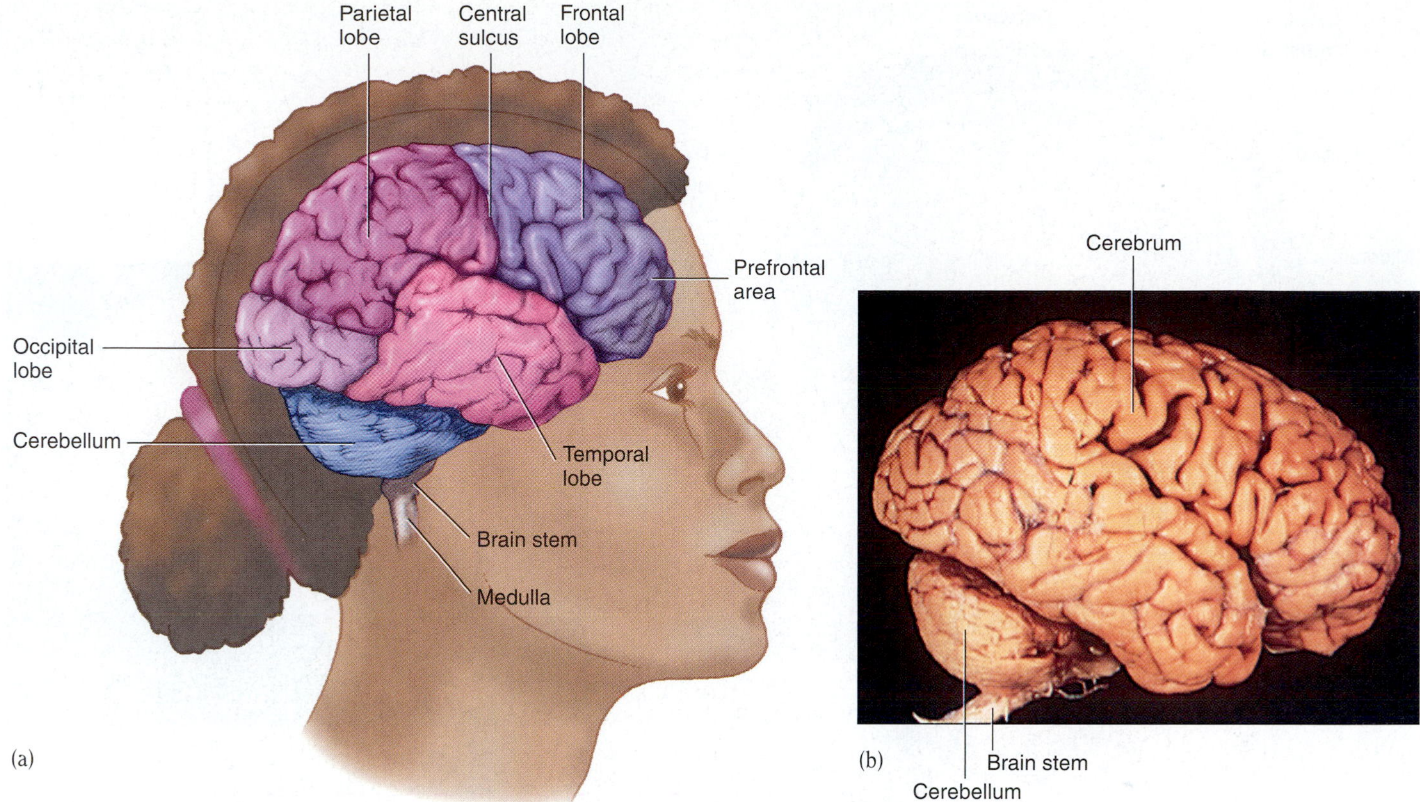

Figure 40–10 The human brain. **(a)** Human brain, lateral view. The diencephalon and part of the brain stem are covered by the cerebrum. **(b)** Photograph of lateral view of the human brain showing the lobes of the cerebrum. *(b, Fred Hossler/Visuals Unlimited)*

sensory and motor areas and are responsible for thought, learning, language, memory, judgment, and personality.

Investigators have been able to map the cerebral cortex, locating the areas responsible for different functions. The **occipital lobes** contain the visual centers. Stimulation of these areas, even by a blow on the back of the head, causes the sensation of light; their removal causes blindness. The centers for hearing are located in the **temporal lobes** of the brain above the ear; stimulation by a blow causes a sensation of noise. Removal of both auditory areas causes deafness. Removal of one does not cause deafness in one ear but rather produces a decrease in the auditory acuity of both ears.

A groove called the **central sulcus** crosses the top of each hemisphere from medial to lateral edge. This groove partially separates the **frontal lobes** from the **parietal lobes.** The **primary motor areas** in the frontal lobes control the skeletal muscles. The **primary sensory areas** in the parietal lobes receive information regarding heat, cold, touch, and pressure from sense organs in the skin.

The size of the motor area in the brain for any given part of the body is proportional not to the amount of muscle but to the complexity of movement involved. Predictably, areas that control the hands and face are relatively large (Fig. 40–12). A similar relationship exists between the sensory area and the sensitivity of the region of the skin from which it receives impulses. In connections between the body and the brain, not only do the fibers cross so that one side of the brain controls the opposite side of the body but another "reversal" makes the uppermost part of the cortex control the lower limbs of the body.

When all the areas of known function are plotted, they cover almost all of the rat's cortex, a large part of the dog's, a moderate amount of the monkey's, and only a small part of the total surface of the human cortex. The remaining cortical areas are the association areas. Somehow the association regions integrate all the diverse impulses reaching the brain into a meaningful unit so that an appropriate response is made. When disease or accident destroys the functioning of one or more association areas, the ability to recognize certain kinds of symbols may be lost. For example, the names of objects may be forgotten, although their functions are remembered and understood.

The white matter of the cerebrum lies beneath the cerebral cortex. Nerve fibers of the white matter connect the cortical areas with one another and with other parts of the nervous system. A large band of white matter, the **corpus callosum,** connects the right and left hemispheres (see Fig. 40–11).

Deep within the white matter of the cerebrum lie the **basal ganglia,** paired groups of nuclei (gray matter). These nuclei play an important role in coordination of movement. The basal ganglia

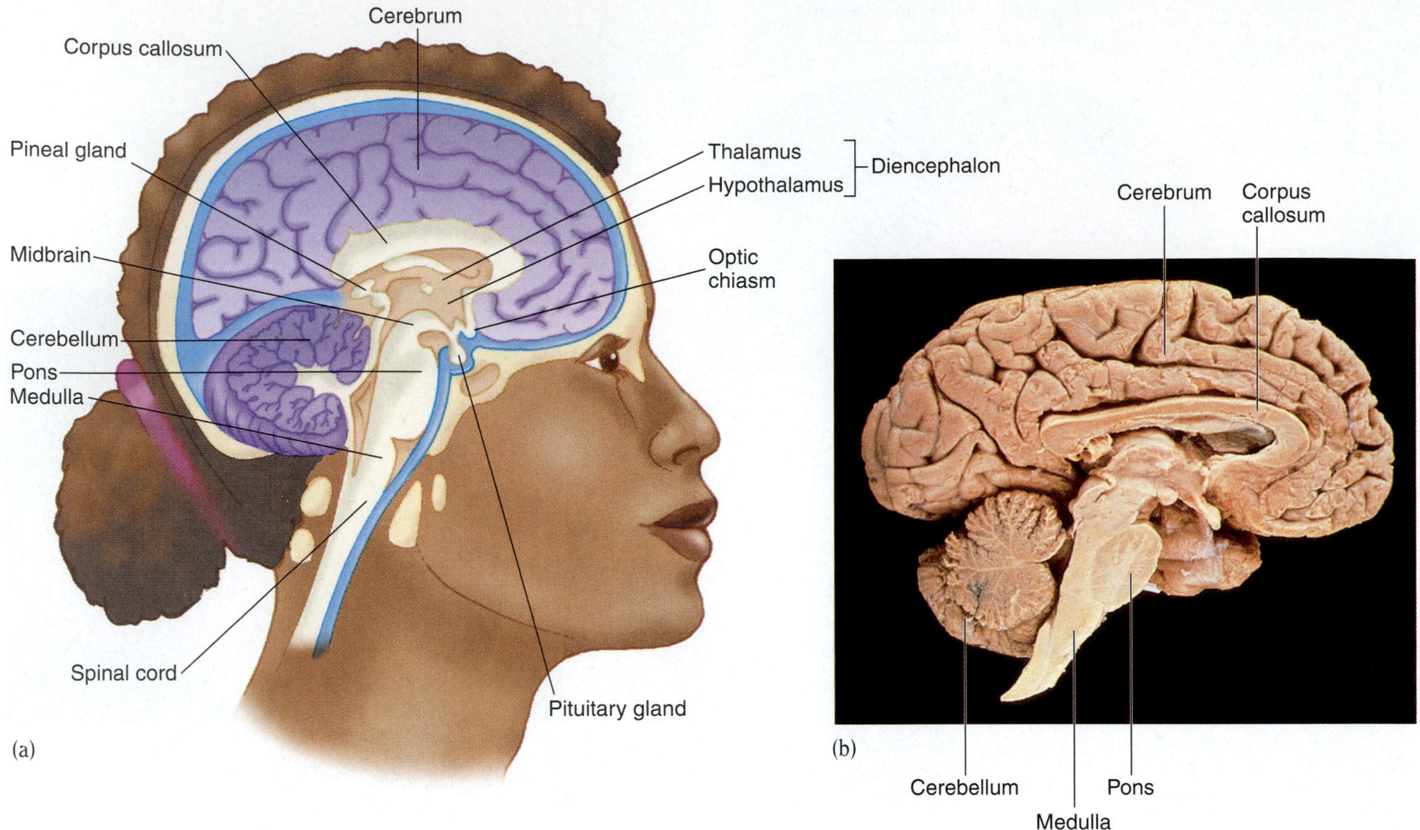

Figure 40–11 Midsagittal section through the human brain. Half of the brain has been cut away, exposing structures normally covered by the cerebrum. Compare the diagram **(a)** with the photograph of the human brain **(b)**. *(b, Science Pictures Limited/Science Photo Library/Photo Researchers, Inc.)*

send signals to the **substantia nigra,** which is located in the midbrain, and they receive inputs back from the substantia nigra. The neurons that project (extend) to the basal ganglia produce **dopamine,** a neurotransmitter essential for sensory-motor coordination. Dopamine helps maintain the balance between inhibition and excitation of neurons involved in motor function. Other neurons from the substantia nigra signal nuclei in the thalamus that in turn relay information to the motor cortex. The substantia nigra neurons that signal the thalamus release the neurotransmitter **gamma-aminobutyric acid (GABA).** Just how these areas work together to coordinate motor function is not yet fully understood.

Process of Science The important role that dopamine plays in motor function was discovered through an interesting series of somewhat unrelated events. During the mid-1950s, the drug reserpine became popular as a major tranquilizer for mental patients. Then, in 1959, investigators noticed that some patients taking reserpine experienced distressing side effects, such as muscle rigidity and persistent tremors (shaking). These symptoms were very similar to those seen in patients with **Parkinson's disease,** one of the most common movement disorders. Victims of Parkinson's disease have a shuffling gait and suffer from tremors at rest. When they attempt purposeful movement, it is slow, shaky, and difficult.

The observation that reserpine caused Parkinson-like symptoms led to studies showing that reserpine greatly reduces the amount of dopamine in two of the basal ganglia within the white matter of the cerebrum. Investigators then discovered that patients with Parkinson's disease have only about 50% of the normal amount of dopamine in their basal ganglia. (About 80% of the brain's dopamine is in the basal ganglia.) Subsequently, investigators discovered that the decrease in dopamine level results from degeneration of dopamine-secreting neurons that originate in the substantia nigra and project to the basal ganglia.

Early attempts to administer dopamine to patients with Parkinson's disease were not successful because dopamine cannot penetrate the blood-brain barrier. However, L-dopa, a substance from which dopamine is synthesized in the body, does penetrate the blood-brain barrier. Its use has dramatically relieved the symptoms of Parkinson's disease in many patients. Unfortunately, the benefits of treatment with L-dopa and other drugs decrease as the disease progresses. Surgical interventions are sometimes used to electrically destroy overactive areas of the basal ganglia.

Investigators have successfully used animal models of Parkinson's disease to study the effects of transplanting

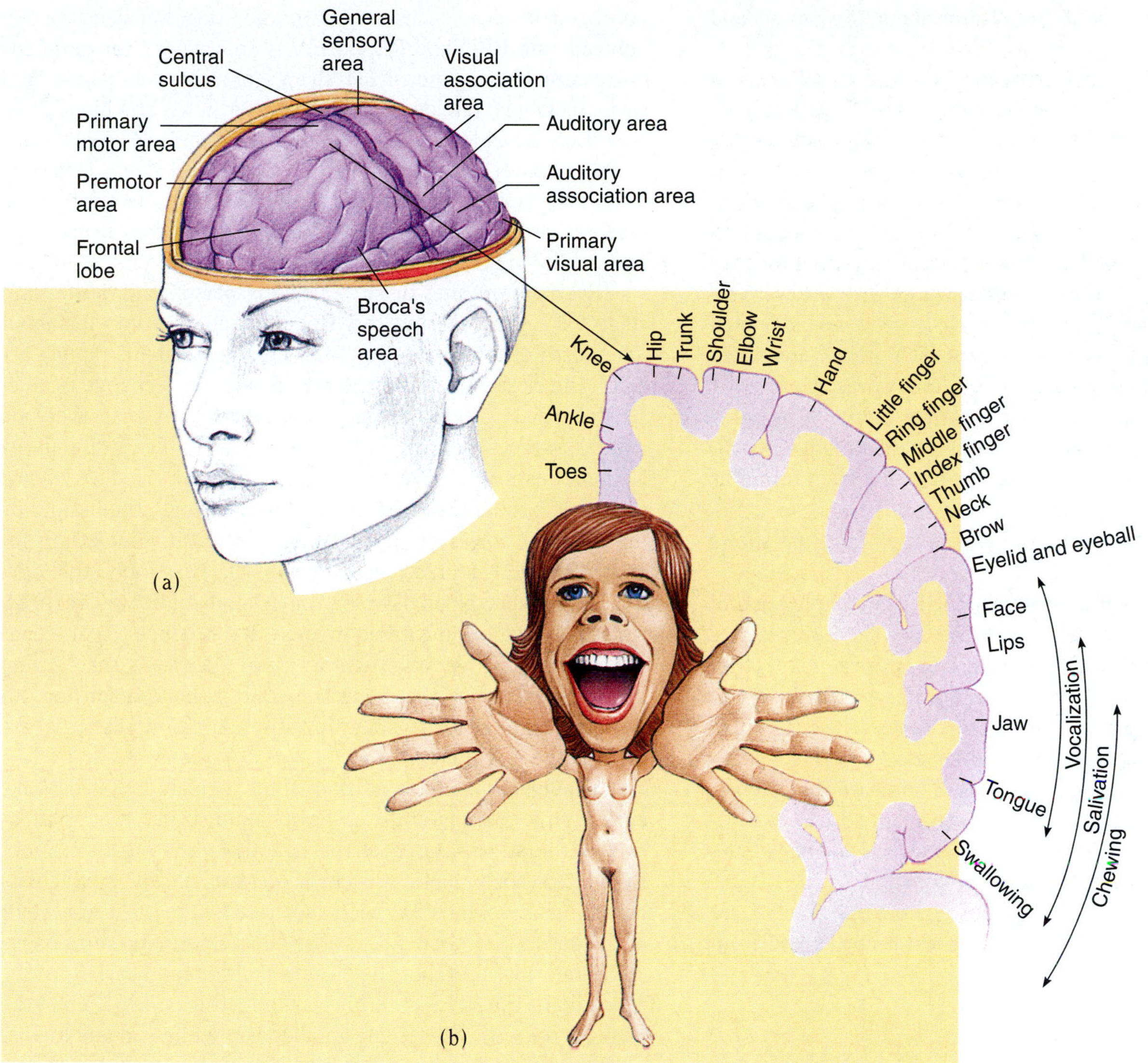

Figure 40–12 Functional areas of the brain. **(a)** Several sensory, motor, and association areas are shown. **(b)** A cross section through the primary motor area showing which area of cerebral cortex controls each body part. The figure shown here, known as a motor homunculus ("little person"), is proportioned to reflect the amount of cerebral cortex that controls each body part. Note that more cortical tissue is devoted to controlling those body structures capable of skilled, complex movement.

dopamine-producing neurons directly into the brain. Embryonic stem cells are a promising source of dopamine-producing neurons that can be transplanted into patients with Parkinson's disease (see *Focus On: Stem Cells,* in Chapter 16).

Dopamine depletion may occur with aging. As a result, even healthy persons experience changes in motor abilities as they age. Body movements and even reflexes slow, and movement becomes more difficult. Studies suggest that L-dopa may be an effective treatment for individuals with these symptoms.

The brain integrates information

Most of the integration of the nervous system takes place within the cerebral cortex. The cerebral cortex integrates information about diverse activities such as arousal and sleep, emotion, and information processing.

Brain activity cycles in a sleep-wake pattern

Brain activity can be studied by measuring and recording the electrical potentials, or "brain waves," generated by thousands of active neurons in various parts of the brain. This electrical activity can be recorded by a device known as an electroencephalograph. To obtain a recording of this electrical activity, called an **electroencephalogram (EEG),** electrodes are taped to different parts of the scalp, and the activity of the cerebral cortex is measured. The EEG shows that the brain is continuously active. The most regular indication of activity appears to be **alpha waves,** which occur rhythmically at the rate of about ten per second (Fig. 40–13).

Alpha waves are generated mainly from the visual areas in the occipital lobes when the person being tested is resting quietly with eyes closed. When the eyes are opened, alpha waves disappear and are replaced by more rapid, irregular waves. As you are reading this biology text, your brain is emitting **beta waves.** These waves have a fast-frequency rhythm and are thought to represent heightened mental activity such as information processing. During sleep, the brain emits waves with a lower frequency and a higher amplitude. The slow, large waves associated with certain stages of sleep are called **delta waves** and **theta waves.** During dreaming, flurries of irregular waves occur.

Certain brain diseases change the pattern of brain waves. Individuals with epilepsy, for example, exhibit a distinctive, recognizable, abnormal wave pattern. The location of a brain tumor or the site of brain damage caused by a blow to the head can sometimes be determined by noting the part of the brain that shows abnormal waves.

The **reticular activating system (RAS)** is a complex neural pathway within the brain stem and thalamus. The RAS receives messages from neurons in the spinal cord and from many other parts of the nervous system and communicates with the cerebral cortex by complex neural circuits. Components of the RAS help regulate **consciousness**—a state of awareness, or of selective attention. When certain neurons of the RAS bombard the cerebral cortex with stimuli, you feel alert and are able to focus your attention on specific thoughts. If the RAS is severely damaged, the victim may pass into a deep, permanent coma.

Sleep is an alteration of consciousness during which there is decreased electrical activity of the cerebral cortex and from which a person can be aroused. Two main stages of sleep are recognized: **non-REM** and **REM.** *REM* is an acronym for *rapid eye movement.* During non-REM sleep, sometimes called normal sleep, metabolic rate decreases, breathing slows, and blood pressure decreases. Slower frequency, higher amplitude waves (theta and delta waves) are characteristic of non-REM sleep. This electrical activity is thought to be generated spontaneously by the cerebral cortex when it is not driven by impulses from other parts of the brain.

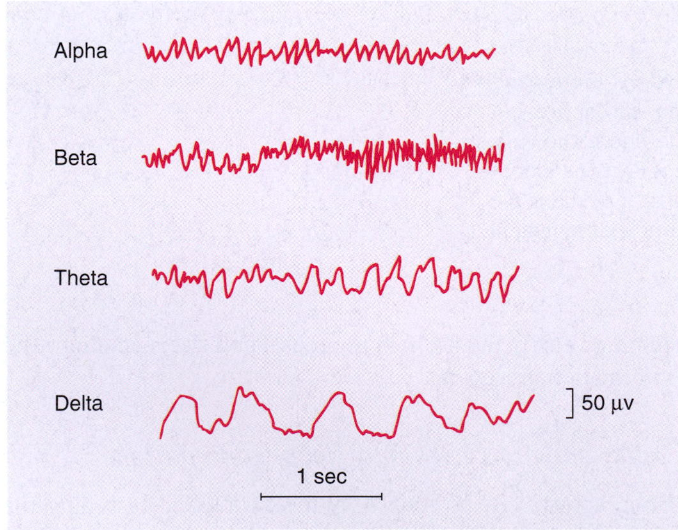

Figure 40–13 EEGs showing electrical activity in the brain. The regular waves characteristic of the relaxed state are called alpha waves. During mental activity, rapid, low-amplitude beta waves are dominant. Recordings made in various stages of sleep show theta and delta waves, which have a lower frequency and greater amplitude.

Every 90 minutes or so, a sleeping person enters the REM stage for a time. During this stage, which accounts for about one-fourth of total sleep time, the eyes move about rapidly beneath the closed but fluttering lids. Brain waves change to a desynchronized pattern. Sleep researchers claim that everyone dreams, especially during REM sleep. In 1998, Allen Braun of the National Institutes of Health and colleagues reported in *Science* that they found differences in blood flow through the brain during different stages of sleep. Comparing positive emission tomography (PET) scans of sleeping subjects, they found that during REM sleep, blood flow in the frontal lobes was reduced, whereas blood flow increased in areas that produce visual scenes and emotion. Other investigators have suggested that during dreams, release of the neurotransmitter **norepinephrine** increases within the RAS, generating stimulating impulses that are fed into the cerebral cortex.

The hypothalamus and brain stem regulate the sleep-wake cycle. Using rats, investigators have demonstrated that stimulation of the *preoptic area* of the hypothalamus results in non-REM sleep. The preoptic nucleus is located near the **suprachiasmatic nucleus,** which has been shown to be the most important of the body's biological clocks. The suprachiasmatic nucleus receives information about the duration of light and dark from the retina of the eye and apparently transmits this information to the preoptic nucleus. The fatigue and decreased physical and mental performance referred to as jet lag occur when we fly to a different time zone where the body is no longer synchronized with the light-dark cycle.

The *raphe nuclei* in the brain stem (lower pons and medulla) also appear to be important in producing REM sleep. Many of the neurons that project from the raphe nuclei release **serotonin,** a neurotransmitter involved in sleep. During REM sleep, neurons in the brain stem also release norepinephrine, which may stimulate heightened activity in certain regions of the brain.

The sleep-wake cycle may be affected by fatigue of the RAS after many hours of activity. Sleep centers are then activated, and their neurons release serotonin. After sufficient rest, the inhibitory neurons of the sleep centers become less excitable, while the excitatory neurons of the RAS become more excitable.

Although birds, mammals, and many other animals have a sleep-wake cycle, neurophysiologists do not know why sleep is necessary. When a person stays awake for unusually long periods, fatigue and irritability result, and even routine tasks cannot be performed well.

Not only is non-REM sleep required, but REM sleep is apparently also necessary. In sleep deprivation experiments performed with human volunteers, lack of REM sleep makes subjects anxious and irritable. When the subjects are permitted to sleep normally again, they spend more time than usual in the REM stage for a period. Many types of drugs, for example, amphetamines, alter sleep patterns and affect the amount of REM sleep. Certain drugs that induce sleep may increase total sleeping time but decrease REM time. When a person stops taking such a drug, several weeks may be required before normal sleep patterns are reestablished.

The limbic system affects emotional aspects of behavior

The **limbic system,** an action system of the brain, includes parts of the cerebrum (parts of the frontal lobe, temporal lobe, hippocampus, and amygdala), parts of the thalamus, and hypothalamus, several nuclei in the midbrain, and the neural pathways that connect these structures. The limbic system affects the emotional aspects of behavior, evaluates rewards, and is important in motivation. This system also plays a role in sexual behavior, biological rhythms, and autonomic responses. Stimulation of certain areas of the limbic system in an experimental animal results in increased general activity and may cause fighting behavior or extreme rage.

Investigators first discovered that the limbic system is an important motivational system when they implanted electrodes in certain areas of the brains of laboratory animals. They found that a rat may press a lever that stimulates the so-called reward area as many as 15,000 times per hour. The rat will forego food and water in favor of self-stimulation until it drops from exhaustion. One of the areas that makes up this motivational system is the substantia nigra in the midbrain. When something happens that feels good, neurons in this area release dopamine. Drugs such as cocaine and amphetamines block the reuptake of dopamine, prolonging the stimulation.

Dopamine neurons are activated by novel stimuli that are associated with reward or pleasure. These neurons signal us about surprising events or stimuli that predict rewards. Dopamine may motivate us to act when something important is happening. These mechanisms may help us understand attention deficit disorder (ADD) and schizophrenia, which involve abnormal amounts of dopamine in the brain. Individuals with these disorders have difficulty filtering out sensory stimuli. The presence of excess dopamine might drive them to divert their attention to so many sensory stimuli that they have difficulty focusing on the most important stimuli.

Using PET scans, Dr. Mark George and his team at the National Institute of Mental Health in Bethesda, Maryland, showed that areas outside the limbic system are also involved in emotion. In 1995 George reported that when his subjects were happy, activity decreased in the temporal-parietal region of the cortex, an area involved in complex planning. During sadness, the left prefrontal area of the cortex, and the **amygdala,** a pair of structures within the limbic system, were activated. In subjects with clinical depression, the left prefrontal area appeared to be shut down, and subjects felt emotionally numb.

Learning involves the storage of information and its retrieval

Learning is the process by which we acquire information as a result of experience. Laboratory experiments have shown that members of every animal phylum can learn. For learning to occur, we must be able to remember what we experience. **Memory** is the process by which information is encoded, stored, and retrieved. Learning and memory can result in changes in behavior.

INFORMATION PROCESSING Two forms of memory are recognized: **implicit** (or declarative) and **explicit** (nondeclarative). **Implicit memory** is unconscious memory for perceptual and motor skills, for example, riding a bicycle. **Explicit memory** involves factual knowledge of people, places, or objects, and it requires conscious recall of the information.

At any moment you are bombarded with thousands of bits of sensory information. At this very moment, your eyes are receiving information about the words on this page, the objects around you, and the intensity of the light in the room. At the same time you may be hearing a variety of sounds—music, your friends talking in the next room, the hum of an air conditioner. Your olfactory epithelium may sense cologne or the smell of coffee. Sensory receptors in your hands may be receiving information regarding the weight and position of your book. Most sensory stimuli are not important to remember and so are filtered out.

When we focus our attention on certain bits of sensory, motor, or other information, that information may be further processed for storage. In both implicit and explicit memory, memory storage involves both **short-term memory,** which lasts only minutes, and **long-term memory,** which may last for a long time.

Memory storage begins with recognition. For stimuli to be recognized, we must relate them to past experience or knowledge. We recognize patterns and begin to encode information. Typically, we can hold only about seven chunks of information (a chunk is some unit such as a word, syllable, or number) at a time in short-term memory.

Short-term memory allows us to recall information for a few minutes. Usually when we look up a phone number, for example, we remember it only long enough to dial. Should we need the same number an hour later, most of us would have to look it up again. One hypothesis of short-term memory suggests that reverberating circuits are involved (see Fig. 39–13). A memory circuit may continue to reverberate for several minutes until it fatigues or until new signals are received that interfere with the old.

When information is selected for long-term storage, the brain rehearses the material and then stores it in long-term memory in association with similar memories. Several minutes are required for a memory to become consolidated within long-term memory. Should a person suffer a brain concussion or undergo electroshock therapy, memory of what happened immediately prior to the incident may be completely lost. This is known as *retrograde amnesia.* When the hippocampus is damaged, short-term memory is unimpaired and the patient can still recall information stored in the past. However, new short-term memories can no longer be converted to long-term memories.

Retrieval of information stored in long-term memory is of considerable interest, especially to students. Some researchers think that once information is consolidated in long-term memory, it remains within the brain permanently. The challenge is to find it when you need it. When you seem to forget something, the problem may be that you have not effectively searched for the memory. Information retrieval can be improved by careful storage. One method is to form strong associations between items when they are being stored.

Where are memories stored? Researchers have worked with the brains of experimental animals for years without finding specific regions where information is stored. Some forms of learning take place in association areas within lower brain regions, the thalamus, for example. Even very simple animals such as flatworms that completely lack a cerebral cortex are capable of some types of learning.

When large areas of the mammalian cerebral cortex are destroyed, information is lost somewhat in proportion to the extent of lost tissue. However, no specific area can be labeled the "memory bank." Rather, memories appear to be stored within the many sensory areas of the brain. For example, visual memories may be stored in the visual centers of the occipital lobes, and auditory memories may be stored in the temporal lobes. Memories are thought to be integrated in many areas of the brain, including association areas of the cerebral cortex, parts of the limbic system (the amygdala and the hippocampus), and the thalamus and hypothalamus.

Wernicke's area in the temporal lobe has been identified as a very important association area for complex thought processes. This area is an important center for language function involved in the recognition and interpretation of words. Neurons within the association areas form highly complex pathways that permit complicated reverberation.

NEUROPHYSIOLOGICAL CHANGES DURING LEARNING

Short-term memory involves brief changes in proteins, resulting in strengthening of synaptic connections. The changes take place in the neurotransmitter receptors of postsynaptic neurons. The receptors are linked by second messengers to ion channels. The second messenger cyclic AMP (cAMP) is thought to facilitate short-term memory by activating a specific protein kinase. By phosphorylating certain molecules, this protein kinase affects specific ion channels.

In long-term memory, gene expression and protein synthesis take place. This process involves slower, but longer lasting, changes in synaptic connections. Long-term memory depends on activated receptors that are linked to G proteins. Cyclic AMP acts as a second messenger. A high level of cyclic AMP activates a protein kinase, which enters the nucleus, leading to gene activation and protein synthesis. In this process, protein kinase phosphorylates a regulatory protein (a transcription factor) known as **CREB** (for cyclic AMP response element binding protein). CREB then turns on the transcription process of certain genes. CREB has been shown to be a signaling molecule in the memory pathway in many animals, including fruit flies and mice. The molecules and processes involved in learning and memory have been highly conserved during evolution.

In some types of learning, changes take place in presynaptic terminals or postsynaptic neurons that permanently enhance or inhibit the transmission of impulses. In some cases, specific neurons may become more sensitive to neurotransmitter.

Unraveling the mechanisms that change the strength of synaptic connections appears to be an important key to understanding the process of learning. Neurons typically transmit action potentials in bursts, and the amount of neurotransmitter released by each action potential may increase or decrease. An increase, referred to as **synaptic enhancement,** is thought to occur as a result of calcium ion accumulation inside the presynaptic terminal. One type of synaptic enhancement, **synaptic facilitation,** can occur after a single action potential. However, synaptic facilitation lasts for only milliseconds. **Potentiation** (which means to strengthen or make more potent) is a longer form of synaptic enhancement that can last for several minutes. It occurs when a presynaptic neuron continues to transmit action potentials at a high rate for a minute or longer.

Process of Science For more than 30 years, neurophysiologists have studied learning in *Aplysia,* a slug-like marine mollusk (a gastropod) (Fig. 40–14). Because its nervous system is relatively simple and because the cellular mechanisms of neural functioning have been conserved during evolution, *Aplysia* has proved to be a good model of memory and learning in other animals, including mammals. Eric Kandel of Columbia University, one of three neurobiologists who shared the 2000 Nobel Prize in physiology or medicine, used *Aplysia* to demonstrate how synaptic changes produce memory and learning.

An important focus of study has been the gill-withdrawal reflex, a relatively simple neural circuit that helps protect the animal from predators. *Aplysia*'s gill is located in its mantle (a fold of tissue), which is normally slightly open, allowing sea water to bathe the gill. The water exits through the animal's siphon. When *Aplysia*'s exposed siphon is touched, a reflex action is initiated resulting in withdrawal of the siphon and gill into the mantle. This behavior is a reflex action that can be modified by experience.

The neural pathway includes sensory neurons that synapse directly with motor neurons. When a sensory neuron is stimulated, a message is transmitted to a motor neuron that stimulates the gill to withdraw. Interneurons also synapse with sensory and motor neurons; they can be activated by the sensory neurons and then signal the motor neurons. Using this model, Kandel and other neurophysiologists have been able to study several neural processes, including habituation, sensitization, and classical conditioning.

Habituation is the decrease in response that occurs after repeated exposure to a harmless stimulus. This process permits animals to disregard constant stimuli. When *Aplysia*'s siphon is repeatedly stimulated, the gill-withdrawal reflex habituates and the gill is no longer withdrawn when the siphon is touched. By studying the mechanism of habituation in this marine mollusk, investigators learned how habituation occurs in many other animals. They found that, with repeated stimulation, sensory neurons release less neurotransmitter, resulting in a decreased num-

(a)

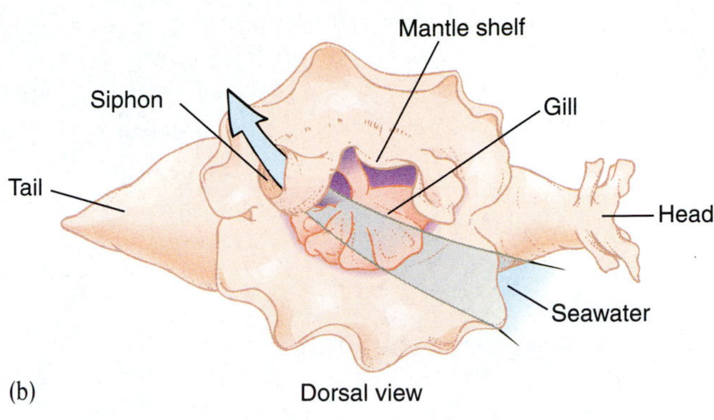

(b)

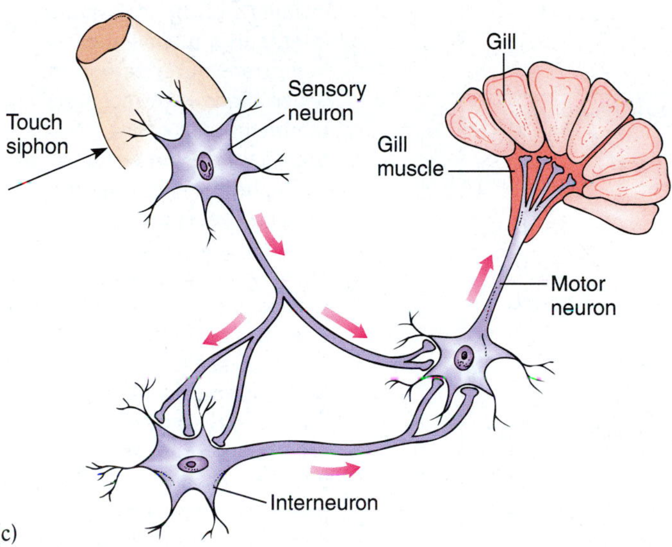

(c)

Figure 40–14 A model for studying learning. **(a)** Neurobiologists have taken advantage of the relatively simple nervous system of the marine mollusk *Aplysia* to study learning. **(b)** Structure of *Aplysia.* Seawater flows in through the slightly open mantle, bathes the gills, then exits through the siphon. **(c)** The neural circuit for the withdrawal reflex in *Aplysia.* Sensory neurons that innervate the siphon are sensitive to touch. These neurons project directly to the motor neurons of the gill muscles. They also project indirectly to the gill muscles by way of interneurons. When the siphon is touched, the siphon and gill are withdrawn into the protective mantle. *(Randy Morse/Tom Stack & Associates)*

ber of action potentials in the motor neurons. The decrease in neurotransmitter release was traced to inactivation of calcium channels in the presynaptic terminals. This leads to a reduction in the number of calcium ions entering the presynaptic terminals in response to repeated action potentials. (Recall from Chapter 39 that the influx of calcium ions stimulates neurotransmitter release.)

Sensitization results in an increased response after experience of an unpleasant, or harmful, stimulus. Stimulating *Aplysia*'s tail with a series of electric shocks sensitizes the gill-withdrawal reflex. This enhanced behavioral response was studied by Kandel and his research team. Sensory neurons in the tail synapse on facilitatory interneurons that end on the siphon sensory neuron and on other interneurons (Fig. 40–15*a*). Note that these facilitatory interneurons synapse on the synaptic terminals of siphon sensory neurons. Sensitization depends on facilitation (enhancement) of neurotransmitter release. Kandel reported that when these interneurons are stimulated, they release serotonin (Fig. 40–15*b*). The serotonin initiates a complex series of reactions involving G-proteins and cAMP. Cyclic AMP activates protein kinase A, which inactivates potassium channels that are important in repolarization of the sensory neuron. The inactivation of these channels provides a longer time for the axon terminal to remain depolarized. As a result more calcium enters the axon terminal and more neurotransmitter is released. This stimulates a stronger response in the postsynaptic motor neuron. Thus, sensitization depends on an increase in neurotransmitter release by the sensory neuron.

Classical conditioning is a form of learning in which an association is formed between some normal response to a normal stimulus, called the unconditioned stimulus (US) and a new stimulus, the conditioned stimulus (CS). The animal learns the association, and the CS then elicits the same response, now called the conditioned response (CR). (Classical conditioning and other types of learning are described in Chapter 50.) *Aplysia*'s gill-withdrawal response can be classically conditioned (Fig. 40–16). A strong electric shock applied to the tail causes a strong unconditioned withdrawal response. In contrast, when the siphon is lightly touched, only a mild response occurs. When the light touch is paired with the electric shock several times, the mollusk *learns* to respond to a light touch (CS) to the siphon with a strong withdrawal response (CR).

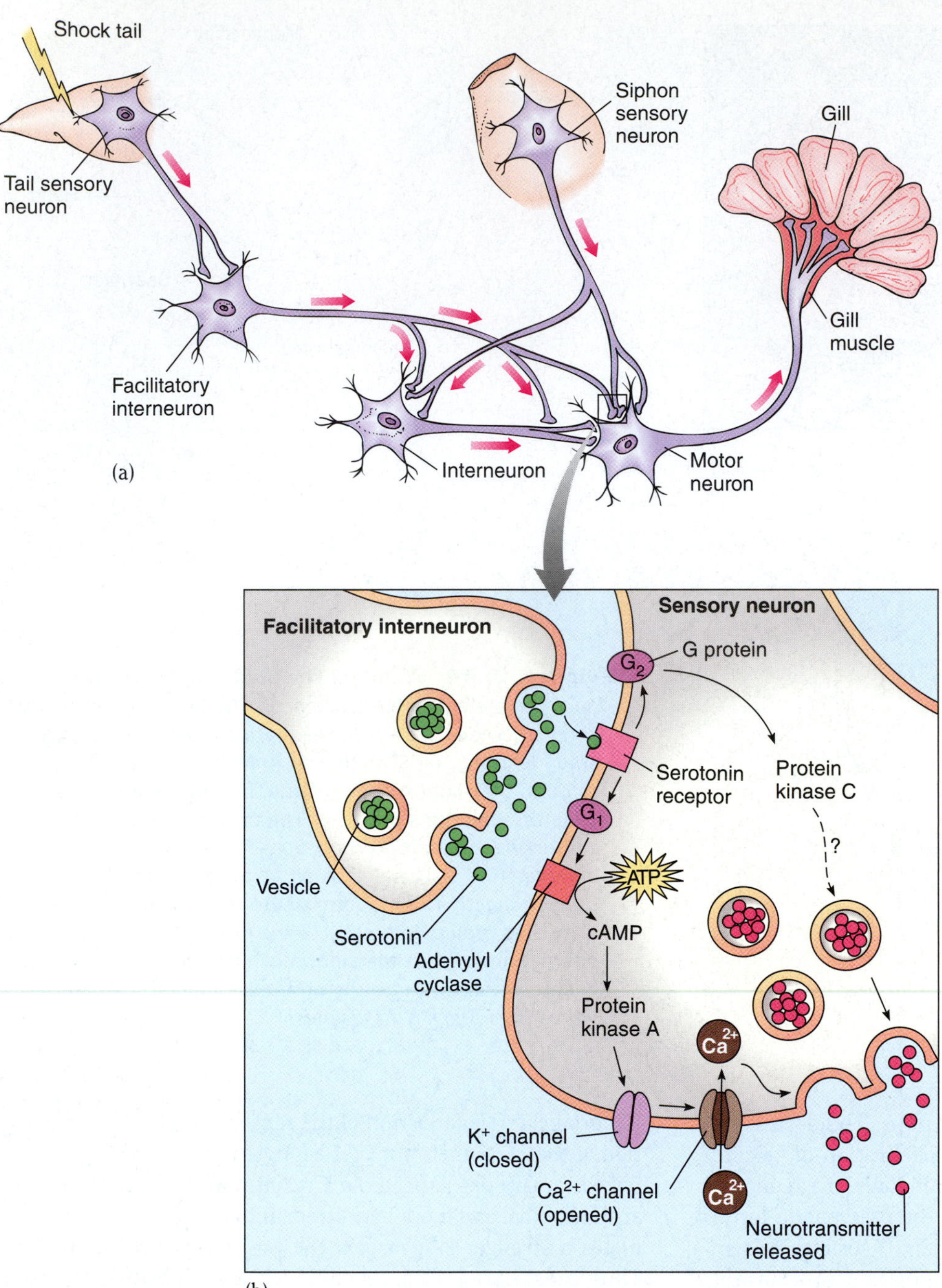

Figure 40–15 Sensitization of the gill-withdrawal reflex in the marine mollusk *Aplysia.* **(a)** Unpleasant stimuli applied to the tail activate facilitatory interneurons. These interneurons signal the presynaptic terminals of siphon sensory neurons. After sensitization, the duration of the presynaptic action potential increases, and larger excitatory postsynaptic potentials are produced. **(b)** A model for the events associated with sensitization of the gill-withdrawal reflex. The axon terminals of the facilitatory interneuron release the neurotransmitter serotonin, which binds with receptors in the membrane of the presynaptic axon terminals of the siphon sensory neuron. The activated receptor activates two G proteins. One activates adenylyl cyclase, which increases cyclic AMP. The cAMP stimulates protein kinase A, which closes K^+ channels. This action promotes passage of Ca^{2+} into the neuron. The second G protein activates protein kinase C, which may potentiate (increase) neurotransmitter release by mobilizing reserve neurotransmitter vesicles.

Classical conditioning is similar to sensitization in that interneurons release serotonin. However, sensitization normally lasts only a few minutes, whereas conditioning can last for weeks, months, or years. One difference may be that in conditioning the sensory neuron is already active when it is signaled by the facilitatory interneuron. If a sensory neuron is already active when serotonin binds with it, it is strongly affected and becomes conditioned. When it fires again, it releases an increased amount of neurotransmitter, resulting in a strong behavioral response. The cellular mechanism is thought to involve a G protein, cAMP, and a protein kinase. The protein kinase may activate genes and protein synthesis.

Recall from the chapter opener that repeated electrical stimulation of neurons causes a functional change at synapses. A series of high-frequency electrical stimuli result in a long-lasting, *increased* response of postsynaptic neurons. The increased strength of the synaptic connections is a response known as **long-term potentiation (LTP).** Low-frequency stimulation of neurons results in a long-lasting *decrease* in the strength of their synaptic connections, called **long-term depression (LTD),** also known as **synaptic depression.** Information storage and forgetting appear to depend on the strengthening and weakening of synaptic connections brought about by LTP and LTD. For example, classical conditioning of *Aplysia*'s siphon-withdrawal reflex

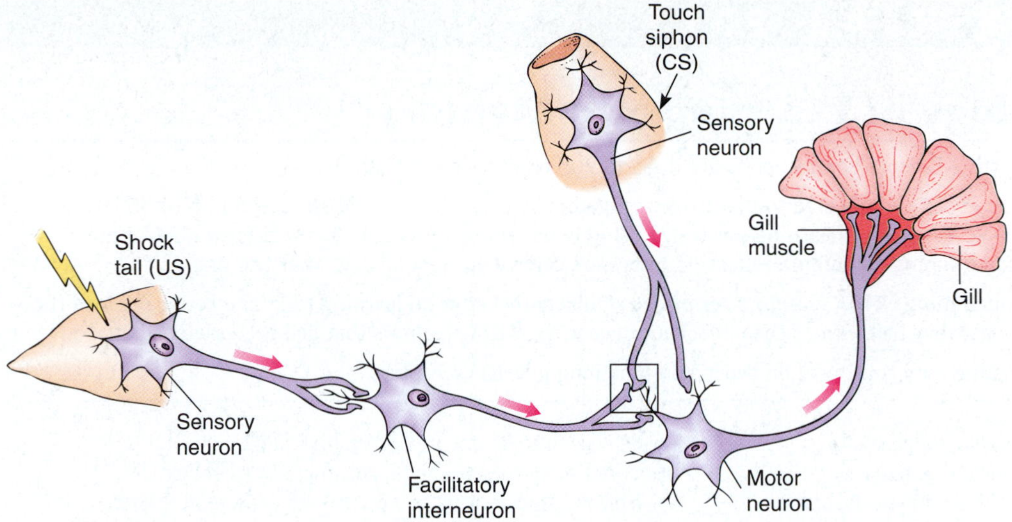

Figure 40–16 Classical conditioning of a withdrawal response in the marine mollusk *Aplysia*. An electric shock to the tail is the unconditioned stimulus (US). A sensory neuron from the tail is stimulated and signals a facilitatory interneuron. The interneuron releases neurotransmitter, which signals the motor neurons to withdraw the gill (the unconditioned response, or UR). A light touch to the siphon is the conditioned stimulus (CS). During conditioning these stimuli (the electric shock and the light touch) are paired. Neurotransmitter from the interneuron then stimulates the already activated sensory neuron from the siphon. The sensory neuron is strongly affected, and the next time it fires it releases a large amount of neurotransmitter. This causes a response as strong as that elicited by the original shock stimulus to the tail. Thus, the light touch (the CS) now elicits an intense response that withdraws the gill (the conditioned response).

has been shown to be mediated by LTP. LTP and LTD have been shown to depend on the activation of NMDA receptors on postsynaptic neurons (see *On the Cutting Edge: The Neurophysiology of Learning and Memory*).

EFFECTS OF EXPERIENCE Many studies have demonstrated neural plasticity, the ability to change in response to environmental stimuli, in rats and other laboratory animals exposed to enriched environments. In contrast with rats housed in standard cages and provided with the basic necessities, those exposed to enriched environments are given toys and other stimulating objects, as well as the opportunity to socially interact with other rats. Animals reared in a complex environment exhibit increased synaptic contacts and may be able to process and remember information more quickly than animals not given such advantages. In 1997, Gerd Kempermann and colleagues reported in the journal *Nature* that mice reared in an enriched environment developed significantly greater numbers of neurons in the hippocampus. These investigators also reported that the experimental mice learned a maze significantly faster than controls.

Early environmental stimulation can also enhance the development of motor areas in the brain. For example, the brains of rats encouraged to exercise become slightly heavier than those of control animals. Characteristic changes occur within the cerebellum, including the development of larger dendrites.

Apparently, during early life certain critical or sensitive periods of nervous system development occur that are influenced by environmental stimuli. For example, when the eyes of young mice first open, neurons in the visual cortex develop large numbers of dendritic spines (structures on which synaptic contact takes place). If the animals are kept in the dark and deprived of visual stimuli, fewer dendritic spines form. If the mice are exposed to light later in life, some new dendritic spines form, but never the number that develop in a mouse reared in a normal environment.

Studies linking the development of the brain with environmental experience indicate that early stimulation is important for the sensory, motor, and intellectual development of children. Such studies have led to the rapidly expanding educational toy market and to widespread acceptance of early education programs. Researchers have also suggested that continuing environmental stimulation is needed to maintain the status of the cerebral cortex in later life.

THE PERIPHERAL NERVOUS SYSTEM INCLUDES SOMATIC AND AUTONOMIC SUBDIVISIONS

The peripheral nervous system (PNS) consists of the sensory receptors, the nerves that link these receptors with the CNS, and

ON THE CUTTING EDGE

The Neurophysiology of Learning and Memory

HYPOTHESIS: NMDA (*N*-methyl-D-aspartate) receptors are important in memory consolidation.

METHOD: Mice were genetically engineered so that the investigators could switch the NMDA receptor off or on in the hippocampus (CA1 region). Learning was tested using two memory tasks (a hidden-platform water maze and a fear-conditioning paradigm). Retention tests were carried out several days after training.

RESULTS: Mice with functioning NMDA receptors performed significantly better on learning tasks and better retained the memory of what they had learned than knockout mice with NMDA receptors that had been switched off.

CONCLUSION: Learning and memory consolidation depend on functioning NMDA receptors.

Learning and memory depend on synaptic plasticity, which involves complex biochemical processes, including actions of neurotransmitters and postsynaptic receptors. Increase in synaptic strength, known as long-term potentiation (LTP), is an important mechanism in learning. Investigators have also described long-term depression (LTD), which is a long-lasting decrease in the strength of synaptic connections. LTP and LTD apparently take place in many parts of the nervous system, including the hippocampus of mammals. Recall that the hippocampus is critically important in forming memories.

Induction of LTP and LTD require activation of NMDA receptors, present on the plasma membranes of postsynaptic neurons. These receptors control the passage of calcium ions into neurons. NMDA receptors belong to a class of receptors known as *ionotropic glutamate receptors.* These receptors respond to the neurotransmitter glutamate by opening ion channels that allow Ca^{2+} and Na^{+} to diffuse into the neuron. Several types of ionotropic glutamate receptors are known. An important difference among them is that some types actively bind NMDA (*N*-methyl-D-aspartate), an artificial ligand. (Recall that a **ligand** is any molecule that binds to a specific receptor.)

When the postsynaptic neuron is at its resting potential, the NMDA ion channels are blocked by Mg^{2+} ions. The channels can open only after the neuron is depolarized and glutamate binds to the receptors. Because the NMDA receptors require both ligand binding and a particular voltage, they are *doubly gated receptors.*

A great deal of evidence now links the NMDA receptor to learning and memory formation, but its role in memory consolidation, a process that can take several days or weeks after initial learning, has been difficult to demonstrate. In 2000, Eiji Shimizu and colleagues reported in *Science** that they tested genetically engineered animals in which they could selectively eliminate and restore NMDA receptor function in neurons of the hippocampus (CA1 region).

Shimizu and colleagues were able to demonstrate that after initial learning took place, functioning NMDA receptors in the hippocampus were necessary for memory consolidation. Without them, the mice did not remember what they had learned.

By changing the time after learning at which NMDA receptors were turned on or off, Shimizu and colleagues showed that these NMDA receptors were not required for memory retrieval. These results suggest that once the hippocampus transfers newly created memories to the cortex, they are permanently stored in the cortex, and the hippocampus is not needed for retrieving the information.

One model for the neurophysiological basis of LTP is illustrated in the figure. A presynaptic neuron releases glutamate, which combines with non-NMDA receptors. The postsynaptic neuron becomes depolarized, and Mg^{2+} moves away from the NMDA receptors, unblocking them. When the NMDA receptor channel opens, Ca^{2+} moves into the cell. The Ca^{2+} appears to be an important trigger for LTP.

Calcium ions act as a second messenger initiating long-term changes. One effect of Ca^{2+} may be to stimulate release of a **retrograde signal** (one that moves backward) from the *postsynaptic* neuron and signals the *presynaptic* neuron. This signal would enhance neurotransmitter release by the presynaptic neuron, thus strengthening the relationship between the two neurons.

The retrograde signal appears to be a soluble gas, perhaps nitric oxide (NO). Calcium ions act on the enzyme NO synthase, which produces NO. This gas passes easily through

* Shimizu, E., Y.P. Tang, C. Rampon, and J.Z. Tsien. "NMDA Receptor-Dependent Synaptic Reinforcement as a Crucial Process for Memory Consolidation." *Science,* Vol. 290, 10 Nov. 2000.

the nerves that link the CNS with the effectors (muscles and glands). The portion of the PNS that helps the body respond to changes in the external environment is the somatic system. The nerves and receptors that maintain homeostasis despite internal changes make up the autonomic nervous system.

The somatic division helps the body adjust to the external environment

The **somatic division** of the nervous system includes the receptors that react to changes in the external environment, the sensory neurons that keep the CNS informed of those changes, and the motor neurons that adjust the positions of the skeletal mus-cell membranes, and because it is very reactive, it is active for only a few seconds. NO may also signal nearby neurons, thus recruiting more neurons in the learning process.

As this brief glimpse of the cellular processes involved in synaptic plasticity indicates, the neurophysiology of learning and memory are extremely complex, and it is likely that other mechanisms for LTP will be discovered through the application of new techniques, including genetic engineering.

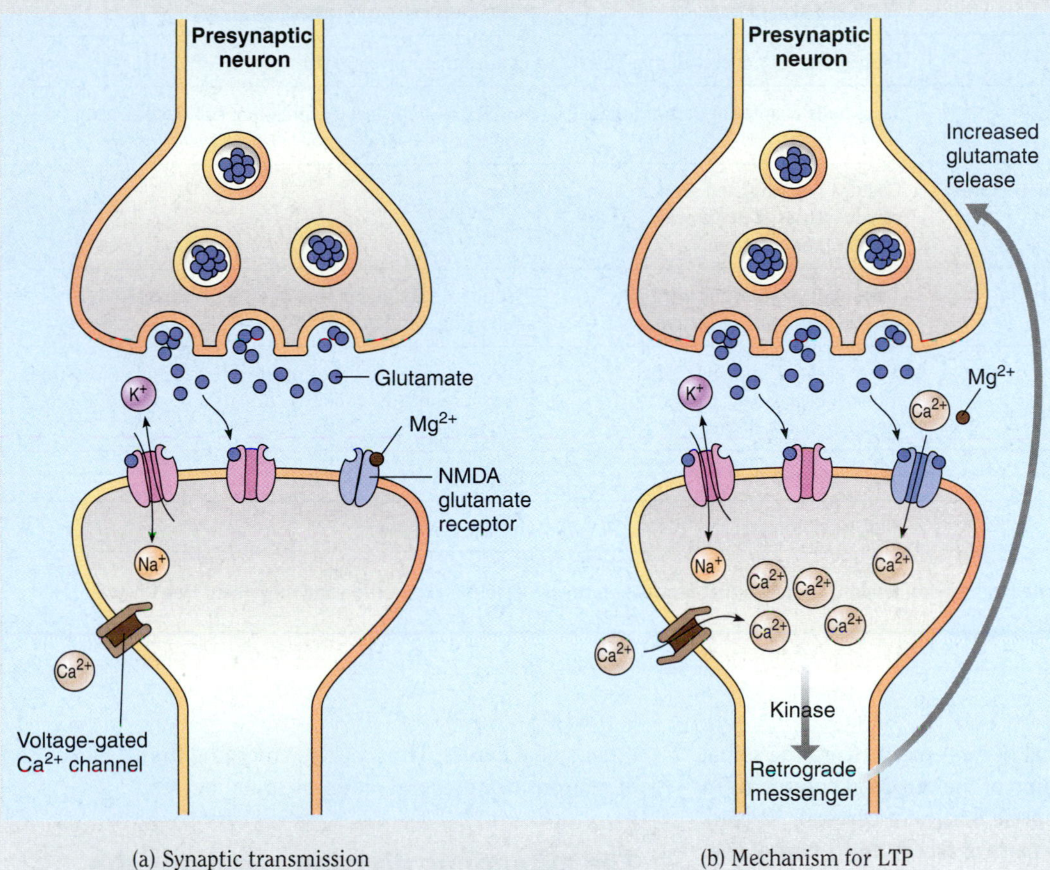

(a) Synaptic transmission

(b) Mechanism for LTP

A proposed model for the mechanism of LTP. **(a)** A single stimulus results in glutamate release from a presynaptic neuron. The glutamate acts mainly on non-NMDA receptors. NMDA receptors are blocked by Mg^{2+}. **(b)** The combination of a series of action potentials and glutamate binding activates NMDA receptors. Calcium channels open, allowing Ca^{2+} to enter the neuron and activate a second messenger system. A retrograde signal is produced that signals the presynnaptic neuron to release more glutamate.

cles that help maintain the body's posture and balance. In mammals, 12 pairs of **cranial nerves** emerge from the brain. They transmit information regarding the senses of smell, sight, hearing, and taste from the special sensory receptors as well as information from the general sensory receptors, especially in the head region. For example, cranial nerve II, the optic nerve, transmits signals from the retina of the eye to the brain. The cranial nerves also bring orders from the CNS to the voluntary muscles that control movements of the eyes, face, mouth, tongue, pharynx, and larynx. Cranial nerve VII, the facial nerve, for example, transmits signals to the muscles used in facial expression and to the salivary glands. The names of the cranial nerves and the structures they innervate are given in Table 40–4.

TABLE 40–4 The Cranial Nerves of Mammals

Number	Name	Origin of Sensory Fibers	Effector Innervated by Motor Fibers
I	Olfactory	Olfactory epithelium of nose (smell)	None
II	Optic	Retina of eye (vision)	None
III	Oculomotor	Proprioceptors* of eyeball muscles	Muscles that move eyeball; muscles that change shape of lens; muscles that constrict pupil
IV	Trochlear	Proprioceptors* of eyeball muscles	Muscles that move eyeball
V	Trigeminal	Teeth and skin of face	Some muscles used in chewing
VI	Abducens	Proprioceptors* of eyeball muscles	Muscles that move eyeball
VII	Facial	Taste buds of anterior part of tongue	Muscles used for facial expression; submaxillary and sublingual salivary glands
VIII	Auditory (vestibulocochlear)	Cochlea (hearing) and semicircular canals (senses of movement, balance, and rotation)	None
IX	Glossopharyngeal	Taste buds of posterior third of tongue and lining of pharynx	Parotid salivary gland; muscles of pharynx used in swallowing
X	Vagus	Nerve endings in many of the internal organs (e.g., lungs, stomach, aorta, larynx)	Parasympathetic fibers to heart, stomach, small intestine, larynx, esophagus, and other organs
XI	Spinal accessory	Muscles of shoulder	Muscles of neck and shoulder
XII	Hypoglossal	Muscles of tongue	Muscles of tongue

*Proprioceptors are receptors located in muscles, tendons, or joints that provide information about body position and movement (see Chapter 41).

In humans, 31 pairs of spinal nerves emerge from the spinal cord. Named for the general region of the vertebral column from which they originate, they comprise 8 pairs of cervical, 12 pairs of thoracic, 5 pairs of lumbar, 5 pairs of sacral, and 1 pair of coccygeal spinal nerves.

Each spinal nerve has a dorsal root and a ventral root (Fig. 40–17). The dorsal root consists of afferent fibers that transmit information from the sensory receptors to the spinal cord. Just before the dorsal root joins with the cord, it is marked by a swelling, the spinal ganglion, which consists of the cell bodies of the sensory neurons. The ventral root is a grouping of the efferent fibers leaving the cord en route to the muscles and glands. Cell bodies of the motor neurons occur within the gray matter of the cord.

Peripheral to the junction of the dorsal and ventral roots, each spinal nerve divides into branches. The autonomic branch innervates the viscera. The dorsal branch serves the skin and muscles of the back. The ventral branch serves the skin and muscles of the sides and ventral part of the body. The ventral branches of several spinal nerves form tangled networks called *plexi* (sing., *plexus*). Within a plexus, the fibers of a spinal nerve may separate and then regroup with fibers that originated in other spinal nerves. Thus, nerves emerging from a plexus consist of neurons from several different spinal nerves.

The autonomic division regulates the internal environment

The **autonomic division** helps maintain homeostasis in the internal environment. For instance, it regulates the heart rate and helps maintain a constant body temperature. The autonomic system works automatically and without voluntary input. Its effectors are smooth muscle, cardiac muscle, and glands. Like the somatic system, it is functionally organized into reflex pathways. Receptors within the viscera relay information via afferent nerves to the CNS. The information is integrated at various levels. Then the decision is transmitted along efferent nerves to the appropriate muscles or glands.

Afferent and efferent neurons of the autonomic division are found within cranial or spinal nerves. For example, afferent fibers of cranial nerve X, the vagus nerve, transmit signals from many organs of the chest and upper abdomen to the CNS. Its efferent fibers transmit signals from the brain to the heart, stomach, small intestine, and several other organs.

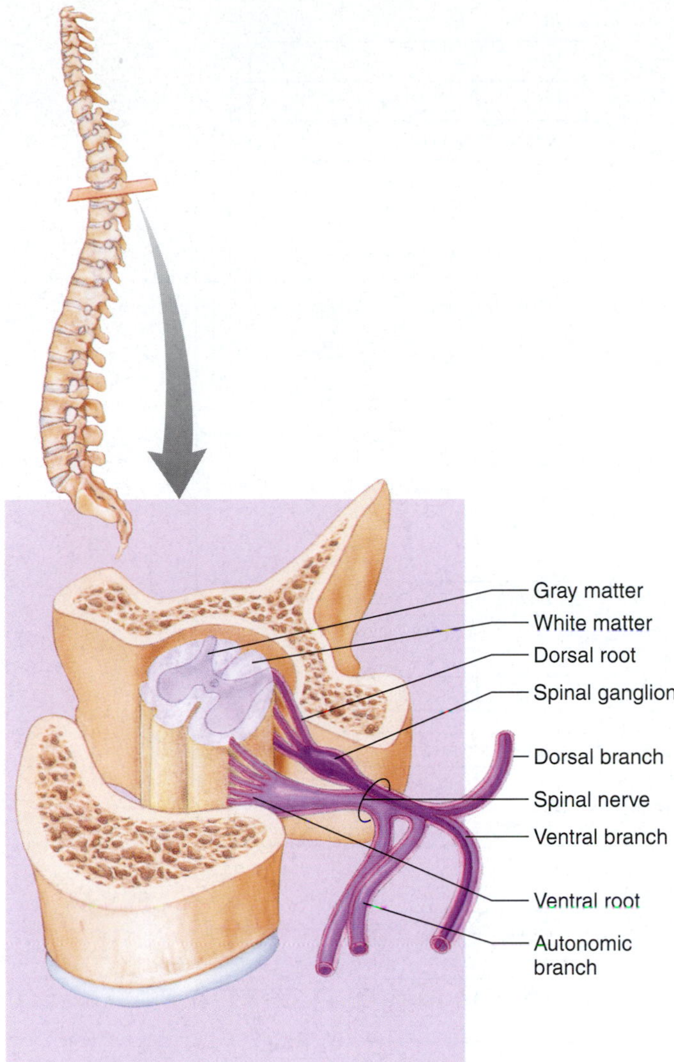

Figure 40–17 Spinal nerve. Dorsal and ventral roots emerge from the spinal cord and join to form a spinal nerve. Each spinal nerve divides into several branches

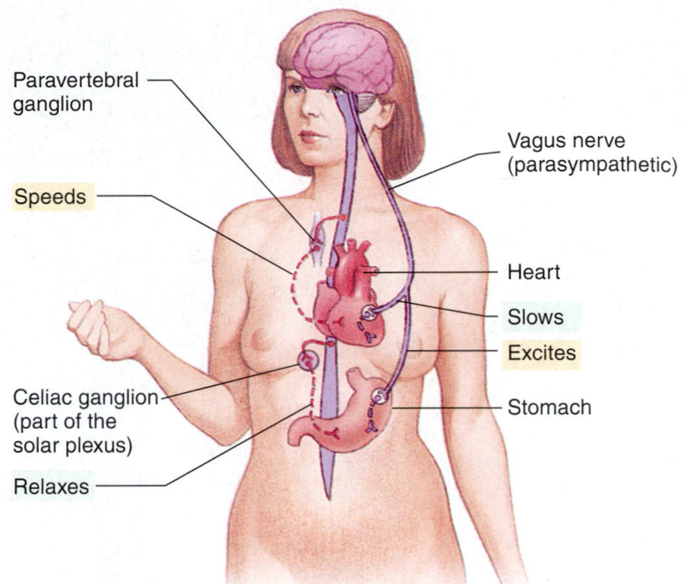

Figure 40–18 Dual innervation of the heart and stomach. Many organs are innervated by both sympathetic and parasympathetic nerves. A sympathetic nerve increases the heart rate whereas the vagus nerve, a parasympathetic nerve, slows the heartbeat. In contrast, sympathetic nerves slow contractions of the stomach and intestine while the vagus nerve stimulates contractions of the digestive organs. Sympathetic nerves are shown in red; postganglionic fibers are shown as dotted lines.

The efferent portion of the autonomic division is subdivided into **sympathetic** and **parasympathetic systems.** Many organs are innervated by both types of nerves. In general, sympathetic and parasympathetic systems have *opposite* effects (Figs. 40–18 and 40–19; Table 40–5). For example, the heart rate is speeded up by messages from sympathetic nerve fibers and slowed by impulses from its parasympathetic nerve fibers. In many cases sympathetic nerves operate to stimulate organs and to mobilize energy, especially in response to stress, whereas the parasympathetic nerves influence organs to conserve and restore energy, particularly during quiet, calm activities.

Instead of using a single efferent neuron, as in the somatic system, the autonomic system uses a relay of two neurons between the CNS and the effector. The first neuron, called the **preganglionic neuron,** has a cell body and dendrites within the CNS. Its axon, part of a peripheral nerve, ends by synapsing with a **postganglionic neuron.** The dendrites and cell body of the postganglionic neuron are in a ganglion outside the CNS. Its axon terminates near or on the effector. The sympathetic ganglia are paired, and a chain of them, the **paravertebral sympathetic ganglion chain,** occurs on each side of the spinal cord from the neck to the abdomen. Some sympathetic preganglionic neurons do not end in these ganglia but instead pass on to **collateral ganglia** in the abdomen, close to the aorta and its major branches. Parasympathetic preganglionic neurons synapse with postganglionic neurons in **terminal ganglia** near or within the walls of the organs they innervate.

The sympathetic and parasympathetic systems also differ in the neurotransmitters they release at the synapse with the effector. Both preganglionic and postganglionic parasympathetic neurons secrete acetylcholine. Sympathetic postganglionic neurons release norepinephrine (although preganglionic sympathetic neurons secrete acetylcholine). Table 40–6 compares the actions of the sympathetic and parasympathetic systems on selected effectors.

The autonomic division got its name from the original belief that it was independent of the CNS, that is, autonomous. Neurobiologists have shown that this is not so and that the hypothalamus and many other parts of the CNS help regulate the autonomic system. Although the autonomic system usually functions automatically, its activities can be consciously influenced. Biofeedback provides a person with visual or auditory evidence concerning the status of an autonomic body function.

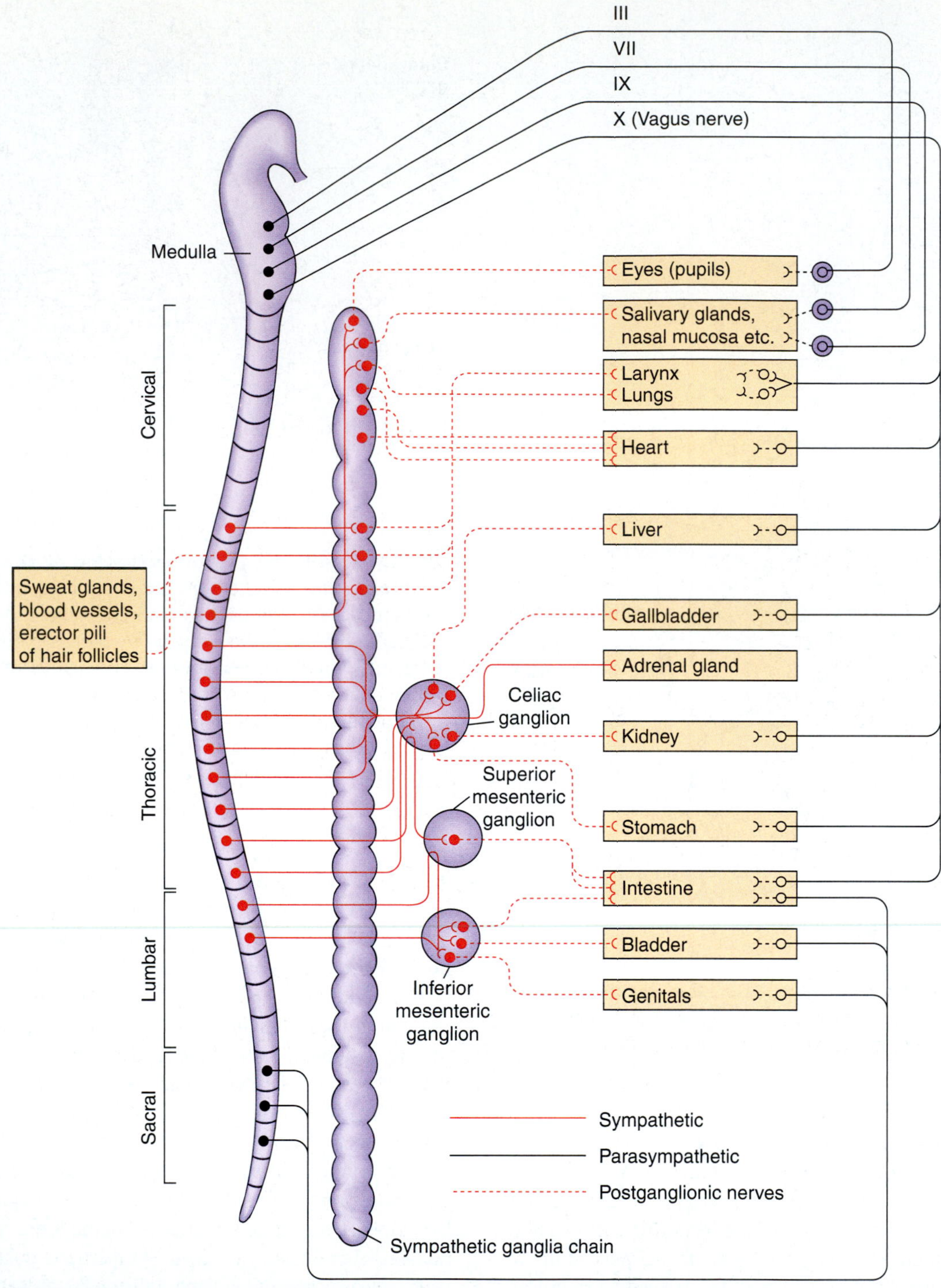

Figure 40–19 The sympathetic and parasympathetic systems. Complex as it appears, this diagram has been greatly simplified. Red lines represent sympathetic nerves, black lines represent parasympathetic nerves, and dotted lines represent postganglionic nerves. See Table 40–6 for specific actions of the nerves.

For example, a tone may be sounded when blood pressure reaches a desirable level. Using such techniques, subjects have learned to control autonomic activities such as blood pressure, brain wave pattern, heart rate, and blood sugar level. Even certain abnormal heart rhythms can be consciously modified.

MANY DRUGS AFFECT THE NERVOUS SYSTEM

About 25% of all prescribed drugs are taken to alter psychological conditions, and almost all the commonly abused drugs affect

TABLE 40–5 Comparison of Sympathetic and Parasympathetic Systems

Characteristic	Sympathetic System	Parasympathetic System
General effect	Prepares body to cope with stressful situations	Restores body to resting state after stressful situation; actively maintains normal body functions
Extent of effect	Widespread throughout body	Localized
Neurotransmitter released at synapse with effector	Norepinephrine (usually)	Acetylcholine
Duration of effect	Lasting	Brief
Outflow from CNS	Thoracic and lumbar nerves from spinal cord	Cranial nerves and sacral nerves from spinal cord
Location of ganglia	Chain and collateral ganglia	Terminal ganglia
Number of postganglionic fibers with which each preganglionic fiber synapses	Many	Few

mood. Many of them act by changing the levels of neurotransmitters within the brain. In particular, levels of norepinephrine, serotonin, and dopamine are thought to influence mood. For example, when excessive amounts of norepinephrine are released in the RAS, we feel stimulated and energetic, whereas low concentrations of this neurotransmitter reduce anxiety.

According to the Substance Abuse and Mental Health Services Administration, an estimated 14.8 million individuals in the United States were current users of illicit drugs in 1999, meaning they used an illicit drug at least once during the 30 days prior to the interview. An estimated 3.6 million were dependent on illicit drugs. Table 40–7 lists several commonly used and abused drugs and gives their effects. Also see *Focus On: Alcohol: The Most Abused of All Drugs* and *Focus On: Crack Cocaine.*

Habitual use of almost any mood-altering drug can result in **psychological dependence,** in which the user becomes emotionally dependent on the drug. When deprived of it, the user craves the feeling of euphoria (well-being) that the drug induces. Some drugs induce **tolerance** after several days or weeks. This means that response to the drug decreases, and greater amounts are required to obtain the desired effect. Tolerance often occurs because the liver cells are stimulated to produce more of the enzymes that metabolize and inactivate the drug. It can also occur owing to a decrease in the number of postsynaptic receptors that bind with the drug (a mechanism known as downregulation).

Drug addiction is a serious societal problem that involves compulsive use of a drug despite negative health consequences and negative effects on the ability to function socially

(text continues on page 882)

TABLE 40–6 Comparison of Sympathetic and Parasympathetic Actions on Selected Effectors*

Effector	Sympathetic Action	Parasympathetic Action
Heart	Increases rate and strength of contraction	Decreases rate; no direct effect on strength of contraction
Bronchial tubes	Dilates	Constricts
Iris of eye	Dilates pupil	Constricts pupil
Sex organs (male)	Constricts blood vessels; ejaculation	Dilates blood vessels; erection
Blood vessels	Generally constricts	No innervation for many
Sweat glands	Stimulates	No innervation
Intestine	Inhibits motility	Stimulates motility and secretion
Liver	Stimulates glycogenolysis (conversion of glycogen to glucose)	No effect
Adipose tissue	Stimulates free fatty acid release from fat cells	No effect
Adrenal medulla	Stimulates secretion of epinephrine and norepinephrine	No effect
Salivary glands	Stimulates thick, viscous secretion	Stimulates a profuse, watery secretion

*Refer to Figure 40–19 as you study this table. Notice that many other examples could be added to this list.

TABLE 40–7 Effects of Some Commonly Used Drugs

Name of Drug	Effect on Mood	Actions on Body	Side Effects/Dangers Associated with Abuse
Antidepressants			
Tricyclic antidepressants (e.g., Elavil, Anafranil)	Elevate mood; relieve depression; also used to treat obsessive-compulsive behavior	Most block reuptake of biogenic amines (especially norepinephrine), increasing their concentration in the synapses	Can cause sedation, weight gain, sexual dysfunction
Selective serotonin reuptake inhibitors (SSRIs) (e.g., Prozac, Zoloft, Celexa)	Relieve depression; used to treat obsessive-compulsive behavior	Block serotonin reuptake	Nausea, headache, insomnia, anxiety
MAO inhibitors (e.g., Parnate, Nardil)	Relieve depression	Block enzymatic breakdown of biogenic amines, increasing their concentration in the synapses	Liver toxicity, excessive CNS stimulation; overdose may affect blood pressure, cause hallucinations
Anti-anxiety Drugs			
Benzodiazepines (e.g., Xanax, Valium, Librium)	Sedation; induce sleep	Bind to GABA receptors on post-synaptic neuron; increase effectiveness of GABA in opening chloride channels, causing hyperpolarization; cause relaxation of skeletal muscles	Drowsiness, confusion; psychological dependence, addiction; effects additive with alcohol
Barbiturates "Downers" (e.g., phenobarbital)	Sedation; induce sleep	Bind to receptor adjacent to GABA receptor on chloride channels; chloride channels open so more chloride ions move in, causing hyperpolarization	Drowsiness, confusion; psychological dependence, tolerance, addiction; overdose can cause severe CNS depression, resulting in coma and death (especially lethal in combination with alcohol)
Antipsychotic Medications			
Phenothiazines (e.g., Thorazine, Mellaril, Stelazine)	Relieve some symptoms of schizophrenia; reduce impulsive and aggressive behavior	Block dopamine receptors	Muscle spasms, shuffling gait, and other symptoms of Parkinson's disease
Newer antipsychotic medications (e.g., Clozaril, Risperdal)	Similar to phenothiazines, but also appear to increase motivation	Block dopamine receptors and several other neurotransmitter receptors	Fewer side effects than with the phenothiazines
Narcotic Analgesics Opiates (e.g., morphine, codeine, Demerol, heroin)	Euphoria; sedation; relieve pain	Mimic actions of endorphins; bind to opiate receptors	Depress respiration; constrict pupils; impair coordination; tolerance, psychological dependence, addiction; convulsions, death from overdose
Cocaine	Euphoria; excitation followed by depression	Stimulates release and inhibits reuptake of norepinephrine and dopamine, leading to CNS stimulation followed by depression; autonomic stimulation; dilation of pupils; local anesthesia	Mental impairment, convulsions, hallucinations, unconsciousness; death from overdose
Marijuana	Euphoria	Impairs coordination; impairs depth perception and alters sense of timing; impairs short-term memory (probably by decreasing acetylcholine levels in the hippocampus); inflames eyes; causes peripheral vasodilation	In large doses, sensory distortions, hallucinations, evidence of lowered sperm counts and testosterone (male hormone) levels

Name of Drug	Effect on Mood	Actions on Body	Side Effects/Dangers Associated with Abuse
Amphetamines			
"Uppers," "pep pills," "crystal meth" (e.g., Dexedrine)	Euphoria, stimulation, hyperactivity	Stimulate release and block reuptake of dopamine and norepinephrine; enhance flow of impulses in RAS, leading to increased heart rate and blood pressure and dilation of pupils	Tolerance, possible physical dependence; hallucinations; increased blood pressure; psychotic episodes; death from overdose
MDMA (3,4-methylenedioxy-methamphetamine) "ecstasy" (most popular of club drugs)	Stimulant, sense of pleasure, self-confidence, feelings of closeness to others; mild hallucinogenic	CNS stimulant; damages serotonin pathways in rat and monkey brains	Hyperthermia (excessive body heat); overdose causes rapid heartbeat, high blood pressure, panic attacks, and seizures; long-term neurological changes; cognitive impairment
Rohypnol (flunitrazepam) "Date rape" drug, "roofies," "roach"	Sedative-hypnotic	Muscle relaxation; amnesia	Physical and psychological dependence; withdrawal may cause seizures; may be lethal when mixed with alcohol and/or other depressants
PCP (phencyclidine) "Angel dust," "ozone"	Feelings of strength, power, invulnerability, and a numbing effect on the mind	Increase in blood pressure and heart rate; shallow respiration; flushing and profuse sweating; numbness of the extremities; muscular incoordination; changes in body awareness, similar to those associated with alcohol intoxication	Psychological dependence, craving; in adolescents, may interfere with hormones related to growth and development; may interfere with learning; can cause hallucinations, delusions, disordered thinking; high doses can cause drop in blood pressure and heart rate, seizures, coma, and death; chronic use can cause memory loss, difficulties with speech and thinking, and depression
LSD (lysergic acid diethylamide)	Overexcitation; sensory distortions, hallucinations	Alters levels of transmitters in brain; potent CNS stimulator; dilates pupils, sometimes unequally; increases heart rate; raises blood pressure	Irrational behavior
Methaqualone (e.g., Quaalude, Sopor)	Hypnotic	Depresses CNS; depresses certain spinal reflexes	Tolerance, physical dependence; convulsions, death
Caffeine	Increases mental alertness; decreases fatigue and drowsiness	Acts on cerebral cortex; relaxes smooth muscle; stimulates cardiac and skeletal muscle; increases urine volume	Very large doses stimulate centers in the medulla (may slow the heart); toxic doses may cause convulsions
Nicotine	Lessens psychological tension	Stimulates sympathetic nervous system; increases dopamine in mesolimbic pathway	Tolerance, physical dependence; stimulates development of atherosclerosis by stimulating synthesis of lipid in arterial wall
Alcohol	Euphoria, relaxation, release of inhibitions	Depresses CNS; impairs vision, coordination, judgment; lengthens reaction time	Physical dependence; damage to pancreas and liver (e.g., cirrhosis), brain damage

FOCUS ON Alcohol: The Most Abused of All Drugs

After tobacco, alcohol is the leading cause of premature death in the United States. It is linked to more than 100,000 deaths every year and costs our society more than $100 billion annually. According to the Substance Abuse and Mental Health Services Administration, an estimated 8.2 million individuals in the United States were dependent on alcohol in 1999. An estimated one in three homes includes someone with a serious drinking problem. Alcohol abuse is not limited to adults. About 4.6 million adolescents, or nearly one of every three high school students, experience negative consequences from alcohol use, including difficulty with parents, poor performance at school, and trouble with the law.

Alcohol abuse results in physiological, psychological, and social impairment for the abuser and has serious negative consequences for family, friends, and society. Alcohol abuse has been linked to the following:

- More than 50% of all traffic fatalities
- More than 50% of violent crimes, and more than 60% of child abuse and spouse abuse cases
- More than 50% of suicides
- More than 15,000 babies born each year with serious birth defects because their mothers drank alcohol during pregnancy; **fetal alcohol syndrome** is the leading cause of preventable mental retardation in the United States
- Recent studies suggest that as few as three drinks per week increase breast cancer risk by 50%

A single drink, that is, 12 oz of beer, 5 oz of wine, or 1.5 oz of 80-proof liquor, results in a blood alcohol concentration of approximately 20 mg/dL (milligrams per deciliter). This represents about 0.5 oz of pure alcohol in the blood. Alcohol accumulates in the blood because absorption occurs more rapidly than do oxidation and excretion. Every cell in the body can take in alcohol from the blood. At first the drinker may feel stimulated, but alcohol actually causes depression of the CNS.

As blood alcohol concentration rises, information processing, judgment, memory, sensory perception, and motor coordination all become progressively impaired. Depression and drowsiness generally occur. Contrary to popular belief, alcohol decreases sexual performance in males. Some individuals become loud, angry, or violent. In most states, a blood alcohol concentration of 80 mg/dL (or 0.08) legally defines driving while intoxicated (DWI). Any state that does not adopt this standard by 2003 will lose a portion of its federal highway funds. A 170-pound man typically reaches this level by drinking four drinks in an hour on an empty stomach. All states have a zero tolerance law for drivers under 21 years old.

Alcohol metabolism occurs at a fixed rate in the liver so only time, not coffee, will decrease its effects. Because alcohol inhibits water reabsorption in the kidneys, more fluid is excreted as urine than is consumed. This results in dehydration that, together with low blood sugar level, may cause the stupor produced by excessive drinking.

In chronic drinkers, cells of the CNS adapt to the presence of the drug. This causes *tolerance* (more and more alcohol is needed to experience the same effect) and physical dependence. Abrupt withdrawal can result in sleep disturbances, severe anxiety, tremors, seizures, hallucinations, and psychoses.

The *A1* allele of the D2 dopamine receptor gene increases the risk for alcoholism. Individuals with this allele have fewer D2 dopamine receptors in their brain. Alcohol, as well as cocaine, heroin, amphetamines, and nicotine, enhances dopamine activity in the *mesolimbic dopamine pathway* of the brain. Individuals with the *A1* allele may compensate for their low dopamine levels by using alcohol and other drugs. However, the mechanisms underlying alcohol addiction are far more complex. Evidence suggests that many neurotransmitters are involved, including glutamate, serotonin, GABA, and opioid peptides.

Treatment for alcohol problems includes various forms of psychotherapy, including relapse prevention therapy in which individuals are encouraged not to consider lapses from abstinence as failure. The group support offered by Alcoholics Anonymous (AA) has proved effective for many struggling with alcohol abuse.

and occupationally. Use of some drugs, such as heroin, tobacco, alcohol, and barbiturates, may also result in physical dependence, in which physiological changes occur that make the user dependent on the drug. For example, when heroin or alcohol is withheld, the addict suffers characteristic withdrawal symptoms. Physical addiction can occur when a drug, for example, morphine, has components similar to substances that body cells normally manufacture on their own. Some highly addictive drugs such as crack cocaine and methamphetamine do not cause serious withdrawal symptoms.

The neurophysiological mechanisms for drug addiction involve a network of neurons that release dopamine. These neurons are located in the midbrain and give rise to the **mesolimbic dopamine pathway** that extends into behavioral control centers in the limbic system. This pathway (important in motivation and experiencing emotion) enables an individual to experience pleasure in response to certain events or substances. Facilitation of neurons in this pathway has been demonstrated in nicotine, heroin, amphetamine, and cocaine addiction.

FOCUS ON Crack Cocaine

Cocaine use is increasing. With the exception of alcoholics, most people seeking treatment for drug abuse are now crack cocaine addicts. Almost 10% of high school seniors report that they have used cocaine, and an estimated 1.5 million individuals in the United States are users.

Cocaine is a CNS stimulant derived from the coca plant. Crack cocaine is a very concentrated and extremely powerful form of cocaine—five to ten times more addictive than other forms of cocaine. This drug is produced in illegal, makeshift labs by converting powdered cocaine into small "rocks" that are up to 80% pure cocaine. Crack is smoked in pipes or added to tobacco or marijuana cigarettes.

Use of crack results in an intense, brief high beginning in about 10 seconds and lasting for 5 to 10 minutes. Physiologically, crack stimulates a massive release of norepinephrine and dopamine in the brain and blocks their reuptake by postsynaptic neurons. Excitation of the sympathetic nervous system occurs, and users report experiencing feelings of self-confidence, power, and euphoria. As the neurotransmitters are depleted, the high is followed by a "crash," a period of fatigue and depression. The abuser experiences an intense craving for another crack "hit" to get more stimulation.

Some abusers spend days smoking crack without stopping to eat or sleep. Although crack rocks can be obtained for as little as $5 to $10 each, many abusers develop habits that cost hundreds of dollars a week. Supporting an expensive drug habit leads many abusers to prostitution, drug dealing, and other forms of crime.

Cocaine use can cause dilated pupils, increased heart rate, elevated blood pressure, respiratory problems, brain seizures, cardiac arrest, and stroke. Many users have suffered fatal reactions to impurities in the drug or have died as a result of accidents related to its use. Cocaine addicts report problems with memory loss, fatigue, depression, insomnia, paranoia, loss of sexual drive, violent behavior, and attempts at suicide.

The effects of cocaine on the developing brain has been the focus of extensive investigation. Pediatrician Ira Chasnoff of the University of Illinois College of Medicine in Chicago reported that children whose mothers used cocaine during pregnancy had significantly more behavioral problems. They were more anxious, aggressive, and impulsive, and had difficulty paying attention and staying focused. They were also depressed. Linda Mayes of Yale University reported that by three to six months of age, infants exposed to cocaine prenatally were more irritable than control infants. By the time they were about one year old, these children had difficulty focusing. Animal studies suggest that when the brain is exposed to high levels of dopamine and serotonin, the brain attempts to compensate, and permanent changes occur that may affect behavior.

No pharmaceutical agent has been developed to treat cocaine overdose or addiction. Addiction is a chronic, relapsing disorder with genetic, physiological, behavioral, and societal components. Effective treatments will need to address all of these aspects of the disorder.

SUMMARY WITH KEY TERMS

I. Two major aspects of neural function are **"hard-wiring"** and **neural plasticity,** more specifically **synaptic plasticity**
   A. Hard-wiring refers to how neurons signal one another, how they connect, and how they carry out basic functions.
   B. Synaptic plasticity refers to the ability of the nervous system to modify synaptic connections in response to experience.

II. Nerve nets and radial nervous systems are typical of radially symmetrical invertebrates. Bilateral nervous systems are characteristic of bilaterally symmetrical animals.
   A. Cnidarians have a **nerve net** consisting of nerve cells scattered throughout the body; no central nervous system (CNS) is present.
   B. Echinoderms typically have a **radial nervous system** consisting of a nerve ring and nerves that extend to various parts of the body.
   C. In a **bilateral nervous system,** nerve cells concentrate to form **nerves, nerve cords, ganglia,** and **brain;** typically, sense organs are concentrated in the head region. An increase in number of neurons, especially association neurons, permits a wide range of responses.
      1. In planarian flatworms the nervous system includes **cerebral ganglia** and, typically, two solid ventral nerve cords connected by transverse nerves.
      2. Annelids and arthropods typically have a ventral nerve cord and numerous ganglia. The cerebral ganglia of arthropods have specialized regions.
      3. Mollusks typically have at least three pairs of ganglia. Octopods and other cephalopod mollusks have the most complex invertebrate nervous systems.

III. The vertebrate nervous system consists of the **central nervous system (CNS)** and **peripheral nervous system (PNS).**
   A. The CNS consists of the brain and dorsal, tubular **spinal cord.**
   B. The PNS consists of sensory receptors and nerves. The PNS is made up of the **somatic division** and the **autonomic division.** Two types of efferent pathways in the autonomic system are the **sympathetic system** and the **parasympathetic system.**

IV. In the vertebrate embryo, the brain and spinal cord arise from the **neural tube.** The anterior end of the tube differentiates into **forebrain, midbrain,** and **hindbrain.**
   A. The hindbrain subdivides into the **metencephalon** and **myelencephalon.**
      1. The myelencephalon develops into the **medulla,** which contains vital centers and other reflex centers. The **fourth ventricle,** the cavity of the medulla, communicates with the central canal of the spinal cord.
      2. The metencephalon gives rise to the **cerebellum,** which is responsible for muscle tone, posture, and equilibrium, and to the **pons,** which connects various parts of the brain.

B. The midbrain is the largest part of the brain in fishes and amphibians. It is their main association area, linking sensory input and motor output. In reptiles, birds, and mammals, the midbrain serves as a center for visual and auditory reflexes.
C. The medulla, pons, and midbrain make up the **brain stem.**
D. The forebrain differentiates to form the **diencephalon** and **telencephalon.**
   1. The diencephalon develops into the **thalamus** and **hypothalamus.**
      a. The thalamus is a relay center for motor and sensory information.
      b. The hypothalamus controls autonomic functions; links nervous and endocrine systems; controls temperature, appetite, and fluid balance; and is involved in some emotional and sexual responses.
   2. The telencephalon develops into the **cerebrum** and **olfactory bulbs.**
      a. In most vertebrates the cerebrum is divided into right and left **hemispheres.**
      b. In fishes and amphibians, the cerebrum mainly integrates incoming sensory information.
      c. In birds, the corpus striatum controls stereotypical but complex behavior patterns, and another part of the cerebrum governs learning.
      d. In mammals, the **neocortex** accounts for a large part of the **cerebral cortex,** and the cerebrum has complex association functions.

V. The human brain and spinal cord are protected by bone and three **meninges**—the **dura mater, arachnoid,** and **pia mater;** brain and spinal cord are cushioned by **cerebrospinal fluid.**
A. The spinal cord transmits impulses to and from the brain and controls many **reflex actions.**
   1. The spinal cord consists of ascending **tracts,** which transmit information to the brain, and descending tracts, which transmit information from the brain. Its gray matter contains nuclei that serve as reflex centers.
   2. A **withdrawal reflex** involves receptors; sensory, association, and motor neurons; and effectors.
B. The human cerebral cortex consists of **gray matter,** which forms folds or **convolutions.** Deep furrows between the folds are called **fissures.**
   1. The cerebrum is functionally divided into **sensory areas** that receive incoming sensory information; **motor areas** that control voluntary movement; and **association areas** that link sensory and motor areas and are responsible for learning, language, thought, and judgment.
   2. The cerebrum consists of lobes including the **frontal lobes, parietal lobes, temporal lobes,** and **occipital lobes.**
   3. The **white matter** of the cerebrum lies beneath the cerebral cortex. The **corpus callosum,** a large band of white matter, connects right and left hemispheres. The **basal ganglia,** a cluster of **nuclei** within the white matter, are important centers for motor function.
C. Brain activity cycles in a sleep-wake pattern that is regulated by the hypothalamus and brain stem.
   1. **Alpha wave** patterns are characteristic of relaxed states, **beta wave** patterns of heightened mental activity, and slower frequency, higher amplitude (**theta** and **delta waves**) of non-REM sleep.
   2. The **reticular activating system (RAS)** is an arousal system; its neurons filter sensory input, selecting which information is transmitted to the cerebrum.
   3. Electrical activity of the cerebral cortex and metabolic rate slows during **non-REM sleep. REM sleep** is characterized by dreaming.
   4. The body's main biological clock (suprachiasmatic nucleus) receives information about light and dark and apparently transmits it to other nuclei that regulate sleep. When sleep centers are activated, they are thought to release **serotonin.**
D. The **limbic system** of the brain affects the emotional aspects of behavior, motivation, sexual behavior, autonomic responses, and biological rhythms.
E. **Learning** is the process by which we acquire information as a result of experience. **Memory** is the process by which information is encoded, stored, and retrieved. Learning and memory can result in changes in behavior.
   1. **Implicit memory** is unconscious memory for perceptual and motor skills. **Explicit memory** is factual memory of people, places, or objects; it requires conscious recall of information.
   2. **Short-term memory** allows us to recall information, such as a telephone number, for a few minutes. Information can be transferred from short-term memory to **long-term memory.**
   3. Memories appear to be stored throughout the cerebrum; they are integrated in many areas of the brain, including the association areas of the cerebral cortex and parts of the limbic system.
   4. **Synaptic plasticity** refers to the ability of the nervous system to modify synapses during learning and remembering.
      a. An increase in neurotransmitter released by a presynaptic neuron as a result of an action potential is **synaptic facilitation.** This type of **synaptic enhancement** lasts less than a second.
      b. Long-term memory storage may involve gene activation and long-term functional changes at synapses. Known as **long-term potentiation (LTP),** such changes involve increased sensitivity to an action potential by a postsynaptic neuron. **Long-term depression (LTD)** is a long-lasting decrease in the strength of synaptic connections.
   5. The marine mollusk *Aplysia* is a model organism used to study memory and learning in animals.
   6. **Habituation** is the decrease in response that occurs after repeated exposure to a harmless stimulus. Habituation results from inactivation of calcium channels in presynaptic terminals in response to repeated action potentials. This results in a decrease in neurotransmitter released by the presynaptic neurons.
   7. **Sensitization** results in an increased response after experience of an unpleasant stimulus. It depends on an increase in neurotransmitter released by sensory (presynaptic) neurons.
   8. **Classical conditioning** is a form of learning in which an association forms between an unconditioned stimulus (US) and a conditioned stimulus (CS). The CS then elicits a conditioned response (CR). The cellular mechanism for conditioning may involve a second messenger, protein kinase, gene activation, and protein synthesis.
   9. The physical structure and chemistry of the brain can be altered by environmental experience.

VI. The peripheral nervous system consists of sensory receptors and nerves, including the **cranial nerves** and **spinal nerves** and their branches. The **somatic division** of the PNS responds to changes in the external environment.

VII. The **autonomic division** regulates the internal activities of the body.
A. The **sympathetic system** permits the body to respond to stressful situations.
B. The **parasympathetic system** influences organs to conserve and restore energy.
C. Many organs are innervated by sympathetic and parasympathetic nerves, which function in a complementary way.

VIII. Some of the types of drugs that affect the nervous system are amphetamines, antianxiety drugs, barbiturates, antipsychotic drugs, antidepressants, narcotic analgesics, and alcohol.
A. Many drugs alter mood by increasing or decreasing the concentrations of specific neurotransmitters within the brain.
B. Habitual use of mood-altering drugs can result in **psychological dependence** or **addiction.** Many drugs induce **tolerance,** in which the body's response to the drug decreases so that greater amounts are needed to obtain the desired effect.

POST-TEST

1. A radially symmetrical animal is likely to have (a) a forebrain (b) a nerve net (c) cerebral ganglia (d) a ventral nerve cord (e) cerebral ganglia and a nerve net
2. In vertebrate embryos the brain develops from the (a) spinal cord (b) sympathetic nervous system (c) parasympathetic nervous system (d) neural tube (e) forebrain
3. Which part of the brain maintains posture, muscle tone, and equilibrium? (a) cerebrum (b) medulla (c) cerebellum (d) neocortex (e) thalamus
4. Which part of the brain controls autonomic functions and regulates body temperature? (a) cerebrum (b) hypothalamus (c) cerebellum (d) pons (e) thalamus
5. In a withdrawal reflex, following reception, a signal is transmitted (a) from a motor neuron to an association neuron in the CNS (b) from an association neuron in the CNS to an afferent neuron (c) from an afferent neuron in the CNS to a motor neuron (d) from a sensory neuron to an association neuron in the CNS (e) from a sensory neuron to an association neuron in the PNS
6. Association areas in the human brain are concentrated in the (a) cerebral cortex (b) medulla (c) ventricle (d) hippocampus (e) meninges
7. The human brain is protected by (a) meninges, cerebrospinal fluid, and skull bones (b) meninges and skull bones only (c) dura mater and fourth ventricle (d) pia mater and skull bones (e) arachnoid, pia mater, cerebrospinal fluid, and ganglia
8. Which of the following is *not* a function of the spinal cord? (a) controls many reflex actions (b) transmits information to the brain (c) transmits information from the brain (d) regulates sleep-wake cycles (e) controls the withdrawal reflex
9. The most prominent part of the amphibian brain is the (a) midbrain (b) medulla (c) cerebellum (d) neocortex (e) cerebrum
10. The visual centers are located in the (a) parietal lobes (b) thalamus (c) occipital lobes (d) limbic lobes (e) frontal lobes
11. As you are answering these questions, what is the predominant type of wave that your brain is emitting? (a) REM (b) alpha (c) theta (d) delta (e) beta
12. Implicit memory refers to (a) short-term memory (b) long-term memory (c) factual knowledge of people, places, or objects (d) unconscious memory for perceptual or motor skills (e) learning that depends on long-term depression (LTD)
13. Long-term potentiation (LTP) (a) is mainly the responsibility of cranial nerve X (b) is a long-lasting increase in synaptic strength (c) is associated with short-term memory (d) is a long-lasting decrease in synaptic strength (e) occurs mainly during REM sleep
14. The heart rate is slowed by (a) sympathetic nerves (b) parasympathetic nerves (c) corpus callosum (d) both sympathetic and parasympathetic nerves (e) the hippocampus working together with sympathetic nerves
15. After taking a mood-altering drug for several weeks, a patient notices that it no longer works as effectively. This is an example of (a) psychological dependence (b) withdrawal (c) addiction (d) tolerance (e) neurotransmitter increases
16. Amphetamines (a) facilitate neurons in the mesolimbic dopamine pathway (b) decrease the amount of dopamine secreted (c) do not result in tolerance (d) are sometimes referred to as "downers" (e) are CNS depressants

REVIEW QUESTIONS

1. Contrast the nervous system of a planarian flatworm with that of *Hydra*.
2. What are some characteristics of the nervous system in a bilaterally symmetrical animal? What are some advantages of this type of system?
3. Compare the flatworm nervous system with that of a vertebrate. In each case, what is the relationship between the type of nervous system and the animal's lifestyle?
4. Identify the parts of the adult vertebrate brain derived from the embryonic forebrain, midbrain, and hindbrain.
5. Give the function of each of the following structures in the human brain: (a) medulla (b) midbrain (c) cerebellum (d) thalamus (e) hypothalamus (f) cerebrum.
6. Compare the midbrain and cerebrum of the fish with those of the mammal.
7. Describe the protective coverings of the human CNS, and give the function of the cerebrospinal fluid.
8. How is information processed? How might short-term memory be adaptive? Where are memories stored?
9. What mechanisms are involved in habituation? sensitization? classical conditioning? What is long-term potentiation (LTP)? What is long-term depression (LTD)?
10. Contrast the functions of the CNS with those of the PNS.
11. Compare the functions of the somatic and autonomic divisions.
12. Contrast the structure and function of the sympathetic system with those of the parasympathetic system.
13. Describe how each of the following drugs affects the CNS: (a) alcohol (b) antipsychotic drugs (c) antidepressants (d) amphetamines (e) MDMA ("ecstasy").
14. Label the diagram. Use Figure 40–11*a* to check your answers.

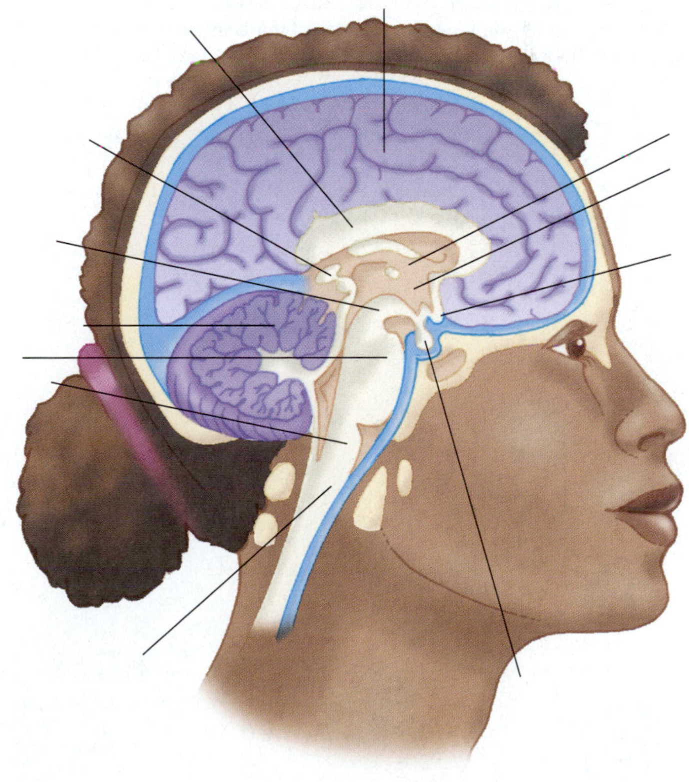

YOU MAKE THE CONNECTION

1. What general trends can you identify in the evolution of the vertebrate brain?
2. In what way does electrical activity of the brain reflect your state of consciousness? Of what use might it be to learn to control your brain wave patterns?
3. Imagine that you have just become a parent. What kind of things could you do to enhance the development of your child's intellectual abilities?
4. Hypothesize a possible relationship between inheritance and intelligence based on gene activation during information processing.

RECOMMENDED READINGS

Gazzaniga, M.S. "The Split Brain Revisited." *Scientific American,* Vol. 279, No. 1, Jul. 1998. Studies on brain organization provide insight into the workings of the mind.

Ho, D.Y., and R.M. Sapolsky. "Gene Therapy for the Nervous System." *Scientific American,* Vol. 276, No. 6, Jun. 1997. Damage from degenerative neurological disease may someday be treated by inserting genes into brain cells.

Kandel, E.R., J.H. Schwartz, and T.M. Jessell. *Essentials of Neural Science and Behavior,* 4th ed. McGraw-Hill, New York, 2000. A well-written, comprehensive exploration of neurobiology.

Karni, A. "Adult Cortical Plasticity and Reorganization." *Science and Medicine,* Vol. 4, No. 1, Jan./Feb. 1997. Studies indicate that the adult brain retains an important degree of plasticity.

Landry, D.W. "Immunotherapy for Cocaine Addiction." *Scientific American,* Vol. 276, No. 2, Feb. 1997. A discussion of cocaine addiction and a possible future treatment.

Malenka, R.C., and R. Nicoll. "Long-Term Potentiation—A Decade of Progress?" *Science,* Vol. 285, 17 Sep. 1999. This neuroscience review discusses current understanding of the molecular mechanisms involved in long-term potentiation.

Maudgil, D.D., and S.D. Shorvon. "Locating the Epileptogenic Focus by MRI." *Science and Medicine,* Vol. 4, No. 5, Sep./Oct. 1997. New imaging techniques provide valuable clinical information regarding lesions in the brain related to epilepsy.

Noble, E.P. "The Gene That Rewards Alcoholism." *Science and Medicine,* Vol. 3, No. 2, Mar./Apr. 1996. A gene that codes for a dopamine receptor has been linked to alcoholism.

Science, Vol. 278, 3 Oct. 1997. A series of articles on "Frontiers in Neuroscience: The Science of Substance Abuse." Discussion of the current view of drug addiction and what we know about neurophysiological mechanisms underlying addiction.

Tsien, J.Z. "Building a Brainier Mouse." *Scientific American,* Vol. 282, No. 4, Apr. 2000. Genetically engineering smart mice has helped researchers identify some of the molecules that are key in learning and memory.

Vander, A., J. Sherman, and D. Luciano. *Human Physiology,* 8th ed. McGraw-Hill, New York, 2001. Chapters 8 and 13 provide a clear introduction to neural function.

Youdim, M.B.H., and P. Riederer. "Understanding Parkinson's Disease." *Scientific American,* Vol. 276, No. 1, Jan. 1997. A discussion of the role of free radicals in the etiology of Parkinson's disease.

- Visit our Web site at **http://www.info.brookscole.com/solomonbergmartin** for links to chapter-related resources on the World Wide Web. Additional on-line materials relating to this chapter can also be found on our Web site.

See chapter activity on BioActive Learner CD for additional help in mastering the chapter's material. Icon location in the chapter's margins shows which topics have tutorials or simulations in the CD.

41

Sensory Reception

Bats are guided by echoes. This California leaf-nosed bat *(Macrotus californicus)* has located an insect. *(Merlin D. Tuttle/Bat Conservation International/Photo Researchers, Inc.)*

LEARNING OBJECTIVES

After you have studied this chapter you should be able to

1. Distinguish among the five kinds of sensory receptors that are classified according to the types of energy they transduce, and give examples of each type of sense organ.
2. Describe how a sensory receptor functions; define energy transduction, receptor potential, and adaptation as part of your answer.
3. Describe the following mechanoreceptors: tactile receptors, proprioceptors, statocysts, and hair cells.
4. Describe the structure and function of lateral line organs.
5. Compare the function of the saccule and utricle with that of the semicircular canals in maintaining equilibrium.
6. Trace the path taken by sound waves through the structures of the ear, and explain how the organ of Corti functions as an auditory receptor.
7. Compare the structure and function of the receptors of taste and smell.
8. Relate the presence of thermoreceptors to the lifestyles of animals that have them.
9. Contrast simple eyes, compound eyes, and vertebrate eyes.
10. Label the structures of the vertebrate eye on a diagram, and give the functions of each of its accessory structures.
11. Describe the events that take place in human vision from the time light passes through the cornea to the integration of visual information in the brain. Compare the two types of photoreceptors and describe the signal transduction pathway as part of your explanation.

Bats, dolphins, and a few other vertebrates can detect distant objects by **echolocation,** sometimes called biosonar. Echolocation works something like radar but uses sound rather than radio waves. For example, a bat emits high-pitched sounds that bounce off objects in its path and echo back. By rapidly responding to the echo, the bat can skillfully avoid obstacles and can capture prey *(see photograph).* The ability to detect sounds emitted by some foraging bats has evolved in certain moths and other insect prey. Some species of bats have adapted, in turn, by emitting very high-frequency sounds that their prey cannot detect.

You may have heard dolphins emit clicking sounds. Humans can hear only a limited number of the wide range of sound frequencies that dolphins emit. Using echolocation, dolphins can determine the size, shape, texture, and density of an object, as well as its location.

Sensory receptors are structures that detect information about changes in the internal or external environment. By informing an animal about its environment, sensory receptors are the link between the animal and the outside world. The kinds of sensory receptors an animal possesses determine just how it perceives its surroundings. We humans live in a world of rich colors, numerous shapes, and varied sounds. But we cannot hear the high-pitched whistles that are audible to dogs and cats or the ultrasonic echoes by which bats navigate. Neither do we ordinarily recognize our friends by their distinctive odors. And although vision is our dominant and most refined sense, we are blind to the ultraviolet (UV) hues that light up the world for insects.

Sensory receptors consist of specialized neuron endings or specialized cells in close contact with neurons. These receptors **transduce** (convert) the energy of the stimulus to electrical signals, the information currency of the nervous system. The transduction mechanisms that couple the stimulus with the opening or closing of ion channels in the plasma membrane of sensory receptors is the focus of a great deal of research. When a sensory receptor is depolarized, an action potential may be initiated in a sensory neuron. As we discussed in Chapter 40, **sensory neurons,** also known as afferent neurons, transmit information from receptors to the central nervous system (CNS).

Sensory receptors, along with other types of cells, make up complex **sense organs:** eyes, ears, nose, and taste buds. A human taste bud, for example, consists of modified epithelial cells that detect chemicals dissolved in saliva. In addition to the five senses of sight, hearing, smell, taste, and touch, neurobiologists recognize balance as a sense. They view touch as a compound sense that allows us to detect pressure, pain, and temperature. In this chapter, we also consider receptors that enable us to sense muscle tension and joint position.

SENSORY RECEPTORS CAN BE CLASSIFIED ACCORDING TO THE TYPE OF ENERGY THEY TRANSDUCE

Animals receive information about their environment in various energy forms. Sensory receptors transduce these types of energy into graded potentials that can result in action potentials. **Mechanoreceptors** transduce mechanical energy—touch, pressure, gravity, stretching, and movement (Table 41–1). Mechanoreceptors convert mechanical forces directly into electrical signals. **Chemoreceptors** transduce certain chemical compounds, and **photoreceptors** transduce light energy. To convert chemical or light energy to electrical energy, these receptors utilize a signal transduction process involving G proteins. **Thermoreceptors** respond to heat and cold. Some fish have well-developed **electroreceptors** that sense differences in electrical potential.

Each kind of sensory receptor is especially sensitive to one particular form of energy. For example, photoreceptor cells (rods and cones) in the eye absorb light energy, while temperature receptors respond to infrared thermal energy (heat). Receptor cells are remarkably sensitive to appropriate stimuli. For example, the photoreceptors of the eye are stimulated by an extremely faint beam of light, and taste receptors are stimulated by a minute amount of a chemical compound.

Sensory receptors can also be classified according to the source of the stimuli to which they respond. **Exteroceptors** receive stimuli from the outside environment, enabling an animal to know and explore the world, search for food, find and attract a mate, recognize friends, and detect enemies. **Interoceptors** are sensory receptors within body organs that detect changes in pH, osmotic pressure, body temperature, and the chemical composition of the blood. We are not usually aware of messages sent to the CNS by these receptors as they work continuously to maintain homeostasis. We become aware of their activity when they enable us to perceive diverse internal conditions such as thirst, hunger, nausea, pain, and orgasm.

RECEPTOR CELLS PRODUCE RECEPTOR POTENTIALS

Sensory receptors perform three important functions: They (1) absorb a small amount of energy from some stimulus; (2) con-

TABLE 41–1 Classification of Receptors by Type of Energy They Transduce

Type of Receptor	Examples	Effective Stimuli
Mechanoreceptors	Tactile receptors	Touch, pressure
	Pacinian corpuscles; Merkel discs; Ruffini corpuscles; Meissner corpuscles	
	Proprioceptors	Movement, body position
	Muscle spindles	Muscle contraction
	Golgi tendon organs	Stretch of a tendon
	Joint receptors	Movement in ligaments
	Statocysts in invertebrates	Gravity
	Lateral line organs in fish	Waves, currents in water
	Vestibular apparatus	
	Saccule and utricle	Gravity; linear acceleration
	Semicircular canals	Angular acceleration
	Hair cells in the organ of Corti of the cochlea	Pressure waves (sound)
Chemoreceptors	Taste buds; olfactory epithelium	Specific chemical compounds
Thermoreceptors	Temperature receptors in blood-sucking insects and ticks; pit organs in pit vipers; nerve endings and receptors in skin and tongues of many animals	Heat
Electroreceptors	Organs in skin of some fishes	Electrical currents in water
Photoreceptors	Eyespots; ommatidia of arthropods; rods and cones in retinas of vertebrates	Light energy

vert the energy of the stimulus into electrical energy, the process known as **transduction;** and (3) produce a **receptor potential,** a depolarization or hyperpolarization of the membrane. With minor variations, this is how all receptors operate.

The function of a receptor potential is to generate action potentials that transmit information to the CNS. However, as we discuss later, each receptor potential does not directly trigger an action potential. When unstimulated, a sensory receptor maintains a resting potential, that is, a difference in charge between the inside and the outside of the cell. When the receptor cell is stimulated and its membrane potential changes, specific types of ion channels in its plasma membrane open or close, altering the permeability of the plasma membrane to various ions. A change in ion distribution can cause a change in voltage across the membrane. If the difference in charge increases, the receptor becomes hyperpolarized. If the potential decreases, the receptor becomes depolarized. The change in membrane potential is the receptor potential.

Like an excitatory postsynaptic potential, or EPSP, (see Chapter 39) the receptor potential is a **graded response** that spreads relatively slowly down the dendrite, fading as it goes. In contrast, according to the *all-or-none law,* the amplitude (size) of each action potential does not vary and therefore has no relation to the magnitude of the stimulus.

The specialized region of the receptor plasma membrane that transduces energy does not generate action potentials. Current generated by receptor potentials flows to a region along the axon where an action potential can be generated. If the receptor membrane is part of a separate cell, receptor potentials stimulate the release of neurotransmitter, which flows across the synapse and binds to sites on the sensory neuron. When the sensory neuron becomes sufficiently depolarized to reach its threshold level, an action potential is generated. The action potential travels along the axon of the sensory neuron to the CNS. In summary,

Stimulus (e.g., light energy) ⟶ transduction into electrical energy ⟶ receptor potential ⟶ action potential

SENSATION DEPENDS ON TRANSMISSION OF A CODED MESSAGE

How do you know whether you are seeing a blue sky, tasting a doughnut, or hearing a note played on a piano? All action potentials are qualitatively the same. Light of the wavelength 450 nm (blue), sugar molecules (sweet), and sound waves of 440 hertz (Hz) (A above middle C) all cause transmission of similar action potentials. Our ability to differentiate stimuli depends on both the sensory receptor itself and on the brain. We can distinguish the color of a blue sky from the scent of cologne; a sweet taste from a light breeze; or the sound of a piano from the heat of the sun because cells of each sensory receptor are connected to specific neurons in particular parts of the brain. Because a receptor normally responds to only one type of stimulus, for example, light or sound, the brain interprets a message arriving from a particular receptor as meaning that a certain type of stimulus occurred (e.g., a flash of color).

Sensation takes place in the brain. Interpretation of the message and the type of sensation depends on which association neurons receive the message. The rods and cones of the eye do not see. When stimulated, they send a message to the brain that interprets the signals and translates them into a rainbow, an elephant, or a child. Artificial (e.g., electrical) stimulation of brain centers can also result in sensation. On the other hand, many sensory messages never give rise to sensations at all. For example, certain chemoreceptors sense internal changes in the body but never stir our consciousness.

When stimulated, a sensory receptor initiates what might be considered a coded message, composed of action potentials transmitted by nerve fibers. This coded message is later decoded in the brain. Impulses from the sensory receptor may differ in the (1) total number of fibers transmitting, (2) specific fibers carrying action potentials and their targets, (3) total number of action potentials passing over a given fiber, and (4) frequency of the action potentials passing over a given fiber. The intensity of a stimulus is coded by the frequency of action potentials fired by sensory neurons during the stimulus. A strong stimulus causes a greater depolarization of the receptor membrane and causes the sensory neuron to fire action potentials with greater frequency than would a weak stimulus. This occurs because as a graded response, the receptor potential can vary in magnitude.

The difference in sound intensity between the gentle rustling of leaves and a clap of thunder depends on the number of neurons transmitting action potentials, as well as the frequency of the action potentials transmitted by each neuron. Just how the sensory receptor initiates different codes and how the brain analyzes and interprets them to produce various sensations are not completely understood.

SENSORY RECEPTORS ADAPT TO STIMULI

Many sensory receptors do not continue to respond at the initial rate, even if the stimulus continues at the same intensity. With time, the response to a continued, constant stimulus decreases. This decrease in frequency of action potentials in a sensory neuron even though the stimulus is maintained is called **sensory adaptation.** Sensory adaptation occurs for two reasons. First, during a sustained stimulus, the receptor sensitivity decreases and it produces a smaller receptor potential (resulting in a lower frequency of action potentials in the sensory neurons). Second, changes take place at synapses in the neural pathway activated by the receptor. For example, the release of neurotransmitter from a presynaptic terminal may decrease in response to a series of action potentials.

Some receptors, such as those for pain or cold, adapt so slowly that they continue to trigger action potentials as long as the stimulus persists. Other receptors adapt rapidly, permitting an animal to ignore persistent unpleasant or unimportant stimuli. For example, when you first pull on a pair of tight jeans, your pressure receptors let you know that you are being squished, and you may feel uncomfortable. Soon, though, these receptors adapt, and you hardly notice the sensation of the tight fit. In the same way, we quickly adapt to odors that at first seem to assault our senses. Sensory adaptation enables an animal to discriminate between unimportant background stimuli that can be ignored and new or important stimuli that require attention.

MECHANORECEPTORS RESPOND TO TOUCH, PRESSURE, GRAVITY, STRETCH, OR MOVEMENT

Mechanoreceptors are activated when they change shape as a result of being mechanically pushed or pulled. Mechanoreceptors in the skin are responsible for touch. Certain mechanoreceptors enable an organism to maintain its body position with respect to gravity (for us, head up and feet down; for a dog, dorsal side up and ventral side down; for a tree sloth, ventral side up and dorsal side down). Other mechanoreceptors are concerned with maintaining postural relations—the position of one part of the body with respect to another. This information is essential for all forms of locomotion and for all coordinated and skilled movements, from spinning a cocoon to completing a reverse one-and-a-half dive with twist.

Mechanoreceptors provide information about the shape, texture, weight, and topographical relations of objects in the external environment. Certain mechanoreceptors affect the operation of internal organs. For example, they supply information about the presence of food in the stomach, feces in the rectum, urine in the bladder, and a fetus in the uterus.

Touch receptors are located in the skin

The simplest mechanoreceptors are free nerve endings in the skin. These nerve endings detect touch, pressure, and pain when stimulated by objects that contact the body surface. Thousands of more specialized tactile (touch) receptors are also located in the skin (Fig. 41–1). **Merkel discs** sense touch and pressure. They adapt slowly, permitting us to know that an object contin-

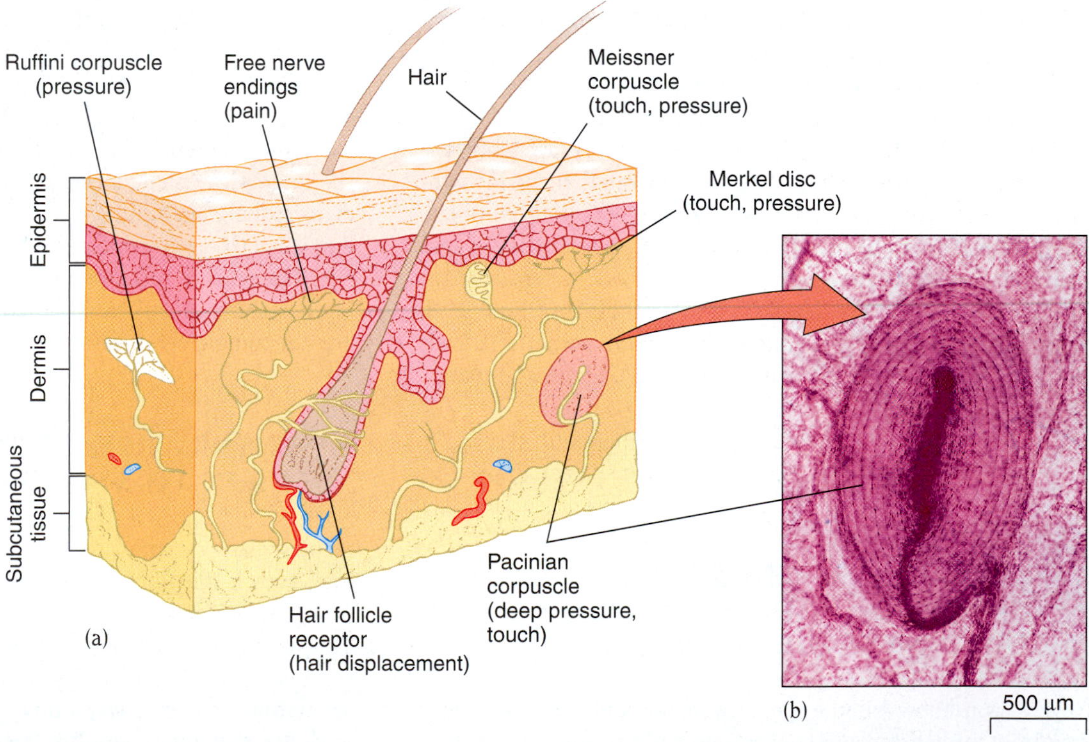

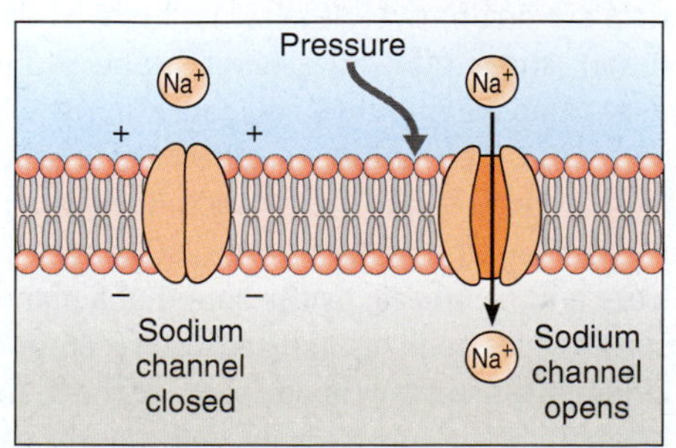

(c) Membrane of the Pacinian corpuscle

Figure 41–1 Sensory reception in human skin. **(a)** This diagrammatic section through the human skin illustrates several mechanoreceptors. Merkel discs, Meissner corpuscles, and Pacinian corpuscles respond to touch and pressure. Ruffini corpuscles are sensitive to pressure. Hair follicle receptors respond to displacement of the hair. Also shown are free nerve endings that respond to pain. **(b)** Pacinian corpuscle, a deep pressure receptor. **(c)** The mechanical forces sensed by these receptors are thought to cause Na^+ channels to open, leading to depolarization of the sensory neuron. *(Ed Reschke)*

ues to touch the skin. Three types of mechanoreceptors in the skin have endings that are encapsulated: Meissner corpuscles, Ruffini corpuscles, and Pacinian corpuscles. **Meissner corpuscles** are sensitive to light touch and pressure and adapt quickly to a sustained stimulus. **Ruffini corpuscles,** which adapt very slowly, inform us of heavy and continuous touch and pressure.

The **Pacinian corpuscle** is sensitive to deep pressure that causes rapid movement of the tissues. It is especially sensitive to stimuli that vibrate. The Pacinian corpuscle consists of a nerve ending surrounded by concentric connective tissue layers interspersed with fluid. Compression causes displacement of the layers, which stimulates the axon. Even though the displacement is maintained under steady compression, the receptor potential rapidly falls to zero, and action potentials cease—an excellent example of sensory adaptation. The Pacinian corpuscle is stimulated only when there is movement of the tissue. The transduction mechanisms that couple touch or pressure on the skin with the opening or closing of ion channels in the plasma membrane of the receptor cell are not yet known.

In many invertebrates as well as vertebrates, tactile receptors lie at the base of a hair or bristle. Tactile hairs may sense the body's orientation in space with respect to gravity. They may also detect air and water vibrations and contact with other objects. The tactile receptor is stimulated indirectly when the hair is bent or displaced. A receptor potential develops, and a few action potentials may be generated. This type of receptor responds only when the hair is moving. Even though the hair may be maintained in a displaced position, the receptor is not stimulated unless there is motion.

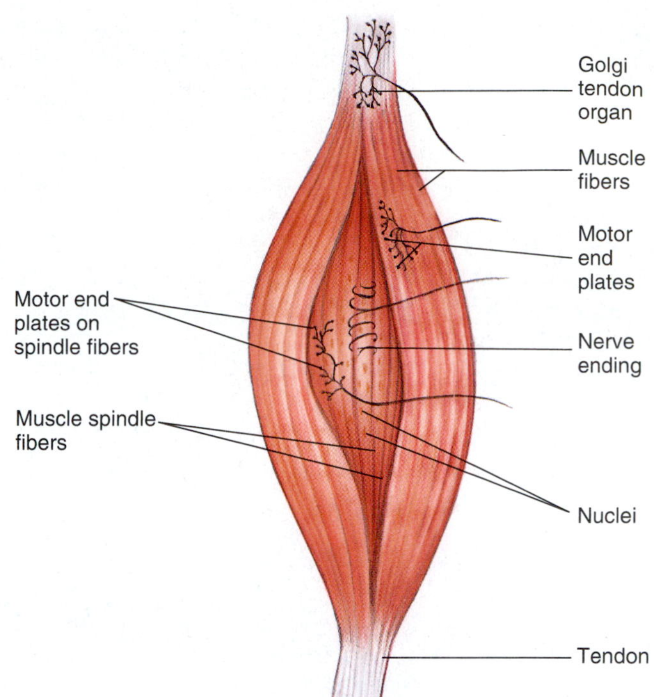

Figure 41–2 Proprioceptors. These sensory receptors respond to changes in movement, tension, and position in muscles and joints. Muscle spindles detect muscle movement. A muscle spindle consists of an elongated bundle of specialized muscle fibers. Golgi tendon organs respond to tension in muscles and their associated tendons.

Proprioceptors help coordinate muscle movement

Proprioceptors are mechanoreceptors within muscles, tendons, and joints that enable an animal to perceive the positions of its arms, legs, head, and other body parts, along with the orientation of its body as a whole. With the help of proprioceptors, humans can carry out activities such as dressing or playing the piano even in the dark. Proprioceptors continually respond to tension and movement in muscles and joints.

Vertebrates have three main types of proprioceptors: **muscle spindles,** which detect muscle movement (Fig. 41–2); **Golgi tendon organs,** which respond to tension in contracting muscles and in the tendons that attach muscle to bone; and **joint receptors,** which detect movement in ligaments. Impulses from the proprioceptors are important in ensuring the harmonious contractions of the several distinct muscles involved in a single movement; without such receptors, complicated, skillful acts would be impossible. These organs are also important in maintaining balance.

Proprioceptors are probably more numerous and more continually active than any of the other sensory receptors, although we are less aware of them than of most of the others. As long as the stimulus is present, receptor potentials are maintained (though not at constant magnitude) and action potentials continue to be generated. Thus, information about position is continuously supplied.

The mammalian muscle spindle, one of the more versatile stretch receptors, helps maintain muscle tone. It consists of a bundle of specialized muscle fibers, in the center of which is a region encircled by sensory nerve endings. These neurons continue to transmit signals for a prolonged period, in proportion to the degree of stretch. Some of the nerve endings also exhibit a strong dynamic response to an increase in length, but only while the length is actually increasing.

Many invertebrates have gravity receptors called statocysts

Every organism is oriented in a characteristic way with respect to gravity. When displaced from this normal position, it quickly adjusts its body to reassume normal orientation. In complex animals, receptors must continually send information to the CNS regarding the position and movements of the body.

Many invertebrates have gravity receptors called **statocysts.** A statocyst is basically an infolding of the epidermis lined with receptor cells, called sensory hair cells, equipped with sensory hairs (not true hairs) (Fig. 41–3). The cavity contains one or more **statoliths,** tiny granules of loose sand grains or calcium carbonate, held together by an adhesive material secreted by cells of the statocyst. Normally the particles are pulled downward by gravity and stimulate the hair cells. When the animal changes

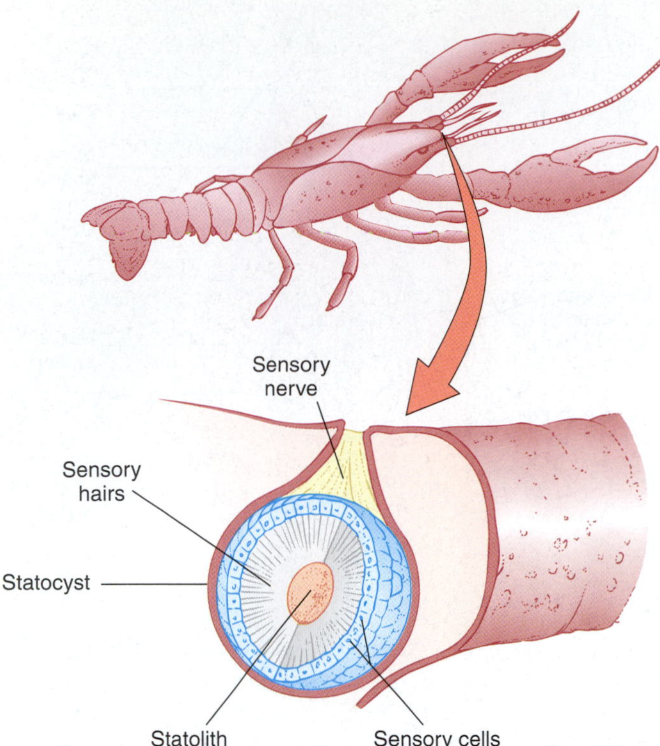

Figure 41-3 A statocyst. Many invertebrates have statocysts, receptors that sense gravitational force and provide information about orientation of the body with respect to gravity.

position, the position of the statolith also changes, stimulating different hair cells. Each sensory cell responds maximally when the animal is at a particular position with respect to gravity. The mechanical displacement results in receptor potentials and action potentials that inform the CNS of the change in position. By "knowing" which hair cells are firing, the animal knows where "down" is and so can correct any abnormal orientation.

In a classic experiment, the function of the statocyst was demonstrated by substituting iron filings for sand grains in the statocysts of crayfish. The force of gravity was overcome by holding magnets above the animals: The iron filings were attracted upward toward the magnets, and the crayfish began to swim upside down in response to the new information provided by their gravity receptors.

Hair cells are characterized by stereocilia

Vertebrate hair cells are mechanoreceptors that detect movement. They are involved in the detection of water movements by the lateral line of fish, in maintaining position and equilibrium, and in hearing. The surface of a vertebrate hair cell typically has a single long **kinocilium** and a number of shorter **stereocilia,** hairlike projections that increase in length from one side of the hair cell to the other. A kinocilium is a true cilium with a 9 + 2 arrangement of microtubules (see Chapter 4). In contrast, stereocilia are not really cilia. They are microvilli that contain actin filaments.

Hair cells have no axons and do not generate action potentials. Mechanical stimulation of the stereocilia causes voltage changes. Mechanical stimulation in one direction causes depolarization and release of neurotransmitter, whereas mechanical stimulation in the opposite direction results in hyperpolarization.

Lateral line organs supplement vision in fish

Lateral line organs of fish and aquatic amphibians detect vibrations in the water. They are thought to supplement vision by informing the animal of obstacles in its way and of moving objects such as prey, enemies, and schooling mates. Researchers have suggested that sharks occasionally attack human swimmers because humans may produce vibrations that are similar to those produced by an injured fish.

Typically the lateral line organ consists of a long canal running the length of the body and continuing into the head (Fig. 41–4). The canal is lined with sensory hair cells. The tips of the stereocilia are enclosed by a **cupula,** a mass of gelatinous material secreted by the hair cells.

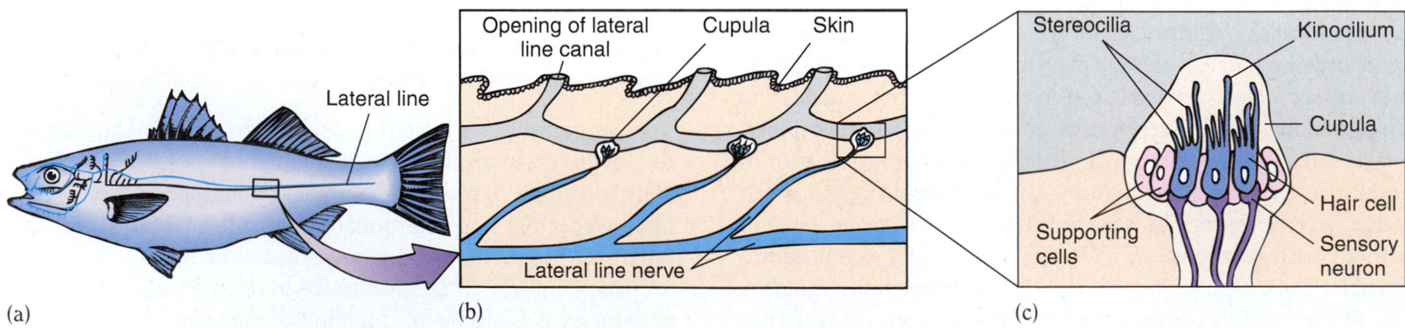

Figure 41-4 Lateral line organ. **(a)** The lateral line organ is a canal that extends the length of the body. **(b)** The canal has numerous openings to the outside environment. **(c)** The receptor cells are hair cells. Each hair cell has stereocilia and a kinocilium that are covered with a gelatinous cupula. The hair cells respond to waves, currents, or other disturbances in the water, informing the fish of obstacles or moving objects.

Waves, currents, and even slight movement in the water cause vibrations in the lateral line organ. The water moves the cupula, causing the stereocilia to bend. This results in changes in the membrane potential of the hair cell, and may result in release of neurotransmitter. If a sensory neuron is sufficiently stimulated, an action potential may be dispatched to the CNS.

Process of Science — Seals use their whiskers to detect water movements

Echolocation has evolved in some species of whales and dolphins, allowing these animals to navigate by sound. Biologists have wondered how other marine mammals find food in dark or murky waters. Studies of seal whiskers have shown that they are extremely sensitive to bending and that seals can identify objects with their whiskers. In 1998 Guido Dehnhardt, then of the University of Bonn, and his colleagues demonstrated that seal whiskers, like the lateral line of fishes, are sensitive to movements of water caused by fish. Some researchers hypothesized that seals could use their whiskers to detect fish at close distances and could then identify them by touch. However, they assumed that seals could not locate prey over long distances because the wake would dissipate too quickly. Dehnhardt's team tested this hypothesis and found that herring and other large fish on which seals feed leave a hydrodynamic trail in the form of a wake that extends as long as 180 meters and lasts for as long as 3 min.

Dehnhardt, now at Ruhr University in Bochum, Germany, reported in *Science* in 2001 that he and his team trained harbor seals to follow hydrodynamic trails generated by a miniature submarine. Using this model, these investigators showed that blindfolded seals used their whiskers to detect and follow such trails even over long distances. These studies suggest that seals and perhaps other marine mammals hunt with their whiskers.

The vestibular apparatus maintains equilibrium

When we think of the ear, we think of hearing. However, in vertebrates its main function is to help maintain equilibrium. Although many vertebrates do not have outer or middle ears, all have inner ears (Fig. 41–5*a*).

The inner ear consists of a complicated group of interconnected canals and sacs, referred to as the **labyrinth,** which includes a membranous labyrinth that fits inside a bony labyrinth. In mammals the membranous labyrinth consists of two saclike chambers—the **saccule** and the **utricle**—and three **semicircular canals.** Collectively, the saccule, utricle, and semicircular canals are referred to as the **vestibular apparatus** (Fig. 41–5*b*). Destruction of these organs leads to a considerable loss of the sense of equilibrium. A pigeon whose vestibular apparatus has been destroyed cannot fly but in time can relearn how to maintain equilibrium using visual stimuli. Proprioceptors, cells sensitive to pressure in the soles of the feet, and vision all contribute to equilibrium in the human.

The saccule and utricle house gravity detectors in the form of small calcium carbonate ear stones called **otoliths** (Fig. 41–6). The sensory cells of the saccule and utricle are hair cells similar to those of the lateral line organ. The stereocilia projecting from the hair cells are covered by a gelatinous cupula in which the otoliths are embedded. The hair cells in the saccule and utricle lie in different planes.

Normally, the pull of gravity causes the otoliths to press against the stereocilia, stimulating them to initiate impulses that are sent to the brain by way of sensory nerve fibers at their bases. When the head is tilted or in linear acceleration (an increase in speed when the body is moving in a straight line), otoliths press on the stereocilia of different cells, deflecting them. Deflection of the stereocilia toward the kinocilium depolarizes the hair cell. Deflection in the opposite direction hyperpolarizes the hair cell. Thus, depending on the direction of movement, the hair cells release more or less neurotransmitter. The brain interprets the neural messages so that the animal is aware of its position relative to the ground regardless of the position of its head.

Information about turning movements, referred to as angular acceleration, is provided by the three semicircular canals. Each canal, a hollow ring connected with the utricle, lies at right angles to the other two and is filled with fluid called **endolymph.** At one of the openings of each canal into the utricle is a small, bulblike enlargement, the **ampulla.** Within each ampulla lies a clump of hair cells called a **crista** (pl., *cristae*), similar to the groups of hair cells in the utricle and saccule. No otoliths are present. The stereocilia of the hair cells of the cristae are stimulated by movements of the endolymph in the canals (Fig. 41–7).

When the head is turned, there is a lag in the movement of the fluid within the canals. The cilia move in relation to the fluid and are stimulated by its flow. This stimulation produces not only the consciousness of rotation but also certain reflex movements in response to it. These reflexes cause the eyes and head to move in a direction opposite that of the original rotation. Because the three canals are in three different planes, head movement in any direction stimulates fluid movement in at least one of the canals.

We humans are used to movements in the horizontal plane but not to vertical movements (parallel to the long axis of the upright body). The motion of an elevator or of a ship pitching in a rough sea stimulates the semicircular canals in an unusual way and may cause seasickness or motion sickness, with resultant nausea or vomiting. When a seasick person lies down, the movement stimulates the semicircular canals in a more familiar way, and nausea is less likely to occur.

Auditory receptors are located in the cochlea

Many arthropods and most vertebrates have sound receptors, but for many of them hearing does not seem to be a sensory priority. It is important in tetrapods, however, and both birds and mammals have a highly developed sense of hearing. Their auditory receptors, located in the **cochlea** of the inner ear, contain mechanoreceptor hair cells that detect pressure waves.

The cochlea is a spiral tube that resembles a snail's shell (Fig. 41–8). If the cochlea were uncoiled, it would be seen to consist

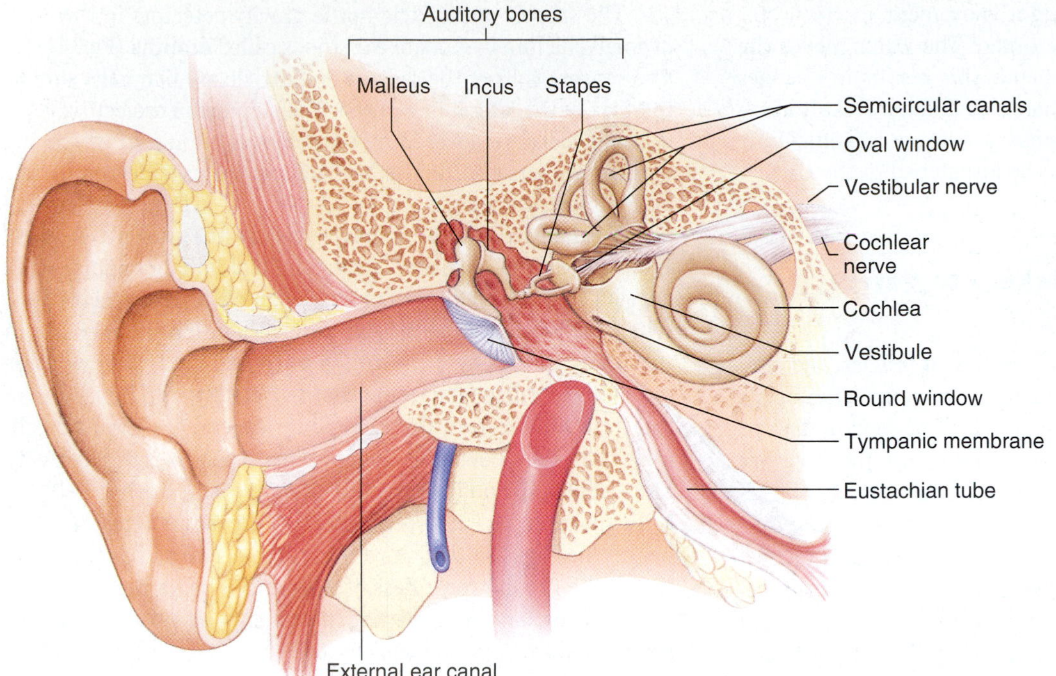

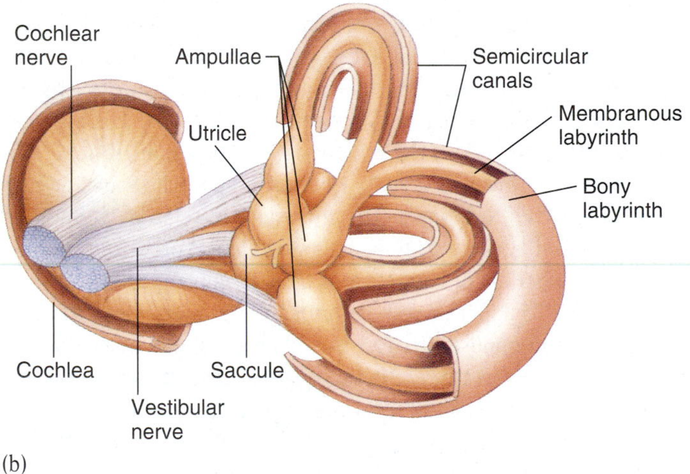

Figure 41–5 The human ear. (a) Human ear, anterior view. Humans have an outer ear, a middle ear, and an inner ear. Many vertebrates do not have outer or middle ears, but all have inner ears that help maintain equilibrium. (b) The inner ear comprises the saccule, utricle, and semicircular canals, collectively referred to as the vestibular apparatus. The utricle and saccule are best seen in the posterior view, which is shown here.

of three canals separated from each other by thin membranes and coming almost to a point at the apex. Two of these canals, or ducts, the **vestibular canal** and the **tympanic canal,** are connected at the apex of the cochlea and are filled with a fluid known as **perilymph.** The middle canal, the **cochlear duct,** is filled with endolymph and contains the auditory organ, the **organ of Corti.**

Each organ of Corti contains about 18,000 hair cells arranged in rows that extend the length of the coiled cochlea. Each hair cell is equipped with stereocilia that extend into the cochlear duct. The hair cells rest on the **basilar membrane,** which separates the cochlear duct from the tympanic canal. Overhanging and in contact with the hair cells of the organ of Corti is another membrane, the **tectorial membrane.**

In terrestrial vertebrates sound waves in the air are transformed into pressure waves in the cochlear fluid. In the human ear, for example, sound waves pass through the **external auditory canal** and cause the **tympanic membrane,** or eardrum (the membrane separating the outer ear and the middle ear), to vibrate. The vibrations are transmitted across the middle ear by three tiny bones—the **malleus, incus,** and **stapes** (or hammer, anvil, and stirrup, so called because of their shapes). The malleus is in contact with the eardrum, and the stapes is in contact with a thin membranous region of the cochlea, called the **oval window.** These bones act as three interconnected levers that help amplify the vibrations. A very small movement in the malleus causes a larger movement in the incus and a very large movement in the stapes. The vibrations pass through the oval window to the fluid in the vestibular canal. If sound waves were conducted directly from air to the oval window, much energy would be lost. The middle ear functions to couple sound waves in the air with the pressure waves conducted through the fluid in the cochlea.

Because liquids cannot be compressed, the oval window could not cause movement of the fluid in the vestibular duct if there were not an escape valve for the pressure. This is provided by the membranous **round window** at the end of the tympanic canal. The pressure wave presses on the membranes separating the three ducts, is transmitted to the tympanic canal, and causes a bulging of the round window. The movements of the basilar membrane produced by these pulsations cause the stereocilia of the organ of Corti to rub against the overlying tectorial membrane. When stereocilia are bent by this contact, ion channels in the plasma membrane of the hair cells open. If depolarization occurs, a receptor potential may be generated. Hair cells release the neurotransmitter *glutamate,* which binds to receptors on sensory neurons that synapse on each hair cell. Glutamate binding may lead to depolarization of sensory neurons. Axons of the sensory neurons join to form the **cochlear nerve,** a component of the vestibulocochlear nerve (cranial nerve VIII; also referred to as the auditory nerve).

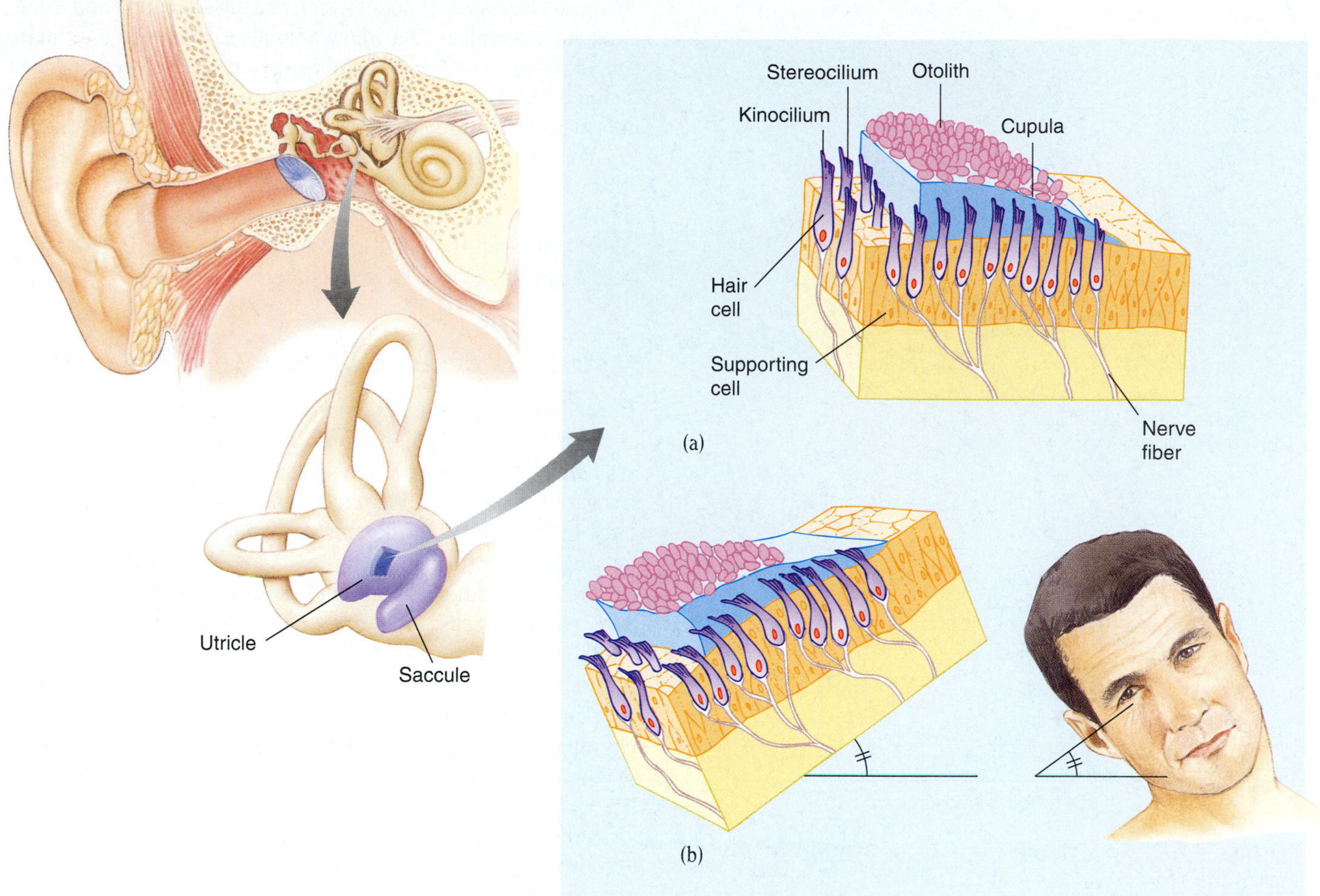

Figure 41–6 Function of the saccule and utricle in maintaining posture. The saccule and utricle sense linear acceleration, allowing us to know our position relative to the ground. Compare the positions of the otoliths and hair cells in **(a)** with those in **(b)**. Changes in head position cause the force of gravity to distort the cupula, which in turn distorts the stereocilia of the hair cells. The hair cells respond by sending impulses along the vestibular nerve (part of the auditory nerve) to the brain.

We can summarize the sequence of events involved in hearing as follows:

> Sound waves enter external auditory canal ⟶ tympanic membrane vibrates ⟶ malleus, incus, and stapes amplify vibrations ⟶ oval window vibrates ⟶ vibrations are conducted through fluid ⟶ basilar membrane vibrates ⟶ hair cells in the organ of Corti are stimulated ⟶ cochlear nerve transmits impulses to brain

Sounds differ in pitch, loudness, and tone quality. **Pitch** depends on frequency of sound waves, or number of vibrations per second, and is expressed as hertz (Hz). Low-frequency vibrations result in the sensation of low pitch, whereas high-frequency vibrations result in the sensation of high pitch. Frequencies greater than 60 Hz result in unequal vibration along the length of the basilar membrane. Sounds of a given frequency set up resonance waves in the cochlear fluid that cause a particular section of the basilar membrane to vibrate. High frequencies are detected by hair cells located near the base of the cochlea, whereas low frequencies are sensed by hair cells near the apex of the cochlea. The brain infers the pitch of a sound from the particular hair cells that are stimulated.

Loud sounds cause resonance waves of greater amplitude (height). The hair cells are more intensely stimulated, and the cochlear nerve then transmits a greater number of impulses per second. Variations in the quality of sound, such as those evident when an oboe, a cornet, and a violin play the same note, depend on the number and kinds of overtones, or harmonics, produced. These provide stimulation to different hair cells in addition to the main stimulation common to all three instruments. Thus, differences in tone quality are recognized in the *pattern* of the hair cells stimulated.

The human ear is equipped to register sound frequencies between about 20 and 20,000 Hz, although individuals vary greatly. Dogs and some other animals can hear sounds of much higher frequencies. The human ear is more sensitive to sounds between

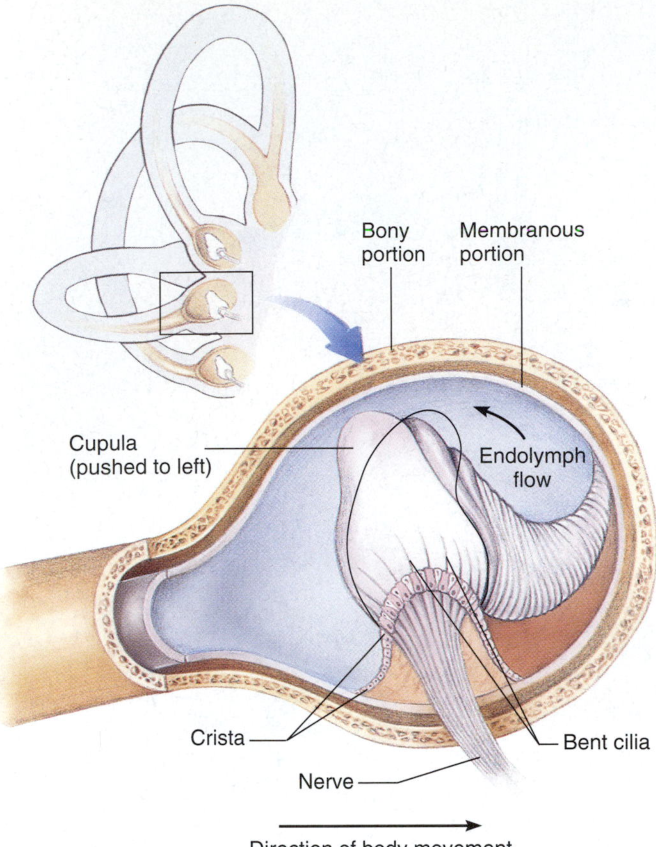

Figure 41–7 Semicircular canals and equilibrium. When the head changes its rate of rotation, endolymph within the ampulla of the semicircular canal distorts the cupula. The stereocilia of the hair cells bend, increasing the frequency of action potentials in sensory neurons. Information is transmitted to the brain via the vestibular nerve.

1000 and 4000 Hz than to higher or lower ones. Within this intermediate range, the ear is extremely sensitive. In fact, when the energy of audible sound waves is compared with the energy of visible light waves, the ear is ten times more sensitive than the eye.

Deafness may be caused by injury to, or malformation of, either the sound-transmitting mechanism of the outer, middle, or inner ear, or the sound-perceiving mechanism of the inner ear. Efferent neurons from the brainstem protect the hair cells of the organ of Corti by dampening their response. Even so, exposure to high-intensity sound, such as heavily amplified music, damages the hair cells of the organ of Corti.

CHEMORECEPTORS ARE ASSOCIATED WITH THE SENSES OF TASTE AND SMELL

Two highly sensitive chemoreceptive systems are the senses of **taste** (gustation) and **smell** (olfaction). These chemical senses, found throughout the animal kingdom, allow animals to detect chemical substances in food, water, and air. Detecting and evaluating such chemical cues allow animals to find food and mates and to avoid predators. Chemoreception is also an important method used by members of the same species to communicate information about food sources, potential mates, and danger.

For terrestrial vertebrates, the sense of smell refers to gaseous substances that reach olfactory receptors through the air. The sense of taste refers to materials dissolved in water (or saliva) in the mouth. For aquatic animals, these distinctions blur. Does a catfish smell or taste the water?

Taste buds detect dissolved food molecules

The ability to discriminate among tastes has survival value. For example, foods with a high caloric value often have a sweet taste, whereas poisons are generally bitter. The organs of taste in insects are sensory hairs. Mammals sense taste with **taste buds** located in the mouth. In humans, taste buds are found mainly in tiny elevations, or papillae, on the tongue. Each of the thousands of taste buds is an oval epithelial capsule containing about 100 taste receptor cells interspersed with supporting cells (Fig. 41–9). Unlike most neurons, both taste and smell receptors are continuously regenerated.

The plasma membrane at the tip of each taste receptor cell has microvilli that extend into a taste pore on the surface of the tongue, where they are bathed in saliva. The taste receptors detect chemical substances dissolved in saliva. Certain molecules, for example, those perceived as sweet, activate a signal transduction process involving a G protein. Adenylyl cyclase activity increases, elevating cyclic AMP levels. A protein kinase is activated that phosphorylates and closes K^+ channels. This decrease in K^+ permeability sets up a depolarizing receptor potential. Action potentials are then generated in sensory neurons that synapse with the taste receptor cell. One sensory neuron can innervate several taste buds.

Traditionally, four basic tastes have been recognized: sweet, sour, salty, and bitter. In 2000, physiologists reported a fifth taste, glutamate. Although the idea of a fifth taste is still somewhat controversial, it was first suggested almost a century ago by the Japanese physiologist Kikunae Ikeda. In 1908, Ikeda identified glutamate as the compound that triggers the taste he called umami, responsible for the savoriness of certain aged cheeses, soy sauce, anchovy, and some seafoods.

Although the greatest sensitivity to each taste occurs in a given area of the human tongue (see Fig. 41–9), not all papillae are restricted to a single category of taste. Recent findings indicate that bitter-sensitive taste receptors are activated by a limited number of bitter compounds, suggesting that taste receptors may be specific for subcategories of tastes.

Flavor depends on the four basic tastes in combination with smell, texture, and temperature. Smell affects flavor because odors pass from the mouth to the nasal chamber. No doubt you have observed that when your nose is congested, food seems to have little "taste." The taste buds are not affected, but the blockage of nasal passages severely reduces the participation of olfactory reception in the composite sensation of flavor.

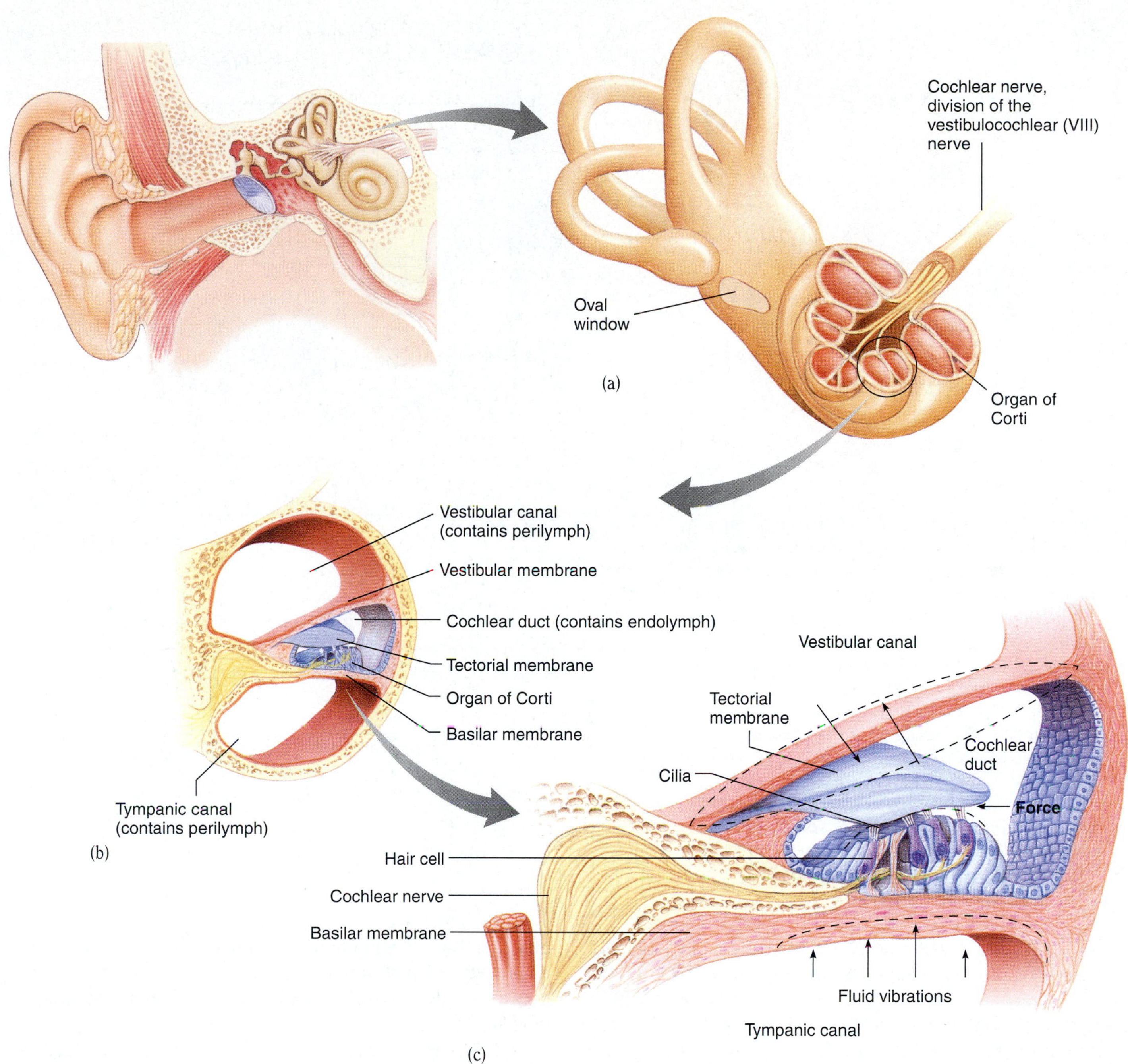

Figure 41–8 The cochlea, organ of hearing. **(a)** Cross section through the cochlea showing the cochlear nerve. **(b)** Enlarged view of the organ of Corti resting on the basilar membrane and covered by the tectorial membrane. **(c)** How the organ of Corti works. Vibrations transmitted by the malleus, incus, and stapes set the fluid in the tympanic canal in motion. These vibrations (indicated by the *dotted lines*) are transmitted to the basilar membrane, and, as the membrane vibrates, hair cells of the organ of Corti rub against the overlying tectorial membrane. This stimulation causes the hair cells to depolarize, resulting in action potentials in the sensory neurons of the cochlear nerve, a branch of the vestibulocochlear (auditory) nerve.

Process of Science Awareness of the genetic component of taste can be traced to 1931 when Arthur L. Fox, a chemist at the Du Pont Company synthesized a compound called phenylthiocarbamide (PTC). Some PTC blew into the air and was inhaled by a colleague who experienced it as very bitter. Because Fox himself could not taste it, he became intrigued and asked other people to taste it. Fox found that about 25% of people were nontasters. Everyone else experienced PTC as bitter. Further research showed that the ability to taste PTC is inherited as a dominant trait.

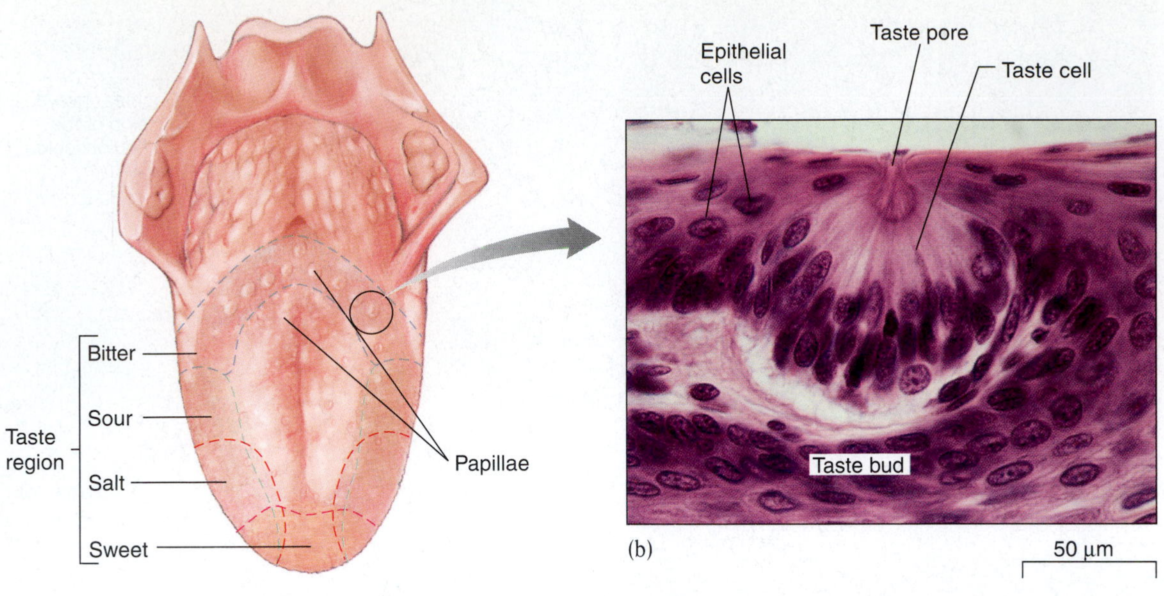

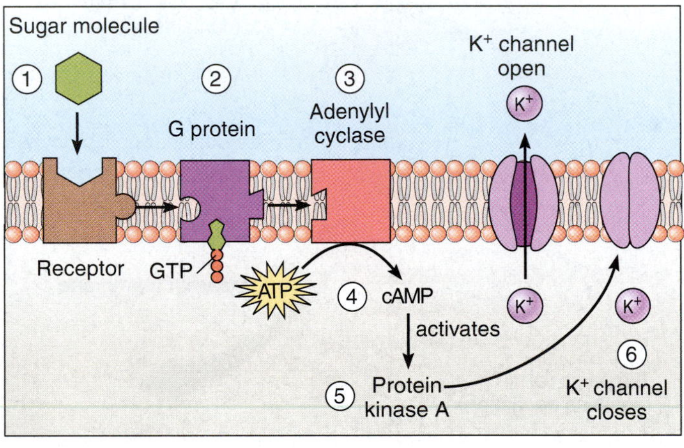

Figure 41–9 Taste buds. Taste receptors are located mainly on the surface of the tongue. **(a)** The surface of the tongue, showing distribution of taste buds that are sensitive to sweet, bitter, sour, and salt. A single taste receptor may respond to more than one category of taste. **(b)** LM of a taste bud, which consists of an epithelial capsule containing several taste receptors. **(c)** A sugar molecule activates a signal transduction process. (1) The sugar binds with a receptor in the plasma membrane of a taste receptor. (2) A G protein is activated. (3) Adenylyl cyclase is activated. (4) ATP is converted to cyclic AMP (cAMP). (5) Cyclic AMP activates a protein kinase. (6) Action of the protein kinase closes K^+ channels. The increase in cAMP also requires the presence of GTP. *(b, Ed Reschke)*

In the 1970s Linda Bartoshuk of Yale University continued taste research using a chemical compound called PROP (6-*n*-propylthiouracil). She found that some tasters were even more sensitive to the bitter taste of PROP. Subsequent studies have suggested that about 25% of the U.S. population are supertasters, 50% are regular tasters, and 25% are nontasters.

In 1997, Adam Drewnowski at the University of Michigan and his team reported that supertasters avoid broccoli, Brussels sprouts, cabbage, and many other vegetables and fruits that have a bitter taste. These foods contain flavonoids and other compounds that are thought to protect against cancer. Continuing investigation of taste preferences and their nutritional consequences may lead to more effective approaches to improving nutrition and to a better understanding of the relative importance of genetics and learning in food selection.

The olfactory epithelium is responsible for the sense of smell

Many animals rely on the sense of smell to find food, identify predators, and find mates. In fact, most invertebrates depend on **olfaction,** the detection of odors, as their main sensory modality. Recent findings indicate that the mechanisms for chemoreception have been highly conserved throughout the animal kingdom.

In terrestrial vertebrates, olfaction occurs in the nasal epithelium. In humans, the **olfactory epithelium** is found in the roof of the nasal cavity (Fig. 41–10). It contains about 100 million olfactory receptor cells with ciliated tips. The long, nonmotile cilia extend into a layer of mucus on the epithelial surface of the nasal passageway. Receptor molecules on the cilia bind with compounds dissolved in the mucus. The other end of each olfactory receptor cell is an axon that projects directly to the brain. These axons make up the **olfactory nerve** (the first cranial nerve), which extends to the olfactory bulb in the brain. From there information is transmitted to the **olfactory cortex,** which is in the limbic system, a part of the brain also associated with emotional behavior. (Odors are often associated with feelings and memories.)

When a molecule binds with a receptor on the cilia of an olfactory receptor cell, a signal transduction process is initiated. A G protein is activated, leading to the synthesis of cyclic AMP,

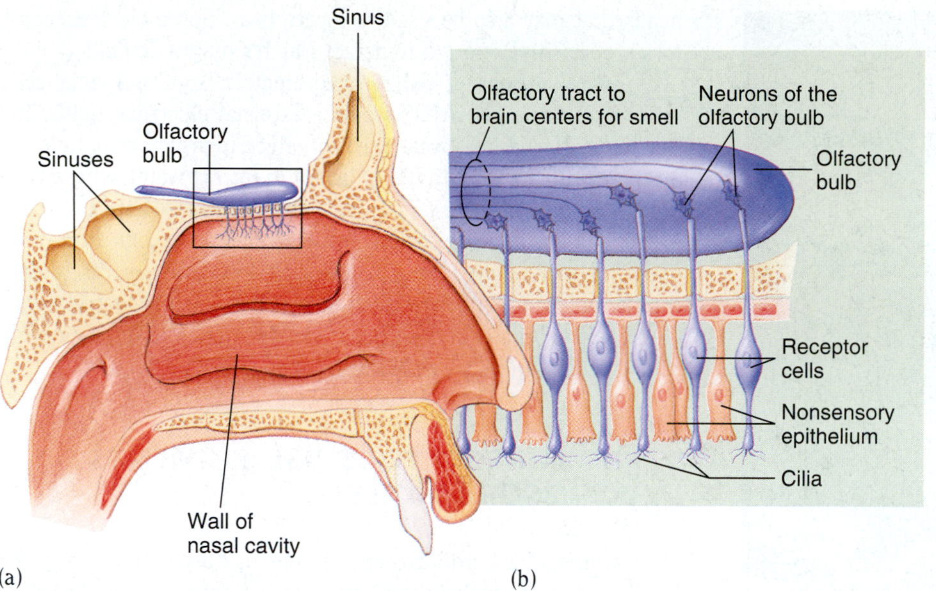

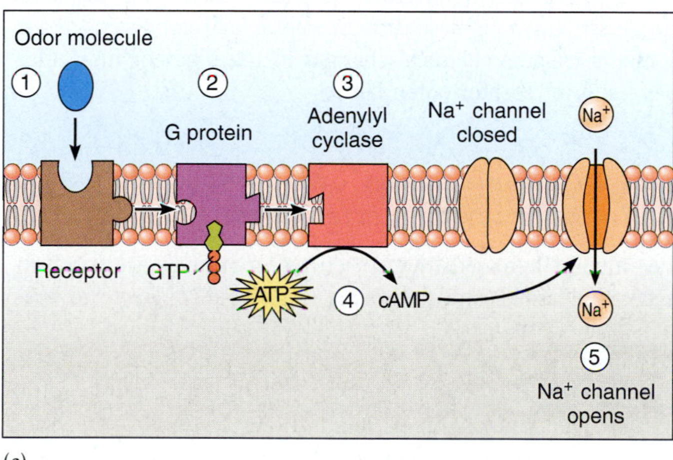

Figure 41-10 Olfactory epithelium. **(a)** Smell depends on thousands of chemoreceptor cells in the roof of the nasal cavity. **(b)** The receptor cells are neurons located in the olfactory epithelium. **(c)** An odor molecule binds to a receptor in the plasma membrane of an olfactory cell, triggering a signal transduction process. A G protein is activated, and a GTP-coupled subunit of the G protein activates adenylyl cyclase. ATP is converted to cyclic AMP, which opens gated channels. Sodium ions enter the cell, leading to depolarization.

which opens gated channels in the plasma membrane. These channels permit Na^+ and other cations to enter the cell, causing depolarization, which is the receptor potential. The number of odorous molecules determines the intensity, which in turn determines the magnitude of the receptor potential.

Humans can detect at least seven main groups of odors: camphor, musk, floral, peppermint, ethereal, pungent, and putrid. About 1000 genes code for 1000 types of olfactory receptors. Each odor consists of several component chemical groups, and each type of receptor may bind with a particular component. The combination of receptors activated determines what odor we perceive. We are capable of perceiving about 10,000 scents.

The olfactory receptors respond to remarkably small amounts of a substance. For example, ionone, the synthetic substitute for the odor of violets, can be detected by most people when it is present in a concentration of only one part to more than 30 billion parts of air. Smell is perhaps the sense that adapts most quickly. The olfactory receptors adapt about 50% in the first second or so after stimulation, so even offensively odorous air may seem odorless after only a few minutes.

Animals within many species communicate with one another with odors. They release **pheromones,** small volatile molecules that are secreted into the environment. For example, female moths release pheromones that attract males, and dogs and wolves use pheromones to mark territory. Mammals have specialized chemoreceptor cells that detect pheromones. These cells make up the **vomeronasal organ** located in the epithelium of the nose. The vomeronasal sensory neurons send signals to areas of the hypothalamus. (Pheromones are discussed in Chapters 47 and 50.)

THERMORECEPTORS ARE SENSITIVE TO HEAT

Heat is another form of radiant energy to which organisms respond. Although not much is known about their specific thermoreceptors, many invertebrates are sensitive to changes in temperature. Mosquitoes, ticks, and other blood-sucking arthropods use thermoreception in their search for an endothermic host. Some have temperature receptors on their antennae that are sensitive to changes of less than 0.5° C. At least two types of snakes, pit vipers and boas, use thermoreceptors to locate their prey (Fig. 41–11).

In mammals, which are endothermic, free nerve endings and specialized receptors in the skin and tongue detect temperature changes in the outside environment. In humans, changes in temperature are detected by at least three types of receptors: cold receptors, warmth receptors, and pain receptors for temperature extremes. (For a discussion of pain, see *Focus On: Pain Perception.*) Thermoreceptors in the hypothalamus detect internal changes in temperature and receive and integrate information from thermoreceptors on the body surface. The hypothalamus then initiates homeostatic mechanisms that ensure a constant body temperature.

Figure 41–11 Thermoreception. The pit organ of this bamboo viper *(Trimersurus stejnegeri)* is a sense organ located between each eye and nostril. The pit organ of many snakes can detect the heat from an endothermic animal up to a distance of 1 to 2 m. *(Carmela Leszczynski/Animals Animals)*

ELECTRORECEPTORS DETECT ELECTRICAL CURRENTS IN WATER

Some predatory species of sharks, rays, and bony fishes can detect the electrical fields generated in the water by the muscle activity of their prey. Their electroreceptors are modified hair cells that may be associated with the lateral line organ. In addition to their lateral line system, sharks have electroreceptors on their heads that may also be used to locate prey. Some electroreceptors are sensitive enough to detect Earth's magnetic field.

Several groups of fishes have electric organs, specialized muscle or nerve cells that produce external electrical fields. In species that produce a weak current, electric organs may help in orientation. This is particularly useful in murky water, where visibility and olfaction are poor. Electroreception also appears to be important in communication, for example, in recognition of a potential mate. Males have a different frequency of discharge than females. A few fishes, such as electric eels or electric rays, have electric organs in their heads capable of delivering powerful shocks that stun prey or predators.

PHOTORECEPTORS USE PIGMENTS TO ABSORB LIGHT

Most animals have photoreceptors that use pigments to absorb light energy. **Rhodopsins** are the photopigments found in the eyes of cephalopod mollusks, arthropods, and vertebrates. Light energy striking a light-sensitive receptor cell containing these pigments triggers chemical changes in the pigment molecules that result in receptor potentials.

Eyespots, simple eyes, and compound eyes are found among invertebrates

The simplest light-sensitive structures in animals are found in certain cnidarians and flatworms (Fig. 41–12). They are **eyespots,** called **ocelli,** that detect light but do not form images. Eyespots are often bowl-shaped clusters of light-sensitive cells within the epidermis. They may detect the direction of the source of light and distinguish light intensity.

Effective image formation requires a more complex **eye,** usually with a **lens.** A lens is a structure that concentrates light on a group of photoreceptors. Vision also requires a brain that

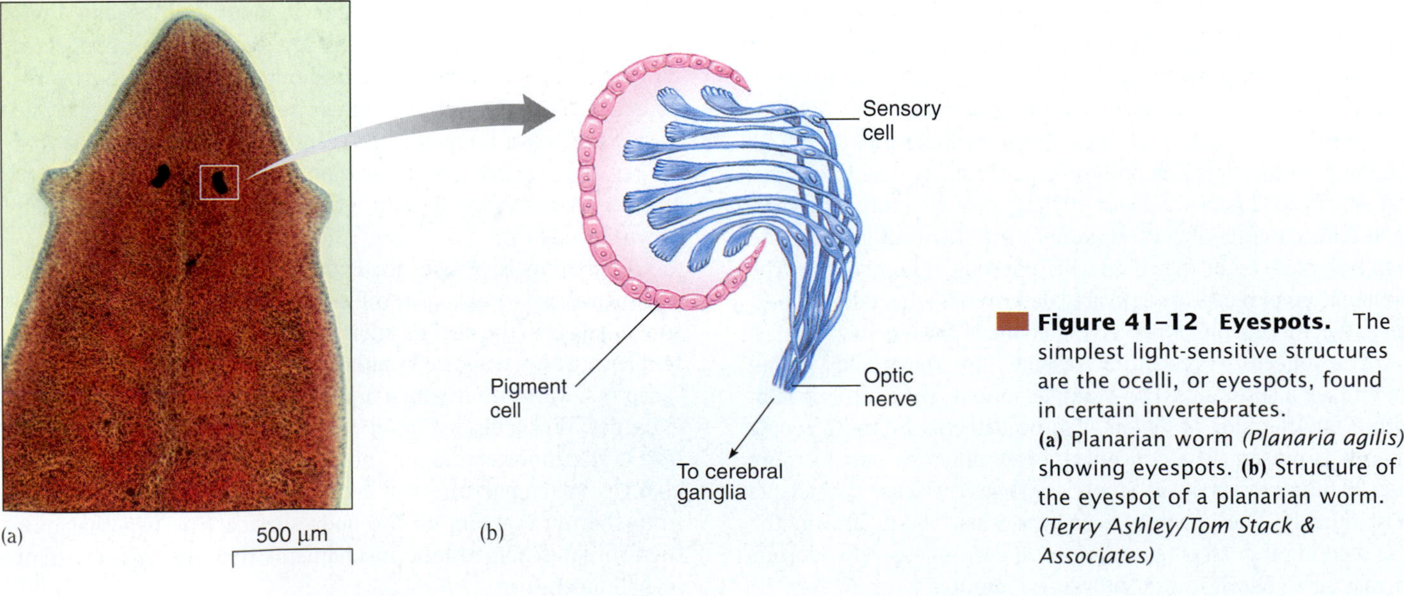

Figure 41–12 Eyespots. The simplest light-sensitive structures are the ocelli, or eyespots, found in certain invertebrates. **(a)** Planarian worm *(Planaria agilis)* showing eyespots. **(b)** Structure of the eyespot of a planarian worm. *(Terry Ashley/Tom Stack & Associates)*

can interpret the action potentials generated by the photoreceptors. The brain must integrate information about movement, brightness, location, position, and shape of the visual stimulus.

Two fundamentally different types of eyes evolved: the camera eye of vertebrates and some mollusks (squids and octopods), and the compound eye of arthropods.

Compound eyes found in crustaceans and insects are structurally and functionally different from vertebrate eyes (Fig. 41–13). The surface of a compound eye appears faceted, which means having many faces, like a diamond. Each **facet** is the convex *cornea* of one of the eye's visual units, called **ommatidia** (sing., *ommatidium*). The number of ommatidia varies with the species. For example, each eye of certain crustaceans has only 20 ommatidia, whereas the eye of a dragonfly has as many as 28,000.

The optical part of each ommatidium includes a biconvex lens and a **crystalline cone.** These structures focus light onto photoreceptor cells, called **retinular cells.** These cells have a light-sensitive membrane made up of microvilli containing rhodopsin. The membranes of adjacent retinular cells may fuse, forming a rod-shaped **rhabdome** that is sensitive to light.

Compound eyes do not perceive form well. Although the lens system of each ommatidium is adequate to focus a small inverted image, there is little evidence that they are actually perceived as images by the organism. However, all the ommatidia together produce a composite image, or **mosaic** picture. Each ommatidium, in gathering a point of light from a narrow sector of the visual field, is in fact sampling a mean intensity from that sector. All of these points of light taken together form a mosaic picture (Fig. 41–13d).

To appreciate the nature of this mosaic picture, we need only look at a newspaper photograph through a magnifying glass; it is a mosaic of many dots of different intensities. The clarity and definition of the picture depend on how many dots there are per unit area—the more dots, the better the picture. So it is with the compound eye. The image perceived by the

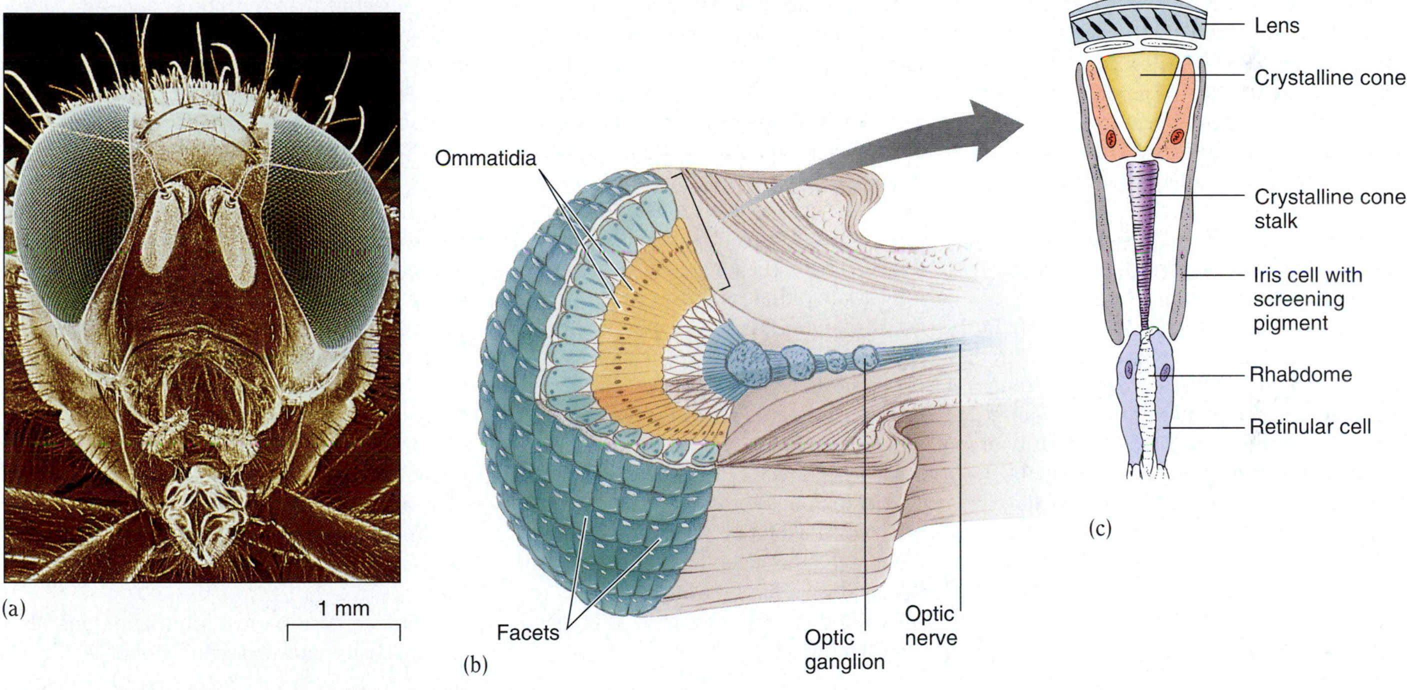

Figure 41–13 Compound eyes. **(a)** SEM of the Mediterranean fruit fly *(Ceratitis capitata)* showing its prominent compound eyes. **(b)** Structure of the compound eye showing several ommatidia. The eye registers changes in light and shade, permitting the animal to detect movement. **(c)** Structure of an ommatidium. The rhabdome is the light-sensitive core of the ommatidium. **(d)** A bee-eye's view of a flower. This photo was taken using an optical device to achieve an approximate simulation of how the bee might see this flower. However, the bee would see UV light rather than red, and circles would be vertically elongated ellipses. *(a, David Scharf/Peter Arnold, Inc.; d, Courtesy of J. Gould, Princeton University)*

FOCUS ON Pain Perception

Pain is a signal that warns an animal to protect the body from injury. A stimulus that causes damage to tissues generally results in the sensation of pain. In humans, pain receptors, called **nociceptors** (from the Latin *nocere,* to injure), are free nerve endings (dendrites) of certain sensory neurons found in almost every tissue. Children born without the ability to perceive pain, or individuals who have learned to dissociate from pain (a result of severe psychological trauma), do not know when they are being injured unless they receive information from other types of sensory receptors (e.g., they may see the blood).

Several types of nociceptors have been identified. Thermal nociceptors respond to temperature extremes (temperatures above 45° C or below 5° C). Mechanical nociceptors respond to strong tactile stimuli such as penetration by sharp objects or pinching. Other nociceptors respond to a variety of stimuli, including certain chemicals.

Stimulation of nociceptors does not always result in the perception of pain. We experience pain when information from nociceptors reaches the brain and is interpreted. Our perception of pain is influenced by past experience, suggestion, and emotions. Pain perception is colored by emotional experience. The intensity of pain we experience is influenced by the particular situation and by how we have learned to deal with pain. A child with a bruised knee may emotionally heighten the feeling of pain, whereas a professional fighter may virtually ignore a long series of well-delivered blows.

When stimulated, nociceptors transmit signals through sensory neurons to interneurons in the spinal cord. Recall from Chapter 40 that some interneurons synapse directly with motor neurons so that a withdrawal reflex can occur before signals are transmitted to the brain. The sensory neurons release the neurotransmitter *glutamate,* as well as several neuropeptides. Of these neuropeptides, the one that has been most studied is **substance P.** The neuropeptides appear to enhance and prolong the actions of glutamate.

The interneuron transmits the message to the opposite side of the spinal cord and then upward through one of five ascending pathways. The principal ascending nociceptive pathway is the spinothalamic tract, which transmits signals to the thalamus, where pain perception begins *(see figure)*. From the thalamus, impulses are sent into the parietal lobes and several other cortical regions, including areas of the limbic system where the emotional aspects of the pain are processed. When the signals reach the cerebrum, the individual becomes fully aware of the pain and can evaluate the situation. How threatening is the stimulus? How intense is the pain? What is the most adaptive response? From the thalamus, messages are also sent to a region in the limbic system, which is the brain's emotional center.

When nociceptors transduce stimuli into action potentials, changes occur in the ion channels and in other parts of the pain pathway that increase sensitivity to the painful stimuli. This process is called **sensitization.**

Neurophysiologists have long known that opiates, such as morphine and codeine, relieve pain, and these compounds and their derivatives are widely used clinically for analgesia (pain control). For example, patient-controlled morphine pumps are routinely used in many hospitals. Opiates work by blocking the release of substance P.

More than ten opiates have now been discovered in the brain, spinal cord, and pituitary gland, including **beta-endorphin** (for "endogenous morphine-like"), **enkephalins,** and **dynorphins.** These endogenous opiates are thought to inhibit nocireceptive neurons in the spinal cord. The endogenous opiates are more effective than morphine, and several are being investigated as potential analgesic drugs.

The brain locates pain on the basis of past experience. Generally, pain at the body surface is accurately projected back to the injured area. For example, when you step on a nail, your brain perceives the pain and then projects it back to the injured foot, so that you feel pain at the site of puncture. Artificial stimulation of the leg nerves may produce a sensation of pain in the foot even though the foot is untouched. In fact, for years after an amputation, a patient may feel **phantom pain** in the missing limb. This occurs because when the severed nerve is stimulated and sends a message to the brain, the brain "remembers" the nerve as it originally was—connected to the missing limb.

Nociceptors in most internal organs are not easily activated. Pain is often not

animal is probably much better in quality than might be suspected from the structure of the compound eye. The nervous system of an insect is apparently capable of image processing similar to that employed to improve the quality of photographs sent to Earth by robot spacecraft.

Arthropod eyes usually adapt to different intensities of light. A sheath of pigmented cells envelops each ommatidium, and screening pigments are present in cells called iris cells and in retinular cells. In nocturnal and crepuscular (active at dusk) insects and many crustaceans, pigment is capable of migrating proximally and distally within each pigmented cell. When the pigment is in the proximal position, each ommatidium is shielded from its neighbor, and only light entering directly along its axis can stimulate the receptors. When the pigment is in the distal position, light striking at any angle can pass through several ommatidia and stimulate many retinal units. As a result, sensitivity is increased in dim light, and the eye is protected from excessive stimulation in bright light. Pigment migration is under neural control in insects and under hormonal control in crustaceans. In some species it follows a daily rhythm.

Although the compound eye can form only coarse images, it compensates by being able to follow flickers to higher frequencies. Flies are able to detect up to about 265 flickers per second. In contrast, the human eye can detect only 45 to 53 flickers per

projected back to the organ that is stimulated. Instead it is *referred* to an area just under the skin that may be some distance from the organ involved. The area to which the pain is referred generally is connected to nerve fibers from the same level of the spinal cord as the organ involved. A person with angina who feels heart pain in the left arm is experiencing **referred pain.** Neurons from both the heart and the arm converge on the same neurons in the central nervous system. The brain interprets the incoming message as coming from the body surface because that is the more common origin. When pain is felt both at the site of the distress and as referred pain, it may seem to spread, or radiate, from the organ to the superficial area.

You may have observed that rubbing the skin around an injury or pressing on the area can alleviate pain. Activating certain large sensory neurons that deliver messages about touch or pressure can stimulate interneurons in the spinal cord that release enkephalins. Thus, these interneurons inhibit neurons in the pain pathway. Clinical methods have been developed for relieving pain by electrical stimulation of large sensory nerve fibers. Stimulation of the skin over a painful area with electrodes has successfully relieved pain in some patients. This procedure is called transcutaneous electrical nerve stimulation (TENS). Electrodes can also be implanted in the brain, allowing the patient to control chronic pain by stimulating the release of endorphins.

In acupuncture, a Chinese therapy for treating pain, needles are inserted to stimulate afferent neurons that inhibit pain signals. Endogenous opioids are also involved in analgesia induced by acupuncture. Evidence suggests that acupuncture needles stimulate nerves deep within the muscles, which in turn stimulate certain neurons in the brain to release endorphins.

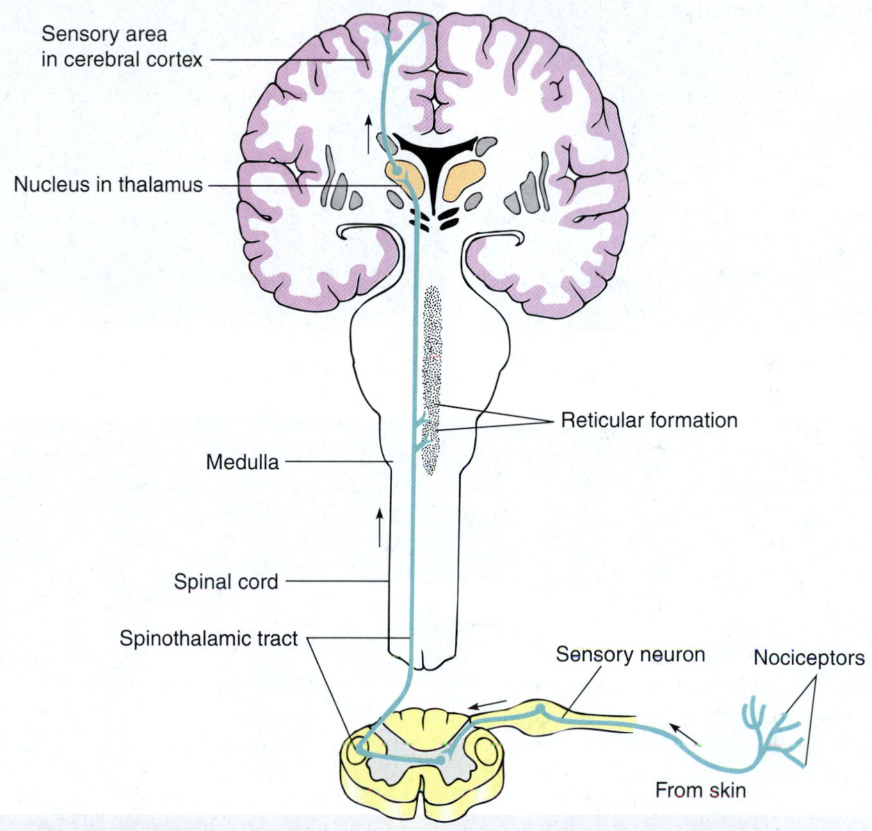

One major pathway for transmission of pain signals. Signals from nociceptors are transmitted by sensory neurons to the spinal cord where they synapse with an ascending neuron in the spinothalamic pathway. Signals are transmitted to the thalamus and then to sensory areas in the parietal lobes of the cerebrum.

second; for us, flickering lights fuse above these values, so we see light provided by an ordinary bulb as steady and the movement in motion pictures as smooth. To an insect, both room lighting and motion pictures must flicker horribly. The advantage of the insect's high critical flicker fusion threshold is that it permits immediate detection of even slight movement by prey or enemy. The compound eye is an important adaptation to the arthropod's way of life.

Compound eyes differ from our eyes in another respect. They are sensitive to wavelengths of light in the range from red to ultraviolet (UV). Accordingly, an insect can see UV light well, and its world of color is very different from ours. Because flowers reflect UV light to various degrees, flowers that appear identically colored to us may appear strikingly dissimilar to insects (see Fig. 35–4).

Vertebrate eyes form sharp images

The position of the eyes in the front of the head of humans and certain other vertebrates permits both eyes to be focused on the same object (Fig. 41–14). The overlap in information they receive results in the same visual information striking the two retinas (light-sensitive areas) at the same time. This **binocular vision** is an important factor in judging distance and depth. Variations of

(a)

(b)

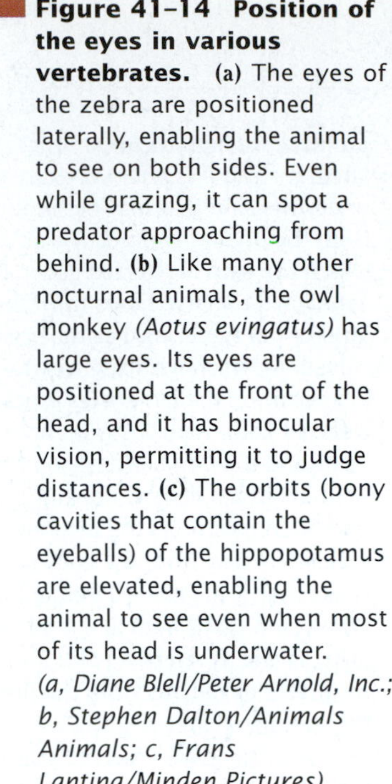

Figure 41–14 Position of the eyes in various vertebrates. **(a)** The eyes of the zebra are positioned laterally, enabling the animal to see on both sides. Even while grazing, it can spot a predator approaching from behind. **(b)** Like many other nocturnal animals, the owl monkey *(Aotus evingatus)* has large eyes. Its eyes are positioned at the front of the head, and it has binocular vision, permitting it to judge distances. **(c)** The orbits (bony cavities that contain the eyeballs) of the hippopotamus are elevated, enabling the animal to see even when most of its head is underwater. *(a, Diane Blell/Peter Arnold, Inc.; b, Stephen Dalton/Animals Animals; c, Frans Lanting/Minden Pictures)*

(c)

eye position in other vertebrates offer different advantages. For example, the eyes of grazing animals are positioned laterally, enabling them to spot a predator approaching from behind.

The vertebrate eye can be compared to a camera. An adjustable lens can be focused for different distances, and a diaphragm, called the **iris,** regulates the size of the light opening, called the **pupil** (Fig. 41–15). The **retina** corresponds to the light-sensitive film used in a camera. Outside the retina is the **choroid layer,** a sheet of cells filled with black pigment that absorbs extra light and prevents internally reflected light from blurring the image. (Cameras are also black on the inside.) The choroid is rich in blood vessels that supply the retina.

The outer coat of the eyeball, called the **sclera,** is a tough, opaque, curved sheet of connective tissue that protects the inner structures and helps to maintain the rigidity of the eyeball. On the front surface of the eye, this sheet becomes the thinner, transparent **cornea,** through which light enters. The cornea serves as a fixed lens that focuses light.

The lens of the eye is a transparent, elastic ball just behind the iris. It bends the light rays coming in and brings them to a focus on the retina. The lens is aided by the curved surface of the cornea and by the refractive properties (ability to bend light rays) of the liquids inside the eyeball. The **anterior cavity** between the cornea and the lens is filled with a watery substance, the **aqueous**

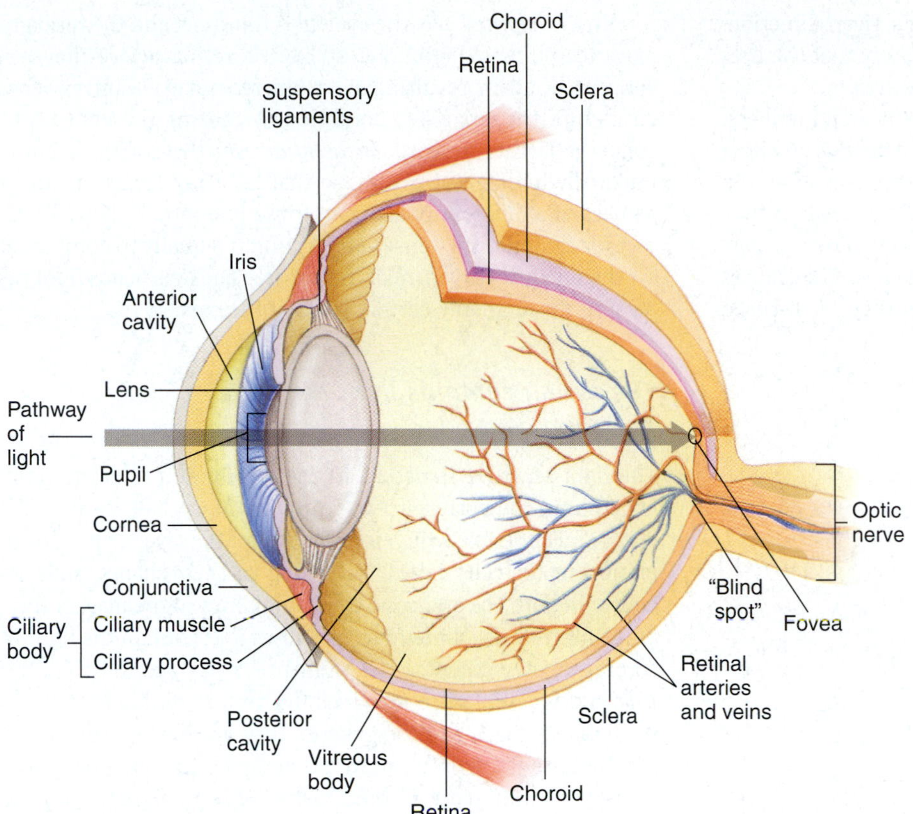

Figure 41–15 Structure of the human eye. Light passes through the human eye to photoreceptor cells in the retina. In this lateral view, the eye is shown partly sectioned along the sagittal plane to show its internal structures.

fluid. The larger **posterior cavity** between the lens and the retina is filled with a more viscous fluid, the **vitreous body.** Both fluids are important in maintaining the shape of the eyeball by providing an internal fluid pressure.

At its anterior margin, the choroid is thick and projects medially into the eyeball to form the **ciliary body,** which consists of ciliary processes and the ciliary muscle. The ciliary processes are glandlike folds that project toward the lens and secrete the aqueous fluid.

We focus a camera by changing the distance between the lens and the film. The eye has the power of **accommodation,** the ability to change focus for near or far vision by changing the shape of the lens (Fig. 41–16). This is accomplished by the **ciliary muscle,** a part of the ciliary body. To focus on objects that are near, the ciliary muscle contracts, causing the elastic lens to assume a rounder shape. To focus on more distant objects, the ciliary muscle relaxes and the lens assumes a flattened (ovoid) shape.

The amount of light entering the eye is regulated by the iris, a ring of smooth muscle that appears as blue, green, gray, or brown depending on the amount and nature of pigment present. The iris is composed of two mutually antagonistic sets of muscle fibers. One set is arranged circularly and contracts to *decrease* the size of the pupil. The other is arranged radially and contracts to *increase* the size of the pupil.

Each eye has six muscles that extend from the surface of the eyeball to various points in the bony socket. These muscles enable

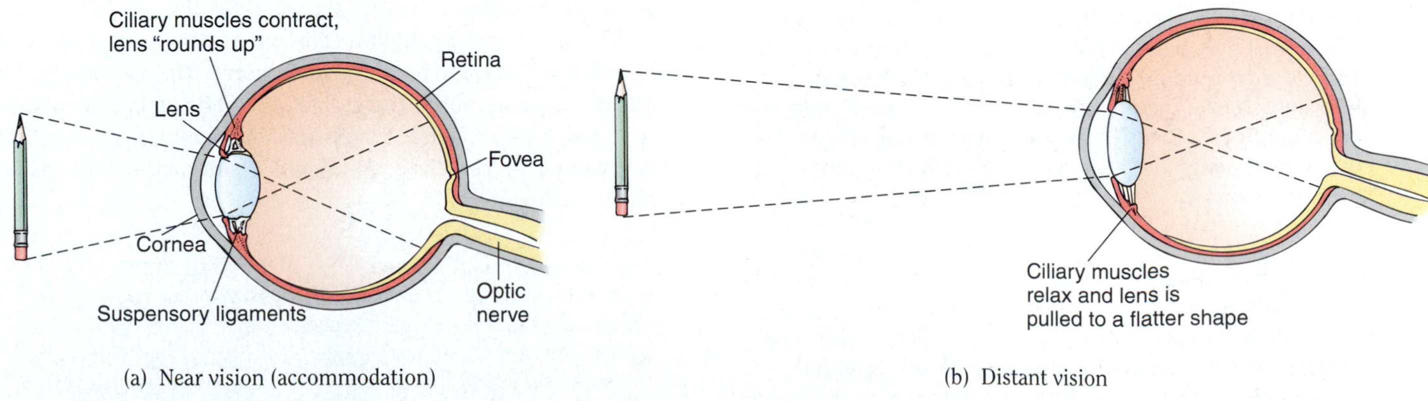

(a) Near vision (accommodation)

(b) Distant vision

Figure 41–16 Focusing on near and far objects.

the eye as a whole to move and be oriented in a given direction. Cranial nerves innervate the muscles in such a way that the eyes normally move together and focus on the same area.

The most common disorders of vision are nearsightedness, farsightedness, and astigmatism (Fig. 41–17). In *nearsightedness (myopia),* the eyeball is elongated. The light rays converge at a point in front of the retina and are diverging again when they reach it. This results in a blurred image. Nearsightedness is corrected by placing a concave lens in front of the eye. This causes the light rays to converge farther back so that they are focused on the retina.

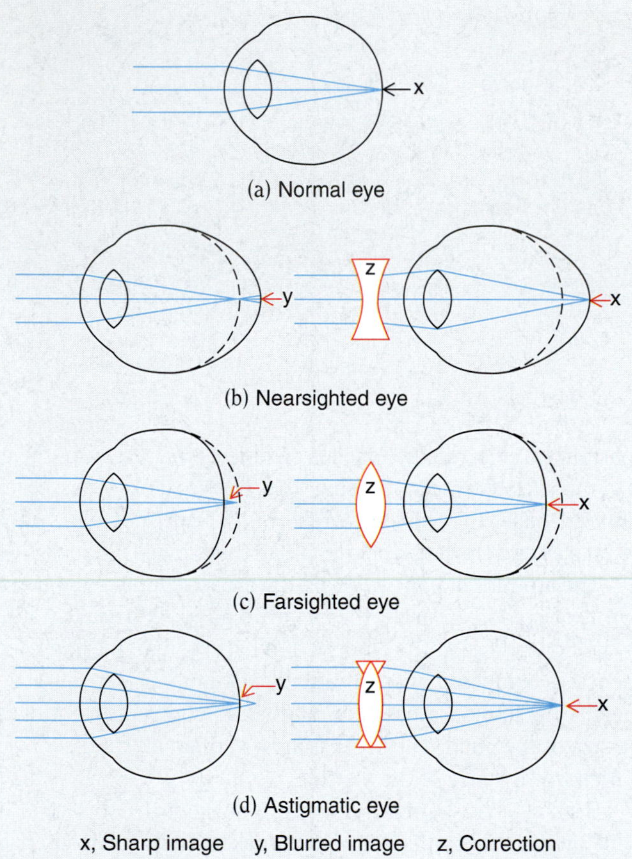

Figure 41–17 Vision disorders. Common abnormalities of the eye result in defects in vision. **(a)** In the normal eye, parallel light rays coming from a point in space are focused as a point on the retina. **(b)** In the nearsighted eye, the eyeball is elongated. Parallel light rays are brought to a focus in front of the retina (on *dotted line,* which represents the position of the retina in a normal eye). A blurred image is formed on the retina. Nearsightedness is corrected with a concave lens. **(c)** In the farsighted eye, the eyeball is shortened and the focal point of the light rays is behind the retina. A convex lens converges the light rays. **(d)** In astigmatism, light rays passing through one part of the lens are focused on the retina, while light rays passing through another area of the lens are not focused on the retina. A cylindrical lens corrects this condition.

In a *farsighted* eye, the eyeball is too short and the retina too close to the lens. Light rays strike the retina before they have converged, again resulting in a blurred image. Convex lenses correct for the farsighted condition by causing the light rays to converge farther forward. In *astigmatism* the cornea is curved unequally in different planes, so that light rays in one plane are focused at a different point from those in another plane. To correct for astigmatism, lenses are ground unequally to compensate for the unequal curvature of the cornea. The lens bends light rays passing through only certain parts of the cornea.

The retina contains light-sensitive rods and cones

The light-sensitive structure in the vertebrate eye is the retina, which lines the posterior two-thirds of the eyeball, covering the choroid. The retina, which is composed of ten layers, contains the photoreceptor cells called, according to their shapes, **rods** and **cones.** In both rod and cone cells, infoldings of the plasma membrane form stacks of membranous discs that contain the light-absorbing photopigments. The human eye has about 125 million rods and 6.5 million cones. Rods function in dim light, allowing us to detect shape and movement. They are not sensitive to colors. Because the rods are more numerous in the periphery of the retina, you can see an object better in dim light if you look slightly to one side of it (allowing the image to fall on the rods).

Cones respond to light at higher levels of intensity, for example daylight, and they allow us to perceive fine detail. Cones are responsible for color vision; they are differentially sensitive to different wavelengths (colors) of light. The cones are most concentrated in the **fovea,** a small depressed area in the center of the retina. The fovea is the region of sharpest vision because it has the greatest density of receptor cells and because the retina is thinner in that area.

Light must pass through several layers of connecting neurons in the retina to reach the rods and cones (Fig. 41–18). The retina consists of five main types of neurons. (1) Photoreceptors (rods and cones) synapse on (2) **bipolar cells,** which make synaptic contact with (3) **ganglion cells.** Two types of lateral interneurons are the (4) **horizontal cells** that receive information from the photoreceptor cells and send it to bipolar cells and (5) **amacrine cells** that receive messages from the bipolar cells and send signals back to the bipolar cells or to ganglion cells (Fig. 41–19).

The axons of the ganglion cells extend across the surface of the retina and unite to form the **optic nerve.** The area where the optic nerve passes out of the eyeball, the **optic disk,** is known as the "blind spot"; because it lacks rods and cones, images falling on it cannot be perceived. A simplified summary of the visual pathway follows:

Light passes through cornea ⟶ through aqueous fluid ⟶ through lens ⟶ through vitreous body ⟶ image forms on retina (photoreceptor cells) ⟶ signals bipolar cells ⟶ signals ganglion cells ⟶ optic nerve transmits signals to thalamus ⟶ integration by visual areas of the cerebral cortex

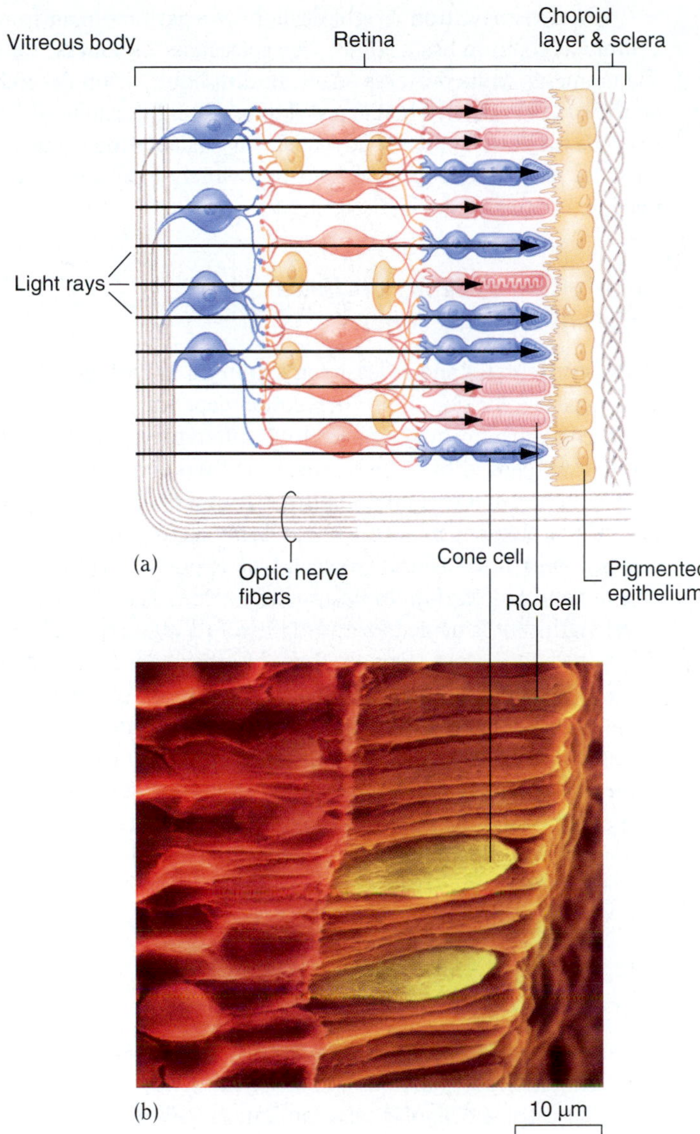

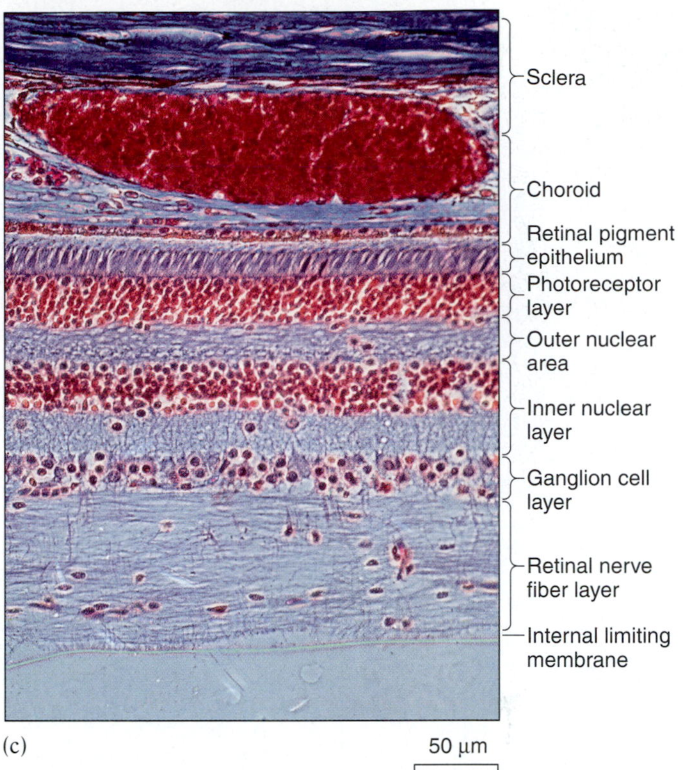

Figure 41–18 Organization of the retina. **(a)** The elaborate interconnections among the various layers of neurons in the retina allow them to interact and to influence one another. The rods and cones are the photoreceptor cells. **(b)** Rods *(red elongated structures)* and two cones *(shorter, thicker, yellow structures)* are seen in this SEM. The elongated rods permit us to see shape and movement, whereas the shorter cones allow us to view our world in color. **(c)** LM of the retina. *(b, Lennart Nilsson, from* The Incredible Machine, *p. 279; c, Manfred Kage/ Peter Arnold, Inc.)*

A chemical change in rhodopsin leads to the response of a rod to light

How is light converted into the neural signals that transmit information about environmental stimuli into pictures in the brain? Rod cells are so sensitive to light that they can respond to a single photon. Rhodopsin in the rod cells and some very closely related photopigments in the cone cells are responsible for the ability to see. Rhodopsin consists of *opsin,* a large protein that is chemically joined with *retinal,* an aldehyde of vitamin A (see Fig. 3–14). Two isomers of retinal exist: the *cis* form, which is folded, and the *trans* form, which is straight.

When it is dark, opsin binds to retinal in the *cis* form. Cyclic GMP (cyclic guanosine monophosphate, a molecule similar to cyclic AMP) opens nonspecific channels that permit passage of Na^+ and other cations into the rod cell (Fig. 41–20). This depolarizes the rod cell, and it releases the neurotransmitter *glutamate*. The glutamate hyperpolarizes the membrane of the bipolar cell so that it does not transmit messages. Note that the photoreceptor is different from other neurons in that the ion channels in its membrane are normally open; it is depolarized and continually releases neurotransmitter. The steady flow of ions into the cell when it is dark is referred to as the *dark current.* Another unusual characteristic of photoreceptors, and also bipolar cells, is that they do not produce action potentials. Their release of neurotransmitter is graded, regulated by the extent of depolarization.

When light strikes rhodopsin, light is transduced. This process can be considered in three stages.

1. Light activates rhodopsin. Light transforms *cis*-retinal to *trans*-retinal. This change in shape causes rhodopsin to change shape and to break down into its components, opsin and retinal. This is the light-dependent process in vision.
2. Rhodopsin is part of a signal transduction pathway. When it changes shape it binds with a G protein called **transducin.** Transducin activates an esterase that hydrolyzes cyclic GMP (cGMP) to GMP, thereby reducing the concentration of cGMP.

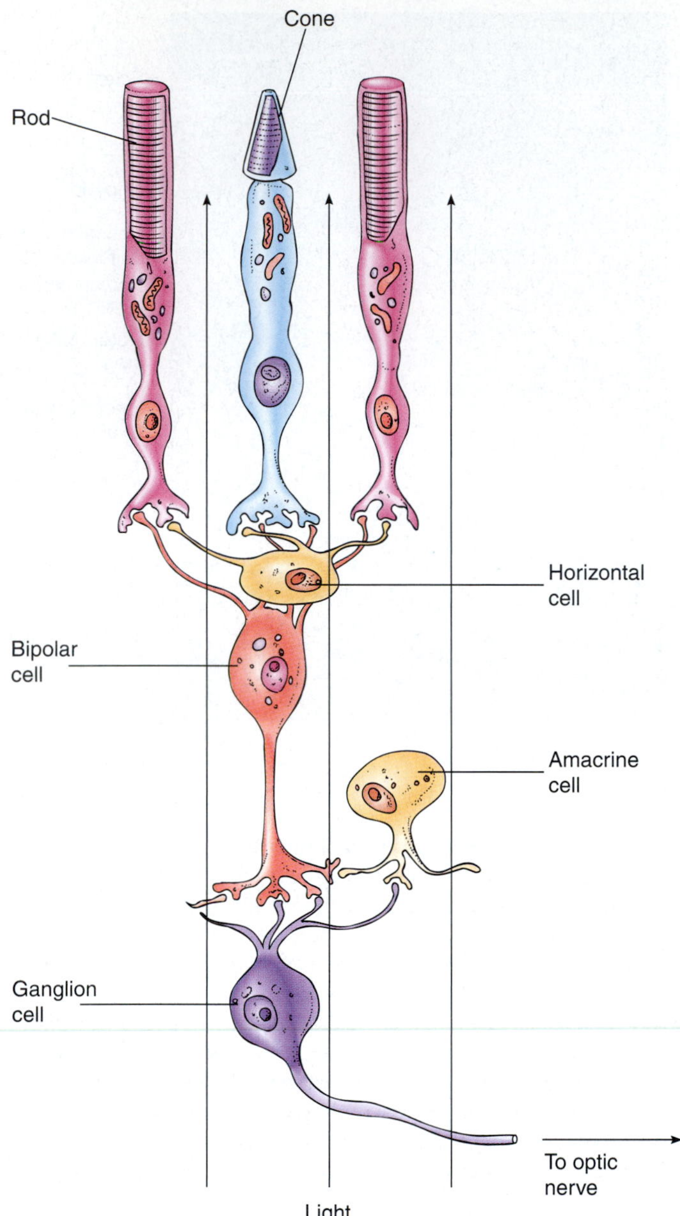

Figure 41–19 Neural pathway in the retina. The photoreceptor cells are located in the back of the retina. They synapse with bipolar and horizontal cells. The bipolar cells synapse with ganglion cells and amacrine cells. Axons of ganglion cells make up the optic nerve.

3. When the concentration of cGMP in the rod decreases, its Na^+ channels begin to close. Fewer cations pass into the rod cell, and it becomes more negative (hyperpolarized). The rod cell then releases less glutamate. Thus, light causes a decrease in neural signals from the rod cells.

The release of glutamate normally hyperpolarizes the membrane of the bipolar cell, so a decrease in glutamate release results in its depolarization. The depolarized bipolar cell increases its release of a neurotransmitter, which typically stimulates the ganglion cell. (Some ganglion cells decrease their firing rate in response to stimulation.)

When we move from bright daylight to a dark room, or from a dimly lit room to bright light, we experience "blindness" for a few moments while our eyes adapt. In dim light, vision depends on the rods. However, when exposed to bright light, rhodopsin in the rods may be completely activated. Dark adaptation occurs as enzymes restore rhodopsin to a form in which it can respond to light.

Color vision depends on three different types of cones

Many invertebrates and at least some animals in each vertebrate class have color vision. Color perception depends on cones. Humans and other primates have three different types of cones, commonly referred to as blue, green, and red cones. Each contains a slightly different photopigment. Although the retinal portion of the pigment molecule is the same as in rhodopsin, the opsin protein differs slightly in each type of photoreceptor. Each type of cone can respond to light within a considerable range of wavelengths but is named for the ability of its pigment to absorb that wavelength more strongly than do other cones. For example, red light can be absorbed by all three types of cones, but those cones most sensitive to red act as red receptors. By comparing the relative responses of the three types of cones, the brain can detect colors of intermediate wavelengths. Color blindness occurs when there is a deficiency of one or more of the three types of cones. This is usually an inherited X-linked condition (see Chapter 10, Fig. 10–15).

Integration of visual information begins in the retina

The size, intensity, and location of light stimuli determine initial processing in the retina. Rods and cones can send signals directly to bipolar cells, and bipolar cells can send signals directly to ganglion cells. However, as illustrated in Figure 41–19, photoreceptor cells can also transmit signals to horizontal cells, and bipolar cells can transmit signals to amacrine cells. Ganglion cells are inhibited by amacrine cells and can be inhibited indirectly by horizontal cells by their action on bipolar cells. Thus, the horizontal and amacrine cells integrate signals laterally.

Bipolar, horizontal, and amacrine cells combine signals from several photoreceptors. Each ganglion cell has a **receptive field,** a specific group of photoreceptors that light must strike for the ganglion cell to be stimulated. Ganglion cells are activated or inhibited, depending on which photoreceptor cells have been stimulated.

The pattern of neuron firing in the retina appears to be very important. The signals sent to ganglion cells depend on spatial patterns and timing of the light striking the retina. Ganglion cells receive signals about specific types of visual stimuli such as color, brightness, and motion and transmit this information about the characteristics of a visual image. Ganglion cells produce action potentials in contrast with rods and cones and most other neurons of the retina that produce only graded potentials.

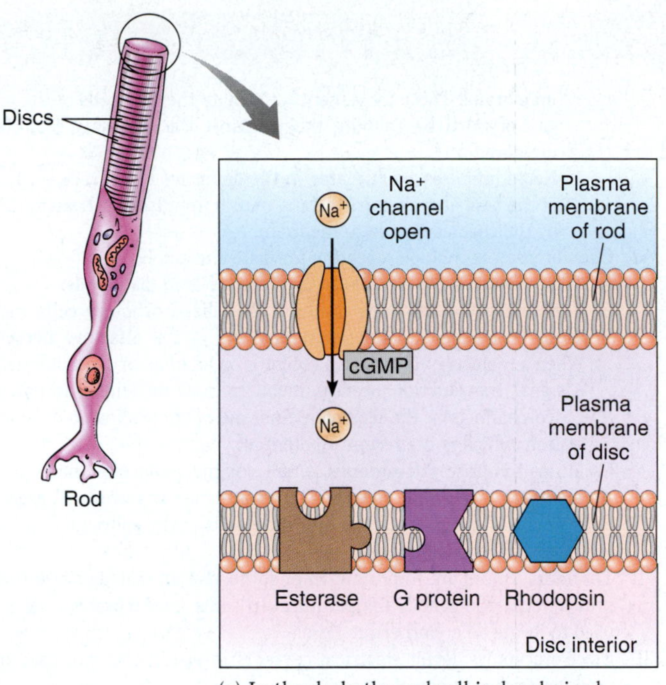

(a) In the dark, the rod cell is depolarized

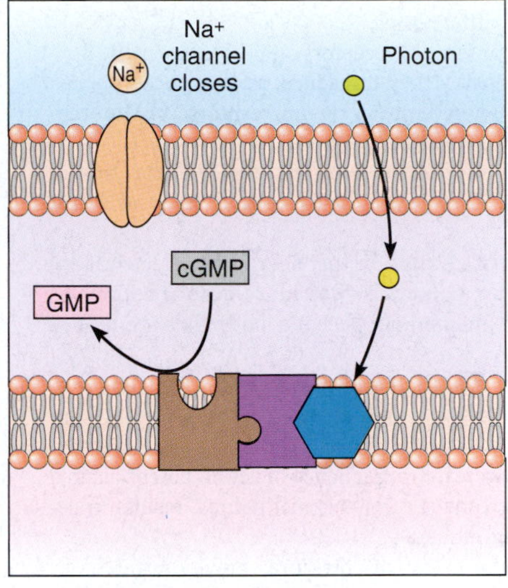

(b) In the light, the rod cell becomes hyperpolarized

Figure 41–20 The biochemistry of vision. **(a)** In the dark, sodium channels are open and the rod cell is depolarized. It releases the neurotransmitter glutamate. **(b)** Light causes rhodopsin to change shape, activating a signal transduction pathway in which activation of a G protein leads to hyperpolarization and a decrease in signals from the rod cell. The discs are infoldings of the plasma membrane. See text for further explanation.

Axons of ganglion cells form the optic nerves that transmit information to the brain by way of complex, encoded signals. The optic nerves cross in the floor of the hypothalamus, forming an X-shaped structure, the **optic chiasm** (Fig. 41–21). Some of the axons of the optic nerves cross over and then extend to the opposite side of the brain. Axons of the optic nerves end in the **lateral geniculate nuclei** of the thalamus. From there, neurons send signals to the **primary visual cortex** in the occipital lobe of the cerebrum. The lateral geniculate nucleus controls which information is sent to the visual cortex. Neurons in the reticular activating system are involved in this integration. Information can be transmitted from the primary visual cortex to other cortical areas for further processing.

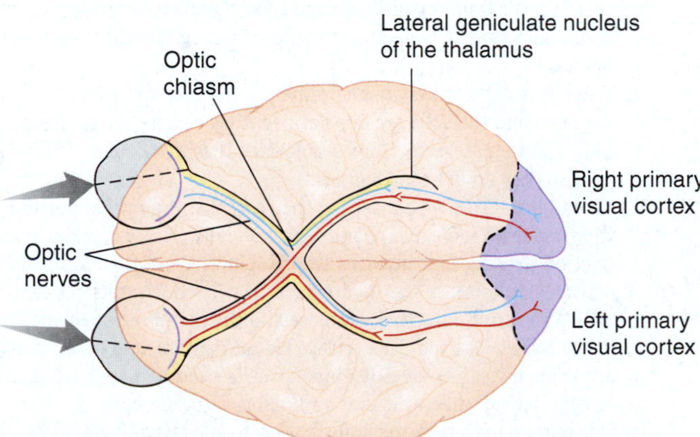

Figure 41–21 Neural pathway for transmission of visual information. Axons of ganglion cells form the optic nerves. The optic nerves cross, forming the optic chiasm. Many optic nerve fibers end in the lateral geniculate nuclei of the thalamus. From there, signals are sent to the visual cortex.

Different types of ganglion cells receive information about different aspects of a visual stimulus. As a result, information from each point in the retina is transmitted in parallel by several different types of ganglion cells. These cells project to different kinds of neurons in the lateral geniculate nucleus and primary visual cortex.

Neurobiologists have not yet discovered all the mechanisms by which the brain makes sense out of the visual information it receives. We do know that a large part of the association areas of the cerebrum are involved in integrating visual input. The neurons of the visual cortex are organized as a map of the external visual field. They are also organized into columns. Each column of neurons receives signals originating from light entering either the right or left eye. Columns are also organized by orientation of light. Neurons in a column receive signals originally triggered by light with the same orientation. However, these signals originate from different locations in the retina.

SUMMARY WITH KEY TERMS

I. **Sensory receptors** are specialized to respond to specific energy stimuli in the environment.
   A. Sensory receptors may be neuron endings or specialized receptor cells in close contact with neurons.
   B. **Sense organs** consist of sensory receptors and accessory cells.
   C. Based on the type of energy they **transduce,** sensory receptors can be classified as **mechanoreceptors, chemoreceptors, photoreceptors, thermoreceptors,** or **electroreceptors.**
   D. **Exteroceptors** receive information from the outside world. **Interoceptors,** sensory receptors within body organs, help maintain homeostasis.

II. Receptor cells absorb energy, transduce that energy into electrical energy, and produce **receptor potentials,** which are depolarizations or hyperpolarizations of the membrane. Receptor potentials are **graded responses.**

III. The ability to experience sensation depends on the transmission of coded signals by sensory receptors and on interpretation of those signals by the brain.

IV. **Sensory adaptation,** the decrease in frequency of action potentials in a sensory neuron even when the stimlulus is maintained, results in decreased response to that stimulus.

V. Mechanoreceptors respond to touch, pressure, gravity, stretch, or movement. They are activated when they change shape as a result of being mechanically pushed or pulled.
   A. The **tactile receptors** in the skin respond to mechanical displacement of hairs or to displacement of the receptor cells themselves. For example, the **Pacinian corpuscle** responds to touch and pressure.
   B. **Proprioceptors,** sensory receptors within muscles, tendons, and joints, enable the animal to perceive orientation of the body and the positions of its parts. **Muscle spindles, Golgi tendon organs,** and **joint receptors** continually respond to tension and movement.
   C. **Statocysts** are gravity receptors found in many invertebrates.
   D. **Hair cells** are vertebrate mechanoreceptors found in the lateral line of fishes, the vestibular apparatus, semicircular canals, and cochlea. Hair cells have cilia-like processes called **stereocilia.**
   E. **Lateral line organs** supplement vision in fish and some amphibians by informing the animal of moving objects or objects in its path.
   F. The vertebrate **inner ear** consists of a **labyrinth** of fluid-filled chambers and canals that help maintain equilibrium. The **vestibular apparatus** in the upper part of the labyrinth consists of the **saccule, utricle,** and **semicircular canals.**
      1. The saccule and utricle contain **otoliths** that change position when the head is tilted or when the body is moving in a straight line. The otoliths stimulate hair cells that send signals to the brain, enabling the animal to perceive the direction of gravity.
      2. The semicircular canals inform the brain about turning movements. Clumps of hair cells, called **cristae,** within each **ampulla** are stimulated by movements of the **endolymph,** a fluid that fills each canal.
   G. In birds and mammals, the **organ of Corti** within the **cochlea** contains auditory receptors.
      1. Sound waves pass through the external auditory canal, cause the **tympanic membrane** (eardrum) to vibrate, and are amplified and transmitted through the middle ear by the **malleus, incus,** and **stapes.**
      2. Vibrations pass through the **oval window** to fluid within the vestibular duct. Pressure waves press on the membranes that separate the three ducts of the cochlea.
      3. The bulging of the **round window** serves as an escape valve for the pressure. The pressure waves cause movements of the **basilar membrane.** These movements stimulate the hair cells of the organ of Corti by rubbing them against the overlying **tectorial membrane.**
      4. Nerve impulses are initiated in the dendrites of neurons that lie at the base of each hair cell, and neural impulses are transmitted by the **cochlear nerve** to the brain.

VI. Chemoreceptors include receptors for taste and smell.
   A. Taste receptors are specialized epithelial cells in **taste buds.**
   B. The **olfactory epithelium** contains specialized olfactory cells with axons that extend to the brain as fibers of the **olfactory nerves.** When a molecule binds with a receptor on an olfactory receptor cell, a signal transduction process involving a G protein is initiated. Gated channels in the plasma membrane open, leading to depolarization, which is the receptor potential.
   C. Animals release **pheromones,** small volatile molecules that signal members of the same species. The mammalian **vomeronasal organ,** made up of specialized chemoreceptor cells in the epithelium of the nose, detects pheromones.

VII. Thermoreceptors are important in endothermic animals because they provide cues about body temperature. In some invertebrates they are used to locate an endothermic host.

VIII. Electroreceptors detect electrical currents in water; they are used by predatory fishes to detect prey.

IX. Photoreceptors have photopigments that absorb light energy. **Rhodopsins** are the photopigments found in the eyes of many animals, including vertebrates.
   A. **Eyespots,** or **ocelli,** found in cnidarians and flatworms, detect light but do not form images.
   B. The **compound eye** of insects and crustaceans consists of visual units called **ommatidia,** which collectively produce a **mosaic** image. Each ommatidium has a transparent **lens** and a **crystalline cone** that focus light onto receptor cells called **retinular cells.**
   C. In the human eye, light enters through the **cornea,** is focused by the lens, and produces an image on the **retina.** The **iris** regulates the amount of light that can enter.
   D. The retina contains the photoreceptor cells: **rods** that function in dim light and form images in black and white; and **cones** that function in bright light and permit color vision.
   E. In addition to the rods and cones, the retina contains **bipolar cells** that send signals to **ganglion cells.** Two types of lateral interneurons integrate information: **Horizontal cells** receive signals from the rods and cones and send signals to bipolar cells, and **amacrine cells** receive signals from bipolar cells and send signals back to bipolar cells and to ganglion cells. Axons of the ganglion cells make up the **optic nerves.**
   F. When it is dark, ion channels in the plasma membranes of rod cells are open and the cells are depolarized; they release the neurotransmitter glutamate, which hyperpolarizes the membranes of bipolar cells so they do not send signals.
   G. When light strikes rhodopsin in the rod cells, the retinal portion of the rhodopsin molecule changes shape and initiates a signal transduction process that involves **transducin,** a G protein. Transducin activates an esterase that hydrolyzes cGMP, reducing its concentration. As a result, ion channels close and the membrane becomes hyperpolarized. The rod cells release less glutamate, and fewer signals are transmitted. As a result, bipolar cells become depolarized, and release neurotransmitter that stimulates ganglion cells.
   H. The optic nerves transmit information to the **lateral geniculate nuclei** in the thalamus. From there neurons project to the **primary visual cortex** and then to integration centers in the cerebral cortex.

POST-TEST

1. Which of the following is *not* true of a sensory receptor? (a) detects a stimulus in the environment (b) converts energy of the stimulus into electrical energy (c) produces a receptor potential (d) produces a graded response (e) interprets sensory stimuli
2. A sensory receptor absorbs energy from some stimulus. The next step is: (a) release of neurotransmitter (b) transmission of an action potential (c) energy transduction (d) transmission of a receptor potential (e) sensory adaptation
3. Interoceptors (a) help maintain homeostasis (b) are located in the skin (c) are photoreceptors (d) adapt rapidly (e) convert absorbed energy directly into action potentials
4. Which of the following are *not* correctly matched? (a) mechanoreceptors—touch, pressure (b) electroreceptors—voltage (c) photoreceptors—light (d) chemoreceptors—gravity (e) nociceptors—extreme temperature
5. A Pacinian corpuscle (a) is a mechanoreceptor (b) responds to heat (c) responds to pressure (d) answers a, b, and c are correct (e) answers a and c only
6. Which of the following is *not* located in the vertebrate inner ear? (a) vestibular apparatus (b) cochlea (c) malleus, incus, and stapes (d) organ of Corti (e) utricle
7. The auditory receptors are located in the (a) utricles (b) organs of Corti (c) vestibular apparati (d) saccule (e) semicircular canals
8. Which of the following is/are associated with informing the brain about turning movements? (a) semicircular canals (b) saccule (c) lymph (d) otoliths (e) utricle
9. Which of the following receptors do *not* have hair cells? (a) lateral line organ (b) saccule (c) utricle (d) semicircular canals (e) rods
10. In the process of hearing, the basilar membrane vibrates. Which event occurs next? (a) tympanic membrane vibrates (b) bones in middle ear amplify and conduct vibrations (c) cochlear nerve transmits impulses to organ of Corti (d) hair cells in organ of Corti are stimulated (e) vibrations conducted to chemoreceptors
11. In olfaction, (a) pheromones stimulate the taste buds (b) the number of odorous molecules determines the magnitude of the receptor potential (c) chemical signals are converted directly into electrical signals (d) a G protein is activated and leads to the closing of gated channels (e) the concentration of cyclic AMP increases, leading to hyperpolarization of the receptor
12. The vomeronasal organ (a) detects pheromones (b) is located at the back of the tongue (c) sends signals directly to the thalamus (d) releases pheromones (e) detects signals from predators
13. Substance P (a) is an opiate receptor (b) inhibits glutamate secretion (c) prolongs the actions of glutamate (d) is responsible for phantom pain (e) three of the preceding answers are correct
14. In the human visual pathway, after light passes through the cornea, it (a) stimulates ganglion cells (b) passes through the lens (c) sends signals through the optic nerve (d) depolarizes horizontal cells (e) hyperpolarizes rod cells
15. Cones (a) are most concentrated in the ganglion area (b) are more numerous than the rods (c) are responsible for bright-light vision (d) are found in all animals with photoreceptors (e) are a type of bipolar cell
16. Rhodopsin (a) is concentrated in the bipolar cells (b) binds with a G protein (c) changes shape when the membrane of the cone is depolarized (d) sends signals to the lateral geniculate nuclei (e) activates substance P

REVIEW QUESTIONS

1. Contrast mechanoreceptors with chemoreceptors.
2. Which sensory receptors permit us to perform actions such as getting dressed or finding our way into bed in the dark? Explain how they work.
3. What is the physiological basis of seasickness?
4. Draw a sequence diagram summarizing the process of hearing.
5. Discuss the mechanism by which the sensory cells of the ear are indirectly stimulated by sound waves.
6. What are otoliths, and what is their role in maintaining equilibrium?
7. Draw a diagram of the human retina, labeling all parts. Diagram the neural pathway in the retina.
8. Contrast the state of a rod cell in darkness and light. Explain what happens when light strikes rhodopsin. (Include a description of the signal transduction pathway in your answer.)
9. How does the human eye adjust to near and far vision?
10. Contrast the function of the insect's compound eye with that of the vertebrate eye.
11. Explain the statement, "Vision happens mainly in the brain."
12. Label the diagrams. Use Figures 41–5 and 41–15 to check your answers.

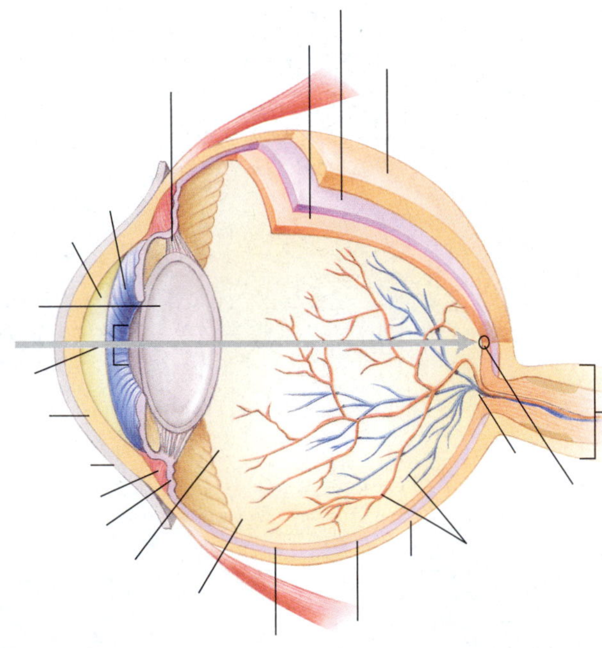

YOU MAKE THE CONNECTION

1. If all neurons transmit the same type of message, how do we know the difference between sound and light? How are we able to distinguish between an intense pain and a mild one? How are these discriminations adaptive?
2. Connoisseurs can recognize many varieties of cheese or wine by "tasting." How can this be, when there are only a few types of taste receptors? How might it be adaptive for organisms to have only four or five basic tastes?
3. Design a synaptic mechanism by which the same neurotransmitter has an excitatory effect on one bipolar cell and an inhibitory effect on another.

RECOMMENDED READINGS

Dehnhardt, G. "Hydrodynamic Trail-Following in Harbor Seals *(Phoca vitulina).*" *Science,* Vol. 293, 6 Jul. 2001. Seals use their whiskers to sense movements in water and can find prey by following hydrodynamic trails. Also see the news summary of this research by Carl Zimmer in the same issue.

Fackelmann, K. "The Bitter Truth: Do Some People Inherit a Distaste for Broccoli?" *Science News,* Vol. 152, Jul. 1997. Twenty-five percent of the U.S. population are supertasters who are especially sensitive to the bitter taste of certain vegetables.

Kandel, E.R., J.H. Schwartz, and T.M. Jessell. *Essentials of Neural Science and Behavior,* 4th ed. McGraw-Hill, New York, 2000. A well-written, comprehensive exploration of neurobiology.

Kirchner, W.H., and W.F. Towne. "The Sensory Basis of the Honeybee's Dance Language." *Scientific American,* Vol. 270, No. 6, Jun. 1994. Bees can hear the sounds produced by the dances they use to communicate.

Matthews, G.G. *Neurobiology: Molecules, Cells, and Systems.* Blackwell Scientific, Malden, MA, 1998. In Part IV the author very clearly discusses the sensory systems in some detail.

Science, Vol. 286, 22 Oct. 1999. Special issue on olfaction. Several articles explore the latest findings in olfactory reception.

Travis, J.T. "Eye-Opening Gene: How Many Times Did Eyes Arise?" *Science News,* Vol. 151, May 1997. An interesting discussion of the controversy surrounding the evolution of eyes.

Vander, A., J. Sherman, and D. Luciano. *Human Physiology,* 8th ed. McGraw-Hill, New York, 2001. Chapters 8 and 13 provide a clear introduction to neural function.

- Visit our Web site at **http://www.info.brookscole.com/solomonbergmartin** for links to chapter-related resources on the World Wide Web. Additional on-line materials relating to this chapter can also be found on our Web site.

See chapter activity on BioActive Learner CD for additional help in mastering the chapter's material. Icon location in the chapter's margins shows which topics have tutorials or simulations in the CD.

42

Internal Transport

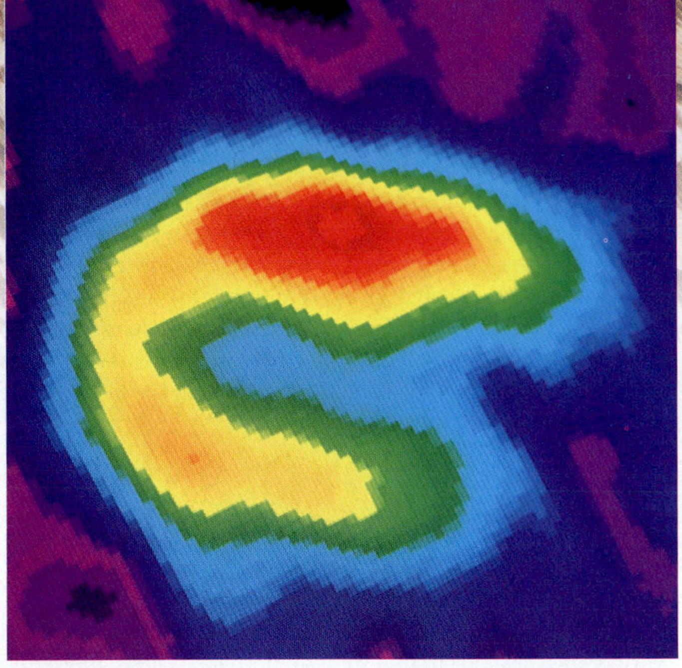

PET scan visualizing the left ventricle of the human heart. The scan measures metabolic activity based on glucose levels in the cardiac muscle tissue, providing information about the condition of the heart. (The color progression from least to most active is blue, green, yellow, red.) *(Ted Horowitz/The Stock Market)*

LEARNING OBJECTIVES

After you have studied this chapter you should be able to

1. Compare and contrast internal transport in animals with no circulatory system, those with an open circulatory system, and those with a closed circulatory system.
2. Relate structural adaptations of the vertebrate circulatory system to each function it performs.
3. Compare the structure and function of red blood cells, white blood cells, and platelets.
4. Summarize in proper sequence the events involved in blood clotting.
5. Compare the structure and function of different types of blood vessels, including arteries, arterioles, capillaries, and veins.
6. Trace the evolution of the vertebrate heart from fish to mammal.
7. Describe the structure and function of the human heart and label a diagram of the heart. Include a description of cardiac muscle and of the heart's conduction system.
8. Trace the events of the cardiac cycle and relate normal heart sounds to these events.
9. Define cardiac output, describe how it is regulated, and identify factors that affect it.
10. Identify factors that determine and regulate blood pressure and compare blood pressure in different types of blood vessels.
11. Trace a drop of blood through the pulmonary and systemic circulations, naming in sequence each structure through which it passes.
12. Identify the risk factors for cardiovascular disease, trace the progress of atherosclerosis, and summarize its possible complications, including angina pectoris and myocardial infarction.
13. Describe the structure and functions of the lymphatic system.

Most cells require a continuous supply of nutrients and oxygen and the removal of waste products. In very small animals, these metabolic needs can be met by simple **diffusion,** the net movement of particles from a region of higher concentration to a region of lower concentration, resulting from random motion. A molecule can diffuse 1 micrometer (μm) in less than 1 millisecond (msec), so diffusion is adequate over microscopic distances. In invertebrates that are only a few cells thick, diffusion is an effective mechanism for distributing materials to and from their cells. No specialized circulatory structures are present in sponges, cnidarians (e.g., jellyfish), ctenophores (comb jellies), flatworms, or nematodes (roundworms). As in all animals, the fluid between the cells, called **interstitial fluid,** or tissue fluid, bathes the cells and provides a medium for diffusion of oxygen, nutrients, and wastes.

The time that diffusion requires increases with the square of the distance over which diffusion occurs. A cell that is 10 μm away from its oxygen (or nutrient) supply can receive oxygen by diffusion in about 50 msec, but a cell that is 1000 μm (1 mm) away from its oxygen supply would have to wait several minutes and could not survive if it had to depend on diffusion alone. In animals that are many cells thick, specialized **circulatory systems** transport oxygen, nutrients, hormones, and other materials to the interstitial fluid surrounding all the cells and remove metabolic wastes. A circulatory system reduces the diffusion distance that needed materials must travel. Typically, a circulatory system interacts with every other organ system in the body.

A circulatory system typically consists of the following: (1) **blood,** a connective tissue consisting of cells and cell fragments dispersed in fluid known as *plasma;* (2) a pumping organ, generally a **heart;** and (3) a system of **blood vessels** or spaces through which the blood circulates.

The human circulatory system, also known as the **cardiovascular system,** is the focus of extensive research because cardiovascular disease is the number one cause of death in the United States and most other industrial societies. The positron-emission tomography (PET) scan shown here visualizes the left ventricle of the human heart. A major risk factor for cardiovascular disease is elevated levels of cholesterol and low-density lipoprotein (LDL) in the blood. In contrast, high-density lipoprotein (HDL) appears to play a protective role, by removing excess cholesterol from the blood and tissues. Cells in the liver and certain other organs bind with the HDL, remove the cholesterol, and use it in synthesizing needed compounds.

In 1997 molecular biologist Monty Krieger of the Massachusetts Institute of Technology and his research team reported that they had identified a gene in mice that codes for an HDL receptor. When these investigators knocked out the mouse gene, cholesterol levels more than doubled. As new receptors and mechanisms involved in lipid transport and metabolism are discovered, they serve as targets for new drugs and other treatments for cardiovascular disease.

SOME INVERTEBRATES HAVE NO CIRCULATORY SYSTEM

As indicated in the chapter introduction, many small, aquatic invertebrates have no circulatory system. In cnidarians, the central gastrovascular cavity (as its name implies) serves as a circulatory organ as well as a digestive organ (Fig. 42–1). The animal's tentacles capture prey and deliver it through the mouth into the cavity, where digestion occurs. The digested nutrients then pass into the cells lining the cavity and through them to cells of the outer layer. As the animal stretches and contracts, movements of the body stir up the contents of the gastrovascular cavity and help distribute nutrients.

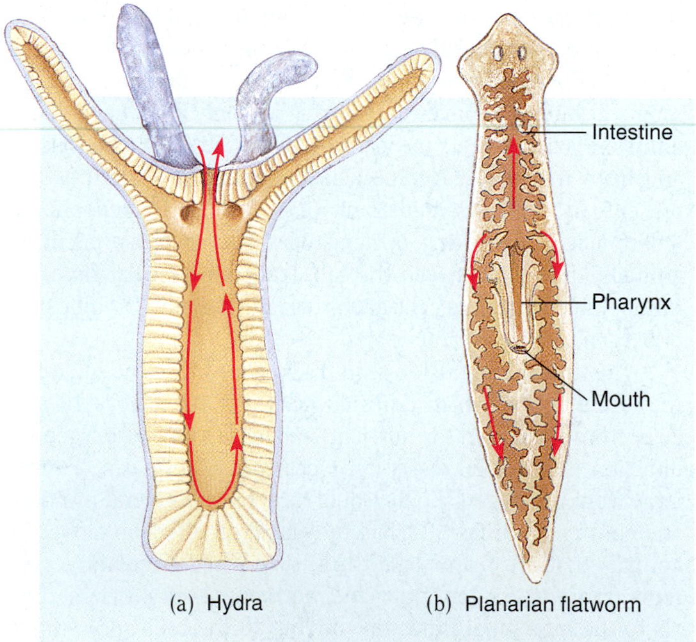

Intestine
Pharynx
Mouth
(a) Hydra
(b) Planarian flatworm

Figure 42–1 Invertebrates with no circulatory system. **(a)** In *Hydra* and other cnidarians, oxygen and nutrients circulate through the gastrovascular cavity and come in contact with the inner layer of body cells. These materials can diffuse the short distance to the outer layer of cells. **(b)** In planarian flatworms, the branched intestine circulates nutrients and oxygen within close proximity to the body cells.

The flattened body of the flatworm permits effective gas exchange by diffusion. Its branched intestine brings nutrients close to all the cells. As in cnidarians, circulation is aided by contractions of the muscles of the body wall, which agitate the intestinal fluid and the tissue fluid. The branching excretory system of planarians provides for internal transport of metabolic wastes that are then expelled from the body.

Fluid in the body cavity of nematodes and other pseudocoelomate animals helps circulate materials. Nutrients, oxygen, and wastes dissolve in this fluid and diffuse through it to and from the cells. Body movements of the animal result in movement of the fluid, facilitating distribution of these materials.

MANY INVERTEBRATES HAVE AN OPEN CIRCULATORY SYSTEM

Arthropods and most mollusks have an **open circulatory system,** in which the heart pumps blood into vessels that have open ends (Fig. 42-2). Blood and interstitial fluid are not distinguishable; they are collectively referred to as **hemolymph.** Hemolymph spills out of the open ends of the blood vessels, filling large spaces, called sinuses, that make up the **hemocoel** (blood cavity), which is not part of the coelom. (In arthropods and mollusks the coelom is reduced.) The hemolymph bathes the cells of the body directly. Blood reenters the circulatory system through openings in the heart (in arthropods) or through open-ended vessels that lead to the gills (in mollusks).

In the open circulatory system of most mollusks, the heart typically consists of three chambers: two atria and a ventricle (Fig. 42–2*a*). The atria receive hemolymph from the gills. Then, the ventricle pumps oxygen-rich hemolymph into blood vessels that conduct it into the large sinuses of the hemocoel. After bathing the body cells, the hemolymph passes into vessels that lead to the gills, where it is recharged with oxygen. The hemolymph is then returned to the heart.

Some mollusks, as well as arthropods, have a hemolymph pigment, **hemocyanin,** containing copper that binds with oxygen. When oxygenated, hemocyanin is blue and imparts a bluish color to the hemolymph of these animals (the original blue bloods!).

In arthropods, a tubular heart pumps hemolymph into blood vessels (arteries) that deliver it to the sinuses of the hemo-

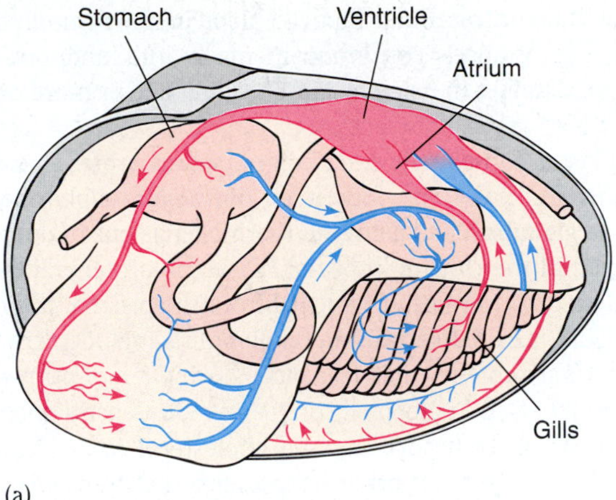

(a)

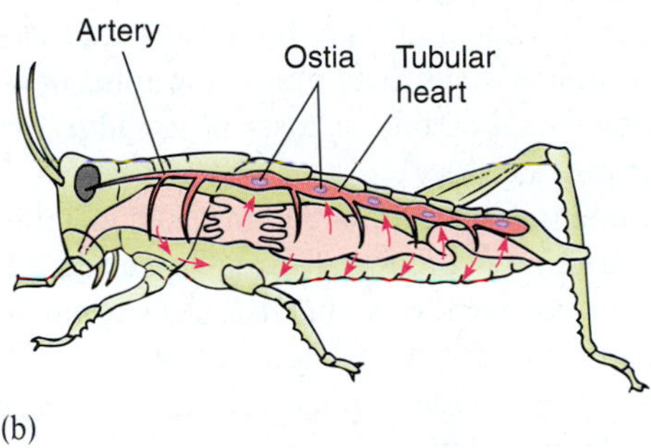

(b)

Figure 42–2 Open circulatory systems. In mollusks and arthropods, a heart pumps the blood into arteries that end in sinuses of the hemocoel. Hemolymph circulates through the hemocoel. **(a)** In most mollusks, hemolymph enters vessels that conduct it to the gills, where it is recharged with oxygen. Blood then returns to the heart. **(b)** In arthropods, a tubular heart pumps hemolymph into arteries that deliver it to the sinuses of the hemocoel. After circulating, hemolymph reenters the heart through ostia (openings) in the heart wall. In crayfish and other crustaceans, gas exchange occurs in the gills.

coel (Fig. 42–2*b*). Hemolymph then circulates through the hemocoel, eventually returning to the pericardial cavity surrounding the heart. Hemolymph enters the heart through tiny openings *(ostia)* that are equipped with valves to prevent backflow. Some insects have accessory "hearts," modified blood vessels that help pump hemolymph through the extremities, particularly the wings. The rate of hemolymph circulation increases when the insect moves. Thus, when an animal is active and most in need of nutrients for fuel, its own movement ensures effective circulation. An open circulatory system cannot provide enough oxygen to maintain the active lifestyle of insects. Indeed, insect hemolymph mainly distributes nutrients and hormones. Oxygen diffuses directly to the cells through a system of air tubes (tracheae) that make up the respiratory system (see Figure 44–2). In crayfish and other crustaceans, gas exchange takes place as hemolymph circulates through the gills.

SOME INVERTEBRATES HAVE A CLOSED CIRCULATORY SYSTEM

Annelids, some mollusks (cephalopods), and echinoderms have a **closed circulatory system.** In them, blood flows through a continuous circuit of blood vessels. The walls of the smallest blood vessels, the **capillaries,** are thin enough to permit diffusion of gases, nutrients, and wastes between blood in the vessels and the interstitial fluid that bathes the cells.

A rudimentary closed circulatory system is found in the proboscis worms (phylum Nemertea). This system consists of a complete network of blood vessels but no heart. Blood flow depends on movements of the animal and on contractions in the walls of the large blood vessels.

Earthworms and other annelids have a complex, closed circulatory system (Fig. 42–3). Two main blood vessels extend lengthwise in the body. The ventral vessel conducts blood posteriorly, and the dorsal vessel conducts blood anteriorly. Dorsal and ventral vessels are connected by lateral vessels in every segment. Branches of the lateral vessels deliver blood to the surface, where it is oxygenated. In the anterior part of the worm, five pairs of contractile blood vessels (sometimes referred to as "hearts") connect dorsal and ventral vessels. Contractions of these paired vessels and of the dorsal vessel, as well as contraction of the muscles of the body wall, circulate the blood. Earthworms have **hemoglobin,** the same red pigment that transports oxygen in vertebrate blood. However, their hemoglobin is not contained within red blood cells but is dissolved in the blood plasma.

Although other mollusks have an open circulatory system, the fast-moving cephalopods, such as the squid and octopus,

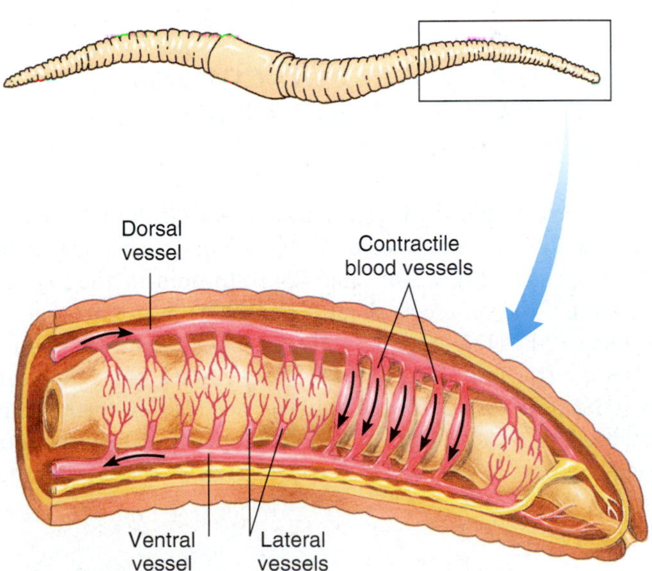

Figure 42–3 Closed circulatory system of the earthworm. Blood circulates through a continuous system of blood vessels. Five pairs of contractile blood vessels deliver blood from the dorsal vessel to the ventral vessel.

require a more efficient means of internal transport. They have a closed system made even more effective by accessory "hearts" at the base of the gills, which speed the passage of blood through the gills.

VERTEBRATES HAVE A CLOSED CIRCULATORY SYSTEM

The circulatory system is basically similar in all vertebrates, from fishes, frogs, and reptiles to birds and mammals. The system consists of heart, blood vessels, blood, lymph, lymph vessels, and associated organs such as the thymus, spleen, and liver. All vertebrates have a ventral, muscular heart that pumps blood into a closed system of blood vessels. Capillaries, the tiniest blood vessels, have very thin walls that permit exchange of materials between blood and interstitial fluid.

The vertebrate circulatory system performs several functions:

1. Transports nutrients from the digestive system and from storage depots to each cell
2. Transports oxygen from respiratory structures (gills or lungs) to the cells
3. Transports metabolic wastes from each cell to organs that excrete them
4. Transports hormones from endocrine glands to target tissues
5. Helps maintain fluid balance
6. Defends the body against invading microorganisms
7. Helps distribute metabolic heat within the body, which helps maintain a constant body temperature in endothermic animals
8. Helps maintain appropriate pH

VERTEBRATE BLOOD CONSISTS OF PLASMA, BLOOD CELLS, AND PLATELETS

In vertebrates, blood consists of a pale yellowish fluid, known as **plasma,** in which red blood cells, white blood cells, and platelets are suspended (Fig. 42–4; Table 42–1). In humans the total circulating blood volume is about 8% of the body weight—5.6 L (6 qt) in a 70-kg (154-lb) person. About 55% of the blood volume is plasma. The remaining 45% is made up of blood cells and platelets. Because cells and platelets are heavier than plasma, they can be separated from it by centrifugation. Plasma does not separate from blood cells in the body because the blood is constantly mixed as it circulates in the blood vessels.

Plasma is the fluid component of blood

Plasma is composed of water (about 92%), proteins (about 7%), salts, and a variety of materials being transported, such as dissolved gases, nutrients, wastes, and hormones. Plasma is in dynamic equilibrium with the interstitial fluid bathing the cells and with the intracellular fluid. As blood passes through the capillaries, substances continuously move into and out of the plasma. Changes in its composition signal one or more organs of the body to restore homeostasis.

Plasma contains several kinds of **plasma proteins,** each with specific properties and functions: **fibrinogen;** alpha, beta, and gamma **globulins;** and **albumin.** Fibrinogen is one of the proteins involved in the clotting process. When the proteins involved in blood clotting have been removed from the plasma, the remaining liquid is called **serum.** Alpha globulins include certain hormones and proteins that transport hormones; prothrombin, a protein involved in blood clotting; and high-density lipoproteins (HDL), which transport fats and cholesterol (see Chapter 45). Beta globulins include other lipoproteins that transport fats and cholesterol, as well as proteins that transport certain vitamins and minerals. The **gamma globulin** fraction contains many types of antibodies that provide immunity to diseases such as measles and infectious hepatitis. Purified human gamma globulin is sometimes used to treat certain diseases or to reduce the possibility of contracting a disease.

Plasma proteins, especially albumins and globulins, help regulate the distribution of fluid between plasma and interstitial fluid. Too large to pass readily through the walls of blood vessels, these proteins contribute to the blood's osmotic pressure, which helps maintain an appropriate blood volume. Plasma proteins (along with the hemoglobin in the red blood cells) are also important acid-base buffers. They help keep the pH of the blood within a narrow range—near its normal, slightly alkaline pH of 7.4.

Red blood cells transport oxygen

Erythrocytes, also called **red blood cells (RBCs),** are highly specialized for transporting oxygen. In most vertebrates except mammals, circulating RBCs have nuclei. For example, birds have large, oval, nucleated RBCs. In mammals, the nucleus is ejected from the RBC as the cell develops. Each mammalian RBC is a flexible, biconcave disc, 7 to 8 μm in diameter and 1 to 2 μm thick. An internal elastic framework maintains the disc shape and permits the cell to bend and twist as it passes through blood vessels even smaller than its own diameter. Its biconcave shape provides a high ratio of surface area to volume, allowing efficient diffusion of oxygen and carbon dioxide into and out of the cell. In an adult human, about 30 trillion RBCs circulate in the blood, approximately 5 million per μL.

Erythrocytes are produced within the red bone marrow of certain bones: vertebrae, ribs, breastbone, skull bones, and long bones. As an RBC develops, it produces great quantities of hemoglobin, the oxygen-transporting pigment that gives vertebrate blood its red color. (Oxygen transport is discussed in Chapter 44.) The life span of a human RBC is about 120 days. As blood circulates through the liver and spleen, phagocytic cells remove worn-out RBCs from the circulation. These RBCs are then disassembled, and some of their components are recycled. In the human body, more than 2.4 million RBCs are destroyed every second, so an equal number must be produced in the bone marrow to replace them. Red blood cell production is regulated by the hor-

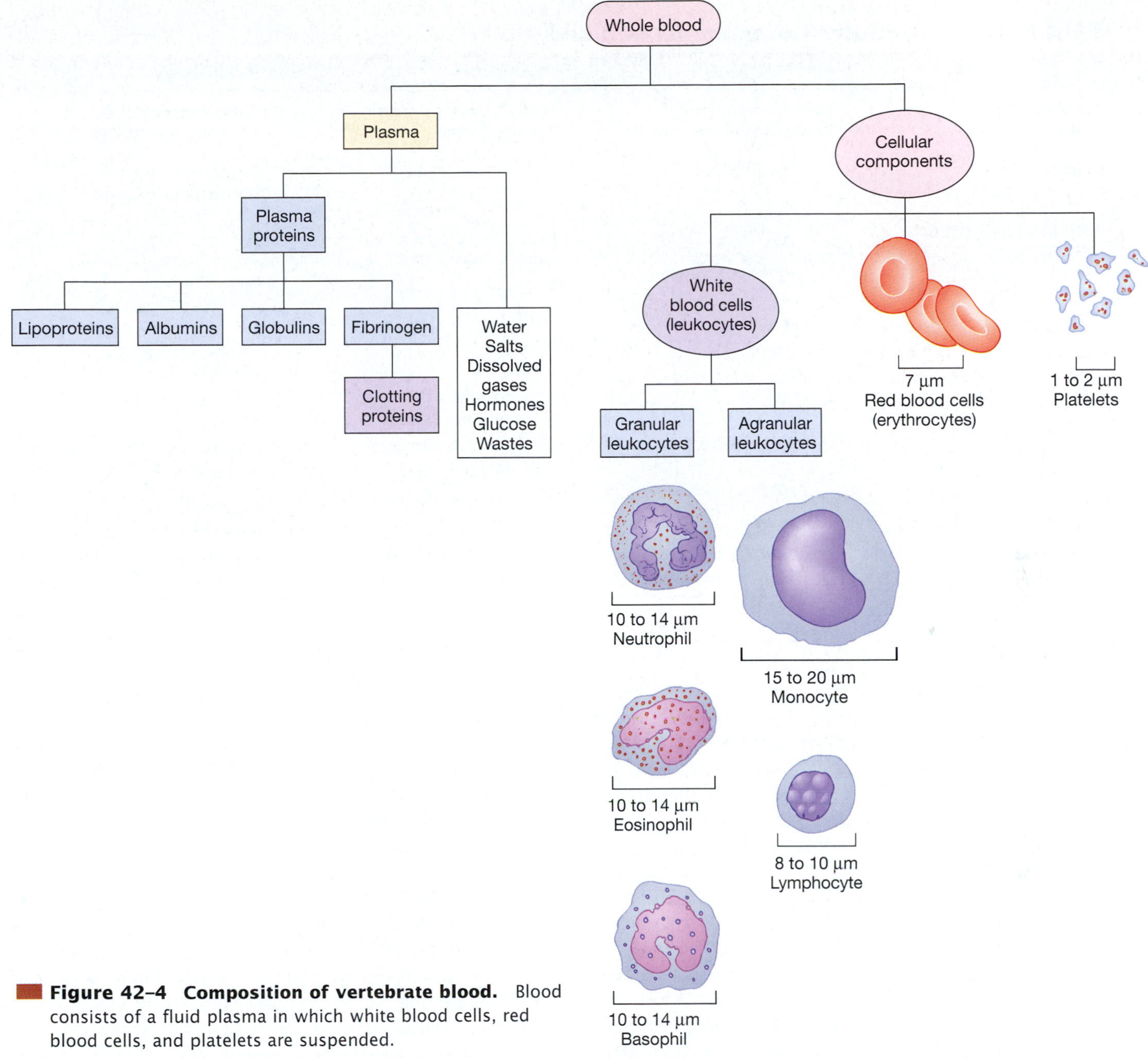

Figure 42–4 Composition of vertebrate blood. Blood consists of a fluid plasma in which white blood cells, red blood cells, and platelets are suspended.

mone **erythropoietin,** which is released by the kidneys in response to a decrease in oxygen.

Anemia is a deficiency in hemoglobin (often accompanied by a decrease in the number of RBCs). When hemoglobin is insufficient, the amount of oxygen transported is inadequate to supply the body's needs. An anemic person may complain of feeling weak and may become easily fatigued. Three general causes of anemia are (1) loss of blood due to hemorrhage or internal bleeding; (2) decreased production of hemoglobin or red blood cells as in iron-deficiency anemia or pernicious anemia (which can be caused by vitamin B_{12} deficiency); and (3) increased rate of RBC destruction—the **hemolytic anemias,** such as sickle cell anemia (see Chapter 15).

White blood cells defend the body against disease organisms

The **leukocytes,** or **white blood cells (WBCs),** are specialized to defend the body against harmful bacteria and other microorganisms. Leukocytes are amoeba-like cells capable of independent movement. Some types routinely slip through the walls of blood vessels and enter the tissues. Human blood contains five kinds of leukocytes, which may be classified as either granular or agranular (see Fig. 42–4). Both types are manufactured in the red bone marrow.

The **granular leukocytes** are characterized by large, lobed nuclei and distinctive granules in their cytoplasm. The three

TABLE 42–1 Cellular Components of Blood

	Normal Range	Function	Pathology
Red Blood Cells (RBCs)	Male: 4.2–5.4 million/μL Female: 3.6–5.0 million/μL	Oxygen transport; carbon dioxide transport	Too few: anemia Too many: polycythemia
Platelets	150,000–400,000/μL	Essential for clotting	Clotting malfunctions; bleeding; easy bruising
White Blood Cells (WBCs)	5000–10,000/μL		
Neutrophils	About 60% of WBCs	Phagocytosis	Too many: may be due to bacterial infection, inflammation, leukemia (myelogenous)
Eosinophils	1%–3% of WBCs	Play some role in allergic response	Too many: may result from allergic reaction, parasitic infestation
Basophils	1% of WBCs	May play role in prevention of inappropriate clotting	
Lymphocytes	25%–35% of WBCs	Produce antibodies; destroy foreign cells	Atypical lymphocytes present in infectious mononucleosis; too many may be due to lymphocytic leukemia, certain viral infections
Monocytes	6% of WBCs	Differentiate to form macrophages	May increase in monocytic leukemia and fungal infections

varieties of granular leukocytes are the neutrophils, eosinophils, and basophils. **Neutrophils,** the principal phagocytic cells in the blood, are especially adept at seeking out and ingesting bacteria. They also phagocytize dead cells, a clean-up task especially demanding after injury or infection. Most of the granules in neutrophils contain enzymes that digest ingested material.

Eosinophils have large granules that stain bright red with eosin, an acidic dye. The lysosomes of these WBCs contain enzymes such as oxidases and peroxidases, suggesting that they function in detoxifying foreign proteins and other substances. Eosinophils increase in number during allergic reactions and during parasitic (e.g., tapeworm) infestations. **Basophils** exhibit deep blue granules when stained with basic dyes. Like eosinophils, these cells play a role in allergic reactions. Basophils do not contain lysosomes. Granules in their cytoplasm contain **histamine,** a substance that dilates blood vessels and makes capillaries more permeable. Basophils release histamine in injured tissues and in allergic responses. Other basophil granules contain **heparin,** an anticoagulant that helps prevent blood from clotting inappropriately within the blood vessels.

Agranular leukocytes lack large, distinctive granules, and their nuclei are rounded or kidney-shaped. Two types of agranular leukocytes are lymphocytes and monocytes. Some **lymphocytes** are specialized to produce antibodies, whereas others attack foreign invaders such as bacteria or viruses directly. Just how they manage these feats is discussed in Chapter 43.

Monocytes are the largest WBCs, reaching 20 μm in diameter. After circulating in the blood for about 24 hours, a monocyte leaves the circulation and completes its development in the tissues. The monocyte greatly enlarges and becomes a **macrophage,** a giant scavenger cell. All of the macrophages found in the tissues develop in this way. Macrophages voraciously engulf bacteria, dead cells, and debris.

In human blood there are normally about 7000 WBCs per μL of blood (only one for every 700 RBCs). During bacterial infections the number may rise sharply, so that a WBC count is a useful diagnostic tool. The proportion of each kind of WBC is determined by a differential WBC count. The normal distribution of leukocytes is indicated in Table 42–1.

Leukemia is a form of cancer in which any one of the various kinds of WBCs multiplies rapidly within the bone marrow. Many of these cells do not mature, and their large numbers crowd out developing RBCs and platelets, leading to anemia and impaired clotting. A common cause of death from leukemia is internal hemorrhaging, especially in the brain. Another frequent cause of death is infection; although the WBC count may rise dramatically, the cells are immature and abnormal and cannot defend the body against disease organisms. *Acute leukemias* have a rapid onset and are characterized by large numbers of immature WBCs, whereas *chronic leukemias* have a more gradual onset and are marked by large numbers of mature WBCs. Some leukemias, particularly acute leukemias in children, can be cured in some cases, and in some other cases complete remissions lasting as long as 15 years or more are possible. Leukemias are treated by radiation therapy, chemotherapy (drug therapy, including steroids and antimitotic drugs), and bone marrow transplants.

Platelets function in blood clotting

In most vertebrates other than mammals, the blood contains small, oval cells called **thrombocytes,** which have nuclei. In mammals, thrombocytes are tiny spherical or disc-shaped bits of cytoplasm that lack nuclei. They are usually referred to as **platelets.** About 300,000 platelets per μL are present in human blood. Platelets are pinched off from very large cells *(megakaryocytes)* in the bone marrow. Thus, a platelet is not a whole cell but a fragment of cytoplasm enclosed by a membrane.

When a blood vessel is cut, it constricts, reducing loss of blood. Platelets stick to the rough, cut edges of the vessel, physically patching the break in the wall. As platelets begin to gather, they release substances that attract other platelets. The platelets become sticky and adhere to collagen fibers in the blood vessel wall. Within about 5 minutes after injury, they form a platelet plug, or temporary clot.

At the same time that the temporary clot forms, a stronger, more permanent clot begins to develop. More than 30 different chemical substances interact in this very complex process. The series of reactions that leads to clotting is triggered when one of the clotting factors in the blood is activated by contact with the injured tissue. In **hemophilia,** one of the clotting factors is absent as a result of an inherited genetic mutation (see Chapter 15).

Prothrombin, a plasma protein manufactured in the liver, requires vitamin K for its production. In the presence of clotting factors, calcium ions, and compounds released from platelets, prothrombin is converted to **thrombin.** Then thrombin catalyzes the conversion of the soluble plasma protein **fibrinogen** to an insoluble protein, **fibrin.** Once formed, fibrin polymerizes, producing long threads that stick to the damaged surface of the blood vessel and form the webbing of the clot. These threads trap blood cells and platelets, which help strengthen the clot. The clotting process is summarized in Figure 42–5.

VERTEBRATES HAVE THREE MAIN TYPES OF BLOOD VESSELS

The vertebrate circulatory system includes three main types of blood vessels: arteries, capillaries, and veins (Fig. 42–6). An **artery** carries blood away from a heart chamber, toward other tissues. When an artery enters an organ, it divides into many smaller branches called **arterioles.** The arterioles deliver blood into the microscopic capillaries. After blood circulates through an organ, capillaries merge to form **veins** that transport the blood back toward the heart. The wall of an artery or vein has three layers (Fig. 42–6*b*). The innermost layer, which lines the blood vessel, consists mainly of **endothelium,** a tissue that resembles squamous epithelium (see Chapter 37). The middle layer consists of connective tissue and smooth muscle cells, and the outer coat is composed of connective tissue rich in elastic and collagen fibers.

The thick walls of arteries and veins prevent gases and nutrients from passing through. Materials are exchanged between

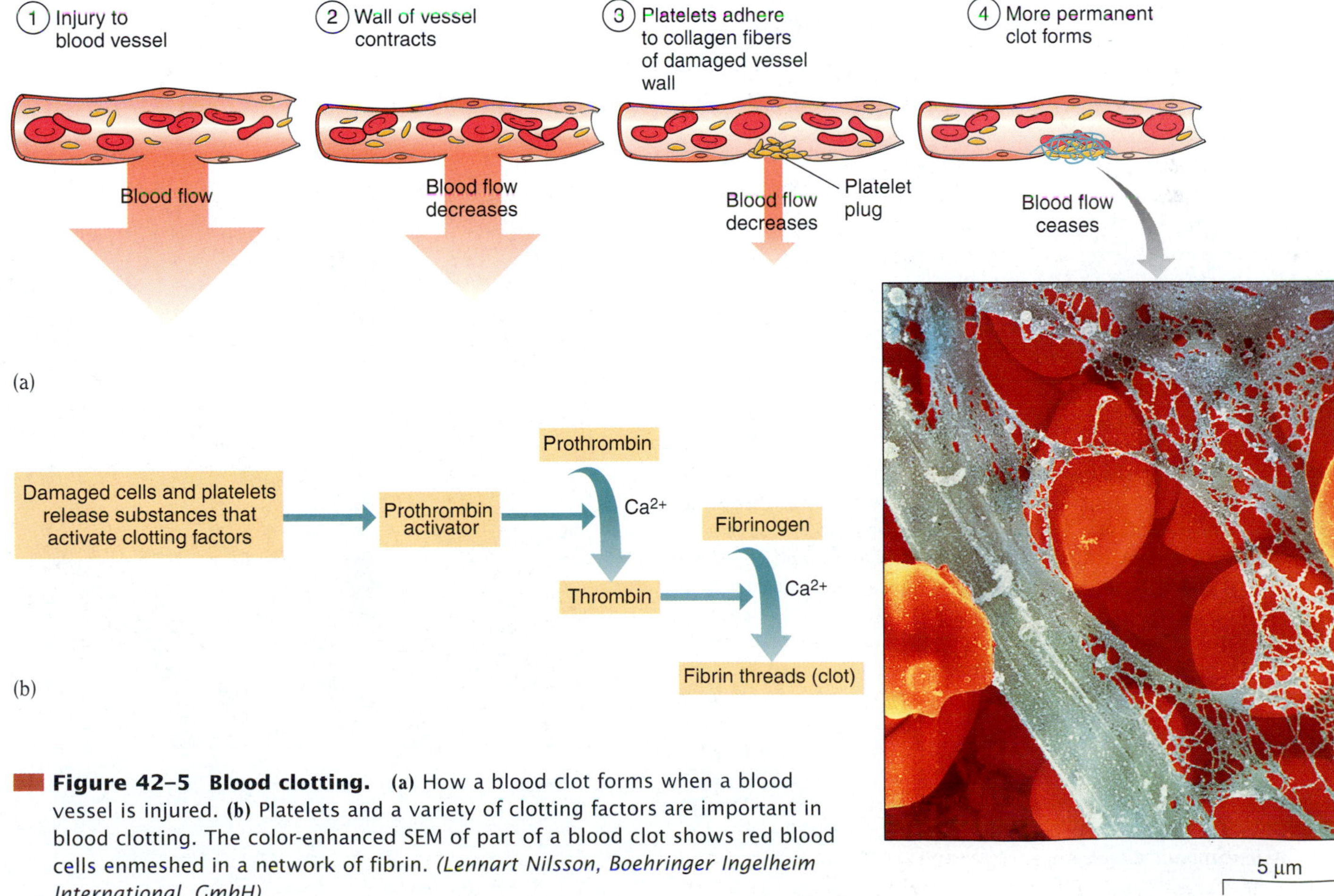

Figure 42–5 Blood clotting. **(a)** How a blood clot forms when a blood vessel is injured. **(b)** Platelets and a variety of clotting factors are important in blood clotting. The color-enhanced SEM of part of a blood clot shows red blood cells enmeshed in a network of fibrin. *(Lennart Nilsson, Boehringer Ingelheim International, GmbH)*

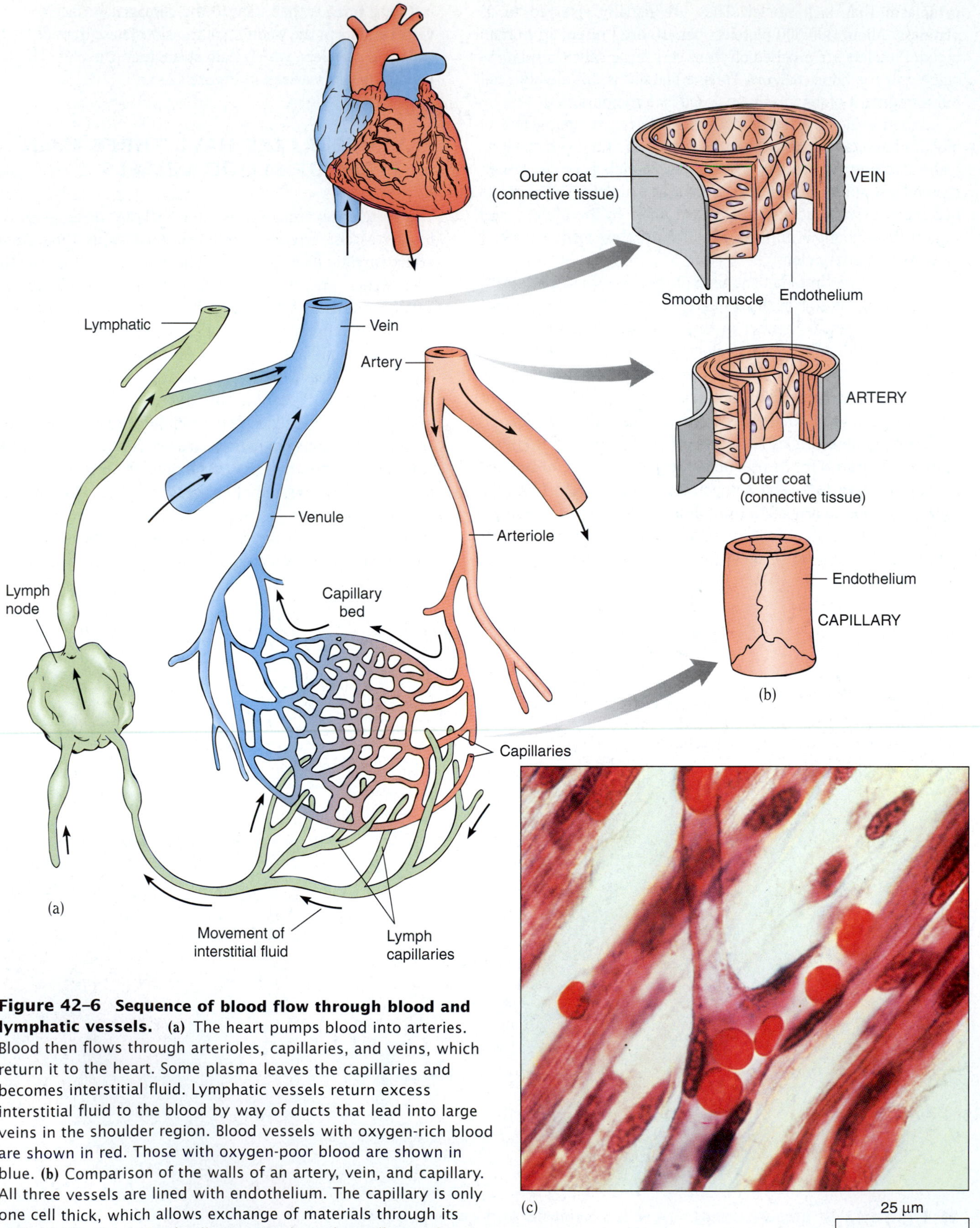

Figure 42–6 Sequence of blood flow through blood and lymphatic vessels. **(a)** The heart pumps blood into arteries. Blood then flows through arterioles, capillaries, and veins, which return it to the heart. Some plasma leaves the capillaries and becomes interstitial fluid. Lymphatic vessels return excess interstitial fluid to the blood by way of ducts that lead into large veins in the shoulder region. Blood vessels with oxygen-rich blood are shown in red. Those with oxygen-poor blood are shown in blue. **(b)** Comparison of the walls of an artery, vein, and capillary. All three vessels are lined with endothelium. The capillary is only one cell thick, which allows exchange of materials through its wall. **(c)** LM of red blood cells passing through capillaries almost in single file. *(c, Lennart Nilsson, Boehringer Ingelheim International, GmbH)*

the blood and interstitial fluid bathing the cells through the capillary walls, which are only one cell thick. Capillary networks in the body are so extensive that at least one of these tiny vessels is located close to every cell in the body. The total length of all capillaries in the body has been estimated to be more than 60,000 miles!

Smooth muscle in the arteriole wall can constrict **(vasoconstriction)** or relax **(vasodilation),** changing the radius of the arteriole. Such changes help maintain appropriate blood pressure and can help control the volume of blood passing to a particular tissue. Changes in blood flow are regulated by the nervous system in response to the metabolic needs of the tissue, as well as by the demands of the body as a whole. For example, when a tissue is metabolizing rapidly, it needs a greater supply of nutrients and oxygen. During exercise, arterioles within skeletal muscles dilate, increasing by more than tenfold the amount of blood flowing to these muscle tissues.

If all the blood vessels were dilated at the same time, there would not be sufficient blood to fill them completely. Normally the liver, kidneys, and brain receive the lion's share of blood. However, if an emergency suddenly occurred requiring rapid action, the blood would be rerouted quickly in favor of the heart and muscles. At such a time the digestive system and kidneys can do with less blood, for they are not critical in responding to the crisis.

The small vessels that directly link arterioles with venules (small veins) are **metarterioles.** The so-called true capillaries branch off from the metarterioles and then rejoin them (Fig. 42–7). True capillaries also interconnect with one another. Wherever a capillary branches from a metarteriole, a smooth muscle cell called a *precapillary sphincter* is present. Precapillary sphincters open and close continuously, directing blood first to one and then to another section of tissue. These sphincters (along with the smooth muscle in the walls of arteries and arterioles) regulate the blood supply to each organ and its subdivisions.

DIFFERENT ADAPTATIONS OF THE HEART AND CIRCULATION HAVE EVOLVED IN EACH VERTEBRATE CLASS

The vertebrate circulatory system became modified in the course of evolution, as the site of gas exchange shifted from gills to lungs and as certain vertebrates became active, endothermic animals with higher metabolic rates. The vertebrate heart has one or two **atria** (sing., *atrium*), chambers that receive blood returning from the tissues, and one or two **ventricles** that pump blood into the arteries (Fig. 42–8). Additional chambers are present in some classes.

The fish heart has one atrium and one ventricle. The atrium pumps blood into the ventricle, which pumps blood into a single circuit of blood vessels. Blood is oxygenated as it passes through capillaries in the gills. After blood circulates through the gill capillaries, its pressure is low, so blood passes very slowly to the other organs. Circulation is helped by the movements of the fish while it is swimming. Blood returning to the heart has a low oxygen content. A thin-walled *sinus venosus* receives blood returning from the tissues and pumps it into the atrium.

In amphibians, blood flows through a double circuit: the **pulmonary circulation** and the **systemic circulation.** Oxygen-rich and oxygen-poor blood are kept somewhat separate. The amphibian heart has two atria and one ventricle. A *sinus venosus* collects oxygen-poor blood returning from the veins and pumps it into the right atrium. Blood returning from the lungs passes directly into the left atrium. Both atria pump into the single ventricle, but oxygen-poor blood is pumped out of the ventricle before oxygen-rich blood enters it. Blood passes into an artery, the *conus arteriosus,* equipped with a fold that helps keep the blood separate. Much of the oxygen-poor blood is directed into the pulmonary circulation, which delivers it to the lungs and skin where it is recharged with oxygen. Oxygen-rich blood is delivered by the

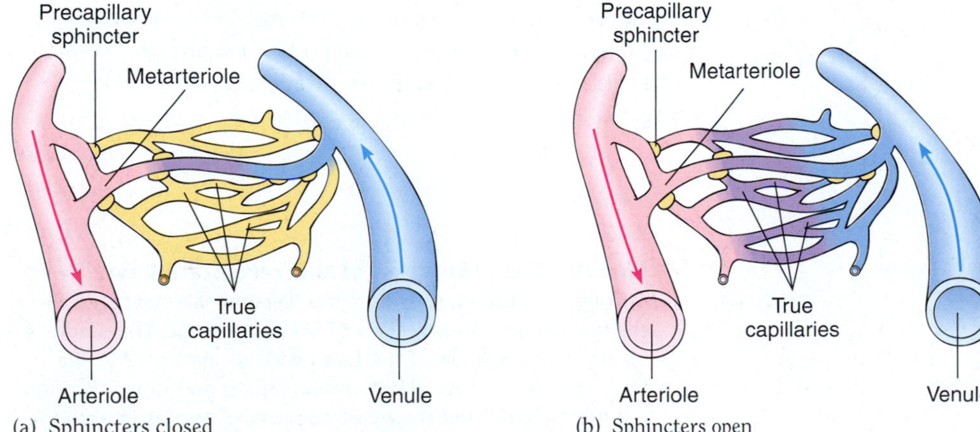

Figure 42–7 Blood flow through a capillary network. As a tissue becomes active, the pattern of blood flow through its capillary networks changes. **(a)** When a tissue is inactive, only its metarterioles are open. **(b)** When the tissue becomes active, decreased oxygen tension in the tissue brings about relaxation of the precapillary sphincters, and the capillaries open. This increases the blood supply, and thus the delivery of nutrients and oxygen to the active tissue.

systemic circulation into arteries that conduct it to the various tissues of the body.

Most reptiles also have a double circuit of blood flow, made more efficient by a wall that partly divides the ventricle. Mixing of oxygen-rich and oxygen-poor blood is minimized by the timing of contractions of the left and right sides of the heart and by pressure differences. In crocodilians (crocodiles and alligators), the wall between the ventricles is complete, so the heart consists of two atria and two ventricles. Thus, a four-chambered heart first evolved among the reptiles.

Unlike birds and mammals, amphibians and reptiles do not ventilate their lungs continuously. Therefore, it would be inefficient to pump blood through the pulmonary circulation continuously. The shunts between the two sides of the heart allow the blood to be distributed to the lungs as needed.

In all birds and mammals (and in crocodilians), the wall between the ventricles is complete, preventing the mixing of oxygen-rich blood in the left chamber with oxygen-poor blood in the right chamber. The conus arteriosus has split and become the base of the **aorta** (the largest artery) and the pulmonary artery. No sinus venosus is present as a separate chamber, but a vestige remains as the *sinoatrial node* (the pacemaker).

Complete separation of the right and left sides of the heart makes it necessary for blood to pass through the heart twice each time it tours the body. As a result, it is possible to maintain higher blood pressures, and materials are delivered to the tissues rapidly and efficiently. Because the blood of birds and mammals contains more oxygen per unit volume and circulates more rapidly than in other vertebrates, the tissues receive more oxygen. As a result, birds and mammals can maintain a higher metabolic rate and a constant, high body temperature even in cold surroundings.

The pattern of blood circulation in birds and mammals can be summarized as follows:

Veins (conduct blood from organs) ⟶ right atrium ⟶ right ventricle ⟶ pulmonary arteries ⟶ capillaries in the lungs ⟶ pulmonary veins ⟶ left atrium ⟶ left ventricle ⟶ aorta ⟶ arteries (conduct blood to organs) ⟶ arterioles ⟶ capillaries

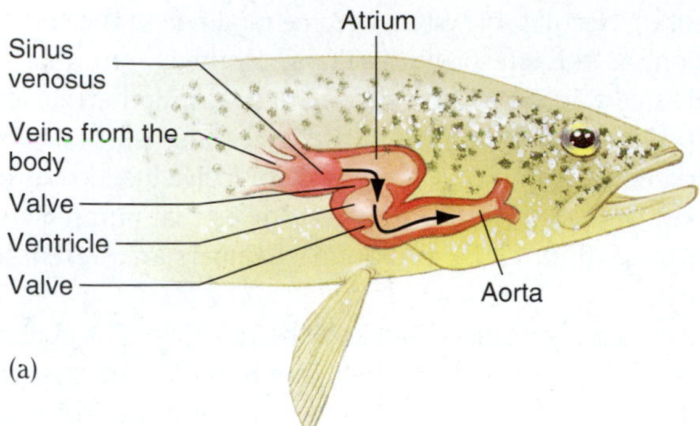

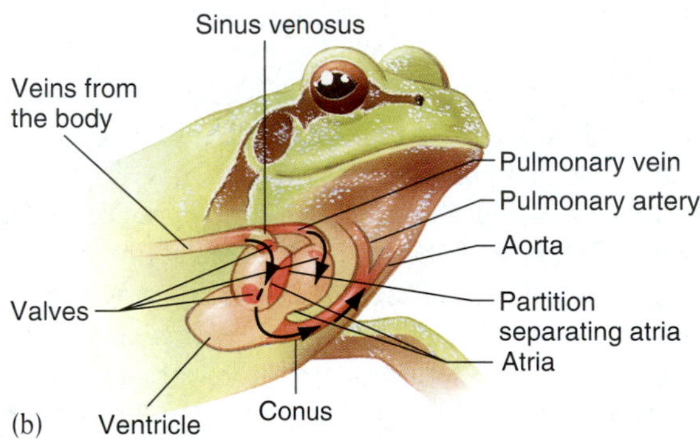

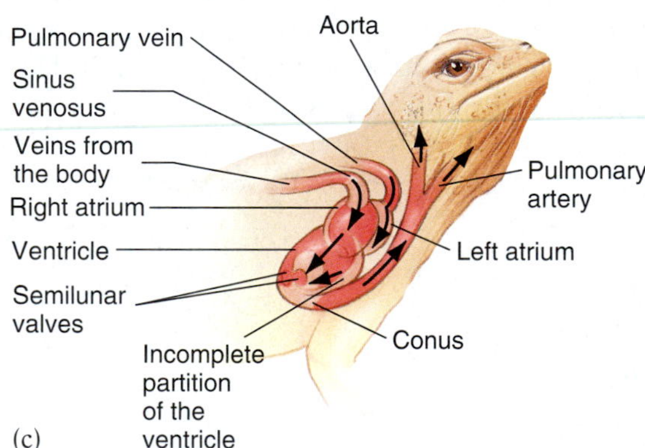

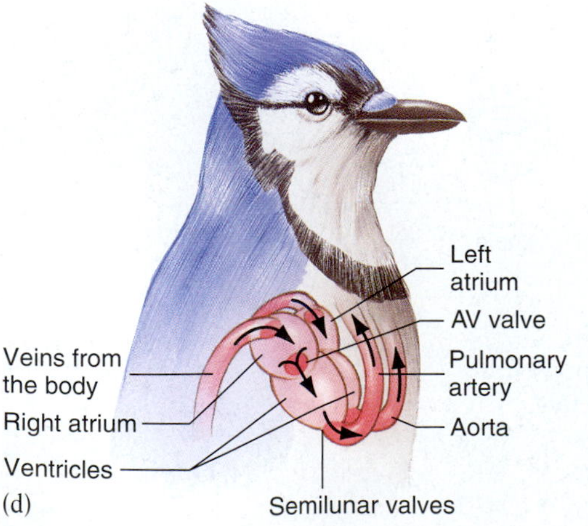

Figure 42–8 Evolution of the vertebrate heart. Through evolution, the heart has become well adapted for the way of life of each class of vertebrates. **(a)** The single atrium and ventricle of the fish heart are part of a single circuit of blood flow. **(b)** In amphibians blood flows through a double circuit and the heart consists of two atria and one ventricle. **(c)** The reptilian heart has two atria and two ventricles; in all but the crocodiles and alligators, the wall separating the ventricles is incomplete so that blood from the right and left chambers mixes to some extent. **(d)** Birds and mammals have two atria and two ventricles, and blood rich in oxygen is kept completely separate from oxygen-poor blood.

THE HUMAN HEART IS WELL ADAPTED FOR PUMPING BLOOD

Not much bigger than a fist and weighing less than a pound, the human **heart** is a remarkable organ that beats about 2.5 billion times in an average lifetime, pumping about 300 million L (80 million gal) of blood (Fig. 42–9). To meet the body's changing needs, the heart can vary its output from 5 to more than 20 L of blood per minute.

The heart is a hollow, muscular organ located in the chest cavity directly under the breastbone. Enclosing it is a tough connective tissue sac, the **pericardium.** The inner surface of the pericardium and the outer surface of the heart are covered by a smooth layer of endothelium. Between these two surfaces is a small **pericardial cavity** filled with fluid, which reduces friction to a minimum as the heart beats.

The wall of the heart is composed mainly of cardiac muscle attached to a framework of collagen fibers. The right atrium and

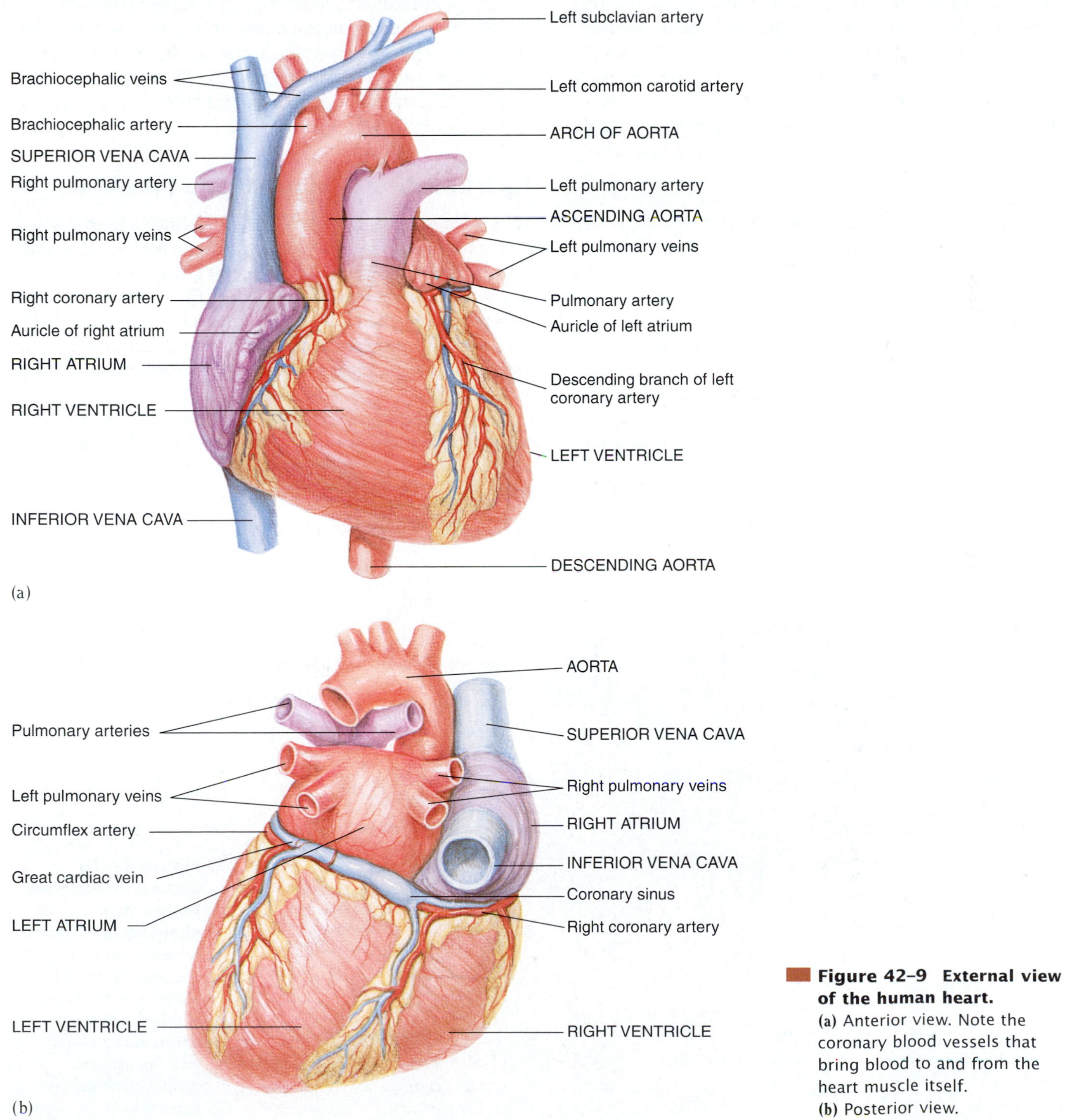

Figure 42–9 External view of the human heart.
(a) Anterior view. Note the coronary blood vessels that bring blood to and from the heart muscle itself.
(b) Posterior view.

ventricle are separated from the left atrium and ventricle by a wall, or **septum.** Between the atria the wall is known as the **interatrial septum;** between the ventricles it is the **interventricular septum.** A shallow depression, the **fossa ovalis,** on the interatrial septum marks the place where an opening, the **foramen ovale,** was located in the fetal heart. In the fetus, the foramen ovale permits the blood to flow directly from right to left atrium so that very little passes to the nonfunctional lungs. At the upper surface of each atrium lies a small muscular pouch called an **auricle.**

To prevent blood from flowing backward, the heart is equipped with valves that close automatically (Fig. 42–10). The valve between the right atrium and right ventricle is called the **right atrioventricular (AV) valve,** or **tricuspid valve.** The **left AV valve** (between the left atrium and left ventricle) is referred to as the **mitral valve,** or **bicuspid valve.** The AV valves are held in place by stout cords, or "heart-strings," the **chordae tendineae.** These cords attach the valves to the papillary muscles that project from the walls of the ventricles.

When blood returning from the tissues fills the atria, blood pressure on the AV valves forces them to open into the ventricles, which then fill with blood. As the ventricles contract, blood is forced back against the AV valves, pushing them closed. Contraction of the papillary muscles and tensing of the chordae tendineae prevent the valves from opening backward into the atria. These valves are like swinging doors that can open in only one direction.

Semilunar valves (named for their flaps, which are shaped like half-moons) guard the exits from the heart. The semilunar valve between the left ventricle and the aorta is the **aortic valve,** and the one between the right ventricle and the pulmonary artery is the **pulmonary valve.** When blood passes out of the ventricles, the flaps of the semilunar valves are pushed aside and offer no resistance to blood flow. But when the ventricles are relaxing and filling with blood from the atria, the blood pressure in the arteries is higher than that in the ventricles. Blood then fills the pouches of the valves, stretching them across the artery so that blood cannot flow back into the ventricle.

Valve deformities are sometimes present at birth, or they may result from certain diseases such as rheumatic fever or syphilis. Inflammation and scarring may narrow the blood passageway by thickening the valves. Sometimes erosion of the valve tissues prevents the flaps from closing tightly, causing blood to leak backward and reducing the efficiency of the heartbeat. Diseased or deformed valves can be surgically repaired or replaced with artificial valves.

Each heartbeat is initiated by a pacemaker

Horror films frequently feature a scene in which a heart cut out of a human body continues to beat. Scriptwriters of these tales actually have some factual basis for their gruesome fantasies, for when carefully removed from the body, the heart does continue to beat for many hours if kept in a nutritive, oxygenated fluid. This is possible because the contractions of cardiac muscle begin within the muscle itself and can occur independently of any nerve supply.

At their ends cardiac muscle cells are joined by dense bands called **intercalated discs** (Fig. 42–11*b, c*). Each disc is a type of

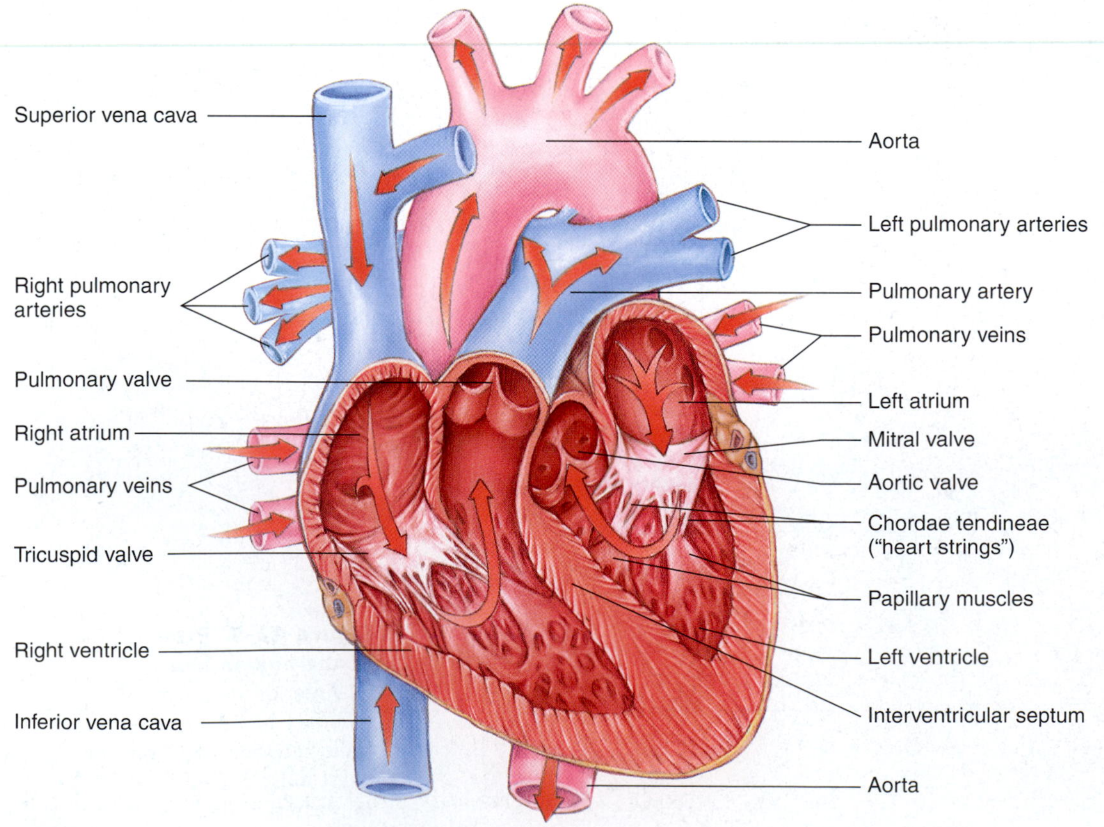

Figure 42–10 Section through the human heart showing the valves. Note the right and left atria, which receive blood, and the right and left ventricles, which pump blood into the arteries. Arrows indicate the pattern of blood flow.

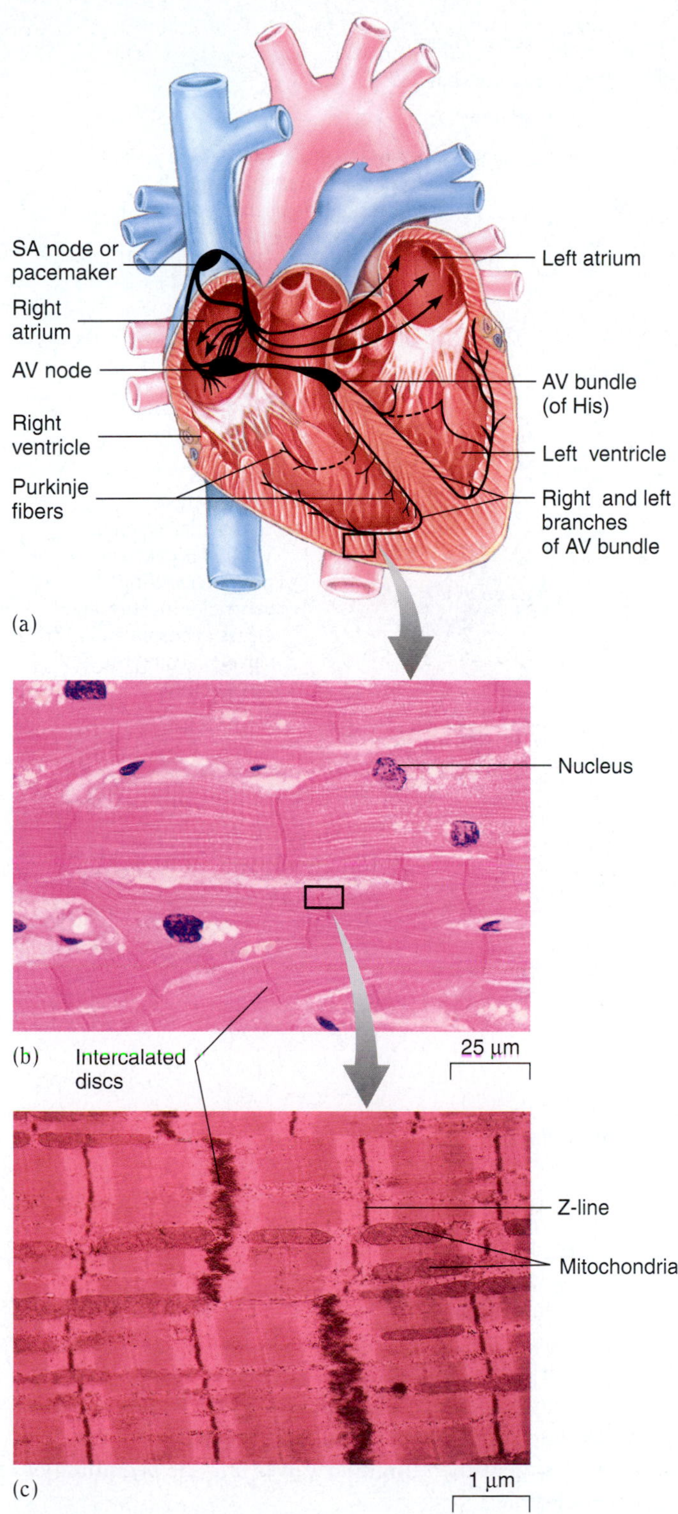

Figure 42–11 Conduction system of the heart. **(a)** The sinoatrial (SA) node initiates each heartbeat. The action potential spreads through the muscle fibers of the atria, producing atrial contraction. Transmission is briefly delayed at the atrioventricular (AV) node before the action potential spreads through specialized muscle fibers into the ventricles. **(b)** LM of cardiac muscle. **(c)** TEM of cardiac muscle. *(b, Ed Reschke; c, Don Fawcett/Visuals Unlimited)*

gap junction (see Chapter 5) in which two cells are connected through pores. This type of junction is of great physiological importance because it offers very little resistance to the passage of an action potential. Ions move easily through the gap junctions, allowing the entire atrial (or ventricular) muscle mass to contract as one giant cell.

Compared with skeletal muscle that has action potentials typically lasting 1 to 2 msec, the action potentials of cardiac muscle are much longer, several hundred milliseconds. Voltage-activated calcium ion channels open during depolarization of cardiac muscle fibers. Entry of Ca^{2+} contributes to the longer depolarization time. Another factor is a type of potassium channel that stays open when the cell is at its resting potential but closes during depolarization, lengthening depolarization by decreasing the permeability of the membrane to K^+. From patch-clamp experiments on isolated cardiac muscle fibers, investigators have found that spontaneous contraction occurs as a result of the combination of a slow decrease in potassium permeability and a slow increase in sodium and calcium permeability.

A specialized conduction system ensures that the heart beats in a regular and effective rhythm. Each beat is initiated by the **pacemaker,** called the **sinoatrial (SA) node** (Fig. 42–11*a*). The SA node is a small mass of specialized cardiac muscle in the posterior wall of the right atrium near the opening of a large vein, the superior vena cava. The action potential in the SA node is triggered mainly by the opening of Ca^{2+} channels. Ends of the SA node fibers fuse with surrounding ordinary atrial muscle fibers so that each action potential spreads through both atria, producing atrial contraction.

One group of atrial muscle fibers conducts the action potential directly to the **atrioventricular (AV) node,** located in the right atrium along the lower part of the septum. Here transmission is delayed briefly, permitting the atria to complete their contraction before the ventricles begin to contract. From the AV node the action potential spreads into specialized muscle fibers called **Purkinje fibers.** These large fibers make up the **AV bundle.** The AV bundle divides, sending branches into each ventricle. When an impulse reaches the ends of the Purkinje fibers, it spreads through the ordinary cardiac muscle fibers of the ventricles.

SA node ⟶ atrial muscle fibers (atria contract) ⟶ AV node ⟶ AV bundle (Purkinje fibers) ⟶ right and left branches (Purkinje fibers) ⟶ ventricular muscle fibers (ventricles contract)

Each minute the heart beats about 70 times. One complete heartbeat takes about 0.8 second and is referred to as a **cardiac cycle.** That portion of the cycle in which contraction occurs is known as **systole;** the period of relaxation is **diastole.** Figure 42–12 shows the sequence of events that occur during one cardiac cycle.

You can measure your heart rate by placing a finger over the radial artery in your wrist or the carotid artery in your neck and

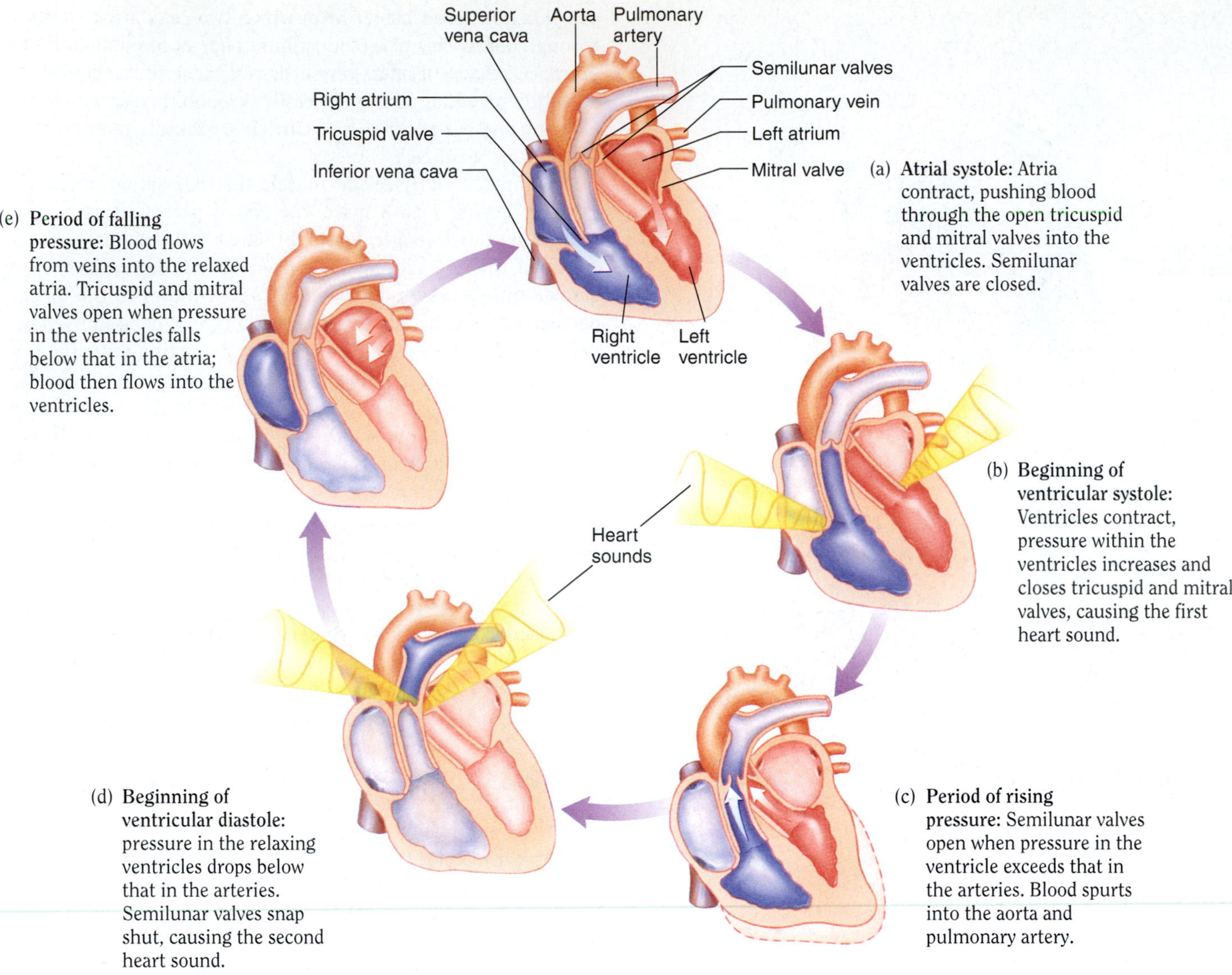

Superior vena cava
Aorta
Pulmonary artery
Semilunar valves
Pulmonary vein
Left atrium
Mitral valve
Right atrium
Tricuspid valve
Inferior vena cava
Right ventricle
Left ventricle
Heart sounds
(a) **Atrial systole:** Atria contract, pushing blood through the open tricuspid and mitral valves into the ventricles. Semilunar valves are closed.
(b) **Beginning of ventricular systole:** Ventricles contract, pressure within the ventricles increases and closes tricuspid and mitral valves, causing the first heart sound.
(c) **Period of rising pressure:** Semilunar valves open when pressure in the ventricle exceeds that in the arteries. Blood spurts into the aorta and pulmonary artery.
(d) **Beginning of ventricular diastole:** pressure in the relaxing ventricles drops below that in the arteries. Semilunar valves snap shut, causing the second heart sound.
(e) **Period of falling pressure:** Blood flows from veins into the relaxed atria. Tricuspid and mitral valves open when pressure in the ventricles falls below that in the atria; blood then flows into the ventricles.

Figure 42–12 The cardiac cycle. The cycle comprises contraction of both atria followed by both ventricles. Arrows indicate the direction of blood flow; dotted lines indicate the change in size as contraction occurs.

counting the pulsations for 1 minute. Arterial pulse is the alternate expansion and recoil of an artery. Each time the left ventricle pumps blood into the aorta, the elastic wall of the aorta expands to accommodate the blood. This expansion moves in a wave down the aorta and the arteries that branch from it. When this pressure wave passes, the elastic arterial wall snaps back to its normal size.

Two main heart sounds can be distinguished

When you listen to the heartbeat with a stethoscope you can hear two main heart sounds, "lub-dup," which repeat rhythmically. These sounds result from the closure of the heart valves. When the valves close they cause turbulence in blood flow that sets up vibrations in the walls of the heart chambers. The first heart sound, lub, is low-pitched, not very loud, and fairly long-lasting. It is caused mainly by the closing of the AV (mitral and tricuspid) valves and marks the beginning of ventricular systole. The lub sound is quickly followed by the higher-pitched, louder, sharper, and shorter dup sound. Heard almost as a quick snap, the dup marks the closing of the semilunar valves and the beginning of ventricular diastole.

The quality of these sounds tells a discerning physician much about the state of the valves. When the semilunar valves are injured, a soft hissing noise ("lub-shhh") is heard in place of the normal sound. This is known as a **heart murmur** and may be caused by any condition that prevents the valves from closing tightly, permitting blood to flow backward into the ventricles during diastole. Damaged mitral and tricuspid valves can also cause heart murmurs.

The electrical activity of the heart can be recorded

As each wave of contraction spreads through the heart, electrical currents flow into the tissues surrounding the heart and onto the body surface. By placing electrodes on the body surface on opposite sides of the heart, the electrical activity can be amplified and recorded either by an oscilloscope or an electrocardiograph. The graph produced is called an **electrocardiogram (ECG** or **EKG).**

An ECG begins with a **P wave,** which is caused by the firing of the SA node and depolarization of the atrial muscle (Fig. 42–13). Then a **QRS complex** appears, reflecting the firing of the AV node and depolarization of the ventricles. The **T wave** results from repolarization of the ventricles. The heart then repeats its pattern of electrical impulses, generating a new P wave, QRS complex, and T wave.

Abnormalities in the ECG indicate disorders in the heart or its rhythm. One class of disorders that can be diagnosed with the help of the ECG is **heart block.** In this condition, transmission of an impulse is delayed or blocked at some point in the conduction system. **Artificial pacemakers** can be implanted in patients with severe heart block. A pacemaker is implanted beneath the skin, and its electrodes are connected to the heart. This device provides continuous rhythmic impulses that avoid the block and drive the heartbeat.

Cardiac output varies with the body's need

The **cardiac output (CO)** is the volume of blood pumped by the left ventricle into the aorta in 1 minute. The volume of blood pumped by one ventricle during one beat is called the **stroke volume.** By multiplying the stroke volume by the number of times the left ventricle beats per minute, the CO can be computed. For example, in a resting adult the heart may beat about 72 times per minute and pump about 70 mL of blood with each contraction.

$$\begin{aligned} \text{CO} &= \text{stroke volume} \times \text{heart rate (number of ventricular contractions per minute)} \\ &= 70 \text{ mL/stroke} \times 72 \text{ strokes/min} \\ &= 5040 \text{ mL/min (about 5 L/min)} \end{aligned}$$

The CO varies dramatically with the changing needs of the body. During stress or heavy exercise, the normal heart can increase its CO fourfold to fivefold, so that 20 to 30 L of blood can be pumped per minute. CO varies with changes in either stroke volume or heart rate.

Stroke volume depends on venous return

Stroke volume depends mainly on venous return, the amount of blood delivered to the heart by the veins. According to **Starling's law of the heart,** if more blood is delivered to the heart by the veins, more blood is pumped by the heart (within physiological limits). When extra amounts of blood fill the heart chambers, the cardiac muscle fibers are stretched to a greater extent and contract with greater force, pumping a larger volume of blood into the arteries. This increase in stroke volume increases the CO (Fig. 42–14).

The release of the neurotransmitter *norepinephrine* by sympathetic nerves also increases the force of contraction of the

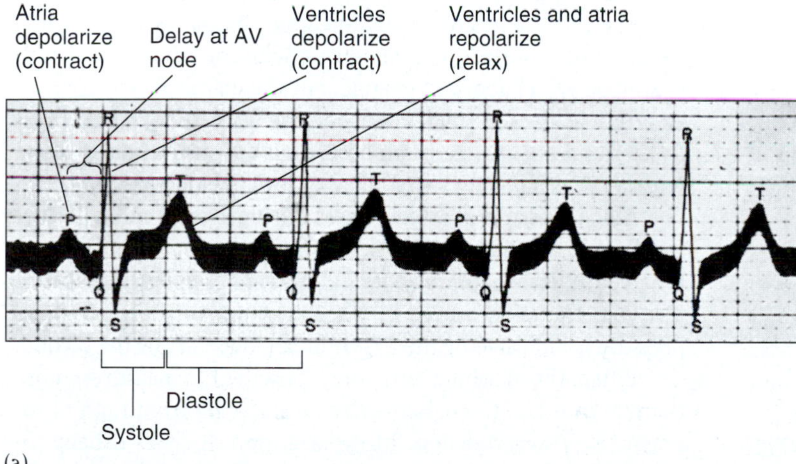

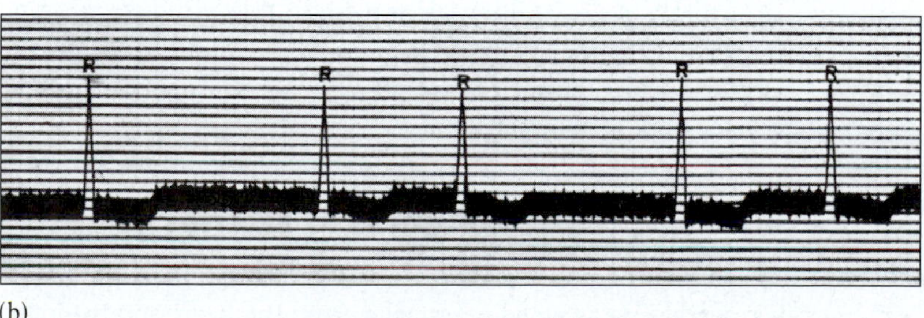

Figure 42–13 Electrocardiograms. **(a)** Tracing from a normal heart. The P wave corresponds to the contraction of the atria, the QRS complex to the contraction of the ventricles, and the T wave to the relaxation of the ventricles. **(b)** Tracing from a patient with atrial fibrillation. The individual muscle fibers of the atria twitch rapidly and independently. There is no regular atrial contraction and no P wave. The ventricles beat independently and irregularly, causing the QRS wave to appear at irregular intervals. *(Courtesy of Dr. Lewis Dexter, Brigham and Women's Hospital, Boston, Massachusetts)*

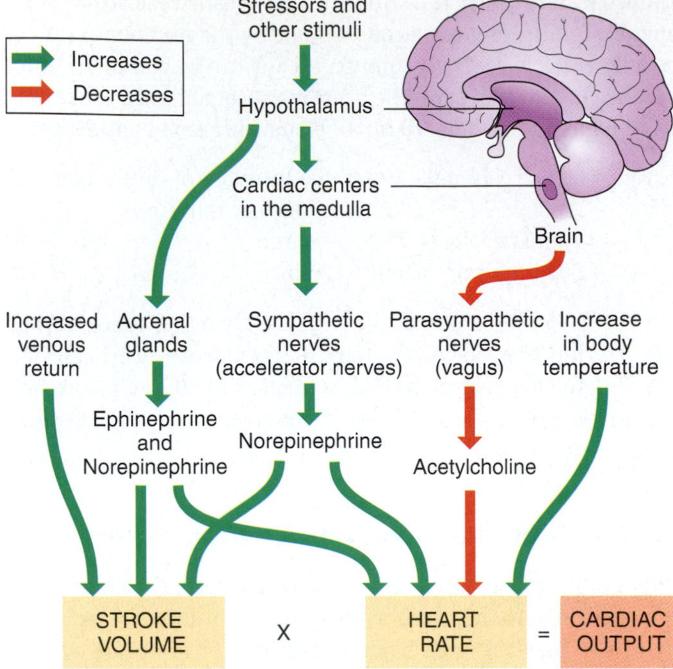

Figure 42–14 Some factors that influence cardiac output.

cardiac muscle fibers. The hormone *epinephrine* released into the blood by the adrenal glands during stress has a similar effect on the heart muscle. When the force of contraction increases, the stroke volume increases, and this in turn increases CO.

Heart rate is regulated by the nervous system

Although the heart is capable of beating independently, its rate is, in fact, carefully regulated by the nervous system and endocrine system (see Fig. 42–14). Sensory receptors in the walls of certain blood vessels and heart chambers are sensitive to changes in blood pressure. When stimulated, they send messages to **cardiac centers** in the medulla of the brain. These cardiac centers maintain control over two sets of autonomic nerves that pass to the SA node. Parasympathetic and sympathetic nerves have opposite effects on heart rate (see Fig. 40–18). Parasympathetic nerves release the neurotransmitter *acetylcholine,* which slows the heart. Acetylcholine slows the rate of depolarization by increasing the membrane's permeability to potassium (Fig. 42–15*a*). Sympathetic nerves release norepinephrine, which speeds the heart rate and, as previously discussed, increases the strength of contraction. Norepinephrine stimulates calcium ion channel opening during depolarization (Fig. 42–15*b*). Both norepinephrine and acetylcholine act indirectly on ion channels. They activate a *signal transduction* process involving a G protein. Norepinephrine binds to *beta-adrenergic receptors,* one of the two main types of adrenergic receptors. These receptors are targeted by *beta blockers,* drugs that block the actions of norepinephrine on the heart and are used clinically in the treatment of hypertension and other types of heart disease.

In response to physical and emotional stressors, the adrenal glands release epinephrine and norepinephrine, which speed the heart. An elevated body temperature also increases heart rate. During fever, the heart may beat more than 100 times per minute. As you might expect, heart rate decreases when body temperature is lowered. This is why a patient's temperature may be deliberately lowered during heart surgery.

BLOOD PRESSURE DEPENDS ON BLOOD FLOW AND RESISTANCE TO BLOOD FLOW

Blood pressure is the force exerted by the blood against the inner walls of the blood vessels. It is determined by CO, blood volume, and the resistance to blood flow (Fig. 42–16*a*). When CO increases, blood flow increases, causing a rise in blood pressure. When CO decreases, blood flow decreases, causing a fall in blood pressure. If the volume of blood is reduced by hemorrhage or by chronic bleeding, the blood pressure drops. On the other hand, an increase in blood volume results in an increase in blood pressure. For example, a high dietary intake of salt causes water retention. This results in an increase of blood volume and leads to higher blood pressure.

Blood flow is impeded by resistance; when the resistance to flow increases, blood pressure rises. **Peripheral resistance** is the resistance to blood flow caused by viscosity of the blood and by friction between the blood and the wall of the blood vessel. In the blood of a healthy person, viscosity remains fairly constant and is only a minor factor influencing changes in blood pressure. More important is the friction between the blood and the wall of the blood vessel. The length and diameter of a blood vessel determine the amount of surface area in contact with the blood. The length of a blood vessel does not change, but the diameter, especially of an arteriole, does. A small change in the diameter of a blood vessel causes a big change in blood pressure. Constriction of blood vessels raises blood pressure; dilation lowers blood pressure.

Blood pressure in arteries rises during systole and falls during diastole. Normal blood pressure (measured in the upper arm) for a young male adult is about 120/80 mm of mercury, abbreviated mm Hg, as measured by the sphygmomanometer. Systolic pressure is indicated by the first number, diastolic by the second.

When the diastolic pressure consistently measures more than 90 mm Hg, the patient may be suffering from high blood pressure, or **hypertension.** In hypertension, there is usually increased vascular resistance, especially in the arterioles and small arteries. The heart's workload increases because it must pump against this greater resistance. If this condition persists, the left ventricle enlarges and may begin to deteriorate in function. Heredity, aging, ethnicity, and obesity may be important in the development of hypertension.

Blood pressure is highest in arteries

As you might guess, blood pressure is greatest in the large arteries, decreasing as blood flows away from the heart and through

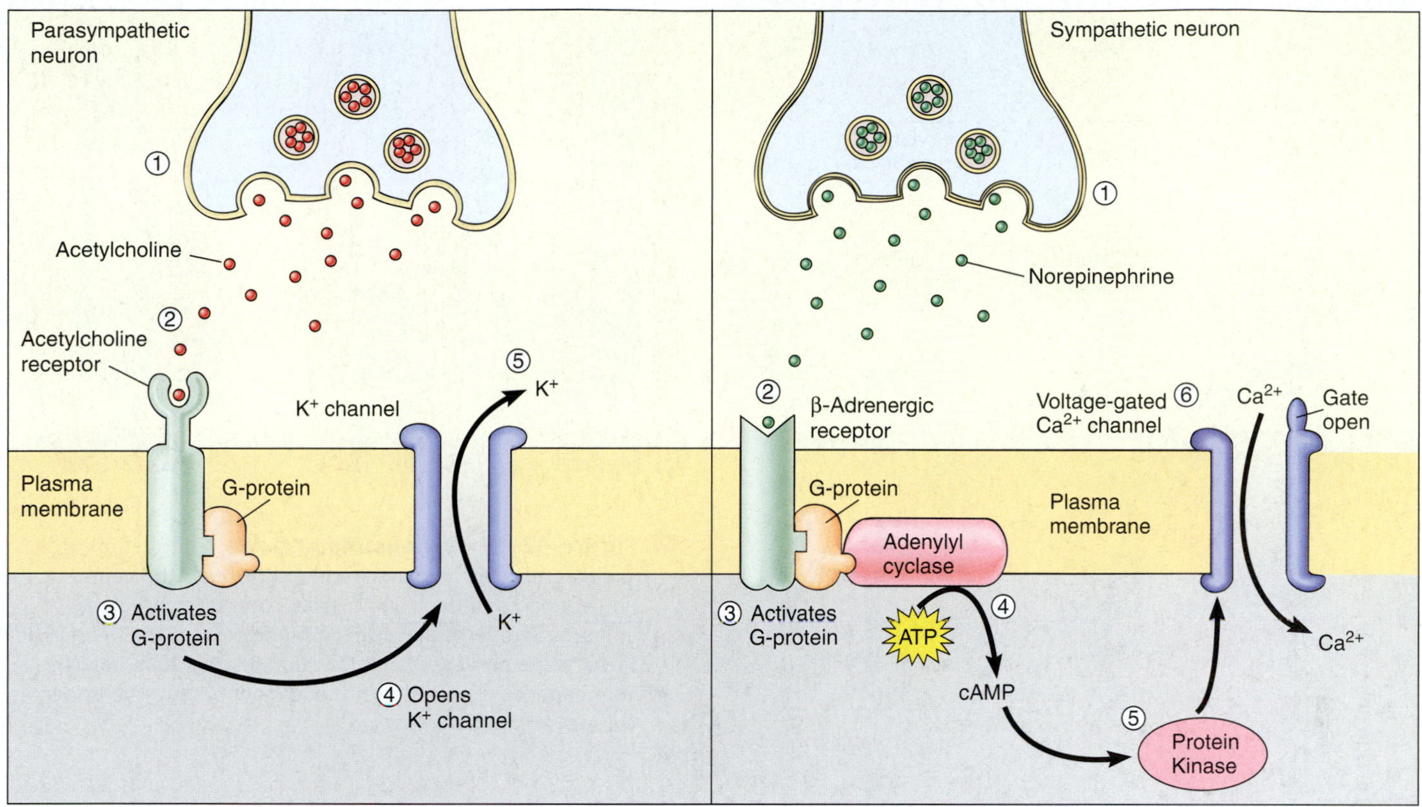

(a) Parasympathetic action on cardiac muscle

(b) Sympathetic action on cardiac muscle

Figure 42–15 Actions of sympathetic and parasympathetic neurons on cardiac muscle cells. Neurotransmitters released by sympathetic and parasympathetic neurons initiate a signal transduction process. **(a)** Parasympathetic neurons release acetylcholine (1), which binds with receptors on the plasma membrane of cardiac muscle (2). The receptor then activates a G protein (3), which can bind with a K^+ channel, causing the channel to open (4). K^+ leaves the cell (5). The membrane becomes hyperpolarized, causing the rate of action potentials to slow. **(b)** Sympathetic neurons release norepinephrine (1), which binds with receptors on the plasma membrane of cardiac muscle (2). The receptor then activates a G protein (3), which activates adenylyl cyclase (4). This enzyme converts ATP to cyclic AMP, which then activates a protein kinase (5). The protein kinase phosphorylates calcium ion channels so that they open more easily when the neuron is depolarized (6). Action potentials occur more rapidly.

the smaller arteries and capillaries (Fig. 42–16b). By the time blood enters the veins, its pressure is very low, even approaching zero. Flow rate can be maintained in veins at low pressure because they are low-resistance vessels. Their diameter is larger than that of corresponding arteries, and there is little smooth muscle in their walls. Flow of blood through veins depends on several factors, including muscular movement, which compresses veins. Most veins larger than 2 mm (0.08 in) in diameter that conduct blood against the force of gravity are equipped with valves to prevent backflow (Fig. 42–17). Such valves usually consist of two cusps formed by inward extensions of the vein wall.

When a person stands perfectly still for a long time, as when a soldier stands at attention or a store clerk stands at a cash register, blood tends to pool in the veins. When fully distended, veins can accept no more blood from the capillaries. Pressure in the capillaries increases, and large amounts of plasma are forced out of the circulation through the thin capillary walls. Within just a few minutes, as much as 20% of the blood volume can be lost from the circulation—with drastic effect. Arterial blood pressure falls dramatically, reducing blood flow to the brain. Sometimes the resulting lack of oxygen in the brain causes fainting, a protective response. Lying in a prone position increases blood supply to the brain. In fact, lifting a person who has fainted to an upright position can result in circulatory shock and even death.

Blood pressure is carefully regulated

Each time you get up from a horizontal position, changes occur in your blood pressure. Several complex mechanisms interact to maintain normal blood pressure so that you do not faint when you get out of bed each morning or change position during the day. When blood pressure falls, sympathetic nerves to the blood

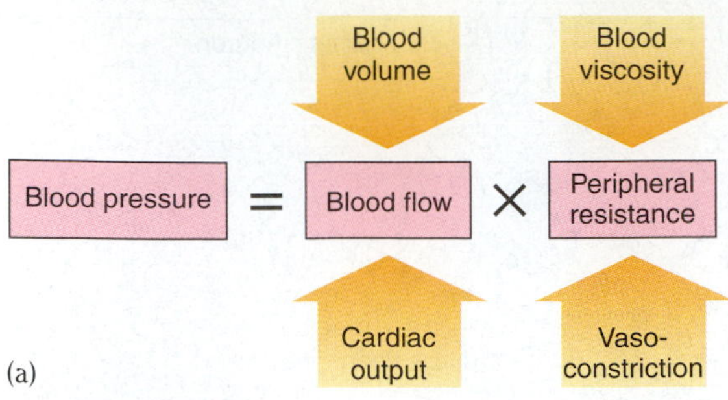

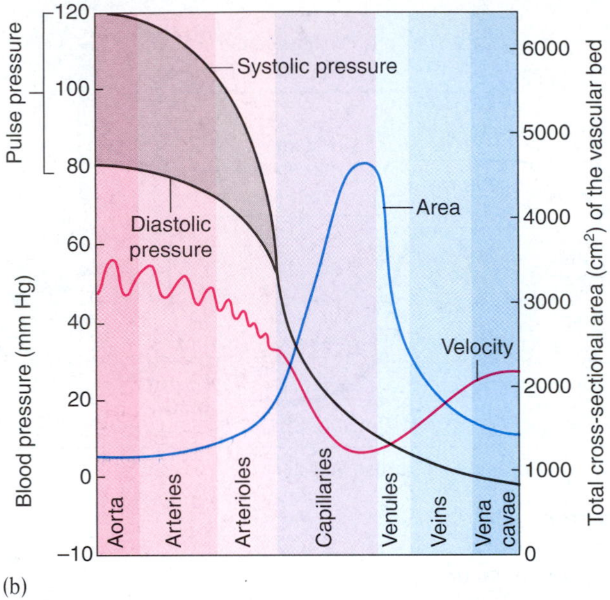

Figure 42–16 Blood pressure. **(a)** Blood pressure depends on blood flow and resistance to that flow. A variety of factors can affect blood flow and resistance. **(b)** Blood pressure in different types of blood vessels. Systolic and diastolic variations in arterial blood pressures are shown. Note that the venous pressure drops below zero (below atmospheric pressure) near the heart.

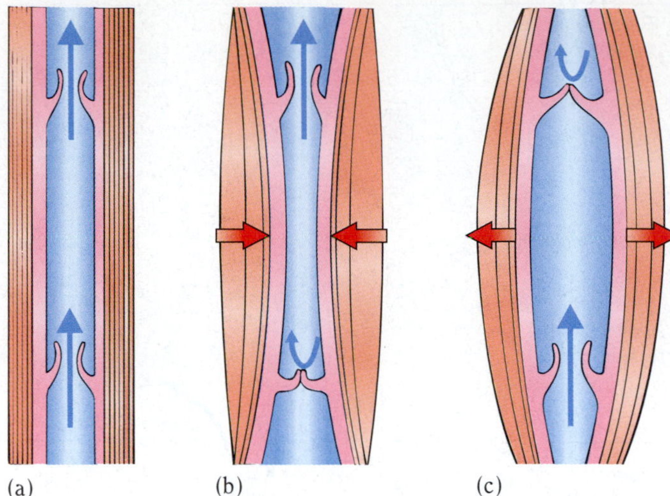

Figure 42–17 Venous blood flow. Contraction of skeletal muscles helps move blood through the veins. **(a)** Resting condition. **(b)** Muscles contract and bulge, compressing veins and forcing blood toward the heart. The lower valve prevents backflow. **(c)** Muscles relax, and the vein expands and fills with blood from below. The upper valve prevents backflow.

vessels stimulate vasoconstriction, causing the pressure to rise again.

The **baroreceptors** present in the walls of certain arteries and in the heart wall are sensitive to changes in blood pressure. When an increase in blood pressure stretches the baroreceptors, messages are sent to the cardiac and vasomotor centers in the medulla of the brain. The cardiac center stimulates parasympathetic nerves that slow the heart, lowering blood pressure. The vasomotor center inhibits sympathetic nerves that constrict arterioles; this action causes vasodilation, which lowers blood pressure. These neural reflexes continuously work in this complementary way to maintain blood pressure within normal limits.

Hormones are also involved in regulating blood pressure. In response to low blood pressure, the kidneys release **renin,** which activates the **renin-angiotensin-aldosterone pathway** (discussed in Chapter 46). Renin acts on a plasma protein (angiotensinogen), initiating a cascade of reactions that produces **angiotensin II,** a hormone that acts as a powerful vasoconstrictor. Angiotensin also acts indirectly to maintain blood pressure by increasing the synthesis and release of the hormone **aldosterone** by the adrenal glands. Aldosterone increases the retention of sodium ions by the kidneys, resulting in greater fluid retention and increased blood volume.

BLOOD IS PUMPED THROUGH PULMONARY AND SYSTEMIC CIRCUITS

One of the main jobs of the circulation is to bring oxygen to all the cells. In most vertebrates other than fish, blood is charged with oxygen in the lungs. As in amphibians and reptiles, birds and mammals have a double circuit of blood vessels: (1) the pulmonary circulation, which connects the heart and lungs; and (2) the systemic circulation, which connects the heart with all of the tissues of the body. This general pattern of circulation may be traced in Figure 42–18.

The pulmonary circulation oxygenates the blood

Blood from the tissues returns to the right atrium of the heart. Partly depleted of its oxygen supply, this oxygen-poor blood, loaded with carbon dioxide, is pumped by the right ventricle into the pulmonary circulation. As it emerges from the heart, the large pulmonary trunk branches to form the two pulmonary arteries, one going to each lung. These are the only arteries that carry oxygen-poor blood.

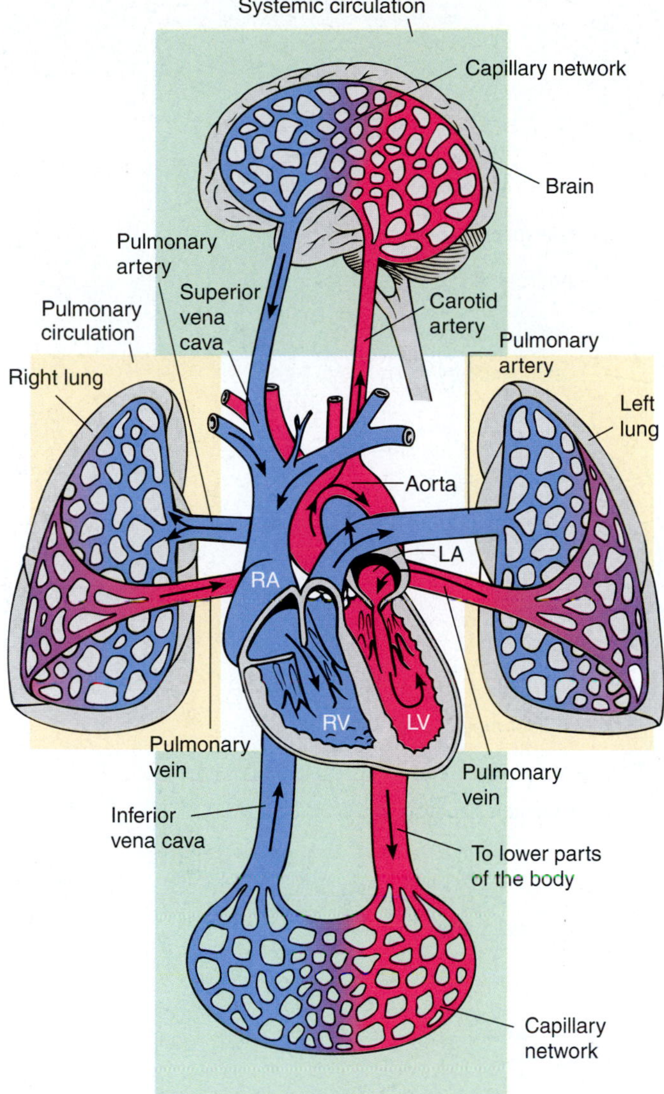

Figure 42–18 Systemic and pulmonary circulation. In this highly simplified diagram, red represents oxygen-rich blood; blue represents oxygen-poor blood.

In the lungs the pulmonary arteries branch into smaller and smaller vessels, which finally give rise to extensive networks of pulmonary capillaries that surround the air sacs of the lungs. As blood circulates through the pulmonary capillaries, carbon dioxide diffuses out of the blood and into the air sacs. Oxygen from the air sacs diffuses into the blood so that, by the time blood enters the pulmonary veins leading back to the left atrium of the heart, it is charged with oxygen. Pulmonary veins are the only veins in the body that carry blood rich in oxygen.

In summary, blood flows through the pulmonary circulation in the following sequence:

Right atrium ⟶ right ventricle ⟶ pulmonary arteries ⟶ pulmonary capillaries (in lungs) ⟶ pulmonary veins ⟶ left atrium

The systemic circulation delivers blood to the tissues

Blood entering the systemic circulation is pumped by the left ventricle into the **aorta,** the largest artery. Arteries that branch off from the aorta conduct blood to all regions of the body. Some of the principal branches include the **coronary arteries** to the heart wall itself, the **carotid arteries** to the brain, the **subclavian arteries** to the shoulder region, the mesenteric artery to the intestine, the **renal arteries** to the kidneys, and the **iliac arteries** to the legs (Fig. 42–19). Each of these arteries gives rise to smaller and smaller vessels, somewhat like branches of a tree that divide until they form tiny twigs. Eventually blood flows into the capillary network within each tissue or organ.

Blood returning from the capillary networks within the brain passes through the **jugular veins.** Blood from the shoulders and arms drains into the **subclavian veins.** These veins and others returning blood from the upper portion of the body merge to form a very large vein that empties blood into the right atrium. In humans this vein is called the **superior vena cava. Renal veins** from the kidneys, **iliac veins** from the lower limbs, **hepatic veins** from the liver, and other veins from the lower portion of the body return blood to the **inferior vena cava,** which delivers blood to the right atrium.

As an example of blood circulation through the systemic circuit, let us trace a drop of blood from the heart to the right leg and back to the heart:

Left atrium ⟶ left ventricle ⟶ aorta ⟶ right common iliac artery ⟶ smaller arteries in leg ⟶ capillaries in leg ⟶ small veins in leg ⟶ common iliac vein ⟶ inferior vena cava ⟶ right atrium

The coronary circulation delivers blood to the heart

In most fishes, amphibians, and reptiles, the heart muscle is spongy and receives oxygen directly from the blood as it passes through the heart chambers. However, in birds and mammals, the walls of the heart are too thick for nutrients and oxygen to diffuse through them to reach all the muscle fibers. Instead, the cardiac muscle is supplied by its own circulation.

In humans, **coronary arteries** branch from the aorta at the point where that vessel leaves the heart. The coronary arteries give rise to a network of blood vessels within the wall of the heart. Nutrients and gases are exchanged through coronary capillaries. Blood from these capillaries flows into **coronary veins,** which join to form a large vein, the **coronary sinus** (see Fig. 42–9). The coronary sinus empties directly into the right atrium; it does not join either of the venae cavae.

Four arteries deliver blood to the brain

Four arteries, two internal carotid arteries and two vertebral arteries (branches of the subclavian arteries), deliver blood to the

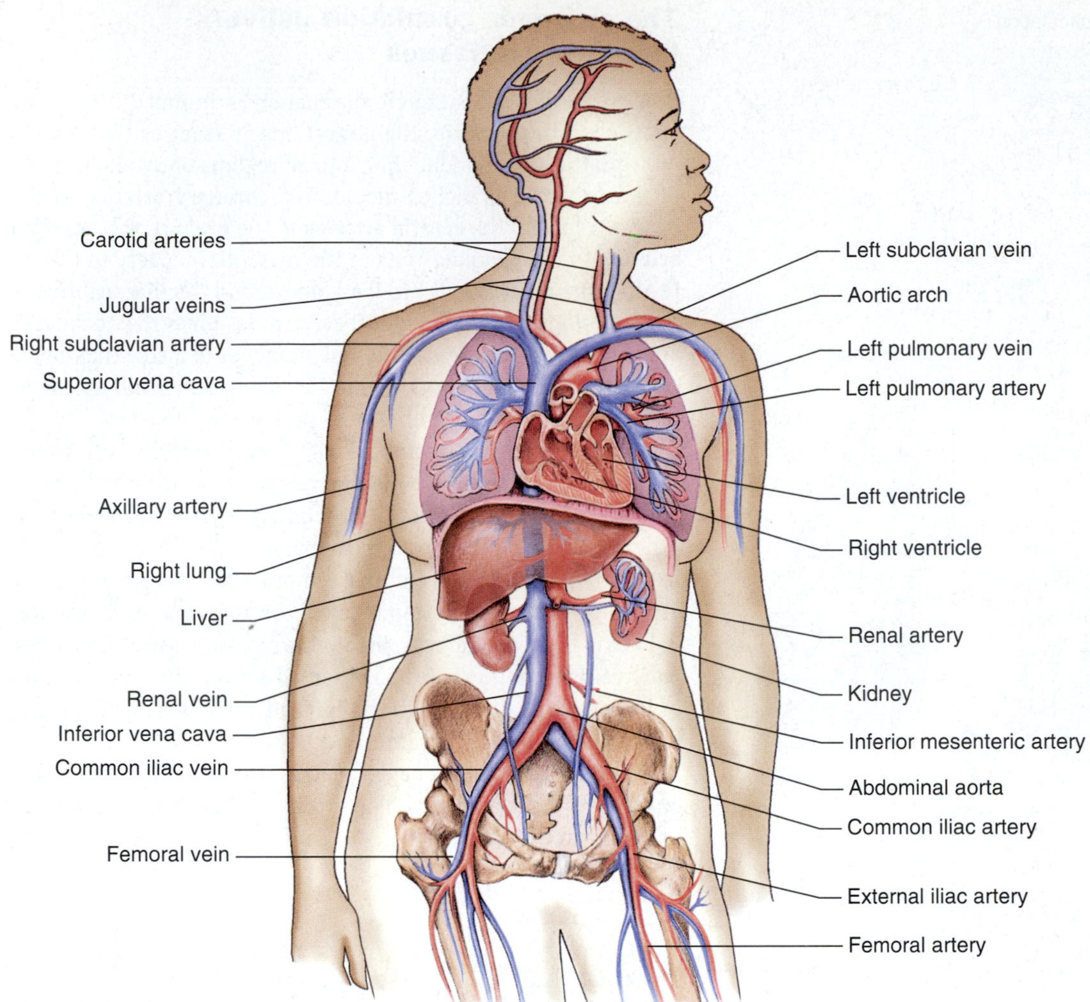

Figure 42–19 Blood circulation through some of the principal arteries and veins of the human body. Blood vessels carrying oxygen-rich blood are red; those carrying oxygen-poor blood are blue.

brain. At the base of the brain, branches of these arteries form an arterial circuit called the **circle of Willis.** In the event that one of the arteries serving the brain becomes blocked or injured in some way, this arterial circuit helps ensure blood delivery to the brain cells via other vessels. Blood from the brain returns to the superior vena cava by way of the internal jugular veins at either side of the neck.

The hepatic portal system delivers nutrients to the liver

Blood almost always travels from artery to capillary to vein to the heart. An exception to this sequence occurs in the **hepatic portal system,** which delivers blood rich in nutrients to the liver. Blood is conducted to the small intestine by the superior mesenteric artery. Then, as it flows through capillaries within the wall of the intestine, blood picks up glucose, amino acids, and other nutrients. This blood passes into the mesenteric vein and then into the **hepatic portal vein.** Instead of going directly back to the heart (as most veins do), the hepatic portal vein delivers nutrients to the liver.

Within the liver, the hepatic portal vein gives rise to an extensive network of tiny blood sinuses. As blood courses through the hepatic sinuses, liver cells remove nutrients and store them. Eventually liver sinuses merge to form hepatic veins, which deliver blood to the inferior vena cava. The hepatic portal vein contains blood that, although laden with food materials, has already given up some of its oxygen to the cells of the intestinal wall. Oxygen-rich blood is supplied to the liver by the hepatic artery.

CARDIOVASCULAR FUNCTION IS ALTERED DURING EXERCISE

Exercise results in marked physiological changes in the body. For example, it alters the rate and distribution of blood flow to the organs. Contracting muscles need more oxygen, so blood flow to the muscles increases dramatically—more than 20 times during aerobic exercise. This increase helps meet not only the greater demand for oxygen but also the greater need for removal of CO_2 and other wastes generated during physical activity. Thus, exercise has a purifying effect on body systems. Although the rate of blood flow to the brain is maintained, blood flow to the digestive organs and the kidneys is markedly reduced, so more blood is available for the muscles.

MAKING THE CONNECTION How does exercise benefit the cardiovascular system? According to the Centers for Disease Control and Prevention, physical inactivity contributes to more than one-third of the approximately 500,000 deaths related to heart disease in the United States each year. Exercise helps reduce stress and favorably modifies risk factors for developing cardiovascular disease (see *Focus On: Cardiovascular Disease*). For example, exercise decreases obesity, decreases hypertension (high blood pressure), and increases the concentration of high-density lipoproteins (HDLs, the "good" lipoproteins that help lower cholesterol levels) in the blood.

Exercise improves fitness. Specifically, exercise improves muscle tone and strength, and enhances muscular and cardiovascular endurance. Your *endurance* determines how long you can continue physical activities, such as running, swimming, and dancing. Endurance requires cardiovascular fitness as well as muscle strength. Endurance can be improved by aerobic exercise, which is any activity that increases the heart rate and requires oxygen. Aerobic exercise includes such activities as rapid walking, in-line skating, and bicycling.

Regular aerobic exercise strengthens the heart muscle and decreases the resting heart rate. Endurance training can result in increased volume of the left ventricle and increased thickening of the left ventricular wall. Such changes increase the amount of blood the heart can pump with each contraction and increases the strength of cardiac muscle contraction. The heart becomes more efficient in its pumping capacity both at rest and during physical activity and does not have to work as hard.

High concentrations of triacylglycerols (triglycerides) and cholesterol in the blood have been associated with heart disease, and exercise reduces their concentrations. At the same time, exercise increases the concentration of HDLs, which appear to protect against heart disease. One mechanism involves the enzyme *lipoprotein lipase*. During exercise, muscle fibers increase their output of lipoprotein lipase, which breaks down triacylglycerols for fuel. As triacylglycerol concentration decreases, low-density lipoproteins (LDLs) in the blood decrease in size. (Recall that LDLs are unhealthy lipoproteins.) Some of the cholesterol they transport is transferred to HDLs. HDLs transport cholesterol to the liver, where it is used to synthesize steroid hormones and other substances.

Another benefit of exercise is increased metabolic rate, which makes it easier to maintain desired body weight and amount of body fat. As a result, more food can be eaten without gaining weight.

In 1996 the National Institute of Health Consensus Development Panel on Physical Activity and Cardiovascular Health reported in the *Journal of the American Medical Association* that children and adults should perform moderate-intensity physical activity for at least 30 minutes on most days each week. The American College of Sports Medicine recommends that exercise should be intense enough to increase the heart rate to 70% to 90% of maximum. You can estimate maximum heart rate by subtracting your age from 220. For example, if you are 20 years old, your maximum heart rate is 200 beats per minute. If you are in good health, and exercise at a level that raises your heart rate to 70% to 90% of maximum, your heart rate will increase to about 160 beats per minute.

If you have not been exercising and are just beginning a program, mild exercise such as walking can raise your heart rate to aerobic levels. As you improve your fitness, you will need to intensify your exercise to reach aerobic range. For example, you might speed up your walking or skating.

Some exercise researchers have proposed that moderate levels of activity, such as walking or gardening, even if it is sporadic, can provide health benefits. Other investigators disagree. Studies indicate that risk of heart disease declines as exercise level increases. As a report by the U.S. surgeon general concluded, the health benefits of intermittent physical activity have not yet been demonstrated.

THE LYMPHATIC SYSTEM IS AN ACCESSORY CIRCULATORY SYSTEM

In addition to the blood circulatory system, vertebrates have an accessory circulatory system, the **lymphatic system** (Fig. 42–20), which has three important functions: (1) to collect and return interstitial fluid to the blood; (2) to defend the body against disease organisms by way of immune mechanisms; and (3) to absorb lipids from the digestive tract. In this section we focus on the first function. Immunity is discussed in Chapter 43, and lipid absorption is discussed in Chapter 45.

The lymphatic system consists of lymphatic vessels and lymph tissue

The lymphatic system consists of (1) an extensive network of **lymphatic vessels,** or simply **lymphatics,** that conduct **lymph,** the clear, watery fluid formed from interstitial fluid, and (2) **lymph tissue,** a type of connective tissue with large numbers of lymphocytes. Lymph tissue is organized into small masses of tissue called **lymph nodes** and **lymph nodules.** The tonsils, thymus gland, and spleen, which consist mainly of lymph tissue, are also part of the lymphatic system.

Tiny "dead-end" capillaries of the lymphatic system extend into almost all the tissues of the body (Fig. 42–21). Lymph

FOCUS ON Cardiovascular Disease

Cardiovascular disease is the number one cause of death in the United States and in most other industrial countries. Most often death results from some complication of **atherosclerosis,** a disease in which the arteries narrow as a result of lipid deposits in their walls. Although it can affect almost any artery, the disease most often develops in the aorta and in the coronary and cerebral arteries. When it occurs in the cerebral arteries, less blood is delivered to the brain. This condition can lead to a **cerebrovascular accident (CVA),** commonly referred to as a stroke.

In atherosclerosis, the endothelial lining of the arterial wall is damaged. The underlying smooth muscle then proliferates, thickening the arterial wall. Lipids in the blood are deposited in the smooth muscle cells of the arterial wall. These cells proliferate, and the inner lining thickens. More lipid, especially cholesterol from low-density lipoproteins, accumulates in the wall, and fibrous tissue forms around the deposits. Eventually calcium is deposited there, contributing to the slow formation of hard plaque that further narrows the diameter of the artery. As the plaque develops, arteries lose their ability to stretch when filled with blood, and they become progressively occluded (blocked), as shown in the figure. Less blood can be delivered to the tissues, which then may become **ischemic,** meaning that they are lacking in blood. Under these conditions the tissue is deprived of an adequate oxygen and nutrient supply.

When a coronary artery becomes narrowed, **ischemic heart disease** can develop. Sufficient oxygen may reach the heart tissue during normal activity, but the increased need for oxygen during exercise or emotional stress results in the pain known as **angina pectoris.** Nitroglycerin pills or a nitroglycerin patch dilates veins, reducing venous return. Cardiac output is lowered so that the heart is not working so hard and requires less oxygen. Nitroglycerin also dilates the coronary arteries slightly, allowing more blood to reach the heart muscle. *Calcium ion channel blockers* are drugs that slow the heart by inhibiting the passage of calcium into cardiac muscle fibers. They also dilate the coronary arteries.

Myocardial infarction (MI), popularly referred to as heart attack, is a very serious, often fatal, consequence of ischemic heart disease. MI often results from a sudden decrease in coronary blood supply. The portion of cardiac muscle deprived of oxygen dies within a few minutes and is then referred to as an **infarct.** MI is the leading cause of death and disability in the United States. Just what triggers the sudden decrease in blood supply is a matter of some debate.

Platelets may adhere to the roughened wall of an artery and initiate clotting. This can produce a **thrombus,** a clot that forms within a blood vessel or within the heart. If a thrombus blocks a sizable branch of a coronary artery, blood flow to a portion of heart muscle is impeded or completely halted. When such blockage, referred to as a *coronary occlusion,* prevents blood flow to a large region of cardiac muscle, the heart may stop beating; that is, **cardiac arrest** may occur, and death can follow within moments. If only a small region of the heart is affected, however, the heart may continue to function. Cells in the region deprived of oxygen die and are replaced by scar tissue.

An episode of ischemia may trigger a fatal arrhythmia such as **ventricular fibrillation,** a condition in which the ventricles contract very rapidly without actually pumping blood. The pulse may stop, and blood pressure may fall precipitously. Ventricular fibrillation has been linked with about 65% of cardiac arrests. The only effective treatment for fibrillation is **defibrillation** with electric shock. The shock appears to depolarize every muscle fiber in the heart so that its timing mechanism can reset.

Patients with progressive cardiovascular disease can be treated with *coronary bypass surgery* in which veins from another location in the patient's body

capillaries join to form larger lymphatics (which we can think of as lymph veins). There are no lymph arteries.

When interstitial fluid enters lymph capillaries, it is referred to as lymph. The lymph is conveyed into lymphatics, which at certain locations empty into lymph nodes. As lymph circulates through the lymph nodes, phagocytes filter out bacteria and other harmful materials. The lymph then flows into lymphatics that conduct it away from the lymph node. Lymphatics from all over the body conduct lymph toward the shoulder region. These vessels join the circulatory system at the base of the subclavian veins by way of ducts: the **thoracic duct** on the left side and the **right lymphatic duct** on the right.

Tonsils are masses of lymph tissue under the lining of the oral cavity and throat. (When enlarged, the pharyngeal tonsils in back of the nose are called **adenoids.**) Tonsils help protect the respiratory system from infection by destroying bacteria and other foreign matter that enter the body through the mouth or nose. Unfortunately, tonsils are sometimes overcome by invading bacteria and become the site of frequent infection themselves.

Some nonmammalian vertebrates such as the frog have lymph "hearts," which pulsate and squeeze lymph along. However, in mammals the walls of the lymphatic vessels themselves pulsate. Valves within the lymph vessels prevent the lymph

are grafted around occluded coronary arteries. The newly positioned blood vessels restore adequate blood flow to the affected area. Another procedure, *coronary angioplasty*, involves inserting a small balloon into an occluded coronary artery. When the balloon is inflated, the plaque in the arterial wall is broken up, increasing the diameter of the vessel. New approaches to treating cardiovascular disease include gene therapy aimed at production of a protein that would dissolve thrombi, as well as metabolic interventions that would shift fatty acid metabolism to glucose metabolism.

Several major modifiable risk factors for cardiovascular disease have been identified:

1. **Elevated cholesterol levels** in the blood, often associated with diets rich in total calories, total fats, saturated fats, and cholesterol.
2. **Hypertension.** The higher the blood pressure, the greater the risk.
3. **Cigarette smoking.** The risk of developing atherosclerosis is two to six times greater in smokers than in nonsmokers and is directly proportional to the number of cigarettes smoked daily. Components of cigarette smoke damage the endothelial lining of blood vessels, leading to atherosclerosis.
4. **Diabetes mellitus,** an endocrine disorder in which glucose is not metabolized normally. The body shifts to fat metabolism, and there is a marked increase in circulating lipids, leading to atherosclerosis.
5. **Physical inactivity.** One in four adults in the United States has a sedentary lifestyle and does not engage in regular exercise.
6. **Obesity** can affect cholesterol levels and increase the risk of hypertension and diabetes.

The risk of developing cardiovascular disease also increases with age. Estrogen hormones are thought to offer some protection in women until after menopause, when the concentration of these hormones decreases. Other probable risk factors currently being studied are hereditary predisposition, stress and behavior patterns, and dietary factors.

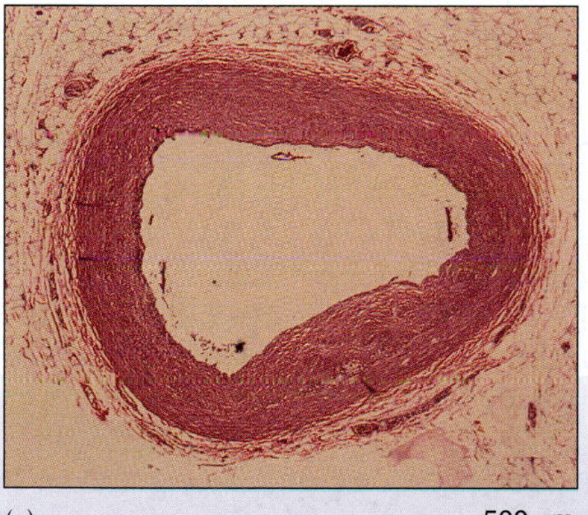

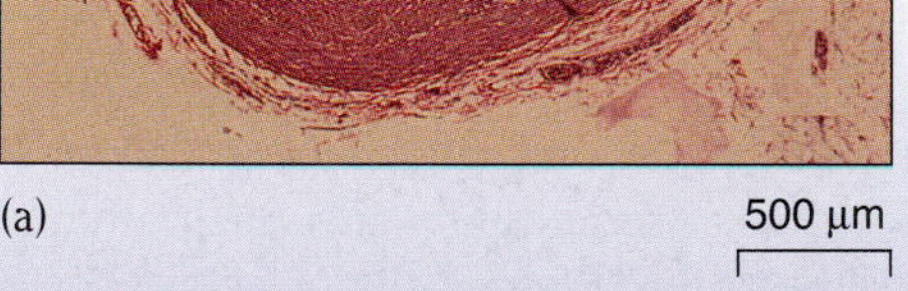

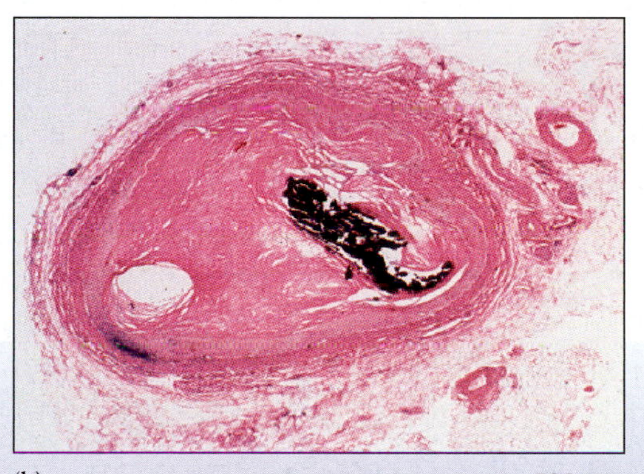

Progression of atherosclerosis. LMs of cross sections through two arteries showing changes that take place in atherosclerosis. **(a)** Normal coronary artery. **(b)** This artery is almost completely blocked with atherosclerotic plaque. *(a, Cabisco/Visuals Unlimited; b, Sloop-Ober/Visuals Unlimited)*

from flowing backward. When muscles contract or when arteries pulsate, pressure on the lymphatic vessels increases lymph flow. The rate at which lymph flows is slow and variable, and the total lymph flow is about 100 mL per hour—very much slower than the 5 L per min of blood flowing in the vascular system.

Lymphomas are a group of cancers that originate in the lymphatic system or bone marrow. The major types, distinguished by the appearance of the cells, are Hodgkin's disease and non-Hodgkin's lymphoma. Advances in radiation treatment and chemotherapy have increased survival rates of both types of lymphoma.

The lymphatic system plays an important role in fluid homeostasis

When blood enters a capillary network, it is under rather high pressure, so some plasma is forced out of the capillaries and into the tissues. Once it leaves the blood vessels, this fluid is called *interstitial fluid,* or *tissue fluid.* It contains no red blood cells or platelets and only a few white blood cells. Its protein content is about one-fourth that found in plasma because proteins are too large to pass easily through capillary walls. However, smaller molecules dissolved in the plasma do pass out with the fluid leaving the blood vessels. Thus, interstitial fluid contains glucose,

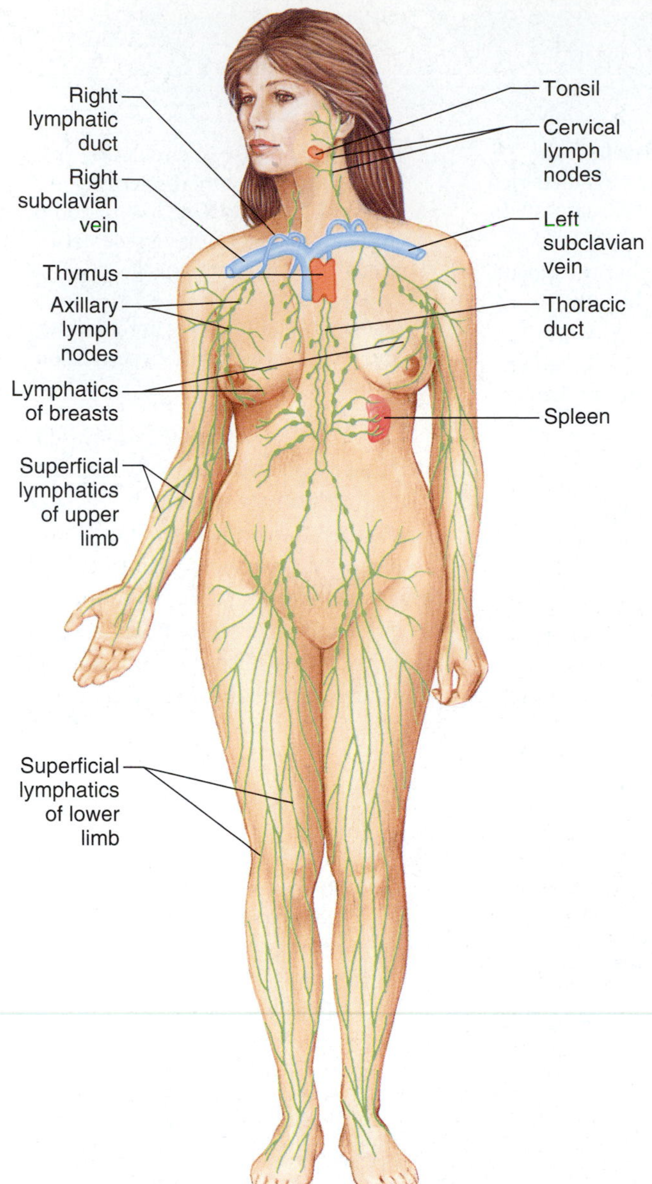

Figure 42–20 Human lymphatic system. Note that while the lymphatic vessels extend into most tissues of the body, the lymph nodes are clustered in certain regions. The right lymphatic duct drains lymph from the upper right quadrant of the body. The thoracic duct drains lymph from other regions of the body.

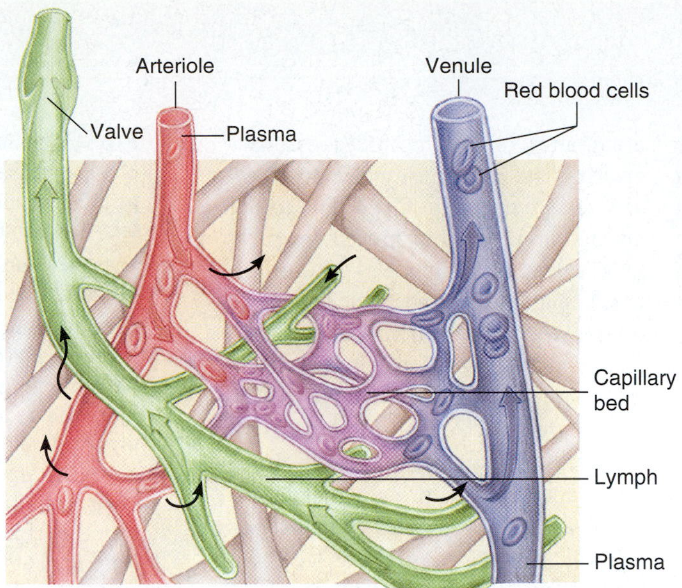

Figure 42–21 Lymph capillaries. Lymph capillaries drain excess interstitial fluid from the tissues. Note that blood capillaries are connected to vessels at both ends, whereas lymph capillaries, shown in green, are "dead-end streets." The arrows indicate direction of flow.

amino acids, other nutrients, and oxygen, as well as a variety of salts. This nourishing fluid bathes all the cells.

The main force *(filtration pressure)* pushing plasma out of the blood is hydrostatic pressure, that is, the blood pressure against the capillary wall (Fig. 42–22). The osmotic pressure of the interstitial fluid adds to the filtration pressure. The principal opposing force is the osmotic pressure of the blood (also called colloid osmotic pressure, or oncotic pressure), which restrains fluid loss from the capillary.

At the venous ends of the capillaries the blood pressure is much lower, and the osmotic pressure of the blood draws fluid back into the capillary. However, not as much fluid is absorbed back into the circulation as is filtered out. Furthermore, protein does not return effectively into the venous capillaries and instead tends to accumulate in the interstitial fluid. These potential problems are so serious that without the lymphatic system, fluid balance in the body would be significantly disturbed within a few hours, and death would occur within about 24 hours. The lymphatic system preserves fluid balance by collecting some of the interstitial fluid (about 10%), including the protein that accumulates in it, and returning it to the circulation.

The walls of the lymph capillaries are composed of endothelial cells that overlap slightly. When interstitial fluid accumulates, it presses against these cells, pushing them inward like tiny swinging doors that can swing in only one direction. As fluid accumulates within the lymph capillary, these cell doors are pushed closed.

Obstruction of the lymphatic vessels causes **edema,** swelling that results from excessive accumulation of interstitial fluid. Lymphatic vessels can be blocked as a result of injury, inflammation, surgery, or parasitic infection. For example, when a breast is removed (mastectomy) because of cancer, lymph nodes in the underarm region may also be removed in an effort to prevent the spread of cancer cells. The disrupted lymph circulation causes the patient's arm to swell. New lymphatic vessels develop within a few months, and the swelling slowly subsides.

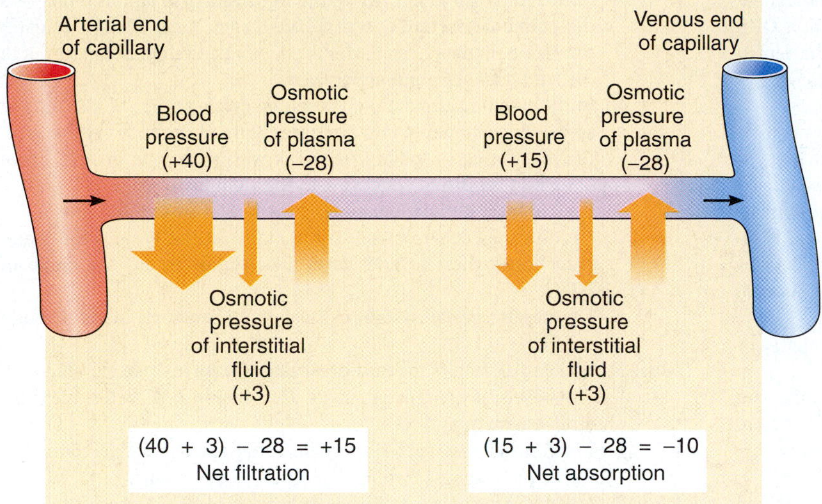

Figure 42–22 Fluid movement between blood and interstitial fluid. Blood pressure (hydrostatic pressure) and osmotic pressures are responsible for fluid movement, and, thus, exchange of dissolved materials, between blood and interstitial fluid. At the arterial end of a capillary, blood pressure forces plasma out of the capillary. The osmotic pressure of the blood is an opposing force acting to draw fluid into the blood. Osmotic pressure of the interstitial fluid contributes to the net filtration pressure but does not change much between arterial and venous ends of the capillary. At the venous end of the capillary, fluid enters the blood because blood pressure is much lower. However, more fluid leaves the blood than returns, and the lymphatic system collects the excess interstitial fluid. The numbers given are hypothetical and represent millimeters of mercury. Net filtration is the total pressure moving fluid out of the capillary. Net absorption refers to the total pressure drawing fluid into the capillary.

SUMMARY WITH KEY TERMS

I. Small, simple invertebrates, such as sponges, cnidarians, and flatworms, depend on **diffusion** for internal transport. Larger animals require a specialized **circulatory system,** which typically consists of **blood,** a **heart,** and a system of blood vessels or spaces through which blood circulates. In all animals **interstitial fluid,** the fluid between cells, is important in bringing oxygen and nutrients into contact with cells.
   A. Arthropods and most mollusks have an **open circulatory system** in which blood flows into a **hemocoel,** bathing the tissues directly.
   B. Some invertebrates and all vertebrates have a **closed circulatory system** in which blood flows through a continuous circuit of blood vessels.

II. The vertebrate circulatory system consists of a muscular heart that pumps blood into a system of **arteries, capillaries,** and **veins.** This system transports nutrients, oxygen, wastes, and hormones; helps maintain fluid balance, appropriate pH, and body temperature; and defends the body against disease.

III. Vertebrate blood consists of liquid **plasma** in which **red blood cells (erythrocytes), white blood cells (leukocytes),** and **platelets** are suspended.
   A. Plasma consists of water, salts, substances in transport, and **plasma proteins,** including **albumins, globulins,** and **fibrinogen.**
   B. Red blood cells transport oxygen and carbon dioxide. In vertebrates, red blood cells produce large quantities of **hemoglobin,** a red pigment that binds with oxygen.
   C. White blood cells defend the body against disease organisms. **Lymphocytes** and **monocytes** are agranular white blood cells; **neutrophils, eosinophils,** and **basophils** are granular white blood cells.
   D. Platelets patch damaged blood vessels and release substances essential for blood clotting. During the clotting process **thrombin** catalyzes the conversion of **fibrinogen** to an insoluble protein, **fibrin.**

IV. Arteries carry blood away from the heart chambers; veins return blood to the heart chambers.
   A. **Arterioles** constrict **(vasoconstriction)** and dilate **(vasodilation)** to regulate blood pressure and distribution of blood to the tissues.
   B. Capillaries are the thin-walled exchange vessels through which materials are transferred between the blood and tissues.

V. The vertebrate heart consists of one or two **atria,** which receive blood, and one or two **ventricles,** which pump blood into the arteries.
   A. The four-chambered hearts of birds and mammals separate oxygen-rich blood from oxygen-poor blood.
   B. The heart is enclosed by a **pericardium** and is equipped with valves that prevent backflow of blood.
      1. The valve between right atrium and ventricle is the **right atrioventricular(AV) valve,** or **tricuspid valve.** The valve between left atrium and ventricle is the **mitral valve.**
      2. **Semilunar valves** guard the exits from the heart.
   C. Cardiac muscle fibers are joined by **intercalated discs.**
   D. The **sinoatrial (SA) node,** or **pacemaker,** initiates each heartbeat. A specialized electrical conduction system ensures that the heart beats in a coordinated manner.
   E. One complete heartbeat makes up a **cardiac cycle.** Contraction occurs during **systole.** The period of relaxation is **diastole.**

F. An **electrocardiogram (ECG or EKG)** begins with a **P wave** caused by depolarization of the SA node and atrial muscle. Then, a **QRS complex** reflects the action potential spreading through the ventricles. The **T wave** occurs during repolarization of the ventricles.

G. **Cardiac output** equals **stroke volume** times heart rate.
   1. Stroke volume depends on **venous return** and on neural messages and hormones, especially epinephrine and norepinephrine.
   2. According to **Starling's law of the heart,** the more blood delivered to the heart by the veins, the more blood the heart pumps.
   3. Heart rate is regulated mainly by the nervous system and is influenced by hormones and body temperature.
   4. The heart rate can be measured by counting the pulse. **Arterial pulse** is the alternate expansion and recoil of an artery.

VI. **Blood pressure** is the force exerted by the blood against the inner walls of the blood vessel.

A. Blood pressure is determined by cardiac output, blood volume, and resistance to blood flow. **Peripheral resistance** is the resistance to blood flow caused by the viscosity of the blood and by friction between the blood and the wall of the blood vessel.

B. Blood pressure is greatest in the arteries and decreases as blood flows through the capillaries.

C. **Baroreceptors** sensitive to changes in blood pressure send messages to the cardiac and vasomotor centers in the medulla of the brain. When informed of an increase in blood pressure, the cardiac center stimulates parasympathetic nerves that slow the heart rate, and the vasomotor center inhibits sympathetic nerves that constrict blood vessels. These actions reduce blood pressure.

D. **Angiotensin** is a hormone that raises blood pressure. **Aldosterone** helps regulate salt excretion, which affects blood volume and blood pressure.

VII. The **pulmonary circulation** connects heart and lungs; the **systemic circulation** connects the heart and the tissues.

A. In the pulmonary circulation, the right ventricle pumps blood into the **pulmonary arteries,** one going to each lung. Blood circulates through pulmonary capillaries in the lung and then is conducted to the left atrium by a **pulmonary vein.**

B. In the systemic circulation, the left ventricle pumps blood into the **aorta,** which branches into arteries leading to the body organs. After flowing through capillary networks within various organs, blood flows into veins that conduct it to the right atrium.

C. The **coronary circulation** supplies the heart muscle with blood.

D. Four arteries deliver blood to the brain. At the base of the brain, branches of these arteries form an arterial circuit, the **circle of Willis.**

E. The **hepatic portal system** circulates nutrient-rich blood through the liver.

VIII. Modifiable risk factors for **cardiovascular disease** include elevated cholesterol levels, hypertension, cigarette smoking, diabetes mellitus, physical inactivity, and obesity.

A. In **atherosclerosis,** arteries become progressively narrowed as they thicken in response to lipid deposition in their walls.

B. Atherosclerosis leads to **ischemic heart disease,** in which the heart muscle does not receive sufficient blood.

C. **Myocardial infarction (MI)** is a serious consequence of ischemic heart disease.

IX. The **lymphatic system** collects interstitial fluid, also called **tissue fluid,** and returns it to the blood. It plays an important role in homeostasis of fluids.

A. **Lymphatic vessels** conduct **lymph,** a clear fluid formed from interstitial fluid, to the shoulder region and return it to the circulatory system.

B. **Lymph nodes** are small masses of lymph tissue that filter the lymph, removing bacteria and harmful materials.

POST-TEST

1. An open circulatory system (a) is found in flatworms (b) typically includes a hemocoel (c) has a continuous circuit of vessels with openings in the capillaries (d) is characteristic of vertebrates (e) is typically found in animals with a two-chambered heart
2. Which of the following is *not* a function of the vertebrate circulatory system? (a) helps maintain appropriate pH (b) transports nutrients, oxygen, and metabolic wastes (c) helps maintain fluid balance (d) produces hemocyanin (e) provides internal defense
3. Lipoproteins (a) are mainly transported in granular leukocytes (b) transport cholesterol (c) have been linked to clotting disorders (d) are associated with platelets (e) are stored in red blood cells
4. Which of the following are most closely associated with oxygen transport? (a) red blood cells (b) platelets (c) neutrophils (d) basophils (e) lymphocytes
5. Which of the following are most closely associated with blood clotting? (a) red blood cells (b) platelets (c) neutrophils (d) basophils (e) lymphocytes
6. In blood clotting (a) thrombin ⟶ prothrombin; fibrinogen ⟶ fibrin (b) prothrombin ⟶ thrombin; fibrin ⟶ fibrinogen (c) prothrombin ⟶ thrombin; fibrinogen ⟶ fibrin (d) clotting factors ⟶ platelets; thrombin ⟶ fibrinogen (e) prothrombin ⟶ thrombin; fibrinogen ⟶ platelets
7. Blood vessels that carry blood away from the heart are (a) arteries (b) arterioles (c) veins (d) capillaries (e) arterioles and arteries
8. Arterioles (a) help regulate blood pressure (b) help regulate distribution of blood to the tissues (c) deliver blood to arteries (d) answers a, b, and c are correct (e) answers a and b only
9. Which sequence most accurately describes a sequence of blood flow? (a) right atrium ⟶ right ventricle ⟶ pulmonary artery (b) right atrium ⟶ left atrium ⟶ left ventricle ⟶ aorta (c) left atrium ⟶ left ventricle ⟶ pulmonary artery (d) left ventricle ⟶ left atrium ⟶ aorta (e) right atrium ⟶ right ventricle ⟶ aorta
10. Which sequence most accurately describes a sequence of blood flow? (a) pulmonary vein ⟶ pulmonary artery ⟶ right atrium (b) pulmonary artery ⟶ left atrium ⟶ left ventricle (c) pulmonary artery ⟶ pulmonary capillaries ⟶ pulmonary vein ⟶ left atrium (d) left ventricle ⟶ aorta ⟶ pulmonary artery (e) pulmonary artery ⟶ pulmonary capillaries ⟶ pulmonary vein ⟶ right atrium
11. A cardiac cycle (a) consists of one ventricular heartbeat (b) includes a systole (c) equals stroke volume times heart rate (d) includes a diastole (e) includes a systole and a diastole
12. Blood pressure is determined by (a) cardiac output (b) peripheral resistance (c) blood volume (d) answers a, b, and c are correct (e) answers b and c only
13. Lymph forms from (a) interstitial fluid (b) blood serum (c) plasma combined with protein (d) fluid released by lymph nodes (e) angiotensins
14. The valve between the right atrium and right ventricle is the (a) mitral valve (b) semilunar valve (c) tricuspid valve (d) pulmonary valve (e) aortic valve
15. Atherosclerosis (a) is associated with thickening of arteries and veins (b) is thought to be caused by high concentrations of low-density lipoprotein (c) can lead to ischemic heart disease (d) answers a, b, and c are correct (e) answers b and c only

REVIEW QUESTIONS

1. Compare how nutrients and oxygen are transported to the body cells in a hydra, planarian, earthworm, insect, and frog.
2. Contrast an open with a closed circulatory system.
3. Describe five functions of the vertebrate circulatory system.
4. List the functions of the main groups of plasma proteins.
5. Contrast the structure and functions of red and white blood cells.
6. Summarize the process by which blood clots.
7. Compare the functions of arteries, capillaries, and veins. How do arterioles function in maintaining homeostasis?
8. Compare and contrast the hearts of a fish, amphibian, reptile, and bird.
9. Define cardiac output and describe factors that influence it.
10. Describe the conduction system of the heart and how the heart is regulated. Include a description of the actions of acetylcholine and norepinephrine.
11. What is the relationship between blood pressure and peripheral resistance? What mechanisms regulate blood pressure?
12. Trace the path of a red blood cell: (a) from the inferior vena cava to the aorta; and (b) from the renal vein to the renal artery.
13. What is the function of the hepatic portal system? How does its sequence of blood vessels differ from that in most other circulatory routes?
14. List five modifiable risk factors associated with the development of cardiovascular disease. Describe the disease process in atherosclerosis, and explain the association between atherosclerosis and ischemic heart disease. What happens in myocardial infarction?
15. What is the role of lipoproteins in the development of cardiovascular disease? How does exercise affect lipoprotein concentration?
16. What is the relationship between plasma, interstitial fluid, and lymph? How does the lymph system help maintain fluid balance?
17. Blood pressure is low in capillaries. Explain how this helps to retain fluid in the circulation.
18. Label the diagram. Use Figure 42-9*a* to check your answers.

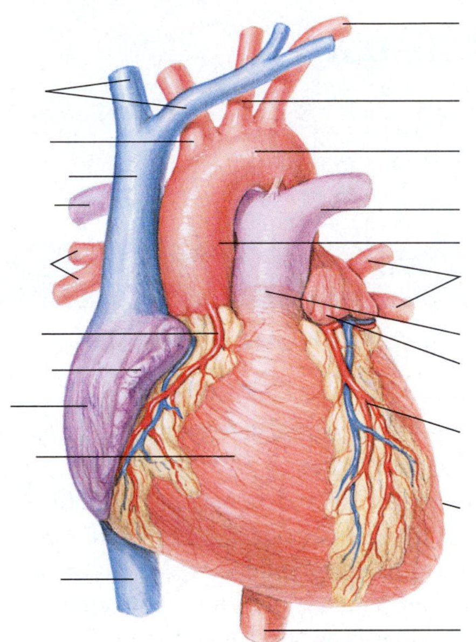

YOU MAKE THE CONNECTION

1. How do the changes that take place during exercise affect the cardiovascular system? Compare the heart function of an adult who does not exercise to that of an athlete.
2. Cartilage lacks blood and lymph vessels. How do you imagine its cells are nourished? What effect, if any, would a lack of blood vessels have on cartilage healing after an injury?
3. How is the heart of the fish specifically adapted to its lifestyle?
4. When the nerves to the heart are cut, the heart rate increases to about 100 contractions per minute. What does this indicate about the regulation of the heart rate?

RECOMMENDED READINGS

Borst, C. "Operating on a Beating Heart." *Scientific American,* Vol. 283, No. 4, Oct. 2000. New surgical techniques make coronary bypass surgery safer and less expensive.

Eisenberg, M.S. "Defibrillation: The Spark of Life." *Scientific American,* Vol. 278, No. 6, Jun. 1998. The author discusses the history of using electricity to restart the heart and compares defibrillation to rebooting a computer that has suddenly frozen.

Lopaschuk, G.D., and W.C. Stanley. "Manipulation of Energy Metabolism in the Heart." *Science & Medicine,* Vol. 4, No. 6, Nov./Dec. 1997. Shifting energy substrate preference away from fatty acid metabolism and toward glucose use appears to be an effective approach to treating cardiovascular disease.

Nabel, E.G. "Cardiovascular Gene Therapy." *Science & Medicine,* Vol. 4, No. 4, Jul./Aug. 1997. Clinical trials of gene therapy that may be used to treat cardiovascular disease are in progress.

National Institutes of Health Consensus Development Panel on Physical Activity and Cardiovascular Health. "Physical Activity and Cardiovascular Health." *Journal of the American Medical Association,* Vol. 276, No. 3, 17 Jul. 1996. An assessment of the relationship between physical activity and cardiac health.

Stamler, J., and J.D. Neaton. "Benefits of Lower Cholesterol." *Scientific American: Science & Medicine,* Vol. 1, No. 2, May/Jun. 1994. Declines in death rate from heart disease appear to be related to lifestyle changes.

- Visit our Web site at **http://www.info.brookscole.com/solomonbergmartin** for links to chapter-related resources on the World Wide Web. Additional on-line materials relating to this chapter can also be found on our Web site.

 See chapter activity on BioActive Learner CD for additional help in mastering the chapter's material. Icon location in the chapter's margins shows which topics have tutorials or simulations in the CD.

43

Internal Defense

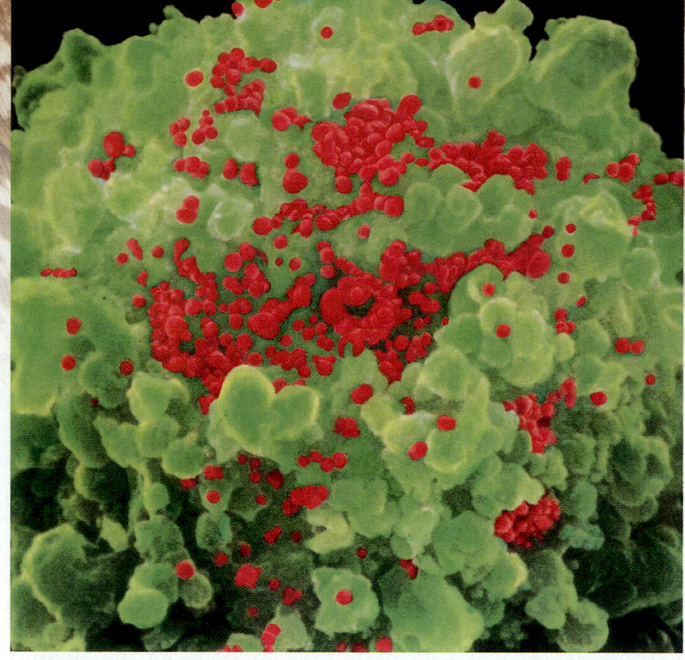

Color-enhanced SEM of a T cell *(green)* infected with human immunodeficiency virus *(red)*. *(NIBSC/Science Photo Library/Photo Researchers, Inc.)*

LEARNING OBJECTIVES

After you have studied this chapter you should be able to

1. Compare in general terms internal defense mechanisms of invertebrates and vertebrates.
2. Distinguish between specific and nonspecific immune responses, and describe nonspecific immune defenses, including mechanical barriers, soluble molecules such as cytokines and the proteins that make up complement, phagocytes, natural killer cells, and inflammation.
3. Describe the principal cells of the immune system and contrast T and B cells with respect to their development and function.
4. Draw the basic structure of an antibody and describe how antibodies recognize antigens.
5. Summarize the mechanisms of antibody-mediated immunity, including the effects of antigen-antibody complexes on pathogens; include a discussion of the complement system.
6. Describe the mechanisms of cell-mediated immunity, including the development of memory cells.
7. Describe the basis of immunological memory; contrast a secondary with a primary immune response.
8. Compare active and passive immunity, giving examples of each.
9. Summarize the immunological basis of graft rejection and explain how the effects of graft rejection can be minimized.
10. Describe the body's response to cancer cells and discuss new approaches to treating cancer.
11. Compare examples of inherited and acquired immunodeficiency; include a discussion of AIDS, emphasizing its cause, risk factors, progress of the disease, and difficulties encountered in developing a vaccine.
12. Describe the events that occur during hypersensitivity reactions, including Rh incompatibility and allergic reactions such as a hayfever response and systemic anaphylaxis.
13. Describe the immunological basis of autoimmune diseases and list possible causes.

Animals have internal defense mechanisms that protect them against disease-causing organisms that enter the body with air, food, and water; during copulation; and through wounds in the skin. Disease-causing microorganisms, including certain viruses, bacteria, fungi, and protozoa, are referred to as **pathogens.** Internal defense depends on the animal's ability to distinguish between *self* and *nonself.* Such recognition is possible because each individual is biochemically unique. Cells have surface proteins different from those on the cells of other species or even other members of the same species. An animal recognizes its own cells and can identify those of other organisms as foreign.

Pathogens manufacture macromolecules that the body identifies as foreign. A single bacterium may have from 10 to more than 1000 distinct macromolecules on its surface. Pathogens may also secrete macromolecules, some of which are toxic to most organisms. When a pathogen invades an animal, its distinctive macromolecules stimulate the animal's defense mechanisms.

An **immune response** involves recognition of foreign macromolecules and a response aimed at eliminating them. The term *immune* is derived from a Latin word meaning "safe." Immune responses depend on communication among cells, or **cell signaling.** Cells of the immune system communicate directly by means of their surface molecules and indirectly by releasing messenger molecules. **Immunology,** the study of internal defense mechanisms, is one of the most rapidly changing and exciting fields of biomedical research today.

Two main types of immune responses are nonspecific and specific immune responses. **Nonspecific immune responses,** also called **innate immunity,** provide general protection against pathogens. These mechanisms prevent most pathogens from entering the body and rapidly destroy those pathogens that do penetrate the outer defenses. For example, the cuticle or skin provides a barrier to pathogens that come in contact with an animal's body. Phagocytosis, another nonspecific defense mechanism, destroys bacteria that invade the body.

Specific immune responses, also referred to as **adaptive** or **acquired immune responses,** are tailor-made to combat specific macromolecules associated with each pathogen. Specific immune responses are directed to the particular type of foreign substance or pathogen that has gained entrance to an animal's body. Any molecule that can be specifically recognized as foreign by cells of the immune system is called an **antigen.** Many macromolecules, including proteins, RNA, DNA, and some carbohydrates, are antigens. An important specific immune response is the production of **antibodies,** highly specific proteins that recognize and bind to specific antigens. In complex animals, specific immune responses include immunological memory, the capacity to respond more effectively the second time foreign molecules invade the body.

Sometimes the body's defenses are outmaneuvered and overcome by pathogens, resulting in disease. Some diseases, as well as certain genetic mutations, prevent or reduce immune function. Human immunodeficiency virus (HIV), the virus that causes acquired immunodeficiency syndrome (AIDS), infects T cells, an important component of the immune system *(see photograph).* At times, the immune system overfunctions, as in allergic reactions, or attacks in ways that are clinically inconvenient, as in Rh incompatibility or destruction of the cells of organ transplants. In about 5% of adults in highly developed countries, certain immune responses are directed against self, resulting in autoimmune disease.

The immune system is a collection of many types of cells and of tissues scattered throughout the body that release hundreds of different compounds. Understanding its complex signaling systems and mechanisms for destroying pathogens has been challenging. Perhaps immunology's greatest accomplishments have been in the development of vaccines that prevent disease and in successful tissue and organ transplantation. More sophisticated research tools, such as gene transfer, have permitted immunologists to expand their knowledge of the cells and molecules that interact to produce immune responses and to develop new approaches to prevention and treatment of disease. Much has been learned, and many challenges lie ahead.

INVERTEBRATES MAKE MAINLY NONSPECIFIC IMMUNE RESPONSES

All invertebrate species that have been studied demonstrate the ability to distinguish between self and nonself. Sponge cells have specific glycoproteins on their surfaces that enable them to recognize their own species. When cells of two different species are mixed together, they reassort according to species. When two different species of sponges are forced to grow in contact with each other, tissue is destroyed along the region of contact. Cnidarians (e.g., corals) also reject grafted tissue and destroy foreign tissue. Invertebrates with a coelom have amoeba-like **phagocytes** that engulf and destroy bacteria and other foreign matter. In mollusks, substances in the hemolymph (blood) enhance phagocytosis.

Antimicrobial peptides have been identified in all multicellular organisms (including plants) that have been investigated. In fact, about 400 of these peptides that inactivate or kill pathogens have been described! For example, the fruit fly *Drosophila* produces at least seven different antimicrobial peptides. Six of them act against bacteria, and one is antifungal.

Certain cnidarians and arthropods (e.g., insects) demonstrate some specificity in their immune responses. In them, and in some echinoderms (e.g., sea stars) and simple chordates (e.g., tunicates), the body appears to remember antigens for a short period; this memory enables the body to respond more effectively when the same pathogens are encountered again. Echinoderms and tunicates are the simplest animals known to have differentiated white blood cells that perform limited immune functions.

VERTEBRATES LAUNCH NONSPECIFIC AND SPECIFIC IMMUNE RESPONSES

Like invertebrates, vertebrates protect themselves against pathogens with both nonspecific and specific immune responses (Fig. 43–1). More sophisticated specific immune responses are possible because vertebrates have a specialized lymphatic system (see Chapter 42). Only vertebrates have **lymphocytes,** white blood cells specialized to carry out immune responses. In the discussion that follows, we focus on the human immune system, with references to those of other animals.

NONSPECIFIC DEFENSE RESPONSES INVOLVE A VARIETY OF MOLECULES AND CELLS

Mechanical and chemical barriers prevent most pathogens from entering the body. An animal's first line of defense is its outer covering and the mucous membranes lining passageways that open to the outside of the body. For example, very few pathogens can penetrate the intact human skin, and glands in the skin produce antimicrobial chemicals. **Lysozyme,** an enzyme found in many tissues and in tears and other body fluids, attacks the cell walls of many gram-positive bacteria.

Microorganisms that enter with food are usually destroyed by the acid secretions and enzymes of the stomach. Pathogens that enter the body with inhaled air may be filtered out by hairs in the nose or trapped in the sticky mucous lining of the respiratory passageways. Once trapped, they may be destroyed by phagocytes.

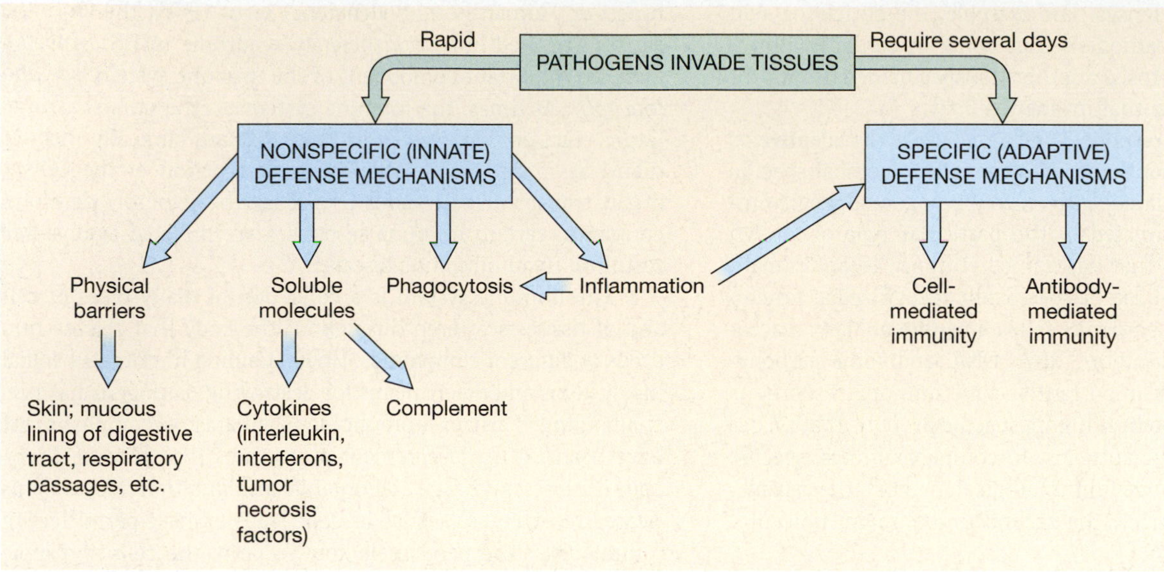

Figure 43–1 Nonspecific and specific immune responses. Nonspecific (innate) immune responses include physical and chemical barriers that prevent entrance of pathogens; soluble peptides and proteins that destroy pathogens; phagocytosis; and inflammation. Cytokines released by cells during inflammation signal specific (adaptive) immune responses: cell-mediated immunity and antibody-mediated immunity.

When pathogens breach the body's outer barriers, other nonspecific defense responses are activated, and the invaders are usually quickly eliminated. Molecules and cells of the nonspecific defense system can be activated by the chemical properties of the pathogen itself. In addition to the mechanical barriers, nonspecific immune responses include soluble molecules that destroy pathogens; phagocytes; and natural killer (NK) cells. The activation and concentration of these molecules and cells produce inflammation, the major characteristic of a nonspecific immune response.

Soluble molecules destroy pathogens

Cells of the immune system secrete a remarkable number of regulatory and antimicrobial peptides and proteins. Similar antimicrobial peptides have been identified in a wide range of animals, from insects to amphibians and mammals, suggesting an early common origin of these molecules. The major groups of soluble molecules of the nonspecific defense system are cytokines and complement.

Cytokines are important signaling molecules

Cytokines are a large group of peptides and proteins (some are glycoproteins) that serve as signals during both nonspecific and specific immune responses. They are also important in regulating many other biological processes such as cell growth, repair, and cell activation. Cytokines can act as *autocrine agents,* affecting the very cells that produce them, or as *paracrine agents* that regulate the activity of nearby cells (see Chapter 47). Some cytokines, like many hormones, circulate in the blood and affect distant tissues. We discuss three groups of cytokines: interferons, interleukins, and tumor necrosis factors.

When infected by viruses or other intracellular parasites (some types of bacteria, fungi, and protozoa), certain cells respond by secreting cytokines called **interferons.** Type I interferons are produced either by **macrophages,** large phagocytic cells, or by *fibroblasts,* cells that produce the fibers of connective tissues . Type I interferons inhibit viral replication and also activate natural killer cells that have antiviral actions. Viruses produced in cells exposed to Type I interferons are not very effective at infecting other cells. Another group, Type II interferons, produced by the specific immune system, enhance the activities of other immune cells. This group of interferons can stimulate macrophages to destroy tumor cells and host cells that have been infected by viruses.

Since their discovery in 1957, interferons have been the focus of much research. Recombinant DNA techniques are now used to produce large quantities of some interferons. The U.S. Food and Drug Administration (FDA) has approved interferons for treating several diseases, including hepatitis B and hepatitis C, genital warts, a type of leukemia, a type of multiple sclerosis, and AIDS-related Kaposi's sarcoma. Interferons are being tested in clinical trials for treatment of HIV infection and several types of cancer.

Interleukins are a diverse group of proteins secreted mainly by macrophages and lymphocytes. These cytokines are numbered according to their order of discovery. They regulate interactions between lymphocytes and other cells of the body, and some interleukins have widespread effects. For example, during infection, **interleukin-1 (IL-1)** can reset the body's thermostat in the hypothalamus, resulting in fever and its symptoms. Just as there are overlaps between the functions of nonspecific and specific immune responses, the cytokines of these subsystems also overlap. For example, cytokines produced by nonspecific cells such as macrophages can activate lymphocytes that are involved in specific immune responses.

Tumor necrosis factors (TNFs) are cytokines that are secreted by macrophages (TNF-alpha) and by lymphocytes called T cells (TNF-beta). TNF can stimulate immune cells to initiate an inflammatory response. TNF also kills tumor cells, offering promise in terms of immunotherapy for cancer patients. Sometimes infection by gram-negative bacteria, such as *Salmonella typhi,* results in the release of large amounts of TNF and other cytokines. This can lead to *septic shock,* a potentially lethal condition that may involve high fever and malfunction of the circulatory system. Thus, cytokines can sometimes have harmful effects. Cytokines that are closely associated with specific immune responses are discussed later in the chapter.

Complement leads to destruction of pathogens

Cytokines produced by phagocytes can activate the complement system. **Complement,** so-named because it *complements* the action of other defense mechanisms, consists of more than 20 proteins present in plasma and other body fluids. Similarities in complement proteins in many species, including horseshoe crabs, sea urchins, tunicates, and mammals, suggest that these molecules evolved millions of years ago and have been conserved.

Normally, complement proteins are inactive until the body is exposed to an antigen. Certain pathogens activate the complement system directly. In other cases, the binding of an antigen and antibody stimulate activation. Complement activation involves a cascade of reactions, with each component acting on the next in the series. Proteins of the complement system then work to destroy pathogens.

Complement proteins can be activated against many antigens, and their actions are nonspecific. Activated complement proteins have four main actions: (1) certain complement proteins lyse the pathogen cell wall; (2) others coat pathogens, making them less "slippery" so that phagocytes (macrophages and neutrophils) can phagocytize them more easily; (3) some complement proteins attract white blood cells to the site of infection; and (4) some complement proteins increase inflammation by stimulating release of **histamine** and other compounds that dilate blood vessels and increase capillary permeability.

Phagocytes and natural killer cells destroy pathogens

Neutrophils (the most common type of white blood cell; see Chapter 42) and macrophages are the phagocytes of the nonspecific immune system. Recall that **phagocytosis** is a type of endocytosis in which cells engulf microorganisms, foreign matter, or other cells (see Chapter 5 and Fig. 5–18). A neutrophil can phagocytize 20 or so bacteria before it becomes inactivated (perhaps by leaking lysosomal enzymes) and dies. A macrophage can phagocytize about 100 bacteria during its lifespan.

Some macrophages wander through the body's tissues, phagocytizing foreign matter (including bacteria) and, when appropriate, releasing antiviral agents (Fig. 43–2). Others stay in one place and destroy bacteria that pass by. For example, air sacs in the lungs contain large numbers of macrophages that destroy foreign matter entering with inhaled air.

Can bacteria counteract the phagocyte's attack? Some bacteria have cell walls or capsules that resist the action of lysosomal enzymes. Other bacteria release enzymes that destroy the membranes of the phagocyte's lysosomes. The powerful lysosomal enzymes then spill out into the cytoplasm and may destroy the phagocyte.

Natural killer (NK) cells are large, granular lymphocytes that originate in the bone marrow. They account for about 15% of circulating lymphocytes. When they were first identified, immunologists thought that NK cells functioned only against tumor cells. However, studies have demonstrated that these cells recognize and are active against a wide variety of targets, including cells infected with viruses, some bacteria, and some fungi. NK cells destroy target cells by both nonspecific and specific (antibody-requiring) processes. NK cells release cytokines as well as enzymes known as **perforins** and **granzymes** that destroy target cells.

NK activity is stimulated by several cytokines. When NK levels are high, resistance to certain cancers may be increased. Psychological stressors are thought to decrease NK cell activity and thus enhance tumor growth.

Inflammation is a protective response

The **inflammatory response (inflammation)** develops within a few hours after pathogen invasion or physical injury (Fig. 43–3). The clinical characteristics of inflammation are *heat, redness, edema,* and *pain.* Inflammation is regulated by proteins in the plasma, by cytokines, by substances released by platelets, by certain white blood cells called **basophils,** and by **mast cells,** large connective tissue cells filled with distinctive granules. *Bradykinin,* a peptide in the plasma, dilates blood vessels and increases capillary permeability. Platelets, basophils, and mast cells release *histamine* and *serotonin,* compounds that dilate blood vessels in the affected area and increase capillary permeability. Blood flow increases to the infected region, bringing great numbers of neutrophils and other phagocytic cells. The increased blood flow makes the skin feel warm and makes skin that contains little pigment appear red. Phagocytes migrate out of the capillaries and into the infected tissues. One of the main functions of inflammation appears to be increased phagocytosis.

Increased capillary permeability allows fluid and antibodies to leave the circulation and enter the tissues. As the volume of interstitial fluid increases, **edema** (swelling) occurs. The edema,

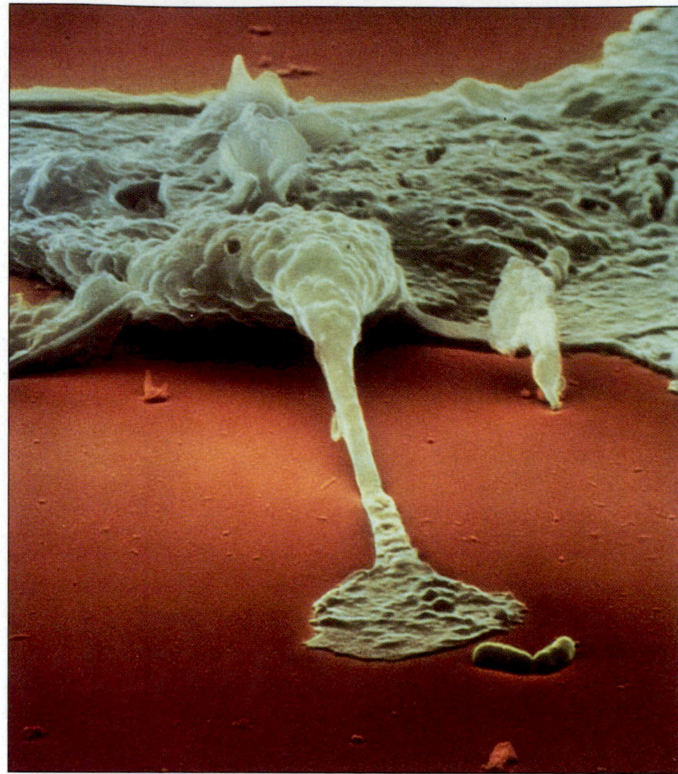

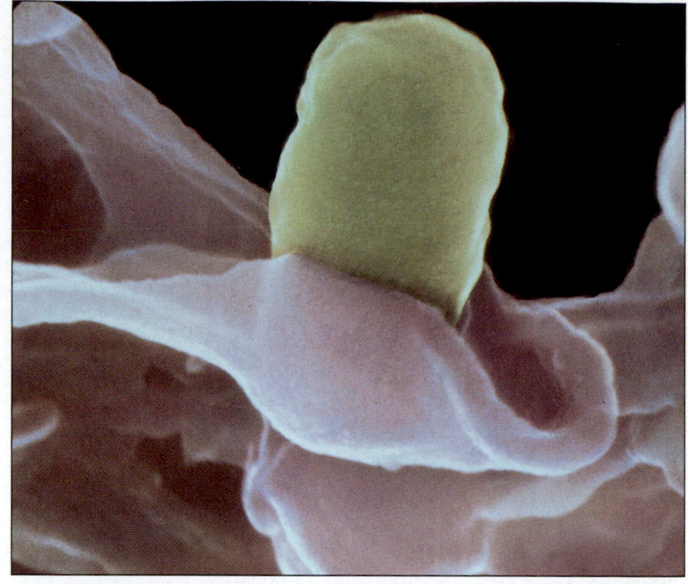

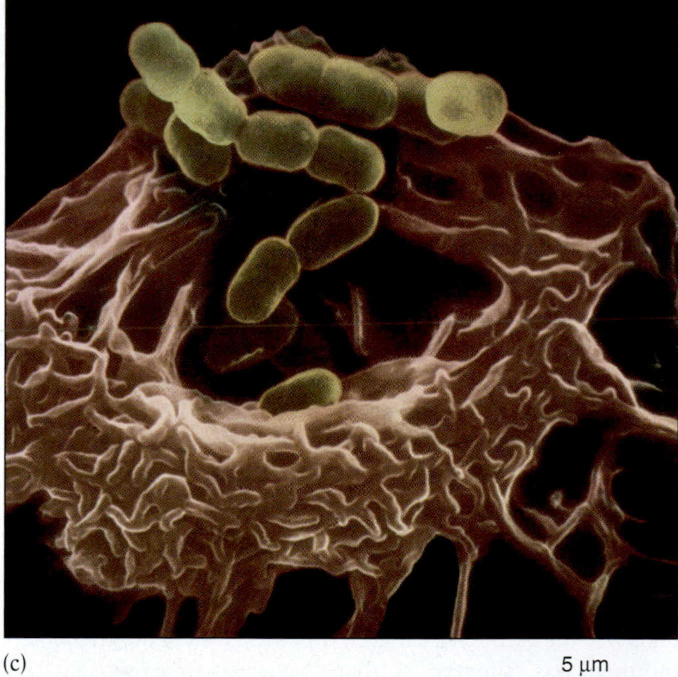

Figure 43–2 Phagocytosis. Color enhanced SEMs of macrophages, which are remarkably efficient warriors. **(a)** A macrophage extends a pseudopod toward an invading *Escherichia coli* bacterium that is already multiplying. **(b)** The bacterium is trapped within the engulfing pseudopod. **(c)** The macrophage takes in the trapped bacteria along with its own plasma membrane. The macrophage plasma membrane will seal over the bacteria, and powerful lysosomal enzymes will destroy them. *(Lennart Nilsson, Boehringer Ingelheim International GmbH)*

along with the action of certain enzymes in the plasma, causes the pain characteristic of inflammation.

Although inflammation is often a local response, sometimes the entire body is involved. **Fever** is a common clinical symptom of widespread inflammatory response. Macrophages and certain other cells release compounds, such as the cytokine IL-1, that reset the body's thermostat in the hypothalamus, resulting in fever. **Prostaglandins,** a group of local hormones derived from fatty acids (see Chapter 47), are also involved in this resetting process.

Fever helps the body fight infection. The increased body temperature interferes with the growth and replication of some microorganisms and may kill some pathogens. Fever also causes lysosomes to break down, destroying cells infected by viruses. In addition, increased temperature promotes the activity of certain lymphocytes (T cells) and the production of antibodies by other lymphocytes, and it increases phagocytosis. A short-term, low fever helps speed recovery.

SPECIFIC IMMUNE RESPONSES INCLUDE ANTIBODY-MEDIATED AND CELL-MEDIATED IMMUNITY

While nonspecific immune responses are destroying pathogens and preventing the spread of infection, the body is mobilizing its specific immune responses. Several days are required to activate specific immune responses, but, once in gear, these mechanisms are extremely effective. Two main types of specific immunity are **antibody-mediated immunity** and **cell-mediated immunity.**

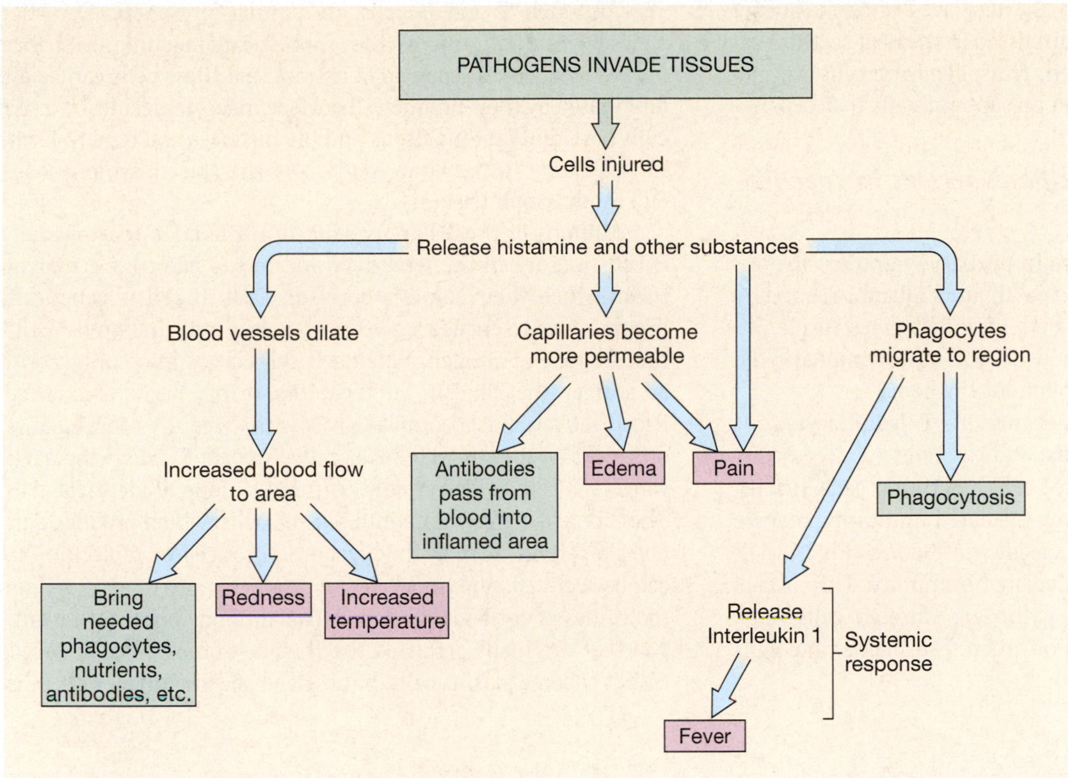

Figure 43-3 Inflammation. During inflammation, phagocytic cells, antibodies, and other needed compounds enter the tissue where pathogen invasion is taking place.

Many types of cells participate in specific immune responses

Among the many cell types that participate in specific immune responses are phagocytes, eosinophils, mast cells, dendritic cells, and lymphocytes. Phagocytes include neutrophils and macrophages. **Eosinophils** are white blood cells that release toxins from their granules to destroy parasitic worms. When basophils move out of the circulation into the tissues, they are called mast cells. These cells release histamine in response to antigens that cause allergic reactions. The main types of lymphocytes are T cells, B cells, and natural killer (NK) cells.

Macrophages are important in nonspecific and specific immune responses

When a macrophage ingests a bacterium, most, but not all, of the bacterial antigens are degraded by lysosomal enzymes. Fragments of the foreign antigens associate with a certain type of self-molecule (MHC, class II, discussed later in this chapter) and are then displayed on the surface of the macrophage. Thus, the macrophage is an **antigen-presenting cell (APC)** that displays foreign antigens as well as its own surface proteins. The foreign antigen–self-molecule combination activates certain T cells.

Macrophages secrete about 100 different compounds, including interferons and enzymes that destroy bacteria. When macrophages are stimulated by bacteria, they secrete interleukins, which activate B cells and certain T cells. Interleukins also promote nonspecific immune responses, causing fever and activating other mechanisms that defend the body against invasion.

Dendritic cells activate T cells

Although **dendritic cells** were first identified in 1973, their small numbers made them difficult to study until more sophisticated techniques became available about 20 years later. Their name stems from their long cytoplasmic processes known as dendrites (not to be confused with the dendrites of neurons). Dendritic cells are specialized to capture antigens. They are strategically stationed in the skin (where they are known as Langerhans cells) and in the linings of the digestive, respiratory, urinary, and vaginal passageways into the body.

When pathogens infect a tissue, dendritic cells recognize molecular patterns characteristic of microbial macromolecules. Dendritic cells then capture the pathogens (or their products) by phagocytosis or by receptor-mediated endocytosis (see Chapter 5). Guided by **chemokines** (a group of cytokines) expressed in the lymph nodes and lymph vessels, dendritic cells make their way to the lymph nodes. As they migrate, dendritic cells mature. Like macrophages, mature dendritic cells break down antigens and display the fragments on the cell surface. These cells also express certain additional signaling molecules on their surfaces. These signaling molecules, along with the displayed antigen, attract and activate the specific T cells (discussed in the next section) capable of responding to the antigen. Thus, dendritic cells are specialized to process, transport, and present antigens to T cells.

Lymphocytes are the principal warriors in specific immune responses

Lymphocytes develop and mature in primary lymph organs, including the bone marrow and the **thymus gland.** Secondary lymph organs, where large numbers of lymphocytes reside, include the spleen, lymph nodes, tonsils, and other lymphatic tissues strategically positioned throughout the body.

Three main types of lymphocytes are **T lymphocytes,** or **T cells; B lymphocytes,** or **B cells;** and NK cells. As already discussed, NK cells kill virally infected cells and tumor cells. B cells, which are responsible for antibody-mediated immunity, mature into **plasma cells** that produce specific antibodies (Fig. 43–4). T cells are responsible for cell-mediated immunity. They attack body cells infected by invading pathogens, foreign cells (e.g., those introduced in tissue grafts or organ transplants) and cells altered by mutation (cancer cells).

Although T and B cells are similar in appearance when viewed with a light microscope, sophisticated techniques such as fluorescence microscopy demonstrate that these cells can be differentiated by their unique cell surface macromolecules. T and B cells have different functions and life histories and tend to locate in (or "home" to) separate regions of the spleen, lymph nodes, and other lymph tissues.

Millions of B cells are produced in the bone marrow daily. B cells mature in the fetal liver and in the adult bone marrow (from which their name is derived). Each B cell is genetically programmed to encode a glycoprotein receptor that binds with a specific type of antigen. When a B cell comes into contact with an antigen that binds to its receptors, it may become activated. B-cell activation is a complex process that requires participation of T cells of a particular type, called *helper T cells.* Once activated, a B cell divides rapidly, forming a clone of identical cells. These B cells differentiate into plasma cells, which produce antibody, a soluble form of the glycoprotein receptor molecule that can be secreted. A plasma cell can produce more than 10 million molecules of antibody per hour! The antibody binds to the antigen that originally activated the B cells. Some activated B cells do not become plasma cells, but instead become **memory B cells,**

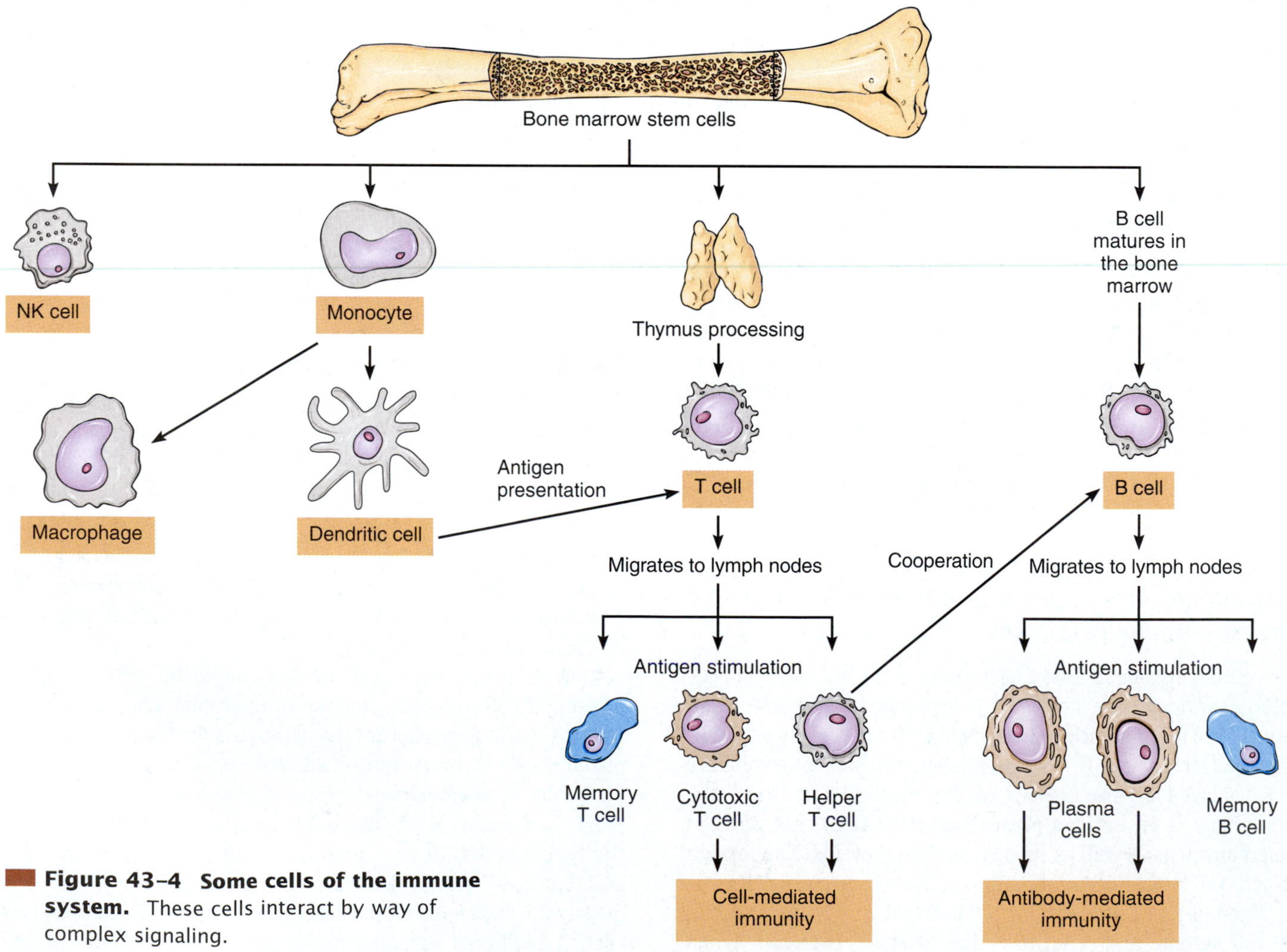

Figure 43–4 Some cells of the immune system. These cells interact by way of complex signaling.

which continue to produce small amounts of antibody after an infection has been overcome.

Like B cells, T cells originate from stem cells in the bone marrow. On their way to the lymph tissues, the future T cells stop off in the **thymus gland** for processing. (The "T" in T cells stands for *thymus-derived.*) The thymus makes T cells **immunocompetent,** that is, capable of immunological response (discussed in a later section). As T cells move through the thymus, they divide many times and develop specific surface proteins with distinctive receptor sites. Only the T cells that have specific receptors are selected to divide. This is a form of **positive selection.**

T cells that react to self-antigens undergo **apoptosis,** or programmed cell death (see Chapter 5). This is a form of **negative selection.** Immunologists estimate that more than 90% of developing T cells are negatively selected. The remaining T cells differentiate and leave the thymus to take up residence in other lymph tissues or to launch immune responses in infected tissues. By selecting only appropriate T cells, the thymus gland ensures that T cells can distinguish between the body's own antigens and foreign antigens.

The differentiation of the majority of T cells within the thymus is thought to take place just before birth and during the first few months of postnatal life. If the thymus is removed before this processing takes place, an animal is not able to develop cellular immunity. If the thymus is removed after that time, cellular immunity is less seriously impaired.

T cells are distinguished by the **T-cell antigen receptor (TCR)** that allows T cells to recognize specific antigens. Two main types of T cells have been identified: CD8 T cells and CD4 T cells. **CD8 T cells,** which have a surface marker (a protein) designated CD8, include **cytotoxic T cells (T_c),** known less formally as *killer T cells.* Cytotoxic T cells recognize and destroy cells with foreign antigens on their surfaces. Among their target cells are virus-infected cells, cancer cells, and foreign tissue grafts. T_c cells kill their target cells by releasing a variety of cytokines and enzymes that lyse cells.

The second main type of T cells are known as **CD4 T cells** because they have a surface marker designated CD4. They are also known as **helper T cells (T_h).** These cells secrete substances that activate or enhance immune responses. Two subsets of helper T cells have been identified: T helper 1 (T_h1) cells and T helper 2 (T_h2) cells. They secrete different types of cytokines and have different functions. T_h1 cells mainly promote cell-mediated immune responses. T_h2 cells stimulate B cells to divide and produce antibodies, and thus function mainly in antibody-mediated immunity. The balance between T_h1 and T_h2 cells appears to be important in effective response to infection. Recent studies suggest that dendritic cells may determine which helper T cell subset is produced. If the relative proportion of T_h1 and T_h2 cells could be shifted to obtain the necessary balance, a more positive outcome might occur in diseases such as AIDS.

T cells produce many cytokines important in immune responses. Some of the cytokines affect T cell development, others B cell development, and still others influence the action of macrophages. T_h1 cells produce IL-2, which stimulates development of NK cells, T cells, and B cells. Efforts have been made to treat people with depressed immune systems with IL-2 to increase their responses to cancer. Unfortunately, this type of therapy tends to be fairly toxic to the patient.

T_h2 cells, B cells, and macrophages produce IL-4. Whereas IL-2 enhances cell-mediated immunity, IL-4 tends to enhance antibody-mediated immunity. Another cytokine, IL-12, is thought to enhance T_h1 action and therefore IL-2 production, as well as the action of cytotoxic T cells. Such interactions of cytokine activity provide mechanisms for checks and balances in this very complex system. Investigators continue to identify cytokines produced by T cells and other cells of the immune system.

Both helper T cells and cytotoxic T cells can suppress immune responses. Both populations of T cells also include memory T cells.

The major histocompatibility complex permits recognition of self

The ability of the vertebrate immune system to distinguish self from nonself depends largely on a group of cell surface proteins, known as **MHC antigens.** These antigens are coded for by a set of closely linked genes known as the **major histocompatibility complex (MHC).** In humans, the MHC is called the **HLA** (human leukocyte antigen) group. These genes are polymorphic (variable). Within the population there are multiple alleles for each locus, sometimes more than 40 alleles for a given gene. As a result, the cell surface proteins for which they code are generally different in each individual. With so many possible combinations, no two people, except identical twins, are likely to have all of the same MHC proteins on their cells. The more closely related two individuals are, the more MHC genes they have in common. Thus, the MHC proteins of an individual are a biochemical "fingerprint."

The MHC is divided into three groups of genes that code for distinct sets of proteins. These proteins differ in terms of tissue distribution and chemical structure. MHC class I antigens are found on most nucleated cells and are important in distinguishing between self and nonself. They bind foreign antigens produced by viruses or foreign cells (e.g., cells of foreign tissue grafts). The foreign antigen–MHC complexes thus formed are displayed on the cell surface and can be recognized by cytotoxic T cells.

MHC class II antigens are found primarily on cells of the immune system, particularly B cells, macrophages, some T cells, and dendritic cells. MHC class II antigens regulate the interactions among T cells, B cells, and antigen-presenting cells. These antigens bind peptide fragments of proteins that have entered the cell via foreign sources such as bacteria and have been degraded. The foreign antigen–MHC complex is displayed on the cell surface and stimulates helper T cells. MHC class III proteins include components of the complement system.

ANTIBODY-MEDIATED IMMUNITY IS A CHEMICAL WARFARE MECHANISM

B cells are responsible for antibody-mediated immunity (also called humoral immunity). A given B cell can produce many copies of one specific antibody. Recall that antibody molecules serve as cell surface receptors that combine with antigens. Only a B cell displaying a matching receptor on its surface can bind a particular antigen. This binding activates the B cell.

In most cases, activation of B cells is a complex process that involves dendritic cells or macrophages and helper T cells (Fig. 43–5). Dendritic cells capture antigens in the tissues and migrate to the lymph nodes or spleen, where they present antigen fragments to helper T cells. Macrophages also display fragments of antigens from pathogens they have engulfed. The foreign antigen forms a complex with MHC class II molecules of the macrophage. This foreign antigen–MHC complex is then displayed on the cell surface.

When an APC displaying a foreign antigen–MHC complex contacts a helper T cell with complementary T cell receptors, a complicated interaction occurs. Multiple chemical signals are sent back and forth between cells. For example, the macrophage secretes interleukins, such as IL-1, that activate helper T cells. T cells do not recognize an antigen that is presented alone. Helper T cells require that the antigen be presented as part of a foreign antigen-MHC class II complex on the surface of an APC.

The antibody receptor of a B cell binds with complementary antigen. Inside the B cell, the antigen is degraded into peptide fragments. The B cell then displays the peptide fragments together with MHC protein class II on its surface. An activated helper T cell binds with the foreign antigen-MHC complex on the B cell. Thus, the B cell can serve as an APC to T cells. An activated helper T cell then releases interleukins, which, together with antigen, activate the B cell.

Once activated, a B cell increases in size. It then divides by mitosis, giving rise to a clone of identical cells (Fig. 43–6). This cell division in response to a specific antigen is known as **clonal selection** (discussed in a later section). Each B cell of the clone makes antibodies specific to the specific antigen that activated the original B cell. It is important to remember that the specificity of the clone is determined *before* the B cell encounters the antigen.

Some cells of the B cell clone mature into plasma cells that secrete the type of antibody specific to the antigen. Unlike T cells, most plasma cells do not leave the lymph nodes. Only the antibodies they secrete pass out of the lymph tissues and make their way via the lymph and blood to the infected area. This sequence is summarized on page 949, right text column.

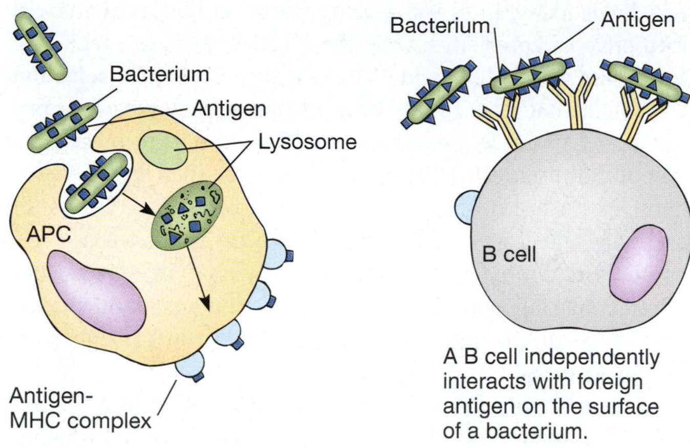

(a) An APC degrades antigen and displays it in combination with MHC class II.

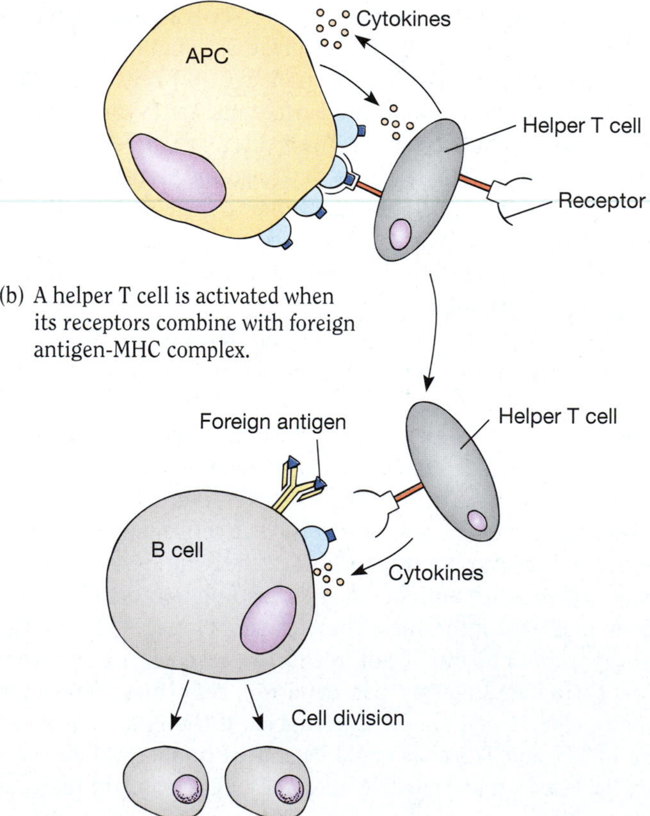

(b) A helper T cell is activated when its receptors combine with foreign antigen-MHC complex.

(c) The activated helper T cell recognizes antigen-MHC complex on a B cell. The T cell secretes cytokines that can activate a B cell. The activated B cell then divides, forming a clone of competent B cells.

Figure 43–5 B-cell activation. Antigen presentation and helper T cells are required. **(a)** An antigen presenting cell (APC), such as a macrophage, takes in a bacterium or foreign antigen, breaks it down, and presents foreign antigens on its surface in combination with MHC antigens. B cells can combine with specific antigens. However, B cells are typically not active until stimulated by helper T cells. **(b)** Helper T cells are activated when their receptors combine with foreign antigen–MHC complexes on an APC. Complex signaling takes place via cytokines secreted by both cells. **(c)** The activated helper T cell secretes cytokines that can activate a B cell. The B cell then divides, forming a clone of identical B cells.

Pathogen bearing foreign antigens invades body ⟶ APC phagocytizes pathogen ⟶ foreign antigen – MHC complex displayed on APC surface ⟶ helper T cell binds with foreign antigen – MHC complex ⟶ activated helper T cell interacts with a B cell that displays the same complex ⟶ B cell activated ⟶ clone of B cells produced ⟶ B cells differentiate, becoming plasma cells ⟶ plasma cells secrete antibodies ⟶ antibodies form complexes with pathogen ⟶ antibody complexes trigger processes leading to pathogen destruction

Some activated B cells do not differentiate into plasma cells but instead become **memory B cells.** In these cells, a "survival gene" is activated that permits them to prevent *apoptosis,* the programmed cell death that is the eventual fate of plasma cells. Memory cells continue to live and produce small amounts of antibody long after an infection has been overcome. This antibody, part of the gamma globulin fraction of the plasma, becomes part of the body's arsenal of chemical weapons. Should the same pathogen enter the body again, this circulating antibody immediately targets it for destruction. At the same time, specific memory cells are stimulated to divide, producing new clones of the appropriate plasma cells. The presence of circulating antibodies can be used clinically to detect previous exposure to specific pathogens. For example, HIV screening tests measure antibody to the virus that causes AIDS.

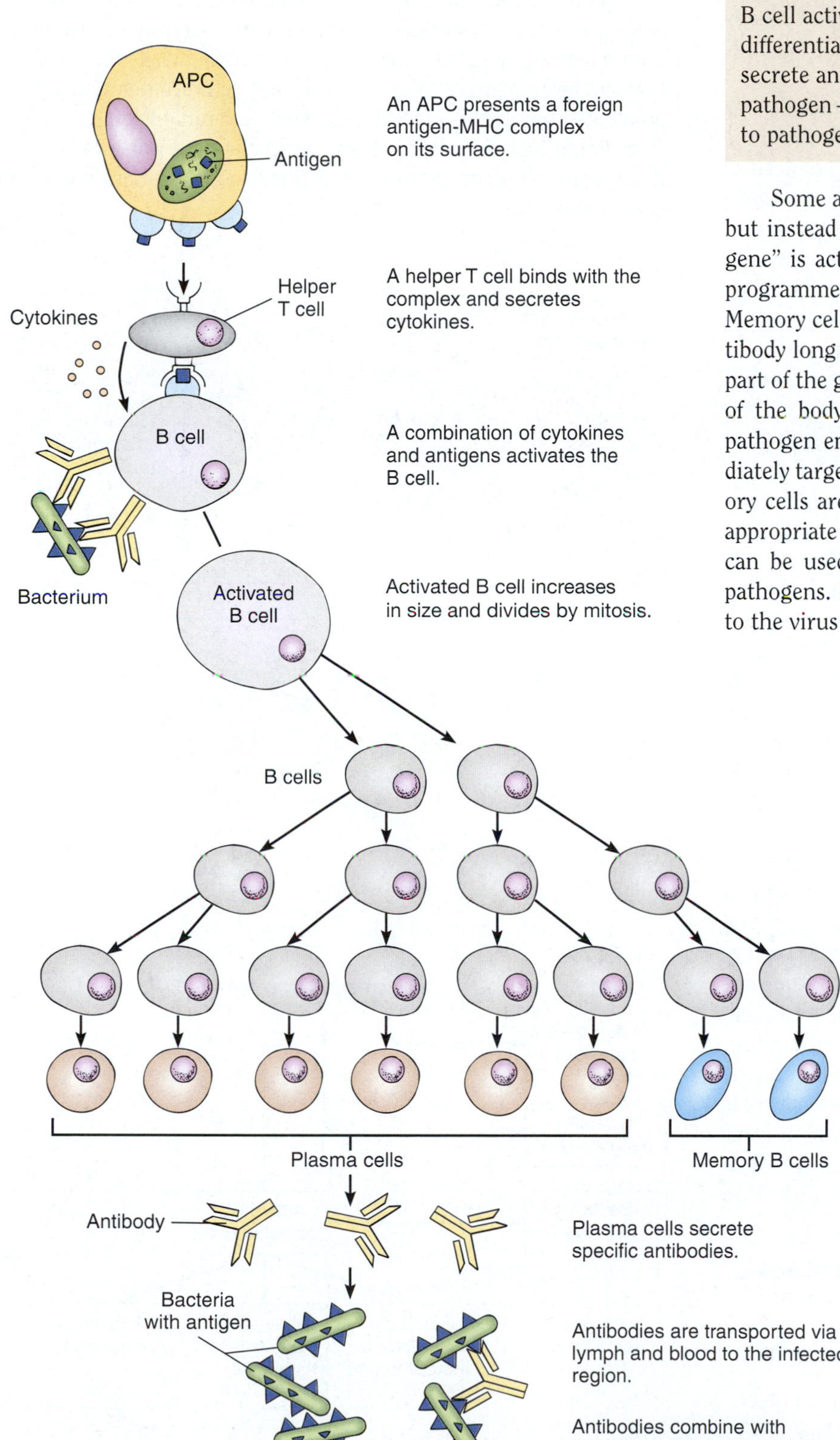

APC
Antigen
An APC presents a foreign antigen-MHC complex on its surface.
Cytokines
Helper T cell
A helper T cell binds with the complex and secretes cytokines.
B cell
Bacterium
A combination of cytokines and antigens activates the B cell.
Activated B cell
Activated B cell increases in size and divides by mitosis.
B cells
A clone of competent B cells is produced.
B cells differentiate into plasma cells and memory B cells.
Plasma cells
Memory B cells
Antibody
Plasma cells secrete specific antibodies.
Bacteria with antigen
Antibodies are transported via lymph and blood to the infected region.
Antibodies combine with antigens on the surface of the pathogen to form antigen-antibody complexes.
Antigen-antibody complex

Figure 43–6 Antibody-mediated immunity. A specific B cell becomes activated when it binds with a specific antigen and when an activated helper T cell releases cytokines. Once activated, the B cell divides, producing a clone. Many of these cells differentiate and become plasma cells that secrete antibodies. The plasma cells remain in the lymph tissues, but the antibodies are transported to the site of infection by the blood or lymph.

A typical antibody consists of four polypeptide chains

An antibody molecule, also called **immunoglobulin** or **Ig,** has two main functions: It combines with antigen, and it activates processes that destroy the antigen that binds to it. For example, the antibody may stimulate phagocytosis. Note that the antibody does not destroy the antigen directly. Rather, it *labels* the antigen for destruction.

The basic structure of the immunoglobulin molecule was clarified by the work of Rodney Porter and Gerald Edelman in the 1960s. Porter used the plant enzyme papain to split Ig molecules into fragments. Based on his findings, Porter developed a working model of the structure of the Ig molecule and was the first to suggest that it is Y-shaped. Two of the fragments retain the ability to bind antigen and are referred to as **Fab** fragments (for antigen-binding fragment). The fragment that interacts with cells of the immune system is the **Fc** fragment (the c indicates that this fragment crystallizes during cold storage). Many cells of the immune system have Fc receptors.

A typical antibody is a Y-shaped molecule in which the two arms of the Y (the Fab portion) bind with antigen (Fig. 43–7). This shape permits the antibody to combine with two antigen molecules and allows formation of **antigen-antibody complexes.** While the arms of the Y bind to antigen, the tail of the Y, the Fc portion, interacts with cells of the immune system, such as phagocytes, or binds with molecules of the complement system.

The antibody molecule consists of four polypeptide chains: two identical long chains called heavy chains, and two identical short chains called light chains. Each chain has a constant segment, a junctional segment, and a variable segment. In the **constant segment,** or **C region,** of the heavy chains, the amino acid sequence is constant within a particular immunoglobulin class. The C region may be thought of as the handle portion of a door key. Like the elongated part of a key that slides into a lock, the amino acid sequence of the **junctional segment,** or **J region,** is somewhat variable. Finally, like the pattern of bumps and notches at the end of a key, the **variable segment,** or **V region,** has a unique amino acid sequence. In B cell receptors, the variable region of the immunoglobulin protrudes from the B cell, whereas the constant region anchors the molecule to the cell.

The V region is the part of the key that is unique for a specific antigen (the lock). At its variable regions, the antibody folds three-dimensionally, assuming a shape that enables it to combine with a specific antigen. When they meet, antigen and antibody fit together somewhat like a lock and key. They must fit in just the right way, though not perfectly, for the antibody to be effective (Fig. 43–8). A given antibody can bind with different strengths, or **affinities,** to different antigens. In the course of an immune response, stronger (higher affinity) antibodies are generated.

In an antigen that is a protein, specific sequences of amino acids make up an **antigenic determinant,** or **epitope.** These amino acids give part of the antigen molecule a specific shape that can be recognized by an antibody or T cell receptor. Usually, an antigen has many different antigenic determinants on its surface. Some have hundreds. These antigenic determinants may differ from one another, so several different kinds of antibodies can combine with a single antigen.

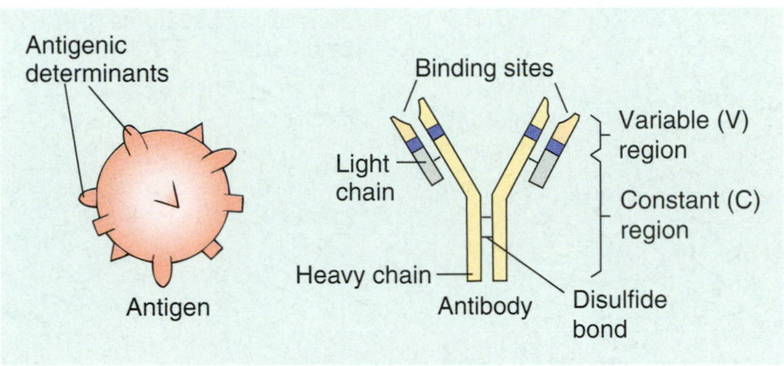

(a)

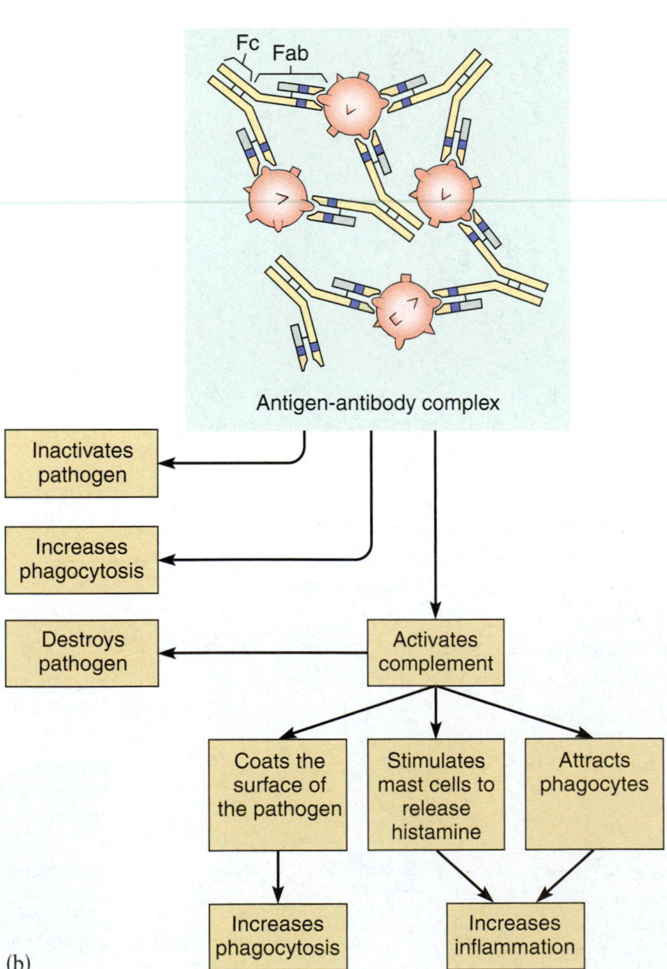

(b)

Figure 43–7 Structure and functions of antibodies. Antibodies combine with antigens, forming antigen-antibody complexes. **(a)** The antibody molecule is composed of two light chains and two heavy chains, joined by disulfide bonds. The constant (C) and variable (V) regions of the chains are labeled. **(b)** The Fab part of the antibody binds to antigen. The Fc part of the molecule binds with cells of the immune system. Antigen-antibody complexes directly inactivate pathogens and increase phagocytosis. They also activate the complement system.

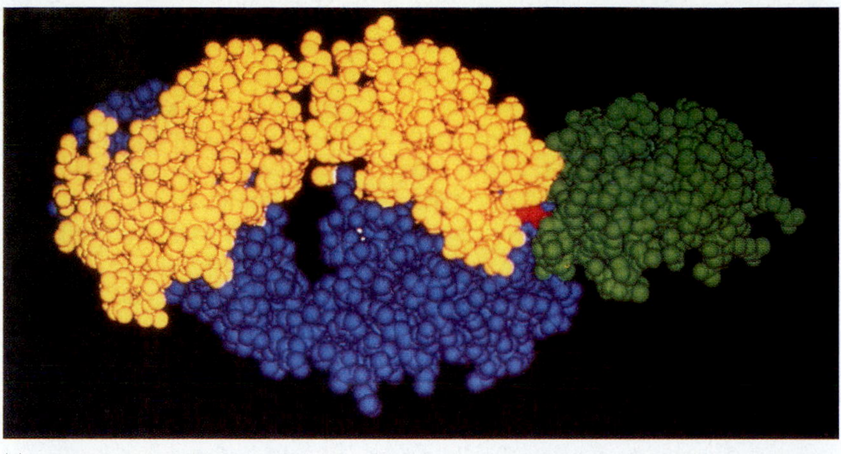

(a)

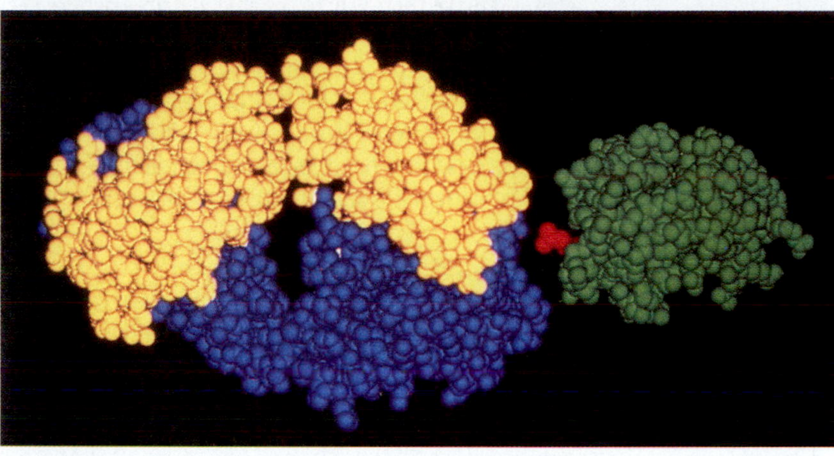

(b)

Figure 43–8 Antigen-antibody complex. The components of an antigen-antibody complex fit together as shown in this computer simulation of their molecular structure. The antigen lysozyme is shown in green, the heavy chain of the antibody is shown in blue, and the light chain is shown in yellow. **(a)** A portion of the antigenic determinant, shown in red, fits into a groove in the antibody molecule. **(b)** The antigen-antibody complex has been pulled apart to show its structure.

Antibodies are grouped in five classes

Antibodies are grouped in five classes defined by unique amino acid sequences in the constant region of the heavy chains. Using the abbreviation Ig for immunoglobulin, the classes are designated IgG, IgM, IgA, IgD, and IgE. In humans, about 75% of the antibodies belong to the **IgG** class; these are part of the gamma globulin fraction of the plasma. IgG and **IgM** interact with macrophages and activate the complement system. They defend against many pathogens carried in the blood, including bacteria, viruses, and some fungi. **IgA,** present in mucus, tears, saliva, and milk, prevents viruses and bacteria from attaching to epithelial surfaces. This immunoglobulin, which defends against inhaled or ingested pathogens, is strategically secreted into the respiratory passageways, digestive tract, urinary tract, and reproductive tract.

IgD has a low concentration in the plasma. Along with IgM, it is an important immunoglobulin on the B cell surface. IgD helps activate B cells following antigen binding. **IgE** also has a low plasma concentration; it can bind to *mast cells,* connective tissue cells that contain potent signaling molecules such as *histamine.* When an antigen binds to IgE on a mast cell, these molecules are released. Histamine triggers many allergy symptoms, including inflammation. IgE is also responsible for immunity to invading parasitic worms.

The binding of antibody to antigen activates other defense mechanisms

Antibodies mark a pathogen as foreign by combining with an antigen on its surface. Generally, several antibodies bind with several antigens, creating a mass of clumped antigen-antibody complexes. The combination of antigen and antibody activates several defense mechanisms:

1. The antigen-antibody complex may inactivate the pathogen or its toxin. For example, when an antibody attaches to the surface of a virus, the virus may lose its ability to attach to a host cell.
2. The antigen-antibody complex stimulates phagocytic cells to ingest the pathogen.
3. Antibodies of the IgG and IgM groups work mainly through the *complement system* (discussed earlier in this chapter). When antibodies combine with a specific antigen on a pathogen, complement proteins destroy the pathogens. IgG molecules have an Fc fragment that binds Fc receptors,

expressed on most immune cells. When the Fc part of an antibody molecule that has bound to a pathogen binds with an Fc receptor on a phagocyte, the pathogen is more easily destroyed.

Process of Science — Monoclonal antibodies are highly specific

Before 1975, the only method for obtaining antibodies for medicine or research was immunizing animals and collecting their blood. Then, Cesar Milstein and Georges Kohler at the Laboratory of Molecular Biology in Cambridge, England, developed **monoclonal antibodies,** identical antibodies produced by cells cloned from a single cell. One way to produce monoclonal antibodies in the laboratory is to inject mice with the antigen of interest, for example, antigens from a particular bacterium. After the mice have produced antibodies to the antigen, their B cells are collected. However, these cells survive in culture for only a few generations. In contrast, cancer cells can live and divide in tissue culture indefinitely. The B cells can be suspended in a culture medium together with lymphoma cells from other mice. (Lymphoma is cancer of lymphocytes.) The B cells and lymphoma cells can be induced to fuse. They form hybrid cells, known as **hybridomas,** that have properties of the two "parent" cells. Hybridomas can be cultured indefinitely (a cancer cell property) and continue to secrete antibodies (a B cell property).

Researchers select hybrid cells that are manufacturing the specific antibody needed and then clone them in a separate cell culture. Cells of this clone secrete large amounts of the specific antibody, thus, the name monoclonal antibodies. Each type of monoclonal antibody is specific for a single antigenic determinant. A recently developed method for producing monoclonal antibodies uses bacteriophages rather than mouse cells.

Because of their purity and specificity, monoclonal antibodies have proved to be invaluable tools in modern biology. For example, a researcher may want to detect a particular molecule even when it is present in very small amounts in a mixture. The reaction of that molecule (the antigen) with a specific monoclonal antibody makes its presence known. Monoclonal antibodies are used in similar ways in various diagnostic tests. For example, the commonly used home pregnancy tests make use of a monoclonal antibody that is specific for human chorionic gonadotropin (hCG), a hormone produced by a developing human embryo (see Chapters 48 and 49). Therapeutic uses for monoclonal antibodies are being developed. For example, Herceptin is a monoclonal antibody used to treat a form of breast cancer.

CELL-MEDIATED IMMUNITY PROVIDES CELLULAR WARRIORS

The T cells and APCs (mainly dendritic cells and macrophages) are responsible for cell-mediated immunity (Fig. 43–9). T cells destroy cells infected with viruses and cells that have been altered in some way, such as cancer cells. They also destroy foreign grafts such as transplanted kidneys. There are thousands of different populations of T cells. Each T cell has more than 50,000 identical receptors (TCRs) that bind to one specific type of antigen. Like B cells, T cells are clonal.

How do T cells know which cells to attack? T cells do not recognize antigens unless they are presented properly. For example, when a virus infects a cell, some of the viral protein is broken down to peptides and displayed with MHC class I molecules on the cell surface of an APC, typically a dendritic cell. Only T cells with receptors that bind to the specific antigen presented together with the MHC receptor become activated. Generally, fewer than 1 in 10,000 T cells can respond. Helper T cell activation is thought to be necessary for activating many aspects of the immune response.

Once activated, a T cell increases in size and gives rise to a clone of helper T cells, cytotoxic T cells, and memory T cells. Cytotoxic T cells make up the cellular infantry. They leave the lymph nodes and make their way to the infected area, where they destroy target cells within seconds after contact.

After a cytotoxic T cell combines with antigen on the surface of the target cell, it secretes granules containing proteins that destroy the target cell. After releasing these cytotoxic substances, the T cell disengages itself from its victim cell and seeks out a new target. This sequence is summarized as follows:

> Virus invades body cell ⟶ foreign antigen–MHC class I complex displayed on cell surface of APC ⟶ specific T cell activated by this complex ⟶ clone of T cells produced; some become cytotoxic T cells ⟶ cytotoxic T cells migrate to area of infection ⟶ cytotoxic T cells release proteins that destroy target cells

Helper T cells and macrophages at the site of infection secrete interleukins that help regulate immune function. For example, some interleukins stimulate activated T cells to divide. Others enhance inflammation, attracting great numbers of macrophages to the site of infection. Experimental evidence suggests that both helper and cytotoxic T cells can, under some conditions, suppress immune responses. They may act by destroying APCs.

THE IMMUNE SYSTEM RECOGNIZES AND RESPONDS TO MILLIONS OF DIFFERENT ANTIGENS

Process of Science

How can the immune system recognize every possible antigen, even those produced by newly mutated viruses never before encountered during the evolution of our species? In the 1920s and 1940s several hypotheses, known as *instructive theories,* attempted to explain antibody diversity. These hypotheses suggested that an antigen serves as a template for an antibody that shapes itself to conform to this template. Thus, the antigen *instructs* the antibody. However, no experimental data were provided, and as the field of molecular biology expanded, these hypotheses could not be reconciled with new knowledge. For example, the central dogma held that information is transmitted from DNA to RNA to protein. The instructive theories required

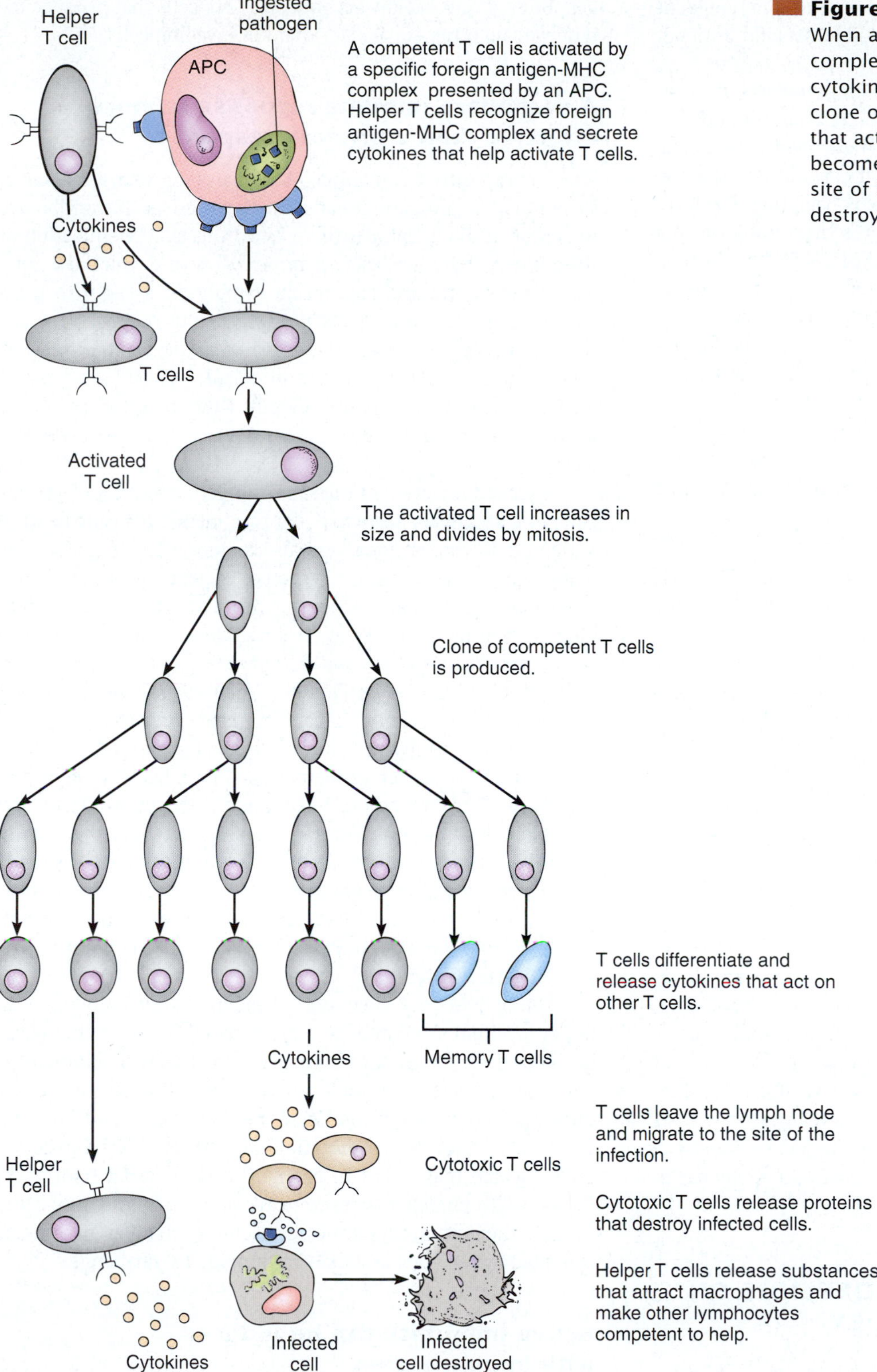
Helper
T cell
Ingested
pathogen
APC
A competent T cell is activated by a specific foreign antigen-MHC complex presented by an APC. Helper T cells recognize foreign antigen-MHC complex and secrete cytokines that help activate T cells.
Cytokines
T cells
Activated
T cell
The activated T cell increases in size and divides by mitosis.
Clone of competent T cells is produced.
T cells differentiate and release cytokines that act on other T cells.
Cytokines
Memory T cells
T cells leave the lymph node and migrate to the site of the infection.
Helper
T cell
Cytotoxic T cells
Cytotoxic T cells release proteins that destroy infected cells.
Helper T cells release substances that attract macrophages and make other lymphocytes competent to help.
Cytokines
Infected
cell
Infected
cell destroyed

Figure 43-9 Cell-mediated immunity. When activated by a foreign antigen–MHC complex presented by an APC and by cytokines, a T cell divides, giving rise to a clone of cells. Helper T cells release cytokines that act on other T cells. Some of these cells become cytotoxic T cells, which migrate to the site of infection and release proteins that destroy invading pathogens.

the information to flow backward, and so these hypotheses were discarded.

In the 1950s Niels Jerne, David Talmadge, and Frank Macfarlane Burnet developed the **clonal selection** hypothesis suggesting that before a lymphocyte ever encounters an antigen, the lymphocyte has specific receptors for that antigen on its surface. When an antigen binds to a matching receptor on a lymphocyte, it activates the lymphocyte, which then divides and gives rise to a clone of cells with identical receptors. A major problem with this hypothesis was that it suggested that our cells must contain millions of separate antibody genes, each coding for an antibody with a different specificity. Although each human cell has a large amount of DNA, it is not enough to provide a different gene to code for each of the millions of possible specific antibody molecules.

In 1965 W.J. Dreyer and J.C. Bennett suggested that the constant (C) region and the variable (V) region of an immunoglobulin are encoded by two separate genes. Their hypothesis was at first rejected by many biologists because it contradicted the prevailing theory that one gene codes for one polypeptide. The technology needed to test Dreyer and Bennett's hypothesis was not immediately available, and it was not until 1976 that Susumu Tonegawa and his colleagues demonstrated that separate genes encode the V and C regions of immunoglobulins. Tonegawa further showed that three separate families of genes code for immunoglobulins and that each gene family contains a large number of DNA segments that code for V regions. Recombination of these DNA segments during the differentiation of B cells is responsible for antibody diversity. Tonegawa was awarded the Nobel prize in 1987 for his work, which transformed the emerging science of immunology.

We now understand that in undifferentiated B cells, gene segments are present for a number of different V regions, for one or more junction (J) regions, and for one or more different C regions (Fig. 43–10). Recombination of these DNA segments can produce an enormous number of potential combinations! Millions of different types of B (and T) cells are produced. By chance, one of those cells may produce just the right antibody to destroy a pathogen that invades the body. To appreciate the capacity for antibody diversity, consider the diverse combinations of things you create in your everyday life. A familiar example is making an ice cream sundae. Imagine the varied combinations that are possible using ten flavors of ice cream, six types of sauce, and 15 kinds of toppings.

Additional sources of antibody diversity are known. For example, the DNA of the mature B cells that codes for the variable regions of immunoglobulins mutates very readily. These somatic mutations produce genes that code for slightly different antibodies.

We have used antibody diversity here as an example, but similar genetic mechanisms account for the diversity of T cell receptors. Although we may actually use only a relatively few types of antibodies or T cells in a lifetime, the remarkable diversity of the immune system prepares it to attack most potentially harmful antigens that may invade the body.

LONG-TERM IMMUNITY DEPENDS ON IMMUNOLOGICAL MEMORY

Memory B and memory T cells are responsible for long-term immunity. They enable the body to launch more effective immune responses, thus providing immunization against disease. Recently, researchers have tracked memory T cells and reported that following an immune response, these cells group strategically in many nonlymphatic tissues, including the lung, liver, kidney, and gut. Many infections occur at such sites, so T cells already stationed there can respond quickly. In response to antigen, these T cells rapidly become cytotoxic cells that produce interferon and other substances that kill invading cells.

A secondary immune response is more effective than a primary response

The first exposure to an antigen stimulates a **primary response.** Injection of an antigen into an animal causes specific antibodies to appear in the blood plasma in 3 to 14 days. After injection of the antigen, there is a brief *latent period* during which the antigen is recognized and appropriate lymphocytes begin to form clones. Then there is a *logarithmic phase,* during which the antibody concentration rises rapidly for several days until it reaches a peak (Fig. 43–11). IgM is the principal antibody synthesized during the primary response. Finally, there is a decline phase, during which the antibody concentration decreases to a very low level.

A second injection of the same antigen, even years later, results in a **secondary response.** Because memory B cells bearing antibodies to that antigen (and also memory T cells) persist for many years, the secondary response is generally much more rapid than the primary response, with a shorter latent period. Much less antigen is necessary to stimulate a secondary response than a primary response, and more antibodies are produced. In addition, the affinity of antibodies is generally much higher. The predominant antibody in a secondary response is IgG.

The body's ability to launch a rapid, effective response during a second encounter with an antigen explains why we do not usually suffer from the same infectious disease several times. For example, most persons contract measles or chickenpox only once. When exposed a second time, the immune system responds quickly, destroying the pathogens before they have time to multiply and cause symptoms of the disease. Booster shots of vaccine are given to elicit a secondary response, thus reinforcing immunological memory.

You may wonder, then, how a person can get influenza (the flu) or a cold more than once. Unfortunately, there are many varieties of these diseases, each caused by a virus with slightly different antigens. For example, more than 100 different viruses cause the common cold, and new varieties of cold and flu virus evolve continuously by mutation (a survival mechanism for them), which may result in changes in their surface antigens. Even a slight change may prevent recognition by memory cells. Because the immune system is so specific, each different antigen is treated by the body as a new immunological challenge.

Active immunity can be induced with immunization

We have been considering **active immunity,** immunity developed following exposure to antigens. When you have chickenpox as a young child, for example, you develop immunity that protects you from contracting it again. Active immunity can be *naturally* or *artificially* induced (Table 43–1). If someone with chickenpox sneezes near you and you contract the disease, you develop active immunity naturally. Active immunity can also be artificially in-

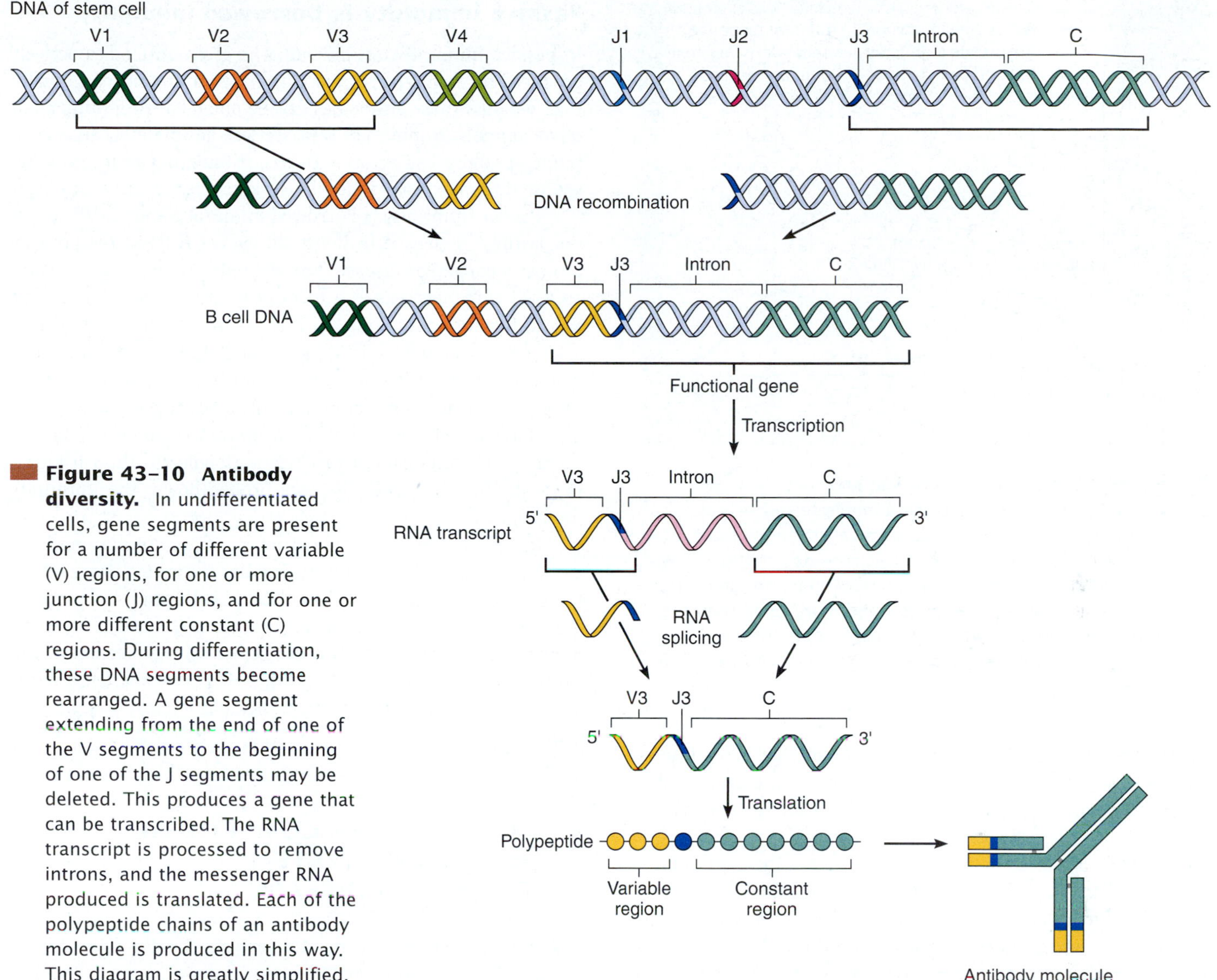

Figure 43–10 Antibody diversity. In undifferentiated cells, gene segments are present for a number of different variable (V) regions, for one or more junction (J) regions, and for one or more different constant (C) regions. During differentiation, these DNA segments become rearranged. A gene segment extending from the end of one of the V segments to the beginning of one of the J segments may be deleted. This produces a gene that can be transcribed. The RNA transcript is processed to remove introns, and the messenger RNA produced is translated. Each of the polypeptide chains of an antibody molecule is produced in this way. This diagram is greatly simplified.

duced by **immunization,** that is, by exposure to a **vaccine.** When an effective vaccine is introduced into the body, the immune system actively develops clones of cells, produces antibodies, and develops memory cells.

Process of Science The term *vaccination* is derived from the first vaccine, prepared in 1796 by Edward Jenner against vaccinia, the cowpox virus. This vaccine provided humans with immunity to smallpox, a deadly disease. Jenner had no knowledge of microorganisms or of immunology, and it remained for Louis Pasteur 100 years later to begin to develop scientific methods for preparing vaccines. Pasteur showed that inoculations with preparations of attenuated (weakened) pathogens could be used to develop immunity against the virulent form of the pathogen. However, it was not until 20th century advances in immunology, for example, Burnet's clonal selection theory in 1957 and the discovery of T and B cells in 1965, that a modern understanding of vaccines was developed. Effective vaccination depends on stimulating the body to launch an immune response against the antigens contained in the vaccine. Memory cells develop, and future encounters with the same pathogen are dealt with swiftly.

Effective vaccines are prepared in a number of ways. A virus may be attenuated (weakened) by successive passage through cells of nonhuman hosts. In the process, mutations adapt the pathogen to the nonhuman host so that it can no longer cause disease in humans. This is how Sabin polio vaccine and measles vaccine are produced. Whooping cough and typhoid fever vaccines are made from killed pathogens that still have the necessary antigens to stimulate an immune response. Tetanus and botulism vaccines are made from toxins secreted by the respective pathogens. The toxin is altered so that it can no longer destroy tissues, but its antigenic determinants are still intact.

Most vaccines consist of the entire pathogen, attenuated or killed, or of a protein from the pathogen. Researchers are investigating a number of approaches that would reduce potential side

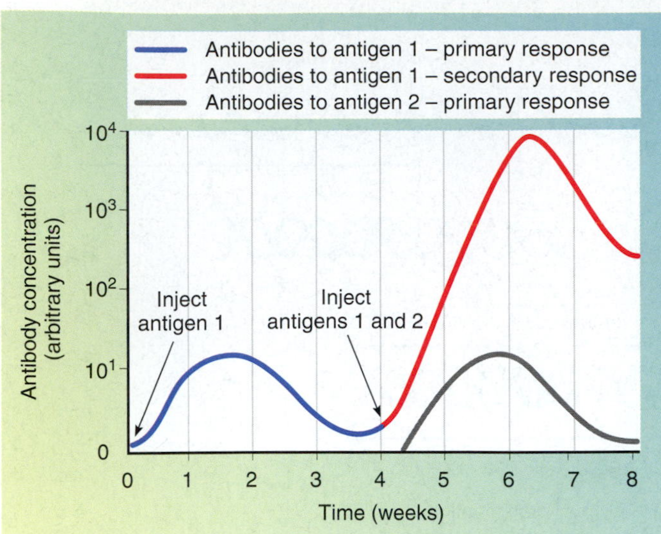

Figure 43–11 Immunological memory. Antigen 1 was injected at day 0, and the immune response was assessed by measuring antibody levels to the antigen. At week 4, the primary response had subsided. Antigen 1 was injected again, together with a new protein, antigen 2. The secondary response was greater and more rapid than the primary response. It was also specific to antigen 1. A primary response was made to the newly encountered antigen 2.

effects. For example, they are designing vaccines consisting of synthetic peptides that are only a small part of the antigen. Another new approach is development of **DNA** (or RNA) **vaccines,** made from a part of the pathogen's genetic material. The DNA of the pathogen is altered so that it transfers genes that specify antigens. When injected into a patient, the altered DNA is taken up by cells and makes its way to the nucleus. The encoded antigens are manufactured and can stimulate both cell-mediated and antibody-mediated immunity. Several DNA vaccines, including vaccines to prevent and treat HIV infection, are in clinical trials.

Passive immunity is borrowed immunity

In **passive immunity,** an individual is given antibodies actively produced by another organism. The serum or gamma globulin that contains these antibodies can be obtained from humans or other animals. Nonhuman sera are less desirable because nonhuman proteins can act as antigens, stimulating an immune response that may result in an illness known as serum sickness.

Passive immunity is borrowed immunity, and its effects are not lasting. It is used to boost the body's defense temporarily against a particular disease. For example, when someone is diagnosed with hepatitis A, a form of viral hepatitis that can be spread through contaminated food or water, persons at risk of infection can be injected with gamma globulin containing antibodies to the hepatitis pathogen. However, such gamma globulin injections offer protection for only a few weeks. Because the body has not actively launched an immune response, it has no memory cells and cannot produce antibodies to the pathogen. Once the injected antibodies are broken down, the immunity disappears.

Pregnant women confer natural passive immunity on their developing babies by manufacturing antibodies for them. These maternal antibodies, of the IgG class, pass through the placenta (the organ of exchange between mother and developing fetus) and provide the fetus and newborn infant with a defense system until its own immune system matures. Babies who are breast-fed continue to receive immunoglobulins, particularly IgA, in their milk.

MANY FACTORS COMPROMISE THE IMMUNE SYSTEM, LEADING TO DISEASE

The immune system is generally very effective in defending us against invading pathogens. However, many factors can compromise immune function, including genetic mutations that impair

TABLE 43–1 Active and Passive Immunity

Type of Immunity	When Developed	Development of Memory Cells	Duration of Immunity
Active			
Naturally induced	After pathogens enter the body through natural encounters (e.g., person with measles sneezes on you)	Yes	Many years
Artificially induced	After immunization with a vaccine	Yes	Many years
Passive			
Naturally induced	After transfer of antibodies from mother to developing baby	No	Few months
Artificially induced	After injection with gamma globulin	No	Few months

the immune system, malnutrition, sleep deprivation, preexisting disease, stress, and virulent pathogens.

Cancer is a failure in immunosurveillance

Cancer has been discussed in several previous chapters including Chapters 16 and 37. Here we focus on how the immune system destroys cancer cells and on new treatment strategies based on immune mechanisms. A few normal cells may be transformed into precancer cells daily in each of us in response to radiation, certain viruses, chemical carcinogens in the environment, and yet unknown factors. Because precancer cells are abnormal, some of their surface proteins are different from those of normal body cells. Such abnormal proteins act as antigens, stimulating an immune response that typically destroys the cancer cells.

Although many components of the immune system help defend against cancer cells, dendritic cells, NK cells, and cytotoxic T cells are central. Dendritic cells recognize cancer cells and process and display their abnormal peptides. The dendritic cells return to lymph nodes and present the cancer antigen to T cells. The matching T cell is activated and forms a clone of cytotoxic T cells. These cells produce interleukins, which attract macrophages and NK cells and activate them. The cytotoxic T cells also produce interferons, which have an antitumor effect (Fig. 43–12). Macrophages produce factors, including TNFs (tumor necrosis factors), that inhibit tumor growth.

Because there are hundreds of subtypes of cancer, the immune system must make varied responses to them. Sometimes, immunosurveillance is not effective and cancer cells multiply unchecked. To treat cancer effectively we must tailor treatment strategies to the particular cancer. Conventional treatment strategies such as chemotherapy and radiation treatment are not specific and destroy normal, as well as cancer, cells. Some new cancer treatments are based on immune strategies and can be precisely targeted to destroy a patient's cancer cells. A few new cancer drugs, like Herceptin, are monoclonal antibodies. Herceptin binds with a receptor that is present in excessive numbers on the cells of about 30% of breast cancers and blocks growth factors that would stimulate proliferation of the cells. Other drugs inhibit the development of blood vessels needed by tumors. More than 50 such angiogenesis inhibitors are currently being tested. These drugs do not cure cancer, but they slow its growth.

One very promising approach is to produce vaccines that stimulate the immune system to attack the cancer cells. (Note that these vaccines differ from traditional vaccines that are used to destroy pathogens *before* they cause disease.) For example, some types of cancer cells are able to evade dendritic cells. If the dendritic cells could be alerted, perhaps they would more effectively find these cancer cells. Several groups of researchers are experimenting with cancer vaccines composed of dendritic cells from a cancer patient mixed with peptides from the patient's cancer cells. The hypothesis is that the dendritic cells would become sensitized to the cancer peptides and then would be more effective in recognizing the cancer cells in the patient. In clinical tests with 18 patients with melanoma (a type of skin cancer; discussed in Chapter 38), a cancer vaccine was injected weekly at first and then less frequently. The results have been promising, with a few patients becoming cancer-free and with tumor reduction in several others. (Cancer vaccine development is also discussed in Part 3 *Career Visions.*)

In 2001 investigators reported that most breast cancer cells produce an excess of cyclin D_1, a protein that helps regulate cell cycles (see Chapter 9), and suggested that this overexpression might play a role in causing breast cancer. Using gene targeting, these researchers produced mice that lack cyclin D_1. The genetically engineered mice are resistant to breast cancer induced by certain oncogenes, suggesting that some forms of human breast cancer could be treated with antibodies to cyclin D_1.

DNA microarrays are being used to detect gene expression patterns in cancer cells. The microarray is a slide or chip that can be dotted with DNA from thousands of genes (see Fig. 16–5). These genes can be used as probes to determine which genes are active in cancer cells. Microarrays are being used to identify cancer subtypes, information that can be used to identify and develop the most effective treatments.

Immunodeficiency disease can be inherited or acquired

Absence or failure of some component of the immune system can result in **immunodeficiency disease,** a condition that causes increased susceptibility to infection. Inherited immunodeficiencies have an estimated incidence of 1 per 10,000 births. *Severe combined immunodeficiency syndromes (SCIDs)* are X-linked and autosomal recessive disorders that profoundly affect both cell-mediated immunity and antibody-mediated immunity, resulting in multiple infections. Babies born with SCID typically die by 2 years of age unless they are maintained in a protective bubble until they can be effectively treated, typically with bone marrow transplants.

In *DiGeorge syndrome,* the thymus is reduced or absent and the patient is deficient in T cells. Children born with this disorder are prone to serious viral infections. Treatment involves transplanting bone marrow or fetal thymus tissue.

Researchers have several important animal models of immunodeficiency. For example, the nude mouse (a hairless mutant mouse) does not develop a functional thymus. Because they are deficient in mature T cells, nude mice can be used to study the effects of compromised cell-mediated immunity, as well as possible treatment. *Stem cell research* and advances in genetic engineering suggest new approaches to treating immunodeficiency. For example, in mouse models, gene transfer into the mouse's stem cells was shown to cure SCID. Based on this work, a French research team announced in 2000 that they had successfully transferred a healthy gene into the bone marrow of two children with SCID.

Worldwide, the leading cause of acquired immunodeficiency in children is protein malnutrition. Lack of protein causes a decrease in T-cell numbers, as well as decreased ability to manufacture antibodies, resulting in increased risk for opportunistic infections. Another important cause of acquired immunodeficiency is chemotherapy administered to cancer patients.

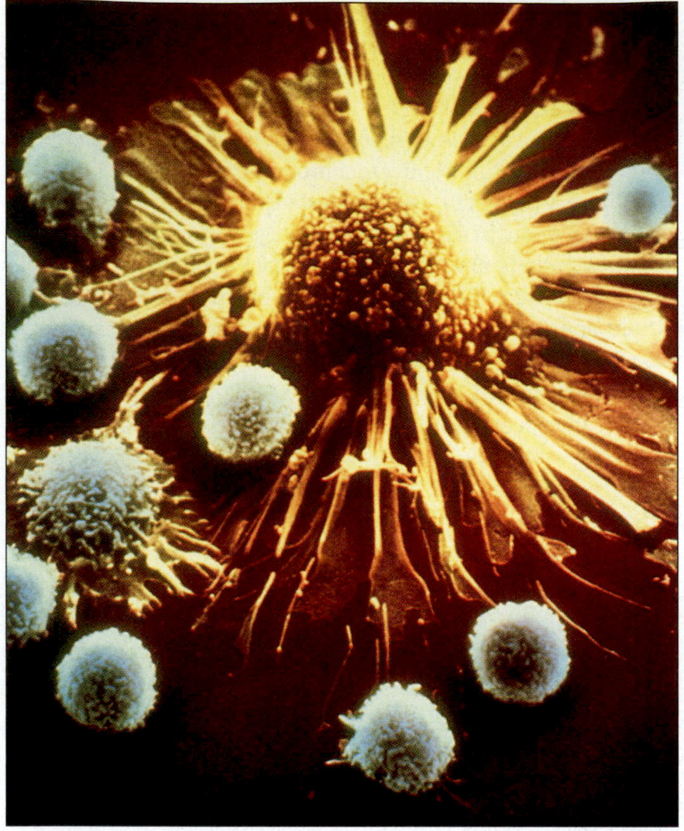

(a) 10 μm

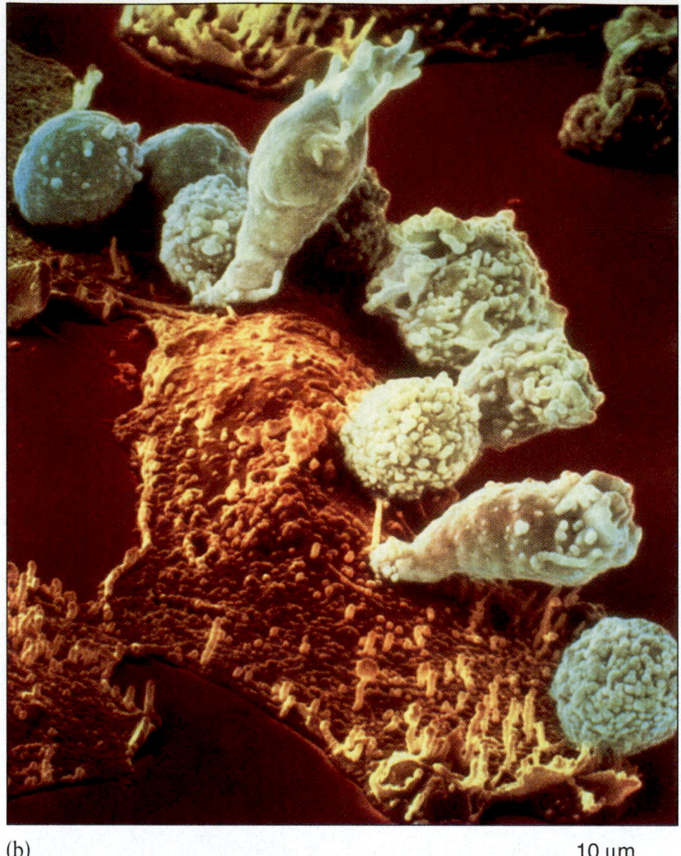

(b) 10 μm

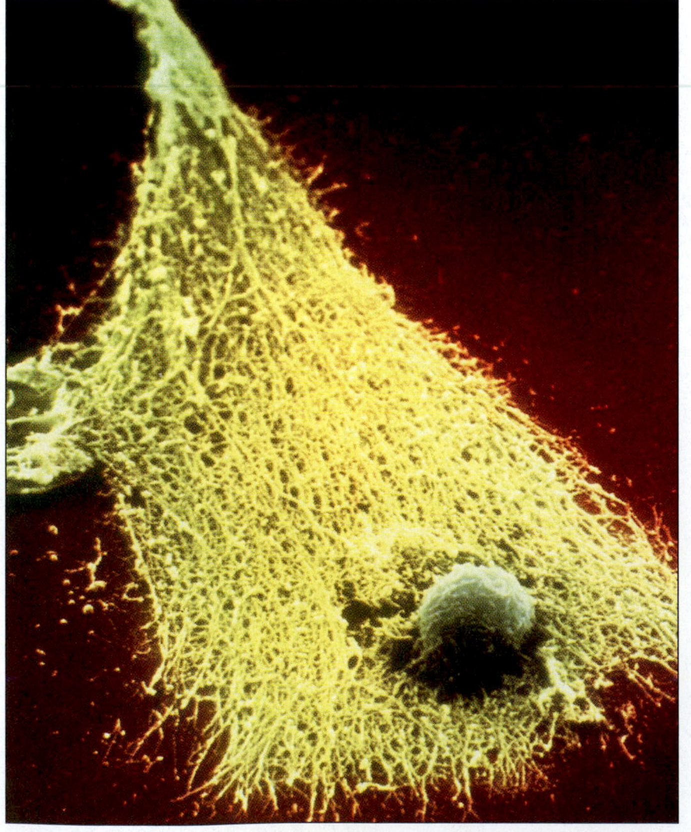

(c) 10 μm

Figure 43–12 Color-enhanced SEMs showing cancer cell destruction. **(a)** An army of cytotoxic T cells surrounds a large cancer cell. The T cells recognize the cancer cell as nonself because it displays altered or unique antigens on its surface. **(b)** Some of the cytotoxic T cells elongate as they chemically attack the cancer cell, breaking down its plasma membrane. **(c)** The cancer cell has been destroyed. Only a collapsed fibrous cytoskeleton remains. *(Lennart Nilsson, Boehringer Ingelheim International GmbH)*

HIV is the major cause of acquired immunodeficiency

Human immunodeficiency virus (HIV) was first isolated in 1983 and was shown to be the cause of **acquired immune deficiency syndrome (AIDS)** in 1984. HIV, a retrovirus, has probably been studied more than any other virus. (Recall from Chapter 23 that a retrovirus is an RNA virus that uses its RNA as a template to make DNA with the help of reverse transcriptase.) Several different strains of the virus are known; HIV-1 is the most virulent form in humans.

AIDS is a serious disease that has claimed more than 23 million lives. Now the fourth highest cause of death globally, AIDS continues to spread through the human population at an alarming rate. Epidemiologists estimate that more than 36 million people worldwide are now infected with HIV, and an estimated 6 million new infections occur each year. These numbers may reflect only a small percentage of the actual number of individuals infected. In 2001, the General Assembly of the United Nations passed a declaration calling AIDS a "global emergency," and outlining measures for mobilizing world resources to combat this disease. The declaration views AIDS as more than a medical issue, stating that it is a political, economic, and human rights threat.

HIV is transmitted mainly during sexual intercourse with an infected person or by direct exposure to infected blood or blood products. This virus is not spread by casual contact. People do not contract the virus by hugging, casual kissing, or using the same bathroom facilities. Friends and family members who live with AIDS patients are not more likely to get the virus.

Most persons currently infected in the United States are men who engage in homosexual and bisexual behavior and individuals who use intravenous drugs. New infections are increasing most rapidly among women and teenagers who contract the virus through heterosexual contact. Heterosexual contact with infected individuals accounts for more than one-third of HIV infections in women and an increasing number of cases in both men and women. Use of a latex condom during sexual intercourse provides some protection against the virus, and use of a spermicide containing nonoxynol-9 may provide additional protection.

An estimated 10% of AIDS patients are children born to infected mothers, but recent drug treatment protocols have reduced the rate of HIV transmission between mother and fetus. Mothers infected with HIV can also transmit the virus to their babies by breastfeeding. Effective blood-screening procedures have been developed to safeguard blood bank supplies, markedly reducing the risk of infection from blood transfusion in highly developed countries.

Dendritic cells appear to be the first cells that HIV targets in the mucosa. When inflammation (due to sexually transmitted disease or other infection) is present in the mucosa (lining) of the cervix or rectum, large numbers of dendritic cells are present. Their presence may increase the risk of HIV infection. Dendritic cells transport HIV from the cervix or rectum to the lymph nodes. HIV has a protein on its outer envelope that attaches to CD4, a protein present on the surface of the helper T cell (CD4 T cell), its main target (chapter opening figure and Fig. 43–13). The virus then enters the helper T cell and destroys it. Over time HIV causes a dramatic decrease in the helper T cell population, which severely impairs the ability to resist infection. Cytotoxic T cells appear to be the main cells that attack HIV, limiting viral replication and delaying the progress of the disease. Too often, the virus eventually wins the battle.

When HIV infects the body, an immune response may be launched. Although most individuals have no symptoms, about 15% of infected individuals experience mild flulike symptoms (fever and aching muscles) for a week or so. Some cells infected by HIV are not destroyed, and the virus may continue to replicate slowly for many years. After a time, a progression of symptoms occurs, including swollen lymph glands, night sweats, fever, and weight loss (Fig. 43–14). Response to exposure with HIV and progression of AIDS depends on a combination of genetic and environmental factors, such as general health and infection with sexually transmitted disease. Some evidence suggests that susceptibility is affected by psychosocial factors, including personality variables and coping styles.

In about one-third of AIDS patients, the virus infects the nervous system, causing *AIDS dementia complex*. These patients exhibit progressive cognitive, motor, and behavioral dysfunction that typically ends in coma and death. As a consequence of immunosuppression, many AIDS patients develop and die from serious opportunistic infections or rare forms of cancer, for example, Kaposi's sarcoma, an endothelial cell tumor that causes purplish spots on the skin.

Researchers throughout the world are searching for drugs that will successfully combat the AIDS virus. Because HIV often infects the central nervous system, an effective drug must cross the blood-brain barrier. **AZT** (azidothymidine, a deoxyribonucleotide analog), the first drug developed to treat HIV infection, can prolong the period prior to the onset of AIDS symptoms. AZT blocks HIV replication by inhibiting the action of reverse transcriptase, the enzyme used by the retrovirus to synthesize DNA. Without producing DNA, the virus cannot incorporate itself into the host cell's DNA. Unfortunately, the reverse transcriptase used by the virus to synthesize DNA makes many mistakes: about 1 in every 2000 nucleotides it incorporates is incorrect. By mutating in this way, viral strains that are resistant to AZT have evolved.

Protease inhibitors block the viral enzyme protease, resulting in viral copies that cannot infect new cells. Protease inhibitors can be used in combination with AZT and other reverse transcriptase inhibitors. Triple-combination treatment has been very effective for many AIDS patients, preventing opportunistic infections and prolonging life. In fact, combination treatment has led to a recent decline in AIDS incidence and mortality in the United States. Unfortunately, people in many parts of the world cannot afford these expensive drugs, and in those poor areas the AIDS epidemic continues to grow. In addition, the emergence of drug-resistant viruses has been a serious problem.

Worldwide containment of the AIDS epidemic will require an effective vaccine that prevents the spread of HIV. Developing such a vaccine has been a most daunting challenge for immunologists. More than three dozen vaccines have been developed and tested clinically, but only one has proved sufficiently effective to progress to large-scale human tests. An effective vaccine would have to overcome the problem created by viral destruction of key cells needed to mount an immune response. Also, because mutations arise at a very high rate, new viral strains with new antigens evolve quickly. A vaccine would not be effective against new antigens and so would quickly become obsolete. Other barriers to the development of a vaccine include the absence of an effective animal model for AIDS and the ethical and practical difficulties associated with using human volunteers to test the vaccine.

While immunologists work to develop a successful vaccine and more effective drugs to treat infected patients, massive educational programs have been launched to slow the spread of HIV. Educating the public that having multiple sexual partners increases the risk of AIDS and teaching sexually active individuals the importance of "safe" sex may help slow the epidemic. Some have suggested that public health facilities offer free condoms to those who are sexually active and free sterile hypodermic needles to those addicted to drugs. The cost of these measures would be

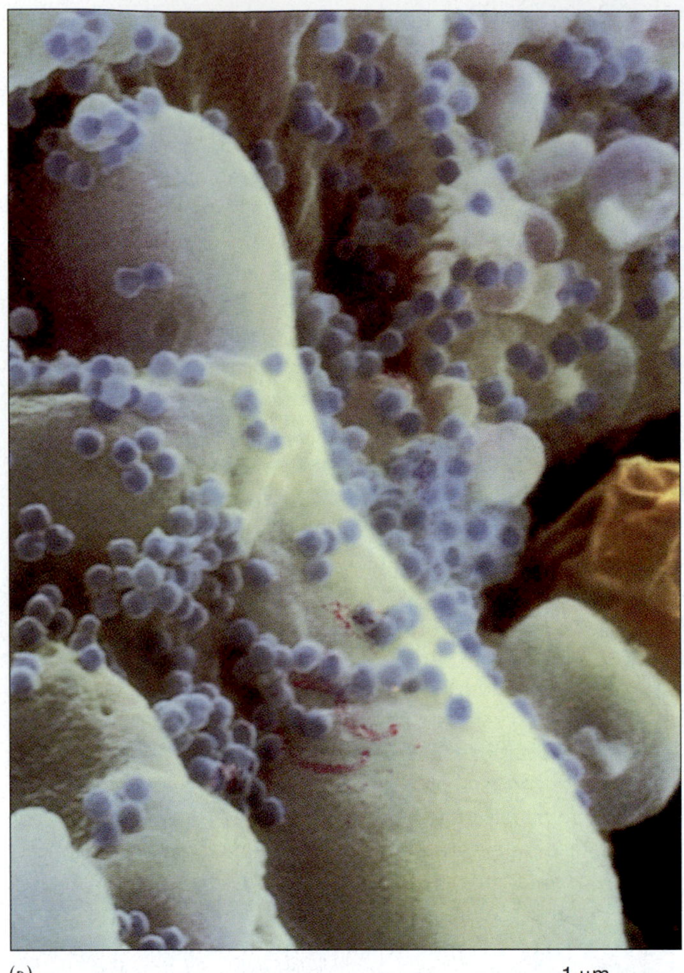

(a) 1 μm

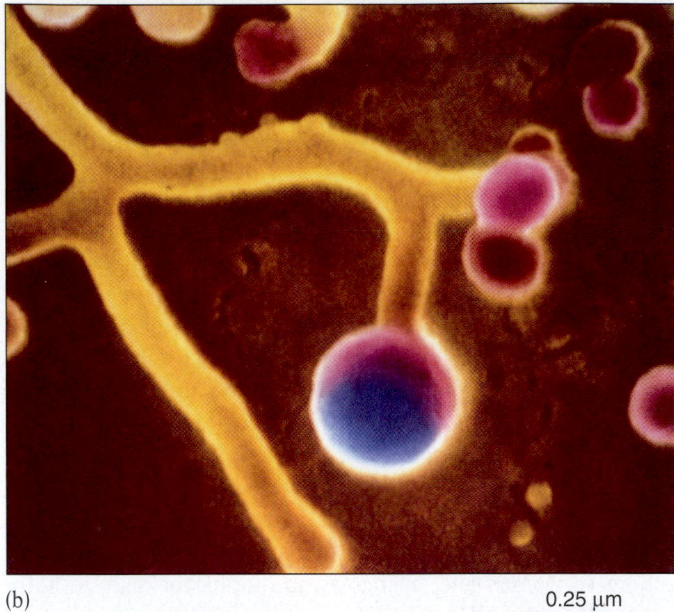

(b) 0.25 μm

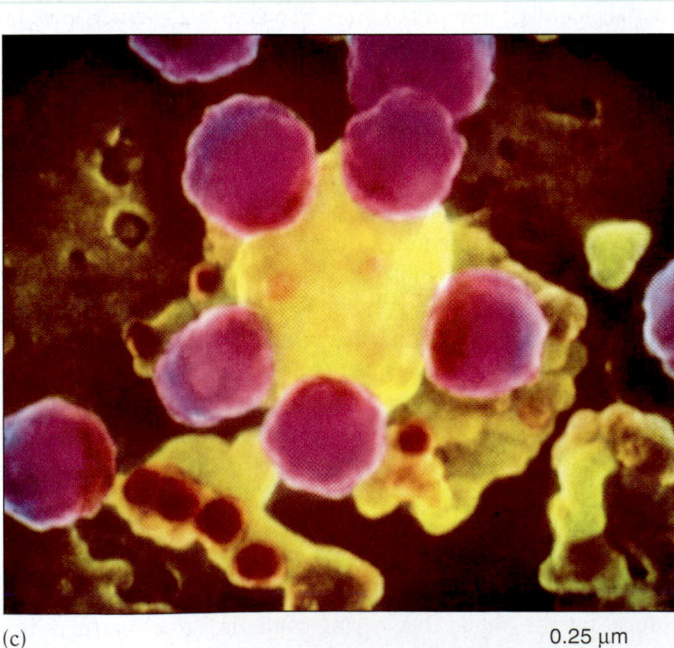

(c) 0.25 μm

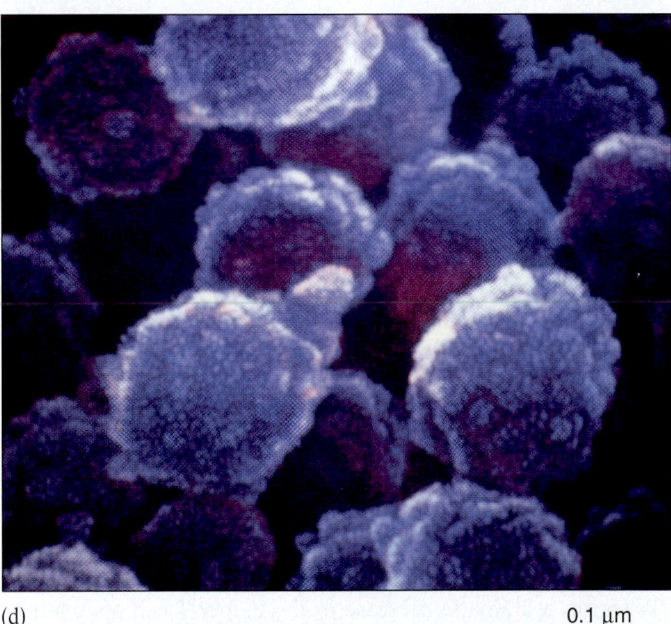

(d) 0.1 μm

Figure 43–13 Color-enhanced SEMs of human immunodeficiency virus (HIV) infection of a helper T cell. **(a)** HIV virus particles *(blue)*, which cause AIDS, attack a helper T cell *(light green)*. **(b)** HIV virus particles budding from the ends of the branched microvilli of a helper T cell. **(c)** An even higher magnification of virus particles budding from a "bleb" (a cytoplasmic extension broader than a microvillus). Note the pentagonal symmetry often evident in biological branching and flowering structures. **(d)** HIV virus particles at extremely high magnification. The surfaces are grainy and the outlines slightly blurred because the preparation was coated with coarse-grained heavy metal (palladium) salt. *(Lennart Nilsson, Boehringer Ingelheim International GmbH)*

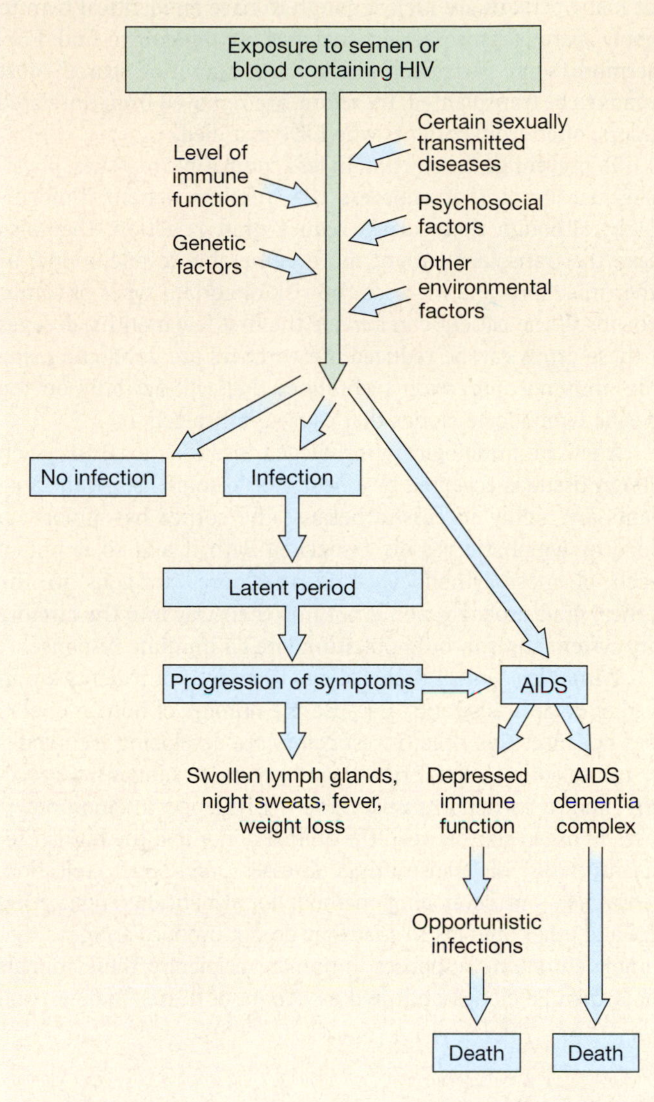

Figure 43–14 Human immunodeficiency virus (HIV) infection. Exposure to semen or blood that contains HIV can lead to AIDS. Although some exposed individuals apparently do not become infected, the risk increases with multiple exposures. Many factors apparently determine whether a person exposed to the AIDS virus contracts the disease.

far less than the cost of medical care for increasing numbers of AIDS patients and the toll in human suffering.

IMMUNE RESPONSES ARE SOMETIMES INCONVENIENT OR HARMFUL

The immune system's ability to distinguish self from nonself sometimes interferes with clinical interventions. For example, efforts to save a patient's life with a blood transfusion or organ transplant can be disastrous if we do not consider the effects of immune function.

Because the immune system is extremely complex, it presents many opportunities for mechanisms to malfunction. **Hypersensitivity** refers to an exaggerated, damaging immune response to an antigen that is normally harmless, as occurs in an allergic reaction. Inappropriately directed or ineffective immune responses can also result in disease states.

Rh incompatibility can result in hypersensitivity

Named for the rhesus monkeys in whose blood it was first found, the Rh system consists of more than 40 kinds of Rh antigens, each referred to as an **Rh factor.** By far the most important of these factors is **antigen D.** About 85% of those United States residents who are of Western European descent are Rh-positive. This means that they have antigen D on the surfaces of their red blood cells (in addition to the antigens of the ABO system and other blood groups). The 15% or so of this population who are Rh-negative have no antigen D. Unlike the situation discussed in Chapter 10 for the ABO blood group, Rh-negative persons do not naturally produce antibodies against antigen D (anti-D). However, they produce anti-D antibodies if they are exposed to Rh-positive blood. The allele coding for antigen D is dominant to the allele for the absence of antigen D. Hence, Rh-negative persons are homozygous recessive, and Rh-positive persons are heterozygous or homozygous dominant.

Although several kinds of maternal-fetal blood type incompatibilities are known, **Rh incompatibility** is probably the most important (Fig. 43–15). If a woman is Rh-negative and the father of the fetus she is carrying is Rh-positive, the fetus may also be Rh-positive, having inherited the D allele from the father. Ordinarily no mixing of maternal and fetal blood occurs; nutrients, oxygen, and other substances are exchanged between these two circulatory systems across the placenta. However, late in pregnancy or during the birth process, a small quantity of blood from the fetus may pass through some defect in the placenta.

The fetus's red blood cells, which bear antigen D, activate the mother's white blood cells, stimulating them to form antibodies to antigen D. If the woman becomes pregnant again, her sensitized white blood cells produce anti-D antibodies that can cross an intact placenta and enter the fetal blood. There they combine with the antigen D molecules on the surface of the fetal red blood cells, causing hemolysis (cells rupture and release hemoglobin into the circulation). The ability to transport oxygen is reduced, and breakdown products of the released hemoglobin damage organs, including the brain. In extreme cases of this disease, known as **erythroblastosis fetalis,** so many fetal red blood cells are destroyed that the fetus may die.

When Rh-incompatibility problems are suspected, fetal blood can be exchanged by transfusion before birth, but this is a risky procedure. More commonly, Rh-negative women are treated just after childbirth (or at termination of pregnancy by miscarriage or abortion) with a preparation of anti-D antibodies known as RhoGAM. These antibodies clear the Rh-positive fetal red blood cells from the mother's blood very quickly, minimizing the chance for her own white blood cells to become sensitized.

The antibodies are also soon eliminated from her body. As a result, if she becomes pregnant again, her blood does not contain the anti-D that could harm her baby.

Graft rejection is an immune response against transplanted tissue

Skin can be successfully transplanted from one part of the same body to another or from one identical twin to another. However, when skin is taken from one person and grafted onto the body of a nontwin, it is rejected and sloughs off. Why? Recall that tissues from the same individual or from identical twins have identical MHC alleles and thus the same MHC antigens. Such tissues are compatible.

Because there are many alleles for each of the MHC genes, it is difficult to find identical matches. When a tissue or organ is taken from a donor and transplanted to the body of a nontwin host, several of the MHC antigens are likely to be different. The host's immune system regards the graft as foreign and launches an immune response called **graft rejection.** T cells attack the transplanted tissue and can destroy it within a week.

> Tissue from donor transplanted into body of recipient ⟶ T cells recognize MHC antigens on transplant cells as foreign ⟶ T cells launch immune response known as graft rejection ⟶ T cells destroy transplant cells

Before transplants are performed, tissues from the patient and from potential donors must be typed and matched as closely as possible. Cell typing is somewhat similar to blood typing but is more complex. If all the MHC antigens are matched, the graft has about a 95% chance of surviving the first year. Unfortunately, not many persons are lucky enough to have an identical twin to supply spare parts, so perfect matches are difficult to find. Furthermore, some parts such as the heart cannot be spared. Most organs to be transplanted, therefore, are removed from unrelated donors, often from patients who have just died.

To prevent graft rejection in less compatible matches, physicians use drugs that suppress the immune system. Unfortunately, although these drugs reduce graft rejection, they also make the transplant patient more vulnerable to pneumonia or other infections and increase the risk of certain types of tumor growths. If the patient can survive the first few months, dosages of these drugs can be reduced. Researchers are developing specific immunosuppression techniques that will act only on the specific lymphocyte clones that cause graft rejection.

A few immunologically privileged locations exist in which foreign tissue is accepted by a host. For example, corneal transplants are highly successful because the cornea has almost no blood or lymphatic vessels associated with it and so is out of reach of most lymphocytes. Furthermore, antigens in the corneal graft probably would not find their way into the circulatory system and so would not stimulate an immune response.

Within the United States alone, thousands of patients are in need of organ transplants. Because the number of human donors does not meet this need, investigators are developing techniques for transplanting animal tissues and organs to humans, a procedure known as *xenotransplantation.* Challenges include differences in organ size between the animal donor and the human recipient, risks of transmitting diseases, and graft rejection. Researchers are developing methods for genetically engineering pigs and other animals so that they do not produce antigens that stimulate immune responses in human recipients. These animals could then be cloned and used as donors of hearts, kidneys, and

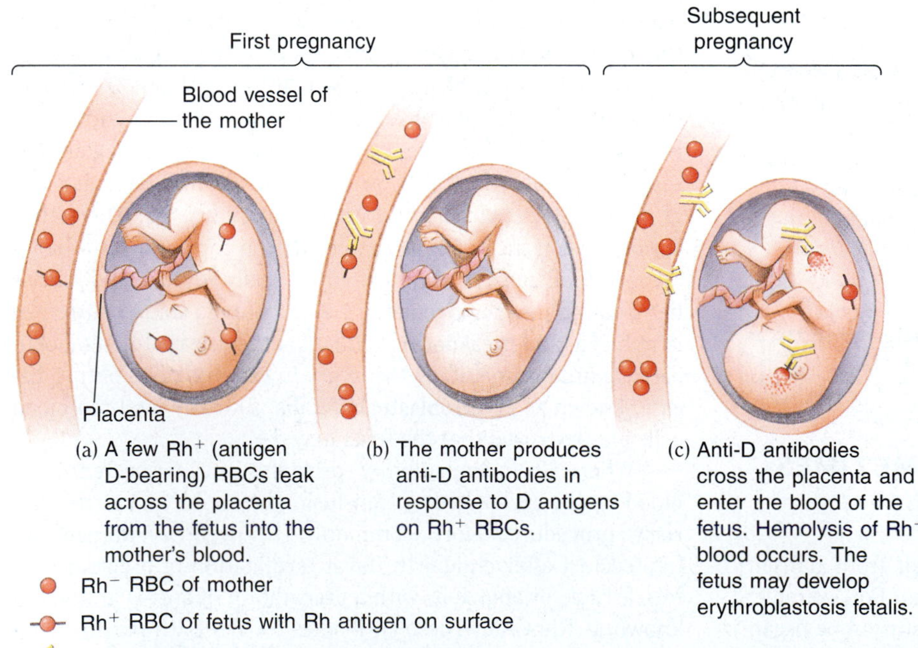
First pregnancy
Subsequent pregnancy
Blood vessel of the mother
Placenta
(a) A few Rh$^+$ (antigen D-bearing) RBCs leak across the placenta from the fetus into the mother's blood.
(b) The mother produces anti-D antibodies in response to D antigens on Rh$^+$ RBCs.
(c) Anti-D antibodies cross the placenta and enter the blood of the fetus. Hemolysis of Rh$^+$ blood occurs. The fetus may develop erythroblastosis fetalis.
Rh$^-$ RBC of mother
Rh$^+$ RBC of fetus with Rh antigen on surface
Anti-D antibody made against Rh$^+$ RBC
Hemolysis of Rh$^+$ RBC

Figure 43–15 Rh incompatibility. When an Rh-negative woman produces Rh-positive offspring, her immune system can be sensitized and produce anti-D antibodies that can cross the placenta and destroy fetal red blood cells.

other organs. Effective artificial organs are also being developed. In 2001, a patient received the first completely implanted artificial heart.

Allergic reactions are directed against ordinary environmental antigens

About 20% of the population of the United States is plagued by an allergic disorder such as allergic asthma or hayfever. A predisposition toward these disorders appears to be inherited. In **allergic reactions,** hypersensitivity results in the manufacture of antibodies against mild antigens, called **allergens,** that normally do not stimulate an immune response. Common environmental agents such as house-dust mites or pollen can trigger allergic reactions in some individuals. Allergic reactions are referred to as *Type I hypersensitivity*. In many kinds of allergic reactions, distinctive *IgE immunoglobulins* are produced.

Let us examine a common allergic reaction, a hayfever response to ragweed pollen (Fig. 43–16). The first step is *sensitization.* Macrophages degrade the allergen and present fragments of it to T cells. The activated T cells then stimulate B cells to become plasma cells and produce IgE. These antibodies attach to receptors on mast cells. Each IgE molecule attaches to a mast cell receptor by its C region end, leaving the V region end free to combine with the ragweed pollen allergen.

The second step is *activation of mast cells.* When a sensitized, allergic person inhales the microscopic pollen, allergen molecules rapidly attach to the IgE on mast cells. This binding of allergen with IgE antibody stimulates the mast cell to release granules filled with chemicals such as *histamine* and *serotonin* that cause *inflammation* (Fig. 43–17). These substances cause blood vessels to dilate and capillaries to become more permeable, leading to edema and redness. Such responses cause the victims' nasal passages to become swollen and irritated. Their noses run, they sneeze, their eyes water, and they feel generally uncomfortable.

A third step may occur in which *the allergic response is prolonged.* Chemical compounds released by the mast cells lure certain white blood cells to leave the circulation and migrate to the inflamed area. These cells then release compounds that damage tissue and prolong the allergic reaction.

> Allergens on pollen ⟶ plasma cells sensitized ⟶ allergen-specific IgE released ⟶ IgE combines with mast cell receptors ⟶ bound IgE combines with allergen ⟶ mast cells release granules containing histamine and other chemicals ⟶ allergic symptoms

In **allergic asthma,** an allergen-IgE response occurs in the bronchioles of the lungs. Mast cells release substances that cause

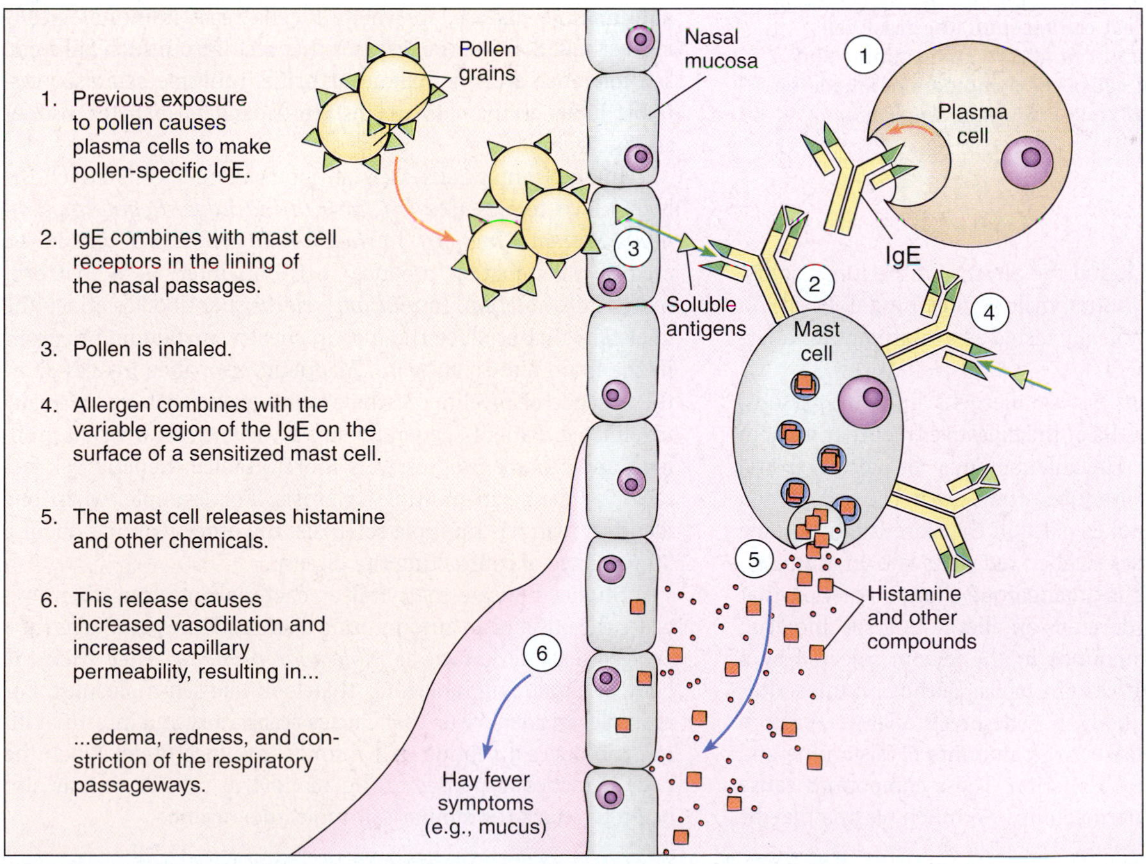

1. Previous exposure to pollen causes plasma cells to make pollen-specific IgE.
2. IgE combines with mast cell receptors in the lining of the nasal passages.
3. Pollen is inhaled.
4. Allergen combines with the variable region of the IgE on the surface of a sensitized mast cell.
5. The mast cell releases histamine and other chemicals.
6. This release causes increased vasodilation and increased capillary permeability, resulting in...
...edema, redness, and constriction of the respiratory passageways.
Pollen grains
Nasal mucosa
Plasma cell
IgE
Soluble antigens
Mast cell
Histamine and other compounds
Hay fever symptoms (e.g., mucus)

Figure 43–16 Allergic response. Pollen causes a common type of allergic response in many people.

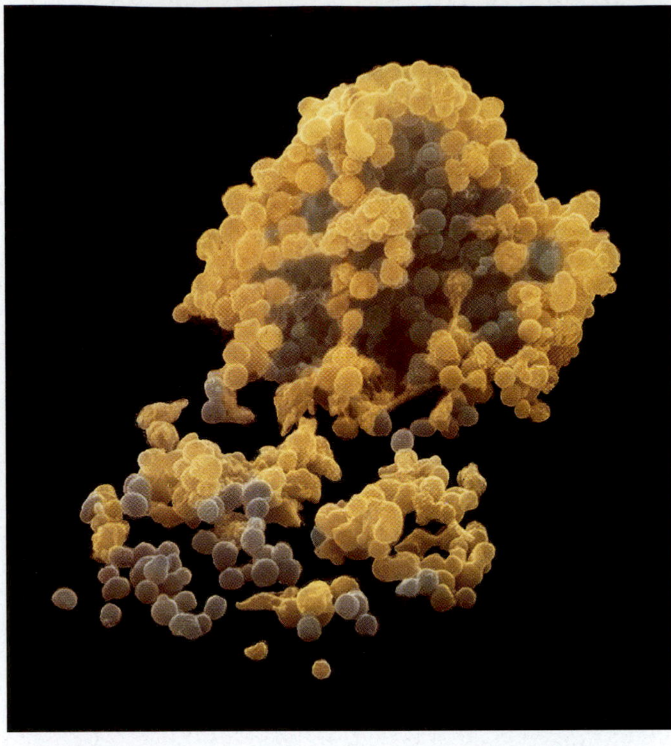

Figure 43–17 Color-enhanced SEM showing release of granules by mast cells. When an allergen combines with IgE bound to a mast cell receptor, the mast cell explosively releases granules filled with histamine and other compounds that cause the symptoms of an allergic response. *(Lennart Nilsson, Boehringer Ingelheim International GmbH)*

smooth muscle to contract, and the airways in the lungs sometimes constrict for several hours, making breathing difficult. An antibody to IgE is being clinically tested as a treatment for allergic asthma.

Certain foods or drugs act as allergens in some persons, causing a reaction in the walls of the digestive tract that leads to discomfort and diarrhea. The allergen may be absorbed and cause mast cells to release granules elsewhere in the body. When the allergen-IgE reaction takes place in the skin, the histamine released by mast cells causes swollen red welts known as **hives.**

Systemic anaphylaxis is a dangerous allergic reaction that can occur when a person develops an allergy to a specific drug such as penicillin, to compounds in the venom injected by a stinging insect, or even to certain foods. Within minutes after the substance enters the body, a widespread allergic reaction takes place. Mast cells release large amounts of histamine and other compounds into the circulation. These compounds cause extreme vasodilation and permeability. So much plasma may be lost from the blood that circulatory shock and death can occur within a few minutes.

The symptoms of allergic reactions are often treated with **antihistamines,** drugs that block the effects of histamines. These drugs compete for the same receptor sites on cells targeted by histamine. When the antihistamine combines with the receptor, it prevents the histamine from binding and thus prevents its harmful effects. Antihistamines are useful clinically in relieving the symptoms of hives and hayfever. They are not completely effective because mast cells release substances other than histamines that also cause allergic symptoms.

In serious allergic disorders, patients are sometimes given a form of immunotherapy known as desensitization. Small amounts of the allergen are injected weekly over a period of months or years. Just how this treatment works is not completely understood, but production of IgG antibody to the allergen or induction of T cell tolerance may be involved.

In an autoimmune disease, the body attacks its own tissues

During their development, complex mechanisms establish self-tolerance (self-recognition) so that lymphocytes do not attack tissues of their own body. However, some lymphocytes remain that have the potential to be **autoreactive,** that is, to launch an immune response against self-tissues. Such autoreactivity can lead to a form of hypersensitivity known as **autoimmunity,** or **autoimmune disease,** in which T cells react immunologically against self. Some of the diseases that result from such failure in self-tolerance are rheumatoid arthritis, multiple sclerosis, systemic lupus erythematosus, insulin-dependent diabetes, psoriasis, and scleroderma.

In autoimmune diseases, antibodies and T cells attack the body's own tissues (see *On the Cutting Edge: Regulating Immune System Function*). In *rheumatoid arthritis,* T cells in the area of inflammation produce IL-15, an interleukin that promotes inflammation. In *multiple sclerosis,* antibodies attack the glial cells that produce the myelin sheath surrounding neurons in the brain and spinal cord. Magnetic resonance images show the absence of myelin sheaths around axons that are normally myelinated. Patients generally suffer weakness and visual problems and become progressively more disabled. Genetic risk factors play a role in multiple sclerosis. For example, when one identical twin has multiple sclerosis, the other twin has about a 30% chance of contracting the disease.

Studies indicate that viral or bacterial infection often precedes the onset of an autoimmune disease. Some pathogens have evolved a tactic known as *molecular mimicry*. They trick the body by producing molecules that look like self-molecules. For example, an adenovirus that causes respiratory and intestinal illness produces a peptide that mimics myelin protein. When the body launches responses to the adenovirus peptide, it may also begin to attack the similar self-molecule, myelin.

ON THE CUTTING EDGE

Regulating Immune System Function

HYPOTHESIS: Tyrosine kinase receptors play an essential role in regulating immune function.

METHOD: Researchers studied immune system function in mice with defective genes for three different tyrosine kinase receptors.

RESULTS: Mutant mice that lack the three tyrosine kinase receptors under investigation develop a severe immune disorder and autoimmunity.

CONCLUSION: Genetic evidence supports the hypothesis that tyrosine receptors are essential in regulating immune function.

The numerous types of cells, regulatory and signaling molecules, and the complex actions of the immune system must be carefully regulated in order to maintain homeostasis. Specific *T cells* and *B cells* must proliferate and respond to challenges by pathogens, but after an effective immune response, the expanded populations of T and B cells must be decreased to normal numbers. Cells that are no longer needed die by *apoptosis* (programmed cell death). Underactivation of critical components would allow pathogens to cause serious disease, and overactivation would result in inflammation and autoimmunity. Either situation can lead to death.

Tyrosine kinase receptors are important in cell signaling and regulation in many organ systems. Researchers Qingxian Lu and Greg Lemke hypothesized that these receptors are also important in regulating immune function. Lu and Lemke studied the immune systems of mice that had mutations in one, two, or three genes encoding tyrosine kinase receptors. The three receptors (Tyro 3, Axl, and Mer) are structurally and functionally related. Lu and Lemke studied the immune system phenotypes of the mutant mice, analyzing their spleens, lymph nodes, B cell and T cell populations, and a variety of signaling molecules. In 2001, these researchers reported in *Science* that the three tyrosine kinase receptors studied are essential in regulating immune function.*

The three tyrosine kinase receptors that were inactivated are not expressed on B cells and T cells. Rather, they are expressed on *antigen-presenting cells (APCs),* including *macrophages* and *dendritic cells.* APCs normally help regulate immune function. In the mutant mice, APCs did not express the tyrosine kinase receptor genes, and without the corresponding functional receptors, the APCs were hyperactive. For example, the mutant APCs produced excessive amounts of a variety of regulatory molecules, including cytokines and tumor necrosis factors that cause inflammation.

The APCs in the mutant mice were unable to regulate B cell and T cell proliferation. Abnormally rapid proliferation of these lymphocytes resulted in their overpopulation in the spleen and lymph nodes, causing these structures to enlarge dramatically. Furthermore, the unregulated proliferation of B and T cells in the triple mutants resulted in the migration of these cells to unusual locations in the body. Populations of B and T cells were found in every adult organ examined, including the brain, heart, lung, kidney, skeletal muscle, and eye.

Dendritic cells normally inhibit T cell proliferation. When dendritic cells are not functioning effectively, and when certain cytokines are overproduced, T cells specific to the body's own antigens may proliferate, resulting in *autoimmunity.* All of the triple mutants developed autoimmunity. Their tissues showed abnormalities similar to those characteristic of human autoimmune disorders, including rheumatoid arthritis and systemic lupus erythematosus. Furthermore, the investigators measured abnormally high levels of circulating antibodies to the mutant animals' DNA, a finding consistent with known symptoms of autoimmune disease.

The overactivation of APCs in the mutant mice suggests that the tyrosine kinase receptors normally help regulate immune function, including lymphocyte proliferation and inflammatory responses. These receptors also help prevent unchecked immune function that can lead to autoimmunity.

Tyrosine receptors are not the only regulators of immune system function. Many other factors operate to maintain homeostasis in the immune system. For example, compounds known as lysophospholipids (produced from compounds in plasma membranes and from compounds in low-density lipoproteins) were recently shown to help regulate T cell proliferation and responses.

* Lu, Q., and G. Lemke. "Homeostatic Regulation of the Immune System by Receptor Tyrosine Kinases of the Tyro 3 Family." *Science,* Vol. 293, 13 Jul. 2001.

SUMMARY WITH KEY TERMS

I. The body defends itself against **pathogens** and other foreign agents. The study of internal defense mechanisms is called **immunology.** An **immune response** involves recognizing foreign macromolecules and mounting a response aimed at eliminating them. Immune responses depend on the ability of an organism to distinguish between self and nonself.
  A. **Nonspecific immune responses,** also called **innate immunity,** provide general and immediate protection against pathogens.
  B. **Specific immune responses,** also known as **acquired** or **adaptive immune responses,** target specific macromolecules associated with a pathogen.
  C. An **antigen** is a molecule specifically recognized as foreign by cells of the immune system. **Antibodies** are highly specific proteins that recognize and bind to specific antigens.

II. Invertebrates depend mainly on nonspecific immune responses such as physical barriers (cuticle, skin, mucous membranes), **phagocytosis,** and **antimicrobial peptides,** soluble molecules that destroy pathogens.

III. Vertebrate nonspecific immune responses include physical barriers, such as the skin and mucous lining of the respiratory and digestive tracts. Should pathogens break through these first-line defenses, other nonspecific defense mechanisms are activated.
  A. Antimicrobial peptides and proteins destroy pathogens.
    1. **Cytokines** are signaling proteins that regulate interactions between cells. Three important groups are interferons, interleukins, and tumor necrosis factors.
      a. **Interferons** inhibit viral replication and enhance activities of immune cells.
      b. **Interleukins** help regulate interactions between lymphocytes and other cells of the body; some have widespread effects.
      c. **Tumor necrosis factors (TNFs)** kill tumor cells and stimulate immune cells to initiate an **inflammatory response.**
    2. Some **complement** proteins lyse the cell wall of the pathogen. Some coat the pathogen, enhancing phagocytosis, whereas others attract white blood cells to the site of infection. Still others increase the inflammatory response.
  B. **Phagocytes,** including **neutrophils** and **macrophages,** phagocytize and destroy bacteria. **Natural killer cells (NK cells)** destroy cells infected with viruses and foreign or altered cells such as tumor cells.
  C. When pathogens invade tissues, they trigger an inflammatory response, which brings needed phagocytic cells and antibodies to the infected area.

IV. Vertebrate specific immune responses include **antibody-mediated immunity** and **cell-mediated immunity.**
  A. Cells of the immune system include phagocytes such as neutrophils and macrophages; **eosinophils** that destroy parasitic worms; **mast cells,** important in allergic reactions; **dendritic cells;** and **lymphocytes—T cells, B cells,** and natural killer cells (NK cells).
    1. Dendritic cells, macrophages, and B cells are **antigen-presenting cells (APCs)** that display foreign antigens as well as their own surface proteins.
    2. B cells (B lymphocytes) are responsible for antibody-mediated immunity. B cells differentiate into **plasma cells,** which produce antibodies. Some activated B cells become **memory B cells,** which continue to produce antibodies after the infection has been overcome.
    3. T cells (T lymphocytes), distinguished by the **T-cell antigen receptor (TCR),** are responsible for cellular immunity. The **thymus gland** confers immunological competence on T cells by making them capable of distinguishing between self and nonself. Among the several types of T cells are the **cytotoxic T cells (CD8 T cells), helper T cells (CD4 T cells),** and **memory T cells.**
  B. Immune responses depend on protein antigens coded for by a group of genes

V. In antibody-mediated immunity, B cells are activated when they combine with antigen. Activation requires an APC (such as a dendritic cell or macrophage) that has a foreign antigen–MHC complex displayed on its surface. A helper T cell that secretes interleukins is also needed.
  A. Activated B cells multiply, giving rise to clones of cells.
  B. Differentiated B cells, called plasma cells, produce specific antibodies, also called **immunoglobulins (Ig),** in response to specific antigens.
    1. A typical antibody is Y-shaped; the two arms of the Y combine with antigen. An antibody molecule consists of four polypeptide chains: two identical heavy chains and two shorter chains, called light chains. Each chain has a **C region (constant segment), J region (junctional segment),** and **V region (variable segment).**
    2. Antibodies are grouped in five classes determined by differences in the constant regions of their heavy polypeptide chains. **IgG** and **IgM** defend against pathogens in the blood. **IgA** prevents pathogens from attaching to epithelial surfaces. **IgD** helps activate B cells after they bind with antigens. **IgE** is important in allergic reactions.
    3. A given antibody can bind with different strengths, or **affinities,** to different antigens. Antibodies with progressively higher affinities are produced during the course of an immune response.
  C. An antibody combines with a specific antigen to form an **antigen-antibody complex,** which may inactivate the pathogen, stimulate phagocytosis, or activate the complement system.
  D. **Monoclonal antibodies** are identical antibodies produced by cells cloned from a single cell. They are used clinically and are important tools in biological research.

VI. In cell-mediated immunity, specific T cells are activated by a foreign antigen–MHC complex on an APC and by helper T cells.
  A. Activated T cells multiply, giving rise to a clone.
  B. Some T cells differentiate to become cytotoxic T cells that migrate to the site of infection and chemically destroy cells infected with viruses.
  C. Some activated T cells become memory T cells; others become helper T cells.

VII. Memory B and memory T cells are responsible for long-term immunity.
  A. The first exposure to an antigen stimulates a **primary response.** Second exposure to the same antigen evokes a **secondary immune response,** which is more rapid and more intense than the primary response.
  B. **Active immunity** develops as a result of exposure to antigens; it may occur naturally after recovery from a disease or be artificially induced by immunization with a **vaccine.**
  C. **Passive immunity** is a temporary condition that develops when an individual receives antibodies produced by another person or animal.

VIII. Many factors can compromise the immune system, leading to disease.
  A. Normally, the immune system destroys precancer cells when they arise; cancer develops when immunosurveillance is not effective. Immune therapy, including cancer vaccines, are being developed to precisely target specific cancers.
  B. **Immunodeficiency disease** can be inherited or acquired.
  C. **Acquired immune deficiency syndrome (AIDS)** is caused by **human immunodeficiency virus (HIV),** a retrovirus. HIV damages the immune system by destroying helper T cells, which severely impairs immunity and puts the patient at risk for opportunistic infections.

IX. The immune system sometimes malfunctions, resulting in hypersensitivity reactions or producing antigens to self-molecules. Such responses can interfere with clinical interventions and also cause disease.
   A. When an Rh-negative woman gives birth to an Rh-positive baby, she may develop anti-D antibodies. **Rh incompatibility** can then occur in future pregnancies.
   B. Transplanted tissues have MHC antigens that stimulate **graft rejection,** an immune response in which T cells destroy the transplant.
   C. In an **allergic reaction,** an **allergen** can stimulate production of IgE, which combines with the receptors on mast cells. When the allergen combines with the IgE, the mast cells release **histamine** and other substances that cause symptoms of allergy such as inflammation. **Systemic anaphylaxis** is a rapid, widespread allergic reaction that can lead to death.
   D. In **autoimmune diseases,** the body reacts immunologically against its own tissues.

POST-TEST

1. A substance that is recognized as foreign by cells of the immune system is a(an) (a) antibody (b) antigen (c) immunoglobulin (d) interferon (e) cytokine
2. Nonspecific (innate) immune responses include (a) inflammation (b) antigen-antibody complexes (c) immunoglobulin action (d) complement and memory T cells (e) interferon and memory B cells
3. Invertebrate defense mechanisms include (a) phagocytosis (b) antimicrobial peptides (c) ability to distinguish between self and nonself (d) answers a, b, and c are correct (e) answers a and b only
4. Cytokines (a) are regulatory nucleic acids (b) prevent the inflammatory response (c) include interferons and interleukins (d) are immunoglobulins (e) include complement proteins
5. Which of the following is *not* an action of complement? (a) enhances phagocytosis (b) enhances inflammatory response (c) coats pathogens (d) stimulates histamine release (e) stimulates allergen release
6. Which of the following cells are antigen-presenting cells? (a) NK cells (b) plasma cells (c) dendritic cells (d) mast cells (e) memory T cells
7. Which of the following cells are especially adept at destroying tumor cells? (a) NK cells (b) plasma cells (c) neutrophils (d) cytotoxic B cells (e) mast cells
8. Which of the following cells become immunologically competent after processing in the thymus gland? (a) NK cells (b) T cells (c) macrophages (d) B cells (e) plasma cells
9. Cells that have a surface marker called CD4 are (a) NK cells (b) cytotoxic T cells (c) helper T cells (d) B cells (e) suppressor cells
10. The major histocompatibility complex (MHC) (a) codes for a group of cell surface proteins (b) codes for proteins that form complexes with certain cytokines (c) is important mainly in allergic reactions (d) suppresses complement release from macrophages (e) is Y-shaped
11. Which sequence most accurately describes antibody-mediated immunity? (1) B cell divides and gives rise to clone (2) antibodies produced (3) cells differentiate, forming plasma cells (4) activated helper T cell interacts with B cell displaying same antigen complex (5) B cell activated (a) $1 \longrightarrow 2 \longrightarrow 3 \longrightarrow 4 \longrightarrow 5$ (b) $3 \longrightarrow 2 \longrightarrow 1 \longrightarrow 4 \longrightarrow 5$ (c) $4 \longrightarrow 5 \longrightarrow 3 \longrightarrow 4 \longrightarrow 1$ (d) $4 \longrightarrow 5 \longrightarrow 1 \longrightarrow 3 \longrightarrow 2$ (e) $4 \longrightarrow 3 \longrightarrow 1 \longrightarrow 2 \longrightarrow 5$
12. A typical antibody (a) has a Y shape (b) has four identical heavy chains and four identical light chains (c) has IgG and IgD components (d) suppresses phagocyte actions (e) attacks cancer cells
13. Immunoglobulin A (a) combines with mast cells in allergic reactions (b) combines with NK cells (c) prevents pathogens from attaching to epithelial surfaces (d) is found mainly on B-cell surfaces (e) is found mainly on T cells
14. When a person is exposed to the same antigen a second time, the response is (a) called a secondary immune response (b) more rapid (c) mediated by dendritic cells (d) answers a, b, and c are correct (e) answers a and b only
15. Graft rejection (a) is an example of passive immunity (b) occurs in mild form after immunization (c) generates monoclonal antibodies (d) does not occur when tissue is transplanted from one identical twin to the other (e) is initiated by the thymus gland
16. In an allergic reaction (a) the body is immunodeficient (b) an allergen binds with IgE (c) helper T cells release histamine (d) allergen stimulates graft rejection (e) mast cells are deactivated
17. HIV (a) is a retrovirus (b) destroys cytotoxic T cells (c) is attacked mainly by B cells (d) answers a, b, and c are correct (e) none of the preceding answers is correct

REVIEW QUESTIONS

1. How does the body distinguish between self and nonself? What evidence is there that invertebrates are capable of making this distinction?
2. Contrast specific and nonspecific immune responses. In what ways do the two types of processes interact?
3. How does inflammation help restore homeostasis?
4. Compare and contrast cell-mediated and antibody-mediated immunity.
5. Describe how antibodies work against pathogens.
6. John is immunized against measles. Jack contracts measles from a playmate in nursery school because his parents failed to have him immunized. Compare the immune responses of the two children. Five years later, John and Jack are playing together when Judy, who has just been diagnosed with measles, sneezes on both of them. Compare their immune responses.
7. Why is passive immunity temporary?
8. What is graft rejection? What is the immunological basis for it?
9. List the immunological events that take place in a common type of allergic reaction such as hayfever.
10. What is an autoimmune disease? Give two examples.
11. Specificity, diversity, and memory are key features of the immune system. Giving specific examples, explain how each of these features is important.

YOU MAKE THE CONNECTION

1. Imagine that you are a researcher developing new HIV treatments. What approaches might you take? What public policy decisions would you recommend that might help slow the spread of AIDS while new treatments or vaccines are being developed?
2. Macrophages can be selectively destroyed in the body by the administration of a certain chemical. What would be the effects of such a loss of macrophages? Which do you think would have a greater effect on the immune system, loss of macrophages or loss of B cells?
3. What are the advantages of having MHC antigens? Disadvantages? What do you think would be the consequences of not having them?

RECOMMENDED READINGS

Abraham, S.N. "Discovering the Benign Traits of the Mast Cell," *Science & Medicine,* Vol. 4, No. 5, Sep./Oct. 1997. A discussion of the role of mast cells in human responses.

Behar, S.M., and S.A. Porcelli. "The Immunology of Inflammatory Arthritis." *Science & Medicine,* Vol. 4, No. 1, Jan./Feb. 1997. A discussion of an abnormal immune response.

Ezzell, C. "Magic Bullets Fly Again." *Scientific American,* Vol. 285, Oct. 2001. Biotechnology and pharmaceutical firms are developing a promising new generation of monoclonal antibodies that target a variety of disorders.

Gura, T. "Innate Immunity: Ancient System Gets New Respect." *Science,* Vol. 291, 16 Mar. 2001. Focuses on recent studies of antimicrobial peptides.

Kuby, J. *Immunology.* W.H. Freeman and Company, New York, 1997. An excellent introduction to immunology.

Langridge, W.H.R. "Edible Vaccines." *Scientific American,* Vol. 283, No. 3, Sep. 2000. In the future, common food plants may be genetically engineered to produce vaccines in their edible parts.

Mackaness, G.B. "Cellular Resistance to Infection." *Journal of Experimental Medicine,* Vol. 116, 1962. This classic study was a major contribution toward understanding cellular immunity and immunological memory.

Roitt, I., J. Brostoff, and D. Male. *Immunology,* 5th ed. Mosby, London, 1998. An excellent immunology textbook.

Science, Vol. 280, No. 5371, 19 Jun. 1998. A special issue on AIDS research.

Science, Vol. 293, 13 Jul. 2001. Several articles on "Vaccines and Immunity" discuss recent findings regarding immune function and how this knowledge might be applied to prevention of infectious disease.

Weiner, D.B., and R.C. Kennedy. "Genetic Vaccines." *Scientific American,* Vol. 281, No. 1, Jul. 1999. An interesting account of how DNA vaccines work and how they are produced.

- Visit our Web site at **http://www.info.brookscole.com/solomonbergmartin** for links to chapter-related resources on the World Wide Web. Additional on-line materials relating to this chapter can also be found on our Web site.

See chapter activity on BioActive Learner CD for additional help in mastering the chapter's material. Icon location in the chapter's margins shows which topics have tutorials or simulations in the CD.

44

Gas Exchange

The white-tailed deer *(Odocoileus virginianus)* and other terrestrial vertebrates ventilate their lungs by breathing air. *(Skip Moody/Dembinsky Photo Associates)*

LEARNING OBJECTIVES

After you have studied this chapter you should be able to

1. Compare the advantages and disadvantages of gas exchange in air with those in water.
2. Describe the following adaptations for gas exchange: body surface, tracheal tubes, gills, and lungs.
3. Identify two respiratory pigments and describe their function.
4. Trace the passage of oxygen through the human respiratory system from nostrils to alveoli.
5. Summarize the mechanics and the regulation of breathing in humans. (Include the role of chemoreceptors.)
6. Describe how oxygen and carbon dioxide are exchanged in the lungs and in the tissues.
7. Explain the role of hemoglobin in oxygen transport, and identify factors that determine and influence the oxygen-hemoglobin dissociation curve.
8. Summarize the mechanisms by which carbon dioxide is transported in the blood.
9. Describe the physiological effects of hyperventilation and of sudden decompression when surfacing too quickly from a deep-sea dive.
10. Describe the defense mechanisms that protect the lungs and the effects on the respiratory system of breathing polluted air.
11. Describe adaptations of diving mammals.

Most animal cells require a continuous supply of oxygen for cellular respiration. Some cells, such as mammalian brain cells, may be damaged beyond repair if their oxygen supply is cut off for only a few minutes. Animal cells must also rid themselves of carbon dioxide. The exchange of gases between an organism and its environment is known as **respiration.** Two phases of respiration are organismic and cellular respiration. During **organismic respiration,** oxygen from the environment is taken up by the animal and delivered to its individual cells. At the same time, carbon dioxide generated during cellular respiration is excreted into the environment. In **aerobic cellular respiration,** which takes place in mitochondria, oxygen is necessary for the citric acid cycle to proceed. Oxygen serves as the final electron acceptor in the mitochondrial electron transport chain (see Chapter 7). Carbon dioxide is produced as a metabolic waste product of cellular respiration.

In small, aquatic organisms such as sponges, hydras, and flatworms, gas exchange occurs entirely by simple **diffusion,** the passive movement of particles (atoms, ions, or molecules) from a region of higher concentration to a region of lower concentration, that is, down a concentration gradient. Most cells are in direct contact with the environment. Dissolved oxygen from the surrounding water diffuses into the cells, while carbon dioxide diffuses out of the cells and into the water. No specialized respiratory structures are needed.

Oxygen diffuses through tissues slowly, however. In an animal more than about 1 mm thick, oxygen cannot diffuse quickly enough through layers of cells to support life. Specialized respiratory structures such as gills or lungs are required to deliver oxygen to the cells or to a transport system and to facilitate the excretion of carbon dioxide. Such respiratory systems, working with circulatory systems, provide the efficient intake and transport of oxygen necessary to support high metabolic rates.

If the air or water supplying oxygen to the cells can be continuously renewed, more oxygen will be available. For this reason animals carry on **ventilation;** that is, they actively move air or water over their respiratory surfaces. Sponges do this by setting up a current of water through the channels of their bodies by means of flagella. Most fishes gulp water, which then passes over their gills. Terrestrial vertebrates have lungs and breathe air *(see photograph);* the diaphragm and other muscles move air in and out of the lungs.

RESPIRATORY STRUCTURES ARE ADAPTED FOR GAS EXCHANGE IN AIR OR WATER

Gills are adapted for gas exchange in water, while the tracheal tubes of insects and lungs of vertebrates are respiratory structures adapted for gas exchange in air. Respiratory surfaces must be kept moist to prevent drying out, and oxygen and carbon dioxide are dissolved in the fluid that bathes the cells of these surfaces. Whether an animal makes its home on land or water, gas exchange takes place across a moist surface.

Animals that carry out gas exchange in water require no special mechanisms for maintaining moisture. In contrast, animals that respire in air struggle continuously with water loss. Adaptations have evolved that keep respiratory surfaces moist and minimize desiccation. For example, the lungs of air-breathing vertebrates are located deep within the body, not exposed like gills. Air is humidified and brought to body temperature as it passes through the upper respiratory passageways, and expired air must again pass through these airways (thus allowing for retention of water) before leaving the body. These adaptations help protect the lungs from the drying and cooling effects of air.

Gas exchange in air has certain advantages over gas exchange in water. Compared with water, air contains a much higher concentration of molecular oxygen. Oxygen also diffuses much faster through air than through water. Another advantage is that less energy is needed to move air than to move water over a gas exchange surface. This is because air is less dense and less viscous.

MANY TYPES OF SURFACES HAVE EVOLVED FOR GAS EXCHANGE

Not only must respiratory structures be moist, they must also have thin walls through which diffusion can easily occur. Respiratory structures are generally richly supplied with blood vessels to facilitate transport and exchange of respiratory gases. Four main types of respiratory surfaces used by animals are: the animal's own body surface, tracheal tubes, gills, and lungs (Fig. 44–1). Some animals use a combination of these adaptations.

The body surface may be adapted for gas exchange

Gas exchange occurs through the entire body surface in many animals, including nudibranch mollusks, most annelids, and some amphibians. All these animals are small, with a high surface-to-volume ratio. They also have a low metabolic rate that requires smaller quantities of oxygen per cell. In aquatic animals, the body surface is kept moist by the surrounding water. In terrestrial animals, the body secretes fluids that keep its surface moist. Many animals that exchange gases across the body surface also have gills or lungs.

How does an animal such as an earthworm exchange gases across its body surface? Gland cells in the epidermis secrete mucus, which keeps the body surface moist while also offering protection. Oxygen, present in air pockets in the loose soil that the earthworm inhabits, dissolves in the mucus. Then, the oxygen diffuses through the body wall and into blood circulating in a network of capillaries just beneath the outer cell layer. Carbon dioxide is transported by the blood to the body surface, and then diffuses out into the environment.

Tracheal tube systems of arthropods deliver air directly to the cells

The relatively inefficient open circulatory system of arthropods cannot supply these active animals with sufficient oxygen. A specialized respiratory system is needed. In insects and some other arthropods (e.g., chilopods, diplopods, some mites, and some spiders), the respiratory system consists of a network of **tracheal tubes,** also called **tracheae** (Fig. 44–2). Air enters the tracheal tubes through a series of up to 20 tiny openings called **spiracles** along the body surface. In some insects, especially large, active insects, muscles help ventilate the tracheae by pumping air in and out of the spiracles. For example, the grasshopper draws air in through the first four pairs of spiracles when the abdomen expands. Then the abdomen contracts, forcing air out through the last six pairs of spiracles.

Once inside the body, the air passes through a system of branching tracheal tubes, which extend to all parts of the animal. The tracheal tubes terminate in microscopic, fluid-filled tracheoles. Gases are exchanged between this fluid and the body cells. The tracheal system supplies oxygen needed to support the high metabolic rates characteristic of many insects.

Gills of aquatic animals are respiratory surfaces

Found mainly in aquatic animals, **gills** are moist, thin structures that extend from the body surface. They are supported by the buoyancy of water but tend to collapse in air. In many animals, the outer surface of the gills is exposed to water, whereas the inner side is in close contact with networks of blood vessels.

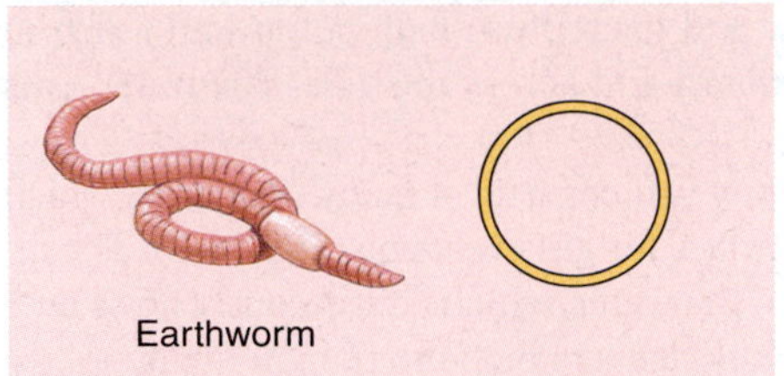

(a) Gas exchange across body surface

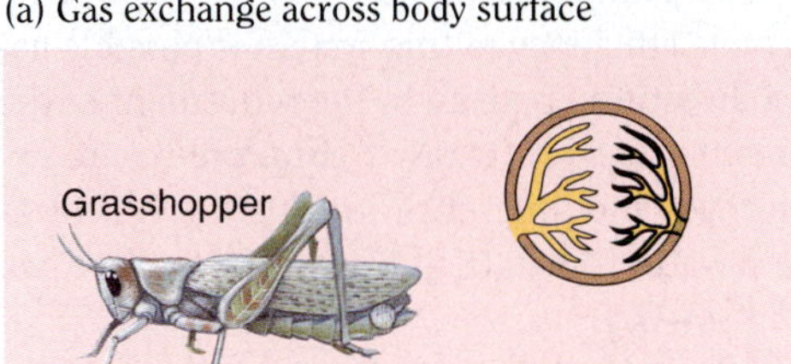

(b) Tracheal tubes

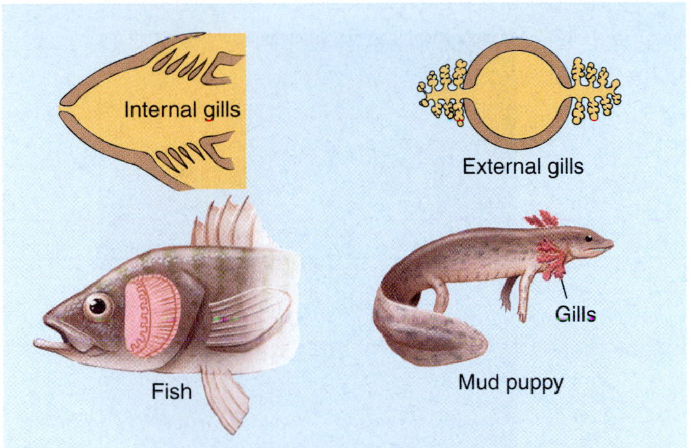

(c) Gills

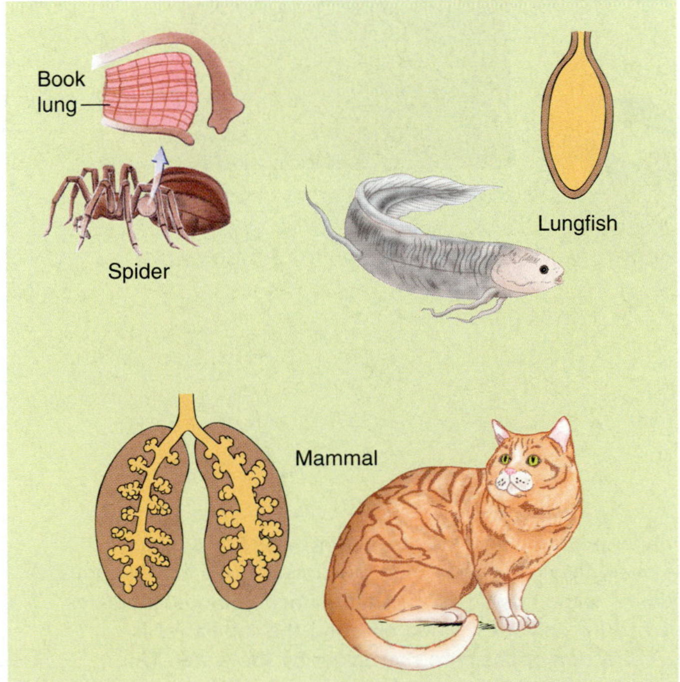

(d) Lungs

Figure 44–1 Adaptations for gas exchange. **(a)** Some animals exchange gases through the body surface. **(b)** Insects and some other arthropods exchange gases through a system of tracheal tubes, or tracheae. **(c)** Gills can be external or internal. **(d)** Lungs are adaptations for terrestrial gas exchange.

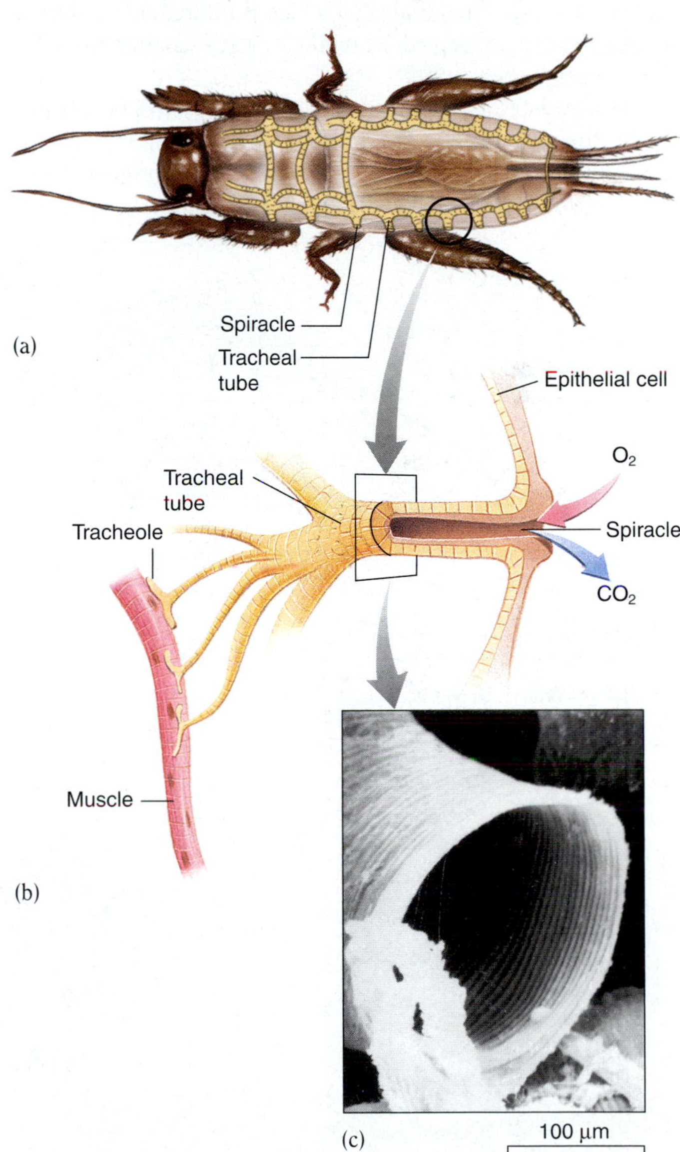

Figure 44–2 Tracheal tubes. **(a)** Air enters the system of tracheal tubes through openings called spiracles. **(b)** Air passes through a system of branching tracheal tubes that conduct oxygen to all cells of the insect. **(c)** SEM of a mole cricket trachea. The wrinkles are a part of a long spiral that wraps around the tube, strengthening the tracheal wall somewhat as a spring strengthens the plastic hoses of many vacuum cleaners. The tracheal wall is composed of chitin. *(c, Courtesy of Dr. James L. Nation and* Stain Technology, *Vol. 58, 1983)*

Sea stars and sea urchins have **dermal gills** that project from the body wall. Their ciliated epidermal cells ventilate the gills by beating a stream of water over them. Gases are exchanged between the water and the coelomic fluid inside the body by diffusion through the gills.

Various types of gills are found in some annelids, aquatic mollusks, crustaceans, fishes, and amphibians. Molluskan gills are folded, providing a large surface for respiration. In clams and other bivalve mollusks and in simple chordates, gills may also be adapted for trapping and sorting food. Rhythmic beating of cilia draws water over the gill area, and food is filtered out of the water as gases are exchanged. In mollusks, gas exchange also takes place through the mantle.

In chordates, gills are usually internal. A series of slits perforates the pharynx, and the gills are located along the edges of these gill slits (see Fig. 30–7*b*). In bony fishes, the fragile gills are protected by an external bony plate, the **operculum.** In some fish, movements of the jaw and operculum help pump water rich in oxygen through the mouth and across the gills. The water exits through the gill slits.

Each gill in the bony fish consists of many **filaments,** which provide an extensive surface for gas exchange (Fig. 44–3). The filaments extend out into the water, which continuously flows over them. A capillary network delivers blood to the gill filaments, facilitating diffusion of oxygen and carbon dioxide between blood and water. The impressive efficiency of this system is possible because blood flows in a direction opposite to the movement of the water. This arrangement, referred to as a **countercurrent exchange system,** maximizes the difference in oxygen concentration between blood and water throughout the area where the two remain in contact (Fig. 44–3*c*).

If blood and water flowed in the *same* direction, that is, *concurrent exchange,* the difference between the oxygen concentrations in blood (low) and water (high) would be very large initially

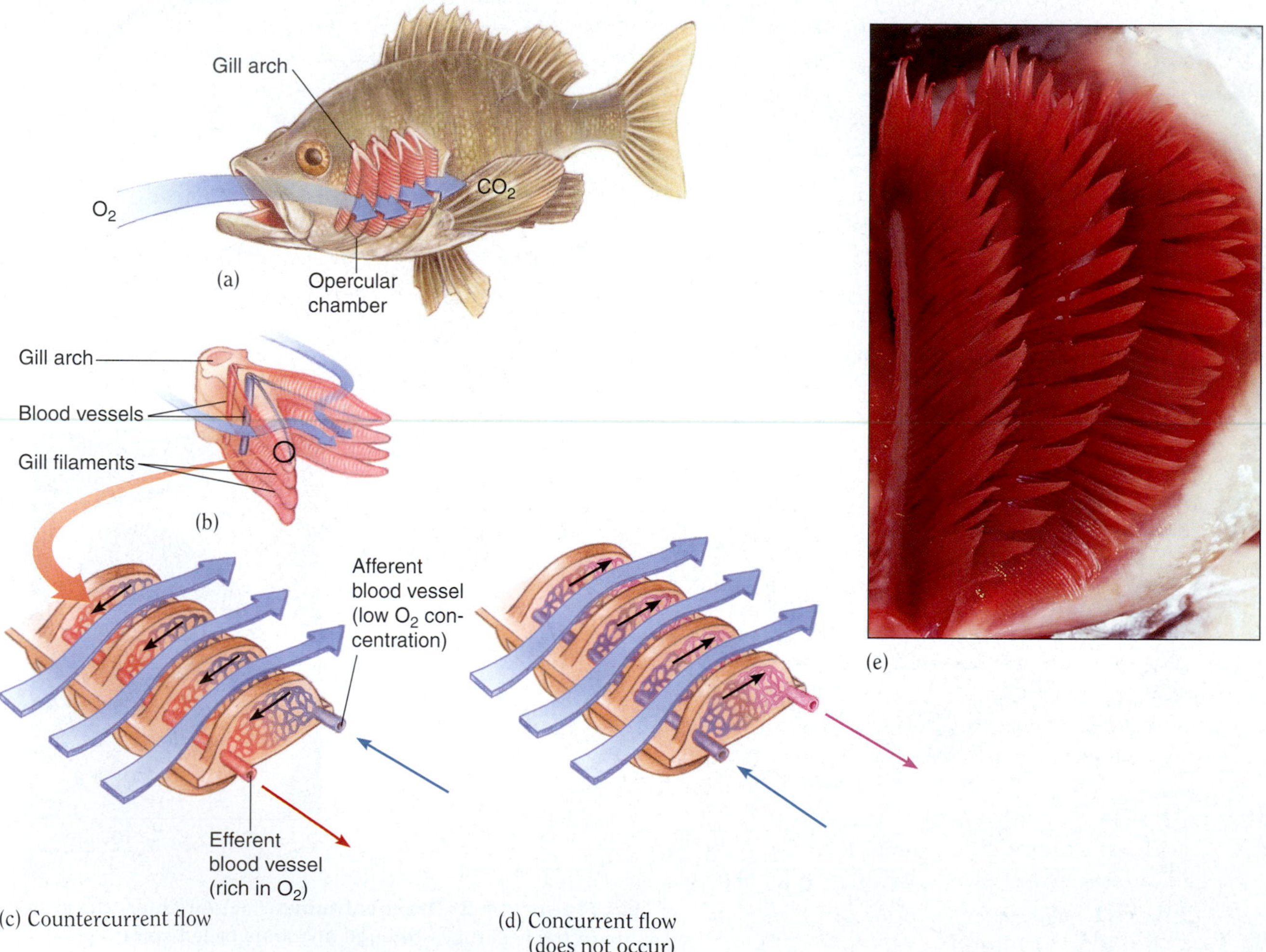

Figure 44–3 Gills in bony fish. **(a)** The gills are located under a bony plate, the operculum, which has been removed in this side view. The gills form the lateral wall of the pharyngeal cavity. **(b)** Each gill consists of a cartilaginous gill arch to which two rows of leaflike gill filaments are attached. As water flows past the gill filaments, blood circulates within them. **(c)** Each gill filament has many smaller extensions rich in capillaries. Blood entering the capillaries is deficient in oxygen. The blood flows through the capillaries in a direction opposite to that taken by the water. This countercurrent exchange system efficiently charges the blood with oxygen. **(d)** If the system were concurrent, that is, if blood flowed through the capillaries in the same direction as the flow of the water, much less of the oxygen dissolved in the water could diffuse into the blood. **(e)** Gills of the salmon. *(e, Bernard Photo Production/Animals Animals)*

and very small at the end. The oxygen concentration in the water would decrease as that in the blood increased. When the concentrations in the two fluids became equal, an equilibrium would be reached, and the net diffusion of oxygen would stop. Only about 50% of the oxygen dissolved in the water could diffuse into the blood.

In the countercurrent exchange system, however, blood low in oxygen comes in contact with water that is partly oxygen-depleted. Then, as the blood flowing through the capillaries becomes more and more oxygen-rich, it comes in contact with water with a progressively higher concentration of oxygen. Thus, all along the capillaries, the diffusion gradient favors passage of oxygen from the water into the gill. A high rate of diffusion is maintained, ensuring that a very high percentage (more than 80%) of the available oxygen in the water diffuses into the blood.

Oxygen and carbon dioxide do not interfere with one another's diffusion, and they simultaneously diffuse in opposite directions. This is because oxygen is more concentrated outside the gills than within, but carbon dioxide is more concentrated inside the gills than outside. Thus, the same countercurrent exchange mechanism that ensures efficient inflow of oxygen also results in equally efficient outflow of carbon dioxide.

Terrestrial vertebrates exchange gases through lungs

Lungs are respiratory structures that develop as ingrowths of the body surface or from the wall of a body cavity such as the pharynx. For example, the **book lungs** of spiders are enclosed in an inpocketing of the abdominal wall. These lungs consist of a series of thin parallel plates of tissue (like the pages of a book) filled with hemolymph (see Fig. 44–1). The plates of tissue are separated by air spaces that receive oxygen from the outside environment through a spiracle. A different type of lung evolved in land snails and slugs. These terrestrial mollusks lack gills. Gas exchange takes place through a lung, which is a vascularized region of the mantle.

Fossil evidence suggests that early lobe-finned fishes had lungs somewhat similar to those of modern lungfish. The Australian lungfish can use either its gills or its lungs, depending on the conditions in its environment. The gills of African and South American lungfishes degenerate with age, so adults must rise to the surface and exchange gases entirely through their lungs.

Remains of early lobe-finned fishes occur extensively in the fossil record. Those found in Devonian strata are thought to be similar to the ancestors of amphibians, and numerous amphibian fossils occur in adjacent ancient strata. Some paleontologists hypothesize that the evolution of lungs occurred as an adaptation to the periodic droughts that occurred in Devonian times and that lungs or lunglike structures may have been present in all early bony fishes. Most modern bony fishes have no lungs, but nearly all of them possess homologous **swim bladders** (see Chapter 30). By adjusting the amount of gas in its swim bladder, the fish can control its buoyancy.

Some amphibians do not have lungs. Among plethodontid (lungless) salamanders, for example, all gas exchange takes place in the pharynx or across the thin, wet skin. However, even though they depend mainly on their body surface for gas exchange, most amphibians have lungs (Fig. 44–4). The lungs of mud puppies are two long, simple sacs richly supplied by capillaries. Frogs and toads have ridges containing connective tissue on the inside of the lungs, thereby increasing the respiratory surface somewhat.

The lungs of most reptiles are rather simple sacs, with only some folding of the wall to increase the surface for gas exchange. Gas exchange is not very efficient and does not supply sufficient oxygen to sustain long periods of activity. In some lizards, turtles, and crocodiles, the lungs are somewhat more complex with subdivisions that give them a spongy texture.

Birds have the most efficient respiratory system of any living vertebrate. Very active, endothermic animals with high metabolic rates, birds require large amounts of oxygen to sustain flight and other activities. Their small, bright red lungs have extensions (usually nine) called **air sacs,** which reach into all parts of the body and even connect with air spaces in some of the bones (Fig. 44–5). The air sacs act as bellows drawing air into the system. Collapse of the air sacs during expiration (exhalation) forces air out. Gas exchange does not take place across the walls of the air sacs.

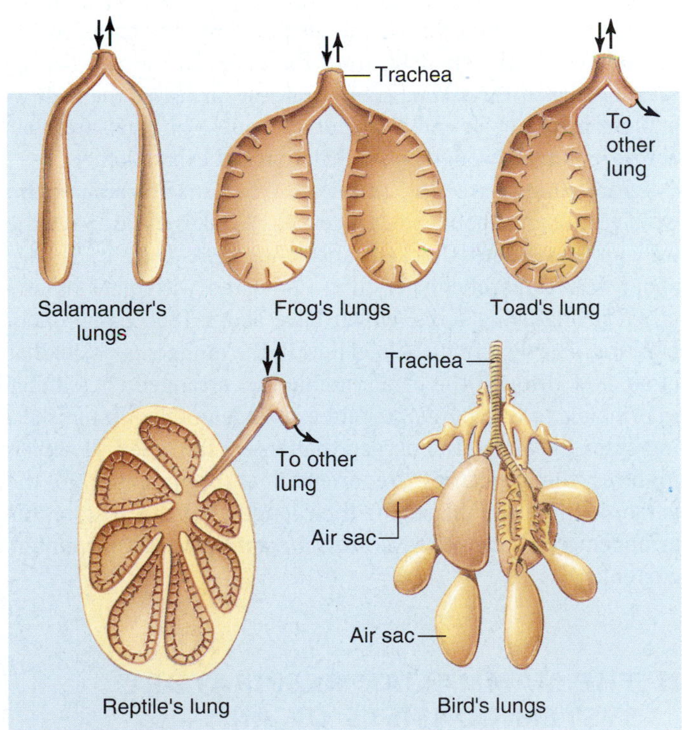

Figure 44–4 Comparison of vertebrate lungs. The surface area of the lung has increased during vertebrate evolution. Salamander lungs are simple sacs. Other amphibians and reptiles have lungs with small ridges or folds that help increase surface area. Birds have an elaborate system of lungs and air sacs. Mammalian lungs (shown in Fig. 44–6, page 975) have millions of alveoli that increase the surface available for gas exchange.

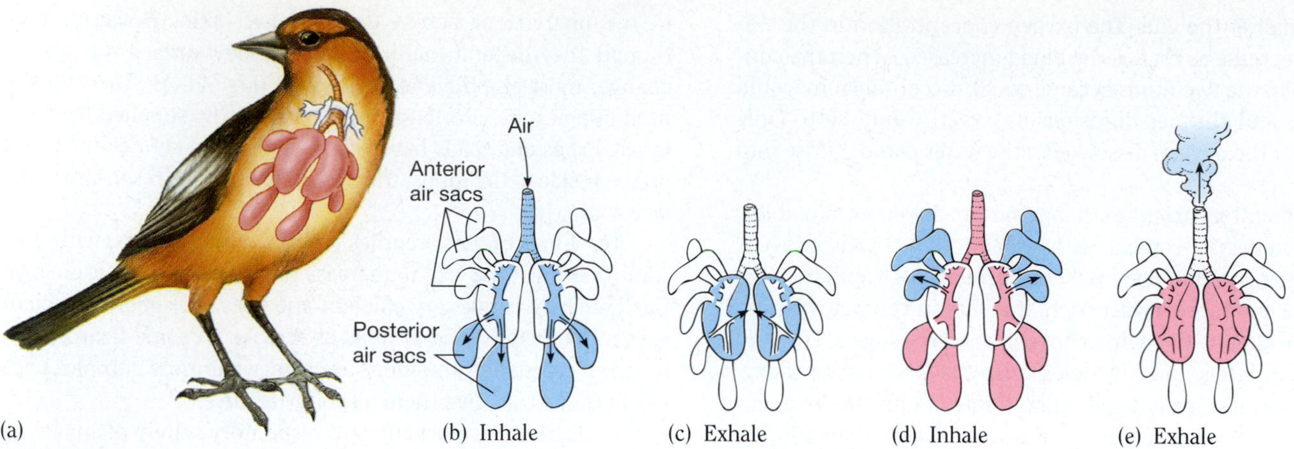

Figure 44–5 How bird lungs function. The bird has a breathing process that requires two cycles of inhalation and exhalation to support a one-way flow of air through the lungs. **(a)** The bird respiratory system includes lungs and extensions called air sacs. **(b)** As the bird inhales, new air flows into the posterior air sacs *(blue)* and partly into the lungs *(not shown)*. **(c)** As the bird exhales, this air is forced into the lungs. **(d)** At the second inhalation, most of the air from the first breath moves into the anterior air sacs. Air from the second inhalation flows into the posterior air sacs *(pink)* and partly into the lungs *(not shown)*. **(e)** At the second exhalation, most of the air from the first inhalation leaves the body, and air from the second inhalation flows into the lungs.

The high-performance bird respiratory system is arranged so that air flows in one direction through the lungs and is renewed during a two-cycle process. Air entering the body passes into the posterior air sacs and the part of the lungs closest to these air sacs. When the bird exhales, that air flows into the lungs. At the second breath, the air flows from the lungs to the anterior air sacs. Finally, at the second exhalation, the air leaves the body as another breath of air enters the lungs. Thus, a bird gets fresh air across its lungs through both inhalation and exhalation.

Bird lungs have tiny, thin-walled tubes, the **parabronchi,** which are open at both ends. Gas exchange takes place across the walls of these tubes. Chickens and other weak-flying birds have about 400 parabronchi per lung compared with pigeons and other strong-flying birds, which have about 1800 parabronchi per lung. The direction of blood flow in the lungs is opposite that of air flow through the parabronchi. This arrangement, similar in principle to the countercurrent exchange in the gills of fishes, increases the amount of oxygen that enters the blood. However, in birds the capillaries are oriented at right angles to the parabronchi rather than along their length. For this reason, this arrangement is referred to as *crosscurrent,* rather than countercurrent.

THE MAMMALIAN RESPIRATORY SYSTEM CONSISTS OF AN AIRWAY AND LUNGS

The respiratory system of mammals consists of the lungs and a series of tubes through which air passes on its journey from the nostrils to the lungs and back (Fig. 44–6). The complex lungs have an enormous surface area. In the following sections we use the human respiratory system as an example.

The airway conducts air into the lungs

A breath of air enters the body through the **nostrils** and flows through the **nasal cavities.** As air passes through the nose, it is filtered, moistened, and brought to body temperature. The nasal cavities are lined with a moist, ciliated epithelium that is rich in blood vessels. Inhaled dirt, bacteria, and other foreign particles are trapped in the stream of mucus that is produced by cells within the epithelium and pushed along toward the throat by the cilia. In this way, foreign particles are delivered to the digestive system, which can more effectively dispose of such materials than can the delicate lungs. A person normally swallows more than a pint of nasal mucus each day, more during an infection or allergic reaction.

The back of the nasal cavities is continuous with the throat region, or **pharynx.** Air finds its way into the pharynx whether one breathes through the nose or mouth. An opening in the floor of the pharynx leads into the **larynx,** sometimes called the "Adam's apple." Because the larynx contains the vocal cords, it is also referred to as the voice box. Cartilage embedded in its wall prevents the larynx from collapsing and makes it hard to the touch when felt through the neck.

During swallowing, a flap of tissue called the **epiglottis** automatically closes off the larynx so that food and liquid enter the esophagus rather than the lower airway. Should this defense mechanism fail and foreign matter enter the sensitive larynx, a cough reflex is initiated, expelling the material. Despite these mechanisms, choking sometimes occurs.

From the larynx, air passes into the **trachea,** or windpipe, which is kept from collapsing by rings of cartilage in its wall. The trachea divides into two branches, the **bronchi** (sing., *bronchus*), each of which connects to a lung.

Both trachea and bronchi are lined by a mucous membrane containing ciliated cells. Many medium-sized particles that have

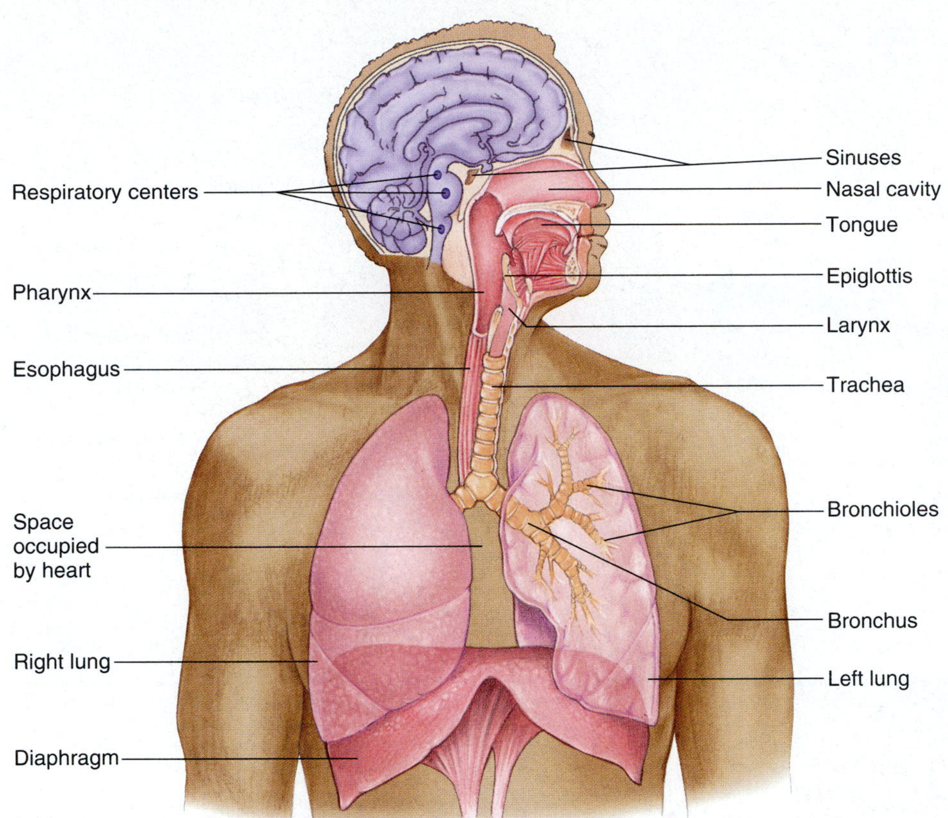

Figure 44–6 The human respiratory system. The muscular diaphragm forms the floor of the thoracic cavity. An internal view of one lung illustrates a portion of its extensive system of air passageways. The respiratory centers in the brain regulate the rate of respiration.

escaped the cleansing mechanisms of nose and larynx are trapped here. Mucus that contains these particles is constantly beaten upward by the cilia to the pharynx, where it is periodically swallowed. This mechanism, functioning as a cilia-propelled elevator of mucus, helps keep foreign material out of the lungs.

Gas exchange occurs in the alveoli of the lungs

The lungs are large, paired, spongy organs occupying the *thoracic* (chest) *cavity*. The right lung is divided into three lobes, the left lung into two lobes. Each lung is covered with a **pleural membrane,** which forms a continuous sac that encloses the lung and becomes the lining of the thoracic cavity. The *pleural cavity* is the space between the pleural membranes. A film of fluid in the pleural cavity provides lubrication between the lungs and the chest wall.

Because the lung consists largely of air tubes and elastic tissue, it is a spongy, elastic organ with a very large internal surface area for gas exchange. Inside the lungs the bronchi branch, becoming smaller and more numerous. These branches give rise to more than 1 million tiny **bronchioles** in each lung. Each bronchiole ends in a cluster of tiny air sacs, the **alveoli** (sing., *alveolus*) (Fig. 44–7). Each human lung contains more than 300 million alveoli with an internal surface area the approximate size of a tennis court. Each alveolus is lined by an extremely thin, single layer of epithelial cells. Gases diffuse freely through the wall of the alveolus and into the capillaries that surround it. Only two thin cell layers, the epithelia of the alveolar wall and the capillary wall, separate the air in the alveolus from the blood.

In summary, the sequence of structures through which air passes after it enters the body is the following:

Nostrils ⟶ nasal cavities ⟶ pharynx ⟶ larynx ⟶ trachea ⟶ bronchi ⟶ bronchioles ⟶ alveoli

Ventilation is accomplished by breathing

Breathing is the mechanical process of moving air from the environment into the lungs and of expelling air from the lungs. Inhaling air is referred to as **inspiration;** exhaling air is **expiration.** The thoracic cavity is closed so that no air can enter except through the trachea. (When the chest wall is punctured, for example, by a gunshot wound, air enters the pleural space and the lung collapses.)

During inspiration, the volume of the thoracic cavity is increased by the contraction of the **diaphragm,** the dome-shaped muscle that forms its floor. When the diaphragm contracts, it moves downward, increasing the volume of the thoracic cavity (Fig. 44–8). During forced inspiration, when a large volume of air is inspired, the *external intercostal muscles* contract as well. This action moves the ribs upward, also increasing the volume of the thoracic cavity. Because the lungs adhere to the walls of the thoracic cavity, when the volume of the thoracic cavity increases,

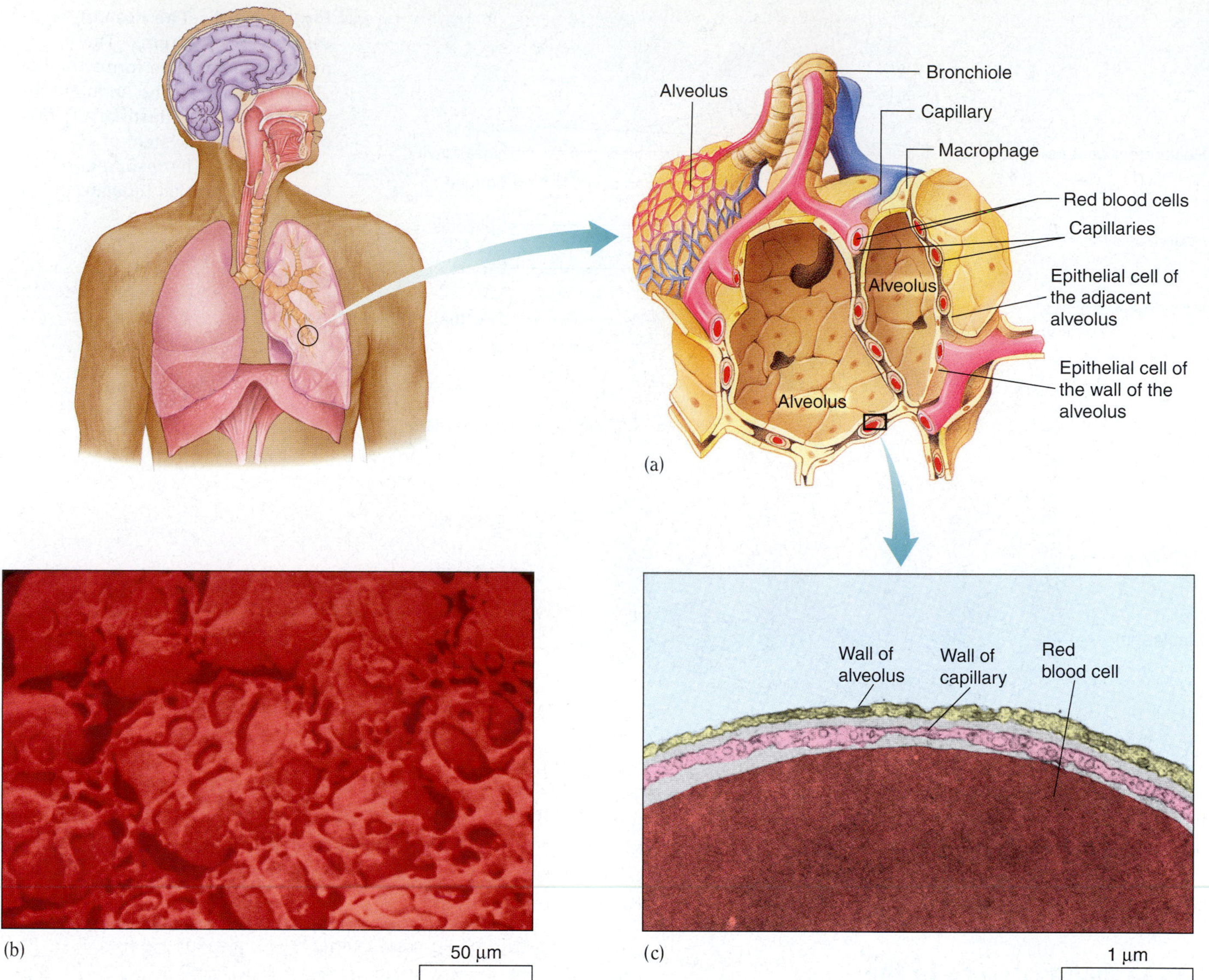

Figure 44–7 Structure of alveoli. **(a)** Gas exchange takes place across the thin wall of the alveolus. Note that the alveolar wall consists of extremely thin squamous epithelium. Extensive capillary networks lie between the walls of the alveoli. **(b)** Color-enhanced SEM showing the capillary network surrounding a portion of several alveoli. **(c)** Color-enhanced TEM of a portion of a capillary and the wall of an alveolus. The dark structure extending through the capillary is a portion of a red blood cell. The wall of the alveolus is visible just above the wall of the capillary. Notice the very short distance oxygen must diffuse to get from the air within the alveolus to the red blood cells that transport it to the body tissues. *(b, R.G. Kessel/ Visuals Unlimited; c, Courtesy of Drs. Peter Gehr, Marianne Bachofen, and Ewald R. Wiebel)*

the space within each lung also increases. The air in the lungs now has more space in which to move about, and the pressure of the air in the lungs falls by 2 or 3 millimeters of mercury (mm Hg) below the pressure of the air outside the body. As a result of this pressure difference, air from the outside rushes in through the respiratory passageways and fills the lungs until the two pressures are equal once again.

Expiration occurs when the diaphragm relaxes. The volume of the chest cavity decreases, increasing the pressure in the lungs to 2 to 3 mm Hg above atmospheric pressure. The millions of distended air sacs partially deflate, expelling the air that was inhaled. The pressure returns to normal, and the lung is ready for another inspiration. Thus, in inspiration, the millions of alveoli fill with air like so many tiny balloons. Then during expiration, the air rushes out of the alveoli, partially deflating them. During exercise or forced expiration, muscles of the abdominal wall and the *internal intercostal muscles* contract, pushing the diaphragm up and the ribs down.

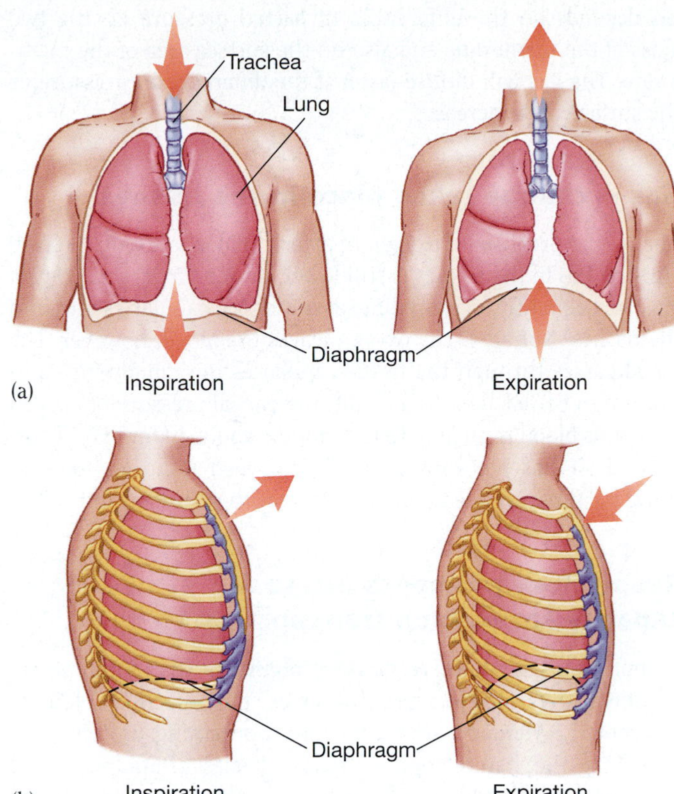

Figure 44–8 Mechanics of breathing. **(a)** Changes in the position of the diaphragm in expiration and inspiration result in changes in the volume of the thoracic cavity. During inspiration, the diaphragm contracts, increasing the volume of the thoracic cavity. When the volume of the thoracic cavity increases, air moves into the lungs. During expiration, the diaphragm relaxes, decreasing the volume of the thoracic cavity. **(b)** During forced inspiration, the elevation of the front ends of the ribs by the external intercostal muscles causes an increase in the front-to-back dimension of the chest and a corresponding increase in the volume of the thoracic cavity. During expiration, the rib cage drops, decreasing the volume of the thoracic cavity.

Some of the work involved in stretching the thorax and the lungs is necessary to stretch elastic connective tissue. Work is also required to overcome the cohesive force of water molecules associated with the respiratory membrane. The forces between the water molecules produce a surface tension that resists stretching. The work of breathing is reduced by **pulmonary surfactant,** a detergent-like phospholipid mixture secreted by specialized epithelial cells in the lining of the alveoli. Pulmonary surfactant intersperses between the water molecules, reducing their cohesive force. This action markedly reduces the surface tension of the water, prevents the alveoli from collapsing, and reduces the energy required to stretch the lungs. Premature infants are often unable to produce sufficient amounts of surfactant and suffer from respiratory distress syndrome (formerly referred to as hyaline membrane disease). In these infants, high surface tension makes it difficult to inflate the lungs, and the alveoli collapse during expiration. Breathing is labored. These infants may be placed on respirators that help them breathe.

The quantity of respired air can be measured

The amount of air moved into and out of the lungs with each normal resting breath is called the **tidal volume.** The normal tidal volume is about 500 mL. The **vital capacity** is the maximum amount of air a person can exhale after filling the lungs to the maximum extent.

Vital capacity is greater than tidal volume because the lungs are not completely emptied of stale air and filled with fresh air with each normal resting breath. The volume of air that remains in the lungs at the end of a normal expiration is the **residual capacity.** Table 44–1 shows the percentages of oxygen and carbon dioxide present in exhaled (alveolar) air compared with inhaled air. Because carbon dioxide is produced during cellular respiration, more of this gas—100 times more, in fact—enters the alveoli from the blood than is present in air inhaled from the environment.

Gas exchange takes place in the alveoli

The respiratory system delivers oxygen to the alveoli, but if oxygen were to remain in the lungs, all the other body cells would soon die. The vital link between alveolus and body cell is the circulatory system. Each alveolus serves as a tiny depot from which oxygen is loaded into blood brought close to the alveolar air by capillaries (Fig. 44–9).

Oxygen molecules efficiently pass by simple diffusion from the alveoli, where they are more concentrated, into the blood in the pulmonary capillaries, where they are less concentrated. At the same time, carbon dioxide moves from the blood, where it is more concentrated, to the alveoli, where it is less concentrated. Each gas diffuses through the single layer of cells lining the alveoli and the single layer of cells lining the capillaries.

Cellular respiration results in the continuous production of carbon dioxide and utilization of oxygen. Consequently, the

TABLE 44–1 Composition of Inhaled Air Compared with Exhaled Air

	Inhaled Air (atmospheric air)	Exhaled Air* (alveolar air)
% Oxygen (O_2)	20.9	14.0
% Carbon Dioxide (CO_2)	0.04	5.60
% Nitrogen (N_2)	79	79

*As indicated, the body uses up about one-third of the inhaled oxygen. The percentage of CO_2 increases more than 100-fold because it is produced during cellular respiration.

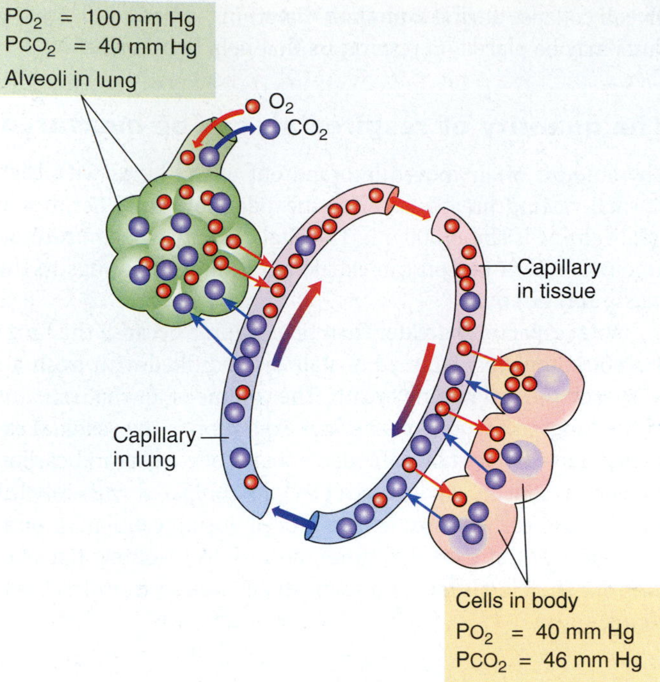

Figure 44-9 Gas exchange in the lungs and tissues. The concentration of oxygen is greater in the alveoli than in the pulmonary capillaries, so oxygen diffuses from the alveoli into the blood. Carbon dioxide is more concentrated in the blood than in the alveoli, so it diffuses out of the capillaries and into the alveoli. In the tissues, oxygen is more concentrated in the blood than in the body cells; it diffuses out of the capillaries into the cells. Carbon dioxide is more concentrated in the cells, so it diffuses out of the cells and moves into the blood. Note the differences in partial pressures of oxygen and carbon dioxide before and after gases are exchanged in the tissues.

concentration of oxygen in the cells is lower than in the capillaries entering the tissues, and the concentration of carbon dioxide is higher in the cells than in the capillaries. Thus, as blood circulates through capillaries of a tissue such as brain or muscle, oxygen moves by simple diffusion from the blood to the cells, while carbon dioxide moves from the cells into the blood.

The factor that determines the direction and rate of diffusion is the pressure or tension of the particular gas. According to **Dalton's law of partial pressures,** in a mixture of gases the total pressure of the mixture is the sum of the pressures of the individual gases. Each gas exerts, independently of the others, a **partial pressure**—the same pressure it would exert if it were present alone. At sea level, the barometric pressure (the pressure of Earth's atmosphere) is able to support a column of mercury 760 mm high. Because oxygen makes up about 21% of the atmosphere, oxygen's share of that pressure is $0.21 \times 760 = 160$ mm Hg. Thus, 160 mm Hg is the partial pressure of atmospheric O_2, abbreviated Po_2. In contrast the partial pressure of atmospheric CO_2 is 0.3 mm Hg, abbreviated Pco_2.

Fick's law of diffusion explains that the amount of oxygen or carbon dioxide that diffuses across the membrane of an alveolus depends on the differences in partial pressure on the two sides of the membrane and also on the surface area of the membrane. The gas will diffuse faster if the difference in pressure or the surface area increases.

Gas exchange takes place in the tissues

The partial pressure of oxygen in arterial blood is about 100 mm Hg. The Po_2 in the tissues is still lower, ranging from 0 to 40 mm Hg. Consequently, oxygen diffuses out of the capillaries and into the tissues. Not all of the oxygen leaves the blood, however. The blood passes through the tissue capillaries too rapidly for equilibrium to be reached. As a result, the partial pressure of oxygen in venous blood returning to the lungs is about 40 mm Hg. Thus, expired air has had only part of its oxygen removed—a good thing for those in need of mouth-to-mouth resuscitation!

Respiratory pigments increase capacity for oxygen transport

Complex animals have **respiratory pigments** that combine reversibly with oxygen and greatly increase the capacity of blood to transport it. **Hemocyanins** are copper-containing proteins dispersed in the hemolymph of many species of mollusks and arthropods. Without oxygen, these pigments are colorless. When oxygen combines with the copper, hemocyanins are blue.

Hemoglobin and **myoglobin** are the most common respiratory pigments in animals. Myoglobin is a form of hemoglobin found in muscle fibers. Hemoglobin is the only type of pigment found in the blood of vertebrates. It is also present in many invertebrate species, including annelids, nematodes, mollusks, and arthropods. In some of these animals, the hemoglobin is dispersed in the plasma rather than confined to blood cells.

Hemoglobin is actually a general name for a group of related compounds, all of which consist of an iron-porphyrin, or heme, group bound to a protein known as a globin. The protein portion of the molecule varies in size, amino acid composition, and physical properties among various species. When combined with oxygen, hemoglobin is bright red; without oxygen, it appears dark red, imparting a purplish color to venous blood.

At rest, the cells of the human body use about 250 mL of oxygen per minute, or about 300 L every 24 hours. With exercise or work this rate may increase as much as 10- or 15-fold. If oxygen were simply dissolved in plasma, blood would have to circulate through the body at a rate of 180 L per minute to supply enough oxygen to the cells at rest. This is because, as we have seen, oxygen is not very soluble in blood plasma. Actually, the blood of a human at rest circulates at a rate of about 5 L per minute and supplies all of the oxygen the cells need. Why are only 5 L per minute rather than 180 L per minute required?

The answer is that hemoglobin transports about 97% of the oxygen. Only about 3% is dissolved in the plasma. In humans and other mammals, inspired oxygen diffuses out of the alveoli and enters the pulmonary capillaries. Plasma in equilibrium with alveolar air can take up only 0.25 mL of oxygen per 100 mL. However, oxygen diffuses into red blood cells and combines with hemoglobin. The properties of hemoglobin permit whole blood to carry some 20 mL of oxygen per 100 mL.

The protein portion of hemoglobin is composed of four peptide chains, typically two α and two β chains, each attached to a heme (iron-porphyrin) ring (see Fig. 3–23*a*). An iron atom is bound in the center of each heme ring. Hemoglobin has the remarkable property of forming a weak chemical bond with oxygen. An oxygen molecule can attach to the iron atom in each heme. In the lung (or gill), oxygen diffuses into the red blood cells and combines with hemoglobin (Hb) to form **oxyhemoglobin (HbO_2).** Because the chemical bond formed between the oxygen and the hemoglobin is weak, the reaction is readily reversible. As blood circulates through tissues where the oxygen concentration is low, the reaction proceeds to the left. Hemoglobin releases oxygen, which diffuses out of the blood and into the tissue cells.

$$Hb + O_2 \rightleftharpoons HbO_2$$

The maximum amount of oxygen that can be transported by hemoglobin is called the **oxygen-carrying capacity.** The actual amount of oxygen bound to hemoglobin is the **oxygen content.** The ratio of O_2 content to O_2 capacity is the **percent O_2 saturation** of the hemoglobin. The percent saturation is highest in the pulmonary capillaries, where the concentration of oxygen is greatest. In the capillaries of the tissues, where there is less oxygen, the oxyhemoglobin dissociates, releasing oxygen. There, the percent saturation of hemoglobin is correspondingly lower. The **oxygen-hemoglobin dissociation curve** shown in Figure 44–10*a* illustrates this relationship. As oxygen concentration increases, there is a progressive increase in the percentage of hemoglobin that is combined with oxygen. The ability of oxygen to combine with hemoglobin and be released from oxyhemoglobin is influenced by several factors in addition to percent O_2 saturation; these include pH, carbon dioxide concentration, and temperature.

Carbon dioxide formed in respiring tissue reacts with water in the plasma to form carbonic acid, H_2CO_3. In this way any increase in the carbon dioxide concentration increases the acidity (lowers the pH) of the blood. Oxyhemoglobin dissociates its oxygen more readily in an acidic environment than in an environment with normal pH. Displacement of the oxygen-hemoglobin dissociation curve by a change in pH is known as the **Bohr effect** (Fig. 44–10*b*). Lactic acid released from active muscles also lowers the pH of the blood and has a similar effect on the oxygen-hemoglobin dissociation curve.

Some carbon dioxide is transported by the hemoglobin molecule. Although it attaches to the hemoglobin molecule in a different way and at a different site than oxygen, the attachment of a carbon dioxide molecule causes the release of an oxygen molecule from the hemoglobin. The effect of carbon dioxide concentration on the oxygen-hemoglobin dissociation curve is important. In the capillaries of the lungs (or gills in fishes), carbon dioxide concentration is relatively low and oxygen concentration is high, so oxygen combines with a very high percentage of hemoglobin. In the capillaries of the tissues, carbon dioxide concentration is high and oxygen concentration is low, so oxygen is

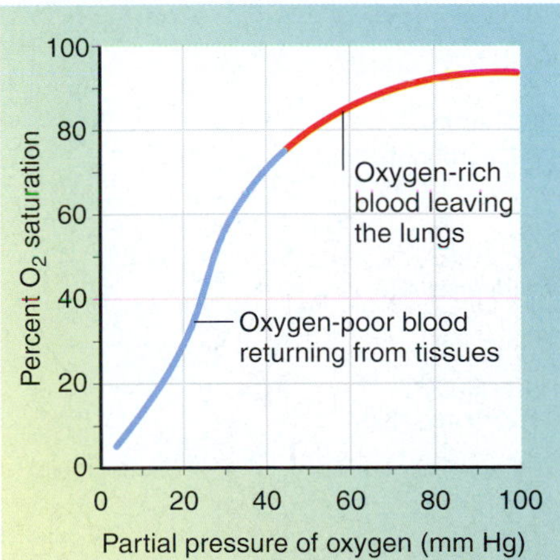

(a) Normal oxygen-hemoglobin dissociation curve

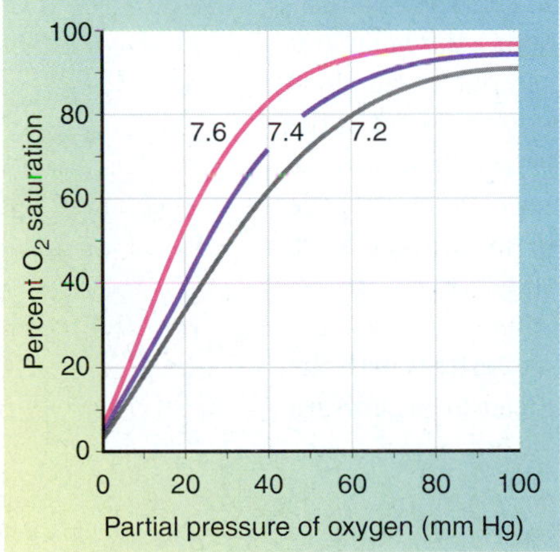

(b) Effect of pH

Figure 44–10 Oxygen-hemoglobin dissociation curves. (a) Normal curve showing the relationship between the partial pressure of oxygen *(horizontal axis)* and the percent oxygen saturation of hemoglobin *(vertical axis)*. Oxygen binds to hemoglobin in the lungs and dissociates from hemoglobin in the tissues. **(b)** Effect of pH on the oxygen-hemoglobin curve (Bohr effect). The normal pH of human blood is 7.4. Find the location on the horizontal axis where the partial pressure of oxygen is 40 and follow the line up through the curves. Notice how the saturation of hemoglobin with oxygen differs among the three curves, even though the partial pressure of oxygen is the same. There is an increased ability to bind oxygen at higher pH (7.6). If the pH is lower (7.2), there is a decreased ability of the hemoglobin to bind oxygen, so oxygen is unloaded. For example, during muscular activity the pH decreases, and more oxygen is unloaded and available for the muscle cells.

readily released from the hemoglobin. The oxygen and carbon dioxide bound to hemoglobin in red blood cells does not contribute to the partial pressures that govern diffusion. The association and dissociation of gases from hemoglobin depend on partial pressures of the plasma and interstitial fluid (which is determined by the gases dissolved in these fluids).

Carbon dioxide is transported mainly as bicarbonate ions

Carbon dioxide is transported in the blood in three ways. About 7% to 10% of carbon dioxide is dissolved in the plasma. Another 20% enters the red blood cells and combines with hemoglobin. Because the bond between the hemoglobin and carbon dioxide is very weak, the reaction is readily reversible. Most of the carbon dioxide (about 70%) is transported in the plasma as **bicarbonate ions** (HCO_3^-).

In the plasma, carbon dioxide slowly combines with water to form carbonic acid. This reaction proceeds much more rapidly inside red blood cells owing to the action of the enzyme **carbonic anhydrase** (Fig. 44–11). Carbonic acid dissociates, forming hydrogen ions and bicarbonate ions.

$$\underset{\text{Carbon dioxide}}{CO_2} + \underset{\text{Water}}{H_2O} \xrightarrow{\text{Carbonic anhydrase}} \underset{\text{Carbonic acid}}{H_2CO_3} \longrightarrow \underset{\text{Hydrogen ion}}{H^+} + \underset{\text{Bicarbonate ion}}{HCO_3^-}$$

Most of the hydrogen ions released from carbonic acid combine with hemoglobin, which is a very effective buffer. Many of the bicarbonate ions diffuse into the plasma. The action of carbonic anhydrase in the red blood cells maintains a diffusion gradient for carbon dioxide to move into and then out of red blood cells. As the negatively charged bicarbonate ions move out of the red blood cells, chloride ions (Cl^-) in the plasma diffuse into the red blood cells to replace them, a process known as the **chloride shift.** In the alveolar capillaries, CO_2 diffuses out of the plasma and into the alveoli. As the CO_2 concentration decreases, the reaction sequence above is reversed.

Any condition (such as pneumonia) that interferes with the removal of carbon dioxide by the lungs can lead to *respiratory acidosis*. In this condition there is an increased concentration of carbonic acid and bicarbonate ions in the blood. When the pH of the blood falls below 7.0, the central nervous system becomes de-

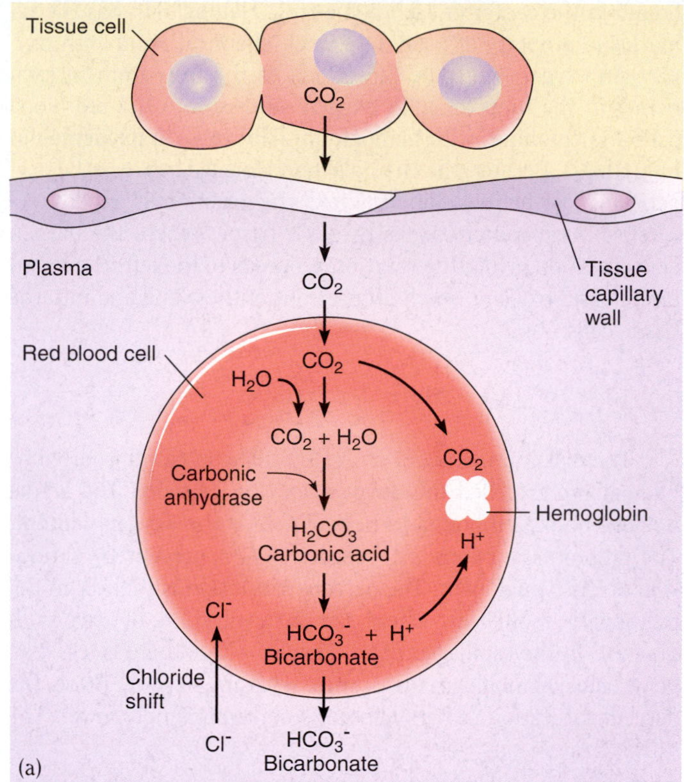

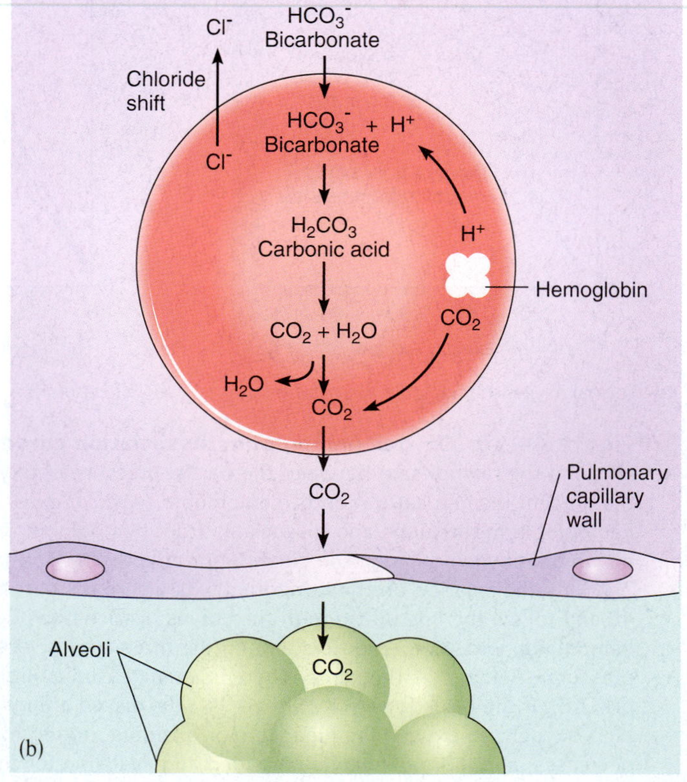

Figure 44–11 Carbon dioxide transport. **(a)** In the tissues, carbon dioxide diffuses from cells into the plasma. Most of the carbon dioxide enters red blood cells, where the enzyme carbonic anhydrase rapidly converts it to carbonic acid. Carbonic acid dissociates, forming bicarbonate and hydrogen ions. As bicarbonate ions move out into the plasma, chloride ions replace them. Hemoglobin combines with hydrogen ions, preventing a decrease in pH. **(b)** In the lungs, carbon dioxide diffuses out of the plasma and into the alveoli, and these processes are reversed. Bicarbonate ions diffuse from the plasma into the red blood cells. H^+ released from hemoglobin combines with bicarbonate ions, forming carbonic acid. Carbon dioxide produced from the carbonic acid diffuses out of the blood and into the alveoli.

pressed. Such depression causes the individual to become disoriented, and if not treated, respiratory acidosis can result in coma and death.

Breathing is regulated by respiratory centers in the brain

Breathing is a rhythmic, involuntary process regulated by **respiratory centers** in the brain stem (see Fig. 44–6). Groups of neurons in the dorsal region of the medulla regulate the basic rhythm of breathing. These neurons send a burst of impulses to the diaphragm and external intercostal muscles, causing them to contract. After several seconds, these neurons become inactive, the muscles relax, and expiration occurs. Respiratory centers in the pons help control the transition from inspiration to expiration. These centers can stimulate or inhibit the respiratory centers in the medulla.

The cycle of activity and inactivity repeats itself so that at rest we breathe about 14 times per minute. A group of neurons in the ventral region of the medulla becomes active only when we need to breathe forcefully. Overdose of certain medications such as barbiturates depresses the respiratory centers and may lead to respiratory failure.

The basic rhythm of respiration can be altered in response to changing needs of the body. When you are engaged in a strenuous game of tennis, you require more oxygen than when studying biology. During exercise the rate of aerobic cellular respiration increases, producing more carbon dioxide. The body must dispose of this carbon dioxide through increased ventilation.

Carbon dioxide concentration is the most important chemical stimulus for regulating the rate of respiration. Specialized **chemoreceptors** in the medulla and in the walls of the aorta and carotid arteries are sensitive to changes in arterial carbon dioxide concentration. When stimulated they send impulses to the respiratory centers, which lead to an increase in breathing rate.

The chemoreceptors in the walls of the aorta, called *aortic bodies,* and those in the walls of the carotid arteries, the *carotid bodies,* are sensitive to changes in hydrogen ion concentration and oxygen concentration, as well as to carbon dioxide levels. Recall that an increase in carbon dioxide concentration results in an increase in hydrogen ions from carbonic acid, lowering the pH of the blood. Even a slight decrease in pH stimulates these chemoreceptors, leading to a faster breathing rate. As carbon dioxide is removed by the lungs, the hydrogen ion concentration in the blood and other body fluids decreases, and homeostasis is reestablished.

Interestingly, oxygen concentration generally does not play an important role in regulating respiration. Only if the partial pressure of oxygen falls markedly do the chemoreceptors in the aorta and carotid arteries become stimulated to send messages to the respiratory centers.

Although breathing is an involuntary process, the action of the respiratory centers can be consciously influenced for a short time by stimulating or inhibiting them. For example, you can inhibit respiration by holding your breath. You cannot hold your breath indefinitely, however, because eventually you feel a strong urge to breathe. Even if you were able to ignore this, you would eventually pass out, and breathing would resume.

Individuals who have stopped breathing because of drowning, smoke inhalation, electric shock, or cardiac arrest can sometimes be sustained by mouth-to-mouth resuscitation until their own breathing reflexes are initiated again. **Cardiopulmonary resuscitation (CPR)** is a method for aiding victims who have suffered respiratory and cardiac arrest. CPR must be started immediately, because irreversible brain damage may occur within about 4 minutes of respiratory arrest. A number of organizations offer training in CPR.

Hyperventilation reduces carbon dioxide concentration

Underwater swimmers and some Asian pearl divers voluntarily **hyperventilate** before going under water. By making a series of deep inhalations and exhalations, they "blow off" CO_2, markedly reducing the carbon dioxide content of the alveolar air and of the blood. As a result, it takes longer before the urge to breathe becomes irresistible.

When hyperventilation is continued for a long period, dizziness and sometimes unconsciousness may occur. This is because a certain concentration of carbon dioxide is needed in the blood to maintain normal blood pressure. (This mechanism operates by way of the vasoconstrictor center in the brain, which maintains the muscle tone of blood vessel walls.) Furthermore, if divers hold their breath too long, the low concentration of oxygen may result in unconsciousness and drowning.

High flying or deep diving can disrupt homeostasis

The barometric pressure decreases at progressively higher altitudes. Because the concentration of oxygen in the air remains at 21%, the partial pressure of oxygen decreases along with the barometric pressure. At an altitude of 6000 m (19,500 ft), the barometric pressure is about 350 mm Hg, the partial pressure of oxygen is about 75 mm Hg, and the hemoglobin in arterial blood is about 70% saturated with oxygen. At 10,000 m (33,000 ft) the barometric pressure is about 225 mm Hg, the partial pressure of oxygen is 50 mm Hg, and arterial oxygen saturation is only 20%. Thus, getting sufficient oxygen from the air becomes an ever-increasing problem at higher altitudes.

When a person moves to a high altitude, the body adjusts over time by producing a greater number of red blood cells. In a person breathing pure oxygen at 10,000 m, the oxygen would have a partial pressure of 225 mm Hg and the hemoglobin would be almost fully saturated with oxygen. Above 13,000 m, however, barometric pressure is so low that even breathing pure oxygen does not permit complete oxygen saturation of arterial hemoglobin.

A person becomes unconscious when the arterial oxygen saturation falls to between 40% and 50%. This level is reached at about 7000 m (23,000 ft) when the person is breathing air, or 14,500 m (47,100 ft) when pure oxygen is used. All high-flying jets have airtight cabins pressurized to the equivalent of the barometric pressure at an altitude of about 2000 m.

Shallow breathing, which occurs in many respiratory diseases, causes **hypoxia,** a deficiency of oxygen. Even *rapid,* shallow breathing results in hypoxia; this is because the tidal volume includes the air in the respiratory passages. When we breathe shallowly, we may not clear out the stale air in the airway and ventilate the lung. Hypoxia causes drowsiness, mental fatigue, headache, and sometimes euphoria. The ability to think and make judgments is impaired, as is the ability to perform tasks requiring coordination. If a jet flying at 11,700 m (over 38,000 ft) underwent sudden decompression, the pilot would lose consciousness in about 30 seconds and become comatose in about 1 minute.

In addition to the problems of hypoxia, a rapid decrease in barometric pressure can cause **decompression sickness** (commonly known as the "bends" because those suffering from it bend over in pain). Whenever the barometric pressure drops below the total pressure of all gases dissolved in the blood and other body fluids, the dissolved gases tend to come out of solution and form gas bubbles. A familiar example of such bubbling occurs each time you uncap a bottle of soda, thus reducing the pressure in the bottle. Carbon dioxide is released from solution and bubbles out into the air. In the body, nitrogen has a low solubility in blood and tissues. When it comes out of solution, the bubbles formed may damage tissues and block capillaries, interfering with blood flow. The clinical effects of decompression sickness are pain, dizziness, paralysis, unconsciousness, and even death.

Decompression sickness is even more common in scuba diving than in high-altitude flying. As a diver descends, the surrounding pressure increases tremendously—1 atmosphere (the atmospheric pressure at sea level, which equals 760 mm Hg) for each 10 m. To prevent the collapse of the lungs, a diver must be supplied with air under pressure, thereby exposing the lungs to very high alveolar gas pressures.

At sea level an adult human has about 1 L of nitrogen dissolved in the body, with about half in the fat and half in the body fluids. After a diver's body has been saturated with nitrogen at a depth of 100 m (325 ft), the body fluids contain about 10 L of nitrogen. To prevent this nitrogen from rapidly bubbling out of solution and causing decompression sickness, the diver must be brought to the surface gradually, with stops at certain levels on the way up. This allows the nitrogen to be expelled slowly through the lungs.

Some mammals are adapted for diving

Some air-breathing mammals can spend rather long periods in the ocean depths without coming up for air. Dolphins, whales, seals, and beavers have structural and physiological adaptations that permit them to dive for food or to elude their enemies (Fig. 44–12). With their streamlined bodies and forelimbs modified as fins or flippers, diving mammals perform impressive aquatic feats. The Weddell seal *(Leptonychotoes weddelli)* can swim under the ice at a depth of 596 m (1955 ft) for more than an hour without coming up for air. The enormous northern elephant seal *(Mirounga angustirostris),* which measures about 5 m (16 to 18 ft) in length and weighs 2 to 4 tons, can plunge even deeper. A female elephant seal carried a depth recorder to a depth of more than 1500 m (almost 5000 ft) and stayed beneath the surface 2 hours.

Figure 44–12 Bottlenose whales *(Hyperoodon ampullatus)* may be the deepest divers on Earth. Researchers have recorded their dives at depths of more than 5000 feet. *(Flip Nicklin/Minden Pictures)*

Turtles and birds that dive depend on oxygen stored in their lungs. Diving mammals do not take in and store extra air before a dive. In fact, seals exhale before they dive. With less air in their lungs they are less buoyant. Their lungs collapse at about 50 to 70 m into their dive and then reinflate as they ascend. This means that their lungs do not function for most of the dive. These adaptations are thought to reduce the chance of decompression sickness, because with less air in the lungs there is less nitrogen in the blood to dissolve during the dive.

Physiological adaptations, including ways to distribute and store oxygen, permit some mammals to dive deeply and remain under water for long periods. Seals have about twice the volume of blood, relative to their body weight, as nondiving mammals. Diving mammals also have high concentrations of *myoglobin,* an oxygen-binding pigment similar to hemoglobin, which stores oxygen in muscles. These animals have up to ten times more myoglobin than do terrestrial mammals. The very large spleen typical of many diving mammals is thought to store oxygen-rich red blood cells. Under pressure (during a dive), the spleen is squeezed and releases these red blood cells into the circulation.

Process of Science Biologists have wondered how diving mammals balance the energy demands of swimming and diving with the need to conserve energy when submerged. In 2000, Terrie M. Williams and her research team reported in *Science* that these diving mammals markedly reduce the energy expended in deep (more than 200 m) diving by gliding. These investigators filmed video sequences of diving seals and whales, showing that they glide most of the way down. Gliding is possible because the buoyancy of the animal decreases as the lungs gradually collapse, decreasing the amount of air in the lungs.

When a mammal dives to its limit, a group of physiological mechanisms known collectively as the **diving reflex** are activated. Metabolic rate decreases by about 20%, which conserves oxygen. Breathing stops and bradycardia (slowing of the heart rate) occurs. The heart rate may decrease to one-tenth of the normal rate, reducing the body's consumption of oxygen and energy. Blood is redistributed; skin, muscles, digestive organs, and other internal organs can survive with less oxygen and receive less blood while an animal is submerged.

The diving reflex is present to some extent in humans, where it may act as a protective mechanism during birth, when an infant may be deprived of oxygen for several minutes. Many cases of near-drownings, especially of young children, have been documented in which the victim had been submerged for as long as 45 minutes in very cold water before being rescued and resuscitated. In many of these survivors there was no apparent brain damage. The shock of the icy water slows the heart rate, increases blood pressure, and shunts the blood to the internal organs of the body that most need oxygen (blood flow in the arms and legs decreases). Metabolic rate decreases so that less oxygen is required.

BREATHING POLLUTED AIR DAMAGES THE RESPIRATORY SYSTEM

Several defense mechanisms protect the delicate lungs from the harmful substances we breathe (Fig. 44–13). The hair around the nostrils, the ciliated mucous lining in the nose and pharynx, and the cilia-mucus elevator of the trachea and bronchi serve to trap foreign particles in inspired air. One of the body's most rapid defense responses to breathing dirty air is **bronchial constriction.** In this process, the bronchial tubes narrow, increasing the chance that inhaled particles will land on the sticky mucous lining. Unfortunately, bronchial constriction narrows the airway so that less air can pass through to the lungs, thus decreasing the amount of oxygen available to body cells. Fifteen puffs on a cigarette during a 5-minute period increases airway resistance as much as threefold, and this added resistance to breathing lasts more than 30 minutes. Chain-smokers and those who breathe heavily polluted air may remain in a state of chronic bronchial constriction.

Figure 44–13 Urban air pollution. Industry spews tons of pollutants into the atmosphere. Harmful gases, such as sulfur dioxide and nitrogen oxides, contribute to acid rain. *(David Nunuk/Photo Researchers, Inc.)*

Neither the smallest bronchioles nor the alveoli are equipped with mucus or ciliated cells. Foreign particles that get through other respiratory defenses and find their way into the alveoli may be engulfed by macrophages. The macrophages may then accumulate in the lymph tissue of the lungs. Lung tissues of chronic smokers and those who work in dirty fossil fuel–burning industries contain large blackened areas where carbon particles have been deposited (Fig. 44–14).

Continued insult to the respiratory system results in disease. Chronic bronchitis, pulmonary emphysema, and lung cancer have been linked to cigarette smoking and breathing polluted air. More than 75% of patients with **chronic bronchitis** have a history of heavy cigarette smoking (see *Focus On: The Effects of Smoking*). In chronic bronchitis, irritation from inhaled pollutants causes the bronchial tubes to secrete too much mucus. Ciliated cells, damaged by the pollutants, cannot effectively clear the mucus and trapped particles from the airways. The body resorts to coughing in an attempt to clear the airways. The bronchioles become constricted and inflamed, and the patient is short of breath.

Victims of chronic bronchitis often develop **pulmonary emphysema,** a disease most common in cigarette smokers. In this disorder, alveoli lose their elasticity, and walls between adjacent alveoli are destroyed. The surface area of the lung is so reduced that gas exchange is seriously impaired. Air is not expelled effectively, and stale air accumulates in the lungs. The emphysema victim struggles for every breath, and still the body does not get enough oxygen. To compensate, the right ventricle of the heart pumps harder and becomes enlarged. Emphysema patients frequently die of heart failure.

Most patients with **chronic obstructive pulmonary disease (COPD),** a condition characterized by obstruction of the airflow, have both chronic bronchitis and emphysema. Asthma also contributes to COPD. Individuals with asthma respond to inhaled stimuli with exaggerated bronchial constriction, and the airway is typically inflamed.

Cigarette smoking is also the main cause of **lung cancer.** More than 40 of the compounds in the tar of tobacco smoke have been shown to cause cancer. These carcinogenic substances irritate the cells lining the respiratory passages and alter their metabolic balance. Normal cells are transformed into cancer cells, which may multiply rapidly and invade surrounding tissues.

FOCUS ON The Effects of Smoking

Smoking is the single most preventable cause of death in our society. In the United States alone, more than 400,000 adults die each year of tobacco-related causes. Tobacco smoke is by far the most important risk factor for lung cancer, the most common lethal cancer in the United States and in the world. In addition, tobacco smoke is an important risk factor for cardiovascular disease and a number of other diseases.

Nonsmokers are also affected by tobacco smoke. Cigarette smoke is a "portable" air pollutant. Smokers move about, exhaling tobacco smoke into the air we all must breathe. This environmental tobacco smoke has been linked to death from lung cancer of about 3000 nonsmokers each year in the United States.

Process of Science Cigarette smoke contains more than 4700 chemical compounds, including compounds that damage the inner lining of blood vessels, leading to development of atherosclerosis, and more than 40 known cancer-causing agents. Understanding the biochemical links between smoking and disease has been a challenging goal for researchers. In 1996, biologist Mikhail Denissenko and his colleagues reported in *Science* that they had demonstrated a direct link between a carcinogen in cigarette smoke and human cancer. The mechanism involves benzo-α pyrene, one of the most potent mutagens and carcinogens known. Denissenko's group used the polymerase chain reaction to show that this carcinogen causes mutations in the *p53* tumor suppressor gene. This gene is mutated in about 60% of lung cancers and also in many other types of cancer. (See *Focus On: Unwelcome Tissues: Cancers* in Chapter 37.)

Nicotine has been shown to be highly addictive. Each year about 35 million U.S. smokers try to quit, but more than 90% resume smoking within one year. A better understanding of the mechanisms leading to and sustaining nicotine addiction may help smokers who want to stop. Gaetano Di Chiara and his colleagues at the University of Cagliari in Italy injected nicotine into the veins of rats and then studied the effects on the brain. In 1996 these researchers reported in *Nature* that nicotine, like morphine, amphetamines, and cocaine, increases dopamine concentration and activates cells in the base of the forebrain, in an area called the *nucleus accumbens.* This area helps integrate emotion, and Di Chiara suggested that dopamine may facilitate learning an association between the pleasurable effects of the drug and other stimuli such as the smell of smoke. Di Chiara's work has been supported and extended. For example, in 2001, Eric Nestler of the University of Texas Southwestern Medical Center reported in *Science* that nicotine, like cocaine and other abused drugs, causes long-lasting changes in the brain. This researcher pointed out that cell signaling and other mechanisms underlying drug addiction are very similar to those of learning and memory. Investigators are also searching for drugs that block the dopamine release caused by nicotine and other addictive drugs.

Some facts about smoking include the following:

- The life of a 30-year-old who smokes 15 cigarettes a day is shortened by an average of more than five years.
- If you smoke more than one pack per day, you are about 20 times more likely to develop lung cancer than is a nonsmoker. According to the American Cancer Society, cigarette smoking causes more than 75% of all lung cancer deaths.
- If you smoke, you double your chances of dying from cardiovascular disease.

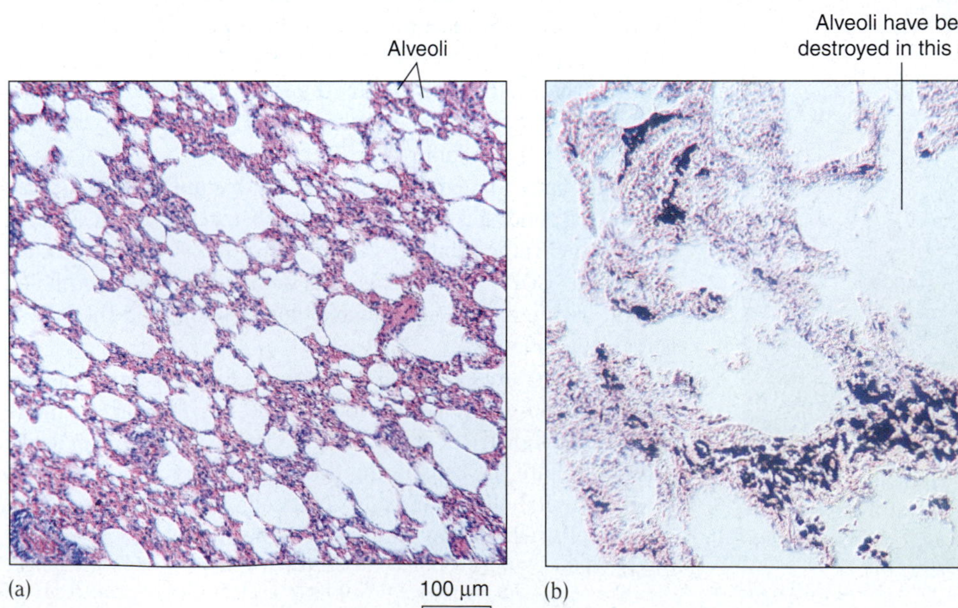

Figure 44–14 Comparison of normal and diseased lung tissue. **(a)** LM of normal lung tissue. **(b)** LM of lung tissue with accumulated carbon particles. Despite the body's defenses, when we inhale smoky, polluted air, especially over a long period, dirt particles enter the lung tissue, and some remain lodged there. *(a, b, Alfred Paseika/Taurus Photos)*

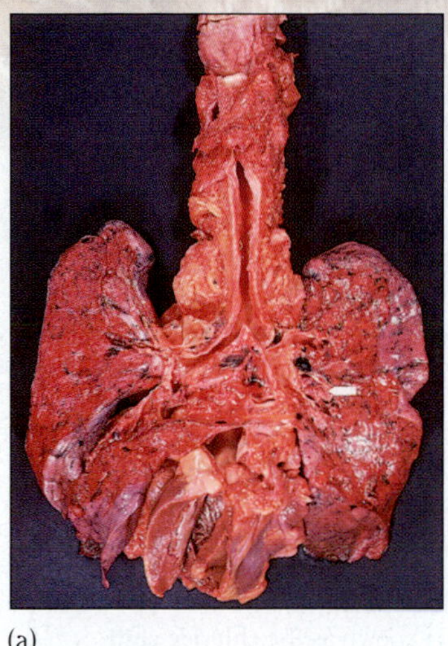

(a)

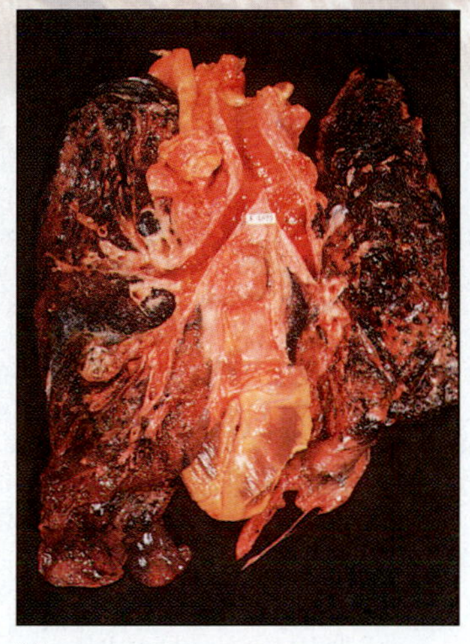

(b)

Effects of cigarette smoking.
(a) Lungs and major bronchi of a nonsmoker.
(b) Lungs and heart of a cigarette smoker. The dark spots in the lung tissue are particles of carbon, tar, and other substances that passed through the respiratory defenses and lodged in the lungs.
(a, b, Martin Rotker/Taurus Photos)

- If you smoke, you are 20 times more likely to develop chronic bronchitis and emphysema than is a nonsmoker.
- If you smoke, you have about 5% less oxygen circulating in your blood (because carbon monoxide binds to hemoglobin) than does a nonsmoker.
- If you smoke when you are pregnant, your baby will weigh about 6 ounces less at birth, and there is double the risk of miscarriage, stillbirth, and infant death.
- Infants whose parents smoke have double the risk of contracting pneumonia or bronchitis in their first year of life.
- Workers who smoke one or more packs of cigarettes per day are absent from their jobs because of illness 33% more often than are nonsmokers.
- When smokers quit smoking, their risk of dying from chronic pulmonary disease, cardiovascular disease, or cancer decreases. (Precise changes in risk depend on the number of years the person smoked, the number of cigarettes smoked per day, the age of starting to smoke, and the number of years since quitting.)
- Nicotine replacement with gum, patches, or nasal spray has been shown to be effective as an aid to smoking cessation with many individuals, especially when used in conjunction with behavioral therapy. More recently certain antidepressants have been used to reduce the craving for nicotine.

Almost every American who takes up smoking is a teenager—3000 every day, more than 1 million every year. The prevalence of smoking among adolescents increased significantly during the 1990s. Ten percent of children who begin smoking start by the fourth grade, and nearly two-thirds start by the tenth grade.

SUMMARY WITH KEY TERMS

I. **Respiration** is the process of gas exchange.
   A. During **organismic respiration** oxygen from the environment is taken up by the animal and delivered to its cells, while carbon dioxide is excreted into the environment.
   B. **Aerobic cellular respiration** is the process by which cells generate ATP through a series of redox reactions in which oxygen serves as the final electron acceptor. Carbon dioxide is produced as a waste product. Aerobic cellular respiration takes place in mitochondria.

II. Gas exchange in air has advantages and disadvantages compared with gas exchange in water.
   A. Air contains a higher concentration of molecular oxygen than does water.
   B. Oxygen diffuses more rapidly through air than through water.
   C. Air is less dense and less viscous than water and so less energy is needed to move air over a gas exchange surface.
   D. Desiccation is a problem for terrestrial animals; these animals have adaptations that protect their respiratory surfaces from drying.

III. Small aquatic animals can exchange gases by diffusion, with no specialized respiratory structures required. Large animals need specialized respiratory structures. Animals carry on **ventilation,** the process of actively moving air or water over respiratory surfaces.
   A. Some invertebrates, including most annelids, and a few vertebrates (many amphibians) exchange gases across the body surface.
   B. In insects and some other arthropods, air enters a network of **tracheal tubes,** or **tracheae,** through openings, called **spiracles,** along the body surface. Tracheal tubes branch and extend to all regions of the body.
   C. **Gills** are moist, thin projections of the body surface found mainly in aquatic animals.
      1. In chordates, gills are usually internal, located along the edges of the gill slits. In bony fishes the gills are protected by an **operculum.**
      2. In bony fishes, a **countercurrent exchange system** maximizes diffusion of oxygen into the blood and diffusion of carbon dioxide out of the blood.

D. Terrestrial vertebrates have **lungs** with some means of ventilating them.
   1. Most fishes do not have lungs but have a homologous swim bladder that permits the fish to control its buoyancy.
   2. Amphibians and reptiles have lungs with only some ridges or folds that increase surface area.
   3. In birds, the lungs have extensions called **air sacs** that act as bellows, drawing air into the system. Air flows in one direction: from the outside into the posterior air sacs, to the lung, through the anterior air sacs and then out of the body. Gas exchange takes place through the walls of the **parabronchi** in the lungs. A crosscurrent arrangement, in which blood flow is at right angles to the parabronchi, increases the amount of oxygen that enters the blood.

IV. Respiratory pigments, such as **hemoglobin** and **hemocyanins,** increase the capacity of blood to transport oxygen.

V. The mammalian respiratory system includes the lungs and a system of airways. A breath of air passes in sequence through the **nostrils, nasal cavities, pharynx, larynx, trachea, bronchi, bronchioles,** and **alveoli.** Each lung occupies a pleural cavity and is covered with a **pleural membrane. Pulmonary surfactant** reduces the cohesive force of the water molecules on the lining of the alveoli, which prevents the alveoli from collapsing and reduces the energy needed to stretch the lungs.
   A. During breathing, the **diaphragm** contracts, expanding the chest cavity. The membranous walls of the lungs move outward along with the chest walls, decreasing the pressure within the lungs. Air from outside the body rushes in through the air passageways and fills the lungs until the pressure equals atmospheric pressure.
   B. **Tidal volume** is the amount of air moved into and out of the lungs with each normal breath. **Vital capacity** is the maximum volume that can be exhaled after the lungs are filled to the maximum extent. The volume of air that remains in the lungs at the end of a normal expiration is the **residual capacity.**
   C. Oxygen and carbon dioxide are exchanged between alveoli and blood by diffusion.
      1. The pressure of a particular gas determines its direction and rate of diffusion.
      2. **Dalton's law of partial pressures** explains that in a mixture of gases, the total pressure of the mixture is the sum of the pressures of the individual gases. Thus, each gas in a mixture of gases exerts a **partial pressure,** the same pressure it would exert if it were present alone. The partial pressure of atmospheric oxygen, Po_2, is 160 mm Hg at sea level.
      3. According to **Fick's law of diffusion,** the greater the difference in pressure on the two sides of a membrane and the larger the surface area, the faster the gas will diffuse across the membrane.
   D. About 97% of the oxygen in the blood is transported as **oxyhemoglobin (HbO_2).**
      1. The maximum amount of oxygen that can be transported by hemoglobin is the **oxygen-carrying capacity.**
      2. The actual amount of oxygen bound to hemoglobin is the **oxygen content.**
      3. The **percent O_2 saturation,** the ratio of oxygen content to oxygen carrying capacity, is highest in pulmonary capillaries, where oxygen concentration is greatest.
      4. The **oxygen-hemoglobin dissociation curve** shows that as oxygen concentration increases, there is a progressive increase in the amount of hemoglobin that combines with oxygen. The curve is affected by pH, temperature, and CO_2 concentration.
      5. Owing to lowered pH caused by carbonic acid, oxyhemoglobin dissociates more readily as carbon dioxide concentration increases. This is the **Bohr effect.**
   E. About 70% of the carbon dioxide in the blood is transported as bicarbonate ions. About 20% combines with hemoglobin, and another 10% is dissolved in plasma.
      1. Carbon dioxide combines with water to form carbonic acid; the reaction is catalyzed by **carbonic anhydrase.** Carbonic acid dissociates, forming bicarbonate ions (HCO_3^-) and hydrogen ions (H^+).
      2. Hemoglobin combines with H^+, buffering the blood. Many bicarbonate ions diffuse into the plasma and are replaced by Cl^- ions; this exchange is known as the **chloride shift.**
   F. **Respiratory centers** in the medulla and pons regulate respiration. These centers are stimulated by **chemoreceptors** sensitive to an increase in carbon dioxide concentration. They also respond to an increase in hydrogen ions and to very low oxygen concentration.
   G. **Hyperventilation** reduces the concentration of carbon dioxide in the alveolar air and in the blood.
   H. As altitude increases, barometric pressure (the pressure of Earth's atmosphere) decreases, and less oxygen enters the blood. This situation can lead to **hypoxia,** a deficiency of oxygen, which can lead to loss of consciousness and death. In addition to hypoxia, rapid decrease in barometric pressure can cause **decompression sickness,** a condition especially common among divers who ascend too rapidly.

VI. Physiological adaptations, including ways to distribute and store oxygen, permit some mammals to dive deeply and remain under water for long periods. Diving mammals have high concentrations of **myoglobin,** a pigment that binds oxygen, in their muscles. The **diving reflex,** a group of physiological mechanisms, including a decrease in metabolic rate, is activated when a mammal dives to its limit.

VII. Inhaling polluted air results in **bronchial constriction,** increased mucous secretion, damage to ciliated cells, and coughing. Breathing polluted air or inhaling cigarette smoke can cause **chronic bronchitis, pulmonary emphysema,** and **lung cancer.**

POST-TEST

1. Breathing is an example of (a) countercurrent exchange (b) cellular respiration (c) ventilation (d) answers a, b, and c are correct (e) answers a and c only
2. Which of the following is a benefit of gas exchange in air compared with water? (a) higher concentration of molecular oxygen (b) oxygen diffuses more slowly in air (c) no energy required for ventilation (d) moist respiratory surface not needed (e) air is denser than water
3. Which of the following adaptations for gas exchange is most characteristic of insects? (a) lungs (b) tracheal tubes (tracheae) (c) parabronchi (d) air sacs (e) gills
4. Which of the following are accurately matched? (a) bony fish/operculum (b) insect/alveoli (c) bird/spiracles (d) mammal/gill filaments (e) shark/dermal gills
5. Tracheal tubes (tracheae) (a) are typically found in mollusks (b) are highly vascular (c) branch and extend to all the cells (d) are characteristic of many mammals (e) end in book lungs
6. The most efficient vertebrate respiratory system is found in (a) amphibians (b) birds (c) reptiles (d) mammals (e) humans
7. In a bird the correct sequence for a breath of air would be (a) anterior air sacs ⟶ posterior air sacs ⟶ lung (b) posterior air sacs ⟶ lung ⟶ anterior air sacs (c) parabronchi ⟶ posterior air sacs ⟶ anterior air sacs (d) posterior air sacs ⟶ alveoli ⟶ anterior air sacs (e) posterior air sacs ⟶ capillaries ⟶ cells
8. Respiratory pigments (a) combine reversibly with oxygen (b) are found only in vertebrates (c) all have a heme (porphyrin) group that combines with oxygen (d) diffuse into the air sacs (e) attach to the alveolar wall

9. Which sequence most accurately describes air flow in the human respiratory system? (a) pharynx ⟶ bronchus ⟶ trachea ⟶ alveolus (b) pharynx ⟶ parabronchi ⟶ alveoli ⟶ bronchioles (c) bronchus ⟶ trachea ⟶ larynx ⟶ lung (d) larynx ⟶ trachea ⟶ bronchus ⟶ bronchiole (e) trachea ⟶ larynx ⟶ bronchus ⟶ alveolus
10. The amount of air moved in and out of the lungs with each normal resting breath is the (a) vital capacity (b) residual capacity (c) vital volume (d) partial pressure (e) tidal volume
11. The greater the difference in pressure and the larger the surface area, the faster a gas will diffuse. This is explained by (a) Dalton's law of partial pressure (b) Fick's law of diffusion (c) the percent saturation (d) the Bohr effect (e) the oxygen-hemoglobin dissociation curve
12. Oxygen in the blood is transported mainly (a) in combination with hemoglobin (b) as bicarbonate ions (c) as carbonic acid (d) dissolved in plasma (e) combined with carbon dioxide
13. The concentration of which of the following substances is most important in regulating the rate of respiration? (a) chloride ions (b) oxygen (c) bicarbonate ions (d) nitrogen (e) carbon dioxide
14. When a diver ascends too rapidly (a) bronchial constriction occurs (b) a diving reflex is activated (c) nitrogen rapidly bubbles out of solution in the body fluids (d) nitrogen hypoxia occurs (e) carbon dioxide bubbles damage the alveoli
15. Pulmonary emphysema (a) can lead to chronic obstructive pulmonary disease (b) is common in cigarette smokers (c) is characterized by loss of elasticity of the alveolar walls (d) answers a, b, and c are correct (e) answers a and b only

REVIEW QUESTIONS

1. Why are specialized respiratory structures necessary in a tadpole but not in a flatworm?
2. Compare ventilation in a sponge with ventilation in a human.
3. Compare gas exchange in the following animals: (a) earthworm (b) grasshopper (c) fish (d) frog (e) bird
4. What is the function of respiratory pigments? How do they work?
5. Why does alveolar air differ in composition from atmospheric air? Explain.
6. What physiological mechanisms bring about an increase in rate and depth of breathing during exercise? Why is such an increase necessary?
7. Under what conditions might it be advantageous for a fish to have lungs as well as gills? What function do the "lungs" of modern fishes serve?
8. How does the countercurrent exchange system increase the efficiency of gas exchange between a fish's gills and blood?
9. What mechanisms does the human respiratory system have for getting rid of inhaled dirt? What happens when so much dirty air is inhaled that these mechanisms cannot function effectively?
10. What is meant by the percent O_2 saturation of hemoglobin? What factors affect the dissociation of oxyhemoglobin?
11. Describe the physiological effects of each of the following: hyperventilation, sudden decompression at 12,000 m altitude, and surfacing too quickly from a scuba dive.

YOU MAKE THE CONNECTION

1. What problems would be faced by a terrestrial animal possessing gills instead of lungs?
2. Aquatic mammals such as whales and dolphins use lungs rather than gills for gas exchange. Propose a hypothesis to explain this.
3. What are the advantages of having millions of alveoli rather than a pair of simple, balloon-like lungs?

RECOMMENDED READINGS

Bartecchi, C.E., T.D. MacKenzie, and R.W. Schrier. "The Global Tobacco Epidemic." *Scientific American,* Vol. 272, No. 5, May 1995. An overview of the worldwide increase in cigarette smoking and of the health effects of tobacco.

Gostin, L.O., P.S. Arno, and A.M. Brandt. "FDA Regulation of Tobacco Advertising and Youth Smoking." *Journal of the American Medical Association,* Vol. 277, No. 5, 5 Feb. 1997. An overview of U.S. policy regarding regulation of tobacco and smoking by minors.

Kessler, D.A., P.S. Barnett, A. Witt, M.R. Zeller, J.R. Mande, and W.B. Schultz. "The Legal and Scientific Basis for FDA's Assertion of Jurisdiction over Cigarettes and Smokeless Tobacco." *Journal of the American Medical Association,* Vol. 277, No. 5, 5 Feb. 1997. An overview of an important shift in regulation of tobacco based on recognition of the addictive qualities of nicotine and on the dangers of secondary smoke.

Kooyman, G.L., and P.J. Ponganis. "The Challenges of Diving to Depth." *American Scientist,* Vol. 85, No. 6, Nov./Dec. 1997. Diving birds and mammals have adaptations for conserving oxygen and responding to pressure.

McAuliffe, K. "Elephant Seals, the Champion Divers of the Deep." *Smithsonian,* Vol. 26, No. 6, Sept. 1995. An overview of research findings about diving mammals.

Moon, R.E., R.D. Vann, and P.B. Bennett. "The Physiology of Decompression Illness." *Scientific American,* Vol. 272, No. 2, Aug. 1995. A discussion of some aspects of decompression sickness.

Wickelgren, I. "Drug May Suppress the Craving for Nicotine." *Science,* Vol. 282, 4 Dec. 1998. A drug that suppresses a rise in dopamine in the brain's reward centers is being tested for use in smoking cessation.

Williams, T.M., R.W. Davis, L.A. Fulman, J. Francis, B.J. LeBoeuf, M. Horning, J. Calambokidis, and D.A. Croll. "Sink or Swim: Strategies for Cost-Efficient Diving by Marine Mammals." *Science,* Vol. 288, 7 Apr. 2000. Gliding during deep diving saves energy.

- Visit our Web site at **http://www.info.brookscole.com/solomonbergmartin** for links to chapter-related resources on the World Wide Web. Additional on-line materials relating to this chapter can also be found on our Web site.

See chapter activity on BioActive Learner CD for additional help in mastering the chapter's material. Icon location in the chapter's margins shows which topics have tutorials or simulations in the CD.

45

Processing Food and Nutrition

Spotted hyenas *(Crocuta crocuta)* and several species of birds of prey gathered around a dead elephant. Photographed in East Africa. *(McMurray Photography)*

LEARNING OBJECTIVES

After you have studied this chapter you should be able to

1. Compare food processing, including ingestion, digestion, absorption, and egestion or elimination, by an animal (such as *Hydra*) that has a single-opening digestive system with an animal (such as a vertebrate) that has a digestive system with two openings.
2. Identify on a diagram or model each of the structures of the human digestive system described in this chapter, and give the function of each.
3. Trace the pathway traveled by an ingested meal in the human digestive system and describe the step-by-step digestion of carbohydrate, protein, and lipid; summarize the functions of the accessory digestive glands of humans and other terrestrial vertebrates.
4. Describe the structural adaptations that increase the surface area of the digestive tract.
5. Compare lipid absorption with absorption of other nutrients.
6. Trace the fate of glucose, lipids, and amino acids after their absorption, and discuss their roles in the body.
7. Discuss the roles of vitamins, minerals, and phytochemicals; distinguish between water-soluble and fat-soluble vitamins.
8. Contrast basal metabolic rate with total metabolic rate; write the basic energy equation for maintaining body weight, and describe the consequences of altering it in either direction.
9. In general terms, describe the effects of malnutrition, including both undernutrition and overnutrition.
10. Summarize what is currently known and hypothesized about the regulation of food intake and energy homeostasis. (Include the roles of leptin, neuropeptide Y, and melanin-concentrating hormone.)

Sponges and baleen whales feed on food particles suspended in the water. Giraffes eat leaves, and antelope graze on grasses. Frogs, lions, and some insects capture other animals. Although their choices of food and feeding mechanisms are diverse, all animals are **heterotrophs,** organisms that must obtain their energy and nourishment from the organic molecules manufactured by other organisms. What they eat, how they obtain food, and how they process and use food are the focus of this chapter.

Nutrients are substances in food that are used as energy sources to run the systems of the body, as ingredients to make compounds for metabolic processes, and as building blocks in the growth and repair of tissues. Obtaining nutrients is of such vital importance that both individual organisms and ecosystems are structured around the central theme of **nutrition,** the process of taking in and using food. An organism's body plan and its lifestyle are adapted to its particular mode of obtaining food. For example, spotted hyenas *(Crocuta crocuta),* which are formidable predators as well as scavengers, have massive jaws that permit them to consume an entire elephant carcass, including bones and hide *(see photograph).* Few nutrients are wasted.

With only slight variations, all animals require the same basic nutrients: minerals, vitamins, carbohydrates, lipids, and proteins. Carbohydrates, lipids, and proteins all can be used as energy sources. Eating too much of any of these nutrients can result in weight gain, whereas eating too few nutrients or an unbalanced diet can result in malnutrition and death. **Malnutrition,** or poor nutritional status, can result from dietary intake that is either below or above required needs. In human populations, both undernutrition (particularly protein deficiency) and obesity (which results from overnutrition) are serious health problems.

Feeding is the selection, acquisition, and ingestion of food. **Ingestion** is the process of taking food into the digestive cavity. In many animals, including vertebrates, ingestion includes taking food into the mouth and swallowing it. Most animals have a specialized digestive system that processes the food they eat. The process of breaking down food is called **digestion.** Because animals eat the macromolecules tailor-made by and for other organisms, they must break down these molecules and refashion them for their own needs. For example, the hyena cannot incorporate the proteins and other complex organic compounds from the elephant carcass directly into its own cells. It must *mechanically digest* its food, and then *chemically digest* it by enzymatic hydrolysis (see Chapter 3). During digestion, complex organic compounds are degraded into smaller molecular components. For example, proteins are broken down into their component amino acids.

Amino acids and other nutrients pass through the lining of the digestive tract and into the blood by **absorption.** Then, they are transported to the body cells by the circulatory system. In the cells they can be used to synthesize proteins and other complex organic compounds required by the animal. Food that is not digested and absorbed is discharged from the body. This process is called **egestion** in simple animals and **elimination** in more complex animals.

Food processing and nutrition are active areas of research. For example, researchers investigating the regulation of digestion are discovering a number of peptide messengers in the digestive tract. Nutritionists are studying a variety of plant compounds that appear to be important in maintaining health. And the food industry continues to search for new fat and sugar substitutes.

ANIMALS ARE ADAPTED TO THEIR MODE OF NUTRITION

Animals can be classified as herbivores, carnivores, or omnivores on the basis of the type of food they typically eat (Fig. 45–1). Animals that feed directly on producers are **herbivores,** or primary consumers. Animals cannot digest the cellulose of plant cell walls, and many adaptations have evolved for extracting nutrients from the plant material they eat. For example, many herbivores, including termites, cows, and horses, have a symbiotic relationship with bacteria that inhabit their digestive tracts. In exchange for food and shelter, the bacteria break down cellulose cell walls, allowing the host's digestive enzymes access to the nutrients within plant cells.

In the cud-chewing ruminants (cattle, sheep, deer, giraffes), the stomach is divided into four chambers. Bacteria inhabiting the first two chambers digest cellulose, splitting some of it into sugars, which are then used by the host and the bacteria themselves. The bacteria produce fatty acids during their metabolism, some of which are absorbed by the animal and serve as an important energy source. Food that is not sufficiently chewed, called cud, is regurgitated into the animal's mouth and chewed again.

Most of what a herbivore eats is not efficiently digested and is eliminated from the body, almost unchanged, as waste. For this reason herbivores must eat large quantities of food to obtain the nourishment they need. Many herbivores—grasshoppers, locusts, elephants, and cattle, for example—spend a major part of their lives eating.

Herbivores are sometimes eaten by flesh-eating **carnivores,** which may also eat one another. Many carnivores (secondary and higher level consumers in ecosystems) are predators, adapted for capturing and killing prey. Some carnivores seize their victims and swallow them alive and whole (Fig. 45–1*d*). Others paralyze, crush, or shred their prey before ingesting it. Carnivorous mammals have well-developed canine teeth for stabbing their prey during combat. The digestive juice of the stomach breaks down proteins, and because meat is more easily digested than plant food, their digestive tracts are shorter than those of herbivores.

Omnivores, such as bears and humans, consume both plants and animals. Earthworms ingest large amounts of soil containing both animal and plant material. The blue whale, the largest animal, is a filter feeder that strains out tiny algae and animals as it swims. Omnivores often possess adaptations that permit them to distinguish among a wide range of smells and tastes and thereby select a variety of foods.

Animals can also be classified according to the mechanisms they use to feed. Many omnivores are **suspension feeders** that remove suspended food particles from the water. Some animals expose a sticky, mucus-coated surface to flowing water; suspended particles adhere to the surface. For example, some echinoderms have tentacles coated with mucus. Others, like bivalve mollusks, filter water. Baleen whales use rows of hard plates (baleen) suspended from the roof of the mouth to filter small crustaceans.

Some animals feed on fluids by piercing and sucking. Mosquitos have highly adapted structures for piercing skin and sucking blood. Birds that feed on pollen and nectar have long bills and tongues. The shape, size, and curve of the beak may be specialized for feeding on a particular type of flower. Bats that feed on nectar have a long tongue and reduced dentition (number of teeth).

Many animals, including carnivores, ingest large pieces of food. Adaptations for this type of feeding include claws, fangs, poison glands, tentacles, and teeth. The beaks of birds and the teeth of many vertebrates are specialized for cutting, tearing, or chewing food.

SOME INVERTEBRATES HAVE A DIGESTIVE CAVITY WITH A SINGLE OPENING

The simplest invertebrates, sponges, have no digestive system at all. Sponges obtain food by filtering microscopic organisms from the surrounding water. Individual cells phagocytize the food particles, and digestion is *intracellular* within food vacuoles. Wastes are egested into the water that continuously circulates through the sponge body.

(a) (b) (c) (d)

Figure 45–1 Adaptations for obtaining and processing food. **(a)** The impressively long "snout" of the herbivorous acorn weevil (*Circulio* sp.) is adapted both for feeding and for making a hole in the acorn through which it deposits an egg. When it has hatched, the larva feeds on the contents of the acorn seed. **(b)** The herbivorous giant panda's *(Ailuropoda melanoleuca)* large, flat teeth and well-developed jaws and jaw muscles are adaptations for grinding high-fiber plant food. **(c)** The mouth *(on left)* of the carnivorous long-nose butterfly fish *(Forcipiger longirostris)* is adapted for extracting small worms and crustaceans from tight spots in coral reefs. **(d)** This carnivorous snake (*Dromicus* sp.) is strangling a lava lizard (*Tropiduris* sp.). *(a, Darwin Dale/Photo Researchers, Inc.; b, Tom McHugh/Photo Researchers, Inc.; c, Carmela Leszczynski/ Animals Animals; d, Frans Lanting/Minden Pictures)*

Cnidarians (such as hydras and jellyfish) and flatworms have a **gastrovascular cavity,** a central digestive cavity with a single opening. Cnidarians capture small aquatic animals with the help of their stinging cells and tentacles (Fig. 45–2*a*). The mouth opens into the gastrovascular cavity. Cells lining this digestive cavity secrete enzymes that break down proteins. Digestion continues *intracellularly* within food vacuoles, and digested nutrients diffuse into other cells. Undigested food particles are egested through the mouth by contraction of the body.

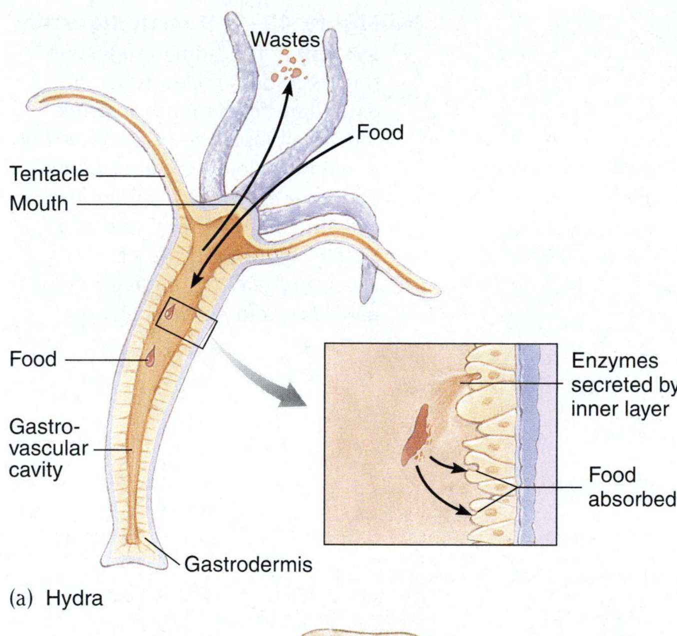

(a) Hydra

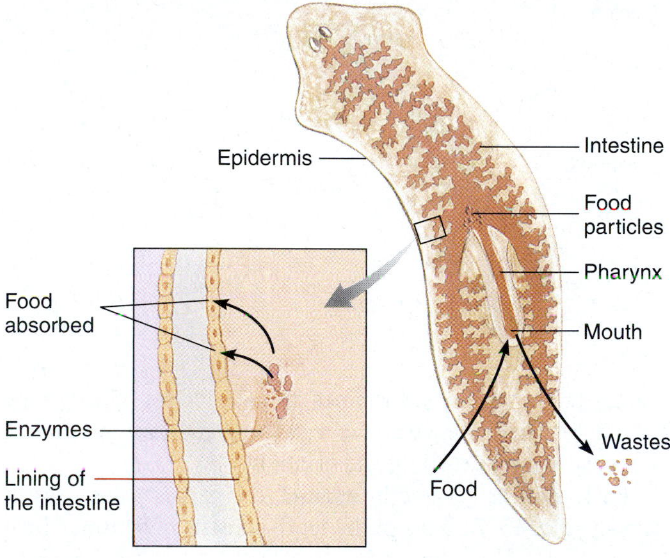

(b) Flatworm

Figure 45–2 Simple invertebrate digestive systems. **(a)** Hydras and **(b)** flatworms (planarians) have digestive tracts with a single opening that serves as both mouth and anus.

Free-living flatworms begin to digest their prey even before ingesting it. They extend the pharynx out through their mouth and secrete digestive enzymes onto the prey (Fig. 45–2*b*). When ingested, the food enters the branched gastrovascular cavity where enzymes continue to digest it. Partly digested food fragments are then phagocytized by cells lining the gastrovascular cavity, and digestion is completed within food vacuoles. As in cnidarians, the flatworm digestive cavity has only one opening, so undigested wastes are egested through the mouth.

MOST ANIMAL DIGESTIVE SYSTEMS HAVE TWO OPENINGS

Most invertebrates, and all vertebrates, have a tube-within-a-tube body plan. The body wall forms the outer tube. The inner tube is a digestive tract with two openings, sometimes referred to as a complete digestive system (Fig. 45–3). Food enters through the mouth, and undigested food is eliminated through the anus. **Motility** refers to the mixing and propulsive movements of the digestive tract. The propulsive activity characteristic of most regions of the digestive tract is **peristalsis,** waves of muscular contraction that push the food in one direction. More food can be taken in while previously eaten food is being digested and absorbed farther down the digestive tract. In a digestive tract with two openings, various regions of the tube are adapted to perform specific functions.

THE VERTEBRATE DIGESTIVE SYSTEM IS HIGHLY SPECIALIZED FOR PROCESSING FOOD

Various regions of the vertebrate digestive tract are specialized to perform specific functions (Fig. 45–4). Food passes in sequence through the following specialized regions:

Mouth ⟶ pharynx (throat) ⟶ esophagus ⟶
stomach ⟶ small intestine ⟶ large intestine ⟶
anus

All vertebrates have accessory glands that secrete digestive juices into the digestive tract. These include the liver, the pancreas, and, in terrestrial vertebrates, the salivary glands.

The wall of the digestive tract is composed of four layers. Although various regions differ somewhat in structure, the layers are basically similar throughout the digestive tract (Fig. 45–5). The **mucosa,** a layer of epithelial tissue and underlying connective tissue, lines the *lumen* (inner space) of the digestive tract. In the stomach and intestine, the mucosa is greatly folded to increase the secreting and absorbing surface. Surrounding the

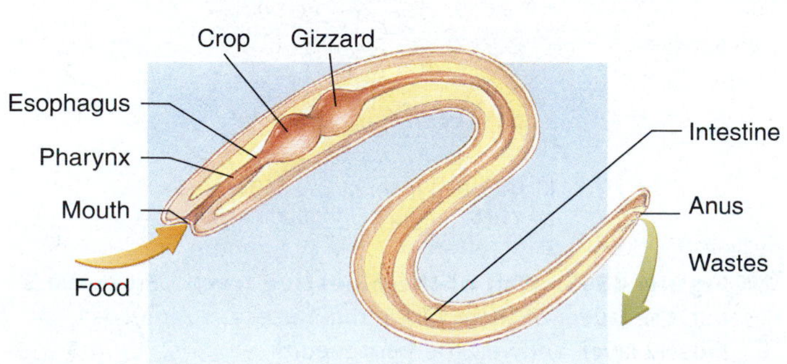

Figure 45–3 Digestive tract with two openings. The earthworm, like most complex animals, has a complete digestive tract extending from mouth to anus. Various regions of the digestive tract are specialized to perform different food processing functions.

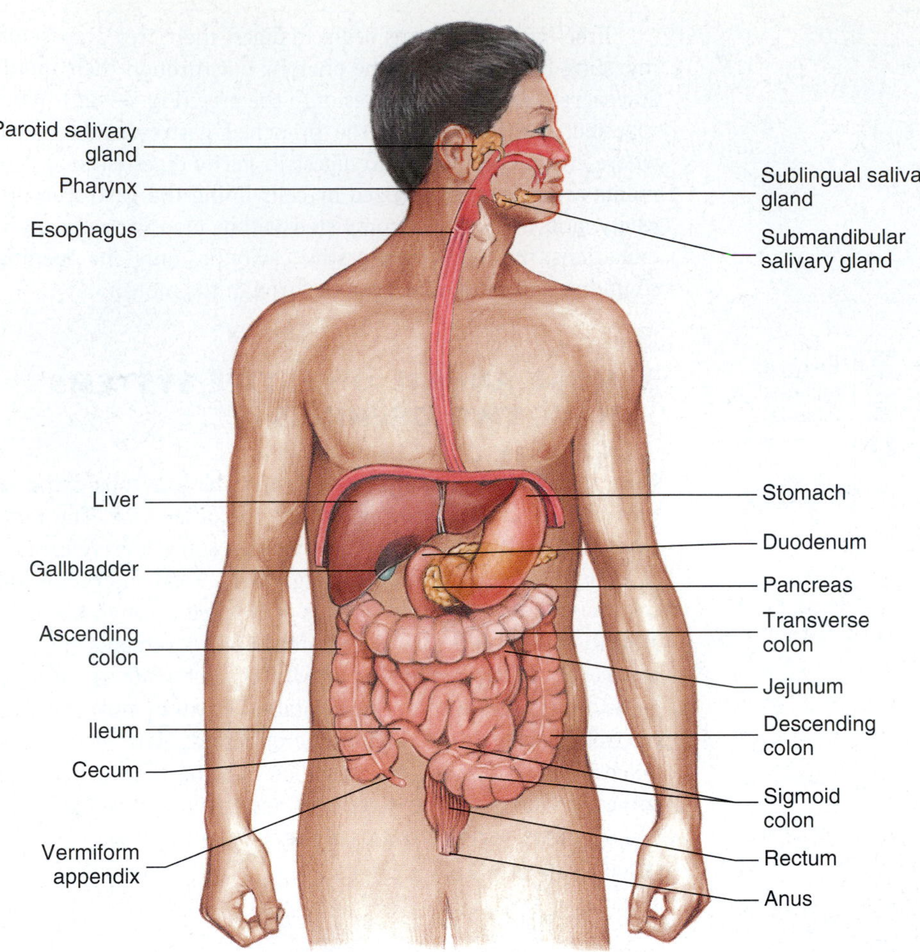

Figure 45–4 Human digestive system. The human digestive tract is a long, coiled tube extending from mouth to anus. The small intestine consists of the duodenum, jejunum, and ileum. The large intestine includes the cecum, colon, rectum, and anus. Locate the three types of accessory glands: the liver, pancreas, and salivary glands.

mucosa is the **submucosa,** a connective tissue layer rich in blood vessels, lymphatic vessels, and nerves.

A **muscle layer,** consisting of two sublayers of smooth muscle, surrounds the submucosa. In the inner sublayer the muscle fibers are arranged circularly around the digestive tube. In the outer sublayer the muscle fibers are arranged longitudinally. Below the level of the diaphragm, the outer connective tissue coat of the digestive tract is called the **visceral peritoneum.** By various folds it is connected to the **parietal peritoneum,** a sheet of connective tissue that lines the walls of the abdominal and pelvic cavities. The visceral and parietal peritonea enclose part of the coelom called the **peritoneal cavity.** Inflammation of the peritoneum, called **peritonitis,** can be very serious because infection can spread along the peritoneum to most of the abdominal organs.

Food processing begins in the mouth

Imagine that you have just taken a big bite of a hamburger. The mouth is specialized for ingestion and for beginning the digestive process. Mechanical digestion begins as you bite, grind, and chew the meat and bun with your teeth. Unlike the simple, pointed teeth of fish, amphibians, and reptiles, the teeth of mammals vary in size and shape and are specialized to perform specific functions. The chisel-shaped **incisors** are used for biting, whereas the long, pointed **canines** are adapted for tearing food (Fig. 45–6). The flattened surfaces of the **premolars** and **molars** are specialized for crushing and grinding.

Each tooth is covered by **enamel,** the hardest substance in the body (Fig. 45–7). Most of the tooth consists of **dentin,** which

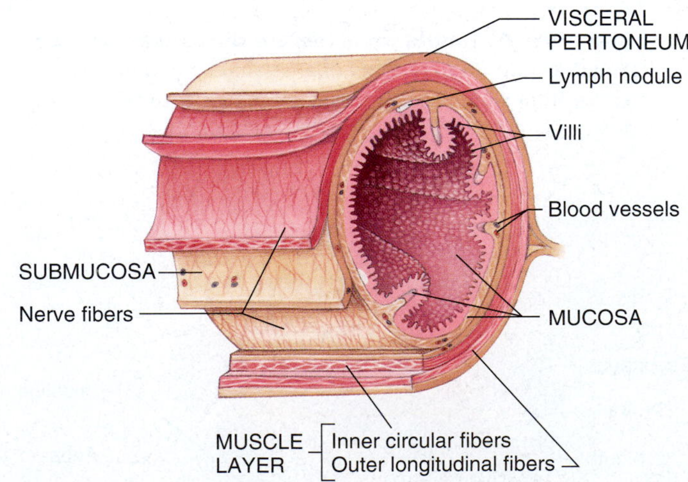

Figure 45–5 Wall of the digestive tract. From inside out, the layers of the wall are the mucosa, submucosa, muscle layer, and visceral peritoneum.

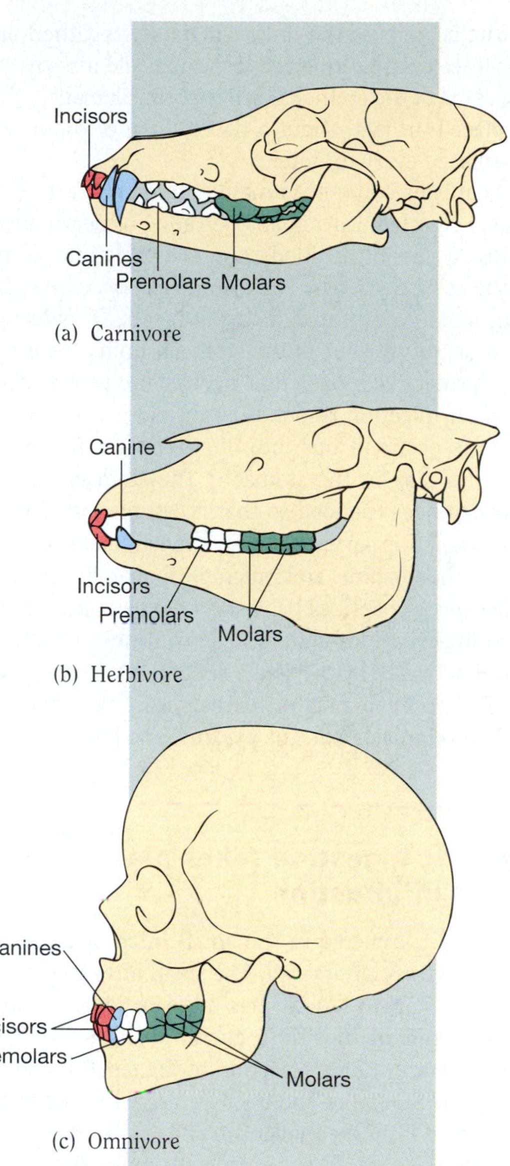

Figure 45–6 Teeth and diet. **(a)** Skull of a coyote showing the pointed incisors and canines, adaptations for ripping flesh. **(b)** In contrast, herbivores, such as horses, have incisors (and sometimes canines) adapted for cutting off bits of vegetation. Canines are absent in some herbivores. The broad, ridged surfaces of the molars are adapted for grinding plant material. **(c)** The teeth of omnivores, such as humans, are adapted for chewing a variety of foods.

resembles bone in composition and hardness. Beneath the dentin is the **pulp cavity,** a soft connective tissue containing blood vessels, lymph vessels, and nerves.

While food is being mechanically disassembled by the teeth, it is also moistened by saliva. Some of its molecules dissolve, enabling you to taste the food. Recall from Chapter 41 that taste buds are located on the tongue and other surfaces of the mouth. Three pairs of **salivary glands** secrete about a liter of saliva into the mouth cavity each day. Saliva contains an enzyme, **salivary amylase,** which initiates the chemical digestion of starch into sugar.

The pharynx and esophagus conduct food to the stomach

After the bite of food has been chewed and fashioned into a lump called a **bolus,** it is swallowed, that is, moved through the **pharynx** and into the **esophagus.** The pharynx, or throat, is a muscular tube that serves as the hallway of the respiratory system as well as the digestive system. During swallowing, the opening to the airway is closed by a small flap of tissue, the **epiglottis.**

Waves of peristalsis sweep the bolus through the pharynx and esophagus toward the stomach (Fig. 45–8). Circular muscle fibers in the wall of the esophagus contract around the top of the bolus, pushing it downward. Almost at the same time, longitudinal muscles around the bottom of the bolus and below it contract, shortening the tube.

When the body is in an upright position, gravity helps move the food through the esophagus, but gravity is not essential. Astronauts are able to eat in its absence, and even if you are standing on your head, food will reach your stomach.

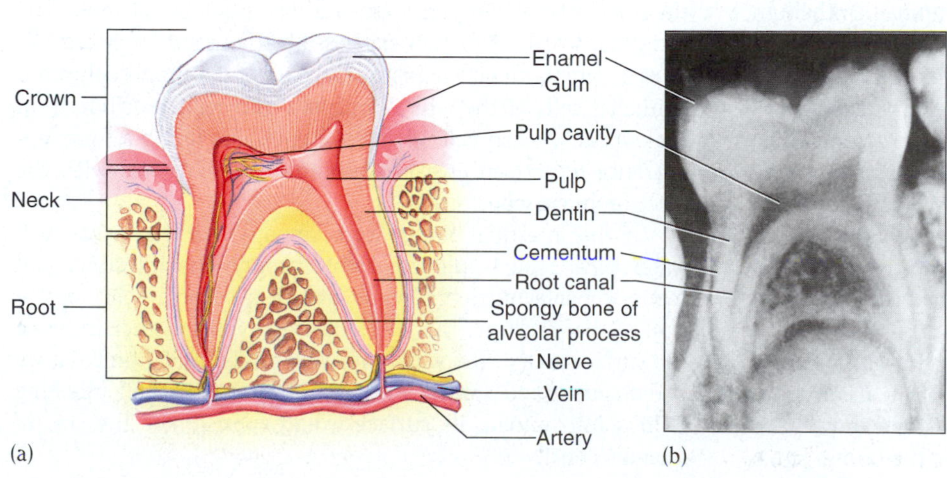

Figure 45–7 Tooth structure. **(a)** Sagittal section through a human lower molar showing the crown, neck, and root. **(b)** X ray of a healthy tooth.

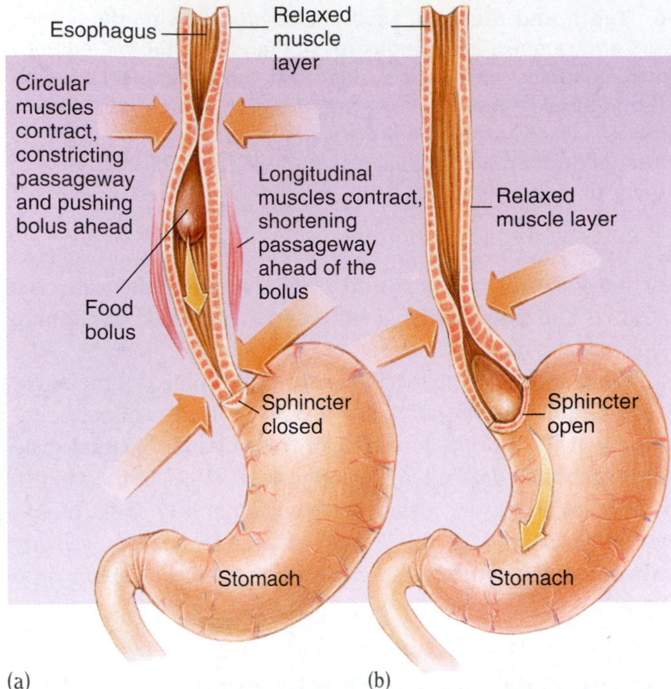

Figure 45–8 Peristalsis. Food is moved through the digestive tract by waves of muscular contractions known as peristalsis. **(a)** A bolus is moved down through the esophagus by peristaltic contractions. **(b)** When the sphincter (ring of muscle) at the entrance of the stomach opens, food enters the stomach.

Food is mechanically and enzymatically digested in the stomach

The entrance to the large, muscular **stomach** is normally closed by a ring of muscle at the lower end of the esophagus. When a peristaltic wave passes down the esophagus, the muscle relaxes, permitting the bolus to enter the stomach (Fig. 45–9). When empty, the stomach is collapsed and shaped almost like a hot dog. Folds of the stomach wall, called **rugae,** give the inner lining a wrinkled appearance. As food enters, the rugae gradually smooth out, expanding the capacity of the stomach to more than a liter.

The stomach is lined with a simple, columnar epithelium that secretes large amounts of mucus. Tiny pits mark the entrances to the millions of **gastric glands,** which extend deep into the stomach wall. **Parietal cells** in the gastric glands secrete hydrochloric acid and a substance known as *intrinsic factor,* needed for adequate absorption of vitamin B_{12}. **Chief cells** in the gastric glands secrete **pepsinogen,** an inactive enzyme precursor. When pepsinogen comes in contact with the acidic gastric juice in the stomach, it is converted to **pepsin,** the main digestive enzyme of the stomach. Pepsin hydrolyzes proteins, converting them to short polypeptides.

Several protective mechanisms prevent the gastric juice from digesting the wall of the stomach. Cells of the gastric mucosa secrete an alkaline mucus that coats the stomach wall and neutralizes the acidity of the gastric juice along the lining. In addition, the epithelial cells of the lining fit tightly together, preventing gastric juice from leaking between them and into the tissue beneath. If some of the epithelial cells become damaged, they are quickly replaced. In fact, about a half million of these cells are shed and replaced every minute!

These mechanisms sometimes malfunction and part of the stomach lining is digested, leaving an open sore, or **peptic ulcer.** Such ulcers often occur in the duodenum and sometimes in the lower part of the esophagus. The bacterium *Helicobacter pylori* has been implicated as a causative factor in ulcers. *H. pylori* infects the mucus-secreting cells of the stomach lining, resulting in a decrease in protective mucus that may lead to peptic ulcers or cancer. *H. pylori* infection responds to antibiotic therapy.

What changes occur in our bite of hamburger during its three- to four-hour stay in the stomach? The stomach churns and chemically degrades the food so that it assumes the consistency of a thick soup; this partially digested food is called **chyme.** Protein digestion then begins, and much of the hamburger protein is degraded to polypeptides. Digestion of the starch in the bun to small polysaccharides and maltose continues until salivary amylase is inactivated by the acidic pH of the stomach. Over a period of several hours, peristaltic waves release the chyme in spurts through the stomach exit, the **pylorus,** and into the small intestine.

Most enzymatic digestion takes place inside the small intestine

Digestion of food is completed in the **small intestine,** and nutrients are absorbed through its wall. The small intestine, which is 5 to 6 m (about 17 ft) in length, has three regions: the **duodenum, jejunum,** and **ileum.** Most chemical digestion takes place in the duodenum, the first portion of the small intestine, not in the stomach. Bile from the liver and enzymes from the pancreas are released into the duodenum and act on the chyme. Then enzymes produced by the epithelial cells lining the duodenum catalyze the final steps in the digestion of the major types of nutrients.

The lining of the small intestine appears velvety because of its millions of tiny finger-like projections, the intestinal **villi** (sing., *villus*) (Fig. 45–10). The villi increase the surface area of the small intestine for digestion and absorption of nutrients. The intestinal surface is further expanded by thousands of **microvilli,** folds of cytoplasm on the exposed surface of the simple columnar epithelial cells of the villi. About 600 microvilli protrude from the surface of each cell, giving the epithelial lining a fuzzy appearance (described as a brush border) when viewed with the electron microscope.

If the intestinal lining were smooth like the inside of a water pipe, food would zip right through the intestine, and many valuable nutrients would not be absorbed. Folds in the wall of the intestine, the villi, and microvilli together increase the surface area of the small intestine by about 600 times. If we could unfold and spread out the lining of the small intestine of an adult human, its surface would approximate the size of a tennis court.

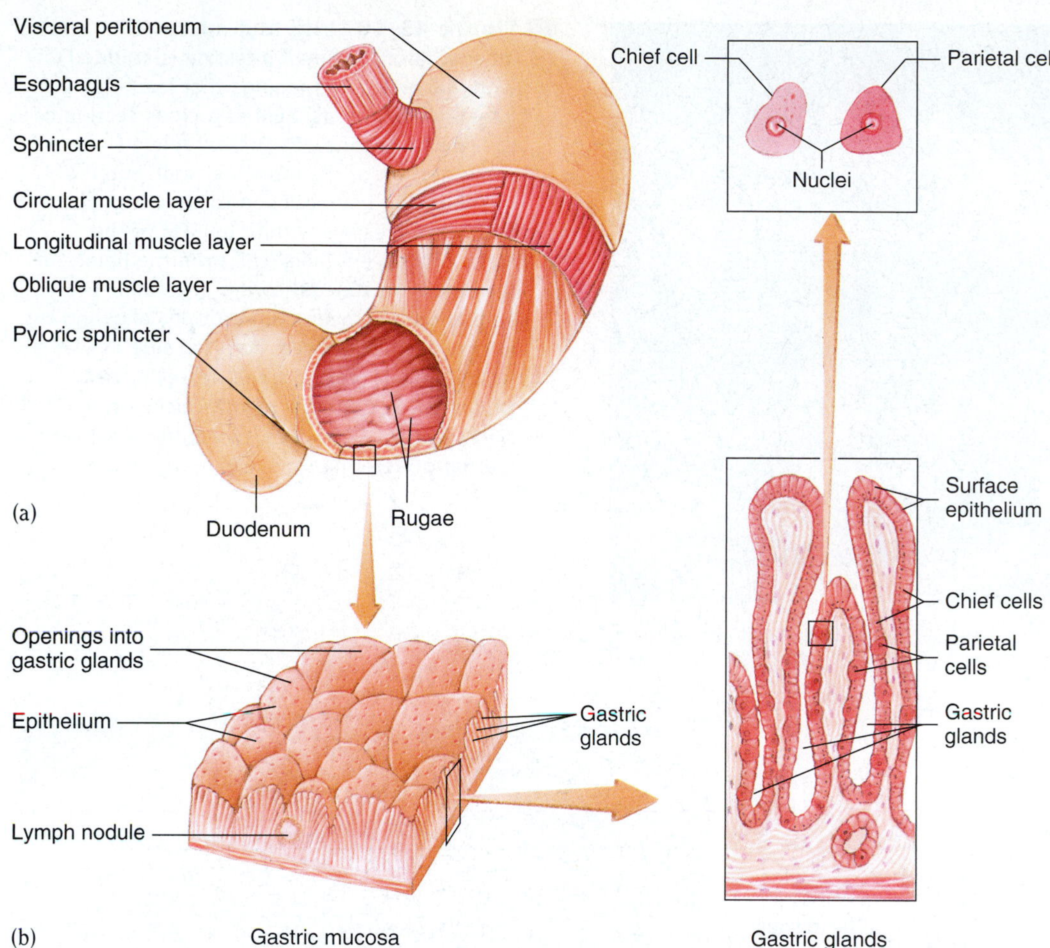

Figure 45–9 Structure of the stomach. From the esophagus, food enters the stomach, where it is mechanically and enzymatically digested. **(a)** The wall of the stomach has been progressively removed to show muscle layers and rugae. **(b)** Stomach lining and gastric glands.

The liver secretes bile, which mechanically digests fats

The **liver,** the largest internal organ and also one of the most complex organs in the body, lies in the upper right part of the abdomen just under the diaphragm (Fig. 45–11). A single liver cell can carry on more than 500 separate, specialized metabolic activities. The liver's food processing functions include the following:

1. Secretes **bile,** which is important in the mechanical digestion of fats.
2. Helps maintain homeostasis by removing or adding nutrients to the blood.
3. Converts excess glucose to glycogen and stores it.
4. Converts excess amino acids to fatty acids and urea.
5. Stores iron and certain vitamins.
6. Detoxifies alcohol and other drugs and poisons.

Bile consists of water, bile salts, bile pigments, cholesterol, salts, and lecithin (a phospholipid). Bile is stored in the pear-shaped gallbladder, which concentrates the bile and releases it into the duodenum as needed. Bile mechanically digests fats by a detergent-like action (discussed in a later section). Because it contains no digestive enzymes, bile does not enzymatically digest food.

The pancreas secretes digestive enzymes

The **pancreas** is an elongated gland that secretes both digestive enzymes and hormones that help regulate the level of glucose in the blood. Among its enzymes are **trypsin** and **chymotrypsin,** which digest polypeptides to dipeptides; **pancreatic lipase,** which degrades fats; **pancreatic amylase,** which breaks down almost all types of carbohydrates, except cellulose, to disaccharides; and **ribonuclease** and **deoxyribonuclease,** which split the nucleic acids ribonucleic acid (RNA) and deoxyribonucleic acid (DNA) to free nucleotides.

Enzymatic digestion occurs as food moves through the digestive tract

Chyme moves through the digestive tract by peristalsis, mixing contractions, and motions of the villi. As chyme passes through the digestive tract, nutrients in the chyme come into contact with enzymes that digest them (Table 45–1 on page 998).

Carbohydrates are digested to monosaccharides

Polysaccharides, such as starch and glycogen, are important components of the food ingested by humans and most other

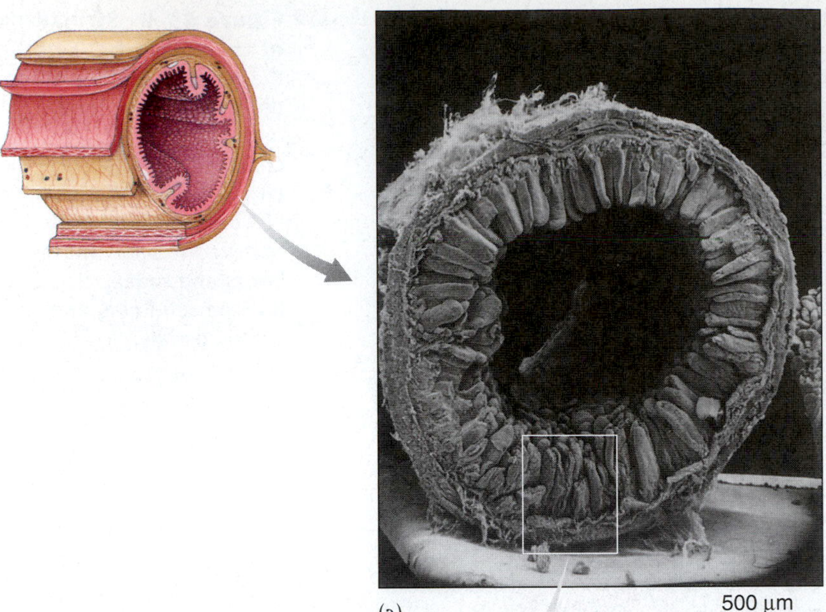

(a)

Figure 45–10 Villi and microvilli. The inner wall of the small intestine is studded with villi and tiny openings that lead into the intestinal glands. **(a)** SEM of a cross section of the small intestine. **(b)** Microscopic view of a small portion of the intestinal wall. Some of the villi have been opened to show the blood and lymph vessels within. **(c)** SEM of the surface of an epithelial cell from the lining of the small intestine, showing microvilli. The epithelium has been cut vertically, allowing the microvilli to be viewed from the side as well as from above. *(a, G. Shih-R. Kessel/Visuals Unlimited; c, Courtesy of J.D. Hoskings, W.G. Henk, and Y.Z. Abdelbaki, from the* American Journal of Veterinary Research, *Vol. 43, No. 10)*

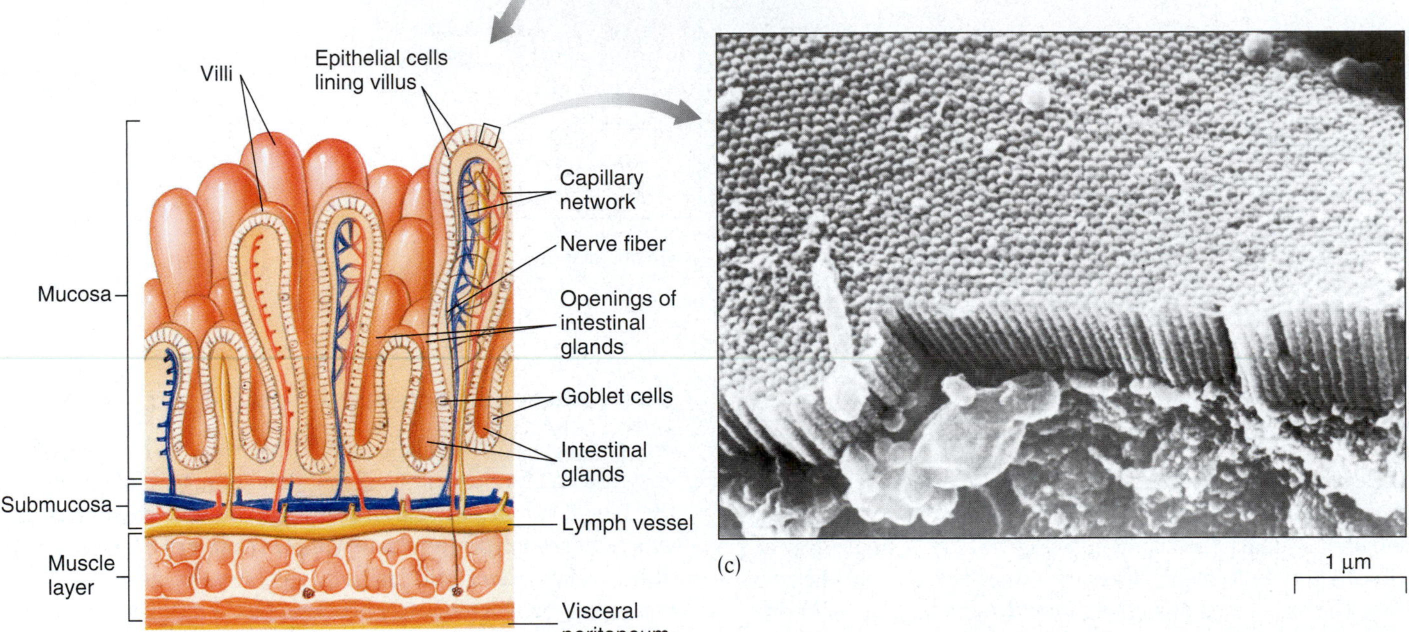

animals. The glucose units of these large molecules are connected by glycosidic bonds linking carbon 4 (or 6) of one glucose molecule with carbon 1 of the adjacent glucose molecule. These bonds are hydrolyzed by **amylases,** enzymes that digest polysaccharides to the disaccharide maltose. Although amylase can split the α-glycosidic linkages present in starch and glycogen, it cannot split the β-glycosidic linkages present in cellulose (see Figs. 3–9 and 3–10 in Chapter 3).

Amylase cannot split the bond between the two glucose units of maltose. Enzymes produced by the cells lining the small intestine break down disaccharides such as maltose to monosaccharides. **Maltase,** for example, splits maltose into two glucose molecules (see Fig. 3–8*a*). Hydrolysis occurs while the disaccharides are being absorbed through the epithelium of the small intestine.

Proteins are digested to amino acids

Several kinds of proteolytic enzymes are secreted into the digestive tract. Each breaks peptide bonds at one or more specific locations in a polypeptide chain. Trypsin, secreted in an inactive form by the pancreas, is activated by an enzyme called enterokinase. The trypsin then activates chymotrypsin and carboxypeptidase, as well as additional trypsin. Pepsin, trypsin, and chy-

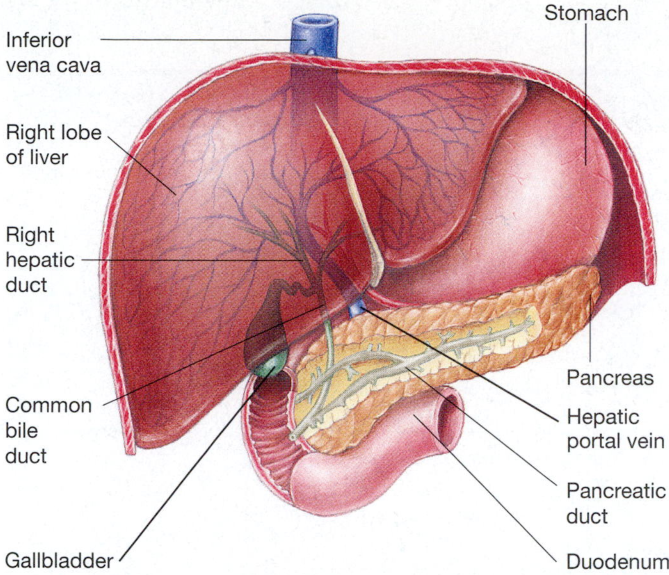

Figure 45–11 The liver and pancreas. The gallbladder stores bile from the liver. Note the ducts that conduct bile to the gallbladder and the duodenum. The stomach has been displaced to expose the pancreas.

motrypsin break certain internal peptide bonds of proteins and polypeptides. Carboxypeptidase removes amino acids with free terminal carboxyl groups from the end of polypeptide chains. **Dipeptidases** released by the duodenum then split the remaining small peptides to amino acids.

Fats are digested to fatty acids and monoacylglycerols

Lipids are usually ingested as large masses of triacylglycerols (also called triglycerides). They are digested mainly within the duodenum by pancreatic lipase. Like many other proteins, lipase is water-soluble, but its substrates are not. Thus, the enzyme can attack only the fat molecules at the surface of a mass of fat. Bile salts act like detergents, reducing the surface tension of fats. Their action, called **emulsification,** breaks large masses of fat into smaller droplets. Emulsification greatly increases the surface area of fat exposed to the action of pancreatic lipase and so increases the rate of lipid digestion.

Conditions in the intestine are usually not optimal for the complete hydrolysis of lipids to glycerol and fatty acids. The products of lipid digestion therefore include monoacylglycerols (monoglycerides) and diacylglycerols (diglycerides) as well as glycerol and fatty acids. Undigested triacylglycerols remain as well, and some of these are absorbed without digestion.

Nerves and hormones regulate digestion

Most digestive enzymes are produced only when food is present in the digestive tract. Salivary gland secretion is controlled entirely by the nervous system, but secretion of other digestive juices is regulated by both nerves and hormones. The wall of the digestive tract contains dense networks of neurons. This so-called *enteric nervous system* continues to regulate many motor and secretory activities of the digestive system even if sympathetic and parasympathetic nerves to these organs are cut. Many neuropeptides present in the brain are also released by neurons in the digestive tract and help regulate digestion. For example, *substance P* stimulates smooth muscle contraction of the digestive tract, while *enkephalin* inhibits it. (The actions of substance P and enkephalin in pain perception are described in Chapter 41.)

At least four hormones—**gastrin, secretin, cholecystokinin (CCK),** and **gastric inhibitory peptide (GIP)**—help regulate the digestive system (Table 45–2). All of these hormones are polypeptides secreted by endocrine cells in the mucosa of certain regions of the digestive tract. Investigators are studying several other messenger peptides that appear to be important in regulating digestive activity.

As an example of the regulation of the digestive system, consider the secretion of gastric juice. Seeing, smelling, tasting, or even thinking about food causes the brain to send neural signals to the gastric glands in the stomach, stimulating them to secrete. In addition, when food distends the stomach, stretch receptors send neural messages to the medulla. The medulla then sends messages to endocrine cells in the stomach wall that secrete the hormone gastrin. Although gastrin is absorbed into the blood, it returns to the stomach, where it stimulates gastric juice release, gastric emptying, and intestinal motility.

Absorption takes place mainly through the villi of the small intestine

Only a few substances—water, simple sugars, salts, alcohol, and certain drugs—are small enough to be absorbed through the stomach wall. Absorption of nutrients is primarily the job of the intestinal villi. As illustrated in Figure 45–10, the wall of a villus consists of a single layer of epithelial cells. Inside each villus is a network of capillaries and a central lymph vessel, called a **lacteal.**

To reach the blood (or lymph), a nutrient molecule must pass through an epithelial cell of the intestinal lining and through a cell of the blood or lymph vessel lining. Absorption occurs by a combination of simple diffusion, facilitated diffusion, and active transport. Because glucose and amino acids cannot diffuse through the intestinal lining, they must be absorbed by active transport. Absorption of these nutrients is coupled with the active transport of sodium (see Chapter 5). Fructose is absorbed by facilitated diffusion.

Amino acids and glucose are transported directly to the liver by the **hepatic portal vein.** In the liver this vein divides into a vast network of tiny blood vessels similar to capillaries. As the nutrient-rich blood moves slowly through the liver, nutrients and certain toxic substances are removed from the circulation.

The products of lipid digestion are absorbed by a different process and different route (Fig. 45–12). Fatty acids and monoacylglycerols combine with bile salts to form soluble complexes called **micelles.** The micelles serve as a reservoir for fatty acids and monoacylglycerols. Only small amounts of free fatty acids

TABLE 45–1 Summary of Digestion

Location	Source of Enzyme	Digestive Process*
Carbohydrate Digestion		
Mouth	Salivary glands	Polysaccharides (e.g., starch) —Salivary amylase→ Maltose + Small polysaccharides 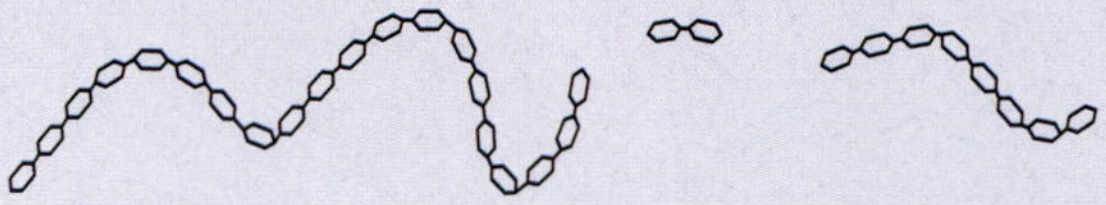
Stomach		Action continues until salivary amylase is inactivated by acidic pH
Small intestine	Pancreas Intestine	Undigested polysaccharides —Pancreatic amylase→ Maltose and small polysaccharides Disaccharides hydrolyzed to monosaccharides as follows: Maltose (malt sugar) —Maltase→ Glucose + Glucose Sucrose (table sugar) —Sucrase→ Glucose + Fructose Lactose (milk sugar) —Lactase→ Glucose + Galactose
Protein Digestion		
Stomach	Stomach (gastric glands)	Protein —Pepsin→ Short polypeptides
Small intestine	Pancreas	Polypeptides (A—A—A—A—A | A—A—A—A—A) —Trypsin, Chymotrypsin→ Polypeptides + Dipeptides (A—A—A A—A) Polypeptides (A—A—A—A) —Carboxypeptidase→ Peptides and free amino acids (A—A—A A A)
	Small intestine	Peptides + Dipeptides (A—A—A A—A) —Peptidases, Dipeptidases→ Free amino acids (A A A)
Lipid Digestion		
Small intestine	Liver	Glob of fat —Bile salts→ Emulsified fat (individual triacylglycerols) 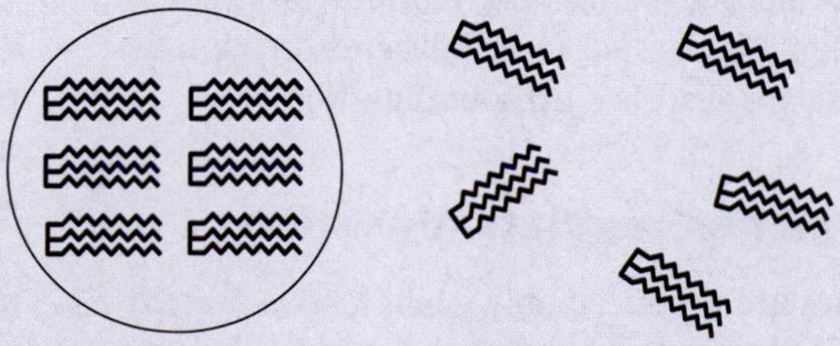
	Pancreas	Triacylglycerol —Pancreatic lipase→ Fatty acids + Glycerol (E)

* ⬭ = monosaccharide; [symbol] = triacylglycerol; [symbol] = fatty acid; A = amino acid; E = glycerol.

TABLE 45–2 Hormonal Control of Digestion

Hormone	Source	Target Tissue	Actions	Factors that Stimulate Release
Gastrin	Stomach (mucosa)	Stomach (gastric glands)	Stimulates gastric glands to secrete pepsinogen	Distention of the stomach by food; certain substances such as partially digested proteins and caffeine
Secretin	Duodenum (mucosa)	Pancreas	Signals secretion of sodium bicarbonate	Acidic chyme acting on mucosa of duodenum
		Liver	Stimulates bile secretion	
Cholecystokinin (CCK)	Duodenum (mucosa)	Pancreas	Stimulates release of digestive enzymes	Presence of fatty acids and partially digested proteins in duodenum
		Gallbladder	Stimulates emptying of bile	
Gastric inhibitory peptide (GIP)	Duodenum (mucosa)	Stomach	Decreases stomach churning, thus slowing emptying	Presence of fatty acids or glucose in duodenum

and monoacylglycerols are present in the intestine. As these molecules are absorbed by diffusion (they are soluble in the lipid of the plasma membrane), they are immediately replaced by molecules from the micelles. This process greatly facilitates absorption of fats. As monoacylglycerols and fatty acids leave a micelle, new fatty acids and monoacylglycerols combine with it. Micelles also dissolve cholesterol and fat-soluble vitamins, facilitating their absorption.

Once free fatty acids and monoacylglycerols enter an epithelial cell in the intestinal lining, they are reassembled as triacylglycerols in the smooth endoplasmic reticulum. The triacylglycerols are packaged into droplets, which become larger as absorbed cholesterol and phospholipids are added. The droplets are then covered with a thin coat of protein. Note that these protein-covered fat droplets, called **chylomicrons,** are processed much as a secreted protein. They are packaged into vesicles in the endoplasmic reticulum, processed in the Golgi complex, and transported to the plasma membrane. After they are released into the interstitial fluid, chylomicrons enter the lacteal (lymph vessel) of the villus.

Chylomicrons are transported by the lymph to the subclavian veins, where the lymph and its contents enter the blood. About 90% of absorbed fat enters the blood circulation in this indirect way. The rest, mainly short-chain fatty acids such as those in butter, are absorbed directly into the blood. After a meal rich in fats, the great number of chylomicrons in the blood may give the plasma a turbid, milky appearance for a few hours.

Most of the nutrients in the chyme are absorbed by the time it reaches the end of the small intestine. What is left (mainly waste) passes through a sphincter, the **ileocecal valve,** into the large intestine.

The large intestine eliminates waste

Undigested material, such as the cellulose of plant foods, along with unabsorbed chyme, passes into the **large intestine** (see Fig. 45–4). Although only about 1.3 m (about 4 ft) long, this part of the digestive tract is referred to as "large" because its diameter is greater than that of the small intestine. The small intestine joins the large intestine about 7 cm (2.8 in) from the end of the large intestine, forming a blind pouch, the **cecum.** The **vermiform appendix** projects from the end of the cecum. (Appendicitis is an inflammation of the appendix.) Herbivores such as rabbits have a large, functional cecum that holds food while bacteria digest its cellulose. In humans, the functions of the cecum and appendix are not known, and these structures are generally considered vestigial organs.

From the cecum to the **rectum** (the last portion of the large intestine), the large intestine is known as the **colon.** The regions of the large intestine are the cecum; ascending colon; transverse colon; descending colon; sigmoid colon; rectum; and anus, the opening for the elimination of wastes.

As the chyme passes slowly through the large intestine, water and sodium are absorbed from it, and it gradually assumes the consistency of normal feces. Bacteria inhabiting the large intestine are nourished by the last remnants of the meal and benefit their host by producing vitamin K and certain B vitamins that can be absorbed and used.

A distinction should be made between elimination and excretion. Elimination is the process of getting rid of digestive wastes—materials that have not been absorbed from the digestive tract and did not participate in metabolic activities. In contrast, excretion is the process of getting rid of *metabolic wastes,* which in mammals is mainly the function of the kidneys and lungs. The large intestine, however, does excrete bile pigments.

When chyme passes through the intestine too rapidly, **defecation** (expulsion of feces) becomes more frequent, and the feces are watery. This condition, called diarrhea, may be caused by anxiety, certain foods, or disease organisms that irritate the intestinal lining. Prolonged diarrhea results in loss of water and salts and leads to dehydration—a serious condition, especially in infants. Constipation results when chyme passes through the

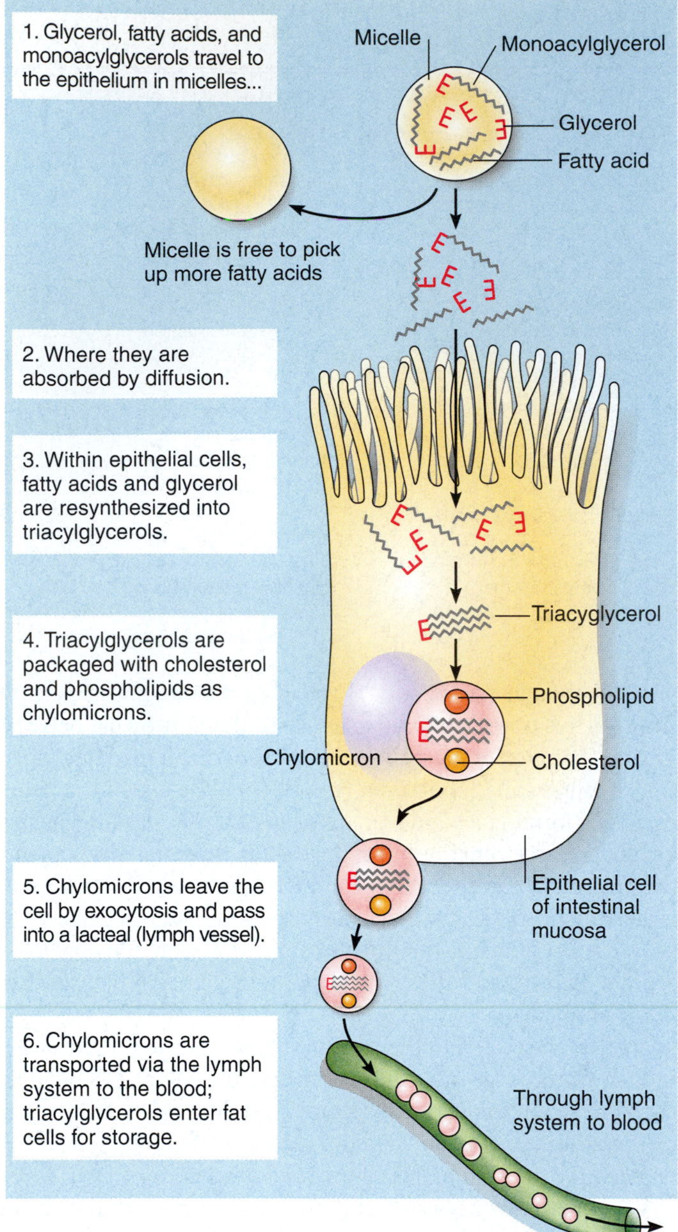

1. Glycerol, fatty acids, and monoacylglycerols travel to the epithelium in micelles...
Micelle
Monoacylglycerol
Glycerol
Fatty acid
Micelle is free to pick up more fatty acids
2. Where they are absorbed by diffusion.
3. Within epithelial cells, fatty acids and glycerol are resynthesized into triacylglycerols.
Triacyglycerol
4. Triacylglycerols are packaged with cholesterol and phospholipids as chylomicrons.
Phospholipid
Chylomicron
Cholesterol
Epithelial cell of intestinal mucosa
5. Chylomicrons leave the cell by exocytosis and pass into a lacteal (lymph vessel).
6. Chylomicrons are transported via the lymph system to the blood; triacylglycerols enter fat cells for storage.
Through lymph system to blood

Figure 45–12 Lipid absorption. Fatty acids and monoacylglycerols combine with bile salts in the small intestine to form micelles, which transport the fatty substances to the epithelial cells lining the small intestine. In the epithelial cells, fatty acids and glycerol combine to form triacylglycerols. These molecules, along with phospholipids and cholesterol, form protein-covered globules called chylomicrons, which are transported to the blood by the lymphatic system.

intestine too slowly. Because more water than usual is removed from the chyme, the feces may be hard and dry. Constipation is often caused by a diet deficient in fiber.

In Western countries, cancer of the colon and rectum accounts for more new cases each year than do any other cancers except lung cancer (Fig. 45–13). Studies suggesting that diets high in fiber content offer some protection against colon cancer have recently been questioned; the factors contributing to colon cancer are not yet known.

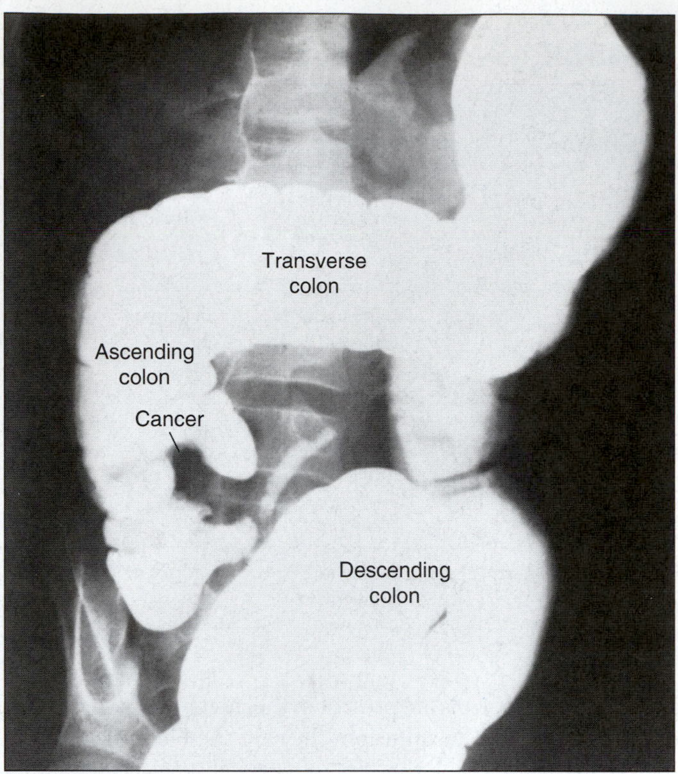

Figure 45–13 Colon cancer. Cancer is evident as a mass that projects into the lumen of the colon in this radiographic view of the large intestine. The large intestine has been filled with a suspension of barium sulfate, which makes irregularities in the wall visible.

ADEQUATE AMOUNTS OF NUTRIENTS ARE NECESSARY TO SUPPORT METABOLIC PROCESSES

All animals require carbohydrates, lipids, proteins, vitamins, and minerals. Although not considered a nutrient in a strict sense, water is a necessary dietary component. Sufficient fluid must be ingested to replace that lost in urine, sweat, feces, and breath. In addition, nutritionists are investigating a host of plant compounds, called *phytochemicals,* that appear to be important in nutrition.

Adequate amounts of essential nutrients are necessary for metabolic processes. Recall from Chapter 1 that **metabolism** refers to all the chemical processes that take place in the body. Metabolic processes include anabolism and catabolism. **Anabolism** includes the synthetic aspects of metabolism such as the production of proteins and nucleic acids. **Catabolism** includes breakdown processes such as hydrolysis. Nutritionists measure the energy value of food in kilocalories, or simply Calories. A Calorie, spelled with a capital C, is a **kilocalorie (kcal),** defined as the amount of heat required to raise the temperature of a kilogram of water 1° C.

Carbohydrates are a major energy source in the omnivore diet

Sugars and starches are the principal sources of energy in the ordinary human diet. However, they are not considered essential nutrients, because the body can obtain sufficient energy from a mixture of proteins and fats. In the average American diet, carbohydrates provide about 48% of the kilocalories ingested daily.

Most carbohydrates are ingested in the form of starch and cellulose, both polysaccharides. (You may want to review the discussion of carbohydrates in Chapter 3.) Nutritionists refer to polysaccharides as **complex carbohydrates.** Foods rich in complex carbohydrates include rice, potatoes, corn, and other cereal grains.

Some nutritionists suggest that we increase our consumption of complex carbohydrates and fiber by eating more fruits, vegetables, and whole grains. **Fiber** is mainly a complex mixture of cellulose and other indigestible carbohydrates. The U.S. diet is low in fiber due to low intake of fruits and vegetables and use of refined flour. Increasing fiber in the diet is thought to have a variety of health benefits, including decreasing cholesterol concentration in the blood. Fiber may also stimulate the feeling of being satisfied (satiety) after food intake and thus could be useful in treating obesity.

When an excess of carbohydrate-rich food is eaten, the liver cells may become fully packed with glycogen and still have more glucose entering them. Liver cells then convert excess glucose to fatty acids and glycerol. These compounds are converted to triacylglycerols and sent to the fat depots of the body for storage.

Lipids are used as an energy source and to make biological molecules

Cells use ingested lipids as fuel, as components of cell membranes, and to make lipid compounds such as steroid hormones and bile salts. Lipids account for about 34% of the kilocalories in the average U.S. diet. Three polyunsaturated fatty acids (linoleic, linolenic, and arachidonic acids) are essential fatty acids that must be obtained in the human diet. Given these and sufficient nonlipid nutrients, the body can make all the lipid compounds (including fats, cholesterol, phospholipids, and prostaglandins) that it needs.

About 98% of the lipids in the diet are ingested in the form of triacylglycerols (triglycerides). (Recall from Chapter 3 that a triacylglycerol is a glycerol molecule chemically combined with three fatty acids; see Fig. 3–12*b*.) Triacylglycerols may be saturated, that is, fully loaded with hydrogen atoms, or their fatty acids may be monounsaturated (containing one double bond) or polyunsaturated (containing two or more double bonds).

In general, animal foods are rich in both saturated fats and cholesterol, whereas most plant foods contain unsaturated fats and no cholesterol. Commonly used polyunsaturated vegetable oils are corn, soy, cottonseed, and safflower oils. Olive, canola, and peanut oils contain large amounts of monounsaturated fats. However, some plant oils, including palm oil and coconut oil, are high in saturated fats. Butter contains mainly saturated fats.

The average U.S. diet provides far more cholesterol than the recommended daily maximum of 300 mg. High cholesterol sources include egg yolks, butter, and meat. The body is not dependent on dietary sources for cholesterol because it is able to synthesize cholesterol from other nutrients.

When needed, stored fats are hydrolyzed to fatty acids and released into the blood. Before these fatty acids can be used by cells as fuel, they must be broken down into smaller compounds and combined with coenzyme A to form molecules of acetyl coenzyme A (acetyl CoA; Fig. 45–14). Acetyl CoA enters the citric acid cycle (see Chapter 7). The conversion of fatty acids to acetyl CoA is accomplished in the liver by a process known as β-oxidation.

For transport to the cells, acetyl coenzyme A is converted into one of three types of *ketone bodies* (four-carbon ketones). Normally, the level of ketone bodies in the blood is low, but in certain abnormal conditions, such as starvation and diabetes mellitus, fat metabolism is tremendously increased. Ketone bodies are then produced so rapidly that their level in the blood becomes excessive, which may cause the blood to become too acidic. Such disruption of normal pH balance can lead to death.

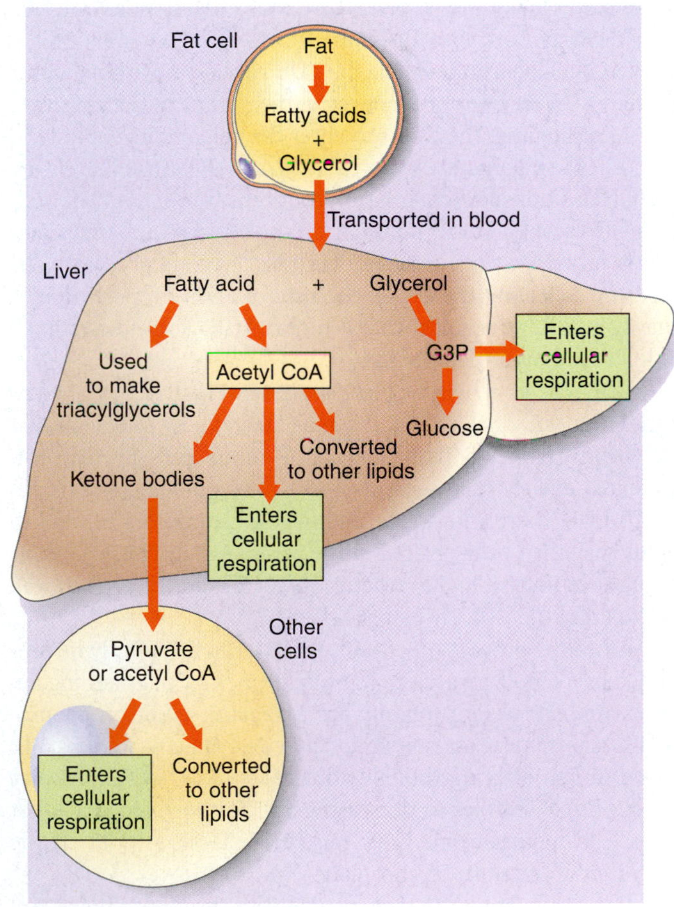

Figure 45–14 How the body uses fat. The liver plays an important role in converting glycerol and fatty acids to compounds that can be used as fuel in cellular respiration. Recall from Chapter 7 that G3P is glyceraldehyde-3-phosphate.

MAKING THE CONNECTION

What is the relationship between fat and cholesterol intake and cardiovascular disease? This issue has been extensively studied because of the role of lipids in atherosclerosis, a progressive disease in which the arteries become clogged with fatty material (see Chapter 42). Research results indicate that the type of fat consumed, as well as other dietary and other lifestyle factors, is important. For example, the Inuits in Greenland consume extremely high-fat diets, but because their dietary fat is rich in **omega-3 fatty acids** (found in fish oils), they have a low incidence of heart disease.

Lipid transport and metabolism are complex. Recall that chylomicrons transport lipids from the intestine to the cells of the body. As chylomicrons circulate in the blood, the enzyme **lipoprotein lipase** breaks down triacylglycerols. The fatty acids and glycerol can then be taken up by the cells. What is left of the chylomicron, a remnant made up mainly of cholesterol and protein, is taken up by the liver.

Cholesterol and triacylglycerols are not transported freely in the blood plasma. In the liver they are bound to proteins and are transported as large molecular complexes, called **lipoproteins.** Some plasma cholesterol is transported by **high-density lipoproteins,** or **HDLs** ("good" cholesterol), but most is transported by **low-density lipoproteins,** or **LDLs** ("bad" cholesterol). LDLs deliver cholesterol to the cells. For LDLs to be taken up by cells, a protein (apolipoprotein B) on the LDL surface must bind with a protein **LDL receptor** on the plasma membrane of the cell.

After binding, the LDL can enter the cell and its cholesterol and other components can be used. When cholesterol levels are high, HDLs transport cholesterol from cells throughout the body to the liver where it is removed from the blood. Thus, HDLs may play a protective role and decrease risk for coronary heart diseases by collecting the excess cholesterol and transporting it to the liver. Studies suggest that a higher HDL level lowers one's risk for cardiovascular disease.

LDL receptors help regulate cholesterol level in the blood by combining with LDL so that it can be taken up by body cells. Recall from Chapter 5 that cholesterol enters cells by receptor-mediated endocytosis (see Fig. 5–20). If the blood contains too much LDL, there will not be enough LDL receptors to bind it; lipids may then be deposited in the arterial wall. This process is thought to involve highly reactive molecules in the arterial wall that oxidize the LDL cholesterol.

A healthy proportion of HDL to LDL can apparently be promoted by a regular exercise program, by diet, by maintaining appropriate body weight (obesity has a negative effect on lipoprotein levels), and by not smoking cigarettes. Omega-3 fatty acids, found in fish oils, are thought to decrease LDL levels and play other protective roles in decreasing the risk for coronary heart disease. Monounsaturated fats raise HDL level, as does stearic acid, the saturated fat in chocolate.

Because polyunsaturated fats decrease the blood cholesterol level, many people now cook with vegetable oils rather than with butter and lard, drink skim milk rather than whole milk, and eat ice milk instead of ice cream. Some use margarine instead of butter. Vegetable oils are partially hydrogenated (some of the carbons accept hydrogen to become fully saturated) to produce margarine. During hydrogenation, some double bonds are transformed from a *cis* arrangement to a *trans* arrangement (see Fig. 3–3*b*), forming **trans fatty acids;** the harder the margarine, the higher the trans fatty acid content. Many processed foods, including cookies, crackers, and potato chips, contain trans fatty acids. Health concerns have been raised regarding trans fatty acids because these fatty acids increase LDL cholesterol in the blood and also lower HDL.

Genetics appears to be important in determining risk for cardiovascular disease. Some individuals with low-fat diets develop high blood cholesterol levels, and some with high-fat diets do not. However, in about one-third of the population, a diet high in saturated fats and cholesterol raises the blood cholesterol level by as much as 25%. Moreover, some individuals with high cholesterol levels do not develop coronary heart disease. Others who seem to have a safe risk profile suddenly drop dead from myocardial infarction (heart attack).

Proteins serve as enzymes and as structural components of cells

Proteins are essential building blocks of cells, serve as enzymes, and are also used to make needed substances such as hemoglobin and myosin. Protein consumption is an index of a country's (or an individual's) economic status, because high-quality protein tends to be the most expensive and least available of the nutrients.

The recommended daily adult intake of protein is about 56 g—only about an eighth of a pound (dry weight). In the United States and other developed countries, most people eat about twice as much protein as they require. In other parts of the world, protein poverty is one of the most pressing health problems. Millions of humans suffer from poor health, disease, and even death as a consequence of protein malnutrition.

Ingested proteins are degraded in the digestive tract to amino acids. Of the 20 or so amino acids important in nutrition, approximately nine (ten in children) cannot be synthesized by humans at all, or at least not in sufficient quantities to meet the body's needs. These, which must be provided in the diet, are referred to as **essential amino acids** (see Chapter 3).

Complete proteins, those that contain the most appropriate distribution of amino acids for human nutrition, are found in eggs, milk, meat, and fish. Some foods, such as gelatin and soybeans, contain a high proportion of protein. However, they either do not contain all the essential amino acids, or they do not contain them in proper nutritional proportions. Most plant proteins are deficient in one or more essential amino acids (see *Focus On: Vegetarian Diets*).

Amino acids circulating in the blood can be taken up by cells and used for the synthesis of proteins. Excess amino acids are removed from the circulation by the liver. In the liver cells, these are deaminated; that is, the amino group is removed (Fig. 45–15). During deamination, ammonia forms from the amino group. Ammonia, which is toxic at high concentrations, is converted to urea and excreted from the body. The remaining carbon

FOCUS ON Vegetarian Diets

Most of the world's population still depends mainly on plants, especially cereal grains (typically rice, wheat, or corn) as the staple food. However, this situation has been changing as people in China and other developing countries have become more affluent and have increased their meat consumption. Meat-based diets are ecologically expensive to produce. About 21 kg of protein in grain, for example, is required to produce just 1 kg of beef protein. Approximately 38% of the world's grain (and about 70% of the grain produced in the United States) is fed to cows, pigs, and chickens. Livestock production has become big business. Animals are no longer raised on family farms. Instead, thousands of animals are confined to small areas, often leading to pollution of nearby waterways by manure. In the Amazon and other parts of the world, forests are being cut down to provide more grazing land for cattle. Thus, meat-based diets exert a negative impact on our environment.

Balanced vegetarian diets offer health benefits in part because they are low in saturated fat and cholesterol and rich in legumes, grains, vegetables, and fruit. Such diets provide more fiber and phytochemicals, as well as more of certain vitamins and minerals, than a typical meat-based diet. Studies of Seventh Day Adventists, a religious group that excludes animal products from their diets, compared with non–Seventh Day Adventists living in the same area indicate that the incidence of cardiovascular disease is reduced by about 50%. Other studies have shown that vegetarians have a lower risk for certain types of cancer.

Vegetarian diets must be nutritionally planned because plant foods typically lack adequate amounts of one or more of the essential amino acids. An important guideline is to select proteins that complement one another. Protein complementation requires knowledge of which amino acids are deficient in each kind of food. For example, if beans and rice are eaten together, all the needed amino acids are provided, because one food contributes what the other lacks. (Rice is limited in the amino acid lysine but is high in methionine; beans are low in methionine, but high in lysine.) Similarly, if dairy products are not excluded from the vegetarian diet, then macaroni can be paired with cheese, or cereal with milk, thereby providing all the essential amino acids. Because the body does not store amino acids, all the essential amino acids should be ingested daily.

Other concerns regarding vegetarian diets include deficiencies of iron, calcium, zinc, and certain vitamins. Dairy products, which are a major source of calcium and vitamin D (in fortified milk), are excluded in certain (vegan) vegetarian diets. Vitamin B_{12} is found almost exclusively in animal products. Vegetarians can ensure adequate amounts of these minerals and vitamins by including supplements or fortified foods in their diet.

chain of the amino acid (called a keto acid) may be converted into carbohydrate or lipid and used as fuel or stored. Thus, even people who eat high-protein diets can gain weight if they eat too much.

Vitamins are organic compounds essential for normal metabolism

Vitamins are organic compounds required in the diet in relatively small amounts for normal biochemical functioning. Many are components of coenzymes (see Chapter 6). Vitamins may be divided into two main groups. **Fat-soluble vitamins,** those that can be dissolved in fat, include vitamins A, D, E, and K. **Water-soluble vitamins** are the B and C vitamins. Table 45–3 provides the sources, functions, and consequences of deficiency for most of the vitamins.

Health professionals debate the advisability of taking large amounts of certain specific vitamins, such as vitamin C to prevent colds or vitamin E to protect against vascular disease. Some studies suggest that vitamin A (found in yellow and green vegetables) and vitamin C (found in citrus fruit and tomatoes) may help protect against certain forms of cancer. We do not yet understand all the biochemical roles played by vitamins or the interactions among various vitamins and other nutrients.

We do know that large overdoses of vitamins, like vitamin deficiency, can be harmful. Moderate overdoses of the B and C vitamins are excreted in the urine, but surpluses of the fat-soluble vitamins are not easily excreted and can accumulate to harmful levels.

Minerals are inorganic nutrients

Minerals are inorganic nutrients ingested in the form of salts dissolved in food and water (Table 45–4). Essential minerals required in amounts of 100 mg or more daily include sodium, chloride, potassium, calcium, phosphorus, magnesium, and sulfur. Several others, such as iron, copper, iodide, fluoride, and selenium, are **trace elements,** minerals that are required in amounts of less than 100 mg per day.

Minerals are needed as components of body tissues and fluids. Salt content (about 0.9% in plasma) is vital in maintaining the fluid balance of the body, and since salts are lost from the body daily in sweat, urine, and feces, they must be replaced by dietary intake. Sodium chloride (common table salt) is the salt needed in largest quantity in blood and other body fluids. A deficiency results in dehydration.

Iron is the mineral most likely to be deficient in the diet. In fact, iron deficiency is one of the most widespread nutritional problems in the world. In most developing countries, an estimated two-thirds of children and women of childbearing age suffer from iron deficiency. In the United States, Europe, and Japan, 10% to 20% of women of childbearing age have this deficiency.

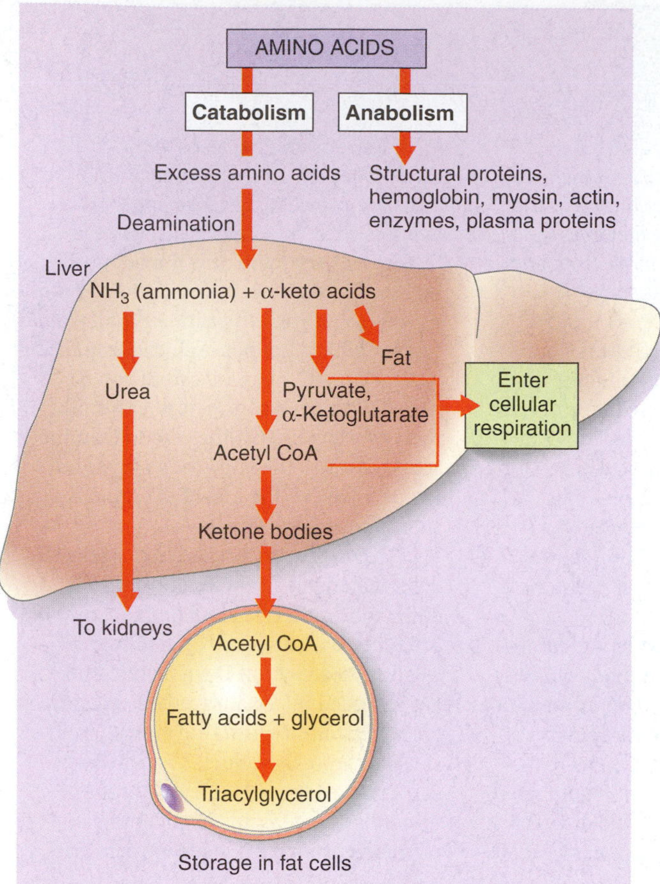

Figure 45–15 How the body uses protein. The liver plays a central role in protein metabolism. Deamination of amino acids and conversion of the amino groups to urea take place there. In addition, many proteins are synthesized in the liver. Note that you can gain weight on a high-protein diet because excess amino acids can be converted to fat.

Antioxidants protect against oxidants

Normal cell processes that require oxygen produce **oxidants,** highly reactive molecules such as free radicals, peroxides, and superoxides. Free radicals, molecules or ions with one or more unpaired electrons, are also generated by ionizing radiation, tobacco smoke, and other forms of air pollution. Oxidants can damage DNA, proteins, and unsaturated fatty acids by snatching electrons. Damage to DNA can cause mutations that lead to cancer, and injury to unsaturated fatty acids can result in damage to cell membranes. Free radicals are thought to contribute to atherosclerosis by causing oxidation of LDL cholesterol. Oxidative damage to the body over many years contributes to the aging process.

Cells have **antioxidants** that destroy free radicals and other reactive molecules. Antioxidants in tissues include certain enzymes, for example, superoxide dismutase, catalase, peroxidase, and glutathione. Their action requires minerals such as selenium, zinc, manganese, and copper. Certain vitamins—vitamin C, vitamin E, and β-carotene—have strong antioxidant activity. Vitamins A and E protect cell membranes from free radicals. In addition, a variety of phytochemicals are potent antioxidants (discussed in next section).

Many antioxidants have more than one action. For example, vitamin E also helps prevent selenium deficiency. We do not know whether antioxidant supplements are useful. Meanwhile nutritionists recommend that we increase the antioxidants in our diet by eating fruits, vegetables, and other foods high in antioxidants.

Phytochemicals play important roles in maintaining health

Results of hundreds of studies indicate that diets rich in fruits and vegetables lower the incidence of cancer and heart disease. In fact, studies comparing diets in various countries suggest that intake of fruits and vegetables may be more important than differences in dietary fat. Yet, nutritionists estimate that in the United States only about 1 in 11 people eats the recommended three servings of vegetables and two of fruit each day. A diet that includes all the essential nutrients does not provide the same health benefits as one rich in fruit and vegetables. The missing ingredients appear to be **phytochemicals,** compounds found in plants that promote health. Some phytochemicals function as antioxidants. In Asian countries where diets are low in fat and high in soy and green tea, the incidence of breast, prostate, and colorectal cancer is low.

Nutritionists are just beginning to intensively investigate phytochemicals. Among the important classes of phytochemicals are the carotenoids, allium compounds, flavonoids, indoles, and isocyanates. Except for carotenoids, yellow-orange pigments that the body can convert into vitamin A, phytochemicals have not been established as essential nutrients. Among the carotenoids are the *lycopenes,* powerful antioxidants that are responsible for the red color of tomatoes.

For hundreds of years garlic has been used medicinally. Nutritionists are just beginning to understand how its *allium compounds* (also found in onions, scallions, and chives) protect against heart disease and cancer. Allium compounds induce enzymes that detoxify carcinogens. They also decrease the production of nitrosamines (cancer-causing compounds produced from certain food preservatives) in the gastrointestinal tract.

More than 4000 *flavonoids* have been identified. These water-soluble pigments in fruits, vegetables, grains, flowers, and seeds are antioxidants. Flavonoids are responsible for the blue color of blueberries and the red colors of raspberries and red cabbage. *Indoles* and *isocyanates* are phytochemicals that enhance the synthesis of enzymes that detoxify cancer-causing agents. Cruciferous vegetables—cabbage, Brussels sprouts, cauliflower, and their relatives—are rich in flavonoids, indoles, and isocyanates. Studies indicate that diets rich in these vegetables reduce the risk of several types of cancer, including colon cancer.

TABLE 45–3 The Vitamins

Vitamins and U.S. RDA*	Actions	Effect of Deficiency	Sources
Fat-soluble			
Vitamin A, retinol 5000 IU†	Converted to retinal; essential for normal vision; essential for normal growth and differentiation of cells; reproduction; immunity	Growth retardation; night blindness; worldwide 250 million children are at risk of blindness from vitamin A deficiency	Liver, fortified milk, yellow and green vegetables such as carrots and broccoli
Vitamin D, calciferol 400 IU	Promotes calcium and phosphorus absorption from digestive tract; essential to normal growth and maintenance of bone	Weak bones; bone deformities: rickets in children, osteomalacia in adults	Fish oils, egg yolk, fortified milk, butter, margarine
Vitamin E, tocopherols 30 IU	Antioxidant; protects unsaturated fatty acids and cell membranes	Increased catabolism of unsaturated fatty acids, so that not enough are available for maintenance of cell membranes; prevention of normal growth; nerve damage	Vegetable oils, nuts; leafy greens
Vitamin K, about 80 mcg‡	Essential for blood clotting	Prolonged blood clotting time	Normally supplied by intestinal bacteria; leafy greens, legumes
Water-soluble			
Vitamin C, ascorbic acid 60 mg	Collagen synthesis; antioxidant; needed for synthesis of some hormones and neurotransmitters; important in immune function	Scurvy (wounds heal very slowly and scars become weak and split open; capillaries become fragile; bone does not grow or heal properly)	Citrus fruit, strawberries, tomatoes leafy vegetables, cabbage
B-complex vitamins			
Vitamin B_1, thiamine 1.5 mg	Active form is a coenzyme in many enzyme systems; important in carbohydrate and amino acid metabolism	Beriberi (weakened heart muscle, enlarged right side of heart, nervous system and digestive tract disorders); common in alcoholics	Liver, yeast, whole and enriched grains, meat, green leafy vegetables
Vitamin B_2, riboflavin 1.7 mg	Used to make coenzymes (e.g., FAD) essential in cellular respiration	Dermatitis, inflammation and cracking at corners of mouth; confusion	Liver, milk, eggs, green leafy vegetables, enriched grains
Niacin 20 mg	Component of important coenzymes (NAD^+ and $NADP^+$); essential to cellular respiration	Pellagra (dermatitis, diarrhea, mental symptoms, muscular weakness, fatigue)	Liver, chicken, tuna, milk, green leafy vegetables, enriched grains
Vitamin B_6, pyridoxine 2 mg	Derivative is coenzyme in many reactions in amino acid metabolism	Dermatitis, digestive tract disturbances; convulsions	Meat, whole grains, legumes, green leafy vegetables
Pantothenic acid 10 mg	Constituent of coenzyme A (important in cellular metabolism)	Deficiency extremely rare	Meat, whole grains, legumes
Folic acid 400 mcg	Coenzyme needed for nucleic acid synthesis and for maturation of red blood cells	A type of anemia; certain birth defects; increased risk of cardiovascular disease; deficiency in alcoholics, smokers, and pregnant women	Liver, legumes, dark green leafy vegetables, orange juice
Biotin 30 mcg	Coenzyme important in metabolism	—	Produced by intestinal bacteria; liver; chocolate, egg yolk
Vitamin B_{12} 6 mcg	Coenzyme important in metabolism; contains cobalt	A type of anemia	Liver, meat, fish

*RDA is the recommended dietary allowance, established by the Food and Nutrition Board of the National Research Council, to maintain good nutrition for healthy adults.

†International Unit: the amount that produces a specific biological effect and is internationally accepted as a measure of the activity of the substance.

‡mcg = micrograms.

TABLE 45–4 Some Important Minerals and Their Functions

Mineral	Functions	Sources; Comments
Calcium	Component of bones and teeth; essential for normal blood clotting; needed for normal muscle and nerve function	Milk and other dairy products, fish, green leafy vegetables; bones serve as calcium reservoir
Phosphorus	Performs more functions than any other mineral; structural component of bone; component of ATP, DNA, RNA, and phospholipids	Meat, dairy products, cereals
Sulfur	Component of many proteins and vitamins	High-protein foods such as meat, fish, legumes, nuts
Potassium	Principal positive ion within cells; influences muscle contraction and nerve excitability	Fruit, vegetables, grains
Sodium	Principal positive ion in interstitial fluid; important in fluid balance; neural transmission	Many foods, table salt; too much ingested in average American diet; excessive amounts may contribute to high blood pressure
Chloride	Principal negative ion of interstitial fluid; important in fluid balance and in acid-base balance	Many foods; table salt
Magnesium	Needed for normal muscle and nerve function	Nuts; whole grains; green, leafy vegetables
Copper	Component of enzyme needed for melanin synthesis; component of many other enzymes; essential for hemoglobin synthesis	Liver, eggs, fish, whole wheat flour, beans
Iodide	Component of thyroid hormones (hormones that increase metabolic rate); deficiency results in goiter (abnormal enlargement of thyroid gland)	Seafood, iodized salt, vegetables grown in iodine-rich soils
Manganese	Necessary to activate arginase, an enzyme essential for urea formation; activates many other enzymes	Whole-grain cereals, egg yolks, green vegetables; poorly absorbed from intestine
Iron	Component of hemoglobin, myoglobin, important respiratory enzymes (cytochromes), and other enzymes essential to oxygen transport and cellular respiration; deficiency results in anemia and may impair cognitive function	Mineral most likely to be deficient in diet. Good sources: meat (especially liver), nuts, egg yolk, legumes, dried fruit
Fluoride	Component of bones and teeth; makes teeth resistant to decay; excess causes tooth mottling	Fish; in areas where it does not occur naturally, fluoride may be added to municipal water supplies (fluoridation)
Zinc	Cofactor for at least 70 enzymes; helps regulate synthesis of certain proteins; needed for growth and repair of tissues; deficiency may impair cognitive function	Meat, milk, yogurt, some seafood
Selenium	Antioxidant (part of a peroxidase that breaks down peroxides)	Seafood, eggs, liver, garlic, mushrooms

Chocolate contains flavonoid phenols, phytochemicals that have potent antioxidant activity. Tea contains flavonoids of the catechin group that act synergistically with vitamins E and C. In mice exposed to nitrosamines, green tea reduced lung cancer by 45%. In a study of more than 35,000 women who drank more than two cups of black tea per day, risk for urinary tract cancer was reduced by 60% and cancers of the gastrointestinal tract by 32%.

ENERGY METABOLISM IS BALANCED WHEN ENERGY INPUT EQUALS ENERGY OUTPUT

The amount of energy liberated by the body per unit time is a measure of the **metabolic rate.** Much of the energy expended by the body is ultimately converted to heat. Metabolic rate may be expressed either in kilocalories of heat energy expended per day or as a percentage above or below a standard normal level.

The **basal metabolic rate (BMR)** is the rate at which the body releases heat as a result of breaking down fuel molecules. BMR is the body's basic cost of living, that is, the rate of energy used during resting conditions. This energy is required to maintain body functions such as heart contraction, breathing, and kidney function. An individual's **total metabolic rate** is the sum of his or her BMR and the energy used to carry on all daily activities. For example, a laborer has a greater total metabolic rate than does an executive whose job requirements do not include a substantial amount of movement and who does not exercise regularly.

An average-sized person who does not engage in any exercise program and who sits at a desk all day expends about 2000 kcal daily. If the food the individual eats each day also contains about 2000 kcal, the body will be in a state of energy balance; that is, energy input will equal energy output. This is an extremely important concept, because body weight remains constant when

$$\text{Energy input} = \text{Energy output}$$

When energy output is greater than energy input, stored fat is burned and body weight decreases. On the other hand, people gain weight when they take in more energy in food than they expend in daily activity, in other words, when

$$\text{Energy input} > \text{Energy output}$$

Obesity is a serious nutritional problem

Obesity, the excess accumulation of body fat, is a serious form of malnutrition that has become a problem of epidemic proportions in affluent societies. More than 50% of U.S. adults are overweight, and more than 33% are clinically obese (defined as having a body mass index [BMI] above 30; discussed later in this section). An estimated 20% of U.S. children and adolescents are obese. Several large studies suggest that an overweight person is at greater risk for heart disease, diabetes mellitus, osteoarthritis, and certain forms of cancer. For example, in the well-known Framingham study of more than 2000 men, those who were 20% overweight had a significantly higher mortality rate from all causes. Obesity contributes to about 300,000 deaths annually in the United States and is the second leading preventable cause of death (second only to smoking). Recently, some researchers have suggested that lack of physical activity may be a confounding factor in many studies, and being physically unfit may be the most important factor in higher mortality.

Body mass index (BMI) is now used worldwide as a measure of body size. BMI is an index of weight in relation to height. It is calculated by dividing the square of the weight (kg^2) by height (m). The English equivalent is 4.89 times the weight (lb) divided by the square of the height (ft^2). A person is considered obese if the BMI is 30 or more. Each of us appears to have a **set point,** or steady state, around which body weight is regulated. When BMI decreases below an individual's set point, energy-conserving mechanisms are activated and energy expenditure decreases.

Obesity can result from an increase in the size of fat cells or from an increase in the number of fat cells, or both. The number of fat cells in the adult is apparently determined mainly by the amount of fat stored during infancy and childhood. When we are overfed early in life, abnormally large numbers of fat cells are formed. Later in life, these fat cells may be fully stocked with excess lipids or may shrink in size, but they are always there. People with such increased numbers of fat cells are thought to be at greater risk for obesity than are those with normal numbers.

Process of Science

In their search for the causes of obesity, many investigators are focusing on the genetic aspects of obesity and on signaling pathways. This research has its roots in the 1950s when a spontaneous recessive mutation occurred in a strain of laboratory mice, resulting in mutant mice that were grossly obese. The increase in adipose tissue in these mice is part of a syndrome that parallels morbid obesity in humans, a condition in which an individual's body weight is 100 lb (45.5 kg) or more above normal. In 1994, Jeffrey M. Friedman's research team at Rockefeller University isolated the gene (*ob* gene) that, when mutated, was responsible for the obese phenotype. Mice with the mutated gene apparently lacked some weight-regulating substance.

In 1995, three different research teams reported in *Science* that injections of the *ob* gene product, later named **leptin,** resulted in weight loss by obese and non-obese mice. All three research groups transferred the *ob* gene into bacteria, which then made large quantities of the protein. When leptin was injected into grossly obese mice, their appetites decreased and their energy use increased, resulting in weight loss, mainly from loss of body fat. Friedman's team fed the same diet to obese mice that were injected with leptin and to a control group of obese mice that did not receive leptin. The mice in the treated group lost 50% more weight than the untreated animals (Fig. 45–16).

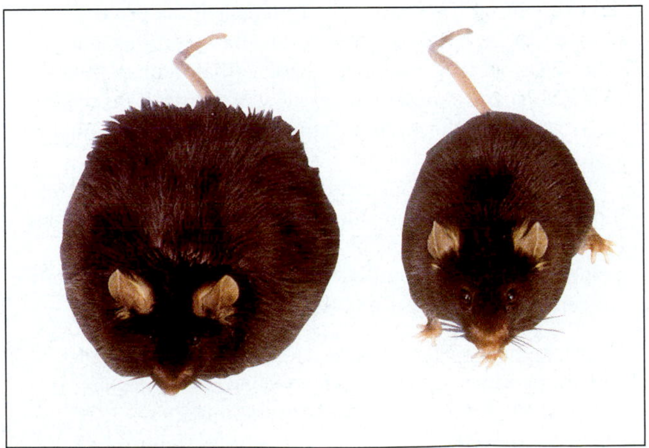

Figure 45–16 Mutant mice before *(left)* and after *(right)* treatment with leptin. *(Courtesy of John Sholtis/ Rockefeller University, New York City)*

Process of Science

ON THE CUTTING EDGE

In Search of the Signaling Molecules that Regulate Food Intake and Energy Homeostasis

HYPOTHESIS: The neuropeptide **melanin-concentrating hormone (MCH)** helps regulate energy homeostasis by responding to fasting and to the hormone leptin.

METHOD: MCH expression was studied in a strain of genetically obese mice before and after a 60-hour fast. Leptin was administered to one group of fasting mice.

RESULTS: After fasting, expression of MCH messenger RNA was increased by 33%. Administration of leptin to fasting mice blunted this increase.

CONCLUSION: MCH is an important signaling molecule in the regulation of energy homeostasis; its expression appears to be regulated by leptin.

Identifying the signaling molecules involved and understanding how they regulate food intake and energy homeostasis are areas of intense research interest. These signaling molecules form complex regulatory pathways involving genetic, neural, and endocrine mechanisms. Nicholas A. Tritos, of the Joslin Diabetes Center and Harvard Medical School, and his team of researchers, including Terry Maratos-Flier, have demonstrated that the neuropeptide MCH is a component of the complex molecular pathways that regulate food intake and energy balance.* (Neuropeptides are a group of neurotransmitters composed of two or more amino acids that also function as signaling molecules in non-neural tissues.) Tritos and his colleagues also showed that MCH is regulated by the hormone leptin.

About the same time that the hormone leptin was identified, Terry Maratos-Flier discovered that MCH is two to three times more active in *ob/ob* mice (which are leptin-deficient).† Maratos-Flier hypothesized that the increased MCH may be a factor in the obesity of these animals. She injected MCH into the brains of rats and observed that they ate more, which supported her hypothesis. When her research group, in collaboration with her husband, Jeffrey Flier, knocked out the MCH gene in mice, the animals ate less and weighed 15% to 25% less. These researchers also discovered that the neurons that produce MCH project to olfactory areas in the cerebral cortex. Based on this and other data, Maratos-Flier hypothesized that neuropeptides like MCH may be present at very low concentrations when we are satiated. However, the smell of tempting food might result in their release.

How do leptin, MCH, and the many other signaling molecules that have been identified interact to regulate food intake? Leptin appears to be produced by fat cells in proportion to body fat mass. When body fat decreases, leptin concentration decreases and the animal eats more. In contrast, when body fat increases, leptin secretion also increases and the animal eats less. Receptors for leptin have been identified in the hypothalamus. The leptin receptor is encoded by the *diabetes (db)* gene. Mice with a *db* mutation produce leptin but cannot respond to it, and so are obese.

Recent studies indicate that the hormone insulin, like leptin, is secreted in response to the amount of adipose tissue in the body. When energy balance is positive, the amount of adipose tissue increases, and secretion of both insulin and leptin increases. Catabolic activities increase, favoring a return to energy homeostasis. A group of peptides called **melanocortins** are also involved in regulation of body weight. Their receptors appear to decrease appetite in response to increased fat stores.

During negative energy balance, the body's adipose tissue decreases, resulting in a decrease in both leptin and insulin secretion. When leptin levels and food intake are low, as during dieting or starvation, the hypothalamus increases secretion of the neurotransmitter **neuropeptide Y (NPY).** This neuropeptide increases appetite and slows metabolism, actions that help restore energy homeostasis.

As researchers discover the genetic and biochemical mechanisms that regulate energy metabolism, they provide potential targets for the development of pharmacological treatments for obesity. When leptin was first discovered, many thought this hormone would be an effective cure for human obesity. (In fact, a biotech company paid more than $20 million for the rights to leptin.) Results of clinical trials have been disappointing, however. Leptin does not work as well in humans as in mice, and pharmaceutical companies are focusing on other strategies, for example, blocking MCH.

The following are among the many targets for drug action that are the focus of research:

1. Reduce appetite so that food intake is decreased.
2. Block absorption of fat.
3. Uncouple metabolism from energy production (see *Focus On: Electron Transport and Heat* in Chapter 7). This strategy results in the dissipation of food energy as heat.
4. Block receptors for molecules that send signals leading to increased energy storage.

*Tritos, N.A., Mastaitis, J.W., Kokkotou, E., and E. Maratos-Flier. "Characterization of Melanin Concentrating Hormone and Preproorexin Expression in the Murine Hypothalamus." *Brain Research*, Vol. 895, 2001.

†Gura, T. "Tracing Leptin's Partners in Regulating Body Weight." *Science*, Vol. 287, 10 Mar. 2000.

Frank Collins's research group at Amgen showed that treated mutant mice became more active and that their metabolic rate increased. Arthur Campfield's team at Hoffman–La Roche, Inc., reported that leptin acts directly on neural networks in the brain. They suggested that leptin is a hormone, produced in adipose tissue, that signals centers in the brain about the status of energy stores. The brain centers then adjust feeding behavior and energy metabolism.

During the past few years, researchers have identified several gene mutations and signaling molecules that affect body weight (see *On the Cutting Edge: In Search of the Signaling Molecules that Regulate Food Intake and Energy Homeostasis*). An *ob* gene has been identified in humans, indicating that this gene has been conserved during evolution. Although some investigators estimate that 40% to 70% of the factors involved in obesity are inherited, there is plasticity in the system that regulates body weight. High-calorie diets, overeating, and underexercising are important factors leading to obesity, and psychosocial factors influence these behaviors. A sedentary lifestyle combined with the ready availability of high-energy foods contributes to the problem. One researcher cleverly described the roots of the obesity problem as the combination of "computer chips and potato chips." For every 9.3 kcal of excess food taken into the body, about 1 g of fat is stored. (An excess of about 140 kcal, less than a typical candy bar, per day for a month results in a 1-lb weight gain.)

Undernutrition can cause serious health problems

While millions of people eat too much, millions of others do not have enough to eat or do not eat a balanced diet. Individuals suffering from undernutrition are weak, easily fatigued, and highly susceptible to infection. Essential amino acids, iron, calcium, and vitamin A are commonly deficient nutrients.

Of all the required nutrients, essential amino acids are the ones most often deficient in the diet. Millions of people suffer from poor health and lowered resistance to disease because of protein deficiency. Children's physical and mental development are retarded when the essential building blocks of cells are not provided in the diet. Because their bodies cannot manufacture antibodies (which are proteins) and cells needed to fight infection, common childhood diseases, such as measles, whooping cough, and chickenpox, are often fatal in children suffering from protein malnutrition.

In young children, severe protein malnutrition results in the condition known as **kwashiorkor.** This term is an African word that means "first-second." It refers to the situation in which a first child is displaced from its mother's breast when a younger sibling is born. The older child is placed on a diet of starchy cereal or cassava that is deficient in protein. Growth becomes stunted, muscles are wasted, and edema develops (as displayed by a swollen belly); the child becomes apathetic and anemic, with an impaired metabolism (Fig. 45–17). Without essential amino acids, digestive enzymes cannot be manufactured, so what little protein is ingested cannot be digested.

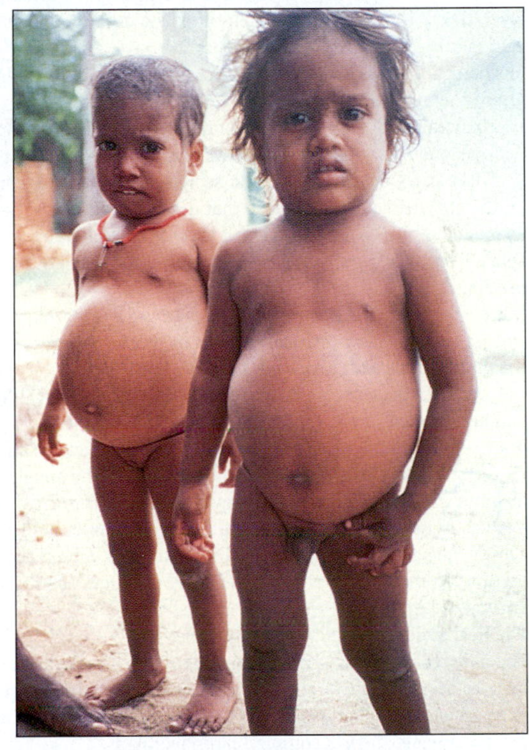

Figure 45–17 Protein deficiency. Millions of children suffer from kwashiorkor, a disease caused by severe protein deficiency. Note the characteristic swollen belly, which results from fluid imbalance. *(P. Pittet/United Nations Food and Agricultural Organization)*

SUMMARY WITH KEY TERMS

I. **Nutrition** is the process of taking in and using food. Animals are **heterotrophs;** they must obtain their **nutrients** from the organic molecules manufactured by other organisms.
   A. **Feeding** is the selection, acquisition, and **ingestion** of food.
   B. The process of breaking down food mechanically and chemically is **digestion.** Nutrients pass through the lining of the digestive tract and into the blood by **absorption.** Food that is not digested and absorbed is discharged from the body by **egestion** or **elimination.**

II. An organism's body plan and lifestyle are adapted to its mode of nutrition. Animals can be classified according to the type of food they typically eat and according to their mechanisms for obtaining food.
   A. **Herbivores** feed directly on producers.
   B. **Carnivores** are adapted for capturing and killing prey.
   C. **Omnivores** eat both plants and animals.
   D. Some animals eat large pieces of food. Others are **suspension feeders** that trap or filter food from water or air surrounding them. Still others suck fluids from other organisms.

III. The simplest invertebrates, the sponges, have no digestive system; they digest food intracellularly. Most other animals have organs or systems specialized for processing food.
   A. In cnidarians and flatworms, food is digested in the **gastrovascular cavity.** This cavity has only one opening that serves as both mouth

and anus. Cnidarians and flatworms have intracellular and extracellular digestion.

B. In more complex invertebrates and in all vertebrates, the digestive tract is a complete tube with an opening at each end. Digestion is mainly extracellular.

IV. In animals with a complete digestive tract, various parts of the tract are specialized to perform specific functions. In vertebrates, food passes in sequence through the mouth, pharynx, esophagus, stomach, small intestine, large intestine, and anus.

A. **Motility** refers to the mixing and propulsive movements of the digestive tract. The propulsive activity characteristic of most regions of the digestive tract is **peristalsis,** waves of muscular contraction that push the food along.

B. The wall of the digestive tract consists of four layers. The **mucosa,** a layer of epithelial and connective tissue, lines the lumen. The **submucosa** consists of connective tissue rich in blood and lymph vessels and nerves. The **muscle layer** consists of an inner sublayer of circularly arranged muscle fibers and an outer longitudinal sublayer. Below the level of the diaphragm, the outer connective tissue coat is called the **visceral peritoneum.**

C. Mechanical digestion and enzymatic digestion of carbohydrates begin in the mouth.

1. Mammalian teeth include **incisors** for biting, **canines** for tearing food, and **premolars** and **molars** for crushing and grinding.
2. Three pairs of **salivary glands** secrete saliva, a fluid containing the enzyme **salivary amylase** that digests starch.

D. As food is swallowed, it is propelled through the **pharynx** and **esophagus.** A **bolus** of food is moved along through the digestive tract by peristalsis.

E. In the **stomach,** food is mechanically digested by vigorous churning, and proteins are enzymatically digested by the action of **pepsin** in the gastric juice. **Rugae** are folds in the stomach wall that expand as the stomach fills with food. **Gastric glands** secrete hydrochloric acid and **pepsinogen**, the precursor of pepsin.

F. After several hours, a soup of partly digested food, called **chyme,** exits the stomach through the **pylorus** and enters the **small intestine** in spurts. The surface area of the small intestine is greatly expanded by folds in its wall, by the intestinal **villi,** and by **microvilli.** Most enzymatic digestion takes place in the **duodenum,** which receives secretions from the liver and pancreas and produces several digestive enzymes of its own.

G. The **liver** produces **bile,** which emulsifies fats.

H. The **pancreas** releases enzymes that digest protein, lipid, and carbohydrate, as well as RNA and DNA. **Trypsin** and **chymotrypsin** digest polypeptides to dipeptides. **Pancreatic lipase** degrades fats, and **pancreatic amylase** digests complex carbohydrates.

I. Activities of the digestive system are regulated by both nerves and hormones. In addition to sympathetic and parasympathetic nerves, digestive activity is regulated by networks of neurons within the intestinal wall. Among the hormones secreted by the digestive tract are **gastrin, secretin, cholecystokinin (CCK),** and **gastric inhibitory peptide (GIP).**

J. Nutrients in chyme are enzymatically digested as they move through the digestive tract.

1. Polysaccharides are digested to the disaccharide maltose by salivary and pancreatic amylases. Maltase in the small intestine splits maltose into glucose, the main product of carbohydrate digestion.
2. Proteins are split by pepsin in the stomach and by proteolytic enzymes in the pancreatic juice. The peptides and dipeptides produced are then split by dipeptidases. The end products of protein digestion are amino acids.
3. Lipids are emulsified by bile salts and then hydrolyzed by pancreatic lipase.

K. Nutrients are absorbed through the thin walls of the intestinal villi. Amino acids and glucose are transported to the liver by the **hepatic portal vein.** Fatty acids and monoacylglycerols combine with bile salts to form soluble complexes called **micelles.** Fatty acids and monacylglycerols leave the micelles and enter epithelial cells in the intestinal lining. There they are reassembled into triacylglycerols and are packaged into droplets that also contain cholesterol and phospholipids. They are covered by a thin coat of protein. Called **chylomicrons,** these lipid droplets are transported by the lymphatic system to the blood circulation.

L. The **large intestine,** which consists of the **cecum, colon, rectum,** and **anus,** is responsible for the elimination of undigested wastes. It also incubates bacteria that produce vitamin K and certain B vitamins.

V. For a balanced diet, humans and other animals require carbohydrates, lipids, proteins, vitamins, and minerals.

A. Most carbohydrates are ingested in the form of polysaccharides—starch and cellulose. Polysaccharides are referred to as **complex carbohydrates. Fiber** is mainly a mixture of cellulose and other indigestible carbohydrates.

1. Carbohydrates are used mainly as fuel.
2. Glucose concentration in the blood is carefully regulated. Excess glucose is stored as glycogen and can also be converted to fat.

B. Lipids are used as fuel, as components of cell membranes, and to synthesize steroid hormones and other lipid substances.

1. Most lipids are ingested in the form of triacyglycerols.
2. Fatty acids are converted to molecules of acetyl coenzyme A and used as fuel. Excess fatty acids are converted to triacylglycerol and stored as fat.
3. Lipids are transported as large molecular complexes called **lipoproteins. Low-density lipoproteins (LDLs)** deliver cholesterol to the cells. **High-density lipoproteins (HDLs)** collect excess cholesterol and transport it to the liver.

C. Proteins serve as enzymes and are essential structural components of cells.

1. The best distribution of **essential amino acids** is found in the complete proteins of animal foods.
2. Excess amino acids are deaminated by liver cells. Amino groups are converted to urea and excreted in urine; the remaining keto acids are converted to carbohydrate and used as fuel or converted to lipid and stored in fat cells.

D. **Vitamins** are organic compounds required in small amounts for many biochemical processes. Many serve as components of coenzymes. **Fat-soluble vitamins** include vitamins A, D, E, and K. **Water-soluble vitamins** are the B and C vitamins.

E. **Minerals** are inorganic nutrients ingested as salts dissolved in food and water. **Trace elements** are minerals required in amounts less than 100 mg per day.

F. **Phytochemicals** are compounds found in plants that promote health. Some phytochemicals function as **antioxidants,** compounds that destroy **oxidants**, free radicals and other reactive molecules that damage DNA and other molecules by snatching electrons.

VI. **Basal metabolic rate (BMR)** is the body's cost of metabolic living.

A. **Total metabolic rate** is the BMR plus the energy used to carry on daily activities.

B. When energy (kilocalories) input equals energy output, body weight remains constant.

C. **Obesity** is a serious nutritional problem in which an excess amount of fat accumulates in adipose tissues. A person gains weight by taking in more energy, in the form of kilocalories, than is expended in activity.

D. Millions of people suffer from undernutrition. Essential amino acids are the nutrients most often deficient in the diet.

E. Researchers are identifying signaling molecules that make up the complex pathways that regulate food intake and energy metabolism. The hormone **leptin** appears to be produced by fat cells in proportion to body fat mass; it signals the brain about the status of energy stores. **Neuropeptide Y (NPY),** a neurotransmitter produced in the hypothalamus, increases appetite and slows metabolism when leptin levels and food intake are low. **Melanin-concentrating hormone (MCH),** a signaling molecule regulated by leptin, appears to stimulate food intake.

POST-TEST

1. The process of taking in and using food is (a) nutrition (b) chemical digestion (c) egestion (d) ingestion (e) absorption
2. Animals that feed mainly on producers are (a) herbivores (b) secondary consumers (c) animals with gastrovascular cavities (d) carnivores (e) secondary consumers and carnivores
3. Teeth adapted for crushing and grinding are (a) incisors (b) canines (c) premolars and molars (d) incisors and premolars (e) canines and molars
4. The layer of tissue that lines the lumen of the digestive tract is the (a) muscle layer (b) visceral peritoneum (c) parietal peritoneum (d) mucosa (e) submucosa
5. Which of the following are accessory digestive glands? (a) salivary glands (b) pancreas (c) liver (d) answers a, b, and c are correct (e) answers a and b only
6. Which of the following is the correct sequence? (a) pharynx ⟶ stomach ⟶ esophagus (b) stomach ⟶ large intestine ⟶ small intestine (c) esophagus ⟶ pharynx ⟶ stomach (d) cecum ⟶ colon ⟶ duodenum (e) pharynx ⟶ esophagus ⟶ stomach
7. Amylase is produced by the (a) liver (b) pancreas (c) salivary glands (d) liver and pancreas (e) pancreas and salivary glands
8. Pepsin is produced by the (a) liver (b) stomach (c) pancreas (d) duodenum (e) salivary glands
9. Which sequence most accurately describes the digestion of protein? (a) polypeptide ⟶ monoacylglycerol ⟶ amino acids (b) polypeptide ⟶ dipeptides ⟶ amino acids (c) protein ⟶ amylose ⟶ amino acid (d) protein ⟶ emulsified peptide ⟶ fatty acids and glycerol (e) protein ⟶ triacylglycerol ⟶ fatty acids and glycerol
10. The surface area of the small intestine is increased by (a) folds in its wall (b) villi (c) microvilli (d) a, b, and c are correct (e) a and b only
11. Lipids are transported from the intestine to the cells of the body by (a) chylomicrons (b) micelles (c) rugae (d) villi (e) leptin
12. Most vitamins are (a) inorganic compounds (b) components of coenzymes (c) used as fuel (d) electrolytes (e) required by herbivores and carnivores but not omnivores
13. When energy input is greater than energy output (a) weight loss occurs (b) weight remains stable (c) weight gain occurs (d) leptin prevents weight gain (e) the *ob* gene is activated
14. A hormone that stimulates gastric glands to secrete pepsinogen is (a) secretin (b) gastric inhibitory peptide (c) gastrin (d) cholecystokinin (CCK) (e) substance P
15. Neuropeptide Y (NPY) (a) increases appetite (b) increases metabolism (c) is encoded by the mutant *ob* gene (d) is a hormone produced by the pancreas (e) decreases bile production

REVIEW QUESTIONS

1. If you were presented with an unfamiliar animal and asked to determine its nutritional lifestyle, how would you go about this task?
2. What are the advantages of a digestive tract with two openings compared to one with a single opening?
3. How are digestive structures and methods of processing food in sponges, hydras, and flatworms adapted to each group's lifestyle? Give specific examples.
4. Trace a bite of food through the human digestive tract, listing each structure through which it passes.
5. Give the functions of the three types of accessory glands that secrete digestive juices in vertebrates. Identify their secretions.
6. The inner lining of the digestive tract is not smooth like the inside of a water pipe. Why is this advantageous? What structures increase its surface area?
7. Summarize the step-by-step digestion of (a) carbohydrates (b) lipids (c) proteins
8. What happens to ingested cellulose in humans? Why?
9. Draw and label an intestinal villus, and explain how its structure is adapted to its function.
10. How does the absorption of fat differ from the absorption of glucose?
11. List the nutrients that must be included in a balanced diet. What are some of the difficulties in planning a nutritionally balanced vegetarian diet? (Include the challenges encountered in obtaining adequate amounts of amino acids in a vegetarian diet.)
12. Why, specifically, is each of the following essential? (a) iron (b) calcium (c) iodide (d) vitamin A (e) vitamin K (f) essential amino acids
13. Draw a diagram to illustrate the fate of each of the following in the body: (a) glucose (b) absorbed amino acids (c) absorbed fat
14. Write an equation to describe energy balance, and explain what happens when the equation is altered in either direction.
15. Label the diagram. Use Figure 45–4 to check your answers.

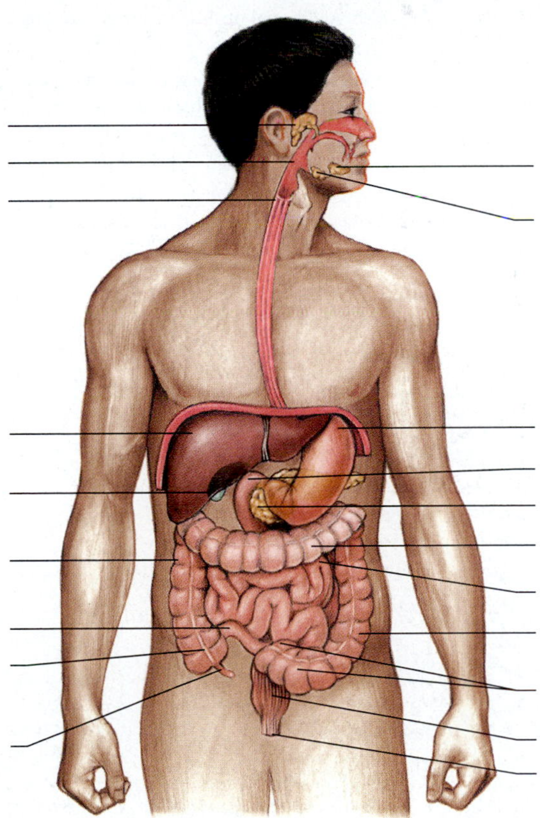

YOU MAKE THE CONNECTION

1. A high percentage of adults suffer from gastrointestinal discomfort, including cramps and diarrhea, when they drink milk. What do you think could cause this condition, known as lactose intolerance?
2. Design an experiment to test the hypothesis that the B vitamin pyridoxine is an essential nutrient.
3. Why are proteolytic enzymes produced in an inactive form?
4. Investigators are unraveling the biochemical pathways that help regulate body weight. How might their work lead to a cure for obesity?

RECOMMENDED READINGS

Borek, C. "Antioxidants and Cancer." *Science and Medicine,* Vol. 4, No. 6. Nov./Dec. 1997. Discussion of the importance of supplemental antioxidant vitamins and minerals.

Brown, L.R., et al. "Eradicating Hunger: A Growing Challenge." *State of the World 2001*. Norton, New York, 2001. A Worldwatch Institute Report on environmental problems and progress. Includes an analysis of our ability to provide adequate food for our increasing population.

"Genetics of Colon Cancer: A Special Report." *Science and Medicine,* Vol. 4, No. 4, Jul./Aug. 1997. Cumulative events, including mutation of oncogenes and inactivation of tumor suppressor genes, appear to cause the progression of colon cancer.

Nature, Vol. 404, 6 Apr. 2000. "Nature Insight: Obesity." Several articles reviewing the current status of research on the molecular biology of obesity.

Science, Vol. 280, No. 5368, 29 May 1998. Several articles on "Regulation of Body Weight." These articles discuss current research on obesity and eating disorders.

Smolin, L.A., and M.B. Grosvenor. *Nutrition: Science and Applications,* 3rd ed. Harcourt College Publishers, Philadelphia, 2000. A readable introduction to nutritional science.

Weigle, D.S., and J.L. Kuijper. "Mouse Models of Human Obesity." *Science and Medicine,* Vol. 4, No. 3. May/Jun. 1997. Genetic variations in signaling molecules involved in regulating body composition account for the inherited portion of body fat.

- Visit our Web site at **http://www.info.brookscole.com/solomonbergmartin** for links to chapter-related resources on the World Wide Web. Additional on-line materials relating to this chapter can also be found on our Web site.

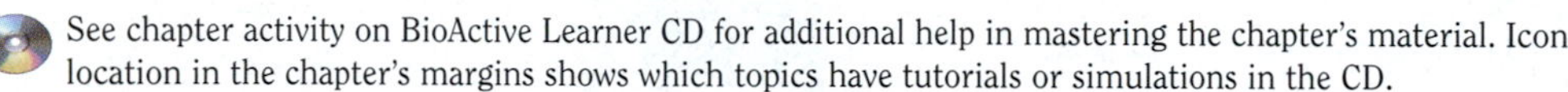

See chapter activity on BioActive Learner CD for additional help in mastering the chapter's material. Icon location in the chapter's margins shows which topics have tutorials or simulations in the CD.

46

Osmoregulation and Disposal of Metabolic Wastes

LEARNING OBJECTIVES

After you have studied this chapter you should be able to

1. Relate the principal functions of excretory systems to specific osmoregulatory challenges posed by various environments.
2. Contrast the advantages and disadvantages of excreting ammonia, uric acid, or urea.
3. Compare nephridial organs and Malpighian tubules as osmoregulatory organs.
4. Relate the function of the vertebrate kidney to the success of vertebrates in a wide variety of habitats.
5. Compare the adaptations to the challenges of osmoregulation that have evolved in freshwater fishes with those of marine bony fishes, sharks, and marine mammals.
6. Label on a diagram the organs of the mammalian urinary system and give the functions of each.
7. Identify on a diagram the principal parts of a nephron (including associated blood vessels) and give the functions of each structure.
8. Trace a drop of filtrate from Bowman's capsule to its release from the body as urine.
9. Describe the hormonal regulation of fluid balance by antidiuretic hormone (ADH), aldosterone, angiotensin II, and atrial natriuretic peptide (ANP).

Most terrestrial animals inhabit areas near water sources. Zebra, wildebeest, and a variety of birds gather to drink at a lake in Ngorongoro Crater in Tanzania. *(McMurray Photography)*

Water, the most abundant molecule both in the cell and on Earth, shapes life and the distribution of organisms on our planet. Water is the medium in which most metabolic reactions take place, and the osmotic and ionic composition of an animal's intracellular and **extracellular fluids** (the interstitial fluid, blood, and other fluid outside the cells) must be maintained within homeostatic limits. Natural selection has resulted in the evolution of a variety of homeostatic mechanisms that regulate the volume and composition of fluids in the internal environment. In this chapter we discuss some of these mechanisms.

Many small animals live in the ocean and obtain their food and oxygen directly from the seawater that surrounds them; they release waste products into the surrounding seawater. In larger, more complex animals and most terrestrial animals, the extracellular fluid serves as an internal sea. Blood plasma, composed mainly of water, transports nutrients, gases, waste products, and other materials throughout the animal body. Interstitial fluid forms from the blood plasma and bathes all the cells. Excess water evaporates from the body surface or is excreted by specialized structures.

Terrestrial animals have a continuous need to conserve water. Its loss from the body must be carefully regulated, and water that is lost must be replaced. Water is ingested with food and drink and is also produced in some metabolic reactions. Most animals require a dependable source of water with which to replenish their body fluids, and so, often inhabit areas near water sources *(see photograph)*.

Two processes that maintain homeostasis of fluids in animals are osmoregulation and excretion of metabolic wastes. **Osmoregulation** is the active regulation of osmotic pressure of body fluids to keep them from becoming too dilute or too concentrated. **Excretion** is the process of ridding the body of metabolic wastes, including excess water. Efficient **excretory systems** have evolved that function in osmoregulation and in disposal of metabolic wastes, excess water and ions, and harmful substances. As we discuss, hormones are important signaling molecules in these regulatory processes.

EXCRETORY SYSTEMS HELP MAINTAIN HOMEOSTASIS

Excretory systems maintain homeostasis by selectively adjusting the concentrations of salts and other substances in blood and other body fluids. Typically, an excretory system collects fluid, generally from the blood or interstitial fluid. It then adjusts the composition of this fluid by selectively returning needed substances to the body fluid. Finally, the adjusted excretory product containing excess or potentially toxic substances is released from the body. In many animals, a major excretory product is **urine,** a watery solution of metabolic wastes and other organic and inorganic substances.

The terms *excretion* and *elimination* are sometimes confused (Fig. 46–1). Food materials that have not been absorbed are eliminated from the body in the feces. Such substances never participated in the animal's metabolism or entered body cells but merely passed through the digestive system.

METABOLIC WASTE PRODUCTS INCLUDE WATER, CARBON DIOXIDE, AND NITROGENOUS WASTES

Metabolic wastes must be excreted so that they do not accumulate and reach concentrations that would disrupt homeostasis. The principal metabolic waste products in most animals are water, carbon dioxide, and **nitrogenous wastes,** those that contain nitrogen. Carbon dioxide is excreted mainly by respiratory structures (see Chapter 44). Water is also lost from respiratory surfaces in terrestrial animals. Excretory organs, such as kidneys, remove and excrete most of the water and nitrogenous wastes.

Nitrogenous wastes include ammonia, uric acid, and urea. Recall that amino acids and nucleic acids contain nitrogen. During the metabolism of amino acids, the nitrogen-containing amino group is removed (in a process known as deamination) and converted to **ammonia** (Fig. 46–2). However, ammonia is highly toxic. Some aquatic animals excrete it into the surrounding water before it can build up to toxic concentrations in their tissues. A few terrestrial animals, including some terrestrial snails and wood lice, vent it directly into the air. But in many animals, humans included, ammonia is converted to some less toxic nitrogenous waste such as uric acid or urea.

Uric acid is produced both from ammonia and by the breakdown of nucleotides from nucleic acids. Uric acid is insoluble in water and forms crystals that can be excreted as a crystalline paste with little fluid loss. This is an important water-conserving adaptation in many terrestrial animals, including insects, certain reptiles, and birds. Also, because uric acid is not toxic and can be safely stored, its excretion is an adaptive advantage for species whose young begin their development enclosed in eggs.

Urea is the principal nitrogenous waste product of amphibians and mammals. It is produced in the liver from ammonia.

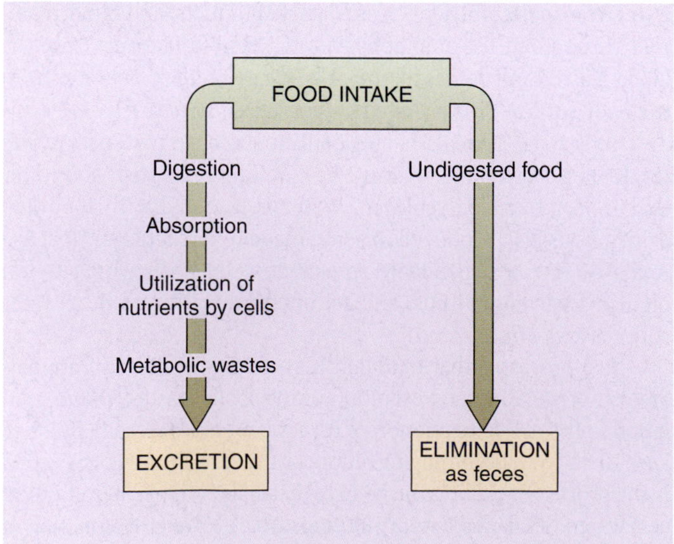

Figure 46–1 Excretion versus elimination. Excretion is the disposal of metabolic wastes. Elimination is the ejection of undigested and unabsorbed food from the body.

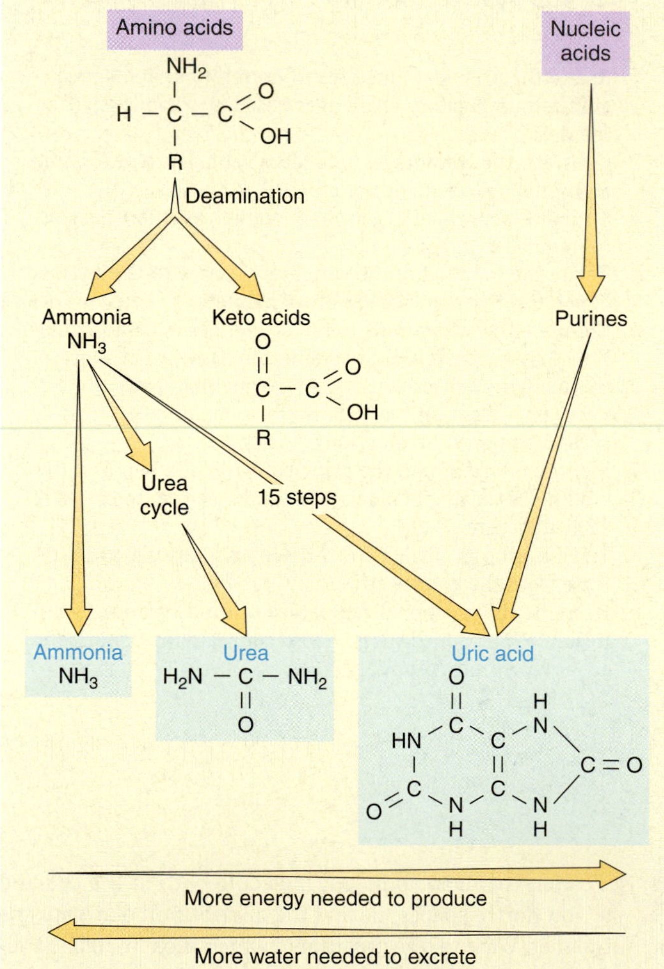

Figure 46–2 Formation of nitrogenous wastes. Deamination of amino acids and metabolism of nucleic acids produce nitrogenous wastes. Ammonia is the first metabolic product of deamination. In many animals, ammonia is converted to urea via the urea cycle. Other animals convert ammonia to uric acid. Energy is required to convert ammonia to urea and uric acid, but less water is required to excrete these wastes.

The sequence of reactions by which urea is synthesized from ammonia and carbon dioxide is known as the **urea cycle.** Like the formation of uric acid, these reactions require specific enzymes and the input of energy by the cells. Urea has the advantage of being far less toxic than ammonia and can accumulate in higher concentrations without causing tissue damage; thus, it can be excreted in more concentrated form. However, because urea is highly soluble, it is dissolved in water, and more water is needed to excrete urea than to excrete uric acid.

INVERTEBRATES HAVE ADAPTATIONS FOR OSMOREGULATION AND METABOLIC WASTE DISPOSAL

The ocean is a stable environment, and its salt concentration does not vary much. The body fluids of most marine invertebrates are in osmotic equilibrium with the surrounding seawater. These animals are known as **osmoconformers** because the concentration of their body fluids varies along with changes in the seawater. Even so, many osmoconforming marine animals regulate some ions in their body fluids.

Coastal habitats, such as estuaries that contain brackish water, are much less stable environments than is the open ocean. Salt concentrations change frequently with shifting tides. Many invertebrates (as well as vertebrates) that inhabit these environments are **osmoregulators** that maintain an optimal salt concentration in their tissues regardless of changes in the salt concentration of their surroundings.

In a coastal environment where fresh water enters the ocean, the water may have a lower salt concentration than the body fluids of the animal. Water osmotically moves into the body, and salt diffuses out. An animal adapted to this environment has excretory structures that remove the excess water. Crabs and some other animals have cells in their gills that remove salts from the surrounding water and transport them into the body fluids. Certain polychaete worms and the blue crab are among the animals that can be osmoconformers or osmoregulators depending on environmental conditions.

Dehydration is a constant threat to terrestrial animals. Because their fluid concentration is higher than that of the air around them, they tend to lose water by evaporation from both the body surface and respiratory surfaces and may also lose water as wastes are excreted. As animals moved onto the land, natural selection favored the evolution of structures and processes that conserve water.

Nephridial organs are specialized for osmoregulation and/or excretion

Marine sponges and cnidarians need no specialized excretory structures. Their wastes pass by diffusion from their cells to the external environment and are washed away by water currents. They expend little or no energy to excrete wastes. When water becomes stagnant and currents do not wash away wastes, aquatic environments, such as coral reefs, are damaged by their accumulation.

Nephridial organs, or nephridia, are osmoregulatory structures that have evolved in many invertebrates, including flatworms, nemerteans, rotifers, annelids, mollusks, and lancelets. Each nephridial organ consists of simple or branching tubes that typically open to the outside of the body through excretory pores, called nephridiopores. Two types of nephridial organs are protonephridia and metanephridia.

In flatworms and nemerteans, metabolic wastes pass through the body surface by diffusion, but these animals also have **protonephridia.** These nephridial organs are composed of tubules with no internal openings. Their enlarged blind ends consist of **flame cells** with brushes of cilia, so named because their constant motion reminded early biologists of flickering flames (Fig. 46–3). The flame cells lie in the interstitial fluid that

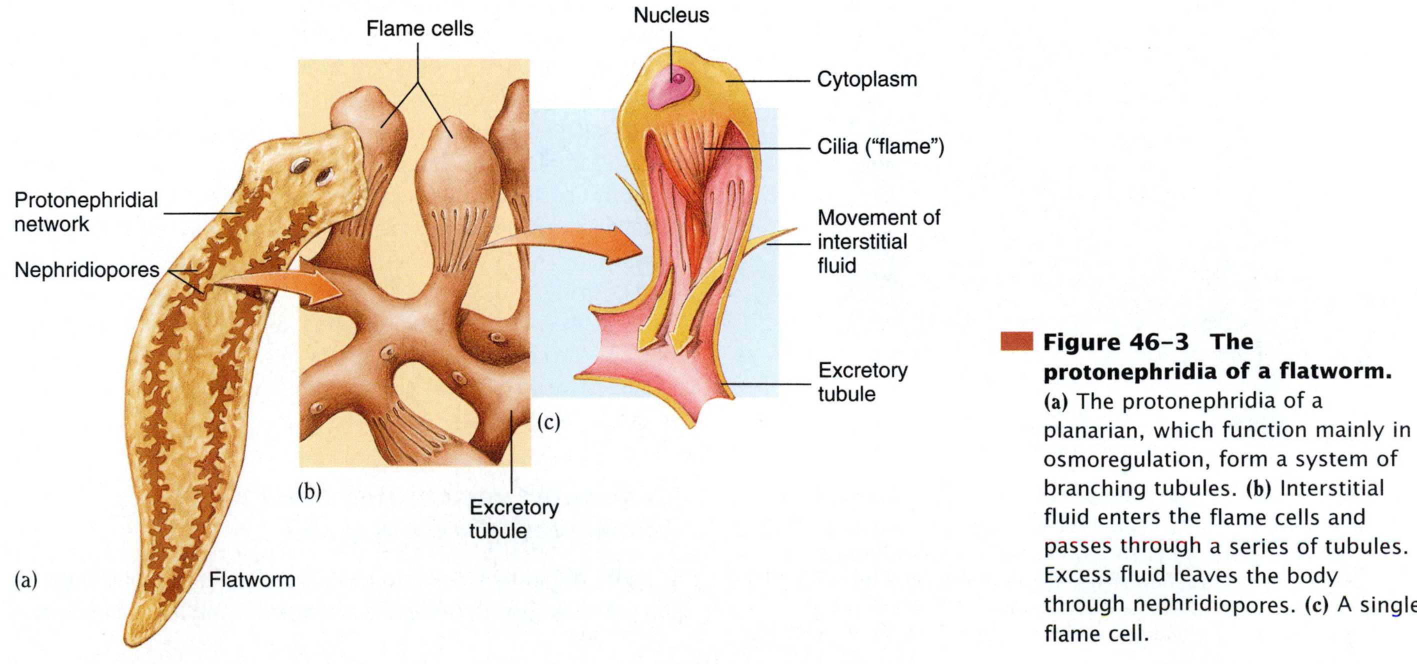

Figure 46–3 The protonephridia of a flatworm. **(a)** The protonephridia of a planarian, which function mainly in osmoregulation, form a system of branching tubules. **(b)** Interstitial fluid enters the flame cells and passes through a series of tubules. Excess fluid leaves the body through nephridiopores. **(c)** A single flame cell.

bathes the body cells. Fluid enters the flame cells, and the beating of the cilia propels the fluid through the tubules. Excess fluid leaves the body through nephridiopores.

Most annelids, as well as mollusks, have more complex nephridial organs called **metanephridia.** Each segment of an earthworm has a pair of metanephridia. A metanephridium is a tubule open at both ends. The inner end opens into the coelom as a ciliated funnel (Fig. 46–4), and the outer end opens to the outside through a nephridiopore. Around each tubule is a network of capillaries. Fluid from the coelom passes into the tubule, bringing with it whatever it contains—glucose, salts, or wastes. As fluid moves through the tubule, needed materials (such as water and glucose) are removed from the fluid by the tubule and are reabsorbed by the capillaries, leaving the wastes behind. In this way, urine is produced that contains concentrated wastes.

Malpighian tubules are an important adaptation for conserving water in insects

The excretory system of insects and spiders consists of several hundred **Malpighian tubules** (Fig. 46–5). Malpighian tubules are slender extensions of the gut wall. Their blind ends lie in the hemocoel (blood cavity) and are bathed in hemolymph. Cells of the tubule wall actively transport uric acid, potassium ions, and some other substances from the hemolymph into the tubule lumen. Other solutes and water follow by diffusion. The Malpighian tubules empty into the gut. Water, some salts, and other solutes are reabsorbed into the hemolymph by a specialized epithelium in the rectum.

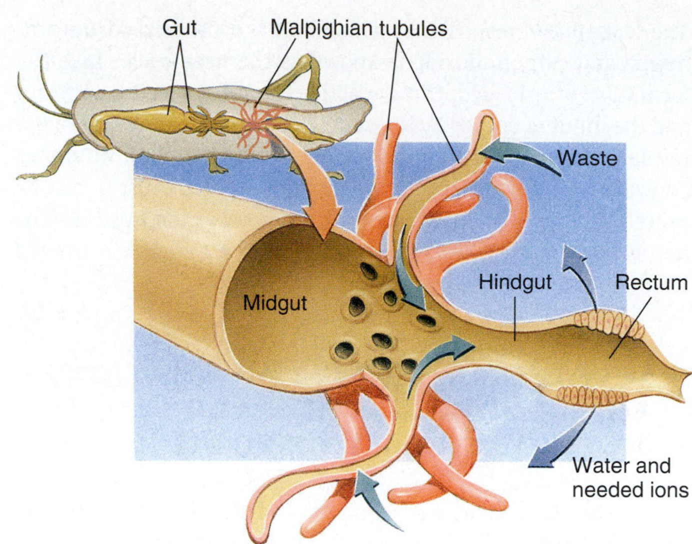

Figure 46–5 The Malpighian tubules of an insect. The slender Malpighian tubules have blind ends that extend into the hemocoel. Their cells transfer uric acid and some ions from the hemolymph to the cavity of the tubule. Water follows by diffusion. The wastes are discharged into the gut. The epithelium lining the rectum (part of the hindgut) actively reabsorbs most of the water and needed ions.

Uric acid, the major waste product, is excreted as a semidry paste with a minimum of water. Because the insect excretory system effectively conserves body fluids, it has contributed significantly to the success of insects in terrestrial environments.

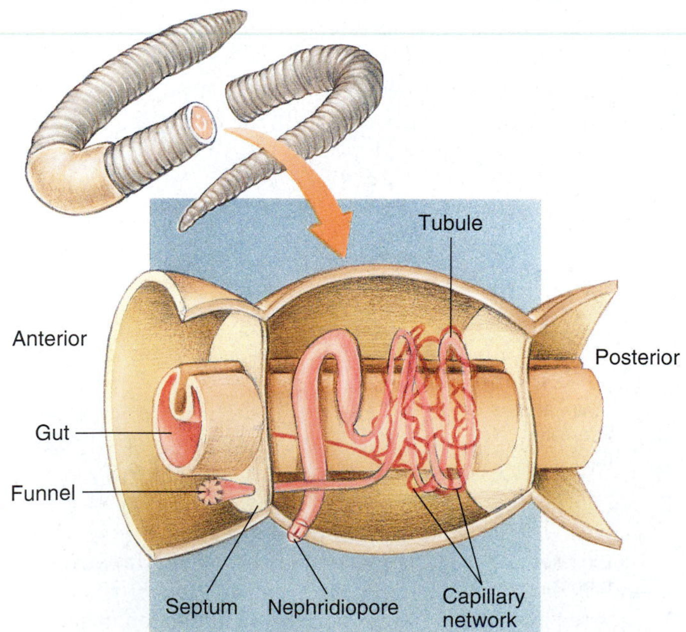

Figure 46–4 A metanephridium of an earthworm. Each metanephridium consists of a ciliated funnel opening into the coelom, a coiled tubule, and a nephridiopore opening to the outside. This three-dimensional internal view shows parts of three segments.

THE KIDNEY IS THE KEY VERTEBRATE ORGAN OF OSMOREGULATION AND EXCRETION

Vertebrates live successfully in a wide range of habitats—in fresh water, the ocean, tidal regions, and on land, even in extreme environments such as deserts. In response to the requirements of these diverse environments, adaptations have evolved for regulating salt and water content and for excreting wastes. An extreme example is the desert-dwelling kangaroo rat, which must carefully conserve water. It obtains most of its water from its own metabolism (recall from Chapter 7 that aerobic respiration produces water), and its kidneys are so efficient that it loses little fluid as urine.

The main osmoregulatory and excretory organ in vertebrates is the **kidney.** In most vertebrates, the skin, lungs or gills, and digestive system also help maintain fluid balance and dispose of metabolic wastes.

Freshwater vertebrates must rid themselves of excess water

As fishes began to move into freshwater habitats about 460 million years ago (mya), there was strong selection for the evolution of adaptations for effective osmoregulation. Because the body fluids of freshwater animals have higher salt concentrations and are thus hypertonic to their surroundings, water passes into them osmotically. As a result, they are in constant danger of becoming waterlogged. Freshwater fishes are covered by scales and a mucous secretion that retards the passage of water into the body. However, water enters through the gills. The kidneys of these fishes have become adapted to excrete large amounts of dilute urine (Fig. 46–6*a*).

Water entry, though, is only part of the challenge of osmoregulation in freshwater fishes. These animals also tend to lose salts to the surrounding fresh water. To compensate, special gill cells have evolved that actively transport salts (mainly sodium chloride) from the water into the body.

Most amphibians are at least semiaquatic, and their mechanisms of osmoregulation are similar to those of freshwater fishes. They, too, produce large amounts of dilute urine. In its urine and through its skin, a frog can lose an amount of water equivalent to one-third of its body weight in one day. Active transport of salt inward by special cells in the skin compensates for loss of salt through skin and in urine.

Marine vertebrates must replace lost fluid

Freshwater fishes have adapted very successfully to their aquatic habitats. Evolution of body fluids more dilute than seawater is one of their chief adaptations. Thus, when some freshwater fishes returned to the sea about 200 mya, their blood and body fluids were less salty than (hypotonic to) their surroundings. They tended to lose water osmotically and to take in salt. To compensate for fluid loss, many marine bony fishes drink seawater (Fig. 46–6*b*). They retain the water and excrete salt by the action of specialized cells in their gills. Very little urine is excreted by the kidneys; the nephrons (microscopic units of the kidney that produce urine) have only small or no capillary clusters (glomeruli) that filter the blood and produce urine in other vertebrates.

Marine cartilaginous fishes (sharks and rays) have different osmoregulatory adaptations that allow them to tolerate the salt concentrations of their environment. These animals accumulate and tolerate urea (Fig. 46–6*c*). Their tissues are adapted to function at concentrations of urea that would be toxic to most other animals. The high urea concentration makes the body fluids slightly hypertonic to seawater, resulting in a net inflow of water. Their well-developed kidneys excrete a large volume of urine. Excess salt is excreted by the kidneys and, in many species, by a rectal salt gland.

The heads of certain reptiles and marine birds have salt glands that excrete salt that has entered the body with ingested seawater. Salt glands are usually inactive; they function only in response to osmotic stress. Thus, only when seawater or salty food is ingested do the salt glands excrete a fluid laden with sodium and chloride.

Whales, dolphins, and other marine mammals ingest seawater along with their food. Their kidneys produce a very concentrated urine, much saltier than seawater. This is an important

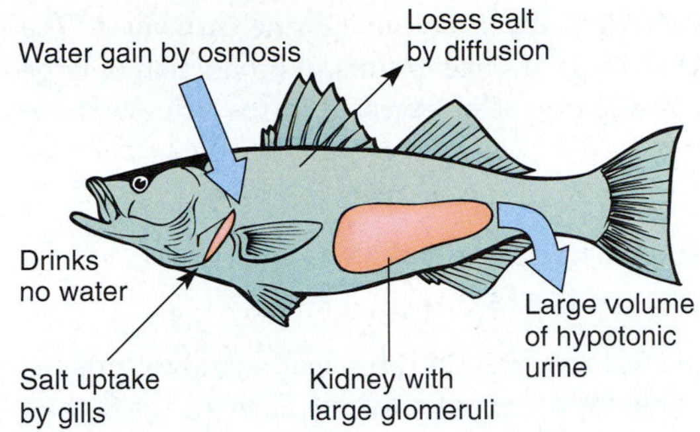

(a) Freshwater fish

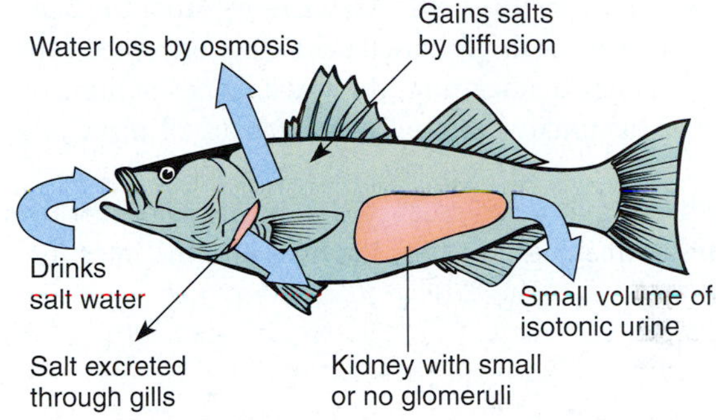

(b) Marine bony fish

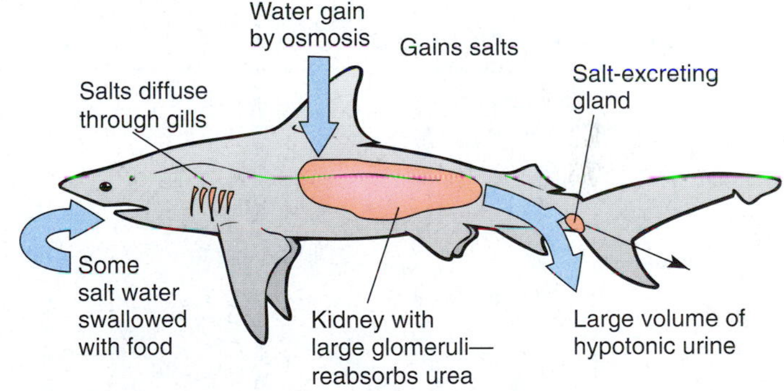

(c) Cartilaginous fish (shark)

Figure 46–6 Osmoregulation in fishes. **(a)** Freshwater fishes live in a hypotonic medium, so water continuously enters the body by osmosis, and salts diffuse out. These fishes excrete large quantities of dilute urine and actively transport salts in through the gills. **(b)** Marine fishes live in a hypertonic medium and therefore lose water by osmosis. They gain salts from the seawater they drink and by diffusion. To compensate, the fish drinks water, excretes the salt, and produces a small volume of urine. **(c)** In contrast, the shark kidney reabsorbs urea in high enough concentration that its tissues become hypertonic to the surrounding medium. As a result, water enters the shark by osmosis. A large quantity of dilute urine is excreted.

physiological adaptation, especially for marine carnivores. The high-protein diet of these animals results in production of large amounts of urea, which must be excreted in the urine without losing much water.

The mammalian kidney helps maintain homeostasis

In mammals, the kidneys are the principal excretory organs. They are responsible for excreting most nitrogenous wastes and for helping to maintain fluid balance by adjusting the salt and water content of the urine. As in other terrestrial vertebrates, the lungs, skin, and digestive system are also important in mammalian osmoregulation and waste disposal (Fig. 46–7). Most carbon dioxide is excreted by the lungs. Water is lost from the body as water vapor in exhaled air. Although primarily concerned with the regulation of body temperature, the sweat glands of humans and some other mammals excrete 5% to 10% of all metabolic wastes.

Most of the bile pigments produced by the breakdown of red blood cells are normally excreted by the liver into the intestine. From the intestine they then pass out of the body with the feces. The liver also produces both urea and uric acid, which are transported by the blood to the kidneys.

The kidneys produce the enzyme *renin* and at least two hormones: *erythropoietin,* which stimulates red blood cell production, and *1,25-dihydroxyvitamin* D_3, which stimulates calcium absorption by the intestine. As we discuss later in this chapter, renin helps regulate fluid balance and blood pressure.

(a)

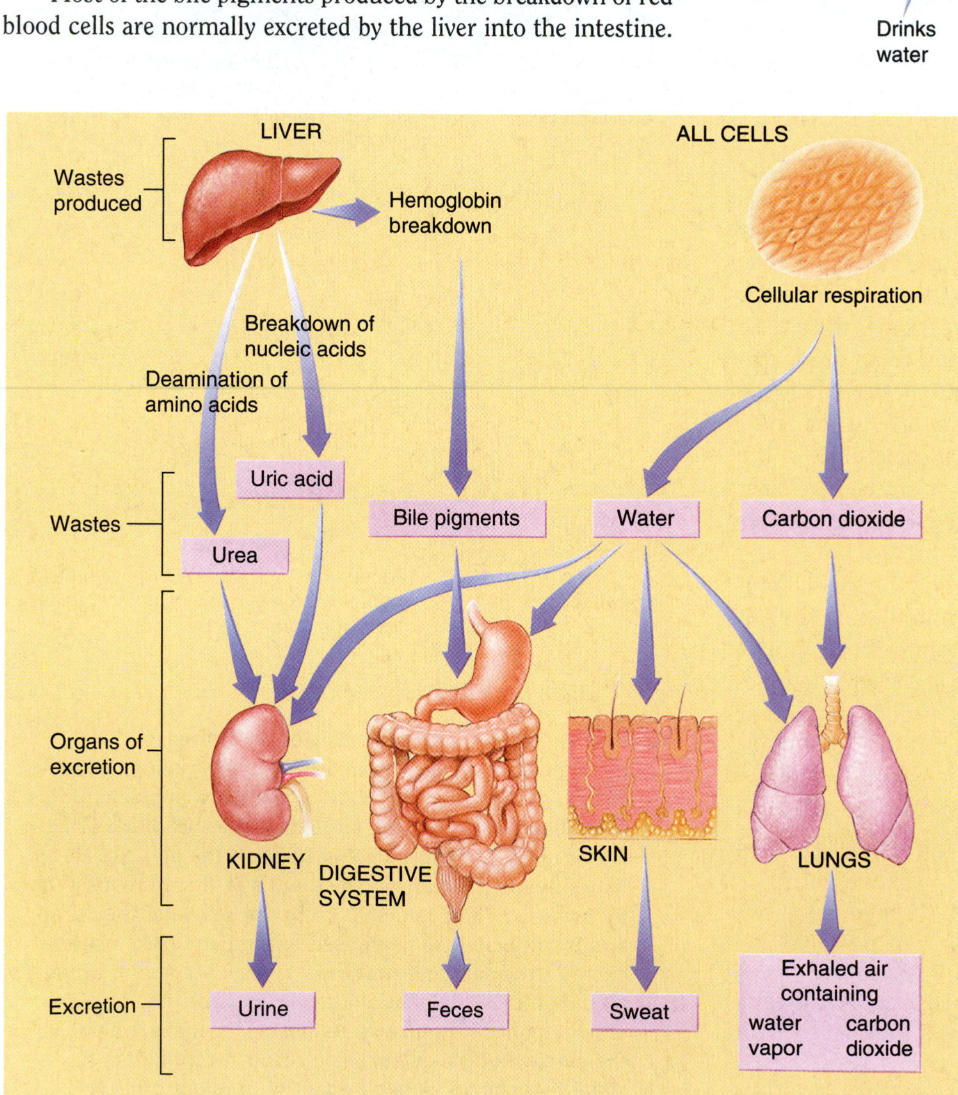

(b)

Figure 46–7 Excretory organs in terrestrial vertebrates. **(a)** The vertebrate kidney conserves water by reabsorbing it. **(b)** Disposal of metabolic wastes in humans and other terrestrial mammals. To conserve water, a small amount of hypertonic urine is produced. Nitrogenous wastes are produced by the liver and transported to the kidneys. All cells produce carbon dioxide and some water during cellular respiration.

THE KIDNEYS, URINARY BLADDER, AND THEIR DUCTS MAKE UP THE URINARY SYSTEM

The mammalian **urinary system** consists of the kidneys, the urinary bladder, and associated ducts. The overall structure of the human urinary system is shown in Figure 46–8. Located just below the diaphragm in the "small of the back," the kidneys look like a pair of giant, dark-red lima beans, each about the size of a fist. Each kidney is covered by a connective tissue capsule. The outer portion of the kidney is called the **renal cortex;** the inner portion is the **renal medulla** (Fig. 46–9). The renal medulla contains eight to ten cone-shaped structures called **renal pyramids.** The tip of each pyramid is a **renal papilla.** Each papilla has several pores, the openings of **collecting ducts.**

As urine is produced, it flows from collecting ducts through a renal papilla and into the **renal pelvis,** a funnel-shaped chamber. Urine then flows into one of the paired **ureters,** ducts that connect the kidney with the **urinary bladder.** The urinary bladder is a remarkable organ capable of holding (with practice) up to 800 mL (about a pint and a half) of urine. Emptying the bladder changes it from the size of a small melon to that of a pecan. This feat is made possible by the smooth muscle and specialized epithelium of the bladder wall, which is capable of great shrinkage and stretching.

During **urination,** urine is released from the bladder and flows through the **urethra,** a duct leading to the outside of the body. In the male, the urethra is lengthy and passes through the penis. Semen, as well as urine, is transported through the male urethra. In the female, the urethra is short and transports only urine. Its opening to the outside is just above the opening of the vagina. The length of the male urethra discourages bacterial invasions of the bladder. This length difference helps explain why bladder infections are more common in females than in males. In summary, urine flows through the following structures:

Kidney (through renal pelvis) ⟶ ureter ⟶ urinary bladder ⟶ urethra

The nephron is the functional unit of the kidney

Each kidney consists of more than one million functional units called **nephrons.** A nephron consists of a cuplike **Bowman's capsule** connected to a long, partially coiled **renal tubule** (Figs. 46–9 and 46–10). Positioned within Bowman's capsule is a cluster of capillaries known as a **glomerulus.**

Three main regions of the renal tubule are the **proximal convoluted tubule,** which conducts the filtrate from Bowman's capsule; the **loop of Henle,** an elongated, hairpin-shaped portion; and the **distal convoluted tubule,** which conducts the filtrate to a collecting duct. Thus, filtrate passes through the following structures:

Bowman's capsule ⟶ proximal convoluted tubule ⟶ loop of Henle ⟶ distal convoluted tubule ⟶ collecting duct

The human kidney has two types of nephrons: the more numerous (85%) cortical nephrons and the more internal

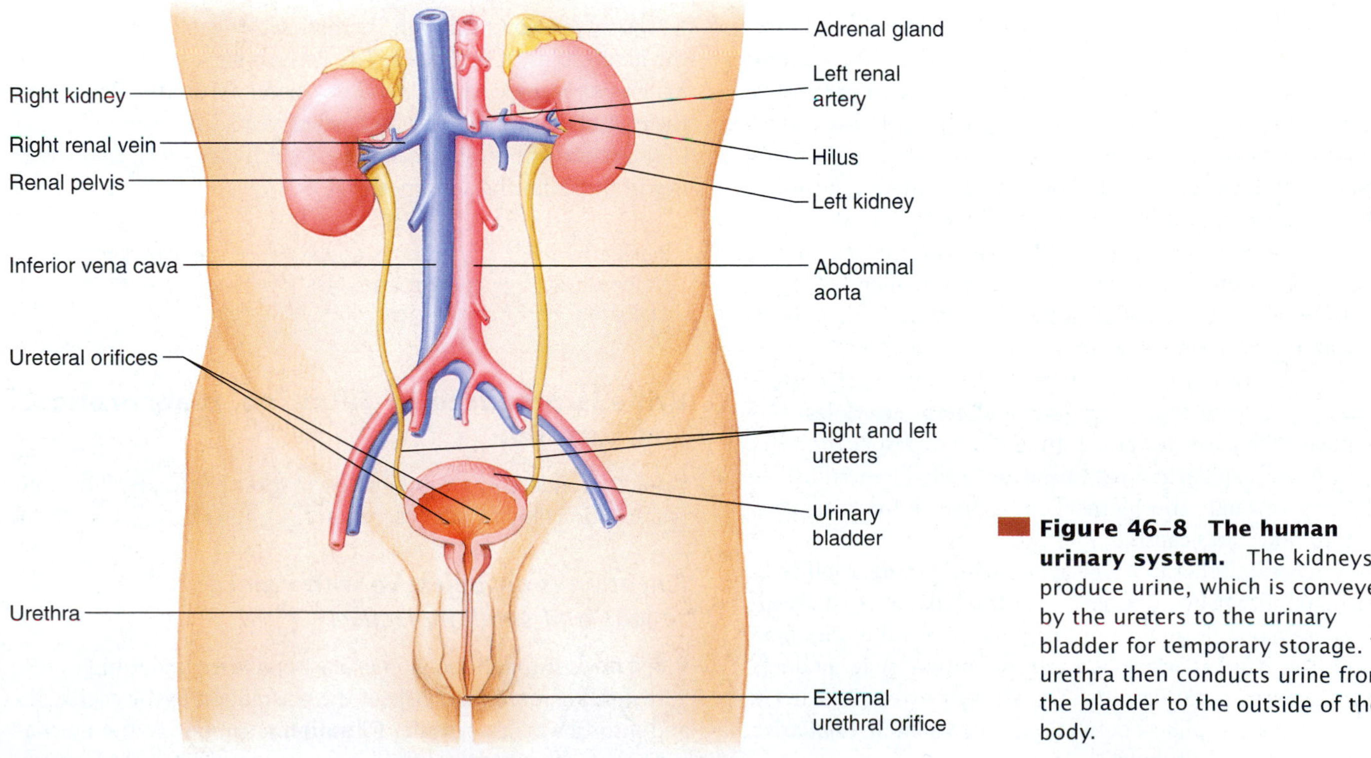

Figure 46–8 The human urinary system. The kidneys produce urine, which is conveyed by the ureters to the urinary bladder for temporary storage. The urethra then conducts urine from the bladder to the outside of the body.

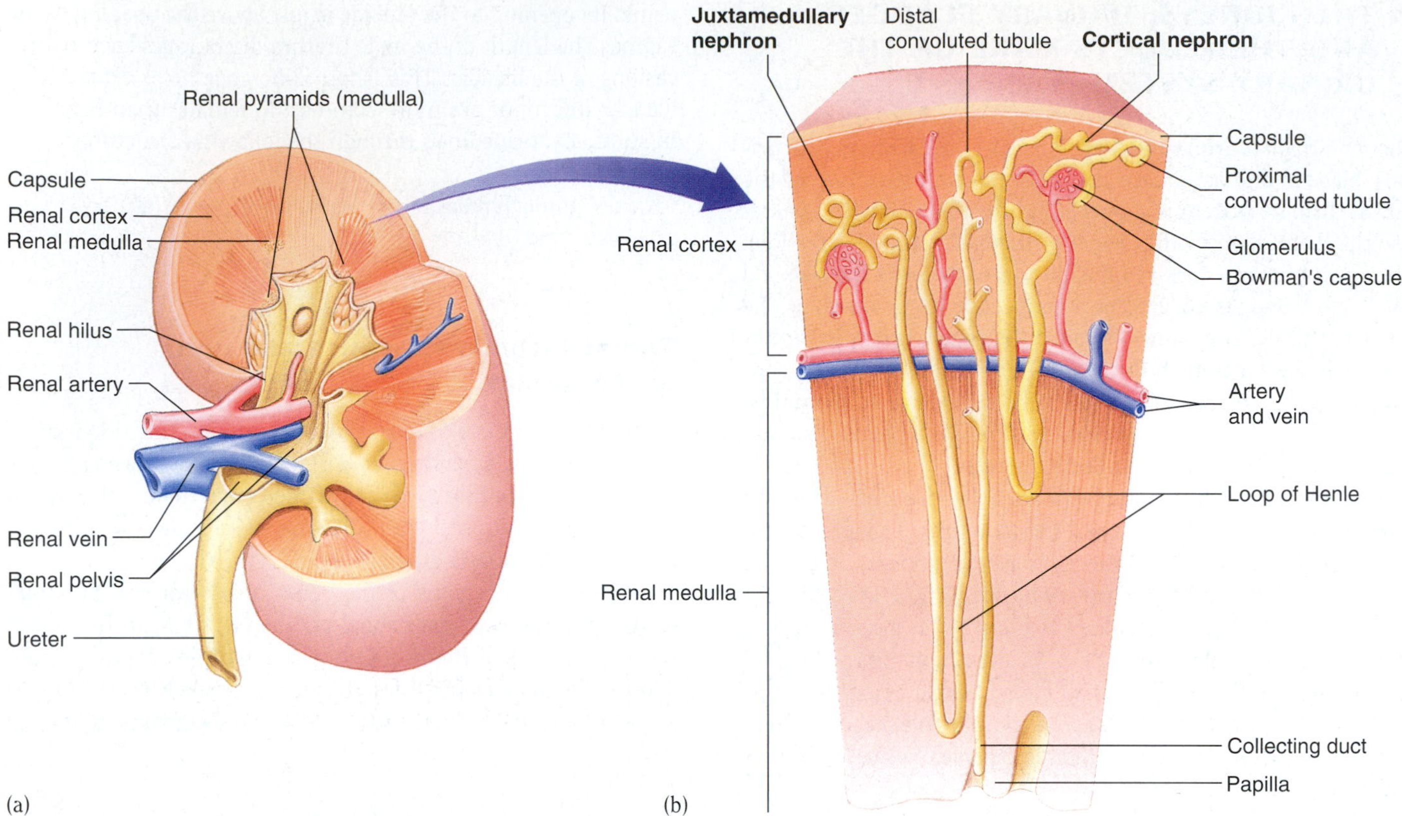

Figure 46–9 Structure of the kidney. **(a)** The kidney is covered by a fibrous capsule. The outer region of the kidney is the cortex; the inner region is the medulla. When urine is produced, it flows into the renal pelvis and leaves the kidney through the ureter. The renal artery delivers blood to the kidney; the renal vein drains blood away from the kidney. **(b)** Longitudinal section showing the location of the two main types of nephrons: the juxtamedullary nephron and the cortical nephron.

juxtamedullary nephrons (see Fig. 46–9*b*). **Cortical nephrons** have relatively small glomeruli and are located almost entirely within the cortex or outer medulla. **Juxtamedullary nephrons** have large glomeruli, and their very long loops of Henle extend deep into the medulla. The loop of Henle consists of a *descending limb* that receives filtrate from the proximal convoluted tubule and an *ascending limb,* through which the filtrate passes on its way to the distal convoluted tubule. The juxtamedullary nephrons contribute to the ability of the mammalian kidney to concentrate urine. Excretion of urine that is hypertonic to body fluids is an important mechanism for conserving water.

Blood is delivered to the kidney by the renal artery. Small branches of the renal artery give rise to **afferent arterioles.** (Afferent means "to carry toward.") An afferent arteriole conducts blood into the capillaries that make up each glomerulus. As blood flows through the glomerulus, some of the plasma is forced into Bowman's capsule.

You may recall that in a typical circulatory route, capillaries deliver blood into veins. Circulation in the kidneys is an exception because blood flowing from the glomerular capillaries next passes into an **efferent arteriole.** Each efferent arteriole conducts blood *away* from a glomerulus. The efferent arteriole delivers blood to a second capillary network, the **peritubular capillaries** surrounding the renal tubule.

As blood flows through the first set of capillaries, those of the glomerulus, it is filtered. The peritubular capillaries receive materials returned to the blood by the renal tubule. Blood from the peritubular capillaries enters small veins that eventually lead to the renal vein. In summary, blood circulates through the kidney in the following sequence:

Renal artery ⟶ afferent arterioles ⟶ capillaries of glomerulus ⟶ efferent arterioles ⟶ peritubular capillaries ⟶ small veins ⟶ renal vein

Urine is produced by filtration, reabsorption, and secretion

Urine is produced by a combination of three processes: filtration, reabsorption, and tubular secretion (Fig. 46–11).

Filtration is not selective with regard to ions and small molecules

Blood flows through the glomerular capillaries under high pressure, forcing more than 10% of the plasma out of the capillaries and into Bowman's capsule. **Filtration** is similar to the mechanism whereby interstitial fluid is formed as blood flows through

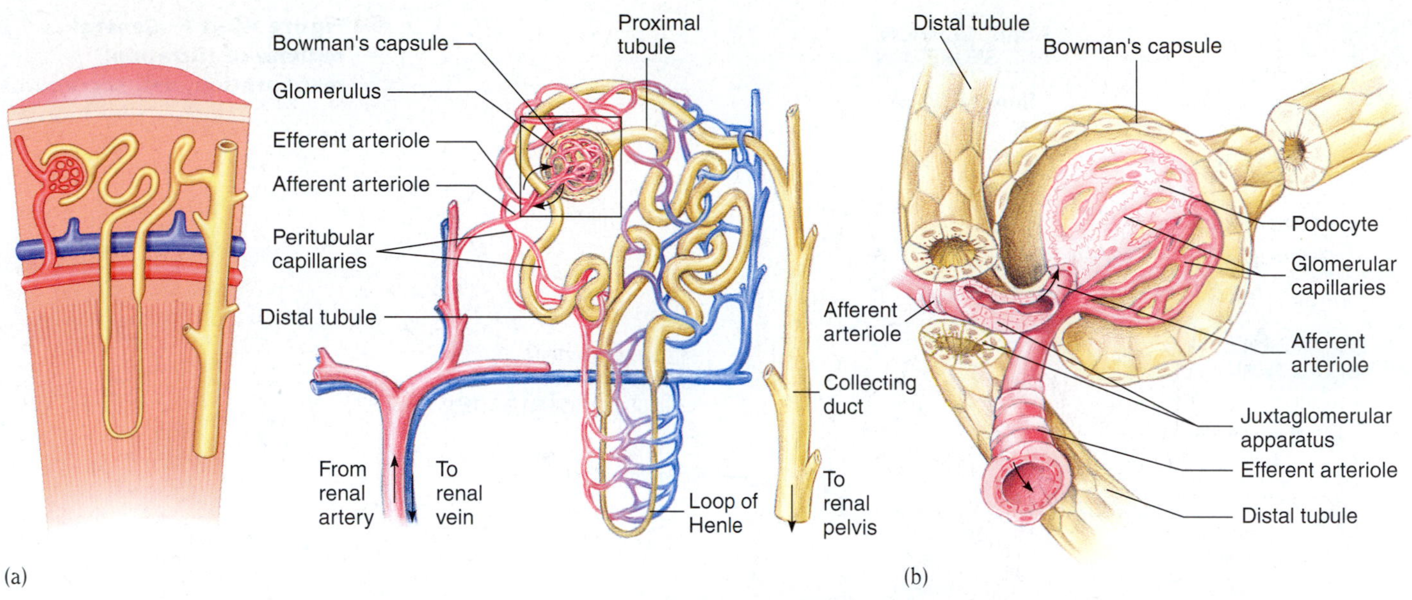

Afferent arteriole

Glomeruli

(c)

100 μm

Figure 46–10 Nephron structure. **(a)** The location and the basic structure of a nephron. Urine forms by filtration of the blood in the glomeruli and by adjustment of the filtrate as it passes through the tubules that drain the glomeruli. **(b)** Detailed view of Bowman's capsule. Note that the distal tubule is actually adjacent to the afferent arteriole. The juxtaglomerular apparatus (discussed later in the chapter) is a small group of cells located in the walls of the tubule and arterioles. **(c)** Low-power SEM of a portion of the kidney cortex, showing glomeruli and associated blood vessels. *(c, CNRI/Science Photo Library/Photo Researchers, Inc.)*

other capillary networks in the body. However, blood flow through glomerular capillaries is at much higher pressure, so more plasma is filtered in the kidney.

Several factors contribute to filtration. First, the hydrostatic blood pressure in the glomerular capillaries is higher than in other capillaries. This high pressure is mainly due to the high resistance to outflow presented by the efferent arteriole, which is smaller in diameter than the afferent arteriole. A second factor that contributes to the large amount of **glomerular filtrate** is the large surface area for filtration provided by the highly coiled glomerular capillaries. A third factor is the great permeability of the glomerular capillaries. Numerous small pores, called *fenestrations,* are present between the endothelial cells that form their walls, making the glomerular capillaries more porous than typical capillaries.

The wall of Bowman's capsule in contact with the capillaries consists of specialized epithelial cells called **podocytes.** These cells have numerous cytoplasmic extensions called foot processes that cover most of the surfaces of the glomerular capillaries (Fig. 46–12). Foot processes of adjacent podocytes are separated by narrow gaps called **filtration slits.** The fenestrated walls of the glomerular capillaries and the podocytes form a **filtration membrane** that permits fluid and small solutes dissolved in the plasma, such as glucose, amino acids, sodium, potassium, chloride, bicarbonate, other salts, and urea, to pass through and become part of the filtrate. This filtration membrane holds back blood cells, platelets, and most of the plasma proteins.

Process of Science Two protein components of the filtration membrane have been identified. In 1998 researchers identified a mutation associated with a congenital kidney disease. Subsequently, the gene was shown to code for a protein named nephrin that is part of the filtration membrane. A second protein, CD2-associated protein, is thought to anchor nephrin to the podocyte cytoskeleton. Mice that do not express CD2-associated protein

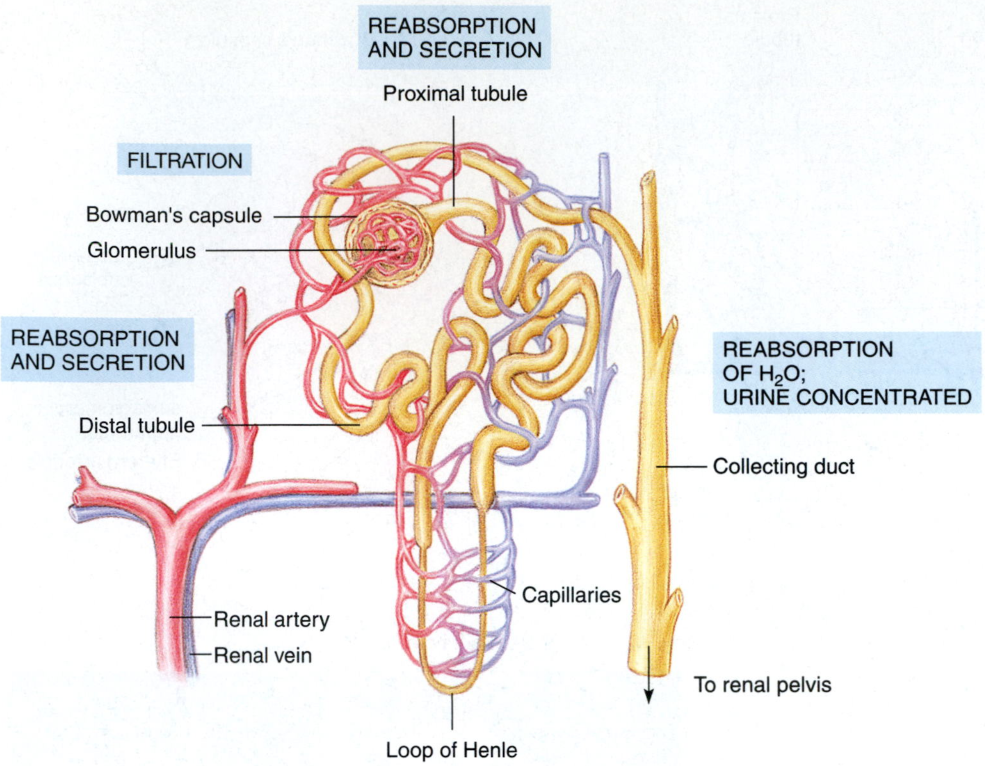

REABSORPTION AND SECRETION

Proximal tubule

FILTRATION

Bowman's capsule

Glomerulus

REABSORPTION AND SECRETION

Distal tubule

REABSORPTION OF H_2O; URINE CONCENTRATED

Collecting duct

Capillaries

Renal artery

Renal vein

To renal pelvis

Loop of Henle

Figure 46-11 General regions of filtration, reabsorption, and secretion.

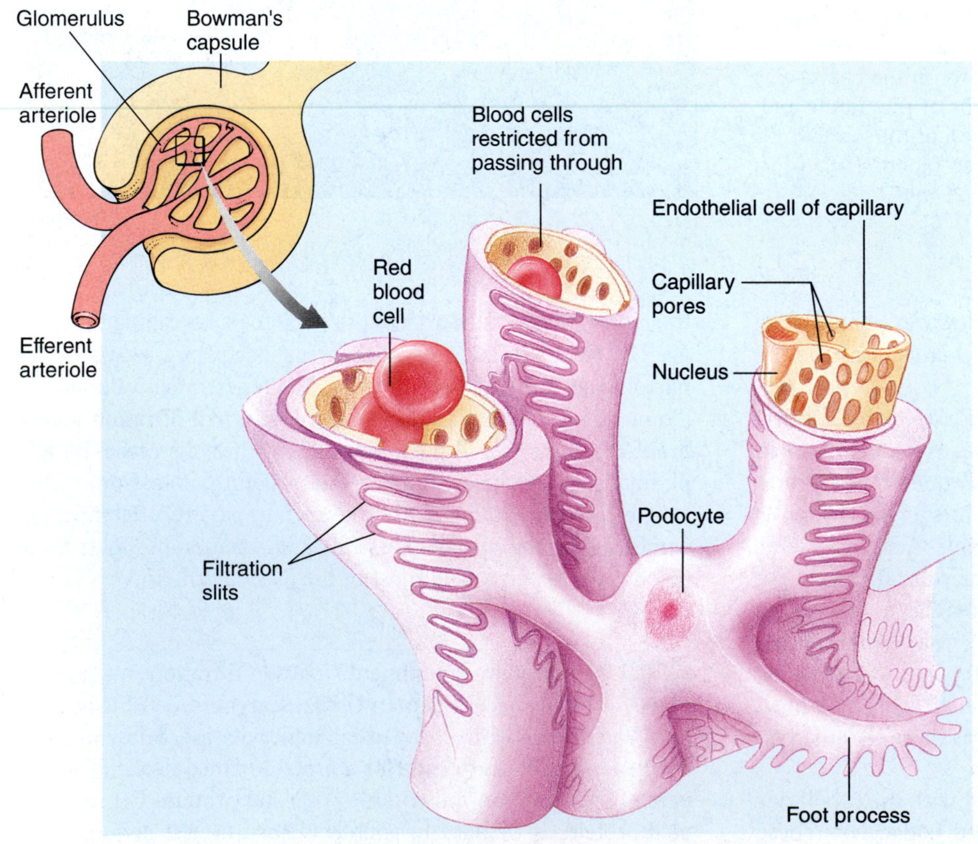

Figure 46-12 Filtration membrane of the kidney. Consisting of podocytes and glomerular capillary walls, this filtration barrier is highly permeable to water and small molecules but restricts the passage of blood cells and large molecules.

FOCUS ON Kidney Disease

Kidney disease ranks fourth in prevalence among major diseases in the United States, affecting more than 13 million people. Kidney function can be impaired by infections, tumors, kidney stones, shock, diabetes, circulatory disease, or exposure to substances such as mercury, lead, or carbon tetrachloride. Many diseases, including congestive heart failure, cause a decrease in the glomerular filtration rate. Sodium builds up in the tissues, resulting in water retention. **Diuretics** are medications commonly used to decrease retention of sodium and water; diuretics increase the urine volume.

A large number of related chronic kidney diseases in which the glomeruli (the filtering units of the kidney) are damaged are referred to as glomerulonephritis. The glomerular damage may result from an autoimmune response. A common symptom of kidney disease is the appearance of protein in the urine.

In chronic kidney disease, a progressive decrease in the number of functioning nephrons may eventually result in kidney failure. A decrease in the rate of filtration occurs, causing loss of homeostatic water and salt balance and an inability to effectively excrete urea and other nitrogenous wastes. Retention of water causes edema (swelling). As the concentration of hydrogen ions increases, body fluids become too acidic (acidosis). Nitrogenous wastes accumulate in the blood and tissues, causing a condition called uremia. If untreated, acidosis and uremia can cause coma and, eventually, death.

Chronic kidney failure is treated by kidney dialysis or by kidney transplant. More than 200,000 individuals in the United States alone undergo kidney dialysis. In dialysis, solutes are exchanged by diffusion across a selectively permeable membrane between solutions of different compositions. A kidney dialysis machine can be used to temporarily restore appropriate blood solute balance to a patient whose kidneys are not functioning.

In one type of dialysis (hemodialysis), blood circulates from one of the patient's arteries through the dialysis machine and is then returned through a vein. A plastic tube is surgically inserted into both an artery and a vein in the patient's arm or leg. These tubes are then connected to a circuit of plastic tubing from a dialysis machine. The patient's blood flows through the tubing, which is permeable to certain substances such as nitrogenous wastes and ions, but not permeable to proteins and blood cells. The tubing is immersed in a solution containing most of the normal blood plasma components in their normal proportions. Nitrogenous wastes diffuse down their concentration gradient from the patient's blood through minute pores in the tubing and into the surrounding solution in the dialysis machine. As the blood circulates repeatedly through the tubing in the machine, dialysis continues, eventually adjusting most of the values of the patient's blood chemistry to normal ranges.

Mechanical dialysis is expensive, clumsy, and inconvenient. Patients typically undergo dialysis three times a week, and the procedure takes 4 to 6 hours. This type of dialysis can also produce serious side effects such as osteoporosis ("brittle bone syndrome," caused by bone calcium loss).

Peritoneal dialysis uses the patient's own peritoneum (the lining of the abdominal cavity), which is a selectively permeable membrane. A plastic bag containing dialysis fluid is attached to the patient's abdomen, and the fluid is allowed to run into the peritoneal cavity. Solutes are exchanged between the dialysis fluid and the patient's blood. After several hours, the dialysis fluid is withdrawn into the bag and discarded. Peritoneal dialysis is convenient, but poses the threat of peritonitis if bacteria enter the body cavity with the dialysis fluid.

Long-term dialysis is not as desirable for the patient as a functioning kidney. With a successful kidney transplant, a patient can live a more normal life with far less long-term expense. At present, more than two-thirds of kidney transplants are successful for several years, although physicians must routinely treat the problems of graft rejection (see Chapter 43). Many recipients of kidney transplants have survived for more than 20 years.

---

appear to have faulty contacts between podocytes and die of renal failure. Researchers have suggested that mutations in genes coding these proteins may increase one's risk for a variety of kidney diseases (see *Focus On: Kidney Disease*).

The total volume of blood passing through the kidneys is about 1200 mL per minute, or about one fourth of the entire cardiac output. The plasma passing through the glomerulus loses more than 10% of its volume to the glomerular filtrate; the rest leaves the glomerulus through the efferent arteriole. The normal glomerular filtration rate amounts to about 180 L (about 45 gal) each 24 hours. This is four and a half times the amount of fluid in the entire body! Common sense tells us that urine could not be excreted at that rate. Within a few moments, dehydration would become life-threatening.

Reabsorption is highly selective

The threat to homeostasis posed by the vast amounts of fluid filtered by the kidneys is avoided by **reabsorption.** About 99% of the filtrate is reabsorbed into the blood through the renal tubules, leaving only about 1.5 L to be excreted as urine during a 24 hour period. Reabsorption permits precise regulation of blood chemistry by the kidneys. Wastes, excess salts, and other materials remain in the filtrate and are excreted in the urine, while needed substances such as glucose and amino acids are returned to the

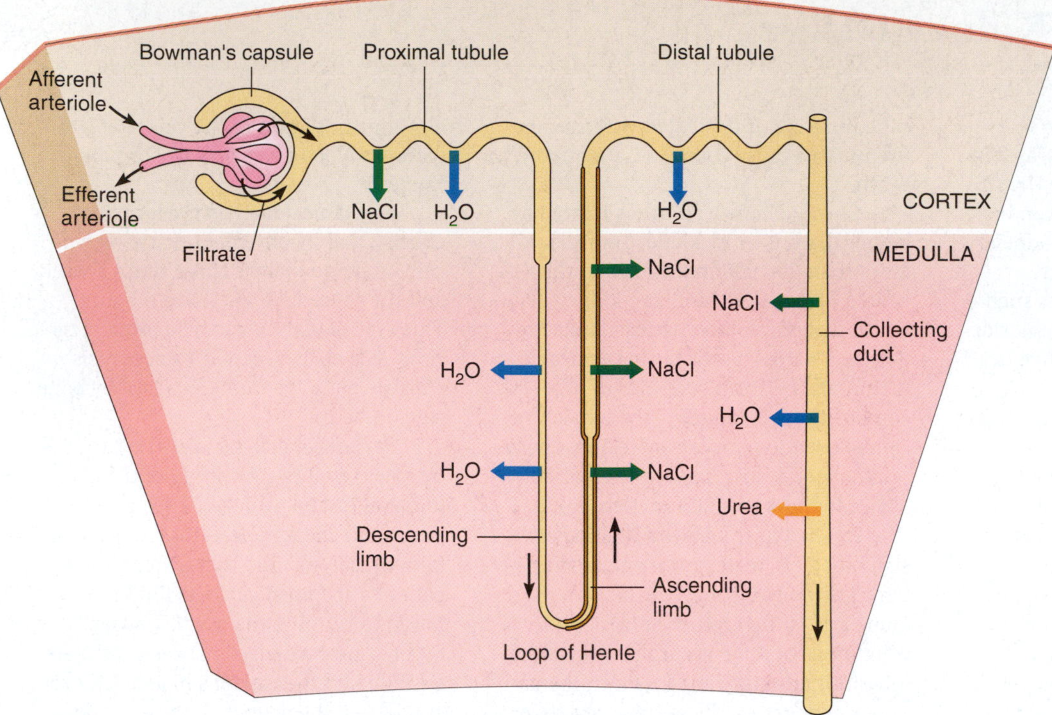

Figure 46–13 Movement of water, ions, and urea through the renal tubule and collecting duct. Water passes out of the descending limb of the loop of Henle, leaving a more concentrated filtrate inside. The heavy outline along the ascending limb indicates that this region is relatively impermeable to water. NaCl diffuses out from the lower (thin) part of the ascending limb. In the upper (thick) part of the ascending limb, NaCl is actively transported into the interstitial fluid. The saltier the interstitial fluid becomes, the more water moves out of the descending limb. This leaves a concentrated filtrate inside, so more salt passes out. Note that this is a positive feedback system. Some urea also moves out into the interstitial fluid through the collecting ducts. Water from the collecting ducts moves out osmotically into this hypertonic interstitial fluid.

blood. Each day the tubules reabsorb more than 178 L of water, 1200 g (2.6 lb) of salt, and about 250 g (0.5 lb) of glucose. Most of this, of course, is reabsorbed many times over.

The simple epithelial cells lining the renal tubule are well adapted for reabsorbing materials. They have abundant microvilli that increase their surface area for reabsorption (and give the inner lining a "brush border" appearance). These cells also contain numerous mitochondria that provide the energy for running the cellular pumps that actively transport materials.

Most (about 65%) of the filtrate is reabsorbed as it passes through the proximal convoluted tubule. Glucose, amino acids, vitamins, and other substances of nutritional value are entirely reabsorbed there. Many ions, including sodium, chloride, bicarbonate, and potassium, are partially reabsorbed. Some of these ions are actively transported; others are reabsorbed by diffusion. Reabsorption continues as the filtrate passes through the loop of Henle and the distal convoluted tubule. Then the filtrate is further concentrated as it passes through the collecting duct that leads to the renal pelvis.

Normally, substances that are useful to the body, such as glucose or amino acids, are reabsorbed from the renal tubules. If the concentration of a particular substance in the blood is high, however, the tubules may not be able to reabsorb all of it. The maximum rate at which a substance can be reabsorbed is called its **tubular transport maximum (Tm).** When that rate is reached, the binding sites are saturated on the membrane proteins that transport the substance. For example, the tubular load of glucose is about 125 mg per minute, and almost all of it is normally reabsorbed. However, in a person with uncontrolled diabetes mellitus, the concentration of glucose in the blood exceeds its Tm (glucose concentration in excess of 320 mg per minute). The excess glucose cannot be reabsorbed and is excreted in the urine (glucosuria). Its presence there is evidence of this disorder (see Chapter 47).

Some substances are actively secreted from the blood into the filtrate

Tubular **secretion** is the passage of substances across the tubule epithelium in a direction opposite to that of reabsorption. Secretion occurs mainly in the region of the distal convoluted tubule. Potassium, hydrogen ions, and ammonium ions, as well as some

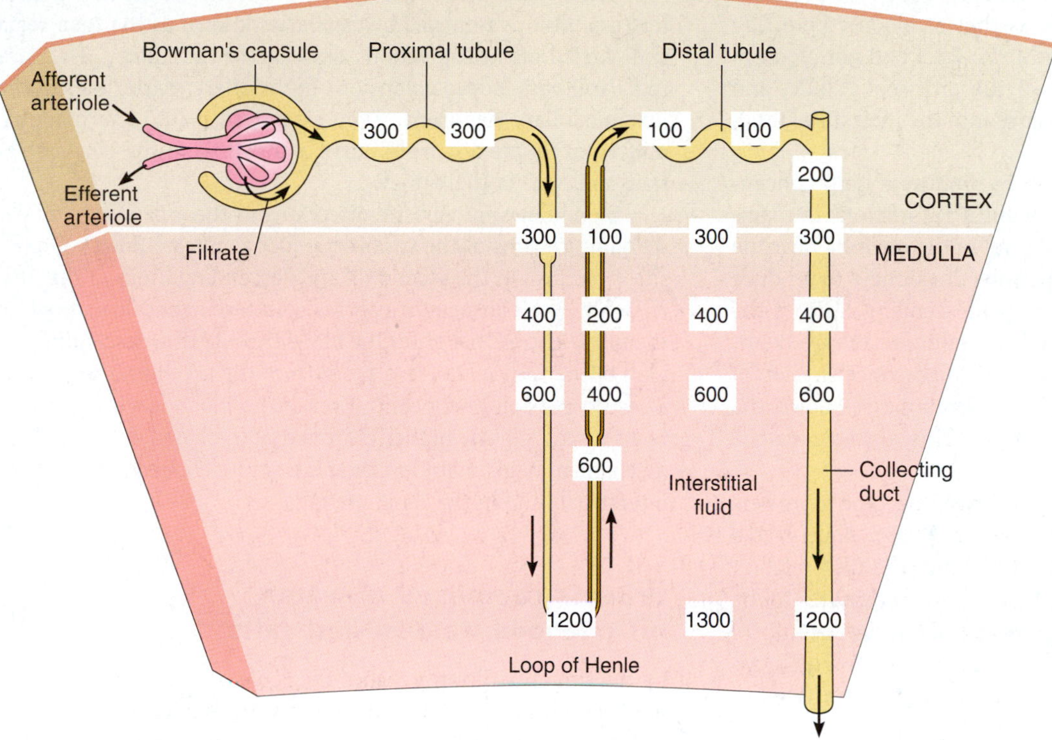

Figure 46–14 Concentration of the filtrate as it moves through the nephron. This figure shows the relative concentration of ions, mainly Na^+ and Cl^-, during formation of a very concentrated urine. Numbers indicate the concentration of salt in the filtrate expressed in milliosmols per liter. The more hypertonic a solution is, the higher its osmotic pressure. The very hypertonic interstitial fluid near the renal pelvis draws water osmotically from the filtrate in the collecting ducts. The heavy outline along the ascending loop indicates that this region is relatively impermeable to water.

organic ions such as creatinine (a metabolic waste), are secreted into the filtrate. Certain drugs, such as penicillin, are also removed from the blood by secretion.

Secretion of hydrogen ions, an important homeostatic mechanism for regulating the pH of the blood, takes place through the formation of carbonic acid. Carbon dioxide, which diffuses from the blood into the cells of the distal tubules and collecting ducts, combines with water to form carbonic acid. This acid then dissociates, forming hydrogen ions and bicarbonate ions. When the blood becomes too acidic, more hydrogen ions are secreted into the urine.

$$CO_2 + H_2O \rightleftharpoons H_2CO_3 \rightleftharpoons H^+ + HCO_3^-$$

Potassium secretion is also an important homeostatic mechanism. When potassium concentration is too high, nerve impulses are not effectively transmitted, and the strength of muscle contraction decreases. The heart can be weakened and even fail. When potassium ions are too highly concentrated, they are secreted from the blood into the renal tubules and then are excreted in the urine. Secretion results partly from a direct effect of the potassium ions on the tubules. In addition, a high potassium ion concentration in the blood stimulates the adrenal cortex to increase its output of the hormone **aldosterone,** which further stimulates secretion of potassium.

Urine becomes concentrated as it passes through the renal tubule

We can survive with limited fluid intake because the kidneys can produce highly concentrated urine—more than four times as concentrated as blood. The osmolarity of human blood is about 300 milliosmols per liter (mOsm/L)[1]. The kidneys can produce urine with an osmolarity of about 1400 mOsm/L.

As the filtrate passes through various regions of the renal tubule, salt (NaCl) is reabsorbed and a salt concentration gradient is established (Figs. 46–13 and 46–14). The gradient is used to produce a concentrated urine.

[1] An osmol is a unit of osmotic pressure equal to the molarity of the solution divided by the number of particles produced when the solute dissolves. For example, glucose dissolves to give only one kind of particle, but NaCl produces two. A milliosmol is 1/1000 of an osmol.

When filtrate flows from Bowman's capsule to the proximal tubule, its osmolarity is the same as that of blood (about 300 mOsm/L). The proximal tubule reabsorbs water and salt. Sodium ions are actively transported out of the proximal tubule, and chloride follows passively. As salt passes into the interstitial fluid, water follows osmotically.

The loop of Henle is specialized to produce a high concentration of sodium chloride in the medulla. It functions to maintain a highly hypertonic interstitial fluid in the medulla near the bottom of the loop, which in turn permits the kidneys to produce a concentrated urine. The walls of the descending limb of the loop of Henle are relatively permeable to water but relatively impermeable to sodium and urea. There is a high concentration of Na^+ in the interstitial fluid, so as the filtrate passes down the loop of Henle, water moves out by osmosis. This concentrates the filtrate inside the loop of Henle.

At the turn of the loop of Henle, the walls become more permeable to salt and less permeable to water. As the concentrated filtrate begins to move up the ascending limb (the thin region), salt diffuses out into the interstitial fluid. This contributes to the high salt concentration in the interstitial fluid in the medulla of the kidney surrounding the loop of Henle. Further along the ascending limb (the thick region), sodium is actively transported out of the tubule.

Because water passes out of the descending limb of the loop of Henle, the filtrate at the bottom of the loop has a high salt concentration. However, because salt (but not water) is removed in the ascending limb, by the time the filtrate moves into the distal tubule, its osmolarity may be the same or even lower than that of blood.

As the filtrate passes through the distal tubule, the filtrate may become even more dilute. The distal tubule is relatively impermeable to water but actively transports salt out into the interstitial fluid. The filtrate passes from the renal tubule into a larger collecting duct that eventually empties into the renal pelvis.

Note that there is a *counterflow* of fluid through the two limbs of the loop of Henle. Filtrate passing down through the descending limb is flowing in a direction opposite the filtrate moving upward through the ascending limb. The filtrate is concentrated as it moves down the descending limb and diluted as it moves up the ascending limb. This countercurrent mechanism helps maintain a high salt concentration in the interstitial fluid of the medulla. The hypertonic interstitial fluid draws water osmotically from the filtrate in the collecting ducts.

The inner medullary collecting ducts are permeable to urea, allowing the concentrated urea in the filtrate to diffuse out into the interstitial fluid. This urea contributes to the high solute concentration of the inner medulla and so helps in the process of concentrating urine.

The collecting ducts pass through the zone of very salty interstitial fluid. As the filtrate moves down the collecting duct, water passes osmotically into the interstitial fluid, where it is collected by capillaries. So much water may leave the collecting ducts that highly concentrated urine can be produced. Because it has a low concentration of water, a hypertonic urine conserves water.

Some of the water that diffuses from the filtrate into the interstitial fluid is removed by capillaries known as the **vasa recta** and carried off in the venous drainage of the kidney. The vasa recta are long, looped extensions of the efferent arterioles of the juxtamedullary nephrons. They extend deep into the medulla, only to negotiate a hairpin curve and return to the cortical venous drainage of the kidney.

Blood flows in opposite directions in the ascending and descending regions of the vasa recta, just as filtrate flows in opposite directions in the ascending and descending limbs of the loop of Henle. As a consequence of this countercurrent flow, much of the salt and urea that enter the blood through the descending region of the vasa recta leave again from the ascending region. As a result, the solute concentration of the blood leaving the vasa recta is only slightly higher than that of the blood entering. This mechanism helps maintain the high solute concentration of the interstitial fluid in the renal medulla.

Urine is composed of water, nitrogenous wastes, and salts

By the time the filtrate reaches the renal pelvis, its composition has been precisely adjusted. Useful materials have been returned to the blood by reabsorption. Wastes and excess materials that entered by filtration or secretion have been retained by the tubules. The adjusted filtrate, called **urine,** is composed of approximately 96% water, 2.5% nitrogenous wastes (mainly urea), 1.5% salts, and traces of other substances, such as bile pigments, that may contribute to the characteristic color and odor.

Healthy urine is sterile and has been used to wash battlefield wounds when clean water was not available. However, urine swiftly decomposes when exposed to bacterial action, forming ammonia and other products. It is the ammonia that produces the diaper rash of infants.

The composition of urine yields many clues to body function and malfunction. **Urinalysis,** the physical, chemical, and microscopic examination of urine, is a very important diagnostic tool that has been used to monitor diabetes mellitus and many other disorders. Urinalysis is also extensively used in drug testing because breakdown products of some drugs can be identified in the urine for several weeks after the drugs are taken.

Kidney function is regulated by hormones

Several hormones regulate urine volume and concentration, and additional hormones that help maintain fluid homeostasis remain to be identified (Table 46–1). The amount of urine produced depends on the body's need to retain or rid itself of water. We have seen that salt reabsorption in the loops of Henle establishes a very salty interstitial fluid that draws water osmotically from the collecting ducts. Permeability of the collecting ducts to water is regulated by **antidiuretic hormone (ADH).** When the body needs to conserve water, the posterior pituitary gland increases its release of ADH (Fig. 46–15). This hormone makes the collecting ducts more permeable to water so that more water

TABLE 46–1 Hormonal Control of Kidney Function

Hormone	Source	Target Tissue	Actions	Factors that Stimulate Release
Antidiuretic hormone (ADH)	Produced in hypothalamus; released by posterior pituitary gland	Collecting ducts	Increases permeability of the collecting ducts to water, increasing reabsorption and decreasing water excretion	Low fluid intake decreases blood volume and increases osmotic pressure of blood; receptors in hypothalamus stimulate posterior pituitary
Aldosterone	Adrenal glands (cortex)	Distal tubules and collecting ducts	Increases sodium reabsorption	Angiotensin II (when blood pressure decreases)
Angiotensin II	Produced from angiotensin I	Blood vessels and adrenal glands	Constricts blood vessels, which raises blood pressure; stimulates aldosterone secretion	Decrease in blood pressure causes renin secretion; renin catalyzes conversion of angiotensinogen to angiotensin I, which is then converted to angiotensin II by ACE
Atrial natriuretic peptide (ANP)	Atrium of heart	Afferent arterioles; collecting ducts	Dilates afferent arterioles; inhibits sodium reabsorption by collecting ducts; inhibits aldosterone secretion	Stretching of atria due to increased blood volume

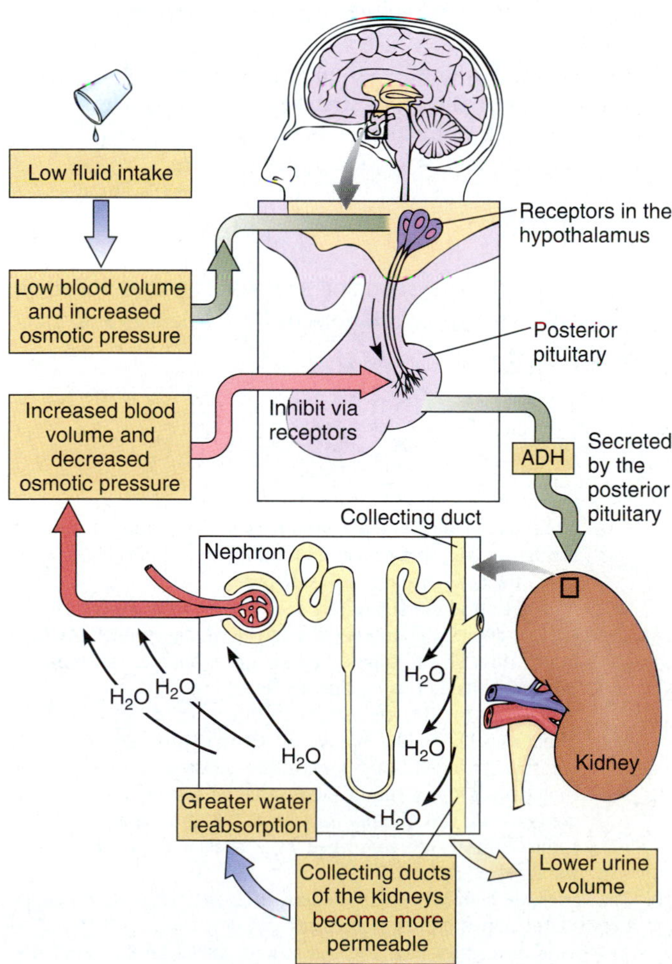

Figure 46–15 Regulation of urine volume by antidiuretic hormone (ADH). When the body is dehydrated, the hormone ADH increases the permeability of the collecting ducts to water. More water is reabsorbed, and only a small volume of concentrated urine is produced.

is reabsorbed and a small volume of concentrated urine is produced.

Secretion of ADH is stimulated by special receptors in the hypothalamus. When fluid intake is low, the body begins to dehydrate, causing the blood volume to decrease. As blood volume decreases, the concentration of salts dissolved in the blood becomes greater, causing an increase in osmotic pressure. Receptors in the hypothalamus are sensitive to this osmotic change and stimulate the posterior lobe of the pituitary to release more ADH. A thirst center in the hypothalamus also responds to dehydration, stimulating an increase in fluid intake.

When one drinks a great deal of water, the blood becomes diluted and its osmotic pressure falls. Release of ADH by the pituitary gland decreases, lessening the amount of water reabsorbed from the collecting ducts. A large volume of dilute urine is produced.

Occasionally, the pituitary gland malfunctions and does not produce sufficient ADH. The resulting condition is called *diabetes insipidus* (not to be confused with the more common disorder, diabetes mellitus). This condition can also result from an acquired insensitivity of the kidney to ADH. In diabetes insipidus, water is not efficiently reabsorbed from the ducts, so a large volume of urine is produced. A person with severe diabetes insipidus may excrete up to 25 quarts of urine each day, a serious loss of water to the body. The affected individual becomes dehydrated and must drink almost continually to offset fluid loss. Diabetes insipidus can often be controlled by injections of ADH or by use of an ADH nasal spray.

Sodium is the most abundant extracellular ion, accounting for about 90% of all positive ions outside cells in the extracellular fluid. Sodium concentration is precisely regulated by the actions of several hormones. Aldosterone, which is secreted by the cortex of the adrenal glands, stimulates the distal tubules and collecting ducts to increase sodium reabsorption. When the adrenal glands of experimental animals are removed, too much sodium is excreted, leading to serious depletion of the extracellular fluid.

Aldosterone secretion can be stimulated by a decrease in blood pressure (which is caused by a decrease in volume of blood and interstitial fluid). When blood pressure falls, cells of the **juxtaglomerular apparatus** secrete the enzyme **renin** and activate the **renin-angiotensin-aldosterone pathway.** The juxtaglomerular apparatus is a small group of cells located in the region where the renal tubule contacts the afferent and efferent arterioles (see Fig. 46–10*b*). Renin acts on a plasma protein (angiotensinogen), converting it to angiotensin I. *Angiotensin converting enzyme (ACE)* converts angiotensin I into its active form, **angiotensin II,** a peptide hormone that stimulates aldosterone secretion. ACE is produced by the endothelial cells of capillary walls, especially by those in the lung.

Angiotensin II not only increases the synthesis and release of aldosterone but also raises blood pressure directly by constricting blood vessels. Angiotensin II also stimulates sodium reabsorption by the proximal convoluted tubules and may stimulate the posterior pituitary to release ADH. All of these actions help restore extracellular fluid volume and normal blood pressure. In individuals with hypertension, *ACE inhibitors* are sometimes used to block the production of angiotensin II.

Decrease in blood volume ⟶ decrease in blood pressure ⟶ cells of juxtaglomerular apparatus secrete renin ⟶ angiotensinogen ⟶ angiotensin I ⟶ angiotensin II ⟶ constricts blood vessels and stimulates aldosterone secretion ⟶ aldosterone increases sodium reabsorption ⟶ blood pressure increases

Process of Science **Atrial natriuretic peptide (ANP),** a hormone produced by the heart, increases sodium excretion and decreases blood pressure. In 1981 Adolpho de Bold and his research team at Queen's University in Ontario reported that when they injected homogenized rat atria into rats, they observed a dramatic increase in excretion of sodium and water by the kidney. During the next few years, the amino acid sequence of ANP was determined, and it was chemically synthesized. Rat, mouse, and human genes for ANP were then cloned, and their nucleotide sequences were determined. ANP is stored in granules in atrial muscle cells and is released into the circulation in response to stretching of the atria by increased blood volume.

ANP dilates afferent arterioles, thereby increasing glomerular filtration rate. It inhibits sodium reabsorption by the collecting ducts directly and also indirectly by inhibiting aldosterone secretion by the adrenal cortex. ANP also reduces plasma aldosterone concentration by inhibiting release of renin. These actions of ANP lower blood volume and blood pressure.

Increase in blood volume ⟶ increase in blood pressure ⟶ atria of heart stretched ⟶ atria release ANP ⟶ directly inhibits sodium reabsorption and inhibits aldosterone secretion, which also inhibits sodium reabsorption ⟶ decrease in blood volume ⟶ blood pressure decreases

The renin-angiotensin system and ANP work antagonistically in regulating fluid balance, salt (electrolyte) balance, and blood pressure.

SUMMARY WITH KEY TERMS

I. **Osmoregulation** is the active regulation of osmotic pressure of body fluids so that homeostasis is maintained. **Excretion** is the process of ridding the body of metabolic wastes. **Excretory systems** help maintain homeostasis by regulating the concentration of body fluids through osmoregulation and excretion of metabolic wastes.

II. The principal waste products of animal metabolism are water; carbon dioxide; and **nitrogenous wastes,** including **ammonia, urea,** and **uric acid.** Urea is synthesized from ammonia and carbon dioxide in a sequence of reactions known as the **urea cycle.**

III. Most marine invertebrates are **osmoconformers**—the salt concentration of their body fluids varies with changes in the seawater. Some marine invertebrates, especially those inhabiting coastal habitats, are **osmoregulators** that maintain an optimal salt concentration despite changes in salinity of their surroundings.
   A. Marine sponges and cnidarians depend on diffusion for excretion of wastes.
   B. Invertebrate mechanisms of osmoregulation and waste disposal include the **protonephridia** of flatworms, **metanephridia** of annelids, and **Malpighian tubules** of insects.
      1. The enlarged blind end of a protonephridium is a **flame cell** with brushes of cilia that propel fluid into a system of tubules. Fluid leaves the body through nephridiopores.
      2. A metanephridium is a tubule open at both ends. Fluid from the coelom enters the tubule through a ciliated funnel. As fluid moves through the tubule, needed materials are reabsorbed by capillaries that surround the tubule.
      3. Malpighian tubules are extensions of the insect gut wall. Their blind ends lie in the hemocoel. Cells of the tubule actively transport uric acid and some other substances from the hemolymph into the tubule, and water follows by diffusion. The contents of the tubule pass into the gut, and water and some solutes are reabsorbed by a specialized epithelium in the rectum.

IV. The vertebrate **kidney,** which functions in excretion and osmoregulation, is vital in maintaining homeostasis.
   A. Aquatic vertebrates have the continuous challenge of osmoregulation. Freshwater fishes take in water osmotically; they excrete a large volume of dilute urine.

B. Marine bony fishes lose water osmotically. They compensate by drinking seawater and excreting salt through their gills; only a small volume of urine is produced.
C. Marine cartilaginous fishes retain large amounts of urea, allowing them to take in water osmotically through the gills. This water can be used to excrete a large volume of urine.
D. Other aquatic vertebrates have specific adaptations for dealing with osmoregulation. Marine birds and reptiles, for example, have salt glands that excrete excess salt.

V. The **urinary system** is the principal excretory system in humans and other vertebrates. The **kidney** is the key organ of the urinary system.
A. In mammals, the kidneys produce urine, which passes through the **ureters** to the **urinary bladder** for storage. During urination, the urine is released from the body through the **urethra.**
B. The outer portion of each kidney is the **renal cortex;** the inner portion is the **renal medulla.** The renal medulla contains eight to ten **renal pyramids.** The tip of each pyramid is a **renal papilla.** As urine is produced, it flows into **collecting ducts,** which empty through a renal papilla into a funnel-shaped chamber, the **renal pelvis.**
C. Each **nephron** consists of a cluster of capillaries, called a **glomerulus,** surrounded by a **Bowman's capsule** that opens into a long, coiled **renal tubule.** The renal tubule consists of a **proximal convoluted tubule, loop of Henle,** and **distal convoluted tubule.**
D. **Cortical nephrons** have small glomeruli and are located almost entirely within the cortex or outer medulla. **Juxtamedullary nephrons** have large glomeruli and long loops of Henle that extend deep into the medulla. These nephrons are important in concentrating urine.
E. Small branches of the **renal artery** deliver blood to **afferent arterioles** that conduct blood to the glomerular capillaries. From the glomerular capillaries, blood flows into an **efferent arteriole** that delivers blood into a second set of capillaries, the **peritubular capillaries** that surround the renal tubule. Blood leaves the kidney through the **renal vein.**
F. Urine formation is accomplished by the **filtration** of plasma, **reabsorption** of needed materials, and **secretion** of a few substances such as potassium and hydrogen ions into the renal tubule.
1. Plasma filters out of the glomerular capillaries and into Bowman's capsule. The permeable walls of the capillaries and specialized epithelial cells, called **podocytes,** which make up the inner wall of Bowman's capsule, serve as a **filtration membrane.** Filtration is nonselective with regard to small molecules; glucose and other needed materials, as well as metabolic wastes, become part of the filtrate.
2. About 99% of the filtrate is reabsorbed from the renal tubules into the blood. Reabsorption is a highly selective process that returns usable materials to the blood but leaves wastes and excesses of other substances to be excreted in the urine. The maximum rate at which a substance can be reabsorbed is its **tubular transport maximum (Tm).**
3. In secretion, certain substances and drugs are actively transported into the renal tubule to become part of the urine.
4. The ability to produce a concentrated urine depends on a high salt and urea concentration in the interstitial fluid of the kidney medulla. The interstitial fluid in the medulla has a concentration gradient in which the salt is most concentrated around the bottom of the loop of Henle. This gradient is maintained, in part, by salt reabsorption from various parts of the renal tubule. There is a counterflow of fluid through the two limbs of the loop of Henle; filtrate becomes concentrated as it moves down the descending loop and diluted as it moves up the ascending loop.
5. Water is drawn by osmosis from the filtrate as it passes through the collecting ducts. This permits the concentration of urine in the collecting ducts.
6. Some of the water that diffuses from the filtrate into the interstitial fluid is removed by a system of capillaries known as the **vasa recta.**
7. **Urine** is a watery solution of nitrogenous wastes, excess salts, and other substances not needed by the body.

G. Urine volume is regulated by **antidiuretic hormone (ADH)** which is released by the posterior lobe of the pituitary gland in response to an increase in osmotic concentration of the blood (caused by dehydration). ADH increases the permeability of the collecting ducts to water. As a result, more water is reabsorbed and only a small volume of urine is produced.
H. **Aldosterone** and **atrial natriuretic peptide (ANP)** work antagonistically.
1. When blood pressure decreases, cells of the **juxtaglomerular apparatus** secrete the enzyme **renin,** which activates a pathway leading to production of **angiotensin II.** This hormone increases release of aldosterone, which increases sodium reabsorption, raising blood pressure.
2. When blood pressure increases, ANP increases sodium excretion and inhibits aldosterone secretion. These actions lower blood pressure.

I. Chronic kidney disease is treated with dialysis or kidney transplant.

POST-TEST

1. The process that maintains homeostasis of body fluids, keeping them from becoming too dilute or too concentrated is called (a) excretion (b) elimination (c) osmoregulation (d) glomerular filtration (e) tubular secretion
2. The main nitrogenous waste product of insects and birds is (a) urea (b) uric acid (c) ammonia (d) carbon dioxide (e) nitrate
3. The main nitrogenous waste product of amphibians and mammals is (a) urea (b) uric acid (c) ammonia (d) carbon dioxide (e) nitrate
4. Osmoconformers (a) maintain an optimal salt concentration despite fluctuations in salt concentration of their surroundings (b) include many animals that inhabit coastal habitats (c) experience variation in concentration of their body fluids along with changes in salinity of the seawater (d) typically have cells in their gills that remove salts from the surrounding water (e) two of the preceding answers are correct
5. Which of the following is *not* a correct pair? (a) protonephridia/flatworm (b) metanephridia/annelid (c) flame cell/flatworm (d) Malpighian tubule/mollusk (e) kidney/vertebrate
6. To compensate for fluid loss, many marine bony fishes (a) accumulate urea (b) have glands that excrete glucose (c) eat a low-protein diet (d) excrete a large volume of hypertonic urine (e) drink seawater
7. Arrange the following structures into an accurate sequence through which urine passes. (1) urethra (2) urinary bladder (3) kidney (4) ureter (a) 4, 3, 2, 1 (b) 3, 4, 2, 1 (c) 1, 2, 3, 4 (d) 4, 2, 1, 3 (e) 3, 1, 2, 4
8. Arrange the following structures into an accurate sequence through which filtrate passes. (1) proximal convoluted tubule (2) loop of Henle (3) collecting duct (4) distal convoluted tubule (5) Bowman's capsule (a) 5, 4, 3, 2, 1 (b) 3, 4, 2, 5, 1 (c) 1, 5, 2, 3, 4 (d) 5, 4, 2, 3, 1 (e) 5, 1, 2, 4, 3
9. The afferent arteriole delivers blood to the (a) renal artery (b) efferent arteriole (c) renal vein (d) capillaries of the glomerulus (e) peritubular capillaries
10. Which of the following does *not* contribute to the process of filtration? (a) high hydrostatic blood pressure in glomerular capillaries (b) large surface area for filtration (c) permeability of glomerular capillaries (d) active transport by epithelial cells lining renal tubules (e) podocytes
11. Tubular transport maximum refers to (a) the maximum concentration of a substance in the plasma that can be reabsorbed by the kidney (b) the most rapid rate at which urine can be transported through the ureter (c) the maximum rate at which a substance can be reabsorbed from the filtrate in the renal tubules (d) the maximum rate at which a

substance can pass through the loop of Henle (e) the maximum rate at which a substance can be secreted into the filtrate

12. Which of the following does *not* contribute to the high salt concentration in the interstitial fluid in the medulla of the kidney? (a) reabsorption of salt from various regions of Bowman's capsule (b) diffusion of salt from the ascending limb of the loop of Henle (c) active transport of sodium from the upper part of the ascending limb (d) counterflow of fluid through the two limbs of the loop of Henle (e) diffusion of urea out of the collecting duct
13. Which of the following is *not* normally present in urine? (a) urea (b) glucose (c) salts (d) water (e) traces of bile pigments
14. Which is *not* true of ADH? (a) released by posterior lobe of the pituitary gland (b) increases water reabsorption (c) secretion increases when osmotic pressure in body increases (d) increases urine volume (e) secretion decreases when you drink a lot of water
15. Aldosterone (a) is released by the posterior pituitary gland (b) decreases sodium reabsorption (c) secretion is stimulated by an increase in blood pressure (d) is an enzyme that converts angiotensin into angiotensin II (e) secretion increases in response to angiotensin II

REVIEW QUESTIONS

1. Compare osmoregulation in flatworms with that in insects.
2. What type of osmoregulatory challenge is faced by marine fishes? By freshwater fishes? What adaptations have evolved that meet these challenges?
3. What are the principal types of nitrogenous wastes? Give examples of animals that excrete each type.
4. Name the structure(s) in the mammalian body associated with each of the following: (a) urea formation (b) urine formation (c) temporary storage of urine (d) conduction of urine out of the body
5. Which part(s) of the nephron is/are associated with the following? (a) filtration (b) reabsorption (c) secretion
6. Contrast filtration, reabsorption, and secretion.
7. List the sequence of blood vessels through which a drop of blood passes as it is conducted to and from a nephron.
8. How is urine volume regulated? Explain. Why must victims of untreated diabetes insipidus drink great quantities of water?
9. What are the actions of the renin-angiotensin-aldosterone pathway?
10. Describe two treatment strategies for kidney failure. What are the advantages and disadvantages of each?

YOU MAKE THE CONNECTION

1. The number of protonephridia in a planarian is related to the salinity of its environment. Planaria inhabiting slightly salty water develop fewer protonephridia, but the number quickly increases when the concentration of salt in the environment is lowered. Explain why.
2. What types of osmoregulatory challenges do humans experience? Explain. What mechanisms do we have to solve these challenges?
3. Why is glucose normally not present in urine? Why is it present in the urine of individuals with diabetes mellitus? Why do you suppose diabetics experience an increased output of urine?
4. The kangaroo rat's diet consists of dry seeds, and it drinks no water. Speculate about the adaptations this animal needs in order to survive.

RECOMMENDED READINGS

McClanahan, L.L., R. Ruibal, and V.H. Shoemaker. "Frogs and Toads in Deserts." *Scientific American,* Vol. 270, No. 3, Mar. 1994. A discussion of some unusual adaptations for life in the desert.

Vander, A., J. Sherman, and D. Luciano. *Human Physiology: The Mechanisms of Body Function,* 8th ed. McGraw-Hill, New York, 2001. Chapter 16 focuses on the kidney and its homeostatic functions.

Wickelgren, I. "Cell Biology: First Components Found for Key Kidney Filter." *Science,* Vol. 286, 8 Oct. 1999. A brief discussion of the discovery and functions of two proteins that are components of the filtration membrane in the nephron.

- Visit our Web site at **http://www.info.brookscole.com/solomonbergmartin** for links to chapter-related resources on the World Wide Web. Additional on-line materials relating to this chapter can also be found on our Web site.

See chapter activity on BioActive Learner CD for additional help in mastering the chapter's material. Icon location in the chapter's margins shows which topics have tutorials or simulations in the CD.

47

Endocrine Regulation

This juvenile Puget Sound king crab *(Lopholithodes mandtii)* is a crustacean that changes color to blend with its background. Color changes are caused by changes in the distribution of pigment in its cells, a process regulated by hormones. *(Shark Song/M. Kazmers/Dembinsky Photo Associates)*

LEARNING OBJECTIVES

After you have studied this chapter you should be able to

1. Describe the sources, general chemical structure, transport, and actions of classical hormones and of related chemical signals.
2. Compare the mechanisms of action of steroid and protein-type hormones; include the role of second messengers, such as cyclic AMP.
3. Identify four functions of hormones in invertebrates, and describe the hormonal regulation of insect development.
4. Identify the principal vertebrate endocrine glands, locate them in the body, and describe the hormones secreted by each.
5. Summarize the regulation of endocrine glands by negative feedback mechanisms, giving specific examples.
6. Describe the mechanisms by which the hypothalamus serves as an important link between nervous and endocrine systems.
7. Compare the functions of the posterior and anterior lobes of the pituitary; describe the actions of their hormones.
8. Describe the actions of growth hormone on growth and metabolism, and contrast the consequences of hyposecretion and hypersecretion.
9. Describe the actions of the thyroid hormones, their regulation, and the effects of thyroid malfunction.
10. Contrast the actions of insulin and glucagon, and describe the disorders associated with impaired function of these hormones or their receptors.
11. Describe the actions of the adrenal glands, including their role in helping the body cope with stress.

A caterpillar becomes a butterfly. A crustacean changes color to blend with its background *(see photograph)*. A young girl develops into a woman. An adult copes with chronic stress. These physiological processes and many other adjustments of metabolism, fluid balance, growth, and reproduction are regulated by the **endocrine system.** This system works closely with the nervous system to maintain the steady state of the body. **Endocrinology,** the study of endocrine activity, is a very active and exciting field of biomedical research.

The endocrine system is a diverse collection of cells, tissues, and organs, including specialized **endocrine glands** that produce and secrete **hormones,** chemical messengers responsible for the regulation of many body processes. The term *hormone* is derived from a Greek word meaning "to excite." Hormones excite, or stimulate, changes in specific tissues.

Endocrine glands have no ducts; they secrete hormones directly into the interstitial fluid or blood. Hormones are typically transported by the blood and produce a characteristic response only after they reach **target cells** and bind with specific receptors. Target cells, the cells influenced by a particular hormone, may be located in another endocrine gland or in an entirely different type of organ, such as a bone or the kidney. Target cells may be located far from the endocrine gland. For example, the vertebrate thyroid gland secretes hormones that stimulate metabolism in tissues throughout the body. Several types of hormones may be involved in regulating the metabolic activities of a particular type of cell. In fact, many hormones produce a synergistic effect in which the presence of one hormone enhances the effects of another.

Modern endocrinology has its roots in experiments performed by German physiologist A.A. Berthold in the 1840s. Berthold removed the testes from young roosters and observed that their combs (a male secondary sex characteristic) did not grow as large as those in normal roosters. He then transplanted testes into some of the birds and observed that the combs grew back to normal size.

Berthold's methods became a model for subsequent studies in endocrinology and are still used by researchers today. To test the hypothesis that a particular tissue is endocrine, an investigator first removes the tissue. A first question is, does removal of the tissue produce deficiency symptoms in the experimental animal? Investigators then replace the suspected endocrine tissue by transplanting similar tissue from another animal. They often transplant the new tissue to a different location in the body to determine whether the effects depend on a signal that moves throughout the body in the blood. As with Berthold's roosters, the changes induced by removing the tissue should be reversed by replacing it.

Investigators extract the suspected compound from the endocrine tissue of one animal and inject it into an experimental animal from which the tissue producing the compound has been removed. Deficiency symptoms should be relieved by replacing the suspected hormone. Researchers then isolate the active compound and determine its chemical structure. Finally, the compound is synthesized in the laboratory and injected into experimental animals. If its effects are those predicted, the researchers have data to support their hypothesis.

Using such procedures, endocrinologists have identified about ten discrete endocrine glands. More recently investigators have identified specialized cells in the digestive tract, heart, kidneys, and many other parts of the body that also release hormones. In addition, some neurons release hormones. As a result of these discoveries, the scope of endocrinology has been broadened to include the production and actions of chemical messengers produced by a wide variety of organs, tissues, and cells.

One of the current areas of active research in endocrinology is the study of the mechanisms of hormone action, which includes characterizing receptors and identifying the molecules involved in *signal transduction.* Some endocrinologists are now using a reverse strategy for discovering new hormones and signaling pathways within the cell. They focus on "orphan" nuclear receptors, those for which ligands (the molecules that bind with them) are not yet known. Some of these "orphan" receptors are receptors for hormones that have not yet been identified. Using this strategy, researchers have identified intracellular signaling pathways for steroids, fatty acids, and several other compounds.

In this chapter we discuss the actions of a variety of hormones. We also examine how overproduction or deficiency of various hormones interferes with normal functioning.

CELLS COMMUNICATE BY CHEMICAL SIGNALS

Cells communicate with neurotransmitters, hormones, and local regulators. We discussed neurotransmitters in Chapters 39 through 41. In this chapter we focus on hormones and local regulators. The same compound may function as a neurotransmitter, hormone, or **local regulator,** a signaling molecule that diffuses through the interstitial fluid and acts on nearby cells. Thus, a neuron, endocrine gland, or some other cell type all may secrete the same chemical message. However, the same chemical message may have different meanings for various target cells.

Pheromones are chemical messengers produced by animals for communication with other animals of the same species. Because pheromones are generally produced by exocrine glands and do not regulate metabolic activities within the animal that produces them, most biologists do not classify them as hormones. Their role in regulating behavior is discussed in Chapter 50.

Traditionally, hormones were defined as chemical messengers secreted by endocrine glands

Endocrine glands differ from **exocrine glands** (such as sweat glands and gastric glands) that release their secretions into ducts. Endocrine glands have no ducts, and secrete their hormones into the surrounding interstitial fluid or into the blood. Typically, hormones diffuse into capillaries and are transported by the blood to target cells (Fig. 47–1*a*). In addition to the discrete classical endocrine glands, specialized cells in many tissues and organs (e.g., the kidneys and heart) also release hormones or hormone-like substances.

The complexity of animal physiology challenges simplistic definitions. As new chemical signals and their modes of action have been discovered, the old definition of a hormone as a substance secreted by an endocrine gland and transported by the blood has become inadequate.

Neurohormones are transported in the blood

Certain neurons, known as **neuroendocrine cells,** are an important link between the nervous and endocrine systems. Neuroendocrine cells produce **neurohormones** that are transported down axons and released into the interstitial fluid. They typically diffuse into capillaries and are transported by the blood (Fig. 47–1*b*). Invertebrate endocrine systems are largely neuroendocrine. In vertebrates, the hypothalamus produces several neurohormones that link the nervous system with the pituitary gland, an endocrine gland that secretes several hormones.

Many endocrinologists include some local regulators as hormones

Certain hormones that are indisputably hormones because they are typically transported by the blood, under some conditions, act as local regulators. In **autocrine regulation,** a hormone, or other regulator, acts on the very cells that produce it (Fig. 47–1*c*). For example, the female hormone estrogen, which functions as a classical hormone, may also exert an autocrine effect

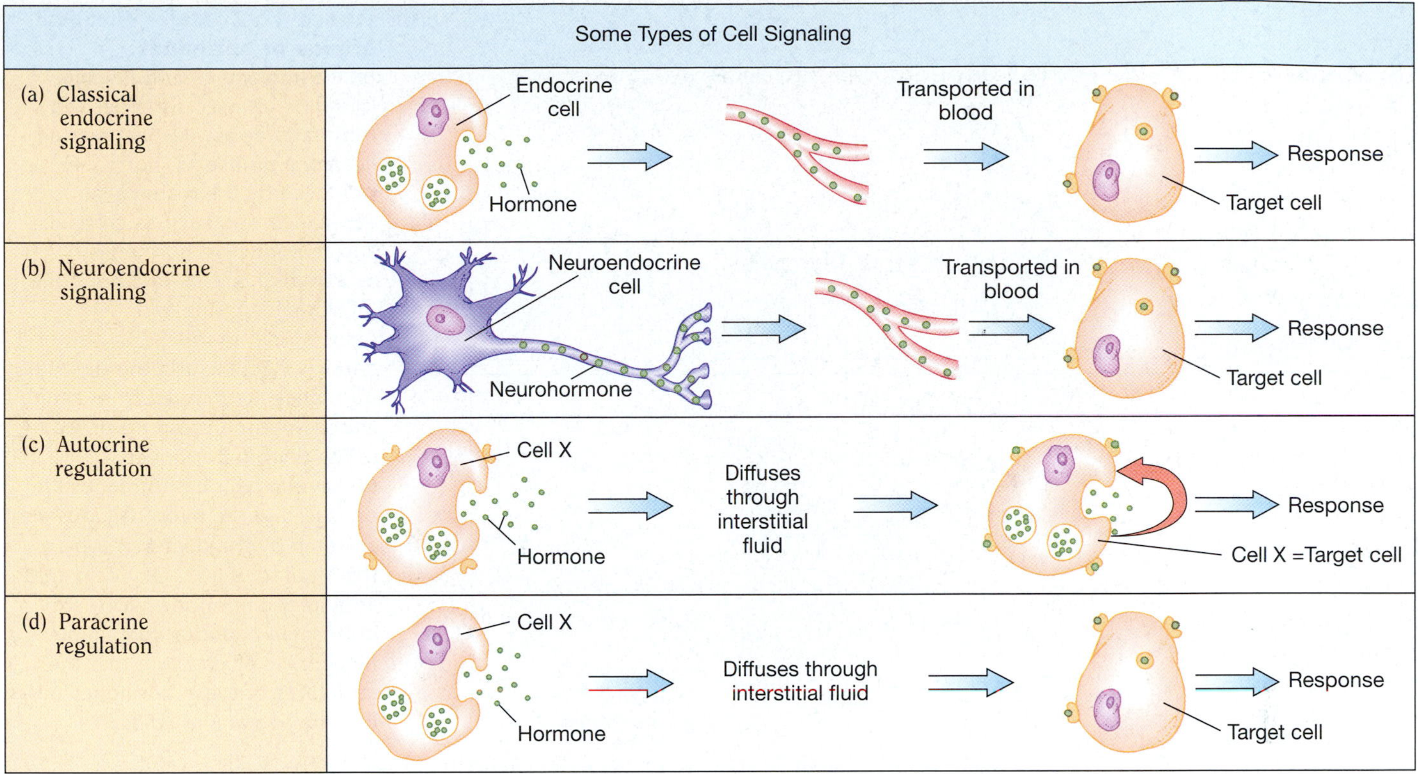

Figure 47–1 Some types of endocrine signaling. **(a)** In classical endocrine signaling, endocrine cells release hormones that are transported to target cells by the blood. **(b)** In neuroendocrine signaling, neurons release neurohormones, which may be transported by blood or may diffuse through interstitial fluid. **(c)** In autocrine regulation, a hormone acts on the very cells that produce it. **(d)** In paracrine regulation, hormones diffuse through the interstitial fluid and act on nearby target cells.

that stimulates additional estrogen secretion. Estrogen can also act on nearby cells, a type of local regulation known as **paracrine regulation** (Fig. 47–1*d*).

Other local regulators include local chemical mediators such as histamine, growth factors, and prostaglandins. Recall from Chapter 43 that **histamine** is stored in mast cells and is released in response to allergic reactions, injury, or infection. Histamine causes blood vessels to dilate and capillaries to become more permeable. More than 50 **growth factors,** typically peptides, stimulate cell division and normal development in specific types of cells. Growth factors have autocrine or paracrine effects, and their mechanisms of action are similar to those of peptide hormones.

Prostaglandins are modified fatty acids released by many different organs, including the lungs, liver, digestive tract, and certain reproductive organs. Although present in very small quantities, prostaglandins affect a wide range of body processes. They are paracrine regulators that act on cells in their immediate vicinity. Prostaglandins modify cyclic adenosine monophosphate (cAMP) levels (discussed later in the chapter), and they interact with other hormones to regulate various metabolic activities.

About 16 prostaglandins are known, and they have different actions on different tissues. Some reduce blood pressure, whereas others raise it. Various prostaglandins dilate the bronchial passageways, inhibit gastric secretion, stimulate contraction of the uterus, affect nerve function, cause inflammation, and affect blood clotting. Those synthesized in the temperature-regulating center of the hypothalamus cause fever. In fact, nonsteroidal antiinflammatory drugs such as aspirin and ibuprofen reduce fever and decrease inflammation by inhibiting prostaglandin synthesis.

Because prostaglandins are involved in the regulation of so many metabolic processes, they have great potential for a variety of clinical uses. At present, prostaglandins are used to induce labor in pregnant women, to induce abortion, and to promote the healing of ulcers in the stomach and duodenum. Prostaglandins may someday be used to treat a wide variety of illnesses, including asthma, arthritis, kidney disease, certain cardiovascular disorders, and some forms of cancer.

HORMONES CAN BE ASSIGNED TO FOUR CHEMICAL GROUPS

Although hormones are chemically diverse, they generally belong to one of four different chemical groups: (1) fatty acid derivatives, (2) steroids, (3) amino acid derivatives, or (4) peptides or proteins (Fig. 47–2).

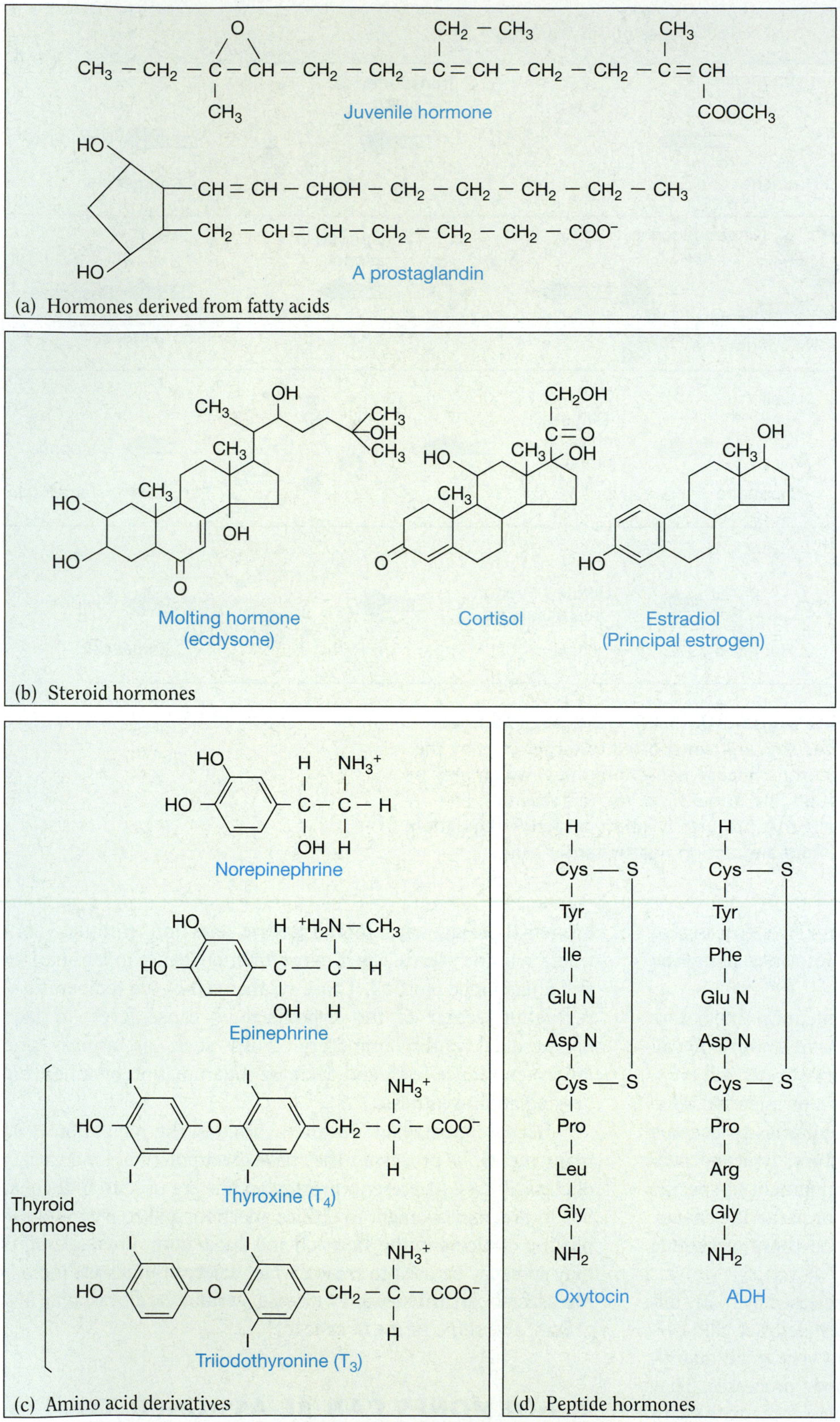

(a) Hormones derived from fatty acids

Juvenile hormone

A prostaglandin

(b) Steroid hormones

Molting hormone (ecdysone)

Cortisol

Estradiol (Principal estrogen)

(c) Amino acid derivatives

Norepinephrine

Epinephrine

Thyroid hormones

Thyroxine (T_4)

Triiodothyronine (T_3)

(d) Peptide hormones

Oxytocin

ADH

Figure 47–2 Major chemical groups of hormones. **(a)** Prostaglandins and juvenile hormone of insects are derived from fatty acids. **(b)** The molting hormone of insects, hormones produced by the mammalian adrenal cortex, such as cortisol, and reproductive hormones, such as estradiol, are steroid hormones. **(c)** Norepinephrine and epinephrine, produced by the adrenal medulla, and the thyroid hormones are derived from amino acids. Note the presence of iodine in the thyroid hormones. **(d)** Several peptide hormones are produced in the hypothalamus and secreted by the posterior lobe of the pituitary gland. Oxytocin and antidiuretic hormone (ADH) are both small peptides containing nine amino acids. Note that the structures of these hormones differ by only two amino acids.

Prostaglandins and the juvenile hormones of insects are **fatty acid derivatives** (Fig. 47–2*a*). Prostaglandins are synthesized from arachidonic acid, a 20-carbon fatty acid. A prostaglandin has a five-carbon ring in its structure.

The *molting hormone* (also called ecdysone) of insects is a **steroid hormone** (Fig. 47–2*b*). In vertebrates, the adrenal cortex, testis, ovary, and placenta secrete steroid hormones synthesized from cholesterol. Examples of steroid hormones are cortisol, se-

FOCUS ON Anabolic Steroids and Other Abused Hormones

An estimated 1 million individuals in the United States, one-half of them adolescents, abuse a group of synthetic hormones known as anabolic steroids, and an unknown number are now abusing other hormones, including human growth hormone (hGH) and erythropoietin.

Anabolic steroids are synthetic androgens (male reproductive hormones) that were developed in the 1930s to prevent muscle atrophy in patients with diseases that prevented them from moving about. In the 1950s, anabolic steroids became popular with professional athletes, who used them to increase muscle mass, physical strength, endurance, and aggressiveness. In truth, their athletic performance was probably enhanced, at least in part, by drug-induced euphoria and increased enthusiasm for training.

As with other hormones, the concentration of steroids circulating in the body is precisely regulated, so use of anabolic steroids interferes with normal physiological processes. Even during short-term use and at relatively low doses, anabolic steroids have a significant effect on mood and behavior. At higher doses, users experience disturbed thought processes, forgetfulness, and confusion, and often find themselves easily distracted. "Steroid rage" refers to the mood swings, unpredictable anger, increased aggressiveness, and irrational behavior exhibited by many users.

Anabolic steroids elevate blood pressure, damage the liver, and increase low-density lipoprotein (LDL) concentration, raising the risk of cardiovascular disease. In adolescents, anabolic hormones cause severe acne and stunt growth by prematurely closing the growth plates in bones. Abuse of these hormones reduces sexual function and can shrink the testes, leading to sterility. These drugs remain in the body for a long time. Their metabolites (breakdown products) can be detected in the urine for up to six months.

When their serious side effects became known in the 1960s, anabolic steroid use became controversial, and in 1973 the Olympic Committee banned their use. They are now prohibited worldwide by amateur and professional sports organizations. However, according to the U.S. Drug Enforcement Administration, a multimillion-dollar black market exists for these synthetic hormones. They are both injected and taken in pill form.

The typical anabolic steroid user is a male (95%) athlete (65%), most often a football player, weight lifter, or wrestler. Surprisingly, though, about 10% of male high school students have used anabolic steroids and about one-third of these students are not even on a high school team. These adolescents use the hormone only to change their physical appearance-to pump up their muscles ("bulk up")-and increase endurance. Many anabolic steroid users have difficulty realistically perceiving their body images. They remain unhappy even after dramatic increases in muscle mass. The use of steroids by youth in the United States has increased steadily and is no longer limited to males. More than 175,000 high school girls have reported using steroids.

Human growth hormone (GH), like anabolic steroids, helps build muscle mass but also causes acromegaly. Erythropoietin increases the concentration of red blood cells. With more red blood cells, the body of an endurance athlete can transport more oxygen, and this enhances performance by up to 10%. Erythropoietin abuse has caused the deaths of several athletes. Reliable tests for detecting abuse of GH and erythropoietin have not yet been developed.

creted by the adrenal cortex, the male hormone testosterone secreted by the testis, and the female hormones progesterone and estrogens produced by the ovary. Synthetic hormones known as anabolic steroids are sometimes abused by athletes (see *Focus On: Anabolic Steroids and Other Abused Hormones*).

Chemically, the simplest hormones are **amino acid derivatives** (Fig. 47–2*c*). The thyroid hormones (T_3 and T_4) are synthesized from the amino acid tyrosine and iodide. Epinephrine (also known as adrenaline) and norepinephrine (also known as noradrenaline), produced by the medulla of the adrenal gland, are also derived from tyrosine. Melatonin is synthesized from the amino acid tryptophan.

The water-soluble **peptide hormones** are the largest hormone group. (Endocrinologists include protein hormones in this group.) Oxytocin and antidiuretic hormone (ADH), produced by neuroendocrine cells in the hypothalamus, are short **neuropeptides.** Neuropeptides are a large group of signaling molecules produced by neurons. Oxytocin and ADH are each composed of nine amino acids (Fig. 47–2*d*). Seven of the amino acids are identical in the two hormones, but the actions of these hormones are quite different. The hormones glucagon (see Fig. 3–20), secretin, adrenocorticotropic hormone (ACTH), and calcitonin are somewhat longer peptides consisting of about 30 amino acids.

Insulin, secreted by the islets of Langerhans in the pancreas, is a small protein consisting of two peptide chains joined by disulfide bonds. Growth hormone, thyroid-stimulating hormone, and the gonadotropic hormones, all secreted by the anterior lobe of the pituitary gland, are large proteins with molecular weights of 25,000 or more.

HORMONE SECRETION IS REGULATED BY NEGATIVE FEEDBACK MECHANISMS

Although present in minute amounts, more than 50 different hormones may be circulating in the blood of a vertebrate at any time. Steroid hormones and thyroid hormones are transported bound to plasma proteins. Peptide hormones are water-soluble and can be transported dissolved in the plasma. Hormone molecules are continuously removed from the circulation by target cells. They are also removed by the liver, which inactivates some hormones, and by the kidneys, which excrete them.

Hormone secretion is typically regulated by **negative feedback mechanisms,** in which a hormone is released in response to some change in a steady state and triggers a response that counteracts the changed condition, restoring homeostasis (see Chapter 37). Information about the amount of hormone or of some other substance in the blood or interstitial fluid is fed back to the gland, which then responds to restore homeostasis. The parathyroid glands, located in the neck of tetrapod vertebrates, provide a good example of how negative feedback works.

The parathyroid glands regulate the calcium concentration of the blood. When the calcium concentration is not within homeostatic limits, nerves and muscles cannot function properly. For example, when insufficient amounts of calcium ions are present, neurons can fire spontaneously, causing muscle spasms. When calcium concentration varies too far from the steady state (either too high or too low), negative feedback mechanisms bring the condition back to the steady state. A decrease in the calcium concentration in the plasma signals the parathyroid glands to release more parathyroid hormone (Fig. 47–3). This hormone increases the concentration of calcium in the blood. (The details of parathyroid action are discussed later in this chapter.) When the calcium concentration rises above normal limits, the parathyroid glands slow their output of hormone. Both responses are negative feedback mechanisms. An increase in calcium concentration results in decreased release of parathyroid hormone, whereas a decrease in calcium leads to increased hormone secretion. In each case, the response counteracts the inappropriate change, thus restoring the steady state.

HORMONES COMBINE WITH SPECIFIC RECEPTORS IN TARGET CELLS

A hormone may be present in the extracellular fluid of many tissues, but is "unnoticed" until it reaches its target cells. How do target cells recognize a hormone? Specific receptors in or on target cells recognize the hormone and bind with it. Researchers have determined that hormone receptors are large proteins or glycoproteins. The receptor site is similar to a lock, and the hormones are similar to different keys. Only the hormone that fits the specific receptor can influence the metabolic machinery of the cell. Receptors are responsible for the specificity of the endocrine system.

Receptors are continuously synthesized and degraded. Their numbers can be increased or decreased by **receptor up-regulation** and **receptor down-regulation.** For example, when the concentration of the hormone insulin is too high for an extended period, insulin receptors are down-regulated. The decrease in number of receptors dampens insulin's ability to stimulate cells to take in glucose.

Receptor up-regulation and down-regulation can be influenced by signals to the genes that code for the receptors. The number of receptors can also be altered by the following mechanisms. In response to continuous high hormone concentration in the extracellular fluid, hormone-receptor complexes can be taken into the cell by receptor-mediated endocytosis (see Chapter 5). The rate of receptor breakdown in the cell increases and

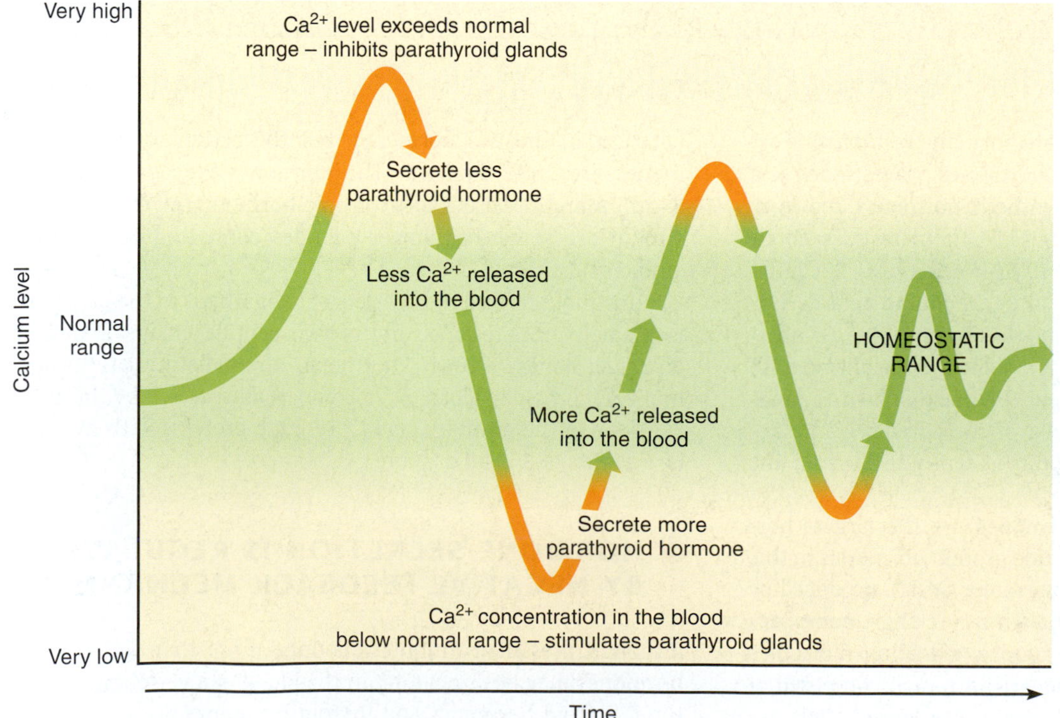

Figure 47–3 Regulation by negative feedback. Should calcium concentration exceed its normal range in the blood, the parathyroid glands are inhibited and slow their release of parathyroid hormone. When the calcium concentration in the blood falls below its normal range, the parathyroid glands are stimulated to release more parathyroid hormone. This hormone acts on target tissues that increase the calcium concentration in the blood, thus restoring homeostasis.

so the number of receptors decreases. Down-regulation suppresses the sensitivity of target cells to the hormone.

In response to low hormone concentrations, receptors stored inside the cell can be inserted into the plasma membrane by exocytosis. A greater number of receptors on the plasma membrane amplifies the hormone's effect on the cell.

SOME HORMONES ENTER THE CELL AND ACTIVATE GENES

Steroid hormones and thyroid hormones are relatively small, lipid-soluble (hydrophobic) molecules that easily pass through the plasma membrane of the target cell. Specific protein receptors in the cytoplasm or in the nucleus bind with the hormone to form a hormone-receptor complex (Fig. 47–4). This complex combines with specific acceptor sites on the nuclear DNA. Their interaction turns specific genes on or off. The hormone–receptor complex either activates or represses transcription of messenger RNA molecules coding for specific proteins. These proteins produce the changes in structure or metabolic activity responsible for the effect of the hormone. Interestingly, some steroid hormone target cells have plasma membrane receptors, as well as intracellular receptors, for these hormones.

MANY HORMONES ACT BY SIGNAL TRANSDUCTION

Because peptide hormones are hydrophilic and not soluble in the lipid layer of the plasma membrane, they do not enter the target cell. Instead, they bind to a specific transmembrane receptor in the plasma membrane. Transmembrane receptors initiate **signal transduction,** that is, they convert an extracellular hormone signal into an intracellular signal that affects some intracellular process (Fig. 47–5). The hormone, which serves as the first messenger, may relay information to a **second messenger,** or intracellular signal. The second messenger then relays information to effector molecules that carry out the action. Some hormones activate a series of molecular events that comprise a *signaling cascade.*

Two main types of cell surface hormone receptors are known: **enzyme-linked receptors** and **G protein–linked receptors.** Most enzyme-linked receptors are transmembrane proteins with a hormone-binding site outside the cell and an enzyme site (typically a protein kinase) inside the cell. G protein-linked receptors are transmembrane proteins that loop back and forth through the plasma membrane seven times. These receptors activate **G proteins,** a group of integral regulatory proteins. The G indicates that they bind **guanosine triphosphate (GTP),** which,

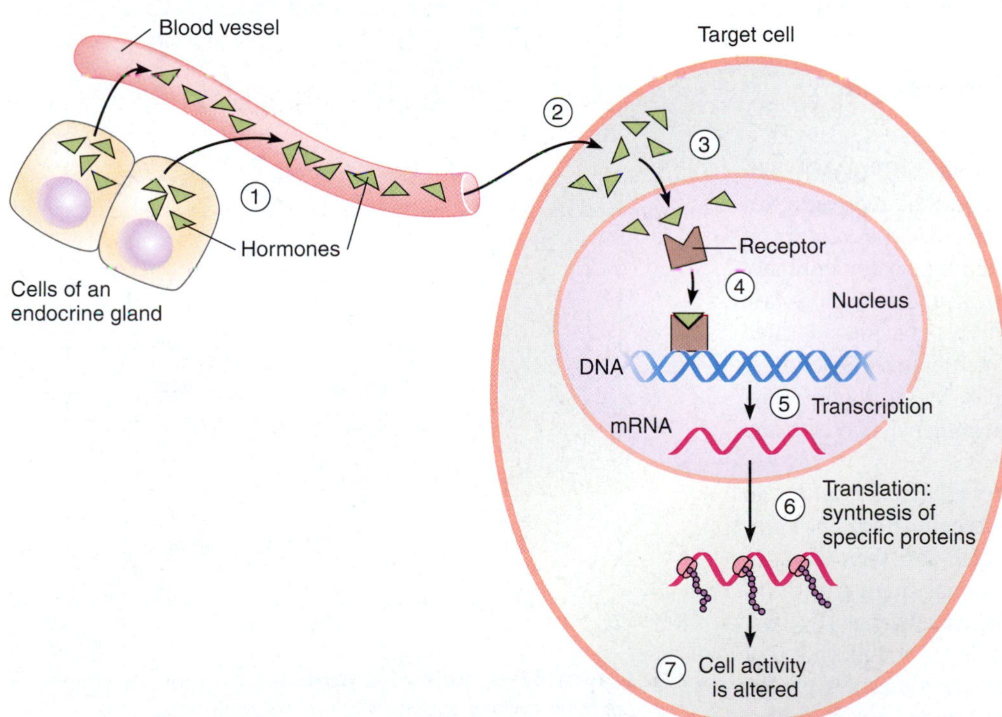

Figure 47–4 Steroid hormones activate genes. (1) Steroid hormones are secreted by an endocrine gland and transported to a target cell. (2) Steroid hormones are small, lipid-soluble molecules that pass freely through the plasma membrane. (3) The hormone passes through the cytoplasm to the nucleus. (4) The hormone combines with a receptor either in the cytoplasm or in the nucleus. Then, the steroid hormone-receptor complex binds with DNA. (5) This activates (or represses) specific genes, leading to mRNA transcription and (6) synthesis of specific proteins. The proteins cause the response recognized as the hormone's action (7).

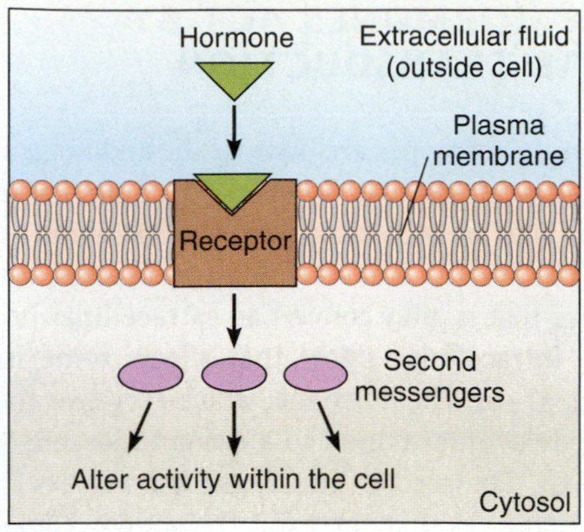

Figure 47–5 Action of hormones that have membrane receptors. Peptide hormones bind with receptors on the plasma membrane of a target cell. The receptor is a signal transducer that converts the hormone signal to an intracellular signal. The signal may be relayed by a second messenger. Typically, a sequence of several signaling molecules relays the message.

like ATP, is an important molecule in energy reactions. When the system is inactive, G protein binds to **guanosine diphosphate** (GDP), which is similar to ADP, the hydrolyzed form of ATP. G proteins are shuttles that transmit a signal between the receptor and a second messenger.

Cyclic AMP is the most common second messenger in animal cells

In the 1960s, Earl Sutherland identified **cyclic AMP (cAMP)** as a second messenger, and for his pioneering work he received a Nobel prize in 1971. When certain hormones bind with receptors on the extracellular surface of the plasma membrane, a membrane-bound enzyme, **adenylyl cyclase,** is activated. Adenylyl cyclase catalyzes the conversion of ATP to cAMP. This enzyme is located on the cytoplasmic side of the plasma membrane. Its action is coupled to the hormone-receptor complex by G proteins (Fig. 47–6). One type of G protein, G_s, stimulates adenylyl cyclase, and another, G_i, inhibits it.

Most hormones bind to a stimulatory receptor, resulting in activation of adenylyl cyclase. After the receptor-hormone complex is formed, the G protein releases GDP and then binds to GTP. This binding produces a conformational change in the G protein that enables it to bind with and activate adenylyl cyclase. By coupling the hormone-receptor complex to an enzyme that generates a signal, many second messenger molecules can be rapidly produced, and the actions of the hormone are amplified.

Once activated, adenylyl cyclase catalyzes the conversion of ATP to cAMP (Fig. 47–7). Cyclic AMP then activates one or more enzymes known as **protein kinases.** Each type of protein kinase phosphorylates (adds a phosphate group to) a specific protein. When the protein is phosphorylated, its function is altered, and it triggers a chain of reactions leading to some metabolic effect. Any increase in cAMP is temporary. Cyclic AMP is rapidly inactivated by enzymes known as **phosphodiesterases,** which convert it to AMP. Thus, the concentration of cAMP depends on the activity of both adenylyl cyclase that produces it and phosphodiesterase that breaks it down.

Protein kinases can inhibit or activate certain cellular proteins. Each protein kinase acts on a different type of protein. Because different types of protein kinases exist in each kind of target cell (and even within different organelles of the same target cell), protein kinases are able to produce a wide variety of responses. For example, activation of one type of protein kinase may have a metabolic effect on the cell, a second protein kinase may affect membrane permeability, and a third may activate genes (Fig. 47–8).

Calcium ions can act as second messengers

Certain receptors are linked to gated ion channels, for example calcium channels. When a hormone binds to the receptor, the calcium channel opens, allowing calcium ions to move into the cell. A G protein may be involved in coupling the hormone-

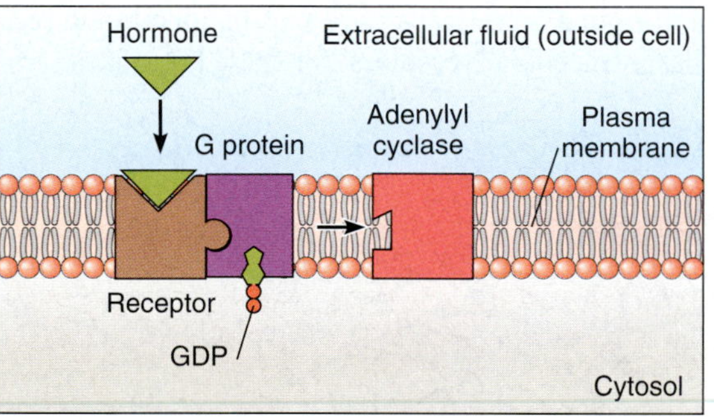

(a)

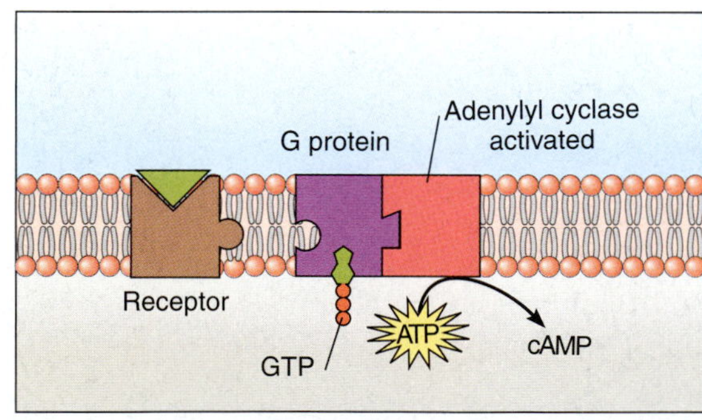

(b)

Figure 47–6 Role of G proteins. G proteins couple adenylyl cyclase activity with membrane receptors. **(a)** When a hormone binds to a receptor on the plasma membrane, the hormone-receptor complex binds to a G protein. **(b)** GDP on the G protein is replaced by GTP. The G protein undergoes a conformational change (change in shape), allowing it to bind with adenylyl cyclase. G_s activates adenylyl cyclase, which catalyzes the conversion of ATP to cAMP. G_i inhibits adenylyl cyclase activity *(not shown).*

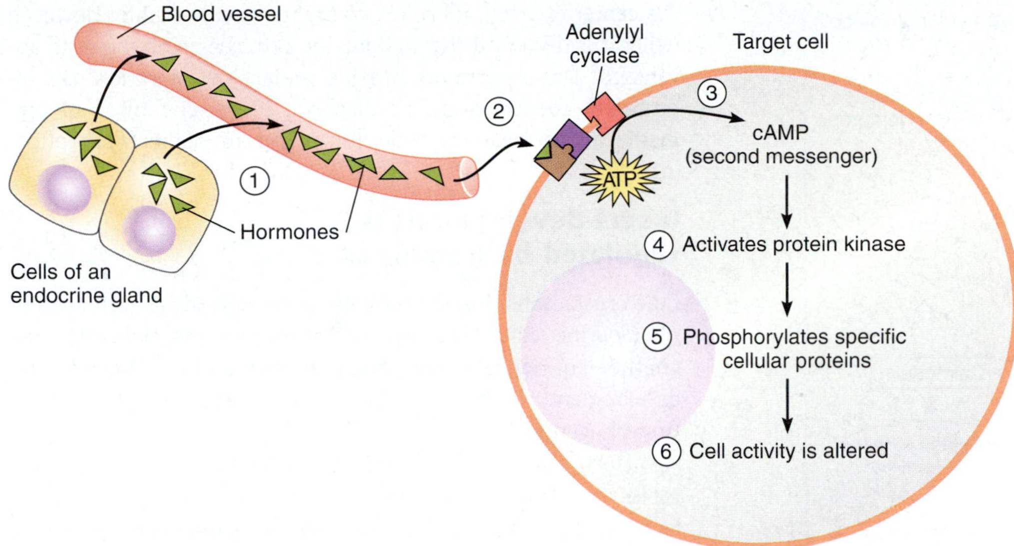

Figure 47–7 Mechanism of action of hormones that use cyclic AMP (cAMP) as a second messenger. (1) Peptide hormones are transported in the blood. (2) They bind with receptors in the plasma membrane of a target cell. (3) The hormone-receptor combination is coupled by a G protein to adenylyl cyclase. The adenylyl cyclase catalyzes the conversion of ATP to cyclic AMP, a second messenger. (4) Cyclic AMP then activates one or more enzymes that (5) phosphorylate proteins. (6) These proteins alter the activity of the cell in some way.

receptor complex to the opening of the calcium channel (Fig. 47–9). The hormone-receptor complex can also increase the cellular concentration of calcium by releasing calcium ions stored in the endoplasmic reticulum. Once the calcium concentration in the cell is increased, calcium ions bind to **calmodulin,** a protein found in all eukaryotic cells that have been examined to date. The calmodulin changes shape and can then activate certain enzymes, including protein kinases. Cellular processes regulated by calmodulin include membrane phosphorylation, neurotransmitter release, and microtubule disassembly.

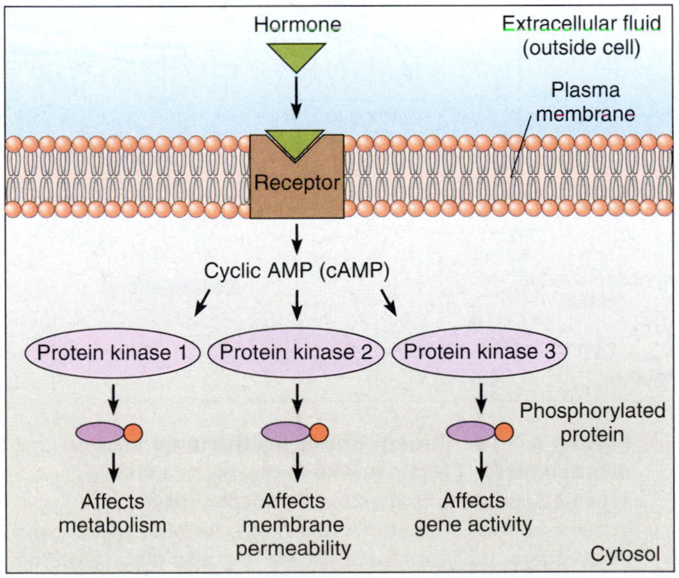

Figure 47–8 Effects of protein kinases. When hormone action increases the amount of cyclic AMP in the cell, protein kinases are activated. Protein kinases phosphorylate proteins, leading to a variety of effects.

Phospholipid products can act as second messengers

Certain hormone-receptor complexes (e.g., ADH-receptor complex) activate a G protein that then activates the membrane-bound enzyme phospholipase C (Fig. 47–10). This enzyme splits a membrane protein (phosphotidylinositol 4,5-bisphosphate, or PIP_2) into two products, **inositol trisphosphate (IP_3)** and **diacylglycerol (DAG),** that act as second messengers. IP_3 stimulates the endoplasmic reticulum to release calcium ions that then act as a third messenger. The calcium ions bind to calmodulin, and together with DAG they activate the enzyme protein kinase C, which phosphorylates specific proteins.

Hormone signals are amplified

Although hormones are present in very small amounts, they effectively regulate many physiological processes. This is in large part the result of **signal amplification,** an increase in signal strength. For example, a single hormone-receptor complex can stimulate the production of many cAMP molecules. In turn, each cAMP can activate a protein kinase that phosphorylates many protein molecules. In this way, a single hormone molecule can activate many proteins.

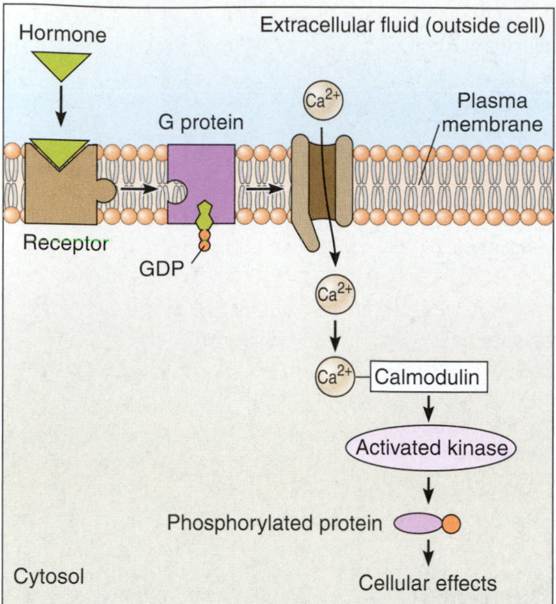

Figure 47–9 Calcium ions as second messengers. Certain receptors are linked by a G protein to gated calcium ion channels. When the gate opens in response to the activated protein, Ca^{2+} enters the cell and binds to calmodulin. The activated calmodulin activates specific protein kinases, enzymes that produce changes in the cell.

INVERTEBRATE ENDOCRINE SYSTEMS ARE MAINLY NEUROENDOCRINE

Among invertebrates, hormones are secreted mainly by neurons rather than by endocrine glands. These neurohormones regulate regeneration in hydras, flatworms, and annelids; molting and metamorphosis in insects; color changes in crustaceans; and growth, gamete production, reproductive behavior, and metabolic rate in other groups. Trends in the evolution of invertebrate endocrine systems include a larger number of both neurohormones and hormones secreted by endocrine glands, and a greater role of hormones in physiological processes.

Color change in crustaceans is regulated by hormones

Crustaceans possess endocrine glands, as well as masses of neuroendocrine cells. Hormonal regulation in crustaceans is complex and affects many activities, including molting, reproduction, heart rate, and metabolism. One of the most interesting and novel activities regulated by hormones is color change (see chapter opening photograph).

Pigment cells of crustaceans are located in the integument beneath the exoskeleton. Pigment may be black, yellow, red, white, or even blue. Color changes are produced by changes in the distribution of pigment granules within the cells. Distribution of pigment is regulated by neurohormones produced by neuroendocrine cells. When the pigment is concentrated near the center of a cell, its color is only minimally visible. However, when it is dispersed throughout the cell, the color shows to advantage. These pigments provide protective coloration. By appropriate concentration or dispersal of specific types of pigments, a crustacean can approximate the color of its background.

Insect development is regulated by hormones

Like crustaceans, insects have endocrine glands as well as neuroendocrine cells. The various hormones interact with one another to regulate reproduction, metabolism, growth, and development, including molting and development of form (morphogenesis).

Hormonal control of development in insects is complex and varies among the many species. Generally, some environmental factor (e.g., temperature change) affects neuroendocrine cells in the brain. When activated, these cells produce a neurohormone referred to as **brain hormone** (**BH,** or ecdysiotropin), which is

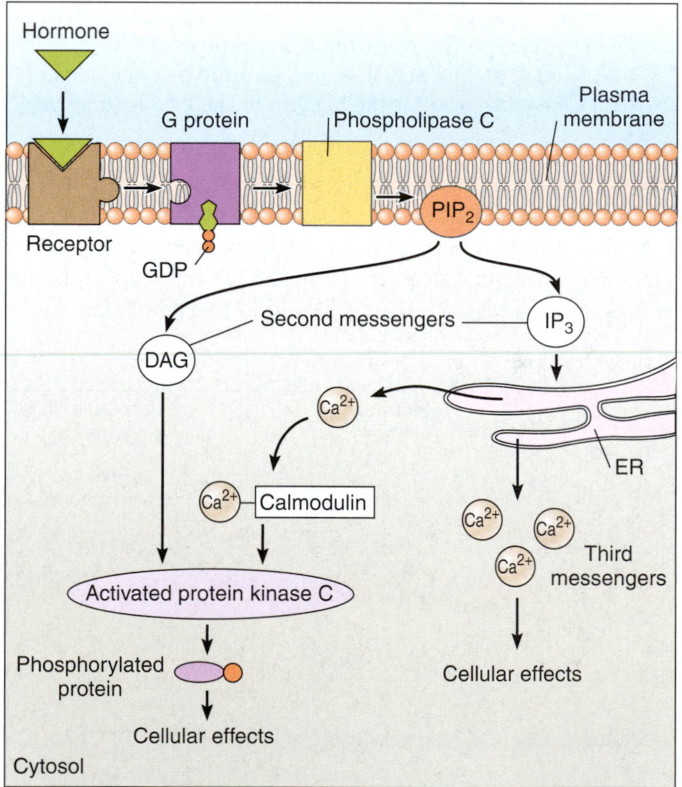

Figure 47–10 Phospholipid products as second messengers. Certain hormone-receptor complexes activate G proteins that activate phospholipase C. This enzyme splits a membrane lipid (PIP_2), forming two second messengers, inositol trisphosphate (IP_3) and diacylglycerol (DAG). IP_3 stimulates the release of Ca^{2+} from the endoplasmic reticulum (ER). Ca^{2+} acts as a third messenger. Calcium ions bind with calmodulin, and together with DAG they activate protein kinase C. The protein kinase phosphorylates certain proteins, leading to various changes in the cell.

transported down axons and stored in the paired **corpora cardiaca** (sing., *corpus cardiacum;* Fig. 47–11). When released from the corpora cardiaca, BH stimulates the **prothoracic glands,** endocrine glands in the prothorax, to produce **molting hormone (MH),** also called **ecdysone.** Molting hormone stimulates growth and molting.

In the immature insect, paired endocrine glands called **corpora allata** (sing., *corpus allatum*) secrete **juvenile hormone (JH).** This hormone suppresses metamorphosis at each larval molt so that the insect increases in size while remaining in its immature state; after the molt, the insect is still in a larval stage. When the concentration of JH decreases, metamorphosis occurs, and the insect is transformed into a pupa (see Chapter 29). In the absence of JH, the pupa molts and becomes an adult. The secretory activity of the corpora allata is regulated by the nervous system, and the amount of JH decreases with successive molts.

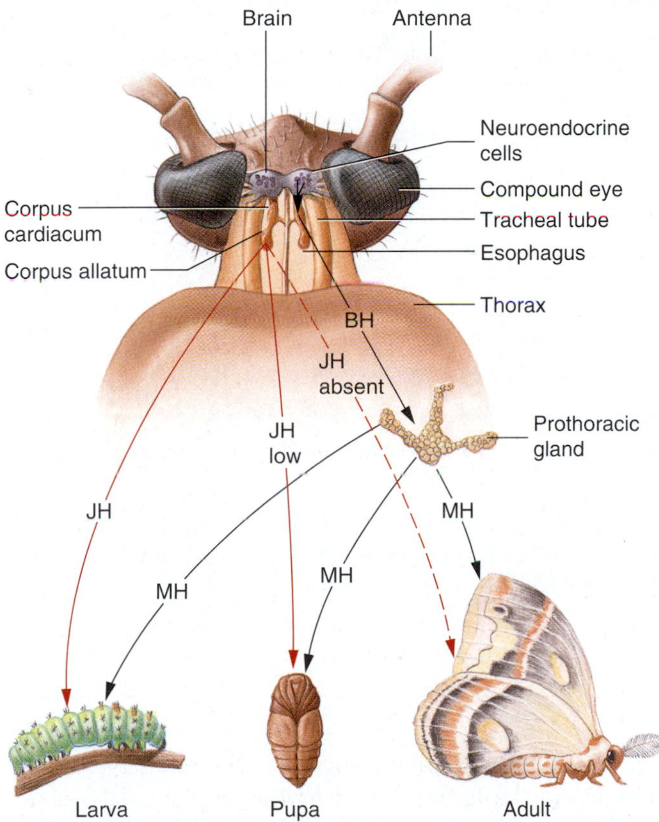

Figure 47–11 Regulation of growth and molting in insects. The dissection of an insect head shows the paired corpora allata and corpora cardiaca, as well as the brain. Neuroendocrine cells in the brain secrete brain hormone (BH), which is stored in the corpora cardiaca. When released, BH stimulates the prothoracic glands to secrete molting hormone (MH), which stimulates growth and molting. In the immature insect, the corpora allata secrete juvenile hormone (JH), which suppresses metamorphosis at each larval molt. Metamorphosis to the adult form occurs when molting hormone acts in the absence of juvenile hormone.

THE HYPOTHALAMUS LINKS THE VERTEBRATE NERVOUS AND ENDOCRINE SYSTEMS

Vertebrate hormones regulate such diverse activities as growth, development, reproduction, metabolic rate, fluid balance, blood homeostasis, and coping with stress. Most vertebrates have similar endocrine glands, though the actions of some hormones may be different in various groups. Table 47–1 gives the sources, target tissues, and physiological actions of some of the major vertebrate hormones. Many of these hormones are regulated by the hypothalamus and pituitary gland. The principal human endocrine glands are illustrated in Figure 47–12.

Homeostasis depends on normal concentrations of hormones

When a disorder or disease process affects an endocrine gland, the rate of secretion may become abnormal. If **hyposecretion** (abnormally reduced output) occurs, target cells are deprived of

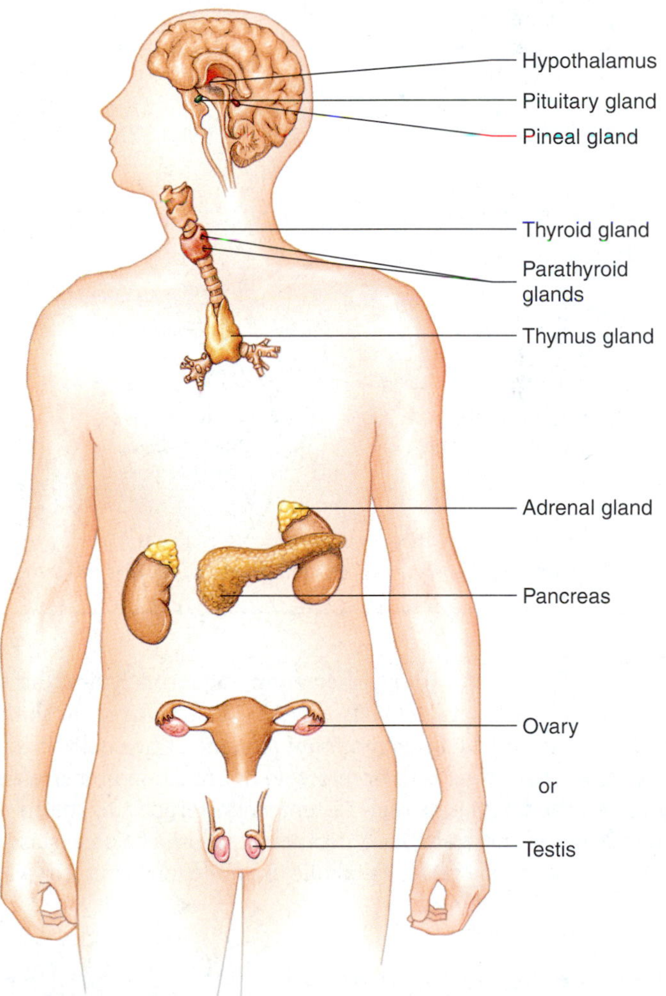

Figure 47–12 The classical human endocrine glands. The hormones produced by these glands and their actions are discussed in this chapter.

TABLE 47–1 Some Endocrine Glands and Their Hormones*

Endocrine Gland and Hormone	Target Tissue	Principal Actions
Hypothalamus	-	
Releasing and inhibiting hormones	Anterior lobe of pituitary	Regulate secretion of hormones by the anterior pituitary
Hypothalamus (production) **Posterior lobe of pituitary** (storage and release)		
Oxytocin	Uterus Mammary glands	Stimulates contraction Stimulates ejection of milk into ducts
Antidiuretic hormone (ADH)	Kidneys (collecting ducts)	Stimulates reabsorption of water; conserves water
Anterior lobe of pituitary		
Growth hormone (GH)	General	Stimulates production of insulin-like growth factors (IGFs); stimulates growth by promoting protein synthesis
Prolactin	Mammary glands	Stimulates milk production
Thyroid-stimulating hormone (TSH)	Thyroid gland	Stimulates secretion of thyroid hormones; stimulates increase in size of thyroid gland
Adrenocorticotropic hormone (ACTH)	Adrenal cortex	Stimulates secretion of adrenal cortical hormones
Gonadotropic hormones* (follicle-stimulating hormone [FSH]; luteinizing hormone [LH])	Gonads	Stimulate gonad function and growth
Thyroid gland		
Thyroxine (T_4) and triiodothyronine (T_3)	General	Stimulate metabolic rate; essential to normal growth and development
Calcitonin	Bone	Lowers blood-calcium level by inhibiting Ca^{2+} release from bone
Parathyroid glands		
Parathyroid hormone	Bone, kidneys, digestive tract	Increases blood-calcium level by stimulating Ca^{2+} release from bone; stimulates calcium reabsorption by kidneys; activates vitamin D, which increases intestinal absorption of calcium

*The gonadotropic hormones (FSH and LH) and the ovaries and testes and their hormones are discussed in Chapter 48.

needed stimulation. If **hypersecretion** (abnormally increased output) occurs, the target cells may be overstimulated. In some endocrine disorders, an appropriate amount of hormone is secreted, but the target cell receptors do not function properly. As a result, the target cells may be unable to respond to the hormone. Any of these abnormalities leads to loss of homeostasis, resulting in predictable metabolic malfunctions and clinical symptoms (Table 47–2).

The hypothalamus regulates the pituitary gland

Most endocrine activity is controlled directly or indirectly by the **hypothalamus,** which links the nervous and endocrine systems both anatomically and physiologically. The **pituitary gland** is connected to the hypothalamus by the pituitary stalk. In response to input from other areas of the brain and from hormones in the blood, neurons of the hypothalamus secrete neurohormones that regulate specific physiological processes.

Because its secretions control the activities of several other endocrine glands, the pituitary gland is known as the master gland of the body. Although it is only the size of a large pea and weighs only about 0.5 g (0.02 oz), the pituitary gland produces and secretes at least seven peptide hormones that exert far-reaching influence over body activities. The human pituitary gland consists of two main parts: the **anterior lobe** and the **posterior lobe.** In some vertebrates, an intermediate lobe is present.

TABLE 47–1 Continued

Endocrine Gland and Hormone	Target Tissue	Principal Actions
Islets of Langerhans of Pancreas	-	
Insulin	General	Lowers glucose concentration in the blood by facilitating glucose uptake and utilization by cells; stimulates glycogen production; stimulates fat storage and protein synthesis
Glucagon	Liver, adipose tissue	Raises glucose concentration in the blood; stimulates glycogen breakdown; mobilizes fat
Adrenal Medulla		
Epinephrine and norepinephrine	Skeletal muscle; cardiac muscle; blood vessels; liver; adipose tissue	Help body cope with stress; increase heart rate, blood pressure, metabolic rate; reroute blood; mobilize fat; raise blood-sugar level
Adrenal Cortex		
Mineralocorticoids (aldosterone)	Kidney tubules	Maintain sodium and potassium balance; increase sodium reabsorption; increase potassium excretion
Glucocorticoids (cortisol)	General	Help body cope with long-term stress; raise blood-glucose level; mobilize fat
Pineal gland		
Melatonin	Hypothalamus	Important in biological rhythms; influences reproductive processes in hamsters and other animals; pigmentation in some vertebrates; may help control onset of puberty in humans
Ovary†		
Estrogens (estradiol)	General; uterus	Develop and maintain sex characteristics in female; stimulate growth of uterine lining
Progesterone	Uterus; breast	Stimulates development of uterine lining
Testis†		
Testosterone	General; reproductive structures	Develops and maintains sex characteristics of males; promotes spermatogenesis; responsible for adolescent growth spurt
Inhibin	Anterior lobe of pituitary	Inhibits FSH release in male

†For more detailed description see Tables 48–1 and 48–2.

The posterior lobe of the pituitary gland releases hormones produced by the hypothalamus

The **posterior lobe** of the pituitary gland, or neurohypophysis, develops from brain tissue. This neuroendocrine organ *secretes* two peptide hormones, **oxytocin** and **antidiuretic hormone,** or **ADH** (also known as *vasopressin*). (The function of ADH in regulating water reabsorption in the kidneys is discussed in Chapter 46.) These hormones are actually *produced* by neuroendocrine cells in two distinct areas of the hypothalamus. Enclosed within vesicles, they are transported down axons that extend into the posterior lobe of the pituitary gland (Fig. 47–13). The vesicles are stored in these axon terminals until the neuron is stimulated. Then the hormones are released and diffuse into surrounding capillaries.

Oxytocin concentration in the blood rises toward the end of pregnancy, stimulating the strong contractions of the uterus needed to expel a baby. Oxytocin is sometimes administered clinically (under the trade name Pitocin) to initiate or speed labor. After birth, when an infant sucks at its mother's breast, sensory neurons signal the pituitary to release oxytocin. The hormone stimulates contraction of cells surrounding the milk glands so that milk is let down into the ducts, from which it can be sucked by the infant. Because oxytocin also stimulates the uterus to contract, breast feeding promotes rapid recovery of the uterus to nonpregnant size.

TABLE 47–2 Consequences of Endocrine Malfunction

Hormone	Hyposecretion	Hypersecretion
Growth hormone	Pituitary dwarfism	Gigantism if malfunction occurs in childhood; acromegaly in adult
Thyroid hormones	Cretinism (in children); myxedema, a condition of pronounced adult hypothyroidism (metabolic rate is reduced by about 40%; patient feels tired and may be mentally slow); dietary iodine defiency can lead to hyposecretion and goiter (abnormal enlargement of the thyroid gland; see figure)	Hyperthyroidism; increased metabolic rate, nervousness irritability; goiter; can be caused by Grave's disease.
Parathyroid hormone	Spontaneous discharge of nerves; spasms; tetany; death	Weak, brittle bones; kidney stones
Insulin	Diabetes mellitus	Hypoglycemia
Hormones of adrenal cortex	Addison's disease (inability to cope with stress; sodium loss in urine may lead to shock)	Cushing's disease (edema gives face a full-moon appearance; fat is deposited about trunk; blood-glucose level rises; immune responses are depressed)

Goiter commonly results from iodine deficiency. *(John Paul Kay/Peter Arnold, Inc.)*

The anterior lobe of the pituitary gland regulates growth and other endocrine glands

The **anterior lobe** of the pituitary develops from epithelial cells rather than neural cells. The anterior lobe functions like a classical endocrine gland in that it receives signals by way of the blood and releases its hormones into the blood. The hypothalamus produces several **releasing hormones** and **inhibiting hormones** that regulate the production and secretion of specific hormones by the anterior pituitary gland. These neurohormones enter capillaries and pass through special portal veins that connect the hypothalamus with the anterior lobe of the pituitary. (These portal veins, like the hepatic portal vein, do not deliver blood to a larger vein directly but connect two sets of capillaries.) Within the anterior lobe of the pituitary, the portal veins divide into a second set of capillaries. The releasing and inhibiting hormones pass through the walls of these capillaries into the tissue of the anterior lobe.

The anterior lobe secretes prolactin, growth hormone, and several **tropic hormones** that stimulate other endocrine glands (Fig. 47–14). **Prolactin** stimulates the cells of the mammary glands to produce milk in a nursing mother.

Growth hormone stimulates protein synthesis

Small children measure themselves periodically against their parents, eagerly awaiting that time when they, too, will be "big." Whether one will be tall or short depends on many factors, including genes, diet, and hormonal balance.

Growth hormone (**GH;** also called *somatotropin*) is referred to as an **anabolic hormone** because it promotes tissue growth. Many of the effects of GH on skeletal growth are indirect. GH stimulates liver cells and cells of many other tissues to produce peptides called **somatomedins,** including **insulin-like growth factors (IGFs).** These growth factors (1) promote the linear growth of the skeleton by stimulating growth of cartilage in growth areas of the bones, and (2) stimulate general tissue growth and increase in size of organs by promoting protein synthesis and other anabolic processes.

Growth hormone stimulates uptake of amino acids by the cells and promotes protein synthesis. GH also affects lipid and carbohydrate metabolism. In adipose tissue, GH decreases glucose uptake and promotes mobilization of fat, raising the level of free fatty acids in the blood. In this protein-sparing process, fatty acids become available for cells to use as fuel. Fat mobilization by GH is also important during fasting or periods of prolonged stress, when the blood-sugar level is low.

Growth is affected by many factors

In adults as well as in growing children, GH is secreted in pulses throughout the day. Secretion of GH is regulated by both a **growth hormone-releasing hormone (GHRH)** and a **growth**

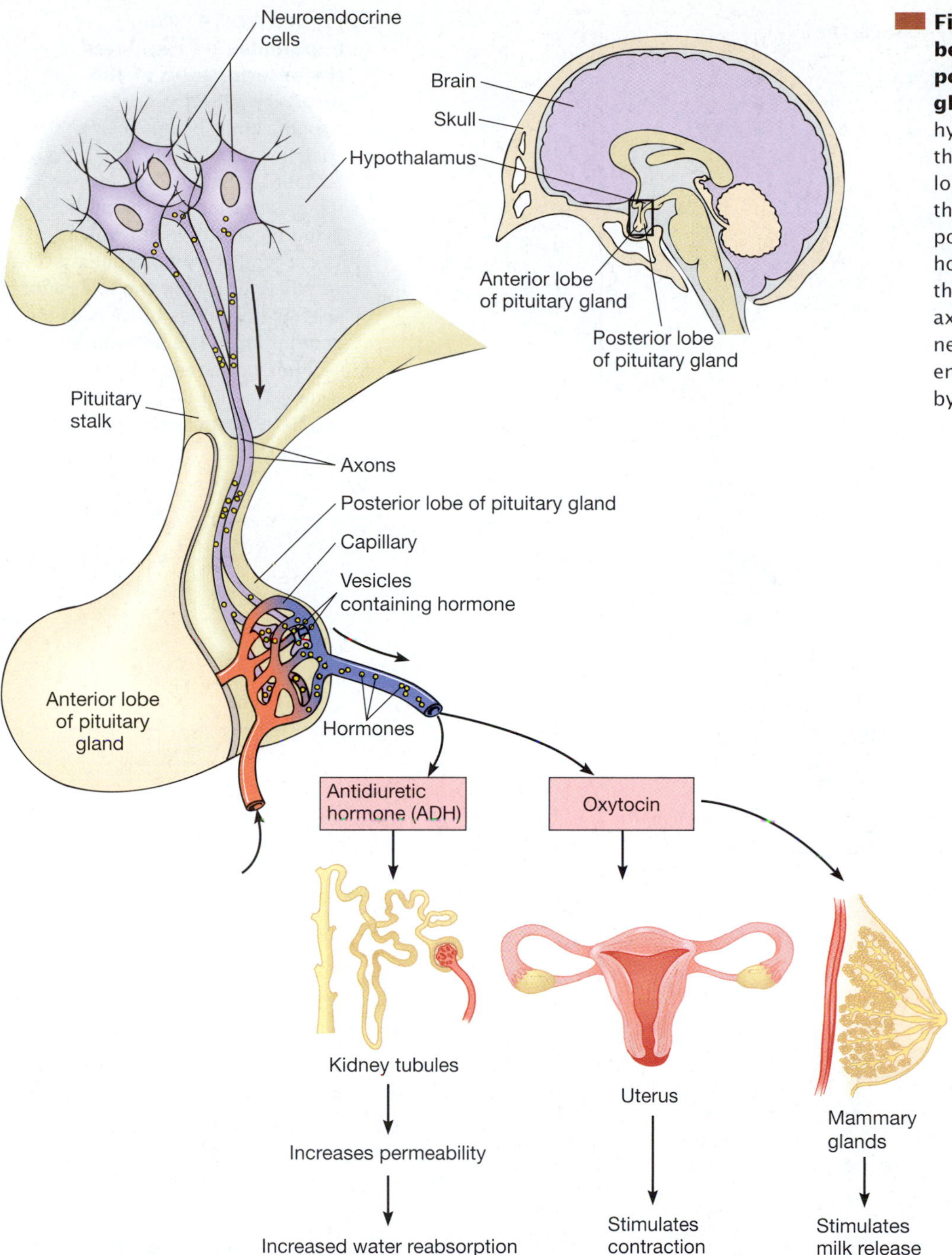

Figure 47–13 The relationship between the hypothalamus and posterior lobe of the pituitary gland. Neuroendocrine cells of the hypothalamus manufacture hormones that are secreted by the posterior lobe of the pituitary. The axons of these neurons extend down into the posterior lobe of the pituitary. Their hormones are packaged in vesicles that are transported through the axons and stored in their ends. When needed, the hormones are secreted, enter the blood, and are transported by the circulatory system.

hormone-inhibiting hormone (GHIH; also called **somatostatin**) released by the hypothalamus. A high level of GH in the blood signals the hypothalamus to secrete GHIH, the inhibiting hormone, and the pituitary release of GH slows. A low level of GH in the blood stimulates the hypothalamus to secrete GHRH, the releasing hormone, which in turn stimulates the pituitary gland to release more GH. Many other factors, including blood-sugar level, decrease in amino acid concentration, and stress, influence GH secretion.

The age-old notions that children need plenty of sleep, a proper diet, and regular exercise in order to grow are supported by research. Secretion of GH increases during exercise, probably because rapid metabolism by muscle cells lowers the blood-sugar level. GH is secreted about 1 hour after the onset of deep sleep and in a series of pulses 2 to 4 hours after a meal.

Emotional support is also necessary for proper growth. Growth may be retarded in children who are deprived of cuddling, playing, and other forms of nurture, even when their needs for food and shelter are met. In extreme cases, childhood stress can produce a form of retarded development known as *psychosocial dwarfism.* Some emotionally deprived children exhibit abnormal sleep patterns, which may be the basis for decreased secretion of GH.

Other hormones also influence growth. Thyroid hormones appear to be necessary for normal GH secretion and function

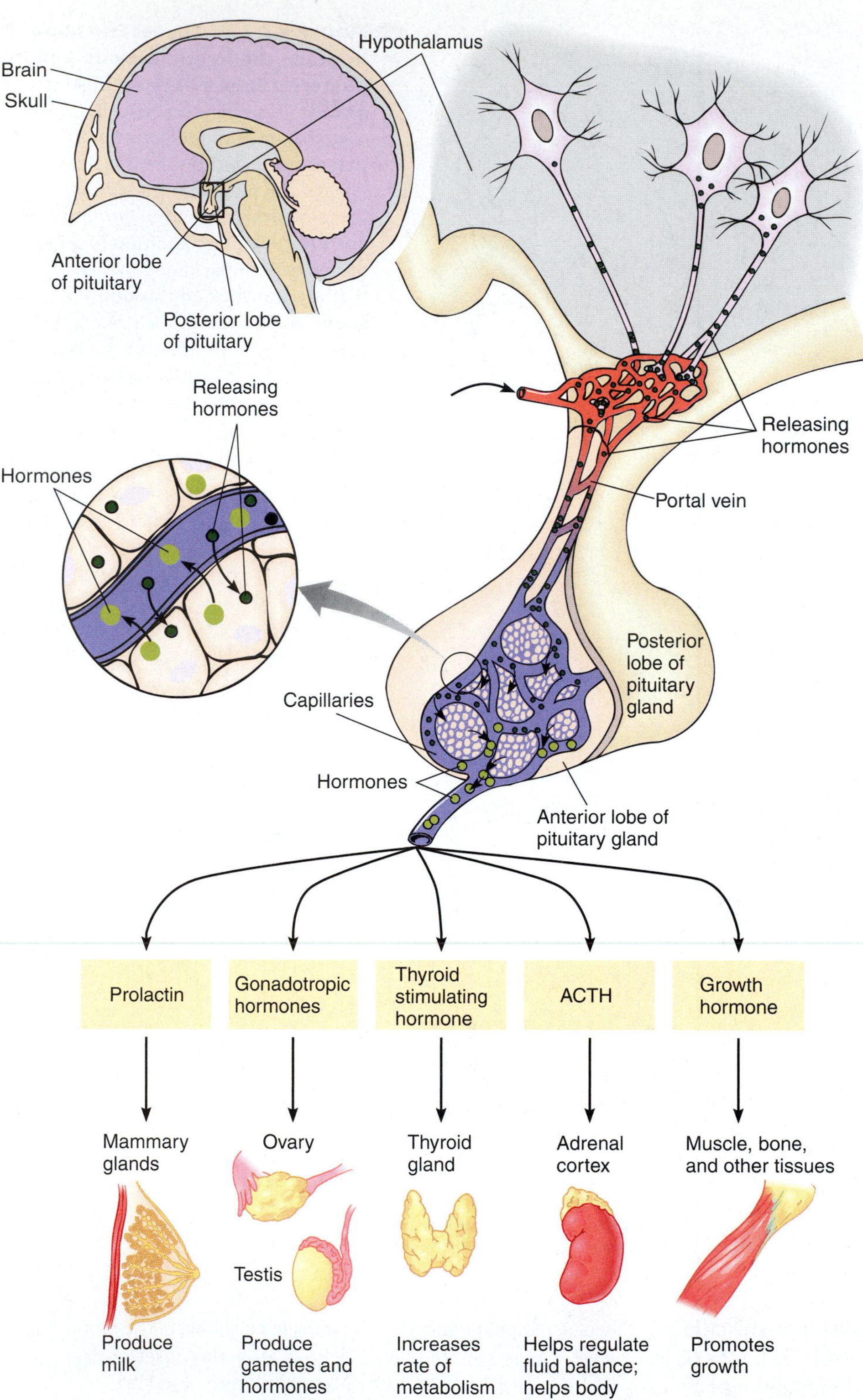

Figure 47-14 The hypothalamus regulates the anterior lobe of the pituitary gland. The hypothalamus secretes several specific releasing and inhibiting hormones that are transported to the anterior lobe of the pituitary gland by way of portal veins. These regulatory hormones stimulate or inhibit the release of specific hormones by cells of the anterior lobe. The anterior lobe secretes hormones that act on a wide variety of target tissues. (ACTH, adrenocorticotropic hormone)

and for normal tissue response to IGFs. Sex hormones must be present for the adolescent growth spurt to occur. However, the presence of sex hormones eventually causes the growth centers within the long bones to ossify, so that further increase in height is impossible even when GH is present.

Inappropriate amounts of growth hormone secretion result in abnormal growth

Have you ever wondered why some people fail to grow normally? **Pituitary dwarfs** are individuals whose pituitary glands do not produce sufficient GH during childhood. Although miniature, a pituitary dwarf has normal intelligence and is usually well-proportioned. If the growth centers in the long bones are still active when this condition is diagnosed, it can be treated clinically by injection with human GH, which can be synthesized by recombinant DNA technology. Growth problems may also result from the malfunction of other mechanisms, such as regulating hormones from the hypothalamus.

An individual may become abnormally tall when the anterior pituitary secretes excessive amounts of GH during childhood. This condition is referred to as **gigantism.** If pituitary malfunction leads to hypersecretion of GH during adulthood, the individual cannot grow taller. Instead, connective tissue thickens, and bones in the hands, feet, and face may increase in diameter. This condition is known as **acromegaly,** which means "large extremities."

Thyroid hormones increase metabolic rate

The **thyroid gland** is located in the neck region, in front of the trachea and below the larynx (see Fig. 47–12). Two of the **thyroid hormones, thyroxine,** also known as $\mathbf{T_4}$**,** and **triiodothyronine,** or $\mathbf{T_3}$**,** are synthesized from the amino acid tyrosine and from iodine. Thyroxine has four iodine atoms attached to each molecule; T_3 has three. In most target tissues, T_4 is converted to T_3, the more active form. We describe the actions of calcitonin, another hormone secreted by the thyroid gland, when we discuss the parathyroid glands.

In vertebrates, thyroid hormones are essential for normal growth and development, and they increase the rate of metabolism in most body tissues. After T_3 binds with its receptor in the nucleus of a target cell, the T_3 receptor complex induces or suppresses synthesis of specific enzymes and other proteins. Thyroid hormones also help regulate the synthesis of proteins necessary for cellular differentiation. For example, tadpoles cannot develop into adult frogs without thyroxine.

Thyroid secretion is regulated by negative feedback mechanisms

The regulation of thyroid hormone secretion depends mainly on a negative feedback loop between the anterior pituitary and the thyroid gland (Fig. 47–15). When the concentration of thyroid hormones in the blood rises above normal, the anterior pituitary secretes less *thyroid-stimulating hormone (TSH):*

High concentration of thyroid hormones ⟶ anterior pituitary secretes less TSH ⟶ thyroid gland secretes less hormone ⟶ homeostasis

Too much thyroid hormone in the blood also affects the hypothalamus, inhibiting secretion of a *TSH-releasing hormone, TRH*. However, the hypothalamus is thought to exert its regulatory effects mainly in certain stressful situations, such as extreme weather change. Exposure to very cold weather may stimulate the hypothalamus to increase secretion of TSH-releasing hormone. This action raises body temperature through increased metabolic heat production. Increased body temperature has a negative feedback effect, limiting further secretion of thyroid hormones.

When the concentration of thyroid hormones decreases, the pituitary secretes more TSH. TSH acts by way of cAMP to promote synthesis and secretion of thyroid hormones and also to promote increased size of the thyroid gland itself. The effect of TSH can be summarized as follows:

Low concentration of thyroid hormones ⟶ anterior pituitary secretes more TSH ⟶ thyroid gland secretes more hormone ⟶ homeostasis

Malfunction of the thyroid gland leads to specific disorders

Extreme hypothyroidism during infancy and childhood results in low metabolic rate and can lead to **cretinism,** a condition of retarded mental and physical development. When diagnosed early enough and treated with thyroid hormones, the effects of cretinism can be prevented.

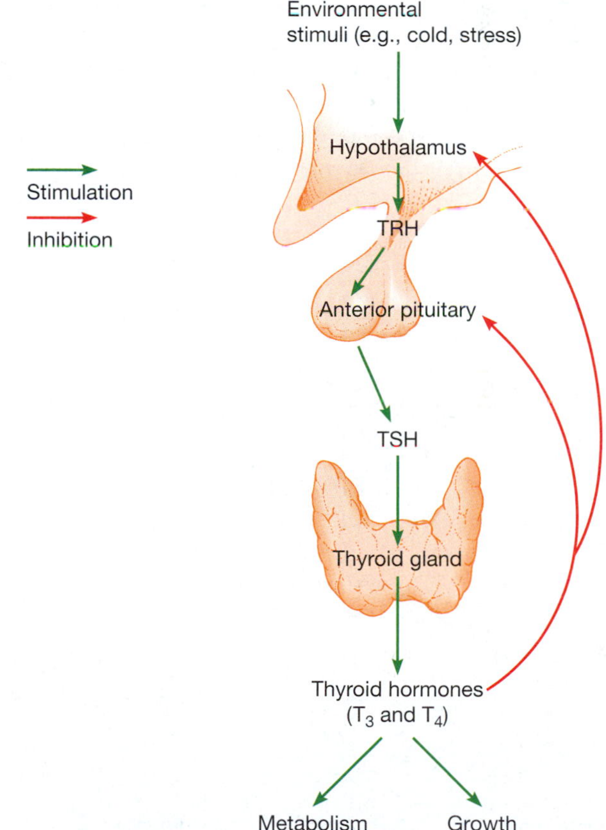

Figure 47–15 Regulation of thyroid hormone secretion by negative feedback. Green arrows indicate stimulation; red arrows indicate inhibition. An increase in concentration of thyroid hormones above normal levels signals the anterior pituitary and hypothalamus to limit thyroid-stimulating hormone (TSH) production. Thus, thyroid hormones limit their own production by negative feedback. (TRH, TSH-releasing hormone)

An adult who feels like sleeping all of the time, has little energy, and is mentally slow or confused may be suffering from hypothyroidism. When there is almost no thyroid function, the basal metabolic rate is reduced by about 40% and the patient develops **myxedema,** characterized by a slowing down of physical and mental activity. Hypothyroidism can be treated readily by oral administration of the missing hormone.

Hyperthyroidism does not cause abnormal growth but does increase metabolic rate by 60% or even more. This increase in metabolism results in the rapid use of nutrients, causing the individual to be hungry and to eat more. But this is not sufficient to meet the demands of the rapidly metabolizing cells, so individuals with this condition often lose weight. They also tend to be nervous, irritable, and emotionally unstable. The most common cause of hyperthyroidism is **Grave's disease,** an autoimmune disease. Abnormal antibodies bind to TSH receptors, activating them. This leads to increased production of thyroid hormones.

An abnormal enlargement of the thyroid gland, or **goiter,** may be associated with either hypersecretion or hyposecretion (see figure in Table 47–2). In Grave's disease, the abnormal antibodies that activate TSH receptors can cause the development of a goiter. Another cause of hyposecretion is dietary iodine deficiency. Without iodine the gland cannot make thyroid hormones, so their concentration in the blood decreases. In compensation, the anterior pituitary secretes large amounts of TSH, which stimulates growth of the thyroid gland, sometimes to gigantic proportions. However, enlargement of the gland cannot increase production of the hormones, because the needed ingredient is still missing. Seafood is a rich source of iodine, and iodine is also added to table salt as a nutritional supplement. In fact, thanks to iodized salt, goiter is no longer common in the United States and other highly developed countries. In other parts of the world, however, hundreds of thousands still suffer from this easily preventable disorder.

The parathyroid glands regulate calcium concentration

The **parathyroid glands** are typically embedded in the connective tissue surrounding the thyroid gland. The parathyroid glands secrete **parathyroid hormone (PTH),** a polypeptide that helps regulate the calcium level of the blood and interstitial fluid (Fig. 47–16). Parathyroid hormone acts by way of a membrane receptor and cAMP to stimulate calcium release from bones and to increase calcium reabsorption from the kidney tubules. It also activates vitamin D, which then increases the amount of calcium absorbed from the intestine.

Calcitonin, a peptide hormone secreted by the thyroid gland, works antagonistically to parathyroid hormone. When the concentration of calcium rises above homeostatic levels, calcitonin is released and rapidly inhibits removal of calcium from bone.

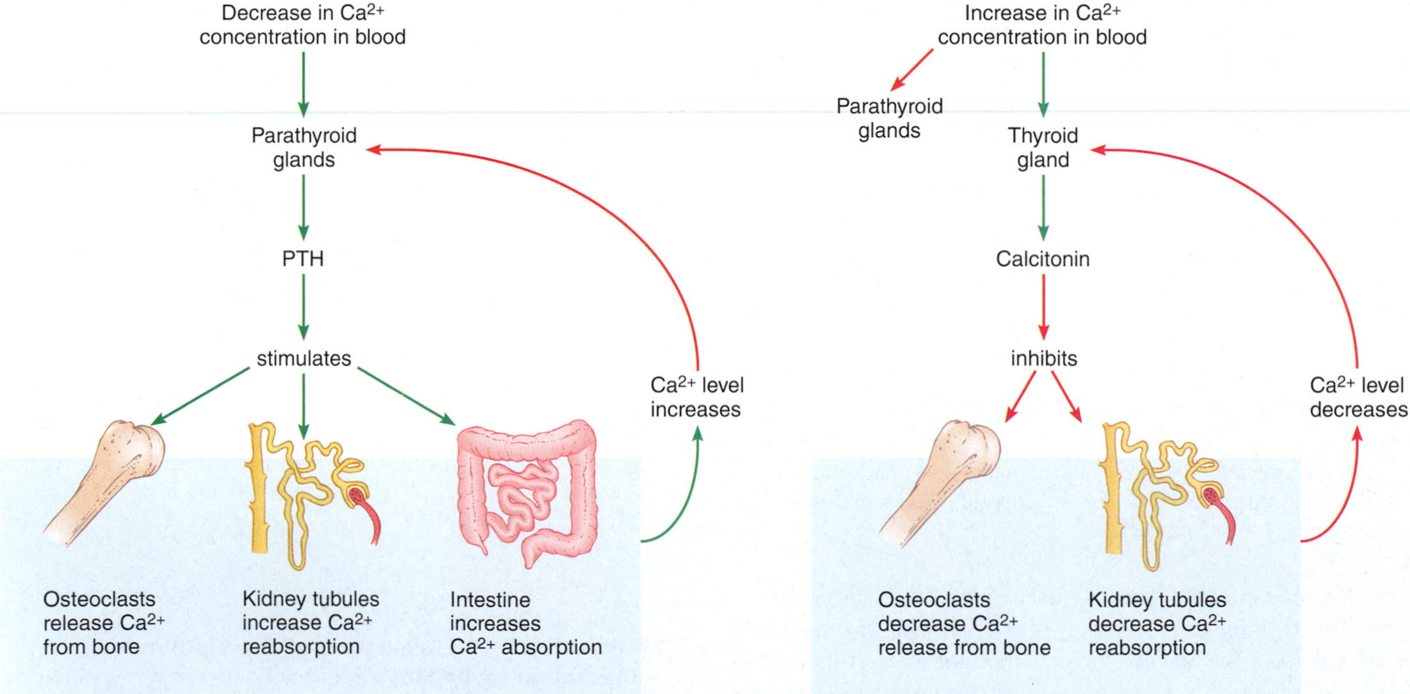

Figure 47–16 Regulation of calcium homeostasis by parathyroid hormone (PTH) and calcitonin. Green arrows indicate stimulation; red arrows indicate inhibition. **(a)** When calcium concentration decreases below the normal range, the parathyroid glands secrete PTH. This hormone stimulates homeostatic mechanisms that restore appropriate calcium concentration. **(b)** When calcium concentration increases above normal limits, the parathyroid glands are inhibited. The thyroid gland secretes calcitonin, which inhibits Ca^{2+} release from bone and decreases Ca^{2+} reabsorption in the kidneys.

The islets of the pancreas regulate glucose concentration

In addition to secreting digestive enzymes (see Chapter 45), the pancreas is an important endocrine gland. Its hormones, **insulin** and **glucagon,** are secreted by cells that occur in little clusters, the **islets of Langerhans,** throughout the pancreas (Fig. 47–17). About 1 million islets are present in the human pancreas. They are composed mainly of **beta cells,** which secrete insulin, and **alpha cells,** which secrete glucagon.

Insulin lowers the concentration of glucose in the blood

When insulin binds to a **tyrosine kinase receptor** on a target cell plasma membrane, it activates the receptor. **Tyrosine kinase** phosphorylates other insulin receptors, increasing their activity. Other proteins, including a large protein called insulin receptor substrate (IRS), are also phosphorylated by the insulin receptor. Activation of IRS initiates a series of phosphorylations that result in activation of transport mechanisms in the plasma membrane and also affect transcription.

Insulin is an anabolic hormone that promotes storage of fuel molecules. It stimulates cells of many tissues, including liver, muscle, and fat cells, to take up glucose from the blood by facilitated diffusion (see Chapter 5). Once glucose enters muscle cells, it is either used immediately as fuel or stored as glycogen. Insulin also inhibits liver cells from releasing glucose. Thus, insulin activity results in *lowering* the glucose level in the blood.

Insulin helps regulate fat and protein metabolism. This hormone reduces the use of fatty acids as fuel and instead stimulates their storage in adipose tissue. Insulin has an anabolic effect on protein metabolism, resulting in a net increase in protein. It promotes protein synthesis by increasing the number of amino acid transporters in the plasma membrane, thus stimulating the transport of certain amino acids into cells. Insulin also promotes transcription and translation.

Glucagon raises the concentration of glucose in the blood

The effects of **glucagon** are opposite to those of insulin. The main action of glucagon is to raise blood-glucose level. It does this by stimulating liver cells to convert glycogen to glucose, a process known as *glycogenolysis*. Glucagon also stimulates *gluconeogenesis,* the production of glucose from other metabolites. Glucagon mobilizes fatty acids and amino acids as well as glucose.

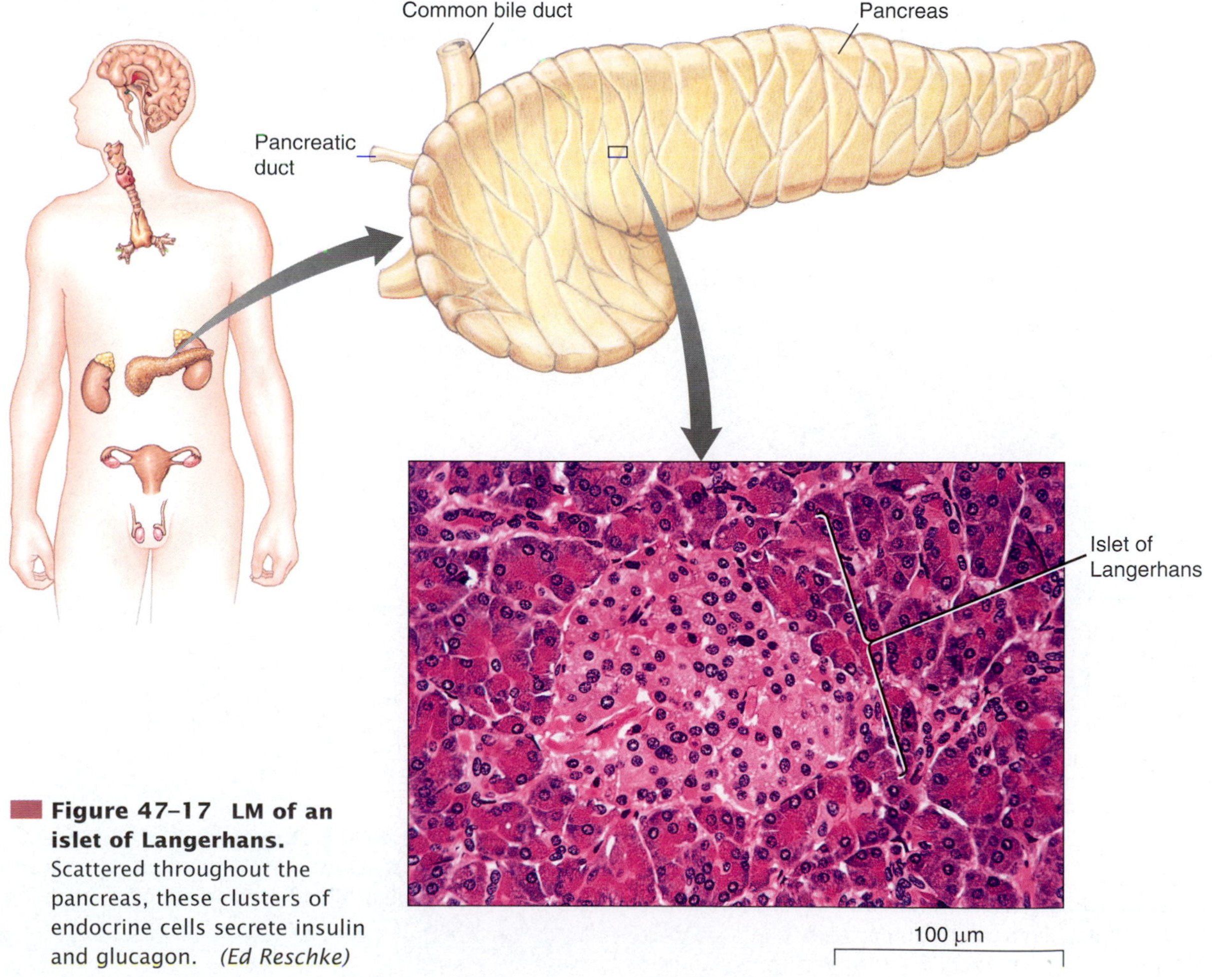

Figure 47–17 LM of an islet of Langerhans. Scattered throughout the pancreas, these clusters of endocrine cells secrete insulin and glucagon. *(Ed Reschke)*

Insulin and glucagon secretion are regulated by glucose concentration

Insulin and glucagon secretion are directly controlled by the concentration of glucose in the blood (Fig. 47–18). After a meal, when the blood-glucose level rises as a result of intestinal absorption, beta cells are stimulated to increase insulin secretion. Then, as cells remove glucose from the blood, decreasing its concentration, insulin secretion decreases.

High blood-glucose concentration ⟶ beta cells secrete insulin ⟶ blood-glucose concentration decreases ⟶ homeostasis

When one has not eaten for several hours, the concentration of glucose in the blood begins to fall. When it decreases from its normal fasting level of about 90 mg of glucose per 100 mL of blood to about 70 mg of glucose, the alpha cells of the islets increase their secretion of glucagon. Glucose is mobilized from storage in the liver cells, and blood-glucose concentration returns to normal:

Low blood-glucose concentration ⟶ alpha cells secrete glucagon ⟶ blood-glucose concentration increases ⟶ homeostasis

The alpha cells respond to the glucose concentration within their own cytoplasm, which reflects the blood-glucose level. When blood-glucose level is high, there is generally a high level of glucose within the alpha cells, and glucagon secretion is inhibited.

Note that these are negative feedback mechanisms and that insulin and glucagon work antagonistically to keep blood-glucose concentration within normal limits. When the glucose level rises, insulin release brings it back to normal; when the glucose concentration falls, glucagon acts to raise it again. The insulin-glucagon system is a powerful, fast-acting mechanism for keeping blood-glucose level within normal limits. Why do you think it is important to maintain a constant blood-glucose level? Recall that brain cells ordinarily are unable to use other nutrients as fuel and so depend on a continuous supply of glucose. As we will discuss, several other hormones also affect blood-glucose concentration.

Diabetes mellitus is a serious disorder of carbohydrate metabolism

Diabetes mellitus, the most common endocrine disorder, is a worldwide health problem, and the World Health Organization reports that its prevalence is increasing. Of the estimated 16 million diabetics in the United States, more than 40,000 die each year as a result of this disorder, making it one of the leading causes of premature death. Diabetes is a leading cause of vascular disease, blindness, nerve disease, kidney disorders, and gangrene of the limbs. (Note that diabetes mellitus is an entirely dif-

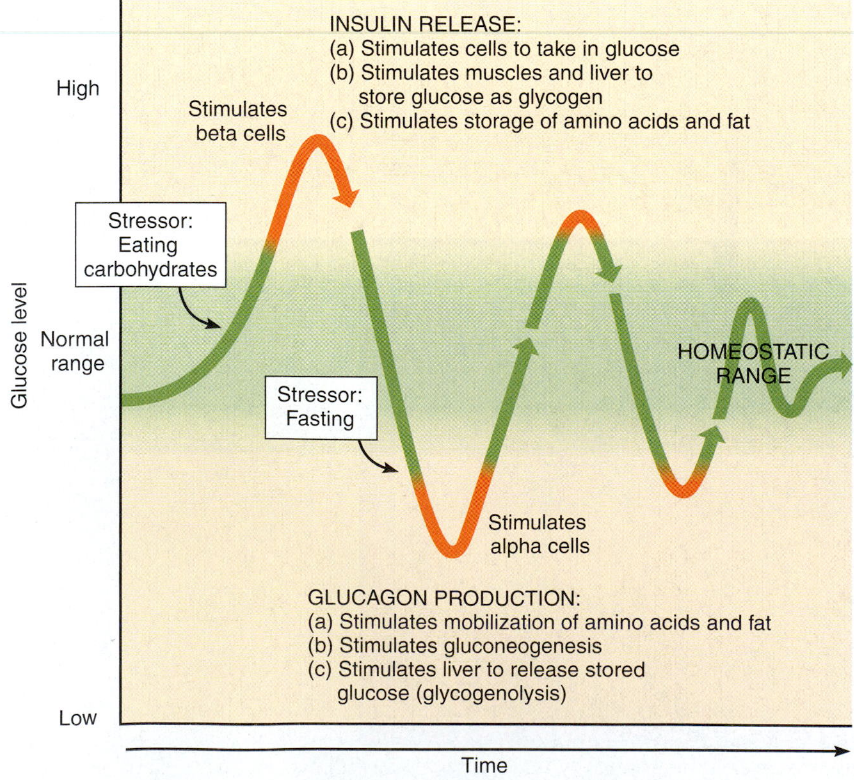

Figure 47–18 Regulation of glucose concentration. Insulin and glucagon work antagonistically to regulate blood-glucose levels.

ferent disorder from diabetes insipidus, which is discussed in Chapter 46.)

About 90% of diabetics have *type 2 diabetes.* This disorder develops gradually, usually in overweight persons older than 40 years of age. In type 2 diabetes, normal (or greater than normal) concentrations of insulin are present in the blood, but insulin receptors on target cells do not bind it, a condition known as **insulin resistance.** The causes of insulin resistance are not fully understood, but there is a strong genetic component. Even if identical twins are raised separately in very different environments, when one identical twin develops insulin resistance, the other one almost always does also. Investigators have also reported an association between increased fat tissue and insulin resistance.

Process of Science In 2001 Claire M. Steppan and her team of researchers at the University of Pennsylvania reported that they had identified a human gene that codes for a hormone they named **resistin** (for insulin resistance). These biologists had first identified the hormone in mice, where it antagonizes the action of insulin. They found that resistin is secreted by fat cells and that its concentration increases significantly in both genetic and diet-induced obesity. Resistin is down-regulated by fasting, as well as by certain drugs used to treat diabetes. Further research will determine whether resistin is a signaling molecule that links diabetes to obesity. When the receptor for resistin is identified, pharmaceutical companies may have a new target for drugs aimed at treating diabetes.

Many type 2 diabetics can keep their blood-glucose levels within normal range by diet management, weight loss, and regular exercise. When this treatment approach is not effective, type 2 diabetics are treated with oral drugs that stimulate insulin secretion and promote its actions.

Insulin-dependent diabetes, referred to as *type 1 diabetes,* usually develops before age 30, often during childhood. In type 1 diabetes there is a marked decrease in the number of beta cells in the pancreas, resulting in insulin deficiency. Daily insulin injections are needed to correct the carbohydrate imbalance that results. Type 1 diabetes is an autoimmune disease in which antibodies mark the beta cells for destruction. This disorder may be caused by a combination of genetic predisposition and infection by a virus.

Similar metabolic disturbances occur in both types of diabetes mellitus: disruption of carbohydrate, fat, and protein metabolism, and electrolyte imbalance.

1. **Decreased use of glucose.** Because the cells of diabetics cannot take up glucose from the blood, it accumulates there, causing hyperglycemia (an abnormally high concentration of glucose in the blood). Instead of the normal fasting concentration of about 90 mg per 100 mL, the level typically exceeds 200 mg per 100 mL and may reach concentrations of more than 1000 mg per mL. When glucose concentration exceeds its **tubular transport maximum (Tm),** the maximum rate at which it can be reabsorbed, glucose is excreted in the urine.
2. **Increased fat mobilization.** Despite the large quantities of glucose in the blood, insulin-dependent cells in a diabetic can take in only about 25% of the glucose they require for fuel. The body turns to fat and protein for energy. The absence of insulin promotes the mobilization of fat stores, providing nutrients for cellular respiration. But unfortunately, the blood lipid level may reach five times the normal level, leading to the development of atherosclerosis (see Chapter 42).

   Increased fat metabolism increases the formation of ketone bodies. These compounds build up in the blood, causing *ketoacidosis,* a condition in which the body fluids and blood become too acidic. When the ketone level in the blood rises, ketones appear in the urine, another clinical indication of diabetes mellitus. If severe, ketoacidosis can lead to coma and death.
3. **Increased protein use.** Lack of insulin also results in increased protein breakdown relative to protein synthesis, so the untreated diabetic becomes thin and emaciated.
4. **Electrolyte imbalance.** When ketone bodies and glucose are excreted in the urine, water follows by osmosis; as a result, urine volume increases. The resulting dehydration causes the diabetic to feel continually thirsty and challenges electrolyte homeostasis. When ketones are excreted, they also take sodium, potassium, and some other cations with them. Loss of these ions in the urine contributes to electrolyte imbalance.

In hypoglycemia the blood-glucose concentration is too low

Hypoglycemia, low blood-glucose concentration, sometimes occurs in people who later develop diabetes. It may be an overreaction by the islets to glucose challenge; that is, too much insulin is secreted in response to carbohydrate ingestion. About 3 hours after a meal, the blood-sugar concentration falls below normal, making the individual feel very drowsy. If this reaction is severe enough, the patient may become uncoordinated or even unconscious.

Serious hypoglycemia can develop if diabetics receive injections of too much insulin or if the islets, because of a tumor, secrete too much insulin. The blood-glucose concentration may then fall drastically, depriving brain cells of their needed fuel supply. These events can lead to **insulin shock,** a condition in which the patient may appear drunk or may become unconscious, suffer convulsions, and even die.

The adrenal glands help the body cope with stress

The paired **adrenal glands** are small, yellow masses of tissue that lie in contact with the upper ends of the kidneys (Fig. 47–19). Each gland consists of a central portion, the **adrenal medulla,** and a larger outer section, the **adrenal cortex.** Although joined anatomically, the adrenal medulla and cortex develop from different types of tissue in the embryo and function as distinctly different glands. Both secrete hormones that help regulate metabolism, and both help the body respond to stress.

The adrenal medulla initiates an alarm reaction

The adrenal medulla is a neuroendocrine gland that is coupled to the sympathetic nervous system. It develops from neural tissue, and its secretion is controlled by sympathetic nerves. The adrenal medulla secretes **epinephrine** and **norepinephrine.** Chemically, these hormones are very similar; they belong to the chemical group known as **catecholamines.** Recall that norepinephrine also serves as a neurotransmitter released by sympathetic neurons and by some neurons in the central nervous system. Most of the hormone output of the adrenal medulla is epinephrine.

Under normal conditions, both epinephrine and norepinephrine are secreted continuously in small amounts. Their secretion is under neural control. When anxiety is aroused, messages are sent from the brain through sympathetic nerves to the adrenal medulla. Acetylcholine released by these neurons triggers the release of epinephrine and norepinephrine.

During a stressful situation, adrenal medullary hormones initiate an *alarm reaction* enabling you to think more quickly, fight harder, or run faster than usual. Metabolic rate increases by as much as 100%. Blood is rerouted to those organs essential for emergency action (Fig. 47–20). Blood vessels going to the brain, muscles, and heart are dilated, while those to the skin and kidneys are constricted. Constriction of blood vessels serving the skin has the added advantage of decreasing blood loss in case of hemorrhage (and explains the sudden paling that comes with fear or rage). At the same time, the heart beats faster. Thresholds in the reticular activating system of the brain are lowered, so you become more alert (see Chapter 40). Strength of muscle contraction increases. The adrenal medullary hormones also raise fatty acid and glucose levels in the blood, ensuring needed fuel for extra energy.

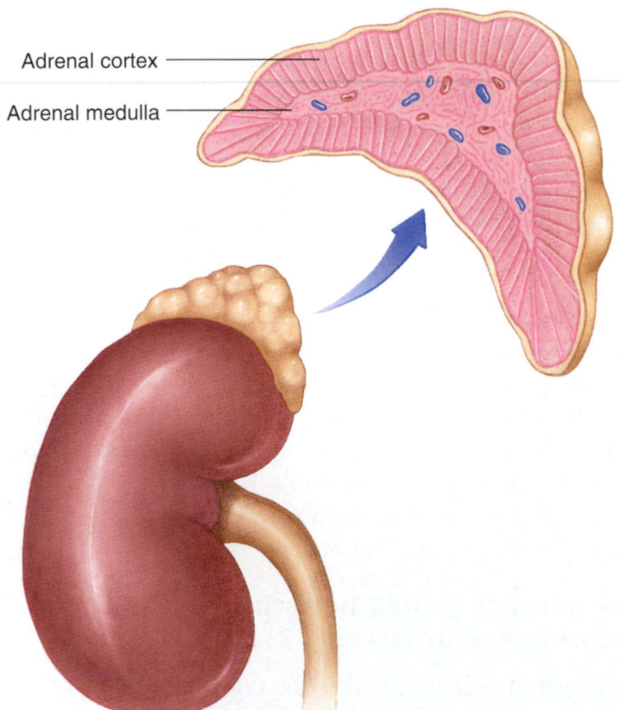

Figure 47–19 Adrenal gland. The adrenal glands are located above the kidneys. Each gland consists of a central adrenal medulla surrounded by an adrenal cortex.

The adrenal cortex helps the body deal with chronic stress

All the hormones of the adrenal cortex are steroids synthesized from cholesterol, which in turn is made in the liver from acetyl coenzyme A. Although more than 30 types of steroids have been isolated from the adrenal cortex, this gland produces only three types of hormones in significant amounts: sex hormone precursors, mineralocorticoids, and glucocorticoids.

Sex hormone precursors, such as androgens (masculinizing hormones), are secreted by the adrenal cortex in both sexes. In the tissues sex hormone precursors are converted to **testosterone,** the principal male sex hormone, and **estradiol,** the principal female sex hormone. In males, androgen production by the adrenal cortex is not significant because the testes produce testosterone. However, in females, the androgen produced by the adrenal cortex accounts for most of the testosterone circulating in the blood.

The principal **mineralocorticoid** is **aldosterone.** Recall from Chapter 46 that aldosterone helps regulate fluid balance by regulating salt balance. In response to aldosterone, the kidneys reabsorb more sodium and excrete more potassium. As a result of increased sodium, extracellular fluid volume increases, which results in greater blood volume and elevated blood pressure.

When the adrenal glands do not produce enough aldosterone, large amounts of sodium are excreted in the urine. Water leaves the body with the sodium (because of osmotic pressure), and the blood volume may be so markedly reduced that the patient dies of low blood pressure.

Cortisol, also called hydrocortisone, accounts for about 95% of the **glucocorticoid** activity of the human adrenal cortex. Cortisol helps ensure adequate fuel supplies for the cells when the body is under stress. Its principal action is to stimulate gluconeogenesis in liver cells. Cortisol helps provide nutrients for glucose production by stimulating transport of amino acids into liver cells (see Fig. 47–20). It also promotes mobilization of fats so that fatty acids are available for conversion to glucose. These actions ensure that glucose and glycogen are produced in the liver, and the concentration of glucose in the blood rises. Thus, the adrenal cortex provides an important backup system for the adrenal medulla, ensuring glucose supplies when the body is under stress and in need of extra energy (see *Focus On: Coping with Stress*).

Almost any type of stress stimulates the hypothalamus to secrete **corticotropin-releasing factor (CRF).** This hormone

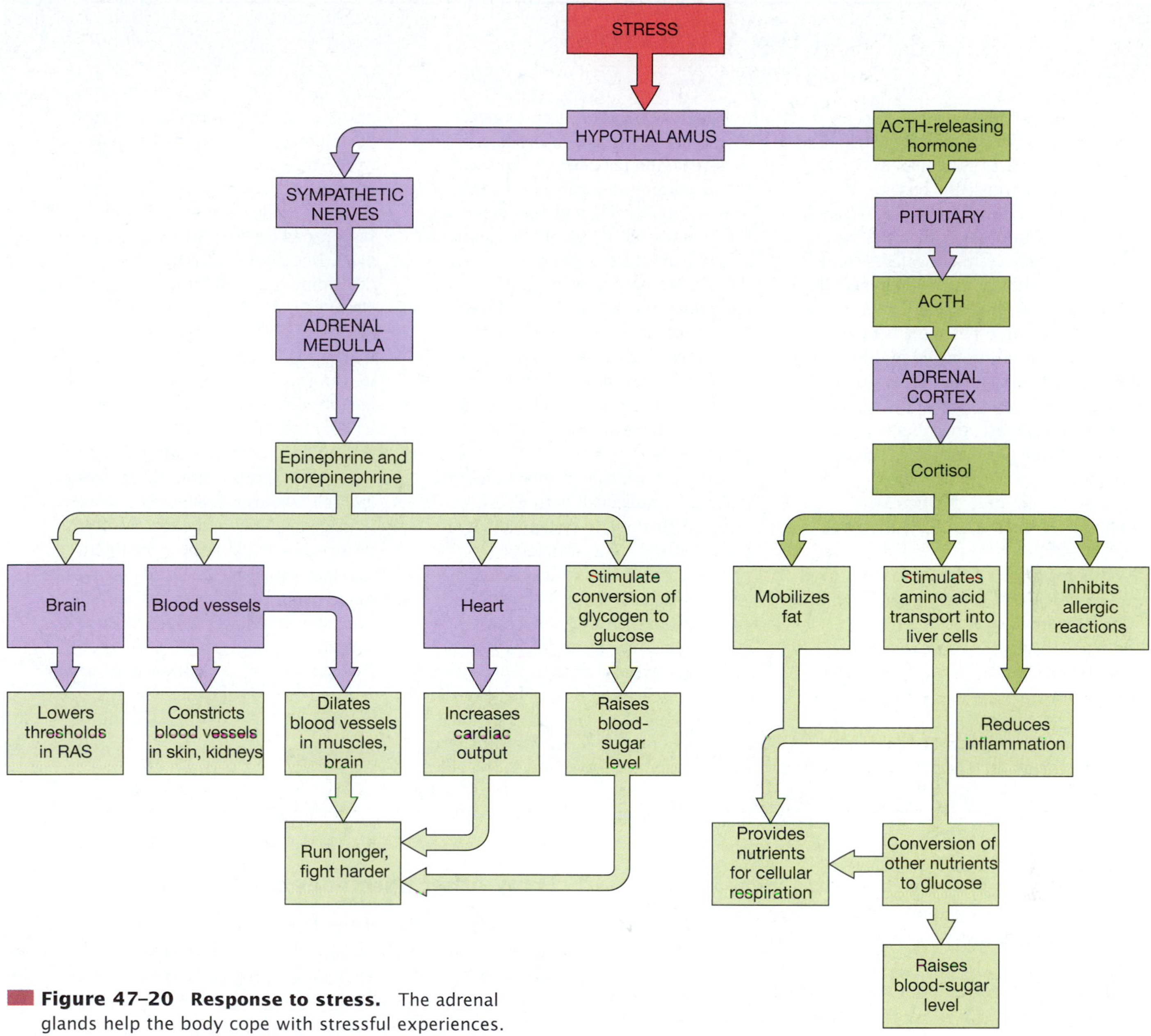

Figure 47–20 Response to stress. The adrenal glands help the body cope with stressful experiences.

stimulates the anterior pituitary to secrete **adrenocorticotropic hormone (ACTH).** ACTH regulates both glucocorticoid and aldosterone secretion. ACTH is so potent that it can result in up to a 20-fold increase in cortisol secretion within minutes. When the body is not under stress, high levels of cortisol in the blood inhibit both CRF secretion by the hypothalamus and ACTH secretion by the pituitary.

Destruction of the adrenal cortex and the resulting decrease in aldosterone and cortisol secretion cause **Addison's disease.** Reduction in cortisol prevents the body from regulating the concentration of glucose in the blood because it cannot synthesize enough glucose. The cortisol-deficient patient also loses the ability to cope with stress. If cortisol levels are significantly depressed, even the stress of mild infections can cause death.

Glucocorticoids are used clinically to reduce inflammation in allergic reactions, infections, arthritis, and certain types of cancer. These hormones inhibit production of prostaglandins (which are mediators of inflammation) by inducing the expression of an inhibitor of phospholipase A, an enzyme required for prostaglandin synthesis. Glucocorticoids also reduce inflammation by decreasing the permeability of capillary membranes, thereby reducing swelling. In addition, they reduce the effects of histamine and so are used to treat allergic symptoms.

When used in large amounts over long periods, glucocorticoids can cause serious side effects. Although they help stabilize lysosome membranes so that they do not destroy tissues with their potent enzymes, the ability of lysosomes to degrade foreign molecules is also decreased. Glucocorticoids decrease

FOCUS ON Coping with Stress

The breakup of a relationship, an infection, or the anxiety of taking a test when you are not prepared all are stressors that arouse the body to action. During stress, the brain and the adrenal glands work together to help the body cope effectively. Neural messages from the brain stimulate the adrenal medulla to release epinephrine and norepinephrine, which prepare the body for fight or flight. As discussed in the main text, the hypothalamus signals the anterior pituitary to signal the adrenal cortex. In response, the adrenal cortex secretes more glucocorticoids such as cortisol. These hormones adjust metabolism to meet the increased demands of the stressful situation (see Fig. 47–20).

Some forms of stress are short-lived. We react, quickly resolving the situation. Chronic stress, such as that from an unhappy marriage or psychologically toxic work situation, is harmful because of the side effects of long-term, elevated levels of glucocorticoids.

Chronic stress has been shown to damage the brain. When rodents and monkeys are subjected to prolonged stress, elevation of glucocorticoids leads to degeneration of neurons, especially in the hippocampus (a part of the brain involved in learning and remembering). Elevated concentration of glucocorticoids also impairs the capacity of neurons in the hippocampus to withstand physiological insult such as reduced blood flow or oxygen to the brain.

Traumatic stress (extreme stress caused by threat of serious injury or death, or witnessing such threats to someone else) can result in Post-traumatic Stress Disorder (PTSD), a condition in which hormone levels are altered. In individuals with PTSD, sympathetic nervous system activity is increased and epinephrine and norepinephrine levels are elevated.

Individuals approach stressful situations in their lives differently. A stressor that may result in chronic, damaging levels of glucocorticoids in one person may be viewed as a challenge by another. One source of such individual differences may be variations in maternal care. Researchers have studied the effects of variations in maternal care in rats during the first 10 days of life. Rats that were licked and groomed more by their mothers showed reduced ACTH and adrenocorticosteroid secretion in response to acute stress when they were older. Does early experience also determine how well humans cope with stress later in life?

We have no control over the maternal care we receive early in life, but we can learn to reduce our psychological and physiological responses to stressors by learning stress management techniques, such as meditation, visual imagery, progressive muscle relaxation, and self-hypnosis. Practicing a stress management technique can result in decreased activity of the sympathetic nervous system and reduced response to norepinephrine. Relaxation training has been shown to lower blood pressure in hypertensive patients, decrease the frequency of migraine headaches, and reduce chronic pain.

interleukin-1 production, thereby blocking cell-mediated immunity and reducing the patient's ability to fight infections. Other side effects include ulcers, hypertension, diabetes mellitus, and atherosclerosis.

Abnormally large amounts of glucocorticoids, whether due to disease or drugs, can result in **Cushing's disease.** In this condition, fat is mobilized from the lower part of the body and deposited about the trunk. Edema gives the patient's face a full-moon appearance. Blood-glucose level rises to as much as 50% above normal, causing *adrenal diabetes*. If this condition persists for several months, the beta cells in the pancreas may "burn out" from secreting excessive amounts of insulin, leading to permanent diabetes mellitus. Reduction in protein synthesis causes weakness and decreases immune responses, so the untreated patient may die of infection.

Many other hormones are known

Many other tissues of the body secrete hormones. The **pineal gland,** located in the brain, produces a hormone called **melatonin,** which influences biological rhythms and the onset of sexual maturity. In humans, melatonin facilitates the onset of sleep. Exposure to light suppresses secretion of melatonin.

Several hormones secreted by the digestive tract regulate digestive processes. The **thymus gland** produces **thymosin,** a hormone that plays a role in immune responses. **Atrial natriuretic factor (ANF),** secreted by the atrium of the heart, promotes sodium excretion and lowers blood pressure (see Chapter 46).

Recently, researchers have demonstrated that adipose tissue also secretes hormones. These include leptin (see Chapter 45) and resistin. In Chapter 48 we discuss the principal reproductive hormones.

SUMMARY WITH KEY TERMS

I. The **endocrine system** consists of endocrine glands, cells, and tissues that secrete **hormones.** This system helps regulate many aspects of metabolism, fluid balance, growth, and reproduction. **Endocrinology** is the study of endocrine activity.

II. Hormones are an important type of chemical signal by which cells communicate with one another.
   A. Classically, hormones were defined as chemical messengers secreted by discrete **endocrine glands,** glands that lack ducts. Hormones are secreted into the interstitial fluid and typically are transported by the blood. They bind with receptors on or in specific **target cells.**
   B. **Neuroendocrine cells** are neurons that secrete **neurohormones,** which are typically transported down axons and then secreted and transported by the blood.
   C. Some **local regulators,** signaling molecules that diffuse through the interstitial fluid and act on nearby cells, are now considered hormones. These include **growth factors,** peptides that stimulate cell division and development, and **prostaglandins,** a group of local hormones that help regulate many metabolic processes
   D. In **autocrine regulation,** a hormone (or other signal molecule) is secreted into the interstitial fluid and then acts on the very cell that produced it. In **paracrine regulation,** a hormone (or other signal molecule) diffuses through interstitial fluid and acts on nearby target cells.

III. Hormones can be assigned to four chemical groups: fatty acid derivatives, steroids, amino acid derivatives, or peptides and proteins.
   A. Prostaglandins and the juvenile hormone of insects are **fatty acid derivatives.**
   B. Hormones secreted by the adrenal cortex, ovary, and testis, as well as the molting hormone of insects are **steroid hormones.**
   C. Thyroid hormones and epinephrine are **amino acid derivatives.**
   D. Antidiuretic hormone (ADH) and glucagon are examples of **peptide hormones.** Insulin is a small protein.

IV. Hormone secretion is typically regulated by **negative feedback mechanisms,** in which a hormone is released in response to some change in a steady state and triggers a response that counteracts the changed condition, thereby restoring homeostasis.

V. Receptors on or in target cells determine the specificity of hormones.
   A. In response to high hormone concentration, **receptor down-regulation** decreases the number of receptors, thereby suppressing the sensitivity of target cells to the hormone.
   B. In response to low hormone concentration, **receptor up-regulation** increases the number of receptors, amplifying the hormone's effect.

VI. Steroid hormones and thyroid hormones are hydrophobic molecules that pass through the plasma membrane and combine with receptors within the target cell; the hormone-receptor complex may activate or repress transcription of messenger RNA coding for specific proteins.

VII. Most hormones are hydrophilic and do not enter target cells. They combine with receptors on the plasma membrane of target cells. Many hormones act via **signal transduction.** An extracellular hormone signal is transduced into an intracellular signal by the receptor.
   A. Most peptide hormones are first messengers that carry out their actions by way of **second messengers,** such as **cyclic AMP (cAMP)** or calcium ions.
      1. The hormone-receptor complex activates a coupling molecule, typically a **G protein.** The G protein either stimulates or inhibits an enzyme that affects the second messenger. For example G proteins stimulate or inhibit **adenylyl cyclase,** the enzyme that catalyzes the conversion of ATP to cAMP.
      2. Second messengers typically act by stimulating the activity of **protein kinases.**
   B. Protein kinases phosphorylate specific proteins that affect the activity of the cell.
   C. Certain hormone-receptor complexes increase the concentration of calcium ions in the cell. Calcium ions bind with **calmodulin,** which activates certain enzymes.
   D. **Inositol triphosphate (IP_3)** and **diacylglycerol (DAG)** are second messengers that increase calcium concentration and activate enzymes.
   E. **Signal amplification** occurs as each hormone-receptor complex stimulates the production of many second messenger molecules. Second messengers, in turn, activate protein kinase molecules that can activate many protein molecules.

VIII. Many invertebrate hormones are secreted by neurons rather than by endocrine glands. These neurohormones help regulate regeneration, molting, metamorphosis, reproduction, metabolism, and behavior.
   A. Pigment distribution in crustaceans is regulated by neurohormones.
   B. Hormones control development in insects.
      1. When stimulated by some environmental factor, neuroendocrine cells in the insect brain secrete **brain hormone (BH).**
      2. BH stimulates the prothoracic glands to produce **molting hormone (ecdysone),** which stimulates growth and molting.
      3. In the immature insect, the **corpora allata** secrete **juvenile hormone,** which suppresses metamorphosis at each larval molt. The amount of juvenile hormone decreases with successive molts.

IX. In vertebrates, hormones regulate growth, reproduction, salt and fluid balance, many aspects of metabolism, and behavior.

X. Nervous and endocrine system regulation are integrated in the **hypothalamus,** which regulates the activity of the **pituitary gland.**
   A. Endocrine disorders can result from **hyposecretion** (abnormally reduced output) or **hypersecretion** (abnormally increased output) of hormones.
   B. The neurohormones **oxytocin** and **antidiuretic hormone (ADH)** are produced by the hypothalamus and released by the **posterior lobe** of the pituitary.
      1. **Oxytocin** stimulates contraction of the uterus and stimulates ejection of milk by the mammary glands.
      2. **ADH** (vasopressin) stimulates reabsorption of water by the kidney tubules.
   C. Secretion of hormones by the **anterior lobe** of the pituitary gland is regulated by **releasing hormones** and **inhibiting hormones** secreted by the hypothalamus. The anterior lobe of the pituitary gland secretes **growth hormone, prolactin,** and several **tropic hormones** that stimulate other endocrine glands.
      1. Growth hormone (GH) is an **anabolic hormone** that stimulates body growth by promoting protein synthesis. GH stimulates the liver to produce **somatomedins,** also called **insulin-like growth factors,** which promote skeletal growth and general tissue growth. Malfunctions in GH secretion can lead to **pituitary dwarfism, gigantism,** or **acromegaly.**
      2. Prolactin stimulates the mammary glands to produce milk.
   D. The **thyroid gland** secretes **thyroid hormones: thyroxine,** or **T_4,** and **triiodothyronine,** or **T_3.** Thyroid hormones stimulate the rate of metabolism.
      1. Regulation of thyroid secretion depends mainly on a negative feedback system between the anterior pituitary gland and the thyroid gland.
      2. Hyposecretion of thyroxine during childhood may lead to **cretinism;** during adulthood it may result in **myxedema. Goiter,** an abnormal enlargement of the thyroid gland, is associated with both hyposecretion and hypersecretion. The most common cause of hyperthyroidism is **Grave's disease,** an autoimmune disease.
   E. The **parathyroid glands** secrete **parathyroid hormone,** which regulates the calcium level in the blood.
      1. Parathyroid hormone increases calcium concentration by stimulating calcium release from bones, increasing calcium reabsorption by kidney tubules, and increasing calcium reabsorption from the intestine.
      2. **Calcitonin,** secreted by the thyroid gland, acts antagonistically to parathyroid hormone.

F. The **islets of Langerhans** in the pancreas secrete **insulin** and **glucagon.**
   1. Insulin stimulates cells to take up glucose from the blood and so lowers blood-glucose concentration.
   2. Glucagon raises blood-glucose concentration by stimulating conversion of glycogen to glucose (glycogenolysis) and by stimulating production of glucose from other nutrients (gluconeogenesis).
   3. Insulin and glucagon secretion are regulated directly by glucose concentration.
   4. In **diabetes mellitus,** either insulin deficiency or **insulin resistance** results in decreased use of glucose, increased fat mobilization, increased protein use, and electrolyte imbalance.

G. The **adrenal glands** secrete hormones that help the body cope with stress.
   1. The **adrenal medulla** secretes **epinephrine** and **norepinephrine.**
   2. The **adrenal cortex** secretes sex hormones; **mineralocorticoids,** such as **aldosterone;** and **glucocorticoids,** such as **cortisol.** Aldosterone increases the rate of sodium reabsorption and potassium excretion by the kidneys. Cortisol promotes gluconeogenesis.
   3. The hormones of the adrenal medulla help the body respond to stress by increasing the heart rate, metabolic rate, and the strength of muscle contraction. These hormones also reroute blood to those organs needed for fight or flight. The adrenal cortex ensures adequate fuel supplies for the rapidly metabolizing cells.

POST-TEST

1. Which of the following is *not* true of endocrine glands? (a) they secrete hormones (b) they have ducts (c) their product is typically transported by the blood (d) they are typically regulated by negative feedback (e) when removed from an experimental animal, the animal exhibits symptoms of deficiency
2. A cell secretes a product that diffuses through the interstitial fluid and acts on nearby cells. This is an example of (a) neuroendocrine secretion (b) autocrine regulation (c) paracrine regulation (d) classical endocrine control (e) peptide hormone function
3. Paracrine regulators that are derived from fatty acids and are found in many different organs are (a) prostaglandins (b) thyroid hormones (c) growth factors (d) anabolic steroids (e) G proteins
4. Which of the following is/are true of steroid hormones? (a) hydrophilic (b) secreted by the posterior pituitary (c) typically work through G proteins and cyclic AMP (d) typically bind with receptor in nucleus and affect transcription (e) two of the preceding answers are correct
5. Which of the following is *not* a correct pair? (a) neurohormone; brain hormone (b) calcium; calmodulin (c) posterior lobe of pituitary; releasing hormone (d) anterior lobe of pituitary; growth hormone (e) hyposecretion; cretinism
6. Which of the following is/are second messengers? (a) hormone-receptor complex (b) calcium ions (c) inositol triphosphate (IP_3) (d) answers a, b, and c are correct (e) only answers b and c are correct
7. Growth hormone (a) is regulated mainly by calcium level (b) stimulates the liver to produce insulin-like growth factors (c) is a catabolic hormone (d) stimulates metabolic rate (e) signals the hypothalamus to produce a releasing hormone
8. Arrange the following events into an appropriate sequence. (1) high thyroid hormone concentration (2) anterior pituitary inhibited (3) homeostasis (4) lower level of thyroid-stimulating hormone (5) thyroid gland secretes less thyroid hormone (a) 1, 2, 4, 5, 3 (b) 5, 4, 3, 2, 1 (c) 1, 2, 5, 4, 3 (d) 4, 5, 2, 3, 1 (e) 1, 4, 2, 5, 3
9. Arrange the following events into an appropriate sequence. (1) blood-glucose concentration increases (2) alpha cells in islets stimulated (3) homeostasis (4) low blood-glucose concentration (5) glucagon secretion increases (a) 1, 2, 3, 5, 4 (b) 5, 4, 2, 1, 3 (c) 1, 2, 5, 4, 3 (d) 4, 2, 5, 1, 3 (e) 4, 5, 1, 2, 3
10. Parathyroid hormone (a) increases glucose level in blood (b) helps body cope with stress (c) increases permeability of kidney tubules to water (d) promotes uptake of amino acids (e) increases calcium concentration in blood
11. An action of cortisol is (a) decreases glucose level in blood (b) helps body cope with stress (c) increases permeability of kidney tubules to water (d) promotes uptake of amino acids (e) increases calcium concentration in blood
12. Which of the following is *not* a correct pair? (a) thyroid gland; calcitonin (b) islets of Langerhans; glucagon (c) posterior lobe of pituitary; oxytocin (d) anterior lobe of pituitary; cortisol (e) adrenal medulla; epinephrine
13. Which of the following occurs in diabetes mellitus? (a) decreased use of glucose (b) decreased fat metabolism (c) decreased protein use (d) increased concentration of thyroid-releasing hormone (e) two of the preceding answers are correct
14. Aldosterone (a) is released by posterior pituitary (b) is an androgen (c) secretion is stimulated by an increase in thyroid-stimulating hormone (d) is an enzyme that converts epinephrine to norepinephrine (e) increases sodium reabsorption
15. Insulin resistance is associated with (a) low insulin secretion by the islets of Langerhans (b) type 1 diabetes (c) impaired function of receptors on target cells (d) hypoglycemia (e) two of the preceding answers are correct

REVIEW QUESTIONS

1. How has the definition of the term *hormone* changed in recent years? Give examples.
2. What are the major chemical groups of hormones? What are some of the important actions of hormones in invertebrates? In vertebrates?
3. What are the roles of receptors and second messengers in hormone action? Describe the mechanism of action of a hormone that uses (a) cyclic AMP as a second messenger (b) calcium ions as second messengers.
4. Why is the hypothalamus considered the link between the nervous and endocrine systems? What is the role of the anterior lobe of the pituitary? The posterior lobe?
5. Describe the actions of (a) prolactin (b) oxytocin (c) thyroid-stimulating hormone.
6. Draw a diagram illustrating the regulation of (a) thyroid hormone secretion and (b) parathyroid hormone secretion.
7. Explain the hormonal basis for (a) acromegaly (b) pituitary dwarfism (c) cretinism (d) hypoglycemia (e) Cushing's disease.
8. Explain the antagonistic actions of insulin and glucagon in regulating blood-glucose level.
9. What is insulin resistance? Describe several physiological disturbances caused by diabetes mellitus.
10. What are the actions of epinephrine and norepinephrine? How is the adrenal medulla regulated?
11. What types of hormones are released by the adrenal cortex, and what are the actions of each type?
12. Explain how the adrenal glands help the body respond to stress.
13. An injection of too much insulin may cause a diabetic to go into insulin shock, in which the patient may appear drunk or may become unconscious, suffer convulsions, and even die. From what you know about the actions of insulin, explain the physiological causes of insulin shock.
14. Label the diagram. Use Figure 47–12 to check your answers.

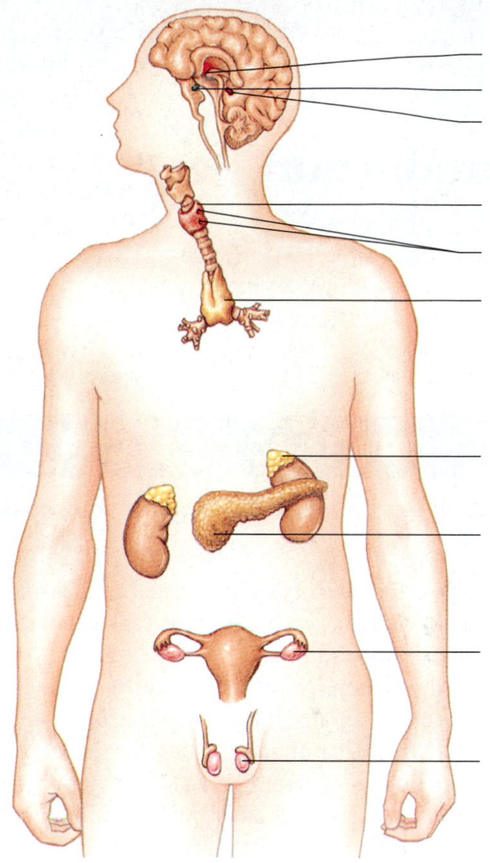

YOU MAKE THE CONNECTION

1. Human males have about the same amount of oxytocin circulating in their blood as do nonpregnant females who are not lactating, but its function in males is unknown. Hypothesize its function, and design an ethical experiment to test your hypothesis. (Hint: Based on its effects on animal behavior, oxytocin has been referred to as the "nurturing hormone.")
2. How do receptors impart specificity within the endocrine system? What might be some advantages of having complex mechanisms for hormone action (e.g., second messengers)?
3. Why do you think it is important to maintain a constant blood-glucose level? Several hormones discussed in this chapter affect carbohydrate metabolism. Why is it important to have more than one? How do they interact?

RECOMMENDED READINGS

Chen, S, A. Casaretto, and F.N. Ziyadeh. "Pathogenesis of Diabetic Nephropathy." *Science and Medicine,* Vol. 7, No. 1, Jan./Feb. 2000. Diabetes mellitus is an important cause of renal failure.

Lowe, W.L., Jr. "Insulin-like Growth Factors." *Scientific American Science & Medicine,* Vol. 3, No. 2, Mar./Apr. 1996. A discussion of the process of establishing the role of insulin-like growth factors and a description of their actions.

Steppan, C.M., S.T. Bailey, S. Bhat, E.J. Brown, R.R. Banerjee, C.M. Wright, H.R. Patel, R.S. Ahima, and M.A. Lazar. "The Hormone Resistin Links Obesity to Diabetes." *Nature,* Vol. 409, 18 Jan. 2001. Investigators report the discovery of resistin, a signaling molecule secreted by fat cells.

Youngren, J.F., and I.D. Goldfine. "The Molecular Basis of Insulin Resistance." *Science & Medicine,* Vol. 4, No. 3, May/Jun. 1997. Insulin resistance is a complex phenotype that may result from defects in insulin receptor signaling.

Zorpette, G. "Sports Medicine, Drug Testing: All Doped Up-and Going for the Gold." *Scientific American,* Vol. 282, No. 5, May 2000. Tests are being developed to detect hormone abuse by athletes.

- Visit our Web site at **http://www.info.brookscole.com/solomonbergmartin** for links to chapter-related resources on the World Wide Web. Additional on-line materials relating to this chapter can also be found on our Web site.

See chapter activity on BioActive Learner CD for additional help in mastering the chapter's material. Icon location in the chapter's margins shows which topics have tutorials or simulations in the CD.

48

Reproduction

Mating nudibranchs (*Chromodoris* sp.). Photographed in Indonesia. *(Bruce Watkins/ Animals Animals)*

LEARNING OBJECTIVES

After you have studied this chapter you should be able to

1. Compare the advantages of asexual and sexual reproduction; describe each mode of reproduction, giving specific examples.
2. Label the structures of the human male reproductive system on a diagram and describe the functions of each.
3. Trace the passage of sperm cells through the human male reproductive system from their origin in the seminiferous tubules to their expulsion from the body in the semen.
4. Describe the endocrine regulation of reproduction in the human male.
5. Label the structures of the human female reproductive system on a diagram and describe the functions of each.
6. Trace the development of a human ovum (egg) and its passage through the female reproductive system until it is fertilized.
7. Describe the endocrine regulation of reproduction in the human female, and identify the important events of the menstrual cycle, such as ovulation and menstruation.
8. Summarize the process of human fertilization.
9. Identify hormones that regulate pregnancy and parturition, and describe their actions.
10. Compare the modes of action, effectiveness, advantages, and disadvantages of the methods of birth control discussed; include sterilization and emergency contraception.
11. Identify common sexually transmitted diseases and describe their symptoms, effects, and treatments.

The ability to reproduce and perpetuate its species is a basic characteristic of living things. The survival of each species requires that its members produce new individuals to replace those that die. The instructions for accomplishing this process are encoded in each organism's nucleic acids.

Many invertebrates, including sponges, cnidarians, and some flatworms and annelids, can reproduce asexually. In **asexual reproduction** a single parent gives rise to offspring that are genetically identical (except for mutation) to the parent. Asexual reproduction is an adaptation of sessile animals that cannot move about to search for mates. For animals that do move about, this method of reproduction is advantageous when the population density is low and mates are not readily available. Some animals reproduce asexually under some conditions and sexually at other times.

Sexual reproduction in animals involves the production and fusion of two types of **gametes**—sperm and eggs. Typically, as exemplified by the mating nudibranchs in the photograph, two different individuals are required. A male parent contributes **sperm** and a female parent contributes an egg, or **ovum** (pl., *ova*). The sperm provides genes coding for some of the male parent's traits, and the egg contributes genes coding for some of the female parent's traits. When sperm and egg unite, a fertilized egg, or **zygote,** forms. The zygote develops into a new organism, similar to both parents but not identical to either. Sexual reproduction typically involves remarkably complex structural, functional, and behavioral processes. In vertebrates, these processes are regulated by hormones secreted by the hypothalamus, pituitary gland, and gonads.

Asexual reproduction is the fastest and most efficient way to reproduce. Compared to asexual reproduction, sexual reproduction is expensive in terms of energy expenditure. It is also less efficient because two cells, rather than just one, are required to make a new organism. Why then do most animals reproduce sexually? This chapter summarizes some major features of animal reproduction, including a discussion of the advantages of sexual reproduction, and then focuses on human reproduction.

ASEXUAL REPRODUCTION IS COMMON AMONG SOME ANIMAL GROUPS

In asexual reproduction, a single parent may split, bud, or fragment to give rise to two or more offspring. Except for mutations, the offspring have hereditary traits identical to those of the parent. Sponges and cnidarians are among the animals that can reproduce by **budding.** A small part of the parent's body separates from the rest and develops into a new individual (Fig. 48–1). Sometimes the buds remain attached and become more or less independent members of a colony.

Oyster farmers learned long ago that when they tried to kill sea stars by chopping them in half and throwing the pieces back into the sea, the number of sea stars preying on the oyster bed doubled! In some flatworms, nemerteans, and annelids this ability to regenerate is part of a method of reproduction known as **fragmentation.** The body of the parent breaks into several pieces; each piece regenerates the missing parts and develops into a whole animal.

Parthenogenesis ("virgin development") is a form of asexual reproduction in which an unfertilized egg develops into an adult animal. The adult is typically haploid. Parthenogenesis is common among insects (especially honeybees and wasps) and other arthropods; it also occurs among some other invertebrate and vertebrate groups, including some species of nematodes, gastropods, fishes, amphibians, and reptiles.

Although a few species appear to reproduce solely by parthenogenesis, in most species episodes of parthenogenesis alternate with periods of sexual reproduction. Parthenogenesis may occur for several generations, followed at some point by sexual reproduction in which males develop, produce sperm, and mate with the females to fertilize their eggs. In some species, parthenogenesis is a means of rapidly producing individuals when conditions are favorable.

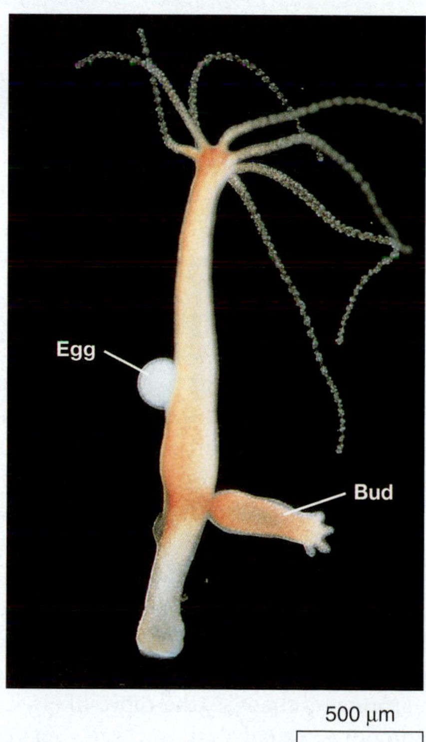

Figure 48–1 Asexual reproduction by budding. A part of *Hydra's* body grows outward and then separates and develops into a new individual. The region of the parent body that buds is not specialized exclusively for reproduction. The *Hydra* shown here is also reproducing sexually as evidenced by the egg *(left). (Richard Campbell/Biological Photo Service)*

SEXUAL REPRODUCTION IS THE MOST COMMON TYPE OF ANIMAL REPRODUCTION

Most animals reproduce sexually by fusion of sperm and egg. The egg is typically large and nonmotile, with a store of nutrients that supports the development of the embryo. The sperm is usually small and motile, adapted to propel itself by beating its long, whiplike flagellum.

Many aquatic animals practice **external fertilization** in which the gametes meet outside the body (Fig. 48–2*a;* also see Fig. 19–5). Mating partners usually release eggs and sperm into the water simultaneously. Gametes live only for a short time, and many are lost in the water; some are eaten by predators. However, so many gametes are released that sufficient numbers of sperm and egg cells do meet to perpetuate the species.

In **internal fertilization,** matters are left less to chance. The male generally delivers sperm cells directly into the body of the female. Her moist tissues provide the watery medium required for the movement of sperm, and the gametes fuse inside the body. Most terrestrial animals, sharks, and aquatic reptiles, birds, and mammals practice internal fertilization (Fig. 48–2*b*).

Hermaphroditism is a form of sexual reproduction in which a single individual produces both eggs and sperm. A few hermaphrodites, such as the tapeworm, are capable of self-fertilization. Earthworms are more typical hermaphrodites. Two animals copulate, and mutual cross-fertilization occurs with each inseminating the other. In some hermaphroditic species, self-fertilization is prevented by the development of testes and ovaries at different times.

Process of Science

Asexual reproduction is an efficient strategy allowing an animal to pass all of its genes to its offspring; yet sexual reproduction, despite the high energy cost of producing gametes and finding mates, is far more common. For years, biologists have struggled with this apparent contradiction. Many biologists explain that sexual reproduction has the biological advantage of promoting genetic variety among the members of a species. Each offspring is the product of a particular combination of genes contributed by both parents, rather than a genetic copy of a single individual. By combining inherited traits of two parents, sexual reproduction gives rise to offspring that may be better able to survive than either parent.

(a)

(b)

Figure 48–2 External and internal fertilization. **(a)** Like many aquatic animals, these spawning frogs *(Rana temporaria)*, practice external fertilization. The female lays a mass of eggs, while the male mounts her and simultaneously deposits his sperm in the water. **(b)** Internal fertilization is practiced by some fishes, aquatic reptiles, birds, and mammals and by most terrestrial animals, such as these lions *(Panthera leo)*. *(a, Carmela Leszczynski/ Animals Animals; b, Fritz Polking/Dembinsky Photo Associates)*

Although they generally agree that sexual reproduction must have some selective advantage, biologists do not agree on the details. The following two major possibilities are being explored:

1. Does sexual reproduction permit beneficial mutations from each parent to come together in offspring that can reproduce and spread these mutations through the population? For example, certain beneficial mutations permit animals to protect themselves from predators and to resist parasites. Sexual reproduction would enable such mutations to spread through the population.
2. Does sexual reproduction remove harmful mutations? Mutations occur constantly, and most mutations are harmful. When animals reproduce asexually, all the offspring inherit all the harmful mutations. As mutations accumulate in a population, individuals carry a bigger and bigger load of harmful genes. In contrast, when animals with different mutations mate, offspring inherit varying numbers and combinations of mutations. Offspring that inherit too many harmful mutations are selected against. They may not live to reproduce, so their harmful mutations are removed from the population.

Wayne Getz, an applied mathematician at the University of California, Berkeley, developed a mathematical model that predicts whether asexual or sexual reproduction will exist within a population. The model assumes that each clone produced by asexual individuals will use resources in the environment in the same way. In contrast, sexual reproduction will result in different gene combinations so that offspring can exploit environmental resources in various ways.

According to the Getz model, if the environment does not change, the genes of animals that reproduce asexually would eventually dominate the population: The clones produced by asexual reproduction would be most successful. However, the more the environment changes, the more likely it is that animals that reproduce sexually will succeed. Sexual reproduction is particularly adaptive in an unstable, changing environment.

Getz suggested that under conditions of moderate change, the two reproductive strategies can coexist, with the asexual reproductive capacity of clones balancing out the survival capacity of the animals that reproduce sexually. This situation occurs, for example, within certain genera of soil mites.

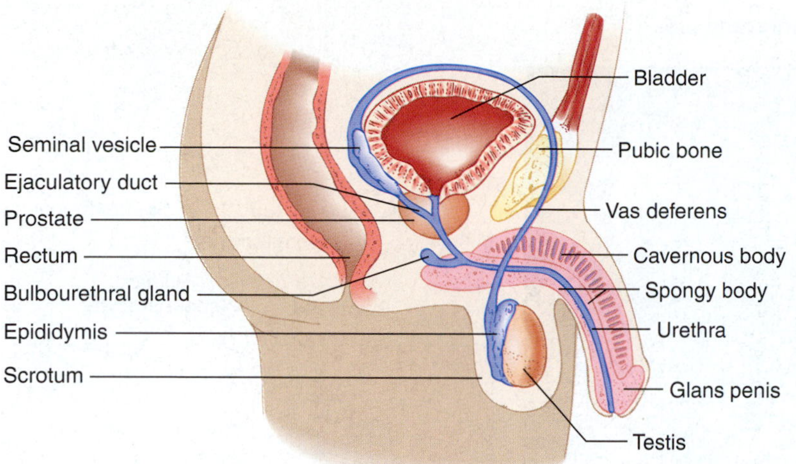

Figure 48–3 Male reproductive system. The scrotum, penis, and pelvic region of the human male are shown in sagittal section to illustrate their internal structure.

Some biologists think the Getz model is too narrow. A broader model might consider the benefit of sexual reproduction in both passing on beneficial mutations *and* ridding the genome of harmful mutations. Competing hypotheses are an expected part of the scientific process because scientific discovery is rarely a straightforward, linear sequence of question-answer, question-answer. In fact, as scientific knowledge expands, many creative hypotheses are discarded as dead ends.

HUMAN REPRODUCTION: THE MALE PROVIDES SPERM

The human male, like other male mammals, has the reproductive role of producing sperm cells and delivering them into the female reproductive tract. The sperm that combines with an egg contributes its genes and determines the sex of the offspring. The male reproductive system is illustrated in Figure 48–3.

The testes produce gametes and hormones

In humans and other vertebrates, **spermatogenesis,** the process of sperm cell production, occurs in the paired male gonads, or **testes** (sing., *testis*). Spermatogenesis takes place within a vast tangle of hollow tubules, the **seminiferous tubules,** within each testis (Fig. 48–4). Spermatogenesis begins with undifferentiated cells, the **spermatogonia** in the walls of the tubules (Fig. 48–5).

The spermatogonia, which are diploid cells, divide by mitosis, producing more spermatogonia. Some enlarge and become **primary spermatocytes,** which undergo **meiosis,** producing haploid gametes. (You may want to review the discussion of meiosis in Chapter 9.) In many animals, gamete production occurs only in the spring or fall, but humans have no special breeding season. In the human adult male, spermatogenesis proceeds continuously, and millions of sperm are produced each day.

Each primary spermatocyte undergoes a first meiotic division, producing **secondary spermatocytes** (Fig. 48–6). In the second meiotic division, each of the two secondary spermatocytes gives rise to two **spermatids.** Four spermatids are produced from the original primary spermatocyte. Each haploid spermatid differentiates into a mature sperm. The sequence is as follows:

Spermatogonium (diploid) ⟶ primary spermatocyte (diploid) ⟶ two secondary spermatocytes (haploid) ⟶ four spermatids (haploid) ⟶ four mature sperm (haploid)

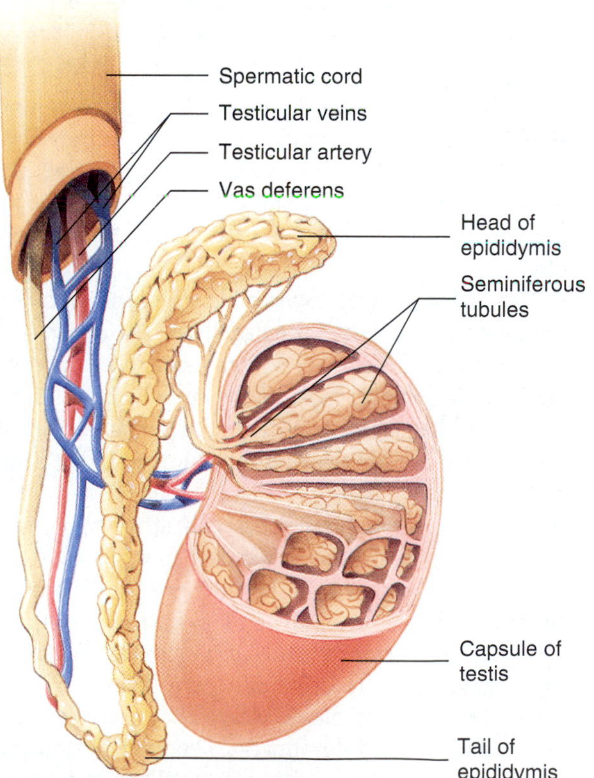

Figure 48–4 Structure of testis, epididymis, and spermatic cord. The organs are shown partly dissected and exposed. The testis is shown in sagittal section to illustrate the arrangement of the seminiferous tubules.

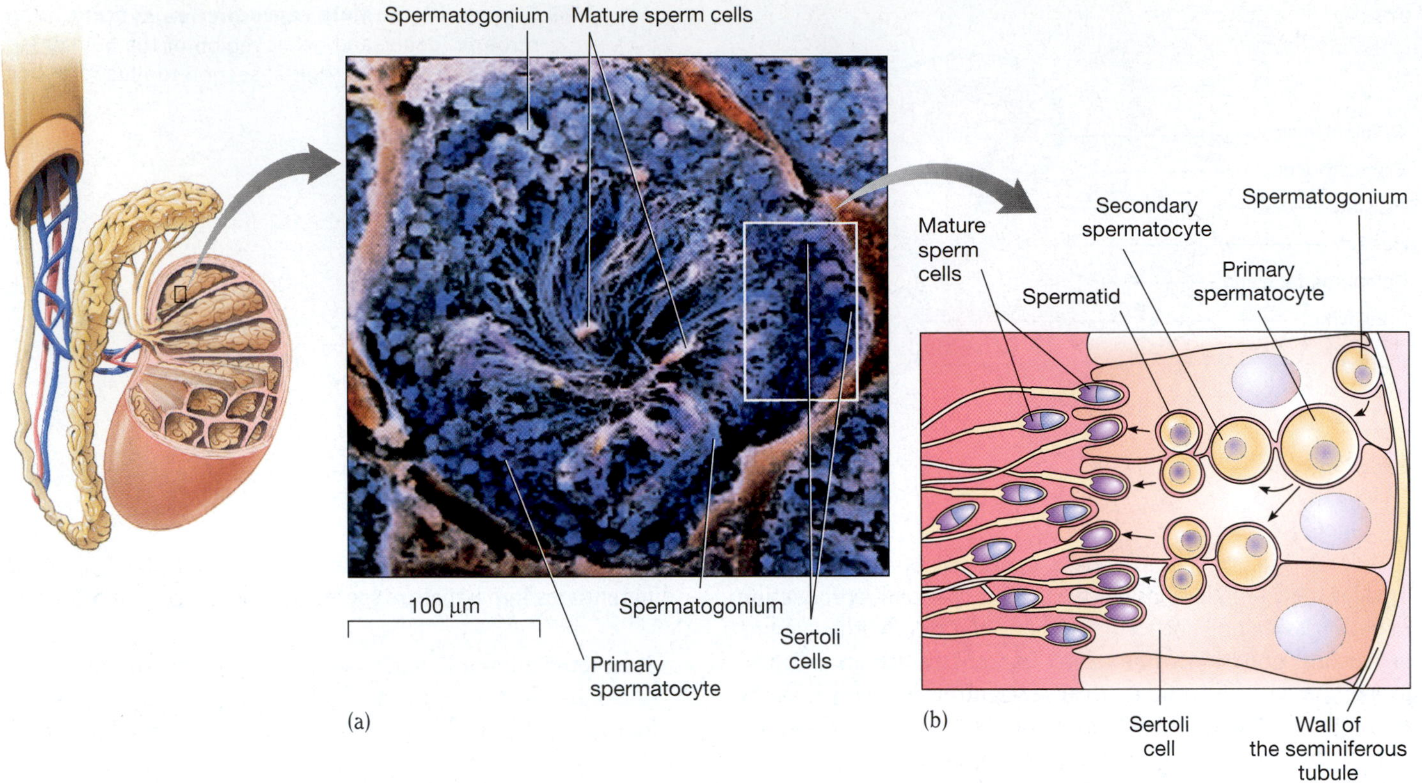

Figure 48–5 Spermatogenesis in the seminiferous tubules. **(a)** Color-enhanced SEM of a transverse section through a seminiferous tubule. **(b)** Sperm cells can be seen in various stages of development between the large, nutritive Sertoli cells. *(a, Custom Medical Stock Photo)*

Each mature sperm consists of a head, a midpiece, and a flagellum (Fig. 48–7). The head consists almost entirely of the nucleus. Part of the nucleus is covered by the **acrosome,** a vesicle that differentiates from the Golgi complex. The acrosome contains proteins, including enzymes that help the sperm penetrate the egg.

Mitochondria, located in the midpiece of the sperm, provide the energy for movement of the flagellum. The sperm flagellum has the typical eukaryotic 9 + 2 arrangement of microtubules. During its development, most of the sperm's cytoplasm is discarded and is phagocytized by the large nutritive **Sertoli cells** that form a ring around the fluid-filled lumen of the seminiferous tubule.

Sertoli cells secrete hormones and other signaling molecules and have several other important functions. Sertoli cells are joined to one another by tight junctions (see Chapter 5) and together form a blood-testis barrier. This barrier prevents the entrance into the tubule of harmful substances that could interfere with spermatogenesis. It also stops sperm from passing out of the tubule and into the blood, where they could stimulate an immune response. The tight junctions between Sertoli cells also form compartments that separate sperm cells in various stages of development.

Human sperm cells cannot develop at body temperature. Although the testes develop within the abdominal cavity of the male embryo, about two months before birth they descend into the **scrotum,** a skin-covered sac suspended from the groin. The scrotum serves as a cooling unit, maintaining sperm below body temperature. In rare cases, the testes do not descend. If this condition is not corrected surgically, or with hormone treatment, the seminiferous tubules eventually degenerate and the male becomes **sterile,** unable to produce offspring.

The scrotum is an outpocketing of the pelvic cavity and is connected to it by the **inguinal canals.** As they descend, the testes pull their blood vessels, nerves, and conducting tubes after them. The inguinal region is a weak place in the abdominal wall. Straining the abdominal muscles by lifting heavy objects sometimes results in tearing the inguinal tissue. A loop of intestine can then bulge into the scrotum through the tear, a condition known as an inguinal hernia.

A series of ducts store and transport sperm

Sperm cells leave the seminiferous tubules of each testis through small tubules that empty into a larger coiled tube, the **epididymis.** There sperm complete their maturation and are stored. During ejaculation, sperm pass from each epididymis into a sperm duct, the **vas deferens** (pl., *vasa deferentia*). The vas deferens extends from the scrotum through the inguinal canal and into the pelvic cavity.

Each vas deferens empties into a short **ejaculatory duct,** which passes through the prostate gland and then opens into the urethra. The single urethra, which at different times conducts urine and semen, passes through the penis to the outside of the body. Thus, the sperm pass in sequence through the following structures:

Seminiferous tubules ⟶ epididymis ⟶ vas deferens ⟶ ejaculatory duct ⟶ urethra ⟶ release from body

The accessory glands produce the fluid portion of semen

As sperm are transported through the conducting tubes, they are mixed with secretions from three types of accessory glands. Approximately 3 mL of **semen** is ejaculated during sexual climax. Semen consists of about 200 million sperm cells suspended in the secretions of these glands.

The paired **seminal vesicles** secrete a nutritive fluid rich in fructose and prostaglandins into the vasa deferentia (see Fig. 48–3). Nutrients in this secretion provide energy for the sperm after they are ejaculated. The single **prostate gland** secretes an alkaline fluid that may be important in neutralizing the acidic environment of the vagina and in increasing sperm cell motility. Prostaglandins in the semen may stimulate contractions of the female uterus, which help move sperm up the female reproductive tract. The prostate gland is a common site of cancer in men older than 50 years of age. Its cause is not known, but prostate cancer is thought to be hormone-related.

During sexual arousal, the paired **bulbourethral glands,** located on each side of the urethra, release a mucous secretion. This fluid lubricates the penis, facilitating its penetration into the vagina.

A major cause of male infertility is insufficient sperm production. When sperm counts drop below 35 million per mL of semen, fertility is impaired, and males with a sperm count lower than 20 million/mL are usually considered sterile. When a couple's attempts to produce a child are unsuccessful, a sperm count and analysis may be performed in a clinical laboratory. Sometimes semen is found to contain large numbers of abnormal sperm or, occasionally, no sperm at all.

In the United States in the 1970s an average healthy young man produced about 100 million sperm per mL of semen. Today that average has dropped to about 60 million. Although the cause of this decrease is not known, low sperm counts have been linked to a variety of environmental factors, including chronic marijuana use, alcohol abuse, and cigarette smoking. Studies show that men who smoke tobacco are more likely than nonsmokers to produce abnormal sperm. Exposure to industrial and environmental toxins such as DDT and PCBs (polychlorinated biphenyls) may also result in low sperm count and sterility. The use of anabolic steroids by athletes to accelerate muscle development can

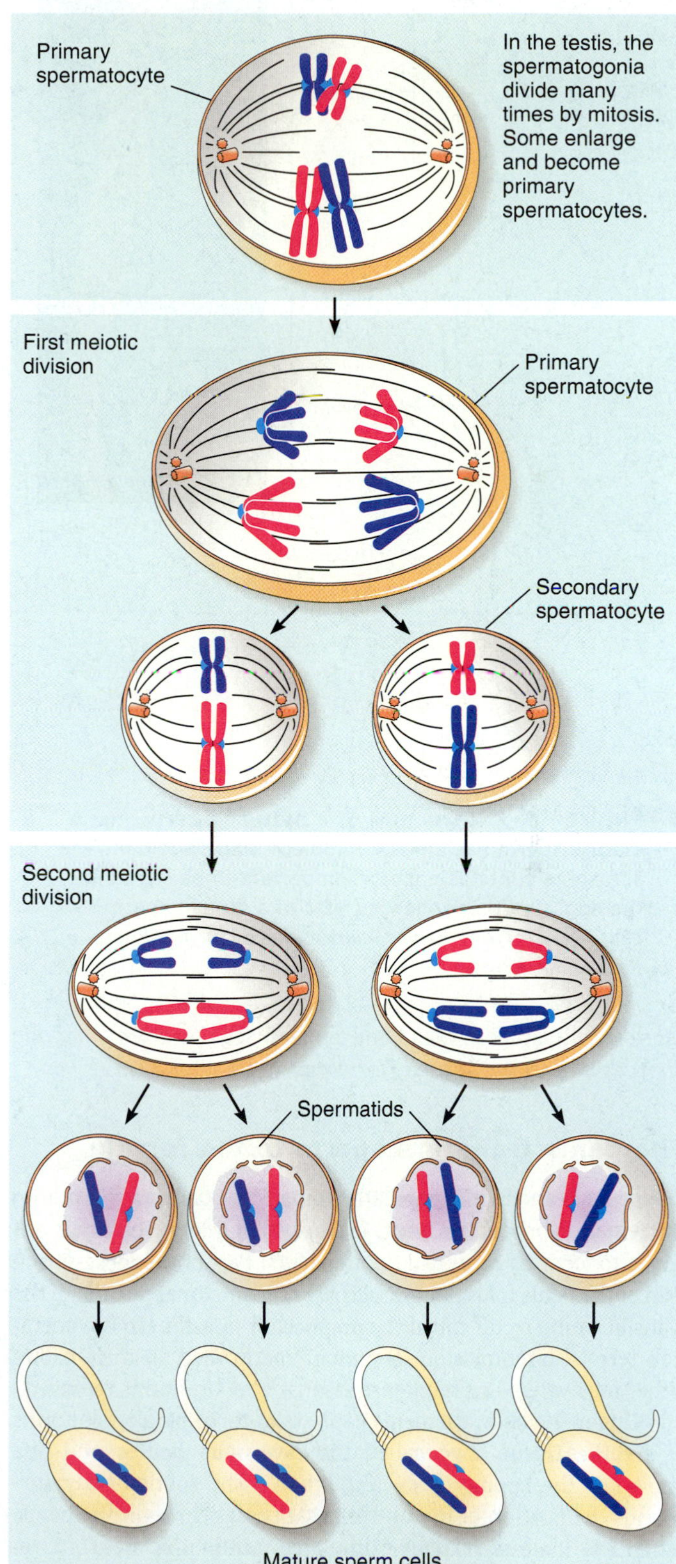

Figure 48–6 Spermatogenesis. Spermatogonia divide by mitosis. Some enlarge and become primary spermatocytes, which undergo meiosis. Each primary spermatocyte gives rise to four spermatids. The spermatids differentiate, becoming mature sperm cells. Note that in this example, four chromosomes are present in the primary spermatocyte (2*n*) and that meiosis produces the haploid (*n*) number (two) in the secondary spermatocytes, spermatids and mature sperm cells.

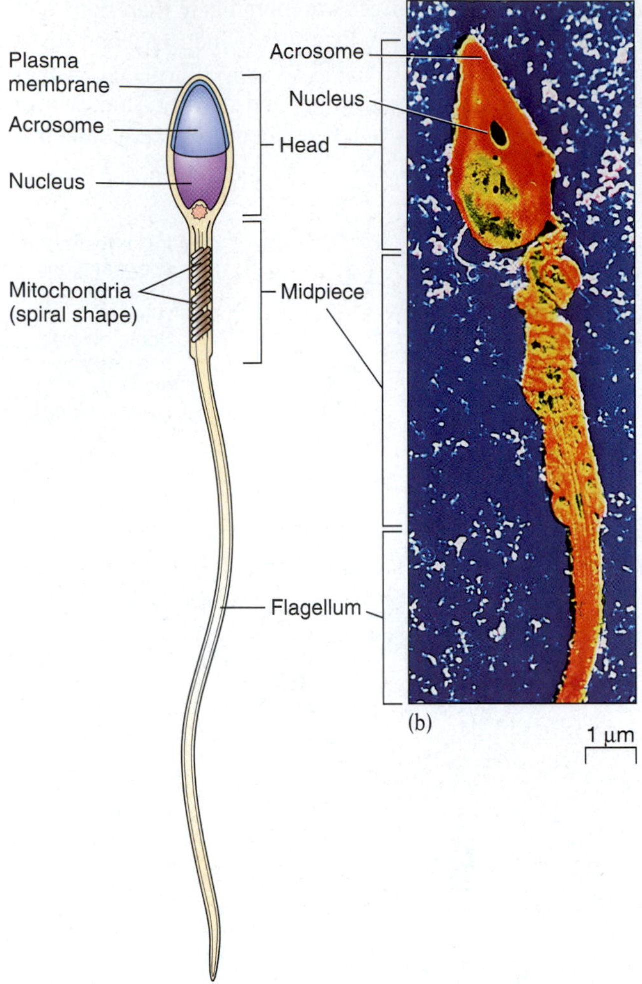

Figure 48–7 Structure of a mature sperm. **(a)** A mature sperm has a head, midpiece, and flagellum. The acrosome contains enzymes important in penetrating the egg. **(b)** Color-enhanced TEM of a human sperm cell. *(b, Dr. Tony Brain/Science Photo Library/Photo Researchers, Inc.)*

cause sterility in both males and females (see *Focus On: Anabolic Steroids and Other Abused Hormones* in Chapter 47).

The penis transfers sperm to the female

The **penis** is an erectile copulatory organ that delivers sperm into the female reproductive tract. It consists of a long shaft that enlarges to form an expanded tip, the **glans.** Part of the loose-fitting skin of the penis folds down and covers the proximal portion of the glans, forming a cuff called the **prepuce,** or foreskin. In the operation termed circumcision (commonly performed on male babies either for hygienic or religious reasons), the foreskin is removed.

Under the skin, the penis consists of three parallel columns of **erectile tissue,** two called the **cavernous bodies** and one known as the **spongy body** (Fig. 48–8). The spongy body surrounds the portion of the urethra that passes through the penis. When the male is sexually stimulated, autonomic neurons release *nitric oxide,* which causes the smooth muscles in the arterial walls to relax. Nitric oxide is part of a signal transduction pathway that involves cyclic guanosine monophosphate (cGMP).

When smooth muscles in the arterial walls relax, the arteries dilate, and blood rushes into the numerous blood vessels of the erectile tissue, causing the tissue to swell. This compresses veins that conduct blood away from the penis, slowing the outflow of blood. Thus, more blood enters the penis than can leave, causing the erectile tissue to become further engorged with blood. The penis becomes erect, that is, longer, larger in circumference, and firm. Although the human penis contains no bone, penis bones do occur in some other mammals, such as bats, rodents, and some primates.

Chronic inability to sustain an erection, termed *erectile dysfunction* (sometimes referred to as impotence), is associated with a variety of physical causes and psychological issues. Erectile dysfunction prevents effective sexual intercourse. This common disorder is now treated with sildenafil (Viagra), which blocks the action of an enzyme that breaks down cGMP.

The hypothalamus, pituitary, and testes interact to regulate male reproduction

When a boy is about ten years old the hypothalamus begins to secrete **gonadotropin-releasing hormone** (**GnRH;** Table 48–1). In an adult male, GnRH is secreted every two hours. This hormone stimulates the anterior pituitary to secrete the gonadotropic hormones **follicle-stimulating hormone (FSH)** and **luteinizing hormone (LH).**

Both FSH and LH are glycoproteins that use cyclic AMP as a second messenger (see Chapters 5 and 47). FSH stimulates Sertoli cells to secrete **androgen-binding protein (ABP)** and other signaling molecules that are necessary for spermatogenesis. LH stimulates the **interstitial cells,** lying between the seminiferous tubules in the testes, to secrete the steroid hormone **testosterone,** which is the principal **androgen,** or male sex hormone. For this reason, LH in the male is also called interstitial cell–stimulating hormone (ICSH).

FSH, LH, and testosterone all directly or indirectly stimulate testosterone secretion and spermatogenesis. A high concentration of testosterone in the testes is required for spermatogenesis. FSH stimulates Sertoli cells to produce ABP, which binds to testosterone and concentrates it in the tubules. Interestingly, developing sperm cells appear to lack receptors for sex steroids, so just how testosterone acts on them is not known.

Reproductive hormone concentrations are regulated by negative feedback mechanisms. Testosterone inhibits mainly LH secretion. It acts on the hypothalamus, decreasing its secretion of GnRH, which results in decreased FSH and LH secretion by the pituitary. Testosterone also directly inhibits the anterior lobe of the pituitary by blocking the normal actions of GnRH on LH synthesis and release.

FSH secretion is inhibited mainly by a peptide hormone, **inhibin,** which is secreted by Sertoli cells. FSH itself stimulates inhibin secretion. Inhibin is transported by the blood to the pituitary gland, where it inhibits FSH secretion (Fig. 48–9). The endocrine regulation of reproductive function is extremely com-

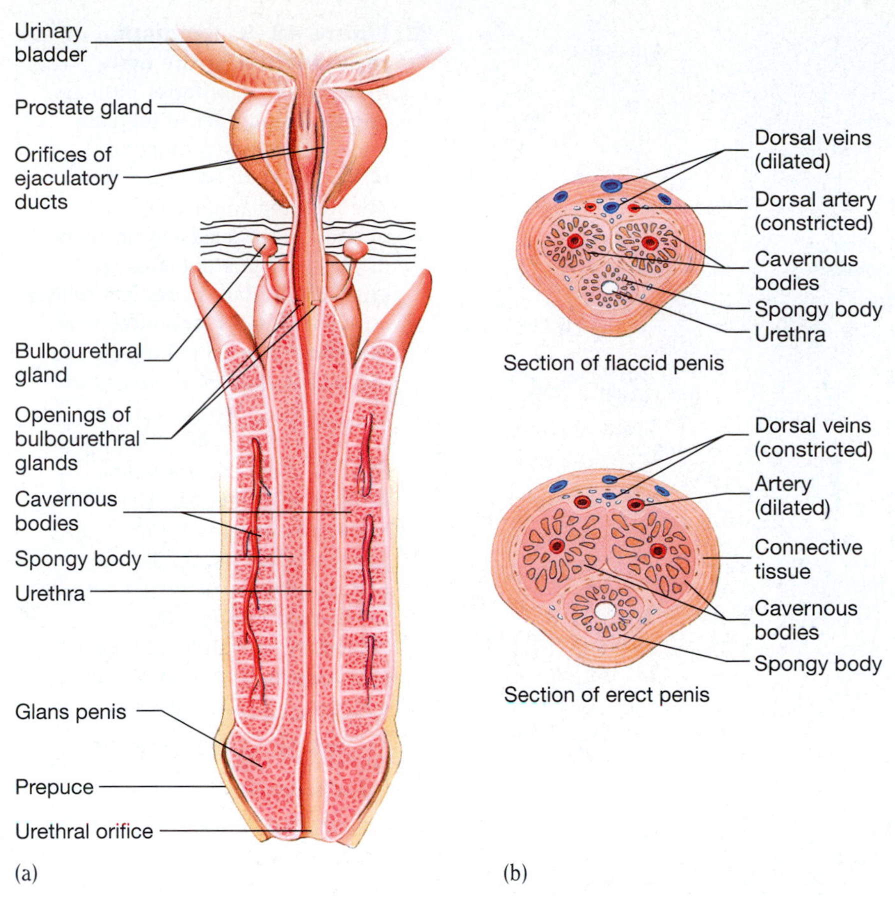

Figure 48-8 Internal structure of the penis. **(a)** Longitudinal section through the prostate gland and penis. Note the three parallel columns of erectile tissue in the penis. **(b)** Cross section through a flaccid and an erect penis. Note that the erectile tissues of the cavernous and spongy bodies are engorged with blood in the erect penis.

TABLE 48–1 Principal Male Reproductive Hormones

Endocrine Gland and Hormones	Principal Target Tissue	Principal Actions
Hypothalamus		
Gonadotropin-releasing hormone (GnRH)	Anterior pituitary	Stimulates release of FSH and LH
Anterior Pituitary		
Follicle-stimulating hormone (FSH)	Testes	Stimulates development of seminiferous tubules; stimulates spermatogenesis
Luteinizing hormone (LH); also called interstitial cell-stimulating hormone (ICSH)	Testes	Stimulates interstitial cells to secrete testosterone
Testes (interstitial cells)		
Testosterone	General	*Before birth:* stimulates development of primary sex organs and descent of testes into scrotum *At puberty:* responsible for growth spurt; stimulates development of reproductive structures and secondary sex characteristics (e.g., male body build, growth of beard, deep voice) *In adult:* responsible for maintaining secondary sex characteristics; stimulates spermatogenesis
Testes (Sertoli cells)		
Inhibin	Anterior pituitary	Inhibits FSH secretion

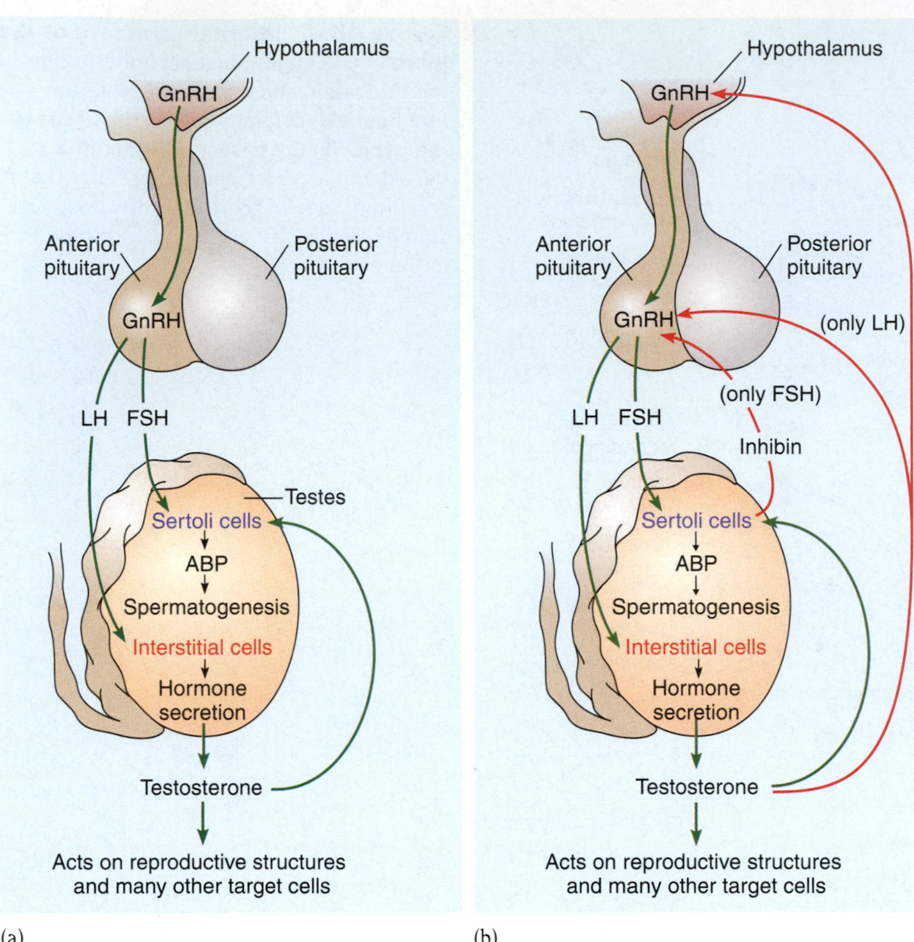

Figure 48–9 Regulation of reproduction in the male. The hypothalamus, anterior pituitary, and testes interact to regulate male reproductive function. **(a)** Overview of hormone action. The hypothalamus secretes gonadotropin-releasing hormone (GnRH), which stimulates the anterior pituitary to secrete follicle-stimulating (FSH) and luteinizing hormone (LH). LH stimulates the interstitial cells in the testes to secrete testosterone. FSH acts on the Sertoli cells, stimulating them to secrete androgen-binding protein (ABP) and other signaling molecules necessary for spermatogenesis. **(b)** Several negative feedback systems operate to regulate hormone level. Testosterone mainly inhibits LH production. It does this by decreasing GnRH secretion by the hypothalamus and inhibiting LH secretion by the pituitary. Inhibin inhibits FSH secretion.

plex, and it is likely that other hormones and signaling molecules will be identified.

Testosterone affects many tissues

Testosterone directly affects muscle and bone. This hormone is responsible for the adolescent growth spurt in males at about age 13. It stimulates growth of the reproductive organs and so is responsible for the male's **primary sex characteristics.**

Testosterone is also responsible for the **secondary sex characteristics** that develop at puberty, including growth of facial and body hair, muscle development, and the increase in vocal cord length and thickness that causes the voice to deepen. Testosterone is necessary for normal sex drive.

In some of its target tissues, testosterone must be converted to other steroids before it can produce an effect. Interestingly, in brain cells, testosterone is converted to *estradiol,* the principal female sex hormone. The implications of this transformation are not yet understood.

What happens when testosterone is insufficient or absent? Insufficient testosterone results in sterility. If a male is **castrated,** that is, the testes are removed before puberty, he is deprived of testosterone and becomes a eunuch. He retains childlike sex organs and does not develop secondary sexual characteristics. If castration occurs after puberty, increased secretion of male hormones by the adrenal glands helps maintain masculinity.

HUMAN REPRODUCTION: THE FEMALE PRODUCES GAMETES AND INCUBATES THE EMBRYO

The female reproductive system produces oocytes (immature gametes), receives the penis and sperm released from it during sexual intercourse, houses and nourishes the embryo during prenatal development, gives birth, and produces milk for the young (lactation). These processes are regulated and coordinated by the interaction of hormones secreted by the hypothalamus, pituitary gland, and ovaries. The principal organs of the female reproductive system are illustrated in Figures 48–10 and 48–11.

The ovaries produce gametes and sex hormones

Like the male gonads, the female gonads, or **ovaries,** produce both gametes and sex hormones. About the size and shape of large almonds, the ovaries are located close to the lateral walls of the pelvic cavity and are held in position by several connective

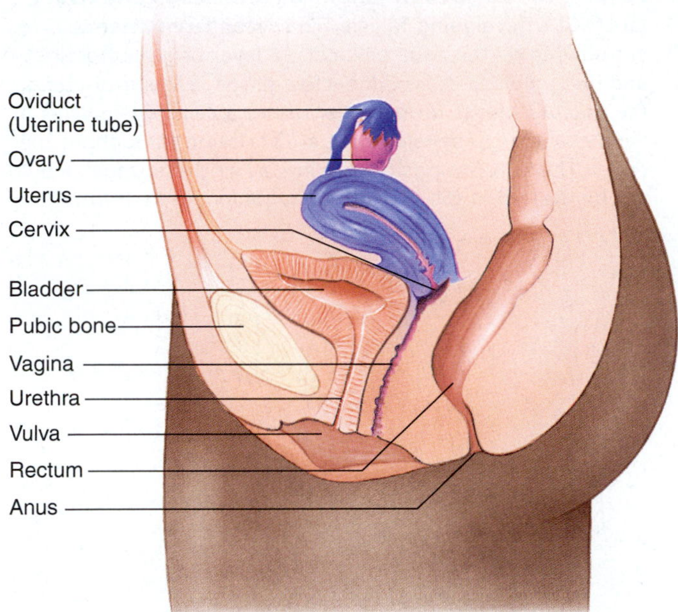

Figure 48–10 Female reproductive system. Midsagittal section through the female pelvis. Note the position of the uterus relative to the vagina.

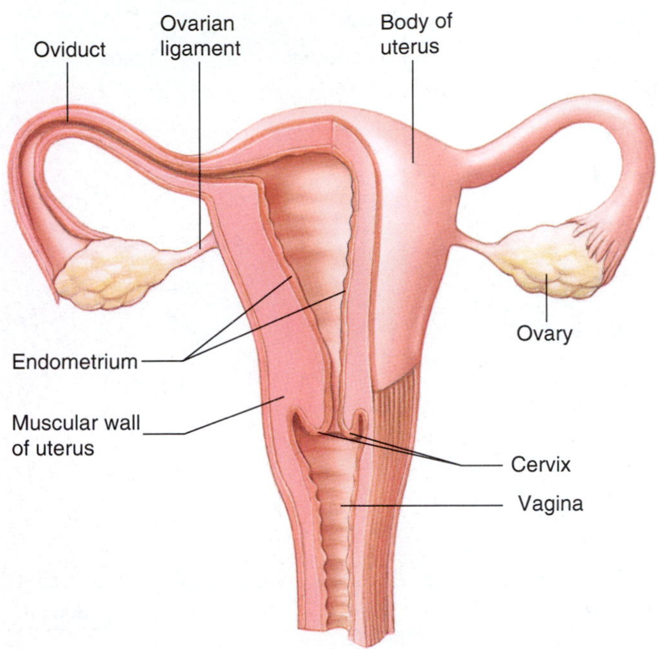

Figure 48–11 Anterior view of the female reproductive system. Some organs have been cut open to expose their internal structure. Connective tissue ligaments anchor the reproductive organs in place.

tissue ligaments. Internally, the ovary consists mainly of connective tissue containing scattered ova in various stages of maturation (Fig. 48–12).

The process of ovum production, called **oogenesis,** begins in the ovaries. Before birth, hundreds of thousands of **oogonia** are present in the ovaries. All of a female's oogonia form during embryonic development. No new oogonia are formed after birth. During prenatal development, the oogonia increase in size and become **primary oocytes.** By the time of birth, they are in the prophase of the first meiotic division. At this stage, they enter a resting phase that lasts throughout childhood and into adult life.

A primary oocyte and the **granulosa cells** surrounding it together make up a **follicle.** The granulosa cells are connected by *tight junctions* that form a protective barrier around the oocyte. With the onset of puberty, a few follicles begin to mature each month in response to FSH secreted by the anterior pituitary gland. As a follicle grows, the granulosa cells proliferate, forming several layers. Connective tissue cells surrounding the granulosa differentiate, forming a layer of **theca cells.**

As the follicle matures, the primary oocyte completes its first meiotic division. The two haploid cells produced are different in size (Fig. 48–13). The smaller one, the first **polar body,** may later divide, forming two polar bodies, but these eventually disintegrate. The larger cell, the **secondary oocyte,** proceeds to the second meiotic division but remains in metaphase II until it is fertilized. When meiosis continues, the second meiotic division gives rise to a single ovum and a second polar body. The polar bodies are small and apparently serve to dispose of unneeded chromosomes with a minimal amount of cytoplasm. The sequence is as follows:

Oogonium (diploid) ⟶ primary oocyte (diploid) ⟶
secondary oocyte + first polar body (both haploid) ⟶
ovum + second polar body (both haploid)

Recall that in the male, each primary spermatocyte gives rise to four functional sperm cells. In contrast, each primary oocyte generates only one ovum.

As an oocyte develops, it becomes separated from its surrounding follicle cells by a layer of glycoproteins called the **zona pellucida.** As the follicle develops, follicle cells secrete fluid, which collects in the **antrum** (space) between them (see Fig. 48–12). The follicle cells also secrete **estrogens,** female sex hormones. The principal estrogen is **estradiol.** Typically, only one follicle fully matures each month. Several others may develop for awhile and then deteriorate by *apoptosis.*

As a follicle matures, it moves closer to the surface of the ovary, eventually resembling a fluid-filled bulge on the ovarian surface. Follicle cells secrete proteolytic enzymes that break down a small area of the ovary wall. During **ovulation,** the secondary oocyte is ejected through the ovary wall and into the pelvic cavity. The portion of the follicle that remains in the ovary develops into the **corpus luteum,** a temporary endocrine gland that secretes estrogen and **progesterone.**

The oviducts transport the secondary oocyte

Almost immediately after ovulation, the secondary oocyte is swept into the funnel-shaped opening of the **oviduct,** or **uterine tube** (also called the *fallopian tube*). Action of cilia on the

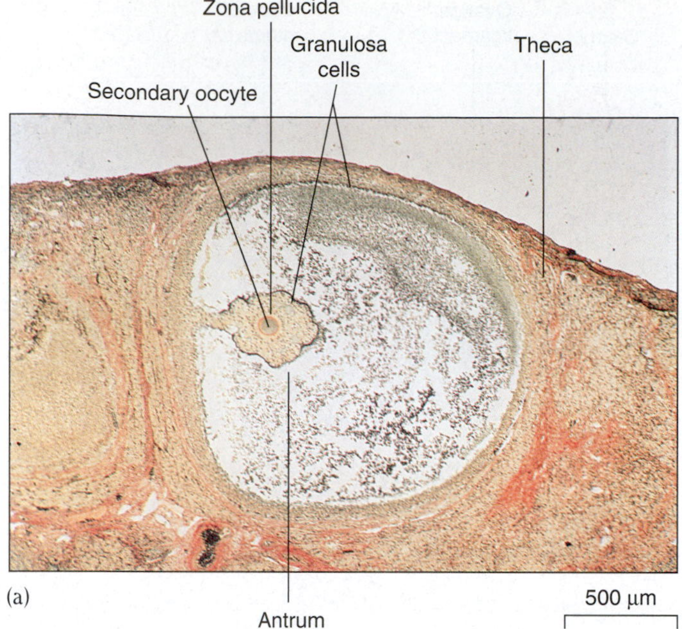

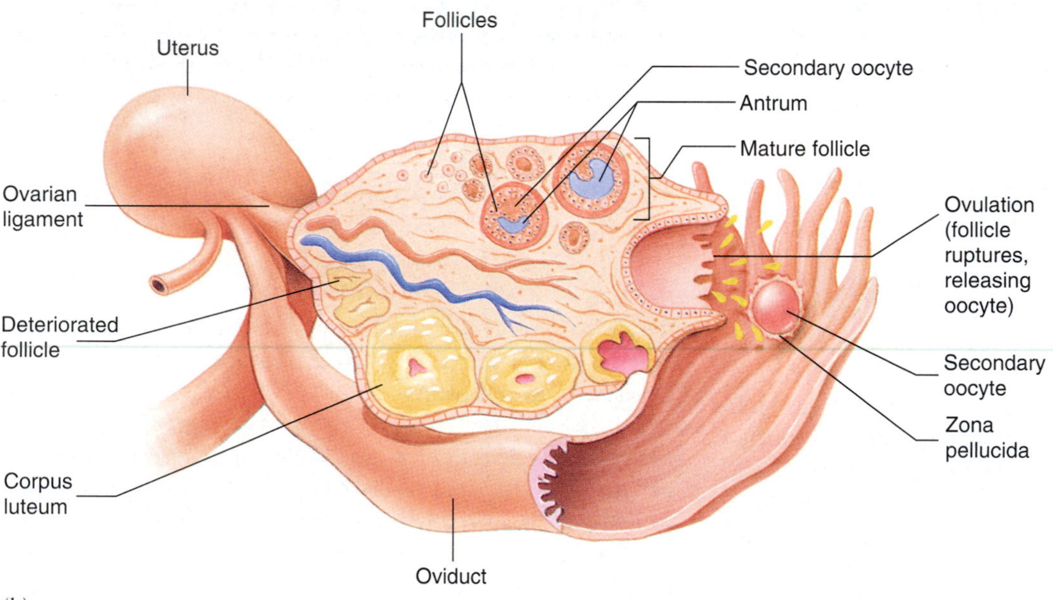

Figure 48–12 Development of follicles in the ovary. **(a)** LM of a developing follicle. The secondary oocyte is surrounded by the zona pellucida (a layer of glycoproteins) and by granulosa cells. Connective tissue cells surrounding the granulosa cells form a layer of theca cells. **(b)** Follicles in various stages of development are scattered throughout the ovary. This is a composite drawing; all of these stages would not be present at the same time. *(a, Biophoto Associates)*

epithelial lining of the oviduct is responsible for both sweeping the secondary oocyte into the oviduct and moving it along toward the uterus. Fertilization takes place within the oviduct. If fertilization does not occur, the secondary oocyte degenerates there.

Scarring of the oviducts (e.g., by pelvic inflammatory disease, a sexually transmitted disease) can block the tubes so that the fertilized ovum cannot pass to the uterus. Sometimes partial constriction of the oviduct results in *tubal pregnancy,* in which the embryo begins to develop in the wall of the oviduct because it cannot progress to the uterus. Oviducts are not adapted to bear the burden of a developing embryo; thus, the oviduct and the embryo it contains must be surgically removed before it ruptures and endangers the life of the mother. Women with blocked oviducts may be infertile.

The uterus incubates the embryo

The oviducts open into the upper corners of the pear-shaped **uterus** (see Fig. 48–11). About the size of a fist, the uterus (or womb) occupies a central position in the pelvic cavity. It has thick walls of smooth muscle and an epithelial lining, the **endometrium,** that thickens each month in preparation for possible pregnancy.

If a secondary oocyte is fertilized, the tiny embryo enters the uterus and implants in the endometrium. There it grows and de-

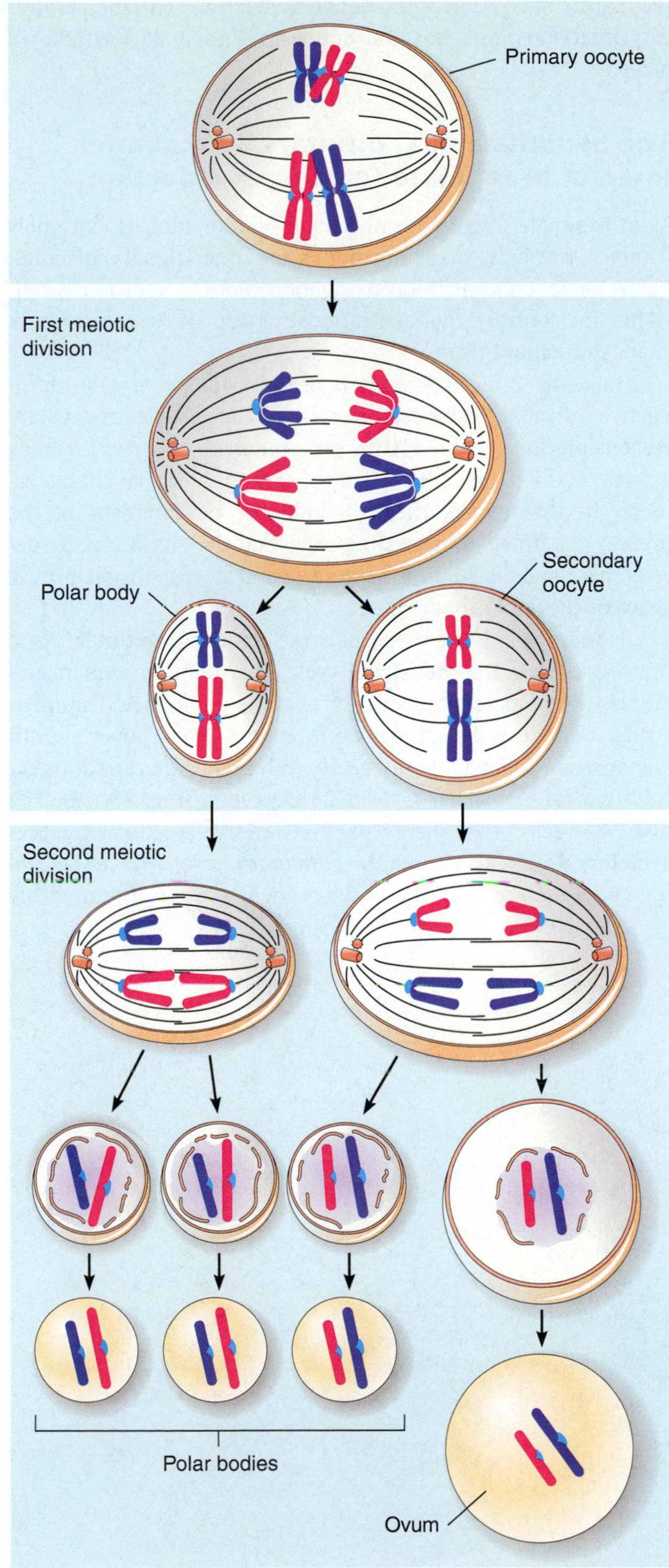

Figure 48–13 Oogenesis. Before birth oogonia divide many times by mitosis. Some oogonia differentiate to become primary oocytes that undergo meiosis. Only one functional ovum is produced from each primary oocyte. The other cells produced are polar bodies that degenerate. The second meiotic division is completed after fertilization. Note that in this example, four chromosomes are present in the primary oocyte ($2n$) and that meiosis produces the haploid *(n)* number (two) in the polar bodies, secondary oocyte, and mature ovum.

velops, sustained by nutrients and oxygen delivered by surrounding maternal blood vessels. If fertilization does not occur during the monthly cycle, the endometrium sloughs off and is discharged in the process known as **menstruation.**

More than 5 million women in the United States are affected by **endometriosis,** a painful disorder in which fragments of the tissue lining the uterus migrate to other areas such as the oviducts or ovaries. Endometriosis causes scarring that can lead to infertility.

The lower portion of the uterus, called the **cervix,** extends slightly into the vagina. The cervix is a common site of cancer in women. Detection is usually possible by the routine Papanicolaou test (Pap smear) in which a few cells are scraped from the cervix during a regular gynecological examination and studied microscopically. When cervical cancer is detected at very early stages of malignancy, the chances that the patient can be cured are good.

The vagina receives sperm

The **vagina** is an elastic, muscular tube that extends from the uterus to the exterior of the body. The vagina serves as a receptacle for sperm during sexual intercourse and as part of the birth canal (see Fig. 48–11). *Vaginismus* is a condition in which, during sexual intercourse, a woman experiences painful involuntary spasms of the outer third of the vaginal muscles. Vaginismus is often associated with a history of sexual abuse.

The vulva are external genital structures

The female external genitalia, collectively known as the **vulva,** include several structures. Liplike folds, the **labia minora,** surround the vaginal and urethral openings (Fig. 48–14). The area enclosed by the labia minora is the **vestibule** of the vagina. Vestibular glands secrete a lubricating mucus into the vestibule. The **hymen** is a thin ring of tissue that forms a border around the entrance to the vagina.

Anteriorly, the labia minora merge to form the prepuce of the **clitoris,** a small erectile structure comparable to the male glans penis. Like the penis, the clitoris contains erectile tissue that becomes engorged with blood during sexual excitement. Rich in nerve endings, the clitoris is highly sensitive to touch, pressure, and temperature and serves as a center of sexual sensation in the female.

External to the delicate labia minora are the thicker **labia majora.** The **mons pubis** is the mound of fatty tissue just above the clitoris at the junction of the thighs and torso. At puberty the

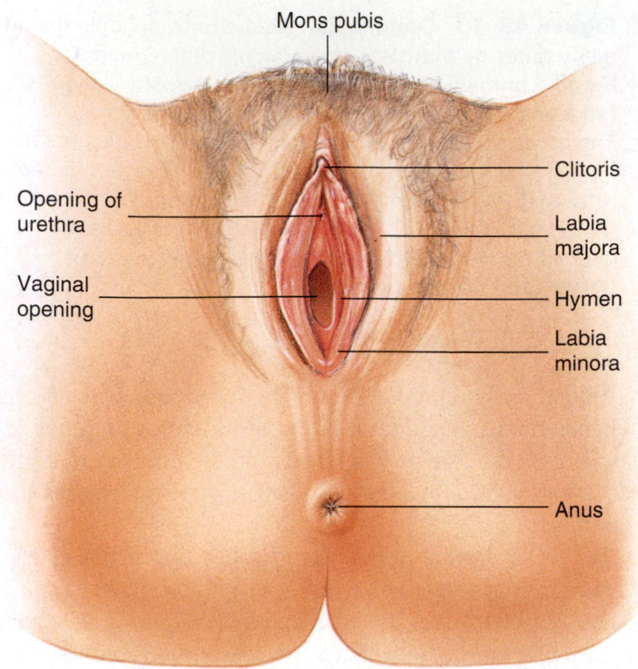

Figure 48–14 External female genital structures. Collectively, the structures shown (excluding the anus) are referred to as the vulva.

mons pubis and labia majora become covered by coarse pubic hair.

The breasts function in lactation

Each breast is composed of 15 to 20 lobes of glandular tissue. The amount of adipose tissue around these lobes determines the size of the breasts and accounts for their softness. Gland cells are arranged in grapelike clusters called **alveoli** (Fig. 48–15). Ducts from each cluster join to form a single duct from each lobe, producing 15 to 20 tiny openings on the surface of each nipple. The breasts are the most common site of potentially deadly cancer in women (see *Focus On: Breast Cancer*).

Lactation is the production of milk for the nourishment of the young. During pregnancy, high concentrations of the female reproductive hormones, estrogen and progesterone, stimulate the breasts to increase in size. For the first couple of days after childbirth, the mammary glands produce a fluid called **colostrum,** which contains protein and lactose but little fat. After birth the hormone **prolactin** stimulates milk production. Recall from Chapter 47 that when a baby suckles, the posterior pituitary releases **oxytocin,** which stimulates ejection of milk from the alveoli into the ducts.

Breastfeeding promotes recovery of the uterus because oxytocin released during breastfeeding stimulates the uterus to contract to nonpregnant size. Breastfeeding offers advantages to the baby as well. It promotes a close bond between mother and child and provides milk tailored to the nutritional needs of the human infant. Breast milk contains antibodies, and breast-fed infants have a lower incidence of diarrhea, ear and respiratory infections, and hospital admissions than do formula-fed babies.

The hypothalamus, pituitary, and ovaries interact to regulate female reproduction

As in the male, regulation of female reproduction is extremely complex, involving many hormones and other signal molecules. In the male, concentration of sex hormones is kept fairly constant. In contrast, concentrations of female sex hormones change in a monthly cycle.

Table 48–2 lists the actions of the principal female reproductive hormones. Like testosterone in the male, estrogens are responsible for growth of the sex organs at puberty, for body growth, and for the development of secondary sexual characteristics. In the female, these include the development of the breasts, the broadening of the pelvis, and the characteristic development and distribution of muscle and fat responsible for the shape of the female body.

Hormones of the hypothalamus, anterior pituitary, and ovaries regulate the **menstrual cycle,** the monthly sequence of events that prepares the body for possible pregnancy. The menstrual cycle runs its course every month from puberty until *menopause* occurs at about age 50. Although wide variations exist, a typical menstrual cycle is 28 days long (Fig. 48–16). The first two weeks of the menstrual cycle are referred to as the **preovulatory phase** (also called the *follicular phase*). The first day of the cycle is marked by the onset of menstruation, the monthly

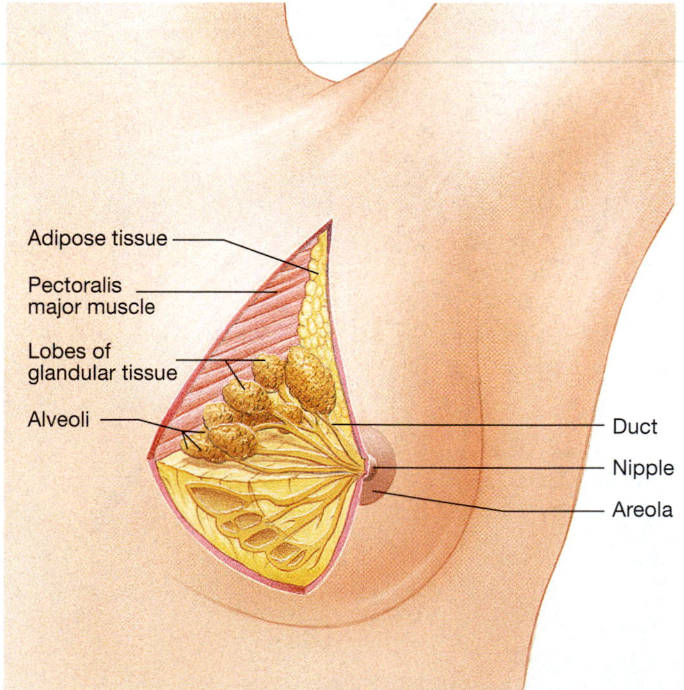

Figure 48–15 Structure of the mature female breast. The breast contains lobes of glandular tissue. The lobes consist of alveoli, clusters of gland cells.

FOCUS ON Breast Cancer

Other than skin cancer, breast cancer is the most common type of cancer among women. It is a leading cause of cancer deaths in women, second only to lung cancer. The causes of breast cancer are not known, but there appears to be a higher risk in women with a family history of the disease. An estimated 10% of breast cancers are familial, and about half of these patients have mutations in a tumor suppressor gene, either *BRCA1* or *BRCA2*. When phosphorylated by a specific protein kinase, *BRCA1* normally works with *BRCA2* and certain other compounds to repair DNA damage.

Recent studies suggest that centrosome abnormalities may be linked to breast cancer. Centrosomes help organize the mitotic spindle and are important for mitosis in animal cells. Many cancer cells have extra centrosomes that increase the error rate in chromosome replication and sorting. Consequently, cancer cells have abnormal chromosome complements (aneuploidy); some have extra chromosomes and some lack chromosomes. (Centrosome abnormalities also have been identified in about 20% of prostate cancers.) Problems with telomeres (the protective caps on chromosomes) can also lead to aneuploidy. Researchers are looking for the causes of centrosome (and telomere) malfunction that appear to lead to cancer.

Investigators are studying a variety of suspected risk factors for breast cancer, including a high-fat diet, obesity, exposure to radiation, and exposure to certain chemicals. Smoking cigarettes increases a woman's risk of dying from breast cancer by at least 25%. Women who smoke two packs or more of cigarettes a day have a 75% greater risk.

About 50% of breast cancers begin in the upper, outer quadrant of the breast. As a malignant tumor grows, it may adhere to the deep tissue of the chest wall. Sometimes it extends to the skin, causing dimpling. Eventually the cancer spreads to the lymphatic system. About two-thirds of breast cancers have metastasized (spread) to the lymph nodes by the time they are first diagnosed. When diagnosis and treatment begin early, 80% of patients survive for five years, and 62% survive for ten years or longer. Untreated patients have a five-year survival rate of only 20%.

Mastectomy (surgical removal of the breast) and radiation treatment are common methods of treating breast cancer. Lumpectomy (surgical removal of only the affected portion of the breast) in conjunction with radiation treatment is thought to be as effective as mastectomy in some cases. Chemotherapy is useful in preventing metastasis, especially in premenopausal patients. A recent development in cancer treatment is the use of *biological response modifiers,* which include substances such as interferons, interleukins, and monoclonal antibodies (see Chapter 43).

About one-third of breast cancers are estrogen-dependent; that is, their growth depends on circulating estrogens. Removing the ovaries in patients with these tumors relieves the symptoms and may cause remission of the disease for months or even years. Developing pharmacological agents that antagonize the action of estrogen receptors has been an important approach in the treatment of breast cancer. One challenge has been to inhibit estrogen in the breast while at the same time retaining its beneficial effects on bone, blood vessels, and the brain.

The synthetic drug *tamoxifen* blocks estrogen receptors and is a potent chemotherapeutic agent for treating breast cancer patients. This drug appears to be effective in preventing cancer development, especially in patients who have had tumors removed surgically and are at high risk for recurrence of breast cancer. However, studies suggest that tamoxifen may increase risk of uterine cancer. The drug *raloxifene* is one member of a more recent generation of selective estrogen receptor modulators. Raloxifene has been used clinically to prevent osteoporosis, particularly in patients unable to take estrogen for hormone replacement therapy. Clinical studies suggest that raloxifene may prevent breast cancer without increasing the risk of uterine cancer.

Because early detection of breast cancer greatly increases the chances of cure and survival, campaigns have been launched to educate women on the importance of self-examination. Mammography, a soft tissue radiological study of the breast, is helpful in detecting very small lesions that might not be identified by routine examination. In mammography, lesions show on an x-ray plate as areas of increased density *(see figure).*

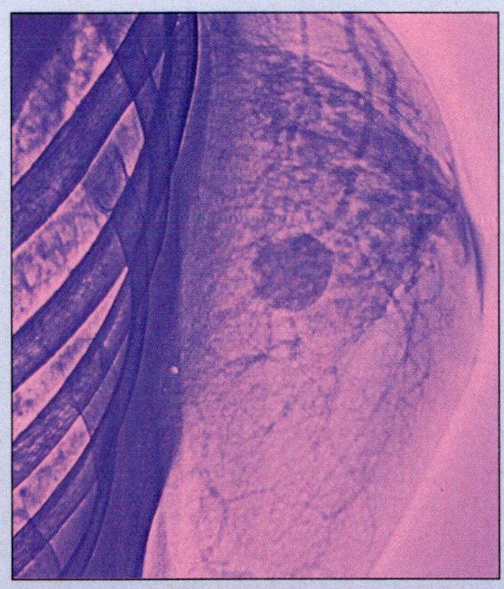

Mammogram showing area of breast cancer. Note the extensive vascularization. *(Visuals Unlimited/SIU)*

discharge through the vagina of blood and tissue from the endometrium. Ovulation occurs on about the 14th day of the cycle. The third and fourth weeks of the menstrual cycle are the **postovulatory phase** (also called the *luteal phase*).

During the *menstrual phase* (the first five days of the preovulatory phase), gonadotropin-releasing hormone (GnRH) is released from the hypothalamus. GnRH stimulates the anterior pituitary to release follicle-stimulating hormone (FSH) and luteinizing hormone (LH) (Fig. 48–17 on page 1074).

During the preovulatory phase, FSH stimulates a few follicles to begin to develop, and stimulates the granulosa cells to multiply and produce estrogen. Some of the estrogen diffuses into the

TABLE 48–2 Principal Female Reproductive Hormones

Endocrine Gland and Hormones	Principal Target Tissue	Principal Actions
Hypothalamus		
Gonadotropin-releasing hormone (GnRH)	Anterior pituitary	Stimulates release of FSH and LH
Anterior Pituitary		
Follicle-stimulating hormone (FSH)	Ovary	Stimulates development of follicles and secretion of estrogen
Luteinizing hormone (LH)	Ovary	Stimulates ovulation and development of corpus luteum
Prolactin	Breast	Stimulates milk production (after breast has been prepared by estrogen and progesterone)
Posterior Pituitary		
Oxytocin	Uterus	Stimulates contraction and stimulates prostaglandin release
	Mammary glands	Stimulates ejection of milk into ducts
Ovaries		
Estrogen (estradiol) (secreted by the granulosa cells of the follicle and by the corpus luteum)	General	Stimulates growth of sex organs at puberty and development of secondary sex characteristics (breast development, broadening of pelvis, distribution of fat and muscle)
	Reproductive structures	Induces maturation; stimulates monthly preparation of the endometrium for pregnancy; makes cervical mucus thinner and more alkaline
Progesterone (secreted mainly by corpus luteum)	Uterus	Completes preparation of endometrium for pregnancy
Inhibin	Anterior pituitary	Inhibits FSH secretion

blood, but estrogen also has an autocrine action (see Chapter 47) on the granulosa cells that produce it and a paracrine effect on nearby granulosa cells. Estrogen stimulates the granulosa cells to multiply (which increases estrogen production).

The amount of estrogen secreted by the granulosa cells is enhanced by the action of LH on the theca cells. LH stimulates the theca cells to proliferate and produce androgens. The androgens diffuse into the granulosa cells, which convert them to estrogen. Estrogen stimulates growth of the endometrium, which thickens and develops new blood vessels and glands.

After the first week of the menstrual cycle, only one follicle continues to develop. Its granulosa cells become sensitive to LH as well as to FSH. This dominant follicle now secretes enough estrogen to cause a rise in the concentration of estrogen in the blood. *Although still at relatively low concentration,* estrogen inhibits secretion of FSH and LH from the pituitary and may also act on the hypothalamus, decreasing secretion of GnRH. In addition, the granulosa cells secrete the hormone *inhibin,* which inhibits mainly FSH secretion. As a result of these negative feedback signals, FSH (and to a lesser extent, LH) concentration decreases.

As its concentration in the blood peaks during the late preovulatory phase, estrogen signals the anterior pituitary to secrete LH. This is a positive feedback mechanism. The surge of LH secreted at the middle of the menstrual cycle stimulates the final maturation of the follicle, and stimulates ovulation.

After the secondary oocyte has been ejected from the ovary, the postovulatory phase begins. LH stimulates development of the corpus luteum, which secretes a large amount of **progesterone** and estrogen, and also inhibin (Fig. 48–17*c*). These hormones stimulate the uterus to continue its preparation for pregnancy. Progesterone stimulates tiny glands in the endometrium to secrete a fluid rich in nutrients.

During the postovulatory phase, the high concentration of progesterone in the blood, along with estrogen, inhibits secretion of GnRH, FSH, and LH. Progesterone is thought to act mainly on the hypothalamus. Inhibin acts on the pituitary to further inhibit FSH secretion. Thus, during the postovulatory phase, FSH and LH concentrations are low and no new follicles develop.

If the secondary oocyte is not fertilized, the corpus luteum begins to degenerate after about eight days. Although the mechanism responsible for corpus luteum degeneration is not completely understood, a decrease in LH may be a factor. In addition, the corpus luteum may become less sensitive to LH.

When the corpus luteum stops secreting progesterone and estrogen, the concentrations of these hormones in the blood fall markedly. As a result, small arteries in the endometrium con-

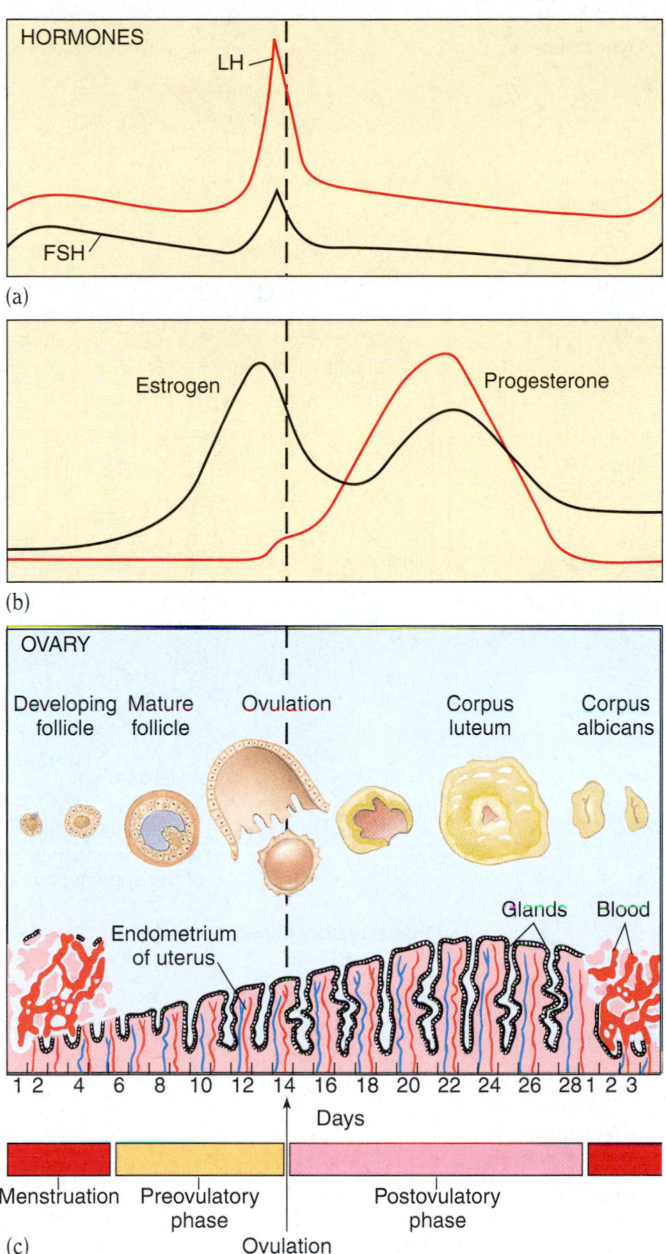

Figure 48-16 Menstrual cycle. When fertilization does not occur, the menstrual cycle repeats itself about every 28 days. The events that take place within the ovary and uterus are precisely coordinated by hormones. Note that estrogen concentration is highest during the preovulatory phase, whereas progesterone concentration is highest during the postovulatory phase. FSH, follicle-stimulating hormone; LH, luteinizing hormone.

strict, reducing the oxygen supply. Menstruation, which marks the beginning of a new cycle, begins as cells die and damaged arteries rupture and bleed. The low levels of estrogen and progesterone are insufficient to inhibit the anterior pituitary, and secretion of FSH and LH increases once again.

Premenstrual syndrome (PMS) is a condition experienced by some women starting several hours to ten days before menstruation and ending a few hours after onset of menstruation. Symptoms include fatigue, anxiety, depression, irritability, headache, edema, and skin eruptions. Its cause is not known, but in many women, symptoms are alleviated with drugs that block *serotonin* reuptake by neurons.

SEXUAL RESPONSE INVOLVES PHYSIOLOGICAL CHANGES

During copulation, also called **coitus** or sexual intercourse in humans, the male deposits semen into the upper end of the vagina. The complex structures of the male and female reproductive systems, and the physiological, endocrine, and psychological processes associated with sexual activity, are adaptations that promote fertilization of the secondary oocyte, and development of the resulting embryo.

Sexual stimulation results in two basic physiological responses: increased blood flow (vasocongestion) to reproductive structures and other tissues such as the skin, and increased muscle tension. During vasocongestion, erectile tissues within the penis and clitoris, as well as in other areas of the body, become engorged with blood.

Sexual response includes four phases: **sexual excitement, plateau, orgasm,** and **resolution.** The *desire* to have sexual activity may be motivated by fantasies or thoughts about sex. This anticipation can lead to (physical) sexual excitement and a sense of sexual pleasure. Physiologically, the excitement phase involves vasocongestion and increased muscle tension. Before the penis can enter the vagina and function in coitus, it must be erect. Penile erection is the first male response to sexual excitement. In the female, vasocongestion occurs in the vagina, clitoris and breasts, and the vaginal epithelium secretes a sticky lubricant. Vaginal lubrication is the female's first response to effective sexual stimulation. During the excitement phase, the vagina lengthens and expands in preparation for receiving the penis.

If erotic stimulation continues, sexual excitement heightens to the plateau phase. Vasocongestion and muscle tension increase markedly. In both sexes, blood pressure increases and heart rate and breathing accelerate.

Sexual intercourse is usually initiated during the plateau phase. The penis creates friction as it is moved inward and outward in the vagina in actions referred to as pelvic thrusts. Physical and psychological sensations resulting from this friction (and from the emotional intimacy experienced) may lead to orgasm, the climax of sexual excitement. In the female, stimulation of the clitoris is important in heightening the sexual excitement that leads to orgasm.

Although it lasts only a few seconds, orgasm is the phase of maximum sexual tension and its release. In both sexes, orgasm is marked by rhythmic contractions of the muscles of the pelvic floor and reproductive structures. These muscular contractions continue at about 0.8-second intervals for several seconds. After the first few contractions, their intensity decreases, and they become less regular and less frequent. Heart rate and respiration more than double, and blood pressure rises markedly, just

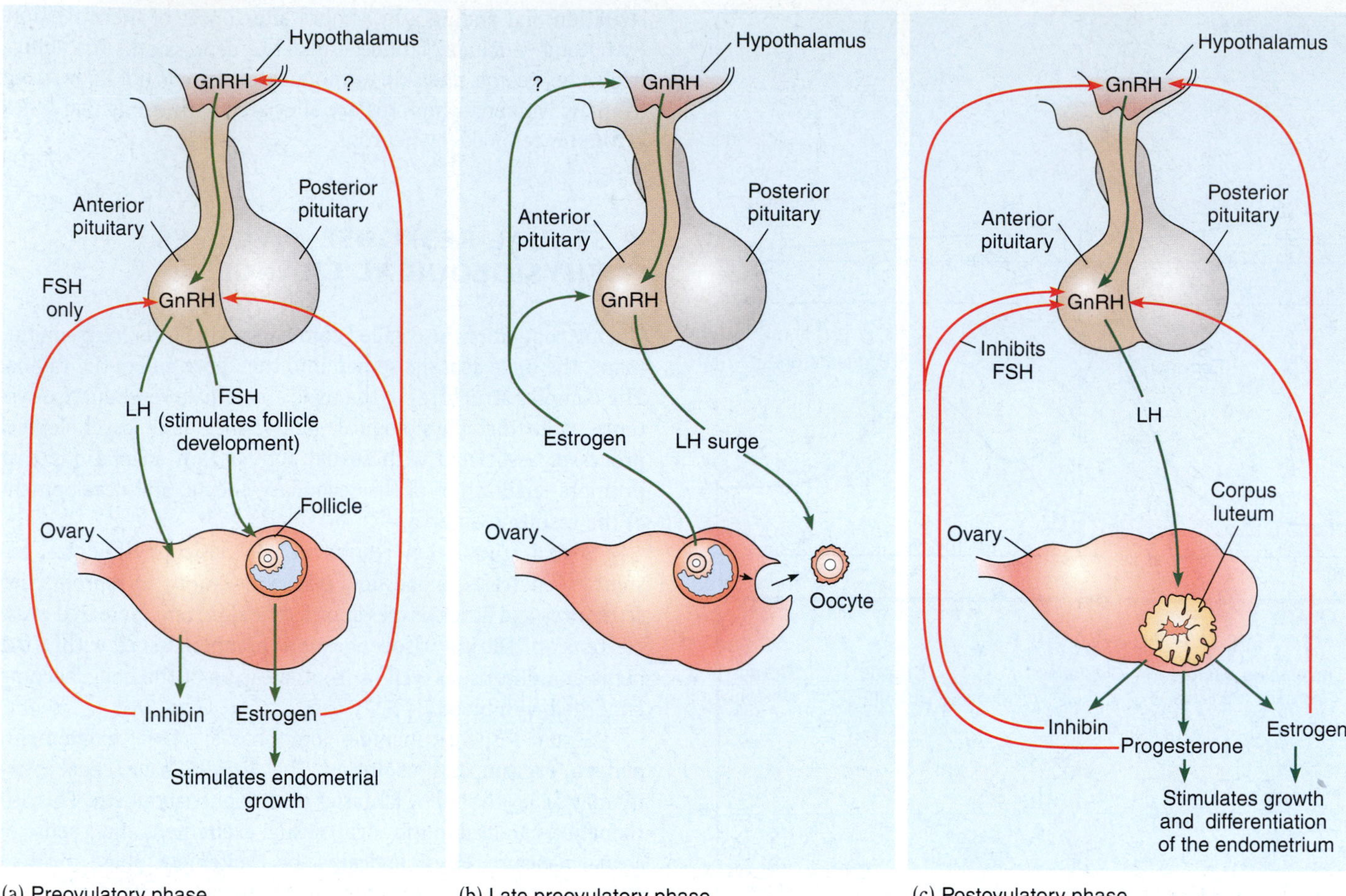

(a) Preovulatory phase (b) Late preovulatory phase (c) Postovulatory phase

Figure 48–17 Regulation of reproduction in the female. Hormones from the hypothalamus, anterior pituitary, and ovary interact to regulate the menstrual cycle. **(a)** Hormonal interactions during the mid-preovulatory phase. **(b)** Hormonal interactions during the late preovulatory phase. **(c)** Hormonal interactions during the postovulatory phase. Note that estrogens have a positive feedback effect on the hypothalamus and pituitary during the late preovulatory phase but a negative feedback effect during the early preovulatory phase and the postovulatory phase. Red arrows indicate inhibition. FSH, follicle-stimulating hormone; GnRH, gonadotropin-releasing hormone; LH, luteinizing hormone.

before and during orgasm. Musculoskeletal contractions occur throughout the body.

In the male, orgasm is marked by the ejaculation of semen from the penis. No fluid ejaculation accompanies orgasm in the female. Orgasm is followed by the resolution phase, a state of well-being during which the body is restored to its unstimulated state.

FERTILIZATION IS THE FUSION OF SPERM AND EGG

The fusion of sperm and egg is the process of **fertilization.** Fertilization and the subsequent establishment of pregnancy together are referred to as **conception.** After ejaculation into the female reproductive tract, sperm remain alive and retain their ability to fertilize an ovum for only a few days. The ovum remains fertile for only about 24 hours after ovulation. Thus, conception is most probable when intercourse takes place on the day of ovulation or the five days preceding ovulation. In a very regular 28-day menstrual cycle, sexual intercourse on days 12 to 16 is most likely to result in fertilization. However, many women do not have regular menstrual cycles, and many factors can cause irregular cycles even in women who are generally regular.

When conditions in the vagina and cervix are favorable, sperm begin to arrive at the site of fertilization in the upper oviduct within a few minutes after ejaculation. However, to pass from the vagina into the uterus, sperm must pass through the cervical mucus. At the time of ovulation when estrogen concentration is high, the cervical mucus has a thin consistency that permits passage of sperm. After ovulation, when progesterone concentration rises, the cervical mucus becomes thick and sticky, blocking entrance of sperm (as well as bacteria that might

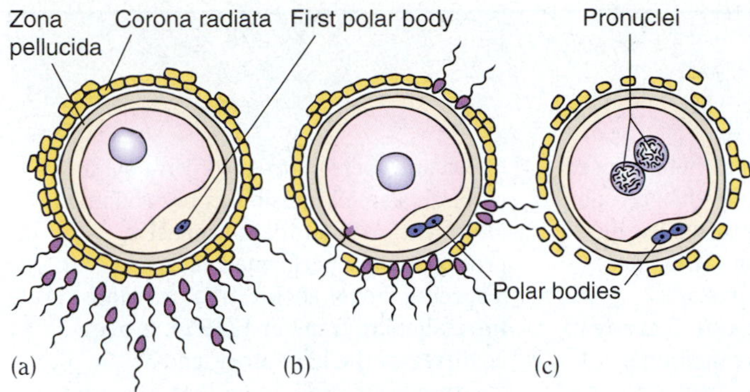

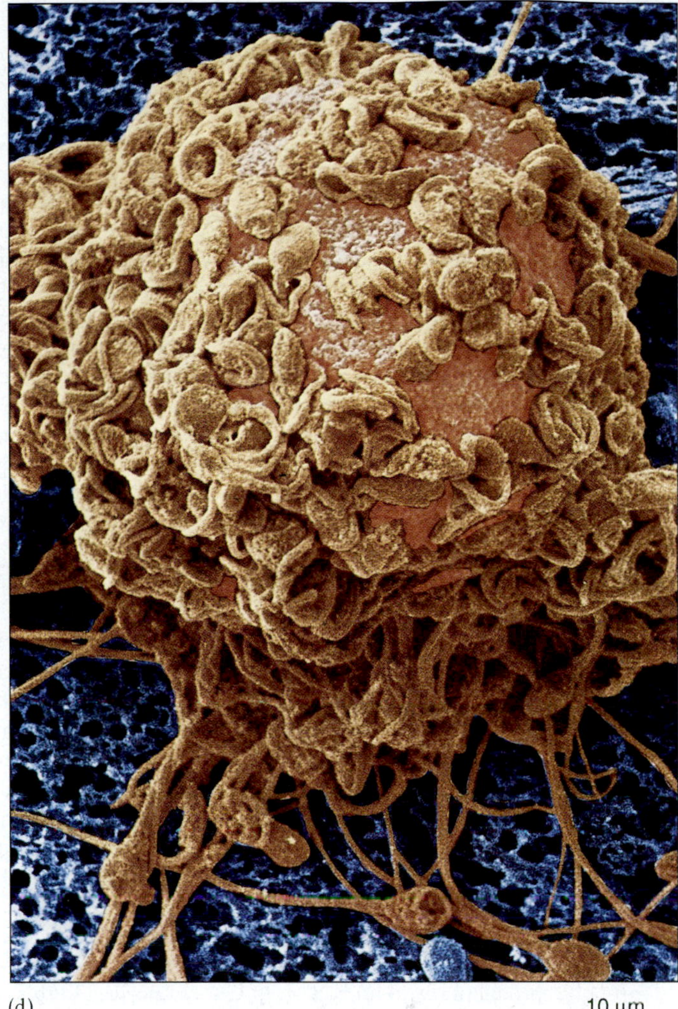

Figure 48–18 Fertilization. **(a)** Each sperm is thought to release a small amount of an enzyme that helps disperse the layer of follicle cells (corona radiata) surrounding the ovum. **(b)** After a sperm cell enters, the secondary oocyte completes its second meiotic division, producing an ovum and a polar body. **(c)** Pronuclei of sperm and ovum combine, producing a zygote with the diploid number of chromosomes. **(d)** Color-enhanced SEM of human sperm cells surrounding a test ovum. Sperm are being tested for viability. *(d, David Scharf/Peter Arnold, Inc.)*

harm the developing embryo). Once sperm enter the uterus, contractions of the uterine wall may help transport them. The sperms' own motility is important, especially in approaching and fertilizing the ovum.

Sperm cannot fertilize a secondary oocyte until they have been in contact with secretions of the female reproductive tract for several hours, a process known as *capacitation*. When a capacitated sperm encounters an egg, openings develop in the sperm acrosome, allowing enzymes to digest a path through the zona pellucida surrounding the secondary oocyte.

As soon as one sperm enters the ovum, a rapid electrical change occurs, followed by a slower chemical change in the plasma membrane of the ovum. These changes prevent the entrance of other sperm. As the fertilizing sperm enters the ovum, it usually loses its flagellum (Fig. 48–18). Sperm entry stimulates the ovum to complete its second meiotic division. The head of the haploid sperm then swells to form the **male pronucleus** and fuses with the **female pronucleus,** forming the diploid nucleus of the zygote. The process of fertilization is described in more detail in Chapter 49. (Also see *Focus On: Novel Origins.*)

If only one sperm is needed to fertilize a secondary oocyte, why are millions ejaculated? Many die as a result of unfavorable pH or phagocytosis by leukocytes and macrophages in the female tract. Only a few hundred succeed in traversing the correct oviduct and reaching the vicinity of the secondary oocyte.

HORMONES REGULATE EARLY DEVELOPMENT

If the secondary oocyte is fertilized, development begins as the embryo is slowly moved to the uterus by the action of cilia lining the oviduct. The embryo cannot enter the uterus for three to four days because estrogen maintains the smooth muscle of the passageway into the uterus in a contracted state. As progesterone concentration increases (due to the action of the corpus luteum), the smooth muscle relaxes, allowing the embryo to pass into the uterus.

When it enters the uterus, the embryo consists of a ball of about 32 cells and is about the same size as the zygote. After floating free in the uterus for about three days, the embryo begins to implant in the thick endometrium on about the seventh day after fertilization (Fig. 48–19). (Development is discussed in Chapter 49.)

Membranes that develop around the embryo secrete **human chorionic gonadotropin (hCG),** a hormone that signals the mother's corpus luteum to continue to function. (The presence of hCG in urine or blood is used as an early pregnancy test.) Concentrations of estrogen and progesterone remain high throughout pregnancy. During the first two months, the corpus luteum secretes almost all of the estrogen and progesterone necessary to maintain the pregnancy. During this time, membranes surrounding the embryo together with uterine tissue form the **placenta,** the organ of exchange between the mother and developing embryo. As the corpus luteum slows its secretion and deteriorates after about three months, the placenta takes over and secretes large amounts of estrogen and progesterone.

FOCUS ON Novel Origins

About 15% of married couples in the United States are affected by infertility, the inability of a couple to achieve conception after using no contraception for at least one year. About 30% of cases involve both male and female factors. Male infertility is often attributed to low sperm count. Among the common causes of female infertility are failure to ovulate, production of infertile eggs (common in older women), and oviduct scarring (often caused by pelvic inflammatory disease) that blocks the passage of the secondary oocyte to the uterus. Women with blocked oviducts can usually produce ova and incubate an embryo normally but require clinical assistance in getting the ovum from the ovary to the uterus.

In the United States, more than 3 million infertile couples consult health care professionals each year. Some are helped with conventional treatment, for example, with hormone therapy that regulates ovulation or with fertility drugs. But more than 40,000 couples need more sophisticated clinical help and turn to high-tech assisted reproductive techniques that have been developed through reproduction and human embryo research. At present these techniques are expensive and the success rate is low, less than one-third.

The most common assisted reproductive procedure is **artificial insemination,** in which a catheter is used to inject sperm directly into the cervix or uterus. Transfer into the uterus is called **intrauterine insemination (IUI).** More than 600,000 IUI procedures are performed annually with a success rate of about 10%. Artificial insemination is indicated when the male partner of a couple desiring a child is sterile or carries a genetic defect. If the male is infertile due to a low sperm count, his sperm can be concentrated, or alternatively, sperm from a donor can be used. Although the sperm donor usually remains anonymous to the couple, his genetic qualifications are screened by physicians.

With **in vitro fertilization (IVF),** a woman takes a fertility drug that induces ovulation of several ova. The ova are removed and fertilized with sperm in laboratory glassware. Embryos can be screened for chromosome or gene abnormalities, and healthy ones can be transferred into the uterus through the cervix. This procedure was first used in England in 1978 to help a couple who had tried unsuccessfully for several years to have a child. Since that time, thousands of "test-tube babies" have been conceived in this way and born to previously infertile women. The success rate of in vitro fertilization is about 20%.

In **gamete intrafallopian transfer (GIFT),** a laparoscope (a fiber-optic instrument) is used to guide the transfer of ova and sperm into a woman's oviduct through a small incision in her abdomen. More than 4000 of these expensive procedures are performed each year with a success rate of about 28%. In **zygote intrafallopian transfer (ZIFT),** ova are fertilized in the laboratory, and a laparoscope is used to guide the transfer of the resulting zygotes into the oviduct. In these procedures, the patient's own ova may be used. However, if she does not produce fertile ova, they can be contributed by a donor **(oocyte donation).**

Another novel procedure is **host mothering.** An embryo is removed from its natural mother and implanted into a female substitute. The foster mother can support the developing embryo either until birth or temporarily until it is implanted again into the original mother or another host. This technique has proved useful to animal breeders. For example, embryos from prize sheep can be temporarily implanted into rabbits for easy shipping by air and then implanted into a host mother sheep, perhaps of inferior quality. Host mothering has the advantage of allowing an animal with superior genetic traits to produce more offspring than would be naturally possible. This procedure is also used to increase the populations of certain endangered species *(see figure).*

Technology is available to freeze the gametes or embryos of many species, including humans, and then transplant them into their donors or into host mothers. Freezing eggs may become popular with young women not yet ready to become parents, but who want to preserve young eggs with lower risk for chromosome abnormality, and reimplant them at a later time.

Newborn bongo with its surrogate mother, an eland. As a young embryo, the bongo was transplanted into the eland's uterus, where it implanted and developed. Bongos are a rare and elusive species inhabiting dense forests in Africa. The larger and more common elands inhabit open areas.

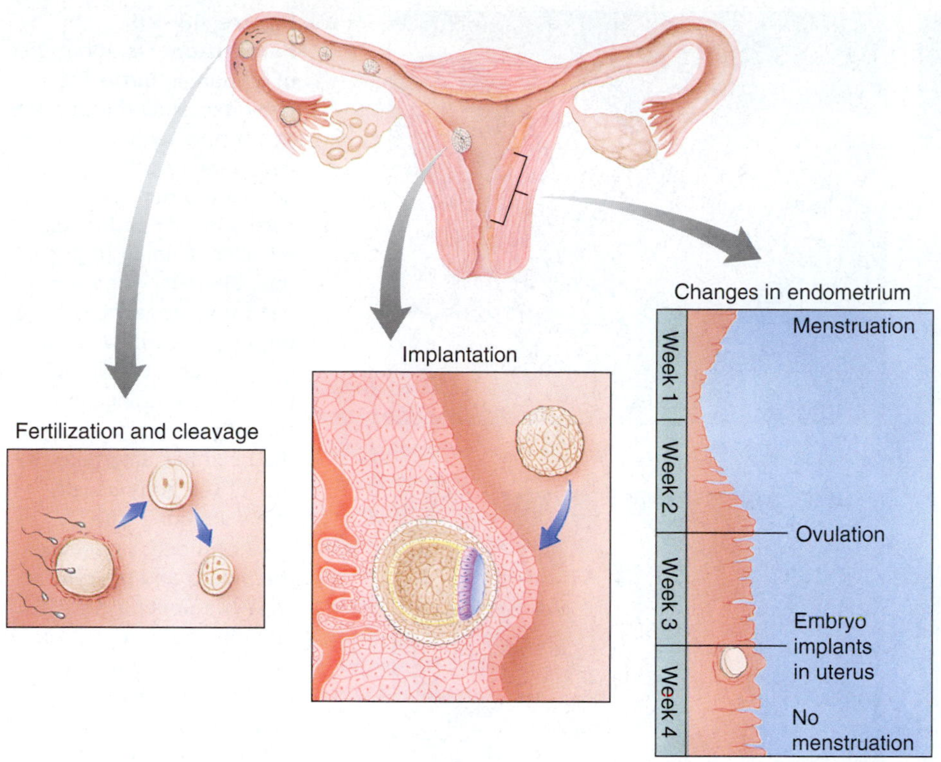

Figure 48–19 Events following fertilization. The menstrual cycle is interrupted when pregnancy occurs. The corpus luteum does not degenerate, and menstruation does not take place. Instead, the wall of the uterus thickens even more, permitting the embryo to develop within it.

Estrogen and progesterone are necessary to maintain pregnancy. Estrogen stimulates development of the uterine wall, including the muscle needed to expel the fetus during delivery. Progesterone inhibits uterine contractions so that the fetus is not expelled too soon. Because these hormones also inhibit FSH and LH, new follicles do not develop and the menstrual cycle stops during pregnancy.

HORMONES REGULATE THE BIRTH PROCESS

A normal human pregnancy is about 38 weeks long, counting from the day of conception, or about 40 weeks from the first day of the last menstrual cycle. The mechanisms that terminate pregnancy and initiate the birth process, called **parturition,** are not fully understood.

Although progesterone inhibits contractions of the uterus, its concentration does not decrease toward the end of pregnancy. However, estrogen concentration continues to increase and exerts several actions that overpower the effects of progesterone. Estrogen stimulates uterine muscle to produce proteins (connexins) that form *gap junctions* between cells (see Chapter 5). Gap junctions permit the uterine muscle fibers to contract together in a coordinated way.

Estrogen stimulates the muscle fibers of the uterine wall to develop additional receptors for the posterior pituitary hormone **oxytocin,** which stimulates uterine contractions. As a result, the uterus becomes more sensitive to oxytocin, and uterine contractions become stronger. Oxytocin also stimulates uterine muscle fibers to secrete *prostaglandins,* which, through paracrine action, powerfully stimulate uterine contraction. Estrogen also stimulates production of enzymes that break down collagen fibers in the wall of the cervix so that the cervix becomes flexible.

A hormone similar to corticotropin-releasing hormone (see Chapter 47) is secreted by the placenta. Investigators hypothesize that this hormone stimulates the adrenal cortex of the fetus to produce androgens. The placenta converts these androgens to estrogen, thus increasing estrogen concentration.

At the end of pregnancy, the stretching of the uterine muscle by the growing fetus combines with the effects of increased estrogen concentration to produce strong uterine contractions. A long series of involuntary contractions of the uterus are experienced as **labor,** which begins when uterine contractions occur every 10 to 15 minutes. Labor can be divided into three stages. During the first stage, which typically lasts about 12 hours, the contractions of the uterus move the fetus toward the cervix, causing the cervix to *dilate* (open) to a maximum diameter of 10 cm (4 in). The cervix also becomes *effaced;* that is, it thins out so that the fetal head can pass through. During the first stage of labor the amnion (the membrane that forms a fluid-filled sac around the embryo/fetus) usually ruptures, releasing about a liter of amniotic fluid, which flows out through the vagina.

During the second stage, which normally lasts between 20 minutes and an hour, the fetus passes through the cervix and the vagina and is born, or "delivered" (Fig. 48–20). With each uterine contraction the woman bears down so that the fetus is expelled by the combined forces of uterine contractions and contractions of abdominal wall muscles.

At birth, the baby is still connected to the placenta by the umbilical cord. Contractions of the uterus squeeze much of the fetal blood from the placenta into the infant. The cord is tied and

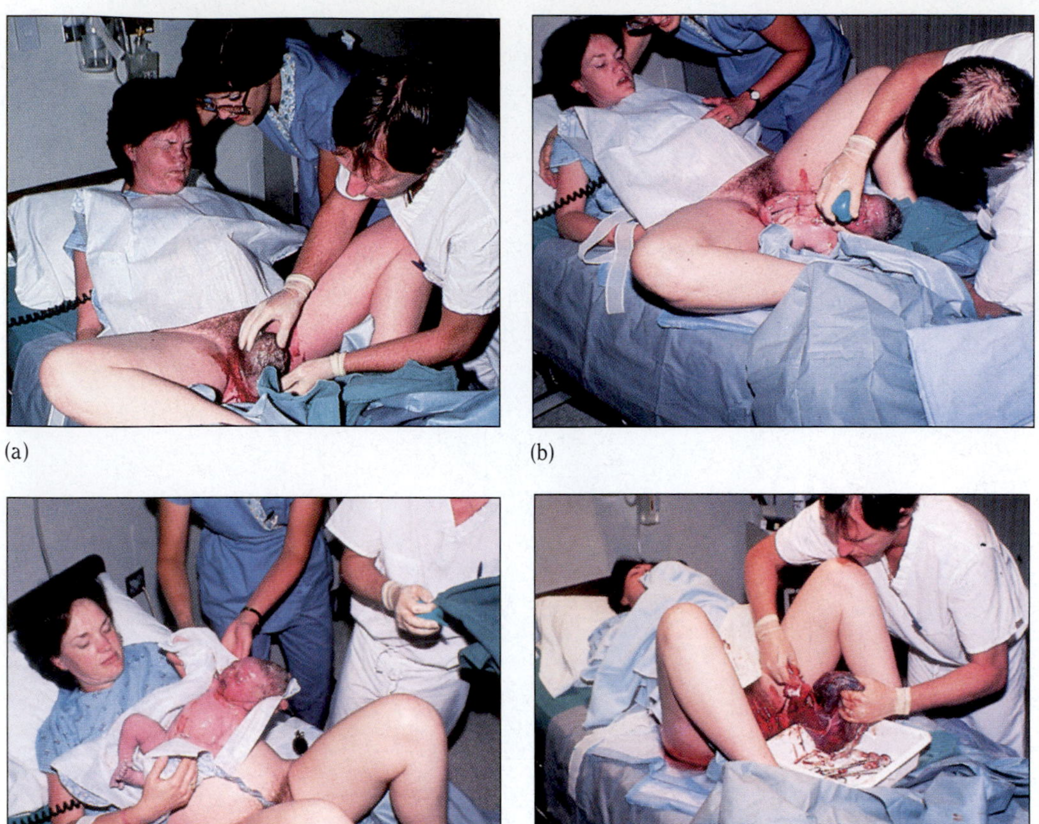

Figure 48–20
Parturition. In about 95% of all human births the baby descends through the cervix and vagina in the head-down position. **(a)** The mother bears down hard with her abdominal muscles, helping to push the baby out. When the head fully appears, the physician or midwife can gently grasp it and guide the baby's entrance into the outside world. **(b)** Once the head has emerged, the rest of the body usually follows readily. The physician gently aspirates the mouth and pharynx to clear the upper airway of amniotic fluid, mucus, or blood. At this time the newborn usually takes its first breath. **(c)** The baby, still attached to the placenta by its umbilical cord, is presented to its mother. **(d)** During the third stage of labor, the placenta is delivered. *(Courtesy of Dan Atchinson)*

cut, separating the child from the mother. (The stump of the cord gradually shrivels until nothing remains but the scar, the **navel.**)

During the third stage of labor, which lasts 10 or 15 minutes after the birth, the placenta and the fetal membranes are loosened from the lining of the uterus by another series of contractions and expelled. At this stage they are collectively called the **afterbirth** because these tissues are expelled from the vagina after the baby has been delivered.

During labor an obstetrician may administer oxytocin or prostaglandins to increase the contractions of the uterus or may assist with special forceps or a vacuum device. In some women, the opening between the pelvic bones is too small to permit the passage of the baby vaginally. In this situation, or if the baby's position prevents normal delivery, the obstetrician may perform a **cesarean section,** an operation in which the baby is delivered through an incision made in the abdominal and uterine walls.

BIRTH CONTROL METHODS ALLOW INDIVIDUALS TO CHOOSE

When a fertile, heterosexually active woman uses no form of birth control, her chances of becoming pregnant during the course of a year are about 90%. Any method for deliberately separating sexual intercourse from reproduction is considered **contraception** (literally, "against conception"). Since ancient times, humans have searched for effective contraceptive methods.

Although many couples worldwide agree that it is best to have babies by choice rather than by chance, the majority either are not educated about contraceptive methods or do not have contraceptives available to them. According to the U.N. Population Division, in underdeveloped countries, 60% of married women do not use modern methods of contraception. Studies indicate that many of these women would use modern birth control methods if they were available and affordable and if someone showed them how to use them.

More than half of the 7 million women who become pregnant each year in the United States did not intend to do so. Young people particularly lack the knowledge and means of protecting themselves from unwanted pregnancy. Only about one-third of sexually active teenagers consistently use birth control. Every year more than 1 million teenagers in the United States become pregnant, and thousands of girls aged 14 or younger have babies.

Scientists have developed a variety of contraceptives with a high percentage of reliability, but the ideal contraceptive has not yet been developed. Women in the United States currently have available four highly effective, reversible methods of contraception: oral contraceptives, injectable contraceptives (Depo-Provera and Lunelle), progestin (a synthetic form of progesterone) implants (Norplant), and intrauterine devices (IUDs). The only male method of contraception currently available is the condom.

Among the issues that must be considered in developing contraceptives are safety, effectiveness, cost, convenience, and ease of use. Risks of cancer, birth defects in case the method fails and the woman becomes pregnant, permanent infertility, and side effects such as menstrual abnormalities must be minimized.

Some researchers predict that the contraceptive methods of the 21st century will control regulatory peptides and the genes that code for them. For example, the genes that code for the pituitary hormones that stimulate the ovaries to release estrogen could be turned off selectively, or other hormone signals could be interrupted. Other approaches being studied are molecular interruption of fertilization and contraceptive vaccines. Hormone therapy that will interfere with sperm production will require an estimated 15 to 20 years to develop. Within 25 years, current methods of contraception will likely be replaced by more sophisticated molecular methods.

Some of the more common methods of birth control are described in the following paragraphs and in Table 48–3 (see also Fig. 48–21). Note that IUDs as well as some types of oral contraceptives do not prevent fertilization; they prevent implantation of the embryo in the uterine wall.

Most hormone contraceptives prevent ovulation

Oral contraceptives, injectable contraceptives, and progestin implants (Norplant) are hormone contraceptives. More than 80 million women worldwide (more than 8 million in the United States alone) use **oral contraceptives.** The most common preparations are combinations of progestin and synthetic estrogen. (Natural hormones are destroyed by the liver almost immediately, but synthetic ones are chemically modified so that they can be absorbed effectively and metabolized slowly.) In a typical regimen, a woman takes one pill each day for about three weeks. Then for one week she takes a sugar pill that allows menstruation to occur because of the withdrawal of the hormones. When taken correctly, these pills are about 99.7% effective in preventing pregnancy.

Most oral contraceptives prevent pregnancy by preventing ovulation. When postovulatory levels of ovarian hormones are maintained in the blood, the body is tricked into responding as though conception has occurred. The pituitary gland is inhibited and does not produce the surge of LH that stimulates ovulation. The chief advantage of oral contraceptives is their high rate of effectiveness. They are also beneficial in regulating menstrual cycles and decreasing menstrual flow.

Studies suggest that women over the age of 35 who smoke or have other risk factors, such as untreated hypertension, should not take oral contraceptives. Women in this category who take oral contraceptives have an increased risk of death from cardiovascular diseases such as stroke and myocardial infarction. The new low-dose oral contraceptive pills appear to be safe for nonsmokers up to the time of menopause. Oral contraceptives are linked to deaths in about 3 per 100,000 users. This compares favorably with the death rate of about 9 per 100,000 pregnancies (Table 48–4 on page 1082).

Injectable contraceptives are becoming increasingly popular. Injectable progestin (Depo-Provera) is injected intramuscularly every three months. Lunelle, a combination of estrogen and progestin, is injected monthly. These injectable contraceptives prevent ovulation by suppressing anterior pituitary function and also thicken the cervical mucus (which makes it more difficult for the sperm to reach the egg).

Implantation of progestin (Norplant) in the woman's body involves inserting several soft, flexible capsules under the skin of the upper arm. There the synthetic hormone is continuously

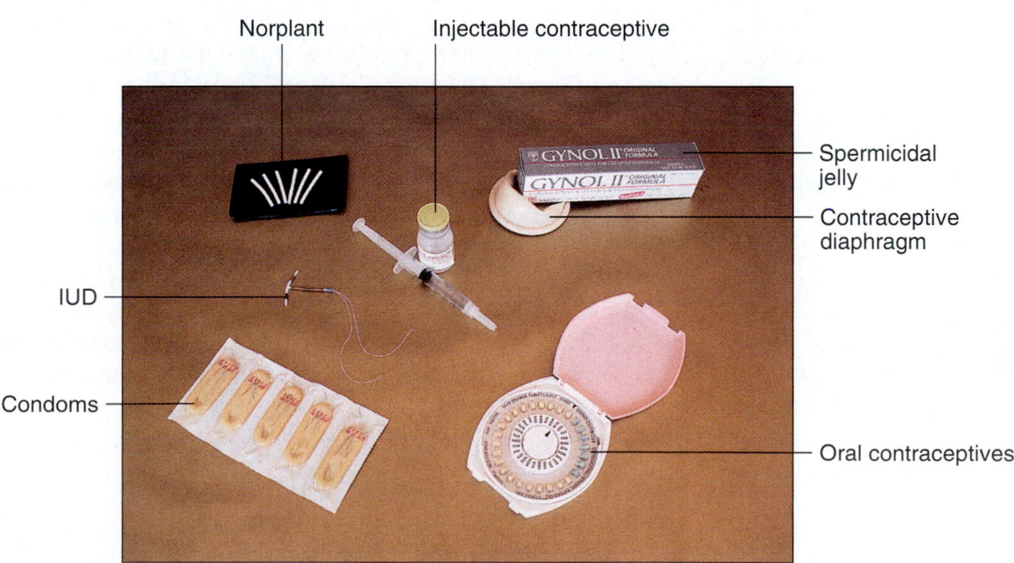

Figure 48–21 Some commonly used contraceptives. The only common male method of contraception is the condom. *(McMurray Photography)*

TABLE 48–3 Birth Control Methods

Method	Failure Rate*	Mode of Action	Advantages	Disadvantages
Oral contraceptives	0.3; 5	Inhibit ovulation; may also affect endometrium and cervical mucus and prevent implantation	Highly effective; regulate menstrual cycle; prevent endometrial and ovarian cancer	Minor discomfort in some women; because of potential for thromboembolism or hypertension, should not be used by women over age 35 who smoke
Injectable contraceptives (Depo-Provera, Lunelle)	About 1	Inhibit ovulation	Effective; long-lasting	Irregular menstrual bleeding, fertility may not return for 6–12 months after contraceptive is discontinued
Progestin implantation (Norplant)	About 1	Inhibits ovulation	Effective; long-lasting	Irregular menstrual bleeding in some women; requires small incision in arm for placement by physician
Intrauterine device (IUD)	1; 1	Not known; probably stimulates inflammatory response and prevents implantation	Provides continuous protection; highly effective for several years	Cramps; increased menstrual flow; spontaneous expulsion; increased risk of pelvic inflammatory disease and infertility; not recommended for women who have not had a child
Spermicides (foams, jellies, creams)	3; 20	Chemically kill sperm	No known side effects; can be used with a condom or diaphram to improve efficacy	Messy to use; must be applied before intercourse

*The lower figure is the failure rate of the method; the higher figure is the rate of method failure plus failure of the user to utilize the method correctly. Based on number of failures per 100 women who use the method per year in the United States.

released and is transported by the blood. The hormone inhibits ovulation and stimulates thickening of the cervical mucus. The capsules are effective for about five years, after which they must be replaced. The most common side effect is irregular menstrual bleeding, which may last up to one year. Injectable contraceptives and progestin implants are very effective, reversible contraceptive procedures. Their failure rate is about 1 in 100.

Use of the intrauterine device has declined in the United States

The **intrauterine device (IUD)** is a small device that must be inserted into the uterus by a medical professional. IUDs are about 99% effective. They are the most widely used method of reversible contraception in the world, used by an estimated 90 million women. Their low use in the United States stems from problems with an early IUD (which was taken off the market in 1974) that caused infections. IUDs in current use have been shown to be safe.

Two types of IUDs presently being used in the United States are the Copper T380 (ParaGard) and the hormone-releasing Progestasert, which continuously releases progestin. Although the Progestasert must be changed yearly, the copper-containing ParaGard can be left in place for up to ten years. Disadvantages of the IUD include uterine cramping and bleeding in a small percentage of women.

The IUD's mode of action is not well understood. The copper or hormone slowly released from an IUD apparently interferes with embryo implantation. In addition, white blood cells mobilized in response to the foreign body in the uterus may produce substances toxic to sperm.

Other common contraceptive methods include the diaphragm and condom

The **contraceptive diaphragm** mechanically blocks the passage of sperm from the vagina into the cervix. It is covered with spermicidal jelly or cream and inserted just prior to sexual intercourse.

TABLE 48–3 *continued*

Method	Failure Rate*	Mode of Action	Advantages	Disadvantages
Contraceptive diaphragm (with jelly)†	3; 14	Diaphragm mechanically blocks entrance to cervix; jelly is spermicidal	No side effects	Must be prescribed and fitted by physician; must be inserted prior to intercourse and left in place for several hours afterward
Condom	2.6; 14	Mechanically prevents sperm from entering vagina	No side effects; some protection against STD, including HIV	Interruption of foreplay to put it on; slightly decreased sensation for male; could break
Rhythm‡	13; 19	Abstinence during fertile period	No known side effects	Not very reliable
Withdrawal (coitus interruptus)	9; 22	Male withdraws penis from vagina prior to ejaculation	No side effects	Not reliable; sperm in the fluid secreted before ejaculation may be sufficient for conception
Sterilization				
Tubal ligation	0.04	Prevents ovum from leaving uterine tube	Most reliable method	Requires surgery; considered permanent
Vasectomy	0.15	Prevents sperm from leaving vas deferens	Most reliable method	Requires surgery; considered permanent
Chance (no contraception)	About 90			

†The failure rate is lower when the diaphragm is used together with spermicides.
‡There are several variations of the rhythm method. For those who use the calendar method alone, the failure is about 35. However, if the body temperature is taken daily and careful records are kept (temperature rises after ovulation), the failure rate can be reduced. When women use some method to help determine the time of ovulation and have intercourse *only* more than 48 hours *after* ovulation, the failure rate can be reduced to about 7.

The **condom** is also a mechanical method of birth control. The only contraceptive device currently sold for men, the condom provides a barrier that contains the semen so that sperm cannot enter the female tract. The latex condom is the only contraceptive that provides some protection against infection with human immunodeficiency virus (HIV) and other sexually transmitted diseases (STDs). Thinner, stronger condoms made of nonlatex material are being developed. A condom for women has been developed but is not widely used.

Emergency contraception is available

Emergency contraception is after-the-fact contraception for rape victims and others who have had unprotected intercourse. In 1997 David A. Kessler, Commissioner of the U.S. Food and Drug Administration, asserted that if physicians advocated and women used emergency contraception, an estimated 2 million unwanted pregnancies and thousands of abortions could be prevented. Emergency contraception is provided in two ways: using increased doses of certain oral contraceptive pills within 72 hours or insertion of a copper IUD within five to seven days. Insertion of a copper IUD within a week of unprotected intercourse is more than 99% effective in preventing pregnancy.

High doses of oral hormone contraceptives can be used as **morning-after pills.** These contraceptives change the endometrium so that the embryo cannot implant in the uterine wall. Taken up to 72 hours after unprotected intercourse, they are about 75% effective in preventing pregnancy.

The drug *RU-486* (mifepristone) binds competitively with progesterone receptors in the uterus but does not activate them. Because the normal actions of progesterone are blocked, the endometrium breaks down and uterine contractions occur. These conditions prevent implantation of an embryo. As discussed later, RU-486 can also be used to abort an implanted embryo.

Sterilization renders an individual incapable of producing offspring

Aside from total abstinence, **sterilization** is the only method of contraception not affected by human error. Sterilization is currently the most popular contraceptive method for couples in which the woman is older than 30 years of age.

TABLE 48–4 Deaths in the United States from Pregnancy and Childbirth and from Some Birth Control Methods

	Death Rate per 100,000
Pregnancy and childbirth	9
Oral contraception	3
Intrauterine device	0.5
Legal abortions	< 1
Illegal abortions performed by medically untrained individuals	About 100

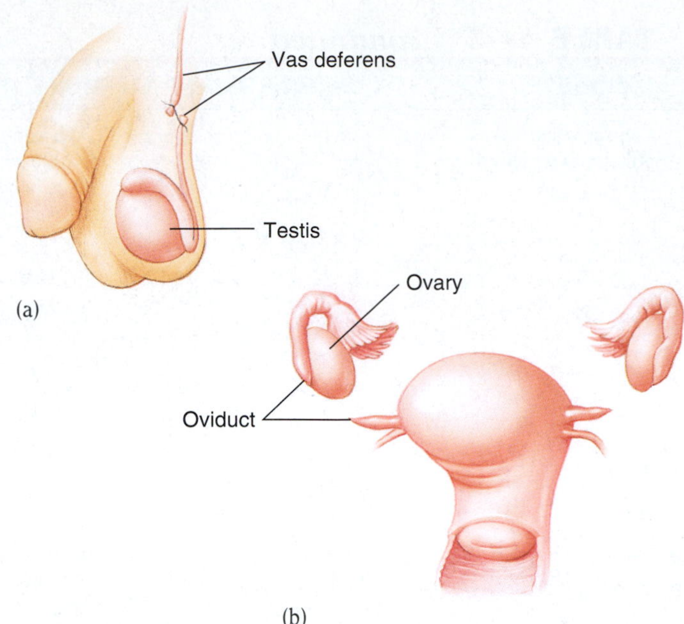

Figure 48–22 Sterilization. **(a)** In vasectomy, the vas deferens (sperm duct) on each side is cut and cauterized. **(b)** In tubal ligation, each oviduct is cut and cauterized so that ovum and sperm can no longer meet.

Male sterilization is performed by vasectomy

An estimated 1 million **vasectomies** are performed each year in the United States. With the use of a local anesthetic, a small incision is made on each side of the scrotum. Then, each vas deferens is cut and its ends sealed so that they cannot grow back together (Fig. 48–22*a*). Because testosterone secretion and transport are not affected, vasectomy does not affect masculinity. Sperm continue to be produced, though at a much slower rate, and are destroyed by macrophages in the testes. No change in the amount of semen ejaculated is noticed because sperm account for very little of the semen volume.

By surgically reuniting the ends of the vasa deferentia, surgeons can successfully reverse male sterilization in about 70% of attempts. Apparently, some sterilized men eventually develop antibodies against their own sperm and remain sterile even after their vasectomies have been surgically reversed. For this reason fertility rates are low (about 30%) for men who undergo vasectomy reversal after ten or more years.

An alternative to vasectomy reversal is the storage of frozen sperm in sperm banks. If the male should decide to father another child after he has been sterilized, he simply "withdraws" his sperm to artificially inseminate his mate. Sperm banks have been established throughout the United States. Not much is known yet about the effects of long-term sperm storage, and there may be an increased risk of genetic defects.

Female sterilization is by tubal ligation

About 41% of married women between the ages of 15 and 44 have chosen sterilization as a means of birth control. **Tubal ligation** is generally performed under general anesthesia by making a small abdominal incision and using a laparoscope to locate and cut the oviducts (Fig. 48–22*b*). The tubes are then cauterized. As in the male, hormone balance and sexual performance are not affected by sterilization.

There are three types of abortion

Abortion is termination of pregnancy that results in the death of the embryo or fetus. Worldwide, an estimated 40 million deliberate abortions are performed each year (more than a million in the United States). Three types of abortions may be distinguished: spontaneous abortions, therapeutic abortions, and those performed as a means of birth control. **Spontaneous abortions** (also called miscarriages) occur without intervention. Spontaneously aborted embryos are frequently abnormal. **Therapeutic abortions** are performed to maintain the health of the mother or when there is reason to suspect that the embryo is grossly abnormal. The third type of abortion, performed as a means of birth control, is the most controversial.

Most first-trimester abortions (those performed during the first three months of pregnancy) are performed by a suction method or by administering drugs that interrupt the pregnancy and induce labor. In the suction method of abortion, the cervix is dilated, and a suction aspirator is inserted into the uterus. The embryo and other products of conception are evacuated. The action of methotrexate, a drug used to treat cancer, is similar to that of RU-486. Both of these drugs can be used to interrupt pregnancy. Prostaglandins may then be used to induce the uterine contractions that expel the embryo.

After the first trimester, abortions are performed mainly because of fetal defects or maternal illness. The method most commonly used is dilation and evacuation. The cervix is dilated, forceps are used to remove the fetus, and suction is used to aspirate the remaining contents of the uterus. Drugs such as prostaglandins are being increasingly used to induce labor in second-trimester abortions.

When abortions are performed by skilled medical personnel, the mortality rate in the United States is less than 1 per 100,000 (see Table 48–4). The death rate from illegal abortions performed by medically untrained individuals is about 100 per 100,000. These statistics can be contrasted with the death rate from pregnancy and childbirth of about 9 per 100,000.

TABLE 48–5 Some Common Sexually Transmitted Diseases

Disease and Causative Organism	Course of Disease	Treatment
Chlamydia (*Chlamydia trachomatis,* a bacterium)	Discharge and burning with urination, or may be asymptomatic; most common cause of nongonococcal urethritis in males; men 15–30 years old with multiple sex partners are most at risk	Antibiotics
Gonorrhea (*Neisseria gonorrhoeae,* a gonococcus bacterium)	Bacterial toxin may produce redness and swelling at infection site; symptoms in males are painful urination and discharge of pus from penis; in about 60% of infected women no symptoms occur in initial states; can spread to epididymis (in males) or uterine tubes and ovaries (in females), causing sterility; can cause widespread pelvic or other infection, plus damage to heart valves, meninges (outer coverings of brain and spinal cord), and joints	Antibiotics
Syphilis (*Treponema pallidum,* a spirochete bacterium)	Bacteria enter body through defect in skin near site of infection and spread throughout the body by lymphatic and circulatory routes; primary chancre (a small, painless ulcer) forms at site of initial infection and heals in about a month: highly infectious at this stage; secondary stage follows, in which a widespread rash and influenza-like symptoms may occur; scaly lesions may occur that teem with bacteria and are highly infectious; latent stage that follows can last 20 years; eventually, lesions called gummae may form, consuming parts of the body surface or damaging liver, bone, or spleen; serious brain damage may occur; death results in 5–10% of untreated cases	Penicillin
Genital herpes (herpes simplex type 2 virus)	Tiny, painful blisters appear on genitals; may develop into ulcers; influenza-like symptoms may occur; recurs periodically; threat to fetus or newborn infant	No effective cure; some drugs may shorten outbreaks or reduce severity of symptoms; medication available that can be taken to prevent outbreak
Human papillomavirus (HPV; about 30 strains of HPV can infect reproductive organs)	Some strains cause abnormal Pap smears and have been linked with cervical cancer; other strains cause genital warts	Immune modifiers; interferon
Trichomoniasis (a protozoon)	Symptoms include itching, discharge, soreness; can be contracted from dirty toilet seats and towels; may be asymptomatic in males	Metronidazole (an antibiotic)
"Yeast" infections (genital candidiasis; *Candida albicans*)	Irritation, soreness, discharge; especially common in females; rare in males	Antifungal drugs
Pelvic inflammatory disease (PID; primarily caused by chlamydia or gonorrhea)	Infection of reproductive organs and pelvic cavity; may lead to sterility (> 15% of cases)	Antibiotics, surgical removal of affected organs
Acquired immune deficiency syndrome (AIDS)* (caused by a retrovirus known as human immunodeficiency virus, HIV)	Influenza-like symptoms: swollen lymph glands, fever, night sweats, weight loss; decreased immunity, leading to pneumonia, rare forms of cancer	Zidovudine (AZT), protease inhibitors, and a variety of experimental drugs

*AIDS was discussed in Chapter 43.

SEXUALLY TRANSMITTED DISEASES ARE SPREAD BY SEXUAL CONTACT

Sexually transmitted diseases (STDs), also called venereal diseases (VD), are, next to the common cold, the most prevalent communicable diseases in the world. The World Health Organization has estimated that more than 250 million people are infected each year with **gonorrhea,** and more than 50 million are infected with **syphilis.**

In the United States, more than 65 million people are currently living with an incurable STD, and more than 15 million new cases of STD are diagnosed each year. According to Planned Parenthood, one of four U.S. teenagers becomes infected with an STD by age 21. Currently, the most common STD in the United

States is **chlamydia;** it is also the most frequently reported infectious disease. Chlamydia is the most common cause of **pelvic inflammatory disease (PID),** an infection of the female reproductive organs that often leads to sterility. Some common STDs are described in Table 48–5 on page 1083. (HIV was discussed in Chapter 43.)

SUMMARY WITH KEY TERMS

I. In **asexual reproduction,** a single parent endows its offspring with a set of genes identical to its own (except for mutations). Asexual reproduction can take place by budding, fragmentation, or parthenogenesis.
   A. In **budding,** a part of the parent's body grows and separates from the rest of the body.
   B. In **fragmentation,** the parent's body may break into several pieces; each piece can develop into a new animal.
   C. In **parthenogenesis,** an unfertilized egg develops into an adult.

II. In **sexual reproduction,** offspring are produced by the fusion of two types of **gametes** (**ovum** [egg] and **sperm**). When sperm and egg fuse, a fertilized egg, or **zygote,** forms.
   A. In **external fertilization,** mating partners typically release eggs and sperm into the water simultaneously.
   B. In **internal fertilization,** the male delivers sperm into the female's body.
   C. In **hermaphroditism,** a single individual produces both eggs and sperm.

III. The human male reproductive system includes the **testes,** which produce sperm and **testosterone;** a series of conducting ducts; accessory glands; and the **penis.**
   A. The testes, housed in the **scrotum,** contain the **seminiferous tubules,** where **spermatogenesis** takes place. The **interstitial cells** in the testes secrete testosterone. **Sertoli cells** are large cells that produce a fluid that nourishes sperm cells; they also produce signaling molecules.
      1. **Spermatogonia** divide by mitosis; some differentiate and become **primary spermatocytes,** which undergo **meiosis.**
      2. The first meiotic division produces two **secondary spermatocytes.** In the second meiotic division, each secondary spermatocyte yields two **spermatids.**
      3. Each spermatid differentiates to form a mature sperm. The head of a sperm consists of the nucleus and a cap, or **acrosome,** containing enzymes that help the sperm penetrate the egg.
   B. Sperm complete their maturation and are stored in the **epididymis** and **vas deferens.**
   C. During ejaculation, sperm pass from the vas deferens to the **ejaculatory duct** and then into the **urethra,** which passes through the penis.
   D. Each ejaculate of semen contains about 200 million sperm suspended in the secretions of the **seminal vesicles** and **prostate gland.** The **bulbourethral glands** release a mucous secretion.
   E. The penis consists of three columns of **erectile tissue,** two **cavernous bodies** and one **spongy body** that surrounds the urethra. When its erectile tissue becomes engorged with blood, the penis becomes erect.
   F. Endocrine regulation of male reproduction involves the hypothalamus, pituitary gland, and testes. The hypothalamus secretes **gonadotropin-releasing hormone (GnRH),** which stimulates the anterior pituitary gland to secrete the gonadotropic hormones: **follicle-stimulating hormone (FSH)** and **luteinizing hormone (LH),** also called **interstitial cell–stimulating hormone (ICSH).**
      1. FSH, LH, and testosterone directly or indirectly stimulate sperm production. LH stimulates the interstitial cells of the testes to produce testosterone.
      2. FSH stimulates the Sertoli cells to produce (1) **androgen-binding protein (ABP),** which binds to testosterone and concentrates it; and (2) **inhibin,** a hormone that inhibits FSH secretion.
      3. Testosterone is responsible for establishing and maintaining male **primary sex characteristics** and **secondary sex characteristics.**

IV. The female reproductive system includes the **ovaries,** which produce oocytes and the steroid hormones **estrogen** and **progesterone;** oviducts (uterine tubes); uterus; vagina; vulva, or external genital structures; and breasts.
   A. **Oogenesis** takes place in the ovaries. **Oogonia** differentiate into **primary oocytes.** A primary oocyte and the **granulosa cells** surrounding it make up a **follicle.**
      1. As the follicle grows, connective tissue cells surrounding the granulosa cells form a layer of **theca cells.**
      2. As the follicle matures, the primary oocyte undergoes the first meiotic division, giving rise to a **secondary oocyte** and a **polar body.**
      3. During **ovulation,** the secondary oocyte is ejected from the ovary and enters one of the paired **oviducts (uterine tubes),** where it may be fertilized.
      4. The part of the follicle remaining in the ovary develops into a **corpus luteum,** a temporary endocrine gland.
   B. The **uterus** serves as an incubator for the developing embryo. The epithelial lining of the uterus, the **endometrium,** thickens each month in preparation for possible pregnancy. The lower part of the uterus, the **cervix,** extends into the vagina.
   C. The **vagina** is the lower part of the birth canal and also receives the penis during sexual intercourse.
   D. The **vulva** include the **labia majora, labia minora, vestibule** of the vagina, **clitoris**, and **mons pubis.**
   E. The breasts function in **lactation,** production of milk for the young. Each breast consists of 15 to 20 lobes of glandular tissue. Gland cells are arranged in **alveoli.** The hormone **prolactin** stimulates milk production; **oxytocin** stimulates ejection of milk from the alveoli into the ducts from which it can be sucked.
   F. Endocrine regulation of female reproduction involves the hypothalamus, pituitary gland, and ovaries. The first day of the **menstrual cycle** is marked by the beginning of menstrual bleeding. **Ovulation** occurs at about day 14 in a typical 28-day menstrual cycle. Events of the menstrual cycle are coordinated by gonadotropic and ovarian hormones.
      1. During the **preovulatory phase, gonadotropin-releasing hormone (GnRH)** from the hypothalamus stimulates the anterior lobe of the pituitary to secrete follicle-stimulating hormone (FSH) and luteinizing hormone (LH).
      2. FSH stimulates follicle development and stimulates the granulosa cells to produce **estrogen.**
      3. LH stimulates theca cells to multiply and to produce androgens. The androgens diffuse into the granulosa cells and are converted to estrogen.
      4. Estrogen stimulates development of the endometrium. After the first week only one follicle continues to develop. Although at relatively low concentration, estrogen inhibits FSH and LH secretion. Granulosa cells produce inhibin, which also inhibits FSH secretion.
      5. During the late preovulatory phase, estrogen concentration peaks, and estrogen signals the anterior pituitary to secrete LH.
      6. LH stimulates final maturation of the follicle and stimulates ovulation.
      7. During the **postovulatory phase,** LH promotes development of the corpus luteum. The corpus luteum secretes progesterone and estrogen, which stimulate final preparation of the uterus for possible pregnancy.

8. During the postovulatory phase, progesterone, along with estrogen, inhibits secretion of GnRH, FSH, and LH.
9. If fertilization does not occur, the corpus luteum degenerates, concentrations of estrogen and progesterone in the blood fall, and menstruation occurs.
10. Estrogen is responsible for the secondary female sex characteristics.

V. Vasocongestion and increased muscle tension are physiological responses to sexual stimulation. The phases of sexual response include **sexual excitement, plateau, orgasm,** and **resolution.**

VI. Human **fertilization** is the fusion of secondary oocyte and sperm to form a zygote. Fertilization and establishment of pregnancy together are called **conception.**

VII. If the secondary oocyte is fertilized, development begins and the embryo implants in the uterus. Membranes that develop around the embryo secrete **human chorionic gonadotropin (hCG),** a hormone that maintains the corpus luteum. During the first 2 to 3 months of pregnancy, the corpus luteum secretes the large amounts of estrogen and progesterone needed to maintain pregnancy. After three months, the **placenta,** the organ of exchange between mother and embryo, assumes this function.

VIII. Several hormones, including estrogen, oxytocin, and prostaglandins, regulate **parturition,** the birth process. Labor can be divided into three stages; the baby is delivered during the second stage.

IX. Effective methods of **contraception** include hormonal methods **(oral contraceptives, injectable contraceptives** [e.g., Depo-Provera], **implantation of progestin,** and **morning-after pills); intrauterine devices; condoms; contraceptive diaphragms;** and **sterilization (vasectomy** or **tubal ligation).** Emergency contraception can be used to prevent unwanted pregnancy after rape or unprotected intercourse.

X. **Spontaneous abortions** (miscarriages) occur without intervention. **Therapeutic abortions** are performed to maintain the mother's health or when the embryo is thought to be grossly abnormal. Abortions are also performed as a means of birth control.

XI. Among the common types of **sexually transmitted diseases (STD)** are **chlamydia, gonorrhea, syphilis, pelvic inflammatory disease (PID), genital herpes,** and **HIV.**

POST-TEST

1. Which of the following is *not* an example of asexual reproduction? (a) budding (b) external fertilization (c) fragmentation (d) parthenogenesis (e) regeneration to form a new individual from a fragment
2. Hermaphroditism (a) is a form of asexual reproduction (b) occurs when an unfertilized egg develops into an adult animal (c) can involve cross-fertilization between two animals (d) typically involves self-fertilization (e) typically involves only one animal
3. The seminiferous tubules (a) are the site of spermatogenesis (b) produce most of the seminal fluid (c) empty directly into the vas deferens (d) are located within the cavernous body (e) receive fluid from the bulbourethral glands
4. Arrange the following stages into the correct sequence. (1) spermatogonium (2) spermatid (3) primary spermatocyte (4) secondary spermatocyte (5) sperm (a) 2, 1, 3, 4, 5 (b) 5, 1, 3, 4, 2 (c) 4, 3, 1, 2, 5 (d) 1, 4, 3, 2, 5 (e) 1, 3, 4, 2, 5
5. Which sequence best describes the passage of sperm? (1) seminiferous tubules (2) vas deferens (3) epididymis (4) ejaculatory duct (5) urethra (a) 3, 1, 2, 4, 5 (b) 1, 3, 2, 4, 5 (c) 5, 4, 2, 3, 1 (d) 1, 3, 4, 2, 5 (e) 3, 1, 4, 2, 5
6. Which of the following characteristics is *not* associated with testosterone? (a) maintains secondary sex characteristics (b) responsible for primary sex characteristics (c) principal androgen (d) protein hormone (e) necessary for spermatogenesis
7. Androgen-binding protein (a) is secreted by Sertoli cells (b) stimulates estrogen production (c) inhibits secretion of FSH (d) inhibits spermatogenesis (e) two of the preceding answers are correct
8. Which of the following cells is haploid? (a) primary oocyte (b) oogonium (c) secondary oocyte (d) corpus luteum (e) follicle cell
9. The corpus luteum (a) is surrounded by ova (b) degenerates if fertilization occurs (c) develops in the preovulatory phase (d) is maintained by prostaglandins (e) serves as a temporary endocrine gland
10. After ovulation the secondary oocyte enters the (a) ovary (b) corpus luteum (c) cervix (d) oviduct (e) vagina
11. The endometrium (a) is the muscle layer of the uterus (b) is thickest during the preovulatory phase (c) is the site of embryo implantation (d) lines the vagina (e) is directly affected by FSH
12. The hormone that reaches its highest level during the postovulatory phase is (a) progesterone (b) estrogen (c) FSH (d) LH (e) testosterone
13. Uterine contraction is strongly stimulated by (a) progesterone (b) FSH (c) LH (d) inhibin (e) oxytocin
14. Which of the following is *not* a hormonal type of contraception? (a) Depo-Provera (b) contraceptive diaphragm (c) injectable progestin (d) progestin implantation (e) oral contraceptives
15. Pelvic inflammatory disease is commonly caused by (a) syphilis (b) gonorrhea (c) genital herpes (d) HIV (e) chlamydia

REVIEW QUESTIONS

1. Compare asexual with sexual reproduction, and describe two specific modes of asexual reproduction.
2. Explain the physiological basis for erection of the penis.
3. Compare the functions of ovaries and testes.
4. Trace the passage of sperm from a seminiferous tubule through the male reproductive system until it leaves the male body during ejaculation. Assuming that ejaculation takes place within the vagina, trace the journey of the sperm until it meets the ovum.
5. What are the actions of testosterone? Give an overview of the endocrine regulation of male reproduction.
6. What is the function of the corpus luteum? Which hormone stimulates its development? Contrast the fate of the corpus luteum when fertilization occurs with its fate when the ovum is not fertilized.
7. Give an overview of the endocrine regulation of female reproduction. What are specific actions of FSH and LH in the female? Of estrogens? Of progesterone?
8. Why are so many sperm produced in the male and so few ova produced in the female?
9. Compare common methods of birth control with regard to their effectiveness, advantages, and disadvantages. Describe emergency contraception.
10. Draw a diagram of the principal events of the menstrual cycle, including ovulation and menstruation. Indicate on which days of the cycle sexual intercourse would most likely result in pregnancy.
11. Describe the actions of hormones during pregnancy and parturition.

12. Label the following diagram of a portion of the male reproductive system. Use Figure 48–4 to check your answers.

13. Label the following diagram of the female reproductive system (anterior view). Use Figure 48–11 to check your answers.

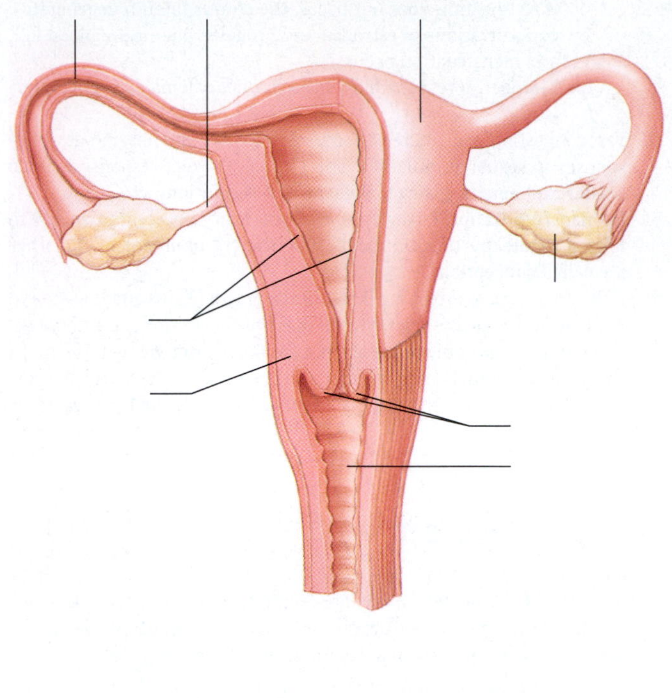

YOU MAKE THE CONNECTION

1. Contrast the biological advantages of hermaphroditism with cross-fertilization and self-fertilization.
2. What would happen if ovulation occurred but no corpus luteum developed?
3. Why do you think a variety of effective methods of male contraception have not yet been developed? If you were a researcher, what are some approaches you might take to developing a method of male contraception?

RECOMMENDED READINGS

Cain, J.M., and M.K. Howett. "Preventing Cervical Cancer." *Science*, Vol. 288, 9 Jun. 2000. A brief Policy Forum review of screening for cervical cancer, the link with human papillomavirus (HPV), and efforts to develop a vaccine against HPV.

Davis, D.L., D. Axelrod, M.P. Osborne, and N.T. Telang. "Environmental Influences on Breast Cancer Risk." *Science & Medicine,* Vol. 4, No. 3, May/Jun. 1997. Exposure to estrogen is a major risk factor for breast cancer.

Duellman, W.E. "Reproductive Strategies of Frogs." *Scientific American,* Vol. 267, No. 1, Jul. 1992. In addition to the egg-to-tadpole progression, frogs have other reproductive strategies such as egg to froglet, egg brooding, and tadpoles in the mother's stomach.

Garnick, M.B., and W.R. Fair. "Combating Prostate Cancer." *Scientific American*, Vol. 279, No. 6, Dec. 1998. Advances in diagnosis and treatment will increase survival time.

Kuiper, G.G.J.M., M. Carlquist, and J. Gustafsson. "Estrogen Is a Male and Female Hormone." *Science & Medicine,* Vol. 5, No. 4, Jul./Aug. 1998. Discovery of a second type of estrogen receptor appears to be related to a new understanding of the actions of estrogen in males.

Marx, J. "Do Centrosome Abnormalities Lead to Cancer." *Science*, Vol. 292, 20 Apr. 2001. Links between centrosome abnormalities and aneuploidy may provide clues to breast and prostate cancer.

Pennisi, E. "Evolutionary Biology: In Search of Biological Weirdness." *Science*, Vol. 290, 10 Nov. 2000. Hypotheses and models about sexual reproduction.

Vander, A., J. Sherman, and D. Luciano. *Human Physiology: The Mechanisms of Body Function,* 8th ed. McGraw-Hill, New York, 2001. Chapter 19 focuses on reproduction.

Wright, K. "Human in the Age of Mechanical Reproduction." *Discover,* May 1998. An interesting discussion of noncoital conception.

Wuethrich, B. "Why Sex? Putting Theory to Test." *Science*, Vol. 281, 25 Sep. 1998. A brief, but excellent, summary of data supporting hypotheses regarding the evolutionary function of sexual reproduction.

- Visit our Web site at **http://www.info.brookscole.com/solomonbergmartin** for links to chapter-related resources on the World Wide Web. Additional on-line materials relating to this chapter can also be found on our Web site.

See chapter activity on BioActive Learner CD for additional help in mastering the chapter's material. Icon location in the chapter's margins shows which topics have tutorials or simulations in the CD.

49

Animal Development

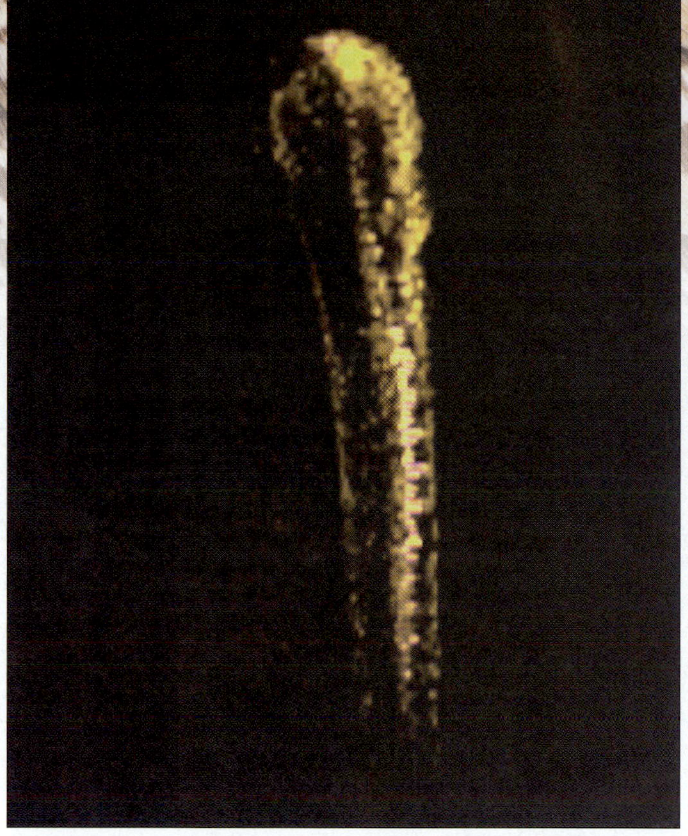

Magnetic resonance image (MRI) of frog embryo. *(Thomas J. Meade, Caltech)*

LEARNING OBJECTIVES

After you have studied this chapter you should be able to

1. Summarize the roles of the following in the development of an animal: cell growth and division; cell determination and cell differentiation; and pattern formation and morphogenesis.
2. Summarize the genetic and physiological functions of fertilization, and describe the four processes involved.
3. Trace the generalized pattern of early development of the embryo from zygote through early cleavage; formation of the morula, blastula, and gastrula; and early organ development.
4. Contrast early development, including cleavage and gastrulation, in the echinoderm (or in amphioxus), the amphibian, and the bird, paying particular attention to the importance of the amount and distribution of the yolk.
5. Summarize the fate of each of the germ layers.
6. Define organogenesis and trace the early development of the vertebrate nervous system.
7. Trace the development of the extraembryonic membranes and placenta (in mammals), giving the functions of each membrane.
8. Describe the general course of early human development, including fertilization, the fates of the trophoblast and inner cell mass, implantation, and the role of the placenta.
9. Contrast postnatal with prenatal life, describing several changes that occur at or shortly after birth that allow the neonate to live independently.
10. List specific steps that a pregnant woman can take to promote the well-being of the developing embryo; describe how the embryo can be affected by nutrients, drugs, cigarette smoking, pathogens, and ionizing radiation.
11. Describe some anatomical and physiological changes that occur with aging, and discuss current hypotheses of aging.

Development includes all the changes that take place during the entire life of an individual. In this chapter, however, we focus mainly on the fertilization of the egg to form a zygote, and the subsequent development of the young animal before birth or hatching. Just how does a microscopic, single-celled zygote give rise to the bones, muscles, brain, and other structures of a complex animal? These are derived from a balanced combination of several fundamental processes: cell division and growth, cell determination and cell differentiation, and pattern formation and morphogenesis.

The zygote divides by mitosis, forming an **embryo** that subsequently undergoes an orderly sequence of cell divisions. In animals, growth (i.e., increase in mass) occurs primarily by an increase in the number of cells but it can also occur by an increase in cell size, as in fat cells. Although the very early embryo usually does not grow, later cell divisions typically do contribute to growth. However, by itself, cell division, which is under the control of a genetic program that interacts with environmental signals (see Chapter 9), would produce only a formless heap of similar cells.

As embryonic development proceeds, certain cells become biochemically and structurally specialized to carry out specific functions, through a process known as cell **differentiation.** In our discussion of the genetic control of development in Chapter 16, we examined how cell differentiation occurs through cell **determination,** a series of molecular events in which the activities of certain genes become altered in ways that cause a cell to become progressively more committed to follow a particular differentiation pathway. This process proceeds even though there may not be any immediate changes in the cell's morphology. We also discussed evidence supporting the principle of **nuclear equivalence,** which states that in most cases neither cell determination nor cell differentiation is accompanied by a loss of genetic information from the cell nucleus. That is, the nuclei of virtually all differentiated cells of an animal contain the same genetic information that was present in the zygote, but each cell type *expresses* a different subset of that information. In fact, nuclear equivalence is the principle that underlies the cloning of organisms (see *Focus On: Mammalian Cloning* in Chapter 16).

Cell differentiation is therefore an expression of changes in the activity of specific genes, and this genetic activity in turn is influenced by a variety of factors inside and outside the cell. It is this **differential gene expression** that is responsible for variations in chemistry, behavior, and structure among cells. Through this process, an embryo develops into an organism composed of more than 200 types of cells, each specialized to perform specific functions. However, not all cells differentiate; some, known as **stem cells,** remain in a relatively undifferentiated state and retain the ability to give rise to a wide variety of cell types (see *Focus On: Stem Cells* in Chapter 16).

Cell differentiation by itself does not explain development. The differentiated cells must become progressively organized, shaping the intricate pattern of tissues and organs that characterizes a multicellular animal. This development of form, known as **morphogenesis,** proceeds through the process of **pattern formation,** a series of steps requiring signaling between cells, changes in the shapes of certain cells, precise cell migrations, interactions with the extracellular matrix, and even **apoptosis** (programmed cell death) of some cells.

In this chapter, we compare and contrast the sequences of developmental events in several different animals. These model organisms have been chosen by researchers because they have certain desirable characteristics. For example, it is advantageous if investigators can study very large numbers of embryos, all developing synchronously. Some embryos are particularly easy to observe because they are transparent. These studies have been facilitated by the use of new technology; for example, the photograph shows a frog embryo in which a technique known as magnetic resonance imaging (MRI) has been used to visualize regions in which a specific developmentally important gene is active. As you read, note the intimate interrelationships among the basic developmental processes, as well as the fundamental similarities among the early developmental events in the animals featured.

FERTILIZATION HAS GENETIC AND PHYSIOLOGICAL CONSEQUENCES

In Chapter 48 we discussed **spermatogenesis** and **oogenesis,** the processes by which meiosis leads to the formation of haploid cells, which differentiate as sperm and eggs, respectively. In **fertilization** a usually flagellated, motile sperm fuses with a much larger, immotile ovum to produce a **zygote,** or fertilized egg. Fertilization has two important genetic consequences: Restoration of the diploid chromosome number and, in mammals and many other animals, determination of the sex of the offspring (see Chapter 10). Fertilization also has profound physiological effects, for it activates the egg by stimulating reactions that permit development to take place.

Fertilization involves four events, some of which may occur simultaneously and do not necessarily follow the same temporal sequence in all animals: (1) the sperm contacts the egg and recognition occurs, (2) the sperm or sperm nucleus enters the egg, (3) the egg becomes activated, and certain developmental changes begin, and (4) the sperm and egg nuclei fuse. Unless otherwise stated, our discussion applies to sea urchins and other echinoderms such as sea stars, which have been studied intensively because they produce very large numbers of gametes, and because fertilization is external.

The first step in fertilization involves contact and recognition

Although eggs are immotile, they are active participants in the fertilization process. An egg is surrounded not only by a plasma membrane but also by one or more external coverings that are important in fertilization. For example, as discussed in Chapter 48, a mammalian egg is enclosed by a thick acellular, **zona pellucida,** which is surrounded by a layer of granulosa cells derived from the follicle in which the egg developed.

The egg coverings not only facilitate fertilization by sperm of the same species but also serve as barriers to interspecific fertilization, a function that is particularly important in species that practice external fertilization. External to the plasma membrane of a sea urchin egg are two acellular layers that interact with sperm: a very thin **vitelline envelope** and, outside of this, a thick glycoprotein layer called the **jelly coat.** Sea urchin sperm become motile when they are released into seawater, and their motility is increased when they reach the vicinity of a sea urchin egg. Motility improves the statistical probability that sperm will encounter the egg, but motility alone may not be sufficient to ensure that actual contact with the egg will occur. In sea urchins, as well as certain fish and some cnidarians, the egg or one of its coverings secretes a chemotactic substance that attracts sperm of the same species. However, for a great many species it has not been possible to demonstrate the existence of a specific chemical that attracts the sperm to the egg.

A sea urchin sperm becomes activated when it contacts the jelly coat. It undergoes an **acrosome reaction,** in which the membranes surrounding the acrosome (the cap at the head of the sperm, see Fig. 48–7) fuse, and pores in the membrane enlarge. Calcium ions from the seawater move into the acrosome, which then swells and begins to disorganize. The acrosome then releases proteolytic enzymes that digest a path through the jelly coat to the vitelline envelope of the egg. If the sperm and egg are of the same species, **bindin,** a species-specific binding protein located on the acrosome, adheres to species-specific bindin receptors located on the vitelline envelope.

Before a mammalian sperm can participate in fertilization, it must first undergo **capacitation,** a maturation process that occurs in the female reproductive tract. During capacitation, which in humans can take several hours, sperm become increasingly motile and capable of undergoing an acrosome reaction when they encounter an egg. There is evidence that mammalian fertilization requires the interaction between sperm and species-specific glycoproteins in the zona pellucida.

Sperm entry is regulated

In sea urchins, once acrosome-vitelline envelope contact occurs, enzymes dissolve a bit of the vitelline envelope in the area of the sperm head. The plasma membrane of the egg is covered with microvilli, several of which elongate to surround the head of the sperm. As this occurs, the plasma membranes of sperm and egg fuse, and a **fertilization cone** is formed which contracts to draw the sperm into the egg (Fig. 49–1).

Fertilization of the egg by more than one sperm, a condition known as **polyspermy**, would result in an offspring with extra sets of chromosomes, which is usually lethal. This is prevented by two reactions, known as the fast and slow blocks to polyspermy. In the fast block to polyspermy the egg plasma membrane becomes depolarized, preventing its fusion with additional sperm. An unfertilized egg is polarized, that is, the cytoplasm is negatively charged relative to the outside, but within a few seconds after sperm fusion, ion channels in the plasma membrane open, permitting positively charged calcium ions to diffuse across the membrane and depolarize the egg.

The slow block to polyspermy, the **cortical reaction,** requires about one to several minutes to complete, but it is a complete block. Binding of the sperm to receptors on the vitelline envelope activates one or more signal transduction pathways (see Chapter 5) in the egg. These events lead to calcium ions stored in the egg endoplasmic reticulum being released into the cytosol. This rise in the level of intracellular calcium causes thousands of cortical granules, membrane-bounded vesicles in the egg cortex (region close to the plasma membrane), to release enzymes, various proteins, and other substances by exocytosis into the space between the plasma membrane and the vitelline envelope (Fig. 49–2). Some of the enzymes dissolve the protein linking the two membranes, allowing the space to enlarge as additional substances released by the cortical granules raise the osmotic pressure, causing an influx of water from the surroundings. Thus the vitelline envelope becomes elevated away from the plasma membrane and forms the fertilization envelope, a hardened covering that prevents entry of additional sperm.

In mammals, a fertilization envelope does not form, but the enzymes released during exocytosis of the cortical granules alter the sperm receptors on the zona pellucida so that no additional sperm can bind to them.

Not all species have these types of blocks to polyspermy. In some, such as salamanders, several sperm usually enter the egg, but only one sperm nucleus fuses with the egg nucleus, and the rest degenerate.

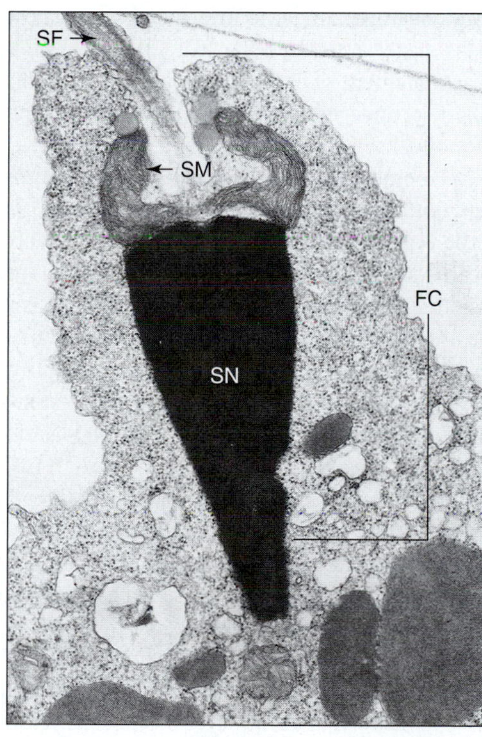

Figure 49–1 Fertilization. A fertilization cone (FC) forms as a sperm enters a sea urchin egg. The TEM shows various sperm organelles in the process of being engulfed by the egg cytoplasm. SN, sperm nucleus; SM, sperm mitochondrion; SF, sperm flagellum. *(Frank J. Longo)*

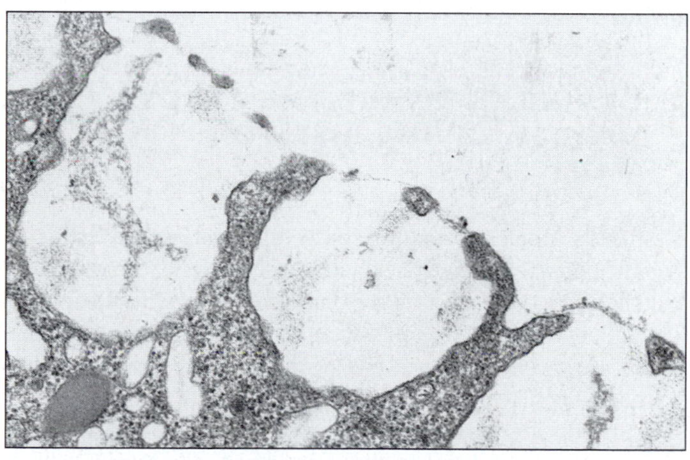

Figure 49–2 The cortical reaction. Three cortical granules are undergoing exocytosis in this TEM of the plasma membrane of a sea urchin egg shortly after fertilization. This release of the contents of the cortical granules initiates the slow block to polyspermy, as it leads to the formation of a hardened fertilization envelope. *(From E. Anderson. Reprinted from* The Journal of Cell Biology, *Vol. 37, 1968, by copyright permission of the Rockefeller University Press)*

Fertilization activates the egg

Release of calcium ions into the egg cytoplasm does more than stimulate the cortical reaction; it also triggers the *activation program,* a series of metabolic changes within the egg. Aerobic respiration increases, certain maternal enzymes and other proteins become active, and within a few minutes after sperm entry, a burst of protein synthesis occurs. In addition, the egg nucleus is stimulated to complete meiosis. (Recall from Chapter 48 that in most animal species, including mammals, at the time of fertilization an egg is actually a secondary oocyte, arrested early in the second meiotic division.)

In some species, an egg can be artificially activated without sperm penetration by swabbing it with blood and pricking the plasma membrane with a needle, by calcium injection, or by certain other treatments. These haploid eggs may go through some developmental stages parthenogenetically, that is, without fertilization (see Chapter 48), but they are unable to complete development. Special mechanisms of egg activation have evolved in those species that are naturally parthenogenetic.

Sperm and egg pronuclei fuse

After the sperm nucleus enters the egg, it is thought to be guided toward the egg nucleus by a system of microtubules that forms within the egg. The sperm nucleus swells, forming the **male pronucleus.** The nucleus formed during the completion of meiosis in the egg becomes the **female pronucleus.** (Recall from Figure 48–13 that the other nucleus formed during meiosis II in the egg becomes the second polar body.) The haploid male and female pronuclei then fuse to form the diploid nucleus of the zygote, and DNA synthesis occurs in preparation for the first cell division.

DURING CLEAVAGE THE ZYGOTE DIVIDES, GIVING RISE TO MANY CELLS

Despite its simple appearance, the zygote is **totipotent,** that is, it has the potential to give rise to all the cell types of the new individual. Because the ovum is very large compared with the sperm, the bulk of the zygote cytoplasm and organelles comes from the ovum. However, the sperm and ovum usually contribute equal numbers of chromosomes.

Shortly after fertilization, the zygote undergoes **cleavage,** a series of rapid mitotic divisions with no period of growth during each cell cycle. For this reason, although the cell number increases, the embryo does not increase in size. The zygote initially divides to form a two-celled embryo. Then, each of these cells undergoes mitosis and divides, bringing the number of cells to four. Repeated divisions increase the number of cells, called **blastomeres,** that make up the embryo. At about the 32-cell stage the embryo is a solid ball of blastomeres called a **morula.** Eventually, anywhere from 64 to several hundred blastomeres form the **blastula,** which is usually a hollow ball with a fluid-filled cavity, the **blastocoel.**

The pattern of cleavage is affected by yolk

Many animal eggs contain **yolk,** a mixture of proteins, phospholipids, and fats that serves as food for the developing embryo. The amount and distribution of yolk vary among different animal groups, depending on the needs of the embryo. Mammalian eggs have very little yolk because the embryo obtains maternal nutritional support throughout most of its development, whereas the eggs of birds and reptiles must contain sufficient yolk to sustain the embryo until hatching. Echinoderm eggs typically need only enough yolk to nourish the embryo until it becomes a tiny larva capable of obtaining its own food.

Most invertebrates and simple chordates have **isolecithal** eggs with relatively small amounts of yolk uniformly distributed through the cytoplasm. Isolecithal eggs divide completely **(holoblastic cleavage).** Cleavage of these eggs can be radial or spiral. Radial cleavage is characteristic of *deuterostomes* such as chordates and echinoderms; spiral cleavage is commonly found in the embryos of *protostomes* such as annelids and mollusks (see Fig. 28–3).

In **radial cleavage,** the first division is vertical and splits the egg into two equal cells. The second cleavage division, also vertical, is at right angles to the first division and separates the two cells into four equal cells. The third division is horizontal, at right angles to the other two, and separates the four cells into eight cells: four above and four below the third line of cleavage. This pattern of radial cleavage occurs in echinoderms and in the cephalochordate amphioxus (Figs. 49–3*a–e* and 49–4).

In **spiral cleavage,** after the first two divisions, the plane of cytokinesis is diagonal to the polar axis (Fig. 49–5). This results in a spiral arrangement of cells, with each cell located above and between the two underlying cells. This pattern is typical of annelids and mollusks.

Many vertebrate eggs are **telolecithal,** meaning that they have large amounts of yolk concentrated at one end of the cell, known as the **vegetal pole.** The opposite, more metabolically active, pole is the **animal pole.** The eggs of amphibians are moderately telolecithal (mesolecithal). Although cleavage is radial and holoblastic, the divisions in the vegetal hemisphere are slowed by the presence of the inert yolk. As a result, the blastula consists of many small cells in the animal hemisphere, and fewer but larger cells in the vegetal hemisphere (Fig. 49–6). The blastocoel is displaced toward the animal pole.

The telolecithal eggs of reptiles and birds have very large amounts of yolk at the vegetal pole and only a small amount of cytoplasm concentrated at the animal pole. The yolk of such eggs never cleaves. Cell division is restricted to the **blastodisc,** the small disc of cytoplasm at the animal pole (Fig. 49–7); this type of cleavage is termed **meroblastic.** In birds and some reptiles, the blastomeres form two layers, separated by the blastocoel cavity: an upper *epiblast* and below that a thin layer of flat cells, the *hypoblast.*

Cleavage may distribute developmental determinants

The pattern of cleavage in a particular species depends not only on yolk but on other factors as well. Recall from Chapter 16 that some organisms have relatively rigid developmental patterns,

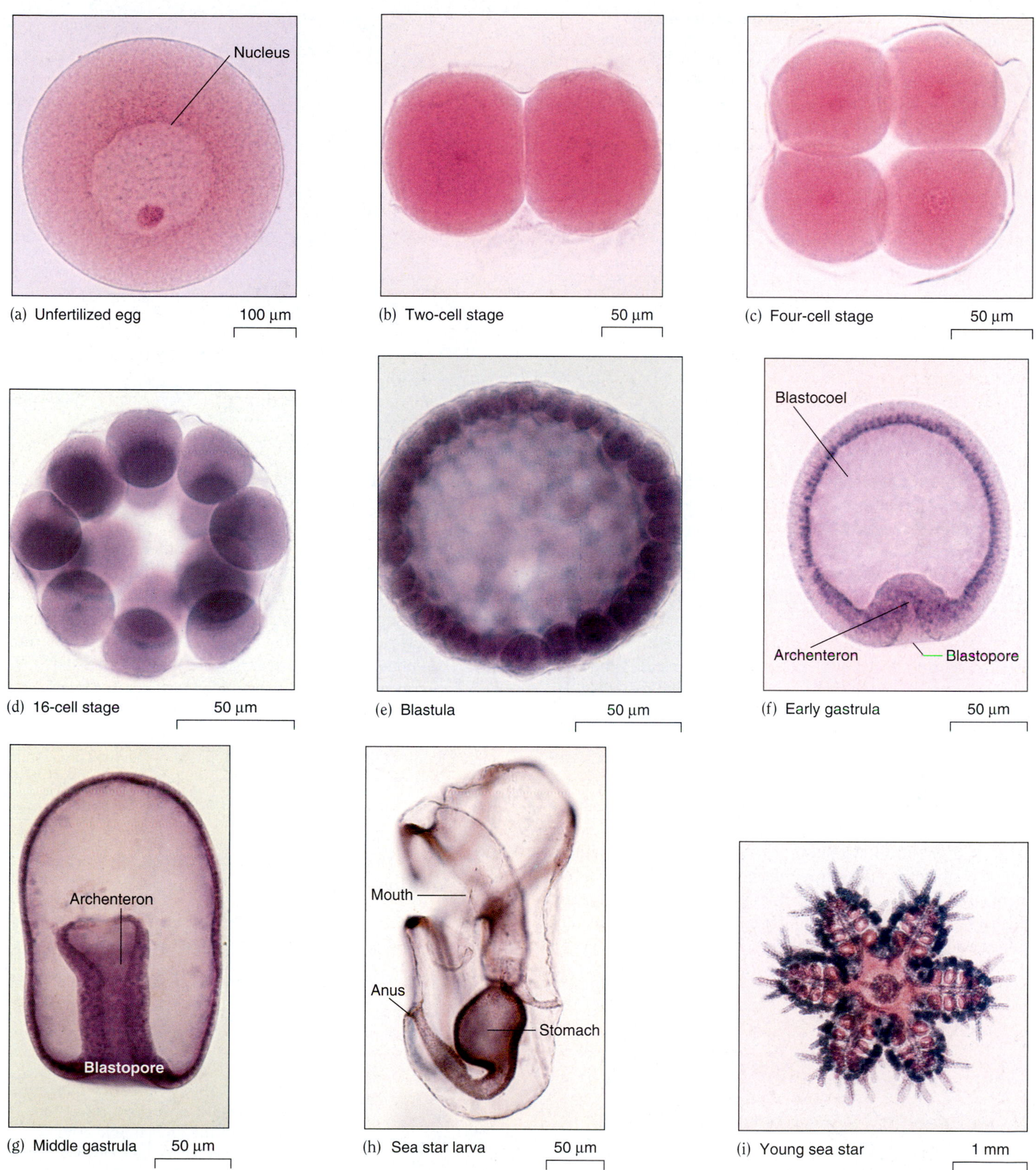

Figure 49–3 LMs showing sea star development. **(a)** The isolecithal egg has a small amount of uniformly distributed yolk. **(b–e)** The cleavage pattern is radial and holoblastic (the entire egg becomes partitioned into cells). **(f, g)** The three germ layers form during gastrulation. The blastopore is the opening into the developing gut cavity, the archenteron. The rudiments of organs are evident in the sea star larva **(h)** and the young sea star **(i)**. All views are side views with the animal pole at the top, except **(c)** and **(i)**, which are top views. Note that the sea star larva is bilaterally symmetrical, but differential growth produces a radially symmetrical young sea star. *(Carolina Biological Supply Company/Phototake-NYC)*

Figure 49–4 Cleavage and gastrulation in amphioxus. As in the sea star, cleavage is holoblastic and radial. The embryos shown are viewed from the side. **(a)** Mature egg with polar body. **(b–e)** Two-, four-, eight-, and 16-cell stages. **(f)** Embryo cut open to show the blastocoel. **(g)** Blastula. **(h)** Blastula cut open. **(i)** Early gastrula showing beginning of invagination at vegetal pole. **(j)** Late gastrula. Invagination is completed and the blastopore has formed.

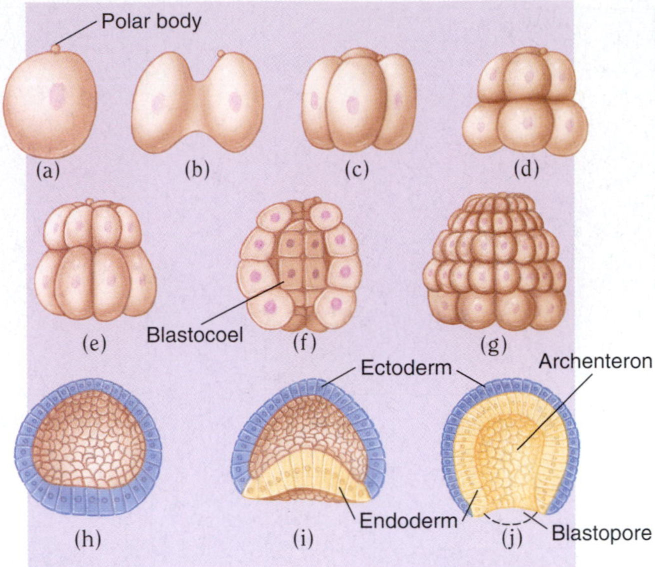

known as **mosaic development.** This is largely a consequence of the unequal distribution of important materials in the cytoplasm of the zygote. Because the zygote cytoplasm is not homogeneous, cytoplasmic developmental determinants portioned out to each new cell during cleavage may be different. Such differences help determine the course of development. At the other extreme are mammals, which have zygotes with very homogenous cytoplasm. They exhibit highly **regulative development,** in which individual cells produced by the cleavage divisions are equivalent, allowing the embryo to develop as a self-regulating whole. Developmental patterns of most animals fall somewhere on a continuum between these two extremes.

In some species the distribution of developmental determinants in the unfertilized egg is maintained in the zygote. In others sperm penetration initiates rearrangement of the cytoplasm. For example, fertilization in the amphibian egg causes a shift of some of the cortical cytoplasm. These cytoplasmic movements can be easily followed because the egg cortex contains dark pigment granules. As a result of this rearrangement, a crescent-shaped region of underlying lighter-colored (gray) cytoplasm becomes evident directly opposite the point on the cell where the sperm penetrated the egg (Fig. 49–8*a*). This **gray crescent** region is thought to contain growth factors and other developmental determinants. The first cleavage bisects the gray crescent, distributing half to each of the first two blastomeres; in this way the position of the gray crescent establishes the future right and left halves of the embryo. As cleavage continues, the gray crescent material becomes partitioned into certain blastomeres. Those that contain parts of the gray crescent eventually develop into the dorsal region of the embryo.

Experiments have confirmed the importance of deteminants in the gray crescent to development. If the first two frog blastomeres are separated experimentally, each develops into a complete tadpole (Fig. 49–8*b*). When the plane of the first division is altered so that the gray crescent is completely absent from one of the cells, that cell does not develop normally (Fig. 49–8*c*).

Cleavage provides building blocks for development

Cleavage partitions the zygote into many small cells that serve as the basic building blocks for subsequent development. At the end

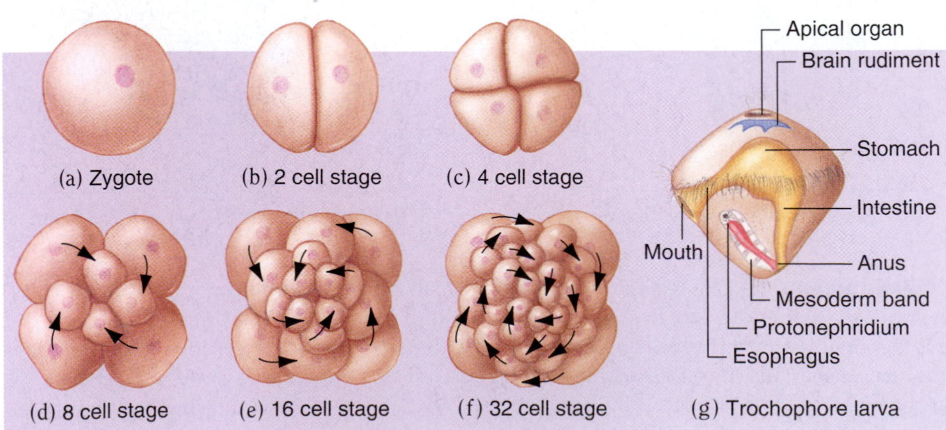

Figure 49–5 Spiral cleavage in an annelid embryo. **(a–f)** Top views of the animal pole. The successive cleavage divisions occur in a spiral pattern as illustrated. **(g)** A typical trochophore larva. The upper half of the trochophore develops into the extreme anterior end of the adult worm; all the rest of the adult body develops from the lower half.

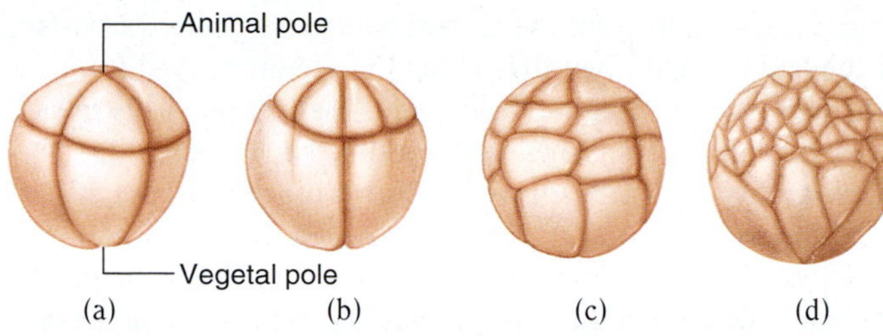

Figure 49–6 Cleavage pattern in a frog egg. **(a–d)** Although cleavage is holoblastic, the large amount of yolk concentrated in the vegetal hemisphere slows cleavage. As a result, fewer cells develop at the vegetal hemisphere than at the animal hemisphere. These embryos are viewed from the side.

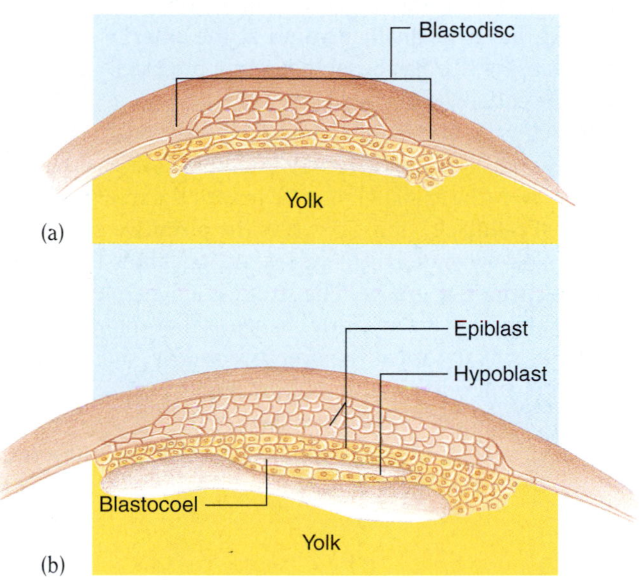

Figure 49–7 Cleavage in a bird embryo. Cleavage is meroblastic; that is, it is restricted to the blastodisc, a small disk of cytoplasm on the upper surface of the egg yolk. **(a)** Early blastodisc formation. This cutaway view shows cells on the blastodisc surface, as well as in the interior. **(b)** The blastodisc splits into two tissue layers, an upper epiblast and a lower hypoblast, separated by the blastocoel.

of cleavage, the small cells that make up the blastula begin to move about with relative ease, arranging themselves into the patterns necessary for further development. Surface proteins are important in helping cells to "recognize" one another and therefore in determining which ones adhere to form tissues.

THE GERM LAYERS FORM DURING GASTRULATION

The process by which the blastula becomes a three-layered embryo, or **gastrula,** is called **gastrulation.** Thus, early development proceeds through the following stages:

Zygote ⟶ early cleavage stages ⟶ morula ⟶ blastula ⟶ gastrula

During gastrulation, the embryo begins to take on a first approximation of its body plan as cells arrange themselves into three distinct **germ layers,** or embryonic tissue layers: the outermost layer, the **ectoderm;** the innermost, the **endoderm;** and the **mesoderm,** which develops between them. Additional cell divisions take place during gastrulation, and the germ layers become established through a combination of processes. Many cells lose their old cell-to-cell contacts and establish new ones through cell recognition and adhesion processes involving complex interactions among the integrins and other plasma membrane proteins, and the extracellular matrix (see Chapters 4 and 5). Many cells undergo cytoskeletal changes, particularly alterations in the distribution of actin microfilaments; these changes in the internal

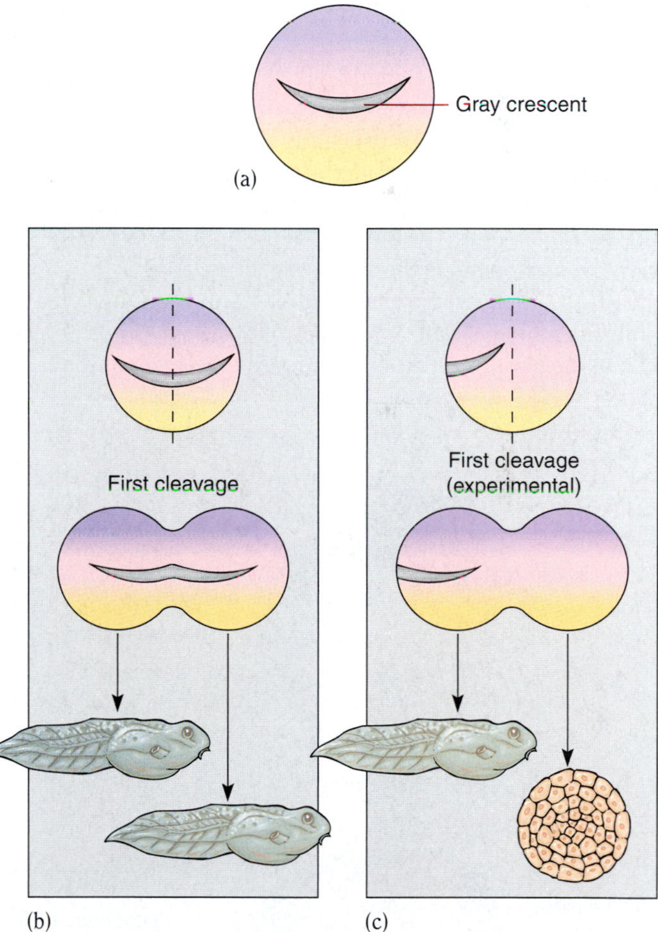

Figure 49–8 Cytoplasmic determinants in frog development. **(a)** The position of the gray crescent in the frog zygote determines the main axes of the body. **(b)** The first division of the zygote partitions the gray crescent into the two daughter cells. If these cells are separated, each can develop into a tadpole. **(c)** If the plane of cleavage is changed experimentally so that only one cell receives the gray crescent, only that cell can develop into a tadpole.

architecture of the cells allow them to change shape and/or undergo specific, directional, amoeboid movements. As a result of these movements, many cells ultimately take up new positions in the interior of the embryo.

Each of the germ layers develops into specific parts of the embryo. As you study the illustrations, keep in mind that a conventional color code accepted by developmental biologists is used to depict the germ layers: endoderm, yellow; mesoderm, red or pink; and ectoderm, blue.

The pattern of gastrulation is affected by the amount of yolk

The simple type of gastrulation that occurs in echinoderms and in amphioxus is illustrated in Figures 49–3*f, g* and 49–4*i, j,* respectively. Gastrulation begins when a group of cells at the vegetal pole undergoes a series of changes in shape, causing that part of the blastula wall to first flatten and then bend inward (invaginate). The invaginated wall eventually meets the opposite wall, obliterating the blastocoel.

We can roughly demonstrate this type of gastrulation by pushing inward on the wall of a partly deflated rubber ball until it rests against the opposite wall. In a similar way the embryo is converted into a double-walled, cup-shaped structure. The new internal wall lines the **archenteron,** the newly formed cavity of the developing gut. The opening of the archenteron to the exterior, the **blastopore,** is the site of the future anus.

This simple type of gastrulation cannot occur in the amphibian embryo because the large yolk-laden cells in the vegetal half of the blastula obstruct any inward movement at the vegetal pole. Instead, cells from the animal pole move down the surface of the embryo and when they reach the region derived from the gray crescent they begin to move into the interior. This inward movement is accomplished as the cells change shape, first becoming flask- or bottle-shaped (so that most of their mass is actually below the surface) and then sinking into the interior as they lose their remaining connections with other cells on the surface. This spot on the embryo's surface, referred to as the **dorsal lip of the blastopore,** is marked by a dimple, shaped like a C lying on its side (Fig. 49–9). As the process continues, the blastopore becomes ring-shaped as cells lateral, and then ventral, to its dorsal lip become involved in similar movements. The yolk-filled cells fill the space enclosed by the lips of the blastopore, forming the yolk plug.

The archenteron forms and is lined on all sides by cells that have moved in from the surface. At first, the archenteron is a narrow slit, but it gradually expands at the anterior end of the embryo, causing the blastocoel to become progressively smaller and to be eventually eliminated.

Although the details differ somewhat, gastrulation in birds is basically similar to amphibian gastrulation. Cells that make up the upper layer (the epiblast) migrate toward the midline to form a thickened cellular region, known as the **primitive streak,** that elongates and narrows as it develops. At its center is a narrow furrow, the **primitive groove.** The streak is a dynamic structure. The cells composing it constantly change as they migrate in from the epiblast, sink inward at the primitive groove, then move out laterally and anteriorly in the interior (Fig. 49–10). The primitive groove is the functional equivalent of the blastopore of the echinoderm, amphioxus, and amphibian embryos. However, a

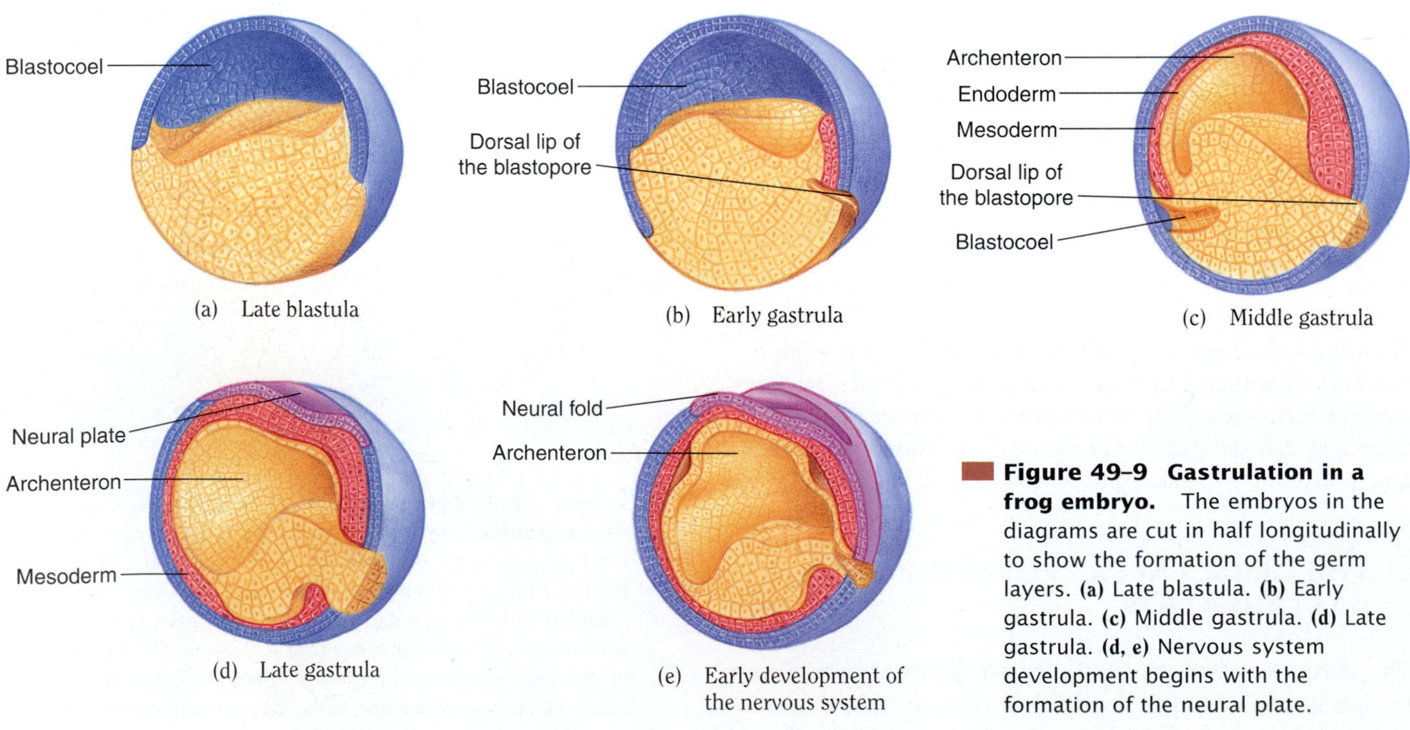

Figure 49–9 Gastrulation in a frog embryo. The embryos in the diagrams are cut in half longitudinally to show the formation of the germ layers. **(a)** Late blastula. **(b)** Early gastrula. **(c)** Middle gastrula. **(d)** Late gastrula. **(d, e)** Nervous system development begins with the formation of the neural plate.

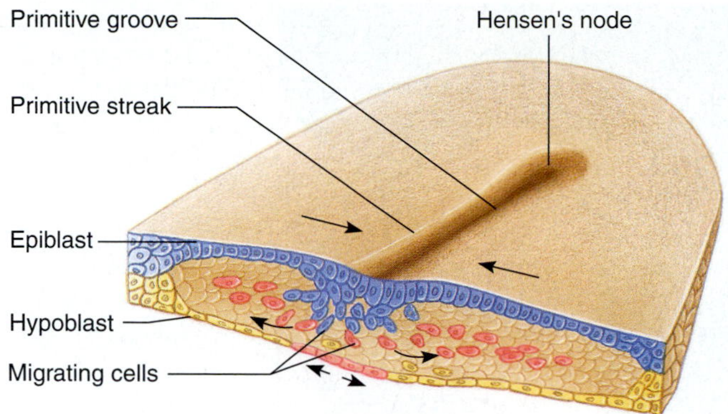

Figure 49–10 Gastrulation in birds. A three-layered embryo is formed as cells move toward the primitive streak, move inward at the primitive groove, and migrate laterally and forward in the interior.

bird embryo contains no cavity that is homologous to the archenteron.

At the anterior end of the primitive streak, cells destined to form the **notochord** (supporting rod of mesodermal, cartilage-like cells that serves as the flexible skeletal axis in all chordate embryos) become concentrated in a thickened knot known as **Hensen's node.** These cells sink into the interior and then move anteriorly just beneath the epiblast, forming a narrow extension from the node. Cells that will form other types of mesoderm move laterally and anteriorly from the primitive streak, between the epiblast (which becomes ectoderm) and the hypoblast. Other cells form the endoderm, displacing the hypoblast cells and causing them to move laterally. These displaced cells will form part of the extraembryonic membranes, discussed in a later section.

ORGANOGENESIS BEGINS WITH THE DEVELOPMENT OF THE NERVOUS SYSTEM

In **organogenesis,** or organ formation, the cells of the three germ layers continue the processes of pattern formation that will lead to the formation of specific structures. The ectoderm eventually forms the outer layer of the skin and gives rise to the nervous system and sense organs. Tissues that eventually line the digestive tract, and organs that develop as outgrowths of the digestive tract (including the liver, pancreas, and lungs) are all of endodermal origin. Skeletal tissue, muscle, and the circulatory, excretory, and reproductive systems all are derived from mesoderm (Table 49–1; see Fig. 16–1).

The notochord, brain, and spinal cord are among the first organs to develop in the early vertebrate embryo (Fig. 49–11; see also Fig. 49–9*d* and *e*). First the notochord, which is mesodermal tissue, grows forward along the length of the embryo as a cylindrical rod of cells. The notochord eventually will be replaced by the vertebral column, although remnants will remain in the discs of cartilage between the vertebrae.

Process of Science The developing notochord has yet another crucial function. Numerous experiments in which researchers have transplanted portions of the notochord mesoderm to other locations in the embryo have demonstrated that it is responsible for causing the overlying ectoderm to thicken and differentiate to form the precursor of the central nervous system, the **neural plate.** Such phenomena, in which certain cells stimulate or otherwise influence the differentiation of their neighbors, are examples of **induction** (see Chapter 16).

It has been shown repeatedly that induction of the neural plate cells does not require direct cell-to-cell contact with the developing notochord cells. It is therefore thought that this induction involves the diffusion of some type of signal molecule, although the chemical signal itself has never been identified. The induction of the neural plate by the notochord is a good example of the importance of the position of a cell in relation to its cellular neighbors. That position is often critical in determining the fate of a cell because it determines its exposure to substances released from other cells.

The cells of the neural plate undergo changes in shape that cause the central cells of the neural plate to move downward and form a depression called the **neural groove;** the cells flanking the groove on each side form **neural folds.** Continued changes in cell shape bring the folds closer together until they meet and fuse, forming the **neural tube.** In this process, the hollow neural tube comes to lie beneath the surface. The ectoderm overlying it will form the outer layer of skin. The anterior portion of the neural tube grows and differentiates into the brain; the remainder of the tube develops into the spinal cord. The brain and the spinal cord remain hollow even in the adult; the ventricles of the brain and

TABLE 49–1 Fate of the Germ Layers Formed at Gastrulation

Germ layer	Structures
Ectoderm	Nervous system and sense organs Outer layer of skin (epidermis) and its associated structures (nails, hair, etc.) Pituitary gland
Mesoderm	Notochord Skeleton (bone and cartilage) Muscles Circulatory system Excretory system Reproductive system Inner layer of skin (dermis) Outer layers of digestive tube and of structures that develop from it, such as part of respiratory system
Endoderm	Lining of digestive tube and of structures that develop from it, such as lining of the respiratory system

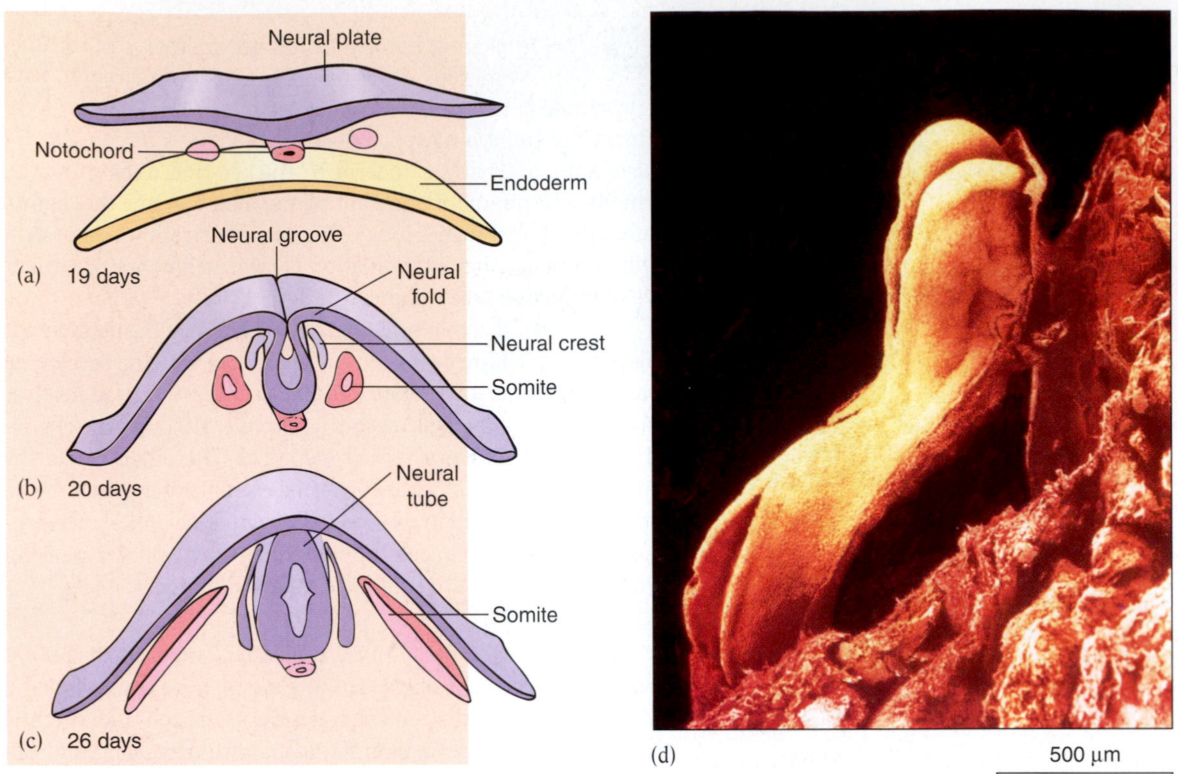

Figure 49–11 Development of the nervous system. **(a–c)** Cross sections of human embryos. **(a)** Approximately 19 days. The neural plate has indented to form a shallow groove flanked by neural folds. **(b)** Approximately 20 days. The neural folds approach one another. The neural crest cells, derived from ectoderm, will migrate in the embryo and give rise to sensory neurons; the somites are blocks of mesoderm that will become the vertebrae and other segmented body parts. The endoderm is not shown. **(c)** Approximately 26 days. The neural folds have joined to form the hollow neural tube, which will give rise to the brain at the anterior end of the embryo and the spinal cord posteriorly. **(d)** The neural folds in the 20-day-old human embryo in this photograph have begun to fuse in the middle, but not in the region of the developing brain *(top)* nor in the posterior portion of the spinal cord. *(d, Lennart Nilsson, Boehringer Ingelheim International GmbH, from* A Child Is Born, *Dell Publishing, 1989, p. 76)*

central canal of the spinal cord are persistent reminders of their embryonic origins.

Various motor nerves grow out of the developing brain and spinal cord, but sensory nerves have a separate origin. When the neural folds fuse to form the neural tube, bits of nervous tissue known as the **neural crest** arise from ectoderm on each side of the tube. **Neural crest cells** migrate downward from their original position and form the dorsal root ganglia of the spinal nerves and the postganglionic sympathetic neurons. From sensory cells in the dorsal root ganglia, dendrites grow out to the sense organs, and axons grow into the spinal cord. Other neural crest cells migrate to various locations in the embryo. They give rise to parts of certain sense organs and nearly all pigment-forming cells in the body. (*On the Cutting Edge: The Evolution of the Vertebrate Head* discusses a fascinating example of evolutionary developmental biology involving neural crest cells.)

As the nervous system develops, other organs also begin to take shape. Blocks of mesoderm known as **somites** form on either side of the neural tube. These will give rise to the vertebrae, muscles, and other components of the body axis. In addition to the skeleton and muscles, mesodermal structures include the kidneys and reproductive structures, as well as the circulatory organs. In fact, the heart and blood vessels are among the first structures to take form, and they must function while still developing.

The development of the four-chambered heart of birds and mammals is a good example of the origin of a complex organ (Fig. 49–12; see Chapter 42). The heart originates through the fusion of paired blood vessels. Venous blood enters the single atrium; passes into the single ventricle, which is located above the atrium; and is then pumped into the embryo. During subsequent development, the atrium undergoes a process of torsion that brings it

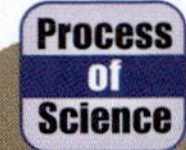

ON THE CUTTING EDGE

The Evolution of the Vertebrate Head

HYPOTHESIS: The vertebrate head evolved, in part, as a result of modifications of conserved (preexisting) *Hox* regulatory genes.

METHOD: Insert regulatory elements of *Hox* genes from the nonvertebrate amphioxus, which lacks a head, into mouse and chick embryos. Each regulatory element is fused to a gene encoding a reporter protein, which produces a dark stain in the tissues in which it is active.

RESULTS: Certain *Hox* gene regulatory elements of amphioxus were shown to be responsible for a characteristic pattern of reporter gene expression in the developing heads of transgenic mouse and chick embryos.

CONCLUSION: *Hox* genes evolved early and have been highly conserved in all animals that have an anteroposterior axis. Modifications of *Hox* gene sequences contributed in part to the evolution of the head in vertebrates.

All vertebrates possess a head, but ancestors of the vertebrates probably lacked well-defined heads. The cephalochordates, close nonvertebrate relatives that may be similar to vertebrate ancestors, lack heads. The lancelet, or amphioxus, is a small, fishlike cephalochordate. Like vertebrates, amphioxus has a notochord, a dorsal tubular nerve cord, pharyngeal slits, and a muscular postanal tail (see Fig. 30–8). Unlike vertebrates, however, amphioxus lacks a well-defined head: It has no jaws; no sense organs such as eyes, ears, or nose; and no brain.

How do vertebrates develop heads? Head development occurs largely as a result of two cell populations in the developing vertebrate embryo: neural crest cells and neurogenic placodes. Cephalochordates do not have either of these sets of cells. *Neural crest cells* separate from the dorsal neural tube and migrate, first in the head region and then along the trunk, to give rise to several cell types, depending on their location along the anteroposterior axis. Examples of some head derivatives of the neural crest are the cartilage and bone of the skull, face, and jaw; connective tissue in the eye and face dermis; sensory nerves and other parts of the peripheral nervous system; and ganglia of the autonomic nervous system. The chick embryo has been the main organism used to study neural crest cell migration and differentiation.

Neurogenic placodes are platelike thickenings of ectoderm that are found only in the head. Depending on their location along the anteroposterior axis, neurogenic placodes differentiate into sensory structures such as the nasal epithelium (the sensory lining of the nose), the lens of the eye, the inner ear, and neurons in the cranial sensory ganglia. Neurogenic placodes have been studied extensively in mouse and chick embryos.

Neural crest cells and neurogenic placodes "know" their location along the anteroposterior axis because different *Hox* genes are expressed at different locations. ***Hox* genes** are a class of developmental genes that help specify the anteroposterior axis during development (see Chapter 16).

Miguel Manzanares, from the MRC National Institute for Medical Research in London, and colleagues studied the evolution of the vertebrate head by introducing *Hox* regulatory elements from amphioxus into the cells of vertebrate embryos.* Each *Hox* regulatory element was fused to a "reporter gene," a gene coding for a protein that produces a dark stain. Thus, mouse and chick cells in which the *Hox* regulatory elements from amphioxus were able to induce transcription were visible because of the dark stain. Certain amphioxus *Hox* regulatory elements were able to control expression of the reporter gene in the hindbrain, neural crest, and neurogenic placode cells in the heads of mouse and chick embryos, each according to its own characteristic pattern.

Evolutionary developmental biologists have recognized that evolution is a conservative process that builds on what has come before rather than starting from scratch. The ability of amphioxus *Hox* gene regulatory elements to respond to their position in a vertebrate embryo and produce a characteristic pattern of gene expression supports the idea that preexisting genes in the common ancestor of amphioxus and vertebrates were conserved. The *Hox* genes and regulatory elements of headless vertebrate ancestors were modified during the course of vertebrate evolution to perform the developmental roles leading to the formation of a head.

*Manzanares, M., H. Wada, N. Itasaki, P.A. Trainor, R. Krumlauf, and P.W.H. Holland. "Conservation and Elaboration of *Hox* Gene Regulation during Evolution of the Vertebrate Head." *Nature,* Vol. 408, 14 Dec. 2000.

to a position above the ventricle, and partitions form that divide the atrium and the ventricle into right and left chambers.

The digestive tract is first formed as a separate foregut and hindgut as the body wall grows and folds, cutting them off from the yolk sac as two simple tubes (Fig. 49–13*a*). As the embryo grows, these tubes, which are lined with endoderm, grow and become greatly elongated. The liver, pancreas, and trachea originate as hollow, tubular outgrowths from the gut. As the trachea grows downward, it gives rise to the paired lung buds, which develop into lungs. The most anterior part of the foregut becomes the pharynx. A series of small outpocketings of the pharynx, the **pharyngeal pouches,** bud out laterally (Fig. 49–13*b*). These

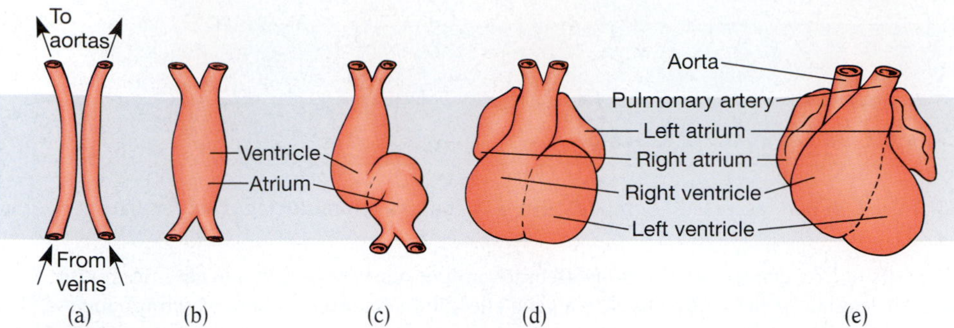

Figure 49–12 Formation of the human heart. These ventral views of successive stages show that the heart forms from the fusion of two blood vessels, and that at first the end that receives blood from the veins is at the bottom. The developing heart twists to carry the atrium to a position above the ventricle, and the chambers then divide to form the four-chambered heart.

pouches meet a corresponding set of inpocketings from the overlying ectoderm, the **branchial grooves.** The arches of tissue formed between the grooves are called **branchial arches.** They contain the rudimentary skeletal, neural, and vascular elements of the face, jaws, and neck.

In fishes and some amphibians, the pharyngeal pouches and branchial grooves meet and form a continuous passage from the pharynx to the outside; these are the gills slits, which function as respiratory organs. In terrestrial vertebrates, each branchial groove remains separated from the corresponding pharyngeal pouch by a thin membrane of tissue, and these structures give rise to organs more appropriate for life on land. For example, the middle ear cavity and its connection to the pharynx, the eustachian tube, are derived from a pharyngeal pouch.

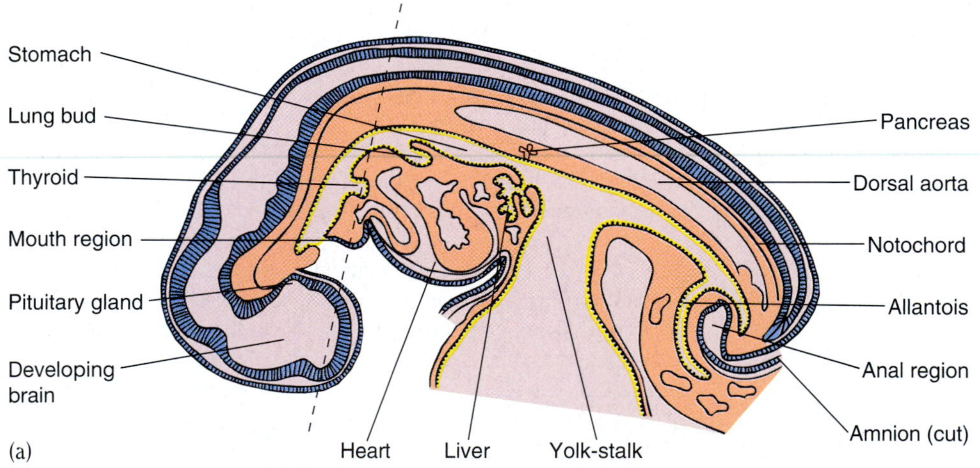

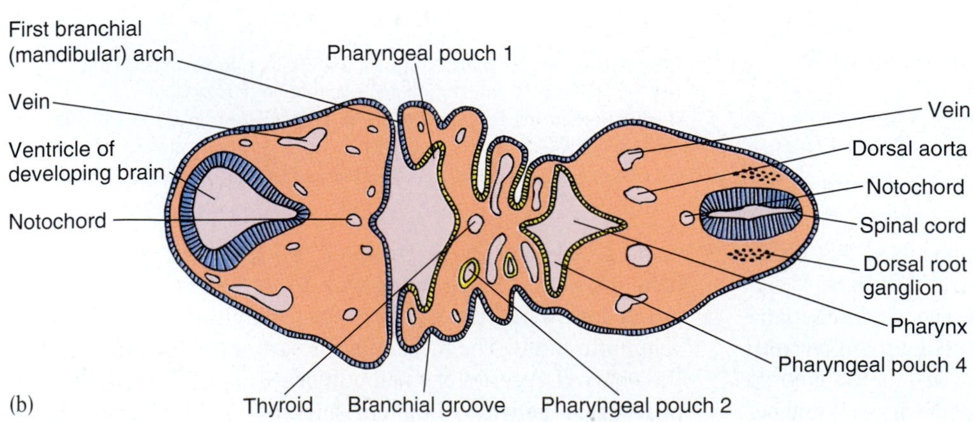

Figure 49–13 Early organ formation. The diagrams illustrate some of the organs that form in a human embryo during the fifth week. **(a)** Sagittal section. Note that the liver and lungs develop as outpocketings from the digestive tract. **(b)** Cross section through the head and neck region (plane indicated by the dotted line in *a*). The embryo is flexed so both the brain and spinal cord are visible. Note the branchial grooves, pharyngeal pouches, and branchial arches.

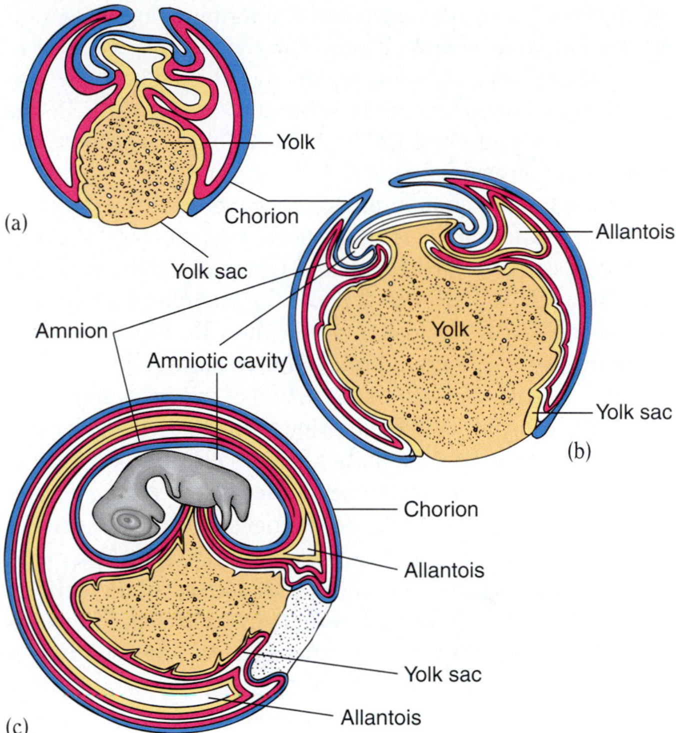

Figure 49–14 Extraembryonic membranes. The formation of the extraembryonic membranes of the chick is illustrated at **(a)** 4 days, **(b)** 5 days, and **(c)** 9 days of development. Each of the membranes develops from a combination of two germ layers. The chorion and amnion form from lateral folds of the ectoderm and mesoderm that extend over the embryo and fuse. The allantois develops from an outpocketing of the gut. The allantois, an elongated sac, and the yolk sac develop from endoderm and mesoderm.

EXTRAEMBRYONIC MEMBRANES PROTECT AND NOURISH THE EMBRYO

Terrestrial vertebrates have four **extraembryonic membranes:** the chorion, amnion, allantois, and yolk sac (Fig. 49–14). Although they develop from the germ layers, these membranes are not part of the embryo itself and are discarded at hatching or birth. The extraembryonic membranes are adaptations to the challenges of embryonic development on land. They protect the embryo, prevent it from drying out, and help in obtaining food and oxygen and eliminating wastes.

The outermost membrane, the **chorion,** encloses the entire embryo. Lying underneath the egg shell in birds and reptiles, it functions as the major organ of gas exchange. As we will see in the next section, its functions have been extended in humans and most other mammals.

Like the chorion, the **amnion** also encloses the entire embryo. The amniotic cavity, the space between the embryo and the amnion, becomes filled with amniotic fluid secreted by the membrane. Recall from Chapter 30 that terrestrial vertebrates are known as *amniotes* because their embryos develop within this pool of fluid. The amniotic fluid prevents the embryo from drying out and permits it a certain freedom of motion. The fluid also serves as a protective cushion that absorbs shocks and prevents the amniotic membrane from sticking to the embryo. In current medical practice a sample of amniotic fluid can be obtained through a procedure known as *amniocentesis* and analyzed for biochemical or chromosomal abnormalities that indicate a possible birth defect (see Fig. 15–12).

The **allantois** is an outgrowth of the developing digestive tract. In reptiles and birds, it stores nitrogenous wastes. In humans, the allantois is small and nonfunctional, except that its blood vessels contribute to the formation of umbilical vessels joining the embryo to the placenta.

In vertebrates with yolk-rich eggs, the **yolk sac** encloses the yolk, slowly digests it, and makes it available to the embryo. A yolk sac, connected to the embryo by a yolk stalk, develops even in mammalian embryos that have little or no yolk. Its walls serve as temporary centers for the formation of blood cells.

HUMAN PRENATAL DEVELOPMENT REQUIRES ABOUT 266 DAYS

The human **gestation period,** the duration of pregnancy, averages 266 days (38 weeks, or about 9 months) from the time of fertilization to the birth of the baby (Table 49–2). We will use this system of timing when discussing the events of early development. Because the time of fertilization is not easily marked, obstetricians usually time the pregnancy by counting from the date of onset of the mother's last menstrual period; by this calculation an average pregnancy is 280 days (40 weeks).

Development begins in the oviduct

Fertilization occurs in the oviduct, and within 24 hours the human zygote has divided to become a two-celled embryo (Fig. 49–15). Cleavage continues as the embryo is propelled along the oviduct by ciliary action.

When the embryo enters the uterus on about the fifth day of development, the *zona pellucida* (its surrounding coat) is dissolved. During the next few days, the embryo floats free in the uterine cavity, nourished by a nutritive fluid secreted by the glands of the uterus. Its cells arrange themselves, forming a blastula, which in mammals is called a **blastocyst** (Fig. 49–16). The outer layer of cells, the **trophoblast,** eventually forms the chorion and amnion that surround the embryo. A little cluster of cells, the **inner cell mass,** projects into the cavity of the blastocyst. The inner cell mass gives rise to the embryo proper.

The embryo implants in the wall of the uterus

On about the seventh day of development the embryo begins the process of **implantation,** in which it becomes embedded in the

TABLE 49–2 Some Important Developmental Events in the Human Embryo

Time from Fertilization	Event
24 hours	Embryo reaches two-cell stage
3 days	Morula reaches uterus
7 days	Blastocyst begins to implant
2.5 weeks	Notochord and neutral plate are formed; tissue that will give rise to heart is differentiating; blood cells are forming in yolk sac and chorion
3.5 weeks	Neural tube forming; primordial eye and ear visible; pharyngeal pouches forming; liver bud differentiating; respiratory system and thyroid gland just beginning to develop; heart tubes fuse, bend, and begin to beat; blood vessels are laid down
4 weeks	Limb buds appear; three primary divisions of brain forming
2 months	Muscles differentiating; embryo capable of movement; gonad distinguishable as testis or ovary; bones begin to ossify; cerebral cortex differentiating; principal blood vessels assume final positions
3 months	Sex can be determined by external inspection; notochord degenerates; lymph glands develop
4 months	Face begins to look human; lobes of cerebrum differentiate; eyes, ears, and nose look more "normal"
Third trimester	A covering of downy hair covers the fetus, then later is shed; neuron myelination begins; tremendous growth of body
266 days (from conception)	Birth

endometrium of the uterus (see Fig. 49–16*a*). The trophoblast cells in contact with the endometrium secrete enzymes that erode an area just large enough to accommodate the tiny embryo. As the embryo slowly works its way down into the underlying connective and vascular tissues, the opening in the endometrium repairs itself. All further development of the embryo takes place *within* the endometrium.

The placenta is an organ of exchange

In placental mammals, the **placenta** is the organ of exchange between mother and embryo (Fig. 49–16*e*). The placenta provides nutrients and oxygen for the fetus and removes wastes, which the mother then excretes. In addition, the placenta is an endocrine organ that secretes hormones (estrogens and progesterone; see Chapter 48) to maintain pregnancy. The placenta develops from both the chorion of the embryo and the uterine tissue of the mother. In early development, the chorion grows rapidly, invading the endometrium and forming finger-like projections known as *chorionic villi*. These structures are used in the chorionic villus sampling technique, which allows the prenatal detection of certain genetic disorders (see Fig. 15–13). The villi become vascularized (infiltrated with blood vessels) as the embryonic circulation develops.

As the human embryo grows, the **umbilical cord** develops and connects the embryo to the placenta (Fig. 49–16*e*). The umbilical cord contains the two umbilical arteries and the umbilical vein. The umbilical arteries connect the embryo to a vast network of capillaries developing within the villi. Blood from the villi returns to the embryo through the umbilical vein.

The placenta consists of the portion of the chorion that develops villi, together with the uterine tissue underlying the villi that contains maternal capillaries and small pools of maternal blood. The fetal blood in the capillaries of the chorionic villi comes in close contact with the mother's blood in the tissues between the villi. However, they are always separated by a membrane through which substances may diffuse or be actively transported. Although certain pathogens have evolved mechanisms that permit them to cross the placenta, maternal and fetal blood do not normally mix in the placenta or any other place.

Several hormones are produced by the placenta. From the time the embryo first begins to implant itself, its trophoblastic cells release **human chorionic gonadotropin (hCG),** which signals the corpus luteum (see Chapter 48) that pregnancy has begun. In response, the corpus luteum increases in size and releases large amounts of progesterone and estrogens. These hormones stimulate continued development of the endometrium and placenta.

Without hCG, the corpus luteum would degenerate and the embryo would be aborted and flushed out with the menstrual flow. The woman would probably not even know that she had been pregnant. If the corpus luteum is removed before about the 11th week of pregnancy, the embryo is spontaneously aborted. After that time, the placenta itself produces enough progesterone and estrogens to maintain pregnancy.

Organ development begins during the first trimester

Gastrulation occurs during the second and third weeks of development. Then the notochord begins to form and induces formation of the neural plate. The neural tube develops, and the forebrain, midbrain, and hindbrain are evident by the fifth week of development. A week or so later the forebrain begins to grow outward, forming the rudiments of the cerebral hemispheres.

The heart begins to develop, and after 3.5 weeks begins to beat spontaneously (see Fig. 49–12). Pharyngeal pouches, branchial grooves, and branchial arches form in the region of the developing pharynx. In the floor of the pharynx, a tube of cells grows downward to form the primordial trachea, which gives rise to the lung buds. The digestive system also gives rise to outgrowths that will develop into the liver, gallbladder, and pancreas. A thin tail becomes evident but does not grow as rapidly as the rest of the body,

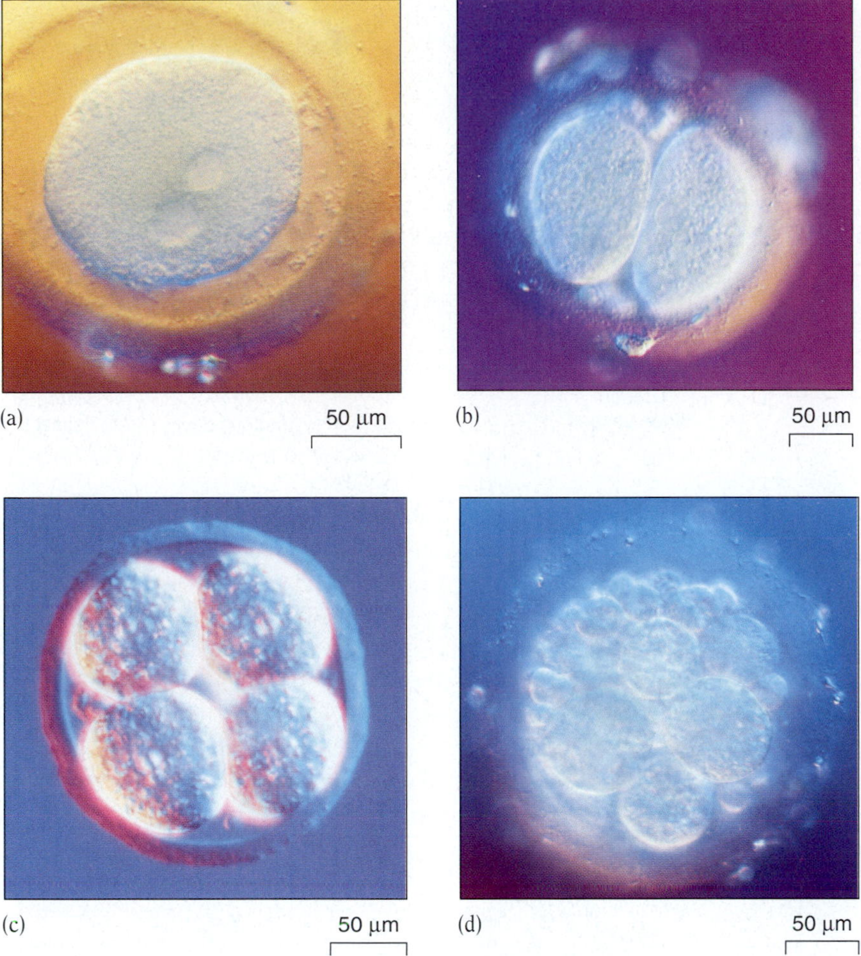

Figure 49–15 Cleavage in a human embryo. **(a)** Male and female pronuclei prior to fusion. **(b)** Two-cell stage. **(c)** Eight-cell stage. **(d)** Cleavage continues, giving rise to a cluster of cells called the morula. *(Lennart Nilsson, from* Being Born, *1992, pp. 14, 15, 17. The Putnam Publishing Group)*

and so becomes inconspicuous by the end of the second month. Near the end of the fourth week the limb buds begin to differentiate; these eventually give rise to arms and legs (Fig. 49–17).

All the organs continue to develop during the second month (Fig. 49–18). Muscles develop, and the embryo becomes capable of movement. The brain begins to send impulses that regulate the functions of some organs, and a few simple reflexes are evident. After the first two months of development, the embryo is referred to as a **fetus** (Fig. 49–19).

By the end of the **first trimester** (the first three months of development) the fetus is about 56 mm (2.2 in) long and weighs about 14 g (0.5 oz). Although small, it has a recognizably human appearance. The external genital structures have differentiated, indicating the sex of the fetus. Ears and eyes approach their final positions. Some of the skeleton becomes distinct, and the notochord has been replaced by the developing vertebral column. The fetus performs breathing movements, pumping amniotic fluid into and out of its lungs, and even makes sucking motions.

Development continues during the second and third trimesters

During the second trimester (months 4 through 6), the fetal heart can be heard with a stethoscope. The fetus moves freely within the amniotic cavity, and during the fifth month the mother usually becomes aware of relatively weak fetal movements ("quickening").

The fetus grows rapidly during the final trimester (months 7 through 9), and final differentiation of tissues and organs occurs. If born at 24 weeks (out of 40 weeks), the fetus has only about a 50% chance of surviving, even with the best of medical care, because its brain is not sufficiently developed to sustain vital functions such as rhythmic breathing, and because the kidneys and lungs are immature.

During the seventh month the cerebrum grows rapidly and develops convolutions. The grasping and sucking reflexes are evident, and the fetus may suck its thumb. Any infant born before 37 weeks of gestation is considered premature. However, if born after 30 weeks, the baby has a good chance of surviving. At birth the average full-term baby weighs about 3000 g (6.6 lb) and measures about 52 cm (20 in) in total length.

More than one mechanism can lead to a multiple birth

Occasionally, the cells of the two-celled embryo separate, and each cell develops into a complete organism. Or sometimes the inner cell mass subdivides, forming two groups of cells, each of which develops independently. Because these cells have identical

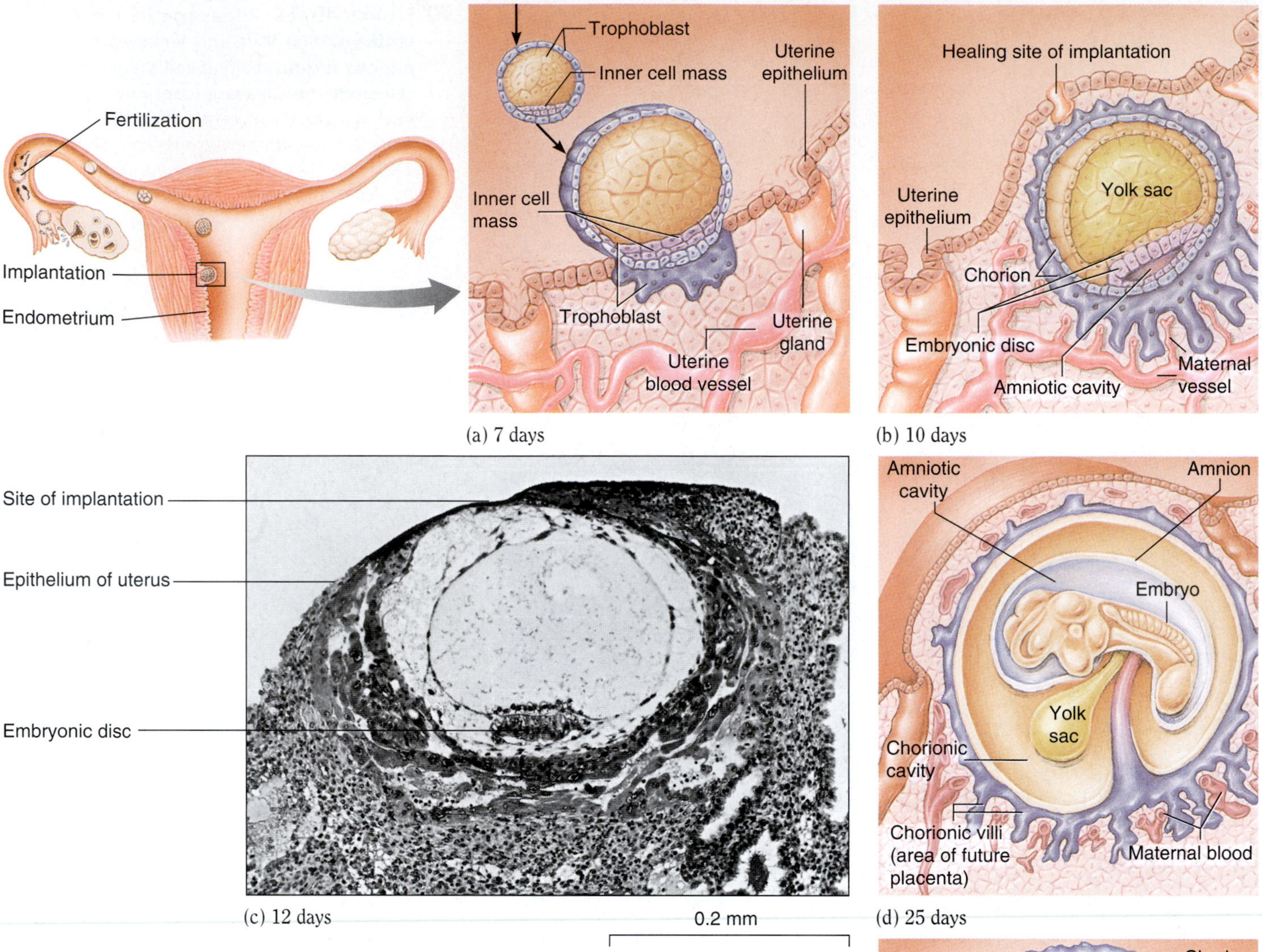

Fertilization
Implantation
Endometrium

Trophoblast
Inner cell mass
Uterine epithelium
Inner cell mass
Trophoblast
Uterine gland
Uterine blood vessel

(a) 7 days

Healing site of implantation
Uterine epithelium
Yolk sac
Chorion
Embryonic disc
Amniotic cavity
Maternal vessel

(b) 10 days

Site of implantation
Epithelium of uterus
Embryonic disc

(c) 12 days

0.2 mm

Amniotic cavity
Amnion
Embryo
Yolk sac
Chorionic cavity
Chorionic villi (area of future placenta)
Maternal blood

(d) 25 days

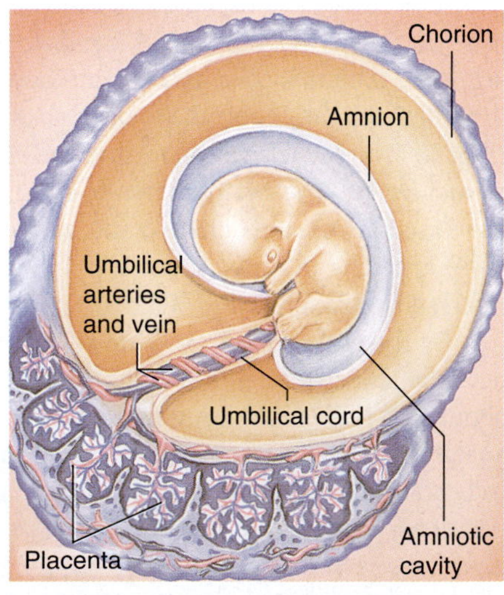

(e) 45 days

Figure 49–16 Implantation and early development in the uterus. **(a)** About seven days after fertilization, the blastocyst drifts to an appropriate site along the uterine wall and begins to implant. The cells of the trophoblast proliferate and invade the endometrium. **(b)** About ten days after fertilization, the chorion has formed from the trophoblast. **(c)** This LM shows an implanted blastocyst at about 12 days after fertilization. **(d)** After 25 days, intimate relationships have been established between the embryo and the maternal blood vessels. Oxygen and nutrients from the maternal blood now satisfy the embryo's needs. Note the specialized region of the chorion that will soon become the placenta. **(e)** At about 45 days the embryo and its membranes together are about the size of a ping-pong ball, and the mother still may be unaware of her pregnancy. The amnion filled with amniotic fluid surrounds and cushions the embryo. The yolk sac has been incorporated into the umbilical cord. Blood circulation has been established through the umbilical cord to the placenta. *(c, Courtesy of Carnegie Institute of Washington)*

sets of genes, the individuals formed are exactly alike—**monozygotic,** or **identical, twins.** Very rarely, the two inner cell masses do not completely separate and so give rise to **conjoined twins,** who are physically attached and usually share one or more body parts.

Dizygotic twins, also called **fraternal twins,** develop when two eggs are ovulated, and each is fertilized by a different sperm. Each zygote has its own distinctive genetic endowment, so the individuals produced are not identical. They may not even be of the same sex. Similarly, triplets (and other multiple births) may be either identical or fraternal.

Before fertility-inducing agents became available in the United States, twins were born once in about 80 births (about 30% of twins are monozygotic), triplets once in 80^2 (or 1 in 6400), and quadruplets once in 80^3 (or 1 in 512,000). However, the Centers for Disease Control and Prevention report that between 1980 and 1995 the rate of twin births increased by about 30%, and the rate of triplet births by over 240%. This increase is attributed to widespread use of drugs to improve fertility.

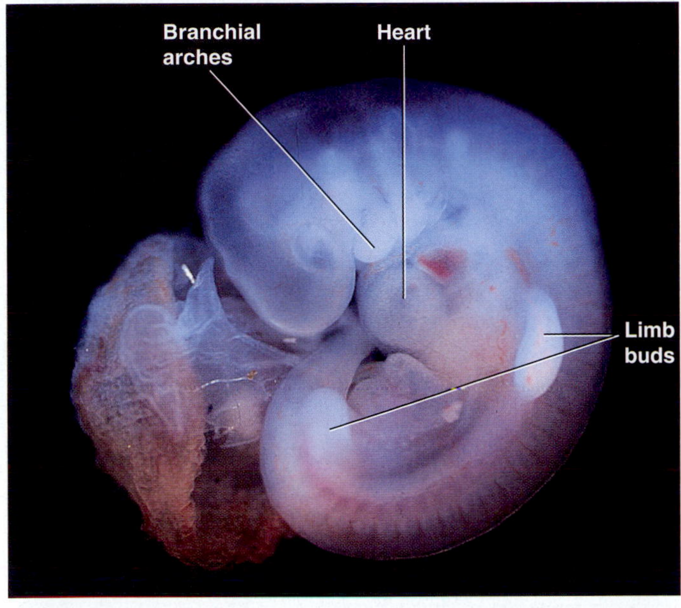

Figure 49–17 A human embryo at 29 days. Developing limb buds are evident, and there is a distinct tail, which will regress during later stages. The heart can be seen below the head near the mouth of the embryo. Branchial arches appear as "double chins." At this stage the embryo is about 7 mm (0.3 in) long. *(Lennart Nilsson, from* A Child Is Born, *Dell Publishing, 1989)*

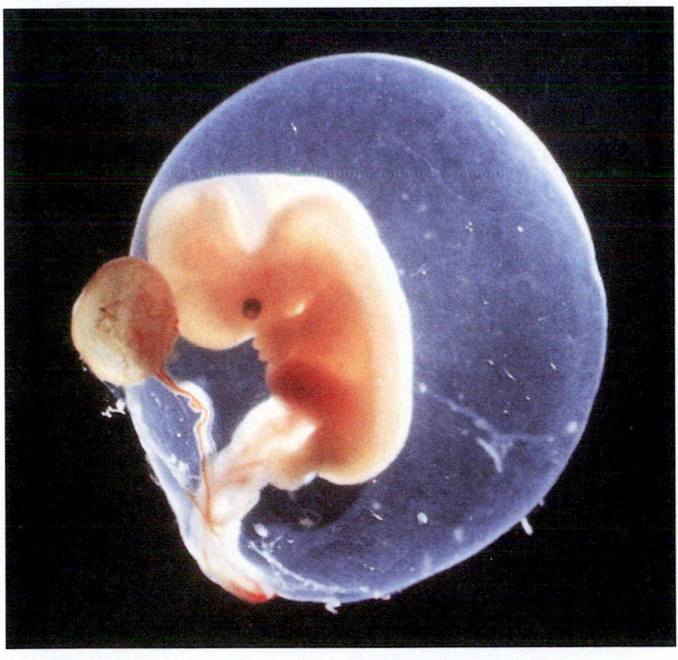

(a)

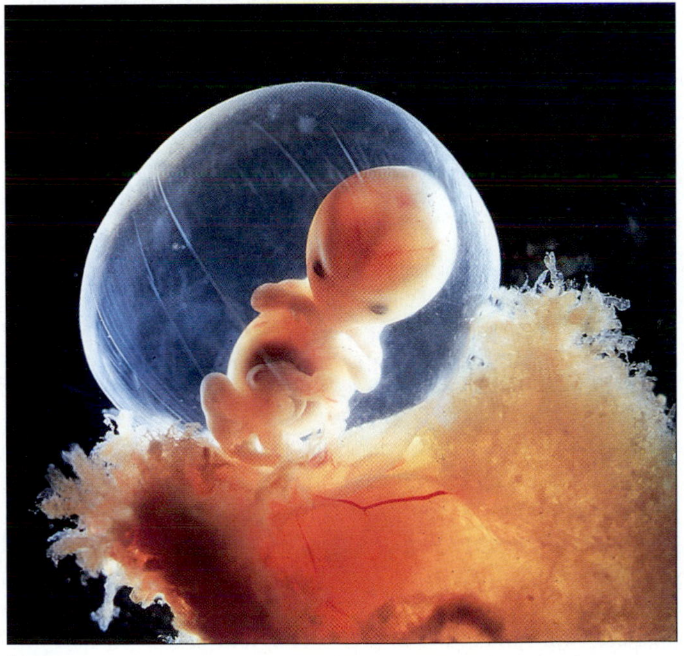

(b)

Figure 49–18 The second month of development. The amnion is prominent as a transparent fluid-filled sac surrounding these embryos. **(a)** Human embryo at 5.5 weeks, 1 cm (0.4 in) long. Limb buds have lengthened, and the eyes have become more evident. The balloon-like object at the left, connected by a stalk, is the yolk sac. **(b)** In its seventh week of development, the embryo is 2 cm (0.8 in) long. The dark-red object inside the embryo is the liver. The large fluffy mass in the lower right-hand corner of the photograph is part of the placenta. *(a, Petit Format/Nestle/Photo Researchers, Inc.; b, Lennart Nilsson, from* A Child Is Born, *Dell Publishing, 1989)*

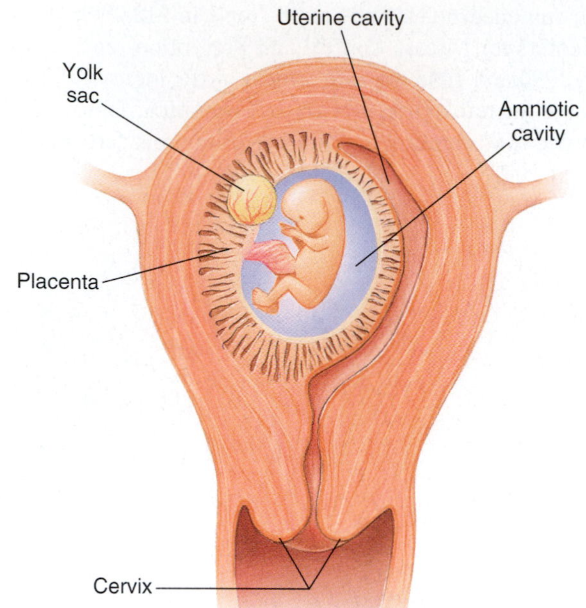

Figure 49–19 Human fetus at ten weeks. Note the position of the fetus within the uterine wall.

A family history of twinning increases the probability of having dizygotic twins. However, giving birth to monozygotic twins does not appear to be influenced by heredity, age of the mother, or other known factors. An estimated two-thirds of multiple pregnancies end in the birth of a single baby; the other embryo(s) may be absorbed within the first 10 weeks of pregnancy, or spontaneously aborted. Ultrasound imaging techniques are used to produce a type of image known as a **sonogram,** which can give valuable information regarding the presence of multiple embryos. This is medically important because multiple births are associated with higher mortality, which is largely a consequence of an increased risk of prematurity and low birth weight.

Environmental factors affect the embryo

We all know that the growth and development of babies are influenced by the food they eat, the air they breathe, the disease organisms that infect them, and the chemicals or drugs to which they are exposed. **Prenatal** development is also affected by these environmental influences. Life before birth is even more sensitive to environmental changes than it is for the fully formed baby. Although there is no direct mixing of maternal and fetal blood, diffusion and various other mechanisms allow many substances—nutrients, drugs, pathogens, and gases—to be transported across the placenta.

Table 49–3 describes some of the environmental influences on development. Some of these are **teratogens,** drugs or other substances that interfere with morphogenesis, causing malformations (Fig. 49–20). Many factors, such as smoking, alcohol use, and poor nutrition contribute to low birth weight, a condition responsible for a great number of infant deaths.

About 5% of newborns (more than 150,000 babies per year) in the United States have a defect of clinical significance. Such birth defects account for about 22% of deaths among newborns. Birth defects may be caused by genetic or environmental factors, or a combination of the two. Genetic factors were discussed in Chapter 15. In this section, we examine some environmental conditions that affect the well-being of the embryo.

Timing is important. Each developing structure has a critical period during which it is most susceptible to unfavorable conditions. Generally, this critical period occurs early in the development of the structure, when interference with cell movements or divisions may prevent formation of normal shape or size, resulting in permanent malformation. Because most structures form during the first three months of embryonic life, the embryo is most susceptible to environmental factors during this early period. During a substantial portion of this time, the mother may not even realize that she is pregnant and so may not take special precautions to minimize potentially dangerous influences.

Physicians are now able to diagnose some defects while the embryo is in the uterus. In some cases, treatment is possible before birth. Amniocentesis and chorionic villus sampling, discussed in Chapter 15, are techniques used to detect certain defects. Ultrasound imaging techniques discussed previously are helpful in diagnosing defects and in determining the position of the fetus. Unlike imaging techniques that use radiation, the available evidence is that ultrasound is harmless to the fetus. New methods currently under development include special types of MRI and high-resolution three-dimensional ultrasound imaging (Fig. 49–21).

THE NEONATE MUST ADAPT TO ITS NEW ENVIRONMENT

Important changes take place after birth. During prenatal life, the fetus received both food and oxygen from the mother through the placenta. Now the newborn's own digestive and respiratory systems must function. Correlated with these changes are several major changes in the circulatory system.

Normally, the **neonate** (newborn infant) begins to breathe within a few seconds of birth and cries within half a minute. If anesthetics have been given to the mother, however, the fetus may also have been anesthetized, and breathing and other activities may be depressed. Some infants may not begin breathing until several minutes have passed. This is one of the reasons that many women request childbirth methods that minimize the use of medication.

The neonate's first breath is thought to be initiated by the accumulation of carbon dioxide in the blood after the umbilical cord is cut. The carbon dioxide stimulates the respiratory centers in the medulla. The resulting expansion of the lungs enlarges its blood vessels (which previously were partially collapsed). Blood from the right ventricle flows in increasing amounts through these larger pulmonary vessels. (During fetal life, blood bypasses the lungs in two ways: by flowing through an opening, the *foramen ovale,*

TABLE 49–3 Environmental Influences on the Embryo

Factor	Example and Effect	Comment
Nutrition	Severe protein malnutrition doubles number of defects, fewer brain cells are produced, and learning ability may be permanently affected; deficiency of folic acid (a vitamin) linked to CNS defects such as spina bifida (open spine), low birth weight	Growth rate mainly determined by rate of net protein synthesis by embryo's cells
Medications	Many medications, even aspirin, affect development of the fetus	
Excessive vitamins	Vitamin D essential, but excessive amounts may result in a form of mental retardation; an excess of vitamins A and K may also be harmful	Vitamin supplements are normally prescribed for pregnant women, but only the recommended dosage should be taken
Thalidomide	Thalidomide, marketed in Europe as a mild sedative, was responsible for serious malformations in more than 7000 babies born in the late 1950's in 20 countries; principal defect was phocomelia, a condition in which babies are born with extremely short limbs, often with no fingers or toes	This teratogenic drug interferes with the development of blood vessels and nerves; most hazardous when taken during fourth to fifth weeks, when limbs are developing; thalidomide has been approved for the treatment of leprosy and certain cancers (e.g., myeloma), and clinical trials are being conducted to test its usefulness in the treatment of other cancers and some of the complications of AIDS*
Accutane (isotretinoin)	Accutane, a synthetic derivitive of vitamin A, is used for the treatment of cystic acne. A woman who takes Accutane during pregnancy has a 1 in 5 chance of having a child with serious malformations of the brain (causing mental retardation), head and face, thymus gland (interfering with immune function), or heart	Although Accutane is marketed with severe restrictions to prevent pregnant women from taking this teratogenic drug, several children are born each year with birth defects attributable to Accutane exposure
Pathogens		
Rubella	Rubella (German measles) virus crosses placenta and infects embryo; interferes with normal metabolism and cell movements; causes syndrome that involves blinding cataracts, deafness, heart malformations, and mental retardation; risk is greatest (about 50%) when rubella is contracted during the first month of pregnancy; risk declines with each succeeding month	Rubella epidemic in the United States in 1963–1965 resulted in about 20,000 fetal deaths and 30,000 infants born with serious defects; immunization is available but must be administered several months before conception
HIV*	HIV can be transmitted from mother to baby before birth, during birth, or by breastfeeding	See discussion of AIDS in Chapter 43
Syphilis	Syphilis is transmitted to fetus in about 40% of infected women; fetus may die or be born with defects and congenital syphilis	Pregnant women are routinely tested for syphilis during prenatal examinations; most cases can be safely treated with antibiotics
Ionizing radiation	When a pregnant woman is subjected to x rays or other forms of radiation, infant has a higher risk of birth defects and leukemia	Radiation was one of the earliest causes of birth defects to be recognized
Recreational and/or abused substances		
Alcohol	When a woman drinks heavily during pregnancy, the baby may be born with fetal alcohol syndrome (FAS), which includes certain physical deformities, and mental and physical retardation; low birth weight and structural abnormalities have been associated with as little as two drinks a day; some cases of hyperactivity and learning disabilities have occurred	Fetal alcohol syndrome is the leading cause of preventable mental retardation in the United States
Cigarette smoking	Cigarette smoking reduces the amount of oxygen available to the fetus because some of the maternal hemoglobin is combined with carbon monoxide; may slow growth and can cause subtle forms of damage	Mothers who smoke deliver babies with lower-than-average birth weights and have a higher incidence of spontaneous abortions, stillbirths, and neonatal deaths; studies also indicate a possible relationship between maternal smoking and slower intellectual development in offspring
Cocaine	Causes constriction of fetal arteries, resulting in retarded development and low birth weight; severe cases may be mentally retarded, have heart defects and other medical problems	Cocaine users frequently abuse alcohol as well, so it is difficult to separate the effects
Heroin	High rates of mortality and prematurity; low birth weight	Infants that survive are born addicted and must be treated for weeks or months

*HIV, human immunodeficiency virus; AIDS, acquired immunodeficiency syndrome.

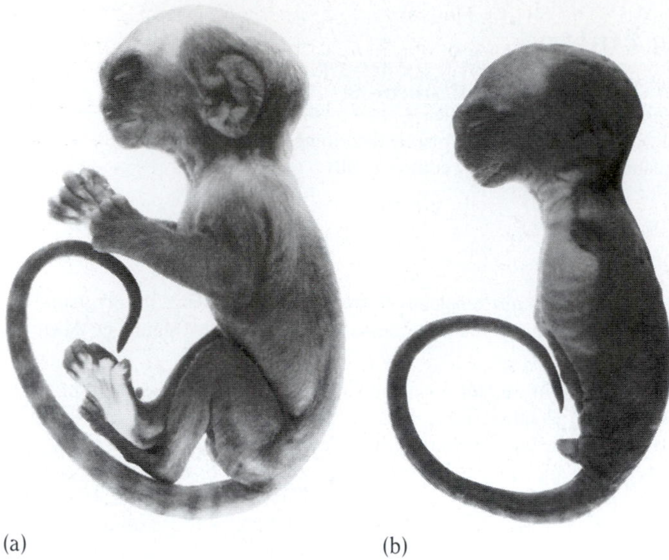

Figure 49–20 Effects of thalidomide. Thalidomide is a potent teratogen in humans and other primates. When administered to the marmoset *(Callithrix jacchus)*, it produces a pattern of developmental defects similar to those found in humans. **(a)** Control marmoset fetus obtained from an untreated mother on day 125 of gestation. **(b)** Fetus of marmoset (same age as control) treated with 25 mg/kg of thalidomide from days 38 to 52 of gestation. The drug suppresses limb formation, at least partly due to the fact that it interferes with the formation of blood vessels. *(Courtesy of W.G. McBride and P.H. Vardy, Foundation 41; from* Development, Growth and Differentiation. *Vol. 25, No. 4, 1983, pp. 361–373)*

which shunts blood from the right atrium to the left atrium, and by flowing through an arterial duct connecting the pulmonary artery and aorta. Both of these routes become closed off after birth.)

THE HUMAN LIFE CYCLE EXTENDS FROM FERTILIZATION TO DEATH

We have examined briefly the development of the embryo and fetus, the birth process, and the adjustments required of the neonate. The human life cycle then proceeds through the stages of infant, child, adolescent, young adult, middle-aged adult, and elderly adult.

Aging is not a uniform process

Development encompasses any biological change that takes place within an organism over time, including the changes commonly called **aging.** Changes during the aging process result in decreased function in the older organism. The declining capacities of the various systems in the human body, although most apparent in the elderly, may begin much earlier in life.

The aging process is far from uniform among different individuals or in various parts of the body. The systems of the body generally decline at different times and rates. On average, between the ages of 30 and 75 a man loses 64% of his taste buds, 44% of the glomeruli in his kidneys, and 37% of the axons in his spinal nerves. His nerve impulses are propagated at a 10% slower rate, the blood supply to his brain is 20% less, his glomerular filtration rate has decreased 31%, and the vital capacity of his lungs has declined 44%. However, the human body has considerable functional capacity in reserve, so bodily functions usually continue to be adequate. Furthermore, there is evidence that some of these declines can be significantly lessened by modifying the lifestyle (e.g., diet and exercise). Women also undergo declines in function, although on average they live about 8 years longer than men.

Although marked improvements in medicine and public health have led to survival to an advanced age of a larger fraction of the total human population, there has been no corresponding increase in the maximum life expectancy. Although relatively little is known about the aging process itself, this is now an active field of scientific investigation, and considerable insight is being gained through genetic studies on model organisms such as the nematode worm, *C. elegans;* the fruit fly, *Drosophila melanogaster;* and the laboratory mouse *(Mus)* (see Chapter 16).

Homeostatic response to stress decreases during aging

Research findings support the idea that most aging occurs because a combination of inheritance and environmental stress makes the individual less able to respond to additional stressors. One major question is whether there exists a genetic program that has evolved to cause aging, or if genetic involvement in the process is more circumstantial. Most of the available evidence favors the latter view.

Certainly some genetically programmed developmental events do seem to be related to the aging process. Cells that nor-

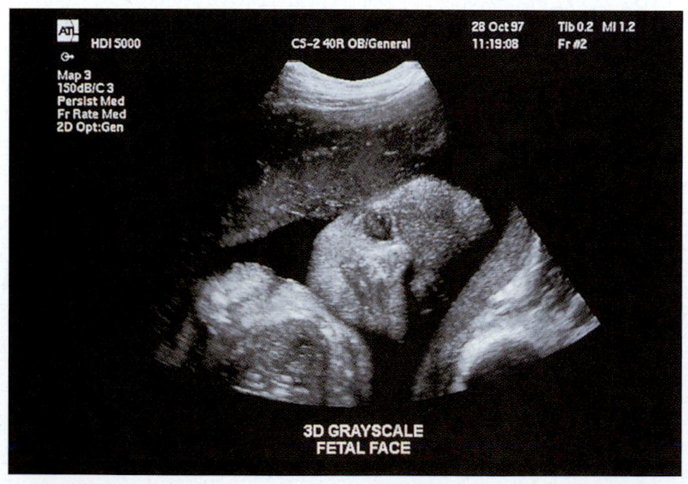

Figure 49–21 Three-dimensional ultrasound image of a human fetus. Note the enhanced soft tissue detail. *(Courtesy of Advanced Technology Laboratories)*

mally stop dividing when they differentiate appear to be more subject to the changes of aging than are those that continue to divide throughout life. Furthermore, it has been hypothesized that certain parts of the body begin to fail to function normally because a genetic program stops key cells from dividing and replenishing themselves. In one model known as **cellular aging,** normal human cells eventually lose their ability to divide when grown in culture. Furthermore, cells taken from an older person can divide fewer times than those from a younger person. Cellular aging appears to be related to the fact that most normal human somatic cells are genetically programmed to lose the ability to produce active telomerase, an enzyme that replicates the DNA of the end caps (telomeres) of the chromosomes (see Chapter 11, *On the Cutting Edge: Telomerase, Cellular Aging, and Cancer*). This loss may help protect the individual against the growth of tumors.

Genetically programmed cell death, *apoptosis,* is an essential developmental mechanism that may play a role in aging (see Chapters 4 and 16). However, it is thought that most aging is not a direct consequence of the genetic program leading to apoptosis; instead, mistakes in the *control* of apoptosis may lead to certain degenerative conditions, such as Alzheimer's disease, or in some of the cell deaths that occur following a heart attack or stroke.

In general, researchers are investigating genes that affect how the body is maintained and repaired in the face of various stressors. Like other life processes, aging may be accelerated by certain environmental influences and may vary because of inherited differences among individuals. Experimental evidence suggests that aging, at least in rats, can be delayed by severe caloric restriction. There is also evidence that premature aging can be precipitated by hormonal changes; by various malfunctions of the immune system, including autoimmune responses; by the accumulation of specific waste products within the cells; by changes in the molecular structure of macromolecules such as collagen; and by damage to DNA by continued exposure to cosmic radiation and x rays.

SUMMARY WITH KEY TERMS

I. Development proceeds as a balanced combination of growth; the molecular events of cell **determination** leading to cell **differentiation;** and **pattern formation** leading to **morphogenesis,** the development of form.

II. Cell differentiation is a consequence of **differential gene expression.**

III. **Fertilization** involves four processes.
   - A. Contact and recognition occur between acellular egg coverings and sperm.
     1. The coverings of echinoderm eggs are the **vitelline envelope** and the **jelly coat;** a **zona pellucida** encloses the mammalian egg.
     2. Upon contact a sperm undergoes an **acrosome reaction,** which facilitates penetration of the coverings.
   - B. Sperm entry is regulated to prevent interspecific fertilization and **polyspermy.** Sea urchin fertilization is followed by a fast block to polyspermy (depolarization of the plasma membrane) and a slow block to polyspermy (the **cortical reaction**).
   - C. Fertilization activates the egg, triggering the events of early development.
   - D. Fusion of sperm and egg **pronuclei** restores the diploid condition.

IV. Development is regulated by the interaction of genes with cytoplasmic factors and environmental factors. **Mosaic development** depends heavily on the distribution of cytoplasmic determinants, whereas the embryo acts as a self-regulating whole in **regulative development.** The developmental patterns of most organisms fall between these two extremes.

V. The zygote undergoes **cleavage,** a series of rapid cell divisions without a growth phase.
   - A. Cleavage leads to the formation of a solid ball of cells (the **morula**) and then usually a hollow ball of cells (the **blastula**).
   - B. The main effect of cleavage is to partition the zygote into many small cells. As cells divide, the distribution of materials in the cytoplasm influences development.
   - C. The **isolecithal** eggs of most invertebrates and simple chordates have evenly distributed **yolk.** They undergo **holoblastic cleavage,** which involves division of the entire egg.
   - D. In the moderately **telolecithal** eggs of amphibians, a concentration of yolk at the **vegetal pole** slows cleavage so that only a few large cells form there, compared to a large number of smaller cells at the **animal pole.**
   - E. The highly telolecithal eggs of reptiles and birds, with a large concentration of yolk at one end, undergo **meroblastic cleavage,** which is restricted to the **blastodisc.**

VI. In **gastrulation,** the basic body plan is laid down as three **germ layers** form: the outer **ectoderm,** the inner **endoderm,** and the **mesoderm** between them; each gives rise to specific structures.
   - A. In the sea star and in amphioxus, cells from the blastula wall invaginate and eventually meet the opposite wall; the new cavity formed, the forerunner of the digestive tube, is the **archenteron,** which has an opening to the exterior, the **blastopore.**
   - B. In the amphibian, invagination at the vegetal pole is obstructed by large, yolk-laden cells; instead, cells from the animal pole move down over the yolk-rich cells and invaginate, forming the **dorsal lip of the blastopore.**
   - C. In the bird, invagination occurs at the **primitive streak,** and no archenteron forms.

VII. **Organogenesis** is the process of organ development. One of the earliest events of organogenesis is the **induction** of nervous system development by the developing **notochord.** The brain and spinal cord develop from the **neural tube.**

VIII. In terrestrial vertebrates, four **extraembryonic membranes—chorion, amnion, allantois,** and **yolk sac**—protect the embryo and help it obtain food and oxygen and eliminate wastes. The amnion is a fluid-filled sac that surrounds the embryo and keeps it moist; it also acts as a shock absorber.

IX. Early human development follows a fairly typical vertebrate pattern.
   - A. Cleavage takes place as the embryo is moved down the oviduct toward the uterus.
   - B. In the uterus, the embryo develops into a **blastocyst** consisting of an outer **trophoblast,** which will give rise to the chorion and amnion, and an **inner cell mass,** which will become the embryo proper. The blastocyst undergoes **implantation** in the endometrium.
   - C. After the first two months of development, the embryo is referred to as a **fetus.**
   - D. In placental mammals, the **umbilical cord** connects the embryo to the **placenta,** the organ of exchange between the maternal and fetal circulation. The placenta is derived from the embryonic chorion and maternal tissue.

E. **Prenatal** development requires 266 days from the time of fertilization; organ development begins during the **first trimester** (three-month period), and growth and refinement of the organs continue in the second and third trimesters. The **neonate** (newborn) must undergo rapid adaptations to independent life.

F. **Monozygotic (identical) twins** arise from a single fertilized egg; **dizygotic (fraternal) twins** arise from fertilization of two different eggs.

X. By controlling environmental factors such as exposure to **teratogens,** diet, smoking, and the intake of alcohol and drugs, a pregnant woman can help ensure the well-being of the embryo as it develops. A **sonogram** (ultrasound image) can diagnose certain defects and also multiple pregnancy.

XI. The **aging** process is marked by a decrease in homeostatic response to stress.

POST-TEST

1. The main function of the acrosome reaction is to (a) activate the egg (b) improve sperm motility (c) prevent interspecific fertilization (d) facilitate penetration of the egg coverings by the sperm (e) cause fusion of the sperm and egg pronuclei
2. The fast block to polyspermy in sea urchins (a) is a depolarization of the egg plasma membrane (b) requires exocytosis of the cortical granules (c) includes the elevation of the fertilization envelope (d) involves the hardening of the jelly coat (e) is a complete block
3. Place the following events of sea urchin fertilization in the proper sequence. (1) fusion of egg and sperm pronuclei (2) DNA synthesis (3) increased protein synthesis (4) release of calcium ions into the egg cytoplasm (a) 4, 3, 1, 2 (b) 3, 2, 4, 1 (c) 2, 3, 1, 4 (d) 1, 2, 3, 4 (e) 4, 1, 2, 3
4. The cleavage divisions of a sea urchin embryo (a) occur in a spiral pattern (b) do not include DNA synthesis (c) do not include cytokinesis (d) are holoblastic (e) do not occur at the vegetal pole
5. Meroblastic cleavage is typical of embryos formed from ______ eggs. (a) moderately telolecithal (b) highly telolecithal (c) isolecithal (d) b and c (e) a, b, and c
6. The primitive groove of the bird embryo is the functional equivalent of the ______ in the amphibian embryo. (a) yolk plug (b) archenteron (c) blastocoel (d) gray crescent (e) blastopore
7. Which of the following are mismatched? (a) endoderm; lining of the digestive tube (b) ectoderm; circulatory system (c) mesoderm; notochord (d) mesoderm; reproductive system (e) ectoderm; sense organs
8. Which of the following has three germ layers? (a) morula (b) gastrula (c) blastula (d) blastocyst (e) trophoblast
9. An unidentified substance (or substances) released from the developing notochord causes the overlying ectoderm to form the neural plate. This phenomenon is known as (a) activation (b) determination (c) induction (d) implantation (e) mosaic development
10. Which of the following is composed of both fetal and maternal tissues? (a) umbilical cord (b) placenta (c) amnion (d) allantois (e) yolk sac
11. The embryo proper of a mammal develops from the (a) trophoblast (b) umbilical cord (c) inner cell mass (d) entire blastocyst (e) yolk sac
12. Which of the following statements about vertebrate organogenesis is *not* true? (a) The notochord, brain, and spinal cord are among the first organs to develop in the early embryo. (b) The developing notochord causes the overlying ectoderm to differentiate into the neural plate. (c) The neural folds meet and fuse, forming the four-chambered heart. (d) Some neural crest cells differentiate into neurons. (e) Blocks of mesoderm known as somites form on either side of the neural tube.
13. On about the seventh day of development, the human embryo (a) implants in the wall of the uterus (b) has a fully developed placenta for obtaining nutrients and oxygen (c) releases human chorionic gonadotropin (d) both a and c (e) a, b, and c

REVIEW QUESTIONS

1. What mechanisms ensure fertilization by only one sperm of the same species?
2. How do the mechanisms of fertilization ensure control of both quality (fertilization by a sperm of the same species) and quantity (fertilization by only one sperm)?
3. Contrast cleavage and gastrulation in sea stars, amphibians, and birds.
4. Trace the formation of the neural tube and explain how cell division, growth, cell differentiation, and morphogenesis interact in the process.
5. Give examples of adult structures that develop from each germ layer.
6. Why do terrestrial vertebrate embryos develop an amnion?
7. Describe human blastocyst formation and implantation.
8. What adaptations must the neonate make immediately after birth?
9. What are some of the nongenetic factors that influence development?
10. What steps can a pregnant woman take to help ensure the safety and well-being of the developing fetus?
11. Describe some of the changes that take place during the aging process.

YOU MAKE THE CONNECTION

1. What is the adaptive value of developing a placenta?
2. For almost 200 years, scientists debated whether an egg or sperm cell contains a completely formed, miniature human (preformation) or if structures develop gradually from a formless zygote (epigenesis). Relate these views to current concepts of development. Can you find any truth in the preformationist view?
3. Not all teratogenic medications are banned by the U.S. Food and Drug Administration. Why?

RECOMMENDED READINGS

Guarente, L., and C. Kenyon. "Genetic Pathways that Regulate Aging in Model Organisms." *Nature,* Vol. 408, 9 Nov. 2000. This review focuses on the current status of research, with a particular emphasis on increasing the maximum lifespan.

Lithgow, G.J., and T.B. Kirkwood. "Mechanisms and Evolution of Aging." *Science,* Vol. 273, 5 Jul. 1996. This review, which provides a perspective on aging research, is one of a series of articles in the same issue devoted to "Patterns of Aging."

Meier, P., A. Finch, and G. Evan. "Apoptosis in Development." *Nature,* Vol. 407, 12 Oct. 2000. This review focuses on the genetic control of programmed cell death, as studied in model organisms, including nematodes *(C. elegans),* fruit flies *(Drosophila)* and laboratory mice *(Mus),* and relates certain human diseases to failures in these controls.

Netting, J. "Telltale Heart." *Science News,* Vol. 160, 7 Jul. 2001. Biologists are uncovering the details of early heart development.

Roush, W. "A Womb with a View." *Science,* Vol. 278, 21 Nov. 1997. A discussion of technological applications that are yielding improved images of developing embryos.

Wright, K. "Thalidomide is Back." *Discover,* April 2000. Thalidomide is becoming an important medical treatment for several serious disorders, including certain types of cancer, amid concerns that controls designed to prevent its use by pregnant women may prove inadequate.

- Visit our Web site at **http://www.info.brookscole.com/solomonbergmartin** for links to chapter-related resources on the World Wide Web. Additional on-line materials relating to this chapter can also be found on our Web site.

See chapter activity on BioActive Learner CD for additional help in mastering the chapter's material. Icon location in the chapter's margins shows which topics have tutorials or simulations in the CD.

50

Animal Behavior

The sand wasp *Philanthus* sp. entering her burrow. *(Darryl T. Gwynne)*

LEARNING OBJECTIVES

After you have studied this chapter you should be able to

1. Contrast proximate and ultimate causes of behavior, and explain how a biologist might decide whether a particular behavior is adaptive.
2. Describe the interactions of heredity, environment, and maturation in animal behavior.
3. Classify a learned behavior as an example of (a) habituation, (b) classical conditioning, (c) operant conditioning, or (d) insight learning.
4. Discuss the adaptive significance of imprinting.
5. Cite examples of biological rhythms, and describe some of the mechanisms responsible for them.
6. Distinguish between directional orientation and navigation, and discuss some biological advantages and disadvantages of migration.
7. Discuss the hypothesis that optimal foraging behavior is adaptive.
8. Contrast a society of social insects with a vertebrate society, and give examples of cultural variation in vertebrate populations.
9. Summarize the modes of communication that animals employ.
10. Describe the concept of a dominance hierarchy, giving at least one example, and discuss its adaptive significance and social function.
11. Distinguish between home range and territory, and describe costs and benefits of territoriality.
12. Describe different types of mating systems and the adaptive value of courtship behavior.
13. Define inclusive fitness and relate this concept to helping behavior.

Suppose your professor provided you with a hypodermic syringe full of poison and instructed you to find a particular type of insect, one that you had never seen before and that was armed with active defenses. You then had to inject the ganglia of your prey's nervous system (about which you had been taught nothing) with just enough poison to paralyze but not kill it. You would have difficulty accomplishing these tasks, but a solitary wasp no larger than the first joint of your thumb does it all with elegance and surgical precision, without instruction.

The sand wasp *Philanthus* sp. captures a bee (or beetle), stings it, and places the paralyzed insect in a burrow excavated in the sand *(see photograph)*. She then lays an egg on her prey, which is devoured alive by the larva that hatches from that egg. From time to time, *Philanthus* returns to her hidden nest to reprovision it until the larva becomes a hibernating pupa in the fall. Her offspring will repeat this behavior, precisely executing each step without ever having seen it done.

An animal's **behavior** is what it *does* and how it does it, usually in response to stimuli in its environment. A dog may wag its tail, a bird may sing, a butterfly may release a volatile sex attractant. Behavior is just as diverse as biological structure and just as characteristic of a given species as its anatomy or physiology. Like its morphology and physiology, an animal's behavior is the product of natural selection on phenotypes and on the genotypes that code for those phenotypes. Thus, an animal's repertoire of behavior is a set of adaptations that equip it for survival in a particular environment.

The capacity for behavior is inherited, but much inherited behavior can be modified by experience. **Learning** involves persistent changes in behavior that result from experience. In considering complex behaviors such as the reproductive behavior of *Philanthus,* we might wonder *why* she behaves as she does, and we might also be interested in *how* she accomplishes her task. Early investigators of animal behavior focused on *how* questions. These questions address **proximate causes,** immediate causes like the genetic, developmental, and physiological processes that permit the animal to carry out the particular behavior.

More recently, biologists have added the *why* perspective, asking questions that address the **ultimate causes.** These questions, which have evolutionary explanations, ask *why* a particular proximate process has evolved. Ultimate considerations address costs and benefits of behavior patterns. When studying ultimate causes, we may ask what the adaptive value of a particular behavior might be. An understanding of behavior requires consideration of both proximate and ultimate causes.

MOST BEHAVIOR IS ADAPTIVE

Whether biologists study behavior in an animal's natural environment or in the laboratory, they must consider that what an animal does cannot be isolated from the way in which it lives. **Behavioral ecology** is the study of behavior in natural environments from an evolutionary perspective. For two decades, behavioral ecology has been the main approach of biologists who study animal behavior. Prior to the emergence of this approach the study of animal behavior was referred to as *ethology,* and this term is sometimes still used to refer to the overall study of animal behavior.

Behavioral ecologists address both the benefits and costs of specific behaviors. A behavior may help an animal obtain food or water, acquire and maintain territory in which to live, protect itself, or reproduce. The benefits typically contribute to **direct fitness,** which is an individual's reproductive success, measured by the number of viable offspring. Reproduction is, of course, a key to evolutionary success. Many behaviors also involve costs. For example, while a parent hunts for food for its offspring, the young may be killed by predators. If the benefits are greater than the costs, the behavior is adaptive.

The ultimate function of a behavior is therefore to increase the probability that the genes of the *individual* animal will be passed to future generations. Certain responses may lead to the death of the individual while increasing the chance that copies of its genes will survive through the enhanced production or survival of its offspring or other relatives. In this chapter, we consider how an animal's behavior contributes to its reproductive success and to the survival of its species.

BEHAVIOR DEPENDS ON THE INTERACTION OF GENES AND ENVIRONMENTAL FACTORS

Early investigators in the field of animal behavior debated about nature versus nurture, that is, the relative importance of genes compared with environmental experience. They defined **innate behavior** (inborn behavior, popularly referred to as instinct) as genetically programmed, and **learned behavior** as behavior that has been modified in response to environmental experience. More recently, behavioral ecologists have recognized that no true dichotomy exists. All behavior has a genetic basis. Even the *capacity* for learned behavior is inherited. However, behavior is modified by the environment in which an animal lives; it is a product of the interaction between genetic capacity and environmental influences. Thus, behavior begins with an inherited framework that experience can modify.

We can think of a range of behaviors from the more rigidly genetically programmed types, through those that, although they have a genetic component, are extensively developed through experience. The wasp *Philanthus,* discussed in the chapter introduction, efficiently carries out a complex, largely genetically programmed, sequence of behaviors. How to dig the burrow, how to cover it, how to kill the bees—these behaviors appear to be genetically determined. Yet some of her behavior is learned. There is no way that her ability to locate the burrow could be genetically programmed. Because a burrow can be dug only in a suitable spot, its location must be learned *after* it is dug. When *Philanthus* covers a nest with sand, she takes precise bearings on the location of the burrow by circling the area a few times before flying off again to hunt.

This behavior of *Philanthus* was studied by the Dutch investigator Niko Tinbergen. He surrounded the wasp's burrow with a circle of pine cones as potential landmarks (Fig. 50–1). Before she returned with another bee, Tinbergen moved or removed the pine cones. The wasp could not find her burrow without them. When Tinbergen moved the pine cones to an area where there was no burrow, the female wasp responded as though the burrow were there. When the investigator completely removed them, the female appeared to be very confused. Only when the experimenter restored the cones to their original location could the wasp find her burrow. When Tinbergen substituted a ring of stones for the cones, the wasp responded as though the nest were in the center of the stones. Thus, *Philanthus* responds to the *arrangement* of the cones, rather than the cones themselves. Tinbergen's findings demonstrate that for *Philanthus* landmark-learning is critical for nest-locating.

Studies of fruit fly courtship and mating have provided interesting examples of interaction between genes and behavior. A ritual consisting of a complex sequence of steps, almost like a dance, must occur before mating takes place. This courtship ritual involves an exchange of visual, auditory, tactile, and chemical signals between the male and female. J.B. Hall at Brandeis University and his colleagues have identified more than a dozen genes controlling these actions, suggesting that courtship behavior is largely inherited and preprogrammed. Despite this, the fruit fly has the capacity to learn from experience, a capacity which, of course, is also inherited.

The interaction between genes and environment has been studied in many vertebrates. Several species of the lovebird *(Agapornis)* differ not only in appearance but in behavior. One species uses its bill to transport small pieces of bark for building a nest. Another species tucks nest-building materials under its rump feathers. In 1962 William Dilger tested the hypothesis that these behaviors are genetically determined by producing hybrids. The hybrid birds appeared confused. They attempted to tuck the material under their feathers, then tried to carry it in their beaks, repeating the pattern several times. Eventually, most of the birds carried the material in their bills, but it took them up to three years to perfect this behavior, and most of them continued to

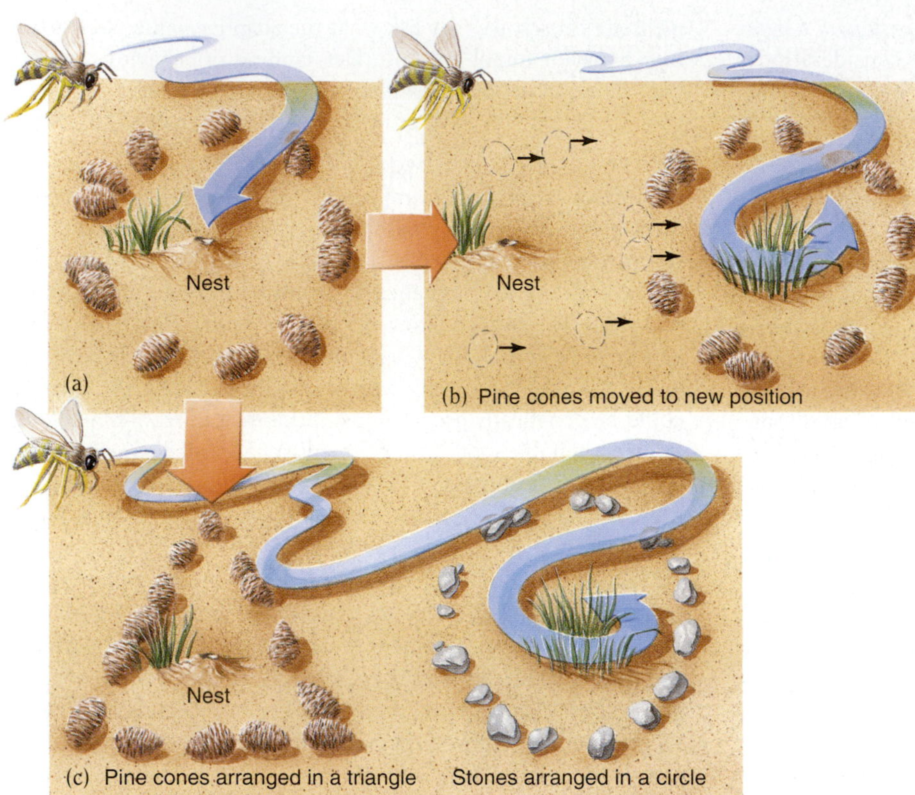

Figure 50–1 Niko Tinbergen's sand wasp experiment. Although the ability of *Philanthus* to learn is quite limited, it is adequate for most natural situations. When the ring of pine cones is moved from position **(a)** to position **(b)**, the *Philanthus* wasp behaves as if her nest were still located at the center because she learned its position in relation to the cones. The wasp responds to the arrangement of the cones, rather than the cones themselves, as shown by the substitution of a ring of stones for cones **(c)**. *(After Tinbergen)*

make futile attempts to tuck material into their feathers. These studies suggest that the method of transporting materials is inherited but somewhat flexible. (Hybrids would likely experience dramatically reduced fitness because of the long delay in breeding onset.)

Behavior depends on physiological readiness

Although behavior involves all body systems, it is influenced mainly by the nervous and endocrine systems. The capacity for behavior depends on the genetic characteristics that govern the development and functions of these systems. Before an animal can exhibit any pattern of behavior, it must be physiologically ready to produce the behavior. For example, breeding behavior does not ordinarily occur among birds or most mammals unless certain concentrations of steroid sex hormones are present in their blood. A human baby cannot walk until its muscles and neurons are sufficiently developed. These states of physiological readiness are themselves produced by a continuous interaction with the environment. The level of sex hormones in a bird's blood may be determined by seasonal variations in day length. The baby's muscles develop with experience, as well as age.

Several factors influence the development of song in male white-crowned sparrows (generally only male songbirds sing a complex song). These birds exhibit considerable regional variation in their song. During early development, young sparrows normally hear adult males sing the distinctive song of their population. Days 10 to 50 after hatching are a critical period for learning the song. When he is several months old, a young male sparrow "practices" the song over several weeks until he eventually sings in the local "dialect."

In laboratory experiments, birds kept in isolation and deprived of the acoustic experience of hearing the song of mature males eventually sing a very poorly developed but recognizable white-crowned sparrow song. When a young white-crowned sparrow is permitted to interact socially with a strawberry finch (which belongs to a different genus of birds), it learns the song of the finch, rather than its own species-specific song. This occurs even if the white-crowned sparrow can hear the song of other sparrows but does not interact with them socially. From these experiments, investigators have concluded that while the white-crowned sparrow is hatched equipped with a rough genetic pattern of its song, social and acoustical stimuli are both important in developing its ability to sing its specific song.

Many behavior patterns depend on motor programs

Many behaviors that we think of as automatic depend on coordinated sequences of muscle actions now referred to as **motor programs.** Some motor programs, for example, walking in newborn gazelles, appear to be mainly innate. Others, such as walking in human infants, have a greater learned component.

A classic example of a motor program in vertebrates is egg-rolling in the European graylag goose. When an egg is removed from the nest of a goose and placed a few centimeters in front of her, she reaches out with her neck and pulls the egg back into the nest (Fig. 50–2). If, while the goose is rolling the egg back toward the nest, the egg veers off to the side, the goose steers it back on course. If the egg is quickly removed, the goose continues her head and neck movements as she persists in moving the now nonexistent egg back toward the nest. Once activated by a simple sensory stimulus, this behavior continues to completion regardless of sensory feedback. There is little flexibility. Ethologists called this behavior a **fixed action pattern (FAP).**

An FAP can be elicited by a **sign stimulus,** or **releaser,** a simple signal that triggers a specific behavioral response. A wooden egg is a sign stimulus that elicits egg-rolling behavior in the graylag goose. Another classic example of a sign stimulus is the red stripe on the ventral surface of a male stickleback fish. The red stripe triggers aggressive behavior by a male whose territory is being invaded. Tinbergen found that crude models painted with a red belly were more likely to be attacked than more realistic models lacking the red belly (Fig. 50–3).

ANIMALS LEARN FROM EXPERIENCE

Recall that learning is a persistent change in behavior due to experience. The capacity to learn appropriate responses to new situations is adaptive, enabling animals to survive as their environment changes. Learning abilities are biased; the information most important to survival appears to be most easily learned. The same rat that may have taken a dozen trials to learn the artificial task of pushing a lever to get a reward learns from one experience to avoid a food that has made it ill. Those who poison rats to get rid of them can readily appreciate the adaptive value of this learning ability. Such quick learning in response to an unpleasant experience forms the basis of warning coloration, which is found in many poisonous insects and brilliantly colored, but distasteful, bird eggs. Once made ill by such a meal, predators quickly learn to avoid them. In the following sections, we consider several types of learning including habituation, imprinting, classical conditioning, operant conditioning, and insight learning.

An animal habituates to irrelevant stimuli

Habituation is a type of learning in which an animal learns to ignore a repeated, irrelevant stimulus, that is, one that neither rewards nor punishes. (The neurophysiological basis of habituation is described in Chapter 40.) Pigeons gathered in a city park learn by repeated harmless encounters that humans are not dangerous to them and behave accordingly. This behavior benefits them. A pigeon intolerant of people might waste energy by flying away each time a human approached and might not get enough to eat. Many African animals habituate to humans on photo safari and to the vans that transport them (Fig. 50–4). Urban humans become habituated to the noise of traffic. In fact, many urban dwellers report that they do not sleep well when they visit a quiet rural area.

Imprinting occurs during an early critical period

Anyone who has watched a mother duck with her ducklings must have wondered how she can "keep track" of such a horde of almost identical little creatures tumbling about in the grass, let alone distinguish them from those belonging to another duck (Fig. 50–5). Although she is capable of recognizing her offspring to an extent, basically they have the responsibility of keeping track of her. The survival of a duckling requires that it quickly learn to discriminate its mother (care-provider) from others.

Imprinting, a type of social learning based on early experience, has been studied in some mammals as well as birds. It occurs during a **critical period,** usually within a few hours or days after birth (or hatching). Konrad Lorenz, an early investigator of this type of learning, discovered that a newly hatched bird imprints on the first moving object it sees—even a human or an inanimate object such as a colored sphere or light. Although the

Figure 50–2 Egg-rolling behavior in the European graylag goose. This behavior is a fixed action pattern (FAP). The goose reaches out by extending her neck and uses her bill to pull the egg back into the nest. If the investigator quickly removes the egg while the goose is in the process of reaching for it or pulling it back, she continues the FAP to completion, as though pulling the non-egg back to the nest.

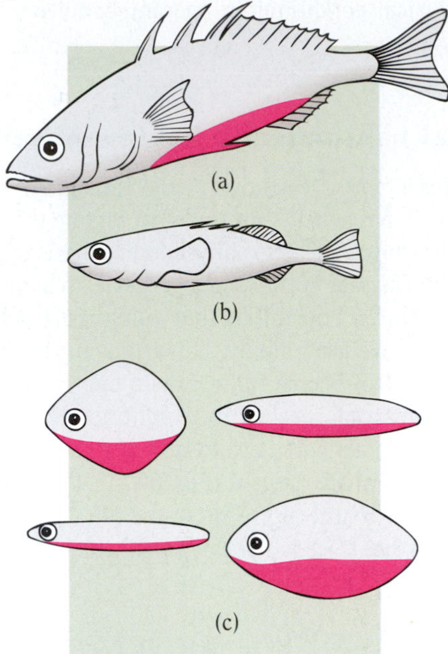

Figure 50–3 A sign stimulus triggers a fixed action pattern. A male stickleback fish **(a)** will not attack a realistic model of another male stickleback if it lacks a red belly **(b)**, but it will attack another model, however unrealistic, that has a red "belly" **(c)**. The aggressive behavior is triggered by the red sign stimulus rather than by recognition based on a combination of features.

process of imprinting is genetically determined, the bird *learns* to respond to a particular animal or object.

Among many types of birds, especially ducks and geese, the older embryos are able to exchange calls with their nest mates and parents through the porous eggshell. When they hatch, at least one parent is normally on hand, emitting the characteristic sounds with which the hatchlings are already familiar. If the parent moves, the chicks follow. This movement plus the sounds produce imprinting. During a brief critical period after hatching, the chicks learn the appearance of the parent.

Imprinting in some mammals depends on scent. Baby shrews, for example, become imprinted on the odor of their mother (or any female nursing them). In many species, the mother also learns to distinguish her offspring during a critical period. The mother in some species of hoofed mammals, such as sheep, will accept her offspring for only a few hours after its birth. If they are kept apart past that time, the young are rejected. Normally, the mother learns to distinguish her own offspring from those of others by olfactory cues.

In classical conditioning, a reflex becomes associated with a new stimulus

In a type of learning called **classical conditioning,** an association is formed between some normal body function and a new stimulus. If you have observed dog or cat behavior, you know that the sound of a can opener at dinner time can captivate a pet's attention. Ivan Pavlov, a Russian physiologist, discovered early in the 20th century that if he rang a bell just before he fed a dog, the dog formed an association between the sound of the bell and the food. Eventually (Fig. 50–6), even when the bell was rung in the absence of food, the dog salivated. Pavlov called the physiologically meaningful stimulus (food, in this case) the *unconditioned stimulus*. The normally irrelevant stimulus (the bell) that became a substitute for it was the *conditioned stimulus*. Because a dog does not normally salivate at the sound of a bell, the association was clearly a learned one. It could also be forgotten. If the bell no longer signaled food, the dog eventually stopped responding to it. Pavlov called this process **extinction.**

In operant conditioning spontaneous behavior is reinforced

In **operant conditioning** (also called *instrumental conditioning*), the animal must do something to gain a reward **(positive reinforcement)** or avoid punishment. Operant conditioning has been studied in many animals, including flatworms, insects, spiders, birds, and mammals. In a typical laboratory experiment, a rat is placed in a cage containing a movable bar. When random actions of the rat result in pressing the bar, a pellet of food rolls down a chute and is delivered to the rat. Thus, the rat is positively reinforced for pressing the bar. Eventually, the rat learns the association and presses the bar to obtain food.

In **negative reinforcement,** removal of a stimulus increases the probability that a behavior will occur. For example, a rat may

Figure 50–4 Habituation. After repeated safe encounters with vans transporting humans on photo safari, many animals, including giraffes, zebras, and lions in the Serengeti, learn to ignore them. Elephants typically ignore the vans unless the driver provokes them by moving too close. In that event an elephant may challenge and even charge the van. Photographed in Tanzania. *(McMurray Photography)*

Figure 50–5 Imprinting. Parent-offspring bonds generally form very early. Some young animals follow the first moving object they encounter. Usually, the object is their mother, but under experimental conditions, young animals have imprinted on humans or even on inanimate objects. *(J.H. Dick/VIREO)*

be subjected to an unpleasant stimulus, such as a low-level electric shock. When the animal presses a bar, this negative reinforcer is removed.

Many variations of these techniques have been developed. A pigeon might be trained to peck at a lighted circle to obtain food, a chimpanzee might learn to perform some task to get tokens that can be exchanged for food, or children might learn to stay quietly in their seats at school to obtain the teacher's praise. Operant conditioning is probably the way that animals learn to perform complex tasks such as perfecting feeding skills.

Operant conditioning plays a role in the development of some behaviors that appear to be genetically programmed. An example is the feeding behavior of gull chicks. Herring gull chicks peck the beaks of the parents, which stimulates the parents to regurgitate partially digested food for them. The chicks are attracted by two stimuli: the general appearance of the parent's beak with its elongated shape and distinctive red spot, and its downward movement as the parent lowers its head. Like the rat's chance pressing of the bar, their initial exploratory pecking behavior is sufficiently functional to get the chicks their first meal, but they waste a lot of energy in pecking. Some pecks are off target and are therefore not rewarded. However, this pecking behavior becomes more efficient over time. Thus, a behavior that might appear to be entirely genetic may be improved by learning (Fig. 50–7).

When presented with food (the unconditioned stimulus), the dog begins to salivate.

A bell (the conditioned stimulus) is rung whenever food is given to the dog. This is repeated a number of times so that an association between the food and the bell is formed.

Eventually the dog salivates at the sound of the bell alone.

Figure 50–6 Classical conditioning. Pavlov's experiment demonstrated that, through classical conditioning, dogs can learn to substitute a new stimulus (the conditioned stimulus) for one that was physiologically meaningful (the unconditioned stimulus).

Insight learning uses recalled events to solve new problems

Perhaps the most complex learning is **insight learning,** which is the ability to adapt past experiences that may involve different stimuli to solve a new problem. A dog can be placed in a blind alley that it must circumvent to reach a reward. The difficulty lies in the fact that the animal must move *away* from the reward to get to it. Typically, the dog flings itself at the barrier nearest the food. Eventually, by trial-and-error, the frustrated dog may find its way around the barrier and reach the reward.

In contrast with the dog, a chimpanzee placed in a similar situation is likely to make new associations between tasks it has learned previously to solve the problem (Fig. 50–8). Primates are especially skilled at insight learning, but some other mammals and a few birds also seem to possess this ability to some degree.

Play may be practice behavior

Perhaps you have watched a kitten pouncing on a dead leaf or practicing a carnivore neck bite or a hind-claw disemboweling stroke on a littermate without causing injury. Many animals, especially young birds and mammals, appear to be practicing adult patterns of behavior while they play (Fig. 50–9). They may improve their ability to escape, kill prey, or perform sexual behavior.

Some investigators have suggested that young animals play just to have "fun." Dolphins appear to play just for pleasure. One play activity that has been studied in bottlenose dolphins is swirling water with their fins, then blowing bubbles to produce rings and helices of air.

Play behavior occurs mainly in birds and mammals. Investigators have proposed many hypotheses for the ultimate causes of play behavior, including exercise, learning to coordinate

Figure 50–7 Operant conditioning. With the experience of being positively reinforced for success, the pelican chick learns to be more accurate in begging for food from a parent. *(H. Cruickshank/VIREO)*

movements, and learning social skills. Of course, there could be some other explanation, yet unknown.

BIOLOGICAL RHYTHMS AFFECT BEHAVIOR

Many types of biological rhythms are known, including daily, monthly, and annual rhythms. Among many animals, including humans, physiological cycles such as body temperature fluctuations and hormone secretion are rhythmic. Human body temperature, for example, follows a typical daily curve. Many behaviors, including activity and sleep, and feeding and drinking are controlled by biological rhythms.

The behavior of many animals, like the activities of many plants (see Chapter 36), appears to be organized around **circadian** (meaning "approximately one day") **rhythms,** which are daily cycles of activity. **Diurnal** animals, such as honeybees and pigeons, are most active during the day. Most bats, moths, and cats are **nocturnal** animals, most active during the hours of darkness. **Crepuscular** animals, like many mosquitoes and fiddler crabs, are busiest at dawn or dusk, or both. Generally, there are ecological reasons for these patterns. If an animal's food is most plentiful in the early morning, for example, its cycle of activity must be regulated so that it becomes active shortly before dawn.

Some biological rhythms of animals reflect the **lunar** (moon) **cycle.** The most striking rhythms are those in marine organisms that are tuned to changes in tides and phases of the moon. For instance, a combination of tidal, lunar, and annual rhythms governs the reproductive behavior of the grunion, a small fish that lives off the Pacific coast of North America. The grunion swarms from April through June on those three or four nights when the highest tides of the year occur. At precisely the high point of the tide, the fish squirm onto the beach and deposit eggs and sperm in the sand. They return to the sea in the next wave. By the time the next tide reaches that portion of the beach 15 days later, the young fishes have hatched in the damp sand and are ready to enter the sea. This synchronization may help protect fish eggs from aquatic predators.

An organism's metabolic processes and behavior are typically synchronized with the cyclic changes in its external environment. Its behavior anticipates these regular changes. The little fiddler crabs of marine beaches often emerge from their burrows at low tide to engage in social activities such as territorial disputes. They must return to their burrows before the tide returns to avoid being washed away. How do the crabs "know" that high tide is about to occur? One might guess that the crabs recognize clues present in the seashore. However, when the crabs are isolated in the laboratory away from any known stimulus that could relate to time and tide, their characteristic behavioral rhythms persist.

Figure 50–8 Insight learning. Confronted with the problem of reaching food hanging from the ceiling, a chimpanzee may stack boxes until it can climb and reach the food. What former experience might the chimp be applying to this new situation?

Figure 50–9 Cheetah cubs playing. Play may serve as a means of practicing behavior that will be used in earnest later in life, possibly in hunting, fighting for territory, or competing for mates. Play may be an example of operant conditioning in action. Photographed in Kenya. *(BIOS/Peter Arnold, Inc.)*

Many biological rhythms are regulated by *internal* timing mechanisms that serve as **biological clocks.** As illustrated by the fiddler crabs, these timing mechanisms do not simply respond to environmental cues, but are capable of sustaining biological rhythms independently. Such internal clocks have been identified in almost every eukaryote, as well as some bacteria. Molecular biologists have demonstrated that biological clocks are controlled by genes. In *Drosophila,* seven genes produce clock proteins that appear to interact in feedback loops in many cell types both inside and outside the nervous system. Similar genes and proteins have been identified in many animal groups, including mammals. The principal clock is located in specific areas of the central nervous system.

In mammals, the master clock is located in the **suprachiasmatic nucleus (SCN)** in the *hypothalamus.* This SCN clock generates approximately 24-hour cycles even without input from the environment. However, the SCN-generated cycles are normally adjusted daily based on visual input received from the retina. Signals from the retina reflecting changes in light intensity are used to adjust internal circadian rhythms to light-dark cycles in the environment. The SCN sends rhythmic signals, in the form of neuropeptides, to the **pineal gland,** an endocrine gland located in the brain. In response, the pineal gland secretes *melatonin,* a hormone important in biological rhythms. Clock genes in the SCN are turned on and off by the very proteins they encode, setting up complex feedback loops that have a 24-hour cycle.

In addition to the master clock, most cells appear to have timing mechanisms that use many of the same clock proteins. The master clock appears to function as a circadian pacemaker that synchronizes these peripheral cellular clocks by complex neural and hormonal signaling. Recent research suggests that complex signaling occurs between the clocks that maintain circadian rhythms and molecules essential for energy metabolism.

MIGRATION INVOLVES INTERACTIONS AMONG BIOLOGICAL RHYTHMS, PHYSIOLOGY, AND THE ENVIRONMENT

Ruby-throated hummingbirds cross the vast distance of the Gulf of Mexico twice each year, and the sooty tern travels across the entire South Atlantic from Africa to reach its tiny island breeding grounds south of Florida. Birds, butterflies, fishes, sea turtles, wildebeest, zebras, and whales are among the many animals that travel long distances. **Migration** is a periodic long-distance movement from one location to another. Many migrations involve astonishing feats of endurance and navigation.

Migration is a response to environmental change

Why do animals migrate? Ultimate causes of migration apparently involve the advantages of moving from an area that seasonally becomes less hospitable to a region more likely to support reproduction or survival. Seasonal changes include shifts in climate, availability of food resources, and safe nesting sites. For example, as winter approaches, many birds migrate to a warmer region.

An interesting example of migration is the annual journey of millions of monarch butterflies *(Danaus plexippus)* from Canada and the continental United States to Mexico, a journey of 2500 km (about 1500 mi) for some. Their dramatic annual migration appears to be related to the availability of milkweed plants on which females lay their eggs. On hatching, the larvae (caterpillars) feed on the leaves of the milkweed plant. In cold regions these plants die in late autumn and grow again with the warmer weather of spring. The wintering destinations of monarchs also seem to be determined by temperature and humidity. Benefits may include the opportunity to winter in an area that offers an abundant food supply, nonfreezing temperatures, and moist air. Thus, the investment made by migrating monarch butterflies in their long journey increases their probability of surviving the winter.

The benefits of migration are not without cost in time, energy, and even survival. Many weeks may be spent each year on energy-demanding journeys. Some animals may become lost or die along the way from fatigue or predation. When in unfamiliar areas, migrating individuals are often at greater risk from predators. In recent years, human activities have interfered with migrations of many kinds of animals. For example, after millions of years of biological success, survival of sea turtles is threatened by fishing and shrimping nets in which they become entangled. As a result, thousands of migrating turtles drown each year. Sea turtles also die as a result of ingesting floating plastic bags they mistake for jellyfish.

Proximate causes of migration include signals from the environment

Some animals migrate when they mature. However, in many animals, certain environmental cues trigger physiological responses that lead to migration. In migratory birds, for example, the *pineal gland* senses changes in day length and then releases hormones that cause restless behavior. The birds show an increased readiness to fly, and to fly for longer periods.

How do migrating animals find their way? **Directional orientation** refers to travel in a specific direction. To travel in a straight line toward a destination requires a sense of direction, or **compass sense.** Many animals use the sun to orient themselves. Because the sun appears to move across the sky each day, an animal must have a sense of time. Biological clocks appear to sense time and regulate circadian rhythms.

Navigation is more complex, requiring both compass sense and **map sense,** which is an "awareness" of location. Navigation involves the use of cues to change direction when necessary to reach a specific destination. When navigating, an animal must integrate information about distance and time, as well as direction.

DNA tests confirm that loggerhead sea turtles that hatch on Florida beaches along the Atlantic swim hundreds of miles across the ocean to the Mediterranean Sea, an area rich in food. Several years later, those that survive to become adults mate, and the females navigate back, often to the same beach, to lay their eggs. Their journey requires both compass and map sense.

Marine biologist Kenneth Lohmann fitted hatchling turtles with harnesses connected to a swivel arm in the center of a large tank (Fig. 50–10). A computer connected to the swivel arm recorded the turtles' swimming movements. Manipulating light and magnetic fields, Lohmann demonstrated that both of these environmental cues are important in turtle migration. The turtles swam toward the east until Lohmann reversed the magnetic field. The turtles reversed their direction to swim toward the new "magnetic east," which was now actually west. Recent work suggests that young turtles use wave direction to help set magnetic direction preference. Some biologists suspect that turtles also use their sense of smell to guide them, particularly to guide the females to the very same beach to lay their eggs. Similarly, adult salmon use the unique odors of different streams to help find their way back to the same stream from which they hatched.

Figure 50–10 Navigation by light and magnetic field. To study turtle navigation, researcher Kenneth Lohmann harnessed leatherback turtles *(Dermochelys coriacea)* like this hatchling and wired them to a computer that recorded their swimming direction. *(Kenneth Lohmann)*

Birds and some other animals that navigate by day rely on the position of the sun (our local star); those that travel at night use the stars to guide them. Working in the 1950s, Franz and Eleonore Sauer hand-reared a number of whitethroats, a species of small European warbler. This ruled out the possibility that the parents had transmitted any information to their offspring. When (and only when) the birds could see the star patterns of the night sky, they attempted to fly in the appropriate migration direction for their species, a direction they had had no opportunity to learn.

Studies by Stephen Emlen in 1975 showed that young indigo buntings *(Passerina cyanea)* learn the constellations, using the position of the North Star as their reference point. (Other stars in the Northern Hemisphere appear to rotate about the North Star.) When Emlen rearranged the constellations in a planetarium sky, birds learned the altered patterns of stars and later attempted to fly in a direction consistent with the artificial pattern. These and other studies suggested that birds have a genetic ability to learn constellation patterns and use them to orient themselves during migration.

Investigators have long observed that some species of birds can navigate even when the sky is overcast and they cannot see the stars. They hypothesized that these birds use Earth's magnetic field to navigate. Studies of garden warblers *(Sylvia borin)* by P. Weindler and colleagues in 1996 indicate that, as these birds make their way from central Europe to Africa each winter, they navigate both by the stars and by Earth's magnetic field. These investigators found that warblers used the stars to determine the general direction of travel, then used magnetic information to refine and correct their course. In addition to birds and sea turtles, honeybees, some fishes, and some other animals also appear to be sensitive to Earth's magnetic field and use it as a guide.

EFFICIENT FORAGING BEHAVIOR CONTRIBUTES TO SURVIVAL

Feeding behavior, or **foraging,** involves locating and selecting food, as well as food gathering and food capture. Some behavioral ecologists study the costs and benefits of searching for and selecting certain types of food, as well as the mechanisms used to locate prey. For example, many camouflage strategies have evolved that make potential prey difficult to detect. As predators experience repeated success in locating a particular prey species, they are thought to develop a *search image,* a constellation of cues that help them identify hidden prey.

Why do grizzly bears spend hours digging Arctic ground squirrels out of burrows while ignoring larger prey such as caribou? It is energetically more efficient to dig for squirrels because the bears' efforts most probably will be rewarded, whereas caribou are more likely to escape, leaving the bears hungry. This is an example of **optimal foraging,** the most efficient way for an animal to obtain food. When animals maximize energy obtained per unit of foraging time, they may maximize reproductive success.

Many factors, such as avoiding predators while foraging, must be considered in determining efficient or optimal strategies.

In habitats where the most preferred food items are abundant and an animal does not have to travel far to obtain them, animals can afford to be very selective. In contrast, the optimal strategy in poor habitats, where it takes longer to find the best food items, is to select less preferred items. Animals may learn to forage efficiently through operant conditioning, that is, by randomly trying various strategies and selecting the one associated with the most rewards (and the fewest hunger pangs at the end of the day).

Foraging is also affected by the forager's risk of predation. Guy Cowlishaw studied foraging behavior in a population of baboons *(Papio cynocephalus ursinus)*. He found that the baboons spent more time foraging in a habitat in which food was relatively scarce than in an area with more abundant food that had a high risk of predation by lions and leopards.

Lions live, and often forage, in social units called *prides*. A pride typically consists of a group of related adult females, their cubs, and unrelated males. Lion behavior has been extensively studied in Serengeti National Park, Tanzania, by Craig Packer and his research team. Packer has radio-collared at least one female in each of 21 prides and has tracked them for several years. During the season when prey is abundant, lions hunt wildebeest, gazelle, and zebra that have migrated into the area. When these herds migrate out of the area during the dry season, these prey become scarce, and lions feed mainly on warthog and Cape buffalo.

Packer has reported that when prey is abundant, the size of the foraging group has little effect on daily food intake. Whether a lion hunts individually or in small (two to four females) or large (five to seven) groups, food is plentiful and is captured by individuals or groups of any size. However, during the season when prey is scarce, lions are more successful if they hunt alone or in large groups (Fig. 50–11). Despite this, Packer's data from radio-collared lions indicated that females typically forage in as large a group as they can, even though this approach decreases foraging efficiency. Thus, selection pressure appears to be stronger to protect cubs and defend territory against larger prides than to forage maximally (alone).

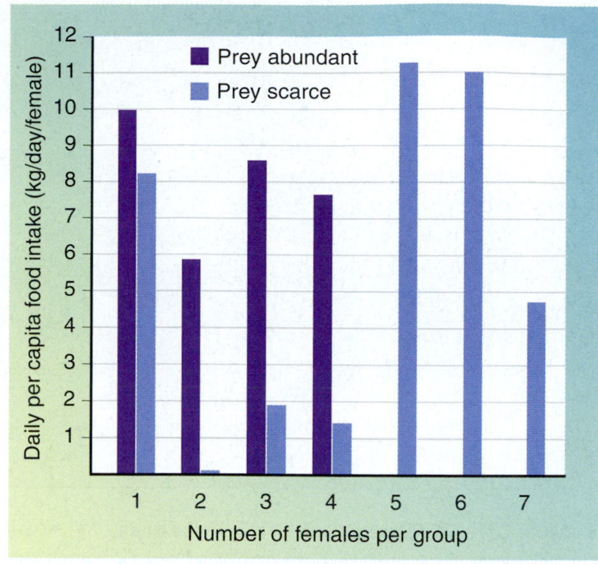

Figure 50–11 Optimal foraging and group size in lions. Researcher Craig Packer and his colleagues observed that during times of prey scarcity, optimal foraging was related to group size: Lions hunted most successfully alone or in large groups of five to seven. Observations of group size suggested that the benefits of optimal defense often outweighed the benefits of optimal foraging. *(Based on data of Craig Packer: Packer, C., D. Scheel, and A.E. Pusey. "Why Lions Form Groups: Food Is Not Enough."* The American Naturalist, *Vol. 136, Jul. 1990)*

SOCIAL BEHAVIOR HAS BOTH BENEFITS AND COSTS

The mere presence of more than one individual does not mean that a behavior is social. Many factors of the physical environment bring animals together in **aggregations,** but whatever interaction they experience may be circumstantial. A light shining in the dark is a stimulus that draws large numbers of moths, and the high humidity under a log may attract wood lice. Although these aggregations may have adaptive value, they are not truly social, because the animals are not responding to one another.

We can define **social behavior** as the interaction of two or more animals, usually of the same species. Many animals benefit from living in groups. By cooperation and division of labor, some insects construct elaborate nests and raise young by mass-production methods. Schools of fishes tend to confuse predators, and so individuals within the school may be less vulnerable to predators than a solitary fish would be (see Fig. 51–2*b*). Zebras can more effectively protect themselves when they are in groups (Fig. 50–12). Zebra stripes appear to be a visual antipredator adaptation. When viewed from a distance, the stripes tend to visually break up the form of the animal so that individuals cannot be distinguished. Thus, a herd of zebras tends to confuse predators, whereas a solitary animal is easy prey to lions, cheetahs, or spotted hyenas.

Social foraging is an adaptive strategy that is used routinely by many animal species. A pack of wolves has greater success in hunting than an individual wolf would have if hunting alone. Among some birds of prey, such as certain hawks, group hunting results in locating prey more quickly.

Some species that engage in social behavior form societies. A **society** is an actively cooperating group of individuals belonging to the same species and often closely related. A hive of bees, a flock of birds, a pack of wolves, and a school of fish are examples of societies. Some societies are loosely organized, whereas others have complex structures. Characteristics of a highly organized society include cooperation and division of labor among animals of different sexes, age groups, or castes. A complex system of communication reinforces the organization of the society. The members tend to remain together and to resist attempts by outsiders to enter the group.

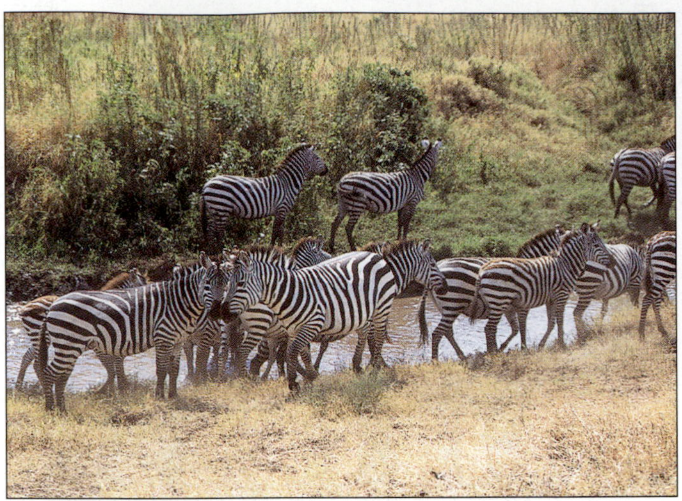

Figure 50–12 **Social behavior in zebras.** The social unit consists of a fairly stable group of females with young and a dominant stallion. Being in a group of many striped zebras may be advantageous to individuals attempting to avoid predators (confusion effect). Photographed in East Africa. *(McMurray Photography)*

Social behavior offers benefits that increase the chances of perpetuating the genes that produce such behavior. However, social behavior also has certain costs. Living together means increased competition for food and habitats, increased risk of attracting predators, and increased risk of transmitting disease. Typically, for animals that exhibit social groupings, the benefits of the society outweigh the costs. For the many animal species that do not form social groups, the costs of forming a society outweigh the benefits.

COMMUNICATION IS NECESSARY FOR SOCIAL BEHAVIOR

One animal can influence the behavior of another only if the two of them can exchange mutually recognizable signals (Fig. 50–13). **Communication** is most evident when an animal performs an act that changes the behavior of another organism. Communication may be important in finding food, as in the elaborate dances of honeybees. Animals may communicate to hold a group together, warn of danger, indicate social status, indicate willingness to accept or provide care, identify members of the same species, or indicate sexual maturity or readiness.

Animals use auditory, visual, tactile, chemical, or electrical signals to transmit information to one another. In many bird species, territorial males announce their presence and willingness to interact socially by singing. Along with their songs, many birds present visual displays. Many birds respond to a particular sound by matching it. Among nonhuman mammals, only bottlenose dolphins are known to match sounds in communicating. These dolphins can respond to the whistle of a member of its own species by imitating and emitting the same sound. Vocal matching is thought to have been important in the evolution of human language.

Orcas (*Orcinus orca;* formerly known as killer whales) communicate by sounds and songs. Members of a given pod (social group) have an average of 12 different calls, which vary in pitch and duration and appear to reflect their "emotional" state. Certain fishes (gymnotids) use electric pulses for navigation and communication, including territorial threat, in a fashion similar to bird vocalization. As sociobiologist Edward O. Wilson has said, "The fish, in effect, sing electrical songs."

Some animals communicate by scent. Dogs and wolves mark territory by frequent urination. Antelopes, deer, and cats rub facial gland secretions on conspicuous objects in their vicinity and urinate on the ground. **Pheromones** are chemical signals

(a)

(b)

Figure 50–13 **Animal communication.** **(a)** A male spring peeper frog *(Hyla crucifer)* calling to locate a mate. **(b)** Communicating with language. Chimpanzee Tatu *(top)* is signing "food" to Washoe *(below)*. Washoe was the first chimp to learn American Sign Language from a human. She then taught other chimps how to sign. *(a, R. Lindholm/ Visuals Unlimited: b, April Ottey, Chimpanzee and Human Communication Institute, Central Washington University)*

secreted into the environment that convey information between members of a species. These small volatile molecules provide a simple, widespread means of communication. Many animals use pheromones to communicate danger, ownership of territory, and availability for mating. For example, female moths release pheromones that attract males. Ants mark their trails with pheromones.

Most pheromones elicit a very specific, immediate, but transitory type of behavior. Others trigger hormonal activities that result in slow but long-lasting responses. Some pheromones may act in both ways. An advantage to pheromone communication is that relatively little energy is needed to synthesize the simple, but distinctive, organic compounds involved. Members of the same species have receptors that fit the molecular configuration of the pheromone; other species usually ignore it or do not detect it at all. Pheromones are effective in the dark, they can pass around obstacles, and they last for several hours or longer. Major disadvantages of pheromone communication are slow transmission and limited information content. Some animals compensate for the latter disadvantage by secreting different pheromones with different meanings.

Pheromones are important in attracting the opposite sex and in sex recognition in many species. Many female insects produce pheromones that attract males, and these chemical signals govern the reproduction of many social insects. We have taken advantage of some sex-attractant pheromones to help control pests such as gypsy moths by luring the males to traps baited with synthetic versions of female pheromones. In honeybees certain fatty acids are mixed with hydrocarbons from wax glands, and this mixture is transferred onto worker bees when they touch the comb. This mixture serves as a pheromone that identifies all the bees that belong to a particular hive. Should a bee from another hive attempt to enter, guard bees sting and may even kill the foreigner.

In vertebrates pheromones affect sexual cycles and reproductive behavior, including choice of a mate, and may play a role in defending territory. When male newts court a female, they release a pheromone into the water that repels other males. The pheromone released by three males is sufficient to keep others away.

Among some mammals, an ovulating female (one physiologically ready to mate) releases pheromones as part of her vaginal secretion. When these chemical odors are detected by males, their sexual interest increases. When the odor of a male mouse is introduced among a group of females, the reproductive cycles of the female mice become synchronized. In some species of mice, the odor of a strange male, a sign of high population density, causes a newly impregnated female to abort.

The extent to which humans are affected by pheromones is currently under study. One interesting finding suggests that an unconsciously perceived body odor is capable of synchronizing the menstrual cycles of women who associate closely (for instance, college roommates or cellmates in prison). A recent study of the effects of human steroids used in commercial fragrances, such as perfumes, suggests that such compounds may act as subtle signals that modulate mood and behavior.

As discussed in Chapter 41, mammals detect pheromones with specialized chemoreceptor cells that make up the **vomeronasal organ** located in the epithelium of the nose. The vomeronasal sensory neurons send signals to the *amygdala* and *hypothalamus,* structures that regulate emotional responses and certain endocrine processes. About 100 genes that are thought to code for pheromone receptors have been identified in the mouse and rat. These receptors initiate **signal transduction** processes that involve G proteins. When neurons of the vomeronasal system are damaged in virgin mice, they do not mate.

DOMINANCE HIERARCHIES ARE SOCIAL RANKINGS

In the spring, female paper-wasps awaken from hibernation and begin to build a nest together. During the early course of construction, a series of squabbles among the females takes place in which the combatants bite one another's bodies or legs. Finally, one of the wasps emerges as dominant. After that, she is rarely challenged. This queen wasp spends more and more time tending the nest and less and less time out foraging for herself. She takes the food she needs from the others as they return.

The queen then begins to take an interest in raising a family—her family. Because she is almost always in the nest, she is able to prevent other wasps from laying eggs in the brood cells by rushing at them, jaws agape. Because of her supreme dominance, the queen can bite any other wasp without serious risk of retaliation.

The other wasps of the nest are further organized into a **dominance hierarchy,** a ranking of social status in which each wasp has more status than the wasps that are lower in rank. Wasps lower in the hierarchy are subordinate to those above them.

Queen → Wasp A → Wasp B → ... Wasp M → Wasp N

Once a dominance hierarchy is established, little or no time is wasted in fighting. When challenged, subordinate wasps exhibit submissive poses that, in turn, inhibit the queen's aggressive behavior. Consequently, few or no colony members are lost through wounds sustained in fighting one another. This ensures greater reproductive success for the colony.

In many species, males and females have separate dominance systems. However, in many monogamous animals, especially birds, the female acquires the dominance status of her mate by virtue of their relationship and the male's willingness to defend his mate.

Like many fishes and some invertebrates, certain coral reef fishes (labrids) are capable of sex reversal. The largest, most dominant individual is always male, and the remaining fishes within his territory are all female. If the male dies or is removed, the most dominant female will become the new male. Should any harm come to him, the next-ranking female will undergo sex reversal, take charge, and protect the group territory. Other types of fishes exhibit the reverse behavior, in which the most dominant fish is always female. In these species, size is less important for aggressive defense than for maximal egg production.

Animals often compete violently for dominance

In establishing dominance, males expend energy in posturing, roaring, leaping about, or sometimes fighting fiercely. These behaviors appear to be a test of male quality. The male with the greatest endurance will likely gain dominance and will have the greatest opportunity to mate and to perpetuate his genes. In many animals, social dominance is a function of aggressiveness (Fig. 50–14). Tremendous energy is often needed for the fights engaged in by some birds and many mammals, including baboons, rams, and elephant seals. Establishing dominance is often strenuous and dangerous.

Aggressiveness is often influenced directly by sex hormones. Among chickens, the rooster is the most dominant; as with most vertebrates, the male reproductive hormone *testosterone* increases aggressiveness. If a hen receives testosterone injections, her place in the dominance hierarchy shifts upward. When male rhesus monkeys are dominant, their testosterone levels are much higher than when they have been defeated. Not only can *estrogen* sometimes reduce dominance and testosterone increase dominance, but dominance may even increase testosterone production. It is not always easy to determine cause and effect. (Studies suggest that the effects of testosterone on human mood and behavior may be different from that in other vertebrates.)

Figure 50–14 Communicating dominance. Social animals signal other members of their group to communicate dominance. This baboon (*Papio* sp.) bares his teeth and screams in an unmistakable show of aggression, a signal that allows him to establish and maintain dominance. *(Gerald Lacz/Peter Arnold, Inc.)*

Dominance affects reproductive success in female chimpanzees

Until 1997 dominance hierarchies were thought to be a "male thing" among chimpanzees. Then, based on data from their 35-year-long field study of chimpanzees in Gombe National Park in Tanzania, Anne Pusey, Jennifer Williams, and Jane Goodall reported that female chimpanzees establish dominance hierarchies. These researchers found that dominant females are more successful reproductively than those with lower social status.

Adult male chimpanzees travel together over the home range of their social community, defending its borders. When challenged by a dominant male, a lower ranking male or female communicates recognition of its lower status by making *pant-grunts.* These "hnn-hnn-hnn" noises signal submission.

An adult female chimpanzee spends about 65% of her time alone with her dependent offspring. Much of her time is spent foraging for food in a core area that may overlap with the core area of another adult. When she meets another female, the two adults often ignore each other. However, sometimes one female pant-grunts to the other.

The research team mapped female dominance hierarchies by recording the direction of pant-grunts between females. They found that the direction was reliably consistent. If female X pant-grunted to female A at any encounter, female X would again pant-grunt to female A the next month or the next year.

The team looked at the reproductive histories of the females and found that rank had a significant effect on reproductive success. Offspring of high-ranking mothers were much more likely to survive to age seven years than offspring of low-ranking mothers. In addition, daughters of high-ranking mothers reached sexual maturity as much as four years earlier. The investigators also found that high-ranking females live longer than low-ranking females.

How does dominance rank affect reproductive success? The research team observed that high-ranking females often grabbed and killed the infants of low-ranking mothers. Infanticide appeared to be a significant threat. The investigators found that female offspring of dominant mothers weighed more than those of low-ranking mothers, and higher weight correlated with reaching sexual maturity at an earlier age. They suggested that better nutrition contributes to the higher survival rates of high-ranking females and their young. Better nutrition may be a consequence of claiming and maintaining a core area with abundant, more nutritious fruit.

How do female chimpanzees achieve social status? Some females apparently become dominant through aggressive behavior. Others achieve high rank by virtue of their mother's status. Does social rank determine success, or does success determine social rank? Or is there another, yet unknown, factor responsible for both? Researchers at Gombe and other primate research sites continue to ask questions and search for answers about the social relationships among our closest relatives.

MANY ANIMALS DEFEND A TERRITORY

Most animals have a **home range,** a geographical area that they seldom leave (Fig. 50–15). Because the animal has the opportunity to become familiar with everything in that range, it has an

Figure 50–15 Home range. The massive Cape buffalo *(Syncerus caffer)* lives in herds of several hundred animals. Each herd, such as this one photographed in the Serengeti, has a fairly constant home range. Buffalo protect herd members, especially calves. *(McMurray Photography)*

advantage over its competitors, predators, and prey in negotiating the terrain and finding food. Some, but not all, animals exhibit **territoriality.** They defend a **territory,** a portion of the home range, often against other individuals of their species and sometimes against individuals of other species. Many species are territorial for only part of the year, often during the breeding season, but others are territorial throughout the year. Territoriality has been positively correlated with the availability of needed resources that occur in small areas that can be defended.

Territoriality is easily studied in birds. Typically, the male chooses a territory at the beginning of the breeding season. This behavior results from high sex hormone concentrations in his blood. The males of adjacent territories fight until territorial boundaries become established. Generally, the dominance of a male is directly associated with how close he is to the center of his territory. Thus, close to home he is like a lion, but when invading some other bird's territory, he may behave more like a lamb. The interplay of dominance values among territorial males may produce a neutral line, an area between their territories in which neither is dominant.

Bird songs announce the existence of a territory and often serve as a substitute for fighting. Furthermore, they announce to eligible females that a propertied male resides in the territory. Typically, male birds take up a conspicuous station, sing, and sometimes display striking patterns of coloration or aerial acrobatics to their neighbors, their rivals, and sometimes their mates.

Studies of lions in the Serengeti show that these animals engage in group territoriality (Fig. 50–16). As discussed previously, a pride consists of several adult females, their dependent young, and a group of immigrant males. The females protect their young and defend the territory against invasion by foreign females. The males defend the territory against strange males. At night males patrol their turf, roaring fiercely to announce their presence.

The costs of territoriality include the time and energy expended in staking out and defending a territory and the risks involved in fighting for it. Benefits often include exclusive rights to food within the territory and greater reproductive success. The males of a lion pride father all the cubs born within that pride. Their ability to defend their territory from invasion by strange males ensures their reproductive success. Among many species, animals that fail to establish territories fail to reproduce. Territoriality also tends to reduce conflict among members of the same species and ensures efficient use of environmental resources by encouraging individuals to spread throughout a habitat.

Usually, territorial behavior is related to the specific lifestyle of the animal and to whatever aspect of its environment is most critical to its reproductive success. For instance, sea birds may range over hundreds of square kilometers of open water but exhibit territorial behavior only at crowded nesting sites on an island. The nesting sites are their scarcest resource and the one for which competition is keenest.

SEXUAL SELECTION RESULTS IN REPRODUCTIVE ADVANTAGE

In many species individuals actively compete for mates. Typically in such populations there is an abundance of males competing for a limited number of receptive females. For a male, reproductive success depends on how many females he can impregnate. Females may have the opportunity to select a sexual partner from among several males. For a female, reproductive success depends on how many eggs she can produce during her reproductive lifetime, the quality of the sperm that fertilize them, and on the survival of her offspring to reproductive age. Recall that an animal's reproductive success is a measure of its *direct fitness*. **Sexual selection,** a type of natural selection, occurs when

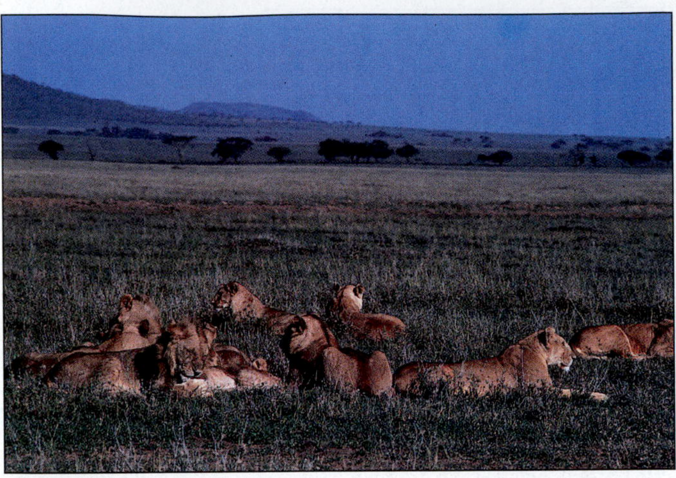

Figure 50–16 Territoriality in lions. Lions *(Panthera leo)* form territorial groups typically consisting of several females and their offspring and accompanied by a coalition of males, generally brothers who are unrelated to the females. Successful breeding appears to depend on having a territory. *(Steve McCutcheon/Visuals Unlimited)*

individuals vary in their ability to compete for mates. This leads to the reproductive advantage that some individuals have over others of the same sex and species.

Dominance and ornamental traits influence mate choice

In many species, success of a male in dominance encounters with other males indicates his quality to the female, and she allows the victorious male to court her. Many studies confirm that males ranking higher in a dominance hierarchy mate more frequently than males that rank lower.

Many exceptions demonstrate the complexity of reproductive behavior among some species. For example, among baboons with a definite dominance hierarchy, males lower in the hierarchy copulated with females as frequently as males of higher status. Investigators observed, though, that dominant males copulated more frequently with females who were in *estrus,* their fertile period. In addition, some lower ranking males develop alternative strategies for attracting females. For instance, a male might win a female's interest by protecting her baby, even though it is not his own.

Among crickets and many other insect species, a courting male offers a gift of food to a prospective mate. Studies show that the larger or higher quality the food offering, the better the chances the male will be accepted.

Females may select their mates based on ornamental displays. Male fishes are often brightly colored (see Fig. 19–13). Among many bird species, males exhibit bright colors and dramatic plumage, and male deer display elaborate antlers. Investigators suggest that the expression of ornamental traits may give the female important information about the male's physical condition. The size of the antlers of red deer *(Cervus elaphus),* for example, reflects proper nutrition and good health.

Among some species of insects, birds, and bats, males gather in a small display area called a **lek,** where they compete for females. When a receptive female appears, males may excitedly display themselves and compete for her attention. Among some species, the female selects a male based on his location in the lek rather than on his appearance. The dominant male may occupy a central position and be chosen by most of the females. The female selects a male, mates with him, and then leaves the lek. The male remains to woo other females.

Courtship rituals ensure that the male is indeed a male and is a member of the same species. Rituals provide the female further opportunity to evaluate him. In some species, courtship may also be necessary as a signal to trigger nest building or ovulation. Courtship rituals can last seconds, hours, or days and often involve a series of fixed action patterns (Fig. 50–17). The first display by the male releases a counterbehavior by the female. This, in turn, releases additional male behavior, and so on until the pair is physiologically ready for copulation. Specific cues enable courtship rituals to function as reproductive isolating mechanisms among species (see Figure 19–3).

An extreme courtship ritual has been described for redback spiders *(Latrodectus hasselti),* which are closely related to the black widow spider. During copulation, the small male spider positions himself above his larger mate's jaws. During 65% of matings, the finale is that the female eats her suitor. The apparent explanation for this behavior is that the risk-taking male is able to copulate for a longer period and thus fertilize more eggs than noncannibalized males and that the female is more likely to reject additional mates.

Sexual selection favors polygynous mating systems

In most species, males make little parental investment in their offspring, apart from providing sperm. Males ensure reproductive success by impregnating many females, increasing the probability that their genes will be propagated in multiple offspring. Thus, sexual selection often favors male **polygyny,** a mating system in which males fertilize the eggs of many females during a breeding season.

In the mating system known as **polyandry,** a female mates with several males. Benefits may include receiving gifts from several males or enlisting several males to help care for the young. Data collected at Jane Goodall's research center at Gombe National Park in Tanzania indicate that female chimpanzees copulate several hundred times. In addition to mating with males of their own group, many females slip off to neighboring communities when they are most fertile and copulate with less familiar males. It is possible that these sexual rendezvous protect against inbreeding. Investigators also suggest that mating with many males provides insurance against infanticide. Males do not aggress against infants of mothers with whom they have copulated. Males in polyandrous species may accept their status because **mate guarding** may be

(a)

(b)

Figure 50–17 Courtship displays. **(a)** The male great frigatebird *(Frigata minor)* inflates his red throat sac in display as part of a courtship ritual. Photographed on Christmas Island in the Pacific. **(b)** Egrets *(Egretta rufescens)* performing a mating dance. *(a, Sid Bart/Photo Researchers, Inc.; b, Arthur Morris/Visuals Unlimited)*

costly, ineffective, or both. Females may be able to range over a wide area, easily escaping guarding behavior of males.

Polyandry and polygyny sometimes occur in the same species. After mating, a female giant water bug attaches a clutch of eggs to the back of her mate. He takes care of the eggs until they hatch. She then may mate with a different male and glue the new clutch of eggs to his back. However, if the male has more space for eggs, he may mate with another female.

Apparently, in most species, it is not certain who fathered the offspring. Raising some other male's offspring is a genetic disadvantage. Thus, males in some species may compromise mate chasing in favor of mate guarding. The male guards his partner after copulation to ensure that she will not copulate with another male. Mate-guarding behavior is likely to occur when the female is receptive and has eggs that might be fertilized by another male. For example, dominant male African elephants guard a female only during the phase of estrus when she is most likely to have a fertile egg. Before that time, or later in estrus after mating has already occurred, younger male elephants of lesser social status can copulate with the female. One cost of mate guarding is the loss of opportunity for a dominant male to mate with other females.

Sexual selection favors males that inseminate many females and produce many offspring. Perhaps for this reason, **monogamy,** a mating system in which a male mates with only one female during a breeding season, is not common. Less than 10% of mammals are monogamous.

Monogamy was long thought to be common among birds because many species form **pair bonds,** stable relationships that ensure cooperative behavior in mating and the rearing of the young (Fig. 50–18). However, genetic evidence shows that some of the offspring are fathered by males other than the one caring for them. For example, DNA tests show that among eastern bluebirds, 15% to 20% of the chicks are fathered by other males. Apparently, females engage in extra-pair copulations. Producing offspring with greater genetic variability may increase chances for survival of at least some of them. Stephen Emlen of Cornell University makes a distinction between social monogamy, in which animals form a pair bond, and genetic (sexual) monogamy.

Monogamy does occur in some species, typically when males are needed to protect and feed the young. For example, the California mouse has been shown to be genetically, as well as socially, monogamous. The offspring require their parents' body heat to survive, and the parents take turns keeping them warm.

Figure 50–18 Pair bond in birds. Black-browed albatrosses *(Diomedea melanophris)* form durable mating pairs and may stay together for life. *(Johnny Johnson/DRK Photo)*

SOME ORGANISMS CARE FOR THEIR YOUNG

Most animals do not invest time and energy in caring for their offspring because the costs of parenting are typically greater than the benefits. The costs include a reduction in the number of offspring that can be produced and the risks taken in protecting the young from predators. The benefit of investment in parental care is the increased probability that each offspring will survive. Natural selection has favored parental care in species in which the benefits to offspring survival outweigh the costs of decreased opportunities to produce additional offspring. This situation occurs in species in which the female produces few young or reproduces only once during a breeding season.

Parental care is an important part of successful reproduction in some invertebrates, including some cnidarians (jellyfish), rotifers, mollusks, and arthropods (crustaceans, insects, spiders, and scorpions; Fig. 50–19*a*). Caring for the young is also common among fishes, reptiles, birds, and mammals (Fig. 50–19*b, c,* and *d*).

Females of many vertebrate animals produce relatively few, large eggs. Because of the time and energy invested in producing eggs and carrying the developing embryo, the female has more to lose than the male if the young do not develop. Thus, females are more likely than males to brood eggs and young, and usually females invest more in parental care. Parental care is especially skewed toward the female in mammals because female mammals, not males, provide milk to nourish their young. Investing time and effort in care of the young is usually less advantageous to a male (assuming that the female can handle the job by herself), for time spent in parenting is time lost from inseminating other females.

In some situations, a male can benefit by helping to rear his own young or even those of a genetic relative. Receptive females may be scarce, breeding territories may be difficult to establish or guard, and gathering sufficient food may require more effort than one parent can provide. In some habitats, the young need protection against predators or cannibalistic males of the same species. For example, among wolves and many other carnivores, males defend the young and the food supply.

Among many species of fishes, the male cares for the young. The ultimate cause for this behavior appears to be that the parenting costs to the males are less than they would be for the females. A male fish must have a territory to attract a female. While guarding his territory, the male guards the fertilized eggs as well. Having eggs also appears to make the male more attractive to potential mates. In contrast, when a female is caring for her eggs or young, she is not able to feed optimally. As a result her body size is smaller and her fertility is reduced.

ELABORATE SOCIETIES ARE FOUND AMONG THE SOCIAL INSECTS

Many insects, such as tent caterpillars, which spin communal nests, cooperate socially. However, the most elaborate insect societies are found among the bees, ants, wasps, and termites. (The first three of these all belong, not coincidentally, to the same order, Hymenoptera.) Insect societies are held together by a complex system of sign stimuli that are keyed to social interaction. As a result, their behaviors tend to be quite rigid.

The social organization of honeybees has been studied more extensively than that of any other social insect. Instructions for the honeybee society are inherited and preprogrammed, and the size and structure of the bee's nervous system permit only a limited range of behavioral variation. Yet these insects are not automatons. Within those limits, the complex bee society can respond with some flexibility to food and other stimuli in the environment.

A honeybee society generally consists of a single adult queen, up to 80,000 worker bees (all female), and, at certain times, a few males called *drones* that fertilize newly developed queens. The queen, the only female in the hive capable of reproduction, deposits about 1000 fertilized eggs per day in the wax cells of a comb.

The composition of a bee society is generally controlled by an antiqueen pheromone secreted by the queen. It inhibits the workers from raising a new queen and prevents development of ovaries in the workers (Fig. 50–20). If the queen dies, or if the colony becomes so large that the inhibiting effect of the pheromone is dissipated, the workers begin to feed some larvae special food that promotes their development into new queens.

The queen bee stores sperm cells from previous matings in a seminal receptacle. If she releases sperm to fertilize an egg as it is laid, the resulting offspring is female; otherwise, it is male. Thus, males develop parthenogenically from unfertilized eggs and are haploid. Because a drone is haploid, each one of his sperm cells has *all* his chromosomes; that is, meiosis does not occur during sperm production. The queen bee stores this sperm throughout her lifetime and uses it to produce worker bees.

The worker bees of a hive are more closely related to one another than would be sisters born of a diploid father. Indeed, they have up to three-quarters of their genes in common. (Assuming the queen bred with a single drone, they share half of the queen's chromosomes and all of the drone's.) As a consequence, they are more closely related to one another than they would be to their own offspring, if they could have any. (A worker bee's offspring would have only one-half of its genes in common with its worker mother.) New queens will also be their sisters. Worker bees are therefore more likely to pass on copies of their genes to the next generation by raising these individuals than they would if they were to produce their own offspring.

Division of labor among worker bees is mostly determined by age. The youngest worker bees serve as nurses that nourish larval bees. After about a week as nurse bees, workers begin to produce wax and build and maintain the wax cells. Older workers are foragers, bringing home nectar and pollen. Most worker bees die at the ripe old age of 42 days.

The most sophisticated known mode of communication among bees is a stereotyped series of body movements referred to as a dance. When a honeybee scout locates a rich source of nectar, it apparently observes the angle between the food, hive, and the sun. The bee then communicates the direction and distance of the food source relative to the hive (Fig. 50–21). If the food supply is within about 50 m (about 55 yd), the scout performs a **round dance,** which generally excites the other bees and causes

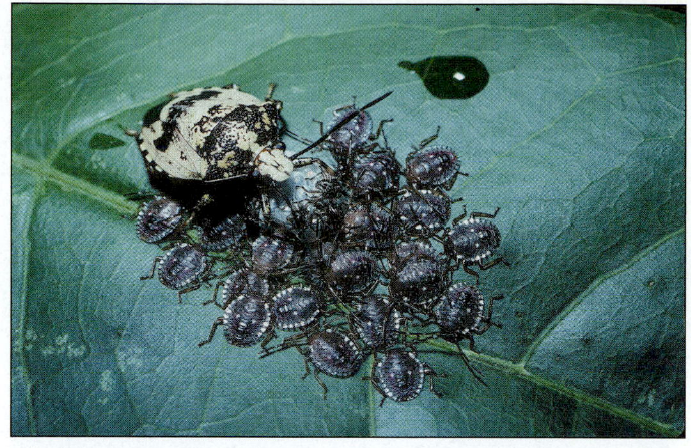

(a)

(b)

(c)

(d)

Figure 50–19 Caring for the young. Parental investment in offspring increases the probability that the young will survive. **(a)** Cocoa shield bug *(Antiteuchus melanoleucus)* female guarding her nymphs. Photographed in the rainforest, Amazonia, Brazil. **(b)** Adult robins invest energy in feeding their young. **(c)** A female crocodile transports hatchlings in her mouth from their nest site to Lake St. Lucia in South Africa. **(d)** Female cheetah *(Acinonyx jubatus)* stands watch as her cubs eat the Thompson's gazelle she has hunted and killed for them. Photographed in Masi Mara, East Africa. *(a, Ken Preston-Mafham/Premaphotos Wildlife; b, Dominique Braud/Dembinsky Photo Associates; c, M. Reardon/Photo Researchers, Inc.; d, McMurray Photography)*

them to fly short distances in all directions from the hive until they find the nectar. If the food is distant, however, the scout performs a **waggle dance,** which follows a figure-eight pattern.

Process of Science

In the 1940s Karl von Frisch pioneered studies in bee communication. He found that the waggle dance conveys information about both distance and direction. The bee typically performs the dance in the hive on the vertical surface of the comb. During the straight run part of the figure eight, the number and frequency of the waggles indicate the distance. The orientation of the movements indicates the direction of the food source in reference to the position of the sun. For example, if the bee dances straight up, the nectar is located directly toward the sun. If the bee dances 40 degrees to the left of the vertical surface of the comb, the nectar is located 40 degrees to the left of a line between the hive and the sun.

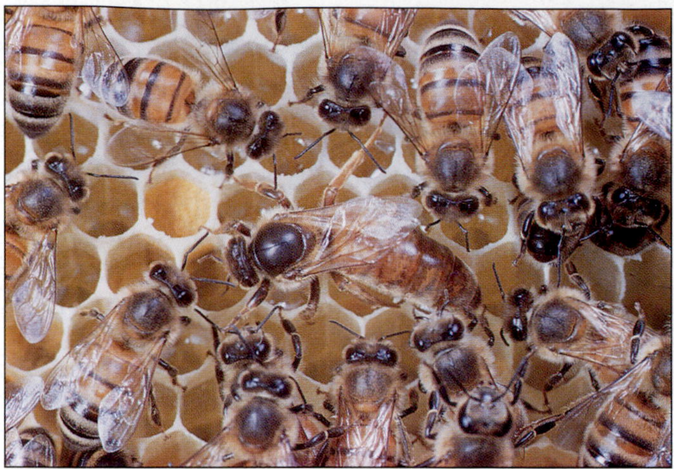

Figure 50–20 Maintaining a complex social structure. Numerous worker honeybees surround their queen *(center)*. Workers constantly lick secretions from the queen bee. These secretions, transmitted throughout the hive, suppress the activity of the workers' ovaries. *(Treat Davidson/Photo Researchers, Inc.)*

Von Frisch thought that through their dances the bee scouts symbolically communicated information about food sources to other foragers. He hypothesized that bees estimated distance from the amount of the energy they expended during a flight. Although human observers can find food sources based on information conveyed by the dances of honeybees, many biologists have been skeptical that bees could have such a complex system of communication. Some suggested that the dance communicated only the general direction of the food source. Others hypothesized that the flower scent on the dancing bee might be the true clue to the source of the nectar, and the dance itself does not really communicate anything to the bees. Although studies using mechanical bees (with no scent) supported von Frisch's conclusions, some biologists remained skeptical.

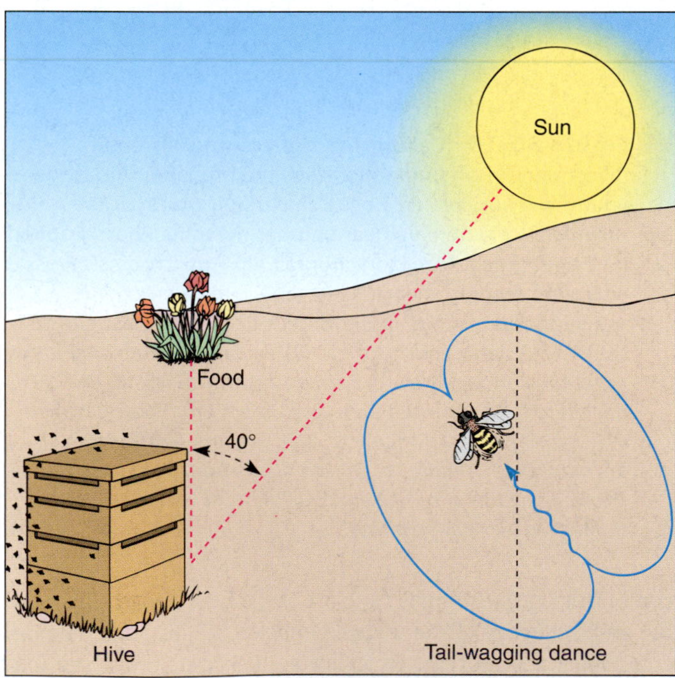

Figure 50–21 The honeybee waggle dance. The scout is waggling upward, indicating that the food is toward the sun, and inclined 40 degrees to the left, which reveals the angle of the food source relative to the sun. The dance takes place inside the hive.

In the mid 1990s, Harold Esch, a neurobiologist at the University of Notre Dame, and a group led by Mandyam Srinivasan at the Australian National University demonstrated that bees visually monitor the distance they travel. Then, in 2001 Esch and his colleagues reported in *Nature* the results of a convincing study in which they tricked foraging honeybees regarding the distance to the food and then observed them communicating the misinformation to other bees. These investigators trained bees to fly through a tunnel to get to a feeder on the other side. They painted a complex pattern inside the tunnel so that the bees calculated that they had traveled 72 m (almost 79 yd) instead of 11 m (12 yd), the actual distance. Based on the dances of these bees, 75% of the other 220 forager bees flew 72 m, the distance communicated, to search for the nectar. These studies clearly demonstrate that the communication of distance by bees depends on visual cues. Their conclusions disprove von Frisch's energy hypothesis.

VERTEBRATE SOCIETIES TEND TO BE RELATIVELY FLEXIBLE

Vertebrate societies usually contain nothing comparable to the physically and behaviorally specialized castes of honeybees or ants. An exception is the naked mole rat *(Heterocephalus glaber)* of southern Africa, a rodent that has a social structure closely resembling that of the social insects. Although in some ways most vertebrate societies seem simpler than insect societies, they are also more flexible. Vertebrate societies share a great range and plasticity of potential behaviors and can effectively modify behavior to meet environmental challenges.

The behavioral plasticity of vertebrates makes possible the transmission of culture in some bird and mammal species. **Culture** can be defined as behavior common to a population that is learned from other members of the group and is transmitted from one generation to another. Culture is not inherited genetically. About 17 behaviors that may be considered cultural have been described among cetaceans (whales, dolphins, porpoises). For example, female orcas teach their offspring to hunt seals according to the custom of their particular group. Orcas also have local dialects that have been documented for at least six generations, suggesting that they are passed on from parent to offspring.

Recent studies of nonhuman primates demonstrate that some groups develop local customs and teach them to their offspring (see *On the Cutting Edge: Cultural Variation Among Chimpanzee Groups*). Many biologists think that studying culture in nonhuman primates will provide insights into the evolutionary origins of human culture. Human society is based to a great extent on the symbolic transmission of culture through spoken and written language.

Process of Science

ON THE CUTTING EDGE

Cultural Variation Among Chimpanzee Groups

HYPOTHESIS: Cultural variations exist among chimpanzee populations. Such customs are learned behaviors that are transmitted from one generation to the next.

METHOD: Investigators gathered data from researchers at seven chimpanzee research sites.

RESULTS: Thirty-nine behaviors were identified that could be considered cultural variations among chimpanzee populations.

CONCLUSION: Chimpanzees develop local customs and teach them to their offspring. These behaviors vary among different populations.

Although researchers have studied primate societies for many years, they have typically focused on a single population at a particular research site. Recently Andrew Whiten of the University of St. Andrews in Scotland and Christophie Boesch of the Max Planck Institute for Evolutionary Anthropology in Leipzig, Germany, asked scientists at seven chimpanzee research sites for a list of behaviors that might be local customs.[*†] Whiten and Boesch then developed a comprehensive list of 65 behaviors that might qualify as cultural behaviors. They asked scientists at each site to indicate whether or not each behavior was present in the population they were studying.

Based on data provided by researchers at the seven research sites, Whiten and Boesch identified 39 chimpanzee behaviors that they considered cultural variations. These behaviors included various ways of using tools, courtship rituals, and grooming techniques *(see figure)*. Each of these local customs was learned from other chimpanzees in the population, and customs varied among different populations. For example, chimpanzees on the western side of the Sassandra-N'Zo River use stone hammers to crack coula nuts. Researchers have photographed mother chimpanzees teaching this behavior to their offspring. Just a few miles away on the eastern side of the river, chimpanzees do not crack nuts even though they are available.

Tetsuro Matsuzawa of the Primate Research Institute at Kyoto University in Japan demonstrated that customs can spread from one population to another.[‡] Matsuzawa studied a population in Guinea that use rock hammers and anvils to crack the hard shells of oil palm nuts. Coula nuts do not grow in this area. Matsuzawa and his team of researchers left coula nuts along with oil palm nuts and rocks in a clearing and videotaped the responses of chimpanzees to the unfamiliar nuts. The researchers observed several chimpanzees picking up the coula nuts, but only one chimpanzee, known as Yo, appeared to know what to do with them. She cracked and ate them. Other adults ignored her behavior, but some young chimps observed her and copied her behavior. Interestingly, Yo had joined this population as an adolescent and may have migrated from an area rich in coula nuts. Was Yo transmitting knowledge that she had learned in a different culture? Is this an example of multiculturalism?

To learn more about how chimpanzees learn from one another, Whiten and his colleagues conducted experiments using artificial fruits. They demonstrated how to open the fruit, a procedure requiring several steps. Many of the observing chimpanzees imitated the researchers, completing the sequence of steps necessary to open the fruit.

As researchers continue to study cultural variation among chimpanzee populations, learning ever more about our closest relatives, it is important to remember that the capacity to learn such behaviors is the product of natural selection. The behaviors that are maintained are adaptive for these animals in their environment.

A demonstration of chimpanzee *(Pan troglodytes)* culture. Mother chimpanzees teach social customs, like use of tools, to their offspring. The chimpanzee in the photograph is inserting a stick into an ant nest within a tree. When the ants climb up the stick, the chimpanzee eats them. This behavior, photographed at a research site in Tanzania, is part of the culture of this chimpanzee population. *(Gunter Ziesler/Peter Arnold)*

[*] Whiten, A., et al. "Cultures in Chimpanzees." *Nature,* Vol. 399, Jun. 17, 1999.

[†] Whiten, A., and C. Boesch. "The Cultures of Chimpanzees." *Scientific American,* Vol. 284, Jan. 2001.

[‡] Matsuzawa, T. (ed.). *Primate Origins of Human Cognition and Behavior.* Springer-Verlag, Tokyo, 2001.

ALTRUISTIC BEHAVIOR CAN BE EXPLAINED BY INCLUSIVE FITNESS

Sociobiology focuses on the evolution of social behavior through natural selection. In his landmark book, *Sociobiology: The New Synthesis,* published in 1975, Edward O. Wilson combined principles of population genetics, evolution, and animal behavior to present a comprehensive view of the evolution of social behavior. Wilson's work influenced the development of behavioral ecology, and many of the concepts discussed in this chapter, such as paternal investment in care of the young, are based on contributions made by sociobiologists.

From the sociobiological perspective, an organism and its adaptations, including its behavior, ensure that genes make more copies of themselves. The cells and tissues of the body support the functions of the reproductive system, and the reproductive system transmits genetic information to succeeding generations.

If an animal's principal evolutionary mandate is to perpetuate its genes, we must wonder why some animals appear to spend time and energy in helping others. We have discussed examples of **cooperative behavior** (sometimes called *mutualism*) such as group hunting in which each animal in the group benefits. In a type of cooperative behavior known as **reciprocal altruism,** one animal helps another with no immediate benefit. However, at some later time the animal that was helped repays the debt.

In **altruistic behavior** there appears to be no payoff: One individual behaves in a way that seems to benefit others rather than itself, with no potential payoff. However, true altruistic behavior would not be adaptive because the animal's response decreases its reproductive success relative to individuals that do not exhibit the altruistic behavior. In 1964 William D. Hamilton offered a plausible explanation. He suggested that evolution does not distinguish between genes transmitted directly from parent to offspring and those transmitted indirectly through close relatives. According to Hamilton, natural selection favors animals that help a relative because the relative's offspring carries some of the helper's genes. **Inclusive fitness** is the sum of direct fitness (which can be measured by the number of genes an animal perpetuates in its offspring) plus indirect fitness (the genes it helps perpetuate in the offspring of its kin).

The potential genetic gain from helping relatives can be measured by the **coefficient of relatedness,** the probability that two individuals inherit the same uncommon allele from a recent common ancestor. The coefficient of relatedness for two animals that are not related is 0, whereas between a parent and its offspring this coefficient would be 0.5, indicating that half of their genes are the same. The coefficient of relatedness for siblings is also 0.5; for first cousins it is 0.125. The higher the coefficient of relatedness, the more likely an animal will help a relative. DNA fingerprinting of lions in the Serengeti National Park and in the Ngorongoro Crater in Tanzania has confirmed that lion brothers that cooperate in prides have a greater probability of perpetuating their genes than those that go off on their own. This is true even if the lion perpetuates his genes only by proxy through his brother. **Indirect selection,** also called **kin selection,** is a form of natural selection that increases inclusive fitness through the breeding success of close relatives.

Among some birds (e.g., Florida jays), nonreproducing individuals aid in the rearing of the young. Nests tended by these additional helpers as well as parents produce more young than do nests with the same number of eggs overseen only by parents. The nonreproducing helpers are close relatives who increase their own biological success by ensuring the successful, though indirect, perpetuation of their genes. Helpers may be prevented from producing their own offspring by limiting factors such as a shortage of mates or territories.

The bee society provides another interesting example of indirect selection. Recall that worker bees do not reproduce, but they nourish larval bees. The worker bees have up to three-quarters of their genes in common with the larval bees and so are more likely to pass on copies of their genes to the next generation by raising these individuals than they would by producing their own offspring.

Process of Science Sentinel behavior involves keeping watch for predators and warning other members of the group that might be foraging when danger threatens. Until recently, biologists thought that sentinels were at greater risk for predation, yet selflessly engaged in risky behavior that benefited their relatives. Their behavior has been considered to be an example of indirect selection. This explanation appears to be valid for some species,

Figure 50–22 Kin selection in prairie dogs. Low-ranking members of this social rodent group act as sentries, risking their own lives by exposing themselves outside their burrows. However, in this way they protect their siblings, and by so doing, ensure that the genes they share in common will be perpetuated in the population. *(Wu Wal Ping/Bruce Coleman, Inc.)*

including prairie dogs and ground squirrels (Fig. 50–22). However, recent studies suggest an opposing view about guarding behavior in other species. P.A. Bednekoff, of the University of British Columbia in Vancouver, developed a model to explain how guarding could result from selfish behavior. Bednekoff observed that for some species there was no evidence that sentinels were at higher risk than others in their group. He suggested that the opposite might be true and that sentinels may actually have an advantage. By detecting predators first, the sentinels could more easily escape.

A study by Clutton-Brock of the University of Cambridge and his colleagues supports Bednekoff's model. These researchers studied sentinel behavior in the African mongoose *(Suricata suricatta)* and reported that during 2000 hours of observation, no sentinel was killed. In fact, they concluded that guarding is a selfish behavior in African mongoose populations because sentinels were able to escape quickly into their nearby burrows. This research is an important reminder that it is risky for biologists to generalize the meanings of complex social behavior from one species to another. Many explanations are possible.

SOCIOBIOLOGY EXPLAINS HUMAN SOCIAL BEHAVIOR IN TERMS OF ADAPTATION

Like Darwin and many other biologists of the past, E.O. Wilson and other sociobiologists suggest that human behavior can be studied in evolutionary terms. Sociobiology is controversial, at least in part because of its possible ethical implications. This approach is sometimes viewed as denying that human behavior is flexible enough to permit substantial improvements in the quality of our social lives. Yet sociobiologists do not disagree with their critics that human behavior is flexible. At least some of the debate therefore seems to rest on the degree to which human behavior is genetically determined and the extent to which it can be modified.

As sociobiologists acknowledge, people can, through culture, change their way of life far more profoundly in a few years than could a hive of bees or a troop of baboons in hundreds of generations of genetic evolution. This capacity to make changes is indeed genetically determined, and that is a great gift. How we use it and what we accomplish with it are not a gift but a responsibility on which our own well-being and the well-being of other species depend.

SUMMARY WITH KEY TERMS

I. **Behavior** is what an animal does and how it does it, usually in response to stimuli in its environment.
   - A. **Proximate causes** of behavior are immediate causes such as genetic, developmental, and physiological processes that permit the animal to carry out a specific behavior. Proximate causes answer *how* questions.
   - B. **Ultimate causes** are the evolutionary explanations for *why* a certain behavior occurs.

II. **Behavioral ecology** is the scientific study of behavior in natural environments from an evolutionary perspective.
   - A. Behavior has costs and benefits. Benefits contribute to **direct fitness,** the animal's reproductive success measured by the number of viable offspring. When benefits outweigh costs, the behavior is adaptive.
   - B. Behavior tends to be adaptive.

III. Behavior results from the interaction of genes **(innate behavior)** and environmental factors. The capacity for behavior is inherited, but **learned behavior** is behavior that has been modified in response to environmental experience.
   - A. Before an organism can show any pattern of behavior, it must be physiologically ready to produce the behavior. Walking and many other behaviors that we view as automatic are **motor programs,** coordinated sequences of muscle actions.
   - B. A **fixed action pattern (FAP)** is an automatic behavior that, once activated by a simple sensory stimulus, continues to completion regardless of sensory feedback.
   - C. An FAP can be triggered by a specific unlearned **sign stimulus,** or **releaser.**
   - D. **Habituation** is a type of learning in which an animal learns to ignore a repeated, irrelevant stimulus.
   - E. **Imprinting** establishes a parent-offspring bond during a critical period early in development.
   - F. In **classical conditioning** an association is formed between some normal body function and a new stimulus.
   - G. In **operant conditioning** an animal learns a behavior to receive **positive reinforcement** or to avoid punishment.
   - H. **Insight learning** is the ability to adapt past experiences that may involve different stimuli to solve a new problem.
   - I. A function of play may be to give a young animal a chance to acquire adult patterns of behavior.

IV. It is adaptive for an organism's metabolic processes and behavior to be synchronized with cyclical changes in the environment.
   - A. In many species, physiological processes and activity follow **circadian rhythms.**
   - B. **Diurnal** animals are most active during the day, **nocturnal** animals at night, and **crepuscular** animals at dawn and/or dusk. Some biological rhythms reflect the **lunar cycle** or the changes in tides due to changes in the phase of the moon.
   - C. Many biological rhythms are regulated by internal timing mechanisms that serve as **biological clocks.** In mammals the master clock is located in the **suprachiasmatic nucleus (SCN)** in the hypothalamus.

V. **Migration** is periodic long-distance travel from one location to another.
   - A. Ultimate causes of migration include the advantage of moving away from an area that seasonally becomes too cold, too dry, or depleted of food to a more hospitable area.
   - B. Proximate causes of migration include physiological responses that are triggered by environmental changes.
   - C. **Directional orientation** refers to travel in a specific direction and requires **compass sense,** a sense of direction. Many migrating animals use the sun to orient themselves.
   - D. **Navigation** requires both compass sense and **map sense,** an awareness of location. Birds that navigate at night use the stars to guide them. Birds and many other animals also use Earth's magnetic field to navigate.

VI. **Optimal foraging,** the most efficient strategy for an animal to obtain food, often enhances reproductive success.

VII. **Social behavior** is adaptive interaction, usually among members of the same species.
   - A. A **society** is a group of individuals of the same species that may work together in an adaptive manner.
   - B. Many animal societies are characterized by a means of communication, cooperation, division of labor, and a tendency to stay together.

VIII. Animal **communication** involves the exchange of mutually recognizable signals.

A. Communication signals can be auditory, visual, electrical, or chemical.
B. **Pheromones** are chemical signals that convey information between members of a species.

IX. A **dominance hierarchy** is a ranking of status within a group in which more dominant members are accorded benefits (such as food or mates) by subordinates, often without overt aggressive behavior.

X. Organisms often inhabit a **home range,** a geographical area that they seldom leave but do not necessarily defend.
A. A defended area within a home range is called a **territory,** and the defensive behavior is **territoriality.**
B. Some animals, such as lions, engage in group territoriality.

XI. **Sexual selection,** a type of natural selection, occurs when individuals vary in their ability to compete for mates. Individuals with reproductive advantages are selected over others of the same sex and species.
A. Mate choice may be influenced by dominance, gifts, ornaments, and courtship displays. Males of some species gather in a **lek,** a small display area where they compete for females.
B. **Courtship rituals** ensure that the male is a member of the same species, and permit the female to assess the quality of the male.
C. Sexual selection often favors male **polygyny,** a mating system in which a male mates with many females. In **polyandry,** a female mates with several males. **Monogamy,** mating with a single partner during a breeding season, is less common.
D. Among many species, males engage in **mate guarding,** especially when the female is most fertile, to prevent other males from fertilizing her eggs.
E. A **pair bond** is a stable relationship between a male and a female that may involve cooperative behavior in mating and in rearing the young.
F. Parental care increases the probability that the offspring will survive. A high investment in parenting is typically less advantageous to the male than to the female.

XII. Insect societies tend to be rigid, with the role of the individual narrowly defined. Division of labor is mainly determined by age.
A. Bees communicate the direction and distance of a food source with a series of body movements referred to as a dance.
B. A **round dance** indicates that the food source is within 50 m; a **waggle dance** communicates information about both distance and direction when the food is farther away.

XIII. Vertebrate societies are more flexible than insect societies. Some bird and mammal species develop **culture,** learned behavior common to a population that is transmitted from one generation to the next. Symbolic transmission of culture is important in human societies.

XIV. Some animals appear to spend time and energy helping others.
A. Some animals engage in **cooperative behavior,** also called mutualism (e.g., hunting), that benefits all. In a type of cooperative behavior known as **reciprocal altruism,** the helper does not immediately benefit but is helped later by the animal it helped.
B. In **altruistic behavior,** one individual appears to behave in a way that benefits others rather than itself.
C. **Inclusive fitness** is the sum of an individual's direct fitness and indirect fitness (number of offspring of kin). This concept suggests that natural selection favors animals that help a relative because this is an indirect way to perpetuate some of the helper's own genes.
D. The **coefficient of relatedness** is the probability that two individuals have the same uncommon allele inherited from a recent common ancestor.
E. **Indirect selection,** also known as **kin selection,** is a type of natural selection that increases inclusive fitness through successful reproduction of close relatives.
F. **Sociobiology** focuses on the evolution of social behavior through natural selection. Sociobiology offers an adaptation explanation for human social behavior.

POST-TEST

1. The contemporary study of behavior in natural environments from the point of view of adaptation is (a) ecology (b) ethology (c) behavioral ecology (d) answers a, b, and c are correct (e) answers a and c only are correct
2. If you were interested in exploring *why* the graylag goose "rolls" a nonexistent egg toward her nest, you might explore (a) proximate causes (b) ultimate causes (c) pheromones (d) answers a, b, and c are correct (e) answers a and b only are correct
3. The responses of an organism to signals from its environment are its (a) behavior (b) learned behavior (c) ultimate behavior (d) releasers (e) motor programs
4. A fixed action pattern (FAP) is elicited by a(an) (a) motor program (b) sign stimulus (c) releaser (d) answers a, b, and c are correct (e) answers b and c only are correct
5. A form of learning in which a young animal forms a strong attachment to a moving object (usually its parent) within a few hours of birth is (a) classical conditioning (b) operant conditioning (c) imprinting (d) insight learning (e) parental investment
6. An animal learns to ignore a repeated, irrelevant stimulus. This is (a) classical conditioning (b) operant conditioning (c) imprinting (d) insight learning (e) habituation
7. Salivation by a student when the noon bell rings is an example of (a) classical conditioning (b) operant conditioning (c) imprinting (d) insight learning (e) habituation
8. Both compass sense and map sense are necessary for (a) navigation (b) migration (c) directional orientation (d) answers a, b, and c are correct (e) answers a and b only are correct
9. In optimal foraging (a) animals always hunt in social groups (b) an animal obtains food in the most efficient way (c) animals can rarely afford to be selective (d) groups of 5 to 7 animals are most successful (e) animals are typically most successful when they hunt alone
10. Chemical signals that convey information between members of a species are (a) pheromones (b) hormones (c) neurotransmitters (d) leks (e) neuropeptides
11. The benefits of territoriality include (a) rights to defend a home range (b) increased reproductive success (c) energy investment in staking out and defending the area (d) monogamy (e) pair bonding
12. Sexual selection (a) is a form of natural selection (b) occurs when animals are very similar in their ability to compete for mates (c) results in animals that have lower direct fitness (d) occurs mainly among animals that practice polygyny (e) occurs mainly among animals that practice polyandry
13. Mate choice is least likely influenced by (a) ornamental displays such as antlers (b) dominance (c) competition in a lek (d) courtship behavior (e) fidelity of the male
14. The round dance of the bee (a) indicates that food is close to the hive (b) communicates direction (c) results in bees flying long distances in all directions (d) indicates that food is distant from the hive (e) indicates both direction and height of food
15. Behavior that appears to have no payoff, that is, an individual appears to act to benefit others rather than itself is known as (a) mutualism (b) altruistic behavior (c) reciprocal altruism (d) inclusive fitness (e) helping behavior
16. Inclusive fitness (a) considers only the genes an animal transmits to its own offspring (b) suggests that natural selection favors animals that help a relative (c) explains territoriality (d) is a form of social behavior (e) is a form of reciprocal altruism
17. Indirect selection (kin selection) (a) increases inclusive fitness (b) is a way of perpetuating genes of nonrelatives (c) accounts for some forms of migration (d) typically involves mate guarding (e) typically involves ornamental displays and use of a lek
18. Culture (a) is common among most animal groups (b) does not vary within the same species (c) is a shared derived character of humans (d) has a strong genetic component in primates (e) is behavior shared by members of a population and is learned from other members of the group

REVIEW QUESTIONS

1. In what ways are the behaviors of *Philanthus,* the sand wasp, adaptive?
2. Why is it adaptive for some species to be diurnal but for others to be nocturnal or crepuscular? What are biological clocks? How do they work in mammals?
3. Behavior capacity is inherited and is modified by learning. Give an example.
4. When Konrad Lorenz kept a graylag goose isolated from other geese for the first week of its life, the goose persisted in following him about in preference to other geese. How can this behavior be explained?
5. How does physiological readiness affect innate behavior? How does it affect learned behavior?
6. What distinguishes an organized society from a mere aggregation of organisms? What role does communication play in animal societies?
7. How does an organism learn its place in a dominance hierarchy? What determines this place? What costs and benefits accrue to dominant individuals in a dominance hierarchy? To subordinate individuals?
8. What is territoriality? What functions does it seem to serve?
9. What is sexual selection? What are some factors that influence mate choice?
10. Under what conditions does sexual selection favor polygynous (or polyandrous) mating systems? Why is monogamy uncommon?
11. What are some advantages of courtship rituals?
12. Why do more females than males invest in caring for their offspring? Explain in terms of costs, benefits, and fitness.
13. Compare and contrast the dance of bees with human language. Compare the society of a social insect with human society.
14. Define inclusive fitness. How is indirect (kin) selection used by sociobiologists to explain the evolution of altruistic behavior? What is the possible role of kin selection in the maintenance of insect and other animal societies?

YOU MAKE THE CONNECTION

1. What might be the adaptive value of sea turtle migration? Consider how this adaptive value has changed if, as a result of human activities, migration now puts sea turtles at greater risk than if they restricted their habitat to a single location. Discuss the possible evolutionary mechanisms by which the behavior of these species may (or may not) adapt to these environmental pressures. What conservation efforts should we take to ensure successful migration?
2. Using what you know about animal learning, propose experiments to test whether it might be possible to teach sea turtles "safer" reproductive behavior. As you do, discuss the following questions: What might be the broader, long-term ecological consequences of your proposals? Is it advisable for humans to attempt to alter the behavior of another species? Is this an ethical way of preserving a species that has become endangered as a result of human activities?
3. Behavioral ecologists have demonstrated that young tiger salamanders can become cannibals and devour other salamanders. Investigators observed that the cannibal salamanders preferred eating unrelated salamanders over their own cousins and preferred eating cousins over their siblings. Explain this behavior based on what you have learned in this chapter.
4. What are some similarities between the transmission of information by heredity and by culture? What are some differences?

RECOMMENDED READINGS

Alcock, J. *Animal Behavior,* 6th ed. Sinauer Associates, Sunderland, MA, 1998. An excellent, basic evolutionary approach to animal behavior.

Bower, B. "Culture of the Sea: Whales and Dolphins Strut Their Social Stuff for Scientists." *Science News,* 28 Oct. 2000. Recent findings suggest that certain cetacean behaviors are cultural.

Dugatkin, L.A. "The Evolution of Cooperation." *Bioscience,* Vol. 47, No. 6, Jun. 1997. The author discusses four evolutionary paths to cooperative behavior.

Dugatkin, L.A. "*Cheating Monkeys and Citizen Bees. The Nature of Cooperation in Humans and Animals.*" The Free Press, New York, 1999. A more detailed discussion of Dugatkin's paper (cited above) on evolutionary paths to cooperative behavior.

Kirchner, W.H., and W.F. Towne. "The Sensory Basis of the Honeybee's Dance Language." *Scientific American,* Vol. 270, No. 6, Jun. 1994. A discussion of honeybee communication.

Lohmann, K.J. "How Sea Turtles Navigate." *Scientific American,* Vol. 266, No. 1, Jan. 1992. Biological compasses and maps guide sea turtles as they migrate across hundreds of miles of ocean.

Packer, C., and A.E. Pusey. "Divided We Fall: Cooperation Among Lions." *Scientific American,* Vol. 276, No. 5, May 1997. A discussion of the costs and benefits of cooperative behavior among lions.

Preston-Mafham, K., and R. Preston-Mafham. "Mating Strategies of Spiders." *Scientific American,* Vol. 279, No. 5, Nov. 1998. Stunning photographs illustrate a brief discussion of mating rituals of some spiders.

Pusey, A., J. Williams, and J. Goodall. "The Influence of Dominance Rank on the Reproductive Success of Female Chimpanzees." *Science,* Vol. 277, 8 Aug. 1997. Data from a 35-year field study of chimpanzee behavior.

Schibler, U., J.A. Ripperger, and S.A. Brown. "Chronobiology—Reducing Time." *Science,* Vol. 293, 20 Jul. 2001. An interesting overview of two research reports on the relationship between circadian rhythms and energy homeostasis published in the same issue.

Seife, C. "Circadian Rhythms: Two Feedback Loops Run Mammalian Clock." *Science,* Vol. 288, 12 May 2000. Proteins that accumulate in the cytoplasm during the day turn off the genes that encode them at night.

Sherman, P.W. and J. Alcock (eds.). *Exploring Animal Behavior: Readings from American Scientist,* 2nd ed. Sinauer Associates, Sunderland, MA, 1998. Compilation of three articles on animal behavior originally published in *American Scientist*.

Tallamy, D.W. "Child Care Among the Insects." *Scientific American,* Vol. 280, No. 1, Jan. 1999. Under some conditions, many insects protect their offspring and help them find food.

Vogel, G. "Primate Abilities: Chimps in the Wild Show Stirrings of Culture." *Science,* Vol. 284, 25 Jun. 1999. An overview of cultural variation in primates.

Whiten, A. "The Cultures of Chimpanzees." *Scientific American,* Vol. 284, No. 1, Jan. 2001. Chimpanzees develop social customs that are passed on from one generation to the next.

Young, M.W. "The Tick-Tock of the Biological Clock." *Scientific American,* Vol. 282, No. 3, Mar. 2000. Specialized clock genes code for proteins that regulate biological rhythms.

- Visit our Web site at **http://www.info.brookscole.com/solomonbergmartin** for links to chapter-related resources on the World Wide Web. Additional on-line materials relating to this chapter can also be found on our Web site.

See chapter activity on BioActive Learner CD for additional help in mastering the chapter's material. Icon location in the chapter's margins shows which topics have tutorials or simulations in the CD.

CAREER VISIONS

Physician Assistant

ELDON LaTRAY

***Eldon LaTray** is a Blackfeet/Chippewa-Cree Native American who lives and works on his native Blackfeet reservation in rural northwest Montana. He practices as a physician assistant at the Browning Service Unit Hospital, part of the Indian Health Service, a federal agency, in Browning, which is located just south of the Canadian border and east of Glacier National Park. Eldon began his undergraduate studies in biology at Montana State University in Bozeman, where he also gained some medical training and research experience. He transferred to Rocky Mountain College in Billings, Montana, to enter the physician assistant program. In 1998, Eldon received dual B.S. degrees in Biology and Physician Assistant Studies. He completed several medical training rotations in Browning, where his family still lives, and returned there after graduation to work full-time. He is trained to offer a variety of primary care services to patients of all ages.*

What exactly is a physician assistant?

A physician assistant, or PA, is a licensed health care provider who practices under the supervision of a physician. Physician assistants in Montana are generally primary care providers, but they also practice in all specialties of medicine. As defined by the American Academy of Physician Assistants, "Physician assistants conduct physical exams, diagnose and treat illnesses, order and interpret tests, counsel on preventive health care, assist in surgery, and, in most states, write prescriptions." Physician assistants must train in programs accredited by the Commission on Accreditation of Allied Health Education Programs.

What influenced your decision to become a physician assistant?

I wanted to come home and be able to practice medicine among my people. Being a PA in Montana allowed me to do that.

Are physician assistants especially important to rural areas?

I can only speak from my own experience; there are so many different PA programs out there. I know that the goal of PA programs in Montana is to put more providers in the rural areas. PAs can provide important services to rural areas. But I hate to generalize. Some schools, for example, are geared to train surgical PAs.

Did your bachelor's degree in biology prepare you well for the PA program?

Definitely. My biology degree provided me with the scientific background I needed to pursue my medical education. An undergraduate degree in biology gives you a lot of flexibility once you graduate; becoming a PA is just what I chose to do.

Is a biology degree specifically required for this program?

No, it's not. You need to meet specific scientific requirements, take certain courses. You're required to take a certain number of credits of biology, chemistry, physics, and psychology, but you don't need a biology degree.

Did you have to have some previous medical experience?

Yes, you need at least a year's medical experience. It could be almost anything. I was a pharmaceutical technician. Many people who become PAs have nursing backgrounds or emergency medicine training. Some are nurses aides, and a lot are lab technicians or respiratory therapists.

In how many medical specialties did you receive training?

The overall emphasis is on general practice, then people choose subspecialties. We were required to do a certain number of different rotations, such as psychiatry, pediatrics, obstetrics and gynecology, orthopedics, and emergency medicine. Then we were able to choose from some elective rotations.

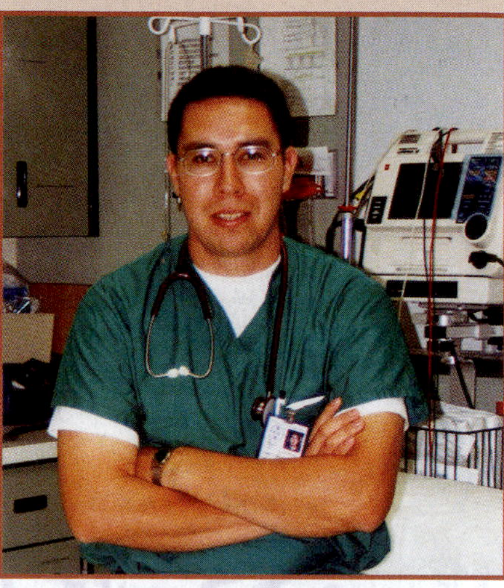

Describe your situation in the Browning Hospital. Are there other PAs or physicians there as well?

The hospital here is an Indian Health Service hospital. It's a federal facility and is pretty well-staffed, far more so than most rural communities the same size in Montana. As a PA, I do primary care and see a lot of walk-in patients with acute problems; they'll see me instead of a physician. I tend to do a lot of orthopedics, because that's what I'm interested in. In addition to being in the clinic, PAs also function in the emergency department.

What are the most common health problems you encounter?

American Indians have the highest rate of Type II diabetes in the world. Some of the other more common pathologies that we deal with are all kinds of arthritis, including rheumatoid arthritis and osteoarthritis. The Blackfeet have a high incidence of rheumatoid arthritis. We also see the simple stuff, like flu-type syndromes, upper respiratory infections or colds, pneumonia, skin infections, and ear infections.

What suggestions would you give a student who wants to be a physician assistant?

Students need a strong, solid scientific background, as well as diverse medical experience, to become a successful PA. They need practical experience as well as interest in the profession.

A field scientist studies plants in a Costa Rican tropical rain forest. Observation is an important part of the scientific process. *(Stephen Ferry/Liaison)*

Part 8

The Interactions of Life: Ecology

Chapters

51

Introduction to Ecology: Population Ecology

A population of Mexican poppies *(Eschscholzia mexicana)* blooms in the desert after the winter rains. Mexican poppies, which thrive on gravelly desert slopes, grow to a height of 40 cm (16 in.). *(Jim Brandenburg/Minden Pictures)*

LEARNING OBJECTIVES

After you have studied this chapter you should be able to

1. Define ecology and distinguish among the following ecological levels: population, community, ecosystem, landscape, and biosphere.
2. Define population density and dispersion, and describe the main types of population dispersion.
3. Explain the four factors (natality, mortality, immigration, and emigration) that produce changes in population size and solve simple problems involving these changes.
4. Define intrinsic rate of increase and carrying capacity and explain the differences between J-shaped and S-shaped growth curves.
5. Distinguish between density-dependent and density-independent factors, and give examples of each.
6. Describe Type I, Type II, and Type III survivorship curves, and explain how life tables and survivorship curves indicate mortality and survival.
7. Define metapopulation and distinguish between source and sink habitats.
8. Summarize the history of human population growth.
9. Explain how highly developed and developing countries differ in population characteristics such as infant mortality rate, total fertility rate, replacement-level fertility, and age structure.
10. Distinguish between people overpopulation and consumption overpopulation.

Living organisms and the physical environment interact in an immense and complicated web of relationships. **Ecology** is the science that studies interactions among organisms **(biotic factors),** and between organisms and their nonliving, physical environment (**abiotic factors,** such as water, temperature, pH, wind, and chemical nutrients). Ecologists formulate hypotheses to explain such phenomena as the distribution and abundance of life on Earth, the ecological role of specific species, the interactions among species in communities, and the importance of ecosystems in maintaining the health of the biosphere. They then test these hypotheses.

The focus of ecology can be local or global, specific or generalized, depending on what questions the scientist is asking and trying to answer. Ecology is the broadest field in biology, with explicit links to evolution and every other biological discipline. Its universality also encompasses subjects that are not traditionally part of biology. Earth science, geology, chemistry, oceanography, climatology, and meteorology are extremely important to ecology, especially when ecologists examine the abiotic environment of planet Earth. Because humans are part of Earth's complex web of life, all our activities, even economics and politics, have profound ecological implications. **Environmental science,** a scientific discipline with ties to ecology, focuses on how humans interact with the environment.

Most ecologists are interested in the levels of biological organization including and above the level of the individual organism: population, community, ecosystem, landscape, and biosphere. Each level has its own characteristic composition, structure, and functioning.

An individual belongs to a **population,** a group consisting of members of the same species that live together in a prescribed area at the same time. The boundaries of the area are defined by the ecologist performing a particular study. A population ecologist might study a population of polar bears or a population of Mexican poppies *(see photograph)* to see how individuals within it live and interact with each other, with other species in their community, and with their physical environment.

Populations are organized into communities. A **community** consists of all the populations of all the different species that live and interact together within the same area. A community ecologist might study how species interact with one another—such as who eats whom, which species compete, and which species cooperate—in a coral reef or in an alpine meadow community. *Ecosystem* is a more inclusive term than *community* because an **ecosystem** links a community to its abiotic environment. Thus, an ecosystem includes not only all the biotic interactions among the organisms of a community but also the interactions between organisms and their abiotic environment. An ecosystem ecologist, for example, might examine how temperature, light, precipitation, and soil factors affect the organisms living in a desert ecosystem or in a coastal bay ecosystem. Ecosystem ecologists focus on such questions as how nutrients cycle through an ecosystem and how energy flows through food webs.

Landscape ecology is a relatively new subdiscipline in ecology that studies the connections in a heterogeneous **landscape** that is composed of interacting ecosystems. Landscape ecologists study the movements of organisms, nutrients, and energy between ecosystems. Consider a simple landscape that contains a forest ecosystem located adjacent to a pond ecosystem. One connection between these two ecosystems might be great blue herons, which eat fish, frogs, insects, crustaceans, and snakes along the shallow water of the pond but often build nests and raise their young in the secluded treetops of the nearby forest. Landscapes, which are typically several to many square kilometers in area, are based on larger land areas than individual ecosystems. We consider landscape ecology as it relates to populations later in the chapter.

The **biosphere** is a global ecological system comprising all of Earth's communities. The biosphere extends from the depths of the ocean and from as deep into the soil as organisms are found to the highest part of the atmosphere in which organisms occur. The organisms of the biosphere depend on one another and on Earth's physical environment—its atmosphere, hydrosphere, and lithosphere. The *atmosphere* is the gaseous envelope surrounding Earth, the *hydrosphere* is Earth's supply of water (liquid and frozen, fresh and salty), and the *lithosphere* is the soil and rock of Earth's crust. Ecologists who study the biosphere examine the complex physical and chemical interrelationships among Earth's organisms, atmosphere, land, and water. The effect of deforestation (the clearance of forests for agriculture or other uses) on increasing carbon dioxide levels is an example of an ecological study at the biosphere level. It is thought that the carbon in forests is released immediately if the trees are burned, or more slowly when unburned parts decay.

In this chapter we begin our study of ecological principles by focusing on the study of populations as functioning systems, ending with a discussion of the human population. Subsequent chapters examine the interactions among different populations within communities (Chapter 52), the dynamic exchanges between communities and their physical environments (Chapter 53), the characteristics of Earth's major biological ecosystems (Chapter 54), and some of the impacts of humans on the biosphere (Chapter 55).

POPULATIONS POSSESS CHARACTERISTIC FEATURES

Populations exhibit characteristics distinctive from those of the individuals of which they are composed. Some of the features discussed in this chapter that are characteristic of populations are population density, population dispersion, birth and death rates, growth rates, survivorship, and age structure.

Although communities are composed of all the populations of all the different species that live together within an area, populations have properties that communities lack. Populations, for example, share a common gene pool (see Chapter 18). As a result, allele frequency changes resulting from natural selection occur in populations. Natural selection, therefore, acts directly to produce adaptive changes in populations and only indirectly affects the community level.

Population ecology considers both the number of individuals of a particular species that are found in an area and the dynamics of the population. **Population dynamics** is the study of changes in populations—how and why those numbers increase or decrease over time. Population ecologists try to determine the processes common to all populations. They study how a population interacts with its environment, such as how individuals in a population compete for food or other resources, and how predation, disease, and other environmental pressures affect the population. Population growth, whether of bacteria, maples, or giraffes, cannot increase indefinitely because of such environmental pressures.

Additional aspects of populations that interest biologists are their reproductive success or failure (i.e., extinction), their evolution, their genetics, and how they affect the normal functioning of communities and ecosystems. Biologists in applied disciplines, such as forestry, agronomy (crop science), and wildlife management, must understand population ecology to manage populations of economic importance, such as forests, field crops, game animals, or fishes. Understanding the population dynamics of endangered and threatened species plays a key role in efforts to prevent their slide to extinction. Knowledge of population ecology helps in efforts to prevent the increase of pest populations to levels that cause significant economic or health impacts.

Density and dispersion are important features of populations

Population size is meaningful only when the boundaries of that population are defined. Consider, for example, the difference between 1000 mice in 100 hectares (250 acres) and 1000 mice in 1 hectare (2.5 acres). Often a population is too large to study in its entirety. Such a population is examined by sampling a part of it and then expressing the population in terms of density. Examples include the number of dandelions per square meter of lawn, the number of water fleas per liter of pond water, or the number of cabbage aphids per square centimeter of cabbage leaf. **Population density,** then, is the number of individuals of a species per unit of area or volume at a given time.

Different environments vary in the population density of any species that they can support. This density may also vary in a single habitat from season to season or year to year. For example, consider red grouse populations in the treeless moors of northwest Scotland at two locations only 2.5 km (1.5 mi) apart (Fig. 51–1). At one location the population density remained stationary during a three-year period, but at the other site it almost doubled in the first two years and then declined to its initial density in the third year. The reason was likely a difference in habitat. The area where the population density increased initially and then decreased had been experimentally burned. Young heather *(Calluna vulgaris)* shoots produced after the burn provided nutritious food for the red grouse. So population density may be determined in large part by biotic or abiotic factors in the environment that are external to the individuals in the population.

The individuals comprising a population often exhibit characteristic patterns of **dispersion,** or spacing, relative to each other. Individuals may be spaced in a clumped, uniform, or random dispersion (Fig. 51–2*a*). Perhaps the most common spacing is **clumped dispersion,** also called **aggregated distribution** or **patchiness,** which occurs when individuals are concentrated in specific parts of the habitat. Clumped dispersion often results from the patchy distribution of resources in the environment. It also occurs among animals owing to the presence of family groups and pairs and among plants owing to limited seed dispersal or asexual reproduction. An entire grove of aspen trees, for example, may originate asexually from a single plant (see Fig.

Figure 51–1 Red grouse. Red grouse *(Lagopus lagopus scoticus)*, which inhabit moorlands, are ground-dwelling game birds whose populations are managed for hunting. Photographed in Deeside, Scotland. *(Roger Wilmshurst/Corbis)*

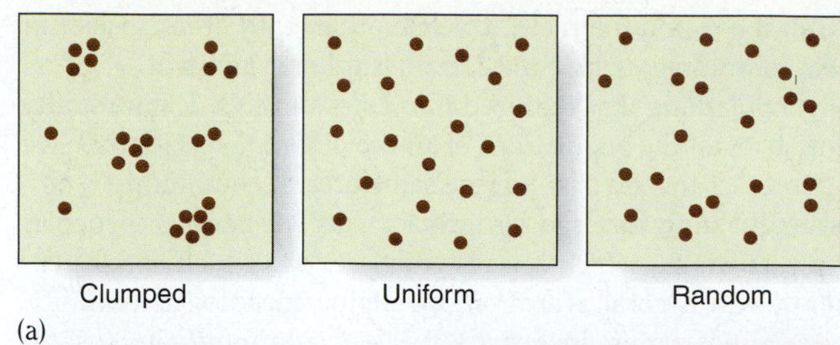

(a)

(b)

(c)

Figure 51–2 Dispersion of individuals within a population. **(a)** Diagram of the spacing of individuals in clumped, uniform, and random dispersion. **(b)** The schooling behavior of certain fish species is an example of clumped dispersion. Shown are bluestripe snappers *(Lutjanus kasmira)*, photographed in Hawaii. This introduced fish, which grows to 30 cm (12 in), may be displacing native fish species in Hawaiian waters. **(c)** Cape gannets *(Morus capensis)* nesting on the coast of South Africa. The birds space their nests more or less evenly. *(b, Andrew G. Wood/Photo Researchers, Inc.; c, Robert Hernandez/Photo Researchers, Inc.)*

35–20*b*). Clumped dispersion may sometimes be advantageous, as social animals derive many benefits from their association. Many fish species, for example, associate in dense schools for at least part of their life cycle, possibly because schooling may reduce the risk of predation for any particular individual (Fig. 51–2*b*). The many pairs of eyes of schooling fish tend to detect predators more effectively than a single pair of eyes of a single fish. When threatened, schooling fish clump together more closely, making it difficult for a predator to single out an individual. If a predator approaches a school of fish, the prey fish retreat around the predator, reforming the school behind the predator.

Uniform dispersion occurs when individuals are more evenly spaced than would be expected from a random occupation of a given habitat. A nesting colony of seabirds, in which the birds are nesting in a relatively homogeneous environment and place their nests at a more or less equal distance from each other, is an example of uniform dispersion (Fig. 51–2*c*). What might this spacing pattern tell us? In this case, uniform dispersion may occur as a result of nesting territoriality. Aggressive interactions among the nesting birds as they peck at one another from their nests cause each pair to place its nest just beyond the reach of nearby nesting birds. Uniform dispersion also occurs when competition among individuals is severe, when plant roots or leaves that have been shed produce toxic substances that inhibit the growth of nearby plants, or when animals establish feeding or mating territories.

Random dispersion occurs when individuals in a population are spaced throughout an area in a manner that is unrelated to the presence of others. Of the three major types of dispersion, random dispersion is least common and hardest to observe in nature, leading some ecologists to question its existence. Trees of the same species, for example, sometimes appear to be distributed randomly in a tropical rain forest. However, an international team of 13 ecologists studied 6 tropical forest plots that were 25 to 52 hectares (62 to 128 acres) in area and reported that most of the 1000 tree species observed were clumped and not randomly dispersed.[1] (Both clumped and uniform dispersion are determined by statistically testing for differences from an assumed random distribution.) Random dispersion may occur infrequently because important environmental factors affecting dispersion usually do not occur at random. Flour beetle larvae in a container of flour are randomly dispersed, for example, but their environment (flour) is unusually homogeneous.

Some populations have different spacing patterns at different ages. Competition for sunlight among same-aged sand pine in a Florida scrub community, for example, resulted in a change over time from either random or clumped dispersion when the plants were young, to uniform dispersion when the plants were old. Sand pine is a fire-adapted plant with cones that do not release their seeds until they have been exposed to high temperatures (45° to 50° C or higher). As a result of seed dispersal and soil conditions following a fire, the seedlings grow back in dense stands that exhibit random or slightly clumped dispersion. Over time, however, many of the more crowded trees tend to die from shading or competition, resulting in uniform dispersion of the surviving trees (Table 51–1).

[1]Conduit, R. et al. "Spatial Patterns in the Distribution of Tropical Tree Species." *Science*, Vol. 288, 26 May 2000.

TABLE 51–1 Dispersion in a Sand Pine Population in Florida

Tree Trunks Examined*	Density (per m^2)	Dispersion
All (alive and dead)	0.16	Random
Alive only	0.08	Uniform

*Adapted from Laessle, A.M. "Spacing and Competition in Natural Stands of Sand Pine." *Ecology*, Vol. 46, 1965, pp. 65–72. Data were collected 51 years following a fire.

MATHEMATICAL MODELS DESCRIBE POPULATION GROWTH

Process of Science

One goal of science is to discover common patterns among separate observations. As mentioned previously, population ecologists wish to understand general processes that are shared by many different populations, so they develop mathematical models based on equations that describe the dynamics of a single population. Population models are not perfect representations of a population, but models help illuminate complex processes. Moreover, mathematical modeling enhances the scientific process by providing a framework with which experimental population studies can be compared. We can test a model and see how it fits or does not fit with existing data. Data that are inconsistent with the model are particularly useful because they demand that we ask how the natural system differs from the mathematical model that we developed to explain it. As more knowledge accumulates from observations and experiments, the model is refined and made more precise.

Population size, whether of sunflowers, elephants, or humans, changes over time. On a global scale, this change is ultimately due to two factors: **natality,** the rate at which individuals produce offspring (the average per capita birth rate), and **mortality,** the rate at which individuals die (the average per capita death rate). In humans the birth rate is usually expressed as the number of births per 1000 people per year and the death rate as the number of deaths per 1000 people per year.

To determine the rate of change in population size, we must also take into account the time interval involved, that is, the change in time. To express change in equations, we employ the Greek letter delta (Δ). In equation (1), ΔN is the change in the number of individuals in the population, Δt the change in time, N the number of individuals in the existing population, b the natality, and d the mortality.

$$(1)\ \Delta N/\Delta t = N\,(b - d)$$

The **growth rate** *(r)*, or rate of change (increase or decrease) of a population on a per capita basis is the birth rate minus the death rate:

$$(2)\ r = b - d$$

As an example, consider a hypothetical human population of 10,000 in which there are 200 births per year (i.e., by convention, 20 births per 1000 people) and 100 deaths per year (i.e., 10 deaths per 1000 people):

$$r = 20/1000 - 10/1000 = 0.02 - 0.01 = 0.01, \text{ or } 1\% \text{ per year}$$

A modification of equation (1) tells us the rate at which the population is growing at a particular instant in time, that is, its instantaneous growth rate *(dN/dt)*. (The symbols *dN* and *dt* are the mathematical differentials of *N* and *t*, respectively; they are not products, nor should the *d* in *dN* or *dt* be confused with the death rate, *d*.) Using differential calculus, this growth rate can be expressed as:

$$(3)\ dN/dt = rN$$

where *N* is the number of individuals in the existing population, *t* the time, and *r* the per capita growth rate.

Because $r = b - d$, if individuals in the population are born faster than they die, *r* is a positive value, and population size increases. If individuals in the population die faster than they are born, *r* is a negative value, and population size decreases. If *r* is equal to zero, births and deaths match, and population size is stationary despite continued reproduction and death.

Dispersal affects the growth rate in some populations

In addition to birth and death rates, **dispersal,** which is movement of individuals among populations, must be considered when examining changes in populations on a *local* scale. There are two types of dispersal: immigration and emigration. **Immigration** occurs when individuals enter a population and thus increase its size. **Emigration** occurs when individuals leave a population and thus decrease its size. The growth rate of a local population must take into account birth rate *(b)*, death rate *(d)*, immigration rate *(i)*, and emigration rate *(e)* on a per capita basis. The per capita growth rate equals the birth rate minus the death rate, plus the immigration rate minus the emigration rate:

$$(4)\ r = (b - d) + (i - e)$$

For example, the growth rate of a human population of 10,000 that has 200 births (by convention, 20 per 1000), 100 deaths (10 per 1000), 10 immigrants (1 per 1000), and 100 emigrants (10 per 1000) in a given year would be calculated as follows:

$$r = (20/1000 - 10/1000) + (1/1000 - 10/1000)$$
$$r = (0.020 - 0.010) + (0.001 - 0.010) = 0.010 - 0.009 = 0.001, \text{ or } 0.1\% \text{ per year}$$

Each population has a characteristic intrinsic rate of increase

The maximum rate at which a population of a given species could increase under ideal conditions, when resources are abundant and its population density is low, is known as its **intrinsic rate of increase *(r_{max})***. Different species have different intrinsic rates of increase. A particular species' intrinsic rate of increase is influenced by several factors. These include the age at which reproduction begins, the fraction of the **life span** (duration of the individual's life) during which the individual is capable of reproducing, the number of reproductive periods per lifetime, and the number of offspring the individual is capable of producing during each period of reproduction. These factors, which we discuss in greater detail later in the chapter, determine whether a particular species has a large or small intrinsic rate of increase.

Generally, large species such as blue whales and elephants have the smallest intrinsic rates of increase, whereas microorganisms have the greatest intrinsic rates of increase. Under ideal conditions (i.e., an environment with unlimited resources), certain bacteria can reproduce by binary fission every 20 minutes. At this rate of growth, a single bacterium would increase to a population of more than 1 billion in just 10 hours!

If we plot the population size versus time, the graph has a J shape that is characteristic of **exponential population growth,** which is the accelerating population growth rate that occurs when optimal conditions allow a constant per capita growth rate (Fig. 51–3). When a population grows exponentially, the larger that population gets, the faster it grows.

Regardless of which species we are considering, whenever a population is growing at its intrinsic rate of increase, population size plotted versus time gives a curve of the same shape. The only variable is time. It may take longer for an elephant population than for a bacterial population to reach a certain size (because elephants do not reproduce as rapidly as bacteria), but both populations will always increase exponentially as long as their per capita growth rates remain constant.

No population can increase exponentially indefinitely

Certain populations may exhibit exponential growth for brief periods. Exponential growth has been experimentally demon-

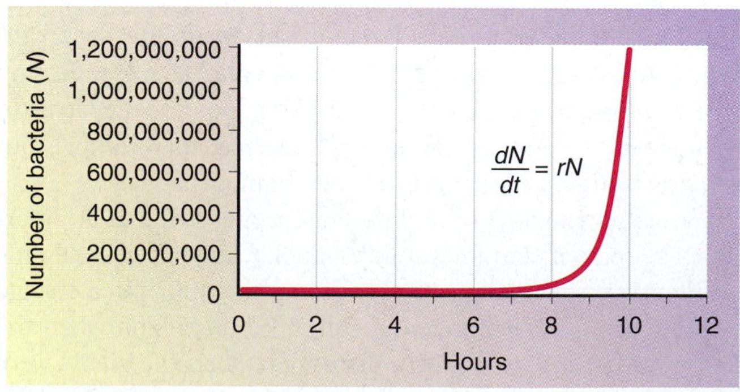

Figure 51–3 Exponential population growth. When bacteria divide every 20 minutes, their numbers increase exponentially. The curve of exponential population growth has a characteristic J shape. The ideal conditions under which bacteria or other organisms reproduce exponentially rarely occur in nature, and when these conditions do occur, they are of short duration.

strated in certain insects and in bacterial and protistan cultures (by continually supplying nutrients and removing waste products). However, organisms cannot reproduce indefinitely at their intrinsic rate of increase because the environment sets limits. These include such unfavorable environmental conditions as the limited availability of food, water, shelter, and other essential resources (resulting in increased competition) as well as limits imposed by disease and predation.

Using the earlier example, bacteria in nature would never be able to reproduce unchecked for an indefinite period because they would run out of food and living space, and poisonous wastes would accumulate in their vicinity. With crowding, bacteria would also become more susceptible to parasites (high population densities facilitate the spread of infectious organisms such as viruses among individuals) and predators (high population densities increase the likelihood of a predator catching an individual). As the environment deteriorated, their birth rate *(b)* would decline and their death rate *(d)* would increase. Conditions might worsen to a point where *d* would exceed *b*, and the population would decrease. The number of individuals in a population, then, is controlled by the ability of the environment to support it. As the number of individuals in a population *(N)* increases, environmental limits act to control population growth.

Over longer periods, the rate of population growth may decrease to nearly zero. This leveling out occurs at or near the limits of the environment to support the population. The **carrying capacity *(K)*** represents, in theory, the largest population that can be maintained for an indefinite period by a particular environment, assuming there are no changes in that environment. In nature, the carrying capacity is dynamic and changes in response to environmental changes. An extended drought, for example, could decrease the amount of vegetation growing in an area, and this change, in turn, would lower the carrying capacity for deer and other herbivores in that environment.

When a population regulated by environmental limits is graphed over longer periods, the curve has a characteristic S shape. The curve shows the population's initial exponential increase (note the curve's J shape at the start, when environmental limits are few), followed by a leveling out as the carrying capacity of the environment is approached (Fig. 51–4). The S-shaped growth curve, also called **logistic population growth,** can be modeled by a modified growth equation called a logistic equation. It describes a population increasing from a small number of individuals to a larger number of individuals that are ultimately limited by the environment. The logistic equation, then, takes the carrying capacity of the environment into account:[2]

$$(5)\ dN/dt = rN[(K - N)/K]$$

Note that part of the equation is the same as equation (3). The added element $[(K - N)/K]$ reflects a decline in growth as a population size approaches its carrying capacity. When the number of organisms *(N)* is small, the rate of population growth is unchecked by the environment, because the expression $[(K - N)/K]$ has a value of almost 1. But as the population *(N)* begins to approach the carrying capacity *(K)*, the growth rate declines because the value of $[(K - N)/K]$ approaches zero.

[2]The logistic model of population growth was developed to explain population growth in continually breeding populations. Similar models exist for populations that have specific breeding seasons.

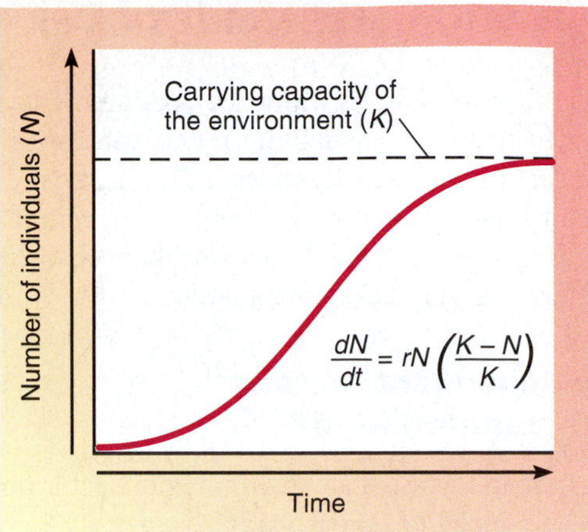

Figure 51–4 Carrying capacity and logistic population growth. In many laboratory studies, exponential population growth slows as the carrying capacity *(K)* of the environment is approached. The logistic model of population growth, when graphed, has a characteristic S-shaped curve.

Although the S curve is an oversimplification of how most populations change over time, it does appear to fit some populations that have been studied in the laboratory, as well as a few that have been studied in nature. For example, Georgyi F. Gause, a Russian ecologist who conducted experiments during the 1930s, grew a population of a single species, *Paramecium caudatum*, in a test tube. He supplied a limited amount of food (bacteria) daily and replenished the growth medium occasionally to eliminate the accumulation of metabolic wastes. Under these conditions, the population of *P. caudatum* increased exponentially at first, but then its growth rate declined to zero, and the population size leveled off (see Fig. 52–4, middle graph).

A population rarely stabilizes at *K* (carrying capacity) but may temporarily rise higher than *K*. It will then drop back to, or below, the carrying capacity. Sometimes a population that overshoots *K* will experience a *population crash,* an abrupt decline from high to low population density. Such an abrupt change is commonly observed in bacterial cultures, zooplankton, and other populations whose resources have been exhausted.

The carrying capacity for reindeer, which live in cold northern habitats, is determined largely by the availability of winter forage. In 1910 a small herd of 26 reindeer was introduced onto one of the Pribilof Islands of Alaska. The herd's population increased exponentially for about 25 years until there were approximately 2000 reindeer, many more than the island could support, particularly in winter. The reindeer overgrazed the vegetation until the plant life was almost wiped out. Then, in slightly over a decade, as reindeer died from starvation, the number of reindeer plunged to eight, one-third the size of the original introduced population. Recovery of subarctic and arctic vegetation after overgrazing by reindeer can take 15 to 20 years, during which time the carrying capacity for reindeer is greatly reduced.

POPULATION SIZE VARIES OVER TIME

Certain natural mechanisms influence population size. Factors that affect population size fall into two categories: density-dependent factors and density-independent factors. These two sets of factors vary in importance from one species to another and, in most cases, probably interact simultaneously in complex ways to determine the size of a population.

Density-dependent factors regulate population size

Sometimes the influence of an environmental factor on the individuals in a population varies with the density or crowding of that population. If a change in population density alters how an environmental factor affects that population, then the environmental factor is said to be a **density-dependent factor.** As population density increases, density-dependent factors tend to slow population growth by causing an increase in death rate and/or a decrease in birth rate. The effect of these density-dependent factors on population growth increases as the population density increases; that is, density-dependent factors affect a larger proportion, not just a larger number, of the population. Density-dependent factors can also affect population growth when population density declines, by decreasing the death rate and/or increasing the birth rate. Density-dependent factors, then, tend to regulate a population at a relatively constant size that is near the carrying capacity of the environment. (Keep in mind, however, that the carrying capacity of the environment frequently changes.) Density-dependent factors are an excellent example of a **negative feedback system** (Fig. 51–5; also see Chapter 37).

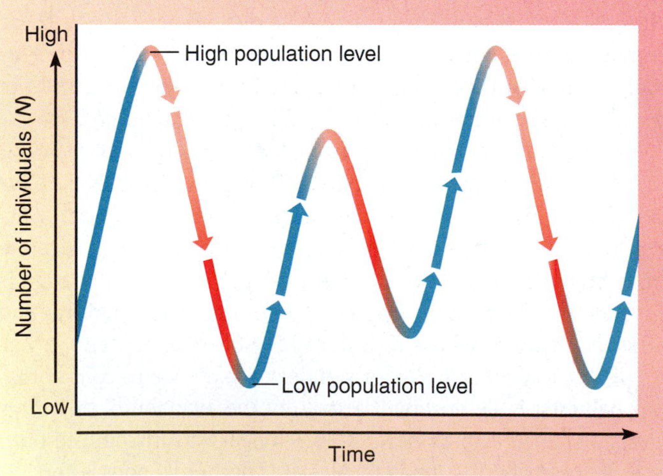

Density-dependent factors are increasingly severe: population peaks and begins to decline.

Density-dependent factors are increasingly relaxed: population bottoms out and begins to increase.

Figure 51–5 Density-dependent factors and negative feedback. When the number of individuals in a population increases, density-dependent factors cause a decline in the population. When the number of individuals in a population decreases, a relaxation of density-dependent factors allows the population to increase.

Predation, disease, and competition are examples of density-dependent factors. As the density of a population increases, predators are more likely to find an individual of a given prey species. When population density is high, the members of a population encounter one another more frequently, and the chance of their transmitting infectious disease organisms increases. As population density increases, so does competition for resources such as living space, food, cover, water, minerals, and sunlight; eventually, it may reach the point where many members of a population fail to obtain the minimum amount of whatever resource is in shortest supply. At higher population densities, density-dependent factors raise the death rate and/or lower the birth rate, inhibiting further population growth. The opposite effect occurs when the density of a population decreases. Predators are less likely to encounter individual prey, parasites are less likely to be transmitted from one host to another, and competition among members of the population for resources such as living space and food declines.

Density-dependent factors may explain what makes certain populations fluctuate cyclically over time

Lemmings are small, stumpy-tailed rodents that are found in colder regions of the Northern Hemisphere (Fig. 51–6). They are herbivores that feed on sedges and grasses. It has long been known that lemming populations have a three- to four-year cyclical oscillation that is often described as "boom or bust." That is, the population increases dramatically and then crashes; the population peaks may be 100 times as high as the low points in the population cycle. Many other populations, such as snowshoe hares and red grouse, also exhibit cyclic fluctuations.

What is the driving force behind these fluctuations? Several hypotheses have been proposed to explain the cyclical periodicity of lemming and other boom-or-bust populations; many hypotheses involve density-dependent factors. One possibility is that the population density of predators, such as long-tailed jaegers (birds that eat lemmings), increases in response to the increasing density of prey. Few jaegers breed when the lemming population is low, whereas when the lemming population is high, most jaegers breed, and the number of eggs per clutch is greater than usual. As more predators consume the abundant prey, the prey population declines. Later, with fewer prey in an area, the population of predators declines (some disperse out of the area, and fewer offspring are produced).

Another explanation for boom-or-bust cycles is that as a prey population becomes more dense it overwhelms its food supply, resulting in population decline. In the lemming example, researchers have recently studied the shape of several lemming population curves during their oscillations; the data suggest that lemming populations crash because they overgraze the plants, not because predators eat them.

Parasites may also interact with their hosts to cause regular cyclic fluctuations. Detailed studies of red grouse (see Fig. 51–1) have shown that even managed populations in wildlife preserves may have significant cyclic oscillations. Reproduction in red grouse is related to the density of parasitic nematodes (round-

Figure 51–6 Lemming. The brown lemming *(Lemmus trimucronatus)* lives in the arctic tundra. Although lemming populations have been studied for decades, much about the cyclic nature of lemming population oscillations and their effects on the rest of the tundra ecosystem is still not well understood. This species ranges from Alaska eastward to the Hudson Bay. *(Tom McHugh/Photo Researchers, Inc.)*

worms) living in adult intestines: Fewer birds breed successfully when adults are infected with worms; thus, a high density of worms leads to a population crash. Hypothesizing that red grouse populations fluctuate in response to the parasites, ecologist Peter Hudson and colleagues at the University of Stirling in the United Kingdom were recently able to reduce or eliminate population fluctuations in several red grouse populations. They accomplished this by catching and orally treating the birds with a chemical that causes worms to be ejected from their bodies.

Competition is an important density-dependent factor

Competition is an interaction among two or more individuals that attempt to use the same essential resource, such as food, water, sunlight, or living space, that is in limited supply. The use of the resource by one of the individuals reduces the availability of that resource for other individuals. Competition occurs both within a given population **(intraspecific competition)** and among populations of different species **(interspecific competition).** We consider the effects of intraspecific competition here; interspecific competition is discussed in Chapter 52.

Individuals of the same species compete for a resource in limited supply by interference competition or by exploitation competition. In **interference competition,** also called **contest competition,** certain dominant individuals obtain an adequate supply of the limited resource at the expense of other individuals in the population; that is, the dominant individuals actively interfere with other individuals' access to resources. In **exploitation competition,** also called **scramble competition,** all the individuals in a population "share" the limited resource more or less equally so that at high population densities none of them obtains an adequate amount. The populations of species in which exploitation competition operates often oscillate over time, and there is always a risk that the population size will drop to zero. In contrast, those species in which interference competition operates experience a relatively small drop in population size, caused by the death of individuals that are unable to compete successfully.

Intraspecific competition among red grouse (see Fig. 51–1) involves interference competition. When red grouse populations are small, the birds are less aggressive, and most young birds are able to establish a feeding territory (an area defended against other members of the same species). However, when the population is large, establishing a territory is difficult because there are more birds than there are territories, and the birds are much more aggressive. Those birds without territories often die from predation or starvation. Thus, birds with territories use a larger share of the limited resource (the territory with its associated food and cover), whereas birds without territories cannot compete successfully.

The moose population on Isle Royale, Michigan, the largest island in Lake Superior, provides a vivid example of exploitation competition (Fig. 51–7) that is similar to the reindeer population on the Pribilof Islands (discussed earlier). Isle Royale differs from most islands in that large mammals can walk to it when the lake freezes over in winter. The minimum distance to be walked is 24 km (15 mi), however, so this movement has happened infrequently. Around 1900, a small herd of moose wandered across the ice of frozen Lake Superior and reached the island for the first time. By 1934 the moose population on the island had increased to about 3000 and had consumed almost all the edible vegetation. In the absence of this food resource, there was massive starvation in 1934. More than 60 years later, in 1996, a similar die-off claimed 80% of the moose after they had again increased to a high density. Thus, exploitation competition for scarce resources can result in dramatic population oscillations.

Figure 51–7 Exploitation competition among moose on Isle Royale in Lake Superior. Individual moose, which forage alone, compete with one another for food during the long, cold winter months. Moose, which inhabit forests with lakes and swamps, stand 1.5 to 2.0 m (5 to 6.5 ft) tall. Males are larger than females and weigh as much as 530 kg (1170 lb). *(Rolf O. Peterson, Michigan Technological University)*

MAKING THE CONNECTION Are there any long-term studies of predator-prey dynamics? In 1949, a few Canadian wolves wandered across frozen Lake Superior and discovered abundant moose prey on Isle Royale. The wolves remained and became established, eating moose as their primary winter prey on the island. Wolves hunt in packs by encircling a moose and trying to get it to run so they can attack it from behind. (The moose is the wolf's largest, most dangerous prey. A standing moose is more dangerous than one that is running because when standing, it can kick and slash its attackers with its hooves.)

Ecologists, such as Rolf Peterson of Michigan Technological University, have studied the effects of both density-dependent and density-independent factors on the Isle Royale populations of moose and wolves. Peterson, who has monitored moose and wolves on Isle Royale since 1970, found that the two populations fluctuated over the years (Fig. 51–8). Generally, as the population of wolves decreased, the population of moose increased. The reverse was also true—as the population of wolves increased, the population of moose decreased. Although the overall effect of wolves was to reduce the population of moose, the wolves did not eliminate moose from the island. Studies indicate that wolves primarily feed on the very old and very young in the moose population. Healthy moose in their peak reproductive years are not eaten.

A new episode in the Isle Royale story began during the late 1980s and early 1990s. The wolf population plunged, from 50 animals in 1980 to a low of 12 animals in 1989, possibly the result of a deadly disease, canine parvovirus. Analysis of their blood revealed the presence of antibodies to canine parvovirus, confirming that the wolves had been exposed to this disease. Inbreeding depression (see Chapter 18) could also be a factor in their decline, as Isle Royale wolves are highly inbred (all are descendants of a single female ancestor). It is not known if the increased wolf population observed from 1998 to 1999 is temporary or the beginning of a long-term recovery. Blood analyses from recently captured wolves indicated they were disease-free.

As expected, the moose population increased as the wolf population declined; in 1995 there were more than 2400 moose on Isle Royale, many more than the island's vegetation could support. The moose overgrazed the island, particularly mountain ash and aspen, their preferred food. Lack of food, in combination with a particularly bad winter (1995–1996), caused hundreds of moose to die; only 500 moose were counted in the 1997 survey. The ratio of moose to wolves was 30:1 in 1999, which should limit the growth rate for moose and allow the overgrazed vegetation (mainly balsam fir) to recover somewhat. Ecologists estimate it will take 5 to 10 years of continued low moose density to allow the fir trees to recover completely.

The effects of density-dependent factors are difficult to assess in nature

Most studies of density dependence have been conducted in laboratory settings where all density-dependent (and density-independent) factors except one are controlled experimentally. But populations in natural settings are exposed to a complex set of variables that continually change. As a result, in natural communities it can be difficult to evaluate the relative effects of different density-dependent factors.

Ecologists from the University of California at Davis noted that few spiders occur on tropical islands inhabited by lizards, whereas more spiders and more species of spiders are found on lizard-free islands. Deciding to study these observations experimentally, David Spiller and Thomas Schoener staked out plots of vegetation (mainly seagrape shrubs) and enclosed some of them with lizard-proof screens (Fig. 51–9*a*). Some of the plots were emptied of all lizards; each control enclosure had approximately nine lizards. Nine web-building spider species were observed in the enclosures. Spiders were counted approximately 30 times from 1989 to 1994 (Fig. 51–9*b*). During the 4.5 years of observations reported here, spider population densities were higher in the lizard-free enclosures than in enclosures with lizards. Moreover, the enclosures without lizards had more species of spiders. Therefore, we might conclude that lizards control spider populations.

But even this relatively simple experiment may be explained by a combination of two density-dependent factors, predation (lizards eat spiders) and interspecific competition (lizards com-

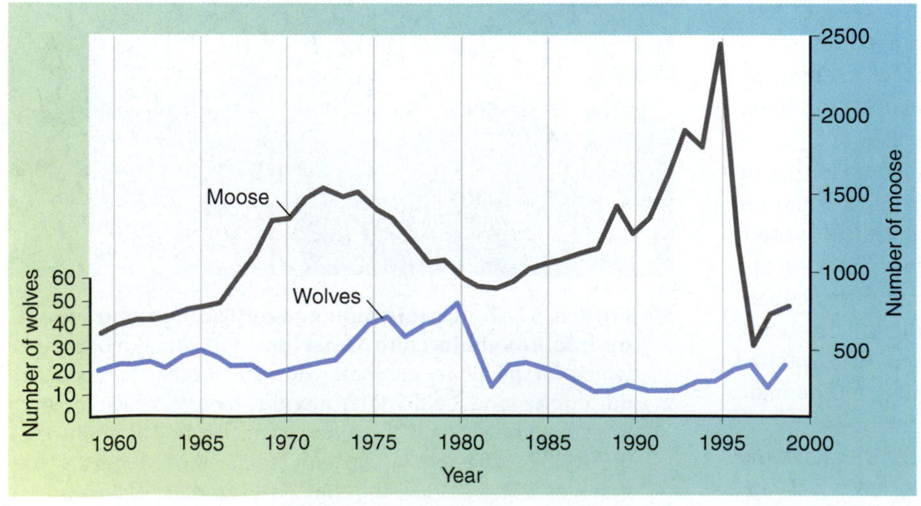

Figure 51–8 Fluctuations in the populations of wolves and moose on Isle Royale. Wolves numbered 25 in 1999, and moose numbered 750 (based on aerial surveys). *(Rolf O. Peterson, Michigan Technological University)*

(a)

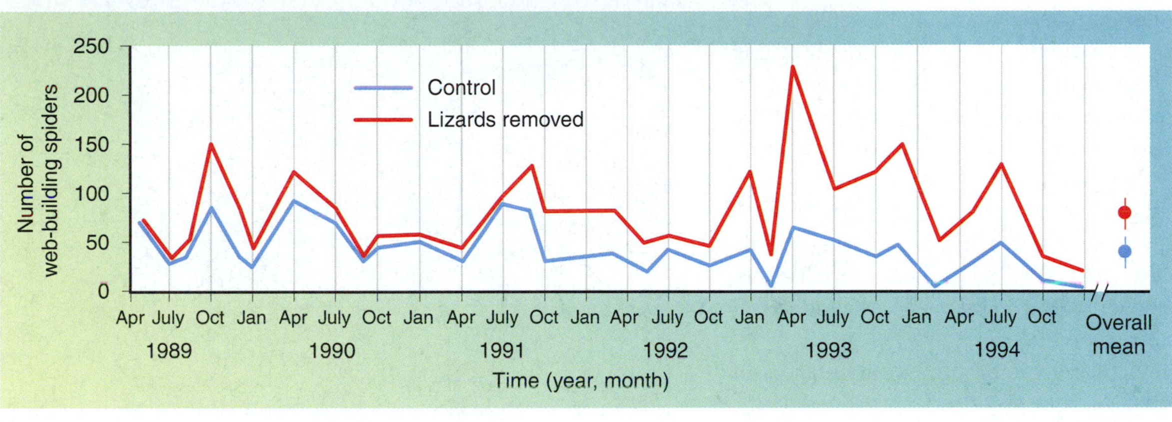

(b)

Figure 51-9 Interaction of density-dependent factors. **(a)** Shown is one of the enclosures used by Spiller and Schoener for their field experiment. Spiders and insects can pass freely into and out of the enclosures, but lizards cannot. **(b)** Graph of the mean number of web-building spiders at each count in enclosures with lizards (control, *blue*) and enclosures with lizards removed *(red)*. A previous experiment indicated that the enclosures themselves have no effect on the number of web-building spiders. *(a, Courtesy of Thomas W. Schoener; b, Spiller, D.A. and T.W. Schoener. "Lizards Reduce Spider Species Richness by Excluding Rare Species."* Ecology, *Vol. 79, No. 2, 1998)*

pete with spiders for insect prey—i.e., both spiders and lizards eat insects). In this experiment, the effects of the two density-dependent factors in determining spider population size cannot be evaluated separately. Additional lines of evidence support the actions of both competition and predation at these sites.

Density-independent factors are generally abiotic

Any environmental factor that affects the size of a population but is not influenced by changes in population density is called a **density-independent factor.** Such factors are typically abiotic. Random weather events that reduce population size serve as density-independent factors. These often affect population density in unpredictable ways. A killing frost, severe blizzard, hurricane, or fire, for example, may cause extreme and irregular reductions in a vulnerable population, regardless of its size, and thus might be considered largely density-independent.

Consider a density-independent factor that influences mosquito populations in arctic environments. These insects produce several generations per summer and achieve high population densities by the end of the season. A shortage of food does not seem to be a limiting factor for mosquitoes, nor is there any shortage of ponds in which to breed. What puts a stop to the skyrocketing mosquito population is winter. Not a single adult mosquito survives winter, and the entire population must grow afresh the next summer from the few eggs and hibernating larvae that survive. Thus, severe winter weather is a density-independent factor that affects arctic mosquito populations.

Density-independent and density-dependent factors are often interrelated. Social animals, for example, are often able to resist dangerous weather conditions by means of their collective behavior, as in the case of sheep huddling together in a snowstorm. In this case it appears that the greater the population density, the better their ability to resist the environmental stress of a density-independent event (i.e., a snowstorm).

SPECIES DIFFER IN REPRODUCTIVE TRAITS

Each species is uniquely suited to its lifestyle. Many years pass before a young magnolia tree flowers and produces seeds, whereas a poppy plant grows from seed, flowers, and dies in a single season. A mating pair of black-browed albatrosses (see Fig. 50–18) produces a single chick every year, but a mating pair of gray-headed albatrosses produces a single chick biennially (every other year).

Species that expend their energy in a single, immense reproductive effort are said to be **semelparous.** Most insects and invertebrates, many plants, and some species of fish exhibit semelparity. Pacific salmon, for example, hatch in fresh water and swim to the ocean, where they live until they mature. Adult salmon swim from the ocean back into the same rivers or streams in which they hatched to spawn (reproduce). After they spawn, the salmon die.

Agaves are semelparous plants that are common in arid tropical and semitropical areas. The thick, fleshy leaves of the agave plant are crowded into a rosette at the base of the stem. Commonly called the century plant because it was mistakenly thought to flower only once in a century, agaves can flower after they are ten years old or so, after which the entire plant dies (Fig. 51–10).

Many species are **iteroparous** and exhibit repeated reproductive cycles—that is, reproduction during several breeding seasons—throughout their lifetimes. Iteroparity is common in most vertebrates, perennial herbaceous plants, shrubs, and trees. The timing of reproduction, earlier or later in life, is a crucial aspect of iteroparity and involves trade-offs. Reproducing earlier in life may mean a reduced likelihood of survival (because an individual is expending energy toward reproduction instead of its own growth), which reduces the potential for later reproduction. On the other hand, reproducing later in life means the individual has less time for additional reproductive events.

Ecologists try to understand the adaptive consequences of various **life history traits,** such as semelparity and iteroparity. Adaptations such as reproductive rate, age at maturity, and **fecundity** (potential capacity to produce offspring), all of which are a part of a species' life history traits, influence an organism's survival and reproduction. The ability of an individual to reproduce successfully, thereby making a genetic contribution to future generations of a population, is called its **fitness** (recall the discussions of fitness in Chapters 18 and 50).

Although many different life histories exist, some ecologists recognize two extremes: *r*-selected species and *K*-selected species. Keep in mind, as you read the following descriptions of *r* selection and *K* selection, that these concepts, although useful, oversimplify most life histories. Many species possess a combination of *r*-selected and *K*-selected traits, as well as traits that cannot be classified as either *r*-selected or *K*-selected. In addition, some populations within a species may exhibit the characteristics of *r*-selection, whereas other populations in different environments may assume *K*-selected traits.

Populations described by the concept of ***r* selection** have traits that contribute to a high population growth rate. Recall that *r* designates the per capita growth rate. Because such organisms have a high *r*, they are known as ***r* strategists** or ***r*-selected species.** Small body size, early maturity, short life span, large broods, and little or no parental care are typical of many *r* strategists, which are usually opportunists found in variable, temporary, or unpredictable environments where the probability of long-term survival is low. Some of the best examples of *r* strategists are insects such as mosquitoes and common weeds such as the dandelion.

Figure 51–10 Semelparity. The century plant (*Agave* sp.) is a succulent with sword-shaped leaves arranged as a rosette around a short stem. Shown is *Agave shawii*, whose leaves grow to 61 cm (2 ft). Note the floral stalk, which can grow more than 3 m (10 ft) tall. Agaves flower once and then die. *(R. Gustafson/Visuals Unlimited)*

In populations described by the concept of ***K* selection,** traits maximize the chance of surviving in an environment where the number of individuals *(N)* is near the carrying capacity *(K)* of the environment. These organisms, called ***K* strategists** or ***K*-selected species,** do not produce large numbers of offspring. They characteristically have long life spans with slow development, late reproduction, large body size, and a low reproductive rate. *K* strategists are found in relatively constant or stable environments, where they have a high competitive ability. Redwood trees are classified as *K* strategists. Animals that are *K* strategists typically invest in parental care of their young.

Tawny owls *(Strix aluco)*, for example, are *K* strategists that pair-bond for life, with both members of a pair living and hunting in adjacent, well-defined territories. Their reproduction is regulated in accordance with the resources, especially the food supply, present in their territories. In an average year, 30% of the birds do not breed at all. If food supplies are more limited than initially indicated, many of those that do breed fail to incubate their eggs. Rarely do the owls lay the maximum number of eggs that they are physiologically capable of laying, and breeding is often delayed until late in the season, when the rodent populations on which they depend have become large. Thus, the behavior of tawny owls ensures better reproductive success of the individual and leads to a stable population at or near the carrying capacity of the environment. Starvation, an indication that the tawny owl population has exceeded the carrying capacity, rarely occurs.

Life tables and survivorship curves indicate mortality and survival

A **life table** can be constructed to show the mortality and survival data of a population or **cohort,** a group of individuals of the same age, at different times during their life span. Insurance companies were the first to use life tables, to calculate the relationship between a client's age and the likelihood of the client surviving to pay enough insurance premiums to cover the cost of the policy. Ecologists construct such tables for animals and plants based on data that rely on a variety of population sampling methods and age determination techniques.

Table 51–2 shows a life table for a cohort of 530 gray squirrels. The first two columns show the units of age (years) and the number of individuals in the cohort that were alive at the beginning of each age interval (the actual data collected in the field by the ecologist). The values for the third column (i.e., the proportion alive at the beginning of each age interval) are calculated by dividing each number in column 2 by 530, the number of squirrels in the original cohort. The values in the fourth column (i.e., the proportion dying during each age interval) are calculated using the values in the third column—by subtracting the number of survivors at the beginning of the next interval from those alive at the beginning of the current interval. For example, the proportion dying during interval 0–1 years = 1.000 − 0.253 = 0.747. The last column, the death rate for each age interval, is calculated by dividing the proportion dying during the age interval (column 4) by the proportion alive at the beginning of the age interval (column 3). For example, the death rate for the age interval 1–2 years is 0.147 ÷ 0.253 = 0.581.

Survivorship is the probability that a given individual in a population or cohort will survive to a particular age. Plotting the log of the number of surviving individuals against age, from birth to the maximum age reached by any individual, produces a **survivorship curve.** Figure 51–11 shows the three main survivorship curves that ecologists recognize.

In Type I survivorship, as exemplified by bison and humans, the young and those at reproductive age have a high probability of surviving. The probability of survival decreases more rapidly with increasing age; mortality is concentrated later in life. Figure 51–12 shows a survivorship curve for a natural population of Drummond phlox, an annual native to East Texas that became widely distributed in the southeastern United States after it escaped from cultivation. Because most Drummond phlox seedlings survive to reproduce, after germination the plant exhibits a Type I survivorship that is typical of annuals.

In Type III survivorship, the probability of mortality is greatest early in life, and those individuals that avoid early death subsequently have a high probability of survival, that is, the probability of survival increases with increasing age. Type III survivorship is characteristic of oysters; young oysters have three free-swimming larval stages before settling down and secreting a shell. These larvae are vulnerable to predation, and few survive to adulthood.

In Type II survivorship, which is intermediate between Types I and III, the probability of survival does not change with age. The probability of death is equally likely across all age groups, resulting in a linear decline in survivorship. This constancy probably results from essentially random events that cause death with little age bias. Although this relationship between age and survivorship is rare, some lizards have Type II survivorship.

The three survivorship curves are generalizations, and few populations exactly fit one of the three. Some species have one

TABLE 51–2 Life Table for a Cohort of 530 Gray Squirrels *(Sciurus carolinensis)*

Age Interval (Years)	Number Alive at Beginning of Age Interval	Proportion Alive at Beginning of Age Interval	Proportion Dying During Age Interval	Death Rate for Age Interval
0–1	530	1.000	0.747	0.747
1–2	134	0.253	0.147	0.581
2–3	56	0.106	0.032	0.302
3–4	39	0.074	0.031	0.418
4–5	23	0.043	0.021	0.488
5–6	12	0.022	0.013	0.591
6–7	5	0.009	0.006	0.666
7–8	2	0.003	0.003	1.000
8–9	0	0.000	0.000	—

Adapted from Table 13.1, p. 150, in Smith, R.L. and T.M. Smith. *Elements of Ecology*, 4th ed. San Francisco, Benjamin/Cummings Science Publishing, 1998.

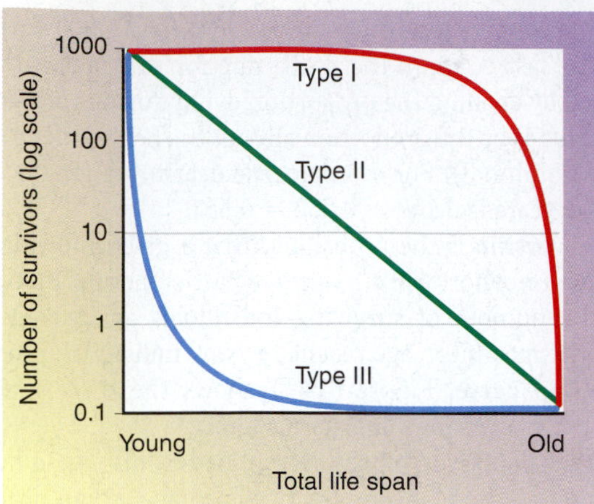

Figure 51–11 Survivorship curves. These curves represent the ideal survivorships of species in which mortality is greatest in old age (Type I), spread evenly across all age groups (Type II), and greatest among the young (Type III). The survivorship of most organisms can be compared to these curves.

type of survivorship curve early in life and another type as adults. Herring gulls, for example, start out with a Type III survivorship curve but develop a Type II curve as adults (Fig. 51–13). The survivorship curve shown in this figure is characteristic of birds in general. Note that most death occurs almost immediately after hatching, despite the protection and care given to the chicks by the parent bird. Herring gull chicks die from predation or attack by other herring gulls, inclement weather, infectious diseases, or starvation following death of the parent. Once the chicks become independent, their survivorship increases dramatically, and death occurs at about the same rate throughout their remaining lives. Few or no herring gulls die from the degenerative diseases of "old age" that cause death in most humans.

MANY SPECIES CONSIST OF A SET OF LOCAL POPULATIONS

The natural environment consists of a landscape in which there are a variety of habitat patches. Consider a forest, for example. The forest landscape is a mosaic of different elevations, temperatures, levels of precipitation, soil moisture, soil types, and other properties. Because each species has its own habitat requirements, this heterogeneity in physical properties is reflected in the different organisms that occupy the various patches in the landscape (Fig. 51–14). Some species occur in very narrow habitat ranges, whereas others have wider habitat distributions.

Population ecologists have discovered that many species are not distributed as one large population across the landscape. Instead, many species exist as a series of local populations that are distributed in distinct habitat patches across the landscape. Each local population has its own characteristic demographic features, such as birth, death, emigration, and immigration rates. A population that is divided into several local populations among which individuals occasionally disperse (emigrate and immigrate) is known as a **metapopulation.** For example, note

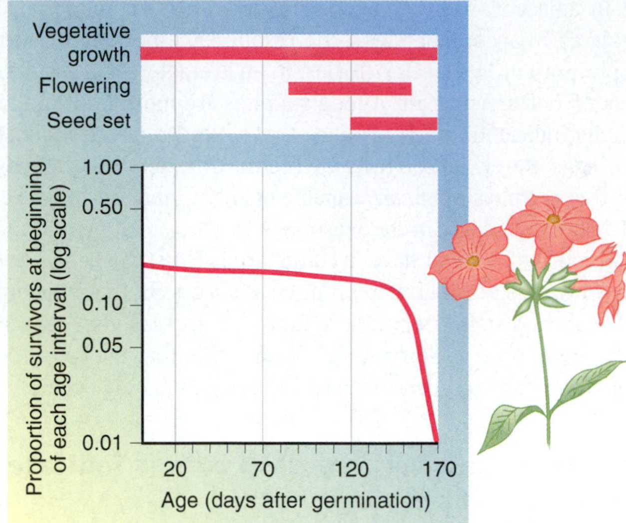

Figure 51–12 Survivorship curve for a Drummond phlox population. Drummond phlox *(Phlox drummondii)* has a Type I survivorship after germination of the seeds. Bars above the graph indicate the various stages in the Drummond phlox life history. Data were collected at Nixon, Texas, in 1974 and 1975. Survivorship on the Y-axis begins at 0.296 instead of 1.000 because the study took into account death during the seed dormancy period prior to germination *(not shown). (Modified from Leverich, W.J. and D.A. Levin. "Age-Specific Survivorship and Reproduction in* Phlox drummondii.*"* The American Naturalist, *Vol. 113, No. 6, Jun. 1979)*

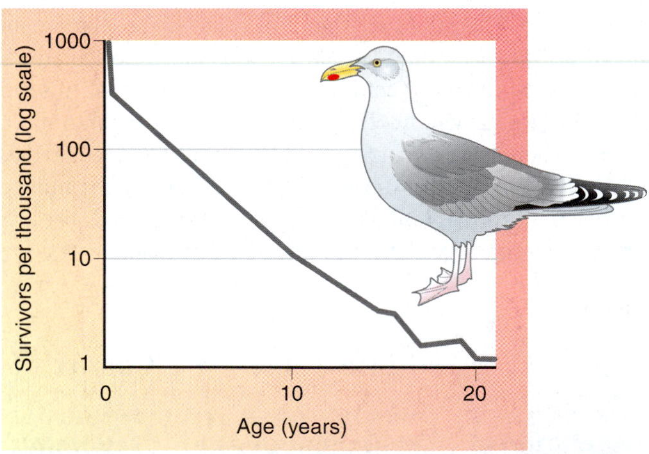

Figure 51–13 Survivorship curve for a herring gull population. Herring gulls have Type III survivorship as chicks and Type II survivorship as adults. Data were collected from Kent Island, Maine, during a five-year period in the 1930s; baby gulls were banded to establish identity. *(After Paynter)*

the various local populations of red oak on the mountain slope in Figure 51–14*b*.

The spatial distribution of a species occurs because different habitats vary in suitability, from unacceptable to preferred. The preferred sites are more productive habitats that increase the likelihood of survival and reproductive success for the individu-

(a)

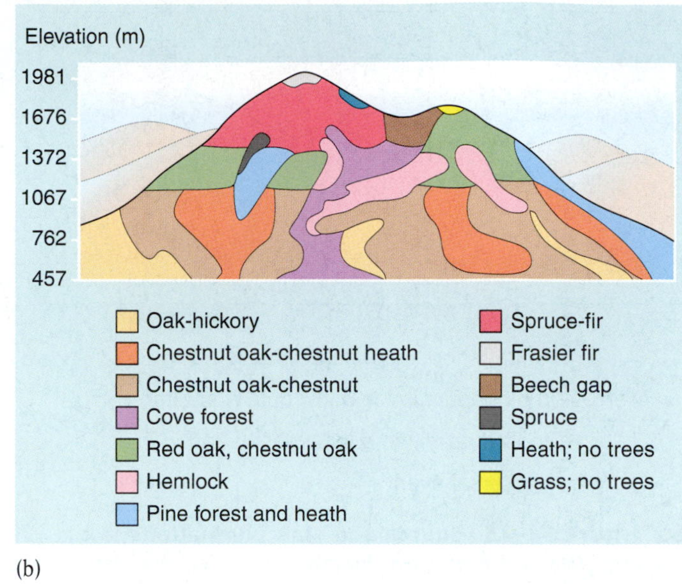

(b)

Figure 51–14 The mosaic nature of landscapes. **(a)** This early spring view of Deep Creek Valley in the Great Smoky Mountains National Park gives some idea of the heterogeneity of the landscape. During the summer, when all the vegetation is a deep green, the landscape appears homogeneous. **(b)** An evaluation of the distribution of vegetation on a typical west-facing slope in the Great Smoky Mountains National Park reveals that the landscape consists of patches. Chestnut oak is a species of oak *(Quercus prinus)*; chestnut heath are areas within chestnut oak forest where the trees are widely scattered and the slopes underneath are covered by a thick growth of laurel (*Kalmia* sp.) shrubs; cove forest is a mixed stand of deciduous trees. *(a, Adam Jones/Photo Researchers, Inc.; b, Adapted from Whittaker, R.H. "Vegetation of the Great Smoky Mountains."* Ecological Monographs, *Vol. 26, 1956)*

als living there. Good habitats, called **sources,** are areas where local reproductive success is greater than local mortality. (*Source* refers to both a habitat and the population living in that habitat.) Source populations generally have greater population densities than less suitable sites, and surplus individuals in the source habitat disperse and find another habitat in which to settle and reproduce.

Individuals living in lower quality habitats may suffer death or, if they survive, poor reproductive success. Lower quality habitats, called **sinks,** are areas where local reproductive success is less than local mortality. (*Sink* refers to both a habitat and the population living in that habitat.) Without immigration from other areas, a sink population declines until extinction occurs. If a local population becomes extinct, individuals from a source habitat may recolonize the vacant habitat at a later time. Source and sink habitats, then, are linked to one another by dispersal (Fig. 51–15).

Metapopulations are becoming more common as humans alter the landscape, fragmenting existing habitats to accommodate homes and factories, agricultural fields, and logging. As a result, the concept of metapopulations, particularly as it relates to endangered and threatened species, has become an important field of study in conservation biology (see Chapter 55).

THE PRINCIPLES OF POPULATION ECOLOGY APPLY TO HUMAN POPULATIONS

Now that we have examined some of the basic concepts of population ecology, we can apply those concepts to the human population. Examine Figure 51–16, which shows the world increase in the human population since the development of agriculture approximately 10,000 years ago. Now look back at Figure 51–3 and compare the two curves. The characteristic J curve of exponential population growth shown in Figure 51–16 reflects the decreasing amount of time it has taken to add each additional billion people to our numbers. It took thousands of years for the human population to reach 1 billion, a milestone that took place around 1800. It took 130 years to reach 2 billion (in 1930), 30 years to reach 3 billion (in 1960), 15 years to reach 4 billion (in 1975), 12 years to reach 5 billion (in 1987), and 12 years to reach

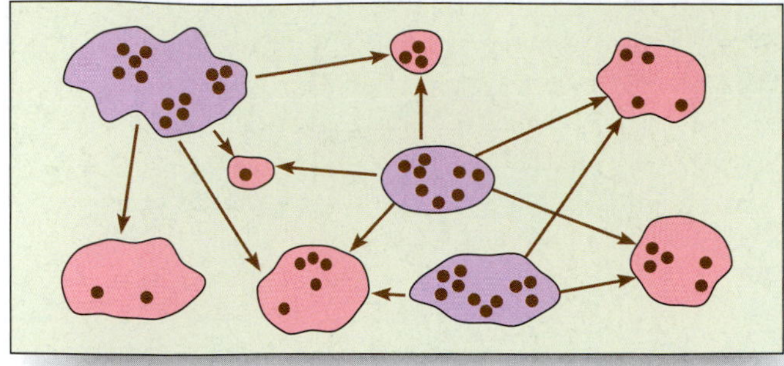

Figure 51–15 Source and sink populations in a hypothetical metapopulation. The local populations in the habitat patches shown here collectively make up the metapopulation. Source habitats provide individuals that emigrate and colonize sink habitats. Studied over time, the metapopulation is shown to exist as a shifting pattern of occupied and vacant habitat patches.

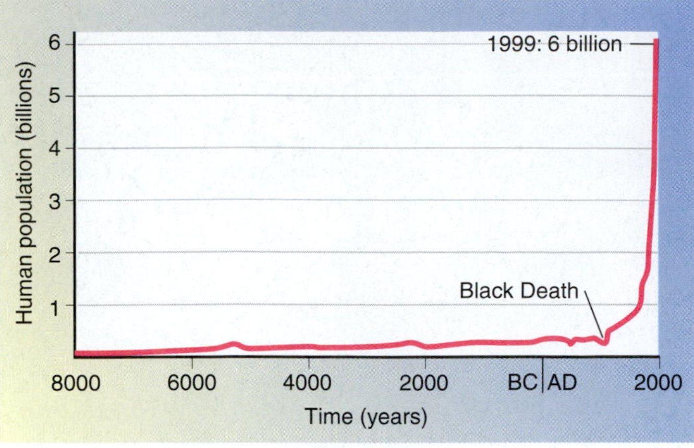

Figure 51–16 Human population growth. During the last 1000 years, the human population has been increasing nearly exponentially. Population experts predict that the population will level out during the 21st century, forming the S curve observed in other species. (Black Death refers to a devastating disease, probably bubonic plague, that decimated Europe and Asia in the 14th century.)

6 billion (in 1999). The United Nations projects that the human population will reach 7 billion by 2013.

Thomas Malthus (1766–1834), a British clergyman and economist, was one of the first to recognize that the human population cannot continue to increase indefinitely (see Chapter 17). He pointed out that human population growth is not always desirable (a view contrary to the beliefs of his day and to those of many people even today) and that the human population is capable of increasing faster than the food supply. He maintained that the inevitable consequences of population growth are famine, disease, and war.

The world population increased by approximately 70 million from 2000 to 2001, reaching 6.14 billion in 2001. This change was not caused by an increase in the birth rate *(b)*. In fact, the world birth rate has actually declined during the past 200 years. The increase in population is due instead to a dramatic decrease in the death rate *(d)*, which has occurred primarily because greater food production, better medical care, and improved sanitation practices have increased the life expectancies of a great majority of the global population. For example, from 1920 to 2000, the death rate in Mexico fell from approximately 40 per 1000 individuals to 4 per 1000, whereas the birth rate dropped from approximately 40 per 1000 individuals to 24 per 1000 (Fig. 51–17).

The human population has reached a turning point. Although our numbers continue to increase, the world per capita growth rate *(r)* has declined over the past several years, from a peak of 2.2% per year in the mid-1960s to 1.3% per year in 2001. Population experts at the United Nations and the World Bank have projected that the growth rate will continue to slowly decrease until zero population growth is attained. Thus, exponential growth of the human population will end, and the J curve will be replaced by the S curve. It is projected that **zero population growth,** the point at which the birth rate equals the death rate $(r = 0)$, will occur toward the end of the 21st century.

The United Nations periodically publishes population projections for the 21st century. The latest (1998) U.N. figures available forecast that the human population will be between 7.7 billion (their low projection) and 11.1 billion (their high projection) in the year 2050, with 9.4 billion thought to be "most likely." Another group that examines population trends is the International Institute for Applied Systems Analysis (IIASA) in Vienna, Austria. IIASA is notable because it takes into account future changes in the death rate as well as future changes in fertility. (Most population projections only examine changes in the birth rate.) In 1996 IIASA projected that the world's population

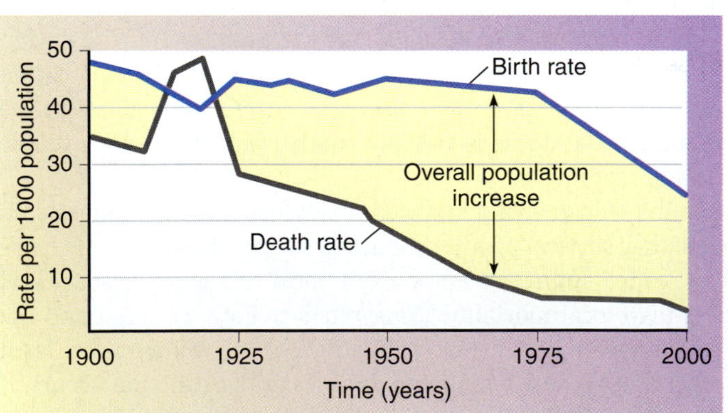

Figure 51–17 Birth and death rates in Mexico, 1900 to 2000. Both birth and death rates declined during the 20th century, but because the death rate declined much more than the birth rate, Mexico has experienced a high growth rate. (The high death rate prior to 1920 was caused by the Mexican Revolution.) *(Population Reference Bureau)*

would peak at 11 billion in 2075. More recently, IIASA informally estimated that the world population may peak at less than 10 billion because of rapid fertility declines in many countries and increased mortality from the spreading AIDS epidemic.

Such population projections are "what if" exercises. Given certain assumptions about future tendencies in natality, mortality, and dispersal, an area's population can be calculated for a given number of years into the future. Population projections indicate the changes that may occur, but they must be interpreted with care because they vary depending on what assumptions have been made. For example, in projecting that the population will be 7.7 billion (their low projection) in 2050, U.N. population experts assume that the average number of children born to each woman in all countries will have declined to 1.5 in the 21st century. In 2001 the average number of children born to each woman was 2.8. If the decline to 1.5 does not occur, the global population could be significantly higher. For example, if the average number of children born to each woman declines only to 2.5 instead of 1.5, the 2050 population will be 11.1 billion (the U.N. high projection). Small differences in fertility, then, produce large differences in population forecasts.

The main unknown factor in any population growth scenario is Earth's carrying capacity. According to Dr. Joel Cohen, who has a joint professorship at Rockefeller University and Columbia University, most published estimates of how many people Earth can support range from 4 billion to 16 billion. These estimates vary so widely because of the assumptions that are made about standard of living, resource consumption, technological innovations, and waste generation. If we want all people to have a high level of material well-being equivalent to the lifestyles common in highly developed countries, then Earth will clearly be able to support far fewer humans than if everyone lives just above the subsistence level. Earth's carrying capacity for the human population, then, is not decided simply by natural environmental constraints. Human choices and values have to be factored into the assessment.

It is also not clear what will happen to the human population if or when the carrying capacity is approached. Optimists suggest that the human population will stabilize because of a decrease in the birth rate. Some experts take a more pessimistic view and predict that the widespread degradation of our environment caused by our ever-expanding numbers will make Earth uninhabitable for humans and other species. These experts contend that a massive wave of human suffering and death will occur. Some experts think the human population has already exceeded the carrying capacity of the environment, a potentially dangerous situation that threatens our long-term survival as a species.

Not all countries have the same growth rate

Although world population figures illustrate overall trends, they do not describe other important aspects of the human population story, such as population differences from country to country. **Human demographics,** the science that deals with human population statistics such as size, density, and distribution, provides interesting information on the populations of various countries. As you probably know, not all parts of the world have the same rates of population increase. Countries can be classified into two groups, highly developed and developing, based on their rates of population growth, degrees of industrialization, and relative prosperity (Table 51–3).

Highly developed countries, such as the United States, Canada, France, Germany, Sweden, Australia, and Japan, have low rates of population growth and are highly industrialized relative to the rest of the world. Highly developed countries have the lowest birth rates in the world. Indeed, some highly developed countries such as Germany have birth rates just below that needed to sustain the population and are thus declining slightly in numbers. Highly developed countries also have low **infant mortality rates** (the number of infant deaths per 1000 live births). The infant mortality rate of the United States was 7.1 in 2001, for example, compared with a world infant mortality rate of 56. Highly developed countries also have longer life expectancies (77 years in the United States versus 67 years worldwide) and higher average GNI PPP per capita ($31,910 in the United States versus $6650 worldwide).[3]

[3]GNI PPP per capita is Gross National Income in purchasing power parity (PPP) divided by midyear population. It indicates the amount of goods and services an average citizen of that particular region or country could buy in the United States.

TABLE 51–3 Comparison of 2001 Population Data in Developed and Developing Countries

	Developed	Developing	
	(Highly Developed) United States	(Moderately Developed) Brazil	(Less Developed) Ethiopia
Fertility rate	2.1	2.4	5.9
Doubling time at current rate (2000)	120 years	48 years	29 years
Infant mortality rate	7.1 per 1000	35 per 1000	97 per 1000
Life expectancy at birth	77 years	68 years	52 years
GNI PPP per capita (U.S. $)	$31,910	$6840	$620
Women using modern contraception	71%	70%	6%

Source: Population Reference Bureau.

Developing countries fall into two subcategories: moderately developed and less developed. Mexico, Turkey, Thailand, and most countries of South America are examples of *moderately developed countries*. Their birth rates and infant mortality rates are higher than those of highly developed countries, but they are declining. Moderately developed countries have a medium level of industrialization, and their average GNI PPP per capita is lower than those of highly developed countries.

Bangladesh, Niger, Ethiopia, Laos, and Cambodia are examples of *less developed countries*. These countries have the highest birth rates, the highest infant mortality rates, the lowest life expectancies, and the lowest average GNI PPP per capita in the world.

One way to represent the population growth of a country is to determine the **doubling time,** the amount of time it would take for its population to double in size, assuming that its current growth rate did not change. (A simplified formula for doubling time [t_d] is $t_d = 70 \div r$. The actual formula involves calculus and is beyond the scope of this text.)

A look at a country's doubling time can identify it as a highly, moderately, or less developed country: The shorter the doubling time, the less developed the country. At rates of growth in 2000, the doubling time is 26 years for Laos, 29 years for Ethiopia, 46 years for Turkey, 70 years for Thailand, 120 years for the United States, and 204 years for France.

It is also instructive to examine **replacement-level fertility,** the number of children a couple must produce to "replace" themselves. Replacement-level fertility is usually given as 2.1 children in highly developed countries and 2.7 children in developing countries. The number is always greater than 2.0 because some children die before they reach reproductive age. Higher infant mortality rates are the main reason that replacement levels in developing countries are greater than in highly developed countries. The **total fertility rate**—the average number of children born to a woman during her lifetime—is 2.8 worldwide, which is well above replacement levels.

The population in many developing countries is beginning to approach stabilization. The fertility rate must decline for the population to stabilize (Table 51–4; also note the general decline in total fertility rate from the 1960s to 2001 in selected developing countries). The total fertility rate in developing countries has decreased from an average of 6.1 children per woman in 1970 to 3.2 in 2001. From 1990 to 2001 fertility rates declined by at least 25% in countries such as Brazil, Indonesia, and Mexico.

Although the fertility rates in these countries have declined, it should be remembered that most still exceed replacement-level fertility. Consequently, the populations in these countries are still increasing. Also, even when fertility rates equal replacement-level fertility, population growth will still continue for some time. To understand why this is so, we now examine the age structure of various countries.

The age structure of a country can be used to predict future population growth

To predict the future growth of a population, it is important to know its **age structure,** which is the number and proportion of people at each age in a population. The number of males and number of females at each age, from birth to death, can be represented in an **age structure diagram.**

The overall shape of an age structure diagram indicates whether the population is increasing, stationary, or shrinking. The age structure diagram of a country with a high growth rate (e.g., Nigeria or Bolivia) is shaped like a pyramid (Fig. 51–18*a*). Because the largest percentage of the population is in the pre-reproductive age group (i.e., 0 to 14 years of age), the probability of future population growth is great. A strong **population growth momentum** exists because when all these children mature, they will become the parents of the next generation, and this group of parents will be larger than the previous group. Thus, even if the fertility rate of such a country declines to replacement level (i.e., couples have smaller families than their parents did), the population will continue to grow for some time. Population growth momentum can have either a positive value (i.e., the population will grow) or a negative value (i.e., the population will decline). However, it is usually discussed in a positive context to explain how the future growth of a population is affected by its present age distribution.

In contrast, the more tapered bases of the age structure diagrams of countries with slowly growing, stable, or declining populations indicate a smaller proportion of the population will become the parents of the next generation (Fig. 51–18*b* and *c*). The age structure diagram of a stable population, one that is neither growing nor shrinking, demonstrates that the number of people at pre-reproductive and reproductive ages are approximately the same. Also, a larger percentage of the population is older, that is, post-reproductive, than in a rapidly increasing population. Many countries in Europe have stable populations.

In a population that is shrinking in size, the pre-reproductive age group is smaller than either the reproductive or post-reproductive group. Germany, Russia, and Bulgaria are examples of countries with slowly shrinking populations.

Worldwide, 30% of the human population is younger than age 15. When these people enter their reproductive years, they

TABLE 51–4 Fertility Changes in Selected Developing Countries

	Total Fertility Rate	
Country	**1960–1965**	**2001**
Bangladesh	6.7	3.3
Brazil	6.2	2.4
China	5.9	1.8
Egypt	7.1	3.5
Guatemala	6.9	4.8
India	5.8	3.2
Kenya	8.1	4.4
Mexico	6.8	2.8
Nepal	5.9	4.8
Nigeria	6.9	5.8
Thailand	6.4	1.8

Source: Population Reference Bureau.

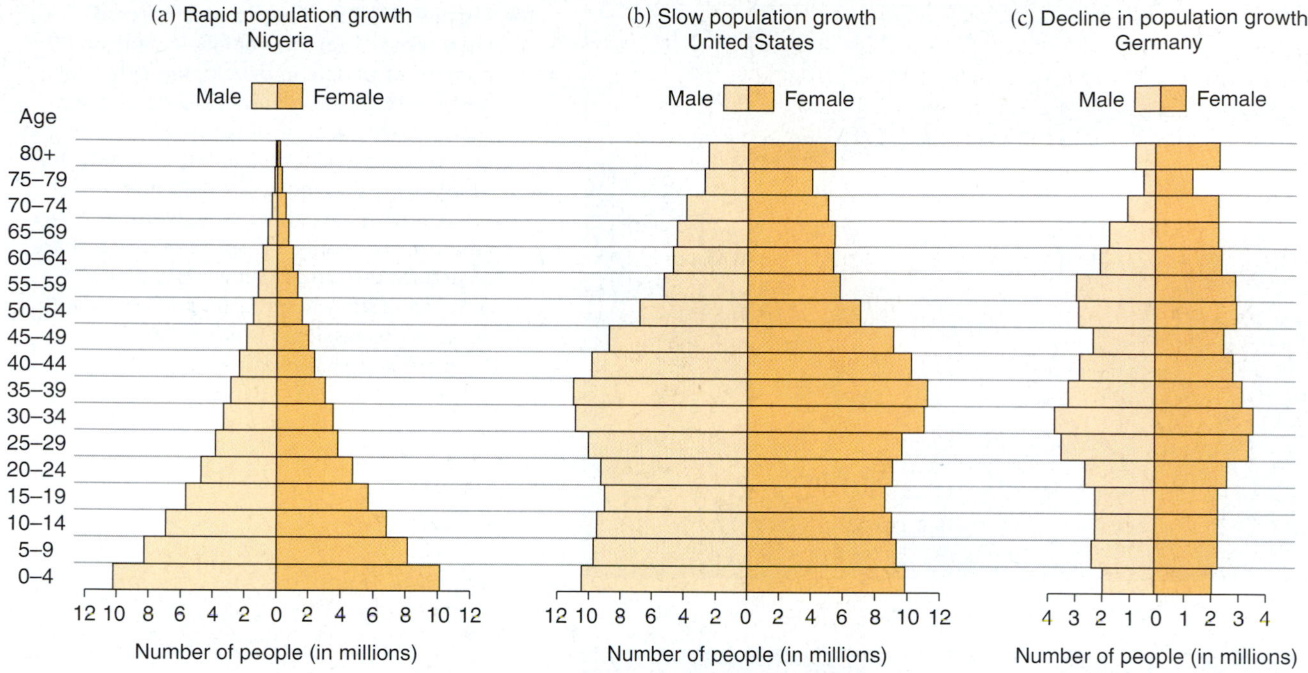

Figure 51–18 Age structure diagrams. These age structure diagrams for **(a)** Nigeria, **(b)** the United States, and **(c)** Germany indicate that less developed countries such as Nigeria have a greater percentage of young people than highly developed countries. As a result, less developed countries are projected to have greater population growth than highly developed countries. *(Reproduced by author using data from the Population Reference Bureau)*

have the potential to cause a large increase in the growth rate. Even if the birth rate does not increase, the growth rate will increase simply because there are more people reproducing.

Most of the world population increase since 1950 has taken place in developing countries as a result of the younger age structure and the higher-than-replacement-level fertility rates of their populations. In 1950, 66.8% of the world's population was in the developing countries in Africa, Asia (minus Japan), and Latin America. Between 1950 and 2001, the world's population more than doubled in size, but most of that growth occurred in developing countries. As a reflection of this, in 2001 the people in developing countries had increased to 80.6% of the world's population. Most of the population increase that will occur during the 21st century will also take place in developing countries, largely the result of their young age structures. These countries, most of which are poor, are least able to support such growth.

Environmental degradation is related to population growth and resource consumption

The relationships among population growth, use of natural resources, and environmental degradation are complex, but we can make two useful generalizations. First, although resources that are essential to an individual's survival may be small, a rapidly increasing population (as in developing countries) tends to overwhelm and deplete a country's soils, forests, and other natural resources (Fig. 51–19*a*). Second, in highly developed nations, individual resource demands are large, far above requirements for survival. To satisfy their desires rather than their basic needs, people in more affluent nations exhaust resources and degrade the global environment through excessive consumption and "throwaway" lifestyles (Fig. 51–19*b*).

Rapid population growth can cause natural resources to be overexploited. For example, large numbers of poor people must grow crops on land that is inappropriate for farming, such as mountain slopes or some tropical rain forests. Although this practice may provide a short-term solution to the need for food, it does not work in the long run, because when these lands are cleared for farming, their agricultural productivity declines rapidly, and severe environmental deterioration occurs.

The effects of population growth on natural resources are particularly critical in developing countries. The economic growth of developing countries is often tied to the exploitation of natural resources, often for export to highly developed countries. Developing countries are faced with the difficult choice of exploiting natural resources to provide for their expanding populations in the short term (i.e., to pay for food or to cover debts) or conserving those resources for future generations. It is instructive to note that the economic growth and development of the United States and of other highly developed nations came about through the exploitation and, in some cases, the destruction of their natural resources.

Although resource issues are clearly related to population size (more people use more resources), an equally, if not more, important factor is a population's resource consumption. People in highly developed countries are extravagant consumers—their use of resources is greatly out of proportion to their numbers. A single child born in a highly developed country such as the United

(a)

(b)

Figure 51-19 People and natural resources. **(a)** The rapidly increasing number of people in developing countries overwhelms their natural resources, even though individual resource requirements may be low. Shown is a typical Indian family, from Ahraura Village, India, with all their possessions. **(b)** People in highly developed countries consume a disproportionate share of natural resources. Shown is a typical American family from Pearland, Texas, with all their possessions. *(a,b, Peter Ginter/Material World)*

States causes a greater impact on the environment and on resource depletion than a dozen or more children born in a developing country. Many natural resources are needed to provide the automobiles, air conditioners, disposable diapers, cell phones, videocassette recorders, computers, clothes, athletic shoes, furniture, boats, and other "comforts" of life in highly developed nations. Thus, the disproportionately large consumption of resources by the United States and other highly developed countries affects natural resources and the environment as much or more than the population explosion in the developing world does.

A country is overpopulated if the level of demand on its resource base results in damage to the environment. If we compare human impact on the environment in developing and highly developed countries, we see that a country can be overpopulated in two ways. **People overpopulation** occurs when the environment is worsening from too many people, even if those people consume few resources per person. People overpopulation is the current problem in many developing nations.

In contrast, **consumption overpopulation** occurs when each individual in a population consumes too large a share of resources. The effect of consumption overpopulation on the environment is the same as that of people overpopulation—pollution and degradation of the environment. Most affluent highly developed nations, including the United States, suffer from consumption overpopulation. Highly developed nations represent only 19.4% of the world's population, yet they consume significantly more than half its resources. According to the Worldwatch Institute, a private research institution in Washington, D.C., highly developed nations account for the lion's share of total resources consumed, as follows: 86% of aluminum used, 76% of timber harvested, 68% of energy produced, 61% of meat eaten, and 42% of fresh water consumed. These nations also generate 75% of the world's pollution and waste.

SUMMARY WITH KEY TERMS

I. **Ecology** is the study of all relationships among organisms and their abiotic environment.
   A. A **population** is all the members of a particular species that live together in the same area.
   B. A **community** is all the populations of different species living in the same area. An **ecosystem** is a community and its environment.
   C. A **landscape** is a large area of terrain (several to many square kilometers) composed of interacting ecosystems.
   D. The **biosphere** is the global ecological system that comprises all the communities on Earth. The biosphere concept includes interactions among all Earth's communities and Earth's atmosphere, lithosphere, and hydrosphere.

II. **Population ecology** is the branch of biology that deals with the number of individuals of a particular species that are found in an area. It includes **population dynamics**—how and why those numbers change over time. Populations have certain properties, such as birth rates and death rates, that individual organisms lack.
   A. **Population density** is the number of individuals of a species per unit of area or volume at a given time.
   B. Population **dispersion** (spacing) may be **clumped** (clustered in specific parts of the habitat), **uniform** (evenly spaced), or **random** (unpredictably spaced).
   C. Population size is affected by the average per capita birth rate *(b)*, average per capita death rate *(d)*, average per capita immigration rate *(i)*, and average per capita emigration rate *(e)*.
      1. The **growth rate *(r)*** is the rate of change (increase or decrease) of a population on a per capita basis.
      2. $r = b - d$ on a global scale (when **dispersal** is not a factor). Populations increase in size as long as the average per capita birth rate **(natality)** is greater than the average per capita death rate **(mortality)**.
      3. $r = (b - d) + (i - e)$ for a local population (where dispersal is a factor).

III. **Intrinsic rate of increase** is the maximum rate at which a species or population could increase in number under ideal conditions.
   A. Although certain populations exhibit an accelerated pattern of growth known as **exponential population growth** for limited periods of time (the J-shaped curve), eventually the growth rate decreases to around zero or becomes negative.
   B. Population size is modified by limits set by the environment.
   C. The **carrying capacity *(K)*** of the environment is the largest population that can be maintained for an indefinite time by a particular environment.
   D. **Logistic population growth,** when graphed, shows a characteristic S-shaped curve. Seldom do natural populations follow the logistic growth curve very closely.

IV. Environmental factors limit population growth.
   A. **Density-dependent factors** regulate population growth by affecting a larger proportion of the population as population density rises. Predation, disease, and competition are examples.
      1. **Intraspecific competition** occurs when individuals of the same species compete for a resource in limited supply, whereas **interspecific competition** occurs when individuals of different species compete.
      2. In **interference competition**, certain dominant individuals obtain an adequate supply of a limited resource at the expense of other individuals in the population. In **exploitation competition**, all the individuals in a population "share" the limited resource more or less equally so that at high population densities, none of them receives an adequate amount.
   B. **Density-independent factors** limit population growth but are not influenced by changes in population density. Hurricanes and fires are examples.

V. Each species has **life history traits** that are uniquely adapted to its reproductive pattern.
   A. **Semelparous** species expend their energy in a single, immense reproductive effort. **Iteroparous** species exhibit repeated reproductive cycles throughout their lifetimes.
   B. Although many different life histories exist, some ecologists recognize two extremes.
      1. An ***r* strategy** emphasizes a high growth rate. These organisms often have small body sizes, high reproductive rates, and short **life spans,** and they typically inhabit variable environments.
      2. A ***K* strategy** results in maintenance of a population near the carrying capacity of the environment. These organisms often have large body sizes, low reproductive rates, and long life spans, and they typically inhabit stable environments.
      3. The two strategies oversimplify most life histories. Many species possess a combination of *r*-selected and *K*-selected traits, as well as traits that cannot be classified as either *r*-selected or *K*-selected.
   C. A **life table** shows the mortality and survival data of a population or **cohort** (a group of individuals of the same age) at different times during their life span.
   D. **Survivorship** is the probability that a given individual in a population or cohort will survive to a particular age. There are three general **survivorship curves**: Type I survivorship, in which mortality is greatest in old age; Type II survivorship, in which mortality is spread evenly across all age groups; and Type III survivorship, in which mortality is greatest among the young.

VI. A **metapopulation** is a set of local populations among which individuals occasionally disperse (by emigration and immigration).
   A. **Source habitats** are preferred sites where local reproductive success is greater than local mortality. Surplus individuals disperse from source habitats.
   B. **Sink habitats** are lower quality habitats where individuals may suffer death or, if they survive, poor reproductive success.
   C. If extinction of a local population living in a sink habitat occurs, individuals from a source habitat may recolonize the vacant habitat at a later time.

VII. The principles of population ecology apply to humans.
   A. The world population increased by 70 million from 2000 to 2001, reaching 6.14 billion in 2001.
      1. Although our numbers continue to increase, the per capita growth rate *(r)* has declined over the past several years, from a peak in 1965 of about 2% per year to a 2001 growth rate of 1.3% per year.
      2. **Demographers,** scientists who study human population statistics, project that the world population will become stationary ($r = 0$, or **zero population growth**) by the end of the 21st century.
   B. **Highly developed countries** have the lowest birth rates, lowest **infant mortality rates,** longest life expectancies, and highest GNI PPP per capita. **Developing countries** have the highest birth rates, highest infant mortality rates, shortest life expectancies, and lowest GNI PPP per capita.
   C. The **age structure** of a population greatly influences population dynamics.
      1. It is possible for a country to have **replacement-level fertility** and still experience population growth if the largest percentage of the population is in the pre-reproductive years.
      2. A young age structure causes a positive **population growth momentum** as the very large pre-reproductive age group matures and becomes parents.

VIII. There are two kinds of overpopulation: people overpopulation and consumption overpopulation.
   A. Developing countries tend to have **people overpopulation,** in which population increase degrades the environment even though each individual uses few resources.
   B. Highly developed countries tend to have **consumption overpopulation,** in which each individual in a slow-growing or stationary population consumes a large share of resources, resulting in environmental degradation.

POST-TEST

1. Population____________ is the number of individuals of a species per unit of habitat area or volume at a given time. (a) dispersion (b) density (c) survivorship (d) age structure (e) demographics
2. Which of the following patterns of cars parked along a street is an example of uniform dispersion? (a) five cars parked next to one another in the middle, leaving two empty spaces at one end and three empty spaces at the other end (b) five cars parked in the pattern: car, empty space, car, empty space, etc. (c) five cars parked in no discernible pattern, sometimes having empty spaces on each side and sometimes parked next to another car
3. This type of dispersion occurs when individuals are concentrated in certain parts of the habitat. (a) clumped (b) uniform (c) random (d) both a and b (e) both b and c
4. The average per capita birth rate, or ____________, increases population size, whereas the average per capita death rate, or ____________, decreases population size. (a) natality; demography (b) exploitation competition; interference competition (c) mortality; natality (d) total fertility rate; mortality (e) natality; mortality
5. The per capita growth rate of a population where dispersal is not a factor is expressed as (a) $i + e$ (b) $b - d$ (c) dN/dt (d) $rN(K - N)$ (e) $(K - N) \div K$
6. The maximum rate at which a population could increase under ideal conditions is known as its (a) total fertility rate (b) survivorship (c) intrinsic rate of increase (d) doubling time (e) age structure
7. When r is a positive number, the population size is (a) stable (b) increasing (c) decreasing (d) either increasing or decreasing, depending on interference competition (e) either increasing or stable, depending on whether the species is semelparous
8. In a graph of population size versus time, a J-shaped curve is characteristic of (a) exponential population growth (b) logistic population growth (c) zero population growth (d) replacement-level fertility (e) population growth momentum
9. The largest population that can be maintained by a particular environment for an indefinite period is known as a (a) semelparous population (b) population undergoing exponential growth (c) metapopulation (d) population's carrying capacity (e) source population
10. Giant bamboos live many years without reproducing, then send up a huge flowering stalk and die shortly thereafter. Bamboo is therefore an example of (a) iteroparity (b) a source population (c) a metapopulation (d) an r strategist (e) semelparity
11. Predation, disease, and competition are examples of ____________ factors. (a) density-dependent (b) density-independent (c) survivorship (d) dispersal (e) semelparous
12. ____________competition occurs within a population and ____________ competition occurs among populations of different species. (a) interspecific; intraspecific (b) intraspecific; interspecific (c) Type I survivorship; Type II survivorship (d) interference; exploitation (e) exploitation; interference
13. Population experts project that during the 21st century, the human population (a) will increase the most in highly developed countries (b) will increase the most in developing countries (c) will increase at similar rates in all countries (d) will decrease in developing countries but stabilize in highly developed countries (e) will increase and then decrease dramatically in all countries
14. A highly developed country has a (a) low doubling time (b) low infant mortality rate (c) high GNI PPP per capita (d) both a and b are correct (e) all three answers (a, b, and c) are correct
15. The continued growth of a population with a young age structure, even after its fertility rate has declined, is known as (a) population doubling (b) iteroparity (c) population growth momentum (d) r selection (e) density dependence

REVIEW QUESTIONS

1. List and define the levels of biological organization studied by ecologists.
2. Give several biological advantages for a clumped dispersion. What are the disadvantages?
3. Define each of the following and explain its effect on population size: (a) natality; (b) mortality; (c) immigration; (d) emigration
4. Distinguish between the J-shaped and S-shaped population growth curves in terms of intrinsic rate of increase and carrying capacity.
5. Draw two graphs to represent the long-term growth of a population of bacteria cultured in a test tube containing a nutrient medium (1) that is replenished, and (2) that is not replenished.
6. Give three examples each of density-dependent and density-independent factors that affect population growth.
7. Draw the three main types of survivorship curves and explain each.
8. What is replacement-level fertility? Why is it greater in developing countries than in highly developed countries?
9. Explain how a single child born in the United States can have a greater effect on the environment and natural resources than a dozen children born in Kenya.
10. The 2001 population of the Netherlands was 16 million, and its land area is 990 square miles. The 2001 population of the United States was 284.5 million, and its land area is 3,615,200 square miles. Which country has the greater population density?
11. The population of India in 2001 was 1033 million, and its growth rate was 1.7% per year. Calculate the 2002 population of India.
12. The world population in 2001 was 6.14 billion, and its annual growth rate was 1.3%. If the 2001 birth rate was 22 per 1000 people in the year 2001, what was the death rate, expressed as number per 1000 people?
13. Complete the following life table for a hypothetical population.

Answers to Review Questions 10 to 13 are included in Appendix A with the Post-Test answers.

Life Table

Age Interval (Years)	Number Alive at Beginning of Age Interval	Proportion Alive at Beginning of Age Interval	Proportion Dying During Age Interval	Death Rate for Age Interval
0–1	1000	____	____	____
1–2	210	____	____	____
2–3	100	____	____	____
3–4	46	____	____	____
4–5	0	____	____	____

YOU MAKE THE CONNECTION

1. How might pigs at a trough be an example of exploitation competition? of interference competition?
2. Explain why the population size of a species that competes by interference competition is often near the carrying capacity, whereas the population size of a species that competes by exploitation competition is often greater than or below the carrying capacity.
3. A female elephant bears a single offspring every two to four years. Based on this information, which survivorship curve do you think is representative of elephants? Explain your answer.
4. In Bolivia, 40% of the population is younger than age 15, and 4% is older than 65. In Austria, 17% of the population is younger than 15, and 15% is older than 65. Which country will have the highest growth rate over the next two decades? Why?

RECOMMENDED READINGS

Bongaarts, J. "Demographic Consequences of Declining Fertility." *Science,* Vol. 282, 16 Oct. 1998. A comprehensive overview of trends in fertility rates in different regions of the world.

Cohen, J.E. *How Many People Can the Earth Support?* W.W. Norton & Company, New York, 1995. An excellent reference on carrying capacity and population.

Ehrlich, P.R., and A.H. Ehrlich. *The Population Explosion*. Simon & Schuster, New York, 1990. Although the population data reported in this book are outdated, the overall message—that overpopulation is our number one environmental problem—is still timely.

Knodel, J. "Deconstructing Population Momentum." *Population Today,* Vol. 27, Mar. 1999. An explanation of how population growth momentum affects population dynamics.

Moore, P.D. "Feeding Patterns on Forest Floors." *Nature,* Vol. 390, 20 Nov. 1997. This article summarizes two reports from the *Journal of Ecology* on the subtle effects of two rainforest mammals (red howler monkeys and tapirs) on the dispersion patterns of certain rainforest plants.

Myers, N. "Consumption: Challenge to Sustainable Development." *Science,* Vol. 276, 4 Apr. 1997. This eminent scholar from Oxford University addresses the issue of overconsumption.

Raven, P.H., and L.R. Berg. *Environment,* 3rd ed. Harcourt College Publishers, Philadelphia, 2001. This environmental science textbook offers a detailed situation analysis of planet Earth, including how humans interact with and affect its physical and living systems.

Ricklefs, R.E., and G.L. Miller. *Ecology,* 4th ed. W.H. Freeman & Company, New York, 1999. This textbook presents the important themes of ecology, such as energy flow, population and community interactions, and mathematical models of ecology, in an accessible manner.

Spiller, D.A., J.B. Losos, and T.W. Schoener. "Impact of a Catastrophic Hurricane on Island Populations." *Science,* Vol. 281, 31 Jul 1998. This fascinating account extends the research on lizard and spider populations on Caribbean islands, which was described in the chapter, to include density independence.

Tuljapurkar, S. "Taking the Measure of Uncertainty." *Nature,* Vol. 387, 19 Jun. 1997. A discussion of the strengths and shortcomings of the methods used by demographers to forecast population.

2001 World Population Data Sheet, Population Reference Bureau, Washington, D.C., 2001. A chart that provides current population data for all countries. Includes birth rates, death rates, infant mortality rates, total fertility rates, and life expectancies, as well as other pertinent information. Visit **www.prb.org,** or call 1-800-877-9881 for details.

- Visit our Web site at **http://www.info.brookscole.com/solomonbergmartin** for links to chapter-related resources on the World Wide Web. Additional on-line materials relating to this chapter can also be found on our Web site.

See chapter activity on BioActive Learner CD for additional help in mastering the chapter's material. Icon location in the chapter's margins shows which topics have tutorials or simulations in the CD.

52

Community Ecology

Community in a rotting log. Such fallen logs, called "nurse logs," shelter plants and other organisms and enrich the soil as they decay. Photographed in Pennsylvania. *(Michael P. Gadomski/Photo Researchers, Inc.)*

LEARNING OBJECTIVES

After you have studied this chapter you should be able to

1. Characterize a community and distinguish between a community and an ecosystem.
2. Define ecological niche, distinguish between an organism's fundamental niche and its realized niche, and give several examples of limiting resources that might affect an organism's ecological niche.
3. Define competition and distinguish between interspecific and intraspecific competition.
4. Summarize the concepts of competitive exclusion, resource partitioning, and character displacement.
5. Define predation and describe the effects of natural selection on predator-prey relationships.
6. Distinguish among mutualism, commensalism, and parasitism, and give examples of each.
7. Summarize the main determinants of species richness in a community, and relate species richness to community stability.
8. Define ecological succession and distinguish between primary and secondary succession.
9. Describe the intermediate disturbance hypothesis.
10. Discuss the two traditional views of the nature of communities, Clements's organismic model and Gleason's individualistic model.

In Chapter 51 we examined the dynamics of single populations, including the ways individual populations change and what factors affect those changes. In the natural world, however, species do not exist as isolated populations. Rather, most populations are the interacting parts of a complex **community,** which consists of populations of different species that live and interact in the same place at the same time.

Communities exhibit characteristic properties that populations lack. These properties, known collectively as *community structure* and *community functioning*, include the number and types of species present, the relative abundance of each species, the interactions among different species, community resilience to disturbances, energy flow throughout the community, and productivity. **Community ecology** is the description and analysis of patterns and processes within the community. Finding common patterns and processes in a wide variety of communities—for example, a pond community, a pine forest community, and a sagebrush desert community—helps ecologists understand community structure and functioning.

Communities are quite complex and therefore exceedingly difficult to study because the large numbers of organisms of many different species interact with one another and are interdependent in a variety of ways. Species compete with one another for food, water, living space, and other resources. (Used in this context, a *resource* is anything from the environment that meets a particular species' needs.) Some organisms kill and eat other organisms. Some species form intimate associations with one another, whereas other species seem only distantly connected. Each organism plays one of three main roles in community life: producer, consumer, or decomposer. Unraveling the many positive and negative, direct and indirect interactions of organisms living together as a community is one of the goals of community ecologists.

Communities vary greatly in size, lack precise boundaries, and are rarely completely isolated. They interact with and influence one another in countless ways, even when the interaction is not readily apparent. Furthermore, there are communities nested within larger communities. A forest is a community, but so is a rotting log in that forest *(see photograph)*. Insects, plants, and fungi invade a fallen tree as it undergoes a series of decay steps. First, wood-boring insects and termites forge tunnels through the bark and wood. Later, other insects, plant roots, and fungi follow and enlarge these openings. Mosses and lichens that establish on the log's surface trap rainwater and extract nutrient minerals, and fungi and bacteria speed decay, thus providing nutrients for other inhabitants. As decay progresses, small mammals burrow into the wood and eat the fungi, insects, and plants.

Organisms exist in an abiotic (nonliving) environment that is as essential to their lives as their interactions with one another. Minerals, air, water, and sunlight are just as much a part of a honeybee's environment, for example, as the flowers that it pollinates and from which it takes nectar. A biological community and its abiotic environment together compose an **ecosystem.**

In this chapter we present some of the questions and challenges that community ecologists face in trying to find common patterns and processes that govern the assembly and persistence of interacting species. Although the living community is emphasized in this chapter, communities and their abiotic environments are inseparably linked. The abiotic components of ecosystems, including energy flow and trophic structure (feeding relationships), nutrient cycling, and climate, are considered in Chapter 53.

COMMUNITIES CONTAIN AUTOTROPHS AND HETEROTROPHS

Sunlight is the source of energy that powers almost all life processes. **Primary producers,** also called *autotrophs* or, simply, *producers,* are organisms that make complex organic molecules from simple inorganic substances (generally carbon dioxide and water), usually using the energy of sunlight. In other words, most producers perform the process of photosynthesis. By incorporating the chemicals they manufacture into their own **biomass** (living material), producers become potential food resources for other organisms. While plants are the most significant producers on land, algae and cyanobacteria are important producers in aquatic environments.

All other organisms in a community are *heterotrophs,* which extract energy from organic molecules produced by other organisms. **Consumers** are heterotrophs that obtain energy and body-building materials by feeding on organic molecules formed by other organisms. Consumers that eat producers are called **primary consumers** or **herbivores** (plant eaters). Deer and grasshoppers are examples of primary consumers. **Secondary consumers** eat primary consumers, whereas **tertiary consumers** eat secondary consumers. Both secondary and tertiary consumers are flesh-eating **carnivores** that eat other animals. Lions and spiders are examples of carnivores. Other consumers, called **omnivores,** eat a variety of organisms, both plant and animal. Bears, pigs, and humans are examples of omnivores.

Some consumers, called **detritus feeders,** or **detritivores,** consume **detritus,** which is organic matter that includes freshly dead or decomposing animal carcasses, leaf litter, and feces. Detritus feeders, such as snails, crabs, clams, and worms, are especially abundant in aquatic environments, where they burrow in the bottom muck and consume the organic matter that collects there. Earthworms are terrestrial detritus feeders, as are termites, certain beetles, snails, mites, and millipedes. Detritus feeders work together with microbial decomposers to destroy dead organisms and waste products. An earthworm, for example, actually eats its way through the soil, digesting much of the organic matter contained there.

Many consumers do not fit readily into a single category of herbivore, carnivore, omnivore, or detritivore. To some degree these organisms modify their food preferences as the need arises; if a consumer is faced with declining levels of one food species, for example, it may switch to other alternatives. Some organisms change food preferences over their lifetime. Tadpoles, which eat algae and plant material, are primary consumers, but adult frogs, which eat earthworms, snails, insects, crayfish, fish, other frogs, and tadpoles, are secondary consumers.

Decomposers (also called **saprotrophs**) include microbial heterotrophs that break down organic material and use the decomposition products to supply themselves with energy. They typically release simple inorganic molecules, such as carbon dioxide and mineral salts, that can then be reused by producers. Most bacteria and fungi are important decomposers. For example, sugar-metabolizing fungi that consume simple carbohydrates, such as glucose and maltose, are the first decomposers to invade dead wood. When these carbohydrates are exhausted in dead wood, other fungi, often aided by termites (with symbiotic zooflagellates in their guts; see Fig. 24–5*a*) complete the digestion of the wood by breaking down cellulose, a complex carbohydrate that is the main component of wood.

IN A COMMUNITY, POPULATIONS INTERACT IN A VARIETY OF WAYS

During the late 1990s, an intricate and fascinating relationship emerged among acorn production, white-footed mice, gypsy moth population growth, and the potential occurrence of Lyme disease in humans (Fig. 52–1). Large-scale experiments conducted in oak forests of the northeastern United States linked bumper acorn crops, which occur every three to four years, to booming mouse populations (the mice eat the acorns). Because the mice also eat gypsy moth pupae, high acorn conditions also lead to low populations of gypsy moths. This outcome helps the oaks, because gypsy moths cause serious defoliation. However, abundant acorns also attract tick-bearing deer. The ticks' hungry offspring feed on the mice, which often carry the Lyme disease-causing bacterium. The bacterium infects the maturing ticks and, in turn, spreads to humans that are bitten by affected ticks. Thus, it may be possible to predict which years pose the greatest potential threat of Lyme disease to humans, based on when oaks are most productive.

Within a community, no species exists independently of other species. As the preceding example shows, the populations of a community interact with and affect one another in complex

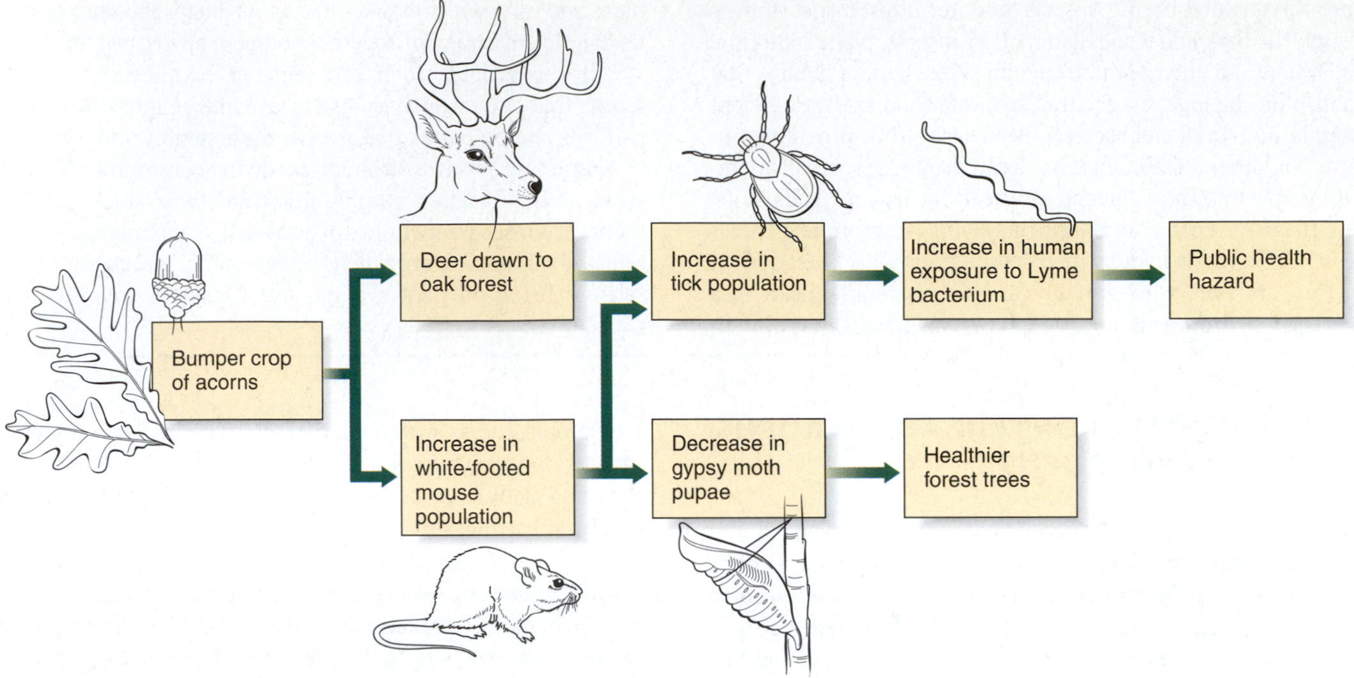

Figure 52-1 Connections to the size of an acorn crop. When there is a bumper crop of acorns, more mice survive and breed in winter, and more deer are attracted to oak forests. Both mice and deer are hosts of ticks that may carry the Lyme disease bacterium to humans. Abundant mice also reduce gypsy moth populations, thereby improving the health of forest trees.

ways that are not always obvious. Three main types of interactions occur among species in a community: competition, predation, and symbiosis. Before we can address these interactions, however, we need to examine the way of life of a given species in its community.

THE NICHE IS A SPECIES' ECOLOGICAL ROLE IN THE COMMUNITY

Every species is thought to have its own ecological role within the structure and function of a community; we call this role its **ecological niche.** Although the concept of ecological niche has been in use in ecology since early in the 20th century, G.E. Hutchinson first described, in 1957, the multidimensional nature of the niche that is accepted today. An ecological niche takes into account all biotic and abiotic aspects of the species' existence, that is, all physical, chemical, and biological factors that the species needs to survive, remain healthy, and reproduce. Among other things, the niche includes the local environment in which a species lives—that is, its **habitat.** A niche also encompasses what a species eats, what eats it, what organisms it competes with, and how it interacts with and is influenced by the abiotic components of its environment, such as light, temperature, and moisture. The niche, then, represents the totality of adaptations by a species to its environment, its use of resources, and the lifestyle to which it is suited. Although a complete description of an organism's ecological niche involves numerous dimensions and is difficult to define precisely, ecologists usually confine their studies to one or a few niche variables, such as feeding behaviors or ability to tolerate temperature extremes.

The ecological niche of a species may be far broader in theory than in actuality. A species is usually capable of using much more of its environment's resources or of living in a wider assortment of habitats than it actually does. The potential ecological niche of a species is its **fundamental niche,** but various factors such as competition with other species may exclude it from part of this fundamental niche. Thus, the lifestyle that a species actually pursues and the resources that it actually uses make up its **realized niche.**

An example may help to make this distinction clear. The green anole, a lizard native to Florida and other southeastern states, perches on trees, shrubs, walls, or fences during the day waiting for insect and spider prey (Fig. 52-2*a*). In the past these little lizards were widespread in Florida. Several years ago, however, a related species, the brown anole, was introduced from Cuba into southern Florida and quickly became common (Fig. 52-2*b*). Suddenly the green anoles became rare, apparently driven out of their habitat by competition from the slightly larger brown anoles. Careful investigation disclosed, however, that green anoles were still around. They were now confined largely to the vegetation in wetlands and to the foliated crowns of trees, where they were less easily seen.

Figure 52–2 Effect of competition on an organism's realized niche. **(a)** The green anole *(Anolis carolinensis)* is the only anole species native to North America. Males are about 12.5 cm (5 in) long; females are slightly smaller. **(b)** The 15.2-cm (6-in) long brown anole *(Anolis sagrei)* was introduced to Florida. **(c, d)** Positions of the two species along a single niche dimension (in this case, habitat). Species 1 represents the green anole, and Species 2 represents the brown anole. **(c)** The fundamental niches of the two lizards overlap. **(d)** The brown anole outcompetes the green anole in the area where their niches overlap, restricting the niche of the green anole. *(a, Ed Kanze/Dembinsky Photo Associates; b, Robert Clay/Visuals Unlimited)*

The habitat portion of the green anole's fundamental niche includes the trunks and crowns of trees, exterior house walls, and many other locations. Once the brown anoles became established in the green anole's habitat, the green anoles were driven from all but wetlands and tree crowns; as a result of competition between species, their realized niche became much smaller than it had been (Fig. 52–2*c*, *d*). Because communities consist of numerous species, many of which compete to some extent, the complex interactions among them produce each species' realized niche.

Limiting resources restrict the ecological niche of a species

A species' structural, physiological, and behavioral adaptations determine its tolerance for environmental extremes. If any feature of an environment lies outside the bounds of its tolerance, then the species cannot live there. Just as you would not expect to find a cactus living in a pond, you would not expect water lilies in a desert.

The environmental factors that actually determine a species' ecological niche can be extremely difficult to identify. For this reason, the concept of ecological niche is largely abstract, although some of its dimensions can be experimentally determined. Any environmental resource that, because it is scarce or unfavorable, tends to restrict the ecological niche of a species is called a **limiting resource.**

Most of the limiting resources that have been studied are simple variables such as the soil's mineral content, temperature extremes, and precipitation amounts. Such investigations have disclosed that any factor exceeding the tolerance of a species, or present in quantities smaller than the required minimum, limits the presence of that species in a community. By their interaction, such factors help define the ecological niche for a species.

Limiting resources may affect only part of an organism's life cycle. For example, although adult blue crabs can live in fresh or slightly brackish water, they cannot become permanently established in such areas because their larvae (immature forms) require salt water. Similarly, the ring-necked pheasant, a popular game bird native to Eurasia, has been widely introduced in North

America but has not become established in the southern United States. The adult birds do well, but the eggs cannot develop properly in the high southern temperatures.

Biotic and abiotic factors may influence a species' ecological niche

In the 1960s, ecologist Joseph H. Connell investigated biotic and abiotic factors that affect the distribution of two barnacle species in the rocky intertidal zone along the coast of Scotland. The intertidal zone is a challenging environment, and organisms in the intertidal zone must be able to tolerate exposure to the drying air during low tides (see Chapter 54). Barnacles are sessile crustaceans whose bodies are covered by a shell of calcium carbonate (see Chapter 29). When the shell is open, feathery appendages extend to filter food from the water.

Along the coast of Scotland, adults of one barnacle species, *Balanus balanoides*, are attached on lower rocks in the intertidal zone than are adults of the other species, *Chthamalus stellatus* (Fig. 52–3). The distributions of the two species do not overlap, despite the fact that immature larvae of both species are found together in the intertidal zone. Connell manipulated the two populations to determine what factors were affecting their distribution. When he removed *Chthamalus* from the upper rocks, *Balanus* barnacles did not expand into the vacant area. Connell's experiments showed that *Chthamalus* is more resistant than *Balanus* to desiccation when the tide retreats. However, when Connell removed *Balanus* from the lower rocks, *Chthamalus* expanded into the lower parts of the intertidal zone. The two species compete for space, and *Balanus*, which is larger and grows faster, is able to outcompete the smaller *Chthamalus* barnacles.

Competition among barnacle species for space was one of the phenomena that Connell's research demonstrated. We now examine some of the principles of competition that Connell and other ecologists have revealed in both laboratory and field studies.

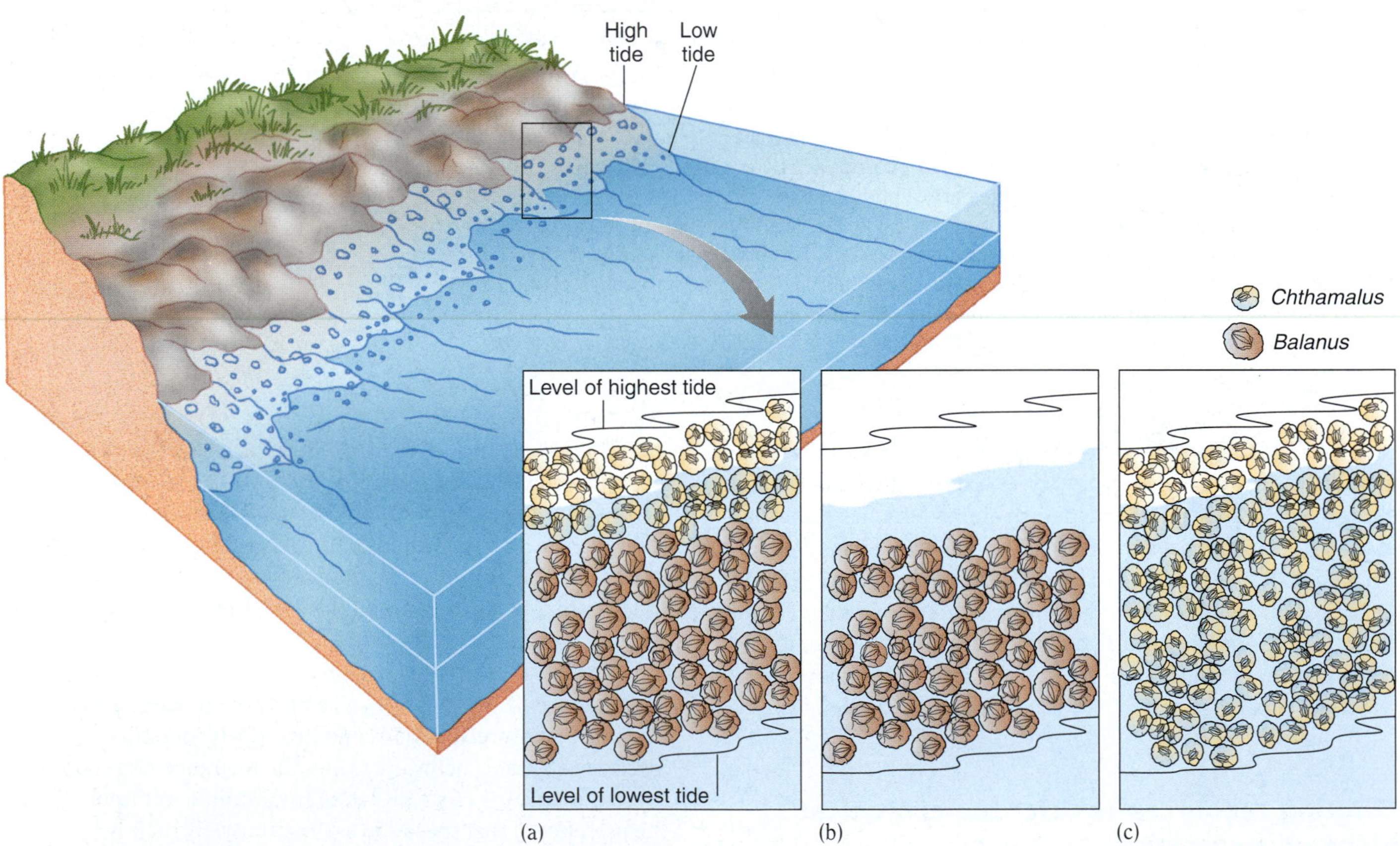

Figure 52-3 Abiotic and biotic factors that affect barnacle distribution. **(a)** Two species of barnacles, *Chthamalus* and *Balanus*, grow in the intertidal zone of a rocky shore in Scotland. **(b)** When *Chthamalus* individuals were experimentally removed, *Balanus* individuals did not expand into their section of the rock. **(c)** When *Balanaus* individuals were experimentally removed, *Chthamalus* individuals spread into the empty area. *(After Connell, J.H. "The Influence of Interspecific Competition and Other Factors on the Distribution of the Barnacle* Chthamalus stellatus." Ecology, *Vol. 42, 1961)*

COMPETITION MAY BE INTRASPECIFIC OR INTERSPECIFIC

Competition occurs when two or more individuals attempt to use the same essential resource, such as food, water, shelter, living space, or sunlight. Because resources are often in limited supply in the environment, their use by one individual decreases the amount available to others (Table 52–1). If a tree in a dense forest grows taller than surrounding trees, for example, it is able to absorb more of the incoming sunlight. Less sunlight is therefore available for nearby trees that are shaded by the taller tree. Competition can occur among individuals within a population **(intraspecific competition)** or between different species **(interspecific competition).** Intraspecific competition was discussed in Chapter 51.

Ecologists traditionally assumed that competition is the most important determinant of both the number of species found in a community as well as the size of each population. Today ecologists recognize that competition is only one of many interacting biotic and abiotic factors that affect community structure. Furthermore, competition is not always a straightforward, direct interaction. A variety of flowering plants, for example, live in a young pine forest and presumably compete with the conifers for such resources as soil moisture and soil nutrient minerals. Their relationship, however, is more complex than simple competition. The flowers produce nectar that is consumed by some insect species that also prey on needle-eating insects, thereby reducing the number of insects feeding on pines. It is therefore difficult to assess the overall effect of flowering plants on pines. If the flowering plants were removed from the community, would the pines grow faster because they were no longer competing for necessary resources? Or would pine growth be inhibited by the increased presence of needle-eating insects (caused by fewer predatory insects)?

Short-term experiments in which one competing plant species is removed from a forest community often have demonstrated an improved growth for the remaining species. However, very few studies have tested the long-term effects on forest species of the removal of single competing species. These long-term effects may be subtle, indirect, and difficult to ascertain; they may lessen or negate the short-term effects of competition for resources.

Competition between species with overlapping niches may lead to competitive exclusion

When two species are similar, as are the green and brown anoles or the two species of barnacles, their fundamental niches may overlap. However, based on experimental and modeling work, many ecologists think that no two species can indefinitely occupy the same niche in the same community because competitive exclusion eventually occurs. In **competitive exclusion,** it is hypothesized that one species excludes another from its niche as a result of interspecific competition. Although it is possible for different species to compete for some necessary resource without being total competitors, two species with absolutely identical ecological niches cannot coexist. Coexistence can occur, however, if the overlap between the two niches is reduced. In the lizard example, direct competition between the two species was reduced as the brown anole excluded the green anole from most of its former habitat.

The initial evidence that interspecific competition contributes to a species' realized niche came from a series of laboratory experiments by the Russian biologist Georgyi F. Gause in the 1930s (see descriptions of other experiments by Gause in Chapter 51). In one study Gause grew populations of two species of protozoa, *Paramecium aurelia* and the larger *Paramecium caudatum,* in controlled conditions (Fig. 52–4). When grown in separate test tubes, that is, in the absence of the second species, the population of each species quickly increased to a level imposed by the resources and remained there for some time thereafter. When grown together, however, only *P. aurelia* thrived, whereas *P. caudatum* dwindled and eventually died out. Under different sets of culture conditions, *P. caudatum* prevailed over *P. aurelia*. Gause interpreted this to mean that although one set of conditions favored one species, a different set favored the other. Nonetheless, because both species were so similar, in time one or the other would eventually triumph over the other. Similar experiments with competing species of fruit flies, mice, certain beetles, and annual plants have supported Gause's results: One species thrives, and the other eventually dies out.

Competition, then, has adverse effects on species that use a limited resource and may result in competitive exclusion of one

TABLE 52–1 Ecological Interactions among Species

Interaction	Effect on Species 1	Effect on Species 2
Predation of Species 2 by Species 1	Beneficial	Harmful
Symbiosis		
Mutualism of Species 1 and Species 2	Beneficial	Beneficial
Commensalism of Species 1 with Species 2	Beneficial	No effect
Parasitism by Species 1 on Species 2	Beneficial	Harmful
Competition between Species 1 and Species 2	Harmful	Harmful

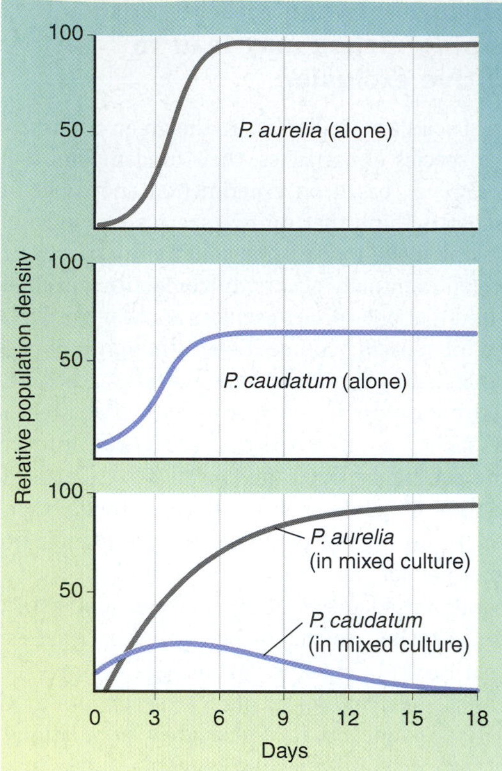

Figure 52–4 Interspecific competition. Competition between two species of *Paramecium* was studied by G.F. Gause. The top and middle graphs show how a population of each species of *Paramecium* flourishes when grown alone. The bottom graph shows how they grow together, in competition with each other. *(Adapted from Gause, G.F.* The Struggle for Existence. *Williams & Wilkins, Baltimore, 1934)*

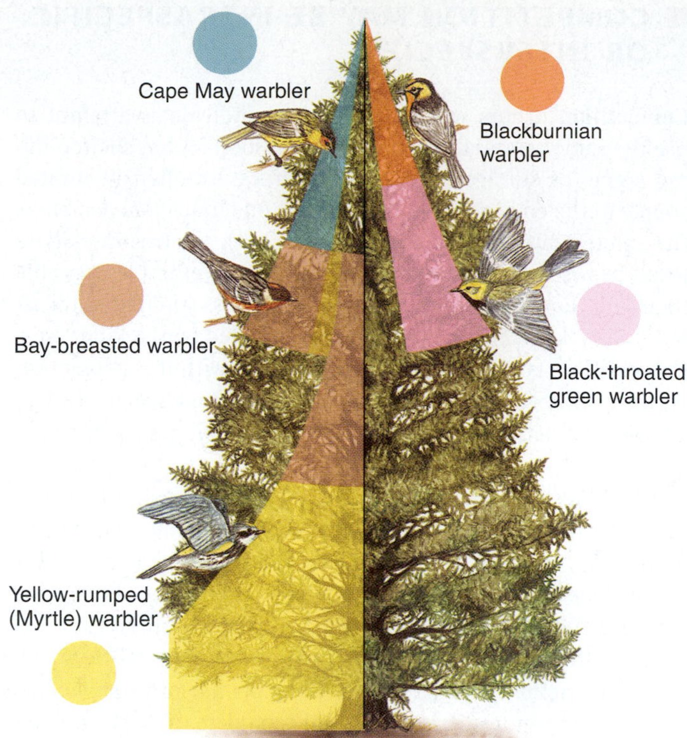

Figure 52–5 Resource partitioning. Each warbler species spends at least half its foraging time in its designated area of a spruce tree, thereby reducing the competition among warbler species. *(After MacArthur, R.H. "Population Ecology of Some Warblers of Northeastern Coniferous Forests."* Ecology, *Vol. 39, 1958)*

or more species. It therefore follows that over time natural selection should favor individuals of each species that avoid or, at least, reduce competition for environmental resources. Such reduction in competition among coexisting species as a result of each species' niche differing from the others in one or more ways is called **resource partitioning.** Resource partitioning is well documented in animals and includes studies in tropical forests of Central and South America that demonstrate little overlap in the diets of fruit-eating birds, primates, and bats that coexist in the same habitat. Although fruits are the primary food for several hundred bird, primate, and bat species, the wide variety of fruits available have allowed fruit eaters to specialize, thereby reducing competition.

Resource partitioning also may include timing of feeding, location of feeding, nest sites, and other aspects of a species' ecological niche. Robert MacArthur's study of five North American warbler species is a classic example of resource partitioning (Fig. 52–5). Although it initially appeared that their niches were nearly identical, MacArthur determined that individuals of each species spend most of their feeding time in different portions of the spruces and other conifer trees they frequent. They also move in different directions through the canopy, consume different combinations of insects, and nest at slightly different times.

Apparent contradictions to the competitive exclusion principle exist. In Florida, for example, native and introduced (nonnative) fish seem to coexist in identical niches. Similarly, botanists have observed closely competitive plant species in the same community. Although such situations seem to contradict the concept of competitive exclusion, the realized niches of these organisms may differ in some way that ecologists do not yet understand, as with the warblers before MacArthur studied them.

Character displacement is an adaptive consequence of interspecific competition

Sometimes populations of two similar species occur together in some locations and separately in others. Where their geographical distributions overlap, the two species tend to differ more in their structural, ecological, and behavioral characteristics than they do where each occurs in separate geographical areas. Such divergence in traits in two similar species living in the same geographical area is known as **character displacement.** It is thought that character displacement reduces competition between two species, since their differences give them somewhat different ecological niches in the same environment.

There are several well-documented examples of character displacement between two closely related species. For example, the flowers of two species of *Solanum* in Mexico are quite simi-

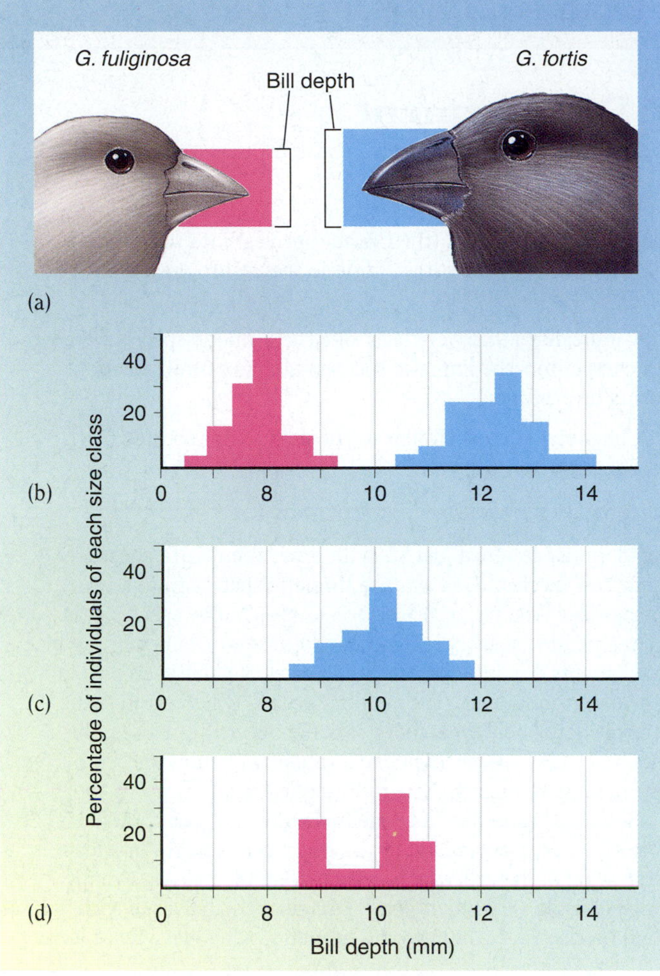

Figure 52–6 Character displacement. (a) Bill depth in two closely related species of Galapagos Island finches, *Geospiza fuliginosa* and *Geospiza fortis*. (b) Where the two species are found on the same island, *G. fuliginosa (pink)* has a smaller average bill depth than *G. fortis (blue)*. Where they occur on separate islands (c, d), the average bill depths of each are similar. *(After Lack, D.* Darwin's Finches. *Cambridge University Press, Cambridge, 1947)*

lar in areas where either one or the other occurs, but in areas where their distributions overlap, the two species differ significantly in flower size and are pollinated by different kinds of bees. In other words, character displacement reduces interspecific competition, in this case for the same animal pollinator.

The bill sizes of Darwin's finches provide another example of character displacement (Fig. 52–6). On large islands in the Galapagos where the medium ground finch *(Geospiza fortis)* and the small ground finch *(Geospiza fuliginosa)* occur together, their bill depths are distinctive: *G. fuliginosa* has a smaller bill depth that enables it to crack small seeds, whereas *G. fortis* has a larger bill depth that enables it to crack medium-sized seeds. However, *G. fortis* and *G. fuliginosa* are also found on separate islands. Where the two species live separately, bill depths tend to be the same intermediate size, perhaps because there is no competition from the other species.

Although these examples of the coexistence of similar species may be explained in terms of character displacement, the character displacement hypothesis has only been demonstrated in field experiments in a few instances (see *On the Cutting Edge: Experimental Evidence of Character Displacement*).

NATURAL SELECTION SHAPES THE BODY FORMS AND BEHAVIORS OF BOTH PREDATOR AND PREY

Predation is the consumption of one species, the *prey,* by another, the *predator* (see Table 52–1). Predation includes animals eating other animals, as well as animals eating plants *(herbivory)*. Predation has resulted in an evolutionary "arms race," with the evolution of predator strategies—more efficient ways to catch prey—and prey strategies—better ways to escape the predator. A predator that is more efficient at catching prey exerts a strong selective force on its prey, which over time may evolve some sort of countermeasure that reduces the probability of being captured. The countermeasure acquired by the prey in turn acts as a strong selective force on the predator. This type of interdependent evolution of two interacting species is known as **coevolution** (see Chapter 35).

We now consider several adaptations that are related to predator-prey interactions. These include predator strategies (pursuit and ambush) and prey strategies (plant defenses and animal defenses). Keep in mind as you read these descriptions that such strategies are not "chosen" by the respective predators or prey. New traits arise randomly in a population as a result of genetic changes. Some new traits may be beneficial, whereas others may be harmful or have no effect. Beneficial strategies, or traits, persist in a population because such characteristics make the individuals that possess them well suited to thrive and reproduce. In contrast, characteristics that make the individuals that possess them poorly suited to their environment tend to disappear in a population.

Pursuit and ambush are two predator strategies

A brown pelican sights its prey—a fish—while in flight. Less than 2 seconds after diving into the water at a speed as great as 72 km/h (45 mph), it has its catch. Orcas (formerly known as killer whales), which hunt in packs, often herd salmon or tuna into a cove so that they are easier to catch. Any trait that increases hunting efficiency, such as the speed of brown pelicans or the intelligence of orcas, favors predators that pursue their prey. Because these carnivores must be able to process information quickly during the pursuit of prey, their brains are generally larger, relative to body size, than those of the prey they pursue.

Ambush is another effective way to catch prey. The yellow crab spider, for example, is the same color as the white or yellow flowers on which it hides (Fig. 52–7). This camouflage keeps unwary insects that visit the flower for nectar from noticing the spider until it is too late. Predators that are able to *attract* prey are particularly effective at ambushing. For example, a diverse group

Process of Science

ON THE CUTTING EDGE

Experimental Evidence of Character Displacement

HYPOTHESIS: Interspecific competition in stickleback fish promotes character displacement.

METHOD: Place a species of stickleback fish that includes a range of phenotypes (the Cranby species) into two divided experimental ponds. Add individuals of a potential competitor species (the limnetic species) to one separated half of each pond.

RESULTS: In the presence of the limnetic species, the Cranby individuals that were least like the limnetic species were more likely to survive and grow. That is, adding a competitor (the limnetic species) alters natural selection on the other species (the Cranby species) by favoring divergence.

CONCLUSION: Evidence was obtained of natural selection against phenotypes most similar to the competitor species. Such natural selection is consistent with the hypothesis of character displacement.

There are many examples in nature of similar species that are markedly different where they are found together, but very alike (yet distinguishable, at least by experts) where each is found alone. For example, two species of threespine stickleback fish, a limnetic species and a benthic species, occur in several lakes of coastal British Columbia. (*Limnetic* organisms live in the open water of a lake; *benthic* organisms live on the bottom.) Where they are found together, individuals belonging to the limnetic species are small, have narrow mouths, and eat zooplankton in the open water. Individuals belonging to the benthic species, on the other hand, are large, have wide mouths, and feed on invertebrates in the sediments or clinging to vegetation along the shoreline. In lakes containing a single stickleback species, the fish are "generalists": They are intermediate in form and eat both plankton and invertebrates.

The hypothesis of character displacement, which has been advanced to explain these observations in threespine sticklebacks and many other organisms, had never been experimentally tested until 1994. Dolph Schluter, a researcher at the University of British Columbia who had previously made extensive observations on natural threespine stickleback populations, designed just such an experiment.* He chose to study two species.† One was the limnetic species from a two-species lake. The other, referred to as the Cranby species, was from a nearby lake where it occurs alone. Prior to beginning his experiments, Schluter performed a series of genetic crosses in the laboratory to make the Cranby species, which exhibits the intermediate phenotype, more variable. After the crosses, the captive Cranby species included a range of phenotypes, from benthic-like to intermediate to limnetic-like.

Schluter began his experiments by dividing two artificial ponds into separate halves. (Two ponds were used to provide a replication of the experiment.) He placed 1800 young Cranby individuals into each half-pond. To one of the halves of each pond he also added 1200 young limnetic individuals; these were

* Schluter, D. "Experimental Evidence that Competition Promotes Divergence in Adaptive Radiation." *Science,* Vol. 266, 4 Nov. 1994.

† Three spine stickleback fish were once thought to be a single species, *Gasterosteus aculeatus*. Although several species are now known to exist, all are currently lumped together in the *G. aculeatus* complex because their taxonomic relationships have not been determined.

of deep-sea fishes called anglerfish possess rodlike luminescent lures close to their mouths to attract prey.

Chemical protection is an effective plant defense against herbivores

Plants cannot escape predators by fleeing, but they possess a number of adaptations that protect them from being eaten. The

Figure 52–7 Ambush. A yellow crab spider *(Misumena vatia)* blends into its surroundings, waiting for an unwary insect to visit the flower. An effective predator strategy, ambush relies on surprising the prey. *(Carmela Leszczynski/Animals Animals)*

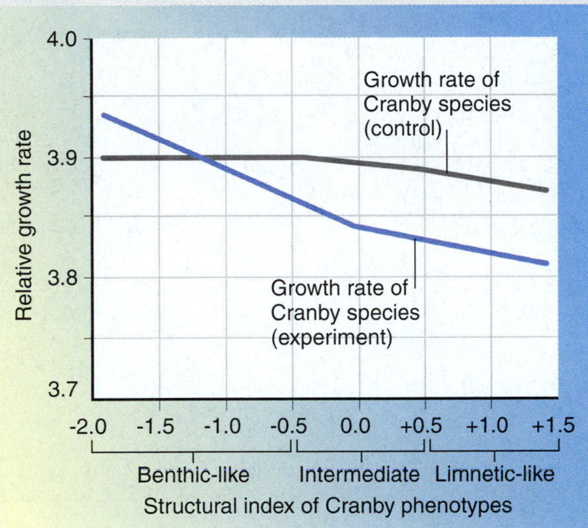

Experimental evidence of character displacement. A comparison of growth rates for the Cranby species alone *(control),* and in the presence of the limnetic species *(experiment).* The structural index reflects the range of Cranby phenotypes, from those most benthic-like (−2.0) to those most limnetic-like (+1.5). Values for relative growth rates were obtained statistically. *(Courtesy of Dolph Schluter)*

the experimental halves. The half-ponds with only Cranby individuals were the control halves.

After 90 days, the fish were harvested and preserved. About 65 randomly chosen Cranby individuals from each group were sized (measured for length), and their phenotypes were noted. There were no significant differences in survival and growth over the range of phenotypes of Cranby individuals when they were grown alone. However, in the presence of the limnetic species, Cranby individuals with a limnetic-like phenotype exhibited significantly lower rates of growth and survival than did the benthic-like and intermediate phenotypes *(see figure)*.

Because only one generation was studied, this experiment investigated only natural selection, the first step in the process of character displacement. The next step, an evolutionary change in gene frequencies over successive generations, will require a long-term evaluation of several generations.

This experiment was both hailed as a landmark and criticized for its design and analysis. For example, some ecologists have argued that the experimental pond sides had much higher population densities (1800 plus 1200 fish) than the control pond sides (only 1800 fish) and that this difference might have affected the results.

During the summer of 1996, Schluter modified the stickleback experiment.‡ He placed the Cranby species (with its range of phenotypes) into both sides of three divided ponds. He then added a limnetic species to one side of each pond and a benthic species to the other side. The total fish densities were the same on both sides at the beginning of the experiment.

After 90 days, the fish were harvested. In the pond side to which the limnetic species had been added, the growth rate of benthic-like phenotypes of the Cranby species was elevated, whereas the growth rate of the limnetic-like phenotypes was depressed. Conversely, in the pond side to which the benthic species had been added, the growth rate of the limneticlike phenotypes was elevated, and the growth rate of the benthic-like phenotypes was depressed.

Further refinements of the experimental design, including studies on multiple generations, are planned and should raise many interesting new questions about character displacement.

‡ Schluter, D. "Ecological Character Displacement in Adaptive Radiation." *American Naturalist*, Vol. 156 Supplement, Oct. 2000.

presence of spines, thorns, tough leathery leaves, or even thick wax on leaves discourages foraging herbivores from grazing. Other plants produce an array of protective chemicals that are unpalatable or even toxic to herbivores. The active ingredients in such plants as marijuana and tobacco affect hormone activity or nerve, muscle, liver, or kidney functions and may discourage foraging by herbivores. Interestingly, many of the chemical defenses in plants are useful to humans. India's neem tree, for example, contains valuable chemicals that are effective against more than 100 species of herbivorous insects, mites, and nematodes. Nicotine from tobacco, pyrethrum from chrysanthemum, and rotenone from the derris plant are other examples of chemicals extracted and used as insecticides. Such plant-derived pesticides are called *botanicals*.

Milkweeds are an excellent example of the biochemical coevolution between plants and herbivores (Fig. 52–8). Milkweeds produce alkaloids and cardiac glycosides, chemicals that are poisonous to all animals except for a small group of insects. During the course of evolution, these insects acquired the ability to either tolerate or metabolize the milkweed toxins. As a result, they can eat milkweeds without being poisoned. These insects avoid competition from other herbivorous insects because few others are able to tolerate milkweed toxins. Predators also learn to avoid these insects, which accumulate the toxins in their tissues and therefore become toxic themselves. The black, white, and yellow coloration of the monarch caterpillar, a milkweed feeder, clearly announces its toxicity to predators that have learned to associate bright colors with illness. Conspicuous colors or patterns, which advertise a species' unpalatability to potential predators, are known as **aposematic coloration** (ap"uh-suh-mat'ik; from the Greek *apo*, away, and *semat*, a mark or sign), or **warning coloration.**

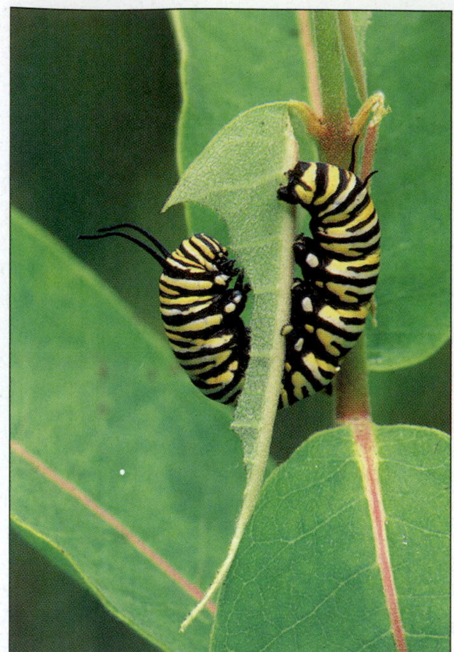

Figure 52–8 Chemical plant defenses. The common milkweed *(Asclepias syriaca)* is protected by its toxic chemicals. Its leaves are poisonous to most herbivores except monarch caterpillars *(Danaus plexippus, shown)* and a few other insects. Monarch caterpillars, which grow to 7 cm (2.75 in), have bright aposematic coloration. Photographed in Michigan. *(Rod Planck/Photo Researchers, Inc.)*

Figure 52–9 Chemical animal defenses. The poison arrow frog advertises its poisonous nature with its conspicuous aposematic coloring, warning away would-be predators. Research suggests that poison arrow frogs obtain the chemicals that make them poisonous from the ants, beetles, and other arthropods they eat. Poison arrow frogs kept in laboratory settings and fed controlled diets do not produce toxins. This poison arrow frog *(Dendrobates tinctorius)*, the smallest of the poison arrow frog species, is 4 cm (1.6 in) long. It is found in French Guiana. *(Michael Fogden/Animals Animals)*

Animals possess a variety of defensive adaptations to avoid predators

Many animals, such as prairie voles and woodchucks, flee from predators by rapidly running to their underground burrows. Others have mechanical defenses, such as the barbed quills of a porcupine and the shell of a pond turtle. Some animals live in groups—for example, a herd of antelope, colony of honeybees, school of anchovies, or flock of pigeons. Because a group has so many eyes, ears, and noses watching, listening, and smelling for predators, this social behavior decreases the likelihood of a predator catching any one of them unaware.

Chemical defenses are also common among animal prey. The South American poison arrow frog (*Dendrobates* sp.) has poison glands in its skin. Its bright aposematic coloration prompts avoidance by experienced predators (Fig. 52–9; also see Fig. 30–18*a*). Snakes and other animals that have tried once to eat a poisonous frog do not repeat their mistake! Other examples of aposematic coloration occur in the striped skunk, which sprays acrid chemicals from its anal glands, and the bombardier beetle, which spews harsh chemicals at potential predators (see Fig. 6–9).

Some animals have **cryptic coloration,** colors or markings that help them hide from predators by blending into their physical surroundings. Certain caterpillars resemble twigs so closely that you would never guess they were animals unless they moved. Pipefish are slender, green fish that are almost perfectly camouflaged in green eelgrass. The smooth-skinned leaf-tailed gecko, discovered in the late 1990s in southern Madagascar, resembles dead leaves not only in color but also in the pattern of leaf veins (Fig. 52–10). Such cryptic coloration has been preserved and accentuated by means of natural selection. These animals are less likely to be captured by predators and, therefore, more likely to live to maturity and produce offspring that also carry the genes for cryptic coloration.

Sometimes a defenseless species (a *mimic*) is protected from predation by its resemblance to a species that is dangerous in some way (a *model*). Such a strategy is known as **Batesian mimicry.** Many examples of this phenomenon exist. For example, a harmless scarlet kingsnake looks so much like a venomous coral snake that predators may avoid it (Fig. 52–11).

In **Müllerian mimicry,** different species, all of which are poisonous, harmful, or distasteful, resemble one another. Although their harmfulness protects them as individual species, their similarity in appearance works as an added advantage. Potential predators can more easily learn a single common aposematic coloration. Viceroy and monarch butterflies are thought to be an example of Müllerian mimicry (see *Focus On: Batesian Butterflies Disproved*).

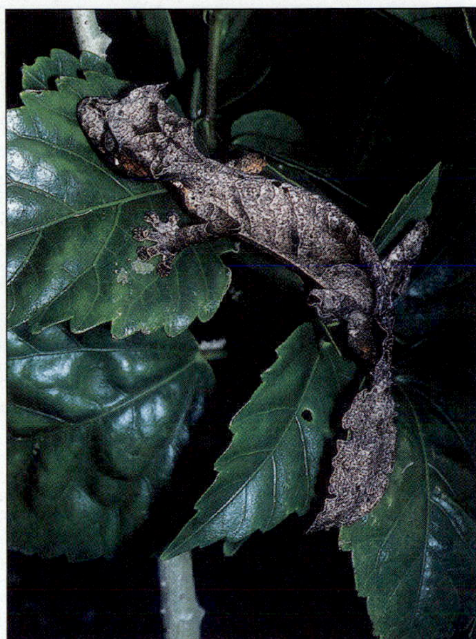

Figure 52–10 Cryptic coloration. The smooth-skinned leaf-tailed gecko *(Uroplatus malama)*, which is about 12 cm (4.7 in) long, hunts for insects by night and sleeps pressed against tree bark by day. It is virtually invisible when it sleeps. Photographed in a rain forest in southern Madagascar. *(Ronald A. Nussbaum/Museum of Zoology, University of Michigan)*

SYMBIOSIS INVOLVES A CLOSE ASSOCIATION BETWEEN SPECIES

Symbiosis is any intimate relationship or association between members of two or more species. Usually, symbiosis involves one species living on or in another species. The partners of a symbiotic relationship, called **symbionts,** may benefit from, be unaffected by, or be harmed by the relationship (see Table 52–1). The thousands, or perhaps even millions, of symbiotic associations all are products of coevolution. Symbiosis takes three forms: mutualism, commensalism, and parasitism.

In mutualism, benefits are shared

Mutualism is a symbiotic relationship in which both partners benefit. Mutualism may be either obligate (essential for the survival of both species) or facultative (when either partner can live alone under certain conditions). The association between **nitrogen-fixing bacteria** of the genus *Rhizobium* and legumes (plants such as peas, beans, and clover) is an example of mutualism (see Fig. 53–8*a*). Nitrogen-fixing bacteria, which live inside nodules on the roots of legumes, supply the plants with all the nitrogen they require to manufacture such nitrogen-containing compounds as chlorophylls, proteins, and nucleic acids. The legumes supply sugars and other energy-rich organic molecules to their bacterial symbionts.

Another example of mutualism is the association between reef-building animals and dinoflagellates called **zooxanthellae** (see Chapters 24, 28, and 54). These symbiotic algae live inside cells of the coral polyp (the coral forms a vacuole around the algal cell), where they photosynthesize and provide the animal with carbon and nitrogen compounds as well as oxygen (Fig. 52–12). Zooxanthellae have a stimulatory effect on the growth of corals, causing calcium carbonate skeletons to form around their bodies much faster when the algae are present. The corals, in turn, supply their zooxanthellae with waste products such as ammonia, which the algae use to make nitrogen compounds for both partners.

(a)

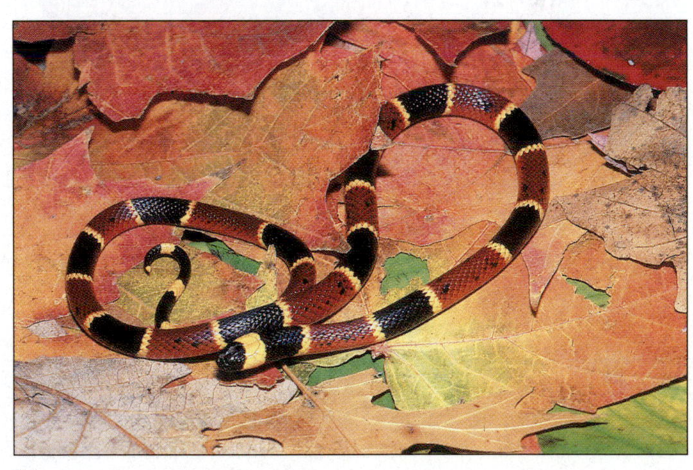

(b)

Figure 52–11 Batesian mimicry. In this example, the scarlet kingsnake *(Lampropeltis triangulum elapsoides)* **(a)** is the mimic and the eastern coral snake *(Micrurus fulvius fulvius)* **(b)** is the model. Note that the red and yellow warning colors touch on the coral snake but do not touch on the harmless mimic. Scarlet kingsnakes are 36 to 51 cm (14 to 20 in) in length, whereas eastern coral snakes are 51 to 76 cm (20 to 30 in). *(a, Suzanne L. and Joseph T. Collins/Photo Researchers, Inc.; b, Suzanne L. Collins/Photo Researchers, Inc.)*

FOCUS ON Batesian Butterflies Disproved

The monarch butterfly *(Danaus plexippus)* is an attractive insect found throughout much of North America *(see figure, left side)*. As a caterpillar, it feeds exclusively on milkweed leaves. The milky white liquid produced by the milkweed plant contains poisons that apparently do not harm the insect but remain in its tissues for life. When a young bird encounters and tries to eat its first monarch butterfly, it sickens and vomits. Thereafter, the bird avoids eating the distinctively marked insect.

Many people confuse the viceroy butterfly *(Limenitis archippus, see figure, right side)* with the monarch. The viceroy, which is found throughout most of North America, is approximately the same size, and the color and markings of its wings are almost identical to those of the monarch. As caterpillars, viceroys eat willow and poplar leaves, which apparently do not contain poisonous substances.

During the past century, it was thought that the viceroy butterfly was a tasty food for birds but that its close resemblance to monarchs gave it some protection against being eaten. In other words, birds that had learned to associate the distinctive markings and coloration of the monarch butterfly with its bad taste tended to avoid viceroys because they were similarly marked. The viceroy butterfly was therefore considered a classic example of Batesian mimicry.

In the journal *Nature* in 1991, ecologists David Ritland and Lincoln Brower of the University of Florida reported the results of an experiment that tested the long-held notion that birds like the taste of viceroys but avoid eating them because of their resemblance to monarchs. They removed the wings of different kinds of butterflies—monarchs, viceroys, and several tasty species—and fed the seemingly identical wingless bodies to red-winged blackbirds. The results were surprising: Monarchs and viceroys were equally distasteful to the birds.

As a result of this work, ecologists are reevaluating the evolutionary significance of different types of mimicry. Rather than being an example of Batesian mimicry, monarchs and viceroys now appear to be an example of Müllerian mimicry, in which two or more different species that are distasteful or poisonous have come to resemble one another during the course of evolution. This likeness provides an adaptive advantage because predators learn quickly to avoid all butterflies with the coloration and markings of monarchs and viceroys. As a result, fewer butterflies of either species die, and more individuals survive to reproduce.

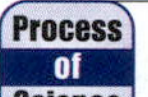

The butterfly study provides us with a useful reminder about the process of science. Expansion of knowledge in science is an ongoing enterprise, and newly acquired evidence helps scientists to reevaluate current models or ideas. Thus, scientific knowledge and understanding is not static, but continually changing.

Müllerian mimicry. Monarch *(left)* and viceroy *(right)* butterflies may be an example of Müllerian mimicry, in which two or more poisonous, harmful, or distasteful organisms resemble each other. *(Thomas C. Emmel)*

Mycorrhizae are mutualistic associations between fungi and the roots of plants. The association is common: About 80% of all plant species are thought to have mycorrhizae. The fungus, which grows around and into the root as well as into the surrounding soil, absorbs essential nutrient minerals, especially phosphorus, from the soil and provides them to the plant. In return, the plant provides the fungus with organic molecules produced by photosynthesis. Plants grow more vigorously in the presence of mycorrhizal fungi (see Figs. 25–15 and 34–10), and they are better able to tolerate environmental stressors such as drought and high soil temperatures. Indeed, some plants cannot maintain themselves under natural conditions if the fungi with which they normally form mycorrhizae are not present.

Commensalism is taking without harming

Commensalism is a type of symbiosis in which one species benefits and the other one is neither harmed nor helped. One example of commensalism is the relationship between social insects and scavengers, such as mites, beetles, or millipedes, that live in their hosts' nests. Certain types of silverfish, for example, move along in permanent association with marching columns of army ants and share the food caught in their raids. The army ants derive no apparent benefit or harm from the silverfish.

Another example of commensalism is the relationship between a host tree and its epiphytes, which are smaller plants, such as orchids, ferns, or Spanish moss, attached to the host's

Figure 52–12 Mutualism. The greenish-brown specks in these polyps of stony coral (*Pocillopora* sp.) are zooxanthellae, algae that live symbiotically within the coral's translucent cells and supply the coral with carbon and nitrogen compounds. In return, the coral provides its zooxanthellae with nitrogen in the form of ammonia. *(P. Parks-OSF/Animals Animals)*

branches (Fig. 52–13). The epiphyte anchors itself to the tree but does not obtain nutrients or water directly from it. Living on the tree enables it to obtain adequate light, water (as rainfall dripping down the branches), and required minerals (washed out of the tree's leaves by rainfall). Thus, the epiphyte benefits from the association, whereas the tree is apparently unaffected. (Epiphytes can harm their host, however, if they are present in a large enough number; in this instance, the relationship is no longer commensalism.)

Parasitism is taking at another's expense

Parasitism is a symbiotic relationship in which one member, the *parasite,* benefits, while the other, the *host,* is adversely affected. The parasite obtains nourishment from its host. A parasite rarely kills its host directly but may weaken it, rendering it more vulnerable to predators, competitors, or abiotic stressors. Some parasites, such as ticks, live outside the host's body; other parasites, such as tapeworms, live within the host. Parasitism is a successful lifestyle; by one estimate, more than two-thirds of all species are parasites, and more than 100 parasites live in or on the human species alone!

Since the 1980s, wild and domestic honeybees in the United States have been dying off. Although habitat loss and pesticide use have contributed to the problem, tracheal mites (Fig. 52–14) and larger varroa mites have been a major reason for the honeybee decline. The number of commercial colonies has fallen by about 50% during the past several decades. Honeybees pollinate up to $10 billion of apples, almonds, and other crops each year and produce about $250 million of honey, so their decline is a major threat to U.S. agriculture. Entomologists are searching for mite-resistant bees among both North American and European honeybees; the entomologists plan to breed the mite-resistant bees with local honeybees to incorporate genetic resistance into the local bees. Farmers are also experimenting with mite-resistant bee species other than honeybees to pollinate crops.

Figure 52–13 Commensalism. Spanish moss *(Tillandsia usneoides)* is a gray-colored epiphyte that hangs suspended from larger plants in the southeastern United States. Spanish moss is not a moss but a flowering plant in the pineapple family. Although it is often 6 m (20 ft) or longer, it is nonparasitic and does not harm the host tree. Photographed in autumn in North Carolina, hanging from a bald cypress whose foliage has turned color. *(Jeff Lepore/ Photo Researchers, Inc.)*

When a parasite causes disease and sometimes the death of a host, it is called a **pathogen.** Crown gall disease, which is caused by a bacterial pathogen, occurs in many different kinds of plants and results in millions of dollars of damage each year to ornamental and agricultural plants. Crown gall bacteria, which also live on organic debris in the soil, enter plants through small wounds such as those caused by insects. They cause galls (tumorlike growths), often at a plant's crown, that is, the area between the stem and roots at or near the surface of the ground. Although plants seldom die from crown gall disease, they weaken, grow more slowly, and often succumb to other pathogens.

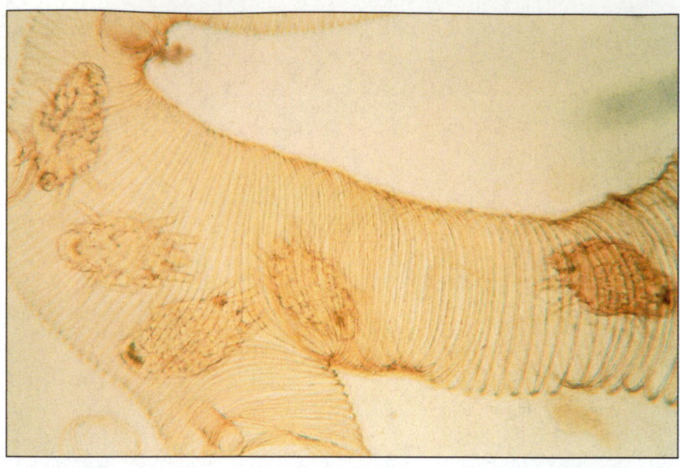

Figure 52-14 Parasitism. Microscopic tracheal mites *(Acarapis woodi)* live in the tracheal tubes of honeybees, clogging their airways so they cannot breathe efficiently. Tracheal mites also suck the bees' circulatory fluid, weakening and eventually killing them. Entomologists think the larger varroa mites *(not shown)*, which also feed on the circulatory fluid, are more devastating to honeybee populations than tracheal mites. Other species of mites may transmit viruses to their honeybee hosts. *(U.S. Department of Agriculture, Agricultural Research Service)*

KEYSTONE SPECIES AFFECT THE CHARACTER OF THE COMMUNITY

Certain species, called **keystone species,** are crucial in determining the nature of the entire community, that is, its species composition and ecosystem functioning. Other species of a community depend on or are greatly affected by the keystone species. Keystone species are usually not the most abundant species in the community. Although present in relatively small numbers, the individuals of a keystone species profoundly influence the entire community because they often affect the amount of available food, water, or some other resource. Identifying and protecting keystone species are crucial goals of conservation biology (see Chapter 55) because if a keystone species disappears from a community, many other species in that community may become more common, more rare, or even disappear.

One problem with the concept of keystone species is that it is difficult to measure all the direct and indirect impacts of a keystone species on an ecosystem. Consequently, most evidence for the existence of keystone species is based on indirect observations rather than on experimental manipulations. For example, consider the fig tree. Because fig trees produce a continuous crop of fruits, they appear to be keystone species in tropical rain forests of Central and South America. Fruit-eating monkeys, birds, bats, and other fruit-eating vertebrates of the forest do not normally consume large quantities of figs in their diets. During that time of the year when other fruits are less plentiful, however, fig trees become important in sustaining fruit-eating vertebrates. It is therefore assumed that, should the fig trees disappear, most of the fruit-eating vertebrates would also be eliminated. In turn, should the fruit-eaters disappear, the spatial distribution of other fruit-bearing plants would become more limited because the fruit-eaters help disperse their seeds. Thus, protecting fig trees in such tropical rainforest ecosystems probably increases the likelihood that monkeys, birds, bats, and many other tree species will survive. The question that begs to be answered is whether this anecdotal evidence of fig trees as keystone species is strong enough for policy makers to grant special protection to fig trees.

Many keystone species are top predators such as the gray wolf. Where wolves were hunted to extinction, the populations of elk, deer, and other larger herbivores increased exponentially. As these herbivores overgrazed the vegetation, many plant species that could not tolerate such grazing pressure disappeared. Smaller animals such as rodents, rabbits, and insects declined in number because the plants that they depended on for food were now less abundant. The number of foxes, hawks, owls, and badgers that prey on these small animals decreased, as did the number of ravens, eagles, and other scavengers that eat wolf kill. Thus, the disappearance of the wolf resulted in communities with considerably less biological diversity.

The reintroduction of wolves to Yellowstone National Park in 1995 has given ecologists a unique opportunity to study the impacts of a keystone species. The wolf's return has already caused substantial changes for other residents in the park. The top predator's effects have ranged from altering relationships among predator and prey species to transforming vegetation profiles. Coyotes are potential prey for wolves, and wolf packs have decimated some coyote populations. A reduction in coyotes has allowed populations of the coyotes' prey, such as ground squirrel and chipmunks, to increase. Scavengers such as ravens, bald eagles, and grizzly bears have benefited from dining on scraps from wolf kills.

The pruning effect that wolves will have on prey populations such as elk should cause broader, long-term impacts. Yellowstone's booming elk population of 35,000 eventually could be reduced by as much as 20%, which would relieve heavy grazing pressure on plants and encourage a more lush and varied plant composition. Richer vegetation should support more herbivores such as beavers and snow hares, which in turn will support small predators such as foxes, badgers, and martens.

COMMUNITIES VARY IN SPECIES RICHNESS AND SPECIES DIVERSITY

Species richness and species diversity vary greatly from one community to another and are influenced by many biotic and abiotic factors. **Species richness** is the number of species present in a community. Tropical rain forests and coral reefs are examples of communities with extremely high species richness. In contrast, geographically isolated islands and mountaintops exhibit low species richness.

Species diversity is sometimes used interchangeably with species richness, but more precisely, species diversity is a mea-

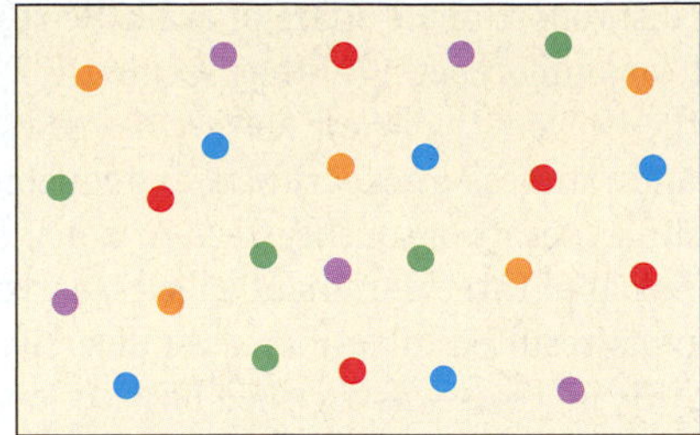

(a) Species evenness and species diversity are high.

(b) Species evenness and species diversity are low.

Figure 52–15 Species evenness. Shown are two hypothetical communities, each with five species. Each circle represents one individual; each color represents a different species. Both communities have the same species richness, but the community in **(a)** has a greater species evenness (and therefore a greater species diversity) than the community in **(b)**.

sure of the relative importance of each species within a community, based on its abundance, productivity, or size. Mathematically, species diversity represents a combination of species richness and **species evenness,** the distribution of individuals of each species in the community. A community in which all species have about the same number of individuals has a greater species evenness than one with both rare and common species (Fig. 52–15). (The actual equations that ecologists use to calculate species evenness and species diversity are too advanced for this text.)

Ecologists seek to explain why some communities have more species than others

What determines the number of species in a community? There seems to be no single conclusive answer, but several explanations appear plausible. These include the abundance of potential ecological niches, geographical isolation, habitat stress, closeness to the margins of adjacent communities, dominance of one species over others, and geological history. Although these and other environmental factors contribute to species richness, there are exceptions and variations in every single explanation. Some explanations vary at different spatial scales: An explanation that seems to work at a large geographical scale (such as a continent) may not work at a smaller local scale (such as a meadow).

In many habitats, species richness is related to the abundance of potential ecological niches. An already complex community offers a greater variety of potential ecological niches than does a simple community (Fig. 52–16). It may become even more complex if species potentially capable of filling those niches evolve or immigrate into the community, because these species create "opportunities" for additional species. Thus, it appears that species richness is self-perpetuating to some degree.

Species richness is inversely related to the geographical isolation of a community. Isolated island communities tend to be much less diverse than are communities in similar environments found on continents. This is due partly to the *distance effect*, the difficulty encountered by many species in reaching and successfully colonizing the island (Fig. 52–17). Also, sometimes species become locally extinct as a result of random events. In isolated habitats such as islands or mountaintops, locally extinct species cannot be readily replaced. Isolated areas are also likely to be small and to possess fewer potential ecological niches.

Generally, species richness is inversely related to the environmental stress of a habitat. Only those species capable of tolerating extreme conditions can live in an environmentally

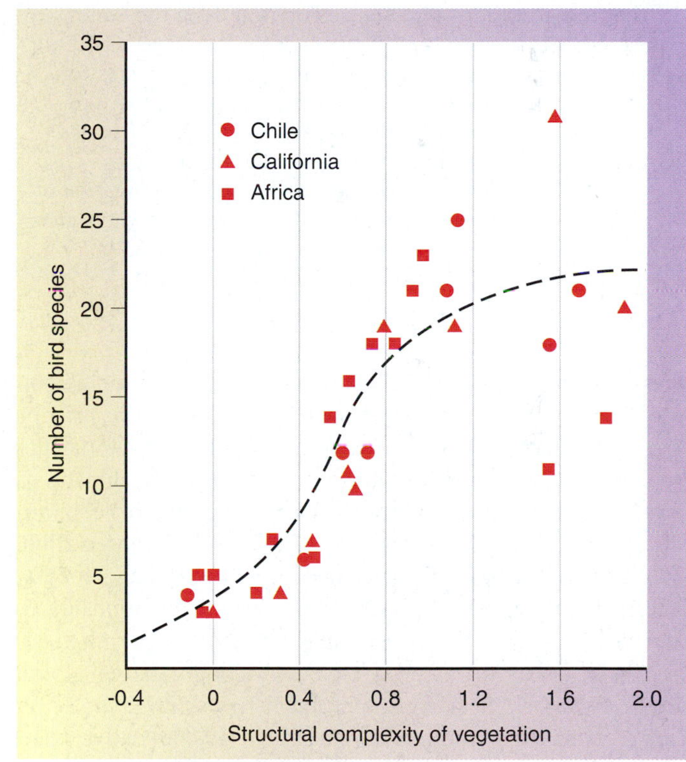

Figure 52–16 Effect of community complexity on species richness. Data were compiled in comparable chaparral habitats (grassy, shrubby, and woody areas) in southern Chile, Africa, and California. The structural complexity of vegetation *(X-axis)* is a numerically assigned gradient of habitats, based on height and density of vegetation, from low complexity (i.e., a grassland) to high complexity (a woodland). A community in which the vegetation is structurally complex generally provides birds with more kinds of food and hiding places than a community with a lower structural complexity. *(After Cody, M.L., and J.M. Diamond, eds.* Ecology and Evolution of Communities. *Harvard University, Cambridge, 1975)*

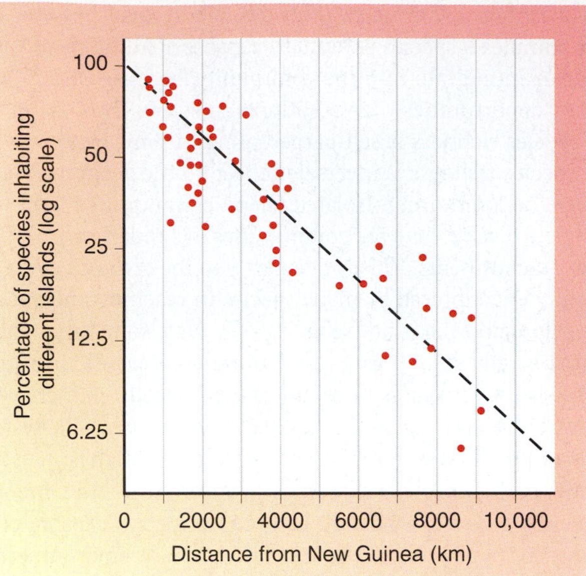

Figure 52–17 The distance effect. This graph shows that the percent of bird species on islands in the South Pacific is related to their distance from the large island of New Guinea, which is a source of colonizing species for these islands. (New Guinea has 100 percent of the bird species living in that region.) Note that species richness declines as the distance from New Guinea increases. *(After J.M. Diamond. "Biogeographic Kinetics: Estimation of Relaxation Times for Avifaunas of Southwest Pacific Islands."* Proceedings of the National Academy of Sciences, *Vol. 69, 1972)*

stressed community. Thus, the species richness of a highly polluted stream is low compared with that of a nearby pristine stream. Similarly, the species richness of high-latitude (farther from the equator) communities exposed to harsh climates is lower than that of lower-latitude (closer to the equator) communities with milder climates (Fig. 52–18). This observation, known as the *species richness–energy hypothesis,* suggests that different latitudes affect species richness due to variations in solar energy (see Fig. 53–12). Greater energy may permit more species to coexist in a given region. Although the equatorial countries of Colombia, Ecuador, and Peru occupy only 2% of Earth's land, they contain a remarkable 45,000 native plant species. The continental United States and Canada, with a significantly larger land area, possess a total of 19,000 native plant species. Ecuador alone contains more than 1300 native species of birds—twice as many as the United States and Canada combined.

Species richness is usually greater at the margins of distinct communities than it is in their centers. This is because an **ecotone,** a transitional zone where two or more communities meet, contains all or most of the ecological niches of the adjacent communities as well as some that are unique to the ecotone (see Chapter 54). This change in species composition produced at ecotones is known as the **edge effect.**

Species richness is reduced when any one species enjoys a position of dominance within a community so that it is able to appropriate a disproportionate share of available resources, thus crowding out, or outcompeting, other species. Ecologist James H. Brown of the University of New Mexico has addressed species composition and richness in experiments conducted since 1977 in the Chihuahuan desert of southeastern Arizona. In one experiment, the removal of three dominant species, all kangaroo rats, from several plots resulted in an increased diversity of other rodent species. This increase was ascribed both to lowered competition for food and also to an altered habitat, because the abundance of grass species increased dramatically after the removal of the kangaroo rats.

Species richness is greatly affected by geological history. Tropical rain forests are thought to be old, stable communities that have undergone relatively few widespread disturbances through Earth's entire history. (In ecology, a **disturbance** is any event in time that disrupts community or population structure.) During this time, myriad species evolved in tropical rain forests. In contrast, glaciers have repeatedly altered temperate and arctic regions during Earth's history. An area recently vacated by glaciers will have low species richness because few species have as yet had a chance to enter it and become established. The idea that older, more stable habitats have greater species richness than habitats subjected to frequent, widespread disturbances is known as the *time hypothesis*.

Species richness may promote community stability

Traditionally, most ecologists assumed that community stability—the ability of a community to withstand disturbances—is a consequence of community complexity. That is, a community with considerable species richness was hypothesized to be more stable than a community with less species richness. According to this view, the greater the species richness, the less critically important any single species should be. With many possible interactions within the community, it appeared unlikely that any single disturbance could affect enough components of the system to make a significant difference in its functioning.

Supporting evidence for this hypothesis can be found in the fact that destructive outbreaks of pests are more common in cultivated fields, which are low-diversity communities, than in natural communities with greater species richness. As another example, the almost complete loss of the American chestnut tree to the chestnut blight fungus had little ecological impact on the moderately diverse Appalachian woodlands of which it used to be a part (see *Focus On: Fighting Chestnut Blight* in Chapter 25).

Ongoing studies by David Tilman of the University of Minnesota and John Downing of the University of Iowa have strengthened the link between species richness and community stability. In their initial study, reported in the journal *Nature* in 1994, they established and monitored 207 plots of Minnesota grasslands for seven years. During the study period, Minnesota's worst drought in 50 years occurred (1987–1988). The ecologists found that those plots with the greatest number of plant species lost less ground cover, as measured by dry weight, and recovered faster than species-poor plots. Later studies by Tilman and his

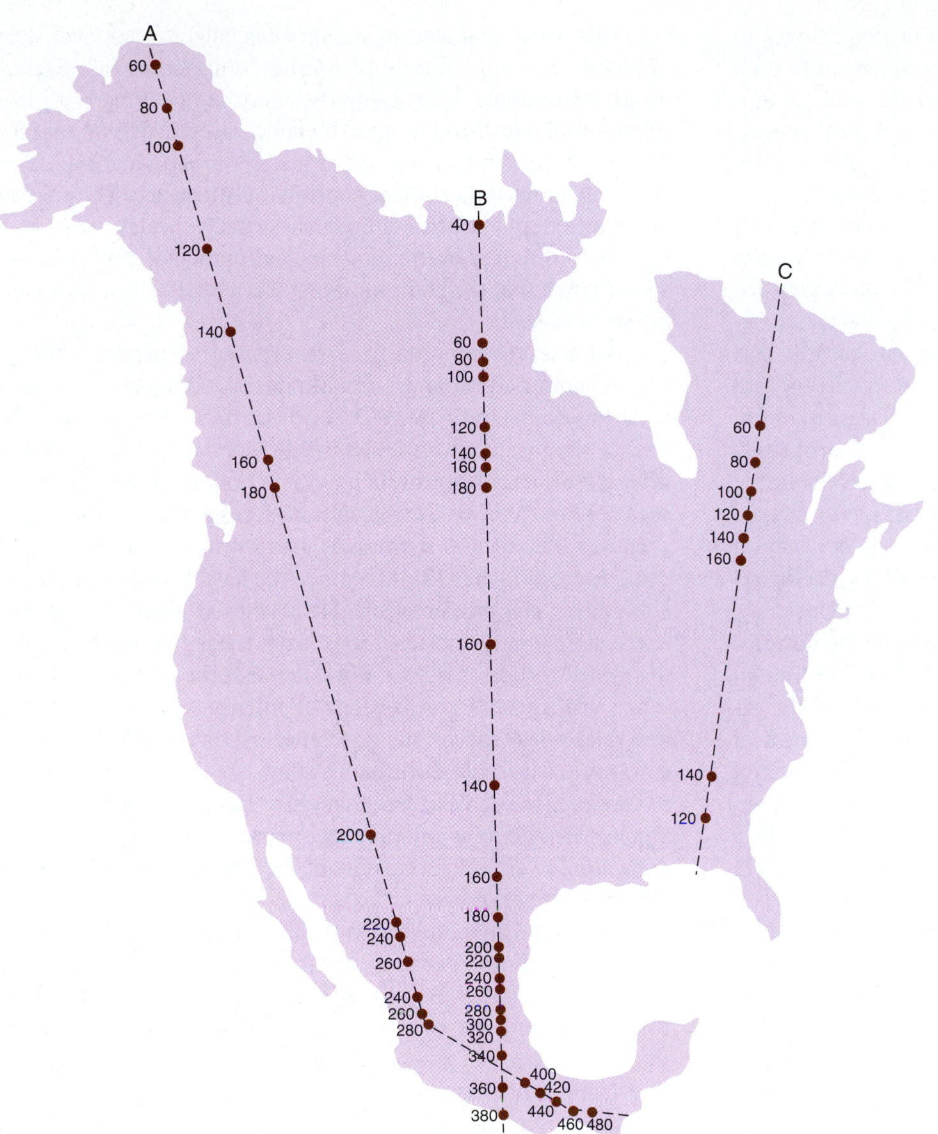

Figure 52–18 Effect of latitude on species richness of breeding birds in North America. The species richness for three north-south transects are shown. Note that the overall number of breeding bird species is greater at lower latitudes, toward the equator, than at higher latitudes. However, this pattern is strongly modified by other factors, such as precipitation and surface features (e.g., mountains). Similar observations have been made in many groups of organisms, from plants to primates, in both terrestrial and marine environments. *(After Cook, R.E. "Variation in Species Density of North American Birds."* Systematic Zoology, *Vol. 18, 1969)*

colleagues supported these conclusions and showed a similar effect of species richness on community stability during non-drought years. Similar work by almost three dozen ecologists at eight grassland sites in Europe, published in 1999 in the journal *Science*, also supports the link between species richness and community stability.

Some scientists do not agree with the conclusions of Tilman and other research groups that the more species the better, in terms of community stability. These critics argue that it is very difficult to separate species number from other factors that could affect productivity, and they suggest it might be better to start with established ecosystems and then study what happens to their productivity when plants are removed.

Another observation that adds a layer of complexity to the species richness–community stability debate is that populations of individual species within a diverse community may vary significantly from year to year. It may seem paradoxical that flux within populations of individual species relates to the stability of the entire community. When one considers all the interactions among the organisms in a community, however, it is obvious that some species benefit at the expense of others. If one species declines in a given year, other species that compete with it may flourish. Thus, it appears that ecosystems with greater species richness are more likely to contain species that are resistant to any given disturbance.

SUCCESSION IS COMMUNITY DEVELOPMENT OVER TIME

A community does not spring into existence full-blown but develops gradually through a series of stages, each dominated by different organisms. The process of community development over time, which involves species in one stage being replaced by different species, is called **succession.** An area is initially colonized by certain early-successional species that are replaced over time by others, which themselves may be replaced much later by still other, late-successional species.

Succession is usually described in terms of the changes in the species composition of an area's vegetation, although each successional stage also has its own characteristic kinds of animals and other species. The time involved in ecological succession is on the order of tens, hundreds, or thousands of years, not the millions of years involved in the evolutionary time scale.

Ecologists distinguish between two types of succession, primary and secondary. **Primary succession** is the change in species composition over time in a habitat that was not previously inhabited by organisms. No soil exists when primary succession begins. Bare rock surfaces, such as recently formed volcanic lava and rock scraped clean by glaciers, are examples of sites where primary succession might take place. The Indonesian island of Krakatoa has provided scientists with a perfect long-term study of primary succession in a tropical rain forest. In 1883 a volcanic eruption destroyed all life on the island. Ecologists have surveyed the ecosystem in the more than 100 years since the devastation to document the return of life forms. As of the 1990s, ecologists had found that the progress of primary succession was extremely slow, in part because of Krakatoa's isolation (recall the distance effect discussed earlier in the chapter). Many species are limited in their ability to disperse over water. Krakatoa's forest, for example, might have only one-tenth the tree species richness of undisturbed tropical rain forest of nearby islands. The lack of plant diversity has in turn limited the number of colonizing animal species. In a forested area of Krakatoa where zoologists would expect more than 100 butterfly species, for example, there are only two species.

Secondary succession is the change in species composition that takes place after some disturbance removes the existing vegetation; soil is already present at these sites. Abandoned agricultural fields or open areas produced by forest fires are common examples of sites where secondary succession occurs. During the summer of 1988, wildfires burned approximately one-third of Yellowstone National Park. This natural disaster provided a valuable chance for ecologists to study secondary succession in areas that had been forests. After the conflagration, gray ash covered the forest floor, and most of the trees, although standing, were charred and dead. Secondary succession in Yellowstone has occurred rapidly since 1988. Less than one year later, in the spring of 1989, trout lily and other herbs sprouted and covered much of the ground. By 1998, a young forest of knee-high to shoulder-high lodgepole pines dominated the area. Douglas fir seedlings also began appearing in 1998. Ecologists continue to monitor the changes in Yellowstone as secondary succession unfolds.

Disturbance influences succession and species richness

Early studies suggested that succession inevitably progressed to a stable and persistent community, known as a *climax community*, which was determined solely by climate. Periodic disturbances, such as fires or floods, were not thought to exert much influence on climax communities. If the climax community was disturbed in any way, it would return in time.

This traditional view of stability has fallen out of favor. The apparent endpoint stability of species composition in a climax forest, for example, is probably the result of how long trees live relative to the human life span. It is now recognized that mature "climax" communities are not in a state of permanent equilibrium but rather in a state of continual disturbance. The species composition and relative abundance of each species varies in a mature community over a range of environmental gradients, despite the fact that the community retains a relatively uniform appearance overall.

Because all communities are exposed to periodic disturbances, both natural and human-induced, ecologists have long tried to understand the effects of disturbance on species richness. A significant advance was the development of the **intermediate disturbance hypothesis** by Joseph Connell. When he examined species richness in tropical rain forests and coral reefs, he proposed that species richness is greatest at moderate levels of disturbance (Fig. 52–19). At a moderate level of disturbance, the community is a mosaic of habitat patches at different stages of succession; a range of sites exists, from those that were recently disturbed to those that had not been disturbed for many years. When disturbances are frequent or intense, only those species best adapted to earlier stages of succession are able to persist, whereas low levels of disturbance allow late-successional species to dominate to such a degree that other species disappear. For example, when periodic wildfires are suppressed in forests, creating a low level of disturbance, some of the "typical" forest herbs decline in number or even disappear.

One of the difficulties with the intermediate disturbance hypothesis is defining precisely what constitutes an "intermediate level of disturbance." Despite this problem, the intermediate disturbance hypothesis has important ramifications for conservation biology because it tells us that we cannot maintain a particular community simply by creating a reserve around it. Both natural and human-induced disturbances will cause changes in species composition, and it may be necessary for some type of intentional human intervention to maintain the species richness of the original community. What kind and how much human intervention are necessary to maintain species richness are controversial.

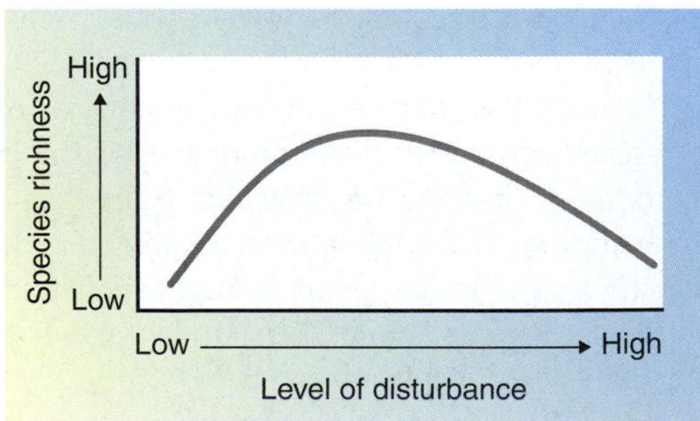

Figure 52–19 Intermediate disturbance hypothesis. Species richness is greatest at an intermediate level of disturbance. *(After Connell, J.H. "Diversity in Tropical Rain Forests and Coral Reefs."* Science, *Vol. 199, 1978)*

ECOLOGISTS CONTINUE TO STUDY COMMUNITY STRUCTURE

Process of Science

One of the big issues in community ecology, from the early 1900s to the present, is the nature of communities. Are communities highly organized systems of predictable species, or are they abstractions produced by the minds of ecologists?

Frederick E. Clements (1874–1945) was struck by the worldwide uniformity of large tracts of vegetation, for example, tropical rain forests in South America, Africa, and Southeast Asia (see Chapter 54). He also noted that even though the species composition of a community in a particular habitat might be different from that of a community in a habitat with a similar climate elsewhere in the world, overall the components of the two communities were usually similar. He viewed communities as something like "superorganisms," whose member species cooperated with one another in a manner that resembled the cooperation of the parts of an individual organism's body. Clements's view was that a community went through certain stages of development, like those of an organism, and eventually reached an adult state; the developmental process was succession, and the adult state was the climax community. This cooperative view of the community, called the **organismic model,** stresses the interaction of the members, which tend to cluster in tightly knit groups within discrete community boundaries (Fig. 52–20*a*).

Opponents of the organismic model, particularly Henry A. Gleason (1882–1975), held that biological interactions are less important in the production of communities than are environmental gradients (such as climate and soil) or even chance. Indeed, the concept of a community is questionable. It might be a classification category with no reality, reflecting little more than the tendency of organisms with similar environmental requirements to live in similar places. This school of thought, called the **individualistic model,** emphasizes species individuality, with each species having its own particular abiotic living requirements. It holds that communities are therefore not interdependent associations of organisms. Rather, each species is independently distributed across a continuum of areas that meets its own individual requirements (Fig. 52–20*b*).

Debates such as this one over the nature of communities are an integral part of the scientific process because they fuel discussion and research that lead to a better understanding of broad scientific principles. Studies testing the organismic and individualistic hypotheses of communities, for example, do not seem to support Clements's interactive concept of communities as discrete units. Instead, most studies favor the individualistic model. As shown in Figure 52–20*c*, tree species in Wisconsin forests are distributed in a gradient from wet to dry environments.

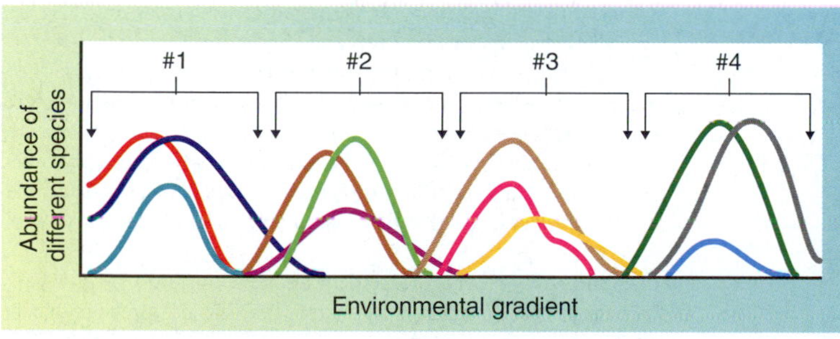

(a) Organismic model

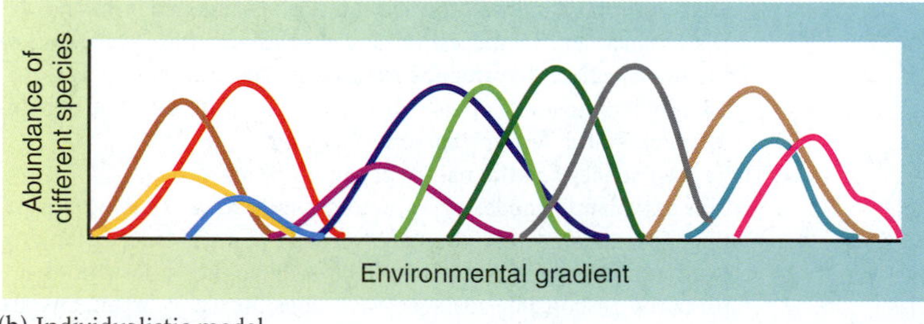

(b) Individualistic model

(c)

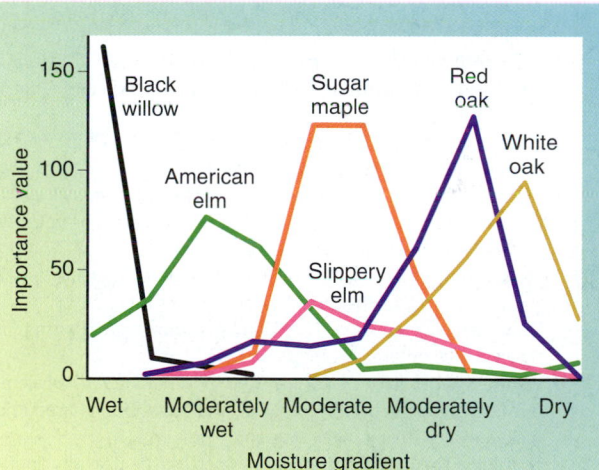

Figure 52–20 Nature of communities. **(a)** The hypothesized organismic model, in which communities are organized as distinct units. Each of the four communities shown (numbered brackets) consists of an assemblage of distinct species (each colored curve represents a single species). The arrows indicate ecotones, regions of transition along community boundaries. **(b)** The hypothesized individualistic model, in which there is a more random assemblage of species along a gradient of environmental conditions. **(c)** Tree species in Wisconsin forests, which are distributed along a moisture gradient, more closely resemble the individualistic model. The "importance value" *(Y-axis)* of each species in a given location combines three aspects (population density, frequency, and size). *(c, Adapted from Curtis, J.T.* The Vegetation of Wisconsin. *University of Wisconsin Press, Madison, Wisconsin, 1959)*

SUMMARY WITH KEY TERMS

I. **Community ecology** is the description and analysis of patterns and processes within the community.
   A. A biological **community** consists of a group of organisms of different species that interact and live together.
   B. A community and its abiotic environment compose an **ecosystem.**

II. The major roles of organisms in communities are those of autotrophs and heterotrophs.
   A. **Producers** are the photosynthetic autotrophs at the base of most food webs. They include plants and algae.
   B. **Consumers** are heterotrophs that feed on other organisms. **Primary consumers,** or **herbivores,** feed on plants; **secondary consumers,** or **carnivores,** feed on primary consumers. **Detritus feeders,** or **detritivores,** feed on **detritus,** freshly dead or decomposing organic material. **Omnivores** eat a variety of foods.
   C. **Decomposers,** or **saprotrophs,** are microbial heterotrophs (bacteria and fungi) that recycle the components of dead organisms and organic wastes.

III. The distinctive lifestyle and role of an organism in a community is its **ecological niche.** An organism's ecological niche takes into account all abiotic and biotic aspects of the organism's existence. An organism's **habitat** (where it lives) is one of the parameters used to describe the niche.
   A. Organisms are potentially able to exploit more resources and play a broader role in the life of their community than they actually do. The potential ecological niche for an organism is its **fundamental niche,** whereas the niche it actually occupies is its **realized niche.**
   B. An organism's **limiting resources** (such as the mineral content of soil, temperature extremes, and amount of precipitation) tend to restrict its realized niche.

IV. **Competition** occurs when two or more individuals attempt to use the same essential resource, such as food, water, shelter, living space, or sunlight.
   A. Competition can occur among individuals within a population **(intraspecific competition)** or between different species **(interspecific competition).**
   B. Interspecific competition may be one of the chief biological determinants of a species' realized niche.
      1. It is thought that **competitive exclusion** prevents two species from occupying the same niche in the same community for an indefinite period.
      2. In competitive exclusion, one species is excluded by another as a result of competition for a limiting resource.
   C. Some species reduce competition by **resource partitioning,** in which they evolve differences in resource use.
   D. Some species reduce competition by **character displacement,** in which their structural, ecological, and behavioral characteristics diverge where their ranges overlap.

V. **Predation** is the consumption of one species (the prey) by another (the predator).
   A. During **coevolution** between predator and prey, the predator evolves more efficient ways to catch prey, and the prey evolves better ways to escape the predator.
   B. Two effective predator strategies are pursuit and ambush.
   C. Plants possess several adaptations that protect them from being eaten, including spines, thorns, tough leathery leaves, and protective chemicals that are unpalatable or toxic to herbivores.
   D. Animals possess many strategies that help them avoid being killed and eaten.
      1. Many animals flee from predators, some have mechanical defenses, and some associate in groups.
      2. Some animals that possess chemical defenses also exhibit **aposematic coloration,** which is also called **warning coloration.**
      3. Some animals exhibit **cryptic coloration** that helps them hide from predators by blending into their surroundings.
      4. In **Batesian mimicry,** a harmless or edible species resembles another species that is dangerous in some way. Predators avoid the mimic as well as the model.
      5. In **Müllerian mimicry,** several different species, all of which are poisonous, harmful, or distasteful, resemble one another. Predators learn to avoid their common aposematic coloration.

VI. **Symbiosis** is any intimate or long-term association between two or more species.
   A. In **mutualism,** both partners benefit. Three examples of mutualism are **nitrogen-fixing bacteria** and legumes, **zooxanthellae** and corals, and **mycorrhizae** (fungi and plant roots).
   B. In **commensalism,** one organism benefits and the other is unaffected. Two examples of commensalism are silverfish and army ants, and epiphytes and larger plants.
   C. In **parasitism,** one organism (the parasite) benefits while the other (the host) is harmed. One example of parasitism is mites that grow in or on honeybees. Some parasites are **pathogens** that cause disease.

VII. Each community may have one or more **keystone species** that are crucial in determining the nature of the entire community. Identifying and protecting keystone species is a goal of conservation biology.

VIII. Community complexity is related to a variety of factors.
   A. Community complexity is expressed in terms of **species richness,** the number of species within a community; **species evenness,** the distribution of individuals of each species in a community; and **species diversity,** a measure of the relative importance of each species within a community. Mathematically, species diversity is a combination of species richness and species evenness.
   B. Species richness is often great when there are many potential ecological niches; when a community is not isolated (the distance effect) or severely stressed; when more energy is available (the species richness-energy hypothesis); in **ecotones** (transition zones between communities); and in communities with long histories without major **disturbances,** events that disrupt community or population structure.

IX. **Succession** is the orderly replacement of one community by another.
   A. **Primary succession** occurs in an area that has not previously been inhabited (e.g., bare rock).
   B. **Secondary succession** begins in an area where there was a preexisting community and well-formed soil (e.g., abandoned farmland).
   C. Disturbance affects succession and species richness. According to the **intermediate disturbance hypothesis,** species richness is greatest at moderate levels of disturbance, which create a mosaic of habitat patches at different stages of succession.

X. There are two views of the nature of communities.
   A. The **organismic model** views a community as a "superorganism" that goes through certain stages of development (succession) toward adulthood (climax). In this view, biological interactions are primarily responsible for species composition, and organisms are highly interdependent.
   B. Most ecologists support the **individualistic model,** which challenges the concept of a highly interdependent community. According to this model, abiotic environmental factors are the primary determinants of species composition in a community, and organisms are largely independent of each other.

POST-TEST

1. An association of populations of different species living together in one area is a(an) (a) organismic model (b) ecological niche (c) ecotone (d) community (e) habitat
2. Ecologically, grasses are classified as ____________ and deer are classified as ____________ (a) herbivores; carnivores (b) primary consumers; secondary consumers (c) detritivores; herbivores (d) primary producers; primary consumers (e) producers; secondary consumers
3. Photosynthetic organisms are known as (a) heterotrophs and primary producers (b) autotrophs and secondary consumers (c) heterotrophs and primary consumers (d) autotrophs and primary producers (e) detritivores and tertiary consumers
4. Monarch and viceroy butterflies are an example of (a) Batesian mimicry (b) character displacement (c) coevolution (d) Müllerian mimicry (e) cryptic coloration
5. A symbiotic association in which organisms are beneficial to one another is known as (a) predation (b) interspecific competition (c) intraspecific competition (d) commensalism (e) mutualism
6. A species' ____________ is the totality of its adaptations, its use of resources, and its lifestyle. (a) habitat (b) ecotone (c) ecological niche (d) competitive exclusion (e) coevolution
7. Competition from other species helps to determine an organism's (a) ecotone (b) fundamental niche (c) realized niche (d) limiting resource (e) ecosystem
8. "Complete competitors cannot coexist" is a statement of the principle of (a) primary succession (b) limiting resources (c) Müllerian mimicry (d) competitive exclusion (e) character displacement
9. The ____________ signifies that species richness is greater where two communities meet than at the center of either community. (a) edge effect (b) fundamental niche (c) character displacement (d) realized niche (e) limiting resource
10. Primary succession occurs on (a) bare rock (b) newly cooled lava (c) abandoned farmland (d) both a and b are examples (e) a, b, and c are examples
11. The tendency for two similar species to differ from one another more markedly in areas where they occur together is known as (a) Müllerian mimicry (b) Batesian mimicry (c) resource partitioning (d) competitive exclusion (e) character displacement
12. An unpalatable species demonstrates its threat to potential predators by displaying (a) character displacement (b) limiting resources (c) cryptic coloration (d) aposematic coloration (e) competitive exclusion
13. An ecologist studying several forest-dwelling, insect-eating bird species is unable to find any evidence of interspecific competition. The most likely explanation is (a) lack of a keystone species (b) low species richness (c) pronounced intraspecific competition (d) coevolution of predator-prey strategies (e) resource partitioning
14. Support for the individualistic model of community structure includes (a) the decline of honeybees due to two species of parasitic mites (b) the identification of fig trees as a keystone species in tropical forests (c) the competitive exclusion of one *Paramecium* species by another (d) the distribution of trees along a moisture gradient in Wisconsin forests (e) the effects of the removal of a dominant rodent species from an Arizona desert
15. Connell's hypothesis to explain the effect of disturbance on species richness is known as the (a) distance effect (b) time hypothesis (c) species richness–energy hypothesis (d) intermediate disturbance hypothesis (e) edge effect

REVIEW QUESTIONS

1. Distinguish a community from an ecosystem.
2. Distinguish between a decomposer and a detritus feeder.
3. Explain the relationship among acorns, gypsy moths, and Lyme disease.
4. Describe how natural selection has affected predator-prey relationships.
5. List the three kinds of symbiosis, give an example of each, and state the effects of the symbiotic interaction on each of the two species involved.
6. Why is an organism's realized niche usually narrower, or more restricted, than its fundamental niche?
7. What is the principle of competitive exclusion?
8. Describe at least two factors that affect species richness in a community.
9. Distinguish between primary and secondary succession and give an example of each.
10. Contrast the organismic and individualistic models of the nature of communities.

YOU MAKE THE CONNECTION

1. In what symbiotic relationships are humans involved?
2. Describe the ecological niche of humans. Do you think our realized niche has changed during the past 1000 years? Why or why not?
3. How is the vertical distribution of barnacles in Scotland an example of competitive exclusion?
4. Which part of the stickleback experiment described in the *Cutting Edge* box exhibited directional selection? Explain your answer.
5. Examine the top and middle graphs in Figure 52–4. Are these examples of exponential or logistic population growth? Where is K in each graph? (You may need to refer to Chapter 51 to answer this question.)
6. Many plants that produce nodules for nitrogen-fixing bacteria are common on disturbed sites. Explain how these plants might simultaneously compete and cooperate with other plant species.

RECOMMENDED READINGS

Baskin, Y. "The Work of Nature." *Natural History,* Feb. 1997. The importance of species richness (particularly keystone species) to community function.

Brodie, E.D., and E.D. Brodie Jr. "Predator-Prey Arms Races." *BioScience,* Vol. 49, No. 7, Jul. 1999. The authors cite research that suggests that predator-prey interactions often do not follow arms race dynamics.

Conniff, R. "Body Beasts." *National Geographic,* Vol. 194, No. 6, Dec. 1998. Excellent photographs highlight this review of various human parasites, from hair follicle mites to head lice.

Doebler, S.A. "The Rise and Fall of the Honeybee." *BioScience,* Vol. 50, No. 9, Sept. 2000. How mite infestations affect bees and the beekeeping industry.

Estes, J.A., M.T. Tinker, T.M. Williams, and D.F. Doak. "Killer Whale Predation on Sea Otters Linking Oceanic and Nearshore Ecosystems." *Science,* Vol. 282, 16 Oct. 1998. A fascinating account of how killer whales have affected the sea otter's keystone role in the offshore oceanic ecosystem of western Alaska.

Findley, R. "Mount St. Helens." *National Geographic,* Vol. 197, No. 5, May 2000. This article examines secondary succession since the eruption of Mount St. Helens in 1980.

Jones, C.G., R.S. Ostfeld, M.P. Richard, E.M. Schauber, and J.O. Wolff. "Chain Reactions Linking Acorns to Gypsy Moth Outbreaks and Lyme Disease Risk." *Science,* Vol. 279, 13 Feb. 1998. Ecologists have unraveled a complex cycle of events that links Lyme disease epidemics and outbreaks of gypsy moths in deciduous forests.

Lasley, E.N. "Having Their Toxins and Eating Them Too." *BioScience,* Vol. 49, No. 12, Dec. 1999. Examines the natural sources of many animals' chemical defenses.

Lovett, R.A. "Mount St. Helens, Revisited." *Science,* Vol. 288, 2 Jun. 2000. Study of the recovery of vegetation since Mount St. Helens erupted in Washington State in 1980 has provided some surprising results.

McClintock, J.B., and B.J. Baker. "Chemical Ecology in Antarctic Seas." *American Scientist,* Vol. 86, May–Jun. 1998. A fascinating account of chemical interactions among species, including the intriguing story of marine amphipods that abduct and carry toxic sea butterflies on their backs to deter predators.

Mlot, C. "The Coyotes of Lamar Valley." *Science News,* Vol. 153, 31 Jan. 1998. The coyote is learning to adapt to the presence of wolves in Yellowstone National Park.

Moffett, M.W. "Ants and Plants." *National Geographic,* Vol. 197, No. 5, May 2000. Ants have evolved a variety of unusual ecological relationships with the trees they live in and depend on.

Raven, P.H., and L.R. Berg. *Environment,* 3rd ed. Harcourt College Publishers, Philadelphia, 2001. This environmental science text offers a detailed situation analysis of planet Earth, including how humans interact with and affect its physical and living systems.

Ricklefs, R.E., and G.L. Miller. *Ecology,* 4th ed. W.H. Freeman & Company, New York, 1999. This textbook presents the important themes of ecology, such as energy flow, population and community interactions, and mathematical models of ecology, in an accessible manner.

Turner, M.G., V.H. Dale, and E.H. Everham III. "Fires, Hurricanes, and Volcanoes: Comparing Large Disturbances." *BioScience,* Vol. 47, No. 11, Dec. 1997. This article compares the recovery of natural communities affected by three large-scale disturbances: the 1980 eruption of Mount St. Helens, the 1988 Yellowstone fires, and Hurricane Hugo in 1989.

Wakelin, D. "Parasites and the Immune System: Conflict or Compromise?" *BioScience,* Vol. 47, No. 1, Jan. 1997. Parasites and their hosts must strike balances between the beneficial and harmful consequences of parasitism.

Zimmer, C. "Attack and Counterattack: The Never-Ending Story of Hosts and Parasites." *Natural History,* Sept. 2000. A fascinating account of how parasites are important as forces of evolution.

- Visit our Web site at **http://www.info.brookscole.com/solomonbergmartin** for links to chapter-related resources on the World Wide Web. Additional on-line materials relating to this chapter can also be found on our Web site.

See chapter activity on BioActive Learner CD for additional help in mastering the chapter's material. Icon location in the chapter's margins shows which topics have tutorials or simulations in the CD.

53

Ecosystems and the Biosphere

Catchment at Hubbard Brook Experimental Forest. Catchments measure the quantity, timing, and quality of water flowing from a forested watershed. *(USDA Forest Service)*

LEARNING OBJECTIVES

After you have studied this chapter you should be able to

1. Compare how matter and energy operate in ecosystems.
2. Summarize the concept of energy flow through a food web.
3. Draw and explain typical pyramids of numbers, biomass, and energy.
4. Distinguish between gross primary productivity and net primary productivity.
5. Diagram the carbon, nitrogen, phosphorus, and hydrological cycles.
6. Summarize the effects of solar energy on Earth's temperatures.
7. Discuss the roles of solar energy and the Coriolis effect in the production of global air and water flow patterns.
8. Define El Niño–Southern Oscillation (ENSO) and describe some of its effects on climate and on organisms.
9. Give three causes of regional precipitation differences.
10. Discuss the effects of fire on certain ecosystems.

Almost completely isolated from everything in the Universe but sunlight, our planet Earth has often been compared to a vast spaceship whose life support system consists of the communities of organisms that inhabit it, plus energy from the sun. These organisms produce oxygen, transfer energy, and recycle nutrients with great efficiency. Yet none of these ecological processes would be possible without the abiotic (nonliving) environment of our spaceship Earth. Much of the climate to which organisms have adapted is produced by the sun, which warms the planet, powers the hydrological cycle (causes precipitation), and drives ocean currents and atmospheric circulation patterns. The sun also supplies the energy that almost all organisms use to carry on life processes.

The science of ecology deals with the abiotic environment as well as with living organisms. Individual communities and their abiotic environments are **ecosystems**. An ecosystem encompasses all the interactions among organisms living together in a particular place, and among those organisms and their abiotic environment. Earth, which encompasses the **biosphere** (all of Earth's communities) and its interactions with Earth's water, soil, rock, and atmosphere, is the largest ecosystem.

Ecologists conduct detailed ecosystem studies in laboratory simulations and in the field to measure such processes as energy flow, the cycling of nutrients, and the effects of natural and human-induced disturbances (e.g., air pollution, tree harvesting, and land-use changes). Some ecosystem studies, such as those performed at the Hubbard Brook Experimental Forest (HBEF), a 3100-hectare reserve in the White Mountain National Forest in New Hampshire, are long-term. Beginning in the early 1960s and continuing to the present, HBEF has been the site of numerous studies that address the hydrology (e.g., precipitation, surface runoff, and groundwater flow), biology, geology, and chemistry of forests and associated aquatic ecosystems. The National Science Foundation (NSF) has designated HBEF as one of its 24 long-term ecological research sites. (NSF provides funding for scientists, graduate students, and undergraduates to conduct well-designed and documented experiments.)

Many of the experiments at HBEF are based on field observations. For example, detailed surveys of the salamander populations in forest communities were originally done in 1970 and were repeated in recent years. Other studies involve manipulative experiments. In 1978, for example, dilute sulfuric acid was added to a small stream in HBEF to study the chemical and biological effects of acidification. This experiment was of practical value because **acid deposition,** a form of air pollution, has acidified numerous lakes and streams in industrialized countries. It is thought that acidification of waterways and soil leads to an increase in the solubility of toxic metals such as aluminum that enter food webs and may contribute to declining populations of fishes and other organisms.

Several researchers have studied the effects on HBEF stream ecosystems of **deforestation,** the clearance of large expanses of forest for agriculture or other uses. When a forest is removed, the total amount of water and nutrient minerals that flows into streams increases drastically. In these studies, concrete dams called *catchments* are constructed across streams to measure the flow of water and chemical components out of the ecosystem. Typically, outflow is measured in two separate ecosystems, one of which serves as a control and one of which is experimentally manipulated *(see figure)*. These studies have demonstrated that deforestation results in decreased soil fertility caused by soil erosion and leaching of essential nutrient minerals. The summer temperatures in streams running through deforested areas are higher than in shady streams running through uncut forests. Many stream dwellers do not fare well in deforested areas, in part because they are adapted to cold temperatures.

Detailed studies such as those at HBEF have provided ecologists with insights into how ecological processes function in individual ecosystems. These experimental manipulations have demonstrated, for example, that terrestrial ecosystems such as forests are strongly linked to downstream aquatic ecosystems. Ecologists compare these data with similar information from other ecosystem studies to develop generalized insights into how ecosystems are structured and how they function. Long-term ecosystem experiments enable ecologists to evaluate and predict the effects of environmental change, including human-induced change.

Ecosystem experiments have also contributed to our practical knowledge about how to maintain water quality, wildlife habitat, and productive forests. **Ecosystem management,** a conservation approach that emphasizes restoring and maintaining the quality of an entire ecosystem rather than the conservation of individual species, makes use of such knowledge. Ecosystem management is based on the understanding that organisms, including humans, are interrelated and interdependent with one another and with their abiotic environment. It takes into account that ecosystems do not recognize political boundaries and may span federal, state, county, and private lands. Ecosystem management also recognizes that economic well-being of human communities is inextricably linked to the sustainable use of natural resources and ecological processes.

This chapter develops three key concepts about ecosystems. First, energy flow in ecosystems is linear. Energy moves through food webs of ecosystems in a one-way direction from the environment to organisms. Once energy has been used by an organism to do biological work, it is unavailable to other organisms because, as work is performed, the energy is changed to heat and given off into the cooler surroundings. Energy cannot be recycled and reused.

Second, matter moves in numerous cycles within an ecosystem, that is, from one organism to another, and from organisms to the abiotic environment and back again. All materials vital to life are continually recycled through ecosystems and so become available to new generations of organisms.

Third, the abiotic environment causes conditions that determine where and how successfully species live. We briefly consider five aspects of the abiotic environment that have helped shape the biotic component of ecosystems: solar radiation, the atmosphere, the ocean, weather and climate, and fire.

THE FLOW OF ENERGY THROUGH ECOSYSTEMS IS LINEAR

The passage of energy in a one-way direction through an ecosystem is known as **energy flow.** Energy enters an ecosystem as radiant energy (sunlight), a tiny portion (far less than 1%) of which is trapped and used by producers during photosynthesis. The energy, now in chemical form, is stored in the bonds of organic (carbon-containing) molecules such as glucose. When these molecules are broken apart by cellular respiration, energy becomes available (in the form of ATP) to do work such as repairing tissues, producing body heat, moving about, or reproducing. As the work is accomplished, energy escapes the organisms and dissipates into the environment as heat. Ultimately, this heat energy radiates into space. Thus, once energy has been used by an organism, it is unavailable for reuse (Fig. 53–1; see also the discussion of the second law of thermodynamics in Chapter 6).

Energy flow describes who eats whom in ecosystems

In an ecosystem, energy flow occurs in **food chains,** in which energy from food passes from one organism to the next in a sequence. Producers form the beginning of the food chain by capturing the sun's energy through photosynthesis. Herbivores (and omnivores) eat plants, obtaining the chemical energy of the producers' molecules as well as building materials from which they construct their own tissues. Herbivores are in turn consumed by carnivores and omnivores, who reap the energy stored in the herbivores' molecules. At every step, decomposers break down organic molecules in the remains (carcasses and body wastes) of all members of the food chain. (See *Focus On: Food Chains and Poisons in the Environment* for a discussion of how certain toxins pass through food chains.)

Simple food chains as just described rarely occur in nature, because few organisms eat just one kind, or are eaten by just one

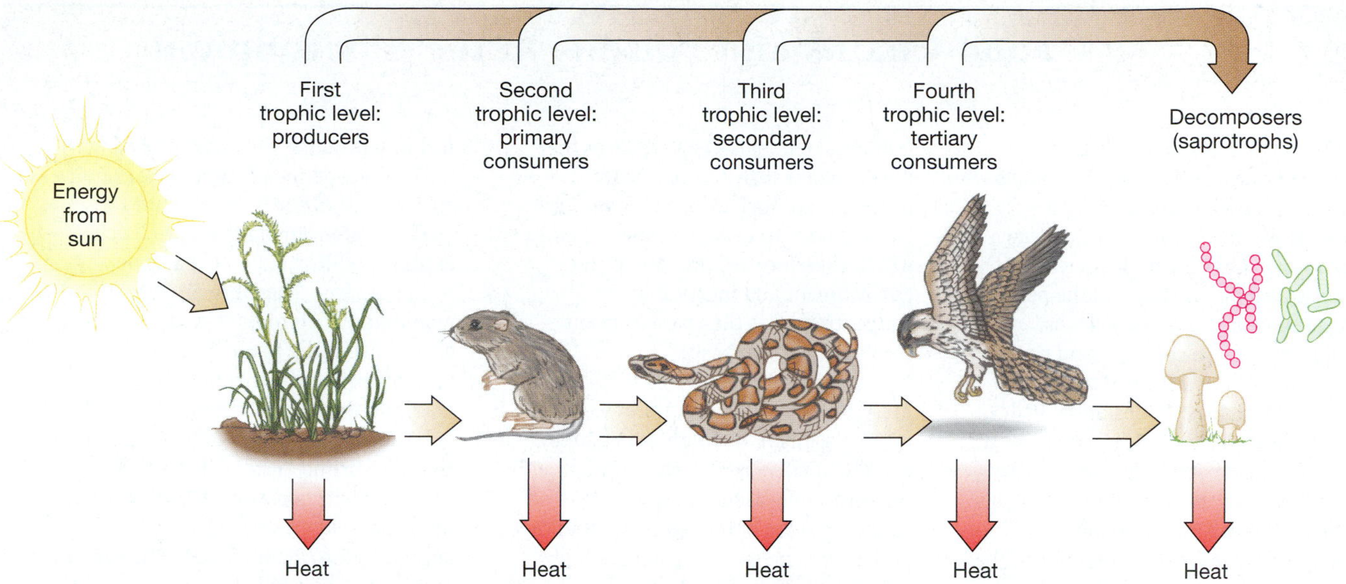

Figure 53–1 One-way energy flow through ecosystems. Energy enters ecosystems from an external source (the sun) and exits as heat loss. As stipulated by the second law of thermodynamics, most of the energy acquired by a given trophic level is released into the environment as heat and is therefore unavailable to the next trophic level.

other kind of organism. More typically, the flow of energy and materials through ecosystems takes place in accordance with a range of food choices for each organism. In an ecosystem of average complexity, hundreds of alternative pathways are possible. Thus, a **food web,** which is a complex of interconnected food chains in an ecosystem, is a more realistic model of the flow of energy and materials through ecosystems (Fig. 53–2).

Food webs are divided into **trophic levels** (from the Greek *tropho,* which means nourishment). The first trophic level is formed by producers (organisms that photosynthesize), the second by primary consumers (herbivores), the third by secondary consumers (carnivores and omnivores), and so on (see Fig. 53–1).

Because food webs are descriptions of "who eats whom," they indicate the negative effects that predators have on their prey. For example, consider a simple food chain: grass ⟶ field mouse ⟶ owl. The owl, which kills and eats mice, obviously exerts a negative effect on the mouse population; in like manner, field mice, which eat grass seeds, reduce the grass population.

A trophic level in a food web also influences other trophic levels to which it is not directly linked. Producers and top carnivores do not exert direct effects on one another, yet each is indirectly affected by the other. In our example, the owls help the producers by keeping the population of seed-eating mice under control. Likewise, the grasses benefit owls by supporting a population of mice on which the owl population feeds. These indirect interactions may be as important as direct, predator-prey interactions are in food web dynamics.

The most important thing to remember about energy flow in ecosystems is that it is linear, or one-way. That is, energy can move along a food web from one trophic level to the next trophic level as long as it is not used to do biological work. Once energy has been used by an organism, however, it is lost as heat and is unavailable to any other organism in the ecosystem.

Ecological pyramids illustrate how ecosystems work

Ecologists sometimes compare trophic levels by determining the number of organisms, the biomass, or the relative energy found at each level. This information is presented graphically as **ecological pyramids.** The base of each ecological pyramid represents the producers, the next level is the primary consumers (herbivores), the level above that is the secondary consumers (carnivores), and so on. The relative area of each bar of the pyramid is proportional to what is being demonstrated.

A **pyramid of numbers** shows the number of organisms at each trophic level in a given ecosystem, with greater numbers illustrated by a larger area for that section of the pyramid. In most pyramids of numbers, each successive trophic level is occupied by fewer organisms. Thus, in African grasslands the number of herbivores, such as zebras and wildebeests, is greater than the number of carnivores, such as lions. Inverted pyramids of numbers, in which higher trophic levels have more organisms than lower trophic levels, are often observed among decomposers, parasites, and herbivorous insects. One tree can provide food for thousands of leaf-eating insects, for example. Pyramids of numbers are of limited usefulness because they do not indicate the biomass of the organisms at each level and they do not indicate the amount of energy transferred from one level to another.

A **pyramid of biomass** illustrates the total biomass at each successive trophic level. **Biomass** is a quantitative estimate of

FOCUS ON Food Chains and Poisons in the Environment

Certain toxic substances, including some pesticides, radioactive isotopes, heavy metals such as mercury, and industrial chemicals such as PCBs, can enter food chains. The problem was first demonstrated by the effects of the pesticide DDT on some bird species. Falcons, pelicans, bald eagles, ospreys, and many other birds are very sensitive to traces of DDT in their tissues. A substantial body of scientific evidence indicates that one of the effects of DDT on these birds is that they lay eggs with extremely thin, fragile shells that usually break during incubation, causing the chicks' deaths. After 1972, the year DDT was banned in the United States, the reproductive success of many birds gradually improved.

The impact of DDT on birds is the result of three characteristics of DDT (and other toxins that cause problems in food webs): its persistence, bioaccumulation, and biological magnification. Some toxins are extremely stable and may take many years to be broken down into less toxic forms. The **persistence** of synthetic pesticides and industrial chemicals is a result of their novel chemical structures. Natural decomposers such as bacteria have not evolved ways to degrade these toxins, which therefore accumulate in the environment and increase in concentration as they pass from one trophic level to another.

When a persistent toxin is not metabolized (broken down) or excreted by an organism, it simply gets stored, usually in fatty tissues. Over time, the organism may accumulate high concentrations of the toxin. The buildup of such a toxin in an organism's body is known as **bioaccumulation.**

Organisms at higher trophic levels in food webs tend to have greater concentrations of bioaccumulated toxins stored in their bodies than those at lower levels. The increase in concentration as the toxin passes through successive levels of the food web is known as **biological magnification.**

As an example of the concentrating characteristic of persistent toxins, consider a food chain studied in a Long Island salt marsh that was sprayed with DDT over a period of years for mosquito control *(see figure)*. The concentration of DDT in water during this period was extremely dilute, 0.00005 parts per million (ppm). The algae and other plankton, which took up and accumulated the toxin, contained 0.4 ppm of DDT. Each shrimp grazing on the plankton ingested and concentrated the pesticide in its tissues, to 0.16 ppm. Eels that ate shrimp laced with pesticide ended up with a pesticide level of 0.28 ppm, whereas other predaceous fishes contained 2.07 ppm of DDT. The top carnivores, ring-billed gulls, had a pesticide value of 75.5 ppm from eating contaminated fishes. Although this example involved a bird at the top of the food chain, it is important to recognize that *all* top carnivores, from fishes to humans, are at risk from biological magnification of persistent toxins. Because of this risk, currently approved pesticides have been tested to ensure they do not persist and accumulate in the environment.

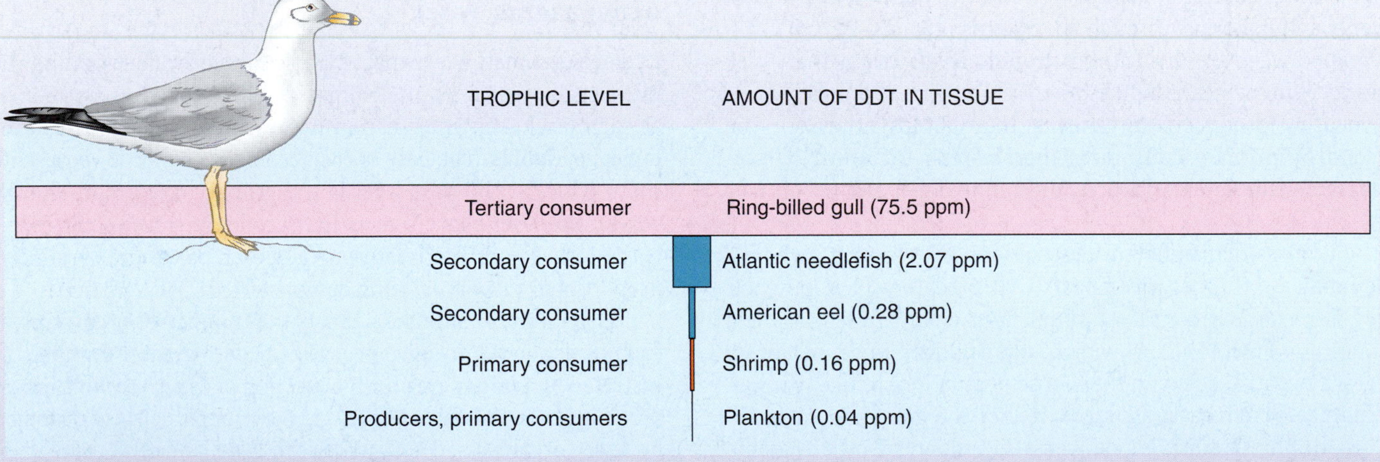

Biological magnification of DDT (expressed as parts per million) in a Long Island salt marsh. Note how the level of DDT increased in the tissues of various organisms as DDT moved through the food chain from producers to consumers. Ring-billed gulls at the top of the food chain had approximately 1 million times more DDT in their tissues than the concentration of DDT in the water (0.00005 ppm). (Plankton consisted of a mixture of phytoplankton and zooplankton.) *(Based on data from Woodwell, G.M., C.F. Worster, Jr., and P.A. Isaacson. "DDT Residues in an East Coast Estuary: A Case of Biological Concentration of a Persistent Insecticide."* Science, *Vol. 156, 12 May 1967)*

Figure 53–2 A deciduous forest food web. This diagram is greatly simplified compared with what actually happens in nature. Groups of species are lumped into single categories such as "spiders"; many species are not included; and numerous links in the web are not shown.

the total mass, or amount, of living material; it indicates the amount of fixed energy at a particular time. Biomass units of measure vary: Biomass may be represented as total volume, dry weight, or live weight. Typically, these pyramids illustrate a progressive reduction of biomass in succeeding trophic levels (Fig. 53–3*a*). Assuming an average biomass reduction of about 90% for each trophic level,[1] 10,000 kg of grass should be able to support 1000 kg of grasshoppers, which in turn support 100 kg of frogs. By this logic, the biomass of frog-eaters (such as snakes)

[1]The 90% reduction in biomass is an approximation; actual biomass reduction from one trophic level to the next varies widely in nature.

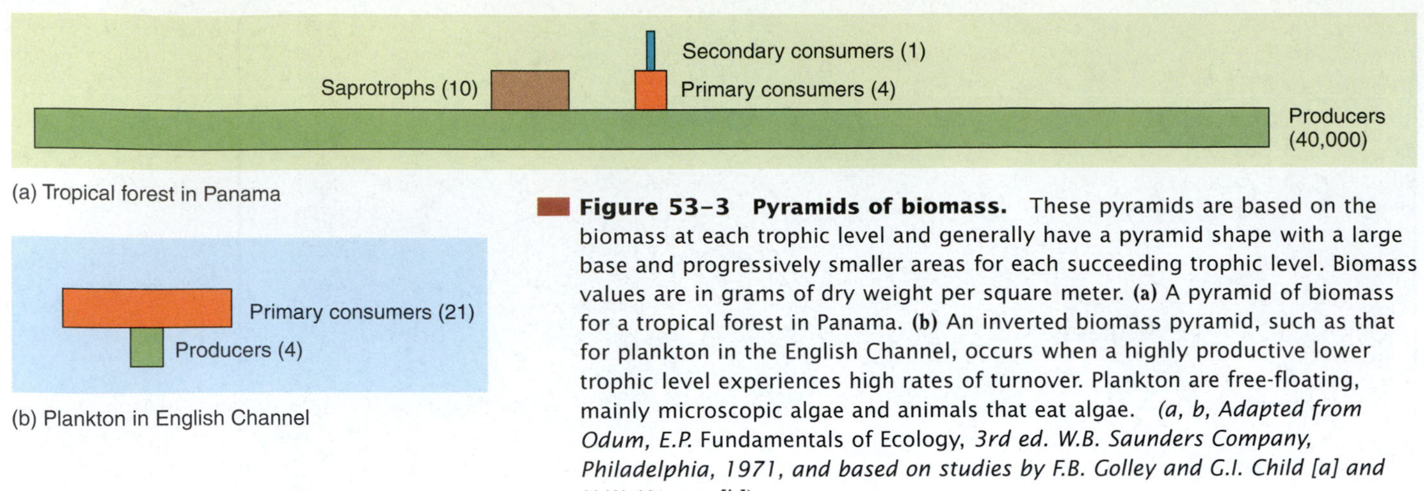

Figure 53–3 Pyramids of biomass. These pyramids are based on the biomass at each trophic level and generally have a pyramid shape with a large base and progressively smaller areas for each succeeding trophic level. Biomass values are in grams of dry weight per square meter. **(a)** A pyramid of biomass for a tropical forest in Panama. **(b)** An inverted biomass pyramid, such as that for plankton in the English Channel, occurs when a highly productive lower trophic level experiences high rates of turnover. Plankton are free-floating, mainly microscopic algae and animals that eat algae. *(a, b, Adapted from Odum, E.P.* Fundamentals of Ecology, *3rd ed. W.B. Saunders Company, Philadelphia, 1971, and based on studies by F.B. Golley and G.I. Child [a] and H.W. Harvey [b])*

could only weigh, at most, about 10 kg. From this brief exercise, you can see that although carnivores do not eat producers, a large producer biomass is required to support carnivores via a food web.

Occasionally we find an inverted pyramid of biomass in which the primary consumers outweigh the producers (Fig. 53–3*b*). In these instances the producers, which are usually unicellular algae that are short-lived and reproduce quickly, are consumed in large numbers by herbivores such as large zooplankton and fish. Thus, although at any point in time relatively few algae are present, the rate of biomass production of the primary consumers is much less than that of the producers.

A **pyramid of energy** indicates the energy content, often expressed as kilocalories per square meter per year, of the biomass of each trophic level. A common method ecologists use to measure energy content is to burn a sample of tissue in a calorimeter; the heat released during combustion is measured to determine the energy content of the organic material in the sample. Energy pyramids, which always have large bases and get progressively smaller through succeeding trophic levels, show that most energy dissipates into the environment when there is a transition from one trophic level to the next. Less energy reaches each successive trophic level from the level beneath it because some of the energy at the lower level is used by those organisms to perform work, while some of it is lost (Fig. 53–4). (Remember, no biological process is 100% efficient.) The second law of thermodynamics explains why there are few trophic levels: Food webs are short because of the dramatic reduction in energy content that occurs at each successive trophic level.

Ecosystems vary in productivity

The **gross primary productivity (GPP)** of an ecosystem is the rate at which energy is captured during photosynthesis.[2] Thus, GPP is the total amount of photosynthetic energy captured in a given period. Of course, plants must respire to provide energy for their life processes, and cellular respiration acts as a drain on

[2]Gross and net primary productivities are referred to as primary because plants occupy the first position in food webs.

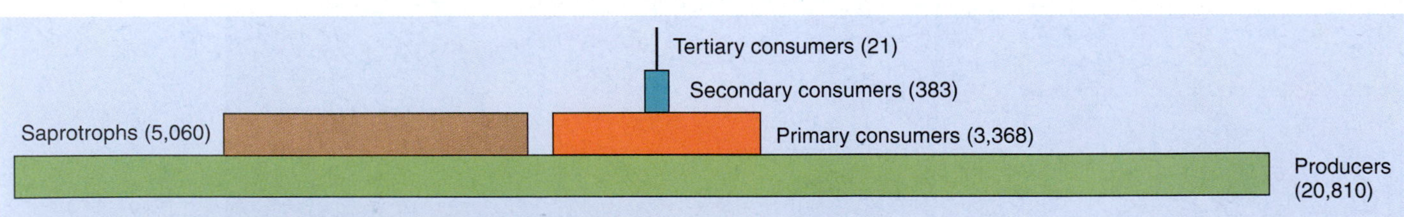

Figure 53–4 Pyramid of energy. A pyramid of energy for Silver Springs, Florida, represents energy flow, the functional basis of ecosystem structure. Energy values are in kilocalories per square meter per year. Note the substantial loss of usable energy from one trophic level to the next. (Energy in producers represents gross primary productivity.) The Silver Spring ecosystem is complex, but tape grass (producers), spiral-shelled snails (primary consumers), young river turtles (secondary consumers), gar (tertiary consumers), and bacteria and fungi (saprotrophs) are representative organisms. When they are young, river turtles are carnivores and consume snails, aquatic insects, and worms; as adults, river turtles are herbivores. *(Based on Odum, H.T. "Trophic Structure and Productivity of Silver Springs, Florida."* Ecological Monographs, *Vol. 27, 1957)*

photosynthetic output. Energy that remains in plant tissues after cellular respiration has occurred is called **net primary productivity (NPP).** That is, NPP is the amount of biomass (the energy stored in plant tissues) found in excess of that broken down by a plant's cellular respiration for normal daily activities. NPP represents the rate at which this organic matter is actually incorporated into plant tissues to produce growth.

Net primary productivity	=	Gross primary productivity	−	Plant respiration
(Plant growth per unit area per unit time)		(Total photosynthesis per unit area per unit time)		(Per unit area per unit time)

Only the energy represented by net primary productivity is available for consumers, and of this energy only a portion is actually used by them. Both GPP and NPP can be expressed as energy per unit area per unit time—for example, kilocalories of energy fixed by photosynthesis per square meter per year—or in terms of dry weight—that is, grams of carbon incorporated into tissue per square meter per year.

Many factors may interact to determine productivity. Some plants are more efficient than others in fixing carbon. Environmental factors are also important. These include the availability of solar energy, nutrient minerals, and water; other climate factors; the degree of maturity of the community; and the severity of human modification of the environment. Many of these factors are difficult to assess.

Ecologists use different methods to measure primary productivity, depending on whether GPP or NPP is being assessed. Methods also vary from one type of ecosystem to another. On land, for example, ecologists might cut, dry, and weigh plants at the end of a growing season to measure NPP. Dry weights are measured because the water content of different species varies considerably. This method would not be useful to measure NPP in the ocean, however, where the main producers are short-lived microscopic algae that are heavily grazed.

Ecosystems differ strikingly in their productivities (Fig. 53–5 and Table 53–1). On land, tropical rain forests have the highest productivity, probably owing to the abundant rainfall, warm temperatures, and intense sunlight. As you might expect, tundra with its short, cool growing season and deserts with their lack of precipitation are the least productive terrestrial ecosystems. In ecosystems with comparable annual temperatures (such as temperate deciduous forest, temperate grassland, and temper-

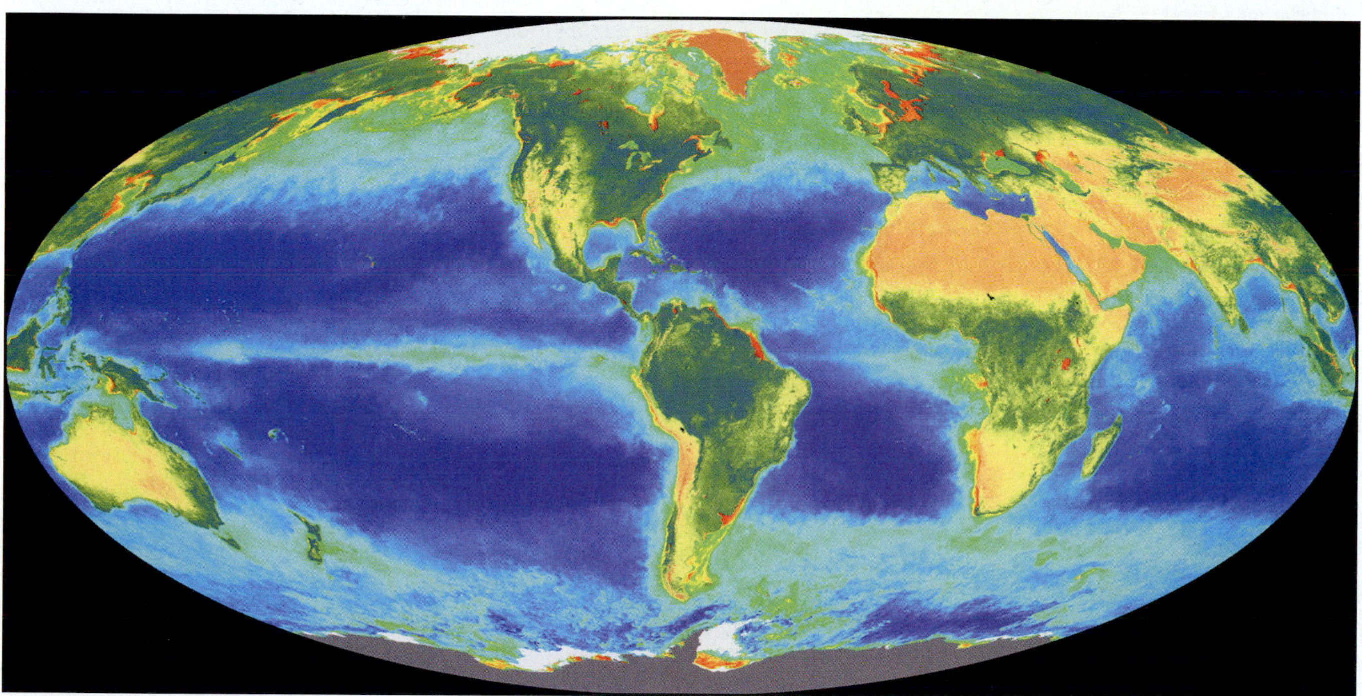

Figure 53–5 A measure of Earth's primary productivity. The data are from a satellite launched as part of NASA's Mission to Planet Earth in 1997. The satellite measured the amount of plant life on land as well as the concentration of phytoplankton (algae) in the ocean. On land, the most productive areas, such as tropical rain forests, are dark green, whereas the least productive ecosystems (deserts) are orange. In the ocean and other aquatic ecosystems, the most productive regions are red, followed by orange, yellow, green, and blue (the least productive). Data are not available for the gray area. *(Provided by the SEAWIFS Project, NASA/Goddard Space Flight Center and ORBIMAGE)*

TABLE 53–1 Net Primary Productivities (NPP) for Selected Ecosystems*

Ecosystem	Average NPP (g dry matter/m²/year)
Algal beds and reefs	2500
Tropical rain forest	2200
Swamp and marsh	2000
Estuaries	1500
Temperate evergreen forest	1300
Temperate deciduous forest	1200
Savanna	900
Boreal (northern) forest	800
Woodland and shrubland	700
Agricultural land	650
Temperate grassland	600
Upwelling zones in ocean	500
Lake and stream	250
Arctic and alpine tundra	140
Open ocean	125
Desert and semidesert scrub	90
Extreme desert (rock, sand, ice)	3

*Based on Whittaker, R.H. *Communities and Ecosystems*, 2nd ed. Macmillan, New York, 1975.

ate desert), water availability (as annual precipitation) directly affects NPP. Availability of essential minerals such as nitrogen and phosphorus can also affect NPP.

Wetlands (swamps and marshes) connect terrestrial and aquatic environments and are extremely productive. The most productive aquatic ecosystems are algal beds, coral reefs, and estuaries. The lack of available nutrient minerals in the sunlit region of the open ocean makes this area extremely unproductive, equivalent to an aquatic desert. Earth's major aquatic and terrestrial ecosystems are discussed in Chapter 54.

The relationship of productivity to biological diversity is complex

Conventional ecological theory once held that the more productive the ecosystem, the greater the biodiversity it supported. Now ecologists are seeing a recurring pattern worldwide: Ecosystems are more diverse as productivity increases, but at some point diversity actually declines with increasing productivity. For example, the resource-poor depths of the Atlantic Ocean's abyssal plain have a higher species richness than the productive shallow waters near the coasts, and intermediate depths exhibit the greatest diversity. Ecologists have little data to help explain the pattern, which holds true with rodents in Israel, birds in South America, and large mammals in Africa. Mathematical ecosystem models, however, suggest that a *patchy distribution of resources* reduces competition and allows the coexistence of a greater variety of organisms (see Chapter 51).

The bad news for global biodiversity is that humans are constantly enriching our environment, for example, with nitrogen and phosphorus inputs from fossil fuels, fertilizers, and livestock. This continual fertilization may make the Earth's ecosystems more and more productive, a shift that some ecologists think could cost the world a substantial loss of biodiversity. Other factors that affect species richness were discussed in Chapter 52.

Humans consume an increasingly greater percentage of global primary productivity

Humans consume far more of Earth's resources than any of the other millions of animal species. Peter Vitousek and colleagues at Stanford University calculated in 1986 how much of the global NPP is appropriated for the human economy. When both direct and indirect human impacts are accounted for, humans are estimated to use 25% of the global annual NPP and 40% of the annual NPP of terrestrial ecosystems. Essentially, humans' use of global productivity is competing with other species' needs for energy. Our use of so much of the world's productivity may contribute to the loss of many species that have unique roles in maintaining functional ecosystems, through extinction or genetic impoverishment. Clearly, at these levels of consumption and exploitation of the Earth's resources, human population growth, which was discussed in Chapter 51, is a serious threat to the planet's ability to support all its occupants.

MATTER CYCLES THROUGH ECOSYSTEMS

Matter moves in numerous cycles from one part of an ecosystem to another—that is, from one organism to another and from living organisms to the abiotic environment and back again. We call these cycles of matter **biogeochemical cycles** because they involve biological, geological, and chemical interactions. With respect to matter, Earth is essentially a **closed system** (see Chapter 6); for all practical purposes, matter cannot escape from Earth's boundaries. The materials used by organisms cannot be "lost," although they can end up in locations outside the reach of organisms for a long period. Usually, materials are reused and often recycled both within and among ecosystems.

We discuss four different biogeochemical cycles of matter—carbon, nitrogen, phosphorus, and water—as representative of all biogeochemical cycles. These four cycles are particularly important to organisms because they involve materials used to make the chemical components of cells. Carbon, nitrogen, and water have gaseous components and so cycle over large distances of the atmosphere with relative ease. Phosphorus, however, is an element that is never in a gaseous phase, and, as a result, only local cycling of phosphorus occurs easily.

CARBON DIOXIDE IS THE PIVOTAL MOLECULE OF THE CARBON CYCLE

Carbon must be available to organisms because proteins, nucleic acids, lipids, carbohydrates, and other molecules essential to life contain carbon. Carbon is present in the atmosphere as the gas carbon dioxide (CO_2), which makes up approximately 0.03% of the atmosphere. It is also present in the ocean and fresh water as dissolved carbon dioxide, that is, carbonate (CO_3^{2-}) and bicarbonate (HCO_3^-); other forms of dissolved inorganic carbon; and dissolved organic carbon from decay processes. Carbon is also present in rocks such as limestone ($CaCO_3$). The global movement of carbon between the abiotic environment, including the atmosphere and ocean, and organisms is known as the **carbon cycle** (Fig. 53–6).

During photosynthesis, plants, algae, and cyanobacteria remove carbon dioxide from the air and fix, or incorporate, it into complex organic compounds such as glucose. Plants use much of the glucose to make cellulose, starch, amino acids, nucleic acids, and other compounds. Thus, photosynthesis incorporates carbon from the abiotic environment into the biological compounds of producers. Many of these compounds are used as fuel for cellular respiration by the producer that made them, by a consumer that eats the producer, or by a decomposer that breaks down the remains of the producer or consumer. The process of cellular respiration returns carbon dioxide to the atmosphere. A similar carbon cycle occurs in aquatic ecosystems between aquatic organisms and dissolved carbon dioxide in the water.

Sometimes the carbon in biological molecules is not recycled back to the abiotic environment for some time. A large amount of carbon is stored in the wood of trees, where it may stay for several hundred years or even longer. In addition, millions of years ago vast coal beds formed from the bodies of ancient trees that were buried and subjected to anaerobic conditions before they had fully decayed. Similarly, the oils of unicellular marine organisms probably gave rise to the underground deposits of oil and natural gas that accumulated in the geological past. Coal, oil, and natural gas, called **fossil fuels** because they formed from the remains of ancient organisms, are vast depositories of carbon compounds, the end products of photosynthesis that occurred millions of years ago. Fossil fuels are nonrenewable resources; that is, Earth has a finite, or limited, supply of them. Although natural forces still form fossil fuels today, they are forming too slowly to replace the fossil fuel reserve we are using.

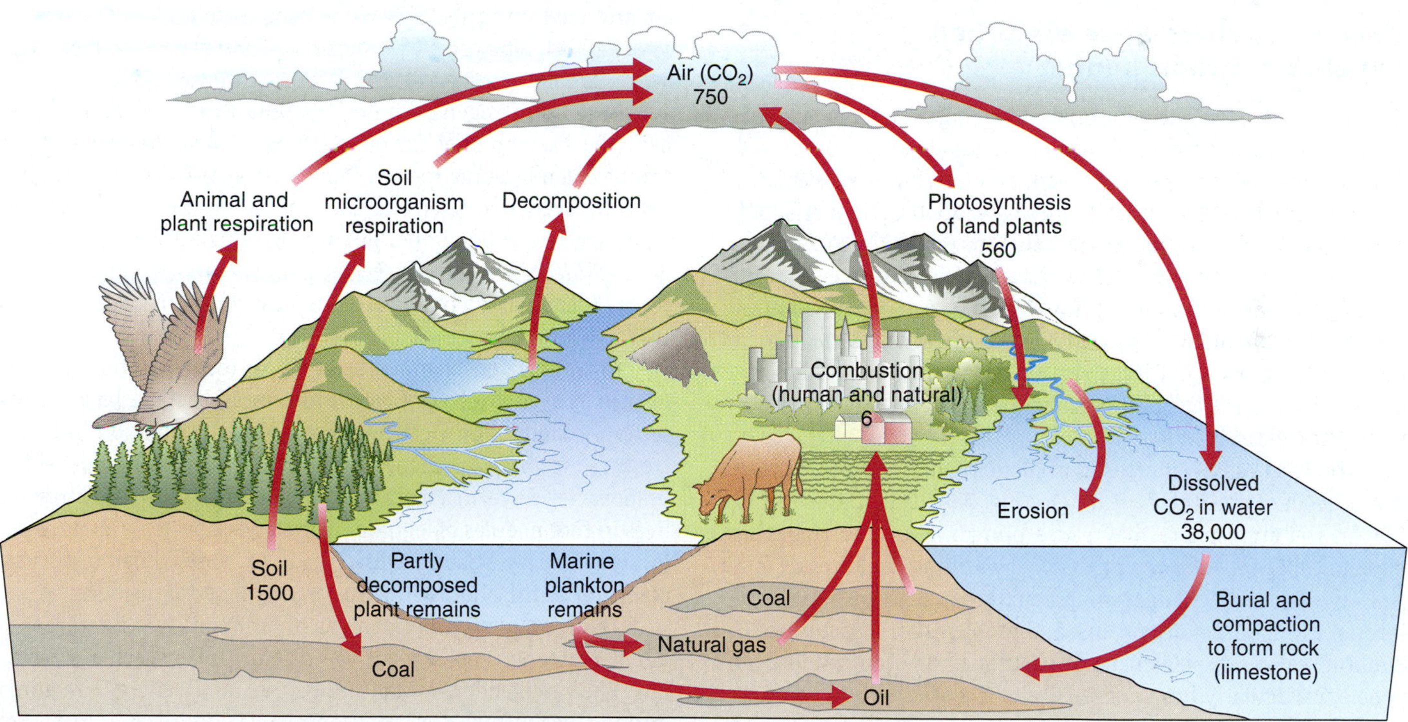

Figure 53–6 A simplified diagram of the carbon cycle. Carbon (as carbon dioxide) enters organisms when primary producers photosynthesize. Carbon returns to the environment by respiration, combustion, and erosion of limestone. The carbon in fossil fuels, limestone rock, and marine animal shells may take millions of years to cycle back to the biotic world. All but a tiny fraction of Earth's estimated 10^{23} g of carbon is buried in sedimentary rocks and fossil fuel deposits. The values shown for some of the active pools in the global carbon budget are expressed as 10^{15} g of carbon. For example, there are 1500×10^{15} g of carbon in the soil. *(Values from Schlesinger, W.H.* Biogeochemistry: An Analysis of Global Change, *2nd ed. Academic Press, San Diego, 1997, and based on several sources)*

The carbon in coal, oil, natural gas, and wood may be returned to the atmosphere by the process of burning, or combustion. In combustion, organic molecules are rapidly oxidized (combined with oxygen), converting them into carbon dioxide and water with an accompanying release of light and heat.

An even greater amount of carbon that is stored for millions of years is incorporated into the shells of marine organisms. When these organisms die, their shells sink to the ocean floor and are covered by sediments, forming seabed deposits thousands of meters thick. The deposits are eventually cemented together to form a sedimentary rock called limestone. Earth's crust is dynamically active, and over millions of years, sedimentary rock on the bottom of the sea floor may lift to form land surfaces. The summit of Mount Everest, for example, is composed of sedimentary rock. When the process of geological uplift exposes limestone, it slowly erodes away by chemical and physical weathering processes. This returns carbon to the water and atmosphere, where it is available to participate in the carbon cycle once again.

Thus, photosynthesis removes carbon from the abiotic environment and incorporates it into biological molecules. Cellular respiration, combustion, and erosion of limestone return carbon to the water and atmosphere of the abiotic environment.

Human activities have disturbed the global carbon budget

Before the Industrial Revolution around 1850, the global carbon cycle was in a steady state. Enormous amounts of carbon moved to and from the atmosphere, ocean, and terrestrial ecosystems, but these movements within the global carbon cycle just about cancelled out one another. Since 1850, our industrial society has required a lot of energy, and we have obtained this energy by burning increasing amounts of fossil fuels—coal, oil, and natural gas. This trend, along with a greater combustion of wood as a fuel and the burning of large sections of tropical forest, has released CO_2 into the atmosphere at a rate greater than the natural carbon cycle can handle.

The level of CO_2 increased particularly dramatically during the last half of the 20th century (see Fig. 55–11), and this rise of CO_2 in the atmosphere may cause human-induced changes in climate called *global warming.* Global warming could result in a rise in sea level, changes in precipitation patterns, death of forests, extinction of organisms, and problems for agriculture. It could force the displacement of thousands or even millions of people, particularly from coastal areas. A more thorough discussion of increasing atmospheric CO_2 and the potential impact of global warming is found in Chapter 55.

BACTERIA ARE ESSENTIAL TO THE NITROGEN CYCLE

Nitrogen is crucial for all organisms because it is an essential part of proteins, nucleic acids, and chlorophyll. Because Earth's atmosphere is about 78% nitrogen gas (N_2), it would appear that there could be no possible shortage of nitrogen for organisms. However, molecular nitrogen is so stable that it does not readily combine with other elements. Therefore, the N_2 molecule must be broken apart before the nitrogen atoms can combine with other elements to form proteins, nucleic acids, and chlorophyll. Chemical reactions that break up N_2 and combine nitrogen with such elements as oxygen and hydrogen require a great deal of energy.

The **nitrogen cycle,** in which nitrogen cycles between the abiotic environment and organisms, has five steps: nitrogen fixation, nitrification, assimilation, ammonification, and denitrification (Fig. 53–7). Bacteria are exclusively involved in all these steps except assimilation. The following paragraphs describe the nitrogen cycle that occurs on land; a similar cycle occurs in aquatic ecosystems.

The first step in the nitrogen cycle, biological **nitrogen fixation,** involves conversion of gaseous nitrogen (N_2) to ammonia (NH_3). This process is called nitrogen fixation because nitrogen is fixed into a form that organisms can use. Nitrogen can also be fixed as nitrate by combustion, volcanic action, lightning discharges, and industrial means; each of these processes supplies enough energy to break apart molecular nitrogen. Nitrogen-fixing bacteria, including cyanobacteria and certain other free-living bacteria, carry on biological nitrogen fixation in soil and aquatic environments. Nitrogen-fixing bacteria employ an enzyme called **nitrogenase** to break up molecular nitrogen and combine the resulting nitrogen atoms with hydrogen.

Because nitrogenase functions only in the absence of oxygen, the bacteria that fix nitrogen must insulate the enzyme from oxygen by some means. Some nitrogen-fixing bacteria live beneath layers of oxygen-excluding slime on the roots of several plant species. Other important nitrogen-fixing bacteria, in the genus *Rhizobium*, live in special swellings, or **nodules,** on the roots of legumes such as beans and peas and some woody plants (Fig. 53–8*a*).

In aquatic environments, most nitrogen fixation is performed by cyanobacteria. Filamentous cyanobacteria have special oxygen-excluding cells called **heterocysts** that function to fix nitrogen (Fig. 53–8*b*). Some water ferns have cavities in which cyanobacteria live, in a manner comparable to the way *Rhizobium* lives in root nodules of legumes. Other cyanobacteria fix nitrogen in symbiotic association with cycads or some other terrestrial plants, or as the photosynthetic partner of certain lichens.

The reduction of nitrogen gas to ammonia by nitrogenase is a remarkable accomplishment by living organisms, achieved without the tremendous heat, pressure, and energy required during the manufacture of commercial fertilizers. Even so, nitrogen-fixing bacteria must consume the energy in 12 g of glucose (or the equivalent) to fix a single gram of nitrogen biologically.

The second step of the nitrogen cycle is **nitrification,** the conversion of ammonia (NH_3) or ammonium (NH_4^+, formed when water reacts with ammonia) to nitrate (NO_3^-). Soil bacteria are responsible for the two-step process of nitrification. First, the soil bacteria *Nitrosomonas* and *Nitrococcus* convert ammonia to nitrite (NO_2^-). Then the soil bacterium *Nitrobacter* oxidizes nitrite to nitrate. The process of nitrification furnishes these bacteria, called nitrifying bacteria, with energy.

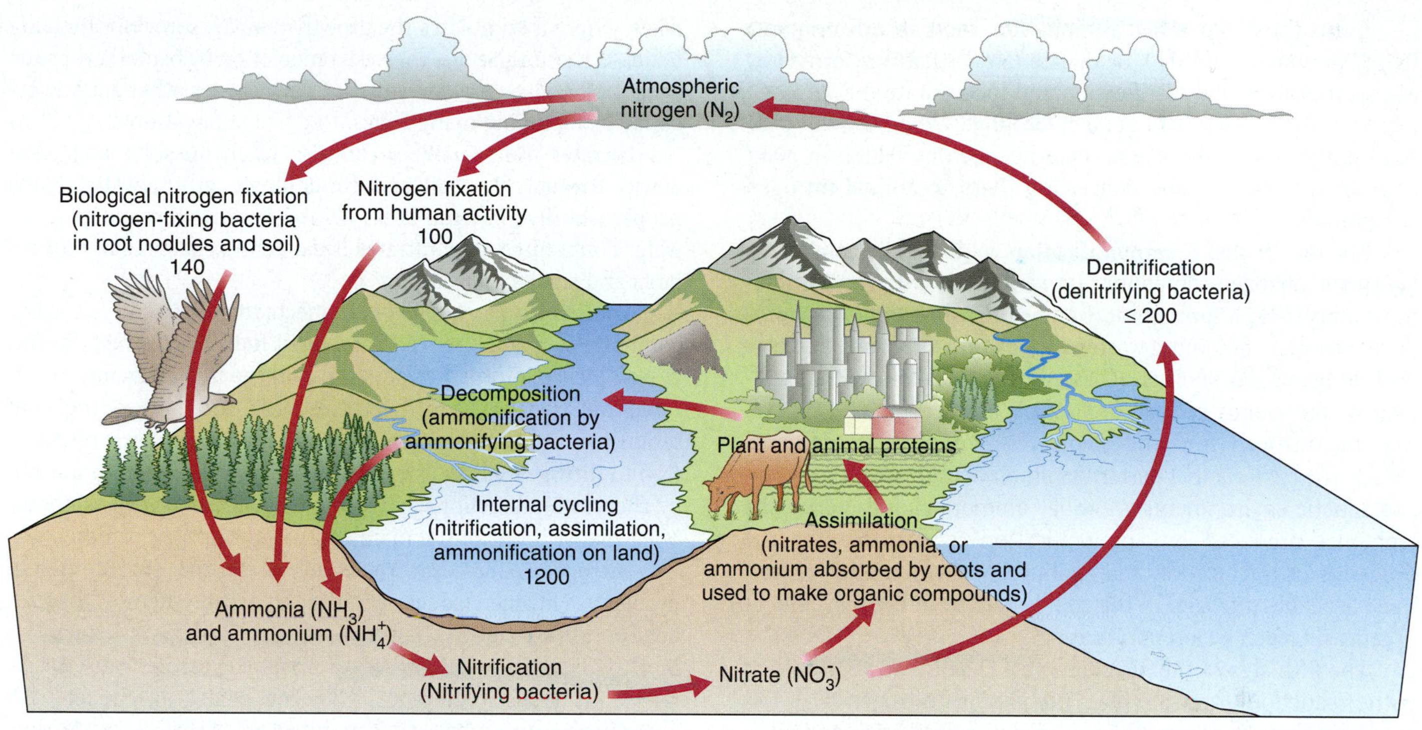

Figure 53–7 A simplified diagram of the nitrogen cycle. Nitrogen-fixing bacteria convert atmospheric nitrogen (N_2) into ammonia (NH_3). Ammonia or ammonium is converted to nitrate (NO_3^-) by nitrifying bacteria in the soil. Plants assimilate nitrate, ammonia, or ammonium, producing proteins and nucleic acids in the process; then animals eat plant proteins and produce animal proteins. Ammonifying bacteria break down the nitrogen compounds of dead organisms, releasing ammonia that can be reused. Nitrogen is returned to the atmosphere by denitrifying bacteria, which convert nitrate to molecular nitrogen. The largest pool of nitrogen, estimated at 3.9×10^{21} g, is in the atmosphere. The values shown for selected nitrogen fluxes in the global nitrogen budget are expressed as 10^{12} g of nitrogen per year and represent terrestrial values. For example, each year 100×10^{12} g of nitrogen is fixed as a result of human activity. *(Values from Schlesinger, W.H.* Biogeochemistry: An Analysis of Global Change, *2nd ed. Academic Press, San Diego, 1997, and based on several sources)*

(a)

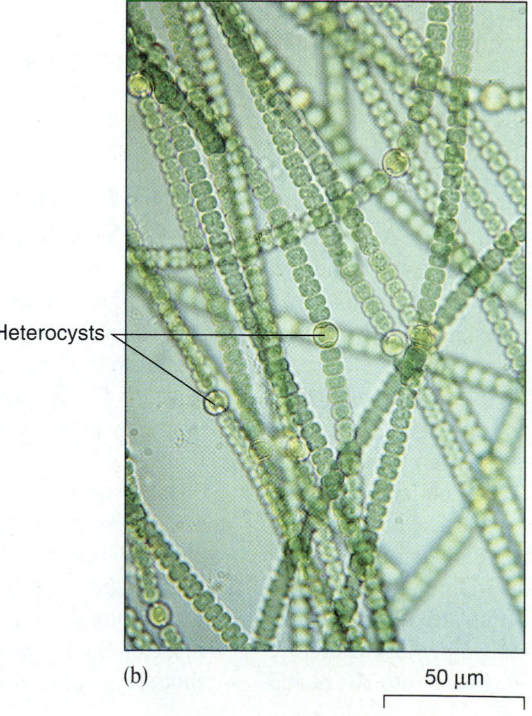

(b)

Figure 53–8 Nitrogen fixation. Nitrogenase, the enzyme involved in biological nitrogen fixation, functions only in the absence of oxygen. **(a)** Root nodules on the roots of a pea plant provide an oxygen-free environment for nitrogen-fixing *Rhizobium* bacteria that live in them. **(b)** LM of *Anabaena*, a cyanobacterium with distinctive oxygen-excluding heterocysts in which nitrogen fixation occurs. *(a, Hugh Spencer/Photo Researchers, Inc.; b, Dennis Drenner)*

In the third step, called **assimilation,** roots absorb ammonia (NH_3), ammonium (NH_4^+), or nitrate (NO_3^-) that was formed by nitrogen fixation and nitrification and incorporate the nitrogen into proteins, nucleic acids, and chlorophyll. When animals consume plant tissues, they assimilate nitrogen by taking in plant nitrogen compounds and converting them to animal nitrogen compounds.

The fourth step is **ammonification,** which is the conversion of organic nitrogen compounds into ammonia (NH_3) and ammonium ions (NH_4^+). Ammonification begins when organisms produce nitrogen-containing wastes such as urea (in urine) and uric acid (in the wastes of birds). These substances, along with the nitrogen compounds in dead organisms, are decomposed, releasing the nitrogen into the abiotic environment as ammonia (NH_3). The bacteria that perform ammonification in both the soil and aquatic environments are called ammonifying bacteria. The ammonia produced by ammonification is available for the processes of nitrification and assimilation. As a matter of fact, most available nitrogen in the soil derives from the recycling of organic nitrogen by ammonification.

The fifth step of the nitrogen cycle is **denitrification,** which is the reduction of nitrate (NO_3^-) to gaseous nitrogen (N_2). Denitrifying bacteria reverse the action of nitrogen-fixing and nitrifying bacteria by returning nitrogen to the atmosphere as nitrogen gas. Denitrifying bacteria are anaerobic and therefore live and grow best where there is little or no free oxygen. For example, they are found deep in the soil near the water table, an environment that is nearly oxygen-free.

Human activities have changed the global nitrogen budget

Human activities have disturbed the balance of the global nitrogen cycle. During the 20th century, humans more than doubled the amount of fixed nitrogen (nitrogen that has been chemically combined with hydrogen, oxygen, or carbon) entering the global nitrogen cycle. The excess nitrogen is seriously altering many terrestrial and aquatic ecosystems.

Large quantities of nitrogen fertilizer, both ammonia and nitrate, are produced from nitrogen gas for agriculture. According to the International Fertilizer Industry Association, 140 million metric tons of fertilizer were used worldwide in 2000; this amount was equal to 23.1 kg (51 lb) of fertilizer per person. The increasing use of fertilizer has resulted in higher crop yields.

In 1997 Peter Vitousek and seven other ecologists published an extensive review of scientific research on the negative environmental impacts of human-produced nitrogen. One serious problem with nitrogen is that it is extremely mobile and is easily transferred from the land to rivers to estuaries to the ocean. Thus, the overuse of commercial fertilizer on the land can cause water quality problems that may help explain long-term declines in many coastal fisheries. The amount of nitrate or ammonium in most aquatic ecosystems is in limited supply and therefore limits the growth of algae. Rain washes fertilizer into rivers and lakes, where it stimulates the growth of algae, some of which are toxic. As these algae die, their decomposition by bacteria robs the water of dissolved oxygen, which in turn causes other aquatic organisms, including many fishes, to die of suffocation.

Nitrates from fertilizer can also leach (dissolve and wash down) through the soil and contaminate groundwater. Many people who live in rural areas drink groundwater, and groundwater contaminated by nitrates is dangerous, particularly for infants and small children.

Another human activity that affects the nitrogen cycle is the combustion of fossil fuels. When fossil fuels are burned, the nitrogen locked in organic compounds in the fuel is chemically altered and transferred to the atmosphere. In addition, the high temperatures of combustion convert some atmospheric nitrogen (N_2) to **nitrogen oxides,** trace gases in the atmosphere produced by chemical interactions between nitrogen and oxygen. Automobile exhaust is one of the main sources of nitrogen oxides.

Nitrogen oxides exacerbate several serious environmental problems. Nitrogen oxides are a necessary ingredient in the production of **photochemical smog,** a mixture of several air pollutants that can injure plant tissues, irritate eyes, and cause respiratory problems in humans. Nitrogen oxides also react with water in the atmosphere to form nitric acid (HNO_3) and nitrous acid (HNO_2). When these and other acids leave the atmosphere as **acid deposition,** they cause the pH of surface waters (lakes and streams) and soils to decrease. Acid deposition has been linked to declining animal populations in aquatic ecosystems. On land, acid deposition alters soil chemistry so that certain essential nutrient minerals such as calcium and potassium wash out of the soil and are therefore unavailable for plants. Nitrous oxide (N_2O), one of the nitrogen oxides, retains heat in the atmosphere (like CO_2) and so promotes global warming. Nitrous oxide also contributes to the depletion of ozone in the stratosphere. (See Chapter 3 for a discussion of acids and pH and Chapter 55 for a discussion of global warming and stratospheric ozone depletion.)

THE PHOSPHORUS CYCLE LACKS A GASEOUS COMPONENT

In the **phosphorus cycle,** phosphorus, which does not exist in a gaseous state and therefore does not enter the atmosphere, cycles from the land to sediments in the ocean and back to the land (Fig. 53–9). As water runs over rocks containing phosphorus, it gradually erodes the surface and carries off inorganic phosphate (PO_4^{3-}).

The erosion of phosphorus rocks releases phosphate into the soil, where it is taken up by roots in the form of inorganic phosphates. Once in cells, phosphates are incorporated into a variety of biological molecules, including nucleic acids, ATP, and the phospholipids that make up cell membranes. Animals obtain most of their required phosphorus from the food they eat, although in some places drinking water may contain a substantial

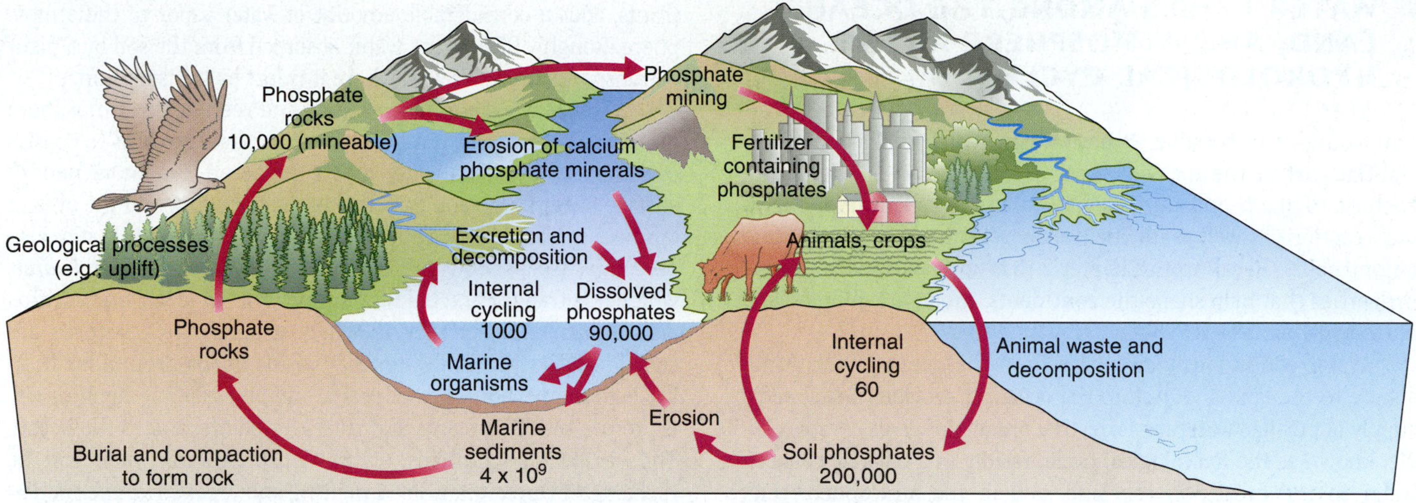

Figure 53-9 A simplified diagram of the phosphorus cycle. Recycling of phosphorus (as phosphate) is slow because no form of phosphorus is gaseous. Phosphate that becomes part of marine sediments may take millions of years to solidify into rock, uplift as mountains, and erode again to become available to organisms. Some values of the global phosphorus budget are given, in units of 10^{12} g phosphorus per year. For example, each year 60×10^{12} g of phosphate cycles from the soil to terrestrial organisms and back to the soil. *(Values from Schlesinger, W.H.* Biogeochemistry: An Analysis of Global Change, *2nd ed. Academic Press, San Diego, 1997, and based on several sources)*

amount of inorganic phosphate. Phosphate released by decomposers becomes part of the pool of inorganic phosphate in the soil that can be reused by plants. Thus, like carbon and nitrogen, phosphorus moves through the food web as one organism consumes another.

Phosphorus cycles through aquatic ecosystems in much the same way that it does through terrestrial ecosystems. Dissolved phosphate enters aquatic ecosystems through absorption by algae and aquatic plants, which are consumed by plankton and larger organisms. These are in turn eaten by a variety of fishes and mollusks. Ultimately, decomposers that break down wastes and dead organisms release inorganic phosphate into the water, making it available for use again by aquatic producers.

Some phosphate in the aquatic food web finds its way back to the land. A tiny fraction of fish and aquatic invertebrates are eaten by sea birds, which may defecate on the land where they roost. Guano, the manure of sea birds, contains large amounts of phosphate and nitrate. Once on land, roots may absorb these minerals. The phosphate contained in guano may enter terrestrial food webs in this way, although the amounts involved are quite small.

Phosphate can be lost for varying time periods from biological cycles. Some phosphate is carried from the land by streams and rivers to the ocean, where it can be deposited on the sea floor and remain for millions of years. The geological process of uplift may someday expose these seafloor sediments as new land surfaces, from which phosphate will be once again eroded. Phosphate deposits, including guano, are also mined for agricultural use in phosphate fertilizers.

Humans affect the natural cycling of phosphorus

You have seen that once phosphorus moves from terrestrial to aquatic ecosystems, its return to land is very slow. Certain human activities accelerate the long-term loss of phosphorus from terrestrial ecosystems. For example, corn grown in Iowa, which contains phosphate absorbed from the soil, may be used to fatten cattle in an Illinois feedlot. Part of the phosphate absorbed by the roots of the corn plants thus ends up in the feedlot wastes, much of which is used as fertilizer and may eventually wash into the Mississippi River. Beef from the Illinois cattle may be consumed by people living far away, in New York City, for instance, where more of the phosphate ends up in human wastes and is flushed down toilets into the New York City sewer system. Sewage treatment rarely removes phosphates, and so they cause water quality problems in rivers and lakes. To compensate for the steady loss of phosphate from their land, farmers must add phosphate fertilizer to their fields. That fertilizer is extracted from the large deposits of phosphate rock in Florida, Tennessee, and several other states.

In natural terrestrial communities, very little phosphorus is lost from the cycle, but few communities today are in a natural state, that is, unaltered in some way by humans. Phosphorus loss from the soil is accelerated by land-denuding practices such as the clearcutting of timber and by erosion of agricultural and residential lands. For practical purposes, phosphorus that washes from the land into the ocean is permanently lost from the terrestrial phosphorus cycle (and from further human use), for it remains in the ocean for millions of years.

WATER MOVES AMONG THE OCEAN, LAND, AND ATMOSPHERE IN THE HYDROLOGICAL CYCLE

Life would be impossible without water, which makes up a substantial part of the mass of most organisms. All species, from bacteria to plants and animals, use water as a medium for chemical reactions as well as for the transport of materials within and among cells. (Recall from Chapter 2 that water has many unique properties that help shape the continents, moderate climate, and allow organisms to survive.)

Water continuously circulates from the ocean to the atmosphere to the land and back to the ocean. It provides a renewable supply of purified water for terrestrial organisms. This complex cycle, known as the **hydrological cycle,** results in a balance between water in the ocean, on the land, and in the atmosphere (Fig. 53–10). Water moves from the atmosphere to the land and ocean in the form of precipitation (rain, sleet, snow, or hail). When water evaporates from the ocean surface and from soil, streams, rivers, and lakes, it eventually condenses and forms clouds in the atmosphere. In addition, **transpiration,** the loss of water vapor from land plants, adds a considerable amount of water vapor to the atmosphere. Roughly 97% of the water absorbed from the soil by a plant is transported to the leaves, where it is lost by transpiration.

Water may evaporate from land and reenter the atmosphere directly. Alternatively, it may flow in rivers and streams to coastal **estuaries,** where fresh water meets the ocean. The movement of surface water from land to ocean is called *runoff,* and the area of land being drained by runoff is called a *watershed.* Water also percolates (seeps) downward in the soil to become *groundwater,* where it is trapped and held for a time. The underground caverns and porous layers of rock in which groundwater is stored are called *aquifers.* Groundwater may reside in the ground for hundreds to many thousands of years, but eventually it supplies water to the soil, to streams and rivers, to plants, and to the ocean. The removal by humans of more groundwater than can be recharged by precipitation or melting snow, called *aquifer depletion,* eliminates groundwater as a water resource.

Regardless of its physical form (solid, liquid, or vapor) or location, every molecule of water eventually moves through the hydrological cycle. Tremendous quantities of water are cycled annually between Earth and its atmosphere. The volume of water

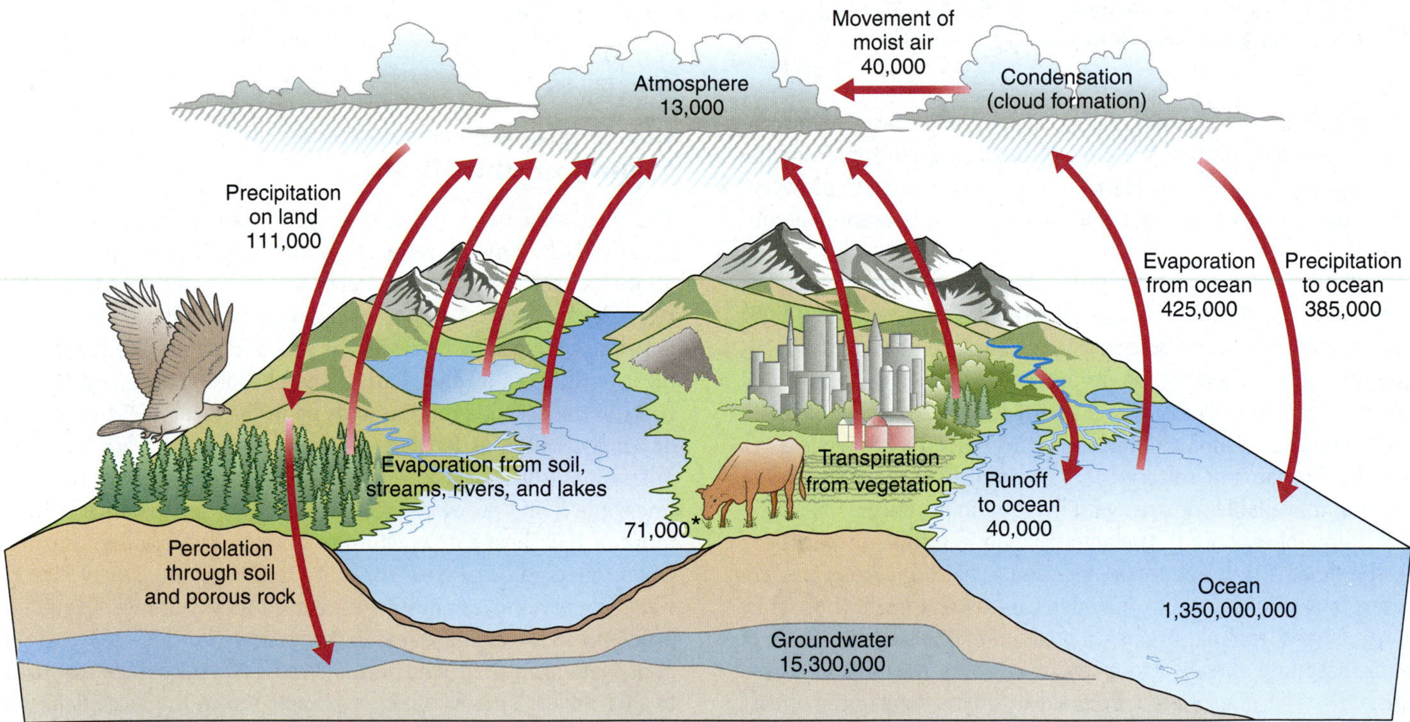

Figure 53–10 A simplified diagram of the hydrological cycle. Tremendous quantities of water are cycled annually between Earth and its atmosphere. An estimated 425,000 km^3 enter the atmosphere by evaporation from the ocean each year. Approximately 90% of this water reenters the ocean directly as precipitation over water; the remainder falls on land. Although some water molecules are unavailable for thousands of years (e.g., locked up in polar ice or underground aquifers), all water molecules eventually pass through the hydrological cycle. The global water budget values shown for pools are expressed as cubic kilometers; values for fluxes (movements associated with arrows) are in cubic kilometers per year. The starred value (71,000 km^3/yr) is the sum of both transpiration and evaporation from soil, streams, rivers, and lakes. *(Values from Schlesinger, W.H.* Biogeochemistry: An Analysis of Global Change, *2nd ed. Academic Press, San Diego, 1997, and based on several sources)*

entering the atmosphere each year is estimated at about 425,000 km^3. Approximately 90% of this water reenters the ocean directly as precipitation over water; the remainder falls on land. As is true of the other cycles, water (in the form of glaciers and polar ice caps) can be lost from the cycle for thousands of years.

ABIOTIC FACTORS INFLUENCE ECOSYSTEMS

We have seen how ecosystems depend on the abiotic environment to supply energy and essential materials (in biogeochemical cycles). Abiotic factors such as solar radiation, the atmosphere, the ocean, weather and climate, and fire also affect ecosystems. For a given abiotic factor, each organism living in an ecosystem has an optimal range in which it can survive and reproduce. Water and temperature are probably the two abiotic factors that most affect organisms in ecosystems.

THE SUN WARMS THE EARTH

The sun makes life on Earth possible. It warms the planet to habitable temperatures. Without the sun's energy, the temperature on planet Earth would approach absolute zero (−273° C), and all water would be frozen, even in the ocean. The sun powers the hydrological cycle, carbon cycle, and other biogeochemical cycles and is the primary determinant of climate. The sun's energy is captured by photosynthetic organisms and used to make organic compounds that are required by almost all forms of life. Most of our fuels, such as wood, oil, coal, and natural gas, represent solar energy captured by photosynthetic organisms. Without the sun, almost all life on Earth would cease (see *Focus On: Life Without the Sun,* on the following page) for an interesting exception).

The sun's energy, which is the product of a massive nuclear fusion reaction, is emitted into space in the form of electromagnetic radiation, especially ultraviolet, visible, and infrared radiation. About one-billionth of the total energy released by the sun strikes the atmosphere, and of this tiny trickle of energy, a minute part operates the biosphere. On average, 30% of the solar radiation that falls on Earth is immediately reflected away by clouds and surfaces, especially snow, ice, and the ocean (Fig. 53–11). The remaining 70% is absorbed by Earth's surface and atmosphere and runs the water cycle, drives winds and ocean currents, powers photosynthesis, and warms the planet. Ultimately, all this energy is lost by the continual radiation of long-wave infrared (heat) energy into space. If heat gains were not exactly balanced by losses, Earth would heat up or cool down.

Temperature changes with latitude

The most significant local variations in Earth's temperature are produced because the sun's energy does not uniformly reach all places. A combination of our planet's roughly spherical shape and the tilted angle of its axis produces a great deal of variation in the exposure of the surface to the energy delivered by sunlight. The sun's rays strike nearly vertically near the equator, concentrating the energy and producing warmer temperatures. Near

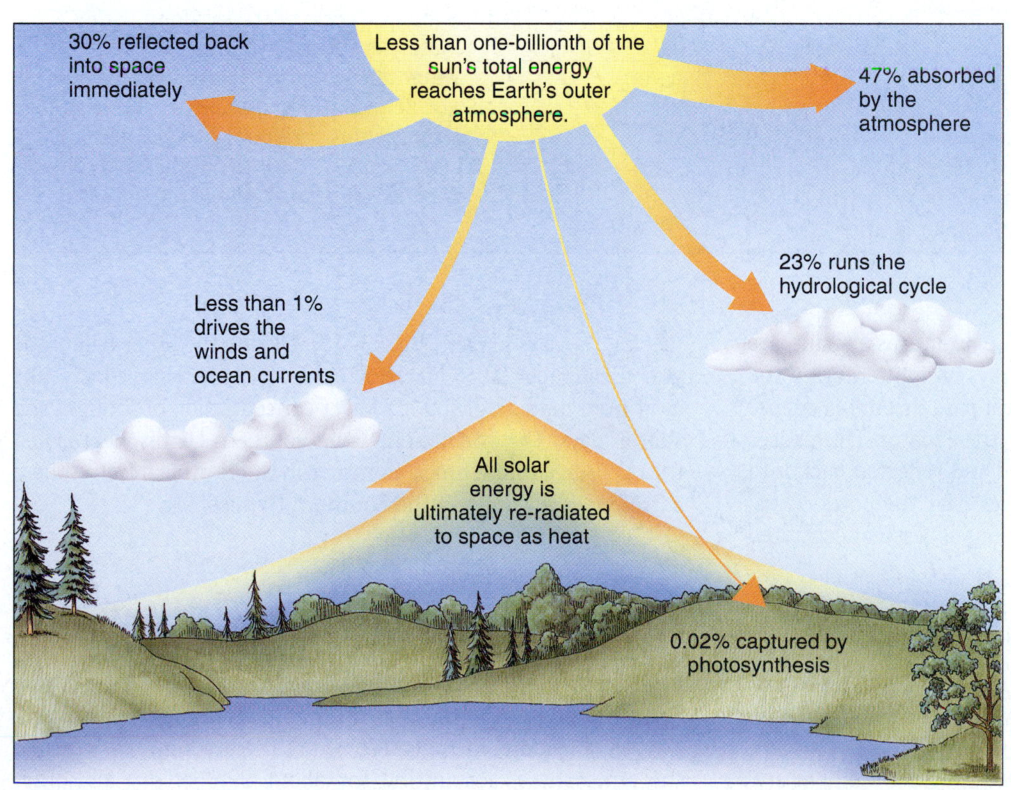

Figure 53–11 The fate of solar radiation that reaches Earth. Most of the energy released by the sun never reaches Earth. The solar energy that does reach Earth warms the planet's surface, drives the hydrological and other biogeochemical cycles, produces our climate, and powers almost all life through the process of photosynthesis.

FOCUS ON Life Without the Sun

In 1977 an oceanographic expedition aboard the research submersible Alvin studied the Galapagos Rift, a deep cleft in the ocean floor off the coast of Ecuador. The expedition revealed, on the floor of the abyss, a series of **hydrothermal vents** where sea water apparently had penetrated and been heated by the hot rocks below (see Chapter 20). During its time within Earth, the water had also become charged with mineral compounds, including hydrogen sulfide (H_2S), which is toxic to most species.

At the tremendous depths of greater than 2500 m (8200 ft) found in the Galapagos Rift, there is no light for photosynthesis. But hydrothermal vents support a rich variety of life forms *(see figure),* in contrast with the surrounding "desert" of the abyssal floor. Many of the species in these oases of life are not found in other habitats. For example, giant blood-red tube worms almost 3 m (10 ft) in length cluster in great numbers around the vents. Other animals around the hydrothermal vents include unique species of clams, crabs, barnacles, and mussels. Since 1977 such hydrothermal vent communities, some as large as football fields, have been identified at many other oceanic sites.

Scientists initially wondered what energy source sustains these organisms. Most deep-sea communities depend on the organic matter that drifts down from the surface waters; in other words, they rely on energy ultimately derived from photosynthesis. Hydrothermal vent communities, however, are too densely clustered and too productive to be dependent on chance encounters with organic material from surface waters. Instead, chemoautotrophic bacteria occupy the base of the food web in these aquatic oases. These bacteria possess enzymes that catalyze the oxidation of hydrogen sulfide to water and sulfur or sulfate. Such chemical reactions are exergonic and provide the energy required to fix CO_2, which is dissolved in the water, into organic compounds. Many of the animals consume the bacteria directly by filter-feeding, but others, such as the giant tube worms, get their energy from bacteria that live in their tissues.

Scientists continue to generate questions about hydrothermal vent communities. How, for example, do the organisms find and colonize vents, which are ephemeral and widely scattered on the ocean floor? How have the inhabitants of these communities adapted to survive the harsh living conditions, including high pressure, high temperatures, and toxic chemicals? As vent community research continues, scientists hope to discover answers to these and other questions.

A hydrothermal vent community. Chemoautotrophic bacteria living in the tissues of these tube worms *(Riftia pachyptila)* extract energy from hydrogen sulfide to manufacture organic compounds. Because these worms lack digestive systems, they depend on the organic compounds provided by the endosymbiotic bacteria, along with materials filtered from the surrounding water and digested extracellularly. Also visible in the photograph are some filter-feeding mollusks *(yellow)* and a crab *(white).* *(D. Foster, Science VU-WHOI/Visuals Unlimited)*

the poles the sun's rays strike more obliquely and, as a result, are spread over a larger surface area. Also, rays of light entering the atmosphere obliquely near the poles must pass through a deeper envelope of air than those entering near the equator. This causes more of the sun's energy to be scattered and reflected back into space, which further lowers temperatures near the poles. Thus, the solar energy that reaches polar regions is less concentrated and produces lower temperatures.

Temperature changes with season

Seasons are primarily determined by Earth's inclination on its axis (23.5 degrees from a line drawn perpendicular to the orbital plane). During half of the year (March 21 to September 22) the Northern Hemisphere tilts toward the sun, concentrating the sunlight and making the days longer (Fig. 53–12). During the other half of the year (September 22 to March 21) the Northern Hemisphere tilts away from the sun, giving it a lower concentration of sunlight and shorter days. The orientation of the Southern Hemisphere is just the opposite at these times. Summer in the Northern Hemisphere corresponds to winter in the Southern Hemisphere.

THE ATMOSPHERE CONTAINS SEVERAL GASES ESSENTIAL TO ORGANISMS

The atmosphere is an invisible layer of gases that envelops Earth. Oxygen (21%) and nitrogen (78%) are the predominant gases in the atmosphere, accounting for about 99% of dry air; other

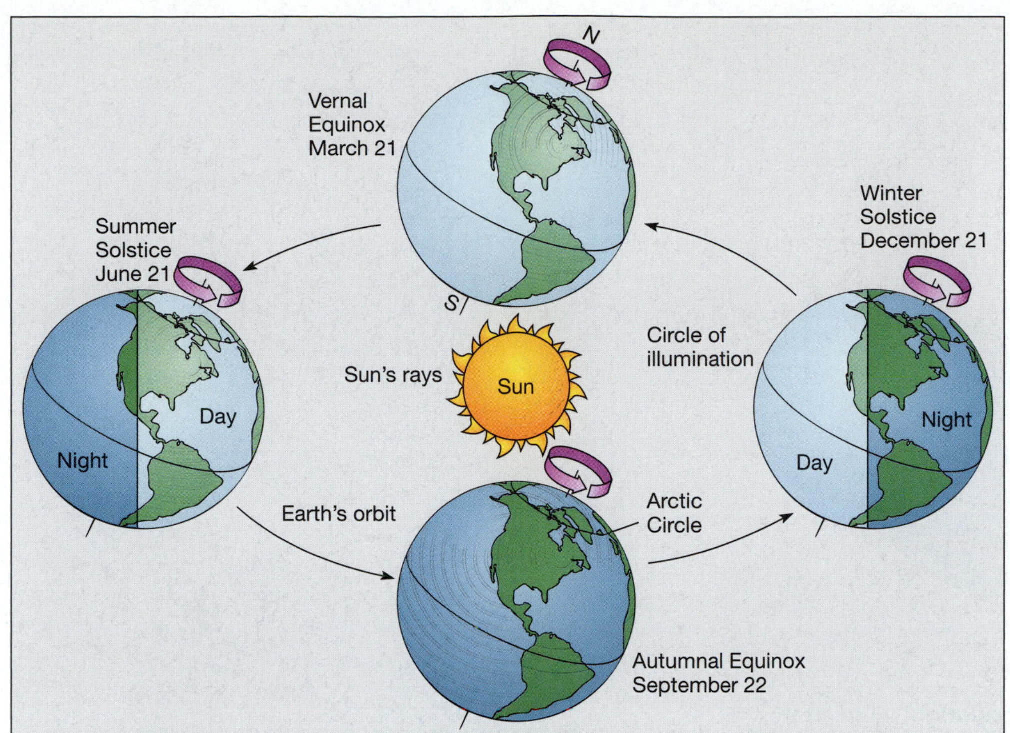

Figure 53–12 Seasonal changes in temperature. Earth's inclination on its axis remains the same as it travels around the sun. Thus, the sun's rays hit the Northern Hemisphere obliquely during winter months and more directly during summer months. In the Southern Hemisphere, the sun's rays are oblique during its winter, which corresponds to the Northern Hemisphere's summer. At the equator, the sun's rays are approximately vertical on March 21 and September 22.

gases, including argon, carbon dioxide, neon, and helium, make up the remaining 1%. In addition, water vapor and trace amounts of various air pollutants, such as methane, ozone, dust particles, pollen, microorganisms and chlorofluorocarbons (CFCs) are present. Atmospheric oxygen is essential to plants, animals, and other organisms that respire aerobically, whereas carbon dioxide is required by plants and other photosynthetic organisms. The atmosphere becomes less dense as it extends outward into space; most of its mass is found near Earth's surface.

The atmosphere performs several ecologically important functions. It protects the surface from most of the sun's ultraviolet radiation and x rays and from lethal amounts of cosmic rays from space. Without this shielding by the atmosphere, life as we know it would cease. Although the atmosphere protects Earth from high-energy radiation, it allows visible light and some infrared radiation to penetrate, and they warm the surface and the lower atmosphere. This interaction between the atmosphere and solar energy is responsible for weather and climate.

Organisms depend on the atmosphere, but they also help maintain and, in certain instances, modify its composition. For example, atmospheric oxygen is thought to have increased to its present level as a result of billions of years of photosynthesis. Today an approximate balance between oxygen-producing photosynthesis and oxygen-using aerobic respiration helps to maintain the level of atmospheric oxygen.

The sun drives global atmospheric circulation

In large measure, differences in temperature caused by variations in the amount of solar energy at different locations on Earth drive the circulation of the atmosphere. The warm surface near the equator heats the air with which it comes into contact, causing this air to expand and rise. As the warm air rises, it flows away from the equator, cools, and sinks again (Fig. 53–13). Much of it recirculates back to the same areas it left, but the remainder splits and flows in two directions, toward the poles. The air chills enough to sink to the surface at about 30 degrees north and south latitudes. Similar upward movements of warm air and its subsequent flow toward the poles occur at higher latitudes, farther from the equator. At the poles, the cold polar air sinks and flows toward the lower latitudes, generally beneath the sheets of warm air that simultaneously flow toward the poles. The constant motion of air transfers heat from the equator toward the poles, and as the air returns, it cools the land over which it passes. This continuous turnover does not equalize temperatures over Earth's surface, but it does moderate them.

The atmosphere exhibits complex horizontal movements

In addition to global circulation patterns, the atmosphere exhibits complex horizontal movements that are commonly referred to as **winds.** The nature of wind, with its turbulent gusts, eddies, and lulls, is complex and difficult to understand or predict. It results in part from differences in atmospheric pressure and from the rotation of Earth.

The gases that constitute the atmosphere have weight and exert a pressure that is, at sea level, about 1013 millibars (14.7 lb/in^2). Air pressure is variable, however, and changes with altitude, temperature, and humidity. Winds tend to blow from areas of high atmospheric pressure to areas of low pressure; the greater the difference between the high and low pressure areas, the stronger the wind.

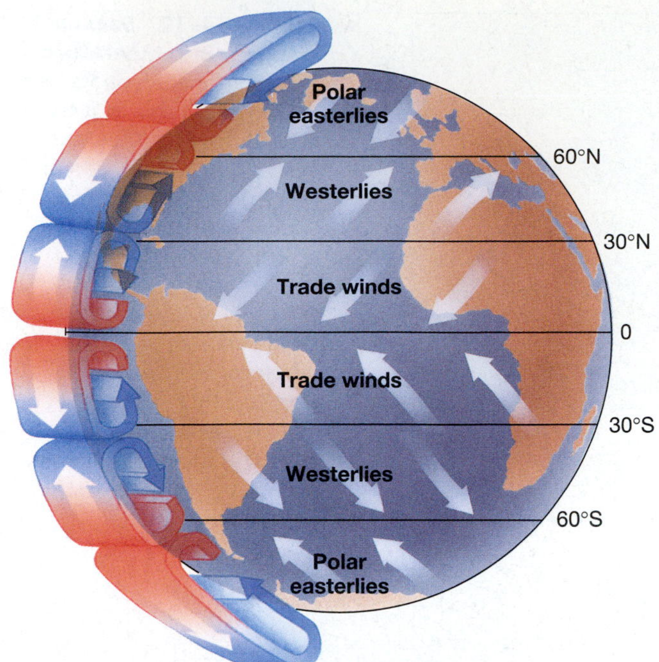

Figure 53–13 Atmospheric circulation. The greatest solar energy input occurs at the equator, heating air most strongly in that area. The air rises and travels poleward *(left)* but is cooled in the process so that much of it descends again around 30 degrees latitude in both hemispheres. At higher latitudes the patterns of air movement are more complex.

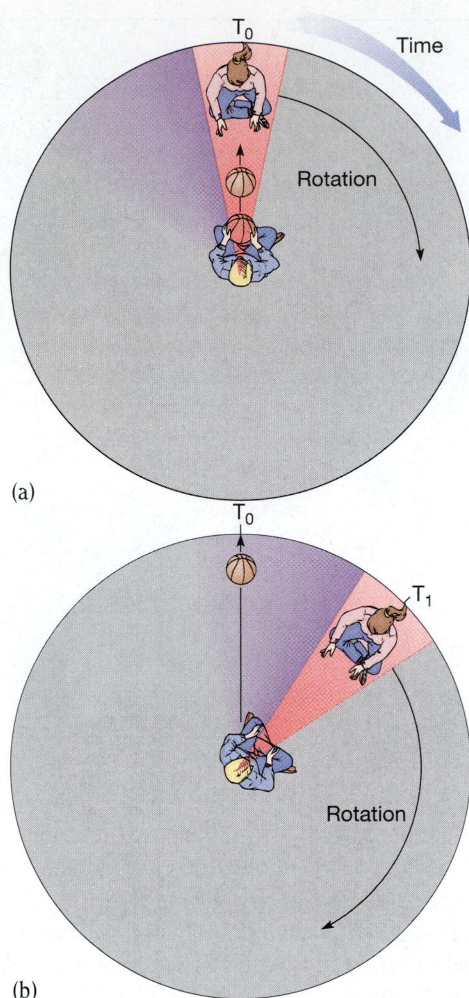

Figure 53–14 The Coriolis effect as demonstrated by a merry-go-round. The center of the merry-go-round *(shown from above)* corresponds to the South Pole, and the outer edge to the equator. **(a)** If, while sitting on the merry-go-round, you throw a ball to a friend (also on the merry-go-round) at time T_0, when the merry-go-round is rotating clockwise, the ball at time T_1 **(b)** appears to curve to the left instead of going straight. Because of Earth's rotation, winds and ocean currents curve to the left (counterclockwise) in the Southern Hemisphere and to the right (clockwise) in the Northern Hemisphere.

Earth's rotation influences the direction that wind blows. Because Earth rotates from west to east, wind swerves to the right in the Northern Hemisphere and to the left in the Southern Hemisphere. This tendency of moving air to be deflected from its path by Earth's rotation is known as the **Coriolis effect.** The Coriolis effect can be visualized by imagining that you and a friend are sitting about 3 m apart on a merry-go-round that is turning clockwise (Fig. 53–14). Suppose you throw a ball directly to your friend. By the time the ball reaches the place where your friend was, he or she is no longer in that spot. The ball will have swerved far to the left of your friend. This is how the Coriolis effect works in the Southern Hemisphere. To visualize how the Coriolis effect works in the Northern Hemisphere, imagine that you and your friend are sitting on the same merry-go-round, only this time it is moving counterclockwise. Now when you throw the ball, it will appear to swerve far to the right of your friend.

THE GLOBAL OCEAN COVERS MOST OF EARTH'S SURFACE

The global ocean is a huge body of salt water that surrounds the continents and covers almost three-fourths of Earth's surface. It is a single, continuous body of water, but geographers divide it into four sections separated by the continents: the Pacific, Atlantic, Indian, and Arctic Oceans. The Pacific Ocean is the largest by far: It covers one-third of Earth's surface and contains more than half of Earth's water.

Surface ocean currents are driven by winds

The persistent prevailing winds blowing over the ocean produce mass movements of surface ocean water known as **ocean currents** (Fig. 53–15). Circular ocean currents generated by the prevailing winds are called *gyres*. For example, in the North Atlantic, the tropical trade winds tend to blow toward the west,

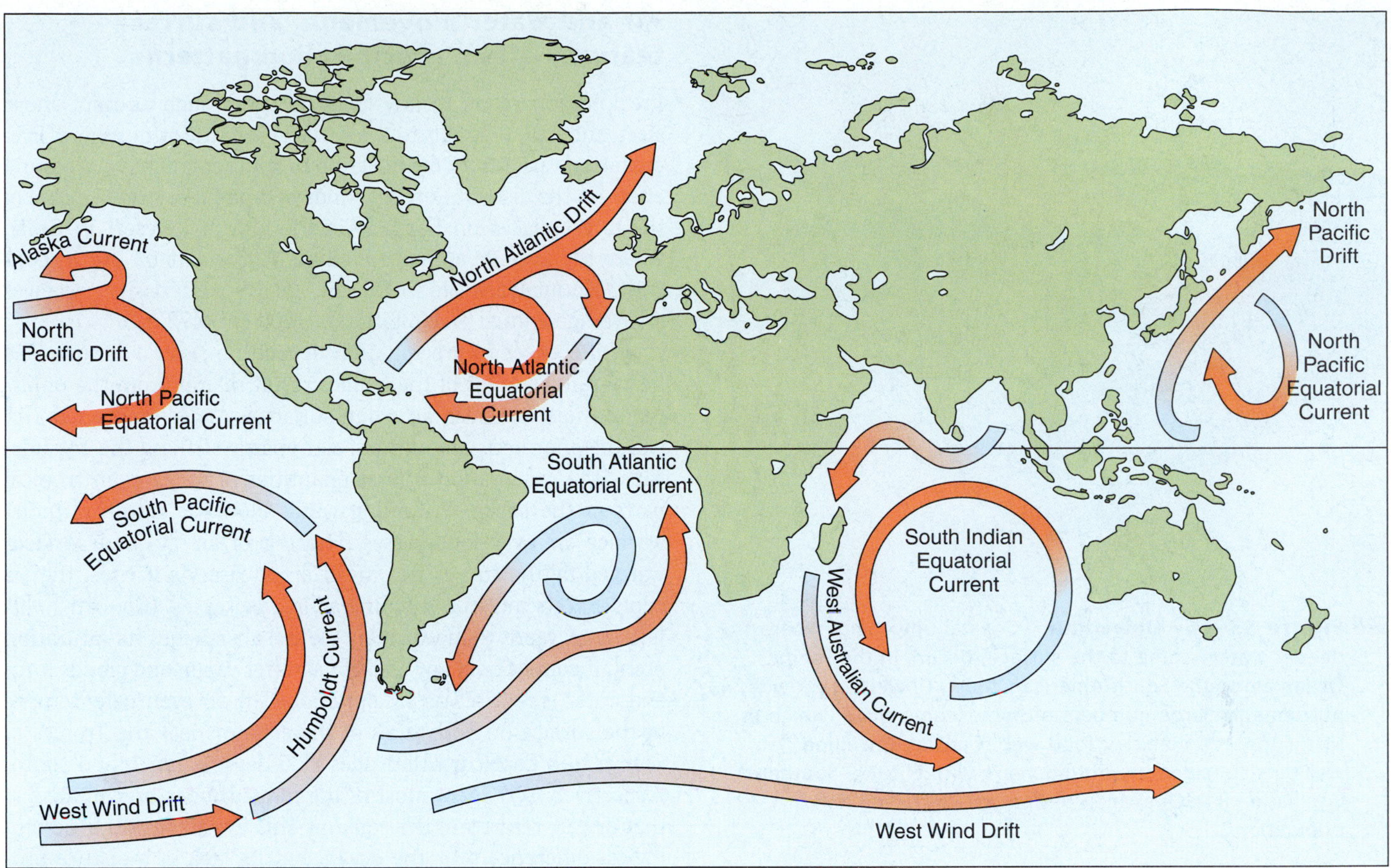

Figure 53-15 Major surface ocean currents. The basic pattern of ocean currents is caused largely by the action of winds. The main ocean current flow—clockwise in the Northern Hemisphere and counterclockwise flow in the Southern Hemisphere—results partly from the Coriolis effect.

whereas the westerlies in the mid-latitudes blow toward the east (see Fig. 53–13). This helps establish a clockwise gyre in the North Atlantic. Thus, surface ocean currents and winds tend to move in the same direction, although there are many variations on this general rule.

The paths that surface ocean currents travel are partly caused by the Coriolis effect. Earth's rotation from west to east causes surface ocean currents to swerve to the right in the Northern Hemisphere, producing a clockwise gyre of water currents. In the Southern Hemisphere, ocean currents swerve to the left, producing a counterclockwise gyre.

The ocean interacts with the atmosphere

The ocean and the atmosphere are strongly linked, with wind from the atmosphere affecting the ocean currents and heat from the ocean affecting atmospheric circulation. One of the best examples of the interaction between ocean and atmosphere is the **El Niño–Southern Oscillation (ENSO)** event. ENSO is a periodic warming of surface waters of the tropical East Pacific that alters both oceanic and atmospheric circulation patterns and results in unusual weather in areas far from the tropical Pacific. Normally, westward-blowing trade winds restrict the warmest waters to the western Pacific (near Australia). Every three to seven years, however, the trade winds weaken and the warm water mass expands eastward to South America, raising surface temperatures in the East Pacific 3° to 4° C above average. Ocean currents, which normally flow westward in this area, slow down, stop altogether, or even reverse and go eastward. The phenomenon is called El Niño (Spanish for "the child") because the warming usually reaches the fishing grounds off Peru just before Christmas. Most ENSOs last from one to two years.

An ENSO event changes biological productivity in parts of the ocean. The warmer sea surface temperatures and accompanying changes in ocean circulation patterns off the west coast of South America prevent colder, nutrient-laden deeper waters from **upwelling** (coming to the surface) (Fig. 53–16). The lack of nutrients in the water results in a severe decrease in the populations of anchovies and many other marine fishes. For example, during the 1982 to 1983 ENSO, one of the worst ever recorded, the anchovy population decreased by 99%. Other species, such as shrimp and scallops, thrive during an ENSO event. Along the Pacific coast of North America, ENSO shifts the distribution of tropical fishes northward and even affects how the salmon run in Alaska.

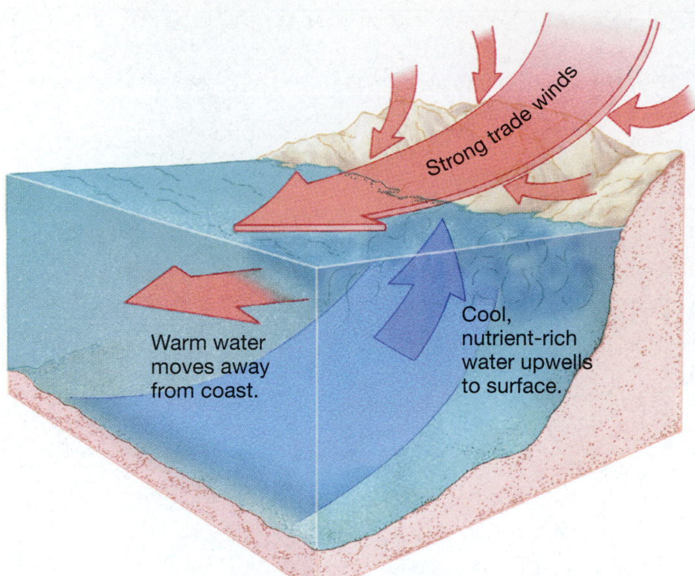

Figure 53-16 Upwelling. Coastal upwelling, where deeper waters come to the surface, occurs in the Pacific Ocean along the South American coast. Upwelling provides nutrients for large numbers of microscopic algae, which in turn support a complex food web. Coastal upwelling weakens considerably during years with El Niño–Southern Oscillation (ENSO) events, temporarily reducing fish populations.

CLIMATE PROFOUNDLY AFFECTS ORGANISMS

Weather refers to the conditions in the atmosphere at a given place and time; it includes temperature, atmospheric pressure, precipitation, cloudiness, humidity, and wind. Weather changes from one hour to the next and from one day to the next.

Climate comprises the average weather conditions plus extremes (records) that occur in a given place over a period of years. The two most important factors that help determine an area's climate are temperature (both average temperature and temperature extremes) and precipitation (both average precipitation and seasonal distribution). Other climate factors include wind, humidity, fog, cloud cover, and lightning-caused wildfires. Unlike weather, which changes rapidly, climate generally changes slowly, over hundreds or thousands of years.

Day-to-day variations, day-to-night variations, and seasonal variations are also important dimensions of climate that affect organisms. Temperature, precipitation, and other aspects of climate are influenced by latitude, elevation, topography, vegetation, distance from the ocean or other large bodies of water, and location on a continent or other landmass.

Earth has many different climates, and because they are relatively constant for many years, organisms have adapted to them. The wide variety of organisms on Earth evolved in part because of the large number of different climates, ranging from cold, snow-covered, polar climates to hot, tropical climates where it rains almost every day.

Air and water movements and surface features affect precipitation patterns

Precipitation refers to any form of water, such as rain, snow, sleet, and hail, that falls to the surface from the atmosphere. Precipitation varies from one location to another and has a profound effect on the distribution and kinds of organisms present. One of the driest places on Earth is in the Atacama Desert in Chile, where the average annual rainfall is 0.05 cm (0.02 in). In contrast, Mount Waialeale in Hawaii, Earth's wettest spot, receives an average annual precipitation of 1200 cm (472 in.).

Differences in precipitation depend on several factors. The heavy-rainfall areas of the tropics result mainly from the uplifting of moisture-laden air. High surface-water temperatures (recall the enormous amount of solar energy striking the equator) cause the evaporation of vast quantities of water from tropical parts of the ocean. Prevailing winds blow the resulting moist warmer air over landmasses. Heating of air by land surfaces warmed by the sun causes moist air to rise. As it rises, the air cools and its moisture-holding ability decreases. (Cool air holds less water vapor than warm air.) When air reaches its saturation point, it cannot hold any additional water vapor, and clouds form and water is released as precipitation. The air eventually returns to the surface on both sides of the equator near the Tropics of Cancer and Capricorn (latitudes 23.5 degrees north and south, respectively). By then, most of the moisture has precipitated so that dry air returns to the equator. This dry air makes little biological difference over the ocean, but its lack of moisture produces some of the great subtropical deserts, such as the Sahara.

Air is also dried during long journeys over landmasses. Near the windward (side from which the prevailing wind blows) coasts of continents, rainfall may be heavy. However, in the temperate zones—the areas between the tropical and polar zones—continental interiors are usually dry because they are far from the ocean that replenishes water vapor in the air passing over it.

Moisture is removed from humid air by mountains, which force air to rise. As it gains altitude, the air cools, clouds form, and precipitation typically occurs, primarily on the windward slopes of the mountains. As the air mass moves down on the other side of the mountain, it is warmed and clouds evaporate, thereby lessening the chance of precipitation of any remaining moisture. This situation exists on the West Coast of North America, where precipitation falls on the western slopes of mountains that are close to the coast. The dry lands on the sides of the mountains away from the prevailing wind (in this case, east of the mountain range) are called **rain shadows** (Fig. 53-17).

Microclimates are local variations in climate

Differences in elevation, in the steepness and direction of slopes, and therefore in exposure to sunlight and prevailing winds, may produce local variations in climate known as **microclimates**, which can be quite different from their overall surroundings. Patches of sun and shade on a forest floor, for example, produce a variety of microclimates for plants, animals, and microorganisms living there. The microclimate of an organism's habitat is of primary importance because that is the climate an organism ac-

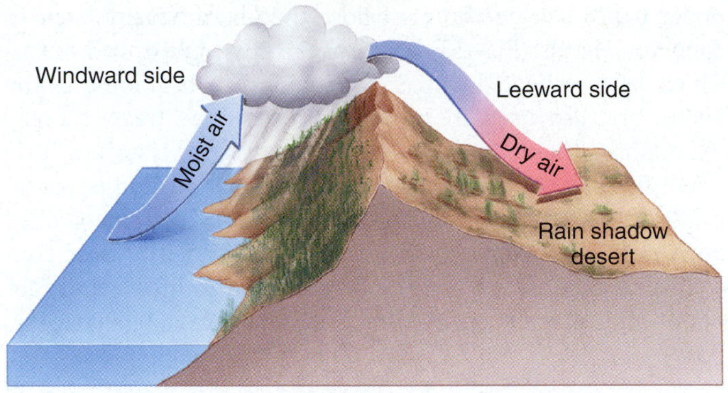

Figure 53–17 Rain shadow. A rain shadow refers to arid or semiarid land that occurs on the leeward side of a mountain. Prevailing winds blow warm, moist air from the windward side of the mountain. Air cools as it rises, releasing precipitation so that dry air descends on the leeward side. Such a rain shadow occurs east of the Cascade Range in Washington State. The west side of Olympic National Park receives more than 500 cm of precipitation annually, whereas the east side gets 40 to 50 cm.

tually experiences and must cope with. (Keep in mind, however, that microclimates are largely affected by the regional climate in which they are located.)

Sometimes organisms modify their own microclimate. For instance, trees modify the local climate within a forest so that in summer the temperature is usually lower, and the relative humidity greater, than outside the forest. The temperature and humidity beneath the litter of the forest floor differ still more; in the summer the microclimate of this area is considerably cooler and moister than the surrounding forest. As another example, many desert-dwelling animals burrow to avoid surface climate conditions that would kill them in minutes. The cooler daytime microclimate in their burrows permits them to survive until night, when the surface cools off and they can come out to forage or hunt.

FIRES ARE A COMMON DISTURBANCE IN SOME ECOSYSTEMS

Wildfires, which are fires started by lightning, are an important ecological force in many geographical areas. Those areas most prone to wildfires have wet seasons followed by dry seasons. Vegetation that grows and accumulates during the wet season dries out enough during the dry season to burn easily. When lightning hits vegetation or ground litter, it ignites the dry organic material, and a fire spreads through the area.

Fires have several effects on organisms. First, combustion frees the nutrient minerals that were locked in dry organic matter. The ashes remaining after a fire are rich in potassium, phosphorus, calcium, and other nutrient minerals essential for plant growth. With the arrival of precipitation, vegetation flourishes following a fire. Second, fire removes plant cover and exposes the soil. This change stimulates the germination and establishment of seeds requiring bare soil, as well as encourages the growth of shade-intolerant plants. Third, fire can cause increased soil erosion because it removes plant cover, leaving the soil more vulnerable to wind and water.

Some plants are fire-adapted. Grasses and some shrubs, for example, have underground stems and buds that are typically unaffected by a fire sweeping over them. After the aerial parts have been killed by fire, the underground parts send up new sprouts. The availability of nutrients in the ash speeds the regrowth of these plants. Fire-adapted trees such as bur oak and ponderosa pine have a thick bark that is resistant to fire. (In contrast, fire-sensitive trees, such as many hardwoods, have a thin bark.) Certain pines such as jack and lodgepole pine depend on fire for successful reproduction, because the heat of the fire opens the cones so that the seeds can be released.

Fires were a part of the natural environment long before humans appeared, and organisms in many terrestrial ecosystems have adapted to fires. African savanna, California chaparral, North American grasslands, and ponderosa pine forests of the western United States are some fire-adapted ecosystems (see Chapter 54). Fire helps maintain grasses as the dominant vegetation in grasslands by removing fire-sensitive hardwood trees.

The influence of fire on plants became even more pronounced once humans evolved. Because humans deliberately and accidentally set fires, fire became a more common occurrence. Humans set fires for many reasons: to provide the grasses and shrubs that many game animals require; to clear the land for agriculture and human development; and to reduce enemy cover in times of war.

Figure 53–18 Fire as a tool of ecological management. Here, a controlled burn helps maintain a ponderosa pine *(Pinus ponderosa)* stand in Oregon. *(Joan Landsberg, U.S. Department of Agriculture Forest Service)*

Humans also try to prevent fires, and sometimes this effort can have disastrous consequences. When fire is excluded from a fire-adapted ecosystem, deadwood and other plant litter accumulate. As a result, when a fire does occur, it is much more destructive. The deadly wildfire in Colorado during the summer of 1994, which claimed the lives of 14 firefighters, was blamed in part on decades of suppressing fires in the region. Prevention of fire also converts grassland to woody vegetation and facilitates the invasion of fire-sensitive trees into fire-adapted forests.

Controlled burning is a tool of ecological management in which the undergrowth and plant litter are deliberately burned under controlled conditions before they have accumulated to dangerous levels (Fig. 53–18 on page 1201). Controlled burns, which were instigated in U.S. national forests beginning in the 1960s, are also used to suppress fire-sensitive trees, thereby maintaining the natural fire-adapted ecosystem. However, little scientific evidence exists on when controlled burning is appropriate or how often it should be carried out. In 2000 a controlled burn near Los Alamos, New Mexico, ignited debate over controlled burn policies when the fire, exposed to high winds, got out of control and burned about 20,200 hectares (50,000 acres).

SUMMARY WITH KEY TERMS

I. **Energy flow** through an ecosystem is linear, from the sun to producer to consumer to decomposer. Much of this energy is converted to heat as it moves from one organism to another and is therefore unusable by organisms occupying the next **trophic level.**
   A. Trophic relationships may be expressed as **food chains** or, more realistically, as **food webs,** which show the many alternative pathways that energy may take among the producers, consumers, and decomposers of an ecosystem.
   B. **Ecological pyramids** typically express the progressive reduction in numbers of organisms, biomass, and energy found in successive trophic levels.
   C. **Gross primary productivity** of an ecosystem is the rate at which energy is captured by photosynthesis. **Net primary productivity** is the energy that remains (as biomass) after plants carry out cellular respiration.
      1. The two most productive ecosystems are algal beds/reefs and tropical rain forests.
      2. The least productive ecosystem is extreme desert (rock, sand, or ice).

II. **Biogeochemical cycles** are the cycling of matter from the abiotic environment to organisms and then back to the abiotic environment.
   A. Carbon dioxide is the important gas of the **carbon cycle.**
      1. Carbon enters plants, algae, and cyanobacteria as CO_2, which is incorporated into organic molecules by photosynthesis.
      2. Cellular respiration, combustion, and erosion of limestone return CO_2 to the water and atmosphere, making it available to producers again.
   B. The **nitrogen cycle** has five steps.
      1. Biological **nitrogen fixation** is the conversion of nitrogen gas to ammonia, an important form of nitrogen for certain plants.
      2. **Nitrification** is the biological conversion of ammonia or ammonium to nitrate, an important form of nitrogen used by plants.
      3. **Assimilation** is the biological conversion of nitrates, ammonia, or ammonium to proteins, chlorophyll, and other nitrogen-containing compounds by plants. The conversion of plant proteins into animal proteins is also considered assimilation.
      4. **Ammonification** is the biological conversion of organic nitrogen to ammonia and ammonium ions.
      5. **Denitrification** is the biological conversion of nitrate to nitrogen gas.
   C. The **phosphorus cycle** has no biologically important gaseous compounds.
      1. Phosphorus erodes from rock as inorganic phosphate and is absorbed from the soil by the roots of plants. Plants incorporate phosphorus into compounds such as nucleic acids and phospholipids.
      2. Animals obtain the phosphorus they need from their diets. Decomposers release inorganic phosphate into the environment.
      3. Phosphorus can be lost from biological cycles for millions of years when it washes into the ocean and is subsequently deposited in sea beds.
   D. The **hydrological cycle,** which continually renews the supply of water so essential to life, involves an exchange of water between the land, atmosphere, and organisms.
      1. Water enters the atmosphere by evaporation and **transpiration** and leaves the atmosphere as precipitation.
      2. On land, water filters through the ground or runs off to lakes, rivers, and the ocean. The underground caverns and porous layers of rock in which groundwater is stored are called aquifers. The movement of surface water from land to ocean is called runoff.

III. The unique planetary environment of Earth makes life possible. Sunlight is the primary (almost sole) source of energy available to the biosphere.
   A. Of the solar energy that reaches Earth, 30% is immediately reflected away and the remaining 70% is absorbed by the atmosphere and surface.
   B. Ultimately, all absorbed solar energy is reradiated into space as infrared (heat) radiation.
   C. A combination of Earth's roughly spherical shape and the tilted angle of its axis concentrates solar energy at the equator and dilutes it at the poles.
      1. The tropics are therefore hotter and less variable in climate than are temperate and polar areas.
      2. Seasons are determined primarily by the inclination of Earth's axis.

IV. The atmosphere protects the surface from most of the sun's ultraviolet radiation and x rays and from lethal amounts of cosmic rays from space. At the same time, visible light and some infrared radiation penetrate to warm the surface and the lower part of the atmosphere.
   A. Atmospheric heat transferred from the equator to the poles produces both a movement of warm air toward the poles and of cool air toward the equator, thus moderating the extremes of global climate.
   B. Winds result in part from differences in atmospheric pressure and from the **Coriolis effect.**

V. The global ocean is a single, continuous body of water that surrounds and covers almost three-fourths of Earth's surface.
   A. Surface **ocean currents** result in part from prevailing winds and the Coriolis effect.
   B. The **El Niño–Southern Oscillation (ENSO)** alters both ocean and atmospheric currents, results in unusual weather patterns, and has significant ecological effects.

VI. An area's **climate** comprises the average **weather** conditions plus extremes that occur there over a period of years.
   A. The two most important factors that help determine an area's climate are temperature (both average temperature and temperature ex-

tremes) and precipitation (both average precipitation and seasonal distribution).

B. Precipitation is greatest where warm air passes over the ocean, absorbing moisture, and then cools, such as when mountains force humid air upward.

C. Deserts develop in the **rain shadows** of mountain ranges or in continental interiors.

D. **Microclimates,** which are local variations in climate, produce a variety of climate conditions for organisms in a given habitat.

VII. Many ecosystems, such as savanna, chaparral, grasslands, and certain forests, contain fire-adapted organisms.

POST-TEST

1. A community and its abiotic environment best defines a(an) (a) biogeochemical cycle (b) biosphere (c) ecosystem (d) food web (e) trophic level
2. The movement of matter is __________ in ecosystems, and the movement of energy is __________. (a) linear; linear (b) linear; cyclic (c) cyclic; cyclic (d) cyclic; linear (e) cyclic; linear or cyclic
3. A complex of interconnected food chains in an ecosystem is called a(an): (a) ecosystem (b) pyramid of numbers (c) pyramid of biomass (d) biosphere (e) food web
4. Which of the following shows the correct flow of energy through ecosystems? (a) sun → primary consumer → secondary consumer → producer (b) sun → producer → secondary consumer → primary consumer (c) sun → secondary consumer → primary consumer → producer (d) sun → producer → primary consumer → secondary consumer (e) sun → primary consumer → producer → secondary consumer
5. The quantitative estimate of the total amount of living material is called (a) biomass (b) energy flow (c) gross primary productivity (d) plant respiration (e) net primary productivity
6. Which of the following equations shows the relationship between gross primary productivity (GPP) and net primary productivity (NPP)? (a) GPP = NPP − photosynthesis (b) NPP = GPP − photosynthesis (c) GPP = NPP − plant respiration (d) NPP = GPP − plant respiration (e) NPP = GPP − animal respiration
7. Which of the following processes increase(s) the amount of atmospheric carbon in the carbon cycle? (a) photosynthesis (b) cellular respiration (c) combustion (d) both a and c (e) both b and c
8. In the carbon cycle, carbon can be found in (a) limestone rock (b) oil, coal, and natural gas (c) living organisms (d) the atmosphere (e) all of the above
9. In the nitrogen cycle, gaseous nitrogen is converted to ammonia during (a) nitrogen fixation (b) nitrification (c) assimilation (d) ammonification (e) denitrification
10. The conversion of ammonia to nitrate, known as __________, is a two-step process performed by soil bacteria. (a) nitrogen fixation (b) nitrification (c) assimilation (d) ammonification (e) denitrification
11. This biogeochemical cycle does not have a gaseous component but cycles from the land to sediments in the ocean and back to the land. (a) carbon cycle (b) nitrogen cycle (c) phosphorus cycle (d) hydrological cycle (e) neither a nor c have a gaseous component
12. Which of the following processes is *not* involved in the hydrological cycle? (a) transpiration (b) evaporation (c) precipitation (d) nitrification (e) condensation
13. The __________, which results from the rotation of Earth, displaces the paths of atmospheric and oceanic currents to the right in the Northern Hemisphere and to the left in the Southern Hemisphere. (a) upwelling (b) prevailing wind (c) ocean current (d) El Niño–Southern Oscillation (e) Coriolis effect
14. The periodic warming of surface waters of the tropical East Pacific that alters both oceanic and atmospheric circulation patterns is known as (a) upwelling (b) prevailing wind (c) ocean current (d) El Niño–Southern Oscillation (e) Coriolis effect
15. A mountain range may produce a downwind arid (a) upwelling (b) rain shadow (c) ocean current (d) microclimate (e) ecological pyramid

REVIEW QUESTIONS

1. Describe energy flow through a food web.
2. What are trophic levels and how are they related to ecological pyramids?
3. Define gross primary productivity (GPP) and net primary productivity (NPP).
4. Why is the cycling of matter essential to the long-term continuance of life?
5. Diagram the carbon cycle, including the following processes: photosynthesis, cellular respiration, combustion, and erosion.
6. List and describe the five steps in the nitrogen cycle.
7. What basic forces determine the circulation of the atmosphere? Describe the general directions of atmospheric circulation.
8. What basic forces produce the main ocean currents? Describe the general directions of these currents.
9. What are some of the factors that produce regional differences in precipitation?

YOU MAKE THE CONNECTION

1. Describe the simplest stable ecosystem that you can imagine.
2. As discussed in the chapter, Earth is a closed system regarding matter. Is Earth a closed or open system regarding energy? Explain.
3. What is the energy source that powers the wind?
4. How do ocean currents affect climate on land?
5. Would the microclimate of an ant be the same as that of an elephant living in the same area? Why or why not?

RECOMMENDED READINGS

Asner, G.P., T.R. Seastedt, and A.R. Townsend. "The Decoupling of Terrestrial Carbon and Nitrogen Cycles." *BioScience*, Vol. 47, No. 4, Apr. 1997. Fertilizer production and land use, such as conversion of forest to agricultural land, are changing natural links between the carbon and nitrogen cycles.

Caraco, N.F. "Disturbance of the Phosphorus Cycle: A Case of Indirect Effects of Human Activity." *Tree*, Vol. 8, No. 2, Feb. 1993. Humans have affected the phosphorus cycle both directly and indirectly.

Carpenter, S.R., et al. "Ecosystem Experiments." *Science*, Vol. 269, 21 Jul. 1995. An overview of how experimental manipulations of entire ecosystems are used to increase our understanding of ecosystem structure and function.

Chahine, M.T. "The Hydrologic Cycle and Its Influence on Climate." *Nature*, Vol. 359, 1 Oct. 1992. A review of our understanding of how the hydrological cycle affects global climate.

Holloway, M. "Uncontrolled Burn." *Scientific American*, Vol. 283, Aug. 2000. Examines the aftermath of the controlled burn that got out of control in New Mexico.

Lawton, J.H. "Ecological Experiments with Model Systems." *Science*, Vol. 269, 21 Jul. 1995. Addresses how model laboratory systems, which are not as complex as natural ecosystems, can help untangle some of the complexities found in nature.

Lutz, R.A. "Deep-Sea Vents." *National Geographic*, Vol. 198, No. 4, Oct. 2000. Knowledge of deep-sea ecosystems has expanded rapidly in the past few decades.

Raven, P.H., and L.R. Berg. *Environment*, 3rd ed. Harcourt College Publishers, Philadelphia, 2001. This environmental science text offers a detailed situation analysis of planet Earth, including how humans interact with and affect its physical and living systems.

Ricklefs, R.E., and G.L. Miller. *Ecology*, 4th ed. W.H. Freeman & Company, New York, 1999. This textbook presents the important themes of ecology, such as energy flow, population and community interactions, and mathematical models of ecology, in an accessible manner.

Schoen, D. "Primary Productivity: The Link to Global Health." *BioScience*, Vol. 47, No. 8, Sep. 1997. Examines an important question that scientists are currently addressing: How terrestrial primary productivity will respond to climate change.

Smil, V. "Global Population and the Nitrogen Cycle." *Scientific American*, Vol. 277, No. 1, Jul. 1997. Humans have had a global impact on the nitrogen cycle, largely due to our increasing production of nitrogen fertilizer to provide food for a growing human population.

Trefil, J. "Nitrogen." *Smithsonian*, Oct. 1997. An excellent overview of the nitrogen cycle and human-produced nitrogen pollution.

- Visit our Web site at **http://www.info.brookscole.com/solomonbergmartin** for links to chapter-related resources on the World Wide Web. Additional on-line materials relating to this chapter can also be found on our Web site.

See chapter activity on BioActive Learner CD for additional help in mastering the chapter's material. Icon location in the chapter's margins shows which topics have tutorials or simulations in the CD.

54

Ecology and the Geography of Life

Black-tailed prairie dog *(Cynomys ludovicianus)*. Prairie dogs never wander far from their burrows, which they use to escape from predators. They grow to 38 cm (15 in) in length, not counting their tails. Photographed in North Dakota. *(Steve Kaufman/Peter Arnold, Inc.)*

LEARNING OBJECTIVES

After you have studied this chapter you should be able to

1. Define biome and briefly describe the nine major terrestrial biomes, giving attention to the climate, soil, and characteristic plants and animals of each.
2. Describe at least one human effect on each of the biomes discussed.
3. Explain the similarities in the changes in vegetation observed with increasing elevation and increasing latitude.
4. Explain the important environmental factors that affect aquatic ecosystems.
5. Distinguish among plankton, nekton, and benthos.
6. Briefly describe the various freshwater, estuarine, and marine ecosystems, giving attention to the environmental characteristics and representative organisms of each.
7. Describe at least one human effect on each of the aquatic ecosystems described.
8. Describe thermal stratification in a temperate lake.
9. Define and describe some of the features of an ecotone.
10. Define biogeography and briefly describe Wallace's biogeographical realms.

In Chapter 53 you learned that climate, particularly temperature and precipitation, provides Earth with many different environments. Natural selection affects an organism's ability to survive and reproduce in a given environment (see Chapter 17). In natural selection, both **abiotic** (nonliving) and **biotic** (living) factors act to eliminate the least-fit individuals in a given population. Over time, succeeding generations of organisms that live in each biome or major aquatic ecosystem become better adapted to local environmental conditions.

To demonstrate how each organism is well adapted to its environment, consider the black-tailed prairie dog, one of several prairie dog species. These small rodents live in the grasslands of western North America, from southern Canada to northern Mexico. Black-tailed prairie dogs are not found in temperate grasslands on other continents, although presumably they could survive there. A species occurs at a given location either because it evolved or dispersed there. Black-tailed prairie dogs evolved in North America and were prevented from dispersing to similar grasslands by biological and geographical barriers.

Black-tailed prairie dogs are superbly adapted to their environment. Their teeth and digestive tracts are modified to eat and easily digest the seeds and leaves of grasses that grow in great profusion on the Great Plains. They have an enlarged pair of incisors, for example, that are used to nibble grasses and gnaw through seeds. Prairie dogs also eat insects and underground roots and tubers.

Black-tailed prairie dogs have many behavioral adaptations that protect them from predators, such as coyotes, eagles, hawks, badgers, and ferrets. For example, prairie dogs are social animals and live in large colonies of about 500 individuals; the eyes of every individual in the colony watch for potential danger, and when they see it, prairie dogs call out to warn the rest of the colony. In fact, they are called "dogs" not because they are related to dogs (they are not), but because their warning calls are similar to a dog's bark.

Black-tailed prairie dogs are burrowing animals *(see photograph)*. When danger approaches—a coyote, perhaps—each prairie dog dives into its underground home. Their burrows have at least two openings and consist of an elaborate network of long tunnels with several chambers. One room may serve as a nursery for the young, one as a sleeping chamber, one as a toilet chamber, and one, which is close to an entrance, as a listening chamber. High piles of excavated soil surround the burrow entrances and help prevent flooding during rainstorms.

Black-tailed prairie dogs survive winter by hibernating in their burrows. Their metabolism slows, and they exist on the stored fat in their bodies. They do not go into deep hibernation, however, and may come out of their burrows to look for food when the weather warms.

Some ranchers view prairie dogs as a nuisance because they dig burrows in fields and pastures. Ranchers fear that their livestock will step into the burrow entrances and injure their legs. (Horses and cattle rarely step into the entrances, however.) Ranchers also think that prairie dogs compete with livestock for grasses. Despite extensive trapping and poisoning, however, these resourceful animals persist.

Why aren't prairie dogs found in other environments, such as deserts or deciduous forests? Natural selection has adapted prairie dogs to the environmental conditions found in temperate grasslands, and prairie dogs can usually tolerate most environmental extremes in that biome. In adapting a population to local conditions, however, natural selection may limit individuals from that population from successfully dispersing and establishing themselves in other places with different environmental conditions.

Like prairie dogs, each species has evolved structural, behavioral, and physiological features that help adapt it to its own particular environment and lifestyle. These features often limit its ability to survive and reproduce elsewhere where different conditions prevail. As we examine Earth's major biomes (large terrestrial ecosystems) and aquatic ecosystems, including the species characteristic of each, think about the variety of adaptations that natural selection has produced in organisms in response to their particular environments.

BIOMES ARE LARGELY DISTINGUISHED BY THEIR DOMINANT FORM OF VEGETATION

A **biome** is a large, relatively distinct terrestrial region characterized by similar climate, soil, plants, and animals regardless of where it occurs. Because it covers such a large geographical area, a biome encompasses a number of interacting ecosystems. In terrestrial ecology, a biome is considered the next level of ecological organization above that of landscape.

Biomes largely correspond to major climate zones, with temperature and precipitation being most important (Fig. 54–1). Near the poles, temperature is generally the overriding climate factor, whereas in temperate and tropical regions, precipitation becomes more significant than temperature (Fig. 54–2). Other abiotic factors to which biomes are sensitive include temperature extremes as well as rapid temperature changes, floods, droughts, strong winds, and fires (see section on fires in Chapter 53). Nine major biomes are discussed: tundra, taiga, temperate rain forest, temperate deciduous forest, temperate grassland, chaparral, desert, savanna, and tropical rain forest (Fig. 54–3). Although we discuss each biome as a distinct entity, biomes intergrade into one another at their boundaries.

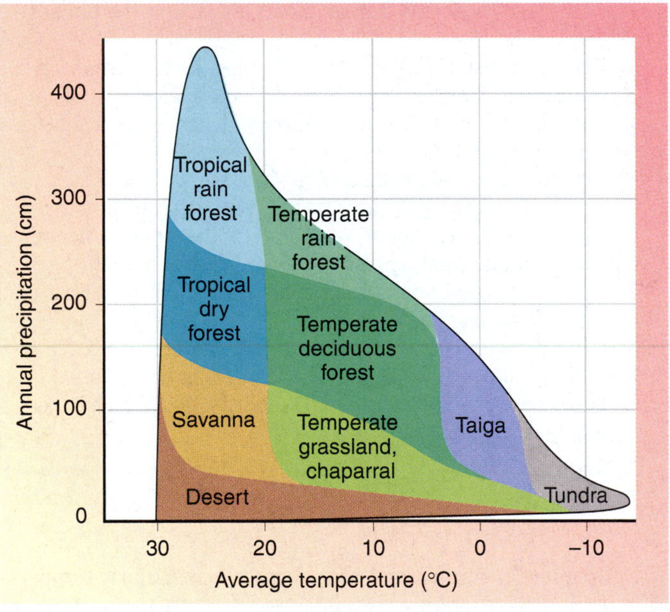

Figure 54–1 Identification of biomes by precipitation and temperature. Factors such as soil type, fire, and seasonality of climate affect whether temperate grassland or chaparral develops. *(Adapted from Whittaker, R.H.* Communities and Ecosystems, *2nd ed. Macmillan, New York, 1975)*

Tundra is the cold, boggy plains of the far north

Tundra (also called **arctic tundra**) occurs in extreme northern latitudes wherever snow melts seasonally (Fig. 54–4; also see Fig. 54–3). The Southern Hemisphere has no equivalent of the arctic tundra because it has no land in the proper latitudes. A similar ecosystem located in the higher elevations of mountains, above the tree line, is called **alpine tundra** to distinguish it from arctic tundra (see *Focus On: The Distribution of Vegetation on Mountains* on page 1210).

Arctic tundra has long, harsh winters and extremely short summers. Although the growing season, with its warmer temperatures, is as short as 50 days (depending on location), the days are long. Above the Arctic Circle the sun does not set at all for many days in midsummer, although the amount of light at midnight is only one-tenth that at noon. There is little precipitation (10 to 25 cm, or 4 to 10 in, per year) over much of the tundra, with most of it falling during summer months.

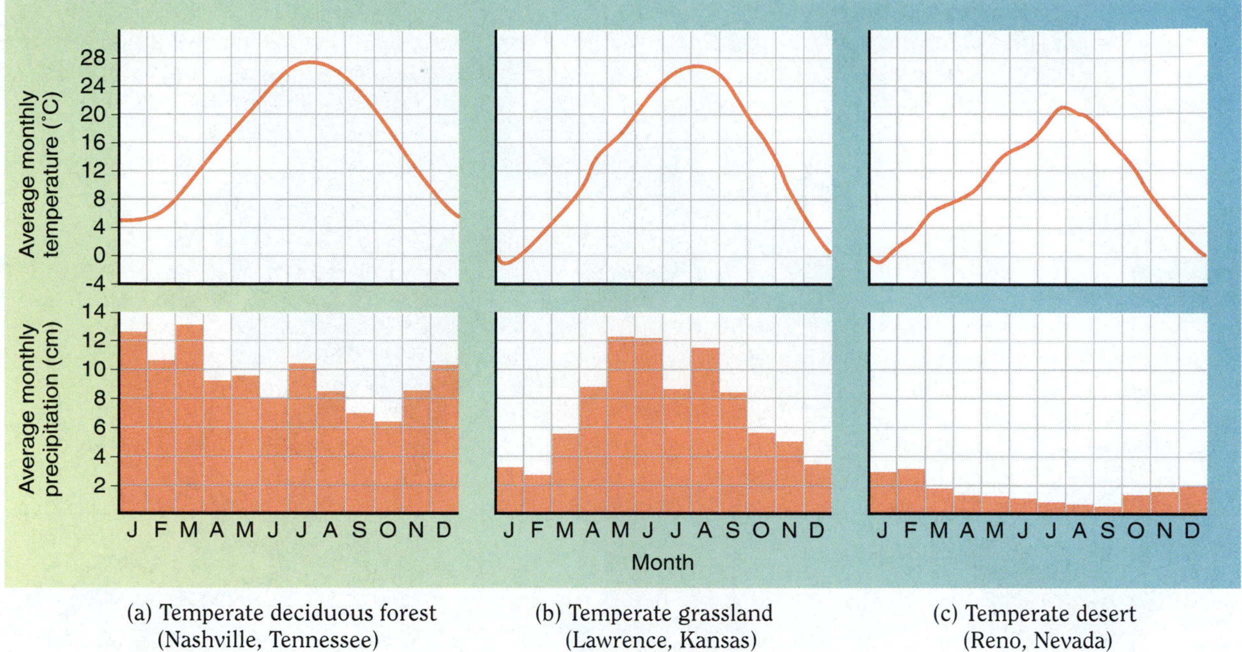

(a) Temperate deciduous forest (Nashville, Tennessee)

(b) Temperate grassland (Lawrence, Kansas)

(c) Temperate desert (Reno, Nevada)

Figure 54–2 Significance of precipitation in temperate biomes. Average monthly temperature *(top)* and precipitation *(bottom)* are given for three temperate biomes in North America. Although temperature is approximately the same in all three locations, precipitation varies a great deal, resulting in **(a)** deciduous forest where precipitation is plentiful, **(b)** grassland where it is less plentiful and more seasonal, and **(c)** desert where it is quite low.

Tundra soils tend to be geologically young, since most were formed only after the last Ice Age.[1] These soils are usually nutrient-poor and have little organic litter (dead leaves and stems, animal droppings, and remains of organisms) in the uppermost layer of soil. Although the soil surface melts during the summer, tundra has a layer of permanently frozen ground called **permafrost** that varies in depth and thickness. Because permafrost interferes with drainage, the thawed upper zone of soil is usually waterlogged during the summer. Permafrost limits the depth to which roots can penetrate, thereby preventing the establishment of most woody species, which tend to be deep-rooting. Limited precipitation, combined with low temperatures, flat topography (surface features), and permafrost, produces a landscape of broad shallow lakes, sluggish streams, and bogs.

Low species richness and low primary productivity characterize tundra. Few plant species occur, but individual species often exist in great numbers. Mosses, lichens (such as reindeer moss), grasses, and grasslike sedges dominate tundra vegetation; most of these short plants are herbaceous perennials that live 20 to 100 years. No readily recognizable trees or shrubs grow except in sheltered locations, although dwarf willows, dwarf birches, and other dwarf trees are common. Tundra plants seldom grow taller than 30 cm (12 in) in open areas.

[1] Glacier ice, which occupied about 29% of Earth's land during the last Ice Age, began retreating about 17,000 years ago. Today, glacier ice occupies about 10% of the land surface.

Year-round animal life of the tundra includes voles, weasels, arctic foxes, gray wolves, snowshoe hares, ptarmigan, snowy owls, musk-oxen, and lemmings (see the discussion of lemming population cycles in Chapter 51). In the summer, caribou migrate north to the tundra to graze on sedges, grasses, and dwarf willow. Dozens of bird species also migrate north in summer to nest and feed on abundant insects. Mosquitoes, blackflies, and deerflies survive the winter as eggs or pupae and occur in great numbers during summer weeks. There are no reptiles or amphibians. (Recall from Chapter 30 that amphibians and reptiles are *ectothermic*. Because their body temperature fluctuates with the temperature of the surrounding environment, reptiles and amphibians are excluded from the cold temperatures of the tundra.)

Tundra regenerates quite slowly after it has been disturbed. Even casual use by hikers can cause damage. Long-lasting injury, likely to persist for hundreds of years, has been done to large portions of the arctic tundra as a result of oil exploration and military use.

Taiga is the evergreen forest of the north

Just south of the tundra is the **taiga,** or **boreal forest,** which stretches across both North America and Eurasia. Taiga is the world's largest biome, covering approximately 11% of Earth's land (Fig. 54–5; also see Fig. 54–3). A biome comparable to the taiga is not found in the Southern Hemisphere because it has no land at the corresponding latitudes. Winters are extremely cold

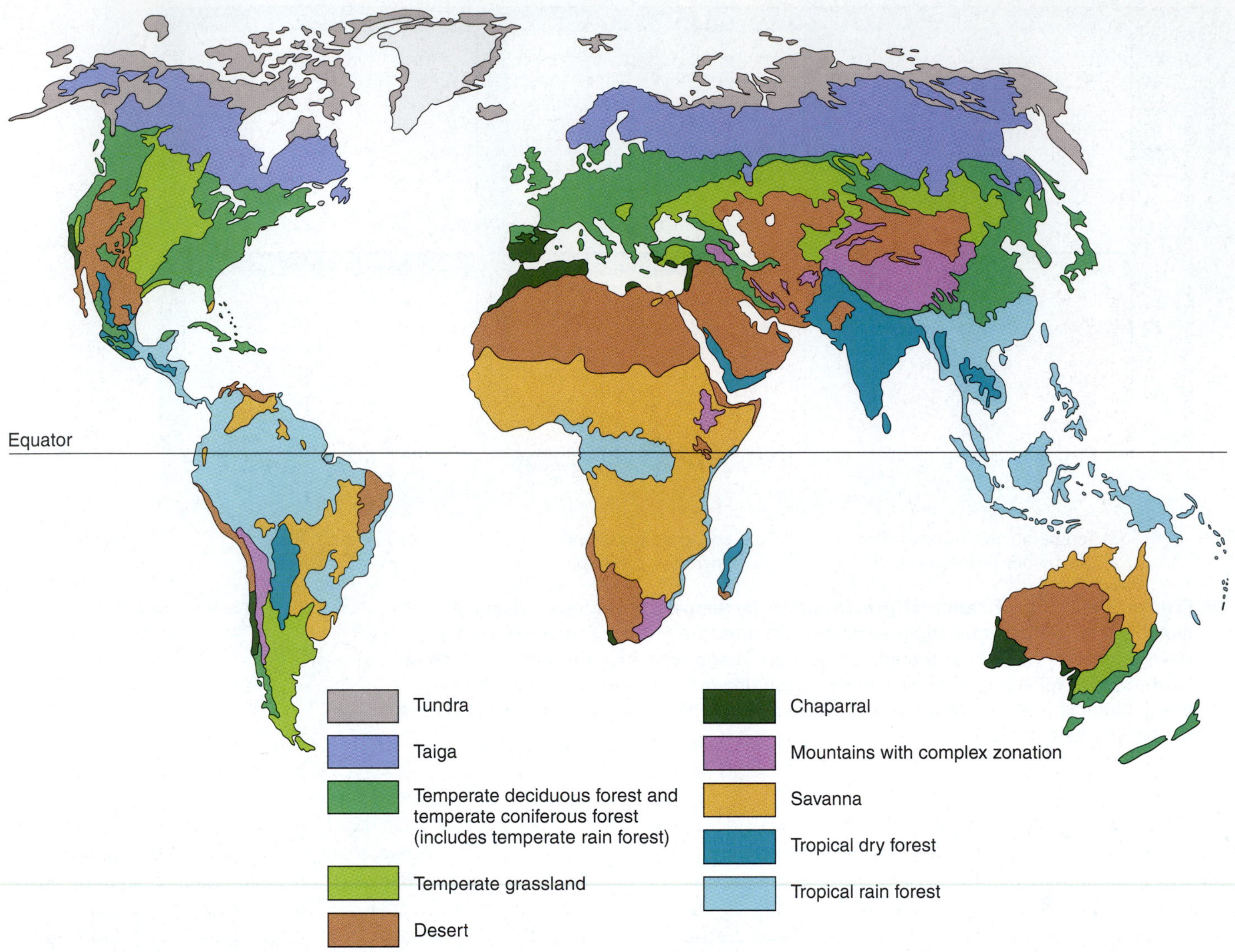

Figure 54–3 The world's major biomes. This highly simplified diagram shows sharp boundaries between biomes. Biomes actually intergrade at their boundaries, sometimes over large areas. (Note that forested mountains such as the Appalachian Mountains are keyed according to their predominant vegetation type, whereas mountains with variable vegetation are keyed as "Mountains with complex zonation.") *(Based on data from the World Wildlife Fund)*

and severe, although not as harsh as in the tundra. The growing season of the boreal forest is somewhat longer than that of the tundra. Taiga receives little precipitation, perhaps 50 cm (20 in) per year, and its soil is typically acidic, mineral-poor, and characterized by a deep layer of partly decomposed conifer needles at the surface. (Conifers are cone-bearing evergreens.) Permafrost is patchy and, where found, is often deep underneath the soil. Taiga contains numerous ponds and lakes in water-filled depressions that were dug by grinding ice sheets during the last Ice Age.

Black and white spruces, balsam fir, eastern larch, and other conifers clearly dominate the taiga, but **deciduous** trees such as aspen or birch, which shed their leaves in autumn, may form striking stands. Conifers have many drought-resistant adaptations, such as needle-like leaves with a minimal surface area to reduce water loss (see Chapter 32). Such an adaptation enables conifers to withstand the "drought" of the northern winter months, when roots cannot absorb water because the ground is frozen. Natural selection also favors conifers in the taiga because, being evergreen, they can resume photosynthesis as soon as warmer temperatures return.

Animal life of the boreal forest includes some larger species such as caribou (which migrate from the tundra to the taiga for winter), wolves, bears, and moose. However, most animal life is medium-sized to small and includes rodents, rabbits, and fur-bearing predators such as lynx, sable, and mink. Most species of birds are seasonally abundant but migrate to warmer climates for winter. Insects are numerous, but there are few amphibians and reptiles except in the southern taiga.

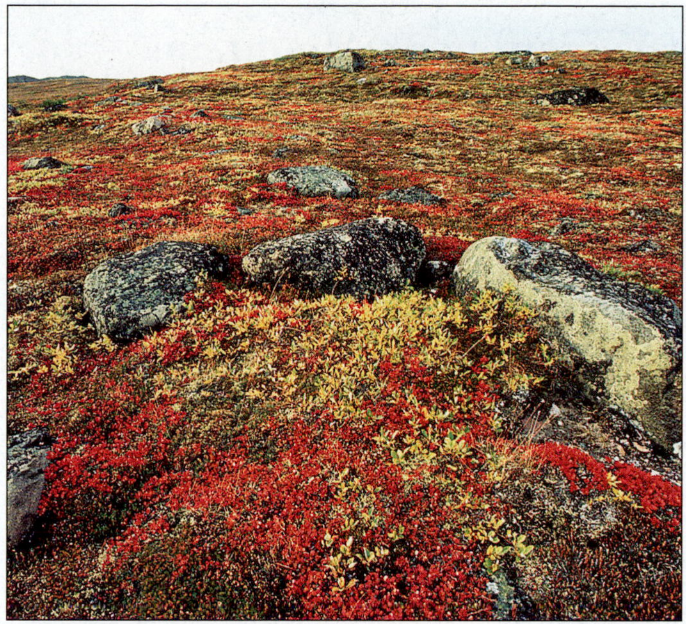

Figure 54–4 Arctic tundra. Because of its short growing season and permafrost, only small, hardy plants grow in the northernmost biome that encircles the Arctic Ocean. Photographed during autumn in Northwest Territories, Canada. *(Eastcott/Momatiuk/Earth Scenes)*

Figure 54–5 Taiga. Taiga is coniferous forest that occurs in cold regions of the Northern Hemisphere adjacent to the tundra. Photographed in Yukon, Canada. *(Beth Davidow/Visuals Unlimited)*

Most of the taiga is not suitable for agriculture because of its short growing season and mineral-poor soil. Taiga, which is being harvested primarily by clearcut logging, is currently the primary source of the world's industrial wood and wood fiber. The annual loss of boreal forests has been estimated to encompass an area twice as large as the Amazonian rain forests of Brazil. About 1 million hectares (2.5 million acres) of Canadian forests are logged annually, and extensive tracts of Siberian forests in Russia are harvested, although exact estimates are unavailable. Alaska's taiga is also at risk because the U.S. government may increase logging on public lands in the future.

Temperate rain forest is characterized by cool weather, dense fog, and high precipitation

A coniferous **temperate rain forest** occurs on the northwest coast of North America. Similar vegetation exists in southeastern Australia and in southwestern South America (see Fig. 54–3). Annual precipitation in this biome is high, from 200 to 380 cm (80 to 150 in), and this is augmented by condensation of water from dense coastal fogs. The proximity of temperate rain forest to the coastline moderates the temperature so that seasonal fluctuation is narrow; winters are mild, and summers are cool. Temperate rain forest has a relatively nutrient-poor soil, although its organic content may be high. Cool temperatures slow the activity of bacterial and fungal decomposers. Thus, needles and large fallen branches and trunks accumulate on the ground as litter that takes many years to decay and release nutrient minerals to the soil.

The dominant vegetation type in the North American temperate rain forest is large evergreen trees such as western hemlock, Douglas fir, Sitka spruce, and western red cedar. Temperate rain forest is rich in epiphytic vegetation, which are smaller plants that grow nonparasitically on the trunks and branches of large trees (Fig. 54–6). Epiphytes in this biome are mainly mosses, lichens, and ferns, all of which also carpet the ground. Deciduous shrubs such as vine maple grow wherever a break in the overlying canopy occurs. Squirrels, woodrats, mule deer, elk, numerous bird species (e.g., jays, nuthatches, and chickadees), and several species of reptiles (e.g., painted turtles and western terrestrial garter snakes) and amphibians (e.g., Pacific giant salamanders and Pacific treefrogs) are common temperate rainforest animals.

Temperate rain forest is one of the richest wood producers in the world, supplying us with lumber and pulpwood. It is also one of the most complex ecosystems in terms of species richness. Care must be taken to avoid overharvesting original old-growth forest, because such an ecosystem takes hundreds of years to develop. When the logging industry harvests old-growth forest, it typically replants the area with a monoculture (a single species) of trees that it can harvest in 40- to 100-year cycles. Thus, the old-growth forest ecosystem, once harvested, never has a chance to redevelop. A small fraction of the original, old-growth temperate rain forest in Washington, Oregon, and northern California remains untouched. These stable forest ecosystems provide biological habitat for many species, including 40 endangered and threatened species.

FOCUS ON The Distribution of Vegetation on Mountains

Hiking up a mountain is similar to traveling toward the North Pole with respect to the major life zones encountered *(see figure)*. This elevation-latitude similarity occurs because, as one climbs a mountain, the temperature drops just as it does when one travels north; the temperature drops about 6° C (11° F) with each 1000-m increase in elevation. The types of species growing on the mountain change as the temperatures change.

The base of a mountain in Colorado, for example, might be covered by deciduous trees, which shed their leaves every autumn. At higher elevations, where the climate is colder and more severe, a coniferous *subalpine forest* resembling northern taiga grows. Spruces and firs are the dominant trees here. Higher still, the forest thins, and the trees become smaller, gnarled, and shrublike. These twisted, shrublike trees, called **krummholz,** are found at their elevational limit (the *treeline*). The exact elevation at which the treeline occurs depends on the latitude and distance from the ocean. In the Rocky Mountains, between 35° and 50° north latitude, the treeline drops 100 m with each 1° latitude northward.

Above the tree line, where the climate is quite cold, a kind of tundra occurs, with vegetation composed of grasses, sedges, and small tufted plants, most of which are hardy perennials. Some alpine plants, for example, buttercups, are lowland species that have adapted to the alpine environment, whereas other plants, for example, mountain douglasia, live exclusively in the mountains. This tundra is called *alpine tundra* to distinguish it from arctic tundra. At the top of the mountain, a permanent ice or snow cap might be found, similar to the nearly lifeless polar land areas.

Important environmental differences exist between high elevations and high latitudes that affect the types of organisms found in each place. Alpine tundra typically lacks permafrost and receives more precipitation than does arctic tundra. High elevations of temperate mountains do not have the great extremes of day length that are associated with the changing seasons in high-latitude biomes. The intensity of solar radiation is greater at high elevations than at high latitudes. At high elevations, the sun's rays pass through less atmosphere, which results in greater exposure to ultraviolet radiation (less is filtered out by the atmosphere) than occurs at high latitudes.

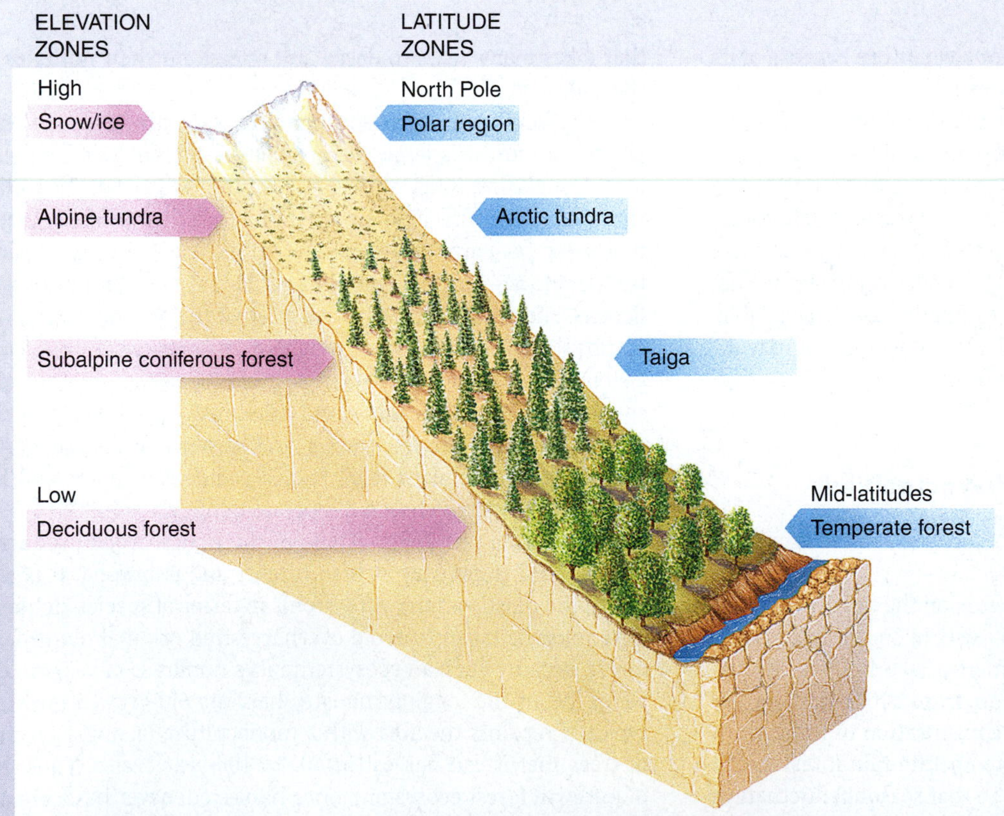

Comparison of elevation and latitude zones. The cooler temperatures at higher elevations of a mountain produce a series of ecosystems similar to those encountered when going from the equator toward the North Pole.

Figure 54–6 Temperate rain forest. Temperate rain forest is characterized by large amounts of precipitation. Note the epiphytes hanging from the branches of coniferous trees. Photographed in Olympic National Park in Washington state. *(Terry Donnelly/Dembinsky Photo Associates)*

Temperate deciduous forest has a canopy of broad-leaf trees

Seasonality (hot summers and cold winters) is characteristic of **temperate deciduous forest,** which occurs in temperate areas where precipitation ranges from about 75 to 126 cm (30 to 50 in) annually (see Fig. 54–3). Typically, the soil of a temperate deciduous forest consists of both a topsoil rich in organic material and a deep, clay-rich lower layer. As organic materials decay, nutrient mineral ions are released. If roots of living trees do not absorb these ions, they leach into the clay, where they may be retained.

Temperate deciduous forests of the northeastern and middle eastern United States are dominated by broad-leaf hardwood trees, such as oak, hickory, maple, and beech, that lose their foliage annually (Fig. 54–7). In southern areas, the number of broad-leaf evergreen trees, such as magnolia, increases. The trees of the temperate deciduous forest form a dense canopy that overlies saplings and shrubs.

Temperate deciduous forests originally contained a variety of large mammals such as puma, wolves, bison, and other species now regionally extinct, plus deer, bears, and many small mammals and birds (e.g., wild turkeys, blue jays, and scarlet tanagers). Both reptiles (e.g., box turtles and rat snakes) and amphibians (e.g., spotted salamanders and wood frogs) abounded, together with a denser and more varied insect life than exists today.

In Europe and North America, logging and land clearing for farms, tree plantations, and cities have removed much of the original temperate deciduous forest. Where it has been allowed to regenerate, temperate deciduous forest is often in a seminatural state, that is, highly modified by humans for recreation, livestock foraging, timber harvest, and other uses. Although these returning forests do not have the biological diversity of virgin stands, many forest organisms have successfully become reestablished.

Worldwide, temperate deciduous forest was among the first biomes to be converted to agricultural use. In Europe and Asia, many soils that originally supported temperate deciduous forest

Figure 54–7 Temperate deciduous forest. The broad-leaf trees that dominate the temperate deciduous forest shed their leaves before winter. Photographed during autumn in Pennsylvania. *(Barbara Miller/Biological Photo Service)*

have been cultivated by traditional agricultural methods for thousands of years without substantial loss in fertility. During the 20th century, however, intensive agricultural practices were adopted; these, along with overgrazing and deforestation, contributed to the degradation of some agricultural lands. The U.N. Environment Program released the first global assessment of soil conditions in 1992. It reported that 1.96 billion hectares (4.84 billion acres) of soil—an area equal to 17% of Earth's total vegetated surface area—have been degraded since the end of World War II. Eleven percent of the Earth's vegetated surface—an area the size of China and India combined—has been degraded so badly that it will be very costly, or in some cases impossible, to reclaim it.

Temperate grasslands occur in areas of moderate precipitation

Summers are hot, winters are cold, fires help to shape the landscape, and rainfall is often uncertain in **temperate grasslands** (see Fig. 54–3). Annual precipitation averages 25 to 75 cm (10 to 30 in). In grasslands with less precipitation, nutrient minerals tend to accumulate in a marked layer just below the topsoil. These nutrient minerals tend to leach out of the soil in areas with more precipitation. Grassland soil contains considerable organic material because surface parts of many grasses die off each winter and contribute to the organic content of the soil, while the roots and rhizomes (underground stems) survive underground. The roots and rhizomes that eventually die also contribute to the soil's organic material. Many grasses are sod formers: Their roots and rhizomes form a thick, continuous underground mat.

Moist temperate grasslands, also known as *tallgrass prairies,* occur in the United States in Iowa, western Minnesota, eastern Nebraska, and parts of other Midwestern states and across Canada's prairie provinces. Although few trees grow except near rivers and streams, grasses, some as tall as 2 m (6.5 ft), grow in great profusion in the deep, rich soil (Fig. 54–8). Before most of this area was converted to arable land, it was covered with herds of grazing animals, particularly bison. The principal predators were wolves, although in sparser, drier areas coyotes took their place. Smaller fauna included prairie dogs and their predators (foxes, black-footed ferrets, and birds of prey such as prairie falcons), western meadowlarks, bobolinks, reptiles (such as gopher snakes and short-horned lizards), and great numbers of insects.

Shortgrass prairies, in which the dominant grasses are less than 0.5 m (1.6 ft) tall, are temperate grasslands that receive less precipitation than the moister grasslands just described but more precipitation than deserts. In the United States, shortgrass prairies occur in the eastern half of Montana, the western half of South Dakota, and parts of other Midwestern states, as well as western Alberta in Canada. Grasses that grow knee high or lower dominate shortgrass prairies. The plants grow in less abundance than in the moister grasslands, and occasionally some bare soil is exposed. Native grasses of shortgrass prairies are drought-resistant.

Figure 54–8 Temperate grassland. This tallgrass prairie in Oklahoma is a preserve owned by the Nature Conservancy. Like other moist temperate grasslands, it is mostly treeless but contains a profusion of grasses and other herbaceous flowering plants. Bison graze on the plants, thereby affecting community structure and diversity. *(Harvey Payne)*

The North American grassland, particularly the tallgrass prairie, was so well suited to agriculture that little of it remains. More than 90% has vanished under the plow, and the remainder is so fragmented that almost nowhere can we see even an approximation of what European settlers saw when they settled in the Midwest. Today, the tallgrass prairie is considered North America's rarest biome. It is not surprising that the American Midwest, Ukraine, and other moist temperate grasslands became the breadbaskets of the world, because they provide ideal growing conditions for crops such as corn and wheat, which are also grasses.

Chaparral is a thicket of evergreen shrubs and small trees

Some hilly temperate environments have mild winters with abundant rainfall, combined with extremely dry summers. Such Mediterranean climates, as they are called, occur not only in the area around the Mediterranean Sea but also in California, western Australia, portions of Chile, and South Africa (see Fig. 54–3). In southern California this environment is known as **chaparral.**[2] Chaparral soil is thin and infertile. Frequent fires occur naturally in this environment, particularly in late summer and autumn.

Chaparral vegetation looks strikingly similar in different areas of the world, even though the individual species are quite different. Chaparral is usually dominated by a dense growth of evergreen shrubs and may contain drought-resistant pine or scrub oak trees (Fig. 54–9). During the rainy winter season the landscape may be lush and green, but the plants lie dormant during the hot, dry summer. Trees and shrubs often have hard, small,

[2] This vegetation type is known as maquis in the Mediterranean region, mallee scrub in Australia, matorral in Chile, and Cape scrub in Africa.

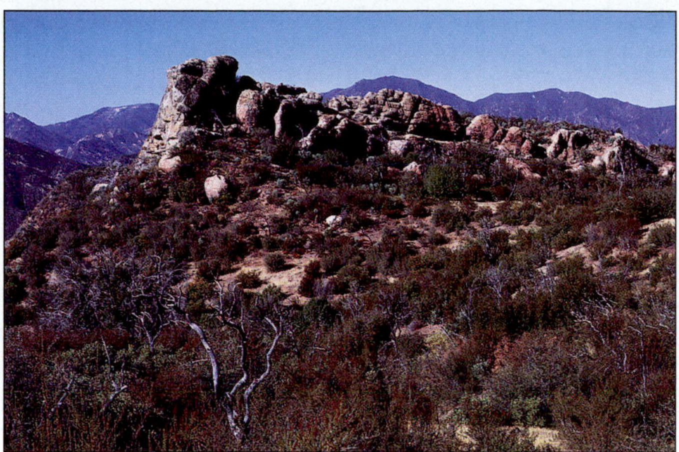

Figure 54–9 Chaparral. Chaparral, which consists primarily of drought-resistant evergreen shrubs and small trees, develops where hot, dry summers alternate with mild, rainy winters. Photographed in the Santa Lucia Mountains, California. *(Edward Ely/Biological Photo Service)*

leathery leaves that resist water loss. Many plants are also fire-adapted and grow best in the months following a fire. Such growth is possible because fire releases nutrient minerals that were tied up in aboveground parts of plants that burned. The underground parts of some plants and the seeds of many others are not killed by the fire, however, and with the new availability of essential nutrient minerals, plants sprout vigorously during winter rains. Mule deer, wood rats, brush rabbits, lizards, and many species of birds (e.g., Anna's hummingbird, scrub jay, and bushtit) are common animals of the chaparral.

Fires, which occur at irregular intervals in California chaparral vegetation, are often quite costly because they consume expensive homes built on the hilly chaparral landscape. Unfortunately, efforts to control naturally occurring fires sometimes backfire. Denser, thicker vegetation tends to accumulate when periodic fires are prevented; then, when a fire does occur, it is much more severe. Removing the chaparral vegetation, whose roots hold the soil in place, can also cause problems—witness the mudslides that sometimes occur during winter rains in these areas.

Deserts are arid ecosystems

Deserts are dry areas found in temperate *(cold deserts)* and subtropical or tropical regions *(warm deserts)* (see Fig. 54–3). In North America there are four distinct deserts. The Great Basin Desert in Nevada, Utah, and neighboring states is a cold desert dominated by sagebrush. The Mojave Desert in Nevada and California is known for its Joshua trees. The Chihuahuan Desert, home of century plants (agaves; see Fig. 51–10), is found in Texas and New Mexico as well as Mexico. The Sonoran Desert, with its many species of cacti, is found in Arizona, California, and Mexico (Fig. 54–10).

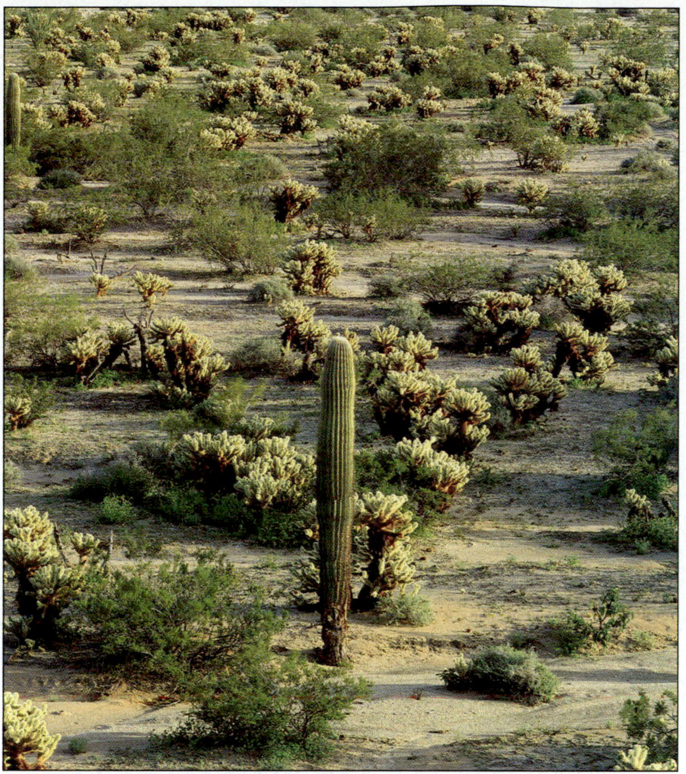

Figure 54–10 Desert. Desert inhabitants are strikingly adapted to the demands of their environment. The warmer deserts of North America, such as the Sonoran Desert shown here, are characterized by summer rainfall. The Sonoran Desert, found in parts of Arizona, California, and Mexico, contains many species of cacti, including the large, treelike saguaro, which grows 15 to 18 m (50 to 60 ft) tall. Photographed in Arizona. *(Willard Clay/Dembinsky Photo Associates)*

The low water vapor content of the desert atmosphere leads to daily temperature extremes of heat and cold, so that a major change in temperature occurs in each 24-hour period. Deserts vary greatly depending on the amount of precipitation they receive, which is generally less than 25 cm (10 in) per year. A few deserts are so dry that virtually no plant life occurs in them. As a result of sparse vegetation, desert soil is low in organic material but often high in mineral content, particularly the salts NaCl, $CaCO_3$, and $CaSO_4$. In some regions, such as areas of Utah and Nevada, the naturally occurring concentration of certain soil minerals reaches toxic levels for many plants.

Desert vegetation includes both perennials and, after a rain, flowering annuals. Plants in North American deserts include cacti, yuccas, Joshua trees, and sagebrushes. Desert plants tend to have reduced leaves or no leaves, an adaptation that conserves water. In cacti such as the giant saguaro, for example, photosynthesis is carried out by the stem, which also expands accordion-style to store water; the leaves are modified into spines, which discourage herbivores. Other desert plants shed their leaves for most of the year, growing only during the brief moist season. Many desert plants possess defensive spines, thorns, or toxins to resist the heavy grazing pressure often experienced in this food- and water-deficient environment.

Desert animals tend to be small. During the heat of the day they remain under cover or return to shelter periodically, whereas at night they come out to forage or hunt. In addition to desert-adapted insects, there are many specialized desert reptiles (such as desert iguanas, desert tortoises, and rattlesnakes) and a few desert-adapted amphibians (such as western spadefoot toads). Mammals include such rodents as the American kangaroo rat, which does not need to drink water but can subsist solely on the water content of its food (primarily seeds and insects). American deserts are also home to jackrabbits, while Australian deserts have kangaroos. Carnivores like the African fennec fox and some birds of prey, especially owls, live on rodents and rabbits. During the driest months of the year, many desert insects, amphibians, reptiles, and mammals tunnel underground, where they remain inactive; this period of dormancy is known as *aestivation*.

Humans have altered North American deserts in several ways. Off-road vehicles damage desert vegetation, which sometimes takes years to recover. Tank crews that trained for World War II in California's Mojave Desert left tracks in 1940 that are still visible today. People who drive across the desert in four-wheel-drive vehicles also inflict environmental damage. When the top layer of desert soil is disturbed, erosion occurs more readily, and less vegetation grows to support native animals. Another problem is that certain cacti and desert tortoises have become rare as a result of poaching. Houses, factories, and farms built in desert areas require vast quantities of water, which must be imported from distant areas. Increased groundwater consumption by many desert cities has caused groundwater levels to drop. Aquifer depletion in U.S. deserts is particularly critical in southern Arizona and southwestern New Mexico.

Savanna is a tropical grassland with scattered trees

The **savanna** biome is a tropical grassland with widely scattered clumps of low trees (Fig. 54–11). Savanna is found in areas of relatively low or seasonal rainfall with prolonged dry periods (see Fig. 54–3). Temperatures in savannas vary little throughout the year, and seasons are regulated by precipitation, not by temperature as they are in temperate grasslands. Annual precipitation is 85 to 150 cm (34 to 60 in). Savanna soil is low in essential nutrient minerals, in part because it is strongly leached. Savanna soil is often rich in aluminum, which resists leaching, and in places the aluminum reaches levels that are toxic to many plants. Although the African savanna is best known, savanna also occurs in South America and northern Australia.

Savanna is characterized by wide expanses of grasses interrupted by occasional trees, such as *Acacia,* which bristles with thorns that provide protection against herbivores. Both trees and grasses have fire-adapted features such as extensive underground root systems that enable them to survive seasonal droughts as well as periodic fires that sweep through the savanna.

The world's greatest assemblage of hoofed mammals occurs in the African savanna. Here live great herds of herbivores, including wildebeests, antelopes, giraffes, zebras, and elephants. Large predators, such as lions and hyenas, kill and scavenge the herds. In areas of seasonally varying rainfall, the herds and their predators may migrate annually.

Figure 54–11 Savanna. Savanna is tropical grassland with widely scattered trees. African savanna formerly supported large herds of grazing animals and their predators; these are swiftly vanishing under pressure from pastoral and agricultural land use. The smaller animals in the foreground are Thomson's gazelles *(Gazella thomsonii)*, and the larger ones near the trees are wildebeests *(Connochaetes taurinus).* Photographed in Tanzania. *(Carlyn Iverson)*

Savanna is rapidly being converted to rangeland for cattle and other domesticated animals, which are replacing the big herds of wild animals. The problem is particularly acute in Africa, which has the most rapidly growing human population of any continent. In some places severe overgrazing by domestic animals has contributed to the conversion of marginal savanna into desert, a process known as **desertification.**[3] In desertification, the reduced grass cover caused by overgrazing allows wind and water to erode the soil, removing the topsoil. Desertification is a progressive process that reduces the productivity of the land, decreasing its ability to support crops or livestock.

There are two basic types of tropical forests

There are many kinds of tropical forests, but ecologists generally classify them as one of two types: tropical dry forests or tropical rain forests. **Tropical dry forests** occur in regions subjected to a wet season and a dry season (usually two to three months each year). Annual precipitation is 150 to 200 cm (60 to 80 in). During the dry season, many tropical trees shed their leaves and remain dormant, much as temperate trees do during the winter. Sometimes a dense understory of grasses and herbs develops during the dry season when the canopy is open. India, Brazil, Thailand, and Mexico are some of the countries that have tropical dry forests (see Fig. 54–3). Tropical dry forests intergrade with savanna on their dry edges and with tropical rain forests on their wet edges. This forest has been fragmented and degraded extensively by logging and overgrazing by domestic animals.

Tropical rain forests occur where temperatures are warm throughout the year and precipitation occurs almost daily (see Fig. 54–3). The annual precipitation of tropical rain forests is 200 to 450 cm (80 to 180 in). Much of this precipitation comes from locally recycled water that enters the atmosphere by transpiration from the forest's own trees.

Tropical rain forests are often located in areas with ancient, highly weathered, mineral-poor soil. Little organic matter accumulates in such soils. Because temperatures are high and soil moisture is abundant year round, decay organisms and detritus-feeding ants and termites decompose organic litter quite rapidly. Nutrient minerals from decomposing materials are quickly absorbed by vast networks of roots and mycorrhizae. Thus, nutrient minerals of tropical rain forests are tied up in the vegetation rather than in the soil. Tropical rain forests are found in Central and South America, Africa, and Southeast Asia. Tropical rain forest is very productive despite the scarcity of nutrient minerals in the soil. Its plants capture considerable energy by photosynthesis, stimulated by abundant solar energy and precipitation. Of all the biomes, the tropical rain forest is unrivaled in species richness. Local factors affecting rainforest species composition include varying soil fertility and topography (e.g., valleys have more plant diversity than hills).

Most trees of tropical rain forests are evergreen flowering plants (Fig. 54–12*a*). A fully developed rain forest has three or more distinct stories of vegetation. The topmost story consists of the crowns of the oldest, tallest trees, some 50 m (164 ft) or more in height; these trees are exposed to direct sunlight and are subject to the warmest temperatures, lowest humidities, and strongest winds. The middle story, which reaches a height of 30 to 40 m (100 to 130 ft), forms a continuous canopy of leaves overhead that lets in little sunlight for the support of the sparse understory. Only 2% to 3% of the light bathing the forest canopy reaches the forest understory, which is exposed to very little wind and is relatively cool and humid. The understory itself consists of both smaller plants specialized for life in shade and seedlings of taller trees. Vegetation of tropical rain forests is not dense at ground level except near stream banks or where a fallen tree has opened the canopy.

Tropical rainforest trees support extensive epiphytic communities of smaller plants such as orchids and bromeliads. Although epiphytes grow in crotches of branches, on bark, or even on the leaves of their hosts, they only use their host trees for physical support, not for nourishment.

Because little light penetrates to the understory, many plants living there are adapted to climb already-established host trees. Lianas (woody tropical vines), some as thick as a human thigh, twist up through the branches of tropical rainforest trees (Fig. 54–12*b*). Once in the canopy, lianas grow from the upper branches of one forest tree to another, connecting the tops of the trees together and providing a walkway for many of the canopy's residents. Lianas and herbaceous vines provide nectar and fruit for many tree-dwelling animals.

Not counting bacteria and other soil-dwelling organisms, about 90% of tropical rainforest organisms live in the middle and upper canopies (Fig. 54–13). Rainforest animals include the most abundant and varied insect, reptile, and amphibian fauna on Earth. Birds, too, are diverse, with some specialized to consume fruits (e.g., parrots) and others to consume nectar (e.g., hummingbirds and sunbirds). Most rainforest mammals, such as sloths and monkeys, live only in the trees and never climb down to the ground surface, although some large ground-dwelling mammals, including elephants, are also found in tropical rain forests.

Unless strong conservation measures are initiated soon, human population growth and agricultural and industrial expansion in tropical countries will spell the end of tropical rain forests by the middle of the 21st century. Biologists know that many rainforest species will become extinct before they have even been identified and scientifically described. Tropical rainforest destruction is discussed in detail in Chapter 55.

AQUATIC ECOSYSTEMS OCCUPY MOST OF EARTH'S SURFACE

Aquatic "biomes" do not exist, in the sense that aquatic ecologists do not distinguish aquatic ecosystems based on the dominant form of vegetation. Aquatic ecosystems are classified primarily on abiotic factors, such as salinity, that help determine an aquatic life zone's boundaries. **Salinity** (the concentration of

[3] Desertification is not restricted to the savanna biome. Temperate grasslands and tropical dry forests can also be degraded to desert.

(a)

(b)

Figure 54–12 Tropical rain forest. **(a)** A broad view of tropical rainforest vegetation along a riverbank in Southeast Asia. Except at riverbanks, tropical rain forest has a closed canopy that admits little light to the forest floor. **(b)** Thick, epiphyte-covered lianas grow into the canopy using a tree trunk for support. This photo was taken in Costa Rica. *(a, Frans Lanting/Minden Pictures; b, Mark Moffett/Minden Pictures)*

dissolved salts, such as sodium chloride, in a body of water) affects the kinds of organisms present in aquatic ecosystems, as does the amount of dissolved oxygen. Water greatly interferes with the penetration of light, so floating aquatic organisms that photosynthesize must remain near the water's surface, and vegetation attached to the bottom can grow only in shallow water. In addition, low levels of essential nutrient minerals often limit the number and distribution of organisms in certain aquatic environments. Other abiotic determinants of species composition in aquatic ecosystems include water depth, temperature, pH, and presence or absence of waves and currents.

Aquatic ecosystems contain three main ecological categories of organisms: free-floating plankton, strongly swimming nekton, and bottom-dwelling benthos. **Plankton** are usually small or microscopic organisms that are relatively feeble swimmers. For the most part, they are carried about at the mercy of currents and waves. They are unable to swim far horizontally, but some species are capable of large vertical migrations and are found at different depths of water at different times of the

Figure 54–13 Studying the rainforest canopy. This construction crane was erected in a tropical rain forest in Panama to study rainforest organisms without harming them. Before cranes were used, biologists were unable to perform many experiments in the middle and upper canopies. *(Photo by N. Guerre/STRI)*

day or at different seasons. Plankton are generally subdivided into two major categories: phytoplankton and zooplankton. **Phytoplankton** (photosynthetic cyanobacteria and free-floating algae) are producers that form the base of most aquatic food webs. **Zooplankton** are nonphotosynthetic organisms that include protozoa, tiny crustaceans, and the larval stages of many animals. **Nekton** are larger, more strongly swimming organisms such as fishes, turtles, and whales. **Benthos** are bottom-dwelling organisms that fix themselves to one spot (sponges, oysters, and barnacles), burrow into the sand (many worms and echinoderms), or simply walk or swim about on the bottom (crawfish, aquatic insect larvae, and brittle stars).

FRESHWATER ECOSYSTEMS ARE CLOSELY LINKED TO LAND AND MARINE ECOSYSTEMS

Freshwater ecosystems include streams and rivers (flowing-water ecosystems), ponds and lakes (standing-water ecosystems), and marshes and swamps (freshwater wetlands). Each type of freshwater ecosystem has its own specific abiotic conditions and characteristic organisms. Although freshwater ecosystems occupy a relatively small portion (about 2%) of Earth's surface, they have an important role in the hydrological cycle: They assist in recycling precipitation that flows as surface runoff to the ocean (see discussion of hydrological cycle in Chapter 53). Large bodies of fresh water also help moderate daily and seasonal temperature fluctuations on nearby land. Freshwater habitats also provide homes for large numbers of species.

Streams and rivers are flowing-water ecosystems

Many different conditions exist along the length of a stream or river (Fig. 54–14). The nature of a **flowing-water ecosystem** changes greatly from its source (where it begins) to its mouth (where it empties into another body of water). Headwater streams (small streams that are the sources of a river) are usually shallow, clear, cold, swiftly flowing, and highly oxygenated. In contrast, rivers downstream from the headwaters are wider and deeper, cloudy (i.e., they contain suspended particulates), not as cold, slower flowing, and less oxygenated. Surrounding forest may shade certain parts of the stream or river, whereas other parts may be exposed to direct sunlight. Along parts of a stream or river, groundwater wells up through sediments on the bottom; this local input of water moderates the water temperature so that summer temperatures are cooler and winter temperatures are warmer than in adjacent parts of the flowing-water ecosystem.

The kinds of organisms found in flowing-water ecosystems vary greatly from one stream to another, depending primarily on the strength of the current. In streams with fast currents, inhabitants may have adaptations such as suckers to attach themselves to rocks so they are not swept away. The larvae of blackflies, for example, attach themselves with suction disks located on the ends of their abdomens. Some stream inhabitants, such as immature

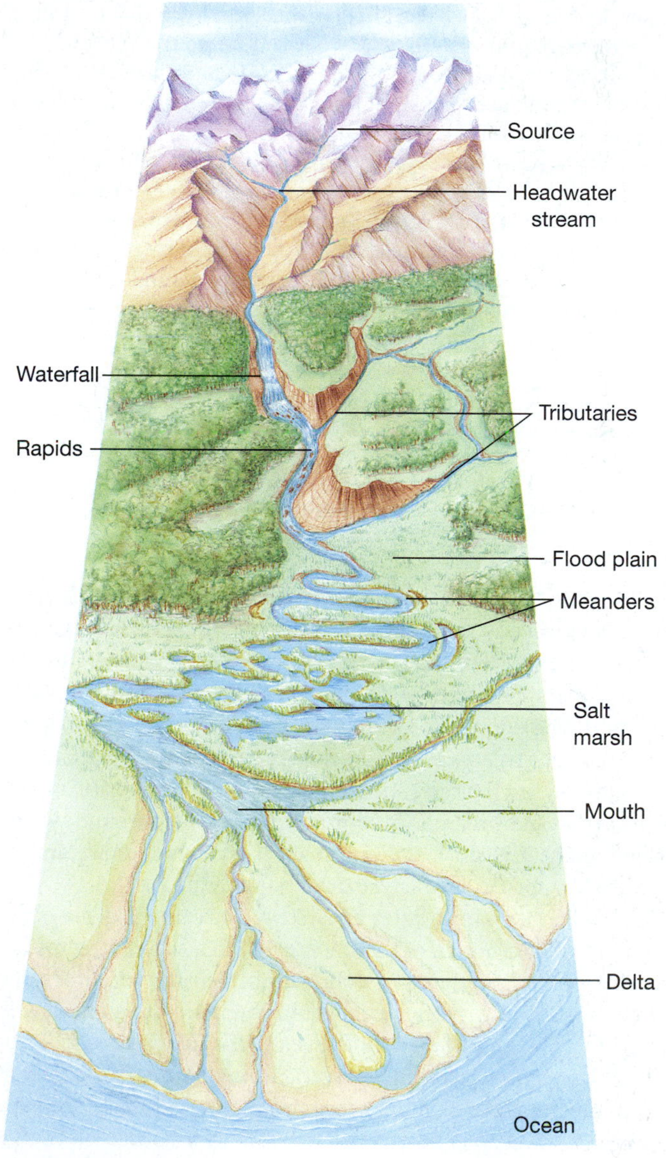

Figure 54–14 Features of a typical river. The river begins at a source, often high in mountains and fed by melting snows or glaciers. Headwater streams flow downstream rapidly, often over rocks (as rapids) or bluffs (as waterfalls). Along the way, smaller streams called tributaries feed into the river, adding to its flow. As the river's course levels out, the river flows more slowly and winds from side to side, forming bends called meanders. The flood plain is the relatively flat area on either side of the river that is subject to flooding. Some flood plains are quite large: The Mississippi River's flood plain is up to 130 km (80 mi) wide. Near the ocean, the river may form a salt marsh where fresh water from the river and salt water from the ocean mix. The delta is a fertile, low-lying plain at the river's mouth that forms from sediments deposited by the slow-moving river as it empties into the ocean.

Process of Science

ON THE CUTTING EDGE

A Detrital Connection Between Terrestrial and Aquatic Ecosystems

HYPOTHESIS: Leaf litter is an important detrital input for secondary productivity in headwater streams.

METHOD: Exclude leaf litter from a forested headwater stream for four years. Collect monthly samples of benthic invertebrates.

RESULTS: Benthic invertebrate abundance, biomass, and secondary productivity in the manipulated stream was significantly less than in a control stream.

CONCLUSION: Leaf litter is an important resource in maintaining the productivity of headwater streams flowing through forests.

One of the long-held assumptions of stream ecology is that a forest provides a stream flowing through it with an important input of energy in the form of detritus (nonliving organic matter). Until recently, however, little direct evidence was available to support this idea. In 1997, J.B. Wallace and others from the University of Georgia and the Virginia Polytechnic Institute and State University published results of a study that linked terrestrial leaf litter to the aquatic organisms in a forested Appalachian headwater stream.*

Wallace and coworkers excluded detritus—specifically, leaf litter—from 180 m of a stream for several years by erecting an overhead canopy and lateral fence. They used a nearby "unmanipulated" stream as a control to minimize the possibility that the changes they observed were due to natural environmental fluctuations instead of their experimental manipulations. They compared the changes in two different habitats in the stream: where the streambed was gravel or sand, and where it was moss-covered bedrock.

Compared to the unmanipulated stream, the abundance, biomass, and secondary production of aquatic invertebrates declined in those portions of the experimental stream where the streambed was gravel or sand *(see figure)*. (**Secondary production** is consumer production—that is, the energy that goes into the production of consumer biomass, either new tissue associated with growth or new individuals.) The decline was observed throughout the food web, from detritivores (animals that eat detritus) to the predators of detritivores. Thus, Wallace's experiment demonstrates how a single change, reduction in the input of leaf litter, propagates through a food web.

Interestingly, excluding terrestrial inputs from the moss-covered portion of the experimental stream caused few changes in the abundance and biomass of invertebrate populations. This result suggests that secondary production in the moss-covered habitat is not as closely linked to detrital inputs from the surrounding forest. Apparently, different food webs exist in

*Wallace, J.B., S.L. Eggert, J.L. Meyer, and J.R. Webster. "Multiple Trophic Levels of a Forest Stream Linked to Terrestrial Litter Inputs." *Science*, Vol. 277, 4 Jul. 1997.

water-penny beetles, have flattened bodies that enable them to slip under or between rocks. The water-penny beetle larva gets its common name from its flattened, nearly circular shape. Alternatively, inhabitants such as the brown trout may be streamlined and muscular enough to swim in the current. Organisms in large, slow-moving streams and rivers do not need such adaptations, although they are typically streamlined, as are most aquatic organisms, to lessen resistance during movement through water. Where current is slow, plants and animals of the headwaters are replaced by species characteristic of ponds and lakes.

Unlike other freshwater ecosystems, streams and rivers depend on land for much of their energy. In headwater streams, up to 99% of the energy input comes from detritus (dead organic material such as leaves) carried from the land into streams and rivers by wind or surface runoff (see *On the Cutting Edge: A Detrital Connection Between Terrestrial and Aquatic Ecosystems*). Downstream, rivers contain more producers and therefore have a slightly lower dependence on detritus as a source of energy than do the headwaters.

Human activities have several adverse impacts on rivers and streams, including water pollution and the effects of dams, which are built to contain the water of rivers or streams. Dams change the nature of flowing-water ecosystems, both upstream and downstream from the dam location. A dam causes water to back up, flooding large areas of land and forming a reservoir, which destroys terrestrial habitat. Below the dam, the once-powerful river is reduced to a relative trickle, altering the flowing-water ecosystem.

Pollution alters the physical environment of a flowing-water ecosystem and changes the biotic component downstream from the pollution source. For example, fertilizer runoff from Midwestern fields and manure runoff from livestock operations in such states as Iowa, Wisconsin, and Illinois eventually find their way into the Mississippi River and, from there, into the Gulf of Mexico. These nutrients have created a huge "dead zone" in the Gulf of Mexico some 17,500 km^2 (7000 mi^2) in area. The dead zone extends from the seafloor up into the water column, sometimes to within a few meters of the surface. It generally persists from April or May, as snowmelt and spring rains flow from the

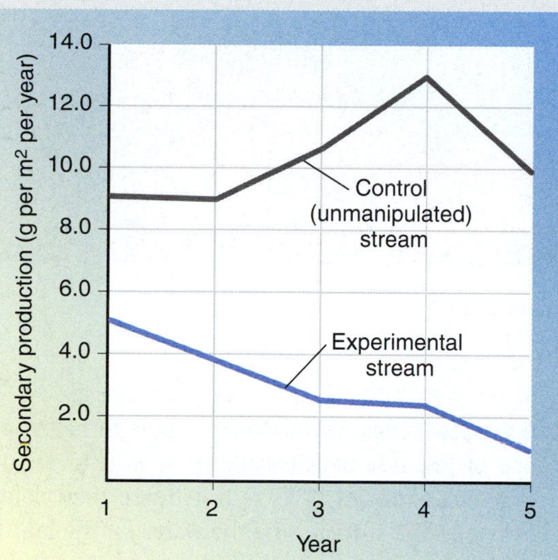

Annual secondary productivity of benthic organisms in gravel and sand portions of control (unmanipulated) and experimental streams. Secondary production is the assimilation of organic matter by the consumers (in this example, the detritivores and predators) in the stream ecosystem. Year 1 measurements were taken before litter was excluded from the experimental stream to determine natural differences between control and experimental streams; litter exclusion occurred during years 2, 3, 4, and 5. *(Data derived from Wallace, J.B., S.L. Eggert, J.L. Meyer, and J.R. Webster. "Multiple Trophic Levels in a Forest Stream Linked to Terrestrial Litter Inputs."* Science, *Vol. 277, 4 Jul. 1997; and Wallace, J.B., S.L. Eggert, J.L. Meyer, and J.R. Webster. "Effects of Resource Limitation on a Detrital-Based Ecosystem."* Ecological Monographs, *Vol. 64, No. 4, 1999)*

different stream habitats, despite the fact that they are in close proximity to each other.

J.L. Meyer, a collaborator in the experiment, determined the relative importance of terrestrial and instream sources of dissolved organic carbon (DOC).[†] DOC serves as food for microorganisms in aquatic food webs and affects properties such as pH and how far ultraviolet radiation penetrates in water. DOC comes from groundwater or from decomposing leaves and wood that fall into streams. Excluding litter from the experimental stream reduced the streamwater concentration of DOC as compared with the control stream. By measuring DOC moving into and out of streams, the experiment showed that a third of DOC in forested streams comes from leaf litter.

Wallace's work has important implications in the field of restoration ecology. Many natural and human-induced disturbances, such as logging, land-use change, and grazing, cause multiple changes in aquatic ecosystems, including a reduction in the supply of leaf litter to stream ecosystems. Wallace's experiment indicates that to restore the biological diversity and productivity in a stream ecosystem, inputs of terrestrial detritus must be established.

[†]Meyer, J.L., J.B. Wallace, and S.L. Eggert. "Leaf Litter as a Source of Dissolved Organic Carbon in Streams." *Ecosystems,* Vol. 1, 1998.

Mississippi River into the Gulf, to September. Only anaerobic bacteria exist in the dead zone because the water does not contain enough dissolved oxygen to support fishes, shrimp, or other aquatic organisms. This low-oxygen condition, known as **hypoxia,** occurs when algae grow rapidly owing to the presence of nutrients in the water. When these algae die, they sink to the bottom and are decomposed by bacteria, whose metabolic activities deplete the water of dissolved oxygen, leaving too little for other aquatic life. Pollution in flowing-water ecosystems has caused hypoxia in more than 60 coastal areas around the world.

Ponds and lakes are standing-water ecosystems

Standing-water ecosystems are characterized by zonation. A large lake has three basic zones: the littoral, limnetic, and profundal zones (Fig. 54–15). Smaller lakes and ponds typically lack a profundal zone. The **littoral zone** is a shallow water area along the shore of a lake or pond. It includes rooted, emergent vegetation, such as cattails and burreeds, plus several deeper-dwelling aquatic plants and algae. The littoral zone is the most productive zone of the lake. Photosynthesis is greatest in the littoral zone, in part because light is abundant and because the littoral zone receives nutrient inputs from surrounding land that stimulate the growth of plants and algae. Animals of the littoral zone include frogs and their tadpoles, turtles, worms, crayfish and other crustaceans, insect larvae, and many fishes such as perch, carp, and bass. Surface dwellers such as water striders and whirligig beetles are found in the quieter areas.

The **limnetic zone** is the open water beyond the littoral zone, that is, away from the shore; it extends down as far as sunlight penetrates to permit photosynthesis. The main organisms of the limnetic zone are microscopic phytoplankton and zooplankton. Larger fishes also spend some of their time in the limnetic zone, although they may visit the littoral zone to feed and reproduce. Owing to its depth, less vegetation grows here than in the littoral zone.

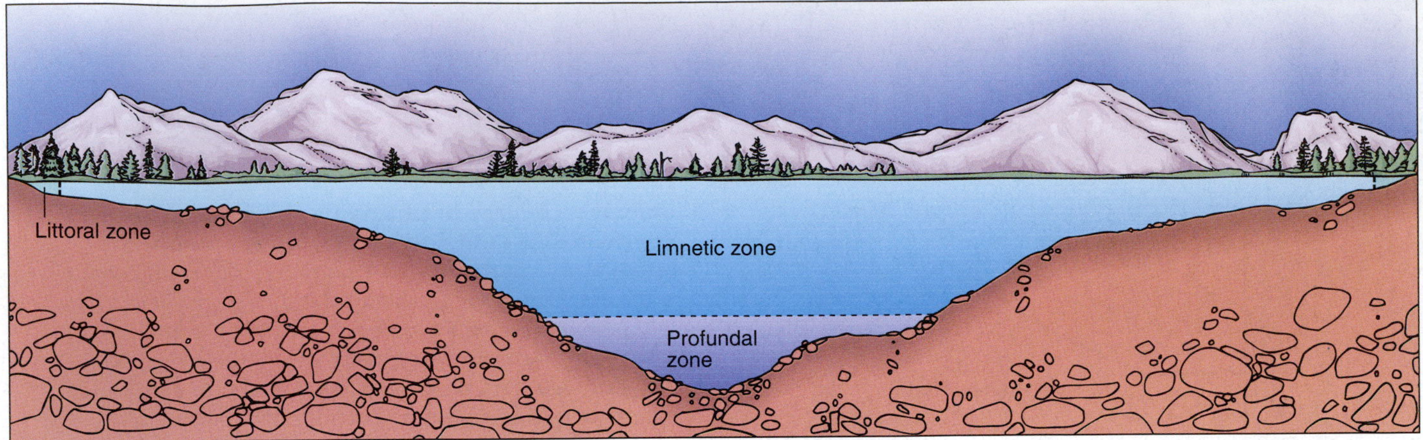

Figure 54–15 Zonation in a large lake. A lake is a standing-water ecosystem surrounded by land. Vegetation around the lake is not drawn to scale.

Beneath the limnetic zone of a large lake is the **profundal zone.** Because light does not penetrate effectively to this depth, plants and algae do not live here. Food drifts into the profundal zone from the littoral and limnetic zones. Bacteria decompose dead plants and animals that reach the profundal zone, liberating nutrient minerals. These minerals are not effectively recycled because no photosynthetic organisms are present to absorb them and incorporate them into the food web. As a result, the profundal zone tends to be both mineral-rich and anaerobic (oxygen deficient), with few organisms other than anaerobic bacteria occupying it.

Thermal stratification occurs in temperate lakes

The marked layering of large temperate lakes caused by light penetration is accentuated by **thermal stratification,** in which the temperature changes sharply with depth (Fig. 54–16*a*). Thermal stratification occurs because the summer sunlight penetrates and warms surface waters, making them less dense.[4] In summer, cool (and therefore more dense) water remains at the lake bottom and is separated from warm (and therefore less dense) water above by an abrupt temperature transition called the **thermocline.** Seasonal distribution of temperature and oxygen (more oxygen dissolves in water at cooler temperatures) affects the distribution of fish in the lake.

In temperate lakes, falling temperatures in autumn cause a mixing of the lake waters called the **fall turnover** (Fig. 54–16*b*). (Since little seasonal temperature variation occurs in the tropics, such turnovers are uncommon there.) As surface water cools to 4° C, its density increases, and eventually it sinks and displaces the less dense, warmer, mineral-rich water beneath. Warmer water then rises to the surface where it, in turn, cools and sinks. This process of cooling and sinking continues until the lake reaches a uniform temperature throughout.

When winter comes, surface water cools to below 4° C, its temperature of greatest density, and, if it is cold enough, ice forms. Ice, which forms at 0° C, is less dense than cold water; thus, ice forms on the surface, and the water on the lake bottom is warmer than the ice on the surface.

In the spring, a **spring turnover** occurs as ice melts and surface water reaches 4° C. Surface water again sinks to the bottom, and bottom water returns to the surface. As summer arrives, thermal stratification occurs once again.

The mixing of deeper, nutrient-rich water with surface, nutrient-poor water during the fall and spring turnovers brings essential nutrient minerals to the surface and oxygenated water to the bottom. The sudden presence of large amounts of essential nutrient minerals in surface waters encourages the development of large algal and cyanobacterial populations, which form temporary **blooms** in the fall and spring.

Increased nutrients, supplied by human activities, stimulate algal growth

Enrichment, the fertilization of a body of water, is caused by the presence of high levels of plant and algal nutrients such as nitrogen and phosphorus. Excess amounts of these nutrients get into waterways from sewage and from fertilizer runoff from lawns and fields. A pond or lake that is enriched is said to be **eutrophic.** The water in a eutrophic pond or lake is cloudy because of the presence of vast numbers of algae and cyanobacteria that are supported by the nutrients. Although eutrophic lakes contain large populations of aquatic animals, the species composition is different than in unenriched lakes. For example, an unenriched lake in the northeastern United States may contain pike, sturgeon, and whitefish in the deeper, colder part of the lake where there is a higher concentration of dissolved oxygen. In eutrophic lakes, on the other hand, the deeper, colder levels of water are depleted of dissolved oxygen because of the greater amount of decomposition on the lake floor. Therefore, fishes such as pike, sturgeon, and whitefish die out and are replaced by fishes such as catfish and carp that can tolerate lower concentrations of dissolved oxygen.

[4] Recall from Chapter 2 that the density of water is greatest at 4° C; both above and below this temperature, water is less dense.

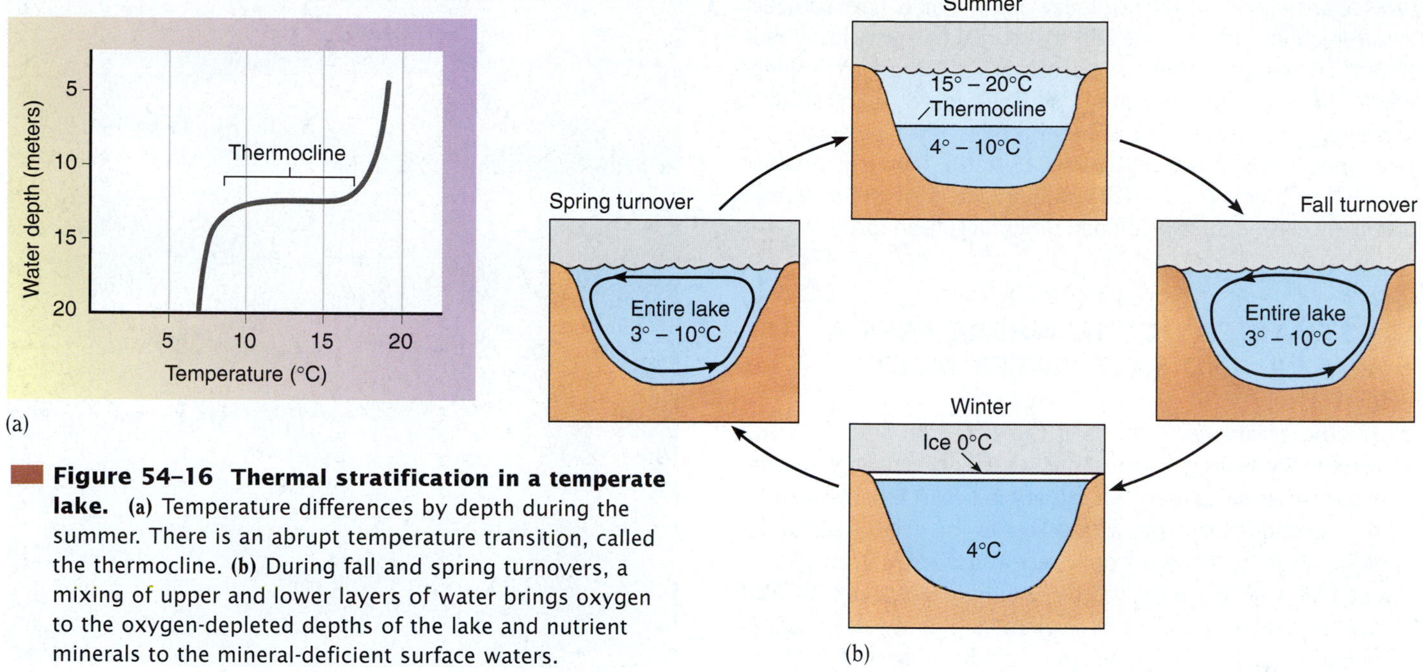

Figure 54–16 Thermal stratification in a temperate lake. **(a)** Temperature differences by depth during the summer. There is an abrupt temperature transition, called the thermocline. **(b)** During fall and spring turnovers, a mixing of upper and lower layers of water brings oxygen to the oxygen-depleted depths of the lake and nutrient minerals to the mineral-deficient surface waters.

Eutrophication is reversible and has declined sharply in North America since the 1970s with the passage of legislation to limit the phosphate content of detergents and with the construction of better sewage treatment plants. According to the 1992 U.S. National Water Quality Inventory Report to Congress, the leading source of water quality impairment of surface waters nationwide is agriculture. Fertilizer runoff, as well as animal wastes and plant residues in waterways, still cause enrichment problems.

Freshwater wetlands are transitional between aquatic and terrestrial ecosystems

Freshwater wetlands, usually covered by shallow water for at least part of the year, have characteristic soils and water-tolerant vegetation. They include marshes, in which grasslike plants dominate, and swamps, in which woody trees or shrubs dominate (Fig. 54–17). Freshwater wetlands also include hardwood bottomland forests (lowlands along streams and rivers that are periodically flooded), prairie potholes (small, shallow ponds that formed when glacial ice melted at the end of the last Ice Age), and peat moss bogs (peat-accumulating wetlands where sphagnum moss dominates).

Wetland plants, which are highly productive, provide enough food to support a wide variety of organisms. Wetlands are valued as a wildlife habitat for migratory waterfowl and many other bird species, beaver, otter, muskrat, and game fishes. Wetlands help control flooding by acting as holding areas for excess water when rivers flood their banks. The floodwater stored in wetlands then drains slowly back into the rivers, providing a steady flow of water throughout the year. Wetlands also serve as groundwater recharging areas. One of their most important roles is to help cleanse and purify water by trapping and holding pollutants in the flooded soil. Such important environmental functions as these are known as **ecosystem services.**

At one time wetlands were considered wastelands, areas to be filled in or drained so that farms, housing developments, and industrial plants could be built on them. Wetlands are also breeding places for mosquitoes and therefore were viewed as a menace to public health. The crucial ecosystem services that wetlands

Figure 54–17 Freshwater swamp. Freshwater swamps are inland areas permanently saturated or covered by water and dominated by trees, such as bald cypress *(shown).* A floating carpet of tiny aquatic plants covers the water surface. Photographed in northeast Texas. *(Gregory G. Dimijian/Photo Researchers, Inc.)*

provide are widely recognized today, and wetlands have some legal protection, but they are still threatened by agriculture, pollution, engineering (dams), and urban and suburban development. In the United States, wetlands have been steadily shrinking by an estimated 47,350 hectares (117,000 acres) per year since 1985. In the contiguous 48 states, more than half of the 89.4 million hectares (221 million acres) of wetlands that originally existed during colonial times have been lost.

ESTUARIES OCCUR WHERE FRESH WATER AND SALT WATER MEET

Where the ocean meets the land there may be one of several kinds of ecosystems: a rocky shore, a sandy beach, an intertidal mud flat, or a tidal estuary. An **estuary** is a coastal body of water, partly surrounded by land, with access to the open ocean and a large supply of fresh water from rivers. Water levels in an estuary rise and fall with the tides, whereas salinity fluctuates with tidal cycles, the time of year, and precipitation. Salinity also changes gradually within the estuary, from unsalty fresh water at the river entrance to salty ocean water at the mouth of the estuary. Because estuaries undergo marked daily, seasonal, and annual variations in temperature, salinity, and other physical properties, estuarine organisms must have a wide tolerance to such changes.

Estuaries are among the most fertile ecosystems in the world, often having much greater productivities than the adjacent ocean or freshwater river (see Table 53–1). This high productivity is the result of four factors. One, the action of tides promotes a rapid circulation of nutrients and helps remove waste products. Two, nutrient minerals are transported from land into streams and rivers that empty into the estuary. Three, a high level of light penetrates the shallow water. Four, the presence of many plants provides an extensive photosynthetic carpet and also mechanically traps detritus, forming the basis of detritus food webs. Most commercially important fishes and shellfish spend their larval stages in estuaries among the protective tangle of decaying stems.

Temperate estuaries usually contain **salt marshes,** shallow wetlands dominated by salt-tolerant grasses (Fig. 54–18). Salt marshes have often appeared to be worthless, empty stretches of land to uninformed people. As a result, they have been used as dumps and become severely polluted or have been filled with dredged bottom material to form artificial land for residential and industrial development. A large part of the estuarine environment has been lost in this way, along with many of its ecosystem services, such as biological habitats, sediment and pollution trapping, groundwater supply, and storm buffering (salt marshes absorb much of the energy of a storm surge and thereby prevent flood damage elsewhere).

Mangrove forests, the tropical equivalent of salt marshes, cover perhaps 70% of tropical coastlines. Like salt marshes, mangrove forests provide valuable ecosystem services. Their interlacing roots are breeding grounds and nurseries for several commercially important fish and shellfish species, such as blue crabs, shrimp, mullet, and spotted sea trout (Fig. 54–19). Mangrove

Figure 54–18 Salt marsh. Salt water and fresh water mix in salt marshes. Cordgrass (*Spartina* sp.) is the dominant vegetation in this salt marsh at Assateague Island National Seashore, Maryland. *(Michael Gadomski/Earth Scenes)*

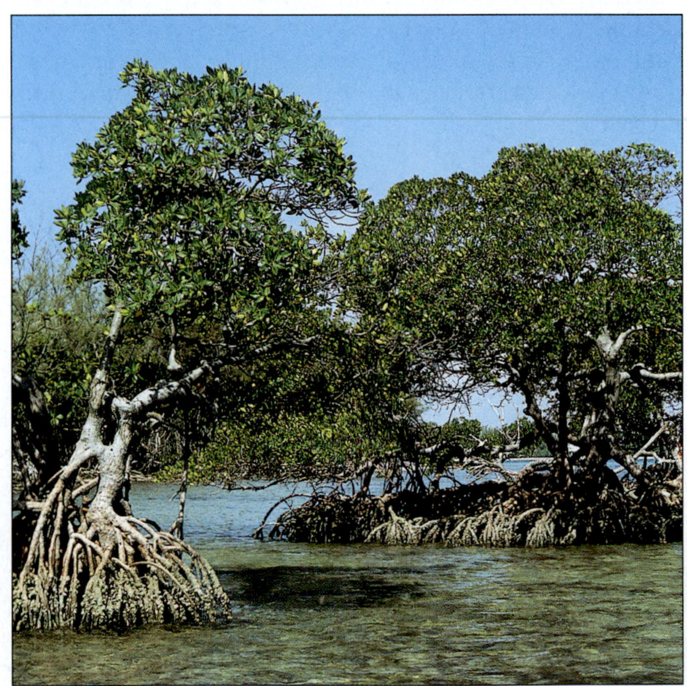

Figure 54–19 Mangrove forest. Mangroves often grow in tropical and subtropical coastal mudflats where tidal waters fluctuate. Red mangroves *(Rhizophora mangle)* have stiltlike roots that support the tree and grow into deeper water as well as into the mudflat exposed by low tide. Many animals live in the complex root system of mangrove forests. Photographed at low tide along the coast of Florida, near Miami. *(Patti Murray/Earth Scenes)*

branches are nesting sites for many species of birds, such as pelicans, herons, egrets, and roseate spoonbills. Mangrove roots stabilize the sediments, thereby preventing coastal erosion and providing a barrier against the ocean during storms. Mangroves are under assault from coastal development, including aquaculture facilities, and unsustainable logging. Some countries, such as the Philippines, Bangladesh, and Guinea-Bissau, have cut down 70% or more of their mangrove forests.

MARINE ECOSYSTEMS DOMINATE EARTH'S SURFACE

Although the ocean and lakes are comparable in many ways, they have many differences. Depths of even the deepest lakes do not approach those of the oceanic abysses, which are extremely deep areas that extend more than 6 km (3.6 mi) below the sunlit surface. Tides and currents profoundly influence the ocean. Gravitational pulls of both sun and moon produce two tides a day throughout the ocean, but the height of those tides varies with season, local topography, and phases of the moon (full moons and new moons cause the highest tides).

The immense and extremely complex marine environment is subdivided into several zones: the intertidal zone, the benthic (ocean floor) environment, and the pelagic (ocean water) environment (Fig. 54–20). The pelagic environment is in turn divided into two provinces—the neritic province and the oceanic province.

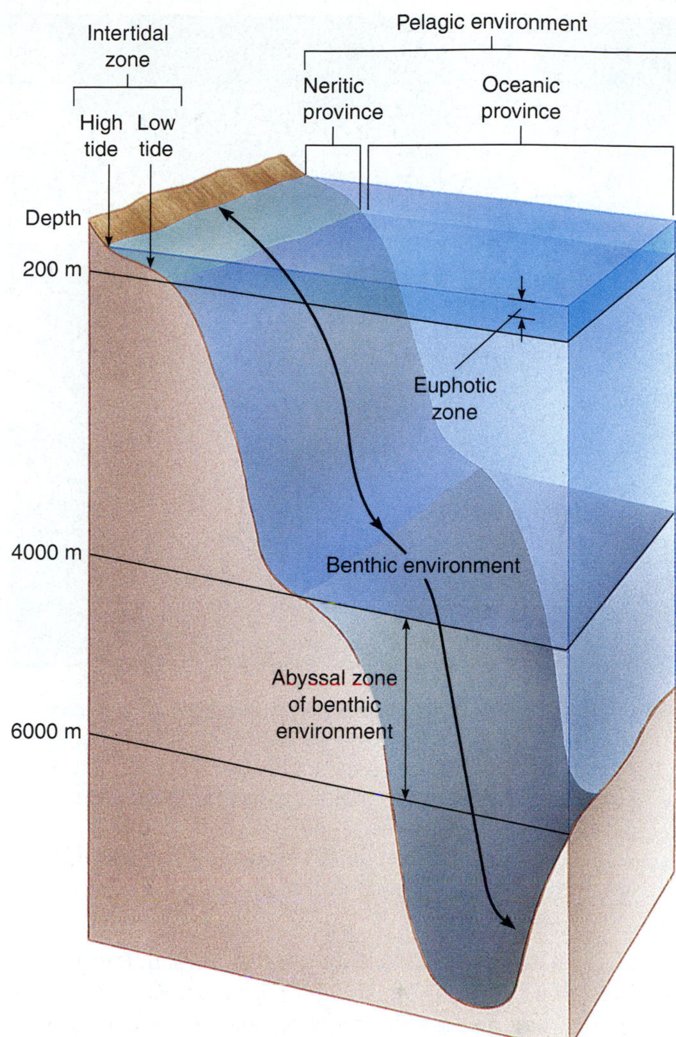

Figure 54–20 Zonation in the ocean. The ocean has three main life zones: the intertidal zone, the benthic environment, and the pelagic environment. The pelagic environment consists of the neritic and oceanic provinces. (The slopes of the ocean floor are not as steep as shown; they are exaggerated to save space.)

The intertidal zone is transitional between land and ocean

The shoreline area between low and high tide is called the **intertidal zone.** Although high levels of light and nutrients, together with an abundance of oxygen, make the intertidal zone a biologically productive environment, it is also a stressful one. If an intertidal beach is sandy, inhabitants must contend with a constantly shifting environment that threatens to engulf them and gives them scant protection against wave action. Consequently, most sand-dwelling organisms, such as mole crabs, are continual and active burrowers. Because they are able to follow the tides up and down the beach, most do not have any notable adaptations to survive drying or exposure.

A rocky shore provides a fine anchorage for seaweeds and invertebrate animals. However, it is exposed to constant wave action when immersed during high tides and to drying and temperature changes when exposed to the air during low tides. A typical rocky-shore inhabitant has some way of sealing in moisture, perhaps by closing its shell, if it has one, plus a powerful means of anchoring itself to rocks. Mussels, for example, have horny, threadlike anchors, and barnacles have special cement glands. Rocky-shore intertidal algae (seaweeds) usually have thick, gummy polysaccharide coats, which dry out slowly when exposed, and flexible bodies not easily broken by wave action (Fig. 54–21). Some rocky-shore community inhabitants hide in burrows or crevices at low tide.

Seagrass beds, kelp forests, and coral reefs are an important part of the benthic environment

The **benthic environment** is the ocean floor. It is divided into zones based on distance from land, light availability, and depth. The benthic environment consists of sediments (mostly sand and mud) in which many marine animals, such as worms and clams, burrow. Bacteria are common in marine sediments, and living bacteria have been found in ocean sediments more than 500 m (1625 ft) below the ocean floor at several different sites in the Pacific Ocean. The **abyssal zone** is that part of the benthic environment that extends from a depth of 4000 to 6000 m (2.5 to 3.7 mi.). *Focus On: Life Without the Sun* describes some of the unusual organisms in hydrothermal vents in the abyssal zone (see Chapter 53).

Figure 54–21 Sea palms (*Postelsia* sp.) in a rocky intertidal zone. These sturdy seaweeds, which are 50 to 75 cm (20 to 30 in) tall, are common on the rocky Pacific coast from Vancouver Island to California. Their bases are firmly attached to the rocky substrate, enabling them to withstand heavy surf action. They were photographed at low tide. *(William E. Ferguson)*

Here we describe benthic communities in shallow ocean waters—seagrass beds, kelp forests, and coral reefs. **Sea grasses** are not true grasses, but are flowering plants that have adapted to complete submersion in salty ocean water (Fig. 54–22). They only occur in shallow water, to depths of 10 m (33 ft.) where they receive enough light to photosynthesize efficiently. Extensive beds of sea grasses occur in quiet temperate, subtropical, and tropical waters; no sea grasses occur in polar waters. Eel grass is the most widely distributed sea grass along the coasts of North America. The most common sea grasses in South Florida and the Caribbean Sea are turtle grass and manatee grass. Sea grasses have a high primary productivity and are therefore ecologically important in shallow marine areas. Their roots and rhizomes help stabilize the sediments, reducing surface erosion. Sea grasses provide food and habitat for many marine organisms. In temperate waters, ducks and geese eat sea grasses, whereas in tropical waters, manatees, green turtles, parrot fish, sturgeon fish, and sea urchins eat them. These herbivores consume only about 5% of the sea grasses. The remaining 95% eventually enters the detritus food web and is decomposed by bacteria when the sea grasses die. The bacteria are in turn consumed by a variety of animals such as mud shrimp, lug worms, and mullet (a type of fish).

Kelps, which may reach lengths of 60 m (200 ft.), are the largest brown algae (see Fig. 24–13*b*). Kelps are common in cooler temperate marine waters of both Northern and Southern Hemispheres. They are especially abundant in relatively shallow waters (depths of about 25 m, or 82 ft.) along rocky coastlines. Kelps are photosynthetic and are therefore the primary food producers for the kelp forest ecosystem. Kelp forests also provide habitats for many marine animals. Tube worms, sponges, sea cucumbers, clams, crabs, fishes (e.g., tuna), and mammals (e.g., sea otters) find refuge in the algal fronds. Some animals eat the fronds, but kelps are mainly consumed in the detritus food web. Bacteria that decompose the kelp remains provide food for sponges, tunicates, worms, clams, and snails. The diversity of life supported by kelp beds almost rivals that found in coral reefs.

Coral reefs, which are built from accumulated layers of calcium carbonate ($CaCO_3$), are found in warm (usually higher than 21° C), shallow seawater. The living portions of coral reefs must grow in shallow waters where light penetrates. Many coral reefs are composed principally of red coralline algae that require light for photosynthesis. Coral animals also require light for the large number of symbiotic dinoflagellates, known as **zooxanthellae,** that live and photosynthesize in their tissues (see Fig. 52–12). Although species of coral without zooxanthellae exist, only those with zooxanthellae build reefs. In addition to obtaining food from the zooxanthellae living inside them, coral animals capture food at night by using their stinging tentacles to paralyze small animals that drift nearby. Coral reefs grow slowly in warm, shallow water, as coral organisms build on the calcareous remains of countless organisms before them. The waters in which coral reefs are found are often poor in nutrients. Other factors favor high productivity, however, including the presence of symbiotic zooxanthellae, warm temperatures, and plenty of sunlight.

Coral reef ecosystems are the most diverse of all marine environments and contain hundreds of species of fishes and invertebrates, such as giant clams, sea urchins, sea stars, sponges, brittle stars, sea fans, and shrimp (Fig. 54–23). The Great Barrier Reef along the northeastern coast of Australia occupies only 0.1% of the ocean's surface, but 8% of the world's fish species live there. Many are brightly colored, which advertises the fact that they are

Figure 54–22 Seagrass beds. Turtle grasses (*Thalassia* sp.) have numerous invertebrates and algal epiphytes attached to their leaves. Note the white sea anemone in the middle of the bed of turtle grass. These underwater meadows are ecologically important for shelter and food for many organisms. Photographed off the coast of Florida. *(Dr. David Campbell)*

Figure 54-23 Coral reef organisms. A panoramic view of a coral reef in the Indian Ocean off the coast of the Maldives shows the many animals that live around coral reefs. *(Denise Tackett/Tom Stack & Associates)*

poisonous. The complex multitude of relationships and interactions that occur at coral reefs is comparable only to the tropical rain forest among terrestrial ecosystems. As in the rain forest, competition is intense, particularly for light and space to grow.

Many unusual relationships occur at coral reefs. Certain tiny fishes, for example, swim over larger fishes and even inside their mouths to remove potentially harmful parasites. Fishes sometimes line up at these cleaning stations, waiting their turn to be serviced. As another example, certain corals have a remarkable defense mechanism that protects them from coral browsers such as the crown-of-thorns sea star. Tiny crabs live among the branches of the coral. When a sea star crawls onto the coral, the crabs swarm over it, nipping off its tube feet and eventually killing it if it does not retreat! (Since the late 1960s, population explosions of the crown-of-thorns sea stars have occurred on many coral reefs, causing extensive devastation of the coral. In most cases, the number of sea stars has eventually declined to normal levels. It is not known what causes these outbreaks, including the role of factors such as pollution.)

Coral reefs are ecologically important because they both provide habitat for a wide variety of marine organisms and protect coastlines from shoreline erosion. They also provide humans with seafood, pharmaceuticals, and recreational/tourism dollars. Although coral formations are important ecosystems, they are being degraded and destroyed. Of 109 countries with large reef formations, 90 are damaging them. According to the first global assessment of reefs jointly published in 1998 by the U.N. Environment Program and several leading conservation organizations,[5] 27% of the world's coral reefs are at high risk. Coral reefs of Southeastern Asia, which contain the most species of all coral reefs, are the most threatened of any region.

In some areas, silt washing downstream from clearcut inland forests has smothered reefs under a layer of sediment. High salinity resulting from the diversion of fresh water to supply the growing human population is thought by some scientists to be killing Florida reefs. Overfishing, pollution from sewage discharge and agricultural runoff, oil spills, boat groundings, fishing with dynamite or cyanide, hurricane damage, disease, coral bleaching, land reclamation, tourism, and the mining of corals for building material are also taking a heavy toll. (Coral bleaching is discussed in Chapter 28.)

Marine biologists are especially concerned about the more than 1 million scuba divers and snorkelers that visit coral reefs each year. Most divers and snorkelers are unaware of how vulnerable coral reefs and their associated organisms are to human interactions. Many reef dwellers are killed or injured simply by touching or squeezing them. When a diver or snorkeler accidentally kicks or grabs the reef, pieces break off. Divers also stir up the bottom sediments, which suffocate the coral animals.

The neritic province consists of shallow waters close to shore

The **neritic province** is open ocean that overlies the continental shelves, that is, the ocean floor from the shoreline to a depth of 200 m (650 ft). Organisms that live in the neritic province all are floaters or swimmers. The upper reaches of the neritic province make up the **euphotic zone,** which extends from the surface to a depth of approximately 100 m. Enough light penetrates the euphotic zone to support photosynthesis.

Large numbers of phytoplankton, particularly diatoms in cooler waters and dinoflagellates in warmer waters, produce food by photosynthesis and are thus the base of food webs. Zooplankton (including tiny crustaceans, jellyfish, comb jellies, protists such as foraminiferans, and larvae of barnacles, sea urchins,

[5] The World Resources Institute, the World Conservation Monitoring Center, and the International Center for Living Aquatic Resources Management.

worms, and crabs) feed on phytoplankton. Zooplankton are consumed by plankton-eating nekton such as herring, sardines, squid, manta rays, and baleen whales. These in turn become prey for carnivorous nekton such as sharks, tuna, dolphins, and toothed whales. Nekton are mostly confined to the shallower neritic waters (less than 60 m, or 195 ft, deep) because that is where their food is.

The oceanic province comprises most of the ocean

The average depth of the world's ocean is 4000 m (2.4 mi). The **oceanic province** is that part of the open ocean that covers the deep ocean basin, that is, the ocean floor at depths more than 200 m. It is the largest marine environment and contains about 75% of the ocean's water. Cold temperatures, high hydrostatic pressure, and an absence of sunlight characterize the oceanic province; these environmental conditions are uniform throughout the year.

Most organisms of the oceanic province depend on **marine snow,** organic debris that drifts down into the **aphotic** ("without light") **region** from the upper, lighted regions. Organisms of this little-known realm are filter-feeders, scavengers, or predators. Many are invertebrates, some of which attain great sizes. The giant squid, for example, measures up to 18 m (59 ft) in length, including its tentacles. Fishes of the oceanic province are strikingly adapted to darkness and food scarcity. For example, the gulper eel has huge jaws that enable it to swallow large prey (Fig. 54–24). (An organism that encounters food infrequently needs to eat as much as possible when it has the chance.) Many animals of the oceanic province have illuminated organs that enable them to see one another for mating or to capture food. Adapted to drifting or slow swimming, they are often characterized by reduced bone and muscle mass.

Human activities are harming the ocean

Because the ocean is so vast, it is hard to visualize that human activities could affect, much less harm, it. Such is the case, however. Development of resorts, cities, industries, and agriculture along coasts alters or destroys many coastal ecosystems, including mangrove forests, salt marshes, seagrass beds, and coral reefs. Coastal and marine ecosystems receive pollution from land, from rivers emptying into the ocean, and from atmospheric contaminants that enter the ocean via precipitation. Disease-causing viruses and bacteria from human sewage contaminate shellfish and other seafood and pose an increasing threat to public health. Millions of tons of trash, including plastic, fishing nets, and packaging materials, end up in coastal and marine ecosystems; some of this trash entangles and kills marine organisms. Less visible contaminants of the ocean include fertilizers, pesticides, heavy metals, and synthetic chemicals from agriculture and industry.

Offshore mining and oil drilling pollute the neritic province with oil and other contaminants. Millions of ships dump oily ballast and other wastes overboard in the neritic and oceanic provinces. Fishing is highly mechanized, and new technologies can remove every single fish in a targeted area of the ocean. Scallop dredges and shrimp trawls are dragged across the benthic environment, destroying entire communities with a single swipe.

Many of these problems have been studied during the past 20 years, and conservation groups and government agencies have made numerous recommendations to protect and manage the ocean's resources. The following are a few ideas that have been proposed:

1. All countries should develop strict policies to protect their coastal and marine ecosystems.
2. Governments should use existing scientific knowledge to develop immediate plans to manage fish stocks; governments

Figure 54–24 Gulper eel *(Saccopharynx lavenbergi)* from the oceanic province. This deepwater fish uses its "trap-door" jaws to swallow prey as large as itself. The tail, of which only a small portion is shown, makes up most of the length of a gulper eel's body. Gulper eels grow to 1.8 m (6 ft). Shown is a live specimen, photographed in an aquarium aboard ship after being captured at a depth of 1500 m (4921 ft) off Southern California. *(Bruce H. Robison, Monterey Bay Aquarium)*

should support scientific research to expand our knowledge in this and other critical fields of marine science.

3. Environmental education should be part of the curriculum in every country and should include a strong marine component.
4. Protected coastal and marine areas should be established and carefully monitored around the world.
5. Reduction of marine pollution from land-based sources (sewage discharges, agricultural pollutants, and industrial effluents) should be a top priority of every government.
6. Pollution from ships and offshore installations should be prohibited, and this prohibition should be effectively enforced.

Most countries recognize the importance of the ocean to life on this planet, but few have the resources or programs to protect and manage the ocean effectively. The actions just listed will require global, regional, and local cooperation, billions of dollars, and years of effort if they are to be realized.

ECOSYSTEMS INTERGRADE AND INTERACT WITH ONE ANOTHER

We have discussed the various terrestrial biomes and aquatic ecosystems as if they were distinct and separate entities, but they intergrade with one another at their boundaries. You learned in Chapter 52 that the transition zone where two communities or biomes meet and intergrade is called an **ecotone.** Ecotones range in size from quite small, such as the area where an agricultural field meets a woodland or where a stream flows through a forest, to continental in scope. For example, at the border between tundra and taiga, an extensive ecotone exists that consists of tundra vegetation interspersed with small, scattered conifers. Such ecotones provide habitat diversity and are often populated by a greater variety and density of organisms than either adjacent ecosystem.

Ecologists who study ecotones look for adaptations that enable organisms to survive there. They also examine the relationship between biological diversity and ecotones, and how ecotones change over time. Long-term studies of ecotones have revealed that they are far from static. The ecotone boundary between desert and semiarid grassland in southern New Mexico, for example, has moved during the past 50 years as the desert ecosystem has expanded into the grassland. Many scientists think that ecotones will show the first measurable responses to global climate change.

ANCIENT ECOSYSTEMS FORM A CONTINUUM WITH MODERN ECOSYSTEMS

The study of the geographical distribution of plants and animals is called **biogeography** (see Chapter 17). Biogeographers search for patterns in geographical distribution and try to explain how such patterns arose, including where populations originated, how they spread, and when. Biogeographers recognize that geological and climate changes such as mountain building, continental drift (see Chapter 17), and periods of extensive glaciation influence the distribution of species. Biogeography is linked to evolutionary history and provides insights into how organisms may have interacted in ancient ecosystems. Studying biogeography helps us relate ancient ecosystems to modern ones.

One of the basic tenets of biogeography is that each species originated only once. The particular place where this occurred is known as the species' **center of origin.** The center of origin is not a single point but the distribution of the population when the new species originated. From its center of origin, each species spread until halted by a barrier of some kind, such as an ocean, desert, or mountain range; unfavorable climate; or the presence of organisms that competed with it for food or shelter.

Most plant and animal species have characteristic geographical distributions. The **range** of a particular species is that portion of the Earth in which it is found. The range of some species may be a relatively small area, as for example, wombats (the marsupial equivalent of groundhogs), which are found only in drier parts of southeastern Australia and nearby islands. Such localized, native species are said to be **endemic,** that is, they are not found anywhere else in the world. In contrast, some species have a nearly worldwide distribution and occur on more than one continent or throughout much of the ocean. Such species are said to be **cosmopolitan.**

One of the early observations of biogeographers is that the ranges of different species do not include everywhere that they *could* survive. Central Africa has elephants, gorillas, chimpanzees, lions, antelopes, umbrella trees, and guapiruvu trees, whereas areas in South America with a similar climate have none of these. These animals and plants, which originated in Africa after continental drift had already separated the supercontinent Pangaea into several land masses, could not expand their range into South America because the Atlantic Ocean was an impassible barrier. Likewise, the ocean was a barrier to South American monkeys, sloths, tapirs, balsa trees, snakewood trees, and many palms, none of which is found in Africa.

Land areas are divided into six biogeographical realms

As the various continents were explored and their organisms studied, biologists observed that the world could be divided into major blocks of vegetation, such as forests, grasslands, and deserts, and that these vegetation types corresponded to specific climates (see Fig. 54–3). The relationship between animal distribution, geography, and climate was not deduced until 1876. At that time, Alfred Wallace, who also discovered the same theory of evolution by natural selection as Charles Darwin (see Chapter 17), divided Earth's land areas into six major biogeographical realms: the Palearctic, Nearctic, Neotropical, Ethiopian, Oriental, and Australian (Fig. 54–25). Each of the six biogeographical realms is separated from the others by a major barrier, such as a mountain range, desert, or ocean, that helps maintain each region's biological distinctiveness. Wallace's classification was quickly embraced by many biologists and is still considered valid, except that human activities, such as the intentional and unintentional introduction of foreign species, are contributing to a homogenization of the biogeographical realms (see Chapter 55).

Refer to Figure 54–25 as we briefly consider some of the characteristic animals in each realm. The *Nearctic* and *Palearctic realms* are more closely related than the other regions, especially in their northern parts where they share many animals such as wolves, hares, and caribou. This similarity may be due to a land bridge that has periodically connected Siberia and Alaska. This bridge was present late in the Pleistocene epoch, about 10,000 years ago. Animals adapted to cold environments could disperse between Asia and North America along this bridge.

The *Neotropical realm* was almost completely isolated from the Nearctic realm and other land masses for most of the past 70 million years. During this time, many marsupial species evolved. The isthmus of Panama, which formed a dry-land connection about 3 million years ago, linked North and South America and provided a route for dispersal. Only three species, the opossum, armadillo, and porcupine, are descendants of animals that survived the northward dispersal from South America, but many species, such as the tapir and llama, are descendants of animals that survived the southward dispersal from North America. Competition from these species caused many of South America's marsupial species to go extinct.

The *Ethiopian realm,* which is separated from other land masses by the Sahara Desert, contains the most varied vertebrates of all six realms. Some overlap exists between the Ethiopian and *Oriental realms* because a land bridge with a moist climate linked Africa to Asia during the Miocene and Pliocene epochs. The Oriental realm has the fewest endemic species of all the tropical realms.

The *Australian realm* has not had a land connection with other regions for more than 85 million years. It has no native placental mammals and is dominated by marsupials and monotremes, including the duck-billed platypus and the spiny anteater. Adaptive radiation of the marsupials during their long period of isolation led to species with ecological niches similar to those of placental mammals of other realms (see Fig. 30–27).

The six biogeographical realms are an example of descriptive biogeography. We now discuss an example of how contemporary research attempts to explain observed distributions of species.

Analysis of mitochondrial DNA is a useful biogeographical tool

Molecular evidence, such as mitochondrial DNA (mtDNA) data, can provide insights about an area's biogeographical history. (Recall from Chapter 21 that mtDNA mutates more rapidly than nuclear DNA and is therefore a sensitive indicator of evolution.) Consider Florida, a peninsula bordered by the Atlantic Ocean on the east and the Gulf of Mexico on the west. During the Pleistocene epoch, recurring periods of glaciation caused the sea level of the Atlantic Ocean and the Gulf of Mexico to lower as snow and ice accumulated on the land. During these periods of glaciation, Florida had a larger land area and experienced drier and cooler conditions. When the glaciers receded, Florida's land area became smaller, and its climate became moister and warmer. The ranges of Florida's plant and animal species presumably varied in response to these changes in climate, but exactly how their ranges were affected was not clear until mtDNA data helped reconstruct Florida's biogeographical history.

During the 1990s the mtDNA of many Florida fish, oyster, terrapin, crab, rodent, and bird species was analyzed. The data revealed that each species has distinctive eastern and western populations. These genetic patterns may be evidence that ancient climate changes in Florida produced some type of east–west barrier running the length of the state. The barrier effectively divided each species into smaller, geographically separate populations that diverged (evolved differences) from each other during their separation.

It should be noted that although mtDNA data indicate an east–west barrier, biogeographers have not yet deduced the nature of the barrier from Florida's climate and ecological history. However, the mtDNA data have helped formulate new hypotheses and provided new directions for future ecological and climate research.

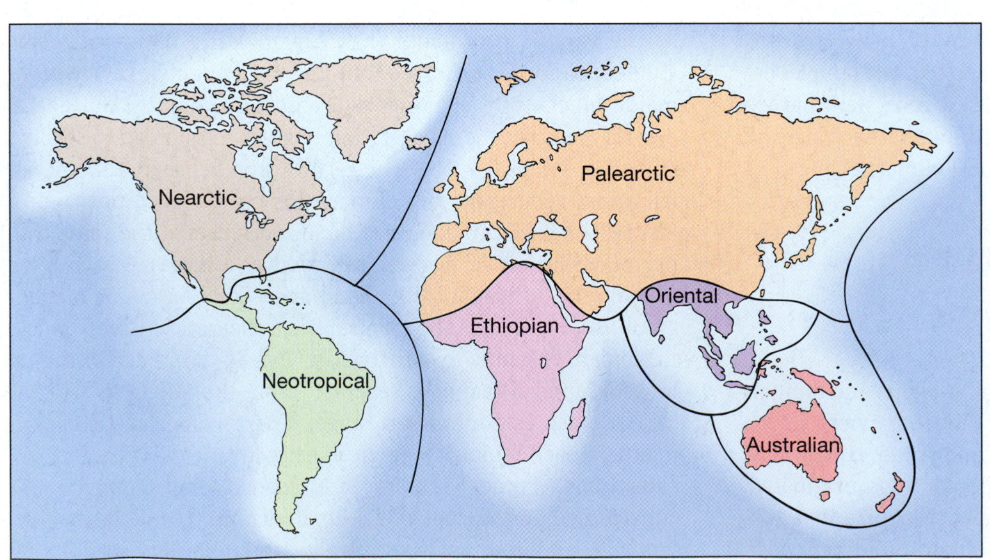

Figure 54–25 Wallace's biogeographical realms. These six biogeographical realms are characterized by the presence of certain unique species whose distributions are the direct outcome of their centers of origin, past dispersals, and the barriers that they encountered.

SUMMARY WITH KEY TERMS

I. A **biome** is a large, relatively distinct terrestrial region with characteristic climate, soil, plants, and animals. Each biome encompasses a number of interacting ecosystems. Temperature and precipitation are important abiotic factors that influence biome distribution.
   A. **Tundra**, the northernmost biome, is characterized by a frozen layer of subsoil **(permafrost),** and low-growing vegetation that is adapted to extreme cold and a short growing season.
   B. The **taiga**, or **boreal forest**, lies south of the tundra and is dominated by coniferous trees that are adapted to the cold winters, short growing season, and acidic, mineral-poor soil.
   C. **Temperate rain forest**, such as occurs on the northwest coast of North America, receives high precipitation and is dominated by large conifers.
   D. **Temperate deciduous forest**, which occurs where precipitation is relatively high and soils are rich in organic matter, is dominated by broad-leaf trees that lose their leaves seasonally.
   E. **Temperate grassland** typically possesses a deep, mineral-rich soil and has moderate but uncertain precipitation. Temperate grassland is well-suited to growing grain crops.
   F. Thickets of small-leaf evergreen shrubs and trees and a climate of wet, mild winters and dry summers characterize **chaparral**.
   G. **Desert**, found in both temperate (cold deserts) and subtropical or tropical regions (warm deserts) with low levels of precipitation, contains organisms with specialized water-conserving adaptations.
   H. Tropical grassland, called **savanna**, has widely scattered trees interspersed with grassy areas. Savanna occurs in tropical areas with low or seasonal rainfall.
   I. Mineral-poor soil and high rainfall that is evenly distributed throughout the year characterize **tropical rain forest**. Tropical rain forest has high species richness and high productivity.

II. In aquatic ecosystems, important environmental factors include **salinity**, the amount of dissolved oxygen, and the availability of light for photosynthesis.
   A. Aquatic life is ecologically divided into **plankton** (free-floating), **nekton** (strongly swimming), and **benthos** (bottom-dwelling).
   B. **Phytoplankton** are photosynthetic algae and cyanobacteria that form the base of the food web in most aquatic communities. **Zooplankton** are nonphotosynthetic organisms that include protozoa, tiny crustaceans, and the larval stages of many animals.

III. Freshwater ecosystems include flowing-water ecosystems (streams and rivers), standing water ecosystems (ponds and lakes), and freshwater wetlands.
   A. In **flowing-water ecosystems** the water flows in a current. Flowing-water ecosystems have few phytoplankton and depend on detritus from the land for much of their energy. The organisms in flowing-water ecosystems vary greatly depending on the current, which is swifter in headwaters than downstream.
   B. Large **standing-water ecosystems** (freshwater lakes) are divided into zones on the basis of water depth.
      1. The marginal **littoral zone** contains both emergent vegetation and algae and is very productive.
      2. The **limnetic zone** is open water away from the shore that extends as far down as sunlight penetrates. Organisms in the limnetic zone include phytoplankton, zooplankton, and larger fishes.
      3. The deep, dark **profundal zone** holds little life other than bacterial decomposers.
   C. **Freshwater wetlands**, lands that are transitional between freshwater and terrestrial ecosystems, are usually covered at least part of the year by shallow water and have characteristic soils and vegetation. They are highly productive areas that perform many valuable **ecosystem services**.

IV. An **estuary** is a coastal body of water, partly surrounded by land, with access to the ocean and a large supply of fresh water from rivers.
   A. Water levels in an estuary rise and fall with the tides.
   B. Salinity fluctuates with tidal cycles, the time of year, and precipitation. Within an estuary, salinity changes gradually, from fresh water at the river entrance to salty ocean water at the estuary's mouth.
   C. Estuaries are very productive, in part because they receive a high input of nutrients from the adjacent land.
   D. One of the important roles of estuaries is to provide a nursery for the young of many aquatic organisms.
   E. Temperate estuaries usually contain **salt marshes**, whereas **mangrove forests** dominate tropical coastlines.

V. Four important marine environments are the intertidal zone, the benthic environment, the neritic province, and the oceanic province.
   A. The **intertidal zone** is the shoreline area between low and high tides. It is a very productive area. Organisms of the intertidal zone possess adaptations to resist wave action and the extremes of being covered by water (high tide) and exposed to air (low tide).
   B. The **benthic environment** is the ocean floor.
      1. **Sea grasses** are flowering plants that have adapted to complete submersion in salty ocean water. They have a high primary productivity and are ecologically important in shallow marine areas.
      2. **Kelps**, the largest brown algae, are common in cooler temperate marine waters, particularly in relatively shallow waters along rocky coastlines. Kelp forests provide food and habitats for many marine animals.
      3. **Coral reefs** are particularly important benthic communities in shallow ocean waters because they have high species richness and are very productive.
   C. The **neritic province** is open ocean from the shoreline to a depth of 200 m. Organisms that live in the neritic province are all floaters or swimmers. Phytoplankton are the base of the food web in the **euphotic zone,** where enough light penetrates to support photosynthesis.
   D. The **oceanic province** is that part of the open ocean that is deeper than 200 m. The uniform environment is one of darkness, cold temperature, and high pressure. Animal inhabitants of the oceanic province are either predators or scavengers that subsist on **marine snow**, detritus that drifts down from other areas of the ocean.

VI. **Biogeography** is the study of the geographical distribution of plants and animals, including where populations came from, how they got there, and when.
   A. Each species originated only once, at its **center of origin.** From its center of origin, each species spread until halted by a physical, environmental, or biological barrier.
   B. The **range** of a particular species is that portion of the Earth in which it is found.
   C. Alfred Wallace divided Earth's land areas into six major biogeographical realms: the Palearctic, Nearctic, Neotropical, Ethiopian, Oriental, and Australian.
      1. Each realm has maintained its biological distinctiveness because it is separated from the others by a mountain range, desert, ocean, or other barrier.
      2. Today, human activities are contributing to a homogenization of the biogeographical realms.

POST-TEST

1. The northernmost biome, known as _______________, typically has little precipitation, a short growing season, and permafrost. (a) chaparral (b) taiga (c) tundra (d) northern deciduous forest (e) boreal forest
2. South of tundra is the _______________, which consists of coniferous forests with many lakes. (a) chaparral (b) taiga (c) alpine tundra (d) northern deciduous forest (e) permafrost
3. Forests of the northeastern and middle eastern United States, which are dominated by broad-leaf hardwood trees that lose their foliage annually, are called (a) temperate deciduous forests (b) tropical deciduous forests (c) northern coniferous forests (d) temperate rain forests (e) tropical rain forests
4. The deepest, richest soil in the world occurs in (a) temperate rain forest (b) tropical rain forest (c) savanna (d) temperate grassland (e) chaparral
5. This biome is characterized by a thicket of evergreen shrubs and small trees found in areas with Mediterranean climates. (a) temperate rain forest (b) tropical rain forest (c) savanna (d) temperate grassland (e) chaparral
6. This biome is a tropical grassland interspersed with widely spaced trees. (a) temperate rain forest (b) tropical rain forest (c) savanna (d) temperate grassland (e) chaparral
7. This biome has the greatest species richness. (a) temperate rain forest (b) tropical rain forest (c) savanna (d) temperate grassland (e) chaparral
8. Organisms in aquatic environments fall into three categories: free-floating _______________, strongly swimming _______________, and bottom-dwelling _______________. (a) nekton; benthos; plankton (b) nekton; plankton; benthos (c) plankton; benthos; nekton (d) plankton; nekton; benthos (e) benthos; nekton; plankton
9. Temperate-zone lakes are thermally stratified, with warm and cold layers separated by a transitional (a) aphotic region (b) thermocline (c) barrier reef (d) ecotone (e) littoral zone
10. Emergent vegetation grows in the _______________ zone of freshwater lakes. (a) littoral (b) limnetic (c) profundal (d) neritic (e) intertidal
11. This coastal body of water has access to the open ocean and a large supply of fresh water from rivers. (a) intertidal zone (b) estuary (c) freshwater wetland (d) neritic province (e) standing-water ecosystem
12. The_______________ is open ocean from the shoreline to a depth of 200 m. (a) benthic environment (b) intertidal zone (c) neritic province (d) oceanic province (e) aphotic region
13. Sea grasses (a) occur in shallow water of the ocean's benthic environment (b) may reach lengths of 60 m (c) contain symbiotic algae known as zooxanthellae (d) are common on rocky shores in the intertidal zone (e) are the main producers in freshwater wetlands
14. The transition zone where two ecosystems or biomes meet and intergrade is called a(an): (a) biosphere (b) aphotic region (c) thermocline (d) biogeographical realm (e) ecotone
15. Which biogeographical realm has been separated from the other land masses for more than 85 million years? (a) Ethiopian (b) Palearctic (c) Nearctic (d) Oriental (e) Australian

REVIEW QUESTIONS

1. What climate and soil factors produce the major biomes?
2. Describe representative organisms of these forest biomes: (a) taiga, (b) temperate deciduous forest, (c) temperate rain forest, (d) tropical rain forest.
3. In which biome do you live? If your biome does not match the description given in this text, explain the discrepancy.
4. Compare tundra with desert and temperate grassland with savanna.
5. Distinguish among plankton, nekton, and benthos, giving examples of each.
6. What environmental factors are most important in determining the adaptations of organisms that live in aquatic environments?
7. Distinguish between freshwater wetlands and estuaries, and between flowing-water and standing-water ecosystems.
8. List and briefly describe the four main marine environments.
9. Which aquatic ecosystem is often compared to tropical rain forests? Why?
10. What is biogeography?

YOU MAKE THE CONNECTION

1. When a black-tailed prairie dog or other small animal dies, other prairie dogs bury it. Develop a hypothesis to explain how this behavior may be adaptive. How would you test this hypothesis?
2. In which biomes would migration be most common? Hibernation? Aestivation? Explain your answers.
3. Develop a hypothesis to explain why animals adapted to the desert are usually small. How would you test your hypothesis?
4. Why do most of the animals of the tropical rain forest live in trees?
5. What would happen to the organisms in a river with a fast current if a dam were built? Would there be any differences in habitat if the dam were upstream or downstream of the organisms in question? Explain your answers.

RECOMMENDED READINGS

Allen, W.H. "Travel Across the Treetops." *BioScience*, Vol. 46, No. 11, Dec. 1996. How construction cranes help ecologists study tropical forests.

Chadwick, D.H. "Kingdom of Coral: Australia's Great Barrier Reef." *National Geographic*, Vol. 199, No. 1, Jan. 2001. A beautiful photo essay of the diversity of life in Australia's Great Barrier Reef.

Dold, C. "Making Room for Prairie Dogs," *Smithsonian*, Mar. 1998. Although millions of prairie dogs still live on the prairie, they are in trouble.

Doubilet, D. "Coral Eden," *National Geographic*, Vol. 195, No. 1, Jan. 1999. A look at some of the spectacular organisms living in one of the richest coral reefs on our planet. (This issue also contains an article on how overfishing, pollution, and disease threaten coral reefs.)

Earle, S. "Does It Matter What We Do to the World's Oceans?" *Environmental Review,* Vol. 6, November 1999. This is the complete text of Dr. Sylvia Earle's marvelous speech to the Ecological Society of America. Her love for the ocean and all living things in it is contagious.

Johnson, S. "Transparent Animals." *Scientific American*, Vol. 282, No. 2, Feb. 2000. Many marine organisms are transparent.

Madigan, M.T., and B.L. Marrs. "Extremophiles." *Scientific American*, Vol. 276, No. 4, Apr. 1997. Some environments once thought to be too harsh for life (such as sea ice, deep-sea vents, acid springs, and salt lakes) contain thriving populations of microorganisms.

Pain, S. "Lair of the Dragon," *New Scientist,* Vol. 161, 27 Mar. 1999. Many fish in the dark depths of the ocean have light-producing organs to help them hunt and find prospective mates.

Raven, P.H., and L.R. Berg. *Environment*, 3rd ed. Harcourt College Publishers, Philadelphia, 2001. This environmental science text offers a detailed situation analysis of planet Earth, including how humans interact with and affect its physical and living systems.

Ricklefs, R.E., and G.L. Miller. *Ecology*, 4th ed. W.H. Freeman & Company, New York, 1999. This textbook presents the important themes of ecology, such as energy flow, population and community interactions, and mathematical models of ecology, in an accessible manner.

Rützler, K., and I.C. Feller. "Caribbean Mangrove Swamps." *Scientific American,* Vol. 247, Vol. 3, Mar. 1996. Discusses the diverse organisms that live around mangrove trees in Belize.

Schultz, J.C., and T. Floyd. "Desert Survivor." *Natural History*, Feb. 1999. The creosote bush, an alien species from South America, dominates the landscape of deserts in the southwestern United States.

Walsberg, G.E. "Small Mammals in Hot Deserts: Some Generalizations Revisited." *BioScience*, Vol. 50, No. 2, Feb. 2000. Much about the biology of desert mammals is still obscure.

Wiley, J.P., Jr. "New Light on Diversity." *Smithsonian*, May 1999. Ecologists studying the diversity of tree species in tropical rain forests examine which plants grow in "holes" in the canopy caused by tree fall.

- Visit our Web site at **http://www.info.brookscole.com/solomonbergmartin** for links to chapter-related resources on the World Wide Web. Additional on-line materials relating to this chapter can also be found on our Web site.

See chapter activity on BioActive Learner CD for additional help in mastering the chapter's material. Icon location in the chapter's margins shows which topics have tutorials or simulations in the CD.

55

Humans in the Environment

LEARNING OBJECTIVES

After you have studied this chapter you should be able to

1. Distinguish among threatened species, endangered species, and extinct species.
2. Discuss at least four causes of declining biological diversity and identify the most important cause.
3. Define conservation biology and compare in situ and ex situ conservation measures.
4. Describe the benefits and shortcomings of the following laws and agreements: the Endangered Species Act; the World Conservation Strategy; the Convention on International Trade in Endangered Species of Wild Flora and Fauna.
5. Discuss the ecosystem services of forests and describe the consequences of deforestation.
6. State at least three reasons why forests are disappearing today.
7. Name at least three greenhouse gases and explain how greenhouse gases contribute to global warming.
8. Describe how global warming may affect sea level, precipitation patterns, organisms (including humans), and food production.
9. Distinguish between surface ozone and stratospheric ozone.
10. Cite the causes and potential effects of ozone destruction in the stratosphere.

Frog deformities. Pollution, ultraviolet radiation, and parasites have been implicated in the recent widespread increase in developmental abnormalities in amphibians. *(Frans Lanting/Minden Pictures)*

In 1995 schoolchildren in Minnesota made an important discovery while they were on a field trip to a local pond. The children found that almost half of the leopard frogs they caught were deformed. (Fewer than 1% of frogs in healthy populations exhibit deformities.) Deformities include frogs with extra legs, extra toes, eyes located on the shoulder or back, deformed jaws, missing legs, missing toes, and missing eyes *(see photograph).* Deformed frogs usually die early, before they can reproduce. Predators can easily catch frogs with extra or missing legs. Since the children's discovery made headlines, 44 states have reported abnormally large numbers of deformities (as high as 60% in some local populations) in 38 amphibian species. Canada has also reported frog deformities.

Many possible causes have been investigated and shown to produce amphibian deformities during development. These include exposure to chemicals such as pesticides, to increased ultraviolet (UV) light due to thinning of the ozone layer, and to parasites. Several pesticides are known to affect normal development in frog embryos. Lab experiments using traces of the pesticide *S*-methoprene and its breakdown products have caused frog deformities that are similar to abnormalities found in ponds in Vermont that are close to where *S*-methoprene is used to control mosquitoes and fleas. Ultraviolet radiation has long been known to cause mutations in deoxyribonucleic acid (DNA), the genetic material that codes for organisms' traits and development. Laboratory tests have also demonstrated that infecting tadpoles with a parasitic trematode (a flatworm) causes the adults that develop from the tadpoles to exhibit deformities like the ones observed in field sites in Santa Clara County, California. No single factor explains the deformities found in all locations. Furthermore, multiple effects, such as habitat loss, disease, and air and water pollution, may interact synergistically with one another to cause deformities. An amphibian that is stressed by pesticide residues, high UV, or drought, for example, may be more susceptible to a parasite.

Amphibians are one example of the kinds of injuries humans are inflicting on biological diversity. The human species *(Homo sapiens)* has been present on Earth for about 800,000 years (see Chapter 21), which is a brief span of time compared with the age of our planet, some 4.6 billion years. Despite our relatively short tenure on Earth, our biological impact on other species has been unparalleled. Our numbers have increased dramatically—the human population reached 6.14 billion in mid-2001—and we have expanded our biological range, moving into almost every habitat on Earth. Wherever we have gone, we have altered the environment and shaped it to meet our needs. In only a few generations we have transformed the face of Earth, placed a great strain on Earth's resources and resilience, and profoundly affected other species. Thus, the impact of humans on the environment merits special study in biology, not merely because we ourselves are humans but also because our impact on the rest of the biosphere has been so extensive.

Many environmental concerns exist today, too many to be considered in a single chapter. The rapidly expanding human population underlies and exacerbates all environmental problems. The increasing population is placing a nonsustainable stress on the environment, as humans consume ever-increasing quantities of food and water, use more and more energy and raw materials, and produce enormous amounts of waste and pollution. Because the human population crisis was discussed in Chapter 51, we conclude this text by focusing our attention on four other serious environmental issues that affect the biosphere: declining biological diversity, deforestation, global warming, and ozone depletion in the stratosphere.

SPECIES ARE DISAPPEARING AT AN UNPRECEDENTED RATE

Extinction, the death of a species, occurs when the last individual member of a species dies (see Chapter 19). A natural biological process, extinction has been greatly accelerated by human activities. The burgeoning human population has forced us to spread into almost all areas of Earth. Whenever humans invade an area, the habitats of many plants and animals are disrupted or destroyed, which can contribute to their extinction. For example, the dusky seaside sparrow, a small bird that was found only in the marshes of the St. Johns River in Florida, became extinct in 1987, largely because of human destruction of its habitat (Fig. 55–1).

Figure 55–1 Extinction. The dusky seaside sparrow *(Ammospiza nigrescens)* became extinct in 1987, largely owing to human destruction of its habitat in Florida. The dusky seaside sparrow was 15 cm (6 in) long. *(U. S. Fish and Wildlife Service)*

Biological diversity, also called **biodiversity,** is the variety of living organisms, from their genes to the ecosystems in which they live. Biological diversity includes **species richness,** the number of different species (see Chapter 52); *genetic diversity*, the genetic variety within a species; and *ecosystem diversity*, the variety of interactions within and among ecosystems.

Biological diversity is currently decreasing at an alarming rate (see *Focus On: Declining Amphibian Populations*). Conservation biologists estimate that species are presently becoming extinct at a rate at least 100 times the natural rate of background extinctions. The U.N. Global Biodiversity Assessment, based on the work of about 1500 scientists from around the world, was released in late 1995. It estimated that more than 31,000 plant and animal species are currently threatened with extinction. More recent surveys indicate that these estimates are probably too conservative. For example, the first World Conservation Union Red List of Threatened Plants, issued in 1998 and based on 20 years of data collection and analysis around the world, lists about 34,000 species of plants currently threatened with extinction.

Some biologists fear that we are entering the greatest period of mass extinction in Earth's history, but the current situation differs from previous periods of mass extinction in several respects. First, its cause is directly attributable to human activities. Second, it is occurring in a tremendously compressed period (just a few decades as opposed to hundreds of thousands of years), much faster than rates of speciation (or replacement). Perhaps even more sobering, larger numbers of plant species are becoming extinct today than in previous mass extinctions. Because plants are the base of terrestrial food webs, extinction of animals that depend on plants cannot be far behind. It is crucial that we determine how the loss of biodiversity affects the stability and functioning of ecosystems, which comprise our life-support system.

According to the U.S. Endangered Species Act, a species is designated as **endangered** when its numbers are so severely

FOCUS ON Declining Amphibian Populations

Over the last several decades, many of the world's frog populations have dwindled or disappeared. North America, Central and South America, and Australia all have experienced dramatic population declines. For example, at least 14 species of Australian rainforest frogs have become endangered or extinct since 1980. In the United States, as many as 38% of its 242 native amphibian species are declining in numbers. In assessing the possible causes of these declines, researchers have observed that the declines are not limited to areas with obvious habitat destruction, such as drainage of wetlands where frogs live, or degradation from pollutants. Some remote, pristine locations also show dramatic declines in amphibians: Populations of all seven native species of frogs and toads in Yosemite National Park have declined. Biologists are not certain what is causing these mysterious declines, and it appears that no single factor is responsible. Potential factors for which there is strong evidence include pollutants, increased UV radiation, infectious diseases, and climate warming.

Agricultural chemicals have been implicated in amphibian declines in California's Sierra Nevada Mountains. Frog populations on the eastern slopes are relatively healthy, but about eight species are declining on the western slopes, where prevailing winds carry residues of 15 different pesticides from the Central Valley, a huge agricultural region. Other areas in which agricultural chemicals may be contributing to amphibian decline include the Eastern Shore of Maryland and Ontario, Canada.

Another possible culprit may be increased UV radiation caused by ozone thinning. Like most other organisms, amphibians possess an enzyme that allows them to repair DNA damage caused by natural UV radiation. Species suffering declines appear to be limited in their ability to repair such cellular damage. Researchers at Oregon State University exposed the eggs of three frog species to natural radiation. Egg survival was high for the Pacific tree frog, which had the greatest enzyme activity and is not in decline. Egg survival was much less (only 45 to 65%) for the western toad and Cascades frog, both of which are declining. Egg survival for these species increased dramatically, however, when eggs were shielded from UV radiation.

Infectious diseases may explain some of the declines. In Australia, data suggest that a fungus called a chytrid (see Chapter 25) is responsible for massive die-offs of more than 12 species, 4 of which are probably now extinct. Laboratory studies in the United States have shown that the chytrid can kill healthy frogs. This fungus may also be responsible for some of the frog declines observed in Central America and the United States.

Golden toads and about 20 other frog species that have disappeared in mountain areas of Costa Rica may have succumbed to climate warming. In recent years increasing global temperatures have reduced moisture levels in the cloud forests of Costa Rica's central highlands. The organisms in these tropical mountain forests depend on the regular formation of clouds and mist, particularly during the dry winter season. As the equatorial Pacific Ocean surface has warmed in recent years, a decline in the frequency of dry season mist has been observed and correlated with declines in the abundance of many animal species, including the frogs that disappeared.

reduced that it is in imminent danger of extinction throughout all or a significant part of its range. (The area in which a particular species is found is its **range.**) Unless humans intervene, an endangered species will probably become extinct. When extinction is less imminent but the population of a particular species is quite small, the species is classified as **threatened.** A threatened species is likely to become endangered in the foreseeable future, throughout all or a significant part of its range. Endangered or threatened species represent a decline in biological diversity, since they have severely diminished genetic diversity. Endangered and threatened species are at greater risk of extinction than species with greater genetic variability, because long-term survival and evolution depend on genetic diversity (see the section on genetic drift in Chapter 18).

Human activities contribute to declining biological diversity

Species become endangered and extinct for a variety of reasons, including the destruction or modification of habitats and the production of pollution. Humans also upset the delicate balance of organisms in a given area by introducing new, foreign species or by controlling native pests or predators. Illegal hunting and uncontrolled commercial harvesting are also factors. Table 55–1 summarizes an extensive study of the human threats to biological diversity in the United States. These data illustrate the overall importance of habitat loss and degradation. Some interesting variations among the groups of organisms are also apparent. Pollution, for example, is a very serious threat to fishes, reptiles, and invertebrates but a relatively minor threat to plants.

To save species, we must protect their habitats

Most species facing extinction today are endangered because of destruction of natural habitats. Building roads, parking lots, and buildings; clearing forests to grow crops or graze domestic animals; and logging forests for timber all take their toll on natural habitats. Draining marshes converts aquatic habitats to terrestrial ones, whereas building dams and canals floods terrestrial habitats (Fig. 55–2). Habitat destruction threatens the survival of species because most organisms require a particular type of environment, and habitat destruction reduces their biological range and ability to survive.

Humans often leave small, isolated patches of natural landscape that are completely surrounded by roads, fields, and buildings. Like a landmass that is surrounded by water, an isolated

TABLE 55–1 Percentages of Imperiled* U.S. Species that Are Threatened by Various Human Activities

Activity	All Species (1880)†	Plants (1055)	Mammals (85)	Birds (98)	Reptiles (38)	Fishes (213)	Invertebrates (331)
Habitat loss/ degradation	85‡	81	89	90	97	94	87
Exotic species	49	57	27	69	37	53	27
Pollution	24	7	19	22	53	66	45
Overexploitation	17	10	45	33	66	13	23

*This includes species classified as imperiled by The Nature Conservancy and all species listed as endangered or threatened under the Endangered Species Act or formally proposed for listing.

†Numbers in parentheses are the total number of species evaluated. The "All Species" category represents about 75% of imperiled species in the United States.

‡Because many of the species are affected by more than one human activity, the percentages in each column do not add up to 100.

Adapted from Table 2 in Wilcove, D.S., et al. "Quantifying Threats to Imperiled Species in the United States." *BioScience*, Vol. 48, No. 8, Aug. 1998.

Figure 55–2 Habitat destruction. This tiny island is located in the Panama Canal. It was once a hilltop in a forest that was flooded when the Panama Canal was constructed. *(Frans Lanting/Minden Pictures)*

habitat that is surrounded by an expanse of unsuitable territory is referred to as an island (recall the discussion of isolated island communities in Chapter 52). Species from the surrounding "developed" landscape may intrude into the isolated habitat. Species that prefer the isolated habitat may occur in greatly reduced numbers or, if they require a large patch of undisturbed habitat, disappear altogether. As a result, habitat fragments often support only a fraction of the species found in the original, unaltered environment. The effects of **habitat fragmentation,** the breakup of large areas of habitat into small, isolated patches, are supported by extensive scientific evidence (see *Focus On: Forest Fragmentation and Declining Bird Populations*).

Even habitats left undisturbed and in their natural state are indirectly modified by human activities that produce acid precipitation and other forms of pollution. Acid precipitation has contributed to the decline of large stands of forest trees and to the biological death of many freshwater lakes, for example, in the Adirondack Mountains and in Nova Scotia. The production of other types of pollutants also adversely affects organisms. Such pollutants include industrial and agricultural chemicals, organic pollutants from sewage, acid wastes seeping from mines, and thermal pollution from the heated wastewater of industrial plants.

To save native species, we must control invasions of nonindigenous species

Biotic pollution, the introduction of a foreign, or nonindigenous, species into an area where it is not native, often upsets the balance among the organisms living in that area. The foreign species may prey on native species or compete with them for food or habitat. Generally, a nonindigenous competitor or predator causes a greater negative effect on local organisms than do native competitors or predators. (Most introduced species lack natural agents, such as diseases, predators, and competitors, that would otherwise control them. Also, without a shared evolutionary history, most native species typically are less equipped to cope with introduced species.) Although nonindigenous species may be introduced into new areas by natural means, humans are usually responsible for such introductions, either knowingly or accidentally.

One of North America's greatest biological threats is the zebra mussel, a native of the Caspian Sea that was probably introduced through ballast water flushed into the Great Lakes by a foreign ship in 1985 or 1986. Since then, the tiny freshwater mussel, which clusters in extraordinary densities, has massed on hulls of boats, piers, buoys, water intake systems and, most damaging of all, on native clam and mussel shells. The zebra mussel's strong appetite for algae, phytoplankton, and zooplankton is also cutting

FOCUS ON

Forest Fragmentation and Declining Bird Populations

Evidence is mounting that the populations of many birds that were once abundant have been dropping steadily for about two decades. Population declines have been noted across the North American continent, particularly in songbirds of forests, shrublands, and grasslands. Many songbirds are *neotropical birds* that spend the winter in Central and South America and the Caribbean and then migrate north to the United States and Canada to breed during the summer. Cerulean warblers, olive-sided flycatchers, yellow-billed cuckoos, eastern wood pewees, and wood thrushes are examples of neotropical birds with declining numbers. Cerulean warblers, for example, have declined by 51% from the 1960s to the 1990s, based on Breeding Bird Survey data.

Neotropical birds are faced with changing environments in both their winter and summer homes, and loss of habitat appears to be an important reason for their decline. These migratory birds are under stress from the burning of tropical rain forests in Central and South America as well as the fragmenting of forest habitat in North America to accommodate suburban development, agriculture, and logging.

Several studies have shown that fragmentation of northern forests increases the likelihood of reproductive failure in neotropical birds, many of which are insect-eaters that nest only in forests. As a result of forest fragmentation, the nests are more likely to be located near a **forest edge,** the often sharp boundary between the forest and residential neighborhoods, farmlands, or clearcut areas, rather than deep in the forest. Nests that are within 100 m (330 ft) of a forest edge are more vulnerable to predation by small predators such as raccoons, opossums, snakes, and feral cats (domestic cats that have gone wild). Blue jays, American crows, and other egg-eating birds do most of their hunting along the forest edge. (See discussion of the edge effect in Chapter 52.)

Nest parasitism by brown-headed cowbirds is also more significant along the forest edge. Female cowbirds lay their eggs in the nests of other bird species and leave all parenting jobs to the hosts *(see figure)*. In a 1989 to 1993 study that monitored 5000 nests of neotropical birds, the majority of nests were parasitized in areas with less than 55% forest cover, and in many cases there were more cowbird eggs in the nests than eggs from the host species. Nest parasitism causes reproductive failure because the host parents provide food for the larger, more aggressive cowbird babies, while their own young starve.

Cowbirds avoid nesting or feeding in intact forests. Before Europeans settled in North America, cowbirds followed the migratory herds of bison across the Great Plains, foraging on seeds and insects stirred up by the bison. With the settlement of the continent by Europeans, many forests were cut to provide agricultural lands, and the cowbird expanded its range into fields and cattle pastures. The cowbirds began parasitizing new bird species that had not yet evolved ways to resist them. (Some birds, such as robins, gray catbirds, and Baltimore orioles, reject nests with cowbird eggs.)

Nest parasitism. Note the speckled cowbird *(Molothrus ater)* egg in the nest of a wood thrush *(Hylocichla mustelina)*. When the wood thrush nestlings hatch, they will have to compete for food with their larger, more aggressive foster nestling. More cowbird eggs are laid in songbird nests along the forest edge than deep within a forest. *(Jeff Lepore/Photo Researchers, Inc.)*

into the food supply of native fishes, mussels, and clams, threatening their survival. By 1994 the zebra mussel had spread from the Great Lakes into the Mississippi River. It is currently found in 20 states.[1] Since the zebra mussel invaded North America, an estimated $5 billion has been spent to repair damage such as clogged pipes (Fig. 55–3).

In the mid-1980s an aggressive aquarium-bred strain of alga known as *Caulerpa* was accidentally released into the Mediterranean Sea when a seaside aquarium cleaned out its tanks. The alga has spread like a dense carpet over the floor of the Mediterranean, crowding out biologically diverse seafloor communities of native sea grasses, sponges, corals, sea fans, anemones, sea stars, and lobsters. *Caulerpa* is also toxic to many Mediterranean

[1] Alabama, Arkansas, Connecticut, Illinois, Indiana, Iowa, Kentucky, Louisiana, Michigan, Minnesota, Mississippi, Missouri, New York, Ohio, Oklahoma, Pennsylvania, Tennessee, Vermont, West Virginia, and Wisconsin.

Figure 55-3 Zebra mussels clog a pipe. Zebra mussels *(Dreissena polymorpha)* have caused billions of dollars in damage in addition to displacing native clams and mussels. *(Illinois Department of Conservation)*

species. Other regions far from the Mediterranean are concerned that *Caulerpa* could cause havoc to their marine ecosystems. The United States, for example, has banned the importation of *Caulerpa* under the Federal Noxious Weed Act.

Islands are particularly susceptible to the introduction of nonindigenous species (Fig. 55–4). For example, Abingdon Island, one of the Galapagos Islands off the coast of South America, was home to an endemic (found nowhere else) giant tortoise. In 1957 several fishermen introduced goats to Abingdon Island, and within five years the Abingdon tortoise was extinct. The goats, with no natural predators on the island, had greatly increased in number and had eaten the tortoises' food. In Hawaii, the introduction of mouplan sheep has imperiled both the mamane tree (because the sheep eat it) and a species of honeycreeper, an endemic bird that relies on the tree for food. (Hawaii's plants evolved in the absence of herbivorous mammals and therefore have no defenses against introduced sheep, pigs, goat, and deer; see discussion of plant adaptations against herbivores in Chapter 52.)

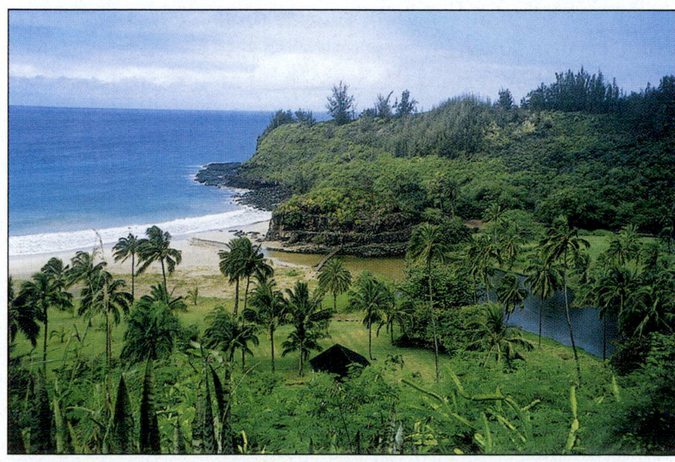

Figure 55-4 Biotic pollution in Hawaii. This view of a tropical paradise on the Hawaiian island of Kauai is a vivid demonstration of biotic pollution. Of all the plant species seen in this photograph, only one is native (the low-growing plant found on the beach). Humans introduced all the other plant species. *(Dr. Linda R. Berg)*

Other human activities affect biodiversity directly or indirectly

Sometimes species become endangered or extinct as a result of deliberate efforts to eradicate or control their numbers, often because they prey on game animals or livestock. Ranchers, hunters, and government agents have decimated populations of large predators such as the wolf, mountain lion, and grizzly bear. Predators of game animals and livestock are not the only animals vulnerable to human control efforts. Some animals are killed because their lifestyles cause problems for humans. The Carolina parakeet, a beautiful green, red, and yellow bird endemic to the southeastern United States, was extinct by 1920, exterminated by farmers because it ate fruit from their trees.

Prairie dogs and pocket gophers were poisoned and trapped so extensively by ranchers and farmers that, between 1900 and 1960, they disappeared from most of their original range. As a result of sharply decreased numbers of prairie dogs, the black-footed ferret, a natural predator of these animals, became endangered. By the winter of 1985–1986, only ten ferrets were known to exist: four in Meeteetse, Wyoming, and six in captivity. A successful captive-breeding program enabled biologists to release black-footed ferrets to the Wyoming prairie beginning in 1991. Black-footed ferrets have successfully reproduced in the wild since being reintroduced, but a recent outbreak of disease (plague) in black-tailed prairie dogs has threatened ferret survival by reducing their prey. Recovery efforts for these charismatic animals continue, although they remain vulnerable to human development of the prairie habitat.

Unregulated hunting, also called overhunting, has caused the extinction of certain species in the past but is now strictly controlled in most countries. The passenger pigeon was one of the most common birds in North America in the early 1800s, but a century of overhunting resulted in its extinction in the early 1900s.

Illegal commercial hunting (poaching) continues to endanger a number of larger animals such as the tiger, cheetah, and snow leopard, whose beautiful furs are quite valuable. Rhinoceroses are slaughtered for their horns (used for ceremonial dagger handles in the Middle East and for purported medicinal purposes in Asian medicine). Bears are killed for their gallbladders (which are used in Asian medicine to treat ailments ranging from indigestion to hemorrhoids). Bush meat—meat from rare primates, elephants, anteaters, and such—is sold to urban restaurants. Although laws protect these animals, demand for their products on the black market has promoted illegal hunting, particularly in impoverished countries where sale of contraband products can support a family for months.

Commercial harvest removes live organisms from the wild. Most organisms that are commercially harvested end up in pet stores. Several million birds are commercially harvested each year for the pet trade, but, unfortunately, many die in transit, and many more die from improper treatment in their owners' homes. At least 40 parrot species are now threatened or endangered, in part because of commercial harvest. Although it is illegal to capture endangered animals from the wild, a thriving black market exists, mainly because collectors in the United States, Europe, and Japan will pay extremely large sums for rare tropical birds (Figure 55–5). Each hyacinth macaw, for example, fetches $7000 to $10,000 in the United States.

Animals are not the only organisms threatened by commercial harvest. A number of unique or rare plants have been collected from the wild so extensively that they are now classified as endangered. These include certain carnivorous plants, cacti, and orchids. On the other hand, carefully monitored and regulated commercial use of animal and plant resources creates an economic incentive to insure that these resources do not disappear.

Where is the problem of declining biological diversity greatest?

Declining biological diversity is a concern throughout the United States but is most serious in the states of Florida, California, and Hawaii, according to a 1995 study by Defenders of Wildlife entitled *Endangered Ecosystems: A Status Report on America's Vanishing Habitat and Wildlife*. Hawaii, for example, has lost hundreds of species and has more species listed as endangered than any other state. At least two-thirds of Hawaii's native forests are gone.

Figure 55–5 Illegal commercial harvesting. These hyacinth macaws *(Anodorhynchus hyacinthus)* were seized in French Guiana in South America as part of the illegal animal trade there. The hyacinth macaw population is much reduced in South America; according to a recent estimate, there are 2500 to 5000 hyacinth macaws living in the wild. Hyacinth macaws, which are the largest parrot species in the world, are typically about 100 cm (40 in) long. *(Jany Sauvanet/NHPA)*

As serious as declining biological diversity is in the United States, it is even more serious abroad. Ecosystem loss and degradation are occurring in many places around the world, but tropical rain forests are being destroyed faster than almost all other ecosystems. Using remote sensing surveys, scientists have determined that approximately 1% of tropical rain forests are being cleared or severely degraded each year. The forests are making way for human settlements, banana plantations, oil and mineral explorations, and other human activities (discussed later in the chapter). Tropical rain forests are home to thousands or even millions of the world's species. Many species in tropical rain forests are endemic; the clearing of tropical rain forests therefore contributes to their extinction.

Conservation biology addresses the issue of declining biological diversity

Conservation biology is the scientific study of how humans impact organisms and of the development of ways to protect biological diversity. Conservation biology is a broad discipline that ranges from safeguarding populations of endangered species to preserving entire ecosystems and landscapes. It includes two problem-solving approaches that are used to save organisms from extinction: in situ and ex situ. **In situ conservation,** which includes the establishment of parks and reserves, concentrates on preserving biological diversity in nature. A high priority of in situ conservation is the identification and protection of sites with a great deal of diversity. With increasing demands on land, however, in situ conservation cannot guarantee the preservation of all types of biological diversity. Sometimes only ex situ conservation can save a species. **Ex situ conservation** involves conserving biological diversity in human-controlled settings. Examples of ex situ conservation include the breeding of captive species in zoos and seed storage of genetically diverse plant crops (see *Focus On: Seed Banks* in Chapter 35).

In situ conservation is the best way to preserve biological diversity

Conservation biologists maintain that protecting animal and plant habitats—that is, conserving and managing the ecosystem as a whole—is the single best way to protect biological diversity (recall the discussion of ecosystem management in the introduction to Chapter 53). Many nations appreciate the need to protect their biological heritage and for this reason have set aside areas for wildlife habitats. Such natural ecosystems offer the best strategy for the long-term protection and preservation of biological diversity. There are currently more than 3000 national parks, marine sanctuaries, wildlife refuges, forests, and other protected areas throughout the world. These encompass some 1 billion hectares, an area almost as large as Canada. Some of these areas have been chosen to protect specific endangered species. The first such refuge, established in 1903 at Pelican Island, Florida, was set aside to protect the brown pelican. Today the National Wildlife Refuge System of the United States includes more than 500 refuges. Although the bulk of the protected land is in Alaska, refuges exist in all 50 states.

Many protected areas have multiple uses that sometimes conflict with the goal of preserving species. National parks may serve recreational needs, for example, whereas national forests may be open for logging, grazing, and mineral extraction. The mineral rights to many refuges are privately owned, and some wildlife refuges have had oil, gas, and other mineral development.

Protected areas are not always effective in preserving biological diversity, particularly in developing countries where biological diversity is greatest, because there is little money or expertise to manage them. Another shortcoming of the world's protected areas is that many are in lightly populated mountain areas, tundra, and the driest deserts, places that often have spectacular scenery but relatively few kinds of species. (Such remote areas are often designated reserves because they are unsuitable for commercial development.) In contrast, ecosystems in which biological diversity is greatest often receive little attention. Protected areas are urgently needed in tropical rain forests, the tropical grasslands and savannas of Brazil and Australia, and dry forests that are widely scattered around the world. Desert organisms are underprotected in northern Africa and Argentina, and the species of many islands and temperate river basins also need protection. These areas are part of what biologists have identified as the world's 25 *biodiversity hotspots*, which collectively comprise 1.4% of Earth's land but contain as many as 44% of all vascular plant species and 35% of all land vertebrates.

Restoring damaged or destroyed habitats is the goal of restoration ecology

Although preserving habitats is an important part of conservation biology, the realities of our world, including the fact that the land-hungry human population continues to increase, dictate a variety of other conservation measures. Scientists can reclaim disturbed lands and convert them into areas with high biological diversity. **Restoration ecology,** in which the principles of ecology are used to help return a degraded environment as close as possible to its former state, is an important part of in situ conservation.

One of the oldest and most famous examples of ecological restoration has been carried out since the 1930s by the University of Wisconsin–Madison Arboretum (Fig. 55–6). Several distinct natural communities have been carefully developed on damaged agricultural land. These communities include a tallgrass prairie, a xeric (dry) prairie, and several types of pine and maple forests native to Wisconsin.

Restoration of disturbed lands not only creates biological habitats but also has additional benefits such as the regeneration of soil damaged by agriculture or mining. The disadvantages of restoration ecology include the time and expense required to restore an area. Nonetheless, restoration ecology is an important aspect of conservation biology, as it is thought that restoration will deter many extinctions.

Ex situ conservation is used in an attempt to save species on the brink of extinction

Zoos, aquaria, and botanical gardens often attempt to save certain endangered species from extinction. Eggs may be collected from nature, for example, or the remaining few animals may be captured and bred in zoos and other research facilities.

Special techniques such as artificial insemination and host mothering are used to increase the number of offspring. Sperm is collected from a suitable male of a rare species and is used to impregnate a female (perhaps located in another zoo in a different city or even in another country). This is an example of **artificial insemination.** In **host mothering,** also called **embryo transfer,** a female of a rare species is treated with fertility drugs, which cause her to produce multiple eggs (see *Focus On: Novel Origins* in Chapter 48). Some of these eggs are collected, fertilized with sperm, and surgically implanted into females of a related but less rare species, who later give birth to offspring of the rare species. In late 1999, for example, the first transfer of a frozen embryo from an endangered African wildcat to a common house cat was announced. Plans are underway to clone endangered species, such as the giant panda, that do not reproduce well in captivity. Another technique involves hormone patches, which are being developed to stimulate reproduction in endangered birds; the patch is attached under the female bird's wing.

A few spectacular successes have occurred in captive breeding programs, in which large enough numbers of a species have been produced to reestablish small populations in the wild. Conservation efforts, for example, have helped the bald eagle make a remarkable comeback. In the mid-1970s, the first eagles bred in captivity were released to the wild. In addition to captive breeding programs, biologists also remove eagle eggs from their nests in the wild, raise the baby eagles in wildlife refuges, and then return them to the wild. (Removal of eggs increases the number of eagles because nesting eagles often lay more eggs to replace those that were removed.) Other efforts also improved the bald eagle's chances, including protection of nesting sites and the banning of chlorinated pesticides (see *Focus On: Food Chains and Poisons in the Environment* in Chapter 53). As a result of continuing efforts on many fronts, the number of nesting pairs in the contiguous United States increased from 417 in 1963 to an estimated 5748 in 1999. In 1994 the bald eagle was removed from the endangered list and transferred to the less critical threatened list. In 1999, the U.S. Fish and Wildlife Service (FWS) proposed that it be removed from the threatened list, an action that should be approved sometime in the early 2000s. Even if the bald eagle is delisted, it will still be protected by the Migratory Bird Treaty Act and the Bald and Golden Eagle Protection Act.

Attempting to save a species on the brink of extinction is usually extremely expensive; therefore, only a small proportion of en-

(a)

(b)

Figure 55-6 Restoring damaged lands. The University of Wisconsin–Madison Arboretum has pioneered restoration ecology. **(a)** The restoration of the prairie was at an early stage in November 1935. The men are digging holes to plant prairie grass sod. **(b)** The prairie as it looks today. This picture was taken at approximately the same location as the 1935 photograph. *(a, Courtesy of the University of Wisconsin–Madison Arboretum; b, Courtesy of Virginia Kline)*

dangered species can be saved. Moreover, zoos, aquaria, and botanical gardens do not have the space to try to save all endangered species. This means that conservation biologists must prioritize which species to attempt to save. Zoos have traditionally focused on large, charismatic animals because the public is more interested in them. Such conservation efforts ignore millions of less glamorous but ecologically important species. However, zoos, aquaria, and botanical gardens serve a useful purpose by educating the public about the value of conservation biology. Clearly, it is more effective to protect and maintain natural habitats by controlling unchecked development so that species do not become endangered in the first place.

Conservation organizations are essential to conservation biology

Conservation organizations are an important part of the effort to maintain biological diversity. They help educate policy makers and the public about the importance of biological diversity. In certain instances they serve as catalysts by galvanizing public support for important biodiversity preservation efforts. They also provide financial support for conservation projects, from basic research to the purchase of land that is a critical habitat for a particular organism or group of organisms.

The World Conservation Union (IUCN)[2] assists countries with conservation biology projects. It and other conservation organizations are currently assessing how effective established wildlife refuges are in maintaining biological diversity. The Minimum Critical Size of Ecosystems Project, a long-term study of the effects of fragmentation on the Amazonian rain forest, is being conducted in Brazil by the World Wildlife Fund[3] and Brazil's National Institute for Amazon Research (Fig. 55–7). In addition, IUCN and the World Wildlife Fund have identified which biomes and ecosystems are not represented by protected areas. IUCN maintains a data bank on the status of the world's species, which is published in *The IUCN Red Data Books*.

The Endangered Species Act provides some legal protection for species and habitats

In 1973 the **Endangered Species Act (ESA)** was passed in the United States, authorizing the FWS to protect endangered and

[2] Formerly called the International Union for Conservation of Nature and Natural Resources, the World Conservation Union still goes by the abbreviation IUCN.

[3] Outside of the United States, Canada, and Australia, the World Wildlife Fund is known as the World Wide Fund for Nature.

Figure 55–7 Minimum Critical Size of Ecosystems Project. When a protected area is set aside, it is important to know what the minimum size of that area must be so that it will not be affected by encroaching species from surrounding areas. Shown are 1-hectare and 10-hectare plots of a long-term study that is being conducted on the effects of fragmentation on Amazonian rain forest in Brazil by the World Wildlife Fund and Brazil's National Institute for Amazon Research. Plots with an area of 100 hectares are also under study, along with identically sized sections of intact forest, which are used as controls. Preliminary data indicate that the smaller forest fragments are unable to maintain their ecological integrity. For example, large trees near forest edges often die or are damaged from exposure to wind, desiccation (from lateral exposure of the forest fragment to sunlight), and invasion by parasitic woody vines. *(R. O. Bierregaard)*

threatened species in the United States and abroad. Many other countries now have similar legislation. The FWS conducts a detailed study of a species to determine if it should be listed as endangered or threatened. Since the passage of the ESA, more than 1200 species in the United States have been listed as endangered or threatened (Table 55–2). The ESA provides legal protection to listed species so that their danger of extinction is reduced. For example, the act makes it illegal to sell or buy any product made from an endangered or threatened species. It further requires officials of the FWS to select critical habitats and design a recovery plan for each species listed. The recovery plan includes an estimate of the current population size, an analysis of what factors contributed to its endangerment, and a list of activities that will be used to help the population recover. As of early 2001, there were 746 recovery plans. Some recovery plans cover more than one species, whereas a few species have two or more recovery plans for different parts of their ranges. The 1995 reintroduction of wolves to Yellowstone National Park is part of that species' recovery plan.

The Endangered Species Act, which was amended in 1982, 1985, and 1988, is considered one of the strongest pieces of environmental legislation in the United States, in part because species are designated as endangered or threatened entirely on biological grounds. Currently, economic considerations cannot influence the designation of endangered or threatened species. Biologists generally agree that as a result of passage of the ESA in 1973, fewer species became extinct than would have had the law never been passed.

The ESA has also been one of the most controversial pieces of environmental legislation. For example, the ESA does not provide compensation for private property owners who suffer financial losses because they cannot develop their land if a threatened or endangered species lives there. The ESA has also interfered with some federally funded development projects.

The ESA was scheduled for congressional reauthorization in 1992 but has been entangled since then in disagreements between conservation advocates and those who support private property rights. Conservation advocates think the ESA does not do enough to save endangered species, whereas those who own land on which rare species live think the law goes too far and infringes on property rights. Some critics—notably business interests and private property owners—view the ESA as an impediment to economic progress. Those who defend the ESA point out that, of 34,000 past cases of endangered species versus development, only 21 cases could not be resolved by some sort of compromise. When the black-footed ferret was reintroduced on the Wyoming prairie, for example, it was classified as an "experimental, nonessential species" so that its reintroduction would not block ranching and mining in the area. Thus, the ferret release program obtained the

TABLE 55–2 Organisms Listed as Endangered or Threatened in the United States, 2001

Type of Organism	Number of Endangered Species	Number of Threatened Species
Mammals	63	9
Birds	78	14
Reptiles	14	22
Amphibians	10	8
Fishes	70	44
Snails	20	11
Clams	61	8
Crustaceans	18	3
Insects	33	9
Spiders	12	0
Flowering plants	564	141
Conifers and cycads	2	1
Ferns and other plants	24	2

Source: U.S. Fish and Wildlife Service.

support of local landowners, support that was deemed crucial to ferret survival in nature.

This type of compromise is essential to the success of saving endangered species because, according to the U.S. General Accounting Office, more than 90% of endangered species live on at least some privately owned lands. Critics of the ESA think the law should be changed so that private landowners are given economic incentives to help save endangered species living on their lands. For example, tax cuts for property owners who are good land stewards could make the presence of endangered species on their properties an asset instead of a liability.

Defenders of the ESA agree that it is not perfect. By mid-2001, only 12 endangered species had recovered enough to be delisted, that is, removed from protection of the ESA (20 additional species had been delisted because new information showed they were not as rare as thought or because they had gone extinct). However, the FWS says that hundreds of listed species are stable or improving; they expect as many as several dozen additional species to be delisted over the next several years. The law is geared more to saving a few popular or unique endangered species rather than the much larger number of less glamorous species that perform valuable **ecosystem services.** Yet it is the less glamorous organisms such as plants, fungi, and insects that play central roles in ecosystems and contribute most to their functioning.

Conservationists would like to see the ESA strengthened in such a way as to manage whole ecosystems and maintain complete biological diversity rather than attempting to save endangered species as isolated entities. This approach offers collective protection to many declining species rather than to single species.

International agreements help protect species and habitats

At the international level, 146 countries participate in the Convention on International Trade in Endangered Species of Wild Flora and Fauna (CITES), which went into effect in 1975. Originally drawn up to protect endangered animals and plants considered valuable in the highly lucrative international wildlife trade, CITES bans hunting, capturing, and selling of endangered or threatened species and regulates trade of organisms listed as potentially threatened. Unfortunately, enforcement of this treaty varies from country to country, and even where enforcement exists, the penalties are not very severe. As a result, illegal trade in rare, commercially valuable species continues.

The goals of CITES often stir up controversy over such issues as who actually owns the world's wildlife, and whether global conservation concerns take precedence over competing local interests. These conflicts often highlight socioeconomic differences between wealthy consumers of CITES products and poor people who trade the endangered organisms. The case of the African elephant bears out these controversies. Listed as an endangered species since 1989 to halt the slaughter of elephants driven by the ivory trade, the species seems to have recovered in southern Africa (Namibia, Botswana, and Zimbabwe). When elephant populations grow too large for their habitat, they root out and knock over so many smaller trees that the forest habitat can support fewer other species. Organizations such as the Humane Society in the United States are developing a birth control vaccine to reduce elephant births. However, the African people living near the elephants want to be able to cull the herd periodically, so they can sell elephant meat, hides, and ivory for profit. In 1997, CITES transferred elephant populations in Namibia, Botswana, and Zimbabwe to a less restrictive, potentially threatened listing to allow trade of stockpiled ivory to Japan. Some conservationists oppose the resumption of any ivory trade, although others think that conservation goals are best attained by cooperating with local hunters and traders rather than by treating these groups as outlaws.

The Convention on Biological Diversity treaty produced by the 1992 Earth Summit has been ratified by more than 180 nations and is now considered binding. (The United States has signed but not ratified it.) Under the conditions of the treaty, each signatory nation must inventory its own biodiversity and develop a **national conservation strategy,** a detailed plan for managing and preserving the biological diversity of that specific country.

DEFORESTATION IS OCCURRING AT AN UNPRECEDENTED RATE

The most serious problem facing the world's forests is **deforestation,** which is defined as the temporary or permanent clearance of forests for agriculture or other uses (Fig. 55–8). The World Commission on Forests, formed following the Earth Summit in 1992, released its first report in 1999 after three years of research and public hearings around the world. It concluded that Earth's forests are shrinking *each year* by 15 million hectares (37 million acres). Causes of forest destruction include fires caused by drought and land clearing practices, expansion of agriculture,

Figure 55–8 Deforestation. Aerial view of clearcut areas in the Gifford Pinchot National Forest in southwest Washington state. *(Gary Braasch/Tony Stone Images)*

construction of roads in forests, tree harvests, and insects and diseases.

When forests are destroyed, they no longer make valuable contributions to the environment or to the people who depend on them. Deforestation results in increased soil erosion and thus decreased soil fertility. Increased sedimentation of waterways, caused by soil erosion, harms downstream aquatic ecosystems by reducing light penetration, covering aquatic organisms, and filling in waterways. Uncontrolled soil erosion, particularly on steep deforested slopes, can cause mud flows that endanger human lives and property and can affect the production of hydroelectric power as silt builds up behind dams. In drier areas, deforestation can lead to the formation of deserts.

Deforestation contributes to the loss of biological diversity. In particular, many tropical species have limited ranges within a forest, so they are especially vulnerable to habitat destruction or modification. Migratory species, such as birds and butterflies, also suffer because of tropical deforestation.

Forests, particularly on hillsides and mountains, provide nearby lowlands with some protection from floods by trapping and absorbing precipitation. When a forest is cut down, the watershed cannot hold water nearly as well, and the total amount of surface runoff flowing into rivers and streams increases. This not only causes soil erosion but puts lowland areas at extreme risk of flooding.

Deforestation may affect regional and global climate changes. Transpiring trees release substantial amounts of moisture into the air. This moisture falls back to the surface in the hydrological cycle (see Chapter 53). When a large forest is removed, rainfall may decline, and droughts may become common in that region. Studies suggest that the local climate has become drier in parts of Brazil where tracts of the rain forest have been burned. Temperatures may also rise slightly in a deforested area because there is less evaporative cooling from the trees.

Deforestation may contribute to an increase in global temperature by causing a release of carbon originally stored in the trees into the atmosphere as carbon dioxide, which enables the air to retain heat. It is thought that the carbon in forests is released immediately if the trees are burned or more slowly when unburned parts decay. If trees are harvested and logs are removed, roughly one-half of the forest carbon remains as dead materials (branches, twigs, roots, and leaves) that decompose, releasing carbon dioxide. When an old-growth forest is harvested, researchers estimate that it takes about 200 years for the replacement forest to accumulate the amount of carbon that was stored in the original forest.

Where and why are forests disappearing?

During the past 1000 years, deciduous forests in temperate areas were largely cleared for housing and agriculture. Today, however, deforestation in the tropics is occurring much more rapidly and over a much larger area. Most of the remaining undisturbed tropical rain forests, which occur in the Amazon and Congo River basins of South America and Africa, are being cleared and burned at a rate unprecedented in human history. Tropical rain forests are also being destroyed at an extremely rapid rate in southern Asia, Indonesia, Central America, and the Philippines.

Exact figures on rates of tropical forest destruction are unavailable. The Food and Agriculture Organization (FAO) of the United Nations, which provides world statistics on forest cover, released its most recent assessment of tropical deforestation in *State of the World's Forests: 1999*. A total of 117 tropical countries were evaluated in this study. The FAO estimated an average annual loss in forests of 0.7% per year from 1990 to 1995, and some areas, such as continental Southeast Asia, experienced a loss of forests estimated at 1.6% per year. If this rate of deforestation—which represents an annual loss of 12.6 million hectares (31.1 million acres)—continues, tropical forests will be all but gone by the first half of the 22nd century.

Several studies show a strong statistical correlation between population growth and deforestation. More people need more food, so the forests are cleared for agricultural expansion. Tropical deforestation is a complex problem, however, that cannot be attributed simply to population pressures. The main causes of deforestation vary from place to place, and a variety of economic, social, and governmental factors interact to cause deforestation. Government policies sometimes provide incentives that favor the removal of forests. For example, the Brazilian government opened the Amazonian frontier for settlement, beginning in the late 1950s, by constructing the Belem-Brasilia Highway, which cut through the Amazon Basin. Sometimes economic conditions encourage deforestation. The farmer who converts more forest to pasture can maintain a larger herd of cattle, which is a good hedge against inflation.

Keeping in mind that tropical deforestation is a complex problem, three agents are thought to be the most immediate causes of deforestation in tropical rain forests: subsistence agriculture, commercial logging, and cattle ranching. Other reasons for the destruction of tropical forests include the development of hydroelectric power, which inundates large areas of forest; mining, particularly when ore smelters burn charcoal produced from rainforest trees; and plantation-style agriculture of crops such as citrus fruits and bananas. The main cause of deforestation in tropical dry forests is fuel wood consumption.

Subsistence agriculture, in which a family produces enough food to feed itself, accounts for perhaps 60% of tropical deforestation. In many developing countries where tropical rain forests are located, the majority of people do not own the land that they live and work on. For example, in Brazil only 5% of the farmers own 70% of the land. Most subsistence farmers have no place to go except into the forest, which they clear to grow food. Land reform in Brazil, Madagascar, Mexico, the Philippines, Thailand, and many other countries would make the land owned by a few available to everyone, thereby easing the pressure on tropical forests by subsistence farmers. This scenario is unlikely, however, because wealthy landowners have more economic and political clout than impoverished land-less peasants.

Subsistence farmers often follow loggers' access roads until they find a suitable spot. They first cut down the trees and allow them to dry, then they burn the area (Fig. 55–9) and plant crops immediately after burning. This is known as **slash-and-burn**

Figure 55–9 Clearing the forest. Tropical rain forest in Brazil is burned to provide agricultural land. This type of cultivation is known as slash-and-burn agriculture. *(Dr. Nigel Smith/Earth Scenes)*

agriculture. Yields from the first crop are often quite high because the nutrients that had been in the trees are now available in the soil. However, soil productivity declines rapidly, and subsequent crops are poor. In a few years the farmer must move to a new part of the forest and repeat the process. Cattle ranchers often claim the abandoned land for grazing, because land that is not fertile enough to support crops can still support livestock.

Slash-and-burn agriculture done on a small scale, with periods of 20 to 100 years between cycles, is sustainable. The forest regrows rapidly after a few years of farming. But when millions of people try to obtain a living in this way, the land is not allowed to lie uncultivated long enough to recover.

About 20% of tropical deforestation is the result of commercial logging, and vast tracts of tropical rain forests, particularly in Southeast Asia, are being harvested for export abroad. Most tropical countries allow commercial logging to proceed at a much faster rate than is sustainable because it supplies them with much-needed revenues. In the final analysis, tropical deforestation does not contribute to economic development; rather, it reduces or destroys the value of an important natural resource.

Approximately 12% of tropical rainforest destruction is carried out to provide open rangelands for cattle. Cattle ranching is particularly important in Central America. Much of the beef raised on these ranches, which are often owned by foreign companies, is exported to fast-food restaurant chains in North America and Europe. After the forests are cleared, cattle can graze on the land for up to perhaps 20 years, after which time the soil fertility is depleted. When this occurs, shrubby plants, known as scrub savanna, take over the range.

CERTAIN ATMOSPHERIC POLLUTANTS MAY AFFECT EARTH'S CLIMATE

Earth's average temperature is based on daily measurements taken at several thousand land-based meteorological stations around the world, as well as data from weather balloons, orbiting satellites, transoceanic ships, and hundreds of sea-surface buoys with temperature sensors. Earth's average surface temperature increased by 0.6° C (1.1° F) during the 20th century, and the 1990s was the warmest decade of the century (Fig. 55–10). The warmest year on record was 1998. The increase in temperature during the 20th century may have been the largest in the last 1000 years. (Although reliable records have been kept only since the mid-19th century, scientists can reconstruct earlier temperatures using indirect climate evidence contained in tree rings, lake and ocean sediments, ancient ice, and coral reefs.)

Other evidence also suggests an increase in global temperature. For example, several studies have documented that spring in the Northern Hemisphere now comes about 6 days earlier than it did in 1959, and autumn is delayed by 5 days. (Spring is determined by when buds of specific plants open, and autumn by when leaves of specific plants turn color and fall.) Since 1949 the United States has experienced an increased frequency of extreme heat stress events, which are extremely hot, humid days during summer months; medical records indicate that heat-related deaths among elderly and other vulnerable people increase during these events. The sea level has risen slightly (0.1 to 0.2 m, or

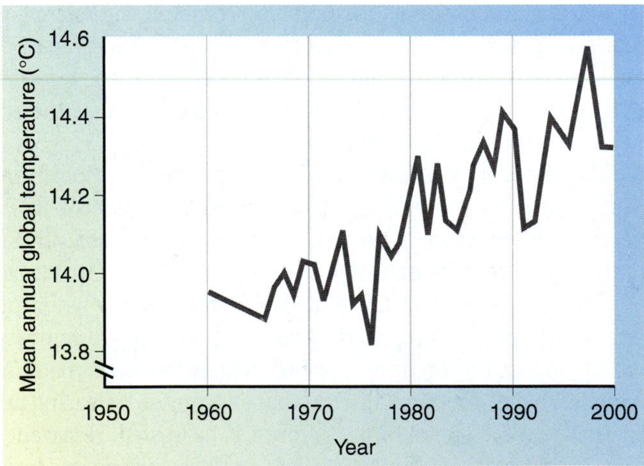

Figure 55–10 Mean annual global temperature, 1960 to 2000. Data are presented as surface temperatures (°C) for 1960, 1965, and every year thereafter. The measurements, which naturally fluctuate, clearly show the warming trend of the last several decades. (The decline in temperature in 1992 and 1993 was due in part to the largest volcanic eruption in the 20th century, Mt. Pinatubo in the Philippines, in June 1991. The eruption injected massive amounts of sulfur into the stratosphere, which reduced the amount of sunlight reaching the planet.) *(Surface Air Temperature Analysis, Goddard Institute for Space Studies, NASA)*

4 to 7.9 in, during the 20th century), mountain glaciers in nonpolar regions have retreated, and extreme weather events such as severe rainstorms have occurred with increasing frequency in certain regions.

Scientists around the world have been researching **global warming** for the past 50 years. As the evidence has accumulated, those most qualified to address the issue have reached a strong consensus that the 21st century will experience significant climate change and that human activities will be responsible for much of this change.

In response to this growing scientific consensus, governments around the world organized the U.N. Intergovernmental Panel on Climate Change (IPCC). With input from hundreds of climate experts, the IPCC provides the most definitive scientific statement about global warming. The IPCC Third Assessment Report, released in January 2001, concluded that human-produced air pollutants continue to change the atmosphere, causing most of the warming observed in the past 50 years. The human influence on climate change can be identified despite questions about how much of the recent warming stems from natural variations. The IPCC report projects a 1.4 to 5.8° C (2.5 to 10.4° F) increase in global temperature by the year 2100, although the warming will probably not be uniform from region to region. Thus, Earth may become warmer during the 21st century than it has been for hundreds of thousands of years.

Process of Science Almost all climate experts agree with the IPCC's assessment that the warming trend has already begun and will continue throughout the 21st century. However, scientists are uncertain over how rapidly the warming will proceed, how severe it will be, and where it will be most pronounced. (Recall that uncertainty and debate are part of the scientific process and that scientists can never claim to know a "final answer.") As a result of these uncertainties about global warming, many people, including policy makers, are confused about what should be done. Yet the stakes are quite high because human-induced global warming has the potential to disrupt Earth's climate for a very long time.

Greenhouse gases cause global warming

Carbon dioxide (CO_2) and certain other trace gases, including methane (CH_4), surface ozone (O_3),[4] nitrous oxide (N_2O), and chlorofluorocarbons (CFCs), are accumulating in the atmosphere as a result of human activities (Table 55–3). The concentration of atmospheric CO_2 has increased from about 288 parts per million (ppm) approximately 200 years ago (before the Industrial Revolution began) to 371 ppm in 2000 (Fig. 55–11). Burning carbon-containing fossil fuels—coal, oil, and natural gas—accounts for about three-fourths of human-made carbon dioxide emissions to the atmosphere. Carbon dioxide is also released by land conversion, such as when tracts of tropical forests are logged or burned. (The burning itself releases carbon dioxide into the atmosphere, and because trees normally remove carbon dioxide from the atmosphere during photosynthesis, tree removal prevents this process.) The levels of the other trace gases associated with global warming are also rising.

[4] Surface ozone (more precisely, ozone in the troposphere) is a greenhouse gas as well as a component of photochemical smog. Ozone in the upper atmosphere, the stratosphere, provides an important planetary service that is discussed later in this chapter.

TABLE 55–3 Increase in Selected Atmospheric Greenhouse Gases, Preindustrial to 2000

Gas	Estimated Preindustrial Concentration	2000 Concentration	Relative Contribution to Warming, 1750 to 2000‖
Carbon dioxide	288 ppm*	371 ppm§	1.6
Methane	848 ppb†	1800 ppb	0.48
Nitrous oxide	285 ppb	312 ppb	0.15
CFC-12	0 ppt‡	538 ppt	0.34 (CFC-12 and CFC-11 combined)
CFC-11	0 ppt	264 ppt	

*ppm = parts per million.

†ppb = parts per billion.

‡ppt = parts per trillion.

§2000 annual average derived from in situ sampling at Mauna Loa, Hawaii. All other data from Mace Head, Ireland, monitoring site.

‖Relative contributions to warming are given in watts per square meter using a measure known as radiative forcing. Estimates are from the U.N. Intergovernmental Panel on Climate Change Third Assessment. The 0.34 is a collective estimate for all chlorine-and bromine-containing substances.

Source: Carbon Dioxide Information Analysis Center, Environmental Sciences Division, Oak Ridge National Laboratory.

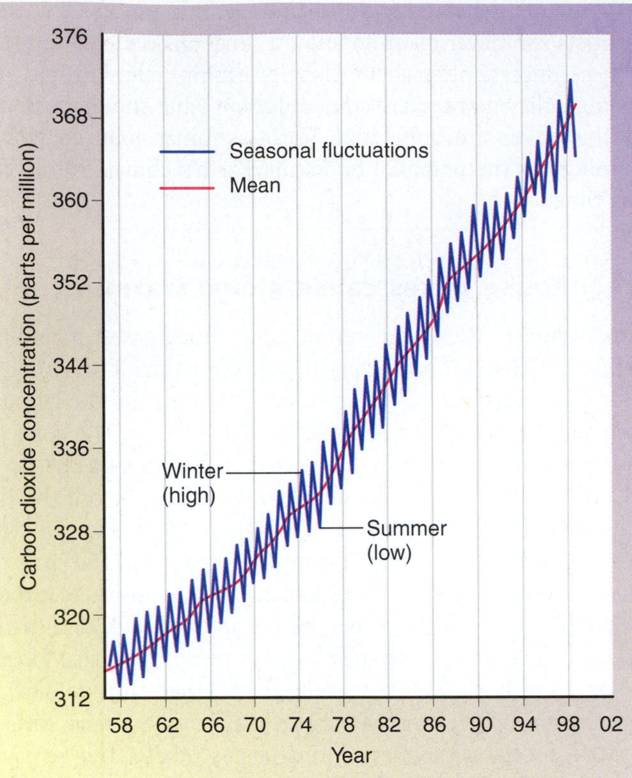

Figure 55–11 Carbon dioxide in the atmosphere, 1958 to 1999. Note the steady increase in the concentration of atmospheric carbon dioxide since 1958, when measurements began at the Mauna Loa Observatory, Hawaii. This location was selected because it is far from urban areas where factories, power plants, and motor vehicles emit carbon dioxide. The fluctuations correspond to the annual cycles of plants in the Northern Hemisphere: winter (a high level of carbon dioxide), when plants are not actively growing and absorbing carbon dioxide, and summer (a low level of carbon dioxide), when they are growing and absorbing carbon dioxide. *(C.D. Keeling and T. P. Whorf, Scripps Institution of Oceanography, University of California, LaJolla, California)*

Global warming occurs because these gases absorb infrared radiation, that is, heat, in the atmosphere. This absorption slows the natural heat flow into space, warming the lower atmosphere. Some of the heat from the lower atmosphere is transferred to the ocean and raises its temperature as well. This retention of heat in the atmosphere is a natural phenomenon that has made Earth habitable for its millions of species. However, as human activities increase the atmospheric concentration of these gases, the atmosphere and ocean continue to warm, and the overall global temperature rises.

Because carbon dioxide and other gases trap the sun's radiation somewhat like glass does in a greenhouse, the natural trapping of heat in the atmosphere is referred to as the **greenhouse effect,** and the gases that absorb infrared radiation are known as **greenhouse gases.** The additional warming that may be produced by increased levels of gases that absorb infrared radiation is known as the **enhanced greenhouse effect** (Fig. 55–12).

Although current rates of fossil fuel combustion and deforestation are high, causing the carbon dioxide level in the atmosphere to increase markedly, scientists think the warming trend will be slower than the increasing level of carbon dioxide might indicate. The reason is that water requires more heat to raise its temperature than gases in the atmosphere do (recall the discussion of the high specific heat of water in Chapter 2). As a result, the ocean takes longer than the atmosphere to absorb heat, and climate scientists think that warming will probably be more pronounced in the second half of the 21st century than in the first half.

What are the probable effects of global warming?

We now consider some of the probable effects of global warming, including changes in sea level; changes in precipitation patterns; effects on organisms, including humans; and effects on agriculture. These changes will persist for many centuries because many greenhouse gases remain in the atmosphere for hundreds of years. Furthermore, even after greenhouse gas concentrations have been stabilized, scientists think Earth's mean surface temperature will continue to rise because the ocean adjusts to climate change on a delayed time scale.

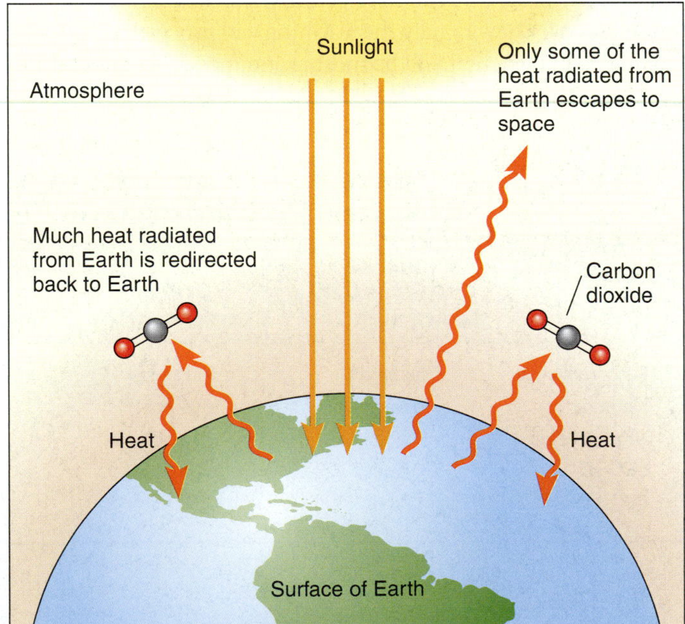

Figure 55–12 Enhanced greenhouse effect. Visible light passes through the atmosphere to Earth's surface. However, the longer wavelength heat (infrared radiation) radiated from the surface is redirected back to the surface by carbon dioxide and other greenhouse gases. Therefore, the buildup of greenhouse gases in the atmosphere results in global warming.

With global warming, the global average sea level is rising

If the overall temperature of the Earth increases by just a few degrees, there could be a major thawing of glaciers and the polar ice caps. During the Antarctic summer that extended from December 1998 to March 1999, for example, the ice shelves along the Antarctic Peninsula shrunk by thousands of square kilometers, an area roughly the size of Rhode Island. (Antarctic ice shelves are thick sheets of floating ice that are fed mainly by glaciers that flow off the land.) This loss of ice coincided with a decades-long trend of atmospheric warming in the Antarctic. A study from the 1970s to the 1990s showed that the area of ice-covered ocean in the Arctic has also retreated, and mountain glaciers around the world are melting at accelerating rates.

In addition to sea-level rise caused by the retreat of glaciers and thawing of polar ice, the sea level will probably rise because of thermal expansion of the warming ocean. Water, like other substances, expands as it warms. As stated previously, during the 20th century, sea level rose between 0.1 and 0.2 m, most of it owing to thermal expansion.

The IPCC estimates that sea level will rise by an additional 0.5 m (20 in) by 2100. Such an increase would flood low-lying coastal areas such as parts of southern Louisiana and South Florida. Coastal areas that are not inundated will be more likely to suffer erosion and other damage from more frequent and more intense weather events such as hurricanes. These likely effects are certainly a cause for concern, particularly since about two-thirds of the world's population lives within 150 km (93 mi) of a coastline.

With global warming, precipitation patterns will change

Global computer simulations of weather changes as global warming occurs indicate that precipitation patterns will change, causing some areas such as mid-latitude continental interiors to have more frequent droughts. At the same time, more intense precipitation events (heavier snow and rainstorms) may cause more frequent flooding in other areas. Changes in precipitation patterns could affect the availability and quality of fresh water in many locations. It is projected that areas that are currently arid or semiarid, such as the Sahel region just south of the Sahara Desert, will have the most troublesome water shortages as the climate changes. Closer to home, water experts predict water shortages in the American West because warmer winter temperatures will cause more precipitation to fall as rain rather than snow; melting snow currently provides 70% of stream flows in the West during summer months.

The frequency and intensity of storms over warm surface waters may also increase. Scientists at the National Oceanic and Atmospheric Administration developed a computer model in 1998 that examined how global warming might affect hurricanes. When the model was run with a sea surface temperature 2.2° C warmer than today, more intense hurricanes resulted. (The question of whether hurricanes will be more *frequent* in a warmer climate remains uncertain.) Changes in storm frequency and intensity are expected because as the atmosphere warms, more water evaporates, which in turn releases more energy into the atmosphere (recall the discussion of water's heat of vaporization in Chapter 2). This energy generates more powerful storms.

With global warming, the ranges of organisms are changing

Biologists have started to examine some of the potential consequences of global warming on organisms. For example, one team of biologists placed infrared heaters above test plots in a Rocky Mountain meadow to mimic future warming. Changes in the kinds of plants growing in the plots were noted in just a few months.

Other researchers determined that populations of zooplankton in the California Current have declined by 80% since 1951, apparently because the current has warmed slightly. (The California Current flows from Oregon southward along the California coast.) The decline in zooplankton has affected the entire ecosystem's food web, and populations of seabirds and plankton-eating fishes have also declined. As in all studies of effects of warming on organisms in nature, researchers are unsure of the relative contributions of human production of greenhouse gases and of natural fluctuations in the climate; despite this uncertainty, the observed changes are real.

As temperatures have risen in the waters around Antarctica during the past two decades, a similar decline in shrimp-like krill has contributed to a reduction in Adélie penguin populations. Because there are fewer krill, the birds are not able to get enough food. Warmer temperatures in Antarctica—during the past 50 years the average annual temperature on the Antarctic Peninsula has increased 2.6° C (5° F)—have also contributed to reproductive failure in Adélie penguins. The birds normally lay their eggs in snow-free rocky outcrops, but the warmer temperatures have caused more snowfall (recall that warmer air can hold more moisture), which melts when the birds incubate the eggs. The melted snow forms cold pools of slush that kill the developing chick embryos.

The first scientific evidence that a species has shifted its geographical range in response to climate change was published in 1996. Camille Parmesan, a biologist at the University of California at Santa Barbara, compared the current geographical range of a western butterfly (the Edith's checkerspot butterfly) to previously recorded observations. She found that the butterfly has shifted its range northward by about 160 km (about 100 mi). Since then, dozens of examples of poleward shifts of plant and animal ranges have been observed, as well as earlier flowering of trees and earlier egg laying in birds.

Each species reacts to changes in temperature differently. Some species will undoubtedly become extinct, particularly those with narrow temperature requirements, those confined to small reserves or parks, and those living in fragile ecosystems. Other species may survive in greatly reduced numbers and ranges. Ecosystems considered most vulnerable to species loss in the short term are polar seas; coral reefs and atolls; mountains, particularly alpine tundra; prairie wetlands; coastal wetlands; tundra; taiga; and tropical forests.

Biologists generally agree that global warming will have an especially severe impact on plants because they cannot move about when environmental conditions change. Although wind and animals disperse seeds, sometimes over long distances, seed dispersal has definite limitations in terms of the speed of dispersal. During past climate warmings, such as during the glacial retreat that took place some 12,000 years ago, tree species are thought to have dispersed from 4 to 200 km (2.5 to 124 mi) per century. If the Earth warms by the projected 1.4 to 5.8° C during the 21st century, the ideal ranges for some temperate tree species (i.e., the environment where they grow the best) may shift northward as much as 480 km (300 mi) (Fig. 55–13). Moreover, soil characteristics, water availability, competition with other plant species, and human alterations of natural habitats all affect the rate at which plants can move into a new area.

Some species may be able to disperse into new environments or adapt to the changing conditions in their present habitats. Also, some species may be unaffected by global warming, whereas others may emerge from it as winners, with greatly expanded numbers and ranges. Those considered most likely to prosper include weeds, pests, and disease-carrying organisms, all of which are generalists that are already common in many different environments.

Global warming will have a more pronounced effect on human health in developing countries

Data linking climate warming and human health problems are accumulating. Scientists hypothesize that the increase in carbon dioxide in the atmosphere and the resultant more frequent and more severe heat waves during summer months will cause an increase in the number of heat-related illnesses and deaths. During the summer 1998 heat wave in Dallas, Texas, for example, more than 100 people died from heatstroke and other heat-related illnesses. India recorded almost 1300 deaths caused by high temperatures during the 1998 summer heat wave.

Human health may also be affected indirectly by climate warming. Mosquitoes and other disease carriers could expand their range into the newly warm areas and spread malaria, dengue fever, yellow fever, Rift Valley fever, and viral encephalitis. As many as 50 to 80 million additional cases of malaria could occur annually in tropical, subtropical, and temperate areas. According to the World Health Organization, during 1998, the warmest year on record, the incidence of malaria, Rift Valley fever, and cholera surged in developing countries.

Highly developed countries are less vulnerable to such disease outbreaks because of better housing (which keeps mosquitoes out-

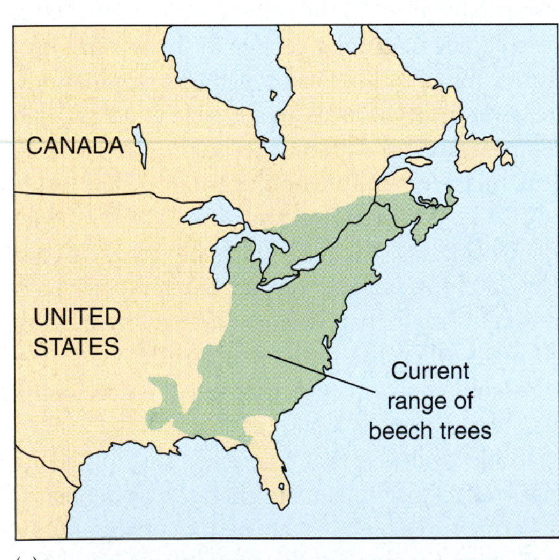

(a)

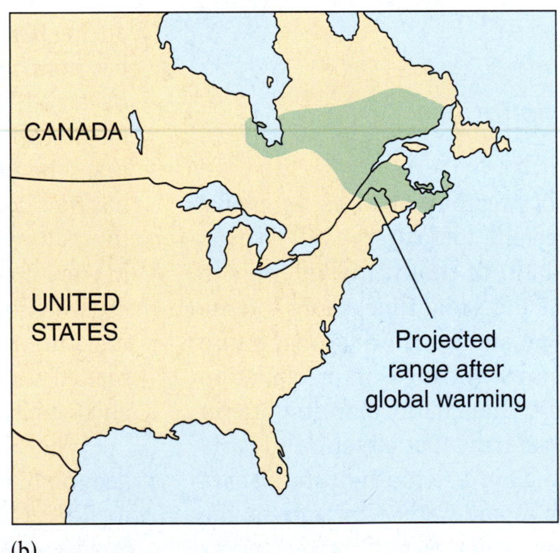

(b)

Figure 55–13 Climate change and beech trees in North America. **(a)** Present geographical range of American beech trees. These large shade trees produce edible beechnuts that are an important food source for squirrels, raccoons, bears, game birds, and many other kinds of wildlife. **(b)** One projected range of beech trees after global warming occurs. This figure is based on a model of climate change developed by the U.S. Geophysical Fluid Dynamics Laboratory (GFDL). According to the GFDL scenario of changed temperature and precipitation patterns, beeches would continue to grow in parts of Maine and Nova Scotia, and they would expand into Quebec and northern Ontario. *(Adapted from Davis, M.B., and C. Zabinski. "Changes in Geographical Range Resulting from Greenhouse Warming: Effects on Biodiversity in Forests."* Global Warming and Biological Diversity, *edited by Peters, R.L., and T.E. Lovejoy. Yale University Press, New Haven, Connecticut, 1992)*

side), medical care, pest control, and public health measures such as water treatment plants. Texas reported a few cases of dengue fever in the late-1990s, for example, whereas nearby Mexico had thousands of cases during that period.

Global warming will probably affect agriculture

Global warming may increase problems for agriculture, which is already beset with the challenge of providing enough food for a hungry world without doing irreparable damage to the environment. Several studies document that the rising sea level will inundate river deltas, which have been important since ancient times as some of the world's best agricultural lands. The Nile River (Egypt), Mississippi River (United States), and Yangtze River (China) are examples of river deltas that have been studied. Certain agricultural pests and disease-causing organisms will probably proliferate. As mentioned earlier, global warming may also increase the frequency and duration of droughts and, in some areas, crop-damaging floods.

On a regional scale, current global warming models forecast that agricultural productivity will increase in some areas and decline in others. Models suggest that Canada and Russia may be able to increase their agricultural productivity in a warmer climate, whereas tropical and subtropical regions where many of the world's poorest people live will be hardest hit by declining agricultural productivity. Central America and Southeast Asia may experience some of the greatest declines in agricultural productivity.

Many actions have been suggested to deal with global warming

Despite certain gaps in our knowledge of global warming, our present understanding of the changing global climate and its potential impact on human society and other species gives us many legitimate reasons to develop strategies to deal with this problem. The amount and severity of global warming depend on how much additional greenhouse gas emissions we add to the atmosphere. Many studies make the assumption that we will be able to stabilize atmospheric carbon dioxide at 550 ppm, which is roughly twice the concentration of atmospheric carbon dioxide that existed in the preindustrial world and about 50% higher than the carbon dioxide currently in the atmosphere.

The international community recognizes that it must stabilize carbon dioxide emissions. At least 174 nations, including the United States, signed the U.N. Framework Convention on Climate Change developed at the 1992 Earth Summit. Its ultimate goal was to stabilize greenhouse gas concentrations in the atmosphere at levels low enough to prevent dangerous human influences on the climate. However, the details of how that goal was to be accomplished were left to future conferences.

At the 1996 U.N. Climate Change Convention held in Geneva, Switzerland, highly developed countries agreed to establish legally binding timetables to cut emissions of greenhouse gases. These timetables for reducing greenhouse gas emissions were decided at a meeting of representatives from 160 countries held in Kyoto, Japan, in December 1997, but will not go into effect until at least 55% of the countries involved have ratified the **Kyoto Protocol,** as the agreement is called. (As of mid-2001, 33 countries had ratified the treaty; the United States has signed but not ratified the agreement.) Although each nation has its own individual emissions target, those goals are similar in many industrial countries: The United States agrees to reduce emissions of carbon dioxide, methane, and nitrous oxide by 7% below its 1990 levels by the year 2012, the European Union by 8%, and Japan by 6%.

The United States signed the Kyoto Protocol in November 1998, but the U.S. Senate must first ratify the treaty before it becomes binding. Some senators are sharply critical of the protocol because of the adverse economic impact it may have. (To curb their production, greenhouse gas emissions will have to be taxed or regulated in some other manner, such as tradable emissions permits.) At the same time, other senators, while supporting the treaty, do not think it is strong enough to prevent dramatic changes in global climate. Most climate scientists agree that to have a significant impact on climate change, the Kyoto Protocol will have to be followed by even more stringent emissions reductions—some 60 to 70% lower than 1990 levels if we want to stabilize greenhouse gas concentrations.

STRATOSPHERIC OZONE CONTINUES TO DECLINE

Ozone (O_3) is a form of oxygen that is a human-made pollutant in the lower atmosphere but a naturally produced, essential part of the stratosphere (see Fig. 20–6). The **stratosphere,** which encircles our planet some 10 to 45 km (6 to 28 mi) above the surface, contains a layer of ozone that shields the surface from much of the ultraviolet radiation coming from the sun (Fig. 55–14). If ozone disappeared from the stratosphere, Earth would become unlivable for most forms of life. Ozone in the lower atmosphere is converted back to oxygen in a few days and so does not replenish the ozone that has been depleted in the stratosphere.

A slight thinning in the ozone layer over Antarctica forms naturally for a few months each year. In 1985, however, the thinning was first observed to be greater than it should have been if natural causes were the only factor inducing it. This increased thinning, which begins each September, is commonly referred to as the "ozone hole" (Fig. 55–15). There, ozone levels decrease by as much as 67% each year. During the 1990s the ozone-thinned area continued to grow, and by 1998 it had reached the record size of 27.5 million km^2 (11 million mi^2), which is larger than the North American continent. In 1999, the ozone-thinned area was slightly smaller—25 million km^2 (10 million mi^2). A less extensive thinning has also been detected in the stratospheric ozone layer over the Arctic.

In addition, worldwide levels of stratospheric ozone have been decreasing for several decades. According to the National Center for Atmospheric Research, ozone levels over Europe and North America have dropped by almost 10% since the 1970s.

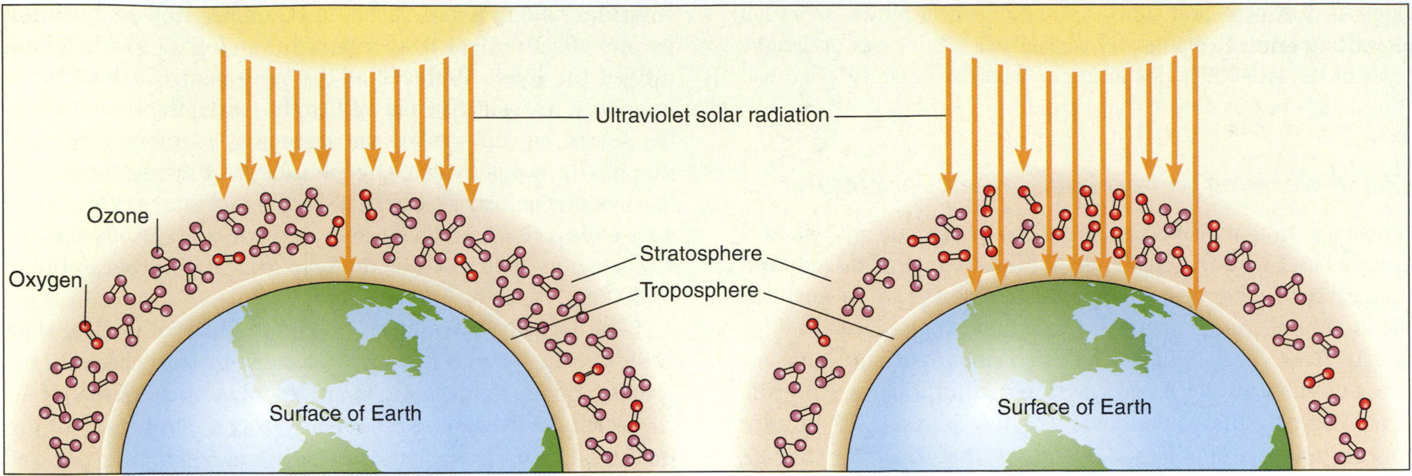

(a) Ozone present at normal levels

(b) Ozone present at reduced levels

Figure 55–14 Ultraviolet radiation and the ozone layer. **(a)** Stratospheric ozone absorbs 99% of incoming ultraviolet radiation, effectively shielding Earth's surface. **(b)** When stratospheric ozone is present at reduced levels, more high-energy ultraviolet radiation penetrates the atmosphere to the surface, where its presence harms organisms.

Certain chemicals destroy stratospheric ozone

Both chlorine- and bromine-containing substances catalyze ozone destruction. The primary chemicals responsible for ozone loss in the stratosphere are a group of chlorine-compounds called chlorofluorocarbons (CFCs). Chlorofluorocarbons have been used as propellants in aerosol cans, as coolants in air conditioners and refrigerators, as foam-blowing agents to make insulation and packaging, and as solvents and cleaners in the electronics industry. Additional compounds that also attack ozone include halons (found in many fire extinguishers); methyl bromide (a pesticide used as a soil and crop fumigant); methyl chloroform (an industrial solvent); and carbon tetrachloride (used in many industrial processes, including the manufacture of pesticides and dyes).

After release into the lower troposphere, CFCs and similar compounds slowly drift up to the stratosphere, where ultraviolet radiation breaks them down, releasing chlorine. Similarly, bromine is released by the breakdown of halons and methyl bromide. The thinning in the ozone layer that was discovered over Antarctica occurs annually between September and November (spring in the Southern Hemisphere). At this time, two important conditions are present: Sunlight returns to the polar region, and the *circumpolar vortex*, a mass of cold air that circulates around the southern polar region and isolates it from the warmer air in the rest of the planet, is well developed. The cold air causes polar stratospheric clouds to form; these clouds contain ice crystals to which chlorine and bromine adhere, making them available to destroy ozone. The sunlight catalyzes the chemical reaction by which chlorine or bromine breaks ozone molecules apart, converting them into oxygen molecules. The chemical reaction by which ozone is destroyed does not alter the chlorine or bromine, and therefore a single chlorine or bromine atom can break down many thousands of ozone molecules. The chlorine and bromine remain in the stratosphere for many years. When the circumpolar vortex breaks up, the ozone-depleted air

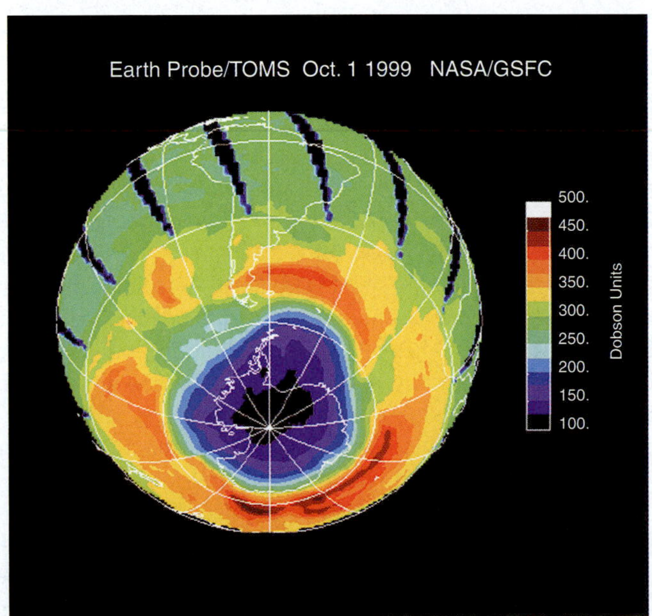

Figure 55–15 Ozone thinning. A computer-generated image of part of the Southern Hemisphere, taken on October 1, 1999, reveals the ozone thinning (black, purple, and blue areas). The ozone-thin area is not stationary, but moves about as a result of air currents. (Dobson units measure stratospheric ozone. They are named after British scientist Gordon Dobson, who studied ozone levels over Antarctica during the late 1950s.) *(NASA/Goddard Space Flight Center)*

spreads northward, diluting ozone levels in the stratosphere over South America, New Zealand, and Australia.

Ozone depletion harms organisms

With depletion of the ozone layer, more ultraviolet radiation reaches the surface. A study conducted in Toronto from 1989 to 1993, for example, showed that wintertime levels of UVB increased by more than 5% each year as a result of lower ozone levels. (UVB is one of three types of UV radiation; see the section on vertebrate skin in Chapter 38.) Excessive exposure to UV radiation is linked to a number of human health problems, including cataracts, skin cancer, and a weakened immune system. The lens of the eye contains transparent proteins that are replaced at a very slow rate. Exposure to excessive UV radiation damages these proteins, and over time, the damage accumulates so that the lens becomes cloudy, forming a cataract. Cataracts can be treated by surgery, but millions of people in developing nations cannot afford the operation and so remain partially or totally blind.

Scientists are concerned that increased levels of UV radiation may disrupt ecosystems. For example, the productivity of Antarctic phytoplankton, the microscopic drifting algae that are the base of the Antarctic food web, has declined from increased exposure to UVB. Research shows that surface UV radiation inhibits photosynthesis in these phytoplankton. Direct damage to natural populations of Antarctic fish has also been documented. Increased DNA mutations in icefish eggs and larvae (young fish) were matched to increased levels of UV radiation; researchers are currently studying if these mutations lessen the animals' ability to survive.

There is also concern that high levels of UV radiation may damage crops and forests, but the effects of UVB radiation on plants are very complex and have not yet been adequately studied. Plants interact with a large number of other species in both natural ecosystems and agricultural ecosystems, and the effects of UV radiation on each of these organisms in turn affect plants indirectly. For example, exposure to higher levels of UVB radiation appears to increase wheat yields by inhibiting fungi that cause disease in wheat. On the other hand, exposure to higher levels of UV radiation appears to decrease cucumber yields by increasing their incidence of disease.

International cooperation will prevent significant depletion of the ozone layer

In 1987, representatives from many countries met in Montreal to sign the **Montreal Protocol,** an agreement that originally stipulated a 50% reduction of CFC production by 1998. Since then, more than 150 countries have signed this agreement. After scientists reported that decreases in stratospheric ozone occurred over the heavily populated mid-latitudes of the Northern Hemisphere in all seasons, the Montreal Protocol was modified to include stricter measures to limit CFC production. Industrial companies that manufacture CFCs quickly developed substitutes.

CFC, carbon tetrachloride, and methyl chloroform production was completely phased out in the United States and other highly developed countries in 1996, except for a relatively small amount exported to developing countries. Existing stockpiles could be used after the deadline, however. Developing countries are on a different timetable and will phase out CFC use by 2005. Methyl bromide will be phased out in highly developed countries, which are responsible for 80% of the global use of that chemical, by 2005. Hydrochlorofluorocarbons (HCFCs) will be phased out in 2030.

Measurements taken in the lower atmosphere between 1994 and 1998 showed that the amount of ozone-depleting chemicals had slowly decreased. Satellite measurements taken in 1997 provided the first evidence that the level of ozone-depleting chemicals were starting to decline in the stratosphere. However, two chemicals—CFC-12 and halon-1211—may have increased and therefore still represent a threat to ozone recovery. Although highly developed countries no longer manufacture CFC-12, it continues to leak into the atmosphere from old refrigerators and vehicle air conditioners discarded in those countries. Moreover, developing countries such as China, India, Korea, and Mexico continue to produce CFC-12 and halon-1211. An international fund known as the Montreal Multilateral Fund is available to help developing countries during their transition from ozone-depleting chemicals to safer alternatives.

CFCs are extremely stable, and those being used today probably will continue to deplete stratospheric ozone for at least 50 years. Scientists expect human-exacerbated ozone thinning to reappear over Antarctica each year, although the area and degree of thinning will gradually decline over time.

ENVIRONMENTAL PROBLEMS ARE INTERRELATED

Several connections have been pointed out among the four environmental problems discussed in this chapter. For example, deforestation, global warming, and ozone depletion will likely cause future reductions in biological diversity. Similarly, any environmental problem you can think of, even if not discussed in this chapter, is related to other environmental concerns. Most of all, however, environmental problems are connected to overpopulation. We have seen that the rate of human population growth is greatest in developing countries but that highly developed nations are overpopulated as well, in the sense that they have a high per capita consumption of resources (see Chapter 51). **Consumption overpopulation**, which is the disproportionately large consumption of resources by people in highly developed countries, affects the environment as much as the population explosion in the developing world does.

Humans share much in common with the fate of other organisms on the planet. We are not immune to the environmental damage we have produced. We differ from other organisms, however, in our capacity to reflect on the consequences of our actions and to alter our behavior accordingly. Humans, both individually and collectively, are able to bring about change. The widespread substitution of constructive change for destructive change is the key to ensuring the biosphere's survival.

SUMMARY WITH KEY TERMS

I. **Biological diversity,** the number and variety of organisms, from their genes to the ecosystems in which they live, is decreasing worldwide.
   A. A species becomes **extinct** when its last individual member dies.
      1. A species whose severely reduced numbers throughout all or a significant part of its **range** put it in imminent danger of extinction is classified as an **endangered species.**
      2. When extinction is less imminent but the population is quite small, a species is classified as a **threatened species.**
   B. Human activities that contribute to a reduction in biological diversity include habitat loss and **habitat fragmentation,** pollution, introduction of nonindigenous species **(biotic pollution)**, pest and predator control, illegal commercial hunting, and commercial harvest. Of these, habitat loss and fragmentation are the most significant.
   C. In the United States, declining biological diversity is most serious in the states of Florida, California, and Hawaii. Worldwide, tropical rain forests are being destroyed faster than almost all other ecosystems.
   D. **Conservation biology** is the study of how humans impact organisms and of the development of ways to protect biological diversity. Conservation biology is a broad discipline that ranges from safeguarding populations of endangered species to preserving entire ecosystems and landscapes.
      1. Efforts to preserve biological diversity in the wild are known as **in situ conservation.** In situ conservation is urgently needed in the world's 25 biodiversity hotspots, which contain as many as 44% of all vascular plant species and 35% of all land vertebrates.
      2. **Ex situ conservation,** such as captive breeding, occurs in human-controlled settings.
   E. The **Endangered Species Act** authorizes the U.S. Fish and Wildlife Service to protect from extinction endangered and threatened species, both in the United States and abroad. Many other countries have similar legislation.

II. Forests provide us with many ecological benefits, including watershed protection, soil erosion prevention, climate moderation, and wildlife habitat.
   A. The greatest problem facing forests today is **deforestation,** the temporary or permanent clearance of forests for agriculture or other uses.
   B. In the tropics, forests are destroyed to provide subsistence farmers with temporary agricultural land; produce timber, particularly for highly developed nations; provide open rangeland for cattle; and supply fuel wood.
      1. **Subsistence agriculture,** in which a family produces enough food to feed itself, accounts for perhaps 60% of tropical deforestation.
      2. About 20% of tropical deforestation is the result of commercial logging, and vast tracts of tropical rain forests, particularly in Southeast Asia, are being harvested for export abroad.

III. Earth's average surface temperature increased by 0.6° C (1.1° F) during the 20th century. The 1990s were the warmest decade of the century.
   A. Carbon dioxide and certain other gases cause the **greenhouse effect,** in which the atmosphere retains heat and warms Earth's surface.
      1. Increased levels of CO_2 and other **greenhouse gases** in the atmosphere are causing concerns about an **enhanced greenhouse effect,** which is additional warming that may be produced by increased levels of gases that absorb infrared radiation.
      2. The combustion of fossil fuels produces pollutants, especially CO_2, nitrous oxide, and surface ozone. Other greenhouse gases are methane and chlorofluorocarbons.
   B. During the 21st century, **global warming** will cause a rise in sea level, changes in precipitation patterns, multiple impacts on natural biological communities, human health problems (particularly in developing countries), and problems for agriculture.

IV. The **ozone (O_3)** layer in the **stratosphere** helps shield Earth's surface from damaging ultraviolet radiation.
   A. The total amount of ozone in the stratosphere is declining (i.e., **ozone depletion**), and large areas of ozone thinning develop over Antarctica and the Arctic each year.
   B. Chlorofluorocarbons and similar chlorine- and bromine-containing compounds attack the ozone layer.
   C. The **Montreal Protocol** is an international agreement to phase out production of ozone-depleting chemicals.

POST-TEST

1. Which of the following statements about extinction is *not* true? (a) extinction is the permanent loss of a species (b) extinction is a natural biological process (c) once a species is extinct, it can never reappear (d) human activities have little impact on extinctions (e) thousands of plant and animal species are currently threatened with extinction
2. An endangered species (a) is severely reduced in number (b) is in imminent danger of becoming extinct throughout all or a significant part of its range (c) usually does not have reduced genetic variability (d) is not in danger of extinction in the foreseeable future (e) both a and b are true
3. The most important reason for declining biological diversity is (a) air pollution (b) introduction of foreign (nonindigenous) species (c) habitat destruction and fragmentation (d) illegal commercial hunting (e) commercial harvesting
4. In situ conservation (a) includes breeding captive species in zoos (b) includes seed storage of genetically diverse crops (c) concentrates on preserving biological diversity in the wild (d) focuses exclusively on large, charismatic animals (e) both a and b are true
5. In ____________, sperm from a male is used to artificially impregnate a female. (a) host mothering (b) artificial insemination (c) in situ conservation (d) biotic pollution (e) seed banks.
6. Which of the following have been linked to the decline of amphibian populations? (a) agricultural chemicals (b) increased UV radiation (c) infectious diseases (d) global climate warming (e) all of these
7. When forests are destroyed (a) soil fertility increases (b) soil erosion decreases (c) some forest organisms decline in number (d) the danger of flooding in nearby lowlands declines (e) carbon dioxide is removed from the atmosphere and stored in forest plants
8. About 60% of tropical deforestation is the result of (a) commercial logging (b) cattle ranching (c) hydroelectric dams (d) mining (e) subsistence agriculture
9. If deforestation of tropical forests continues at present estimated rates, the forests will (a) disappear in a few hundred years (b) disappear in the next 20 years (c) disappear by the first half of the 22nd century (d) regrow and replenish themselves (e) persist into the foreseeable future
10. Which of the following gases contributes to both global warming and thinning of the ozone layer? (a) CO_2 (b) CH_4 (c) surface O_3 (d) CFCs (e) N_2O
11. The first scientific evidence that a species had shifted its geographical range in response to climate change involved a(an) (a) tree (b) butterfly (c) mosquito (d) coral reef dweller (e) soil-dwelling fungus
12. Global warming occurs because (a) carbon dioxide and other greenhouse gases react chemically to produce excess heat (b) Earth has too many greenhouses and other glassed buildings (c) volcanic eruptions produce large quantities of sulfur and other greenhouse gases (d) carbon dioxide and other greenhouse gases trap infrared radiation in the atmos-

phere (e) carbon dioxide and other greenhouse gases allow excess heat to pass out of the atmosphere

13. What gas is a human-made pollutant in the lower (surface) atmosphere but a natural and beneficial gas in the stratosphere? (a) CO_2 (b) CH_4 (c) O_3 (d) CFCs (e) N_2O
14. Where is stratospheric ozone depletion most pronounced? (a) over Antarctica (b) over the equator (c) over South America (d) over North America and Europe (e) over Alaska and Siberia
15. The Montreal Protocol is an agreement to (a) phase out greenhouse gas emissions (b) curtail CFC production (c) design a recovery plan for each endangered species (d) curtail deforestation (e) prevent the selling of products made from endangered or threatened species

REVIEW QUESTIONS

1. Which organism is more likely to become extinct, an endangered species or a threatened species? Explain your answer.
2. How does habitat fragmentation contribute to declining biological diversity?
3. Which type of conservation measure, in situ or ex situ, will help the greatest number of species? Why?
4. State at least three ecosystem services that forests provide.
5. Discuss two reasons for tropical deforestation.
6. Describe the enhanced greenhouse effect and list three or more greenhouse gases.
7. What are some of the significant problems that may be caused by global warming during the 21st century?
8. What environmental service does the stratospheric ozone layer provide?
9. Describe how ozone depletion occurs and discuss some of the consequences of the thinning ozone layer.

YOU MAKE THE CONNECTION

1. If half of the world's biological diversity were to disappear, how would it affect your life?
2. Because new species will eventually evolve to replace those that humans are driving to extinction, why is declining biological diversity such a threat to us?
3. Why does the most recent version of the World Conservation Strategy include stabilizing the human population as one of its goals?
4. Why might captive breeding programs that reintroduce species into natural environments fail?
5. If you were given the task of developing a policy for the United States to deal with global climate change during the next 50 years, what would you propose? Explain your answer.
6. It has been suggested that the wisest way to "use" fossil fuels would be to leave them in the ground. How would this affect global warming? Energy supplies?
7. This statement was overheard in an elevator: "CFCs cannot be the cause of ozone depletion over Antarctica because there are no refrigerators in Antarctica." Criticize the reasoning behind this statement.
8. It has been suggested that a more descriptive name for *Homo sapiens* would be "*Homo dangerous*." Explain this specific epithet, based on what you have learned in this chapter.

RECOMMENDED READINGS

Ashley, M.V. "Molecular Conservation Genetics." *American Scientist,* Vol. 87, Jan.–Feb. 1999. The tools of molecular genetics are helpful in efforts to conserve species.

Climate Change Impacts on the United States: The Potential Consequences of Climate Variability and Change. U.S. Global Change Research Program. Cambridge University Press, New York, 2001. This massive report outlines the impacts of climate changes expected in the United States in the next 100 years.

Cohn, J.P. "Working Outside the Box." *BioScience,* Vol. 50, No. 7, July 2000. Increasingly, zoos are focusing on preserving natural habitats and animals in the wild.

Epstein, P.R. "Is Global Warming Harmful to Health?" *Scientific American,* Vol. 283, No. 2, Aug. 2000. Signs of future troubles with infectious diseases are starting to be recognized.

Hesman, T. "Greenhouse Gassed." *Science News,* Vol. 157, 25 Mar. 2000. Although higher carbon dioxide levels make plants grow faster, they produce more starch and less protein and are therefore not as nutritious to the animals that eat them.

Houlahan, J.E., et al. "Quantitative Evidence for Global Amphibian Population Declines." Nature, Vol. 404, 13 Apr. 2000. This paper examines the results of more than 900 studies of amphibian declines that were conducted between 1950 and 1997.

Lanza, R.P., B.L. Dresser, and P. Damiani. "Cloning Noah's Ark." *Scientific American,* Vol. 283, No. 5, Nov. 2000. Examines the potential of cloning endangered species.

Novacek, M.J., Editor. *The Biodiversity Crisis: Losing What Counts.* New Press, New York, 2001. Numerous highly respected scientists have written the chapters of this book, published by the American Museum of Natural History.

Raven, P.H., and L.R. Berg. *Environment,* 3rd ed. Harcourt College Publishers, Philadelphia, 2001. This environmental science text offers a detailed situation analysis of planet Earth, including how humans interact with and affect its physical and living systems.

Weidensaul, S. "The Rarest of the Rare." *Smithsonian,* Nov. 2000. Considers the captive breeding program at the Smithsonian's Conservation and Research Center in Virginia.

Wuethrich, B. "How Climate Change Alters Rhythms of the Wild." *Science,* Vol. 287, 4 Feb. 2000. Examines some of the many reported effects of global warming on animal species.

- Visit our Web site at **http://www.info.brookscole.com/solomonbergmartin** for links to chapter-related resources on the World Wide Web. Additional on-line materials relating to this chapter can also be found on our Web site.

See chapter activity on BioActive Learner CD for additional help in mastering the chapter's material. Icon location in the chapter's margins shows which topics have tutorials or simulations in the CD.

CAREER VISIONS

Environmental Consultant

SCOTT SMYERS

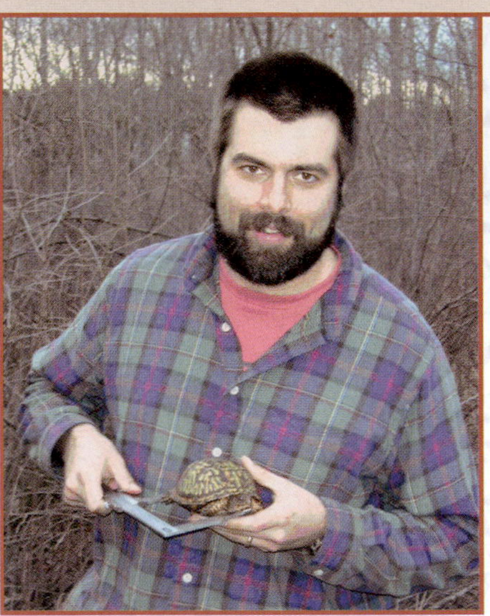

***Scott Smyers,** who was born in Pittsburgh and grew up in Massachusetts, works as a field biologist at Oxbow Associates, in Acton, Massachusetts, a wetlands and wildlife consulting company specializing in protection of rare amphibians and reptiles. Scott received a B.A. in Environmental Science with a biology focus from the University of Massachusetts at Lowell in 1993. He worked five years as an environmental consultant before returning to school and earning his M.S. in Biology from the University of Louisiana at Lafayette in 2000. Scott has studied the behavior and ecology of amphibian and reptile species native to the eastern United States for more than five years. He has recently published a part of his master's research thesis, which explored behavioral interactions within and between two species of juvenile salamanders. Scott is passionate about his work and finds it particularly challenging in Massachusetts, where environmental regulations are very stringent. He provides consultation services for a variety of projects, such as residential developments, roadway improvements, and airport expansions, yet continues to conduct independent ecological research.*

What was the focus of your undergraduate coursework?

I earned an environmental science degree with a focus in biology, which meant that I had to take a certain number of required upper-level biology classes, in addition to those required by the core curriculum. Many of my environmental science core courses—geology, oceanography, environmental geochemistry, hydrogeology—had to do with water and glacial geology.

How and when did you begin working in environmental consulting?

When I was a senior in college, I participated in internships at the state department of environmental protection and at an environmental consulting firm. After I graduated, I applied for full-time positions as an environmental consultant of wetlands, and I ended up getting a job a couple of weeks after graduation, mostly because of my internships, according to the person who hired me.

In general, what is the function of an environmental consultant?

The function of an environmental consultant, from my perspective, is to provide our clients with our expertise on the ecology and biology of an ecological system that may become either directly or indirectly affected by a proposed project or activity. We then intertwine that assessment with the actual laws and regulations governing the particular project. We advise our clients as to how they can proceed on a project, within the laws and regulations, and have a minimal impact to the sensitive area or sensitive system or organism that may be involved.

What types of studies does Oxbow Associates specialize in?

At Oxbow Associates, we are wetlands and wildlife consultants who provide services to regulatory authorities and the private sector alike. We help them either by conducting inventories to find what rare species or habitats may be on their project site or by determining the limit of wetland resource areas as defined by regulations. We then help clients plan their project around those areas, by considering the specific sensitive natures of each area that may be under the authority of local, state, or federal statutes or regulations. We really specialize in the rare amphibians and reptiles in the wetlands of the U.S. Northeast, a niche that sets us apart from other wetlands and environmental consultants.

Who are Oxbow Associates' clients?

For some of the jobs, we work for the regulatory authorities, such as towns that hire us to review another consultant's work. We also often work for developers or builders, whom we encourage to take a proactive role in trying to identify what rare species or sensitive areas could be within their project layout. This effort streamlines the regulatory process substantially.

What are your specific responsibilities?

I contact clients and field requests that come in. After I take on a project, I consult with my coworkers and prepare a proposal that details the services I believe will be required and how much I believe they will cost. I obtain maps of the area we're interested in working in and use computerized site data available through the Internet. I tend to be out in the field quite a bit on projects, often half a day every day, which I really enjoy. My field work includes trapping animals; radiotelemetry, where we monitor the locations of animals that have transmitters on them; wetlands delineation; and general inventories of large sites. I also write a lot of reports.

What steps should a biology major take if he or she wishes to pursue work in environmental consulting?

I would recommend doing as well as you can in all your biology classes. Take as many courses as possible on ecology-related topics. Take opportunities to do field work and go different places. If you work hard at it and take the opportunities when they come, environmental consulting can be a really exciting and rewarding career. You use all areas of biology to understand natural systems, because the disciplines are connected.

Appendix A

Post-Test Answers

CHAPTER 1

1. a 2. a 3. e 4. c 5. a 6. b 7. a
8. c 9. c 10. b 11. a 12. b 13. b 14. d
15. c 16. b 17. b

CHAPTER 2

1. c 2. d 3. e 4. a 5. b 6. c 7. d 8. a
9. e 10. a 11. d 12. e

CHAPTER 3

1. a 2. d 3. b 4. e 5. e 6. d 7. c 8. c
9. a 10. d 11. a 12. e 13. c 14. b

CHAPTER 4

1. b 2. a 3. d 4. d 5. d 6. a 7. e 8. d
9. a 10. a 11. b 12. a 13. a 14. e 15. d

CHAPTER 5

1. d 2. c 3. c 4. c 5. a 6. b 7. a 8. c
9. b 10. b 11. b 12. e 13. c 14. e

CHAPTER 6

1. a 2. c 3. c 4. d 5. b 6. d 7. b 8. d
9. e 10. e 11. a 12. e 13. c 14. b 15. a

CHAPTER 7

1. c 2. a 3. d 4. e 5. a 6. a 7. b 8. c
9. c 10. a 11. c 12. d 13. e 14. b 15. c

CHAPTER 8

1. a 2. d 3. c 4. a 5. a 6. c 7. a 8. e
9. e 10. a 11. c 12. b 13. c 14. a 15. c

CHAPTER 9

1. d 2. a 3. d 4. e 5. c 6. a 7. e 8. d
9. b 10. d 11. c 12. b

CHAPTER 10

1. c 2. d 3. c 4. c 5. d 6. b 7. e 8. c
9. e 10. c 11. b 12. a

Chapter 10 Review Questions

1. a. all yellow b. $\frac{1}{2}$ yellow: $\frac{1}{2}$ green c. all yellow
   d. $\frac{3}{4}$ yellow: $\frac{1}{4}$ green
2. 150
3. The short-winged condition is recessive. Both parents are heterozygous.
4. The blue-eyed man is homozygous recessive; his brown-eyed wife is heterozygous.
5. Repeated matings of the roan bull and the white cow will yield an approximate 1 : 1 ratio of roan-to-white offspring. Repeated matings among roan offspring will yield red, roan, and white offspring in an approximate 1 : 2 : 1 ratio. The mating of two red individuals will yield only red offspring.
6. There are 36 possible outcomes when a pair of dice is rolled. There are six ways of obtaining a seven: 1,6; 6,1; 2,5; 5,2; 3,4; and 4,3. There are five ways of rolling a six: 1,5; 5,1; 2,4; 4,2; and 3,3. There are also five ways of rolling an eight: 2,6; 6,2; 3,5; 5,3; and 4,4.
7. The genotype of the brown spotted rabbit is *bbSS*. The genotype of the black, solid rabbit is *BBss*. An F_1 rabbit would be black, spotted *(BbSs)*. The F_2 is expected to be 9/16 black, spotted *(B_S_)*, 3/16 black, solid *(B_ss)*, 3/16 brown, spotted *(bbS_)*, and 1/16 brown, solid *(bbss)*.
8. The F_1 offspring are expected to be black, short-haired. There is a 1/16 chance of brown and tan, long-haired offspring in the F_2 generation.
9. Yes
10. Pleiotropic
11. The rooster is *PpRr;* hen A is *PpRR;* hen B is *Pprr;* and hen C is *PPRR*.
12. a. Both parents are heterozygous for a single locus. b. One parent is heterozygous and the other parent homozygous recessive (for a single locus). c. Both parents are heterozygous (for two loci) d. One parent is heterozygous and the other is homozygous recessive (for two loci).

13. The expected types of F_2 plants are: 1/16 of the plants produce 2 kg fruits; $\frac{1}{4}$ of the plants produce 1.75 kg fruits, $\frac{3}{8}$ of the plants produce 1.5 kg fruits, $\frac{1}{4}$ of the plants produce 1.25 kg fruits, and 1/16 of the plants produce 1 kg fruits.
14. All male offspring of this cross are barred, and all females are nonbarred, thus allowing the sex of the chicks to be determined by their phenotypes.
15. This is a two-point test cross involving linked loci. The parental class of offspring are *Aabb* and *aaBb;* the recombinant classes are *AaBb* and *aabb*. There is 4.6% recombination, which corresponds to 4.6 map units between the loci.
16. If genes B and C are 10 map units apart, gene A is in the middle. If genes B and C are 2 map units apart, gene C is in the middle.

CHAPTER 11

1. d 2. d 3. b 4. e 5. a 6. d 7. c 8. e 9. b 10. a 11. c

CHAPTER 12

1. e 2. d 3. c 4. b 5. a 6. e 7. c 8. a 9. b 10. d 11. e

CHAPTER 13

1. a 2. c 3. a 4. e 5. c 6. a 7. c 8. a 9. b 10. c 11. a 12. b 13. c

CHAPTER 14

1. a 2. b 3. e 4. d 5. a 6. b 7. a 8. b 9. c 10. a 11. b

CHAPTER 15

1. d 2. e 3. a 4. c 5. c 6. a 7. e 8. d 9. a 10. a 11. d 12. c 13. e 14. e 15. d

CHAPTER 16

1. c 2. d 3. a 4. b 5. a 6. c 7. b 8. d 9. e 10. b 11. c 12. d

CHAPTER 17

1. d 2. b 3. c 4. b 5. c 6. b 7. a 8. c 9. d 10. b 11. e 12. b 13. b 14. b 15. d

CHAPTER 18

1. b 2. b 3. b 4. d 5. b 6. b 7. c 8. d 9. a 10. e 11. e 12. d 13. b 14. c 15. b

Chapter 18 Review Questions

1. 0.6 2. 0.25; 0.5 3. 0.8 4. 0.4; 0.6 5. No. If it were at genetic equilibrium, the genotype frequency of *Aa* would be $2pq = 0.48$. 6. Frequency of $t = 0.55$; frequency of $T = 0.45$ 7. 0.3 8. $T = 0.6$; $t = 0.4$ 9. $T = 0.8$; $t = 0.2$ 10. 0.42

CHAPTER 19

1. e 2. b 3. a 4. a 5. c 6. b 7. d 8. b 9. b 10. c 11. b 12. e 13. c 14. e 15. d

CHAPTER 20

1. a 2. c 3. b 4. d 5. c 6. d 7. c 8. a 9. d 10. a 11. b 12. c 13. b 14. c 15. c

CHAPTER 21

1. c 2. a 3. e 4. d 5. a 6. b 7. b 8. e 9. c 10. a 11. b 12. b 13. d 14. c 15. c

CHAPTER 22

1. b 2. c 3. a 4. d 5. b 6. b 7. e 8. c 9. a 10. a 11. a 12. a 13. b 14. e 15. a

CHAPTER 23

1. a 2. e 3. a 4. b 5. c 6. a 7. d 8. b 9. b 10. a 11. b 12. a 13. d 14. c 15. e 16. c 17. b 18. e 19. d 20. a 21. b 22. d

CHAPTER 24

1. c 2. a 3. c 4. d 5. b 6. d 7. d 8. e 9. b 10. b 11. e 12. d 13. a 14. e 15. c

CHAPTER 25

1. a 2. b 3. e 4. d 5. d 6. a 7. b 8. c 9. a 10. c 11. c 12. c 13. e 14. b 15. a

CHAPTER 26

1. d 2. a 3. b 4. c 5. d 6. b 7. a 8. e 9. a 10. c 11. d 12. c 13. e 14. c 15. d 16. b 17. d 18. e 19. a 20. b 21. b 22. e 23. d 24. c 25. e

CHAPTER 27

1. e 2. b 3. c 4. d 5. a 6. e 7. d 8. d 9. c 10. b 11. d 12. a 13. c 14. a 15. c

CHAPTER 28

1. d 2. d 3. e 4. b 5. e 6. a 7. c 8. c
9. b 10. a 11. a 12. b 13. a 14. c 15. c

CHAPTER 29

1. b 2. c 3. a 4. b 5. e 6. b 7. d 8. e
9. c 10. b 11. c 12. d 13. a 14. d 15. b
16. b 17. a 18. e 19. e

CHAPTER 30

1. d 2. b 3. a 4. a 5. b 6. e 7. c 8. b
9. a 10. d 11. c 12. c 13. a 14. d 15. c

CHAPTER 31

1. c 2. a 3. d 4. e 5. b 6. c 7. c 8. b
9. e 10. c 11. b 12. a 13. c 14. b 15. d

CHAPTER 32

1. c 2. b 3. d 4. d 5. b 6. e 7. c 8. a
9. d 10. d 11. a 12. e 13. c 14. c 15. a

CHAPTER 33

1. e 2. b 3. d 4. c 5. d 6. e 7. b 8. a
9. c 10. d 11. e 12. c 13. b 14. a 15. b

CHAPTER 34

1. c 2. b 3. b 4. d 5. b 6. e 7. a 8. c
9. c 10. d 11. d 12. a 13. d 14. b 15. e

CHAPTER 35

1. e 2. c 3. d 4. e 5. c 6. a 7. c 8. b
9. c 10. b 11. d 12. c 13. a 14. d 15. a

CHAPTER 36

1. d 2. b 3. e 4. c 5. e 6. a 7. d 8. a
9. a 10. b 11. c 12. c 13. e 14. e 15. b

CHAPTER 37

1. b 2. b 3. a 4. c 5. b 6. c 7. a 8. e
9. c 10. e 11. a 12. d 13. b 14. d 15. a
16. a 17. e

CHAPTER 38

1. d 2. a 3. c 4. c 5. a 6. b 7. d 8. b
9. a 10. e 11. a 12. c 13. e 14. b 15. d

CHAPTER 39

1. c 2. e 3. a 4. b 5. e 6. d 7. e 8. e
9. a 10. c 11. c 12. c 13. c 14. b 15. d

CHAPTER 40

1. b 2. d 3. c 4. b 5. d 6. a 7. a 8. d
9. a 10. c 11. e 12. d 13. b 14. b 15. d
16. a

CHAPTER 41

1. e 2. c 3. a 4. d 5. e 6. c 7. b 8. a
9. e 10. d 11. b 12. a 13. c 14. b 15. c
16. b

CHAPTER 42

1. b 2. d 3. b 4. a 5. b 6. c 7. e 8. e
9. a 10. c 11. e 12. d 13. a 14. c 15. d

CHAPTER 43

1. b 2. a 3. d 4. c 5. e 6. c 7. a 8. b
9. c 10. a 11. d 12. a 13. c 14. e 15. d
16. b 17. a

CHAPTER 44

1. c 2. a 3. b 4. a 5. c 6. b 7. b 8. a
9. d 10. e 11. b 12. a 13. e 14. c 15. d

CHAPTER 45

1. a 2. a 3. c 4. d 5. d 6. e 7. e 8. b
9. b 10. d 11. a 12. b 13. c 14. c 15. a

CHAPTER 46

1. c 2. b 3. a 4. c 5. d 6. e 7. b
8. e 9. d 10. d 11. c 12. a 13. b 14. d
15. e

CHAPTER 47

1. b 2. c 3. a 4. d 5. c 6. e 7. b
8. a 9. d 10. e 11. b 12. d 13. a 14. e
15. c

CHAPTER 48

1. b 2. c 3. a 4. e 5. b 6. d 7. a
8. c 9. e 10. d 11. c 12. a 13. e 14. b
15. e

CHAPTER 49

1. d 2. a 3. a 4. d 5. b 6. e 7. b
8. b 9. c 10. b 11. c 12. c 13. d

CHAPTER 50

1. c 2. b 3. a 4. e 5. c 6. e 7. a
8. a 9. b 10. a 11. b 12. a 13. e 14. a
15. b 16. b 17. a 18. e

CHAPTER 51

1. b 2. b 3. a 4. e 5. a 6. c 7. b
8. a 9. d 10. e 11. a 12. b 13. b 14. e
15. c

Chapter 51 Review Questions 10 to 13

10. Netherlands. (The density of the Netherlands is 16 million ÷ 990 mi^2 = 16,161 people/mi^2. The density of the United States is 284.5 million ÷ 3,615,200 mi^2 = 78 people/mi^2.)

11. The 2002 population of India will be 1050.5 million.

12. The 2002 world death rate was 9 per 1000 people.

13.

TABLE 51–A Life Table

Age Interval (Years)	Number Alive at Beginning of Age Interval	Proportion Alive at Beginning of Age Interval	Proportion Dying During Age Interval	Death Rate for Age Interval
0–1	1000	1.000	0.790	0.790
1–2	210	0.210	0.110	0.524
2–3	100	0.100	0.054	0.540
3–4	46	0.046	0.046	1.000
4–5	0	0.000	0.000	

CHAPTER 52

1. d 2. d 3. d 4. d 5. e 6. c 7. c 8. d
9. a 10. d 11. e 12. d 13. e 14. d 15. d

CHAPTER 53

1. c 2. d 3. e 4. d 5. a 6. d 7. e 8. e
9. d 10. b 11. c 12. d 13. e 14. d 15. b

CHAPTER 54

1. c 2. b 3. a 4. d 5. e 6. c 7. b 8. d
9. b 10. a 11. b 12. c 13. a 14. e 15. e

CHAPTER 55

1. d 2. e 3. c 4. c 5. b 6. e 7. c 8. e
9. c 10. d 11. b 12. d 13. c 14. a 15. b

Appendix B

Periodic Table of the Elements

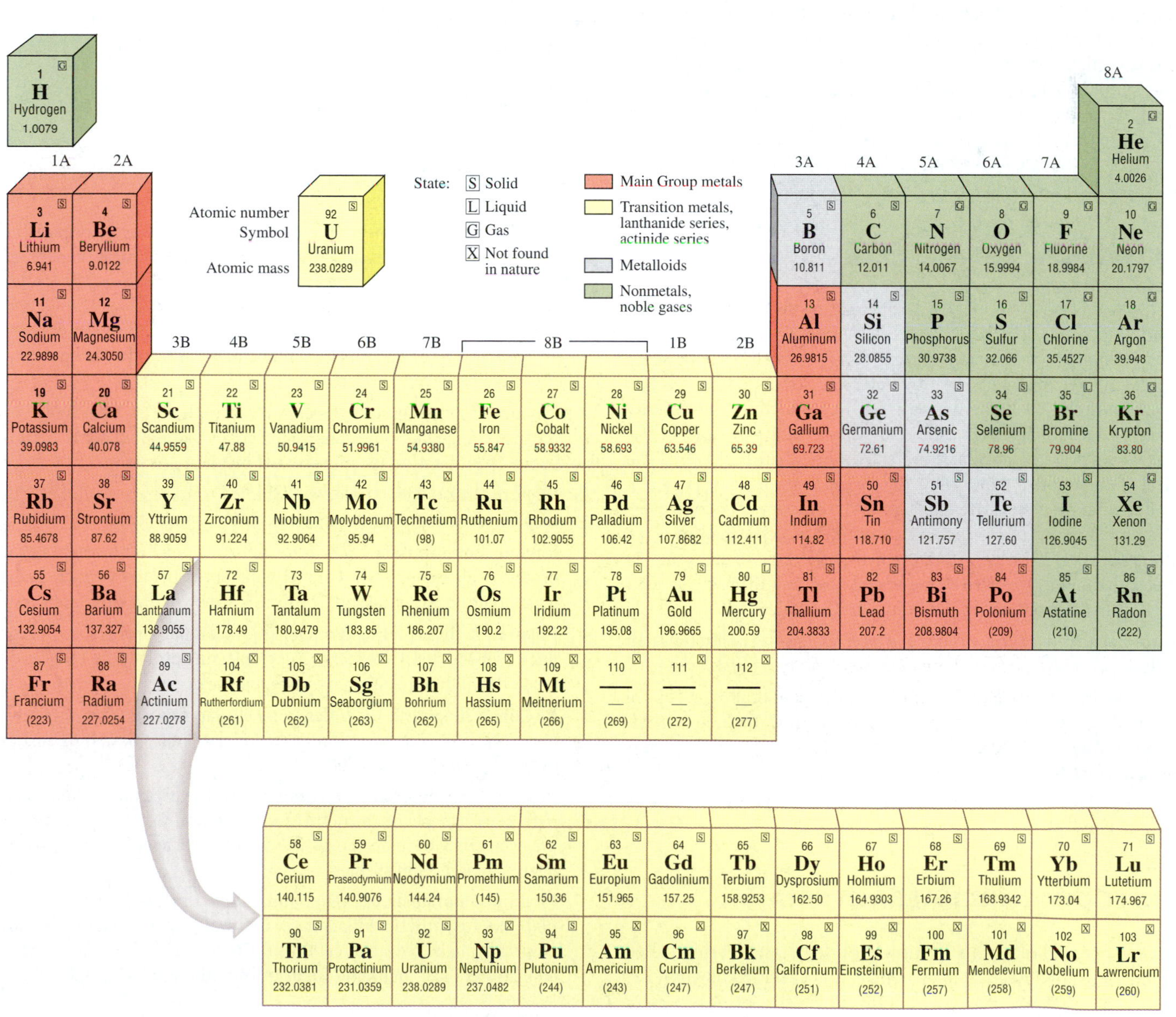

Appendix C

The Classification of Organisms

The system of cataloging organisms used in this book is described in Chapter 1 and in Part 5. In this edition of *Biology,* we use a six-kingdom classification: Eubacteria, Archaebacteria, Protista, Fungi, Plantae, and Animalia. Prokaryotic organisms, distinguished from eukaryotes by their smaller ribosomes and absence of a nuclear envelope and other membranous organelles, are classified in kingdoms Eubacteria and Archaebacteria. Because many biologists prefer the three-domain approach, we also discuss the domains: **Archaea** (which corresponds to kingdom Archaebacteria), **Eubacteria** (or simply **Bacteria**), and **Eukarya** (eukaryotes). We have omitted many groups, especially extinct ones, to simplify the vast number of diverse categories of living organisms and their relationships to one another. Note that we have omitted the viruses from this survey, because they do not fit into any of the six kingdoms.

KINGDOM EUBACTERIA

Very large, diverse group of prokaryotic organisms (lack nuclear envelopes, mitochondria, and other membranous organelles; ribosomes smaller than in eukaryotes). Typically unicellular, but some form colonies or filaments. Mainly heterotrophic, but some groups are photosynthetic or chemosynthetic. Reproduction is primarily asexual by fission. Bacteria are nonmotile or move by beating flagella. When present, flagella are solid (rather than the 9 + 2 type typical of eukaryotes). More than 10,000 species.

Bacterial nomenclature and taxonomic practices are controversial and changing. In one system, eubacteria are classified in three main groups: (1) bacteria with a gram-negative type of cell wall, (2) bacteria with a gram-positive type of cell wall, and (3) bacteria with no cell walls (mycoplasmas). See Chapter 23, Table 23–3.

Gram-negative bacteria. Thin cell wall containing peptidoglycan. Include *proteobacteria, chlamydias, spirochetes, cyanobacteria.*

Gram-positive bacteria. Thick cell wall of peptidoglycan. All are nonphotosynthetic, and many produce spores. Include *actinomycetes, lactic acid bacteria, mycobacteria, streptococci, staphylococci, clostridia.*

Bacteria that lack a rigid cell wall. Extremely small bacteria bounded by a plasma membrane. *Mycoplasmas.*

KINGDOM ARCHAEBACTERIA

Prokaryotes with cell walls lacking peptidoglycan. Distinguished by their ribosomal RNA, lipid structure, and specific enzymes. Archaebacteria are found in extreme environments—hot springs, sea vents, dry and salty seashores, boiling mud, and near ash-ejecting volcanoes. Three main groups (considered phyla by some microbiologists) are *methanogens, extreme halophiles,* and *extreme thermophiles.*

Methanogens. Anaerobes that produce methane gas from simple carbon compounds.

Extreme Halophiles. Inhabit saturated salt solutions.

Extreme Thermophiles. Grow at 70°C or higher, some thrive above the boiling point.

KINGDOM PROTISTA

Primarily unicellular or simple multicellular eukaryotic organisms that do not form tissues and that exhibit relatively little division of labor. Most modes of nutrition occur in this kingdom. Life cycles may include both sexually and asexually reproducing phases and may be extremely complex, especially in parasitic forms. Locomotion is by cilia, flagella, amoeboid movement, or by other means. Flagella and cilia have 9 + 2 structure.

Phylum Zoomastigina. *Flagellates.* Single cells that move by means of flagella. Some free-living; many symbiotic; some pathogenic. Reproduction usually asexual by binary fission.

Phylum Rhizopoda. *Amoebas.* Shelled or naked unicellular protists whose movement is associated with pseudopodia.

Phylum Foraminifera. *Foraminiferans.* Unicellular protists that produce calcareous tests (shells) with pores through which cytoplasmic projections extend, forming a sticky net to entangle prey.

Phylum Actinopoda. *Actinopods.* Unicellular protists that produce axopods (long, filamentous cytoplasmic projections) that protrude through pores in their siliceous shells.

Phylum Ciliophora. *Ciliates.* Unicellular protists that move by means of cilia. Reproduction is asexual by binary fission or sexual by conjugation. About 7200 species.

Phylum Apicomplexa. *Apicomplexans.* Parasitic unicellular protists that lack specific structures for locomotion. At some stage in life cycle, they develop spores (small infective agents). Some pathogenic. About 3900 species.

Phylum Dinoflagellata. *Dinoflagellates.* Unicellular (some colonial), photosynthetic, biflagellate. Cell walls, composed of overlapping cell plates, contain cellulose. Contain chlorophylls *a* and *c* and carotenoids, including fucoxanthin. About 2100 species.

Phylum Bacillariophyta. *Diatoms*. Unicellular (some colonial), photosynthetic. Most nonmotile, but some move by gliding. Cell walls composed of silica rather than cellulose. Contain chlorophylls *a* and *c* and carotenoids, including fucoxanthin. About 5600 species.

Phylum Chrysophyta. *Golden algae*. Unicellular (some colonial), photosynthetic, biflagellate (some lack flagella). Cells covered by tiny scales of either silica or calcium carbonate. Contain chlorophylls *a* and *c* and carotenoids, including fucoxanthin. About 500 species.

Phylum Euglenophyta. *Euglenoids*. Unicellular, photosynthetic, two flagella (one of them very short). Flexible outer covering. Contain chlorophylls *a* and *b* and carotenoids. About 1000 species.

Phylum Chlorophyta. *Green algae*. Unicellular, colonial, siphonous, and multicellular forms. Some motile and flagellated. Photosynthetic; contain chlorophylls *a* and *b* and carotenoids. About 7000 species.

Phylum Rhodophyta. *Red algae*. Most multicellular (some unicellular), mainly marine. Some (coralline algae) have bodies impregnated with calcium carbonate. No motile cells. Photosynthetic; contain chlorophyll *a*, carotenoids, phycocyanin, and phycoerythrin. About 4000 species.

Phylum Phaeophyta. *Brown algae*. Multicellular, often quite large (kelps). Photosynthetic; contain chlorophylls *a* and *c* and carotenoids, including fucoxanthin. Biflagellate reproductive cells. About 1500 species.

Phylum Myxomycota. *Plasmodial slime molds*. Spend part of life cycle as a thin, streaming, multinucleate plasmodium that creeps along on decaying leaves or wood. Flagellated or amoeboid reproductive cells; form spores in sporangia. About 500 species.

Phylum Acrasiomycota. *Cellular slime molds*. Vegetative (nonreproductive) form unicellular; move by pseudopods. Amoeba-like cells aggregate to form a multicellular pseudoplasmodium that eventually develops into a fruiting body that bears spores. About 70 species.

Phylum Oomycota. *Water molds*. Consist of branched, coenocytic mycelia. Cellulose and/or chitin in cell walls. Produce biflagellate asexual spores. Sexual stage involves production of oospores. Some parasitic. About 580 species.

KINGDOM FUNGI

All eukaryotic, mainly multicellular organisms that are heterotrophic with saprotrophic or parasitic nutrition. Body form often a mycelium, and cell walls consist of chitin. No flagellated stages except in chytrids. Reproduce by means of spores, which may be produced sexually or asexually. Cells usually haploid or dikaryotic, with brief diploid period following fertilization.

Phylum Chytridiomycota. *Chytridiomycetes* or *chytrids*. Parasites and decomposers found principally in fresh water. Motile cells (gametes and zoospores) contain a single, posterior flagellum. Reproduce both sexually and asexually. About 1000 species.

Phylum Zygomycota. *Zygomycetes*. Produce sexual resting spores called zygospores, and asexual spores in a sporangium. Hyphae are coenocytic. Many are heterothallic (two mating types). About 800 species.

Phylum Ascomycota. *Ascomycetes* or *sac fungi*. Sexual reproduction involves formation of ascospores in little sacs called asci. Asexual reproduction involves production of spores called conidia, which pinch off from conidiophores. Hyphae usually have perforated septa. About 30,000 species.

Phylum Basidiomycota. *Basidiomycetes* or *club fungi*. Sexual reproduction involves formation of basidiospores on a basidium. Asexual reproduction uncommon. Heterothallic. Hyphae usually have perforated septa. About 25,000 species.

Phylum Deuteromycota. *Deuteromycetes* or *imperfect fungi*. Sexual stage has not been observed. Most reproduce only by conidia. About 25,000 species.

KINGDOM PLANTAE

Multicellular eukaryotic organisms with differentiated tissues and organs. Cell walls contain cellulose. Cells frequently contain large vacuoles; photosynthetic pigments in plastids. Photosynthetic pigments are chlorophylls *a* and *b*, and carotenoids. Nonmotile. Reproduce both asexually and sexually, with alternation of gametophyte (n) and sporophyte ($2n$) generations.

Phylum Bryophyta. *Mosses*. Nonvascular plants that lack xylem and phloem. Marked alternation of generations with dominant gametophyte generation. Motile sperm. The gametophytes generally form a dense green mat consisting of individual plants. About 9000 species.

Phylum Hepatophyta. *Liverworts*. Nonvascular plants that lack xylem and phloem. Marked alternation of generations with dominant gametophyte generation. Motile sperm. The gametophytes of certain species have a flat, liver-like thallus; other species are more mosslike in appearance. About 6000 species.

Phylum Anthocerotophyta. *Hornworts*. Nonvascular plants that lack xylem and phloem. Marked alternation of generations with dominant gametophyte generation. Motile sperm. The gametophyte is a small, flat green thallus with scalloped edges. Spores are produced on an erect, hornlike stalk. About 100 species.

Phylum Pterophyta. *Ferns*. Vascular plants with a dominant sporophyte generation. Generally homosporous. Gametophyte is free-living and photosynthetic. Reproduce by spores. Motile sperm. About 11,000 species.

Phylum Psilotophyta. *Whisk ferns*. Vascular plants with a dominant sporophyte generation. Homosporous. Stem is distinctive because it branches dichotomously; plant lacks true roots and leaves. The gametophyte is subterranean and nonphotosynthetic and forms a mycorrhizal relationship with a fungus. Motile sperm. About 12 species.

Phylum Sphenophyta. *Horsetails*. Vascular plants with hollow, jointed stems and reduced, scalelike leaves. Although modern representatives are small, some extinct species were treelike. Homosporous. Gametophyte a tiny photosynthetic plant. Motile sperm. About 15 species.

Phylum Lycophyta. *Club mosses*. Sporophyte plants are vascular with branching rhizomes and upright stems that bear microphylls. Although modern representatives are small, some extinct species were treelike. Some homosporous, others heterosporous. Motile sperm. About 1000 species.

Phylum Coniferophyta. *Conifers*. Heterosporous vascular plants with woody tissues (trees and shrubs) and needle-shaped or scale-like leaves. Most are evergreen. Seeds are usually borne naked on the surface of cone scales. Nutritive tissue in the seed is haploid female gametophyte tissue. Nonmotile sperm. About 550 species.

Phylum Cycadophyta. *Cycads*. Heterosporous, vascular, dioecious plants that are small and shrubby or larger and palmlike. Produce naked seeds in conspicuous cones. Flagellated sperm. About 140 species.

Phylum Ginkgophyta. *Ginkgo*. Broad-leaved deciduous trees that bear naked seeds directly on branches. Dioecious. Contain vascular tissues. Flagellated sperm. The ginkgo tree is the only living representative. One species.

Phylum Gnetophyta. *Gnetophytes*. Woody shrubs, vines, or small trees that bear naked seeds in cones. Contain vascular tissues. Possess many features similar to flowering plants. About 70 species.

Phylum Anthophyta. *Flowering plants* or *angiosperms*. Largest, most successful group of plants. Heterosporous; dominant sporophytes with extremely reduced gametophytes. Contain vascular tissues. Bear flowers, fruits, and seeds (enclosed in a fruit; seeds contain endosperm as nutritive tissue). Double fertilization. More than 235,000 species.

KINGDOM ANIMALIA

Multicellular eukaryotic heterotrophs with differentiated cells. In most animals, cells are organized to form tissues; tissues are organized to form organs; and tissues and organs form specialized body systems that carry on specific functions. Most animals have a well-developed nervous system and can respond rapidly to changes in their environment. Most are capable of locomotion during some time in their life cycle. Most animals are diploid and reproduce sexually with eggs and sperm. A flagellated haploid sperm unites with a large, nonmotile haploid egg to form a zygote that undergoes cleavage.

Phylum Porifera. *Sponges*. Mainly marine. Body bears many pores through which water circulates. Food is filtered from the water by collar cells (choanocytes). Solitary or colonial. Asexual reproduction by budding; external sexual reproduction in which sperm are released and swim to internal egg. Larva is motile. About 10,000 species.

Phylum Cnidaria. *Hydras, jellyfish, sea anemones, corals*. Radial symmetry. Tentacles surrounding mouth. Stinging cells (cnidocytes) contain stinging structures called nematocysts. Polyp and medusa forms. Planula larva. Solitary or colonial. Marine, with a few freshwater forms. About 10,000 species.

Phylum Ctenophora. *Comb jellies*. Biradial symmetry. Free-swimming; marine. Two tentacles and eight longitudinal rows of cilia resembling combs; animal moves by means of these bands of cilia. About 100 species.

Phylum Platyhelminthes. *Flatworms*. Acoelomate (no body cavity); region between body wall and internal organs filled with tissue. Planarians are free-living; flukes and tapeworms are parasitic. Body dorsoventrally flattened; cephalization; three tissue layers. Simple nervous system with ganglia in head region. Excretory organs are protonephridia with flame cells. About 20,000 species.

Coelomates

These animals have a true coelom, a body cavity completely lined with mesoderm. Two evolutionary branches of coelomates are protostomes and deuterostomes.

Protostomes

Coelomates with spiral, determinate cleavage; mouth typically develops from blastopore. Two branches of protostomes are Lophotrochozoa and Ecdysozoa.

Lophotrochozoa

Phylum Nemertea. *Proboscis worms* (also called ribbon worms). Long, dorsoventrally flattened body with complex proboscis used for defense and for capturing prey. Definite organ systems. Complete digestive tract. Circulatory system with blood. Functionally acoelomate, but do have a small true coelom in the proboscis. About 900 species.

Phylum Mollusca. *Snails, clams, squids, octopods*. Unsegmented, soft-bodied animals usually covered by a dorsal shell. Have a ventral, muscular foot. Most organs located above foot in visceral mass. A shell-secreting mantle covers the visceral mass and forms a mantle cavity, which contains gills. Trochophore and/or veliger larva. About 50,000 species.

Phylum Annelida. *Segmented worms*. Polychaetes, earthworms, leeches. Both body wall and internal organs are segmented. Body segments separated by septa. Some have nonjointed appendages. Setae used in locomotion. Closed circulatory system; metanephridia; specialized regions of digestive tract. Trochophore larva. About 15,000 species.

Phylum Brachiopoda. *Lamp shells*. One of the lophophorate phyla. Marine; body enclosed between two shells. About 350 species.

Phylum Phoronida. One of the lophophorate phyla. Tube-dwelling marine worms. About 12 species.

Phylum Bryozoa. One of the lophophorate phyla. Mainly marine; sessile colonies produced by asexual budding. About 5000 species.

Ecdysozoa

Phylum Rotifera. *Wheel animals*. Microscopic, wormlike animals. Anterior end has ciliated crown that looks like a wheel when the cilia beat. Posterior end tapers to a foot. Characterized by pseudocoelom (body cavity not completely lined with mesoderm). Constant number of cells. About 1800 species.

Phylum Nematoda. *Roundworms. Ascaris,* hookworms, pinworms. Slender, elongated, cylindrical worms; covered with cuticle. Characterized by pseudocoelom. Free-living and parasitic forms. About 15,000 species.

Phylum Arthropoda. *Arachnids (spiders, mites, ticks), crustaceans (lobsters, crabs, shrimp), insects, centipedes, millipedes*. Segmented animals with paired, jointed appendages and a hard exoskeleton made of chitin. Open circulatory system with dorsal heart. Hemocoel occupies most of body cavity, and coelom is reduced. More than 1 million species.

Deuterostomes

Coelomates with radial, indeterminate cleavage. Blastopore develops into anus, and mouth forms from a second opening.

Phylum Echinodermata. *Sea stars, sea urchins, sand dollars, sea cucumbers.* Marine animals that have pentaradial symmetry as adults but bilateral symmetry as larvae. Endoskeleton of small, calcareous plates. Water vascular system; tube feet for locomotion. About 7000 species.

Phylum Hemichordata. *Acorn worms.* Marine animals with an anterior muscular proboscis, connected by a collar region to a long wormlike body. The larval form resembles an echinoderm larva. About 85 species.

Phylum Chordata. *Subphylum Urochordata (tunicates), subphylum Cephalochordata (lancelets), subphylum Vertebrata (fishes, amphibians, reptiles, birds, mammals).* Notochord; pharyngeal gill slits; dorsal, tubular nerve cord, and postanal tail present at some time in life cycle. About 48,000 species.

Appendix D

Careers in Biology

The following organizations, professional associations, and World Wide Web sites provide career information upon request.

GENERAL

American Association for the Advancement of Science

1200 New York Avenue, NW
Washington, DC 20005
202-326-6400
Web Site: http://www.aaas.org
E-mail: info@aaas.org

American Institute of Biological Sciences

Communications Office
1444 Eye Street NW, Suite 200
Washington, DC 20005
202-628-1500 (ext. 261)
Web Site: http://www.aibs.org/careers/index.html
E-mail: cmoulton@aibs.org

Association for Tropical Biology

Executive Director, W. John Kress
Department of Botany MRC-166
National Museum of Natural History
Smithsonian Institute
Washington, DC 20560
202-357-3392
(Fax) 202-786-2563
Web Site: http://atb.botany.ufl.edu
E-mail: kressj@nmnh.si.edu

Association for Women in Science (AWIS)

1200 New York Avenue, NW
Suite 650
Washington, DC 20005
202-326-8940
(Fax) 202-326-8960
Web Site: http://www.awis.org
E-mail: awis@awis.org

Biology Careers Page

Web Site: http://furman.careerhighway.com

Bio Online

Web Site: http://www.bio.com
E-mail: jobs@bio.com

Bioscience Links

Web Site: http://mcb.harvard.edu/BioLinks.html

Federation of American Societies for Experimental Biology

9650 Rockville Pike
Bethesda, MD 20814-3998
301-530-7020
(Fax) 301-571-0699
Web Site: http://www.faseb.org
E-mail: careers@faseb.org

JobHunt: Science, Engineering, and Medicine

Web Site: http://www.job-hunt.org/science.html

National Academy of Science, National Academy of Engineering, Institute of Medicine, National Research Council: A Career Planning Center

Web Site: http://www.nas.edu/careers

Pursuit

Pursuing careers in science, engineering, and mathematics for persons with disabilities
Web Site: http://rehab.uiuc.edu/career/library.html

Women in Technology International (WITI)

6345 Balboa Boulevard, #257
Encino, CA 91316
818-334-WITI
(Fax) 818-342-9891
Web Site: http://www.witi.com

AGRICULTURE/AGRONOMY

Careers in Agriculture and Agri-Food
Web Site: http://www.cfa-fca.ca/careers/index1.html

American Society of Agronomy

667 South Segoe Road
Madison, WI 53711
608-273-8080 (ext. 314)
Web Site: http://www.agronomy.org
E-mail: Lmalison@agronomy.org (Leann Malison)

ART AND COMMUNICATIONS

Association of Medical Illustrators

2965 Flowers Road South, Suite 105
Atlanta, GA 30341
770-454-7933
(Fax) 770-458-3314
Web Site: http://www.medical-illustrators.org
E-mail: assnhq@mindspring.com

Bio Communications Association

115 Stoneridge Drive
Chapel Hill, NC 27514-9737
(Phone/Fax) 919-967-8246
Web Site: http://www.bca.org
E-mail: BCAoffice@aol.com

Medical Library Association

65 East Wacker Place, Suite 1900
Chicago, IL 60602-4805
312-419-9094
(Fax) 312-419-8950
Web Site: http://www.mlanet.org
E-mail: mlapd1@mlahq.org

National Association of Science Writers

P.O. Box 294
Greenlawn, NY 11740
516-757-5664
(Fax) 516-757-0069
Web Site: http://www.nasw.org
E-mail: diane@nasw.org

BIOCHEMISTRY

American Society for Biochemistry and Molecular Biology (ASBMB)

9650 Rockville Pike
Bethesda, MD 20814-3996
Web Site: http://www.faseb.org/asbmb
E-mail: aps@scisoc.org

BIOMEDICAL ENGINEERING

Biomedical Engineering Society

8401 Corporate Drive, Suite 110
Landover, MD 20785-2224
301-459-1999
(Fax) 301-459-2444
Web Site: http://mecca.org/BME/BMES/society/index.htm
E-mail: bmes@netcom.com

BIOMEDICAL SCIENCE

Employment Links for the Biomedical Scientist

Web Site: http://www.his.com/~graeme/employ.html

BIOPHYSICS

The Biophysical Society

9650 Rockville Pike
Bethesda, MD 20814
301-530-7114
(Fax) 301-530-7133
Web Site: http://www.biophysics.org/biophys/society/biohome.htm
E-mail: society@biophysics.faseb.org

BOTANY

American Phytopathological Society

3340 Pilot Knob Road
St. Paul, MN 55121-2097
612-454-7250
(Fax) 612-454-0766
Web Site: http://www.apsnet.org

American Society of Plant Physiologists

15501 Monona Drive
Rockville, MD 20855-2768
301-251-0560
(Fax) 301-279-2996
Web Site: http://www.aspp.org
E-mail: aspp@aspp.org

Botanical Society of America

Kim Hiser, Business Manager
1735 Neil Avenue
Columbus, OH 43210-1293
614-292-3519
Web Site: http://www.botany.org

Mycological Society of America

Attention: Linda Hartwick
Allen Marketing and Management
810 E. 10th Street
Lawrence, KS 66044
785-843-1235
(Fax) 785-843-1274
Web Site: http://www.erin.utoronto.ca/~w3msa
E-mail: lhardwick@allenpress.com

Phycological Society of America

Blackwell Science, Inc.
350 Main Street, Commerce Place
Malden, MA 02148
718-388-8250
(Fax) 781-388-8270
Web Site: http://www.psaalgae.org
E-mail: csjournal@blacksci.com

CELL BIOLOGY

American Society for Cell Biology (ASCB)

8120 Woodmont Avenue, Suite 750
301-347-9300 (ext. 2755)
(Fax) 301-347-9310
Web Site: http://www.ascb.org
E-mail: ascbinfo@ascb.org

Cell & Molecular Biology Online
Web Site: http://www.cellbio.com

DENTISTRY

American Association of Dental Schools
202-667-9433
(Fax) 202-667-0642
Web Site: http://www.aads.jhu.edu
E-mail: adea@adea.org

American Dental Association
211 E. Chicago Avenue
Chicago, IL 60611
312-440-2500
(Fax) 312-440-2800
Web Site: http://www.ada.org
E-mail: publicinfo@ada.org

EDUCATION

National Association of Biology Teachers (NABT)
12030 Sunrise Valley Drive, Suite 110
Reston, VA 20191
703-264-9696
(Fax) 703-264-7778
Web Site: http://www.nabt.org
E-mail: office@nabt.org

ENTOMOLOGY

Entomological Society of America
9301 Annapolis Road
Lanham, MD 20706-3115
301-731-4535
(Fax) 301-731-4538
Web Site: http://www.entsoc.org
E-mail: esa@entsoc.org

ENVIRONMENT/ECOLOGY

Ecological Society of America
1707 H Street NW, Suite 400
Washington, DC 20006
202-833-8773
(Fax) 202-833-8775
Web Site: http://www.sdsc.edu/projects/ESA/esa.htm
E-mail: esahq@esa.org

Environmental Careers Organization (ECO)
179 South Street
Boston, MA 02111
617-426-4375
(Fax) 617-423-0998
Web Site: http://www.eco.org

Midcontinent Ecological Science Center: Explore Science Careers
Midcontinent Ecological Science Center
4512 McMurry Avenue
Fort Collins, CO 80525-3400
970-226-9100
(Fax) 970-226-9230
Web Site: http://www.mesc.nbs.gov/science-careers.html

National Wildlife Federation
11100 Wildlife Center Drive
Reston, VA 20190-5362
703-438-600
Web Site: http://www.nwf.org
E-mail: jobopp@nwf.org

Nature Conservancy
4245 N. Fairfax Drive, Suite 100
Arlington, VA 22203
1-800-628-6860
Web Site: http://nature.org

ENVIRONMENTAL LAW

Environmental Defense Fund
257 Park Avenue South
New York, NY 10010
1-800-684-3322
Web Site: http://www.edf.org

Environmental Law Institute
1616 P Street, NW, Suite 200
Washington, DC 20036
202-939-3800
(Fax) 202-939-3868
Web Site: http://www.eli.org

FORENSICS

American Academy of Forensic Sciences
The Forensic Sciences Foundation, Inc.
410 North 21st Street, Suite 203
Colorado Springs, CO 80904-2798
719-636-1100
(Fax) 719-636-1993
Web Site: http://www.aafs.org
E-mail: Membship@aafs.org

GENETICS

Genetics Society of America
9650 Rockville Pike
Bethesda, MD 20814-3889
301-571-1825
Web Site: http://www.nsgc.org
E-mail: estrass@genetics.faseb.org

National Society of Genetic Counselors
NSGC Executive Office
233 Canterbury Drive
Wallingford, PA 19086-6617
610-872-7608
Web Site: http://members.aol.com/nsgcweb/nsgchome.htm
E-mail: nsgc@aol.com

Careers in Biotechnology
Web Site: http://www.accessexcellence.org

MARINE BIOLOGY

American Society of Limnology and Oceanography
Web Site: http://www.aslo.org

Careers & Jobs in Marine Biology & Oceanography
Web Site: http://www-marine.stanford.edu/HMS web/careers.html

Scripps Research Institute
10550 North Torrey Pines Road
La Jolla, CA 92037
858-784-1000
Web Site: http://www.scripps.edu

Virginia Institute of Marine Science
Sea Grant Advisory Board Program
Gloucester Point, VA 23062-1346
804-684-7164
Web Site: http://www.vims.edu/adv/ed/careers

Woods Hole Oceanographic Institution
Information Office
Co-op Building, MS #16
Woods Hole, MA 02543
508-548-1400
(Fax) 508-457-2034
Web Site: http://www.whoi.edu
E-mail: information@whoi.edu

MATHEMATICAL BIOLOGY (BIOINFORMATICS)

The Society for Mathematical Biology
Web Site: http://www.smb.org

MEDICINE

American Academy of Family Physicians
11400 Tomahawk Creek Parkway
Leawood, KS 66211-2672
913-906-6000
Web Site: http://www.aafp.org
E-mail: fp@aafp.org

American Academy of Pediatrics
141 Northwest Point Boulevard
Elk Grove Village, IL 60007-1098
847-434-4000
(Fax) 847-434-8000
Web Site: http://www.aap.org
E-mail: pedscareer@aap.org

American Association of Immunologists
9650 Rockville Pike
Bethesda, MD 20814-3994
301-530-7178
(Fax) 301-571-1816
Web Site: http://www.aai.org
E-mail: infoaai@aai.faseb.org

American Medical Association
515 North State Street
Chicago, IL 60610
312-464-5000
Web Site: http://www.ama-assn.org

Clinical Immunology Society
611 E. Wells Street
Milwaukee, WI 53202
414-224-8095
(Fax) 414-272-6070
Web Site: http://www.clinimmsoc.org

Research! America (Medical Research)
908 King Street, Suite 400E
Alexandria, VA 22314
703-739-2577
Web Site: http://www.researchamerica.org
E-mail: researcham@aol.com

MICROBIOLOGY

American Society for Microbiology
1752 N Street
Washington, DC 20036
202-737-3600
Web Site: http://www.asmusa.org
E-mail: FellowshipsCareerInformation@asmusa.org

MICROSCOPY

Microscopy Society of America
Bostrom Corp.
230 East Ohio, Suite 400
Chicago, IL 60611
312-644-1527
(Fax) 312-644-8557
Web Site: http://www.msa.microscopy.com
E-mail: BusinessOffice@MSA.Microscopy.Com

NURSING

American Nurses Association

600 Maryland Avenue, SW
Suite 100W
Washington, DC 20024-2571
1-800-274-4ANA
Web Site: http://www.ana.org

National Association for Practical Nurse Education and Service

1400 Spring Street
Suite 300
Silver Spring, MD 20910
301-588-2491
Web Site: http://www.aoa.dhhs.gov/aoa/die/bo.html
E-mail: napnes@bellatlantic.net

NUTRITION

American Dietetic Association

216 West Jackson Boulevard
Chicago, IL 60606-6995
312-899-0040
(Fax) 312-899-1979
Web Site: http://www.eatright.org/careers.html
E-mail: network@eatright.org

Institute of Food Technologists

Professional Development Department
221 North LaSalle Street, Suite 300
Chicago, IL 60601-1291
312-782-8424
(Fax) 312-782-8348
Web Site: http://www.ift.org

PHARMACEUTICALS

American Association of Colleges of Pharmacy

1426 Prince Street
Alexandria, VA 22314-2841
703-729-2330
(Fax) 703-836-8982
Web Site: http://www.aacp.org
E-mail: angieaacp@aol.com

American Pharmaceutical Association

2215 Constitution Avenue, NW
Washington, DC 20037-2985
202-628-4410
(Fax) 202-783-2351
Web Site: http://www.aoa.dhhs.gov/aoa/dir/49.html

PHYSICAL THERAPY

American Physical Therapy Association

1111 North Fairfax Street
Alexandria, VA 22314
703-684-2782
(Fax) 703-684-7343
Web Site: http://www.apta.org

PHYSIOLOGY

American Physiological Society

9650 Rockville Pike
Bethesda, MD 20814
301-530-7160
Web Site: http://www.biophysics.org
E-mail: mfrank@aps.faseb.org

PSYCHIATRY/PSYCHOLOGY

American Psychiatric Association

Division of Public Affairs
1400 K Street, NW
Washington, DC 20005
1-888-357-7924
(Fax) 202-682-6850
Web Site: http://www.psych.org
E-mail: apa@psych.org

American Psychological Association

750 First Street, NE
Washington, DC 20002
202-336-5500
Web Site: http://www.apa.org
E-mail: education@apa.org

PUBLIC HEALTH

American Public Health Association

1015 15th Street, NW
Washington, DC 20005-2605
202-789-5600
(Fax) 202-789-5661
Web Site: http://www.apha.org

TAXONOMY

American Society of Plant Taxonomists

Web Site: http://www.sysbof.org

Association of Systematic Collections

1725 K Street, NW, Suite 601
Washington, DC 20006-1401
202-835-9050
(Fax) 202-835-7334
Web Site: http://www.ascoll.org
E-mail: asc@ascoll.org

ZOOLOGY

American Association of Zoo Keepers

Topeka Zoological Park
635 S.W. Gage Boulevard
Topeka, KS 66606
913-272-5821 or 913-273-1980
Web Site: http://www.aazk.org

American Society of Mammalogists

Monte L. Bean Life Science Museum
Brigham Young University
Provo, UT 84602-0200
Web Site: http://wkuwebl.wku.edu/~asm

American Zoo and Aquarium Association

8403 Colesville Road
Suite 710
Silver Spring, MD 20910-3314
301-562-0777
(Fax) 562-0888
Web Site: http://www.aza.org

Society for Integrative and Comparative Biology (SICB) (formerly: American Society of Zoologists)

1313 Dolley Madison Boulevard
Suite 402
McLean, VA 22101
703-790-1745
(Fax) 703-790-2672
Web Site: http://www.sicb.org
E-mail: sicb@burkinc.com

Appendix E

Understanding Biological Terms

Your task of mastering new terms will be greatly simplified if you learn to dissect each new word. Many terms can be divided into a prefix, the part of the word that precedes the main root, the word root itself, and often a suffix, a word ending that may add to or modify the meaning of the root. As you progress in your study of biology, you will learn to recognize the more common prefixes, word roots, and suffixes. Such recognition will help you analyze new terms so that you can more readily determine their meaning and will also help you remember them.

PREFIXES

a-, ab- from, away, apart (abduct, move away from the midline of the body)
a-, an-, un- less, lack, not (asymmetrical, not symmetrical)
ad- (also **af-, ag-, an-, ap-**) to, toward (adduct, move toward the midline of the body)
allo- different (allometric growth, different rates of growth for different parts of the body during development)
ambi- both sides (ambidextrous, able to use either hand)
andro- a man (androecium, the male portion of a flower)
anis- unequal (anisogamy, sexual reproduction in which the gametes are of unequal sizes)
ante- forward, before (anteflexion, bending forward)
anti- against (antibody, proteins that have the capacity to react against foreign substances in the body)
auto- self (autotroph, organism that manufactures its own food)
bi- two (biennial, a plant that takes two years to complete its life cycle)
bio- life (biology, the study of life)
circum-, circ- around (circumcision, a cutting around)
co-, con- with, together (congenital, existing with or before birth)
contra- against (contraception, against conception)
cyt- cell (cytology, the study of cells)
di- two (disaccharide, a compound made of two sugar molecules chemically combined)
dis- apart (dissect, cut apart)
ecto- outside (ectoplasm, outer layer of cytoplasm)
end-, endo- within, inner (endoplasmic reticulum, a network of membranes found within the cytoplasm)
epi- on, upon (epidermis, upon the dermis)
ex-, e-, ef- out from, out of (extension, a straightening out)
extra- outside, beyond (extraembryonic membrane, a membrane that encircles and protects the embryo)
gravi- heavy (gravitropism, growth of a plant in response to gravity)
hemi- half (cerebral hemisphere, lateral half of the cerebrum)
hetero- other, different (heterozygous, having unlike members of a gene pair)
homeo- unchanging, steady (homeostasis, reaching a steady state)
homo-, hom- same (homologous, corresponding in structure; homozygous, having identical members of a gene pair)
hyper- excessive, above normal (hypersecretion, excessive secretion)
hypo- under, below, deficient (hypotonic, a solution whose osmotic pressure is less than that of a solution with which it is compared)
in-, im- not (incomplete flower, a flower that does not have one or more of the four main parts)
inter- between, among (interstitial, situated between parts)
intra- within (intracellular, within the cell)
iso- equal, like (isotonic, equal osmotic concentration)
macro- large (macronucleus, a large, polyploid nucleus found in ciliates)
mal- bad, abnormal (malnutrition, poor nutrition)
mega- large, great (megakaryocyte, giant cell of bone marrow)
meso- middle (mesoderm, middle tissue layer of the animal embryo)
meta- after, beyond (metaphase, the stage of mitosis after prophase)
micro- small (microscope, instrument for viewing small objects)
mono- one (monocot, a group of flowering plants with one cotyledon, or seed leaf, in the seed)
oligo- small, few, scant (oligotrophic lake, a lake deficient in nutrients and organisms)
oo- egg (oocyte, cell that gives rise to an egg cell)
paedo- a child (paedomorphosis, the preservation of a juvenile characteristic in an adult)
para- near, beside, beyond (paracentral, near the center)
peri- around (pericardial membrane, membrane that surrounds the heart)
photo- light (phototropism, growth of a plant in response to the direction of light)
poly- many, much, multiple, complex (polysaccharide, a carbohydrate composed of many simple sugars)
post- after, behind (postnatal, after birth)
pre- before (prenatal, before birth)
pseudo- false (pseudopod, a temporary protrusion of a cell, i.e., "false foot")
retro- backward (retroperitoneal, located behind the peritoneum)
semi- half (semilunar, half-moon)
sub- under (subcutaneous tissue, tissue immediately under the skin)
super, supra- above (suprarenal, above the kidney)
sym- with, together (sympatric speciation, evolution of a new species within the same geographical region as the parent species)

syn- with, together (syndrome, a group of symptoms that occur together and characterize a disease)
trans- across, beyond (transport, carry across)

SUFFIXES

-able, -ible able (viable, able to live)
-ad used in anatomy to form adverbs of direction (cephalad, toward the head)
-asis, -asia, -esis condition or state of (euthanasia, state of "good death")
-cide kill, destroy (biocide, substance that kills living things)
-emia condition of blood (anemia, a blood condition in which there is a lack of red blood cells)
-gen something produced or generated or something that produces or generates (pathogen, an organism that produces disease)
-gram record, write (electrocardiogram, a record of the electrical activity of the heart)
-graph record, write (electrocardiograph, an instrument for recording the electrical activity of the heart)
-ic adjective-forming suffix that means *of* or *pertaining to* (ophthalmic, of or pertaining to the eye)
-itis inflammation of (appendicitis, inflammation of the appendix)
-logy study or science of (cytology, study of cells)
-oid like, in the form of (thyroid, in the form of a shield, referring to the shape of the thyroid gland)
-oma tumor (carcinoma, a malignant tumor)
-osis indicates disease (psychosis, a mental disease)
-pathy disease (dermopathy, disease of the skin)
-phyll leaf (mesophyll, the middle tissue of the leaf)
-scope instrument for viewing or observing (microscope, instrument for viewing small objects)

SOME COMMON WORD ROOTS

abscis cut off (abscission, the falling off of leaves or other plant parts)
angi, angio vessel (angiosperm, a plant that produces seeds enclosed within a fruit or "vessel")
apic tip, apex (apical meristem, area of cell division located at the tips of plant stems and roots)
arthr joint (arthropods, invertebrate animals with jointed legs and segmented bodies)
aux grow, enlarge (auxin, a plant hormone involved in growth and development)
blast a formative cell, germ layer (osteoblast, cell that gives rise to bone cells)
brachi arm (brachial artery, blood vessel that supplies the arm)
bry grow, swell (embryo, an organism in the early stages of development)
cardi heart (cardiac, pertaining to the heart)
carot carrot (carotene, a yellow, orange, or red pigment in plants)
cephal head (cephalad, toward the head)
cerebr brain (cerebral, pertaining to the brain)
cervic, cervix neck (cervical, pertaining to the neck)
chlor green (chlorophyll, a green pigment found in plants)
chondr cartilage (chondrocyte, a cartilage cell)
chrom color (chromosome, deeply staining body in nucleus)
cili small hair (cilium, a short, fine cytoplasmic hair projecting from the surface of a cell)
coleo a sheath (coleoptile, a protective sheath that encircles the stem in grass seedlings)
conjug joined together (conjugation, a sexual phenomenon in certain protists)
cran skull (cranial, pertaining to the skull)
decid falling off (deciduous, a plant that sheds its leaves at the end of the growing season)
dehis split (dehiscent fruit, a fruit that splits open at maturity)
derm skin (dermatology, study of the skin)
ecol dwelling, house (ecology, the study of organisms in relation to their environment, i.e., "their house")
enter intestine (enterobacteria, a group of bacteria that include species that inhabit the intestines of humans and other animals)
evol to unroll (evolution, descent with modification, or gradual directional change)
fil a thread (filament, the thin stalk of the stamen in flowers)
gamet a wife or husband (gametangium, the part of a plant, protist, or fungus that produces reproductive cells)
gastr stomach (gastrointestinal tract, the digestive tract)
glyc, glyco sweet, sugar (glycogen, storage form of glucose)
gon seed (gonad, an organ that produces gametes)
gutt a drop (guttation, loss of water as liquid "drops" from plants)
gymn naked (gymnosperm, a plant that produces seeds that are not enclosed with a fruit, i.e., "naked")
hem blood (hemoglobin, the pigment of red blood cells)
hepat liver (hepatic, of or pertaining to the liver)
hist tissue (histology, study of tissues)
hydr water (hydrolysis, a breakdown reaction involving water)
leuk white (leukocyte, white blood cell)
menin membrane (meninges, the three membranes that envelop the brain and spinal cord)
morph form (morphogenesis, development of body form)
my, myo muscle (myocardium, muscle layer of the heart)
myc a fungus (mycelium, the vegetative body of a fungus)
nephr kidney (nephron, microscopic unit of the kidney)
neur, nerv nerve (neuromuscular, involving both the nerves and muscles)
occiput back part of the head (occipital, back region of the head)
ost bone (osteology, study of bones)
path disease (pathologist, one who studies disease processes)
ped, pod foot (bipedal, walking on two feet)
pell skin (pellicle, a flexible covering over the body of certain protists)
phag eat (phagocytosis, process by which certain cells ingest particles and foreign matter)
phil love (hydrophilic, a substance that attracts, i.e., "loves," water)
phloe bark of a tree (phloem, food-conducting tissue in plants that corresponds to bark in woody plants)
phyt plant (xerophyte, a plant adapted to xeric, or dry, conditions)
plankt wandering (plankton, microscopic aquatic protists that float or drift passively)
rhiz root (rhizome, a horizontal, underground stem that superficially resembles a root)
scler hard (sclerenchyma, cells that provide strength and support in the plant body)
sipho a tube (siphonous, a type of tubular body form found in certain algae)
som body (chromosome, deeply staining body in the nucleus)
sor heap (sorus, a cluster or "heap" of sporangia in a fern)

spor seed (spore, a reproductive cell that gives rise to individual offspring in plants, protists, and fungi)
stom a mouth (stoma, a small pore, i.e., "mouth," in the epidermis of plants)
thigm a touch (thigmotropism, plant growth in response to touch)
thromb clot (thrombus, a clot within a blood vessel)
tropi turn (thigmotropism, growth of a plant in response to contact with a solid object, as when a tendril "turns" or wraps around a wire fence)
visc pertaining to an internal organ or body cavity (viscera, internal organs)
xanth yellow (xanthophyll, a yellowish pigment found in plants)
xyl wood (xylem, water-conducting tissue in plant, the "wood" of woody plants)
zoo an animal (zoology, the science of animals)

Appendix F

Abbreviations

The biological sciences use a great many abbreviations and with good reason. Many technical terms in biology and biological chemistry are both long and difficult to pronounce. Yet it can be difficult for beginners, when confronted with something like NADPH or EPSP, to understand the reference. Here are some of the common abbreviations used in biology for your ready reference.

A Adenine
ABA Abscisic acid
ACTH Adrenocorticotropic hormone
AD Alzheimer's disease
ADA Adenosine deaminase
ADH Antidiuretic hormone
ADP Adenosine diphosphate
AIDS Acquired immunodeficiency syndrome
AMP Adenosine monophosphate
amu Atomic mass unit (dalton)
APC Anaphase-promoting complex *or* antigen presenting cell
ATP Adenosine triphosphate
AV node or valve Atrioventricular node or valve (of heart)
B lymphocyte or B cell Lymphocyte responsible for antibody-mediated immunity
BH Brain hormone (of insects)
BMR Basal metabolic rate
bya Billion years ago
C Cytosine
C_3 Three-carbon pathway for carbon fixation (Calvin cycle)
C_4 Four-carbon pathway for carbon fixation (Hatch-Slack pathway)
CAM Crassulacean acid metabolism
cAMP Cyclic adenosine monophosphate
CAP Catabolite gene activator protein
CD4 T cell Helper T cell (T_h); has a surface marker designated CD4
CD8 T cell T cell with a surface marker designated CD8; includes cytotoxic T cells
Cdk cyclin-dependent protein kinase
cDNA Complementary deoxyribonucleic acid
CFCs Chlorofluorocarbons
CFTR Cystic fibrosis transmembrane conductance regulator
CITES The Convention on International Trade in Endangered Species of Wild Flora and Fauna
CNS Central nervous system
CoA Coenzyme A
COPD Chronic obstructive pulmonary disease
CP Creatine phosphate
CPR Cardiopulmonary resuscitation
CR Conditioned response
CS Conditioned stimulus
CSF Cerebrospinal fluid
CVS Cardiovascular system
DAG Diacylglycerol
DNA Deoxyribonucleic acid
DOC Dissolved organic carbon
E_A Activation energy (of an enzyme)
ECG Electrocardiogram
ECM Extracellular matrix
EEG Electroencephalogram
EKG Electrocardiogram
EM Electron microscope or micrograph
ENSO El Niño—Southern Oscillation
EPSP Excitatory postsynaptic potential (of a neuron)
ER Endoplasmic reticulum
ES cells Embryonic stem cells
F_1 First filial generation
F_2 Second filial generation
Fab portion The part of an antibody that binds to an antigen
Factor VIII Blood clotting factor (absent in hemophiliacs)
FAD/$FADH_2$ Flavin adenine dinucleotide (oxidized and reduced forms, respectively)
FAP Fixed action pattern
Fc portion The part of an antibody that interacts with cells of the immune system
FSH Follicle-stimulating hormone
G Guanine
G_1 phase First gap phase (of the cell cycle)
G_2 phase Second gap phase (of the cell cycle)
G3P Glyceraldehyde-3-phosphate
G protein Cell signaling molecule that requires GTP
GA_3 Gibberellin
GABA Gamma-aminobutyric acid
GH Growth hormone (somatotropin)
GnRH Gonadotropin-releasing hormone
GTP Guanosine triphosphate
HBEF Hubbard Brook Experimental Forest
HCFCs Hydrochlorofluorocarbons
hCG Human chorionic gonadotropin
HD Huntington disease
HDL High-density lipoprotein
HFCs Hydrofluorocarbons
hGH Human growth hormone
HIV Human immunodeficiency virus
HLA Human leukocyte antigen
IAA Indole acetic acid (natural auxin)
Ig Immunoglobulin, as in IgA, IgG, etc.
IGF Insulin-like growth factor
IP_3 Inositol triphosphate

IPCC United Nations Intergovernmental Panel on Climate Change
IPSP Inhibitory postsynaptic potential (of a neuron)
IUCN World Conservation Union
IUD Intrauterine device
JH Juvenile hormone (of insects)
kb Kilobase
LDH Lactic dehydrogenase enzyme
LDL Low-density lipoprotein
LH Luteinizing hormone
LM Light microscope or micrograph
LSD Lysergic acid diethylamide
LTP Long-term potentiation
MAO Monoamine oxidase
MAPs Microtubule-associated proteins
MHC Major histocompatibility complex
MI Myocardial infarction
MPF Mitosis-promoting factor
MRI Magnetic resonance imaging
mRNA Messenger RNA
MSAFP Maternal serum α-fetoprotein
mtDNA Mitochondrial DNA
MTOC Microtubule organizing center
mya Million years ago
9 + 2 structure Cilium or flagellum (of a eukaryote)
9 × 3 structure Centriole or basal body (of a eukaryote)
n*, 2*n The chromosome number of a gamete and of a zygote, respectively
NAD⁺/NADH Nicotinamide adenine dinucleotide (oxidized and reduced forms, respectively)
NADP⁺/NADPH Nicotinamide adenine dinucleotide phosphate (oxidized and reduced forms, respectively)
NAG *N*-acetyl glucosamine
NK cell Natural killer cell
NMDA *N*-methyl-D aspartate (an artificial ligand)
NSF National Science Foundation
P generation Parental generation
P53 A tumor suppressor gene
P680 Reaction center of photosystem II
P700 Reaction center of photosystem I
PABA Para-aminobenzoic acid
PCR Polymerase chain reaction
PEP Phosphoenolpyruvate
Pfr Phytochrome (form that absorbs far red light)
PGA Phosphoglycerate
PID Pelvic inflammatory disease
PKU Phenylketonuria
PNS Peripheral nervous system
pre-mRNA Precursor messenger RNA (in eukaryotes)
Pr Phytochrome (form that absorbs red light)
PTH Parathyroid hormone
RAS Reticular activating system
RBC Red blood cell (erythrocyte)
REM sleep Rapid eye movement sleep
RFLP Restriction fragment length polymorphism
RNA Ribonucleic acid
rRNA Ribosomal RNA
Rubisco Ribulose bisphosphate carboxylase/oxygenase
RuBP Ribulose bisphosphate
S phase DNA synthetic phase (of the cell cycle)
SA node Sinoatrial node (of heart)
SCID Severe combined immunodeficiency
SEM Scanning electron microscope or micrograph
snRNP Small nuclear ribonucleoprotein complex
ssp Subspecies
STD Sexually transmitted disease
T Thymine
T lymphocyte or T cell Lymphocyte responsible for cell-mediated immunity
T_c lymphocyte Cytotoxic T cell
T_h lymphocyte Helper T cell (CD4 T cell)
TATA box Base sequence in eukaryotic promoter
TCA cycle Tricarboxylic acid cycle (synonym for citric acid cycle)
TCR T-cell antigen receptor
TEM Transmission electron microscope or micrograph
Tm Tubular transport maximum
TNF Tumor necrosis factor
tRNA Transfer RNA
U Uracil
UPE Upstream promoter element
UR Unconditioned response
US Unconditioned stimulus
UV light Ultraviolet light
WBC White blood cell (leukocyte)

Glossary

abiotic factors Elements of the nonliving, physical environment that affect a particular organism. Compare with *biotic factors*.

abscisic acid (ab-sis'ik) A plant hormone involved in dormancy and responses to stress.

abscission (ab-sizh'en) The normal (usually seasonal) fall of leaves or other plant parts, such as fruits or flowers.

abscission layer The area at the base of the petiole where the leaf will break away from the stem. Also known as abscission zone.

absorption (ab-sorp'shun) The movement of nutrients and other substances through the wall of the digestive tract and into the blood or lymph.

absorption spectrum A graph of the amount of light at specific wavelengths that has been absorbed as light passes through a substance. Each type of molecule has a characteristic absorption spectrum. Compare with *action spectrum*.

accessory fruit A fruit composed primarily of tissue other than ovary tissue, e.g., apple, pear. Compare with *aggregate, simple,* and *multiple fruits*.

acetyl coenzyme A (acetyl CoA) (as'uh-teel) A key intermediate compound in metabolism; consists of a two-carbon acetyl group covalently bonded to coenzyme A.

acetyl group A two-carbon group derived from acetic acid (acetate).

acetylcholine (ah''see-til-koh'leen) A common neurotransmitter released by cholinergic neurons, including motor neurons.

achene (a-keen') A simple, dry fruit with one seed in which the fruit wall is separate from the seed coat, e.g., sunflower fruit.

acid A substance that is a hydrogen ion (proton) donor; acids unite with bases to form salts. Compare with *base*.

acidic solution A solution in which the concentration of hydrogen ions (H^+) exceeds the concentration of hydroxide ions (OH^-). An acidic solution has a pH less than 7. Compare with *basic solution* and *neutral solution*.

acid precipitation Precipitation that is acidic as a result of both sulfur and nitrogen oxides forming acids when they react with water in the atmosphere.

acclimitization Adjustment to seasonal changes.

acoelomate (a-seel'oh-mate) An animal lacking a body cavity (coelom). Compare with *coelomate* and *pseudocoelomate*.

acquired immunodeficiency syndrome (AIDS) A serious, potentially fatal disease caused by the human immunodeficiency virus (HIV).

acromegaly (ak''roh-meg'ah-lee) A condition characterized by overgrowth of the extremities of the skeleton, fingers, toes, jaws, and nose. It may be produced by excessive secretion of growth hormone by the anterior pituitary gland.

acrosome reaction (ak'roh-sohm) A series of events in which the acrosome, a caplike structure covering the head of a sperm cell, releases proteolytic (protein-digesting) enzymes and undergoes other changes that permit the sperm to penetrate the outer coverings of the egg.

actin (ak'tin) The protein of which microfilaments are composed. Actin, together with the protein myosin, is responsible for muscle contraction.

actin filaments Thin filaments composed mainly of the protein actin; actin and myosin filaments make up the myofibrils of muscle fibers.

actinopods (ak-tin'o-podz) Protozoa characterized by axopods that protrude through pores in their shells. See *radiolarians*.

action potential An electric signal resulting from depolarization of the plasma membrane in a neuron or muscle cell. Compare with *resting potential*.

action spectrum A graph of the effectiveness of light at specific wavelengths in promoting a light-requiring reaction. Compare with *absorption spectrum*.

activation energy (E_A) The kinetic energy required to initiate a chemical reaction.

activator protein A positive regulatory protein that stimulates transcription when bound to DNA. Compare with *repressor protein*.

active immunity Immunity that develops as a result of exposure to antigens; it may occur naturally after recovery from a disease or be artificially induced by immunization with a vaccine. Compare with *passive immunity*.

active site A specific region of an enzyme (generally near the surface) that accepts one or more substrates and catalyzes a chemical reaction. Compare with *allosteric site*.

active transport All forms of transport of a substance across a membrane that do not rely on the potential energy of a concentration gradient for the substance being transported and therefore require an additional energy source (often ATP); includes carrier-mediated active transport, endocytosis, and exocytosis. Compare with *diffusion* and *facilitated diffusion*.

adaptation (1) An evolutionary modification that improves an organism's chances of survival and reproductive success; (2) A decline in the response of a receptor subjected to repeated or prolonged stimulation.

adaptive immune responses See *specific immune responses*.

adaptive radiation The evolution of a large number of related species from an unspecialized ancestral organism.

adaptive zone A new ecological opportunity that was not exploited by an ancestral organism; used by evolutionary biologists to explain the ecological paths along which different taxa evolve.

addiction Physical dependence on a drug, generally based on physiological changes that take place in response to the drug; when the drug is withheld, the addict may suffer characteristic withdrawal symptoms.

adenine (ad′eh-neen) A nitrogenous purine base that is a component of nucleic acids and ATP.

adenosine triphosphate (ATP) (a-den′oh-seen) An organic compound containing adenine, ribose, and three phosphate groups; of prime importance for energy transfers in cells.

adhering junction A type of anchoring junction between cells; connects epithelial cells.

adipose tissue (ad′i-pohs) Tissue in which fat is stored.

adrenal cortex (ah-dree′nul kor′teks) The outer region of each adrenal gland; secretes steroid hormones, including mineralocorticoids and glucocorticoids.

adrenal glands (ah-dree′nul) Paired endocrine glands, one located just superior to each kidney; secrete hormones that help regulate metabolism and help the body cope with stress.

adrenal medulla (ah-dree′nul meh-dull′uh) The inner region of each adrenal gland; secretes epinephrine and norepinephrine.

adrenergic neuron (ad-ren-er′jik) A neuron that releases norepinephrine or epinephrine as a neurotransmitter. Compare with *cholinergic neuron*.

adventitious (ad″ven-tish′us) Of plant organs, such as roots or buds, that arise in an unusual position on a plant.

aerobe Organism that grows or metabolizes only in the presence of molecular oxygen. Compare with *anaerobe*.

aerobic (air-oh′bik) Growing or metabolizing only in the presence of molecular oxygen. Compare with *anaerobic*.

aerobic respiration See *respiration*.

afferent (af′fer-ent) Leading toward some point of reference. Compare with *efferent*.

afferent neurons Neurons that transmit action potentials from sensory receptors to the brain or spinal cord. Compare with *efferent neurons*.

age structure The number and proportion of people at each age in a population. Age structure diagrams represent the number of males and females at each age, from birth to death, in the population.

aggregate fruit A fruit that develops from a single flower with many separate carpels, e.g., raspberry. Compare with *simple, accessory,* and *multiple fruits*.

aggregated distribution See *clumped dispersion*.

agnathans (ag-na′thanz) Jawless fishes; historical class of vertebrates, including lampreys, hagfishes, and many extinct forms.

albinism (al′bih-niz-em) A hereditary inability to form melanin pigment, resulting in light coloration.

AIDS See *acquired immunodeficiency syndrome*.

albumin (al-bew′min) A class of protein found in most animal tissues; a fraction of plasma proteins.

aldehyde An organic molecule containing a carbonyl group bonded to at least one hydrogen atom. Compare with *ketone*.

aldosterone (al-dos′tur-ohn) A steroid hormone produced by the vertebrate adrenal cortex; stimulates sodium reabsorption. See *mineralocorticoids*.

algae (al′gee) (sing. *alga*) An informal group of unicellular, or simple multicellular, photosynthetic protists that are important producers in aquatic ecosystems; includes dinoflagellates, diatoms, euglenoids, golden algae, green algae, red algae, and brown algae.

allantois (a-lan′toe-iss) An extraembryonic membrane of reptiles, birds, and mammals that stores the embryo's nitrogenous wastes; most of the allantois is detached at hatching or birth.

allele frequency The proportion of a specific allele in the population.

alleles (al-leels′) Genes governing variation of the same character that occupy corresponding positions (loci) on homologous chromosomes; alternative forms of a gene.

allelopathy (uh-leel′uh-path″ee) An adaptation in which toxic substances secreted by roots or shed leaves inhibit the establishment of competing plants nearby.

allergen A substance that stimulates an allergic reaction.

allergy A hypersensitivity to some substance in the environment, manifested as hay fever, skin rash, asthma, food allergies, etc.

all-or-none law The principle that neurons transmit an impulse in a similar way no matter how weak or strong the stimulus; the neuron either transmits an action potential (all) or does not (none).

allopatric speciation (al-oh-pa′trik) Speciation that occurs when one population becomes geographically separated from the rest of the species and subsequently evolves. Compare with *sympatric speciation*.

allopolyploid (al″oh-pol′ee-ploid) A polyploid whose chromosomes are derived from two species. Compare with *autopolyploid*.

allosteric regulators Substances that affect protein function by binding to allosteric sites.

allosteric site (al-oh-steer′ik) (1) A regulatory site located on a protein that is separate from the functional site. The binding of a specific regulator to the allosteric site alters the conformation and function of the protein; (2) A site on an enzyme other than the active site, to which a specific substance (other than the normal substrate) can bind, thereby changing the shape and activity of the enzyme. Compare with *active site*.

alpha (α) helix A regular, coiled type of secondary structure of a polypeptide chain, maintained by hydrogen bonds. Compare with *beta (β)-pleated sheet*.

alpine tundra An ecosystem located in the higher elevations of mountains, above the tree line and below the snow line. Compare with *tundra*.

alternation of generations A type of life cycle characteristic of plants and a few algae and fungi in which they spend part of their life in a multicellular n gametophyte stage and part in a multicellular $2n$ sporophyte stage. The gametophyte develops from a spore and produces gametes; the sporophyte develops from a zygote and produces spores.

altruistic behavior Behavior in which one individual helps another, seemingly at its own risk or expense.

alveolus (al-vee′o-lus) (pl. *alveoli*) (1) An air sac of the lung through which gas exchange with the blood takes place; (2) A saclike unit of some glands, e.g., mammary glands.

Alzheimer's disease (AD) A progressive, degenerative brain disorder characterized by amyloid plaques and neurofibrillary tangles.

amino acid (uh-mee′no) An organic compound containing an amino group ($—NH_2$) and a carboxyl group ($—COOH$); may be joined by peptide bonds to form the polypeptide chains of protein molecules.

amino group A weakly basic functional group; abbreviated $—NH_2$.

aminoacyl-tRNA (uh-mee″no-ace′seel) Molecule consisting of an amino acid covalently linked to a transfer RNA.

aminoacyl-tRNA synthetase One of a family of enzymes, each responsible for covalently linking an amino acid to its specific transfer RNA.

ammonification (uh-moe''nuh-fah-kay'shun) The conversion of nitrogen-containing organic compounds to ammonia (NH_3) by certain soil bacteria (ammonifying bacteria); part of the nitrogen cycle.

amniocentesis (am''nee-oh-sen-tee'sis) Sampling of the amniotic fluid surrounding a fetus to obtain information about its development and genetic makeup. Compare with *chorionic villus sampling*.

amnion (am'nee-on) In terrestrial vertebrates, an extraembryonic membrane that forms a fluid-filled sac for the protection of the developing embryo.

amniotes Terrestrial vertebrates: reptiles, birds, and mammals; animals whose embryos are enclosed by an amnion.

amoeba (a-mee'ba) (pl. *amoebas*) A unicellular protozoon that moves by means of pseudopodia.

amphibians Members of vertebrate class that includes salamanders, frogs, and caecilians.

amphipathic molecule (am''fih-pa'thik) A molecule containing both hydrophobic and hydrophilic regions.

ampulla Any small saclike extension, e.g., the expanded structure at the end of each semicircular canal of the ear.

amylase (am'-uh-laze) Starch-digesting enzyme, e.g., human salivary amylase or pancreatic amylase.

amyloplasts See *leukoplasts*.

anabolic steroids Synthetic androgens that increase muscle mass, physical strength, endurance, and aggressiveness, but cause serious side effects; these drugs are often abused.

anabolism (an-ab'oh-lizm) The aspect of metabolism in which simpler substances are combined to form more complex substances, resulting in the storage of energy, the production of new cellular materials, and growth. Compare with *catabolism*.

anaerobe Organism that grows or metabolizes only in the absence of molecular oxygen. See *facultative anaerobe* and *obligate anaerobe*. Compare with *aerobe*.

anaerobic (an''air-oh'bik) Growing or metabolizing only in the absence of molecular oxygen. Compare with *aerobic*.

anaerobic respiration See *respiration*.

anagenesis Progressive evolutionary changes in a single lineage over long periods. Also called phyletic evolution. Compare with *cladogenesis*.

anaphase (an'uh-faze) The stage of mitosis, and of meiosis I and II, in which the chromosomes move to opposite poles of the cell; anaphase occurs after metaphase and before telophase.

anaphylaxis (an''uh-fih-lak'sis) An acute allergic reaction following sensitization to a foreign substance or other substance.

ancestral characters See *shared ancestral characters*.

androgen (an'dro-jen) Any substance that possesses masculinizing properties, such as a sex hormone. See *testosterone*.

androgen-binding protein (ABP) A protein produced by Sertoli cells in the testes; binds and concentrates testosterone.

anemia (uh-nee'mee-uh) A deficiency of hemoglobin or red blood cells.

aneuploidy (an'you-ploy-dee) Any chromosomal aberration in which there are either extra or missing copies of certain chromosomes.

angiosperms (an'jee-oh-spermz'') The traditional name for flowering plants, a very large (about 235,000 species), diverse phylum of plants that form flowers for sexual reproduction and produce seeds enclosed in fruits; include monocots and dicots.

angiotensin I (an-jee-o-ten'sin) A polypeptide produced by the action of renin on a plasma protein (angiotensinogen).

angiotensin II A peptide hormone formed by the action of angiotensin-converting enzyme on angiotensin I; stimulates aldosterone secretion by the adrenal cortex.

animal pole The non-yolky, metabolically active pole of a vertebrate or echinoderm egg. Compare with *vegetal pole*.

anion (an'eye-on) A particle with one or more units of negative charge, such as a chloride ion (Cl^-) or hydroxide ion (OH^-). Compare with *cation*.

anisogamy (an''eye-sog'uh-me) Sexual reproduction involving motile gametes of similar form but dissimilar size. Compare with *isogamy* and *oogamy*.

annelid (an'eh-lid) A member of phylum Annelida; segmented worm such as earthworm.

annual plant A plant that completes its entire life cycle in one year or less. Compare with *perennial* and *biennial*.

antenna complex The currently accepted arrangement of chlorophyll, accessory pigment molecules, and pigment-binding proteins into light-gathering units in the thylakoid membranes of photoautotrophic eukaryotes. See *reaction center* and *photosystem*.

antennae (sing. *antenna*) Sensory structures characteristic of some arthropod groups.

anterior Toward the head end of a bilaterally symmetrical animal. Compare with *posterior*.

anther (an'thur) The part of the stamen in flowers that produces microspores and, ultimately, pollen grains.

antheridium (an''thur-id'ee-im) (pl. *antheridia*) In plants, the multicellular male gametangium (sex organ) that produces sperm cells. Compare with *archegonium*.

anthropoid (an'thra-poid) A member of a suborder of primates that includes monkeys, apes, and humans.

antibody (an'tee-bod''ee) A specific protein (immunoglobulin) that recognizes and binds to specific antigens; produced by plasma cells.

antibody-mediated immunity A type of specific immune response in which B cells differentiate into plasma cells and produce antibodies that bind with foreign antigens, leading to the destruction of pathogens.

anticodon (an''tee-koh'don) A sequence of three nucleotides in transfer RNA that is complementary to, and combines with, the three nucleotide codon on messenger RNA, thus helping to specify the addition of a particular amino acid to the end of a growing polypeptide.

antidiuretic hormone (ADH) (an''ty-dy-uh-ret'ik) A hormone secreted by the posterior lobe of the pituitary that controls the rate of water reabsorption by the kidney.

antigen (an'tih-jen) Any molecule, usually a protein or large carbohydrate, that can be specifically recognized as foreign by cells of the immune system.

antigen-antibody complex The combination of antigen and antibody molecules.

antimicrobial peptides Soluble molecules that destroy pathogens.

anti-oncogene A gene (also known as a tumor suppressor gene) whose normal role is to block cell division in response to certain growth inhibiting factors; when mutated, may contribute to the formation of a cancer cell. Compare with *oncogene*.

antioxidants Certain enzymes (e.g., catalase and peroxidase), vitamins, and other substances that destroy free radicals and other reactive molecules.

anus (ay'nus) The distal end and outlet of the digestive tract.

aorta (ay-or'tah) The largest and main systemic artery of the vertebrate body; arises from the left ventricle and branches to distribute blood to all parts of the body except the lungs.

aphotic region (ay-fote'ik) The lower layer of the ocean (deeper than 100 m or so) where light does not penetrate.

apical dominance (ape'ih-kl) The inhibition of lateral buds by a shoot tip.

apical meristem (mehr'ih-stem) An area of dividing tissue, located at the tip of a shoot or root, that gives rise to primary tissues; apical meristems cause an increase in the length of the plant body. Compare with *lateral meristems*.

apicomplexans A group of parasitic protozoa that lack structures for locomotion and that produce sporozoites as infective agents; malaria is caused by an apicomplexan. Also called *sporozoa*.

apoenzyme (ap''oh-en'zime) Protein portion of an enzyme; requires the presence of a specific coenzyme to become a complete functional enzyme.

apomixis (ap''uh-mix'us) A type of reproduction in which fruits and seeds are formed asexually.

apoplast A continuum consisting of the interconnected, porous plant cell walls, along which water moves freely. Compare with *symplast*.

apoptosis (ap-uh-toe'sis) Programmed cell death; apoptosis is a normal part of an organism's development and maintenance. Compare with *necrosis*.

aposematic coloration The conspicuous coloring of a poisonous or distasteful organisms that enables potential predators to easily see and recognize it. Also called warning coloration. Compare with *cryptic coloration*.

arachnids (ah-rack'nids) Eight-legged arthropods such as spiders, scorpions, ticks, and mites.

arachnoid The middle of the three meningeal layers that cover and protect the brain and spinal cord; see *pia mater* and *dura mater*.

archaebacteria (ar''kuh-bak-teer'ee-uh) Prokaryotic organisms with a number of features, such as the absence of peptidoglycan in their cell walls, that set them apart from the rest of the bacteria. Compare with *eubacteria*.

archaic *Homo sapiens* Regionally diverse descendants of *Homo erectus* that lived in Africa, Asia, and Europe from about 800,000 to 100,000 years ago; considered by some paleoanthropologists to be a separate species, *Homo heidelbergensis*.

archegonium (ar''ke-go'nee-um) (pl. *archegonia*) In plants, the multicellular female gametangium (sex organ) that contains an egg. Compare with *antheridium*.

archenteron (ark-en'ter-on) The central cavity of the gastrula stage of embryonic development that is lined with endoderm; primitive digestive system.

arctic tundra See tundra.

Ardipithecus ramidus The earliest known hominid; an australopithecine that lived about 4.4 million years ago. See *australopithecines*.

arterial pulse See *pulse, arterial*.

arteriole (ar-teer'ee-ole) A very small artery. Vasoconstriction and vasodilation of arterioles help regulate blood pressure.

artery A thick-walled blood vessel that carries blood away from a heart chamber and toward the body organs. Compare with *vein*.

arthropod (ar'throh-pod) An invertebrate that belongs to phylum Arthropoda; characterized by a hard exoskeleton, a segmented body, and paired, jointed appendages.

artificial insemination The impregnation of a female by artificially introducing sperm from a male.

artificial selection The selection by humans of traits that are desirable in plants or animals, and breeding only those individuals that possess the desired traits.

ascocarp (ass'koh-karp) The fruiting body of an ascomycete.

ascomycete (ass''koh-my'seat) Member of a phylum of fungi characterized by the production of nonmotile asexual conidia and sexual ascospores.

ascospore (ass'koh-spor) One of a set of sexual spores, usually eight, contained in a special spore case (an ascus) of an ascomycete.

ascus (ass'kus) A saclike spore case in ascomycetes that contains sexual spores called ascospores.

asexual reproduction Reproduction in which there is no fusion of gametes and in which the genetic makeup of parent and of offspring is usually identical. Compare with *sexual reproduction*.

assimilation (of nitrogen) The conversion of inorganic nitrogen (nitrate, NO_3^-, or ammonia, NH_3) to the organic molecules of living things; part of the nitrogen cycle.

association areas Areas of the brain that link sensory and motor areas; responsible for thought, learning, memory, language abilities, judgment, and personality.

association neuron See *interneuron*.

assortative mating Sexual reproduction in which individuals pair nonrandomly, i.e., select mates on the basis of phenotype.

asters Clusters of microtubules radiating out from the poles in dividing cells that have centrioles.

astrocyte A type of glial cell; some are phagocytic; others regulate the composition of the extracellular fluid in the central nervous system.

atherosclerosis (ath''ur-oh-skle-row'sis) A progressive disease in which lipid deposits accumulate in the inner lining of arteries, leading eventually to impaired circulation and heart disease.

atom The smallest quantity of an element that can retain the chemical properties of that element.

atomic mass The total number of protons and neutrons in an atom; expressed in atomic mass units or daltons.

atomic mass unit (amu) The approximate mass of a proton or neutron; also called dalton.

atomic number The number of protons in the atomic nucleus of an atom, which uniquely identifies the element to which the atom corresponds.

ATP See *adenosine triphosphate*.

ATP synthase Large enzyme complex that catalyzes the formation of ATP from ADP and inorganic phosphate by chemiosmosis; contains a transmembrane channel through which protons diffuse down a concentration gradient; located in the inner mitochondrial membrane, the thylakoid membrane of chloroplasts, and the plasma membrane of bacteria.

atrial natriuretic peptide (ANP) A hormone released by the atrium of the heart; helps regulate sodium excretion and lowers blood pressure.

atrioventricular (AV) node (ay''tree-oh-ven-trik'you-lur) Mass of specialized cardiac tissue that receives an impulse from the sinoatrial node (pacemaker) and conducts it to the ventricles.

atrioventricular (AV) valve (of the heart) A valve between each atrium and its ventricle that prevents backflow of blood. The right AV valve is the tricuspid valve, the left AV valve is the mitral valve.

atrium (of the heart) (ay'tree-um) A heart chamber that receives blood from the veins.

australopithecines Early hominids that lived between about 4.4 and 1.25 million years ago, based on fossil evidence. Includes several species in two genera, *Ardipithecus* and *Australopithecus*.

Australopithecus afarensis Hominids that lived between about 3.6 and 3.0 million years ago, e.g., Lucy, discovered at Hadar, Ethiopia, in 1974. *A. afarensis* may have arisen from *A. anamensis*.

Australopithecus africanus Hominids that lived between about 3.0 and 2.5 million years ago. *A. africanus* may have arisen from *A. afarensis*.

Australopithecus anamensis Hominids that lived between about 3.9 and 4.2 million years ago. May have arisen from *Ardipithecus ramidus;* had an upright posture and was bipedal.

autocrine regulation A type of regulation in which a signaling molecule (e.g., a hormone) is secreted into interstitial fluid and then acts on the cells that produce it. Compare with *paracrine regulation*.

autogenous model The idea that eukaryotes arose from prokaryotes by the proliferation of internal membranes, derived from the plasma membrane, to form cellular compartments. Compare with *endosymbiont theory*.

autoimmune disease (aw''toh-ih-mune') A disease in which the body produces antibodies against its own cells or tissues. Also called autoimmunity.

autonomic nervous system (aw-tuh-nom'ik) The portion of the peripheral nervous system that controls the visceral functions of the body, e.g., regulates smooth muscle, cardiac muscle, and glands, thereby helping to maintain homeostasis. Its divisions are the sympathetic and parasympathetic nervous systems. Compare with *somatic nervous system*.

autopolyploid A polyploid whose chromosomes are derived from a single species. Compare with *allopolyploid*.

autoradiography Method for detection of radioactive decay; radiation causes the appearance of dark silver grains in special x-ray film.

autosome (aw'toh-sohm) A chromosome other than the sex (X and Y) chromosomes.

autotroph (aw'toh-trof) An organism that synthesizes complex organic compounds from simple inorganic raw materials; also called producer or primary producer. Compare with *heterotroph*. See *chemoautotroph* and *photoautotroph*.

auxin (awk'sin) A plant hormone involved in various aspects of growth and development, such as stem elongation, apical dominance, and root formation on cuttings, e.g., indole acetic acid (IAA).

avirulent Unable to cause disease in a host. Compare with *virulent*.

Avogadro's number The number of units (6.02×10^{23}) present in one mole of any substance.

axillary bud A bud in the axil of a leaf. Compare with *terminal bud*.

axon (aks'on) The long extension of the neuron that transmits nerve impulses away from the cell body. Compare with *dendrite*.

axopods (aks'o-podz) Long, filamentous cytoplasmic projections characteristic of actinopods.

B cell (B lymphocyte) A type of white blood cell responsible for antibody-mediated immunity. When stimulated, B cells differentiate to become plasma cells that produce antibodies. Compare with *T cell*.

bacillus (bah-sill'us) (pl. *bacilli*) A rod-shaped bacterium. Compare with *coccus, spirillum, vibrio,* and *spirochete*.

background extinction The continuous, low-level extinction of species that has occurred throughout much of the history of life. Compare with *mass extinction*.

bacteria (bak-teer'ee-uh) A general term for two groups of unicellular, prokaryotic microorganisms, the archaebacteria and eubacteria. Most bacteria are decomposers, but some are parasites and others are autotrophs.

bacteriophage (bak-teer'ee-oh-fayj) A virus that can infect a bacterium (literally, "bacteria eater"). Also called phage.

balanced polymorphism (pol''ee-mor'fizm) The presence in a population of two or more genetic variants that are maintained in a stable frequency over several generations.

bark The outermost covering over woody stems and roots; consists of all plant tissues located outside the vascular cambium.

baroreceptors (bare''oh-ree-sep'torz) Receptors within certain blood vessels that are stimulated by changes in blood pressure.

Barr body A condensed and inactivated X chromosome appearing as a distinctive dense spot in the nucleus of certain cells of female mammals.

basal body (bay'sl) Structure involved in the organization and anchorage of a cilium or flagellum. Structurally similar to a centriole; each is in the form of a cylinder composed of nine triplets of microtubules (9×3 structure).

basal metabolic rate (BMR) The amount of energy expended by the body at resting conditions, when no food is being digested and no voluntary muscular work is being performed.

base (1) A substance that is a hydrogen ion (proton) acceptor; bases unite with acids to form salts. Compare with *acid*. (2) A nitrogenous base in a nucleotide or nucleic acid. See *purines* and *pyrimidines*.

basement membrane The thin, noncellular layer of an epithelial membrane that attaches to the underlying tissues; composed of tiny fibers and polysaccharides produced by the epithelial cells.

base-substitution mutation A change in one base pair in DNA. See *missense mutation* and *nonsense mutation*.

basic solution A solution in which the concentration of hydroxide ions (OH^-) exceeds the concentration of hydrogen ions (H^+). A basic solution has pH greater than 7. Compare with *acidic solution* and *neutral solution*.

basidiocarp (ba-sid'ee-o-karp) The fruiting body of a basidiomycete, e.g., a mushroom.

basidiomycete (ba-sid''ee-o-my'seat) Member of a phylum of fungi characterized by the production of sexual basidiospores.

basidiospore (ba-sid'ee-o-spor) One of a set of sexual spores, usually four, borne on a basidium of a basidiomycete.

basidium (ba-sid'ee-um) The clublike spore-producing organ of basidiomycetes that bears sexual spores called basidiospores.

basilar membrane The multicellular tissue in the inner ear that separates the cochlear duct from the tympanic canal; the sensory cells of the organ of Corti rest on this membrane.

Batesian mimicry (bate'see-un mim'ih-kree) The resemblance of a harmless or palatable species to one that is dangerous, unpalatable, or poisonous. Compare with *Müllerian mimicry*.

behavioral ecology The scientific study of behavior in natural environments from the evolutionary perspective.

behavioral isolation A prezygotic reproductive isolating mechanism in which reproduction between similar species is prevented because each group possesses its own characteristic courtship behavior; also called sexual isolation.

bellwether species An organism that provides an early warning of environmental damage. Examples include lichens, which are very sensitive to air pollution, and amphibians, which are sensitive to a wide variety of environmental stressors.

benthos (ben'thos) Bottom-dwelling sea organisms that fix themselves to one spot, burrow into the sediment, or simply walk about on the ocean floor.

berry A simple, fleshy fruit in which the fruit wall is soft throughout, e.g., tomato, banana, grape.

beta (β) oxidation Process by which fatty acids are converted to acetyl CoA before entry into the citric acid cycle.

beta (β)-pleated sheet A regular, folded, sheetlike type of protein secondary structure, resulting from hydrogen bonding between two different polypeptide chains or two regions of the same polypeptide chain. Compare with *alpha (α) helix*.

biennial plant (by-en ee-ul) A plant that takes two years to complete its life cycle. Compare with *annual* and *perennial*.

bilateral symmetry A body shape with right and left halves that are approximately mirror images of one another. Compare with *radial symmetry*.

bile The fluid secreted by the liver; emulsifies fats.

binary fission (by'nare-ee fish'un) Equal division of a cell or organism into two; a type of asexual reproduction.

binomial system of nomenclature (by-nome'ee-ul) System of naming a species by the combination of the genus name and a specific epithet.

bioaccumulation The buildup of a persistent toxic substance, such as certain pesticides, in an organism's body.

biodiversity See *biological diversity*.

biogenic amines A class of neurotransmitters that includes norepinephrine, serotonin, and dopamine.

biogeochemical cycle (bye''o-jee''o-kem'ee-kl) Process by which matter cycles from the living world to the nonliving, physical environment and back again, e.g., the carbon cycle, the nitrogen cycle, and the phosphorus cycle.

biogeography The study of the past and present geographical distributions of organisms.

bioinformatics A new scientific field that uses powerful computers to store, retrieve, and compare biological information. Much of bioinformatics is concerned with DNA sequences within human DNA and between genomes of different species. Also called biological computing.

biological clocks Mechanisms by which activities of organisms are adapted to regularly recurring changes in the environment. See *circadian rhythm*.

biological computing See *bioinformatics*.

biological diversity The variety of living organisms, from their genes to the ecosystems in which they live; includes species richness, genetic diversity, and ecosystem diversity. Also called biodiversity.

biological magnification The increased concentration of toxic chemicals, such as PCBs, heavy metals, and certain pesticides, in the tissues of organisms at higher trophic levels in food webs.

biological species concept See *species*.

biomass (bye'o-mas) A quantitative estimate of the total mass, or amount, of living material in a particular ecosystem.

biome (by'ohm) A large, relatively distinct terrestrial region characterized by a similar climate, soil, plants, and animals, regardless of where it occurs on Earth.

bioremediation A method to clean up a hazardous waste site that uses microorganisms to break down toxic pollutants, or plants to selectively accumulate toxins so they can be easily removed from the site.

biosphere All of Earth's living organisms.

biotic factors Elements of the living world that affect a particular organism, that is, its relationships with other organisms. Compare with *abiotic factors*.

biotic potential See *intrinsic rate of increase*.

bipedal Walking on two feet.

bipolar cell A type of neuron in the retina of the eye; receives input from the photoreceptors (rods and cones) and synapses on ganglion cells.

biramous appendages Appendages with two jointed branches at their ends; characteristic of crustaceans.

bivalent (by-vale'ent or biv'ah-lent) See *tetrad*.

blade (1) The thin, expanded part of a leaf; (2) The flat, leaflike structure of certain multicellular algae.

blastocoel (blas'toh-seel) The fluid-filled cavity of a blastula.

blastocyst The mammalian blastula. See *blastula*.

blastodisc A small disc of cytoplasm at the animal pole of a reptile or bird egg; cleavage is restricted to the blastodisc (meroblastic cleavage).

blastopore (blas'toh-pore) The primitive opening into the body cavity of an early embryo that may become the mouth (in protostomes) or anus (in deuterostomes) of the adult organism.

blastula (blas'tew-lah) In animal development, a hollow ball of cells produced by cleavage of a fertilized ovum. Known as a blastocyst in mammalian development.

blood A fluid, circulating connective tissue that transports nutrients and other materials through the bodies of many types of animals.

blood pressure The force exerted by blood against the inner walls of the blood vessels.

bloom The sporadic occurrence of huge numbers of algae in freshwater and marine ecosystems.

body mass index (BMI) An index of weight in relation to height; calculated by dividing the square of the weight (square kilograms) by height (meters).

Bohr effect Increased oxyhemoglobin dissociation due to lowered pH; occurs as carbon dioxide concentration increases.

bolting The production of a tall flower stalk by a plant that grows vegetatively as a rosette (growth habit with a short stem and a circular cluster of leaves).

bond energy The energy required to break a particular chemical bond.

bone tissue Principal vertebrate skeletal tissue; a type of connective tissue.

boreal forest (bor'ee-uhl) See *taiga*.

bottleneck A sudden decrease in a population size due to adverse environmental factors; may result in genetic drift; also called genetic bottleneck or population bottleneck.

Bowman's capsule A double-walled sac of cells that surrounds the glomerulus of each nephron.

brachiopods (bray'kee-oh-pods) The phylum of solitary marine invertebrates possessing a pair of shells, and internally, a pair of coiled arms with ciliated tentacles; one of the lophophorate phyla.

brain A concentration of nervous tissue that controls neural function; in vertebrates, the anterior, enlarged portion of the central nervous system.

brain stem The part of the vertebrate brain that includes the medulla, pons, and midbrain.

branchial Pertaining to the gills or gill region.

branching evolution See *cladogenesis*.

bronchiole (bronk'ee-ole) Air duct in the lung that branches from a bronchus; divides to form air sacs (alveoli).

bronchus (bronk'us) (pl. *bronchi*) One of the branches of the trachea and its immediate branches within the lung.

brown alga One of a phylum of predominantly marine algae that are multicellular and contain the pigments chlorophyll *a* and *c*, and carotenoids, including fucoxanthin.

bryophytes (bry'oh-fites) Nonvascular plants including mosses, liverworts, and hornworts.

bryozoans Animals belonging to phylum Bryozoa, one of the three lophophorate phyla; form sessile colonies by asexual budding.

bud An undeveloped shoot that can develop into flowers, stems, or leaves. Buds are enclosed in bud scales.

bud scale A modified leaf that covers and protects a dormant bud.

bud scale scar Scar on a twig left when a bud scale abscises from the terminal bud.

budding Asexual reproduction in which a small part of the parent's body separates from the rest and develops into a new individual; characteristic of yeasts and certain other organisms.

buffer A substance in a solution that tends to lessen the change in hydrogen ion concentration (pH) that otherwise would be produced by adding an acid or base.

bulb A globose, fleshy, underground bud that consists of a short stem with fleshy leaves, e.g., onion.

bundle scar Marks on a leaf scar left when vascular bundles of the petiole break during leaf abscission.

bundle sheath cells Tightly packed cells that form a sheath around the veins of a leaf.

bundle sheath extension Support cells that extend from the bundle sheath of a leaf vein toward the upper and/or lower epidermis.

buttress root A bracelike root at the base of certain trees that provides upright support.

C_3 plant Plant that carries out carbon fixation solely by the Calvin cycle. Compare with *C_4 plant* and *CAM plant*.

C_4 plant Plant that fixes carbon initially by the Hatch-Slack pathway, in which the reaction of CO_2 with phosphoenolpyruvate is catalyzed by PEP carboxylase in leaf mesophyll cells; the products are transferred to the bundle sheath cells, where the Calvin cycle takes place. Compare with *C_3 plant* and *CAM plant*.

calcitonin (kal-sih-toh'nin) A hormone secreted by the thyroid gland that rapidly lowers the calcium content in the blood.

callus (kal'us) Undifferentiated tissue formed on an explant (excised tissue or organ) in plant tissue culture.

calmodulin A calcium-binding protein; when bound it alters the activity of certain enzymes or transport proteins.

calorie The amount of heat energy required to raise the temperature of 1g of water 1°C; equivalent to 4.184 joules. Compare with *kilocalorie*.

Calvin cycle Cyclic series of reactions in the chloroplast stroma in photosynthesis; fixes carbon dioxide and produces carbohydrate. See *C_3 plant*.

calyx (kay'liks) The collective term for the sepals of a flower.

cambium See *lateral meristems*.

Cambrian explosion A span of 40 million years, from about 565 to 525 million years ago, during which many new animal groups appeared in the fossil record.

CAM plant Plant that carries out crassulacean acid metabolism; carbon is initially fixed into organic acids at night in the reaction of CO_2 and phosphoenolpyruvate, catalyzed by PEP carboxylase; during the day the acids break down to yield CO_2, which enters the Calvin cycle. Compare with *C_3 plant* and *C_4 plant*.

cAMP See *cyclic AMP*.

cancer cells See *malignant*.

CAP See *catabolite gene activator protein*.

capillaries (kap'i-lare-eez) Microscopic blood vessels in the tissues that permit exchange of materials between cells and blood.

capillary action The ability of water to move in small diameter tubes as a consequence of its cohesive and adhesive properties.

capping See *mRNA cap*.

capsid Protein coat surrounding the nucleic acid of a virus.

capsule (1) The portion of the moss sporophyte that contains spores; (2) A simple, dry, dehiscent fruit that opens along many sutures or pores to release seeds; (3) A gelatinous coat that surrounds some bacteria.

carbohydrate Compound containing carbon, hydrogen, and oxygen, in the approximate ratio of C:2H:O, e.g., sugars, starch, and cellulose.

carbon cycle The worldwide circulation of carbon from the abiotic environment into living things and back into the abiotic environment.

carbon fixation reactions Reduction reactions of photosynthesis in which carbon from carbon dioxide becomes incorporated into organic molecules, leading to the production of carbohydrate; requires ATP and NADPH.

carbonyl group A polar functional group consisting of a carbon attached to an oxygen by a double bond; found in aldehydes and ketones.

carboxyl group A weakly acidic functional group; abbreviated —COOH.

carcinogen (kar-sin'oh-jen) An agent that causes cancer or accelerates its development.

cardiac cycle One complete heart beat.

cardiac muscle Involuntary, striated type of muscle found in the vertebrate heart. Compare with *smooth muscle* and *skeletal muscle*.

cardiac output The volume of blood pumped by the left ventricle into the aorta in one minute.

cardiovascular disease Disease of the heart or blood vessels; the leading cause of death in most industrial societies.

carnivore (kar'ni-vor) An animal that feeds on other animals; flesh-eater; also called secondary or tertiary consumer. Secondary consumers eat primary consumers (herbivores), whereas tertiary consumers eat secondary consumers.

carotenoids (ka-rot'n-oidz) A group of yellow to orange plant pigments synthesized from isoprene subunits; include carotenes and xanthophylls.

carpel (kar'pul) The female reproductive unit of a flower; carpels bear ovules. Compare with *pistil*.

carrier-mediated active transport Transport across a membrane of a substance from a region of low concentration to a region of high concentration; requires both a transport protein with a binding site for the specific substance and an energy source (often ATP).

carrier-mediated transport Any form of transport across a membrane that uses a membrane-bound transport protein with a binding site for a specific substance; includes both facilitated diffusion and carrier-mediated active transport.

carrying capacity The largest population that a particular habitat can support and sustain for an indefinite period, assuming there are no changes in the environment.

cartilage A flexible skeletal tissue of vertebrates; a type of connective tissue.

Casparian strip (kas-pare′ee-un) A band of waterproof material around the radial and transverse walls of endodermal root cells.

catabolism The aspect of metabolism in which complex substances are broken down to form simpler substances; catabolic reactions are particularly important in releasing chemical energy stored by the cell. Compare with *anabolism*.

catabolite gene activator protein (CAP) A positively acting regulator that becomes active when bound to cAMP; active CAP stimulates transcription of the lactose operon and other operons that code for enzymes used in catabolic pathways. Also known as cyclic AMP receptor protein (CRP).

catalyst (kat′ah-list) A substance that increases the speed at which a chemical reaction occurs without being used up in the reaction. Enzymes are biological catalysts.

catecholamine (cat″eh-kole′-ah-meen) A class of compounds including dopamine, epinephrine, and norepinephrine; these compounds serve as neurotransmitters and hormones.

cation A particle with one or more units of positive charge, such as a hydrogen ion (H^+) or calcium ion (Ca^{2+}). Compare with *anion*.

cDNA library A collection of recombinant plasmids that contain complementary DNA (cDNA) copies of mRNA templates. The cDNA, which lacks introns, is synthesized by reverse transcriptase. Compare with *genomic DNA library*.

cell The basic structural and functional unit of life, which consists of living material bounded by a membrane.

cell cycle Cyclic series of events in the life of a dividing eukaryotic cell; consists of mitosis, cytokinesis, and the stages of interphase.

cell determination See *determination*.

cell differentiation See *differentiation*.

cell fractionation The technique used to separate the components of cells by subjecting them to centrifugal force. See *differential centrifugation* and *density gradient centrifugation*.

cell plate The structure that forms during cytokinesis in plants, separating the two daughter cells produced by mitosis.

cell signaling Mechanisms of communication between cells. Cells can signal one another with secreted signaling molecules, or a signaling molecule on one cell can combine with a receptor on another cell. Examples include the synaptic signaling of neurons and endocrine signaling. See *signal transduction*.

cell theory The theory that the cell is the basic unit of life, of which all living things are composed, and that all cells are derived from preexisting cells.

cell wall The structure outside the plasma membrane of certain cells; may contain cellulose (plant cells), chitin (most fungal cells), peptidoglycan and/or lipopolysaccharide (most bacterial cells), or other material.

cellular respiration See *respiration*.

cellular slime mold A phylum of fungus-like protists whose feeding stage consists of unicellular, amoeboid organisms that aggregate to form a pseudoplasmodium during reproduction.

cellulose (sel′yoo-lohs) A structural polysaccharide composed of beta glucose subunits; the main constituent of plant primary cell walls.

Cenozoic era A geological era that began about 65 million years ago and extends to the present time.

center of origin The geographical area where a given species originated.

central nervous system (CNS) In vertebrates, the brain and spinal cord. Compare with *peripheral nervous system (PNS)*.

centrifuge A device used to separate cells or their components by subjecting them to centrifugal force.

centriole (sen′tree-ohl) One of a pair of small, cylindrical organelles lying at right angles to each other near the nucleus in the cytoplasm of animal cells and certain protist and plant cells; each centriole is in the form of a cylinder composed of nine triplets of microtubules (9×3 structure).

centromere (sen′tro-meer) A specialized constricted region of a chromatid; contains the kinetochore. In cells at prophase and metaphase, sister chromatids are joined in the vicinity of their centromeres.

cephalization The evolution of a head; the concentration of nervous tissue and sense organs at the front end of the animal.

cephalochordates Members of the chordate subphylum that includes the lancelets.

cerebellum (ser-eh-bel′um) A convoluted subdivision of the vertebrate brain concerned with the coordination of muscular movements, muscle tone, and balance.

cerebral cortex (ser-ee′brul kor′tex) The outer layer of the cerebrum composed of gray matter and consisting mainly of nerve cell bodies.

cerebrospinal fluid (CSF) The fluid that bathes the central nervous system of vertebrates.

cerebrum (ser-ee′brum) A large, convoluted subdivision of the vertebrate brain; in humans, it functions as the center for learning, voluntary movement, and interpretation of sensation.

chaos The tendency of a simple system to exhibit complex, erratic dynamics; used by some ecologists to model the state of flux displayed by some populations.

chaparral (shap″uh-ral′) A biome with a Mediterranean climate (mild, moist winters and hot, dry summers). Chaparral vegetation is characterized by drought-resistant, small-leaved evergreen shrubs and small trees.

chaperones See *molecular chaperones*.

character displacement The tendency for two similar species to diverge (become more different) in areas where their ranges overlap; reduces interspecific competition.

chelicerae (keh-lis′er-ee) The first pair of appendages in certain arthropods; clawlike appendages located immediately anterior to the mouth and used to manipulate food into the mouth.

chemical bond A force of attraction between atoms in a compound. See *covalent bond, hydrogen bond,* and *ionic bond*.

chemical compound Two or more elements combined in a fixed ratio.

chemical evolution The origin of life from nonliving matter.

chemical formula A representation of the composition of a compound; the elements are indicated by chemical symbols with subscripts to indicate their ratios. See *molecular formula, structural formula,* and *simplest formula*.

chemical symbol The abbreviation for an element; usually the first letter (or first and second letters) of the English or Latin name.

chemiosmosis Process by which phosphorylation of ADP to form ATP is coupled to the transfer of electrons down an electron transport chain; the electron transport chain powers proton pumps that produce a proton gradient across the membrane; ATP is formed as protons diffuse through transmembrane channels in ATP synthase.

chemoautotroph (kee''moh-aw'toh-trof) Organism that obtains energy from inorganic compounds and synthesizes organic compounds from inorganic raw materials; includes some bacteria. Compare with *photoautotroph, photoheterotroph,* and *chemoheterotroph*.

chemoheterotroph (kee''moh-het'ur-oh-trof) Organism that uses organic compounds as a source of energy and carbon; includes animals, fungi, and many bacteria. Compare with *photoautotroph, photoheterotroph,* and *chemoautotroph*.

chemoreceptor (kee''moh-ree-sep'tor) A sensory receptor that responds to chemical stimuli.

chemotroph (kee'moh-trof) Organism that uses organic compounds or inorganic substances, such as iron, nitrate, ammonia, or sulfur, as sources of energy. Compare with *phototroph*. See *chemoautotroph* and *chemoheterotroph*.

chiasma (ky-az'muh) (pl. *chiasmata*) An X-shaped site in a tetrad (bivalent) usually marking the location where homologous (nonsister) chromatids previously underwent crossing-over.

chimera (ky meer'' uh) An organism composed of two or more kinds of genetically dissimilar cells.

chitin (ky'tin) A nitrogen-containing structural polysaccharide that forms the exoskeleton of insects and the cell walls of many fungi.

chlorophyll (klor'oh-fil) A group of light-trapping green pigments found in most photosynthetic organisms.

chlorophyll-binding proteins About 15 different proteins associated with chlorophyll molecules in the thylakoid membrane.

chloroplasts (klor'oh-plastz) Membranous organelles that are the sites of photosynthesis in eukaryotes; occur in some plant and algal cells.

cholinergic neuron (kohl''in-air'jik) A nerve cell that secretes acetylcholine as a neurotransmitter. Compare with *adrenergic neuron*.

chondrichthyes (kon-drik'-thees) The class of cartilaginous fishes that includes the sharks, rays, and skates.

chondrocytes Cartilage cells.

chordates (kor'dates) Deuterostome animals that, at some time in their lives, possess a cartilaginous, dorsal skeletal structure called a notochord; a dorsal, tubular nerve cord; pharyngeal gill grooves; and a postanal tail.

chorion (kor'ee-on) An extraembryonic membrane in reptiles, birds, and mammals that forms an outer cover around the embryo, and in mammals contributes to the formation of the placenta.

chorionic villus sampling (CVS) (kor''ee-on'ik) Study of extraembryonic cells that are genetically identical to the cells of an embryo, making it possible to assess its genetic makeup. Compare with *amniocentesis*.

choroid layer A layer of cells filled with black pigment that absorbs light and prevents reflected light from blurring the image that falls on the retina; the layer of the eyeball outside the retina.

chromatid (kroh'mah-tid) One of the two identical halves of a duplicated chromosome; the two chromatids that make up a chromosome are referred to as sister chromatids.

chromatin (kro'mah-tin) The complex of DNA and protein that makes up eukaryotic chromosomes.

chromoplasts Pigment-containing plastids; usually found in flowers and fruits.

chromosomes Structures in the cell nucleus that are composed of chromatin and contain the genes. The chromosomes become visible under the microscope as distinct structures during cell division.

chylomicrons (kie-low-my'kronz) Protein-covered fat droplets produced in the intestinal cells; they enter the lymphatic system and are transported to the blood.

chytrid See *chytridiomycete*.

chytridiomycete (ki-trid''ee-o-my'seat) A member of a phylum of fungi characterized by the production of flagellated cells at some stage in their life history. Also called chytrid.

ciliate (sil'e-ate) A unicellular protozoon covered by many short cilia.

cilium (sil'ee-um) (pl. *cilia*) One of many short, hairlike structures that project from the surface of some eukaryotic cells and are used for locomotion or movement of materials across the cell surface; composed of two single microtubules surrounded by nine double microtubules (9 + 2 structure), covered by a plasma membrane.

circadian rhythm (sir-kay'dee-un) An internal rhythm that approximates the 24-hour day. See *biological clocks*.

circulatory system The body system that functions in internal transport and protects the body from disease.

cisternae (sing. *cisterna*) Stacks of flattened membranous sacs that make up the Golgi complex.

citrate (citric acid) A six-carbon organic acid.

citric acid cycle Series of chemical reactions in aerobic respiration in which acetyl coenzyme A is completely degraded to carbon dioxide and water with the release of metabolic energy that is used to produce ATP; also known as the Krebs cycle and the tricarboxylic acid (TCA) cycle.

clade A taxon containing a common ancestor and all the taxa descended from it; a monophyletic group.

cladistics An approach to classification based on recency of common ancestry rather than degree of structural similarity. Also called phylogenetic systematics. Compare with *phenetics* and *evolutionary systematics*.

cladogenesis A branching type of evolution in which two or more populations of an ancestral species split and diverge; also called branching evolution. Compare with *anagenesis*.

cladogram A branching diagram that illustrates taxonomic relationships based on the principles of cladistics.

class A taxonomic category made up of related orders.

classical conditioning A type of learning in which an association is formed between some normal response to a stimulus and a new stimulus, after which the new stimulus elicits the response.

cleavage Series of mitotic cell divisions, without growth, that converts the zygote to a multicellular blastula.

cleavage furrow A constricted region of the cytoplasm that forms and progressively deepens during cytokinesis of animal cells, thereby separating the two daughter cells.

cline Gradual change in phenotype and genotype frequencies among contiguous populations that is the result of an environmental gradient.

clitoris (klit'o-ris) A small, erectile structure at the anterior part of the vulva in female mammals; homologous to the male penis.

cloaca (klow-a'ka) An exit chamber in some animals that receives digestive wastes and urine; may also serve as an exit for gametes.

clonal selection Lymphocyte activation in which a specific antigen causes activation, cell division, and differentiation only in cells that express receptors with which the antigen can bind.

clone (1) A population of cells descended by mitotic division from a single ancestral cell; (2) A population of genetically identical organisms asexually propagated from a single individual. Also see *DNA cloning*.

cloning The process of forming a clone.

closed circulatory system A type of circulatory system in which the blood flows through a continuous circuit of blood vessels; characteristic of annelids, cephalopods, and vertebrates. Compare with *open circulatory system*.

closed system An entity that does not exchange energy or matter with its surroundings. Compare with *open system*.

club mosses A phylum of seedless vascular plants with a life cycle similar to ferns.

clumped dispersion The spatial distribution pattern of a population in which individuals are more concentrated in specific parts of the habitat. Also called aggregated distribution and patchiness. Compare with *random dispersion and uniform dispersion*.

cnidarians (ni-dah'ree-anz) Phylum of animals that have stinging cells called cnidocytes, two tissue layers, and radial symmetry; include hydras and jellyfish.

cnidocytes Stinging cells characteristic of cnidarians.

coated pit A depression in the plasma membrane, the cytosolic side of which is coated with the protein clathrin; important in receptor-mediated endocytosis.

cochlea (koke'lee-ah) The structure of the inner ear of mammals that contains the auditory receptors (organ of Corti).

coccus (kok'us) (pl. *cocci*) A bacterium with a spherical shape. Compare with *bacillus, spirillum, vibrio,* and *spirochete*.

codominance (koh''dom'in-ants) Condition in which two alleles of a locus are expressed in a heterozygote.

codon (koh'don) A triplet of mRNA nucleotides. The 64 possible codons collectively constitute a universal genetic code in which each codon specifies an amino acid in a polypeptide, or a signal to either start or terminate polypeptide synthesis.

coelacanths A genus of lobe-finned fish that have survived to the present day.

coelom (see'lum) The main body cavity of most animals; a true coelom is lined with mesoderm. Compare with *pseudocoelom*.

coelomate (seel'oh-mate) Animal possessing a true coelom. Compare with *acoelomate* and *pseudocoelomate*.

coenocyte (see'no-site) An organism consisting of a multinucleate cell, i.e., the nuclei are not separated from one another by septa.

coenzyme (koh-en'zime) An organic cofactor for an enzyme; generally participates in the reaction by transferring some component, such as electrons or part of a substrate molecule.

coenzyme A (CoA) Organic cofactor responsible for transferring groups derived from organic acids.

coevolution The reciprocal adaptation of two or more species that occurs as a result of their close interactions over a long period.

cofactor A nonprotein substance needed by an enzyme for normal activity; some cofactors are inorganic (usually metal ions); others are organic (coenzymes).

cohesive Having the property of sticking together.

cohort A group of individuals of the same age.

colchicine A drug that blocks the division of eukaryotic cells by binding to tubulin subunits, which make up the microtubules that comprise the major component of the mitotic spindle.

coleoptile (kol-ee-op'tile) A protective sheath that encloses the young stem in certain monocots.

collagens (kol'ah-gen) Proteins found in the collagen fibers of connective tissues.

collecting duct A tube in the kidney that receives filtrate from several nephrons and conducts it to the renal pelvis.

collenchyma (kol-en'kih-mah) Living cells with moderately but unevenly thickened primary cell walls; collenchyma cells help support the herbaceous plant body.

commensalism (kuh-men'sul-iz-m) A type of symbiosis in which one organism benefits and the other one is neither harmed nor helped. Compare with *mutualism* and *parasitism*.

commercial harvest The collection of commercially important organisms from the wild. Examples include the commercial harvest of parrots (for the pet trade) and cacti (for houseplants).

community An association of different species living together in a defined habitat with some degree of interdependence. Compare with *ecosystem*.

community ecology The description and analysis of patterns and processes within the community.

compact bone Dense, hard bone tissue found mainly near the surfaces of a bone.

companion cell A cell in the phloem of flowering plants that is responsible for loading and unloading sugar into the sieve tube member for translocation.

competition The interaction among two or more individuals that attempt to use the same essential resource, such as food, water, sunlight, or living space. See *interspecific* and *intraspecific competition*. See *interference* and *exploitation competition*.

competitive exclusion The concept that no two species with identical living requirements can occupy the same ecological niche indefinitely. Eventually, one species will be excluded by the other as a result of interspecific competition for a resource in limited supply.

competitive inhibitor A substance that binds to the active site of an enzyme, thus lowering the rate of the reaction catalyzed by the enzyme. Compare with *noncompetitive inhibitor*.

complement A group of proteins in blood and other body fluids that are activated by an antigen-antibody complex, and then destroy pathogens.

complementary DNA (cDNA) DNA synthesized by reverse transcriptase, using RNA as a template.

complete flower A flower that possesses all four parts: sepals, petals, stamens, and carpels. Compare with *incomplete flower*.

compound eye An eye, such as that of an insect, composed of many light-sensitive units called ommatidia.

concentration gradient A difference in the concentration of a substance from one point to another, as for example, across a cell membrane.

condensation synthesis A reaction in which two monomers are combined covalently through the removal of the equivalent of a water molecule. Compare with *hydrolysis*.

cone (1) In botany, a reproductive structure in many gymnosperms that produces either microspores or megaspores; (2) In zoology, one of the conical photoreceptive cells of the retina that is particularly sensitive to bright light and, by distinguishing light of various wavelengths, mediates color vision. Compare with *rod*.

conidiophore (kah-nid'e-o-for'') A specialized hypha that bears conidia.

conidium (kah-nid'e-um) (pl. *conidia*) An asexual spore that is usually formed at the tip of a specialized hypha called a conidiophore.

conifer (kon'ih-fur) Any of a large phylum of gymnosperms that are woody trees and shrubs with needle-like, mostly evergreen, leaves and with seeds in cones.

conjugation (kon''jew-gay'shun) (1) A sexual process in certain protists that involves exchange or fusion of a cell with another cell; (2) A mechanism for DNA exchange in bacteria that involves cell-to-cell contact.
connective tissue Animal tissue consisting mostly of intercellular substance (fibers scattered through a matrix) in which the cells are embedded, e.g., bone.
consanguineous mating A mating between close relatives.
conservation biology A multidisciplinary science that focuses on the study of how humans impact organisms and on the development of ways to protect biological diversity.
constitutive gene A gene that is constantly transcribed.
consumer See *heterotroph*.
consumption overpopulation A situation in which each individual in a human population consumes too large a share of resources; results in pollution, environmental degradation, and resource depletion. Compare with *people overpopulation*.
contest competition See *interference competition*.
continental drift The theory that continents were once joined together and later split and drifted apart.
contraception Any method used to intentionally prevent pregnancy.
contractile root (kun-trak'til) A specialized type of root that contracts and pulls a bulb or corm deeper into the soil.
contractile vacuole A membrane-bounded organelle that is found in certain freshwater protists, such as *Paramecium,* and that appears to have an osmoregulatory function; it periodically fills with water, then contracts to expel the contents into the surroundings.
control group In a scientific experiment, a group in which the experimental variable is kept constant. The control provides a standard of comparison used to verify the results of the experiment.
controlled mating A mating in which the genotypes of the parents are known.
convergent circuit (kun-vur'jent) A neural pathway in which a postsynaptic neuron is controlled by signals coming from two or more presynaptic neurons. Compare with *divergent circuit*.
convergent evolution (kun-vur'jent) The independent evolution of structural or functional similarity in two or more distantly related species, usually as a result of adaptations to similar environments.
corepressor Substance that binds to a repressor protein, converting it to its active form, which is capable of preventing transcription.
Coriolis effect (kor''e-o'lis) The tendency of moving air or water to be deflected from its path to the right in the Northern Hemisphere and to the left in the Southern Hemisphere. Caused by the direction of Earth's rotation.
cork cambium (kam'bee-um) A lateral meristem that produces cork cells and cork parenchyma; cork cambium and the tissues it produces make up the outer bark of a woody plant. Compare with *vascular cambium*.
cork cell A cell in the bark that is produced outwardly by the cork cambium; cork cells are dead at maturity and function for protection and reduction of water loss.
cork parenchyma (par-en'kih-mah) One or more layers of parenchyma cells produced inwardly by the cork cambium.
corm A short, thickened underground stem specialized for food storage and asexual reproduction, e.g., crocus, gladiolus.
cornea (kor'nee-ah) The transparent covering of an eye.
corolla (kor-ohl'ah) A collective term for the petals of a flower.
corpus callosum (kah-loh'sum) In mammals, a large bundle of nerve fibers interconnecting the two cerebral hemispheres.
corpus luteum (loo'tee''um) The temporary endocrine tissue in the ovary that develops from the ruptured follicle after ovulation; secretes progesterone and estrogen.
cortex (kor'tex) (1) The outer part of an organ, such as the cortex of the kidney; compare with *medulla*. (2) The tissue between the epidermis and vascular tissue in the stems and roots of many herbaceous plants.
cortical reaction Process occurring after fertilization that prevents additional sperm from entering the egg; also known as the slow block to polyspermy.
cosmopolitan species Species that have a nearly worldwide distribution and occur on more than one continent or throughout much of the ocean. Compare with *endemic species*.
cotransport The active transport of a substance from a region of low concentration to a region of high concentration by coupling its transport to the transport of a substance down its concentration gradient.
cotyledon (kot''uh-lee'dun) The seed leaf of a plant embryo, which may contain food stored for germination.
cotylosaurs The first reptiles; also known as stem reptiles.
countercurrent exchange system A biological mechanism that enables maximum exchange between two fluids. The two fluids must be flowing in opposite directions and have a concentration gradient between them.
coupled reactions A set of reactions in which an exergonic reaction provides the free energy required to drive an endergonic reaction; energy coupling generally occurs through a common intermediate.
covalent bond The chemical bond involving shared pairs of electrons; may be single, double, or triple (with one, two, or three shared pairs of electrons, respectively). Compare with *ionic bond* and *hydrogen bond*.
covalent compound A compound in which atoms are held together by covalent bonds; covalent compounds consist of molecules. Compare with *ionic compound*.
cranial nerves The ten to twelve pairs of nerves in vertebrates that emerge directly from the brain.
cranium The bony framework that protects the brain in vertebrates.
crassulacean acid metabolism See *CAM plant*.
creatine phosphate An energy-storing compound found in muscle cells.
cretinism (kree'tin-izm) A chronic condition due to lack of thyroid secretion during fetal development and early childhood; results in retarded physical and mental development if untreated.
cri-du-chat A human genetic disease caused by loss of part of the short arm of chromosome 5 and characterized by mental retardation, a cry that sounds like a kitten mewing, and death in infancy or childhood.
cristae (kris'tee) (sing. *crista*) Shelflike or finger-like inward projections of the inner membrane of a mitochondrion.
Cro-Magnons Prehistoric humans *(Homo sapiens)* with modern features (tall, erect, lacking a heavy brow) who lived in Europe some 30,000 years ago.
cross bridges The connections between myosin and actin filaments in muscle fibers; formed by the binding of myosin heads to active sites on actin filaments.
crossing-over The breaking and rejoining of homologous (nonsister) chromatids during early meiotic prophase I that results in an exchange of genetic material.

CRP See *catabolite gene activator protein.*

cryptic coloration Colors or markings that help some organisms hide from predators by blending into their physical surroundings. Compare with *aposematic coloration.*

cryptochrome A proteinaceous pigment that strongly absorbs blue light: implicated in resetting the biological clock in plants, fruit flies, and mice.

ctenophores (ten′oh-forz) Phylum of marine animals (comb jellies) whose bodies consist of two layers of cells enclosing a gelatinous mass. The outer surface is covered with comblike rows of cilia, by which the animal moves.

cuticle (kew′tih-kl) (1) A noncellular, waxy covering over the epidermis of the aerial parts of plants that reduces water loss; (2) The outer covering of some animals, such as roundworms.

cyanobacteria (sy-an″oh-bak-teer′ee-uh) Prokaryotic photosynthetic microorganisms that possess chlorophyll and produce oxygen during photosynthesis. Formerly known as blue-green algae.

cycad (sih′kad) Any of a phylum of gymnosperms that live mainly in tropical and semitropical regions and have stout stems (to 20 m in height) and fernlike leaves.

cyclic AMP (cAMP) A form of adenosine monophosphate in which the phosphate is part of a ring-shaped structure; acts as a regulatory molecule and second messenger in organisms ranging from bacteria to humans.

cyclic AMP receptor protein See *catabolite gene activator protein.*

cyclic electron transport In photosynthesis, the cyclic flow of electrons through Photosystem I; ATP is formed by chemiosmosis, but O_2 and NADPH are not produced. Compare with *noncyclic electron transport.*

cyclins Regulatory proteins whose levels oscillate during the cell cycle; activate cyclin-dependent protein kinases.

cystic fibrosis A genetic disease with an autosomal recessive inheritance pattern; characterized by secretion of abnormally thick mucus, particularly in the respiratory and digestive systems.

cytochromes (sy′toh-kromz) Iron-containing heme proteins of an electron transport system.

cytokines Signaling proteins that regulate interactions between cells in the immune system. Important groups include interferons, interleukins, tumor necrosis factors, and chemokines.

cytokinesis (sy″toh-kih-nee′sis) Stage of cell division in which the cytoplasm divides to form two daughter cells.

cytokinin (sy″toh-ky′nin) A plant hormone involved in various aspects of plant growth and development, such as cell division and delay of senescence.

cytoplasm The plasma membrane and cell contents with the exception of the nucleus.

cytosine A nitrogenous pyrimidine base that is a component of nucleic acids.

cytoskeleton The dynamic internal network of protein fibers that includes microfilaments, intermediate filaments, and microtubules.

cytosol The fluid component of the cytoplasm in which the organelles are suspended.

cytotoxic T cell T lymphocyte that destroys cancer cells and other pathogenic cells on contact. Also known as CD8 T cell and killer T cell.

dalton See *atomic mass unit (amu).*

day-neutral plant A plant whose flowering is not controlled by variations in day length that occur with changing seasons. Compare with *long-day, short-day,* and *intermediate-day plants.*

deamination (dee-am-ih-nay′shun) The removal of an amino group (—NH_2) from an amino acid or other organic compound.

decarboxylation A reaction in which a molecule of CO_2 is removed from a carboxyl group of an organic acid.

deciduous A term describing a plant that sheds leaves or other structures at regular intervals; e.g., during autumn. Compare with *evergreen.*

decomposers Microbial heterotrophs that break down dead organic material and use the decomposition products as a source of energy. Also called saprotrophs or saprobes.

deductive reasoning The reasoning that operates from generalities to specifics and can make relationships among data more apparent. Compare with *inductive reasoning.* See *hypothetico-deductive approach.*

deforestation The temporary or permanent removal of forest for agriculture or other uses.

dehydrogenation (dee-hy″dro-jen-ay′shun) A form of oxidation in which hydrogen atoms are removed from a molecule.

deletion (1) A chromosome abnormality in which part of a chromosome is missing, e.g., cri-du-chat; (2) The loss of one or more base pairs from DNA, which can result in a frameshift mutation.

demographics The science that deals with human population statistics, such as size, density, and distribution.

denature (dee-nay′ture) To alter the physical properties and three-dimensional structure of a protein, nucleic acid, or other macromolecule by treating it with excess heat, strong acids, or strong bases.

dendrite (den′drite) A branch of a neuron that receives and conducts nerve impulses toward the cell body. Compare with *axon.*

dendritic cells A set of immune cells present in many tissues that capture antigens and present them to T cells.

dendrochronology (den″dro-kruh-naal′uh-gee) A method of dating using the annual rings of trees.

denitrification (dee-nie″tra-fuh-kay′shun) The conversion of nitrate (NO_3^-) to nitrogen gas (N_2) by certain bacteria (denitrifying bacteria) in the soil; part of the nitrogen cycle.

dense connective tissue A type of tissue that may be irregular, as in the dermis of the skin, or regular, as in tendons.

density-dependent factor An environmental factor whose effects on a population change as population density changes; density-dependent factors tend to retard population growth as population density increases and enhance population growth as population density decreases. Compare with *density-independent factor.*

density gradient centrifugation Procedure in which cellular components are placed in a layer on top of a density gradient, usually made up of a sucrose solution and water. Cell structures migrate during centrifugation, forming a band at the position in the gradient where their own density equals that of the sucrose solution.

density-independent factor An environmental factor that affects the size of a population but is not influenced by changes in population density. Compare with *density-dependent factor.*

deoxyribonucleic acid (DNA) Double-stranded nucleic acid; contains genetic information coded in specific sequences of its constituent nucleotides.

deoxyribose Pentose sugar lacking a hydroxyl (—OH) group on carbon-2′; a constituent of DNA.

depolarization (dee-pol''ar-ih-zay'shun) A decrease in the charge difference across a plasma membrane; may result in an action potential in a neuron or muscle cell.

derived characters See *shared derived characters*.

dermal tissue system The tissue that forms the outer covering over a plant; the epidermis or periderm.

dermis (dur'mis) The layer of dense connective tissue beneath the epidermis in the skin of vertebrates.

desert A temperate or tropical biome in which lack of precipitation limits plant growth.

desertification The degradation of once-fertile land into nonproductive desert; caused partly by soil erosion, deforestation, and overgrazing by domestic animals.

desmosomes (dez'moh-somz) Button-like plaques, present on two opposing cell surfaces, that hold the cells together by means of protein filaments that span the intercellular space.

determinate growth Growth of limited duration, as for example, in flowers and leaves. Compare with *indeterminate growth*.

determination The developmental process by which one or more cells become progressively committed to a particular fate. Determination is a series of molecular events usually leading to differentiation. Also called cell determination.

detritivore (duh-try'tuh-vore) An organism, such as an earthworm or crab, that consumes fragments of freshly dead or decomposing organisms; also called detritus feeder.

detritus (duh-try'tus) Organic debris from decomposing organisms.

detritus feeder See *detritivore*.

deuteromycetes (doo''ter-o-my'seats) An artificial grouping of fungi characterized by the absence of sexual reproduction but usually having other traits similar to ascomycetes; also called imperfect fungi.

deuterostome (doo'ter-oh-stome) Major division of the animal kingdom in which the anus develops from the blastopore; includes the echinoderms and chordates. Compare with *protostome*.

development All the progressive changes that take place throughout the life of an organism.

diabetes mellitus (mel'i-tus) The most common endocrine disorder; in Type I diabetes, there is a marked decrease in the number of beta cells in the pancreas resulting in insulin deficiency. In the more common Type II diabetes, insulin receptors on target cells do not bind with insulin (insulin resistance). Both types result in hyperglycemia and decreased use of glucose by cells.

diacylglycerol (DAG) (di''as-il-glis'er-ol) A lipid consisting of glycerol combined chemically with two fatty acids; also called diglyceride. Can act as a second messenger that increases calcium concentration and activates enzymes. Compare with *monoacylglycerol* and *triacylglycerol*.

dialysis The diffusion of certain solutes across a selectively permeable membrane.

diaphragm In mammals, the muscular floor of the chest cavity; contracts during inhalation, expanding the chest cavity.

diastole (di-ass'toh-lee) Phase of the cardiac cycle in which the heart is relaxed. Compare with *systole*.

diatom (die'eh-tom'') A usually unicellular alga that is covered by an ornate, siliceous shell consisting of two overlapping halves; an important component of plankton in both marine and fresh waters.

dichotomous branching (di-kaut'uh-mus) In botany, a type of branching in which one part always divides into two more or less equal parts.

dicot (dy'kot) One of the two classes of flowering plants; dicot seeds contain two cotyledons, or seed leaves. Compare with *monocot*.

differential centrifugation Separation of cellular particles according to their mass, size, or density. In differential centrifugation the supernatant is spun at successively higher revolutions per minute.

differential gene expression The expression of different subsets of genes at different times and in different cells during development.

differentiated cell A specialized cell; carries out unique activities, expresses a specific set of proteins, and usually has a recognizable appearance.

differentiation (dif''ah-ren-she-ay'shun) Development toward a more mature state; a process changing a young, relatively unspecialized cell to a more specialized cell. Also called cell differentiation.

diffusion The net movement of particles (atoms, molecules, or ions) from a region of higher concentration to a region of lower concentration (i.e., down a concentration gradient), resulting from random motion. Compare with *facilitated diffusion* and *active transport*.

digestion The breakdown of food to smaller molecules.

diglyceride See *diacylglycerol*.

dihybrid cross (dy-hy'brid) A genetic cross that takes into account the behavior of alleles of two loci. Compare with *monohybrid cross*.

dikaryotic (dy-kare-ee-ot'ik) Condition of having two nuclei per cell (i.e., $n + n$), characteristic of certain fungal hyphae. Compare with *monokaryotic*.

dimer An association of two monomers (e.g. a disaccharide or a dipeptide).

dinoflagellate (dy''noh-flaj'eh-late) A unicellular, biflagellate, typically marine alga that is an important component of plankton; usually photosynthetic.

dioecious (dy-ee'shus) Having male and female reproductive structures on separate plants; compare with *monoecious*.

dipeptide See *peptide*.

diploid (dip'loyd) The condition of having two sets of chromosomes per nucleus. Compare with *haploid* and *polyploid*.

diplomonads Small, mostly parasitic zooflagellates with one or two nuclei, no mitochondria, and one to four flagella

direct fitness An individual's reproductive success, measured by the number of viable offspring it produces. Compare with *inclusive fitness*.

directed evolution See *in vitro evolution*.

directional selection The gradual replacement of one phenotype with another due to environmental change that favors phenotypes at one of the extremes of the normal distribution. Compare with *stabilizing selection* and *disruptive selection*.

disaccharide (dy-sak'ah-ride) A sugar produced by covalently linking two monosaccharides (e.g., maltose or sucrose).

disomy The normal condition in which both members of a chromosome pair are present in a diploid cell or organism. Compare with *monosomy* and *trisomy*.

dispersal The movement of individuals among populations. Compare with *migration*.

dispersion The pattern of distribution in space of the individuals of a population relative to their neighbors; may be clumped, random, or uniform.

disruptive selection A special type of directional selection in which changes in the environment favor two or more variant phenotypes at the expense of the mean. Compare with *stabilizing selection* and *directional selection*.

distal Remote; farther from the point of reference. Compare with *proximal*.

distal convoluted tubule The part of the renal tubule that extends from the loop of Henle to the collecting duct. Compare with *proximal convoluted tubule*.

disturbance In ecology, any event that disrupts community or population structure.

divergent circuit A neural pathway in which a presynaptic neuron stimulates many postsynaptic neurons. Compare with *convergent circuit*.

diving reflex A group of physiological mechanisms, such as decrease in metabolic rate, that are activated when a mammal dives to its limit.

division A taxonomic category below that of kingdom, comparable to a phylum; often used in classifying plants, fungi, and certain protists.

dizygotic twins Twins that arise from the separate fertilization of two eggs; commonly known as fraternal twins. Compare with *monozygotic twins*.

DNA See *deoxyribonucleic acid*.

DNA cloning The process of selectively amplifying DNA sequences so their structure and function can be studied.

DNA-dependent RNA polymerase See *RNA polymerase*.

DNA ligase Enzyme that catalyzes the joining of the 5′ and 3′ ends of two DNA fragments; essential in DNA replication and used in recombinant DNA technology.

DNA methylation A process in which gene inactivation is perpetuated by enzymes that add methyl groups to DNA.

DNA microarray A diagnostic test involving thousands of DNA molecules placed on a glass slide or chip.

DNA polymerases Family of enzymes that catalyze the synthesis of DNA from a DNA template, by adding nucleotides to a growing 3′ end.

DNA provirus Double-stranded DNA molecule that is an intermediate in the life cycle of an RNA tumor virus (retrovirus).

DNA replication The process by which DNA is duplicated; ordinarily a semiconservative process in which a double helix gives rise to two double helices, each with an "old" strand and a newly synthesized strand.

DNA sequencing Procedure by which the sequence of nucleotides in DNA is determined.

domain (1) A structural and functional region of a protein; (2) A taxonomic category that includes one or more kingdoms.

dominance hierarchy A linear "pecking order" into which animals in a population may organize according to status; regulates aggressive behavior within the population.

dominant allele (al-leel′) An allele that is always expressed when it is present, regardless of whether it is homozygous or heterozygous. Compare with *recessive allele*.

dopamine A neurotransmitter of the biogenic amine group.

dormancy A temporary period of arrested growth in plants or plant parts such as spores, seeds, bulbs, and buds.

dorsal (dor′sl) Toward the uppermost surface or back of an animal. Compare with *ventral*.

dosage compensation Genetic mechanism by which the expression of X-linked genes is made equivalent in XX females and XY males.

double fertilization A process in the flowering plant life cycle in which there are two fertilizations; one fertilization results in the formation of a zygote, whereas the second results in the formation of endosperm.

doubling time The amount of time it takes for a population to double in size, assuming that its current rate of increase does not change.

Down syndrome An inherited condition in which individuals have abnormalities of the face, eyelids, tongue, and other parts of the body, and are physically and mentally retarded; usually results from trisomy of chromosome 21.

drupe (droop) A simple, fleshy fruit in which the inner wall of the fruit is hard and stony, e.g., peach, cherry.

duodenum (doo″o-dee′num) The portion of the small intestine into which the contents of the stomach first enter.

duplication An abnormality in which a set of chromosomes contains more than one copy of a particular chromosomal segment; the translocation form of Down syndrome is an example.

dura mater The tough, outer meningeal layer that covers and protects the brain and spinal cord. Also see *arachnoid* and *pia mater*.

dynamic equilibrium The condition of a chemical reaction when the rate of change in one direction is exactly the same as the rate of change in the opposite direction, i.e., the concentrations of the reactants and products are not changing, and the difference in free energy between reactants and products is zero.

ecdysone (ek′dih-sone) See *molting hormone*.

Ecdysozoa A branch of the protostomes that includes animals that molt, such as the rotifers, nematodes, and arthropods.

echinoderms (eh-kine′oh-derms) Phylum of spiny-skinned marine deuterostome invertebrates characterized by a water vascular system and tube feet; include sea stars, sea urchins, and sea cucumbers.

echolocation Determination of the position of objects by detecting echos of high-pitched sounds emitted by an animal; a type of sensory system used by bats and dolphins.

ecological niche See *niche*.

ecological pyramid A graphical representation of the relative energy value at each trophic level. See *pyramid of biomass, pyramid of energy,* and *pyramid of numbers*.

ecological succession See *succession*.

ecology (ee-kol′uh-jee) A discipline of biology that studies the interrelations among living things and their environments.

ecosystem (ee′koh-sis-tem) The interacting system that encompasses a community and its nonliving, physical environment. Compare with *community*.

ecosystem management A conservation focus that emphasizes restoring and maintaining ecosystem quality rather than the conservation of individual species.

ecosystem services Important environmental services, such as clean air to breathe, clean water to drink, and fertile soil in which to grow crops, that ecosystems provide.

ecotone The transition zone where two communities or biomes meet and intergrade.

ectoderm (ek′toh-derm) The outer germ layer of the early embryo; gives rise to the skin and nervous system. Compare with *mesoderm* and *endoderm*.

ectotherm An animal whose temperature fluctuates with that of the environment; may use behavioral adaptations to regulate temperature; sometimes referred to as cold-blooded. Compare with *endotherm*.

edge effect The ecological phenomenon in which ecotones between adjacent communities often contain a greater number of species or greater population densities of certain species than either adjacent community.

effector A muscle or gland that contracts or secretes in direct response to nerve impulses.

efferent (ef′fur-ent) Leading away from some point of reference. Compare with *afferent*.

efferent neurons Neurons that transmit action potentials from the brain or spinal cord to muscles or glands. Compare with *afferent neurons*.

ejaculation (ee-jak′′yoo-lay′shun) A sudden expulsion, as in the ejection of semen from the penis.

electrolyte A substance that dissociates into ions when dissolved in water; the resulting solution can conduct an electrical current.

electron A particle with one unit of negative charge and negligible mass, located outside the atomic nucleus. Compare with *neutron* and *proton*.

electron configuration The arrangement of the electrons around the atom. In a Bohr model the electron configuration is depicted as a series of concentric circles.

electron microscope A microscope capable of producing high-resolution, highly magnified images through the use of an electron beam (rather than light). Transmission electron microscopes (TEMs) produce images of thin sections; scanning electron microscopes (SEMs) produce images of surfaces.

electron shell Group of orbitals of electrons with similar energies.

electron transport system A series of chemical reactions during which hydrogens or their electrons are passed along an electron transport chain from one acceptor molecule to another, with the release of energy.

electronegativity A measure of an atom's attraction for electrons.

electrophoresis, gel See *gel electrophoresis*.

electroreceptor A receptor that responds to electrical stimuli.

element A substance that cannot be changed to a simpler substance by a normal chemical reaction.

elimination Ejection of undigested food from the body. Compare with *excretion*.

El Niño—Southern Oscillation (ENSO) (el nee′nyo) A recurring climatic phenomenon that involves a surge of warm water in the Pacific Ocean and unusual weather patterns elsewhere in the world.

elongation (in protein synthesis) Cyclic process by which amino acids are added one by one to a growing polypeptide chain. See *initiation* and *termination*.

embryo (em′bree-oh) (1) A young organism before it emerges from the egg, seed, or body of its mother; (2) Developing human until the end of the second month, after which it is referred to as a fetus; (3) In plants, the young sporophyte produced following fertilization and subsequent development of the zygote.

embryo sac The female gametophyte generation in flowering plants.

embryo transfer See *host mothering*.

emigration A type of migration in which individuals leave a population and thus decrease its size. Compare with *immigration*.

enantiomers (en-an′tee-oh-merz) Two isomeric chemical compounds that are mirror images.

endangered species A species whose numbers are so severely reduced that it is in imminent danger of extinction throughout all or part of its range. Compare with *threatened species*.

Endangered Species Act A U.S. law that authorizes the U.S. Fish and Wildlife Service to protect from extinction all endangered and threatened species in the United States and abroad.

endemic species Localized, native species that are not found anywhere else in the world. Compare with *cosmopolitan species*.

endergonic reaction (end′er-gon′′ik) A nonspontaneous reaction; a reaction requiring a net input of free energy. Compare with *exergonic reaction*.

endocrine gland (en′doh-crin) A gland that secretes hormones directly into the blood or tissue fluid instead of into ducts. Compare with *exocrine gland*.

endocrine system The body system that helps regulate metabolic activities; consists of ductless glands and tissues that secrete hormones.

endocytosis (en′′doh-sy-toh′sis) The active transport of substances into the cell by the formation of invaginated regions of the plasma membrane that pinch off and become cytoplasmic vesicles. Compare with *exocytosis*.

endoderm (en′doh-derm) The inner germ layer of the early embryo; becomes the lining of the digestive tract and the structures that develop from the digestive tract—liver, lungs, and pancreas. Compare with *ectoderm* and *mesoderm*.

endodermis (en′′doh-der′mis) The innermost layer of the plant root cortex. Endodermal cells have a waterproof Casparian strip around their radial and transverse walls that ensures that water and minerals can enter the root xylem only by passing through the endoderm cells.

endolymph (en′doh-limf) The fluid of the membranous labyrinth and cochlear duct of the ear.

endomembrane system See *internal membrane system*.

endometrium (en′′doh-mee′tree-um) The uterine lining.

endoplasmic reticulum (ER) (en′doh-plaz′′mik reh-tik′yoo-lum) An interconnected network of internal membranes in eukaryotic cells enclosing a compartment, the ER lumen. Rough ER has ribosomes attached to the cytosolic surface; smooth ER, a site of lipid biosynthesis, lacks ribosomes.

endorphins (en-dor′finz) Neuropeptides released by certain brain neurons; block pain signals.

endoskeleton (en′′doh-skel′eh-ton) Bony and/or cartilaginous structures within the body that provide support. Compare with *exoskeleton*.

endosperm (en′doh-sperm) The 3*n* nutritive tissue that is formed at some point in the development of all angiosperm seeds.

endospore A resting cell formed by certain bacteria; highly resistant to heat, radiation, and disinfectants.

endosymbiont (en′′doe-sim′bee-ont) An organism that lives inside the body of another kind of organism. Endosymbionts may benefit their host (mutualism) or harm their host (parasitism).

endosymbiont theory The theory that certain organelles such as mitochondria and chloroplasts originated as symbiotic prokaryotes that lived inside other, free-living, prokaryotic cells. Compare with *autogenous model*.

endothelium (en-doh-theel′ee-um) The tissue that lines the cavities of the heart, blood vessels, and lymph vessels.

endotherm (en′doh-therm) An animal that uses metabolic energy to maintain a constant body temperature despite variations in environmental temperature; e.g., birds and mammals. Compare with *ectotherm*.

endotoxin A poisonous substance in the cell walls of gram-negative bacteria. Compare with *exotoxin*.

end product inhibition See *feedback inhibition*.

energy The capacity to do work; can be expressed in kilojoules or kilocalories.

energy of activation See *activation energy*.

enhanced greenhouse effect See *greenhouse effect*.

enhancers Regulatory DNA sequences that can be located long distances away from the actual coding regions of a gene.

enkephalins (en-kef'ah-linz) Neuropeptides released by certain brain neurons that block pain signals.

enterocoely (en'ter-oh-seely) The process by which the coelom forms as a cavity within mesoderm produced by outpocketings of the primitive gut (archenteron); characteristic of many deuterostomes. Compare with *schizocoely*.

enthalpy The total potential energy of a system; sometimes referred to as the heat content of the system.

entropy (en'trop-ee) Disorderliness; a quantitative measure of the amount of the random, disordered energy that is unavailable to do work.

environmental resistance Unfavorable environmental conditions, such as crowding, that prevent organisms from reproducing indefinitely at their intrinsic rate of increase.

enzyme (en'zime) An organic catalyst (usually a protein) that accelerates a specific chemical reaction by lowering the activation energy required for that reaction.

enzyme-substrate complex The temporary association between enzyme and substrate that forms during the course of a catalyzed reaction; also called ES complex.

eosinophil (ee-oh-sin'oh-fil) A type of white blood cell whose cytoplasmic granules absorb acidic stains; functions in parasitic infestations and allergic reactions.

epidermis (ep-ih-dur'mis) (1) An outer layer of cells that covers the body of plants and functions primarily for protection; (2) The outer layer of vertebrate skin.

epididymis (ep-ih-did'ih-mis) (pl. *epididymides*) A coiled tube that receives sperm from the testis and conveys it to the vas deferens.

epiglottis A thin, flexible structure that guards the entrance to the larynx, preventing food from entering the airway during swallowing.

epinephrine (ep-ih-nef'rin) Hormone produced by the adrenal medulla; stimulates the sympathetic nervous system.

epistasis (ep'ih-sta-sis) Condition in which certain alleles of one locus can alter the expression of alleles of a different locus.

epithelial tissue (ep-ih-theel'ee-al) The type of animal tissue that covers body surfaces, lines body cavities, and forms glands; also called epithelium.

epoch The smallest unit of geological time; a subdivision of a period.

equilibrium See *dynamic equilibrium, genetic equilibrium,* and *punctuated equilibrium*.

era One of the main divisions of geological time; eras are subdivided into periods.

erythroblastosis fetalis (eh-rith'row-blas-toe''-sis fi-tal'is) Serious condition in which Rh^+ red blood cells (which bear antigen D) of a fetus are destroyed by maternal anti-D antibodies.

erythrocyte (eh-rith'row-site) A vertebrate red blood cell; contains hemoglobin, which transports oxygen.

erythropoietin (eh-rith''row-poy'ih-tin) A peptide hormone secreted mainly by kidney cells; stimulates red blood cell production.

ES complex See *enzyme-substrate complex*.

essential nutrient A nutrient that must be provided in the diet because the body cannot make it or cannot make it in sufficient quantities to meet nutritional needs, e.g., essential amino acids and essential fatty acids.

ester linkage Covalent linkage formed by the reaction of a carboxyl group and a hydroxyl group, with the removal of the equivalent of a water molecule; the linkage includes an oxygen atom bonded to a carbonyl group.

estivation A state of torpor caused by lack of food or water during periods of high temperature. Compare with *hibernation*.

estrogens (es'troh-jens) Female sex hormones produced by the ovary; promote the development and maintenance of female reproductive structures and of secondary sex characteristics.

estuary (es'choo-wear-ee) A coastal body of water that connects to an ocean, in which fresh water from the land mixes with salt water.

ethology (ee-thol'oh-jee) The study of animal behavior under natural conditions from the point of view of adaptation.

ethyl alcohol A two-carbon alcohol.

ethylene (eth'ih-leen) A gaseous plant hormone involved in various aspects of plant growth and development, such as leaf abscission and fruit ripening.

eubacteria (yoo''bak-teer'ee-ah) Prokaryotes other than the archaebacteria.

euchromatin (yoo-croh'mah-tin) A loosely coiled chromatin that is generally capable of transcription. Compare with *heterochromatin*.

euglenoids (yoo-glee'noids) A group of mostly freshwater, flagellated, unicellular algae that move by means of an anterior flagellum and are usually photosynthetic.

eukaryote (yoo''kar'ee-ote) An organism whose cells possess nuclei and other membrane-bounded organelles. Includes protists, fungi, plants, and animals. Compare with *prokaryote*.

euphotic zone The upper reaches of the ocean, in which enough light penetrates to support photosynthesis.

eustachian tube (yoo-stay'shee-un) The auditory tube passing between the middle ear cavity and the pharynx in vertebrates; permits the equalization of pressure on the tympanic membrane.

eutrophic lake A lake enriched with nutrients such as nitrate and phosphate and consequently overgrown with plants or algae.

evergreen A term describing a plant that sheds leaves over a long period, so that some leaves are always present. Compare with *deciduous*.

evolution Any cumulative genetic changes in a population from generation to generation. Evolution leads to differences in populations and explains the origin of all of the organisms that exist today or have ever existed.

evolutionary systematics An approach to classification that considers both evolutionary relationships and the extent of divergence that has occurred since a group branched from an ancestral group. Compare with *cladistics* and *phenetics*.

excitatory postsynaptic potential (EPSP) A change in membrane potential that brings a neuron closer to the firing level. Compare with *inhibitory postsynaptic potential (IPSP)*.

excretion (ek-skree'shun) The discharge from the body of a waste product of metabolism (not to be confused with the elimination of undigested food materials). Compare with *elimination*.

excretory system The body system in animals that functions in osmoregulation and in the discharge of metabolic wastes.

exergonic reaction (ex'er-gon''ik) A reaction characterized by a release of free energy. Also called spontaneous reaction. Compare with *endergonic reaction*.

exocrine gland (ex'oh-crin) A gland that excretes its products through a duct that opens onto a free surface such as the skin (e.g., sweat glands). Compare with *endocrine gland*.

exocytosis (ex''oh-sy-toh'sis) The active transport of materials out of the cell by fusion of cytoplasmic vesicles with the plasma membrane. Compare with *endocytosis*.

exon (1) A protein-coding region of a eukaryotic gene; (2) The RNA transcribed from such a region. Compare with *intron*.

exoskeleton (ex''oh-skel'eh-ton) An external skeleton, such as the shell of mollusks or outer covering of arthropods; provides protection and sites of attachment for muscles. Compare with *endoskeleton*.

exotoxin A poisonous substance released by certain bacteria. Compare with *endotoxin*.

explicit memory Factual knowledge of people, places, or objects; requires conscious recall of the information.

exploitation competition An intraspecific competition in which all the individuals in a population "share" the limited resource equally, so that at high-population densities, none of them obtains an adequate amount. Also called scramble competition. Compare with *interference competition*.

exponential population growth The accelerating population growth rate that occurs when optimal conditions allow a constant per capita growth rate. When the increase in number versus time is plotted on a graph, exponential growth produces a characteristic J-shaped curve. Compare with *logistic population growth*.

ex situ conservation Conservation efforts that involve conserving biological diversity in human-controlled settings, such as zoos. Compare with *in situ conservation*.

exteroceptor (ex'tur-oh-sep''tor) One of the sense organs that receives sensory stimuli from the outside world, such as the eyes or touch receptors. Compare with *interoceptor*.

extinction The elimination of a species; occurs when the last individual member of a species dies.

extracellular matrix (ECM) A network of proteins and carbohydrates that surrounds many animal cells.

extraembryonic membranes Multicellular membranous structures that develop from the germ layers of a terrestrial vertebrate embryo but are not part of the embryo itself. See *chorion, amnion, allantois,* and *yolk sac*.

F_1 generation (first filial generation) The first generation of hybrid offspring resulting from a cross between parents from two different true-breeding lines.

F_2 generation (second filial generation) The offspring of the F_1 generation.

facilitated diffusion The passive transport of ions or molecules by a specific carrier protein in a membrane. As in simple diffusion, net transport is down a concentration gradient, and no additional energy has to be supplied. Compare with *diffusion* and *active transport*.

facilitation A process in which a neuron is brought closer to its threshold level by stimulation from various presynaptic neurons.

facultative anaerobe An organism capable of carrying out aerobic respiration but able to switch to fermentation when oxygen is unavailable; e.g., yeast. Compare with *obligate anaerobe*.

FAD/FADH$_2$ Oxidized and reduced forms, respectively, of flavin adenine dinucleotide, a coenzyme that transfers electrons (as hydrogen) in metabolism, including cellular respiration.

fallopian tube See *oviduct*.

family A taxonomic category made up of related genera.

fatty acid A lipid that is an organic acid containing a long hydrocarbon chain, with no double bonds (saturated fatty acid), one double bond (monounsaturated fatty acid), or two or more double bonds (polyunsaturated fatty acid); components of triacylglycerols, and phospholipids, as well as monoacylglycerols and diacylglycerols.

fecundity The potential capacity of an individual to produce offspring.

feedback inhibition A type of enzyme regulation in which the accumulation of the product of a reaction inhibits an earlier reaction in the sequence; also known as end product inhibition.

fermentation An anaerobic process by which ATP is produced by a series of redox reactions in which organic compounds serve both as electron donors and terminal electron acceptors.

fern One of a phylum of seedless vascular plants that reproduce by spores produced in sporangia usually borne on leaves in sori; ferns have an alternation of generations between the dominant sporophyte (fern plant) and the gametophyte (prothallus).

fertilization The fusion of two n gametes; results in the formation of a $2n$ zygote. Compare with *double fertilization*.

fetus The unborn human offspring from the third month of pregnancy to birth.

fiber (1) In plants, a type of sclerenchyma cell; fibers are long, tapered cells with thick walls. Compare with *sclereid.* (2) In animals, an elongated cell such as a muscle or nerve cell. (3) In animals, the microscopic, threadlike protein and carbohydrate complexes scattered through the matrix of connective tissues.

fibrin An insoluble protein formed from the plasma protein fibrinogen during blood clotting.

fibroblasts Connective tissue cells that produce the fibers and the protein and carbohydrate complexes of the matrix of connective tissues.

fibronectins Glycoproteins of the extracellular matrix that bind to integrins (receptor proteins in the plasma membrane).

fibrous root system A root system consisting of several adventitious roots of approximately equal size that arise from the base of the stem. Compare with *taproot system*.

Fick's law of diffusion A physical law governing rates of gas exchange in animal respiratory systems; states that the rate of diffusion of a substance across a membrane is directly proportional to the surface area and to the difference in pressure between the two sides.

filament In flowering plants, the thin stalk of a stamen; the filament bears an anther at its tip.

first law of thermodynamics The law of conservation of energy, which states that the total energy of any closed system (any object plus its surroundings, i.e., the universe) remains constant. Compare with *second law of thermodynamics*.

fitness See *direct fitness*.

fixed action pattern (FAP) An innate behavior triggered by a sign stimulus.

flagellum (flah-jel'um) (pl. *flagella*) A long, whiplike structure extending from certain cells and used in locomotion. (1) Eukaryote flagella are composed of two central single microtubules surrounded by nine double microtubules (9 + 2 structure), all covered by a plasma membrane. (2) Prokaryote flagella are filaments rotated by special structures located in the plasma membrane and cell wall.

flame cells Collecting cells that have cilia; part of the osmoregulatory system of flatworms.

flavin adenine dinucleotide See *FAD/FADH$_2$*.

flowering plants See *angiosperms*.
flowing-water ecosystem A river or stream ecosystem.
fluid-mosaic model The currently accepted model of the plasma membrane and other cell membranes, in which protein molecules float in a phospholipid bilayer.
fluorescence The emission of light of a longer wavelength (lower energy) than the light originally absorbed.
follicle (fol′i-kl) (1) A simple, dry, dehiscent fruit that splits open at maturity along one suture to liberate the seeds; (2) A small sac of cells in the mammalian ovary that contains a maturing egg; (3) The pocket in the skin from which a hair grows.
follicle-stimulating hormone (FSH) A gonadotropic hormone secreted by the anterior lobe of the pituitary gland; stimulates follicle development in the ovaries of females and sperm production in the testes of males.
food chain The series of organisms through which energy flows in an ecosystem. Each organism in the series eats or decomposes the preceding organism in the chain. See *food web*.
food web A complex interconnection of all of the food chains in an ecosystem.
foram See *foraminiferan*.
foramen magnum The opening in the vertebrate skull through which the spinal cord passes.
foraminiferan (for″am-in-if′er-an) A marine protozoon that produces a shell, or test, that encloses an amoeboid body. Also called foram.
forebrain In the early embryo, one of the three divisions of the developing vertebrate brain; subdivides to form the telencephalon, which gives rise to the cerebrum, and the diencephalon, which gives rise to the thalamus and hypothalamus. Compare with *midbrain* and *hindbrain*.
forest decline A gradual deterioration (and often death) of many trees in a forest; may be caused by a combination of factors, such as acid precipitation, toxic heavy metals, and surface-level ozone.
fossil Parts or traces of an ancient organism usually preserved in rock.
fossil fuel Combustible deposits in Earth's crust that are composed of the remnants of prehistoric organisms that existed millions of years ago, e.g., oil, natural gas, and coal.
founder cell A cell from which a particular cell lineage is derived.
founder effect Genetic drift that results from a small population colonizing a new area.
fovea (foe′vee-ah) The area of sharpest vision in the retina; cone cells are concentrated here.
fragile site A weak point at a specific location on a chromosome where part of a chromatid appears to be attached to the rest of the chromosome by a thin thread of DNA.
fragile X syndrome A human genetic disorder caused by a fragile site that occurs near the tip on the X chromosome; effects range from mild learning disabilities to severe mental retardation and hyperactivity.
frameshift mutation A mutation that results when one or two nucleotide pairs are inserted into or deleted from the DNA. The change causes the mRNA transcribed from the mutated DNA to have an altered reading frame such that all codons downstream from the mutation are changed.
free energy The maximum amount of energy available to do work under the conditions of a biochemical reaction.
free radicals Toxic, highly reactive compounds that have unpaired electrons that can bond with other compounds in the cell, interfering with normal function.
frequency-dependent selection Selection in which the relative fitness of different genotypes is related to how frequently they occur in the population.
freshwater wetlands Land that is transitional between freshwater and terrestrial ecosystems and is covered with water for at least part of the year; e.g., marshes and swamps.
frontal lobes In mammals, the anterior part of the cerebrum.
fruit In flowering plants, a mature, ripened ovary. Fruits contain seeds and usually provide seed protection and dispersal.
fruiting body A multicellular structure that contains the sexual spores of certain fungi; refers to the ascocarp of an ascomycete and the basidiocarp of a basidiomycete.
fucoxanthin (few″koh-zan′thin) The brown carotenoid pigment found in brown algae, golden algae, diatoms, and dinoflagellates.
functional genomics The study of the roles of genes in cells.
functional group A group of atoms that confers distinctive properties on an organic molecule (or region of a molecule) to which it is attached, e.g., hydroxyl, carbonyl, carboxyl, amino, phosphate, and sulfhydryl groups.
fundamental niche The potential ecological niche that an organism could occupy if there were no competition from other species. Compare with *realized niche*.
fungus (pl. *fungi*) A heterotrophic eukaryote with chitinous cell walls and a body usually in the form of a mycelium of branched, threadlike hyphae. Most fungi are decomposers; some are parasitic.
G protein One of a group of proteins that bind GTP and are involved in the transfer of signals across the plasma membrane.
G_1 phase The first gap phase within the interphase stage of the cell cycle; G_1 occurs before DNA synthesis (S phase) begins. Compare with *S* and *G_2 phases*.
G_2 phase Second gap phase within the interphase stage of the cell cycle; G_2 occurs after DNA synthesis (S phase) and before mitosis. Compare with *S* and *G_1 phases*.
gallbladder A small sac that stores bile.
gametangium (gam″uh-tan′gee-um) Special multicellular or unicellular structure of plants, protists, and fungi in which gametes are formed.
gamete (gam′eet) A sex cell; in plants and animals, an egg or sperm. In sexual reproduction, the union of gametes results in the formation of a zygote. The chromosome number of a gamete is designated *n*. Species that are not polyploid have haploid gametes and diploid zygotes.
gametic isolation (gam-ee′tik) A prezygotic reproductive isolating mechanism in which sexual reproduction between two closely related species cannot occur because of chemical differences in the gametes.
gametogenesis The process of gamete formation. See *spermatogenesis* and *oogenesis*.
gametophyte generation (gam-ee′toh-fite) The *n*, gamete-producing stage in the life cycle of a plant. Compare with *sporophyte generation*.
gamma-aminobutyric acid (GABA) A neurotransmitter that has an inhibitory effect.
ganglion (gang′glee-on) (pl. *ganglia*) A mass of neuron cell bodies.
ganglion cell A type of neuron in the retina of the eye; receives input from bipolar cells.
gap junction Structure consisting of specialized regions of the

plasma membrane of two adjacent cells; contains numerous pores that allow the passage of certain small molecules and ions between them.

gastrin (gas'trin) A hormone released by the stomach mucosa; stimulates the gastric glands to secrete pepsinogen.

gastrovascular cavity A central digestive cavity with a single opening that functions as both mouth and anus; characteristic of cnidarians and flatworms.

gastrula (gas'troo-lah) A three-layered embryo formed by the process of gastrulation.

gastrulation (gas-troo-lay'shun) Process in embryonic development during which the three germ layers (ectoderm, mesoderm, and endoderm) form.

gel electrophoresis Procedure by which proteins or nucleic acids can be separated on the basis of size and charge as they migrate through a gel in an electrical field.

gene A segment of DNA that serves as a unit of hereditary information; includes a transcribable DNA sequence (plus associated sequences regulating its transcription) that yields a protein or RNA product with a specific function. Most eukaryotic genes are in chromosomes.

gene amplification The developmental process in which certain cells produce multiple copies of a gene by selective replication, thus allowing for increased synthesis of the gene product. Compare with *nuclear equivalence* and *genomic rearrangement*.

gene flow The movement of alleles between local populations due to the migration of individuals; can have significant evolutionary consequences.

gene locus See *locus*.

gene pool All the alleles of all the genes present in a freely interbreeding population.

generation time The time required for the completion of one cell cycle.

gene replacement therapy Any of a variety of methods designed to correct a disease or alleviate its symptoms through the introduction of genes into the affected person's cells. Also known as gene transfer therapy.

gene transfer therapy See *gene replacement therapy*.

genetic bottleneck See *bottleneck*.

genetic code See *codon*.

genetic counseling Medical and genetic information provided to couples who are concerned about the risk of abnormality in their children.

genetic drift A random change in allele frequency in a small breeding population.

genetic engineering Manipulation of genes, often through recombinant DNA technology.

genetic equilibrium The condition of a population that is not undergoing evolutionary change, i.e., in which allele and genotype frequencies do not change from one generation to the next. See *Hardy-Weinberg principle*.

genetic polymorphism (pol''ee-mor'fizm) The presence in a population of two or more alleles for a given gene locus.

genetic probe A single-stranded nucleic acid (either DNA or RNA) that can be used to identify a complementary sequence by hydrogen-bonding to it.

genetic recombination See *recombination, genetic*.

genetic screening A systematic search through a population for individuals with a genotype or karyotype that might cause a serious genetic disease in them or their offspring.

genome (jee'nome) Originally, all the genetic material in a cell or individual organism. The term is used more than one way, depending on context: e.g., an organism's haploid genome is all the DNA contained in one haploid set of its chromosomes, and its mitochondrial genome is all the DNA in a mitochondrion. See *human genome*.

genomic DNA library A collection of recombinant plasmids in which all the DNA in the genome is represented. Compare with *cDNA library*.

genomic rearrangement A physical change in the structure of one or more genes that occurs during the development of an organism and leads to an alteration in gene expression; compare with *nuclear equivalence* and *gene amplification*.

genotype (jeen'oh-type) The genetic makeup of an individual. Compare with *phenotype*.

genotype frequency The proportion of a particular genotype in the population.

genus (jee'nus) A taxonomic category made up of related species.

germination Resumption of growth of an embryo or spore; occurs when a seed or spore sprouts.

germ layers In animals, three embryonic tissue layers: endoderm, mesoderm, and ectoderm.

germ line cell In animals, a cell that is part of the line of cells that will ultimately undergo meiosis to form gametes. Compare with *somatic cell*.

germplasm Any plant or animal material that may be used in breeding; includes seeds, plants, and plant tissues of traditional crop varieties and the sperm and eggs of traditional livestock breeds.

gibberellin (jib''ur-el'lin) A plant hormone involved in many aspects of plant growth and development, such as stem elongation, flowering, and seed germination.

gills (1) The respiratory organs characteristic of many aquatic animals, usually thin-walled projections from the body surface or from some part of the digestive tract; (2) The spore-bearing, platelike structures under the caps of mushrooms.

ginkgo (ging'ko) A member of an ancient gymnosperm group that consists of a single living representative (*Ginkgo biloba*), a hardy, deciduous tree with broad, fan-shaped leaves and naked, fleshy seeds (on female trees).

gland See *endocrine gland* and *exocrine gland*.

glial cells (glee'ul) In nervous tissue, cells that support and nourish neurons.

globulin (glob'yoo-lin) One of a class of proteins in blood plasma, some of which (gamma globulins) function as antibodies.

glomerulus (glom-air'yoo-lus) The cluster of capillaries at the proximal end of a nephron; the glomerulus is surrounded by Bowman's capsule.

glucagon (gloo'kah-gahn) A hormone secreted by the pancreas that stimulates glycogen breakdown, thereby increasing the concentration of glucose in the blood. Compare with *insulin*.

glucose A hexose aldehyde sugar that is central to many metabolic processes.

glutamate An amino acid that functions as the major excitatory neurotransmitter in the vertebrate brain.

glyceraldehyde-3-phosphate (G3P) Phosphorylated 3-carbon compound that is an important intermediate in glycolysis and in the Calvin cycle.

glycerol A three-carbon alcohol with a hydroxyl group on each carbon; a component of triacylglycerols and phospholipids, as well as monoacylglycerols and diacylglycerols.

glycocalyx (gly''koh-kay'lix) A coating on the outside of an animal cell, formed by the polysaccharide portions of glycoproteins and glycolipids associated with the plasma membrane.

glycogen (gly'koh-jen) The principal storage polysaccharide in animal cells; formed from glucose and stored primarily in the liver and, to a lesser extent, in muscle cells.

glycolipid A lipid with covalently attached carbohydrates.

glycolysis (gly-kol'ih-sis) The first stage of cellular respiration, literally the "splitting of sugar." The metabolic conversion of glucose into pyruvate, accompanied by the production of ATP.

glycoprotein (gly'koh-pro-teen) A protein with covalently attached carbohydrates.

glycosidic linkage Covalent linkage joining two sugars; includes an oxygen atom bonded to a carbon of each sugar.

glyoxysomes (gly-ox'ih-somz) Membrane-bounded structures in cells of certain plant seeds; contain a large array of enzymes that convert stored fat to sugar.

gnetophyte (nee'toe-fite) One of a small phylum of unusual gymnosperms that possess some features similar to flowering plants.

goblet cells Unicellular glands that secrete mucus.

goiter (goy'ter) An enlargement of the thyroid gland.

golden alga A member of a phylum of algae, most of which are biflagellated, unicellular, and contain pigments, including chlorophyll *a* and *c* and carotenoids, including fucoxanthin.

Golgi complex (goal'jee) Organelle composed of stacks of flattened, membranous sacs. Mainly responsible for modifying, packaging, and sorting proteins that will be secreted or targeted to other organelles of the internal membrane system or to the plasma membrane; also called Golgi body or Golgi apparatus.

gonad (goh'nad) A gamete-producing gland; an ovary or a testis.

gonadotropin-releasing hormone (GnRH) A hormone secreted by the hypothalamus that stimulates the anterior pituitary to secrete the gonadotropic hormones: follicle-stimulating hormone (FSH) and luteinizing hormone (LH).

gonadotropic hormones (go-nad-oh-troh'pic) Hormones produced by the anterior pituitary gland that stimulate the testes and ovaries; include follicle-stimulating hormone (FSH) and luteinizing hormone (LH).

graded potential A local change in electrical potential that can vary in magnitude depending on the strength of the applied stimulus.

gradualism The idea that evolution occurs by a slow, steady accumulation of genetic changes over time. Compare with *punctuated equilibrium*.

graft rejection An immune response directed against a transplanted tissue or organ.

grain A simple, dry, one-seeded fruit in which the fruit wall is fused to the seed coat, making it impossible to separate the fruit from the seed, e.g., corn and wheat kernels.

granulosa cells In mammals, cells that surround the developing oocyte and are part of the follicle; produce estrogens and inhibin.

granum (pl. *grana*) A stack of thylakoids within a chloroplast.

gravitropism (grav''ih-troh'pizm) Growth of a plant in response to gravity.

gray crescent The grayish area of cytoplasm that marks the region where gastrulation begins in an amphibian embryo.

gray matter Nervous tissue in the brain and spinal cord that contains cell bodies, dendrites, and unmyelinated axons. Compare with *white matter*.

green alga A member of a diverse phylum of algae that contain the same pigments as plants (chlorophylls *a* and *b* and carotenoids).

greenhouse effect The natural global warming of Earth's atmosphere caused by the presence of carbon dioxide and other greenhouse gases, which trap the sun's radiation in much the same way that glass does in a greenhouse. The additional warming that may be produced by increased levels of greenhouse gases that absorb infrared radiation is known as the enhanced greenhouse effect.

greenhouse gases Trace gases in the atmosphere that allow the sun's energy to penetrate to Earth's surface but do not allow as much of it to escape as heat.

gross primary productivity The rate at which energy accumulates (is assimilated) in an ecosystem during photosynthesis. Compare with *net primary productivity*.

ground state The lowest energy state of an atom.

ground tissue system All tissues in the plant body other than the dermal tissue system and vascular tissue system; consists of parenchyma, collenchyma, and sclerenchyma.

growth factors A group of more than 50 extracellular peptides that signal certain cells to grow and divide.

growth hormone (GH) A hormone secreted by the anterior lobe of the pituitary gland; stimulates growth of body tissues; also called somatotropin.

growth rate The rate of change of a population's size on a per capita basis.

guanine (gwan'een) A nitrogenous purine base that is a component of nucleic acids and GTP.

guard cell One of a pair of epidermal cells that adjust their shape to form a stomatal pore for gas exchange.

guttation (gut-tay'shun) The appearance of water droplets on leaves, forced out through leaf pores by root pressure.

gymnosperm (jim'noh-sperm) Any of a group of seed plants in which the seeds are not enclosed in an ovary; gymnosperms frequently bear their seeds in cones. Includes four phyla: conifers, cycads, ginkgoes, and gnetophytes.

habitat The natural environment or place where an organism, population, or species lives.

habitat fragmentation The division of habitats that formerly occupied large, unbroken areas into smaller pieces by roads, fields, cities, and other human land-transforming activities.

habitat isolation A prezygotic reproductive isolating mechanism in which reproduction between similar species is prevented because they live and breed in different habitats

habituation (hab-it''yoo-ay'shun) A type of learning in which an animal becomes accustomed to a repeated, irrelevant stimulus and no longer responds to it.

hair cell A vertebrate mechanoreceptor found in the lateral line of fishes, the vestibular apparatus, semicircular canals, and cochlea.

half-life The period of time required for a radioisotope to change into a different material.

haploid (hap'loyd) The condition of having one set of chromosomes per nucleus. Compare with *diploid* and *polyploid*.

"hard-wiring" Refers to how neurons signal one another, how they connect, and how they carry out basic functions such as regulating heart rate, blood pressure, and sleep-wake cycles.

Hardy-Weinberg principle The mathematical prediction that allele frequencies do not change from generation to generation in a large population in the absence of microevolutionary processes (mutation, genetic drift, gene flow, natural selection).

Hatch-Slack pathway See C_4 *plant*.
haustorium (hah-stor′ee-um) (pl. *haustoria*) In parasitic fungi, a specialized hypha that penetrates a host cell and obtains nourishment from the cytoplasm.
Haversian canals (ha-vur′zee-un) Channels extending through the matrix of bone; contain blood vessels and nerves.
heat The total amount of kinetic energy in a sample of a substance.
heat energy The thermal energy that flows from an object with a higher temperature to an object with a lower temperature.
heat of vaporization The amount of heat energy that must be supplied to change one gram of a substance from the liquid phase to the vapor phase.
helicases Enzymes that unwind the two strands of a DNA double helix.
heliotropism The ability of leaves or flowers of certain plants to follow the sun by aligning themselves either perpendicular or parallel to the sun′s rays; also called solar tracking.
helper T cell T lymphocyte that activates B lymphocytes and can stimulate cytotoxic T cell production. Also known as CD4 T cell.
hemichordates A phylum of sedentary, wormlike deuterostomes.
hemizygous (hem″ih-zy′gus) Possessing only one allele for a particular locus; a human male is hemizygous for all X-linked genes. Compare with *homozygous* and *heterozygous*.
hemocoel Blood cavity characteristic of animals with an open circulatory system.
hemocyanin A hemolymph pigment that transports oxygen in some mollusks and arthropods.
hemoglobin (hee′moh-gloh″bin) The red, iron-containing protein pigment in blood that transports oxygen and carbon dioxide and aids in regulation of pH.
hemolymph (hee′moh-limf) The fluid that bathes the tissues in animals with an open circulatory system, e.g., arthropods and most mollusks.
hemophilia (hee″-moh-feel′ee-ah) A hereditary disease in which blood does not clot properly; the form known as hemophilia A has an X-linked, recessive inheritance pattern.
Hensen′s node See *primitive streak*.
hepatic (heh-pat′ik) Pertaining to the liver.
hepatic portal system The portion of the circulatory system that carries blood from the intestine through the liver.
herbivore (erb′uh-vore) An animal that feeds on plants or algae. Also called primary consumer.
hermaphrodite (her-maf′roh-dite) An organism that possesses both male and female sex organs.
heterochromatin (het″ur-oh-kroh′mah-tin) Highly coiled and compacted chromatin in an inactive state. Compare with *euchromatin*.
heterocyst (het′ur-oh-sist″) An oxygen-excluding cell of cyanobacteria that is the site of nitrogen fixation.
heterogametic A term describing an individual that produces two classes of gametes with respect to their sex chromosome constitutions. Human males (XY) are heterogametic, producing X and Y sperm. Compare with *homogametic*.
heterospory (het″ur-os′pur-ee) Production of two types of n spores, microspores (male) and megaspores (female). Compare with *homospory*.
heterothallic (het″ur-oh-thal′ik) Pertaining to certain algae and fungi that have two mating types; only by combining a plus strain and a minus strain can sexual reproduction occur. Compare with *homothallic*.
heterotroph (het′ur-oh-trof) An organism that cannot synthesize its own food from inorganic raw materials and therefore must obtain energy and body-building materials from other organisms. Also called consumer. Compare with *autotroph*. See *chemoheterotroph* and *photoheterotroph*.
heterozygote advantage A phenomenon in which the heterozygous condition confers some special advantage on an individual that either homozygous condition does not (i.e., *Aa* has a higher degree of fitness than does *AA* or *aa*).
heterozygous (het-ur′oh-zye′gus) Possessing a pair of unlike alleles for a particular locus. Compare with *homozygous*.
hexose A monosaccharide containing six carbon atoms.
hibernation Long-term torpor in response to winter cold and scarcity of food. Compare with *estivation*.
histamine (his′tah-meen) Substance released from mast cells that is involved in allergic and inflammatory reactions.
hindbrain In the early embryo, one of the three divisions of the developing vertebrate brain; subdivides to form the metencephalon, which gives rise to the cerebellum and pons, and the myelencephalon, which gives rise to the medulla. Compare with *forebrain* and *midbrain*.
histones (his′tones) Small, positively charged (basic) proteins in the cell nucleus that bind to the negatively charged DNA. See *nucleosomes*.
holdfast The basal structure for attachment to solid surfaces found in multicellular algae.
holoblastic cleavage A cleavage pattern in which the entire embryo cleaves; characteristic of eggs with little or moderate yolk (isolecithal or moderately telolecithal), e.g., the eggs of echinoderms, amphioxus, and mammals. Compare with *meroblastic cleavage*.
homeobox A DNA sequence of approximately 180 base pairs found in many homeotic genes and some other genes that are important in development; genes containing homeobox sequences code for certain transcription factors.
homeodomain A functional region of certain transcription factor proteins; consists of approximately 60 amino acids specified by a homeobox DNA sequence and includes a recognition alpha helix, which can bind to specific DNA sequences and affect their transcription.
home range A geographical area that an animal seldom or never leaves.
homeostasis (home″ee-oh-stay′sis) The balanced internal environment of the body; the automatic tendency of an organism to maintain such a steady state.
homeotic gene (home″ee-ah′tik) A gene that controls the formation of specific structures during development. Such genes were originally identified through insect mutants in which one body part is substituted for another.
hominid (hah′min-id) Any of a group of extinct and living humans. Also called hominine.
hominoid (hah′min-oid) The apes and hominids.
Homo erectus An extinct hominid that lived between 2 million and 200,000 years ago; made stone tools and may have worn clothing, built fires, and lived in caves.
Homo ergaster An extinct hominid that is recognized by some paleoanthropologists as separate from *Homo erectus;* other paleoanthropologists do not recognize *H. ergaster* as a separate species and instead consider it to be the African line of *H. erectus*.
homogametic Term describing an individual that produces gametes with identical sex chromosome constitutions. Human

females (XX) are homogametic, producing all X eggs. Compare with *heterogametic*.

Homo habilis An extinct hominid that lived between about 2.3 and 1.5 million years ago; lived in Africa and fashioned crude stone tools.

Homo heidelbergensis See *archaic Homo sapiens*.

homologous chromosomes (hom-ol'ah-gus) Chromosomes that are similar in morphology and genetic constitution. In humans there are 23 pairs of homologous chromosomes; one member of each pair is inherited from the mother, and the other from the father.

homologous features See *homology*.

homology Similarity in different species that results from their derivation from a common ancestor. The features that exhibit such similarity are called homologous features. Compare with *homoplasy*.

Homo neanderthalensis Primitive humans who lived in Europe and western Asia from about 230,000 to 30,000 years ago; possessed sophisticated stone tools, cared for the sick and elderly, and buried their dead; also called Neandertals.

homoplastic features See *homoplasy*.

homoplasy Similarity in the characters in different species that is due to convergent evolution, not common descent. Characters that exhibit such similarity are called homoplastic features. Compare with *homology*.

Homo sapiens The modern human species.

homospory (hoh''mos'pur-ee) Production of one type of n spore that gives rise to a bisexual gametophyte. Compare with *heterospory*.

homothallic (hoh''moh-thal'ik) Pertaining to certain algae and fungi that are self-fertile. Compare with *heterothallic*.

homozygous (hoh''moh-zy'gous) Possessing a pair of identical alleles for a particular locus. Compare with *heterozygous*.

hormone An organic chemical messenger in multicellular organisms that is produced in one part of the body and often transported to another part where it signals cells to alter some aspect of metabolism.

hornwort A type of spore-producing, nonvascular, thallose plant with a life cycle similar to mosses.

horsetail A phylum of seedless vascular plants with a life cycle similar to ferns.

host mothering The introduction of an embryo from one species into the uterus of another species, where it implants and develops; the host mother subsequently gives birth and may raise the offspring as her own; also called embryo transfer.

human chorionic gonadotropin (hCG) A hormone secreted by cells surrounding the early embryo; signals the mother's corpus luteum to continue to function.

human genetics The science of inherited variation in humans.

human genome The totality of genetic information in human cells; includes the DNA content of both the nucleus and mitochondria. See *genome*.

human immunodeficiency virus (HIV) The retrovirus that causes AIDS (acquired immunodeficiency syndrome).

human leukocyte antigen (HLA) See *major histocompatibility complex*.

humus (hew'mus) Organic matter in various stages of decomposition in the soil; gives soil a dark brown or black color.

Huntington disease A genetic disease that has an autosomal dominant inheritance pattern and causes mental and physical deterioration.

hybrid The offspring of two genetically dissimilar parents.

hybrid breakdown A postzygotic reproductive isolating mechanism in which, although an interspecific hybrid is fertile and produces a second (F_2) generation, the F_2 has defects that prevent it from successfully reproducing.

hybrid inviability A postzygotic reproductive isolating mechanism in which the embryonic development of an interspecific hybrid is aborted.

hybridization (1) Interbreeding between members of two different taxa; (2) Interbreeding between genetically dissimilar parents; (3) In molecular biology, complementary base pairing between nucleic acid (DNA or RNA) strands from different sources.

hybrid sterility A postzygotic reproductive isolating mechanism in which an interspecific hybrid cannot reproduce successfully.

hybrid vigor The genetic superiority of an F_1 hybrid over either parent, due to the presence of heterozygosity for a number of different loci.

hybrid zone An area of overlap between two closely related populations, subspecies, or species, in which interbreeding occurs.

hydration Process of association of a substance with the partial positive and/or negative charges of water molecules.

hydrocarbon An organic compound composed solely of hydrogen and carbon atoms.

hydrogen bond A weak attractive force existing between a hydrogen atom with a partial positive charge and an electronegative atom (usually oxygen or nitrogen) with a partial negative charge. Compare with *covalent bond* and *ionic bond*.

hydrological cycle The water cycle, which includes evaporation, precipitation, and flow to the ocean. The hydrological cycle supplies terrestrial organisms with a continual supply of fresh water.

hydrolysis Reaction in which a covalent bond between two subunits is broken through the addition of the equivalent of a water molecule; a hydrogen atom is added to one subunit and a hydroxyl group to the other. Compare with *condensation synthesis*.

hydrophilic Interacting readily with water; having a greater affinity for water molecules than they have for each other. Compare with *hydrophobic*.

hydrophobic Not readily interacting with water; having less affinity for water molecules than they have for each other. Compare with *hydrophilic*.

hydroponics (hy''dra-paun'iks) Growing plants in an aerated solution of dissolved inorganic minerals; i.e., without soil.

hydrostatic skeleton A type of skeleton found in some invertebrates in which contracting muscles push against a tube of fluid.

hydroxide ion An anion (negatively charged particle) consisting of oxygen and hydrogen; usually written OH^-.

hydroxyl group (hy-drok'sil) Polar functional group; abbreviated —OH.

hyperpolarize To change the membrane potential so that the inside of the cell becomes more negative than its resting potential.

hypertonic A term referring to a solution having an osmotic pressure (or solute concentration) greater than that of the solution with which it is compared. Compare with *hypotonic* and *isotonic*.

hypha (hy'fah) (pl. *hyphae*) One of the threadlike filaments composing the mycelium of a water mold or fungus.

hypocotyl (hy'poh-kah''tl) The part of the axis of a plant embryo or seedling below the point of attachment of the cotyledons.

hypothalamus (hy-poh-thal'uh-mus) Part of the mammalian brain that regulates the pituitary gland, the autonomic system, emotional responses, body temperature, water balance, and appetite; located below the thalamus.

hypothesis A testable statement about the nature of an observation or relationship. Compare with *theory* and *principle*.

hypothetico-deductive approach Emphasizes the use of deductive reasoning to test hypotheses. Compare with *hypothetico-inductive approach*. See *deductive reasoning*.

hypothetico-inductive approach Emphasizes the use of inductive reasoning to discover new general principles. Compare with *hypothetico-deductive approach*. See *inductive reasoning*.

hypotonic A term referring to a solution having an osmotic pressure (or solute concentration) less than that of the solution with which it is compared. Compare with *hypertonic* and *isotonic*.

hypotrichs A group of dorsoventrally flattened ciliates that exhibit an unusual creeping-darting locomotion.

illuviation The deposition of material leached from the upper layers of soil into the lower layers.

imaginal discs Paired structures in an insect larva that develop into specific adult structures during complete metamorphosis.

imago (ih-may'go) The adult form of an insect.

imbibition (im''bi-bish'en) The absorption of water by a seed prior to germination.

immigration A type of migration in which individuals enter a population and thus increase its size. Compare with *emigration*.

immune response Process of recognizing foreign macromolecules and mounting a response aimed at eliminating them. See *specific* and *nonspecific immune responses; primary* and *secondary responses*.

immunoglobulin (im-yoon''oh-glob'yoo-lin) See *antibody*.

imperfect flower A flower that lacks either stamens or carpels. Compare with *perfect flower*.

imperfect fungi See *deuteromycetes*.

implantation The embedding of a developing embryo in the inner lining (endometrium) of the uterus.

implicit memory The unconscious memory for perceptual and motor skills, e.g., riding a bicycle.

imprinting (1) The expression of a gene based on its parental origin. (2) A type of learning by which a young bird or mammal forms a strong social attachment to an individual (usually a parent) or object within a few hours after hatching or birth.

inborn error of metabolism A metabolic disorder cause by the mutation of a gene that codes for an enzyme needed for a biochemical pathway.

inbreeding The mating of genetically similar individuals. Homozygosity increases with each successive generation of inbreeding. Compare with *outbreeding*.

inbreeding depression The phenomenon in which inbred offspring of genetically similar individuals have lower fitness (e.g., decline in fertility and high juvenile mortality) than noninbred individuals.

inclusive fitness The total of an individual's direct and indirect fitness; includes the genes contributed directly to offspring and those contributed indirectly by kin selection. Compare with *direct fitness*. See *kin selection*.

incomplete dominance A condition in which neither member of a pair of contrasting alleles is completely expressed when the other is present.

incomplete flower A flower that lacks one or more of the four parts: sepals, petals, stamens, and/or carpels. Compare with *complete flower*.

independent assortment, principle of The genetic principle, first noted by Gregor Mendel, that states that the alleles of unlinked loci are randomly distributed to gametes.

indeterminate growth Unrestricted growth, as for example, in stems and roots. Compare with *determinate growth*.

index fossils Fossils restricted to a narrow period of geological time and found in the same sedimentary layers in different geographical areas.

indoleacetic acid See *auxin*.

induced fit Conformational change in the active site of an enzyme that occurs when it binds to its substrate.

inducer molecule A substance that binds to a repressor protein, converting it to its inactive form, which is unable to prevent transcription.

inducible operon An operon that is normally inactive because a repressor molecule is attached to its operator; transcription is activated when a specific inducer molecule binds to the repressor, making it incapable of binding to the operator, e.g., the lactose operon of *Escherichia coli*. Compare with *repressible operon*.

induction The process by which the differentiation of a cell or group of cells is influenced by interactions with neighboring cells.

inductive reasoning The reasoning that uses specific examples to draw a general conclusion or discover a general principle. Compare with *deductive reasoning*. See *hypothetico-inductive approach*.

infant mortality rate The number of infant deaths per 1000 live births. (A child is an infant during its first two years of life.)

inflammatory response The response of body tissues to injury or infection, characterized clinically by heat, swelling, redness, and pain, and physiologically by increased dilation of blood vessels and increased phagocytosis.

inflorescence A cluster of flowers on a common floral stalk.

ingestion The process of taking food (or other material) into the body.

inhibin A hormone that inhibits FSH secretion; produced by Sertoli cells in the testes and by granulosa cells in the ovaries.

inhibitory postsynaptic potential (IPSP) A change in membrane potential that takes a neuron farther from the firing level. Compare with *excitatory postsynaptic potential (EPSP)*.

initiation (of protein synthesis) The first steps of protein synthesis, in which the large and small ribosomal subunits and other components of the translation machinery bind to the 5′ end of mRNA. See *elongation* and *termination*.

initiation codon The codon AUG, which serves as the signal to begin translation of messenger RNA; also called a start codon. Compare with *termination codon*.

innate behavior Behavior that is inherited and typical of the species; also called instinct.

innate immune responses See *nonspecific immune responses*.

inner cell mass The cluster of cells in the early mammalian embryo that gives rise to the embryo proper.

inorganic compound A simple substance that does not contain a carbon backbone. Compare with *organic compound*.

inositol triphosphate (IP_3) A second messenger that increases intracellular calcium concentration and activates enzymes.

insight learning A complex learning process in which an animal adapts past experience to solve a new problem that may involve different stimuli.

in situ conservation Conservation efforts that concentrate on preserving biological diversity in the wild. Compare with *ex situ conservation*.

instinct See *innate behavior*.

instrumental conditioning See *operant conditioning*.

insulin (in′suh-lin) A hormone secreted by the pancreas that lowers blood-glucose concentration. Compare with *glucagon*.

insulin-like growth factors (IGF) Somatomedins; proteins that mediate responses to growth hormone.

insulin resistance A condition in which insulin is present, but insulin receptors on target cells do not bind with it; this condition is present in Type 2 diabetes.

insulin shock A condition in which the blood glucose concentration is so low that the individual may appear intoxicated, or may become unconscious, and even die; can be caused by the injection of too much insulin or by certain metabolic malfunctions.

integrins Receptor proteins that bind to specific proteins in the extracellular matrix and to membrane proteins on adjacent cells; transmit signals into the cell from the extracellular matrix.

integral membrane protein A protein that is tightly associated with the lipid bilayer of a biological membrane; a transmembrane integral protein spans the bilayer. Compare with *peripheral membrane protein*.

integration The process of summing (adding and subtracting) incoming neural signals.

integumentary system (in-teg′′yoo-men′tar-ee) The body's covering, including the skin and its nails, glands, hair, and other associated structures.

integuments The outer cell layers that surround the megasporangium of an ovule; develop into the seed coat.

intercellular substance In connective tissues, the combination of matrix and fibers in which the cells are embedded.

interference competition Intraspecific competition in which certain dominant individuals obtain an adequate supply of the limited resource at the expense of other individuals in the population. Also called contest competition. Compare with *exploitation competition*.

interferons (in′′tur-feer′on) Cytokines produced by animal cells when challenged by a virus; prevent viral reproduction and enable cells to resist a variety of viruses.

interkinesis The stage between meiosis I and meiosis II. Interkinesis is usually brief; the chromosomes may decondense, reverting at least partially to an interphase-like state, but DNA synthesis and chromosome duplication do not occur.

interleukins A diverse group of cytokines produced mainly by macrophages and lymphocytes.

intermediate-day plant A plant that flowers when it is exposed to days and nights of intermediate length but does not flower when the daylength is too long or too short. Compare with *long-day, short-day,* and *day-neutral plants*.

intermediate disturbance hypothesis In community ecology, the idea that species richness is greatest at moderate levels of disturbance, which create a mosaic of habitat patches at different stages of succession.

intermediate filaments Cytoplasmic fibers that are part of the cytoskeletal network and are intermediate in size between microtubules and microfilaments.

internal membrane system The group of membranous structures in eukaryotic cells that interact through direct connections or by means of vesicles; includes the endoplasmic reticulum, outer membrane of the nuclear envelope, Golgi complex, lysosomes, and the plasma membrane; also called endomembrane system.

interneuron (in′′tur-noor′on) A nerve cell that carries impulses from one nerve cell to another and is not directly associated with either an effector or a sensory receptor. Also known as an association neuron.

internode The region on a stem between two successive nodes. Compare with *node*.

interoceptor (in′tur-oh-sep′′tor) A sense organ within a body organ that transmits information regarding chemical composition, pH, osmotic pressure, or temperature. Compare with *exteroceptor*.

interphase The stage of the cell cycle between successive mitotic divisions; its subdivisions are the G_1 (first gap), S (DNA synthesis), and G_2 (second gap) phases.

interspecific competition The interaction between members of different species that vie for the same resource in an ecosystem (e.g., food or living space). Compare with *intraspecific competition*.

interstitial cells (of testis) The cells between the seminiferous tubules that secrete testosterone.

interstitial fluid The fluid that bathes the tissues of the body; also called tissue fluid.

intertidal zone The marine shoreline area between the high tide mark and the low tide mark.

intraspecific competition The interaction between members of the same species that vie for the same resource in an ecosystem (e.g., food or living space). Compare with *interspecific competition*.

intrinsic rate of increase The theoretical maximum rate of increase in population size occurring under optimal environmental conditions. Also called biotic potential.

intron A non–protein-coding region of a eukaryotic gene and also of the pre-mRNA transcribed from such a region. Introns do not appear in mRNA. Compare with *exon*.

invertebrate An animal without a backbone (vertebral column); invertebrates account for about 95% of animal species.

in vitro Occurring outside a living organism (literally "in glass"). Compare with *in vivo*.

in vitro evolution Test tube experiments that demonstrate that RNA molecules in the RNA world could have catalyzed the many different chemical reactions needed for life. Also called directed evolution.

in vivo Occurring in a living organism. Compare with *in vitro*.

ion An atom or group of atoms bearing one or more units of electrical charge, either positive (cation) or negative (anion).

ion channels Channels for the passage of ions through a membrane; formed by specific membrane proteins.

ionic bond The chemical attraction between a cation and an anion. Compare with *covalent bond* and *hydrogen bond*.

ionic compound A substance consisting of cations and anions, which are attracted by their opposite charges; ionic compounds do not consist of molecules. Compare with *covalent compound*.

ionization The dissociation of a substance to yield ions, e.g., the ionization of water yields H^+ and OH^-.

iris The pigmented portion of the vertebrate eye.

iron-sulfur world hypothesis The hypothesis that simple organic molecules that are the precursors of life originated at hydrothermal vents in the deep-ocean floor. Compare with *prebiotic broth hypothesis*.

irreversible inhibitor A substance that permanently inactivates an enzyme. Compare with *reversible inhibitor*.

islets of Langerhans (eye'lets of Lahng'er-hanz) The endocrine portion of the pancreas that secretes glucagon and insulin, hormones that regulate the concentration of glucose in the blood.

isogamy (eye-sog'uh-me) Sexual reproduction involving motile gametes of similar form and size. Compare with *anisogamy* and *oogamy*.

isogenic A term describing a strain of organisms in which all individuals are genetically identical and homozygous at all loci.

isolecithal egg An egg containing a relatively small amount of uniformly distributed yolk. Compare with *telolecithal egg*.

isomer (eye'soh-mer) One of two or more chemical compounds having the same chemical formula, but different structural formulas, e.g., structural and geometric isomers and enantiomers.

isoprene units Five-carbon hydrocarbon monomers that make up certain lipids such as carotenoids and steroids.

isotonic (eye''soh-ton'ik) A term applied to solutions that have identical concentrations of solute molecules and hence the same osmotic pressure. Compare with *hypertonic* and *hypotonic*.

isotope (eye'suh-tope) An alternate form of an element with a different number of neutrons, but the same number of protons and electrons. See *radioisotopes*.

iteroparity The condition of having repeated reproductive cycles throughout a lifetime. Compare with *semelparity*.

jelly coat One of the acellular coverings of the eggs of certain animals, such as echinoderms.

joint The junction between two or more bones of the skeleton.

joule A unit of energy, equivalent to 0.239 calorie.

juvenile hormone (JH) An arthropod hormone that preserves juvenile structure during a molt. Without it, metamorphosis toward the adult form takes place.

juxtaglomerular apparatus (juks''tah-glo-mer'yoo-lar) A structure in the kidney that secretes renin in response to a decrease in blood pressure.

***K* selection** A reproductive strategy recognized by some ecologists in which a species typically has a large body size, slow development, long life span, and does not devote a large proportion of its metabolic energy to the production of offspring. Compare with *r selection*.

karyotype (kare'ee-oh-type) The chromosomal constitution of an individual. Representations of the karyotype are generally prepared by photographing the chromosomes and arranging the homologous pairs according to size, centromere position, and pattern of bands.

keratin (kare'ah-tin) A horny, water-insoluble protein found in the epidermis of vertebrates and in nails, feathers, hair, and horns.

ketone An organic molecule containing a carbonyl group bonded to two carbon atoms. Compare with *aldehyde*.

keystone species A species whose presence in an ecosystem largely determines the species composition and functioning of that ecosystem.

kidney The paired vertebrate organ important in excretion of metabolic wastes and in osmoregulation.

killer T cell See *cytotoxic T cell*.

kilobase (kb) 1000 bases or base pairs of a nucleic acid.

kilocalorie The amount of heat required to raise the temperature of 1 kg of water 1°C; also called Calorie, which is equivalent to 1000 calories.

kilojoule 1000 joules. See *joule*.

kinases Enzymes that catalyze the transfer of phosphate groups from ATP to acceptor molecules. Protein kinases activate or inactivate proteins through the addition of phosphates at specific locations.

kinetic energy Energy of motion. Compare with *potential energy*.

kinetochore (kin-eh'toh-kore) The portion of the chromosome centromere to which the mitotic spindle fibers attach.

kingdom A broad taxonomic category made up of related phyla; many biologists currently recognize six kingdoms of living organisms.

kin selection A type of natural selection that favors altruistic behavior toward relatives (kin), thereby ensuring that, although the chances of an individual's survival are lessened, some of its genes will survive through successful reproduction of close relatives; increases inclusive fitness.

Klinefelter syndrome Inherited condition in which the affected individual is a sterile male with an XXY karyotype.

Koch's postulates A set of guidelines used to demonstrate that a specific pathogen causes specific disease symptoms.

Krebs cycle See *citric acid cycle*.

krummholz The gnarled, shrublike growth habit found in trees at high elevations, near their upper limit of distribution.

labyrinth The system of interconnecting canals of the inner ear of vertebrates.

labyrinthodonts The first successful group of tetrapods.

lactation (lak-tay'shun) The production or release of milk from the breast.

lacteal (lak'tee-al) One of the many lymphatic vessels in the intestinal villi that absorb fat.

lactic acid (lactate) A three-carbon organic acid.

lagging strand A strand of DNA that is synthesized as a series of short segments, called Okazaki fragments, which are then covalently joined by DNA ligase. Compare with *leading strand*.

lamins Proteins attached to the inner surface of the nuclear envelope that provide a type of skeletal framework.

landscape A large land area (several to many square kilometers) composed of interacting ecosystems.

landscape ecology The subdiscipline in ecology that studies the connections in a heterogenous landscape.

large intestine The portion of the digestive tract of humans (and other vertebrates) consisting of the cecum, colon, rectum, and anus.

larva (pl. *larvae*) An immature form in the life history of some animals; may be unlike the parent.

larynx (lare'inks) The organ at the upper end of the trachea that contains the vocal cords.

lateral meristems Areas of localized cell division on the side of a plant that give rise to secondary tissues. Lateral meristems, including the vascular cambium and the cork cambium, cause an increase in the girth of the plant body. Compare with *apical meristem*.

leaching The process by which dissolved materials are washed away or carried with water down through the various layers of the soil.

leader sequence Noncoding sequence of nucleotides in mRNA that is transcribed from the region that precedes (is upstream to) the coding region.

leading strand Strand of DNA that is synthesized continuously. Compare with *lagging strand.*

learning A change in the behavior of an animal that results from experience.

lectins Sugar-binding proteins that were originally extracted from plant seeds, where they are found in large quantities; also occur in many other organisms.

legume (leg'yoom) (1) A simple, dry fruit that splits open at maturity along two sutures to release seeds; (2) Any member of the pea family, e.g., pea, bean, peanut, alfalfa.

lek A small territory in which males compete for females.

lens The oval, transparent structure located behind the iris of the vertebrate eye; bends incoming light rays and brings them to a focus on the retina.

lenticels (len'tih-sels) Porous swellings of cork cells in the stems of woody plants; facilitate the exchange of gases.

leptin A hormone produced by adipose tissue that signals brain centers about the status of energy stores.

leukocytes (loo'koh-sites) White blood cells; colorless amoeboid cells that defend the body against disease-causing organisms.

leukoplasts Colorless plastids; include amyloplasts, which are used for starch storage in cells of roots and tubers.

lichen (ly'ken) A compound organism composed of a symbiotic fungus and an alga or cyanobacterium.

life history traits Significant features of a species' life cycle, particularly traits that influence survival and reproduction.

life span The maximum duration of life for an individual of a species.

life table A table showing mortality and survival data by age of a population or cohort.

ligament (lig'uh-ment) A connective tissue cable or strap that connects bones to each other or holds other organs in place.

ligand A molecule that binds to a specific site in a receptor or other protein.

light-dependent reactions Reactions of photosynthesis in which light energy absorbed by chlorophyll is used to synthesize ATP and usually NADPH. Includes *cyclic electron transport* and *noncyclic electron transport*.

lignin (lig'nin) A substance found in many plant cell walls that confers rigidity and strength, particularly in woody tissues.

limbic system In vertebrates, an action system of the brain. In humans, plays a role in emotional responses, motivation, autonomic function, and sexual response.

limiting resource An environmental resource that, because it is scarce or unfavorable, tends to restrict the ecological niche of an organism.

limnetic zone (lim-net'ik) The open water away from the shore of a lake or pond extending down as far as sunlight penetrates. Compare with *littoral zone* and *profundal zone*.

linkage The tendency for a group of genes located on the same chromosome to be inherited together in successive generations.

lipase (lip'ase) A fat-digesting enzyme.

lipid Any of a group of organic compounds that are insoluble in water but soluble in nonpolar solvents; lipids serve as energy storage and are important components of cell membranes.

lipoprotein (lip-oh-proh' teen) A large molecular complex consisting of lipids and protein; transports lipids in the blood. High-density lipoproteins (HDLs) transport cholesterol to the liver; low-density lipoproteins (LDLs) deliver cholesterol to many cells of the body.

littoral zone (lit'or-ul) The region of shallow water along the shore of a lake or pond. Compare with *limnetic zone* and *profundal zone*.

liver A large, complex organ that secretes bile, helps maintain homeostasis by removing or adding nutrients to the blood, and performs many other metabolic functions.

liverworts A phylum of spore-producing, nonvascular, thallose or leafy plants with a life cycle similar to mosses.

local hormones See *local regulators*.

local regulators Prostaglandins (a group of local hormones), growth factors, cytokines, and other soluble molecules that act on nearby cells by paracrine regulation or act on the cells that produce them (autocrine regulation).

locus The place on the chromosome at which the gene for a given trait occurs, i.e., a segment of the chromosomal DNA containing information that controls some feature of the organism; also called gene locus.

logistic population growth Population growth that initially occurs at a constant rate of increase over time (i.e., exponential) but then levels out as the carrying capacity of the environment is approached. When the increase in number versus time is plotted on a graph, logistic growth produces a characteristic S-shaped curve. Compare with *exponential population growth*.

long-day plant A plant that flowers in response to shortening nights; also called short-night plant. Compare with *short-day, intermediate-day,* and *day-neutral plants*.

long-night plant See *short-day plant*.

long-term potentiation (LTP) Long-lasting increase in the strength of synaptic connections that occurs in response to a series of high frequency electrical stimuli. Compare with *long-term synaptic depresssion (LTD)*.

long-term synaptic depression (LTD) Long-lasting decrease in the strength of synaptic connections that occurs in response to low-frequency stimulation of neurons. Compare with *long-term potentiation (LTP)*.

loop of Henle (Hen'lee) The U-shaped loop of a mammalian kidney tubule, which extends down into the renal medulla.

loose connective tissue A type of connective tissue that is widely distributed in the body; consists of fibers strewn through a semifluid matrix.

lophoporate phyla Three related invertebrate protostome phyla, characterized by a ciliated ring of tentacles that surrounds the mouth.

Lophotrochozoa A branch of the protostomes that includes the nemerteans (proboscis worms), mollusks, annelids, and the lophophorate phyla.

lumen (loo'men) (1) The space enclosed by a membrane, such as the lumen of the endoplasmic reticulum or the thylakoid lumen; (2) The cavity or channel within a tube or tubular organ, such as a blood vessel or the digestive tract.

lung An internal respiratory organ that functions in gas exchange; enables an animal to breathe air.

luteinizing hormone (LH) (loot'eh-ny-zing) Gonadotropic hormone secreted by the anterior pituitary; stimulates ovulation and maintains the corpus luteum in the ovaries of females; stimulates testosterone production by interstitial cells in the testes of males.

lymph (limf) The colorless fluid within the lymphatic vessels that is derived from blood plasma and resembles it closely in composition; contains white blood cells; ultimately lymph is returned to the blood.

lymph node A mass of lymph tissue surrounded by a connective tissue capsule; manufactures lymphocytes and filters lymph.

lymphatic system A subsystem of the cardiovascular system; returns excess interstitial fluid (lymph) to the circulation; defends the body against disease organisms.

lymphocyte (lim′foh-site) White blood cell with nongranular cytoplasm that is responsible for immune responses. See *B cell* and *T cell*.

lysis (ly′sis) The process of disintegration of a cell or some other structure.

lysogenic conversion The change in properties of bacteria that results from the presence of a prophage.

lysosomes (ly′soh-somes) Intracellular organelles present in many animal cells; contain a variety of hydrolytic enzymes.

lysozyme An enzyme found in many tissues and in tears and other body fluids; attacks the cell wall of many gram-positive bacteria.

macroevolution Large-scale evolutionary change over long time spans. Compare with *microevolution*.

macromolecule A very large organic molecule, such as a protein or nucleic acid.

macronucleus A large nucleus found, along with one or several micronuclei, in ciliates. The macronucleus regulates metabolism and growth. Compare with *micronucleus*.

macronutrient An essential element that is required in fairly large amounts for normal growth. Compare with *micronutrient*.

macrophage (mak′roh-faje) A large phagocytic cell capable of ingesting and digesting bacteria and cellular debris. Macrophages are also antigen-presenting cells.

major histocompatibility complex (MHC) A group of membrane proteins, present on the surface of most cells, that are slightly different in each individual. In humans, the MHC is called the HLA (human leukocyte antigen) group.

malignant The term used to describe cancer cells, tumor cells that are able to invade tissue and metastasize.

malnutrition Poor nutritional status; can result from dietary intake that is either below or above required needs.

Malpighian tubules (mal-pig′ee-an) The excretory organs of many arthropods.

mammals The class of vertebrates characterized by hair, mammary glands, a diaphragm, and differentiation of teeth.

mandible (man′dih-bl) (1) The lower jaw of vertebrates; (2) Jaw-like, external mouthparts of insects.

mangrove forest A tidal wetland dominated by mangrove trees, in which the salinity fluctuates between that of sea water and fresh water.

mantle In the mollusk, a fold of tissue that covers the visceral mass and that usually produces a shell.

marine snow The organic debris (plankton, dead organisms, fecal material, etc.) that "rains" into the dark area of the oceanic province from the lighted region above; the primary food of most organisms that live in the ocean's depths.

marsupials (mar-soo′pee-uls) A subclass of mammals, characterized by the presence of an abdominal pouch in which the young, which are born in a very undeveloped condition, are carried for some time after birth.

mass extinction The extinction of numerous species during a relatively short period of geological time. Compare with *background extinction*.

mast cell A type of cell found in connective tissue; contains histamine and is important in allergic reactions.

maternal effect genes Genes of the mother that are transcribed during oogenesis and subsequently affect the development of the embryo. Compare with *zygotic genes*.

matrix (may′triks) (1) In cell biology, the interior of the compartment enclosed by the inner mitochondrial membrane; (2) In zoology, nonliving material secreted by and surrounding connective tissue cells; contains a network of microscopic fibers.

matter Anything that has mass and takes up space.

maxillae Appendages used for manipulating food; characteristic of crustaceans.

mechanical isolation A prezygotic reproductive isolating mechanism in which fusion of the gametes of two species is prevented by morphological or anatomical differences.

mechanoreceptor (meh-kan′oh-ree-sep′′tor) A sensory cell or organ that perceives mechanical stimuli, e.g., touch, pressure, gravity, stretching, or movement.

medulla (meh-dul′uh) (1) The inner part of an organ, such as the medulla of the kidney; compare with *cortex*. (2) The most posterior part of the vertebrate brain, lying next to the spinal cord.

medusa A jellyfish-like animal; a free-swimming, umbrella-shaped stage in the life cycle of certain cnidarians. Compare with *polyp*.

megaphyll (meg′uh-fil) Type of leaf found in horsetails, ferns, gymnosperms, and angiosperms; contains multiple vascular strands (i.e., complex venation). Compare with *microphyll*.

megaspore (meg′uh-spor) The n spore in heterosporous plants that gives rise to a female gametophyte. Compare with *microspore*.

meiosis (my-oh′sis) Process in which a $2n$ cell undergoes two successive nuclear divisions (meiosis I and meiosis II), potentially producing four n nuclei; leads to the formation of gametes in animals and spores in plants.

melanin A dark pigment present in many animals; contributes to the color of the skin.

melanin concentrating hormone (MCH) A neuropeptide signaling molecule that helps regulate energy homeostasis.

melanocortins A group of peptides that appear to decrease appetite in response to increased fat stores.

melatonin (mel-ah-toh′ nin) A hormone secreted by the pineal gland that plays a role in setting circadian rhythms.

memory cell B or T lymphocyte that permits rapid mobilization of immune response on second or subsequent exposure to a particular antigen.

meninges (meh-nin′jeez) (sing. *meninx*) The three membranes that protect the brain and spinal cord: the dura mater, arachnoid, and pia mater.

menopause The period (usually occurring between 45 and 55 years of age) in women when the recurring menstrual cycle ceases.

menstrual cycle (men′stroo-ul) In the human female, the monthly sequence of events that prepares the body for pregnancy; the cycle begins with the first day of menstruation; the first two weeks of the cycle are the preovulatory phase; ovulation takes place on about the 14th day of the cycle; the third and fourth weeks of the cycle are the postovulatory phase.

menstruation (men-stroo-ay′shun) The monthly discharge of blood and degenerated uterine lining in the human female; marks the beginning of each menstrual cycle.

meristem (mer′ih-stem) A localized area of mitotic cell division in the plant body. See *apical meristem* and *lateral meristems*.

meroblastic cleavage Cleavage pattern observed in the telolecithal eggs of reptiles and birds, in which cleavage is restricted to a small disc of cytoplasm at the animal pole. Compare with *holoblastic cleavage*.

mesenchyme (mes'en-kime) A loose, often jelly-like connective tissue containing undifferentiated cells; found in the embryos of vertebrates and the adults of some invertebrates.

mesoderm (mez'oh-derm) The middle germ layer of the early embryo; gives rise to connective tissue, muscle, bone, blood vessels, kidneys, and many other structures. Compare with *ectoderm* and *endoderm*.

mesophyll (mez'oh-fil) Photosynthetic tissue in the interior of a leaf; sometimes differentiated into palisade mesophyll and spongy mesophyll.

Mesozoic era That part of geological time extending from roughly 248 to 65 million years ago.

messenger RNA (mRNA) RNA that specifies the amino acid sequence of a protein; transcribed from DNA.

metabolic pathway A series of chemical reactions in which the product of one reaction becomes the substrate of the next reaction.

metabolic rate Energy production of an organism per unit time. See *basal metabolic rate*.

metabolism The sum of all the chemical processes that occur within a cell or organism: the transformations by which energy and matter are made available for use by the organism. See *anabolism* and *catabolism*.

metamorphosis (met''ah-mor'fuh-sis) Transition from one developmental stage to another, such as from a larva to an adult.

metanephridia (sing. *metanephridium*) The excretory organs of annelids and mollusks; each metanephridium consists of a tubule open at both ends; at one end a ciliated funnel opens into the coelom, and the other end of the tube opens to the outside of the body.

metaphase (met'ah-faze) The stage of mitosis, and of meiosis I and II, in which the chromosomes line up on the equatorial plane of the cell. Occurs after prophase and before anaphase.

metapopulation A population that is divided into several local populations among which individuals occasionally disperse.

metastasis (met-tas'tuh-sis) The spreading of cancer cells from one organ or part of the body to another.

methyl group A nonpolar functional group; abbreviated $—CH_3$.

microclimate Local variations in climate produced by differences in elevation, in the steepness and direction of slopes, and in exposure to prevailing winds.

microevolution Small-scale evolutionary change due to changes in allele or genotype frequencies that occur within a population over successive generations. Compare with *macroevolution*.

microfilaments Thin fibers composed of actin protein subunits; form part of the cytoskeleton.

microfossils Ancient traces (fossils) of microscopic life.

microglia Phagocytic glial cells found in the CNS.

micronucleus One or more smaller nuclei found, along with the macronucleus, in ciliates. The micronucleus is involved in sexual reproduction. Compare with *macronucleus*.

micronutrient An essential element that is required in trace amounts for normal growth. Compare with *macronutrient*.

microphyll (mi'kro-fil) Type of leaf found in club mosses; contains one vascular strand (i.e., simple venation). Compare with *megaphyll*.

microsphere A protobiont produced by adding water to abiotically formed polypeptides.

microspore (mi'kro-spor) The *n* spore in heterosporous plants that gives rise to a male gametophyte. Compare with *megaspore*.

microtubules (my-kroh-too'bewls) Hollow cylindrical fibers composed of tubulin protein subunits; major components of the cytoskeleton and found in mitotic spindles, cilia, flagella, centrioles, and basal bodies.

microtubule-associated proteins (MAPs) Include structural proteins that help regulate microtubule assembly and cross-link microtubules to other cytoskeletal polymers; and motors, such as kinesin and dynein, that use ATP to produce movement.

microtubule-organizing center (MTOC) The region of the cell from which microtubules are anchored and possibly assembled. The MTOCs of many organisms (including animals, but not flowering plants or most gymnosperms) contain a pair of centrioles.

microvilli (sing. *microvillus*) Minute projections of the plasma membrane that increase the surface area of the cell; found mainly in cells concerned with absorption or secretion, such as those lining the intestine or the kidney tubules.

midbrain In vertebrate embryos, one of the three divisions of the developing brain. Also called mesencephalon. Compare with *forebrain* and *hindbrain*.

middle lamella The layer composed of pectin polysaccharides that serves to cement together the primary cell walls of adjacent plant cells.

midvein The main, or central, vein of a leaf.

migration The periodic or seasonal movement of an organism (individual or population) from one place to another, usually over a long distance. See *dispersal*.

mineralocorticoids (min''ur-al-oh-kor'tih-koidz) Hormones produced by the adrenal cortex that regulate mineral metabolism and, indirectly, fluid balance. The principal mineralocorticoid is aldosterone.

minerals Inorganic nutrients ingested as salts dissolved in food and water.

missense mutation A type of base-substitution mutation that causes one amino acid to be substituted for another in the resulting protein product. Compare with *nonsense mutation*.

mitochondria (my''toh-kon'dree-ah) (sing. *mitochondrion*) Intracellular organelles that are the sites of oxidative phosphorylation in eukaryotes; include an outer membrane and an inner membrane.

mitochondrial DNA (mtDNA) DNA present in mitochondria that is transmitted maternally, from mothers to their offspring. Mitochondrial DNA mutates more rapidly than nuclear DNA.

mitosis (my-toh'sis) The division of the cell nucleus resulting in two daughter nuclei, each with the same number of chromosomes as the parent nucleus; mitosis consists of four phases: prophase, metaphase, anaphase, and telophase. Cytokinesis usually overlaps the telophase stage.

mitosis-promoting factor (MPF) A cyclin-dependent protein kinase that controls the transition of a cell from the G_2 stage of interphase to mitosis; formerly known as maturation promoting factor.

mitotic spindle Structure consisting mainly of microtubules that provides the framework for chromosome movement during cell division.

mitral valve See *atrioventricalar valve*.

mobile genetic element See *transposon*.

mole The atomic mass of an element or the molecular mass of a compound, expressed in grams; one mole of any substance has 6.02×10^{23} units (Avogadro's number).

molecular anthropology The branch of science that compares genetic material from individuals of regional human populations to help unravel the origin and migrations of modern humans.

molecular chaperones Proteins that help other proteins fold properly. Although not dictating the folding pattern, chaperones make the process more efficient.

molecular clock analysis A comparison of the DNA nucleotide sequences of related organisms to estimate when they diverged from one another during the course of evolution.

molecular formula The type of chemical formula that gives the actual numbers of each type of atom in a molecule. Compare with *simplest formula* and *structural formula*.

molecular mass The sum of the atomic masses of the atoms that make up a single molecule of a compound; expressed in atomic mass units (amu) or daltons.

molecule The smallest particle of a covalently bonded element or compound that has the composition and properties of a larger part of the substance.

mollusks A phylum of coelomate protostome animals characterized by a soft body, visceral mass, mantle, and foot.

molting The shedding and replacement of an outer covering such as an exoskeleton.

molting hormone A steroid hormone that stimulates growth and molting in insects. Also called ecdysone.

monoacylglycerol (mon''o-as''-il-glis'er-ol) Lipid consisting of glycerol combined chemically with a single fatty acid. Also called monoglyceride. Compare with *diacylglycerol* and *triacylglycerol*.

monocot (mon'oh-kot) One of the two classes of flowering plants; monocot seeds contain a single cotyledon, or seed leaf. Compare with *dicot*.

monoclonal antibodies Identical antibody molecules produced by cells cloned from a single cell.

monocyte (mon'oh-site) A type of white blood cell; a large phagocytic, nongranular leukocyte that enters the tissues and differentiates into a macrophage.

monoecious (mon-ee'shus) Having male and female reproductive parts in separate flowers or cones on the same plant; compare with *dioecious*.

monogamy A mating system in which a male animal mates with a single female during a breeding season.

monoglyceride See *monoacylglycerol*.

monohybrid cross A genetic cross that takes into account the behavior of alleles of a single locus. Compare with *dihybrid cross*.

monokaryotic (mon''o-kare-ee-ot'ik) The condition of having a single *n* nucleus per cell, characteristic of certain fungal hyphae. Compare with *dikaryotic*.

monomer (mon'oh-mer) A molecule that can be linked with other similar molecules; two monomers join to form a dimer, whereas many can be linked to form a polymer. Monomers can be small (e.g., sugars or amino acids) or they can be large (e.g., tubulin or actin proteins).

monophyletic group (mon''oh-fye-let'ik) A group made up of organisms that evolved from a common ancestor. Compare with *polyphyletic group* and *paraphyletic group*.

monosaccharide (mon-oh-sak'ah-ride) A simple sugar that cannot be degraded by hydrolysis to a simpler sugar.

monosomy The condition in which only one member of a chromosome pair is present and the other is missing. Compare with *trisomy* and *disomy*.

monotremes (mon'oh-treems) Egg-laying mammals such as the duck-billed platypus of Australia.

monounsaturated fatty acid See *fatty acid*.

monozygotic twins Genetically identical twins that arise from the division of a single fertilized egg; commonly known as identical twins. Compare with *dizygotic twins*.

morphogen Any chemical agent thought to be responsible for the processes of cell differentiation and pattern formation that lead to morphogenesis.

morphogenesis (mor-foh-jen'eh-sis) The development of the form and structures of an organism and its parts; proceeds by a series of steps known as pattern formation.

mortality The rate at which individuals die; the average per capita death rate.

morula (mor'yoo-lah) An early embryo consisting of a solid ball of cells.

mosaic development A highly invariant developmental pattern in which the fate of each blastomere becomes determined at a very early stage. Compare with *regulative development*.

mosses A phylum of spore-producing nonvascular plants with an alternation of generations in which the dominant *n* gametophyte alternates with a 2*n* sporophyte that remains attached to the gametophyte.

motor neuron An efferent neuron that transmits impulses away from the central nervous system to skeletal muscle.

motor unit All the skeletal muscle fibers that are stimulated by a single motor neuron.

mRNA cap An unusual nucleotide, 7-methylguanylate, that is added to the 5′ end of a eukaryotic messenger RNA. Capping enables eukaryotic ribosomes to bind to mRNA.

mucosa (mew-koh'suh) See *mucous membrane*.

mucous membrane A type of epithelial membrane that lines a body cavity that opens to the outside of the body, eg., the digestive and respiratory tracts; also called mucosa.

mucus (mew'cus) A sticky secretion composed of covalently linked protein and carbohydrate; serves to lubricate body parts and trap particles of dirt and other contaminants. (The adjectival form is spelled *mucous*.)

Müllerian mimicry (mul-ler'ee-un mim'ih-kree) The resemblance of dangerous, unpalatable, or poisonous species to one another so that they are more easily recognized by potential predators, which learn to avoid all of them after tasting one. Compare with *Batesian mimicry*.

multiple alleles (al-leels') Three or more alleles of a single locus (in a population), such as the alleles governing the ABO series of blood types.

multiple fruit A fruit that develops from many ovaries of many separate flowers, e.g., pineapple. Compare with *simple, aggregate,* and *accessory fruits*.

muscle (1) A tissue specialized for contraction; (2) An organ that produces movement by contraction.

mutagen (mew'tah-jen) Any agent capable of producing mutations.

mutation Any change in DNA; may include a change in the nucleotide base pairs of a gene, a rearrangement of genes within the chromosomes so that their interactions produce different effects, or a change in the chromosomes themselves.

mutualism (1) In ecology, a symbiotic relationship in which both partners benefit from the association. Compare with *parasitism* and *commensalism*. (2). In animal behavior, cooperative behavior in which each animal in the group benefits.

mycelium (my-seel'ee-um) (pl. *mycelia*) The vegetative body of most fungi and certain protists (water molds); consists of a branched network of hyphae.

mycorrhizae (my''kor-rye'zee) Mutualistic associations of fungi

and plant roots that aid in the plant's absorption of essential minerals from the soil.
mycotoxins Poisonous chemical compounds produced by fungi, e.g., aflatoxins that harm the liver and are known carcinogens.
myelin sheath (my'eh-lin) The white fatty material that forms a sheath around the axons of certain nerve cells, which are then called myelinated fibers.
myocardial infarction (MI) Heart attack; serious consequence occurring when the heart muscle receives insufficient oxygen.
myofibrils (my-oh-fy'brilz) Tiny threadlike structures in the cytoplasm of striated and cardiac muscle that are responsible for contractions of the cell; contain myofilaments.
myofilament (my-oh-fil'uh-ment) One of the filaments making up a myofibril; the structural unit of muscle proteins in a muscle cell. See *myosin filament* and *actin filament*.
myoglobin (my'oh-glob''bin) A hemoglobin-like, oxygen-transferring protein found in muscle.
myosin (my'oh-sin) A protein that, together with actin, is responsible for muscle contraction.
myosin filaments Thick filaments composed mainly of the protein myosin; actin and myosin filaments make up the myofibrils of muscle fibers.
n The chromosome number of a gamete. The chromosome number of a zygote is $2n$. If an organism is not polyploid, the n gametes are haploid and the $2n$ zygotes are diploid.
NAD^+/NADH Oxidized and reduced forms, respectively, of nicotinamide adenine dinucleotide, a coenzyme that transfers electrons (as hydrogen), particularly in catabolic pathways, including cellular respiration.
$NADP^+$/NADPH Oxidized and reduced forms, respectively, of nicotinamide adenine dinucleotide phosphate, a coenzyme that acts as an electron (hydrogen) transfer agent, particularly in anabolic pathways, including photosynthesis.
nanoplankton Extremely minute (< 50 μm in length) algae that are major producers in the ocean because of their great abundance; part of phytoplankton.
nastic movement A temporary, reversible movement of a plant organ in response to external stimuli; movement is caused by changes in the turgor of certain cells.
natality The rate at which individuals produce offspring; the average per capita birth rate.
natural selection The mechanism of evolution proposed by Charles Darwin; the tendency of organisms that possess favorable adaptations to their environment to survive and become the parents of the next generation. Evolution occurs when natural selection results in changes in allele frequencies in a population.
natural killer cell (NK cell) A large, granular lymphocyte that functions in both nonspecific and specific immune responses; releases cytokines and proteolytic enzymes that target tumor cells and cells infected with viruses and other pathogens.
Neandertal See *Homo neanderthalensis*.
necrosis Uncontrolled cell death that causes inflammation and damages other cells. Compare with *apoptosis*.
nectary (nek'ter-ee) A gland or other structure that secretes nectar.
negative feedback mechanism A homeostatic mechanism in which a change in some condition triggers a response that counteracts, or reverses, the changed condition, restoring homeostasis, e.g., how mammals maintain body temperature. Compare with *positive feedback mechanism*.
nekton (nek'ton) Free-swimming aquatic organisms such as fish and turtles. Compare with *plankton*.
nematocyst (nem-at'oh-sist) A stinging structure found within cnidocytes (stinging cells) in cnidarians; used for anchorage, defense, and capturing prey.
nematodes The phylum of animals commonly known as roundworms.
nemerteans The phylum of animals commonly known as ribbon worms; possess a proboscis for capturing prey.
neo-Darwinism See *synthetic theory of evolution*.
neonate Newborn individual.
neoplasm See *tumor*.
nephridial organ (neh-frid'ee-al) The excretory organ of many invertebrates; consists of simple or branching tubes that usually open to the outside of the body through pores; also called nephridia.
nephron (nef'ron) The functional, microscopic unit of the vertebrate kidney.
neritic province (ner-ih'tik) Ocean water that extends from the shoreline to where the bottom reaches a depth of 200 m. Compare with *oceanic province*.
nerve A bundle of axons (or dendrites) wrapped in connective tissue that conveys impulses between the central nervous system and some other part of the body.
nerve net A system of interconnecting nerve cells found in cnidarians and echinoderms.
nervous tissue A type of animal tissue specialized for transmitting electrical and chemical signals.
nest parasitism A behavior practiced by brown-headed cowbirds and certain other bird species in which females lay their eggs in the nests of other bird species and leave all parenting jobs to the hosts.
net primary productivity The energy that remains in an ecosystem (as biomass) after cellular respiration has occurred; net primary productivity equaly gross primary productivity minus respiration. Compare with *gross primary productivity*.
neural crest (noor'ul) The group of cells along the neural tube that migrate and form structures of the peripheral nervous system and certain other structures.
neural plasticity The ability of the nervous system to change in response to experience.
neural plate See *neural tube*.
neural transmission See *transmission, neural*.
neural tube The hollow, longitudinal structure in the early vertebrate embryo that gives rise to the brain and spinal cord. The neural tube forms from the neural plate, a flattened, thickened region of the ectoderm that rolls up and sinks below the surface.
neuroendocrine cells Neurons that produce neurohormones.
neurohormones Hormones produced by neuroendocrine cells; transported down axons and released into interstitial fluid; typically diffuse into capillaries and are transported by the blood; common in invertebrates; in vertebrates, neurohormones are produced by the hypothalamus.
neuron (noor'on) A nerve cell; a conducting cell of the nervous system that typically consists of a cell body, dendrites, and an axon.
neuropeptide The group of peptides produced in neural tissue that function as signaling molecules; many are neurotransmitters.
neuropeptide Y A signaling molecule produced by the hypothalamus that increases appetite and slows metabolism; helps restore

energy homeostasis when leptin levels and food intake are low.

neurotransmitter A chemical signal used by neurons to transmit impulses across a synapse.

neutral solution A solution of pH 7; there are equal concentrations of hydrogen ions (H^+) and hydroxide ions (OH^-). Compare with *acidic solution* and *basic solution*.

neutral variation Variation that does not appear to confer any selective advantage or disadvantage to the organism.

neutron (noo'tron) An electrically neutral particle with a mass of one atomic mass unit (amu) found in the atomic nucleus. Compare with *proton* and *electron*.

neutrophil (new'truh-fil) A type of granular leukocyte important in immune responses; a type of phagocyte that engulfs and destroys bacteria and foreign matter.

niche (nich) The totality of an organism's adaptations, its use of resources, and the lifestyle to which it is fitted in its community. How an organism uses materials in its environment as well as how it interacts with other organisms; also called ecological niche. See *fundamental niche* and *realized niche*.

nicotinamide adenine dinucleotide See NAD^+/NADH.

nicotinamide adenine dinucleotide phosphate See $NADP^+$/NADPH.

nitric oxide (NO) A gaseous signaling molecule; a neurotransmitter.

nitrification (nie''tra-fuh-kay'shun) The conversion of ammonia to nitrate by certain bacteria (nitrifying bacteria) in the soil; part of the nitrogen cycle.

nitrogen cycle The worldwide circulation of nitrogen from the abiotic environment into living things and back into the abiotic environment.

nitrogen fixation The conversion of atmospheric nitrogen (N_2) to ammonia (NH_3) by certain bacteria; part of the nitrogen cycle.

nitrogenase (nie-traa'jen-ase) The enzyme responsible for nitrogen fixation under anaerobic conditions.

nociceptors (no'sih-sep-torz) Pain receptors; free endings of certain sensory neurons whose stimulation is perceived as pain.

node The area on a stem where each leaf is attached. Compare with *internode*.

nodules Swellings on the roots of plants, such as legumes, in which symbiotic nitrogen-fixing bacteria *(Rhizobium)* live.

noncompetitive inhibitor A substance that lowers the rate at which an enzyme catalyzes a reaction but does not bind to the active site. Compare with *competitive inhibitor*.

noncyclic electron transport In photosynthesis, the linear flow of electrons through Photosystems I and II; results in the formation of ATP (by chemiosmosis), NADPH, and O_2. Compare with *cyclic electron transport*.

nondisjunction Abnormal separation of sister chromatids or of homologous chromosomes caused by their failure to disjoin (move apart) properly during mitosis or meiosis.

nonpolar covalent bond Chemical bond formed by the equal sharing of electrons between atoms of approximately equal electronegativity. Compare with *polar covalent bond*.

nonpolar molecule Molecule that does not have a positively charged end and a negatively charged end; nonpolar molecules are generally insoluble in water. Compare with *polar molecule*.

nonsense mutation A base substitution mutation that results in an amino acid–specifying codon being changed to a termination (stop) codon; when the abnormal mRNA is translated, the resulting protein is usually truncated and nonfunctional. Compare with *missense mutation*.

nonspecific immune responses Mechanisms such as physical barriers (e.g., the skin) and phagocytosis that provide immediate and general protection against pathogens. Also called innate immunity. Compare with *specific immune responses*.

norepinephrine (nor-ep-ih-nef'rin) A neurotransmitter that is also a hormone secreted by the adrenal medulla.

Northern blot hybridization A technique in which RNA fragments, previously separated by gel electrophoresis, are transferred to a nitrocellulose membrane; a specific radioactive genetic probe is then allowed to hybridize to complementary fragments, thus marking their locations. Compare with *Southern blot hybridization* and *Western blotting*.

notochord (no'toe-kord) The flexible, longitudinal rod in the anteroposterior axis that serves as an internal skeleton in the embryos of all chordates and in the adults of some.

nuclear area Region of a bacterial cell that contains DNA but is not enclosed by a membrane.

nuclear envelope The double membrane system that encloses the cell nucleus of eukaryotes.

nuclear equivalence The concept that, with few exceptions, the nuclei of all differentiated cells of an adult organism are genetically identical to each other and to the nucleus of the zygote from which they were derived. Compare with *genomic rearrangement* and *gene amplification*.

nuclear pores Structures in the nuclear envelope that allow passage of certain materials between the cell nucleus and the cytoplasm.

nucleolus (new-klee'oh-lus) (pl. *nucleoli*) Specialized structure in the cell nucleus formed from regions of several chromosomes; site of assembly of the ribosomal subunits.

nucleoplasm The contents of the cell nucleus.

nucleoside triphosphate Molecule consisting of a nitrogenous base, a pentose sugar, and three phosphate groups, e.g., adenosine triphosphate (ATP).

nucleosomes (new'klee-oh-somz) Repeating units of chromatin structure, each consisting of a length of DNA wound around a complex of eight histone molecules. Adjacent nucleosomes are connected by a DNA linker region associated with another histone protein.

nucleotide (noo'klee-oh-tide) A molecule composed of one or more phosphate groups, a five-carbon sugar (ribose or deoxyribose), and a nitrogenous base (purine or pyrimidine).

nucleus (new'klee-us) (pl. *nuclei*) (1) The central region of an atom, containing the protons and neutrons; (2) A cellular organelle in eukaryotes that contains the DNA and serves as the control center of the cell; (3) A mass of nerve cell bodies in the central nervous system.

nut A simple, dry fruit that contains a single seed and is surrounded by a hard fruit wall.

nutrients The chemical substances in food that are used as components for synthesizing needed materials and/or as energy sources.

nutrition The process of taking in and using food (nutrients).

obesity Excess accumulation of body fat; a person is considered obese if the body mass index (BMI) is 30 or more.

obligate anaerobe Organism that grows only in the absence of oxygen. Compare with *facultative anaerobe*.

occipital lobes Posterior areas of the mammalian cerebrum; interpret visual stimuli from the retina of the eye.

oceanic province That part of the open ocean that overlies an ocean bottom deeper than 200 m. Compare with *neritic province*.

Okazaki fragment One of many short segments of DNA, each 100 to 1000 nucleotides long, that must be joined by DNA ligase to form the lagging strand in DNA replication.

olfactory epithelium Tissue containing odor-sensing neurons.

oligodendrocyte A type of glial cell that forms myelin sheaths around neurons in the CNS.

ommatidium (om''ah-tid'ee-um) (pl. *ommatidia*) One of the light-detecting units of a compound eye, consisting of a lens and a crystalline cone that focus light onto photoreceptors called retinular cells.

omnivore (om'nih-vore) An animal that eats a variety of plant and animal materials.

oncogene (on'koh-jeen) An abnormally functioning gene implicated in causing cancer. Compare with *proto-oncogene* and *anti-oncogene*.

onycophorans (on''ih-kof'or-anz) Phylum of rare, tropical, caterpillar-like animals, structurally intermediate between annelids and arthropods, possessing an annelid-like excretory system, and claw-tipped short legs.

oocytes (oh'oh-sites) Meiotic cells that give rise to egg cells (ova).

oogamy (oh-og'uh-me) The fertilization of a large, nonmotile female gamete by a small, motile male gamete. Compare with *isogamy* and *anisogamy*.

oogenesis (oh''oh-jen'eh-sis) Production of female gametes (eggs) by meiosis. Compare with *spermatogenesis*.

oospore A thick-walled, resistant spore formed from a zygote during sexual reproduction in water molds and certain algae.

open circulatory system A type of circulatory system in which the blood bathes the tissues directly; characteristic of arthropods and many mollusks. Compare with *closed circulatory system*.

open system An entity that can exchange energy and matter with its surroundings. Compare with *closed system*.

operant conditioning A type of learning in which an animal is rewarded or punished for performing a behavior it discovers by chance; also called instrumental conditioning.

operator site One of the control regions of an operon; the DNA segment to which a repressor binds, thereby inhibiting the transcription of the adjacent structural genes of the operon.

operculum In bony fishes, a protective flap of the body wall that covers the gills.

operon (op'er-on) In prokaryotes, a group of structural genes that are coordinately controlled and transcribed as a single message, plus their adjacent regulatory elements.

optimal foraging The process of obtaining food in a manner that maximizes benefits and/or minimizes costs.

orbital Region in which electrons occur in an atom or molecule.

order A taxonomic category made up of related families.

organ A specialized structure, such as the heart or liver, made up of tissues and adapted to perform a specific function or group of functions.

organ of Corti (kor'tie) The structure within the inner ear of vertebrates that contains receptor cells that sense sound vibrations.

organ system An organized group of tissues and organs that work together to perform a specialized set of functions, e.g., the digestive system or circulatory system.

organelle One of the specialized structures within the cell, such as the mitochondria, Golgi complex, ribosomes, or contractile vacuole; many organelles are membrane-bounded.

organic compound A compound composed of a backbone made up of carbon atoms. Compare with *inorganic compound*.

organism Any living system composed of one or more cells.

organismic respiration See *respiration*.

organogenesis The process of organ formation.

orgasm (or'gazm) The climax of sexual excitement.

origin of replication A specific site on the DNA where replication can begin.

osmoconformer An animal in which the salt concentration of body fluids varies with changes in surrounding seawater. Compare with *osmoregulator*.

osmoregulation (oz''moh-reg-yoo-lay'shun) The active regulation of the osmotic pressure of body fluids so that they do not become excessively dilute or excessively concentrated.

osmoregulator An animal that maintains an optimal salt concentration in its body fluids despite changes in salinity of its surroundings. Compare with *osmoconformer*.

osmosis (oz-moh'sis) The net movement of water (the principal solvent in biological systems) by diffusion through a selectively permeable membrane from a region of higher concentration of water (a hypotonic solution) to a region of lower concentration of water (a hypertonic solution).

osmotic pressure The pressure that must be exerted on the hypertonic side of a selectively permeable membrane to prevent diffusion of water (by osmosis) from the side containing pure water.

osteichthyes (os''tee-ick'thees) Historically, the vertebrate class of bony fishes. Biologists now divide bony fishes into three classes: Actinopterygii, the ray-finned fishes; Actinistia, the coelacanths; and Sarcopterygii, the lobe-finned fishes.

osteoblast (os'tee-oh-blast) A type of bone cell that secretes the protein matrix of bone. Also see *osteocyte*.

osteoclast (os'tee-oh-clast) Large, multinucleate cell that helps sculpt and remodel bones by dissolving and removing part of the bony substance.

osteocyte (os'tee-oh-site) A mature bone cell; an osteoblast that has become embedded within the bone matrix and occupies a lacuna.

osteon (os'tee-on) The spindle-shaped unit of bone composed of concentric layers of osteocytes organized around a central Haversian canal containing blood vessels.

otoliths (oh'toe-liths) Small calcium carbonate crystals in the saccule and utricle of the inner ear; sense gravity and are important in static equilibrium.

outbreeding The mating of individuals of unrelated strains. Compare with *inbreeding*.

ovary (oh'var-ee) (1) In animals, one of the paired female gonads responsible for producing eggs and sex hormones; (2) In flowering plants, the base of the carpel that contains ovules; ovaries develop into fruits after fertilization.

oviduct (oh'vih-dukt) The tube that carries ova from the ovary to the uterus, cloaca, or body exterior. Also called fallopian tube or uterine tube.

oviparous (oh-vip'ur-us) Bearing young in the egg stage of development; egg-laying. Compare with *viviparous* and *ovoviviparous*.

ovoviviparous (oh'voh-vih-vip''ur-us) A type of development in which the young hatch from eggs incubated inside the mother's body. Compare with *viviparous* and *oviparous*.

ovulation (ov-u-lay'shun) The release of an ovum from the ovary.

ovule (ov'yool) The structure (i.e., megasporangium) in the plant ovary that develops into the seed following fertilization.

ovum (pl. *ova*) Female gamete of an animal.

oxaloacetate Four-carbon compound; important intermediate in the citric acid cycle and in the C_4 and CAM pathways of carbon fixation in photosynthesis.

oxidants Highly reactive molecules such as radicals, peroxides, and superoxides that are produced during normal cell processes that require oxygen; can damage DNA and other molecules by snatching electrons.

oxidation The loss of one or more electrons (or hydrogen atoms) by an atom, ion, or molecule. Compare with *reduction*.

oxidative phosphorylation (fos''for-ih-lay'shun) The production of ATP using energy derived from the transfer of electrons in the electron transport system of mitochondria; occurs by chemiosmosis.

oxygen-carrying capacity The maximum amount of oxygen that can be transported by hemoglobin.

oxygen debt The oxygen necessary to metabolize the lactic acid produced during strenuous exercise.

oxygen-hemoglobin dissociation curve A curve depicting the percentage saturation of hemoglobin with oxygen, as a function of certain variables such as oxygen concentration, carbon dioxide concentration, or pH.

oxyhemoglobin Hemoglobin that has combined with oxygen.

oxytocin (ok''see-tow'sin) Hormone secreted by the hypothalamus and released by the posterior lobe of the pituitary gland; stimulates contraction of the pregnant uterus and the ducts of mammary glands.

ozone A blue gas, O_3, with a distinctive odor that is a human-made pollutant near Earth's surface (in the troposphere) but a natural and essential component of the stratosphere.

P generation (parental generation) Members of two different true-breeding lines that are crossed to produce the F_1 generation.

P680 Chlorophyll *a* molecules that serve as the reaction center of Photosystem II, transferring photoexcited electrons to a primary acceptor; named by their absorption peak at 680 nm.

P700 Chlorophyll *a* molecules that serve as the reaction center of Photosystem I, transferring photoexcited electrons to a primary acceptor; named by their absorption peak at 700 nm.

pacemaker (of the heart) See *sinoatrial (SA) node*.

Pacinian corpuscle (pah-sin'-ee-an kor'pus-el) A receptor located in the dermis of the skin that responds to pressure.

paedomorphosis Retention of juvenile or larval features in a sexually mature animal.

pair bond A stable relationship between animals of opposite sex that ensures cooperative behavior in mating and rearing the young.

paleoanthropology (pay''lee-o-an-thro-pol'uh-gee) The study of human evolution.

Paleozoic era That part of geological time extending from roughly 570 to 248 million years ago.

palindromic Reading the same forward and backward; some DNA sequences are palindromic because the base sequence of one strand is the reverse of the base sequence in its complement; for example, the complement of 5′-GAATTC-3′ is 3′-CTTAAG-5′.

palisade mesophyll (mez'oh-fil) The vertically stacked, columnar mesophyll cells near the upper epidermis in certain leaves. Compare with *spongy mesophyll*.

pancreas (pan'kree-us) Large gland located in the vertebrate abdominal cavity. The pancreas produces pancreatic juice containing digestive enzymes; also serves as an endocrine gland, secreting the hormones insulin and glucagon.

panspermia The idea that life did not originate on Earth, but began elsewhere in the galaxy and drifted through space to Earth.

parabronchi (sing. *parabronchus*) Thin-walled ducts in the lungs of birds; gases are exchanged across their walls.

paracrine regulation A type of regulation in which a signal molecule (e.g., certain hormones) diffuses through interstitial fluid and acts on nearby target cells. Compare with *autocrine regulation*.

paraphyletic group A group of organisms made up of a common ancestor and some, but not all, of its descendants. Compare with *monophyletic group* and *polyphyletic group*.

parapodia (par''uh-poh'dee-ah) (sing. *parapodium*) Paired, thickly bristled paddle-like appendages extending laterally from each segment of polychaete worms.

parasite A heterotrophic organism that obtains nourishment from the living tissue of another organism (the host).

parasitism (par'uh-si-tiz''m) A symbiotic relationship in which one member (the parasite) benefits and the other (the host) is adversely affected. Compare with *commensalism* and *mutualism*.

parasympathetic nervous system A division of the autonomic nervous system concerned with the control of the internal organs; functions to conserve or restore energy. Compare with *sympathetic nervous system*.

parathyroid glands Small, pea-sized glands closely adjacent to the thyroid gland; their secretion regulates calcium and phosphate metabolism.

parathyroid hormone (PTH) A hormone secreted by the parathyroid glands; regulates calcium and phosphate metabolism.

parenchyma (par-en'kih-mah) Highly variable living plant cells that have thin primary walls; function in photosynthesis, the storage of nutrients, and/or secretion.

parsimony The principle based on the experience that the simplest explanation is most probably the correct one.

parthenogenesis (par''theh-noh-jen'eh-sis) The development of an unfertilized egg into an adult organism; common among honey bees, wasps, and certain other arthropods.

parturition (par''to-rish'un) The birth process.

partial pressure (of a gas) The pressure exerted by gas in a mixture, which is the same pressure it would exert if alone. For example, the partial pressure of atmospheric oxygen (P_{O_2}) is 160 mm Hg at sea level.

passive immunity Temporary immunity that depends on the presence of immunoglobulins produced by another organism. Compare with *active immunity*.

passive ion channel A channel in the plasma membrane that permits the passage of specific ions such as Na^+, K^+, or Cl^-.

patch clamp technique A method that allows researchers to study the ion channels of a tiny patch of membrane by tightly sealing a micropipette to the patch and measuring the flow of ions through the channels.

patchiness See *clumped dispersion*.

pathogen (path'oh-gen) An organism, usually a microorganism, capable of producing disease.

pattern formation See *morphogenesis*.

pedigree A chart constructed to show an inheritance pattern within a family through multiple generations.

peduncle The stalk of a flower or inflorescence.

pellicle A flexible outer covering of protein; characteristic of certain protists, e.g., ciliates and euglenoids.

penis The male sexual organ of copulation in reptiles, mammals, and a few birds.

pentose A sugar molecule containing five carbons.

people overpopulation A situation in which there are too many people in a given geographical area; results in pollution, environmental degradation, and resource depletion. Compare with *consumption overpopulation*.

pepsin (pep'sin) An enzyme produced in the stomach that initiates digestion of protein.

peptide (pep'tide) A compound consisting of a chain of amino acid groups linked by peptide bonds. A dipeptide consists of two amino acids, a polypeptide of many.

peptide bond A distinctive covalent carbon-to-nitrogen bond that links amino acids in peptides and proteins.

peptidoglycan (pep''tid-oh-gly'kan) A modified protein or peptide possessing an attached carbohydrate; component of the eubacterial cell wall.

perennial plant (purr-en'ee-ul) A woody or herbaceous plant that grows year after year, i.e., lives more than two years. Compare with *annual* and *biennial*.

perfect flower A flower that has both stamens and carpels. Compare with *imperfect* flower.

pericentriolar material Fibrils surrounding the centrioles in the microtubule organizing centers in cells of animals and other organisms possessing centrioles; chemically similar to the material in the microtubule organizing centers of organisms lacking centrioles.

pericycle (pehr'eh-sy''kl) A layer of meristematic cells typically found between the endodermis and phloem in roots.

periderm (pehr'ih-durm) The outer bark of woody stems and roots; composed of cork cells, cork cambium, and cork parenchyma, along with traces of primary tissues.

period An interval of geological time that is a subdivision of an era. Each period is divided into epochs.

peripheral membrane protein A protein associated with one of the surfaces of a biological membrane. Compare with *integral membrane protein*.

peripheral nervous system (PNS) In vertebrates, the nerves and receptors that lie outside the central nervous system. Compare with *central nervous system (CNS)*.

peristalsis (pehr''ih-stal'sis) Rhythmic waves of muscular contraction and relaxation in the walls of hollow tubular organs, such as the ureter or parts of the digestive tract, that serve to move the contents through the tube.

permafrost Permanently frozen subsoil characteristic of frigid areas such as the tundra.

peroxisomes (pehr-ox'ih-somz) Membrane-bounded organelles in eukaryotic cells containing enzymes that produce or degrade hydrogen peroxide.

persistence A characteristic of certain chemicals that are extremely stable and may take many years to be broken down into simpler forms by natural processes.

petal One of the parts of the flower attached inside the whorl of sepals; petals are usually colored.

petiole (pet'ee-ohl) The part of a leaf that attaches to a stem.

pH The negative logarithm of the hydrogen ion concentration of a solution (expressed as moles per liter). Neutral pH is 7, values less than 7 are acidic, and those greater than 7 are basic.

phage See *bacteriophage*.

phagocytosis (fag''oh-sy-toh'sis) Literally, "cell eating"; a type of endocytosis by which certain cells engulf food particles, microorganisms, foreign matter, or other cells.

pharmacogenomics A new field of gene-based medicine in which drugs are personalized to match a patient's genetic makeup.

pharynx (fair'inks) Part of the digestive tract. In complex vertebrates it is bounded anteriorly by the mouth and nasal cavities and posteriorly by the esophagus and larynx; the throat region in humans.

phenetics (feh-neh'tiks) An approach to classification based on measurable similarities in phenotypic characters, without consideration of homology or other evolutionary relationships. Compare with *cladistics* and *evolutionary systematics*.

phenotype (fee'noh-type) The physical or chemical expression of an organism's genes. Compare with *genotype*.

phenotype frequency The proportion of a particular phenotype in the population.

phenylketonuria (PKU) (fee''nl-kee''toh-noor'ee-ah) An inherited disease in which there is a deficiency of the enzyme that normally converts phenylalanine to tyrosine; results in mental retardation if untreated.

pheromone (fer'oh-mone) A substance secreted by an organism to the external environment that influences the development or behavior of other members of the same species.

phloem (flo'em) The vascular tissue that conducts dissolved sugar and other organic compounds in plants.

phosphate group A weakly acidic functional group that can release one or two hydrogen ions.

phosphodiester linkage Covalent linkage between two nucleotides in a strand of DNA or RNA; includes a phosphate group bonded to the sugars of two adjacent nucleotides.

phosphoenolpyruvate (PEP) Three-carbon phosphorylated compound that is an important intermediate in glycolysis and is a reactant in the initial carbon fixation step in C_4 and CAM photosynthesis.

phosphoglycerate (PGA) Phosphorylated three-carbon compound that is an important metabolic intermediate.

phospholipids (fos''foh-lip'idz) Lipids in which there are two fatty acids and a phosphorus-containing group attached to glycerol; major components of cell membranes.

phosphorus cycle The worldwide circulation of phosphorus from the abiotic environment into living things and back into the abiotic environment.

phosphorylation (fos''for-ih-lay'shun) The introduction of a phosphate group into an organic molecule. See *kinases*.

photoautotroph An organism that obtains energy from light and synthesizes organic compounds from inorganic raw materials; includes plants, algae, and some bacteria. Compare with *photoheterotroph, chemoautotroph,* and *chemoheterotroph*.

photoheterotroph An organism that is able to carry out photosynthesis to obtain energy but is unable to fix carbon dioxide and therefore requires organic compounds as a carbon source; includes some bacteria. Compare with *photoautotroph, chemoautotroph,* and *chemoheterotroph*.

photolysis (foh-tol'uh-sis) The photochemical splitting of water in the light-dependent reactions of photosynthesis, catalyzed by a specific enzyme.

photon (foh'ton) A particle of electromagnetic radiation; one quantum of radiant energy.

photoperiodism (foh''teh-peer'ee-o-dizm) The physiological response (such as flowering) of plants to variations in the length of daylight and darkness.

photophosphorylation (foh''toh-fos-for-ih-lay'shun) The production of ATP in photosynthesis.

photoreceptor (foh''toh-ree-sep'tor) (1) A sense organ specialized to detect light; (2) A pigment that absorbs light before triggering a physiological response.

photorespiration (foh''toh-res-pur-ay'shun) The process that reduces the efficiency of photosynthesis in C_3 plants during hot spells in summer; consumes oxygen and produces carbon dioxide through the degradation of Calvin cycle intermediates.

photosynthesis The biological process that captures light energy and transforms it into the chemical energy of organic molecules (e.g., carbohydrates), which are manufactured from carbon dioxide and water; performed by plants, algae, and certain bacteria.

photosystem One of two photosynthetic units, consisting of chlorophyll molecules, accessory pigments, proteins, and associated electron acceptors, responsible for capturing light energy and transferring excited electrons; photosystem I best absorbs and uses light of about 700 nm, whereas photosystem II best absorbs and uses light of about 680 nm.

phototroph (foh'toh-trof) Organism that uses light as a source of energy. Compare with *chemotroph*. See *photoautotroph* and *photoheterotroph*.

phototropism (foh''toh-troh'pizm) The growth of a plant in response to the direction of light.

phycocyanin (fy''koh-sy-ah'nin) A blue pigment found in cyanobacteria and red algae.

phycoerythrin (fy''koh-ee-rih'thrin) A red pigment found in cyanobacteria and red algae.

phyletic evolution See *anagenesis*.

phylogenetic systematics See *cladistics*.

phylogenetic tree A branching diagram that shows lines of descent among a group of related species.

phylogeny (fy-loj'en-ee) The complete evolutionary history of a group of organisms.

phylum (fy'lum) A taxonomic grouping of related, similar classes; a category beneath the kingdom and above the class.

phytochemicals Compounds found in plants that play important roles in preventing certain diseases; some function as antioxidants.

phytochrome (fy'toh-krome) A blue-green, proteinaceous pigment involved in a wide variety of physiological responses to light; occurs in two interchangeable forms depending on the ratio of red to far-red light.

phytoplankton (fy''toh-plank'tun) Microscopic floating algae and cyanobacteria that are the base of most aquatic food webs. Compare with *zooplankton*. See *plankton* and *nanoplankton*.

pia mater (pee'a may'ter) The inner membrane covering the brain and spinal cord; the innermost of the meninges; also see *dura mater* and *arachnoid*.

pigment A substance that selectively absorbs light of different wavelengths.

pili (pie'lie) (sing. *pilus*) Hairlike structures on the surface of many bacteria. Function in conjugation or attachment.

pineal gland (pie-nee'al) Endocrine gland located in the brain.

pinocytosis (pin''oh-sy-toh'sis) Cell drinking; a type of endocytosis by which cells engulf and absorb droplets of liquids.

pioneer The first organism to colonize an area and begin the first stage of succession.

pistil The female reproductive organ of a flower; consists of either a single carpel or two or more fused carpels. See *carpel*.

pith The innermost tissue in the stems and roots of many herbaceous plants; primarily a storage tissue.

pituitary gland (pi-too'ih-tehr''ee) An endocrine gland located below the hypothalamus; secretes several hormones that influence a wide range of physiological processes.

placenta (plah-sen'tah) The partly fetal and partly maternal organ whereby materials are exchanged between fetus and mother in the uterus of placental mammals.

placoderms (plak'oh-durms) A group of extinct jawed fishes.

plankton Free-floating, mainly microscopic aquatic organisms found in the upper layers of the water; composed of phytoplankton and zooplankton. Compare with *nekton*.

planula larva (plan'yoo-lah) A ciliated larval form found in cnidarians.

plasma The fluid portion of blood in which red blood cells, white blood cells, and platelets are suspended.

plasma membrane The selectively permeable surface membrane that encloses the cell contents and through which all materials entering or leaving the cell must pass.

plasma cell Cell that secretes antibodies; a differentiated B lymphocyte (B cell).

plasma proteins Proteins such as albumins, globulins, and fibrinogen that circulate in the blood plasma.

plasmids (plaz'midz) Small circular double-stranded DNA molecules that carry genes separate from the main DNA of a cell.

plasmodesmata (sing. *plasmodesma*) Cytoplasmic channels connecting adjacent plant cells and allowing for the movement of molecules and ions between cells.

plasmodial slime mold (plaz-moh'dee-uhl) A fungus-like protist whose feeding stage consists of a plasmodium.

plasmodium (plaz-moh'dee-um) A multinucleate mass of living matter that moves and feeds in an amoeboid fashion.

plasmolysis (plaz-mol'ih-sis) The shrinkage of cytoplasm and the pulling away of the plasma membrane from the cell wall when a plant cell (or other walled cell) loses water, usually in a hypertonic environment.

plastids (plas'tidz) A family of membrane-bounded organelles occurring in photosynthetic eukaryotic cells; include chloroplasts, chromoplasts, and amyloplasts and other leukoplasts.

platelets (playt'lets) Cell fragments in vertebrate blood that function in clotting; also called thrombocytes.

platyhelminthes The phylum of acoelomate animals commonly known as flatworms.

plesiomorphic characters See *shared ancestral characters*.

pleiotropic (ply''oh-troh'pik) A term referring to an allele that affects a number of characteristics of an individual.

pleural membrane (ploor'ul) The membrane that lines the thoracic cavity and envelops each lung.

ploidy The number of chromosome sets in a nucleus or cell. See *haploid, diploid,* and *polyploid*.

plumule (ploom'yool) The embryonic shoot apex, or terminal bud, located above the point of attachment of the cotyledon(s).

pluripotent (ploor-i-poh'tent) A term describing a stem cell that is able to divide to give rise to all tissues of the body. Compare with *totipotent*.

pneumatophore (noo-mat'uh-for'') Roots that extend up out of the water in swampy areas and are thought to provide aeration between the atmosphere and submerged roots.

polar body A small *n* cell produced during oogenesis in female animals that does not develop into a functional ovum.

polar covalent bond Chemical bond formed by the sharing of electrons between atoms that differ in electronegativity; the end of the bond near the more electronegative atom has a partial negative charge, the other end has a partial positive charge. Compare with *nonpolar covalent bond*.

polar molecule Molecule that has one end with a partial positive charge and the other with a partial negative charge; polar molecules are generally soluble in water. Compare with *nonpolar molecule*.

polar nucleus In flowering plants, one of two n cells in the embryo sac that fuse with a sperm during double fertilization to form the $3n$ endosperm.
pollen grain The immature male gametophyte of seed plants (gymnosperms and angiosperms) that produces sperm capable of fertilization.
pollen tube In gymnosperms and flowering plants, a tube or extension that forms after germination of the pollen grain and through which male gametes (sperm cells) pass into the ovule.
pollination (pol′′uh-nay′shen) In seed plants, the transfer of pollen from the male to the female part of the plant.
polyadenylation (pol′′ee-a-den-uh-lay′shun) That part of eukaryotic mRNA processing in which multiple adenine-containing nucleotides (a poly A tail) are added to the 3′ end of the molecule.
polyandry A mating system in which a female mates with several males during a breeding season. Compare with *polygyny*.
polygyny A mating system in which a male animal mates with many females during a breeding season. Compare with *polyandry*.
poly A tail See *polyadenylation*.
polygenic inheritance (pol′′ee-jen′ik) Inheritance in which several independently assorting or loosely linked nonallelic genes modify the intensity of a trait or contribute to the phenotype in additive fashion.
polymer (pol′ih-mer) A molecule built up from repeating subunits of the same general type (monomers), such as a protein, nucleic acid, or polysaccharide.
polymerase chain reaction (PCR) A method by which a targeted DNA fragment can be amplified in vitro to produce millions of copies.
polymorphism (pol′′ee-mor′fizm) (1) The existence of two or more phenotypically different individuals within a population; (2) The presence of more than one allele for a given locus in a population.
polyp (pol′ip) A hydra-like animal; the sessile stage of the life cycle of certain cnidarians. Compare with *medusa*.
polypeptide See *peptide*.
polyphyletic group (pol′′ee-fye-let′ik) A group made up of organisms that evolved from two or more different ancestors. Compare with *monophyletic group* and *paraphyletic group*.
polyploid (pol′ee-ployd) The condition of having more than two sets of chromosomes per nucleus. Compare with *diploid* and *haploid*.
polyribosome A complex consisting of a number of ribosomes attached to an mRNA during translation; also known as a polysome.
polysaccharide (pol-ee-sak′ah-ride) A carbohydrate consisting of many monosaccharide subunits, e.g., starch, glycogen, and cellulose.
polysome See *polyribosome*.
polyspermy The fertilization of an egg by more than one sperm.
polytene A term describing a giant chromosome consisting of many (usually > 1000) parallel DNA double helices. Polytene chromosomes are typically found in cells of the salivary glands and some other tissues of certain insects, such as the fruit fly, *Drosophila*.
polyunsaturated fatty acid See *fatty acid*.
pons (ponz) The white bulge that is the part of the brain stem between the medulla and the midbrain; connects various parts of the brain.
population A group of organisms of the same species that live in a defined geographical area at the same time.
population bottleneck See *bottleneck*.
population crash An abrupt decline in the size of a population.
population density The number of individuals of a species per unit of area or volume at a given time.
population dynamics The study of changes in populations, such as how and why population numbers change over time.
population ecology That branch of biology that deals with the numbers of a particular species that are found in an area and how and why those numbers change (or remain fixed) over time.
population genetics The study of genetic variability in a population and of the forces that act on it.
population growth momentum The continued growth of a population after fertility rates have declined, as a result of a population's young age structure.
poriferans Sponges; members of phylum Porifera.
positive feedback mechanism A homeostatic mechanism in which a change in some condition triggers a response that intensifies the changing condition. Compare with *negative feedback mechanism*.
posterior Toward the tail end of a bilaterally symmetrical animal. Compare with *anterior*.
postsynaptic neuron A neuron that transmits an impulse away from a synapse. Compare with *presynaptic neuron*.
postzygotic barrier One of several reproductive isolating mechanisms that prevent gene flow between species after fertilization has taken place; e.g., hybrid inviability, hybrid sterility, and hybrid breakdown. Compare with *prezygotic barrier*.
potential energy Stored energy; energy that can do work as a consequence of its position or state. Compare with *kinetic energy*.
potentiation A form of synaptic enhancement (increase in neurotransmitter release) that can last for several minutes; occurs when a presynaptic neuron continues to transmit action potentials at a high rate for a minute or longer.
preadaptation A novel evolutionary change in a preexisting biological structure that enables it to have a different function; feathers, which evolved from reptilian scales, represent a preadaptation for flight
prebiotic broth hypothesis The hypothesis that simple organic molecues that are the precursors of life originated and accumulated at Earth's surface, in shallow seas or on rock or clay surfaces. Compare with *iron-sulfur world hypothesis*.
Precambrian time All of geological time before the Paleozoic era, encompassing approximately the first 4 billion years of Earth's history.
predation Relationship in which one organism (the predator) kills and devours another organism (the prey).
pre-mRNA RNA precursor to mRNA in eukaryotes; contains both introns and exons.
premise Information supplied from which conclusions can be drawn in the process of deductive reasoning.
pressure-flow hypothesis The mechanism by which dissolved sugar is thought to be transported in phloem; caused by a pressure gradient between the source (where sugar is loaded into the phloem) and the sink (where sugar is removed from phloem).
prenatal A term referring to the time before birth.
presynaptic neuron A neuron that transmits an impulse to a synapse. Compare with *postsynaptic neuron*.

prezygotic barrier One of several reproductive isolating mechanisms that interfere with fertilization between male and female gametes of different species; e.g., temporal isolation, habitat isolation, behavioral isolation, mechanical isolation, and gametic isolation. Compare with *postzygotic barrier*.

primary consumer See *herbivore*.

primary growth An increase in the length of a plant that occurs at the tips of the shoots and roots due to the activity of apical meristems. Compare with *secondary growth*.

primary mycelium A mycelium in which the cells are monokaryotic and haploid; a mycelium that grows from either an ascospore or a basidiospore. Compare with *secondary mycelium*.

primary producer See *autotroph*.

primary response The response of the immune system to first exposure to an antigen. Compare with *secondary response*.

primary structure (of a protein) The complete sequence of amino acids in a polypeptide chain, beginning at the amino end and ending at the carboxyl end. Compare with *secondary, tertiary,* and *quaternary protein structure*.

primary succession An ecological succession that occurs on land that has not previously been inhabited by plants; no soil is present initially. See *succession*. Compare with *secondary succession*.

primer See *RNA primer*.

primitive groove See *primitive streak*.

primitive streak Dynamic, constantly changing structure that forms at the midline of the blastodisc in birds, mammals, and some other vertebrates, and is active in gastrulation. Cells from the surface enter the interior through the furrow at its center (primitive groove), which is the functional equivalent of the blastopore. The anterior end of the primitive streak is Hensen's node.

primosome A complex of proteins responsible for synthesizing the RNA primers required in DNA synthesis.

principle A scientific theory that has withstood repeated testing and has the highest level of scientific confidence. Compare with *hypothesis* and *theory*.

prion An infectious agent that consists only of protein.

producer See *autotroph*.

product Substance formed by a chemical reaction. Compare with *reactant*.

product rule The rule for combining the probabilities of independent events by multiplying their individual probabilities. Compare with *sum rule*.

profundal zone (pro-fun'dl) The deepest zone of a large lake, located below the level of penetration by sunlight. Compare with *littoral zone* and *limnetic zone*.

progesterone (pro-jes'ter-own) A steroid hormone secreted by the ovary (mainly by the corpus luteum) and placenta; stimulates the uterus (to prepare the endometrium for implanation) and breasts (for milk secretion).

progymnosperm (pro-jim'noh-sperm) An extinct group of plants that are thought to have been the ancestors of gymnosperms.

prokaryote (pro-kar'ee-ote) A cell that lacks a nucleus and other membrane-bounded organelles; includes the bacteria, members of kingdoms Eubacteria and Archaebacteria. Compare with *eukaryote*.

promoter The nucleotide sequence in DNA to which RNA polymerase attaches to begin transcription.

prophage (pro'faj) Bacteriophage nucleic acid that is inserted into the bacterial DNA.

prophase The first stage of mitosis and of meiosis I and meiosis II. During prophase the chromosomes become visible as distinct structures, the nuclear envelope breaks down, and a spindle forms. Meiotic prophase I is complex and includes synapsis of homologous chromosomes and crossing-over.

proplastids Organelles that are plastid precursors; may mature into various specialized plastids, including chloroplasts, chromoplasts, or leukoplasts.

proprioceptors (pro''pree-oh-sep'torz) Receptors in muscles, tendons, and joints that respond to changes in movement, tension, and position; enable an animal to perceive the position of its body.

prop root An adventitious root that arises from the stem and provides additional support for a plant such as corn.

prostaglandins (pros''tah-glan'dinz) Derivatives of unsaturated fatty acids that produce a wide variety of hormone-like effects; synthesized by most cells of the body; sometimes called local hormones.

prostate gland A gland in male animals that produces an alkaline secretion that is part of the semen.

protein A large, complex organic compound composed of covalently linked amino acid subunits; contains carbon, hydrogen, oxygen, nitrogen, and sulfur.

prothallus (pro-thal'us) (pl. *prothalli*) The free-living, n gametophyte in ferns and other seedless vascular plants.

protist (pro'tist) One of a vast kingdom of eukaryotic organisms, primarily unicellular or simple multicellular; mostly aquatic.

protobionts (pro''toh-by'ontz) Assemblages of organic polymers that spontaneously form under certain conditions. Protobionts may have been involved in chemical evolution.

proton A particle present in the nuclei of all atoms that has one unit of positive charge and a mass of one atomic mass unit (amu). Compare with *electron* and *neutron*.

protonema (pro''toh-nee'mah) (pl. *protonemata*) In mosses, a filament of n cells that grows from a spore and develops into leafy moss gametophytes.

protonephridia (pro''toh-nef-rid'ee-ah) (sing. *protonephridium*) The flame-cell excretory organs of flatworms and some other simple invertebrates.

proto-oncogene A gene that normally promotes cell division in response to the presence of certain growth factors; when mutated it may become an oncogene, possibly leading to the formation of a cancer cell. Compare with *oncogene*.

protostome (pro'toh-stome) A major division of the animal kingdom in which the blastopore develops into the mouth, and the anus forms secondarily; includes the annelids, arthropods, and mollusks. Compare with *deuterostome*.

protozoa (proh''toh-zoh'a) (sing. *protozoon*) An informal group of unicellular, animal-like protists, including amoebas, foraminiferans, actinopods, ciliates, flagellates, and apicomplexans. (The adjectival form is *protozoan.*)

provirus (pro-vy'rus) A part of a virus, consisting of nucleic acid only, that has been inserted into a host genome. See *DNA provirus*.

proximal Closer to the point of reference. Compare with *distal*.

proximal convoluted tubule The part of the renal tubule that extends from Bowman's capsule to the loop of Henle. Compare with *distal convoluted tubule*.

proximate causes (of behavior) The immediate causes of behavior, such as genetic, developmental, and physiological processes that permit the animal to carry out a specific behavior. Compare with *ultimate causes of behavior*.

pseudocoelom (sue''doh-see'lom) A body cavity between the mesoderm and endoderm; derived from the blastocoel. Compare with *coelom*.

pseudocoelomate (sue''doh-seel'oh-mate) An animal possessing a pseudocoelom. Compare with *coelomate* and *acoelomate*.

pseudoplasmodium (sue''doe-plaz-moh'dee-um) In cellular slime molds, an aggregation of amoeboid cells that forms a spore-producing fruiting body during reproduction.

pseudopodium (sue''doe-poe'dee-um) (pl. *pseudopodia*) A temporary extension of an amoeboid cell that is used for feeding and locomotion.

puff In a polytene chromosome, a decondensed region that is a site of intense RNA synthesis.

pulmonary circulation The part of the circulatory system that delivers blood to and from the lungs for oxygenation. Compare with *systemic circulation*.

pulse, arterial The alternate expansion and recoil of an artery.

pulvinus (pul-vy'nus) A special structure, often located at the base of the petiole, that functions in leaf movement by changes in turgor.

punctuated equilibrium The idea that evolution proceeds with periods of little or no genetic change, followed by very active phases, so that major adaptations or clusters of adaptations appear suddenly in the fossil record. Compare with *gradualism*.

Punnett square The grid structure, first developed by Reginald Punnett, that allows direct calculation of the probabilities of occurrence of all possible offspring of a genetic cross.

pupa (pew'pah) (pl. *pupae*) A stage in the development of an insect, between the larva and the imago (adult); a form that neither moves nor feeds, and may be in a cocoon.

pure line See *true-breeding line*.

purines (pure'eenz) Nitrogenous bases with carbon and nitrogen atoms in two attached rings, e.g., adenine and guanine; components of nucleic acids, ATP, GTP, NAD^+, and certain other biologically active substances. Compare with *pyrimidines*.

pyramid of biomass An ecological pyramid that illustrates the total biomass, as, for example, the total dry weight, of all organisms at each trophic level in an ecosystem.

pyramid of energy An ecological pyramid that shows the energy flow through each trophic level of an ecosystem.

pyramid of numbers An ecological pyramid that shows the number of organisms at each trophic level in an ecosystem.

pyrimidines (pyr-im'ih-deenz) Nitrogenous bases, each composed of a single ring of carbon and nitrogen atoms, e.g., thymine, cytosine, and uracil; components of nucleic acids. Compare with *purines*.

pyruvate (pyruvic acid) A three-carbon compound; the end product of glycolysis.

quadrupedal (kwad'roo-ped''ul) Walking on all fours.

quantitative trait A trait that shows continuous variation in a population (e.g., human height) and typically has a polygenic inheritance pattern.

quaternary structure (of a protein) The overall conformation of a protein produced by the interaction of two or more polypeptide chains. Compare with *primary, secondary,* and *tertiary protein structure*.

***r* selection** A reproductive strategy recognized by some ecologists, in which a species typically has a small body size, rapid development, short life span, and devotes a large proportion of its metabolic energy to the production of offspring. Compare with *K selection*.

radial cleavage The pattern of blastomere production in which the cells are located directly above or below one another; characteristic of early deuterostome embryos. Compare with *spiral cleavage*.

radial symmetry A body plan in which any section through the mouth and down the length of the body divides the body into similar halves. Jellyfish and other cnidarians have radial symmetry. Compare with *bilateral symmetry*.

radicle (rad'ih-kl) The embryonic root of a seed plant.

radioactive decay The process in which a radioactive element emits radiation and, as a result, its nucleus changes into the nucleus of a different element.

radioisotopes Unstable isotopes that spontaneously emit radiation; also called radioactive isotopes.

radiolarians Those actinopods that secrete elaborate shells of silica (glass).

radula (rad'yoo-lah) A rasplike structure in the digestive tract of chitons, snails, squids, and certain other mollusks.

rain shadow An area that has very little precipitation, found on the downwind side of a mountain range. Deserts often occur in rain shadows.

random dispersion The spatial distribution pattern of a population in which the presence of one individual has no effect on the distribution of other individuals. Compare with *clumped dispersion* and *uniform dispersion*.

range The area where a particular species occurs.

ray A chain of parenchyma cells (one to many cells thick) that functions for lateral transport in stems and roots of woody plants.

ray-finned fishes A class (Actinopterygii) of modern bony fishes; contains about 95% of living fish species.

reabsorption The selective removal of certain substances from the glomerular filtrate by the renal tubules and collecting ducts of the kidney, and their return into the blood.

reactant Substance that participates in a chemical reaction. Compare with *product*.

reaction center The portion of a photosystem that includes chlorophyll *a* molecules capable of transferring electrons to a primary electron acceptor, which is the first of several electron acceptors in a series; the reaction center of Photosystem I is P700 and of Photosystem II is P680. See *antenna complex* and *photosystem*.

realized niche The lifestyle that an organism actually pursues, including the resources that it actually uses. An organism's realized niche is narrower than its fundamental niche because of interspecific competition. Compare with *fundamental niche*.

receptacle The end of a flower stalk where the flower parts (sepals, petals, stamens, and carpels) are attached.

reception Process of detecting a stimulus.

receptor down-regulation The process by which some hormone receptors decrease in number, thereby suppressing the sensitivity of target cells to the hormone. Compare with *receptor up-regulation*.

receptor-mediated endocytosis A type of endocytosis in which extracellular molecules become bound to specific receptors on the cell surface and then enter the cytoplasm enclosed in vesicles.

receptor up-regulation The process by which some hormone receptors increase in number, thereby increasing the sensitivity of the target cells to the hormone. Compare with *receptor down-regulation*.

recessive allele (al-leel') An allele that is not expressed in the heterozygous state. Compare with *dominant allele*.

recombinant DNA Any DNA molecule made by combining genes from different organisms.

recombination, genetic The appearance of new gene combinations. Recombination in eukaryotes generally results from meiotic events, either crossing-over or shuffling of chromosomes.

red alga A member of a diverse phylum of algae that contain the pigments chlorophyll *a,* carotenoids, phycocyanin, and phycoerythrin.

red blood cell (RBC) See *erythrocyte*.

redox reaction (ree'dox) The chemical reaction in which one or more electrons are transferred from one substance (the substance that becomes oxidized) to another (the substance that becomes reduced). See *oxidation* and *reduction*.

red tide A red or brown coloration of ocean water caused by a population explosion, or bloom, of dinoflagellates.

reduction The gain of one or more electrons (or hydrogen atoms) by an atom, ion, or molecule. Compare with *oxidation*.

reflex action An automatic, involuntary response to a given stimulus that generally functions to restore homeostasis.

refractory period The brief period that must elapse after the response of a neuron or muscle fiber, during which it cannot respond to another stimulus.

regulative development The very plastic developmental pattern in which each individual blastomere retains totipotency. Compare with *mosaic development*.

regulon A group of operons that are coordinately controlled.

renal (ree'nl) Pertaining to the kidney.

renal pelvis The funnel-shaped chamber of the kidney that receives urine from the collecting ducts; urine then moves into the ureters.

renin (reh'nin) An enzyme released by the kidney in response to a decrease in blood pressure, which activates a pathway leading to production of angiotensin II, a hormone that increases aldosterone release.

replacement-level fertility The number of children a couple must produce to "replace" themselves. The average number is greater than two because some children die before reaching reproductive age.

replication fork Y-shaped structure produced during the semiconservative replication of DNA.

replication See *DNA replication*.

repolarization The process of returning membrane potential to its resting level.

repressible operon An operon that is normally active but can be controlled by a repressor protein, which becomes active when it binds to a corepressor; the active repressor binds to the operator, making the operon transcriptionally inactive, e.g., the tryptophan operons of *Escherichia coli* and *Salmonella*. Compare with *inducible operon*.

repressor protein A negative regulatory protein that inhibits transcription when bound to DNA; some repressors require a corepressor to be active; some other repressors become inactive when bound to an inducer molecule. Compare with *activator protein*.

reproduction The process by which new individuals are produced. See *asexual reproduction* and *sexual reproduction*.

reproductive isolating mechanisms The reproductive barriers that prevent a species from interbreeding with another species; as a result, each species' gene pool is isolated from other species. See *prezygotic barrier* and *postzygotic barrier*.

reptiles A class of vertebrates characterized by dry skin with horny scales and adaptations for terrestrial reproduction; include turtles, snakes, and alligators; reptiles are not a monophyletic group.

residual capacity The volume of air that remains in the lungs at the end of a normal exhalation.

resin A viscous organic material that certain plants produce and secrete into specialized ducts; may play a role in deterring disease organisms or plant-eating insects.

resolution See *resolving power*.

resolving power The ability of a microscope to show fine detail, defined as the minimum distance between two points at which they can be seen as separate images; also called resolution.

resource partitioning The reduction of competition for environmental resources such as food that occurs among coexisting species as a result of each species' niche differing from the others in one or more ways.

respiration (1) Cellular respiration is the process by which cells generate ATP through a series of redox reactions. In aerobic respiration the terminal electron acceptor is molecular oxygen; in anaerobic respiration the terminal acceptor is an inorganic molecule other than oxygen. (2) Organismic respiration is the process of gas exchange between a complex animal and its environment, generally through a specialized respiratory surface, such as a lung or gill.

respiratory centers Centers in the medulla and pons that regulate breathing.

resting potential The membrane potential (difference in electrical charge between the two sides of the plasma membrane) of a neuron in which no action potential is occurring. The typical resting potential is about -70 millivolts. Compare with *action potential*.

restoration ecology The scientific field that uses the principles of ecology to help return a degraded environment as closely as possible to its former undisturbed state.

restriction enzyme One of a class of enzymes that cleave DNA at specific base sequences; produced by bacteria to degrade foreign DNA; used in recombinant DNA technology.

restriction fragment length polymorphism (RFLP) analysis A technique that permits assessment of the degree of relatedness among individuals within a population; individuals are compared on the basis of the different patterns of DNA fragments generated when their DNA is cut with the same restriction enzyme.

restriction map A physical map of DNA in which sites cut by specific restriction enzymes serve as landmarks.

reticular activating system (RAS) (reh-tik'yoo-lur) A diffuse network of neurons in the brain stem responsible for maintaining consciousness.

retina (ret'ih-nah) The innermost of the three layers (retina, choroid layer, and sclera) of the eyeball, which is continuous with the optic nerve and contains the light-sensitive rod and cone cells.

retrovirus (ret'roh-vy''rus) An RNA virus that uses reverse transcriptase to produce a DNA intermediate, known as a DNA provirus, in the host cell. See *DNA provirus*.

reverse transcriptase An enzyme produced by retroviruses that catalyzes the production of DNA using RNA as a template.

reversible inhibitor A substance that forms weak bonds with an enzyme, temporarily interfering with its function; a reversible inhibitor can be competitive or noncompetitive. Compare with *irreversible inhibitor*.

Rh factors Red blood cell antigens, known as D antigens, first identified in *Rhesus* monkeys. Persons possessing these antigens are Rh^+ those lacking them are Rh^- See *erythroblastosis fetalis*.

rhizome (ry'zome) A horizontal underground stem that bears leaves and buds and often serves as a storage organ and a means of asexual reproduction, e.g., iris.

rhodopsin (rho-dop'sin) Visual purple; a light-sensitive pigment found in the rod cells of the vertebrate eye; a similar molecule is employed by certain bacteria in the capture of light energy to make ATP.

ribonucleic acid (RNA) A family of single-stranded nucleic acids that function mainly in protein synthesis.

ribosomal RNA (rRNA) See *ribosomes*.

ribosomes (ry'boh-sohms) Organelles that are part of the protein synthesis machinery of both prokaryotic and eukaryotic cells; consist of a larger and smaller subunit, each composed of ribosomal RNA (rRNA) and ribosomal proteins.

ribozyme (ry'boh-zime) A molecule of RNA that has catalytic properties.

ribulose bisphosphate (RuBP) A five-carbon phosphorylated compound with a high energy potential that reacts with carbon dioxide in the initial step of the Calvin cycle.

ribulose bisphosphate carboxylase See *Rubisco*.

RNA polymerase An enzyme that catalyzes the synthesis of RNA from a DNA template. Also called DNA-dependent RNA polymerase.

RNA primer The sequence of about five RNA nucleotides that are synthesized during DNA replication to provide a 3′ end to which DNA polymerase can add nucleotides. The RNA primer is later degraded and replaced with DNA.

RNA world A model that proposes that, during the evolution of cells, RNA was the first informational molecule to evolve, followed at a later time by proteins and DNA.

rod One of the rod-shaped, light-sensitive cells of the retina that are particularly sensitive to dim light and mediate black and white vision. Compare with *cone*.

root cap A covering of cells over the root tip that protects the delicate meristematic tissue directly behind it.

root graft The process of roots from two different plants growing together and becoming permanently attached to one another.

root hair An extension, or outgrowth, of a root epidermal cell. Root hairs increase the absorptive capacity of roots.

root pressure The pressure in xylem sap that occurs as a result of the active absorption of mineral ions followed by the osmotic uptake of water into roots from the soil.

root system The underground portion of a plant that anchors it in the soil and absorbs water and dissolved minerals.

rough ER See *endoplasmic reticulum*.

Rubisco The common name of ribulose bisphosphate carboxylase, the enzyme that catalyzes the reaction of carbon dioxide with ribulose bisphosphate in the Calvin cycle.

rugae (roo'jee) Folds, such as those in the lining of the stomach.

runner See *stolon*.

S phase Stage in interphase of the cell cycle during which DNA and other chromosomal constituents are synthesized. Compare with *G_1 and G_2 phases*.

saccule The structure within the vestibule of the inner vertebrate ear that along with the utricle houses the receptors of static equilibrium.

salinity The concentration of dissolved salts (e.g., sodium chloride) in a body of water.

salivary glands Accessory digestive glands found in vertebrates and some invertebrates; in humans there are three pairs.

salt An ionic compound consisting of an anion other than a hydroxide ion and a cation other than a hydrogen ion. A salt can be formed by the reaction between an acid and a base.

salt marsh A wetland dominated by grasses in which the salinity fluctuates between that of sea water and fresh water; salt marshes are usually located in estuaries.

saltatory conduction The transmission of a neural impulse along a myelinated neuron; ion activity at one node depolarizes the next node along the axon.

saprobe See *decomposer*.

saprotroph (sap'roh-trof) See *decomposer*.

sarcolemma (sar''koh-lem'mah) The muscle cell plasma membrane.

sarcomere (sar'koh-meer) A segment of a striated muscle cell located between adjacent Z-lines that serves as a unit of contraction.

sarcoplasmic reticulum The system of vesicles in a muscle cell that surrounds the myofibrils and releases calcium in muscle contraction; a modified endoplasmic reticulum.

saturated fatty acid See *fatty acid*.

savanna (suh-van'uh) A tropical grassland containing scattered trees; found in areas of low rainfall or seasonal rainfall with prolonged dry periods.

scaffolding proteins Nonhistone proteins that help maintain the structure of a chromosome.

schizocoely (skiz'oh-seely) The process of coelom formation in which the mesoderm splits into two layers, forming a cavity between them; characteristic of protostomes. Compare with *enterocoely*.

Schwann cells Supporting cells found in nervous tissue outside the central nervous system; produce the myelin sheath around peripheral neurons.

sclera (skler'ah) The outer coat of the eyeball; a tough, opaque sheet of connective tissue that protects the inner structures and helps maintain the rigidity of the eyeball.

sclereid (skler'id) In plants, a sclerenchyma cell that is variable in shape but typically not long and tapered. Compare with *fiber*.

sclerenchyma (skler-en'kim-uh) Cells that provide strength and support in the plant body, are often dead at maturity, and have extremely thick walls; includes fibers and sclereids.

scramble competition See *exploitation competition*.

scrotum (skroh'tum) The external sac of skin found in most male mammals that contains the testes and their accessory organs.

secondary consumer See *carnivore*.

secondary growth An increase in the girth of a plant due to the activity of the vascular cambium and cork cambium; secondary growth results in the production of secondary tissues, i.e., wood and bark. Compare with *primary growth*.

secondary mycelium A dikaryotic mycelium formed by the fusion of two primary hyphae. Compare with *primary mycelium*.

secondary response The rapid production of antibodies induced by a second exposure to an antigen several days, weeks, or even months after the initial exposure. Compare with *primary response*.

secondary structure (of a protein) A regular geometric shape produced by hydrogen bonding between the atoms of the uniform polypeptide backbone; includes the alpha helix and the beta-pleated sheet. Compare with *primary, tertiary,* and *quaternary protein structure*.

secondary succession An ecological succession that takes place after some disturbance destroys the existing vegetation; soil is already present. See *succession*. Compare with *primary succession*.

second law of thermodynamics The physical law that states that the total amount of entropy in the universe continually increases. Compare with *first law of thermodynamics*.

second messenger A substance within a cell that relays a message and (usually) triggers a response to a hormone combined with a receptor at the cell's surface, e.g., cyclic AMP and calcium ions.

secretory vesicles Small cytoplasmic vesicles that move substances from an internal membrane system to the plasma membrane.

seed A plant reproductive body composed of a young, multicellular plant and nutritive tissue (food reserves), enclosed by a seed coat.

seed coat The outer protective covering of a seed.

seed fern An extinct group of seed-bearing woody plants with fernlike leaves; seed ferns probably descended from progymnosperms and gave rise to cycads and possibly ginkgoes.

segregation, principle of The genetic principle, first noted by Gregor Mendel, that states that two alleles of a locus become separated into different gametes.

selectively permeable membrane A membrane that allows some substances to cross it more easily than others. Biological membranes are generally permeable to water, but restrict the passage of many solutes.

self-incompatibility A genetic condition in which the pollen cannot effect fertilization in the same flower or in flowers on the same plant.

semelparity The condition of having a single reproductive effort in a lifetime. Compare with *iteroparity*.

semen The fluid composed of sperm suspended in various glandular secretions that is ejaculated from the penis during orgasm.

semicircular canals The passages in the vertebrate inner ear containing structures that control the sense of equilibrium (balance).

semiconservative replication See *DNA replication*.

semilunar valves Valves between the ventricles of the heart and the arteries that carry blood away from the heart; aortic and pulmonary valves.

seminal vesicles (1) In mammals, glandular sacs that secrete a component of seminal fluid; (2) In some invertebrates, structures that store sperm.

seminiferous tubules (sem-ih-nif′er-ous) Coiled tubules in the testes in which spermatogenesis takes place in male vertebrates.

senescence (se-nes′cents) The aging process.

sensory neuron A neuron that transmits an impulse from a receptor to the central nervous system.

sensory receptor A cell (or part of a cell) specialized to detect specific energy stimuli in the environment.

sepal (see′pul) One of the outermost parts of a flower, usually leaflike in appearance, that protect the flower as a bud.

septum (pl. *septa*) A cross wall or partition, e.g., the walls that divide a hypha into cells.

sequencing See *DNA sequencing*.

serotonin A neurotransmitter of the biogenic amine group.

Sertoli cells (sur-tole′ee) Supporting cells of the tubules of the testis.

sessile (ses′sile) Permanently attached to one location, e.g., coral animals.

setae (sing. *seta*) Bristle-like structures that aid in annelid locomotion.

set point A normal condition maintained by homeostatic mechanisms.

sex-influenced trait A genetic trait that is expressed differently in males and females.

sex-linked gene A gene carried on a sex chromosome. In mammals almost all sex-linked genes are borne on the X chromosome, i.e., are X-linked.

sexual dimorphism Marked phenotypic differences between the two sexes of the same species.

sexual isolation See *behavioral isolation*.

sexual reproduction A type of reproduction in which two gametes (usually, but not necessarily, contributed by two different parents) fuse to form a zygote. Compare with *asexual reproduction*.

sexual selection A type of natural selection that occurs when individuals of a species vary in their ability to compete for mates; individuals with reproductive advantages are selected over others of the same sex.

shade avoidance The tendency of plants that are adapted to high light intensities to grow taller when they are closely surrounded by other plants.

shared ancestral characters Traits that were present in an ancestral species that have remained essentially unchanged; suggest a distant common ancestor. Also called plesiomorphic characters. Compare with *shared derived characters*.

shared derived characters Homologous traits found in two or more taxa that are present in their most recent common ancestor but not in earlier common ancestors. Also called synapomorphic characters. Compare with *shared ancestral characters*.

shoot system The above-ground portion of a plant, such as the stem and leaves.

short-day plant A plant that flowers in response to lengthening nights; also called long-night plant. Compare with *long-day, intermediate-day,* and *day-neutral plants*.

short-night plant See *long-day plant*.

sickle cell anemia An inherited form of anemia in which there is abnormality in the hemoglobin beta chains; the inheritance pattern is autosomal recessive.

sieve tube members Cells that conduct dissolved sugar in the phloem of flowering plants.

signaling molecule See *cell signaling*.

signal transduction A process in which a cell converts and amplifies an extracellular signal into an intracellular signal that affects some function in the cell. Also see *cell signaling*.

sign stimulus Any stimulus that elicits a fixed action pattern in an animal.

simple fruit A fruit that develops from a single ovary. Compare with *aggregate, accessory,* and *multiple fruits*.

simplest formula A type of chemical formula that gives the smallest whole number ratio of the component atoms. Compare with *molecular formula* and *structural formula*.

sink habitat A lower quality habitat in which local reproductive success is less than local mortality. Compare with *source habitat*.

sinoatrial (SA) node The mass of specialized cardiac muscle in which the impulse triggering the heartbeat originates; the pacemaker of the heart.

skeletal muscle The voluntary striated muscle of vertebrates, so-called because it usually is directly or indirectly attached to some part of the skeleton. Compare with *cardiac muscle* and *smooth muscle*.

slash-and-burn agriculture A type of agriculture in which tropical rain forest is cut down, allowed to dry, and burned. The crops that are planted immediately afterwards thrive because the ashes provide nutrients; in a few years, however, the soil is depleted and the land must be abandoned.

small intestine Portion of the vertebrate digestive tract that extends from the stomach to the large intestine.

small nuclear ribonucleoprotein complexes (snRNP) Aggregations of RNA and protein responsible for binding to pre-mRNA in eukaryotes and catalyzing the excision of introns and the splicing of exons.

smooth ER See *endoplasmic reticulum.*

smooth muscle Involuntary muscle tissue that lacks transverse striations; found mainly in sheets surrounding hollow organs, such as the intestine. Compare with *cardiac muscle* and *skeletal muscle.*

social behavior Interaction of two or more animals, usually of the same species.

sociobiology The branch of biology that focuses on the evolution of social behavior through natural selection.

sodium-potassium pump Active transport system that transports sodium ions out of, and potassium ions into, cells.

soil erosion The wearing away or removal of soil from the land; although soil erosion occurs naturally from precipitation and runoff, human activities (such as clearing the land) accelerate it.

solar tracking See *heliotropism.*

solute A dissolved substance. Compare with *solvent.*

solvent Substance capable of dissolving other substances. Compare with *solute.*

somatic cell In animals, a cell of the body not involved in formation of gametes. Compare with *germ line cell.*

somatic nervous system That part of the vertebrate peripheral nervous system that keeps the body in adjustment with the external environment; includes sensory receptors on the body surface and within the muscles, and the nerves that link them with the central nervous system. Compare with *autonomic nervous system.*

somatomedins See *insulin-like growth factors.*

somatotropin See *growth hormone.*

sonogram See *ultrasound imaging.*

soredium (sor-id'e-um) (pl. *soredia*) In lichens, a type of asexual reproductive structure that consists of a cluster of algal cells surrounded by fungal hyphae.

sorus (soh'rus) (pl. *sori*) In ferns, a cluster of spore-producing sporangia.

source habitat A good habitat in which local reproductive success is greater than local mortality. Surplus individuals in a source habitat may disperse to other habitats. Compare with *sink habitat.*

Southern blot hybridization A technique in which DNA fragments, previously separated by gel electrophoresis, are transferred to a nitrocellulose membrane. A specific radioactive genetic probe is then allowed to hybridize to complementary fragments, marking their locations. Compare with *Northern blot hybridization* and *Western blotting.*

speciation Evolution of a new species.

species According to the biological species concept, one or more populations whose members are capable of interbreeding in nature to produce fertile offspring and do not interbreed with members of other species.

species diversity A measure of the relative importance of each species within a comunity; represents a combination of species richness and species evenness.

species evenness The distribution of individuals of each species in the community.

species richness The number of species in a community.

specific immune responses Defense mechanisms that target specific macromolecules associated with a pathogen. Includes cell-mediated immunity and antibody-mediated immunity. Also known as acquired or adaptive immune responses.

specific epithet The second part of the name of a species; designates a specific species belonging to that genus.

specific heat The amount of heat energy that must be supplied to raise the temperature of 1 g of a substance 1°C.

sperm The motile, n male reproductive cell of animals and some plants and protists; also called spermatozoan.

spermatid (spur'ma-tid) An immature sperm cell.

spermatocyte (spur-mah'toh-site) A meiotic cell that gives rise to spermatids and ultimately to mature sperm cells.

spermatogenesis (spur''mah-toh-jen'eh-sis) The production of male gametes (sperm) by meiosis and subsequent cell differentiation. Compare with *oogenesis.*

spermatozoan (spur-mah-toh-zoh'un) See *sperm.*

sphincter (sfink'tur) A group of circularly arranged muscle fibers, the contractions of which close an opening, e.g., the pyloric sphincter at the exit of the stomach.

spinal cord In vertebrates, the dorsal, tubular nerve cord.

spinal nerves In vertebrates, the nerves that emerge from the spinal cord.

spindle See *mitotic spindle.*

spine A leaf that is modified for protection, such as a cactus spine.

spiracle (speer'ih-kl) An opening for gas exchange, such as the opening of a trachea on the body surface of an insect.

spiral cleavage A distinctive spiral pattern of blastomere production in an early protostome embryo. Compare with *radial cleavage.*

spirillum (pl. *spirilla*) A long, rigid, helical bacterium. Compare with *spirochete, vibrio, bacillus,* and *coccus.*

spirochete A long, flexible, helical bacterium. Compare with *spirillum, vibrio, bacillus,* and *coccus.*

spleen An abdominal organ located just below the diaphragm that removes worn-out blood cells and bacteria from the blood and plays a role in immunity.

spongy mesophyll (mez'oh-fil) The loosely arranged mesophyll cells near the lower epidermis in certain leaves. Compare with *palisade mesophyll.*

spontaneous reaction See *exergonic reaction.*

sporangium (spor-an'jee-um) (pl. *sporangia*) A spore case, found in plants, certain protists, and fungi.

spore A reproductive cell that gives rise to individual offspring in plants, fungi, and certain algae and protozoa.

sporophyll (spor'oh-fil) A leaflike structure that bears spores.

sporophyte generation (spor'oh-fite) The $2n$, spore-producing stage in the life cycle of a plant. Compare with *gametophyte generation.*

sporozoa See *apicomplexans.*

sporozoite The infective sporelike state in apicomplexans.

stabilizing selection Natural selection that acts against extreme phenotypes and favors intermediate variants; associated with a population well-adapted to its environment. Compare with *directional selection* and *disruptive selection.*

stamen (stay'men) The male part of a flower; consists of a filament and anther.

standing-water ecosystem A lake or pond ecosystem.

starch A polysaccharide composed of alpha glucose subunits; made by plants for energy storage.

start codon See *initiation codon*.
stasis Long periods in the fossil record in which there is little or no evolutionary change.
statocyst (stat'oh-sist) An invertebrate sense organ containing one or more granules (statoliths); senses gravity and motion.
statoliths (stat'uh-liths) Granules of loose sand or calcium carbonate found in statocysts.
stele The cylinder in the center of roots and stems that contains the vascular tissue.
stem cell A relatively undifferentiated cell capable of repeated cell division. At each division at least one of the daughter cells usually remains a stem cell, whereas the other may differentiate as a specific cell type.
sterilization A procedure that renders an individual incapable of producing offspring; the most common surgical procedures are vasectomy in the male and tubal ligation in the female.
stereocilia Hairlike projections of hair cells; microvilli that contain actin filaments.
steroids (steer'oids) Complex molecules containing carbon atoms arranged in four attached rings, three of which contain six carbon atoms each and the fourth of which contains five; e.g., cholesterol and certain hormones, including the male and female sex hormones of vertebrates.
stigma The portion of the carpel where pollen grains land during pollination (and before fertilization).
stipe A short stalk or stemlike structure that is a part of the body of certain multicellular algae.
stipule (stip'yule) One of a pair of scalelike or leaflike structures found at the base of certain leaves.
stolon (stow'lon) An above-ground, horizontal stem with long internodes; stolons often form buds that develop into separate plants, e.g., strawberry; also called runner.
stomach Muscular region of the vertebrate digestive tract, extending from the esophagus to the small intestine.
stomata (sing. *stoma*) Small pores located in the epidermis of plants that provide for gas exchange for photosynthesis; each stoma is flanked by two guard cells, which are responsible for its opening and closing.
stop codon See *termination codon*.
stratosphere The layer of the atmosphere between the troposphere and the mesosphere. It contains a thin ozone layer that protects life by filtering out much of the sun's ultraviolet radiation.
stratum basale (strat'um bah-say'lee) The deepest sublayer of the human epidermis, consisting of cells that continuously divide. Compare with *stratum corneum*.
stratum corneum The most superficial sublayer of the human epidermis. Compare with *stratum basale*.
strobilus (stroh'bil-us) (pl. *strobili*) In certain plants, a conelike structure that bears spore-producing sporangia.
stroke volume The volume of blood pumped by one ventricle during one contraction.
stroma A fluid space of the chloroplast, enclosed by the chloroplast inner membrane and surrounding the thylakoids; site of the reactions of the Calvin cycle.
stromatolite (stroh-mat'oh-lite) A column-like rock that is composed of many minute layers of prokaryotic cells, usually cyanobacteria.
structural formula A type of chemical formula that shows the spatial arrangement of the atoms in a molecule. Compare with *simplest formula* and *molecular formula*.
structural isomer One of two or more chemical compounds having the same chemical formula, but differing in the covalent arrangement of their atoms, e.g., glucose and fructose.
style The neck connecting the stigma to the ovary of a carpel.
subsidiary cell In plants, a structurally distinct epidermal cell associated with a guard cell.
substance P A peptide neurotransmitter released by certain sensory neurons in pain pathways; signals the brain regarding painful stimuli; also stimulates other structures including smooth muscle in the digestive tract.
substrate A substance on which an enzyme acts; a reactant in an enzymatically catalyzed reaction.
succession The sequence of changes in the species composition in a community over time. See *primary succession* and *secondary succession*.
sucker A shoot that develops adventitiously from a root; a type of asexual reproduction.
sulcus (sul'kus) (pl. *sulci*) A groove, trench, or depression, especially one occurring on the surface of the brain, separating the convolutions.
sulfhydryl group Functional group abbreviated —SH; found in organic compounds called thiols.
sum rule The rule for combining the probabilities of mutually exclusive events by adding their individual probabilities. Compare with *product rule*.
summation The process of adding together excitatory postsynaptic potentials (EPSPs).
suppressor T cell T lymphocyte that suppresses the immune response.
supraorbital ridge (soop''rah-or'bit-ul) The prominent bony ridge above the eye socket; ape skulls have prominent supraorbital ridges.
surface tension The attraction that the molecules at the surface of a liquid may have for each other.
survivorship The probability that a given individual in a population or cohort will survive to a particular age; usually presented as a survivorship curve.
survivorship curve A graph of the number of surviving individuals of a cohort, from birth to the maximum age attained by any individual.
suspensor (suh-spen'sur) In plant embryo development, a multicellular structure that anchors the embryo and aids in nutrient absorption from the endosperm.
swim bladder The hydrostatic organ in bony fishes that permits the fish to hover at a given depth.
symbiosis (sim-bee-oh'sis) An intimate relationship between two or more organisms of different species. See *commensalism, mutualism,* and *parasitism*.
sympathetic nervous system A division of the autonomic nervous system; its general effect is to mobilize energy, especially during stress situations; prepares the body for fight-or-flight response. Compare with *parasympathetic nervous system*.
sympatric speciation (sim-pa'trik) The evolution of a new species within the same geographical region as the parental species. Compare with *allopatric speciation*.
symplast A continuum consisting of the cytoplasm of many plant cells, connected from one cell to the next by plasmodesmata. Compare with *apoplast*.
synapomorphic characters See *shared derived characters*.
synapse (sin'aps) The junction between two neurons or between a neuron and an effector (muscle or gland).

synapsis (sin-ap′sis) The process of physical association of homologous chromosomes during prophase I of meiosis.

synaptic enhancement An increase in neurotransmitter release thought to occur as a result of calcium ion accumulation inside the presynaptic neuron.

synaptic plasticity The ability of synapses to change in response to certain types of stimuli. Synaptic changes occur during learning and memory storage.

synaptonemal complex The structure, visible with the electron microscope, produced when homologous chromosomes undergo synapsis.

syngamy (sin′gah-mee) The union of the gametes in sexual reproduction.

synthetic theory of evolution The synthesis of previous theories, especially of Mendelian genetics, with Darwin's theory of evolution by natural selection to formulate a comprehensive explanation of evolution; also called neo-Darwinism.

systematics The scientific study of the diversity of organisms and their evolutionary relationships. Taxonomy is an aspect of systematics. See *taxonomy*.

systemic anaphylaxis A rapid, widespread allergic reaction that can lead to death.

systemic circulation The part of the circulatory system that delivers blood to and from the tissues and organs of the body. Compare with *pulmonary circulation*.

systole (sis′tuh-lee) The phase of the cardiac cycle when the heart is contracting. Compare with *diastole*.

T cell (T lymphocyte) The type of white blood cell responsible for a wide variety of immune functions, particularly cell-mediated immunity. T cells are processed in the thymus. Compare with *B cell*.

T tubules Transverse tubules; system of inward extensions of the muscle fiber plasma membrane.

taiga (tie′gah) The northern coniferous forest biome found primarily in Canada, northern Europe, and Siberia; also called boreal forest.

taproot system A root system consisting of a prominant main root with smaller lateral roots branching off it; a taproot develops directly from the embryonic radicle. Compare with *fibrous root system*.

target cell or tissue A cell or tissue with receptors that bind a hormone.

TATA box A component of a eukaryotic promoter region; consists of a sequence of bases located about 30 base pairs upstream from the transcription initiation site.

taxon A formal taxonomic group at any level, e.g., phylum or genus.

taxonomy (tax-on′ah-mee) The science of naming, describing, and classifying organisms; see *systematics*.

Tay-Sachs disease A serious genetic disease in which abnormal lipid metabolism in the brain causes mental deterioration in affected infants and young children; inheritance pattern is autosomal recessive.

tectorial membrane (tek-tor′ee-ul) The roof membrane of the organ of Corti in the cochlea of the ear.

telolecithal egg An egg with a large amount of yolk, concentrated at the vegetal pole. Compare with *isolecithal egg*.

telophase (teel′oh-faze or tel′oh-faze) The last stage of mitosis and of meiosis I and II when, having reached the poles, chromosomes become decondensed, and a nuclear envelope forms around each group.

temperate deciduous forest A forest biome that occurs in temperate areas where annual precipitation ranges from about 75 cm to 125 cm.

temperate grassland A grassland characterized by hot summers, cold winters, and less rainfall than is found in a temperate deciduous forest biome.

temperate rain forest A coniferous biome characterized by cool weather, dense fog, and high precipitation, e.g., the north Pacific coast of North America.

temperate virus A virus that can become integrated into the host DNA as a prophage.

temperature The average kinetic energy of the particles in a sample of a substance.

temporal isolation A prezygotic reproductive isolating mechanism in which genetic exchange is prevented between similar species because they reproduce at different times of the day, season, or year.

tendon A connective tissue structure that joins a muscle to another muscle, or a muscle to a bone. Tendons transmit the force generated by a muscle.

tendril A leaf or stem that is modified for holding or attaching onto objects.

tension-cohesion model The mechanism by which water and dissolved inorganic minerals are thought to be transported in xylem; water is pulled upward under tension due to transpiration while maintaining an unbroken column in xylem due to cohesion; also called transpiration-cohesion model.

teratogen Any agent capable of interfering with normal morphogenesis in an embryo, thereby causing malformations; examples include radiation, certain chemicals, and certain infectious agents.

terminal bud A bud at the tip of a stem. Compare with *axillary bud*.

termination (of protein synthesis) The final stage of protein synthesis, which occurs when a termination (stop) codon is reached, causing the completed polypeptide chain to be released from the ribosome. See *initiation* and *elongation*.

termination codon Any of the three codons in mRNA that do not code for an amino acid (UAA, UAG, or UGA). This stops translation at that point. Also known as a stop codon. Compare with *initiation codon*.

territoriality Behavior pattern in which one organism (usually a male) stakes out a territory of its own and defends it against intrusion by other members of the same species and sex.

tertiary consumer See *carnivore*.

tertiary structure (of a protein) (tur′she-air″ee) The overall three-dimensional shape of a polypeptide that is determined by interactions involving the amino acid side chains. Compare with *primary, secondary,* and *quaternary protein structure*.

test A shell.

test cross The genetic cross in which either an F_1 individual, or an individual of unknown genotype, is mated to a homozygous recessive individual.

testis (tes′tis) (pl. *testes*) The male gonad that produces sperm and the male hormone testosterone; in humans and certain other mammals the testes are located in the scrotum.

testosterone (tes-tos′ter-own) The principal male sex hormone (androgen); a steroid hormone produced by the interstitial cells of the testes; stimulates spermatogenesis and is responsible for primary and secondary sex characteristics in the male.

tetrad The chromosome complex formed by the synapsis of a pair of homologous chromosomes (i.e., four chromatids) during meiotic prophase I; also known as a bivalent.

tetrapods (tet'rah-podz) Four-limbed vertebrates: the amphibians, reptiles, birds, and mammals.

thalamus (thal'uh-mus) The part of the vertebrate brain that serves as a main relay center, transmitting information between the spinal cord and the cerebrum.

thallus (thal'us) (pl. *thalli*) The simple body of an alga, fungus, or nonvascular plant that lacks root, stems, or leaves, e.g., a liverwort thallus or a lichen thallus.

theca cells The layer of connective tissue cells that surrounds the granulosa cells in an ovarian follicle; stimulated by luteinizing hormone (LH) to produce androgens which are converted to estrogen in the granulosa cells.

theory A widely accepted explanation supported by a large body of observations and experiments. A good theory relates facts that appear to be unrelated; it predicts new facts and suggests new relationships. Compare with *hypothesis* and *principle*.

therapsids (ther-ap'sids) A group of mammal-like reptiles of the Permian period; gave rise to the mammals.

thermal stratification The marked layering (separation into warm and cold layers) of temperate lakes during the summer. See *thermocline*.

thermocline (thur'moh-kline) A marked and abrupt temperature transition in temperate lakes between warm surface water and cold deeper water. See *thermal stratification*.

thermodynamics Principles governing energy transfer (often expressed in terms of heat transfer). See *first law of thermodynamics* and *second law of thermodynamics*.

thermoreceptor A sensory receptor that responds to heat.

thigmomorphogenesis (thig''moh-mor-foh-jen'uh-sis) An alteration of plant growth in response to mechanical stimuli such as wind, rain, hail, and contact with passing animals.

thigmotropism (thig'moh-troh'pizm) Plant growth in response to contact with a solid object, such as the twining of plant tendrils.

threatened species A species in which the population is small enough for it to be at risk of becoming extinct throughout all or part of its range, but not so small that it is in imminent danger of extinction. Compare with *endangered species*.

threshold level The potential that a neuron or other excitable cell must reach for an action potential to be initiated.

thrombocytes See *platelets*.

thylakoid lumen See *thylakoids*.

thylakoids (thy'lah-koidz) An interconnected system of flattened, saclike membranous structures inside the chloroplast; the thylakoid membranes contain chlorophyll and the electron transport chain and enclose a compartment, the thylakoid lumen.

thymine (thy'meen) A nitrogenous pyrimidine base found in DNA.

thymus gland (thy'mus) An endocrine gland that functions as part of the lymphatic system; important in the ability to make immune responses.

thyroid gland An endocrine gland that lies anterior to the trachea and releases hormones that regulate the rate of metabolism.

thyroid hormones Hormones, including thyroxin, secreted by the thyroid gland; stimulate rate of metabolism.

tidal volume The volume of air moved into and out of the lungs with each normal resting breath.

tight junctions Specialized structures that form between some animal cells, producing a tight seal that prevents materials from passing through the spaces between the cells.

tissue A group of closely associated, similar cells that work together to carry out specific functions.

tissue culture The growth of tissue or cells in a synthetic growth medium under sterile conditions.

tissue fluid See *interstitial fluid*.

tolerance A decreased response to a drug over time.

tonoplast The membrane surrounding a vacuole.

topoisomerases (toe-poe-eye-sahm'er-ases) Enzymes that relieve twists and kinks in a DNA molecule by breaking and rejoining the strands.

torpor An energy-conserving state of low metabolic rate and inactivity. See *estivation* and *hibernation*.

torsion The twisting of the visceral mass characteristic of gastropod mollusks.

total fertility rate The average number of children born to a woman during her lifetime.

totipotent (toh-ti-poh'tent) A term describing a cell or nucleus that contains the complete set of genetic instructions required to direct the normal development of an entire organism. Compare with *pleuripotent*.

trace element An element required by an organism in very small amounts.

trachea (tray'kee-uh)(pl. *tracheae*) (1) Principal thoracic air duct of terrestrial vertebrates; windpipe; (2) One of the microscopic air ducts (or tracheal tubes) branching throughout the body of most terrestrial arthopods and some terrestrial mollusks.

tracheal tubes See *trachea*.

tracheid (tray'kee-id) A type of water-conducting and supporting cell in the xylem of vascular plants.

tract A bundle of nerve fibers within the central nervous system.

transcription The synthesis of RNA from a DNA template.

transcription factors DNA-binding proteins that regulate transcription in eukaryotes; include positively acting activators and negatively acting repressors.

transduction (1) The transfer of a genetic fragment from one cell to another, e.g., from one bacterium to another, by a virus. (2) In the nervous system, the conversion of energy of a stimulus to electrical signals.

transfer RNA (tRNA) RNA molecules that bind to specific amino acids and serve as adapter molecules in protein synthesis. The tRNA anticodons bind to complementary mRNA codons.

transformation (1) The incorporation of genetic material into a cell, thereby changing its phenotype; (2) The conversion of a normal cell to a malignant cell.

transgenic organism A plant or animal that has foreign DNA incorporated into its genome.

translation The conversion of information provided by mRNA into a specific sequence of amino acids in a polypeptide chain; process also requires transfer RNA and ribosomes.

translocation (1) The movement of organic materials (dissolved food) in the phloem of a plant; (2) Chromosome abnormality in which part of one chromosome has become attached to another; (3) Part of the elongation cycle of protein synthesis in which a transfer RNA attached to the growing polypeptide chain is transferred from the A site to the P site.

transmembrane protein An integral membrane protein that spans the lipid bilayer.

transmission, neural The conduction of a neural impulse along a neuron or from one neuron to another.

transpiration The loss of water vapor from the aerial surfaces of a plant (i.e., leaves and stems).

transpiration-cohesion model See *tension-cohesion model.*

transport vesicles Small cytoplasmic vesicles that move substances from one membrane system to another.

transposable element See *transposon.*

transposon (tranz-poze'on) A DNA segment that is capable of moving from one chromosome to another or to different sites within the same chromosome; also called a transposable element or mobile genetic element.

transverse tubules See *T tubules.*

triacylglycerol (try-ace''il-glis'er-ol) The main storage lipid of organisms, consisting of a glycerol combined chemically with three fatty acids; also called triglyceride. Compare with *monoacylglycerol* and *diacylglycerol.*

tricarboxylic acid (TCA) cycle See *citric acid cycle.*

trichocyst (trik'oh-sist) A cellular organelle found in certain ciliates that can discharge a threadlike structure that may aid in trapping and holding prey.

trichome (try'kohm) A hair or other appendage growing out from the epidermis of a plant.

tricuspid valve See *atrioventricular valve.*

triglyceride See *triacylglycerol.*

triose A sugar molecule containing three carbons.

triplet A sequence of three nucleotides that serves as the basic unit of genetic information.

triplet code The sequences of three nucleotides that compose the codons, the units of genetic information in mRNA that specify the order of amino acids in a polypeptide chain.

trisomy (try'sohm-ee) The condition in which each chromosome has two copies, except for one, which is present in triplicate. Compare with *monosomy* and *disomy.*

trochophore larva (troh'koh-for) A larval form found in mollusks and many polychaetes.

trophic level (troh'fik) Each sequential step of matter and energy in a food web, from producers to primary, secondary, or tertiary consumers; each organism is assigned to a trophic level based on its primary source of nourishment.

trophoblast (troh'foh-blast) The outer cell layer of a late blastocyst which, in placental mammals, gives rise to the chorion and to the fetal contribution to the placenta.

tropic hormone (trow'pic) A hormone, produced by one endocrine gland, that targets another endocrine gland.

tropical dry forest A tropical forest where enough precipitation falls to support trees but not enough to support the lush vegetation of a tropical rain forest; often occurs in areas with pronounced rainy and dry seasons.

tropical rain forest A lush, species-rich forest biome that occurs in tropical areas where the climate is very moist throughout the year. Tropical rain forests are also characterized by old, infertile soils.

tropism (troh'pizm) In plants, a directional growth response that is elicited by an environmental stimulus.

tropomyosin (troh-poh-my'oh-sin) A regulatory muscle protein involved in contraction.

true-breeding line A genetically pure strain of organism, i.e., one in which all individuals are homozygous for the traits under consideration; also called a pure line.

tube feet Structures characteristic of echinoderms; function in locomotion and feeding.

tuber A thickened end of a rhizome that is fleshy and enlarged for food storage, e.g., white potato.

tubular transport maximum (Tm) The maximum rate at which a substance can be reabsorbed from the renal tubules of the kidney.

tumor A mass of tissue that grows in an uncontrolled manner; a neoplasm.

tumor necrosis factors (TNF) Cytokines that can kill tumor cells and can stimulate immune cells to initiate an inflammatory response.

tumor suppressor gene See *anti-oncogene.*

tundra (tun'dra) A treeless biome between the taiga in the south and the polar ice cap in the north that consists of boggy plains covered by lichens and small plants. Characterized by harsh, very cold winters and extremely short summers. Also called arctic tundra. Compare with *alpine tundra.*

tunicates Chordates belonging to subphylum Urochordata; sea squirts.

turgor pressure (tur'gor) Hydrostatic pressure that develops within a walled cell, such as a plant cell, when the osmotic pressure of the cell's contents is greater than the osmotic pressure of the surrounding fluid.

Turner syndrome An inherited condition in which only one sex chromosome (an X chromosome) is present in cells; karyotype is designated XO; affected individuals are sterile females.

tyrosine kinase An enzyme that phosphorylates the tyrosine part of proteins.

tyrosine kinase receptor A plasma membrane receptor that phosphorylates the tyrosine part of proteins; when a ligand binds to the receptor, the conformation of the receptor changes and it may phosphorylate itself as well as other molecules; important in immune function and serves as a receptor for insulin.

ultimate causes (of behavior) Evolutionary explanations for why a certain behavior occurs. Compare with *proximate causes of behavior.*

ultrasound imaging A technique in which high-frequency sound waves (ultrasound) are used to provide an image (sonogram) of an internal structure.

ultrastructure The fine detail of a cell, generally only observable by use of an electron microscope.

umbilical cord In placental mammals, the organ that connects the embryo to the placenta.

uniform dispersion The spatial distribution pattern of a population in which individuals are regularly spaced. Compare with *random dispersion* and *clumped dispersion.*

unsaturated fatty acid See *fatty acid.*

upstream promoter elements (UPE) Components of a eukaryotic promoter, found upstream of the RNA polymerase-binding site; the strength of a promoter is affected by the number and type of UPEs present.

upwelling An upward movement of water that brings nutrients from the ocean depths to the surface. Where upwelling occurs, the ocean is very productive.

uracil (yur'ah-sil) A nitrogenous pyrimidine base found in RNA.

urea (yur-ee'ah) The principal nitrogenous excretory product of mammals; one of the water-soluble end products of protein metabolism.

ureter (yur'ih-tur) One of the paired tubular structures that conducts urine from the kidney to the bladder.

urethra (yoo-ree'thruh) The tube that conducts urine from the bladder to the outside of the body.

uric acid (yoor'ik) The principal nitrogenous excretory product of insects, birds, and reptiles; a relatively insoluble end product of protein metabolism; also occurs in mammals as an end product of purine metabolism.

urinary bladder An organ that receives urine from the ureters and temporarily stores it.

urinary system The body system in vertebrates that consists of kidneys, urinary bladder, and associated ducts.

urochordates A subphylum of chordates; includes the tunicates.

uterine tube (yoo'tur-in) See *oviduct*.

uterus (yoo'tur-us) The hollow, muscular organ of the female reproductive tract in which the fetus undergoes development.

utricle The structure within the vestibule of the vertebrate inner ear that, along with the saccule, houses the receptors of static equilibrium.

vaccine (vak-seen') A commercially produced, weakened or killed antigen associated with a particular disease that stimulates the body to make antibodies.

vacuole (vak'yoo-ole) A fluid-filled, membrane-bounded sac found within the cytoplasm; may function in storage, digestion, or water elimination.

vagina The elastic, muscular tube, extending from the cervix to its orifice, that receives the penis during sexual intercourse and serves as the birth canal.

valence electrons The electrons in the outer electron shell, known as the valence shell, of an atom; in the formation of a chemical bond an atom can accept electrons into its valence shell, or donate or share valence electrons.

van der Waals forces Weak attractive forces between atoms; caused by interactions among fluctuating charges.

vas deferens (vas def'ur-enz) (pl. *vasa deferentia*) One of the paired sperm ducts that connects the epididymis of the testis to the ejaculatory duct.

vascular cambium A lateral meristem that produces secondary xylem (wood) and secondary phloem (inner bark). Compare with *cork cambium*.

vascular tissue system The tissues specialized for translocation of materials throughout the plant body, i.e., the xylem and phloem.

vasoconstriction Narrowing of the diameter of blood vessels.

vasodilation Expansion of the diameter of blood vessels.

vector (1) Any carrier or means of transfer; (2) Agent, e.g., a plasmid or virus, that transfers genetic information; (3) Agent that transfers a parasite from one host to another.

vegetal pole The yolky pole of a vertebrate or echinoderm egg. Compare with *animal pole*.

vein (1) A blood vessel that carries blood from the tissues toward a chamber of the heart; compare with *artery*. (2) A strand of vascular tissue that is part of the network of conducting tissue in a leaf.

veliger larva The larval stage of many marine gastropods (snails) and bivalves (e.g., clams); often is a second larval stage that develops after the trochophore larva.

ventilation The process of actively moving air or water over a respiratory surface.

ventral Toward the lowermost surface or belly of an animal. Compare with *dorsal*.

ventricle (1) A cavity in an organ; (2) One of the several cavities of the brain; (3) One of the chambers of the heart that receives blood from an atrium.

vernalization (vur''nul-uh-zay'shun) The induction of flowering by a low temperature treatment.

vertebrates A subphylum of chordates; possess a bony vertebral column; include fishes, amphibians, reptiles, birds, and mammals.

vesicle (ves'ih-kl) Any small sac, especially a small spherical membrane-bounded compartment, within the cytoplasm.

vessel element A type of water-conducting cell in the xylem of vascular plants.

vestibular apparatus Collectively, the saccule, utricle, and semicircular canals of the inner ear.

vestigial (ves-tij'ee-ul) Rudimentary; an evolutionary remnant of a formerly functional structure.

vibrio A spiral-shaped bacterium that has the form of a short helix. Compare with *spirillum, spirochete, bacillus,* and *coccus*.

villus (pl. *villi*) A multicellular, minute, elongated projection, from the surface of an epithelial membrane, e.g., villi of the mucosa of the small intestine.

viroid (vy'roid) A tiny, naked infectious particle consisting only of nucleic acid.

virulent Able to cause disease in a host. Compare with *avirulent*.

virus A tiny pathogen composed of a core of nucleic acid usually encased in protein and capable of infecting living cells; a virus is characterized by total dependence on a living host.

viscera (vis'ur-uh) The internal body organs, especially those located in the abdominal or thoracic cavities.

visceral mass The concentration of body organs (viscera) located above the foot in mollusks.

vital capacity The maximum volume of air a person can expire after filling the lungs to the maximum extent.

vitamin A complex organic molecule required in very small amounts for normal metabolic functioning.

vitelline envelope An acellular covering of the eggs of certain animals (e.g., echinoderms), located just outside the plasma membrane.

viviparous (vih-vip'er-us) Bearing living young that develop within the body of the mother. Compare with *oviparous* and *ovoviviparous*.

voltage-activated ion channels Ion channels in the plasma membrane of neurons that are regulated by changes in voltage. Also called voltage-gated channels.

vomeronasal organ In mammals, an organ in the epithelium of the nose, made up of specialized chemoreceptor cells that detect pheromones.

vulva The external genital structures of the female.

warning coloration See *aposematic coloration*.

water mold A fungus-like protist with a body consisting of a coenocytic mycelium that reproduces asexually by forming motile zoospores and sexually by forming oospores.

water potential Free energy of water; the water potential of pure water is zero and that of solutions is a negative value. Differences in water potential are used to predict the direction of water movement (always from a region of less negative water potential to a region of more negative water potential).

water vascular system Unique hydraulic system of echinoderms; functions in locomotion and feeding.

wavelength The distance from one wave peak to the next; the energy of electromagnetic radiation is inversely proportional to its wavelength.

weathering processes Chemical or physical processes that help form soil from rock; during weathering processes, the rock is gradually broken into smaller and smaller pieces.

Western blotting A technique in which proteins, previously separated by gel electrophoresis, are transferred to paper. A specific labeled antibody is generally used to mark the location of a particular protein. Compare with *Southern blot hybridization* and *Northern blot hybridization*.

whisk ferns One of a phylum of seedless vascular plants with a life cycle similar to ferns.

white matter Nervous tissue in the brain and spinal cord that contains myelinated axons. Compare with *gray matter*.

wild type The phenotypically normal (naturally occurring) form of a gene or organism.

wobble The ability of some tRNA anticodons to associate with more than one mRNA codon; in these cases the 5′ base of the anticodon is capable of forming hydrogen bonds with more than one kind of base in the 3′ position of the codon.

work Any change in the state or motion of matter.

X-linked gene A gene carried on an X chromosome.

x-ray diffraction A technique for determining the spatial arrangement of the components of a crystal.

xylem (zy′lem) The vascular tissue that conducts water and dissolved minerals in plants.

XYY karyotype Chromosome constitution that causes affected individuals (who are fertile males) to be unusually tall, with severe acne.

yeast A unicellular fungus (ascomycete) that reproduces asexually by budding or fission and sexually by ascospores.

yolk sac One of the extraembryonic membranes; a pouchlike outgrowth of the digestive tract of embryos of certain vertebrates (e.g., birds) that grows around the yolk and digests it. Embryonic blood cells are formed in the mammalian yolk sac, which lacks yolk.

zero population growth Point at which the birth rate equals the death rate. A population with zero population growth does not change in size.

zona pellucida (pel-loo′sih-duh) The thick, transparent covering that surrounds the plasma membrane of a mammalian ovum.

zooflagellate A unicellular, nonphotosynthetic protozoon that possesses one or more long, whiplike flagella.

zooplankton (zoh′′oh-plank′tun) The nonphotosynthetic organisms present in plankton, e.g., protozoa, tiny crustaceans, and the larval stages of many animals. See *plankton*. Compare with *phytoplankton*.

zoospore (zoh′oh-spore) A flagellated motile spore produced asexually by certain algae, chytrids, water molds, and other protists.

zooxanthellae (zoh′′oh-zan-thel′ee) (sing. *zooxanthella*) Endosymbiotic, photosynthetic dinoflagellates found in certain marine invertebrates; their mutualistic relationship with corals enhances the corals' reef-building ability.

zygomycetes (zy′′gah-my′seats) Fungi characterized by the production of nonmotile asexual spores and sexual zygospores.

zygosporangium (zy′gah-spor-an′gee-um) A thick-walled sporangium containing a zygospore.

zygospore (zy′gah-spor) A sexual spore produced by a zygomycete.

zygote The $2n$ cell that results from the union of n gametes in sexual reproduction. Species that are not polyploid have haploid gametes and diploid zygotes.

zygotic genes Genes that are transcribed after fertilization, either in the zygote or in the embryo. Compare with *maternal effect genes*.

zygotic segmentation genes In *Drosophila,* genes transcribed in the embryo that are responsible for controlling formation of the segmented body.

Index

Note: **Boldface** numbers indicate pages on which the index term is defined; "il" following the page number indicates an illustration, "t" a table, "f" a focus box, and "n" a footnote.

Scientific Measurement

Some Common Prefixes

		Examples
kilo	1000	a kilogram is 1000 grams
centi	0.01	a centimeter is 0.01 meter
milli	0.001	a milliliter is 0.001 liter
micro (μ)	one-millionth	a micrometer is 0.000001 (one-millionth) of a meter
nano (n)	one-billionth	a nanogram is 10^{-9} (one-billionth) of a gram
pico (p)	one-trillionth	a picogram is 10^{-12} (one-trillionth) of a gram

The relationship between mass and volume of water (at 20°C)

$1 \text{ g} = 1 \text{ cm}^3 = 1 \text{ mL}$

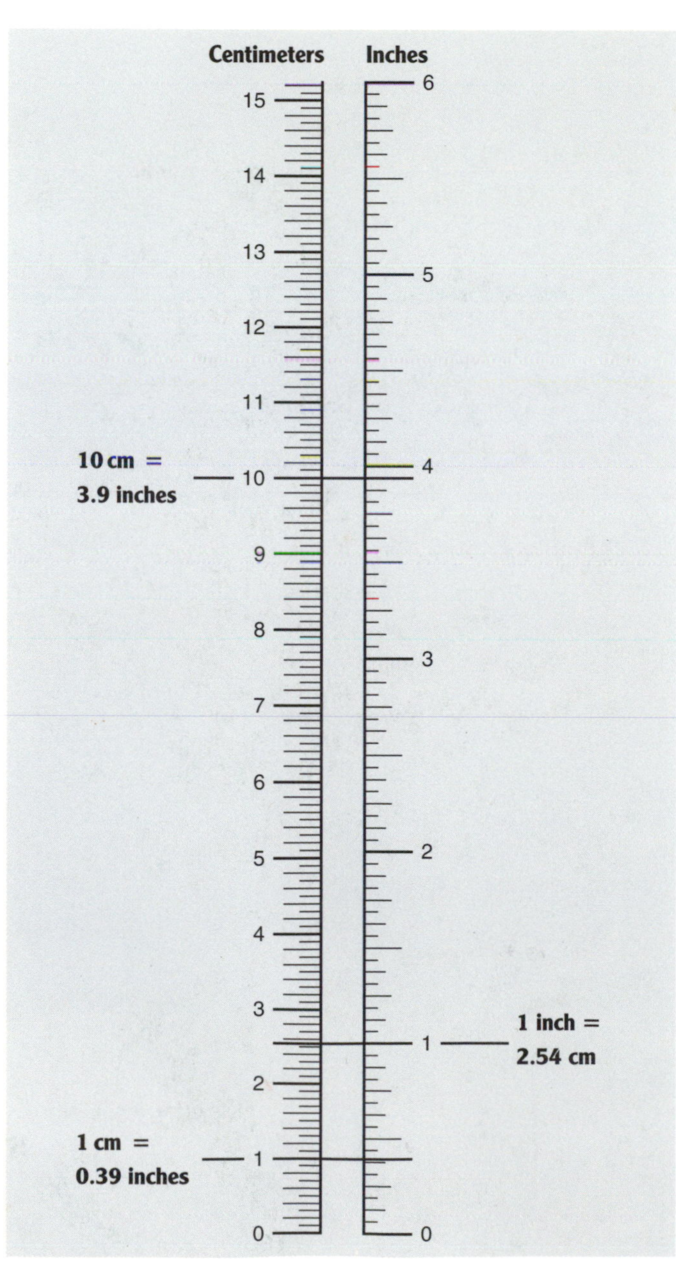

Some Common Units of Length

Unit	Abbreviation	Equivalent
meter	m	approximately 39 in
centimeter	cm	10^{-2} m
millimeter	mm	10^{-3} m
micrometer	μm	10^{-6} m
nanometer	nm	10^{-9} m

Length Conversions

1 in = 2.5 cm	1 mm = 0.039 in
1 ft = 30 cm	1 cm = 0.39 in
1 yd = 0.9 m	1 m = 39 in
1 mi = 1.6 km	1 m = 1.094 yd
	1 km = 0.6 mi

To convert	*Multiply by*	*To obtain*
inches	2.54	centimeters
feet	30	centimeters
centimeters	0.39	inches
millimeters	0.039	inches

Standard Metric Units

		Abbreviation
Standard unit of mass	gram	g
Standard unit of length	meter	m
Standard unit of volume	liter	L